Physics: The Nature of Things

Susan M. Lea
San Francisco State University

John Robert Burke
San Francisco State University

Brooks/Cole Publishing Company

West Publishing Company

I(T)P® An International Thomson Publishing Company

Pacific Grove • Albany • Belmont • Bonn • Boston • Cincinnati • Detroit
Johannesburg • London • Madrid • Melbourne • Mexico City
New York • Paris • Singapore • Tokyo • Toronto • Washington

PROOFREADING:	Elliot Simon, Simon and Associates
TEXT DESIGN:	Geri Davis, Quadrata, Inc.
COMPOSITION:	G&S Typesetters, Inc.
ART:	Scientific Illustrators, Inc.
COVER IMAGE:	Leo de Wys Inc./De Wys/Sipa/Fritz

PHOTO CREDITS FOLLOW THE INDEX.

West's Commitment to the Environment: In 1906, West Publishing Company began recycling materials left over from the production of books. This began a tradition of efficient and responsible use of resources. Today, 100% of our legal bound volumes are printed on acid-free, recycled paper consisting of 50% new fibers. West recycles nearly 27,700,000 pounds of scrap paper annually—the equivalent of 229,300 trees. Since the 1960s, West has devised ways to capture and recycle waste inks, solvents, oils, and vapors created in the printing process. We also recycle plastics of all kinds, wood, glass, corrugated cardboard, and batteries, and have eliminated the use of polystyrene book packaging. We at West are proud of the longevity and the scope of our commitment to the environment.

West pocket parts and advance sheets are printed on recyclable paper and can be collected and recycled with newspapers. Staples do not have to be removed. Bound volumes can be recycled after removing the cover.

Production, Prepress, Printing and Binding by West Publishing Company.

Copyright 1997   By West Publishing Company
610 Opperman Drive
P.O. Box 64526
St. Paul, MN 55164-0526
1-800-328-9352

All rights reserved

Printed in the United States of America

04 03 02 01 00 99 98 97   8 7 6 5 4 3 2 1 0

Library of Congress Cataloging-in-Publication Data
Lea, Susan.
    Physics : the nature of things / Susan Lea, John Burke.
      p.   cm.
    Includes bibliographical references and index.
    ISBN 0-314-05273-9 student edition (alk. paper)
    ISBN 0-314-07012-5 annotated instructor's edition (alk. paper)
    1. Physics.   I. Burke, John (John Robert)   II. Title.
QC21.2.L43   1997
530—dc20   96-13354
                                                                                         CIP

British Library Cataloging in Publication Data
A catalogue record for this book is available from the British Library.

Systematic Use of Color in this Text

(Components of vectors are shown in a lighter shade of the same color)

Part I: Newtonian Mechanics
- Path of a particle
- Unit vector
- Position vector
- Displacement vector
- Velocity vector
- Acceleration vector
- Force vector
- Moving frame box

Part II: Conservation laws
- Momentum vector
- Gravitational field vector
- Angular momentum vector
- Torque vector

Part III: Continuous systems
- Angular velocity vector
- Angular acceleration vector
- Streamlines

Part IV: Oscillatory and Wave Motion
- Sound wave phasors
- Light wave phasors
- Light rays

Part V: Thermodynamics
- Adiabat
- Isotherm
- Isochor
- Isobar

Part VI: Electromagnetic Fields
- Positive charge
- Negative charge
- Electric field vector
- Magnetic field vector
- Equipotential surface
- Electric dipole
- Magnetic dipole
- Electric current
- Electric displacement vector
- Gaussian surface
- Amperian curve

Part VII: Electrodynamics
- Poynting vector

Part VIII: Twentieth Century Physics
- World line
- Photon
- α-decay
- β-decay

Unit Systems and Conversion Factors

SI Units

Fundamental units	mass	kg	kilogram
	length	m	meter
	time	s	second
	electric current	A	ampere
Supplementary unit	angle	rad	radian
Some derived units	force	$N = kg \cdot m/s^2$	newton
	energy	$J = N \cdot m$	joule
	power	$W = J/s$	watt
	frequency	$Hz = 1/s$	hertz
	pressure	$Pa = N/m^2$	pascal
	charge	$C = A \cdot s$	coulomb
	electric potential	$V = J/C$	volt
	magnetic field	$T = N/A \cdot m$	tesla
	magnetic flux	$Wb = T \cdot m^2$	weber
	capacitance	$F = C/V$	farad
	resistance	$\Omega = V/A$	ohm

Selected British Units

length	1 inch ≡ 2.540 cm
	1 foot = 0.3048 m
	1 mile = 1.609 km
mass	1 pound mass (used in the U.K.) = 0.4536 kg
	1 slug (mass unit used in the U.S.) = 14.59 kg
	1 ton = 2240 lb mass (British or long ton)
	1 U.S. ton = 2000 lb mass (short ton)
energy	1 British thermal unit = 1.055×10^3 J
power	1 horse power = 745.7 W
force	1 pound (called pound-weight in the U.K.) = 4.448 N
pressure	1 lb/in.² = 6.895×10^3 Pa

Selected cgs/Gaussian Units

length	1 cm = 10^{-2} m
mass	1 g = 10^{-3} kg
energy	1 erg = 10^{-7} J
force	1 dyne = 10^{-5} N
pressure	1 dyn/cm² = 0.1 Pa
magnetic field	1 gauss corresponds to 10^{-4} T

Selected Units Used in Astronomy

length	1 astronomical unit = 1.50×10^{11} m
	1 light-year = 9.46×10^{15} m
	1 angstrom = 10^{-10} m
mass	1 solar mass = 1.99×10^{30} kg
power	1 solar luminosity = 3.90×10^{26} W

Miscellaneous Units

time	1 y ≈ $\pi \times 10^7$ s
	1 d = 86400 s
length	1 nautical mile = 1.852 km
speed	1 mph = 0.4470 m/s
	1 knot = 0.5145 m/s
mass	1 u (atomic mass unit) = 1.660×10^{-27} kg
	1 metric ton = 1000 kg
energy	1 calorie = 4.18 J
	1 electron volt = 1.60×10^{-19} J
	1 kilowatt-hour = 3.60×10^6 J
	1 kiloton of TNT = 4.2×10^{12} J
pressure	1 atmosphere = 1.013×10^5 Pa
	1 torr = 133.3 Pa
	1 cm Hg = 1.333×10^3 Pa
	1 in. Hg = 3.386×10^3 Pa
volume	1 liter = 10^{-3} m³
	1 U.S. gallon = 3.785×10^{-3} m³
area	1 acre = 4.05×10^3 m²
	1 barn = 10^{-28} m²
angle	1° = 1.745×10^{-2} rad
	1' = 1 minute of arc = $\frac{1}{60}$°
	1" = 1 second of arc = $\frac{1}{60}$'

"This dread and darkness of the mind cannot be dispelled by the sunbeams, the shining shafts of day, but only by an understanding of the outward form and inner workings of nature."

LUCRETIUS

To the special people in my life:

    my father and mother, my husband Michael and my daughter Jennifer.

                                                   Thank you.

                                                   Susan Lea

To my father, whose thirst for knowledge was an inspiration.

                                                   John Burke

ABOUT THE AUTHORS

Susan Lea is a professor of Physics and Astronomy at San Francisco State University, where she has taught since 1981. Born in Wales, she received her undergraduate degree from Cambridge University, with 1st class honors in applied mathematics and theoretical physics. She did her graduate work at the University of California, Berkeley, receiving a Ph.D in Astrophysics. She worked extensively with data from x-ray satellite missions, including Uhuru, HEAO 1 and the Einstein Observatory. She and her husband own and operate a software company offering optical ray tracing software. She has published extensively in the astronomical journals, but her first refereed paper (in an engineering journal) was on the theory of loudspeaker design! She began teaching physics at the age of 16 (in high school), and hasn't stopped since.

Dr. Lea's interests include flying (she holds a flight instructor certificate with airplane and instrument ratings), horse riding and music.

Professor of Physics at San Francisco State University since 1972, Dr. Burke has enjoyed sharing his love of science with young people deciding on their careers. As a voracious young consumer of science fiction and serious studies of space exploration, Dr. Burke's own path was set by visits to dad's job at then new particle accelerators and by Fred Hoyle's popular astronomy books. "It was so cool to know we could explore atoms or picture the Earth four billion years ago, melted by meteorite bombardment, its core forming from liquid iron dribbling inward." Undergraduate work at Caltech and graduate work in astrophysics at Harvard led to a research specialty in physics of the interstellar medium, with occasional forays into acoustics, economics, and relativity. It was also at Harvard that Dr. Burke's interest in physics eduation bloomed. "I had the opportunity to study teaching with outstanding masters of the craft. Concern for how people come to understand, how they fit science into their lives, and how they learn to think with precision have since guided all my work."

Of course, it's not all work. On occasion "J.R.B." can be caught taking in an early music concert, trekking a wilderness, climbing the odd mountain, or taking his plane into some out-of-the-way airport.

PREFACE

TO THE INSTRUCTOR

Our book's title is taken from De Rerum Natura, a work by Lucretius, a Roman writer of the first century AD[1] who tried to persuade his readers by using logical arguments based on observation and experience. This approach is still in style—modern physicists employ the same methods. Like Lucretius, both research physicists and physics students struggle to understand "the nature of things."

GOALS

A primary goal of this book is to help science students develop the kinds of logical thinking that they will need to understand physics. These skills are useful in physics and other disciplines as well. Students often find physics the most difficult of the sciences because, even in the introductory courses, it demands much more than the memorization of facts. To study physics successfully, students need to learn to think like physicists. Students must move beyond being hunters and gatherers of formulae to solve problems—they must become, like physicists, creative problem solvers. In this book we have tried to help students develop the logical reasoning and analytical skills that enable a physicist to practice his or her art.

Citizens in a modern technological society need to be scientifically literate. That only a small, elite group of bright students survives introductory physics and goes on to become aerodynamic engineers or physics professors is no longer acceptable. We hope to make physics accessible to all those who choose to take a physics course. We make it accessible not by watering it down, but by giving students the tools they need to grab hold of the subject and make it their own. Physics is fascinating and fun—at least we think so—and we have tried to convey some of our own enthusiasm for the subject. Examples such as the motion of a "hot-dog" skier (Example 3.5, Exercise 3.2) show the power of physics as a tool for understanding the world and, at the same time, spark students' interest.

Intended for a course that requires calculus as a prerequisite or co-requisite, this text uses calculus throughout, in derivations, examples, and problems. In the first few chapters calculus is used sparingly, mostly in optional sections, so that those students who are just starting calculus will not be overwhelmed. Later in the book, more familiarity is assumed. An interlude following Chapter 7 discusses the use of integration in physics and presents a five-step plan for setting up integrals. Basic knowledge of algebra, geometry, and trigonometry is assumed. Appendix I includes some basic relations from these disciplines as a reminder and reference for students.

This book can be used by students with widely varying levels of ability. Each chapter stresses the basic concepts first. By including or excluding the *Digging Deeper boxes*, the optional *Math Topic boxes, Optional sections* (marked with an ✱), and the *Advanced and Challenge Problems*, the instructor can tailor the text to her or his own students. *Instructor marginal notes (in blue)* indicate which optional topics are used later in the book, and also explain the reason for some of our choices of topic and organization. We have also given references for some of our sources.

ORGANIZATION

The order of topics in the book is largely traditional, but is organized to allow a large range of sequencing options. For example, introducing angular momentum of a particle in Part II offers the option of foregoing rigid body dynamics in favor of a faster move to the twentieth century. The chapter on oscillatory motion could be used any time after the discussion of energy (Chapter 8). We have included optics in the section on wave motion, to stress the unity of such wave phenomena as interference. However, Chapters 16–18 on optics could easily be covered after E&M if desired. Part V on thermodynamics is self contained and could be studied any time after basic mechanics. The first three sections of Chapter 34 (relativity) could be covered after Chapter 3, and Section 34.4 could be introduced after Chapter 8. The chapters on modern physics tend to be more qualitative, because of the level of mathematical sophistication required for a detailed treatment. They are designed to serve as the culmination of a two- or three-semester survey or as an impedance-matching introduction to a standard course on modern physics. These chapters emphasize the conservation principles developed in Part II.

---

[1] Lucretius based his book on earlier work by the Greek philosopher Democritus.

Throughout the book we stress *two major themes: conceptual understanding and a consistent approach to problem solving*. The material in the book is divided into eight parts, each introducing a unified body of concepts: Newtonian Mechanics; Conservation Laws; Continuous Systems; Oscillations and Waves; Thermodynamics; Electromagnetic Fields; Electrodynamics; and Twentieth Century Physics. This division helps the students organize their knowledge. The introduction to each part explains the theme to be covered and provides some historical perspective. We begin each chapter with a discussion of the opening photograph, frequently raising a question that we answer within the chapter. Just as each chapter begins with a physical situation to introduce the concepts of the chapter, each topic within the chapter is introduced with a conceptual discussion before the mathematics is presented. In this way we emphasize that working with the concepts is the first essential step in solving a problem. Then the mathematics is used to complete the solution. Similarly, we place a great deal of emphasis on using diagrams to help conceptualize problems and plan their solution. We encourage students to use diagrams as graphical tools to aid their understanding and to help make the transition from a verbal presentation to a mathematical model. Unlike many texts, we not only tell students to use diagrams, we *always* do it ourselves.

PROBLEM SOLVING

Two *Interludes* in the early parts of the text help lay the groundwork for a systematic approach to problem solving. In the first interlude, following Chapter 3, we lay out *our basic four-part problem-solving strategy*. The major stages of each problem solution— **MODEL** , **SETUP** , **SOLVE** , and **ANALYZE** — are identified and discussed at this point. These steps are used and labelled in every example throughout the book. Seeing the method at work in each example better enables students to apply a similar approach in their own solutions.

The second Interlude, following Chapter 7, shows students how to set up problem solutions using *integration*. The method involves five steps. The first four steps are a procedure for describing a physical process or system in terms of differential elements and transforming a sum over such elements to a standard mathematical form. Only at the final step does the actual evaluation of an integral occur. This final step is the one that students learn in their calculus classes. In each example requiring integration we use this method, with the steps clearly labelled.

Throughout the book we present *Solution Plans*. These are problem-solving strategies that show the logical steps necessary in certain specific classes of problems. Each plan is explicitly laid out in flow-diagram form. The method for analyzing dynamical systems with Newton's laws (Chapter 5, p. 167) provides a good example of a Solution Plan. A table in the appendix lists all the plans for easy reference. These Solution Plans will help the students develop the skills they need to solve problems in physics, and help them to go beyond that hunter-gatherer, "find the right equation and stuff in," stage. As students become more proficient they will be able to adapt these problem-solving strategies to their personal style.

The Solution Plans can also be valuable teaching tools, allowing you to identify precisely where students have difficulties. For example, using the plan in Chapter 5, we found that an astonishingly large number of students are convinced that they can't analyze a system with strings unless they know the value of the tension before carrying out the algebra. Once these difficulties have been identified, it is much easier to confront them and, ultimately, eliminate them.

The careful use of *vectors* is stressed throughout. In particular we introduce vectors as the primary descriptive tool in kinematics, using geometrical addition (Sections 1.4–1.6), and then solve one-dimensional problems as a special case of one-component vectors (Section 2.3). Not only does this approach stress the importance of vectors from the beginning, but it makes the meaning of signs in one-dimensional motion obvious. (An instructor's marginal note on page 52 explains how this material can be presented in other sequences.) In addition to boldface type, we have used the "*arrow-over*" *notation* so that equations in the book will look the same as the equations you write on the blackboard, or the students write in their notes. We have avoided the use of "magic" minus signs (as in the spring force) that are not explicitly tied to a coordinate choice or stated sign convention.

Beginning students often focus on finding "the answer" without first framing any expectation of what the magnitude, units or other characteristics of the answer might be. As scientists, instructors know the importance of estimation as a problem-solving strategy. It can be difficult to integrate this strategy into teaching, however, especially if students don't see it used regularly in their text. We introduce students to these valuable skills by using *back-of-the-envelope calculations* to estimate results, or to decide what is or is not important in a given situation. These methods are also used to estimate the reasonableness of an answer or to figure out the basic physics behind a complicated event like a thunderstorm. The envelope symbol (✉) alerts the students whenever we use these techniques in examples or discussions. Some problems show this symbol to indicate that the students should use these techniques in their solution, and that an exact answer is not expected.

EXAMPLES, QUESTIONS, AND PROBLEMS

Each chapter starts by emphasizing the basic concept, then developing it through a carefully graded series of *Examples*. All Examples consistently use the four-part problem-solving strategy presented in the first Interlude, and show the appropriate free-body diagram or other illustration at each step. While we have attempted to keep the introductory examples straightforward, and to assure that they demonstrate a steady gradual increase in difficulty throughout a chapter or part, twenty *Study Problems,* spread throughout the book, emphasize the use of the problem-solving method in detail with interesting and sometimes intricate problems. The inclusion of these problems should help to alleviate the complaint that the "examples didn't prepare me to do the problems."

The text offers many opportunities for students to test their knowledge and their ability to use the material. Within each chapter, *Exercises* allow students to practice with ideas they have just learned. Abbreviated solutions—not just answers—are given at the end of each chapter, so students can get real feedback after they work an exercise.

The end-of-chapter material includes a carefully structured array of problems for student review or assignment by an instructor. *Review Questions* emphasize conceptual understanding and can be answered by a quote or paraphrase of material from the chapter; *Basic Skill Drill* is a set of problems that test student's knowledge of fundamental mathematical relations and the meaning of terms introduced in the chapter; an extensive set of *Questions and Problems* include practical applications and conceptual questions as well as the usual "textbook exercises." Symbols preceding each problem identify the level of difficulty, and also indicate the conceptual problems. Many of the problems are sorted by chapter sections, but numerous *Additional Problems* are included that may require use of material from several sections, or even from previous chapters. *Computer problems* give the students an opportunity to hone their computer skills—an increasingly important component of education. Most of these problems can be solved using a spreadsheet program, or one of the simple programs on the supplementary computer disk available with the text. Students with more advanced computer skills will have an opportunity to incorporate these skills into their physics problem solving. *Challenge problems* introduce the more capable students to interesting and stimulating exercises that require advanced problem-solving skills. *Part Problems,* found at the end of each of the eight parts of the text, give students an opportunity to synthesize their understanding and to see how each topic builds on and enhances what went before.

OTHER HELPFUL FEATURES

Math Toolboxes appear throughout the text. Each one presents a set of techniques that are necessary tools for doing physics. They are located in the text where the techniques are first needed. For examples, see the Math Toolbox on the properties of the scalar product (p. 229) or the one called *How to Solve a Differential Equation* on p. 1005.

Digging Deeper boxes and *Math Topic* boxes present ideas that are not essential but that provoke interest, give greater depth to a point in the text, or simply point out a delightful consequence of the physical principles. See for example *More on Cyclotrons* (p. 928), *Use of Calculus in Circular Motion* (p. 98), and *How Do Fish Survive the Winter?* (p. 686).

Essays, some by guest authors, address interesting sidelights or more advanced topics. We happily remember the student who suddenly remarked "I get it!" after reading the bicycle essay. By applying Newton's laws to a subject he enjoyed, he finally made sense of it all.

Definitions and equations are color-coded to help students recognize their level of importance. Despite the emphasis throughout the text on problem-solving as a reasoning process, some things must be memorized to be used efficiently. Anything in a gold box is fundamental and should be memorized!!! Level 2 equations, in tan boxes, are important and will often be useful in solving problems. Level 3 equations, unboxed and unnumbered, are intermediate or less important results that need not be memorized. Occasionally we need to refer to intermediate results in order to guide students through a problem solution or derivation. Such results are given lower case Roman numerals. Any reference to these equations is local (within a page or so of the original statement).

Marginal notes (in black) alert the students to common errors, point out important features and special cases, give additional references, refer to previously discussed, related issues, and add clarifying commentary.

Instructor's Marginal Notes (in blue) appear throughout the *Instructor's Annotated Edition*. In this special version of the text, these marginal notes signal the location of related material, explain why a particular approach is used, cite references to the physics literature, provide suggestions and comments on possible changes in the sequence of topics and so forth. Many of the *Instructor's Marginal Notes* in the text are the result of "dialogues" that are carried on between reviewers of the manuscript and ourselves through many drafts of the text. In the *Instructor's Annotated Edition*, the Contents (on page *xi*) includes instructor's notes that comment on various organizational and content features of the text.

Our book has *more art* than other texts presenting the same material. We don't just tell students that good problem solving starts by making a drawing as conceptual link from the physical situation to the correct mathematical model, we consistently follow this practice. To help reinforce the importance of using and understanding graphical models, *color is consistently used in all illustrations* throughout the book. Acceleration is always blue, for example. See the color key that appears on page ii in the front of the book.

ACCURACY

The authors and publisher recognize that errors in quantitative material can undermine the effectiveness of a text. A great deal of attention and effort has been invested to assure that all of the quantitative material in the text (and the solutions manuals) is correct and accurate. Accuracy checking went on throughout preparation of the manuscript, as well as during production of the physical book. During the years that manuscript was being written and developed, many people were involved in assuring accuracy.

- Dozens of physics professors reviewed numerous drafts of the manuscript. All were asked to review the examples, exercises and problems. Many reviewers focused specifically on this quantitative material, at the publisher's request.
- The authors solved every end-of-chapter problem and checked each to be sure that it did not make unstated assumptions or rely upon unstated information.
- Jon Celesia of San Francisco State University carefully reviewed the final manuscript, checking for unstated assumptions, unclear explanations, and any possible inaccuracies.

During the year-long process of drawing all the art and setting all the type, numerous checks were performed.

- The authors proofread every syllable and symbol through two (in some cases three) stages of proof.

- One independent proofreader was hired to read the entire first stage proof. Another professional proofreader checked both stages of proof, and all subsequent revisions.
- Barbara Uchida of the College of San Mateo checked every one of the in-chapter Examples and Exercises for accuracy and clarity.
- Every end-of-chapter problem was solved by a team of physics professors and graduate students. These solutions were then reviewed by an independent accuracy checker before going to the authors for final approval.

Because students remember best what they learned first, we have taken pains to keep the concept discussions accurate. Even if topics must be expanded on later, students should never have to unlearn anything. Extensive review of the manuscript by dozens of teaching colleagues and consultations with several authorities on specific topics have helped to ensure that all concepts are correctly presented.

A final source of quality assurance for both quantitative and non-quantitative material has been students. During the development of the manuscript, many of the Examples, Exercises and end-of-chapter problems were tested with students in class and in homework assignments. The first half of the book has been used by students at San Francisco state University and at the University of California at Davis. The results have been very gratifying, both for the instructors and the students.

SUPPLEMENTS

A carefully prepared package of supplements has been created to support both the instructor and the student. Contact your local sales representative for information about the complete list of print and electronic supplements available. Among them are the following:

The Solutions Manuals have been written by the authors in conjunction with a team of graduate students and fellow physics professors. The Solutions Manuals have been carefully checked by the authors at least three times and by an independent accuracy checker for clarity, consistency, and accuracy. A clearly designed format featuring accurate, professionally rendered art makes the Solutions Manuals more accessible and useful. The Solutions Manuals also include a section of recommended readings ("For Further Reading") for reference.

Student's Solutions Manual—provides complete solutions to selected odd-numbered end-of-chapter problems including solutions for every odd-numbered Basic Skill Drill problem.

Instructor's Solutions Manual—contains complete solutions to all of the odd-numbered end-of-chapter problems and answers to all of the even-numbered problems.

The Test Bank, prepared by Darry S. Carlstone of the University of Central Oklahoma, includes over 3000 questions in multiple-choice format. The test bank is available in hard copy and on disk with a computerized test generator that allows instructors to modify, write, and display test questions. The testing program has outstanding graphics capability and a full range of physics symbols. IBM and Macintosh versions available.

The Optional Student Program Disk, created by Susan Lea and Michael Lampton in conjunction with Donnelly Software, contains data files for use with computer problems in the text (see above). In addition, some small programs for demonstrations and for problems on specific topics are also included. The files are in ASCII and can be imported into any spreadsheet program. The text may be ordered with or without the Student Program Disk.

Acetate Transparencies of numerous important illustrations from the text are available in full color.

ACKNOWLEDGEMENTS

This textbook represents not only the work of the authors, but also the extraordinary efforts and contributions of a number of people to whom we owe thanks.

This project would not have been possible without the support, encouragement, and help of our editor, Richard Mixter. His guidance and insight have been invaluable. He has kept us going through the difficult times, and shared our joy in the happy times. He has held our hands as we learned the ins and outs of publishing. Without him, this book would never have been completed.

Keith Dodson, the developmental editor, has also provided important guidance through his perceptive and thoughtful analyses of reviews and manuscript. We are grateful to the publishing team in Eagan, Minnesota, especially to Tamborah Moore and Emily Autumn for their outstanding work in the production of this complex book, and to Ann Hillstrom and Ellen Stanton for their insight and effort in marketing. Our thanks go to George Morris of Scientific Illustrators for creatively rendering the hundreds of pieces of art in the text, beautifully, clearly and correctly, and to Denee Reiton Skipper for her elegant and effective page layouts. Patricia Burke deserves recognition for her invaluable help with the illustration program. We should also like to acknowledge the contributions of those photographers and scientists who have allowed us to use their pictures, and especially to Tom Pantages who took many photographs to our specifications. We appreciate the help of Chuck and Janet Donnelly of Donnelly Software with the optional student program disk. (Any errors in the programs, however, are our responsibility.)

We also wish to thank our colleagues for their help: Jon Celesia for checking every word of the manuscript, Barbara Uchida for checking all of the examples and exercises, J. David Jackson for helpful discussions on topics in E&M, Edwin F. Taylor for brilliant thought experiments that clarified our thinking and inspired several end-of-chapter problems, Jim Lockhart and Shirley Chiang for searching out errors in the preliminary edition and the solutions manual, Alma Zook for taking data on vibrating strings especially for this book, Peter Linde for compiling the gas data in Figure 19.6, and our colleagues at SFSU for their continuing support. We especially appreciate the contributions of our guest essayists and their commitment to giving students a perspective on a wide range of careers in applied physics.

An important group of people has been helpful in working end-of-chapter problems to check both problems and solutions for accuracy and clarity. Special thanks go to Chris Kelly, Shuleen Martin, Russ Patrick, Peter Salzman, and Ladye Wilkinson. We appreciate the herculean efforts of Jeremy Hayhurst and his team at Chrysalis Productions, who turned a mountain of manuscript into the two Solutions Manuals. Lauren Fogel, at West, also deserves our thanks for her steadfast work on coordinating the entire Solutions Manual project, and keeping the authors going when we thought no more was possible. And thanks for the brownies, Lauren!

The development of this book has greatly benefitted from the contributions of numerous reviewers who offered their various perspectives and insights. We are sincerely grateful for their ideas and suggestions and offer each of them our thanks: Bill Adams, Baylor University; Edward Adelson, Ohio State University; Clifton Albergotti, University of San Francisco; S. N. Antani, Edgewood College; Paul Baum, CUNY, Queen's College; David Boness, Seattle University; Peter Border, University of Minnesota; Nick Brown, California State Polytechnic University, San Luis Obispo; Michael Browne, University of Idaho; Joseph J. Boyle, Miami-Dade Community College; Anthony Buffa, California State Polytechnic University, San Luis Obispo; Lou Cadwell, Providence College; Bob Camley, University of Colorado at Colorado Springs; D. S. Carlstone, University of Central Oklahoma; Colston Chandler, University of New Mexico; Edward Chang, University of Massachusetts; William Cochran, Youngstown State University; James R. Conrad, Contra Costa College; Roger Crawford, LA Pierce College; John E. Crew, Illinois State University; Gordon Emalbie, University of Alabama, Huntsville; Lewis Ford, Texas A&M University; David Gavenda, University of Texas at Austin; Edward F. Gibson, California State University, Sacramento; Gerald Hart, Moorhead State University; Scott Hildreth, Chabot College; Richard Hilt, Colorado College; Stanley Hirschi, Central Michigan University; Laurent Hodges, Iowa State University; C. Gregor Hood, Tidewater Community College; Ruth H. Howes, Ball State University; John Hubisy, North Carolina State University; Alvin W. Jenkins, North Carolina State University; Darrell Huwe, Ohio University; Larry Johnson, Northeast Louisiana University; Karen Johnston, North Carolina State University; John King, University of Central Oklahoma; Leonard Kleinman, University of Texas at Austin; Claude M. Laird, University of Kansas; Robert Larson, St. Louis Community College; Michael Lieber, University of Arkansas; David Markowitz, University of Connecticut; L. C. McIntyre, Jr., University of Arizona; Howard Miles, Washington State University; Lewis Miller, Canada College; David Mills, College of the Redwoods; Matthew J. Moelter, California State University, Sacramento; Richard Mould, SUNY Stony Brook; Raymond Nelson, U.S. Military Academy; Jack Noon, University of Central Florida; Aileen O'Donoghue, St. Lawrence University; Harry Otteson, Utah State University; Rob Parsons, Bakersfield College; Eric Peterson, Highland Community College; R. Jerry Peterson, University of Colorado; Ronald Poling, University of Minnesota; Richard Reimann, Boise State University; Charles W. Scherr, University of Texas at Austin; Arthur Schmidt, Northwestern University; Achin Sen, Eastern Washington University; John Shelton, College of Lake County; Stanley J. Shepherd, Pennsylvania State University; Gregory Snow, University of Michigan; Kevork Spartalian, University of Vermont; Richard Swanson, Sandhills Community College; Chuck Taylor, University of Oregon; Carl T. Tomizuka, University of Arizona; Sam Tyagi, Drexel University; Gianfranco Vidali, Syracuse University; James S. Walker, Washington State University; Arthur West, Shoreline Community College; Gary Williams, University of California, Los Angeles; John G. Wills, University of Indiana at Bloomington; Lowell Wood, University of Houston.

We also thank all the students at SFSU on whom we have tried out our ideas over the years, and who have used various early editions of this book.

To Andy Crowley, who started with us on this project many years ago: we thank you for your faith in us, and wish you well in your present ventures.

Finally, we owe an enormous debt of thanks to our families, who have endured the enormous piles of paper that have littered our homes for years and who have tolerated our long hours and grouchiness for lack of sleep. We would especially like to thank Jennifer Lampton who gracefully consented to be the guinea-pig on whom we tested our ideas and explanations to see if they made sense. To those friends we haven't seen for ages, perhaps we'll see you soon, and we thank you, too, for your patience.

IN CONCLUSION

In writing this book we have been guided by our students. We have listened to their complaints, watched how they work and noted where they have difficulties. We have also been cognizant of recent research on physics education, which, for the most part, supports our own observations. Thus, this book is written for the student. No book can make physics easy for everyone, but we can show students an approach that works. Our problem-solving strategy has been tested and approved by hundreds of our students, and it has increased their exam scores dramatically. We are confident it can work for your students as well.

CONTENTS IN BRIEF

Preface v
Prologue xxv

PART ONE: NEWTONIAN MECHANICS

CHAPTER 0   The Roots of Science   2
CHAPTER 1   Introducing the Language of Physics   18
CHAPTER 2   Kinematics   51
CHAPTER 3   Advanced Kinematic Models   83
Interlude 1: Solving Physics Problems   116
CHAPTER 4   Force and Newton's Laws   125
Essay 1: Newton's Discoveries and Their Impact   162
CHAPTER 5   Using Newton's Laws   164

PART TWO: CONSERVATION LAWS

CHAPTER 6   Linear Momentum   198
CHAPTER 7   Work and Kinetic Energy   224
Interlude 2: Using Integration in Physics Problems   248
CHAPTER 8   Conservation of Energy   257
Essay 2: The Gravitational Field   290
CHAPTER 9   Angular Momentum   294
Essay 3: Orbits   329
CHAPTER 10   Collisions   332

PART THREE: CONTINUOUS SYSTEMS

CHAPTER 11   Rigid Bodies in Equilibrium   362
CHAPTER 12   Dynamics of Rigid Bodies   390
Essay 4: The Bicycle   426
CHAPTER 13   Fluids   431

PART FOUR: OSCILLATORY AND WAVE MOTION

CHAPTER 14   Oscillatory Motion   470
CHAPTER 15   Introduction to Wave Motion   495
CHAPTER 16   Sound and Light Waves   523
CHAPTER 17   Interference and Diffraction   562
CHAPTER 18   Geometrical Optics   597
Essay 5: Ray Tracing with a Computer   634

PART FIVE: THERMODYNAMICS

CHAPTER 19   Temperature and Thermal Energy   643
CHAPTER 20   Thermodynamics of Real Substances   676
Essay 6: Low Temperatures and their Measurement   698
CHAPTER 21   Heat Transfer   700
CHAPTER 22   Entropy and the Second Law of Thermodynamics   718
Essay 7: Entropy, Evolution, and the Arrow of Time   751

PART SIX: ELECTROMAGNETIC FIELDS

Overview of Electromagnetism   755
CHAPTER 23   Charge and the Electric Field   764
CHAPTER 24   Static Electric Fields   790
CHAPTER 25   Electric Potential Energy   813
CHAPTER 26   Introduction to Electric Circuits   843
CHAPTER 27   Capacitance and Electrostatic Energy   875
CHAPTER 28   Static Magnetic Fields   898
CHAPTER 29   Static Magnetic Fields: Applications   923

PART SEVEN: ELECTRODYNAMICS

CHAPTER 30   Dynamic Fields   955
CHAPTER 31   Introduction to Time-Dependent Circuits   987
CHAPTER 32   Introduction to Alternating Current Circuits   1016
CHAPTER 33   Electromagnetic Waves   1039

PART EIGHT: TWENTIETH-CENTURY PHYSICS

CHAPTER 34   Relativity and Space-Time   1069
Essay 8: General Relativity: A Geometric Theory of Gravity   1103
CHAPTER 35   Light and Atoms   1110
Essay 9: The Scanning Tunneling Microscope   1142
CHAPTER 36   Atomic Nuclei   1145
CHAPTER 37   Particle Physics   1177

Epilogue   1199

Index   I-1

CONTENTS

Preface v
Prologue xxv
    Why Do Physics?   xxv
    So, What is Physics?   xxv
    What Are the Aims of This Text?   xxvi
    Suggestions for Using the Text   xxvi

The Universe: An Overview   xxvii
    The Everyday Scale   xxviii
    The Solar System   xxix
    The Universe of Stars   xxix
    The World as Atoms   xxx
    The Subatomic World   xxxi
    Summary Chart   xxxi

PART ONE
NEWTONIAN MECHANICS

CHAPTER 0   The Roots of Science   2

0.1   Why Study History?   3
    0.1.1   Substance and Structure   3
    0.1.2   Simplicity   4
    0.1.3   What is a Satisfactory Explanation?   4
    0.1.4   Aristotle and the Nature of Motion   5
0.2   Kepler, Planetary Motion, and Physical Law   6
    0.2.1   Scientific Revolutions   6
    0.2.2   The Earth Moves   6
    0.2.3   Kepler's Laws   6
    DIGGING DEEPER: ELLIPSES   9
0.3   Galileo and Experimental Science   10
    0.3.1   Composition of Motion   10
    0.3.2   The Inclined Plane Experiments   11
0.4   The Nature of Physics   12
    0.4.1   The Nature of Experiment   12
    0.4.2   Universal Law   12
    0.4.3   Perspective and Point of View   13
    0.4.4   Theory and Experiment   14
    0.4.5   How Nature Works versus Why Nature Works   14
    0.4.6   Ideal versus Real: Modeling and the Role of Mathematics   15

CHAPTER SUMMARY 15 ● SOLUTION TO EXERCISE 16
BASIC SKILLS 16 ● QUESTIONS AND PROBLEMS 17

CHAPTER 1   Introducing the Language of Physics   18

1.1   A Model of Space and Time   19
    1.1.1   Space   19
    1.1.2   Time   20
    1.1.3   How Good Is the Cartesian Model?   21
1.2   The International System of Units   21
    1.2.1   Metrology, the Science of Measurement   21
    1.2.2   The Origin of the Metric System   22
    1.2.3   The SI Unit of Time—The Second   22
    1.2.4   The SI Unit of Length—The Meter   23
    1.2.5   The SI Unit of Angle—The Radian   24
    DIGGING DEEPER: HOW FAR IS A SECOND?
    HOW QUICK IS A METER?   25
    1.2.6   The SI Unit of Mass—The Kilogram   26
1.3   Using SI   26
    1.3.1   Significant Figures   26
    1.3.2   Units Conversion and SI Prefixes   28
    1.3.3   Dimensional Analysis and Estimation   29
1.4   Vectors and Scalars   31
    1.4.1   The Basic Distinction   31
    1.4.2   Displacement   32
    1.4.3   Vector Addition   33
    1.4.4   The Zero Vector   33
    1.4.5   Displacement and Distance   34
    1.4.6   Multiplication of a Vector by a Scalar   34
    1.4.7   Subtraction of Vectors   34
1.5   The Position Vector   35

1.6 Vector Algebra 37
   1.6.1 Components 37
   1.6.2 Unit Vectors 39
   1.6.3 Choosing Coordinate Systems 40
   1.6.4 Vector Algebra 41

CHAPTER SUMMARY 42 • SOLUTIONS TO EXERCISES 43
BASIC SKILLS 44 • QUESTIONS AND PROBLEMS 46

CHAPTER 2 Kinematics 51

2.1 Speed and Velocity 52
   2.1.1 Average Speed 52
   2.1.2 Average Velocity 54
   2.1.3 Instantaneous Velocity and the Concept of Limit 55
   2.1.4 ✻ Calculus as a Kinematic Model 58
2.2 Average and Instantaneous Acceleration 58
   2.2.1 The Acceleration Vector 58
   2.2.2 Velocity and Acceleration in Component Notation 62
2.3 Linear Motion 63
   2.3.1 Position, Velocity, and Acceleration in Linear Motion 63
   2.3.2 The Acceleration Due to Gravity 64
   2.3.3 Galileo's Law Using Formal Calculus 66
   DIGGING DEEPER: GRAVITY 66
   2.3.4 Interpreting Graphs of Position or Velocity versus Time 68
   2.3.5 Uniformly Accelerated Linear Motion 69

CHAPTER SUMMARY 74 • SOLUTIONS TO EXERCISES 74
BASIC SKILLS 76 • QUESTIONS AND PROBLEMS 78

CHAPTER 3 Advanced Kinematic Models 83

3.1 Projectile Motion 84
   3.1.1 Problems with Given Initial Conditions 85
   3.1.2 Problems with a Specific Goal 86
   3.1.3 Simultaneous Motion 89
   DIGGING DEEPER: FREE FALL 90
   3.1.4 The Shape of a Projectile's Trajectory 91
   DIGGING DEEPER: PROJECTILES AND ORBITS 92

3.2 Circular Motion 93
   3.2.1 Describing Circular Motion 93
   3.2.2 Uniform Circular Motion 94
   3.2.3 Speed and Instantaneous Velocity in Circular Motion 95
   MATH TOPIC: FORMAL EVALUATION OF THE LIMIT 97
   3.2.4 Instantaneous Acceleration in Circular Motion 97
   MATH TOPIC: USE OF CALCULUS IN CIRCULAR MOTION 98
   3.2.5 Rigid Body Rotation 99
   3.2.6 Using Linear and Circular Motion to Model General Motion 101
3.3 Relative Motion 102
   3.3.1 Navigation 102
   3.3.2 In What Frame is a Problem Easiest? 105
   3.3.3 Rolling Without Slipping 106

CHAPTER SUMMARY 106 • SOLUTIONS TO EXERCISES 107
BASIC SKILLS 108 • QUESTIONS AND PROBLEMS 109

Interlude 1: Solving Physics Problems 116

I1.1 Top-Down Reasoning 116
I1.2 Steps in Building a Problem Solution 117
I1.3 Study Problems 118
   Study Problem 1: *Lunch at Noon?* 118
   Study Problem 2: *Slugger José's Pop Fly* 119
   Study Problem 3: *Mars or Bust!* 122

SUMMARY 124 • SOLUTIONS TO EXERCISES 124

CHAPTER 4 Force and Newton's Laws 125

4.1 Force 126
4.2 Force in the Newtonian Model 127
   4.2.1 What Kinds of Force Occur? 127
   4.2.2 Forces are Vectors 129
   4.2.3 Force Vectors Occur in Pairs 130
4.3 Newton's Second Law 132
   4.3.1 The Relation Between Force and Acceleration 132
   4.3.2 Mass 133
   4.3.3 The SI Unit of Force—The Newton 134

4.4 Weight 134
4.5 Practical Expressions for Spring and Friction Forces 136
    4.5.1 The Elastic Force Exerted by a Spring 136
    4.5.2 Friction 139
4.6 Motion of a Single Object Subject to Several Forces 140
    4.6.1 The Method of Free-Body Diagrams 140
    4.6.2 Examples Involving Linear Motion 141
    Study Problem 4: Hotdog Skiing on Spring Snow 143
4.7 Dynamics of Circular Motion 145
4.8 Newton's Laws of Motion 148
    4.8.1 The First Law of Motion and Inertial Frames of Reference 148
    4.8.2 The Second Law of Motion 148
    DIGGING DEEPER: INERTIAL AND NONINERTIAL REFERENCE FRAMES 149
    4.8.3 Newton's Third Law 149
    4.8.4 What Is Mass? 150
    4.8.5 The Structure of Newtonian Mechanics 150

CHAPTER SUMMARY 151 • SOLUTIONS TO EXERCISES 152
BASIC SKILLS 154 • QUESTIONS AND PROBLEMS 156

Essay 1: Newton's Discoveries and Their Impact 162

CHAPTER 5 Using Newton's Laws 164

5.1 A First Example 165
5.2 Solution Method for Systems of Particles 165
    5.2.1 Identifying Particles 165
    5.2.2 Free-Body Diagrams 166
    5.2.3 Analysis of the Free-Body Diagrams 166
    5.2.4 Connecting the Particles Back into a System 166
    5.2.5 Using the Solution Method 166
5.3 ✻ Strings 170
    5.3.1 Why the Ideal String Model Works 170
    5.3.2 Strings Subject to Weight and Friction Forces 172
    5.3.3 Pulley Systems 173
5.4 The Law of Universal Gravitation 175
    5.4.1 The Gravitational Force Law 175
    DIGGING DEEPER: GRAVITATIONAL FORCE DUE TO A UNIFORM SPHERE 178
    DIGGING DEEPER: FORCE AT A DISTANCE 179
    Study Problem 5: A Trip Around the Moon 179
    5.4.2 Finding the Mass of the Earth 181
    5.4.3 ✻ Newton's Gravity and Galileo's Law of Falling Objects 181
    5.4.4 ✻ Are there Different Kinds of Mass? 182
    5.4.5 ✻ Kepler's Laws 183
5.5 ✻ The Beanstalk 183

CHAPTER SUMMARY 185 • SOLUTIONS TO EXERCISES 186
BASIC SKILLS 187 • QUESTIONS AND PROBLEMS 188

PART ONE: PROBLEMS 196

PART TWO
CONSERVATION LAWS

CHAPTER 6 Linear Momentum 198

6.1 Linear Momentum 199
    6.1.1 Momentum of a Particle 199
    6.1.2 Momentum Transfer 200
    6.1.3 Exchange of Momentum Between Particles 204
    6.1.4 Conservation of Momentum 205
6.2 Using Conservation of Momentum 206
    6.2.1 General Solution Plan 206
    6.2.2 Problems Involving Mass Flow 208
    Study Problem 6: Delivering the Mail 212
    DIGGING DEEPER: ACCELERATION OF THE ROCKET 212
6.3 ✻ Formal Proof of Momentum Conservation 214

CHAPTER SUMMARY 216 • SOLUTIONS TO EXERCISES 217
BASIC SKILLS 217 • QUESTIONS AND PROBLEMS 218

CHAPTER 7 Work and Kinetic Energy 224

7.1 Energy and Its Transfer 225
    7.1.1 Work and Kinetic Energy in One-Dimensional Motion 225
    7.1.2 Work in More Than One Dimension 227
    MATH TOOLBOX: PROPERTIES OF THE SCALAR PRODUCT 229
    7.1.3 Work Done When More Than One Force Acts 230
    7.1.4 Formal Proof of the Work-Energy Theorem for a Particle 232
    7.1.5 Kinetic Energy of a System of Particles 233
7.2 Power and Simple Machinery 234
    7.2.1 Power 234
    DIGGING DEEPER: THE HORSEPOWER 235
    7.2.2 Simple Machines 237
    7.2.3 Energy Transmission by Machines 238

CHAPTER SUMMARY 239 • SOLUTIONS TO EXERCISES 240
BASIC SKILLS 241 • QUESTIONS AND PROBLEMS 242

Interlude 2:
Using Integration in Physics Problems 248

- I2.1 The Integral of a Function as the Limit of a Sum 248
- I2.2 General Method for Evaluating Physical Integrals 250
  MATH TOPIC: INTEGRALS AS AREAS 251

SUMMARY 254 • SOLUTION TO EXERCISE 254
BASIC SKILLS 254 • QUESTIONS AND PROBLEMS 255

CHAPTER 8 Conservation of Energy 257

- 8.1 Elastic Potential Energy 258
  - 8.1.1 Stored Energy in a Compressed Spring 258
  - 8.1.2 Potential Energy 259
  DIGGING DEEPER: CALCULATION OF WORK DONE ON THE FREIGHT CARS 260
  - 8.1.3 Conservation of Energy in the Mass-on-Spring System 261
- 8.2 Gravitational Potential Energy 263
  - 8.2.1 Practical Description for Use Near a Planetary Surface 263
  - 8.2.2 An Exact Expression for Gravitational Potential Energy 266
- 8.3 Conservation of Mechanical Energy 269
  - 8.3.1 Conservative Forces 269
  - 8.3.2 Conservation of Total Mechanical Energy 270
  MATH TOPIC: WORK DONE AROUND A CLOSED PATH 271
- 8.4 Potential Energy in Systems of Particles 273
- 8.5 Internal Energy 274
  - 8.5.1 Thermal Energy 274
  - 8.5.2 Conservation of Energy 275
    Study Problem 7: The Egg Factory 276
  - 8.5.3 What Does Make a Car Go Up a Hill? 278

CHAPTER SUMMARY 278 • SOLUTIONS TO EXERCISES 279
BASIC SKILLS 280 • QUESTIONS AND PROBLEMS 281

Essay 2: The Gravitational Field 290

- E2.1 The Gravitational Field of the Earth 290
- E2.2 The Field of a Two-Particle System 292
- E2.3 Is the Gravitational Field Real? 293

CHAPTER 9 Angular Momentum 294

- 9.1 Angular Momentum of a Particle 295
  - 9.1.1 What Is Angular Momentum? 295
  - 9.1.2 Angular Momentum as a Vector Product 297
  MATH TOOLBOX: PROPERTIES OF THE CROSS PRODUCT 299
- 9.2 Torque 301
  - 9.2.1 What Is Torque? 301
  - 9.2.2 Work and Power in Rotating Systems 304
- 9.3 The Center of Mass 306
  - 9.3.1 Center of Gravity and Center of Mass 306
  - 9.3.2 Definition of the Center of Mass 307
  - 9.3.3 Motion of the Center of Mass 309
  - 9.3.4 The Center of Mass Reference Frame 310
- 9.4 Conservation of Angular Momentum 313
  - 9.4.1 A System of Two Particles 313
  - 9.4.2 Proof that Angular Momentum is Conserved 313
  DIGGING DEEPER: THE STRONG FORM OF NEWTON'S THIRD LAW 314
    Study Problem 8: The Two Skaters 315

CHAPTER SUMMARY 318 • SOLUTIONS TO EXERCISES 319
BASIC SKILLS 320 • QUESTIONS AND PROBLEMS 322

Essay 3: Orbits 329

- E3.1 Energy and the Semimajor Axis 329
- E3.2 Angular Momentum and Eccentricity 331

CHAPTER 10 Collisions 332

- 10.1 What Is a Collision? 333
- 10.2 Collisions Between Two Particles 334
  - 10.2.1 Elastic Collisions 334
  - 10.2.2 Elastic Collisions in the CM Reference Frame 337
  - 10.2.3 Inelastic Collisions 338
- 10.3 Models for Elastic and Inelastic Collisions 341

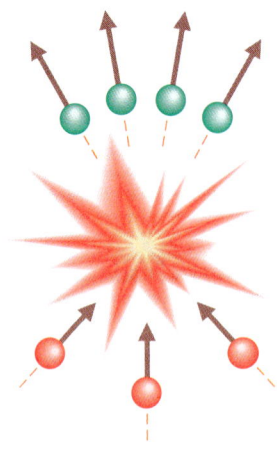

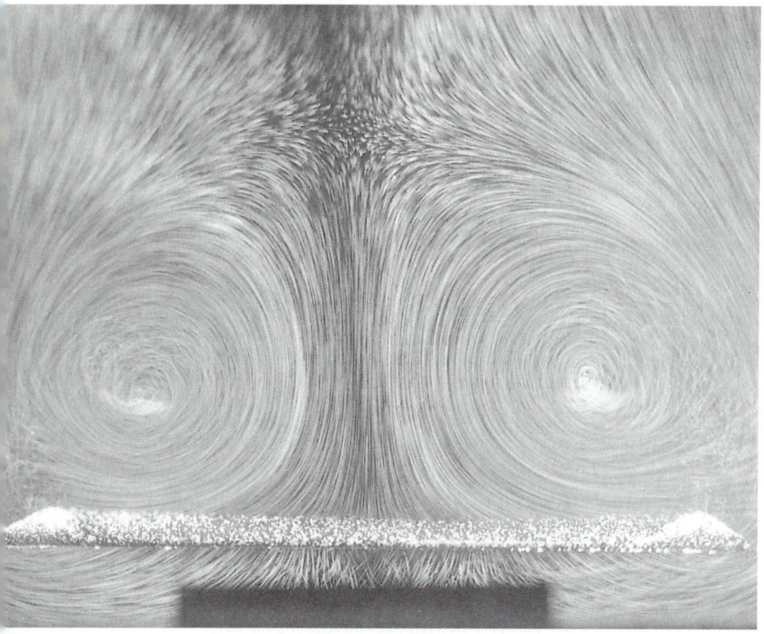

10.4 ✲ Some Applications of Collision Theory 345
10.4.1 Atomic and Subatomic Particles 345
10.4.2 When Molecules Collide 347
10.4.3 Gravitational Collisions 349

Chapter Summary 350 • Solutions to Exercises 351
Basic Skills 352 • Questions and Problems 354

Part Two: Problems 359

PART THREE CONTINUOUS SYSTEMS

CHAPTER 11 Rigid Bodies in Equilibrium 362

11.1 Static Equilibrium 363
11.1.1 Conditions for Equilibrium 363
11.1.2 Stability 366
11.1.3 Couples 367
11.1.4 Three Forces in Equilibrium 368
11.2 Dynamic Equilibrium 369
11.3 Equilibrium of Systems 371
11.4 The Ladder Problem 372
11.5 The Center of Mass of Extended Bodies 374
11.6 ✲ Bridges 376

Chapter Summary 378 • Solutions to Exercises 379
Basic Skills 381 • Questions and Problems 382

CHAPTER 12 Dynamics of Rigid Bodies 390

12.1 Rotational Kinematics 391
12.1.1 Body Coordinates 391
12.1.2 Angular Velocity and Angular Acceleration 392
12.1.3 Rotation About a Fixed Axis 393
Math Topic: Vector Relations In Circular Motion 396
12.2 Rotational Kinetic Energy and Angular Momentum 396
12.2.1 Energy 396
12.2.2 Angular Momentum 397
12.3 Dynamic Behavior of Rigid Bodies 400
12.3.1 General Solution Method 400
Study Problem 9: *Rockfall!* 404
12.3.2 ✲ Limitations of the Rigid Body Model 406
12.4 Application of the Conservation Laws 406
12.5 Calculation of Rotational Inertia 409
12.5.1 Integration 409
12.5.2 The Parallel Axis Theorem 410
Math Topic: General Proof of the Parallel Axis Theorem 411
12.6 ✲ Precession 413

Chapter Summary 415 • Solutions to Exercises 416
Basic Skills 417 • Questions and Problems 418

Essay 4: The Bicycle 426
Digging Deeper: Average Torque on a Bicycle Crank 429
Digging Deeper: Air Resistance 430

CHAPTER 13 Fluids 431

13.1 What Is a Fluid? 433
13.2 Basic Properties of Fluids 434
13.2.1 Density 434
13.2.2 Pressure 435
Digging Deeper: Von Guericke's Demonstration 437
13.2.3 What Causes Pressure? 438
13.3 Fluids in Equilibrium 440
13.3.1 Variation of Pressure in a Fluid at Rest 440
13.3.2 Hydrostatic Equilibrium 442
Digging Deeper: The Hydrostatic Paradox 443
13.3.3 The Barometer 443
13.3.4 The Atmosphere 443

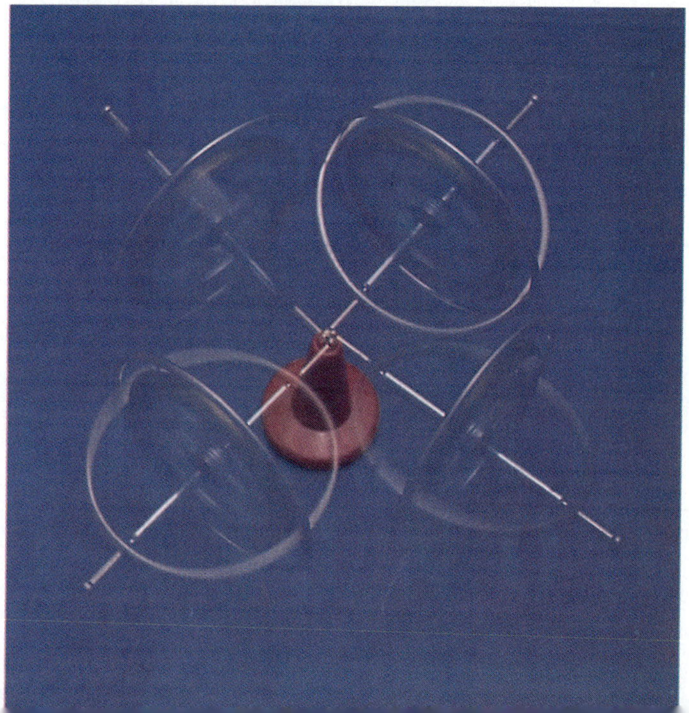

13.4 Archimedes' Principle   445
   13.4.1 Buoyant Force   445
      DIGGING DEEPER: THE SHIP IN DRY DOCK   446
      Study Problem 10: *An Accident in the Lock*   448
13.5 Fluid Dynamics   449
   13.5.1 Streamlines   449
   13.5.2 The Equation of Continuity   450
   13.5.3 Bernoulli's Law   451
   13.5.4 Solving Problems in Fluid Dynamics   453
13.6 Incompressible Flow of Air   454
   13.6.1 Airflow Over a Wing   454
   13.6.2 Ram Pressure   455

CHAPTER SUMMARY 456 • SOLUTIONS TO EXERCISES 457
BASIC SKILLS 459 • QUESTIONS AND PROBLEMS 460

PART THREE: PROBLEMS   468

PART FOUR
OSCILLATORY
AND WAVE MOTION

CHAPTER 14   Oscillatory Motion   470

14.1 Simple Harmonic Motion   471
   14.1.1 Restoring Forces   471
   14.1.2 Equation of Motion for an Oscillating Particle   472
   14.1.3 Analogy with Circular Motion   472
   MATH TOOLBOX: HARMONIC FUNCTIONS   474
   14.1.4 Using the Solution for Simple Harmonic Motion   475
14.2 The Pendulum   477
   14.2.1 The Simple Pendulum   477
   14.2.2 The Physical Pendulum   478
14.3 Energy in Oscillatory Motion   480
      Study Problem 11: *Bungee Jumping!*   482

14.4 ✱ The Effect of External Forces   484
   14.4.1 Forced Oscillations and Resonance   484
   DIGGING DEEPER: THE AMPLITUDE OF A FORCED OSCILLATION   485
   DIGGING DEEPER: A DRIVING FORCE   486
   14.4.2 Damping   486

CHAPTER SUMMARY 489 • SOLUTIONS TO EXERCISES 489
BASIC SKILLS 490 • QUESTIONS AND PROBLEMS 491

CHAPTER 15   Introduction to Wave Motion   495

15.1 Mechanical Waves   496
   15.1.1 What Causes Mechanical Waves?   496
   15.1.2 The Language of Wave Theory   498
15.2 Mathematical Description of a Wave Disturbance   501
   15.2.1 The Wave Function   501
   15.2.2 Harmonic Waves   502
   15.2.3 The Wave Equation   505
   MATH TOPIC: DEMONSTRATION THAT $f(x \pm vt)$ SOLVES THE WAVE EQUATION   506
   15.2.4 ✱ Derivation of the Wave Equation for a String   506
15.3 Energy Transmission by Harmonic Waves   507
15.4 Superposition   508
   15.4.1 Reflections of Waves at a Boundary   508
   15.4.2 Standing Waves   510
   DIGGING DEEPER: STANDING WAVES AND MUSICAL NOTES   512
15.5 ✱ Reflection and Transmission of Waves at a Junction of Two Strings   513

CHAPTER SUMMARY 515 • SOLUTIONS TO EXERCISES 517
BASIC SKILLS 517 • QUESTIONS AND PROBLEMS 518

CHAPTER 16   Sound and Light Waves   523

16.1 Sound   524
   16.1.1 Sound Waves in a Tube   525
   DIGGING DEEPER: DERIVATION OF EQN. (16.2)   528
   16.1.2 The Speed of Sound   528
   16.1.3 Standing Sound Waves   530
   DIGGING DEEPER: THE WAVE EQUATION FOR SOUND   531
16.2 Light   532
   16.2.1 The Electromagnetic Spectrum   532
   DIGGING DEEPER: THE SPEED OF LIGHT   534
   DIGGING DEEPER: HOW DOES THE EYE DETECT LIGHT? (SUZANNE MCKEE)   536
   16.2.2 The Speed of Light   536
16.3 Energy in Sound and Light   537
   16.3.1 Wave Fronts and Rays   537
   16.3.2 Power in Sound Waves   538
   DIGGING DEEPER: HOW DOES THE EAR DETECT SOUND?   540
   16.3.3 The Inverse Square Law   541
16.4 The Doppler Effect   542
   16.4.1 Source Moving with Respect to the Air   542
   16.4.2 Observer Moving with Respect to the Air   543
   16.4.3 Source and Observer Moving with Respect to the Air   544
   16.4.4 The Doppler Effect for Light   545

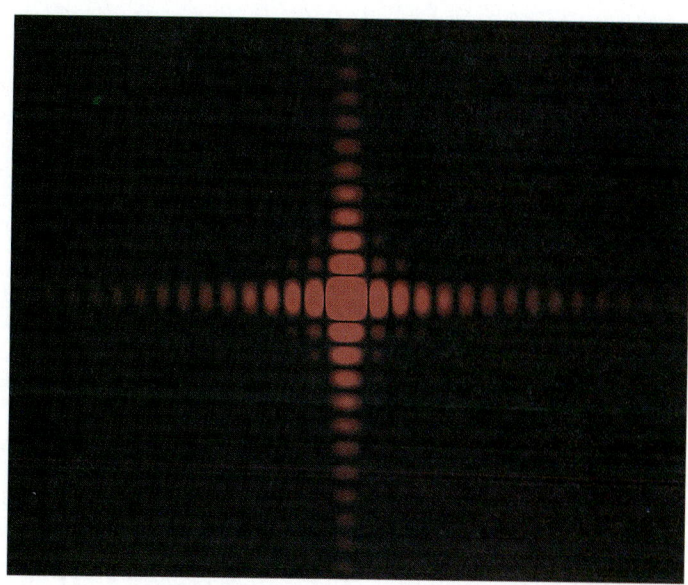

16.5 Reflection and Refraction of Sound and Light   547
   16.5.1 Plane Waves   547
   16.5.2 Reflection of Plane Waves   548
   16.5.3 Refraction of Plane Waves   549
   16.5.4 Total Internal Reflection   550
   16.5.5 Dispersion   552

CHAPTER SUMMARY 553 • SOLUTIONS TO EXERCISES 554
BASIC SKILLS 555 • QUESTIONS AND PROBLEMS 556

CHAPTER 17 Interference and Diffraction   562

17.1 Interference   563
   17.1.1 Superposition of Two Harmonic Wave Functions   563
   17.1.2 Beats   564
   DIGGING DEEPER: ENERGY REDISTRIBUTION IN BEATS   566
   DIGGING DEEPER: PHASE SPEED AND GROUP SPEED   566
   17.1.3 Interference Between Two Spatially Separated Wave Sources   567
   17.1.4 Coherence   570
   17.1.5 Thin-Film Interference   571
17.2 The Michelson Interferometer and the Michelson–Morley Experiment   573
17.3 Diffraction   576
   17.3.1 The Rectangular Aperture   577
   17.3.2 The Circular Aperture   578
   17.3.3 Resolution   578
   17.3.4 Combined Diffraction and Interference: The Double Slit   579
17.4 Intensity in Interference Patterns   580
   17.4.1 Energy Redistribution in the Two-Slit Interference Pattern   580
   DIGGING DEEPER: CLOSELY SPACED SOURCES   581
   17.4.2 Phasors   581
   17.4.3 Interference of Multiple Sources   583
   17.4.4 Gratings   585
   17.4.5 Intensity in Diffraction Patterns   586
17.5 X Ray Diffraction   587

CHAPTER SUMMARY 589 • SOLUTIONS TO EXERCISES 590
BASIC SKILLS 591 • QUESTIONS AND PROBLEMS 592

CHAPTER 18 Geometrical Optics   597

18.1 Images Formed by Plane Surfaces   598
   18.1.1 Images in a Plane Mirror   599
   18.1.2 Objects, Images and Sign Conventions   600
   18.1.3 Images Formed by Plane Refracting Surfaces   602
18.2 Images Formed by Curved Surfaces   603
   18.2.1 Spherical Mirrors   603
   DIGGING DEEPER: WAVEFRONTS   606
   18.2.2 Spherical Refracting Surfaces   607
18.3 Lenses   608
   18.3.1 Optical Surfaces in Series   608
   18.3.2 Thin Lenses   610
18.4 Ray Tracing and Images of Extended Objects   612
   18.4.1 Images in Plane Mirrors   612
   DIGGING DEEPER: LEFT, RIGHT, AND PARITY   613
   18.4.2 Images in Curved Mirrors   613
   18.4.3 Images Formed by Thin Lenses   615
   18.4.4 Visibility of Images   617
18.5 Compound Optical Systems   618
   18.5.1 The Eye   618
   18.5.2 The Simple Magnifier   618
   18.5.3 Microscopes   619
   18.5.4 Telescopes   620
      Study Problem 12: F-Stops   621
   18.5.5 Astronomical Telescopes—Reflectors   622
18.6 Aberrations   624

CHAPTER SUMMARY 625 • SOLUTIONS TO EXERCISES 626
BASIC SKILLS 627 • QUESTIONS AND PROBLEMS 628

Essay 5: Ray Tracing Using a Computer   634
(Dr. Michael Lampton)

PART FOUR: PROBLEMS   639

PART FIVE THERMODYNAMICS

CHAPTER 19 Temperature and Thermal Energy 643

19.1 Temperature 644
- 19.1.1 Thermodynamic Systems 644
- 19.1.2 Thermal Equilibrium and the Definition of Temperature 645
- 19.1.3 Thermometers and Temperature Scales 646
- Digging Deeper: Why a Degree? 647

19.2 Temperature in an Ideal Gas 649
- 19.2.1 The Ideal Gas Law 649
- 19.2.2 Molecules and Moles 651
- 19.2.3 The Relationship Between Temperature and Internal Kinetic Energy in an Ideal Gas 651
- 19.2.4 Internal energy of a Monatomic Ideal Gas 652

19.3 The First Law of Thermodynamics 653
- 19.3.1 Heat, Work, and the first Law of Thermodynamics 653
- 19.3.2 Work Done by an Ideal Gas 655

19.4 The *P-V* Diagram 656
- 19.4.1 Representation of a Thermodynamic State in a *P-V* Diagram 656
- 19.4.2 Representation of Processes in the *P-V* Diagram 656

19.5 Specific Heats of an Ideal Gas 658
- 19.5.1 Heat Transfer and Specific Heat 658
- 19.5.2 Specific Heat of a Monatomic Ideal Gas at Constant Volume 658
- 19.5.3 Specific Heat of a Monatomic Ideal Gas at Constant Pressure 659
- 19.5.4 The Relation Between c_p and c_v 660

19.6 Adiabatic Processes in an Ideal Gas 661
- Study Problem 13: A Cycle in the *P-V* Diagram 663

19.7 ✲ Equipartition of Energy 665
- 19.7.1 Modes of Energy Storage 665
- 19.7.2 Energy Stored in Each Mode 666
- 19.7.3 Specific Heats of Polyatomic Ideal Gases 666
- 19.7.4 Brownian Motion 667

Chapter Summary 668 ● Solutions to Exercises 669
Basic Skills 670 ● Questions and Problems 671

CHAPTER 20 Thermodynamics of Real Substances 676

20.1 The Behavior of Real Gases 677
- 20.1.1 The van der Waals Equation of State 677
- Digging Deeper: Molecular Forces 677
- 20.1.2 Isotherms of the van der Waals Equation 679

20.2 Change of Phase 680
- 20.2.1 Phase Transitions at Constant Pressure 680
- Digging Deeper: Why Is Temperature Constant During a Phase Transition? 681
- 20.2.2 Latent Heat 681

20.3 Thermal Expansion 683
- 20.3.1 Linear Expansion 683
- Digging Deeper: Thermal Expansion of Solids 684
- 20.3.2 Volume Expansion 684
- Digging Deeper: How Do Fish Survive the Winter? 686
- Study Problem 14: *A Thermal Switch* 686

20.4 Calorimetry 687
- 20.4.1 Specific Heat of Real Substances 687
- Digging Deeper: The Calorie 688
- 20.4.2 Heat Capacity 689
- 20.4.3 Calorimetry 689

20.5 Thermometry 690
- 20.5.1 Temperature Standards 690
- 20.5.2 Practical Thermometers 691

Chapter Summary 692 ● Solutions to Exercises 693
Basic Skills 694 ● Questions and Problems 694

Essay 6: Low Temperatures and their Measurement 698
(J. M. Lockhart)

CHAPTER 21 Heat Transfer 700

21.1 Conduction 701
- 21.1.1 Heat Transfer Along a Rod 701
- Digging Deeper: Why is Conductive Heat Flux Proportional to the Temperature Gradient? 703
- 21.1.2 Steady State Heat Flow 703
- 21.1.3 Thermal Resistance 705

21.2 Convection 707
- Digging Deeper: Conditions for Convection 708

21.3 Radiation 709
- 21.3.1 The Nature of Thermal Radiation 709
- 21.3.2 Radiation and Heat Transport 710

21.4 An Empirical Approach to Heat Transfer: Newton's Law of Cooling 711

Chapter Summary 712 ● Solutions to Exercises 713
Basic Skills 714 ● Questions and Problems 714

CHAPTER 22 Entropy and the Second Law of Thermodynamics 718

- 22.1 Why is a Second Law of Thermodynamics Necessary? 719
- 22.2 Heat Engines 720
  - 22.2.1 Efficiency of Engines 720
  - 22.2.2 The Otto Cycle 721
- 22.3 The Carnot Cycle 723
  - 22.3.1 A Reversible Cycle 723
  - 22.3.2 Refrigerators 724
- 22.4 Entropy 726
  - 22.4.1 The Carnot Cycle and Entropy Change 726
  - 22.4.2 Entropy Change in an Arbitrary Process 727
  - 22.4.3 Entropy as a State Variable 728
  - 22.4.4 Entropy Change and Reversibility 729
- 22.5 A Limit on Efficiency 732
- 22.6 The Significance of Absolute Zero 733
  - 22.6.1 The Thermodynamic Temperature Scale 733
  - 22.6.2 Absolute Zero 734
- 22.7 ✷ Statistical Mechanics 734
  - 22.7.1 The Boltzmann Factor 734
  - DIGGING DEEPER: WHAT IS A DISTRIBUTION? 736
  - 22.7.2 The Maxwell–Boltzmann Distribution 736
  - 22.7.3 Mean Free Path and the Establishment of Equilibrium 739
  - DIGGING DEEPER: THE DRUNKARD'S WALK 741
  - 22.7.4 A Microscopic View of Entropy 741
  - 22.7.5 Entropy and Equilibrium 742

CHAPTER SUMMARY 743 • SOLUTIONS TO EXERCISES 744
BASIC SKILLS 745 • QUESTIONS AND PROBLEMS 746

Essay 7: Entropy, Evolution, and the Arrow of Time 751

PART FIVE: PROBLEMS 753

PART SIX ELECTROMAGNETIC FIELDS

Overview of Electromagnetism 755

- VI.1 Magnetic Field 755
- VI.2 Electric Charge 757
- VI.3 The Electrical Structure of Matter 759
- VI.4 Electric Field 760
- VI.5 Moving Charge as the Source of Magnetic Field 760
- VI.6 Magnetic Force on Moving Charges 761
- VI.7 SI Units for Charge and Current 762
- VI.8 Unity of the Electromagnetic Field 762
- VI.9 Electromagnetic Waves 762

CHAPTER 23 Charge and the Electric Field 764

- 23.1 Electric Charge 765
  - 23.1.1 Charge and Matter 765
  - 23.1.2 The Forces Between Charges 766
  - 23.1.3 The Strength of the Electric Force 767
  - 23.1.4 Triboelectricity 768
  - 23.1.5 Conductors and Insulators 768
- 23.2 The Electric Field of a Point Charge 769
- 23.3 The Principle of Superposition 771
  - 23.3.1 Superposition of Fields at a Point 771
  - DIGGING DEEPER: MEASURING THE ELECTRON CHARGE 772
  - 23.3.2 Field Line Diagrams 773
  - 23.3.3 Calculation of Fields as a Function of Position 775
    Study Problem 15: *Two Unequal Charges* 776
- 23.4 Gauss' Law 777
  - 23.4.1 The Relation Between Charge and Field Lines 777
  - 23.4.2 Electric Flux 779
  - DIGGING DEEPER: FLUX 779
  - 22.4.3 Gauss' Law for the Electric Field 781
  - DIGGING DEEPER: FORMAL PROOF OF GAUSS' LAW 782

CHAPTER SUMMARY 783 • SOLUTIONS TO EXERCISES 784
BASIC SKILLS 785 • QUESTIONS AND PROBLEMS 786

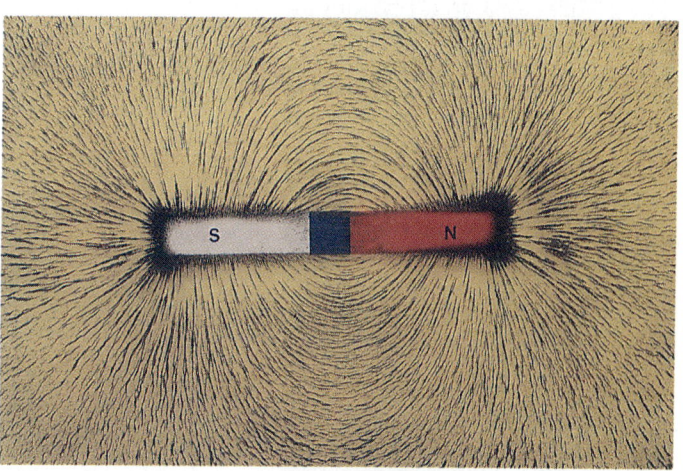

CHAPTER 24 Static Electric Fields 790

24.1 Using Gauss' Law to Calculate Electric Field 791
24.2 The Electric Field Due to a Linear Charge Distribution 793
    24.2.1 An Infinitely Long, Uniformly Charged Filament 793
    24.2.2 A Finite, Uniformly Charged Filament 794
    24.2.3 Comparison of Exact and Approximate Calculations 796
    24.2.4 More Complicated Linear Charge Distributions 796
24.3 Electric Field Due to Surface and Volume Charge Distributions 797
    24.3.1 Surface Charge Distributions 797
    24.3.2 Volume Charge Distributions 799
24.4 Motion of Charges in an Electric Field 800
24.5 The Dipole 802

CHAPTER SUMMARY 805 • SOLUTIONS TO EXERCISES 806
BASIC SKILLS 807 • QUESTIONS AND PROBLEMS 808

CHAPTER 25 Electric Potential Energy 813

25.1 Potential Energy of a Pair of Point Charges 814
    25.1.1 Work Done by the Coulomb Force 814
    25.1.2 The Electric Force as a Conservative Force 815
    25.1.3 Potential Energy of a Pair of Charges 815
25.2 Electric Potential 817
    25.2.1 Potential Energy of a Charge in an Arbitrary Electric Field 817
    25.2.2 Electric Potential 817
    25.2.3 Calculation of Field from Potential 820
25.3 Potential Energy of a System of Charges 821
    25.3.1 The Principle of Superposition 821
    25.3.2 Potential Energy of Systems of Charges 822
    Study Problem 16: *The Collapsing Square* 824
25.4 Equipotential Surfaces 825
    25.4.1 The Relation Between Equipotential Surfaces and Field Lines 825
    25.4.2 Equipotential Surfaces for a System of Point Charges 826
    MATH TOPIC: ELECTRIC FIELD AS THE GRADIENT OF POTENTIAL 826

25.5 Potential Due to a Continuous Distribution of Charge 828
25.6 The Behavior of Conductors 829
    25.6.1 Response of a Conductor to an Electric Field 829
    25.6.2 Conductors and Electric Potential 831
    25.6.3 The Relation Between Field and Surface Charge Density on a Conductor 832
    25.6.4 The Relation Between the Shape of a Conductor and the Electric Field at Its Surface 832

CHAPTER SUMMARY 834 • SOLUTIONS TO EXERCISES 835
BASIC SKILLS 836 • QUESTIONS AND PROBLEMS 837

CHAPTER 26 Introduction to Electric Circuits 843

26.1 Basic Circuit Behavior 844
    26.1.1 Electric Current 844
    26.1.2 Batteries and Electromotive Force 844
    26.1.3 Resistance 846
    26.1.4 Energy Relations in a Simple Circuit 847
    26.1.5 Safety Considerations 848
26.2 A Model for Current and Resistance 849
26.3 Series and Parallel Combinations of Resistors 853
    26.3.1 Resistors in Series 853
    DIGGING DEEPER: NONOHMIC DEVICES 854
    26.3.2 Resistors in Parallel 856
    26.3.3 Combined Series and Parallel Circuits 857
26.4 Kirchhoff's Rules 858
    Study Problem 17: *A Compound Circuit* 860
26.5 ✹ Electrical Measurement 862
    26.5.1 An Ammeter 862
    26.5.2 Voltmeters 863
    26.5.3 The Wheatstone Bridge 864

CHAPTER SUMMARY 865 • SOLUTIONS TO EXERCISES 866
BASIC SKILLS 867 • QUESTIONS AND PROBLEMS 868

CHAPTER 27 Capacitance and Electrostatic Energy 875

27.1 Capacitance 876
    27.1.1 The Parallel Plate Capacitor 876
    27.1.2 Calculating Capacitance 877
    DIGGING DEEPER: GROUNDING A DC CIRCUIT 879
27.2 Energy Storage in Capacitors 880
    27.2.1 Charging a Capacitor 880
    27.2.2 Capacitors in Parallel 881
    27.2.3 Capacitors in Series 882
    27.2.4 Series and Parallel Combinations 882
27.3 Dielectrics and Practical Capacitors 884
    27.3.1 The Dielectric Constant 884
    27.3.2 ✹ Polarization and Susceptibility 887
    27.3.3 ✹ Electric Displacement 888
27.4 Energy in the Electric Field 889
    27.4.1 Electrostatic Energy Density in Vacuum 889
    27.4.2 ✹ Electrostatic Energy of Two Point Charges 890
    27.4.3 ✹ The Classical Electron Radius and Renormalization 891

CHAPTER SUMMARY 892 • SOLUTIONS TO EXERCISES 893
BASIC SKILLS 894 • QUESTIONS AND PROBLEMS 894

CHAPTER 28 Static Magnetic Fields 898

- 28.1 Magnetic Force 899
- 28.2 Current as the Source of Magnetic Fields 901
  - 28.2.1 The Biot-Savart Law 901
  - 28.2.2 The Magnetic Field Produced by a Straight Wire Segment 902
  - 28.2.3 The Magnetic Field of Loops and Coils 903
  - DIGGING DEEPER: MAGNETIC MOMENT OF AN ARBITRARY PLANAR LOOP 906
- 28.3 Integral Laws for Static Magnetic Fields 907
  - 28.3.1 Gauss' Law for the Magnetic Field 907
  - 28.3.2 Circulation and Ampère's Law 907
  - DIGGING DEEPER: DEMONSTRATION OF AMPÈRE'S LAW 909
  - 28.3.3 Finding Magnetic Fields with Ampère's Law 910
  - 28.3.4 Summary of the Integral Laws for Static Fields 914
    Study Problem 18: *An Electron Beam* 914

CHAPTER SUMMARY 916 • SOLUTIONS TO EXERCISES 917
BASIC SKILLS 918 • QUESTIONS AND PROBLEMS 919

CHAPTER 29 Static Magnetic Fields: Applications 923

- 29.1 Motion of Charged Particles in a Magnetic Field 924
  - 29.1.1 Motion Perpendicular to a Uniform Magnetic Field 924
  - 29.1.2 Practical Applications of Circular Particle Motion 925
  - 29.1.3 Motion in Combined Electric and Magnetic Fields 927
  - DIGGING DEEPER: MORE ON CYCLOTRONS 928
- 29.2 Forces on Current-Carrying Wires 931
  - 29.2.1 Force on a Wire Segment 931
  - 29.2.2 Force and Torque on Current Loops 934
  - DIGGING DEEPER: MAGNETIC FORCES AND NEWTON'S THIRD LAW 936
- 29.3 The Hall Effect 937
- 29.4 Magnetic Materials 938
  - 29.4.1 Atomic Model of Magnetization 938
  - DIGGING DEEPER: MAGNETIC MOMENT AND ANGULAR MOMENTUM 941
  - 29.4.2 ✱ The Magnetic Field Intensity $\vec{H}$ 941
  - DIGGING DEEPER: MAGNETIC RESONANCE IMAGING 943

CHAPTER SUMMARY 943 • SOLUTIONS TO EXERCISES 944
BASIC SKILLS 945 • QUESTIONS AND PROBLEMS 947

PART SIX: PROBLEMS 952

PART SEVEN ELECTRODYNAMICS

CHAPTER 30 Dynamic Fields 955

- 30.1 Induced EMF 956
  - 30.1.1 Faraday's Law 956
  - 30.1.2 Lenz's Law 958
  - 30.1.3 Induced Electric Field 959
  - 30.1.4 Sign Conventions 960
- 30.2 Motional EMF 961
  - 30.2.1 EMF in Circuits with Moving Boundaries 961
  - 30.2.2 Generators and Motors 962

- 30.3 The Nature of EMF 964
  - 30.3.1 Potential Difference and EMF 964
  - 30.3.2 ✱ Mathematical Properties of EMF and Potential Difference 965
  - 30.3.3 ✱ EMF and Choice of Reference Frame 966
  - DIGGING DEEPER: MAGNETIC FORCE AND EMF 966
- 30.4 Calculation of Induced Electric Field 967
  - DIGGING DEEPER: THE COMPLETE MATHEMATICAL STATEMENT OF FARADAY'S LAW 967
    - Study Problem 19: The Betatron 970
- 30.5 Eddy Currents 973
  - DIGGING DEEPER: FORCES DUE TO EDDY CURRENTS 974
- 30.6 The Ampère–Maxwell Law 975
  - DIGGING DEEPER: CONTINUITY OF TOTAL CURRENT 976

CHAPTER SUMMARY 977 ● SOLUTIONS TO EXERCISES 978
BASIC SKILLS 979 ● QUESTIONS AND PROBLEMS 980

CHAPTER 31 Introduction to Time-Dependent Circuits 987

- 31.1 Resistor-Capacitor Circuits 988
  - 31.1.1 Discharging a Capacitor 988
  - MATH TOOLBOX: SIGN CONVENTIONS 990
  - 31.1.2 The Solution Method 991
  - 31.1.3 Charging a Capacitor 992
- 31.2 Inductance 993
  - 31.2.1 Self-Inductance 993
  - 31.2.2 Energy Storage in an Inductor 995
  - 31.2.3 Mutual Inductance 997
- 31.3 Inductor Circuits 999
  - 31.3.1 The LR Circuit 999
  - 31.3.2 The LC Circuit 1001
  - 31.3.3 The LRC Circuit 1003
  - MATH TOOLBOX: HOW TO SOLVE A LINEAR DIFFERENTIAL EQUATION 1005
- 31.4 ✱ Multiloop Circuits 1006

CHAPTER SUMMARY 1007 ● SOLUTIONS TO EXERCISES 1008
BASIC SKILLS 1009 ● QUESTIONS AND PROBLEMS 1010

CHAPTER 32 Introduction to Alternating Current Circuits 1016

- 32.1 Single-Element Circuits 1017
  - 32.1.1 Voltage and Current 1017
  - 32.1.2 Power 1019
  - 32.1.3 Reactance and Phase Shift 1020
- 32.2 Two-Component Circuits 1021
  - 32.2.1 Steady State Response 1021
  - 32.2.2 Power 1024
  - 32.2.3 Transient Response 1025
- 32.3 Circuit Analysis Using Phasors 1025
  - 32.3.1 Phasors 1025
  - 32.3.2 Phasor Representation of a Series Circuit 1025
  - 32.3.3 Phasor Representation of a Parallel Circuit 1027
- 32.4 The LRC Circuit 1029

CHAPTER SUMMARY 1032 ● SOLUTIONS TO EXERCISES 1033
BASIC SKILLS 1034 ● QUESTIONS AND PROBLEMS 1035

CHAPTER 33 Electromagnetic Waves 1039

- 33.1 Plane Electromagnetic Waves 1040
  - 33.1.1 Origin and Structure of a Plane EM Wave 1040
  - 33.1.2 The Wave Equation for $\vec{E}$ and $\vec{B}$ 1041
- 33.2 Energy and Momentum Transport by EM Waves 1044
  - 33.2.1 Energy Density and the Poynting Vector 1044
  - 33.2.2 Momentum Density and Radiation Pressure 1046
  - DIGGING DEEPER: THE MOMENTUM OF LIGHT 1047
  - 33.2.3 Energy Transport in Circuits 1048
  - DIGGING DEEPER: OBLIQUE INCIDENCE 1049
- 33.3 Polarization 1050
  - 33.3.1 Linear Polarization 1050
  - 33.3.2 Polarization by Reflection 1053
  - DIGGING DEEPER: POLARIZATION IN NATURE 1054
  - 33.3.3 Circular Polarization 1056
- 33.4 ✱ Electromagnetic Oscillations and Microwaves 1057
  - 33.4.1 Cavity Oscillators 1057
  - 33.4.2 Waveguides 1058
  - DIGGING DEEPER: SUPERPOSITION OF REFLECTING WAVES 1060

CHAPTER SUMMARY 1060 ● SOLUTIONS TO EXERCISES 1061
BASIC SKILLS 1062 ● QUESTIONS AND PROBLEMS 1063

PART SEVEN: PROBLEMS 1066

PART EIGHT
TWENTIETH-CENTURY PHYSICS

CHAPTER 34 Relativity and Space-Time 1069

- 34.1 Special Relativity 1070
  - 34.1.1 What is a Relativity Theory? 1070
  - 34.1.2 Einstein's Postulates 1072
  - 34.1.3 Time Dilation 1073
  - 34.1.4 Length Contraction 1074
  - 34.1.5 Simultaneity 1075
- 34.2 Space-time 1077
  - 34.2.1 Representation of Space-time in a Single Reference Frame 1077
  - DIGGING DEEPER: DEFINING COORDINATES IN A REFERENCE FRAME 1078
  - 34.2.2 Space-time Interval 1079
- 34.3 The Lorentz Transformation 1081
  - 34.3.1 Coordinate Transformation 1081
  - DIGGING DEEPER: DERIVATION OF THE LORENTZ TRANSFORMATION 1082
    - Study Problem 20: *The Student's Revenge* 1083
  - 34.3.2 Velocity Transformation 1085
  - 34.3.3 Acceleration in Special Relativity 1087
- 34.4 Relativistic Dynamics 1088
  - 34.4.1 Momentum 1088
  - DIGGING DEEPER: RELATIVISTIC MOMENTUM 1089
  - 34.4.2 Mass and Energy 1090
  - 34.4.3 The Energy-Momentum Invariant 1092

CHAPTER SUMMARY 1094 • SOLUTIONS TO EXERCISES 1096
BASIC SKILLS 1097 • QUESTIONS AND PROBLEMS 1098

Essay 8: General Relativity: A Geometric Theory of Gravity 1103

CHAPTER 35 Light and Atoms 1110

- DIGGING DEEPER: THE ORIGINS OF THE QUANTUM IDEA 1111
- 35.1 Photons 1112
  - 35.1.1 The Photoelectric Effect 1112
  - 35.1.2 The Compton Effect 1116
  - 35.1.3 ✱ The Planck Radiation Law 1118
- 35.2 Bohr's Atomic Model 1119
  - 35.2.1 The Structure of Atoms 1119
  - 35.2.2 Balmer's Spectrum and Bohr's Atom 1121
  - 35.2.3 The Correspondence Principle 1126
- 35.3 Electron Waves 1126
  - 35.3.1 De Broglie's Hypothesis 1126
  - 35.3.2 Schrödinger's Picture of the Hydrogen Atom 1128
  - 35.3.3 The Pauli Exclusion Principle and Chemistry 1130
- 35.4 Quantum Mechanics 1132
  - 35.4.1 The Heisenberg Uncertainty Principle 1132
  - 35.4.2 The Meaning of the Wave Function 1133

CHAPTER SUMMARY 1135 • SOLUTIONS TO EXERCISES 1136
BASIC SKILLS 1137 • QUESTIONS AND PROBLEMS 1138

Essay 9: The Scanning Tunneling Microscope 1142
(Shirley Chiang)

CHAPTER 36 Atomic Nuclei 1145

- 36.1 Basic Nuclear Structure 1146
  - 36.1.1 Charge and Mass 1146
  - 36.1.2 The Size of Nuclei 1149
  - 36.1.3 Nucleons 1150
  - 36.1.4 Nuclear Forces 1151
  - DIGGING DEEPER: WHY THE NEUTRON IS NECESSARY 1151
  - 36.1.5 Binding Energy 1152

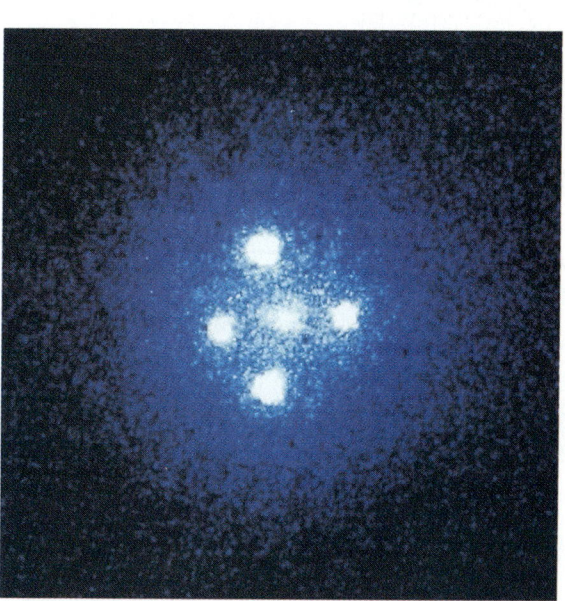

DIGGING DEEPER: WE ARE THE CHILDREN OF THE STARS 1168
36.3.3 Fission 1169

CHAPTER SUMMARY 1171 • SOLUTIONS TO EXERCISES 1172
BASIC SKILLS 1172 • QUESTIONS AND PROBLEMS 1173

CHAPTER 37 Particle Physics 1177

37.1 Particle Creation and Fundamental Forces 1178
    37.1.1 Creation and Destruction 1178
    37.1.2 Virtual Particles and Fundamental Forces 1179
    37.1.3 Feynman Diagrams 1179
    37.1.4 The π Meson 1180
    37.1.5 Isospin 1181
37.2 Subnuclear Particles and the Quark Model 1182
    37.2.1 The Population Explosion 1182
    37.2.2 Strangeness and Quarks 1183
    37.2.3 Proton Structure and the Reality of Quarks 1186
37.3 The Standard Model 1187
    37.3.1 Electroweak Unification 1187
    37.3.2 Quantum Chromodynamics 1189
    37.3.3 Conservation Laws for the Strong and Weak Forces 1190
    37.3.4 Limitations of the Standard Model 1191
37.4 Characteristics of Modern Particle Theories 1192
    37.4.1 Symmetries and Groups 1192
    MATH TOPIC: GROUP PROPERTIES ILLUSTRATED BY AN EXAMPLE 1192
    37.4.2 Renormalization 1193
    37.4.3 Spontaneous Symmetry Breaking 1193

CHAPTER SUMMARY 1194 • SOLUTIONS TO EXERCISES 1195
BASIC SKILLS 1195 • QUESTIONS AND PROBLEMS 1196

PART EIGHT: PROBLEMS 1197

Epilogue 1199

Appendix I Mathematics A-1

Appendix II Symbols A-7

Appendix III Solutions to Selected Problems A-11

Index I-1

36.2 Natural Radioactivity 1153
    36.2.1 Conservation Laws and Quantum Numbers 1153
    36.2.2 α Decay 1154
    36.2.3 β Decay 1156
    36.2.4 Antiparticles and Positron (β^+) Decay 1157
    DIGGING DEEPER: NEUTRINOS 1158
    36.2.5 γ Decay 1160
    36.2.6 The Law of Decay 1161
    DIGGING DEEPER: γ-RAY IMAGING 1163
    36.2.7 Radioactive Series 1165
36.3 Nuclear Reactions 1166
    36.3.1 Transmutation by Neutron Bombardment 1166
    36.3.2 Energy Generation in Stars 1167

> *... just for the fun of doing Physics.*
> MARIA GOEPPERT-MEYER

> *To see the world for a moment as something rich and strange is the private reward of many a discovery.*
> EDWARD M. PURCELL

> *... If there turn out to be any practical applications, that's fine and dandy. But we think it's important that the human race understands where sunlight comes from.*
> WILLIAM FOWLER

PROLOGUE

WHY DO PHYSICS?

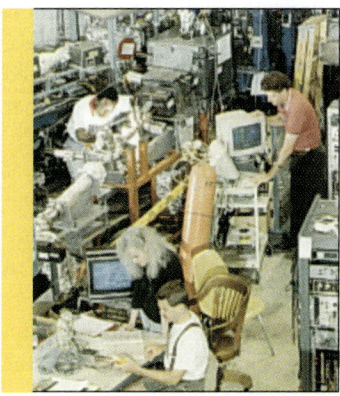

Reasons for doing physics are nearly as diverse as the people who do it. For the professional, the challenge of teasing secrets from nature is a calling, an exciting occupation, and often a source of profound personal satisfaction. Physicists often view their discoveries as major additions to human culture, not unlike great symphonies or epic poems. Physics is, at the same time, a very practical science—basic to the design of your alarm clock, the computer that handles your bank account, and whatever transportation system gets you to work and school. Most students take a physics course because of this practical aspect.

Every citizen in a modern society needs to have some scientific understanding. The scientific way of thinking about our world has become an integral part of modern culture, interwoven with theories of politics and justice and with the economic structure of our society. Most scientists believe that a scientific worldview liberates the mind and that technological progress will continue to be beneficial. Critics of science argue that it diminishes traditional, humanistic ways of thinking without offering a valid, alternative view, and that technology has left us with problems of pollution, atomic bombs, global warming, to name a few. As both an individual and a citizen, you will need to judge these issues for yourself, and this physics course offers an introduction to the necessary scientific reasoning.

Whatever your reason for studying physics, you will acquire powerful skills that you can use professionally as well as in developing your personal philosophy. Perhaps you will also come to share some of the physicist's deep fascination with the beauty and logic of the universe *and* to enjoy solving its puzzles. Welcome to the enterprise!

SO, WHAT IS PHYSICS?

The name is derived from an ancient Greek word meaning *the nature of things that move of themselves.* Through physics, we strive to discover the fundamental structure of the universe and the rules by which it operates. This structure turns out to be both *simple* and *complex!* It is *simple* because only a small number of rules are needed to explain the world around us. It is *complex* because of the large numbers of objects that interact. We have a good set of rules for the behavior of everyday objects and can understand those rules in terms of atoms and only two kinds of interaction. For atoms we have yet deeper *levels of description* that involve three kinds of interaction. We're pretty sure we haven't reached bottom yet!

Occasionally someone (who should know better) declares to the world that we now know it all . . . then someone else will discover nuclear energy, or semiconductors, or lasers! Physics is a dynamic subject, and physicists continually test the limits of current ideas, probe for exciting new phenomena, attempt to explain puzzling phenomena when they are discovered, and strive to create new ideas that provide deeper or more wide-ranging explanations. The fun of science is in this dynamic quest.

New scientific ideas, as Einstein put it, are "free creations of the human mind," as fresh and unpredictable as any other creative endeavor. But any theoretical picture must be consistent with the actual behavior of the world. So, scientific ideas experience evolutionary pressure as intense as do biological species in a jungle—with similar results: some stable, well-adapted, broad-ranging ideas thrive, while certain variant ideas test the limits of survival. Most of the variants become extinct, but occasionally one proves highly adaptive, takes over the whole environment, and establishes a new level of description. Physics research is the process of creativity, skepticism, and competition that drives this evolution.

Good ideas, unlike dinosaurs, don't always become fossils when a new one takes over their habitat. Most often, a long-lived old idea remains the easiest to learn and use where it is valid, even though it is recognized as a *special case* of the newer and more penetrating idea. For example, a mechanical engineer works almost entirely with mechanical principles obeyed by everyday objects; a metallurgist uses atomic physics to develop stronger metals. Neither would probably ever work with subatomic physics or the theory of relativity.

So, what *is* physics? It is at least three things: a set of ideas describing the universe at various levels of detail; a set of methods for using these ideas to understand the world about us; and a dynamic, evolutionary process for testing, extending, and refining those ideas and methods. The study of physics calls on us to employ a peculiar way of thinking—that of viewing familiar events as the sum of many parts, each governed by the principles of physics and interacting with one another. The term *natural philosophy*, used until recently in Britain, describes physics well: it is a method that has evolved for thinking successfully about the natural world.

What Are the Aims of this Text?

Fortunately, you don't need to master the whole of physics to achieve your purposes in this introductory course. Our main aim here is to help you learn how to become a *natural philosopher*—to understand the structure of physics and to be able to apply it to the world. Like most introductory physics texts, we shall work primarily with classical physics. These ideas, developed largely before 1900, describe most systems on the everyday scale of existence and still find broad application. Though everyday events are familiar and we can study them at a level consistent with your mathematical experience, don't make the error of thinking them trivial. It took 2000 years to get *everyday* physics right, and you will find it a challenge to figure out just how the basic rules work. Once you've met the challenge though, you'll have a method for using physics, for further study of science, or for deciding whether a political candidate takes sound positions on technical issues.

At the beginning of the twentieth century, physicists discovered that phenomena involving strong gravity, objects moving near the speed of light, small numbers of atoms, or low temperature are not well described by classical ideas. The last part of the text introduces you to the modern ideas that have resolved these difficulties and provides a framework for appreciating discoveries at the current frontiers of physics.

We know you will find your study of physics challenging. We hope you will also find it fascinating and rewarding. Good luck!

Suggestions for Using the Text

We have divided the text into eight parts. The chapters in each part form a conceptual unit that will prove useful in organizing your knowledge. We suggest that you read each chapter before attending a lecture on the material. You will understand the lecture better and also be able to ask your instructor about anything that was not clear. Be sure to work the exercises.

Complete solutions are given at the end of the chapter. Peek for hints, but don't just copy them; that doesn't do you much good. The chapter summaries review the major ideas.

The lists of concepts and goals indicate the ideas and methods you should understand after reading the chapter. A wise way to use them is to scan the list as you begin reading so that you know which terms to look for. When you have finished the chapter, go back and be sure you know what each item is about. Then you are ready to tackle the problem set.

The problem set is divided into two parts: *Basic Skills* and *Questions and Problems.* The *Basic Skills* section includes review questions and a basic skill drill. The review questions bring out the main points of the chapter and should be answered with a short quote or paraphrase. The skill drill tests your knowledge of the most fundamental concepts in the chapter. We suggest that you answer all the questions in *Basic Skills,* whether or not your instructor assigns them.

We have provided questions and problems for each chapter section, as well as additional problems for the whole chapter. They are rated according to the following scheme:

CONCEPTUAL ❖
These questions involve primarily verbal and/or graphical discussion. These questions are not necessarily easy!

BASIC ♦
These problems are mostly calculations (more than 10% of the total effort), but ones that involve only a single physical principle from the current chapter.

INTERMEDIATE ♦♦
These problems (except those in the *Additional Problems* category) rely on ideas from the current chapter or ideas encountered so frequently before that they are now taken for granted.

ADVANCED ♦♦♦
Advanced problems may involve subtleties that go beyond the examples and exercises, require more difficult mathematics, take more than one page to complete, or involve ideas from previous chapters. These problems usually involve more than one physical principle.

COMPUTER PROBLEMS are intended to be used with a simple computer program or spreadsheet. Some may be solved graphically, or with a calculator and patience.

CHALLENGE PROBLEMS, at the end of each problem set, require an intricate or subtle argument and/or an expert level of computational skill.

The *Additional Problems* may involve concepts from one or more sections of the chapter, or even from different chapters. The text is divided into eight parts, and you will find a problem set at the end of each part. These problem sets involve material such as might be asked on comprehensive examinations.

THE UNIVERSE: AN OVERVIEW

Small children quickly learn that the world is made up of definite objects with identifiable properties: soft blankets, hard floors, hot water, cold ice. They also learn that certain behaviors are predictable: push your cup off the table and it falls to the floor! As adults, we notice that changes occur because the objects interact with each other. To model this world, we need to classify the kinds of objects that exist and the ways in which they interact. Physicists do this systematically, distilling intuitive experience into a precise and succinct set of ideas, then probing far beyond common experience with carefully designed experiments.

In daily life, we interact with a wide variety of objects more or less similar in size to our own bodies. A description on this scale is completely adequate for a study of mechanics and yields precise methods for problems as diverse as the design of machines or the maneuvering of spacecraft. However, on the everyday scale we find no explanation of why such a huge variety of objects exists or of the reasons for their interactions. Better understanding comes

SEE APPENDIX IA FOR A DISCUSSION OF SCIENTIFIC NOTATION.

from looking at different size scales—different magnifications. For both very large and very small systems, we find a simpler, if weirder, description, although the everyday description, which we shall study first, remains an important and useful approximation. Physicists can now shed light on phenomena with size scales ranging from 10^{-37} meter to 10^{26} meters and can discuss events that occurred as early as 10^{-45} second after a beginning some 10^{10} years ago or that will occur as late as some 10^{100} years in the future. Touring the universe on different length scales will allow us to sample the ideas physicists now use.

The Everyday Scale

■ **New York City.** *We are very familiar with size scales ranging from 1 millimeter to about 10 kilometers—that is, from roughly the size of a grain of sand to the size of a city.*

■ *We are familiar with the sensation of force. Your muscles ache after carrying your physics books around all day. In Part I we'll begin to study the forces we experience in our daily lives.*

■ *A jet aircraft is a good example of modern technology. To build one, you must understand mechanics (Part I), to understand how it flies is an exercise in fluid mechanics (Part III), its engine is a thermodynamic machine (Part V), and plotting its course is an exercise in kinematics (Part I). This picture also shows interesting optical effects due to refraction of sunlight through the jet engine exhaust (Part IV).*

The Solar System

■ At a size scale of 10^7 meters, things begin to look different. The Earth's surface now appears curved. Gravitational attraction by the Earth is the dominant interaction, accelerating the space shuttle in its orbit around the Earth. The Hubble telescope (Part IV), being refurbished on this shuttle mission, offers us a view of the universe we live in.

■ On a scale of 10^9 meters, the Earth's spherical shape is obvious. This view is from Earth's closest natural companion, the Moon, a rocky body similar to Earth but without atmosphere or native life-forms. Both the Earth and the Moon exert gravitational forces on each other and on the spacecraft used to reach the Moon.

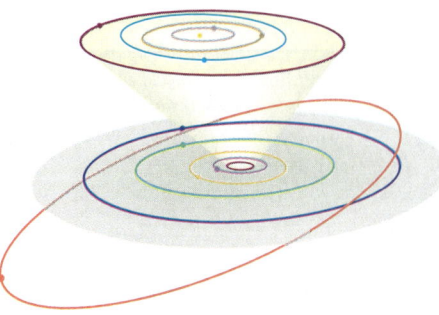

■ At a scale of 10^{12} meters, the Earth has faded into insignificance, and there is no evidence of the magnificent detail we are so familiar with on Earth. The Sun's gravitational pull holds planets, comets, and sundry other debris in orbits that form the solar system. Isaac Newton's study of the solar system led him to discover that every object exerts a gravitational attraction on every other object (Part I).

The Universe of Stars

Outside the solar system, there is no trace of human existence. On very large scales, the interactions between objects are simplified, and a single force—the gravitational force—dominates.

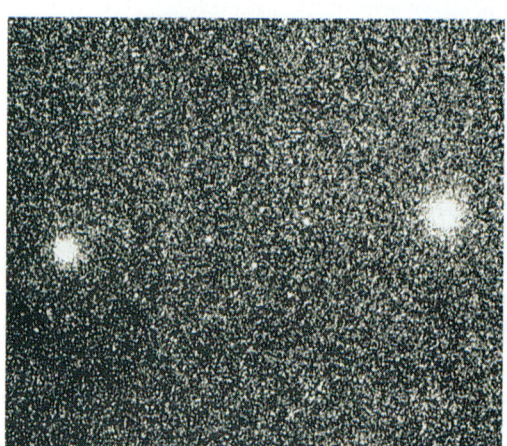

■ The distance to the nearest star is some 30 000 times the size of the solar system. A cube around the Sun with sides of 10^{18} meters contains about 10 000 stars so distant from one another that their individual gravitational attractions have negligible influence on the motions of the other stars.

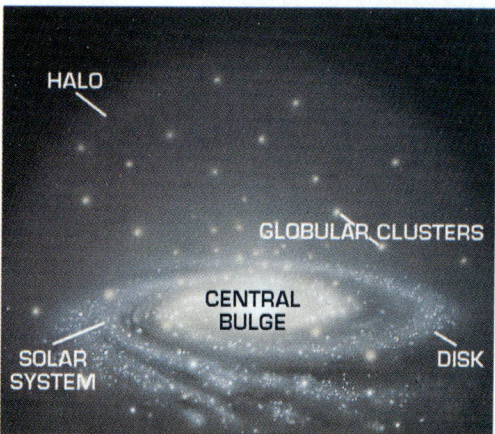

■ We belong to a galaxy of roughly a thousand billion stars called the Milky Way, a system that is about 10^{21} meters in size. All the stars in a galaxy (plus about ten times as much material that we don't see visually) together exert enough gravitational force to hold individual stars in orbit around the center of the galaxy.

■ On a scale of 10^{24} meters, galaxies cluster together, moving under their mutual gravitational attraction.

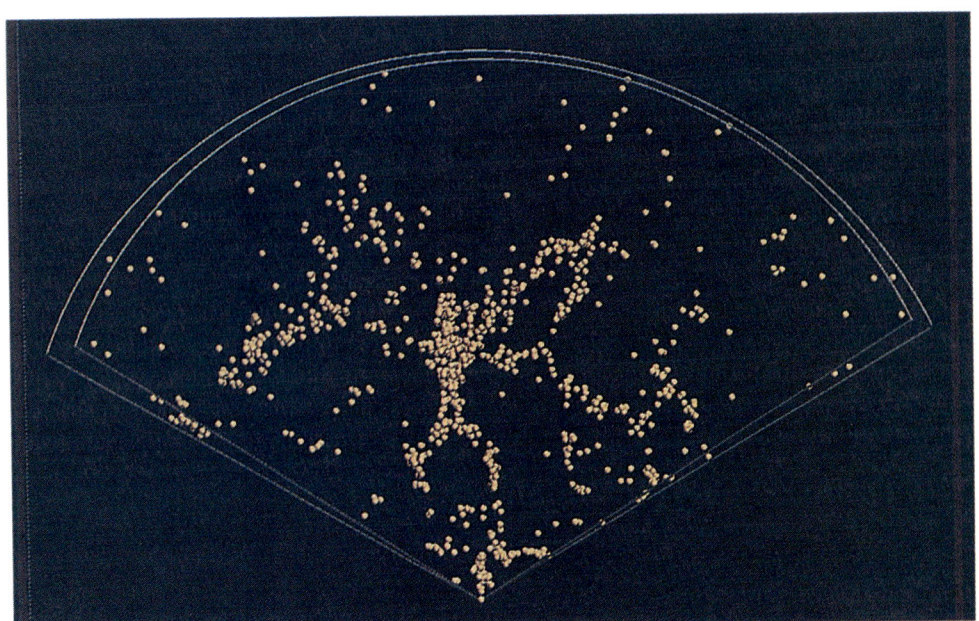

■ Here we see a plot of galaxy positions within a slice of the universe. The distribution is clumpy, but on size scales larger than about 10^{27} meters, the structure seems to average out. This uniform universe is expanding; all the galaxies are rushing away from each other. Albert Einstein's concept of gravity as variations in the geometry of space and time (Part VIII) explains this expansion, but current observations cannot yet determine whether the expansion will stop or continue forever. In the past, the part of the universe we can see must have been much smaller. Cosmological theories suggest that, between 10 and 20 billion years ago, all the stars and galaxies we can see were squeezed into a volume the size of a single atomic nucleus.

The World as Atoms

As with the large scales of astronomy, our description of the world changes radically when we look at very small size scales. Again, we find odd and wonderful things and a small number of fundamental interactions.

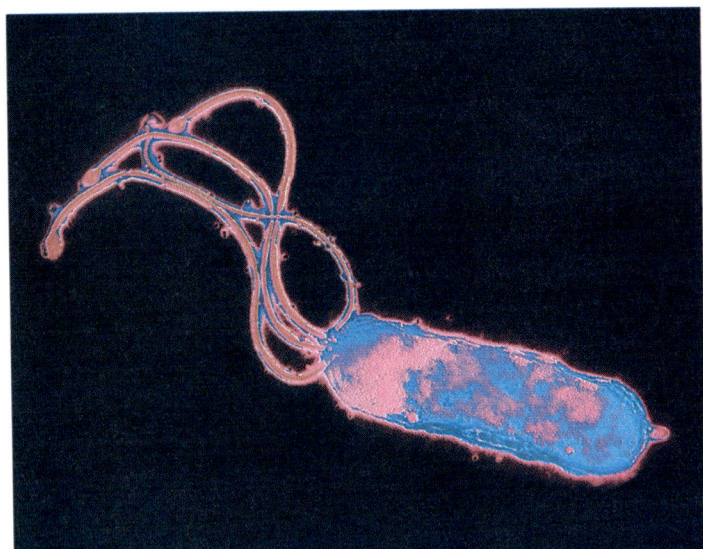

■ Single cells of living creatures are several micrometers (10^{-6} meter) in size. Although we cannot see them with our own eyes, we can still comprehend their behavior with concepts from the everyday world.

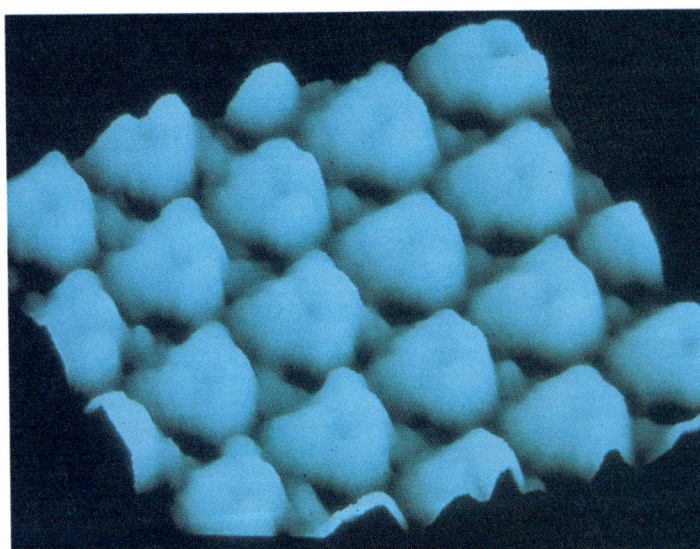

■ At a scale of 10^{-9} meter, we observe the atomic nature of matter. Fluid forces result from collisions between rapidly moving atoms of gas or liquid. Forces between solid bodies in contact result from the forces between individual atoms in the surfaces of the bodies. The atoms consist of electrons, with negative electric charge, surrounding small, positively charged nuclei. Interatomic forces are electromagnetic forces between these charged pieces of the atoms. All of the kinds of force we experience on the everyday scale are either gravitational or result from electromagnetic interactions between atoms (Part VI). This photo shows benzene molecules.

The Subatomic World

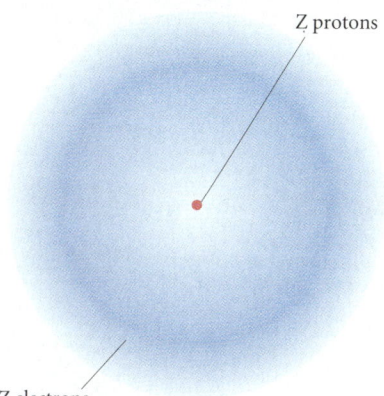

■ Nearly all the mass of an atom is concentrated in the nucleus, pointlike compared with the size of the atom, 10^{-10} meter. The volume of the atom is filled by much less massive electrons, which form a cloud around the nucleus. (Part VIII)

■ On a scale of 10^{-15} meter, we become aware of the electrically neutral neutrons and positively charged protons that comprise the nucleus. The protons repel each other electrically and are held together by the strong nuclear force. A second nuclear force, the weak nuclear force, causes some nuclei to change form.

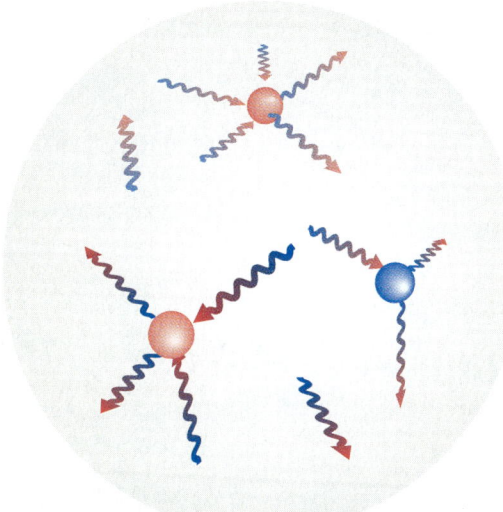

■ Nuclear particles themselves have structure. A proton consists of three particles called quarks, which exert strong nuclear forces on each other by exchanging particles called gluons, and exert electromagnetic forces by exchanging photons. The quarks also exert weak forces through the exchange of particles. In the 1970s, these particles were shown to be cousins of the photons. In this sense, there is but one "electroweak" kind of force, rather than separate electromagnetic and weak nuclear forces. Theorists are now trying to show that the electroweak and strong nuclear forces are just different aspects of one force. Yet more intriguing is the possibility that gravity and this unified force may be aspects of a single interaction. An experimental test of these ideas is far beyond current techniques. Because only these most fundamental particles could exist at the beginning of the universe, the way they behave may be responsible for the way the universe is today. In this way, the smallest and largest scales are intimately connected.

Summary Chart

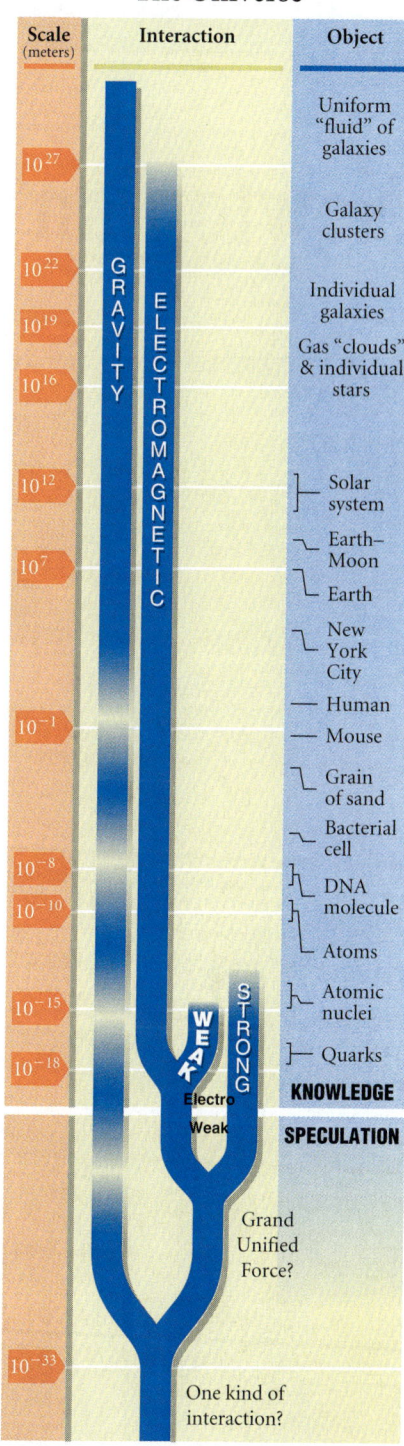

■ The most important forms of material substance are listed for each size scale. For the fundamental types of force, solid lines denote scales at which the force is of major importance. Fuzzy lines indicate scales at which a particular kind of force is present but relatively unimportant.

Newtonian Mechanics

PART ONE

Physical principles now taken for granted in civil or mechanical engineering are based on the work of Isaac Newton in the seventeenth century. Newton's work itself built on a tradition that had been evolving for some 22 centuries, and each of the major achievements in this tradition highlights concepts and methods that a beginning student must also master to understand physics. Our discussion in this part parallels this historical development.

Today physicists describe the world with theories that are tested and applied through careful experimental technique. At the beginning of the seventeenth century, Johannes Kepler's work in astronomy set the standard for mathematical precision in theory. At about the same time, Galileo Galilei used careful experiments to find an accurate description of motion and to show that only change of motion requires explanation. In Chapter 0 we trace the evolution of physics through the work of these first modern physicists.

A complete theory of mechanics includes methods for describing both the motion of objects and the reasons why objects move as they do. Building such a theory on the work of Kepler and Galileo, Newton achieved four major advances: invention of the calculus; a careful definition of force as the cause of change in motion; recognition of mass as the essential mechanical property of objects; and discovery of the universal gravitational attraction. In Chapters 1–3 we discuss concepts of space, time, measurement, and the motion of simple objects. Describing motion involves both graphical and geometrical methods as well as calculus. In Chapter 4 we develop the intuitive models of force and mass that underlie Newtonian physics. Chapters 4 and 5 show how Newtonian theory is used in practical problems. The interlude between Chapters 3 and 4 outlines a very important practical approach to using physics.

When you understand the material in this part, you will have mastered both a powerful way of thinking about the everyday world and some practical methods for predicting a good deal of its behavior.

CHAPTER 0
The Roots of Science

Concepts

Natural law
Substance versus structure
Composition of motion
Perspective
Experiment and theory
Modeling

Goals

Appreciate the goals, methods, and scope of physics.
Be able to apply Kepler's laws and Galileo's law of falling objects.

Astronaut David Scott releases a hammer and a feather on the Moon to demonstrate that they fall from rest side by side.

*If I have seen farther,
it is by standing on the shoulders of giants.*

ISAAC NEWTON

*I*n 1971, during the *Apollo 15* Moon landing, astronaut David Scott dropped a hammer and a feather so television viewers could watch them fall side by side. If you try the same experiment in your backyard, the feather will fall in a complicated fluttering motion, and much more slowly than the hammer. The surrounding air here on Earth has a much bigger influence on the feather than on the hammer and accounts for the complexity of the feather's motion. Inside a cylinder with the air pumped out, the hammer and the feather fall together.

Scott's simple experiment surprised no one. Why then did it rate time on interplanetary television? The experiment links three major ideas developed together at the beginning of the seventeenth century by Galileo Galilei:

- An experiment is the way to obtain definite answers to interesting physical questions.
- Physics on the Moon is the same as physics on the Earth.
- Gravity causes both heavy and light objects to fall from rest in a vacuum at the same rate.

None of these ideas is obvious. Before Galileo, they were considered either unimportant, wrong, or even dangerous! A large proportion of adults in our own society still do not accept them on an intuitive level. Physicists believe them because they have proved valuable in understanding nature and because they have survived numerous experimental tests. The technology developed from these ideas made the trip to the Moon possible.

0.1 Why Study History?

How can a seemingly simple question—how do objects fall?—be one of the deepest questions of science? Albert Einstein spoke of the reason when he described his childhood reaction to playing with a toy compass. The strange behavior of the compass, he concluded, could only be understood in terms of a deep structure underlying reality. At this deep level, a small number of fundamental rules govern the behavior of very many particles of matter and give rise to the complex world we inhabit.

People have had all of history to discover these rules, but a student cannot spend 2000 years mulling over the behavior of hammers and feathers. You have roughly until the first examination! What you need is a sense of the deep structure, ways to decide which facts and methods relate to a particular problem, and techniques for *creating procedures* to solve problems. This is the real challenge in learning physics.

Like any other skill, learning a new method of thought requires both practice and knowing just what to practice. We look back at history to observe how ideas developed and to guide the evolution of our own thinking. Historical figures are important precisely because they overcame major conceptual barriers and showed the rest of us where to go.

0.1.1 Substance and Structure

The philosopher Alfred Whitehead once remarked that societies typically share agreements so basic that they are rarely mentioned, much less questioned. For physicists, the most fundamental such assumption is that the universe is rational and follows consistent rules that people can discover. That nature in and of itself has at least some fixed properties seems obvious because our culture conditions us to think so. But, in a world seemingly full of random, unpredictable activity, people haven't always believed that reliable rules exist; this idea had to be invented.

Belief in order probably evolved from astronomy. Archaeological evidence suggests that people recorded the phases of the Moon as long as 100 000 years ago (■ Figure 0.1). Elaborate prehistoric structures such as Stonehenge in England (■ Figure 0.2) apparently served as astronomical observatories, which allowed the societies that built them to predict changes in the seasons from the orderly sequence of heavenly motions. Stonehenge was built and expanded over a period of three centuries. Its size and complexity show that early peoples placed a high value indeed on astronomy. Though they usually interpreted heavenly bodies as gods who could, by acts of will, control events on Earth, those peoples recognized and *used* the predictable order of the heavens. They realized that the heavens follow *rules*.

■ Figure 0.1
Bone carvings from as early as 100 000 years ago show regular patterns of crescent shapes. Alexander Marshack has argued that these depict the phases of the Moon and form a lunar calendar. The sample shown is thought to be 32 000 years old.

■ Figure 0.2
Stonehenge, a late Stone Age monument in Britain. Sight lines between pairs of stones point toward sunrise or toward moonrise at significant times in the cycles of these bodies' motions. An outer ring of 56 chalk pits could have been used to compute and predict cycles of lunar and solar eclipses. Early peoples' interest in astronomy ultimately led to the concept of natural law—regular order in the universe.

This was important knowledge that enabled such societies to plant their crops at the right time.

Around 600 B.C., the Ionian philosophers conceived the idea that fixed rules—natural laws—might hold on Earth as well as in the heavens. To appreciate the magnitude of the Ionians' advance, try to think as the very early Greeks did. For example, try to imagine a flame as alive and willful. Why the Ionians made the transition from merely observing to asking why and how is not known, since we have only later philosophers' reports of their thoughts. From our perspective, the idea of natural law appears suddenly as an unquestioned foundation. This event marks the beginning of the western scientific tradition.

Thales of Miletus founded the Ionian school, named after the region in which he lived, now the Mediterranean coast of Turkey. He considered the question, Of what substance is the universe made? and suggested water as the answer. His suggestion is quite reasonable: the Greeks viewed the world as being alive; water is essential to all forms of life; and water takes on solid, liquid, and vaporous forms. The mixture of river and ocean water at the mouths of major rivers even appears to form new land.

However, Thales' idea cannot be right, for water is wet and is thus unbelievable as the basis of dry things such as desert sand. Nor could Thales explain other opposite properties of things such as hotness and coldness.

Thales' followers made numerous attempts to imagine more abstract universal substances as the answer to his question but finally recognized that no answer is possible. If the universal substance has any properties, everything should have just those properties and no others. But if the universal substance has no properties, it cannot explain why anything has any properties.

Pythagoras (ca. 500 B.C.) found a solution to this dilemma. He discovered simple numerical relationships among harmonic notes on stringed instruments and suggested that such mathematical relationships form an underlying structure that explains nature. The properties of matter arise not from the nature of a universal substance, but from the mathematical structure of its arrangement.

Pythagoras and the Ionians did not leave us with any final answers. Their legacy is the two questions about substance and structure they were the first to raise. At the frontiers of research, we still wrestle with the same questions:

1. Of what sorts of things is the universe composed?
2. What are the rules that govern their behavior, and what structures do they form?

THESE QUESTIONS STILL FORM THE BASIS FOR RESEARCH IN PHYSICS. IN THIS TEXT WE'LL DISCUSS THE ANSWERS THAT CURRENTLY SEEM BEST.

0.1.2 Simplicity

We have also inherited a taste for simple and elegant answers like the single substance of the Ionians or Pythagoras' simple ratios of harmonious strings. Of course, what is simple and elegant is a matter of taste, of culture, and of the mathematical knowledge available.

Historically, scientific experience strongly supports a belief in simplicity and elegance. Whenever accepted answers to either of the two questions above have grown overly complex, further progress has replaced those answers with simpler schemes. For example, in the nineteenth century, chemists recognized nearly a hundred chemical elements. In the 1930s, physicists could explain them all as structures built from three kinds of particle: protons, electrons, and neutrons. Since that time, numerous similar kinds of particle have been discovered and their explanation found in terms of a few, more basic kinds. Einstein's *deep structure* has several levels, each explained simply by the next. Yet simpler and deeper levels may be found as our understanding grows.

0.1.3 What Is a Satisfactory Explanation?

That events have causes seems basic to human thought. A person in a hypnotic trance can be instructed to open a window when the hypnotist scratches an ear. Later, released from the trance, the person will open the window on cue. Asked why, the person becomes puzzled and then feels a need to explain, saying perhaps that the room is hot. Having opened the window for no apparent reason is too disconcerting.

One can think of a number of different kinds of causes. For example, if we ask why the bus stopped at Fifth Street, we might come up with the following answers.

1. Because friction on the tires slowed it to a stop.
2. Because the operator put on the brakes.
3. Because there was a passenger waiting to get on.
4. Because the bus schedule says it will stop.
5. Because the taxpayers paid for a public transit system.

FRICTION IS DISCUSSED IN CHAPTER 4.

All of these answers are reasonable and each conveys part, but not all, of the truth. They are arranged according to increasing content of purpose or will and decreasing content of mechanical action and specific detail. Which answer provides a satisfying response depends on your view of the question, its context, and your reason for being interested. An engineer working for the bus system cares about causes 1 and maybe 2. A small child, learning about the world, cares about 2 and 3. An adult passenger cares about 3 and 4; and a sociologist, politician, or taxpayer cares about 5.

A physical theory is an attempt to understand, at least in part, the structure and substance of the universe. If a theory is correct, then it must be able to *explain* a number of physical phenomena—that is, to determine their causes. The example of the bus points out in homely fashion that the things we think require explanation and the causes we find satisfying depend strongly on our overall cultural outlook.

0.1.4 Aristotle and the Nature of Motion

Aristotle (ca. 350 B.C.), whose writings embody the Greek synthesis, gave one of the first careful discussions of the nature of causes. He was also the first philosopher to present a complete, coherent view of the universe.

In Aristotle's view, the heavens, as the realm of the gods, had to consist of an ideal substance and uniformly rotating spherical shapes. A very clever, though intricate, mechanism of spheres carried the stars, the Sun, the Moon, and the planets through their various motions (■ Figure 0.3). Inside the sphere of the Moon, each of the four fundamental substances—fire, air, water, and earth—had its natural sphere. These four *elements,* disturbed and mixed by the rotation of the system, accounted for the different forms of matter that occur on Earth. Wood, for example, would be a mixture of fire and earth. As the wood burns, the fire is released and earth (ashes) is left behind.

Aristotle's physics was part of a general theory of change based on the predominantly biological Greek worldview. In addition to their actual states, systems were supposed to have potential states. For example, a young child has the potential to become an adult. Potential adulthood was seen as the cause of *natural* change as the child matures. Similarly, according to the theory, a material object, composed mostly of earth, falls because of its potential state: at rest in its proper place—on the central sphere of the universe. Flame, being fire, rises to achieve its proper place at the sphere below the Moon. The image of a flame rejoicing at its release emphasizes the great poetic beauty of this scheme.

The theory also allowed for unnatural or *violent* change. An accident might interrupt the development of a child. An outside mover (such as a horse) might cause a heavy object (such as a loaded cart) to move horizontally instead of vertically. Such *unnatural* motions need *external* causes.

Aristotle's ideas survived through 20 centuries. It's not hard to see why. All of us, as young children, shared the idea of natural motions. We cannot feel any motion of the Earth, and we *see* the Sun move across the sky. Aristotle was convincing because he was building on common, natural notions, and because his theory unified all of human knowledge and experience in one grand structure.

Aristotle's system suffered from two weaknesses that seemed minor but which ultimately led to its downfall. His scheme of uniformly rotating spheres was unable to provide even crude *numerical* predictions of the planets' motions. The worst problem was *retrograde motion:* each of the five visible planets occasionally appears to stop, turn around and go backward, stop again, and then continue its journey across the sky (■ Figure 0.4). Aristotle's scheme could provide only a rough, qualitative account of such motions.

Projectile motion was the second difficulty. An arrow released from the bow, being no longer affected by an external mover, should adopt its natural motion and fall straight to

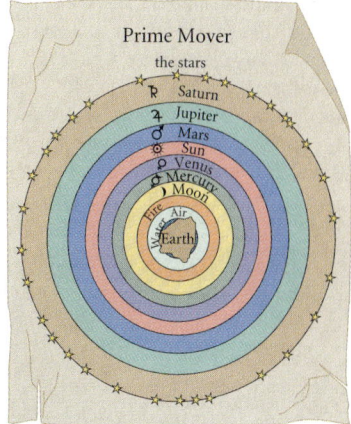

■ FIGURE 0.3
Aristotle's universe. Each planet—the Sun and the Moon were included—rides on a set of spheres that cause the planet's motion with respect to the stars. The sphere of stars rotates once per day, carrying the whole apparatus around Earth. Stationary at the center are the spheres of four basic elements. Earth is not exactly spherical because of the processes of change and mixture in the elements, which oppose their natural desire to inhabit their proper spheres.

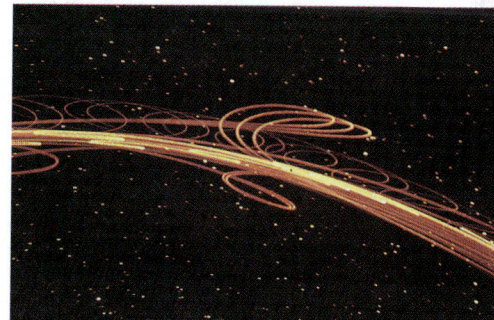

■ FIGURE 0.4
Retrograde motion of the planets. Viewed against the stars, the planets move generally eastward, but each occasionally reverses its direction, traveling westward for a time before resuming eastward motion. Mars, for example, reverses each time it is in the opposite direction in the sky from the Sun. This photograph from the Munich Planetarium simulates the motions of Mars, Venus, and Mercury over a period of 17 years.

WE'LL DISCUSS GALILEO'S RESOLUTION OF THIS DIFFICULTY IN §0.3. A MODERN ANALYSIS IS GIVEN IN CHAPTER 3.

KEPLER'S WORK EXEMPLIFIES MODERN SCIENTIFIC UNDERSTANDING. IT IS BASED ON OBSERVATIONS UNIFIED BY A PRECISE, MATHEMATICAL STATEMENT AND IS GROUNDED IN A COHERENT, SATISFYING, QUALITATIVE PICTURE. KEPLER'S LAWS THEMSELVES ILLUSTRATE PROPORTION AS A FEATURE OF PHYSICAL RELATIONSHIPS.

■ FIGURE 0.5
A painting of Aristotle's universe by Piero de Puccio from the fourteenth century. This model, with its stationary Earth and ideal heavens, was incorporated into the political, religious, and social ideas of medieval Europe.

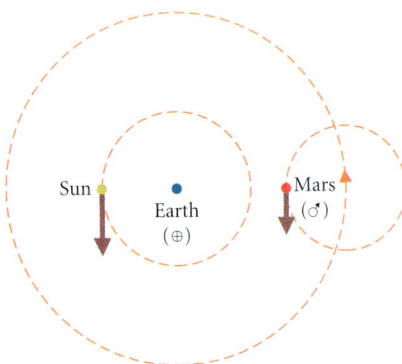

■ FIGURE 0.6
Motion of Mars according to Ptolemy. Mars moves on the small circle, whose center is carried around Earth on the large circle. Retrograde motion occurs when Mars is closest to Earth and the two motions oppose each other. This happens only when the Sun and Mars are on opposite sides of Earth. Ptolemy could offer no explanation why. The most careful refinements of Ptolemy's model can follow actual motions of the planets with moderate accuracy for no more than 200 years.

Earth. Instead, it follows a curved path. The archer clearly causes the arrow's motion but cannot act on the arrow in flight. Aristotle himself recognized this problem but never produced a convincing explanation.

0.2 Kepler, Planetary Motion, and Physical Law

0.2.1 Scientific Revolutions

Progress in physical theory has been marked by a number of upheavals separated by comparatively long periods of less dramatic change. Once a theory has been put forward and found satisfactory, it is used to explain more and more phenomena and to ask new and more detailed questions. Upheavals occur because some phenomena cannot fit within the accepted scheme of thought without major conceptual change. Often, a theory's downfall is caused by phenomena once considered minor or that were unknown before the advent of the theory.

Just such an upheaval occurred in European thought during the sixteenth and seventeenth centuries. Medieval Europe had accepted Aristotle's ideas of the structure of the universe and had fused them into its religious and social beliefs (■ Figure 0.5). At the same time, its practical interests demanded a precise understanding of projectile motion and of the astronomy used to construct the religious calendar. These two topics were just those that Aristotle's physics could not describe properly. The resulting conceptual revolution imposed major alterations on European society, as well as a new view of physics, and forms one of the most interesting episodes in European history.

0.2.2 The Earth Moves

Numerical accuracy in predicting the location of the planets in the sky has always been of value to astronomers. Aristotle's successors kept his spheres primarily as a poetic image, but they retained his ideas of a mechanical model for the heavens and of uniform rotation as the proper form of heavenly motion. Ptolemy of Alexandria (second century A.D.) produced such a model, which accounted for all the observations known at his time. Each planet was viewed as moving on a small circle whose center moves around Earth on a larger circle. The odd reversals of the planet's retrograde motion occur when the planet is nearest Earth and the motion on the small circle opposes the motion on the large circle (■ Figure 0.6). We now know models like Ptolemy's can only represent observations with reasonable accuracy for about 200 years.

By the beginning of the sixteenth century, Ptolemy's original computations were hopelessly out of date, and several attempts at recomputation of the scheme had failed to improve its accuracy. Astronomers had concluded that some fundamental alteration was required. The solution, when it came, was startling. Nicholas Copernicus (1473–1543, ■ Figure 0.7) noticed that much of the Ptolemaic scheme's complexity could be eliminated at a stroke by supposing that Earth and the other planets move around the Sun (■ Figure 0.8). The odd motions of the other planets would no longer be a mystery but instead the result of the Earth and the planets passing each other in their trips around the Sun.

The initial arguments in favor of the Copernican scheme were primarily its simplicity and consequent beauty. But it was considered physically absurd! A moving Earth was impossible in Aristotle's scheme of the universe. The entire structure of cause and effect and, in medieval Europe, God's plan for the universe required a stationary Earth. Besides, if the Earth moved, there would be giant winds as it sped through the air! The powerful elite of Europe could tolerate such an idea only as a convenient scheme for astronomical arithmetic but could not acknowledge its claim to be physical reality. Copernicus' ideas waited half a century for others to press that claim.

0.2.3 Kepler's Laws

Johannes Kepler (1571–1630; ■ Figure 0.9) trained in astronomy under Michael Mästlin, one of the few professional astronomers who believed in the Copernican model. Kepler, a rather intense and mystically oriented young man, was attracted to the Copernican model by its

■ FIGURE 0.7
Nicholas Copernicus (1473–1543). Copernicus studied at the University of Crakow and at several universities in northern Italy, where he became convinced that the Sun was stationary. He developed this idea into a complete astronomical theory. The first printed copy of his book, *On the Revolutions of the Heavenly Spheres,* was shown to him on his deathbed.

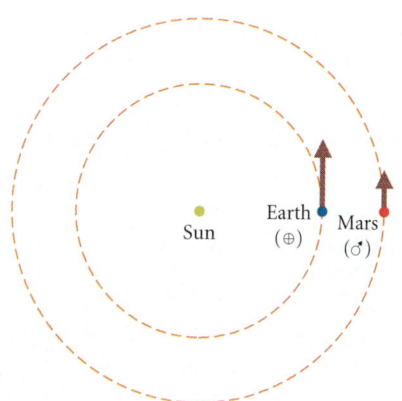

■ FIGURE 0.8
Motion of Mars according to Copernicus. Both Mars and Earth revolve about the Sun. Mars *appears* to move backward at the time shown, because Earth is passing it at higher speed. Retrograde motion when the planet appears opposite the Sun is a natural consequence of this model.

aesthetic purity. Coming from a family of meagre means, Kepler seemed destined to be a school teacher, but his career was abruptly changed by a most remarkable idea. While teaching an elementary geometry lesson on circumscribed polygons, he conceived a connection between the existence of six planets in the Copernican scheme and the five possible regular solids (■ Figure 0.10). If spheres are inscribed within and circumscribed about the regular solids in a nested arrangement, six spheres are required and their radii are very nearly proportional to the distances of the planets from the Sun (■ Figure 0.11). This is just how Pythagoras had expected mathematics to structure the world, and Kepler was exultant. His students must have been perplexed!

■ FIGURE 0.9
Johannes Kepler (1571–1630). As an assistant to Tycho Brahe and later as Imperial Mathematician to the Holy Roman Emperor, Kepler discovered the elliptical motion of the planets about the Sun. Kepler's insistence on mathematical precision in theory and on a physical explanation for astronomical phenomena set the stage for Newton's discovery of the law of gravitation.

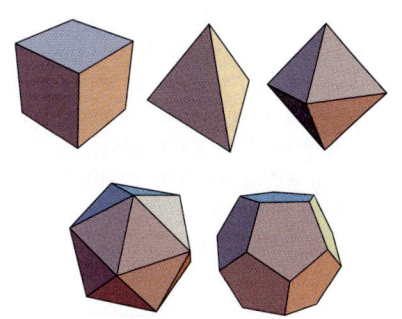

■ FIGURE 0.10
The Pythagorean solids. There are five possible solid figures formed from identical plane faces, each of which is a regular plane figure:

tetrahedron	4 equilateral triangles
cube	6 squares
octahedron	8 equilateral triangles
dodecahedron	12 pentagons
icosahedron	20 equilateral triangles

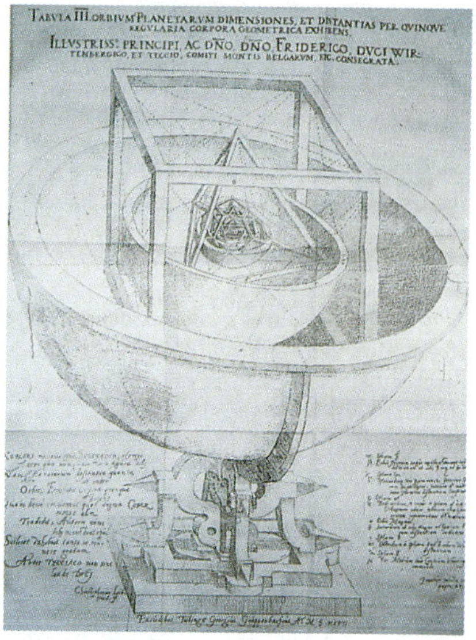

■ FIGURE 0.11
Kepler's solar system. Spheres nested among the five Pythagorean solids have radii very nearly in the same proportion as the distances from the sun of the six planets known to Kepler. The order of the solids going inward from Saturn is cube, tetrahedron, dodecahedron, icosahedron, octahedron. Though this model inspired Kepler to further work, it remains only a curiosity since it has no physical basis and cannot account for planets discovered since Kepler's time.

■ FIGURE 0.12
Tycho Brahe (1546–1601). This Danish astronomer built astronomical instruments of unprecedented size and precision, and used them to make careful observations of stellar and planetary positions. The quadrant is the brass arc. The astronomer at right looks through a movable sight on the quadrant and a hole in the wall (top left) to determine the angle of an astronomical object above the horizon. The painting within the arc represented Tycho Brahe overseeing the activities at his observatory. Kepler discovered his laws of planetary motion by analyzing Brahe's observations.

Kepler's construction works with 95% accuracy and earlier philosophers would have been satisfied, but Kepler's curiosity was only whetted. Another radical idea was taking form in his mind: numerical inaccuracy isn't just inconvenient, it means something is wrong. He left his teaching position and went to work with Tycho Brahe, who had recently made highly reliable observations of the planet Mars (■ Figure 0.12). Using the Copernican model, Kepler determined Mars' orbit with 100 times more accuracy than any preceding work. This alone would have justified the Copernican theory, but Kepler wasn't satisfied. The theory still did not properly represent the observed data. He realized that the ancient ideal of uniform circular motion, considered obvious by Aristotle, Ptolemy, and Copernicus, was incorrect and would have to be abandoned.

The task Kepler then set himself was extraordinarily difficult. With no theory for the actual shape of a planet's orbit, no idea of the rule governing its speed in the orbit, and no mathematics adequate for describing arbitrary shapes, he had to find an orbit that fit the observations. After seven years of labor, he finally obtained simple statements for the orbit shape and speed law. The results, now known as Kepler's first two laws, achieved totally unprecedented accuracy and stand today as research tools for understanding the motion of bodies in orbit (■ Figure 0.13).

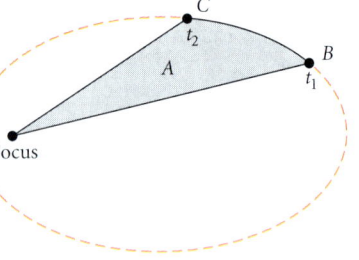

■ FIGURE 0.13
Between times t_1 and t_2, the planet moves from point B to point C. The area swept out by the radius vector is A, shown shaded. Kepler's second law refers to this shaded area.

KEPLER'S BOOK IS LIKE A DIARY OF HIS EFFORTS AND PROVIDES FASCINATING INSIGHT INTO THE MIND OF A WORKING SCIENTIST. KOESTLER'S BIOGRAPHY OF KEPLER —*THE WATERSHED*—GIVES A VERY READABLE VERSION OF THIS STORY.

KEPLER'S LAWS

1. A planet revolves around the Sun on an elliptical path with the Sun at one focus.
2. The line from the Sun to the planet sweeps out equal areas in equal times.

Kepler's cherished model of nested spheres and polyhedrons received no support from his great achievement, yet he could not forget his Pythagorean ideals. In his final work, *The Harmony of the World*, he laid out a series of mystical ideas, only one of which proved more than a historical curiosity. That one, now known as Kepler's third law, gives an important relation between the size of a planet's orbit and its period—the time required for each revolution around the orbit.

Digging Deeper

Ellipses

The ellipse is defined as the set of points P for which the sum of the distances r_1 and r_2 in ■ Figure 0.14 is constant. Points F_1 and F_2 are the foci of the ellipse. Point C is the center. The semimajor axis is a, and the semiminor axis is b. The eccentricity is e, the distance between the center and a focus divided by the semimajor axis. For a point at the end of the major axis, $r_1 = a(1 + e)$, $r_2 = a(1 - e)$, and their sum is

$$r_1 + r_2 = 2a.$$

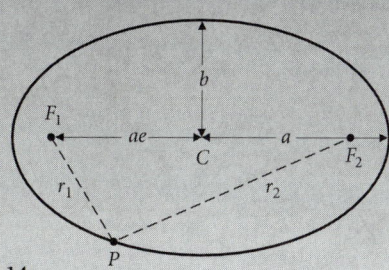

■ **Figure 0.14**
An ellipse.

Since the sum is constant, this formula is true for every point on the ellipse.

This property can be used to draw an ellipse: Put thumbtacks at the foci and tie a loop of string loosely around the tacks. Now place a pencil in the loop and pull the string taut. As you move the pencil around, it will trace out an ellipse. When you vary the distance between the tacks and the length of the loop, you find that the eccentricity and semimajor axis of the ellipse change.

Ellipses have many curious and useful properties. One, which we shall discuss in Part IV, is that light rays emitted by a source at one focus of an elliptical mirror all pass through the other focus. This property is the basis of an experimental oven for heating solid samples. The oven was used on a *Spacelab* shuttle mission to investigate crystallization of silicon in free fall (■ Figures 0.15 and 0.16).

An elliptically shaped dome also focuses sound waves. Standing at one focus, you could eavesdrop on a whispered conversation at the other focus (Jearl Walker, *The Amateur Scientist*, *Scientific American*, October 1978:179).

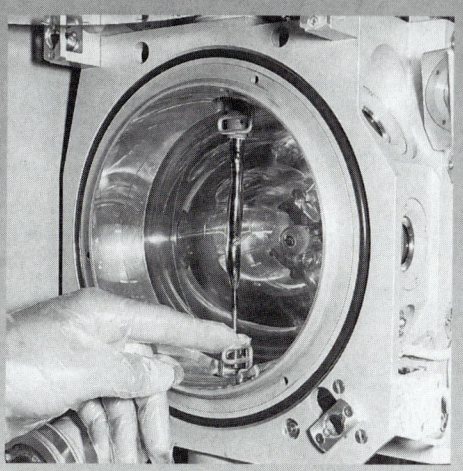

■ **Figure 0.15**
The mirror heating facility, used in a *Spacelab* experiment on crystallization of silicon, uses two half-ellipsoidal mirrors to concentrate light onto a thin section of silicon rod. An ellipsoid is the shape formed by rotating an ellipse around its major axis.

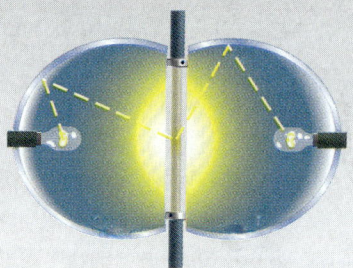

■ **Figure 0.16**
Schematic diagram of the mirror heating apparatus. A high-intensity lamp is placed at the focus of each partial ellipsoid. The other focus of each is at the center of the silicon rod.

---

3. The period T of a planet is proportional to the $\frac{3}{2}$ power of its orbit's semimajor axis a.

$$T \propto a^{3/2}.$$

Kepler's laws stand as the first fully modern description of a physical system. Although Kepler himself did not correctly understand *why* the planets behave this way, both the laws themselves and their very small inaccuracies proved important to Newton (Chapter 5) in developing a physical explanation of planetary motion.

Kepler stated these laws for the planets. In fact they correctly describe the motion of any objects orbiting another object, e.g., satellites orbiting the Earth.

THE LABELS (MODEL, SETUP, SOLVE, ANALYZE) INDICATE THE STEPS IN THE PROBLEM-SOLVING METHOD LAID OUT IN THE FIRST INTERLUDE.

SOLUTIONS TO THE EXERCISES ARE AT THE END OF THE CHAPTER.

GALILEO IS CREDITED WITH THE INVENTION OF EXPERIMENTAL SCIENCE. THAT CAREFUL EXPERIMENTAL DESIGN CAN ISOLATE IMPORTANT PHENOMENA FROM THE WHOLE OF EXPERIENCE WAS AS FOREIGN A CONCEPT TO ANCIENT SCIENCE AS IT IS ESSENTIAL TO MODERN IDEAS.

■ FIGURE 0.17
Galileo Galilei (1564–1642). Galileo revolutionized astronomical observation through his use of the telescope, and he revolutionized physics with several well-chosen experiments and a style of theoretical argument based on experiment. His work set the standard for future physical thought. His writings offended established religious doctrine, and he was held under house arrest for the last decade of his life.

EXAMPLE 0.1 ♦ The semimajor axis of the Martian orbit is 1.5 times larger than that of Earth's orbit. How long is the Martian *year*?

MODEL One year means the time Earth takes to go once around the Sun. The *Martian year* means, similarly, the period of Mars in its orbit.

SETUP According to Kepler's third law,

$$\frac{\text{Martian year}}{\text{Earth year}} = \frac{T_m}{T_e} = \left(\frac{a_m}{a_e}\right)^{3/2}.$$

We are given $a_m/a_e = 1.5$.

SOLVE

$$\frac{T_m}{1\text{ y}} = (1.5)^{3/2} \quad \text{so} \quad T_m = 1.8 \text{ y}.$$

ANALYZE The time we calculated is the period as observed by an astronomer looking down on the solar system from outside, and is called the *sidereal* period. The time for Mars to return to the same relative position as viewed from a moving Earth is called the *synodic* period. This distinction is the same as the difference between the time required to run once around a track and the time required to run one lap more than your rival. ■

EXERCISE 0.1 ♦ Halley's comet is in an orbit about the Sun with a period of 76 years. What is the semimajor axis of its orbit?

0.3 GALILEO AND EXPERIMENTAL SCIENCE

The new astronomy of Copernicus and Kepler required a new physics, which emerged from the work of Galileo Galilei (1564–1642; ■ Figure 0.17). After a short sketch of his life, we shall consider his important ideas individually.

Galileo began his career as professor of mathematics at the University of Padua, where he established himself as an inventor of precision instruments and an expert in theories of fortification and the use of artillery. These subjects were a standard part of the mathematics curriculum, reflecting the intense military activity of the time. In 1610, Galileo received reports of telescopes made in Holland and was able to duplicate the invention. Not content merely to look at distant ships, he turned his telescope to the sky. His observations of sunspots, lunar mountains, and the satellites of Jupiter destroyed Aristotle's model of the heavens and lent great support to the Copernican Sun-centered theory.

Movement of the Earth presented the Catholic Church with great ideological difficulties, and Galileo's vigorous advocacy of Copernicanism led to a trial. He was required to renounce his belief, forbidden to write about astronomy, and kept under house arrest. He spent the remainder of his life preparing his ideas on physics for publication.

0.3.1 Composition of Motion

Before Galileo, neither the cause nor the description of projectile motion was well understood. Aristotle thought an arrow should fall straight downward unless some cause acts on it continuously. Later philosophers decided that an archer must impress some sort of internal *cause of motion* into the arrow. According to this theory, the arrow should rise along a straight line until the internal cause is used up, and then it should fall straight down. Anyone who has caught a baseball knows that projectiles do not fall straight down at the ends of their paths.

Galileo devised clever experiments to study projectile motion, and he made a startling discovery:

> A projectile combines horizontal motion at constant speed with a completely independent vertical motion.

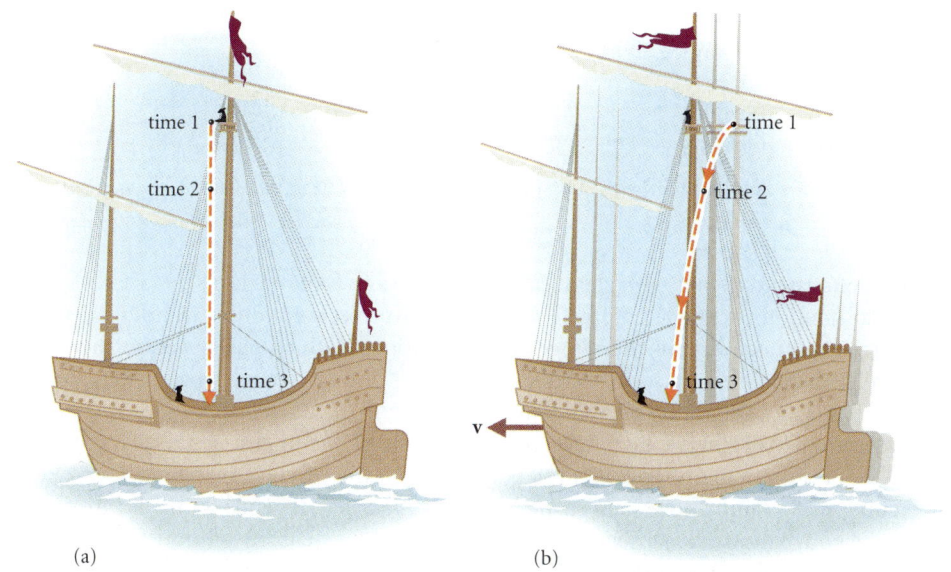

(a)   (b)

■ FIGURE 0.18
Galileo's thought experiment. Sailors on a ship see a cannonball fall directly down along the ship's mast (a), while an observer on shore sees it following the curved path of a projectile (b). Galileo concluded that a projectile's horizontal motion depends on who is observing the projectile, is independent of the vertical motion, and requires no physical cause. These facts contradict Aristotle's most basic assumptions about the causes of motion and the structure of the universe.

The surprising concept is two motions *occurring simultaneously*. For 2000 years people had thought that impossible, even meaningless!

Galileo investigated both vertical and horizontal motions with experiments described in the next section. To make composite motion believable, however, he devised a *thought experiment*, a logical argument that allows you to draw a profound conclusion from common experience. Imagine yourself in a cabin on a large ship at sea, cruising on a calm day. Your pet bird flies about normally. The wine poured by the steward falls normally into your glass. Everything in the cabin appears to occur in the same way as if the ship were not moving at all. The same is true for a cannonball dropped from the top of the ship's mast; it falls directly downward along the mast.

To an observer on shore, the cannonball would be a projectile with both forward and downward motion (■ Figure 0.18). The projectile's horizontal motion is just that of the ship; its vertical motion is the same as if it had no horizontal motion. Galileo then generalized this observation to include projectiles given horizontal motion by something other than a ship.

In Galileo's thought experiment, the horizontal motion of the cannonball is a consequence of the observer's point of view and requires no physical cause. This is a death blow to Aristotle's entire explanation for the causes of motion and to his model of the universe. Galileo's work established the *principle of inertia*:

REMEMBER THIS WHEN YOU READ THE MODERN ANALYSIS OF PROJECTILE MOTION IN CHAPTER 3!

ALTHOUGH GALILEO COULD NOT HAVE DONE SO, YOU MAY FIND THE ARGUMENT CLEARER IF YOU IMAGINE YOURSELF FLYING IN A JET AIRPLANE RATHER THAN CRUISING IN A SLOWLY MOVING SHIP.

OFTEN THIS PRINCIPLE IS DESCRIBED AS A RESISTANCE OF THE OBJECT TO CHANGE IN ITS MOTION. THIS RESISTANCE IS GIVEN THE NAME INERTIA.

> The motion of an object requires no cause. Only change in an object's motion requires a physical explanation.

0.3.2   The Inclined Plane Experiments

Galileo carried out a number of experiments to *measure* the nature of vertical motion. An object dropped vertically falls sufficiently rapidly that timing it by eye, even with a modern stopwatch, is nearly impossible. One can tell little more than that the object moves faster at the bottom of its path than at the top. Thus Galileo's major problem was to time a falling object reliably. Part of his solution was to consider a slower motion, that of brass spheres rolling down a hard wooden inclined plane, which he argued theoretically should be of the same character as vertical fall, only slower. He checked his theory by timing the motion of a ball on ramps of different angles and comparing the results with his predictions. There is some controversy about how he measured times, which he reports to an accuracy of $\frac{1}{64}$ second without describing the method. One likely possibility is that he used his musical skill (■ Figure 0.19). Clicks of the ball rolling over guitar strings stretched across the inclined planes

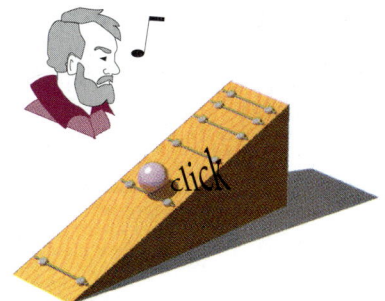

■ FIGURE 0.19
Galileo's experiment on speed of descent. Guitar strings are placed across an inclined plane so the rolling ball crosses them at equal time intervals, corresponding to the notes in a tune hummed by the experimenter. The ball rolls a distance proportional to the square of the elapsed time.

SECTION 0.3 • GALILEO AND EXPERIMENTAL SCIENCE

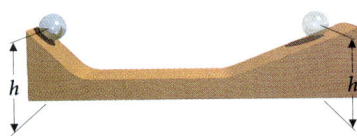

■ **Figure 0.20**
Galileo's experiment on quantity of motion. The ball starting on one inclined plane returns to its original height h regardless of the angles of the inclines or the distance between them. The ball possesses a *quantity of motion* determined by its initial height.

This expression seems vague to a modern ear, but was used by both Galileo and Newton.

We'll do this in Part II.

could be compared with notes in a melody hummed during the experiment! However Galileo measured times, he obtained a clear-cut result:

> **Galileo's Law of Fall**
> The distance an object falls from rest is proportional to the square of the elapsed time.

Galileo's law, together with the idea of decomposing motion into separate vertical and horizontal motions, forms a complete *description* of projectile motion without providing a theoretical understanding of the *causes* of such motions. Galileo performed further experiments with inclined planes to investigate the nature of those causes. A ball rolling down an inclined plane and then up a second plane, as in ■ Figure 0.20, returns to the same height it started from. This remains essentially true regardless of the angles of the two planes and regardless of the length of horizontal roll between the planes. Galileo interpreted these results by arguing that a falling body gains a *quantity of motion* that depends on the height through which the body falls. In the absence of friction, none of this quantity of motion is used up by moving horizontally. The quantity of motion is used to move vertically, and so the body returns to its original height.

Galileo's conclusions provide an illuminating example of how concepts develop in physics. The idea of uniform horizontal motion proceeding without cause comes very close to the modern idea, but Galileo didn't quite go far enough. He conceived of an ideal unforced motion as a circle about the Earth. The modern concept that unforced motion occurs in a straight line (Chapter 4) draws heavily on Galileo's ideas but required further philosophical progress that took a half-century to occur. Galileo's quantity of motion became another starting point for modern ideas, although we now recognize the need to distinguish carefully among several different quantities of motion.

0.4 The Nature of Physics
0.4.1 The Nature of Experiment

Physics is an experimental science that prides itself in getting *close to* reality through laboratory testing of theory. This tradition arose with the experiments of Galileo. The point of an experiment is to learn about nature by creating an idealized situation, different from common experience. An experiment requires careful design to make all effects except the one of interest negligible. For example, we may learn about free fall from balls rolling on an inclined plane, but only if we use accurately constructed spherical balls of hard metal rolling on hard, polished wood to reduce the effects of friction.

How can we be certain that the experimental process of dissecting nature into component parts is ultimately correct? We can't! Belief in experimental science depends on one's overall worldview. The Greek view of nature, which we have described as biological, saw the world as a whole, resulting from the interweaving of many inseparable processes. In that worldview, Galileo's experiments would be meaningless. In contrast, sixteenth-century Europeans had already made great progress in practical engineering and the use of machines. Such activities encourage a view in which whole systems can be understood as the combined action of many distinct parts. With a worldview based on mechanical imagery, people readily accepted a picture of nature operating on distinct principles that could be tested separately. The spectacular achievements of experimental science certainly are a strong argument for its methods, but do not guarantee success. Experimental science is itself a grand experiment still in progress.

0.4.2 Universal Law

To Aristotle, the Earth seemed greatly different from the heavens, of different substance and subject to different rules. In an era when men have walked on the Moon, we know this is not true; scientists now routinely assume that physical laws discovered on Earth apply to the most

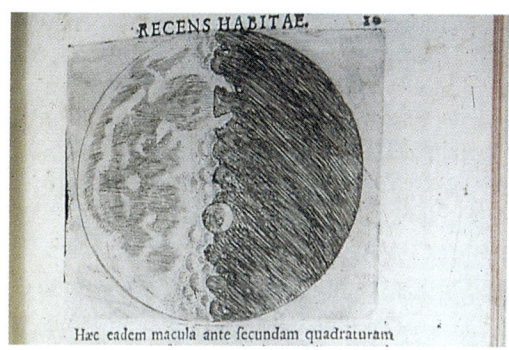

■ FIGURE 0.21
Galileo's drawing of the Moon from *The Starry Messenger* (1610). Note the ridges and mountaintops still lit by the Sun while their bases are in the dark on the night side of the Moon.

bizarre and distant astronomical object. This concept of universally correct natural laws began with the Copernican view of the Earth as one of several planets, but we find its first modern use in the work of Kepler and Galileo.

For example, consider two of Galileo's conclusions about the Moon. Through the telescope, he saw sunlight shining from mountaintops on the surface of the Moon (■ Figure 0.21) and, using geometry, he found the lunar mountains to be comparable in size to those on Earth.

To convince skeptics that the Moon has a rough surface, Galileo pointed out that the diffuse light we see from the Moon is similar to sunlight reflected from a stucco wall but that it does not at all resemble the bright, concentrated reflection from a mirror (■ Figure 0.22). Galileo assumed that light behaves in the same way on the Moon as it does on Earth, and he concluded that the lunar material is similar to that of the Earth.

0.4.3 Perspective and Point of View

One of the most important and practical attributes of physicists is flexibility in choosing a point of view for describing a physical system. Such flexibility arose first in painting. Early artists painted a two-dimensional world. To represent a three-dimensional scene on a two-dimensional canvas, one must understand how distance affects the appearance and position of an object in the picture and how the scene changes when the artist's point of view changes. These effects of *perspective* were first incorporated into paintings by artists in the fourteenth and fifteenth centuries (■ Figure 0.23 and 0.24). Similar ideas entered science with the work of Galileo and Kepler.

Galileo's thought experiment about horizontal motion relies on his use of the ship's deck and the ocean shore as equally valid points of view for describing the motion of a cannonball. Kepler simplified his calculations by imagining the solar system from the point of view of an astronomer *on Mars*. At a time when no one had ever seen Mars as anything other than a

■ FIGURE 0.22
An application of universal law. Aristotle held that the Moon must be a perfectly smooth sphere of ideal substance. Galileo pointed out that light from the Moon resembles sunlight reflected from a rough stucco wall rather than from a mirror. The distant Moon is of similar material and reflects light in the same way as an ordinary stucco wall.

■ FIGURE 0.23
Thirteenth-century painting. Spatial relations among the figures in the painting are unimportant to the artist, who is emphasizing their religious significance.

■ FIGURE 0.24
Painting by Raphael from the fifteenth century. Note the care with which he represents anatomical features of the figures and the spatial relations among them.

reddish orange point of light in the sky, it was remarkable to think of the planet as another place, like Earth, where astronomers might work. These changes in point of view are just the kind that prove useful in physics. Often you can simplify the description of a physical system by *imagining* yourself moving in some particular way, looking at a system from some particular place, or focusing your attention on some particular size scale.

0.4.4 Theory and Experiment

Galileo's experiments and Kepler's astronomy proved Aristotle's theories wrong. Did they prove anything right? Certainly, any *believable* ideas about physics must predict the elliptical motion of planets or that balls fall a distance proportional to time squared—at least as very good approximations. Belief, however, is not proof. Experiments cannot prove the exact truth of any physical concept, but do put ever tighter constraints on the ideas that deserve belief.

New theories are often, but not always, guided by new experimental results. In Einstein's view, the invention of a new physical concept is a creative act of the sort associated with artists, writers, and musicians. We do not know how to describe this creative process very accurately, but we do know how to test its results. Once we have a believable set of basic concepts and physical laws, we may derive conclusions just as geometrical results follow from a few axioms about points and lines. The conclusions from a good physical idea should be consistent with known experiments and should suggest new experiments and more precise tests. The best theories connect old concepts in new and unexpected ways and offer powerful methods for understanding more subtle properties of nature.

Consistency with experiment and usefulness in understanding nature are the properties of a good physical theory. The word *truth* is conspicuously absent. Aristotle's theory met all the tests of a great physical theory and was believed to be absolutely true for 2000 years. It collapsed when confronted with experimental solutions to the two research problems that it had made important. The work of Kepler and Galileo laid the foundation for Newtonian physics, thought absolutely true for 250 years. In the twentieth century, we have learned that Newtonian physics is not exact but stands as an excellent approximation. Absolute truth is elusive. We continue to seek greater depth in our understanding, greater elegance in our theories, and greater precision in our experiments. Whether truth can be achieved in some approximate sense by this process is unanswerable. We believe in physics because we know we can organize our knowledge and employ it to describe the behavior of nature with great accuracy using only a small number of fundamental ideas.

0.4.5 How Nature Works versus Why Nature Works

The scientific revolution of the sixteenth century marked a great change in the kind of question science could be expected to answer. The science of Aristotle was concerned primarily with the reasons *why* nature is as we see it. Detailed and accurate descriptions of phenomena were of secondary interest. The glory of this worldview was its success in placing the whole range of natural phenomena and human affairs into one grand, coherent scheme. Galileo and Kepler and those who followed reversed the emphasis of science, making an accurate description of *how* nature works the focus of interest.

Physics has not abandoned the idea of causes, but it has dramatically changed its view of what is a satisfactory explanation. No longer is the idea of a natural place in the universe an acceptable reason for why a stone falls. The kind of explanation we now require is a well-defined law of gravitational attraction that allows us to compute the stone's path. But why does the law of gravity have the mathematical form it does? Accurate description of gravitational attraction does not yield an ultimate cause, a final reason that needs no further explanation. Physics cannot offer that.

We no longer live in the intellectually tidy world of the Middle Ages. Each individual faces the task of building a personal worldview from the many disparate ideas the world offers. Several authors have discussed the role that physics plays in this task.

SEE, FOR EXAMPLE, *GOD AND THE NEW PHYSICS* BY PAUL DAVIES OR *THE VIEW FROM PLANET EARTH* BY VINCENT CRONIN.

0.4.6 Ideal versus Real: Modeling and the Role of Mathematics

Let us look more closely at how a physical theory allows us to make predictions. Suppose you throw a rock off a cliff and wish to predict its motion. Galileo's description of projectile motion is clearly the correct starting point. However, that description applies to an ideal object whose size and shape are unimportant and for which air resistance is negligible. The theory applies to the ideal object and not directly to the actual rock, but calculations from the theory give a fairly accurate description of the rock's actual motion. We say that a point object falling in a vacuum is a good *model* of the rock.

To get a more accurate answer, you must include the effects of air acting on an object of finite size. With a powerful enough computer, you could even include details of the shape of the rock. At some level, you would decide that including further detail in the model does not usefully improve your description of the rock's motion.

Such modeling is the essence of physics. Research physicists continually attempt to increase the accuracy and level of detail with which we can model the universe. Applications of physics involve modeling interesting physical systems with known concepts, as in the example of the rock. To study physics, we begin with Newtonian mechanics and concentrate on learning how to model systems and apply theory. Thus in every example and problem solution, the first step is to construct the correct model.

Mathematics by its nature deals with ideal situations: points, lines, equations with exact solutions. As such, it is a language well suited for describing the ideal models of physical theory. With Aristotle, the vision of nature ruled by mathematical structures became submerged in verbal logic. The mathematical vision reemerged with Kepler and Galileo, whose results have simple, elegant mathematical statements. This has proved true of every significant physical theory since—mathematics would appear to be the language of physical theory.

IN USING PHYSICS, WE MODEL REAL THINGS WITH APPROPRIATE IDEAL OBJECTS; THE ACCURACY OF OUR PREDICTIONS DEPENDS ON THE COGENCY OF OUR MODEL.

IN EACH EXAMPLE, THIS STEP IS IDENTIFIED BY THE LABEL "MODEL."

Chapter Summary

Where Are We Now?

We have summarized how early thinkers came to the conclusions that are the starting point for Newtonian physics. The ideas that were disproved in the seventeenth century are often the very ones students bring to their first course in physics. The story of how these early ideas were superceded is relevant to your own thinking about how and why things happen.

What Did We Do?

The vision of a universe subject to natural law was invented by Greek philosophers who also posed basic questions about its substance and structure. Aristotle gave us a comprehensive view of the universe that was accepted for 2000 years but ultimately proved unable to account for the motions of planets or of projectiles on Earth. The solution to those two problems required belief in a moving Earth; acceptance of mathematical accuracy and experimental test as standards of truth; and revision of ideas about the relation of causes to motion and the relation of Earth to the heavens.

Today, physicists assume that nature follows consistent rules, and they attempt to determine what those rules are. Physical law is assumed to be universal, not varying in space or time. Physicists create *models* of the universe, idealizations whose behavior can be predicted by mathematics and verified by experiment. In this book we'll show you how to use such models to advantage. In so doing, you will notice that problems are often more easily solved by looking at them from the right perspective or frame of reference.

Practical Applications

Although we haven't yet said why, Kepler's laws of planetary motion also apply to satellites in orbit about Earth and are the starting point for understanding space shuttle navigation. Galileo's law of falling bodies finds numerous applications, from skydiving to the design of pile drivers. The intellectual skill of modeling has proved fruitful in fields as diverse as medicine and economics.

Solution to Exercise

0.1 Kepler's third law applies to anything in orbit around the Sun, not just to the major planets for which Kepler derived it. Thus

$$\frac{\text{Axis of comet's orbit}}{\text{Axis of Earth's orbit}} = \frac{a_h}{a_e} = \left(\frac{T_h}{T_e}\right)^{2/3},$$

$$\frac{a_h}{a_e} = (76)^{2/3} = 18.$$

The radius of Earth's orbit, a_e, is called 1 astronomical unit (AU). So, the semimajor axis of Halley's comet's orbit is about 18 AU. Compare this result with the 5.2-AU semimajor axis of Jupiter's orbit (inside front cover) and the 19-AU value for Uranus.

The $T \propto a^{3/2}$ law applies to objects orbiting any gravitating center—for example, to the satellites of Jupiter. Each such system has its own proportionality constant.

Basic Skills

Review Questions

§0.1 WHY STUDY HISTORY?

- Why was the concept of natural law an important invention?
- Why is it impossible for a single universal substance to explain the variety of the universe?
- How did Pythagoras propose to explain structure?
- What does it mean to say a physical theory is *simple*?
- What are some criteria for deciding whether an explanation of cause is satisfactory?
- In Aristotle's theory what causes change to occur?

§0.2 KEPLER, PLANETARY MOTION, AND PHYSICAL LAW

- What brings about a scientific revolution?
- How does the Copernican model explain retrograde motion of the planets?
- What are Kepler's three laws?

§0.3 GALILEO AND EXPERIMENTAL SCIENCE

- Describe Galileo's thought experiment and explain how it shows that horizontal motion at constant speed needs no physical cause.
- In what sense is projectile motion a composite of two motions occurring simultaneously?
- What is the *principle of inertia*? How does it follow from Galileo's thought experiment?
- What is Galileo's *law of fall*?

§0.4 THE NATURE OF PHYSICS

- What is the goal of an experiment in physical science?
- Explain how Galileo's conclusion that the Moon has a rough surface depends on the idea of universal law.
- What is *perspective* in painting?
- Why is flexibility in point of view a useful technique of physics?
- Does an experiment ever prove anything right? Or wrong?
- Describe the relation between physical theory and experimental test.
- Can physics ever explain *why* the universe is as it is?
- What is a physical *model*, and how does a model relate to mathematics?

Basic Skill Drill

§0.2 KEPLER, PLANETARY MOTION, AND PHYSICAL LAW

1. Think about proportion in daily life. The amount you pay for goods is generally proportional to the amount you buy. For example, carrots may cost $0.33 per pound. With packaged goods, the constant of proportionality may not be the same for each package. Suppose you are trying to decide whether to buy the giant economy size of Brand X shampoo, offering 750 cm³ for $4.50, or the convenient travel package of Brand Y, offering 120 cm³ for $0.80. For which package is the price per unit of shampoo less?

2. The volumes of different spheres are proportional to the cubes of their radii. How are their surface areas related to their radii? If one sphere has 3.4 times the radius of another, what are the ratios of their volumes and of their surface areas?

3. Refer to the discussion of ellipses (*Digging Deeper*, §0.2.3). Use the string and thumbtack method to construct an ellipse with a semimajor axis of $a = 12$ cm and eccentricity $e = 0.30$. How far apart should you put the thumbtacks? How long should you make the loop of string?

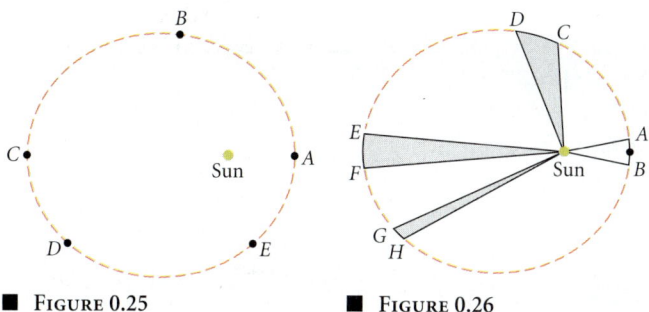

■ Figure 0.25

■ Figure 0.26

4. Suppose you were to draw a diagram of planetary orbits to scale with the semimajor axis of Earth's orbit (1 AU) shown as 5.0 cm. Using numbers from the examples and exercises in this chapter, find what lengths would represent the semimajor axes of the Martian orbit and that of Halley's comet. If the eccentricity of the Martian orbit is 0.093, how far apart would the center of the Martian orbit and the Sun appear in your diagram? If 1 AU is 93 million miles, and the Moon's orbit about Earth has a radius of $\frac{1}{4}$ million miles, could you draw the Moon's orbit to scale on your diagram?

5. A comet moves in the elliptical orbit shown in ■ Figure 0.25. At which point in the orbit is the comet moving most slowly? Most rapidly?

6. Use Kepler's third law to find the semimajor axis of Venus' orbit if its period is $P = 0.62$ y.

§0.3 GALILEO AND EXPERIMENTAL SCIENCE

7. A person leaning from the window of a railroad train drops an apple. If air resistance is negligible, where should the person look to see the apple hit the ground?

8. Your heart beats twice during the time a large stone is falling a given distance from rest. How many heartbeats will occur while a similar stone falls twice that distance from rest?

9. A rock falling down an old mine shaft falls one-quarter of the way to the bottom in 1 s. How far does it fall in 2 s?

Questions and Problems

§0.1 WHY STUDY HISTORY?

10. ❖ Using Aristotle's four elements, how might you explain the difference between desert sand and a fertile soil, between a stone and a block of wood? How might Aristotle explain why a stone sinks while a block of wood floats on water?

§0.2 KEPLER, PLANETARY MOTION, AND PHYSICAL LAW

11. ❖ Which of the shaded regions CD, EF, or GH most closely represents a time interval in a comet's motion equal to that between points A and B in ■ Figure 0.26 above?

12. ◆ The asteroid Vesta has a period of 3.63 y. What is the semimajor axis of its orbit?

13. ◆ Io and Callisto are satellites of Jupiter. Their periods are 1.77 d and 16.7 d. What is the ratio of their distances from Jupiter?

14. ◆◆ The Moon's distance from Earth is approximately 60 Earth radii, and it takes 27 d to orbit Earth. A satellite just above Earth's atmosphere orbits in 84 min. Are these data consistent with Kepler's third law for objects orbiting Earth? (That is, $T^2 \propto a^3$?) What radius orbit is required for a communications satellite to take exactly 1 d for an orbit? Such a satellite appears to remain *stationary* above a point on the equator and is in *geosynchronous* orbit.

15. ◆◆ The eccentricity of the Martian orbit is $e \approx 0.093$. A scale drawing of the orbit is made with a semimajor axis of 5 cm and a pencil line width of 0.05 cm. Can you detect the difference of the drawing from a circle? (*Hint:* Which points on the ellipse have the greatest and least distances from the center of the ellipse, and what is the difference in those distances?)

§0.3 GALILEO AND EXPERIMENTAL SCIENCE

16. ❖ Two boxes of nails, one twice as heavy as the other, slide off the back of a moving truck at the same time. Which box hits the ground first?

17. ❖ A cannon ball is dropped from the mast of a sailing ship. Describe the ball's motion as seen by observers in the following positions. **a.** Standing on the deck beside the mast. **b.** Sitting on the top of the mast. **c.** Floating over the ship in a hot-air balloon. **d.** Standing on shore looking at right angles to the ship's motion. **e.** Standing on shore looking parallel to the ship's motion. In your answers, refer explicitly to horizontal and vertical components of the motion.

§0.4 THE NATURE OF PHYSICS

18. ❖ Discuss how David Scott's lunar experiment (§ 0.1) tests the notion of universal law.

19. ❖ Describe why it is easier to visualize the path of an expedition to the Moon as it would be seen from a location outside of the plane of the Earth–Moon system than as it is actually *seen* from Earth's surface.

Additional Problems

20. ◆◆◆ Figure 0.21 shows one of Galileo's sketches of the half-full Moon. A schematic view of the observation is shown in ■ Figure 0.27. The spot labeled A is one that Galileo interpreted as sunlight shining on a mountain top whose base was in darkness. If the distance d is taken as $\frac{1}{10}$ lunar radius, and the radius of the Moon is approximately 1740 km, estimate the height h of the mountain. How does your result compare to the height of Mt. Everest?

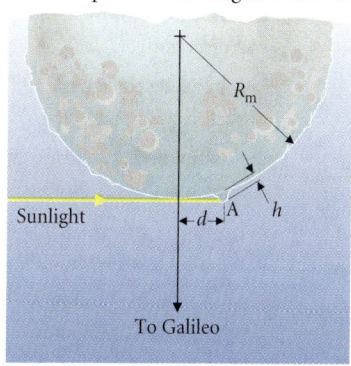

■ Figure 0.27 Schematic diagram of Galileo's observation of a lunar mountaintop illuminated by the Sun. The view is perpendicular to that in his drawing in Figure 0.21.

Painting by Chesley Bonestell shows Long Island as seen by a rocket ship 25 miles above New Jersey. Bonestell's paintings are an amazing depiction of future space ventures that have inspired many young people (including one of the authors!) to enter science. In the twenty-first century, supersonic "scramjets" may well carry passengers from Los Angeles to New York in about an hour. Approaching New York, the pilot will need accurate information about the positions of the scramjet and any other air traffic in the area. Today's pilots require the same sort of information, though their needs for accuracy and rapid response are less critical.

CHAPTER 1
Introducing the Language of Physics

Concepts

Cartesian model of space and time
Reference frame
SI units
Significant figures
Physical dimensions
Scalar
Vector
Displacement
Position vector
Coordinate system
Vector components
Unit vector

Goals

Be able to:

Use the SI units of distance, time, mass, and angle.

Keep track of precision and accuracy in numerical calculations.

Use estimation techniques to find rough answers to numerical questions.

Describe a vector by its magnitude and direction.

Recognize displacement and position as vector quantities.

Describe a vector in terms of components and unit vectors.

Do vector arithmetic using either plane trigonometry or components.

Philosophy is written in this grand book, the universe, which stands continually open to our gaze. But the book cannot be understood unless one first learns to comprehend the language and read the letters in which it is composed. It is written in the language of mathematics, and its characters are triangles, circles and other geometric figures without which it is humanly impossible to understand a word of it; without these, one wanders about in a dark labyrinth.

GALILEO GALILEI

Everything moves—galaxies, jetliners, atoms, people! A careful description of their motion is basic to controlling jetliners and getting people where they want to go, or to understanding atoms and how they make up galaxies. Learning how to describe motion was the first step in the long process of understanding how the universe and the things in it behave. In Chapter 0 we traced this process through the work of Galileo and Kepler. After them, it took another half-century of mathematical progress and debate on the principles of mechanics before a complete theory was possible. It was Isaac Newton, a professor of mathematics at Cambridge University, who created this theory, published in his book *The Principia*, in 1687.[1] Newton's theory proved so successful that no deviation from its laws was discovered for more than 200 years. During the twentieth century, we have found more far-reaching theories, but Newtonian theory stands as the starting point in a study of physics and as the correct approximation to use for a huge variety of practical applications. In Chapters 1–5 we shall study Newtonian theory and a number of those applications.

Galileo, Kepler, and Newton established mathematics as the language of physics, a language we'll begin to apply in this chapter. Physical processes occur in space and proceed through time. To model them, we set up conventions for measuring positions, lengths, time intervals, and angles. To understand motion, we need mathematical tools that allow us to describe quantities with direction as well as size. So let's begin!

1.1 A Model of Space and Time

1.1.1 Space

Suppose someone asks you for directions to the local grocery store. Your response might be, "Turn left at the gas station, go three blocks, and turn right at the second stoplight." That is, you tell them how to locate the store relative to other objects that have an unchanging relation to the store—a reference frame.

> A *reference frame* is a set of objects not moving with respect to each other that act as a background for describing the position and motion of other objects.

The surface of the Earth is a familiar reference frame, and we know what it means to locate the grocery store or to say that the view in Chesley Bonestell's painting is from 25 miles above New Jersey. The interior of a jetliner is just as natural a reference frame to the passengers in it as the room you're in right now is to you. There are many possible choices of reference frame, but usually only a few are convenient to use.

We perceive the objects in a reference frame as occupying different places in the world, giving us reference points in something—space—which exists whether or not any objects are present. A simple experiment shows that space has three dimensions. You can move your hand independently in three different directions: up–down; left–right; forward–backward. Space, then, is an abstract, three-dimensional set of points in which we assume, for now, that familiar rules of Euclidean geometry apply.

Choosing a reference frame is the first step in describing this abstract space. The objects that define the frame, like the gas station and second stop light in the grocery store example, mark a few reference positions we choose to consider *fixed*. We can use meter sticks to measure the distance of any object from one of the reference positions. With a large number of meter sticks arranged in a rigid, rectangular structure (■ Figure 1.1), we can read the position of any object or any point directly from the meter sticks.

This model isn't restricted to the room in Figure 1.1. We can easily imagine the grid of meter sticks extending without limit to fill all of space, thereby assigning three position readings, or *coordinates,* to each point in space. This model is called a Cartesian coordinate system

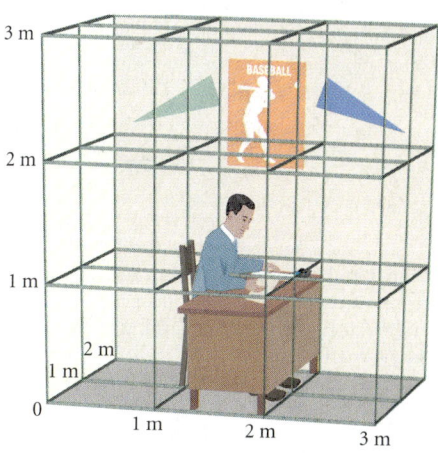

■ **Figure 1.1**
We imagine the space in a library study area filled with a grid of meter sticks. The student in the middle of the room is "pierced" by the grid line 1 m up from the floor and 1 m in from the corner marked zero. The head is between the 1-m and 2-m marks on each line. Increasing the number of meter sticks in each dimension would provide greater precision in measuring positions of things. Such a three-dimensional set of meter sticks serves as a *model* of space.

---

[1] The full title is *Philosophiae Naturalis Principia Mathematica,* or *The Mathematical Principles of Natural Philosophy.*

RECALL HOW DIFFICULT IT WAS FOR COPERNICUS TO MOVE THE CENTER OF THE UNIVERSE FROM EARTH TO THE SUN. HERE WE GO FURTHER AND SAY THERE IS NO CENTER. FOR CONVENIENCE IN CALCULATING, WE CAN DEFINE AN ORIGIN ANYWHERE.

after Rene Descartes (1596–1650), who invented this idea of describing points in space by numbers. In a conventional diagram (■ Figure 1.2), the Cartesian coordinates are named x, y, and z. The imaginary meter sticks are not shown.

The Cartesian model expresses an important property of space: it is uniform. Space has no lumps or bumps, no special directions, and no special center. Mt. Everest is a lump on the Earth, but our imaginary meter sticks pass through the mountain as though it weren't there. In Newtonian physics, space has no bump corresponding to Mt. Everest. Furthermore, this description of space is not affected by objects moving in the space. As the space shuttle moves around Earth, it has no effect on the imaginary meter sticks.

Galileo's relativity principle (§0.3.1) states that a second reference frame that moves at constant speed in a fixed direction with respect to the first is equally valid for describing physics. If you choose the cabin of an airliner as your reference frame, an imaginary set of meter sticks fixed to the airplane is just as good a model of space as a set of meter sticks fixed to the surface of Earth, even though a "single point" in the moving plane's model of space is in different places with respect to the Earth at different times. It's the relationships between different points that are important.

René Descartes, French mathematician and inventor of analytic geometry, lecturing at the Swedish court.

1.1.2 Time

The concept of time has varied considerably throughout history and among cultures. Many ancient cultures viewed time as something repetitive: a pattern of world ages of predetermined length. As modern science emerged, this notion of cycles was rejected and replaced by a sense of time as a uniform progression without structure and unaffected by events. Newton defined time as follows:

> Absolute, true and mathematical time, of itself and from its own nature, flows equably without relation to anything external.

In Newtonian physics, time flows equally for all observers, independent of their reference frame.

You probably have a strong intuition about time, and you know that time is measured out by a good clock. We can add time to the Cartesian model of space by attaching an imaginary clock at each intersection of the meter sticks and by taking care that the clocks are synchronized (■ Figure 1.3). Now, if an *event* occurs, say a flashbulb flashes, we can give it a time coordinate—the clock reading when it happens—in addition to the three spatial coordinates that tell where it happens. It is events happening at specific places and times that are the fundamental things in physics.

We experience a crucial, intuitive difference between time and space: we can move freely in either direction in all three dimensions of space, but we can only move forward in time. The *arrow of time* points into the future, defined as the direction (1) in which we have no memory, (2) in which a popped balloon explodes, and (3) in which the universe expands. It is a fascinating and unsolved question as to why these three *arrows of time* are the same.

Our ideas of cause and effect are closely related to the arrow of time. Aristotle believed that events could be caused by the need to achieve some future state: a child grows because it is to become an adult. Modern scientists have rejected such final causes. Rather, we believe the child grows because of its genetic background and interaction with the environment. It has been fruitful to build our descriptions around the *principle of causality*:

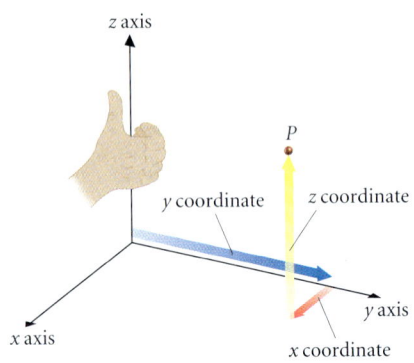

■ **FIGURE 1.2**
Cartesian coordinates. The z-axis is perpendicular to the plane containing the x- and y-axes. By convention, the direction of the z-axis (up versus down here) is chosen using the *right-hand rule*, illustrated by the hand in the figure. Curl the fingers of your right hand in the x–y plane from x toward y, and your thumb points in the direction of z. Because of this rule, the coordinate system is described as "right-handed." The coordinates of the point P are distances measured parallel to the axes, as shown. For this point, all three coordinates are positive.

SEE THE ESSAY FOLLOWING CHAPTER 22 FOR A FURTHER DISCUSSION.

> An event occurs earlier in time than any other event that it causes.

A pebble dropped into a pond is an extremely plausible cause for outwardly moving water waves on the surface of the pond. Inward-moving waves that would meet at the center and eject a pebble from the pond could only result from such an extremely intricate set of causes that we believe such a thing will never happen. We expect physics to explain events with relatively simple causes that occur before their effects.

1.1.3   How Good Is the Cartesian Model?

In Aristotle's concept of space, there was a definite center—the proper place for Earth—with different natural places for other elements. This notion is definitely incorrect. We know that the Cartesian model is an improvement because physical theories based on it describe the world with mathematical precision. As but one example, Newtonian mechanics is completely adequate for sending men to the Moon or guiding spacecraft to the outer planets of the solar system.

Is the Cartesian model a final picture of space and time? No. Classical physics is successful when gravity is not too strong and when the objects we describe are large enough to contain many atoms, are not too cold, and move slowly with respect to each other. Twentieth-century theories, which do describe strong gravity, atomic systems, and rapid motion, require revision rather than rejection of the Cartesian model. For now, we shall restrict ourselves to the large and useful class of problems where classical physics is accurate and the Cartesian model of space is appropriate.

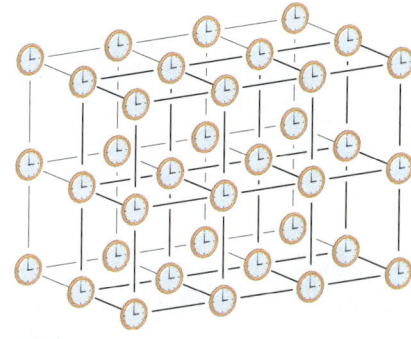

■ FIGURE 1.3
To describe events in physics fully, we also need to measure time; that is, we add clocks to our grid of meter sticks. In Newtonian physics, we can arrange for all the clocks to read the same time. (But see Chapter 34 for Einstein's view!)

HERE "SLOWLY" MEANS RELATIVE SPEED MUCH LESS THAN THE SPEED OF LIGHT.

1.2   THE INTERNATIONAL SYSTEM OF UNITS

1.2.1   Metrology, the Science of Measurement

Measurement is a common practical activity. To buy the right amount of lumber for a doghouse or to get to class on time, you need practical measurements of length and time. In the last section we considered the meaning of space and time and ended up talking about meter sticks and clock readings. This illustrates an important fact about physics: our models of the physical world are always intimately connected with measurement.

Every good physical measurement shares a number of common features. Suppose, for example, that you measure a sailboat's length as 10 meters. What does that mean? First, it is a comparison of the boat's length with that of another object, the meter stick.

> Every physical measurement is a comparison of two similar physical quantities.

Second, we accept the meter stick as a valid device for measuring the boat's length. At the factory, the meter stick was marked by a machine, itself adjusted by comparison with a standard length.

> To be valid, a measuring device must be compared against a widely accepted *standard*.

Third, we trust the manufacturer to have been careful in producing the meter stick, and we believe that the length of the stick doesn't change significantly between manufacture and use.

> The measurement must be sufficiently *accurate*. Also, the procedure must be *stable* so that we know how to compare different measurements made at different times.

Accuracy describes how much a measurement might differ from another measurement made with greater care. Whether the accuracy is sufficient depends on the purpose of the measurement. You probably wouldn't care if the boat were *really* 10.05 meters long, but you'd certainly be disgruntled if it were really only 9 meters long and your so-called meter stick were faulty!

A fourth feature of good measurement is adequate precision.

> The *precision* of a measurement is the smallest amount of the measured quantity that can reliably be distinguished.

Meter sticks are usually marked in thousandths of a meter, which is much more precise than needed for measuring the boat's length, but inadequate for finding the diameter of an engine part. Greater precision requires a more carefully manufactured device (■ Figure 1.4).

The process of comparing a particular measuring instrument against a standard is called *calibration*. The stability of the equipment determines how frequently calibration is necessary. Our meter stick was compared only once against the factory standard, yet we don't hesitate to use it for many measurements. In an ideal world, an instrument, once calibrated, could be used for any number of measurements. So long as we don't need extreme accuracy, our meter stick meets this ideal. Your watch, in contrast, needs occasional adjustment. In precision measurement, changes in instrumental adjustments, aging of components, and so forth make frequent recalibration essential. Good measurement involves a compromise among trusting a calibration, need for accuracy, and making every measurement a direct comparison against standards.

Physicists attempt to describe the universe correctly in a way that has the same meaning for all. At the same time, advancing technology creates ever-increasing practical needs for accuracy and precision. To meet these goals, measurement standards and calibration techniques must combine permanence, uniformity, convenience, and as great a precision as the state of science allows. *Metrology* is the name for this study of measurement technique and the development of precise standards and methods of measurement. It is a fundamental part of experimental science.

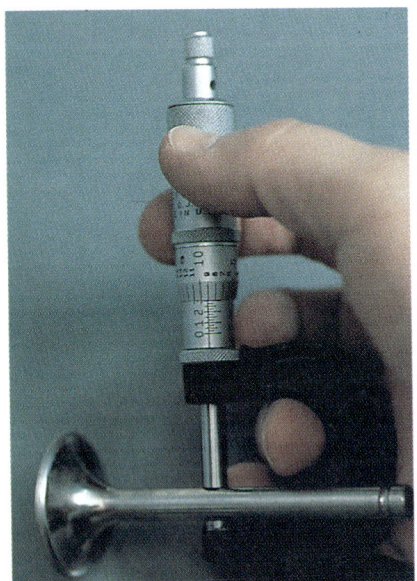

■ FIGURE 1.4
Precision micrometer for measuring small dimensions. The knurled knob at the top turns on a screwhead so that, as it rotates, the cylindrical rod retracts to open a gap. The object to be measured is placed in the gap and the rod is closed on the object. The friction clutch allows every measurement to be made with the same pressure on the object. The scales allow reading to a precision of $\frac{1}{25}$ turn, or 1/1 000 inch.

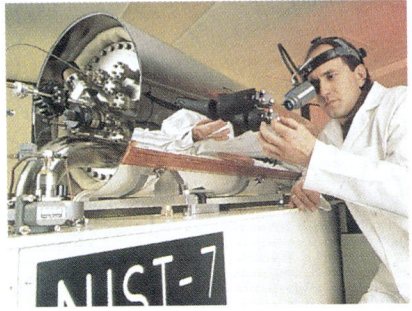

Cesium frequency standard at the Boulder laboratories of the National Institute of Standards and Technology (NIST). This apparatus is used to calibrate precise clocks.

1.2.2 The Origin of the Metric System

Who chooses how we calibrate a meter stick? Communication among scientists and cooperation among manufacturers both demand a common standard. The modern metric system began shortly after the French Revolution, when Napoleon decided to eliminate obviously inconvenient standards inherited from the past. Later, in 1875, the International Bureau of Weights and Measures (Bureau Internationale des Poids et Mesures, or BIPM) was established to develop an international set of standards. Today, the BIPM oversees the International System of Units (le Système Internationale d'Unités), abbreviated SI. This system was adopted by physicists worldwide in 1960. The United States is one of only three countries that still allow official use of nonmetric units.

THE OTHER TWO ARE BURMA AND LIBERIA. SINCE 1988, THE UNITED STATES HAS BEEN MOVING TOWARD ADOPTION OF THE METRIC SYSTEM FOR COMMERCE.

Metrology is a very active science, and the definition of a system of units, no matter how well done, cannot remain static. As greater understanding of physics and greater precision in technology develop, the BIPM oversees increasingly precise definitions of standards.

1.2.3 The SI Unit of Time—The Second

The SI unit of time is the *second*, abbreviated s. Its definition provides an excellent example of how standards evolve. The units of time originally derive from the motions of the Earth. The Babylonians (ca. 2000 B.C.) divided things into 60 parts or an even fraction of 60 parts, such as 12. Hipparchus of Alexandria (ca. 150 B.C.) adopted this system to give us day and night divided into 12 hours, with 60 minutes in each hour and 60 seconds in each minute. We no longer define time by Earth's rotation because it is slowing down (■ Figure 1.5)! Gravi-

THAT IS, ACCORDING TO THE OLD STANDARD, EVERYTHING ELSE WAS SPEEDING UP!

tational forces exerted by the Sun and the Moon produce the ocean tides, and, loosely speaking, friction between moving ocean water and the sea floor causes the slowdown. Keeping our watches synchronized with the cycle of day and night requires the insertion of an extra leap second into the civil time standard every few years.

Until the 1960s, observing planetary motion in the solar system provided the most accurate method of measuring time, but this method is much less precise than current laboratory techniques. According to atomic physics, electromagnetic radiation from atoms in very dilute gases should be extremely regular. Recent advances in electronics allow measurement of time intervals defined by such radiation to a precision of 1 part in 10^{14} (1-second error in approximately 3 million years)! These are the most precise measurements now possible, and atomic time standards promise to remain for the foreseeable future. Consequently, in SI the second is now defined by radio waves from cesium atoms.

We don't notice changes in the definition of the second in daily life because, whenever the BIPM decides on a change, the new standard is chosen to preserve as closely as possible the physical value of the unit defined by the old standard. The second according to the cesium standard is defined so that, as accurately as anyone can tell, it equals the standard second based on Earth rotation in 1900. The language in the definition may be unfamiliar to you, but the important idea is that a specific type of radiation from a specific kind of atom was chosen.

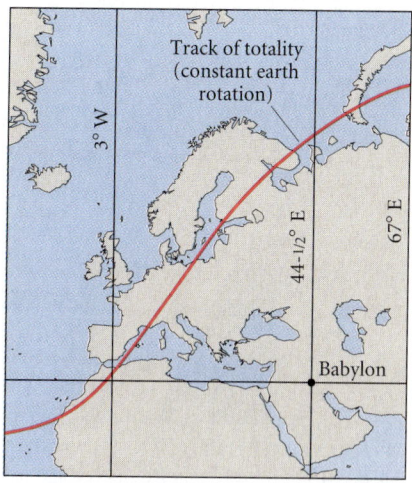

■ **Figure 1.5**
Track predictions for the total solar eclipse of April 15, 136 B.C. The curved line shows the calculation of those places on Earth where a total solar eclipse should have been seen, assuming Earth's rotation rate has been constant between 136 B.C. and now. The eclipse should have been visible from Algeria. In fact, it was seen in Babylon, indicating that Earth's rotation rate has slowed down.

> The *second* is the duration of 9 192 631 770 periods of the radiation corresponding to the transition between the two hyperfine levels of the ground state of the cesium 133 atom (adopted 1967).

In 1990, a day contained 24 hours plus 1.35×10^{-3} seconds.

1.2.4   The SI Unit of Length—The Meter

Standardization of the length unit has as intricate a history as that of time. The SI unit of length is the meter, abbreviation m; it equals 3.28 ft, roughly the distance from your nose to the tip of your outstretched index finger. French scientists originally defined the *meter* to be 10^{-7} of the distance from the equator to the North Pole along the meridian through Paris. Such a definition is inconvenient to use, and so the *standard* actually chosen for the meter was the distance between two finely ruled lines on a bar of platinum–iridium metal kept in Paris. Other nations could obtain copies of the bar to act as national metric standards.

Mechanical methods for comparing lengths are much less precise than methods using light waves. Furthermore, the speed of light in vacuum is now believed to be a universal constant, the same in all experiments. Thus, the most precise length standard now possible is the distance light travels in a specified time:

> The *meter* is the distance traveled by light in vacuum during a time interval of 1/299 792 458 second (adopted 1983).

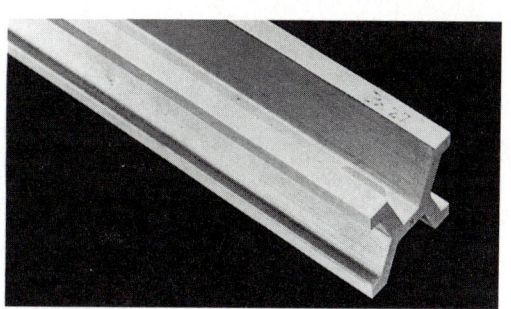

National standard meter bar kept at the NIST, Washington, D.C. Until the 1960s, the meter was defined as the distance between two lines machined on a similar bar kept at the BIPM in Paris.

Light source currently used to realize the definition of the meter. Measuring length with light waves is much more precise than mechanical comparisons. Cleverly devised electronics is used to relate the period of this light source to the period of the cesium standard. From the period and the speed of light, the wavelength is determined in meters. (See Chapter 16.)

■ Figure 1.6
Laser ranger used for mapping. It measures direction to a point using two angles, and distance by timing a laser pulse reflected from the object at the point. These measurements correspond to *spherical polar coordinates,* θ, ϕ, and r.

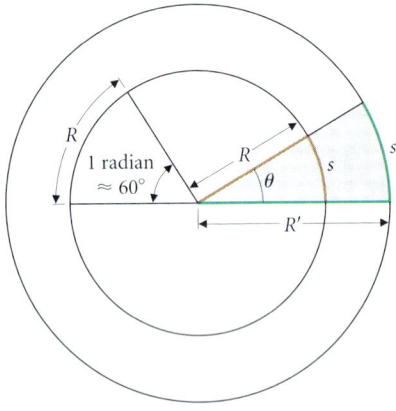

■ Figure 1.8
Radian measure is used to measure angles in physics. The angle θ measured in radians is the arc length s divided by the radius R of the circle. This definition is independent of the radius of the circle: $s/R = s'/R'$.

The time interval is chosen so that new and old definitions of the meter are as close to each other as possible. Measurements based on this standard are precise to a few parts in 10^{10}.

1.2.5 The SI Unit of Angle—The Radian

A laser ranger used by the U.S. Geological Survey (U.S.G.S.) for mapping (■ Figure 1.6) measures distance to an object by timing a laser pulse reflected from the object. Digital readouts from two scales give direction to the object as angles in the vertical and horizontal planes.
■ Figure 1.7 shows a close-up of one of the angular scales on an older and simpler surveying device. The scales divide the circular arc into 360 *degrees*. The degrees are subdivided into 60 *minutes* of arc, each of which contains 60 *seconds*.

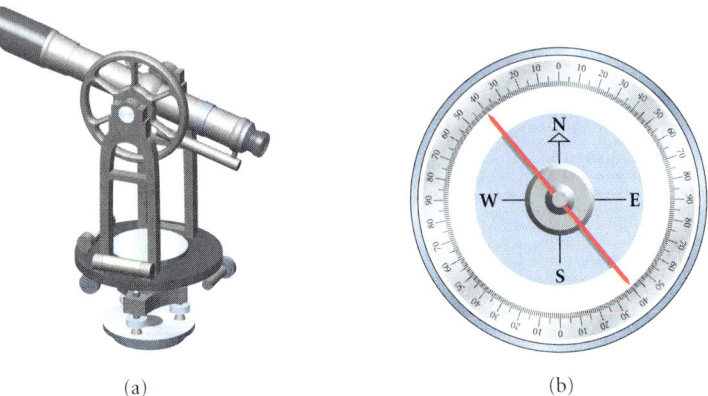

(a) (b)

■ Figure 1.7
This theodolite (a) measures angles in the same way as the laser ranger but was built before digital readouts were available. The scale (b) that measures the angle ϕ in the horizontal plane shows the 360° angle of a full circle divided into four 90° quadrants.

Though angle is very different from length, it is measured as the ratio of two lengths, an arc length and the radius of the scale (■ Figure 1.8). The measured angle does not depend on the radius of the scale, since the length of the arc is proportional to its radius. In Figure 1.8, the shaded portion of the circle is a fraction f of the whole circle, where

$$f = \frac{\text{angle (degrees)}}{360°} = \frac{\text{arc length}}{\text{circumference}} = \frac{\text{arc length}}{2\pi \times \text{radius}}.$$

So

$$\text{angle (degrees)} = \frac{360°}{2\pi} \frac{\text{arc length}}{\text{radius}}.$$

Degrees have a long history and are widely used, but are an awkward unit for physics. The number $(360/2\pi)$ has no physical significance and would needlessly complicate many calculations. We may define a unit of angle by cutting a circle into any number of pieces, not necessarily 360. We obtain a particularly convenient unit, the SI unit of angle, or *radian* (abbreviated rad), by choosing 2π pieces.

THE SI SUPPLEMENTARY UNIT OF ANGLE.

An angle of 1 *radian* corresponds to an arc length equal to the radius of the circle.

$$\text{Angle (radians)} \equiv \frac{\text{arc length}}{\text{radius}} = \frac{s}{r} \qquad (1.1)$$

CHECK HOW YOUR CALCULATOR HANDLES ANGLES. IT WILL PROBABLY EXPECT DEGREES UNLESS YOU TELL IT OTHERWISE. SPREADSHEET PROGRAMS USUALLY EXPECT RADIANS.

NOTE: In this text you should assume an angle is expressed in radians unless stated otherwise. The unit name is often omitted for angles in radian measure.

Digging Deeper

How Far Is a Second? How Quick Is a Meter?

The number 1/299 792 458 in the definition of the meter results from the history of the SI. What if we were free to choose a more convenient number, say 1 exactly? We could do this by defining new, non-SI units either for distance or for time: 1 "second" of distance ≡ distance light travels in one SI second ≡ 299 792 458 m; and 1 "meter" of time ≡ time light takes to travel one SI meter ≡ (1/299 792 458) s.

In addition to being good fun, this idea is useful. Theoretical physicists often find their results easier to interpret when expressed with the speed of light, c, equal to unity: $c \equiv 1$. For example, in Einstein's relativity theory, which models space and time as a unified, four-dimensional space-time, it is convenient to measure space and time in the same units.

How far is a "second" of length? Light goes around Earth in $\frac{1}{7}$ s. It reaches the Moon in approximately $\frac{5}{6}$ s.

How quick is a "meter" of time? One heartbeat is about 1 s, the time it takes for light to travel about 300 000 000 m.

A "second" is a large distance, and a "meter" is a short time.

EXAMPLE 1.1 ♦ Express an angle of 1° in terms of radians.

MODEL The arc length corresponding to 1° is 1/360 of the circumference.

SETUP Thus $s = (2\pi r/360)$, and so, from eqn. (1.1),

SOLVE $$1° \text{ (measured in radians)} = \frac{s}{r} = \frac{2\pi r/360}{r} = \frac{\pi}{180} = 1.745 \times 10^{-2} \text{ rad}.$$

ANALYZE The degree is a small fraction of a radian. For rough calculations, with $\pi \approx 3$, we may use $1° \approx \frac{1}{60}$ rad. ∎

EXERCISE 1.1 ♦ Show that 1.00 rad = 57.3°. What facts about equilateral triangles make it easy to remember that 1 rad ≈ 60°? ∎

Spherical polar coordinates (■ Figure 1.9) are an alternative way of assigning numbers to points in space, directly measurable with the U.S.G.S. laser ranger. Cylindrical polar coordinates (■ Figure 1.10) are another alternative that keeps the z-coordinate of Cartesian coordinates but uses distance and angle in the x–y plane. Relations among these alternative coordinates are explored in the problems.

Many mathematics texts use the names θ and ϕ in the opposite sense from that used in physics; i.e., they use ϕ for the angle from the z-axis.

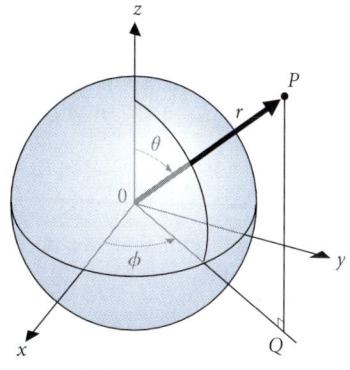

■ **Figure 1.9**
Spherical coordinates: r is the distance from the origin to P and θ is the angle between the z-axis and line OP. Line PQ is perpendicular to the x–y plane, and ϕ is the angle between OQ and the x-axis. So $z = r \cos\theta$, $x = r \sin\theta \cos\phi$, $y = r \sin\theta \sin\phi$.

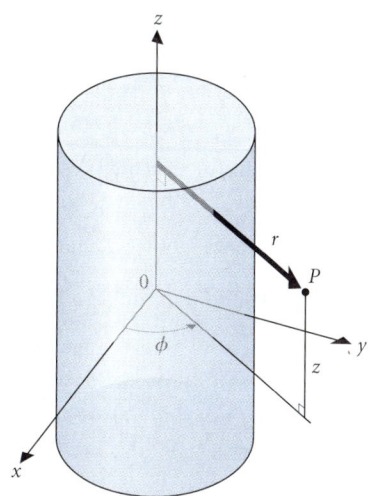

■ **Figure 1.10**
Cylindrical polar coordinates. The angle ϕ is defined as for spherical polars (Figure 1.9). The z-coordinate is defined as in cartesian coordinates (Figure 1.2). The third coordinate is r (sometimes called ρ), the perpendicular distance from the z-axis to the point P.

1.2.6 The SI Unit of Mass—The Kilogram

Roughly speaking, mass is the quantity in Newtonian physics that describes the amount of matter in an object. We shall have to wait until Chapter 4 to study the role of mass in physical laws, but we describe its unit now so as to complete the set of fundamental SI units.

The SI unit of mass is the *kilogram,* equal to 1000 grams. The original metric standard of mass defined the gram (abbreviation g) as the mass of 10^{-6} cubic meter (1 cubic centimeter or 1 milliliter) of pure water under specified conditions of pressure and temperature. Like the original definition of the meter, this definition of mass proved less precise than one based on a carefully manufactured object.

OFFICIAL SI DEFINITION OF THE KILOGRAM.

> The *kilogram* is equal to the mass of the international prototype of the kilogram (adopted 1889).

The international prototype is a finely polished cylinder of platinum–iridium alloy kept at the BIPM in Paris. This is the only standard still based on a particular physical object. The accuracy of mass measurements depends on such factors as atmospheric pressure and humidity, dust collecting on the prototype, accuracy of the laboratory equipment, and so on. The mass of an object compared with the prototype can be known at best to 1 part in 10^8.

1.3 USING SI

Suppose you ask the contractor building your house how much sand is necessary for the concrete in the foundation and get the answer "four." How could you interpret that? You would probably prefer this answer: "between 4.2 and 4.3 metric tons, using the volume estimated from the building plans and a standard formula for the mixture." Every physical quantity, like the amount of sand, is expressed as a numerical multiple of a unit and has both a history of how its value was measured or calculated and some uncertainty in its numerical value. To use SI, we have to learn how to talk about each of these features of physical quantities.

1.3.1 Significant Figures

Only rarely in physics are numerical values known exactly. Even when a number (such as π or the speed of light) is exact, it is usually unnecessary to write out the number to many decimal places. In calculations, we do not want either to lose information or to waste effort by retaining a false and misleading precision in our results. So we need an efficient and convenient set of rules for deciding how much precision to keep as we proceed through a calculation.

A measured physical quantity is only known within some range of experimental uncertainty. For example, if you find a length to be between the values 3.3755×10^7 m and 3.3765×10^7 m, a standard statement of your result is

$$\text{length} = (3.3760 \pm 0.0005) \times 10^7 \text{ m}.$$

This is a careful statement of the result and its uncertainty, more careful than we need for most purposes in this text. So, we shall use the abbreviated statement

$$\text{length} = 3.376 \times 10^7 \text{ m},$$

and *understand* that the uncertainty is 0.0005×10^7 m.

> RULE 1. Unless the uncertainty in a number is explicitly stated, it is assumed to be half a unit in the last decimal place quoted in the number.
>
> The *accuracy* of a number is the number of significant figures quoted. The *precision* is the uncertainty in the number.

National Standard Kilogram #20 at NIST. The kilogram is a platinum cylinder housed under the double bell jar.

The length 3.376×10^7 m has an *accuracy* of four significant figures and a *precision* of 0.0005×10^7 m.

When two numbers are multiplied or divided, the result is also uncertain. For example, 4.62×10^{-5} m is shorthand for a physical value in the range $(4.615$ to $4.625) \times 10^{-5}$ m. The area of a line $\ell = 3.376 \times 10^7$ m long and $w = 4.62 \times 10^{-5}$ m wide is the product of these two numbers, $A = \ell w$, and lies in the range

$$A_{min} = \ell_{min} w_{min} = (3.3755 \times 4.615) \times 10^2 \text{ m}^2 = 15.5779325 \times 10^2 \text{ m}^2$$

to

$$A_{max} = \ell_{max} w_{max} = (3.3765 \times 4.625) \times 10^2 \text{ m}^2 = 15.6163125 \times 10^2 \text{ m}^2.$$

Most of the decimals in these numbers are useless! The average of the two numbers is 15.5971×10^2 m², and the two calculated values differ from the average by ± 1.9 m², so we might write the area as $(15.60 \pm 0.02) \times 10^2$ m². In shorthand form, the area is

$$A = 15.6 \times 10^2 \text{ m}^2 = 1.56 \times 10^3 \text{ m}^2.$$

We should keep three significant figures. The implied uncertainty of 0.05×10^2 m² is slightly too large, but writing four significant figures would claim more accuracy than we actually have. The original numbers have four and three significant figures, respectively. This is an example of a second general rule.

REMEMBER: THE IMPLIED UNCERTAINTY IS ONE-HALF OF THE LAST DIGIT SHOWN.

> **RULE 2.** The number of significant figures in a product or quotient equals the smallest number of significant figures in any of the numbers being multiplied or divided.

A different rule holds for addition or subtraction. Common sense says that if you are timing two runners with your wristwatch and their times differ by a hundredth of a second, you won't notice the difference. For example, if 1.00×10^{-2} s is added to 2.32×10^2 s, the answer is still 2.32×10^2 s. The smaller time is completely insignificant given the precision with which we know the large number. On the other hand, if we add 3.37 s to 0.042 s, the sum lies in the range

$$(3.365 + 0.0415) \text{ s} = 3.4065 \text{ s to } (3.375 + 0.0425) \text{ s} = 3.4175 \text{ s}.$$

The best representation of this range according to Rule 1 is 3.41 s. These two examples illustrate the general rule.

NOTE THAT 3.4175 S IS SLIGHTLY OUTSIDE THE RANGE IMPLIED BY WRITING 3.41 S WITH THREE SIGNIFICANT FIGURES. REMEMBER, THE SIGNIFICANT FIGURES RULES ARE ALWAYS A COMPROMISE.

> **RULE 3.** The precision of a sum or difference is the same as the least precise of the numbers being added or subtracted.

When rounding an answer to the proper number of significant figures, round up if the first nonsignificant figure is 5 or greater, and round down if it's less. Roundoff errors build up more slowly if you keep extra decimal places until the end of the calculation.

Zeros are not written to the right of the decimal point unless they are significant. For example, the speed of light is exactly 2.99792458×10^8 m/s and so, correct to three figures, is 3.00×10^8 m/s. The zeros in a number like 300 kg = 3×10^2 kg are not significant. The corresponding number with three significant figures is 3.00×10^2 kg.

Significant figures are a shorthand replacing a lengthier and more specific statement of accuracy. For nearly all our purposes in this text the significant figure rules are adequate. In your own work you must use your judgment to be sure you produce sensible results. Some methods for improving on the significant figure rules are explored in the problems for this section.

A MORE EXTENSIVE TREATMENT OF UNCERTAINTIES IS USUALLY GIVEN IN THE LABORATORY PORTION OF THE COURSE.

> **EXERCISE 1.2** ♦ Evaluate the following expression showing explicitly each use of the significant figure rules. What is the uncertainty in the answer according to Rule 1?
>
> $$A = \frac{(3.37 \text{ m} + 0.42 \text{ m}) \times (6.23 \times 10^6 \text{ m}^2 - 5.4 \times 10^4 \text{ m}^2)}{27.94 \text{ m} - 27.63 \text{ m}}.$$

1.3.2 Units Conversion and SI Prefixes

SEE THE INSIDE FRONT COVER FOR DEFINITIONS OF BRITISH UNITS.

Any physical quantity can be expressed as the product of a numerical value (a dimensionless number) and a unit. The physical quantity does not depend on the units used to measure it, but the dimensionless number does. For example, the 200-in. telescope at Mt. Palomar is now called the 5-m telescope. Its diameter is 2.00×10^2 in. = 5.08 m. Changing the units in which something like the telescope diameter is expressed is called *units conversion*. The conversion is no more than a multiplication by unity, expressed as a ratio between two physically equal quantities. For the telescope, we use the definition of the inch: 1 in. $\equiv 2.54 \times 10^{-2}$ m (exactly).

THE QUANTITY 2.54×10^{-2} M IS PHYSICALLY EQUAL TO 1 IN. EVEN THOUGH THE RATIO IS NUMERICALLY 2.54×10^{-2} M/IN.

$$2.00 \times 10^2 \text{ in.} = (2.00 \times 10^2 \text{ in.})\left(\begin{array}{c}\text{physical}\\ \text{unity}\end{array}\right)$$

$$= (2.00 \times 10^2 \text{ in.})\left(\frac{2.54 \times 10^{-2} \text{ m}}{1 \text{ in.}}\right) = 5.08 \text{ m}.$$

The *names* of units cancel just like algebraic quantities. Since "inch" appears in both numerator and denominator of the above expression, it cancels, leaving "meter" as the only remaining unit name.

In SI a set of standard prefixes express base units divided by or multiplied by powers of 10 (● Table 1.1). For example, a typical bacterial cell has a diameter of 1 μm = 1 *micro*meter = 10^{-6} m. The radius of Earth is about 6.4 Mm = 6.4 *mega*meter = $6.4 \times 10^{+6}$ m.

The prefixes *deci*, *deka*, and *hekto* are seldom used in physics, and *centi* is used almost exclusively for the centimeter: 1 cm $\equiv 10^{-2}$ m. The other prefixes correspond to increases or decreases by successive factors of 1000. The common practice in engineering is to use prefixes so that the numbers expressing physical quantities are between 1 and 1000. Among physicists, the use of prefixes versus scientific notation is mostly a matter of taste, with no general convention.

TABLE 1.1

Power of 10	Prefix	Symbol
24	yotta	Y
21	zetta	Z
18	exa	E
15	peta	P
12	**tera**	**T**
9	**giga**	**G**
6	**mega**	**M**
3	**kilo**	**k**
2	hekto	h
1	deka	da
−1	deci	d
−2	**centi**	**c**
−3	**milli**	**m**
−6	**micro**	**μ**
−9	**nano**	**n**
−12	**pico**	**p**
−15	femto	f
−18	atto	a
−21	zepto	z
−24	yocto	y

*The prefixes printed in bold are the ones most commonly used. Memorize them!

EXAMPLE 1.2 ♦ A *Saturn V* rocket was 0.111 km long at launch. What was its length in micrometers? In feet?

MODEL First we work out the units conversion, then we apply it to the given quantity.

SETUP
$$1 \text{ km} = 10^3 \text{ m} = (10^3 \text{ m})\left(\frac{1 \text{ } \mu\text{m}}{10^{-6} \text{ m}}\right) = 10^9 \text{ } \mu\text{m} \text{ (exactly)}.$$

SOLVE
$$0.111 \text{ km} = (0.111 \text{ km})\left(\frac{10^9 \text{ } \mu\text{m}}{1 \text{ km}}\right) = 1.11 \times 10^8 \text{ } \mu\text{m}.$$

The *Saturn V* was 1.11×10^8 μm long.

SETUP One inch is defined to be 2.54 cm, so 1 ft $\equiv$ 12 in. $\equiv$ 12(2.54 cm) = 0.3048 m, exactly. Thus

$$0.111 \text{ km} = (0.111 \text{ km})\left(\frac{10^3 \text{ m}}{1 \text{ km}}\right)\left(\frac{1 \text{ ft}}{0.3048 \text{ m}}\right) = 364 \text{ ft}.$$

ANALYZE The conversion factors are exact, so the number of significant figures is determined by the accuracy of the given length. Notice that 364 ft is about 120 yd, or the length of a football field, including the end zones. ∎

EXAMPLE 1.3 ♦ Light travels at 3.00×10^8 m/s. Express its speed in miles per hour (mph).

MODEL There are two conversions to consider: meters to miles and seconds to hours.

REFER TO THE DISCUSSION IN §1.3.1. THE SPEED OF LIGHT HAS BEEN ROUNDED TO THREE SIGNIFICANT FIGURES.

SETUP From the data given on the inside front cover, 1 mile ≡ 5280 ft, and from the previous example, 1 ft = 0.3048 m.

SOLVE

$$3.00 \times 10^8 \text{ m/s} = \left(3.00 \times 10^8 \frac{\text{m}}{\text{s}}\right)\left(\frac{1 \text{ ft}}{0.3048 \text{ m}}\right)\left(\frac{1 \text{ mi}}{5280 \text{ ft}}\right)\left(\frac{60 \text{ s}}{1 \text{ min}}\right)\left(\frac{60 \text{ min}}{1 \text{ h}}\right)$$
$$= 0.671 \times 10^9 \text{ mph}.$$

ANALYZE Equivalently, the speed of light is 671 million mph. That's fast!

EXERCISE 1.3 ♦♦ Show that $1 \text{ y} = \pi \times 10^7 \text{ s}$. To how many significant figures is this expression correct?

1.3.3 Dimensional Analysis and Estimation

One day at lunch, so the story goes, Enrico Fermi asked the question, "Where are they?" Fermi loved to make rough estimates of obscure quantities, and his companions quickly realized what he was asking. Having estimated that a technical civilization could expand across the galaxy within 500 million years, Fermi wondered why a 10-billion-year-old galaxy isn't filled by some such civilization! His fascinating question remains unanswered. It is one of the best examples of how even a crude estimate can expose the essence of a problem.

Estimating the answer to a complicated problem without directly solving it is a favorite technique of physicists, who often cover paper napkins or the *back of an envelope* with such calculations while discussing problems over lunch or during coffee breaks. Like Fermi's question, such estimates are often good for isolating the important parts of a problem and deciding where to begin work on a complete solution. Let's look at a few useful techniques for making such estimates.

One such technique, dimensional analysis, is based on the concept of *physical dimensions.* Time, length, and mass are distinct kinds of physical quantities; each is called a physical dimension. As we study physical laws, we shall see that the dimensions of all other physical quantities can be expressed in terms of these three *fundamental* dimensions.

The dimension of a physical quantity is symbolized by enclosing the quantity in square brackets. Then symbols for the fundamental dimensions are

[time] ≡ *T*; [length] ≡ *L*; and [mass] ≡ *M*.

Speed is found by dividing a length by a time, so [speed] = *L/T*. Another example is the volume flow rate of water through a pipe, the volume that emerges from the pipe per unit time: [flow rate] = L^3/T.

Any equation relating physical variables must express equality between two quantities having the same dimension—you are unlikely to believe that five bananas equals six monkeys or that 76 m/s = 35 kg. Furthermore, you cannot add or subtract quantities of different

A Saturn V was 0.111 km long.

WE'LL USE THIS TECHNIQUE AND, WHEN WE DO, YOU'LL SEE THE ENVELOPE SYMBOL: ✉.

THE NAME *DIMENSION* AS USED HERE MEANS SOMETHING VERY DIFFERENT FROM THE THREE DIMENSIONS OF SPACE.

Enrico Fermi (1901–1954). An Italian physicist and one of the pioneers of nuclear physics, he was famous for his back-of-the-envelope estimates.

dimensions. You may, however, multiply or divide quantities of different dimension, as we did in the case of speed. These facts give rise to several useful techniques. Treating names of dimensions like algebraic symbols allows you to determine the dimension of a quantity from a relation that defines it, or sometimes to guess the relation from the known dimensions. Checking the dimensions of the answer to a complicated problem is one of the best methods for finding blunders. These techniques are typical of dimensional analysis and are illustrated in the following examples.

SEE §0.2 FOR A DISCUSSION OF KEPLER'S LAWS.

EXAMPLE 1.4 ♦ What are the dimensions of the proportionality constant in Kepler's third law (§0.2)?

MODEL Kepler's third law states that the square of the period T of a planet's orbit—that is, the time to complete one orbit—is proportional to the cube of its semimajor axis a.

SETUP Equivalently,
$$T^2 = Ca^3,$$
where C is a constant, a has the dimension of length, $[a] = L$, and T has the dimension of time, $[T] = T$.

SOLVE Both sides of the equation have equal dimensions.
$$[T^2] = T^2 = [Ca^3] = [C]L^3 \Rightarrow [C] = T^2/L^3$$

WE COULD AVOID THE PROBLEM BY USING A DIFFERENT SYMBOL, MAYBE P, FOR THE PERIOD. BUT THIS DILEMMA WILL CROP UP AGAIN, SO WE MAY AS WELL CONFRONT IT NOW.

ANALYZE Unfortunately, the symbol T is doing double duty, so we have to decide from context which meaning is intended. ∎

EXAMPLE 1.5 ♦♦ The density of a fluid is the mass contained in each cubic meter of the fluid—that is, the mass per unit volume. By comparing physical dimensions, find a relation between the amount of mass per unit time that flows through a pipe, the density ρ, and volume flow rate R.

MODEL The equation relating the three quantities must have the same physical dimension on each side.

SETUP We have [density] $= M/L^3$; [mass flow rate] $= M/T$; and [volume flow rate] $= L^3/T$. We write an equation in the form
$$\text{mass flow rate} = \rho^m R^n,$$
where m and n are exponents to be found. Next we look at the dimensions of both sides.
$$[\text{mass flow rate}] = \frac{M}{T} = [\rho]^m[R]^n = \left(\frac{M}{L^3}\right)^m\left(\frac{L^3}{T}\right)^n.$$

SOLVE We obtain the same dimensions on both sides of the equation if $m = n = 1$. Thus
$$\text{mass flow rate} = \rho R = (\text{density})(\text{volume flow rate}).$$

ANALYZE This method does not allow us to find dimensionless constants like π or 2 that may be in the correct formula. Therefore, we can only say that mass flow rate is *proportional* to the product of density and volume flow rate. ∎

In rough estimates, it is usually sufficient to find the *order of magnitude* of the desired result.

> The *order of magnitude* of a number is the power of 10 closest to the number.

Any number in the range $0.3 \times 10^n <$ number $\leq 3.0 \times 10^n$ is of order of magnitude 10^n. The symbol $\sim$ is frequently used to indicate that two numbers are of the same order of magnitude. It is read "is of the order of." Thus $6 \sim 10$; $20 \sim 10$; and $700 \sim 1000$. A typical person ~ 1 m high (i.e., closer to 1 m than 0.1 m or 10 m) and has a mass that $\sim 10^2$ kg (i.e., much more than 10 kg but much less than 10^3 kg).

WE CHOOSE 3 AS THE DIVIDING POINT OF THE DECADE BECAUSE LOG 3 ≈ 0.5.

EXAMPLE 1.6 ♦♦♦ Estimate the number of nitrogen molecules from Julius Caesar's last breath that you inhale each time you breathe.

MODEL At first glance, this question seems completely unanswerable! But then you can ask what you do know or what you can look up about people and nitrogen in the Earth's atmosphere. After 2000 years, it is reasonable to suppose that gas molecules have completely mixed throughout the atmosphere and that each breath you take is a fair sample of the atmosphere's molecules.

SETUP Then, the share N of Caesar's molecules you take in with a breath is

$$N \sim \frac{\text{volume of a breath}}{\text{volume of the atmosphere}} \times (\text{number of } N_2 \text{ molecules Caesar exhaled}).$$

The number of molecules exhaled is the mass of nitrogen in full lungs divided by the mass of a nitrogen molecule. You'll have to look up the volume of a typical breath, the density of nitrogen in the atmosphere, and the molecular mass. These numbers are 2×10^{-3} m³, 0.8 kg/m³, and 4.6×10^{-26} kg, respectively. The volume of the atmosphere is roughly the Earth's surface area $4\pi R_e^2$ times a typical height of about 10 km. From the inside front cover, $R_e = 6.4 \times 10^3$ km.

SOLVE

$$N \sim \frac{\binom{\text{volume of}}{\text{breath}}}{\binom{\text{surface area}}{\text{of Earth}}\binom{\text{height of}}{\text{atmosphere}}} \times \frac{\binom{\text{volume of}}{\text{breath}}\binom{\text{density of}}{\text{nitrogen}}}{\binom{\text{mass of nitrogen}}{\text{molecule}}}$$

$$\sim \frac{(2 \times 10^{-3} \text{m}^3)^2 (0.8 \text{ kg/m}^3)}{(4\pi)(6.4 \times 10^6 \text{ m})^2 (1 \times 10^4 \text{ m})(4.6 \times 10^{-26} \text{ kg})} = 14.$$

ANALYZE Clearly you don't breathe in very many of Caesar's molecules. On the other hand, it can be shown that only one breath in a million will contain none. Apart from a slightly macabre amusement, what do we gain from this calculation? Back-of-the-envelope work is supposed to pin down a rough order of magnitude and point out what would be needed to improve the estimate. If you care about a better estimate, you can probably trust the mixing hypothesis, but you had best find out what fraction of the Earth's nitrogen is tied up in soil, plants, and so on. But then why just the last breath? Hmmm—0.5 breaths per second for 40 y . . . The most important lesson is to turn your imagination loose. You can estimate the size of the most bizarre quantities, if you wish. ∎

1.4 VECTORS AND SCALARS

1.4.1 The Basic Distinction

If you are flying from New York to Boston, you want to know that the distance between the two cities is 300 km. You also want to know which way to fly—northeast. The information you need about the two cities is described by a *vector* (■ Figure 1.11).

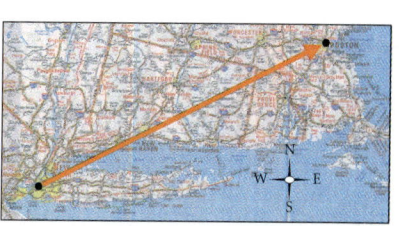

■ **FIGURE 1.11**
Boston is 300 km northeast of New York. This information is encoded mathematically by a *vector*—an arrow that points from New York to Boston.

A *vector* is a quantity with both magnitude and direction.

THESE FIRST DESCRIPTIONS TRY ONLY TO CAPTURE THE ESSENTIAL DIFFERENCE BETWEEN VECTOR AND SCALAR QUANTITIES. AS WE LEARN MORE PHYSICS, WE SHALL ALSO LEARN MORE ABOUT THE MATHEMATICS OF VECTORS AND SCALARS AND THEREBY DEVELOP SHARPER DEFINITIONS.

You also want to know whether Boston is cooler or hotter than New York. The temperature at each city at any given time is completely described by its numerical size and is an example of a *scalar*.

> A *scalar* is a quantity with size (or magnitude) but without any associated direction.

Our discussion so far has been about calculation with scalars. Understanding physics also requires visual, geometrical, and algebraic tools for working with vectors. In the rest of the chapter we introduce two vector quantities, position and displacement, and with them develop some basic tools for vector arithmetic: addition, subtraction, and multiplication by a number.

1.4.2 Displacement

When an object moves, its change of position is described by a *displacement* vector. The vector in Figure 1.11 is your displacement during the trip to Boston. It is represented by an arrow with its tail at the starting point—New York—and its head at the destination—Boston. Since displacement vectors are easy to visualize, we use a story involving several displacements to illustrate general properties, accepted notation, and rules for arithmetic with any kind of vector.

Jim and Alice hike from point P along a level trail and stop near a lake for lunch at point Q (■ Figure 1.12). Their displacement, $\vec{D}_{PQ}$ is the straight line segment between the two points with *direction* from the starting point P (arrow tail) to the end point Q (arrow head). The size, or *magnitude*, of the displacement is the length of the line segment—the straight line distance between P and Q. The arrow represents the vector geometrically. Algebraically, it is written as a letter with an arrow over it, or printed in boldface type, or both. Its magnitude is written with absolute value signs, $|\vec{D}_{PQ}|$, or, for convenience, as D_{PQ} whenever it won't be confused with another quantity. The magnitude of a vector is a scalar—that is, a number without direction—and is always positive.

EXAMPLE 1.7 ♦ Describe the hikers' displacement vector $\vec{D}_{PQ}$ in Figure 1.12.

MODEL To describe a vector, we must give both its magnitude and its direction.

SETUP From the figure caption, Q is 3 km east of P, and at the same altitude.

SOLVE $$\vec{D}_{PQ} = (3 \text{ km, east and horizontal}).$$

ANALYZE Alice and Jim probably walked a good deal farther than 3 km, since the trail they followed was *not* a straight line between points P and Q. The distance they walked is the *path length* between the two points. Do not confuse path length with displacement. (See §1.4.5.)

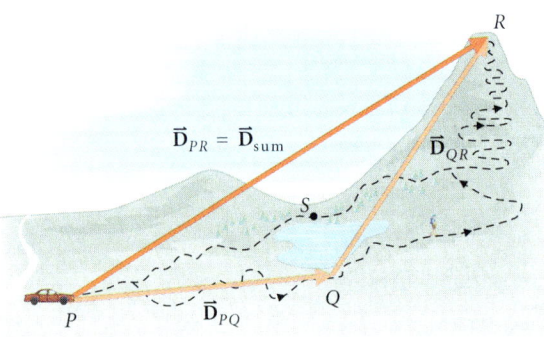

■ **FIGURE 1.12**
Alice and Jim start at P and hike to point Q, 3 km east of P and at the same altitude. There they stop for lunch. After lunch, the hike leads up the mountain to R. They camp at S for the night and return to the car at P in the morning. Although the hikers walked along a trail with switchbacks, the displacement *vectors* are straight lines between start and end of a trip segment.

1.4.3 Vector Addition

Invigorated by lunch, the hikers decide to climb a mountain, and arrive in midafternoon at the summit, point R (Figure 1.12). The displacement since lunch is $\vec{D}_{QR}$, and the total displacement since morning is $\vec{D}_{PR}$, the *sum* of the two individual vectors.

$$\vec{D}_{PR} = \vec{D}_{PQ} + \vec{D}_{QR}.$$

After reaching the summit, the hikers descend and camp for the night at point S. The displacement for the full day is the sum of three vectors (■ Figure 1.13):

$$\vec{D}_{\text{full day}} \equiv \vec{D}_{PS} = \vec{D}_{PQ} + \vec{D}_{QR} + \vec{D}_{RS}.$$

These examples motivate the basic definition for the sum of vectors.

> Each vector in a sum has its tail at the head of the preceding vector. The sum extends from the tail of the first vector to the head of the last vector.

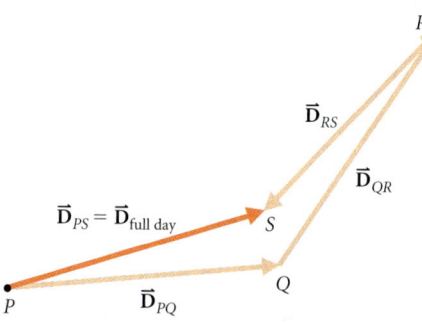

■ FIGURE 1.13
Adding three vectors corresponding to the three hikes (start to lunch, PQ, lunch to summit, QR, and summit to camp, RS) gives the hikers' total *displacement* $\vec{D}_{PS}$ for the day.

Our first example shows how to use this rule with two vectors in a horizontal plane.

EXAMPLE 1.8 ◆ Add the displacements $\vec{D}_1 = (3 \text{ km, west})$ and $\vec{D}_2 = (6 \text{ km}, 60°$ south of east).

MODEL In ■ Figure 1.14 we see the vectors added—that is, with the tail of $\vec{D}_2$ located at the head of $\vec{D}_1$.

SETUP The sum $\vec{D}_1 + \vec{D}_2$ extends from the tail of $\vec{D}_1$ to the head of $\vec{D}_2$. In this case, the vectors form a 30°, 60°, 90° right triangle.

SOLVE $\vec{D}_1 + \vec{D}_2 = (3\sqrt{3} \text{ km, south}) = (5 \text{ km, south}).$

ANALYZE The magnitude of the answer has one significant figure because the magnitudes of the original vectors are also given to one significant figure. ■

EXERCISE 1.4 ◆ Find the sum of the two displacements $\vec{D}_1 = (4 \text{ km, south})$ and $\vec{D}_2 = (4 \text{ km, west}).$

THE TERM RESULTANT IS A SYNONYM FOR THE "SUM" OF VECTORS.

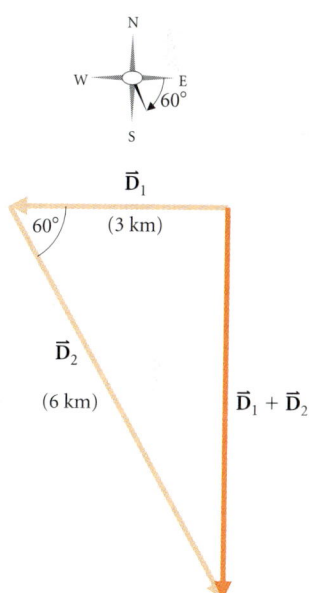

■ FIGURE 1.14
The vector addition in Example 1.8. Note that the sum is not always the longest vector.

1.4.4 The Zero Vector

The hikers return to their car in the morning. The displacement for the full trip is the sum of four vectors (■ Figure 1.15).

$$\vec{D}_{\text{trip}} = \vec{D}_{PQ} + \vec{D}_{QR} + \vec{D}_{RS} + \vec{D}_{SP}.$$

Since $\vec{D}_{\text{trip}}$ begins and ends at P, it has no length and is an example of the zero vector.

$$\vec{D}_{\text{trip}} = 0.$$

None of the individual displacements is zero, but because they fit together in a closed figure, their *sum* is zero.

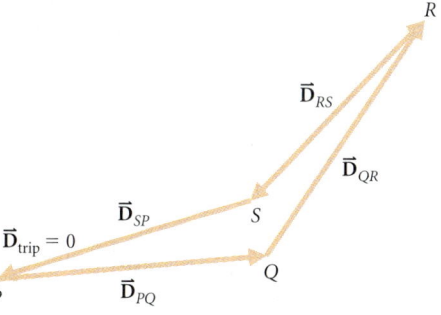

■ FIGURE 1.15
In the morning, the hikers return to their car. The sum of the four displacement vectors is an arrow that begins and ends at P: it has no length. This is an example of the zero vector.

1.4.5 Displacement and Distance

The hikers certainly won't boast about a trip with zero *displacement*. Friends are much more impressed by tales about the great *distance* they hiked. Their displacement is the vector sum that describes their trip. The sum of the lengths of the individual vectors is an *estimate* of the distance s they hiked:

$$s \approx |\vec{D}_{PQ}| + |\vec{D}_{QR}| + |\vec{D}_{RS}| + |\vec{D}_{SP}|$$

THIS IS ONLY AN ESTIMATE BECAUSE THE HIKERS DIDN'T WALK IN STRAIGHT LINES BETWEEN P AND Q, OR BETWEEN Q AND R, AND SO ON.

and is certainly not zero! A more accurate estimate of distance could be obtained by dividing the trip into shorter segments and adding the lengths of those vectors. We'll return to these distinctions in the next chapter, when we discuss speed and velocity.

We have encountered an important fact about the language of physics. A common word, displacement, has been given a precise definition, and the quantity it defines has a property ($\vec{D}_{trip} = 0$) we wouldn't have expected from common usage of the word. In physics, we avoid the multiple meanings of words that produce richness and flexibility in literature and speaking. No one abandons that richness in everyday speech, but we need to recognize that physicists use language with precise and specific meanings that must be carefully respected.

DON'T GET CAUGHT SHORT; MEMORIZE THE DEFINITIONS—IN PICTURES, IN ENGLISH, AND IN A FORMULA, IF ANY.

EXERCISE 1.5 ♦♦ A yacht tacking upwind moves $\frac{1}{4}$ km in a direction 30° east of north, comes about, and then moves $\frac{1}{4}$ km in a direction 30° west of north. What is the boat's displacement? What is the distance sailed?

TO SAIL UPWIND, A BOAT FOLLOWS A ZIG-ZAG PATH IN A MANEUVER CALLED *TACK-ING*. THE CHANGE FROM ZIG TO ZAG IS ABRUPT AND IS CALLED *COMING ABOUT*.

1.4.6 Multiplication of a Vector by a Scalar

Let us return briefly to our hikers to develop definitions for multiplication of a vector by a scalar and for subtraction of vectors.

During their hike, Alice and Jim take a side trip before returning to the main trail. They walk together 1 km northeast, a displacement $\vec{D}_1$ shown in ■ Figure 1.16a. Jim stops and Alice walks 2 more kilometers *in the same direction*. Her total displacement vector $\vec{D}_2$ (Figure 1.16b) is three times as long as $\vec{D}_1$. We may also think of Alice's total displacement as three repetitions of the same 1-km displacement.

$$\vec{D}_2 = \vec{D}_1 + \vec{D}_1 + \vec{D}_1.$$

Both of these statements mean that $\vec{D}_2 = 3\vec{D}_1$. We conclude that:

> Multiplication of a vector by a positive number multiplies its magnitude by that number without changing its direction.

To think of Alice's 3-km walk as three repetitions of the *same* displacement $\vec{D}_1$ points out an important mathematical property of vectors. Each of the three 1-km displacements is defined by its end points, which are different for each vector. Yet each displacement has the same magnitude (1 km) and the same direction as the others; the three vectors are the same despite being associated with different locations in space. In a diagram used to calculate with vectors, the vectors may be moved without any change in their meaning, so long as their magnitudes and directions are not altered. This idea is crucial for adding vectors other than displacements (e.g., velocities, Chapter 2; forces, Chapter 4; electric fields, Chapter 23). Such vectors are added tail to head but, unlike displacements, may have to be moved and placed tail to head.

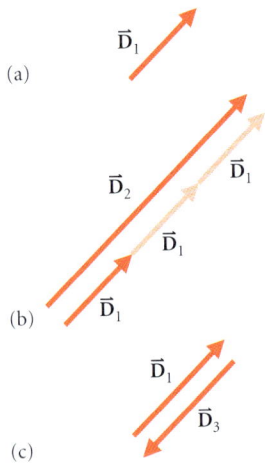

■ **FIGURE 1.16**
Multiplying a vector by a number. (a) The displacement vector $\vec{D}_1$. (b) The expression $\vec{D}_2 \equiv 3\vec{D}_1$ means that $\vec{D}_2$ has the same direction as $\vec{D}_1$, but three times the length. (c) Here $\vec{D}_3 \equiv -\vec{D}_1$ has the same length as $\vec{D}_1$ but is opposite in direction.

1.4.7 Subtraction of Vectors

Instead of continuing on with Alice, Jim returns to the starting point of the side trip with a displacement $\vec{D}_3$ (Figure 1.16c). Jim's total displacement $\vec{D}_t$ is zero, since he is back where he started. It is also the sum of his two displacements $\vec{D}_1$ and $\vec{D}_3$.

$$\vec{D}_t = \vec{D}_1 + \vec{D}_3 = 0.$$

So, we can write $\vec{D}_3 = -\vec{D}_1$ or $\vec{D}_3 = (-1)\vec{D}_1$: multiplication by -1 reverses the direction of a vector.

We have learned all we can from Jim's and Alice's trip. The rules we found are valid for any vector $\vec{A}$, not just for the displacements we've been thinking about. In fact, these rules become part of the definition of a vector.

> Rule for multiplying any vector $\vec{A}$ by a scalar c:
> The vector $c\vec{A}$ means a vector of magnitude $|c||\vec{A}|$ with direction the same as $\vec{A}$ if c is positive and with the opposite direction from $\vec{A}$ if c is negative.
>
> Rule for subtraction of two vectors:
> To subtract one vector from another, reverse its direction and then add. See Figure 1.17.
>
> $$\vec{A} - \vec{B} = \vec{A} + (-\vec{B}).$$

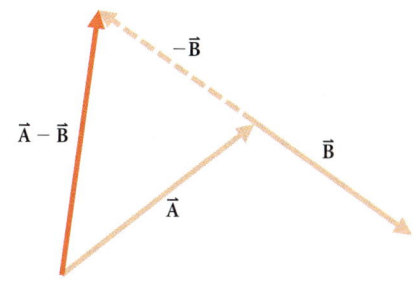

■ **FIGURE 1.17**
Subtracting vectors: $\vec{A} - \vec{B} = \vec{A} + (-\vec{B})$. Note that $-\vec{B}$ is in the opposite direction from $\vec{B}$, and is shown dashed with its tail on the head of $\vec{A}$.

■ Figure 1.18 illustrates an alternative statement of the rule for addition or subtraction of vectors, the *parallelogram* rule:

> Two vectors are placed with their tails together to form two sides of a parallelogram (Figure 1.18a). The sum of the vectors is the vector along the diagonal of the parallelogram starting at the tails of the two vectors. The difference lies along the other diagonal, with its tail at the head of the subtracted vector (Figure 1.18b).

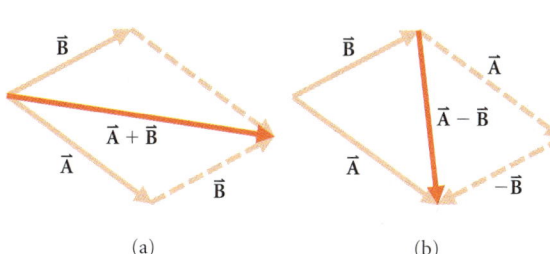

■ **FIGURE 1.18**
The parallelogram rule for adding and subtracting vectors. Draw the two vectors tail to tail, and complete the parallelogram. The sum is the long diagonal of the parallelogram starting at the two tails, and the difference is the diagonal from head to head. To get the correct direction of $\vec{A} - \vec{B}$, notice that $\vec{A} = \vec{B} + (\vec{A} - \vec{B})$, so $\vec{A} - \vec{B}$ must have its head at the head of $\vec{A}$.

EXAMPLE 1.9 ◆ If $\vec{A} = (3 \text{ km, east})$ and $\vec{B} = (4 \text{ km, north})$ (■ Figure 1.19a), find $\vec{A} - \vec{B}$.

MODEL To subtract the vectors, we first reverse the direction of $\vec{B}$ and then add.

SETUP $-\vec{B} = (4 \text{ km, south})$. Placing $(-\vec{B})$ at the head of $\vec{A}$, we form a 3-4-5 right triangle (Figure 1.19b).

SOLVE The angle θ is found from $\tan \theta = \frac{4}{3}$. Then $\theta = 53°$.

$$\vec{A} - \vec{B} = \vec{A} + (-\vec{B}) = (5 \text{ km, } 53° \text{ south of east}).$$

ANALYZE The dashed lines in Figure 1.19b show the parallelogram (rectangle in this case) formed by $\vec{A}$ and $-\vec{B}$. The vector $\vec{C} = \vec{A} - \vec{B}$ is along the diagonal, according to the parallelogram rule.

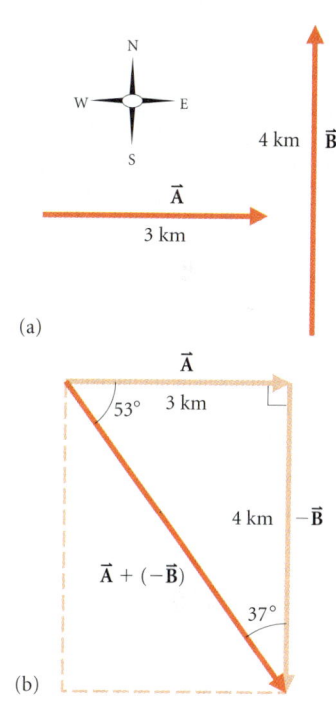

■ **FIGURE 1.19**
(a) Vectors $\vec{A}$ and $\vec{B}$. (b) Triangle showing the addition of $\vec{A}$ and $-\vec{B}$. The dashed lines show that the result may also be found from the parallelogram rule.

1.5 THE POSITION VECTOR

We represented the hiker's displacements in the last section by geometrical lines between well-defined points on the mountainous terrain surrounding the hikers. The surface of the Earth is the *frame of reference* in which the displacements are measured. As we saw in the Cartesian model of space, a frame of reference is a chosen set of points that are fixed with respect to each other. The reference frame has a specific point, the origin, with respect to which positions of all the other points of the frame are measured, and a coordinate system to use in measuring. In our examples the coordinate axes are chosen along the mutually perpendicular directions north–south, east–west, and up–down.

EARTH'S SURFACE IS NOT A PERFECT REFERENCE FRAME, SINCE IT IS BEING SLOWLY BUT CONTINUALLY DISTORTED BY GEOLOGICAL PROCESSES. THESE EFFECTS ARE EASILY MEASURABLE BUT OF SMALL ENOUGH MAGNITUDE THAT THEY CAN BE NEGLECTED FOR OUR PRESENT DISCUSSION. HIKERS OCCASIONALLY FIND ONE OF THE NUMEROUS BRASS MARKERS SET IN THE GROUND BY THE U.S. GEOLOGICAL SURVEY. ACCURATE SURVEYING REQUIRES REGULAR MONITORING OF THE RELATIVE POSITIONS OF THESE MARKERS.

The location of a point in the reference frame is described by the *position vector* $\vec{r}$, which extends from the origin to the point. For example, in Figure 1.20 the origin is chosen at Joe's gas station, and the position of the driver at point i on the freeway is described by

$$\vec{r}_i = (3 \text{ km}, 45° \text{ west of north}).$$

A position vector is similar to a displacement: its magnitude is the length of a straight line segment between two points, and its direction is defined by the two points. Displacement, however, is the *change in position* of an object and refers to two points occupied by a single object at two different times. The position vector is defined by two different points at the same time. In Figure 1.20 the driver undergoes a displacement $\vec{D}$ between two points labeled i for initial and f for final. With respect to origin O, the driver has initial position $\vec{r}_i$ and final position $\vec{r}_f$. Applying the parallelogram rule, we find

$$\vec{D} = \vec{r}_f - \vec{r}_i \equiv \Delta \vec{r}. \tag{1.2}$$

The displacement of an object is the *change* of its position vector. The symbol Δ means the change in a quantity and always means a final value of a quantity minus its initial value.

EXAMPLE 1.10 ♦♦ An airplane flies at constant altitude from its initial position 11 km northeast of Chicago O'Hare airport to its final position 22 km due south of Chicago O'Hare. Find its displacement.

MODEL The positions of the airplane are given from an origin at Chicago O'Hare airport (Figure 1.21). The displacement $\vec{D}$ is the change of the position vector.

$$\vec{D} = \vec{r}_f - \vec{r}_i.$$

SETUP The vectors form an oblique triangle that we can solve with rules from plane trigonometry (Appendix IB). Since we know the lengths of the two position vectors and the angle ϕ between them, we can find the magnitude of the displacement with the cosine rule:

$$|\vec{D}| = \sqrt{r_i^2 + r_f^2 - 2 r_i r_f \cos \phi},$$

where $\phi = 180° - 45° = 135°$ (Figure 1.21).

SOLVE

$$|\vec{D}| = \sqrt{\{11^2 + 22^2 - 2(11)(22)[\cos(135°)]\} \text{ (km)}^2}$$
$$= (11)\sqrt{(1 + 4 + 2\sqrt{2})} \text{ km}$$
$$= 31 \text{ km}.$$

SETUP Angle θ describes the displacement's direction. Now we know the magnitude D and the angle opposite it, as well as the length of the side opposite angle θ. So, we can use the sine rule (Appendix IB):

$$\frac{\sin \theta}{r_i} = \frac{\sin \phi}{D} \Rightarrow \frac{\sin \theta}{11 \text{ km}} = \frac{\sin(135°)}{31 \text{ km}} = \frac{1}{31\sqrt{2} \text{ km}}.$$

SOLVE
$$\sin \theta = \frac{11 \text{ km}}{31\sqrt{2} \text{ km}} = 0.25 \Rightarrow \theta = 15°.$$

Thus $\vec{D} = (31 \text{ km}, 15° \text{ west of south})$.

ANALYZE Complete the parallelogram in Figure 1.21 to verify the vector subtraction. ■

EXERCISE 1.6 ♦♦ If in Figure 1.20 $\vec{D} = (4 \text{ km, north})$, find the position vector $\vec{r}_f$ of the driver at the interchange.

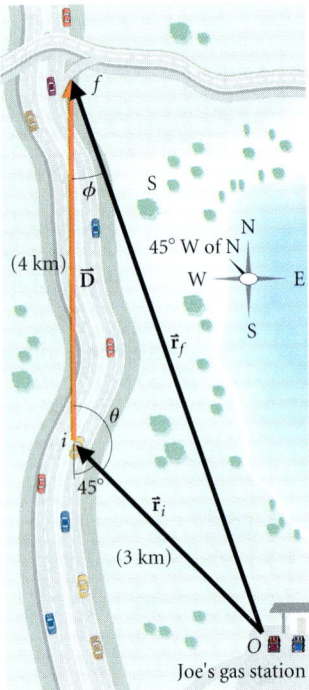

■ **FIGURE 1.20**
The position vector. First we pick an origin: Joe's gas station in this example. The position vector points from the origin to the car on the freeway. Two such vectors are shown here: $\vec{r}_i$, the initial position; and $\vec{r}_f$, the final position.

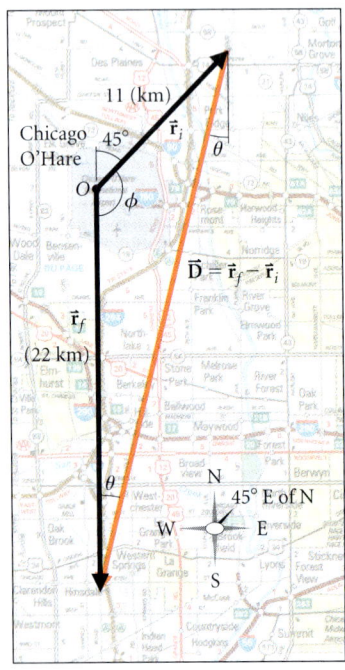

■ **FIGURE 1.21**
Displacement equals change in position. Here we subtract the position vectors of the airplane, with Chicago O'Hare airport as the origin. Note we have used the parallelogram rule (Figure 1.18).

1.6 Vector Algebra

1.6.1 Components

When adding or subtracting vectors that do not form a right triangle, as in finding the displacement of the airplane in Example 1.10, calculations are often simplified by choosing a convenient coordinate system and *resolving* (breaking up) the vectors into *components* that describe how much the vector points in the direction of each coordinate axis. To see how the method works, look at the two ways we have to describe the position of an object in a plane: by its position vector $\vec{r}$ and by its coordinates (x, y). ■ Figure 1.22 shows the relation between the two descriptions. Constructing lines from the tip of $\vec{r}$ perpendicular to each coordinate axis, we define two *component vectors*, $\vec{r}_x$ and $\vec{r}_y$, whose vector sum is $\vec{r}$. Because the component vectors lie along the coordinate axes, they are completely described by the coordinates. For example, the magnitude of $\vec{r}_x$ is $|x|$, and the algebraic sign of x tells whether $\vec{r}_x$ points along the positive x-axis or in the opposite direction. The coordinates (x, y) are also called the *components* of the position vector $\vec{r}$.

We can find components for any vector in a similar way. Constructing a perpendicular from the tip of the vector to a coordinate axis creates a right triangle with the vector along the hypotenuse. For vectors in a plane, the other two sides of the triangle give the desired components (■ Figure 1.23). Our next example shows how to find them.

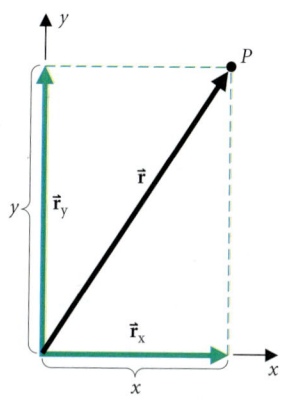

■ **Figure 1.22**
Any vector is the sum of its component vectors, each parallel to one of the coordinate axes. Here $\vec{r} = \vec{r}_x + \vec{r}_y$. The coordinates of a point are the components of its position vector.

> **EXAMPLE 1.11** ♦ Peter jogs 0.500 km along a straight country road in a direction 20.0° south of east. Find the components of Peter's displacement along coordinate axes oriented toward the north and toward the east.
>
> **MODEL** Peter's displacement $\vec{D}$ is a vector from point O to point A in ■ Figure 1.24. We model it as the sum of two vectors, one pointing east (from O to B) and one pointing south (from B to A). Each component vector describes how far Peter has gone in the corresponding direction. We define coordinates with the origin at O, the x-axis to the east, and the y-axis to the north.
>
> **SETUP** First we find the lengths of lines OB and BA, which are the magnitudes of the components. Then we'll worry about the signs, which indicate directions.
>
> $$\frac{OB}{OA} = \cos(20.0°),$$
>
> so
>
> $$OB = OA \cos(20.0°) = |\vec{D}|\cos(20.0°)$$
> $$= (0.500 \text{ km})(0.940) = 0.470 \text{ km}.$$
>
> Similarly,
>
> $$\frac{BA}{OA} = \sin(20.0°),$$

Remember, in such a diagram you are free to move the vector so that its tail is at the origin.

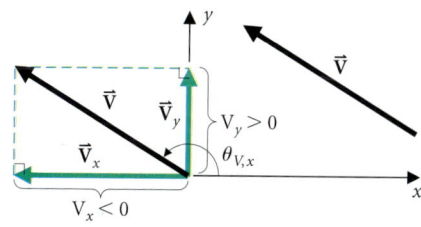

■ **Figure 1.23**
Components for an arbitrary vector in a plane. The vector $\vec{V}$ may be drawn with its tail at the coordinate origin. Components are then found by constructing perpendiculars from the arrow head to the coordinate axes. This vector has a negative x-component; its component vector points in the negative x-direction.

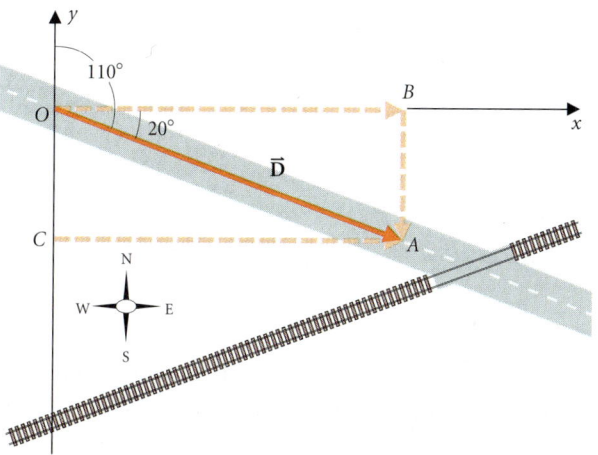

■ **Figure 1.24**
Peter jogs along the road from O to A. His displacement has an eastward (x) component equal to the length OB and positive; its northward (y) component has magnitude equal to $OC = AB$, and is negative.

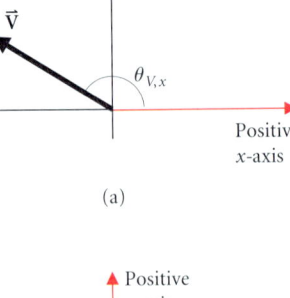

(a)

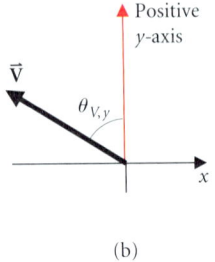

(b)

■ **FIGURE 1.25**
The angle $\theta_{V,x}$ is the angle between the vector and the *positive x*-axis. Similarly, $\theta_{V,y}$ is the angle between the vector and the positive *y*-axis. These angles are used to calculate the components (eqn. 1.3).

so
$$BA = OA\sin(20.0°) = |\vec{D}|\sin(20.0°)$$
$$= (0.500\text{ km})(0.342) = 0.171\text{ km}.$$

SOLVE The component vector $\vec{D}_x$ points east, in the positive *x*-direction, so the component D_x is positive: $D_x = 0.470$ km. On the other hand, the component vector $\vec{D}_y$ points south, in the *negative y*-direction, so the component D_y is negative: $D_y = -0.171$ km.

ANALYZE Since the direction of $\vec{D}$ is 110° east of north, and the cosine of an angle between 90° and 180° is negative, we can find an alternative expression for D_y:

$$|\vec{D}|\cos(110°) = (0.500\text{ km})(-0.342) = -0.171\text{ km} = D_y.$$ ■

Example 1.11 illustrates a general rule for finding components.

The *component* of a vector along a given coordinate axis equals the magnitude of the vector multiplied by the cosine of the angle between the vector and that axis. The correct angle is defined in ■ Figure 1.25.

$$V_x = |\vec{V}|\cos\theta_{V,x}. \qquad V_y = |\vec{V}|\cos\theta_{V,y}. \tag{1.3}$$

Often you will want the component of a vector along some direction other than a coordinate axis. You can use the same general rule to find it. For example, suppose Peter wants to know the component of his displacement parallel to the railroad track that lies in the direction 20.0° north of east (■ Figure 1.26). Peter's displacement vector $\vec{D}$ makes an angle of 40.0° with the track, so the desired component is $(0.500\text{ km})\cos(40.0°) = 0.388\text{ km} = 388$ m.

Because a vector's components along any set of coordinate axes completely define the vector, a common way of referring to a vector is to write its components, like coordinates, in parentheses: $\vec{V} = (V_x, V_y)$.

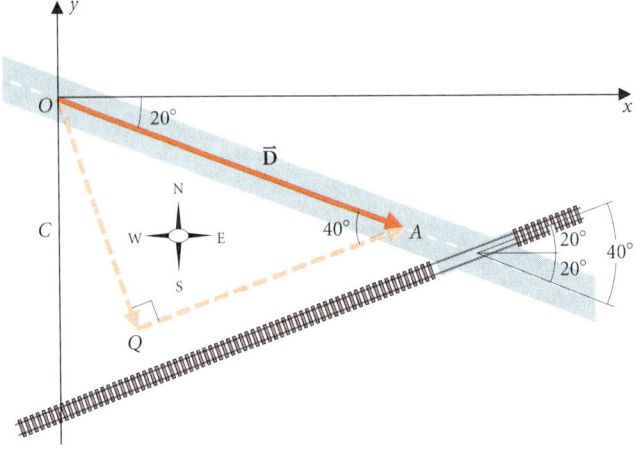

■ **FIGURE 1.26**
The angle between Peter's displacement vector and the railroad track is 40°; *QA* and *OQ* give the components along and perpendicular to the railroad track.

Vector arithmetic can often be simplified once the vectors are written in component form. When a vector $\vec{A} = (A_x, A_y)$ is multiplied by a number n, each component is multiplied by n (■ Figure 1.27).

$$n\vec{A} = (nA_x, nA_y).$$

When two vectors are added, their components are added (■ Figure 1.28).

$$\vec{C} = \vec{A} + \vec{B} \Rightarrow C_x = A_x + B_x; \quad C_y = A_y + B_y.$$

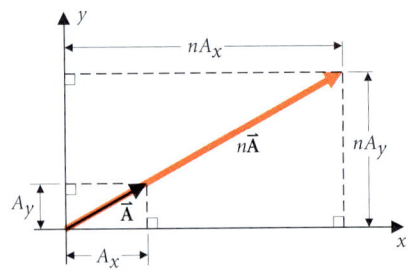

■ **FIGURE 1.27**
When a vector $\vec{A}$ is multiplied by a number n, each component is multiplied by n.

EXAMPLE 1.12 ◆◆ Calculate the displacement of the aircraft that flies from an initial position 11 km northeast of the Chicago O'Hare airport to a final position 22 km due south (cf. Example 1.10 and Figure 1.21) using vector components.

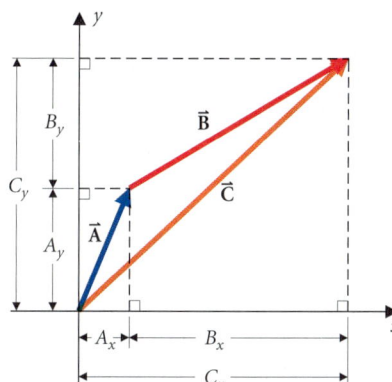

■ **Figure 1.28**
The vectors $\vec{A}$, $\vec{B}$, and $\vec{C} = \vec{A} + \vec{B}$ form a triangle. Perpendicular lines from the vertices to the x-axis produce line segments A_x, B_x, and C_x that represent the x-components. The segments A_x and B_x divide C_x into two pieces. The same is true for the y-components. When adding vectors, we may just add the components: $C_x = A_x + B_x$. Remember to take proper account of the signs! All the components shown here are positive, but that will not always be the case.

MODEL To find the airplane's displacement, we subtract its initial position from its final position. First, we express the two positions in terms of their components and then subtract the components.

SETUP We choose the x-axis to be east and the y-axis to be north, with the origin at the airport (■ Figure 1.29).

$$\text{Initial position } \vec{r}_i: x_i = y_i = (11 \text{ km})\cos(45°) = 7.78 \text{ km}.$$
$$\text{Final position } \vec{r}_f: x_f = 0; y_f = -22 \text{ km}.$$
$$\text{Displacement } \vec{D} = \Delta\vec{r} = \vec{r}_f - \vec{r}_i.$$

SOLVE $\quad D_x = x_f - x_i = 0 - 7.8 \text{ km} = -7.8 \text{ km}$

and $\quad D_y = y_f - y_i = -22 \text{ km} - 7.8 \text{ km} = -30 \text{ km}.$

ANALYZE Each answer has two significant figures, so we should write 30 km as 3.0×10^1 km.

We could stop here, since the vector is known when its components are known. Let's check the previous answer.

$$|\vec{D}| = \sqrt{(D_x)^2 + (D_y)^2} = [(-7.8 \text{ km})^2 + (-30 \text{ km})^2]^{1/2} = 31 \text{ km}.$$

The angle of the displacement from south (θ in Figure 1.29) is given by

$$\theta = \tan^{-1}\frac{|D_x|}{|D_y|} = \tan^{-1}\left(\frac{7.8}{30}\right) = 15°.$$

These are, of course, the same results as before.

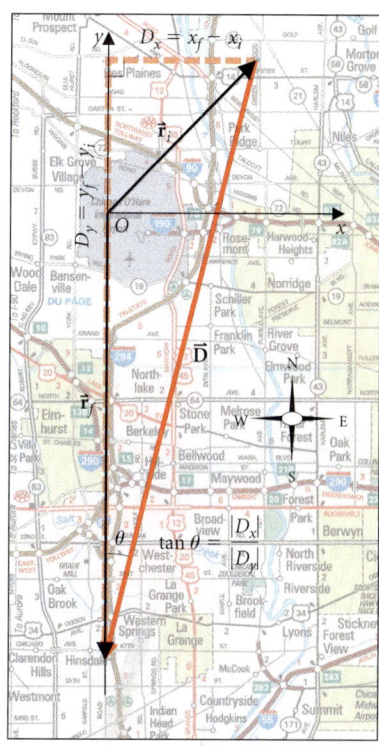

■ **Figure 1.29**
In Example 1.10 we computed the displacement of the airplane from its change in position. Here we do that calculation using components. The magnitude is found from the components using the Pythagorean theorem. The ratio of the components gives the direction of the vector: $\tan\theta = D_x/D_y$.

EXERCISE 1.7 ♦♦ The sailboat *Stars and Stripes* sails 15.0 km at an angle of 32° from the upwind direction. In the same time period, *Australian Pride* sails 17.2 km at an angle of 43° from upwind. Which boat, if either, has made more progress in the upwind direction?

To find vector components in three dimensions (■ Figure 1.30), we use the same techniques and follow the same general rule we derived for two dimensions. We explore some of the details in the problems.

1.6.2 Unit Vectors

Dividing any vector $\vec{a}$ by its own magnitude produces a new, dimensionless vector with the same direction but with magnitude unity. This new vector is called a unit vector and is written with the symbol ∧.

$$\hat{a} \equiv \frac{\vec{a}}{|\vec{a}|}, \quad \text{and} \quad \vec{a} = |\vec{a}|\hat{a}. \tag{1.4}$$

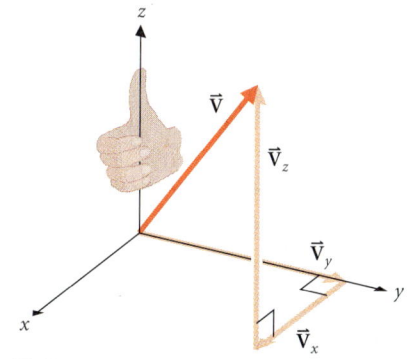

■ **Figure 1.30**
Vector components in three dimensions. With the tail of $\vec{V}$ at the origin, V_x has a magnitude equal to the perpendicular distance from the tip of $\vec{V}$ to the y–z plane. The other components are found similarly.

EXAMPLE 1.13 ♦ What are the x- and y-components of the unit vector in the direction of the aircraft's displacement $\vec{D}$ (Figure 1.29)?

MODEL In Example 1.12 we found the magnitude of $\vec{D}$ and its components. We use eqn. (1.4) to calculate the unit vector.

SETUP
$$\hat{D} \equiv \frac{\vec{D}}{D} = \frac{(D_x, D_y)}{D}$$

SOLVE
$$= \frac{(-7.8 \text{ km}, -30 \text{ km})}{31 \text{ km}} = (-0.25, -0.97).$$

ANALYZE Check that this is a unit vector. Its magnitude is
$$\sqrt{0.25^2 + 0.97^2} = 1.0,$$
and it is dimensionless.

THE NAMES *i*, *j*, AND *k* ARE TRADITIONAL EXCEPTIONS TO THE RULE OF NAMING UNIT VECTORS AFTER THE VECTOR THAT DESCRIBES THEIR DIRECTION.

Unit vectors serve to label directions. The most common unit vectors are
$$\hat{i} \equiv \hat{x}; \qquad \hat{j} \equiv \hat{y}; \qquad \text{and} \qquad \hat{k} \equiv \hat{z}.$$

They lie along the Cartesian coordinate axes and are used when writing a vector in terms of its components. For the aircraft's displacement vector in Example 1.12,
$$\vec{D} = \vec{D}_x + \vec{D}_y = D_x \hat{i} + D_y \hat{j} = (-7.8 \text{ km})\hat{i} + (-30 \text{ km})\hat{j}.$$

The corresponding unit vector (Example 1.13) is
$$\hat{D} = (-0.25)\hat{i} + (-0.97)\hat{j}.$$

A vector equals the sum of its component vectors. Each component vector is the component multiplied by the corresponding unit vector.

This rule extends directly to three dimensions. For the arbitrary vector $\vec{V}$ represented in Figure 1.30 as the sum of component vectors,
$$\vec{V} = \vec{V}_x + \vec{V}_y + \vec{V}_z = V_x \hat{i} + V_y \hat{j} + V_z \hat{k}.$$

1.6.3 Choosing Coordinate Systems

Nothing in the behavior of a physical system depends on how you choose to label it—that is, on which directions you choose for the coordinate axes. Thus, since the component method is supposed to simplify problems, you should choose a coordinate system for any problem so as to make component calculations as easy as possible.

EXAMPLE 1.14 ♦♦ During a boat race, a yacht sails 1.50 km northeast, turns 28° to the left, and sails another 1.25 km. What is its displacement from the start?

MODEL ■ Figure 1.31 shows a diagram of the boat's displacements. Neither of the vectors is due east or north, so if we chose these "obvious" directions for our coordinate axes, each vector would have two components. However, if we choose the x-axis to be along the first vector (northeast), that vector has only one component. In addition, since we are told that the yacht turns 28°, it is convenient to find the components of $\vec{D}_2$ in this system. We choose the y-axis to be perpendicular to the x-axis, as shown in the figure.

SETUP With our chosen coordinate system, the components of the vectors are
$$\vec{D}_1 = (1.50 \text{ km})\hat{i} \qquad \text{and} \qquad \vec{D}_2 = (1.25 \text{ km})(\hat{i} \cos 28° + \hat{j} \sin 28°).$$

The total displacement is the sum of these two vectors.
$$\vec{D} = \vec{D}_1 + \vec{D}_2 = (1.50 \text{ km})\hat{i} + (1.25 \text{ km})(\hat{i} \cos 28° + \hat{j} \sin 28°).$$

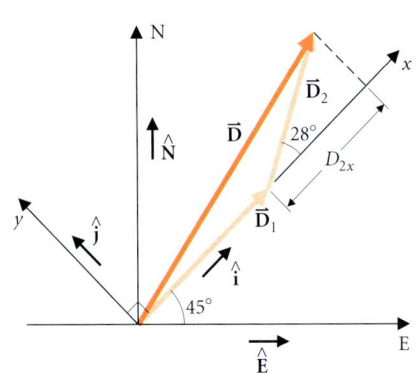

■ **FIGURE 1.31**
The displacement of the boat is most easily calculated using components in the x- and y-directions. The directions labeled x and y do not have to be east and west, or horizontal and vertical. You can choose to orient the axes any way that you want!

SOLVE Collecting terms gives

$$\vec{D} = \hat{i}[1.50 \text{ km} + (1.25 \text{ km})\cos 28°] + \hat{j}[(1.25 \text{ km})\sin 28°]$$
$$= (2.6 \text{ km})\hat{i} + (0.59 \text{ km})\hat{j}.$$

ANALYZE The magnitude of this displacement is

$$|\vec{D}| = \sqrt{D_x^2 + D_y^2} = \sqrt{(2.6 \text{ km})^2 + (0.59 \text{ km})^2} = 2.7 \text{ km}.$$

EXERCISE 1.8 ♦♦ Solve Example 1.14 using coordinate axes in the directions east and north. Compare both method and answer with Example 1.14.

Coordinate systems and components of vectors form a powerful set of tools for doing vector mathematics. It is sometimes easy to forget that the choice of a coordinate system is no more than an arbitrary assignment of names to particular directions and has no physical significance. We may choose the system entirely for convenience in calculation without affecting the basic properties of vectors or of the physical system. Of course, changing the coordinate system may change the numerical values of the vector components or the angles the vector makes with the axes. These are not fundamental properties of the vector, like its magnitude or the physical objects it points toward. Numerical quantities such as a vector's magnitude that are independent of coordinate choice are called *invariants*. A scalar has to be an invariant. Vector components are numerical quantities but *not* scalars.

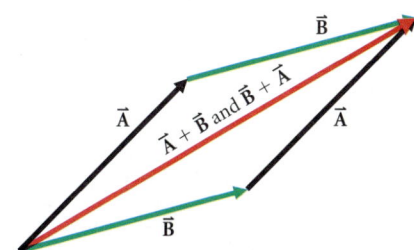

■ **FIGURE 1.32**
Vector addition commutes: $\vec{A} + \vec{B} = \vec{B} + \vec{A}$. The top sides of the parallelogram represent $\vec{A} + \vec{B}$, while the bottom sides represent $\vec{B} + \vec{A}$. Both sums are represented by the diagonal of the parallelogram.

1.6.4 Vector Algebra

Often you will need to manipulate expressions involving vectors algebraically, rather than calculating directly with components or from a vector triangle. From our work with components, we know that the basic rules for addition and scalar multiplication are the same as for ordinary numbers. We may illustrate these rules geometrically without reference to any coordinate system.

BUT, NEVER TRY TO ADD A VECTOR TO A SCALAR!

COMMUTATIVE RULE FOR ADDITION The order in which vectors are added makes no difference (■ Figure 1.32).

$$\vec{A} + \vec{B} = \vec{B} + \vec{A}.$$

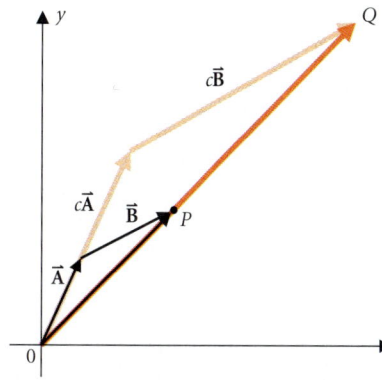

■ **FIGURE 1.33**
The distributive law for vector addition and multiplication by a scalar.
$$\overrightarrow{OP} = \vec{A} + \vec{B};$$
$$\overrightarrow{OQ} = c\vec{A} + c\vec{B}.$$
$\overrightarrow{OQ}$ also equals $c(\vec{A} + \vec{B})$.

DISTRIBUTIVE RULE FOR SCALAR MULTIPLICATION It does not matter in which order we do addition and scalar multiplication (■ Figure 1.33).

$$c(\vec{A} + \vec{B}) = c\vec{A} + c\vec{B}.$$

EXERCISE 1.9 ♦ Simplify the following expression.

$$\vec{D} = \frac{6(\vec{A} + 4\vec{B} - 2\vec{C}) + 4[3\vec{C} - (2\vec{A} + 5\vec{B})]}{8}.$$

Chapter Summary

Where Are We Now?

In this chapter we have introduced the basic tools we'll need to study physics. We have described a model for thinking about space and time, and discussed the SI units for expressing physical quantities. We developed the concept of vectors and rules for calculating with them. With these tools, we are ready to study the motion of simple objects and changes in their motion.

What Did We Do?

In Newtonian physics, space is without structure and time flows uniformly. To model space and time, first choose a *reference frame,* a set of objects fixed with respect to each other, and imagine a grid of meter sticks for measuring position and clocks for measuring time.

A physical measurement is a comparison of two similar physical quantities. A valid measurement procedure uses a stable device, accurately calibrated against a widely accepted standard. The Bureau Internationale des Poids et Mesures is responsible for defining standards and the methods of applying them.

The SI (Système Internationale) units for the fundamental dimensions are those of length (the meter), time (the second), and mass (the kilogram). The unit for angle is the radian.

Rules for significant figures provide a convenient, though only moderately accurate, way to keep track of uncertainty in the results of numerical calculations.

> The uncertainty in a number is assumed to be half a unit in the last decimal place quoted.
>
> The result of multiplication or division is as accurate as the least accurate of the numbers multiplied or divided.
>
> The result of addition or subtraction is as precise as the least precise of the numbers added or subtracted.

To convert the expression of a physical quantity from one unit to another, multiply by unity expressed as the ratio of the new unit to its value in terms of the old unit. The SI prefixes are useful for expressing measurements in a convenient form—for example, distances between cities in kilometers rather than meters.

Physical quantities have a *dimension* that can be expressed in terms of three fundamental dimensions: mass, length, and time. All terms in an equation must have the same physical dimension. Frequently, one may estimate the order of magnitude of a calculation by considering only the physical dimensions of known information in the problem.

A vector is a quantity that has both magnitude and direction. Change of position, or displacement, is a good intuitive example of a vector. All vectors obey the arithmetic rules we demonstrated for displacement. In contrast, a scalar quantity has magnitude only, and its value does not depend on the coordinate system used. Temperature is an example of a scalar quantity.

Two vectors are added when the tail of the second lies at the head of the first. The sum is the third side of the resulting triangle, with direction from the tail of the first vector to the head of the second. Multiplying a vector by a scalar multiplies the magnitude of the vector. If the scalar is negative, the vector's direction is also reversed.

The position vector $\vec{r}$ of an object has its tail at the origin and its head at the location of the object. The displacement of an object is the change of its position.

$$\vec{D} = \Delta \vec{r} = \vec{r}_f - \vec{r}_i.$$

It is often convenient to express vectors in component form.

$$\vec{V} = V_x \hat{i} + V_y \hat{j} + V_z \hat{k}.$$

The unit vectors $\hat{\mathbf{i}}, \hat{\mathbf{j}}$, and $\hat{\mathbf{k}}$ point in the direction of the coordinate axes and have magnitude unity. The components V_x, V_y, and V_z equal the magnitude of $\vec{\mathbf{V}}$ multiplied by the cosine of the angle between the vector and the corresponding coordinate axis. The orientation of coordinate axes is arbitrary and may be chosen to make calculations simpler. The rules of formal algebra for addition and scalar multiplication of vectors are identical to the rules for algebra with numerical quantities.

$$c\vec{\mathbf{V}} = cV_x\hat{\mathbf{i}} + cV_y\hat{\mathbf{j}} + cV_z\hat{\mathbf{k}}$$
$$\vec{\mathbf{V}} + \vec{\mathbf{U}} = (V_x + U_x)\hat{\mathbf{i}} + (V_y + U_y)\hat{\mathbf{j}} + (V_z + U_z)\hat{\mathbf{k}}.$$

A compilation of vector relations is given in Appendix IC.

Practical applications

In this chapter we have learned some fundamental tools for use in the rest of our work. Reference frames, coordinate systems, and measurement of position underlie the practice of surveying and navigation. Precision metrology is essential to competitive manufacturing technology.

Solutions to Exercises

1.1 (■ Figure 1.34). Angle ϕ is 1 rad when the arc AB equals the radius of the circle. Using eqn. (1.1) with arc length equal to radius,

$$1 \text{ rad} = \frac{360°}{2\pi} \times \frac{\text{radius}}{\text{radius}} = \frac{180°}{\pi} = 57.3°.$$

The chord AB is almost the same length as the arc AB, so triangle OAB is almost an equilateral triangle with three equal angles of 60°. Thus ϕ is almost equal to 60°.

1.2 First evaluate the sum and differences using Rule 3. Note that $5.4 \times 10^4 = 0.054 \times 10^6$, so that 6.23×10^6 is the less precise number.

$$A = \frac{(3.79 \text{ m}) \times (6.18 \times 10^6 \text{ m}^2)}{0.31 \text{ m}} = 7.56 \times 10^7 \text{ m}^2.$$

Next, notice that the denominator is accurate only to two significant figures. So,

$$A = 7.6 \times 10^7 \text{ m}^2.$$

By rule 1, the uncertainty is $0.05 \times 10^7 \text{ m}^2$.

1.3 Converting a year into seconds,

$$1 \text{ y} = (1 \text{ y})\left(\frac{365.25 \text{ d}}{1 \text{ y}}\right)\left(\frac{24 \text{ h}}{1 \text{ d}}\right)\left(\frac{60 \text{ min}}{1 \text{ h}}\right)\left(\frac{60 \text{ s}}{1 \text{ min}}\right) = 3.1558 \times 10^7 \text{ s}.$$

Now, $\pi \times 10^7 \text{ s} = 3.1416 \times 10^7 \text{ s}$, so the discrepancy between the two numbers is

$$\delta = 3.1558 \times 10^7 \text{ s} - 3.1416 \times 10^7 \text{ s} = 0.0142 \times 10^7 \text{ s}.$$

The agreement is within 0.5%, but only to one significant figure.

1.4 The vector triangle (■ Figure 1.35) is an isosceles, right-angled triangle. The total displacement has magnitude $4\sqrt{2}$ km and direction 45° west of south (or 225° clockwise from north).

1.5 ■ Figure 1.36 shows the vector triangle. It is an isosceles triangle with 30° angles. The total displacement (AC in the figure) has direc-

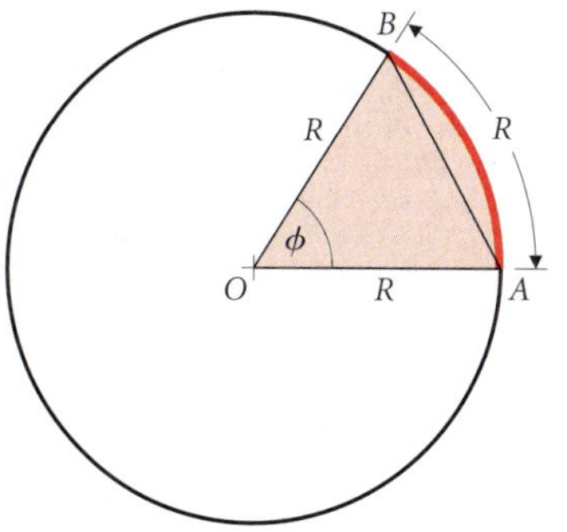

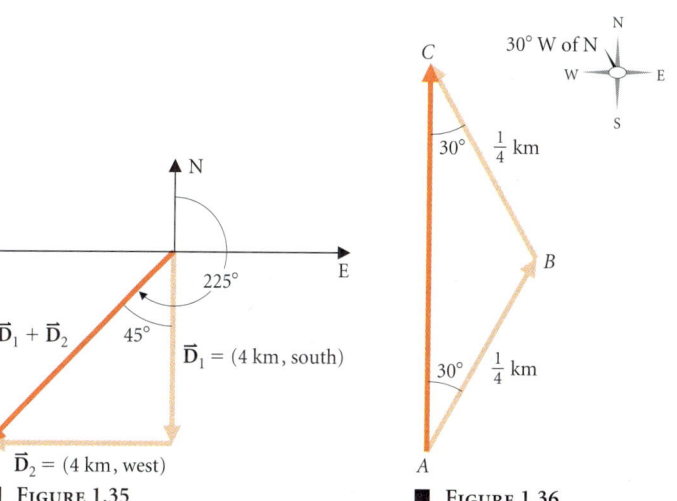

■ Figure 1.34

■ Figure 1.35

■ Figure 1.36

tion north and magnitude $2(\frac{1}{4}\text{ km})\cos 30° = \frac{1}{2}(\sqrt{3}/2)\text{ km} = \sqrt{3}/4\text{ km}$ = 0.43 km. The displacement is thus
$$\vec{D} = (0.43\text{ km, north}).$$
The distance traveled is the sum of the two $\frac{1}{4}$-km distances, and equals $\frac{1}{2}$ km, 0.07 km more than the magnitude of the displacement.

1.6 We apply the law of cosines and law of sines to the vector triangle in Figure 1.20, with $|r_i| = 3$ km, $|D| = 4$ km, and $\theta = 180° - 45° = 135°$.

$$\begin{aligned}r_f &= [r_i^2 + D_1^2 - 2r_iD_1\cos\theta]^{1/2} \\ &= [(3\text{ km})^2 + (4\text{ km})^2 - 2(3\text{ km})(4\text{ km})\cos(135°)]^{1/2} \\ &= [9 + 16 - 24(-\sqrt{2}/2)]^{1/2}\text{ km} = 6.48\text{ km}.\end{aligned}$$

$$\frac{\sin\phi}{r_i} = \frac{\sin\theta}{r_f} \Rightarrow$$

$$\sin\phi = \frac{r_i\sin\theta}{r_f} = \frac{(3\text{ km})\sin(135°)}{6.48\text{ km}} = 0.327.$$

So, $\phi = 19°$. Taking proper account of the significant figure rules, the answer is: $\vec{r}_f = (6\text{ km}, 20°\text{ west of north})$.

1.7 We are asked to compare the components of the two sailboats' displacements along the upwind direction. These are:

Stars and Stripes: $D_\text{upwind} = (15.0\text{ km})\cos 32° = 13\text{ km}$.
Australian Pride: $D_\text{upwind} = (17.2\text{ km})\cos 43° = 13\text{ km}$.

To the accuracy of the given angles, there is no difference in the two boats' progress upwind. It takes a good helmsman to steer to within 1°, so two significant figures in the angle is the best we can expect.

1.8 We calculate the components of the two displacements in the east and north directions. To avoid confusion with our previous solution, we'll call the unit vectors $\hat{E}$ and $\hat{N}$. The angle $\vec{D}_2$ makes with east is $45° + 28° = 73°$.

$$\begin{aligned}\vec{D}_1 &= (1.50\text{ km})(\hat{E}\cos 45° + \hat{N}\sin 45°) \\ &= (1.06\text{ km})\hat{E} + (1.06\text{ km})\hat{N}.\end{aligned}$$

and
$$\begin{aligned}\vec{D}_2 &= (1.25\text{ km})(\hat{E}\cos 73° + \hat{N}\sin 73°) \\ &= (0.365\text{ km})\hat{E} + (1.20\text{ km})\hat{N}.\end{aligned}$$

Their sum is $\vec{D} = \hat{E}(1.4\text{ km}) + \hat{N}(2.3\text{ km})$.

The magnitude of $\vec{D}$ is
$$|\vec{D}| = \sqrt{(1.43\text{ km})^2 + (2.26\text{ km})^2} = 2.7\text{ km},$$
and its angle from north is $\phi = \tan^{-1}(1.43/2.26) = 32°$.

This same angle computed from the results of Example 1.14 is
$$\phi = 45° - \tan^{-1}(0.59/2.60) = 45° - 13° = 32°.$$

We obtain the same result in both coordinate systems, but this solution (using the E–N system) involves a few more steps.

1.9 First collecting terms containing the individual vectors, we have
$$\vec{D} = \frac{\vec{A}(6 - 8) + \vec{B}(24 - 20) + \vec{C}(-12 + 12)}{8}.$$

then
$$\vec{D} = \frac{-2\vec{A} + 4\vec{B} + 0}{8} = -\frac{\vec{A}}{4} + \frac{\vec{B}}{2}.$$

Basic Skills

Review Questions

§1.1 A MODEL OF SPACE AND TIME

- What is a *reference frame*?
- Describe the Cartesian model of space.
- What does it mean to say that space is uniform?
- What does Galileo's relativity principle imply about the structure of space?
- What was Newton's definition of time?
- What is required in order to include time in the Cartesian model?
- What are the *arrow of time* and the *principle of causality*?
- Is the Cartesian model completely accurate?

§1.2 THE INTERNATIONAL SYSTEM OF UNITS

- Describe four features of every good physical measurement.
- What is the BIPM?
- What is wrong with using Earth's rotation as a standard of time?
- Describe in general terms the current definition of the second.
- What is the current definition of the meter? Why is it better than the original definition?
- What is the definition of the radian? Why is it superior to the degree as a unit of angle in physics?
- What is the current definition of the unit of mass?

§1.3 USING SI

- In general, can we express the value of a physical quantity exactly? Why or why not?
- Explain the three rules for significant figures.
- What are the conventions for deciding whether zeros are counted as significant?
- Describe the method for converting the units of a physical quantity.
- What are the most commonly used SI prefixes, their symbols, and the multipliers they stand for?
- Why is a rough estimate often useful?
- What is a *physical dimension*?
- What must be true of any equation relating physical variables?
- What is meant by *order of magnitude*?

§1.4 VECTORS AND SCALARS

- What is the distinction between vectors and scalars?
- What is a *displacement vector*?
- State the rule for vector addition.
- What is the *zero vector*?
- Describe the difference between displacement and distance.
- In what way does the language of physics differ from everyday usage of words?

- Describe scalar multiplication of a vector by (1) a positive scalar, and (2) a negative scalar.
- Describe the basic rule for subtraction of vectors.
- State the parallelogram rules for addition and for subtraction of vectors.

§1.5 THE POSITION VECTOR

- Define the position vector of an object.
- How is a displacement similar to, and different from, a position vector?

§1.6 VECTOR ALGEBRA

- What is the general rule for determining the component of a vector in any given direction?
- What is a *unit vector,* and how are unit vectors usually named?
- What are the traditional names for unit vectors along the x-, y-, and z-coordinate axes?
- How is a vector expressed in terms of components and unit vectors?
- Does a physics problem place any restriction on your choice of direction for the coordinate axes? What criterion *should* you use in choosing coordinates?
- How do the algebraic rules for addition and scalar multiplication of vectors compare to the corresponding rules for numerical quantities?

Basic Skill Drill

§1.2 THE INTERNATIONAL SYSTEM OF UNITS

1. Estimate the precision and accuracy you can achieve when measuring the length of a piece of paper with a meter stick.
2. Express the following angles in radians. For each case, draw a sketch of the angle at the center of a circle and state the result as a ratio of arc length to radius. Your calculator is taboo until you've got the answer. Then be sure you know how to make your calculator generate the correct result.
(a) 90° **(b)** 120° **(c)** 180°
Similarly, express the following angles in degrees.
(a) $\pi/6$ **(b)** $2\pi/3$ **(c)** $3\pi/4$

§1.3 USING SI

3. What are the precision and accuracy of each of the following quantities? State each rule or convention you follow.
(a) 6.724×10^{-4} s
(b) 300 kg **(c)** 0.00139
(d) 0.0300 m
4. Perform each of the indicated operations and give the result with the correct number of significant figures. State which rules you use.
(a) $(6.307 \times 10^{-3}$ m$) \times (4.10$ m$)^2 \times (0.2913$ kg/m$^3)$
(b) 2.63×10^{-2} kg $+ 0.497$ kg $+ 26.34$ kg **(c)** $(724$ m$)/(0.21$ s$)$
(d) $(33.719$ m$)/(2.7196$ s $- 2.7184$ s$)$
5. How many μs are there in 1 Ts?
6. An Angstrom unit (symbol Å) is 10^{-10} m. How many μm are there in 1 Å?
7. What is the physical dimension of each of the following quantities? **(a)** the density of a piece of metal **(b)** the angle between two wooden beams **(c)** the area of a farmer's field **(d)** the volume of a milk carton
8. Estimate the order of magnitude of each of the following quantities. **(a)** the mass of your car **(b)** the height of the physics building

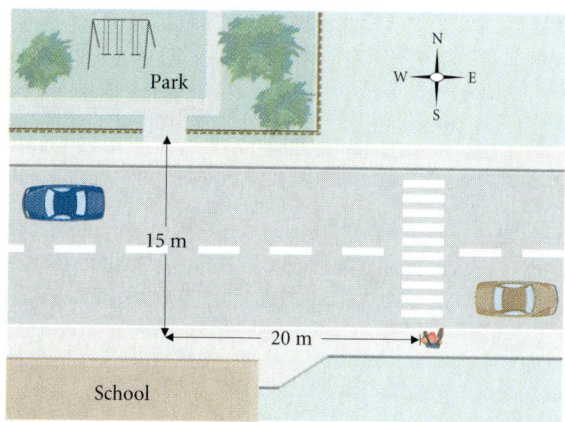

■ FIGURE 1.37

on your campus **(c)** the distance you travel to work **(d)** the amount of time you sleep each night

§1.4 VECTORS AND SCALARS

9. A child wishes to cross the road from school to the park, but the crosswalk is 20 m down the street as shown in ■ Figure 1.37. Describe the child's displacement vector after crossing the street at the crosswalk and walking to the park.
10. For the vectors $\vec{A}$ and $\vec{B}$ in ■ Figure 1.38, sketch the vector sum $2\vec{A} - \vec{B}$. Which of the vectors labeled (a) through (e) best represents your sketch?

§1.5 THE POSITION VECTOR

11. London is 350 km from Paris toward direction 330° (clockwise from north). With Paris as the origin, draw the position vector of London.
12. Here are instructions to find buried treasure: "Starting from the well, walk six paces south, seven paces east, and fifteen paces north. Dig down three meters." Making a reasonable estimate of the length of a pace in meters, draw a diagram showing the position vector of the treasure from the well.

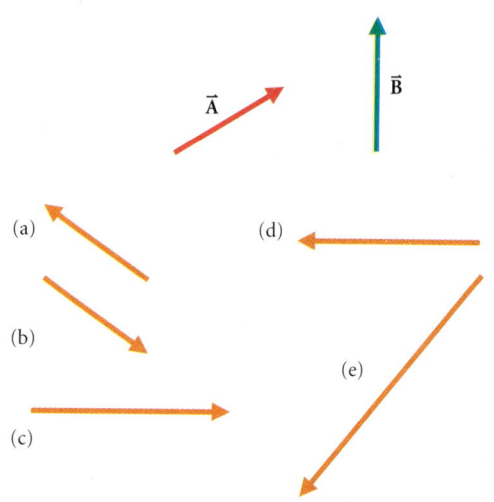

■ FIGURE 1.38

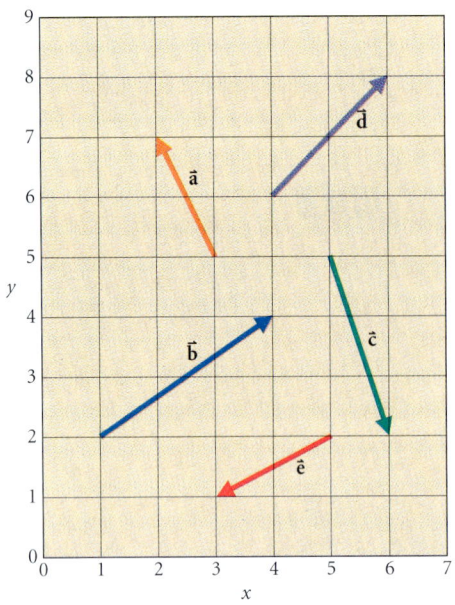

■ FIGURE 1.39

§1.6 VECTOR ALGEBRA

13. Estimate the *x*- and *y*-components of the vectors shown in ■ Figure 1.39. Draw in the unit vector corresponding to each vector shown.

14. Find the components of the unit vector $\hat{a}$ if $\vec{a}$ has components (1 m, 2 m).

15. Two vectors are given by $\vec{A} = 3\hat{i} + 2\hat{j}$ and $\vec{B} = -\hat{i} + \hat{j}$. Draw a careful diagram of an *x*–*y* coordinate system with scales on the axes. Draw the vectors $\vec{A}$ and $\vec{B}$ in your diagram and *construct* their sum. Calculate the components of the sum and compare them with measured values from your diagram.

Questions and Problems

§1.1 A MODEL OF SPACE AND TIME

16. ❖ In the nineteenth century, ship captains at anchor in port would set their clocks against a *time-ball* that was released from the top of a mast precisely at noon (according to the local astronomical observatory). A similar practice in many towns was to mark noon by sounding a steam whistle at the local factory. What assumption underlies these methods for "synchronizing" clocks? Do these methods differ *in principle* from the best current method, which uses coded radio signals from Global Positioning System satellites? If so, how?

17. In the Cartesian model, we envision lines of meter sticks that extend forever parallel to each other. The only way to test whether the model is a real picture of the world is by surveying. To see what it would mean for the Cartesian model to be inaccurate, imagine an intelligent race of ants who think the surface of the Earth is flat. The ants survey a grid of north–south lines and east–west lines on the surface. **(a)** ❖ Describe what happens to the ants' north–south lines as they are extended farther and farther northward from the equator. **(b)** ◆◆ Supposing the ants to be about 2×10^{-3} m in size, how far north of the Earth's equator would the ants have to extend their grid before noticing that two lines 1 m apart at the equator had come closer together by an amount equal to the size of an ant's body? **(c)** ❖ Would the ants find their Cartesian model of the Earth's surface adequate for the 10-m-square area they inhabit around their Capitol hill?

§1.2 THE INTERNATIONAL SYSTEM OF UNITS

18. ❖ Why, do you suppose, do American manufacturers who sell machine parts in the international market generally support establishment of the metric system in American commerce?

19. ❖ Estimate the precision and accuracy you can achieve with each of the following measurement procedures. In each case, comment on whether the accuracy and precision are likely to be adequate for the purpose of the measurement.
(a) measuring the length of a room (≈ 4.5 m) with a meter stick
(b) measuring the length of the boat described in the text with a meter stick (*Hint:* Can you simply lay your meter stick along a straight line or must you use a more contrived procedure?) **(c)** measuring, with a wristwatch, the time your friend takes to run a quarter-mile lap around the track **(d)** using a wristwatch to measure the lap time for an Olympic champion runner

20. ❖ The measurement of a physical quantity always involves an operational procedure that describes how to perform the measurement and generate its numerical outcome. Hand in hand with the procedure goes a theoretical understanding of why the procedure actually measures what it purports to measure. This philosophy is abstract, particularly when applied to what the standard of time means. Area is much more familiar to us than atomic oscillations. Let's observe the same philosophy at work in what we mean by a measurement of area. Consider the area within the curve shown in ■ Figure 1.40. Your theoretical resource is the definition of area for an ideal rectangle. Propose a procedure for measuring the area within the curve and carry out the procedure. How could you improve your method? Can your method be made arbitrarily precise? If not, what limits the precision of the method? To what extent does the concept of area have meaning apart from a detailed story about how it is measured?

21. ◆ **(a)** A telescope is designed to transmit laser pulses to the Moon and to detect the signal reflected from mirrors left there by *Apollo* astronauts. If the time between the transmission of a pulse and the reception of the reflection is 2.433 s, what is the measured dis-

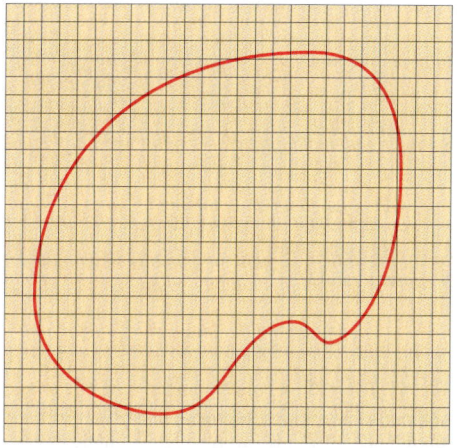

■ FIGURE 1.40

■ FIGURE 1.42

tance from the telescope to the mirror on the lunar surface? **(b)** If the timing system measures time intervals with an uncertainty of 1.5×10^{-10} s, what is the corresponding uncertainty in the measured distance to the Moon?

22. ♦ A surveying instrument can measure the angle between two distant objects, each at a distance of 452.7 m, with an uncertainty of 1.5×10^{-6} rad. What is the corresponding uncertainty in the separation of the two objects?

23. ♦♦ Eratosthenes measured the circumference of the Earth by noting that the Sun was at an angle of 7° south of the vertical at Alexandria (■ Figure 1.41) at the same time that, at Syene, 800 km south of Alexandria, the Sun was observed to be exactly overhead. (Assume Syene is directly south of Alexandria.) Based on these data, what is the circumference of the Earth in kilometers?

24. ♦♦ Suppose the polar axis of a spherical polar coordinate system were along the *y*-axis instead of the *z*-axis of a Cartesian system. Make a sketch depicting these polar coordinates and find expressions for their values at a point in terms of the Cartesian coordinates, and vice versa.

25. ♦♦ Suppose the polar axis of a cylindrical polar coordinate system were along the *x*-axis instead of the *z*-axis of a Cartesian system. Make a sketch depicting these polar coordinates and find expressions for their values at a point in terms of the Cartesian coordinates, and vice versa.

26. ♦♦ Find expressions for the spherical polar coordinates of a point in terms of cylindrical polar coordinates, and vice versa.

§1.3 USING SI

27. ❖ The number on the bridge in ■ Figure 1.42 is intended to give its distance in miles from the Kansas–Oklahoma border. **(a)** What is the precision in meters claimed for the bridge's position? **(b)** Is this precision necessary to avoid confusion in the labeling of different bridges? **(c)** If the bridge is 100 m long, how closely parallel to the border must it be for the marking to make equal sense at both ends? Is such an accurate alignment likely? **(d)** The bridge has the same label on both sides. Does this make sense? **(e)** Is it worth the paint to quote the distance to four significant figures? (See D. L. Mathieson, Why round the answer? *The Physics Teacher*, Oct. 1990:471.)

28. ❖ Pun lovers will enjoy translating the following sentences: A cheerleader uses 10^6 phones; a rock star uses 10^{-6} phones; 10^{-2} pedes will certainly bug you; this is a 10^{12} bulls question.

29. ♦ Express 1 mm/μs in SI units without prefixes.

30. ♦ How many mg are there in a metric ton (1000 kg)?

31. ♦ What is your height in meters?

32. ♦ What is your mass in kilograms?

33. ♦ Express a day in milliseconds.

34. ♦ How many km are there in a mile?

35. ♦ Speeds are sometimes jokingly quoted in furlongs per fortnight. A furlong is 660 ft and a fortnight is 14 d. What is the speed of light (3.0×10^8 m/s) in furlongs/fortnight?

36. ♦ Measuring distance in terms of time, the nearest star is about 4 light-years away; that is, light takes 4 years to reach us from the star. What is its distance in meters?

37. ♦♦ A stainless steel sphere 3.0 cm in diameter has a mass of 0.11 kg. What is the mass of a sphere 5.0 cm in diameter?

38. ♦♦ Gasoline in Europe costs roughly $1.50 per liter. What would a cubic meter of gasoline cost you? Roughly how far could you drive your car using a cubic meter of gas? (If you don't own a car, ask a friend or family member what their car's fuel consumption rate is.)

39. ♦♦ ✉ Suppose the gasoline economy of your car is 35 miles per gallon. What is the physical dimension of gasoline economy? If your car were to scoop in gasoline from a stream in the middle of the road (!) instead of carrying it in a tank, what cross-sectional area should the scoop have? Could you pull the car with a rope of the same cross-section? (See A. A. Bartlett, Physics and the measurement of automobile performance, *The Physics Teacher*, Oct. 1988:433.)

40. ♦♦♦ ✉ Estimate the order of magnitude of the following quantities. State each step in your reasoning and what data you think it necessary to look up in tables. **(a)** the volume of gasoline burned annually by private automobiles in the United States **(b)** the height of a fence you could build around Tennessee using the stone from the Great Pyramids **(c)** the mass of hair swept annually from the floors of U.S. barbershops **(d)** the mass of the Earth **(e)** the total length of the interstate highway system **(f)** the number of professional disk jockeys in the United States

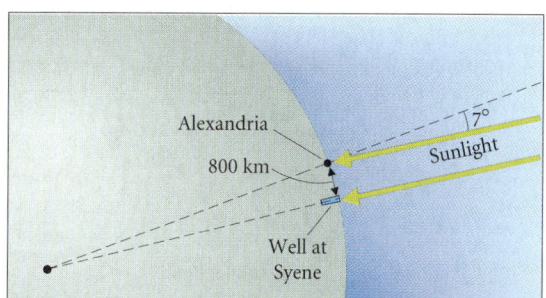

■ FIGURE 1.41

41. ♦♦♦ Suppose you wish to add (and/or subtract) several quantities, each with the same precision. (a) How does the difference between the greatest and least possible values of the sum increase with the number of quantities added or subtracted? (b) How many quantities can you add or subtract before the sum is actually one decimal place less precise than given by Rule 3, §1.3.1. (c) Quantities usually vary randomly within their uncertainty range, some being actually greater than and some less than their mean value. Thus, in a sum, discrepancies tend to cancel, and the *probable* range in values of the sum grows more slowly than the difference between least and greatest *possible* values. The probable range of a sum is known to increase as the square root of the number of quantities added or subtracted. Revise your estimates of part (b) accordingly.

42. ♦♦♦ The *fractional uncertainty* in a number is defined as the number's uncertainty divided by the number itself. (a) From Rule 1, find the range of fractional uncertainties in numbers with N significant figures. (b) Assuming the fractional uncertainties in the numbers A and B are small enough that their product can be neglected, show that the fractional uncertainty in the product AB is approximately equal to the sum of fractional uncertainties in A and B. (c) Estimate how many numbers with equal fractional uncertainty may be multiplied or divided in an expression before the accuracy of the result is one significant figure less than given by Rule 2. (d) The fractional uncertainty in a product actually grows with the square root of the number of factors. Revise your estimate from part (c) accordingly.

43. ♦♦♦ Draw a schematic graph to demonstrate the following rule.

The uncertainty in value of a function of a single independent variable is the absolute value of the function's derivative multiplied by the uncertainty in the independent variable.

$$\Delta f(x) = \left|\frac{df}{dx}\right| \Delta x.$$

Use this rule to find expressions for uncertainty in the following.

(a) $f_1(x) = x^3$ (b) $f_2(x) = \sin x$ (c) $f_3(x) = \dfrac{x^2 + 5}{2x^2 - 1}$

In each case, evaluate the uncertainty in the function if $x = \pi/3 \pm 2.0 \times 10^{-2}$.

44. ♦♦♦ Use the rule from the previous problem to find expressions for uncertainty in the functions e^x and $\ln x$. Use your results to develop significant figure rules for exponentials and logarithms.

§1.4 VECTORS AND SCALARS

45. ❖ Can vectors with different physical dimensions be added? Why or why not?

46. ❖ A spiral staircase makes one complete turn of radius 3 m while climbing 5 m. Describe your displacement vector when you have climbed (a) 5 m and (b) 7.5 m measured vertically from the bottom.

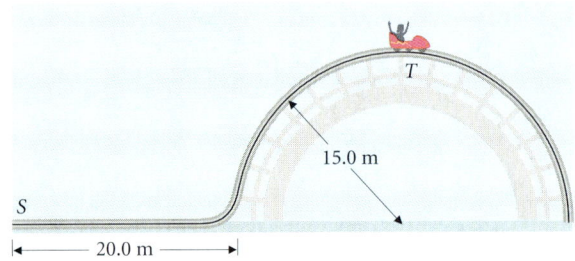

■ FIGURE 1.43

47. ❖ You ride a roller coaster that consists of a straight line segment and a circular segment as shown in ■ Figure 1.43. What is your displacement from the start S when you are at the top of the loop, T?

48. ❖ (a) What must be true of two vectors $\vec{a}$ and $\vec{b}$ if the length of their sum equals the sum of their lengths? (b) What is true of $\vec{a}$ and $\vec{b}$ if the sum of the squares of their lengths equals the square of the length of their sum?

49. ❖ Peter and Mary start in the center of a large lawn and run 41 m and 55 m, respectively, each in a straight line, but not necessarily in the same direction. (a) What are the maximum and minimum possible distances between Peter and Mary? Explain your reasoning, using a sketch of their displacements. (b) What is the distance between them if they run in directions at right angles to each other?

50. ❖ An astronaut rides into orbit on the space shuttle, makes 110 orbits, and lands back on the runway at Kennedy Space Center. Describe the astronaut's net displacement vector at the end of the trip.

51. ♦ Which of the vectors shown in ■ Figure 1.44 most nearly equals $\frac{1}{2}\vec{A} - \vec{B}$?

52. ♦ Using ■ Figure 1.45, construct the vectors $\vec{A} - \vec{B}$ and $\vec{B} - \frac{1}{2}\vec{A}$.

53. ♦ An express bus travels 3 km east to its first stop, then 5 km north to the second stop. At the second stop, what is the total displacement of the bus from its starting point?

54. ♦ A delivery truck drives 1.0 km north, 0.50 km east, 1.0 km south, 2.0 km west, 0.50 km northwest, and 0.75 km northeast. Draw a vector diagram showing the truck's journey, and find graphically the magnitude and direction of the truck's displacement vector at the end of the trip.

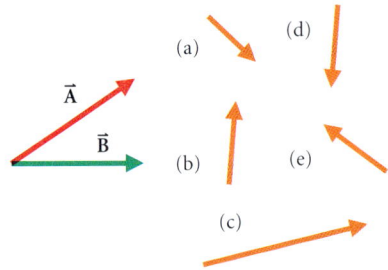

■ FIGURE 1.44

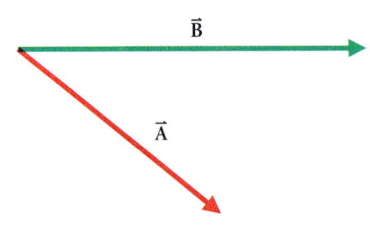

■ FIGURE 1.45

55. ♦♦ (a) Two vectors $\vec{a}$ and $\vec{b}$ have equal magnitude: $|\vec{a}| = |\vec{b}| = a$. Their directions are described by their angles from north, θ_a and θ_b. Use the parallelogram rule to show that the direction of their sum $\vec{s} \equiv \vec{a} + \vec{b}$ is

$$\theta_s = (\theta_a + \theta_b)/2.$$

What is the magnitude of $\vec{s}$?
(b) What are the magnitude and direction of the difference $\vec{d} = \vec{a} - \vec{b}$?

§1.5 THE POSITION VECTOR

56. ♦ What is the displacement of an airplane initially 80 km northwest of the Cincinnati airport and later 80 km northeast of the airport? Draw a diagram showing the initial and final position vectors and the displacement.
57. (a) ♦ An (x, y) coordinate system is set up with origin O. Points P_1, P_2, and P_3 have coordinates (0, 2.0 m), (1.0 m, 0), and (2.0 m, 3.0 m), respectively. Find the length and direction of the position vector for each point. Give the directions as angles counterclockwise from the x-axis.
(b) ♦♦ The point (0, 1.0 m) on the y-axis is now used as the origin instead of the conventional point with coordinates (0, 0). Find the position vectors of P_1, P_2, and P_3 with respect to this new origin.
58. ♦♦ On a road trip, Cindy is initially 71 km west of a highway intersection. Cindy's displacement during the next 2 h is 1.00×10^2 km southeast. (a) What is Cindy's position vector, taking the intersection as the origin? (b) Taking Cindy's initial position as the origin, find the position vectors of the intersection and of Cindy's final position.
59. ♦♦ The displacement of an airplane flying from San Francisco International airport (SFO) to San Jose International (SJC) is 30 nautical miles, 36° south of east. Flying from SJC to Tracy, the plane's displacement is 36 nautical miles, 52° east of north. What is Tracy's position vector, if SFO is taken as origin? (A nautical mile is slightly larger than a statute mile. Give your answer in nautical miles. All numbers are known to two significant figures.)
60. ♦♦ Barry wishes to meet his friend Abigail at the park 3.0 km northwest of Barry's house. Abigail's house is 4.0 km southwest of Barry's house. What is Abigail's displacement between home and the park?
61. ♦♦ The road from Marysville to Petersburg goes through Morriston. Morriston is 5.0 km northwest of Marysville and 7.0 km south of Petersburg. With Marysville as origin, what is the position of Petersburg?

§1.6 VECTOR ALGEBRA

62. ♦ Draw the vector (1 m, 3 m, 2 m) in a three-dimensional diagram.
63. ♦ Measure the components of the vectors $\vec{A}$ and $\vec{B}$ in ■ Figure 1.46. Construct the vectors $\frac{1}{2}\vec{A} + \vec{B}$ and $\vec{B} - \vec{A}$ and compute their components. Compute the magnitudes and directions of the two constructed vectors from their components and compare with measurements from the figure.
64. ♦♦ A train goes between three stops A, B, and C, in that order. From A to B the train travels 75 km south and between B and C it travels 1.0×10^2 km northwest. (a) Show the locations of A, B, and C on a diagram with clearly labeled coordinate axes. (b) Calculate the components of the train's displacement between A and C. (c) What are the magnitude and direction of the displacement?
65. ♦♦ Solve Problem 54 by calculating components of the truck's total displacement.

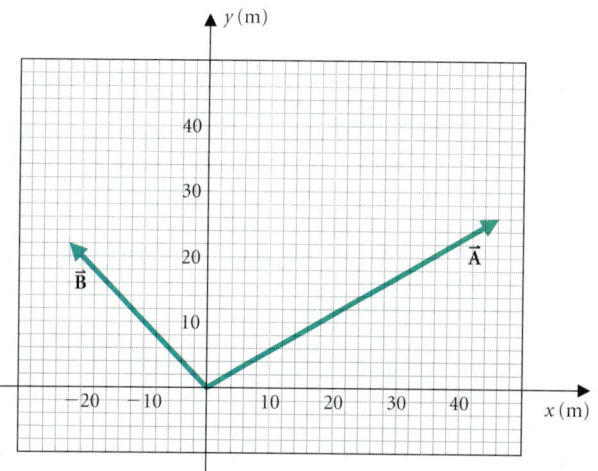

■ FIGURE 1.46

66. ♦♦ A yacht sails overnight on a course of 60° (clockwise from north). Observations at dawn show the yacht to be 1.00×10^2 nautical miles farther north than it was at evening twilight. What was the yacht's displacement between twilight and dawn?
67. ♦♦ Using the Pythagorean theorem for plane triangles, show that the magnitude of a three-dimensional vector $\vec{a}$ is given by the square root of the sum of the squares of its components.

$$|\vec{a}| = \sqrt{a_x^2 + a_y^2 + a_z^2}.$$

Find the magnitudes of the vectors: $\vec{a}$, with components (1 m, 2 m, 3 m); $\vec{b}$, with components (2 m, 2 m, 2 m); and the difference $\vec{a} - \vec{b}$.
68. ♦♦ For the vectors $\vec{a}$ and $\vec{b}$ in ■ Figure 1.47, estimate the x- and y-components and x'- and y'-components. (All vectors and axes lie in a single plane.)
69. ♦♦ Use a diagram to demonstrate that the commutative rule for addition is true for more than two vectors:

$$\vec{A} + \vec{B} + \vec{C} = \vec{C} + \vec{B} + \vec{A}.$$

With a similar figure, demonstrate the associative rule:

$$(\vec{A} + \vec{B}) + \vec{C} = \vec{A} + (\vec{B} + \vec{C}).$$

Develop a short algebraic argument, using this associative rule and the commutative rule for two vectors, to derive the commutative rule for three vectors.

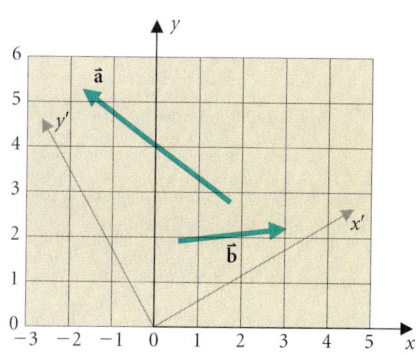

■ FIGURE 1.47

Additional Problems

70. ❖ Comment on the similarities and differences among the concepts of *standard*, *unit*, and *physical dimension*.

71. ◆◆ You measure the diameter of a copper sphere using vernier calipers and obtain the value 2.33 ± 0.02 cm. What is your best estimate of the sphere's volume? What are the maximum and minimum values? Using the significant figure rules, you report your measurement as 2.33 cm. What value do you obtain for the volume? What is the implied uncertainty? How does it compare with the value obtained from the measured uncertainty in the diameter?

72. ◆◆ ✻ Express the acceleration of gravity at the Earth's surface **(a)** in units where length is expresed in *seconds*, and **(b)** in units where time is expressed in *meters*. (One meter of time is the time light requires to travel a meter.)

73. ◆◆◆ ✻ Researchers in general relativity often choose to define Newton's gravitation constant,

$$G = 6.67 \times 10^{-11} \frac{m^3}{kg \cdot s^2},$$

to be $G \equiv 1$ at the same time as they take $c \equiv 1$. **(a)** Express the kilogram in terms of seconds in such a unit system. **(b)** Express the kilogram in terms of meters.

74. ❖ Suppose you lived in the nineteenth century, several miles from the local factory, and you decide to improve on the factory whistle method for setting your clock that was described in Problem 16. You have a steam whistle of your own but no way to communicate with the factory foreman except by steam whistle or by walking to the factory and talking with the foreman. Describe how to cooperate with the foreman to correct for the travel time of sound between you and the factory.

75. ◆◆◆ Coral samples 500 million years old, taken deep below the surface of certain South Sea island reefs, show approximately 400 daily growth patterns within each annual growth pattern. Assuming the length of the year has remained unchanged, how many seconds shorter was Earth's rotation 500 million years ago? Express the rate of change of the day's length in seconds per day, and as a dimensionless number (i.e., seconds per second or days per day).

76. ◆◆◆ The two coordinate systems in ■ Figure 1.48 are rotated by an angle θ with respect to each other. **(a)** By finding their x- and y-components, express the unit vectors $\hat{x}'$ and $\hat{y}'$ in terms of $\hat{i}$ and $\hat{j}$.
(b) Components of an arbitrary vector $\vec{V}$ can be found in either the "primed" or "unprimed" coordinates; either coordinate choice describes the same vector.

$$V_{x'}\hat{x}' + V_{y'}\hat{y}' = \vec{V} = V_x\hat{i} + V_y\hat{j}.$$

Use this fact and the result of part (a) to obtain simultaneous equations for the components $V_{x'}$ and $V_{y'}$ in terms of V_x and V_y. Solve the equations for $V_{x'}$ and $V_{y'}$. These relations express the *transformation* between the two coordinate systems.

If you have done Problem 68, check to see whether your results there agree with your results in this problem.

77. ◆◆◆ Show that a right circular cylinder with a given volume has a minimum surface area when its height equals its diameter. The standard kilogram was made in this shape to minimize errors due to contamination of its surface.

78. ◆◆◆ Show by considering N vectors along the sides of a regular polygon that

$$\sum_{n=0}^{N-1} \cos \frac{2\pi n}{N} = 0 \quad \text{and} \quad \sum_{n=0}^{N-1} \sin \frac{2\pi n}{N} = 0.$$

Challenge Problem

79. ✉ *Fermi's problem.* Assume that the maximum speed of travel achievable by an interstellar colonization mission is 0.001 times the speed of light, and that the average distance between stars is 10 light-years (ly). Model an expanding galactic civilization as occupying a spherical volume whose size increases because freshly colonized worlds develop mature local economies and then launch new colonization missions. From the history of the Earth, estimate the time required for a civilization to become mature. Does spacecraft speed or development time place a stronger limit on the expansion speed of the civilization? Estimate the time required for the sphere to achieve a radius of 3×10^4 ly—the size of the galaxy. Where are they?! (A light-year is the distance traveled by light in 1 y.)

Essay Question/A Point to Ponder

- You wake up on April 1st and read a newspaper headline claiming that the universe and everything in it has shrunk by an order of magnitude overnight, but no other change has occurred. Write a letter to the editor criticizing the claim. Consider how you determine the *length* of an object in daily life and whether you could *detect* that shrinkage had occurred. Then, you had best consider the definition of the meter, and whether it would allow you to detect a shrinkage. Are there any common experiences depending on speed that you could use to detect a shrinkage?

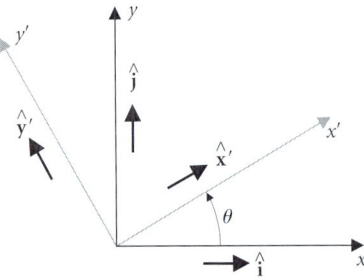

■ Figure 1.48

Kinematics—the subject of this chapter—allows us to describe the motion of the car as it negotiates the curves and the straightaways, changing both its speed and its direction.

But all that moveth doeth mutation love.
EDMUND SPENSER

CHAPTER 2
Kinematics

CONCEPTS
Speed
Average velocity
Limit
Instantaneous velocity
Acceleration
Linear motion
Uniform acceleration

GOALS
Be able to:
Estimate velocity and acceleration from an object's path.

Relate position, velocity, and acceleration in one-dimensional motion.

Solve problems involving uniformly accelerated linear motion.

THE NAME COMES FROM THE GREEK ROOT *KINE* FOR MOTION, AS IN *CINEMA*—A PLACE TO WATCH MOVIES.

FOR NOW, WE DEFINE A PARTICLE AS AN OBJECT WHOSE SIZE AND SHAPE ARE NOT IMPORTANT. WE'LL REFINE THE DEFINITION FURTHER IN §5.2.1.

How far is the race car from the finish line? How fast is it moving? When will it finish the race? We can obtain a complete record of the car's motion by measuring its position at regular time intervals with a camera in a helicopter overhead. But how can an automotive engine designer use that information to find the car's speed? A scientist with experiments on board a spacecraft bound for Mars (■ Figure 2.1) has a different problem. Radio signals from the craft give its direction and speed away from us. How can the scientist determine where it is? These are questions of *kinematics*, the description of motion. Both scientists rely on kinematic methods to find the information they need from the available data.

Both the space scientist and the engine expert can answer their questions by modeling each vehicle as a point-like object—a particle. The same model is effective in many of the problems we shall encounter. So, we begin this chapter with definitions of the basic kinematic quantities for a particle—velocity and acceleration.

Despite this simplification, the race car's motion is quite complicated as it swoops around the curves and continually changes its speed. Even so, we can describe its motion as a combination of simpler motions. In this chapter we concentrate on the model of uniformly accelerated linear motion; in Chapter 3 we'll show how to combine different models to describe the car's motion. Together with Kepler's laws from Chapter 0 and the simple harmonic oscillations described in Chapter 14, we'll then have a complete set of kinematic models for use in all branches of physics, from space science to elementary particle physics.

2.1 SPEED AND VELOCITY

■ Figure 2.2 represents a short segment of the road the race car follows through the mountains. A photographer in a helicopter overhead has taken photos of the car at $\frac{1}{4}$-s intervals and its position in each photo is plotted as a dot on the map. The race starts at point *A* where the timing clock is set to zero. Several positions we will refer to in later examples are marked with letters, and position vectors are shown for two of them, with *A*, the starting point, as origin. We have claimed that such a record is a complete description of the race car's motion, but a good deal seems to be missing. The rate at which things are happening—the car's speed and its acceleration around the turns—are what the driver senses most vividly. Our task in this section is to make the connection between the driver's intuitive sense of motion and the formal description of position versus time.

2.1.1 Average Speed

Speed, measured by the speedometer in a car, is familiar to nearly everyone. A speedometer reading tells how far the car would travel at constant speed in 1 h (e.g., 60 km/h). But we rarely travel a full hour at constant speed, and speedometer readings may vary a great deal. We commonly use an average value of speed.

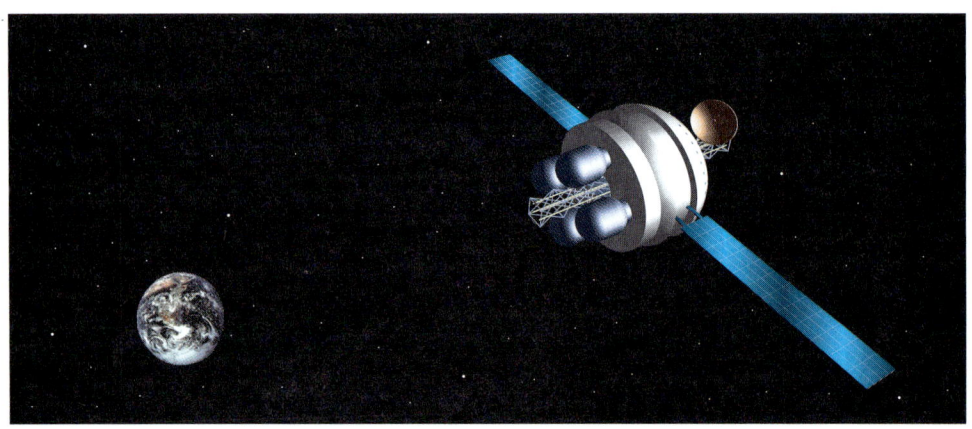

■ FIGURE 2.1
Using radio signals returned from the spacecraft, scientists can measure the rate at which distance to the spacecraft is increasing. Kinematic methods are used to derive the spacecraft's position from these data.

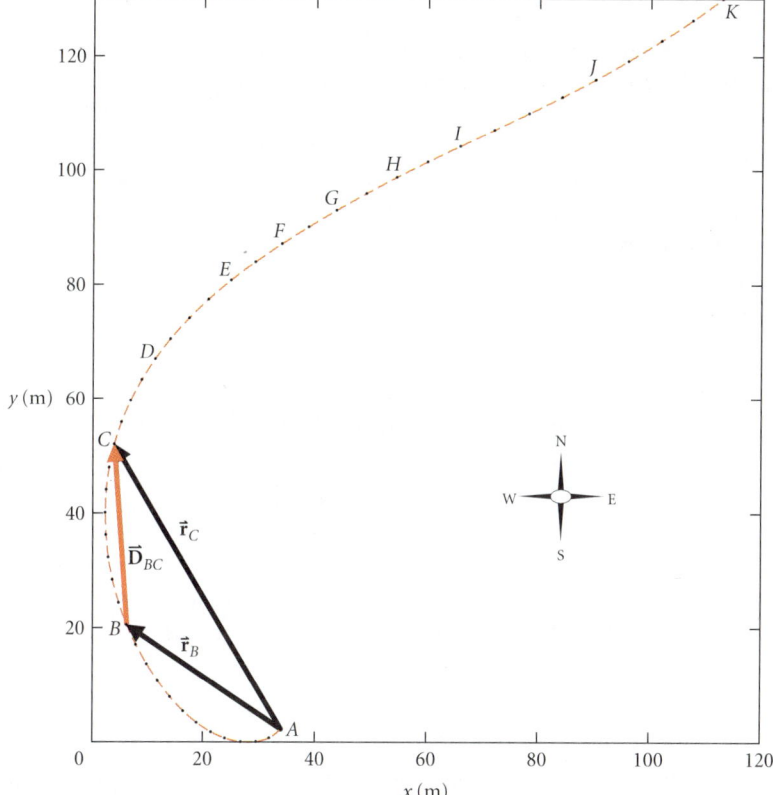

FIGURE 2.2a
Map of the race car's path along a curved but level portion of a road during a time interval of 11.0 s from the start at point A to point K.

The *average speed* of an object during a time interval Δt is the distance ℓ the object travels during that time interval divided by the time interval.

$$S_{av} \equiv \frac{\ell}{\Delta t}. \qquad (2.1)$$

NOTE THAT ℓ = TOTAL DISTANCE TRAVELED, NOT THE MAGNITUDE OF THE DISPLACEMENT.

Average speed would be the same as the speedometer reading only if that reading were constant during the entire time interval Δt.

EXAMPLE 2.1 ♦ The measured distance along the road between points A and K is 0.194 km. What is the race car's average speed over this stretch of road?

MODEL We use the definition of average speed (eqn. 2.1).

SETUP According to the information in Figure 2.2, the car was at point A at time 0.0 s and at point K at time 11.0 s. Thus the time interval $\Delta t = t_K - t_A = 11.0$ s. Using the given distance $\ell = 0.194$ km and converting seconds to hours, we have

SOLVE
$$S_{av} = \frac{\ell}{\Delta t} = \left(\frac{0.194 \text{ km}}{11.0 \text{ s}}\right)\left(\frac{3600 \text{ s}}{1 \text{ h}}\right) = 63.5 \text{ km/h}.$$

ANALYZE Is the answer reasonable? Since 1 km is about $\frac{5}{8}$ mile, 63.5 km/h is about 40 mph. That's not an unreasonable speed for this winding road. Check to see if the car's speed is constant as it moves between points A and K. ■

Suppose you wish to plan a trip consisting of several legs, each with a different average speed. The following example shows a proper approach: consider the distance traveled and the corresponding time interval separately for each leg.

EXAMPLE 2.2 ♦♦ Judy hopes to drive 60 km in 1 h. The first half of the distance is along a very pleasant country road, where she drives at an average speed of 45 km/h. At what average speed must Judy drive for the remaining distance?

MODEL The relationship we need is the definition of average speed, eqn. (2.1): $S_{av} = \ell/\Delta t = 60$ km/h for the whole trip. Since we are given the average speed for the first 30 km, we can find the time Δt_1 Judy took to drive that distance. The time remaining for the second 30 km is $\Delta t_2 = 1$ h $- \Delta t_1$.

SETUP At 45 km/hr, the first 30 km of the journey takes a time interval Δt_1 given by

$$\Delta t_1 = \frac{\ell_1}{S_1} = \frac{30 \text{ km}}{45 \text{ km/h}} = \tfrac{2}{3} \text{ h}.$$

The time available for the second half of the trip is $\Delta t_2 = \tfrac{1}{3}$ h, and so the required speed is

SOLVE
$$S_2 = \frac{\ell_2}{\Delta t_2} = \frac{30 \text{ km}}{\tfrac{1}{3} \text{ h}} = 90 \text{ km/h}.$$

ANALYZE Don't be misled into using an incorrect definition of average. Suppose Judy tried to find the speed that averages with 45 km/h to give 60 km/h.

$$\frac{45 \text{ km/h} + S}{2} = 60 \text{ km/h} \implies S = \cancel{75 \text{ km/h}}.$$

At 75 km/h, she would take 0.4 h to cover the second 30 km, and the total time for the trip would be over 1 h. The method fails because she doesn't spend equal *time* driving the two distances. Remember the definitions and use them! ■

EXERCISE 2.1 ♦♦ Suppose instead that Judy drives only the first third of the distance at 45 km/h. What average speed is required for the rest of her trip? ■

2.1.2 Average Velocity

IN EVERYDAY SPEECH, THE WORDS *SPEED* AND *VELOCITY* ARE OFTEN USED TO MEAN THE SAME THING. IN PHYSICS, SPEED IS A SCALAR QUANTITY WHILE VELOCITY ALWAYS MEANS A VECTOR; THAT IS, IT INCLUDES INFORMATION ABOUT DIRECTION.

Average speed gives only a rough description of a road trip, telling us nothing about the direction of the trip or how tortuous the road was. A first step toward including more information is to define a vector, called *average velocity*, that relates the displacement, or change in position, of an object to the time interval in which it occurs:

> The *average velocity* of an object during a time interval Δt is the object's displacement $\vec{D}$ during that interval divided by the interval.
>
> $$\vec{v}_{av} \equiv \frac{\vec{D}}{\Delta t} = \frac{\Delta \vec{r}}{\Delta t}. \tag{2.2}$$

EXAMPLE 2.3 ♦♦ Find the average velocity of the race car between points B and C in Figure 2.2.

MODEL To use the definition of average velocity, we need the car's displacement between the two points, which we can measure from the figure.

SETUP The displacement $\vec{D}_{BC}$ is drawn in the figure. We measure its length and direction:

$$\vec{D}_{BC} = \Delta \vec{r}_{BC} \equiv \vec{r}_C - \vec{r}_B = (32 \text{ m}, 4° \text{ west of north}).$$

The corresponding time interval is:

$$\Delta t_{BC} = t_C - t_B = 5.0 \text{ s} - 3.0 \text{ s} = 2.0 \text{ s}.$$

SOLVE To find the average velocity, we divide the displacement vector by the time interval:

$$\vec{v}_{BC} = \frac{\vec{D}_{BC}}{\Delta t_{BC}} = \frac{(32 \text{ m}, 4° \text{ west of north})}{2.0 \text{ s}} = (16 \text{ m/s}, 4° \text{ west of north}).$$

ANALYZE Because the length of the displacement vector is shorter than the car's path length between B and C, the magnitude of the average velocity is less than the average speed.

EXERCISE 2.2 ♦♦ A horse racing around a square field 1.0 km on a side runs each side of the field in 75 s. What are the average speed and average velocity of the horse for the first 75 s? The first 150 s? (The horse begins at the southwest corner of the field running toward the north.)

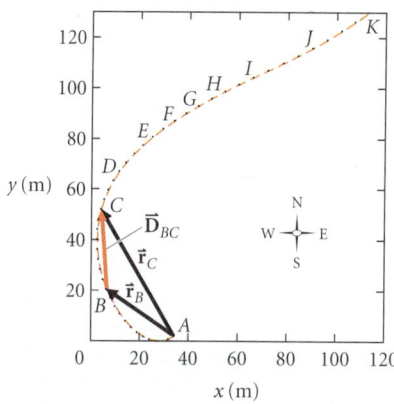

■ **FIGURE 2.2b**
The race car's path. Positions occupied by the car are represented at quarter-second intervals by dots on the path. Position vectors are shown extending from point A to points B and C on the path. The displacement vector extends between these two points.

When the horse in Exercise 2.2 and the car in Example 2.3 finish their races and return to their starting points, they each have zero net displacement. Their average velocities for an entire race are the same, zero! This odd result emphasizes the differences between average speed and average velocity. Average velocity successfully combines direction with speed for small displacements but is progressively less useful when applied to larger portions of a complicated motion.

2.1.3 Instantaneous Velocity and the Concept of Limit

An average speed of 60 km/h for a road trip from San Francisco to Los Angeles only hints at the pleasant cruise through ranching country, the struggle through a terrible snowstorm in the mountains, and the final descent into Los Angeles freeway traffic. Just as this little tale is made vivid by including detail, the description of any complicated motion is made more precise by considering the displacements and velocities occurring over ever smaller time intervals. Our next task is to develop this method.

Let us return to our race car example and study how the car's average velocity depends on the size of time intervals considered. In ■ Figure 2.3 we have drawn four displacements of

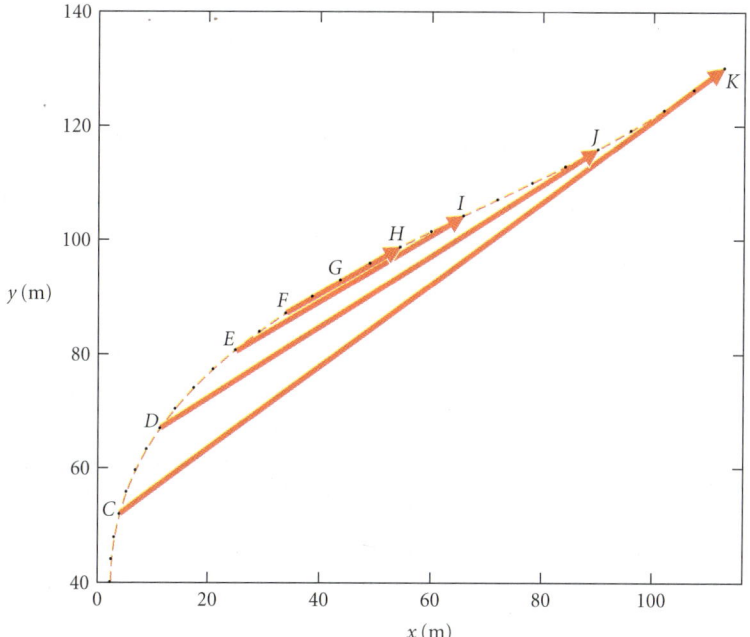

■ **FIGURE 2.3**
Displacements of the race car during time intervals centered at $t = 8.0$ s when the car is at point G. The average velocities corresponding to these displacements form a sequence of vectors whose limit is the instantaneous velocity at G. With $|\vec{D}_{FH}| = 24$ m and $t_{FH} = 1.00$ s, the limit has magnitude 24 m/s.

A STATIC DIAGRAM CANNOT FULLY CAPTURE THE LIMITING PROCESS. TRY TO IMAGINE A MOVIE IN WHICH THE AVERAGE VELOCITY EVOLVES CONTINUOUSLY TOWARD THE LIMIT AS Δt BECOMES SMALL. FIGURE 2.3 GIVES YOU A FEW FRAMES FROM THAT MOVIE.

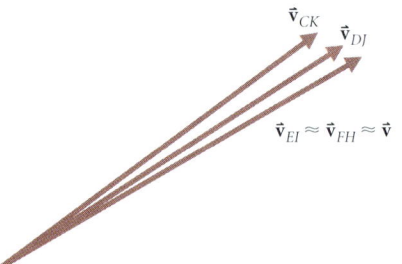

■ **FIGURE 2.4**
Instantaneous velocity at point G as the limit of average velocities. Average velocities for time intervals of 2.0 s and shorter, in this case, are already so close to the limiting value that they are not distinguishable on the graph. The limit is 24 m/s at 60° east of north. We'll use this result in Example 2.6.

IT IS USUAL TO DROP THE WORD *INSTANTA-NEOUS* AND USE SIMPLY *VELOCITY* AND *SPEED* FOR THESE QUANTITIES. WE SHALL INCLUDE *INSTANTANEOUS* IN THIS SECTION AND DROP IT AFTERWARD.

the car during time intervals centered on $t = 8.0$ s, when the car is at point G. The time intervals range from 6 s (interval CK) to 1 s (interval FH). The corresponding average velocity vectors $\vec{v}_{av} = \vec{D}/\Delta t$ are drawn in ■ Figure 2.4. As the time interval decreases, the differences between successive average velocities decrease dramatically in both magnitude and direction: the average velocity approaches a limiting value $\vec{v}$. The average velocities for the 2-s interval EI and smaller intervals differ from $\vec{v}$ by less than the plotting error in this graphical procedure. The limit $\vec{v}$ is called the *instantaneous velocity*.

> The *instantaneous velocity* of a particle at a time t is the limit of its average velocity during a time interval including t, as the size of the interval approaches zero.
>
> $$\vec{v} \equiv \lim_{\Delta t \to 0} \vec{v}_{av} = \lim_{\Delta t \to 0} \frac{\Delta \vec{r}}{\Delta t} = \frac{d\vec{r}}{dt}. \quad (2.3)$$
>
> The magnitude of an object's instantaneous velocity is its *instantaneous speed*.

When you read a speedometer or judge motion by the appearance of objects near the road, what you perceive is instantaneous speed.

■ Figure 2.5 illustrates a simple rule for the direction of an object's instantaneous velocity at any point on its path, here point D on the race car's path. Smaller and smaller time intervals correspond to displacements between closer and closer points. In the limit, the end points converge on point D, and the corresponding displacement lies along a line that touches the curve only at D—the tangent line.

> The direction of an object's instantaneous velocity is tangent to its path.

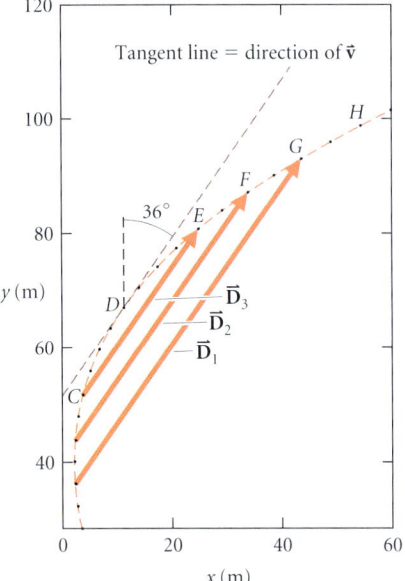

■ **FIGURE 2.5**
Instantaneous velocity is tangent to the path. Displacement vectors for ever smaller time intervals extend between points ever closer to D and so are ever closer to being parallel to the tangent line at D. The limit is in the direction of the tangent line.

EXAMPLE 2.4 ♦♦ Estimate the instantaneous velocity of the race car at point D in Figure 2.2.

MODEL "Estimate" means that an approximate calculation is good enough. Since our information is given graphically, an *exact* calculation isn't possible (maybe not even meaningful, see §2.1.4). We need a method that is as accurate as the given data. By analogy with our calculation for point G, we might be willing to accept the average velocity for a 1-s interval as a good enough approximation. However, the displacement of roughly 12 m that occurs during 1 s is fairly large, and we can see from the figure that the straight line between D and E does not approximate the curved path well. Instead, we find the direction and magnitude of the instantaneous velocity separately.

SETUP The direction of $\vec{v}$ at D is measured from a tangent line drawn in the figure: 36° east of north. The instantaneous speed of the race car, the magnitude of $\vec{v}$, is the limit of the magnitudes of the average velocities. We may find the limit by plotting a graph of $|\vec{v}_{av}|$ versus the corresponding time interval Δt and extrapolating the graph to $\Delta t = 0$. The car is at point D at time $t = 6.0$ s. For each time interval Δt, we find the position of the car (Figures 2.2, 2.5) at times 6.0 s $\pm \Delta t/2$ and measure the displacement between those two positions. Using this procedure, we estimate that we can determine D to ± 0.5 m. Then we calculate the average velocity during the corresponding time interval. The results are listed in ● Table 2.1, and plotted in ■ Figure 2.6.

TABLE 2.1 Data for Example 2.4

Time Interval Δt (s)	|Displacement| $\|\vec{D}\|$ (m)	|Average Velocity| $\|\vec{v}_{av}\|$ (m/s)
8.00	134 ± 0.5	16.8 ± 0.05
7.00	119	17.0 ± 0.05
6.00 (B to I)	104	17.3 ± 0.1
5.00	88	17.6 ± 0.1
4.00	71	17.8 ± 0.1
3.00	54	17.9 ± 0.15
2.50	45	18.0 ± 0.2
2.00 (C to E)	36	18.0 ± 0.25
1.50	27	18.1 ± 0.35
1.00	9	18 ± 0.5

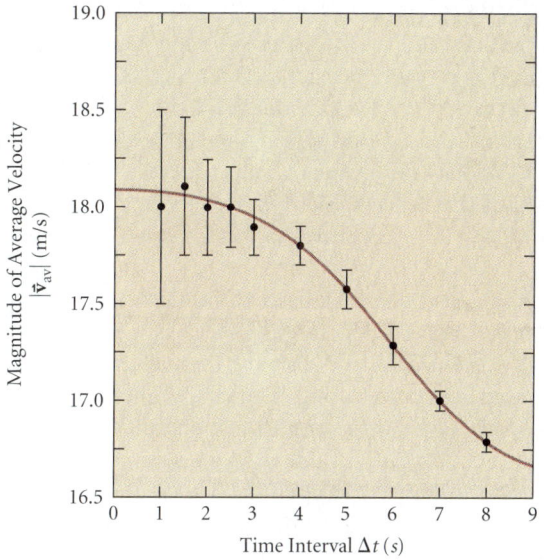

■ **FIGURE 2.6**
Instantaneous speed at point D. Each data point in the graph is the average velocity during a time interval centered on $t = 6.0$ s, when the car is at point D. In this graph the process of taking a limit $\Delta t \to 0$ is accomplished by drawing a smooth curve through the measured values and extrapolating the curve to $\Delta t = 0$. Measurement uncertainty for small Δt leads to an uncertainty of $\Delta|\vec{v}_{av}| \approx \pm 0.25$ m/s.

THE LIMITING VECTOR PROVIDES THE MOST POWERFUL IMAGE, BUT THE ARITHMETIC IS MOST EFFICIENT IF DIRECTION AND MAGNITUDE ARE DONE SEPARATELY.

SOLVE The average velocity approaches a limit of

$$\vec{v}(D) = (18.1 \text{ m/s}, 36° \text{ east of north}).$$

ANALYZE The uncertainty in the speed is about 0.25 m/s. We may estimate it either from the tabulated data or from the graph. Notice that the average speed of the car is almost constant between points C and E but that the direction of its motion changes. ■

The following exercise shows that the value of $\vec{v}$ at a point is independent of how the intervals converge on the point.

EXERCISE 2.3 ♦♦ We may also obtain the velocity at point D using displacements that begin at point D (● Table 2.2). Using these data, plot a graph similar to Figure 2.6 and show that the result for the instantaneous speed of the race car at point D is the same as that obtained in Example 2.4. (*Note:* The two values may be considered the same if they differ by less than about three times the uncertainty.) ■

TABLE 2.2 Data for Exercise 2.3

Time Interval Δt (s)	|Displacement| $\|\vec{D}\|$ (m)
4.0	93
3.0	67
2.0	42
1.0 (C to D)	19
0.75	14
0.5	9

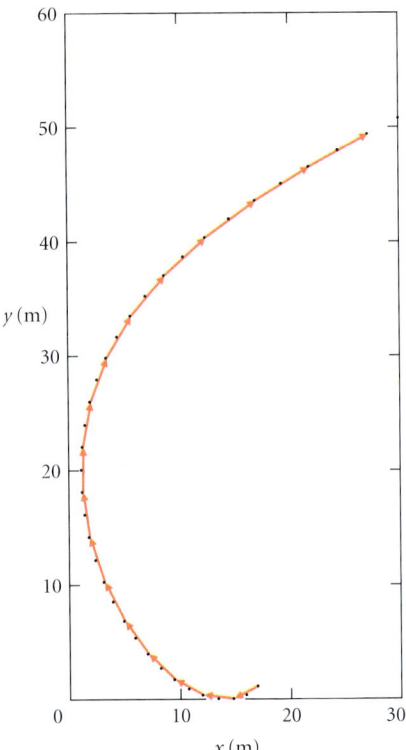

■ **Figure 2.7**
The race car's path modeled as a sequence of $\frac{1}{2}$-s displacements. Within this accuracy, we can barely distinguish between a smooth path of the car and a path consisting of lots of straight lines.

See also §0.4.6 for a discussion of modeling.

Recall that, for the remainder of the text, the word *velocity* alone means instantaneous velocity.

✷2.1.4 Calculus as a Kinematic Model

Our intuitive notion of the race car's path is *smooth*; that is, we expect displacements during ever smaller time intervals to be themselves ever smaller in magnitude and to form an ever better approximation to the curve. In the examples, with measurements $\frac{1}{4}$ s apart, it is the intuitive concept of smoothness that leads us to draw the car's path as a smooth curve passing through the measured positions. No matter how cleverly we measure, there will always be some finite time interval between measurements. What time interval is short enough? Even displacements over a $\frac{1}{2}$-s interval *look like* a good approximation to the actual path of the car, and we might be content to model the path as a sequence of $\frac{1}{2}$-s segments (■ Figure 2.7).

Such terms as "look like" or "approximation" are sloppy, and we should not accept them uncritically. The error we can accept in a description depends on our purpose. Describing the car's path with $\frac{1}{2}$-s intervals is probably adequate for a tire company engineer interested in the performance of an experimental rubber compound. However, the car moves a distance about equal to its length in $\frac{1}{2}$ s, and such a description is probably too crude for the mechanical engineer who is testing vibration in an advanced suspension system design. Half-second time intervals are certainly too crude for the mathematician's ideal world, where describing motion of a point requires time intervals that approach zero in the strict limiting sense of differential calculus.

Our definitions of instantaneous velocity and speed are based on the standard ideal definitions of the calculus. Indeed, this text (or any other physics text) takes great pains to express concepts in as precise a mathematical form as possible. Someone using physics needs not only these mathematical tools, but also a healthy sense of their role in modeling reality. Our graphical method of calculating instantaneous velocity for the race car illustrates the tension between ideal definitions and real-world data. But that tension isn't removed by imagining infinitely precise data. Any model of the world will fail if we ask it to describe distances or time intervals approaching a limit that is strictly zero.

What it means for a model to fail is illustrated by a baseball in flight, destined to be a home run. Modeling the ball as a point moving without air resistance is a good approximation that predicts a very smooth path but overestimates the distance the ball moves by several meters. An improved model, including air resistance acting on a ball of finite size, eliminates that error and still predicts a smooth path. If, however, you care about an error in position comparable to the ball's diameter, then you must also include random variations of air density and wind speed in your model. Good luck! You could attempt to solve the problem on a computer using finite time intervals. If you were to succeed, you would predict a path with small, random jiggles.

Should you care? It's unlikely that you'd ever need to include random wind variations when calculating a baseball's path; but you should care about what calculus does and does not do for you. Motion of a sphere through a uniform wind can be solved using calculus. The solution gives a useful answer but one less accurate than the mathematics might lead you to believe. It is the exact solution to an inexact model.

So, as we use calculus to express physical ideas, try to develop the skill of reasoning about *small pieces of a problem* (*differential elements*) such as the small displacements of the race car, applying calculus as if it were exact, and keeping a wary eye on how good your model is. Choosing proper models and mathematical techniques and judging an appropriate level of approximation are at the heart of using physics to answer questions about the world.

2.2 Average and Instantaneous Acceleration

2.2.1 The Acceleration Vector

While you are riding in a jetliner at 300 m/s, calmly eating your lunch, the plane encounters turbulence and suddenly speeds up by 3 m/s. Your coffee spills in your face, and your lunch tray slides onto your lap. The change in velocity, *acceleration,* causes the disruption of your

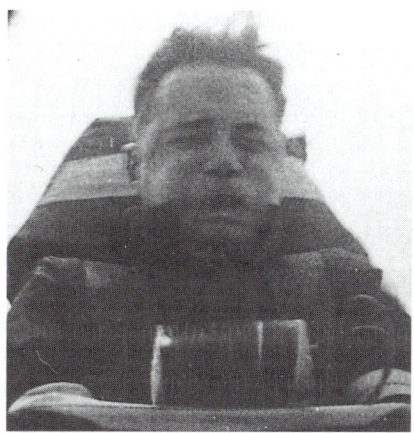

■ FIGURE 2.8
Colonel Stapp's wild ride. Colonel John P. Stapp carried out a long series of experiments on the effects of acceleration on the human body. Stapp's rocket sled was decelerated by a pool of water at up to 120 times the acceleration of gravity!

■ FIGURE 2.9
A carnival ride. The thrill of acceleration justifies the cost of a ticket to the ride.

lunch. Larger accelerations can be quite painful (■ Figure 2.8). Less rapid velocity changes produce the thrilling sensations of a carnival ride (■ Figure 2.9). Humans can detect even tiny accelerations. Small, fluid-filled chambers in our ears contain nerve fibers sensitive to motion of the fluid. Unconsciously, our brains use signals from these nerves to maintain balance when standing or to keep our eyes pointed in a fixed direction as our heads turn. Our next task is to describe acceleration carefully. The definitions follow the same mathematical *pattern* as those for average and instantaneous velocity.

YOU CANNOT FEEL *VELOCITY* DIRECTLY, BUT YOU ARE ABLE TO DETECT EVEN SMALL *ACCELERATIONS*.

> The *average acceleration* of an object during a time interval Δt is the change in the object's velocity divided by the time interval.
>
> $$\vec{a}_{av} \equiv \frac{\Delta \vec{v}}{\Delta t} = \frac{\vec{v}_f - \vec{v}_i}{\Delta t}. \qquad (2.4)$$

EXAMPLE 2.5 ♦ During launch, the space shuttle gains a vertical velocity of 121 km/h in 4.0 s. What is the average vertical acceleration of the shuttle?

MODEL To compute the average acceleration during the given time interval, we need to know the velocity of the shuttle at both ends of the interval.

SETUP The shuttle's initial velocity is $\vec{v}_i = 0$, and its final velocity is $\vec{v}_f = $ (121 km/h, up). We use these values in eqn. (2.4).

SOLVE Thus the average acceleration is

$$\vec{a}_{av} = \frac{\vec{v}_f - \vec{v}_i}{\Delta t} = \frac{(121 \text{ km/h, up}) - 0}{4.0 \text{ s}} = \left(30.3 \frac{\text{km}}{\text{h}\cdot\text{s}}, \text{up}\right).$$

We can tidy up the units by converting hours to seconds and kilometers to meters.

$$\vec{a}_{av} = \left[\left(30.3 \frac{\text{km}}{\text{h}\cdot\text{s}}\right)\left(\frac{10^3 \text{ m}}{1 \text{ km}}\right)\left(\frac{1 \text{ h}}{3600 \text{ s}}\right), \text{up}\right] = (8.4 \text{ m/s}^2, \text{up}).$$

REMEMBER: THE ZERO VECTOR HAS NEITHER MAGNITUDE NOR DIRECTION.

ANALYZE The units of acceleration look strange at first. The shuttle's speed increases by 30 km/h, or 8.4 m/s, every second. Acceleration is measured in units of speed/time, that is, $(\text{m/s})/\text{s} = \text{m/s}^2$.

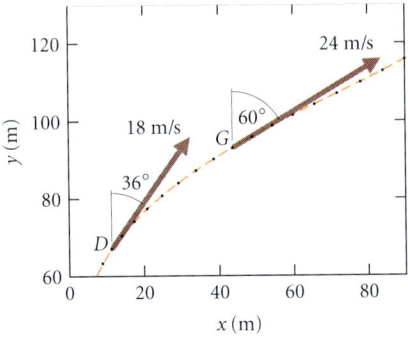

FIGURE 2.10
Race car velocities at points *D* and *G*. The velocities calculated in §2.1 are shown tangent to the car's path. We need the difference $\Delta \vec{v} \equiv \vec{v}_G - \vec{v}_D$ to find the average acceleration between *D* and *G*.

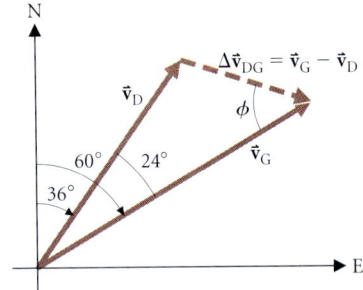

FIGURE 2.11
The velocity change $\Delta \vec{v}$ is computed from a triangle with the two vectors $\vec{v}_D$ and $\vec{v}_G$ drawn with tails coincident.

REMEMBER: MOVING A VECTOR PARALLEL TO ITSELF DOESN'T CHANGE THE VECTOR.

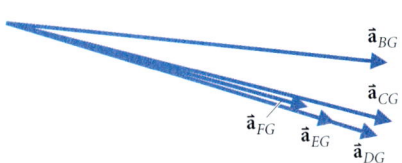

FIGURE 2.12
Instantaneous acceleration at point *G*. Average acceleration vectors are plotted for decreasing time intervals around *G*. (We calculated $\vec{a}_{DG}$ in Example 2.6.) Within graphical accuracy, the average $\vec{a}_{FG}$ for $\Delta t = 0.5$ s is indistinguishable from the limit. In this example the average acceleration vector rotates past, and then returns to, the limit as the time interval decreases.

EXAMPLE 2.6 ♦♦ Estimate the average acceleration of the race car between points *D* and *G*, (■ Figure 2.10), using velocity values determined in §2.1.

MODEL Once again we use eqn. (2.4). Here the velocity vectors do not lie along a single line.

SETUP We found $\vec{v}_D$ in Example 2.4, and $\vec{v}_G$ is shown in Figure 2.4. We draw the two velocity vectors with their tails together (■ Figure 2.11). The angle between the vectors is $\theta = 60° - 36° = 24°$. The change of velocity is their difference $\Delta \vec{v}_{DG} \equiv \vec{v}_G - \vec{v}_D$. From the law of cosines,

$$|\Delta \vec{v}_{DG}| = (v_D^2 + v_G^2 - 2v_D v_G \cos \theta)^{1/2}$$
$$= [(18 \text{ m/s})^2 + (24 \text{ m/s})^2 - 2(18 \text{ m/s})(24 \text{ m/s})\cos(24°)]^{1/2}$$
$$= 11 \text{ m/s}.$$

From the law of sines,

$$\frac{\sin \phi}{|\vec{v}_D|} = \frac{\sin \theta}{|\Delta \vec{v}_{DG}|}.$$

So $\sin \phi = \dfrac{|\vec{v}_D|}{|\Delta \vec{v}_{DG}|} \sin \theta = \dfrac{18 \text{ m/s}}{11 \text{ m/s}} \sin(24°) = 0.67 \Rightarrow \phi = 42°.$

The angle between $\Delta \vec{v}_{DG}$ and north is $60° + \phi = 102°$. Now we are ready to apply eqn. (2.4). The car takes $\Delta t = 2.0$ s to move between points *D* and *G*, so

SOLVE
$$\vec{a}_{av} = \frac{\Delta \vec{v}_{DG}}{\Delta t} = \frac{(11 \text{ m/s}, 102° \text{ east of north})}{2.0 \text{ s}}$$
$$= (5.5 \text{ m/s}^2, 102° \text{ east of north}).$$

ANALYZE Notice that the acceleration is almost perpendicular to $\vec{v}_D$. It changes the speed slightly, but the direction of the velocity substantially. ■

EXERCISE 2.4 ♦♦ The horse in Exercise 2.2 ran around a square field at a speed of 13 m/s, running northward for 75 s and then eastward for 75 s. What is the horse's average acceleration during the time interval from $t = 37.5$ s to $t = 112.5$ s? Where does the acceleration occur?

Like average velocity, average acceleration only gives a reasonable description for very short time intervals. Ultimately, we desire the limit for time intervals approaching zero.

> The *instantaneous acceleration* of a particle at a time *t* is the limit of its average acceleration during time intervals including *t*, as the size of the intervals approaches zero.
>
> $$\vec{a} \equiv \lim_{\Delta t \to 0} \vec{a}_{av} = \lim_{\Delta t \to 0} \frac{\Delta \vec{v}}{\Delta t} = \frac{d\vec{v}}{dt}. \quad (2.5)$$

Let's use the definition to calculate the race car's instantaneous acceleration at point *G* (■ Figure 2.12 and ● Table 2.3). The average acceleration during the 2.0-s time interval *DG* was obtained in Example 2.6. Average acceleration vectors during time intervals *BG* (5.0 s) to *FG* (0.5 s) may be obtained similarly. The behavior of the average acceleration vectors as they approach the limit is similar to the limiting behavior of average velocity vectors (Table 2.3).

There are some important differences between use of the word acceleration in physics and in everyday speech. In physics, acceleration refers to *any* change of velocity: magnitude, direction, or both. The space shuttle's motion in Example 2.5 comes closest to the everyday meaning: its velocity is fixed in direction but increases in magnitude (its speed increases). *Deceleration* commonly refers to a decrease in speed. In both cases, the acceleration vector is

TABLE 2.3 Average Acceleration Vectors near Point G

Interval	Δt (s)	$|\vec{a}_{av}|$ (m/s^2)	Direction East of North
BG	5.00	5.4	96°
CG	3.00	5.6	104°
DG	2.00	5.4	106°
EG	1.00	4.8	106°
FG	0.50	4.4	105°
—	0.25	4.1	105°

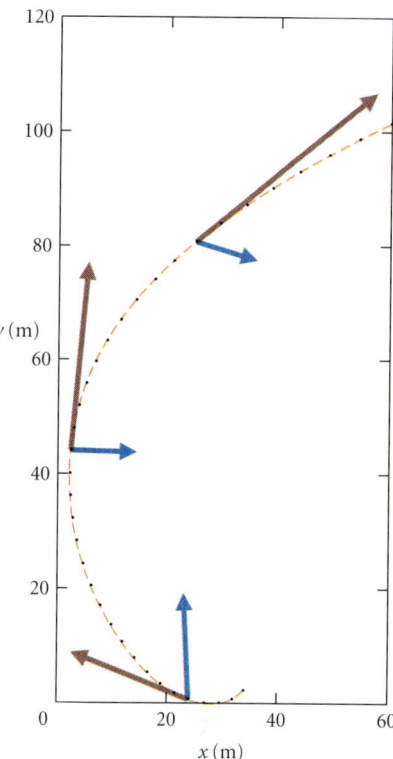

■ FIGURE 2.15
Velocity and acceleration vectors along the race car's path.

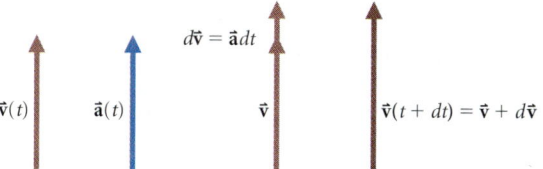

■ FIGURE 2.13
Acceleration parallel to velocity changes the magnitude of the velocity—that is, the speed, but not the direction.

either parallel to or directly opposite the velocity vector (■ Figure 2.13). We shall study this special case in detail in the last section of this chapter.

Alternatively, if the speed remains constant while the direction changes, then the acceleration vector is perpendicular to the velocity vector (■ Figure 2.14). Uniform circular motion, a special case with acceleration always perpendicular to velocity, is a major topic in Chapter 3.

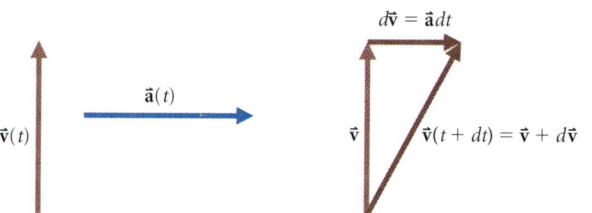

■ FIGURE 2.14
Acceleration perpendicular to velocity changes the direction of the velocity, but not its magnitude.

The race car's velocity changes in both magnitude and direction; its acceleration vector has components both along and perpendicular to the velocity (■ Figure 2.15).

An acceleration vector may have two components: one parallel to or opposite the velocity vector, which means that speed is changing, and one perpendicular to the velocity vector, which means that the direction of motion is changing.

EXAMPLE 2.7 ♦♦ Explain qualitatively the relation between velocity and acceleration at point G (Figure 2.10). How rapidly is the race car's speed changing at point G?

MODEL The path at G is curving to the right, meaning that the direction of the car's velocity is changing toward the right—hence, the acceleration has a component perpendicular to $\vec{v}_G$ and to the right. The acceleration component parallel to $\vec{v}_G$ indicates that the car is speeding up.

SETUP From Figure 2.10, $\vec{v}_G$ = (24 m/s, 60° east of north), and from Table 2.3, $\vec{a}_G$ = (4.1 m/s^2, 105° east of north). Thus the angle between the two vectors is 105° − 60° = 45° (■ Figure 2.16).

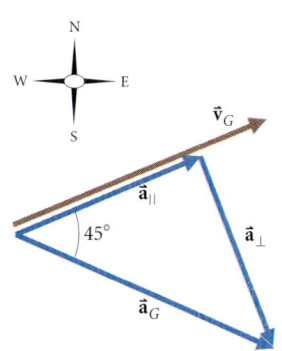

■ FIGURE 2.16
Velocity and acceleration at point G. Since $\vec{a}_G$ has a component parallel to $\vec{v}_G$, the race car is speeding up. The component $\vec{a}_\perp$ perpendicular to $\vec{v}_G$ indicates that $\vec{v}$ is changing direction to the right at point G.

HERE WE USE EQN. (1.3) TO CALCULATE THE COMPONENT.

SOLVE The rate of increase in speed equals the acceleration component parallel to $\vec{v}$.

$$\frac{d|\vec{v}|}{dt} = a_{\parallel} = (4.1 \text{ m/s}^2)\cos 45° = 2.9 \text{ m/s}^2.$$

ANALYZE We might expect the car to be speeding up, since the road is straightening out ahead of G. Since 1 m/s is about 2 mph (check this!), the car's speed is increasing at about 6 mph each second. The 1995 Porsche Carrera can achieve this acceleration at speeds up to 70 mph.

JERK, THE RATE OF CHANGE OF ACCELERATION, IS SOMETIMES IMPORTANT IN ENGINEERING DESIGN.

We could define quantities to describe average and instantaneous changes of acceleration, and so on in a never-ending sequence. For a known motion, such quantities are easily calculated but are rarely of interest in basic physics. As we shall discuss in Chapter 4, acceleration is directly related to the causes of change in motion, and so we do not need any further definitions.

2.2.2 Velocity and Acceleration in Component Notation

In rectangular coordinates, a particle's position as a function of time is expressed as

$$\vec{r}(t) = x(t)\hat{i} + y(t)\hat{j} + z(t)\hat{k}. \tag{2.6}$$

IN CALCULUS, CARTESIAN UNIT VECTORS ACT LIKE ANY OTHER CONSTANT. BUT BE CAREFUL! THE UNIT VECTORS IN OTHER COORDINATE SYSTEMS ARE NOT ALWAYS CONSTANT.

The three coordinates $x(t)$, $y(t)$, and $z(t)$ are ordinary functions of the single variable t. The Cartesian unit vectors have constant magnitude and direction, so the derivative of $\vec{r}$ is expressed

$$\vec{v}(t) \equiv \frac{d\vec{r}}{dt} = \hat{i}\frac{dx}{dt} + \hat{j}\frac{dy}{dt} + \hat{k}\frac{dz}{dt}. \tag{2.7}$$

The velocity components are

$$v_x = \frac{dx}{dt}, \quad v_y = \frac{dy}{dt}, \quad \text{and} \quad v_z = \frac{dz}{dt}.$$

Differentiating again gives an expression for the acceleration.

$$\vec{a} \equiv \frac{d\vec{v}}{dt} = \hat{i}\frac{dv_x}{dt} + \hat{j}\frac{dv_y}{dt} + \hat{k}\frac{dv_z}{dt}. \tag{2.8}$$

The acceleration components are

$$a_x = \frac{dv_x}{dt} = \frac{d^2x}{dt^2}, \quad a_y = \frac{dv_y}{dt} = \frac{d^2y}{dt^2}, \quad \text{and} \quad a_z = \frac{dv_z}{dt} = \frac{d^2z}{dt^2}. \tag{2.9}$$

Whenever it is possible to model a particle's motion by expressing the coordinates as algebraic functions of time, these calculus expressions are the most useful tools for computing velocity and acceleration.

EXAMPLE 2.8 ♦♦ The position of a particle is described by the vector

$$\vec{r}(t) = \hat{i}[(1.0 \text{ m/s}^3)t^3 - (5.0 \text{ m/s})t] + \hat{j}\{(2.0 \text{ m})\cos[(3.0 \text{ rad/s})t]\} + \hat{k}(8.0 \text{ m}).$$

Find the particle's velocity and acceleration at $t = 1.5$ s.

MODEL We just plug into eqns. (2.7) and (2.8).

SETUP The velocity components are

$$v_x = \frac{dx}{dt} = \frac{d}{dt}[(1.0 \text{ m/s}^3)t^3 - (5.0 \text{ m/s})t] = (3.0 \text{ m/s}^3)t^2 - 5.0 \text{ m/s},$$

$$v_y = \frac{dy}{dt} = \frac{d}{dt}\{(2.0 \text{ m})\cos[(3.0 \text{ rad/s})t]\} = -(6.0 \text{ m/s})\sin[(3.0 \text{ rad/s})t],$$

and $\quad v_z = \frac{dz}{dt} = \frac{d}{dt}(8.0 \text{ m}) = 0.$

The acceleration components are

$$a_x = \frac{dv_x}{dt} = \frac{d}{dt}[(3.0 \text{ m/s}^3)t^2 - 5.0 \text{ m/s}] = (6.0 \text{ m/s}^3)t,$$

$$a_y = \frac{dv_y}{dt} = \frac{d}{dt}\{-(6.0 \text{ m/s})\sin[(3.0 \text{ rad/s})t]\} = -(18 \text{ m/s}^2)\cos[(3.0 \text{ rad/s})t],$$

and $\quad a_z = \frac{dv_z}{dt} = \frac{d}{dt}(0) = 0.$

SOLVE Evaluating these expressions at $t = 1.5$ s, we have

$$v_x = (3.0 \text{ m/s}^3)(1.5 \text{ s})^2 - 5.0 \text{ m/s} = 1.8 \text{ m/s},$$

and $\quad v_y = -(6.0 \text{ m/s})\sin[(3.0 \text{ rad/s})(1.5 \text{ s})] = -(6.0 \text{ m/s})\sin(4.5 \text{ rad}).$

So $\quad v_y = 5.9$ m/s, and $v_z = 0.$

$a_x = (6.0 \text{ m/s}^3)(1.5 \text{ s}) = 9.0 \text{ m/s}^2.$

$a_y = -(18 \text{ m/s}^2)\cos[(3.0 \text{ rad/s})(1.5 \text{ s})]$

$\quad = -(18 \text{ m/s}^2)\cos(4.5 \text{ rad}) = +3.8 \text{ m/s}^2.$

$a_z = 0.$

ANALYZE We may write the results as

$$\vec{v} = \hat{i}(1.8 \text{ m/s}) + \hat{j}(5.9 \text{ m/s}) \quad \text{and} \quad \vec{a} = \hat{i}(9.0 \text{ m/s}^2) + \hat{j}(3.8 \text{ m/s}^2).$$

The vectors are shown in ■ Figure 2.17. Because the z-component of $\vec{r}$ is constant, both $\vec{v}$ and $\vec{a}$ lie in the x–y plane.

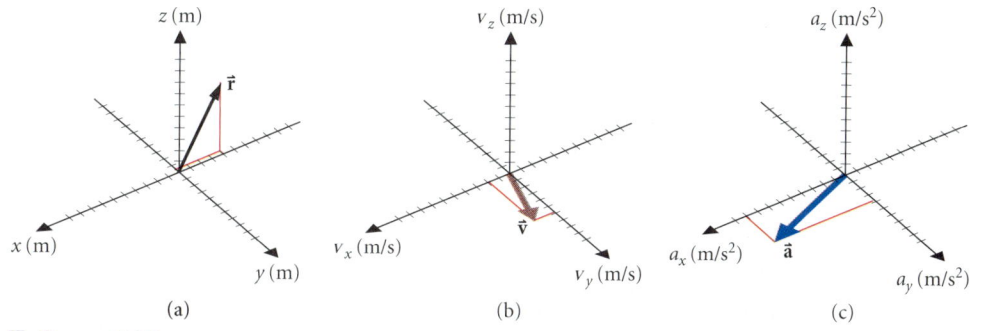

■ **FIGURE 2.17**
(a) The position vector at time $t = 1.5$ s has components $x = -4.1$ m, $y = -0.4$ m, and $z = 8.0$ m.
(b) The velocity vector has no z-component because z remains constant and equal to 8.0 m.
(c) Similarly, the acceleration vector has no z-component.

2.3 LINEAR MOTION

2.3.1 Position, Velocity, and Acceleration in Linear Motion

Imagine Galileo dropping a cannonball from the Leaning Tower of Pisa to demonstrate his theories about falling bodies (■ Figure 2.18). To the accuracy we may tell by eye, the ball falls vertically; that is, its displacement and velocity vectors point downward. Since the ball's velocity vector has constant direction, its acceleration vector also lies along the vertical line. Careful *choice* of a coordinate system greatly simplifies description of such a *linear* motion.

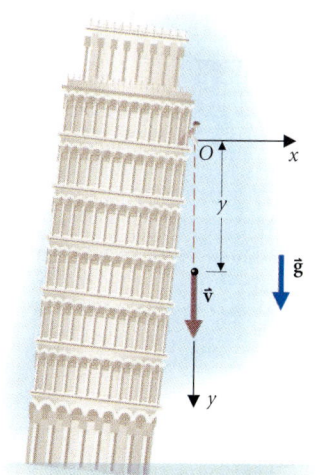

■ **FIGURE 2.18**
A cannonball falls from the Leaning Tower of Pisa. The origin of the reference frame is at the experimenter's hand, with y increasing downward. Galileo describes such an experiment, and one of his students included it as a fact in his biography. Most modern scholars believe Galileo only thought about the outcome. Too bad! It's a story so nice that it ought to be true.

Choosing the line of motion as a coordinate axis means that the ball's position vector has only one nonzero component, as do its velocity and acceleration. Next, we need to choose the name, origin, and positive direction of the coordinate. It is wise to base these choices on significant features of the problem. Here, we'll *choose* the starting point—Galileo's hand—as the origin; *name* the coordinate y; and let the "positive" direction be downward, the way the ball moves. Finally, we *choose* an event at which to set the clock reading at zero. The instant Galileo releases the ball is the most convenient time to start the clock.

With these choices, the vertical vectors have only y-components, positive meaning downward and negative meaning upward. We know the ball moves downward. We also know it moves faster near the ground than when just released; it accelerates downward. Thus displacement, velocity, and acceleration of the ball are all represented by positive numbers—their y-components.

> REMEMBER: In each problem you may freely choose the location of the origin, the coordinate name and its positive direction, and the event when the clock starts. Your choices determine how *words* like positive, negative, or zero express the *ideas* of up and down or before and after. Once made, keep your choices in mind both for translating meaning into algebra and for extracting meaning from your results.

2.3.2 The Acceleration Due to Gravity

Galileo discovered that a ball drops from rest a distance proportional to the square of the time of fall. With our choice of conventions, the ball's time of fall equals the clock reading t, and the distance fallen equals the coordinate y. Thus Galileo's experimental law is expressed

$$y \propto t^2.$$

Then the velocity cannot be constant—that would give distance proportional to time. (For constant velocity, $\vec{v}_{av} \equiv \vec{v} = \vec{D}/t \Rightarrow \vec{D} = \vec{v}t$.) Does the acceleration vary? Let us use kinematics to make a prediction and then compare our prediction with experimental fact. Since graphs are an extremely valuable visual tool in kinematic problems, we shall first practice by doing the analysis entirely with graphical methods that were available in Galileo's time.

IN THE NEXT SUBSECTION WE'LL DO THE DERIVATION USING CALCULUS.

To learn the method, consider first the simpler case of an object being lowered at constant velocity. In a graph of v_y versus time (■ Figure 2.19), the velocity component is represented by a horizontal line at the constant value v_o. The displacement y ($= v_o t$) is represented by a straight line in a graph of $y(t)$. The two graphs are closely related. The vertical line AB represents the displacement at clock reading t. The same magnitude, $v_o t$, is represented by the area beneath the velocity curve, shown shaded in the figure.

> The ordinate in a graph of displacement versus time and the area beneath the curve in a graph of velocity versus time are equally good representations of the object's displacement.

BE ALERT TO THE USAGE HERE. IT IS AWKWARD TO SAY VELOCITY COMPONENT OR DISPLACEMENT COMPONENT TIME AFTER TIME, SO THE WORD COMPONENT IS USUALLY LEFT UNSTATED, ASSUMING YOU WILL REMAIN AWARE WHAT IS MEANT.

Now, let's apply the method to the falling cannonball. The simplest guess we can make is uniform (i.e., constant) acceleration. Let's see if it works. The acceleration is represented by a

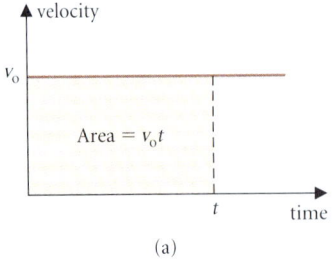
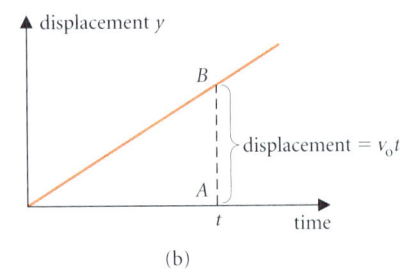

■ FIGURE 2.19
Graphs of velocity and displacement versus time for an object moving with constant velocity. (Check to see that the area under the velocity curve has units of distance.)

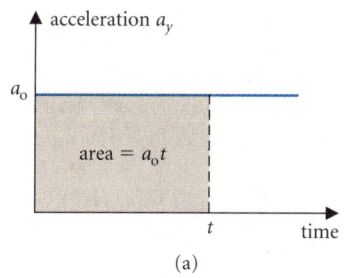
(a)
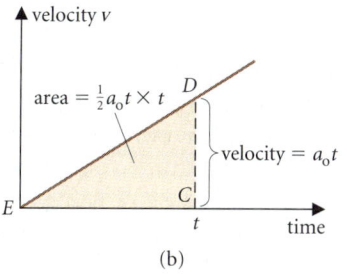
(b)
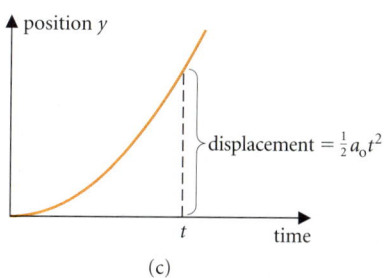
(c)

FIGURE 2.20
Graphs for uniformly accelerated motion from rest. (Check to see that the area under the acceleration curve has units of velocity.)

horizontal line in a graph of $a_y(t)$ (■ Figure 2.20a). In the graph of velocity versus time (Figure 2.20b), $v_y(t) = a_o t$ is represented by a straight line of constant slope; it passes through zero at $t = 0$ since the ball starts from rest ($v_y = 0$ when $t = 0$). The ordinate of the velocity curve at a time t (line CD) and the corresponding area under the acceleration curve (shaded) both represent the ball's velocity.

As before, the area under the velocity curve represents the ball's displacement (■ Figure 2.21). Between any two closely spaced times t and $t + \Delta t$, the velocity is almost constant, and the displacement increases by an amount $\Delta y = v_y(t)\, \Delta t$. The area of the heavily shaded, skinny rectangle in Figure 2.21 represents one such displacement. The total displacement is the sum of all the displacements for intervals Δt between $t = 0$ and $t = t_f$. The sum of the areas of all the corresponding rectangles is the total area under the velocity curve.

The ball's displacement is represented by the area of triangle CDE in Figure 2.20b.

$$\text{Area} = \tfrac{1}{2} \times \text{base} \times \text{altitude} = \tfrac{1}{2} \times \text{time} \times (\text{velocity})$$
$$= \tfrac{1}{2} \times \text{time} \times (\text{acceleration} \times \text{time}).$$

That is, displacement $= \tfrac{1}{2} \times$ acceleration $\times$ time^2.

Uniform acceleration does successfully model Galileo's law of falling bodies, which we now state in its modern form:

> Bodies released from rest near the Earth's surface fall with uniform acceleration, so long as air resistance is negligible. This uniform acceleration is given the traditional symbol g and the name *acceleration due to gravity*.

$$a_y \equiv g \ (\text{constant}).$$
$$v_y = gt.$$
$$y = \tfrac{1}{2}gt^2.$$

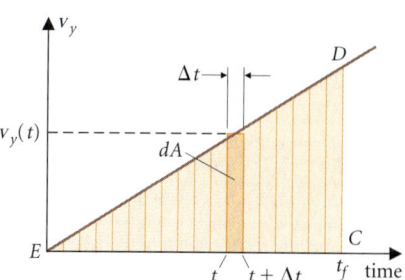

FIGURE 2.21
The area under the velocity curve is composed of skinny rectangles, with area $\Delta A = v(t)\, \Delta t = \Delta y$. Thus the sum of all the rectangles gives both the total area under the curve and the total displacement.

EXAMPLE 2.9 ♦ If Galileo's cannonball (Figure 2.18) requires 2.02 s to fall 20.0 m from its release point to the base of the tower, what is the measured value of g at the town square of Pisa?

MODEL We need only insert the given values for distance and time into Galileo's law.

SETUP $\qquad y = \tfrac{1}{2}gt^2 \ \Rightarrow \ g = 2y/t^2.$

SOLVE
$$g_{\text{Pisa}} = \frac{2y}{t^2} = \frac{2(20.0 \text{ m})}{(2.02 \text{ s})^2} = 9.80 \text{ m/s}^2.$$

ANALYZE The value of g depends on location, though $g = 9.8$ m/s^2 is accurate to two significant figures anywhere on the Earth's surface. Greater accuracy at a specific location requires a measurement. (■ Figure 2.22)

EXERCISE 2.5 ♦ The acceleration due to gravity on the Moon is $\tfrac{1}{6}$ that on Earth. If an astronaut drops a tool from rest, how long does it take to fall 20.0 m to the Moon's surface?

FIGURE 2.22
The National Institute of Standards and Technology experiment for measuring g. A ball is released to fall within an evacuated cylinder. The position of the ball is measured continuously with a laser interferometer (cf. Chapter 17) and its acceleration is thus derived.

2.3.3 Galileo's Law Using Formal Calculus

To use the definitions of instantaneous velocity and acceleration, we have to differentiate a particle's position and velocity vectors. In linear motion, a vector is completely described by one component, an ordinary numerical function of time. The rules for calculus of ordinary functions apply without change to vector components. (See §2.2.2.)

With this in mind, let us see how to apply calculus to the connection between Galileo's law and uniform acceleration.

Statement of Galileo's result for position: $\quad y = \frac{1}{2}gt^2 \quad$ (with g = constant).

Velocity is the derivative of displacement with respect to time.

$$v_y = \frac{dy}{dt} = \frac{d}{dt}(\tfrac{1}{2}gt^2) = gt.$$

Acceleration is the derivative of velocity (the second derivative of position).

$$a_y = \frac{dv_y}{dt} = \frac{d}{dt}(gt) = g.$$

That is, the acceleration is constant.

Using integration, we can work in the opposite sense and derive Galileo's law assuming uniform acceleration.

The derivative of velocity is acceleration: $\quad dv_y/dt = g.$

Integrate both sides over a time interval t:

$$\int_0^t \frac{dv_y}{dt'}dt' = \int_0^t g\, dt'.$$

Evaluate the integral.

$$v_y(t')\Big|_0^t = gt'\Big|_0^t$$

Because the cannonball starts from rest, $v_y(0) = 0$ and

$$v_y(t) = v_y(t) - v_y(0) = gt.$$

This is the same result we obtained from the graphical method. Since an integral is represented by the area under a graph of the integrated function, this method is no more than an algebraic version of the graphical method.

> IN MATHEMATICS, THE VARIABLE INSIDE AN INTEGRAL IS CALLED A *DUMMY VARIABLE* BECAUSE THE RESULT DOES NOT DEPEND ON THE SYMBOL USED. THE SAME IS TRUE IN PHYSICS APPLICATIONS, BUT THE VARIABLES t AND t' HAVE DIFFERENT PHYSICAL INTERPRETATIONS. THE RESULT OF THE CALCULATION WILL BE A FORMULA FOR THE VELOCITY AT TIME t. THE VALUE OF THAT VELOCITY DEPENDS ON THE ACCELERATION AT *EARLIER* TIMES LABELED t'.

Digging Deeper

GRAVITY

Galileo's discovery that all objects fall freely with the same acceleration g is of fundamental importance. In Chapter 4 we'll see how it influenced Newton as he developed his laws of motion. Einstein used it to create a picture of space-time like an elastic jello, warped by gravity. All free bodies move in the same way through space-time, and we interpret them as accelerating equally. These consequences of Galileo's simple law are so fascinating that experimental physicists continually attempt to test it more accurately. Galileo's measurements were probably accurate within a few percent. The most recent test supports the law to an accuracy of 1 part in 10^{13}.

Though g is the same for all bodies at any one place, it does not have exactly the same value at different locations on the Earth's surface. The Earth's rotation, the altitude above sea level, the presence of a nearby mountain range, or even of a nearby body of mineral ore can influence the value of g at any given location. These variations typically amount to a few tenths of a percent.

The expression for position follows from a second integration. The derivative of position is velocity.

$$\frac{dy}{dt} = v_y(t).$$

Substitute the known expression for $v_y(t)$ and integrate over the time interval 0 to t.

$$\int_0^t \frac{dy}{dt'} dt' = \int_0^t v_y(t')\, dt' = \int_0^t (gt')\, dt'.$$

Evaluate the integrals (remember g = constant).

$$y(t')\Big|_0^t = \tfrac{1}{2}gt'^2\Big|_0^t$$

Since the cannonball starts from the origin, $y(0) = 0$, and

$$y(t) = y(t) - y(0) = \tfrac{1}{2}gt^2.$$

Once again, we find the familiar result.

Formal use of integrals is equivalent to the graphical method. When an object's acceleration varies and is described by an integrable algebraic function, the formal approach is the easiest method for calculations. Then, a combined use of graphs for qualitative understanding together with formal calculus is the most powerful approach.

Airborne equipment is often used to measure variations in g. Such measurements are useful for geological mapping and location of mineral deposits.

EXAMPLE 2.10 ♦♦ A sensitive accelerometer is placed inside a falling cannonball to measure the effect of air resistance on the ball's motion. Data from the accelerometer show that the magnitude of the ball's acceleration is reduced by an amount αt^2, where $\alpha = 1.5 \times 10^{-3}$ m/s^4. Assuming g at Pisa is 9.800 m/s^2, what errors are made in the ball's position and velocity after 2.00 s by neglecting air resistance?

MODEL We continue to use the coordinate system with the positive y-axis downward. Then the acceleration of the ball is $a_y = g - \alpha t^2$. We may find the position and velocity by integrating. Neglecting air resistance means setting $\alpha = 0$.

SETUP First we find expressions for velocity and position including air resistance.

$$v_y = \int_0^t a_y(t')\, dt' = \int_0^t (g - \alpha t'^2)\, dt' = gt - \alpha \frac{t^3}{3}.$$

$$y = \int_0^t v_y(t')\, dt' = \int_0^t \left(gt' - \frac{\alpha}{3}(t')^3\right) dt' = \tfrac{1}{2}gt^2 - \frac{\alpha}{12}t^4.$$

SOLVE Neglecting α, we would make an error δv_y in the calculated velocity, where

$$\delta v_y = -\frac{\alpha}{3}t^3 = -\frac{1.5 \times 10^{-3} \text{ m/s}^4}{3}(2.00 \text{ s})^3 = -4.0 \times 10^{-3} \text{ m/s}.$$

With $v \approx gt = (9.8 \text{ m/s}^2)(2.0 \text{ s}) = 20 \text{ m/s}^2$, the fractional error is $\delta v/v \approx -2 \times 10^{-4}$. The error in position is

$$\delta y = -\frac{\alpha}{12}t^4 = -\frac{1.5 \times 10^{-3} \text{ m/s}^4}{12}(2.00 \text{ s})^4 = -2 \times 10^{-3} \text{ m}.$$

With $y \approx \tfrac{1}{2}gt^2 = \tfrac{1}{2}(9.8 \text{ m/s}^2)(2.0 \text{ s})^2 = 20$ m, the fractional error is $\delta y/y \approx -1 \times 10^{-4}$.

ANALYZE Galileo could safely neglect air resistance on the cannonball. ∎

IT DOESN'T MAKE ANY DIFFERENCE WHETHER WE USE THE UNCORRECTED OR CORRECTED VALUE FOR v_y IN CALCULATING THE FRACTIONAL ERROR, SINCE THEY DIFFER BY SO LITTLE. TRY IT AND SEE!

EXERCISE 2.6 ♦♦ Using radar observations of the starship *Enterprise* made from Origin Base, the mathematical function that best describes its position is found to be

$$x(t) = \frac{At_0^2}{t^2 - 4t_0^2}$$

THE WARP SCALE IS A NONSTANDARD SPEED SCALE USED EXCLUSIVELY IN TV AND MOVIE STUDIOS.

during the interval from $t = 0$ to $t = t_o$, when the *Enterprise* goes to warp speed and radar contact is lost. What is the *Enterprise*'s acceleration just before it makes a transition to warp speed? Is it approaching or receding from Origin base? Is it slowing down or speeding up just before entering warp drive?

2.3.4 Interpreting Graphs of Position or Velocity versus Time

So far we have used graphs to find displacement from velocity or velocity from acceleration by computing the area under a given graph. Alternatively, we may obtain velocity and acceleration from a given graph of position. Since velocity and acceleration are the first and second derivatives of position, we can estimate them from the slope and curvature of the position graph.

VELOCITY EQUALS THE SLOPE OF A GRAPH OF POSITION VERSUS TIME.

■ Figure 2.23 is a graph of position versus time for an oscillating object suspended from the roof by a spring. The tangent line drawn at point A has the same slope as the curve, equal to the derivative of position y with respect to time. That derivative equals the velocity of the object. Reading from the graph,

$$v_y(A) = \frac{dy}{dt} \text{ at } A = \text{slope of tangent line} = \frac{\Delta y}{\Delta t} = \frac{-18 \text{ cm}}{1 \text{ s}} = -18 \text{ cm/s}.$$

At B, the curve has a positive slope, $v_y(B) = +18$ cm/s. At C, the tangent line is horizontal; the velocity is zero.

■ **FIGURE 2.23**
Graph of position versus time for an object suspended from the roof by a spring and oscillating vertically.

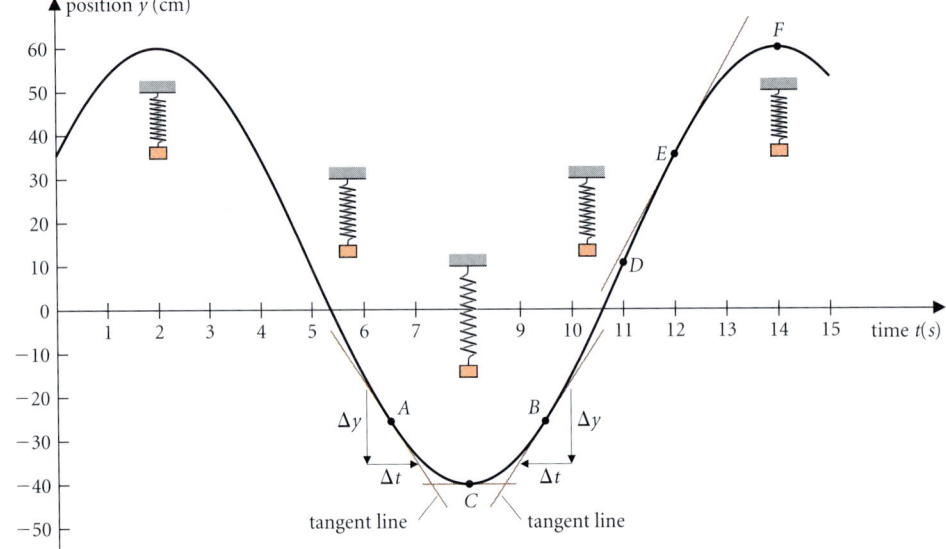

Since acceleration is the rate of change of velocity, it is represented by change in the slope of the tangent line—that is, the curvature of the graph. Both the particle's speed and the choice of scale on the axes also influence the geometrical curvature of the graph, so only a rough, qualitative estimate of acceleration is possible. Near point C, the velocity is changing from negative to positive values, so acceleration is positive, and the $y(t)$ line curves upward. At E and F, downward curvature corresponds to negative acceleration.

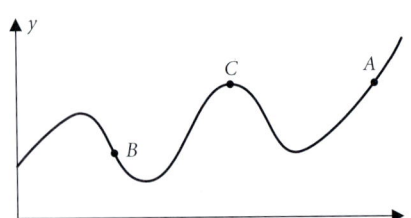

■ **FIGURE 2.24**
Graph of position versus time for Exercise 2.7.

EXERCISE 2.7 ♦ ■ Figure 2.24 represents the one-dimensional motion of an object. Qualitatively compare its velocities and its accelerations at times A, B, and C.

You can find the acceleration of an object with much greater precision from a graph of velocity versus time. Since acceleration is the first derivative of velocity, it is given by the slope of the velocity graph. ■ Figure 2.25 is the velocity graph for the oscillating object whose posi-

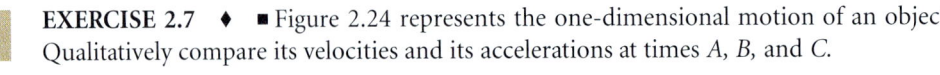

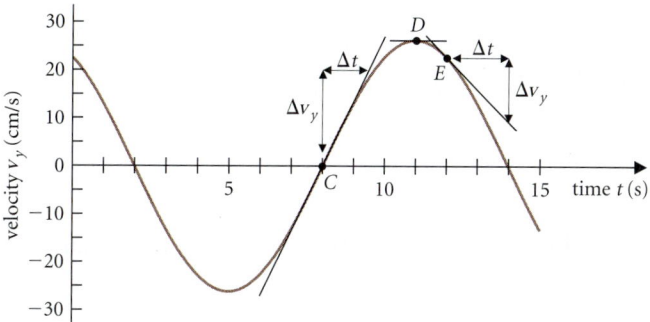

■ **Figure 2.25**
Graph of velocity versus time for the motion shown in Figure 2.23. Estimating the acceleration is best done by finding the slope of a velocity versus time graph. At C, the slope is $\Delta v_y/\Delta t = (20 \text{ m/s})/(1.5 \text{ s}) = 13 \text{ m/s}^2$.

tion is represented in Figure 2.23. At C, the slope of the graph is +13 cm/s²; at D, the slope is zero; and at E, it is −7 cm/s². These results correspond well with the qualitative conclusions from Figure 2.23.

EXAMPLE 2.11 ♦♦ ■ Figure 2.26 shows the graph of velocity versus time for an object in linear motion. Sketch the corresponding graph for acceleration versus time.

MODEL Acceleration is the slope of the velocity curve.

SETUP Before time A, the velocity curve has a constant slope ≈ 20 cm/s², corresponding to constant acceleration. In the intervals between B and C and after E, the velocity is constant and so the acceleration is zero.

SOLVE These facts are recorded in ■ Figure 2.27. The variation of acceleration in the intervals A to B and C to E is sketched qualitatively.

ANALYZE The peak acceleration at D is the same as before time A, since the tangent at D is parallel to the curve to the left of A. ■

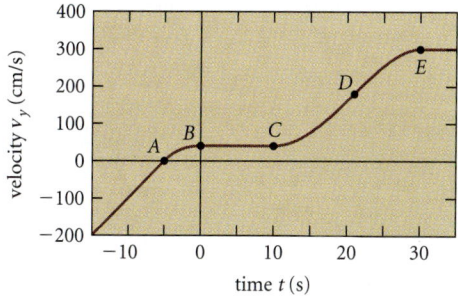

■ **Figure 2.26**
Graph of velocity versus time for Example 2.11.

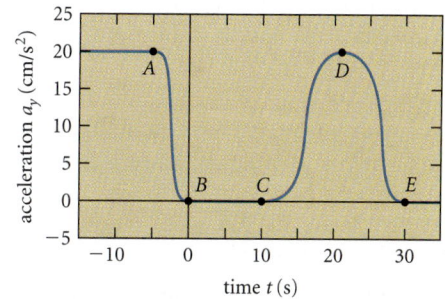

■ **Figure 2.27**
Graph of acceleration versus time resulting from the solution of Example 2.11. The acceleration is the slope of the velocity curve (Figure 2.26).

2.3.5 Uniformly Accelerated Linear Motion

Galileo's law for free fall from rest is a very special case that we need to generalize. For example, after pulling away from the platform, a subway train moves at a relatively low speed until it leaves the station, then it accelerates to a much higher speed for the trip between stations. The train's acceleration is reasonably uniform, but the initial velocity isn't zero, and we might not want our coordinate origin where the acceleration begins. We may apply the graphical method for any initial values of position and velocity. We keep the choice $t = 0$ for the initial clock reading (■ Figure 2.28). The acceleration curve is again a horizontal line, and the area under it after time interval t represents the change of velocity during the interval, $\Delta v = a_o t$. The line CD in the velocity graph also represents this velocity change.

Velocity at t = initial velocity + area under acceleration graph

$$v = v_i + a_o t. \tag{2.10}$$

The area under the velocity graph represents the object's displacement (change in position).

Displacement at t ≡ position at time t − initial position
= area under velocity graph
= area $CEFG$ + area CDE.

$$\Delta y \equiv y - y_i = v_i t + \tfrac{1}{2} t(a_o t).$$

$$\Delta y = v_i t + \tfrac{1}{2} a_o t^2. \tag{2.11}$$

Expressions for velocity and displacement completely describe the motion of an object under uniform acceleration but are not always best for problem solving. We shall derive two more relations that are sometimes more useful.

The average velocity is defined as displacement divided by time interval.

$$v_{av} \equiv \frac{\Delta y}{t} = \frac{v_i t + \tfrac{1}{2} a_o t^2}{t} = v_i + \tfrac{1}{2} a_o t = \tfrac{1}{2} v_i + \tfrac{1}{2}(v_i + a_o t).$$

$$v_{av} = \tfrac{1}{2}(v_i + v_f). \tag{2.12}$$

An object moving with this average velocity would have the same displacement as the accelerated object. Its velocity curve would be the straight line IHJ in Figure 2.28b. Areas under the actual velocity curve EHD and the average velocity curve IHJ must be equal, as they represent the same displacement. This requires that the triangles HIE and HDJ have equal areas and thus be congruent. Then point H, when instantaneous and average velocities are equal, is at the middle of the time interval.

$$v_{av} \text{ (between 0 and } t\text{)} = v(t/2).$$

Our final result is a relation that does not involve the time interval. When instantaneous acceleration is uniform, its constant value equals its average value during any time interval. Then

$$a_{av} \equiv \frac{\Delta v}{t} = a_o, \quad \text{so} \quad t = \frac{\Delta v}{a_o} \equiv \frac{v_f - v_i}{a_o}.$$

Combining this expression for t with eqn. (2.12), we find:

$$\Delta y \equiv v_{av} t = \tfrac{1}{2}(v_f + v_i) \frac{v_f - v_i}{a_o},$$

or

$$\Delta y = \frac{v_f^2 - v_i^2}{2 a_o}. \tag{2.13}$$

These equations are collected in ● Table 2.4. The following examples and exercises illustrate their use in problem solving. Note carefully in each case how coordinates are *chosen* for convenience and how the given information and the desired answers lead to the *choice* of the most useful relations.

> **EXAMPLE 2.12** ♦♦ Slugger José throws a baseball directly upward at a speed of 20 m/s. Use the value $g \approx 10$ m/s^2 to find when the baseball reaches its maximum height and when José catches it again.

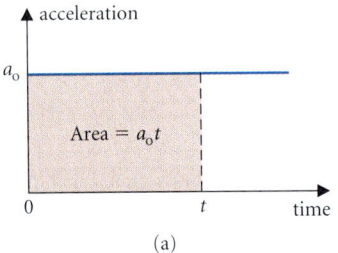
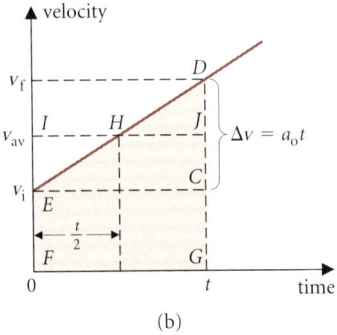
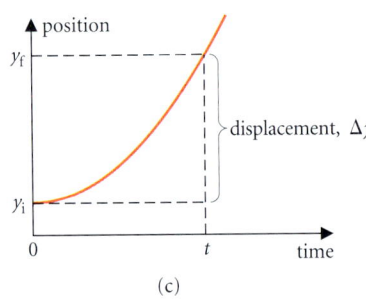

■ **FIGURE 2.28**
The graphical method applied to uniformly accelerated motion without special initial conditions. (a) The acceleration curve appears the same as in the previous special case. (b) It is now the *change* in velocity ordinate $\Delta v = v_f - v_i$ that equals the area $a_o t$ between acceleration curve and time axis. Triangle IHE is constructed equal to triangle JHD, so that the areas of the triangle EDC and the rectangle $EIJC$ are equal. (c) The displacement equals the area between velocity curve and time axis.

REMEMBER: $a^2 - b^2 = (a - b)(a + b)$.

TABLE 2.4	Equations for Uniformly Accelerated Linear Motion	
Velocity as a function of time:	$v = v_i + a_o t$	(2.10)
Displacement as a function of time:	$\Delta y \equiv y - y_i = v_i t + \frac{1}{2} a_o t^2$	(2.11)
Average velocity:	$v_{av} = \frac{1}{2}(v_f + v_i)$	(2.12)
Displacement in terms of speed:	$2 a_o \Delta y = v_f^2 - v_i^2$	(2.13)

WARNING: These equations are safe when used as directed, but have disastrous side effects when applied in cases of nonuniform acceleration.

MODEL In ■ Figure 2.29a a stroboscope shows the positions of a freely falling ball at equal time intervals, and Figure 2.29b illustrates the ball's instantaneous velocity at each of these times. Upward motion of José's baseball is similarly slowed by gravity. The ball's speed decreases to zero at the highest point and increases again as the ball descends. If we choose the positive y-axis to be directly upward, the ball's initial velocity is positive and the acceleration due to gravity is negative.

SETUP Since we are given the initial velocity of the ball, the time to reach maximum height may be obtained from the velocity relation, eqn. (2.10), with $a_o = -g$. We set the velocity at the top equal to zero.

$$v_{top} = 0 = v_i + a_o t_{top} = v_i - g t_{top}.$$

SOLVE
$$t_{top} = \frac{v_i}{g} = \frac{20 \text{ m/s}}{10 \text{ m/s}^2} = 2 \text{ s}.$$

SETUP When José catches the ball, it has returned to its initial position; so, we set the displacement equal to zero in eqn. (2.11).

$$\Delta y_{catch} = 0 = v_i t_{catch} - \tfrac{1}{2} g t_{catch}^2.$$

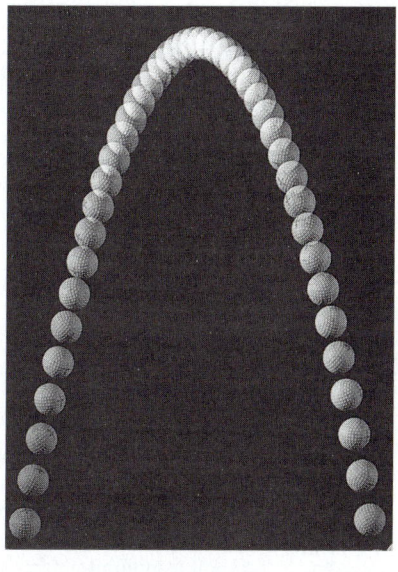

(a)

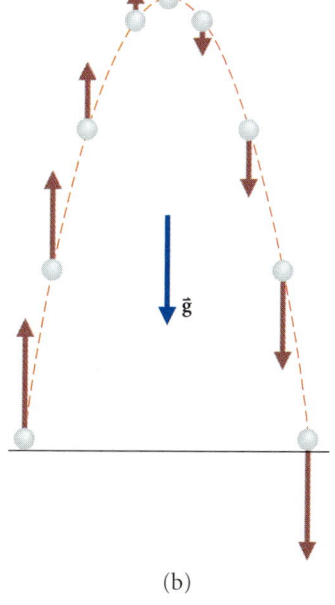

(b)

■ **FIGURE 2.29**
(a) A stroboscope reveals the position of a freely falling golfball at equal time intervals. The ball is given a small horizontal motion so that the images of the ball rising do not overlap those of the ball falling. Decreasing displacements as the ball approaches its highest point indicate decreasing speed. The speed increases again after the ball starts downward. (b) Schematic of the golfball's motion showing a velocity vector at each of several selected positions. The ball's acceleration is $\vec{g}$ at all times.

A MOVIE OF THE BASEBALL'S MOTION WOULD LOOK THE SAME IF RUN BACKWARDS, SO THIS FEATURE IS OFTEN CALLED *TIME-REVERSAL SYMMETRY*. SIMILAR KINDS OF SYMMETRICAL BEHAVIOR ARE COMMON IN PHYSICS AND GIVE RISE TO POWERFUL PROBLEM-SOLVING METHODS. SUCH SYMMETRY IS AN ESSENTIAL FEATURE OF ELEMENTARY PARTICLE BEHAVIOR (CHAPTER 37).

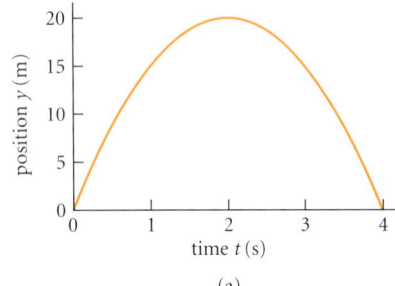

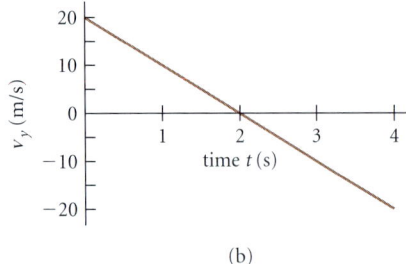

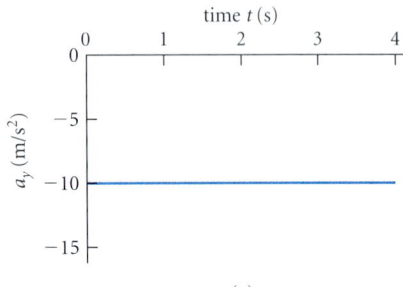

■ **FIGURE 2.30**
(a) Graph of the baseball's position versus time indicating the *time-reversal symmetry* of the ball's motion. The downward portion of the ball's motion is just what you would see by running a movie of the upward motion backward. (b) Velocity of the baseball versus time. The single velocity component is zero at the top of the ball's path ($t = 2$ s) and negative on the way down ($t > 2$ s). (c) The acceleration is constant throughout the motion.

SOLVE The equation has two solutions, but $t_\text{catch} = 0$ only repeats what we already know: the ball started upward at $t = 0$. The other solution gives the time when José catches the ball:

$$t_\text{catch} = 2\frac{v_i}{g} = 2\frac{20 \text{ m/s}}{10 \text{ m/s}^2} = 4 \text{ s}.$$

ANALYZE As ■ Figure 2.30 illustrates, the rising portion of the baseball's motion is symmetric with the falling portion. Using this symmetry is often the quickest way to answer a question. Here, for example, since the ball returns to the same height it started from, the baseball's total time of flight is just twice the time for it to reach its maximum height, so $t_\text{top} = 2$ s gives $t_\text{catch} = 4$ s.

Notice that v_y is zero at the top of the ball's path, but its acceleration is *not* zero there. The acceleration has the constant value $a_y = -g$ throughout the motion. ■

EXERCISE 2.8 ♦ What is the baseball's speed when José catches it? How high does it go? ■

EXAMPLE 2.13 ♦♦ *The speeder's dilemma.* An automobile is speeding along a boulevard at 29 m/s when a traffic signal 51 m ahead turns yellow (see ■ Figure 2.31). If the driver's reaction time is 0.50 s, what uniform acceleration is required for the car to stop at the signal?

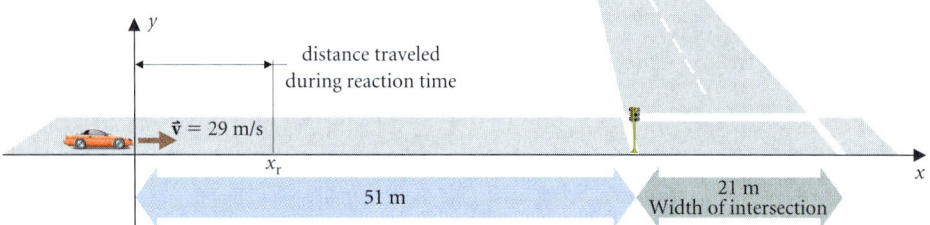

■ **FIGURE 2.31**
The speeder's dilemma. An unwise driver is going too fast for the road. Can the driver decelerate rapidly enough to stop for the red light or accelerate rapidly enough to pass through the intersection before the light turns red?

MODEL We must break the problem into two parts. First, during the reaction time, the car moves at constant velocity. Then the brakes are applied and the car slows to a stop. Assuming the brakes cause uniform acceleration, we plot a graph of the car's position as a function of time (■ Figure 2.32). The graph is a straight line (indicating constant ve-

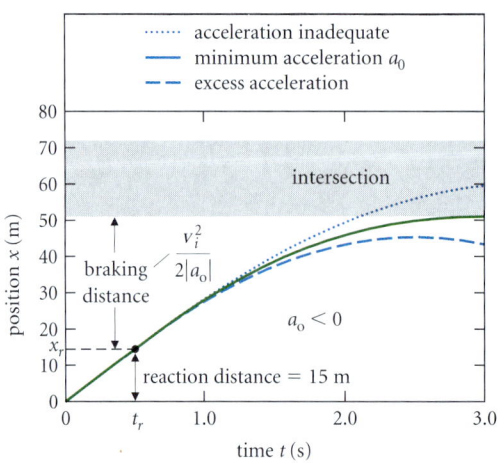

■ **FIGURE 2.32**
Position versus time graph if the speeder tries to stop. The graph is a straight line during the reaction time and then becomes parabolic. The graph for minimum necessary acceleration is just tangent to the edge of the region labeled "intersection."

locity) until the driver reacts ($t = t_r$). Afterward, the graph curves downward indicating the car's deceleration. We choose the x-axis along the road in the direction of the car's motion, with origin at the car's initial position; positive then means *toward the right* in the figure, and the signal is at $x = +51$ m. We start the clock when the signal turns yellow. We are not interested in the time required to stop the car, only that its coordinate when it stops be less than 51 m. So, we plan to work with displacements.

SETUP First we find the distance to the traffic signal at the end of the reaction time, then the acceleration that will stop the car within that distance. During the reaction interval, the acceleration is zero. From eqn. (2.11) with $a_o = 0$, the car travels

$$\Delta x = x - 0 = vt = (29 \text{ m/s})(0.50 \text{ s}) = 14.5 \text{ m},$$

so the remaining distance is $x_f - x = 51 \text{ m} - 14.5 \text{ m} = 36.5 \text{ m}$. To find the required braking acceleration, we use eqn. (2.13), which relates speed and distance.

$$a_o = \frac{v_f^2 - v_i^2}{2\Delta x}.$$

Since the car is to stop, $v_f = 0$. The initial velocity is $v_i = +29$ m/s, so

$$a_o = \frac{0 - (29 \text{ m/s})^2}{2(36.5 \text{ m})} = -12 \text{ m/s}^2.$$

ANALYZE The minus sign indicates that the acceleration points to the left, as we would expect, but its magnitude is larger than the acceleration of gravity, g. Even the best cars have a maximum braking acceleration of about 10 m/s² or 1g. Unless the car has rockets in its front end, the driver should perhaps try to make it through the intersection. ∎

THE ACCELERATION DUE TO GRAVITY, g, IS OFTEN USED AS A UNIT OF ACCELERATION, PARTICULARLY WHEN ESTIMATING THE EFFECT OF ACCELERATION ON PEOPLE. DON'T CONFUSE THE SYMBOL WITH g FOR GRAM!

EXERCISE 2.9 ♦♦ In Example 2.13, if the yellow light is on for 2.0 s and the road intersection is 21 m wide, what acceleration is necessary for the car to pass through the intersection before the light turns red? ∎

Often we can use the results for uniform acceleration to make very informative estimates in situations where the variation of acceleration is much less important than its order of magnitude. One such example involves estimating the effects of automobile accidents.

EXAMPLE 2.14 ♦ A car with an initial speed of 30 m/s skids off the road and runs into a wall. The front 1.5 m of the car is crumpled in the accident. Estimate the acceleration experienced by the driver, who had wisely buckled her seatbelt.

MODEL Even though the acceleration in this accident is unlikely to be constant, we may *estimate* its magnitude using the relations in Table 2.4. We choose the y-axis to be along the direction of the car's motion.

SETUP We are given speed and distance, so the appropriate relation is eqn. (2.13), with $v_i = +30$ m/s and $v_f = 0$.

SOLVE
$$a \approx \frac{v_f^2 - v_i^2}{2\Delta y} = \frac{-(30 \text{ m/s})^2}{3.0 \text{ m}} = -300 \text{ m/s}^2 = -30g.$$

ANALYZE Colonel Stapp's experiments (cf. Figure 2.8) showed that a human body can sustain accelerations of magnitude 35g without permanent damage, so the driver is probably not seriously injured. ∎

EXERCISE 2.10 ♦ Joe, riding in the passenger seat, doesn't believe in shoulder harnesses and his forehead impacts the padded dash, which compresses 2.0 cm. Estimate the acceleration of Joe's head, and comment. ∎

These estimates don't have to be terribly accurate for us to get the point!

Chapter Summary

Where Are We Now?

We have defined the quantities *velocity* and *acceleration* to describe how a particle's position changes with time. We applied these definitions to motion in a straight line. Now we are ready to tackle less restricted models in the next chapter.

What Did We Do?

The *average velocity* of an object in a time interval Δt is the displacement occurring in that interval divided by Δt. *Average speed* is the total distance traveled divided by the time interval. *Instantaneous velocity* is defined as the limit of average velocity as the time interval approaches zero.

$$\vec{v}_{av} \equiv \frac{\Delta \vec{r}}{\Delta t} \quad \text{and} \quad \vec{v} \equiv \lim_{\Delta t \to 0} \vec{v}_{av} \equiv \frac{d\vec{r}}{dt}.$$

Instantaneous speed is the magnitude of the instantaneous velocity vector. *Acceleration* is the rate of change of instantaneous velocity. Both average and instantaneous acceleration are defined with respect to velocity in the same way that average and instantaneous velocity are defined in terms of displacement:

$$\vec{a}_{av} \equiv \frac{\Delta \vec{v}}{\Delta t} \quad \text{and} \quad \vec{a} \equiv \lim_{\Delta t \to 0} \vec{a}_{av} \equiv \frac{d\vec{v}}{dt}.$$

A graphical method shows that Galileo's experimental result for falling bodies corresponds to constant downward acceleration. Extending the graphical method gives a general formalism for uniformly accelerated linear motion, summarized by the equations in Table 2.4. For linear motion, the graphical method is equivalent to the use of integration.

Since velocity and acceleration are the first and second derivatives of position, you may compare velocity and acceleration at different times using the slope and curvature of a graph of position versus time.

Practical Applications

A clear description of motion is necessary for engineers studying race car performance, for scientists tracking spacecraft, and even for people planning a trip. As we shall see, Newton discovered that the instantaneous acceleration of a body is directly related to the forces acting on the body. Consequently, it is acceleration (approximately $3g$ or 30 m/s^2) that makes astronauts feel stressed during a space shuttle launch or determines whether a person is injured in an auto crash ($\gtrsim 30g!$). Aircraft are designed to withstand a maximum acceleration of about $4g$ upward or $1.5g$ downward. Maximum acceleration and braking deceleration are major considerations in designing a car for safe operation.

Solutions to Exercises

2.1 Driving 20 km at 45 km/h requires a time

$$\Delta t_1 = \frac{20 \text{ km}}{45 \text{ km/h}} = \tfrac{4}{9} \text{ h}.$$

The average speed for the remaining 40 km should be

$$S_{av} = \frac{40 \text{ km}}{\tfrac{5}{9} \text{ h}} = 72 \text{ km/h}.$$

Rounding to one significant figure, $S_{av} = 70$ km/h.

2.2 The average speed for the first 75 s is found from the given distance and time.

$$S_{av} = (1.0 \times 10^3 \text{ m})/(75 \text{ s}) = 13 \text{ m/s}.$$

Since, during this interval, the motion is along a straight line, $|\vec{v}_{av}| = S_{av}$, and the average velocity is $\vec{v}_{av} = (13 \text{ m/s, north})$.

Over the 150-s interval, the horse runs twice the distance in twice the time and so has the same average speed. However, the average velocity vector for 150 s (■ Figure 2.33) is:

$$\vec{v}_{av} = \frac{\text{displacement}}{\text{time interval}}$$
$$= \frac{(1.0 \text{ km, north}) + (1.0 \text{ km, east})}{150 \text{ s}}$$
$$= \frac{(1.4 \times 10^3 \text{ m, northeast})}{150 \text{ s}}$$
$$= (9.4 \text{ m/s, northeast}).$$

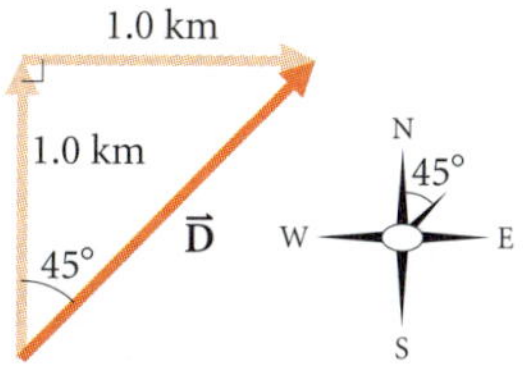

■ **FIGURE 2.33**

2.3 First we use the given data to calculate the average velocity during each time interval: $|\vec{v}_{av}| = |\vec{D}|/\Delta t$.

Time Interval Δt (s)	$|\text{Displacement}|$ $|\vec{D}|$ (m)	$|\text{Average velocity}|$ $|\vec{v}_{av}|$
4.0	93 ± 0.5	23.3 ± 0.12
3.0	67	22.3 ± 0.2
2.0	42	21.0 ± 0.25
1.0 (C to D)	19	19.0 ± 0.5
0.75	14	18.7 ± 0.6
0.5	9	18 ± 1

The results are plotted in ■ Figure 2.34. From the graph, the value for the limiting speed is 18.3 ± 0.4 m/s. The result using symmetric time intervals was 18.1 ± 0.25 m/s. The difference is 0.2 m/s, less than half the sum of the estimated uncertainties in the two results. Thus the two values are consistent.

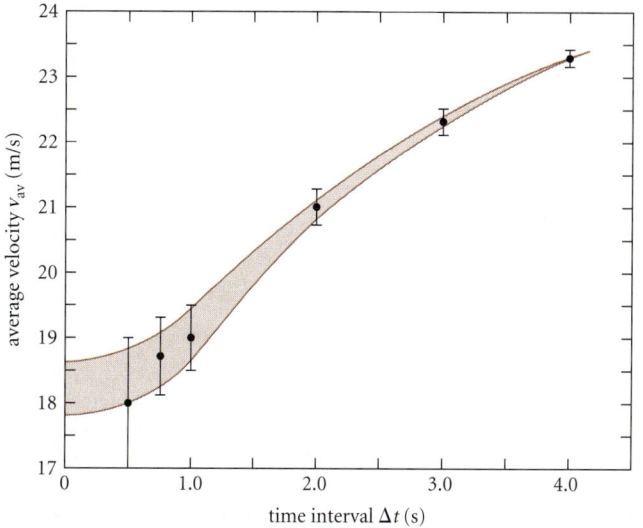

■ **FIGURE 2.34**

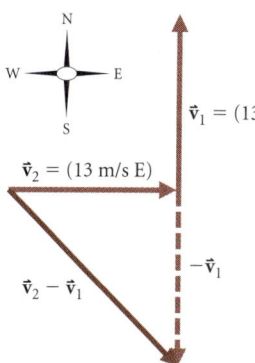

■ **FIGURE 2.35**
Vector diagram showing the subtraction of $\vec{v}_1$ from $\vec{v}_2$ (Exercise 2.4). The change in velocity is
$$\Delta \vec{v} = \vec{v}_2 - \vec{v}_1$$
$$= (13 \text{ m/s, east}) - (13 \text{ m/s, north}).$$

2.4 The average acceleration is the change in velocity divided by the time interval (■ Figure 2.35).

$$\vec{a}_{av} = \frac{\Delta \vec{v}}{\Delta t} = \frac{(13 \text{ m/s, east}) - (13 \text{ m/s, north})}{75 \text{ s}}$$
$$= \frac{(13\sqrt{2} \text{ m/s, southeast})}{75 \text{ s}} = (0.25 \text{ m/s}^2, \text{ southeast}).$$

All of the velocity change occurs as the horse rounds the corner of the field during a time interval left unmentioned. The average acceleration we computed doesn't say much about what the horse must do at the corner.

2.5 The distance the tool falls is $\Delta y = \frac{1}{2} g_{\text{Moon}} t^2$. So, the required time is

$$t = \sqrt{\frac{2 \Delta y}{g_{\text{Moon}}}} = \sqrt{\frac{2(20.0 \text{ m})}{(9.80 \text{ m/s}^2)/6}} = 4.9 \text{ s}.$$

2.6 To obtain an expression for acceleration, we need to differentiate the given expression for position twice. To find the required value, plug in the given time $t = t_o$.

$$v_x(t) \equiv \frac{dx}{dt} = \frac{-At_o^2(2t)}{(t^2 - 4t_o^2)^2}.$$

Note: During the interval $0 \leq t \leq t_o$, both $x(t)$ and $v_x(t)$ have negative values. Thus the *Enterprise*'s position is in the direction labeled *negative* and its velocity points in the *negative* direction—away from Origin Base.

$$a_x(t) \equiv \frac{dv_x}{dt} = \frac{(-2)(-At_o^2)(2t)^2}{(t^2 - 4t_o^2)^3} + \frac{-2At_o^2}{(t^2 - 4t_o^2)^2} = \frac{At_o^2(6t^2 + 8t_o^2)}{(t^2 - 4t_o^2)^3}.$$

Note: $a_x < 0$ at any time $t < 2t_o$, so $\vec{a}$ is parallel to $\vec{v}$; the *Enterprise* is speeding up. Plugging in $t = t_o$,

$$a_x(t_o) = \frac{At_o^2(6t_o^2 + 8t_o^2)}{(t_o^2 - 4t_o^2)^3} = \frac{14At_o^4}{(-3)^3 t_o^6} = -0.52 \frac{A}{t_o^2}.$$

Remember: A negative value for a_x does not necessarily imply deceleration. It means that the acceleration vector points in the negative x-direction.

2.7 At point A, the velocity is positive (■ Figure 2.36). The acceleration is very small, since the curve is almost straight at point A. At

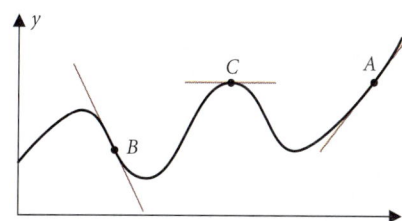

■ **FIGURE 2.36**
Solution to Exercise 2.7.

point B, the velocity is large and negative while the acceleration is small and positive. At C, the velocity is zero and the acceleration is large and negative.

2.8 Since the baseball returns to its original height and the rising and falling portions of its motion are symmetric (cf. Figure 2.30), we may conclude immediately that the speed of the ball when caught is 20 m/s. Computing the result from the formula for velocity as a function of time with $a = -g$, we have

$$v_{\text{strike}} = v_i - gt_{\text{strike}} \approx 20 \text{ m/s} - (10 \text{ m/s}^2)(4 \text{ s}) = -20 \text{ m/s}.$$

Since we chose the upward direction to be positive, the negative velocity component means the ball is moving downward. The speed is 20 m/s.

We may obtain the height at the top of the path in two ways. From the example, $t_{\text{top}} = 2$ s, so

$$\Delta y = y_{\text{top}} - 0 = v_i t - \tfrac{1}{2} g t_{\text{top}}^2;$$

$$y_{\text{top}} = (20 \text{ m/s})(2 \text{ s}) - \tfrac{1}{2}(10 \text{ m/s}^2)(2 \text{ s})^2 = 20 \text{ m}.$$

Alternatively, we could use eqn. (2.13).

$$\Delta y = y_{\text{top}} - 0 = \frac{v_{\text{top}}^2 - v_i^2}{-2g} \approx \frac{0 - (20 \text{ m/s})^2}{(-2)(10 \text{ m/s})} = 20 \text{ m}.$$

2.9 Again, we have two motions to consider (■ Figure 2.37): constant speed during the reaction interval followed by accelerated motion through the intersection.

The coordinate of the far side of the intersection is 51 m + 21 m = 72 m. In this case, the driver wants the car's coordinate to exceed 72 m before $t = 2.0$ s. Thus, for the second time interval, we apply a relation among acceleration, distance, and time. For the time variable t in eqn. (2.11), we must use time since the acceleration began:

$$\Delta t \equiv t - t_r = 2.0 \text{ s} - 0.50 \text{ s} = 1.5 \text{ s}.$$

$$\Delta x = x_f - x_i = v_i \Delta t + \tfrac{1}{2} a_o (\Delta t)^2.$$

Again, + means to the right and v_i is positive; $x_f = 72$ m and x_i, the car's coordinate at the end of the reaction time, = 14.5 m.

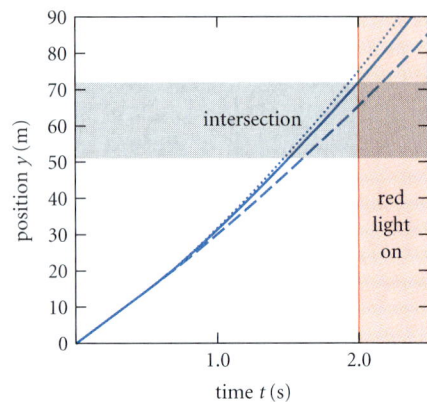

■ **FIGURE 2.37**
Position versus time graph if the driver decides to make the light. Most of the reaction time involves making a decision and so is assumed to be the same as for the other case. Here the curve is to reach the top of the intersection region before entering the "red light on" region.

$$a_o = \frac{2(\Delta x - v_i \Delta t)}{(\Delta t)^2}$$

$$= \frac{2[(72 \text{ m} - 14.5 \text{ m}) - (29 \text{ m/s})(1.5 \text{ s})]}{(1.5 \text{ s})^2} = 12 \text{ m/s}^2.$$

The magnitude of this acceleration is also greater than g. The poor driver can't stop without rockets in the front of the car and can't get through the intersection without rockets in the back. Perhaps speeding wasn't a wise idea!

2.10 ✉ A first estimate, $a \approx (30 \text{ m/s})^2/(4 \times 10^{-2} \text{ m}) = 2 \times 10^3 g$, is terrifyingly large, but overstates the effect of the impact. The car will still be crumpling when Joe's head reaches the dash. The relative speed of head and dash will be less than the initial speed of the car. Even if the relative speed is only 10 m/s, Joe's acceleration of 200g is still frightening!

Basic Skills

Review Questions

§2.1 SPEED AND VELOCITY

- Define *average speed*. Explain why average speed is not the same as a single speedometer reading.
- Describe a proper method for calculating average speed on a journey with several distinct legs.
- Define *average velocity*. Explain why there is, in general, no direct relationship between average speed and the magnitude of average velocity.
- In what special case is average speed equal to the magnitude of average velocity? Why does this relation hold in the special case when it is wrong in general?
- Why does average velocity become a more accurate description of motion when calculated for smaller time intervals?
- Define *instantaneous velocity*. Describe how average velocity vectors approach a limit.
- Explain why an object's instantaneous velocity is tangent to the object's path.
- What is the relation between a car's instantaneous velocity and the speedometer reading?

§2.2 AVERAGE AND INSTANTANEOUS ACCELERATION

- Why is acceleration an important quantity in physics?
- Define *average acceleration*.
- Define *instantaneous acceleration*.
- How does use of the word *acceleration* differ in physics and in everyday speech? In what special case are both usages the same?
- What change results from an acceleration component (1) along the line of an object's velocity and (2) perpendicular to the object's velocity?
- Why is there no need to define terms for the rate of change of acceleration and so on?

§2.3 LINEAR MOTION

- Describe how kinematic relationships simplify mathematically in the case of linear motion.
- What *choices* are you free to make in setting up coordinates for a problem in linear motion?
- Do the words *positive, negative,* or *zero* have any meaning independent of your choice of coordinate system? Why or why not?
- How is displacement represented in a graph of position versus time? In a graph of velocity versus time?
- How is velocity represented in a graph of velocity versus time? In a graph of acceleration versus time?
- State Galileo's law of falling bodies in modern terms.
- What is the *acceleration due to gravity*? Does it vary in magnitude over the Earth's surface? By a large amount?
- In deriving position from acceleration by formal integration, how does your knowledge of initial position and velocity enter the calculation?
- Describe why the velocity of an object in linear motion is represented by the slope of the graph of position versus time.
- Explain why the curvature of a position versus time graph gives a qualitative estimate of acceleration.
- How could you determine acceleration from a graph of velocity versus time?
- In describing the rise and fall of a baseball thrown vertically upward, explain why you do not have to describe the upward and downward parts of the motion separately.
- State in words the meaning of each symbol in eqns. (2.10) to (2.13).
- What is the *Physicist General*'s warning regarding eqns. (2.10) to (2.13)?

Basic Skill Drill

§2.1 SPEED AND VELOCITY

1. A person wants to drive between Salina and Green River, Utah, a distance of 170 km, in 2.0 h. At what average speed should the person drive?

2. An astronaut rides into orbit on the space shuttle, makes 110 orbits, and lands back on the runway at the Kennedy Space Center. What is the astronaut's average velocity for this trip?

3. A ball is rolling on a plane surface along the direction 150° from the *x*-axis. The ball's speed is 3.0 cm/s. What are the *x*- and *y*-components of the ball's velocity?

4. In ■ Figure 2.38, at what time is the instantaneous speed largest? When is it smallest? Is it ever zero?

5. Estimate the instantaneous velocity at points *A*, *B*, and *C* in Figure 2.38.

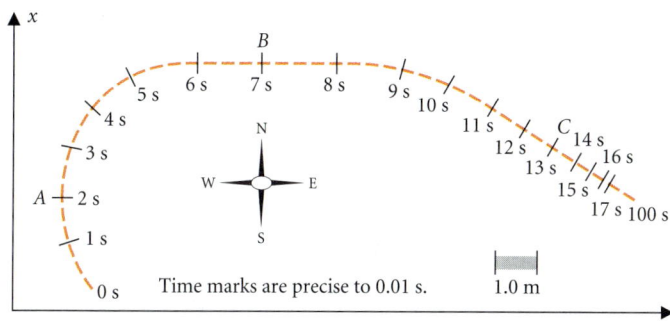

■ FIGURE 2.38

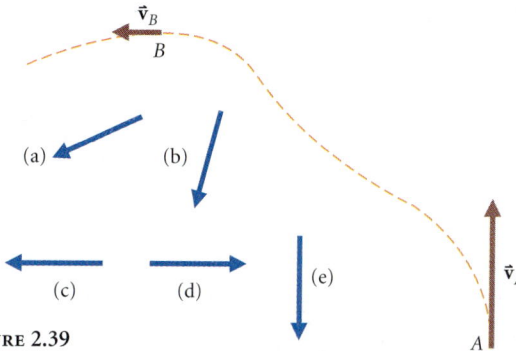

■ FIGURE 2.39

§2.2 AVERAGE AND INSTANTANEOUS ACCELERATION

6. In a road test, a Toyota Celica automobile accelerates in a straight line from rest to a speed of 27 m/s in 8.0 s. What is the magnitude of the car's average acceleration during the test?

7. A particle follows the path shown in ■ Figure 2.39. Its velocity is shown at points *A* and *B*. Which of the vectors (a)–(e) best represents the average acceleration of the particle between *A* and *B*?

§2.3 LINEAR MOTION

8. Explain why the following sentence is *false*: In uniformly accelerated linear motion, the speed of a particle always increases.

9. Explain why the following sentence is *false*: In uniformly accelerated linear motion, the average velocity is always one-half the final velocity.

10. You drop a rock over the edge of a cliff into a river 30.0 m below. How much time passes before the rock hits the water? (Neglect air resistance.)

11. ■ Figure 2.40 is the graph of acceleration versus time for an object in linear motion. Plot the corresponding graphs for velocity and position, assuming the object starts from rest at the origin at $t = 0$.

12. ■ Figure 2.41 is the graph of position versus time for an object in linear motion. Draw the corresponding graph of velocity versus time.

13. The green car in ■ Figure 2.42 has a speed of 1.00×10^2 km/h at point *A* and 80.0 km/h at point *D*. What is the car's acceleration between *A* and *D*? (Assume the acceleration is constant.)

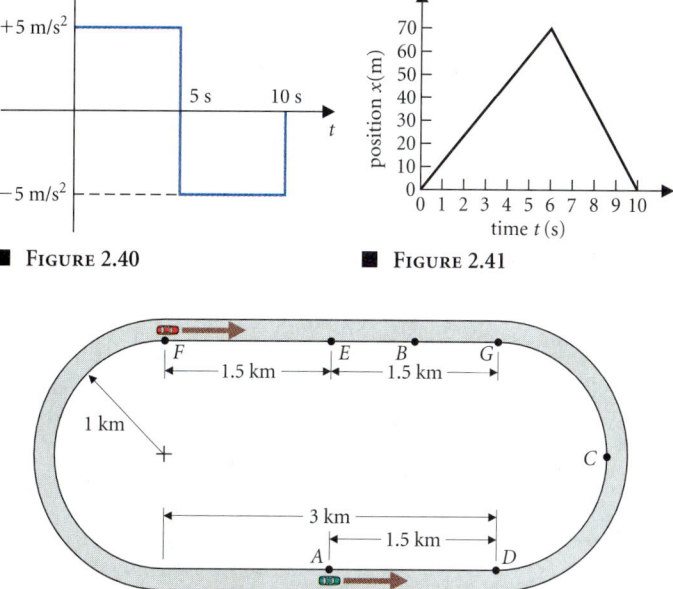

■ FIGURE 2.40 ■ FIGURE 2.41

■ FIGURE 2.42

BASIC SKILLS 77

Questions and Problems

§2.1 SPEED AND VELOCITY

14. ♦ A tortoise walks 10.0 m in 15 min. What is its average speed?

15. ♦ A carbon nucleus, being accelerated in the Heavy Ion Linear Accelerator, moves 3.0 m in 1.5 μs. What is its average speed?

16. ♦ If you walk 0.500 km northeast in 15 min, what is your average velocity for the walk?

17. ♦ A car drives 10 km west in 10 min, turns and goes 5 km north in 3 min. What is the car's average velocity for the trip?

18. ♦ A racetrack is an oval with straight sides 3.0 km long and circular ends, each a semicircle of radius 1.0 km. (Figure 2.42) A car travels clockwise around the track in 7.0 min. What is the average velocity of the car during one complete lap? Assuming constant speed, what is the instantaneous velocity at points A and B?

19. ♦♦ When making a long journey on the highway by car, the Smith family drives at 88 km/h for 3.0 h, then takes a half-hour break for gas and coffee. After the break, they drive another 3.0 h at the same speed. What is their average speed for the entire journey?

20. ♦♦ The leading car in a race comes out of the final bend at a speed of 161 km/h, 1.0 s ahead of the number two car. The final straight is 2.00 km long. What speed must the second car have as it comes out of the bend if it is to win the race? Assume that both cars maintain constant speed on the straightaway.

21. ♦♦ An express bus travels 3.0 km east to its first stop, then 5.0 km north to the second stop. If the bus takes 5.0 min to reach the first stop and 6.0 min between the first and second stops, what is its average velocity? (Neglect the time interval during which the bus is stopped.)

22. ♦♦ To set a track club record, Melinda must run 10.0 km at an average speed of 12 km/h. Having lost sleep the night before, she finishes the first 8.0 km at an average speed of only 9.6 km/h. Is it still possible for Melinda to match the record? If so, at what speed must she run the last 2 km?

§2.2 AVERAGE AND INSTANTANEOUS ACCELERATION

23. ❖ A particle moves along the path shown as a dashed line in ■ Figure 2.43. When the particle is at point A, its speed is increasing. The particle's acceleration at point A is best represented by which of the vectors shown? Explain why your choice is correct and the others are incorrect.

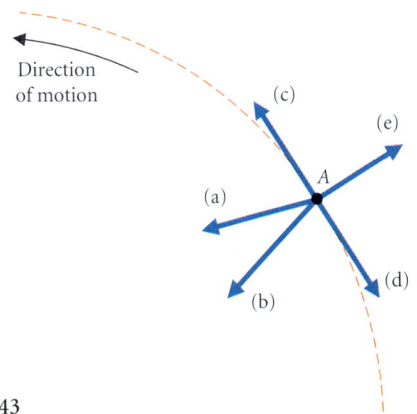

■ FIGURE 2.43

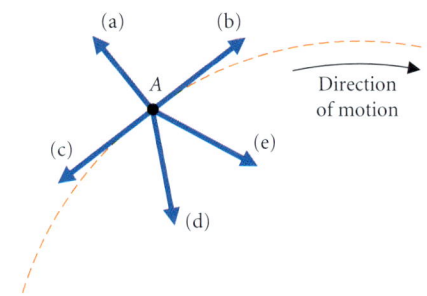

■ FIGURE 2.44

24. ❖ A particle moves along the path shown as a dashed line in ■ Figure 2.44. When the particle is at point A, its speed is decreasing. The particle's acceleration at point A is best represented by which of the vectors shown? Explain why your choice is correct and the others are incorrect.

25. ♦ During its takeoff roll, a Cessna 152 aircraft reaches a speed of 25 m/s in 5.0 s. What is the magnitude of the plane's average acceleration?

26. ♦ A car traveling east at 55 km/h accelerates through a turn and ends up, 6.0 s later, going northeast at 68 km/h. What is its average acceleration during the turn?

27. ♦♦ Each of ■ Figures 2.45a, b, and c shows the path of a particle drawn to scale. The time interval between dots in each figure is 0.10 s

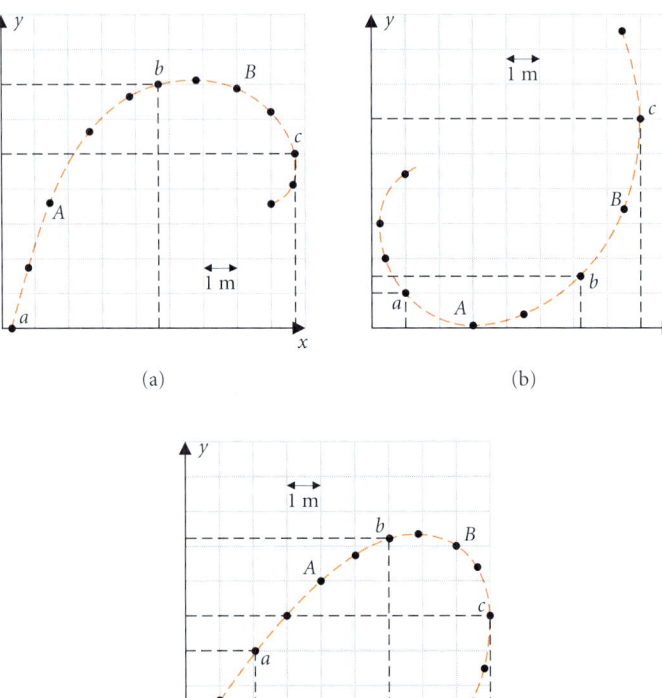

■ FIGURE 2.45

78 CHAPTER 2 • KINEMATICS

TABLE 2.5   Velocity Components Measured at Several Clock Readings

t (s)	v_x (m/s)	v_y (m/s)
1.00	13.00	−13.00
1.50	13.00	−8.25
2.00	12.00	−1.00
2.25	11.13	3.56
2.50	10.00	8.75
2.75	8.63	14.56
3.00	7.00	21.00
3.50	3.00	35.75
4.00	−2.00	53.00

and the spatial scale is 1 division equals 1.0 m. For each diagram, draw displacement vectors for the intervals from a to b and from b to c, and determine the x- and y-components of each displacement. Compute the components of the average velocities $\vec{v}_{ab}$ and $\vec{v}_{bc}$ and draw the vectors. Construct the difference $\vec{v}_{bc} - \vec{v}_{ab}$ and assume it is an adequate approximation for the difference $\Delta \vec{v}_{AB} \equiv \vec{v}_B - \vec{v}_A$. Use your result to find the average acceleration $\vec{a}_{AB}$.

28. ♦♦ • Table 2.5 gives measured values for the components of an object's velocity at several different clock readings. **(a)** Find components of the average acceleration for each of the following time intervals: 1.00–4.00 s; 1.50–3.50 s; 2.00–3.00 s; and 2.25–2.75 s. What is the value of the instantaneous acceleration at $t = 2.50$ s? **(b)** Find the magnitude of the velocity and the angle it makes with the x-axis at $t = 2.25$ s and at $t = 2.75$ s. Calculate the average acceleration for this time interval using the laws of sines and cosines and compare with the result from part (a). **(c)** Repeat the calculation of part (b) for the time interval 2.50–3.00 s. How do these results compare with those of parts (a) and (b)? **(d)** Repeat part (b) for the time interval 2.50–2.75 s. **(e)** Find the rate at which the object's speed is changing at $t = 2.50$ s and at $t = 2.25$ s by using your result for the instantaneous acceleration. Then find the change of speed during the interval $t = 2.25$–2.50 s directly from the data in Table 2.5. Compare your results.

§2.3 LINEAR MOTION

29. ❖ ■ Figure 2.46 is a graph of velocity versus time for a particle in one-dimensional motion. Which of the following statements about the graph is correct? Explain your reasoning; in particular, explain what is wrong with the incorrect statements.
(a) The shaded area represents the distance traveled during the time interval Δt.

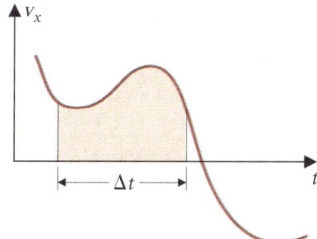

■ FIGURE 2.46

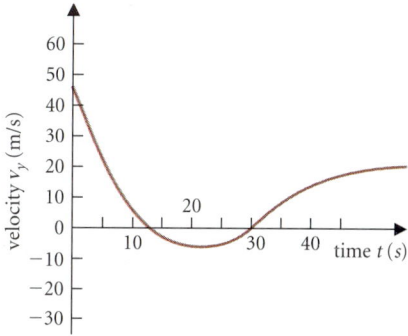

■ FIGURE 2.47

(b) The shaded area represents the acceleration during the time interval Δt.
(c) The acceleration is constant during the time interval Δt.
30. ❖ ■ Figure 2.47 shows the velocity as a function of time for an object in linear motion. Make plots showing **(a)** its displacement and **(b)** its acceleration as functions of time.
31. ❖ Draw the graph of acceleration versus time for the particle in Figure 2.41. Can the position graph for a real particle take the sharp bend shown at $t = 6.0$ s?
32. ❖ Does the acceleration due to gravity vary by a large amount within a few hundred meters of the Earth's surface? How can you tell?
33. ❖ What is meant by the "slope" of a position versus time graph at a given point? Can it be measured with a protractor?
34. ❖ Objects A and B are thrown directly upward. Object A's initial speed is four times that of B. Neglecting air resistance, the time until A strikes the ground is
(a) half as long as the time for B.
(b) four times as long as the time for B.
(c) one-fourth as long as the time for B.
(d) twice as long as the time for B.
(e) the same as for B.
35. ♦ ✉ A brick falls onto the surface of a sandy beach at a speed of 20 m/s. When it is removed, the brick leaves a hole 2×10^{-2} m deep. What uniform acceleration, caused by the sand, would stop the brick in 2 cm?
36. ♦ Galileo, while rolling a ball down an inclined plane, observes that it rolls 1 cubit (the distance from elbow to fingertip) as he hums ten beats of his favorite tune. How far has the ball rolled when Galileo has hummed 20 beats?
37. ♦ Leaving point F with a speed of 80.0 km/h, the racecar in Figure 2.42 accelerates at 2.0 m/s². How long does it take to reach point E, a distance of 1.5 km? What is the car's velocity at E?
38. ♦ During the first 2.5 s out of the starting gate, a skier slides 10.0 m. Assuming uniform acceleration, find the time when the skier reaches a speed of 30.0 m/s.
39. ♦ A road test of a Dodge Neon automobile shows that it can be stopped from a speed of 32 m/s in a distance of 58 m. What is the car's acceleration, assumed to be uniform, during the test? What is the time required to stop, and what is the car's average speed during the test?
40. ♦ During a road test, the test car accelerates from rest through a distance 0.40 km in 18 s. What is the acceleration of the car, assumed to be uniform? How fast is it going at the end of the test?
41. ♦ A police car, accelerating uniformly, passes the 10th Street off-ramp at a speed of 30.0 m/s. At the 12th Street ramp, 0.200 km from 10th, its speed is 50.0 m/s. What is the car's acceleration?

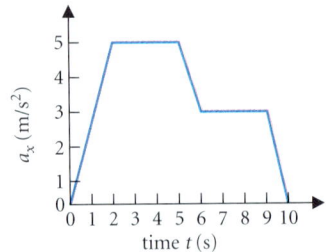

■ FIGURE 2.48

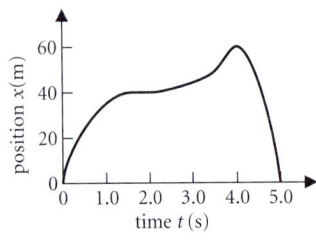

■ FIGURE 2.49

42. ♦♦ ■ Figure 2.48 is the graph of acceleration versus time for an object in linear motion. Plot the corresponding graphs of velocity and position, assuming the object starts from rest at the origin at $t = 0$.

43. ♦♦ ■ Figure 2.49 shows the graph of position versus time for an object in linear motion. Sketch the corresponding graphs of velocity and acceleration versus time.

44. ♦♦ While standing at the edge of the flat roof of an office building 40 m high, you throw a ball upwards at 10 m/s. When does it hit the ground? What is its speed when it passes you on the way down? What is its speed when it hits the ground?

45. ♦♦ Johnny's bunk bed is 2.0 m above Jimmy's. Instead of sleeping, the two are amusing themselves by playing catch. Each in turn throws the ball upward at a speed of 7.0 m/s and the other then catches it. Find the times between each boy's throw and the other boy's catch, and the two speeds of the ball as it is caught.

46. ♦♦ You are driving along a highway at 80.0 km/h. Exactly 10 m in front of you is a truck 15 m long driving at 60.0 km/h. Your car is 5.0 m long. If you wish to pass the truck in 3.0 s, how fast must you accelerate? What will your speed be when you pass the front end of the truck?

47. ♦♦ (a) The tunnel between two nearby subway stations is 0.300 km long. The subway train is required to keep its speed at 3.0 m/s whenever any part of it is within a station. The 50.0 m-long train accelerates uniformly from its initial speed of 3.0 m/s to reach a desired speed of 30.0 m/s midway between the stations. What minimum acceleration is required? What is the train's average speed from the time when it begins to enter the tunnel until it is completely out of the tunnel at the second station? (b) Construct graphs of velocity and position versus time for the front end of the subway train.

48. ♦♦ The automobile in Example 2.13 (§2.3.5) accelerates or decelerates with a maximum magnitude of $\frac{1}{2}g = 4.9$ m/s^2. Find the minimum initial speed that will allow the automobile to clear the intersection before the signal turns red, and the maximum initial speed that will allow the automobile to stop before reaching the signal. Make a schematic graph showing the car's position as a function of time for each case. Show on the graph the location of the intersection and the time when the light turns red.

49. ♦♦ The red car traveling clockwise around the track in Figure 2.42 leaves point E at a speed of 240 km/h. It must decelerate to 80.0 km/h before reaching the turn at point G. Maximum braking yields a deceleration of 8.0 m/s^2. What is the minimum distance from G at which the driver can apply the brakes and still be able to decelerate sufficiently? In this case, find the total time taken to travel from E to G.

50. ♦♦ If the position of an object in linear motion is given in terms of clock reading by the formula $x = (3.0$ m/s$^2)t^2 - (6.0$ m/s$^3)t^3$, find the object's acceleration at $t = 2.0$ s.

51. ♦♦ If the velocity of an object in linear motion is given in terms of clock reading by the formula $v = (15.0$ m/s$^2)t - (7.0$ m/s$^3)t^2$, find the object's acceleration at $t = 0.50$ s. If the object was at the origin at $t = 0$, find its position at $t = 0.50$ s.

52. ♦♦♦ The x-coordinate of an object in linear motion along the x-axis is described by the function: $x(t) = (12.35$ m/s$^2)t^2 - (7.45$ m/s$^3)t^3$ between $t = 0$ and $t = 1.0$ s. What is the maximum speed reached by the object, and when does it occur?

Additional Problems

53. ♦♦ Two heavy balls are released from different heights, one 2.2 s after the other. If the two balls hit the ground simultaneously, 4.0 s after the first is released, from what heights were they dropped?

54. ♦♦ Two friends stand on the roof of a 20.0-m-high building. Annie throws a baseball directly upward at 10.0 m/s. At the same time, Laurie throws a ball directly downward at the same speed. How much later than Laurie's ball does Annie's ball strike the ground? Compare the speeds at which the two balls strike the ground. Do you find the result surprising? Why or why not?

55. ♦♦ A repairman, working at the top of an elevator shaft, 60.0 m above the elevator car, is startled by the sound of the elevator starting upward and drops his wrench. If the elevator car accelerates upward at $0.10g$ for 3.00 s and then runs at constant speed, how far does the wrench fall before striking the elevator car?

56. ♦♦ Clarice is standing at the edge of a steep cliff when a boulder comes loose from the edge and falls directly downward toward the lake below. Clarice hears the sound of the boulder hitting the water 7.0 s later. If the speed of sound is 340 m/s, how high is the cliff?

57. ♦♦♦ Simon Stevinus of the University of Leyden, who we believe first actually did the experiment of dropping two objects from a tower, released the objects from a height of 19.5 m. The noises made by the objects striking the ground occurred $\frac{1}{16}$ s apart. Assume that air resistance has a negligible effect on the object that strikes first, and estimate the following. (a) The height difference between the two objects just as the first strikes the ground. (Hint: How fast are they going? Neglect the difference between their speeds.) (b) The difference in the two objects' average velocities between release and striking the ground. (c) The effect of air resistance on the acceleration of the object that strikes the ground second. (Hint: Use the model of Example 2.10 and find α.)

58. ♦♦♦ Juggler Jim can catch and rethrow a ball and shift attention to the next ball, all in time Δt. How high must he throw each ball in order to keep N balls in the air? Evaluate your answer for $N = 10$ and $\Delta t = 0.2$ s. Is the result reasonable? Estimate how many balls Jim can actually handle at one time.

59. ♦♦♦ Near the start of a ski race, two cameras videotape a racer as she passes. The first tape shows that the skier moved 10.0 m in the time interval $t = 0-2.0$ s. The second camera, downhill from the first, recorded a displacement of 26 m in the interval from 4.0–6.0 s. What is the skier's acceleration, assumed to be uniform, and what is the time interval from the start of the race to the beginning of the first camera's record? If the cameras begin recording when the skier passes through an infrared light beam, what is the distance between the triggering beams for the two cameras?

60. ♦♦♦ You are driving along a two-lane highway at 80.0 km/h (■ Figure 2.50). An approaching car, also with speed 80.0 km/h, is 0.250 km ahead of you. You wish to pass a recreational vehicle going 60.0 km/h, whose front end is 50.0 m ahead of you. How fast must you accelerate to pass the RV and pull back into the right-hand lane

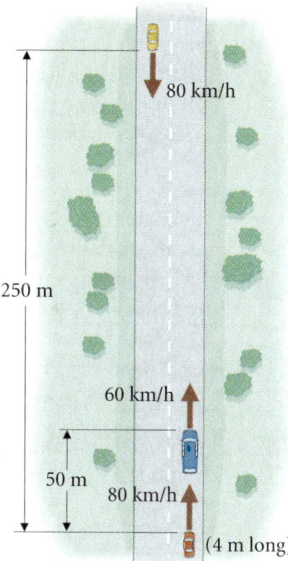

Figure 2.50

before the approaching car can reach you? (Assume that each car is 4.0 m long.)

61. ♦♦♦ Nanette has forgotten her lunch bag (definitely a no-no). However, her commuter train will pass under a pedestrian bridge, and her well-practiced husband waits on the bridge, 15 m above the train windows, and drops the bag when she is 32 m from the bridge. She deftly catches the lunch bag. What is the train's speed?

62. ♦♦♦ A sports car speeding north on a four-lane highway at 40.0 m/s (!) meets a police car traveling south at 20.0 m/s. If the police car can both decelerate and accelerate at 5.0 m/s² (the driver skillfully taking no time to switch direction), when does it catch the speeder? At what speed does the police car catch up with the speeder?

63. ♦♦♦ A body moves along a straight line with an acceleration that increases linearly with time: $a = a_o t/t_o$, where a_o and t_o are constants. Find the body's position and velocity as functions of time assuming it starts from rest at the origin at $t = 0$.

64. ♦♦♦ Show how to derive eqns. (2.10) and (2.11) for uniformly accelerated linear motion using formal integration. Comment on how you include the initial position and velocity of the object.

65. ❖ ✳ The *Apollo* spacecraft was 4 m long. The distance from the Earth to the Moon is 3.84×10^5 km. In calculating *Apollo*'s orbit to the Moon, would you expect improved accuracy from modeling the spacecraft as anything more complicated than a particle? Why or why not?

66. ❖ ✳ The Lunar Excursion Module (LEM) had *feet* of diameter 1 m. Lunar craters range in size from 10^{-1} m to 10^5 m in diameter. In programming the landing controls system, could you reasonably model the LEM as a particle? Suppose the LEM's motion in response to its controls is best modeled as a sequence of small displacements. How large can these individual displacements be and still allow control accurate enough to ensure the LEM's safety?

67. ❖ ✳ The space shuttle tracking system predicts the position of the shuttle orbiter with an accuracy that varies between 30 m and 100 m. Its orbital radius is slightly larger than the radius of the Earth. What, approximately, is the ratio of the position error to the size of the shuttle's orbit? The orbiter itself is approximately 37 m long and 17 m high. Is it reasonable to model the shuttle as a particle for the purpose of computing its orbit?

Computer Problem

68. The file named CH2P68 on your supplementary computer disk contains values of the *x*-coordinate as a function of time for three objects that move along the *x*-axis. Using a spreadsheet program, calculate the average velocity for each object over 2.0-s, 1.5-s, 1.0-s, and 0.50-s intervals. Determine the instantaneous velocity at each time. To how many significant figures can you determine the limit? Using your results for velocity, use a similar procedure to calculate the acceleration. Plot your results, and comment.

69. A Styrofoam sphere 10.00 cm in diameter is dropped from rest and falls through the beams of three photodetectors placed respectively 10.00, 20.00, and 30.00 cm directly below the center of the sphere. The recorded data are as follows:

Time t (s)	Event
0.0000	sphere released
0.1010	beam 1 interrupted
0.1751	beam 1 on again; beam 2 interrupted
0.2262	beam 2 on again; beam 3 interrupted
0.2677	beam 3 on again

First show that these data are *not* consistent with uniform acceleration. Then, using the air resistance model of Example 2.10, determine values of g and α from the given data. (*Hint:* Show that the data should lie on a straight-line graph of y/t^2 versus t^2.)

Challenge Problems

70. An experimental nozzle for an ink jet printer system is set up to eject ink drops horizontally through two photodetector systems set 0.150 m apart (■ Figure 2.51). As a drop passes through one of the light beams, the signal output from that beam decreases and is recorded versus time by a computer. (A pulse from the circuit that activates the nozzle starts the timer.) The signal recorded for the passage of a single drop is graphed in the figure. Assume the drop travels along an exactly horizontal line through the centers of the two light beams, and assume the beams are circular. (*Hint:* Draw sketches

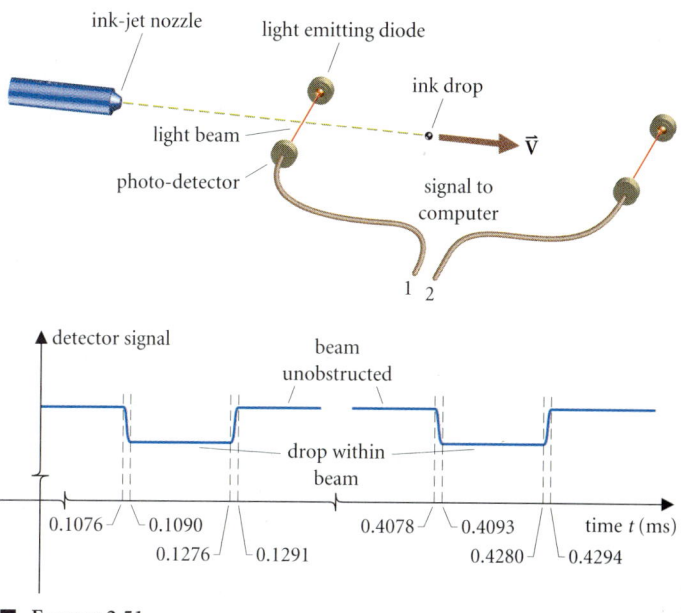

■ **Figure 2.51**

showing where the drop is with respect to the light beams at each of the times given on the graph.) **(a)** First, assuming constant speed for the drop, determine its speed. **(b)** Next, find the diameter of the drop and the diameters of the two light beams. **(c)** How accurately do the data determine your results for parts (a) and (b)? **(d)** If the deceleration of the drop due to air resistance were $10g$, could you detect the deceleration from your data?

71. An object undergoes an acceleration given by $a(t) = a_o \cos \omega t$, where a_o and ω are constants. At $t = 0$, the object is at rest at the origin. Find its velocity and displacement as functions of time t.

72. The following expressions fit the data for the velocity of the object in Table 2.5.

$$v_x(t) = (10.00 \text{ m/s}) + (5.00 \text{ m/s}^2)t - (2.00 \text{ m/s}^3)t^2$$
$$v_y(t) = (-15.00 \text{ m/s}) - (3.00 \text{ m/s}^2)t + (5.00 \text{ m/s}^3)t^2.$$

(a) Using these expressions, calculate the acceleration of the object as a function of time; evaluate your result at $t = 2.50$ s; and compare your results with those of Problem 28. **(b)** Obtain a formula for the speed of the object as a function of time and calculate its rate of change directly. Compare with the parallel component of acceleration.

73. (a) Daphne's dune buggy will carry her rowboat at a speed v_b over the beach. She can row the boat at a speed $v_r = v_b/2$. On a day with no wind or ocean currents, she wishes to get from home to Paradise Island (see ■ Figure 2.52) by driving to the shore and rowing to the island. Show that, for the path that gets Daphne to the island in minimum time,

$$\frac{\sin \theta}{\sin \phi} = \frac{v_b}{v_r} = \frac{\text{speed on land}}{\text{speed on water}} \quad \text{(regardless of numerical values).}$$

(b) (Computer problem.) For the specific value $v_b/v_r = 2$, determine the direction θ at which she should leave home to get to the island in minimum time.

Essay Question/Point to Ponder

Sometimes, the median strip of a freeway is planted with a hedge of large flowering bushes. Tests show that a car slamming into such a hedge will crunch through but emerge with negligible speed. Apart from beauty, why is such a hedge a good idea?

Home Activities

1. Using a partner (to avoid bonking an innocent person), drop a basketball and a baseball from a moderate height. Can you detect a difference in the time of fall, timing them separately with a stopwatch? Dropping them together?

2. With the help of friends, measure your average speed running a 30-m dash from a standing start. Obtain separate results for intervals 0–5 m, 5–10 m, and 10–30 m. What is your average speed running full tilt? Over what distance do you accelerate to that speed? What is your acceleration from a standing start?

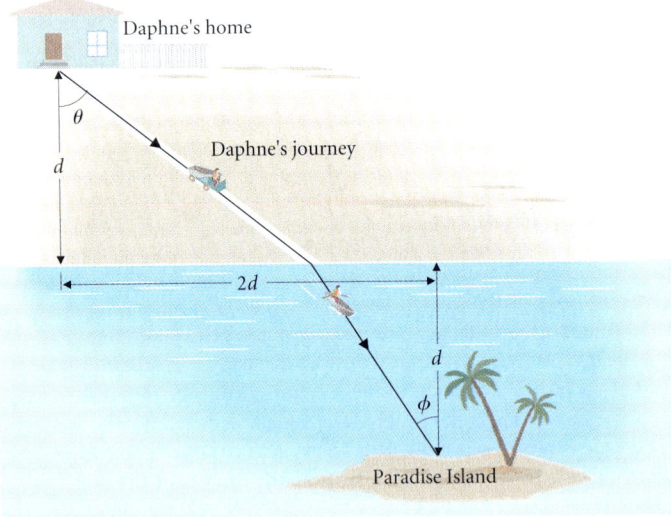

■ **Figure 2.52**

A stunt skier leaps from a cliff, performs a somersault, and lands on the steep snow slope below the cliff. The stunt requires careful coordination of the skier's rotation with her motion as a projectile.

The wind was flapping a temple flag, and two monks were having an argument about it. One said the flag was moving, the other that the wind was moving; and they could come to no agreement on the matter. They argued back and forth. Eno, the Patriarch said, "It is not that the wind is moving; it is not that the flag is moving; it is that your honorable minds are moving."

PLATFORM SUTRA

CHAPTER 3
Advanced Kinematic Models

Concepts

Projectile motion
Angular speed
Period
Frequency
Uniform circular motion
Relative motion of reference frames
Rolling without slipping

Goals

Be able to:

Compute the trajectory of a particle projected at any angle and allowed to fall freely under gravity.

Compute the velocity and acceleration of a particle moving in a circle.

Relate velocities between reference frames in relative motion.

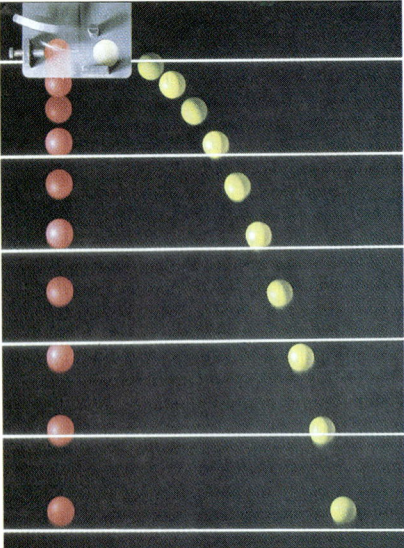

■ **FIGURE 3.1**
Two balls are released simultaneously, one dropped from rest and the other projected horizontally. A stroboscopic flash simultaneously records the positions of both balls at successive times. The vertical motions of the two balls are identical; the second ball also moves horizontally at constant velocity.

WE'LL RETURN TO THIS IMAGE IN §3.3.

■ **FIGURE 3.2**
Streams of water projected from a fountain flow in beautiful parabolic arcs. Because the continuous stream of water eliminates air from its path, each portion of the stream moves as a projectile free of air resistance.

The skier is *very* interested in landing upright—that is, coordinating her body rotation with her plunge back to the snow slope. She does this quickly and automatically, or ends up in a hospital, but we have plenty of time to think about her task. First we must imagine her task divided into simpler pieces, and consider her rotation separately from her motion as a particle. Then we can combine the separate solutions for the two pieces to describe her actual motion. This method won't give us an exact description because people have strange, variable shapes, but it's good enough to tell a conservative stunt from a totally hare-brained one.

This method for analyzing a ski stunt illustrates an important technique in physics: imagining big problems as several small problems that fit together rather like pieces in a jigsaw puzzle. Most interesting physical systems, like the skier, involve complex motions that may often be modeled as combinations of simpler motions. What we need is a toolkit of basic results—kinematic models—that we may fit together to answer the bigger problems we'll encounter. Fortunately, only a small number of such models prove to be necessary. In Chapter 2 we developed the model of uniformly accelerated linear motion. Using that model together with circular motion to describe the skier's adventure is an example of how we fit the models together. We shall also study relations among uniformly moving reference frames, essential in aviation and marine navigation, and important for describing systems of particles.

3.1 PROJECTILE MOTION

A projectile is any particle given an initial velocity and then allowed to fall freely under gravity. As we saw in Chapter 0, Galileo gave the first accurate description of projectile motion:

> In the absence of air resistance, a projectile moves horizontally with constant velocity. Its vertical motion is the same as if it had no horizontal motion (■ Figure 3.1).

The combined vertical and horizontal motions produce a parabolic path in a vertical plane, beautifully illustrated by the water fountain in ■ Figure 3.2. The arc of water acts like a stream of individual water particles, each projected from the fountain with the same velocity and following the same path. Air resistance is small, since each particle of water finds the air already pushed out of the way by the particles that have gone before. Imagine watching a single blob of water in a stream. Now run beside the jet at the same horizontal speed as the water. You would see the blob rise and fall, just like a baseball thrown upward with the same vertical speed. From your new point of view, the water's horizontal motion has been eliminated without affecting the *independent* vertical motion.

In ■ Figure 3.3 we have drawn velocity and acceleration vectors at several points on the path of the water. The vertical velocity component $\vec{v}_v$ is uniformly accelerated downward at $\vec{g}$, and the horizontal component $\vec{v}_h$ is constant. The velocity

$$\vec{v} = \vec{v}_v + \vec{v}_h$$

is tangent to the path. Between points A and C, the water blob slows until at C its speed reaches a minimum equal to the horizontal speed v_h and the vertical velocity v_v is zero. After C, the water goes faster, gaining downward velocity.

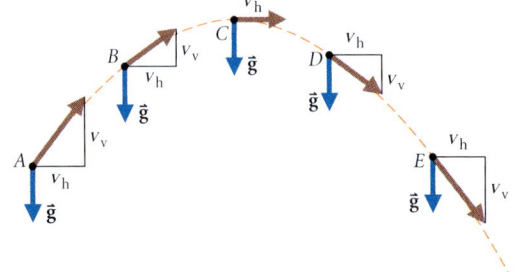

■ **FIGURE 3.3**
At each point on the path of a water element, the element's velocity is tangent to its path. The horizontal velocity component v_h is the same at each point ($a_h = 0$), while the vertical component v_v is accelerated downward with the acceleration due to gravity $\vec{g}$.

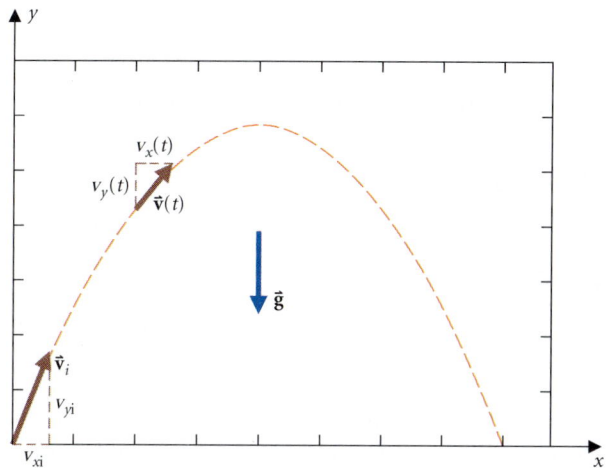

■ **FIGURE 3.4**
Projectile motion occurs in a vertical plane. Choosing vertical and horizontal coordinate axes simplifies the analysis of projectile motion, since the horizontal velocity component is unaccelerated and the vertical component is accelerated uniformly. Here we also choose the coordinate origin at the point where the projectile is launched.

Since a projectile moves in a plane, we need only two coordinates. Choosing x as the horizontal coordinate and y as the vertical coordinate with positive direction upward (■ Figure 3.4) takes advantage of the independence of horizontal and vertical motions. Then, in the x-direction we have constant velocity, and in the y-direction we have uniformly accelerated linear motion. We take the coordinate origin where the water emerges from the fountain. Its initial velocity $\vec{v}_i \equiv (v_{xi}, v_{yi})$. Following a particular blob of water, choose a clock reading of $t = 0$ when the blob emerges from the fountain. Expressions for the position and velocity of the water blob at later times follow from our results for linear motion in Chapter 2 (Table 2.4).

x-DIRECTION, ZERO ACCELERATION, $a_x = 0$:

$$v_x(t) = v_{xi} \qquad \Delta x \equiv x - x_i = v_{xi} t. \tag{3.1a}$$

y-DIRECTION, CONSTANT ACCELERATION, $a_y = -g$:

$$v_y(t) = v_{yi} - gt \qquad \Delta y \equiv y - y_i = v_{yi} t - \tfrac{1}{2} g t^2. \tag{3.1b}$$

The analysis is finished! Problems involving projectile motion can all be solved using similar relations appropriate to your choice of coordinates. Such problems differ primarily in the kind of information you know and the kind of answers you wish to obtain. In the rest of this section, we work examples of the major types of projectile problems and examine their logic plans.

3.1.1 Problems with Given Initial Conditions

In problems of this type, we are given the initial velocity of the projectile and asked to say something about where it goes.

EXAMPLE 3.1 ♦♦ Mary, standing at first base, throws a softball with a speed of 15 m/s at an angle of 25° above the horizontal. The ball leaves Mary's hand 1.2 m above the ground. How far away does the ball land?

MODEL We assume the softball field is flat. We choose the x-axis to be horizontal, the positive y-axis to be directly up, and the origin to be on the ground immediately beneath Mary's hand as she releases the ball (■ Figure 3.5). We also choose to start the clock as Mary releases the ball. The ball is unaccelerated in the x-direction, while in the y-direction it has constant acceleration: $a_y = -g$. We want to find the x-coordinate when $y = 0$.

■ **FIGURE 3.5**
We choose the origin on the ground beneath Mary's hand. We are told $|\vec{v}_i| = 15$ m/s, $\theta_i = 25°$, and $y_i = 1.2$ m. We want to find the x-coordinate of the ball when it reaches $y = 0$.

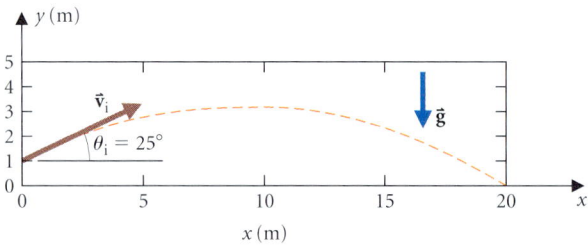

SETUP Here's what we know.

$$x_i = 0 \qquad v_{xi} = v_i \cos\theta_i = (15 \text{ m/s})\cos(25°) = 13.6 \text{ m/s}$$
$$y_i = 1.2 \text{ m} \qquad v_{yi} = v_i \sin\theta_i = (15 \text{ m/s})\sin(25°) = 6.33 \text{ m/s}.$$

Using eqns. (3.1a and b), we can find the time at which $y = 0$, and then determine how far the ball travels horizontally in that time. First the vertical motion:

$$y - y_i = v_{yi}t - \tfrac{1}{2}gt^2 \Rightarrow 0 - y_i = v_{yi}t - \tfrac{1}{2}gt^2.$$

This is a quadratic equation for the time t. It has two possible solutions. Rearranging:

$$t^2 - \frac{2v_{yi}}{g}t - \frac{2y_i}{g} = 0 \Rightarrow t = \frac{2v_{yi}/g \pm \sqrt{(2v_{yi}/g)^2 + 8y_i/g}}{2} = \frac{v_{yi}}{g}\left(1 \pm \sqrt{1 + \frac{2gy_i}{v_{yi}^2}}\right).$$

Then we use this result in eqn. (3.1a) for the horizontal coordinate.

$$x - x_i = x = v_{xi}t.$$

SOLVE Putting in the numbers,

$$t = \frac{6.33 \text{ m/s}}{9.8 \text{ m/s}^2}\left[1 \pm \sqrt{1 + \frac{2(9.8 \text{ m/s}^2)(1.2 \text{ m})}{(6.33 \text{ m/s})^2}}\right]$$
$$= (0.646 \text{ s})(1 \pm 1.26) = 1.46 \text{ s or } -0.17 \text{ s}.$$

Since Mary threw the ball at $t = 0$, the positive root, $t = 1.46$ s, is the answer we want. Then

$$x = v_{xi}t = (13.6 \text{ m/s})(1.46 \text{ s}) = 19.9 \text{ m}.$$

We should round the answer to two significant figures: the ball lands 20 m (2.0×10^1 m) away.

ANALYZE Now what about that negative time solution? If the ball had been thrown from ground level so as to pass Mary's hand at $t = 0$ with velocity $\vec{v}_i$, it would have left the ground at $t = -0.17$ s—that is, 0.17 s before our clock started. ■

3.1.2 Problems with a Specific Goal

In these problems we do not know all of the initial information, but we are told something about where the projectile ends up or a point through which it passes.

EXAMPLE 3.2 ◆◆ Wilhelm Tell, legendary archer and hero of Swiss independence, was forced to shoot at an apple placed on his son's head. If Tell must stand 13 m from his son and he launches the arrow with a speed of $v_i = 58$ m/s, at what angle θ_i above the horizontal should he point the arrow? Assume the arrow starts at the same height as the apple and neglect air resistance.

MODEL The goal is for the arrow to hit the apple. Achieving the goal requires a synchronization of the vertical and horizontal motions of the arrow: it must reach the horizontal position of the apple at the *same time* that it reaches the vertical position of the apple.

SETUP In ■ Figure 3.6 we show the quantities we know and the angle θ_i we want to find. In terms of the initial speed v_i and launch angle θ_i, the arrow's velocity components at launch are

$$v_{xi} = v_i \cos\theta_i \quad \text{and} \quad v_{yi} = v_i \sin\theta_i. \tag{i}$$

EQUATIONS ARE GIVEN ROMAN NUMBERS LIKE (i) WHEN THEY ARE NOT FUNDAMENTAL BUT WE NEED TO REFER TO THEM IN A SOLUTION.

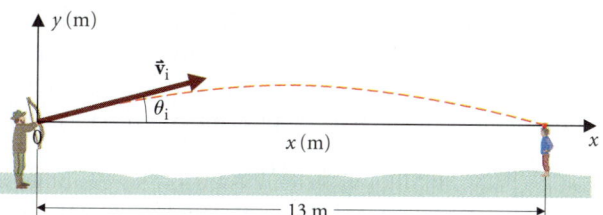

■ **FIGURE 3.6**
Wilhelm Tell wants to know the angle θ_i at which he should shoot his arrow. We know the initial speed $v_i = 58$ m/s and the distance the arrow should go, 13 m.

The basic relations (eqns. 3.1a and b) allow us to find the time of flight from either the vertical or the horizontal motion. The arrow starts at the same height as the apple. So, taking the origin where the arrow is launched, its y-coordinate is zero both leaving the bow and hitting the apple. From eqn. (3.1b) with $y_i = 0$, the arrow is at $y = 0$ at times t where

$$0 = y = v_{yi}t - gt^2/2 = t(v_{yi} - gt/2).$$

There are two possible solutions: $t = 0$ is the time that the arrow was first at $y = 0$—that is, when it left the bow. The second solution, $t = 2v_{yi}/g$, is the time when the arrow reaches $y = 0$ (the apple's height) again.

Next we require that the arrow also be at the apple's x-coordinate, $x = d$, at the *same* time. Using eqn. (3.1a) with $x_i = 0$ and $t = 2v_{yi}/g$,

$$d = x_f = v_{xi}t = v_{xi}(2v_{yi}/g).$$

Substituting relations (i) for the velocity components in terms of θ_i, we have

$$d = \frac{2v_i^2 \sin\theta_i \cos\theta_i}{g} = \frac{v_i^2 \sin 2\theta_i}{g}.$$

WE USED THE TRIG IDENTITY $2\sin\theta\cos\theta = \sin 2\theta$. SEE APPENDIX IB.

SOLVE So

$$\sin 2\theta_i = \frac{gd}{v_i^2} = \frac{(9.8 \text{ m/s}^2)(13 \text{ m})}{(58 \text{ m/s})^2} = 3.8 \times 10^{-2}.$$

Then $2\theta_i = 2.2°$, so $\theta_i = 1.1°$. Tell should aim the arrow 1.1° above the horizontal.

ANALYZE There is a second possible solution: $\sin 2\theta$ also equals 3.8×10^{-2} for $2\theta = 180° - 2.2° = 177.8°$, or $\theta = 89°$. This is not a useful alternative in this problem: the landing arrow might be rather painful for the younger Tell.

Since the arrow pierces the apple at the same height it started from, the arrow's upward motion is symmetric with its downward motion. This suggests an alternative method for finding the flight time. The arrow is at its greatest height when $v_y = 0$, or, from eqn. (3.1b) for v_y, at time $t_{top} = v_{yi}/g$. Thus the time required to return to the original height is $t = 2t_{top} = 2v_{yi}/g$. ∎

THE ALTERNATIVE METHOD WORKS ONLY WHEN $\Delta y = 0$.

As a by-product of Example 3.2, we found a general result for the *range* of a projectile in terms of its initial speed and launch angle.

The *range* of a projectile is the horizontal distance it travels while returning to its original height. If the projectile's initial speed and launch angle are v_i and θ_i, its range R is

$$R = \frac{v_i^2 \sin 2\theta_i}{g}. \tag{3.2}$$

REMEMBER: EQUATION (3.2) APPLIES ONLY WHEN THE PROJECTILE LANDS AT THE SAME HEIGHT IT STARTED FROM. IT WOULD THEREFORE NOT APPLY TO EXAMPLE 3.1. TRY IT AND SEE!

"Trajectory" means the path of a projectile.

This plan is good for many, but not all, problems. Remember the steps, but try to be flexible.

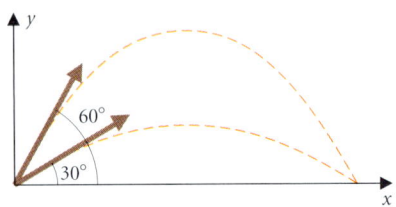

FIGURE 3.7
Projectile paths for two pairs of complementary launch angles. Such paths achieve the same range but take different times.

EXERCISE 3.1 ♦♦ Show that the maximum range Tell can achieve is $R = v_i^2/g$ and that it occurs when he releases the arrow at $\theta_i = 45°$.

Equation (3.2) shows that any range less than the maximum can be achieved in two ways. Launching at a small angle θ_1 leads to a low trajectory. The launch angle $\theta_2 = 90° - \theta_1$ gives the same range ($\sin 2\theta_2 = \sin(180° - 2\theta_1) = \sin 2\theta_1$) with a higher trajectory that takes a longer time (■ Figure 3.7).

The logic plan of Example 3.2 is common to a large variety of projectile problems:

- Identify an important *event* in the problem. Satisfying the goal will usually mean that some specific event occurs. Tell's goal is for the arrow to arrive at the apple.
- Find the *time* at which that event should occur, by thinking about either the vertical or the horizontal coordinate. Often one will provide a much easier method than the other. In this case, both are about equally useful.
- Substitute your expression for time into the equation for the other coordinate and obtain an equation you can solve for the desired *answer*.

Can you see this plan at work in the next example?

EXAMPLE 3.3 ♦♦ At an air show, a passenger attempts to hit a target on the ground with a bag of flour dropped from a plane. If the plane flies at 32 m/s at a constant altitude of 220 m, when and where must the bag be released? (Neglect air resistance.)

MODEL Seen from the ground, the bag has the same horizontal velocity as the plane and travels in an arc. It remains directly below the airplane.

Let's choose the origin at the point where the bag is dropped. This is a daring move, since the launch point is what we are asked to find, but we can readily calculate where the projectile hits in these coordinates and finish the problem by demanding that the strike occur on target (the *event*). We take the x-direction to be horizontal and the positive y-direction to be downward (■ Figure 3.8) so the bag travels through positive values of y. With these coordinates, $a_y = +g$. Since the passenger simply lets go of the flour bag, its initial vertical velocity component v_{yi} is zero and its initial horizontal velocity component is that of the plane, $v_{xi} = 32$ m/s.

SETUP We find first the *time* when the bag lands by setting $y = h$ in the equation for the y-coordinate.

$$h = \Delta y = y - y_i = y - 0 = v_{yi}t + \tfrac{1}{2}gt^2 = \tfrac{1}{2}gt^2.$$

SOLVE
$$t = \sqrt{2h/g} = \sqrt{2(220 \text{ m})/(9.8 \text{ m/s}^2)} = 6.7 \text{ s}.$$

During the time of fall, the bag's horizontal displacement is

$$\Delta x = x - 0 = v_{xi}t = (32 \text{ m/s})(6.7 \text{ s}) = 210 \text{ m}.$$

A successful bag-bomber releases the flour 6.7 s and 210 m before reaching the target.

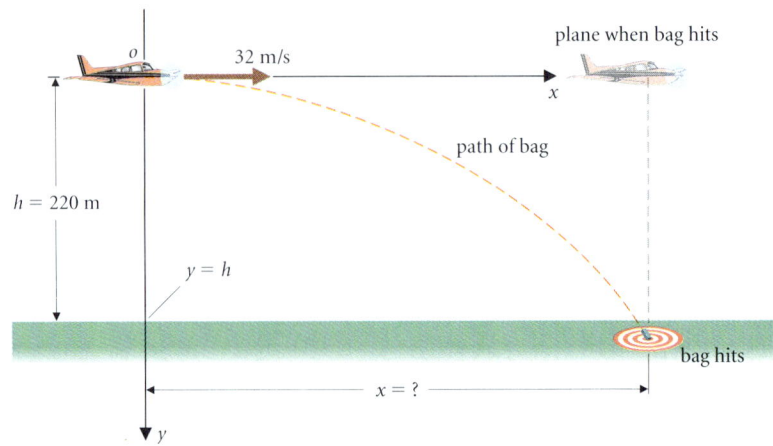

FIGURE 3.8
The origin of coordinates is chosen to be the initial position of an airplane dropping a bag of flour. The vertical axis is chosen as positive downward, so in this case $a_y = +g$.

ANALYZE As an alternative, we could let the coordinate system move with the airplane so that the bag falls straight down. Then the *target* moves from $x = 200$ m to $x = 0$ during the 6.7-s time interval. The calculations are the same as in the original method. Techniques like this are explored in §3.3.

3.1.3 Simultaneous Motion

A third kind of situation involves the simultaneous motion of two particles. Usually, such problems require the two objects to meet at the same place at the same time and then ask how to launch a projectile to achieve that goal. In our next example, note first how the algebra again follows the general solution plan, then notice how you could miss a major part of the question, not spelled out in the statement, by quitting when the algebra seems finished.

EXAMPLE 3.4 ♦♦ A monkey is hanging from a branch in a tree 3.0 m away from a zookeeper and 2.0 m above the keeper's hand (■ Figure 3.9). The keeper knows that the monkey always lets go of the branch just as a banana is thrown to it. At what angle must the keeper throw the banana for the monkey to catch it?

MODEL The monkey falls straight down while the banana behaves as a projectile. The zookeeper can control the initial speed and direction of the banana, which are the unknowns we want to find. The goal is for banana and monkey to be at the same position (coordinates x_1, y_1) at the same time, t.

SETUP First, we write equations for the x- and y-components of position for both monkey and banana. We choose the x-axis to be horizontal, the y-axis to be directly upward, and the origin to be the keeper's hand. Then $a_y = -g$. Setting the expressions for the two positions equal gives algebraic relations between the unknown initial velocity of the banana and the unknown time at which the monkey gets its meal.

MONKEY initial conditions: $\quad x_i = d, \quad y_i = h, \quad v_{xi} = v_{yi} = 0.$
x-motion (eqn. 3.1a): $\quad x_m(t) = d = \text{constant}.$
y-motion (eqn. 3.1b): $\quad y_m(t) = h - \frac{1}{2}gt^2.$
BANANA initial conditions: $\quad x_i = 0, \quad y_i = 0.$
$\quad v_{xi} = v_i \cos \theta_i, \quad v_{yi} = v_i \sin \theta_i.$
x-motion (eqn. 3.1a): $\quad x_b(t) = (v_i \cos \theta_i)t.$
y-motion (eqn. 3.1b): $\quad y_b(t) = (v_i \sin \theta_i)t - \frac{1}{2}gt^2.$

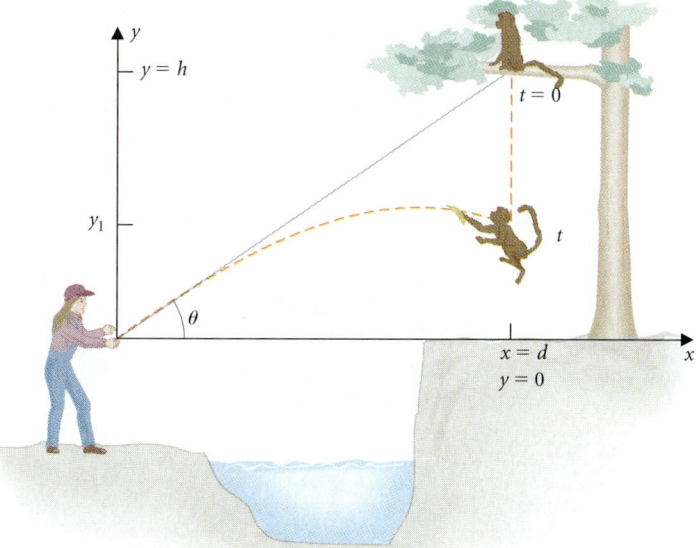

■ **FIGURE 3.9**
A monkey has the curious habit of dropping from its perch just as the zookeeper throws it a banana. For the monkey to catch the banana, the keeper must throw it directly toward the monkey's position in the tree.

Setting $x_b(t)$ equal to $x_m(t)$ gives one relation,

$$t = d/(v_i \cos \theta_i), \quad \text{(i)}$$

while setting $y_m(t) = y_b(t)$ gives a second,

$$h - \tfrac{1}{2}gt^2 = (v_i \sin \theta_i)t - \tfrac{1}{2}gt^2.$$

The term $\tfrac{1}{2}gt^2$ on each side of the equation may be cancelled, leaving

$$t = h/(v_i \sin \theta_i). \quad \text{(ii)}$$

TRIAL We have two expressions for the same time, t. They must give the same result. So setting (i) equal to (ii), we have

$$\frac{d}{v_i \cos \theta_i} = \frac{h}{v_i \sin \theta_i}.$$

The speed cancels out, so we may solve this equation for θ_i.

SOLVE $\quad \tan \theta_i = h/d = (2.0 \text{ m})/(3.0 \text{ m}) = 0.667 \Rightarrow \theta_i = 33.7°.$

Since $\tan \theta_i = h/d$, the keeper must throw the banana directly at the monkey (Figure 3.9).

ANALYZE But, we aren't finished. We have to ask whether the banana actually gets to the monkey, or does one of them hit the ground first?

SETUP The ground beneath the monkey's tree is at $y = 0$. Using expression (ii) for t, the monkey's position when it catches the banana is

$$y_m = h - \tfrac{1}{2}gt^2 = h - \frac{gh^2}{2v_i^2 \sin^2 \theta_i} = h\left(1 - \frac{gh}{2v_i^2 \sin^2 33.7°}\right) = h(1 - 1.62gh/v_i^2).$$

SOLVE For the monkey to catch the banana before reaching the ground, y_m should be greater than zero. This requires $v_i^2 > 1.62gh = 1.62(9.8 \text{ m/s}^2)(2.0 \text{ m}) = 32 \text{ m}^2/\text{s}^2$. The keeper must throw the banana at a speed $v_i > 5.6$ m/s.

ANALYZE We could equally well have used expression (i) for t. Of course, the requirement on v_i is the same. Check it and see.

Digging Deeper

Free Fall

The keeper in Example 3.4 aims the banana directly toward the monkey. Since both fruit and monkey fall with the same acceleration, there is no change in their velocity *with respect to each other*. The monkey always sees the banana coming straight at itself. This fact can be seen in the algebraic solution where g cancels from expression (ii) for the mealtime t.

Galileo's law assures us that all freely falling objects accelerate in the same way; any change in their relative velocity has to be caused by something other than gravity, which seems (but only seems) to be absent. Freely falling astronauts in the space shuttle can ignore gravity for days; its presence is quickly brought to the monkey's attention when the monkey collides with the ground.

A reference frame attached to a freely falling object is accelerated. At this stage in your study, you should avoid the use of accelerated reference frames. Your car, for example, provides a psychologically natural, but puzzling, reference frame. When it accelerates away from a stoplight you feel "thrown backward." What does the throwing? Nothing! (You might ponder why a helium balloon appears "thrown forward" at the same time—try it!) Most accelerated reference frames introduce such paradoxes, but a freely falling frame simplifies things. This fact was the starting point for Einstein's theory of gravity (Essay 8).

3.1.4 The Shape of a Projectile's Trajectory

The previous examples illustrate how to solve problems using a description of a particle's horizontal and vertical positions as functions of time. Many problems, including Examples 3.1, 3.2, and 3.4, are not really concerned with the time a projectile spends in its motion, but instead ask where it will land or if it will hit something. For such applications, it is useful to have an equation that relates the vertical position of the projectile directly to its horizontal position without reference to time—that is, $y(x)$.

Since the projectile moves horizontally at constant velocity, the time t is directly proportional to the projectile's horizontal displacement. Starting the clock when the projectile leaves the origin, the projectile's horizontal displacement is (eqn. 3.1a) $x = v_{xi}t$. So $t = x/v_{xi}$.

We can substitute this result wherever time appears in the equation for vertical motion (eqn. 3.1b), and so eliminate time from the equations. With the positive y-axis upward, $a_y = -g$, and we have

$$y(x) = v_{yi}t - \tfrac{1}{2}gt^2$$
$$= v_{yi}\left(\frac{x}{v_{xi}}\right) - \frac{g}{2}\left(\frac{x}{v_{xi}}\right)^2 = \left(\frac{v_{yi}}{v_{xi}}\right)x - \left(\frac{g}{2v_{xi}^2}\right)x^2.$$

In terms of the projectile's initial speed and launch angle, we have

$$y(x) = (\tan\theta_i)x - \left(\frac{g}{2v_i^2}\right)(\sec^2\theta_i)x^2. \qquad (3.3)$$

See Appendix IB for the equation of a parabola. The correspondence becomes more obvious if you put the origin at the peak of the projectile's path (Problem 3.34).

Remember: $\sec\theta = 1/\cos\theta$.

Equation (3.3) is a quadratic function of the form $y = ax - bx^2$, which is the equation of a parabola.

The next example describes a fourth type of projectile problem. We return to our intrepid skier to ask where she is going to land. Both her projectile path and the ski slope are described by algebraic functions $y(x)$. The landing spot is at the intersection of the two curves. From the horizontal coordinate of the landing spot, we find the time of landing—a reversal of the previous logic plan.

EXAMPLE 3.5 ♦♦ The skier leaves the snow with velocity 11 m/s, 23° below the horizontal, and lands on the 55° slope below (■ Figure 3.10). Where and when does she land? What is her velocity when she hits the snow?

MODEL Choosing coordinates with the origin at the point the skier leaves the snow, her trajectory and the snow surface are each described by an alegbraic equation $y(x)$. At the point where she hits the snow, the skier's y-coordinate and that of the surface are the same. Setting them equal gives an equation for the x-coordinate of the impact point.

SETUP The skier's initial velocity makes an angle 23° *below* the x-axis, so $\theta_i = -23°$. Using the subscript hd for our "hotdog" skier, we apply eqn. (3.3).

$$y_{hd} = x_{hd}\tan(-23°) - \frac{gx_{hd}^2}{2v_i^2\cos^2(-23°)}. \qquad (i)$$

The slope also makes a negative angle with the x-axis. It is described by the equation for a straight line, $y = mx + c$, with $m = \tan(-55°) = -1.43$ and, since the origin is on the slope, $c = 0$. Thus, using the subscript s for coordinates on the slope,

$$y_s = -1.43 x_s. \qquad (ii)$$

SOLVE Then $y_{hd} = y_s$ when $x_{hd} = x_s \equiv x_\ell$, the x-coordinate of the landing point. Setting (i) equal to (ii), with $\tan(-23°) = -0.424$ and $\cos(-23°) = 0.921$,

$$-0.424 x_\ell - \frac{gx_\ell^2}{2v_i^2(0.921)^2} = -1.43 x_\ell.$$

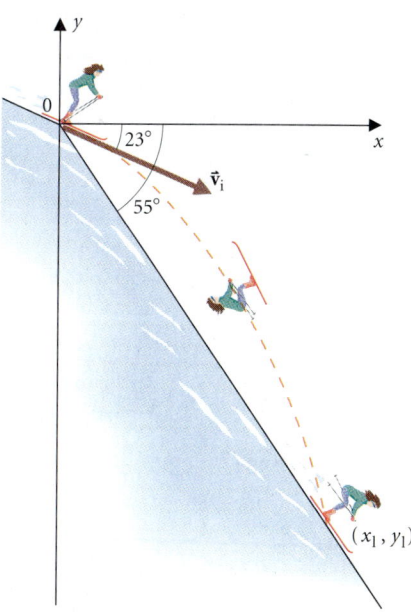

■ **FIGURE 3.10**
Schematic drawing of the somersaulting skier's path.

Are you worried about that minus sign? Look at the figure—the velocity clearly has a negative y-component.

We don't care about the solution $x_\ell = 0$ (the takeoff point), so

$$x_\ell = \frac{1.69 v_i^2}{g}(1.43 - 0.424) = \frac{1.69(11 \text{ m/s})^2}{9.8 \text{ m/s}^2}(1.0) = 21 \text{ m}.$$

Then from eqn. (ii), $\quad y_\ell = (-1.43)(21 \text{ m}) = -30 \text{ m}.$

The distance along the slope from the jump point is

$$d = \sqrt{(21 \text{ m})^2 + (30 \text{ m})^2} = 37 \text{ m}.$$

We may use eqn. (3.1a) to find the time at impact.

$$t_\ell = \frac{x_\ell}{v_{xi}} = \frac{x_\ell}{v_i \cos(23°)} = \frac{21 \text{ m}}{(11 \text{ m/s})(0.92)} = 2.1 \text{ s}.$$

Digging Deeper

Projectiles and Orbits

According to Galileo, in the absence of air resistance, projectiles travel on parabolic trajectories. On the other hand, a projectile is, for a short time, orbiting Earth and, according to Kepler's laws, should follow an elliptical path. How can both claims be correct? ■ Figure 3.11 shows Earth and the trajectory of a projectile. The dotted ellipse is the path the particle would follow if Earth were a point and didn't get in the way. The projectile path is one end of this very long, skinny ellipse and is approximately a parabola so long as the projectile moves a very small distance compared with the radius of the Earth. Then we can neglect any variation of g with height and any difference between the Earth's surface and a plane.

Newton gave a very nice image for using projectile motion to understand the orbits of satellites such as the space shuttle or the Moon. Imagine there is no atmosphere and we fire a sequence of projectiles with ever-increasing horizontal velocity from the top of a very high mountain (■ Figure 3.12). Faster projectiles land ever farther from the base of the mountain, both because of their greater horizontal velocity and because the curvature of the Earth increases the distance they have to fall. If a projectile moves rapidly enough, the Earth curves away as fast as the projectile falls; the projectile never reaches the surface and thus becomes a satellite.

How fast does the satellite have to go? We must wait until Chapter 5 for a careful calculation, but a quick estimate gets the right answer. Gravity is described near Earth's surface by g, and a typical dimension for a satellite orbit is Earth's radius R. The product gR has dimensions L^2/T^2, or speed squared. So, a satellite's speed should be of the order of

$$v \sim \sqrt{gR} = \sqrt{(9.8 \text{ m/s}^2)(6.4 \times 10^6 \text{ m})} = 8 \text{ km/s},$$

or about 20 000 mph.

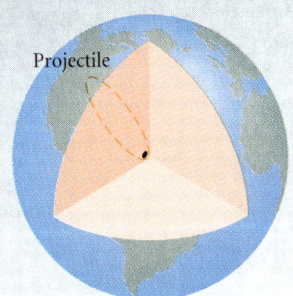

■ **Figure 3.11**
Neglecting air resistance, motion of an object near the ground is a special case of satellite motion—rudely interrupted by Earth's surface. If the Earth were condensed to a point so that it didn't get in the projectile's way, the projectile would follow a long, skinny ellipse with the Earth's center at one focus. The end of such a skinny ellipse is very nearly a parabola.

■ **Figure 3.12**
Newton's explanation of satellites. Projectiles launched horizontally from an imaginary mountaintop, above the atmosphere, travel a greater distance around Earth, the greater their launch speed. With a great enough speed, a projectile "falls" all the way around Earth: it has become a satellite.

The skier's landing velocity has components:

$$v_x = v_{xi} = (11 \text{ m/s})(0.92) = 10 \text{ m/s},$$

and
$$v_y = v_{yi} - gt_\ell = v_i \sin(-23°) - gt_\ell$$
$$= -(11 \text{ m/s})(0.39) - (9.8 \text{ m/s}^2)(2.1 \text{ s}) = -25 \text{ m/s}.$$

Thus her velocity is

$$\vec{\mathbf{v}} = (10 \text{ m/s})\hat{\mathbf{i}} - (25 \text{ m/s})\hat{\mathbf{j}}.$$

ANALYZE The skier's speed is large, but her velocity component perpendicular to the snow is about 6 m/s, the same as for a vertical jump from only about 1.8 m. (See Problem 31.)

3.2 CIRCULAR MOTION

We began this chapter wondering about a skier somersaulting above a steep snow slope and left her in the last section landing some 37 m down the slope. Whether she lands upright is a question of rotational motion, which is our next subject. Modeling the skier is not, however, a good starting point. She begins with a slow, almost unnoticeable rotation, tucks up to spin rapidly in the somersault, then straightens out to slow her rotation for the landing. An airplane propeller (■ Figure 3.13) behaves more simply: in cruise flight, the pilot keeps the propeller's rotation rate constant. Even so, its different parts have different velocities. What we need to study first is a particle moving around a circle at a uniform rate. That model is the foundation on which we may build.

■ **FIGURE 3.13**
In cruise flight, the pilot of this airplane sets the rotation rate of the propeller, which then remains constant. Even so, the propeller tips move much faster than points near the center.

3.2.1 Describing Circular Motion

Imagine yourself on the carnival ride shown in ■ Figure 3.14—whirling around on the inside surface of the *Rotor*. Your speed is constant, but your velocity, continuously changing in direction, is not. This acceleration and the thrill of terror when the floor is removed are the reasons why you buy a ticket.

WE WILL LEARN IN CHAPTER 4 WHY THE PEOPLE DON'T FALL OUT.

■ **FIGURE 3.14**
In this carnival ride—called the *Rotor*—people are whirled around on the inside of a drum, and then the floor is dropped out, leaving the people *stuck* to the vertical wall.

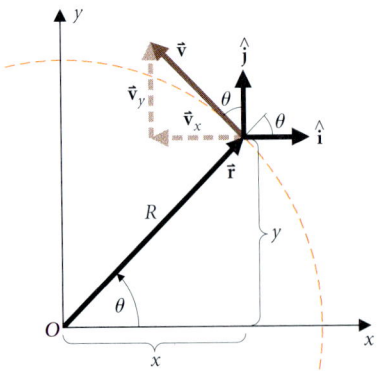

FIGURE 3.15
Coordinates for the carnival ride seen from the top. The origin O is at the center of a rider's circular path. The position vector has components $x = R\cos\theta$ and $y = R\sin\theta$. The rider's velocity is tangent to the path (§2.1.3), and has components $v_x = -v\sin\theta$ and $v_y = v\cos\theta$.

THE UNITS OF ω ARE RADIANS PER SECOND (rad/s).

THE UNITS OF α ARE RADIANS PER SECOND PER SECOND (rad/s^2).

REMEMBER: ANGLES ARE MEASURED IN RADIANS UNLESS IT IS CLEARLY STATED OTHERWISE.

The description of any circular motion, including a trip in the *Rotor*, follows the outline set out in §2.1: describe the particle's position vector as a function of time and then work out its velocity and acceleration. Because circles are so symmetric, we may obtain exact expressions for velocity and acceleration and their relation to the position vector.

As the *Rotor* turns, each rider follows a circular path in a horizontal plane. The most convenient coordinate system has its origin at the center of that circle (■ Figure 3.15). A rider's position vector $\vec{r}$ has constant length equal to the radius of the circle, and its direction is determined by the angle θ from the x-axis. The velocity vector $\vec{v}$ is tangent to the circle and thus perpendicular to $\vec{r}$, as shown in the figure. In components, the position and velocity vectors are

$$\vec{r} = R[(\cos\theta)\hat{\imath} + (\sin\theta)\hat{\jmath}] \quad \text{and} \quad \vec{v} = |\vec{v}|[-(\sin\theta)\hat{\imath} + (\cos\theta)\hat{\jmath}]. \quad (3.4)$$

Like linear motion, circular motion is described by a single coordinate, here an angle $\theta(t)$. The rate at which θ (measured in radians) changes with time is called *angular speed* and given the standard symbol ω.

$$\omega \equiv \lim_{\Delta t \to 0} \frac{\Delta\theta}{\Delta t} = \frac{d\theta}{dt}.$$

Angular acceleration, with standard symbol α, is the rate of change of angular speed.

$$\alpha \equiv \lim_{\Delta t \to 0} \frac{\Delta\omega}{\Delta t} = \frac{d\omega}{dt}.$$

The pattern of these definitions is the same as for linear motion.

3.2.2 Uniform Circular Motion

The simplest case of circular motion, the basic kinematic model we set out to find, is that of uniform motion, in which the angular speed is constant and the angular acceleration is zero. In that case, for any time interval Δt the change in the particle's angular position is $\Delta\theta = \omega\,\Delta t$, or

$$\theta - \theta_i = \omega(t - t_i). \quad (3.5)$$

So long as we are describing only one particle (you in the *Rotor*), we may start the clock at $t = 0$ when the particle is at $\theta = 0$—that is, on the line chosen as the x-axis. Then, the particle's angular position as a function of time becomes

$$\theta(t) = \omega t. \quad (3.6)$$

If you want to consider more than one particle on the same circle, such as other terrorized *Rotor* riders, each could be assigned a value of θ_i and have a position given by a personal version of eqn. (3.5).

Uniform circular motion is *periodic*; that is, each time the particle travels completely around the circle, its motion is an identical repetition of every other trip around the circle. The motion repeats after a time interval called the *period*:

> The *period* T of a particle in uniform circular motion is the time interval required for the particle to go once around the circle.

In each interval T, θ increases by 2π, so from eqn. (3.6),

$$T \equiv 2\pi/\omega. \tag{3.7}$$

An alternative description of a periodic motion is the number of complete cycles per unit of time:

> The *frequency f* of a particle in uniform circular motion is the number of complete trips it makes about the circle per unit of time.

The product of frequency with any time interval gives the number of turns completed during that interval. In a time interval of one period, the particle makes one turn, so $fT = 1$, and

$$f = 1/T = \omega/2\pi. \tag{3.8}$$

Recall that angle is dimensionless, being the ratio of two lengths (arc length and radius). For this reason, some authors use s^{-1} as the unit of angular speed, omitting the word "radians." "Cycles" or "complete trips" are also dimensionless, leaving s^{-1} as the unit of frequency. To avoid ambiguity, the SI unit of frequency has a special name, *hertz* (symbol Hz), honoring the German physicist Heinrich Hertz (1857–1894).

$$1 \text{ Hz} \equiv 1 \text{ cycle/s}.$$

HERTZ WAS THE FIRST TO PRODUCE AND DETECT RADIO WAVES IN THE LABORATORY. (SEE CHAPTER 33.)

EXAMPLE 3.6 ♦ If the *Rotor* spins you around with an angular speed $\omega = 5.0$ rad/s, what are your frequency f and period T? Express the frequency both in SI units and in revolutions per minute (rpm). How many revolutions are completed in 15 s?

MODEL We model a person in the *Rotor* as a particle in uniform circular motion.

SETUP We apply eqns. (3.7) and (3.8).

SOLVE The period is

$$T = \frac{2\pi}{\omega} = \frac{2\pi \text{ rad}}{5.0 \text{ rad/s}} = 0.40\pi \text{ s} = 1.3 \text{ s}.$$

The frequency is

$$f = \frac{1}{T} = \frac{1}{(0.40)\pi \text{ s}} = 0.80 \text{ Hz}.$$

"Revolutions" is just another name for cycles or complete trips and requires no unit conversions, so

$$f = (0.80 \text{ /s})\frac{60 \text{ s}}{1 \text{ min}} = 48 \text{ /min} \equiv 48 \text{ rpm}.$$

The number of revolutions completed in 15 s is

$$N = \frac{\Delta t}{T} = \frac{15 \text{ s}}{1.3 \text{ s}} = 12.$$

ANALYZE We could also calculate the number of revolutions as $N = f\Delta t$. ∎

3.2.3 Speed and Instantaneous Velocity in Circular Motion

As a particle moves around a circle with constant angular speed, we might expect its linear speed to be constant as well. Since the radius of the circle and the particle's angular speed are the significant quantities describing the motion and their product has units of meters per second, we expect the speed to be of the order of ωR. Let's check and see.

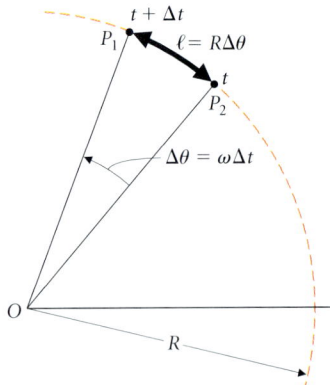

FIGURE 3.16
In a time interval Δt, a particle moves through an angle $\Delta\theta$ and travels a distance ℓ. From the definition of radian measure, $\ell = R\,\Delta\theta$. Divide this by Δt to find the average speed: $S_{av} = \ell/\Delta t = R\,\Delta\theta/\Delta t = R\omega$.

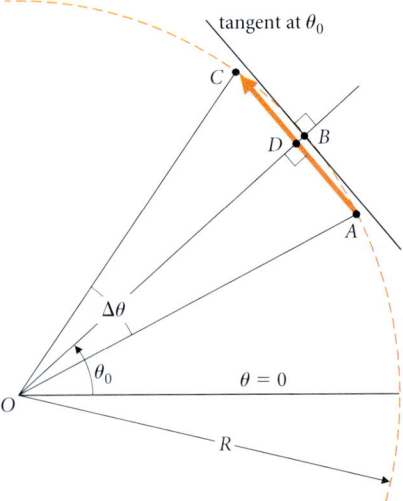

FIGURE 3.17
A particle's displacement during a time interval Δt is the chord $\Delta\vec{r}$ between points A and C. The tangent at B is in the direction of the particle's instantaneous velocity at B.

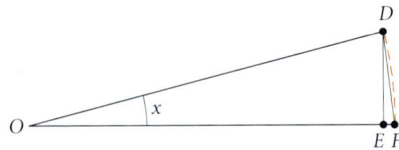

FIGURE 3.18
A *skinny triangle* illustrates the rule that arc length and chord length are almost equal for very small angles. As the angle x decreases, the distance EF also decreases, and the straight lines DE and DF become almost identical with the arc DF. Thus $\sin x \approx x \approx \tan x$, where x is measured in radians.

In a time interval Δt, the particle moves through an angle $\Delta\theta = \omega\,\Delta t$ (■ Figure 3.16). The distance the particle has traveled is the arc length between points P_1 and P_2 in the figure. From the definition of radian measure, arc length and angle are proportional.

$$\ell = R\,\Delta\theta = R\omega\,\Delta t.$$

The average speed S_{av} is the distance divided by the time interval.

$$S_{av} = \frac{\ell}{\Delta t} = \frac{R\omega\,\Delta t}{\Delta t} = \omega R.$$

The average speed is independent of the time interval considered, and has the expected value.

In §2.1 we defined instantaneous velocity as the limit of average velocities during ever decreasing time intervals, and instantaneous speed as the magnitude of the instantaneous velocity. Thus our next task is to calculate the average velocity for an arbitrary time interval, and then take the limit as Δt approaches zero.

During a time interval Δt, the particle undergoes a displacement $\Delta\vec{r}$ centered on point B (■ Figure 3.17) where $\theta(t) = \theta_o$. Triangle OAC is an isosceles triangle, and the line ODB bisects it, so that OD is perpendicular to AC. The tangent to the circle at B is also perpendicular to the radius ODB. So regardless of the size of the angular displacement $\Delta\theta$, $\Delta\vec{r}$ and the corresponding average velocity $\Delta\vec{r}/\Delta t$ are both parallel to the tangent line at the middle of the displacement—that is, at θ_o. Thus the instantaneous velocity is tangent to the circle, as we asserted in eqn. (3.4).

The instantaneous speed is the limiting magnitude of the average velocity.

$$|\vec{v}| = \lim_{\Delta t \to 0}\left|\frac{\Delta\vec{r}}{\Delta t}\right| = \lim_{\Delta t \to 0}\frac{|\Delta\vec{r}|}{\Delta t}.$$

Taking this limit boils down to the following geometrical argument. From Figure 3.17 we see that chord AC ($\equiv |\Delta\vec{r}|$) and arc AC have nearly the same length. So

$$|\vec{v}| \approx |\Delta\vec{r}|/\Delta t \approx AC/\Delta t = \text{average speed} = \omega R.$$

The *skinny triangle* in ■ Figure 3.18 shows perhaps more compellingly that the approximation arc $\approx$ chord length becomes ever more precise for ever smaller angles. Thus we have the following rule:

> The *velocity* of an object in *uniform circular motion* is tangent to the circle and has a magnitude equal to the product of the radius of the circle and the angular speed.
>
> $$\vec{v} = (\omega R,\ \text{tangent to the circle}) \qquad (3.9)$$

EXAMPLE 3.7 ◆ A ferris wheel with radius 11 m turns with a period of 4.5 min. What is the instantaneous velocity of a rider at the top of the wheel and of a rider at the bottom? (■ Figure 3.19)

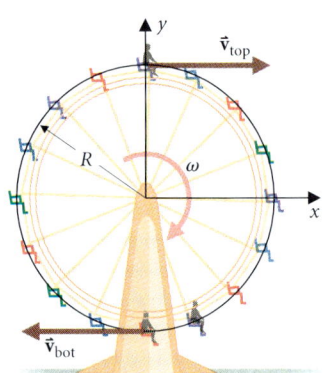

FIGURE 3.19
Each rider on the ferris wheel undergoes uniform circular motion with speed ωR. The velocity of a rider at the top is $\vec{v}_{top} = \omega R\hat{\imath}$.

CHAPTER 3 • ADVANCED KINEMATIC MODELS

Math Topic

FORMAL EVALUATION OF THE LIMIT

To obtain a formula for the magnitude of the average velocity, notice that in Figure 3.17 half the chord AC is one side of the right triangle OAD:

$$\tfrac{1}{2}|\Delta\vec{r}| = R\,\sin(\Delta\theta/2),$$

where $\Delta\theta = \omega\,\Delta t$. The magnitude of the average velocity is

$$|\vec{v}_{av}| = \frac{|\Delta\vec{r}|}{\Delta t} = \frac{2R\,\sin(\omega\,\Delta t/2)}{\Delta t}.$$

If we rearrange the expression so that Δt appears in the combination $x \equiv \omega\,\Delta t/2$, then

$$|\vec{v}| = \lim_{\Delta t \to 0} \omega R \frac{\sin(\omega\,\Delta t/2)}{\omega\,\Delta t/2} = \omega R \lim_{x \to 0} \frac{\sin x}{x}.$$

Both x and $\sin x$ approach zero, and L'Hôpital's Rule applies.

$$\lim_{x \to 0} \frac{\sin x}{x} = \lim_{x \to 0} \frac{d/dx(\sin x)}{d/dx(x)} = \lim_{x \to 0} \cos x = 1.$$

The limit is unity, as we found from the geometrical approach. From the **skinny triangle** in Figure 3.18,

$$\sin x = \frac{ED}{OD} \approx \frac{FD}{OD} = x = \frac{FD}{OF} \approx \frac{ED}{OE} = \tan x.$$

This shows yet again that our limit is unity; but we also have the very useful approximate relation

$$\sin x \approx x \approx \tan x,$$

which is accurate to about 1% for $x \lesssim 0.1$ rad $\approx 6°$. The series expansions in Appendix IB may be used to improve this approximation if necessary.

MODEL Each rider undergoes uniform circular motion.

SETUP The angular speed $\omega = 2\pi/T$ (eqn. 3.7), so the speed of any rider (eqn. 3.9) is

$$v = \omega R = \frac{2\pi}{T}R = \left(\frac{2\pi}{4.5\ \text{min}}\right)\left(\frac{1\ \text{min}}{60\ \text{s}}\right)(11\ \text{m}) = 0.26\ \text{m/s}.$$

The direction of the instantaneous velocity is tangent to the circle—in this case, parallel to the ground.

SOLVE The velocity vector at the top points to the right in Figure 3.19, while the velocity at the bottom points to the left. Drawing the x-axis horizontal with positive to the right,

$$\vec{v}_{\text{top}} = (0.26\ \text{m/s})\hat{\mathbf{i}}$$

and

$$\vec{v}_{\text{bot}} = (0.26\ \text{m/s})(-\hat{\mathbf{i}}).$$

ANALYZE During the ride each person's velocity vector changes direction continuously. All riders have the same constant speed. ∎

3.2.4 Instantaneous Acceleration in Circular Motion

The continuously changing direction of an object's velocity in circular motion means that the object is continuously accelerated. To calculate the acceleration, we follow the same steps used to calculate velocity. ■ Figure 3.20 shows the velocities on the *Rotor* at the end points A and C of the same arbitrary interval used in the discussion of velocity (Figure 3.17). We shall find the particle's instantaneous acceleration at point B, the midpoint of the arc.

Instantaneous acceleration is the limiting value of average acceleration, which, in turn, is the change in instantaneous velocity, divided by time interval.

$$\vec{a} = \lim_{\Delta t \to 0} \vec{a}_{av} = \lim_{\Delta t \to 0} \frac{\Delta \vec{v}}{\Delta t} = \lim_{\Delta t \to 0} \frac{\vec{v}_C - \vec{v}_A}{\Delta t}.$$

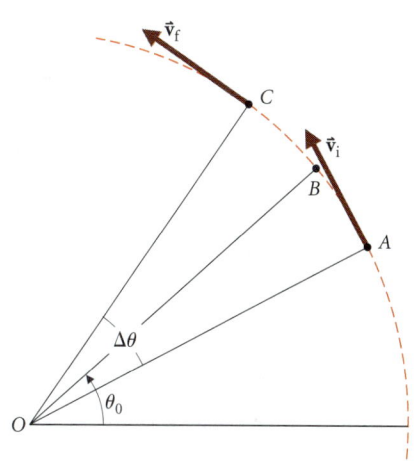

■ **FIGURE 3.20**
The particle's instantaneous velocity, always tangent to the circle, changes direction as the particle moves around the circle.

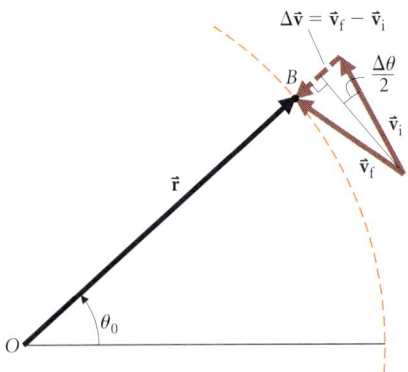

FIGURE 3.21
The change of velocity $\Delta\vec{v}$ is constructed from a triangle where the initial and final velocities are placed tail to tail. The construction is placed in the figure to show that $\Delta\vec{v}$ is parallel to the line from point B toward the center of the circle.

The velocity triangle in ■ Figure 3.21 is similar to the triangle OAC in Figure 3.17, and we analyze it identically. The difference $\Delta\vec{v}$ is parallel to BO—that is, from B toward the center of the circle—so the average acceleration is directed toward the center, regardless of the value of $\Delta\theta$. Thus in the limit as $\Delta\theta \to 0$, the direction of the instantaneous acceleration is toward the center. The magnitude of $\vec{a}_{av}$ is

$$|\vec{a}_{av}| = \frac{|\Delta\vec{v}|}{\Delta t} = \frac{2|\vec{v}|\sin(\Delta\theta/2)}{\Delta\theta/\omega} = \omega|\vec{v}|\frac{\sin(\Delta\theta/2)}{\Delta\theta/2}.$$

So
$$|\vec{a}| = \lim_{\Delta t \to 0}\frac{|\Delta\vec{v}|}{\Delta t} = \omega(\omega R)\lim_{\Delta\theta \to 0}\frac{\sin(\Delta\theta/2)}{\Delta\theta/2} = \omega^2 R = \frac{(\omega R)^2}{R} = \frac{v^2}{R}.$$

> The *acceleration* of an object in *uniform circular motion* is toward the center of the circle and has magnitude equal to the radius of the circle multiplied by the square of the angular speed.
>
> $$\vec{a} = (\omega^2 R, \text{ toward the center of the circle}). \quad (3.10)$$
>
> Alternatively, its magnitude equals the linear speed squared divided by radius.
>
> $$\vec{a} = (v^2/R, \text{ toward the center of the circle}). \quad (3.11)$$

Math Topic

Use of Calculus in Circular Motion

Since velocity is the derivative of the position vector (eqn. 2.3), we have

$$\vec{v}(t) = \frac{d\vec{r}(t)}{dt} = \frac{d}{dt}\left[R(\hat{i}\cos\theta + \hat{j}\sin\theta)\right]$$
$$= R\left(\hat{i}\frac{d}{dt}\cos\theta + \hat{j}\frac{d}{dt}\sin\theta\right).$$

Using the chain rule to differentiate the sine and cosine functions,

$$\vec{v}(t) = R(-\hat{i}\sin\theta + \hat{j}\cos\theta)\frac{d\theta}{dt}$$
$$= (\omega R)(-\hat{i}\sin\theta + \hat{j}\cos\theta) \equiv (\omega R)\hat{v}.$$

So we verify the general form for the velocity in eqn. (3.4) and the result of §3.2.3 for the speed.

Differentiating again to obtain an expression for the acceleration,

$$\vec{a}(t) = \frac{d\vec{v}(t)}{dt} = \omega R(-\hat{i}\cos\theta - \hat{j}\sin\theta)\frac{d\theta}{dt} + R\frac{d\omega}{dt}(-\hat{i}\sin\theta + \hat{j}\cos\theta)$$
$$= -\omega^2\vec{r}(t) + R\alpha\hat{v} = \omega^2 R(-\hat{r}) + R\alpha\hat{v}.$$

The first term verifies our previous result for uniform circular motion. This component of the acceleration corresponds to change in the direction of the velocity vector and points inward toward the center of the circle. The second term gives the parallel component of acceleration in a nonuniform circular motion and corresponds to change in speed.

In uniform circular motion, each differentiation multiplies the vector by ω and rotates it by 90° (■ Figure 3.22).

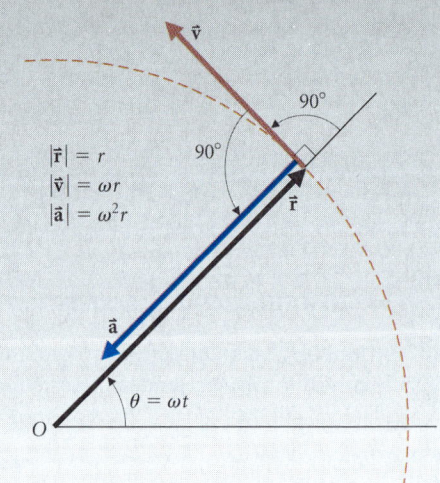

FIGURE 3.22
In uniform circular motion, each differentiation of the position vector multiplies the vector by ω and rotates it through 90°.

Like position and velocity, acceleration in uniform circular motion has constant magnitude but continuously changes direction—always pointing from the particle toward the center of the circle. You can feel this acceleration by twirling a large ball around your head like the contestant in ■ Figure 3.23. You must continuously pull inward! It is your pull that causes the inward acceleration.

EXAMPLE 3.8 ♦ Part of the thrill of an amusement park ride is the acceleration you experience. Riders on the *Wild Whirl* are carried in a circle of radius R at speed V (■ Figure 3.24). Those braving the *Polar Bear* (■ Figure 3.25) start from rest and accelerate along a straight track of length $2R$, reaching speed V as they enter the mouth of the immense, angry bear. Compare the magnitudes of acceleration experienced by passengers on the two rides.

MODEL We model the *Wild Whirl* as an example of uniform circular motion (eqn. 3.11), and the *Polar Bear* as an example of uniformly accelerated linear motion over a comparable distance, for which we may use the relations in Table 2.4.

SETUP AND SOLVE On the *Wild Whirl*, the acceleration (eqn. 3.11) has magnitude

$$|\vec{a}_{ww}| = V^2/R.$$

For the *Polar Bear*, we need the acceleration given speed and distance (eqn. 2.13).

$$|\vec{a}_{pb}| = \frac{v_f^2 - v_i^2}{2(\Delta y)} = \frac{V^2 - 0^2}{2(2R)} = \frac{V^2}{4R}.$$

ANALYZE The acceleration on the *Polar Bear* is only one-quarter that on the *Wild Whirl*. We hope it's correspondingly cheaper.

■ **FIGURE 3.23**
To hold the enormous ball in a circular path around his body, heavyweight champion Francis Brebner must pull inward with all his strength.

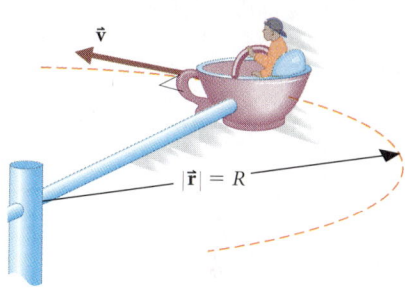
■ **FIGURE 3.24**
Riders on the *Wild Whirl* travel in a circle.

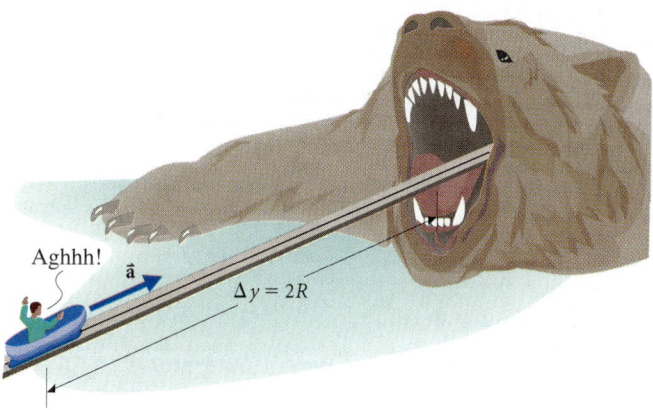
■ **FIGURE 3.25**
Riders on the *Polar Bear* accelerate in a straight line. Which riders experience the greater acceleration?

3.2.5 Rigid Body Rotation

As a rigid body like an aircraft propeller rotates at a constant rate, each small piece of the body undergoes uniform circular motion, taking the same time to rotate once around. Each piece has the same period T and angular speed ω, but has a linear velocity tangent to its own circular path. Its acceleration is toward the center of its path. The magnitudes of velocity and acceleration for a piece of the body are given by eqns. (3.9) and (3.10), with the radius equal to the distance of the piece from the object's rotation axis. ■ Figure 3.26 illustrates this variation, showing velocities and accelerations for four different pieces of a wheel.

EXAMPLE 3.9 ♦♦ Compare the velocities and accelerations of San Francisco and Mombasa, Kenya, due to the rotation of the Earth. Compare your values for these accelerations with the acceleration due to gravity.

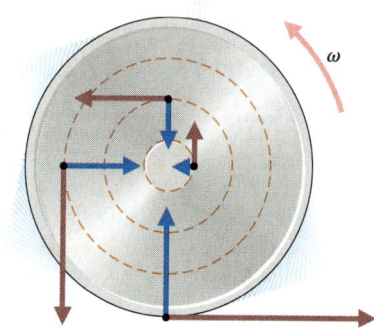
■ **FIGURE 3.26**
Velocities and accelerations for four different pieces of a uniformly rotating wheel. Each part of the wheel undergoes uniform circular motion with the same angular speed ω. The magnitudes of a part's velocity and acceleration are both proportional to the radius of its circular path and are oriented tangent to its path and toward its center, respectively. These directions are at any one time different for different pieces of the wheel.

■ **FIGURE 3.27**
Every point on Earth rotates once a day in a circular path about Earth's axis. The radius of the circular path is greater near the equator than near the poles. The velocity vectors are tangent to the circular paths, and the acceleration vectors point toward the center of each circle (*not* toward the center of Earth!) Thus Mombasa experiences a greater acceleration than San Francisco. Both accelerations are of order 0.3% of *g* (cf. §2.3.2).

MODEL Each city undergoes uniform circular motion with a period of 1 d due to the Earth's rotation. To find the magnitude of velocity and acceleration for the two cities, we need their distances from Earth's rotation axis, and for that we need their latitudes (■ Figure 3.27), and the radius of the Earth, $R_E = 6.4 \times 10^3$ km (inside front cover).

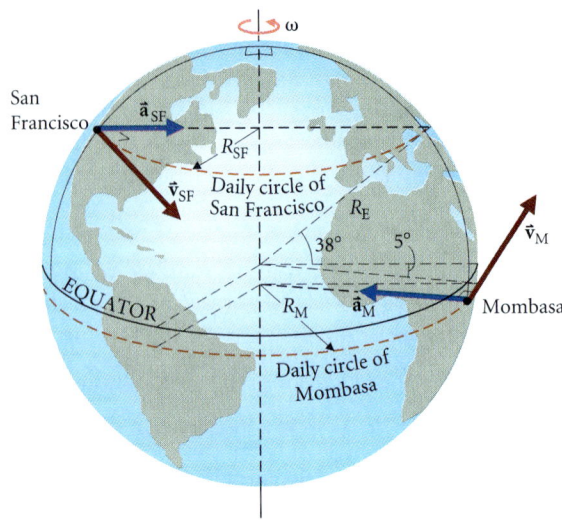

SETUP From an atlas, we find

CITY	LATITUDE (λ)
San Francisco	38° N
Mombasa	5° S

Since 1 d = (24 h)(60 min/h)(60 s/min) = 86 400 s, the angular speed of the Earth is

$$\omega = \frac{2\pi \text{ rad}}{1 \text{ d}} = \frac{2\pi \text{ rad}}{86\ 400 \text{ s}} = 7.27 \times 10^{-5} \text{ rad/s}.$$

Each city travels along a circle of radius $r = R_E \cos \lambda$. Thus the speed of Mombasa is

$$|\vec{v}_M| = \omega R_M = \omega R_E \cos \lambda = (7.27 \times 10^{-5} \text{ rad/s})(6.4 \times 10^6 \text{ m})\cos(5°)$$
$$= 0.46 \text{ km/s}.$$

Mombasa's acceleration has magnitude

$$|\vec{a}_M| = \omega|\vec{v}_M| = (7.27 \times 10^{-5} \text{ rad/s})(0.464 \text{ km/s}) = 3.4 \times 10^{-2} \text{ m/s}^2.$$

The corresponding values for San Francisco are

$$|\vec{v}_{SF}| = (7.27 \times 10^{-5} \text{ rad/s})(6.4 \times 10^6 \text{ m})\cos(38°) = 0.37 \text{ km/s}$$

and $|\vec{a}_{SF}| = \omega|\vec{v}_{SF}| = (7.27 \times 10^{-5} \text{ rad/s})(0.367 \text{ km/s}) = 2.7 \times 10^{-2} \text{ m/s}^2.$

ANALYZE Using ωR to compute the speed, we obtain an answer in rad·m/s. But we know that the units of speed are just m/s! Since radians are dimensionless, we can drop them.

The differences between the results for the two cities are much less significant than their order of magnitude. The speeds are over 800 mph! We remain completely unaware of this speed or of the much larger speed of the Earth with respect to the Sun. It is acceleration, not velocity, that we feel.

The acceleration values are about 0.3% of *g*. Whatever effects these accelerations may have, they are small compared with the effects of Earth's gravity (Chapter 5 and Essay 8). ■

EXERCISE 3.2 ♦ The skier in Example 3.5 intends to do two complete somersaults before landing upright on the snow slope. She is airborne for 2.1 s. Find the average angular speed she must attain.

3.2.6 Using Linear and Circular Motion to Model General Motion

In Chapter 2 we remarked that for an object moving on an arbitrary path, the acceleration component parallel to the velocity gives the rate of change of the object's speed and that the acceleration component $\vec{a}_\perp$ perpendicular to the velocity corresponds to change in the object's direction of motion. In Chapter 2 we couldn't express this second relation in any simple form, but we can now that we know about circular motion. During a very short time interval, an arbitrary piece of path can be modeled as a portion of a circle (■ Figure 3.28). To find the radius of the circle and rate of change in the direction of motion, we apply the relations derived for circular motion.

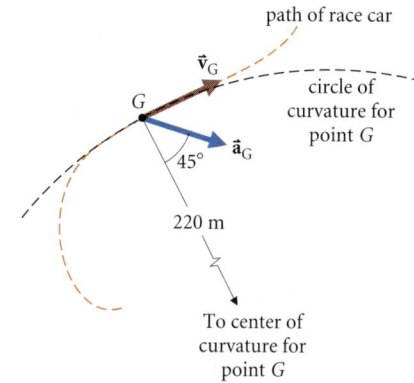

■ **FIGURE 3.28**
Example 3.10. Path of the race car from Chapter 2. From the car's speed and the components of its acceleration at point G, we may find the rate of change of the car's direction and the radius of curvature of its path. The circle of curvature is tangent to the path at G and represents the car's path well for points near G.

> Radius of curvature: $R = \dfrac{v^2}{|\vec{a}_\perp|}$.
>
> Change of direction: $\dfrac{\Delta\theta}{\Delta t} = \dfrac{v}{R} = \dfrac{|\vec{a}_\perp|}{v}$.

THESE RELATIONS ARE TRUE EVEN WHEN THE SPEED IS NOT CONSTANT.

EXAMPLE 3.10 ♦ A car traveling along the path shown in Figure 3.28 has speed $v_G = 24$ m/s at point G. Its acceleration is 4.1 m/s², and the angle between the two vectors is 45°. Calculate the radius of curvature of the path and the rate of change of the car's direction.

MODEL The radius of curvature and the rate of change of direction are determined by the perpendicular component of acceleration $|\vec{a}_\perp|$. We find them from the expressions for circular motion. It doesn't matter that the car's speed is also changing.

SETUP The perpendicular component of acceleration is

$$|\vec{a}_\perp| = |\vec{a}|\sin 45° = (4.1 \text{ m/s}^2)(0.707) = 2.9 \text{ m/s}^2.$$

SOLVE

$$R_G = \dfrac{v_G^2}{|\vec{a}_\perp|} = \dfrac{(24 \text{ m/s})^2}{2.9 \text{ m/s}^2} = 0.20 \text{ km},$$

and

$$\dfrac{\Delta\theta}{\Delta t} = \dfrac{v_G}{R_G} = \dfrac{24 \text{ m/s}}{200 \text{ m}} = 0.12 \text{ rad/s}.$$

THIS IS THE SAME CAR WE STUDIED IN CHAPTER 2. SEE ESPECIALLY EXAMPLE 2.7, §2.2.1, WHERE WE CALCULATED THE RATE AT WHICH SPEED IS CHANGING.

ANALYZE The predicted change in direction of the velocity vector over 0.5 s is

$$\Delta\theta = (0.12 \text{ rad/s})(180°/\pi \text{ rad})(0.50 \text{ s}) = 3.4°,$$

as compared with 3° measured by drawing tangents in Figure 2.2a. ■

EXERCISE 3.3 ♦♦ A roller coaster passes the bottom of a drop at a speed of 20 m/s. If the radius of curvature of the track at the bottom is 20 m, what acceleration do the passengers experience? (■ Figure 3.29)

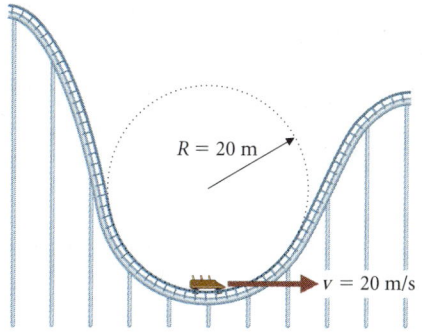

■ **FIGURE 3.29**
At the bottom of a drop, a roller coaster's path is approximated by a circle with a radius of 20 m. What acceleration do the passengers experience at the bottom?

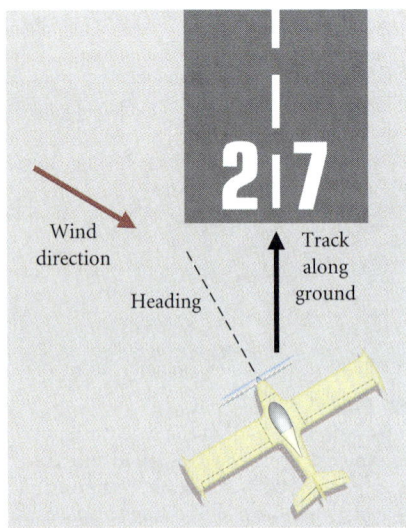

■ FIGURE 3.30
In a *crosswind* the pilot landing a plane has to point the plane at an angle to the runway so that the plane moves parallel to the runway.

3.3 RELATIVE MOTION

The pilot of the airplane in ■ Figure 3.30 isn't crazy; with a strong wind blowing across the runway, the pilot has to point the plane into the wind to keep the plane's motion along the ground parallel to the runway. This is a common type of problem in navigating airplanes and boats. The pilot wants to follow a particular path over the Earth's surface, but the design of a plane or boat only allows control over the motion with respect to the air or water. In the examples we have worked out so far, only the motion of a single object has been of interest, and we have been able to use coordinates stationary with respect to the Earth's surface. That is, we needed only a single reference frame, and the Earth's surface provided it. In navigation problems, one has to imagine more than one reference frame and to relate positions, velocities and accelerations between different frames. For navigators, this is an everyday activity; in physics you will often find it a necessary tool.

3.3.1 Navigation

Let us imagine a balloon and an airplane both flying over the countryside on a windy day. First let's consider what we know about balloons and airplanes. After release, the balloon rises to a certain altitude, and then, without propulsion, moves horizontally with the wind. The motion of the balloon reveals the motion of the surrounding air. The airplane's engine and flight controls allow the pilot to determine the plane's velocity through the air. Thus, the motion of both balloon and plane are most simply described using the air as a reference frame. A good method for visualizing reference points fixed in the air is to imagine a large swarm of balloons (■ Figure 3.31). The plane's displacement could then be visualized by watching it pass balloons. A pilot uses an airspeed indicator, clock, and compass to measure this displacement.

Now, let us imagine the swarm of balloons drifting with the wind, as if encased in a transparent plastic sheet sliding over a map of the ground below (■ Figure 3.32). Meanwhile the airplane flies eastward with respect to the balloons. The total displacement of the airplane over the ground is the sum of its displacement through the air and the displacement of the air over the ground (■ Figure 3.33). This principle underlies all relative motion problems.

■ FIGURE 3.31
Balloons drifting with the wind provide an image of fixed points in an *air reference frame*.

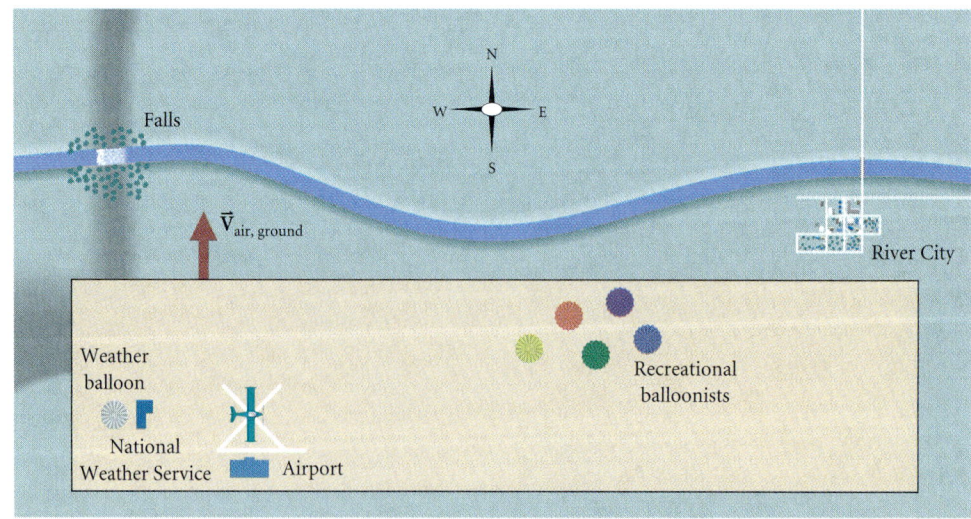

■ FIGURE 3.32
Model of reference frames in relative motion. We imagine the reference frame of the air as a *box* that moves over the ground with velocity $\vec{v}_{\text{air, ground}}$. The airplane moves through this box in an eastward direction.

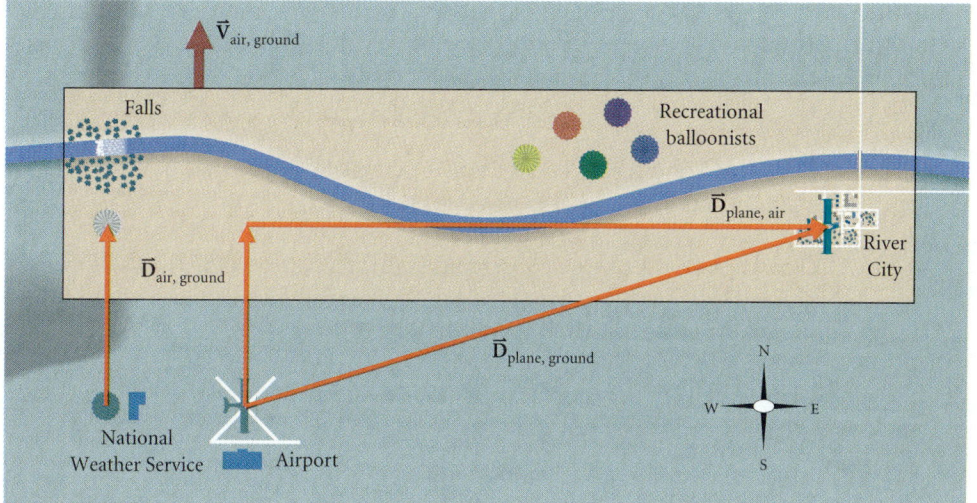

■ **FIGURE 3.33**
After a time interval Δt, the air reference frame has undergone a displacement $\vec{D}_{air,\ ground}$. The displacement of the airplane over the ground is the vector sum of its displacement with respect to the air and the displacement of the air over the ground.

EXAMPLE 3.11 ♦♦ An airplane is headed due east. The pilot determines that the plane moves a distance of 35 km *through the air* in 15 min. During the same time interval, observers on the ground note that a weather balloon moves due north a distance of 11 km over the ground. Find the total displacement of the airplane with respect to the ground. (*Note: Heading* is the direction an airplane is pointed—that is, the direction of its motion through the air.)

MODEL The situation is precisely that described in Figures 3.32 and 3.33. The pilot knows her heading and speed through the air, while the information about the balloon describes the motion of the air over the ground.

SETUP The displacement of the air reference frame is $\vec{D}_{air,\ ground} = (11\ \text{km, north})$ and the displacement of the airplane with respect to the air is $\vec{D}_{plane,\ air} = (35\ \text{km, east})$. The required displacement is their vector sum.

SOLVE
$$\vec{D}_{plane,\ ground} = \vec{D}_{plane,\ air} + \vec{D}_{air,\ ground}.$$

In one sense, our work is finished, since we have the northward and eastward components of the displacement. However, we should follow the custom of navigation problems and give the magnitude and direction.

$$|\vec{D}_{plane,\ ground}| = \sqrt{(11\ \text{km})^2 + (35\ \text{km})^2} = 37\ \text{km}.$$

The angle of the displacement from north is

$$\theta = \tan^{-1}[(35\ \text{km})/(11\ \text{km})] = \tan^{-1}(3.2) = 73°.$$

The custom in navigation is to label directions in degrees clockwise from north, so

$$\vec{D}_{plane,\ ground} = (37\ \text{km, toward}\ 073°).$$

ANALYZE The leading zero is used when expressing angles in navigation to avoid possible confusion between, for example, 15° (015) and 150° (150). Note that "073°" means 73° east of north.

The principle involved in this example—that the displacements form a vector triangle—is quickly stated, and the necessary calculations are very short. Yet, the problem is not easy. It is the *process* of visualizing the situation, organizing the information, and recognizing the method of approach that provides the challenge. Together with its introductory discussion, this is the first example that clearly distinguishes process from calculation, but *every* example attempts to teach something about both. We present a complete and systematic approach to problem solving in the *Interlude* following this chapter.

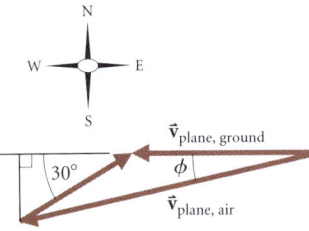

FIGURE 3.34
Velocity vector triangle for an airplane landing in a crosswind. The wind velocity and the plane's velocity with respect to the air have equal and opposite components perpendicular to the runway.

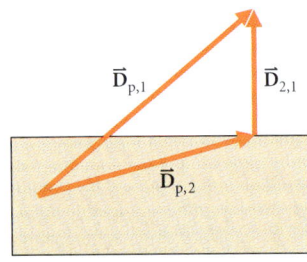

FIGURE 3.35
General version of the rule for displacements: drawing the moving frame as a shaded box, we may state the rule for combining displacements as, Displacement equals displacement *in* the box plus displacement *of* the box.

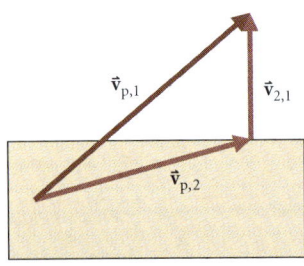

FIGURE 3.36
General version of the rule for velocities: Velocity equals velocity *in* the box plus velocity *of* the box.

AS AN AID TO REMEMBERING THESE RELATIONS, NOTE THAT THE CENTER SUBSCRIPTS "CANCEL."

Since the displacements in Example 3.11 all occur in the same time interval Δt, the rule for relating displacements becomes a rule for relating velocities. The average velocity vectors form a triangle similar to the triangle of displacement vectors.

$$\vec{v}_{\text{plane, ground}} = \vec{v}_{\text{plane, air}} + \vec{v}_{\text{air, ground}}.$$

EXERCISE 3.4 ♦ Find the average velocity of the plane in Example 3.11 with respect to the air, $\vec{v}_{\text{plane, air}}$, and with respect to the ground, $\vec{v}_{\text{plane, ground}}$.

EXAMPLE 3.12 ♦♦ A Cessna 152 aircraft lands at an airspeed of 31 m/s. If the runway direction is 270° (Figure 3.30) and the wind is blowing *from* 240° at 13 m/s, in what direction should the plane head to fly parallel to the runway?

MODEL The velocity of the airplane over the ground equals its velocity through the air plus the velocity of the air over the ground.

SETUP We know the direction and magnitude of $\vec{v}_{\text{air, ground}}$, the direction of $\vec{v}_{\text{plane, ground}}$, and the magnitude of $\vec{v}_{\text{plane, air}}$. We want to find the direction of $\vec{v}_{\text{plane, air}}$. The vectors satisfy the addition rule (■ Figure 3.34).

$$\vec{v}_{\text{plane, ground}} = \vec{v}_{\text{plane, air}} + \vec{v}_{\text{air, ground}}.$$

SOLVE Let's solve this using components. Since $\vec{v}_{\text{plane, ground}}$ is to point along the runway, the wind velocity and the plane's velocity through the air must have equal and opposite components perpendicular to the runway. Thus, if ϕ is the angle of the plane's heading to the runway,

$$v_{\text{air, ground}} \sin(30°) = v_{\text{plane, air}} \sin \phi.$$
$$(13 \text{ m/s})\sin(30°) = (31 \text{ m/s})\sin \phi.$$

The speed units cancel. Solving for $\sin \phi$,

$$\sin \phi = 13(0.50)/31 = 0.21 \quad \text{and} \quad \phi = 12°.$$

ANALYZE Using the components parallel to the runway, we could find the airplane's groundspeed when it touches down. Because the parallel component of wind velocity is opposite the plane's parallel component of velocity through the air, the groundspeed on landing is less than 31 m/s. That's why aircraft generally land upwind. ■

These examples have emphasized that position, displacement, and velocity are intimately connected with a reference frame. Whenever we speak of an object's velocity, we should have in mind the reference frame in which the velocity is measured. Galileo's discovery (cf. §0.3) that reference frames moving at constant velocity with respect to each other are equally valid assures us that we may choose a reference frame for convenience. In problems such as Example 3.12, one reference frame (the air) is convenient for expressing the motion of the plane, while another (the ground) is convenient for expressing where the plane ends up. To solve such problems, we need the rules for comparing displacements and velocities between reference frames that we used in Examples 3.11 and 3.12. Since the reference frames won't always be air and ground, we should state the rules for two arbitrary reference frames labeled 1 and 2.

> *Displacements:* The displacement $\vec{D}_{p,1}$ of a particle measured in reference frame 1 equals the sum of its displacement $\vec{D}_{p,2}$ measured in frame 2 and the relative displacement $\vec{D}_{2,1}$ of frame 2 with respect to frame 1 (■ Figure 3.35).
>
> $$\vec{D}_{p,1} = \vec{D}_{p,2} + \vec{D}_{2,1}. \quad (3.12)$$
>
> *Velocities:* The velocity $\vec{v}_{p,1}$ of a particle measured in reference frame 1 equals the sum of its velocity $\vec{v}_{p,2}$ measured in frame 2 and the relative velocity $\vec{v}_{2,1}$ of frame 2 with respect to frame 1 (■ Figure 3.36).
>
> $$\vec{v}_{p,1} = \vec{v}_{p,2} + \vec{v}_{2,1}. \quad (3.13)$$

In Examples 3.11 and 3.12, reference frame 1 is the ground, reference frame 2 is the air, and the relative velocity $\vec{v}_{2,1}$ is the velocity of the air over the ground.

3.3.2 In What Frame Is a Problem Easiest?

In navigation the use of different reference frames is essential. In other cases, problems that can be solved perfectly well in one reference frame turn out to be easier or even already solved when set in a different frame. Projectile motion provides a good example: running beside the projectile, you see uniformly accelerated vertical motion. To develop our taste for *frame jumping*, let us analyze an example involving the linear motion of two trains. As our example involves acceleration, we note first that acceleration of an object is the same in two frames, so long as their relative velocity $\vec{v}_{2,1}$ is constant.

$$\vec{a}_{p,1} = \frac{d}{dt}\vec{v}_{p,1} = \frac{d}{dt}(\vec{v}_{p,2} + \vec{v}_{2,1}) = \frac{d\vec{v}_{p,2}}{dt} + \frac{d\vec{v}_{2,1}}{dt}.$$

Since $\vec{v}_{2,1}$ is constant, its time derivative is zero, so

$$\vec{a}_{p,1} = \vec{a}_{p,2} + 0 = \vec{a}_{p,2}.$$

ALBERT EINSTEIN SHOWED IN 1905 THAT THESE EXPRESSIONS AREN'T EXACT. THE CORRECTIONS ARE ONLY IMPORTANT WHEN THE RELATIVE SPEED OF THE FRAMES IS NEAR THE SPEED OF LIGHT. (SEE CHAPTER 34.)

EXAMPLE 3.13 ♦♦ The *Overnight Flyer* is speeding across the Kansas plains at 41 m/s. Unknown to the *Flyer*'s engineer, a freight train is ahead on the same track moving at 22 m/s in the same direction. Suddenly, the engineer sees the lights of the freight 0.50 km ahead and applies the *Flyer*'s maximum braking deceleration of 0.41 m/s². Determine whether disaster occurs.

MODEL The particle whose motion we care about is the *Overnight Flyer*. We are given information in the Earth reference frame, which we call frame 1. If we choose frame 2 moving with the freight train, (■ Figure 3.37), then the freight train is at rest in frame 2. The problem becomes, Does the *Flyer* stop before hitting the stationary freight train? The motion is one-dimensional. We choose to place the *x*-axis along the railroad tracks, so that all the vectors have a single *x*-component.

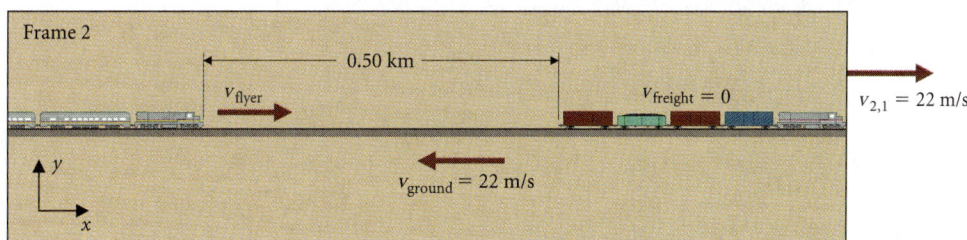

■ **FIGURE 3.37**
Potential train disaster shown in the reference frame of the slow freight train. In this frame, the freight is stationary, and the ground moves to the left. Can the *Overnight Flyer* stop in time?

SETUP The *x*-component of the *Flyer*'s velocity in frame 2 (eqn. 3.13) is

$$v_{x,2} = v_{x,1} - v_{2,1} = 41 \text{ m/s} - 22 \text{ m/s} = 19 \text{ m/s}.$$

SOLVE The distance needed to stop (eqn. 2.13) is

$$d_{min} = \frac{\Delta(v^2)}{2a} = \frac{0 - (19 \text{ m/s})^2}{2(-0.41 \text{ m/s}^2)} = 440 \text{ m}.$$

ANALYZE Whew! The trains don't crash! The choice of a moving reference frame turns the problem into a straightforward plug-in to eqn. (2.13).

EXERCISE 3.5 ♦♦ Solve Example 3.13 by working in the Earth reference frame. (*Hint:* Find the time when the *Flyer*'s speed is reduced to 22 m/s.) Which method do you think is easier?

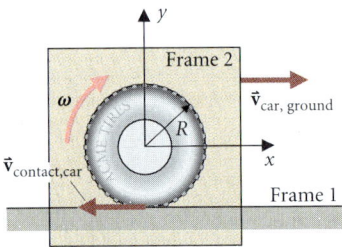

FIGURE 3.38
A wheel rolling without slipping. In the reference frame of the wheel's center, the point of contact moves to the left. The velocity of the wheel center with respect to the ground is to the right. The sum of these two velocities gives zero, the velocity of the point of contact with respect to the ground.

WE SHALL NEED THIS RESULT IN CHAPTER 12.

3.3.3 Rolling Without Slipping

We usually try not to leave a trail of smoking rubber behind our cars. That is, we want the tires to roll along the road without their surfaces skidding over the pavement.

> *Rolling without slipping* means that the relative velocity of the road and the piece of tire in contact with the road is zero.

■ Figure 3.38 shows how this requirement becomes a relation between linear speed of the car and the angular speed of the tire. The velocity of the car reference frame 2 with respect to the ground is

$$\vec{v}_{car,ground} = v_{car}\hat{i}.$$

In the car reference frame, the axle is at rest and the surface of the tire is in uniform circular motion. The velocity of the piece of the tire in contact with the road is backwards and is given by the formula for uniform circular motion (eqn. 3.9).

$$\vec{v}_{contact,car} = -\omega R \hat{i}.$$

The velocity of the contact point with respect to the ground, given by eqn. (3.13), is zero.

$$\vec{v}_{contact,ground} = 0 = \vec{v}_{contact,car} + \vec{v}_{car,ground} = v_{car}\hat{i} - \omega R \hat{i}.$$

So, the criterion for rolling without slipping is

$$v_{car} = \omega R. \tag{3.14}$$

■ **EXERCISE 3.6** ♦ What is the speed of the top of the tire with respect to the road?

Chapter Summary

Where Are We Now?

Using projectile motion as an example, we have learned how to fit models together to solve problems in kinematics. We are now ready to discuss what causes the accelerations we have learned to describe.

What Did We Do?

According to Galileo, projectile motion is freely falling motion in a moving reference frame. Neglecting air resistance, the projectile's horizontal velocity component remains constant, while the independent vertical component is subject to constant downward acceleration $\vec{g}$. So we analyze the vertical motion with the results of Chapter 2 for uniformly accelerated linear motion. Many common projectile problems may be solved with the following plan:

1. Determine a significant event in the problem.
2. Find an expression for the time when that event occurs.
3. Use that time to determine the unknowns.

The shape of the projectile's path is a parabola.

Uniform circular motion is another example of accelerated motion. The particle's speed is constant, but the direction of its velocity vector changes continuously. With the origin at the center of the circle, the angle of the position vector $\vec{r}$ increases uniformly with time, at angular speed $\omega \equiv d\theta/dt$. The particle's instantaneous velocity is

$$\vec{v} = (\omega R, \text{ tangent to the circle}).$$

Its acceleration is $\vec{a} = (\omega^2 R,$ toward the center of the circle).

If the speed v and radius of curvature R of a particle's noncircular path are known at a point, the acceleration component perpendicular to the velocity is v^2/R, the same relation as in circular motion.

If an object has velocity $\vec{v}_{p,2}$ measured in a reference frame, called 2, which itself moves at velocity $\vec{v}_{2,1}$ with respect to a frame called 1, then the object's velocity with respect to frame 1 is

$$\vec{v}_{p,1} = \vec{v}_{p,2} + \vec{v}_{2,1}.$$

If a wheel rolls without slipping, the point of contact between the wheel and the surface it rolls on is instantaneously at rest with respect to the surface. A wheel of radius R and angular speed ω rolls with linear speed $v = \omega R$.

Practical Applications

The model for projectile motion is needed for design of such things as gymnastic equipment, rescue line launchers for Coast Guard ships, baseball pitching machines, etc. Concepts of circular motion are applicable in the design of highways, centrifuges, and all kinds of rotating machinery. Relative motion of objects and the comparison of velocities between reference frames are crucial concepts in marine, air, and space navigation.

Solutions to Exercises

3.1 The distance Tell shoots with a launch angle θ_i is

$$\text{Range} = (v_i^2/g)\sin 2\theta_i.$$

The sine function has a maximum value of 1.0 when its argument $2\theta_i = 90°$. Thus, the maximum range is v_i^2/g and occurs for $\theta_i = 45°$.

3.2 After turning through 4π rad, the skier returns to her original orientation with skis 23° below the horizontal. She then has to turn 32° further to have her skis parallel to the snow slope. She has 2.1 s to complete the turn. So, the necessary average angular speed is

$$\omega = \frac{4\pi + (32°)(\pi/180°)}{2.1 \text{ s}} = 6.2 \text{ rad/s}.$$

3.3 We know from experience that the roller coaster reaches maximum speed at the bottom. It is no longer falling and speeding up, and it is not yet rising and slowing down. Instantaneously, its speed isn't changing, and it has no horizontal acceleration. The direction of its velocity *is* changing—from having a downward component to having an upward component. So, the coaster is accelerating upward with magnitude

$$|\vec{a}| = v^2/\text{radius} \approx (20 \text{ m/s})^2/(20 \text{ m}) = 20 \text{ m/s}^2$$

and $\vec{a} \approx (2g,$ upward).

3.4 We are to put the results of Example 3.11 into the definition of average velocity, using a time interval of 15 min (0.25 h). ■ Figure 3.39 shows the displacements and the corresponding velocities.

$$\vec{v}_{\text{av, air}} = \frac{\vec{D}_{\text{plane}}}{\Delta t} = \frac{(35 \text{ km, east})}{0.25 \text{ h}} = (140 \text{ km/h, east}).$$

$$\vec{v}_{\text{av, ground}} = \frac{\vec{D}_{\text{total}}}{\Delta t} = \frac{(37 \text{ km, 73° east of north})}{0.25 \text{ h}}$$

$$= (150 \text{ km/h, toward 073°}).$$

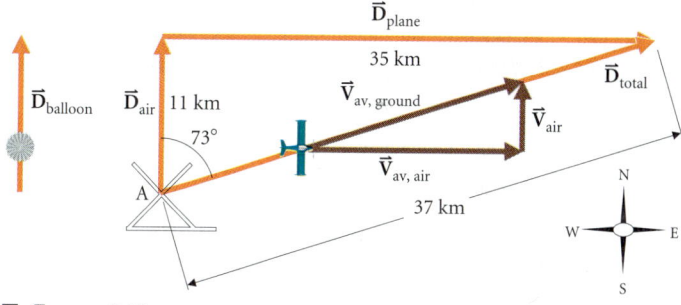

■ **FIGURE 3.39**
Average velocity is found by dividing displacement by time interval. Thus the average velocities of both the airplane and the air form a triangle similar to the triangle of their displacements.

The speed over the ground is greater than the speed through the air. The pilot is happy to have a *tailwind* helping her toward her destination.

3.5 The *Flyer*'s speed at time t is

$$v(t) = v_i + at = 41 \text{ m/s} - (0.41 \text{ m/s}^2)t.$$

This speed equals 22 m/s, the speed of the freight, at time

$$t = (-19 \text{ m/s})/(-0.41 \text{ m/s}^2) = 46.3 \text{ s}.$$

At this time, the position of the freight train, taking the origin at the point where the engineer sights the freight, is

$$x_f = x_{fi} + v_f t$$
$$= 500 \text{ m} + (22 \text{ m/s})(46.3 \text{ s}) = 1.52 \text{ km}.$$

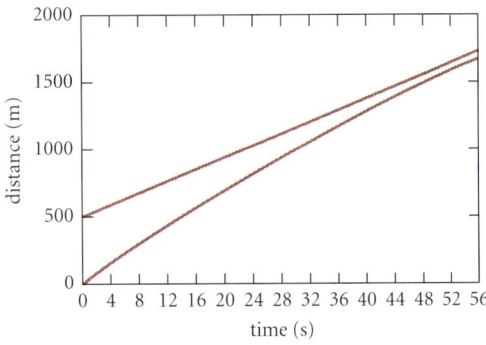

FIGURE 3.40
Graph showing the positions of the trains relative to the Earth reference frame. The Flyer's position graph never intersects that of the freight train, which proves that no collision occurs.

At the same time, the position of the Flyer is

$$x_F = x_{Fi} + v_{Fi}t + \tfrac{1}{2}at^2$$
$$= 0 + (41 \text{ m/s})(46.3 \text{ s}) - \tfrac{1}{2}(0.41 \text{ m/s}^2)(46.3 \text{ s})^2$$
$$= 1.46 \text{ km}.$$

Since the position of the Flyer is to the left of the freight when they have equal speeds, there is no collision. ■ Figure 3.40 shows graphs of the positions of both trains as functions of time. Since both 1.52 and 1.46 should be rounded to 1.5, we lose the answer in the significant figure rules with this method!

3.6 The velocity of a rubber particle at the top of the tire is in the same direction as the velocity of the car with respect to the road. Thus the velocity of the particle with respect to the road is the sum of two equal velocities; its speed is $2v_{car}$.

Basic Skills

Review Questions

§3.1 PROJECTILE MOTION

- Describe the vertical and horizontal motions of a projectile in the absence of air resistance.
- Describe the motion of a projectile when you are running at the same horizontal speed as the projectile.
- Outline a common plan for solving projectile problems. Does it always apply?
- What is the *range* of a projectile?
- Describe how to express mathematically (i) the requirement that two projectiles meet and (ii) the requirement that a projectile land on a given surface.
- For a given launch speed and target, how many different ways can you launch a projectile to strike the target? Do the different paths require the same time?

§3.2 CIRCULAR MOTION

- For an object in uniform circular motion, define *angular speed, period,* and *frequency*. What are the units of angular speed and of frequency?
- In uniform circular motion, what feature of an object's velocity is variable and what feature is constant?
- What is the *skinny triangle* approximation, and why is it useful for computing the speed of an object in uniform circular motion?
- In uniform circular motion, what feature of an object's acceleration is variable? What feature is constant? How does the magnitude of acceleration depend on angular speed and radius? On speed and radius?
- For an object undergoing uniform circular motion, how are the directions of its position, velocity, and acceleration vectors related?
- How does the velocity vary among different pieces of a solid object rotating at constant angular speed?
- Explain how the model of uniform circular motion is useful in describing motion on an arbitrary path.

§3.3 RELATIVE MOTION

- Explain why it is helpful to make use of different reference frames when solving navigation problems.
- What is the rule relating displacements of an object with respect to two different reference frames and the relative displacement of the frames? What is the similar rule relating velocities?
- Why would one consider transforming a problem into a different reference frame when it is possible to solve it in a given frame?
- What is the relation between the acceleration of an object as measured in one reference frame and its acceleration as measured in a second frame moving at constant velocity with respect to the first?
- Explain why two reference frames are useful in describing rolling without slipping. What is the relation between the linear and angular speeds of a wheel that is rolling without slipping?

Basic Skill Drill

§3.1 PROJECTILE MOTION

1. A long fly ball is hit into the outfield. Neglecting air resistance, decide when the acceleration of the baseball is the least. Explain why your choice is correct and the others are incorrect.
(a) While the ball is rising.
(b) When the ball is at its highest point.
(c) While the ball is moving downward.
(d) None of the above: the acceleration is constant.
(e) None of the above: the ball is in free fall so its acceleration is zero.

2. A projectile is launched from ground level at an angle of 30.0° above the horizontal with a speed of 20.0 m/s. (a) Where does it hit the ground? (b) What is its maximum height above the ground? (c) What is its minimum speed and where does it occur? (d) What is its velocity just before impact? (e) When does it reach maximum height? (f) At what time does it hit the ground? (g) When is its horizontal position 10.0 m away from the launch point? (h) At what times is the projectile 2.50 m above the ground?

§3.2 CIRCULAR MOTION

3. A go-kart travels around a circular track of radius 10.0 m in 10.0 s.
(a) What are the speed, frequency, and angular speed of the go-kart?
(b) What is the magnitude of its acceleration?

4. The flywheel in an automobile engine has a radius of $R = 0.10$ m. Find the speed of a point on the edge of the flywheel when the engine is operating at a rotation rate of 4.0×10^3 rpm.

5. An amusement park ride has a track shaped as shown in ■ Figure 3.41: two semicircles of radius 5 m, joined at the end of a diameter C and lying in a horizontal plane. A car travels along the track at 10 m/s. What are the velocity and acceleration of the car at points A, B, C, D, and E? (*Hint:* What is the curvature of the track at point C?)

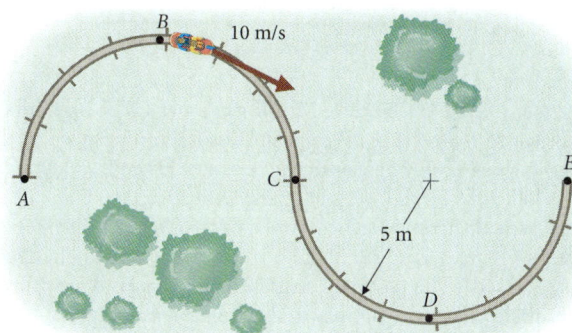

■ **Figure 3.41**

§3.3 RELATIVE MOTION

6. You are driving along the road at 30 m/s and spot a police car ahead of you traveling at 25 m/s. What is the relative velocity of the police car with respect to you? Is your answer different if the police car is behind you?

7. You and a friend set out together on a hike. You both walk 3.0 miles north. After lunch, you decide to walk to Clearlake, another 2.0 miles to the northeast, and your friend decides to walk to Hifalls, 1.0 mile to the northwest. When you arrive at your respective destinations, what is **(a)** your displacement? **(b)** your friend's displacement? **(c)** your friend's displacement relative to you? (Use Cartesian coordinates with x eastward and y northward.)

8. A man is able to walk at 3 m/s on a level surface. Hurrying to catch a plane, he strides along a passenger conveyor belt that moves at 2 m/s with respect to the airport floor. What is the man's speed with respect to the floor when he steps off the end of the conveyor? Back from his trip, the man childishly decides to walk in the *wrong* direction on the conveyor. What is his speed with respect to the floor?

9. Raindrops are falling straight down at a speed of 3 m/s, and you are running at a speed of 7 m/s. At what angle do the raindrops hit you in the face?

10. On a day when the wind is blowing toward the northeast at 30.0 km/h, a plane heads northwest at an airspeed of 60.0 km/h. What is the plane's velocity with respect to the ground?

Questions and Problems

§3.1 PROJECTILE MOTION

11. ❖ Two brothers each hit a golf ball at the same speed, but the first brother hits his ball at an angle of 60° with the horizontal, and the second brother hits his at 30° with the horizontal. Which ball goes further? Which one hits the ground first? (Neglect air resistance.)

12. ❖ A mother and daughter are playing golf. They each make shots at the same angle, but the daughter gives her ball an initial speed 10% greater than the mother's. Which ball goes further? Which one hits the ground first? (Neglect air resistance.)

13. ♦ A projectile fired from ground level at 25 m/s hits the ground 31.25 m away. At what angle was it fired?

14. ♦ A projectile fired from ground level at 22.5° above the horizontal hits the ground 141 m away. With what speed was it fired?

15. ♦ As a research rocket is launched from the White Sands Missile Range, the launch tower begins to topple over. The rocket motors are immediately shut off. Instruments on board the rocket show that it had reached a speed of 1.00×10^2 m/s when the motors were shut off. The rocket is recovered 1.00 km downrange. What was the angle of the launch tower when the rocket left it?

16. ♦♦ A golfer can drive a ball a maximum distance of 110 m over flat terrain. With what speed does the ball leave the golf club? (Ignore effects due to the air.)

17. ♦♦ Whammo the Magnificent is launched from a cannon and is to land in a net 10.0 m below the launch point (■ Figure 3.42). If Whammo's initial velocity components are 20.0 m/s upward and 10.0 m/s horizontally, how long is he in the air? Where should the net be placed? Does he miss the wall?

18. ♦♦ A ski jump is designed so that the skier leaves the jump moving horizontally and then lands $h = 10.0$ m below and $D = 20.0$ m beyond the edge of the jump. Find **(a)** The time Δt the skier will remain in the air. **(b)** The speed v_i required to travel a distance D horizontally. **(c)** The vertical component of the skier's velocity upon landing. **(d)** The angle θ of the landing ramp so that the skier lands parallel to the ramp.

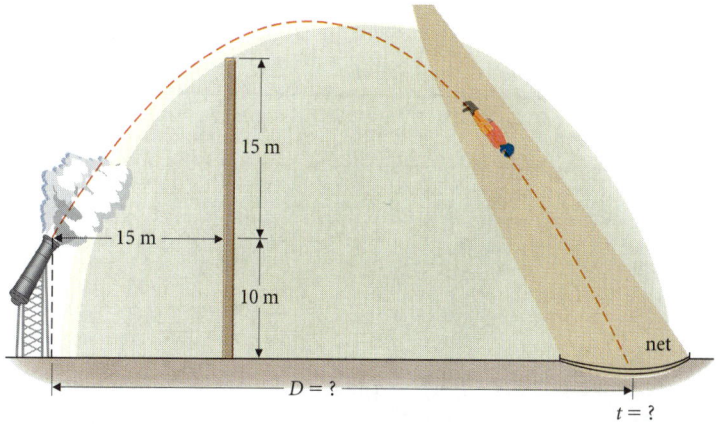

■ **Figure 3.42**

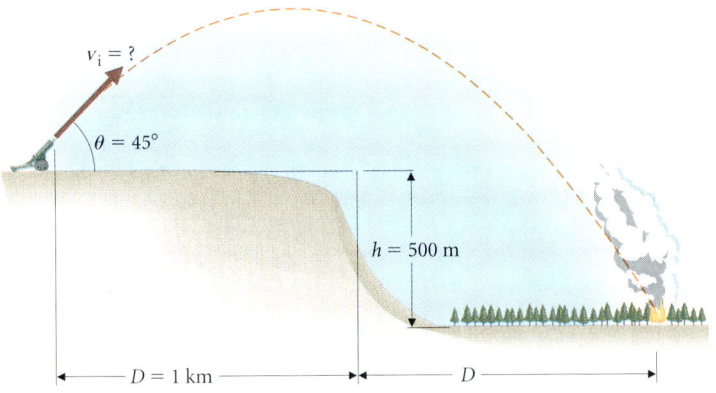

FIGURE 3.43

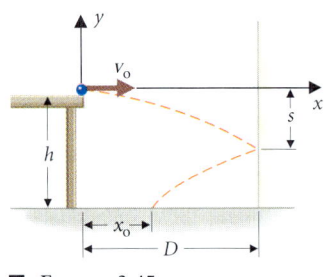

FIGURE 3.45

19. ◆◆ A cannon is used by the Department of Forestry to launch containers of fire retardant chemical onto a fire (■ Figure 3.43). The cannon is a distance $D = 1.0$ km from a cliff $h = 0.50$ km high, and the fire is 1.0 km from the base of the cliff. If the retardant capsules are launched at an angle of 45°, what initial speed v_i is necessary?

20. ◆◆ A cannon is used by a ski resort to fire artillery shells at a cornice hanging over the ski slopes (■ Figure 3.44). (If cornices are allowed to grow, there is a danger of avalanche.) The shells are launched at an angle of 45° above the horizontal at a distance $D = 1.0$ km from the cornice. What initial speed v_i is necessary if the containers are to hit the snow at a height of $h = 0.50$ km above the ground?

21. ◆◆ In a circus act, Whammo the Magnificent is propelled from a cannon through a ring of fire. The center of the ring is 20.0 m horizontally from the cannon and 10.0 m above it. If Whammo is launched at a 45° angle, what initial speed is necessary to pass through the center of the ring? Show that Whammo is moving horizontally as he passes through the ring.

22. ◆◆ In a circus act, a clown jumps from a "burning building" 10.0 m high into a net 3.0 m from the bottom of the building. If the clown jumps upward at an angle of 30° from the horizontal, what initial speed is necessary for her to land in the net?

23. ◆◆ In the movie *The Gods Must Be Crazy*, a bottle is thrown from a cliff overlooking the Indian Ocean. If the cliff is 0.50 km high, and the bottle is thrown horizontally at an initial speed of 10.0 m/s, what are the bottle's position and velocity 0.50 s later? Neglecting effects of air resistance, find the distance from the bottom of the cliff at which the bottle hits the water. Find its velocity just before impact.

24. ◆◆ A tennis player standing 12 m from a net 1.0 m high hits a ball so as just to clear the net. What must be the initial speed of the ball if it is hit at a 10.0° angle and at a height of 1.0 m?

25. ◆◆ A ball rolls off a horizontal table with speed v_o. The ball bounces elastically from a vertical wall a horizontal distance D from the table. (■ Figure 3.45) ("Elastically" means that v_y is unchanged and v_x is reversed.) The ball then strikes the floor a distance x_o from the table, as shown in the figure. **(a)** Write expressions for x- and y-coordinates of the ball as functions of time (valid for the period before the ball hits the wall). **(b)** Find s and the value of v_y when the ball hits the wall. **(c)** Find a formula for x_o and solve for v_o such that $x_o = 0$.

26. ◆◆ In a circus act, clowns are shot from a cannon at a constant speed $v_o = 10.4$ m/s. The angle of launch may be varied. What is the maximum distance D that a clown may travel, if the landing pad is at the same height as the mouth of the cannon? The circus master wants to shoot two clowns from his cannon, one after the other, and have them land on the same landing pad a distance $D/2$ away at the same time. How can it be done? How long should he wait to launch the second clown?

27. (a) ◆◆ In an English carnival game, you throw a ball to hit a coconut on a stand 1.0 m above the point at which you release the ball and 4.0 m away (■ Figure 3.46). If you throw the ball at 10.0 m/s, at what angle θ should you throw it? **(b)** ◆◆◆ Show that the vertical velocity of the ball when it hits the coconut is nearly zero for the shallow launch and downward for the steep launch.

28. ◆◆◆ Whammo the Magnificent is to be propelled from a cannon at a speed of 20 m/s and at an angle of 45°. He is to land on a net that

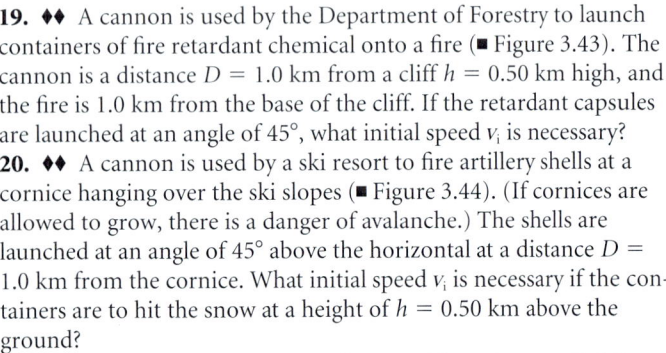

FIGURE 3.44

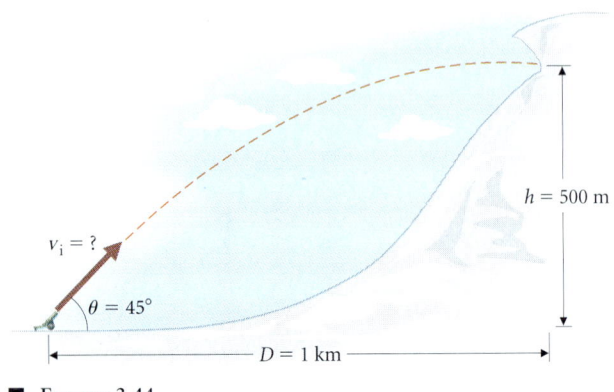

FIGURE 3.46

is stretched upward at an angle of 30° from a point 10 m below the cannon's mouth. Where does Whammo hit the net?

29. ♦♦♦ If the monkey in Example 3.4 stays in the tree on a branch 2.0 m high, at what angle must the zookeeper throw the banana (at 8.0 m/s) for it to reach the monkey 3.0 m away?

30. ♦♦♦ A tennis player standing 12 m from a net 1.0 m high hits a ball so as just to land inside the back line, 12 m beyond the net. The initial height of the ball is 0.50 m. Find the initial velocity and angle required for the ball just to clear the net.

31. ♦♦♦ In Example 3.5, a skier leaves a 23° ski slope at 11 m/s and lands on a 55° slope below. Use coordinates y' and x' perpendicular and parallel to the 55° ski slope. Show that neither the x' nor the y' component of the acceleration is zero. What is the skier's maximum distance from the snow and her velocity component $v_{y'}$ perpendicular to the snow surface at landing?

32. ♦♦♦ On the Moon, mail may be sent in cannisters launched by mass drivers. The settlements of Rimtown and Crater City are separated by a steep cliff (■ Figure 3.47). The Rimtown mass driver has a constant launch speed v_i and a maximum range $2D/\sqrt{3}$, where D is the horizontal distance to the cliff edge. (*Range* here still means the horizontal distance a projectile would go between points at the same height.) What is the launch speed from the Rimtown mass driver? At what minimum distance d from the cliff can the postmaster of Crater City locate the reception point for mail from Rimtown?

33. ♦♦♦ The postmaster at Rimtown launches mail packages for Copernicus Spaceport at 30°. Thieves from Renegade Outpost plan to intercept a mail cannister at the top of its trajectory by launching a cannister-catching projectile from their stronghold in the wilderness of Mare Malefactorum (■ Figure 3.48). The thieves can choose a launch speed of their catcher of up to twice the launch speed v_i of the postmaster's cannon at Rimtown. Find an expression for the launch angle they should use as a function of launch speed. Is it possible for them to wait until the postmaster shoots and still carry out their evil scheme?

34. ♦♦♦ Equation (3.3) gives the equation for a projectile trajectory with the coordinate origin at the launch point of the projectile. Rewrite the equation in terms of coordinates with the origin at the high point of the trajectory, and show that the equation has the form $y'(x') = -\text{const}(x')^2$, the elementary form for a parabola.

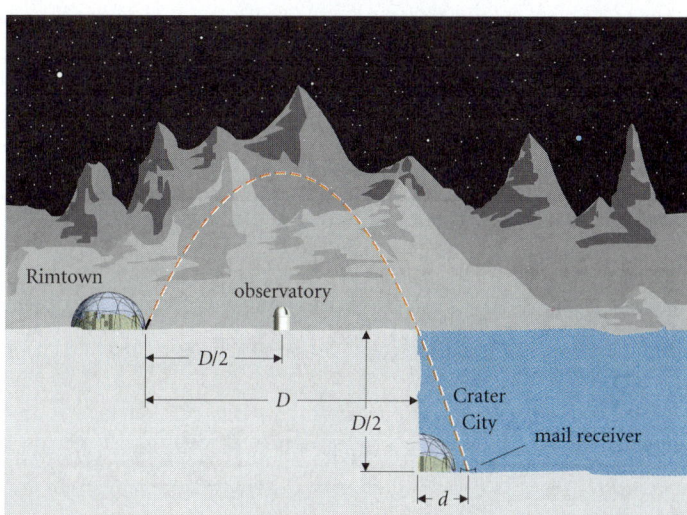

■ FIGURE 3.47

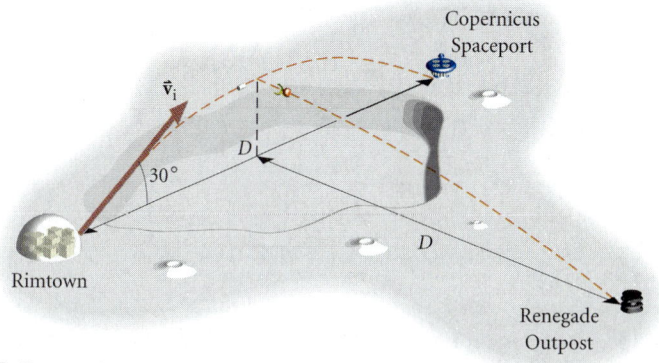

■ FIGURE 3.48

§3.2 CIRCULAR MOTION

35. ❖ The angular speed of San Francisco, due to Earth's rotation, is (a) greater than, (b) less than, or (c) the same as the angular speed of Nome, Alaska. Explain your reasoning.

36. ❖ Explain why the following sentence is *false:* The acceleration of a particle undergoing uniform circular motion is always the same.

37. ❖ Explain why the following sentence is *false:* The *average* acceleration of a particle during one period of uniform circular motion equals the particle's speed squared divided by the radius of the circle.

38. ❖ Explain why the following sentence is *false:* The *average* speed of a particle during one period of a uniform circular motion equals the magnitude of the particle's *average* velocity during the same time interval.

39. ❖ Explain why the following sentence is *false:* As a wheel rotates uniformly, the velocity of a point on its rim remains constant.

40. ❖ Test cars A and B are each driven at constant speed around a circular track of radius R. If A requires half as much time as B to go once around the track, then the magnitude of A's acceleration is (a) the same as that of B. (b) twice that of B. (c) half that of B. (d) four times that of B. (e) one fourth that of B.
Explain your reasoning.

41. ❖ A particle undergoes uniform circular motion (■ Figure 3.49). Which vector best represents the particle's *average* velocity during the time when it moves from point A to point B? Which vector best represents its average acceleration? Explain your reasoning.

42. ❖ (a) Two go-karts, A and B, travel around two circular tracks of radius 5 m and 10 m, respectively, each in 10 s. Does A or B have the greater acceleration? Which has the greater speed? (b) If instead the go-karts travel around the two tracks at the same speed, which has the greater acceleration, A or B? Which takes less time to complete one revolution?

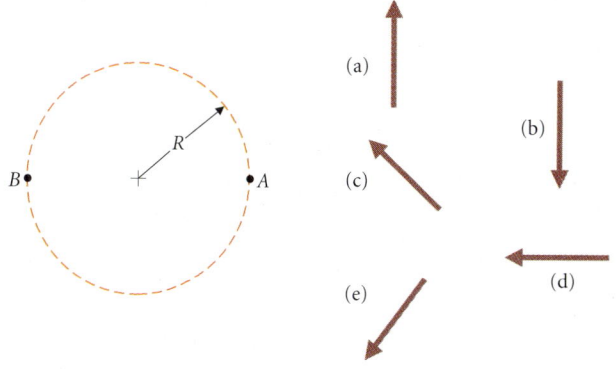

■ FIGURE 3.49

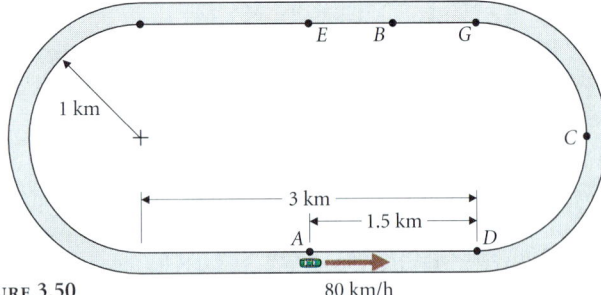

FIGURE 3.50

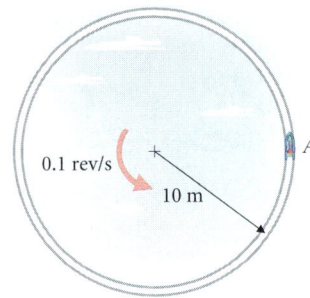

FIGURE 3.51

43. ♦ A toy train moves around a circular track at 0.50 m/s. If the radius of the track is 1.0 m, how long does it take to complete one circuit? What is the angular speed ω of the train's motion?

44. ♦ A machine contains an arm 0.30 m long, which rotates about a pivot at one end with an angular speed of 1.5 rad/s. What is the average speed of the revolving end of the arm? What is the rotation frequency in revolutions per minute (rpm)? In Hz?

45. ♦ The Train Grand Vitesse cruises through the French countryside at 216 km/h. For passenger comfort, horizontal acceleration is limited to $0.050g$. What is the smallest allowable radius of a curve in the tracks?

46. ♦ A particle undergoes uniform circular motion on a circle of radius 2.0 m and with angular frequency $\omega = 10.0$ rad/s. Draw the particle's velocity vectors when it is at two points A and B, 90° apart ($\Delta t_{AB} = \pi/2\omega$). Draw a diagram showing $\Delta \vec{v} \equiv \vec{v}_A - \vec{v}_B$ and compute the particle's *average* acceleration during the time interval Δt_{AB}.

47. ♦ A car moves counterclockwise on the race track in ■ Figure 3.50 at a constant speed of 80 km/h. What is the instantaneous velocity of the car at point C? What is its instantaneous acceleration at C? At D?

48. ♦ A car traveling along the highway at a constant speed of 85 km/h negotiates a turn with a radius of curvature 0.75 km. What is the car's acceleration?

49. ♦ The propeller on a Cessna 210 aircraft has a diameter of 2.08 m. What is the speed of the propeller tip on take-off when the propeller rotates at 2625 rpm? Compare your answer with the speed of sound at sea level, about 330 m/s.

50. ♦♦ A particle moves around a circle of radius $r = 0.14$ m in the x–y plane. Its angular speed is $\omega = 3.7$ rad/s. It is on the x-axis at $t = 0$. (The origin is at the center of the circle.) Write expressions for the particle's position, velocity, and acceleration at $t = 0.40$ s and $t = 1.3$ s.

51. ♦♦ A particle is moving on a circle with velocity

$$\vec{v} = -(1.0 \text{ m/s})\{\sin[(4.5 \text{ rad/s})t]\hat{i} + \cos[(4.5 \text{ rad/s})t]\hat{j}\}.$$

What is its angular speed? Write expressions for its position and acceleration as functions of time.

52. ♦♦ A sled is moving on a circular track of radius 10.0 m on the snow at 0.10 rev/s. Suddenly the sled is stopped at point A in ■ Figure 3.51. A package that was on the sled continues to move at the velocity the sled had just before it was stopped. Describe the position of the package relative to the center of the circle after 2.0 s.

53. ♦♦ The space shuttle circles the Earth in 89.5 min in an orbit of radius 6.63×10^6 m. What is the shuttle's acceleration? Compare this value with the acceleration due to gravity at the Earth's surface. If the radius of Earth is 6.38×10^6 m, find n in the following expression:

$$\left(\frac{r_{\text{Earth}}}{r_{\text{shuttle orbit}}}\right)^n = \frac{a_{\text{shuttle}}}{g}.$$

§3.3 RELATIVE MOTION

54. ❖ One day, just after a storm, the river near John's house is flowing at 5 km/h. John can row his boat at 3 km/h with respect to the water. His neighbor warns him that he can't get across because the river is moving faster than he can row. Is his neighbor correct? Why or why not?

55. ❖ Railroad wheels have a flange to keep the wheels from sliding sideways off the rails (■ Figure 3.52). The wheel rolls without slipping. Discuss qualitatively the path of a point on the rim of the flange with respect to the ground, and sketch the path.

56. ❖ Do you get wetter by running or by walking a short distance in a rainstorm (cf. Problem 9)? Does your answer change if a wind is causing the rain to fall at an angle?

57. ❖ A sailboat moves substantially more slowly through the water when close-hauled (heading nearly opposite the relative wind) than when running with the wind (heading in the same direction as the wind). Explain why the wind seems to blow much harder when the boat is close-hauled and much more gently when the boat is running with.

58. ♦ Snowflakes are falling straight down at 0.50 m/s in the calm air between the trees. In a nearby open meadow, you observe that the snowflakes are moving at 65° to the horizontal. What is the wind speed in the meadow?

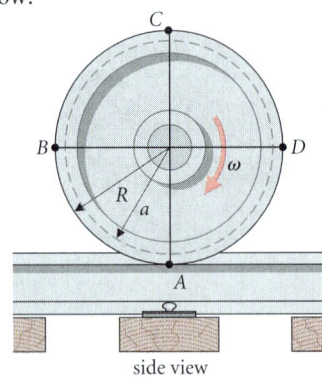

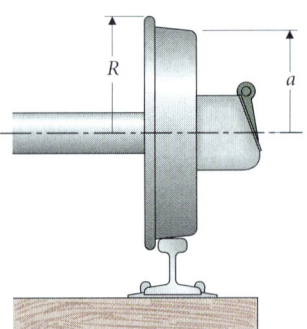

FIGURE 3.52

59. ♦ The current in a river is flowing northwest at a speed of 10.0 km/h, and a boat is observed, moving southward at 20.0 km/h with respect to the shore. What is the boat's speed with respect to the water?

60. ♦ Joe cruises northward in his speedboat for 35 min at 19 m/s, then turns eastward for 9.0 min at 38 m/s and, finally turns northeastward into a narrow estuary, slowing to 5.0 m/s for 21 min. During this entire episode, Lucia cruises northward in her boat at 21 m/s. What is Lucia's average velocity relative to Joe?

61. ♦♦ John, who can row at 3.0 km/h relative to water heads his boat directly across a river 1.0 km wide, flowing at 2.0 km/h. How far downstream from its starting point is the boat when it reaches the opposite shore? How far has the boat moved with respect to the shore? How much time does it take to get across?

62. ♦♦ On a day when the wind is from the west at 7.5 m/s, a cruise liner is steaming northward at 15 m/s. What wind velocity, relative to the ship, would be measured by onboard instruments?

63. ♦♦ A car has wheels of radius 0.33 m. When the car is traveling at 82 km/h, what is the rotation rate of the wheels in revolutions per minute (rpm)?

64. ♦♦ A blimp with an airspeed of 42 km/h heads in direction 060° (clockwise from north). If the wind blows over the water at 18 km/h toward direction 180°, what velocity must a speedboat have to remain directly under the blimp?

65. ♦♦ The town of Clinch City is a distance $D = 4.00 \times 10^2$ km north of Aces, while Badwater is $D\sqrt{2} = 566$ km northwest of Aces. On a day when the wind is blowing toward the northwest at 60.0 km/h, an airplane flies from Aces to Badwater and then to Clinch City. If the airplane's airspeed is $v = 3.00 \times 10^2$ km/h, find the total time for the trip.

66. ♦♦ What are the velocities with respect to the ground and accelerations of the railroad wheel in Figure 3.52 at the points labeled A, B, C, and D? Draw vectors illustrating your results for the velocities. Does it matter whether you find accelerations with respect to the ground or with respect to the center of the wheel?

67. ♦♦ **(a)** A pilot wishes to fly from San Francisco to Sacramento, 138 km northeast. The wind is blowing from the north at 24 km/h, and the plane has an airspeed of 240 km/h. What heading must the pilot take, and how long will it take to get to Sacramento? **(b)** How long does it take the pilot to get back to San Francisco? Compare the round-trip time with the time required with no wind.

68. ♦♦ The skipper of a racing sailboat measures the boat's speed through the water at 8 km/h; the boat's heading (the direction it is pointed) is toward 060° (clockwise from north). After 30 min, the captain observes that the boat has moved 4 km directly toward a lighthouse bearing due north (i.e., in direction 000°). Use these data to make a diagram showing the velocity $\vec{v}_b$ of the boat through the water and $\vec{v}_g$, the velocity of the boat with respect to the lighthouse. Determine $\vec{v}_w$, the velocity of the water with respect to the lighthouse.

69. ♦♦ Mei-Lin, a champion swimmer, is able to swim at 3.0 m/s for 2.0 h before tiring. On a vacation at Paradise Island, she decides to swim directly downstream to Treasure Cove and back. The current in the river flows at 0.50 m/s toward Treasure Cove, which is 10.0 km downstream. Is Mei-Lin's decision wise? Is your answer different if the current flows in the other direction at the same speed?

70. ♦♦ An aircraft flies from San Francisco to New York, a distance of 5700 km, in 5.5 h. The airspeed of the plane is 8.00×10^2 km/h. How large a tailwind does the aircraft experience that day? Assuming that the wind doesn't change, how long does it take to fly back to San Francisco?

71. ♦♦♦ Chuck steers his airplane due north over the ground and directly away from a radio beacon. During an hour the measured distance from the beacon increases by 72 nautical miles. Meanwhile, the plane's heading is toward 20° east of north, and the airspeed indicator reads 91 knots (nautical miles per hour). What wind velocity should Chuck report to Air Traffic Control?

Additional Problems

72. ❖ Black Bart, having stolen the famous god's eye sapphire from a safe on the train, is trapped on the roof of the caboose. Just as the train enters a sharp turn, Bart throws the sapphire directly upward. Explain why the jewel does not land back on the train, but is caught instead by Bart's partners beside the track.

73. ♦♦ Just as you throw a package horizontally from a cliff 20.0 m high, your friend, who is rowing toward the cliff at 1.0 m/s, is 10.0 m away. With what velocity should you throw the package if it is to land in your friend's boat?

74. ♦♦ **(a)** A pitching machine projects two baseballs directly upward. The second ball is launched 0.500 s after the first, which is still rising. Immediately after the second ball is launched, is the distance between the balls increasing, decreasing, or not changing? Is their relative velocity changing? Explain your reasoning. **(b)** If the launch speed of each baseball is $v_o = 35.0$ m/s, determine when and where they collide.

75. ♦♦ At the beginning of the movie *The Gods Must Be Crazy*, an airline pilot flying high over the Namibian desert at 250 m/s drops a Coke bottle. Assuming air resistance produces a horizontal acceleration of $5.0g$ without affecting the vertical acceleration of the bottle, find its position and velocity with respect to the plane and with respect to the ground 1.4 s after it is dropped.

76. ♦♦ In a carnival game you are to blow a small bean from a tube and win a prize if the bean lands in a cup; the value of the prizes increases rapidly with the cup's distance from you. The cup is at the same height as the tube's mouth. Sneakily, the ventilation system is set up to produce a draft that blows any bean above height h off course. If you can blow the bean at speed v_o, how far away is the most valuable prize you can win without correcting for the draft?

77. ♦♦ A bicycle is moving along a level street at 5.0 m/s. If the wheels have a radius of 0.50 m, what is their angular speed? Find an expression for the speed of a point on the rim, with respect to the ground, as a function of time. (Assume the point is in contact with the ground at $t = 0$.)

78. ♦♦ Joe is lounging in his speed boat, hoping that his friend, Lucia, will also show up on the lake. At $t = 0$, she appears from behind the headland 0.50 km due north and heading due east at 25 m/s. Joe's boat can maintain an acceleration of 0.50 m/s². In what direction should he head to intercept Lucia?

79. ♦♦ David, who is 1.6 m tall and is terrified of the approaching giant, Goliath, puts a rock in his sling and swings it with a frequency of 10.0 Hz in a horizontal circle of radius 0.20 m. To his chagrin, the rock lands at the feet of the amused giant. How far away from David was the giant when the rock was released?

80. ♦♦ Two airplanes are flying in the same direction. The first is at an altitude of 1000 m and has a speed of 80 m/s. The second is at an altitude of 500 m and has a speed of 60 m/s. A passenger in the first airplane drops a bag of flour so as to hit the second airplane. What must be the location of the first airplane with respect to the second when the bag is dropped?

81. ♦♦ Joe Skyswimmer is in circular orbit around planet Barsoom at a radius of 1.00×10^4 km with a speed of 1.00×10^4 m/s. He wishes to intercept the spaceship of Lily Organic, 1.00×10^2 km ahead of him and also traveling at 1.00×10^4 m/s in the same orbit. Joe plans to remain on the circular path and to reach Lily within 2.00 min to help ward off an attack by Dirk Vapour. What component of acceleration parallel to his track is required? What is his total acceleration just after he starts the engines? What is his total acceleration just before he reaches Lily?

82. ♦♦♦ James Bond plans to jump from a stationary blimp and land on the SS *Smersh* using a paraglider that glides downward at an angle of 10° from the horizontal and at a speed of 10 m/s. From what altitude and distance should Bond jump to arrive in 5 min if the *Smersh* is cruising at 5 m/s directly away from the blimp? Directly toward the blimp? Perpendicular to the direction to the blimp?

83. ♦♦♦ Petra, who can run at 10.0 m/s is playing ball with her trusty dog Pinscher, who can run at 15.0 m/s. They start from the positions shown in ■ Figure 3.53 just as the ball bounces from the wall at 5.0 m/s. Who gets to the ball first?

84. ♦♦♦ A ball is projected directly upward with an initial speed v_o, bounces elastically from a roof inclined at 45°, and later strikes a table a horizontal distance x_1 from its starting point (■ Figure 3.54). (In an elastic collision, the velocity component parallel to the surface is unchanged and the component perpendicular to the surface is reversed.) **(a)** Write expressions for y and v_y as functions of time (valid for the period of time before the ball hits the roof). **(b)** Find v_y when the object hits the roof, and find the velocity components after impact. **(c)** Find a formula for x_1 and solve for v_o such that $x_1 = 2D$.

85. ♦♦♦ A hard steel ball is dropped from rest at position $y = 2h$ and bounces from a hard surface tilted at angle ϕ from the horizontal (■ Figure 3.55). In the collision, the ball's velocity component perpendicular to the surface is reversed, and its velocity component parallel to the surface is unchanged. Find x_1—that is, where the ball strikes the ground ($y = 0$).

86. ♦♦ A pitching machine throws a baseball at speed v_o and angle θ to the horizontal (■ Figure 3.56). When the ball strikes the vertical wall at $x = D$, its horizontal velocity component is reversed and its vertical velocity component is unchanged. If the ball is to land in a basket at height h above the launch point, how must v_o and θ be related?

87. ♦♦♦ An emergency rescue package is to be dropped alongside a sinking yacht by a Coast Guard plane that dives at 45° to the horizontal at a speed of 40.0 m/s. The minimum altitude at which the plane can safely release the package is 0.100 km. When the plane is at the proper point to release the package, the yacht appears at an angle ϕ below the plane's longitudinal axis. Find ϕ. (Neglect any effects of air resistance on the package.)

88. ♦♦♦ Tullio, skiing down a slope 30° from the horizontal, is carefully controlling his speed at 15 m/s while turning in a circle of radius 45 m around a tree. Give vertical and horizontal components of Tullio's velocity and acceleration at the top, middle, and bottom of the turn. (Careful, this isn't a 2-d question!)

89. ♦♦♦ A pilot needs to plan a flight starting at Atlantis and visiting Bodie (1.00×10^2 km due north of Atlantis) and Camelot (141 km northwest of Atlantis). The wind is blowing due east with speed 70.7 km/h. The plane's speed through the air is 1.00×10^2 km/h. If the pilot wishes to complete both visits in the least time, should Bodie or Camelot be visited first? (*Hint:* Let 100 km = D. Then 141 km = $D\sqrt{2}$. Similarly, if v_p = 100 km/h, then $v_w = v_p/\sqrt{2}$.)

90. ♦♦♦ Coast Guard radar detects a suspicious boat 12 km to the east moving at 35 km/h, northwest. A cutter capable of making 55 km/h is dispatched to intercept the boat. What direction should the cutter head? How long does it take to intercept the boat?

91. ♦♦♦ A sailboat can sail upwind in a maneuver called tacking. Suppose the boat can maintain a speed through the water of $|\vec{v}_s| = v_m \sin\theta$, where θ is the angle between the direction of the boat and

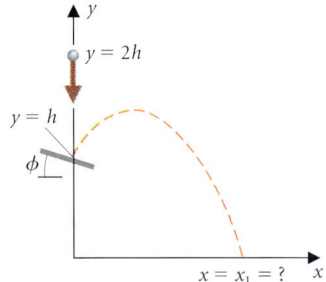

■ FIGURE 3.55

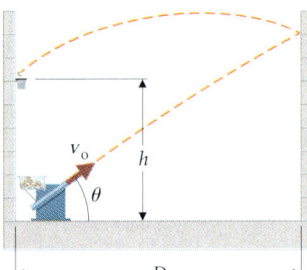

■ FIGURE 3.56

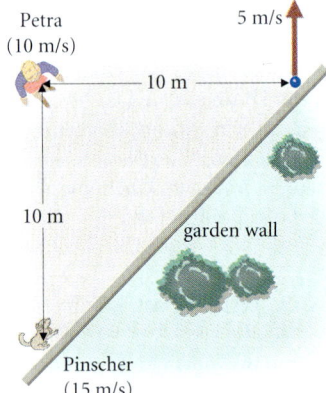

■ FIGURE 3.53

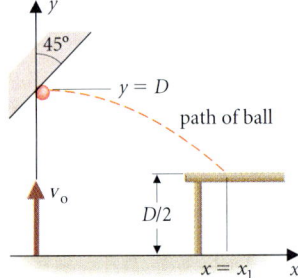

■ FIGURE 3.54

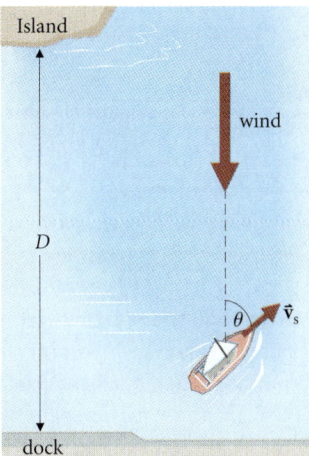

■ **FIGURE 3.57**

the direction of the wind, and v_m is a constant. At what angle should the sailboat tack to travel a distance D upwind in minimum time? (See ■ Figure 3.57)

Computer Problems

92. (a) Using the dimensionless variables $Y = gy/v_i^2$ and $X = gx/v_i^2$, plot the trajectory of a projectile for a launch angle of $-23°$. Then redo the first part of Example 3.5 by plotting the ski slope on the same graph and finding the intersection. (You may make the plot without a computer if you prefer.) **(b)** Introduce a dimensionless velocity variable $\vec{V} = \vec{v}/v_i$, and find a formula for its components as a function of X. Re-do the second part of Example 3.5 by substituting the result of part (a) for X to find V for the skier at impact.

93. Use a spreadsheet program to calculate the position of a projectile as a function of time. **(a)** With an initial speed of 14 m/s and a launch angle of 45°, calculate the time that the projectile is airborne. Using a timestep of 0.1 s, fill a column of your spreadsheet with times between 0 and your calculated time. Using eqns. (3.1a and b), fill four more columns with the values of v_x, v_y, x, and y at each time. **(b)** Next find a numerical algorithm that allows you to compute the position from the values of v_x and v_y. Hint: $y(t + \Delta t) = y(t) + v_y \Delta t$. Which value of v_y gives values for y closest to the values calculated from the algebraic formula? Try $v_y(t)$, $v_y(t + \Delta t)$, and $[v_y(t) + v_y(t + \Delta t)]/2$. **(c)** Now assume that air resistance contributes an acceleration $\vec{a}_{ar} = -\alpha \vec{v}$. Try $\alpha = 0.1$ /s. Set up columns in the spreadsheet for a_x, a_y, v_x, v_y, x, and y. Calculate a_x and a_y using values of v_x and v_y from the previous timestep. Then use the algorithm you found in part (b) to calculate v_x, v_y, x, and y at the next timestep. You may have to modify the algorithm slightly for the first one or two timesteps, until you have enough "history." How does air resistance change the maximum height that the projectile reaches? What effect does it have on the range? Try different values of α. How large can air resistance be without changing the position noticeably?

94. Modify eqn. (3.3) to allow for a nonzero height y_i at launch. Find the horizontal distance traveled by the projectile before it reaches $y = 0$. (Your answer should be the solution to a quadratic equation. Express your answer in terms of the dimensionless variable $\alpha = gy_i/v_i^2$. Evaluate α for $y_i = 15$ m and $v_i = 22$ m/s. Use a spreadsheet program to calculate the horizontal distance traveled for launch angles between 0 and 90° in 1° increments. At what angle is the horizontal distance maximized?

Challenge Problems

95. Light from the stars arrives near Earth traveling at speed $c = 3.0 \times 10^8$ m/s with respect to the Sun, regardless of the direction from which it arrives. It is the velocity of light from a star relative to the Earth that determines the apparent direction of the star. Use the relative velocity law (only qualitatively correct applied to light) to estimate how the apparent direction of stars depends on the angle between their true direction and Earth's velocity. How would this aberration vary throughout the year, and how might you use it to measure the speed of Earth in its orbit? (This effect was first noticed by the English astronomer James Bradley in the eighteenth century.)

96. A projectile is to be launched so as to pass over a hemispherical mountain tangent to the highest point on the mountain (■ Figure 3.58). Show that this can only be accomplished if the projectile is launched at a distance D from the center of the mountain such that $D \geq R\sqrt{2}$. Find the necessary velocity at launch as a function of D. (Hint: Consider the radius of curvature of the trajectory at the top and how it compares with the radius of the sphere.)

97. A gymnast runs toward a padded horse, leaps to land on the horse, pivoting on her hands to enter a double somersault and finally to land on a padded mat. The gymnast's path is modeled by the two projectile paths shown in ■ Figure 3.59. **(a)** If she lands with a speed of 8.0 m/s at an angle of 62° below the horizontal, as in the figure, at what angle ϕ does she leave the horse? **(b)** At what horizontal distance Δx_2 from the horse does she land? **(c)** If she leaves the ground at 8.0 m/s and lands on the horse at an angle of 45°, as shown, at what distance Δx_1 from the horse and at what angle ψ must she jump? **(d)** Assuming the landing takes 0.15 s and the pivot on the horse takes 0.30 s, what is her average acceleration during each maneuver? **(e)** What is her average angular speed during each of the two projectile phases of her motion?

98. ❖ On a night when the Moon is nearly overhead at Mexico City, it is near the horizon at the same time for observers in Wellington, New Zealand. Observers at both locations observe the direction to the Moon against the background stars and later find that their measurements differ by approximately 1°. Explain why this is the case and use the skinny triangle approximation to compare the distance to the Moon with the Earth's radius. The Moon, seen directly overhead, appears to be $\approx \frac{1}{2}°$ in angular diameter. What is its radius compared with that of the Earth?

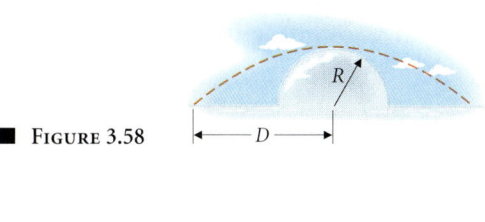

■ **FIGURE 3.58**

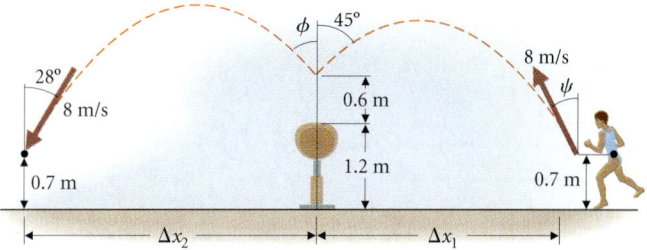

■ **FIGURE 3.59**

INTERLUDE 1
Solving Physics Problems

I1.1 Top-Down Reasoning

To understand how physical systems behave, you must learn both the principles and facts of physics and the methods for applying those ideas. To help you with this task, we outline here an approach to problem solving that uses proven techniques organized so they will work most effectively for you. We apply this approach in examples and exercises throughout the text; you should find it useful in building your personal approach to this subject.

Studies of how people approach physics problems show definite differences between beginners and experienced physicists. Beginners typically attempt to build details and learned equations into a big picture. With experience, a physicist uses a *top-down* approach, analyzing the big picture first and getting down to details last. The beginner's approach of manipulating equations often works deceptively well in simple exercises but fails frustratingly with even moderately complex problems. *Real-world* problems are almost always complex: they involve several different physical principles operating simultaneously and are understood by a process of progressive simplification. Our goal is to learn how to tackle real-world problems, so we need to make the top-down method work.

Top-down reasoning occurs in a great variety of activities. If you were designing a new house, for example, you would probably begin by deciding on your needs and your budget and by sketching the floor plan and exterior appearance. Only later would you get down to designing windows or calculating the amount of lumber required. Computer programs are another good example, typically using a main program to interact with the user and calling on subroutines to solve simultaneous equations or to prepare a graph for printing. These examples illustrate a common plan: First gain a clear idea of the overall goal. Then outline the tasks needed to meet that goal. Finally, determine in detail how to complete each task.

Computer subroutines or building codes for windows provide a good model for how to organize your physics knowledge. Designed to carry out a specific sort of task, each is made to be flexible so the task can fit into a larger whole. Similarly, with each new set of concepts and methods in physics, first learn how to use each idea to solve problems in which it is the only new concept. Then make the idea into a *subroutine* by asking what kind of situation a method applies to, what kind of information is needed to apply it, and what kind of answers it returns. Here's an example.

Subroutine: Uniform Circular Motion

Applies to motion on a curved path. Relates instantaneous speed, radius of the curve, and the acceleration component toward the center of the curve.

$$v = \omega R; \quad |\vec{a}| = \omega^2 R = v^2/R.$$

The method we outline in §I1.2 is a sound starting point but not a rigid recipe. With training and practice, you should grow away from it and develop your own personal style of top-down reasoning. One can design a house from a recipe, but an inspiring design calls on the architect for a personal sense of elegance and style and a joy in the process of design. These qualities too come with experience, but there is no known recipe. We shall try to share our sense of style and our delight in physics and hope you come to develop your own.

I1.2   Steps in Building a Problem Solution

Our method is organized into four groups of steps: modeling the physical system; setting up the solution; solving the equations; and analyzing the results. Steps in a group interlock with each other and may merge together in some problems.

Group 1:   Modeling the Physical System

A. Reading and Context

Read the problem carefully to identify the relevant objects involved, their physical properties, what you are given, and what you are asked for. Context is *very* important. Your problem solution will apply to an ideal model that includes some features of your actual system and excludes others; it is context—what you know, whence the problem arises, and what you want to learn—that guides your choice of model. For example, in navigation problems a point particle is a perfectly adequate model of an airplane, but, if you are asked how much load the plane can carry, your model must include detailed information about the shape and size of the plane's wings.

B. Visualization

Draw a diagram of the system, showing and labeling all the features that are important in your model. Visualize how you expect the system to behave—think of it as a movie, not a static picture. Tell yourself a story about what it does.

C. Identification of Central Concepts

Decide which physical principles explain the system's behavior. A simple exercise may involve only one principle, but a complex problem may involve many. Avoid committing to the first idea that comes into your head! Spend a few minutes to ensure you have identified the most important set of concepts.

Group II: Setting Up the Solution

D. Outline of Plan

Write down how you plan to solve the problem. This forces you to examine the logic and consistency of your model. It may be helpful to summarize the plan in a flow diagram, showing how the *subroutines* fit together to make a solution.

E. Construction of Equations

It is rare that a formula can simply be lifted from the text. The mathematical statement of a physical principle is the starting point, but you will have to identify reference frames, set up coordinate systems, and so on. Be sure that you have clearly identified the physical meaning of each algebraic symbol that you use.

F. Trial Solutions

Test your plan to see if the mathematics works, but avoid large amounts of computation. If the math doesn't work, refine the previous steps to determine what is missing.

Group III: Solving the Equations

G. Solution of Equations

Solve for the quantity you want algebraically, with symbols. Show all your steps in the solution. This makes it easier for you to check for errors and for a reader to help you overcome difficulties. Avoid using numbers until the end of a solution.

Group IV: Analyzing the Results

H. Consistency Check

Always check your answer for consistency.
- Does it have the right physical dimensions?
- Does it have a reasonable magnitude?
- Does it reduce to a solution you already know in some limit? (See *Digging Deeper* in §3.1.4 for an example of this.)
- Does it surprise you? Double check it. Can you use your result to develop your intuition further?

I. Comments

If you used any approximations, do they seem valid? What did you learn from the method or the result? Does the method seem useful for other problems?

I1.3 Study Problems

Examples that are particularly well suited for illustrating the solution method are presented occasionally throughout the text as *Study Problems*. Here, we give three such examples that use ideas developed in Chapters 0–3.

The first problem involves reference frames moving relative to one another.

Study Problem 1 ◆◆ Lunch at Noon?

John and Maya are good friends who live on opposite sides of a river 1.0 km wide. John owns a rowboat and can row it in still water at a speed of 3.0 km/h. The river flows at a speed of 1.0 km/h. John plans to row to Maya's house directly across the river and arrive for lunch at noon. When should he leave, and in what direction should he point the boat?

I. Modeling the Physical System

A. Reading and Context. Because the river moves with respect to the shore, the boat has different velocities with respect to river and shore. The problem involves their relation. The speed at which John can row is the speed of the boat with respect to the water, and we are given the speed of the water with respect to the shore. Since it is not stated otherwise, it is reasonable to assume that the river's velocity is constant and parallel to its shores. John wants the velocity of the boat with respect to the shore to be perpendicular to the shore—*directly across*. The required answers are the direction of the boat's velocity vector with respect to the river, which is what the rower may control, and the time interval required to cross the river. (The problem actually asks for the departure time after giving the expected arrival time.)

B. Visualization. ■ Figure I1.1a shows the situation at a time when John is partway across the river. He must point the boat upriver to compensate for the water's motion; the angle θ is one of the required answers. We take the speed of the water to be the same everywhere in the river. This is an adequate, if not precise, *model* of the river.

C. Identification of Central Concepts. Since we are concerned with velocities of a boat with respect to two different reference frames, we may apply the general principle, expressed as eqn. (3.13), to conclude that the velocity of the boat with respect to the river and the velocity of the river add to give the velocity of the boat with respect to the shore: $\vec{v}_{b,r} + \vec{v}_{r,s} = \vec{v}_{b,s}$. Figure I1.1b shows the vector addition. We draw the vector $\vec{v}_{b,s}$ perpendicular to the riverbanks, a fact we deduced in the discussion of context.

II. Setup of Solution

D. Outline of Plan. Solving the vector triangle in Figure I1.1b gives θ as well as the magnitude of $\vec{v}_{b,s}$. From $v_{b,s}$ and the width of the river, W, the required time interval follows immediately.

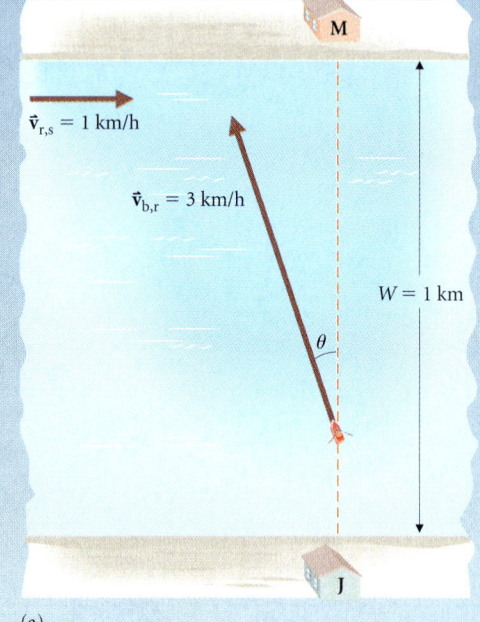

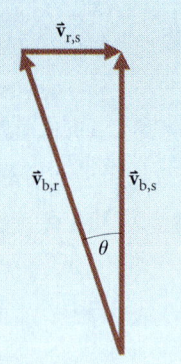

■ **Figure I1.1**
(a) John is rowing to Maya's house (M) for lunch. The river flows with velocity $\vec{v}_{r,s}$ to the right, so John rows upstream at angle θ. His track across the river is perpendicular to the banks. (b) Vector triangle that shows the velocity addition. Velocity of the boat with respect to the river + velocity of the river with respect to the banks = velocity of the boat with respect to the banks, or:

$\vec{v}_{b,r} + \vec{v}_{r,s} = \vec{v}_{b,s}.$

E. Construction of Equations.

From trigonometry,

$$\sin \theta = \frac{v_{r,s}}{v_{b,r}}. \tag{i}$$

From the Pythagorean theorem,

$$v_{b,s}^2 = v_{b,r}^2 - v_{r,s}^2. \tag{ii}$$

The required time interval is

$$\Delta t = W/v_{b,s}. \tag{iii}$$

F. Trial Solutions.
We have three equations for three unknown quantities; no trials are necessary.

III. Solution

G. Solution of Equations.
From eqn. (i):

$$\theta = \sin^{-1}\left(\frac{v_{r,s}}{v_{b,r}}\right) = \sin^{-1}\left(\frac{1.0 \text{ km/h}}{3.0 \text{ km/h}}\right) = \sin^{-1}(\tfrac{1}{3}) = 19°.$$

From eqn. (ii):

$$v_{b,s} = \sqrt{v_{b,r}^2 - v_{r,s}^2} = \sqrt{(3.0 \text{ km/h})^2 - (1.0 \text{ km/h})^2} = \sqrt{8.0} \text{ km/h}.$$

The required time is (eqn. iii):

$$\Delta t = \frac{1.0 \text{ km}}{\sqrt{8.0} \text{ km/h}} = (0.354 \text{ h})\frac{60 \text{ min}}{1 \text{ h}} = 21 \text{ min}.$$

To arrive at noon, John should start at 11:39 and aim his boat 19° upstream.

IV. Analysis

H. Consistency Check.
To perform the same trip in still water, John would require $\tfrac{1}{3}$ h or 20 min. The time is increased only 5% by the flow of the river. From John's point of view, the important result is the 19° angle. Notice that with $v_{r,s} = 0$ we would obtain $\theta = 0$, the expected result for rowing in still water.

I. Comments.
John wants to arrive by boat in the shortest possible time. How do we know this plan achieves that? Suppose John is willing to walk and can do so faster than he can row on water. Could he reduce his time by including some walking on Maya's side of the river?

EXERCISE I1.1 ♦ If John were to point the rowboat directly across the stream, how far would he have to walk along the shore to get to Maya's house?

The second study problem involves uniformly accelerated linear motion. There is more than one successful plan for this question, so we shall take this opportunity to illustrate how trial solutions interact with the formation of plans.

Study Problem 2 ♦♦ Slugger José's Pop Fly

José, superslugger of the sandlot, hits a baseball directly upward next to the wall of an apartment building. José's kid sister observes the game for her physics project. Standing at an apartment window 1.00 m square, she sees the ball start upward. The ball reaches the bottom of the window 1.50 s later and passes by the window in 0.050 s. How high is the window above the batter? What is the initial speed of the ball? At what time does the ball reach the top of the window on the way down?

■ FIGURE I1.2
José hits the ball upward with an initial speed v_i. It passes the window, at height h, where his sister sees it go by. We choose coordinates with the origin at the batter and the y-axis directly upward.

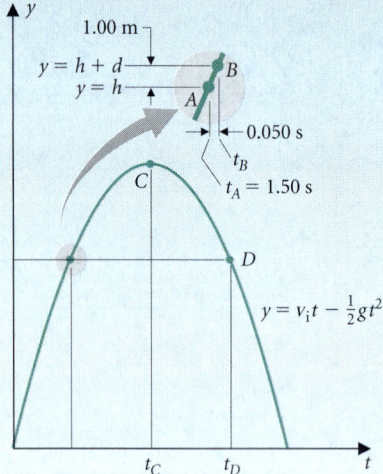

■ FIGURE I1.3
Sketch of the ball's position, described by its y-coordinate, as a function of time t since José hit it. Important events in the problem are marked. *A*: Ball passes the bottom of the window on the way up. *B*: Ball passes the top of the window on the way up. *C*: Ball is at its highest point. *D*: Ball passes the top of the window on the way down. Given information about positions and times is marked. We need to find $y_A = h$, v_i, and t_D.

I. Modeling the Physical System

A. Reading and Context. We neglect air resistance and assume the baseball undergoes one-dimensional motion under the constant acceleration due to gravity. The apartment building serves as a reference frame. We are given information about the location and motion of the baseball during a particular time interval and are asked to derive further details of its motion. (To measure the $\frac{1}{20}$-s time interval, José's sister must be a sharp student!)

B. Visualization. ■ Figure I1.2 illustrates the situation with the upward direction chosen as positive and with the origin at the batter. ■ Figure I1.3 is a graph of the baseball's position versus time in this coordinate system. Each of the events we care about is represented as a point on the graph. We want to find y_A, v_i and t_D.

C. Identification of Central Concepts. The central concept is one-dimensional motion under the constant acceleration due to gravity.

II. Setup of Solution

D. Outline of Plan. Since the motion is one of constant acceleration, our resources are eqns. (2.10) through (2.13). Equation (2.11) gives the position of the baseball as a function of time. With our choice of origin, the starting position of the baseball is $y_i = 0$. The acceleration due to gravity is downward, represented by a negative y-component. The ball starts at the origin, so its position equals its displacement. The initial upward velocity v_i is the only unknown quantity. Equation (2.11) becomes

$$y(t) = v_i t - \tfrac{1}{2}gt^2, \qquad (i)$$

as noted on the graph. (This is convenient to do first, even if it is logically part of step E: Construction of equations. Our plan is NOT rigid.)

One way to proceed from here would be to plug given data into this equation for position versus time and generate equations to be solved for the required unknowns. But the given information allows us to calculate the average velocity of the ball as it passes the window: $v_{av} = d/\Delta t = (1.00 \text{ m})/(0.050 \text{ s}) = 20$ m/s. Can we make use of that? Aha! In uniformly accelerated linear motion, the average velocity equals the instantaneous velocity at the middle of the time interval. Which is easier, the quadratic equations offered by the first plan or the less straightforward second plan? It's time for a trial solution.

E. Construction of Equations. First plan. We evaluate eqn. (i) at points A and B:

$$y(t_A) \equiv h = v_i t_A - \tfrac{1}{2}g t_A^2. \qquad (ii)$$

$$y(t_B) \equiv h + d = v_i t_B - \tfrac{1}{2}g t_B^2. \qquad (iii)$$

COMMENT: Two equations in two unknowns.

Second plan. Use the definition of average velocity: $v_{av} = d/(t_B - t_A)$.

Middle of the time interval: $t_m \equiv (t_B + t_A)/2$.

$$v_{av} = v(t_m) = v_i - g t_m = v_i - g(t_B + t_A)/2. \qquad (iv)$$

COMMENT: One equation in one unknown; add eqn. (ii) for h.

F. Trial Solution. First plan. Subtracting eqn. (ii) from eqn. (iii):

$$d = v_i(t_B - t_A) - \tfrac{1}{2}g(t_B^2 - t_A^2).$$

Solving for v_i:

$$v_i = \frac{d}{(t_B - t_A)} + \tfrac{1}{2}g(t_B + t_A). \qquad (v)$$

This equation says $v_i = v_{av} + gt_m$, which is eqn. (iv). Plan 1 is plan 2 in disguise. Let's get the answer.

III. Solution

G. Solution of Equations.

$$t_B = t_A + 0.050 \text{ s} = 1.50 \text{ s} + 0.050 \text{ s} = 1.55 \text{ s}.$$

From its definition:

$$t_m = (t_B + t_A)/2 = (1.55 \text{ s} + 1.50 \text{ s})/2 = 1.525 \text{ s}.$$

From eqn. (v):

$$v_i = \frac{1.00 \text{ m}}{0.050 \text{ s}} + (9.8 \text{ m/s}^2)(1.525 \text{ s}) = 35 \text{ m/s}.$$

> REMEMBER TO KEEP AN ADDITIONAL FIGURE HERE. WE'LL ROUND OFF AT THE END.
>
> DO WE KNOW g TO MORE THAN TWO FIGURES? NOT UNLESS WE KNOW WHERE JOSÉ LIVES!

We find the height h from eqn. (ii):

$$h = (35 \text{ m/s})(1.5 \text{ s}) - \tfrac{1}{2}(9.8 \text{ m/s}^2)(1.5 \text{ s})^2 = 41 \text{ m}.$$

We still need to find the time t_D at which the ball reappears, so we return to planning.

D. Outline of Plan.
Again, there are two possible plans. We could rely on the formula for position versus time, plug in the height of the top of the window, and solve the resulting quadratic equation. One solution would be $t = 1.55$ s when the ball disappears on the way up, the other would be the desired time. We leave this approach as an exercise and instead use the fact that the baseball's motion is symmetric. Finding the time t_C when the baseball reaches the top of its path gives us the time interval during which the ball is above the window.

E/G. Construction/Solution of Equations.
The time at which the ball reaches the top of its path equals the time required for its upward velocity to become zero. We use eqn. (2.10) with $v(t_C) = 0$.

$$0 = v(t_C) = v_i - gt_C;$$

$$t_C = \frac{v_i}{g} = \frac{35 \text{ m/s}}{9.8 \text{ m/s}^2} = 3.57 \text{ s}.$$

Thus the ball takes 2.02 s after passing by the window to reach the top and takes 2.02 s more to come back down. The ball reappears at:

$$t_{\text{return}} = 3.57 \text{ s} + 2.02 \text{ s} = 5.6 \text{ s}.$$

> NOW WE ROUND TO TWO SIGNIFICANT FIGURES.

IV. Analysis of Result

H. Consistency Check.
The building is rather substantial if one of its apartments is approximately 40 m (or about 120 ft) above ground, but the size is consistent with any number of apartment buildings in, say, New York. Is it reasonable to hit a ball that high? Professional players hitting a home run send a baseball much greater horizontal distances and high enough to leave the stadium.

I. Comments.
Our choices of coordinates were made to maximize convenience. An alternative would be to choose the origin at the bottom or the top of the window. Reasonable people will differ in deciding which choice is more convenient.

At most points in the arithmetic the 0.05-s time interval the baseball spends passing the window is insignificant. We could have approximated $g \approx 10 \text{ m/s}^2$, since that introduces about the same error as neglecting the 0.05 s:

$$1\% \approx \frac{0.05 \text{ s}}{5 \text{ s}}; \quad \frac{0.2 \text{ m/s}^2}{9.8 \text{ m/s}^2} \approx 2\%.$$

However, in the calculation of v_i from the average velocity of the ball passing the window, the 0.05 s is an essential quantity. This is an example of the following rule:

> A small quantity may be neglected in a term where it is added to a much larger quantity, but it may not be neglected in a term where it is a multiplicative factor.

EXERCISE I1.2 ♦ Find the time t_D from solving the equation: $y(t_D) = h + d$.

Our third example is a rather challenging question for which nearly the entire effort lies in careful visualization of the system's motion in time.

Study Problem 3 ♦♦♦ Mars or Bust!

The National Aeronautics and Space Administration (NASA) is considering a manned expedition to Mars early in the twenty-first century. The expedition will follow a path that requires the least rocket fuel: half an elliptical orbit around the Sun, tangent to Earth's orbit at departure and tangent to the orbit of Mars at arrival. Approximate the orbits of Earth and Mars as circles in the same plane and find: 1. How much time is needed to get to Mars? 2. Where should Mars be in its orbit at launch?

THIS IS A BACK-OF-THE-ENVELOPE PROBLEM.

I. Modeling the Physical System

SEE §0.2.3 FOR KEPLER'S LAWS.

A. Reading and Context. As originally stated, Kepler's laws apply to planets, but now we know that they apply to any object orbiting the Sun, such as our Mars-bound spacecraft. So it is a bit redundant of the problem statement to tell us the spacecraft's path is elliptical. The orbits of Earth and Mars are also ellipses, but their deviations from circular shape are small; Kepler needed the most accurate measurements possible in his day to determine their shape. We are also told to neglect the small angle between the planes of the two planets' orbits, lest the problem become one for professional astronomers. Our calculation will be good enough for estimating food supplies and presenting a rough budget to Congress. NASA will use none of these assumptions in programming the guidance computers for the mission! Since no dimensions are given, we shall have to look up any necessary data about the solar system (see inside front cover).

B. Visualization. ■ Figure I1.4 shows the relation of the three orbits looking perpendicularly to their (approximately) common plane. Point A is the position of Earth and spacecraft at launch and point B is the position of Mars and spacecraft at arrival. Kepler's third law tells us qualitatively that things farther from the Sun take greater time to complete an orbit. So, Mars completes less than half an orbit (arc CB), and the Earth completes more than half an orbit (arc AD) while the mission is en route. The angle α shown in the figure is the position of Mars relative to Earth at launch.

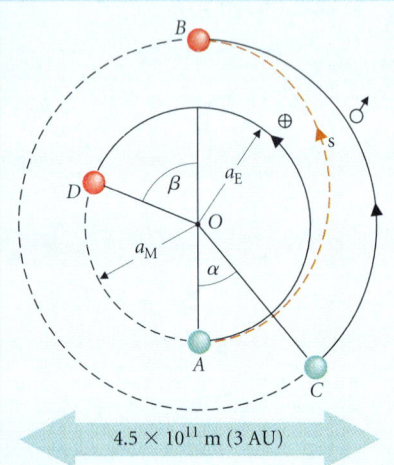

■ **FIGURE I1.4**
The path of a spacecraft to Mars. The spacecraft leaves Earth at point A, and reaches Mars at point B. Mars is at C when the spacecraft is launched, and the Earth is at D when the spacecraft reaches Mars. The Sun is at O. The angle α specifies the relative positions of Earth and Mars at launch. The Earth travels through an angle $\pi + \beta$ during the spacecraft's journey, while Mars moves through the angle $\pi - \alpha$. (The symbol ♂ represents Mars and the symbol ⊕ represents Earth.)

C. Identification of Central Concepts. Kepler's third law states a proportion between the periods of the orbits and their semimajor axes. The semimajor axis of the Earth's orbit, called the astronomical unit, or AU, is a convenient length unit for this problem. Then, from the inside front cover, the orbit of Mars has a semimajor axis $a_M = 1.52$ AU. For the spacecraft, we find a_s from Figure I1.4. The period of the Earth's orbit, the year, is a convenient time unit. Then Kepler's third law gives the orbital period in years:

$$\left(\frac{T}{1\text{ y}}\right)^2 = \left(\frac{a}{1\text{ AU}}\right)^3. \tag{i}$$

Since we are taking the planets' orbits to be circular, we may also assume that Mars and Earth move with constant angular speed ω, and so we can use our subroutine for uniform circular motion. Knowing the transit time of the spacecraft, we can find the angles through which the planets move (eqn. 3.5):

$$\Delta\theta = \omega\, \Delta t. \tag{ii}$$

II. Setup

D. Outline of Plan. Kepler's third law gives us the periods of planets and spacecraft in years. Half the spacecraft period is the transit time. From the ratio of transit time to orbital period, we find the angle swept out by each planet during the transit, and hence the angles α and β.

E. Construction of Equations. From Figure I1.4, we see that the major axis of the spacecraft orbit is given by:

$$2a_s = a_E + a_M. \tag{iii}$$

The periods of the three orbits, in years, are given by eqn. (i):

$$T_E = 1 \text{ y}.$$

$$T_M = (1 \text{ y})(a_M/a_E)^{3/2}.$$

$$T_s = (1 \text{ y})(a_s/a_E)^{3/2}.$$

The angular speeds of Earth and Mars are (eqn. 3.8):

$$\omega_E = 2\pi/T_E = 2\pi \text{ rad/y}.$$

$$\omega_M = 2\pi/T_M.$$

The spacecraft reaches Mars in half an orbit, taking a time $\Delta t_s = T_s/2$. In this time Mars moves through an angle $\Delta\theta_M$, where (eqn. ii):

$$\Delta\theta_M = \omega_M\, \Delta t_s = (2\pi/T_M)(T_s/2) = \pi\,(T_s/T_M).$$

Referring to Figure I1.4:
$$\Delta\theta_M = \pi - \alpha.$$

F. Trial Solution. A trial solution is not necessary here.

III. Solution

G. Solution of Equations. From eqn. (iii), $a_s = (1.00 \text{ AU} + 1.52 \text{ AU})/2 = 1.26 \text{ AU}$. Then $T_s = (1.26)^{3/2} \text{ y} = 1.41 \text{ y}$. The transit time to Mars is then $\frac{1}{2}T_s = 0.71 \text{ y}$.
From either the inside front cover or Kepler's law, we find the period of Mars:

$$T_M = (1.52)^{3/2} \text{ y} = 1.87 \text{ y}.$$

So, the relative location of Mars at launch is given by:

$$\alpha = \pi(1 - T_s/T_M)$$
$$= \pi[1 - (1.41 \text{ y})/(1.87 \text{ y})] = 0.26\pi \text{ rad} = 46°.$$

IV. Analysis

H. Consistency Check. There is not much to check, since the method of doing the algebra virtually guarantees correct units. The result is daunting for astronauts who think of volunteering, but comparable to the periods of the two planets.

I. Comments. A trip to Mars could be mounted with current rocket technology. The problems we need to solve involve life-support systems and human medical and social/psychological health. Of course, the will and daring of Earth's people are also at issue. You could well find yourself in a career where this kind of astronautics is a real personal concern.

Summary

Where Are We Now?

We have presented proven problem-solving techniques in a format you can use to develop your personal method for applying physics.

What Did We Do?

Our approach to problem solving is summarized in an outline:

I. Modeling the physical system
    A. Reading and context
    B. Visualization
    C. Identification of central concepts
II. Setup of solution
    D. Outline of plan
    E. Construction of equations
    F. Trial solution
III. Solution
    G. Solution of equations
IV. Analysis of results
    H. Consistency check
    I. Comments

Practical Applications

Top-down reasoning is a practical mental habit in a wide variety of activities—physics, computer science, economic modeling, planning a business partnership, constructing the family budget.

Solutions to Exercises

I1.1 Pointing the boat directly across the stream results in a velocity component of 3.0 km/h perpendicular to the banks. John will get across in 20 min and will have drifted downstream a distance $(\tfrac{1}{3}\text{ h})(1.0\text{ km/h}) = \tfrac{1}{3}$ km.

I1.2 We have $h + d = v_i t_D - \tfrac{1}{2} g t_D^2$.

Rearranging:

$$t_D^2 - (2v_i/g) t_D + 2(h + d)/g = 0.$$

From the quadratic formula:

$$t_D = (v_i/g) \pm \sqrt{(v_i/g)^2 - 2[(h + d)/g]}$$

Choosing the minus sign, the solution for t_D is less than v_i/g. This is the time the ball passes the top of the window on the way up, which is t_B. We need the $+$ sign:

$$t_D = \frac{v_i}{g}\left[1 + \sqrt{1 - \frac{2g(h+d)}{v_i^2}}\right]$$

$$= \frac{35\text{ m/s}}{9.8\text{ m/s}^2}\left\{1 + \left[1 - \frac{2(9.8\text{ m/s}^2)(42\text{ m})}{(35\text{ m/s})^2}\right]^{1/2}\right\} = 5.6\text{ s.}$$

This is the same result that we found in Study Problem 2.

Nature and Nature's Laws lay hid in Night.
God said "Let Newton Be" and all was Light.

ALEXANDER POPE

CHAPTER 4
Force and Newton's Laws

CONCEPTS

Force
Mass
Gravitation/weight
Normal force
Friction force
Tension and spring force
Force pairs
Balance of forces
Hooke's law
Inertial reference frame
Newton's laws of motion

GOALS

Be able to:

Recognize the different types of force.

Identify the force pairs acting in a system.

Distinguish mass from weight.

Apply Newton's laws to the motion of a particle.

WE'LL ASK YOU TO ANSWER THE QUESTION IN §4.7

A daring cyclist can ride around a vertical curved wall. The trick cannot be done if the cyclist travels too slowly or on a straight wall. Why? This is a question of *dynamics,* the study of forces and the accelerations they cause. Isaac Newton (1642–1727, ■ Figure 4.1) developed the first successful theory of dynamics. Its success rests on his three basic achievements:

- A clear description of force, free of inaccurate preconceptions
- A precise statement of the relation between force and acceleration
- Recognition of gravitation as a universal process described by a simple law

In this chapter and the next, we shall study Newton's theory and learn to apply it to a wide range of practical problems.

Newton published his discoveries in the form of three laws of motion and a law of universal gravitation. These laws form a beautiful, logical, and closed system whose structure we shall discuss at the end of this chapter. To learn the theory, though, we need to look at its pieces one at a time. Our starting point is Newton's first law:

> An object on which no net force is exerted remains at rest or moves along a straight line at constant speed. (That is, its velocity is constant.)

Galileo discovered that change of velocity, rather than velocity itself, requires physical explanation. In the first law, Newton gave the name *force* to the process that provides that explanation.

Everyone has experience with forces, and some intuition about them. If you pull a rowboat onto the beach or carry a heavy object, muscle sensations give you a direct *feel* for the forces you are exerting. But these are complicated situations in which you are not the only object exerting force. To build our intuition into a clear concept of force, we need to reason carefully. This was Newton's first step, and it will be ours. We therefore begin this chapter with a discussion of the nature of force.

■ FIGURE 4.1
Isaac Newton (1642–1727), **English mathematician and physicist.** Newton was a prolific scientist. Among his many accomplishments, he invented the calculus, recognized that sunlight is a composition of pure colors, stated the laws of mechanics clearly for the first time, and discovered the law of universal gravitation.

4.1 FORCE

The word *force* has several common meanings: force of character, a forceful personality, forced to work overtime. All of these uses describe people having will or causing action and are close to the original, pre-Newtonian meaning of force in physics. Originally, people thought any motion required purpose from an active mover such as a person or a horse. This idea makes sense when you're tugging on a rowboat stuck in the sand at the beach (■ Figure 4.2). But we have to think more carefully: the boat doesn't accelerate, so Newton's law tells us that the *total* force exerted on the boat is zero. (This is not what you want to hear while straining on the towrope!) The boat won't accelerate because the *beach pulls back.*

The boat does, however, move very slowly as you tug. Finally it breaks loose, and the beach then exerts a much smaller force. The boat *accelerates* rapidly, and you must be quick to avoid eating sand!

Experience with the rowboat raises three questions we must answer to understand the meaning of *force* in physics:

WHAT KINDS OF FORCE OCCUR?

Both your muscles and the sand can exert force. What other kinds of force occur, and how can we describe the processes that give rise to the forces?

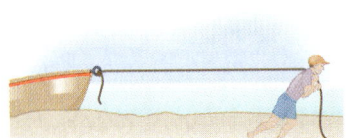

■ FIGURE 4.2
Wet sand exerts force on your stuck rowboat, which counteracts the force you exert via the towrope. When you stop pulling, the sand also stops.

HOW DO DIFFERENT FORCES COMBINE?

Two forces acting on the stuck rowboat can counteract each other to give a net force of zero. When several different forces occur, how do they combine to produce the net force exerted on an object?

■ FIGURE 4.3
The objects in this nautical scene are pushing and pulling on each other—that is, exerting forces. Only a small number of kinds of force occur in mechanics, and examples of each kind occur in this scene.

What Happens When Force Occurs?

Net force exerted on an object causes acceleration. What is the precise relation between force and acceleration; how does the same amount of force affect different objects?

Let us address these questions in this order.

4.2 Force in the Newtonian Model
4.2.1 What Kinds of Force Occur?

At the deepest level we know of only three kinds of interaction among elementary particles. Here, however, we want a less sophisticated and more practical answer: how can we describe the pushes and pulls exerted by ordinary objects such as the boats, people, and fish in Figures 4.2 and ■ 4.3? These figures illustrate all the kinds of force encountered in mechanics. All of the objects portrayed are unaccelerated, which means they are experiencing zero total force. This is very different from experiencing *no force*. Each object is subject to two or more forces that balance to give a net force of zero. We start with the familiar forces exerted by our muscles. Then, as we look at different objects in the pictures, we can recognize other kinds of force. For each object, we first account for all the forces we know about. If they don't balance, we conclude that another type of force is occurring. The resulting list is quite short.

The successful fisherman in Figure 4.3 is holding an object composed of a spring scale, a rope, and a prize fish. The boater's arm exerts muscle force $\vec{F}$ upward on the top of the scale. Balancing this upward force is a downward force $\vec{W}$, weight (■ Figure 4.4). Exerting muscle forces to balance weight is one of our most common experiences, and we normally speak of gravity pulling downward on things. A more precise statement is that a downward gravitational force is exerted on each object.

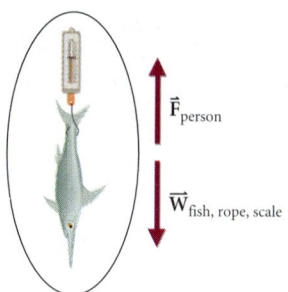

■ FIGURE 4.4
The fisherman's arm is exerting upward force on the system of scale, rope, and fish. Since the system isn't accelerating, this force must be balanced. Weight, the gravitational force exerted by the Earth on the system, is the balancing force.

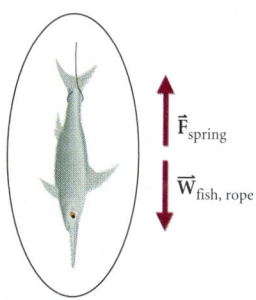

■ FIGURE 4.5
Now we think of the rope and fish together as an object separate from the spring. Gravity acts on the rope and fish, and this weight has to be balanced by a force that the spring exerts.

> *Weight* means the gravitational force being exerted on an object.

Of the three fundamental interactions, gravitation is the only one that appears in our practical list of mechanical forces (● Table 4.1, page 128).

As we look closer (■ Figure 4.5), we notice that the rope and fish are supported by the spring scale. The spring exerts an upward force $\vec{F}_{spring}$ on the rope and fish that balances the downward force due to gravity. Because it is stretched, we can see that the spring is exerting a force. The spring is said to be elastic, and the spring force is one kind of *elastic* force. The upward force the spring exerts is in the opposite direction from the displacement of the spring's lower end.

TABLE 4.1   A Practical List of the Forces That Occur in Classical Mechanics

Name	When It Occurs	Comments	Relation to Fundamental Forces
Gravitation (weight)	Gravitational force occurs between any two bodies. The gravitational force of the Earth on objects near it is what we call weight.	Unless one of the two bodies is of astronomical size, gravitational forces can be detected only by careful measurement. We usually neglect the tiny forces between ordinary bodies.	Gravitation is a fundamental interaction.
Elastic (springs, normal forces, tension in ropes, strings, cables)	When an object's shape is distorted, it exerts an elastic force with direction opposite the distortion.	Except for springs, the distortion is usually not obvious. Normal forces are exerted perpendicular to surfaces in contact.	Balanced electromagnetic forces (Parts VI and VII) hold the atoms of a solid in place. In a distorted solid, these forces no longer balance. The imbalance gives rise to the forces we call elastic.
Friction	Friction occurs between surfaces in contact; it opposes relative sliding motion.	Friction forces depend on the nature of the surfaces, their relative motion, and the normal forces exerted between them. Friction forces are exerted parallel to the surfaces.	Friction forces arise from chemical bonds and the meshing of surface irregularities, both of which involve electromagnetic interaction among atoms.
Fluid (buoyancy, thrust, drag, lift)	Fluid forces are exerted between different elements of fluid or between a fluid and a solid body immersed in or moving through it.	Buoyancy forces are exerted directly upward; thrust is exerted forward and drag backward along the axis of boat or plane; lift is exerted perpendicular to the long axis of a plane. We'll discuss fluid forces in more detail in Chapter 13.	The atoms in fluids are in continual random motion. They exert electromagnetic forces when they collide and bounce off each other or the atoms of a solid object.

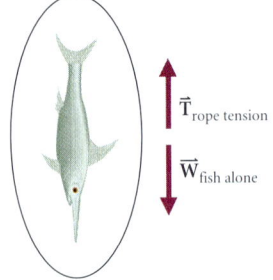

■ FIGURE 4.6
The rope in turn exerts a tension force on the fish that balances the fish's weight.

SINCE BOTH FRICTION AND NORMAL FORCES ONLY OCCUR WHEN SURFACES ARE IN CONTACT, THEY ARE OFTEN REFERRED TO AS PARALLEL AND PERPENDICULAR COMPONENTS OF A SINGLE "CONTACT" FORCE.

The rope directly supports the fish (■ Figure 4.6). The rope force, called a *tension* force, is also an elastic force, but the stretch of the rope is much smaller than that of the coiled spring and is not visible to the fisherman or to us.

A third kind of elastic force is exerted by the gangplank (■ Figure 4.7). The plank is curved, showing the distortion that gives rise to the force. This kind of force, exerted perpendicular, or *normal,* to the surfaces in contact (the plank and the fisherman's shoes) is called a *normal* force. Notice that the dock in Figure 4.3 also exerts a normal force to support the photographer, but the distortion of the dock is too small for us to see.

The tension force exerted by the rope used to pull on a stuck rowboat (■ Figure 4.8) is balanced by a *friction* force exerted by the sand.

Friction is the name for forces between surfaces in contact that are exerted parallel to the surfaces.

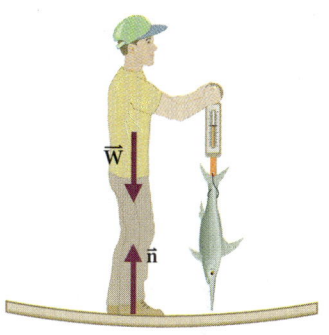

■ **Figure 4.7**
The gangplank exerts a *normal force* $\vec{\mathbf{n}}$ to balance the weight of the fisherman, as well as that of scale, rope, and fish. The gangplank is not too sturdy and bends noticeably, illustrating that normal forces arise from distortion of the surfaces that exert them.

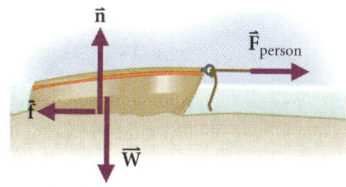

■ **Figure 4.8**
The force exerted by the sand on the rowboat is directed parallel to the sand surface. It is therefore classified as a *friction* force.

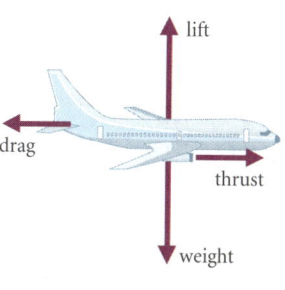

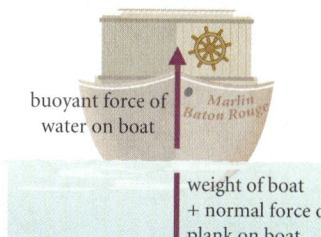

■ **Figure 4.9**
Fluid forces act on the boat and airplane. These forces arise from the impact of fluid molecules on the boat and plane. For now, we describe such forces by their common names. The water exerts an upward force on the boat (similar to normal force), which is called *buoyant force*. Air flowing over the wings of the plane exerts *lift* and *drag*. (Buoyant force on the plane is much less than lift and is not shown.) The engines produce a *thrust* on the plane.

Friction is a nuisance when it keeps your rowboat stuck or ruins your car's engine after an oil leak. On the other hand, friction forces keep your car from sliding off the road, hold knots together, and allow you to walk.

Fluids such as water or air exert contact forces similar to normal and friction forces. We won't study these processes until Chapter 13, but we shall occasionally need to refer to fluid forces by name. ■ Figure 4.9 illustrates the various kinds of fluid forces. The list of mechanical forces is now complete and is summarized in Table 4.1.

4.2.2   Forces Are Vectors

Forces are exerted in definite directions, so we might expect forces to be described by vectors. This idea must be, and is, amply confirmed by experience. The following examples suggest two simple experiments that illustrate vector addition of forces.

EXAMPLE 4.1 ♦♦   Teams of three, four, and five people compete in a three-way game of tug-of-war (■ Figure 4.10). Assuming that each of the 12 contestants is equally strong and that forces add like vectors, what angle θ between the three- and four-person teams is required for a fair contest?

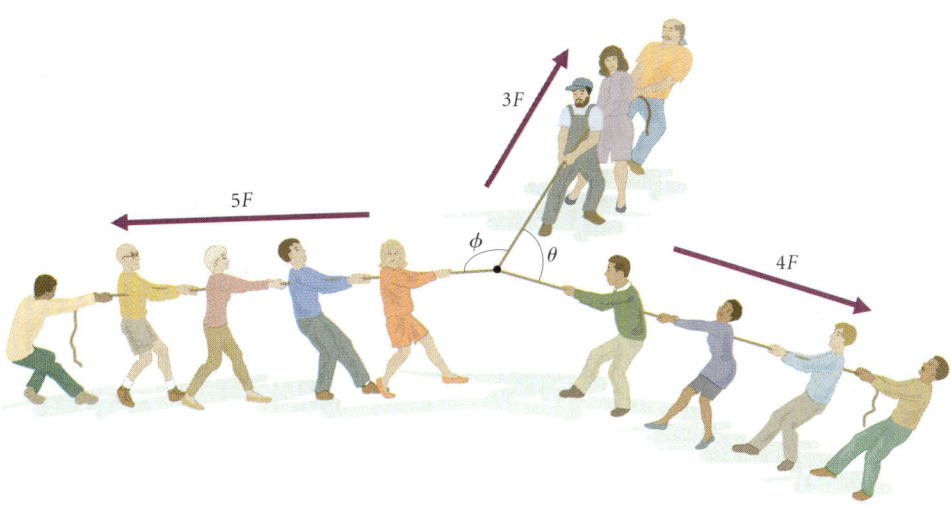

■ **Figure 4.10**
Three teams compete in a game of tug-of-war. We can find the conditions for a fair contest by recognizing that forces follow the rules of vector addition.

An abnormal way to obtain normal force? The Jesus Christ lizard, an inhabitant of the Costa Rican jungle, runs on water to avoid crocodiles and piranha. The lizard strikes the water rapidly and vigorously with its feet. The water pushes back, exerting the force needed to balance the lizard's weight.

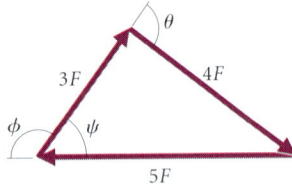

FIGURE 4.11
The net force acting on the knot connecting the ropes has to be zero, or the knot would accelerate. Added tail to head, the three force vectors form a triangle.

MODEL In a *fair* game of tug-of-war, an equal effort by each contestant results in balanced forces (until one team tires or someone slips!). In the three-way game, forces of magnitudes 3, 4, and 5 times the force $|\vec{F}|$ exerted by a single person act on the knot connecting the ropes and should add vectorially to zero (■ Figure 4.11).

SETUP The triangle is solved either by applying the law of cosines or by noticing that the vector sum forms a classic 3-4-5 right triangle.

SOLVE So, $\theta = 90°$ for a fair contest.

ANALYZE Suppose one of the contestants in the five-person team moved to the three-person team? What would the angle be then?

EXERCISE 4.1 ♦♦ In the tug-of-war game, what is the angle ϕ between the three-person and five-person rope? Check that force components perpendicular and parallel to the five-person rope balance separately.

EXAMPLE 4.2 ♦♦ Unable to break your rowboat free of the sand (cf. Figure 4.2), you have a clever idea: tie the rope to a tree and push sideways (■ Figure 4.12). If the angle ψ in the figure is 10°, how much greater is the force exerted on the boat by the rope than the force you can exert directly?

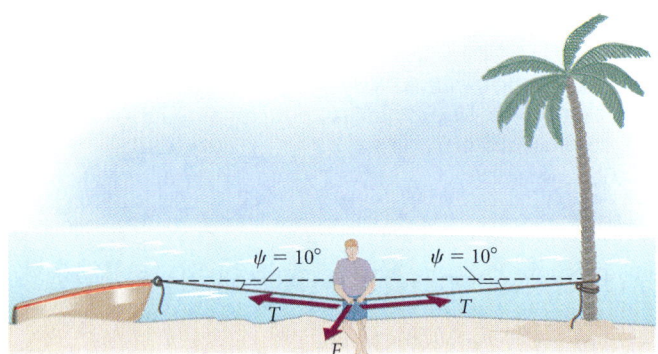

FIGURE 4.12
A clever way to multiply the force you can exert with a rope is to push sideways on it. Because the tension forces exerted by the rope very nearly balance each other, they each have greater magnitude than the force $\vec{F}$ you exert directly.

MODEL Part of our intuition about ropes is that they *transmit* forces; that is, the rope exerts forces of the same magnitude T wherever it acts. Because the rope isn't accelerating, the forces acting on the piece of rope where you push are in balance.

SETUP We express force balance in component form (■ Figure 4.13):

$$x\text{-components:} \sum F_x = 0 = -T \cos \psi + T \cos \psi,$$
$$y\text{-components:} \sum F_y = 0 = F - 2T \sin \psi.$$

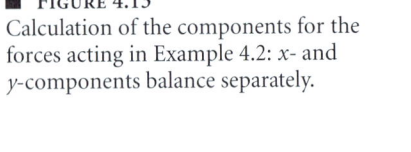

FIGURE 4.13
Calculation of the components for the forces acting in Example 4.2: x- and y-components balance separately.

SOLVE The first equation confirms that the rope tension T is the same on both sides of your hands; the second gives the required relation:

$$\text{Force on boat} = T = \frac{F}{2\sin(10°)} = 3F.$$

ANALYZE By being clever, you can triple the force exerted on the boat. This effect explains why a strong wind can rip power cables from their connections. The result should also remind you to use a strong rope lest it break and you sprawl on your face in the sand!

4.2.3 Force Vectors Occur in Pairs

After a day at the beach, you row your boat back to the dock, where your younger brother Anton foolishly tries to jump ashore (■ Figure 4.14). The rowboat accelerates backward and Anton finds himself swimming. To understand why Anton ends up in the water, let's look at

FIGURE 4.14
Brother Anton leaps from a boat. To jump, Anton exerts a force $\vec{F}_{\text{A on B}}$ on the boat. The boat exerts an equal and opposite force, $\vec{F}_{\text{B on A}}$, on Anton. Unfortunately for Anton, the boat accelerates away from the dock.

130   CHAPTER 4 • FORCE AND NEWTON'S LAWS

the forces involved. When you jump, you rely on the normal and friction forces exerted by the ground to accelerate you. With your muscles, you exert forces on the ground, distorting it ever so slightly, and the ground reacts by exerting the forces you desire. The Earth is enormous, and you don't notice its tiny acceleration caused by the forces you exert. The rowboat is not so large, and when Anton jumps, the forces he exerts on the boat cause it to accelerate out from under him. This situation illustrates several basic facts about forces:

- Forces always occur in pairs and always involve a pair of objects interacting with each other. (One process occurs; two forces are exerted.)
- The forces in each pair are equal in magnitude and opposite in direction. Each object exerts one of the forces and experiences the other.
- Since they result from a single process, the forces in each pair are of the *same kind*: either both normal, both friction, both gravitational, and so on.

IN OUR NARRATIVE, THE PROCESS IS "ANTON JUMPS."

Force is a description of the process by which two objects interact. A force is not itself an object, nor does it have any existence independent of the interacting objects. Newton himself had a difficult time in giving up the notion of force as a quality or substance possessed by objects and also in recognizing two-object interactions as the fundamental fact. Only after making this conceptual leap was he able to develop his theory. The nature of force pairs is usually summarized as Newton's third law:

IF NEWTON HAD TO WRESTLE WITH SOMETHING, THAT MEANS IT'S DIFFICULT. DON'T BE SURPRISED IF YOU TOO HAVE TO WRESTLE WITH THIS ONE A BIT.

> When an object A exerts a force on object B, object B exerts an equal and opposite force on object A.

NEWTON'S SECOND LAW CONCERNS WHAT FORCES *DO* RATHER THAN WHAT THEY *ARE*; IT WILL MAKE ITS DEBUT SHORTLY.

EXAMPLE 4.3 ❖ A block of stone rests on the ground. What pairs of forces occur in the interaction between the block and the Earth?

MODEL For each force exerted *on* the block, a paired force is exerted *by* the block. The other object involved in the interaction is the Earth.

SETUP First we determine the kinds of force the Earth exerts on the block, using the list in Table 4.1. Then we use Newton's third law to describe the paired forces.

FORCES ACTING ON THE BLOCK (■ FIGURE 4.15):

- Gravitation: The Earth pulls on the block with the weight force $\vec{W}_{b,e}$.
- Elastic: There are no strings, springs, cables, and so forth. The surfaces of the block and the Earth are in contact. The Earth's surface is stressed and reacts by exerting the normal force $\vec{n}_{b,e}$.
- Friction: There is no force component acting parallel to the ground, attempting to make the rough surfaces move over each other. Consequently, no friction force is exerted on the block.

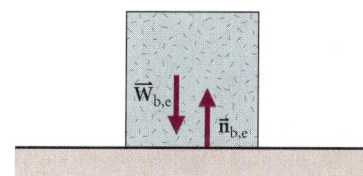

(a) Forces acting on the block

SOLVE

Paired Forces

	Force exerted on block		Paired force exerted on Earth	
Type		Exerted by		Exerted by
Normal	$\vec{n}_{b,e}$	Earth's surface	$\vec{n}_{e,b} = -\vec{n}_{b,e}$	block's surface
Gravity	$\vec{W}_{b,e}$	pull of entire Earth on block	$\vec{W}_{e,b} = -\vec{W}_{b,e}$	pull of block on entire Earth

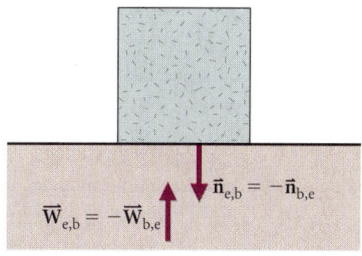

(b) Forces acting on the earth

■ **FIGURE 4.15**
Pairs of forces exerted on each other by the Earth and a heavy block. (a) The Earth exerts an upward normal force *and* a downward gravitational force on the block. (b) The forces exerted on the Earth are equal and opposite the forces exerted on the block. Each force exerted on the Earth is of the same kind (normal or gravitational) as its pair acting on the block.

ANALYZE The pairing of normal forces makes intuitive sense. We know the block pushes down on the Earth's surface; the Earth pushes back. Most people are surprised to learn that the block pulls upward on the Earth; it is certainly *not* intuitive. In astronomy, it becomes clear that gravity follows the third law. For Newton, recognizing this was a crucial step in his discovery of classical mechanics. ■

EXERCISE 4.2 ❖ While climbing Mt. Katahdin, Sally is standing on a steep rock slope. Sketch figures showing the forces being exerted on Sally and their corresponding paired forces. (*Hint:* Which force keeps Sally from sliding down the mountain?)

4.3 Newton's Second Law
4.3.1 The Relation Between Force and Acceleration

The essay following the chapter describes Newton's discovery of the laws of motion.

We have described the kinds of force that arise in mechanics, and we know from Newton's first law that force is the explanation for acceleration. Our next logical step is to determine the precise relation between force and acceleration. Newton was an accomplished experimentalist as well as a great theorist, and he carefully tested his ideas against laboratory experiments and astronomical observation. Rather than describing Newton's careful but complex tests, we will employ straightforward thought experiments that illustrate the issues.

■ **Figure 4.16**
Thought experiment using the Air Force rocket sled. (a) Technicians strap Colonel John P. Stapp into his nine-rocket experimental sled at Holloman Air Force Base on December 10, 1954 for the 632-mph test run. The experiment tested the responses of the human body to rapid acceleration (cf. Figure 2.8). (b) Here we imagine measuring the sled's acceleration as a function of the number of standard rockets used to propel it. (c) When the number of rockets is doubled, the acceleration also doubles. Such experiments show that an object's acceleration is proportional to the force acting on the object. Double the force, and the resulting acceleration doubles.

The vector sum of N identical rocket forces is just N times the force exerted by a single rocket. (See §1.4.6.)

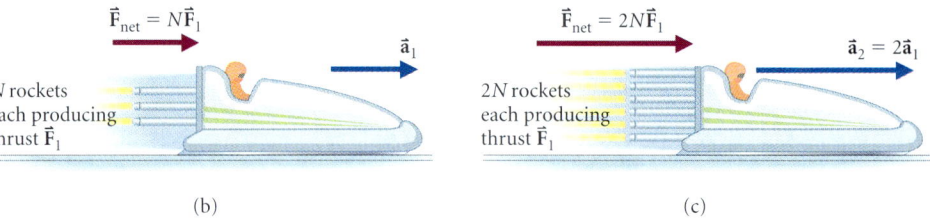

(b)   (c)

Imagine running a series of trials using a rocket sled (■ Figure 4.16) with a large number of small, identical rockets mounted on the back. In each trial, a different number of rockets are ignited, and the sled's acceleration is measured. The measured acceleration is directly proportional to the number of rockets firing—that is, to the net force exerted on the sled.

> The acceleration of any object is directly proportional to the net force exerted on the object:
> $$\vec{a} \propto \vec{F}_{\text{net}}.$$

■ **Figure 4.17**
Thought experiment with the rocket sled. In each trial, one standard rocket is used to propel the sled. The sled is loaded with different numbers of standard lead bricks, and its acceleration is measured as a function of the number of bricks. Such experiments show that an object's acceleration is inversely proportional to the object's total mass. Double the total mass (sled plus bricks) and the acceleration is reduced by one-half.

To test how acceleration depends on the nature of the accelerating body, imagine using the rocket sled with the same number of rockets ignited each time, but loaded in each trial with a different number of lead bricks (■ Figure 4.17). We expect a sled with many bricks to accelerate less than a sled with few bricks. Experiments confirm the simplest such relation, an inverse proportion:

> The acceleration produced by a given force is inversely proportional to the amount of material being accelerated:
> $$|\vec{a}| \propto \frac{1}{\text{amount of material}}.$$

"Amount of material" is a clear concept when comparing two different numbers of identical lead bricks but is fuzzy when comparing, for example, the material of the sled to that of the bricks. The term "amount of material" is replaced by the technical term *mass*.

Together, these two results comprise Newton's second law:

THE UNIT OF MASS—THE KILOGRAM— WAS INTRODUCED IN §1.2.6.

> An object's acceleration is directly proportional to the net force exerted on the object and inversely proportional to its mass:
>
> $$\vec{a} \propto \vec{F}_{net}/m.$$
>
> In SI units, the constant of proportionality is unity:
>
> $$\vec{a} = \vec{F}_{net}/m. \tag{4.1}$$

REMEMBER: IF TWO VECTORS ARE PROPORTIONAL, THEY ARE IN THE SAME DIRECTION AND HAVE MAGNITUDES THAT ARE PROPORTIONAL.

Newton's second law is the basic rule of dynamics. We can now begin using the rule, learning first how to measure forces and masses and then how to solve practical problems.

4.3.2 Mass

Replacing such expressions as "amount of material" or "quantity of matter" with the new term *mass* was a necessary step in Newton's work.[1] We have a good deal of intuition about mass: no one wants to use a bowling ball in a baseball game. Newton's second law captures this intuition: mass is proportional to amount of material; greater mass means less acceleration for the same effort. Thus the second law allows us to go beyond intuition; indeed, it enables us to define and to investigate quantitative properties of mass. In effect, mass *is* the quantity that works in Newton's law.

For example, using Newton's law, we can test whether an object's mass depends on speed. ■ Figure 4.18 illustrates an ideal version of such an experiment. A real-world example is the response of the space shuttle to its engines. The shuttle's mass shows no variation with speed, location in orbit, or direction in which the engines are fired, and is accurately calculated by adding the separately measured masses of each of the shuttle's parts.

In Newtonian physics,

1. The mass of an object is a constant property of the object, independent of the object's velocity, position, or any other of its properties.
2. The mass of a composite object is the sum of the masses of its parts.

One way to compare the masses of objects is to subject them to known forces and observe their accelerations. In ■ Figure 4.19, a Skylab astronaut, Alan Bean, is measuring his mass using Newton's second law. The chair is mounted on springs and is oscillating. The more massive an astronaut is, the more slowly the spring forces can accelerate him, and the longer the time required for each oscillation. Once the oscillation time has been measured for a known mass, an unknown mass can be found by comparing oscillation times. The Skylab astronauts made regular measurements to determine their loss of body mass during their stay in orbit.

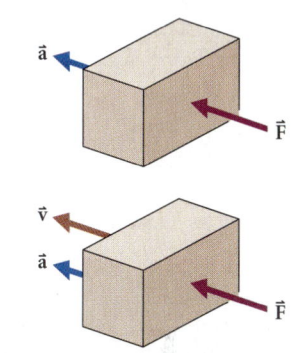

■ **FIGURE 4.18**
Thought experiment to test the dependence of mass on speed. A force $\vec{F}$ exerted on an object at rest causes acceleration $\vec{a}$. An equal force exerted on the same object when moving with velocity $\vec{v}$ causes the same acceleration $\vec{a}$ (provided that the speed is much less than the speed of light). In Newtonian theory, an object's mass is independent of its velocity.

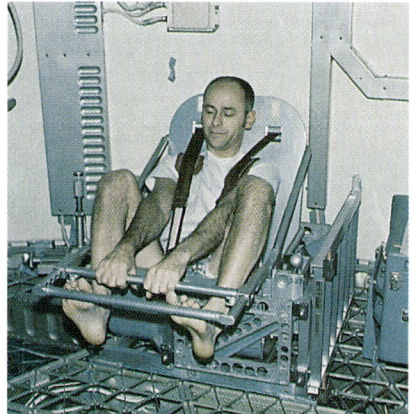

■ **FIGURE 4.19.**
Skylab astronaut Alan Bean measuring his mass. The chair is spring mounted, and its oscillation frequency is determined by the mass of its occupant (see Chapter 14). This relation is calibrated with known masses, and the astronaut in orbit then determines his mass by timing oscillations. Data on changes in the Skylab astronauts' masses have been used in research on the medical effects of long-term space voyages.

EXAMPLE 4.4 ♦♦ With all rockets ignited, the acceleration of Colonel Stapp's sled (Figure 4.16a) is measured at 90.0 m/s². When the sled is loaded with ten lead bricks, in trial two, it accelerates at 67.9 m/s². In trial three, with only the colonel on the sled, the acceleration is 76.6 m/s². What is the ratio of his mass to that of a lead brick?

---

[1] Getting the words right is important, although, with hindsight, it seems easy. During the decades before Newton, the concepts we now distinguish as force, energy, impulse, and momentum (cf. Ch. 6–9) were all regularly confused and interchangeably called force or velocity by various authors. Eliminating this confusion and giving careful definitions of force, mass, and velocity was a necessary step in the discovery of the second law.

MODEL The force F acting on the sled is the same in each trial. So the measured acceleration in each trial is inversely proportional to the combined mass of the sled and the load used in that trial. The force F and the mass of the sled are unknown, but with three measured values, we hope to eliminate them.

SETUP Name the mass of the sled m_s, the mass of a brick m_b, and the mass of the colonel m_c. Then, expressing Newton's law (eqn. 4.1) in the form $F = ma$,

$$F = m_s a_1 \quad \text{(i)} \qquad F = (m_s + 10m_b)a_2 \quad \text{(ii)} \qquad F = (m_s + m_c)a_3. \quad \text{(iii)}$$

TRIAL Equating expressions (i) and (ii), we can eliminate F to find m_b in terms of m_s:

$$m_s a_1 = (m_s + 10m_b)a_2 \quad \Rightarrow \quad m_s(a_1 - a_2) = m_b(10a_2). \quad \text{(iv)}$$

Similarly, equating expressions (i) and (iii) for F:

$$m_s a_1 = (m_s + m_c)a_3 \quad \Rightarrow \quad m_s(a_1 - a_3) = m_c a_3. \quad \text{(v)}$$

SOLVE Now, if we divide eqn. (v) by eqn. (iv), the mass of the sled cancels out, and we obtain:

$$\frac{m_c a_3}{m_b(10a_2)} = \frac{(a_1 - a_3)}{(a_1 - a_2)}.$$

Thus: $\dfrac{m_c}{m_b} = \dfrac{10a_2(a_1 - a_3)}{a_3(a_1 - a_2)} = \dfrac{10(67.9 \text{ m/s}^2)(90.0 \text{ m/s}^2 - 76.6 \text{ m/s}^2)}{(76.6 \text{ m/s}^2)(90.0 \text{ m/s}^2 - 67.9 \text{ m/s}^2)} = 5.37.$

ANALYZE A commercial lead brick's mass is about 13 kg, so the colonel's mass is 70 kg. Notice that you do not need to know the value of the force F to find the ratios of the masses, only that it has some constant value.

4.3.3 The SI Unit of Force—The Newton

Newton's second law relates the physical dimension of force to the fundamental dimensions of mass, length, and time:

$$[\text{Acceleration}] = \frac{[\text{force}]}{[\text{mass}]}$$

or,

$$[\text{Force}] = [\text{mass}] \times [\text{acceleration}] = ML/T^2.$$

Because of this relation, the dimension of force is said to be *derived* from the fundamental dimensions. In SI, the unit for such a derived quantity is defined by replacing the dimension symbols M, L, and T with the corresponding SI units. The resulting unit of force is named the newton, symbol N. (The SI units named after people are not capitalized, but their abbreviations are.)

A force of 1 *newton* exerted on a mass of 1 kilogram causes an acceleration of 1 meter per second squared.

$$1 \text{ newton} = (1 \text{ kg}) \times (1 \text{ m/s}^2)$$
$$1 \text{ N} = 1 \text{ kg} \cdot \text{m/s}^2.$$

EXERCISE 4.3 ♦ A rope exerts a force F on an object of mass 0.70 kg, causing it to accelerate at 0.30 m/s^2. A second object, subject to an equal force F, accelerates at 2.1 m/s^2. Compute the magnitude of the force and the mass of the second object.

4.4 WEIGHT

Suppose an apple falls in Newton's garden (■ Figure 4.20). The gravitational force the Earth exerts on the apple—its weight—causes the apple to accelerate downward with the local acceleration due to gravity $\vec{g}$. Using Newton's second law to interpret the apple's behavior, we obtain a practical relation between the weight of the apple—or of any other body—and its mass:

■ **FIGURE 4.20**
A legend that ought to be true but probably isn't holds that Newton thought of universal gravitation while watching an apple fall in his garden. According to Galileo, the apple's acceleration is $\vec{g}$. This acceleration has to be the result of force exerted on the apple, and the only force acting is weight $\vec{W}$. Using these observations in Newton's second law, we obtain a practical formula for weight.

WE YIELD TO POPULAR USAGE AND SPEAK OF OBJECTS AS "HAVING" WEIGHT. BEWARE! WEIGHT MEANS A FORCE THAT OCCURS; IT IS NOT SOMETHING THE OBJECT CAN POSSESS. IT'S YOUR JOB TO AVOID BEING FLUMMOXED.

> The weight $\vec{W}$ of an object at any point equals its mass multiplied by the local acceleration due to gravity.
>
> $$\vec{W} = m\vec{g}. \quad (4.2)$$

AN OBJECT WITH A WEIGHT ON EARTH OF 1 N WEIGHS ABOUT $\frac{1}{4}$ LB.

Equipment used until recently by the National Institute of Standards and Technology to compare masses against that of the prototype kilogram (cf. §1.2.6). The machine compares the weight forces acting on the two masses. The theoretical idea that weight is proportional to mass is built into the operational definition of mass.

We derived this formula from an experiment in which the object accelerates freely, but it is valid in any circumstance. The same gravitational force acts, whether or not it is balanced by other forces. Thus a good name for $\vec{g}$ might be the gravitational force per unit mass.

You must distinguish carefully between weight and mass, because popular usage confuses them. Mass is a scalar *and* a constant property of an object independent of its location. Weight is a vector and depends on the local value of the acceleration due to gravity. At any specific place, weight is proportional to mass.

The SI operational definition of mass involves the comparison of weights. If you buy a bag of potatoes, the storekeeper puts them on a scale that *senses* their weight. Let us consider a few examples to illustrate the distinction between mass and weight and to learn how it is that the storekeeper determines the mass of the potatoes while sensing their weight.

EXAMPLE 4.5 ♦ The acceleration due to gravity is 9.8 m/s² at Las Vegas, Nevada. What is the weight of a 3.0-kg bag of potatoes in Las Vegas? What is the weight of the same bag of potatoes after shipment to a colony on the Moon, where the acceleration due to gravity is 1.6 m/s²?

MODEL AND SETUP We use eqn. (4.2) for the weight of the bag, using the appropriate value for g.

SOLVE In Las Vegas, the bag's weight has magnitude:

$$W = mg = (3.0 \text{ kg})(9.8 \text{ m/s}^2) = 29 \text{ N}.$$

On the Moon, $\quad W_{\text{Moon}} = mg_{\text{Moon}} = (3.0 \text{ kg})(1.6 \text{ m/s}^2) = 4.8 \text{ N}.$

ANALYZE The bag of potatoes weighs less on the Moon, though its mass is unchanged.

EXERCISE 4.4 ❖ A champion weight lifter can lift four boxes of nails on the surface of the Earth. How many boxes can he lift when he is on the surface of the Moon?

EXAMPLE 4.6 ❖ On the surface of the Earth, a champion shot-putter can accelerate a shot horizontally at 3 m/s². What horizontal acceleration can the champ produce when performing on the Moon?

MODEL The champ's arm strength is the same on the Moon as on the Earth, so he exerts forces of equal magnitude on the shot in both performances. (■ Figure 4.21)

SETUP The mass of a shot is also the same, so the resulting acceleration is the same:

SOLVE $\quad a_{\text{Moon}} = a_{\text{Earth}} = F/m = 3 \text{ m/s}^2.$

ANALYZE It is irrelevant that the shot has only one-sixth as much weight when on the Moon.

■ **FIGURE 4.21**
A shot put champion performs on the Earth and then on the Moon. If the spacesuit doesn't cramp the champ, the force exerted on the shot is the same in both trials. The mass of the shot is also the same in both trials and thus so is the acceleration. (Would the shot put distance record be the same on both worlds?)

EXAMPLE 4.7 ❖ Would a grocer's scale in Las Vegas and one in the lunar colony give the same reading for the same bag of potatoes? (cf. Example 4.5)

MODEL The potatoes weigh only one-sixth as much on the Moon as at Las Vegas. However, all grocer's scales read the same number of kilograms in each place (■ Figure 4.22). To calibrate the scales, an officer from the local department of weights and measures comes around from time to time and weighs standard objects, adjusting the scale to give the proper readings for those standards. Notice that although the scales on the Earth and Moon are marked in mass units (kilograms), the markings have to be different for the calibration to work. Scales use the fact that weight is proportional to mass. By comparing weights, the scales compare the mass of the potatoes to the mass of a standard object m_o:

$$\frac{|\vec{W}|_p}{|\vec{W}|_o} = \frac{m_p g}{m_o g} = \frac{m_p}{m_o}, \text{ independent of the value of } g.$$

ANALYZE Both the potatoes and the standard objects weigh less on the Moon. The ratio of their weights is the same in both locations and equals the ratio of their masses. ■

■ **FIGURE 4.22**
Grocer's scales are used to determine the mass of a potato sack, first at the Las Vegas spaceport and later at the lunar colony. *Elastic* force exerted by the spring inside each scale supports the sack of potatoes, and the rotation of the dial is proportional to this elastic force. Hence, the rotation of the dial is proportional to the weight of the potato sack. On the Moon, the potato sack causes only one-sixth the dial rotation as it would on Earth, but the dial markings are changed accordingly. The scales are checked occasionally with a standard mass.

 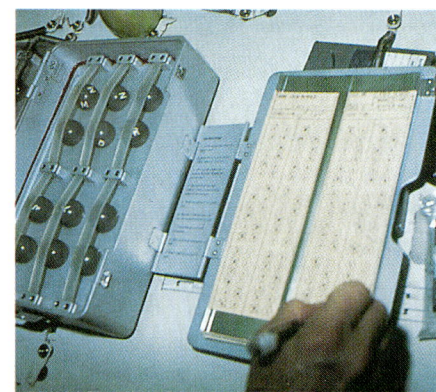

One of the experiments in Spacelab I *tested how accurately astronauts could sense the relative masses of several metal balls by accelerating them back and forth by hand. The astronauts' ability to sense mass differences was about half as sensitive in orbit as on the ground. Like a grocer's scale, we too rely somewhat on a sense of weight to judge mass. (a) Spacelab astronaut Owen Garriott performing the experiment. (b) Close-up showing the set of experimental objects.*

4.5 Practical Expressions for Spring and Friction Forces

4.5.1 The Elastic Force Exerted by a Spring

When a metal cylinder is hung on the end of a spring (■ Figure 4.23a), the spring stretches an amount s_1 from its relaxed length ℓ. In Figure 4.23b, two more identical metal cylinders are attached to the spring, and its stretch s_3 is three times the amount s_1. The spring stretches by an amount proportional to the number of identical cylinders supported. In each situation, the spring balances the weight of the cylinders being supported, so the force exerted by the spring is also proportional to the number of cylinders supported. It follows that the force exerted by the spring is proportional to the stretch of the spring.

With a stiff enough spring, such as one from an automobile suspension, we can do the experiment upside down. Piling objects on top of such a spring compresses it, and we find that the compression force exerted by the spring is proportional to the distance it is squeezed. The constant of proportionality is the same as that found when the spring is in tension.

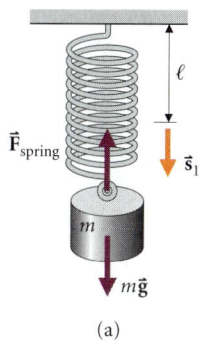

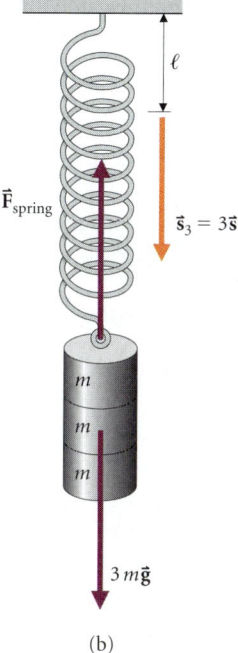

■ FIGURE 4.23
Hooke's law. The force an ideal spring exerts is proportional to the distortion of the spring. The direction of the force exerted by the spring is opposite the displacement of the end of the spring exerting the force: $\vec{F} = -k\vec{s}$.

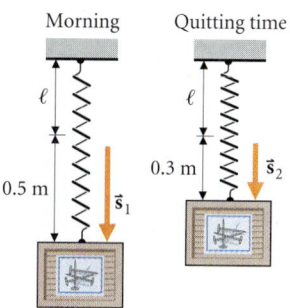

■ FIGURE 4.24
During the day, the spring's stretch was reduced from 0.50 m to 0.30 m. What fraction of the nails has been used?

The results of such experiments are described by Hooke's law, after Robert Hooke (1635–1703):

> The force exerted by a spring is proportional to the distortion of the spring. With one end fixed, the force at the free end is exerted in the direction opposite from the displacement $\vec{s}$:
>
> $$\vec{F}_{spring} = -k\vec{s}. \quad (4.3)$$

The *spring constant k* is a property of the particular spring used. A variety of springs with a wide range of constants are used in practical systems such as your bathroom scale or an automobile suspension. Designing such a system is standard mechanical engineering. Alternatively, for any given spring, k may be determined experimentally.[2]

EXAMPLE 4.8 ♦♦ During the workday, carpenters use nails from a large box. To measure usage, the box is suspended from a spring twice a day. At the beginning of the day, the spring stretches 0.50 m. At the end of the day, the spring stretches only 0.30 m. What fraction of the nails has been used? (■ Figure 4.24)

MODEL Three ideas come together in this problem: the stretch of the spring is proportional to the force it exerts (eqn. 4.3); that force must balance the weight exerted on the nails; and the weight is proportional to the mass, hence the number, of nails in the box. We ignore the mass of the box because we expect it to be much less than that of the nails. The forces are shown in ■ Figure 4.25.

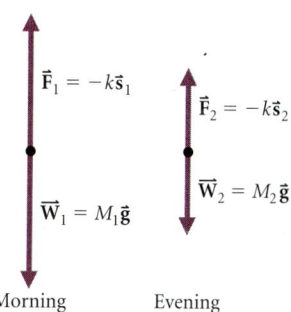

■ FIGURE 4.25
Diagrams showing the forces exerted on the box of nails in the morning and evening.

---

[2] Hooke's law is a reasonable approximation for small distortions of real springs. For very large distortions, the law clearly fails: If you attempt to use the spring from the bathroom scale to suspend an elephant, the spring will be pulled into a straight wire and will probably snap. If the elephant steps on the automobile spring, the coils will be squeezed into contact. In this text, whenever we call something a spring, we are assuming its distortion remains small enough for Hooke's law to apply.

SETUP At the beginning of the day, the weight is: $W_1 = F_1 = ks_1$.
At the end of the day the weight is: $W_2 = F_2 = ks_2$.

Dividing these expressions gives: $\dfrac{W_2}{W_1} = \dfrac{ks_2}{ks_1} = \dfrac{s_2}{s_1}$.

SOLVE Since weight is proportional to mass, the fraction of nails used is:

$$\dfrac{M_1 - M_2}{M_1} = \dfrac{W_1 - W_2}{W_1} = 1 - \dfrac{W_2}{W_1} = 1 - \dfrac{s_2}{s_1}$$

$$= 1 - \dfrac{0.30 \text{ m}}{0.50 \text{ m}} = 1 - 0.60 = 0.40.$$

ANALYZE Thus 40% of the nails have been used.

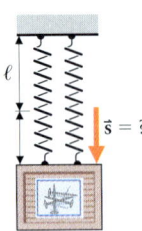

■ FIGURE 4.26
If two identical springs are used to suspend the full box of nails, what is their stretch s?

EXERCISE 4.5 ❖ If two springs identical to the one in Example 4.8 are used to suspend the nails (■ Figure 4.26), how much does each stretch at the start of the day?

EXAMPLE 4.9 ❖❖ Peter, spending a winter day working on his physics project, uses a compressed spring to push a wooden block over the surface of a frozen lake (■ Figure 4.27). Peter can accelerate the block from rest at $a = 1.3$ m/s^2, and he notices that the spring is compressed by 0.48 cm. If the block's mass is 140 kg, what is the spring constant?

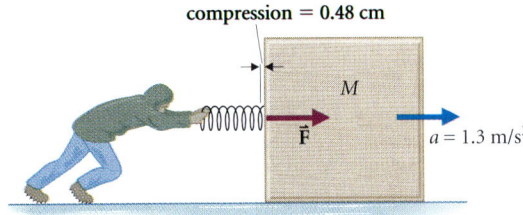

■ FIGURE 4.27
Peter's spring experiment. Wearing spiked shoes so that his feet don't slide, Peter uses a spring to accelerate a wooden block over the (frictionless) ice.

MODEL First we identify all the forces exerted on the block. Ice is a code word in physics problems for an ideal, frictionless surface, so the problem statement is instructing us to ignore friction. The weight and normal forces are exerted in the vertical direction; the spring force is exerted in the horizontal direction. The block remains on the lake surface, so there is no acceleration component in the vertical direction. We conclude that the weight and normal forces balance each other and the spring force F exerted on the sliding block is the only force contributing to its acceleration.

SETUP With the x-axis horizontal and the y-axis vertical, we have:

$$\Sigma F_x = ma_x$$
$$F = ks = ma_x.$$

SOLVE Using Hooke's law, the spring constant is:

$$k = \dfrac{F}{s} = \dfrac{ma_x}{s} = \dfrac{(140 \text{ kg})(1.3 \text{ m/s}^2)}{0.48 \times 10^{-2} \text{ m}} = 3.8 \times 10^4 \dfrac{\text{kg} \cdot \text{m/s}^2}{\text{m}} = 3.8 \times 10^4 \text{ N/m}.$$

ANALYZE Peter is using a rather feeble spring compared with one you might find in a bathroom scale (See Exercise 4.7).

EXERCISE 4.6 ❖❖ George, the football player, agrees to push Peter's block over the ice and manages to compress the spring by 0.80 cm. What acceleration does George achieve?

EXERCISE 4.7 ❖ Estimate the spring constant of a bathroom scale.

4.5.2 Friction

Consider trying to slide a chair over a rug. Applying a small force to the chair seems to produce no effect—friction prevents the chair from moving. If you push hard enough, the chair begins to slide, and then a somewhat smaller force keeps it sliding. Once it stops, the larger force is required to start it again. Of course, the whole task is easier if your buddy gets out of the chair first and thus reduces the force pressing the chair surface against the rug (■ Figure 4.28)! This whimsical example illustrates several important properties of friction:

- Friction acts differently between stationary surfaces (static friction) and sliding surfaces (kinetic friction).
- Static friction forces between two surfaces occur so as to balance forces that would otherwise cause the surfaces to slide.
- Up to a limit, static friction between surfaces has whatever magnitude is necessary to prevent the surfaces from sliding.
- The limit on static friction increases as the normal force between the surfaces is increased.

Friction is notoriously variable, and there is a vast amount of experimental lore for engineers who need precise information for machine design. We need a simple, approximate relation, reasonably true to experiment, that we can use to investigate the significance of friction in mechanical systems. Fortunately, a simple relation works quite well:

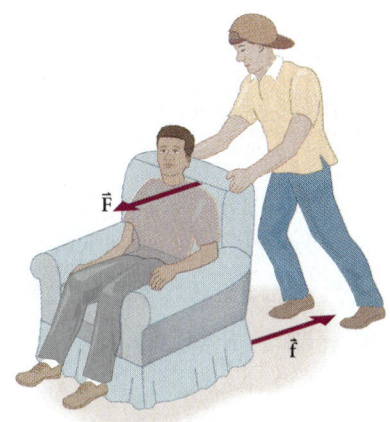

■ **Figure 4.28**
A small force will not budge your friend's chair at all. The force is balanced by static friction and the chair remains stationary. Once you push hard enough to start the chair sliding, a lesser force is needed to keep it sliding. The chair is easier to start and easier to keep sliding if you first evict your buddy from the chair!

THE MORE YOU PUSH, THE MORE STATIC FRICTION PUSHES BACK ... UP TO A POINT.

REMEMBER: $a \propto b$ MEANS $a =$ CONSTANT $\times b$.

> The magnitude of static friction between two surfaces is less than or equal to a limiting value. The limit is approximately proportional to the magnitude of the normal forces exerted by the surfaces.
>
> $$f_s \leq \mu_s n. \qquad (4.4)$$

The dimensionless coefficient μ_s is called the *coefficient of static friction*. It depends on the character of the two surfaces—the material they are made of, their cleanliness, their degree of polish, and so forth. For any application of eqn. (4.4), the coefficient μ_s has to be determined experimentally.

> REMEMBER: 1. Equation (4.4) is a relation between the *magnitudes* of two forces, f_s and n, not a relation between vectors. The friction and normal forces are perpendicular.
> 2. Equation (4.4) is an inequality. It doesn't give you the magnitude of the static friction force, but a *limit* that static friction cannot exceed.

Kinetic friction is also approximately proportional to the magnitude of the normal force between two sliding surfaces:

$$f_k = \mu_k n. \qquad (4.5)$$

The *coefficient of kinetic friction* μ_k also depends on the character of the surfaces and needs to be measured for use in any practical application. For a specific pair of surfaces, μ_k is usually less than μ_s.

A microscopic view of two surfaces in contact shows how friction forces arise (■ Figure 4.29). Irregularities in the surfaces act like sets of teeth that mesh together and make it difficult for the two surfaces to slide over each other. The rougher two surfaces are, the greater the friction forces that can occur between them. Lubricants work because small particles of lubricant fill up the spaces between the surface irregularities and prevent them from meshing together.

This rough-surface model of friction explains eqns. (4.4) and (4.5) in a qualitative way. Normal force causes the bumps and lumps on contacting surfaces to mesh. In the absence of

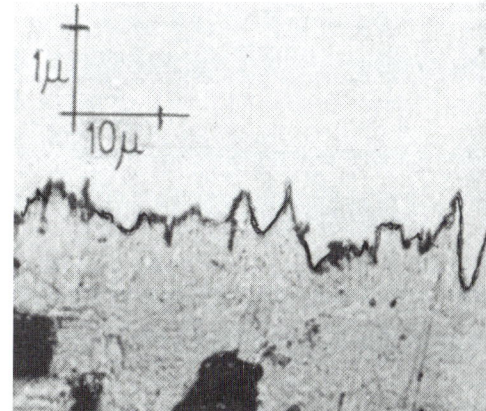

■ **Figure 4.29**
Even the surface of a polished metal piece is rough on the scale of a thousand atomic diameters. The section was cut so as to exaggerate the height of the irregularities ten times. When two such surfaces are in contact, the irregularities mesh and give rise to friction.

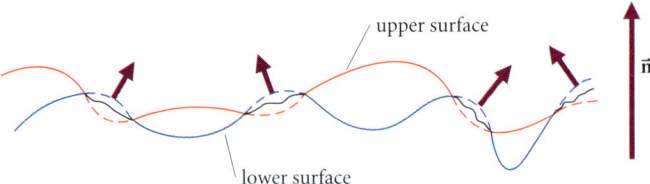

■ **Figure 4.30**
Rough-surface model of normal forces. Two surfaces in contact are shown exerting normal forces but no friction on each other. Contact forces (arrows) arise where surface irregularities touch and are distorted. (Dashed lines show the undistorted shapes of the surfaces.) The contact forces add to give the total normal force $\vec{n}$ on the upper surface. An equal and opposite force is exerted on the lower surface.

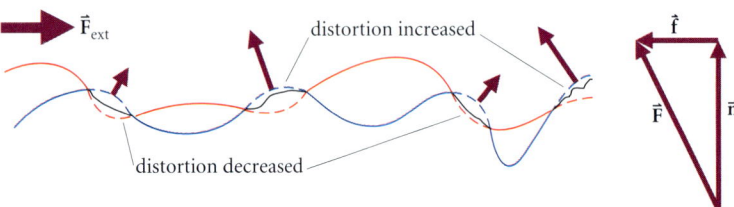

■ **Figure 4.31**
Rough-surface model of static friction. An external force attempts to slide the upper surface to the right. Some surface distortions are increased and some decreased. The contact forces now add to a total force $\vec{F}$ with components $\vec{n}$ perpendicular to the surface and $\vec{f}$ parallel to the surface.

An ice skater seems to glide without resistance over the surface of ice. The ice is slick because it self-lubricates. Apparently, a thin surface layer remains liquid until temperatures reach several degrees below the bulk freezing temperature (0° C).

MEMORIZE AND PRACTICE THIS METHOD; IT WORKS EVERY TIME.

other force components parallel to the surfaces, the microscopic normal forces that occur where the surface bumps make contact add together to form the total normal force perpendicular to the surface (■ Figure 4.30). When an external force $\vec{F}_{ext}$ is applied parallel to the surfaces (■ Figure 4.31), the lumps are bent parallel to the overall surface, and the microscopic contact forces have, on average, components parallel to the surface. These components add to make the observed friction force. A sufficiently large external force causes a large enough distortion to remove the meshing and the surfaces slide. How deeply the bumps and rough spots mesh depends on how strongly they are pressed together, which explains the dependence of friction on normal force. Kinetic friction may be modeled as the effect of many collisions between the many irregularities when the surfaces slide past each other.

The rough-surface model tells only part of the story of friction. Chemical bonding between surface atoms also contributes to friction. While important for understanding lubrication, chemical bonding forces do not change the qualitative predictions of the rough-surface model.

4.6 MOTION OF A SINGLE OBJECT SUBJECT TO SEVERAL FORCES

Will your parked car slide down the hill? How rapidly can a champion skier accelerate? In these situations, several forces, acting in different directions, are exerted simultaneously on a single object—the car or the skier. To apply Newton's second law to such problems, we need a systematic method.

4.6.1 The Method of Free-Body Diagrams

The first step is to decide whether you can reasonably model the object of interest as a particle. This step is important because Newton's second law is stated for an abstract particle, and we haven't yet learned how to use it for anything more complicated. We'll comment further on this in the next chapter; but for now, if you don't care about the size, shape, orientation, or internal structure of the object, then it's a particle.

The second step is to model the rest of the universe (not as awesome a task as it sounds). According to Newton's theory, the *only* way any object can influence another mechanically is by exerting force. So, if you work down the list of forces in Table 4.1 and determine all the forces being exerted on your object, those forces are a *complete model* of how the rest of the universe influences your object.

Next construct a diagram representing the force vectors with arrows; show the object as a point labeled by its mass. Because we are modeling the object as a particle isolated from the rest of the universe and free to move subject to the forces exerted on it, the diagram is called a *free-body diagram*.

Now analyze the free-body diagram. There are four steps:

1. State clearly any information you have about the object's acceleration.
2. Choose the most convenient coordinate axes. Choosing one axis to be parallel to the acceleration is usually wise.
3. State Newton's second law for each coordinate direction:

$$F_{net,x} = \sum F_x = ma_x, \qquad F_{net,y} = \sum F_y = ma_y, \qquad F_{net,z} = \sum F_z = ma_z,$$

and substitute in what you know about forces and acceleration.
4. Verify that you have the same number of equations as unknowns.
   If so, solve.
   If not, either (a) you may be able to solve for some combination of the unknowns, (b) you've missed something in the earlier steps, or (c) the problem can't be solved.

Let's look at a few examples.

REMEMBER: (1) IF IT TOUCHES YOUR OBJECT, ASSUME IT EXERTS FORCE. (2) IF IT ISN'T LISTED IN TABLE 4.1, IT ISN'T A MECHANICAL FORCE.

OFTEN, ONE OR MORE OF THE FORCES BEING EXERTED ARE UNKNOWN QUANTITIES. IT IS UNIMPORTANT AT THIS STAGE WHETHER YOU HAVE COMPLETE INFORMATION ABOUT SOME OF THE FORCES. IT IS CRUCIAL, HOWEVER, THAT YOU NOT LEAVE ANY OUT OF YOUR LIST.

IS THE ACCELERATION ZERO? DO YOU KNOW ITS DIRECTION?

4.6.2 Examples Involving Linear Motion

EXAMPLE 4.10 ♦♦ Sporty drives over the top of a hill onto a 47-m-long icy downslope at an angle of 30° with the horizontal and begins to slide. If Sporty has slowed the car nearly to rest before it reaches the ice (■ Figure 4.32), how much time does the car take to reach the bottom?

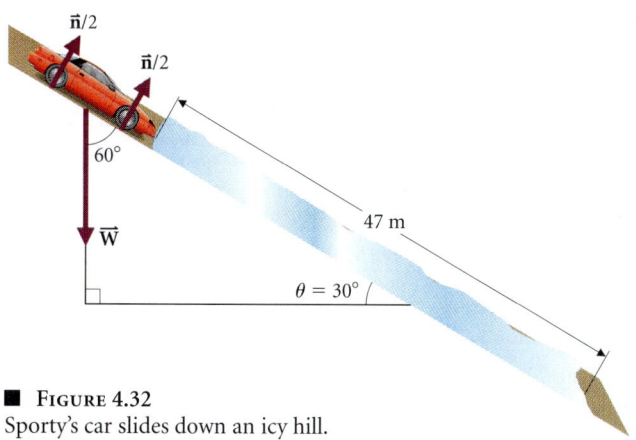

■ FIGURE 4.32
Sporty's car slides down an icy hill.

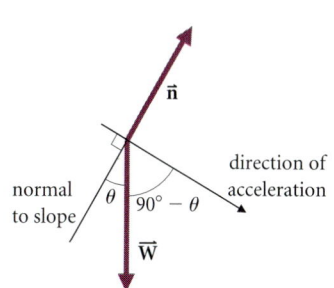

■ FIGURE 4.33
Free-body diagram for Sporty's car as it appears before coordinates are chosen. The normal force is at right angles to the slope. The weight is vertical, and makes an angle θ with the normal. The car accelerates down the slope.

MODEL We begin by drawing the free-body diagram for the car (■ Figure 4.33), which is sliding with negligible friction on the *icy* surface. The forces acting are weight and normal force. Then we apply Newton's second law.

SETUP The car has no acceleration perpendicular to the road, since it remains on the surface. So the normal force exactly balances the component of the car's weight perpen-

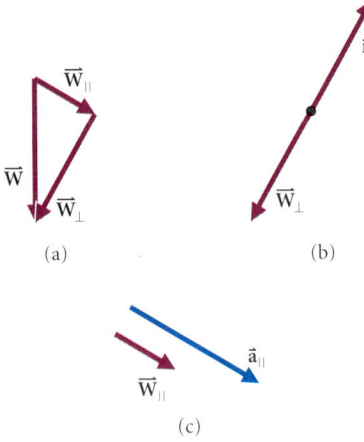

FIGURE 4.34
(a) The weight of Sporty's car resolved into components. (b) The normal force balances the component of weight perpendicular to the slope. (c) The component of weight down the slope is unbalanced.

HAVING FOUND THE ACCELERATION, WE HAVE A KINEMATICS PROBLEM TO SOLVE.

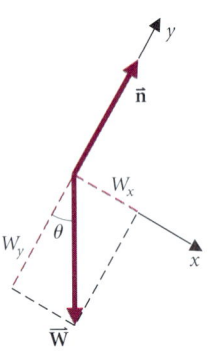

FIGURE 4.35
The free-body diagram as usually drawn after the coordinates have been chosen and the force components resolved.

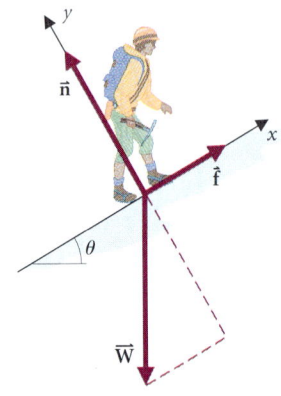

FIGURE 4.36
The climber has no acceleration, so the forces exerted on her balance.

dicular to the road (■ Figure 4.34). The component of weight parallel to the road accelerates the car down the hill (Figure 4.34c). Choosing coordinates with the x-axis downward along the hill makes maximum use of these facts (■ Figure 4.35).

$$\vec{F}_{net} = \vec{n} + \vec{W} = m\vec{a}.$$

The components of the forces are:

Force	x-component	y-component
$\vec{n}$	0	$+n$
$\vec{W} = m\vec{g}$	$mg \sin \theta$	$-mg \cos \theta$

Then applying Newton's second law in each coordinate direction we have:

$$\sum F_x = ma_x. \qquad \sum F_y = ma_y.$$
$$mg \sin \theta = ma_x. \quad (i) \qquad n - mg \cos \theta = 0. \quad (ii)$$

SOLVE From eqn. (i), we have:

$$a_x = g \sin \theta.$$

The car has constant acceleration down the plane, and so the time taken may be found from the displacement Δx (eqn. 2.11 with $v_i = 0$):

$$\Delta x = \tfrac{1}{2} a_x t^2.$$

Thus,

$$t = \sqrt{\frac{2\Delta x}{a_x}} = \sqrt{\frac{2\Delta x}{g \sin \theta}} = \sqrt{\frac{2(47 \text{ m})}{(9.8 \text{ m/s}^2) \sin 30°}} = 4.4 \text{ s}.$$

ANALYZE Notice that for $\theta = 0$ (horizontal surface), $a_x = 0$ and for $\theta = 90°$ (vertical surface), $a_x = g$. These are the results we would expect.

Also note that we didn't need to use the equation for force balance in the y-direction; that equation determines the magnitude of the normal force. ■

Since the magnitude of a static friction force can take any value up to a maximum $f_{max} = \mu_s n$, problems involving friction often require finding a maximum or minimum value of some quantity, such as the slope of a mountainside in the following example.

EXAMPLE 4.11 ♦♦ A rock climber's shoes have a coefficient of friction $\mu_s = 1.4$ when used on granite. Using only friction, what is the steepest slope the climber can walk up without slipping?

MODEL We begin by drawing a free-body diagram showing the forces acting on the climber (■ Figure 4.36). In the diagram, we've chosen the x-direction parallel to the slope and the y-direction perpendicular to the slope. The friction force exerted on the climber by the rock balances the component of her weight parallel to the slope, while the normal force exerted by the rock balances the perpendicular component of her weight. On steeper slopes, a greater friction force is required. On the steepest possible slope, the static friction reaches its maximum value:

$$f_s = f_{max} = \mu_s n.$$

The climber walks carefully and smoothly, minimizing acceleration.

SETUP She is not accelerating, so:

$$\sum F_x = f_x - W \sin \theta = 0 \quad \text{and} \quad \sum F_y = n - W \cos \theta = 0.$$

TRIAL We may solve these equations for $f_s = W \sin \theta$ and $n = W \cos \theta$. We do not need to know the climber's weight to find θ, as it cancels out in the ratio f_s/n.

SOLVE

$$\mu_s \geq \frac{f_s}{n} = \frac{W \sin\theta}{W \cos\theta} = \tan\theta.$$

Since f_s/n has a maximum value of $\mu_s = 1.4$, $\tan\theta \leq 1.4$. The steepest slope the climber can walk up is $\tan^{-1} 1.4 = 54°$.

ANALYZE Because the contact forces $\vec{f}$ and $\vec{n}$ together balance the climber's weight, the ratio of their magnitudes, f/n, equals the ratio of the two components of the single force $\vec{W}$. This ratio is independent of the magnitude of $\vec{W}$ and depends only on the angle θ. ∎

EXERCISE 4.8 ♦♦ A mountain goat is standing on a slope of 30°. If the animal doesn't slide, what is the minimum coefficient of friction between the goat's hooves and the rock?

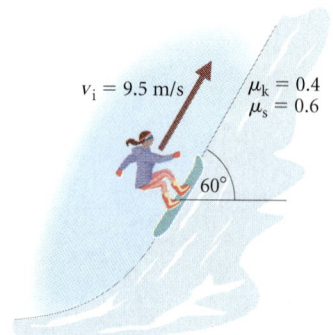

FIGURE 4.37
Melinda's initial conditions.

Study Problem 4 ♦♦ *Hotdog Skiing on Spring Snow*

Melinda starts directly up a snowslope inclined at angle 60° with the horizontal on her skiboard (■ Figure 4.37). Her initial speed is $v_i = 9.5$ m/s. The coefficients of kinetic and static friction between skiboard and snow are $\mu_k = 0.40$ and $\mu_s = 0.60$. (Sticky!) Determine whether Melinda slides back to her starting point after she comes to rest and, if so, how much time she remains on the slope.

Modeling the System

This physical situation is similar to that in Example 4.10, with the addition of friction. The kinetic friction force that opposes Melinda's motion is proportional to the normal force between skiboard and slope and is constant since the normal force is constant. Thus Melinda moves up the slope with constant deceleration. Her velocity eventually reaches zero at some point, where she will stick unless the slope is sufficiently steep (cf. Example 4.11). Then she will slide down again.

Plan

The problem statement implies that we should consider motion both up and down the slope. First we describe the forces acting on Melinda in each case. Newton's law $\vec{F} = m\vec{a}$ gives her acceleration. To find the net force $\vec{F} = \vec{W} + \vec{f} + \vec{n}$, we'll need the relation $f = \mu_k n$ between the magnitudes of friction and normal forces. Once we have $\vec{a}$, we can find the time for Melinda, moving up the slope, to decelerate to zero velocity. At the top, we use $f_s \leq \mu_s n$ to determine whether she slides down again. If she does, the distance traveled up the slope is also necessary to compute the time she takes sliding back down. Since acceleration is constant in each case, Table 2.4 provides the necessary relations.

Construction of Equations

■ Figure 4.38 is Melinda's free-body diagram. Whether Melinda slides upward or downward, the normal force balances the perpendicular component of her weight.

Case 1. Sliding Upward:

x-components: $\sum F_x = ma_{x,1}$ y-components: $\sum F_y = 0$
$\qquad\qquad\qquad W_x - f_1 = ma_{x,1}$ $n + W_y = 0$
$\qquad\qquad -mg\sin\theta - \mu_k n = ma_{x,1}$ (i) $n - mg\cos\theta = 0$ (ii)

From eqn. (ii): $\qquad n = mg\cos\theta$,

and substituting this result into eqn. (i) gives

$$a_{x,1} = -g(\sin\theta + \mu_k \cos\theta). \qquad\qquad (iii)$$

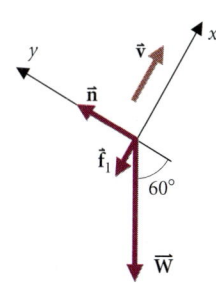

FIGURE 4.38a
Free-body diagram for Melinda sliding upward.

Once we have the acceleration, we may use eqns. (2.10) and (2.12) to calculate the distance traveled up the slope and the time required:

$$\Delta t = \frac{v_f - v_i}{a_{x,1}} \qquad \text{(iv)}$$

$$s = v_{av}\Delta t = \tfrac{1}{2}(v_f + v_i)\Delta t. \qquad \text{(v)}$$

When Melinda comes to rest at the top, the snow exerts static friction on her skiboard. If she is to slide down again, the component of her weight down the plane has to exceed the maximum static friction $f_{max} = \mu_s n$. She slides down if

$$mg \sin\theta > \mu_s n = \mu_s mg \cos\theta$$

or, as we saw in the previous example, if $\tan\theta > \mu_s$.

If she slides down, kinetic friction opposes her motion. That is, the direction of the friction force is *up* the slope (■ Figure 4.38b).

Case 2. Sliding Downward:

x-components: $\sum F_x = ma_{x,2}$ y-components: $\sum F_y = 0$
$W_x + f_2 = ma_{x,2}$ $n + W_y = 0$
$-mg \sin\theta + \mu_k n = ma_{x,2}$ $n - mg \cos\theta = 0$

Again, we solve these equations for the acceleration:

$$a_{x,2} = g(\mu_k \cos\theta - \sin\theta). \qquad \text{(vi)}$$

Starting from rest, the time to travel down the slope (cf. eqn. 2.11) is:

$$\Delta t = \sqrt{2\Delta x/a_{x,2}}. \qquad \text{(vii)}$$

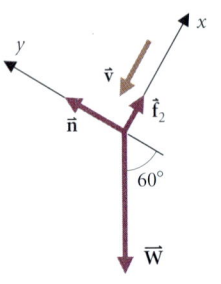

■ FIGURE 4.38b
Free-body diagram for Melinda sliding downward. Friction opposes the motion.

Solution

Since the final upward velocity is zero and the initial velocity is v_i, we may obtain the time and distance for the upward motion using eqns. (iii) and (iv):

$$t_{up} = \frac{-v_i}{-g(\sin\theta + \mu_k \cos\theta)} = \frac{v_i}{g(\sin 60° + \mu_k \cos 60°)}$$

$$= \left(\frac{9.5 \text{ m/s}}{9.8 \text{ m/s}^2}\right)\left(\frac{1}{\sqrt{3}/2 + 0.40/2}\right) \qquad \text{(viii)}$$

$$= 0.91 \text{ s}.$$

Then using this result in eqn. (v):

$$s = \tfrac{1}{2}v_i t_{up} = \tfrac{1}{2}(9.5 \text{ m/s})(0.909 \text{ s}) = 4.3 \text{ m}. \qquad \text{(ix)}$$

Now, $\tan 60° = \sqrt{3} > \mu_s = 0.60$, so Melinda doesn't stick. Since she slides the same distance downward as upward $\Delta x = -s$, and the required time from eqns. (vi), (vii), and (ix) is:

$$t_{down} = \sqrt{\frac{-2s}{a_{x,2}}} = \sqrt{\frac{2(-v_i^2)/[2g(\sin\theta + \mu_k \cos\theta)]}{g(\mu_k \cos\theta - \sin\theta)}}$$

$$= \frac{v_i}{g}\sqrt{\frac{1}{\sin^2\theta - \mu_k^2 \cos^2\theta}}$$

$$= \frac{9.5 \text{ m/s}}{(9.8 \text{ m/s}^2)\sqrt{[\sin^2 60° - (0.40 \cos 60°)^2]}}$$

$$= 1.2 \text{ s}.$$

Analysis

It takes more time for Melinda to come down than to go up. She travels the same distance but at a lower average speed. In both directions, friction opposes sliding. On the way down, friction partially counteracts gravity. Thus her acceleration has a smaller magnitude than on the upward trip when friction and gravity act in the same direction.

On a lesser slope ($\tan\theta < \mu_s$), Melinda would be caught by static friction at the top of her path. She could avoid sticking by jumping the skiboard at the top or by starting out not quite straight up the slope, so her velocity never becomes zero. Our model of friction supposes an abrupt transition from static friction to kinetic friction. According to this model, any motion is sufficient to avoid static friction. Measurement of the minimum speed actually needed to keep Melinda from sticking on the lesser slope would make an interesting test of the model!

4.7 DYNAMICS OF CIRCULAR MOTION

A particle in uniform circular motion at speed v has acceleration (eqn. 3.11):

Uniform circular motion $\vec{a} = (v^2/R,$ toward the center of the circle$)$.

Thus an object can undergo uniform circular motion only if the total force acting on it is always directed toward the center of its circular path. For this reason, the name *centripetal*, or center directed, is often used to describe the net force in circular motion.[3]

> REMEMBER: 1. It is the *total* force that has to be center directed. In your free-body diagram, it may be that *no individual force* is directed toward the center.
>
> 2. Nothing named "centripetal force" should **ever** appear in a free-body diagram. It is not another kind of force; what it describes is the effect of one or more of the forces in the diagram.

EXAMPLE 4.12 ♦♦ Peter, continuing his physics experiments on the frozen lake, rides a sled off a ramp onto the ice (■ Figure 4.39). He grabs the end of a rope and hangs on through a 90° turn before letting go. If Peter comes off the ramp at a speed of $v_o = 9.6$ m/s, and the rope has length $\ell = 8.5$ m, compare Peter's weight with the force he exerts on the rope.

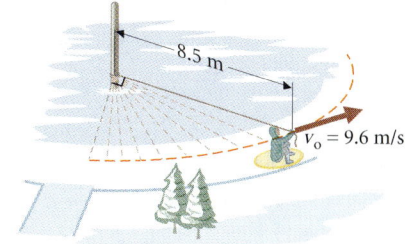

■ FIGURE 4.39
Peter's experiment with circular motion.

MODEL So long as Peter holds onto the rope, he travels along a circle with radius equal to the rope's length: $R = \ell$. Since the lake is frozen, we neglect friction. Thus the forces acting on Peter are his weight, the normal force, and the rope tension. The free-body diagram (■ Figure 4.40) is drawn in a vertical plane through Peter and through the center of the circle. Peter's acceleration in uniform circular motion is toward the center of the circle with magnitude v_o^2/R. The net force exerted on Peter must have the correct direction and magnitude to cause this acceleration. We are asked for the force "Peter exerts on the rope." According to Newton's third law, this is the pair of the force T that the rope exerts on Peter, and it has the same magnitude. We are to compare this force with Peter's weight $W = mg$.

SETUP In the free-body diagram, take coordinate axes vertical and inward along the rope. The vertical forces, the weight and normal force, balance each other, and the horizontal rope tension force causes Peter's acceleration:

$$T = \sum F_x = ma_x = \frac{mv_o^2}{\ell}.$$

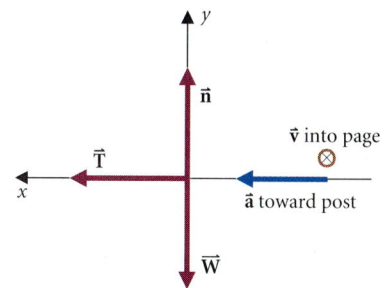

■ FIGURE 4.40
Free-body diagram for Peter. The diagram is drawn in a vertical plane passing through the post and through Peter's position. The unbalanced tension force accelerates Peter toward the post. In these coordinates, $\vec{a} = a\hat{\imath}, \vec{v} = v_o\hat{k}$.

---

[3] The term *centrifugal*, or center fleeing, also exists, and it has a legitimate use in the study of accelerated reference frames. However, it is often misused to refer to a nonexistent outward force not caused by any object. Beware, your intuition may betray you here; Newton's did for quite some time.

FIGURE 4.41
Peter's summertime experiment.

SEE §3.2.6.

SOLVE We want the following ratio:

$$\frac{T}{W} = \frac{mv_o^2/\ell}{mg} = \frac{v_o^2}{g\ell} = \frac{(9.6 \text{ m/s})^2}{(9.8 \text{ m/s}^2)(8.5 \text{ m})} = 1.1.$$

ANALYZE Peter's task is equivalent to hanging from a vertical rope after gaining 10% in mass. To complete a 90° turn (one-quarter of a circle), he has to hang on for a time interval $\Delta t = (2\pi R/4)/v_o = \pi \ell/(2v_o) = 1.4$ s.

EXERCISE 4.9 ❖ Experimenting at a summer picnic, ■ Figure 4.41, Peter runs from the dock, swings on a rope along a horizontal circle over the river and then drops into the water. Draw a free-body diagram showing the forces acting on Peter as he swings on the rope. Are any of the individual forces directed toward the center of Peter's path? Explain why the rope cannot be horizontal, no matter how fast Peter leaps from the dock.

A particle moving along an arbitrary path with instantaneous speed v has an acceleration component perpendicular to the path that describes how its direction changes: $a_\perp = v^2/R$ is given by the same formula as acceleration in uniform circular motion. In the next example, we use this result to analyze how a car negotiates a turn and to illustrate why an understanding of friction is important in studying highway safety.

EXAMPLE 4.13 ❖❖ A car, traveling along a level road, enters a turn with a radius of curvature $R = 1.0 \times 10^2$ m. The coefficient of friction between the road and the tires is $\mu_s = 0.50$. What is the maximum speed at which the car can negotiate the turn?

MODEL When a car turns a corner on a level road, friction is the only force acting horizontally on the car. It is therefore the friction exerted by the road on the car that accelerates it around the turn. Because the tires roll without slipping, we are concerned with static friction, and it is the limit on static friction that sets a maximum speed for rounding the turn.

SETUP ■ Figure 4.42 shows two views of the car, an end view and one looking down on it. The end view shows the horizontal and vertical forces. Since the car is not accelerating vertically:

$$0 = \sum F_y = n - W \Rightarrow n = W = Mg. \quad \text{(i)}$$

The top view shows the horizontal frictional force $\vec{f}_s$ acting on the car. Since friction is the only unbalanced force acting, it equals the car's mass times its acceleration:

$$f_s = \sum F_x = Ma_x = Mv^2/R. \quad \text{(ii)}$$

The maximum speed is that which requires maximum possible friction $f_{max} = \mu_s n$. Combining this result with eqns. (i) and (ii), we have:

$$Mv_{max}^2/R = \mu_s n = \mu_s Mg.$$

FIGURE 4.42
A car rounds a turn on a level road. (a) The forces exerted on the car act in the vertical plane of the end view. That is the plane of the free-body diagram (b). (c) From above, we see the relation of the forces to the car's motion around the turn. Friction exerted by the road on the tires is unbalanced and accelerates the car.

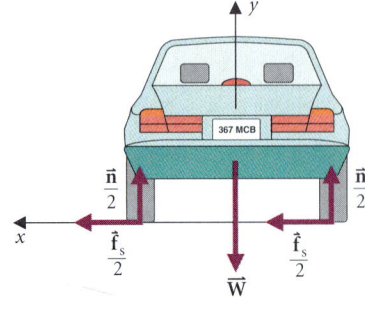

(a) End view

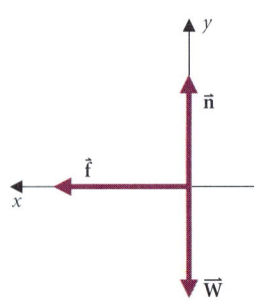

(b) Free-body diagram

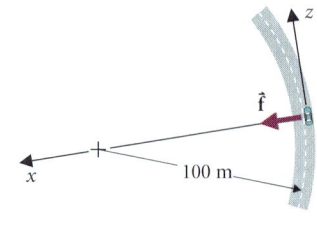

(c) Top view

SOLVE The mass of the car cancels out, and

$$V_{max} = \sqrt{\mu_s gR} = \sqrt{0.50(9.8 \text{ m/s}^2)(1.0 \times 10^2 \text{ m})} = \sqrt{490} \text{ m/s} = 22 \text{ m/s}.$$

ANALYZE The maximum speed depends on the road conditions via the coefficient of friction. On a wet road, the coefficient of friction between the tires and the road is reduced, and the car cannot turn as rapidly as on dry pavement.

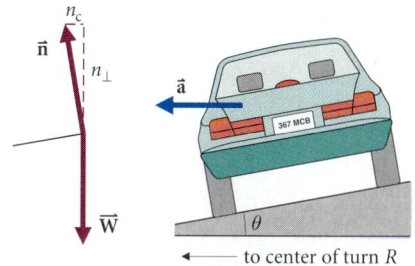

FIGURE 4.43
Car rounding a banked turn. Here the car is moving at a speed that requires no friction force on the car's tires. The vertical component of normal force balances the car's weight, and the horizontal component accelerates the car around the turn. At any other speed, friction forces arise either to prevent the car sliding outward or to prevent it sliding inward. See Problem 87.

Roads designed for high-speed traffic have banked turns (■ Figure 4.43). Then both the friction and normal forces exerted by the road on the car have horizontal components that together cause the necessary acceleration. No friction is necessary, and you can round the turn even on an icy road at the proper speed for a given bank angle. On the icy road though, you'll slide down if you try to go slower!

Airplanes also make turns by banking. The lift force, due to contact forces of moving air on the wing, acts at right angles to the wing chord (■ Figure 4.44). When the aircraft banks, the pilot maneuvers to obtain greater lift than necessary for level flight (■ Figure 4.45); the vertical component of lift balances the airplane's weight, and the horizontal component accelerates the plane.

EXAMPLE 4.14 ◆◆ An aircraft of mass 2.0×10^3 kg flying at 1.0×10^2 m/s banks at a 30° angle. What is the magnitude of the lift force? What is the radius of the turn? How long does the aircraft take to complete a 360° turn?

MODEL Figure 4.45 is the free-body diagram. The lift vector is inclined 30° from the vertical toward the center of the turn. In a properly banked turn, the vertical forces balance. The horizontal component of lift provides the centripetal force.

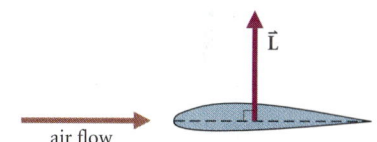

FIGURE 4.44
Lift, the result of normal forces exerted by air moving around an airplane wing, acts perpendicular to the wing chord.

SETUP
$$0 = \sum F_y \qquad \sum F_x = ma_x$$
$$0 = L_2 \cos \theta_b - W$$
$$L_2 \cos \theta_b = Mg. \quad (i) \qquad L_2 \sin \theta_b = \frac{Mv^2}{r}. \quad (ii)$$

SOLVE From eqn. (i):

$$L_2 = \frac{Mg}{\cos \theta_b} = \frac{(2.0 \times 10^3 \text{ kg})(9.8 \text{ m/s}^2)}{\cos 30°} = 2.3 \times 10^4 \text{ N}.$$

Substituting into (ii):

$$\frac{Mv^2}{r} = \frac{Mg}{\cos \theta_b} \sin \theta_b = Mg \tan \theta_b$$

and
$$r = \frac{v^2}{g \tan \theta_b} = \frac{(1.0 \times 10^2 \text{ m/s})^2}{(9.8 \text{ m/s}^2)\tan 30°} = 1800 \text{ m}.$$

The time for a 360° turn equals the period of the circular motion;

$$T = 2\pi r/v = 2\pi(1770 \text{ m})/(1.0 \times 10^2 \text{ m/s}) = 110 \text{ s} = 1.8 \text{ min}.$$

ANALYZE The lift exceeds the weight by 15%.

When an aircraft banks steeply, the pilot and passengers *feel* heavier even though their actual weight is not changed. To minimize this feeling, airliners make shallow banks. An airliner making a *standard-rate turn* completes a 360° turn in 4 min.

What does it mean *to feel* heavier? In level flight, each passenger's weight is balanced by the normal force exerted by the seat, exactly as if the person were at rest on the ground. The person's muscles tense to maintain an upright posture. This muscle tension and the pressure on our bottoms is what we sense when we speak of feeling our weight. When the airplane banks, the seat has to exert enough normal force to balance weight and to accelerate the

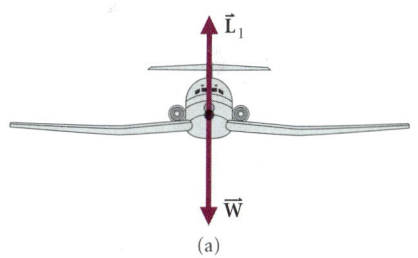

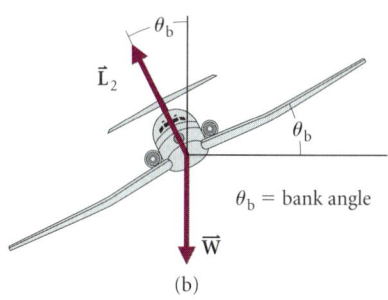

θ_b = bank angle

FIGURE 4.45
In level flight (a), the lift balances an aircraft's weight. In a banked turn (b), the horizontal component of lift is unbalanced and accelerates the plane. The normal component of lift balances the plane's weight.

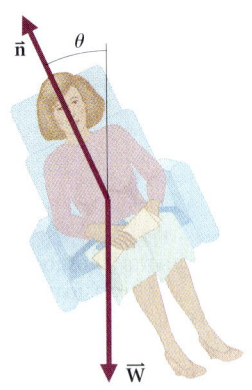

■ **Figure 4.46**
Forces on a passenger in a banked turn. Normal force exerted by the seat balances the passenger's weight and accelerates her around the turn. The passenger feels as if her weight were $|\vec{n}|$.

passenger around the turn (■ Figure 4.46). The increased normal force and the larger muscle tension in response to it is perceived as "feeling heavier."

EXERCISE 4.10 ◆◆ If an airliner flying at 2.00×10^2 m/s makes a standard-rate turn ($T = 4.00$ min), what bank angle is required? What is the percentage increase in the *perceived* weight of the passengers?

We are now ready to answer the question posed at the beginning of the chapter: how can a motorcyclist ride on a curved vertical wall?

EXERCISE 4.11 ◆◆◆ If a motorcyclist rides on the interior of a cylinder 10.0 m in radius and the coefficient of static friction between the cycle's tires and the wall is $\mu_s = 0.65$, discuss why there is a minimum speed at which the cyclist can perform the stunt. Calculate the value of this speed. (*Hint:* Start by drawing the free-body diagram. Which force points toward the center of the circle and which points upward? What is the relation between them?)

4.8 Newton's Laws of Motion

4.8.1 The First Law of Motion and Inertial Frames of Reference

> An object on which no net force acts remains at rest or moves in a straight line at constant speed.

Why should we bother to state Newton's first law? After all, we can derive it easily from the second law: if the total force acting on a body is zero, its acceleration is zero. That is, it moves at constant velocity, which is what the first law says.

We have already noted in §4.1 that the first law reaffirms Galileo's relativity principle: motion at constant velocity is natural and requires no explanation. There is a deeper significance to this law. In Chapter 2, we saw that motion must always be described relative to a particular reference frame—for example, the velocity of a boat with respect to a river. Newton's first law defines a particular kind of reference frame, called an inertial reference frame, in which the rest of Newton's laws apply.

> An *inertial reference frame* in Newtonian physics is one in which Newton's first law holds. That is, in an inertial reference frame, all observed accelerations result from forces exerted by identifiable objects.

Newton's first law thus serves three functions in the theory: it acts as a definition of inertial reference frame; it claims that such frames actually exist; and it specifies that the rest of the theory applies only in an inertial frame.

4.8.2 The Second Law of Motion

> The acceleration of an object is proportional to the total force acting on the object and inversely proportional to the mass of the object.
>
> $$\vec{a} = \vec{F}_{net}/m$$

Digging Deeper

Inertial and Noninertial Reference Frames

Consider people riding on a merry-go-round (■ Figure 4.47). The horses, and the other parts of the merry-go-round are a set of objects with fixed relations to each other. So you might think them a perfectly valid reference frame for measuring velocities and accelerations. But if someone drops a marble, it accelerates rapidly off the edge of the merry-go-round without any apparent force to cause the acceleration. We must either decide that Newton's second law is faulty or that something is wrong with the reference frame.

Since the merry-go-round is rotating, every point on it is accelerating in uniform circular motion. Acceleration of any kind turns out to make a reference frame noninertial.

The surface of the Earth is only approximately an inertial reference frame because of the Earth's daily rotation, but you don't notice this effect in most laboratory experiments. A reference frame defined by the positions of the stars is much more nearly inertial and is completely adequate for problems such as tracking spacecraft in the solar system. However, astronomers can detect acceleration of the stars within our local galaxy and must use very distant, extragalactic objects as references for the most accurate work.

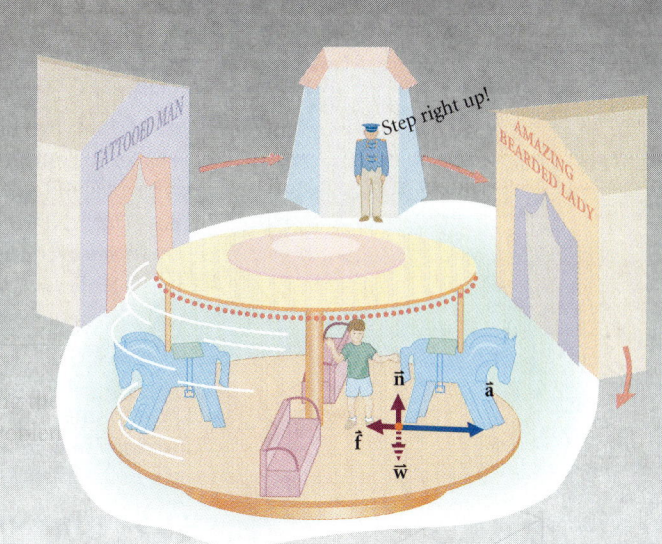

■ **Figure 4.47**
Merry-go-round reference frame. All the universe twirls about the chairs and horses that define the reference frame. Meanwhile, a child's marble accelerates in the opposite direction from the net force acting on it.

Experiments still cannot tell us whether there is an **exactly** inertial reference frame, though the most distant objects in the universe define an extremely good practical standard.

The second law ties together the kinematic description of how objects move, the nature of objects as described by their mass, and force as the cause of acceleration. It is the central relation in Newton's theory and appears explicitly in every application. Writing it in its three possible forms, we can emphasize three different aspects of this relation.

The first version of the law, $\vec{a} = \vec{F}_{net}/m$, emphasizes applications in which we know the mass of an object and the force acting on it and can thus compute the object's path from its acceleration. This is the most common application of the law and the form in which the theory predicts the behavior of nature.

The traditional form of the law, $\vec{F}_{net} = m\vec{a}$, has a slightly different emphasis; one determines the net force acting by knowing an object's mass and acceleration. In this form, the law may be used to identify the kinds of force that exist, along the lines we followed in §4.1. If the theory is to be useful, there should not be a bewildering number of different kinds of force, and those that do exist should have fairly simple descriptions.

In its third possible form, $m = |\vec{F}_{net}|/|\vec{a}|$, the second law acts as a definition of mass, and suggests a procedure for measuring mass by applying a known force and observing an object's acceleration. Thus the second law is both a statement about nature and a research plan for learning physics.

4.8.3 Newton's Third Law

> Forces occur in pairs. The two forces in each pair are of the same kind, have equal magnitude and opposite direction, and act on different members of a pair of objects.

The third law recognizes force as a process of interaction between objects. Its importance will become clearer as we study systems of particles and conservation principles in Chapters 5–10.

4.8.4 What Is Mass?

Experiments establish mass as a precise form of the intuitive concept "amount of matter"; beyond that, they do not tell us what mass *is*. In a sense they cannot, for Newtonian physics accepts the concept of mass as fundamental. In this way, mass is like the universal substance of the ancient Greek philosophers. Alone, it explains nothing. Only with a structure or set of rules—here Newton's laws—does the concept of mass have meaning.

To say a concept is fundamental may do little for your personal intuition, but it does suggest a plan of action: practice working with the whole theory and expect intuition to grow in the process. Not coincidentally, this is also the path of scientific progress. As theory is applied, unexpected connections are found and still more fundamental insights are discovered. For example, the central role of mass in Newtonian physics led Antoine Lavoisier (1743–1794) to introduce mass measurements in chemistry. His experiments demonstrated the role of oxygen in combustion and led to a revolution in chemical theory.

4.8.5 The Structure of Newtonian Mechanics

We are to understand mass by gaining experience using the whole of Newtonian mechanics. The same is true of force, reference frame, or any other concept we have discussed. Like any successful scientific theory, Newtonian mechanics provides a model of nature that is fully understood only as a whole.

You now have some experience working with mechanics and answering real-world questions, but you have encountered the theory one piece at a time. We close this chapter with a short summary of how it all fits together (■ Figure 4.48).

The job of a physical theory is to explain the world by accurately describing both the things that exist in it and the rules that govern their behavior. With Newton's theory, we describe the world as made up of small particles and assume that understanding the motion of these particles explains the world. Thus Newton's theory has three aspects: the nature of motion and the space–time arena in which it occurs; the nature of objects as collections of particles with a single fundamental property, mass; and the nature of force, the process that changes motion. These three aspects are tied together by Newton's second law, which is a precise relation among quantitative measurements of the three aspects. The concept of an *inertial reference frame* precedes the rest of the theory. The first law declares that such frames exist and that the rest of the theory is valid when applied in such a frame.

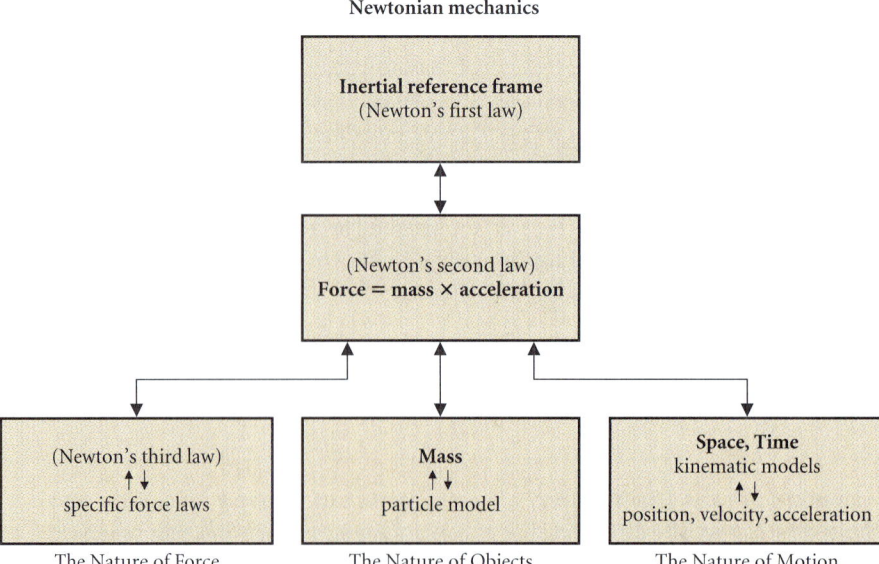

■ FIGURE 4.48
The logical structure of Newtonian mechanics. Three aspects of the theory—the nature of force, the nature of objects, and the nature of motion—are tied together by Newton's second law. The entire theory applies in an inertial reference frame, whose existence is claimed by Newton's first law.

Where Are We Now?

Newton's laws of motion give us a highly successful model for the mechanical behavior of matter. We have described the nature of force, identified the kinds of force that occur in mechanics, and developed practical expressions for describing them. These force laws, kinematics, and Newton's three laws of motion are our basic tools for applying Newtonian physics. We have developed a systematic procedure for solving problems involving forces acting on a single object. Our next step is to extend this procedure to more complex systems.

What Did We Do?

Everyone has intuitive experience with force, but several important facts about force seem to go against that experience: nonliving things exert forces; force changes velocity but is not required for motion at constant velocity; objects exert and experience forces but do not possess force within them; gravitational force acts at a distance between objects that are not in contact.

Newton's first law restates Galileo's relativity principle and gives the name *force* to the process that causes change of velocity:

> An object on which no net force acts remains at rest or moves along a straight line at constant speed.

Newton's second law defines mass as the essential mechanical property of an object and states the precise relation among net force, mass, and acceleration:

> The acceleration of an object is proportional to the total force acting on the object and inversely proportional to the mass of the object.
>
> $$\vec{\mathbf{a}} = \vec{\mathbf{F}}_{net}/m.$$

Newton's third law emphasizes that force occurs when pairs of objects interact:

> Forces occur in pairs. The two forces in each pair are of the same kind, have equal magnitude and opposite direction, and act on different members of a pair of objects.

Using the first law to analyze stationary systems, we obtained a list of forces that occur in mechanics (Table 4.1). With the second law, we were able to find practical expressions for describing these forces.

1. Gravity
   An object's weight means the gravitational force exerted on the object. The weight of an object of mass m is defined by the relation
   $$\vec{\mathbf{W}} = m\vec{\mathbf{g}}.$$
   At any one place and time, the acceleration due to gravity $\vec{\mathbf{g}}$ is fixed and an object's weight is proportional to its mass. Weight is a vector and depends on an object's position. It is to be distinguished carefully from the object's mass, which is both a scalar and a fixed property of the object.

2. Spring Forces
   A spring exerts a force when it is compressed or stretched from its relaxed length. The direction of the force is opposite the distortion of the spring. The magnitude of the force is proportional to the amount the spring is stretched or compressed.
   $$|\vec{\mathbf{F}}| = ks.$$

3. String Forces
   The *tension* force exerted by a string is similar in nature to that of a stretched spring except that the stretch of the string is too small to be measured conveniently.

4. Normal Forces

   When two surfaces are in contact, they exert *normal* forces on each other. The forces act normally (perpendicular) to the surfaces and, like string forces, usually result from distortions too small to be measured conveniently.

5. Friction Forces

   When two rough surfaces come in contact, they may exert *friction* forces on each other parallel to the surfaces. Friction forces result from microscopic irregularities (roughness) of the surface and from chemical bonds between atoms at the surfaces. Friction opposes relative motion between the surfaces.

   Kinetic friction is proportional in magnitude to the normal force acting between two surfaces moving relative to one another: $f_k = \mu_k n$.

   Static friction may have any magnitude up to a maximum that is proportional to the normal force acting: $f_s \leq \mu_s n$.

We apply Newton's laws to the motion of an object using the method of *free-body diagrams*. The sum of force components in each coordinate direction equals the object's mass times the component of its acceleration in that direction.

The total force acting on an object in uniform circular motion acts centripetally—it changes with time so as always to point toward the center of the circle.

Newton's laws, kinematics, and the force laws form a complete logical system whose concepts are learned by practice and experience using the theory. The first law declares the existence of at least one inertial reference frame in which the other two laws are valid. The second law defines mass, provides a research plan for studying the kinds of force that occur, and is a mathematical relation for predicting the motion of objects under known forces.

Practical Applications

The Newtonian model is the key to understanding most physical situations we encounter in our everyday lives. Single-particle dynamics already allows practical calculations about highway design, airplane flight, amusement park rides, skiers, climbers, and so on. The methods we develop in this chapter are the first step in studying more complicated systems.

Solutions to Exercises

4.1 The force triangle in Figure 4.11 is a right triangle, so $\sin\psi = \frac{4}{5}$ and $\psi = 53°$. However, the angle we're asked for is $\phi = 180° - \psi = 127°$.

COMPONENTS PERPENDICULAR TO THE FIVE-PERSON ROPE

$$F_\perp = 3F\sin\psi - 4F\sin(90° - \psi) = 3F(\tfrac{4}{5}) - 4F(\tfrac{3}{5}) = 0.$$

COMPONENTS PARALLEL TO THE FIVE-PERSON ROPE

$$\begin{aligned} F_\parallel &= -5F + 3F\cos\psi + 4F\cos(90° - \psi) \\ &= F[-5 + 3\cos(53°) + 4\cos(37°)] \\ &= F(-5 + 1.8 + 3.2) = 0.0. \end{aligned}$$

4.2 See ■ Figure 4.49; gravity, normal, and friction forces each form a pair. The normal force is perpendicular to the mountain slope, and the friction force is parallel to the slope. Both these forces have vertical components that together support Sally's weight. Alternatively, by resolving the forces into components both along and perpendicular to the slope, we can see that the friction and normal force each support one component of Sally's weight.

Of the two forces in each pair, one acts on Sally, and the other acts on the Earth. The forces in each pair have equal magnitude and

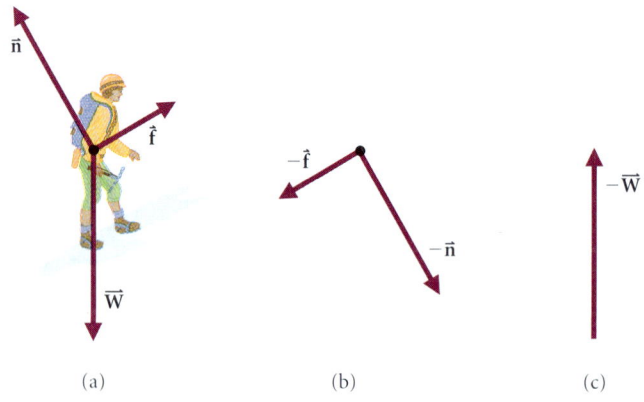

■ **FIGURE 4.49**
(a) Free-body diagram for Sally. (b) Two of the forces Sally exerts on the mountain. (c) The third paired force is Sally's gravitational attraction for the entire Earth.

opposite direction; thus the friction force acting on the mountain points downslope.

■ FIGURE 4.50
On the Moon, the force $\vec{F}$ the weight lifter can exert is the same as on Earth. Thus the weight he can support is also the same. The mass is six times larger.

■ FIGURE 4.51
Each of the two springs exerts half the force needed to balance the box's weight.

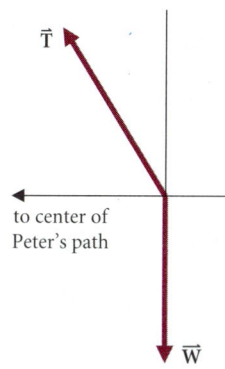

■ FIGURE 4.52
Free-body diagram for Peter's summertime experiment.

4.3 We find the magnitude of F from the data about the first object:
$$|\vec{F}| = m_1|\vec{a}| = (0.70 \text{ kg})(0.30 \text{ m/s}^2) = 0.21 \text{ N}.$$
Since the same force accelerates the second object seven times more rapidly, the second object has one-seventh the mass of the first: $m_2 = 0.10$ kg.

4.4 The champ's strength on the Moon is the same as his strength on Earth, so he can lift a weight equal to that of four boxes of nails on Earth (■ Figure 4.50). On the Moon, gravity is about one-sixth as strong as on Earth, so $6 \times 4 = 24$ boxes weigh as much as four on Earth. The champ can lift 24 boxes of nails on the Moon.

4.5 The weight of the nails is balanced by the forces exerted by the two identical springs (■ Figure 4.51). Each provides a force $F = W_1/2$. Since the force exerted by each spring is proportional to its stretch, half the force requires half the stretch. Each spring stretches 0.25 m.

4.6 George, compressing the spring 0.80 cm, exerts a force $0.80/0.48 = 1.7$ times as large as the force Peter is able to exert. Acting on the same block, this force causes an acceleration 1.7 times as large as Peter achieves. Thus $a_{\text{George}} = 1.7(1.3 \text{ m/s}^2) = 2.2 \text{ m/s}^2$.

4.7 ✉ When you step on your scale in the morning it compresses a distance of the order of 1 mm. An adult human being's mass is of the order of $M \sim 100$ kg. So, your weight $W = Mg$ is of the order of 10^3 N and is supported by a spring force of equal magnitude: $ks = W$. The spring constant of the scale is of the order of $k = W/s \sim (10^3 \text{ N})/(10^{-3} \text{ m}) = 10^6$ N/m.

4.8 Using the result of Example 4.11, force balance requires
$$\mu_s \geq \frac{f}{n} = \frac{W \sin \theta}{W \cos \theta} = \tan \theta = \tan 30° = \frac{\sqrt{3}}{3} = 0.58.$$

4.9 ■ Figure 4.52 shows Peter's free-body diagram. The forces acting on Peter are weight (vertically downward) and tension (along the rope). Neither is directed horizontally inward toward the center of Peter's path, but their sum is. The rope can never be horizontal, because the rope tension could not have a vertical component to balance Peter's weight.

4.10 A passenger's perception depends on the normal force exerted on the passenger (Figure 4.46). The centripetal acceleration has magnitude $\omega^2 r$ (eqn. 3.10), where $\omega = 2\pi/T$ (eqn. 3.8), and the radius of the turn is $r = v/\omega$ (eqn 3.9).

$$\Sigma F_y = 0 \qquad\qquad \Sigma F_x = M\omega^2 r$$
$$n \cos \theta_b - Mg = 0 \qquad n \sin \theta_b = M\omega v. \quad \text{(ii)}$$
$$n \cos \theta_b = Mg. \quad \text{(i)}$$

Dividing eqn. (ii) by eqn. (i), we have:
$$\tan \theta_b = \frac{\omega v}{g} = \left(\frac{2\pi}{4.00 \text{ min}}\right)\left(\frac{1 \text{ min}}{60 \text{ s}}\right)\left(\frac{2.00 \times 10^2 \text{ m/s}}{9.80 \text{ m/s}^2}\right) = 0.534,$$

and $\qquad \theta_b = 28.1°$.

To find the increase in perceived weight, we need n. From eqn. (i):
$$n = Mg/\cos \theta_b = 1.13 \, Mg.$$
The passengers feel as if they were about 13% heavier.

4.11 ■ Figure 4.53 is a free-body diagram for the motorcycle and rider, modeled as a single particle. Static friction, exerted by the cylinder walls on the motorcycle tires, balances the weight of cycle and rider. The normal force acting on the tires causes the centripetal acceleration of cycle and rider. If the rider tries the stunt at too low a speed, the normal force will be correspondingly small, and the maximum possible friction will be too small to balance the weight. (On a straight wall, there is no horizontal acceleration, no normal force arises no matter what the speed, and the stunt cannot be done.) The minimum speed for the stunt is that for which maximum friction can just balance the weight.

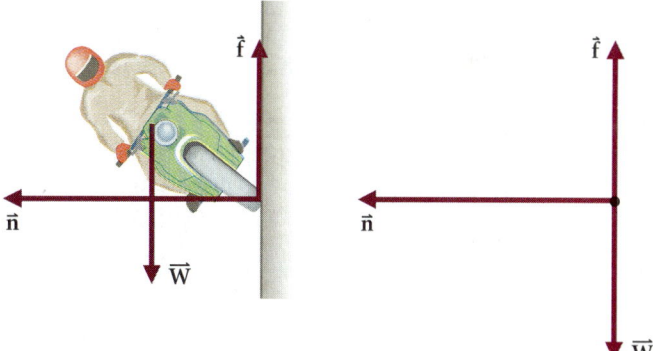

■ FIGURE 4.53
Forces acting on the stunt motorcyclist. Friction exerted by the wall balances the cyclist's weight. Normal force accelerates the cyclist in a circular path. "Right," you say, "this particle model is OK, maybe, but the cyclist is going to tip over!" Nope, but the cyclist must ride at an angle above the horizontal as shown. In Chapter 12, we'll find out why.

VERTICAL COMPONENTS	HORIZONTAL COMPONENTS
$\Sigma F_y = 0$	$\Sigma F_x = Ma_x$
$f - Mg = 0$	$n = Mv^2/R$

At the minimum speed, friction is at its limit: $\mu_s n = f_{\max} = Mg$. Thus $\mu_s M v_{\min}^2/R = Mg$; so:
$$v_{\min} = \left(\frac{gR}{\mu_s}\right)^{1/2} = \left(\frac{(9.8 \text{ m/s}^2)(10.0 \text{ m})}{0.65}\right)^{1/2} = 12 \text{ m/s}.$$
The minimum speed is 12 m/s, or 27 mph.

SOLUTIONS TO EXERCISES 153

Basic Skills

Review Questions

- State Newton's first law. What is its relation to Galileo's discoveries?

§4.1 FORCE

- Why might a pre-Newtonian physicist find the concept of beach sand exerting force on a rowboat surprising?
- State three questions we must answer to understand the meaning of force in physics.

§4.2 FORCE IN THE NEWTONIAN MODEL

- State why zero force is different from no force.
- What is the definition of *weight* we shall use?
- Why is the force exerted by a spring called *elastic*?
- What is the similarity and what is the difference between a spring force and a rope tension force?
- What is a *normal force*? Is a normal force elastic? Why or why not?
- What is the definition of a *friction force*?
- What are the names of the common fluid forces?
- Describe how the situations in Examples 4.1 and 4.2 show that forces follow the rules of vector arithmetic.
- What three facts about force pairs are summarized by Newton's third law?
- What basic idea of Newtonian physics do these three facts reflect?
- What is the force that forms a pair with the weight of a block at rest on the ground?

§4.3 NEWTON'S SECOND LAW

- How does the acceleration of an object depend on the net force exerted on the object?
- What concept does the technical term *mass* represent?
- How does the acceleration of an object depend on its mass?
- How does the mass of an object depend on its velocity? How does the mass of a composite object depend on the mass of its parts?
- What is the definition of the newton?
- Write Newton's second law expressed in SI units.

§4.4 WEIGHT

- State a practical formula for the weight of an object.
- State two ways in which the weight of an object differs from its mass.
- Describe why a grocer's scale determines the mass of an object while responding to the object's weight.

§4.5 PRACTICAL EXPRESSIONS FOR SPRING AND FRICTION FORCES

- State Hooke's law for the force exerted by a spring. How do you determine the direction of the force from the distortion of the spring?
- State four basic facts about static friction forces.
- State a practical formula that describes the maximum value of static friction. Is this always the correct expression for the static friction force in a particular situation?
- State a practical formula for kinetic friction.
- Describe qualitatively why friction forces occur.

§4.6 MOTION OF A SINGLE OBJECT SUBJECT TO SEVERAL FORCES

- Why are the forces acting on an object a complete model of how it interacts with the rest of the universe?
- What is a *free-body diagram*?
- State four steps in the analysis of a free-body diagram.

§4.7 DYNAMICS OF CIRCULAR MOTION

- What must be true of the net force acting on an object that undergoes uniform circular motion?
- Define the word *centripetal*. Does it describe a particular kind of force, like friction or spring force?
- What kind of force accelerates a car around a turn on a level road?
- Describe what it means to *feel* heavier than normal.

§4.8 NEWTON'S LAWS OF MOTION

- What is an *inertial reference frame*?
- ✱ What may cause an otherwise useful reference frame to be noninertial?
- ✱ Do we know that an inertial reference frame actually exists?
- Explain how each of the three algebraic forms of Newton's second law corresponds to a distinct application of the law.
- What are the three aspects of Newtonian mechanics that are tied together by the second law?

Basic Skill Drill

§4.2 FORCE IN THE NEWTONIAN MODEL

1. Identify the forces acting on the mountaineer in ■ Figure 4.54 and those acting on the spherical mooring buoy in ■ Figure 4.55.

2. An elephant stands on a three-legged stool at the circus (■ Figure 4.56). Identify the forces acting on the elephant and on the stool, and identify the force that forms a pair with each.

3. Two forces, $\vec{F}_1 = 2.0$ N at 30° to the x-axis and $\vec{F}_2 = 4.0$ N at 75° to the x-axis, act on an object. What third force $\vec{F}_3$ acting on the object will balance the first two?

■ FIGURE 4.54

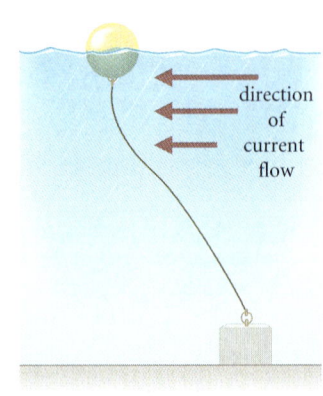

■ FIGURE 4.55

■ FIGURE 4.56

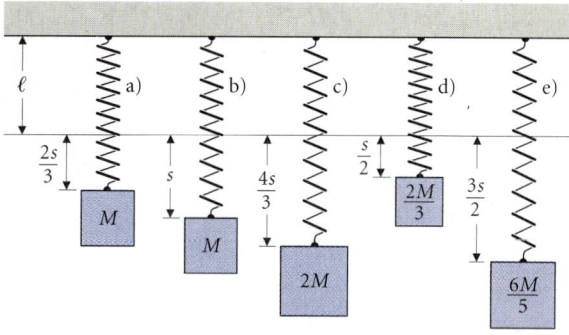

■ FIGURE 4.58

§4.3 NEWTON'S SECOND LAW

4. (a) High-speed cameras record the launches of identical standard arrows from two experimental bows. Measured accelerations of the two arrows are 1.32×10^3 m/s^2 and 1.45×10^3 m/s^2, respectively. What is the ratio of the forces exerted by the two bows? **(b)** The more forceful of the two bows from part (a) is now used to launch an experimental arrow, whose acceleration is measured as 1.62×10^3 m/s^2. What is the ratio of the experimental arrow's mass to that of a standard arrow? **(c)** If the mass of a standard arrow is 0.056 kg, what force is exerted by each of the experimental bows?

5. What kind of physical quantity would be expressed as 1 Gg·cm/ps^2? Express the quantity in terms of its usual unit.

§4.4 WEIGHT

6. What is your mass in kilograms? How much do you weigh in newtons?

7. A table is constructed so as to withstand the weight of six full boxes of nails (on Earth). If the table is taken to Mars ($g_{\text{Mars}} = 0.38 g_{\text{Earth}}$), how many boxes of nails may safely be placed on the table?

8. A mechanical arm designed to remove nail boxes from a warehouse table on Earth with a horizontal acceleration of $a = 2.0$ m/s^2 is used in a Martian colony warehouse. What is the acceleration of a nail box on Mars when the machine exerts the same horizontal force on the box as it does on Earth? State your reasoning.

§4.5 PRACTICAL EXPRESSIONS FOR SPRING AND FRICTION FORCES

9. A carnival attraction claims to measure how much you love your companion. In fact, it measures how far you can compress a spring (■ Figure 4.57). If the spring constant of the device is $k = 2.7 \times 10^5$ N/m and you can push on it with a force whose magnitude is equal to half your weight, to what position x can you compress the spring? Write *vector* expressions for the force you exert on the spring, the force the spring exerts on you, and the force the spring exerts on its mounting. State the relation of each of these results to the formula $\vec{F}_{\text{spring}} = -k\vec{s}$.

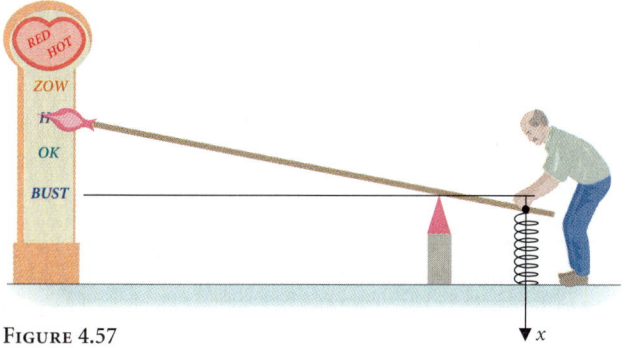

■ FIGURE 4.57

10. The springs in ■ Figure 4.58 all have unstretched length ℓ. Which has the greatest spring constant? Which has the smallest?

11. A large crate sits (at rest) on a concrete floor while you push on it with a horizontal force F. How big is the friction force exerted by the crate on the floor? State your reasoning.

12. A bulldozer is used to slide a 1200-kg boulder over a rock surface. The coefficient of static friction between the boulder and the surface is $\mu_s = 0.60$, and the coefficient of kinetic friction is $\mu_k = 0.35$. What force must the bulldozer exert to start the boulder sliding? What force will keep the boulder sliding?

§4.6 MOTION OF A SINGLE OBJECT SUBJECT TO SEVERAL FORCES

13. An engine bumps into a freight car of mass M and gives it an acceleration a. What normal force is exerted on the car by the engine? (Neglect friction.)

14. During lunch, furniture movers leave a desk tilted 10° from the horizontal. A book of mass 0.50 kg rests on top of the desk. Find the normal force and the friction force exerted on the book by the desk. For practice, go through each step in the method of free-body diagrams and state your reasoning in detail.

§4.7 DYNAMICS OF CIRCULAR MOTION

15. A car of mass $M = 1500$ kg rounds an unbanked turn on a mountain road at a speed of $V = 21$ m/s. If the radius of curvature of the road is $R = 320$ m, find the magnitudes of the forces acting on the car and describe their direction and nature.

16. A cord can support a maximum tension of 1.0×10^4 N before it breaks. The cord is used to swing a 15-kg stone in a circular path of radius 1.0 m over an icy surface. What cord tension is necessary as a function of angular frequency? What is the maximum angular frequency such that the cord will not break?

Questions and Problems

§4.2 FORCE IN THE NEWTONIAN MODEL

17. ❖ Identify the pairs of forces acting in the system shown in (a) Figure 4.54, (b) Figure 4.55, and (c) Figure 4.67 (see Problem 59; treat the coach as a single object).

18. ❖ What pushes a high-jumper off the ground? What pushes a car up the hill?

19. ❖ Three blocks are at rest on a ramp (■ Figure 4.59). On which block is the friction force largest? What is its magnitude? Explain your answer.

20. ❖ The block in ■ Figure 4.60 is at rest. Which of the force vectors shown acting on the block is (are) incorrect? Explain your reasoning and state what correction is needed.

21. ❖ Decide whether the following statement is true or false and explain your reasoning: When you support a heavy object, you exert an upward force on the object; the force that pairs with the force you exert is a normal force exerted by the object on your hands.

22. ♦ Two forces are exerted on an object:

$$\vec{F}_1 = \frac{(75\ N)(\hat{i} - \hat{j})}{\sqrt{2}} \quad \text{and} \quad \vec{F}_2 = \frac{(150\ N)(\hat{i} + \hat{j})}{\sqrt{2}}.$$

What third force $\vec{F}_3$ is needed to balance the first two?

23. ♦ Igor the strong man can sustain a force of 2.5×10^3 N on a rope. He challenges any two persons to a tug-of-war contest. If both opponents pull on ropes at a 45° angle to Igor's rope, what force must each sustain to win the match?

24. ♦♦ Susanne falls asleep leaning against a wire stretched between two posts. If she exerts a force of magnitude F on the wire, and the tension in the wire is measured as $10.0F$, what is the angle between the wire segments on opposite sides of Susanne?

§4.3 NEWTON'S SECOND LAW

25. ❖ Two objects, the first of mass M_1 and the second of mass M_2, are subjected to equal forces. The measured acceleration of the second object is five times that of the first. What is the ratio of the object's masses?

26. ❖ Two objects are observed to accelerate at the same rate when a force F is applied to the first and $4F$ is applied to the second. What is the ratio of the objects' masses?

27. ♦ A machine is designed to exert a force of 20.0 N on objects and to measure their resulting acceleration. Object 1 is given an acceleration of 15.0 m/s^2, and object 2 is accelerated at 10.0 m/s^2. What acceleration results when objects 1 and 2 are fastened together and tested?

28. ♦ An object of mass 15 kg has an acceleration of 10.0 m/s^2 at an angle of 30° with the x-axis. What are the components of the force acting on the object? If a second force of 150 N at an angle of 150° with the x-axis is exerted in addition, what is the object's new acceleration?

29. ♦ Forces with magnitudes 3.0 N and 4.0 N act at right angles on a particle of mass 2.0 kg. What is the magnitude of the particle's acceleration?

30. ♦♦ Suppose force, speed, and time were taken as fundamental units. What would be the physical dimensions of the meter and kilogram as derived units in such a system?

31. ♦♦ In one version of the *British* unit system, the mass unit is called a pound (1 lb ≡ 0.454 kg), and the force unit is also called a pound, defined as the weight of 1 pound-mass (at 45° north latitude, mean sea level). Write Newton's second law in this form of British units and give the value of the constant of proportionality.

The *British Engineering System* (used only in the United States) keeps the pound as a force unit and defines the slug as the mass that accelerates at 1 ft/s^2 when acted on by a force of 1 lb. How many pounds (mass) make up 1 slug? How many kilograms are there in 1 slug?

32. ♦♦ A car of mass 9.7×10^2 kg stops from a speed of 38 m/s in a distance of 95 m. What is the magnitude of the total force acting on the car?

§4.4 WEIGHT

33. ❖ A book lies on the surface of your dining room table. Which of the following statements about the book's weight is correct? Explain why the others are wrong.
(a) Its weight is an unvarying property of the book, no matter where in the universe the book is located.
(b) The force that forms a pair with the book's weight is the normal force exerted by the table on the book.
(c) The weight of the book is directly proportional to its mass.
(d) If the book were accelerated upward, its weight would increase.

34. ❖ A batter hits a pop fly directly upward. Which of the following best describes how the weight of the baseball varies between the time it is hit and when it is caught? Explain why your chosen answer is correct and the others are incorrect.
(a) The ball is weightless on the way up, but weight takes over at the top and draws the ball back down.
(b) Its weight is overcome by the ball's upward force while the ball is rising. At the top, weight dominates upward force and draws the ball downward.
(c) The ball's weight is constant and, except for air resistance, is the only force acting on the ball.
(d) The ball is weightless during the entire time it is in the air.
(e) Weight is irrelevant; the ball falls back because of the acceleration of gravity.

35. ❖ An empty coffee cup of mass M rests on a table. What is the normal force exerted on the cup by the table? What happens when coffee of mass m is poured into the cup?

36. ♦♦ A bathroom scale, calibrated for use on Earth, is imported to the lunar colony. If its scale is marked in kilograms, what is its read-

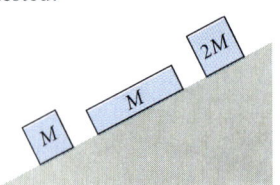
■ FIGURE 4.59

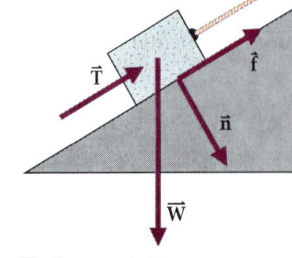
■ FIGURE 4.60

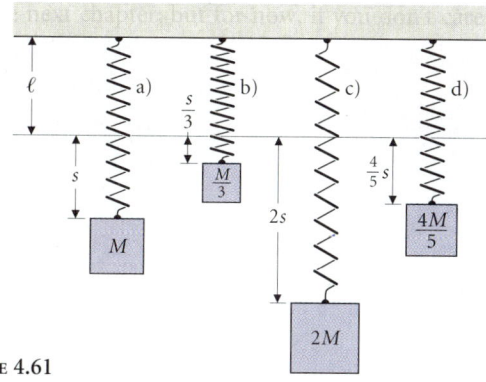

■ FIGURE 4.61

ing when used by a 105-kg person? If its scale is marked in newtons, what would it read? How do these results correspond to the person's weight on the moon?

§4.5 PRACTICAL EXPRESSIONS FOR SPRING AND FRICTION FORCES

37. ❖ A mountaineer proposes to climb a steep crack in a granite rock face, as in Figure 4.54. If the inside surfaces of the crack have no bumps or ledges for the climber to stand on, explain how the climber can ascend.

38. ❖ Two identical, massless springs of constant k and length ℓ are attached end to end. What is the spring constant of the combined spring? What may you conclude about how a spring constant depends on the unstretched length of the spring? The quantity $k\ell \equiv S$ is called the stiffness of a spring. What do you suppose stiffness depends on?

39. ❖ All the springs shown in ■ Figure 4.61 have unstretched length ℓ. Do they all have the same spring constant, or is one different from the rest? If so, which?

40. ♦ An explorer's sled has a mass of 110 kg and rests on a snowy, horizontal surface. If the explorer must exert a horizontal force of 550 N to start the sled sliding, what is the coefficient of static friction between the sled's runners and the surface?

41. ♦♦ A 2500-kg crate of machine tools is at rest on a horizontal surface. The coefficient of static friction between the crate and the surface is $\mu_s = 0.65$. The crate has just been brought to rest by a spring with spring constant $k = 6.43 \times 10^4$ N/m. If the spring is compressed by an amount $s = 2.00 \times 10^{-2}$ m, what static friction force acts on the crate?

42. ♦♦ A box of mass 15 kg sits on the floor of a truck. When the truck pulls away from the stop light with an acceleration $a = 3.0$ m/s², what force accelerates the box? Find the magnitude of the force and describe its direction.

43. ♦♦♦ Three identical massless springs of constant k are attached as shown in ■ Figure 4.62. What is the spring constant of the combined system?

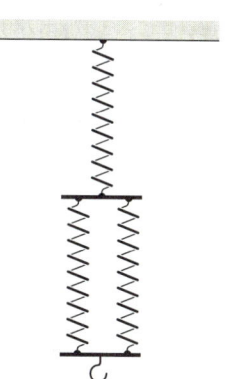

■ FIGURE 4.62

§4.6 MOTION OF A SINGLE OBJECT SUBJECT TO SEVERAL FORCES

44. ❖ A block is sliding to the right over a level surface. Which of the following statements is correct, and why are the others false?
(a) The friction forces acting on the block and on the surface both point right.
(b) The friction force on the block points left and the friction force on the surface points right.
(c) The friction forces acting on the block and on the surface both point left.
(d) The friction force on the block points right and the friction force on the surface points left.

45. ❖ A weather balloon carries an instrument of mass m aloft. If the balloon rises at constant speed, what is the tension in the (massless) cable suspending the instrument from the balloon?

46. ❖ A person is snowshoeing on a spherical mountain covered with hard snow. If the coefficient of friction between snow and snowshoes is μ_s, how far from the summit (what angle) can the person stand without sliding?

47. ❖ A block of mass M rests on the floor. An upward tension force is exerted on the block by a string held by a person. What happens to the normal force exerted by the floor as the person pulls harder on the string? What happens if the person releases the tension in the string?

48. ❖ A ball of mass M is attached to the front wall of a railroad car by a string (■ Figure 4.63). Assume that the floor of the car is smooth and exerts no horizontal forces on the ball. When the train pulls away from the station, the ball moves toward the back of the car and the string goes taut. Why? What happens after the train reaches and maintains a constant final speed?

49. ❖ A block of mass M sits on the floor and is attached to the ceiling by a taut string. Can you find the normal force and the tension in the string? Discuss.

50. ♦ A 0.100-kg lead fishing sinker is suspended on a fishing line from the end of a pole (■ Figure 4.64). At the start of a cast, the end of the pole is first lowered with an acceleration of $0.50g$ and then raised upward at an initial acceleration of $0.70g$. What is the tension in the fishing line in each of these cases?

51. ♦ A woman of mass 65 kg stands inside an elevator on a bathroom scale calibrated to read in newtons. Calculate the scale reading

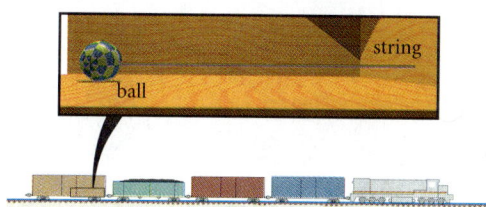

■ FIGURE 4.63

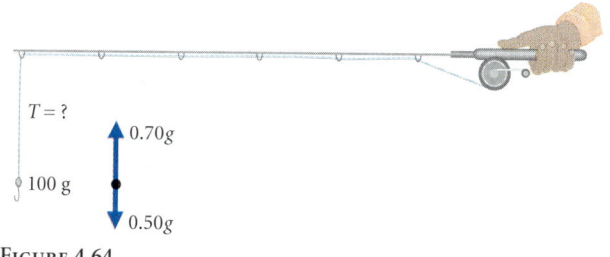

■ FIGURE 4.64

QUESTIONS AND PROBLEMS 157

in each of the following situations and explain in terms of forces acting *on the scale* why it reads as it does: **(a)** elevator stationary; **(b)** elevator accelerating upward at 2.0 m/s^2; **(c)** elevator accelerating downward at 2.0 m/s^2; **(d)** elevator descending with constant velocity; and **(e)** the elevator in free fall after the cable breaks.

52. ♦ A force $\vec{F}$ is exerted horizontally on an object of mass M suspended from the roof by an ideal rope. What is the angle of the rope from the vertical as a function of $|\vec{F}|$? At what angle is $|\vec{F}|$ equal to the weight of the object? At what angle is the rope tension equal to twice the weight of the object?

53. ♦ A glider (aircraft without an engine) of mass M is descending at constant speed at an angle of 15° with the horizontal. What are the drag and lift forces acting on the glider?

54. ♦ A toboggan slides down a 30° slope. The coefficient of kinetic friction between the snow and the toboggan is 0.1. Find the acceleration of the toboggan.

55. ♦ A balloon is tethered to a building to advertise a sale. The forces on the balloon are an upward force due to its buoyancy, the tension in the string, and a horizontal pressure force due to the wind. (The weight of the balloon is small and may be ignored.) If the buoyant force is 12 N and the wind force is 15 N, what is the tension in the string?

56. ♦♦ An orange crate of mass M is at rest on a horizontal surface a distance $3\ell/2$ from a wall. The crate is attached to the wall by a spring of constant k and unstretched length ℓ. Find the friction force acting on the crate and the minimum possible coefficient of friction between the surface and the crate.

57. ♦♦ Maria is attempting to slide a 95-kg crate of dishes up a ramp into a moving van (■ Figure 4.65). The ramp is at an angle of 27° with the horizontal, and the coefficient of static friction between the ramp and the crate is $\mu_s = 0.30$. If Maria pulls on the rope with a force of 650 N, what static friction force acts on the crate? Does the crate move?

58. ♦♦ Each of the systems in ■ Figure 4.66 remains at rest, and in each case the magnitude F of the force is $Mg/2$. What is the normal force acting on the block in each case, and what is the minimum necessary coefficient of static friction?

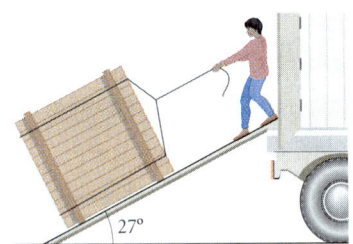

■ FIGURE 4.65

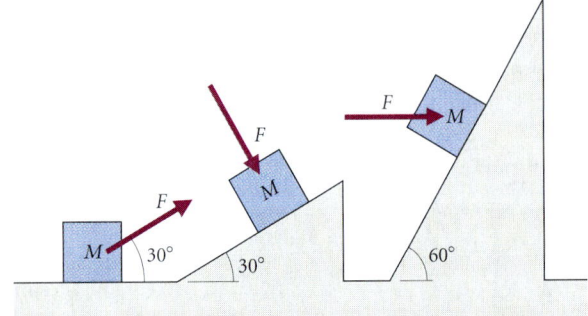

■ FIGURE 4.66

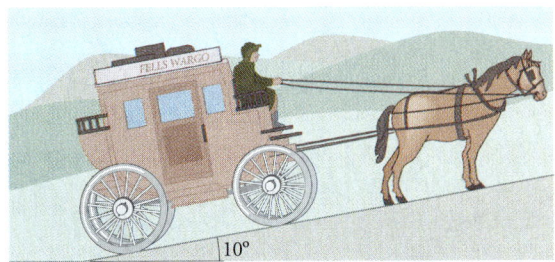

■ FIGURE 4.67

59. ♦♦ A horse is holding a stagecoach on a slope with the brakes off (■ Figure 4.67) while the movie director gets the bad guys ready to shoot the scene. Identify the forces acting on the stagecoach. If the coach has a mass of 1100 kg, how hard must the horse pull?

60. ♦♦ An elevator whose mass is 1.0 metric ton (1.0 Mg) is mounted on a cable that can safely support 3.0 metric tons. What is the minimum distance in which the elevator can safely be stopped from a speed of 2.0 m/s?

61. ♦♦ Melinda has a decorative pendant hanging from the rearview mirror of her car. At what angle to the vertical will the pendant hang in each of the following situations? Explain your answers. **(a)** The car is driving at constant speed v. **(b)** The car has an acceleration of magnitude a on a level road. **(c)** The car is sliding down a frictionless icy slope making angle θ with the horizontal.

62. ♦♦ A block of unknown mass is loaded on a light sled at rest on a 15° slope. It is found that the maximum force that can be exerted downslope on the sled before it slides is 220 N. A force upslope of 727 N is required to start the sled sliding. Find the mass of the block (plus sled) and the coefficient of static friction.

63. ♦♦ A hot-air balloon of total mass $M = 755$ kg is descending at a constant speed of 5.0 m/s. After 25 kg of ballast are dropped overboard, the balloon rises at 5.0 m/s. What is the magnitude of the drag force that acts on the balloon at a speed of 5.0 m/s? What is the upward acceleration of the balloon immediately after the ballast is dropped?

64. ♦♦ Two springs of constants k and k' and negligible unstretched length are connected and stretched between two posts a distance ℓ apart. Where is the point of connection of the springs?

65. ♦♦ Springs 1 and 2 in ■ Figure 4.68 have equal spring constants k but unstretched lengths $\ell_1 = 2\ell$ and $\ell_2 = \ell/2$. Is there a friction force acting on the mass M? If so, what are its magnitude and direction?

66. ♦♦ A ball of mass M and negligible size is at rest on a 30° inclined plane (■ Figure 4.69). If the spring has a constant k and an unstretched length $3\ell/2$, find the friction force exerted on the object.

67. ♦♦ A Nissan Maxima automobile, of mass $M = 1.40 \times 10^3$ kg, brakes to a stop from a speed $v = 113$ km/h in a distance $s = 56.4$ m. What kind of force decelerates the car? What are the magnitude and direction of the force?

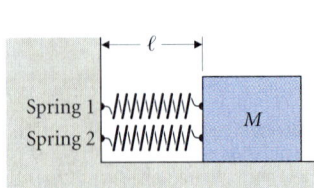

■ FIGURE 4.68

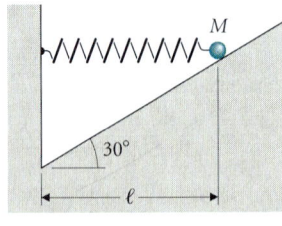

■ FIGURE 4.69

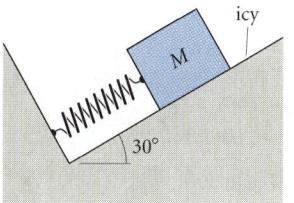

■ Figure 4.70

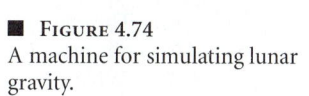

■ Figure 4.71

68. ♦♦ A cargo container of mass $M = 2.0 \times 10^3$ kg rests on an ice-covered ramp inclined at 30° (■ Figure 4.70). What is the compression of the spring if its spring constant is $k = 6.0 \times 10^4$ N/kg? If the ice melts and the coefficient of friction between the dry surface and the container is $\mu_s = 0.50$, find the maximum and minimum possible values of the spring's compression.
69. ♦♦ A television camera for a surveillance system is hung from the roof by two cables (■ Figure 4.71). If the tension in the left cable is $T_2 = 12$ N, find the tension T_1 and the weight of the camera.
70. ♦♦ A sign of mass $m = 17.7$ kg is suspended by two strings (■ Figure 4.72). Find the tension in each of the strings.
71. ♦♦ A block of mass M is pressed against a vertical surface by a spring of unstretched length ℓ (■ Figure 4.73). If the coefficient of static friction between the block and the surface is μ_s, what is the minimum spring constant k_{min} that will keep M in place? If instead $k = 2mg/(\mu_s d)$, find the normal force and the friction force on the block.

■ Figure 4.72

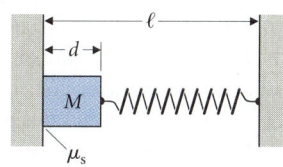

■ Figure 4.73

72. ♦♦ At Alta, Utah, a rope tow pulls skiers from the lodge to the base of one of the ski lifts. The ground has a small slope of 2.0°. Once set in motion, the skier glides downslope at constant speed with no further force from the rope. What is the coefficient of kinetic friction between the skis and the snow? What force must be exerted on a 95-kg skier by the rope to tow the skier upward at constant speed?
73. ♦♦ NASA developed an apparatus (■ Figure 4.74) to enable astronauts to practice maneuvering under reduced gravitational acceleration. Discuss how the system works and find what inclination θ of the surface is needed to simulate lunar gravity of $\frac{1}{6}$ Earth gravity.
74. ♦♦ During its landing roll, 12% of an airplane's weight is supported by lift on the wings when the pilot applies the brakes. The wheels skid. If the coefficient of kinetic friction between the skidding wheels and the ground is 0.50, find the deceleration of the airplane.
75. ♦♦♦ A small ball of mass M is attached to two identical springs of constant k, which are attached to the floor and the roof (■ Figure

■ Figure 4.74
A machine for simulating lunar gravity.

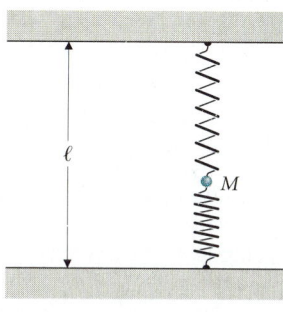

■ Figure 4.75

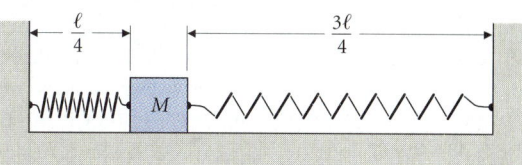

■ Figure 4.76

4.75). The springs have unstretched length $\ell/2$. At what position will the ball remain at rest?
76. ♦♦♦ In ■ Figure 4.76 the springs each have constant $k = 250$ N/m and an unstretched length $\ell/2 = 0.50$ m. What is the friction force acting on the stationary block? If the mass of the block is $M = 75$ kg, what is the minimum possible coefficient of friction between the block and the surface?

If the x-coordinate measures displacement of the block from the midpoint between the walls, find the friction force $\vec{f}(x)$ as a function of x. If $\mu_s = 0.25$, what is the maximum value of $|x|$ at which the block can remain at rest?

For the situation shown in Figure 4.76, suppose that the coefficient of friction is actually less than the minimum value for no acceleration. Discuss qualitatively what happens when the system is released.
77. ♦♦♦ An object of mass m is suspended using a string of length ℓ and a spring of constant k ($< 2mg/\ell$) and unstretched length $\ell/2$ (■ Figure 4.77). Find the tension in the string. What happens if $k > 2mg/\ell$?

§4.7 DYNAMICS OF CIRCULAR MOTION

78. ❖ A tetherball is swinging around its pole on the end of a rope when the rope breaks. Describe the resulting motion of the ball.

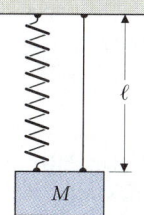

■ Figure 4.77

79. ❖ An airplane is in straight and level flight at constant speed with four forces acting: lift (normal force of air on the wings); drag (friction force of air on the entire plane); thrust from the engine; and, of course, weight. If the pilot suddenly operates the controls in a way that increases the lift force without changing any of the others, describe the effect on the plane's flight path.

80. ♦ A car rounds a turn of radius 225 m at 65 km/h. What is the minimum value of the coefficient of friction between the tires and the road?

81. ♦ An experimental design for a weed cutter uses a metal ball of mass 5.0 g on the end of a string 12 cm long, which spins at a frequency of 33 Hz. What is the tension in the string when the cutter is operating? Criticize the design of this machine on grounds of practicality and safety.

82. ♦ Jasmine enters a freeway exit at a speed of 15 m/s. The exit lane is level and has a radius of curvature of 53 m. The coefficient of static friction between the road and the tires is $\mu_s = 0.80$. What friction force acts on Jasmine's 950-kg car?

83. ♦♦ A car of mass 1.0×10^3 kg, traveling at 32 m/s, drives through a dip in the road that has a circular cross section of radius $R = 11$ m. What is the normal force on the car at the bottom of the dip?

84. ♦♦ In his experiment (Exercise 4.9, Figure 4.41), Peter swings on a rope of length $\ell = 8.0$ m. Find the angle of the rope from the vertical as a function of Peter's speed. Assume he tucks his body tightly, so that you can neglect his size, and that he moves horizontally in a circular path. If Peter's mass is 63 kg and his speed is 8.0 m/s, what are the tension in the rope and the time for him to swing through one-quarter circle?

85. ♦♦ A fairground attraction called *The Rotor* (shudder!) is a spinning drum with a movable floor that is dropped down as the drum speeds up (see Figure 3.14). The people inside are held up on the walls by friction. If the minimum coefficient of friction expected between people's clothes and the wall is 0.50 (DON'T WEAR SILK!), what rotation rate (in revolutions per second—i.e., hertz) is required before the floor is dropped? (Assume the drum has a radius of 5.0 m.)

86. ♦♦ A skier encounters a bump whose top is shaped like a sphere of radius R. If R is very small, the skier will leave the ground. Why? if the skier has a speed $V = 12$ m/s, find the minimum bump radius such that the skier will stay on the snow.

87. ♦♦♦ Cars can negotiate turns in a road at much higher speed if the roadway is inclined, or banked, rather than horizontal (Figure 4.43). (a) If a road turns in a circle of radius $R = 1.0$ km and is banked at an angle $\theta = 5.0°$, at what speed v_1 does the car require no friction between the tires and the road? (b) If the coefficient of static friction between the tires and the road is $\mu_s = 0.40$, what is the maximum speed v_{max} at which the car may make the turn? How does this compare with the maximum speed for a level road? (c) What happens if the car's speed is less than v_1? Under what conditions is there a minimum speed at which the car can round the turn? Compare your results with the solution to Exercise 4.11.

88. ♦♦♦ An acrobatic plane flies in a vertical circle of radius R and constant speed v_0. (This is quite a plane!) If the drag force on the plane is assumed to be constant, describe how the pilot must control engine thrust and lift to achieve this feat. For what value of R will the pilot require zero lift at the top of the loop? What is the required lift in this case at the bottom of the loop?

89. ♦♦♦ An airplane flying at 310 km/h executes a turn using a 55° bank angle. (a) By how much must the lift be increased if the circle is to be made in a horizontal plane? (b) What is the radius of the turn? (c) Describe the aircraft's flight path qualitatively if the lift is *not* changed from its value in level flight.

§4.8 NEWTON'S LAWS OF MOTION

90. ❖ Are any of the following *exactly* inertial reference frames? Are any of them approximately inertial? Explain your answers. (a) The inside of a car during a typical journey. (b) The inside of a physics lab in your university. (c) The inside of the *Turn of the Century* roller coaster. (d) The inside of a Mariner spacecraft en route to Mars (well after launch and before reaching Mars).

Additional Problems

91. ❖ Devise an experiment using one spring, a clock, and a circular, frictionless track of unmeasured radius that allows you to compare the mass of an object with the international prototype kilogram. (It is extremely unlikely you will be permitted to test this idea.)

92. ❖ A magician rapidly jerks a tablecloth from under the dishes set on the table, leaving the dishes in place. Pulling the tablecloth off slowly dumps the dishes on the floor. Explain.

93. ❖ Climbing ropes are designed to stretch up to 80% of their normal length. Explain why this is important to a rock climber taking a long fall.

94. ❖ When making a turn, a skier desires to reduce the normal force acting on her skis to nearly zero. One method of doing this is called *down unweighting*, in which the skier rapidly pulls her body into a crouch. On a 30° slope, how rapidly must she accelerate her body toward the slope to unweight the skis? In another method, called *up unweighting*, the skier jumps away from the snow. Assuming it takes $\frac{1}{3}$ s to turn her skis, at what speed must she jump to accomplish the turn? Are these maneuvers more difficult or less if the skier skis over a bump and performs them at the top?

95. ♦♦ ✉ A baseball pitcher throws the ball at a speed of 45 m/s. Assuming the pitcher exerts constant force on the ball during the throw, estimate the magnitude of that force.

96. ♦♦ Racing aircraft make *figure eight* turns around pylons at a speed of 150 m/s. If the pilot would pass out when the normal force exerted by her seat exceeds six times her weight, what is the maximum bank angle? What is the smallest achievable turning radius?

97. ♦♦ Igor the Incredible, the daring sport parachutist, dives from a plane holding his knees to his chin so as to reach a terminal speed of 1.0×10^2 m/s. Two hundred meters above the ground (!), he "pops" his parachute, which opens in 0.50 s and which then decelerates him to a *gentle* landing at 5.0 m/s. (a) If Igor's mass is 85 kg, what is the drag force acting on him before the parachute opens? (b) Neglecting any effect of the parachute before it is fully open, find the force it exerts on Igor (assumed constant).

98. ♦♦♦ The force acting on an object of mass $M = 4.2$ kg varies with time according to the formula

$$\vec{F}(t) = \hat{i}\,[(6.7 \text{ N/s})t - (48.2 \text{ N/s}^2)t^2]$$
$$+ \hat{j}\,[(-4.1 \text{ N/s})t + (76.3 \text{ N/s}^2)t^2].$$

If the object is at rest at the origin at $t = 0$, what are its velocity and position as functions of time?

99. ♦♦♦ (a) Merilee stands 5.0 m from the center of a merry-go-round that is rotating with frequency $f = 0.20$ Hz. What is the minimum coefficient of static friction between her shoes and the floor that will allow her to stand without sliding off the merry-go-round? Is the required coefficient greater or smaller if Merilee walks radially inward or outward? Why? (b) *The Sphere* is a frictionless, hemispherical bowl of radius 5.0 m that rotates at a variable frequency f. Find Merilee's position as a function of $\omega = 2\pi f$ when she rides *The Sphere*.

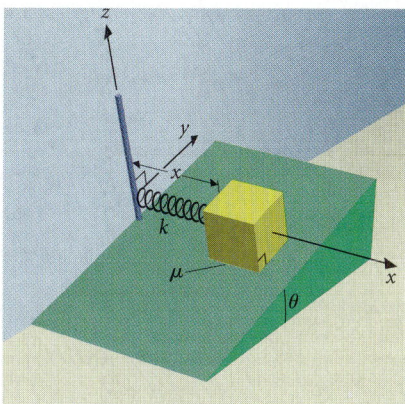

■ FIGURE 4.78

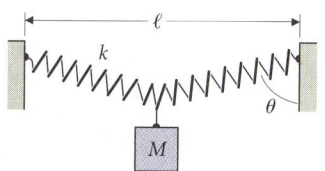

■ FIGURE 4.80

100. ♦♦♦ A particle of mass $M = 0.42$ kg moves along a circular path of radius 3.0 m with an angular position given by $\theta(t) = (2.7 \text{ rad/s}^2)t^2 + (3.0 \text{ rad/s})t$. Find the force acting on the particle at $t = 2.0$ s.

101. ♦♦♦ If an explorer pulls on a sled of mass 110 kg with a rope at an angle θ with the horizontal, what force $F(\theta)$ is needed to start the sled moving? What angle results in the minimum value of $F(\theta)$ and what is that value? How does your result change if the sled is on a 30° slope? (Take $\mu_s = 0.40$.)

102. ♦♦♦ A pilot wishes to fly *over the ground* on a circular track of radius 1.0 km. The airspeed of the plane is 220 km/h and the wind is blowing at 30 km/h. What bank angle is required when the airplane is flying **(a)** directly upwind and **(b)** directly downwind? Find the lift force required in each case if the vertical acceleration is zero.

103. ♦♦♦ A block of mass M rests on an inclined plane and is attached to a wall by a spring of constant k and unstretched length ℓ (■ Figure 4.78). If the coefficient of friction between the block and the plane is μ, what are the minimum and maximum distances from the wall at which the block can remain at rest?

Computer Problems

In Problems 104, 105, and 106, you need to analyze the free-body diagrams in the usual way. You will obtain equations that cannot be solved algebraically. Use a computer or calculator to evaluate the functions on both sides of your equation in 1° or 2° increments for angles between 0° and 90°. Compare the values to find the solution for the angle.

104. An object with mass M is suspended by a string of length ℓ. A spring of unstretched length ℓ and constant k is also attached to the mass. The spring is massless and its left end is free to slide without friction on a vertical rod (■ Figure 4.79). What is the angle θ between the string and the vertical? Take $k = 210$ N/m, $\ell = 2.0$ m, and $M = 5.0$ kg.

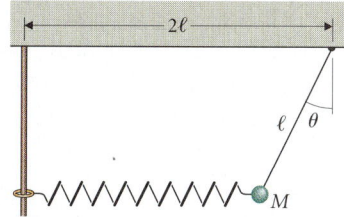

■ FIGURE 4.79

105. A massless spring of constant $k = 78.5$ N/m and unstretched length $\ell = 0.275$ m is fixed at two points at the same height and separation ℓ (■ Figure 4.80). A mass $M = 0.453$ kg is suspended from the middle of the spring. Find the angle θ. (*Hint:* If a spring has constant k, half the spring has constant $2k$; see Problem 38.)

106. A mass is suspended using a string and a spring of constant k and unstretched length ℓ (■ Figure 4.81). If $k = 200.0$ N/m, $\ell = 1.00$ m, and $M = 5.00$ kg, find the tension in the string and the stretch of the spring. (*Hint:* First find the angle the string makes with the vertical.)

Challenge Problems

107. A particle of mass M is suspended by two strings (■ Figure 4.82). If the horizontal string is suddenly cut, does the other string's tension increase, decrease, or remain the same immediately after the cut?

108. The drag force acting on an object moving through a gas follows an expression of the form $C_D \rho v^2$, where C_D is the drag coefficient, ρ is the density of the gas, and v is the speed of the object. A reasonable model of how the density of the Earth's atmosphere varies with height is $\rho = \rho_\circ \exp(-h/h_\circ)$. **(a)** Assume the space shuttle has a constant upward acceleration $a_\circ$ and find an expression for the drag force acting on the shuttle as a function of time. **(b)** Find an expression for the maximum drag force acting on the shuttle. **(c)** ✉ Argue that C_D is, in order of magnitude, given by the cross-sectional area of the shuttle. Make reasonable estimates of other quantities in your expression for the maximum drag and give a numerical estimate of its value.

Essay Question/Point to Ponder

Newton's laws take mass as a fundamental property. That is, according to Newtonian mechanics, the only way to change an object's mass is by adding material to or removing material from the object. Contrast that with each of the following properties of objects: color, surface hardness, shape, economic value. For each property, describe a change in the object's environment that will alter the property. Can you think of any property of objects, other than mass, that cannot be changed without the removal or addition of material?

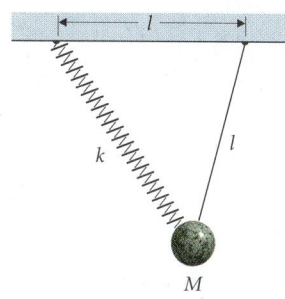

■ FIGURE 4.81

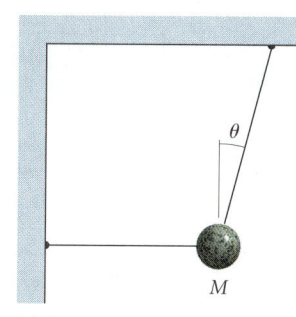

■ FIGURE 4.82

We hold these truths to be self-evident...
THOMAS JEFFERSON

ESSAY 1
Newton's Discoveries and Their Impact

The nineteenth century Japanese artist Hōshū, in a series of prints portraying great men of Western culture, chose Isaac Newton to represent science (■ Figure E1.1). Hōshū's print captures the legend of Newton discovering the law of universal gravitation while sitting in his garden watching an apple fall. Newton's discovery actually took place over a period of 20 years and is a prime example of how science progresses. During those two decades, Newton refined and built on inherited ideas, struggled with error, and finally achieved powerful results. His achievement has had an enormous impact on both our science and our general culture—this is reflected in Hōshū's caption: "Isaac Newton. Very great head of school."

By 1632, Galileo had laid the foundations of mechanics, and Kepler had accurately described planetary motion. Together, they had left a challenge: to explain the physical causes of behavior on Earth and throughout the solar system. By the last third of the century, many people were attempting to meet this challenge, but little progress had been made. Among the key pieces of the puzzle that were coming together was the calculation by Christiaan Huygens (1629–1695) of an object's acceleration in uniform circular motion. Huygens combined his result with Kepler's third law to show that accelerations of the planets decrease as the squares of their distances from the Sun (■ Figure E1.2):

$$a \propto 1/r^2.$$

But no one knew how to go further.

As a young man, Newton had invented the calculus, discovered his third law from a careful study of collisions, and found an original way to re-derive Huygens' result. But Newton did not know what cause would produce any relation among accelerations of the different planets, let alone the inverse distance squared relation. The stumbling block was confusion about the nature of force and acceleration. Newton had not yet fully understood force as a two-body process. At first he thought of force as a property of the object experiencing the force. Also, following Huygens, he viewed the acceleration of an object in circular motion as a "centrifugal endeavor", or center-fleeing tendency, of the object.

You'll have no trouble finding the same stumbling blocks in your own intuition. Holding a heavy object, you gain a clear but incorrect impression that it possesses weight as some sort of internal property. Similarly, when you catch a baseball, it seems to possess some sort of *oomph* that you have to stop. When your car rounds a tight turn, you feel "thrown against the door." In fact, you go straight until the car door presses on you and accelerates you around the turn. What you actually sense when lifting an object is your own muscle stress. Since Newton's time, we have learned to distinguish *momentum* (a fancier name than "oomph," cf. Chapter 6) from the forces that change it. To go forward in his exploration of the natural world, Newton first had to recognize these distinctions.

■ FIGURE E1.1
The Japanese artist Hōshū's representation of Newton deducing the law of universal gravitation from the fall of an apple. Hōshū chose this legendary version of Newton's discovery to represent Western science.

Ironically, a letter from his bitter rival, Robert Hooke, was the necessary catalyst. Prompted by Hooke, Newton realized that there is no such thing as centrifugal endeavor. Rather, each planet accelerates directly toward the Sun because of a force acting in that direction. It is the Sun exerting an attractive force on each planet that holds the planet in orbit.

Such an attraction was foreign to physical ideas of the time. It seems instead to have come from chemistry. During the 20 years after his first work in mechanics, Newton experimented in alchemy, a mixture of laboratory know-how and mystical notions. He satisfied himself that chemical phenomena were explained by general attractive and repulsive tendencies among chemical species. In fact, many of his observations are remarkably similar to modern ideas of electrical forces. Newton's chemical work had thus primed him to recognize a very general principle of attraction. This is one of the clearest examples of how research in seemingly totally unrelated fields can bear remarkable fruit.

Newton's next step was to make an analogy with physics on the Earth. He knew that forces on Earth always come in pairs, and he supposed that his third law should be a general rule about forces anywhere. If the Sun pulls on the Earth, the Earth also pulls on the Sun!

Since the Sun and each planet exert mutually attractive forces, Newton reasoned that planets also attract each other. He knew that telescopic observations of Jupiter and Saturn had shown them to deviate slightly from their elliptical paths when passing each other. Newton was able to show that the deviations are exactly what would result from an attraction between the planets varying inversely with the interplanetary distance squared (■ Figure E1.3). It was this line of reasoning that led Newton to the idea of a universal attraction.

The final step was to test whether the universal attraction was the same as gravity on Earth. If both are caused by Earth's attraction, the fall of an apple should be related to the Moon's acceleration in its orbit. Their accelerations toward Earth should be in an inverse ratio to the squares of their distances from the Earth's center. Newton made the comparison and found the two ratios to agree "pretty nearly."

■ **Figure E1.2**
A planet's elliptical orbit around the Sun appears circular on the scale of the figure. Christiaan Huygens discovered the relation for acceleration in circular motion:

$$a = v^2/r.$$

In terms of the planet's period T,

$$a = \frac{(2\pi r/T)^2}{r} = \frac{4\pi^2 r}{T^2}.$$

According to Kepler's third law, a planet's period squared is proportional to the cube of its semimajor axis (the radius of a circular orbit): $T^2 \propto r^3$. So, for the acceleration we have

$$a \propto 4\pi^2 r/r^3 \propto 1/r^2.$$

The acceleration is inversely proportional to the square of the orbital radius.

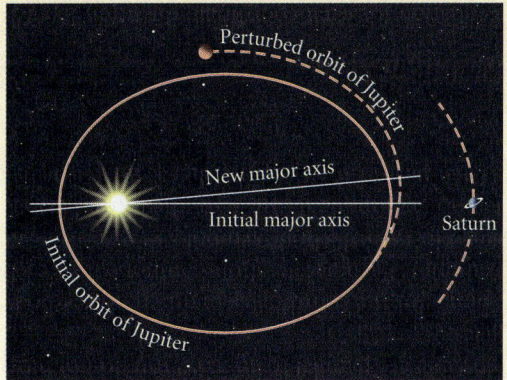

■ **Figure E1.3**
When Saturn and Jupiter pass each other in their orbits, their attraction slightly alters the elliptical shape of their orbits. The pull of Saturn on Jupiter reduces Jupiter's acceleration toward the Sun, thereby decreasing the curvature of Jupiter's orbit. The result is a slight rotation of the major axis of Jupiter's orbit, shown here greatly exaggerated. Newton showed that attraction by Saturn proportional to inverse distance squared was just sufficient to account for an observed slow rotation of Jupiter's semimajor axis.

Newton's concurrent discovery of universal gravitation and the three laws of motion transformed science. These laws give both an explanation of planetary motions in terms of a single law of attraction and a complete prescription for doing physics on Earth. Physics thus ceased to be a dispute over basic principles and became a program of applying Newton's laws in research and in practical engineering.

Newton's work also had a profound effect on European civilization. He believed that his laws of motion and gravitation were not suppositions, but truths about the world established by observation and experiment. Understanding of the world was to follow with mathematical precision by deduction from these truths. The success of this philosophy in physics appeared as a revelation in the eighteenth century. Newtonianism became the intellectual framework of ethics, economics, politics, and the arts. For Americans, the most compelling example of Newtonianism is the theory behind the Declaration of Independence, where the Aristotelian divine right of kings is replaced by the *natural* right of a people to alter their government and to institute the form of government they find most suitable to their needs.

"What attracted me into science as a career? . . . It deals in facts and truth, its conclusions are unambiguous, and it tests them against reality."

JAMES P. HOGAN *ENDGAME ENIGMA*

CHAPTER 5
Using Newton's Laws

CONCEPTS

Systems as collections of particles
Equilibrium
Conservation of string
Element of string
Newton's law of universal gravitation
The gravitational constant

GOALS

Be able to:

Use the free-body method to analyze dynamic systems.

Compute the gravitational force exerted by one particle on another.

Discuss qualitatively how the law of gravitation explains motions in the solar system.

Beanstalk *viewed from the vicinity of the launching platform. Beanstalks exist today only in speculative studies and science fiction novels, but they may be built in the next century if strong enough cable can be developed. A large satellite, like the communication satellites of today, is in geosynchronous orbit, revolving around Earth once every 24 hours and so remaining always* above *the same place on the Earth's equator. A cable runs from the Earth's surface to the satellite and outward to the launching platform. The platform, which also revolves once a day, moves at greater speed than it would in a free orbit around the Earth, fast enough to escape from Earth if it were to break loose. Tension in the cable, in addition to the Earth's gravitational force, accelerates the platform.*

Cargo containers use ordinary electric motors to move away from Earth to the launching platform, from which they are placed into orbits to the Moon or nearby space colonies simply by being released from the platform at the proper moment. Incoming vessels dock gently at the platform with minimal use of rockets. As returning containers descend the beanstalk, their electric motors brake their descent (like cars on the trolley line) and return energy to the system.

In 1960, the Russian engineer Y. Artsutanov envisioned an interplanetary launching system, called a *beanstalk,* that would be as efficient as an electric trolley. Though his Earth-based beanstalk may remain the stuff of science fiction because it requires stronger cables than we can now foresee building, smaller stalks could well provide the most efficient mode of travel among space colonies, and some of us may live to see the first such colonies. For now, the beanstalk is a marvelous fantasy that is fun to think about.

Systems far less exotic than beanstalks consist of objects exerting forces on each other. Since Newton's laws are stated for a single particle, we must learn how to apply them to such collections of particles. We begin by analyzing practical systems on Earth.

Newton's law of universal gravitation explains gravity on Earth, and governs the motions of objects in space. It explains why Kepler's laws apply to satellites of any astronomical body. Satellite technology and space exploration are practical applications of the law of gravity. Finally, we shall be able to investigate some of the requirements for building a beanstalk.

■ **FIGURE 5.1**
Anchored to the ice by a short length of rope, the rescuer lowers an injured climber in a sled. The rope to the sled passes through a "belay plate," which allows the rescuer to control the sled's descent.

5.1 A First Example

The climber in ■ Figure 5.1, who is slowly lowering an injured friend down an ice sheet, needs to know how much tension each rope must support. In Chapter 4 we solved problems involving forces acting on a single mountain climber; now let's consider the dynamic system formed by rescuer and casualty, which we can analyze as two simpler systems connected by a rope. After getting the results we can think about our reasoning and develop a reliable general method for analyzing complex systems.

Several forces act on the injured climber and sled (■ Figure 5.2): the normal force exerted by the ice surface, weight, and rope tension. Similarly (■ Figure 5.3) normal force, weight, and two tension forces act on the rescuer. As in Chapter 4, we model the rope between the climbers as an ideal string that exerts forces of equal magnitude T_1 on each climber. As usual, we neglect any friction forces exerted by the ice. Since the rescue climber is stationary, and the injured climber is being lowered at constant speed, all the acceleration components are zero.

We choose a coordinate system for each person with the x-axis upward along the slope, and the y-axis perpendicular to the slope (Figures 5.2 and 5.3). Then we apply Newton's second law to each person separately.

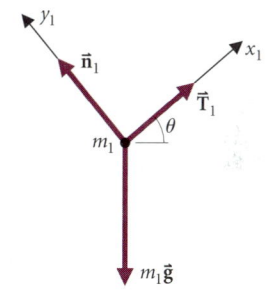

■ **FIGURE 5.2**
Free-body diagram for the injured climber.

THE INJURED CLIMBER

x_1-components: $\sum F_{x_1} = m_1 a_{x_1}$.
$T_1 - m_1 g \sin \theta = 0$.
y_1-components: $\sum F_{y_1} = m_1 a_{y_1}$.
$n_1 - m_1 g \cos \theta = 0$.

THE RESCUER

x_2-components: $\sum F_{x_2} = m_2 a_{x_2}$.
$T_2 - T_1 - m_2 g \sin \theta = 0$.
y_2-components: $\sum F_{y_2} = m_2 a_{y_2}$.
$n_2 - m_2 g \cos \theta = 0$.

The resulting set of equations has the solution:

$$n_1 = m_1 g \cos \theta, \qquad n_2 = m_2 g \cos \theta,$$
$$T_1 = m_1 g \sin \theta, \qquad T_2 = (m_1 + m_2) g \sin \theta,$$

where we used the result for T_1 in the expression for T_2. As we might have expected, the rope holding the rescuer to the mountainside supports the x-components of both climbers' weights.

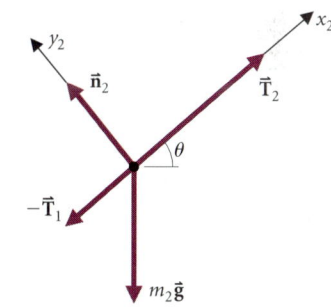

■ **FIGURE 5.3**
Free-body diagram for the rescuer.

5.2 Solution Method for Systems of Particles

5.2.1 *Identifying Particles*

In analyzing the mountain rescue, we focus on two objects, the rescuer and the injured climber, representing each as a particle. The rest of the system (ropes, ice, climbing gear, and Earth) is left out of focus and is represented by the forces acting on the two people. We make three judgments in choosing the two people as the significant particles in the system: that any internal motions of the people do not alter the system's behavior within the degree of accuracy we care about; that we can model the rest of the system well enough to describe the forces it exerts on the people; and that we can accept knowledge of these forces and the motions that result from them as an adequate analysis of the problem. Making these judgments is the first step in attacking textbook problems or analyzing real-world systems.

5.2.2 Free-Body Diagrams

Each significant part of the system is subject to Newton's laws. The second step in analyzing a system is to draw a free-body diagram for each part.

> REMEMBER: In a *free-body diagram* a part of a system is represented as an isolated particle. The rest of the system is not shown but is represented by all the forces it exerts on the part.

5.2.3 Analysis of the Free-Body Diagrams

IN CHAPTER 4 WE SHOWED HOW TO ANALYZE A FREE-BODY DIAGRAM FOR A SINGLE PARTICLE.

Recall that Newton's second law relates vectors. To apply the second law, choose a coordinate system and take vector components along each coordinate axis. For each free-body diagram, choose the coordinates to simplify your problem as much as possible. For example, if you know the direction of a particle's acceleration, choose an axis along that direction; otherwise, choose axes so that as many force vectors as possible are parallel to an axis. You do not need to use the same coordinates for different particles.

It is important, however, to use *different names* for the coordinates in different free-body diagrams (subscripts 1 and 2 in the rescue example). Each acceleration component of each particle is a distinct variable and must be assigned its own algebraic symbol.

THE WORD *EQUILIBRIUM* IS DERIVED FROM THE LATIN WORDS *AEQUUS* MEANING EQUAL AND *LIBRA,* A BALANCE OR PAIR OF SCALES.

For the special case in which all the acceleration components are zero, the forces balance and we say that the system is *in equilibrium.* Each part of the system either is at rest or has constant velocity.

5.2.4 Connecting the Particles Back into a System

Applying Newton's second law gives you one equation for each coordinate direction in each free-body diagram, but you may still have more unknowns than equations. To find the remaining relations you need, look at the connections between the particles and consider what you know about their motions. For instance, in the rescue example we noted that the climbers are not accelerating and that the rope connecting them exerts a tension of equal magnitude on each climber. When you finish this step, you should have an equal number of equations and unknowns, and you can then proceed with the algebra.

The solution method we have described has six steps (■ Figure 5.4). These six steps show how to apply the general problem-solving method to this particular class of problem. The following examples illustrate the use of this method.

5.2.5 Using the Solution Method

> **EXAMPLE 5.1** ♦ Two identical freight cars of mass M, connected by a coupling rod, are pushed by an engine with a force $\vec{F}$ (■ Figure 5.5). What forces are exerted by the coupling?
>
> **MODEL** Figure 5.5 illustrates the system. The two freight cars are the significant particles. The coupling rod connects them together and, like string, transmits forces. Since the rod is stiff, it can push as well as pull. We neglect the rod's mass. We model the engine by the force $\vec{F}$ it exerts.
> **STEP I**
>
> **STEP II** Next we draw the free-body diagrams for the cars (■ Figure 5.6). The forces exerted by the coupling are labeled $\vec{C}_1$ and $\vec{C}_2$.
>
> **STEP III** Since the problem is one-dimensional, the cars accelerate horizontally. We choose the x_1- and x_2-axes in this direction.

■ FIGURE 5.4   Method for Analyzing Dynamic Systems

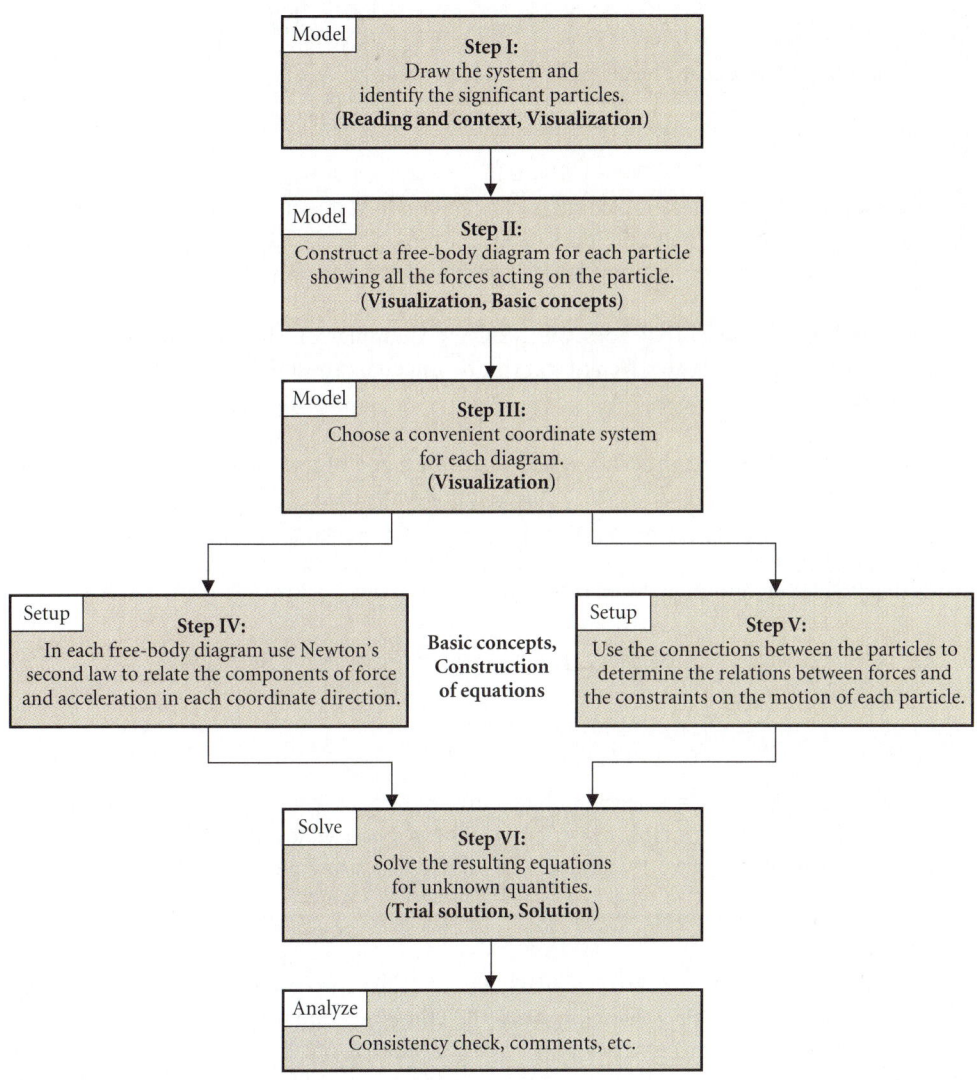

SETUP
STEP IV Applying Newton's second law to the two free-body diagrams gives:

$$\sum F_{x_1} = F - C_1 = Ma_{x_1} \quad \text{and} \quad \sum F_{x_2} = C_2 = Ma_{x_2}.$$

STEP V The two cars are connected by the coupling rod, which is represented by the force components C_2 and $-C_1$ it exerts on the two cars. According to Newton's third law, each car exerts an equal and opposite force on the coupling. If the coupling's mass is negligible, the product of its mass and its acceleration is also negligible, and the net force $C_1 - C_2$ acting on it is zero. Thus $C_1 = C_2 \equiv C$. Like an ideal string, the ideal coupling exerts forces of equal magnitude at each end.

The rod's length is constant, so the cars have equal velocities at all times and therefore equal accelerations: $a_{x_1} = a_{x_2} \equiv a_x$.

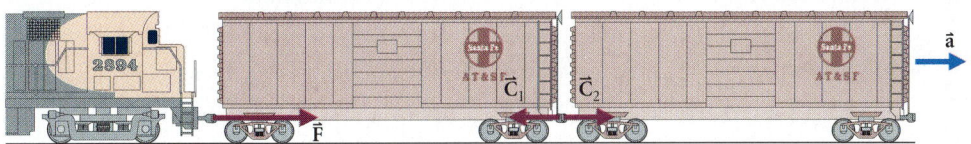

■ FIGURE 5.5
Railroad engine pushing two freight cars. The coupling between the freight cars is compressed and exerts the force needed to accelerate the freight car on the right.

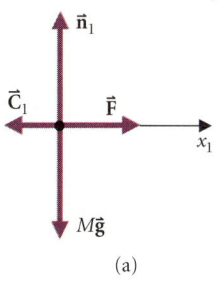
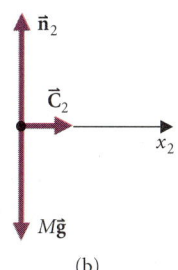

■ FIGURE 5.6
Free-body diagrams for (a) the left and (b) the right freight cars.

IF WE HAD ASSUMED THE COUPLING WAS UNDER TENSION, WE WOULD HAVE FOUND THE RESULT $T = -F/2$.

SOLVE
STEP VI   Substituting the results of Step V into the force equations, we have

$$F - C = Ma_x \quad \text{and} \quad C = Ma_x.$$

Combining these equations gives

$$F = 2Ma_{x_1} \quad \text{or} \quad a_{x_1} = F/(2M)$$
$$\text{and} \quad C = F/2.$$

ANALYZE   The acceleration could have been obtained directly by thinking of the system as a single object of mass $2M$ upon which a force F acts, but to find the compression force C exerted by the coupling, we had to consider the freight cars individually.

EXERCISE 5.1   ❖   How does the result of Example 5.1 change if the engine pulls instead of pushes? If the freight cars have unequal masses? If there are more than two cars?

WHILE CONSERVATION OF STRING IS AN EXTREMELY USEFUL PRINCIPLE, IT DOES NOT HAVE THE FUNDAMENTAL IMPORTANCE OF THE GREAT CONSERVATION LAWS WE SHALL DESCRIBE IN PART II. WE USE THIS TERM TONGUE-IN-CHEEK, BUT PERHAPS THE NAME WILL HELP YOU REMEMBER THE IDEA.

In Example 5.1 the freight cars are connected by a rod of fixed length, so their positions always differ by a fixed amount equal to the length ℓ of the rod:

$$x_2 = x_1 + \ell.$$

Thus their velocities and their accelerations are also related:

$$v_{x_2} = \frac{dx_2}{dt} = \frac{d}{dt}(x_1 + \ell) = \frac{dx_1}{dt} + \frac{d\ell}{dt} = v_{x_1} + 0 = v_{x_1},$$

and

$$a_{x_2} = \frac{dv_{x_2}}{dt} = \frac{dv_{x_1}}{dt} = a_{x_1}.$$

A similar relation holds whenever two objects are connected by a rod or a taut string of fixed length. We shall call this principle *conservation of string* because, for the velocities to differ, the length ℓ of string would have to change. (Of course, the principle fails to hold if the string becomes slack.)

EXAMPLE 5.2   ♦♦   Maria, of mass m, has fallen into a crevasse while descending a steep, icy glacier and is hanging from the climbing rope tied to her father, Joachim (■ Figure 5.7). Joachim, of mass M, decides to pull her out by sliding down the glacier. Assuming that both Joachim and the rope slide without friction, find his acceleration down the glacier.

■ **FIGURE 5.7**
Maria and her father in trouble on the glacier. Maria has fallen into a crevasse, and her father expects to rescue her by sliding down the glacier.

MODEL
STEP I   The significant particles are the two climbers, daughter and father, of masses m and M.

STEP II   ■ Figure 5.8 shows the free-body diagrams for the climbers.

STEP III   Joachim slides along the glacier slope, so we choose the x_1-axis down the slope and the y_1-axis perpendicular to the slope. Maria moves vertically within the crevasse, so we choose x_2 horizontal and y_2 vertical.

SETUP   We analyze the free-body diagrams in the usual way.
STEP IV

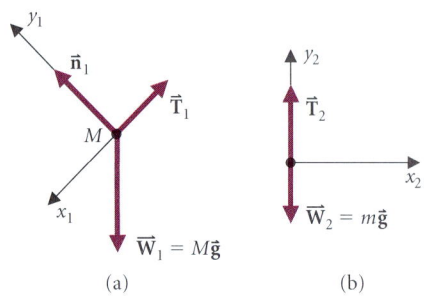

■ **FIGURE 5.8**
Free-body diagrams for (a) Joachim and (b) Maria.

JOACHIM

x_1-components: $\sum F_{x_1} = Ma_{x_1}.$
$\quad Mg \sin\theta - T_1 = Ma_{x_1}.$ (i)

y_1-components: $\sum F_{y_1} = Ma_{y_1}.$
$\quad n - Mg \cos\theta = 0.$ (ii)

MARIA

x_2-components: none

y_2-components: $\sum F_{y_2} = ma_{y_2}.$
$\quad T_2 - W_2 = T_2 - mg = ma_{y_2}.$ (iii)

STEP V   Father and daughter are connected by a massless, frictionless rope, along which the tension has constant magnitude: $T_1 = T_2 \equiv T$.

We find a relation between the two accelerations by using *conservation of string*. Because the rope's length is fixed and it remains taut, Joachim slides down the glacier the same distance as Maria rises. So Joachim's speed down the glacier equals Maria's upward speed at all times, and their accelerations have equal magnitudes.

$$\text{Conservation of string: } |\vec{a}_1| = |\vec{a}_2| \Rightarrow a_{x_1} = a_{y_2}.$$

SOLVE
STEP VI Eliminating T and a_{y_2} gives a relation for the acceleration a_{x_1}. From eqn. (iii):

$$T = m(g + a_{y_2}) = m(g + a_{x_1}). \quad \text{(iv)}$$

Substituting this result in eqn. (i) gives:

$$Mg \sin\theta - (ma_{x_1} + mg) = Ma_{x_1},$$

and rearranging gives

$$a_{x_1} = \frac{g(M \sin\theta - m)}{M + m}.$$

From eqn. (iv), the tension in the rope is:

$$T = m(g + a_{x_1}) = mg\left(1 + \frac{M \sin\theta - m}{M + m}\right) = g\frac{Mm}{M + m}(1 + \sin\theta).$$

ANALYZE Provided that $M \sin\theta > m$, the tension is greater than mg, which is the requirement for Maria to accelerate upward. In this case, the tension is also less than $Mg \sin\theta$, the requirement for Joachim to accelerate downward. For an angle $\theta = 30°$, Joachim has to have at least twice the mass of his daughter for the rescue to work.

We could think of the two climbers as a single object of mass $(m + M)$ subject to forces with magnitudes W_2 and W_{1,x_1}. All other forces are balanced. This approach enables us to make a quick derivation of the acceleration but yields no information about the rope tension. ■

EXERCISE 5.2 ♦♦ Verify that $mg < T < Mg \sin\theta$ if $M \sin\theta > m$.

EXERCISE 5.3 ♦♦ At what minimum angle θ can Joachim succeed in his rescue attempt if his mass is 90 kg and Maria's mass is 75 kg? Describe what happens if θ is only 30°; what is Maria's acceleration in this case?

EXAMPLE 5.3 ♦♦ Two arc lights with equal mass $m = 10.0$ kg, connected by a spring with negligible mass, are attached to the roof and wall of a stage by wires (■ Figure 5.9). If the two lights are at the same height and the angle $\theta = 30.0°$, find the tension in the spring and in each of the wires. What is the angle ϕ?

MODEL
STEP I There are three significant particles in this system: the two lights and the spring.

STEP II ■ Figure 5.10 shows the free-body diagrams. Weight, a wire tension force, and a spring tension force act on each light. The spring stretches horizontally between the lights and exerts horizontal forces on them.

The spring has negligible mass, so we neglect its weight. Hooke's law is not useful here because we are given neither the spring constant nor any information about the stretch of the spring. The spring force F_s will emerge as a result of analyzing the other two free-body diagrams.

STEP III In each of our free-body diagrams most of the forces are in the horizontal or vertical directions, so we choose a coordinate system for each diagram with axes along those directions.

REMEMBER: THIS IS A RELATION BETWEEN ACCELERATION COMPONENTS. THE SIGNS ARE IMPORTANT. SINCE MARIA RISES WHEN JOACHIM MOVES DOWNHILL, $+a_{x_1} = +a_{y_2}$.

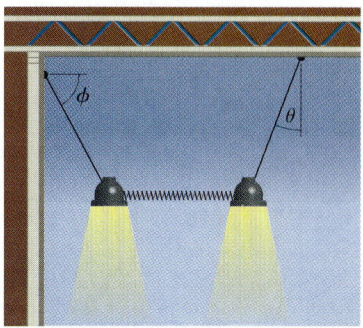
■ **FIGURE 5.9**
Arc lights suspended by ropes and a spring.

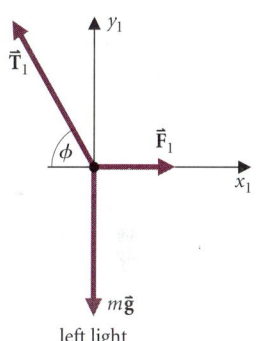
left light

spring

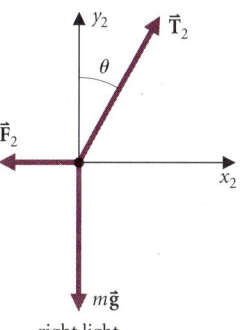
right light

■ **FIGURE 5.10**
Free-body diagrams for the arc lights and the spring.

SETUP
STEP IV All of the objects remain at rest so, applying Newton's second law, we have:

$$\text{SPRING: } \sum F_x = 0 \Rightarrow F_{sR} - F_{sL} = 0 \Rightarrow F_{sR} = F_{sL} \equiv F_s.$$

LEFT LIGHT:

x_1-components: $\sum F_{x_1} = 0.$
$-T_1 \cos\phi + F_1 = 0.$ (i)

y_1-components: $\sum F_{y_1} = 0.$
$T_1 \sin\phi - mg = 0.$ (iii)

RIGHT LIGHT:

x_2-components: $\sum F_{x_2} = 0.$
$-F_2 + T_2 \sin\theta = 0.$ (ii)

y_2-components: $\sum F_{y_2} = 0.$
$T_2 \cos\theta - mg = 0.$ (iv)

STEP V According to Newton's third law, the lights exert forces on the spring that are equal and opposite to the forces the spring exerts on them: $\vec{F}_1 = -\vec{F}_{sL}; \vec{F}_2 = -\vec{F}_{sR}$. Since $F_{sR} = F_{sL} = F_s$, then also $F_1 = F_2 = F_s$.

SOLVE
STEP VI From eqns. (i) and (iv):

$$T_1 \cos\phi = F_s \quad \text{and} \quad T_2 \cos\theta = mg.$$

From eqns. (iii) and (ii):

$$T_1 \sin\phi = mg \quad \text{and} \quad T_2 \sin\theta = F_s.$$

Dividing gives us: $\tan\phi = mg/F_s$ and $\cot\theta = mg/F_s.$

Thus $\cot\theta = \tan\phi$ and so $\phi = 90° - \theta$. Since $\theta = 30.0°$, $\phi = 60.0°$. Then

$$F_s = mg \tan\theta = (10.0 \text{ kg})(9.80 \text{ m/s}^2)\tan 30.0° = 56.6 \text{ N},$$

$$T_1 = mg/\sin\phi = (98.0 \text{ N})/\sin 60.0° = 113 \text{ N},$$

and $\quad T_2 = mg/\cos\theta = (98.0 \text{ N})/\cos 30.0° = 113 \text{ N}.$

ANALYZE With hindsight, we can find a method that requires less algebra. Each wire has to exert a vertical component of force that balances the weight of the light it supports. Once we realize that the wires also exert equal but opposite horizontal force components, we see that the wire tensions have to be equal in magnitude and at the same angle to the vertical. From these conclusions, the numerical answers follow in two steps.

Note that after solving each of these first three examples we found a shortcut. Not really obvious at all, these shortcuts become clear only after careful thought. With experience, you will get better at seeing them. For the time being, the six-step method is the safest approach. ∎

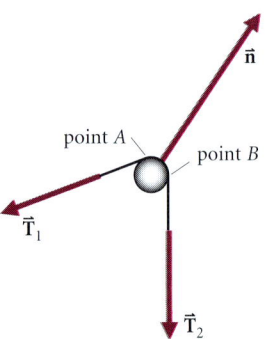

■ **FIGURE 5.11**
Ideal string passing over a support. Because the region of contact between the string and the support is curved, there is no obvious way to determine the direction of the normal force $\vec{n}$.

✳ 5.3 STRINGS

So far we have considered ropes and cables as *ideal strings*: massless, frictionless connections that transmit tension forces. Ideal string is a useful model because it fits a wide range of real experience. But the variation of cable tension in a suspension bridge is an important feature of the bridge's design. When we cannot neglect a cable's mass, the friction acting on it, or the way load is placed on it, we have to imagine the cable itself as a system of particles. This analysis follows the general six-step method, but the identification of significant pieces has a flavor all its own that we investigate in this section.

5.3.1 Why the Ideal String Model Works

Let us first use Newton's laws to verify that tension is constant along an ideal string. Consider a string, with negligible mass, that passes over a frictionless support (■ Figure 5.11). The two objects that the string connects exert tension forces on its ends, and the support exerts a normal force on the string. Because the surface is curved, though, we don't know the direction of $\vec{n}$. Analysis of a free-body diagram for the entire string (■ Figure 5.12) generates only two equations for the magnitudes of the three unknown forces and the unknown direction of $\vec{n}$. We reach Step VI of the standard method with an insoluble problem—a clue that we haven't chosen the significant particles carefully enough.

The entire string has a complex shape that is important to the relationships among the three forces, and so it cannot be regarded as a "particle." We need to imagine the string as a system of smaller pieces. Imaginary cuts at points A and B in ■ Figure 5.13 separate the two straight sections from the curved section around the support. At the imaginary cuts, points A and B, atoms in the string are tugging on each other, and Newton's third law guarantees that the tension forces they exert on opposite sides of a *cut* are equal and opposite, as shown.

The forces acting on each of the straight pieces of string have to balance (■ Figure 5.14). Otherwise, the string, having negligible mass, would accelerate very rapidly, which we know it does not do. The *massless* string segments have negligible weight, so the two tension forces acting on each segment balance—they are equal and opposite. This shows both that each segment is indeed straight and that the magnitude of tension in each segment is the same at both ends.

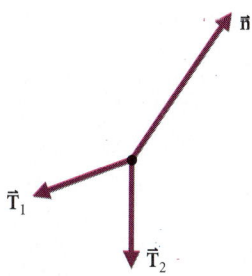

■ Figure 5.12
Free-body diagram for the string. Without knowing the direction of $\vec{n}$ relative to the two tension forces, we haven't enough information to determine both $\vec{n}$ and a relation between the magnitudes of $\vec{T}_1$ and $\vec{T}_2$.

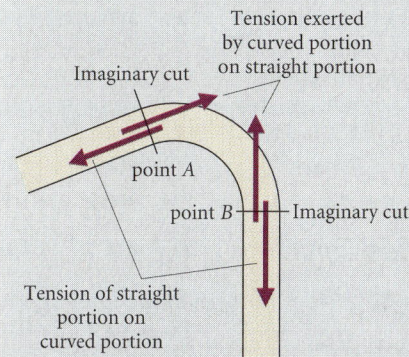

■ Figure 5.13
Imaginary *cuts* at points A and B separate the straight portions of the string from the curved portion. Molecules of string on opposite sides of the cuts attract each other, giving rise to tension forces. Newton's third law tells us that the tension forces exerted by the string segments on each other are equal and opposite.

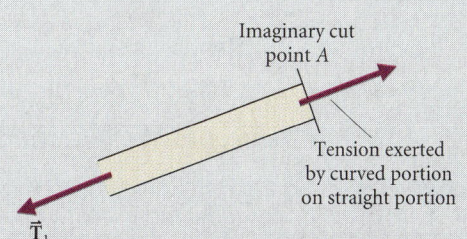

■ Figure 5.14
The *massless* string segment has negligible weight, so the tension forces on its ends must balance. They lie along the same line, so the string has to be straight.

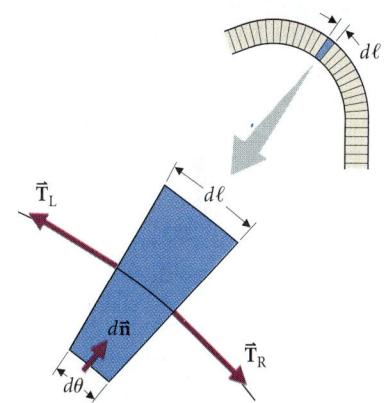

■ Figure 5.15
To analyze the curved portion of the string, we imagine dicing it into a large number of small pieces. One such piece, magnified, bends through the element of angle $d\theta$, so the tension forces acting on it are not quite aligned. The normal force $d\vec{n}$ is perpendicular to the surface *at the element* and so has a known direction.

AGAIN, WE ARE NEGLECTING THE ELEMENT'S MASS AND SETTING THE PRODUCT OF MASS AND ACCELERATION TO ZERO.

To study the curved part of the string, we imagine it as a large number of very small elements of length $d\ell$ (■ Figure 5.15). The normal force $d\vec{n}$ exerted on a single element of string is perpendicular to the support (■ Figure 5.16). The two tension forces do not quite lie along the same line, because the element bends around the support by an angle $d\theta$ between its two ends. Analyzing the free-body diagram in the usual way, we find:

x-components: $\sum F_x = T_R \cos(d\theta/2) - T_L \cos(d\theta/2) = 0$.

y-components: $\sum F_y = dn - T_R \sin(d\theta/2) - T_L \sin(d\theta/2) = 0$.

From the equation for the x-components, we find that the tension is the same at both ends of the element:

$$T_R = T_L.$$

Since the whole string consists of individual elements connected together, it follows that there is no change of tension between the ends of the entire curved section. This completes our demonstration that tension is constant along an ideal string.

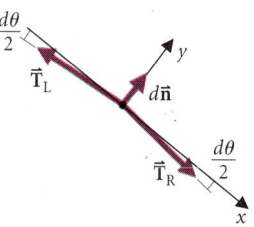

■ Figure 5.16
Free-body diagram for an element of string.

■ **Figure 5.17**
A footbridge on a mountain trail in Nepal.

■ **Figure 5.18**
The Humber Bridge is the world's longest single-span suspension bridge. It crosses the Humber River near Hull, England.

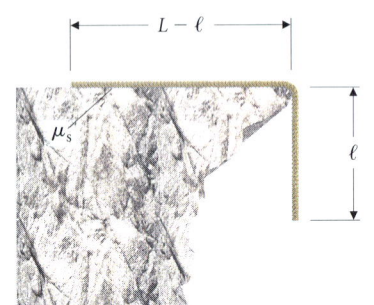

■ **Figure 5.19**
Uniform rope hanging over an edge.

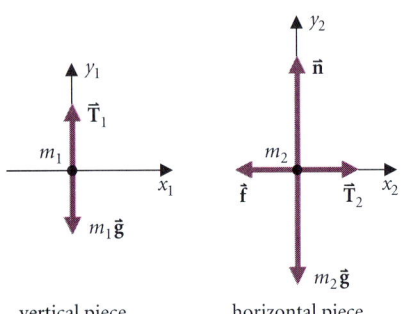

■ **Figure 5.20**
Free-body diagrams for the two pieces of rope.

5.3.2 Strings Subject to Weight and Friction Forces

A massive cable cannot stretch horizontally in a straight line, since horizontal tension forces, no matter how large, cannot balance the cable's (vertical) weight. Instead, the cable hangs in a graceful curve called a catenary or chain curve. The behavior of hanging cables is important in bridge design. Walking the steep curve of a mountain footbridge (■ Figure 5.17) is awkward, but a bridge with a shallower curve would require much greater cable tension and would be correspondingly more difficult to build. In a highway suspension bridge (■ Figure 5.18), a level roadway hangs from the cable. The cable's weight is small compared with the weight of roadway and traffic, and the cable hangs in a parabola (see Problem 85).

All cable problems require insight to decide how to divide the system into particles (Step I). The mathematical challenge occurs when solving the resulting equations (Step VI). Cables or ropes hanging vertically or lying on a horizontal surface are less beautiful than bridge cables, but more readily analyzed.

EXAMPLE 5.4 ♦♦ A uniform rope of total mass M and length L hangs partway over a smooth edge (■ Figure 5.19). If the coefficient of static friction between the rope and the rock surface is $\mu_s = 0.60$, what is the maximum fraction of the rope that can hang without pulling the remainder over the edge?

MODEL
STEP I The edge of the cliff divides the rope into two pieces, the hanging piece of length ℓ and the remainder lying on the horizontal surface. We model the system with these two pieces as the significant particles.

STEP II ■ Figure 5.20 shows free-body diagrams for the two pieces.

STEP III All the forces are horizontal or vertical, so we choose coordinates along these directions.

SETUP
STEP IV We apply Newton's second law, with all the accelerations zero:

HANGING PIECE	HORIZONTAL PIECE
y_1-components: $\sum F_{y_1} = 0.$	x_2-components: $\sum F_{x_2} = 0.$
$T_1 - m_1 g = 0.$ (i)	$T_2 - f = 0.$ (ii)
	y_2-components: $\sum F_{y_2} = 0.$
	$n - m_2 g = 0.$ (iii)

STEP V Since the edge is "smooth," the magnitude of the tension is constant in the small piece of rope at the bend, and $T_1 = T_2 = T$.

In a uniform rope the mass of a piece of rope is proportional to its length. Thus:

$$m_1 = (M/L)\ell$$

and

$$m_2 = (M/L)(L - \ell).$$

Because we are asked for the *maximum* length that can hang, static friction is at its limit: $f = \mu_s n$.

SOLVE
STEP VI From eqn. (i), $T = m_1 g$, and from eqn. (iii), $n = m_2 g$. Substituting these results into eqn. (ii):

$$m_1 g = T = f = \mu_s n = \mu_s m_2 g.$$

Expressing the masses in terms of lengths, we find:

$$\ell = \mu_s (L - \ell).$$

So, the maximum fraction of hanging rope ℓ/L is given by:

$$\frac{\ell}{L} = \frac{\mu_s}{\mu_s + 1} = \frac{0.60}{1.60} = \frac{3}{8}.$$

ANALYZE We needed to analyze only two *particles* to find a relatively simple relation between the two portions of the rope. The following exercise illustrates what to do when rope properties vary or when we need to find how tension varies along the rope.

EXERCISE 5.4 ♦♦ A rope with a significant amount of mass hangs vertically. Measuring x upward from the end of the rope, draw a free-body diagram for the portion of the rope below point x. Analyze the diagram to find the tension in the rope at x.

5.3.3 Pulley Systems

To lift a 100-kg object, you must somehow exert an upward force of 980 N on it. Without help, you are likely to injure yourself; you will certainly become quickly exhausted; and you will not be able to hold the object steady for any length of time. A pulley system (■ Figure 5.21) allows you to lift an object while exerting a much smaller force in a more convenient direction and with greatly improved control over the object's motion. We'll now investigate how this is achieved.

In a single pulley system (■ Figure 5.22), the rope exerts equal tension force at its two ends. At one end, tension balances the force you exert: $\vec{T} = -\vec{F}$. At the other end, tension balances the object's weight, $\vec{T} = -\vec{W}$. Thus, $\vec{F} = \vec{W}$. The single pulley is useful because it allows you to pull downward rather than upward, but it doesn't reduce the necessary force. With two pulleys, we can build a more useful machine.

■ **FIGURE 5.21**
A person lifting a metal punch with a hoist.

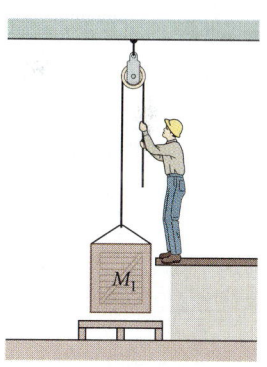

(a)

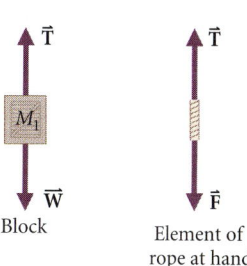

(b)

■ **FIGURE 5.22**
(a) Object being lifted with the aid of a single pulley. (b) Free-body diagrams.

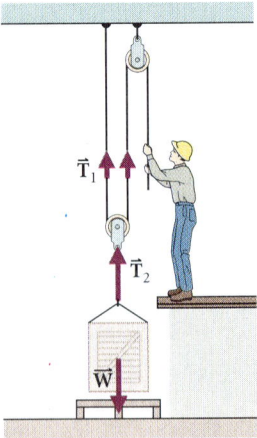

■ **FIGURE 5.23**
Object being supported with a system of two pulleys. Two segments of rope lead away from the lower pulley, so the required tension is only one-half the weight of the load.

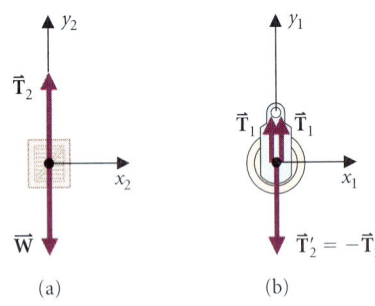

■ **FIGURE 5.24**
Free-body diagrams for (a) the crate and (b) the pulley.

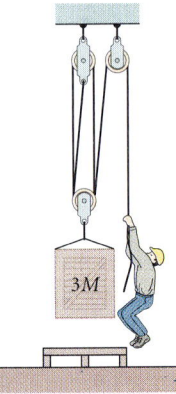

■ **FIGURE 5.25**
Heroic worker saving a crate of electronics with a 4-to-1 pulley system. Does he succeed?

EXAMPLE 5.5 ♦♦ What force F is required to support a crate of mass $M = 175$ kg using the system of two pulleys shown in ■ Figure 5.23?

MODEL
STEP I There are two significant pieces in this system: the second pulley, which is supported by the rope and from which the crate hangs, and the crate itself. (If you need to know the force required of the roof, then the fixed pulley is also significant.)

STEP II ■ Figure 5.24 shows the free-body diagrams for the two particles.

STEP III All forces are vertical, so we choose the y-axis to be directly upward.

SETUP
STEP V The tension has constant magnitude T_1 along the continuous rope running around the two pulleys. All accelerations are zero. By looking at the piece of rope where the force is applied, we see that $T_1 = F$.

STEP IV The pulley is supported by two parallel lengths of rope, each with a tension T_1. The total upward force on the pulley is $2T_1$. The rope supporting the crate pulls downward on the pulley with a tension T_2. Now apply Newton's second law:

$$\text{Pulley:} \sum F_{y_1} = 0 \Rightarrow T_2 = 2T_1.$$

The crate is supported by the rope from the pulley with tension T_2.

$$\text{Crate:} \sum F_{y_2} = 0 \Rightarrow T_2 = W = Mg.$$

SOLVE
STEP VI Combining these relations, $F = T_1 = T_2/2 = Mg/2$:

$$F = (175 \text{ kg})(9.81 \text{ m/s}^2)/2 = 858 \text{ N}.$$

ANALYZE With two pulleys, the force you need to support the crate is only half its weight because each parallel length of rope leading away from the second pulley exerts the same upward tension. This result illustrates a useful generalization:

When only one continuous rope is used, the reduction factor in the required force equals the number of times the rope leads away from the pulley supporting the load. ■

In the next example we look at a pulley system that is not in equilibrium. Watch carefully how *conservation of string* is applied in Step V of the solution process.

EXAMPLE 5.6 ♦♦ A heroic worker of mass $M = 110$ kg leaps onto the rope of a pulley system trying to keep a crate of electronic instruments (mass $3M = 330$ kg) from crashing to the floor of the factory (■ Figure 5.25). What are the accelerations of worker and crate?

MODEL
STEP I The two pulleys fixed to the roof serve to redirect the rope. The movable pulley and the crate are tied together and act as one particle, unless we care about tension in the rope connecting them. The worker is a second significant particle.

STEP II Free-body diagrams for the crate/pulley and the worker are shown in ■ Figure 5.26. We recognized constant tension in the ideal rope here instead of in Step V. (It's OK to do part of Step V here as long as you remember to think carefully about what you're doing.)

STEP III The crate and the worker both move in the vertical direction, and we choose coordinates with the y-axis upward for each.

SETUP Newton's second law gives two equations.
STEP IV
Worker: $\sum F_{y_1} = m_1 a_{y_1}$.   Crate: $\sum F_{y_2} = m_2 a_{y_2}$.
$T - Mg = Ma_{y_1}$.   (i) $4T - 3Mg = 3Ma_{y_2}$.   (ii)

STEP V We use conservation of string to find a relation between a_{y_1} and a_{y_2}. If the crate rises by an amount δy_2, then *each* of the four rope segments supporting the pulley decreases in length by δy_2. The decreased length $4\delta y_2$ represents rope that passes over the pulley to the worker, who moves *downward* by $4\delta y_2$. That means the resulting displacement of the worker is four times that of the crate and in the opposite direction. So:

$$\delta y_1 = -4\delta y_2. \quad \text{(Note: The minus sign is IMPORTANT.)}$$

Differentiating this relation twice gives relations between the velocities and between the accelerations of the crate and the worker:

$$v_{y_1} = -4v_{y_2} \quad \text{and} \quad a_{y_1} = -4a_{y_2}. \quad \text{(iii)}$$

The minus sign in these relations expresses the fact that if one acceleration is upward, the other is downward. It DOES NOT mean that we know the actual direction of either acceleration as yet!

SOLVE
STEP VI The two accelerations and the tension are unknown, and we have three relations among them. From eqn. (i):

$$T = M(g + a_{y_1}).$$

Substituting this result for T into eqn. (ii), along with eqn. (iii) for a_{y_2}, we have:

$$4(Mg + Ma_{y_1}) - 3Mg = 3M(-a_{y_1}/4).$$

Rearranging: $\quad Mg = Ma_{y_1}(-\tfrac{3}{4} - 4) = -(\tfrac{19}{4})Ma_{y_1}.$

Thus, $\quad a_{y_1} = -4g/19 = -4(9.81 \text{ m/s}^2)/19 = -2.07 \text{ m/s}^2.$

ANALYZE The minus sign indicates that the worker accelerates downward. Whether he succeeds in saving the crate depends on its velocity when he grabs the rope. The accelerations don't depend on the initial velocity, but the displacements do, and that is what determines the fate of the crate.

According to the string counting rule described in Example 5.5, a crate of mass $4M$ would be in equilibrium with the worker of mass M. We could thus predict the qualitative result of this example: the crate, with mass less than $4M$, accelerates upward. ∎

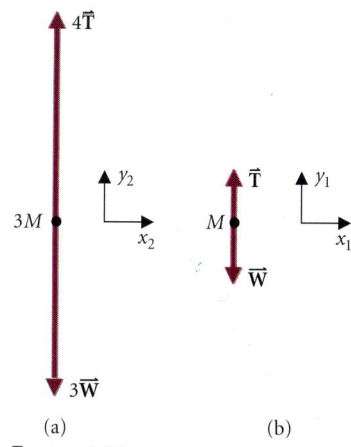

■ **FIGURE 5.26**
Free-body diagrams for (a) crate/pulley and (b) worker.

5.4 THE LAW OF UNIVERSAL GRAVITATION
5.4.1 The Gravitational Force Law

Newton's study of planetary motion convinced him that every particle in the universe exerts a gravitational force on every other particle. Planets are nearly spherical and are very small compared with the distances between them,[1] so regarding them as particles is a good approximation. From the behavior of the planets, Newton could deduce the law for gravitational force between arbitrary pairs of particles. Let's see how.

Newton knew from Kepler's third law that the acceleration of a planet toward the Sun decreases with the inverse square of its distance from the Sun. Furthermore, the acceleration doesn't depend on the mass of the planet—giant Jupiter follows the same rule as tiny Mercury. Applying Newton's second law to the planet:

$$a_{\text{planet}} = \frac{F_{\text{Sun on planet}}}{\text{mass of planet}} \propto \frac{1}{r^2} \text{ independent of the mass.}$$

SEE THE ESSAY ON NEWTON'S DISCOVERY OF THE LAW OF UNIVERSAL GRAVITATION FOR MORE DETAIL. FOR THE PROOF OF $a \propto 1/r^2$, SEE FIGURE E1.2.

---

[1] For example, Earth has a radius of 6400 km and is 150 million km from the Sun. Furthermore, Newton was able to show that a spherical object acts gravitationally as if all its mass were at its center.

Thus the gravitational force acting on the planet is proportional to its mass:

$$F_{\text{Sun on planet}} \propto \frac{m_{\text{planet}}}{r^2}.$$

According to Newton's third law, the planet also pulls on the Sun. That force obeys the same gravitational force law and so depends in the same way on distance and mass:

$$F_{\text{planet on Sun}} \propto \frac{M_{\text{Sun}}}{r^2}.$$

Since the magnitudes of the two paired forces are equal, both of these results can be true only if the magnitude of each force is proportional to the product of the two masses:

$$F_{\text{Sun on planet}} = F_{\text{planet on Sun}} \propto \frac{M_{\text{Sun}} \times m_{\text{planet}}}{r^2}.$$

Newton's *law of universal gravitation* states that the same result holds for any pair of particles:

■ FIGURE 5.27
Newton's law of universal gravitation. Two particles exert attractive gravitational force on each other. The forces lie along the line connecting the particles. Their magnitude is proportional to the product of the two masses and inversely proportional to the square of the distance between them.

Two particles attract each other with forces whose magnitudes are proportional to the product of the particles' masses and inversely proportional to the square of the distance between them (■ Figure 5.27).

$$F_G = |\vec{F}_{2 \text{ on } 1}| = |\vec{F}_{1 \text{ on } 2}| \propto \frac{m_1 m_2}{r^2}, \quad \text{or} \quad F_G = G\frac{m_1 m_2}{r^2}. \quad (5.1)$$

The constant of proportionality, Newton's gravitational constant G, is determined experimentally:

$$G = 6.67 \times 10^{-11} \text{ N} \cdot \text{m}^2/\text{kg}^2.$$

EXAMPLE 5.7 ♦♦ The space shuttle is in a circular orbit 200 km above the Earth's surface. Use Newton's result that the spherical Earth attracts like a particle at its center to find the speed of the shuttle in its orbit.

MODEL The gravitational force acting on the shuttle produces its acceleration. In circular motion, the acceleration has magnitude v^2/r and points toward the center of the circle, the same direction as the force (■ Figure 5.28).

SETUP We use eqn. (5.1), and apply Newton's second law:

$$\vec{F}_G = M_{\text{shuttle}} \vec{a}.$$

$$G\frac{M_{\text{Earth}} M_{\text{shuttle}}}{r^2} = M_{\text{shuttle}} \frac{v^2}{r}.$$

We need to look up data for the mass and radius of the Earth (inside front cover): $R_{\text{Earth}} = 6400$ km, so the radius of the shuttle orbit is 200 km larger: $r = 6600$ km.

SOLVE
$$v^2 = G\frac{M_{\text{Earth}}}{r} = (6.67 \times 10^{-11} \text{ N} \cdot \text{m}^2/\text{kg}^2)\left(\frac{5.98 \times 10^{24} \text{ kg}}{6600 \text{ km}}\right)$$
$$= 6.0 \times 10^7 \text{ N} \cdot \text{m/kg} = 6.0 \times 10^7 \text{ (m/s)}^2,$$

and $v = 7.8 \times 10^3$ m/s.

ANALYZE The result is independent of the shuttle's mass. Any object in orbit at the same radius would have the same speed.

Earth is not a perfect sphere. Its polar radius and its equatorial radius differ in the third significant figure, thus limiting our result to two significant figures. ■

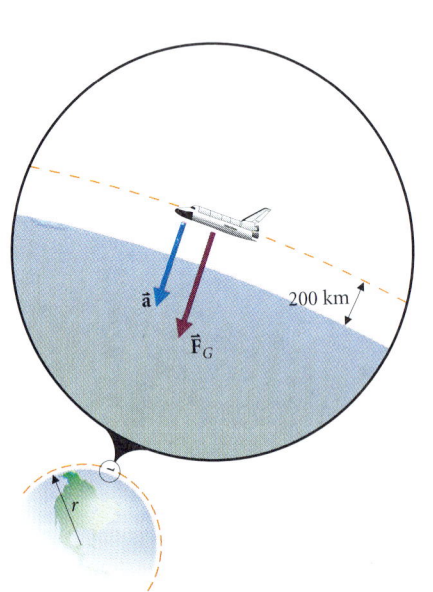

■ FIGURE 5.28
The space shuttle in a circular orbit, 200 km above the Earth's surface. Gravitational force accelerates it in its orbit.

EXAMPLE 5.8 ♦♦♦ Suppose that a hole were drilled to the center of the Earth and an evacuated pipe were inserted in the hole. How long would it take for a test probe dropped down the pipe to reach the center of the Earth (■ Figure 5.29)?

MODEL The object is attracted toward the center by gravity and accelerates according to Newton's second law. We shall make our estimate using dimensional analysis.

SETUP Dimensionally, acceleration is a length divided by a time squared, so:

$$\text{force/mass} = \text{acceleration} = L/T^2.$$

Dimensionally, the law of universal gravitation is $[\text{force}] = [G]M^2/L^2$. So:

$$[\text{force}]/M = [G]M/L^2.$$

Together, the two laws yield:

$$\frac{[G]M}{L^2} = \frac{L}{T^2},$$

or

$$T = \sqrt{\frac{L^3}{[G]M}}.$$

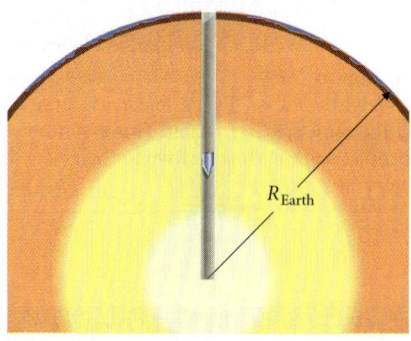

■ **FIGURE 5.29**
A probe is dropped into a hole drilled to the center of the Earth. How long does it take to reach the center?

MODEL Now we replace the dimensions in this result with corresponding quantities in the problem. Since the gravitational acceleration of the object is independent of its mass, the relevant mass is that of Earth. The radius of Earth is both the distance to fall *and* the appropriate scale of distance in the law of gravitation: this must be the relevant length.

SOLVE Thus the time τ we desire is *roughly*:

$$\tau \sim \left(\frac{R_{\text{Earth}}^3}{GM_{\text{Earth}}}\right)^{1/2} = \left[\frac{(6.4 \times 10^6 \text{ m})^3}{(6.7 \times 10^{-11} \text{ m}^3/\text{kg}\cdot\text{s}^2)(6.0 \times 10^{24} \text{ kg})}\right]^{1/2}$$

$$= 800 \text{ s, or about 10 min.}$$

ANALYZE A careful calculation, assuming the Earth to be of uniform density, shows that this result is too small by a factor of $2/\pi \approx 0.6$. This discrepancy is of similar magnitude to the error made in assuming a constant density for Earth. The answer obtained from dimensional analysis is as good as can be obtained without very careful attention to detail.

EXERCISE 5.5 ♦♦ Compare the time for the probe to fall to the center of the Earth with the time for the space shuttle (Example 5.7) to complete one orbit.

EXAMPLE 5.9 ♦ Compare the gravitational force on an astronaut in orbit 200 km above the Earth with her weight on the surface.

MODEL The mass of the astronaut is the same whether on Earth or in orbit. Thus, the gravitational force acting on her changes entirely because of the change in distance from the Earth's center.

SETUP Using eqn. (5.1):

$$\frac{W_{\text{orbit}}}{W_{\text{surface}}} = \frac{M_{\text{astronaut}}M_{\text{Earth}}/r_{\text{orbit}}^2}{M_{\text{astronaut}}M_{\text{Earth}}/r_{\text{Earth}}^2}$$

SOLVE

$$= \left(\frac{r_{\text{Earth}}}{r_{\text{orbit}}}\right)^2 = \left(\frac{6400 \text{ km}}{6600 \text{ km}}\right)^2 = 0.94.$$

ANALYZE The gravitational force on an astronaut in orbit is 6% less than when she is on the surface.

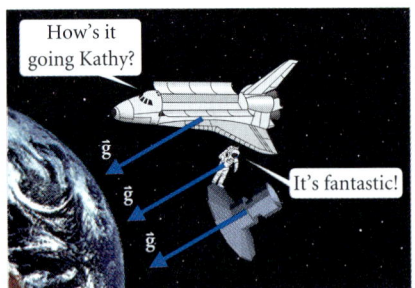

FIGURE 5.30
"Weightlessness." Every object accelerates at the same rate in free fall. Things, and people, seem to float relative to their environment.

Hold it! Everybody knows an astronaut is *weightless* in orbit. Popular use of the term *weightless* refers to the absence of any perception of weight. Standing on the Earth's surface, the astronaut's muscles and skeleton have to exert substantial forces to keep her heavy body upright and unaccelerated. Aloft (■ Figure 5.30), the Earth's gravitational attraction accelerates the astronaut and the shuttle in nearly identical orbits. No skeletal or muscular forces are needed to keep her body from crumpling to the floor of the shuttle.

A much less confusing term for the astronaut's experience is *free fall*. The astronaut and any other loose objects appear to float around the shuttle because the shuttle and everything in it is falling freely toward the Earth with the same acceleration.

EXERCISE 5.6 ♦ Considering the Earth, Moon, and Sun as particles, compare the magnitude of the gravitational force exerted on the Moon by the Earth with that exerted by the Sun.

Digging Deeper

GRAVITATIONAL FORCE DUE TO A UNIFORM SPHERE

The law of universal gravitation describes the forces occurring between two particles. From the point of view of an apple, however, Earth certainly does not seem like a particle, though we can model it as a large uniform sphere.

Newton showed that a uniform sphere exerts the same gravitational force as a point mass at its center. To prove this result, he imagined Earth diced into a huge number of small particles, and he used calculus to add the forces exerted on the apple by all of the particles.

There are many ways to cut a sphere into little bits. For example, we could slice it perpendicular to the diameter *AX* (■ Figure 5.31), cut each slice into rings, and then slice each ring up into pieces of equal size. The gravitational force exerted on the apple by each piece of a ring has the same magnitude, since each piece is the same distance from the apple. The net force due to a pair of pieces at opposite ends of a diameter of the ring points toward the ring's center. Thus the whole ring attracts the apple toward its center. Since the centers of all the rings lie along the sphere's diameter *AX*, the net force points along that diameter, toward the center of the sphere.

What about the magnitude of the force? Gravitational attraction is much larger when particles are closer together, so you might expect the ground directly below the apple to overwhelm the influence of more distant parts of the Earth and make the attraction seem to come from a point closer than the Earth's center. But there is a lot more distant material than nearby stuff! In the interlude after Chapter 7, we shall discuss how to set up Newton's integral and prove that the whole

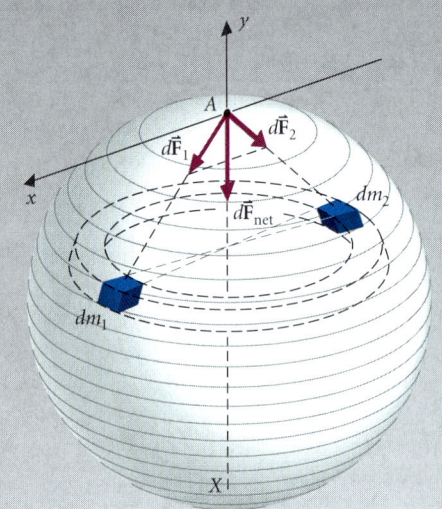

FIGURE 5.31
Two particles located symmetrically on opposite sides of an elementary mass ring exert forces $d\vec{F}_1$ and $d\vec{F}_2$ on the apple at *A*. Their vector sum $d\vec{F}_{net}$ points toward the center of the ring along the sphere's diameter *AX*.

sphere attracts like a point at its center. We'll also do a similar calculation for electric forces in Chapter 24. For now, we ask you to believe the result.

Planets and stars are not actually uniform. They are composed of layered spherical shells, with the denser material near their centers. Newton's theorem holds separately for each layer: each shell attracts another body in the same way as a point mass at the center of the shell. Consequently, the entire body behaves gravitationally like a point mass at its center.

Digging Deeper

Force at a Distance

Gravity is strange. The other mechanical forces occur when objects touch each other. What physical mechanism accounts for attraction between two objects that aren't touching? Newton answered that he didn't care to make any hypotheses about the nature of gravity, but instead chose to accept it as observed fact.

Einstein proposed a theory in 1915, known as general relativity, which improves on the accuracy of Newton's work and offers a "mechanism" for gravitational force. In Einstein's picture, "empty" space has elastic properties, somewhat like the surface of a trampoline. The Sun distorts surrounding space, much as a person standing on the trampoline distorts its surface. The Earth orbits the Sun in response to this distortion of the space through which it travels.

A force acting at a distance? A distortion of space (and time)? Gravity *is* strange!

Study Problem 5 ♦♦ *A Trip Around the Moon*

In Jules Verne's novel *A Trip from the Earth to the Moon* (1881), travelers in a capsule are shot from a huge cannon in Florida into an orbit that rounds the Moon and returns to Earth (■ Figure 5.32). Their dog, *Satellite,* who unfortunately fails to survive the launch, is jettisoned through an airlock and floats eerily just outside the capsule for the rest of the trip. Meanwhile, the passengers stand on the floor nearest the Earth and prepare to pass the mysterious point where the Moon's gravity becomes larger than the Earth's. They expect floor and ceiling to interchange.

Determine what is right and what is wrong in this story. If the Moon's mass is $\frac{1}{81}$ that of Earth, and its distance from Earth is $D = 60.0$ Earth-radii, where is the *mysterious point*? Compare the gravitational force of the Earth on the capsule at the *mysterious point* with its value on the Earth's surface.

Reading and Context

Once the capsule leaves the Earth's atmosphere on its way to the Moon, the only significant forces acting on it and its occupants are the gravitational forces exerted by the Earth, Moon, and Sun. The two major features of the story, one of which is wrong, are the canine companion and the interchange of floor and ceiling at the mysterious equal-gravity point. (Of course, there is plenty wrong with the idea of shooting people to the Moon from a cannon!)

The quantitative questions give information on the masses and distances of the Earth and Moon but do not mention the Sun. Indeed, the force the Sun exerts on the voyagers is essentially constant throughout the journey and so may be ignored in the calculations, which thus involve the gravitational forces of Earth and Moon only.

Visualization and Identification of Central Concepts

Once launched, the capsule falls freely along an orbit in the Earth–Moon system. The occupants and nearby objects appear to float since each object's weight causes it to undergo the same acceleration along the orbit. Just as in the space shuttle, the voyagers float inside the capsule *throughout* the trip, not only at some mysterious point. The luckless dog floats beside the capsule. There is no *floor* or *ceiling* inside the spacecraft, since objects do not fall in a way that defines up and down.

At the mysterious point, the gravitational force exerted by the Earth on any object is equal and opposite that exerted by the Moon, and the sum of these forces is zero.

■ **Figure 5.32**
Travelers to the Moon peer from their cannon-launched spacecraft at their unfortunate dog, who died at launch and whose body accompanies the spacecraft on its journey. From Jules Verne's novel *A Trip from the Earth to the Moon* (1881).

Compare this with Exercise 5.6. The Sun–Earth distance equals the Sun–Moon distance to two significant figures (Exercise 5.6). That's why the force exerted by the Sun is constant.

Outline of Plan and Construction of Equations

The desired point lies on the line of centers of Earth and Moon where the force *vectors* have opposite directions (■ Figure 5.33). The magnitudes of $\vec{F}_{Earth}$ and of $\vec{F}_{Moon}$ on an object of mass m at a distance x from the Moon are given by Newton's law of gravitation (eqn. 5.1):

$$|\vec{F}_{Earth}| = \frac{GmM_{Earth}}{(D-x)^2} = \frac{GmM_{Moon}}{x^2} = |\vec{F}_{Moon}|.$$

Solution of Equations

The mass m cancels out, as expected. The most efficient way to solve the quadratic equation for x is to take the square root:

$$\frac{x^2}{(D-x)^2} = \frac{M_{Moon}}{M_{Earth}} \;\Rightarrow\; \frac{x}{D-x} = \sqrt{\frac{M_{Moon}}{M_{Earth}}} = \sqrt{\frac{1}{81}} = \frac{1}{9}.$$

So, $9x = D - x$; and $x = D/10 = 60.0 R_{Earth}/10 = 6.0 R_{Earth}$. The distance of the point from the center of the Moon is 6 Earth-radii.

The ratio of gravitational force exerted by the Earth on the capsule at the mysterious point to that exerted when the capsule is on the Earth's surface is the inverse ratio of the respective distances from the center of the Earth, squared:

$$\frac{F_{point}}{F_{surface}} = \frac{r^2_{surface}}{r^2_{point}} = \frac{(R_{Earth})^2}{[(60.0-6.0)R_{Earth}]^2}$$

$$= \frac{1}{54.0^2} = 3.43 \times 10^{-4}.$$

Analysis

Jules Verne was fascinated by the idea of zero gravity, but that doesn't occur at the *mysterious point*, since the Sun's gravitational attraction is *not* balanced. There is nothing at all mysterious about the point, and space travelers passing through it wouldn't notice anything. Yet the point is not uninteresting: What would happen to a space station placed at rest there? Would it stay, unaccelerated toward either Earth or Moon? No. The station would have to orbit the Earth once a month to stay in line between the Earth and Moon, but the force needed to produce the corresponding acceleration does not occur. In the eighteenth century, Jean J. L. Lagrange showed that there are two places where an object can stably maintain its position relative to the Earth and Moon. These *Lagrange points* are beloved of present-day space colony enthusiasts.

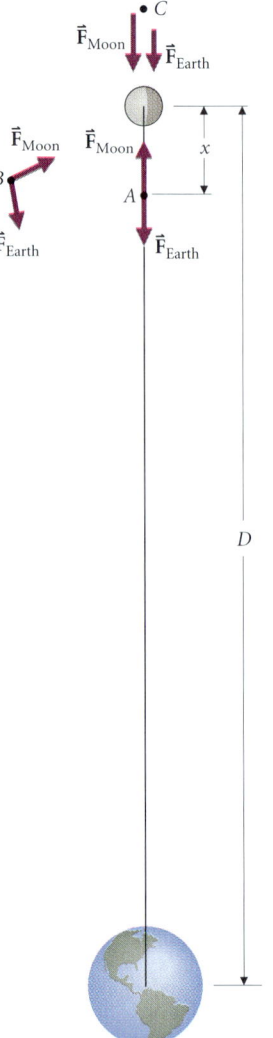

■ FIGURE 5.33
The Earth's gravitational acceleration can balance that of the Moon only at a point such as A where the vectors $\vec{F}_{Earth}$ and $\vec{F}_{Moon}$ are in opposite directions. The point has to be nine times closer to the Moon than to Earth to compensate for the Earth's greater mass.

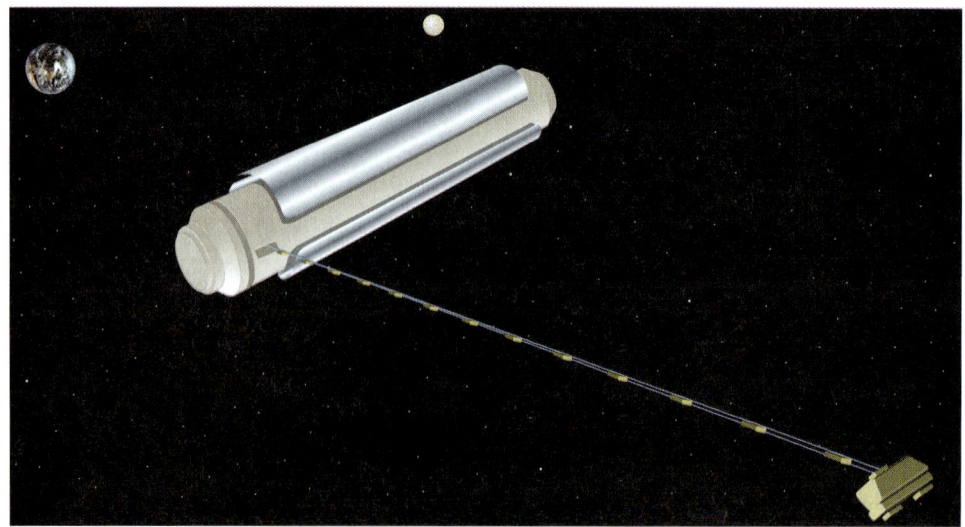

View of the Earth–Moon system from a space colony at the L5 point. There are two points where a space colony can maintain a fixed, stable relation with the Earth and Moon. These Lagrange points lie along the orbital path of the Moon at equal angles ahead of and behind the Moon.

5.4.2 Finding the Mass of the Earth

The gravitational forces between everyday objects are very small compared with other forces. Imagine what life would be like if your body exerted large attractive forces on things like your breakfast cereal. While struggling to hold the cereal bowl on the table, you would be unable to dodge a blob of milk pulled from the bowl! In fact, gravitational force is noticeable only because we live near a very massive object—the Earth.

How can we measure the mass of the Earth? The Earth's radius is well known, but geophysics can provide only a rough idea of rock densities in Earth's interior (■ Figure 5.34). The most precise approach is to measure the acceleration of gravity near the Earth, to obtain the product of Earth's mass and Newton's gravitational constant G. Since the force acting on a mass m near the surface is given both by $F = mg_{\text{surface}}$ and $F = GM_{\text{Earth}}m/R_{\text{Earth}}^2$, we have:

$$GM_{\text{Earth}} = g_{\text{surface}} R_{\text{Earth}}^2 = (9.8 \text{ N/kg})(6.4 \times 10^6 \text{ m})^2 = 4.0 \times 10^{14} \text{ N} \cdot \text{m}^2/\text{kg}.$$

The gravitational constant G has to be obtained separately by measuring the gravitational forces between objects with known mass. The first experiment capable of detecting gravitational force between small objects was conducted by Henry Cavendish (1731–1810). His experiment was called "weighing the Earth."

In a Cavendish balance (■ Figure 5.35) two small lead spheres are at opposite ends of a rod suspended from a thin quartz fiber. Small horizontal forces exerted on the suspended spheres cause the fiber to twist through an angle proportional to the applied force. The twist angle is observed from the displacement of a light beam reflected off a mirror attached to the fiber. When the balance is used to measure the gravitational constant, the deflecting force is exerted by two larger lead spheres placed a known distance from the two suspended spheres. To eliminate systematic errors in the experiment, the large spheres are placed first on one side of the suspended spheres and then on the opposite side. From the resulting change in the fiber's twist, it is possible to calculate the gravitational attraction between one pair of lead spheres.

EXAMPLE 5.10 ♦♦ In a Cavendish experiment, lead spheres of radii 1.0 cm and 10.0 cm are placed so that their surfaces are 1.0 cm apart. The masses of the spheres are 47.8 g and 47.8 kg, respectively. The gravitational force between the spheres is found to be 1.05×10^{-8} N. What is the measured value of the gravitational constant G?

MODEL The measured force between the pair of spheres can be expressed using the law of universal gravitation, eqn. (5.1).

SETUP The distance between the *centers* of the two spheres is $d = 10.0 \text{ cm} + 1.0 \text{ cm} + 1.0 \text{ cm} = 0.120 \text{ m}$. Thus, we have:

$$F_G = G \frac{m_1 m_2}{d^2} \Rightarrow G = \frac{d^2 F_G}{m_1 m_2}.$$

SOLVE $G = \dfrac{(0.120 \text{ m})^2 (1.05 \times 10^{-8} \text{ N})}{(47.8 \times 10^{-3} \text{ kg})(47.8 \text{ kg})} = 6.62 \times 10^{-11} \text{ N} \cdot \text{m}^2/\text{kg}^2.$

ANALYZE The measured value of G in this experiment differs from the accepted value by 0.05×10^{-11} N·m²/kg², or 0.7%. ■

✱ 5.4.3 Newton's Gravity and Galileo's Law of Falling Objects

According to Galileo, an apple should fall from rest with constant acceleration $\vec{g}$. How does this fit with Newton's law of universal gravitation? The apple's weight, which causes its acceleration, is given both by the law of universal gravitation and by our practical formula from Chapter 4:

$$W_{\text{apple}} = G \frac{M_{\text{Earth}} M_{\text{apple}}}{r^2} = M_{\text{apple}} g.$$

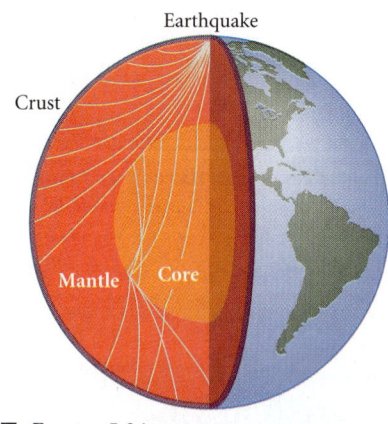

■ **FIGURE 5.34**
Seismologists learn about the Earth's interior from the bending of earthquake waves. Here the Earth's liquid core is shown bending waves and concentrating them on the far side of the Earth from the earthquake. Modeling Earth's interior is a complex research problem. Knowing the mass of Earth from gravitational measurements is a valuable constraint on proposed models. (After B. Gutenberg, "Internal constitution of the Earth," 1928.)

THE CONSTANT OF PROPORTIONALITY BETWEEN FORCE AND ANGLE IS DETERMINED IN A SEPARATE EXPERIMENT BY TIMING OSCILLATIONS OF THE ROD AS IT ROTATES BACK AND FORTH, ALTERNATELY TWISTING AND UNTWISTING THE FIBER. SEE CHAPTER 14.

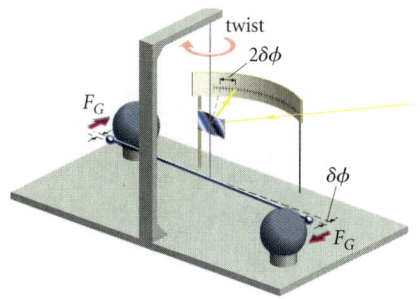

■ **FIGURE 5.35**
Schematic of a Cavendish balance. Gravitational attraction of the large lead spheres acting on the small spheres twists the quartz fiber suspending the small spheres. A laser beam reflected from the mirror attached to the fiber is deflected by the mirror's rotation. The magnitude of the attraction between a pair of spheres is deduced from the measured rotation of the mirror.

Thus, the gravitational acceleration depends on the distance r from the Earth's center:

$$g = G\frac{M_{\text{Earth}}}{r^2}. \quad (5.2)$$

THE NOTATION $g(h)$ MEANS g AT HEIGHT h. DON'T CONFUSE IT WITH g TIMES h.

Galileo's observations were made very near the surface of the Earth, so we shouldn't be surprised if his conclusions don't hold exactly much higher up. By approximating eqn. (5.2), we can find a useful formula for estimating how rapidly g varies with height. First, let h be the distance of the apple above the Earth's surface and R be the radius of the Earth (■ Figure 5.36). Then:

$$r = R + h = R(1 + h/R),$$

and substituting into eqn. (5.2):

$$g(h) = G\frac{M_{\text{Earth}}}{R^2(1 + h/R)^2} = \frac{g_{\text{surface}}}{(1 + h/R)^2} \quad (5.3a)$$

where $g_{\text{surface}} = GM_{\text{Earth}}/R^2$. This expression is very nearly the same as g_{surface} so long as $h \ll R$. A useful approximation in this case is obtained by using the binomial theorem (Appendix IB): $(1 + x)^n \approx 1 + nx$ with $n = -2$ and $x = h/R \ll 1$.

$$g(h) \approx g_{\text{surface}}\left(1 - \frac{2h}{R}\right) \quad (5.3b)$$

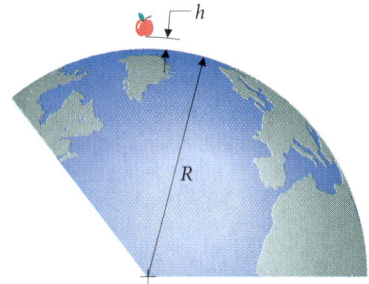

■ FIGURE 5.36
The weight of an apple a distance h above the Earth's surface depends on h, which is greatly exaggerated for clarity.

The second term in eqn. (5.3b) is a correction to Galileo's law.

EXAMPLE 5.11 ♦ At what height above the surface has g decreased by 1%?

MODEL Since we are interested in a small variation in g, we may use eqn. (5.3b). Then the fractional change of g is $\delta g/g_{\text{surface}} = -2h/R$.

SETUP We need $2h/R$ to be 0.01, or 1%:

SOLVE
$$h = \frac{0.01R}{2} = \frac{0.01}{2}(6400 \text{ km}) = 32 \text{ km}.$$

ANALYZE This is about four times the height of Mt. Everest. The correction to Galileo's law is very small ($\ll 1\%$) for everyday phenomena near the Earth's surface. ■

EXERCISE 5.7 ♦♦ How far above the surface of the Moon has the acceleration due to gravity decreased by 1%? Would Galileo have discovered his law for freely falling objects on the Moon? ■

✱ 5.4.4 Are There Different Kinds of Mass?

Mass plays two roles in Newtonian physics: it produces gravitational attraction, and it relates acceleration to the net force exerted on a body. It is not obvious or logically necessary that one quantity should do both things. How do we decide that there aren't two different kinds of mass, one for each of the two roles?

If there were more than one kind of mass, then we should be able to devise an experiment to detect the difference. For example, let's suppose that each object has a gravitational mass M_G that responds to attraction and that is different from the object's inertial mass M_I, which determines its response to forces. Then, using Newton's second law and the law of universal gravitation:

$$\text{(second law)} \quad M_{I,\text{apple}}\,a_{\text{apple}} = \frac{GM_{G,\text{Earth}}}{R_{\text{Earth}}^2}M_{G,\text{apple}} \quad \text{(gravitation law)}$$

The acceleration of the apple would be:

$$a_{\text{apple}} = \frac{M_{\text{G,apple}}}{M_{\text{I,apple}}} g.$$

The acceleration of another falling object, say, Galileo's cannonball, would be:

$$a_{\text{ball}} = \frac{M_{\text{G,ball}}}{M_{\text{I,ball}}} g.$$

If their ratios M_G/M_I were not the same, the apple and the cannonball would have different accelerations.

Experiments conducted by Galileo, by Newton, and by modern physicists all find equal accelerations for every kind of object. Thus, M_G/M_I is the same for all objects, and a sensible choice of units then makes M_G and M_I numerically equal for every object. This is a remarkable fact that distinguishes gravity from the other fundamental forces. Einstein made it the starting point for his theory linking gravity to the structure of space and time.

✻ 5.4.5 Kepler's Laws

One of Newton's most convincing triumphs was the derivation of Kepler's three laws of planetary motion from the law of universal gravitation. Because they are consequences of universal gravitation, Kepler's laws apply not only to planets in the solar system, but equally well to satellites of Earth or of other planets, or to distant astronomical bodies revolving about each other.

Kepler's first law: A satellite moves along an ellipse with the primary body at one focus.

Kepler's second law: The radius vector from the primary to a satellite sweeps out equal areas in equal times.

Kepler's third law: The square of a satellite's period in its orbit is proportional to the cube of its semimajor axis.

Program "Kepler" on your supplementary computer disk shows how to trace orbits using Newton's laws; use it to convince yourself that Kepler's laws are correct.

SEE §0.2 FOR A DISCUSSION OF ELLIPSES AND KEPLER'S LAWS.

✻ 5.5 THE BEANSTALK

In the science fiction of Arthur Clarke and Charles Sheffield, a century or so from now the human race has colonies on the Moon, on several planetary surfaces, and in large artificial structures in independent orbits around the Sun. Rockets would be a very inefficient method of commuting between such colonies. It would be cleverer to attach long cables that rotate with the colonies and so have very large speeds at their outer ends. Containers released from the ends of such cables would have speeds sufficient to follow orbits that rendezvous with other colonies. Such *beanstalks* could operate entirely on electric power from solar panels.

The most spectacular beanstalks would be on planets. Getting away from Earth, for example, is quite a feat. A speed of 11 km/s at departure is needed to make a trip to the Moon; exploratory spacecraft to the other planets need even greater speeds. Even if our descendants develop the necessary engineering skill to build a beanstalk, they may be wary. Should the cable snap, tens of thousands of kilometers of failed cable could fall to Earth, wreaking havoc comparable to several nuclear bombs!

WE USED EQNS. (3.7) AND (3.10).

EXAMPLE 5.12 ♦♦ Use the law of universal gravitation to compute the radius of a geosynchronous orbit—one in which the satellite revolves once a day and so remains always above the same spot on the equator.

MODEL We assume the orbit is circular with speed v and radius R_s. The Earth's gravitational force on the satellite produces the centripetal acceleration $\omega^2 R_s$, and the angular speed $\omega = 2\pi/T$, where $T = 1$ day is the orbital period.

SETUP Letting M be the mass of Earth and m the mass of the satellite, we apply the law of gravitation (eqn. 5.1) and Newton's second law:

$$\frac{GMm}{R_s^2} = m\frac{v^2}{R_s} = m\left(\frac{2\pi}{T}\right)^2 R_s.$$

SOLVE The satellite mass cancels out, and we find:

$$R_s = \left(\frac{GMT^2}{4\pi^2}\right)^{\frac{1}{3}}$$

$$= \left[\frac{(6.7 \times 10^{-11} \text{ m}^3/\text{kg}\cdot\text{s}^2)(6.0 \times 10^{24} \text{ kg})(8.6 \times 10^4 \text{ s})^2}{4\pi^2}\right]^{\frac{1}{3}}$$

$$= 4.2 \times 10^4 \text{ km}.$$

ANALYZE The geosynchronous station on a beanstalk can be arbitrarily large, since no cable tension is needed to hold it in orbit. Tension is necessary to support the cable extending toward Earth and to help accelerate the launching platform and the cable extending toward it. ∎

THE RESULT FOR V_{esc} IS EASIEST TO DEMONSTRATE USING ENERGY METHODS (CHAPTER 8).

EXAMPLE 5.13 ♦♦♦ Given that the speed needed by an object at distance R to escape from Earth is $v_{esc} = \sqrt{2GM/R}$, calculate the distance between the geosynchronous station and the launching platform of the beanstalk. What is the tension in the cable at the launching platform if the mass of the platform is $m_p = 6 \times 10^{12}$ kg?

MODEL ■ Figure 5.37 is a schematic of the beanstalk system showing the forces acting on the launching platform. To find the tension $\vec{F}$, we need to know the acceleration of the launching platform and the gravitational force $\vec{F}_G$ acting on it. Both of these depend on the distance R_p of the platform from the center of the Earth. The platform rotates once a day with the Earth, so its speed is $v_p = 2\pi R_p/T$, with $T = 1$ day.

SETUP We want the speed of the launching platform v_p to equal escape speed v_{esc}:

$$v_p = \frac{2\pi R_p}{T} = \sqrt{\frac{2GM}{R_p}} = v_{esc}.$$

SOLVE
$$R_p = \left(\frac{GMT^2}{2\pi^2}\right)^{\frac{1}{3}} = 2^{1/3} R_s$$
$$= (1.26)(4.2 \times 10^4 \text{ km}) = 5.3 \times 10^4 \text{ km}.$$

ANALYZE We used the result for R_s from the previous example. The distance from station to launching platform is $R_p - R_s = 1.1 \times 10^4$ km.

SETUP The total force on the platform, gravitational force plus the cable tension F, provides the centripetal force.

$$F + \frac{GMm_p}{R_p^2} = m_p \frac{v_p^2}{R_p}.$$

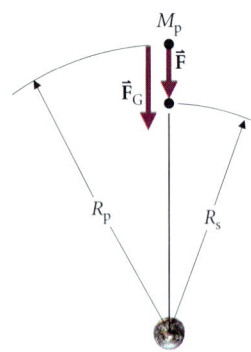

■ FIGURE 5.37
Free-body diagram for the launching platform, located at $r = R_p$. Both gravitational force $\vec{F}_G$ and cable tension $\vec{F}$ act inward on the launching platform. The platform travels a circular path of radius R_p in 1 d.

SOLVE From the first part of the calculation, $GM/R_p = v_p^2/2$. So:

$$F = m_p \frac{v_p^2}{R_p}\left(1 - \tfrac{1}{2}\right) = \frac{m_p}{2}\frac{(2\pi R_p/T)^2}{R_p}$$

$$= \frac{2\pi^2 m_p R_p}{T^2} = \frac{2\pi^2(6.0 \times 10^{12}\text{ kg})(5.3 \times 10^7\text{ m})}{(8.6 \times 10^4\text{ s})^2}$$

$$= 8.5 \times 10^{11}\text{ N}.$$

ANALYZE The given mass of the launching platform corresponds to a structure about 3 km on a side. A much smaller structure could handle the task of releasing and launching cargo vessels but wouldn't be large enough to anchor the 50 000 km of cable! This technology can't be built on a small scale. It would take an impressive solar system economy to maintain such structures and keep them profitable.

Chapter Summary

Where Are We Now?

We have drawn together ideas from all the preceding chapters to see how Newton's laws and kinematics together describe the motion of mechanical systems.

What Did We Do?

To apply Newton's laws to systems, we suggest a standard method with six logical steps (Figure 5.4). Model the system as a collection of objects, each with its own free-body diagram. Model the influence of the rest of the system on each object by the forces acting on the object, as illustrated in the free-body diagram. Newton's second law relates each object's acceleration to the total force acting on the object. The structure of the system constrains its motion and provides relations among the forces acting. Equations derived from Newton's law and from the system structure should form a soluble set if the system has been aptly modeled.

Newton discovered the law of universal gravitation by studying the motions of the planets. The force exerted by one body of mass M_1 on another of mass M_2 separated by distance r is $F = GM_1 M_2/r^2$ and points along the line joining the bodies. An object's mass both produces gravitational attraction and responds to force acting on the object. The term *weightless* as applied to an object in space means that the gravitational force on the object accelerates the object in its orbit and need not be balanced by normal forces exerted by floors and so on. Galileo's law of fall is a consequence of Newton's laws of motion and the law of universal gravitation in the limit of small displacement from the Earth's surface. Cavendish's experiment determines the value of the gravitational constant and, consequently, the mass of the Earth.

Practical Applications

The dynamics of systems is a necessary input to the design of all kinds of machinery and transportation systems. Understanding accelerations produced by gravitational force is essential to the design of space missions, and to modeling astrophysical systems such as our own galaxy.

Solutions to Exercises

5.1 If the engine pulls on the cars with a force $-\vec{F}$, the acceleration of the system is reversed in direction but still has magnitude $F/2M$. The coupling still exerts forces of magnitude $F/2$ on each car, but the forces arise from tension in the rod rather than compression.

If the engine is pushing with force $\vec{F}$ on two cars of unequal masses M_1 and M_2, the solution in the example is altered as follows:
$$F - C = M_1 a_{x_1}, \quad C = M_2 a_{x_2}, \quad a_{x_1} = a_{x_2} \equiv a_x.$$
Thus: $\quad a_x = \dfrac{F}{M_1 + M_2} \quad$ and $\quad C = \dfrac{M_2 F}{M_1 + M_2}.$

If there are more than two cars, the acceleration of the train is $F/$(total mass of all cars). Each coupling exerts a force that causes all the cars to its right to accelerate at this rate:
$$\text{Compression} = F \dfrac{\text{mass of cars to the right}}{\text{total mass of all cars}}.$$

5.2 From the solution to Example 5.2, we have:
$$T = g\dfrac{mM}{m+M}(1 + \sin\theta) = mg\dfrac{M + M\sin\theta}{m + M}.$$
If $M\sin\theta > m$, then: $\quad T > mg\dfrac{M+m}{m+M} = mg,$
and $\quad T = (M\sin\theta)g\dfrac{m + m\sin\theta}{(m+M)\sin\theta}$
$$< (M\sin\theta)g\dfrac{m + m\sin\theta}{m\sin\theta + m} = Mg\sin\theta.$$
So, $mg < T < Mg\sin\theta$.

5.3 The condition for Joachim to accelerate down the glacier $(a_{x_1} > 0)$ is $M\sin\theta > m$. Thus the minimum angle is:
$$\theta_{\min} = \sin^{-1}\left(\dfrac{m}{M}\right) = \sin^{-1}\left(\dfrac{75 \text{ kg}}{90 \text{ kg}}\right) = \sin^{-1}(0.83) = 56°.$$

If the angle is only 30°, Maria will accelerate downward into the crevasse, pulling her father over the edge. Maria's acceleration is:
$$a_m = g\dfrac{m - M\sin\theta}{m + M}$$
$$= (9.8 \text{ m/s}^2)\dfrac{75 \text{ kg} - (90 \text{ kg})\sin 30°}{75 \text{ kg} + 90 \text{ kg}}$$
$$= 1.8 \text{ m/s}^2.$$

5.4 The forces acting on a segment of the rope are its weight and the tension exerted by the rope above it (■ Figure 5.38). In equilibrium, the two forces balance, so $T(x) = gm(x)$. For a uniform rope, $m(x) = Mx/L$, where M is the total mass of the rope and L its length; so $T(x) = gMx/L$. The tension in a uniform, hanging rope varies linearly from Mg at the top to zero at the bottom.

5.5 The time for the shuttle to complete an orbit is the circumference of the orbit divided by the known speed:
$$T_{\text{shuttle}} = \dfrac{2\pi(6.6 \times 10^6 \text{ m})}{7.8 \times 10^3 \text{ m/s}} = 5.3 \times 10^3 \text{ s}.$$

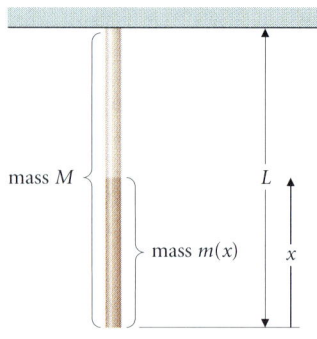

■ **Figure 5.38a**
Uniform rope hanging vertically. We imagine a length x of the rope cut off to form a free particle.

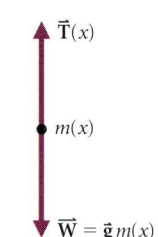

■ **Figure 5.38b**
Free-body diagram for the rope particle.

The shuttle period is some seven times longer than the time to fall to the center. It would be fairer to compare with the time to fall through a hole to the opposite side of the Earth and return. That time is clearly of the same order as the shuttle period.

5.6 We use the gravitational force law, eqn. (5.1). From the data on the inside front cover, the distance of the Moon from the Earth averages $r_{\text{EM}} \approx 3.8 \times 10^5$ km. The distance of the Earth from the Sun averages $r_{\text{ES}} \approx 1.5 \times 10^8$ km. Working roughly, we may neglect any changes in the distance of the Moon from the Sun and simply use the Sun–Earth distance for the Sun–Moon distance. Then:
$$\dfrac{F_{\text{Sun on Moon}}}{F_{\text{Earth on Moon}}} \approx \dfrac{M_{\text{Sun}}/r_{\text{SM}}^2}{M_{\text{Earth}}/r_{\text{EM}}^2}$$
$$= \left(\dfrac{M_{\text{Sun}}}{M_{\text{Earth}}}\right)\left(\dfrac{r_{\text{EM}}}{r_{\text{SM}}}\right)^2$$
$$= \left(\dfrac{2 \times 10^{30} \text{ kg}}{6 \times 10^{24} \text{ kg}}\right)\left(\dfrac{4 \times 10^5 \text{ km}}{1.5 \times 10^8 \text{ km}}\right)^2$$
$$= 2.$$

How can this be? If the Sun's pull is twice as strong, why doesn't it rip the Moon away from the Earth? Remember that the Sun is also pulling on the Earth. The Sun's gravitational pull accelerates both the Moon and the Earth along nearly the same annual orbit (■ Figure 5.39). The Earth's weaker pull causes small changes in the Moon's path around the Sun. Seen from the Earth, those small changes are the orbital motion of the Moon around the Earth.

5.7 We may use eqn. (5.3) with appropriate values for the Moon. The Moon's radius is 1.74×10^3 km (inside front cover). The gravitational acceleration has decreased by 1% when $2h/R = 0.01$, or:
$$h = 0.005R = 0.005(1.74 \times 10^3 \text{ km}) = 8.70 \text{ km}.$$

This value is considerably smaller than the corresponding value for Earth (32 km, Example 5.11) but large enough that Galileo would have drawn the same conclusion.

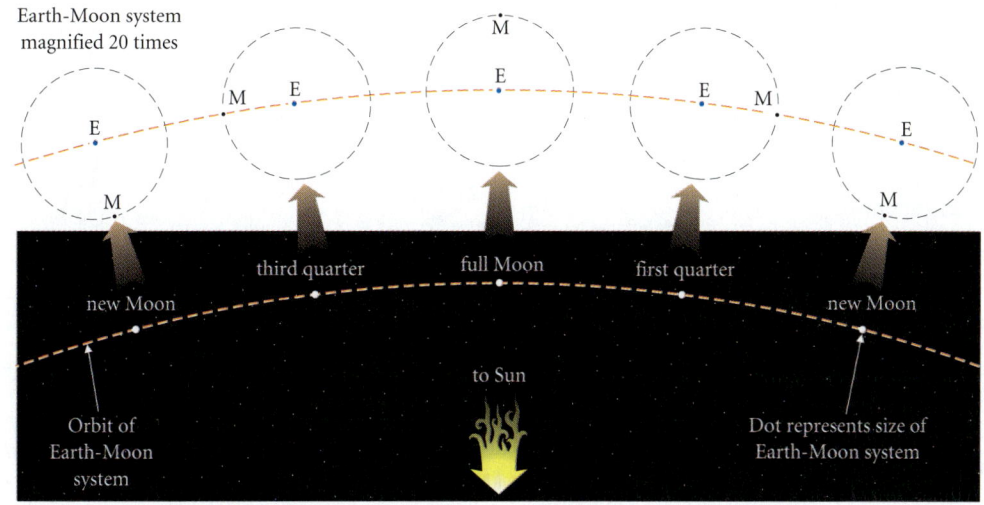

Figure 5.39
Orbit of the Earth–Moon system around the Sun compared with the orbit of the Moon about the Earth. Positions of the Earth and Moon are shown for a period of 1 month. On this scale, the entire Earth–Moon system lies within the 1-mm dot used to represent its position. The primary influence of the Sun on both bodies is to accelerate them together in this annual orbit around the Sun. The influence of the Earth on the Moon is only noticeable if we look at finer detail. Magnified views (20×) are shown at $\frac{1}{4}$-month intervals.

Basic Skills

Review Questions

§5.2 SOLUTION METHOD FOR SYSTEMS OF PARTICLES

- What three judgments must you make when deciding how to model a system as a collection of significant particles?
- How does the free-body diagram for a part of a system represent the part; how does it represent the rest of the system?
- What criteria should you use in choosing coordinates for your free-body diagrams? Are you required to use the same axes for all diagrams? Why or why not?
- Why should you use different names for the coordinate axes in different free-body diagrams?
- How do you obtain relations among unknowns beyond those found from analyzing the free-body diagrams?
- What check should you make on the numbers of unknowns and equations *before* doing any algebra?
- What is meant by conservation of string? Describe how conservation of string is applied.

✱ §5.3 STRINGS

- Why can we not analyze a curved piece of ideal string with a single free-body diagram?
- When we imagine a string cut into pieces, how does the tension compare on opposite sides of a cut? Which physical law gives the answer to this question?
- Explain why the forces acting on opposite ends of a straight piece of ideal string balance.
- Why must we analyze a curved piece of string using a large number of very small elements? Why do we know the direction of normal force acting on an element of string when we cannot tell the direction of the total normal force on a finite piece of curved string?
- Why is it not possible to stretch a massive cable exactly horizontally?
- List three advantages of using a pulley system versus lifting an object directly.
- Describe the easiest way to determine the reduction in required force you can achieve using an ideal pulley system with parallel ropes.
- Describe how conservation of string is used to relate accelerations of the particles in Example 5.6. What does the minus sign mean in the relation between the accelerations? When did we *choose* this meaning for the sign?

§5.4 THE LAW OF UNIVERSAL GRAVITATION

- State Newton's law of universal gravitation and give the value of the gravitational constant.
- Explain why astronauts in orbit are popularly described as *weightless*. How is this usage inconsistent with the definition of weight used in this text? Explain why we have a sense of weight in the first place and why the astronaut in orbit would not have the usual sensations of weight.
- What was Jules Verne's "mysterious point"? Why is it not particularly mysterious? Why is it interesting?
- Describe how a Cavendish balance can measure the gravitational force between two small objects. Why was Cavendish's experiment called "weighing the Earth"?
- ✱ Explain how Galileo's law of falling objects follows from Newton's law of universal gravitation. Is Galileo's law exact? If not, how inexact is it?

- ✲ Why might we think there is more than one kind of mass?
- ✲ Describe how Galileo's experiment on the law of falling objects discourages belief in different gravitational and inertial kinds of mass.

✲ **§5.5 THE BEANSTALK**

- Why would a beanstalk system be useful? Why might one be dangerous?
- Roughly how long a cable is needed for an Earth-based beanstalk?

Basic Skill Drill

§5.2 SOLUTION METHOD FOR SYSTEMS OF PARTICLES

1. A car is towing another car of equal mass, $m = 850$ kg, with a rope. If the lead car accelerates at 2.0 m/s², what is the tension in the rope? What is the force exerted on the lead car's tires by the road?

2. Two fruit boxes are at rest on icy ramps with inclinations of 20° and 30° as shown in ■ Figure 5.40. The boxes are connected by a cord that passes over a frictionless rod. If the mass of the box on the left is $M = 42$ kg, what is the mass m of the other box?

✲ **§5.3 STRINGS**

3. Draw a free-body diagram for an element of a massive cable and use your diagram to explain why the cable hangs in a curve between two points of suspension.

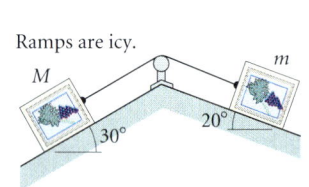

Ramps are icy.

■ FIGURE 5.40

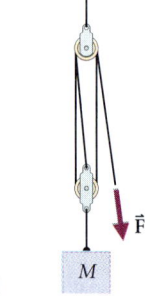

■ FIGURE 5.41

4. A mass M is suspended from a frictionless, massless pulley system as shown in ■ Figure 5.41. All the cables are approximately parallel. What downward force F is required to maintain the system in equilibrium?

§5.4 THE LAW OF UNIVERSAL GRAVITATION

5. ✉ Estimate the gravitational force you exert on your friend sitting in a chair 3 m away. Compare the result with your friend's weight.

6. Compare the acceleration of the apple falling in Newton's garden with the acceleration of the Moon. (You'll need some data from the inside front cover.)

7. An asteroid, 1.0 km in radius, has the same density as Earth. What is the mass of the asteroid? What is the gravitational acceleration on its surface? How far above the asteroid's surface has the gravitational acceleration decreased to half its surface value?

Questions and Problems

§5.2 SOLUTION METHOD FOR SYSTEMS OF PARTICLES

8. ❖ Explain why victory at a game of tug-of-war goes to the team that can sustain the greater *contact* force on the ground, and why force exerted on the rope does *not* explain victory.

9. ❖ Three crates, each of mass M, are stacked on top of each other as shown in ■ Figure 5.42. Identify the normal forces acting and compare their magnitudes.

10. ❖ Is it possible for the system shown in ■ Figure 5.43 to be in equilibrium if the inclined planes are frictionless? Why or why not?

11. ❖ A block of mass M rests on a ramp inclined at angle θ (■ Figure 5.44). There is a coefficient of friction μ_s between the block and the ramp. The ramp itself rests on a horizontal, frictionless surface and is held away from a wall by a spring with spring constant k. What is the compression of the spring if the system is in equilibrium? (*Hint:* This is a conceptual problem; there is a visualization that eliminates the need for equations.)

■ FIGURE 5.43

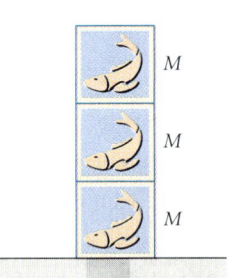

■ FIGURE 5.42

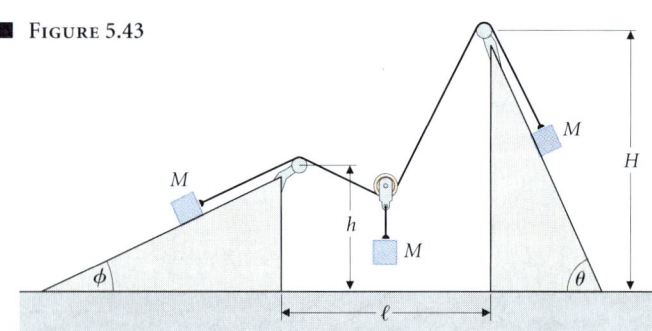

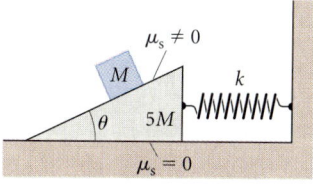
■ FIGURE 5.44

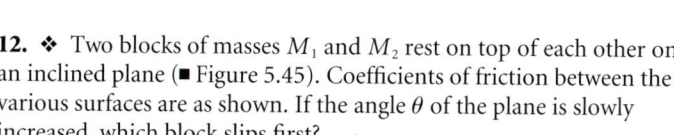
■ FIGURE 5.45

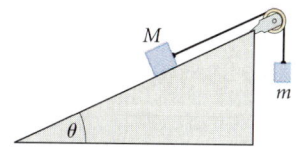
■ FIGURE 5.50

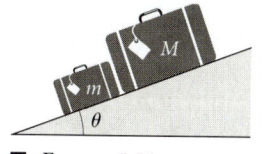
■ FIGURE 5.51

12. ❖ Two blocks of masses M_1 and M_2 rest on top of each other on an inclined plane (■ Figure 5.45). Coefficients of friction between the various surfaces are as shown. If the angle θ of the plane is slowly increased, which block slips first?

13. ◆ (a) An orange crate of mass M is at rest on an icy surface. One cord connects the crate to a wall. A second cord passes over a frictionless rod and suspends another crate of equal mass (■ Figure 5.46). What is the tension in each cord? (b) If the cord between the orange crate and the wall is replaced by a spring with spring constant k and unstretched length ℓ, how far is the orange crate from the wall?

14. ◆ Atwood's machine provided an early method for testing Newton's laws. It consists of two objects of masses M and $m < M$ suspended with ideal string over a massless, frictionless pulley (■ Figure 5.47). Find the acceleration of each object.

15. ◆ A box of mass M sits on a table and is connected by a rope passing over a smooth support to another box of mass m that hangs over the edge of the table. What is the acceleration of M (a) ignoring friction and (b) if the coefficient of kinetic friction between box and table is $\mu_k = 0.7$?

16. ◆ Find the tension in each string in the system of two blocks shown in ■ Figure 5.48.

17. ◆ Two masses are suspended using a string and a spring with spring constant k (■ Figure 5.49). Find the tension in the string and the stretch of the spring.

18. ◆◆ An object of mass m is hung from a rope that passes over a pulley at the top of a ramp and is attached to a block of mass M (■ Figure 5.50). (a) Assume m is large enough that the block accelerates up the ramp. Find an expression for the block's acceleration. (b) From your result for part (a), determine the minimum ratio m/M such that the block, once moving, accelerates up the ramp. Evaluate this minimum ratio if $\theta = 30°$ and the coefficient of kinetic friction between the ramp and the block $\mu_k = 0.3$. (c) If the coefficient of static friction is $\mu_s = 0.4$ and the system is released from rest, find the range of possible values of m/M such that the system remains at rest.

19. ◆◆ (a) Two suitcases of different mass are at rest on a ramp (■ Figure 5.51). What is the minimum coefficient of static friction between the cases and the ramp if the system is in equilibrium? (b) Suppose there is no friction between the upper suitcase and the ramp. What minimum coefficient of friction between the ramp and the lower case is needed for equilibrium?

20. ◆◆ A concrete block of mass $M = 1.5 \times 10^4$ kg is at rest on a dock. The coefficient of static friction between block and dock is $\mu_s = 0.90$. A cable is used to lift a mooring anchor from the harbor as shown in ■ Figure 5.52. What is the maximum mass m of the anchor that can be held in the position shown? If m has its maximum value, what is the tension in each cable?

21. ◆◆ An object of mass $m = 0.0100$ kg slides uniformly around a circle of radius $r = 0.100$ m on a frictionless table (■ Figure 5.53). A frictionless string connects the object to a suspended object of mass $M = 0.100$ kg, which remains at rest. At what speed v does m slide?

22. ◆◆ A sphere of mass M is suspended from a heavy shackle of mass m, which is itself suspended by two cords (■ Figure 5.54). Find the tension in each of the three cords.

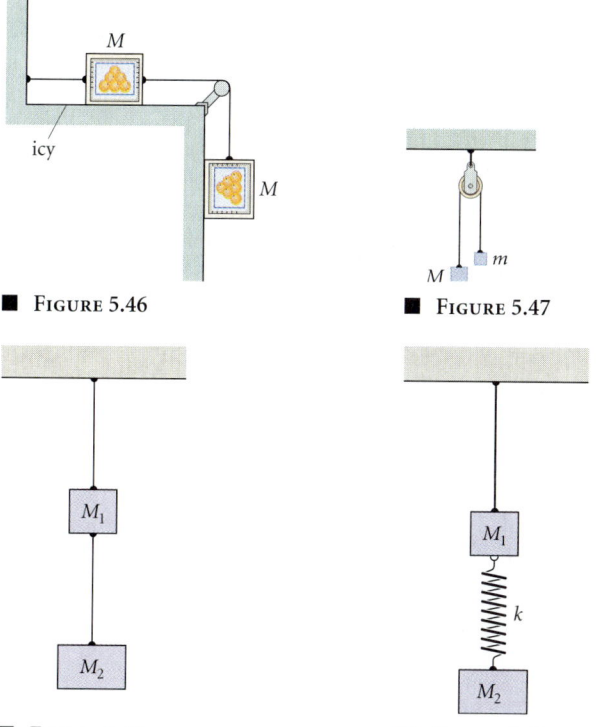
■ FIGURE 5.46

■ FIGURE 5.47

■ FIGURE 5.48

■ FIGURE 5.49

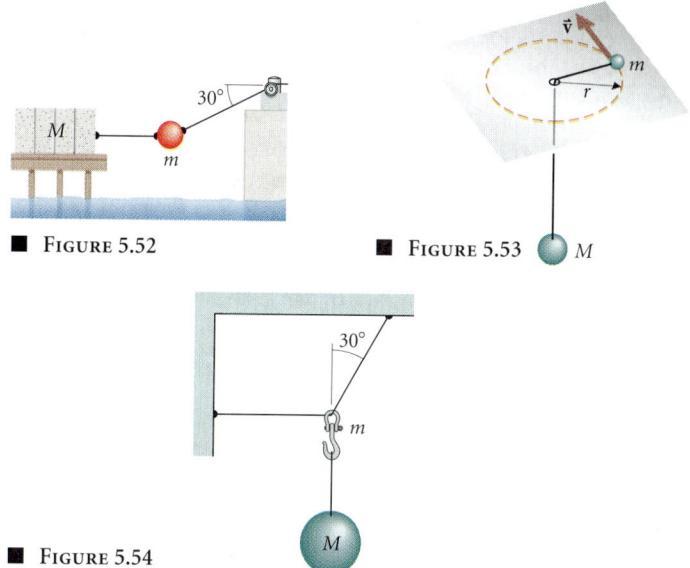
■ FIGURE 5.52

■ FIGURE 5.53

■ FIGURE 5.54

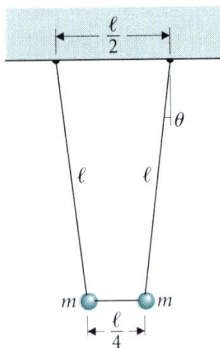

Figure 5.55

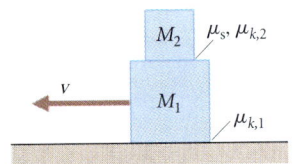

Figure 5.56

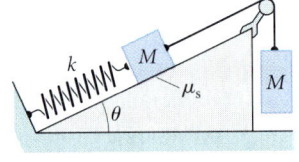

Figure 5.59

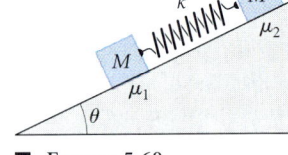

Figure 5.60

23. ♦♦ Identical balls of $m = 10.0$ kg are suspended from points a distance $\ell/2$ apart with strings of length $\ell = 1.0$ m (■ Figure 5.55). The balls are connected by a string of length $\ell/4$. Find the tension in each of the strings.

24. ♦♦ An ice block (mass $m = 10.0$ kg) is attached by a rope to a wooden box of mass $M = 15.0$ kg. The coefficient of kinetic friction between the box and the floor is $\mu_k = 0.50$. If the system is set moving at $v = 10.0$ m/s with the block in front of the box, how long does it take for the system to come to rest?

25. ♦♦ A block of mass M_2 sits on top of a larger block of mass $M_1 = 5.0 M_2$ (■ Figure 5.56). The coefficient of static friction between the small block and the large one is $\mu_s = 0.40$, the coefficient of kinetic friction is $\mu_{k,2} = 0.30$, and the coefficient of kinetic friction between the large block and the floor is $\mu_{k,1} = 0.50$. If the large block initially moves with speed v, does the small block slide with respect to the large one? Find the acceleration of each block.

26. ♦♦ Two blocks of equal mass M sit on a plane inclined at 30° to the horizontal (■ Figure 5.57). The coefficient of friction between the lower block (A) and the plane is 0.50, and between the upper block (B) and the plane is 0.30. (a) Find the acceleration of each block, and the normal force exerted on A by B. (b) What happens if the two blocks are reversed so that B leads the way down the plane?

27. ♦♦ An ice block of mass 10.0 kg is attached by a rope to a wooden crate of mass 20.0 kg on a ramp that makes an angle $\theta = 20.0°$ with the horizontal (■ Figure 5.58). The coefficient of kinetic friction between the crate and the ramp is $\mu_k = 0.400$. (a) What is the acceleration of the ice block? (b) What happens when the ice block reaches the horizontal surface? When the crate reaches the horizontal surface? (c) What would happen if the ice block were behind the crate on the ramp?

28. ♦♦ A block of mass M rests on an inclined plane. It is attached to a wall by a spring with spring constant k. A string connects the block over a frictionless support to a second block of mass M that hangs freely (■ Figure 5.59). If the coefficient of static friction between the block and the ramp is μ_s, what are the maximum and minimum values of the spring stretch that allow the system to remain in equilibrium?

29. ♦♦♦ Two blocks, both of mass M and connected by a spring with spring constant k, slide on a ramp inclined at angle θ to the horizontal. The blocks are made of different materials and so have different coefficients of kinetic friction with the surface of the ramp (■ Figure 5.60). Describe the motion of the system assuming the spring compression s remains constant. Find s in the case where $\tan\theta > \mu_1 > \mu_2$.

✻ §5.3 STRINGS

30. ❖ A massive cable is hanging in the shape of a chain curve. Draw a free-body diagram for an arbitrary element of the cable and use it to determine whether the horizontal component of tension varies along the cable.

31. ❖ Neglecting any mass of the material of the spring and, consequently, any weight of the spring material, show that each turn of a coiled spring exerts a force on the one above it equal in magnitude to the tension exerted by the entire spring on the object it supports.

32. ♦ Two strings, each at an angle of 43.20° with the vertical, support an object with mass $m = 1.275$ kg. The tension in each string is measured at a position 1.25 m from the suspended object, with the result $T = 8.579$ N for each string. If $g = 9.810$ m/s², are the strings ideal within the accuracy of the given data? Explain your reasoning.

33. ♦ Is the pulley system in ■ Figure 5.61 ideal within the accuracy of the given data? Explain your reasoning. (Take $g = 9.81$ m/s².)

34. ♦♦ Two spring scales are tied together by a length of string. You hold one scale in your hand; the other is suspended from the string. The top scale reads 1.854 N and the suspended scale reads 1.756 N. What is the mass of the string? (Take $g = 9.810$ m/s².)

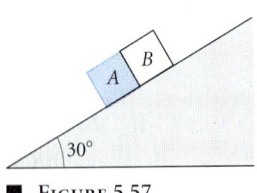

Figure 5.57

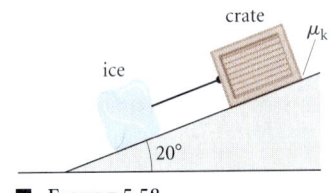

Figure 5.58

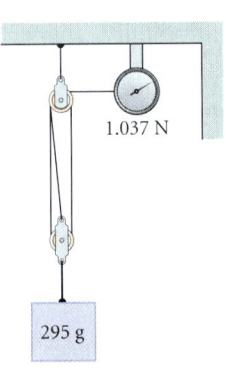

Figure 5.61

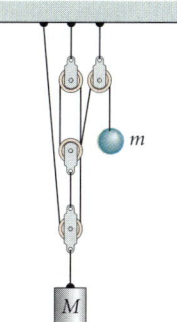
■ FIGURE 5.62

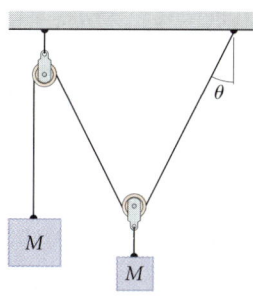
■ FIGURE 5.63

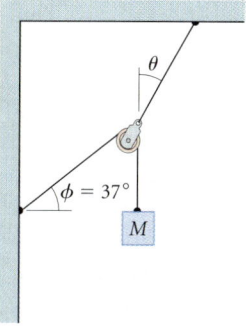
■ FIGURE 5.64

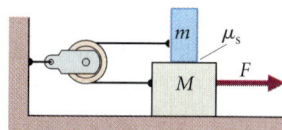
■ FIGURE 5.65

35. ♦♦ A uniform rope of length $\ell = 5.0$ m with mass $M = 2.5$ kg is thrown over a horizontal tree branch. The rope hangs in equilibrium so long as neither end is more than 0.70 m lower than the other. At this limit, what is the change in rope tension caused by friction between rope and tree?

36. ♦♦ A climbing rope has mass $m = 15.0$ kg and total length 50.0 m. A climber is pulling on the rope that runs over a rounded edge and hangs vertically. With exactly half the rope hanging, the climber needs to pull with a force of 125 N to raise the rope at constant speed. What is the change in rope tension caused by friction between rope and rock?

37. ♦♦ What force F is required to support the block in Figure 5.41 if each pulley has mass $M/5$?

38. ♦♦ (a) If the pulleys in the system shown in ■ Figure 5.62 are massless and frictionless, what mass m must the ball have if it is to support the cylinder of mass M? (b) If $m = M$, find the acceleration of each of the objects.

39. ♦♦ Two objects of equal mass M are suspended from the pulley system shown in ■ Figure 5.63. If the pulleys are massless, frictionless, and of negligible size, find the angle θ when the system is in equilibrium.

40. ♦♦ A box of mass $M = 5.7$ kg is suspended by a rope over a pulley (■ Figure 5.64). Find the tension in the rope supporting the box and in the rope supporting the pulley.

41. ♦♦ A physics professor uses the ideal pulley system shown in Figure 5.41 for a demonstration. Instead of a crate, the pulley supports a light canvas chair in which the professor sits while pulling downward with force F on the rope. What force F is required, expressed as a fraction of the professor's weight?

42. ♦♦ A box of mass $m = 5.0$ kg rests on a larger box of mass $M = 25$ kg, which is at rest on a frictionless surface (■ Figure 5.65). The coefficient of static friction between the two boxes is $\mu_s = 0.45$. An ideal string runs between the two boxes around a fixed pulley, as shown in the figure. What is the maximum force F that can be exerted on the lower box without causing the boxes to slide? How would your answer change if the simple pulley were replaced with the pulley system shown in Figure 5.41, mounted on the wall with the pulleys attached to M and the rope end attached to m? What if the pulleys were connected to m and the rope to M?

43. ♦♦♦ Suppose the pulleys in Figure 5.62 each have mass $M/2$. What mass m of the ball is then required to support the cylinder, which has mass M?

44. ♦♦♦ Three objects are suspended from massless, frictionless pulleys (■ Figure 5.66). Find the acceleration of each object and the tension in the string.

45. ♦♦♦ An object of unknown mass M is suspended from the pulley system in ■ Figure 5.67. The string has a total length $2\ell = 5.7$ m and is attached to the floor by a spring with spring constant $k = 6.3 \times 10^4$ N/m and negligible relaxed length. The pulleys are frictionless and of negligible size, and the spring is stretched by an amount $s = \alpha\ell$, with $\alpha = 0.32$. Find the angles θ and ϕ and the mass M in terms of k and α. Evaluate your results using the given data.

■ FIGURE 5.66

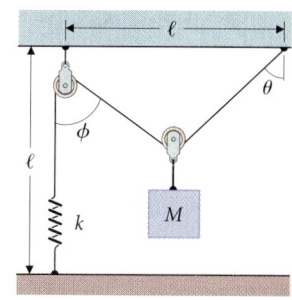
■ FIGURE 5.67

§5.4 THE LAW OF UNIVERSAL GRAVITATION

46. ❖ What force or forces act on the space shuttle as it orbits the Earth? Identify the force that forms a Newton's third law pair with each.

47. ❖ Is it possible to shield one object from the gravitational attraction of another? On what features of the law of gravitation and on what rules for adding forces do you base your answer?

48. ❖ As Alice fell down the rabbit hole in *Alice in Wonderland,* she wondered: . . . "if I shall fall right *through* the earth! How funny it'll seem to come out among the people that walk with their heads downwards!" How should you address Alice's concerns?

49. ❖ During her fall (cf. Problem 48) Alice picked a jar of marmalade from a shelf on the wall of the hole. She was afraid to drop it "for fear of killing somebody underneath." Would her response be different if she had read this chapter?

50. ◆ The Trojan asteroids follow the same orbit around the Sun as does Jupiter, but they lead Jupiter in the orbit by an angle of 60°. Jupiter's mass is approximately 10^{-3} times the mass of the Sun. Compare the force of Jupiter on a Trojan asteroid with that of the Sun. (Check the inside front cover for any data you need.)

51. ◆ Mars has a mass 0.11 times that of Earth and a radius 0.54 times that of Earth. Compare the weight of a colonist on Mars to the colonist's weight when visiting Earth. (Assume the colonist doesn't overeat during the luxury cruise from the home planet!) What is the gravitational acceleration on the surface of Mars?

52. ◆◆ **(a)** How much is the acceleration due to gravity reduced in a jet airplane flying at a 10-km altitude compared with its value on the runway at Tampa, Florida? **(b)** At what height above the Earth's surface does the acceleration due to gravity have half the value it has at the surface? What is the period of a satellite in orbit at that altitude?

53. ◆◆ Treating Earth's orbit about the Sun as a circle of radius 1.5×10^{11} m, use the observed period of the Earth's motion to find the mass of the Sun.

54. ◆◆ The lead spheres in a Cavendish balance experiment have diameters of 2.67 cm and 5.78 cm. When they are placed with their surfaces 0.05 cm apart, the force they exert on each other is measured as 4.70×10^{-9} N. What is the corresponding experimental value of G?

55. ◆◆ The acceleration of gravity at the surface of an asteroid is measured as 7.3×10^{-3} m/s². If the asteroid's radius is 10.0 km, what is its density? (Assume the asteroid is spherical to sufficient accuracy.)

56. ◆◆ Jupiter's satellite Ganymede has an almost circular orbit of radius 1.07×10^9 m and period 7.155 d. Use these data to find the mass of Jupiter.

57. ◆◆ Calculate the orbital period of a satellite just above the Earth's atmosphere. (Neglect the difference between the radius of the Earth and the orbital radius of the satellite.)

58. ◆◆◆ Assume a person who jumps leaves a planetary surface with roughly the same speed no matter on which planet the jump occurs. If Estrellita can jump 1 m high on Earth, how high can she jump on the Moon? Neglecting any variation of g, how high could she jump on an asteroid with density $\rho = 5.5 \times 10^3$ kg/m³ and radius 1.0 km? Is it OK to neglect variation of g above the asteroid's surface?
✉ Assume, as an approximation, that being able to jump to a height equal to an asteroid's radius against a constant gravitational acceleration equal to that at the surface means that you can escape completely from the asteroid. Estimate the radius of the smallest asteroid on which Estrellita may safely leap from place to place.

∗ §5.5 THE BEANSTALK

59. ❖ Explain why the tension in a beanstalk cable increases inward from the launching platform toward the geosynchronous station. Explain why tension also increases upward from the ground toward the station. Thus show that the geosynchronous station is at the point of maximum tension, despite the fact that no *net* tension is required to act on the station itself.

60. ◆◆ Consider a beanstalk system designed for a space colony. Such colonies would follow independent orbits about the Sun with the same radius as the Earth's orbit, and spacecraft launch speeds of 3 km/s would be adequate for travel between them. Suppose a typical colony is a cylindrical structure several kilometers long and a kilometer in radius, and rotates so that points on its circumference have a centripetal acceleration of 9.8 m/s². How long a beanstalk is needed to achieve the necessary launch speed?

61. ◆◆◆ The mass of the Moon is $\frac{1}{81}$ that of the Earth and the rotational period of the Moon is 27.5 d. Calculate the dimensions of a lunar beanstalk. Compare these dimensions with the distance from the Moon to the *mysterious point* where the gravitational accelerations of Earth and Moon balance each other. Based on this calculation, do you think a lunar beanstalk would prove practical? Why or why not? (*Note:* the expression $v_{esc} = \sqrt{2GM/r}$ works for the Moon as well as for the Earth.)

Additional Problems

62. ❖ A brick of mass M is suspended from the roof by a spring. Another brick of equal mass hangs by a string from the bottom of the first brick. If the string is cut, what is the acceleration of the upper brick immediately after the cut?

63. ◆◆ A freight train consists of three engines, each with mass 62 Mg, and 100 freight cars, each with mass 15 Mg. The engines are pulling the train up a 1.0% slope (i.e., tangent of the angle is 0.010) at constant speed. Neglecting friction forces acting on the freight cars, find the forces acting on the nth freight car. (Cars are labeled 1, 2, 3, etc., starting at the back end of the train.) If the static coefficient of friction between the engines' wheels and the track is $\mu_s = 0.50$, what is the maximum slope the train can ascend?

64. ❖ ◆◆ The strings and pulleys in ■ Figure 5.68 are massless and frictionless. What is the force exerted by the spring on the floor, and what is the minimum necessary coefficient of static friction between the surfaces of the box and the ramp for the system to remain in equilibrium?

65. ◆◆ A 30.0-kg monkey climbs hand over hand on a rope that passes over a frictionless support and is tied to a 20.0-kg bunch of

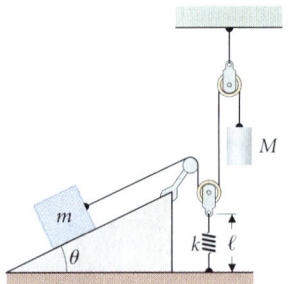

■ FIGURE 5.68

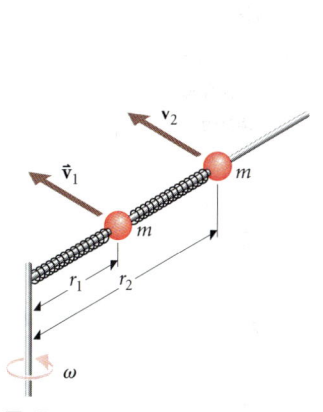

■ FIGURE 5.69

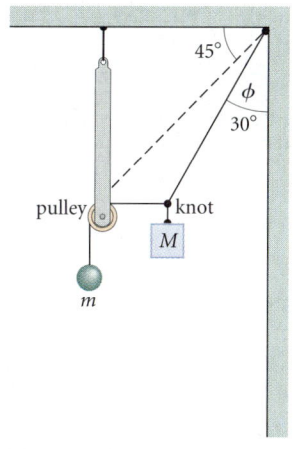

■ FIGURE 5.70

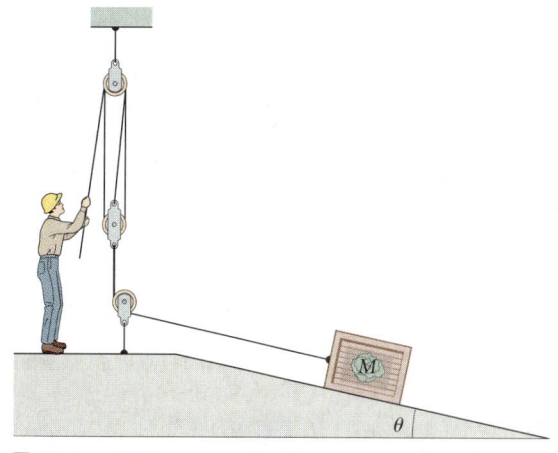

■ FIGURE 5.71

bananas initially at rest. Describe what happens (a) if the monkey simply clutches the rope tightly, (b) if the monkey "climbs" the rope in such a way as to remain the same height above the ground, (c) if the monkey climbs so as to accelerate upward at $g/3$. (d) If the monkey was originally 10.0 m above the bananas, which strategy would get it to the bananas most rapidly? How much time is required in each case? (e) Can the monkey reach the bananas starting from 10 m below them? If so, describe the necessary strategy and find the time required to reach the fruit.

66. ◆◆ Two objects have masses $m_1 = 1.000$ kg and m_2 unknown. They are connected by a stretched spring on a frictionless surface and released from rest. Immediately after release, the objects' accelerations are measured as $a_1 = 2.643$ m/s^2 and $a_2 = 1.769$ m/s^2. What is the value of m_2? If the relaxed length of the spring is 0.243 m and its length at release is 0.327 m, what is its spring constant?

67. ◆◆ Two beads of equal mass m are mounted on a frictionless rod that is rotating at uniform angular speed ω in a horizontal plane (■ Figure 5.69). Two springs, each with spring constant k and negligible relaxed length connect the outer bead to the inner, and the inner bead to the axis of rotation, as shown. Find the distance from the axis of rotation at which each of the beads circles. Is there a solution for any set of values of m, k, and ω?

68. ✽ ◆◆ The pulley and strings in ■ Figure 5.70 are ideal, and the string leading from m to the knot is horizontal when $\phi = 30°$, as shown. Find the ratio of m to M for the system to be in equilibrium. Describe qualitatively what will happen as the ratio m/M is increased or decreased. Is it possible for ϕ to exceed 45°?

69. ✽ ◆◆ A pulley system and a ramp are used as a combined machine (■ Figure 5.71). Assuming the pulley system is ideal and the ramp is frictionless, find the force $\vec{F}$ required to raise a vegetable crate of mass M up the ramp. Show that the system's mechanical advantage, the ratio W/F of the crate's weight to the required force, is the product of the mechanical advantages you could achieve using the pulley alone and using the ramp alone.

70. ✽ ◆◆ In the system shown in Figure 5.68, the middle pulley is attached to a spring of negligible unstretched length and spring constant k, and remains at rest. Find the acceleration of each block and the height of the middle pulley above the ground, if the ramp surface is frictionless and the pulleys are massless.

71. ◆◆ Later in his rescue attempt, Joachim (Example 5.2, §5.2.5) digs his ice axe into the surface. The ice exerts a force on the axe equal to Joachim's weight and directed up the glacier. What then are the accelerations of the two climbers? If Joachim has mass $M = 82$ kg, Maria's mass is $m = 55$ kg, the glacier's slope is $\theta = 51°$, and Maria is initially 15 m below the edge of the crevasse, how soon after he begins to slide should Joachim dig in his ice axe?

72. ✽ ◆◆◆ A loop of cable has accidentally been released by a member of the space station construction crew. The loop forms a circle that rotates with angular frequency $\omega = 10$ rad/s. If the loop has radius $R = 10$ m and mass $M = 10^2$ kg, what is the tension in the cable?

73. ◆◆◆ In a laboratory test two objects of mass $M = 1.000$ kg are suspended by a string passing over a frictionless rod (■ Figure 5.72). A *rider* of mass $m_1 = 6.00$ g is placed on the left object and a *rider* of mass $m_2 = 3.00$ g is placed on the right object. The system is released from rest. After 15.0 s, the 6-g rider is *gently* removed. The system later comes to rest after a total time interval of 30.0 s, when the left object has descended a total distance of 1.241 m. (a) Determine the acceleration due to gravity. Are you on Earth, the Moon, or Mars? (b) Verify that removal of the rider did not disturb the position or velocity of the system within the accuracy of the experiment. (*Hint:* Assume that the final position and velocity after the first 15 s are the initial position and velocity for the second 15 s, and confirm the time and distance for the system to come to rest.)

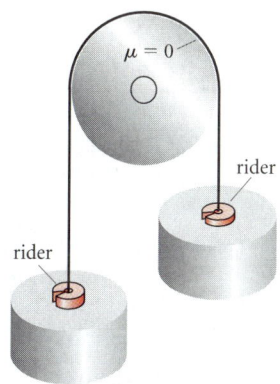

■ FIGURE 5.72

74. ✳ ♦♦♦ **(a)** If the rope in Example 5.4 (§5.3.2, Figure 5.19) is hanging with the maximum amount over the edge, what is the tension in the rope as a function of position? **(b)** If less than the maximum amount is hanging over the edge, how much of the portion on the rock surface can you lift before the rope slides?

75. ♦♦♦ ✉ Assume the Earth to be a perfect sphere of radius 6400 km except for Mt. Everest, which is a cone 5 km high with a base 30 km in radius. Compare the gravitational force of Mt. Everest on a space shuttle 100 km directly overhead with the force on the shuttle due to the entire Earth. Assume that Earth and Mt. Everest have the same uniform density. Based on this calculation, do you think you would make a major error by modeling the Earth as a collection of particles the size of Mt. Everest?

76. ✳ ♦♦♦ A lead sphere and a copper sphere have equal gravitational masses as determined by measuring their weights on a sensitive electronic balance. The two spheres are released simultaneously from rest at the top of a 5-m-high evacuated tube. Supposing the ratio of inertial to gravitational mass for lead were 0.999 times the corresponding ratio for copper, which sphere would reach the bottom of the tube first? How much earlier? Is it important that the gravitational masses of the spheres be equal? If the gravitational mass of each sphere is 10.0 g, what difference in air resistance on the two spheres would mask the supposed difference in their inertial masses? Would it be wiser to make the spheres of equal radius?

77. ♦♦♦ Newton proved that a spherical shell of mass surrounding an object exerts no *net* force on the object. **(a)** Use this result, and the theorem that a sphere attracts objects outside itself as if it were a point, to find how the acceleration due to gravity would vary in a drill hole bored through the center of the Earth. Let g_o be the value at the surface, and assume the Earth to be a uniform sphere. **(b)** Suppose, slightly less impractically, that the tunnel were drilled between Los Angeles and Stockholm, ■ Figure 5.73, to carry magnetically levitated trains of mass m. Show that the component of a train's weight along the tunnel is given by:

$$F_x = -mg_o(x/R).$$

(We will show in Chapter 14 that the time for the train to go from Stockholm to Los Angeles is $t = \pi\sqrt{R/g_o}$, the same time as required by a near-Earth satellite to travel half an orbit.) **(c)** Express the quoted result for *t* in terms of the density of the Earth. How much time would it take to travel through an asteroid with the same density as the Earth but with a radius of 10.0 km? THAT might be practical in 50 years!

78. ♦♦♦ Three identical stars of mass *M* move in circular orbits of radius *R* while always remaining at the corners of an equilateral triangle. Show that this is a possible motion according to Newton's laws and find an expression for its period.

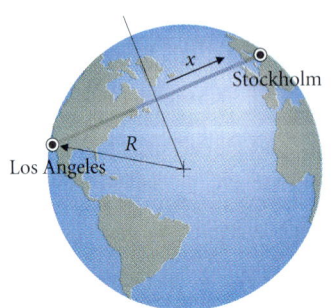

■ **Figure 5.73**

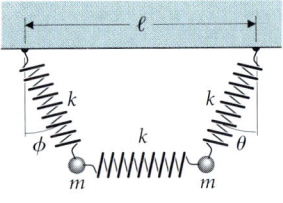

■ **Figure 5.74**

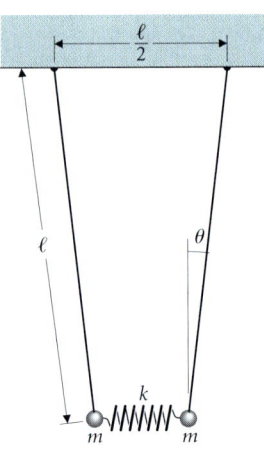

■ **Figure 5.75**

Computer Problems

79. Two identical balls of mass *m* are suspended by three identical springs, each having spring constant *k* and relaxed length $\ell/2$ (■ Figure 5.74). Find the angles θ and ϕ. Take $k = 250.0$ N/m, $\ell = 75.00$ cm, and $m = 10.00$ kg.

80. Two objects with equal masses $m = 10.0$ kg are suspended from points a distance $\ell/2$ apart with strings of length $\ell = 1.00$ m (■ Figure 5.75). The objects are connected by a spring with constant $k = 5.00$ N/m and negligible relaxed length. Find the angle θ from the vertical at which the objects hang, and find the tension in the spring and in each string.

Challenge Problems

81. A tightrope walker with mass 85 kg stands in the middle of a cable of mass 31 kg stretched between two buildings 72 m apart. The cable under the performer is 0.20 m lower than its suspension points. **(a)** First assuming the mass of the cable to be concentrated at its center, estimate the tension of the cable at its suspension points. **(b)** Make a second estimate assuming the mass of the cable to be concentrated in three equal particles equally spaced along the cable. **(c)** ✉ Prepare a graph for plotting estimated tension versus $1/n$, where *n* is the number of mass points used to model the cable. Plot your results from parts (a) and (b) and draw a straight line through your plotted points. From the intercept of this line with the tension axis, estimate the limit of the tension as $n \to \infty$.

82. A wedge of mass *M* is free to slide on a horizontal, frictionless surface (■ Figure 5.76). A box of mass $m = M/5$ is free to slide, also without friction, along the wedge. Find the accelerations of the box and the wedge. If $d = 10.0$ m, find the speed of the box when

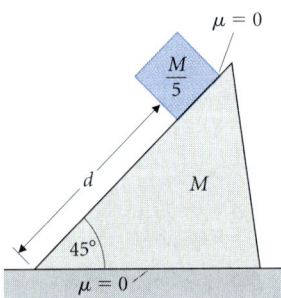

■ **Figure 5.76**

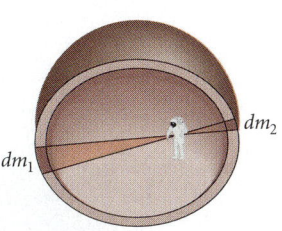

■ FIGURE 5.77

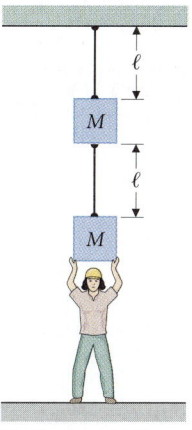

■ FIGURE 5.78

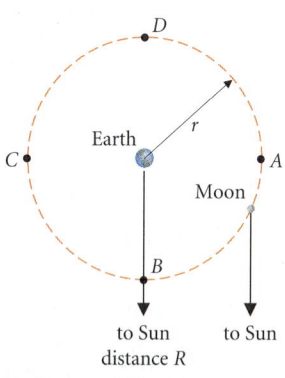

■ FIGURE 5.79

it reaches the bottom of the wedge. (*Hint:* The box remaining on the wedge is a constraint on the relative velocity of the two objects. Use this fact to obtain a relation among the acceleration components.)

83. *Newton's theorem.* Suppose you were inside a large, uniform, thin, spherical shell of mass, perhaps an asteroid hollowed out to produce an interstellar colony ship. Newton's theorem claims that the net gravitational force exerted on you by the shell is exactly zero.
(a) Imagine two tiny cones with their apices at your body (■ Figure 5.77). These cones intersect the shell, defining mass elements dm_1 and dm_2. Explain why the gravitational forces these elements exert on you balance exactly. **(b)** Use a large number of such cones that divide the entire shell into mass elements to prove Newton's theorem.

84. ✉ The (otherwise) ideal strings in the system of ■ Figure 5.78 will break if their tension exceeds $3Mg/2$. If the upper box is suspended first and then the second is gently lowered, the upper string breaks. If the lower box is simply dropped, the lower string breaks and the upper one does not. **(a)** Gentle lowering means that once the lower string becomes taut, the person gradually reduces the upward force she exerts on the lower box. Describe what happens to the tensions in the two strings during this process and explain why the upper string breaks. **(b)** Suppose instead the lower box is released from rest and falls a distance ℓ before the lower string becomes taut. Also suppose that each string can stretch a small amount δs before breaking. Argue that the force required to decelerate the box is of an approximate magnitude $F \sim Mg(\ell/\delta s)$, and explain why this means the lower string will snap. **(c)** Similarly, show that the displacement of the upper block caused by tension in the snapping string is of the order of $\delta s_{ub} \sim (\delta s)^2/\ell$, and argue that this fact explains why the upper string does *not* snap.

✱ **85.** In a typical suspension bridge, the mass of the suspended roadway is much greater than the mass of the cables. We may model the bridge with a massless cable supporting a uniform roadway with mass per unit length λ. Apply Newton's second law to a small element of the cable that makes angles θ and $\theta + d\theta$ with the horizontal at its two ends. Show that $T \cos \theta$ is constant and $T \sin \theta$ varies linearly along the cable. Obtain an expression for $\tan \theta = dy/dx$, and integrate to find $y(x)$. Show that the cable hangs in a parabola. How does the sag of the cable depend on the distance between the towers?

✱ **86. (a)** Find the tension at a distance x from the end of a hanging chain by considering forces on each of its N links. **(b)** Draw a free-body diagram for an element of vertically hanging rope with mass λ per unit length. By analyzing the diagram, obtain an equation for the tension $T(x)$ at a distance x above the end of the rope. Solve the resulting differential equation for $T(x)$. Compare your result with part (a) and with the result of Exercise 5.4.

87. Suppose the Moon's orbit around the Earth would be circular in the absence of any tidal effects due to the Sun. At each point *A*, *B*, *C*, and *D* in ■ Figure 5.79 *estimate* the tidal acceleration of the Earth–Moon system, which is the difference between the Moon's acceleration due to the Sun and the Earth's acceleration due to the Sun, in terms of r, R, G, and M_{Sun}. Show that the magnitude of the tidal acceleration is twice as great at *B* and *D* as at *A* and *C*. Discuss qualitatively the effect of the tidal acceleration on the curvature of the Moon's path and the resulting effect on the overall shape of the Moon's orbit around the Earth.

Part I Problems

1. ❖ At time $t = 0$, a car traveling with speed 50 km/h passes a motorcycle, which has just turned onto the same road. The graphs of speed v versus time t for both vehicles are shown in ■ Figure I.1. Which of the following statements are true and which false at time $t = 20$ s? Explain your reasoning.
   (a) The motorcycle is behind the car.
   (b) The motorcycle is passing the car.
   (c) The motorcycle is in front of the car.
   (d) Both vehicles have the same acceleration.
   (e) The car is accelerating faster than the motorcycle.
2. ◆◆ An object moves in a circle of radius R with a speed that increases with time, $v = v_o + at$. Find the average speed of the particle during the first revolution.
3. ◆◆ A constant force $\vec{F} = (6.7\text{ N})\,\hat{i}$ is exerted on a 1.3-kg object, which starts from rest. What are the displacement and velocity of the object after 12 s?
4. ◆◆ A car traveling at 30.0 km/h accelerates smoothly to a speed of 60.0 km/h in 30.0 s. What is the minimum coefficient of friction between the tires and the road?
5. ◆◆◆ An object of mass M moves in one dimension subject to a force with x-component $F_x = F_o - bt$, where b is a constant. At $t = 0$, the object is at the origin with velocity component $v_x = v_o$. What are the position and velocity of the object as functions of time?
6. ◆◆◆ An airplane's engine has stopped, and the pilot decides to make an emergency landing on a grassy slope inclined at 10.0° to the horizontal. The wind is blowing directly up the incline at a speed of 20.0 km/h, and the landing speed of the airplane is 100.0 km/h, *with respect to the air*. Should the pilot land upslope or downslope to minimize the distance along the ground required to stop the airplane? Assume a coefficient of friction $\mu = 0.30$ between the tires and the grass. (The friction is rolling friction. You may assume that it behaves like kinetic friction.)
7. ◆◆◆ *Distance to Mars according to Copernicus:* Opposition of a planet is the time when it appears in the sky directly opposite from the direction to the Sun. Quadrature is the time when the directions from Earth to the Sun and to the planet are at right angles. The average time between Mars' appearance at opposition to its next appearance at quadrature is 110 d. Take the period of Mars in its orbit as 1.8 y. (a) Draw a diagram of the orbits of Earth and Mars around the Sun, showing the planets' positions with Mars at opposition and 110 d later with Mars at quadrature. (Model the orbits as perfect circles.) (b) From your diagram, derive a formula for the radius of Mars' orbit in terms of the planets' periods, the time between opposition and quadrature, and the radius of Earth's orbit (the astronomical unit). (c) Use the given data to find the radius of Mars' orbit in astronomical units.
8. (a) ❖ Taking the rotation of the Earth into account, explain why the measured acceleration of falling objects at the Earth's equator differs from the acceleration measured at the North Pole. (b) ◆◆ If the measured acceleration of a freely falling object with respect to the Earth's surface is called the *effective gravitational acceleration* $\vec{g}_{eff}$, find the difference between $\vec{g}$ and $\vec{g}_{eff}$ as a function of latitude. (Assume the Earth is a perfect sphere.) (c) ◆◆ How rapidly would the Earth

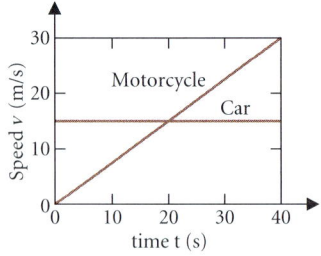

■ Figure I.1

■ Figure I.2

have to rotate for $\vec{g}_{eff}$ to be zero at the equator? How rapidly would an asteroid with the same density as the Earth but with a radius of 10 km have to rotate for there to be zero $\vec{g}_{eff}$ at its equator?
9. ◆◆ Astronaut Toichiro has made an error and is floating away from the space station tethered to a 350-kg piece of equipment by a 0.100-km safety line. Studying a movie of his predicament after the rescue, Toichiro notices that he and the equipment were rotating about a point 19 m from the equipment. Explain how Toichiro can use this observation to compute his mass when space-suited. What is his mass?
10. ◆◆◆ NASA trains astronauts in a KC 135 airplane fondly known as the *Vomit Comet* (■ Figure I.2). The aircraft dives to achieve a speed of 840 km/h and pulls up into a parabolic arc. The pilot adjusts the controls to provide thrust equal to drag and zero lift while the plane is in the arc. If the initial velocity is at 45° to the horizontal, how long do the astronauts feel *weightless*?

Challenge Problem

11. Ulf is a glider pilot. During a competition, he takes off from Stuttgart to fly to Munich, 1.00×10^2 km away. There is no wind. A thermal over Stuttgart carries him vertically upward at 1.0 m/s. Once he leaves the thermal, he must fly directly to Munich using a straight-line descent. His airspeed depends on his glide angle: $v_a = v_o \tan \alpha$ with $v_o = 6.0 \times 10^3$ km/h. How high should he go in the thermal to reach Munich in minimum time? (*Note:* The angle α will be very small. It is important to make a suitable approximation. The problem is much harder if you do not.)

Conservation Laws

PART TWO

In dynamics we have studied the motion of particles in response to the forces that occur between the particles in a system. This approach grew from Galileo's experiments with falling bodies and from Newton's discovery of the relation among force, mass, and acceleration. Another, equivalent, way of thinking about motion also began with Galileo's inclined-plane experiments. In one of his experiments, a ball rolling to and fro between two inclined planes continually returns to the same height it started from, as if it possesses some property that remains constant—is *conserved*—in the absence of outside influences. This idea of a physical property possessed by a body because of its motion is essential in modern formulations of physics. In developing this concept, physicists have learned that there is more than one such conserved property and that they are important in physical processes other than the motion of particles. The conservation of these quantities is a principle that unifies the many different branches of physics and remains unaltered as we revise and refine our understanding of physical laws.

Our next goal is to describe the three kinds of property possessed by a moving body, called linear momentum, energy, and angular momentum, and to discover how to use them to increase our ability to solve physics problems.

Each of the three properties obeys a conservation law: the total amount of each in the universe remains constant. To understand what is meant by conservation laws, imagine pouring water from a bottle into a pan and then freezing it. The water does not disappear in this process, though it does move from place to place and change its form. But water is not an absolutely conserved substance; for example, it can be produced from, or changed into, other substances in chemical reactions. In contrast, in any one reference frame momentum and energy are absolutely conserved: no process can alter their total amount.

The study of conserved properties enriches the concept of force. As we shall see, objects exchange energy and momentum as a result of the forces that occur between them. Newton's third law explains force as an interaction between pairs of objects. At the most fundamental level, we understand that interaction as the transfer of particles that carry energy and momentum. Force both explains, and is explained by, the transfer of conserved quantities.

THE NUMERICAL VALUES ASSIGNED TO MOMENTUM AND ENERGY DO DEPEND ON THE REFERENCE FRAME WE USE; SO FOR NOW, WE SHALL STICK TO A SINGLE REFERENCE FRAME IN EACH EXAMPLE. IT IS POSSIBLE TO DEFINE A COMBINATION OF ENERGY AND MOMENTUM THAT IS THE SAME IN ALL REFERENCE FRAMES. WE'RE NOT READY TO DO THIS YET, BUT WE'LL DO IT IN CHAPTER 34.

Rarefied glowing gas and sunlight reflected from small, solid dust particles form the beautiful tail of Halley's comet, shown here during its close approach to the Sun in 1986. The nucleus of Halley's comet was photographed by the European Space Agency's Giotto spacecraft from a distance of 600 km. The nucleus is an extremely dark, dirty snowball roughly 16 km × 8 km × 8 km. Heated by the Sun, the snowball boils off the gas and dust that form the comet we see. Astronomers estimate that kilometer-sized objects strike the Earth about once every 500 000 years with the effect of a hundred thousand nuclear bombs. An even larger collision may well have killed off the dinosaurs 65 million years ago. Your chance of dying from an asteroid impact this year is about 1 in 2 million.

What if we knew an impact were coming a few years hence? That is the stuff of science-fiction thrillers but also, because we could do something about it, the subject of serious study. Warding off such a disaster would involve a lot of the physics we will study in the next several chapters. We will look back on occasion to see what it would take to avoid going the way of the dinosaurs. First of all, we would have to change the impactor's linear momentum (Example 6.4).

CHAPTER 6
Linear Momentum

Concepts
Momentum of a particle
Impulse
Total momentum of a system
Conservation law

Goals
Be able to:
Evaluate the linear momentum of a system of particles.
Recognize when and how to apply conservation laws to solve problems.

*Since 'tis Nature's law to change,
Constancy alone is strange.*

John Wilmot, Earl of Rochester

What caused the demise of the dinosaurs? Some scientists believe that a large meteor collided with Earth, wreaking havoc on the environment for decades. We know that objects in the solar system do collide. Impact craters on the Moon and Mars provide evidence of past calamities, and in July 1994, a comet collided with Jupiter. Suppose we found a comet on a collision course with Earth? Do we have the ability to avert disaster? Probably. In this chapter we begin to study the most fundamental principles of physics—the conservation laws—and we'll see how powerful they can be in solving problems like changing a comet's orbit.

Catching a softball at a neighborhood picnic is much easier than catching a baseball thrown by a major league pitcher. You wouldn't want to stop a bowling ball no matter who threw it! These intuitive impressions suggest that a moving body possesses some physical property that depends both on its mass and on its speed. Mathematically, the simplest such quantity is directly proportional to both mass and speed. Direction is also important: it is much easier to withstand a glancing impact from an object than to stop it completely. Thus one important property of motion is the product of an object's mass and its velocity vector. Since the mass of a particle is a scalar, its product with the velocity is also a vector in the same direction as the velocity. Newton originally formulated his laws of motion in terms of this property, which he called "quantity of motion" and which we now call *linear momentum*. In this chapter we shall show how linear momentum is used in the study of dynamic systems.

6.1 LINEAR MOMENTUM

6.1.1 Momentum of a Particle

> The *linear momentum* $\vec{\mathbf{p}}$ of a particle in a given reference frame is its mass multiplied by its velocity measured in that reference frame:
>
> $$\vec{\mathbf{p}} = m\vec{\mathbf{v}}. \qquad (6.1)$$

WHEN THE WORD *MOMENTUM* IS USED ALONE, IT IS UNDERSTOOD THAT LINEAR MOMENTUM IS MEANT.

Newton's first law states that the momentum of a particle is constant in time unless an external force acts on the particle. That is, a particle's momentum cannot change unless the particle interacts with some other object. Newton's second law describes how momentum changes when forces do act. Since a *particle*'s mass does not depend on time:

IF AN OBJECT'S MASS CHANGES, WE CANNOT LEGITIMATELY DESCRIBE IT AS A PARTICLE. BUT, SEE §6.2.2.

> $$\vec{\mathbf{F}} = m\vec{\mathbf{a}} = m\frac{d\vec{\mathbf{v}}}{dt} = \frac{d}{dt}(m\vec{\mathbf{v}}) = \frac{d\vec{\mathbf{p}}}{dt}. \qquad (6.2)$$
>
> The rate at which a particle's linear momentum changes is equal to the net force acting on it.

This is Newton's original version of the law, and it remains correct for complex systems as well as for particles.

The unit of momentum is the unit of mass times the unit of velocity: kg·m/s. There is no special SI name or symbol for this unit; it is frequently expressed as 1 newton-second: kg·m/s = kg·(m/s²)·s = N·s.

EXAMPLE 6.1 ♦ A centrifuge cage of mass 230 kg moves on a circle of radius 1.0 m at a speed of 4.0 m/s. Describe its momentum (■ Figure 6.1).

MODEL In uniform circular motion, the cage's velocity continuously changes, remaining tangent to the circle at all times. Its momentum at any time is also tangent to the circle.

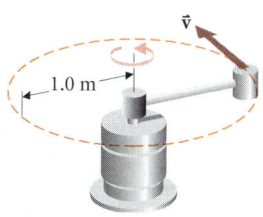

■ FIGURE 6.1
In this centrifuge, the cage moves around a circle of radius 1.0 m at a speed of 4.0 m/s.

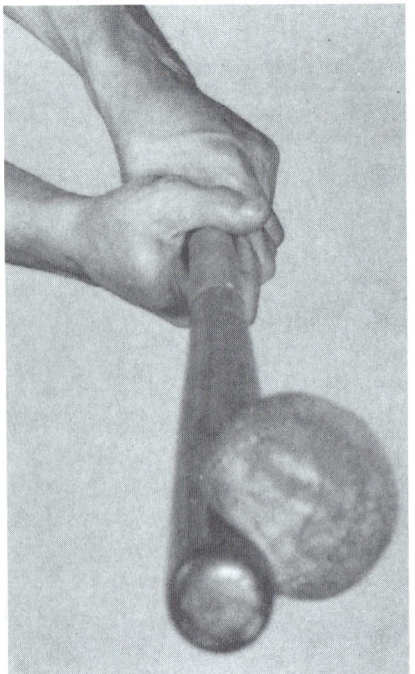

■ FIGURE 6.2
A baseball hit by a bat experiences a very large force that lasts for a very short time interval. This *impulsive* force deforms the baseball as shown in this photograph, which was taken with a high-speed camera.

SETUP The magnitude of the cage's momentum is the product of its mass and speed.

SOLVE
$$|\vec{p}| = m|\vec{v}| = mv = (230 \text{ kg})(4.0 \text{ m/s}) = 9.2 \times 10^2 \text{ kg} \cdot \text{m/s}.$$
$$\vec{p} = (920 \text{ N} \cdot \text{s}, \text{ tangent to the circle}).$$

ANALYZE The direction of the change in momentum is toward the center of the circle, the direction in which the force acts.

EXERCISE 6.1 ♦ A particle of mass 5.0 kg has a speed of 3.0 m/s. What is the magnitude of its momentum?

EXAMPLE 6.2 ♦ A constant force of 15.2 N acts for 7.55 s on a particle initially at rest. What is the change in the particle's momentum?

MODEL The rate of change of the particle's momentum equals the force acting on the particle.

SETUP The change of momentum is the product of its rate of change and the time interval.

SOLVE
$$\vec{F} = \frac{d\vec{p}}{dt}.$$

So
$$|\Delta\vec{p}| = \left|\frac{d\vec{p}}{dt}\right|\Delta t = |\vec{F}|\Delta t = (15.2 \text{ N})(7.55 \text{ s}) = 115 \text{ N} \cdot \text{s}.$$

Thus $\Delta\vec{p} = (115 \text{ N} \cdot \text{s}$, in the direction of $\vec{F})$.

ANALYZE Change of momentum is often the most important thing we wish to know about an interaction.

6.1.2 Momentum Transfer

Example 6.2 shows that, when a constant force acts on a particle, the resulting change in momentum equals the force multiplied by the time during which it acts.

> Force acting over time is the process that transfers momentum. When a constant force acts on a particle, the amount of momentum transferred, or *impulse delivered*, is the force $\vec{F}$ multiplied by the time interval Δt during which it acts:
> $$\vec{I} \equiv \vec{F}\Delta t = \Delta\vec{p}. \tag{6.3}$$

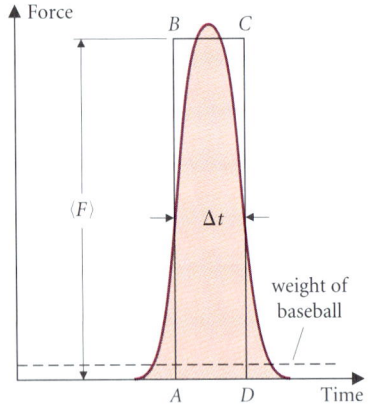

■ FIGURE 6.3
The magnitude of the force acting on the baseball as a function of time. The magnitude of the force peaks rapidly as the ball deforms and then decreases as the ball rebounds. The impulse delivered to the ball is represented by the area under the graph, shown shaded. In Example 6.3 we estimate the time scale, $\Delta t \sim 0.5$ ms, of the collision and the average magnitude, $\langle F \rangle \sim 4 \times 10^4$ N, of the force. [The area of the rectangle $ABCD$ equals the area under the $F(t)$ curve.] The force is 10 000 times greater than the ball's weight.

Impulse is particularly useful for describing rapid interactions such as batting a baseball. The change from fastball to home run means a great change in the baseball's momentum. Yet the ball is in contact with the bat for only a very brief time. For a large change in momentum to occur in a short time interval, the rate of momentum transfer—the force—must be very large indeed. ■ Figure 6.2 shows the deformation this force causes in the baseball, and ■ Figure 6.3 illustrates how the magnitude of force varies in time during the collision. The time interval Δt during which the force acts is negligible compared with other time scales in the ball game, but the force is so large that the product $F \Delta t$ causes a significant change in the ball's motion. In this and similar situations, the momentum change is what we observe and what we care about; the magnitude of the force and the time interval during which it occurs have to be estimated. A force of this nature is called *impulsive*.

EXAMPLE 6.3 ♦♦ A pitcher throws a 150-g baseball at 41 m/s. The ball leaves the bat with a velocity of 52 m/s at an angle of 30° to the incoming pitch (■ Figure 6.4).
(a) What impulse did the bat deliver to the ball?
(b) Estimate the magnitude of the force acting on the ball and the time interval during which it occurs.

Part (a)

MODEL The impulse is equal to the observed change in the ball's momentum.

SETUP The change in momentum is:

$$\Delta \vec{p} = \vec{p}_f - \vec{p}_i = m(\vec{v}_f - \vec{v}_i).$$

■ Figure 6.5 shows the vector subtraction. With the x-direction along the direction of the incoming pitch, the x- and y-components of $\Delta \vec{v}$ are:

$$\Delta v_x = -v_f \cos 30° - v_i = -(v_f\sqrt{3}/2 + v_i)$$

and

$$\Delta v_y = v_f \sin 30° - 0 = v_f/2.$$

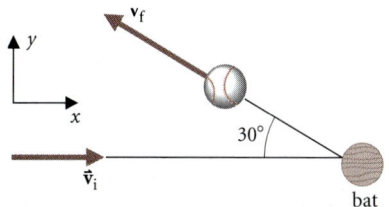
■ **FIGURE 6.4**
Velocity of a baseball before and after striking the bat. The change of the ball's momentum equals the impulse delivered by the bat.

SOLVE The impulse $\Delta \vec{p}$ has magnitude:

$$|\Delta \vec{p}| = \sqrt{(\Delta p_x)^2 + (\Delta p_y)^2} = \sqrt{(m\,\Delta v_x)^2 + (m\,\Delta v_y)^2}$$

$$= m\sqrt{(v_f\sqrt{3}/2 + v_i)^2 + (v_f^2/4)} = m\sqrt{v_f^2 + v_i^2 + \sqrt{3}v_f v_i}$$

$$= (0.15 \text{ kg})\sqrt{(52 \text{ m/s})^2 + (41 \text{ m/s})^2 + \sqrt{3}(52 \text{ m/s})(41 \text{ m/s})}$$

$$= 13 \text{ kg} \cdot \text{m/s}.$$

Its direction makes angle θ with the negative x-axis:

$$\tan \theta = \left|\frac{\Delta p_y}{\Delta p_x}\right| = \left|\frac{\Delta v_y}{\Delta v_x}\right| = \frac{v_f/2}{v_f\sqrt{3}/2 + v_i}$$

$$= \frac{v_f}{v_f\sqrt{3} + 2v_i} = \frac{52 \text{ m/s}}{(52 \text{ m/s})\sqrt{3} + 2(41 \text{ m/s})} = 0.30$$

$$\Rightarrow \theta = \tan^{-1}(0.30) = 17°.$$

$$\vec{I} \equiv \Delta \vec{p} = (13 \text{ N} \cdot \text{s at } 17° \text{ with the negative x-axis}).$$

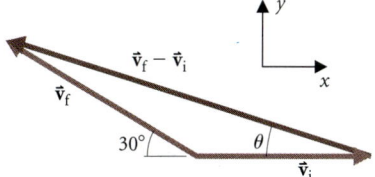
■ **FIGURE 6.5**
Vector triangle for finding the change in the baseball's momentum.

ANALYZE The angle between the impulse and the horizontal is about half the angle of the final momentum. The bat has to deliver the vertical component of the final momentum, as well as enough horizontal impulse to absorb the ball's original momentum and to transfer its final momentum component toward the outfield. The impulse almost reverses the initial momentum, so $|\vec{I}| \sim 2|\vec{p}| \sim 2mv_i = 2(0.15 \text{ kg})(41 \text{ m/s}) = 10 \text{ N} \cdot \text{s}$, and our answer is reasonable.

Part (b)

MODEL We know the baseball isn't squashed flat by its impact with the bat; it compresses and re-expands by some small fraction of its diameter. A baseball's diameter is of the order of 10 cm, so we can estimate that it compresses about 1 cm in the collision (see Figure 6.2). The ball's speed relative to the bat decelerates to zero, then increases again to its final value. Its average speed is of the same order as its initial and final speeds.

So, the impulse is delivered in roughly the time required for the baseball to move 1 cm inward and outward again at its speed before the collision.

SETUP The collision time is of the order of:

$$\tau_{\text{coll}} \sim \frac{2(\text{compression})}{\text{speed}} \sim \frac{2(0.01 \text{ m})}{40 \text{ m/s}} = 0.5 \text{ ms}.$$

SOLVE The magnitude of the collision force is then:

$$F \sim \frac{|\vec{I}|}{\tau_{coll}} \sim \frac{10 \text{ N·s}}{5 \times 10^{-4} \text{ s}} = 2 \times 10^4 \text{ N}.$$

ANALYZE The baseball's weight is roughly 2.5 N, about 10^{-4} of the force exerted by the bat. Little wonder the baseball distorts! The equal and opposite impulse on the bat occasionally shatters it.

EXERCISE 6.2 ♦ In a billiard game, the player's cue contacts the white ball and sets it moving at 10 m/s. What impulse did the cue impart to the ball? (Assume a mass of 200 g for the ball and give only the magnitude of the impulse.)

EXAMPLE 6.4 ♦♦♦ Suppose that in 2005, astronomers discover a comet the same size as Halley's that will strike the Earth in the year 2010. Missiles, carrying the necessary nuclear bombs, can be dispatched to arrive at the comet 1 y before its impact with Earth. By exploding the bombs it is possible to change the comet's course. Estimate the minimum impulse the explosion must deliver to a 10^{15}-kg comet for it to miss Earth.

SEE THE ILLUSTRATION AT THE BEGINNING OF THE CHAPTER.

EARTH'S GRAVITATIONAL PULL AFFECTS THE COMET'S PATH ONLY FOR A VERY SHORT TIME BEFORE IMPACT, AND IS IMPORTANT IN DETERMINING THE DIRECTION OF THE IMPULSE BUT NOT ITS MAGNITUDE.

MODEL A comet like Halley's would follow an elongated elliptical orbit. For the year prior to the rendezvous with Earth, we may approximate the comet's path as a straight line. If Earth is to be saved, the comet's path has to be deflected by at least the radius of Earth: $\Delta r_\perp \sim R_{Earth}$ (■ Figure 6.6). The minimum impulse has to produce the corresponding change in momentum $\Delta \vec{p}$.

SETUP The explosion should give the comet a component of velocity perpendicular to its original path of magnitude at least $\Delta v \sim \Delta r_\perp / \Delta t \sim R_{Earth}/(1 \text{ y})$.

The necessary impulse Δp equals the mass multiplied by Δv:

$$I_{min} = m \Delta v \sim m R_{Earth}/(1 \text{ y}).$$

SOLVE
$$I \sim \frac{(10^{15} \text{ kg})(6 \times 10^6 \text{ m})}{\pi \times 10^7 \text{ s}} \sim 10^{14} \text{ N·s}.$$

IN PART II, PROBLEM 11, WE OUTLINE HOW TO ESTIMATE THE NUMBER OF BOMBS REQUIRED.

ANALYZE This impulse corresponds to 10^{13} hits with a baseball bat! One way to deliver this impulse is to vaporize some of the comet's material with nuclear bombs and produce a jet of hot gas that acts like a rocket motor.

Our first description of impulse (eqn. 6.3) was strictly valid only for a constant force acting during a sharply defined time interval. Then the product $F \Delta t$ gives the area beneath a graph of the force versus time (■ Figure 6.7). As we saw in Chapter 2, the area beneath a graph represents an integral of the function that is graphed. We can show that the precise definition of impulse is an integral of the force over time, $\int \vec{F} \, dt$:

$$\Delta \vec{p} = \vec{p}_f - \vec{p}_i = \int_{t_i}^{t_f} \frac{d\vec{p}}{dt} \, dt = \int_{t_i}^{t_f} \vec{F}(t) \, dt \equiv \vec{I}.$$

This equation serves both as the precise definition of impulse and as a statement of the *impulse-momentum theorem*.

> The change of a particle's momentum during any time interval equals the impulse delivered to the particle during that interval by the net force acting on the particle:
>
> $$\Delta \vec{p} = \vec{I} = \int \vec{F} \, dt. \qquad (6.4)$$

The impulse delivered to the baseball in Example 6.3 corresponds to the shaded area in Figure 6.3. The estimates we made for the collision time and for the force acting on the base-

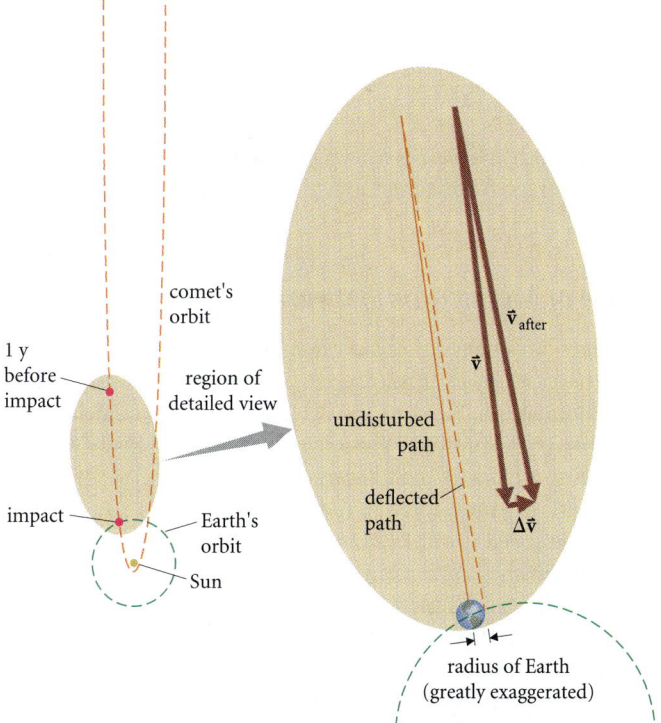

FIGURE 6.6
The orbit of a comet is a highly elongated ellipse. We may approximate the comet's path roughly as a straight line during the year before it is scheduled to strike Earth. To save life on Earth, the path must be displaced a distance greater than the Earth's radius.

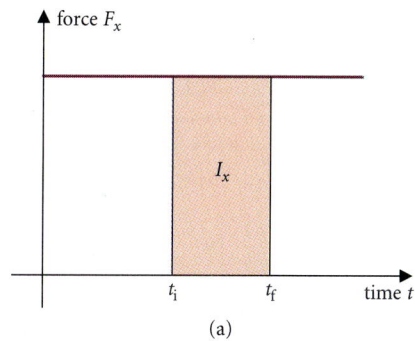

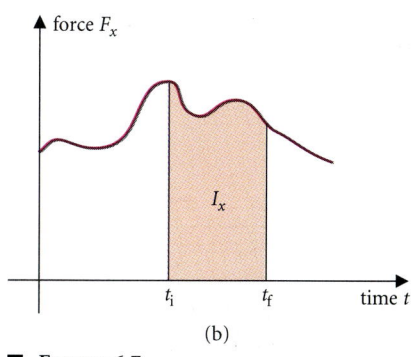

ball are the dimensions of a rectangle with an area equal to the shaded area under the graph of $F(t)$. In other words, we defined the average force acting to be the impulse divided by the interaction time.

EXAMPLE 6.5 ♦♦ A force $\vec{F} = [(3.30 \text{ N/s})t - (4.25 \text{ N/s}^2)t^2]\hat{i}$ acts on a particle of mass $m = 2.72$ kg during the time interval $t = 0$ to $t = 1.29$ s. The particle is at rest at $t = 0$. Find the impulse delivered by the force, and the final velocity of the particle.

MODEL Since the force acts in a constant direction, the resulting impulse is also in the same direction. Its magnitude, and the change in the particle's momentum, are given by eqn. (6.4).

■ **FIGURE 6.7**
(a) A constant force, here in the x-direction, imparts an impulse $I_x = F_x \Delta t$ in the time interval $\Delta t = t_f - t_i$, where I_x is represented by the area under the $F_x(t)$ curve between t_i and t_f. (b) When F_x varies in time, the impulse is still represented by the area under the $F_x(t)$ curve: $I_x = \int F_x \, dt$.

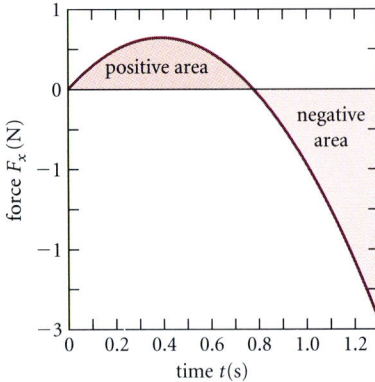

FIGURE 6.8
Graph of $F_x(t)$ between $t = 0$ and $t = 1.29$ s. The force accelerates the particle first in the positive x-direction, then in the negative x-direction. The impulse is the area under the curve. In this example, the total area between 0 and 1.29 s is negative. The particle is moving in the negative x-direction after 1.29 s.

SETUP
$$I_x = \int_0^{1.29 \text{ s}} F_x(t) \, dt = \int_0^{1.29 \text{ s}} [(3.30 \text{ N/s})t - (4.25 \text{ N/s}^2)t^2] dt.$$

SOLVE
$$I_x = (3.30 \text{ N/s})\frac{t^2}{2} - (4.25 \text{ N/s}^2)\frac{t^3}{3} \Big|_0^{1.29 \text{ s}}$$
$$= (3.30 \text{ N/s})\frac{(1.29 \text{ s})^2}{2} - (4.25 \text{ N/s}^2)\frac{(1.29 \text{ s})^3}{3} = -0.295 \text{ N·s}.$$

ANALYZE The negative value of I_x means that the net impulse acts in the negative x-direction. We can check this by plotting F_x as a function of time (■ Figure 6.8). After $t = 0.78$ s, F_x is negative, and the negative contribution to I_x after that time dominates the initial positive contribution.

SETUP The impulse changes the particle's momentum:
$$p_{x,f} - p_{x,i} = I_x.$$

Since the particle is at rest at $t = 0$, $p_{x,i} = 0$. Thus:
$$p_{x,f} = mv_{x,f} = I_x.$$

SOLVE
$$v_{x,f} = \frac{I_x}{m} = \frac{-0.295 \text{ N·s}}{2.72 \text{ kg}} = -0.109 \text{ m/s}.$$

ANALYZE At 1.29 s, the particle is moving in the negative x-direction at 0.109 m/s. The magnitude of the average force acting may be estimated as $I/\Delta t = (0.30 \text{ N·s})/(1.3 \text{ s}) = 0.23$ N. ∎

6.1.3 Exchange of Momentum Between Particles

A particle's momentum can change if some other object exerts a force on it. According to Newton's third law, the particle, in turn, exerts a reaction force on the other object, which changes that object's momentum. Suppose the two objects are a stranded astronaut of mass m_s and his partner (mass m_r), who jumps to the rescue with initial velocity $\vec{v}_i$ (■ Figure 6.9). When the two meet, the rescuing astronaut's hand exerts a normal force $\vec{n}(t)$ on her comrade's shoulder, which responds with a paired force $-\vec{n}(t)$ acting on the rescuer. The speed v_r of the rescuer decreases and the speed v_s of her stranded comrade increases until $v_s = v_r = v_f$, their common final speed. Then the rescuer's muscles relax and there are no more accelerations.

Finding the normal forces as functions of time and computing the resulting motion of the astronauts using the methods of Chapters 4 and 5 would be quite difficult and would require detailed knowledge of the interaction. Fortunately, the ideas of momentum and impulse allow us to find $\vec{v}_f$ without knowing such details.

The change in momentum of the stranded astronaut is:
$$\Delta \vec{p}_s = \vec{I}_s = \int \vec{n}(t) \, dt.$$

The change in the rescuer's momentum is:
$$\Delta \vec{p}_r = \vec{I}_r = \int [-\vec{n}(t)] \, dt = -\int \vec{n}(t) \, dt = -\vec{I}_s.$$

We discuss the properties of force pairs in §4.2.3.

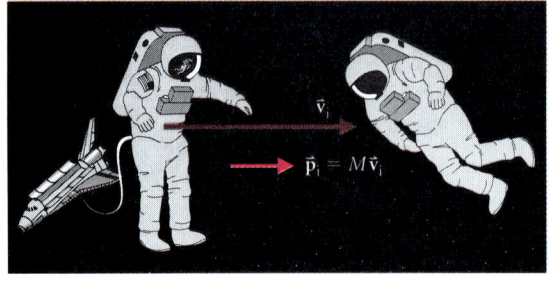
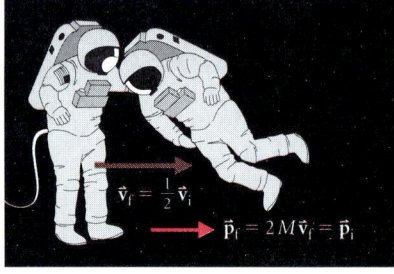

(a) (b)

FIGURE 6.9
(a) An astronaut leaps to the rescue of her stranded colleague. They move away from the spacecraft until the safety line becomes taut. (b) On contact, the two astronauts exert normal forces on each other that vary with time in a complex way. The normal forces exchange equal and opposite amounts of momentum between the astronauts. Their total momentum remains unchanged by the encounter.

Thus $\Delta \vec{p}_r = -\Delta \vec{p}_s$, regardless of how the normal force $\vec{n}(t)$ varies. From the definition of linear momentum, we have:

$$\Delta \vec{p}_r = m_r \Delta \vec{v}_r = -m_s \Delta \vec{v}_s = -\Delta \vec{p}_s.$$

Assuming for simplicity that the two have equal mass, $m_r = m_s$:

$$\vec{v}_f - 0 = -(\vec{v}_f - \vec{v}_i) \quad \text{and} \quad \vec{v}_f = \vec{v}_i/2.$$

The two astronauts move together with half the initial velocity of the rescuer.

■ **EXERCISE 6.3** ♦ Find the final velocity of the astronauts if $m_s = 2m_r/3$. ■

We introduced this example by asking what happens when an individual object's momentum changes. The answer, required by Newton's third law, is that some other object's momentum changes by an equal and opposite amount. The normal forces acting between the astronauts remove momentum from the rescuer and transfer exactly the same amount of momentum to her comrade. If we think of the two astronauts as forming a system, we find that the sum of their linear momentum vectors—their total momentum—is not changed by their collision. With $m_r = m_s = m$, we may calculate their total momentum both *before* and *after* their interaction.

Before: $\vec{P} = \vec{p}_{r,i} + \vec{p}_{s,i} = m\vec{v}_i + 0 = m\vec{v}_i.$

After: $\vec{P} = \vec{p}_{r,f} + \vec{p}_{s,f} = m\vec{v}_i/2 + m\vec{v}_i/2 = m\vec{v}_i.$

NOTE: WE ARE USING LOWERCASE $\vec{p}$ FOR THE MOMENTUM OF A PARTICLE AND UPPERCASE $\vec{P}$ FOR THE TOTAL MOMENTUM OF A SYSTEM.

6.1.4 Conservation of Momentum

The astronaut example illustrates two general, fundamental facts about the momentum of a system of particles. The first is a definition of what we mean by the total momentum of a system.

> The *total linear momentum* of a system is the sum of the linear momentum vectors of all the particles composing the system:
>
> $$\vec{P} = \vec{p}_1 + \vec{p}_2 + \vec{p}_3 + \cdots = \sum_{\text{all particles}} \vec{p}_{\text{particle}}.$$

The second principle is the *law of conservation of momentum:*

> No process can occur that creates or destroys linear momentum.

Forces acting between particles in a system simply redistribute the total momentum among those particles. Only *external* forces, those exerted by objects outside the system, result in the transfer of momentum into or out of the system. Thus the conservation law is usually restated for an ideal, *isolated* system—a system upon which no external forces are exerted:

> The total linear momentum of an isolated system of particles is conserved; the system's total momentum cannot change with time.

The definition of total momentum is just what we would expect. Finding a total by adding amounts of stuff possessed by individual particles is what you do with more familiar properties of things—such as the mass of apples you buy in the market. That momentum is a vector rather than a scalar doesn't change the principle.

EXAMPLE 6.6 ♦♦ At the ice rink a skater of mass 85 kg falls and slides across the ice at 6.0 m/s. The skater bumps into an innocent bystander (of mass 67 kg), and together they slide on across the ice. What is their speed as they slide together?

MODEL The interaction we care about is between the skater and the bystander, and the two of them form the appropriate system. Strictly speaking, this system is not isolated; gravity pulls downward on the two people and the ice exerts contact forces on them. However, the people's weight and the normal forces exerted on them by the ice balance, so there is no *net* vertical force on the system. As usual, we model the ice as a frictionless surface. That is, we neglect the only external force acting horizontally on the system. Thus the net external force is negligible, and we may treat the system as if it were isolated. The system therefore has constant total momentum.

All velocities in this situation are parallel to the initial velocity of the skater. So we choose the x-axis to be along that direction, and we need only consider x-components.

SETUP Initially, the bystander is at rest, so the total momentum of the system equals that of the fallen skater alone:

$$P_{x,i} = m_1 v_{x,i}.$$

After the collision, the two slide on together with velocity $\vec{v}_f$ and their momentum has x-component:

$$P_{x,f} = (m_1 + m_2) v_{x,f}.$$

The total momentum is conserved:

$$P_{x,i} = P_{x,f}.$$

SOLVE
$$m_1 v_{x,i} = (m_1 + m_2) v_{x,f},$$

so
$$v_{x,f} = \left(\frac{m_1}{m_1 + m_2}\right) v_{x,i} = \left(\frac{85 \text{ kg}}{85 \text{ kg} + 67 \text{ kg}}\right)(6.0 \text{ m/s}) = 3.4 \text{ m/s}.$$

ANALYZE As with the astronauts, we do not need to consider the collision of the two skaters in detail; momentum is conserved regardless of the details.

Since both people together have slightly less than twice the mass of the fallen skater, they should slide on at somewhat more than half the initial speed, as we found. ∎

EXERCISE 6.4 ❖ The two people in Example 6.6 ultimately come to rest, either because the friction acting on them isn't exactly zero, or because they come to the edge of the rink. Thus their momentum goes away. Where does it go?

Note the contrast between the addition of momentum vectors in the conservation law and our previous examples of vector addition. Adding force vectors describes how distinct processes acting on a single particle combine their effects. The individual vectors, or their sum, may change over time. The sum of momentum vectors for an isolated system describes a physical property of that system, shared by many particles. Momentum can transfer from particle to particle within the system, but the total is constant over time.

6.2 USING CONSERVATION OF MOMENTUM
6.2.1 General Solution Plan

Many problems are most easily solved using conservation principles. Problems in this class follow a pattern that allows you to recognize them and to apply a standard method for planning a solution. Specifically, such problems do not involve details about how a system evolves in time. Instead, they ask you to compare two states of the system, *just before* something happens and *just after*. The skater collision that we discussed in §6.1 is a typical example. We knew the initial velocity of the fallen skater; we were asked for the final velocity of the two people; and no one asked how much time the collision required. Moreover, we did not need to know

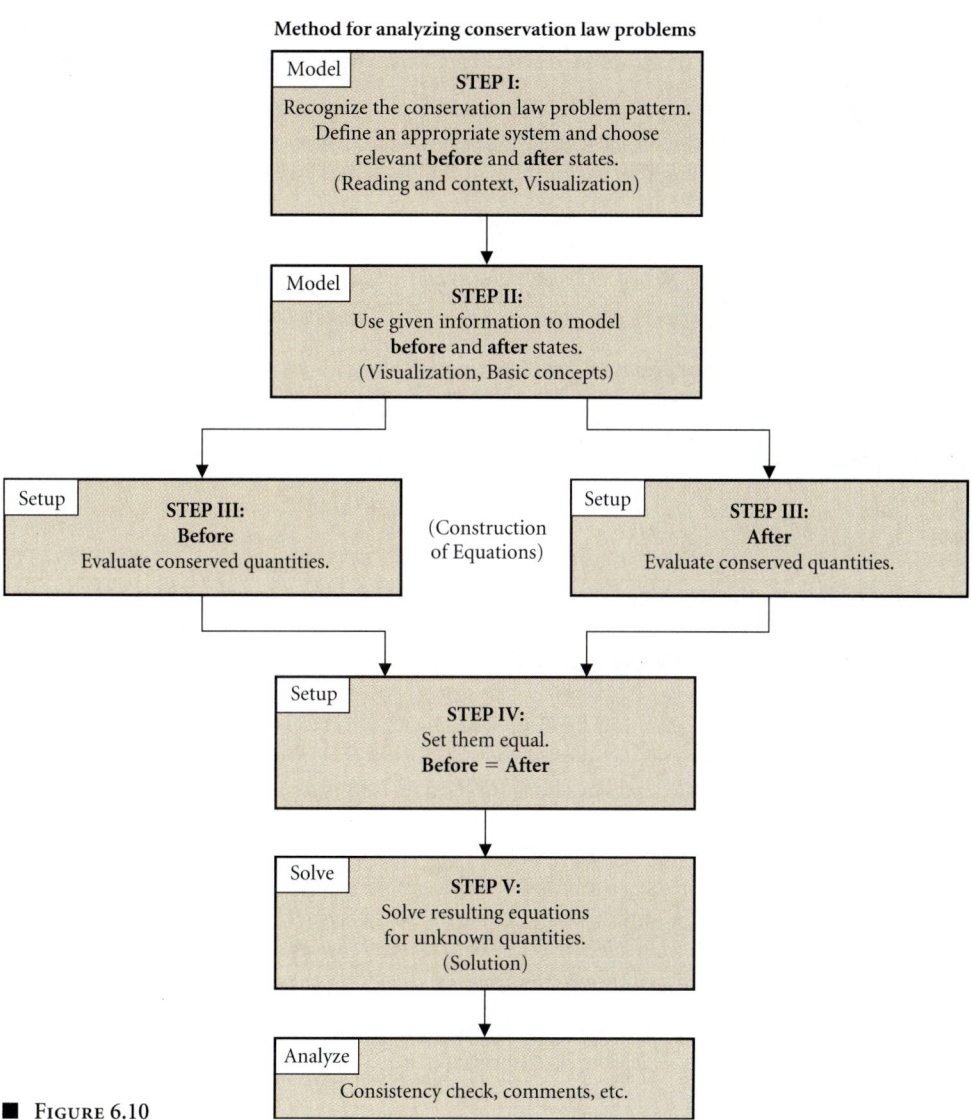

■ FIGURE 6.10

anything about the forces acting between the skaters. The law of conservation of momentum provided a relation, independent of those details, between the *before* and *after* states of the two-skater system. ■ Figure 6.10 presents a formal method for planning solutions of conservation law problems. We'll use the next example to illustrate the method.

EXAMPLE 6.7 ◆◆ Two vehicles of equal mass with equal speeds of 25 m/s approach an intersection (■ Figure 6.11), where they collide and stick together. What is the final velocity of the combined wreckage after the collision?

MODEL
STEP I The question gives information about the vehicles before they collide and asks about their combined state after the collision. Nothing is asked about the collision itself, particularly how much time it requires. This is the pattern of a conservation law problem. The two relevant states are just before and just after the crumpling of the metal.

STEP II As when a bat strikes a baseball, the forces crumpling the cars in a serious accident are much larger than the weight and friction forces acting on the cars, so we can neglect friction during the collision. The weight of each vehicle is balanced by the normal force. We model the system as isolated and recognize the problem as a momentum conservation problem in two (horizontal) dimensions.

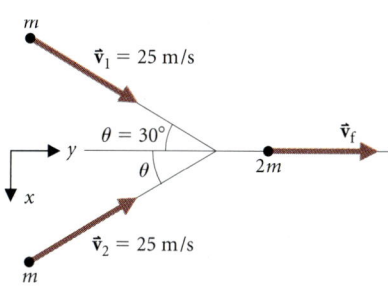

■ **FIGURE 6.11**
Velocity vectors of two colliding vehicles and of their wreckage after the crash. Momentum is conserved during the collision. $|\vec{v}_1| = |\vec{v}_2| = v = 25$ m/s.

SECTION 6.2 • USING CONSERVATION OF MOMENTUM   207

So, the *before* state contains two particles with known momenta, and the *after* state contains one particle, the combined wreckage, with known mass and unknown velocity.

SETUP We evaluate the momenta in a coordinate system that simplifies the calculation.
STEP III Since the vehicles have equal mass and equal speed v, there is symmetry about the line bisecting the angle between their initial velocities. We choose this line to be the y-coordinate axis.

BEFORE

$$\vec{P}_{bef} = mv(\sin\theta\,\hat{i} + \cos\theta\,\hat{j}) + mv(-\sin\theta\,\hat{i} + \cos\theta\,\hat{j})$$
$$= 2mv\cos\theta\,\hat{j}.$$

AFTER

$$\vec{P}_{aft} = 2m\vec{v}_f.$$

STEP IV Set the initial and final momenta equal:

$$2mv\cos\theta\,\hat{j} = 2m\vec{v}_f.$$

SOLVE
STEP V

$$\vec{v}_f = v\cos\theta\,\hat{j}$$
$$= (25 \text{ m/s})(\cos 30°)\,\hat{j}$$
$$= (22 \text{ m/s})\,\hat{j}.$$

ANALYZE In this problem the symmetry arises because the two colliding cars have equal mass and equal speed. Think about how you might choose coordinates for a similar problem with cars of differing mass and speed. ∎

EXAMPLE 6.8 ♦♦ A 1450-kg Honda attempting to stop on an icy road bumps into a 1940-kg Cadillac stopped at the toll booth. Just before impact, the Honda's speed is 2.30 m/s. Afterward, the Cadillac slides forward at 1.80 m/s. What is the final velocity of the Honda?

MODEL At these speeds, car bumpers are designed to prevent damage to the vehicles,
STEP I so we may assume that the two cars in the system move independently after the collision. As usual, we neglect friction between the cars and the icy road. The collision is one-dimensional, and we choose the x-axis to be along the road.

STEP II In the *before* state, the Honda moves toward the stationary Cadillac. Its velocity has x-component $v_i = +2.30$ m/s. Afterward, the x-components of the cars' velocities are u_H and u_C.

SETUP   **BEFORE**   **AFTER**
STEP III x-component of momentum $P_x = m_H v_i$ $P_x = m_H u_H + m_C u_C$

STEP IV Set equal: $\quad m_H v_i = m_H u_H + m_C u_C.$

SOLVE
STEP V

$$u_H = v_i - \frac{m_C}{m_H}u_C$$
$$= 2.30 \text{ m/s} - \frac{1940 \text{ kg}}{1450 \text{ kg}}(1.80 \text{ m/s})$$
$$= -0.11 \text{ m/s}.$$

ANALYZE The minus sign indicates that the Honda rebounds backward after the collision. ∎

6.2.2 Problems Involving Mass Flow

So far, we have considered systems containing a fixed set of particles. But there are many interesting problems in which mass enters and/or leaves a region. For example, as a rocket burns its fuel, it expels the exhaust and so reduces its mass. In these problems, the choice of system is not always obvious. We can often make progress by considering *before* and *after* states separated by a small time interval.

FIGURE 6.12
The engine must exert force on the train to maintain its speed as grain is loaded from the hopper.

EXAMPLE 6.9 ♦♦ A train moves at a constant speed of 1.0 m/s underneath a hopper, which disgorges grain at the rate of 1.0×10^3 kg/s (■ Figure 6.12). What force, beyond that needed to balance rolling friction, must the engine exert to keep the train moving?

MODEL
STEPS I and II We define the system to be the train, including the grain it carries. Its mass increases at a rate $dm/dt = 10^3$ kg/s. The falling grain has vertical momentum but no horizontal momentum. Once it is on the train, it has horizontal momentum but no vertical momentum. The train exerts normal and friction forces on the grain to produce the necessary change in momentum. The grain exerts equal and opposite forces on the train, which would slow it down unless the engine exerts a compensating force. The impulse delivered by this force in a time interval Δt must be just enough to account for the increase in momentum of the grain-plus-train system in that interval. Choose the *before* state at the beginning of the interval Δt and the *after* state at its end. We choose the x-axis to be along the rails, parallel to the train's velocity.

SETUP
STEPS III and IV If M_i is the initial mass of the system, its final mass is:

$$M_f = M_i + (dm/dt)\Delta t.$$

The train's speed v remains constant, so the initial and final momentum components of the system are:

$$P_{x,i} = M_i v \quad \text{and} \quad P_{x,f} = [M_i + (dm/dt)\Delta t]v.$$

Their difference is equal to the impulse $F_x \Delta t$ delivered by the engine:

$$I_x = \Delta P_x.$$

So
$$F_x \Delta t = P_{x,f} - P_{x,i}$$
$$= [M_i + (dm/dt)\Delta t]v - M_i v = (dm/dt)(\Delta t)v.$$

SOLVE
STEP V
$$F_x = (dm/dt)v$$
$$= (1.0 \text{ m/s})(1.0 \times 10^3 \text{ kg/s})$$
$$= 1.0 \times 10^3 \text{ N}.$$

ANALYZE The engine probably does not notice this miniscule extra force needed from the engine! ■

The arithmetic in Example 6.9 is very short once we visualize the momentum flow and decide to analyze states separated by a short time interval. Formal calculus gives us a nice way to remember the result:

$$\vec{F} = \frac{d\vec{P}}{dt} = \frac{d(m\vec{v})}{dt} = m\frac{d\vec{v}}{dt} + \vec{v}\frac{dm}{dt}. \qquad (6.5)$$

For a particle, m is constant, the second term vanishes, and we retrieve Newton's second law in the form $\vec{F} = m\vec{a}$. In Example 6.9, the velocity of the system is constant, the first term vanishes, dm/dt is constant, and we retrieve the result of the example. In the following examples and exercises, observe carefully how the system is defined, and whether eqn. (6.5) is useful.

CAREFUL DEFINITION OF THE SYSTEM UNDER CONSIDERATION IS THE CRUCIAL STEP IN ANALYZING VARIABLE-MASS SYSTEMS.

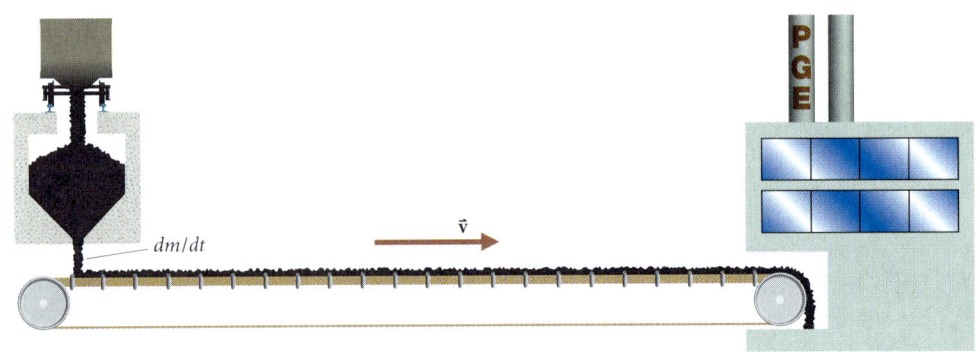

■ **FIGURE 6.13**
A force is required to keep the conveyor moving as it transports coal to the electric plant.

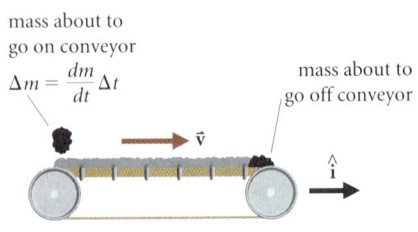

(a) *Before*

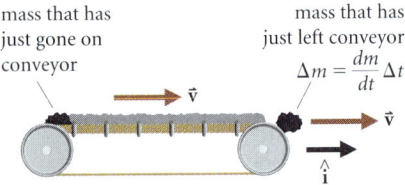

(b) *After*

■ **FIGURE 6.14**
A careful definition of system is needed to see why the force is necessary. The system consisting of the belt, the coal on it, and the coal that falls onto the belt during the time interval Δt is the system to which the belt motor delivers an impulse in the interval Δt. This system gains momentum $\Delta \vec{p} = (\Delta m)\vec{v}$ as the coal that falls onto the belt gains velocity $\vec{v}$.

A Scout sounding rocket used for research at high altitude.

EXAMPLE 6.10 ❖ A conveyor belt carries coal at a rate of dm/dt from a delivery chute into a power plant (■ Figure 6.13). The conveyor is supported on frictionless rollers and driven by a motor at one end of the belt. What force must the motor exert to transport the coal?

MODEL Why must the motor exert any force at all? We have assumed frictionless rollers and the material on the belt moves with constant velocity $\vec{v}$. Furthermore, a system consisting of the conveyor belt and the coal on it has the same mass and the same linear momentum at any given time. But, the objects in this system do not remain the same, so it is difficult to use.

Instead, let's define our system to be the belt plus the coal on it at time t, plus the coal that falls onto it between time t and $t + \Delta t$ (■ Figure 6.14). We choose the x-axis parallel to the belt. At time t, the mass Δm of coal that is about to fall on the belt has no x-component of momentum. At time $t + \Delta t$, this coal is on the belt and has gained momentum $\Delta p_x = (\Delta m)v_x$. The momentum of the rest of the coal in the system has not changed even though a part of it, Δm, has fallen off the belt.

SETUP In time Δt, the motor must transfer momentum,

$$\Delta \vec{p} = (\Delta m)v_x \hat{i} = \left(\frac{dm}{dt}\Delta t\right)\vec{v}$$

to bring this coal up to speed (cf. Example 6.9).

SOLVE The force required of the motor is:

$$\vec{F} = \frac{\Delta \vec{p}}{\Delta t} = \left(\frac{dm}{dt}\right)\vec{v}.$$

ANALYZE The key point in this argument is tracking the momentum in a system consisting of the belt, the coal on the belt, *and* the coal that falls onto the belt in the interval Δt. This system's momentum increases between the *before* and *after* states shown in Figure 6.14. The impulse delivered by the conveyor belt increases the momentum of the system.

EXERCISE 6.5 ◆ What force is required to transport 1.5 metric tons of coal per minute at a speed of 3.0 m/s?

EXERCISE 6.6 ❖ If the coal falls off the conveyor belt onto a second belt moving at the same speed, what force is needed to transport coal on the second belt?

The force necessary to accelerate the coal on a conveyor belt is a minor feature of the belt's design. By contrast, force resulting from the acceleration of exhaust gases is the basic principle of rocket propulsion. The rocket exerts force on the exhaust, and the exhaust exerts an equal and opposite force on the rocket. The rocket's total mass decreases continuously as it exhausts spent fuel, so it is useful to apply momentum conservation to *before* and *after* states that are separated by a differential time interval dt.

EXAMPLE 6.11 ♦♦♦ A rocket is designed so that 10.0% of its initial mass is structure and payload and the other 90.0% is fuel. If the rocket moves on a straight line and exhausts burned fuel at a speed $v_{ex} = 2.0$ km/s with respect to the rocket, what final speed does it achieve starting from rest? (The rocket is launched from a space station, so we may ignore external forces.)

MODEL
STEP I As the rocket fuel burns, the exhaust leaves the back of the rocket at high speed. The rocket pushes on the exhaust gases to accelerate them. According to Newton's third law, the gases also accelerate the rocket. Since no external forces act on the rocket-and-fuel system, its momentum is conserved.

STEP II We define the system to be the rocket, including the fuel that it contains, at time t—that is, in the *before* state. A time dt later, in the *after* state, a small amount dM of the fuel has burned and left the rocket as exhaust (■ Figure 6.15). This material is still part of the system, and we must include its momentum in our calculation.

SETUP
STEP III We choose the x-axis to be along the direction of the rocket's velocity. Then all vectors have x-components only. At time t, the rocket's velocity is $v(t)\hat{i}$ and its mass is $m(t)$. At time $t + dt$, the velocity is $[v(t) + dv]\hat{i}$ and its mass is $m(t) + dm$. The exhaust gas has a velocity $-v_{ex}\hat{i}$ with respect to the rocket. Thus its velocity with respect to the space station, our approximately inertial reference frame, is:

$$[v + (-v_{ex})]\hat{i}.$$

	BEFORE	AFTER
p_x:	mv	$(m + dm)(v + dv) + dM(v - v_{ex})$

STEP IV The mass dM of fuel burned in the time interval dt reduces the rocket's mass: $dm = -dM$. Momentum *before* equals momentum *after*:

$$mv = (m + dm)(v + dv) + (-dm)(v - v_{ex}).$$

SOLVE
STEP V
$$mv = mv + v\, dm + m\, dv + dm\, dv + v(-dm) - v_{ex}(-dm).$$
$$0 = m\, dv + v_{ex}\, dm + dm\, dv.$$

As we let the differential time interval approach zero, the product $dm\, dv$ of two differential quantities is negligible compared with the other terms, so:

$$0 = m\, dv + v_{ex}\, dm.$$

Then: $\qquad dv = -v_{ex}(dm/m).$ \hfill (i)

Remember: $dm < 0$, so $dv > 0$. To find the final speed $v_{burnout}$, we integrate eqn. (i):

$$\int_{launch}^{burnout} dv = \int_{launch}^{burnout} -v_{ex}\frac{dm}{m}.$$

$$v\Big|_{launch}^{burnout} = -v_{ex}\ln m\Big|_{launch}^{burnout}$$

$$v_{burnout} - v_{launch} = -v_{ex}(\ln m_{burnout} - \ln m_{launch}).$$

SUBROUTINE: RELATIVELY MOVING REFERENCE FRAMES.

(a) *Before*: Rocket at time t

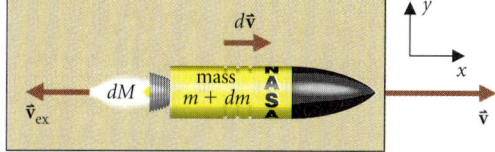

(b) *After*: Rocket at time $t + dt$

■ **FIGURE 6.15**
Conservation of momentum is applied to the system consisting of the rocket and the fuel it carries at time t. (a) In the *before* state the rocket has mass m and speed v. (b) After the interval dt, the rocket has burned a mass $dM\, (= -dm)$ of fuel and exhausted it at a relative velocity $\vec{v}_{ex}$; the rocket's velocity is changed by an amount $d\vec{v}$.
 Puzzle: Does the exhaust gas move backward or forward in the inertial frame of the space station?

With $v_{launch} = 0$, we have:

$$v_{burnout} = -v_{ex} \ln\left(\frac{m_{burnout}}{m_{launch}}\right). \tag{6.6}$$

At burnout, the mass of the rocket shell plus payload is 10% of the total mass at launch, so:

$$v_{burnout} = -v_{ex} \ln \tfrac{1}{10} = v_{ex} \ln 10 = 4.6 \text{ km/s}.$$

ANALYZE The logarithmic dependence of final speed on mass ratio makes it very difficult to achieve speeds much greater than exhaust speed with a single rocket. Once fuel is used, it is inefficient to drag along empty fuel tanks. Most designs for large rockets use staging; that is, empty structures are dropped once used (■ Figure 6.16).

■ **FIGURE 6.16**
After dropping its solid fuel boosters, the space shuttle continues into orbit using the main engines.

Study Problem 6 ♦♦♦ *Delivering the Mail*

A lunar postmaster at Copernicus Spaceport has developed a clever method using one launch to deliver mail to both Rimtown and the observatory (■ Figure 6.17). Explosive bolts are mounted between two parts of the mail canister and are detonated by a timer when the canister is over Rimtown, at the top of its trajectory. The explosion delivers a horizontal impulse to each part of the canister. Find the magnitude of this impulse and the ratio of masses of the two parts of the canister so that one part falls straight down and the other continues onward to the observatory. The postmaster's mass driver has a maximum range of $2D_1$. Rimtown is a distance D_1 away, and the distance to the observatory is $D_2 > 2D_1$.

Modeling the Physical System

The problem involves two phases of projectile motions: before and after the explosion of the bolts divides up the canister's momentum between the two pieces. Thus the main concepts are projectile motion and conservation of momentum. Before the bolts explode, the canister's velocity is determined by its trajectory from the spaceport. After the explosion, the velocities of the two parts have to satisfy the condition that the pieces are to arrive at specific places. In

---

Digging Deeper

ACCELERATION OF THE ROCKET

To find the rocket's velocity as a function of time, we need to know its mass as a function of time. Replacement of $m_{burnout}$ by $m(t)$ gives $v(t)$. For a specific example, suppose the fuel burns at a constant rate $dm/dt = -\mu$. Then:

$$m(t) - m(0) = \int_0^t \frac{dm}{dt} dt = \int_0^t -\mu \, dt = -\mu t.$$

So, $m(t) - m_i = -\mu t \Rightarrow m(t) = m_i - \mu t,$

where m_i is the initial mass of the rocket, including fuel. From eqn. (6.6):

$$v(t) = v_{ex} \ln\left(\frac{m_i}{m_i - \mu t}\right).$$

As the rocket burns off fuel, its mass decreases while the force exerted on it by the exhaust remains constant, and so its acceleration ($a_x = F_x/m$) increases. Thus maximum acceleration occurs just before the rocket motor shuts off, when m has its minimum value. To derive an expression for the acceleration, divide eqn. (i) in Example 6.11 by dt:

$$a_x = \frac{dv}{dt} = \frac{-v_{ex}}{m(t)}\frac{dm}{dt} = \frac{-v_{ex}(-\mu)}{m_i - \mu t} = \frac{\mu}{m_i - \mu t}v_{ex}.$$

If the rocket in Example 6.11 burns its fuel in 300 s, then $\mu = (0.9 m_i)/(300 \text{ s})$, and the final acceleration, just before the rocket turns off at $t = 300$ s, is:

$$a_x \text{ (at 300 s)} = \frac{(0.9 m_i)/(300 \text{ s})}{m_i - 0.9 m_i}(2 \text{ km/s}) = 60 \text{ m/s}^2$$

or about $6g$. This acceleration would be uncomfortable but tolerable for an astronaut in good health.

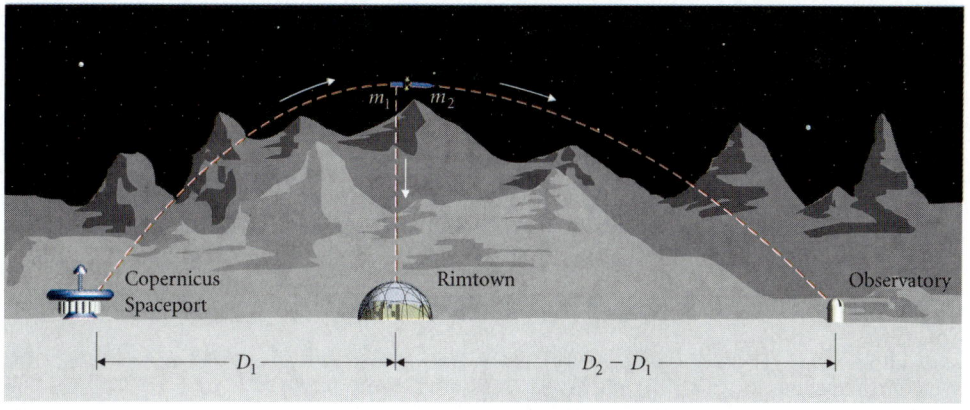 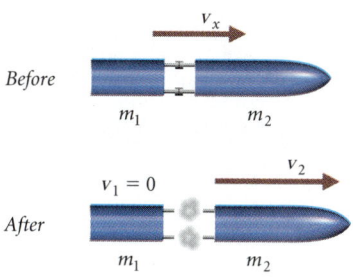

■ **FIGURE 6.17**
(a) A mail delivery system on the Moon uses projectiles launched by an electromagnetic mass driver. To deliver loads of mail to two locations with a single shot, the postmaster designs the mail canister to separate into two parts over Rimtown. One part descends vertically to Rimtown and the other reaches the observatory, which is beyond the range of a single launch. (b) The separation of the mail canister occurs very rapidly compared with the time required for the canister to follow its trajectory. Thus we can analyze the separation using conservation of momentum and neglect any motion of the canister during the separation process.

addition, we do not need to know *when* either canister is at a specific place, only that it gets there. The bolts explode rapidly, and we may neglect any distance the canister travels during the explosion. Since the impulse is horizontal (parallel to the *x*-axis) and is imparted when the canister is moving horizontally, we need only apply conservation of momentum in the *x*-direction.

Because the maximum range of the mass driver is $2D_1$, the postmaster will have to launch at a 45° angle to ensure that the canister reaches the top of its trajectory over Rimtown, a distance D_1 away.

SEE EXERCISE 3.1.

Construction of Equations

We use the standard method (Figure 6.10) to apply conservation of momentum during the explosion, plus the plan for projectile motions with a specific goal (§3.1.2) to describe the motions after the explosion. First we obtain the initial speed from its relation to maximum projectile range (eqn. 3.2):

$$2D_1 = \text{maximum range} = \frac{v_o^2}{g},$$

so

$$v_o = \sqrt{2gD_1}.$$

The velocity of the canister before separation is horizontal, with:

$$v_x = v_o \cos\theta_i = v_o\sqrt{2}/2 = \sqrt{gD_1}. \qquad (i)$$

After separation, one part of the canister (mass m_1) must have zero velocity ($v_1 = 0$) so that it falls straight down to Rimtown.

Evaluate momentum:

BEFORE	AFTER
$P_x = (m_1 + m_2)v_x.$	$P_x = m_1v_1 + m_2v_2$
	$= 0 + m_2v_2.$

Set them equal: $(m_1 + m_2)v_x = m_2v_2.$

Solve for v_2:
$$v_2 = v_x\left(1 + \frac{m_1}{m_2}\right). \qquad (ii)$$

The specific event in the projectile problem is that the second piece of the canister lands at the observatory. The time it falls is determined by its height and is the same time it took to rise to that height. From eqn. (3.1b):

$$t_{\text{fall}} = t_{\text{rise}} = \frac{v_{y,i}}{g} = v_o \frac{\sin\theta_i}{g} = \sqrt{2D_1 g}\left(\frac{\sqrt{2}/2}{g}\right) = \sqrt{D_1/g}. \quad \text{(iii)}$$

During that time, the second piece moves a horizontal distance (eqn. 3.1a):

$$\Delta x = v_2 t_{\text{fall}} = D_2 - D_1. \quad \text{(iv)}$$

Solution

We combine eqns. (i), (ii), (iii), and (iv):

$$D_2 - D_1 = v_x\left(1 + \frac{m_1}{m_2}\right)\sqrt{\frac{D_1}{g}}$$

$$= \sqrt{gD_1}\left(1 + \frac{m_1}{m_2}\right)\sqrt{\frac{D_1}{g}}$$

$$= D_1\left(1 + \frac{m_1}{m_2}\right),$$

which yields
$$\frac{m_1}{m_2} = \frac{D_2}{D_1} - 2.$$

The impulse delivered by the explosion equals the change in momentum of m_1. (The bolts impart an equal and opposite impulse to m_2.) Its magnitude is:

$$|\vec{\mathbf{I}}| = |\Delta\vec{\mathbf{p}}_1| = m_1|\Delta\vec{\mathbf{v}}_1| = m_1|0 - v_x| = m_1\sqrt{gD_1}.$$

Analysis

Since the mass ratio must be a positive number, this method works only if $D_2 > 2D_1$. The method necessarily increases the range of mail delivery. Adding extra propulsion to one part of the canister after it has traveled part of its trajectory is similar to the use of staging for rockets. Numerous variations on the method are possible by varying the timing or direction of the impulse; the same sort of calculation works for any such case, but the details can be a good deal more complicated.

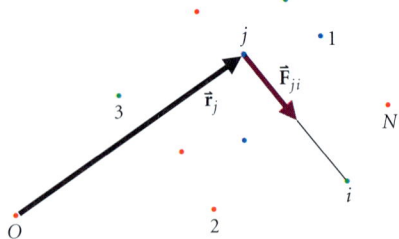

FIGURE 6.18
A general system may be viewed as a large number of distinct particles that exert forces on each other and perhaps experience forces exerted by objects outside the system. We assume only that the particles and forces obey Newton's laws and that the particles are numbered 1, 2, 3, ..., N, where N is the total number of particles. The linear momentum of any such system changes only if external forces act on it.

✱ 6.3   FORMAL PROOF OF MOMENTUM CONSERVATION

Our picture of momentum and its transfer is a generalization from the special case of the two astronauts. To establish this picture firmly, we need to prove the law of conservation of momentum for a general system. We model such a system as a large number of particles (■ Figure 6.18), labeled with numbers, 1, 2, 3, ..., N, where N is the total number of particles in the system. The particles and the forces they exert obey Newton's laws. The particle with number j has mass m_j, velocity $\vec{\mathbf{v}}_j$, and momentum $\vec{\mathbf{p}}_j = m_j\vec{\mathbf{v}}_j$. The total force acting on particle j is $\vec{\mathbf{F}}_j$, which is the sum of forces exerted by objects outside the system, $\vec{\mathbf{F}}_{j,\text{ext}}$, and forces exerted by particles within the system, $\vec{\mathbf{F}}_{j,\text{int}}$. Then, Newton's second law, applied to an arbitrary particle number j, reads:

$$\frac{d\vec{\mathbf{p}}_j}{dt} = \text{sum of forces acting on particle } j \equiv \vec{\mathbf{F}}_j = \vec{\mathbf{F}}_{j,\text{int}} + \vec{\mathbf{F}}_{j,\text{ext}}.$$

The forces exerted by particles within the system may be written as:

$$\vec{F}_{j,\text{int}} = \sum_{\substack{i=1 \\ i \neq j}}^{N} \vec{F}_{ji},$$

where $\vec{F}_{ji}$ stands for the force exerted on particle number j by particle number i. The term with $i = j$ is not included in the sum, because a particle cannot exert a force on itself.

Now we express the rate at which forces change the system's total momentum:

$$\frac{d\vec{P}}{dt} \equiv \frac{d}{dt}\left(\sum_{j=1}^{N} \vec{p}_j\right)$$

$$= \sum_{j=1}^{N} \frac{d\vec{p}_j}{dt} = \sum_{j=1}^{N} \vec{F}_j$$

$$= \sum_{j=1}^{N} \vec{F}_{j,\text{ext}} + \sum_{j=1}^{N}\left(\sum_{\substack{i=1 \\ i \neq j}}^{N} \vec{F}_{ji}\right).$$

In words, this expression reads:

$$\frac{d\vec{P}}{dt} = \frac{\text{sum of external forces}}{\text{acting on particles}} + \frac{\text{sum of internal forces}}{\text{acting on particles}}.$$

The first term is the total external force:

$$\vec{F}_{\text{ext}} \equiv \sum_{j=1}^{N} \vec{F}_{j,\text{ext}}.$$

Then,

$$\frac{d\vec{P}}{dt} = \vec{F}_{\text{ext}} + \frac{\text{sum of internal forces}}{\text{acting on particles}}.$$

The sum of internal forces is the vector sum of every force acting on every particle. Newton's third law tells us that these forces occur in pairs, and since we are summing over all the particles in the system, we are adding both forces in each pair. Thus the sum of internal forces is a sum of zeros, and equals zero.

$$\frac{d\vec{P}}{dt} = \vec{F}_{\text{ext}} + \vec{F}_{\text{int}} = \vec{F}_{\text{ext}} + 0.$$

We have shown that the rate of change of a system's total momentum equals the total external force acting on the system:

$$\boxed{\frac{d\vec{P}}{dt} = \vec{F}_{\text{ext}}.} \qquad (6.7)$$

In the special case of an isolated system, there is no external force and the system's total momentum cannot change, which proves the second statement of the law of conservation of momentum. When external forces do act on a system, any change of its momentum results from an impulse delivered by the external forces. No momentum is created or destroyed; it is transferred to or from objects outside the system.

Equation 6.7 has one more very important consequence. It is the first of three theorems that tell us why it makes sense to model things like trucks or people as particles. That is, the result tells us why everything we've done so far is correct. So long as internal forces keep a system together well enough that we would want to think of it as a single object, we can forget its internal structure and internal forces and apply Newton's second law (eqn. 6.7) using only external forces.

READ THE SYMBOL AS "FORCE ON j DUE TO i."

The globular cluster Omega Centauri. A globular cluster is a real-world illustration of the abstract system in Figure 6.18. Containing some 10^5 stars orbiting each other in a roughly spherical system, this globular cluster has remained an intact satellite of our galaxy for some 10 billion years.

WE SHALL DISCUSS THE OTHERS IN CHAPTER 9.

Chapter Summary

Where Are We Now?

We have described the first conserved quantity of motion: linear momentum. Conservation of momentum follows directly from Newton's laws of motion and provides us with another powerful method for solving problems that complements the dynamic methods of Chapter 5.

What Did We Do?

A particle's linear momentum vector is the product of the particle's mass and its velocity:

$$\vec{\mathbf{p}} = m\vec{\mathbf{v}}.$$

Force acting over time changes a particle's momentum. Newton's first law states that a particle's momentum remains constant unless a net force acts on it. Newton's second law states that the force acting on a particle equals the rate of change of its momentum:

$$\vec{\mathbf{F}} = \frac{d\vec{\mathbf{p}}}{dt}.$$

The impulse-momentum theorem states that the change of a particle's momentum in an interaction equals the impulse delivered by the net force acting on the particle—the integral of the force over the total time of the interaction:

$$\vec{\mathbf{p}}_f - \vec{\mathbf{p}}_i = \Delta\vec{\mathbf{p}} = \int \vec{\mathbf{F}}\, dt = \vec{\mathbf{I}}.$$

Impulse is a particularly useful concept for rapid interactions that involve large, variable forces.

When two particles interact, Newton's third law requires that any momentum transferred to one particle by the interaction be transferred away from the other particle. Because the forces acting on the particles form a third-law pair, the impulses they deliver are equal and opposite.

The law of conservation of momentum:

> No process can occur that creates or destroys linear momentum.
>
> The total linear momentum of an isolated system (i.e., one with $\vec{\mathbf{F}}_{ext} = 0$) cannot change with time.

Problems involving conservation principles can be recognized because they require the comparison of two states of a system at different times and do not involve details of the system's motion between those times. A method for solving these problems is given in Figure 6.10.

Conservation of momentum can be applied to a system with variable mass. When both mass and velocity change, it is usually necessary to consider states of the system separated by a differential amount of time and to include the changes in momentum of any mass that enters or leaves the system during the time interval.

In problems such as the behavior of an exploding projectile in which a very rapid transition occurs between two different phases of motion, momentum ideas are useful for describing the relation between the two phases.

Study Hint: Memorize this method and practice it.

Practical Applications

The rocket equation (6.6) is our most important practical result. It is the starting point for rocket design. Momentum conservation enables us to evaluate forces in a material transport system such as a conveyor belt. The forces are significant for high-speed systems with a high mass-transfer rate. Further applications of momentum conservation involve the other quantities of motion, which we shall discuss in the next three chapters.

Solutions to Exercises

6.1 The magnitude of the particle's momentum is the product of its mass and speed:
$$|\vec{p}| = mv = (5.0 \text{ kg})(3.0 \text{ m/s}) = 15 \text{ N·s}.$$

6.2 The impulse equals the change in momentum of the billiard ball, which in turn equals the mass of the ball multiplied by the change in its velocity:
$$|\Delta\vec{p}| = m|\Delta\vec{v}| = (0.2 \text{ kg})(10 \text{ m/s} - 0) = 2 \text{ N·s}.$$

6.3 The result $\Delta\vec{p}_r = -\Delta\vec{p}_s$ does not depend on the astronauts' mass ratio. The rescuer's momentum changes by an amount:
$$\Delta\vec{p}_r = m_r \Delta\vec{v}_r = m_r(\vec{v}_f - \vec{v}_i),$$
and the stranded astronaut's momentum changes by:
$$\Delta\vec{p}_s = m_s \Delta\vec{v}_s = m_s\vec{v}_f - 0.$$
Setting $\Delta\vec{p}_s = -\Delta\vec{p}_r$: $m_s\vec{v}_f = m_r(\vec{v}_i - \vec{v}_f).$
Then: $\vec{v}_f = m_r\vec{v}_i/(m_r + m_s) = \vec{v}_i m_r/(m_r + 2m_r/3) = \tfrac{3}{5}\vec{v}_i.$

6.4 The object exchanging momentum with the sliding skaters is the Earth itself. We don't notice any effect of the momentum transfer from the skaters to the Earth, because the resulting change in the Earth's velocity is minuscule:
$$|\Delta\vec{p}_{\text{skaters}}| = |\Delta\vec{p}_{\text{Earth}}| \quad \text{and} \quad m_{\text{skaters}}|\Delta\vec{v}_{\text{skaters}}| = m_{\text{Earth}}|\Delta\vec{v}_{\text{Earth}}|.$$
Thus:
$$|\Delta\vec{v}_{\text{Earth}}| = (m_{\text{skaters}}/m_{\text{Earth}})|\Delta\vec{v}_{\text{skaters}}|$$
$$\sim (100 \text{ kg})(3 \text{ m/s})/(6 \times 10^{24} \text{ kg}) \sim 10^{-22} \text{ m/s}$$
$$\sim \text{an atomic diameter in 60 000 y!}$$

6.5 This is a plug-in to the result of Example 6.10.
$$F = \frac{dm}{dt}v = \frac{1.5 \text{ ton}}{1 \text{ min}}(3 \text{ m/s}) = \frac{1.5 \times 10^3 \text{ kg}}{60 \text{ s}}(3 \text{ m/s}) = 75 \text{ N}.$$

6.6 As in Example 6.10, let the system be the coal on the second belt at time t plus the coal that falls onto the belt between t and $t + \Delta t$. The momentum of the system doesn't change during the interval Δt because the incoming coal already has the necessary x-component of momentum. Thus, ignoring friction at the belt rollers, no force is needed.

Basic Skills

Review Questions

- Give intuitive reasons for believing *quantity of motion* should depend on both the mass and the velocity of an object.

§6.1 LINEAR MOMENTUM

- Define *linear momentum*.
- What do Newton's first and second laws tell us about the linear momentum of a particle?
- Describe how to calculate the impulse delivered by a constant force.
- Describe qualitatively how to compute the impulse delivered by a variable force.
- What happens to a particle as a result of an impulse being delivered to it?
- When two particles interact, what does Newton's third law tell us about the momentum transferred between them?
- Describe how to compute the linear momentum of a system of particles given the momentum of each particle comprising the system. How does this rule support the notion of linear momentum as a real physical property possessed by the particles in the system?
- What is true of the total momentum of an isolated system?

§6.2 USING CONSERVATION OF MOMENTUM

- What are two characteristics of problems that are readily solved using a conservation principle?
- Describe the method for using conservation laws in problem solutions.
- Explain why a conveyor belt has to deliver momentum to the material it carries despite the fact that the momentum of the material on the belt at any time is the same.
- If the *system* for a rocket at some time t is all the mass contained *within the rocket* at that time, describe the system an element of time dt later.
- Explain why large rockets use staging to achieve large speeds.

Basic Skill Drill

§6.1 LINEAR MOMENTUM

1. Compare the momentum of a Volkswagen Beetle moving at a speed of 30 km/h with that of a second Beetle with speed of 80 km/h.

2. A 1000-kg car is traveling northeast at 70 km/h. What is the momentum of the car?

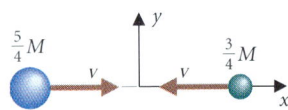

■ FIGURE 6.19

3. Which of the following is the total momentum of the system of two particles shown in ■ Figure 6.19? Explain your reasoning and what is wrong with the answers you don't choose.
(a) $2Mv$ (b) zero   (c) $\frac{1}{2}Mv\hat{i}$ (d) $2Mv\hat{i}$ (e) $\frac{1}{2}Mv$

4. A 10-metric-ton truck and an 850-kg Volkswagen are both traveling at a speed of 80 km/h. Each has brakes that can exert a force of 10^3 N. How much time is required to stop each of the vehicles?

5. An astronaut of mass 100 kg fires a maneuvering pistol and changes speed from 1 m/s to 3 m/s without changing direction. What impulse did the pistol deliver?

6. Four particles of equal mass are placed at the corners of a square. Each is given a velocity of 5 m/s along a diagonal toward the center of the square. What is the total momentum of the system?

7. In a race, three horses of mass 390 kg, 350 kg, and 420 kg are galloping along a straight stretch with speeds of 12.0 m/s, 10.0 m/s, and 9.0 m/s, respectively. What is the momentum of the system of three horses?

§6.2 USING CONSERVATION OF MOMENTUM

8. Decide whether the following sentence is true or false, and explain your reasoning: If the momentum of one part of an isolated system changes by $\Delta\vec{p}$, the momentum of the rest of the system *must* change by $-\Delta\vec{p}$.

9. Two skaters, initially at rest, push off against each other and move apart. One skater (mass 85 kg) moves off at 9.0 m/s. What is the speed of the other skater (mass 63 kg)?

10. A hockey player of mass 85 kg is initially stationary on the frictionless ice. The player delivers to a hockey puck an impulse $\Delta\vec{p} = 24$ N·s at an angle of 35° with the x-axis. Describe the motion of the player after hitting the puck.

11. A 75-kg astronaut moves toward a stationary 1248-kg satellite at 2.67 m/s and grabs onto it. What is the speed of the satellite and astronaut after they connect?

12. A Tomahawk rocket achieves a speed of 2.7×10^3 m/s with a propellant fraction of $m_{fuel}/m_i = 0.80$. What is the speed of the exhaust with respect to the rocket?

Questions and Problems

§6.1 LINEAR MOMENTUM

13. ❖ Compare the momentum of a 170-kg motor scooter moving down the highway at 60 km/h with that of an 850-kg Volkswagen moving at the same speed.

14. ❖ Three dogs of equal mass are at the corners of an equilateral triangle. Each dog runs with speed v clockwise along a side toward the next corner. What is the total momentum of the system of three dogs?

15. ❖ Five identical particles are at the vertices of a pentagon (■ Figure 6.20). The pentagon rotates, so each of the particles undergoes uniform circular motion at speed v. Which of the following best describes the total linear momentum $\vec{P}$ of the system? Explain your reasoning and state what is wrong with the answers you reject.
(a) $5mv$ (b) $mv[(3\cos 75°)\hat{i} - (2\sin 75°)\hat{j}]$ (c) zero
(d) oriented in the $-\hat{i}$ direction   (e) none of the above

16. ♦ An airplane of mass 2000 kg is traveling east at 300 km/h. What is its momentum?

17. ♦ A bowling ball of mass 10 kg is bowled down the alley at 5 m/s. What is its momentum?

18. ♦ A baseball of mass $m = 0.15$ kg falls directly downward at speed $v = 40$ m/s. What impulse $\vec{I}$ must a spectator catching the ball deliver to bring the ball to rest? If the ball is stopped in 0.01 s, what is the average force $\vec{F}_{av}$ exerted by the spectator on the ball?

19. ♦ A system consists of two balls, one of mass $\frac{1}{2}$ kg moving at 8 m/s in the positive x-direction, and one of mass $\frac{1}{3}$ kg moving at 9 m/s in the negative x-direction. What is the momentum of the system?

20. ♦ A cue strikes a pool ball at rest and remains in contact with the ball for 15 ms. If the mass of the ball is 0.20 kg and its final speed is 1.5 m/s, what was the average force exerted on the ball by the cue?

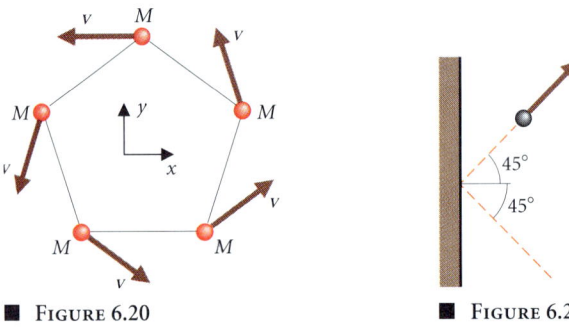

■ FIGURE 6.20   ■ FIGURE 6.21

21. ♦ What is the impulse required to stop a 0.250-kg cricket ball moving at 50.0 m/s?

22. ♦ What is the impulse required to stop a 10-kg bowling ball moving at 5 m/s?

23. ♦ A golfer putts his ball across the green. The ball's mass is 100 g and its speed after it is struck is 1 m/s. What is the magnitude of the impulse the golfer imparted to the ball?

24. ♦ Geoffrey propels a 120-g handball at a wall so that it hits with a speed of 10.0 m/s at a 45° angle to the wall. It bounces off with the same speed (■ Figure 6.21). What impulse did the wall impart to the ball? What impulse did the ball deliver to the wall?

25. ♦ A truck towing a 1000-kg car that is initially at rest, exerts a force of 1000 N for 30 s. Ignoring friction, what is the final speed of the towed car?

26. ♦ Two cars approach each other head-on. One has a mass of 1000 kg and a speed of 20 m/s. The other car has a mass of 800 kg and a speed of 25 m/s. What is the total momentum of the system of two cars?

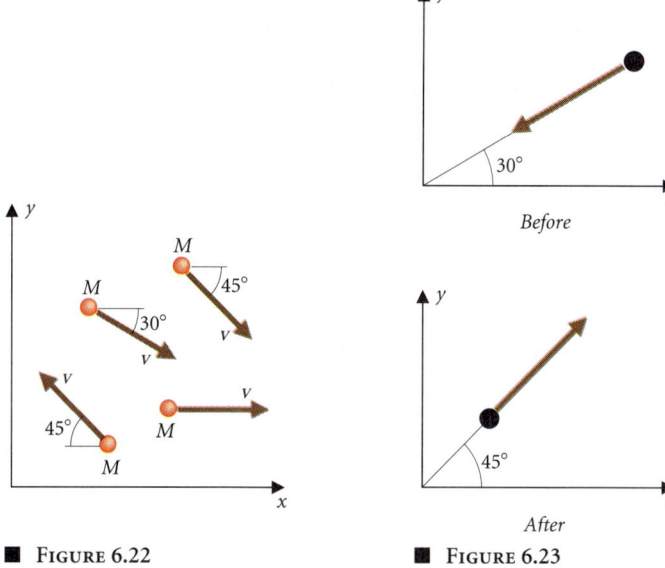

■ FIGURE 6.22

■ FIGURE 6.23

27. ◆ All the particles in ■ Figure 6.22 have the same mass M and the same speed v, but they move in different directions, as shown. What is the total momentum of the system?

28. ◆ Linda (mass 40.0 kg) is standing on an ice-covered pond when her dog (mass 15 kg) leaps from the shore at 3.0 m/s into Linda's arms. What is the speed of Linda and her dog after the catch?

29. ◆◆ A 100-kg astronaut moving at 3 m/s fires a maneuvering pistol and receives an impulse of 800 N·s at right angles to his direction of motion. What is his final velocity?

30. ◆◆ A 0.50-kg hockey puck sliding across the ice with a speed of 30.0 m/s at a 30° angle with the x-axis is hit by a player and moves away with a speed of 20.0 m/s at a 45° angle with the axis (■ Figure 6.23). What impulse did the player impart to the puck?

31. ◆◆ The force whose magnitude is graphed in ■ Figure 6.24 acts on a particle of mass 10 kg. If the particle is initially moving at 5 m/s, what is its final speed? (The force and the initial velocity have the same direction.)

32. ◆◆ The force whose magnitude is graphed in ■ Figure 6.25 acts on a particle of mass 15 kg. If the particle is initially moving at 2 m/s, what is its final speed? (The force and the initial velocity have the same direction.)

33. ◆◆ ✉ A car crashes into a wall head-on at 30 m/s and is brought to rest in 0.1 s. Estimate the average force exerted on a 90-kg test dummy by the seat belt.

34. ◆◆ A spacecraft accelerometer consists of a 0.100-kg ball mounted between two springs with spring constant $k = 1.00$ N/m (■ Figure 6.26). The ball slides without friction in a shaft. Explain how the device can measure the component of acceleration parallel to its shaft. If the ball is observed to be displaced 0.625 cm from its equilibrium position for 31.5 s, what impulse is delivered to it by the springs? What is the change in the component of the spacecraft's velocity along the shaft?

35. ◆◆ A conveyor belt in a factory is 100 m long and carries 100 wheels per minute into the assembly room. If each wheel has a mass of 20 kg and remains on the belt for 5 min, what is the total momentum of all the wheels on the belt?

36. ◆◆ An electric toy locomotive of mass 500 g, moving along the tracks at 0.5 m/s, bumps into a flat car of mass 100 g and latches onto it. What is the speed of engine and car after they connect? (Assume that no power is applied to the engine during the collision.)

37. ◆◆ A 0.6-kg steel ball falls from a height of 5 m into a box of sand. (a) What impulse does the ball deliver to the sand? (b) ✉ If the impression in the sand is 1 cm deep, estimate the magnitude of the force, assumed constant, that acts on the ball during its deceleration.

38. ◆◆◆ How much impulse is required to change the speed of a 1000-kg car from 10 m/s to 30 m/s? If the force exerted on the car is parallel to its velocity and increases linearly as $F = (1000 \text{ N/s})t$, how much time is required?

39. ◆◆◆ Three particles each of mass M lie at the corners of an equilateral triangle of side L (■ Figure 6.27). Compute the gravitational forces exerted on each particle by the other two, and show explicitly that the sum of the forces acting on the three masses is zero. (This result confirms that $\vec{F}_{int} = \Sigma \vec{F}_{ij} = 0$ for this system.)

§6.2 USING CONSERVATION OF MOMENTUM

40. ❖ When a cannon is fired, it recoils (moves backward). Why?

41. ❖ A railroad car containing an angry bull is standing on the tracks, but the brakes are not set. The bull, mass M, initially standing at one end of the car, charges to the other end and crashes into the wall. What is the final velocity of the railroad car if its mass is m and the bull runs at speed V?

42. ❖ ✉ Your friend tells you a tale of African safaris in the nineteenth century in which the hunters carried rifles so powerful that one bullet would stop a charging elephant and even throw it backwards. Could this story be true? Explain how you decide whether the story is reasonable or unreasonable.

43. ❖ Which of the following statements correctly describe why a rocket is able to accelerate in outer space? Why do you think the other statements are incorrect?

(a) Hot exhaust gases exert force on the structure of the rocket engine.

(b) There is no friction in outer space, so no force is needed to accelerate the rocket.

(c) Hot exhaust gases press on the air behind the rocket.

(d) The total momentum of the rocket and exhaust gases remains constant.

(e) There is no gravity in outer space, so no force is needed to accelerate the rocket.

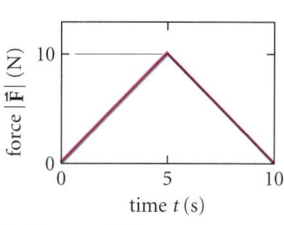

■ FIGURE 6.24

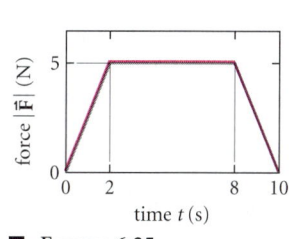

■ FIGURE 6.25

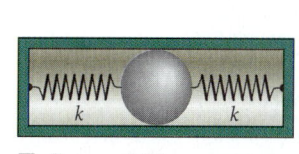

■ FIGURE 6.26

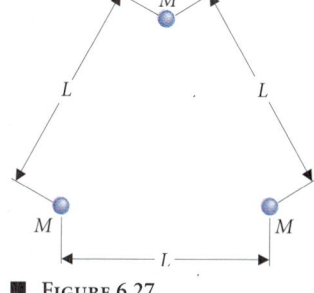

■ FIGURE 6.27

44. ❖ Ice skaters Gregory, of mass M_G, and Natasha, of mass M_N, are initially at rest. Gregory then pushes Natasha and the two separate. Which of the following statements is correct? Explain why you choose your answer, and what is wrong with the answers you reject.
(a) Gregory has the greater speed because he is the one who did the pushing.
(b) Gregory and Natasha both have the same speed.
(c) Natasha has the greater speed because she is the one who was pushed.
(d) Whoever is the more massive has the greater speed.
(e) Whoever is the less massive has the greater speed.
45. ❖ Is the total momentum of the exhausted gas from a rocket (fired in space) less than, greater than, or equal to its mass times the rocket's exhaust velocity? Why?
46. ❖ Several firefighters are often required to change the direction that an operating firehose is pointing? Why?
47. ❖ If a hose through which water is flowing develops a kink as shown in ■ Figure 6.28, the kink grows larger with time. Explain.
48. ❖ Two trucks, one of mass 6000 kg and the other of mass 10 000 kg and traveling at right angles to each other, collide and lock together (■ Figure 6.29). The larger truck's speed is 15 km/h and the smaller truck's speed is 25 km/h. In which of the directions shown in Figure 6.29 do they slide after the collision? State the reasons for your choice and why you rejected the others.

■ FIGURE 6.28

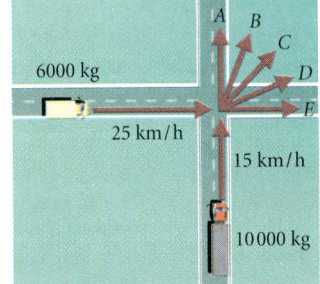

■ FIGURE 6.29

49. ❖ Two dogs are together on the surface of an icy lake and are desperate to get to shore. The dumb dog howls, but the smart dog kicks the dumb dog. Explain why this might be a smart thing to do.
50. ❖ Two boats, each of mass M, float at rest on the surface of a frictionless, still lake. A woman, of mass $m < M$, originally standing still in boat 1, suddenly jumps to boat 2 and stands still in the second boat. You may neglect any up-and-down bobbing motion of the boats. Which of the following statements correctly describes the motion of the boats relative to the lake as a result of the jump? Explain why your choice is correct and why the others are wrong.
(a) Both boats are standing still.
(b) Boat 1 is moving; boat 2 is standing still.
(c) Boat 2 is moving; boat 1 is standing still.
(d) Both boats are moving.
51. ❖ If a process is developed to increase the exhaust velocity of a rocket by a factor of 2, and its fuel to payload ratio is not changed, by what factor does the rocket's final speed change?
52. ❖ An astronaut is in space far from any large masses. She must return to her ship, which is nearby, but she has no tether line and her rocket pack has failed. She does have a tool kit with her. She throws the kit directly opposite from the direction to her ship. Which of the following statements are correct, and why?
(a) She stays where she is because she is much more massive than the tool kit.
(b) She follows the tool kit because of the velocity she has imparted to the kit.
(c) She moves in a direction opposite from that of the moving kit, since the total linear momentum vector was zero before she threw the kit.
(d) The impulse provided by the force exerted by her arms caused the kit to move.
53. ❖❖ If the astronaut in Problem 52 has mass $M = 65$ kg, her tool kit has mass $m = 11$ kg, and she throws the kit with a relative speed of 5.0 m/s, find her speed and the speed of the kit with respect to the spaceship.
54. ◆ In a factory that manufactures lead shot, a conveyor belt transports completed shot to the packaging room at 1 m/s. Shot falls onto the belt at a rate of 50 kg/s. What force is required to move the belt? (Ignore friction.)
55. ◆ A fireboat is spraying water on a burning barge at the rate of 500 gallons per minute. The hoses are aimed at an angle of 45° to the horizontal. If 1 gallon of water has a mass of 5 kg, and the speed of the water as it leaves the hose is 45 m/s, what is the horizontal component of the force required to keep the boat stationary?
56. ◆ A concrete truck pumps concrete at a rate of 12 kg/s with a speed of 0.25 m/s. What force must the pump exert?
57. ◆ An ice skater of mass 60.0 kg catches a 1.0-kg bunch of flowers thrown toward her at a speed of 5.0 m/s and ends up at rest on the ice. What was her velocity before catching the flowers?
58. ◆ At launch, the space shuttle's engines develop 3.0×10^7 N of thrust. If the speed of the exhaust gases is 15 km/s, what mass of gas is exhausted per second?
59. ◆◆ A hunter standing in a boat at rest on the lake shoots a rifle horizontally. The bullet has a mass of 50 g and travels at 350 m/s relative to the barrel. The hunter and boat together have a mass of 300 kg. Ignoring any frictional force between the boat and the water, what is the speed of the boat after the rifle is fired?
60. ◆◆ A cat chasing a catnip mouse jumps onto a rug on a polished floor. What happens? If the cat is moving horizontally at 8 m/s before landing on the rug, and at 6 m/s afterwards, find the mass of the rug plus mouse in cat-masses.

FIGURE 6.30

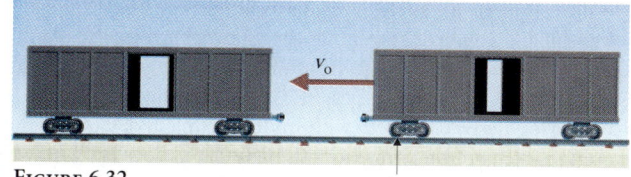

FIGURE 6.32

61. ♦♦ Two 25-g wads of bubble gum are moving along the positive x-axis (■ Figure 6.30) with speeds $v_1 = 3.5$ m/s and $v_2 = 2.2$ m/s. What is the total momentum of the system? What is the speed of the combined wad after the two collide and stick?

62. ♦♦ In a billiard game, one ball moving at 5.0 m/s strikes a second ball, which is at rest. The second ball moves off at 4.3 m/s at an angle of 30° to the direction of the first ball's original velocity. What is the *velocity* of the first ball after the collision?

63. ♦♦ Three equal-mass clowns skating in an ice show collide and stick together all at the same time. If their velocities before the collision are equal in magnitude and arranged as in ■ Figure 6.31 with $\theta = 120°$, what is the velocity of the combined clowns after the collision? What is this velocity if $\theta = 60°$?

64. ♦♦ Lucy, who is 1.30 m tall and 25 cm wide, runs through the rain at 5.2 m/s. The rain is falling straight down at a uniform rate such that at any time a cubic meter of air contains 1.2×10^3 raindrops, each with a mass of 0.13 g. Find the horizontal force acting on Lucy due to her impact with the raindrops.

65. ♦♦ A freight train consists of cars 15 m long, each capable of holding 2.0×10^4 kg of iron ore. Ore is loaded into the cars at a rate of 1.0×10^3 metric tons per hour from a hopper as the train moves slowly underneath. At what speed should the train move? Neglecting friction, what force is required to keep the train moving?

66. ♦♦ The single-stage Black Brant IIIA rocket produces a total impulse of 4.40×10^5 N·s during the 7.5 s that its engine burns. The maximum speed achieved is 1880 m/s. The launch mass is 326 kg, of which 40 kg is payload. Find the mass ratio m_i/m_f and the speed of the exhaust with respect to the rocket.
✱ Find the maximum acceleration of the rocket.

67. ♦♦ A sounding rocket launched from the Earth's surface is to achieve a final speed of 1000 m/s. If the exhaust speed of the spent fuel is 2000 m/s, what fraction of the rocket's total mass at launch must be fuel? (Assume that the engine burns rapidly enough that you may ignore any effects due to Earth's gravity during the burn.)

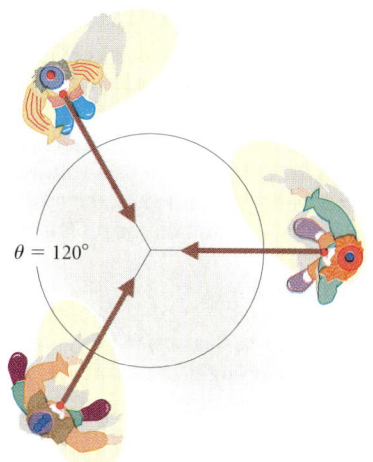

FIGURE 6.31

68. ♦♦ Two railroad boxcars with equal mass can roll on horizontal rails with the same effective coefficient of rolling friction μ_r. Initially, one car is stationary and the other is rolling toward it (■ Figure 6.32). If it didn't collide, the moving car would roll a distance d beyond the collision point before stopping. How far do the two cars actually roll after the collision before stopping?

69. ♦♦ If a rocket's final mass m_f is increased by a factor of 2 while the total mass at launch (m_i) is not changed, by how much does the final speed achieved by the rocket change?

70. ♦♦ A cart is sitting on a road when a dog jumps into it with a velocity of 3.0 m/s making an angle of 45° with the horizontal. What forces act on the cart? On the dog? Ignoring friction, if the dog has a mass of 20.0 kg and the cart has a mass of 30.0 kg, with what speed does the cart move along the road after the dog has jumped into it?

71. ♦♦♦ A spacecraft in orbit about the Earth must increase its speed by 41% to go to the Moon. How much fuel must be carried if the spacecraft is to have a mass of 3.0×10^3 kg upon reaching the Moon and the exhaust speed v_{ex} is 0.20 times the orbital speed? Assume that the engine burn is rapid enough that you may ignore any acceleration of the spacecraft due to Earth's gravity during the burn.

72. ♦♦♦ Sand falls into a container on one side of a balance at a rate of 40 g/s from a height of 30 cm, and leaks out of the bottom of the container at the same rate. If 10 g of sand are in the container at any time, what mass must be placed on the other side of the balance for equilibrium?

73. ♦♦♦ In a multistage rocket, parts of the rocket structure are dropped behind when their fuel content has been used. In a two-stage rocket, suppose the first stage has a total mass 5 times that of the second stage. In each stage, fuel is 90% of the initial mass; 10% is payload plus structure. Compare the speed of the second stage at burnout with what it would be if the first stage were not dropped.

74. ♦♦♦ A lunar mail canister is fired from Copernicus Base (see Study Problem 6) at speed $v_o = \sqrt{2gD}$ at an angle θ with the horizontal. At the top of its trajectory, explosive bolts fire, which cause one-half of the canister to return to the launch point. Where does the other half of the canister land?

75. ♦♦♦ If a rocket's exhaust speed is increased by a factor of 2, how much more payload can it carry to the same final speed?

Additional Problems

76. ❖ Explain why the lunar postmaster's scheme (Study Problem 6) will not work unless the distance to the observatory is greater than the maximum range the mass driver can achieve without the scheme.

77. ❖ When a box is placed on a scale, the reading is 4.0 kg. Make a sketch showing roughly how the scale reading varies as a function of time when the box is dropped onto the scale.

78. ❖ An astronaut swings a 10-kg rivet hammer at 5 m/s onto a rivet on the space station. The astronaut's feet and other hand hold him to the station. What impulse must the station deliver to the hammer when it strikes the rivet? Draw a series of sketches showing qualitatively the impulses delivered to the astronaut, to the space station,

and to the hammer (**a**) as the hammer is lifted back, (**b**) as the hammer is propelled forward, and (**c**) as the hammer strikes the rivet.

79. ❖ Discuss the following newspaper editorial, which ridiculed early experimental work by Robert H. Goddard on rocket propulsion: "That Professor Goddard, with his 'chair' in Clark College and the countenancing of the Smithsonian Institution, does not know the relation of action to reaction, and of the need to have something better than a vacuum against which to react—to say that would be absurd. Of course, he seems only to lack the knowledge ladled out daily in high schools."

80. ❖ Theodora, relaxing on a raft in the middle of a lake, plans to move the raft by turning a propeller to blow air into the raft's sail, which is set perpendicular to the air flow from the fan. Decide whether the method will actually propel the raft. A good analogy is to imagine Theodora throwing things at the sail. Contrast the effect of throwing mud at a limp sail versus throwing superballs at a rigid board. Which of these seems to you a better model of how air blown by the fan would behave? Does the scheme work if the sail is *not* perpendicular to the air flow? Which way does the raft move in this case? Would it be smarter to turn the fan around to blow air toward the rear of the boat? (See the Letters section in *The Physics Teacher*, October 1986, page 392, for a lively discussion of the problem.)

81. ◆◆ A circus performer jumps on a springboard and is projected directly upward at speed v_o. At height $h = v_o^2/(4g)$ her identical twin is standing on a platform and the two clasp each other as they pass. What height do the twins reach after they join?

82. ◆◆ Three equal-mass astronauts are conferring about a construction problem on the space station. Because one of them has an intercom failure, they are holding their helmets in contact. They push off from each other to return to work, one of them with velocity $\vec{V}$, and a second with velocity ($0.80V$ at an angle of $60°$ to $\vec{V}$). What is the velocity of the third astronaut?

83. ◆◆ A firehose projects a horizontal stream of water at a vertical wall. If the hose delivers $0.20 \text{ m}^3/\text{s}$ at a speed of 25 m/s, what mechanical force does the water stream exert on the wall? (Assume the water comes to rest at the wall and does not rebound.)

84. ◆◆ A machine designed for quality testing in a handball factory propels the 0.15-kg balls toward a wall with a velocity of 9.5 m/s at an angle of $60°$ from the normal to the wall surface (■ Figure 6.33). A typical ball rebounds at $70°$ from the normal with a speed of 9.0 m/s.

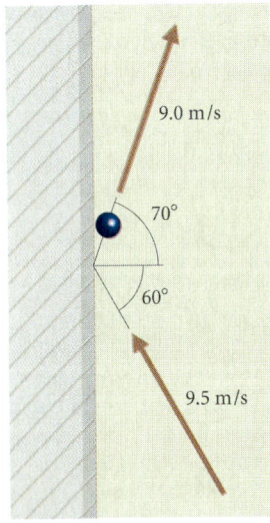

■ FIGURE 6.33

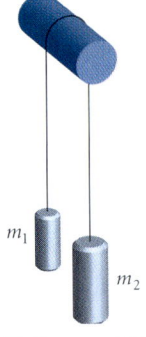

■ FIGURE 6.34

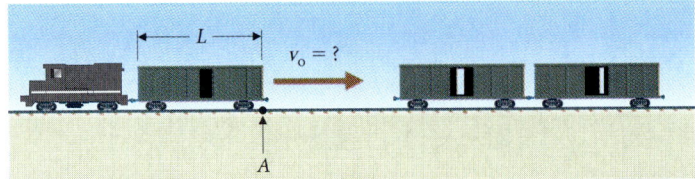

■ FIGURE 6.35

What impulse does a typical ball deliver to the wall? If the ball is in contact with the wall for 4.9 ms, what average force acts on the wall during a ball's impact? If the balls are fired at a rate of one ball per second, what is the average force acting on the wall over an interval of 1 s?

85. ◆◆ Two ice skaters, each of mass $M = 75$ kg, are initially at rest on a frictionless ice surface. One skater throws a 0.65-kg basketball to the other, who catches it. (**a**) If the skater imparts an impulse of $I = 20.0$ N·s to the basketball, find the speed of each skater after the ball is thrown but before the second skater catches it. (**b**) What are the skaters' speeds after the ball is caught? (**c**) What difference (if any) does it make if the skaters, instead of being stationary, are originally gliding parallel to each other and perpendicular to the line between them?

86. ◆◆◆ A circus star of mass 75 kg is to be launched from ground level by a spring-loaded device so that she clears a wall 3.0 m high and 4.0 m away and lands on the ground 5.0 m behind the wall. What is the minimum impulse the spring must deliver to achieve the desired trajectory?

87. ◆◆◆ Two cylinders with masses m_1 and $m_2 > m_1$ are suspended by an ideal string over a frictionless support (■ Figure 6.34). If the objects are released from rest, find their total momentum as a function of time, and show that this momentum equals the impulse delivered by the net external force acting on the system.

88. ◆◆◆ A space transport vehicle of mass M carries N robot asteroid mining units, each of mass m. Each robot unit is designed to launch itself from the transport at a relative velocity $\vec{v}_r$. If the transport is originally at rest with respect to the asteroid, which procedure will result in the greater final speed of the transport, launching the robot units all at the same time or one at a time? Find the final speed of the transport in each case.

89. ◆◆◆ A switch engine operator working at a railroad yard intends to connect three boxcars of equal mass and equal length L for transport to a factory. The cars have a coefficient of rolling friction μ_r on the horizontal rails of the yard. If the operator releases a freight car at point A (■ Figure 6.35) with speed v_o, how far will it roll before coming to rest? The second car is released from A with the same speed v_o; where do the two connected cars come to rest? The third car is again released with speed v_o. What should that speed be for the rear end of the third car to come to rest at point A?

90. ◆◆◆ ✉ In a crash safety test, two 1200-kg cars, moving in opposite directions at 30 m/s, collide head-on. Estimate the magnitude of the force acting on the cars, and compare your result with an estimate of the mass that can be supported on the roof of a car without crushing it. (*Hint:* A person can sit on the roof. Could an elephant?)

Computer Problem

91. The magnitude of a force that acts in the *x*-direction is measured at intervals of 0.200 s, beginning at clock reading $t = 0.000$ s and ending at $t = 4.000$ s. The measured values are listed in file CH6P91 on the supplementary computer disk. The force acts on a particle that

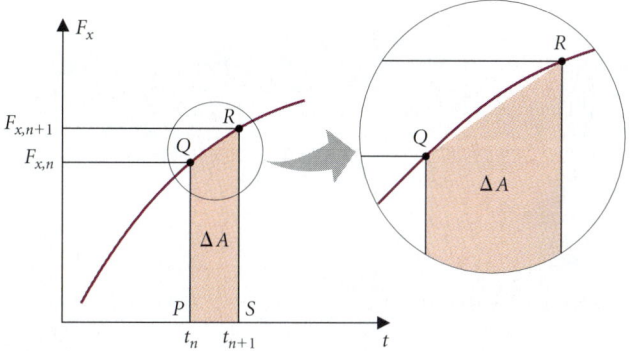

FIGURE 6.36

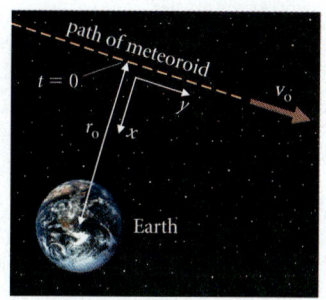

FIGURE 6.37

is initially at rest. Evaluate the x-component of the particle's momentum as a function of time. Use the following method. (a) Derive the trapezoidal rule. We have values of the function at equally spaced values of time: $F_{x,n} = F_x(n\,\Delta t)$ for integer values of n from $n = 0$ to $n = N = 20$. Approximating the curve $F_x(t)$ as a straight line for the small time interval Δt (■ Figure 6.36), the area under the curve equals the area of the trapezoid $PQRS$:

$$\Delta A \approx \frac{(\Delta t)(F_{x,n+1} + F_{x,n})}{2}.$$

Thus show that the total area under the curve between $t = 0$ and $t = N\,\Delta t = 4.000$ s is:

$$A = \int_0^{N\Delta t} F_x(t)\,dt \approx \frac{\Delta t}{2}\left(F_{x,0} + 2\sum_{n=1}^{n=N-1} F_{x,n} + F_{x,N}\right).$$

(b) Use a spreadsheet program to calculate the impulse delivered to the particle between $t = 0$ and $t = N\,\Delta t$ for each value of N from 1 to 20, and plot the particle's momentum as a function of time.

Challenge Problems

92. The force acting on an object varies in time according to the formula:

$$\vec{F}(t) = F_o(\hat{\mathbf{i}}e^{-\alpha t} + \hat{\mathbf{j}}e^{-\beta t}), \quad \text{for } t \geq 0.$$
$$\vec{F}(t) = 0, \quad \text{for } t < 0.$$

What is the impulse delivered to the object by the force between $t = -\infty$ and $t = +\infty$?

93. The force acting on an object varies in time according to the formula:

$$\vec{F}(t) = \frac{F_o}{1 + t^2/t_o^2}\hat{\mathbf{i}}.$$

What is the impulse delivered to the object by the force between $t = -\infty$ and $t = +\infty$?

94. A meteoroid of mass m passes Earth with a speed v_o so rapid that the meteoroid's path is closely approximated as a straight line whose least distance from Earth is r_o (■ Figure 6.37). Show that the gravitational force acting on the meteoroid as a function of time is:

$$\vec{F}(t) = \frac{GMm}{(r_o^2 + v_o^2 t^2)^{3/2}}(r_o\hat{\mathbf{i}} - v_o t\hat{\mathbf{j}}),$$

where M is the mass of the Earth and $t = 0$ when the meteoroid is at its minimum distance r_o from the center of the Earth. Find the impulse delivered to the meteoroid by Earth's gravity, and find the angle by which the direction of the meteoroid's velocity is altered.

[*Hints:* Use a symmetry argument to show that the impulse has no y-component. Use the substitution $t \equiv (r_o \tan\theta)/v_o$ to simplify the integral for the x-component of the impulse.]

95. A sounding rocket with an exhaust speed of 2000 m/s is to achieve a final speed of 1000 m/s (cf. Problem 67). Assume that the rocket motor fires for 100 s. What effect on the rocket's speed does gravitational acceleration have during the 100-s burn? What initial mass ratio is needed for its final speed actually to be 1000 m/s? In what time interval must the fuel be burned to limit the effect of the gravitational acceleration on the necessary mass ratio to less than 5%? Compare your answer with a burn time of 7.5 s for the Black Brant IIIA rocket (Problem 66). Is it reasonable to ignore gravity in that problem?

96. An angry bull is at one end of a railroad car at rest on the tracks. The car's brakes are not set, but there is an effective coefficient of rolling friction μ_r between the freight car's wheels and the tracks. The car's length is L and the bull runs at speed v to the opposite end and crashes into the wall. Find the speed of the freight car after the bull crashes into the wall and the distance the freight car rolls before stopping. (Assume the stunned bull lies slumped on the floor of the car.)

97. An hourglass with a total mass of 10.5 kg contains 10.0 kg of sand (■ Figure 6.38). When the glass is started, the first grains of sand fall 20.0 cm. The last grains, at the end of the hour, fall 5.0 cm. The hourglass is placed on a sensitive scale. Describe how the reading of the scale varies during the hour the sand is falling. Give values for the scale readings before the sand begins to fall; when it has started falling, but the first grains have not yet reached the bottom; while the sand is piling up on the bottom; and (whew!) between the time the last grains leave the upper chamber and when they land in the bottom chamber.

FIGURE 6.38

A Boeing 777 jetliner rotates its nose skyward and climbs into the air.

CHAPTER 7
Work and Kinetic Energy

CONCEPTS

Work
Kinetic energy
Work-energy theorem
Dot product
Power
Machine
Mechanical advantage
Efficiency

GOALS

Be able to:
Compute the work done by a known force acting through a known displacement.

Compute the kinetic energy of a particle or system of particles.

Find the mechanical advantage of simple machines.

*Most of the change we think we see in life
Is due to truths being in and out of favor.*

ROBERT FROST

The takeoff of a jet airliner is an impressive sight. The airliner rotates its nose upward and its powerful engines thrust it into the sky—or do they? The engines produce a smaller upward force component than does aerodynamic lift acting on the plane's wings (■ Figure 7.1). It is the air itself that pushes the plane upward. Though the engines don't *push* the plane into the sky, they are what *causes* it to climb. An engine causes forces to occur by transforming and transferring *energy*. We shall now begin to study these processes.

In Chapter 6 we found that linear momentum is a conserved physical property. That is, momentum is possessed by a moving body; it can be transferred between bodies as forces occur; and its total amount in an isolated system cannot change. *Energy*, like momentum, is a conserved physical property. Unlike momentum, however, energy exists in a number of different forms, so its description has both a greater richness and complexity. In one of Galileo's inclined-plane experiments, described in Chapter 0, a metal sphere rolls back and forth, down one ramp along a horizontal surface and then up another ramp. In today's language, the sphere possesses energy that alternates between gravitational and *kinetic* forms. Friction slowly transforms the energy into thermal form—the apparatus gets ever so slightly warmer as the ball slows down.

Work is the name given to a *process* of energy exchange and transformation. Our job now is to describe this process together with several forms of energy and to demonstrate that energy is a conserved property. In this chapter we shall describe the kinetic form of energy—energy of motion—and work, the transfer process. In Chapter 8 we continue this discussion, which culminates with the law of conservation of energy, one of the most fundamental principles of physics.

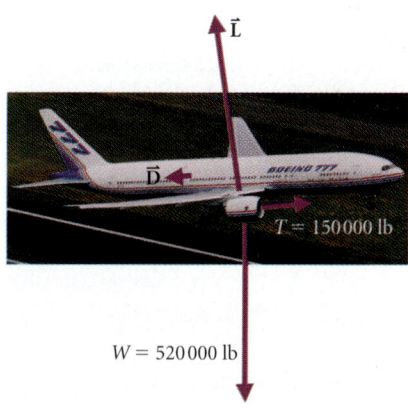

■ **FIGURE 7.1**
Free-body diagram for the departing jetliner. The values given for the magnitudes of the forces acting are for a Boeing 777 making a normal departure at sea level. The aircraft's path slopes at 15° with the horizontal.

7.1 ENERGY AND ITS TRANSFER

7.1.1 Work and Kinetic Energy in One-Dimensional Motion

In Chapter 6 we showed how force results in momentum transfer—a result called the impulse-momentum theorem. When a force $\vec{F}$ acts on a particle during a *time interval dt*, the resulting impulse $\vec{F}\,dt$ causes a change $d\vec{p} = \vec{F}\,dt$ in the particle's momentum. The particle undergoes a displacement $d\vec{s} = \vec{v}\,dt$ during the same time interval. We can obtain an alternative description of the process by asking what happens when a force acts on a particle undergoing a *spatial* displacement. We're looking for a physical property of the moving object—its energy—that depends on its mass and velocity and that is changed by the force acting through the displacement.

The simplest case we can use to test this idea is that of a constant force $\vec{F}$ acting on a particle of mass m with initial velocity $\vec{v}_i$ parallel to $\vec{F}$ (■ Figure 7.2). Taking the x-axis in the direction of $\vec{F}$, the force $\vec{F}$ causes a constant acceleration $\vec{a}$ with x-component:

$$F_x/m = a_x = dv_x/dt.$$

In uniformly accelerated linear motion, the final speed $\vec{v}_f$ of the particle is found from its initial speed $\vec{v}_i$, its displacement Δx, and its acceleration a_x, according to eqn. (2.13):

$$v_f^2 = v_i^2 + 2a_x\,\Delta x.$$

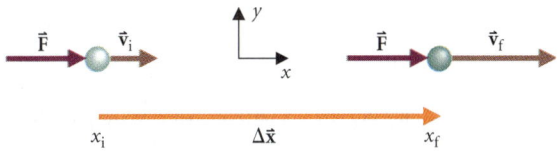

■ **FIGURE 7.2**
A constant force $\vec{F}$ acts on a particle moving with velocity $\vec{v}$ parallel to the force. We take the x-axis in the common direction of $\vec{F}$ and $\vec{v}$. As the particle undergoes a displacement Δx, the force delivers energy to the particle, measured by the work done, $W = F_x\,\Delta x$. The transfer of energy results in a change in the kinetic energy possessed by the particle.

So
$$F_x\,\Delta x = ma_x\,\Delta x = m(v_f^2 - v_i^2)/2 = \tfrac{1}{2}mv_f^2 - \tfrac{1}{2}mv_i^2$$

$$F_x\,\Delta x = \Delta(\tfrac{1}{2}mv^2). \tag{7.1}$$

The force acting through the displacement Δx changes the quantity $\tfrac{1}{2}mv^2$—the kinetic energy possessed by the object.

> The *kinetic energy* K of a particle in a given reference frame is one-half its mass m multiplied by the square of its speed v measured in that reference frame:
> $$K = \tfrac{1}{2}mv^2. \tag{7.2}$$

The expression $F_x\,\Delta x$ on the left side of eqn. (7.1) is the amount of energy transferred to the object in the process—the work done by the force.

> In one-dimensional motion, the *work* done by a constant force with one component F_x on a particle that undergoes a displacement Δx is the product of the force and displacement:
> $$W = F_x\,\Delta x.$$

Notice that the change of kinetic energy can be an increase or a decrease. If the force is parallel to the displacement, the object speeds up: work is positive and kinetic energy increases. A force in the opposite direction from the displacement slows the object down: its kinetic energy decreases; energy is removed from the object; the work is negative.

The unit for both work and kinetic energy is the newton-meter (N·m). It is named the joule (symbol J) after James P. Joule (1818–1889), an English physicist who made important early contributions to the idea of energy conservation.

> A force of 1 newton acting through a displacement of 1 meter does 1 joule of work:
> $$1\ \mathrm{N\cdot m} \equiv 1\ \mathrm{J}.$$

From our very special case of one-dimensional motion, we have found a first statement of the *work-energy theorem for a particle*.

> The work done on a particle equals the change in its kinetic energy:
> $$W = \Delta K \equiv K_f - K_i. \tag{7.3}$$

Next we shall look at some examples that help us become familiar with the concepts of work and energy. Then we can develop a statement of the theorem that is no longer restricted to constant forces and motion along a straight line.

EXAMPLE 7.1 ♦ A spacetug (■ Figure 7.3) exerts a constant force of 1.50×10^6 N on a spaceship of mass 2.25×10^5 kg over a distance of 12 km. The ship is initially at rest. How much work is done by the force? What is the change in the ship's kinetic energy?

MODEL We compute work from force and displacement. The change in kinetic energy follows from the work-energy theorem. We choose the x-axis to be along the direction of the tug's motion.

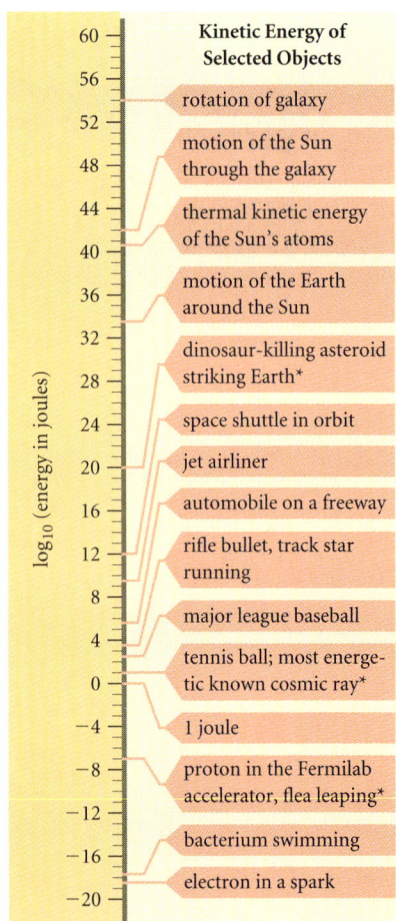

Kinetic energies of selected objects. The rotational energy of our galaxy is 76 orders of magnitude greater than the smallest energy shown on this logarithmic scale. Entries marked * are taken from a similar table by H. Keller in *The Physics Teacher*, **30** *Nov. 1992:*455.

REMEMBER: CONSTANT FORCE IS A SPECIAL CASE. WE'LL GIVE A GENERAL DEFINITION OF WORK IN §7.1.4.

IN CONTRAST WITH LINEAR MOMENTUM, KINETIC ENERGY IS A SCALAR QUANTITY. BOTH QUANTITIES HAVE THE SAME DEPENDENCE ON MASS, BUT THEY HAVE DIFFERENT DEPENDENCE ON SPEED.

SETUP The work is the product of the force and the displacement.

SOLVE
$$W = F_x \Delta x = (1.50 \times 10^6 \text{ N})(12 \times 10^3 \text{ m})$$
$$= 1.80 \times 10^{10} \text{ N·m} = 1.80 \times 10^{10} \text{ J}.$$

The ship's kinetic energy increases by 1.80×10^{10} J.

ANALYZE The ship was initially at rest, so its final kinetic energy equals the change of kinetic energy—that is, 1.80×10^{10} J.

EXERCISE 7.1 ♦ Show that $1 \text{ kg·m}^2/\text{s}^2 = 1 \text{ N·m}$.

EXAMPLE 7.2 ♦♦ A ball of mass 1.2 kg is thrown upward with an initial speed of 13 m/s. How high does it go?

MODEL Gravitational force acts downward on the ball, which is moving upward. So, gravitational force removes energy from the ball, which reaches its highest point when its speed is zero—when all its kinetic energy has been removed.

SETUP We choose the positive y-axis to be upward. Then the ball's weight has only one component: $F_y = -mg$. The displacement Δy is positive, so the work done is negative. In the equation for the work-energy theorem, set the final kinetic energy K_f equal to zero.

SOLVE From the definition of work, $W = F_y \Delta y = -mg \Delta y$. The work-energy theorem gives:
$$-mg \Delta y = W = K_f - K_i = 0 - \tfrac{1}{2}mv_i^2 = -\tfrac{1}{2}mv_i^2,$$

since the final kinetic energy is zero. Thus:
$$\Delta y = \frac{v_i^2}{2g} = \frac{(13 \text{ m/s})^2}{2(9.8 \text{ m/s}^2)} = 8.6 \text{ m}.$$

ANALYZE This type of problem is one of the first we learned to solve. The work-energy theorem places it in a new perspective and gives a new method for solving it.

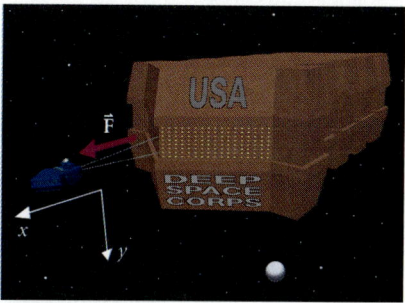

■ **FIGURE 7.3**
A spacetug exerts force on a space craft, thus transferring energy to the craft. The force and the velocity are both in the x-direction.

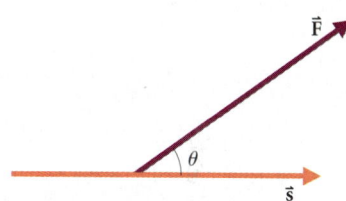

■ **FIGURE 7.4**
The work done by a force acting at an angle θ with the displacement of a particle is the magnitude of the force multiplied by the component of the displacement parallel to the force. That component equals the magnitude of the displacement multiplied by the cosine of the angle θ.

7.1.2 Work in More Than One Dimension

How can we express the work done when the force acting on a particle isn't parallel to the particle's displacement? Uniform circular motion gives us an important clue. In uniform circular motion, a particle has constant speed and thus constant kinetic energy; no energy is transferred to the particle, which means that no work is done on the particle. A force acts and the particle undergoes a displacement, but the force acts toward the center of the circle and the particle is moving tangent to the circle, *perpendicular* to the force.

We have discovered that work should be defined as the product of force with the particle's displacement *in the direction of the force*. This product is zero in uniform circular motion. The same rule fits our results for linear motion: if force and displacement are parallel, the work is the product of their magnitudes; when force and displacement are in opposite directions, work is minus the product of their magnitudes. In general, the component of displacement $\vec{s}$ in the direction of the force $\vec{F}$ is $s_\parallel = s \cos \theta$, where θ is the angle between $\vec{F}$ and $\vec{s}$ (■ Figure 7.4): $W = Fs \cos \theta$.

SEE §3.2.6: ACCELERATION PARALLEL TO VELOCITY CHANGES SPEED; ACCELERATION PERPENDICULAR TO VELOCITY CHANGES DIRECTION. KINETIC ENERGY DEPENDS ONLY ON SPEED AND NOT ON DIRECTION.

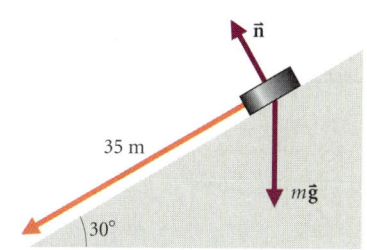

■ **FIGURE 7.5**
A hockey puck slides down an icy slope. How much work does gravity do on the puck?

EXAMPLE 7.3 ♦ A hockey puck of mass 0.50 kg slides 35 m down an icy slope inclined at 30° with the horizontal (■ Figure 7.5). How much work does gravity do on the puck? If it starts with a speed of 9.0 m/s, what is the puck's speed at the bottom of the slope?

MODEL The displacement $\vec{s}$ of the puck is at a 60° angle to the gravitational force, and we have to include the angle in calculating work done on the puck. From the work-energy theorem, we find the final kinetic energy and thus the final speed.

SETUP The puck's weight is $\vec{F}_g = m\vec{g}$, so the work done by gravity is:

$$W = F_g s \cos\theta = mgs \cos 60°$$

SOLVE
$$= (0.50 \text{ kg})(9.8 \text{ m/s}^2)(35 \text{ m})(\tfrac{1}{2}) = 86 \text{ J}.$$

SETUP The change in the puck's kinetic energy is:

$$\Delta K \equiv \Delta(\tfrac{1}{2}mv^2) = \tfrac{1}{2}mv_f^2 - \tfrac{1}{2}mv_i^2,$$

where $v_i = 9.0$ m/s is the initial speed of the puck and v_f is the final speed, whose value we wish to find.

SOLVE Since Δ(kinetic energy) = work done by gravity:

$$\tfrac{1}{2}m(v_f^2 - v_i^2) = W.$$
$$v_f^2 = v_i^2 + 2W/m = (9.0 \text{ m/s})^2 + 2(86 \text{ J})/(0.50 \text{ kg})$$
$$v_f = \sqrt{424 \text{ m}^2/\text{s}^2} = 21 \text{ m/s}.$$

ANALYZE Compare this method with that of Example 4.10 (§4.6.2), where we used Newton's second law to calculate the acceleration down the slope. Note that the normal force is perpendicular to the puck's displacement and so does no work on the puck. ∎

EXAMPLE 7.4 ❖ A laborer carries a 50-kg bag a distance of 100 m from the north end to the south end of a warehouse. How much work does the laborer do while walking at constant speed?

MODEL Assuming the floor of the warehouse to be level, the displacement of the bag is horizontal. The force exerted by the laborer on the bag balances the weight of the bag and so is vertical. Since the laborer's force is exerted perpendicular to the bag's displacement, the component of displacement parallel to the force is zero. The laborer does zero work on the bag!

ANALYZE Should the boss refuse to pay a salary since the laborer does no work? Absurd! In physics, work is carefully defined by its relation to kinetic energy. Simply exerting force requires the laborer's muscles to use the day's lunch and thus counts as work in the common usage of the word. The boss should pay, and you should be aware of the difference between the technical and common usages of the word *work*. ∎

A laborer carries a heavy bag horizontally between two locations. Does the laborer deserve pay despite having done zero work?

The relation among work, force, and displacement is a kind of multiplication that is used frequently in physics. It is called the scalar (or dot) product of two vectors.

The *scalar (dot) product* of two vectors $\vec{a}$ and $\vec{b}$ is the product of their magnitudes multiplied by the cosine of the angle between them (■ Figure 7.6):

$$\vec{a} \cdot \vec{b} = ab \cos\theta.$$

This product is a scalar since it is constructed from the lengths of vectors and an angle between them, none of which depends on the choice of coordinate system. The scalar product is most commonly called the *dot product*, after the "dot" used to express it symbolically.

The *work* done by the constant force $\vec{F}$ acting on a particle that undergoes a displacement $\vec{s}$ is the dot product of the force with the displacement:

$$W = \vec{F} \cdot \vec{s}. \tag{7.4}$$

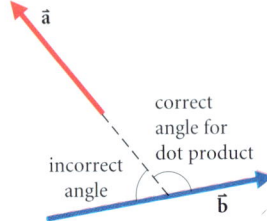

■ **FIGURE 7.6**
The dot product of two vectors $\vec{a}$ and $\vec{b}$ is $ab \cos\theta$. The angle θ is the angle between the two vectors when they are placed tail to tail. This figure shows how to choose the correct angle when the vectors are not placed tail to tail.

Math Toolbox

Properties of the Scalar Product

I: The dot product does not depend on the order of multiplication The dot product of two vectors, in either order, is the same product of three ordinary numbers:

$$\vec{a} \cdot \vec{b} = ab \cos \theta = ba \cos \theta = \vec{b} \cdot \vec{a}.$$

II: Component Description of the Dot Product In ■ Figure 7.7 the vector $\vec{a}$ is shown as the sum of component vectors $\vec{a}_{\|b}$ and $\vec{a}_{\perp b}$. Letting $\vec{b}$ define the positive direction, the component of $\vec{a}$ parallel to $\vec{b}$ is:

$$a_{\|b} = a \cos \theta.$$

The dot product $\vec{a} \cdot \vec{b}$ is the magnitude of $\vec{b}$ multiplied by the component of $\vec{a}$ parallel to $\vec{b}$.

■ Figure 7.8 shows $\vec{b}$ as the sum of component vectors perpendicular and parallel to $\vec{a}$; we see that:

$$b_{\|a} = b \cos \theta.$$

Equivalently, $\vec{a} \cdot \vec{b}$ equals the magnitude of $\vec{a}$ multiplied by the component of $\vec{b}$ parallel to $\vec{a}$.

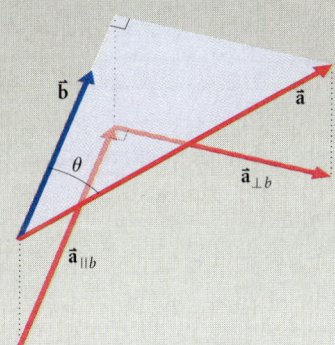

■ **Figure 7.7**
The vector $\vec{a}$ is the sum of two vectors, one perpendicular to, and the other parallel to, a second vector $\vec{b}$.

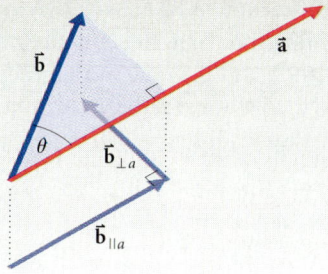

■ **Figure 7.8**
The vector $\vec{b}$ may also be resolved into components perpendicular and parallel to $\vec{a}$.

For example, the components of a vector in a coordinate system are the dot products of the vector with the unit vectors of the system:

$$\vec{a} \cdot \hat{i} = |\hat{i}| \text{ (component of } \vec{a} \text{ in direction of } \hat{i})$$
$$= (1)(a_x)$$
$$= a_x.$$

III: Perpendicular Vectors The dot product of two perpendicular vectors is zero since $\cos 90° = 0$. This fact provides a useful test:

Two nonzero vectors are perpendicular if and only if their dot product is zero.

IV: Parallel Vectors Parallel vectors have a zero angle between them, so their dot product is just the product of their lengths. The dot product of a vector with itself gives a useful expression for the vector's length:

$$a^2 = \vec{a} \cdot \vec{a}.$$

V: Dot Products of Unit Vectors The unit vectors of a coordinate system have unit magnitude and are perpendicular to each other. Thus, from properties III and IV:

$$\hat{i} \cdot \hat{i} = \hat{j} \cdot \hat{j} = \hat{k} \cdot \hat{k} = 1; \text{ and } \hat{i} \cdot \hat{j} = \hat{j} \cdot \hat{k} = \hat{k} \cdot \hat{i} = 0.$$

VI: Distributive Law for the Dot Product Suppose $\vec{a}$ is the sum of $\vec{b}$ and $\vec{c}$, and we wish to consider the dot product of $\vec{a}$ with an arbitrary vector $\vec{d}$ (■ Figure 7.9):

$$\vec{a} \cdot \vec{d} = |\vec{d}| \text{ (component of } \vec{a} \text{ parallel to } \vec{d}) = d(OP).$$

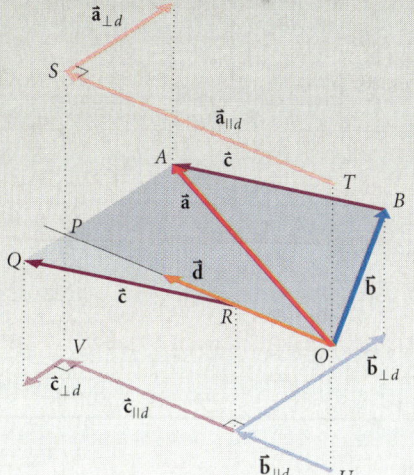

■ **Figure 7.9**
The dot product of a vector sum $\vec{a} = \vec{b} + \vec{c}$ with a vector $\vec{d}$ is the magnitude of $\vec{d}$ multiplied by the component of the sum parallel to $\vec{d}$. The components of the vectors in the sum add to give the overall component parallel to $\vec{d}$. The individual components are the lines OR and RP. Their sum is OP, the component of the vector sum parallel to $\vec{d}$. In the diagram, $\vec{c}$ has been moved to the line RQ to show more clearly that RP is its component parallel to $\vec{d}$. The components of $\vec{a}$, $\vec{b}$ and $\vec{c}$ are drawn in parallel planes above and below the main diagram.

continued

In §1.6.1 we showed that the component of a sum of vectors equals the sum of the components of the individual vectors:

Component of $\vec{a}$ parallel to $\vec{d}$
$$= \text{component of } \vec{b} \text{ parallel to } \vec{d}$$
$$+ \text{component of } \vec{c} \text{ parallel to } \vec{d}.$$

From Figure 7.9:
$$\vec{b} \cdot \vec{d} = d(OR),$$
$$\vec{c} \cdot \vec{d} = d(RP),$$
and
$$OP = OR + RP.$$
Thus, $d(OP) = d(OR) + d(RP);$
or $\vec{a} \cdot \vec{d} = \vec{b} \cdot \vec{d} + \vec{c} \cdot \vec{d}.$

So, the dot product follows the **distributive law**:
$$(\vec{b} + \vec{c}) \cdot \vec{d} = \vec{b} \cdot \vec{d} + \vec{c} \cdot \vec{d}.$$

VII: Components and the Dot Product We may now express the dot product of two vectors in terms of their components. We show the method in two dimensions for clarity. The result generalizes easily to three dimensions. Expressing the two vectors $\vec{a}$ and $\vec{b}$ in terms of their components:
$$\vec{a} \cdot \vec{b} = (a_x \hat{i} + a_y \hat{j}) \cdot (b_x \hat{i} + b_y \hat{j}).$$

Applying the distributive law:
$$\vec{a} \cdot \vec{b} = a_x \hat{i} \cdot (b_x \hat{i} + b_y \hat{j}) + a_y \hat{j} \cdot (b_x \hat{i} + b_y \hat{j}).$$

Using the distributive law again, with property I:
$$\vec{a} \cdot \vec{b} = a_x b_x (\hat{i} \cdot \hat{i}) + a_x b_y (\hat{i} \cdot \hat{j})$$
$$+ a_y b_x (\hat{j} \cdot \hat{i}) + a_y b_y (\hat{j} \cdot \hat{j}).$$

Using the values for dot products of unit vectors:
$$\vec{a} \cdot \vec{b} = a_x b_x (1) + a_x b_y (0) + a_y b_x (0) + a_y b_y (1).$$
$$\vec{a} \cdot \vec{b} = a_x b_x + a_y b_y.$$

Generalizing to three dimensions:
The dot product of two vectors equals the sum of the products of their x-, y-, and z-components.
$$\vec{a} \cdot \vec{b} = a_x b_x + a_y b_y + a_z b_z.$$

EXERCISE 7.2 ♦♦ Write expressions for the position $\vec{r}$ and velocity $\vec{v}$ of a particle in uniform circular motion as functions of time. Show by computing their dot product that $\vec{r}$ and $\vec{v}$ are perpendicular. (See Math Topic "Use of calculus in circular motion" §3.2.4.)

VIII: Differentiating a Dot Product The component expression for the dot product is a powerful tool for proving theorems about vectors. For example, consider the derivative of a dot product of two vectors that lie in the x–y plane:

$$\frac{d}{dt}(\vec{a} \cdot \vec{b}) = \frac{d}{dt}(a_x b_x + a_y b_y)$$
$$= \frac{da_x}{dt} b_x + a_x \frac{db_x}{dt} + \frac{da_y}{dt} b_y + a_y \frac{db_y}{dt}$$
$$= \frac{da_x}{dt} b_x + \frac{da_y}{dt} b_y + a_x \frac{db_x}{dt} + a_y \frac{db_y}{dt}$$
$$= \frac{d\vec{a}}{dt} \cdot \vec{b} + \vec{a} \cdot \frac{d\vec{b}}{dt}.$$

This rule for differentiating a dot product is similar to the rule for differentiating the product of two ordinary functions.

EXERCISE 7.3 ♦♦ Demonstrate the identity:
$$\frac{d}{dt}(v^2) = 2\vec{v} \cdot \frac{d\vec{v}}{dt}.$$

Use this result and the expression for $\vec{v}$ from Exercise 7.2 to verify that the speed is constant in uniform circular motion.

7.1.3 Work Done When More Than One Force Acts

The work-energy theorem states that the change in a particle's kinetic energy equals the work done by the net force acting on the object. When several forces act on an object at the same time, we would expect the total energy they transfer to be the sum of the energy transfers by each of the individual forces. That is, we should be able to compute the work done by any force as if that force were acting alone and then find the total work by adding the results. For example, where two forces $\vec{F}_1$ and $\vec{F}_2$ occur, the net force is the vector sum:

$$\vec{F}_{net} = \vec{F}_1 + \vec{F}_2.$$

The total work done over a displacement $\vec{s}$ is:

$$W = \vec{F}_{net} \cdot \vec{s} = (\vec{F}_1 + \vec{F}_2) \cdot \vec{s}.$$

The distributive law for the dot product (property VI) allows us to expand the sum into two terms:

$$W = \vec{F}_1 \cdot \vec{s} + \vec{F}_2 \cdot \vec{s} = W_1 + W_2,$$

where W_1 and W_2 are the amounts of work done by the individual forces. As expected, their sum equals the net work done. We could extend the demonstration easily to any number of individual forces:

$$W_{net} = \sum_i W_i.$$

EXAMPLE 7.5 ♦ Two astronauts push symmetrically on a satellite (■ Figure 7.10). They each exert a force of 215 N at 30° to the direction of the satellite's motion. If the satellite moves 8.00 m, how much work does each astronaut do?

MODEL Each astronaut does an amount of work computed from the force exerted individually by that astronaut. Because their action is symmetric, the work done by each is the same. We choose the y-axis to be in the direction of the satellite's motion.

SETUP AND SOLVE $W_1 = \vec{F}_1 \cdot \vec{s} = (215 \text{ N})(\cos 30°)(8.00 \text{ m}) = 1.49 \times 10^3 \text{ J}.$

ANALYZE The net force exerted on the satellite is $\vec{F}_{net} \equiv \vec{F}_1 + \vec{F}_2$. The x-components of the two forces cancel and their y-components add:

$$\vec{F}_{net} = 2F \cos 30° \hat{j} = 2(215 \text{ N})\sqrt{3}/2 \hat{j}.$$

The total work done by the two astronauts is:

$$W = \vec{F}_{net} \cdot \vec{s} = (215\sqrt{3} \text{ N})\hat{j} \cdot (8.00 \text{ m})\hat{j}$$
$$= (2.98 \times 10^3 \text{ N·m})(\hat{j} \cdot \hat{j}) = 2.98 \times 10^3 \text{ J}.$$

This equals the sum of the separate amounts of work done by the individual astronauts. ■

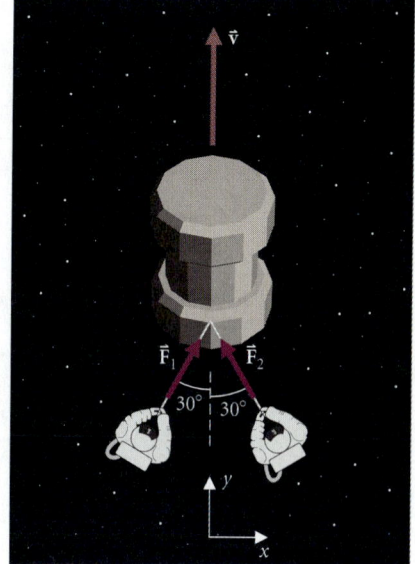

■ FIGURE 7.10
Two astronauts exert forces of equal magnitude on a satellite.

EXAMPLE 7.6 ♦♦ A horse pulls a sleigh and rider of total mass 210 kg along the snow (■ Figure 7.11). The coefficient of kinetic friction between sleigh and snow is $\mu_k = 0.20$, and the traces are at an angle of 10° with the horizontal. How much work does the horse do in pulling the sleigh 250 m at constant speed?

MODEL Since the sleigh moves at constant speed, the net force acting on it is zero. The total work done on the sleigh is zero, but the horse cares about the work *it* is asked to do. Thus we need to find the force the horse exerts, which is equal to the tension in the traces. We choose the y-axis to be vertical and the x-axis to be horizontal.

SETUP From the free-body diagram for the sleigh (■ Figure 7.12), we have:

Vertical components $\sum F_y = n + T \sin \theta - mg = 0.$
Horizontal components $\sum F_x = T \cos \theta - f = 0.$
Kinetic friction $f = \mu_k n.$

SOLVE Combining these equations:

$$T \cos \theta = f = \mu_k n = \mu_k(mg - T \sin \theta).$$

Thus: $$T = \frac{\mu_k mg}{\cos \theta + \mu_k \sin \theta}.$$

The work done by the horse is:

$$W = \vec{T} \cdot \vec{s} = Ts \cos \theta = \frac{\mu_k mg}{\cos \theta + \mu_k \sin \theta} s \cos \theta$$

$$= \frac{(0.20)(210 \text{ kg})(9.8 \text{ m/s}^2)(250 \text{ m})\cos 10°}{\cos 10° + (0.20)\sin 10°} = 9.9 \times 10^4 \frac{\text{kg·m}^2}{\text{s}^2}.$$

$$= 9.9 \times 10^4 \text{ J}.$$

■ FIGURE 7.11
How much work must the horse do to draw the sleigh 250 m over the snow?

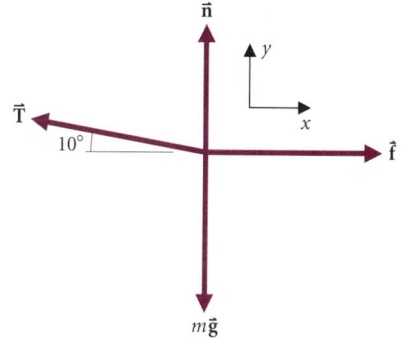

■ FIGURE 7.12
Free-body diagram for the sleigh.

ANALYZE Gravity and the normal force both act perpendicular to the displacement, so neither does work on the sleigh. The friction force is directed opposite the displacement and so does negative work on the sleigh. This negative work adds to the work done by the horse to give the total of zero. Energy added to the sleigh by the horse is removed by friction, and there is no change of kinetic energy.

7.1.4 Formal Proof of the Work-Energy Theorem for a Particle

THE MATHEMATICAL TECHNIQUES NEEDED FOR ARBITRARY PATHS AND VARIABLE FORCES ARE DISCUSSED IN INTERLUDE 2.

In our use of the work-energy theorem so far, we have only considered examples where the object moves along a straight line subject to constant forces. In some important cases, such as circular motion, an object's velocity changes direction continuously; in other cases, the force may change with time in both magnitude and direction. We shall now give a formal proof of the work-energy theorem valid for motion along an arbitrary path, subject to arbitrarily variable forces.

Suppose a particle of mass m is subject to a force whose value as a function of time is $\vec{F}(t)$.[1] During any finite displacement of the particle, the force is not constant, and we *cannot* choose a single value of $\vec{F}$ to calculate the work done. But, we can imagine the particle's path divided into a huge number of small elementary displacements $d\vec{s}$. Each individual elementary displacement is small enough that we can neglect any variation of the force while the particle undergoes that displacement (■ Figure 7.13). Then, the energy transferred to the particle during the elementary displacement is:

$$dW = \vec{F} \cdot d\vec{s}.$$

The displacement of the particle during a time interval dt is:

$$d\vec{s} = \vec{v}(t)\, dt.$$

So, the work done on the particle in a time interval dt is:

$$dW = \vec{F}(t) \cdot \vec{v}(t)\, dt.$$

Using Newton's second law, we may replace the force acting on the particle with its mass times its acceleration:

$$\vec{F}(t) = m\vec{a}(t) = m\, d\vec{v}/dt.$$

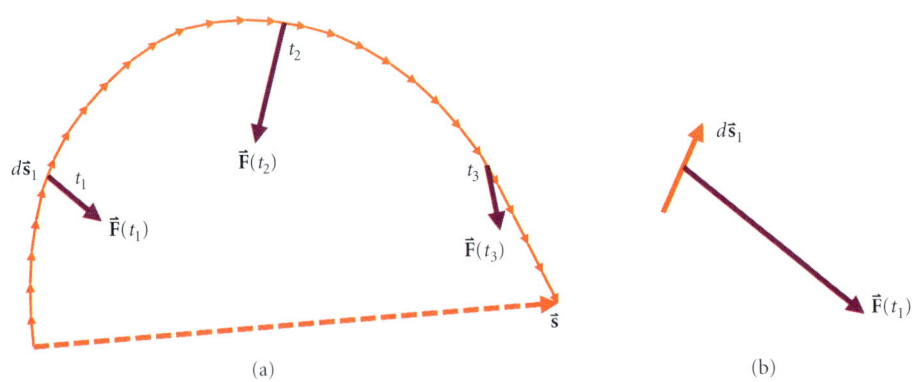

■ **FIGURE 7.13**
(a) The path of a particle may be divided into a large number of elementary displacements $d\vec{s}$. (b) During an elementary displacement, variation of the force may be neglected, and the work done on the particle during the displacement may be expressed as $dW = \vec{F} \cdot d\vec{s}$.

---

[1] The force acting on an object can vary for many reasons. Gravitational force on a satellite depends on the satellite's position; air resistance on a parachutist depends on the person's speed; the force you can exert in a game of tug-of-war decreases as you tire out. Regardless of the reason for its variation, we can imagine measuring the force at frequent times and giving our results as the function $F(t)$.

Then:
$$dW = \vec{F} \cdot d\vec{s} = m\frac{d\vec{v}}{dt} \cdot \vec{v}\, dt.$$

As in Exercise 7.3, we may express $\vec{v} \cdot (d\vec{v}/dt)$ as a single time derivative:

$$\frac{d\vec{v}}{dt} \cdot \vec{v} = \tfrac{1}{2}\frac{d}{dt}(v^2).$$

REMEMBER: $v^2 = \vec{v} \cdot \vec{v}$.

Thus:
$$dW = m\,\tfrac{1}{2}\frac{d}{dt}(v^2)\, dt = \frac{d}{dt}(\tfrac{1}{2}mv^2)\, dt.$$

WE MAY MOVE m INSIDE THE DERIVATIVE SINCE IT IS CONSTANT IN TIME.

The total work done over a finite displacement is the sum of these differential elements of work:

$$W = \int dW = \int_{\text{path of particle}} \vec{F} \cdot d\vec{s}. \tag{7.5a}$$

The overall change of kinetic energy is the sum of the differential changes that occur on each piece of the particle's path:

$$W = \int_{t_i}^{t_f} \frac{d}{dt}(\tfrac{1}{2}mv^2)\, dt = \tfrac{1}{2}mv_f^2 - \tfrac{1}{2}mv_i^2. \tag{7.5b}$$

Thus we arrive at the work-energy theorem for a particle:

> The work done by the net force acting on a particle that undergoes a displacement equals the net change in the particle's kinetic energy during that displacement.

7.1.5 Kinetic Energy of a System of Particles

In Chapter 6 we found it possible to define the total linear momentum of a system of particles simply by adding the momentum vectors of the individual particles in the system. This procedure makes sense because we view momentum, like mass, as a physical property possessed by individual particles. Kinetic energy is also such a physical property, and we would expect the total kinetic energy of a system to be a simple sum of the individual particle energies. To illustrate this fact, let's look at a system that can be viewed either as a single object or as a system of two objects and apply the work-energy theorem to it from both points of view.

EXAMPLE 7.7 ♦♦ Two astronauts, each of mass M, are returning from a construction assignment (■ Figure 7.14). They are connected by an ideal cord and are initially at rest. One of the astronauts is connected to the space station by an ideal cord. To start the astronauts back to the station, a motor exerts a constant tension force T_1 on this cord as it reels in an amount of cord s. Find the net work done on each astronaut during this process and so find the kinetic energy of each. Show that the sum of the two kinetic energies equals the total work done on them as a system.

MODEL We may think of the astronauts either as separate particles or as a system of two connected particles. We want to find the work done and kinetic energy changes in the two views in terms of the known quantities M, s, and T_1.

SETUP To compute the net work done on either of the individual astronauts, we must first solve a dynamics problem to find the tension T_2 in the connecting cord. From the free-body diagram for the second astronaut (■ Figure 7.15):

$$\sum F_{x_2} = Ma_{x_2} \Rightarrow T_2 = Ma_{x_2}.$$

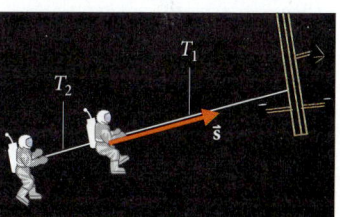

■ FIGURE 7.14
Two astronauts are pulled back to the space station. A motor exerts tension on their tether cord until they are displaced a distance s. Energy is transferred from the motor via the cord to the first astronaut, and some is then transmitted to the second astronaut via their connecting cord. The energy is transformed into kinetic energy of the astronauts. The system's kinetic energy is the sum of the astronauts' individual kinetic energies.

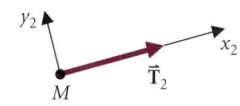

■ FIGURE 7.15
Free-body diagram for the second astronaut.

FIGURE 7.16
Free-body diagram for the first astronaut.

For the first astronaut (■ Figure 7.16):

$$\sum F_{x_1} = Ma_{x_1} \Rightarrow T_1 - T_2 = Ma_{x_1}.$$

The cord connecting the two astronauts ensures that they have equal acceleration:

$$a_{x_1} = a_{x_2} \equiv a.$$

SOLVE From these relations, we find:

$$T_1 = Ma_{x_1} + Ma_{x_2} = 2Ma \Rightarrow a = T_1/(2M) \quad \text{and} \quad T_2 = T_1/2.$$

The net work done on astronaut 2, and hence his final kinetic energy, is:

$$W_2 = T_2 s = T_1 s/2.$$

The net work done on astronaut 1, equal to her final kinetic energy, is:

$$W_1 = (T_1 - T_2)s = T_1 s/2.$$

The total work done on the system is:

$$W_T = T_1 s = W_1 + W_2.$$

This is equal to the sum of the two kinetic energies, as required.

ANALYZE If we think of this system as a single object of mass $2M$ subject to the force T_1, then the total work $T_1 s$ should transfer the total kinetic energy of the system. The cord from the station transfers all of the energy to astronaut 1. The cord connecting her to her colleague removes some of the energy from her and transfers it to him. The total energy transfer is thus parceled out in the form of the individual kinetic energies of the two astronauts. The image of energy as physically real stuff that can flow about and change form but cannot be created or destroyed seems odd at first, but is a powerful tool for visualizing how systems interact. ■

7.2 Power and Simple Machinery

7.2.1 Power

When a horse pulls a sleigh and rider along a snowy path, we may want to know how quickly the horse can accomplish this task. The rate at which a horse, or a person, or a motor does work is referred to as *power*. Practical devices for doing work are frequently limited much more by their maximum power than by the total work they can do. A person running full tilt up a flight of stairs produces more power than a working draft horse, but only for a minute or so. The same person hiking in the mountains produces far less power than when running up stairs, but can hike all day, expending a good deal more energy than in a minute of stair climbing.

The unit of power is named the watt (symbol W) after James Watt (1736–1819), a British engineer who invented the first truly successful steam engine.

The *power* delivered by a system is the rate at which the system does work:

$$P = dW/dt. \tag{7.6}$$

One *watt* is the power produced by a device that does 1 joule of work in 1 second:

$$1 \text{ W} = 1 \text{ J/s}.$$

Digging Deeper

THE HORSEPOWER

Watt originally defined a unit of power called the *horsepower*, which he intended to be the power produced by a typical working horse; 1 horsepower lifts 550 lb at a rate of 1 ft/s:

$$1 \text{ hp} = 746 \text{ W}.$$

This is actually substantially more power than a typical horse can produce for a long period of time. Either Watt used rather phenomenal horses, or he measured the power they could produce over a short time interval (■ Figure 7.17).

■ **FIGURE 7.17**
This painting shows horses and an early steam engine working at a mine head. Watt measured the work output of such horses in order to determine the horsepower and so compare the power of a steam engine with that of a horse.

EXAMPLE 7.8 ♦ If a horse does 9.9×10^4 J of work pulling a sled 250 m in 2.2 min, what average power does the horse produce?

MODEL The required average power is the work done by the horse divided by the time interval in which it is done.

SETUP AND SOLVE
$$P = \frac{W}{\Delta t} = \frac{9.9 \times 10^4 \text{ J}}{(2.2)(60 \text{ s})} = 750 \text{ J/s} = 0.75 \text{ kW}.$$

ANALYZE The horse could probably do this task but most likely would not be able to produce this amount of power for an extended period. ■

EXERCISE 7.4 ♦ How much power do you need (a) to run up a flight of stairs 10 m high in 10 s and (b) to climb a mountain 1 km high in 3 h? ■

Let's think for a moment about moving into a new apartment at the top of a long flight of stairs. If you've packed your belongings into large, heavy boxes, you'll carry each one slowly up the stairs and rest often; your power output will be small, even though you are exerting a large force on each box as you lift it. If, instead, you carry small items up one at a time, you can move much faster but you will waste most of your energy lifting your own body. Again, your useful power output is small. The best way to use your ability as a power source is to pack your items in medium-sized boxes that you can carry up at a steady rate. Your power output as a function of box mass is plotted in ■ Figure 7.18. The figure is an example of a

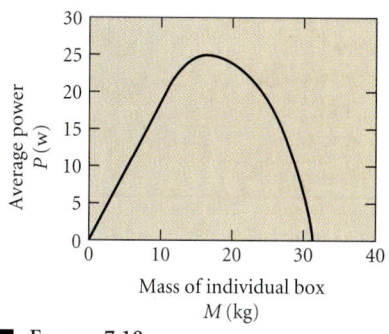

■ **FIGURE 7.18**
The average power a student can deliver over a period of several hours depends on the load to be carried on each trip up the stairs. A student with this schematic power curve would be most effective lifting 17-kg boxes.

SEE THE ESSAY ON THE BICYCLE FOR A DISCUSSION OF HOW A BICYCLE IS DESIGNED TO MATCH A HUMAN POWER SOURCE.

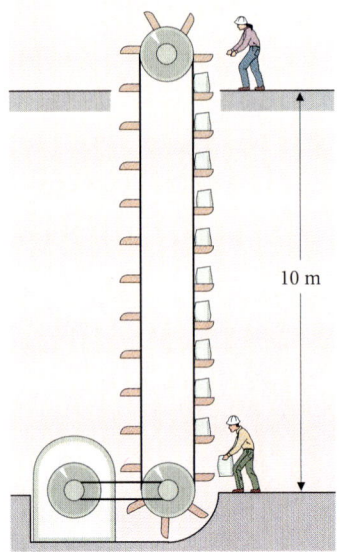

FIGURE 7.19
A belt-drive system lifts bags of flour directly from the warehouse floor to a storage area 10 m above. The belt is kept moving by a motor.

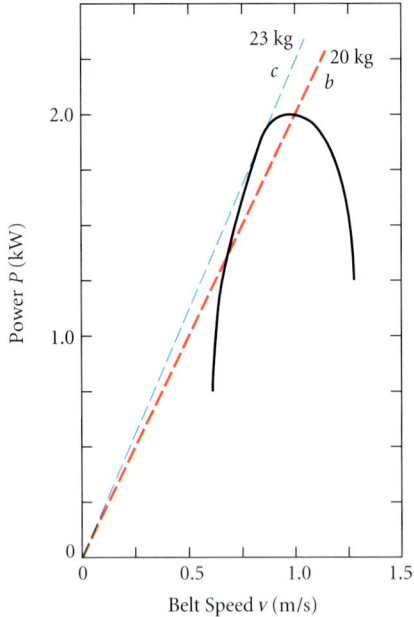

FIGURE 7.20
The power curve for the motor peaks at a belt speed of 1.0 m/s. The straight lines labeled b and c show the power required to lift bags of mass 20 kg and 23 kg as functions of belt speed. Line b passes through the peak of the motor's power curve; with 20-kg bags, the motor can operate at maximum power. Line c is tangent to the power curve. For bags with mass greater than 23 kg, the system cannot operate at all.

power curve. Similar power curves describe how the measured output of a power source depends on a parameter such as load or operating speed.

EXAMPLE 7.9 A motor in a warehouse is used to turn a belt that lifts bags of flour from the warehouse floor to a storage area 10.0 m above the floor (■ Figure 7.19). The power output of the motor versus the speed of the belt is plotted in ■ Figure 7.20.
(a) ◆◆ What is the maximum rate (in kilograms per second) at which the system can deliver flour to the storage area?
(b) ◆◆ What mass should each bag have if the maximum delivery rate is to be achieved?
(c) ◆◆◆ What is the maximum bag-mass that the system can handle?

MODEL The system has to exert a force on each bag equal to its weight and has also to displace the bag vertically a distance $H = 10$ m. Thus the work done on a bag of mass M is $W_{bag} = MgH$ (cf. Example 7.2). The power required from the system is the work to lift each bag multiplied by the number of bags per second delivered to the storage area. From Figure 7.19, we see that there are ten bags on the belt at any one time. All ten bags are delivered to the storage area in the time for the belt to move a distance H at speed v: $\Delta t = H/v$.

$$P_{req} = \frac{(10 \text{ bags})(W_{bag})}{\Delta t} = \frac{10MgH}{H/v} = 10Mgv.$$

Thus P_{req} is proportional to the belt speed v.

SETUP Ten bags are delivered to the upper floor in time $\Delta t = H/v$. Thus the rate at which flour is delivered is:

$$\frac{dm}{dt} = \frac{(10 \text{ bags})(M)}{\Delta t} = \frac{10Mv}{H} = \frac{P_{req}}{gH}.$$

SOLVE A Flour is transported at the maximum rate when the required power P_{req} equals the maximum power output of the motor: 2.0 kW (Figure 7.20). Thus the maximum rate is:

$$\left(\frac{dm}{dt}\right)_{max} = \frac{P_{max}}{gH} = \frac{2.0 \times 10^3 \text{ W}}{(9.8 \text{ m/s}^2)(10.0 \text{ m})} = 20 \text{ kg/s}.$$

SETUP At the maximum delivery rate, the required power equals the maximum available: $P_{req} = 10Mvg = P_{max}$.

SOLVE B Since the maximum power, $P_{max} = 2.0$ kW, occurs at belt speed $v = 1.0$ m/s:

$$M = \frac{2.0 \times 10^3 \text{ W}}{10(1.0 \text{ m/s})(9.8 \text{ m/s}^2)} = 20 \text{ kg}.$$

SETUP For a given bag mass, the power required by the system is proportional to the belt speed and is represented by a straight line on a graph of P versus v. Points where this line intersects the power curve indicate values of v at which the required power equals the motor's power output—that is, speeds at which the system can operate. The slope of the line is proportional to the mass of an individual flour bag. Too great a mass results in a line too steep to intersect the power curve at all. The limiting mass corresponds to line c, which is tangent to the power curve.

SOLVE C Reading off the graph, line c has:

$$\text{Slope} = \frac{\text{rise}}{\text{run}} = \frac{1.9 \text{ kW}}{0.85 \text{ m/s}} = 10M_{max}g.$$

$$M_{max} = \frac{1.9 \times 10^3 \text{ W}}{10(0.85 \text{ m/s})(9.8 \text{ m/s}^2)} = 23 \text{ kg}.$$

ANALYZE Check the units:

$$\frac{W}{m^2/s^3} = \frac{(J/s)\cdot s^3}{m^2} = \frac{N\cdot m\cdot s^2}{m^2} = \frac{kg\cdot(m/s^2)\cdot s^2}{m} = kg.$$

This system would not win any prizes for inspired engineering design, but it does illustrate why the details of a practical system have to be carefully matched to the characteristics of its power source.

EXERCISE 7.5 ◆ The formula for power required from the system in Example 7.9 is the product of the speed v and the force ($10Mg$) exerted by the system. Check the generality of this result. Specifically, (a) show that the unit of power is the product of the units of force and velocity; and, (b) using an intermediate result from the proof of the work-energy theorem, show that the power delivered to a particle with velocity $\vec{v}$ by a force $\vec{F}$ is:

$$P = \vec{F} \cdot \vec{v}. \tag{7.7}$$

7.2.2 Simple Machines

We are surrounded by machines. Most of us ride to work, entrust our clothes to a washer, and use elevators. Similar mechanical devices are essential to the mining and manufacturing industries that support a modern economy. A machine serves to multiply forces, to transmit energy, and generally to match an energy source to its task. In the rest of this section we shall study a few simple machines and describe how well they perform.

A machine's input is where it connects to the energy source that causes it to function. The output is where the machine performs the task that is its purpose.

> The *mechanical advantage* of a simple machine is the magnitude of the force it exerts at its output divided by the magnitude of force exerted on it at its input:
>
> $$MA = \frac{F_O}{F_I}. \tag{7.8}$$

EXAMPLE 7.10 ◆◆ A ramp is often considered as a machine whose purpose is the lifting of heavy objects. What is the mechanical advantage of a frictionless ramp whose angle with the horizontal is $\theta = 45°$?

MODEL The ramp must operate together with a cable if anything is to be lifted (■ Figure 7.21). It is the combination of ramp and cable that we treat as a machine.

SETUP The input force $\vec{F}_I$ produces a tension in the cable, which exerts a force $\vec{T}$ on the crate being lifted. The crate moves at constant velocity, so the net force on it is zero: $\vec{T}$ together with the normal force $\vec{n}$ balance the crate's weight. We choose coordinates with x-axis parallel to the ramp, and y-axis perpendicular to the ramp. The equation for force balance is:

$$\sum \vec{F} = \vec{T} + \vec{n} + m\vec{g} = 0. \tag{i}$$

The x-component of this equation is:

$$\sum F_x = T - mg \sin \theta = 0. \tag{ii}$$

SOLVE From eqn. (ii), the input force is:

$$|F_I| = |T| = mg \sin \theta.$$

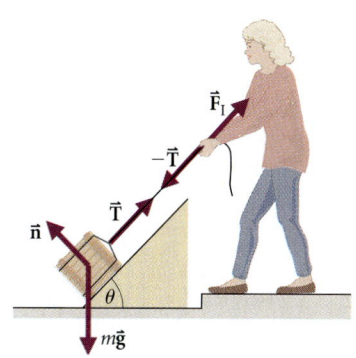

■ **FIGURE 7.21**
A frictionless ramp forms a simple machine. Its input is the cable used to pull objects up the ramp. Its output is the sum of cable tension and normal force exerted by the ramp, which together balance the object's weight.

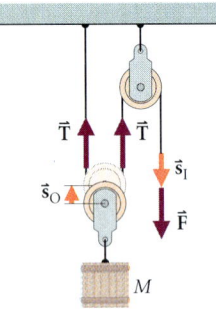

FIGURE 7.22
A pulley system. Energy is transferred to the machine by the force exerted at the input end of the cord. Energy flows through the machine and is transferred to the load at the output end of the machine.

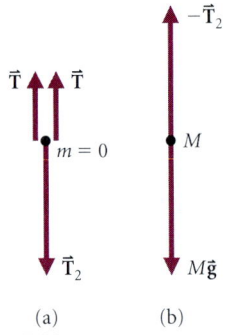

FIGURE 7.23
Free-body diagrams for (a) the movable pulley and (b) the crate in Figure 7.22.

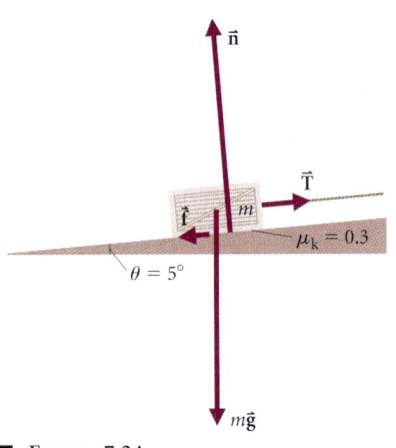

FIGURE 7.24
(a) A ramp with friction is a machine that is not perfectly efficient.

The output force is the force lifting the crate, here the vector sum of the tension in the rope and the normal force. From eqn. (i), the output force is:

$$|F_O| = |\vec{T} + \vec{n}| = |-m\vec{g}| = mg.$$

Then the mechanical advantage is:

$$MA \equiv \frac{F_O}{F_I} = \frac{mg}{mg \sin \theta} = \csc \theta = \csc 45° = 1.4.$$

ANALYZE The shallower the ramp's slope, the greater is the mechanical advantage, but the further the crate has to be dragged.

7.2.3 Energy Transmission by Machines

A second function of machines is to transmit energy. For example, the frictionless pulley system shown in ■ Figure 7.22 lifts the crate without acceleration, and the forces on the crate are balanced. From the free-body diagrams for the crate and the movable pulley (■ Figure 7.23), the tension in the rope is $T = Mg/2$. The input force balances the tension in the rope, so $F_I = T = Mg/2$. The output force of the machine, $F_O = T_2$, balances the crate's weight, so $F_O = Mg$. The system's mechanical advantage is:

$$MA = \frac{F_O}{F_I} = \frac{Mg}{Mg/2} = 2.$$

To compare input and output work, we need a relation between the displacements. If the input end moves downward an amount s_I, the length of string between the fixed pulley and the fixed end decreases by s_I. Each of the two vertical segments of rope supporting the movable pulley decreases in length by $s_I/2$. That is, the output end of the machine is displaced upward by $s_O = s_I/2$.

Work input: $W_I = \vec{F}_I \cdot \vec{s}_I = (Mg/2)s_I = Mgs_I/2.$

Work output: $W_O = \vec{F}_O \cdot \vec{s}_O = Mg(s_I/2) = Mgs_I/2.$

The input work equals the output work. The energy input is transmitted through the pulley system and is transferred to the crate at the end of the machine.

The pulley system and the ramp we have considered are *ideal* machines; in our models of them, we have neglected any processes such as friction that divert input energy from the output end of the machine. A real machine diverts energy by transforming some of the input to nonuseful forms. Minimizing this energy *dissipation* is a major factor in machine design. The standard measure of energy dissipation is a machine's efficiency.

> The *efficiency* of a machine is the ratio of its work output to the work input:
> $$e \equiv \frac{W_O}{W_I}. \tag{7.9}$$

EXAMPLE 7.11 ♦♦ What is the efficiency of a ramp at an angle $\theta = 5°$ with the horizontal if the coefficient of friction between ramp and object is $\mu_k = 0.3$?

MODEL The tension pulling an object up the ramp balances both the component of gravity down the ramp and the friction force, which opposes the object's motion (■ Figure 7.24). As before, the output force is the sum of all the forces balancing the object's weight.

SETUP We choose the x-axis to be parallel to the slope. Then with an object of mass m on the ramp, the force balance equations are:

$$\sum F_x = T - mg \sin \theta - f = 0$$

and
$$\sum F_y = n - mg \cos\theta = 0.$$

The relation between f and n is $f = \mu_k n$.
Combining these relations, we find:
$$T = mg \sin\theta + \mu_k mg \cos\theta.$$

The work done by this input force in pulling the rope a distance $s_I = s$ up the ramp is:
$$W_I = \vec{F}_I \cdot \vec{s}_I = Ts = mgs(\sin\theta + \mu_k \cos\theta).$$

The crate also moves a distance s up the ramp. Thus the work output of the ramp system is (Figure 7.24b):
$$W_O = \vec{F}_O \cdot \vec{s}_O = mgs \cos(90° - \theta) = mgs \sin\theta.$$

SOLVE The efficiency of the ramp is:
$$e_{ramp} = \frac{W_O}{W_I} = \frac{mgs \sin\theta}{mgs(\sin\theta + \mu_k \cos\theta)} = \frac{1}{1 + \mu_k \cot\theta}$$
$$= \frac{1}{1 + 0.3 \cot 5°} = 0.2.$$

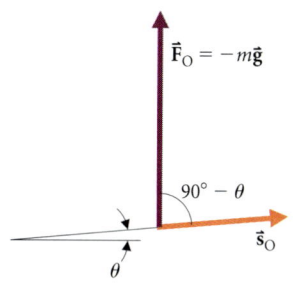

■ **FIGURE 7.24**
(b) The angle between the output force and the displacement is $90° - \theta$.

NOTICE THAT e IS LESS THAN UNITY WHENEVER $\mu_k > 0$.

ANALYZE The efficiency is a dimensionless number and is independent of the object being pulled up the ramp or the distance it is raised. An efficiency of 20% is poor but well within the range of real systems. ■

EXERCISE 7.6 ♦ Treating an airliner as a machine similar to a ramp and using the data in Figure 7.1, find the efficiency of the airliner at lifting its mass into the air. ■

Chapter Summary

Where Are We Now?

We have begun the study of energy by identifying one of its forms—kinetic energy possessed by moving bodies—and the process by which it is transferred—work. Together with power, the rate of doing work, these concepts are the foundation upon which a technological society is built. We are ready to identify other forms of energy and to derive the law of energy conservation.

What Did We Do?

The kinetic energy of a particle equals one-half its mass multiplied by its speed squared:
$$K = \tfrac{1}{2}mv^2.$$

Forces transfer energy by doing work—that is, by having components that act parallel to a displacement of the particle. The work done by a constant force $\vec{F}$ during a displacement $\vec{s}$ is expressed by the dot product $\vec{F} \cdot \vec{s}$.

When several different forces act on a particle at the same time, the total work done is the sum of the amounts of work done by each individual force. These ideas are summarized by the *work-energy theorem* for a particle.

The work done by the total force acting on a particle that undergoes a displacement equals the net change in the particle's kinetic energy during that displacement:
$$W = \int_{\text{path of particle}} \vec{F} \cdot d\vec{s} = \Delta(\tfrac{1}{2}mv^2).$$

Power is the rate of doing work. Practical devices are often limited by their maximum power output rather than by the total amount of work they can do.

The unit of work and of kinetic energy is the joule, equal to the work done by a force of 1 newton acting through a displacement of 1 meter. The unit of power is the watt, a work output of 1 joule per second.

$$1 \text{ J} = 1 \text{ N·m} \quad \text{and} \quad 1 \text{ W} = 1 \text{ J/s}.$$

Machines multiply force and transmit energy. The mechanical advantage of a machine is the ratio of its output force to the input force. The efficiency of a machine is the ratio of output work to input work.

Mathematical Results

The dot product $\vec{a} \cdot \vec{b}$ of two vectors $\vec{a}$ and $\vec{b}$ is the product of their magnitudes multiplied by the cosine of the angle between them.

$$\vec{a} \cdot \vec{b} = ab \cos \theta.$$

In components:
$$\vec{a} \cdot \vec{b} = a_x b_x + a_y b_y + a_z b_z.$$

The dot product of two vectors is a scalar, independent of the coordinate system used.

The derivative of a dot product is:

$$\frac{d}{dt}(\vec{a} \cdot \vec{b}) = \frac{d\vec{a}}{dt} \cdot \vec{b} + \vec{a} \cdot \frac{d\vec{b}}{dt}.$$

Practical Applications

Every technological process involves transformation of energy from one form to another, and the provision of energy in a useful form is one of the basic activities of every society. With the concepts of work and kinetic energy, we have begun a study of energy that will continue throughout the text.

Power is an equally important concept for practical design. Cars require adequate power to go up steep hills, and amplifiers need power to coax sound from speaker systems. Energy and power requirements are important constraints on the design of most useful devices or in estimating the cost of most construction projects.

Simple machines find diverse uses in our everyday lives, and also act as components in more complex machines.

Solutions to Exercises

7.1 $1 \text{ kg·m}^2/\text{s}^2 = 1(\text{kg·m/s}^2)\cdot\text{m} = 1 \text{ N·m}$.

7.2 This is an exercise in applying the mathematical rules for dot products. We begin with the expressions for $\vec{r}$ and $\vec{v}$ from Chapter 3 (eqns. 3.4 and 3.6):

$$\vec{r}(t) = R[\hat{i}(\cos \omega t) + \hat{j}(\sin \omega t)]$$

and
$$\vec{v}(t) = \omega R[\hat{i}(-\sin \omega t) + \hat{j}(\cos \omega t)].$$

Then:
$$\vec{r} \cdot \vec{v} = \{R[\hat{i}(\cos \omega t) + \hat{j}(\sin \omega t)]\}$$
$$\cdot \{\omega R[-\hat{i}(\sin \omega t) + \hat{j}(\cos \omega t)]\}.$$

Factoring out the magnitudes of the two vectors, we find:
$$\vec{r} \cdot \vec{v} = \omega R^2 \{[\hat{i}(\cos \omega t) + \hat{j}(\sin \omega t)]$$
$$\cdot [-\hat{i}(\sin \omega t) + \hat{j}(\cos \omega t)]\}.$$

Now we apply the distributive law to expand the dot product into products of unit vectors:
$$\vec{r} \cdot \vec{v} = \omega R^2 [-(\cos \omega t \sin \omega t)(\hat{i} \cdot \hat{i}) - (\sin^2 \omega t)(\hat{j} \cdot \hat{i})$$
$$+ (\cos^2 \omega t)(\hat{i} \cdot \hat{j}) + (\sin \omega t \cos \omega t)(\hat{j} \cdot \hat{j})]$$
$$= \omega R^2 [-(\cos \omega t)(\sin \omega t) + 0 + 0 + (\sin \omega t)(\cos \omega t)]$$
$$= 0.$$

Since the dot product of the two vectors is zero, they are perpendicular.

7.3 Using property VIII of the dot product:

$$\frac{d}{dt}(v^2) = \frac{d}{dt}(\vec{v} \cdot \vec{v}) = \frac{d\vec{v}}{dt} \cdot \vec{v} + \vec{v} \cdot \frac{d\vec{v}}{dt} = 2\vec{v} \cdot \frac{d\vec{v}}{dt}.$$

In the last step, property I was used to reverse the factors in the first term. Differentiating the expression for $\vec{v}$ in Exercise 7.2, we find:

$$\frac{d\vec{v}}{dt} = (-\omega^2 R)[\hat{i}(\cos \omega t) + \hat{j}(\sin \omega t)].$$

Then: $\vec{v} \cdot \frac{d\vec{v}}{dt} = \{\omega R[-\hat{i}(\sin \omega t) + \hat{j}(\cos \omega t)]\}$
$\cdot \{-\omega^2 R[\hat{i}(\cos \omega t) + \hat{j}(\sin \omega t)]\}.$

From here algebra nearly identical to that in Exercise 7.2 shows that the dot product is zero. Thus, v^2 doesn't change with time.

7.4 When running up stairs, you accelerate at the beginning and decelerate at the end, but you run at a more or less constant speed once started. Your kinetic energy is essentially constant during the running phase, and so the work required to lift your body is just that required to balance the negative work done by gravity on your body. Assuming you have a mass of 80 kg, the power required to run upstairs is:

$$P = W/T = Mgh/T$$
$$= (80 \text{ kg})(10 \text{ m/s}^2)(10 \text{ m})/(10 \text{ s}) = 800 \text{ W} \approx 1 \text{ hp}.$$

The power required to climb the mountain is:

$$P_m = (80 \text{ kg})(10 \text{ m/s}^2)(1000 \text{ m})/(3 \times 3600 \text{ s}) = 70 \text{ W}.$$

You can adjust these values to your own mass by multiplying by $M_{you}/(80 \text{ kg})$. The human body is roughly 20% efficient in its conversion of chemical (food) energy to mechanical energy. So, five times these power rates have to be supplied by your metabolism, and four-fifths of the total has to be eliminated by your cooling system. Running upstairs is definitely not *no sweat!*

7.5 (a) The product of the units of force and velocity is:

$$[F][v] = N \cdot \frac{m}{s} = \frac{N \cdot m}{s} = \frac{J}{s} = W.$$

(b) In the proof of the work-energy theorem, we showed that the work done by the force $\vec{F}$ during a time interval dt is $dW = \vec{F} \cdot \vec{v}dt$. The corresponding power input to the particle is:

$$P = \frac{dW}{dt} = \frac{\vec{F} \cdot \vec{v}dt}{dt} = \vec{F} \cdot \vec{v}.$$

7.6 If the jetliner were perfectly efficient, the work done by the input force—engine thrust—would equal the work done by the output force—the gravitational work on the jetliner. In that case, we would have:

(Engine thrust)(displacement) = (weight)(vertical displacement).

Or, $\dfrac{\text{Weight}}{\text{Engine thrust}} = \dfrac{\text{displacement}}{\text{vertical displacement}} = \csc(15°) = 3.9.$

This is the mechanical advantage for a frictionless ramp at 15°, the climb angle of the jetliner. From the given data, the actual ratio of these forces is:

$$MA = F_O/F_I = (5.2 \times 10^5 \text{ lb})/(1.5 \times 10^5 \text{ lb}) = 3.5.$$

The engine thrust that would be needed with perfect efficiency is a factor $e = 3.5/3.9 = 0.90$ less than the thrust that is actually needed. From this point of view, the aircraft's efficiency is 90%.

This calculation ignores a great deal. First is the energy diverted in converting from energy stored in the fuel to mechanical work done. Second is the fact that the useful payload is the passengers and baggage. Lifting the mass of the aircraft and fuel doesn't count. Most important is that the task of the machine is to move people from place to place, not just to lift them. Thus fuel use per passenger mile is a more useful measure of aircraft "efficiency." Less tangible, but equally important, is the time required for a trip. For a machine with a socially defined purpose, physical notions of efficiency are useful input but not decisive criteria.

Basic Skills

Review Questions

- State some similarities and differences between linear momentum and energy.
- What is *work*? How is it related to energy? How is it different from energy?

§7.1 ENERGY AND ITS TRANSFER

- How is the kinetic energy of a particle calculated from its mass and speed?
- Describe, for motion in one dimension, the work done by a force $\vec{F}$ acting through a displacement $\vec{s}$.
- State the work-energy theorem for a particle moving in one dimension.
- What is the unit for both work and energy? Express this unit in terms of newtons and meters.
- Describe how to compute the work done on a particle by a force that acts at an angle to the particle's displacement. State your answer first in terms of components of $\vec{F}$ and $\vec{s}$ and then in terms of their dot product.
- Explain why the dot product of two vectors forms a scalar quantity.
- State eight mathematical properties of the dot product.
- When more than one force is acting on an object, explain how to find the total work done on the object.
- Describe qualitatively how to compute the work done on a particle by a variable force.
- Describe how to compute the total kinetic energy of a system of particles.

§7.2 POWER AND SIMPLE MACHINERY

- What is the definition of *power*? Why is power a significant feature of a practical device?

- What is the unit of power? Express this unit in terms of newtons, meters, and seconds.
- What is a *power curve*?
- What is the purpose of a simple machine?
- What is *mechanical advantage*?
- Explain the motivation for thinking of energy as being transmitted by simple machines.
- What is the *efficiency* of a machine? Why are real machines not perfectly efficient?

Basic Skill Drill

§7.1 ENERGY AND ITS TRANSFER

1. A car of mass 1.0×10^3 kg and a truck of mass 1.0×10^4 kg both move at 12 m/s. Find the kinetic energy of each vehicle and compare the values. Compare the car's energy with that of an identical car moving at 35 m/s.

2. A car, skidding to a halt, decelerates in a straight line at $a = 8.0 \text{ m/s}^2$. If the car's mass is $m = 950$ kg, what is the friction force acting on the car's tires? If the car requires 3.0 s to stop, what was its initial kinetic energy? What is the car's displacement during the 3.0-s interval? How much work is done on the car by the friction force? Show that the work-energy theorem is satisfied.

3. A car of mass 1200 kg moving at 28 m/s runs into a haystack and is brought to rest. How much work does the haystack do on the car? How much work does the car do on the haystack?

4. A block and tackle in an auto body shop rolls along a horizontal rail fixed to the roof. To move the system 15 m from one side of the shop at constant speed, a mechanic has to pull on its chain with a force of 35 N at an angle of 10° with the vertical. How much work does the mechanic do to transport the system? How much work is done on the system by **(a)** its weight, **(b)** the normal force exerted by the rail, and **(c)** friction?

5. Estimate $\vec{a} \cdot \vec{b}$ for the vectors $\vec{a}$ and $\vec{b}$ in ■ Figure 7.25. First use estimates of the vectors' components. Repeat the estimate using their lengths and the angle between them. Are your two estimates consistent?

6. A system consists of ten 0.50-kg balls, each moving radially outward from a common center in a symmetrical pattern, all at a speed of 12 m/s. What is the kinetic energy of the system?

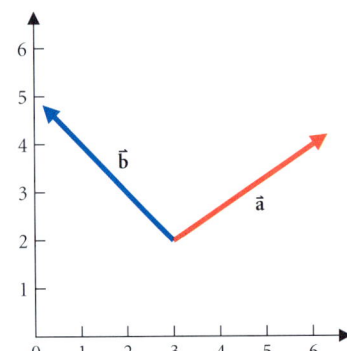

■ FIGURE 7.25

■ FIGURE 7.26

§7.2 POWER AND SIMPLE MACHINERY

7. At launch the space shuttle's rocket motors produce a thrust force of magnitude 2.7×10^7 N. The shuttle is held fixed on the launch pad for about $\frac{1}{2}$ s until full thrust is achieved. (■ Figure 7.26). How much power do the shuttle motors produce during this period?

8. Compare the power required to accelerate an automobile to a speed of 25 m/s in 5 s and in 10 s. Also compare the power required to accelerate to 25 m/s versus 40 m/s in 10 s.

9. A Porsche Carrera can achieve a maximum speed of 257 km/h. The maximum power produced by the engine is 270 hp. How much friction plus drag force does the car experience at 257 km/h?

10. How much work must a person do to lift a 10-kg box a height of 10 m using a machine whose mechanical advantage is 10? Compare both the necessary force and displacement with those needed to lift the box without mechanical assistance.

Questions and Problems

§7.1 ENERGY AND ITS TRANSFER

11. ❖ Describe similarities and differences between the work-energy theorem and the impulse-momentum theorem.

12. ❖ Compare the kinetic energies of two identical baseballs, one with a speed of 30 m/s and one with a speed of 20 m/s.

13. ❖ An athlete swings a hammer of mass 2 kg around on the end of a line, spinning faster and faster, until the hammer is let go with a speed of 10 m/s. If the hammer was originally at rest, how much work does the athlete do on the hammer? Neglect air resistance.

14. ❖ A projectile is launched from ground level at a 45° angle. How much work does gravity do on the projectile between its launch and when it hits the ground?

15. ❖ What are the possible directions of a vector $\vec{a}$, if $\hat{k} \cdot \vec{a} = 0$ and $\hat{i} \cdot \vec{a} = \hat{j} \cdot \vec{a}$?

16. ❖ If you know $\vec{a}$ and $\vec{a} \cdot \vec{b}$, you can easily determine the component of $\vec{b}$ parallel to $\vec{a}$. Is it possible to define a scalar *asterisk product* $\vec{a} * \vec{b}$ so that $\vec{a}$ and $\vec{a} * \vec{b}$ determine the component of $\vec{b}$ *perpendicular* to $\vec{a}$? If not, what additional information is necessary to determine $\vec{b}_{\perp a}$?

17. ❖ An object accelerates because of a force exerted on it. The work done by the force is:
(a) always positive.   (b) always negative.   (c) never zero.
(d) sometimes zero.   (e) always zero.
Explain why you chose your answer and give a counter example for each case you reject.

18. ❖ Identify the forces acting on a car rounding a curve. Which of these forces do work?

19. ❖ Identify the forces acting on the ball of a roulette wheel. Which forces do work? Why does the ball eventually fall into one of the slots?

20. ♦ Express the kinetic energy of a particle in terms of the particle's momentum.

21. ♦ A 250-g softball is pitched at a speed of 28 m/s. What is the ball's kinetic energy?

22. ♦ What is the kinetic energy of an air molecule with a speed of 0.50 km/s and a mass of 4.6×10^{-26} kg?

23. ♦ A particle of mass 10 kg has velocity $\vec{v} = (10 \text{ m/s})\hat{i}$. What is its kinetic energy?

24. ♦ A particle of mass 20 kg has velocity $\vec{v} = (20 \text{ m/s})\hat{i} + (10 \text{ m/s})\hat{j}$. What is its kinetic energy?

25. ♦ A particle of mass 4.8 kg has velocity $\vec{v} = (2.2 \text{ m/s})\hat{i} + (4.9 \text{ m/s})\hat{j} + (2.7 \text{ m/s})\hat{k}$. What is its kinetic energy?

26. ♦ A particle of mass $m = 2$ kg is initially at rest on a frictionless surface. A constant force of magnitude $F = 4$ N then acts on it while it moves a distance $s = 9$ m. Find the final speed of the particle.

27. ♦ A hockey player pushes a puck of mass 0.50 kg across the ice using a constant force of 10.0 N over a distance of 0.50 m. How much work does the hockey player do? If the puck was initially stationary, what is its final speed? (Ignore friction.)

28. ♦ If a constant force $\vec{F} = (3.0 \text{ N})\hat{i} + (5.0 \text{ N})\hat{j}$ acts on a particle that undergoes a displacement $\vec{s} = (4.0 \text{ m})\hat{i} + (1.0 \text{ m})\hat{j}$, how much work is done on the particle?

29. ♦ If $\vec{a} = (1, 2)$ and $\vec{b} = (2, -1)$, what is $\vec{a} \cdot \vec{b}$?

30. ♦ Use the dot product to find the angle between the vectors $(1, 2)$ and $(-1, 3)$.

31. ♦ Use the dot product to find the angle between the vectors $(3.0, 4.0)$ and $(4.0, 3.0)$.

32. ♦ If $\vec{a}$ is (2 km, east) and $\vec{b}$ is (3 km, northwest), what is $\vec{a} \cdot \vec{b}$?

33. ♦ What is the dot product of $(3, 4, 5)$ and $(1, 2, 3)$?

34. ♦ Prove $c (\vec{a} \cdot \vec{b}) = (c\vec{a}) \cdot \vec{b} = \vec{a} \cdot (c\vec{b})$, where c is a scalar.

35. ♦ Prove that $(\vec{a} - \vec{b}) \cdot (\vec{a} + \vec{b}) = a^2 - b^2$.

36. ♦ Find the magnitudes of the vectors $\vec{a} = (1.0, 2.0, 3.0)$ and $\vec{b} = (2.0, 2.0, 1.0)$.

37. ♦ What angle does the vector $(1, 2, 2)$ make with the x-axis?

38. ♦ Use the dot product to find the angle between the vectors $(2.0, 2.0, 1.0)$ and $(1.0, 1.0, 2.0)$.

39. ♦ During rush hour, there are about 7000 cars on the San Francisco Bay Bridge, each traveling at about 80 km/h. If the average mass of a car is 800 kg, what is the total kinetic energy of the system consisting of the bridge and all the cars?

40. ♦ A system consists of three balls with these masses and velocities:

Ball 1. $M = 20.0$ g $\vec{v} = (10.0 \text{ m/s})\hat{i}$
Ball 2. $M = 50.0$ g $\vec{v} = (10.0 \text{ m/s})(\hat{i} + \hat{j})$
Ball 3. $M = 30.0$ g $\vec{v} = (5.00 \text{ m/s})\hat{j}$

Find the kinetic energy of the system.

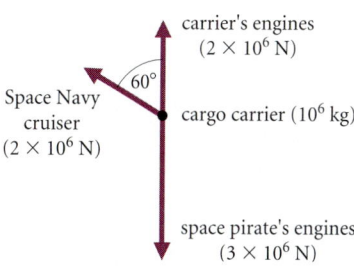

■ **FIGURE 7.27**

41. (a) ♦ In deep space, space pirates attack an interplanetary cargo carrier of mass 10^6 kg by pulling on it with a force of 3×10^6 N. The carrier's captain resists by using her engines to exert a force of 2×10^6 N in the opposite direction. The pirates succeed in pulling the carrier across 100 km of space. How much work is done by the pirate ship? How much work do the carrier's engines do? What is its final speed? (b) ♦♦ A Space Navy cruiser detects the conflict and rushes to the rescue. The cruiser exerts a force of 2×10^6 N at an angle of 60° to the line of conflict (■ Figure 7.27). Does it succeed in rescuing the cargo carrier? How much work does the cruiser do in 1 min?

42. ♦♦ A jet airplane's engines produce constant thrust. If the jet is to take off before the end of the runway, what minimum fraction of takeoff speed must it have when it reaches the middle of the runway?

43. ♦♦ (a) A spaceship of mass 1.0×10^4 kg moving at 1.0×10^3 km/s accelerates by firing its thrusters for 1.0 s. Assume that (i) the thrusters exert a constant force of 1.0×10^9 N, and (ii) you may ignore any change in the speed of the spaceship during the thrust. Estimate the final speed of the spaceship. (b) Find the error made by using assumption (ii).

44. ♦♦ A mass driver 5.0×10^2 m long is designed to launch a 10.0-kg projectile at a speed of 5.0 km/s. Use the work-energy theorem to find the constant force the driver exerts on the projectile.

45. ♦♦ Prove the identity: $\vec{v} = |\vec{v}|(\vec{v} \cdot \hat{i}, \vec{v} \cdot \hat{j}, \vec{v} \cdot \hat{k})$.

46. ♦♦ For the vectors $\vec{a}$ and $\vec{b}$ in ■ Figure 7.28, estimate x_1- and y_1-components and x_2- and y_2-components. Compute $\vec{a} \cdot \vec{b}$ using each of your estimates.

47. ♦♦ Three vectors lie along the sides of a triangle (■ Figure 7.29). Use the law of cosines to show: $\vec{a} \cdot \vec{b} = \frac{1}{2}(c^2 - a^2 - b^2)$. Derive the same result from the fact that $\vec{c} = \vec{a} + \vec{b}$.

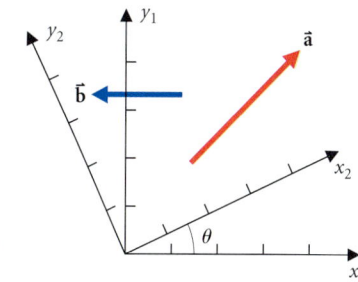

■ **FIGURE 7.28**

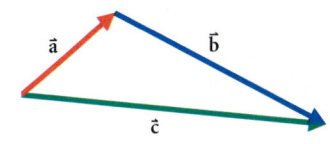

■ **FIGURE 7.29**

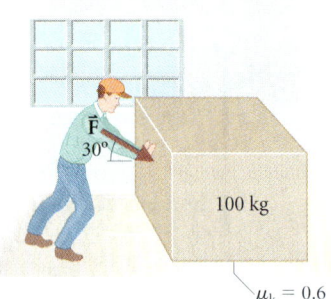

FIGURE 7.30

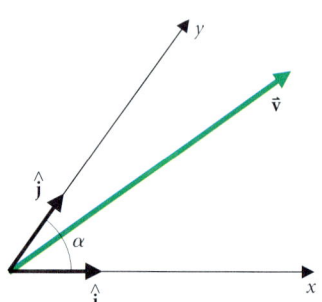

FIGURE 7.33

48. ♦♦ A custodian pushes a box across the floor. The force he exerts makes a 30° angle with the horizontal (■ Figure 7.30). The box has a mass of 100 kg and a coefficient of kinetic friction between it and the floor of 0.6. If the box moves at constant speed once set in motion, how much work must the custodian do to push it 100 m?

49. ♦♦ A tractor pulling a plow exerts a horizontal force of 1000 N (on average). A field measures 1 km × 2 km with furrows along the long dimension 1 m apart. How much work does the tractor do to plow the field? If the field had a 3° slope parallel to the furrows, would the amount of work needed be less, more, or the same?

50. ♦♦ In attempting to jump-start a car of mass 1100 kg, two brothers each push with an equal force of magnitude 670 N parallel to the car's velocity. The car moves at a constant speed of 2.8 m/s for 55 m, when it starts. The car is on a road that slopes downhill at 0.7° with the horizontal. Find: **(a)** The friction force that acts on the car. **(b)** The work done by each brother. **(c)** The work done by gravity. **(d)** The work done by friction.

51. ♦♦ Three boxes of equal mass M are initially at rest on a smooth floor (■ Figure 7.31). A force $\vec{F}$ is applied to the system until it has moved a distance s. **(a)** Show that the coefficient of static friction $\mu_s \geq F/(3Mg)$ is necessary if box 3 is to accelerate together with the other two. **(b)** Show that the work-energy theorem applies to the system as a whole and to each individual box.

52. ♦♦ Two horses on either side of a canal pull a canal boat of mass 1.0×10^4 kg at constant speed for a distance of 10.0 km (■ Figure 7.32). One horse exerts a force of 3.0×10^2 N at an angle of 20° to the canal, and the other exerts 5.0×10^2 N. Find the work done by each horse and the work done by friction between the boat and the water.

53. ♦♦♦ In Figure 7.28 the angle between the x_2- and x_1-axes is θ. **(a)** Find expressions for $\hat{x}_2 \equiv \hat{i}_2$ and $\hat{y}_2 \equiv \hat{j}_2$ in terms of $\hat{x}_1 \equiv \hat{i}_1$, $\hat{y}_1 \equiv \hat{j}_1$, and θ. **(b)** Use the result of part (a) to express the components of an arbitrary vector $\vec{A}$ in coordinate system 1 in terms of its components in system 2. **(c)** Use the result of part (b) to show explicitly that $\vec{A} \cdot \vec{B}$ is the same in both coordinate systems for arbitrary vectors $\vec{A}$ and $\vec{B}$.

54. ♦♦♦ Even with nonperpendicular coordinate axes (■ Figure 7.33), a vector $\vec{v}$ can be split into components along the two axes: $\vec{v} = v_x \hat{i} + v_y \hat{j}$. Find $\hat{i} \cdot \vec{v}$ and $\hat{j} \cdot \vec{v}$ in terms of these components.

55. ♦♦♦ **(a)** A block of mass M slides with uniform angular speed ω on a circular track of radius R with coefficient of kinetic friction μ_k (■ Figure 7.34). A string, attached to the block, provides the necessary force to balance friction and accelerate the block. Find the work done by the string during one revolution of the block and show that it is minus the work done by friction. (*Hint:* Show first that the conclusion is true for each small displacement of the system.) **(b)** Describe how a person's hand can maintain the tension in the string and its angle to the radius of the circle. Show that the work done by the hand on the string during one revolution does not depend on the angle θ. Also show that this work exactly balances the negative work done by friction.

56. ♦♦♦ If $\vec{r}$ is the position vector of a particle and $\vec{v}$ is its velocity, show that

$$\vec{v} \cdot \vec{r} = \frac{d}{dt}(\tfrac{1}{2}r^2) \quad \text{and} \quad \vec{v} \cdot \hat{r} = \frac{dr}{dt}.$$

§7.2 POWER AND SIMPLE MACHINERY

57. ❖ Show that the efficiency of two machines operating in tandem is the product of the efficiencies of the two individual machines.

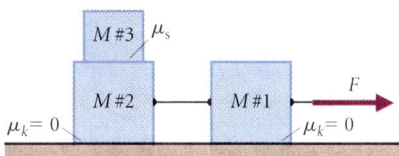

FIGURE 7.31

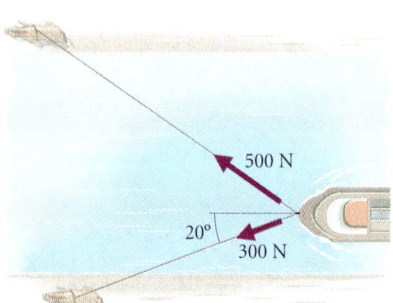

FIGURE 7.32

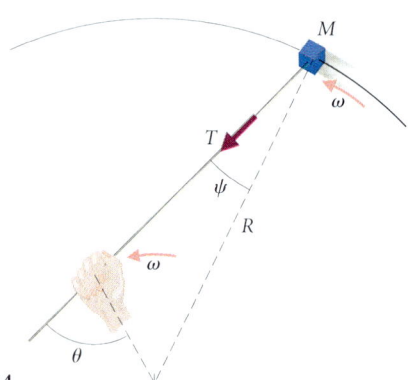

FIGURE 7.34

244 CHAPTER 7 • WORK AND KINETIC ENERGY

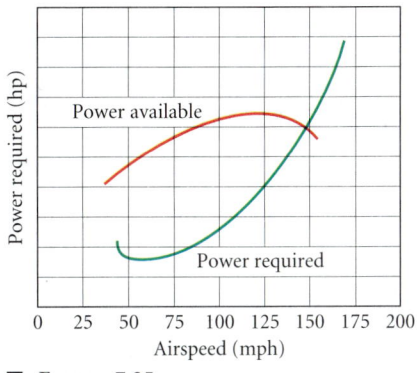

■ FIGURE 7.35

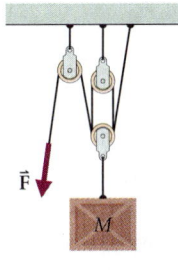

■ FIGURE 7.36

("In tandem" means that the output of the first machine is the input to the second.)

58. ❖ ■ Figure 7.35 shows curves of power required in level flight and total power available for a typical light aircraft. At what speed does the aircraft achieve its greatest rate of climb?

59. ❖ Why does the space shuttle turn to the east as rapidly as possible after launch? (*Hint:* Does the Earth's rotation make any difference?)

60. ◆ How much power must an automobile engine exert to accelerate a 1000-kg automobile to 20 m/s from a standing start in 10 s? Give your answer in watts and in horsepower.

61. ◆ A dock worker uses the ideal pulley system shown in ■ Figure 7.36 to lift boxes from the dock onto a ship, a height of 20.0 m. Each box has a mass of 5.00×10^2 kg. How much work must be done to lift each box? If the worker lifts ten boxes per hour, what is the worker's average power output? (*Hint:* Do you have to analyze the pulley's operation to answer this question?)

62. ◆ A railroad engine pulls a train of mass 1.2×10^6 kg up an average slope of 1.2° for a distance of 55 km. If the speed of the train is 42 km/h, what is the average power produced by the engine? How much total work does the engine do?

63. ◆ A perfectly efficient machine is used to move a 50-kg load. The input force moves a distance of 10 m as the load rises 1 m. What is the mechanical advantage of the machine? What is the magnitude of the input force?

64. ◆ A machine is used to drag a crate of mass 1.0×10^2 kg across a floor with a coefficient of friction = 0.70. The input force is 35 N. What is the displacement of the input force for each meter that the crate moves?

65. ◆ A machine draws a power of 10.0 W when lifting a 6.00-kg load at 0.120 m/s. What is the efficiency of the machine?

66. ◆ The input force to a perfectly efficient machine is observed to move through 30 cm when the load moves 10 cm. What is the mechanical advantage of the machine?

67. ◆ A force of 10 N is required to lift a 5-kg load with machine X. The input force moves a distance of 6 m as the load rises 1 m. Find the mechanical advantage and the efficiency of machine X.

68. ◆◆ A 950-kg car, initially moving at 24 m/s, skids to a halt with an acceleration of 8.0 m/s². Find its speed as a function of time. Use the expression $P = Fv$ to determine the rate kinetic energy is removed from the car. Give numerical values for this power at $t = 0.0$ s, $t = 1.5$ s, and $t = 3.0$ s.

69. ◆◆ A motor can produce a maximum power of 2.7 kW. The motor exerts tension on a cable used to pull a sled at 3.3 m/s over a horizontal surface with a coefficient of kinetic friction $\mu_k = 0.36$. How massive a sled can the motor pull?

70. ◆◆ Find the mechanical advantage of the pulley system shown in Figure 7.36.

71. ◆◆ A 100-car freight train has a total mass of 1500 metric tons. Model the effects of friction on the wheels of the cars with a *coefficient of rolling friction* $\mu_r = 5.0 \times 10^{-3}$. What power must a freight engine provide to pull the train over level ground at a speed of 45 km/h?

72. ◆◆ A father drags his child, riding in a sled, up a snowslope. The child and the sled together have a mass of 58 kg. The slope makes an angle of 21° with the horizontal, and the rope is at a 12° angle to the slope. The coefficient of friction between the slope and the sled is 0.23. How much work must the father do to drag the sled 0.20 km along the slope? If he does this five times in an hour, what is his average power output?

73. ◆◆ A Honda Accord LX, of mass 1200 kg, brakes to a stop from 35 m/s in a distance of 7.0×10^1 m. How much work is done to stop the car? If the car decelerates uniformly, find the power dissipated by the brakes as a function of time.

74. ◆◆ Find an expression for the work required to lift a cubic meter of water from a depth h below the ground to the surface. (The density of water is 1.0×10^3 kg/m³.) If a motor can produce maximum power of 8.0 kW and it is desired to pump water at a rate of 5.0×10^{-2} m³/s, how deep a well can the motor service? If the pump discharges the water at the surface with a speed of 2.5 m/s, what error is made in calculating the maximum depth by neglecting the kinetic energy of the water?

75. ◆◆ The brake system on the San Francisco cable cars uses a wedge pushed into the cable slot (■ Figure 7.37). If the apex angle of the wedge is 10°, how much larger is the braking force compared with simply pushing a flat block onto the ground with the same force F; that is, what is the mechanical advantage of the wedge as a braking machine? (Assume that the coefficient of friction is the same in both cases.)

76. ◆◆ The pulley system in Figure 7.22 is used in tandem with a 45° ramp. (Instead of being attached directly to the load, the rope from the lower pulley passes around a smooth peg at the top of the ramp to a load sitting on the ramp.) Compute the mechanical advantage of this machine directly and as a product of the individual mechanical advantages of the separate ramp and pulley systems.

77. ◆◆ What is the efficiency of the pulley system in ■ Figure 7.38 when lifting a 125-kg load if each pulley has a mass of 1.20 kg?

78. ◆◆ What is the efficiency of the machine shown in ■ Figure 7.39 when raising a 50.0-kg load? The coefficient of friction between the

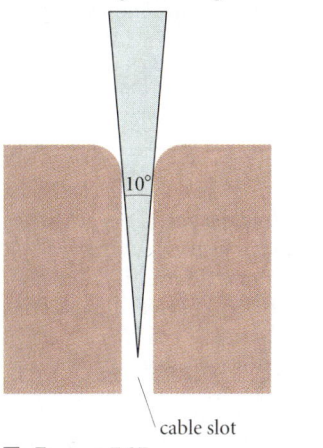

■ FIGURE 7.37

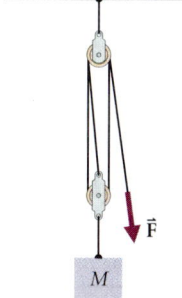

■ FIGURE 7.38

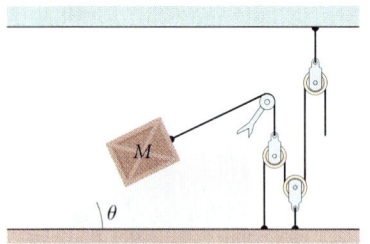

Figure 7.39

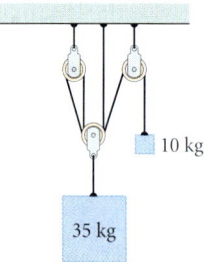

Figure 7.40

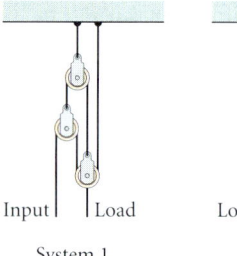
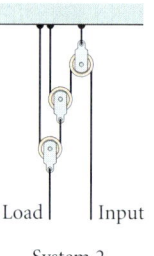

Figure 7.41

crate and the ramp is 0.400, each pulley has a mass of 1.00 kg, and the angle $\theta = 30.0°$.

79. ◆◆ Suppose the pulleys in ■ Figure 7.40 are massless but not frictionless. It is found that an object of mass $m = 10.0$ kg slowly lifts an object of mass $M = 35$ kg. What is the efficiency of the pulley system? If the 10.0-kg object falls 4.0 m, how much work is done by the friction forces in the pulleys?

80. ◆◆◆ Determine the mechanical advantage for each pulley system shown in ■ Figure 7.41. If the pulleys are frictionless but each has a mass $m = 0.50$ kg, what is the efficiency of each system when lifting 120.0-kg objects?

Additional Problems

81. ❖ "Friction acting on an object *always* removes energy from the object." Give a counterexample to this statement and explain why your example shows the statement to be false.

82. ❖ An object is tied to one end of an ideal string, and the other end is tied to the roof. The object is released from rest with the string taut, and it swings downward until the string is vertical. Explain why the following statement is *false:* The string tension force does the same amount of work on the object as does gravity.

83. ❖ Suppose that the object in Problem 82 is instead released from rest at the point where the string is tied to the roof. Decide whether the following statement is true or false and explain your reasoning: After the ball has finally come to rest again, the work done on the ball by the string tension force is minus the amount of work done on it by gravity.

84. ❖ Two blocks of masses M_1 and M_2 are initially at rest on a frictionless surface and are connected by a stretched spring. At any instant after the blocks are released, their kinetic energies have the same ratio. Explain why this is so, and give the value of the ratio. (*Hint:* Refer to the result of Problem 20.)

85. ◆◆ The position of a particle moving in one dimension is described by:

$$x(t) = x_o(t/t_o)^{1.5}.$$

The particle's kinetic energy is proportional to:
(a) t^2; **(b)** t; **(c)** $t^{1/2}$; **(d)** $t^{1.5}$; **(e)** t^0; **(f)** t^3.
The power delivered by the force accelerating the particle is proportional to which of the above answers? Explain your reasoning.

86. ◆◆ A skier enters a flat section of the run at 18 m/s. If the coefficient of friction between snow and skis is 0.050, compute how far the skier can slide. Neglect air resistance. Compute the result first using the work-energy theorem, and then using Newton's second law. Compare the two solutions.

87. ◆◆ A 3200-kg Mercedes Benz truck traveling on a German autobahn goes out of control and hits an 1100-kg Volkswagen; the VW is parked at a rest stop. The wreckage comes to rest 23 m from the VW's parking space. The coefficient of friction between the wreckage and the road is 0.58. You are an insurance investigator, and wish to find: **(a)** the speed of the wreckage immediately after the collision, and **(b)** the speed of the truck before the collision in km/h. **(c)** Assuming the collision took 0.50 s (from the time the two vehicles first touched until the system first reached a uniform velocity of all its parts), what was the average acceleration of the VW during the collision? **(d)** Is it reasonable to neglect friction forces acting on the system *during* the collision? **(e)** How much energy was transformed by the process of crumpling metal during the collision?

88. ◆◆ A rope tow pulls a skier of mass 75 kg up a 12° slope at a constant speed of 3.0 m/s. When the skier has gained a 15-m altitude, how much work has gravity done on the skier? If the power delivered by the motor has to increase by 490 W in order to lift the skier, what is the coefficient of sliding friction between the skis and the snow?

89. ◆◆ What is the mechanical advantage of the *machine* consisting of an ideal rope stretched between the load and a fixed object (■ Figure 7.42)? How does the mechanical advantage vary with displacement d of the input point P?

90. ◆◆◆ A crate of mass $2M$ slides on a surface with a coefficient of kinetic friction μ_k. The crate is connected by a string over a frictionless pulley to a suspended mass M (■ Figure 7.43). If the sliding block starts from rest, find its speed when it reaches the pulley. Show that the final kinetic energy of the system equals the work done by gravity plus the work done by friction.

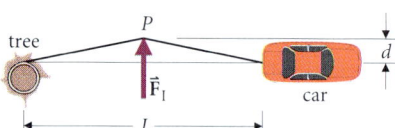

Figure 7.42

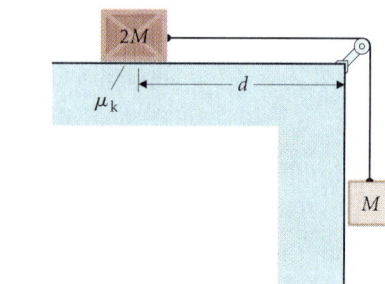

Figure 7.43

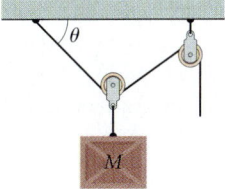

■ FIGURE 7.44

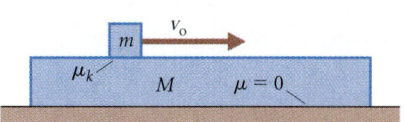

■ FIGURE 7.45

91. ♦♦♦ What is the mechanical advantage of the machine shown in ■ Figure 7.44? Show that at any value of θ the input work needed to lift the mass a distance dh is $Mg\,dh$. Argue that the total work done to lift the block a height h is Mgh.

92. ♦♦♦ Redesigning the flour warehouse (Example 7.9), you decide to have the motor run a conveyor belt that lifts 40-kg flour sacks. The belt is to be at an angle in the range 9° to 10° with the horizontal. **(a)** What range of mechanical advantage is available to you? **(b)** What combinations of belt angle and number of sacks on the belt at any given time makes the optimal match to the motor?

93. ♦♦♦ In the *skew* coordinate system shown in Figure 7.33, the quantities

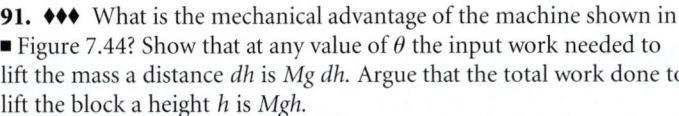

are called metric coefficients. Show that for two arbitrary vectors $\vec{a}$ and $\vec{b}$:

$$\vec{a} \cdot \vec{b} = g_{xx}a_xb_x + g_{xy}a_xb_y + g_{yx}a_yb_x + g_{yy}a_yb_y.$$

This is a generalization of the Pythagorean theorem to a skew coordinate system. What are the metric coefficients in the usual, rectangular Cartesian coordinates?

Computer Problem

94. During a test of an experimental aircraft, the engine power and the airspeed are measured at 0.10-s intervals. See file CH7P94 on the supplementary computer disk. Use these data to calculate the engine thrust at each tabulated time t. Calculate the total work done by the engine in two ways: **(a)** Numerically integrate the power over time, using the trapezoidal rule (Problem 6.91) or another numerical algorithm. **(b)** Compute the displacement of the aircraft, and hence the work done, during each 0.10-s interval. Sum the differential amounts of work to obtain the total work done.

Compare the two results. Which scheme do you think provides the most accurate estimate of W, and why?

Challenge Problems

95. (a) A long block of mass M is initially at rest on a frictionless surface (■ Figure 7.45). A small block with mass m and initial speed v_o slides on top of the larger block. The coefficient of friction between the blocks is μ_k. At the time when the small block ceases to slide, find the displacement of each block and the work done by friction on each. Verify that the total work done equals the change in the kinetic energy of the system. Check your results using conservation of momentum. **(b)** If the length of the lower block is L, find the maximum initial speed v_{max} of the small block such that it ceases sliding before falling off the lower block. For $v_o > v_{max}$, find the work done by friction and the speeds of the blocks just as the top block falls off. Verify that the frictional work equals the change in the kinetic energy of the system.

96. ✉ Consider the design of a human-powered helicopter. The pilot/engine turns pedals that drive an overhead rotor blade. Model the system by assuming that the rotor blade propels air of density ρ downward at speed v throughout a cylinder of diameter D. If the mass of the system is M, what rate of momentum transfer to the air is required for the craft to fly? Express your result as a relation among ρ, v, D, and M. Find an expression for the minimum power the pilot must produce by calculating the rate at which the rotors transfer kinetic energy to the air. If the mass of the system is $M = 200$ kg, the length of the rotor blades is $D = 10$ m, and the density of air is $\rho = 1.23$ kg/m³, decide whether the design is feasible.

97. A 25-g bullet with a speed of 350 m/s penetrates a block of wood, traveling 23 cm in 1.5 ms. **(a)** Estimate the average force that acts on the bullet using, first, the work-energy theorem and, second, the impulse-momentum theorem. Are your two estimates consistent? **(b)** Show that the two estimates are consistent if the force on the bullet is constant. **(c)** Determine from the given data whether the force acting on the bullet increased, decreased, or remained constant in magnitude as the bullet slowed.

When a mathematician engaged in investigating physical actions and results has arrived at his conclusions, may they not be expressed in common language as fully, clearly, and definitely as in mathematical formulae?

MICHAEL FARADAY

INTERLUDE 2
Using Integration in Physics Problems

CONCEPTS

Representation of an integral as the limit of a sum

Representation of an integral by an area

Differential element

Physical integral

GOAL

Be able to:

Apply the five-step process for expressing a physical integral as a mathematical integral.

We have already encountered a number of problems for which integration is a useful technique. For example, force describes the rate at which momentum is exchanged between particles. One statement of this fact is Newton's second law: $\vec{F} = d\vec{p}/dt$. By integrating, we found an equivalent relation, the impulse-momentum theorem:

$$\Delta \vec{p} = \int \vec{F}(t)\,dt \equiv \int [\hat{i}F_x(t) + \hat{j}F_y(t) + \hat{k}F_z(t)]\,dt.$$

Since the Cartesian unit vectors are constants, they may be factored from the integral:

$$\begin{aligned}\Delta \vec{p} &= \int \vec{F}(t)\,dt \\ &\equiv \hat{i}\left[\int F_x(t)\,dt\right] + \hat{j}\left[\int F_y(t)\,dt\right] + \hat{k}\left[\int F_z(t)\,dt\right].\end{aligned} \quad (I2.1)$$

If we know the force acting on a particle as a function of time, then each component of $\Delta \vec{p}$ is the integral of a known function of a single variable t. This mathematical problem is very similar to those you see in your calculus classes. With the study of work and energy, however, we begin to encounter problems that use integration but are not so directly expressed as familiar calculations.

For example, the energy transferred to a particle by a variable force is described by the integral (eqn. 7.5):

$$W = \int dW = \int \vec{F}\cdot d\vec{s}. \quad (I2.2)$$

This definition uses the language of integration but is not yet expressed in a familiar form that is easy to evaluate. You cannot look it up in an integral table! There is a common, five-step method for approaching all such integration problems. In this interlude we set out that method and illustrate it with a few examples.

I2.1 THE INTEGRAL OF A FUNCTION AS THE LIMIT OF A SUM

Recall that the area under a graph of a function represents the integral of the function. Thus, in ■ Figure I2.1, the shaded area represents the integral of the function f over the interval from a to b:

$$A = \int_a^b f(x)\,dx.$$

We can visualize an integral in terms of areas (■ Figure I2.2). First divide the integral a to b into N pieces and construct two sets of rectangles, colored red and green in the figure. The height of a red rectangle is the maximum value of the function in an interval and the height of a green rectangle is the minimum value. So, for any N, the sum of red rectangles is bigger than the integral, and the sum of green rectangles is less. If both these sums have the same limit as $N \to \infty$, the integral exists and equals the common limit.

The limit $N \to \infty$ is an ideal. When applying integrals, it is useful for several reasons to step back from the ideal. To evaluate an integral numerically, you would slice the integration interval into a finite, though large, number of pieces. By choosing a large enough number of pieces, you can ensure that the difference between the integral and the sum of either set of approximating rectangles is less than the precision you need. Available computer time limits the size of the intervals and hence the precision you can achieve. If you are applying the impulse-momentum theorem, you can know the force $\vec{F}(t)$ only at the times it was measured. There is no point in dividing the integration interval more precisely than your measurements. Finally, none of us is very good at thinking about infinitely small things. When figuring out how to apply integration to some system or process, we need to picture and reason about

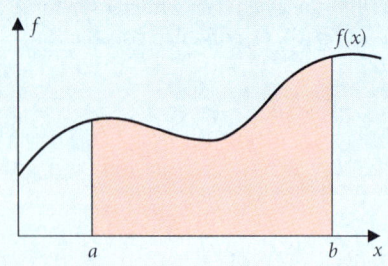

■ FIGURE I2.1
The integral of the function $f(x)$ over the interval (a, b) is represented as the area between a graph of the function and the x-axis, shown shaded. Area *below* the x-axis would count as negative and subtract from the integral, as expected for negative function values.

248

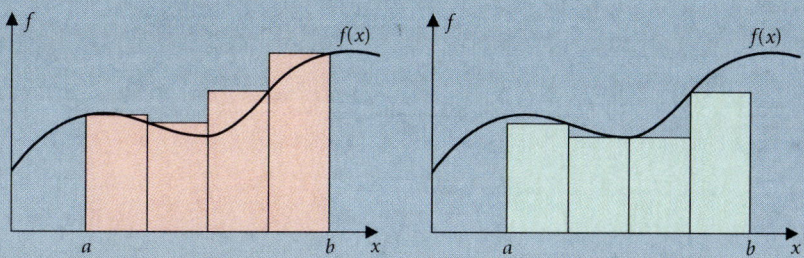

finite, if very small, things. We can then think of an integral as being the sum of a large number of such *differential elements.*

■ Figure I2.3 shows a differential element of area dA labeled by its coordinate x. The element is approximately a rectangle whose height equals the value $f(x)$ of the function at x and whose width is the differential length dx along the coordinate axis. Its area is $dA = f(x)\,dx$ and we may write:

$$\int f(x)\,dx = \int dA = A.$$

This expression for the integral is similar to the definition of work. You should read the expression $\int dA$ as the instruction "add up a large number of differential elements of area." Similarly, $\int dW$ means "add up a large number of differential energy transfers."

NOTICE: In each of these expressions, there are *two* kinds of differential elements: elements of coordinate dx and elements of area dA in one case, elementary displacements $d\vec{s}$ and elements of work dW in the other case.

Here is the main idea: we must learn to model physical systems and processes as large numbers of such *differential elements,* find the corresponding elementary contributions to the quantity we are calculating, and compute their sum. We shall use the name *physical integral* for the resulting sum.

A *physical integral* is the sum of quantities evaluated for each differential element of a physical process or system.

Let us first do an example with the definition of impulse in eqn. (I2.1) and using area to represent the necessary integral.

EXAMPLE I2.1 ♦♦ A particle of mass 6.00 kg is moving along the x-axis with a speed of 3.00 m/s. At time $t = 0.0$ s, a force in the x-direction begins to act on it. Measurements of the force are best fit by a linear increase from 0.0 N at $t = 0.0$ s to 10.0 N at $t = 5.0$ s followed by a linear decrease to 0.0 N in the next 7.0 s. What is the speed of the particle at $t = 15.0$ s?

MODEL Since $\vec{v} = \vec{p}/m$, we may find the speed by first finding the momentum. The force is in the x-direction, so we need only the first term in the impulse-momentum theorem (eqn. I2.1). The necessary integral is an area under a graph of F_x versus time. We desire a result after the force has ceased acting, so we need the area corresponding to the time interval $t = 0.0$ s to 12.0 s. (The force and the acceleration are both zero after $t = 12.0$ s, and the velocity remains constant.)

SETUP The graph (■ Figure I2.4) is a triangle with height 10.0 N and base 12.0 s.

SOLVE The area under the graph is $I_x = \frac{1}{2}(10.0\text{ N})(12.0\text{ s}) = 60.0$ N·s. Since $I_x = \Delta p_x = m(v_{x,f} - v_{x,i})$, we have:

$$v_{x,f} - v_{x,i} = \Delta p_x/m = (60.0\text{ N·s})/(6.00\text{ kg}) = 10.0 \text{ m/s}.$$

So
$$v_{x,f} = v_{x,i} + \Delta v_x = 3.00 \text{ m/s} + 10.0 \text{ m/s} = 13.0 \text{ m/s}.$$

■ **FIGURE I2.2**
The interval (a, b) is divided into N segments. The function f is bounded by step functions, illustrated in red and green. In each segment, the red step function has a value equal to the maximum value of the true function f in that segment; the value of the green function equals the minimum value of f. A sequence of such step functions is defined by increasing the number N of segments.

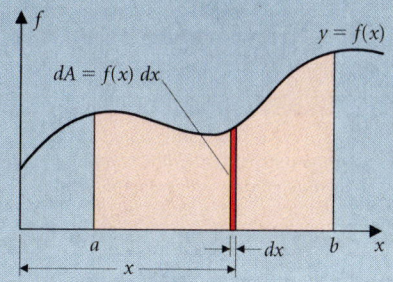

■ **FIGURE I2.3**
In physics problems, it is often useful to step back from the ideal limit $N \to \infty$. We visualize the differential element dx as small but finite, as shown here. An element of area $dA = f(x)\,dx$, and the integral is the sum of these differential areas.

WE USE THE TERM *PHYSICAL INTEGRAL* TO EMPHASIZE THAT THE USUAL MATHEMATICAL EXPRESSION YOU LEARN ABOUT IN CALCULUS DOESN'T APPEAR UNTIL STEP V OF THE INTEGRATION PROCESS (SEE FIGURE I2.5).

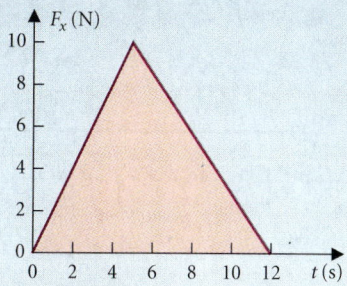

■ **FIGURE I2.4**
The force exerted on a particle is in the x-direction. Its magnitude increases linearly from 0 to 10.0 N during the first 5.0 s, and then decreases again to zero at $t = 12.0$ s.

SECTION I2.1 • THE INTEGRAL OF A FUNCTION AS THE LIMIT OF A SUM

The five-step integration method

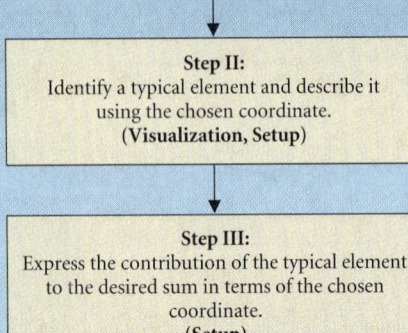

Step I:
Choose appropriate coordinates to describe the system or process and model it as a collection of differential elements. Each element should correspond to a differential change in one coordinate.
(Reading and context, Visualization)

↓

Step II:
Identify a typical element and describe it using the chosen coordinate.
(Visualization, Setup)

↓

Step III:
Express the contribution of the typical element to the desired sum in terms of the chosen coordinate.
(Setup)

↓

Step IV:
Determine the appropriate limits of integration.
(Visualization, Setup)

↓

Step V:
Evaluate the resulting integrals.
(Solution)

↓

Consistency check, comments, etc.

■ FIGURE I2.5

BE ALERT; THIS SEEMS EASY, BUT CHOOSING NONTYPICAL ELEMENTS IS ONE OF THE MOST COMMON STUDENT HANG-UPS. DON'T BE THE ONE WHO GETS CAUGHT.

ANALYZE The result we have found is true at $t = 12.0$ s and at all times thereafter, including $t = 15.0$ s. We could have found the answer by writing a function $F_x(t)$ and integrating, but in this example, evaluating the area geometrically is the easier approach. ■

I2.2 GENERAL METHOD FOR EVALUATING PHYSICAL INTEGRALS

We shall now use two examples to introduce the five-step integration method (■ Figure I2.5). The first four steps transform the integral from a sum of differential elements to a standard mathematical form. The fifth step calls on the powerful analytic or numerical methods that exist for evaluating integrals. The first example is deliberately chosen to have familiar and straightforward calculations so as to exhibit the steps clearly. Then we present an example of work done by a variable force to show how the integration process is used in a problem that does not at first look like a simple integral.

EXAMPLE I2.2 ◆◆ A particle with initial velocity $\vec{v} = (3.0 \text{ m/s})\hat{\imath} + (5.0 \text{ m/s})\hat{k}$ and mass $m = 0.70$ kg is acted upon by a force $\vec{F}(t) = \hat{\jmath}[(2.0 \text{ N/s})t] + \hat{k}[(3.0 \text{ N/s}^2)t^2]$ between $t = 0.0$ s and $t = 2.0$ s. Find the velocity of the particle at $t = 2.0$ s.

MODEL This is another application of eqn. I2.1, but now in three dimensions. We need to compute the y- and z-components of the impulse. Since $F_x = 0$, the x-component of the impulse vanishes. In ■ Figures I2.6 and I2.7 we represent the given force components as functions of time.

STEP I Choose appropriate *coordinates* to describe the system or process and model it as a collection of differential elements. A differential element should correspond to a differential change in *one* of the coordinates.

The force acts over a given time interval $\Delta t = 2.0$ s. The relevant coordinate is the time t. We may think of the process as a collection of differential momentum transfers occurring during differential time intervals.

STEP II Identify a *typical* element and describe it using the chosen coordinate.

A typical element is a differential interval dt somewhere between $t = 0$ s and 2 s; one such element is shown in Figure I2.7. *Typical* means that the element you visualize should not be at a special time, such as one end or the exact center of the time interval.

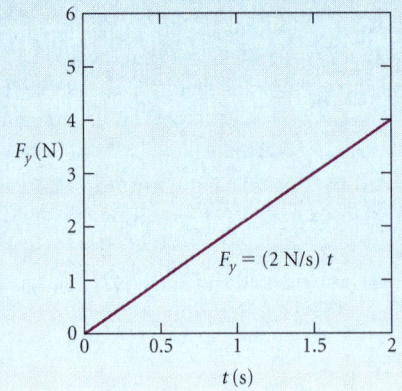

■ FIGURE I2.6
The y-component of the force exerted on a particle increases linearly with time according to the formula $F_y(t) = (2.0 \text{ N/s})t$.

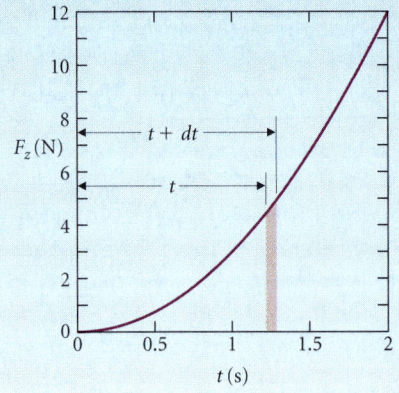

■ FIGURE I2.7
The z-component of the force exerted on the particle increases according to the formula $F_z(t) = (3.0 \text{ N/s}^2)t^2$. (The curve is a parabola.) A typical differential impulse $dI_z = F_z(t) \, dt$ is represented by the shaded rectangle.

SETUP
STEP III *Express the contribution* of the typical element to the desired sum in terms of the chosen coordinate. It is important that all dependence on the coordinate be shown *explicitly*.

The desired contribution is the differential impulse $\vec{F}(t)\,dt$. We are given an explicit expression for the force: $F_y(t) = (2.0\text{ N/s})t;\ F_z(t) = (3.0\text{ N/s}^2)t^2$. Thus the differential impulse has components:

$$dI_y = F_y(t)\,dt = (2.0\text{ N/s})t\,dt$$

and
$$dI_z = F_z(t)\,dt = (3.0\text{ N/s}^2)t^2\,dt.$$

STEP IV Determine the appropriate *limits of integration*.

We are given the velocity at $t = 0.0$ s, and are asked for the velocity at 2.0 s. These are the appropriate time limits for the integration.

$$p_{y,f} - p_{y,i} = I_y = \int F_y(t)\,dt \qquad p_{z,f} - p_{z,i} = I_z = \int F_z(t)\,dt$$

$$= \int_0^{2.0\text{ s}} (2.0\text{ N/s})t\,dt. \qquad\qquad = \int_0^{2.0\text{ s}} (3.0\text{ N/s}^2)t^2\,dt.$$

SOLVE *Evaluate* the resulting integrals.
STEP V The integrals of powers of t are standard:

$$p_{y,f} - p_{y,i} = \frac{t^2}{2}(2.0\text{ N/s})\Big|_{0\text{ s}}^{2.0\text{ s}} = 4.0\text{ N}\cdot\text{s}$$

and
$$p_{z,f} - p_{z,i} = \frac{t^3}{3}(3.0\text{ N/s}^2)\Big|_{0\text{ s}}^{2.0\text{ s}} = 8.0\text{ N}\cdot\text{s}.$$

Lastly, we use these relations to calculate the final velocity:

$$v_{y,f} = p_{y,f}/m = (4.0\text{ N}\cdot\text{s})/(0.70\text{ kg}) = 5.7\text{ m/s}$$

and $v_{z,f} = p_{z,f}/m = (8.0\text{ N}\cdot\text{s})/(0.70\text{ kg}) + v_{z,i} = 11\text{ m/s} + 5.0\text{ m/s} = 16\text{ m/s}.$

So the final velocity is:

$$\vec{v} = (3\text{ m/s})\hat{i} + (6\text{ m/s})\hat{j} + (16\text{ m/s})\hat{k}.$$

ANALYZE We could have done this example by calculating $\vec{a}(t) = \vec{F}/m$ and integrating each acceleration component to find the corresponding change in velocity component. The only difference is that we would divide by the mass at the beginning rather than at the end. ∎

Math Topic

INTEGRALS AS AREAS

Before the invention of electronic computers, graphical methods served as a research technique for evaluating integrals of functions that were numerically defined or whose integrals did not have a known analytic form. Undergraduates were often hired to plot graphs carefully and then cut them out so that their area could be measured by weighing the graph paper! This method is still valuable for evaluating the integral of a function displayed by a chart recorder, though as computers take over, its use becomes ever rarer.

Computing the integral from the measured mass of paper is an exercise in units conversion! The desired area is proportional to the measured mass. For that matter, the integral is proportional to the area under the curve rather than being equal to the area. For a given integral, the actual value of the area under the curve (in m²) depends on the choice of scales on the axes used to plot the function. For this reason, when speaking carefully, we say that the area under a curve **represents** an integral rather than calling them equal.

You probably could have done Example I2.2 without referring to any general plan, although it illustrates that the plan is the basis of all integration problems. However, we have deliberately chosen the next example—work done by a variable force—to illustrate the techniques you typically have to consider in each of the steps.

EXAMPLE I2.3 ♦♦♦ A railroad freight car of mass M and initial speed v_o rolls without friction toward the positive x-direction (■ Figure I2.8). The car is brought to rest by a spring bumper with spring constant k on a second freight car with its brakes locked. Find the final compression s_f of the spring.

MODEL We are asked to find the compression s_f of the spring when the car has come to rest, that is, when its kinetic energy is zero. According to the work-energy theorem, the (negative) work done by the spring on the freight car is just enough to remove the car's initial kinetic energy $K_i = \tfrac{1}{2}Mv_o^2$.

The force exerted by the spring depends on its compression, and hence on the position of the freight car. Since the force acting on the car is variable, we have to compute differential energy transfers $dW = \vec{F} \cdot d\vec{s}$ occurring during differential displacements $d\vec{s}$. The total work done by the spring is the sum of these differential transfers: $W = \int dW$ (eqn. I2.2).

STEP I Choose appropriate coordinates to describe the system or process and model it as a collection of differential elements, each of which corresponds to a differential change in one of the coordinates.

The process is one of energy transfer as the freight car moves along a straight line, which we take as the x-axis (Figure I2.8). We choose the origin at the end of the relaxed spring, with positive direction to the right. The differential elements are the displacements $d\vec{s}$ of the car as it moves along the axis.

REMINDER: "TYPICAL" MEANS NOT IN A SPECIAL PLACE—NOT IN THE MIDDLE, NOT AT EITHER END. DON'T BLOW THIS ONE JUST BECAUSE IT SEEMS SIMPLE IN THE FIRST EXAMPLES.

SETUP STEP II Identify a *typical* element and describe it using the chosen coordinate. The typical element is a differential displacement of the car,

$$d\vec{s} = \hat{\imath}\, dx,$$

identified by its (arbitrary) coordinate x and its length $|dx|$. Since the car moves to the right, $dx > 0$.

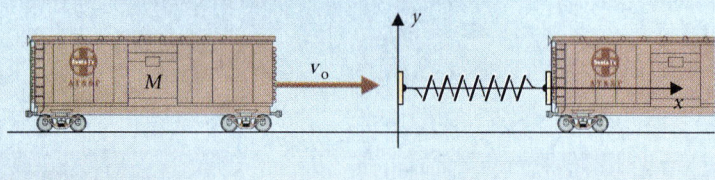

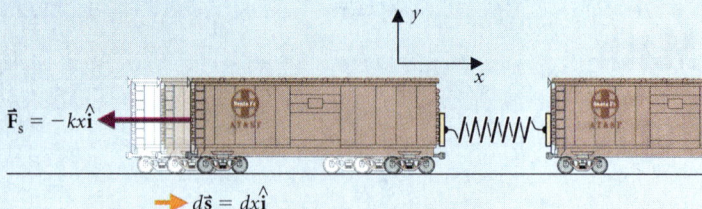

■ **FIGURE I2.8**
(a) A freight car rolling on level tracks encounters a second car with its brakes locked. The rolling car compresses the spring bumper on the stationary car by an amount s_f. The spring does (negative) work on the freight car while bringing it to a stop. (b) A typical element of work done by the spring occurs as the freight car moving to the right undergoes the displacement $d\vec{s} = \hat{\imath}\, dx$. The car's x-coordinate equals the spring's compression, and the spring force acts in the direction opposite the compression.

STEP III Express the contribution of the typical element to the desired sum in terms of the chosen coordinate.

It is *important* that all dependence on the coordinates be shown *explicitly*. When you come to evaluate the integral, you want nothing in your expression for the integrand but constants and the integration variable.

The desired quantity is the differential work done by the spring force acting through the displacement $d\vec{s}$. With our coordinate choice, x equals the compression of the spring, and the spring exerts force (eqn. 4.3) toward the left (the negative x-direction):

$$\vec{F} = (kx)(-\hat{i}).$$

A typical element of work done on the freight car is:

$$dW = \vec{F} \cdot d\vec{s} = (-kx\hat{i}) \cdot (dx\hat{i}) = -kxdx.$$

(Check the sign: as x increases the work done is negative, so the car slows down.)

STEP IV Determine the appropriate limits of integration.

The limits of integration are from $x = 0$, when the freight car contacts the spring, to $x = s_f$, when the car comes to rest. It is important to determine which is the lower and which the upper limit. There is no general rule. We noticed in Step II that a typical displacement has a positive value of dx, implying that the car goes from smaller values of x to larger values. The lower limit is $x = 0$, and $x = s_f$ is the upper limit. The total work done by the spring is then:

$$W = \int_{\text{path of car}} \vec{F} \cdot d\vec{s} = \int_0^{s_f} -kxdx = -k \int_0^{s_f} xdx.$$

SOLVE Evaluate the resulting integral.

STEP V The problem is now reduced to a straightforward mathematical exercise, integration of a polynomial:

$$W = -kx^2/2 \Big|_0^{s_f} = -ks_f^2/2.$$

Applying the work-energy theorem with $K_f = 0$:

$$W = \Delta K = K_f - K_i.$$

$$-\tfrac{1}{2}ks_f^2 = 0 - \tfrac{1}{2}Mv_o^2$$

or

$$s_f = v_o \sqrt{M/k}.$$

ANALYZE Though we derived this result as an exercise in learning how to do integrals, it has a much wider significance. According to Newton's third law, the force exerted by the car on the spring is equal and opposite to the force exerted by the spring. What we have found, expressed in terms of the properties and final state of the spring, is:

Work done on spring = $\tfrac{1}{2}$(spring constant) × (final distortion)^2.

■ Figure I2.9 is a graph representing the integral as an area. We could have made any of several other reasonable choices for the coordinate origin and direction in Step I. Such choices do not affect the final result for W. ∎

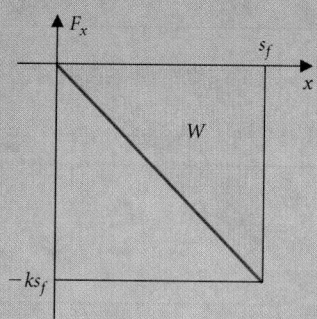

■ FIGURE I2.9
The x-component of force exerted by the spring is graphed versus x for the coordinate choice used in Example I2.3. The work done on the freight car is represented by the area beneath the line $y = F_x(x)$.

THIS OFTEN HAPPENS IN STEP II. YOU MAKE COMMITMENTS TO THE MEANING OF YOUR SYMBOLS THAT YOU NEED TO REMEMBER IN LATER STEPS.

CHECK THAT THE INTEGRAND INVOLVES ONLY CONSTANTS (k) AND x.

EVALUATION OF THE INTEGRAL MAY REQUIRE A CAREFUL CHOICE OF METHOD.

WE'LL DEVELOP THIS IDEA FURTHER IN CHAPTER 8.

GOOD HABIT: ANY IMPORTANT RESULT WILL STICK WITH YOU BETTER IF YOU TRANSLATE IT INTO WORDS, AS WE DO HERE.

EXERCISE I2.1 ♦♦♦ A ball is attached to a spring and the spring is initially stretched to a length $3L/2$. What is the work done on the ball by the spring as it contracts to its relaxed position with length L?

Summary

Physical quantities are frequently defined in terms of integrals, which may be thought of as sums of very large numbers of differential elements. We have developed a general technique for evaluating such physical integrals, summarized in Figure I2.5. Follow this procedure carefully and you will soon become proficient at setting up integrals.

Solution to Exercise

I2.1 **STEP I** Use a coordinate system with the origin at the relaxed position of the spring and with x increasing to the right.

STEP II ■ Figure I2.10 shows a typical displacement $d\vec{s}$, which points toward the left. It's convenient to use the same expression as before for the displacement: $d\vec{s} = \hat{i}\, dx$. So $dx < 0$, since the path is traversed from larger values of x to smaller.

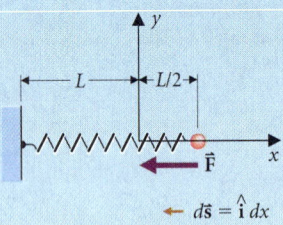

■ **FIGURE I2.10**
Here we choose the origin to be the relaxed position of the spring and the positive direction of the x-axis to be toward the position of the ball. This choice simplifies the expression for F_x.

STEP III With this choice of origin, the stretch of the spring equals the x-coordinate: $s = x$. Since the spring is stretched, the force on the ball points toward the origin:

$$\vec{F} = kx(-\hat{i}).$$

Thus the differential work done by the spring during the displacement $d\vec{s}$ is:

$$dW = \vec{F} \cdot d\vec{s} = (-kx\hat{i}) \cdot (\hat{i}\, dx) = -kx\, dx.$$

STEP IV The beginning of the path is at $x = L/2$ and the end is at $x = 0$. Since the path is traversed toward decreasing x, $L/2$ is the lower limit:

$$W = \int_{L/2}^{0} (-kx)\, dx.$$

STEP V
$$W = (-kx^2/2)\Big|_{L/2}^{0} = kL^2/8.$$

We found a similar result in Example I2.3. Whether the distortion s of the spring is a stretch or a compression, the work done by the spring has magnitude:

$$|W| = \tfrac{1}{2}ks^2.$$

Basic Skills

Review Questions

§I2.1 THE INTEGRAL OF A FUNCTION AS THE LIMIT OF A SUM

- Describe how to define an integral as the limit of a sum of rectangular areas.
- Why is the limit $N \to \infty$ an impractical ideal in some applications of integration?
- What is a *differential element*?
- Why is it possible to think of a *physical integral* as a sum of differential elements?

§I2.2 GENERAL METHOD FOR EVALUATING PHYSICAL INTEGRALS

- What are the five steps of the general method?
- In Step II, why is it important to describe a *typical element*?
- In Step III, why is it important to show all dependence on coordinates explicitly?
- In Example I2.3, describe how the choice of coordinates made in Steps I and II influences the calculation of limits of integration in Step IV.

Basic Skill Drill

1. The density of fluid in a cylindrical test tube varies along the length of the tube. Describe the differential elements you would consider using to calculate the total mass of fluid.

2. You wish to find the work done on a particle that moves along a circular path of radius R. Describe a differential element of the particle's path using angular position on the circle as a coordinate.

Questions and Problems

3. ❖ If you were to compare two graphs of the same function $F_x(t)$ whose scales on each axis differ by a factor 2 (■ Figure I2.11), how would the areas beneath the graphs differ? Explain why the different graphs give consistent values for the integral

$$\int_{t_1}^{t_2} F_x(t)\, dt.$$

4. ❖ Refer to Example I2.3. Consider two alternative choices of coordinates: (a) origin at the stationary freight car with x increasing to the right, and (b) origin at the end of the spring with x increasing to the left. In each case, give an expression for the typical differential work dW and state the limits of integration. Draw graphs similar to Figure I2.9 and describe why the different graphs represent the same total work.

5. ❖ Describe three different ways to imagine a circular disk *sliced* into elementary strips of area.

6. ❖ Use the representation of an integral as an area and the idea of differential elements to argue qualitatively why the following expressions are correct:

$$\frac{d}{dw}\int_a^w f(x)\, dx = f(w)$$

and

$$\frac{d}{dw}\int_w^b f(x)\, dx = -f(w).$$

7. ❖ Assume that the formula for the volume of a sphere of radius r is known. Discuss how to find the formula for the surface area of a sphere by modeling a spherical shell as the difference of two spheres.

8. ❖❖ Refer to Example I2.1. Write expressions for the force as a function of time in the intervals 0–5 s and 5–12 s. Integrate the force analytically and so re-derive the result of the example.

9. ❖❖ A body moves along a straight line with an acceleration that increases linearly with time: $a(t) = Ct$. Find the particle's displacement and velocity as functions of time.

10. ❖❖ Consider a triangle as a collection of horizontal strips of height dy (■ Figure I2.12). Calculate the area of the triangle by integrating—that is, by summing the areas of the differential strips.

11. ❖❖ A right circular cone stands with its base on the x–y plane. The radius of its base is R and its height is h. Consider the cone as a collection of disks of height dz and calculate the volume of the cone by integrating the volumes of the differential disks.

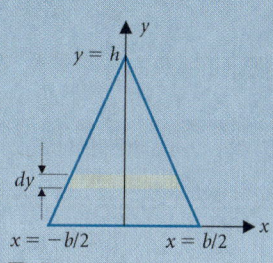

■ FIGURE I2.12

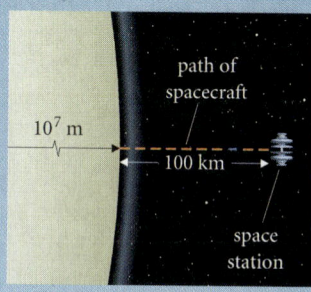

■ FIGURE I2.13

12. ❖❖ Consider a sphere of radius R as a collection of disks of thickness dz parallel to the x–y plane. Calculate the volume of the sphere by integrating the volumes of the differential disks.

13. ❖❖❖ Consider a circle of radius R in the x–y plane as a collection of strips of width dx, and calculate the area of the circle by integrating the areas of the differential strips.

14. ❖❖❖ A spaceship with mass 1.0×10^7 kg rises from the surface of a planet with mass 1.0×10^{24} kg and radius 1.0×10^7 m. Assuming that the ship moves along a radial line as shown in ■ Figure I2.13, how much work must its engines do **(a)** just to lift the ship to the height of a space station 1.0×10^2 km above the surface; and **(b)** once there, to accelerate the ship from rest to the speed of the space station? (Ignore any rotation of the planet.)

15. ❖❖❖ A particle moves in a circle of radius R with a speed that increases with time according to the formula $v = ct^3$. Find the average speed of the particle during the first revolution.

16. ❖❖❖ A block of mass M is initially at rest on a frictionless surface as shown in ■ Figure I2.14. The block is attached to a string that passes over a frictionless pulley a height h above the block and a horizontal distance d away. A person pulls downward on the string with a constant force of magnitude F, which is less than the weight of the block. Find the speed of the block when it is directly beneath the pulley.

17. ❖❖❖ A block of mass M is supported by a frictionless pulley system as shown in ■ Figure I2.15. A person pulls slowly on the rope so as to lift the block. Compute the tension $T(y)$ in the rope as a function of y. Show that the work done by the person balances the negative work done on the block by gravity.

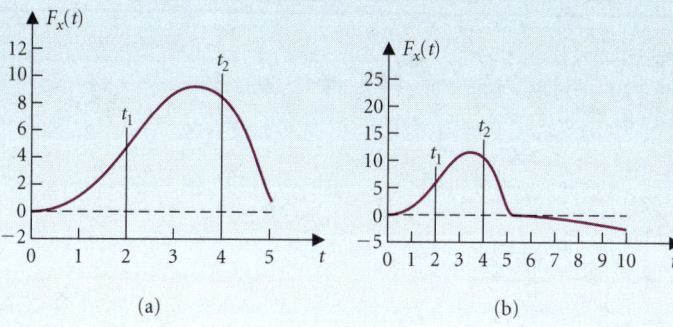

■ FIGURE I2.11

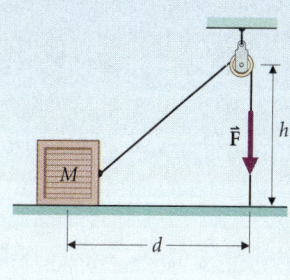

■ FIGURE I2.14

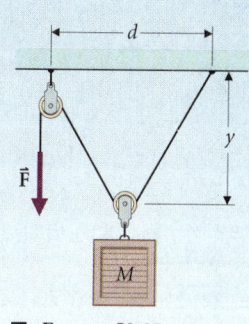

■ FIGURE I2.15

18. ♦♦♦ A bead slides along a frictionless horizontal wire lying at a 45° angle to the x-axis, as shown in ■ Figure I2.16. In addition to any normal forces exerted by the wire, the bead is subject to an external force that depends on position according to the formula:

$$\vec{F} = F_o \left(\frac{x}{x_o}\right)^2 \hat{\imath}.$$

If the bead has a speed v_o as it leaves the origin, find its speed as it leaves the right end of the wire.

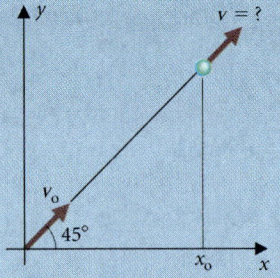

■ FIGURE I2.16

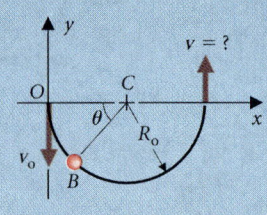

■ FIGURE I2.17

Challenge Problems

19. Archimedes (287–212 B.C.) calculated the value of π by finding the limit of the perimeters of polygons inscribed in a circle. (a) Explain how this method is equivalent to integrating the path length around a circle. (b) Show that the perimeter of a regular polygon of n sides inscribed within a circle of radius R is $p_i(n) = (2nR)\sin(180°/n)$. Show that the perimeter of a circumscribed polygon is $p_c(n) = (2nR)\tan(180°/n)$. (c) Using known values of the trig functions for 30° and 45° and the half-angle relations from Appendix IA, evaluate the perimeters of both kinds of polygon for $n = 4, 6, 8$ and 12. (d) Plot a graph of $[p(n)/2R]$ versus n^{-1} for each kind of polygon. Extrapolate your graphs smoothly to the limit $n^{-1} \to 0$ and estimate π. (e) (Computer problem) Use a spreadsheet program to evaluate $p_i(n)/(2R)$ and $p_c(n)/(2R)$. Using the average of these two values to estimate π, how large an n do you need to compute π to (a) three significant figures and (b) four significant figures?

20. A particle with mass $m = 1.2 \times 10^{-13}$ kg moves along the x-axis, and is at rest at $t = 0$. The force F_x acting on the particle is measured at three times:

Time t (s)	Force F_x (10^{-10} N)
0.0	0.0
2.5	8.0
5.0	4.0

Find the coefficients a and b for which the formula $F_x(t) = at + bt^2$ fits these data. Estimate the time at which the force returns to zero, and apply the impulse-momentum theorem to determine the particle's speed at that time.

21. A bead slides on a frictionless wire bent into a horizontal semicircle of radius R_o, as in ■ Figure I2.17. In addition to any normal forces exerted by the wire, the bead is subject to an external force that points directly away from the origin and depends on distance r from the origin according to the formula:

$$\vec{F} = F_o \left(\frac{r}{R_o}\right)^2 \hat{r}.$$

If the bead has a speed v_o as it leaves the origin, find its speed as it leaves the right end of the wire. (Hint: When the bead is at a point B on the wire, describe its position using an angle $\theta = \angle OCB$.)

Energy is Eternal Delight.
WILLIAM BLAKE

CHAPTER 8
Conservation of Energy

CONCEPTS

Elastic energy
Potential energy
Gravitational energy
Conservative force
Conservation of mechanical energy
Internal energy

GOALS

Be able to:
Compute the total mechanical energy of a given system.

Identify when changes occur in a system's internal energy.

Use conservation of energy in problem solving.

Riding the Kumba roller coaster is a thrilling, modern version of Galileo's experiment on quantity of motion. Released from high above the ground, the coaster car converts gravitational energy into kinetic energy, much to the delight of the passengers. Alas, friction ultimately converts the energy to thermal form and the ride ends.

257

GALILEO'S EXPERIMENT IS DESCRIBED IN §0.3.2.

One of Galileo's inclined-plane experiments—in this modern version—is so popular that people stand in line to take part and then scream with delight during the experimental trials. The roller coaster, like Galileo's metal sphere, rolls along its track, losing kinetic energy as it gains height above the Earth's surface and regaining kinetic energy as it drops. Energy, like momentum, is a conserved quantity, but, unlike momentum, it can occur in different forms. The roller coaster interchanges kinetic and gravitational energy, while friction slowly transforms the coaster's total mechanical energy into less obvious forms.

Part of the delight in studying physics is discovering new ways in which systems can store and transform energy. In this chapter we shall study transformations among kinetic energy and the gravitational and elastic forms of potential energy. In isolated systems of particles, the sum of these three *mechanical* forms of energy is conserved. Transformation of energy between mechanical forms and various *internal* forms describes how the roller coaster finally comes to rest and how a motor can set it going in the first place.

THESE FORMS OF ENERGY INCLUDE THERMAL AND SOUND ENERGY; SEE §8.5.

8.1 Elastic Potential Energy
8.1.1 Stored Energy in a Compressed Spring

Experience tells us that a compressed spring can do many things that a relaxed spring cannot—accelerate an object, power a music box, or keep your antique clock running. As a compressed spring expands it exerts a force on an object attached to it, does work, and accelerates the object. The object gains kinetic energy. If we are to think of this energy as a real physical quantity, it shouldn't simply appear from nowhere; the energy must have been stored in the spring.[1] The compressed spring contains energy in a form we shall call *elastic* energy. In Interlude 2 we calculated the work $W = \frac{1}{2}ks^2$ done on a spring to compress it by an amount s. The energy stored in the spring equals the work done on it:

$$U = W = \tfrac{1}{2}ks^2. \tag{8.1}$$

SEE INTERLUDE 2, EXAMPLE I2.3, FOR A CALCULATION OF THE WORK DONE.

As the spring re-expands, it does an equal amount of work, and the stored energy is retrieved.

EXAMPLE 8.1 ♦ A spring has a spring constant $k = 5.10 \times 10^3$ N/m and it is compressed by 0.140 m. How much work can the spring do on an object placed on its free end?

MODEL The compressed spring exerts a force on an object placed on its free end.

SETUP The work available is given by eqn. (8.1).

SOLVE
$$W = \tfrac{1}{2}ks^2 = \tfrac{1}{2}(5.10 \times 10^3 \text{ N/m})(0.140 \text{ m})^2$$
$$= 50.0 \text{ N}\cdot\text{m} = 50.0 \text{ J}.$$

ANALYZE Notice that the units are correct. ∎

Using a system of two colliding freight cars, we can show how elastic energy may be distributed among different objects. A freight car rolling without friction with speed v_o collides with a second car at rest on the tracks with its brakes locked. The spring bumper is compressed by an amount s_f as the incoming car is brought to rest. The system gains elastic energy equal to the initial kinetic energy of the incoming car:

$$\tfrac{1}{2}ks_f^2 = W = \Delta K = \tfrac{1}{2}mv_o^2 \equiv E_o.$$

---

[1] The Earth gains momentum equal and opposite to that gained by the object, so no net momentum results. But, the Earth's mass is huge compared with that of the object, so its velocity change is tiny and its kinetic energy change is completely negligible (see Problem 8.21).

Table 8.1 Energy in the Freight Car/Spring System

State	Kinetic Energy, K		Elastic Energy, U	Total
	Car 1	Car 2	Spring Compression	
Initial state:	$E_o = \tfrac{1}{2}mv_o^2$	none	none	E_o
Spring fully compressed:	none	none	$E_o = U = \tfrac{1}{2}ks_f^2$	E_o
Spring re-expanded:	$mv_o^2/4$	$mv_o^2/4$	none	E_o

If the brakes hold, the incoming car rebounds and regains its initial kinetic energy (cf. Example 8.1). However, if we release the brakes, both cars move apart and the energy is divided equally between them (Example 8.2):

$$K_{1f} = K_{2f} = \tfrac{1}{4}ks_f^2 = \tfrac{1}{2}E_o.$$

The total energy remains constant regardless of how it is shared.

These results are summarized in ● Table 8.1. The first freight car does work on the spring to compress it and thereby transfers energy to the spring in the form of elastic energy. This elastic energy is then converted back to kinetic form as it is transferred in equal amounts to each of the two cars. The total amount of energy is conserved, though it is transformed twice, divided, and shared.

DIG DEEPER

8.1.2 Potential Energy

In the spring examples we have just discussed, kinetic energy is changed into a different form called elastic energy:

REMEMBER: ELASTIC REFERS TO THE ABILITY OF SOME SOLIDS TO UNDERGO REPEATED DISTORTIONS OF SHAPE AND STILL RECOVER THEIR ORIGINAL FORM.

> A spring stretched or compressed by an amount s contains *elastic energy* U given by:
>
> $$U = \tfrac{1}{2}ks^2, \qquad (8.2)$$
>
> where k is the spring constant.

A standard name for forms of energy such as this is *potential energy*.[2] It is important to understand that potential energy is just as real as kinetic energy and that there is more than one form of potential energy.

Let us investigate elastic energy further to discover some properties that are true of potential energy in general. In Interlude 2 we computed the potential energy stored in the spring from the work it did to decelerate the freight car. The spring does this work while changing between two states—an initial, relaxed state at the instant the car first encounters the spring, and a final, compressed state at the instant the car comes to rest. The potential energy in the initial state is zero. The potential energy in the final state equals the work done *on* the spring by the freight car, which is the negative of the work done *by* the spring between the initial and final states:

$$U_f - U_i = -W_{i \to f}.$$

---

[2] The name derives from the time of Newton, when physicists were first developing these ideas and believed kinetic energy to be the only real form of energy. An object attached to a compressed spring was said to have the potential (a concept from Aristotle's theory of change) to achieve a state with kinetic energy. Aristotle's description of the process has been abandoned, but the name remains.

Digging Deeper

CALCULATION OF WORK DONE ON THE FREIGHT CARS

EXAMPLE 8.2 ♦♦♦ After the first car comes to rest, both freight cars are then released simultaneously and roll without friction. If each car has mass m, find the kinetic energy of each when the spring reaches its equilibrium length.

MODEL Once the second car is released, both cars are free to roll and the spring does work on both of them. Since both cars move, the length of the spring and hence the spring force depends on both of their positions. Their velocities are related because the momentum of the system—two cars and a spring—is conserved. With both cars initially at rest, the system starts with no linear momentum, and gains none, since no *external* forces act in the x-direction. The resulting motion is symmetric with $\vec{p}_2 = -\vec{p}_1$, and since the cars have equal mass, $\vec{v}_2 = -\vec{v}_1$. To find the kinetic energies of both cars, we need only compute the (now positive) work done by the spring on one of them. The other car's motion is symmetric, and the work done on it is the same.

SETUP For x_2 we choose the positive direction to the right and the origin at the right end of the compressed spring. Similarly, for x_1 we choose the positive direction to the left and the origin at the left end of the compressed spring. (*Note:* the positive values of x_1 and x_2 correspond to opposite directions.) Since the speeds of the cars are always equal, the magnitudes of their displacements are also equal: $x_1 = x_2$. The compression of the spring at any instant (■ Figure 8.1b) is:

$$s = s_f - (x_1 + x_2) = s_f - 2x_1,$$

and the spring force acting on the first car is:

$$\vec{F}_1 = ks\,\hat{\mathbf{i}}_1 = k(s_f - 2x_1)\,\hat{\mathbf{i}}_1 \text{ (i.e., to the left).}$$

The work done on car 1 is:

$$W_1 = \int \vec{F}_1 \cdot d\vec{s}_1 = \int [k(s_f - 2x_1)\hat{\mathbf{i}}_1] \cdot (dx_1\hat{\mathbf{i}}_1).$$

The motion begins when $x_1 = 0$, and the spring reaches its equilibrium length when $s = 0$, that is, when $x_1 = x_2 = s_f/2$. So, applying the work-energy theorem:

SOLVE
$$K_1 - 0 = K_1 = W_1 = \int_0^{s_f/2} k(s_f - 2x_1)\,dx_1$$
$$= k(s_f x_1 - x_1^2)\Big|_0^{s_f/2}$$
$$= \frac{ks_f^2}{4}.$$

ANALYZE Each freight car ends up with half of the energy $\frac{1}{2}ks_f^2$ that was stored in the spring. Cars of differing mass would divide the energy unequally (see Problem 21).

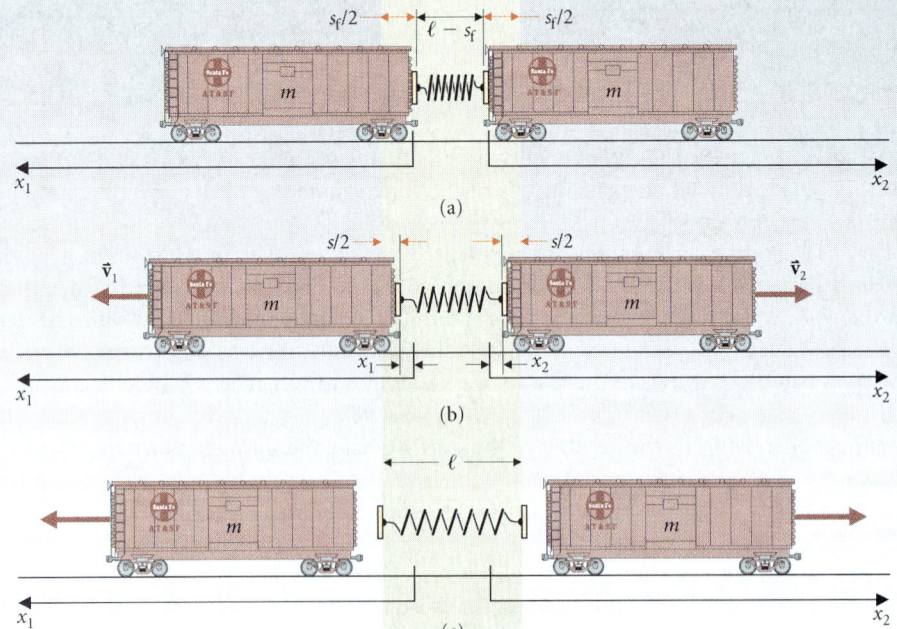

■ **FIGURE 8.1**
The brakes on the second freight car are released after the first car comes to rest. The spring now accelerates both cars in opposite directions. A coordinate is assigned to each car, with the positive sense chosen in the direction that car accelerates. (a) The initial compression of the spring is s_f. (b) At a later time, the spring has re-expanded an amount $x_1 + x_2$. Its compression $s = s_f - (x_1 + x_2)$. (c) After the spring reaches its relaxed length, the cars separate at constant speed.

Similarly, we can compare any two states of the spring: the work done by the spring as the freight car compresses it from s_A to s_B is:

$$W_{A \to B} = \int_{s_A}^{s_B} -(kx\hat{\mathbf{i}}) \cdot (dx\hat{\mathbf{i}}) = k\int_{s_A}^{s_B} -x\, dx = -kx^2/2 \Big|_{s_A}^{s_B}$$

$$= k(-\tfrac{1}{2}s_B^2 + \tfrac{1}{2}s_A^2) \equiv -U(B) + U(A) = -\Delta U_{A \to B}.$$

> The change of a spring's potential energy between any two states A and B is the negative of the work done *by* the spring between the two states:
>
> $$\Delta U_{A \to B} = -W_{A \to B}. \qquad (8.3)$$

Equation (8.3) is a general statement of the relation between work and potential energy. It embodies the idea that a spring can use stored energy to do work. For its stored energy to increase, the spring has to do negative work—that is, extract energy from other objects that do work on it. When a spring does work, it loses potential energy; when external forces do work on it, the spring gains potential energy.

Our results illustrate three features common to all forms of potential energy.

| PROPERTY 1 Net positive work done by forces in a system like the car-and-spring corresponds to a decrease in the system's potential energy.

It is equally important that change of potential energy does not depend on the way a system changes its state. In the calculation of $W_{A \to B}$ for the spring, nothing shows that the spring was compressed by a moving freight car. Using a locomotive to cause a slow compression of the same spring would cause the same changes in potential energy as the rapid compression by the freight car.

| PROPERTY 2 The system's potential energy depends only on the state of the system and not on the way it came to be in that state.

A third important fact is that work is related to a *change* in potential energy. Though it is conceptually important to think of energy as a real physical quantity contained in the spring, the only thing we can measure is how much the energy changes when the spring does something. We can write a formula for the spring's potential energy (eqn. 8.2) because there is a *reference state*—the relaxed spring—to which we can assign zero energy.

| PROPERTY 3 Only *changes* in potential energy are measurable. To assign numerical values to a system's potential energy, we must choose a *reference state* to which we assign zero potential energy.

8.1.3 Conservation of Energy in the Mass-on-Spring System

With the work-energy theorem and these properties of potential energy, we can now show that energy is conserved in a system of object-and-spring. In a change between any two states A and B of the system, the work-energy theorem requires:

Work done by spring = change in kinetic energy of the object,

or $\qquad W_{A \to B} = K(B) - K(A).$

But the work done by the spring equals the decrease in its potential energy (eqn. 8.3):

$$W_{A \to B} = -[U(B) - U(A)] = U(A) - U(B).$$

Thus: $\qquad U(A) - U(B) = K(B) - K(A),$

or $\qquad U(A) + K(A) = U(B) + K(B).$

The total energy of the system (potential plus kinetic) does not change as the system evolves between any two states A and B.

STUDY HINT: THE WHOLE LOGIC OF ENERGY CONSERVATION IS LAID OUT IN THIS SECTION.

■ **FIGURE 8.2**
The energy of a mass-on-spring system remains constant. The energy changes in form from kinetic to potential and back as the spring's length oscillates and the velocity of the mass changes.

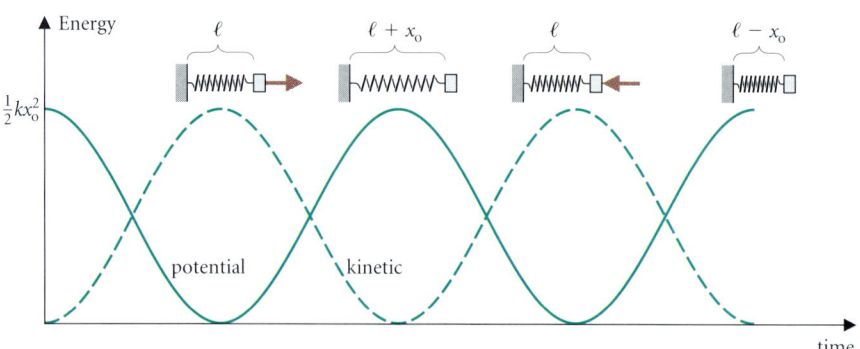

It is important to consider a complete system in understanding this conservation principle. The object alone is subject to an external spring force and its kinetic energy is not constant. The spring alone is subject to the force exerted by the object and so has variable potential energy. No forces external to the system of object *and* spring do work, and the energy (kinetic plus potential) of the *system* is conserved. ■ Figure 8.2 illustrates these relations for a system in which the mass is connected to the spring. Energy is transformed from potential energy to kinetic energy and back as the system evolves.

In this system only two forms of energy are significant. In the rest of the chapter we discuss how additional forms of energy fit into the overall conservation scheme.

REMEMBER: IN CONSERVATION LAW PROBLEMS WE COMPARE DIFFERENT STATES OF A COMPLETE SYSTEM (FIGURE 6.10).

EXAMPLE 8.3 ♦♦ A spring with spring constant $k = 120$ N/m and relaxed length $L = 0.32$ m is attached to a wall, and a ball of mass $m = 0.84$ kg on the end of the spring is free to slide without friction on a horizontal floor (■ Figure 8.3). The ball is pulled out until the spring has length $5L/4$ and then released. What is the maximum speed attained by the ball? What is the length of the spring when the ball is next at rest?

MODEL Both from context and from its pattern, this is a conservation law problem, so we follow the standard method. The relevant states are named by their descriptions: *release, maximum speed,* and *next at rest.*

Initially, the spring is stretched and the system possesses potential energy. When the ball is released, the spring accelerates it inward until it passes the point where the spring is at its relaxed length L. In this state, the spring extension is zero, and there is no potential energy; all of the system's energy is in the form of kinetic energy of the ball, which has its maximum speed. The spring then decelerates the ball until it is brought to rest. At this point, the spring is at maximum compression and all the energy of the system is again potential energy (see the following table).

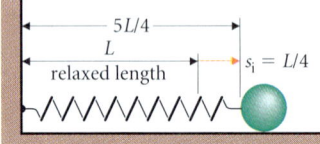
(a) Release

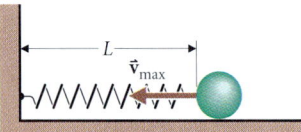
(b) Maximum speed

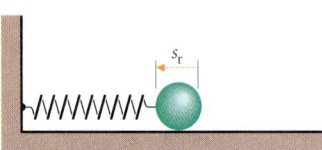
(c) Next at rest

■ **FIGURE 8.3**
A ball attached to a spring slides on a frictionless floor. (a) The ball is released when the spring's length is $5L/4$. (b) The ball has its maximum speed as it passes through the position where the spring's length equals the relaxed length L. (c) The ball next comes to rest when the spring is compressed an amount s_r.

SETUP The ball-and-spring system has zero initial kinetic energy since the ball is released from rest. The spring has initial potential energy $\frac{1}{2}ks_i^2 = \frac{1}{2}k(L/4)^2$ since it is extended by an amount $L/4$. The total energy of the system is:

$$E = E_i = K_i + U_i = 0 + kL^2/32 = (120 \text{ N/m})(0.32 \text{ m})^2/32 = 0.38 \text{ J}.$$

It is the same in all three states. Let the compression of the spring when *next at rest* be s_r and its *maximum speed* be v_{max}.

STATE OF SYSTEM	KINETIC ENERGY	POTENTIAL ENERGY	TOTAL
Release	none	$\frac{1}{2}k(L/4)^2$	$kL^2/32 = 0.38$ J
Ball at max. speed	$\frac{1}{2}mv_{max}^2$	none	$\frac{1}{2}mv_{max}^2 = 0.38$ J
Ball next at rest	none	$\frac{1}{2}ks_r^2$	$\frac{1}{2}ks_r^2 = 0.38$ J

Maximum speed (set total energy in the second line of the table equal to total energy in the first line):

$$\tfrac{1}{2}mv_{max}^2 = kL^2/32. \tag{i}$$

Next at rest (total energy in third line of the table equals total energy in first line):

$$\tfrac{1}{2}ks_r^2 = kL^2/32. \tag{ii}$$

SOLVE From eqn. (i), we find:

$$v_{max} = \sqrt{\frac{kL^2}{16m}} = \frac{L}{4}\sqrt{\frac{k}{m}} = \frac{0.32 \text{ m}}{4}\sqrt{\frac{120 \text{ N/m}}{0.84 \text{ kg}}} = 0.96 \text{ m/s}.$$

From eqn. (ii): $\qquad s_r^2 = L^2/16 \;\Rightarrow\; s_r = L/4.$

So, the spring's length is:

$$\ell = L - s_r = L - L/4 = 3L/4 = 3(0.32 \text{ m})/4 = 0.24 \text{ m}.$$

ANALYZE Let's check the units of the speed:

$$m\left(\frac{\text{N/m}}{\text{kg}}\right)^{1/2} = m\left(\frac{\text{kg}\cdot\text{m/s}^2}{\text{m}\cdot\text{kg}}\right)^{1/2} = \text{m/s},$$

as expected. Incidentally, since the units of $\sqrt{m/k}$ are seconds, this expression gives the time scale for significant changes in displacement or speed of the mass-on-spring system: $T \sim (m/k)^{1/2}$.

EXERCISE 8.1 ♦♦ At what two spring lengths are the kinetic and potential energies of the system in Example 8.3 equal?

8.2 GRAVITATIONAL POTENTIAL ENERGY

8.2.1 Practical Description for Use near a Planetary Surface

A ball thrown upward rises to a maximum height and then falls again. Just like an object compressing and rebounding from a spring, the ball loses kinetic energy as it rises and regains the same amount of kinetic energy as it falls. According to this analogy with the mass-on-spring system, the ball should be part of a system that gains potential energy from the kinetic energy of the rising ball and transforms the potential energy back into kinetic form as the ball falls. The other object in this system is the Earth, which exerts gravitational force on the ball.

The gravitational potential energy for the system of Earth and an object moving near its surface must possess the three properties of potential energy listed in §8.1. To find the potential energy function, according to Property 1, we should first compute the work done by gravity on an object that moves between two arbitrary points near the Earth's surface.

EXAMPLE 8.4 ♦♦♦ A bead of mass m slides on a frictionless wire from point A at the top of the wire to point B at the bottom. (■ Figure 8.4) Find the work done by gravity on the bead.

MODEL The gravitational force on a particle moving near the Earth's surface, $\vec{F}_g \equiv m\vec{g}$, is essentially constant. We choose the y-axis to be upward. Since the gravitational force acts downward, $\vec{F}_g = -mg\hat{j}$.

SETUP The work done on the bead during a differential displacement $d\vec{s}$ is then:

$$dW = \vec{F}_g \cdot d\vec{s} = -mg\hat{j} \cdot d\vec{s} = -mg\,dy,$$

where dy is the y-component of the displacement vector $d\vec{s}$. Since the bead is moving downward, dy is negative and therefore dW is positive. As the bead accelerates, we expect gravity to do positive work on it. The total work done in going from A to B is:

$$W_{A\to B} = \int_A^B -mg\,dy = -mg\int_A^B dy, \quad \text{since } g \text{ is constant.}$$

So: $\qquad W_{A\to B} = -mg(y_B - y_A) = -mg\,\Delta y.$

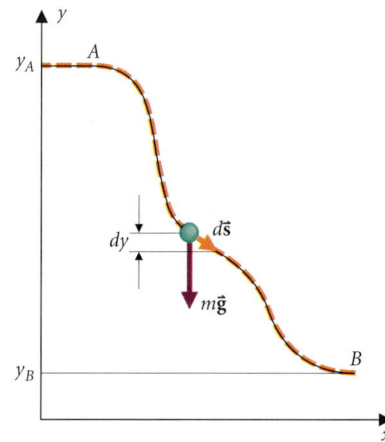

■ **FIGURE 8.4**
The gravitational force does work on a bead sliding along a frictionless wire. Because the gravitational force acts vertically, only the vertical component of the bead's displacement contributes to the work done on the bead: $dW = -mg\,dy$.

THIS EXPRESSION HOLDS NEAR THE SURFACE OF THE EARTH, WHERE WE MAY REASONABLY TAKE g TO BE CONSTANT.

THE POTENTIAL ENERGY BELONGS TO THE SYSTEM OF PARTICLE-AND-EARTH, BUT IT IS BOTH AWKWARD AND UNCONVENTIONAL TO SAY THAT EACH TIME. SO, WE SPEAK LOOSELY AS IF THE ENERGY BELONGS TO THE PARTICLE ALONE.

ANALYZE The net work done by gravity on the bead equals its weight, mg, multiplied by the decrease in its height. This result is independent of the shape of the wire.

EXERCISE 8.2 ❖ How much work is done on the bead by the forces the frictionless wire exerts on it?

Not only is the calculation in Example 8.4 independent of the shape of the wire, it doesn't even depend on the fact that the example concerns a bead sliding on a wire. We used only the expression for the gravitational force and the coordinates of points A and B. The same result applies to any object changing height for any reason. Thus, from Property 1:

$$-\Delta U \equiv W_{A \to B} = -mg\,\Delta y = -\Delta(mgy).$$

The height y of an object is independent of how the object got to that height, so Property 2 is satisfied if U is proportional to height. Property 3 requires that we express U as a difference from the energy in some reference state. Thus we are led to the following practical definition:

> A particle with mass m near the surface of the Earth has *gravitational potential energy* equal to its weight multiplied by its height above a suitably chosen reference level.
>
> $$U \equiv mgy. \qquad (8.4)$$

The definition makes intuitive sense: rolling heavy barrels horizontally is easy, while lifting them onto a shelf is exhausting (■ Figure 8.5). Only the lifting of the barrels requires energy input and results in an increase of potential energy. Conversely, one glance at a barrel about to fall and you know what it means to release gravitational potential energy.

Since we have described the work done on an object by gravitational forces as a decrease in the object's gravitational potential energy, we know immediately that kinetic plus gravitational energy is conserved in motions such as the rise and fall of the ball or the sliding of the bead. The proof follows exactly the same steps as the demonstration for spring energy.

> If only the gravitational force does work on a particle, the sum of its kinetic energy and gravitational potential energy is conserved.

EXAMPLE 8.5 ◆ A ball is thrown directly upward with an initial speed of $v = 11$ m/s. How high does it rise?

MODEL We neglect air resistance, as usual, and treat the ball as a particle whose kinetic plus gravitational energy is conserved.

SETUP Initially, the ball has a kinetic energy $\frac{1}{2}mv^2$, where m is its mass. If we choose the reference level at the ball's initial position, then we describe its initial potential energy as zero. At the top of its trajectory, at unknown height h, the ball's potential energy is mgh, and its kinetic energy is zero. Since energy is conserved, we use the standard form for conservation law problems:

BEFORE		**AFTER**	
(*immediately after throw*)		(*top of trajectory*)	
Kinetic energy	Gravitational energy	Kinetic energy	Gravitational energy
$\frac{1}{2}mv^2$	none	none	mgh

Set the total energies equal: $\quad \frac{1}{2}mv^2 = mgh.$

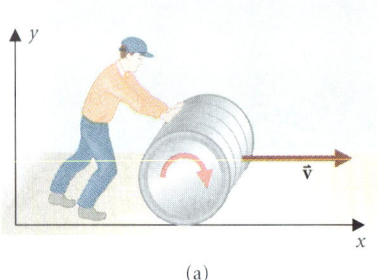

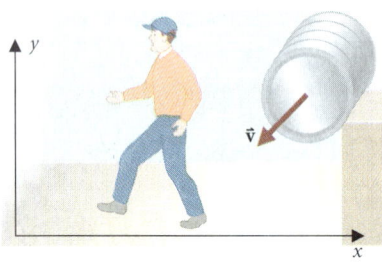

■ **FIGURE 8.5**
(a) Rolling a barrel along a level floor is pretty easy. (b) Lifting a barrel onto a shelf is exhausting. (c) A barrel falling off a shelf is dangerous! It makes sense to say that an object has greater potential energy the higher it is above the ground.

SOLVE Thus: $$h = \frac{v^2}{2g} = \frac{(11\text{ m/s})^2}{2(9.8\text{ m/s}^2)} = 6.2\text{ m}.$$

ANALYZE We have now solved this problem several different times with different methods (cf. Exercise 2.8, Example 7.2). The energy method is the quickest and the easiest. ∎

EXAMPLE 8.6 ♦♦♦ At the bottom of "The Big Drop" (point Q in ■ Figure 8.6) the Kumba roller coaster is 2 m higher than the ticket booth, and the track is best approximated by a circle with a radius of 23 m in a horizontal plane. If the mass of the coaster and passengers is 1600 kg, they started from rest at point P, a height of 43 m above the ticket booth, and friction is negligible, what normal force is the track exerting on the coaster at point Q?

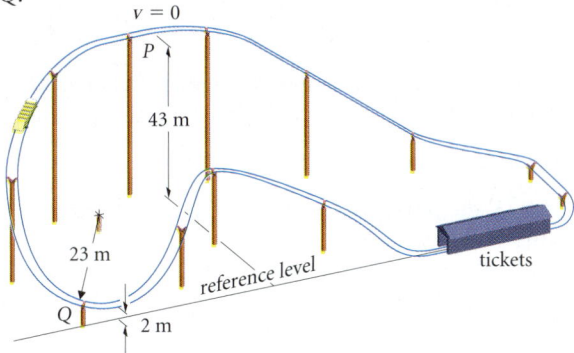

■ **FIGURE 8.6**
Diagram of a roller coaster ride. The car's velocity is essentially zero at the apex of the track (point P), 43 m above the ticket booth. At point Q, the track forms part of a horizontal circle with radius 23 m. Point Q is 2 m above the reference level.

MODEL The normal force causes the car's centripetal acceleration as it follows the circular path. We may use Newton's second law to find $\vec{n}$, once we know the acceleration. In circular motion, the acceleration has magnitude v^2/r. We are given $r = 23$ m, so we need v. The coaster's speed was zero when it was 43 m above the ticket booth, so we can use conservation of energy to find its kinetic energy—and hence its speed—at height 2 m.

SEE §3.2 FOR ACCELERATION IN CIRCULAR MOTION.

SETUP With the reference level at the ticket booth, we can evaluate the coaster's total energy both at its start and 2 m above the booth, and set them equal.

	POINT P		POINT Q		
	Potential	Kinetic	Potential	Kinetic	
	mgh_P	$+\ 0$	$=\ mgh_Q$	$+\ \tfrac{1}{2}mv_Q^2.$	(i)

■ Figure 8.7 is the free-body diagram for the coaster, with the x-axis horizontal and the y-axis directly upward. The coaster's acceleration is toward the center of the circle: $a_x = v_Q^2/r$; $a_y = 0$. So, applying Newton's second law:

$$\sum F_x = ma_x. \qquad\qquad \sum F_y = ma_y.$$
$$n\cos\phi = mv_Q^2/r. \quad\text{(ii)} \qquad n\sin\phi - mg = 0. \quad\text{(iii)}$$

SOLVE From conservation of energy (eqn. i):

$$v_Q^2 = 2g(h_P - h_Q). \tag{iv}$$

Dividing eqn. (iii) by eqn. (ii) and using the result (iv), the direction of the normal force is given by:

$$\tan\phi = \frac{n\sin\phi}{n\cos\phi} = \frac{gr}{v_Q^2} = \frac{r}{2(h_P - h_Q)} = \frac{23\text{ m}}{2[(43\text{ m}) - (2\text{ m})]} = 0.280.$$

So, $\phi = 15.7°$.

From eqn. (iii), the normal force has magnitude:

$$n = \frac{mg}{\sin\phi} = \frac{mg}{\sin(15.7°)} = \frac{(1600\text{ kg})(9.8\text{ m/s}^2)}{0.270} = 5.8 \times 10^4\text{ N}.$$

Thus $\vec{n} = (5.8 \times 10^4$ N, at an angle 16° above the horizontal).

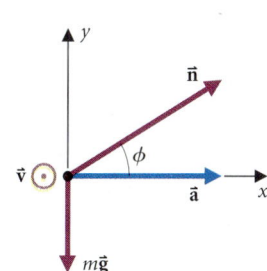

■ **FIGURE 8.7**
Free-body diagram for the roller coaster car. The forces acting on the car are its weight and the normal force exerted by the track. We are neglecting any frictional force. The car's velocity $\vec{v}$ is outward from the plane of the diagram. (The symbol ⊙ indicates the tip of the arrow pointing at you.) The car's acceleration is to the right and is horizontal, along our choice of x-axis.

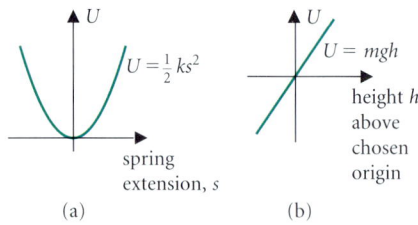

FIGURE 8.8
Graphs of potential energy versus position for elastic energy stored in a spring and for an object near the Earth's surface. The elastic energy function, $U(s) = \frac{1}{2}ks^2$, has a minimum at $s = 0$, where the spring is relaxed. This minimum provides a natural choice of reference state for the spring. The graph of gravitational energy is a straight line, $U(h) = mgh$, with no special feature to guide a choice of reference state.

BUT, DON'T MOVE THE REFERENCE LEVEL IN THE MIDDLE OF A CALCULATION.

SEE §5.4 FOR AN EXTENSIVE DISCUSSION OF THE GRAVITATIONAL FORCE.

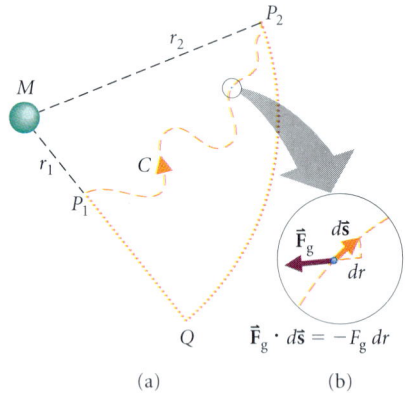

FIGURE 8.9
(a) The potential energy of a system depends only on its state and not on how the system came to be in that state. Thus we have to check that the work done by the Earth's gravitational force on a particle moving between two arbitrary points P_1 and P_2 does not depend upon the particle's path. (b) The particle undergoes displacement $d\vec{s}$ along an element of the arbitrary path C.

ANALYZE We can obtain an alternative expression for n from eqns. (ii) and (iii):

$$n^2 = n^2 \cos^2 \phi + n^2 \sin^2 \phi = (mv_Q^2/r)^2 + (mg)^2,$$

and

$$n = mg\sqrt{1 + (v_Q^2/gr)^2}.$$

In this form we can see that $n \to mg$ as $v_Q \to 0$, as expected.

The major new feature in this problem is the meshing of techniques for solving dynamics and conservation law problems. ∎

These examples illustrate an odd feature in the practical formula for gravitational potential energy: the value assigned to gravitational energy depends on the choice of reference level. We are left with the question:

- How should we choose the reference level?

This point is a common source of confusion in working with gravitational energy. The method used in the examples suggests the best practical answer to the question:

- Put the reference level wherever it makes the calculations easiest!

According to Property 3, the physically significant feature of potential energy is how much it *changes* during a process. Choosing a different reference level simply adds a constant amount to the numerical value of a particle's potential energy at every location. This constant cancels when energy changes are computed.

EXERCISE 8.3 ♦ Re-do Example 8.5, choosing the reference level at a height $d = 10.0$ m above the launch point of the ball.

We could also add an arbitrary constant to the definition of a spring's elastic potential energy with a similar lack of physical significance. In ▪ Figure 8.8 we have plotted the potential energy functions both for the spring and for an object near the Earth's surface. Our choice of reference state determines which point on the energy (vertical) axis is labeled as the origin. The spring's energy function has a minimum when the spring is relaxed, and this is a natural and intuitive choice for the reference state. The gravitational energy function is a straight line with no minimum and hence no obvious choice of origin.

This odd ambiguity is not a feature of gravity, but of the practical approximation we are using. When we use the exact form of Newton's gravitation law, we do find a natural choice for the reference state.

8.2.2 An Exact Expression for Gravitational Potential Energy

Near the Earth's surface, representing gravity as a constant force is a good practical approximation. But the magnitude of the gravitational force on a particle of mass m actually varies with position according to Newton's formula:

$$F_g = GMm/r^2.$$

We need the corresponding expression for gravitational energy to understand the energy requirements for a trip to the Moon or for the launching of a communications satellite.

The first step is to compute the work done by the gravitational force acting on a particle of mass m that moves along some arbitrary path C from position P_1 at a distance r_1 from a body of mass M to position P_2 at a distance r_2 from the body (▪ Figure 8.9). The element of work done on the particle by the gravitational force during a displacement $d\vec{s}$ along C is:

$$dW = \vec{F}_g \cdot d\vec{s} = -\frac{GMm}{r^2}\hat{r} \cdot d\vec{s}.$$

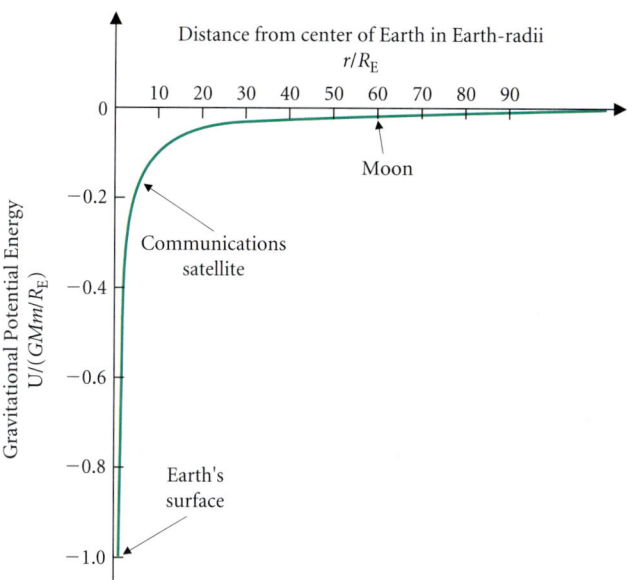

■ **FIGURE 8.10**
Exact form of the gravitational potential energy. The Earth's gravitational influence becomes negligible at very great distances from the planet. The potential energy of an object at a large distance is thus nearly constant, so very large separation is a natural choice of reference state for gravitational potential energy. The scale on the vertical axis is the object's potential energy divided by the magnitude of its energy when on the Earth's surface. The horizontal axis gives distance as a multiple of the Earth's radius.

The product $\hat{r} \cdot d\vec{s}$ is the component of the displacement $d\vec{s}$ in the direction of $\hat{r}$—that is, the change of the particle's distance from the origin (Figure 8.9); $\hat{r} \cdot d\vec{s} = dr$. Thus the total work done on the particle is:

$$W_{1\to 2} = \int_{P_1}^{P_2} dW = \int_{r_1}^{r_2} -\frac{GMm}{r^2}\, dr$$

$$= (GMm)\frac{1}{r}\bigg|_{r_1}^{r_2} = (GMm)\left(\frac{1}{r_2} - \frac{1}{r_1}\right).$$

As we found near the Earth's surface, the work depends only on the initial and final distances from the Earth. It takes work to lift an object upward but not to carry it at constant height. On a path from P_1 to P_2 that goes radially outward from P_1 to Q, and then around a circle from Q to P_2, work is done along $P_1 Q$ (the vectors $\vec{F}_g$ and $\vec{s}$ are antiparallel), but not along QP_2 (the displacement $d\vec{s}$ is perpendicular to $\vec{F}_g$ at each point). The same work is done along the arbitrary path from P_1 to P_2, since it results in the same change of distance from Earth. So, the expression for $W_{1\to 2}$ satisfies Properties 2 and 3 for potential energy. From Property 1:

$$W_{1\to 2} = (GMm)\left(\frac{1}{r_2} - \frac{1}{r_1}\right) \equiv -\Delta U_{1\to 2} = -[U(r_2) - U(r_1)].$$

Finally, we choose the reference state. With the conventional choice, the particle is at a great distance from the central body ($r \to \infty$)—that is, where the influence of the gravitational force is negligible and potential energy changes very slowly with distance (■ Figure 8.10). With $r_1 = r$ and $r_2 \to \infty$, $GMm/r_2 \to 0$, $U(r_2) \equiv 0$, and we find:

> The gravitational potential energy of two objects with masses M and m separated by a distance r is:
>
> $$U(r) = -GMm/r. \tag{8.5}$$

The negative sign in this definition reflects the fact that gravity is an attractive force. An object gains kinetic energy as it falls toward the Earth, and so loses potential energy. If we choose to call the object's potential energy zero at large distances, then it is negative for all finite values of r.

SECTION 8.2 • GRAVITATIONAL POTENTIAL ENERGY

EXAMPLE 8.7 ♦ How much energy is required to lift a kilogram of payload from the Earth to the radius of the Moon's orbit?

MODEL To lift the payload, we have to increase its potential energy. In other words, the work we do on the payload is the negative of the work gravity does on it. We assume the kinetic energy is unchanged.

SETUP We use eqn. (8.5) to calculate the change in potential energy:

$$\frac{\Delta U}{m} = GM\left(\frac{1}{R_E} - \frac{1}{R_o}\right),$$

where R_E is the radius of the Earth, and R_o is the radius of the Moon's orbit.

SOLVE Using values from the inside front cover:

$$\frac{\Delta U}{m} = (6.7 \times 10^{-11} \text{ m}^3/\text{kg} \cdot \text{s}^2)(6.0 \times 10^{24} \text{ kg})\left(\frac{1}{6.4 \times 10^6 \text{ m}} - \frac{1}{3.8 \times 10^8 \text{ m}}\right)$$

$$= 6.2 \times 10^7 \text{ J/kg}.$$

ANALYZE The quantity $1/R_o$ is much less than $1/R_E$; almost all the energy is used to get to a distance of a few Earth-radii from the planet. It takes little more energy to get the payload to the Moon than to the orbit of a communications satellite, even though the satellite is ten times closer.

EXAMPLE 8.8 ♦♦ What speed must a spacecraft have when leaving the top of the atmosphere if it is to escape from Earth?

MODEL By escaping, we mean that the craft rises to a distance much greater than the Earth's radius. As the craft leaves the atmosphere, its kinetic energy is positive and its gravitational energy is negative. As it moves outward, the gravitational force slows it down, reducing its kinetic energy and increasing its potential energy. If the total (kinetic plus potential) energy of the spacecraft is negative, gravitational force eventually pulls it back to Earth. If the total energy is positive, the craft escapes Earth with positive kinetic energy. With the minimum initial speed for escape, the probe's speed approaches zero as the distance approaches *infinity*. The total energy is zero.

SETUP We use the standard form for conservation law problems. The *before* state is leaving the top of the atmosphere, and the *after* state is arrival at *infinity*. The height of the atmosphere makes a negligible difference to the initial radius.

Before		After	
Kinetic energy	Gravitational energy	Kinetic energy	Gravitational energy
$\frac{1}{2}mv_{esc}^2$	$-GMm/R_E$	none	none

Once the craft leaves the atmosphere, only gravitational forces act on the Earth–spacecraft system, so its total mechanical energy is the same in both the initial and final states:

$$\frac{1}{2}mv_{esc}^2 - \frac{GMm}{R_E} = 0.$$

SOLVE So: $v_{esc} = \sqrt{2GM/R_E}$

$$= \sqrt{2(6.7 \times 10^{-11} \text{ m}^3/\text{kg} \cdot \text{s}^2)(6.0 \times 10^{24} \text{ kg})/(6.4 \times 10^6 \text{ m})}$$

$$= 1.1 \times 10^4 \text{ m/s}.$$

ANALYZE For comparison, the speed of a jet airliner is about 300 m/s, or about one-fortieth the escape speed.

> **EXERCISE 8.4** ♦♦ A mass driver on the Moon sends rock to a space colony by firing it along a tangent to the Moon's surface. What speed is required for the rock to escape to infinity?

8.3 CONSERVATION OF MECHANICAL ENERGY

8.3.1 Conservative Forces

Gravity and spring forces have a special character that allows us to relate them to potential energy. That character is described by the first two features of potential energy that we discovered for the spring, and any force that shares it is called conservative:

> A force is *conservative* if the work it does on a particle depends only on the initial and final positions of the particle.

Usually a conservative force is a function of position only and not of velocity or of time. The spring force is conservative, since the work done by a spring depends only on the initial and final compressions of the spring. Gravitational forces are conservative since the work done by gravity on a particle depends only on the particle's initial and final heights. The friction force is an example of a nonconservative force. The work it does on a particle depends on the total path length between the initial and final positions, as well as other factors like the slope of the path and the material over which sliding occurs.

To clarify the connection between conservative forces and potential energy, let us see how to define a potential energy with the properties listed in §8.1 for any conservative force. ■ Figure 8.11 shows two points A and B and two paths a particle might follow under the influence of a conservative force $\vec{F}$ that points generally in the direction from A toward B. Suppose we choose point A as the reference state. As a particle moves from A to B, the conservative force does positive work on the particle, and its potential energy decreases by the same amount. Since A is the reference state, $U(A) = 0$, and so the potential energy function we desire is:

$$U(B) - U(A) = U(B) \equiv -W_{A \to B}. \tag{8.6}$$

The value of a potential energy function must depend only on the particle's location and not on how it got there. That is why only conservative forces correspond to potential energies; if $W_{A \to B}$ were to depend on the path between A and B, the definition of $U(B)$ would not make sense. We would have to specify the path in order to have a unique value for $U(B)$.

According to Property 1, the work done by the conservative force on a particle that moves from point B to a third point C is given by:

$$W_{B \to C} = U(B) - U(C).$$

Let's verify that the potential energy defined in eqn. (8.6) satisfies this relation. The potential energy at C is the work done by the conservative force along *any* path from A to C (eqn. 8.6). If we choose a path that goes first from A to B and then from B to C (■ Figure 8.12), then the work done is the sum of the works done on each of the two parts of the path:

$$W_{A \to C} = W_{A \to B} + W_{B \to C},$$

or

$$-U(C) = -U(B) + W_{B \to C}.$$

Rearranging this equation, we obtain Property 1: the work done is equal to the decrease in potential energy.

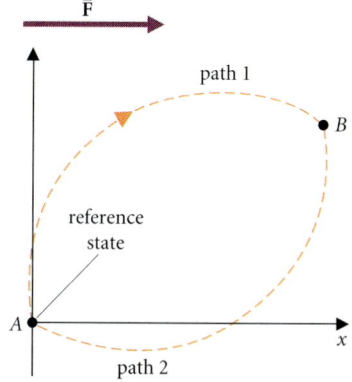

■ **FIGURE 8.11**
If the force $\vec{F}$ acting on a particle is *conservative*, the work it does on a particle moving from point A to point B, $W_{A \to B}$, is independent of the particle's path between the two points. A closed path beginning at A and returning to A follows arbitrary path 1 from A to B and arbitrary path 2 back to A. The net work done by the conservative force $\vec{F}$ around this closed path is zero. This property serves as an alternative definition of conservative force.

IN THE DEFINITION OF CONSERVATIVE FORCE, $\vec{F}$ REFERS TO A MATHEMATICAL FUNCTION OF POSITION THAT DESCRIBES THE FORCE THAT WOULD ACT ON AN OBJECT AT DIFFERENT POSITIONS IT MIGHT OCCUPY. THE "WORK DONE ALONG DIFFERENT PATHS" REFERS TO PROCESSES THAT MIGHT OCCUR.

NOTE THAT $\vec{F}$ NEED NOT BE, AND IN GENERAL IS NOT, CONSTANT.

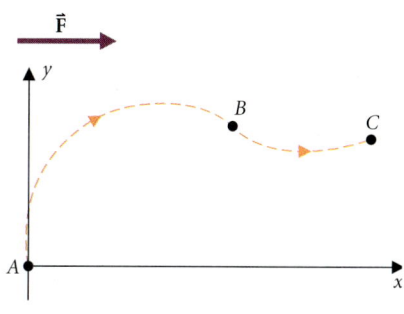

■ **FIGURE 8.12**
If the particle moves from A to B and then continues to point C, the work done on it by the conservative force $\vec{F}$ is the sum of works done on the two parts of the path.

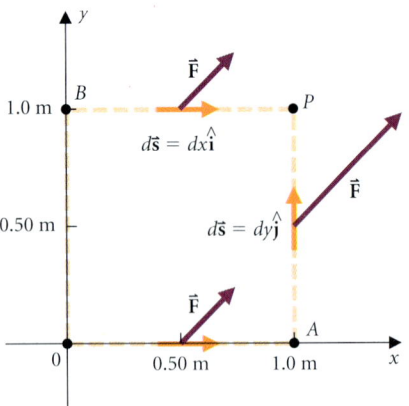

FIGURE 8.13
The force increases in magnitude as x increases. Thus the work done along path OAP is greater than the work done along path OBP. This force is not conservative.

WE'LL INCLUDE NONCONSERVATIVE FORCES IN §8.5.

EXAMPLE 8.9 ♦♦ Find the work done by the force $\vec{F} = (1.0 \text{ N/m})(\hat{i} + \hat{j})x$ between the origin and the point P ($x = 1.0$ m, $y = 1.0$ m) along two paths (■ Figure 8.13).

Path 1: along the x-axis to A ($x = 1.0$ m, $y = 0$), then along the line $x = 1.0$ m to P.
Path 2: along the y-axis to B ($x = 0$, $y = 1.0$ m), then along the line $y = 1.0$ m to P.
Can you determine whether the force is conservative?

MODEL We calculate the work done as $W = \int \vec{F} \cdot d\vec{s}$ along the two paths.

SETUP Path 1: Along OA, $d\vec{s} = dx\hat{i}$, and along AP, $d\vec{s} = dy\hat{j}$. Thus the work done is:

$$W = \int_O^A (1.0 \text{ N/m})(\hat{i} + \hat{j})x \cdot dx\hat{i} + \int_A^P (1.0 \text{ N/m})(\hat{i} + \hat{j})x \cdot dy\hat{j}.$$

The x-coordinate varies between 0 and 1.0 m along OA but has the fixed value 1.0 m between A and P. Along AP, y varies between 0 and 1.0 m. Remember: $\hat{i} \cdot \hat{i} = 1$; $\hat{i} \cdot \hat{j} = 0$. Then:

SOLVE
$$W = (1.0 \text{ N/m}) \left(\int_0^{1.0 \text{ m}} x \, dx + \int_0^{1.0 \text{ m}} (1.0 \text{ m}) \, dy \right)$$

$$= (1.0 \text{ N/m}) \left(\tfrac{1}{2}x^2 \Big|_0^{1.0 \text{ m}} + (1.0 \text{ m})y \Big|_0^{1.0 \text{ m}} \right)$$

$$= (1.0 \text{ N/m})[\tfrac{1}{2}(1.0 \text{ m})^2 + (1.0 \text{ m})^2] = 1.5 \text{ N·m}.$$

SETUP Path 2: Along OB, $d\vec{s} = dy\hat{j}$, and along BP, $d\vec{s} = dx\hat{i}$. Thus the work done is:

$$W = \int_O^B (1.0 \text{ N/m})(\hat{i} + \hat{j})x \cdot dy\hat{j} + \int_B^P (1.0 \text{ N/m})(\hat{i} + \hat{j})x \cdot dx\hat{i}.$$

SOLVE The x-coordinate $= 0$ everywhere along OB and varies between 0 and 1.0 m along BP:

$$W = (1.0 \text{ N/m}) \left(\int_0^{1.0 \text{ m}} 0 \, dy + \int_0^{1.0 \text{ m}} x \, dx \right)$$

$$= (1.0 \text{ N/m}) \left(0 + \tfrac{1}{2}x^2 \Big|_0^{1.0 \text{ m}} \right)$$

$$= (1.0 \text{ N/m})[\tfrac{1}{2}(1.0 \text{ m})^2] = 0.5 \text{ N·m}.$$

Since the amounts of work done along the two paths are not the same, the force is not conservative.

ANALYZE If the two values had been the same, we could not have decided that the force is conservative, since we would not have proved the general case: the work done must be the same on *any* path between O and P, not just the two chosen here. ■

8.3.2 Conservation of Total Mechanical Energy

We have now seen that there is a form of potential energy corresponding to each kind of conservative force that acts on a particle and that, when only one of these conservative forces acts, the sum of kinetic and potential energy is conserved. We are now ready to demonstrate a more general theorem that holds when several conservative forces act simultaneously in a system. Each force does work on a particle that moves between two points A and B, and as we showed in Chapter 7, the total work done is the sum of the works done by the individual forces:

$$W_{A \to B} = \sum_{\text{all forces}} W_j.$$

Math Topic

WORK DONE AROUND A CLOSED PATH

A conservative force is frequently defined as a force that does zero net work on a particle that moves around a closed path—a path that returns to its starting point:

The force $\vec{F}$ is conservative if and only if

$$\oint \vec{F} \cdot d\vec{s} = 0,$$

when the integral is taken around any closed path.

This statement is completely equivalent to the requirement that the work done between any two points be independent of the path. To see why, we'll calculate the work done along an arbitrary path beginning and ending at point A in Figure 8.11. The path consists of two parts: path 1 from A to B and path 2 returning from B to A. The net work W_{net} around the closed path is:

$$W_{net} = W_{A \to B \text{ on path 1}} + W_{B \to A \text{ on path 2}}.$$

Now, the work done going from A to B on path 2 is the negative of the work done if the particle goes in the opposite direction from B to A along the same path. The direction of the force is unchanged at each point, but the direction of the particle's displacement is reversed. The corresponding elements of work have the same magnitude but opposite sign:

$$W_{B \to A \text{ on path 2}} = -W_{A \to B \text{ on path 2}}.$$

So: $\quad W_{net} = W_{A \to B \text{ on path 1}} - W_{A \to B \text{ on path 2}}.$

For a conservative force, the work between A and B is the same regardless of path, so W_{net} is zero. The two definitions are equivalent, and it is useful to have both.

Each individual amount of work W_j done by a conservative force changes the corresponding potential energy U_j:

$$W_j = U_j(A) - U_j(B).$$

REMEMBER: WORK DONE *BY* THE SYSTEM DECREASES ITS POTENTIAL ENERGY.

If we define total potential energy U as the sum of the individual potential energies, we find that the total work done by conservative forces changes the total potential energy:

$$W_{A \to B} = \sum W_j \equiv \sum [U_j(A) - U_j(B)] = \sum U_j(A) - \sum U_j(B)$$
$$\equiv U(A) - U(B).$$

The sum of the total kinetic energy and the total potential energy is called the total mechanical energy of a system. The work-energy theorem now allows us to prove that it is conserved:

$$U(A) - U(B) = W_{A \to B} = K(B) - K(A),$$

or
$$U(A) + K(A) = U(B) + K(B).$$

The total mechanical energy (kinetic plus potential) of a system does not change. This is the law of conservation of mechanical energy:

> In the absence of nonconservative forces, the total mechanical energy of an isolated system is conserved.

EXAMPLE 8.10 ♦♦♦ A spring launcher is to be used for training baseball outfielders to catch high fly balls. The spring has relaxed length $L = 0.18$ m and spring constant $k = 8.5 \times 10^3$ N/m. (a) If the balls are to be launched at a 45° angle, what compression of the spring is required to send a baseball of mass $M = 160$ g a distance $D = 0.10$ km? (b) What maximum height does the baseball reach?

MODEL Air resistance, which we neglect, is the only nonconservative force acting on the baseball before it strikes the fielder's glove. Thus the total mechanical energy of the launcher–Earth–baseball system is conserved. Since we aren't told the fielder's height, we are invited to neglect any height difference between the launcher and the fielder. Similarly, we may neglect any change of the baseball's height as the launcher's spring expands.

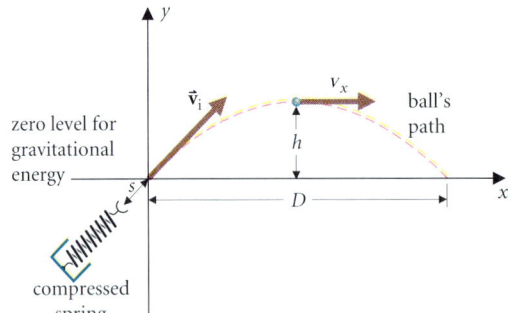

FIGURE 8.14
A spring launcher for baseballs. The spring is initially compressed an amount s. As it expands, it does work on the ball, giving it a speed v_i and corresponding kinetic energy K_i. As the ball travels, part of its kinetic energy is converted to potential energy until it reaches height h. Gravitational energy is converted back to kinetic energy as the ball descends.

We take the ball's position leaving the launcher as the reference level for gravitational energy (■ Figure 8.14).

Part (a)

SETUP The given 45° launch angle results in maximum range for a particular launch speed. The range is given by eqn. (3.2):

$$R_{max} = v_i^2/g.$$

If this range equals D, then $v_i^2 = Dg$. The ball's initial kinetic energy $K_i = \tfrac{1}{2}Mv_i^2$ is provided by potential energy stored in the launching spring. If the compression of the spring is s, we have (eqn. 8.2):

$$\tfrac{1}{2}ks^2 = \tfrac{1}{2}Mv_i^2.$$

SOLVE $\quad s = \sqrt{\dfrac{Mv_i^2}{k}} = \sqrt{\dfrac{MDg}{k}} = \sqrt{\dfrac{(0.16 \text{ kg})(100 \text{ m})(9.8 \text{ m/s}^2)}{8.5 \times 10^3 \text{ N/m}}} = 0.14 \text{ m}.$

ANALYZE The spring compression is almost 80% of the relaxed length—large but not impossible. From experience, we know that greater spring compression is needed if we want to launch bigger baseballs or to send them farther. We find that the necessary compression is directly proportional to the square root of both mass and distance.

Part (b)

SETUP Convenient choices of *before* and *after* states for this part are *just leaving the launcher*, and *maximum height*. The horizontal component of velocity $v_x = v_i \cos 45° = v_i/\sqrt{2}$ remains constant in projectile motion, and the y-component is zero at the top of the trajectory:

BEFORE		**AFTER**	
(*just released*)		(*maximum height*)	
Kinetic energy	Potential energy	Kinetic energy	Potential energy
$\tfrac{1}{2}Mv_i^2$	none	$\tfrac{1}{2}Mv_x^2$	Mgh

SOLVE Set the total mechanical energy in the *before* state equal to that in the *after* state:

$$\tfrac{1}{2}Mv_i^2 = \tfrac{1}{2}Mv_x^2 + Mgh.$$

Put in the known expressions for the initial speed and v_x:

$$\tfrac{1}{2}Mv_i^2 = \tfrac{1}{2}M[v_i^2/2] + Mgh.$$

$$Mv_i^2/4 = MDg/4 = Mgh.$$

So, $h = D/4 = 25$ m or, in a form familiar from previous problems, $h = v_i^2/4g$.

ANALYZE Finding the baseball's maximum height with energy methods provides a second example of problems that use energy conservation together with other kinds of physical information. ■

The law of conservation of mechanical energy has useful and interesting applications in problems where nonconservative forces are negligible. The most accurate such applications are to planetary motion, maneuvering of spacecraft, and atomic physics. The existence of nonconservative forces makes the law an incomplete statement for everyday applications. The remedy for this incompleteness lies in a description of all possible forms of energy, a task that remains at the heart of physical research.

SEE §8.5 FOR MORE ON THIS TOPIC.

8.4 POTENTIAL ENERGY IN SYSTEMS OF PARTICLES

How can we evaluate the total mechanical energy of a system of several particles and springs? The change in the potential energy of a particle is independent of the path it actually moves along, and depends only on the end points of the path. Similarly, the potential energy of a more complex system should depend only on its state and not how it got into that state. We may compute the potential energy of a system by imagining a convenient procedure for putting the system together and then computing the work required to do so. That is, we use the most convenient path for calculating the work done on each particle in the system.

For practical problems near the Earth, a simpler method is equivalent. We already know that the kinetic energies of the particles simply add to give a total kinetic energy of the system. The same rule applies to the potential energies of springs, and for particles near the Earth, we may simply add the gravitational energies of the various particles.

EXAMPLE 8.11 ◆◆ In an engineering model, two test objects, each of mass M, are hung one underneath the other, each supported by a spring with spring constant k and negligible equilibrium length (■ Figure 8.15). If the reference level for gravitational energy is at the point of suspension, find the potential energy of the system.

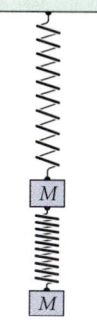

■ FIGURE 8.15
What is the potential energy of this system? (The relaxed length of each spring is negligible compared with its length in the equilibrium shown here.)

MODEL We shall compute the work required to assemble the system from isolated parts. First, we investigate the equilibrium state of the system.

SETUP We draw the free-body diagrams for the two test objects (■ Figure 8.16) and use them to calculate the stretch in each spring.

Lower block: $\sum F_y = 0$.   Upper block: $\sum F_y = 0$.
$F_2 - Mg = 0$. (i) $F_1 - Mg - F_2 = 0$. (ii)

From eqn. (i), $F_2 = Mg$, and from eqn. (ii), $F_1 = Mg + F_2 = 2Mg$. Each spring force has magnitude ks, so the spring stretches are given by:

$$ks_1 = 2Mg \Rightarrow s_1 = 2Mg/k \quad \text{and} \quad ks_2 = Mg \Rightarrow s_2 = Mg/k.$$

Now we may proceed to compute the energy of the system. Imagine the sequence of steps shown in ■ Figure 8.17, beginning with springs unstretched and objects at the level of the support point, where their gravitational potential energy is zero.

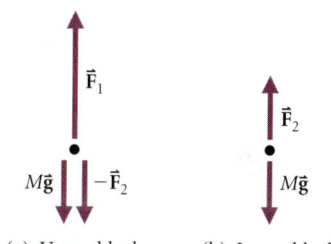

(a) Upper block   (b) Lower block

■ FIGURE 8.16
Free-body diagrams for the two objects. The lower spring exerts equal and opposite forces of magnitude F_2 on the two objects.

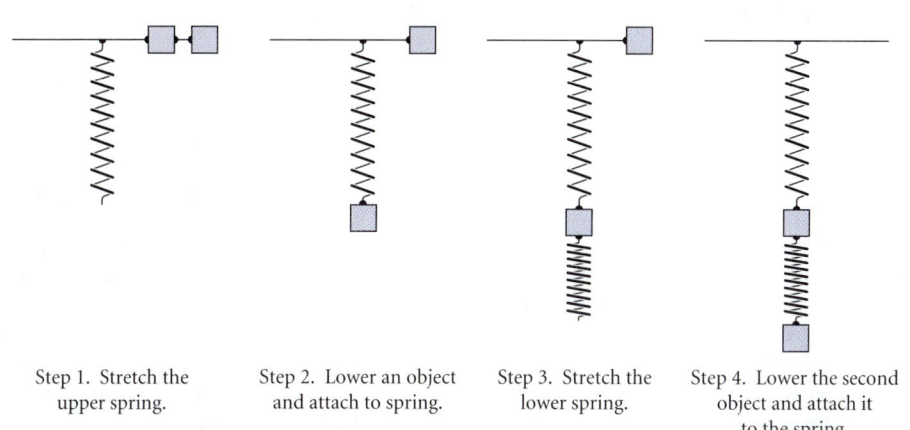

Step 1. Stretch the upper spring.

Step 2. Lower an object and attach to spring.

Step 3. Stretch the lower spring.

Step 4. Lower the second object and attach it to the spring.

■ FIGURE 8.17
We may assemble the system shown in Figure 8.15 from objects with zero potential energy in a sequence of four steps. The work done in this sequence of steps gives the system's potential energy, no matter how the system actually got into the given state.

SOLVE	Step	Work Done
	1. Stretch the upper spring to extension s_1.	$\frac{1}{2}ks_1^2$
	2. Lower the first object and attach it to the spring.	$-Mgs_1$
	3. Stretch the lower spring to extension s_2.	$\frac{1}{2}ks_2^2$
	4. Lower the second object and attach it to the spring.	$-Mg(s_1 + s_2)$

Once Step 4 is accomplished, the system may be released and it will be in equilibrium. The total work required to assemble the system, equal to the system's potential energy, is:

$$W \equiv U = \tfrac{1}{2}ks_1^2 - Mgs_1 + \tfrac{1}{2}ks_2^2 - Mg(s_1 + s_2)$$
$$= \frac{2(Mg)^2}{k} - \frac{2(Mg)^2}{k} + \frac{(Mg)^2}{2k} - \frac{3(Mg)^2}{k} = -\frac{5}{2}\left[\frac{(Mg)^2}{k}\right].$$

ANALYZE As expected, U is the sum of the individual potential energies of the springs and objects.

8.5 INTERNAL ENERGY

8.5.1 Thermal Energy

The most common example of a *nonconservative* force is kinetic friction. When a box slides across the floor, the frictional force acts in the direction opposite from the box's velocity and reduces the box's kinetic energy. Where does the energy go?

RECALL THE ROUGH-SURFACE MODEL IN §4.5.2.

The frictional force results from small irregularities in the surfaces of the box and floor. As the box slides across the floor, banging these irregularities into one another, the vibrations of molecules near the surfaces increase. Kinetic energy is transferred from the motion of the box to random motions of the atoms that make up the box and floor. Such energy is called thermal energy.[3]

WORK MAY ALSO BE DONE IN DEFORMING THE SURFACE, IF THE BOX SCRATCHES THE FLOOR.

> *Thermal energy* is energy, either kinetic or potential, that is randomly distributed among the molecules that make up an object.

For now, the crucial feature of thermal energy is that its production is not reversible. The energy put into compressing a spring is recovered when the spring re-expands. The work dissipated as heat when a box is pushed across a rough surface remains as molecular energy and cannot be retrieved by dragging the box back to its original position. The box will certainly not transform thermal energy back into kinetic form and spontaneously slide along the floor!

Thermal energy is our first example of internal energy:

> *Internal energy* is energy contained within an object being modeled as a particle.

---

[3] When mechanical energy is converted into thermal form, it is conventional to say that "energy is dissipated as heat." This usage, though standard, is doubly unfortunate. Dissipation conveys the notion of disappearance, which is quite wrong. Transformation of mechanical energy into thermal form does make a system appear to lose energy. Thermal energy does tend to spread out and become unnoticeable, but it does **not** disappear. Secondly, the word "heat" is used very differently from the sense here, once the serious study of thermal energy begins (Chapters 19–22). Not much harm is done if you keep in mind that "dissipated as heat" is a poor translation of "transformed into thermal energy."

8.5.2 Conservation of Energy

Throughout this chapter we have been converting the work-energy theorem for a particle into the law of conservation of energy. Originally, we recognized work as the process of exchanging kinetic energy with a particle. Then we recognized work done by conservative forces as the process of transforming a system's energy among the kinetic and potential forms. We are now ready to include nonconservative forces in a full statement of the law of energy conservation.

If we distinguish the work done by conservative forces from that done by nonconservative forces during a process, then the total work done by all forces is:

$$W_{total} = W_{noncon} + W_{con}.$$

This total work equals the change in the system's kinetic energy (eqn. 7.5):

$$W_{noncon} + W_{con} = \Delta K.$$

The net work done by conservative forces changes the system's potential energy (eqn. 8.3):

$$W_{noncon} = \Delta K - W_{con} = \Delta K - (-\Delta U) = \Delta(K + U).$$

> The net work done by the nonconservative forces in an isolated system equals the change in the system's total mechanical energy.
>
> $$W_{noncon} = \Delta(K + U). \tag{8.7}$$

THERE IS ONLY ONE WORK-ENERGY THEOREM, BUT WE CAN EXPRESS IT IN DIFFERENT FORMS. EQUATION (8.7) IS AN ALTERNATIVE VERSION OF EQN. (7.5) IN WHICH WORK DONE BY CONSERVATIVE FORCES IS REPLACED BY THE CONSEQUENT CHANGE IN POTENTIAL ENERGY.

Equation (8.7) shows how nonconservative forces modify the conservation of mechanical energy. Often, the only tool we have to compute the work done by nonconservative forces is to notice the change of total mechanical energy.

The work done by nonconservative forces is related to internal energy in the same way that work done by conservative forces is related to potential energy:

> The net work done by nonconservative forces during any process equals the decrease in the system's internal energy U_{int}.
>
> $$W_{noncon} = -\Delta U_{int}. \tag{8.8}$$

Then we have $-\Delta U_{int} = \Delta(K + U)$, which gives the final form of the *law of conservation of energy*:

> The total energy of an isolated system—equal to the kinetic energy of its particles, the potential energy associated with conservative forces acting within the system, and internal energy—is conserved.
>
> $$\Delta(K + U + U_{int}) = 0. \tag{8.9}$$

A system's internal energy is usually much more difficult to locate and describe than are the kinetic and potential forms of energy. Sometimes, as in an experimental auto design or a wind-up toy, we may look "inside" to find a spinning flywheel or a compressed spring. We would call such energy "internal" only because of the way we chose to model the system as "particles." More commonly, internal energy is in thermal form or in the chemical form stored in a jetliner's fuel tanks or an animal's body. We shall describe thermal energy more fully in Chapters 19–22. For now, our only two methods for computing internal energy changes are use of eqn. (8.7) and calculation of the work done by kinetic friction. So, the final form of the energy conservation law does not increase our ability to calculate. It does give more precision to the idea of mechanical energy being transformed to internal energy, and it also emphasizes the importance of energy as a conserved quantity.

> **EXAMPLE 8.12** ♦ A 6.4-kg bundle of sheets falls 25 m down a laundry chute in a hotel, reaching the hamper at the bottom with a speed of 7.4 m/s. How much work was done by air drag as the bundle fell? Estimate the average value of the drag force on the bundle.
>
> **MODEL** We may compare the total mechanical energy of the bundle at the top and bottom of the shaft. The difference is the work done by the nonconservative drag force.
>
> **SETUP** We take the reference level for gravitational energy to be at the bottom of the chute.
>
> | | **BEFORE** (Top of the chute) | **AFTER** (Bottom of the chute) |
> |---|---|---|
> | Kinetic energy | 0 | $\frac{1}{2} m v_f^2$ |
> | Gravitational energy | mgh | 0 |
> | Total mechanical energy | mgh | $\frac{1}{2} m v_f^2$ |
>
> The difference is the work done by the nonconservative forces (eqn. 8.7):
>
> **SOLVE**
> $$W_{\text{noncon}} = E_f - E_i = \tfrac{1}{2} m v_f^2 - mgh$$
> $$= (6.4 \text{ kg})[\tfrac{1}{2}(7.4 \text{ m/s})^2 - (9.8 \text{ m/s}^2)(25 \text{ m})]$$
> $$= 175 \text{ J} - 1568 \text{ J} = -1.4 \times 10^3 \text{ J}.$$
>
> **SETUP** We may express the work done in terms of the magnitude of the average drag force, which acts antiparallel to the bundle's velocity.
>
> $$W_{\text{noncon}} = -F_{\text{av}} h \implies F_{\text{av}} = -W_{\text{noncon}}/h.$$
>
> **SOLVE**
> $$F_{\text{av}} = (1.4 \times 10^3 \text{ J})/(25 \text{ m}) = 56 \text{ N}.$$
>
> **ANALYZE** This is 90% of the bundle's weight! Of course, the force is not constant during the drop, but increases as the bundle's speed increases. Despite the 1600-J decrease in gravitational potential energy, only 180 J ends up as kinetic energy. ■

Study Problem 7 ♦♦♦ *The Egg Factory*

Egg shipping crates of mass $m = 25$ kg slide down a ramp a vertical distance $h = 5.0$ m (■ Figure 8.18). The coefficient of kinetic friction between a crate and the ramp is $\mu_k = 0.50$. A spring with spring constant k and relaxed length ℓ is at the bottom of the ramp, a distance $d = 9.2$ m from the top. How would you design the system so that the spring pushes the crate back up the ramp and the crate comes to rest at the end of the relaxed spring, where it will be removed by the trucker? Find the maximum compression of the spring when a crate is brought to rest.

Modeling the System

We need to find a suitable spring constant k that brings the egg crate to a gentle stop and pushes it back up to the end of the relaxed spring. Since the problem asks for relations among various states of the crate–ramp–Earth–spring system and does not inquire about times required for its motion, we use the conservation laws.

Energy is the central concept in this problem. Three forms of mechanical energy are present: kinetic and gravitational energy of a crate and elastic energy of the spring. Because of friction, total mechanical energy is not conserved, but the work-energy theorem relates the work done by friction to the change in the total mechanical energy of the system.

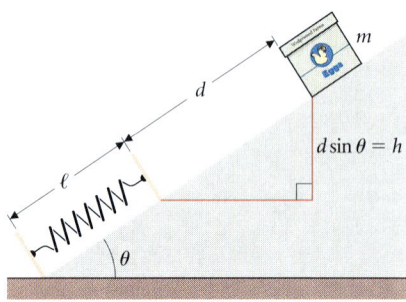

FIGURE 8.18
Egg shipping crates are allowed to slide down a ramp where they impact a spring bumper, which re-expands and leaves the crates at rest for workers to remove. In state 1 of this system the crate has just been released from rest at the top of the ramp.

Once the crate is brought to rest, it reaccelerates up the plane only if the force the spring exerts is large enough to overcome both the component of gravity down the plane and static friction. In that case, determining how far the crate returns up the plane again involves mechanical energy and frictional work.

Plan

For convenience, we choose the reference level for gravitational energy at the end of the relaxed spring. We need to consider the three states of the system shown in Figures 8.18 and ■ 8.19.

State 1: Crate begins sliding at the top of the ramp.
State 2: Crate is brought to rest near the bottom of the ramp. The spring is at its maximum compression, s_f.
State 3: Crate is at rest again at the end of the relaxed spring.

An equation for the spring constant k and the maximum compression of the spring can be found either by comparing states 1 and 2 or by comparing states 1 and 3. By eliminating s_f, we can find the required k.

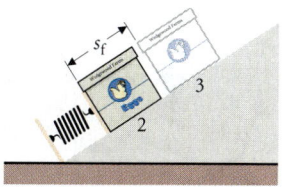

■ **FIGURE 8.19**
In state 2 of the egg crate handling system, the spring is at its maximum compression. In state 3, the spring has re-expanded to its relaxed length, leaving the crate at rest on the ramp.

Construction of Equations

The energies in the three states are as follows.

	State 1	State 2	State 3
Kinetic energy	none	none	none
Gravitational energy	$mgd \sin\theta$	$-mgs_f \sin\theta$	none
Elastic energy	none	$\tfrac{1}{2}ks_f^2$	none

Notice that in state 2 the crate is below the reference level, so its gravitational energy is negative.

The work done by friction depends on the total distance traveled by the crate. Between states 1 and 2, the crate travels down the slope a distance $d + s_f$. Between states 2 and 3, the crate travels back up the slope a distance s_f. The total distance traveled between states 1 and 3 is $(d + s_f)$(down) $+ s_f$(up) $= d + 2s_f$. Since $\vec{f}$ is antiparallel to the displacement throughout the motion, the work done by friction is $W_{\text{noncon}} = (-f_k)$(distance traveled), where $f_k = \mu_k n$ and $n = mg\cos\theta$.

Comparing states 1 and 2:

$$W_{\text{noncon}} = \Delta(K + U) = K_2 + U_2 - K_1 - U_1.$$

$$-\mu_k mg\cos\theta(d + s_f) = 0 + (\tfrac{1}{2}ks_f^2 - mgs_f \sin\theta) - 0 - mgd\sin\theta.$$

Rearranging gives:

$$\tfrac{1}{2}ks_f^2 - (\sin\theta - \mu_k \cos\theta)mg(s_f + d) = 0. \tag{i}$$

Comparing states 1 and 3:

$$W_{\text{noncon}} = \Delta(K + U) = K_3 + U_3 - K_1 - U_1.$$

$$-\mu_k mg\cos\theta(d + 2s_f) = 0 + 0 - 0 - mgd\sin\theta. \tag{ii}$$

SEE, FOR EXAMPLE, STUDY PROBLEM 4.

Solution

First we solve eqn. (ii) for s_f: $\quad d + 2s_f = \dfrac{d}{\mu_k}\tan\theta.$

Now $\sin\theta = h/d = (5.0 \text{ m})/(9.2 \text{ m}) = 0.54 \Rightarrow \theta = 32.9°.$

Thus: $\quad s_f = \dfrac{d}{2}\left(\dfrac{\tan\theta}{\mu_k} - 1\right) = \dfrac{9.2 \text{ m}}{2}\left(\dfrac{\tan 32.9°}{0.50} - 1\right) = 1.36 \text{ m}.$

SECTION 8.5 • INTERNAL ENERGY

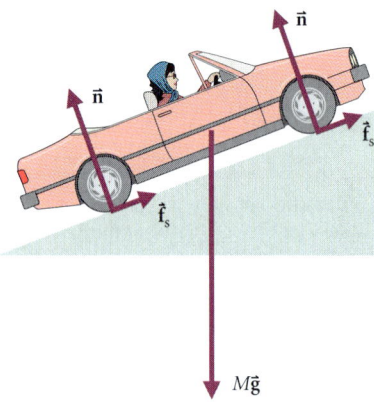

■ FIGURE 8.20
Forces acting on a car climbing a hill. The only force acting upslope on the car is the static friction force the road exerts on the car's tires, and this force does no work.

THE SAME IS TRUE OF YOU WALKING UP THE HILL. THE INCREASED GRAVITATIONAL ENERGY COMES FROM YOUR BREAKFAST.

SEE CHAPTER 22 FOR A DISCUSSION OF ENGINE DESIGN.

Then we solve eqn. (i) for k:

$$k = \frac{2(\sin\theta - \mu_k \cos\theta)mg(s_f + d)}{s_f^2}$$

$$= \frac{2[\sin 33° - (0.50)\cos 33°](25 \text{ kg})(9.8 \text{ m/s}^2)[(1.4 \text{ m}) + (9.2 \text{ m})]}{(1.36 \text{ m})^2}$$

$$= 350 \text{ N/m}.$$

The spring constant is 350 N/m and its maximum compression is 1.4 m.

Analysis

The spring must be longer than 1.4 m, since it compresses that much, and its size extends the length of the device considerably. We shall not inquire whether the eggs survive the impact with the spring! Otherwise, the design we have given does the assigned job. You might enjoy designing a better system!

> **EXERCISE 8.5** ♦♦ What value of μ_s is required so that the spring exerts enough force at maximum compression to push the crate back up the ramp, but the crate then remains at rest at the end of the relaxed spring?

8.5.3 What Does Make a Car Go Up a Hill?

In Chapter 4 we remarked that the only force acting upslope on a car climbing a hill is the static friction force exerted by the road on the car's tires (■ Figure 8.20). The road pushes the car up the hill, but only gravity does work on the car, and the work it does is negative! The normal force is perpendicular to the car's displacement. Because the tires roll without slipping, the static friction force does not act through any displacement. Any increase in the car's gravitational potential energy must come from the transformation of internal energy.

That's no surprise. Everybody knows that burning gasoline in the car's engine *causes* the car to go up the hill. Exactly how it does so is not so easy to understand. The engine is designed so that burned gasoline exerts mechanical force and delivers some of its thermal energy to the car in mechanical form. The internal forces that hold the car together transmit the energy needed to lift each part of the car. But none of this succeeds unless the tires can push on the road and thereby cause the road to push back. The car will not go up an icy hill, no matter how clever its engine design.

Chapter Summary

Where Are We Now?

We have identified the property observed by Galileo in his inclined-plane experiments—total mechanical energy. Galileo's observation captures the idea of energy conservation: energy continually changes in form but is neither created nor destroyed. By recognizing chemical and thermal forms of energy, we have stated the law of conservation of energy, perhaps the most fundamental principle in physics.

What Did We Do?

We developed the idea of potential energy by considering a spring that is compressed as it removes kinetic energy from a freight car. This energy, stored as elastic energy, $U = \frac{1}{2}ks^2$, can later be recovered by allowing the spring to expand.

The system of the Earth and a small particle near its surface has gravitational potential energy. A practical formula for this energy is the particle's weight multiplied by its height above a reference level. Since only changes in potential energy are physically significant, the reference level can be chosen arbitrarily. For convenience in doing mathematics, we assign values to gravitational energy itself. The arbitrary choice of reference level always cancels out of any problem solution. An exact calculation for the gravitational potential energy of an object of mass m at a distance r from the center of the Earth (mass M) is $U = -GMm/r$. The natural choice of reference state is at large separation, $r \to \infty$.

Gravity and spring forces are two examples of conservative forces. The work done on a particle by a conservative force depends only on the end points of the particle's path. Since potential energy should depend only on the state of a system and not how it got there, only such conservative forces are associated with potential energy.

In systems where more than one particle is free to move, the total kinetic energy of the system is the sum of the individual particles' kinetic energies. The potential energy stored in springs is also the sum of individual spring energies. For practical problems with objects near the Earth's surface, we may neglect any gravitational interaction among the particles and add their individual gravitational energies. It is always possible to compute the potential energy of a system by finding the work necessary to assemble the system starting from an imaginary state of zero potential energy.

The total mechanical energy of a system is the sum of its kinetic energy and all its forms of potential energy. Conservation of energy requires that the change in the total mechanical energy of a system equal the work done on that system by nonconservative forces:

$$W_{noncon} = \Delta(K + U).$$

Friction is the most common nonconservative force in mechanical problems. The energy it removes from a mechanical system is transformed into thermal energy (one form of internal energy). If we include internal energy together with mechanical energy, then the system's total energy is conserved.

Practical Applications

Every practical activity involves the transformation of energy from one form to another. Your home computer may receive some of its power from a hydroelectric power plant. The plant uses gravitational potential energy of mountain water to do work on its electrical generators. The gravitational energy came from the sunlight that evaporated the water from the oceans, and so on. This flow of energy is important whether you are designing the computer or the power plant, or predicting the weather! As we continue our study of physics, we will find energy to be an increasingly rich and important topic.

Solutions to Exercises

8.1 If the system's kinetic and potential energies are equal, then they are both equal to half the total. Let s be the compression (or extension) of the spring; then:

$$\tfrac{1}{2}ks^2 = \tfrac{1}{2}(kL^2/32) \Rightarrow s = \pm L/\sqrt{32} = \pm(L\sqrt{2})/8.$$

So, the two possible spring lengths are $L(1 \pm \sqrt{2}/8)$.

8.2 Since the wire is frictionless, it exerts only normal forces perpendicular to its surface. The bead moves along the wire, so its displacement is perpendicular to the normal forces. The wire does *no* work on the bead.

8.3 The only changes required by a redefinition of the reference level of potential energy are in the table entries for gravitational energy; the outline of procedure is identical to that in Example 8.5:

Before		After	
Kinetic energy	Gravitational energy	Kinetic energy	Gravitational energy
$\tfrac{1}{2}mv^2$	$-mgd$	none	$mg(h-d)$

Set total energies equal: $\frac{1}{2}mv^2 - mgd = mg(h - d)$. As before, we obtain $h = v^2/(2g) = 6.2$ m. Both the table entries for gravitational energy are negative in value, because the ball remains below the reference level. The change in potential energy is, of course, the same as in the example.

8.4 Since the derivation of escape velocity depends only on the initial and final energies of the rock, independent of the direction of the initial velocity, the necessary speed is:

$$V_{esc} = \sqrt{2GM_{Moon}/R_{Moon}}.$$

Using values from the inside front cover:

$$V_{esc} = \sqrt{2(6.7 \times 10^{-11} \text{ kg·m/s}^2)(7.4 \times 10^{22} \text{ kg})/(1.7 \times 10^6 \text{ m})}$$
$$= 2.4 \times 10^3 \text{ m/s}.$$

The direction of the velocity does influence the shape of the trajectory. It is a hyperbola (for $v > v_{esc}$), whereas for a rocket shot off the Moon along a radius the trajectory would be a straight line.

8.5 ▪ Figure 8.21 is the free-body diagram for the crate at rest on the end of the relaxed spring.

$$\Sigma F_x = 0. \qquad \Sigma F_y = 0.$$
$$f - mg \sin \theta = 0; \qquad n - mg \cos \theta = 0.$$

Thus: $f = mg \sin \theta \leq \mu_s n = \mu_s mg \cos \theta \;\Rightarrow\; \mu_s \geq \tan 33° = 0.65$.

▪ Figure 8.22 is the free-body diagram for the crate on the end of the compressed spring. If the crate is to accelerate up the ramp, we need:

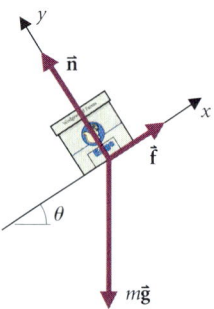

■ **FIGURE 8.21**
Free-body diagram for the crate at rest at the end of the relaxed spring ($F_s = 0$).

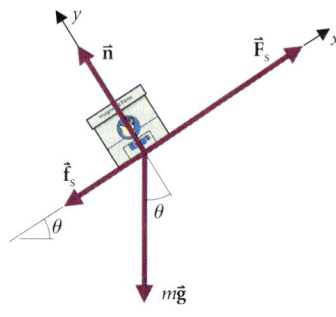

■ **FIGURE 8.22**
Free-body diagram for the crate on the end of the compressed spring.

$$F_s = ks \geq f_{max} + mg \sin \theta = \mu_s mg \cos \theta + mg \sin \theta.$$

Thus:

$$\mu_s \leq \frac{ks}{mg \cos \theta} - \tan \theta = \frac{(350 \text{ N/m})(1.36 \text{ m})}{(25 \text{ kg})(9.8 \text{ m/s}^2)\cos 33°} - \tan 33°$$
$$= 1.7.$$

Thus we need $0.65 < \mu_s < 1.7$.

Basic Skills

Review Questions

§8.1 ELASTIC POTENTIAL ENERGY

- What common experiences indicate that a compressed spring contains energy?
- List the three properties of potential energy and describe how the definition of elastic potential energy satisfies them. Describe how the definition of elastic energy turns the work-energy theorem into a statement of energy conservation for a mass–spring system.

§8.2 GRAVITATIONAL POTENTIAL ENERGY

- Describe the analogy between a ball rising and falling and a mass oscillating on a spring.
- Explain why the work done by gravity on an object depends only on its mass and the change in its height above the Earth's surface.
- What is the practical definition for gravitational potential energy near the Earth's surface?
- Is there any fixed rule for choosing the reference level for gravitational potential energy? Why or why not? What is a good practical rule?
- What are the similarities and differences between the exact calculation of potential energy and the derivation of the practical formula?
- What natural choice of reference state arises in the exact calculation of gravitational potential energy? What does negative potential energy mean?
- What does *escape speed* mean?

§8.3 CONSERVATION OF MECHANICAL ENERGY

- What is a *conservative force*?
- Give an example of a conservative force and a nonconservative force.
- Explain why only conservative forces correspond to potential energies.
- State the law of conservation of mechanical energy.

§8.4 POTENTIAL ENERGY IN SYSTEMS OF PARTICLES

- We can find the potential energy of a system by computing the work necessary to assemble it in any way we care to imagine. Explain why we don't have to consider the actual way the system got into its final state.

§8.5 INTERNAL ENERGY

- What is *thermal energy*? What are some ways it differs from the kinetic or potential energy of a large body?
- What is meant by *internal energy*?
- State the law of conservation of energy.

Basic Skill Drill

§8.1 ELASTIC POTENTIAL ENERGY

1. Compare the energy stored in a spring compressed by 1 cm with that stored in the same spring when it is stretched 2 cm.

2. If 1 kg of apples are weighed on a spring scale with a spring constant $k = 10$ N/m, how much energy is stored in the spring?

3. A pinball machine has balls of mass 100 g. The springs have a spring constant of 90 N/m and a relaxed length of 10 cm. To what length must the spring be compressed if it is to give a ball a speed of 2 m/s?

4. A system consists of an object that slides on a frictionless, horizontal surface and is connected to an ideal spring. Which of the following statements about the system are true, and which are false? Explain your reasoning.
  (a) The maximum potential energy occurs when the mass passes through its equilibrium position.
  (b) The maximum kinetic energy occurs when the mass passes through its equilibrium position.
  (c) The maximum value of the potential energy equals the maximum value of the kinetic energy.
  (d) The maximum value of the kinetic energy occurs when the potential energy is negative.
  (e) The minimum value of the potential energy occurs at the same time as the kinetic energy has its maximum value.

§8.2 GRAVITATIONAL POTENTIAL ENERGY

5. A 45-kg drum of chemicals, initially on a shelf, falls to the floor of a warehouse 3.2 m below. The top of the highest shelf is 12.8 m above the floor. What is the gravitational potential energy of the drum before and after the fall? Evaluate the energies using each of three choices of reference level: the floor, the initial position of the drum, and the highest shelf. Show that the change of potential energy during the fall is the same for each reference level, and find the speed at which the drum strikes the floor.

6. What is the gravitational potential energy of the Earth–Moon system? Treat the Earth and Moon as particles. The Earth's mass is 6.0×10^{24} kg, the Moon's mass is $\frac{1}{81}$ that of the Earth, and the distance between them is 3.8×10^5 km.

§8.3 CONSERVATION OF MECHANICAL ENERGY

7. A lunar colonist, working on a tall radio transmission tower, drops a wrench. As the wrench falls, its *total energy*
(a) increases;   **(b)** decreases;   **(c)** is necessarily zero;
(d) is constant but not necessarily zero.
Explain why you chose your answer and what is wrong with the others.

8. A 180-g baseball is launched upward by a spring, with spring constant $k = 1.0 \times 10^3$ N/m, that is compressed by an amount

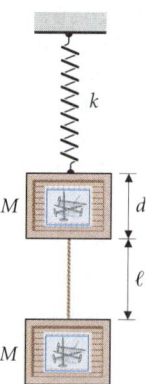

■ **FIGURE 8.23**

$s = 21$ cm. How high does the baseball rise above the launcher? On the way down, at what speed does the ball strike a player's glove held at the same height as the launcher?

9. A sounding rocket with a payload of mass $M = 500$ kg is fired directly upward and reaches a height of 3 km. How much energy from the fuel is transferred to the rocket's payload during the launch?

§8.4 POTENTIAL ENERGY IN SYSTEMS OF PARTICLES

10. (a) Two cubical nail boxes, each of mass $M = 25$ kg and sides $d = 15$ cm, are connected by a rope $\ell = 30$ cm long. The upper crate is attached to the roof by a spring with spring constant $k = 4.0 \times 10^3$ N/m and relaxed length ℓ (■ Figure 8.23). Suppose a box is at the reference level for gravitational energy when its top is against the roof. Evaluate the total potential energy of the system. **(b)** What is the potential energy of the system if the rope and spring are interchanged?

§8.5 INTERNAL ENERGY

11. A rope is pulled horizontally to drag a box of mass 100 kg across the floor at 1 m/s. At what power is mechanical energy transformed into thermal energy? (The coefficient of friction between box and floor is $\mu_k = 0.7$).

12. A 75-kg firefighter, starting from rest 5.0 m above the floor of the firehouse, slides down a rope and strikes the floor at a speed of 6.0 m/s. How much mechanical energy is transformed into thermal energy during the slide? What average friction force acts on the firefighter?

Questions and Problems

§8.1 ELASTIC POTENTIAL ENERGY

13. ❖ If the spring constant in your bathroom scale were doubled, would the energy it stores as you stand on it increase, decrease, or stay the same? By what factor does the energy change?

14. ◆ A car of mass 1.0×10^3 kg is supported on four springs, each of relaxed length 0.20 m and spring constant $k = 1.0 \times 10^5$ N/m. What is the energy of compression stored in the springs?

15. ◆ The launcher in a pinball machine consists of a spring of relaxed length $\ell = 10.0$ cm and constant $k = 2.5 \times 10^3$ N/m. The player fires balls of mass 0.10 kg by compressing the spring to 0.50ℓ

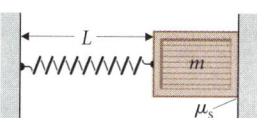

■ **FIGURE 8.24**

with the ball on the end of the spring. If the ball is then released, with what speed does it leave the spring?

16. ◆ Use the results of Example 8.3 to verify the relation $v_{max} \sim s_r/T$, which we would expect from dimensional analysis.

17. ◆◆ A spring presses a block of wood with mass m against a wall (■ Figure 8.24). The spring has relaxed length $\ell > L$, and the coeffi-

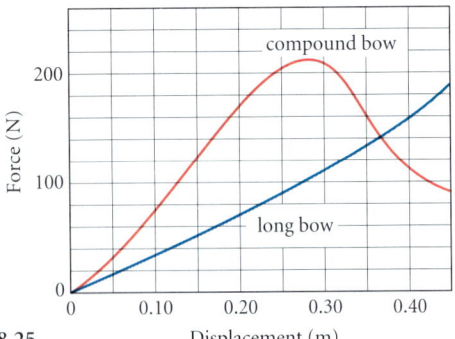

Figure 8.25

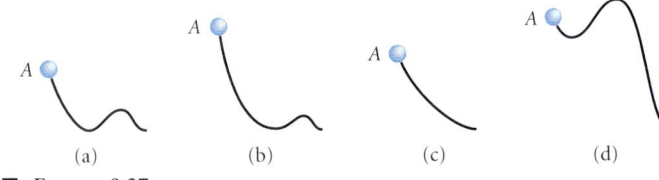

Figure 8.27

cient of friction between the block and the wall is μ_s. What minimum spring constant k is necessary for equilibrium? How much energy is stored in the spring?

18. ◆◆ A block of mass $M = 10.7$ kg and initial speed $v = 5.06$ m/s is decelerated by a spring with $k = 2.17 \times 10^3$ N/m. What is the potential energy in the spring when the block is moving at half its initial speed? What is the compression of the spring at that time?

19. ◆◆ ✉ Spring-mounted pads are to be placed around a hockey rink to cushion the impact of fallen players sliding into the walls. Estimate the spring constant and maximum compression necessary to limit a player's maximum acceleration to $5g$. Is the design practical? Why or why not?

20. ◆◆ ✉ ■ Figure 8.25 is a graph of the force required versus the displacement of the bowstring for two types of archery bows, a longbow and a modern compound bow. Estimate, using measurements from the graph, the potential energy stored in each kind of bow for displacements of 20, 30, and 40 cm. What *spring constant* would you assign the longbow? Is the concept of spring constant applicable to the compound bow? Why or why not? (Courtesy of T. Southworth, Potential energy analysis of a bow, *The Physics Teacher*, Jan. 1990: 42–43.)

21. ◆◆◆ A spring of constant k is compressed by an amount s_f between two freight cars of masses M and m. If the cars are free to roll on frictionless, horizontal rails (■ Figure 8.26) and are released at the same time, find their final kinetic energies by computing the work done by the spring on each object (cf. Example 8.2). Compare the sum of the resulting energies with the initial energy stored in the spring. Discuss how the two cars share energy and momentum in the limit $M \gg m$. (cf. footnote 1)

22. ◆◆◆ An improved bumper design for freight cars acts like a spring with the following relation between force and distortion:

$$|\vec{F}_s| = \beta s^3.$$

What features of Examples I2.3 and 8.2 would be different with this new bumper? If a freight car's mass is $M = 2.5 \times 10^4$ kg, and the initial speed of car 1 is $v_o = 1.6$ m/s, for what value of β is the maximum compression of the bumper $s_f = 10.7$ cm? (Show details in your setup of the necessary integral.) Give an expression for potential energy stored in the bumper as a function of s.

§8.2 GRAVITATIONAL POTENTIAL ENERGY

23. ❖ Beads are released at point A on each of the four frictionless wires in ■ Figure 8.27. Which bead has the greatest speed at the bottom of its wire? Explain your reasoning.

24. ❖ Ball A is attached to a string of length ℓ and is released from rest (■ Figure 8.28). Ball B is released from rest and falls straight down. Which of the following statements for what happens at point P are correct, and which are incorrect? Explain your reasoning and state what is wrong with the incorrect statements.

(a) Ball A has greater speed than B because it takes more time to fall.

(b) Ball A has the same speed as B.

(c) Ball A has lesser speed than B because the string slows it down.

(d) The two balls have the same velocity.

(e) We don't have enough information to compare their speeds.

25. ❖ In the discussion of both practical and exact expressions for gravitational potential energy, which features depend on being near the Earth, and which would be the same near any other planet? Describe your reasoning.

26. ◆ A window cleaner drops a bucket of water from his perch on the side of a building 32 m high. At what speed does the bucket strike the ground?

27. ◆ A bead slides on a frictionless wire shaped as shown in ■ Figure 8.29. If the bead is released from rest at point A, what is its speed when it leaves the wire at point E?

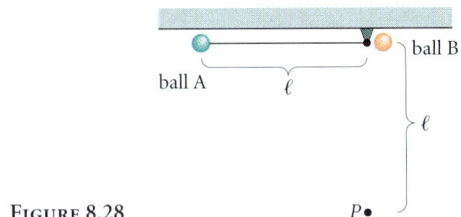

Figure 8.28

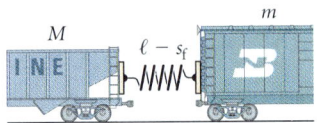

Figure 8.26

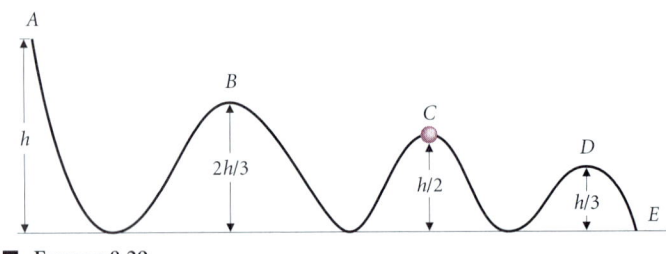

Figure 8.29

28. ♦ The California Water Project pumping plants lift water 1200 m between Bakersfield and Gorman, California. If the project delivers 2.0×10^6 m³ of water per day, what is the average power required by the pumping plant?

29. (a) ♦ A ski lift carries 700 people per hour to a point 300 m higher than the base of the lift. Neglecting friction, what power is required of the motors to operate the lift? Assume the average person has a mass of 70 kg. (Assume also that the lift moves slowly.) (b) ♦♦ If the ski lift has a speed of 1 m/s, does this change the answer appreciably? Discuss.

30. ♦♦ What is the potential energy of the Earth, considered as a satellite orbiting the Sun? Compare your result with the kinetic energy of the Earth in its orbit.

31. ♦♦ A bead sliding on the frictionless wire shown in Figure 8.29 is released from rest at point A. What is the speed of the bead at points B, C, and D?

32. ♦♦ A bead slides toward the left on the frictionless wire shown in Figure 8.29. What speed must it have at point D if it is to come to rest at point B?

33. ♦♦ Certain sources of celestial x radiation are believed to be powered by material falling onto the surface of a very compact star. If the star has a mass of 10^{30} kg and a radius of 10 km, at what rate must material fall onto the star to account for an observed power of 10^{30} W? Assume perfect efficiency in the conversion of energy from gravitational potential energy to x rays.

34. ♦♦♦ A meteor approaching the Sun radially from distant parts of the solar system, strikes the Earth. What is the meteor's speed at the top of the atmosphere? How much energy is transformed into thermal energy when a meteor of mass $m = 10$ kg strikes the Earth's atmosphere and burns up before reaching the surface?

§8.3 CONSERVATION OF MECHANICAL ENERGY

35. ❖ Are the following forces conservative? Why or why not? (a) Drag force on an airplane: $|\vec{F}_D| \propto \rho v^2$, where ρ is air density and v is the airplane's speed relative to the air. (b) Magnetic force on a particle moving in a plane: $\vec{F}$ proportional to the particle's speed and perpendicular both to its velocity $\vec{v}$ and to a constant magnetic field vector $\vec{B}$. (c) Electric force on a particle due to a charged wire: $\vec{F} \propto \hat{r}/r$, where r is the distance of the particle from the wire.

36. ❖ An object of mass M is hanging vertically, suspended by a massless spring of relaxed length ℓ and spring constant k. Which of the following statements are correct, and which are incorrect? Explain your reasoning and state what is wrong with the incorrect statements.
  (a) In equilibrium the spring is longer than its relaxed length because of the gravitational force on the object.
  (b) If the object is pulled down, the potential energy stored in the spring increases.
  (c) If the object is pulled down, the object's gravitational potential energy increases.
  (d) If the object is pulled down and released, its kinetic energy will be zero when the spring returns to its relaxed length.

37. ♦♦ A box of mass M is attached to the roof by a spring of constant k and relaxed length ℓ. Initially, the spring is compressed to a length of $\ell/2$ (■ Figure 8.30). If the box is released, at what distance below the roof will the box first be brought to rest by the spring?

38. ♦♦♦ An object of mass M is connected to a fixed point by an ideal string of length ℓ that allows the object to circle freely in a vertical plane (■ Figure 8.31). As the object passes the top of the circle, the tension in the string is momentarily zero. What is the string tension as the object passes the bottom of the circle?

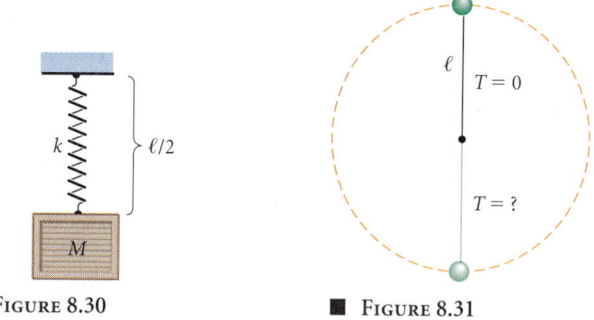

■ FIGURE 8.30

■ FIGURE 8.31

39. ♦♦♦ Find the potential energy function for the following forces.
(a) $\vec{F} = k(r^2 - r_o^2)\hat{r}$, where r is distance from the origin and $\hat{r}$ is a vector that points from the origin to the point at which the force acts.
(b) $\vec{F} = k \sin(\alpha x)\hat{i}$.

§8.4 POTENTIAL ENERGY IN SYSTEMS OF PARTICLES

40. ♦ Five crates of machine tools, each with mass 3.74×10^2 kg, are on shelves in a warehouse. Two of the crates are on a shelf 2.40 m above the warehouse floor; two are on a 1.20-m-high shelf, and one is on a shelf 3.60 m high. With the floor as reference level, what is the total potential energy of the crates?

41. ♦ Five springs, each with negligible relaxed length, are connected between two parallel rods, so that when the rods are pulled apart, the springs are parallel to each other and perpendicular to the rods. If three of the springs have spring constant $k = 617$ N/m and the other two have $k = 445$ N/m, what is the elastic potential energy of the system when Igor the Incredible pulls the rods 1.73 m apart?

42. ♦♦ An egg crate of mass 25 kg is supported on a frictionless slope by a spring with relaxed length $\ell = 1.0$ m and spring constant $k = 1.2 \times 10^3$ N/m (■ Figure 8.32). Take the reference level for potential energy so that the crate has zero gravitational energy when its lower edge is at the end of the relaxed spring, and evaluate the total potential energy of the system.

43. ♦♦ In ■ Figure 8.33, the spring has negligible relaxed length and spring constant $k = \sqrt{3} Mg/(3\ell)$. Verify that the system is in equilibrium and evaluate its total potential energy. Take the reference level for potential energy to be at the roof.

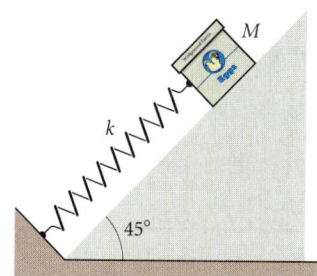

■ FIGURE 8.32

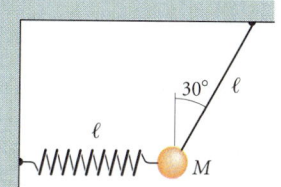

■ FIGURE 8.33

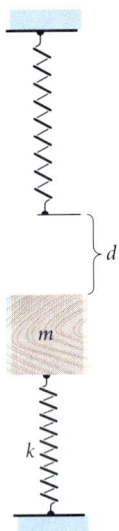

■ FIGURE 8.34

■ FIGURE 8.36

44. ♦♦ A balsa wood cube of mass m is at rest on a spring with spring constant k compressed by an amount s_i. The cube is released so that the spring projects it upward to strike a relaxed spring with the same spring constant, a distance d above the original position of the cube (■ Figure 8.34). Find the height above the cube's original position where it first comes to rest **(a)** if $s_i > [2mgd/k]^{1/2}$ and **(b)** if $s_i < [2mgd/k]^{1/2}$.

45. ♦♦ Ore cars with total mass $M = 1.0 \times 10^4$ kg start from rest and roll without friction down the track to an unloading area 1.0×10^1 m below (■ Figure 8.35). They are stopped by a spring bumper and held with a clamp while they are unloaded. What spring constant is needed if the bumper compresses by 1.0 m when stopping an ore car? After a car is unloaded, its remaining mass is 1.0×10^3 kg. After the clamp is released, the empty car returns back up the track. With what speed does the car reach the top?

46. ♦♦♦ Considering the Earth and Moon as particles, evaluate their gravitational potential energy by calculating the work to bring them from infinite distance to their present separation. Show in detail how you compute the necessary integral. What is the present total energy of the Earth–Moon system? How much energy would have to be transformed into thermal energy to form the Earth–Moon system from two infinitely separated bodies?

47. ♦♦♦ Three stars of mass M orbit each other, undergoing uniform circular motion while remaining at the vertices of an equilateral triangle of side L. Treat the individual stars as particles.
(a) Compute the gravitational potential energy of this system by calculating the work you would have to do to assemble the system one star at a time.
(b) Compare your result with the stars' total kinetic energy.

48. ♦♦♦ Two masses, m_1 and $m_2 > m_1$, are suspended by an ideal string over a frictionless support (■ Figure 8.36). Find how much the potential energy of the system changes when the object of mass m_2 descends a distance h. What is the corresponding change in the kinetic energy of the system? Find the work done on each object by string tension and the work done by nongravitational forces on the entire system. Show that your conclusions about kinetic and potential energy changes are consistent with your calculations of work.

49. ♦♦♦ In ■ Figure 8.37 each spring has a relaxed length $\ell/2$, but the spring constants k_1 and k_2 are different. Evaluate the potential energy of the system as a function of x and find the value of x where the potential energy is minimum. Show that the system is in equilibrium at this value of x.

50. ♦♦♦ In ■ Figure 8.38 each spring has a relaxed length $\ell/2$. If the cube has zero gravitational energy when its top surface touches the roof, evaluate the potential energy of the system as a function of x and find the value of x where the potential energy is a minimum. Show that the system is in equilibrium at the same value of x.

51. ♦♦♦ A block of mass m is at rest at the top of a wedge-shaped object of mass M (■ Figure 8.39). All surfaces are frictionless. At the bottom of the wedge is an ideal spring with spring constant k and relaxed length ℓ. The block is released and slides down the wedge. Find the compression of the spring when the block comes to rest with respect to the wedge.

§8.5 INTERNAL ENERGY

52. ❖ A large truck and a compact car are both driving down the same hill at the same speed. The truck's brakes are more likely to smoke. Why?

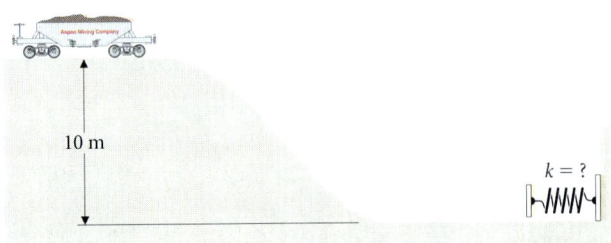

■ FIGURE 8.35

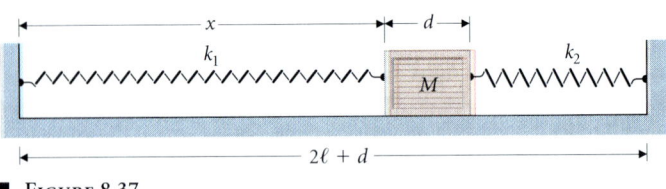

■ FIGURE 8.37

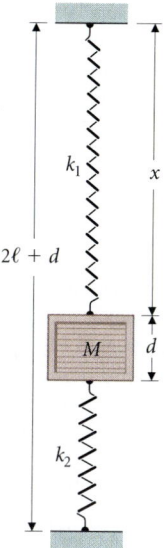

■ FIGURE 8.38

■ FIGURE 8.41

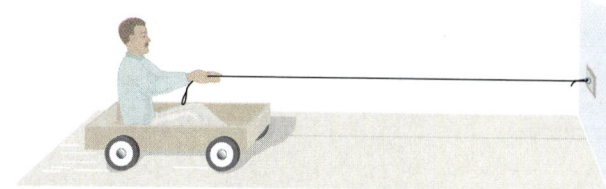

■ FIGURE 8.42

53. ❖ If a car moves downhill at constant velocity, what happens to its kinetic and gravitational potential energies? If there is a net change in the car's total energy, explain where the energy comes from or goes to.

54. ❖ After winning a rope-climbing contest, Sonia climbs back down the rope at constant speed. Describe how each of the following change during this process: Sonia's potential energy, her kinetic energy, her internal energy. Peter, who was second in the contest, decides to drop the last 2 m to the gym floor. Describe how his potential, kinetic, and internal energies change during the drop. What happens to these energies as he comes to rest on the floor?

55. ❖ Whammo the Magnificent, famous human projectile, reaches a height of 15 m and then falls into a net (■ Figure 8.40). Air resistance is small but not negligible. At what point or points is the sum of Whammo's potential and kinetic energy the least? Explain why the answers you reject are wrong.
(a) just after launch
(b) at the top of the trajectory
(c) just before hitting the net
(d) equally small at launch and at the top
(e) equally small at top and just before hitting the net
(f) none of the above

56. ❖ In a circus stunt, a clown who can jump 1 m high repeatedly tries to grab a delicious-looking candy stick suspended 2 m above her. In a last attempt, she lands on a spring-loaded platform, then leaps up and easily gets the candy stick. Explain why the stunt works.

57. ❖ When you run up stairs, what force acts to oppose your weight? How much work does this force do? What forces actually *cause* your body to rise?

58. ❖ Lucy compresses the spring in the base of her pogo stick (■ Figure 8.41) by jumping on the foot bars. The spring then propels Lucy and stick into the air, allowing her to hop along. Identify the forces acting on the Lucy–pogo stick system. Which forces do work? How does the system gain kinetic energy?

59. ❖ Identify the forces acting on the man-plus-cart system in ■ Figure 8.42. Which forces do work? What is the source of the cart's increasing kinetic energy?

60. ❖ A motor mounted on a cart of mass 50 kg turns the cog wheel, which winds up the rope and pulls the cart toward the wall (■ Figure 8.43). Identify the forces acting on the cart. Which forces do work? What is the source of the cart's kinetic energy?

61. ❖ Isaac pulls a toy truck attached to a string across a table at constant speed. Identify the forces acting on the truck. Which of these forces do work on the truck? What happens to the work Isaac does on the string?

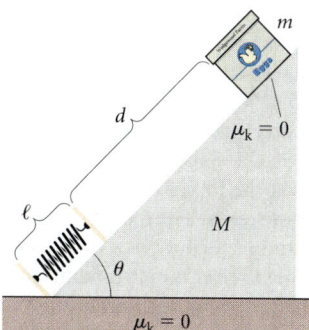

■ FIGURE 8.39

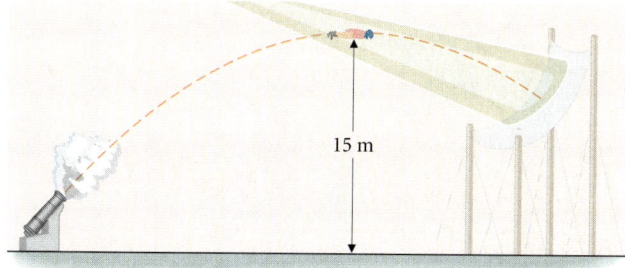

■ FIGURE 8.40

■ FIGURE 8.43

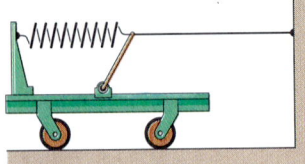

■ FIGURE 8.44

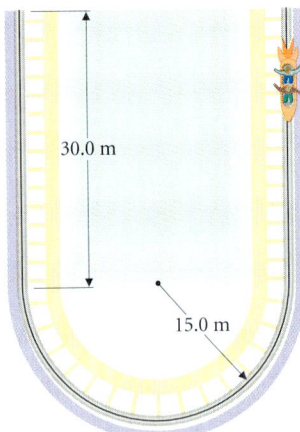

■ FIGURE 8.45

62. ◆ A car of mass 850 kg is descending a 20° slope. How much energy must be transformed into thermal energy to keep the car at constant speed during a descent of 1.0 km (measured along the slope)?

63. ◆ A blob of putty of mass $m = 0.20$ kg is dropped on a table from a height of $h = 2.0$ m above the table. How much energy is transformed into thermal energy when the blob strikes the table?

64. (a) ◆ A parachute allows a 0.45-metric-ton supply package to fall from a great height at a constant speed. What happens to the package's potential energy during the fall? If a package falls from 3150 m, how much energy is dissipated? Ignoring the effects of air resistance on the package itself, find the speed at which the package would hit the ground if the chute failed to open. (b) ◆◆ Air resistance acts on the package even without the parachute. As a result, the package reaches a terminal speed of 115 m/s. What is the rate at which energy is dissipated in a fall at terminal speed?

65. ◆◆ The spring in ■ Figure 8.44 is stretched 6.3 cm and then released. If the spring constant k is 1200 N/m and the cart has a mass of 15 kg, find the final kinetic energy of the cart. Identify all the forces acting on the cart. Which forces do work?

66. ◆◆ If the spring constant in Study Problem 7 were reduced by a factor of 2, find where the egg crates would first come to rest.

67. ◆◆ The spring in Figure 8.30 has relaxed length ℓ. The block is released and after a long time reaches equilibrium. Describe the equilibrium state. How much energy is transformed into thermal energy before the system reaches equilibrium?

68. ◆◆ An amusement park ride has a U-shaped track with vertical sides 30.0 m high and a semicircular bottom of radius 15 m (■ Figure 8.45). A motor and ratchet pulls the car up one side of the track and releases it at the top. (a) If the car's mass is 720 kg and it reaches a height of 27 m on the far side of the track, what is the magnitude of the average friction force acting on the car? (b) Assume that the friction force has a constant magnitude equal to this average and find how far the car travels up the first side of the track on the return trip. What total distance does it travel before coming to rest at the bottom of the track?

69. ◆◆◆ ✉ A child slides down the bannister of a spiral staircase, which slopes at a 30° angle and completes a circle of radius 5 m. The coefficient of friction between the child's pants and the bannister is 0.5. Estimate the speed at which the child flies off the bannister by applying Newton's laws, as if the bannister were straight. Explain qualitatively why the bannister's curvature affects the result, and estimate how important the correction is in the given situation.

70. ◆◆◆ The drag force on an airplane is proportional to its speed squared. When the plane is flying level at a speed of 1.0×10^2 m/s, the engine thrust is equal to one-half the weight of the plane. If the plane then descends at an angle of 30° to the horizontal without changing the engine thrust, show that its speed during the descent is 140 m/s.

Additional Problems

71. ❖ Verify that momentum is conserved throughout the interaction of the two freight cars described in Examples I2.3 and 8.2. Describe the momentum exchanges in each phase of the interaction and determine where the original momentum of freight car 1 ends up.

72. ❖ Experienced hikers prefer to step over a log in the trail rather than stepping up on the log and back down. Why?

73. ❖ ✉ A pole vaulter boasts that with a long and springy enough pole he could set a record height of 9 m. Given that a winning time in the 100-m run ~ 10 s, is the vaulter's boast reasonable? Explain your reasoning.

74. ◆◆ If a wrench of mass 1 kg is dropped by a construction worker from a crane of height 100 m, what impulse does the wrench impart to the ground when it hits? Assume the wrench doesn't bounce. What difference would it make if the wrench bounces?

75. ◆◆ Lydia's garage is 5.0 m higher than the road where her driveway begins. One day she decides to shut off the engine at the bottom of the driveway and coast to a stop in front of the garage door. Considering only mechanical energy, find the minimum speed she should have at the bottom in order to reach the garage. Will this speed be great enough in fact? If not, why not?

76. ◆◆ A ski-jumper slides down a frictionless ramp and lands on a slope inclined 45° to the horizontal (■ Figure 8.46). If the start is a distance h_o above the end of the ramp, where does the jumper land?

77. ◆◆ A 250-g baseball is thrown upward at a speed of 8.7 m/s from the top of a 43-m-high building. Later, the ball hits the ground with a measured speed of 25 m/s. What is the change in the ball's total mechanical energy between the times just after its release and just before it strikes the ground? Can you account for the change?

78. ◆◆ An 850-kg car is coasting down a slope of 10° with the horizontal at a speed of 25 m/s. If the engine has to deliver 25 kW of power to drive at 25 m/s on level ground, at what rate are the car's brakes transforming energy on the slope?

79. ◆◆ A stunt car is at the top of the looped track shown in ■ Figure 8.47 when its engine quits. The loop has radius $R = 9.8$ m and the car's speed at the top of the loop is 9.8 m/s. (a) Use Newton's second law to show that the car barely remains on the track at the top of the loop. (b) If the car coasts without friction to the bottom of the loop, what acceleration is required to stop it in the 25-m straight section at the bottom of the loop?

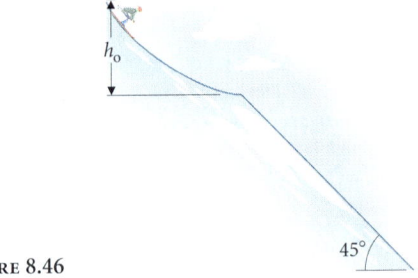

■ FIGURE 8.46

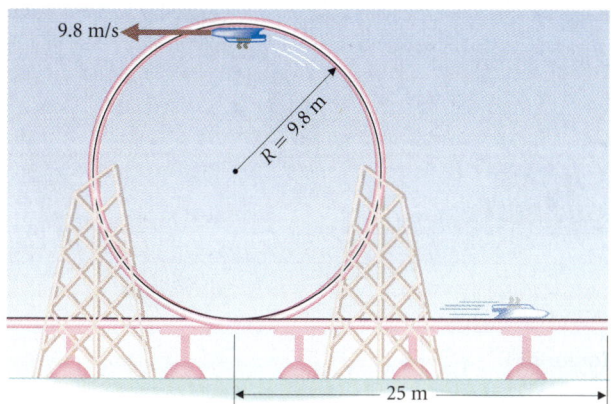

■ FIGURE 8.47

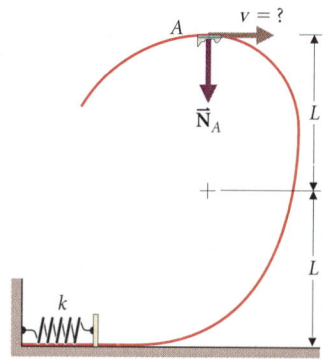

■ FIGURE 8.49

80. ◆◆ A 12.5-kg packing crate slides down a loading ramp at a constant speed of 1.33 m/s. If the ramp makes an angle of 5.5° with the horizontal, how much power is dissipated by nonconservative forces? Find the coefficient of friction between the crate and the ramp.

81. ◆◆ A downhill skier of mass 87 kg gets into a tuck at a speed of 65 km/h and descends a 17° slope 0.85 km long. Passing through the finish line at the bottom, he is clocked at 84 km/h. How much energy was dissipated by friction and wind drag? Find the average value of the (friction plus drag) force acting on the skier.

82. (a) ◆◆ At what point between the Earth and the Moon is the net gravitational force on an object zero? (cf. Study Problem 5.) Call this point P, and assume you may reasonably take an object's gravitational energy to be zero at P. An Apollo spacecraft is sent toward the Earth from the Moon in such a way that it leaves point P going *very slowly*. Find an expression for the speed at which the spacecraft strikes Earth's atmosphere in terms of G and Earth's radius R_E and mass M_E. (The Moon's distance from Earth is $60R_E$ and its mass is $\frac{1}{81}$ of the Earth's mass.) Recalling that $g = GM_E/(R_E^2)$, give a numerical value for the spacecraft's speed. (b) ◆◆ Ignoring the presence of the Moon, what would the spacecraft's gravitational energy be at point P? How large an error is made in ignoring this energy in your answer to part (a)? (c) ◆◆◆ ✉ What effect does the Moon have on the gravitational energy of the spacecraft between point P and the Earth's surface? Estimate how large an error is made by ignoring the Moon's gravity in your answer to part (a).

83. ◆◆◆ A block of mass M slides on a frictionless, horizontal surface and is attached to the wall by a spring. An ideal string is attached to the block and runs over a frictionless rod to a second block of mass m (■ Figure 8.48). In equilibrium, m hangs a distance ℓ below the rod. An external force is applied to pull m down to a distance 1.5ℓ below the rod. (a) What work is done by the external force to pull m down? (b) If m is now released, how far below the rod is it when it next comes to rest? (c) What is the upward speed of m when it is a distance $s < 1.5\ell$ below the rod? (d) What is the tension in the string when m is at distance s below the rod?

84. ◆◆◆ A roller coaster car with mass m moves with negligible friction along the track shown in ■ Figure 8.49. As the coaster passes point A, the normal force exerted on it by the track is measured as $N_A = (2.0)mg$. Near A, the track is circular with radius L. When the car reaches the bottom of the track, it is brought to rest by a spring bumper with spring constant k. (a) Use Newton's second law to find the speed of the car at point A. (b) Give expressions for the kinetic and potential energies of the car at point A. (c) Give an expression for the amount the spring must compress to stop the car. (d) If $m = 3.0 \times 10^2$ kg, $L = 15$ m, and $k = 1.0 \times 10^6$ N/m, find the spring compression and comment on whether the design is reasonable.

85. ◆◆◆ Suppose you were an astronaut working on the (very odd) dumbbell-shaped asteroid shown in ■ Figure 8.50. How would your gravitational potential energy vary with distance above the surface? Would any version of the practical formula be useful? If so, for roughly what range of distances above the surface?

86. ◆◆◆ A large block of mass M slides on a frictionless surface with an initial speed of v_0. On top of the block is a small box of mass m. The coefficients of static and kinetic friction between box and block are μ_s and μ_k. The sliding block encounters an ideal spring with constant k (■ Figure 8.51). (a) Assuming for now that the box doesn't slide, what is the maximum compression of the spring? (b) Still assuming no relative motion, what is the acceleration of block and box at the instant of maximum compression? (c) What is the maximum value of k for which it remains true that the box does not slide? (d) Suppose k is just slightly greater than the value found in part (c), so that the box begins to slide just as the spring reaches maximum compression. What then are the accelerations of box and block?

87. ◆◆◆ A particle of mass m is suspended from the roof on a string

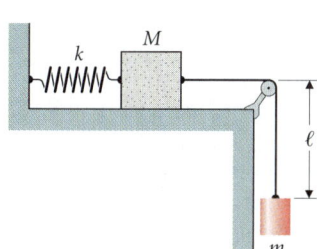

■ FIGURE 8.48

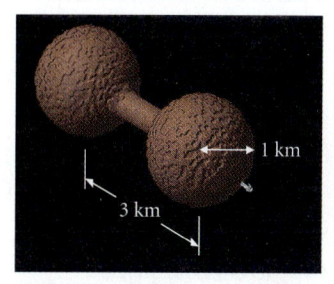

■ FIGURE 8.50

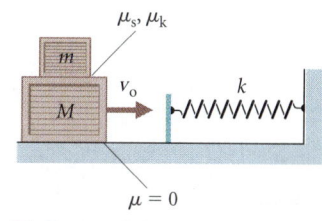

■ FIGURE 8.51

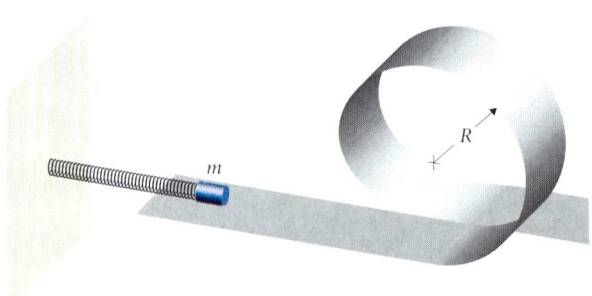

Figure 8.52

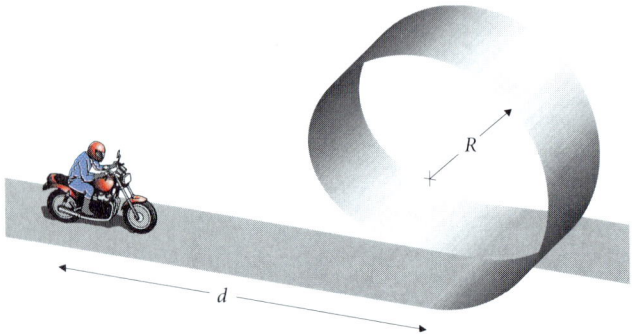

Figure 8.54

of length L. A second string is attached to the particle and pulled horizontally, moving the particle very slowly until it is hanging at an angle θ from the vertical. Calculate the work done by tension in the horizontal string using the definition: $W = \int \vec{F} \cdot d\vec{s}$. Check your answer against the change in the particle's energy.

88. ◆◆◆ A bungi jumper of mass 65 kg leaps from a hot-air balloon with loose bungi cords of length 18 m attached to her legs. The cords stretch and bring her to rest momentarily 36 m below the balloon. Modeling the cords as an ideal spring, find their spring constant. The jumper oscillates up and down for a while before finally coming to rest at a constant distance below the balloon. What is that distance? Does her energy remain constant during this process? If not, find how much energy is transformed by the stretching and contraction of the bungi cords. (Assume that the balloon remains at a fixed height.)

89. ◆◆◆ A block of mass m is accelerated by a spring of constant k and initial compression s_o. The block slides on a frictionless track with a circular loop (■ Figure 8.52). Find the speed of the block at the top of the loop as a function of the initial compression s_o. What initial compression is needed if the block is not to leave the track at the top of the loop?

90. ◆◆◆ A roller coaster car is released from a height h_o above the bottom of a track on which it rolls without friction (■ Figure 8.53). Find the speed of the car at the top of the loop as a function of h_o. What value of h_o is required if the car is to remain on the track?

91. ◆◆◆ Evil Kobold rides a stunt motorcycle whose engine produces a constant power P (■ Figure 8.54). At what distance d must Evil begin the stunt if he begins coasting at the bottom of the loop and is to remain on the track at the top? Neglect work done by nonconservative forces.

92. ◆◆◆ A ski-jumper starts from rest and slides down a frictionless, hemispherical snow dome of radius R (■ Figure 8.55). Find the point where the jumper leaves the surface.

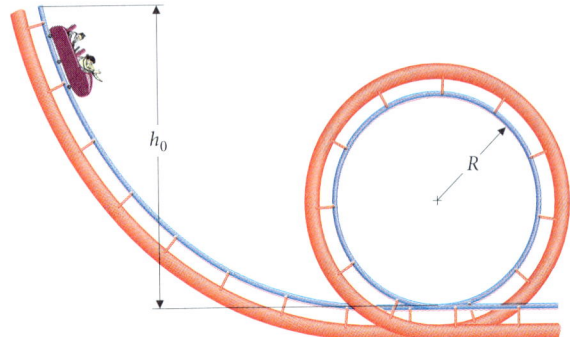

Figure 8.53

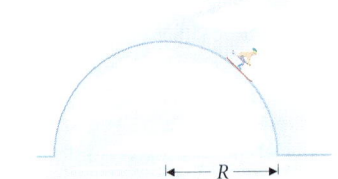

Figure 8.55

Computer Problem

93. The magnitude of force needed to stretch a bungi cord used to secure a load on a truck is measured as a function of the cord's length (see file CH8P93 on the supplementary computer disk). (The relaxed length of the cord is 2.0 m.) **(a)** Plot the magnitude of the force as a function of the cord's stretch. Over what range of lengths does the cord obey Hooke's law? Use the data to estimate the effective spring constant k over this range. How does k change as the bungi length changes? **(b)** Now use the trapezoidal rule or another numerical algorithm to calculate the potential energy stored in the cord as a function of its stretch. Plot your results as a function of s and as a function of s^2. Over what range does the energy increase as s^2? Use the data to estimate k. Does your value agree with your estimate in part (a)? How does the ratio of the two estimates change as s changes?

Challenge Problems

94. What is the gravitational potential energy of a uniform sphere with radius R and mass M? Obtain the answer by finding the work you would have to do to add a spherical shell of thickness dr to a sphere of radius $r < R$, and imagine assembling the entire sphere out of such shells, one at a time. Using numbers from the inside front cover, estimate the gravitational potential energy of the Sun. ✉ Suppose that transformation of gravitational potential energy is what causes the Sun to shine, and estimate the length of time the Sun has been shining. Compare your result with the Sun's age of 4.5 billion years and discuss whether gravitational energy is believable as the source of the Sun's energy emission.

95. In the Earth's reference frame, a block of mass m slides toward the right with initial speed v_o on a belt that moves to the left with constant speed v_1 (■ Figure 8.56). The coefficient of kinetic friction is μ_k. **(a)** Find the time intervals required and the displacements the block undergoes in coming to rest with respect to the Earth and later with respect to the belt. **(b)** Calculate the work done by kinetic friction on the block and verify that the work-energy theorem is satis-

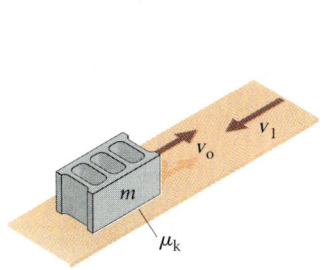

Figure 8.56

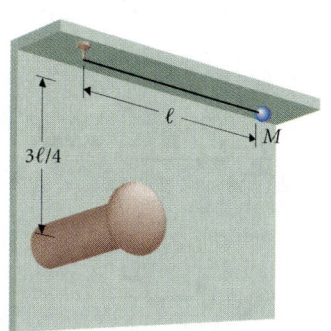

Figure 8.57

fied. **(c)** What forces act on the belt? Calculate the work done by each. **(d)** What is the net energy transferred to the system of block-and-belt? How much energy is transformed into thermal energy? **(e)** Show that the amount of thermal energy transformed is: $\Delta U = -\hat{\mathbf{f}}_k \cdot \vec{\mathbf{s}}_{rel}$, where $\hat{\mathbf{f}}_k$ is the friction force acting on the block and $\vec{\mathbf{s}}_{rel}$ is the block's displacement relative to the belt. **(f)** Now analyze the motion of the block in the reference frame moving with the belt and comment on the features of energy flow that change between reference frames and those that remain unchanged.

96. When a ball bounces on a surface, the ratio of its upward speed just *after* bouncing to its downward speed just *before* bouncing is called the *coefficient of restitution* ϵ. Assume that ϵ does not depend on impact speed, that the ball is initially dropped from rest at height h_o, and that the ball continues to bounce up and down along a vertical line. Find the height to which the ball rises after the nth bounce. What is the time interval between the initial release and the ball's arrival at the top of the nth bounce? Explain how you could measure the coefficient of restitution by measuring the time interval after which the ball has just ceased bouncing. Would your calculations change if the ball has some horizontal velocity? For a steel ball bouncing on a steel surface, $\epsilon = 0.980$. How long will a steel ball continue to bounce if it is dropped from a height $h_o = 0.20$ m? *Hint:* First prove that

$$\sum_{n=0}^{N} \epsilon^n = \frac{1 - \epsilon^{N+1}}{1 - \epsilon}.$$

and let $N \to \infty$.

97. What is the predicted speed of the small block in Figure 8.39 just before it encounters the spring?

98. A particle of mass M is suspended from the roof by an ideal string of length $\ell = 1.31$ m and is released from rest with the string horizontal (■ Figure 8.57). A horizontal peg perpendicular to the plane of the particle's motion has a radius of $\ell/10$, and its center is a distance $3\ell/4$ directly below the point of suspension so that the string wraps around the peg as the particle moves. Describe the motion of the particle, and find its speed when it strikes the peg.

> *Why sometimes I've believed as many as six impossible things before breakfast.*
> THE WHITE QUEEN

ESSAY 2
The Gravitational Field

Magritte's painting of a city on a floating mountain invites our imaginations into a strange, surreal universe. In the real world, there is no way to levitate a mountain; some enormous structure would be needed to support it. But how can the Earth pull the mountain down without reaching up to grasp it?

Newton's critics thought gravitational attraction to be unacceptably mystical, and Newton himself treated it as an observed fact for which he had no explanation. The modern idea of a *gravitational field* offers you a way to visualize gravity and to develop your understanding of it.

E2.1 THE GRAVITATIONAL FIELD OF THE EARTH

In Newton's theory, the gravitational force between two objects is proportional to the product of their masses. If one of the objects is Earth (mass M), the force on a small object of mass m is:

$$\vec{F}_g = -\frac{GMm}{r^2}\hat{r},$$

where $\hat{r}$ is a unit vector that points outward from the center of the Earth toward the object. If no other force acts on the small object, it accelerates toward the Earth with the *acceleration due to gravity*:

$$\vec{g} = \frac{\vec{F}_g}{m} = -\frac{GM}{r^2}\hat{r}.$$

This, as we know from Galileo, is the same for any small body. Another name for $\vec{g}$ is the *gravitational field strength*. With the reference point at *infinity*, the gravitational potential energy of the small body is:

$$U = -\frac{GMm}{r}.$$

The ratio of this potential energy to the mass of the object,

$$\phi \equiv \frac{U}{m} = -\frac{GM}{r},$$

is also the same for every small body at the same distance r from the center of the Earth. The quantity ϕ is called the *gravitational potential*.

Once the potential and the field strength at a particle's position are known, we can simply multiply them by the particle's mass to find the particle's potential energy and the force acting on it. The quantities $\vec{g}$ and ϕ each have a specific value at a given location, regardless of which object is there to experience the force or share energy with the Earth. Indeed, the field strength and potential exist whether or not *any* object is present. We say that the Earth surrounds itself with a *gravitational field*. Gravitational force and energy describe the interaction of objects with this field.

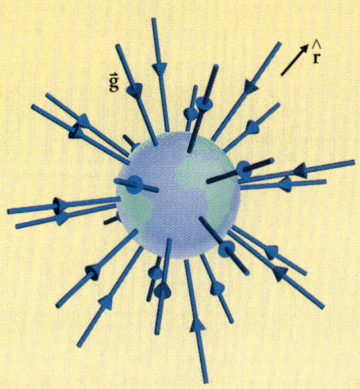

■ **Figure E2.1**
Earth surrounds itself with a gravitational field that accounts for the gravitational forces that act on bodies near the Earth. *Field lines* give one representation of the field. At each point, the acceleration due to gravity, or *gravitational field strength*, is parallel to the field line through that point. Thus the field lines show the direction in which a body falling freely near the Earth would accelerate.

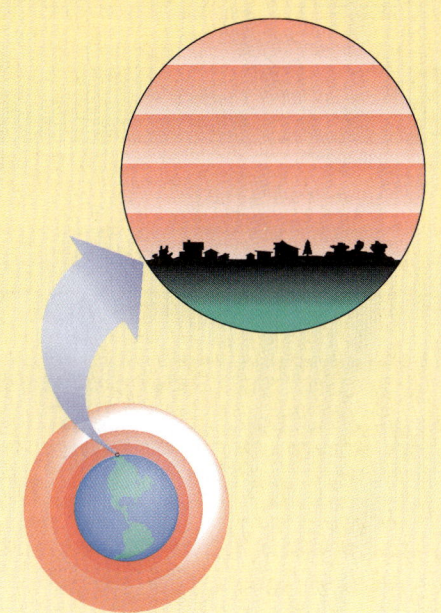

■ **Figure E2.2**
Equipotential surfaces give a second representation of the gravitational field. The surfaces are spheres centered on Earth, since the gravitational potential depends only on distance from Earth's center. The gravitational potential on successive surfaces in this diagram changes by one-sixth of the value at the Earth's surface. The closer spacing near the Earth illustrates the increasing rate of change of ϕ and consequently the increasing gravitational field strength. On small scales near Earth, the equipotential surfaces appear as planes parallel to the surface. The evenly spaced planes represent the uniform change of potential with height over distances small compared with Earth's radius.

Diagrams of the field strength $\vec{g}$ and the potential ϕ are an effective way to visualize the gravitational field. ■ Figure E2.1 shows gravitational *field lines* pointing radially inward toward the Earth. A field line is constructed tangent to $\vec{g}$ at each point so as to show the direction of the field. Since $\vec{g}$ points *downward*, the field lines also run radially toward Earth's center. The lines get closer together near Earth's surface, corresponding to increasing field strength. Outside the Earth, the number of field lines is a constant, and the surface area they pass through is proportional to r^2:

$$\text{Field lines per unit area} = \frac{\text{total field lines}}{\text{total area}} \propto \frac{1}{r^2} \propto |\vec{g}|.$$

This illustrates a general relation:

> In a gravitational field diagram, the number of field lines per unit area passing through a surface is proportional to the field strength $|\vec{g}|$ on that surface.

The field lines enter the Earth and come to an end. Newton showed that the gravitational acceleration at any point at radius r within a spherical distribution of mass equals that due to the sphere of mass interior to the point. Thus the field strength $|\vec{g}|$ decreases to zero at Earth's center and so must the density of field lines. The number of field lines that terminate in any given volume is proportional to the amount of mass in that volume. Each piece of mass contributes its own set of field lines, which together make up the field of the entire Earth.

■ Figure E2.2 shows a second representation of the field, using *equipotential surfaces,* surfaces on which the potential ϕ has a constant value. For the Earth alone, these surfaces are easy to describe: ϕ depends only on the distance from the center, and the surfaces are spheres. Seen on a small scale, these concentric spheres appear as equally spaced planes parallel to Earth's surface. This corresponds to the practical formula for potential near Earth's surface:

$$\phi_{\text{near surface}} = g \times (\text{height above a chosen reference level}).$$

TECHNICALLY, THERE ARE AN INFINITE NUMBER OF FIELD LINES, BUT WE CAN DRAW ONLY A FINITE NUMBER IN ANY DIAGRAM. OFTEN A VERY FEW LINES GIVE AN ADEQUATE QUALITATIVE PICTURE. THE MORE LINES YOU CHOOSE TO DRAW, THE MORE PRECISE THE RESULTING PICTURE.

SINCE EARTH'S DENSITY IS NOT CONSTANT, $|\vec{g}|$ ACTUALLY INCREASES SLIGHTLY WITHIN THE EARTH BEFORE DECREASING TO ZERO AT THE CENTER.

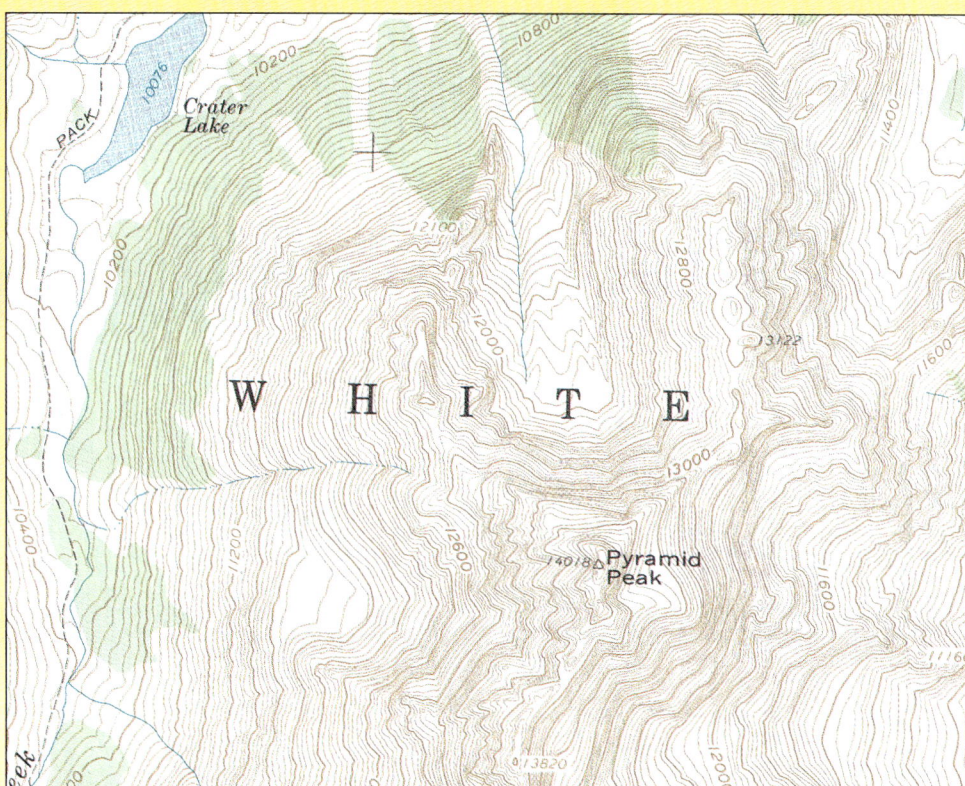

■ FIGURE E2.3
Portion of the United States Geological Survey map of the Maroon Bells, near Aspen, Colorado. The contour lines are lines of constant ground elevation above sea level. The contours are also lines of constant gravitational potential, the intersections of equipotential surfaces with the ground. Practical use of such maps often depends on this depiction of both terrain and the gravitational field.

Equipotential surfaces have a concrete application in a geologist's contour map (■ Figure E2.3) where *contour lines* on the map form a picture of the terrain. One may think of these contours as the intersections of regularly spaced equipotential surfaces with the ground. Understanding water flow is one of the major uses of such maps precisely because of this relation between contours and the gravitational field. Water flows downhill—toward lesser gravitational potential.

E2.2 THE FIELD OF A TWO-PARTICLE SYSTEM

■ Figure E2.4 is a diagram of the gravitational field produced by a two-body system similar to the Earth and Moon. The gravitational forces of the Earth and Moon on any small body add as vectors, and so do their contributions to the total acceleration due to gravity. The contributions of the Earth and Moon to the potential energy of any object are scalar quantities, and so their gravitational potentials add numerically. The field lines and equipotential surfaces in the figure are computed using these rules. Notice that near either body, the field is spherically symmetric like the field of a single object. Because gravitational attraction decreases with distance squared, even a relatively small body like the Moon is the dominant source of gravitational field close to its surface. Far from the system, the field is also spherically symmetric, like the field of a particle having the total mass of the system. Only at points not very near either body but not very far from the system as a whole do the details of the system's structure influence the shape of the field lines.

We can now describe how a system's gravitational potential energy is stored. When the Moon's distance from Earth increases, the system's energy increases, but neither the Earth nor the Moon is altered. It is the details of the gravitational field that change. The gravitational energy is stored throughout space in the field. On a much smaller scale, your contribution to the Earth's gravitational field changes as you climb a hill. All that effort goes into ever-so-slight a change in the field pattern of Earth-plus-you!

E2.3 Is the Gravitational Field Real?

In some ways, a field is the complete opposite of a particle. While a particle is, ideally, point-like, the gravitational field pervades all of space and exists everywhere at the same time. Can this field be real? The argument we have given for the existence of gravitational field is suggestive but not really a proof. The field is an abstract but elegant and very powerful way to interpret gravity—one that Einstein believed has a better claim to reality than particles. Both field and particle models are indispensible in our present theories.

As you walk out the door, try to picture the world permeated by glowing blue field lines and red equipotential surfaces (■ Figure E2.5)! Such a picture is certainly odd—more surreal than Magritte—but, in a way, very, concretely real.

Points to Ponder

- From the work-energy theorem, explain why the equipotential surfaces are always perpendicular to the field lines.
- Do the field lines represent the path a particle would take if released in a gravitational field? Why or why not?
- Can you locate the *mysterious point* of Study Problem 5 in Figure E2.4? (*Hint:* Could a field line pass through the mysterious point?)

■ **Figure E2.4**
Qualitative diagram of the gravitational field surrounding the Earth and Moon. (For clarity, we represent the mass ratio of Earth to Moon as 5 rather than the actual ratio of 81. Five times as many sample lines converge toward Earth as converge toward the Moon.) The figure shows field lines (blue) in a single plane and the intersections of the equipotential surfaces (red) with that plane. The diagram would be identical in any plane containing the line through the objects. Notice that each object is the dominant influence on the field very near itself—Figure E2.2 would fit entirely within the three surfaces nearest the Earth in this figure. At great distances from such a system—roughly ten times the separation of the objects—the field is also spherically symmetric, as if the system were a single pointlike object. The diagram hints at this trend even at distances comparable to the separation. The system's gravitational energy is stored throughout space in the details of the field pattern.

■ **Figure E2.5**
If the gravitational field were visible, everything surrounding you would appear to be sliced by equipotential surfaces and pierced by field lines that literally weigh you down, wherever you may be. Such a vision, worthy of Magritte's surrealism, represents Earth's very real influence on the space surrounding it.

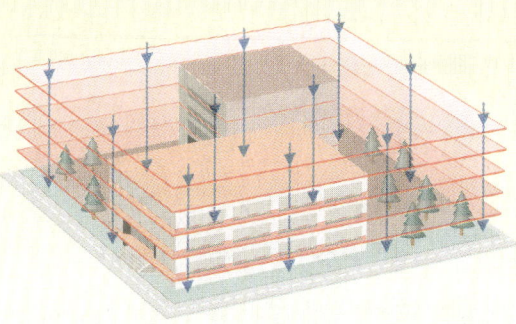

A pair of skaters gliding over the ice possess a third conserved quantity of motion, in addition to linear momentum and energy. Their spectacular spins emphasize its importance.

CHAPTER 9
Angular Momentum

Concepts

Angular momentum of a particle or system

Cross product of vectors

Torque

Energy and power in angular motion

Center of gravity

Center of mass

Force-torque diagram

CM reference frame

Goals

Be able to:

Describe the cross product of two vectors geometrically and in terms of components.

Compute the angular momentum about any center of a particle or system of particles.

Compute the torque produced by a given force about a given center.

Locate the center of mass of a set of particles.

> *We can't return, we can only look behind*
> *From where we came*
> *And go round and round and round*
> *In the circle game.*
>
> Joni Mitchell

Two skaters glide toward each other across the ice, join hands as they meet, and spin around their joined hands. Next, the skaters pull together, their angular speed increases, and their motion becomes blurred to the eye (■ Figure 9.1). If they let go while they are close together, they glide apart much faster than they approached, but if they stretch their arms first, they separate at their initial speeds of approach. Their final state depends on their separation when they release, but not on such details as how long they wait before pulling together or how many turns they make before letting go. From our study of energy and linear momentum, we know that a relation between initial and final states of a system, independent of how the system interacts, is the hallmark of a conservation law. The skaters possess a third conserved physical property, *angular momentum,* related to their spinning motion. In this chapter we shall define the *angular momentum* of a system and *torque,* the rate at which a force transfers angular momentum. We shall find that the angular momentum of an isolated system is conserved. Then, having described the last of the three conserved quantities in mechanics, we can show why the particle model applies to real objects. Finally, we shall study some basic methods for maneuvering spacecraft.

■ FIGURE 9.1
The skaters rotate after joining hands. As they pull themselves closer together, their angular speed increases till they appear blurred to the eye. Their actions illustrate the conservation of a third mechanical property, angular momentum.

9.1 ANGULAR MOMENTUM OF A PARTICLE

9.1.1 What Is Angular Momentum?

Our goal is to describe a conserved property—angular momentum—in rotating systems. As usual, it is simplest to think of a system as a collection of particles and describe first the angular momentum of a single particle. Orbital motion of a particle—a planet around the Sun, or the space shuttle around the Earth—is an example that leads us directly to the description we need. A planet could follow a very complex orbit, or even crash into the Sun, without violating conservation of energy or of linear momentum. But according to Newton's laws, an object that is not launched directly toward the Sun orbits around the Sun rather than crashing into it. There is a third conservation law, which we are seeking, that follows from Newton's laws and explains why planets follow regular elliptical paths. We are looking for a combination of the planet's physical properties that remains constant throughout the orbit.

According to Kepler's second law, the rate at which a planet's position vector sweeps out area in the orbital plane is a constant. As a planet moves along its elliptical orbit (■ Figure 9.2) during a small time interval Δt, it travels a distance $\Delta s = v\,\Delta t$, and its radius vector $\vec{r}(t)$ sweeps out the element of area ΔA. This elementary area is approximately a triangle with base $|\vec{r}|$ and height $\Delta s \sin\phi$, where ϕ is the angle between $\vec{r}$ and the planet's velocity $\vec{v}$:

$$\Delta A \approx \tfrac{1}{2} rv\,\Delta t \sin\phi.$$

Thus the constant rate at which the planet sweeps out area is:

$$\frac{dA}{dt} \equiv \lim_{\Delta t \to 0} \frac{\Delta A}{\Delta t} = \tfrac{1}{2} rv \sin\phi.$$

KEPLER DISCOVERED HIS SECOND LAW FROM HIS ANALYSIS OF TYCHO BRAHE'S OBSERVATIONS OF MARS. SEE CHAPTER 0.

This result leads us to the definition of a general physical property. As with momentum and energy, such a property should be proportional to a particle's mass—a planet ought to have more of it than a baseball following the same orbit. Therefore, we consider the expression:

$$m\frac{dA}{dt} = \tfrac{1}{2} mrv \sin\phi.$$

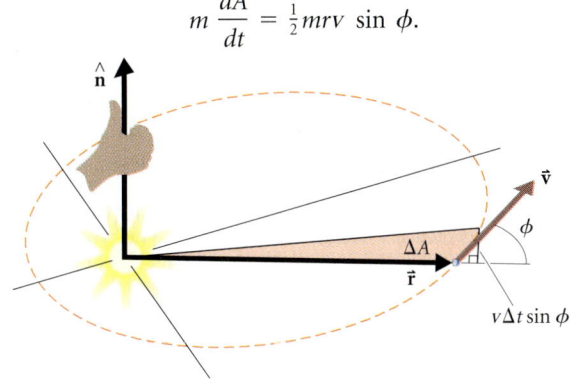

■ FIGURE 9.2
According to Kepler's second law, a planet revolving about the Sun sweeps out area at a constant rate $\Delta A/\Delta t$. The area swept out in time Δt is approximately a triangle with base $|\vec{r}|$ and altitude $|\vec{v}|\Delta t \sin\phi$. The vector $\hat{n}$ perpendicular to the orbital plane is a conventional way to describe the direction of rotation. To choose between the two possible perpendicular directions, curl the fingers of your right hand in the direction the planet moves and choose $\hat{n}$ parallel to your thumb.

THE NUMERICAL FACTOR OF 2 IS INTRODUCED HERE SO THAT L HAS ITS USUAL FORM. IF $m\,dA/dt$ IS CONSTANT, THEN SO IS $2m\,dA/dt$.

The product $p \equiv mv$ is the magnitude of the planet's linear momentum. Thus Kepler's second law suggests we define the planet's angular momentum as a particular combination of its position and momentum:

$$L \equiv rp \sin \phi = 2m \frac{dA}{dt} = \text{constant}.$$

This expression cannot be complete because angular momentum is not a scalar. For example, two identical satellites initially at the same position but with opposite velocities travel the same orbit in opposite directions, and should have opposite angular momenta, yet the product $rp \sin \phi$ is the same for both.

Our planetary example tells us how to assign direction to an angular momentum *vector* $\vec{L}$. The planet's orbital plane is fixed in space, and we want to describe its orientation with a single vector. We could pick a vector lying in the orbital plane (■ Figure 9.3), but it would also lie in many other planes. However, a vector perpendicular to the orbital plane defines that plane's orientation uniquely. So, we define $\vec{L}$ to be perpendicular to the orbital plane—that is, always perpendicular to both $\vec{r}$ and $\vec{v}$. There are two directions perpendicular to the plane corresponding to the two senses in which a planet could move around the same orbit. We use a *right-hand rule* to relate the direction of $\vec{L}$ to the sense of revolution:

ODDLY, $\vec{L}$ IS IN A DIRECTION TOWARD WHICH THE PLANET NEVER MOVES.

RIGHT-HAND RATHER THAN LEFT-HAND RULES ARE NEARLY UNIVERSAL. COMPARE THIS RULE WITH THE CONVENTIONAL CHOICE OF DIRECTION FOR THE z-AXIS IN A THREE-DIMENSIONAL COORDINATE SYSTEM (FIG. 1.2).

> Curl the fingers of your *right* hand so your fingertips point in the direction of the planet's motion. Then the planet's angular momentum $\vec{L}$ lies along the direction of your thumb (Figure 9.2).

From Kepler's law we now have the complete definition for the angular momentum of a particle about the point we choose as the origin for position vectors:

> A particle at position $\vec{r}$ from point O and with linear momentum $\vec{p}$ has *angular momentum about O*, $\vec{L}_O$, with magnitude equal to the product of the magnitudes of $\vec{r}$ and $\vec{p}$ and the sine of the angle between them.
>
> $$|\vec{L}_O| \equiv |\vec{r}||\vec{p}|\sin \phi. \qquad (9.1)$$
>
> The direction of $\vec{L}_O$ is perpendicular to both $\vec{r}$ and $\vec{p}$, and parallel to the thumb of your right hand when your fingers curl from $\vec{r}$ toward $\vec{p}$ (■ Figure 9.4).

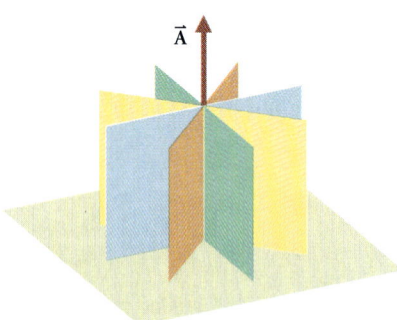

■ FIGURE 9.3
Any given vector $\vec{A}$ is contained in an infinite number of planes that intersect along the line of the vector. But the plane perpendicular to the vector and passing through its tail is unique. Thus a vector describes the orientation of a plane perpendicular to the vector.

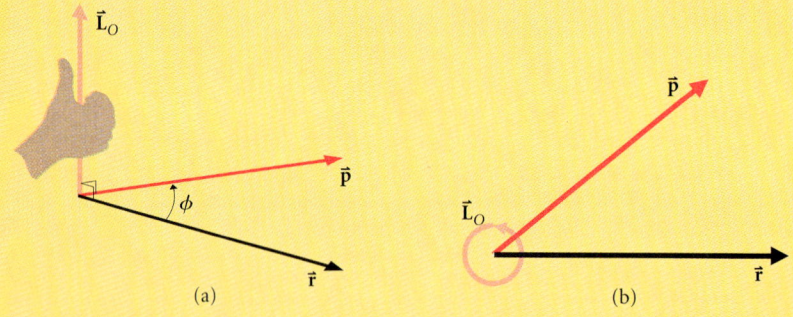

■ FIGURE 9.4
(a) The angular momentum of a particle about point O, $\vec{L}_O$, is defined by the particle's position $\vec{r}$ and linear momentum $\vec{p}$. When $\vec{r}$ and $\vec{p}$ are placed tail to tail and your right hand is placed so the fingers curl from $\vec{r}$ toward $\vec{p}$, your thumb gives the direction of $\vec{L}_O$, which is perpendicular to both $\vec{r}$ and $\vec{p}$. (b) An alternative diagram for $\vec{L}_O$. Here $\vec{r}$ and $\vec{p}$ lie in the plane of the diagram and $\vec{L}_O$ points out of the page, represented by the symbol $\hat{\odot}$, which should remind you of an arrowhead pointing out of the page. The symbol $\hat{\otimes}$ represents a vector pointing into the page. It should remind you of an arrow's tail feathers.

The unit of angular momentum is $\text{kg} \cdot \text{m}^2/\text{s} = \text{J} \cdot \text{s}$ (joule·second). It has no special SI name.

EXAMPLE 9.1 ♦ A particle with mass $m = 10.0$ kg is on the x-axis at $x = 2.50$ m. Its velocity is 25.0 m/s at an angle of 45° to the y-axis (■ Figure 9.5). What is the particle's angular momentum $\vec{L}_O$ about the origin?

MODEL We substitute the given data into the definition of angular momentum. We draw the vectors tail to tail (■ Figure 9.6), both for applying the right-hand rule and for computing the angle between them.

SETUP From Figure 9.6, $\phi = 135°$. For the fingers of a right hand to curl from $\vec{r}$ toward $\vec{p}$, the thumb has to point in the positive z-direction. Using eqn. (9.1) we find:

SOLVE
$$\vec{L}_O = \hat{k}|\vec{r}||\vec{p}|\sin\phi = \hat{k}(2.50 \text{ m})(10.0 \text{ kg})(25.0 \text{ m/s})\sin(135°)$$
$$= (442 \text{ kg}\cdot\text{m}^2/\text{s})\hat{k} = (442 \text{ J}\cdot\text{s})\hat{k}.$$

ANALYZE Until you are very confident with the right-hand rule, it may be useful to make a two-dimensional diagram like Figure 9.6 before applying the rule. ■

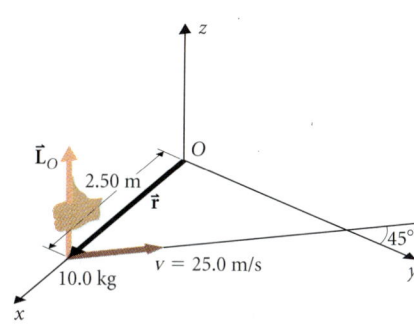

■ **FIGURE 9.5**
What is the particle's angular momentum $\vec{L}_O$ about the origin?

9.1.2 Angular Momentum as a Vector Product

The way in which the position vector $\vec{r}$ and momentum $\vec{p}$ combine to form angular momentum $\vec{L}_O$ occurs in very many physical applications. The pattern describes a second type of vector multiplication called the cross product.

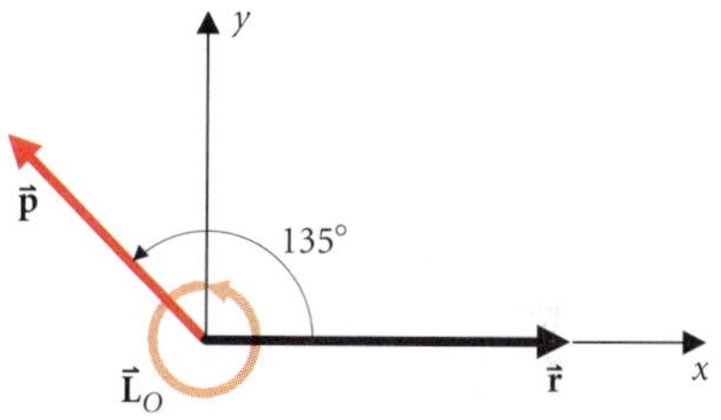

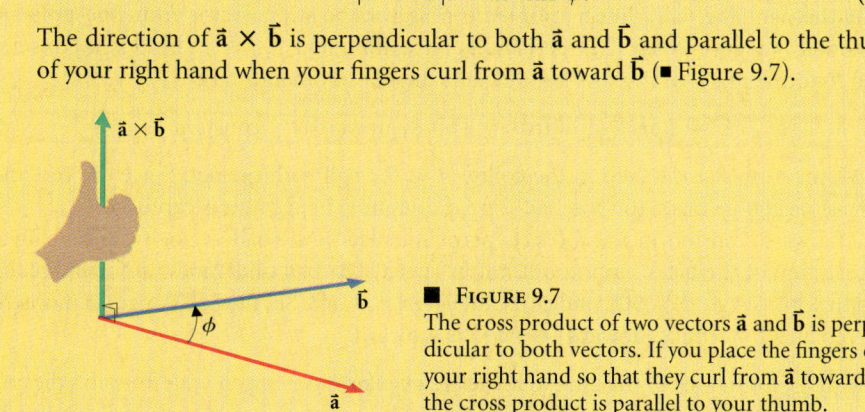

■ **FIGURE 9.6**
A two-dimensional diagram for computing the angular momentum in Example 9.1. The angular momentum vector points out of the page.

The *cross product* of two vectors $\vec{a}$ and $\vec{b}$, written $\vec{a} \times \vec{b}$, is a vector whose magnitude equals the product of the magnitudes of $\vec{a}$ and $\vec{b}$ and the sine of the angle between them:
$$|\vec{a} \times \vec{b}| = ab \sin\phi. \quad (9.2)$$

The direction of $\vec{a} \times \vec{b}$ is perpendicular to both $\vec{a}$ and $\vec{b}$ and parallel to the thumb of your right hand when your fingers curl from $\vec{a}$ toward $\vec{b}$ (■ Figure 9.7).

■ **FIGURE 9.7**
The cross product of two vectors $\vec{a}$ and $\vec{b}$ is perpendicular to both vectors. If you place the fingers of your right hand so that they curl from $\vec{a}$ toward $\vec{b}$, the cross product is parallel to your thumb.

Thus the angular momentum of a particle about point O is:

$$\vec{L}_O = \vec{r} \times \vec{p}. \quad (9.3)$$

Two useful ways to evaluate angular momentum, or any quantity defined as a cross product, are illustrated in ■ Figure 9.8. The factor $|\vec{r}|\sin\phi$ equals $r_{\perp p}$, the component of $\vec{r}$ perpen-

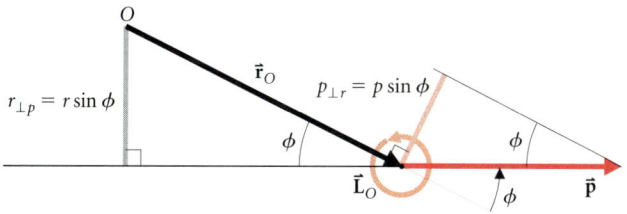

■ **FIGURE 9.8**
The magnitude of $\vec{a} \times \vec{b}$ is the product of the two vector magnitudes and the sine of the angle between them. An angular momentum calculation is shown as an example. The product $r \sin\phi$ is $r_{\perp p}$, the component of $\vec{r}$ perpendicular to $\vec{p}$. Geometrically, $r_{\perp p}$ is the line drawn from O perpendicular to the momentum vector. Similarly, the product $p \sin\phi$ is $p_{\perp r}$, the component of $\vec{p}$ perpendicular to $\vec{r}$. So $|\vec{L}| = |\vec{r} \times \vec{p}| = rp \sin\phi = r(p_{\perp r}) = (r_{\perp p})p$.

REMEMBER: THE DOT PRODUCT IS A SCALAR, BUT THE CROSS PRODUCT IS A VECTOR.

THE COMPONENT OF A VECTOR PARALLEL TO ANOTHER CAN BE EITHER POSITIVE OR NEGATIVE. HERE, "PERPENDICULAR COMPONENT" MEANS THE (POSITIVE) LENGTH OF THE COMPONENT.

dicular to $\vec{p}$. Equivalently, the product $|\vec{p}|\sin\phi$ is $p_{\perp r}$, the component of $\vec{p}$ perpendicular to $\vec{r}$. Thus:

$$|\vec{L}_O| \equiv L_O = (r_{\perp p})p = (p_{\perp r})r. \qquad (9.4)$$

The dot product of two vectors and the magnitude of their cross product have similar geometrical descriptions; each multiplies one vector's magnitude by a component of the other vector. The dot product $\vec{a} \cdot \vec{b}$ is $|\vec{a}|$ times the component of $\vec{b}$ *parallel* to $\vec{a}$, while the magnitude of $\vec{a} \times \vec{b}$ equals $|\vec{a}|$ times the component of $\vec{b}$ *perpendicular* to $\vec{a}$. In addition, the direction of the cross product is perpendicular to the plane in which the two vectors lie, while the dot product is a scalar and has no direction.

EXERCISE 9.1 ♦ For the particle in Example 9.1, compute $r_{\perp p}$ and $p_{\perp r}$. Show that the equations $|L_O| = r(p_{\perp r}) = (r_{\perp p})p$ give the same angular momentum obtained in the example.

EXAMPLE 9.2 ❖ In the carnival game shown in ■ Figure 9.9, you win a teddy bear if you can throw a baseball (mass m, speed v) into a net. The net is attached to a vertical pole by a rod with length R, negligible mass, and a frictionless hinge, so that the baseball will swing in a circle and knock your prize from the shelf. Find the angular momentum $\vec{L}$ of the baseball about the hinge. Show that $\vec{L}$ remains constant (until the ball strikes the prize) and find the relation between $\vec{L}$ and the baseball's angular speed ω in the circle.

MODEL We assume the baseball is thrown rapidly enough that we can neglect any vertical motion. We divide the motion into three segments: (1) ball is moving toward the net; (2) ball is captured by net; and (3) ball and net swing around pole in a circle.
(1) The ball moves horizontally with constant linear momentum of magnitude mv, until it hits the net. The ball's linear momentum and its position vector from the pole are horizontal (■ Figure 9.10). The perpendicular component $r_{\perp p} \equiv r\sin\phi$ is a constant equal to the length R of the rod. So, the ball's angular momentum is:

$$\vec{L} = (pr_{\perp p}, \text{upward}) = (mvR, \text{upward}) = \text{constant}.$$

(2) Momentum is conserved in the collision of the ball with the net. Since the rod and net have negligible mass, the baseball's speed does not change as it is captured.
(3) Afterward, tension in the rod acts perpendicular to the ball's velocity, and changes the direction of the ball's motion but not its speed. The ball undergoes uniform circular motion with $r = R$, $\phi = 90°$, and constant speed $v = \omega R$. So, the ball's angular momentum remains $\vec{L} = (mvR, \text{upward}) = (mR^2\omega, \text{upward})$.

ANALYZE It may seem startling to associate *angular* momentum with the ball's motion along a straight line. However, if you imagine yourself standing at the pole, you would have to turn your head to keep looking straight at the ball. In this sense, the ball is rotating about the pole. Angular momentum measures the feature of this *rotation* that remains the same when the ball's motion becomes circular.
Our reasoning illustrates two general facts: a particle in uniform linear motion has constant angular momentum about any chosen point; a particle in uniform circular motion has constant angular momentum about the center of the circle. ■

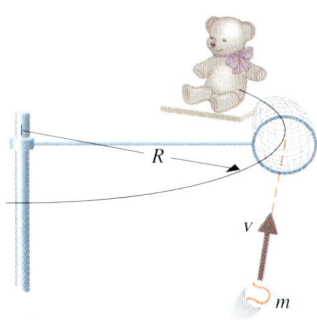

■ **FIGURE 9.9**
Diagram of a carnival game. You win a teddy bear if you throw a baseball into the net.

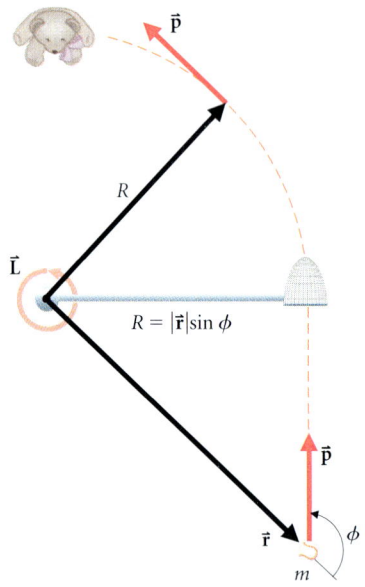

■ **FIGURE 9.10**
The carnival game seen from above. The baseball's position and momentum vectors lie in the horizontal plane of the diagram. As the ball travels along a straight line toward the net, the component of its position perpendicular to its momentum is a constant equal to the distance R of the net from the pole. After striking the net, the ball moves in a circle of radius R about the pole.

Because the definition of angular momentum involves a particle's position vector, the value assigned to angular momentum depends on the choice of origin. This is similar to the dependence of gravitational potential energy on choice of reference level. We are similarly free to choose the origin to make our calculations as easy as possible. For example, with the Sun as the origin, the magnitude of a planet's angular momentum is twice the rate the line from the Sun to the planet sweeps out area. If we used any other origin, we wouldn't obtain this simple relation. The description is clearest when the origin is at the Sun, the object that is exerting force on the planet.

Math Toolbox

PROPERTIES OF THE CROSS PRODUCT

Be sure you know and understand the eight properties of the cross product that are laid out here. We shall use them frequently. (*Note:* The vectors $\vec{a}$, $\vec{b}$, and $\vec{c}$ represent arbitrary vectors.)

I. The cross product of two parallel vectors is zero, because the angle between them is zero:

If $\vec{a} \parallel \vec{b}$, then $|\vec{a} \times \vec{b}| = ab \sin(0) = 0$.

Neither vector has a component perpendicular to the other. In particular, the cross product of any vector with itself is zero.

II. Each of the unit vectors in a rectangular x–y–z coordinate system is a cross product of the other two.
Each of the vectors in ■ Figure 9.11 has magnitude 1 and the angle between any two of them is 90°, so $\sin \phi$ is 1. Thus the cross product of any two of the vectors has unit magnitude. Now the direction of $\hat{i} \times \hat{j}$ is along $\hat{k}$, since the right-hand rule for this product is the same as the rule we used to define a right-handed coordinate system in Chapter 1. Thus:

$$\hat{k} = \hat{i} \times \hat{j}. \quad (9.5a)$$

With your thumb along $\hat{i}$, your fingers curl from $\hat{j}$ to $\hat{k}$ as shown by the arrow labeling angle ϕ_{yz} in Figure 9.12.

$$\hat{i} = \hat{j} \times \hat{k}. \quad (9.5b)$$

Similarly: $\hat{j} = \hat{k} \times \hat{i}. \quad (9.5c)$

(Figure 9.14 gives a way to remember these equations.)

III. Reversing the order of multiplication reverses the direction of a cross product (■ Figure 9.12):

$$\vec{b} \times \vec{a} = -\vec{a} \times \vec{b}. \quad (9.6)$$

Curling one's fingers from $\vec{b}$ to $\vec{a}$ rather than from $\vec{a}$ to $\vec{b}$ gives the opposite thumb direction. The two products have the same magnitude:

$$|\vec{a} \times \vec{b}| = |\vec{b} \times \vec{a}| = ab \sin \phi.$$

IV. For any constant C, $(C\vec{a}) \times \vec{b} = C(\vec{a} \times \vec{b})$. (9.7) (See problem 20.)

V. A cross product represents an area.
■ Figure 9.13 shows a piece of a plane surface in the shape of a parallelogram with sides defined by the vectors $\vec{a}$ and $\vec{b}$. If $|\vec{a}|$ is the base of the parallelogram, $|\vec{b}| \sin \phi$ is its altitude. The product $ab \sin \phi \equiv |\vec{a} \times \vec{b}|$ is the area of the parallelogram. The direction of $\vec{a} \times \vec{b}$, perpendicular to the surface, specifies its orientation. Thus $\vec{a} \times \vec{b}$ completely represents the piece of surface.

VI. Distributive law: The cross product of a vector with a sum of vectors equals the sum of its cross products with the individual vectors in the sum:

$$\vec{a} \times (\vec{b} + \vec{c}) = \vec{a} \times \vec{b} + \vec{a} \times \vec{c}. \quad (9.8)$$

VII. The cross product in terms of components
It is often useful to compute the cross product of two vectors directly from their components. Here we obtain the result for

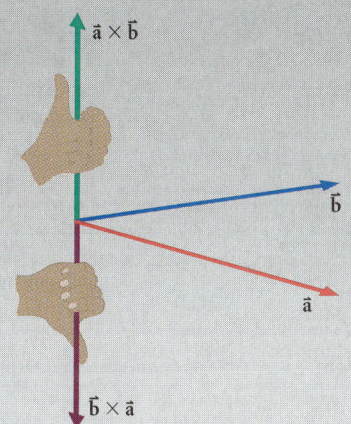

■ **FIGURE 9.12**
Reversing the order of factors in a cross product does not change the magnitude of the result. However, reversing the direction your fingers curl reverses the direction of your right thumb: $\vec{b} \times \vec{a} = -\vec{a} \times \vec{b}$.

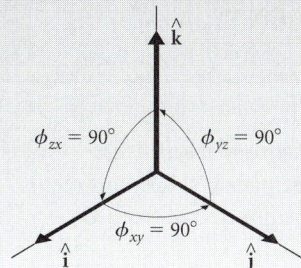

■ **FIGURE 9.11**
The unit vectors $\hat{i}$, $\hat{j}$, and $\hat{k}$ are each equal to a cross product of the other two. The order of the product matters!

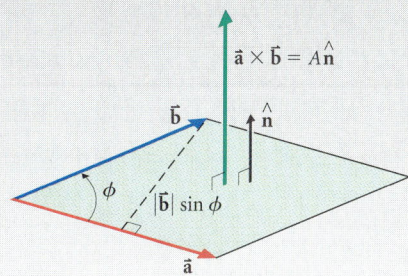

■ **FIGURE 9.13**
The product $\vec{a} \times \vec{b}$ represents the area of a parallelogram bounded by vectors $\vec{a}$ and $\vec{b}$.

continued

two vectors in the x-y plane, and leave proof of the result for three dimensions as a problem. Let

$$\vec{a} = a_x \hat{i} + a_y \hat{j} \quad \text{and} \quad \vec{b} = b_x \hat{i} + b_y \hat{j},$$

then $\vec{a} \times \vec{b}$ is perpendicular to the x-y plane—that is, in the z-direction.

$$\vec{a} \times \vec{b} = (a_x \hat{i} + a_y \hat{j}) \times (b_x \hat{i} + b_y \hat{j})$$

(use Property VI to expand out)

$$= (a_x \hat{i} + a_y \hat{j}) \times b_x \hat{i} + (a_x \hat{i} + a_y \hat{j}) \times b_y \hat{j}$$

(use Property IV to expand out again)

$$= a_x b_x (\hat{i} \times \hat{i}) + a_y b_x (\hat{j} \times \hat{i})$$
$$+ a_x b_y (\hat{i} \times \hat{j}) + a_y b_y (\hat{j} \times \hat{j})$$

(use properties I and III, and eqn. 9.5a)

$$= \hat{k}(a_x b_y - a_y b_x).$$

The general result for two vectors each having x-, y- and z-components (see Problem 9.31) is:

$$\vec{a} \times \vec{b} = \hat{i}(a_y b_z - a_z b_y) + \hat{j}(a_z b_x - a_x b_z)$$
$$+ \hat{k}(a_x b_y - a_y b_x). \quad (9.9)$$

One helpful aid for remembering which terms are positive and which negative is to look at the pattern of subscripts and unit vector names in this formula. Think of $\hat{i} a_y b_z$ as "xyz" and imagine the letters to be written on a cylinder (■ Figure 9.14).

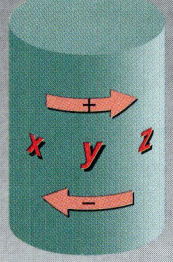

■ **FIGURE 9.14**
Device for remembering the signs of terms in the component representation of the cross product. The term $\hat{i} a_y b_z$, or "xyz", goes around the cylinder in the + direction, and so is a positive term. The term $\hat{k} a_y b_x$, or "zyx", is negative.

The subscript patterns in the positive terms go around the cylinder one way (call this the positive way). The subscript patterns in the negative terms go around the cylinder in the opposite (negative) sense. Another popular method is to think of the cross product as a 3×3 determinant:

$$\vec{a} \times \vec{b} = \begin{vmatrix} \hat{i} & \hat{j} & \hat{k} \\ a_x & a_y & a_z \\ b_x & b_y & b_z \end{vmatrix} \quad (9.10)$$

EXAMPLE 9.3 ♦ The object in Example 9.1, located on the x-axis at $x = 2.50$ m with velocity 25.0 m/s at 45° to the y-axis, has angular momentum $\vec{L}_O = (442 \text{ J·s})\hat{k}$ about the origin. What is its angular momentum $\vec{L}_A$ about the point A on the y-axis at $y = -2.50$ m?

MODEL ■ Figure 9.15 shows the object and its position vector $\vec{r}_A$ from A.

SETUP We compute the particle's angular momentum about point A from the definition using eqn. (9.3), or equivalently, eqn. (9.1). The length of the position vector $\vec{r}_A$ from point A to the particle's position is found from the 45° right triangle OAP: $|\vec{r}_A| = (2.50 \text{ m})\sqrt{2}$. The two vectors $\vec{r}$ and $\vec{p}$ are at right angles.

SOLVE Thus:

$$|\vec{L}_A| = |\vec{r}_A \times \vec{p}| = |\vec{r}_A| |\vec{p}| \sin \phi$$
$$= (2.50\sqrt{2} \text{ m})(10.0 \text{ kg})(25.0 \text{ m/s})\sin(90°) = 884 \text{ J·s}.$$

Using the right-hand rule, $\vec{L}_A$ is parallel to the z-axis: $\vec{L}_A = (884 \text{ J·s})\hat{k}$.

ANALYZE The particle's angular momentum about A has a greater magnitude than its angular momentum about O, because the particle's position vector from P is both longer and more perpendicular to $\vec{p}$ than its position vector from O. ■

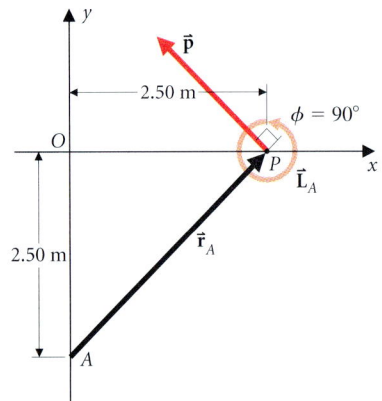

■ **FIGURE 9.15**
The particle's angular momentum about point A is $\vec{L}_A$. This demonstrates that $\vec{L}$ depends on the origin chosen. In any problem choose the origin that makes the calculations easiest.

EXERCISE 9.2 ❖ (a) About what points P does the particle of Examples 9.1 and 9.3 have angular momentum $\vec{L}_P$ equal to $\vec{L}_O$? (b) About what points is the angular momentum zero? (c) About what points is the angular momentum in the z-direction? (d) The negative z-direction?

EXAMPLE 9.4 ♦♦ A race car of mass 1.5×10^3 kg is rounding a tight turn with center at point O. At a certain time, the car's position is $\vec{r}_O = (85 \text{ m})(\hat{i} - \hat{j})$ and its velocity is $\vec{v} = (43 \text{ m/s})(\hat{i} + \hat{j})$. Find the car's angular momentum about O.

MODEL First, we draw the vectors (■ Figure 9.16), then we use the definition of angular momentum in the form of eqn. (9.3). Using the right-hand rule, we find that $\vec{L}$ is in the $+z$-direction.

SETUP The car's linear momentum is its mass times its velocity: $\vec{p} = m\vec{v}$. Then:

$$\vec{L}_O = \vec{r}_O \times \vec{p} = \vec{r}_O \times (m\vec{v})$$

SOLVE
$$= [(85 \text{ m})(\hat{i} - \hat{j})] \times [(1.5 \times 10^3 \text{ kg})(43 \text{ m/s})(\hat{i} + \hat{j})]$$

USE PROPERTY IV:
$$= (5.5 \times 10^6 \text{ kg} \cdot \text{m}^2/\text{s})(\hat{i} - \hat{j}) \times (\hat{i} + \hat{j})$$

USE PROPERTY VI:
$$= (5.5 \times 10^6 \text{ J} \cdot \text{s})(\hat{i} \times \hat{i} - \hat{j} \times \hat{i} + \hat{i} \times \hat{j} - \hat{j} \times \hat{j})$$

USE PROPERTIES II AND III:
$$= (5.5 \times 10^6 \text{ J} \cdot \text{s})[0 - (-\hat{k}) + \hat{k} - 0]$$
$$= (1.1 \times 10^7 \text{ J} \cdot \text{s})\hat{k}.$$

ANALYZE This calculation differs from ordinary algebra only in the rules for cross products of unit vectors. Products of the vectors with themselves vanish and $-\hat{j} \times \hat{i} = +\hat{i} \times \hat{j}$. ■

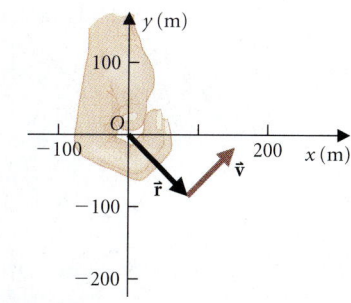

■ **FIGURE 9.16**
The position and velocity vectors are perpendicular, as you can readily check by computing their dot product. The right-hand rule gives the direction of $\vec{L}$ as out of the page, parallel to $\hat{k}$.

9.2 Torque

9.2.1 What Is Torque?

We know that forces transfer energy and linear momentum. Anyone who has tried to open a heavy door in a hurry has learned intuitively how forces transfer angular momentum. If you apply a moderate force $\vec{F}_1$ to the door of a department store (■ Figure 9.17) in the usual way at the edge opposite the hinges, the door accelerates rapidly and you can rush through. If you make a mistake and apply an equal force $\vec{F}_2$ near the hinges, you get a painful surprise. Even though $\vec{F}_3$ is applied to the proper side of the door, you wouldn't expect it to do anything except compress the door against the hinges. The successful strategy is to exert force *perpen-*

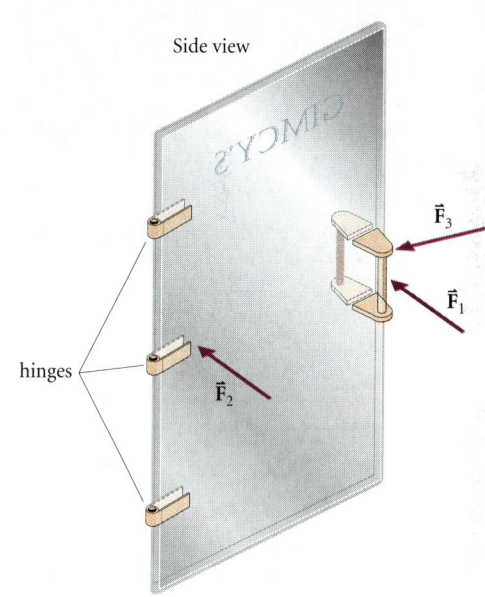

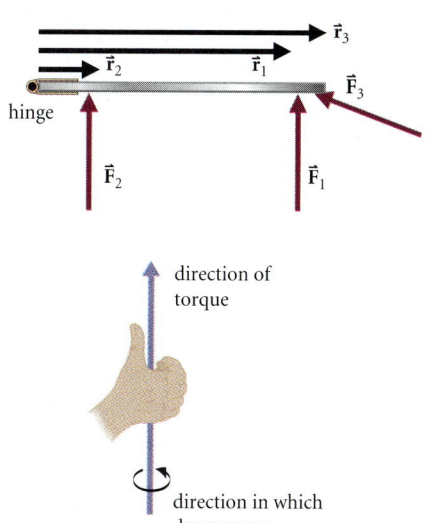

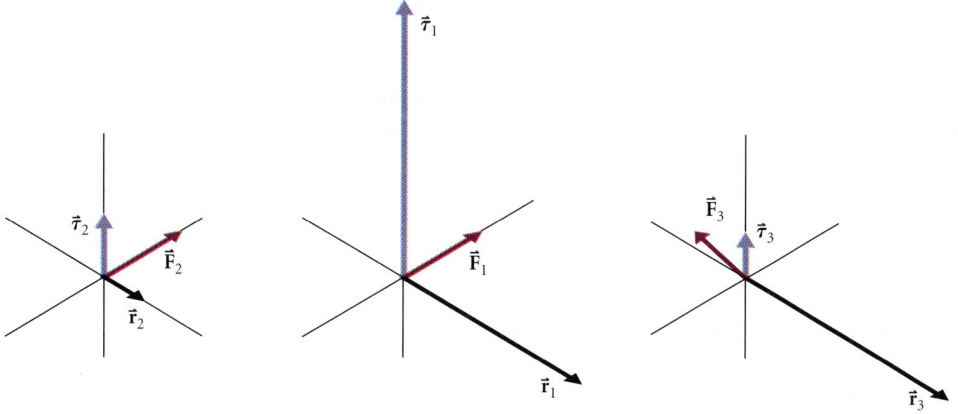

■ **FIGURE 9.17**
From experience, we know that the easiest way to open a door is with force applied perpendicular to the door as far as possible from the hinges. Thus $\vec{F}_1$ is effective in opening the door, while $\vec{F}_2$ and $\vec{F}_3$ are not. Torque, the cross product of position with force, describes the effectiveness of a force in causing the door to rotate. The direction of torque and the sense of the rotation it causes are related by a right-hand rule: thumb parallel to torque, rotation in the sense your fingers curl.

Torque on a wine press. Biologist Jim Duncan stands on the press to ensure adequate friction, while physicist Susan Lea exerts torque on a long rod. By pushing perpendicular to the rod at its end, Dr. Lea produces maximum torque for a given effort.

PICTURES OF AN ARROW COMING OUT OF THE PAGE $\hat{\odot}$ OR GOING INTO THE PAGE $\hat{\otimes}$ ARE OFTEN USED TO DRAW VECTORS IN A TWO-DIMENSIONAL DIAGRAM. WITH A HAT, THE SYMBOLS REPRESENT UNIT VECTORS INTO AND OUT OF THE DIAGRAM.

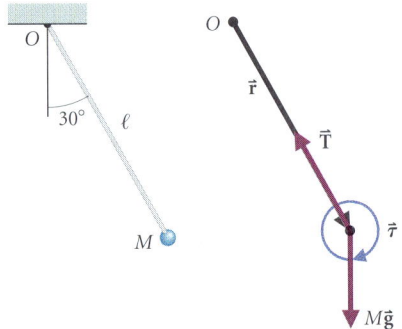

■ FIGURE 9.18
Gravitational force produces a torque on a pendulum. The tension $\vec{T}$ causes no torque about the suspension point O.

dicular to the door at maximum distance from the hinges. *Torque*, the product of perpendicular force component and distance, is what causes the door to rotate out of your way.

The magnitude of one vector multiplied by the perpendicular component of another is one of the descriptions of a cross product given in §9.1: $|\vec{a} \times \vec{b}| = b(a_{\perp b})$. So, our intuitive sense of torque suggests that we define torque as a cross product:

> The *torque* $\vec{\tau}_O$ exerted about point O when a force $\vec{F}$ is exerted at position $\vec{r}$ from O is equal to the cross product of $\vec{r}$ with $\vec{F}$.
>
> $$\vec{\tau}_O \equiv \vec{r} \times \vec{F}. \qquad (9.11)$$

If you put the thumb of your right hand parallel to a torque, your fingers curl in the sense of rotation the torque tends to cause (Figure 9.17).

Like work, torque is a product of distance and force, and its unit is:

$$(1 \text{ newton})(1 \text{ meter}) = 1 \text{ N} \cdot \text{m}.$$

Unlike work, however, torque describes an interaction happening at a particular instant of time, not the result of a process acting over a time interval. Thus, although torque and work have the same physical dimensions, torque is not the same thing as work, and we do not think of its unit as a joule. The unit of torque has no special SI name.

EXAMPLE 9.5 ♦ A pendulum consists of a ball of mass $M = 0.50$ kg suspended by a rod of length $\ell = 0.50$ m and with negligible mass. When the pendulum is at an angle $\theta = 30°$ from the vertical, what is the torque about the pivot point exerted on it?

MODEL The forces acting on the pendulum ball are shown in ■ Figure 9.18. They act at position $\vec{r}$ from the pivot point.

SETUP Tension acts along the rod and exerts no torque about the pivot O, since $\vec{T}$ is antiparallel to $\vec{r}$. The gravitational force does exert a torque about O:

$$\vec{\tau}_O = \vec{r} \times (M\vec{g}).$$

SOLVE From the figure:

$$\vec{r} \times (M\vec{g}) = (\ell M g \sin \theta, \text{ into the page}) \ (\hat{\otimes})$$
$$= \hat{\otimes}(0.50 \text{ m})(0.50 \text{ kg})(9.8 \text{ m/s}^2)\sin 30°.$$
$$\vec{\tau}_O = (1.2 \text{ N}\cdot\text{m})\hat{\otimes} \quad \text{(into the page)}.$$

ANALYZE As the ball accelerates, its angular momentum about the pivot changes, consistent with the idea that nonzero torque changes angular motion. Also note that θ, and hence τ, changes with time.

EXERCISE 9.3 ❖ Find the torque about the Sun that is exerted on a planet due to the gravitational interaction between the two objects.

Our intuitive definition of torque suggests that angular momentum is transferred only when torque occurs. We can verify this suggestion by calculating the rate at which a particle's angular momentum changes. We'll need the derivative of a cross product of two vectors.

EXERCISE 9.4 ♦♦♦ Show that:

$$\frac{d}{dt}(\vec{A} \times \vec{B}) = \frac{d\vec{A}}{dt} \times \vec{B} + \vec{A} \times \frac{d\vec{B}}{dt}.$$

(*Hint:* Express the cross product in components before differentiating.)

The rate of change of a particle's angular momentum is:

$$\frac{d}{dt}\vec{L}_O = \frac{d}{dt}(\vec{r} \times \vec{p}) = \frac{d\vec{r}}{dt} \times \vec{p} + \vec{r} \times \frac{d\vec{p}}{dt}.$$

REMEMBER: WHEN USING THIS RESULT, BE CAREFUL NOT TO REVERSE THE ORDER OF THE VECTORS IN ANY TERM.

Now $d\vec{r}/dt$ is the velocity $\vec{v}$ of the particle and $\vec{p} = m\vec{v}$, so the first term in $d\vec{L}_O/dt$ is zero because it is the cross product of two parallel vectors. In the second term, we may use Newton's second law to replace the rate of change of the particle's momentum by the force acting on the particle. Thus:

WE DERIVED THIS RELATION IN §6.1.

$$\frac{d}{dt}\vec{L}_O = \vec{r} \times \frac{d\vec{p}}{dt} = \vec{r} \times \vec{F} = \vec{\tau}_O. \qquad (9.12)$$

The rate of change of a particle's angular momentum about a point equals the torque about that point acting on the particle.

In short, torque transfers angular momentum. Compare eqn. (9.12) with the relation $\vec{F} = d\vec{p}/dt$, which shows that force transfers linear momentum.

EXAMPLE 9.6 ♦♦ A lecture demonstration uses a small rocket motor of mass 0.19 kg, fed by compressed air, and mounted on a massless rod 0.21 m long that rotates about a pivot at point O (■ Figure 9.19). If the device starts from rest and develops a constant thrust of 0.50 N for 11 s, what is its final angular speed?

MODEL Despite the fact that the rocket's thrust changes direction continuously, the torque it exerts about the pivot is constant (Figure 9.19b). This constant torque produces a constant rate of change of the rocket's angular momentum about O.

SETUP The torque is:

$$\vec{\tau}_O = \vec{r} \times \vec{F} = |\vec{r}||\vec{F}|\sin(90°)\hat{\mathbf{o}} \quad \text{(out of page)}.$$

REMEMBER: TO FIND THE DIRECTION, PLACE $\vec{r}$ AND $\vec{F}$ TAIL TO TAIL, AND CURL YOUR FINGERS FROM $\vec{r}$ TO $\vec{F}$.

And,

$$\frac{d}{dt}\vec{L}_O = \vec{\tau}_O = \text{constant}.$$

So:

$$\Delta\vec{L}_O = (\vec{\tau}_O)\Delta t = rF\,\Delta t\,\hat{\mathbf{o}}. \qquad (i)$$

Since the rocket starts from rest, its initial angular momentum is zero, and so its final angular momentum equals $\Delta\vec{L}_O$. In terms of the rocket's angular speed ω, its momentum has magnitude $p = m\omega r$, and its final angular momentum (Figure 9.19) is:

$$\vec{L}_O = \vec{r} \times \vec{p} = |\vec{r}||\vec{p}|\sin(90°)\hat{\mathbf{o}}$$
$$= r(m\omega r)\hat{\mathbf{o}}. \qquad (ii)$$

■ **FIGURE 9.19**
(a) A rocket motor on the end of a light rod moves in a circle of radius 20 cm. (b) The torque $\vec{\tau}_O$ is the same no matter what the position of the rocket motor, since $\vec{F} \perp \vec{r}$. Tension in the rod causes no torque about O. (c) The final angular momentum is $\vec{L}_O = \vec{r} \times \vec{p}$.

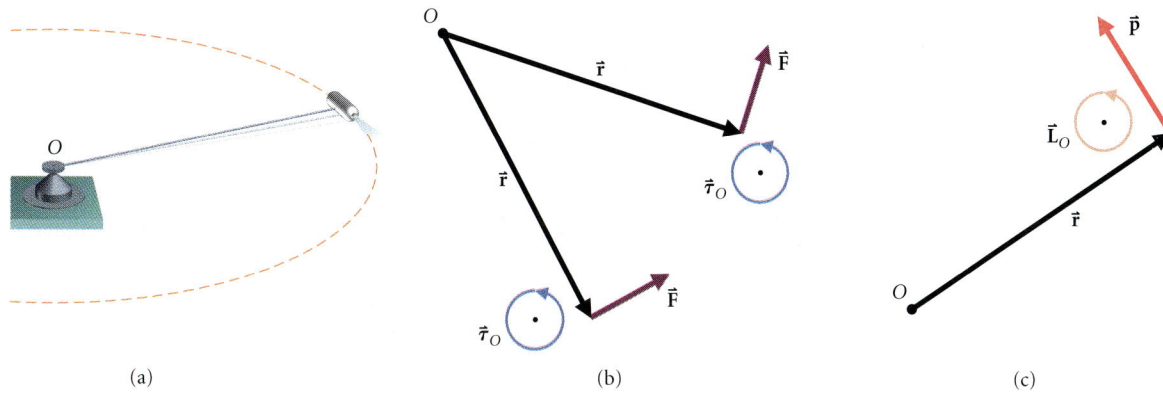

(a)   (b)   (c)

SOLVE Combining eqns. (i) and (ii),

$$m\omega r^2 = rF\,\Delta t,$$

or:
$$\omega = \frac{F\,\Delta t}{mr} = \frac{(0.50\text{ N})(11\text{ s})}{(0.19\text{ kg})(0.21\text{ m})}$$
$$= 140\,\frac{\text{N}\cdot\text{s}}{\text{kg}\cdot\text{m}} = 140\text{ rad/s}.$$

ANALYZE Check the units:

$$\frac{\text{N}\cdot\text{s}}{\text{kg}\cdot\text{m}} = \frac{\text{kg}\cdot\text{m}}{\text{s}^2}\cdot\frac{\text{s}}{\text{kg}\cdot\text{m}} = \frac{1}{\text{s}}.$$

> COMPARE THE DISCUSSION OF LINEAR IMPULSE (§6.1.2).

The change in angular momentum caused by the rocket motor is the integral of the torque exerted by the motor over time, an *angular impulse*:

$$\Delta\vec{\mathbf{L}} \equiv \int \vec{\boldsymbol{\tau}}\,dt = \vec{\boldsymbol{\tau}}_{\text{av}}\,\Delta t.$$

Often, in situations where large torques act for very short periods of time, angular impulse is a more useful concept than the torque that produces it, just as linear impulse is frequently easier to use than the force that produces it.

9.2.2 Work and Power in Rotating Systems

> RECALL THE DISCUSSION OF SIMPLE MACHINES IN §7.2.

The linear machines we studied in Chapter 7 need to be restarted or reset frequently: the load arrives at the top of a ramp; pulleys run out of rope. Because rotating machines are easily designed to function continuously, they have been of immense technical and economic importance since the beginnings of the Industrial Revolution. The earliest practical sources of power—wind and water—were captured by windmills and rotating water wheels. Until the invention of electric motors, the most practical way to transmit power throughout a factory was via long, rotating, overhead shafts. Today, nearly every mechanical device of any complexity uses revolving machinery somewhere in its structure.

For rotating systems, torque and angular displacement prove to be the most useful concepts for expressing work and power transmission. From a simple case, we can find the necessary relations. The thin disk in ■ Figure 9.20 rotates about its center, point O, in the plane of the figure. A force $\vec{\mathbf{F}}$, also in the plane of the figure, acts on the disk at a position $\vec{\mathbf{r}}$ with respect to O.

In a time interval dt, the force does an amount of work $dW = \vec{\mathbf{F}}\cdot d\vec{\mathbf{s}}$, where $d\vec{\mathbf{s}}$ is the displacement of the point where the force acts. Each piece of the disk is in circular motion about O with the same angular speed ω. So the displacement is perpendicular to $\vec{\mathbf{r}}$ and has magnitude $r\,d\theta$, where $d\theta = \omega\,dt$ is the magnitude of the disk's angular displacement during the interval dt. Then, expressing the dot product in terms of the force component parallel to the displacement:

$$dW = \vec{\mathbf{F}}\cdot d\vec{\mathbf{s}} = (F_{\|s})|d\vec{\mathbf{s}}| = (F_{\|s})r\,d\theta.$$

Since $d\vec{\mathbf{s}}$ is perpendicular to $\vec{\mathbf{r}}$, the force component parallel to $d\vec{\mathbf{s}}$ is perpendicular to $\vec{\mathbf{r}}$:

$$|F_{\|s}| = |F_{\perp r}|.$$

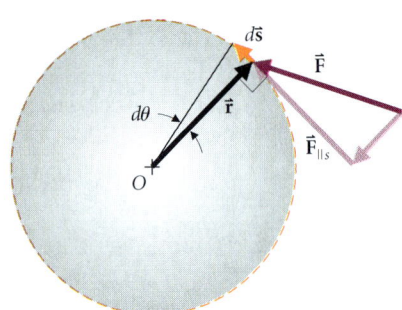

■ **FIGURE 9.20**
A force $\vec{\mathbf{F}}$ acts on a disk. The torque about O has magnitude $rF_{\|s}$. Every point of the disk rotates through the same angle in the same time Δt and so has the same angular speed ω.

> WE USED EQN. (9.11) IN THE THIRD STEP.

Then: $|dW| = |F_{\perp r}|r\,d\theta = |\vec{\mathbf{r}}\times\vec{\mathbf{F}}|\,d\theta = |\vec{\boldsymbol{\tau}}|\,d\theta.$

> The work done by a force acting on a rotating system is the torque multiplied by the angular displacement of the system.
>
> $$|dW| = \tau\,d\theta. \tag{9.13}$$

The power delivered by the force is

$$P = \frac{dW}{dt} = \frac{\tau\, d\theta}{dt} = \tau\omega. \quad (9.14)$$

The power delivered by a force acting on a rotating system equals the torque multiplied by the angular speed of the system.

COMPARE THIS RESULT WITH THE EXPRESSION $P = \vec{F} \cdot \vec{v}$.

WHEN YOU USE THE RIGHT-HAND RULE TO FIND TORQUE, THE POWER IS POSITIVE IF YOUR FINGERS CURL IN THE SENSE OF THE ROTATION. IN CHAPTER 12, WE WILL DEVELOP A FORMAL WAY TO KEEP TRACK OF SIGNS IN THESE EXPRESSIONS.

■ **EXERCISE 9.5** ♦ Show that the product $\tau\omega$ has units of power.

EXAMPLE 9.7 ♦♦ A polishing tool of mass $m = 1.5$ kg is attached to a motor by a horizontal massless rod of length $R = 0.10$ m. A motor turns the rod at a frequency $f = 33$ rpm, so that the tool slides in uniform circular motion over the optical glass surface it is polishing. The coefficient of kinetic friction between the tool and the surface is 0.32. How much power must the motor develop to drive the tool?

MODEL We focus attention on the polishing tool and draw a free-body diagram showing the forces acting on it (■ Figure 9.21). In uniform circular motion, the tool has constant angular momentum about the center of the circle, so the net torque acting on the tool is zero. The motor must exert a torque, via the horizontal force $\vec{H}$, that balances the torque due to kinetic friction. The tension in the rod causes the centripetal acceleration: it exerts no torque and does no work.

SETUP We analyze the free-body diagram in the usual way, using the coordinate system shown in Figure 9.21. We need only the z-components:

$$\Sigma F_z = ma_z.$$
$$n - mg = 0 \Rightarrow n = mg.$$

Then the kinetic friction force has magnitude $f = \mu_k n = \mu_k mg$, and its direction is perpendicular to $\vec{r}$. The torque due to this frictional force has magnitude:

$$\tau = |\vec{r} \times \vec{f}| = Rf = R\mu_k mg.$$

The motor exerts a torque of equal magnitude and opposite direction ($\vec{r} \times \vec{f} = -\vec{r} \times \vec{H}$). Thus the power developed by the motor, given by eqn. (9.14), is:

$$P = \tau\omega = R\mu_k mg\omega.$$

SOLVE
$$P = (0.10 \text{ m})(0.32)(1.5 \text{ kg})(9.8 \text{ m/s}^2)\left(33 \frac{\text{rev}}{\text{min}}\right)\left(\frac{2\pi \text{ rad}}{1 \text{ rev}}\right)\left(\frac{1 \text{ min}}{60 \text{ s}}\right)$$
$$= 0.52\pi \frac{\text{kg} \cdot \text{m}^2}{\text{s}^2} = 1.6 \text{ W}.$$

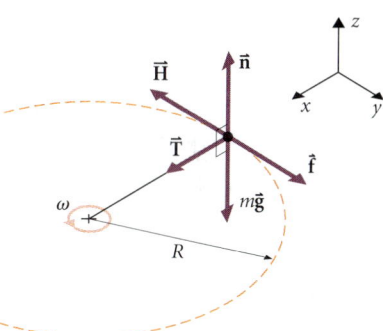

■ **FIGURE 9.21**
A tool on the end of a rotating arm. The motor must provide a torque to overcome the effects of friction. The force exerted by the rotating arm has two components: a tension $\vec{T}$ that provides the centripetal force and a horizontal force $\vec{H}$ that opposes the friction force $\vec{f}$.

ANALYZE In this example it is convenient, but not essential, to express the power in terms of torque rather than force. In Exercise 9.6, by contrast, there is no other practical way to express the power output from the engine. In Essay 3 we shall see that torque is an essential tool for understanding a bicycle's gear system. Even in such simple machines as levers, torque balance is the key idea for computing mechanical advantage. ■

EXERCISE 9.6 ♦♦♦ A competition dragster has a mass $m = 1.8 \times 10^3$ kg. At a certain time, its speed is $v = 15$ m/s, its acceleration is $a = 27$ m/s², and its engine is turning at 5.0×10^3 rpm. What minimum torque must the engine exert on the dragster's drive train? Why can you only find a minimum necessary torque from the given information rather than the actual torque exerted by the engine? ■

9.3 THE CENTER OF MASS
9.3.1 Center of Gravity and Center of Mass

If someone asks you to balance a horizontal golf club on your fingertip, you know what to do. The club won't balance with your fingertip centered, so you move your finger toward the heavy end. Finally, you find the balance point—the *center of gravity* (CG)—and the club is fairly easy to support.

Any object has a single center of gravity, though it is trickier to locate for, say, a pumpkin than for the club. The pumpkin can be supported in any orientation—stem up, stem horizontal, or with your finger on the stem (■ Figure 9.22). You must move your finger around in two dimensions to keep the pumpkin's weight over your fingertip. Once the pumpkin is balanced, imagine piercing it with a thin vertical wire running upward from your finger and then repeating the procedure for several different balance points. All the wires would meet at a single point inside the pumpkin, called its *center of gravity,* where the entire pumpkin's weight seems to act.

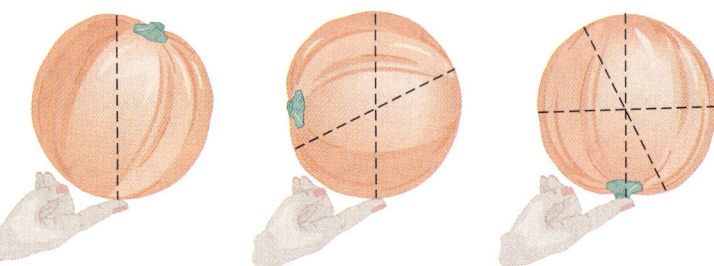

■ FIGURE 9.22
Supporting a pumpkin with your finger. Vertical wires passed through different support points all meet at a single point, the center of gravity of the pumpkin.

REMEMBER: JUST BECAUSE THE WEIGHT AND NORMAL FORCES ARE EQUAL AND OPPOSITE, THEY ARE NOT A THIRD-LAW PAIR!

Balancing an object is a double task. You have to counteract the object's weight to keep it from accelerating downward, and you have to keep the object from rotating off your finger. As we have learned, rotation involves angular momentum, which can change only if a torque acts. "Balancing" means that both zero net force and zero net torque about any origin act on the balanced object (■ Figure 9.23). The balanced object's weight, acting at the center of gravity, and the normal force exerted by your finger are equal and opposite forces acting along the same line through the center of gravity. Neither force produces torque about the CG.

The idea of torque balance allows us to give a precise definition of the center of gravity. When we say that an object's weight *seems to act at the CG,* we mean that the actual distribution of weight acting on the object causes no torque about the CG:

> The *center of gravity* of an object is the unique point about which weight forces acting on the object cause zero torque, regardless of how the object is oriented.

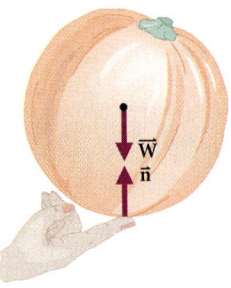

■ FIGURE 9.23
The pumpkin's weight *appears to act* at the pumpkin's center of gravity. Applying normal force directly below the CG results in balanced forces that lie along the same vertical line and cause no net torque about the CG.

A FORCE-TORQUE DIAGRAM IS A FREE-BODY DIAGRAM MODIFIED TO INCLUDE THE TORQUES.

To apply this definition, we need to augment the free-body diagrams we have used since Chapter 4. The linear acceleration caused by the net force acting on a body is the same regardless of where the forces act on the body, so a free-body diagram need only represent the body as a point particle. Torque does depend on where a force acts, so we need to use force-torque diagrams that show where forces act relative to a point chosen as the origin of the diagram.

> A *force-torque diagram* represents a part of a system as an isolated object. The rest of the universe is modeled by both the forces *and torques* it exerts on the part.

■ Figure 9.24 is a force-torque diagram for a balanced, nonuniform rod. For simplicity, we model the rod as two unequal masses m and M connected by a massless, unbendable fiber,

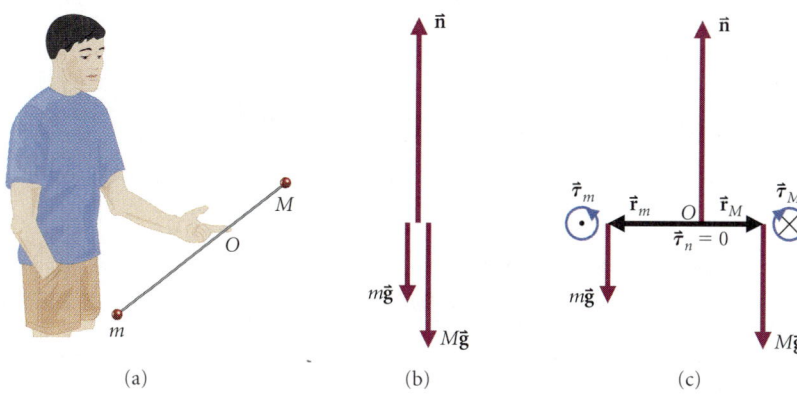

■ **FIGURE 9.24**
(a) A nonuniform rod is modeled as two particles at the ends of a massless, unbendable fiber. (b) A free-body diagram shows that the supporting force has to balance the total weight of the rod. (c) A force-torque diagram shows the torques about the support point caused by the particles' weights. Use the right-hand rule to verify the directions of the torques.

and we choose the origin O to be at the rod's center of gravity. The supporting fingertip exerts normal force at O but exerts no torque about O because $\vec{r}_O = 0$ for this force. The two weight forces do contribute to the total torque.

$$\vec{\tau}_O = \vec{\tau}_{O,M} + \vec{\tau}_{O,m}$$
$$= Mg|\vec{r}_M|\hat{\otimes} \text{ (into the page)} + mg|\vec{r}_m|\hat{\odot} \text{ (out of the page)}$$
$$= g(Mr_M - mr_m)\hat{\otimes}.$$

The torque acting on a balanced rod is zero, so:

$$g(Mr_M - mr_m) = 0 \quad \text{or} \quad r_M/r_m = m/M.$$

We know from experience that the greater mass must be closer to the balance point. Now we can be more precise: the distance of each object from the balance point is in inverse proportion to its mass.

The requirement for torque balance is independent of the convenient, but arbitrary, choice of the center of gravity as the origin of the force-torque diagram. The following exercise illustrates this point.

EXERCISE 9.7 ♦ Show that the total torque about the particle of mass M is also zero. Be sure to include the normal force at the support point O.

9.3.2 Definition of the Center of Mass

When we determined the rod's center of gravity by applying torque balance, the factor g canceled from the equation, leaving a relation among masses and distances. Thus center of gravity is the intuitive version of a more fundamental concept: *center of mass* (CM). We can think of an object's center of gravity as the average location of its weight. Its center of mass is the average location of its mass. If the acceleration due to gravity is constant, an object's weight is distributed in the same way as its mass, and the two average locations are identical. Unless we are designing an experiment that can detect variations in $\vec{g}$ near the Earth or building a space station on an asteroid, where $\vec{g}$ varies noticeably throughout the structure, there is no practical need to distinguish the centers of gravity and mass. Your intuition about one should carry over to the other. When you do need to distinguish the two, it is the center of mass that depends only on the object's structure and the center of gravity that depends also on any peculiarities of the local gravitational field.

Next, let us consider torque balance about the center of mass in a general system containing an arbitrary number of particles. As in Chapter 6, we model such a system as a set of isolated particles labeled with the numbers $1, \cdots, N$, where N is the total number of particles (■ Figure 9.25). Choosing the origin at the center of mass, the position of the jth particle with

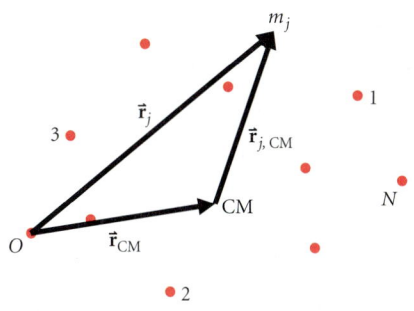

■ **FIGURE 9.25**
A system may be modeled as a collection of particles numbered $1, \ldots, N$. Particle j has mass m_j. Its position is described by $\vec{r}_{j,\text{CM}}$ from the center of mass and $\vec{r}_j$ from the origin O of a convenient reference frame. (See also Figure 6.18.) From the vector triangle, $\vec{r}_{j,\text{CM}} = \vec{r}_j - \vec{r}_{\text{CM}}$.

FROM NOW ON, WE WILL FREQUENTLY USE THE ABBREVIATION CM FOR CENTER OF MASS.

mass m_j is $\vec{r}_{j,CM}$. The total torque exerted by gravity on the system is the sum of the torques exerted on each particle:

$$0 = \vec{\tau} = \sum_j \vec{\tau}_j = \sum_j \vec{r}_{j,CM} \times (m_j \vec{g}) = \left[\sum_j (m_j \vec{r}_{j,CM})\right] \times \vec{g},$$

where we use the distributive law to factor out the constant vector $\vec{g}$. Now we can draw a general conclusion about the center of mass of any system of particles:

> With the center of mass as the origin, the sum of the particle masses m_j multiplied by their position vectors $\vec{r}_{j,CM}$ is zero.
>
> $$0 = \sum_j m_j \vec{r}_{j,CM}. \qquad (9.15)$$

We may use eqn. (9.15) to derive a practical formula for finding the position $\vec{r}_{CM}$ of a system's CM with respect to an origin other than the CM. If the position of an individual particle, number j, with respect to that origin is $\vec{r}_j$ (Figure 9.25), then:

$$\vec{r}_{CM} + \vec{r}_{j,CM} = \vec{r}_j \quad \Rightarrow \quad \vec{r}_{j,CM} = \vec{r}_j - \vec{r}_{CM}.$$

Substituting in eqn. (9.15),

$$0 = \sum_j m_j (\vec{r}_j - \vec{r}_{CM}) = \sum_j m_j \vec{r}_j - \sum_j m_j \vec{r}_{CM}.$$

Since $\vec{r}_{CM}$ does not depend on the particle number j, we may factor it out of the second term and solve for it:

$$\vec{r}_{CM} \equiv \frac{\sum m_j \vec{r}_j}{\sum m_j} = \frac{\sum m_j \vec{r}_j}{M}, \qquad (9.16)$$

where $M \equiv \sum m_j$ is the total mass of the system. This formula for computing $\vec{r}_{CM}$ is known as a *mass-weighted average* and makes precise the statement that the CM is the average location of the system's mass. Thus, eqn. (9.16) serves as the *definition* of the center of mass.

EXAMPLE 9.8 ♦ Four swimmers with masses 35 kg, 41 kg, 54 kg, and 63 kg sit on the corners of a square diving raft with 2.0-m sides (■ Figure 9.26). Where is the CM of the group of swimmers?

MODEL The *particles* in this system are the four people. Since we are given their positions in the horizontal plane with no information about their sizes, we cannot say anything about the vertical (z) position of their CM.

SETUP We work out the x- and y-components of $\vec{r}_{CM}$ separately, using the coordinate system shown in Figure 9.26.

SOLVE We use eqn. (9.16), one component at a time.

$$x_{CM} = \frac{\sum_{j=1}^{4} m_j x_j}{\sum_{j=1}^{4} m_j}$$

$$= \frac{(54 \text{ kg})(0 \text{ m}) + (63 \text{ kg})(0 \text{ m}) + (35 \text{ kg})(2.0 \text{ m}) + (41 \text{ kg})(2.0 \text{ m})}{35 \text{ kg} + 41 \text{ kg} + 54 \text{ kg} + 63 \text{ kg}}$$

$$= \frac{152 \text{ kg} \cdot \text{m}}{193 \text{ kg}} = 0.79 \text{ m}.$$

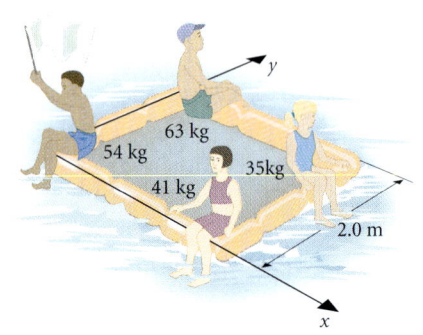

■ **FIGURE 9.26**
Four swimmers sit at the corners of a raft. Where is their center of mass?

Similarly:

$$y_{CM} = \frac{[(54 \text{ kg}) + (41 \text{ kg})](0 \text{ m}) + [(63 \text{ kg}) + (35 \text{ kg})](2.0 \text{ m})}{193 \text{ kg}} = 1.0 \text{ m}.$$

ANALYZE Even though the data were given to two significant figures, we should give the answers to only one figure: $x_{CM} = 0.8$ m, $y_{CM} = 1$ m. People's sizes and the accuracy with which they can "sit at the corners of the raft" make a second figure meaningless. ∎

9.3.3 Motion of the Center of Mass

When the center of mass of a system of particles is not at rest, we may express its velocity directly by differentiating eqn. (9.16):

$$\vec{v}_{CM} \equiv \frac{d}{dt}\vec{r}_{CM} = \frac{d}{dt}\left[\frac{\sum m_j \vec{r}_j}{M}\right] = \frac{\sum m_j(d\vec{r}_j/dt)}{M} \equiv \frac{\sum m_j \vec{v}_j}{M}.$$

THE DERIVATIVE ONLY ACTS ON THE QUANTITIES THAT CHANGE WITH TIME, NAMELY THE POSITIONS $\vec{r}_j$.

Since $m_j \vec{v}_j$ is the momentum $\vec{p}_j$ of the jth particle:

$$\vec{v}_{CM} = (\sum \vec{p}_j)/M = \vec{P}/M.$$

$$\vec{P} = M\vec{v}_{CM}. \tag{9.17}$$

Since $\vec{P}$ changes only if external forces act, the same must be true of $\vec{v}_{CM}$.

> If the net external force acting on a system is zero, its center of mass moves at constant velocity.

We have also shown (§6.3) that the net force acting on a system equals the rate of change of its total momentum. Thus if the system's mass is constant, we have:

$$\vec{F}_{net,\,ext} = \frac{d\vec{P}}{dt} = \frac{d}{dt}(M\vec{v}_{CM}) = M\frac{d\vec{v}_{CM}}{dt}.$$

$$\vec{F}_{net,\,ext} = M\vec{a}_{CM}. \tag{9.18}$$

Now we see why the center of mass is a fundamental feature of a system: Newton's first and second laws apply to a system as if it were a single particle located at the CM. Until now, we have modeled complex objects—people, rockets, automobiles—as particles but have not really justified such a bold step. These theorems show why the particle model works, and they also identify the center of mass of a complex object as the proper location for the particle we use as its model.

For example, when the external forces acting on a system are entirely due to gravity, the total external force is:

$$\vec{F}_{ext} = \sum \vec{F}_{g,j} = \sum (m_j \vec{g}) = (\sum m_j)\vec{g} = M\vec{g}.$$

Comparing this with eqn. (9.18), the acceleration of the system is $\vec{a}_{CM} = \vec{g}$. In the absence of air resistance, a system's CM falls just like a particle, regardless of what the system's individual pieces are doing.

EXAMPLE 9.9 ❖ A lunar mail canister en route from Copernicus Base to Rimtown, a distance D away, explodes into two pieces of equal mass at the top of its trajectory. Immediately after the breakup, both parts are moving horizontally. Describe the subsequent motion of the system's center of mass. If one part lands at distance $2D$ from Copernicus Base, where does the other part land?

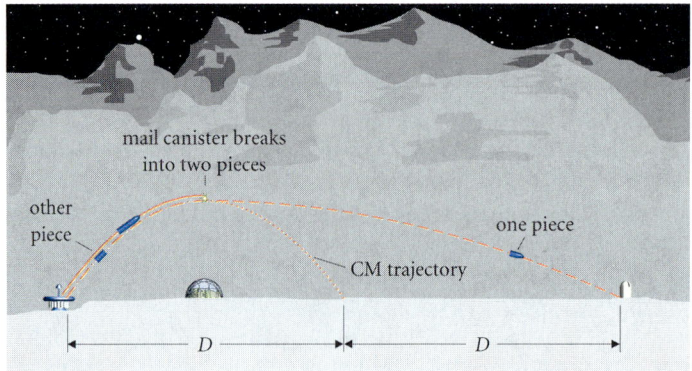

FIGURE 9.27 A lunar mail canister breaks into two pieces at the top of its trajectory. The center of mass continues to move along the planned trajectory.

MODEL This problem is similar to Study Problem 6 in §6.2, where a mail canister was split up by a spring. After the canister breaks up, we may still consider the two pieces as parts of the same system. The total mass of the system, and the total gravitational force on it, are unchanged. Thus the system's CM continues to move along the same trajectory that the unbroken canister would have followed. Since the two pieces of the broken canister are of equal mass, their center of mass is always halfway between them. So if one part lands twice as far away as the unbroken canister would have landed, then the other piece lands at the launch point (■ Figure 9.27).

ANALYZE Beware! This argument relies on the fact that both parts of the canister and the center of mass are moving horizontally immediately after the breakup, and so all land on the ground at the same time. If this were not so, we could use this argument to compute the CM's motion only until the first piece hit the ground. The ground exerts a large force on that piece to stop it. This force is also an external force acting on the system as a whole and so accelerates its center of mass. ■

REFERENCE FRAMES AND THEIR RELATIVE MOTION ARE DISCUSSED IN §§1.1 AND 3.3.

9.3.4 The Center of Mass Reference Frame

A ballet dancer performing the grande jeté (■ Figure 9.28) illustrates both the simple motion of a system's center of mass and how interesting and complex the system's motion about the center of mass can be. By moving her body appropriately, the dancer appears to float horizontally, as her center of mass follows a parabola. Those same body motions, carried out while floating inside the space shuttle, would require exactly the same sequence of muscle forces, though the result would be less beautiful without the parabolic CM motion.

Physical situations are always described with respect to some frame of reference. Astronauts working with a satellite would naturally take the nearby space shuttle to define their reference frame; an insurance agent investigating a traffic accident would use the highway surface. What makes such a reference frame natural is that the measurement equipment and people observing events are at rest in it. Physicists often refer to such a natural frame as the *laboratory frame* (even if there is no laboratory to be seen!). As the ballet dancer illustrates, the motion of particles within a system are often best described with respect to the system's center of mass—that is, in a reference frame with its origin stationary at the system's CM. We shall need to describe systems in this *center of mass reference frame* as well as in a *laboratory frame* and to transform physical quantities between the two descriptions. We developed the necessary tools in Chapter 3:

■ **FIGURE 9.28**
A ballet dancer performs a grand jeté. The dancer's CM moves along a parabolic trajectory typical of a projectile. By moving the rest of her body with respect to her CM, the dancer creates a visual illusion of floating.

The velocity $\vec{v}_j$ of a particle in the laboratory frame equals its velocity $\vec{v}_{j,\text{CM}}$ in the center of mass frame plus the velocity of the center of mass with respect to the laboratory frame:

$$\vec{v}_j = \vec{v}_{j,\text{CM}} + \vec{v}_{\text{CM}}. \qquad (9.19)$$

We already used the corresponding relation among position vectors (Figure 9.25) in the derivation of eqn. (9.16).

We may use eqn. (9.19) to calculate the linear momentum of a system in its CM frame. The linear momentum of particle j in this frame is:

$$\vec{p}_{j,\text{CM}} = m_j \vec{v}_{j,\text{CM}} = m_j(\vec{v}_j - \vec{v}_{\text{CM}}) = \vec{p}_j - m_j \vec{v}_{\text{CM}}.$$

The total linear momentum of the system in the CM frame is:

$$\vec{P}_{\text{CM}} = \sum \vec{p}_{j,\text{CM}} = \sum (\vec{p}_j - m_j \vec{v}_{\text{CM}}) = \vec{P}_{\text{lab}} - \left(\sum m_j\right) \vec{v}_{\text{CM}} = \vec{P}_{\text{lab}} - M\vec{v}_{\text{CM}}.$$

But $\vec{P}_{\text{lab}} = M\vec{v}_{\text{CM}}$ (eqn. 9.17), so $\vec{P}_{\text{CM}} = 0$.

Because the total momentum vanishes in the CM frame, it is sometimes called the *center of momentum* reference frame.

EXAMPLE 9.10 ♦♦ Laura, competing for her club's bocci championship, launches a ball of mass $3m = 1.20$ kg with speed $v_0 = 3.2$ m/s toward the target ball of mass $m = 0.40$ kg (■ Figure 9.29). What is the velocity of the two balls' center of mass? What is the velocity of each ball with respect to the center of mass? What is the momentum of each in the CM frame?

MODEL Assuming Laura has aimed accurately, this is a one-dimensional problem.

SETUP With the x-direction parallel to $\vec{v}_0$, all vectors have only x-components. The CM velocity $\vec{v}_{\text{CM}}$ may be found from the total momentum and total mass of the balls:

SOLVE Using eqn. (9.17):

$$\vec{v}_{\text{CM}} = \frac{\vec{P}}{M} = \frac{3mv_0 + 0}{4m}\hat{i} = \frac{3v_0}{4}\hat{i}$$

$$= \frac{3(3.2 \text{ m/s})}{4}\hat{i} = (2.4 \text{ m/s})\hat{i}.$$

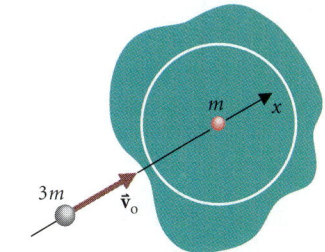

(a) Laboratory frame

SETUP The velocities of the balls in the CM frame are found from eqn. (9.19) by subtracting $\vec{v}_{\text{CM}}$ from the given velocities:

SOLVE

$$\vec{v}_{3m,\text{CM}} = \vec{v}_{3m} - \vec{v}_{\text{CM}} = (v_0 - 3v_0/4)\hat{i} = (v_0/4)\hat{i} = (0.80 \text{ m/s})\hat{i},$$

and $\quad \vec{v}_{m,\text{CM}} = \vec{v}_m - \vec{v}_{\text{CM}} = (0 - 3v_0/4)\hat{i} = (-3v_0/4)\hat{i} = -(2.4 \text{ m/s})\hat{i}.$

The momenta of the balls in the CM frame are:

$$\vec{p}_{3m,\text{CM}} = 3m\vec{v}_{3m,\text{CM}} = (3mv_0/4)\hat{i}$$
$$= (1.20 \text{ kg})(0.80 \text{ m/s})\hat{i} = (0.96 \text{ kg} \cdot \text{m/s})\hat{i},$$

and $\quad \vec{p}_{m,\text{CM}} = m\vec{v}_{m,\text{CM}} = (-3mv_0/4)\hat{i}$
$$= (0.40 \text{ kg})(-2.4 \text{ m/s})\hat{i} = -(0.96 \text{ kg} \cdot \text{m/s})\hat{i}.$$

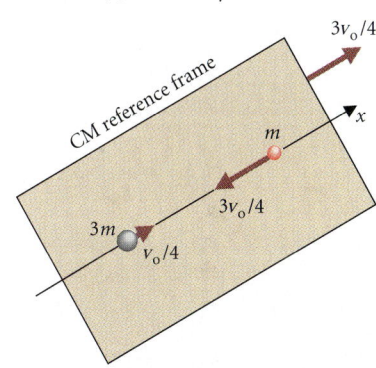

(b) CM reference frame

■ **FIGURE 9.29**
A bocci ball of mass $3m$ is launched at speed v_0 toward a target ball of mass m. (a) The *laboratory frame* is the playing field. (b) The *CM reference frame* moves at speed $3v_0/4$.

ANALYZE The balls' momenta in the CM frame are equal and opposite, as expected since the total momentum in the CM frame is always zero. ■

The system's kinetic energy in the lab frame may also be written in terms of the velocity of the CM:

$$K_{\text{lab}} = \tfrac{1}{2} \sum m_j |\vec{v}_j|^2 = \tfrac{1}{2} \sum m_j |\vec{v}_{j,\text{CM}} + \vec{v}_{\text{CM}}|^2.$$

Expanding the squared factor:

$$K_{\text{lab}} = \tfrac{1}{2} \sum m_j (v_{j,\text{CM}}^2 + 2\vec{v}_{j,\text{CM}} \cdot \vec{v}_{\text{CM}} + v_{\text{CM}}^2)$$
$$= \tfrac{1}{2} \sum m_j v_{j,\text{CM}}^2 + \left(\sum m_j \vec{v}_{j,\text{CM}}\right) \cdot \vec{v}_{\text{CM}} + \tfrac{1}{2}\left(\sum m_j\right) v_{\text{CM}}^2.$$

The sum in the middle term is the system's momentum in its CM frame, which is zero, as we have just shown. So:

$$K_{lab} = \tfrac{1}{2}\sum m_j v_{j,CM}^2 + \tfrac{1}{2}Mv_{CM}^2 \equiv K_{int} + \tfrac{1}{2}Mv_{CM}^2. \qquad (9.20)$$

The result contains a term that looks like the kinetic energy of a particle moving with the CM. The other, *internal* term is the energy of the system's motion with respect to the CM, and is independent of the CM's motion with respect to the lab. This separation of the energy into *internal* and *particle* terms works because the system's total internal momentum is zero.

Using the particle model for a system works well when the change in its internal properties is negligible. Conversely, when there are significant changes in the system's internal state, the particle model is inadequate.

EXAMPLE 9.11 ♦♦ A diatomic molecule, two identical atoms of mass m bound together a distance ℓ apart, is confined in a laboratory apparatus. At a certain time, one of the two atoms is observed to have speed v_o and the other, moving in the opposite direction, speed $v_o/3$ (■ Figure 9.30). Find the molecule's center of mass velocity and the velocity of each atom in the CM frame. What is the molecule's internal energy? What is the molecule's internal angular momentum (its angular momentum about its CM, measured in the CM frame)?

MODEL We model the molecule as a dumbbell—two identical particles connected by a massless, unbendable fiber of length ℓ.

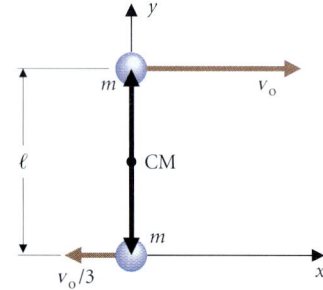

■ **FIGURE 9.30**
A diatomic molecule as observed in the laboratory frame.

SETUP Since the lab velocities of the two particles have only x-components at this instant, the CM velocity is in the x-direction. We use eqn. (9.17):

SOLVE
$$\vec{v}_{CM} = \frac{\vec{P}}{M} = \frac{m(-v_o/3)\hat{\imath} + mv_o\hat{\imath}}{2m} = (v_o/3)\hat{\imath}.$$

SETUP The particles' velocities in the CM frame are found by subtracting $\vec{v}_{CM}$ from their lab velocities (eqn. 9.19). ■ Figure 9.31 shows the results:

SOLVE
$$\vec{v}_{1,CM} = \vec{v}_1 - \vec{v}_{CM} = (v_o - v_o/3)\hat{\imath} = 2v_o\hat{\imath}/3.$$
$$\vec{v}_{2,CM} = \vec{v}_2 - \vec{v}_{CM} = (-v_o/3 - v_o/3)\hat{\imath} = -2v_o\hat{\imath}/3.$$

SETUP From eqn. (9.20) the internal energy is:

SOLVE
$$K_{int} = \tfrac{1}{2}\sum_{j=1}^{2} m_j v_{j,CM}^2$$
$$= \tfrac{1}{2}m\left[\left(\frac{2v_o}{3}\right)^2 + \left(\frac{-2v_o}{3}\right)^2\right]$$
$$= m\left(\frac{4v_o^2}{9}\right)$$

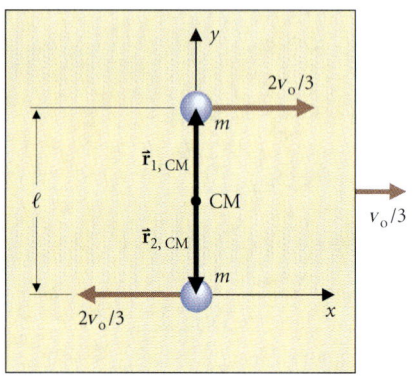

CM reference frame

■ **FIGURE 9.31**
The molecule's CM frame moves to the right at speed $v_o/3$.

SETUP The internal angular momentum is the sum of the two particles' angular momenta in the CM frame about the CM as origin.

SOLVE We obtain the direction using the right-hand rule:
$$\vec{L}_{int} = \vec{r}_{1,CM} \times (m\vec{v}_{1,CM}) + \vec{r}_{2,CM} \times (m\vec{v}_{2,CM})$$
$$= |\vec{r}_{1,CM}||m\vec{v}_{1,CM}|(\sin 90°)(-\hat{k}) + |\vec{r}_{2,CM}||m\vec{v}_{2,CM}|(\sin 90°)(-\hat{k})$$
$$= 2(\ell/2)m(2v_o/3)(1)(-\hat{k})$$
$$= (-2mv_o\ell/3)\hat{k}.$$

WE ASK YOU TO DEMONSTRATE THE DECOMPOSITION OF ANGULAR MOMENTUM INTO PARTICLE AND INTERNAL PORTIONS IN PROBLEM 87.

ANALYZE Changes in a molecule's internal angular momentum can be directly observed in the molecule's emission and absorption of light. ■

9.4 CONSERVATION OF ANGULAR MOMENTUM

9.4.1 A System of Two Particles

An object's angular momentum can change only when a torque acts on the object. Some other object must exert a force to produce that torque, and Newton's third law ensures that the second object will also experience force and torque. The interaction results in an exchange of linear and angular momentum.

For example, suppose two astronauts with equal mass and equal but opposite velocities (■ Figure 9.32) are working near a space station. The two are moving with a speed of 3.0 m/s along parallel lines a distance $2\ell = 15.0$ m apart, and they simultaneously grab the ends of a light cable of length 2ℓ. In the space station frame (Figure 9.32), the astronauts have equal and opposite linear momentum vectors before grabbing the cable, so their total linear momentum is zero and their CM is stationary. What can we conclude about their motion after they grab the cable?

The astronauts exert forces on each other via the cable. According to Newton's third law, the forces are equal and opposite. They also act along the same line, here the line of the cable. Then (■ Figure 9.33) for any chosen origin O, the component of $\vec{r}$ perpendicular to $\vec{F}$, $r_\perp \equiv OP$, is the same for each force. Thus the torque about O exerted on astronaut 2 by astronaut 1, $\vec{\tau}_{21}$, is exactly equal and opposite to the torque $\vec{\tau}_{12}$ exerted on astronaut 1 by astronaut 2. These torques produce equal and opposite changes in the angular momentum of each astronaut, but they do not change the total. The system's angular momentum remains constant.

The astronauts' linear momentum is also conserved, so their CM remains stationary in the space station frame. The center of the rope remains fixed and each astronaut moves in a circle of radius ℓ around it. Let ω be their angular speed in that circle.

The total angular momentum of the system is the sum of individual values for each astronaut. Before the meeting, they each move with constant velocity, while afterward they each move on a circle of radius ℓ with constant speed $\omega\ell$. Tension in the cable provides the forces that change their velocities. We use the standard pattern for applying conservation laws:

	BEFORE	AFTER
Evaluate angular momentum:	$M(\vec{r}_1 \times \vec{v}_1) + M(\vec{r}_2 \times \vec{v}_2)$	$2M\omega\ell^2\hat{k}$
	$= M[-\ell\hat{j} \times v\hat{i}) + \ell\hat{j} \times (-v\hat{i})]$	
	$= 2Mv\ell\hat{k}.$	
Set equal:	$2Mv\ell\hat{k} = 2M\omega\ell^2\hat{k}.$	
	$\omega = v/\ell = (3.0 \text{ m/s})/(7.50 \text{ m}) = 0.40 \text{ rad/s}.$	

In the final motion, the astronauts spin with frequency $f = \omega/2\pi = 0.20/\pi$ revolutions per second = 3.8 rpm. Their kinetic energy remains constant in the encounter, since the cable tension changes only the directions and not the magnitudes of their velocities.

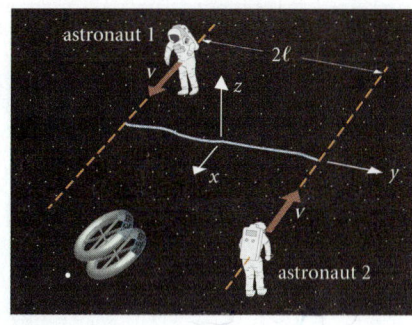

■ FIGURE 9.32
Two astronauts grab the ends of a cable as they pass each other. In the final motion, they spin about the center of the cable.

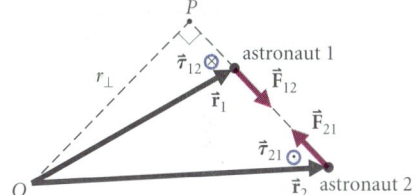

■ FIGURE 9.33
The pair of forces $\vec{F}_{12}$ and $\vec{F}_{21}$ produce canceling torques about any point O. The perpendicular component of position, $r_\perp = OP$, is the same for both forces.

9.4.2 Proof that Angular Momentum is Conserved

From the example of astronauts grabbing a cable, we learn *why* angular momentum is conserved. The astronauts exert equal and opposite torques, ensuring that a change in one astronaut's angular momentum is offset by an opposite change in the other's. The total angular momentum remains fixed. Generalizing this principle to any system of particles follows the same reasoning we applied to linear momentum in §6.1:

> The *total angular momentum* of a system is the sum of the angular momentum vectors of all the particles composing the system:
> $$\vec{L}_{\text{system}} \equiv \sum_{\text{all particles}} \vec{L}_{\text{particle}}. \tag{9.21}$$

Digging Deeper

THE STRONG FORM OF NEWTON'S THIRD LAW

We know from experience that the internal torques in a system sum to zero: bodies left to themselves do not spontaneously start rotating! According to the strong form of Newton's law, torque balance happens for each particle pair. This is true for all the kinds of mechanical force we have studied.

Often you have to look carefully to see that the strong form works in mechanics. The normal forces, exerted on each other by the two blocks shown in ■ Figure 9.34, clearly do not lie along the line joining the centers of the two blocks. Does that mean they violate the strong form? No, because Newton's laws apply to particles, and we can't model the blocks as particles if we care **where** the normal forces act on them. Instead, imagine the blocks as made up of smaller pieces (Figure 9.35b). The normal forces these pieces exert on each other obey the strong form of the third law perfectly well.

We know it is useful to model a block as a single particle when we are only interested in its acceleration as a whole. Questions involving angular momentum and torque may require a more detailed model. Chapters 11 and 12 develop such a model for rigid objects.

Energy and linear and angular momentum reside in electromagnetic fields as well as in particles (Parts V and VI), and Newton's third law does not always hold for the particles alone. The conservation laws remain valid when transfers of conserved quantities between fields and particles are included.

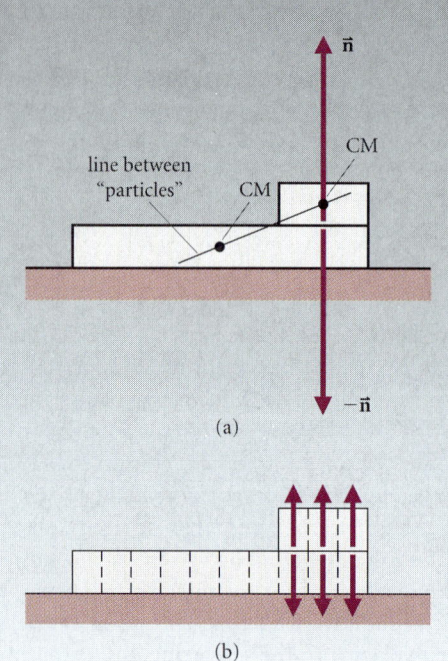

■ **FIGURE 9.34**
(a) The strong form of Newton's law seems in question for the normal forces exerted by two blocks on each other, if we treat the blocks as particles. The forces do not lie along the line of centers of the blocks. (b) If we model the blocks as a large number of smaller particles, however, the normal forces acting between pairs of these particles do obey the strong form of the third law.

The *law of conservation of angular momentum* states:

> No process can occur that creates or destroys angular momentum.

For an isolated system, the law states:

> The total angular momentum of an isolated system of particles is conserved; the system's total angular momentum cannot change in time.

If the system is not isolated, the external forces acting on it produce torques that can transfer angular momentum to or from the system:

$$\frac{d}{dt}\vec{L}_{\text{system}} = \sum \vec{\tau}_{j,\text{ext}} = \vec{\tau}_{\text{ext}}. \quad (9.22)$$

The rate of change of a system's angular momentum equals the total external torque acting on the system.

The proofs of these theorems are very similar to the proofs of the theorems regarding change of a system's linear momentum (§6.3). According to Newton's third law, the forces between a pair of particles are equal and opposite. To prove that angular momentum is conserved, we also need to know that the forces act *along the same line* between the particles. This is an additional assumption, known as the strong form of Newton's third law. It is true for all the mechanical forces we have encountered. With this assumption, the torques a pair of particles exert on each other are also equal and opposite, so that the net torque exerted on the system by any pair of particles is zero (■ Figure 9.35), and internal forces cannot change the angular momentum of the system.

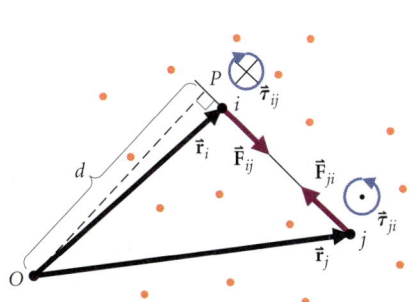

■ FIGURE 9.35
The *strong form* of Newton's third law: the pair of forces $\vec{F}_{ji}$ and $\vec{F}_{ij}$, which the jth and ith particles exert on each other, lie along the line between the particles; $r_\perp = OP = d$ is the same for both forces, and so the torques they exert about O are equal and opposite.

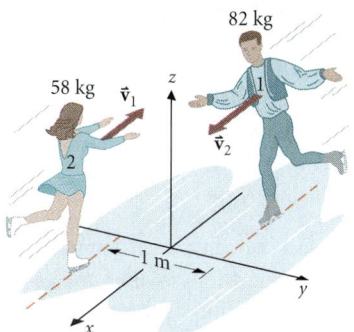

■ FIGURE 9.36
Two skaters approaching each other at speed $v = 1.0$ m/s. They are separated by 1.0 m when they meet and join hands. The velocity of their CM is unchanged, and their angular momentum about the CM is conserved.

Study Problem 8   ♦♦♦   *The Two Skaters*

Two ice skaters with masses of 82 kg and 58 kg approach each other at equal speeds of 1.0 m/s. Their parallel paths are 1.0 m apart (■ Figure 9.36). If they link hands as they pass by, describe how they move afterwards. If they pull on each other and reduce their separation to 0.75 m, what happens? (Ignore friction.)

Modeling the Physical System

Figure 9.36 shows the skaters as seen by the spectators shortly before they join hands. Each skater's weight is balanced by a normal force exerted by the ice, and we are neglecting frictional forces. Thus the two skaters form an isolated system; the net external force acting on them is negligible. So, their total linear and angular momentum are each conserved and their center of mass moves with constant velocity.

The most efficient description of the skaters' motion after they link hands is a constant linear motion of their center of mass together with rotation about the center of mass. Their angular speed is found from their *internal* angular momentum. The key principle in the last part of the question is that this internal angular momentum is conserved when the skaters pull together.

THE SPECTATORS DEFINE THE LABORATORY FRAME.

SEE EXAMPLE 9.11, WHERE WE DEFINED INTERNAL ANGULAR MOMENTUM.

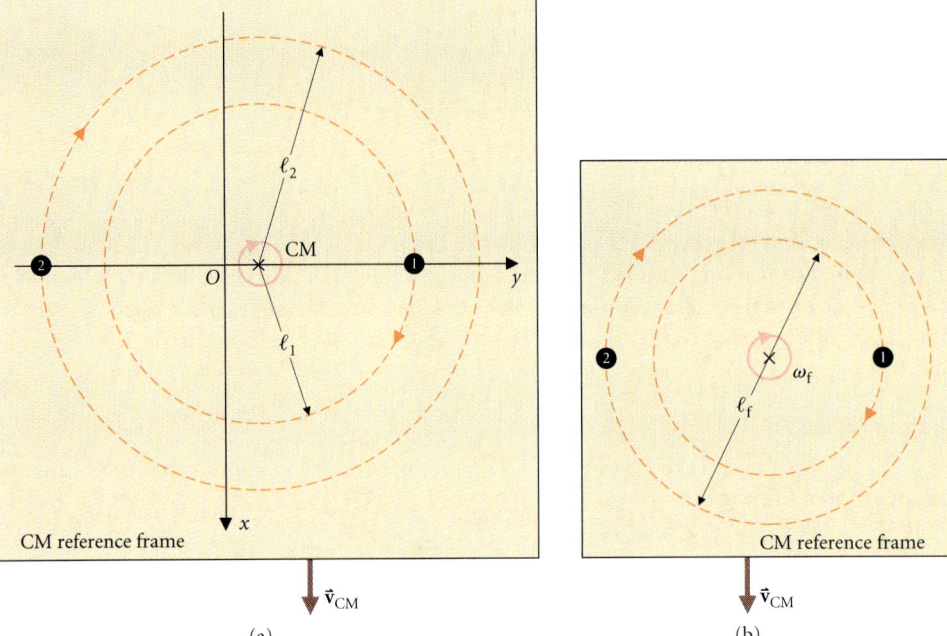

■ **FIGURE 9.37**
(a) The skaters in the CM reference frame.
(b) When the skaters reduce their separation, their angular momentum is conserved because no external torques about the CM act on the system. Thus they spin faster as their separation decreases.

Setup of Solution

We choose a coordinate system with the x-axis parallel to one skater's velocity, the y-axis along the line joining them as they meet, and the origin halfway between them (Figures 9.36 and ■ 9.37). The position of the center of mass as they meet is given by eqn. (9.16):

$$x_{CM} = \frac{\sum m_i x_i}{\sum m_i} = 0$$

and $\quad y_{CM} = \dfrac{\sum m_i y_i}{\sum m_i} = \dfrac{(82\ \text{kg})(0.50\ \text{m}) + (58\ \text{kg})(-0.50\ \text{m})}{82\ \text{kg} + 58\ \text{kg}} = 0.086\ \text{m}.$

The skaters' distances from the CM are:

$\ell_1 = 0.50\ \text{m} - 0.086\ \text{m} = 0.414\ \text{m} \quad$ and $\quad \ell_2 = 0.50\ \text{m} + 0.086\ \text{m} = 0.586\ \text{m}.$

We may calculate the center of mass velocity $\vec{v}_{CM}$ directly from eqn. (9.17):

$$\vec{v}_{CM} = \frac{m_1 \vec{v}_1 + m_2 \vec{v}_2}{m_1 + m_2} = \frac{(82\ \text{kg})(1.0\ \text{m/s})\hat{i} + (58\ \text{kg})(-1.0\ \text{m/s})\hat{i}}{82\ \text{kg} + 58\ \text{kg}}$$
$$= (0.17\ \text{m/s})\hat{i}.$$

Figure 9.37 shows the skaters in their CM reference frame at the instant they join hands. After joining, the skaters move on the circles with radii ℓ_1 and ℓ_2, as shown. The line between the skaters rotates at some angular speed ω, and this is the common angular speed of the skaters around their circular paths. From the relation among angular speed, linear speed, and radius in circular motion:

$$v = \omega r \quad \Rightarrow \quad \omega = \frac{v_{1,CM}}{\ell_1} = \frac{v_{2,CM}}{\ell_2}. \tag{i}$$

We find the two speeds in the CM frame from the skaters' velocities immediately after joining (eqn. 9.19):

$$\vec{v}_{1,CM} = \vec{v}_{1,\text{lab}} - \vec{v}_{CM} = (1.0\ \text{m/s} - 0.17\ \text{m/s})\hat{i} = (0.83\ \text{m/s})\hat{i}.$$
$$\vec{v}_{2,CM} = \vec{v}_{2,\text{lab}} - \vec{v}_{CM} = (-1.0\ \text{m/s} - 0.17\ \text{m/s})\hat{i} = -(1.17\ \text{m/s})\hat{i}.$$

Solution

From eqn. (i), the skaters' angular speed is:

$$\omega = \frac{v_{1,CM}}{\ell_1} = \frac{0.83 \text{ m/s}}{0.414 \text{ m}} = 2.0 \text{ rad/s}.$$

We can check this result using the second skater:

$$\omega = \frac{v_{2,CM}}{\ell_2} = \frac{1.17 \text{ m/s}}{0.586 \text{ m}} = 2.0 \text{ rad/s}.$$

After they join hands, the skaters' center of mass continues to move in the positive x-direction at 0.17 m/s. The skaters spin clockwise around their CM with an angular speed of 2.0 rad/s.

Setup of Solution

In the CM frame, the skaters' angular momentum about the CM is conserved as they pull together (Figure 9.37b). Its initial value is:

$$\vec{L}_{CM} = \vec{L}_1 + \vec{L}_2 = m_1 \vec{r}_1 \times \vec{v}_{1,CM} + m_2 \vec{r}_2 \times \vec{v}_{2,CM}$$
$$= (m_1 \ell_1 v_{1,CM} + m_2 \ell_2 v_{2,CM})(-\hat{k}).$$

Using eqn. (i), we may express $\vec{L}$ in terms of ω:

$$|\vec{L}_{CM}| = (m_1 \ell_1^2 + m_2 \ell_2^2)\omega.$$

It is their total separation that the skaters change, so it is useful to express the angular momentum in terms of $s \equiv \ell_1 + \ell_2$. Since ℓ_1 and ℓ_2 are the distances of the skaters from the CM, by eqn. (9.15) we have:

$$\frac{\ell_1}{\ell_2} = \frac{m_2}{m_1}.$$

Thus:

$$s = \ell_1 + \ell_2 = \ell_1(1 + m_1/m_2).$$

So:

$$\ell_1 = \frac{m_2}{m_1 + m_2} s \quad \text{and} \quad \ell_2 = \frac{m_1}{m_1 + m_2} s.$$

Then:

$$|\vec{L}_{CM}| = \left[\frac{m_1 m_2^2}{(m_1 + m_2)^2} + \frac{m_2 m_1^2}{(m_1 + m_2)^2}\right] s^2 \omega = \left(\frac{m_1 m_2}{m_1 + m_2}\right) s^2 \omega.$$

After they pull together, the skaters spin with angular speed ω_f. We apply the conservation law for angular momentum:

	BEFORE	AFTER
Evaluate angular momentum in CM reference frame. Set equal:	$\dfrac{m_1 m_2}{m_1 + m_2} s_i^2 \omega$	$\dfrac{m_1 m_2}{m_1 + m_2} s_f^2 \omega_f$
	$\omega s_i^2 = \omega_f s_f^2.$	

Solution

$$\omega_f = \omega \frac{s_i^2}{s_f^2} = (2.0 \text{ rad/s})\left(\frac{1.0 \text{ m}}{0.75 \text{ m}}\right)^2 = 3.6 \text{ rad/s}.$$

Analysis

Is this feat possible? Each skater has to pull with a force of magnitude $F = ma = m\omega^2 r \sim$ (80 kg)(4 rad/s)^2(0.4 m) = 500 N to provide the necessary centripetal force—not impossible for a good athlete.

This problem illustrates how useful the CM reference frame is for analyzing systems. To describe the effect of the skaters pulling together in the lab frame would require an elaborate tracing of the skaters' positions as functions of time. That complexity results in a stunning performance for the spectators who define the laboratory frame. It also obscures much of the power of the conservation principle and presents a daunting physics problem.

WORK AND ENERGY IN A SIMILAR SYSTEM ARE DISCUSSED IN PROBLEM 102.

Chapter Summary

Where Are We Now?

With angular momentum we have completed the list of conserved mechanical properties: linear and angular momentum and energy. All three are conserved in isolated systems upon which no external forces act. These conservation laws are equivalent to Newton's laws; they give us an alternative way of viewing physical situations and provide powerful computational tools.

What Did We Do?

Kepler's second law of planetary motion is a conservation law for a vector quantity whose magnitude is proportional to the rate at which the line from the Sun to a planet sweeps out area, and whose direction is defined by the constant orientation of the planet's orbital plane. This vector quantity is the *angular momentum* of a particle.

> The angular momentum of a particle about a point O is the cross product of the particle's position vector from O with its linear momentum:
> $$\vec{L}_O = \vec{r} \times \vec{p}.$$

The angular momentum of a system of particles is the vector sum of the individual angular momenta of the particles.

Transfer of angular momentum to a particle results from the net torque caused by forces acting on the particle.

> A force $\vec{F}$ exerted at position $\vec{r}$ from a point O exerts a torque about that point given by:
> $$\vec{\tau}_O = \vec{r} \times \vec{F}.$$

The rate of change of a system's angular momentum about any point equals the total torque acting on the system about that point:

$$\frac{d\vec{L}}{dt} = \vec{\tau}_{\text{net}}.$$

The power transmitted by a rotating mechanical system equals the torque acting multiplied by the system's angular speed.

$$P = \tau\omega.$$

A system's center of mass is located at the mass-averaged position of the particles:

$$\vec{r}_{\text{CM}} \equiv \frac{\sum m_j \vec{r}_j}{M}.$$

The center of mass accelerates according to Newton's second law for a particle, which is the reason why we can often model complex objects as particles. The CM reference frame is uniquely useful for describing the structure of a system. The system's *internal* energy and angular momentum are evaluated in the CM frame.

The strong form of Newton's third law, valid in mechanics, states that the forces any two particles exert on each other lie along the line between the particles. This implies that a system of particles exerts no total torque on itself. As a consequence:

> The angular momentum of an isolated system is conserved—that is, constant in time.

Practical Applications

Angular momentum is useful in calculating the evolution of systems, like the two skaters, where circular motions occur. One of the most important applications is to the orbits of planets and spacecraft. The size and shape of a spacecraft's elliptical orbit are determined by its energy and angular momentum per unit mass. To navigate a spacecraft, one first decides which ellipse goes to the desired destination and then fires rockets to produce the appropriate changes in energy and angular momentum.

Angular momentum also has great importance in the study of atomic and subatomic phenomena, where Newton's laws cease to be useful.

Mathematical Results

We defined the cross product of two vectors:

> The cross product of two vectors $\vec{a}$ and $\vec{b}$ has magnitude equal to the product of magnitudes $|\vec{a}|$ and $|\vec{b}|$ with the sine of the angle between the vectors. The direction of $\vec{a} \times \vec{b}$ is along the thumb of your right hand when you curl your fingers from $\vec{a}$ toward $\vec{b}$.

The order of the factors in a cross product is important, since:

$$\vec{b} \times \vec{a} = -\vec{a} \times \vec{b}.$$

The cross product of two vectors forms a convenient representation for the area of a parallelogram with the two vectors as sides.

The cross product obeys a distributive law

$$\vec{a} \times (\vec{b} + \vec{c}) = (\vec{a} \times \vec{b}) + (\vec{a} \times \vec{c}),$$

from which we obtain the component representation of the cross product:

$$\vec{a} \times \vec{b} = \hat{i}(a_y b_z - a_z b_y) + \hat{j}(a_z b_x - a_x b_z) + \hat{k}(a_x b_y - a_y b_x).$$

Solutions to Exercises

9.1 ■ Figure 9.38 shows the lengths $r_\perp$ and $p_\perp$. From the figure:

$$r_\perp = |\vec{r}|\sin 45° = |\vec{r}|\sqrt{2}/2$$

and

$$p_\perp = |\vec{p}|\cos 45° = |\vec{p}|\sqrt{2}/2.$$

Then: $r_\perp p = r p_\perp = |\vec{r}| |\vec{p}|\sqrt{2}/2 = |\vec{r}| |\vec{p}|\sin 135°$
$= |\vec{L}_A|$, as required.

9.2 (a) With any choice of reference point in the x–y plane (■ Figure 9.39), both $\vec{r}$ and $\vec{p}$ are in the x–y plane, and the resulting angular momentum $\vec{L}$ is in the z-direction. The magnitude of $\vec{L}$ equals the magnitude of the particle's momentum multiplied by the component of $\vec{r}$ perpendicular to $\vec{p}$. That component is the same for any choice of reference point (e.g., P) along the line OP parallel to $\vec{p}$ through the origin O. **(b)** The angular momentum about a point is zero if there is no component of $\vec{r}$ perpendicular to $\vec{p}$. That is true for points along the line parallel to $\vec{p}$ through the particle's position (e.g., B).

(c) For reference points in the region below the line through the particle, MN in Figure 9.39, the right-hand rule gives $\vec{L}$ in the positive z-direction. **(d)** With reference points above that line, $\vec{L}$ is in the negative z-direction.

9.3 Since the gravitational force the Sun exerts on a planet lies along the radius from the Sun to the planet, it causes no torque about the Sun. The gravitational force does cause torque about other reference points. The planet's angular momentum about a reference point other than the Sun varies as a result of this torque. The same physical behavior has very different descriptions with different choices of origin.

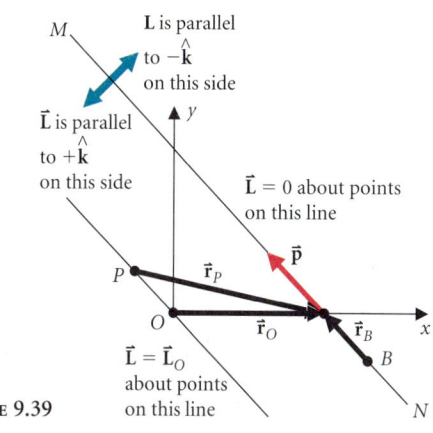

■ FIGURE 9.39

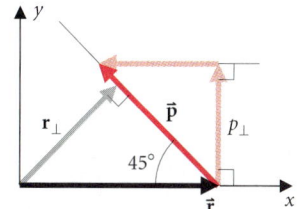

■ FIGURE 9.38

9.4 First we express the cross product in terms of components, then differentiate. Note that the unit vectors are constants.

$$\frac{d}{dt}(\vec{A} \times \vec{B}) = \frac{d}{dt}[\hat{i}(A_yB_z - A_zB_y) + \hat{j}(A_zB_x - A_xB_z)$$
$$+ \hat{k}(A_xB_y - A_yB_x)]$$
$$= \hat{i}\left(\frac{dA_y}{dt}B_z + A_y\frac{dB_z}{dt} - \frac{dA_z}{dt}B_y - A_z\frac{dB_y}{dt}\right)$$
$$+ \hat{j}\left(\frac{dA_z}{dt}B_x + A_z\frac{dB_x}{dt} - \frac{dA_x}{dt}B_z - A_x\frac{dB_z}{dt}\right)$$
$$+ \hat{k}\left(\frac{dA_x}{dt}B_y + A_x\frac{dB_y}{dt} - \frac{dA_y}{dt}B_x - A_y\frac{dB_x}{dt}\right).$$

Now group together terms with derivatives of $\vec{A}$, and terms with derivatives of $\vec{B}$.

$$\text{RHS} = \hat{i}\left(\frac{dA_y}{dt}B_z - \frac{dA_z}{dt}B_y\right) + \hat{j}\left(\frac{dA_z}{dt}B_x - \frac{dA_x}{dt}B_z\right)$$
$$+ \hat{k}\left(\frac{dA_x}{dt}B_y - \frac{dA_y}{dt}B_x\right) + \hat{i}\left(A_y\frac{dB_z}{dt} - A_z\frac{dB_y}{dt}\right)$$
$$+ \hat{j}\left(A_z\frac{dB_x}{dt} - A_x\frac{dB_z}{dt}\right) + \hat{k}\left(A_x\frac{dB_y}{dt} - A_y\frac{dB_x}{dt}\right)$$
$$= \frac{d\vec{A}}{dt} \times \vec{B} + \vec{A} \times \frac{d\vec{B}}{dt}, \text{ as required.}$$

9.5 The units of $\tau\omega$ are $(N \cdot m)(rad/s) = N \cdot m/s = J/s = W$, the unit of power.

9.6 Assuming complete efficiency, the power output of the engine equals the rate the car's kinetic energy is increasing. The engine's power output equals its torque multiplied by its angular speed, and in linear motion, the rate the car's kinetic energy increases is given by the force acting on it multiplied by its speed. Using Newton's second law to write the force in terms of mass and acceleration:

$$\tau = \frac{P}{\omega} = \frac{Fv}{\omega} = \frac{mav}{\omega}$$
$$= \frac{(1.8 \times 10^3 \text{ kg})(27 \text{ m/s}^2)(15 \text{ m/s})}{(5.0 \times 10^3 \text{ rev/min})(2\pi \text{ rad/rev})(1 \text{ min})/(60 \text{ s})}$$
$$= 1.4 \times 10^3 \text{ N} \cdot \text{m}.$$

Some energy is transformed into thermal form because of friction in the drive train and tires and the air resistance acting on the car. The

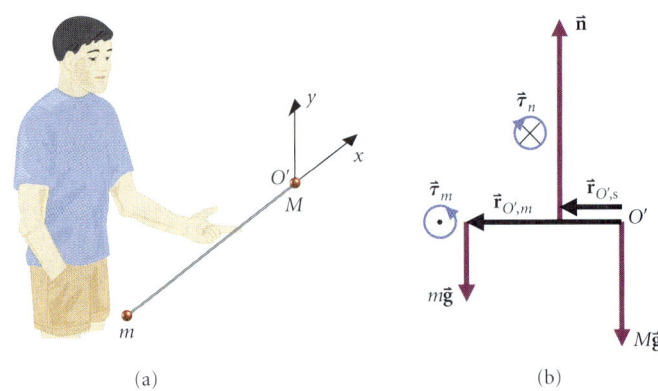

■ **FIGURE 9.40**
Since we know the system is balanced, the torque about O' is also zero.

engine must supply this energy via the drive train as well, and so has to exert greater torque than we calculated. The thermal energy dissipated directly by the engine is roughly twice the amount finally transformed into kinetic energy. The question avoids the issue of overall engine efficiency by asking about mechanical energy output from the engine and not about chemical energy input.

9.7 In ■ Figure 9.40 the origin of the force-torque diagram is at O', the position of the large particle. The positions of the smaller particle and the support are:

$$\vec{r}_{O'm} = -(r_m + r_M)\hat{i} \quad \text{and} \quad \vec{r}_{O's} = -r_M\hat{i}.$$

The torque caused by the smaller particle's weight is:

$$\vec{\tau}_{O'm} = [-(r_m + r_M)\hat{i}] \times [-mg\hat{j}] = [(r_m + r_M)mg]\hat{k}.$$

The normal force at the support point balances the total weight:

$$\vec{n} = (M + m)g\hat{j},$$

and it causes a torque:

$$\vec{\tau}_{O'n} = (-r_M\hat{i}) \times [(m + M)g\hat{j}] = -r_M(m + M)g\hat{k}.$$

The larger particle's weight acts at O' and causes no torque about O'. Since $r_m = r_M(M/m)$, the total torque about O' is:

$$\vec{\tau}_{O'} \equiv \vec{\tau}_{O'm} + \vec{\tau}_{O'n} = \left[r_M\left(\frac{M}{m} + 1\right)mg - r_M(m + M)g\right]\hat{k} = 0,$$

as required.

Basic Skills

Review Questions

- What fact about the skaters' interaction in Figure 9.1 indicates that a conservation law is involved?

§9.1 ANGULAR MOMENTUM OF A PARTICLE

- What fact about planetary motion requires a third conservation law for its explanation?
- What quantity is conserved according to Kepler's second law?

- How do we know that angular momentum is not a scalar? What conventions are used to assign it a direction? State the definition of *angular momentum*.
- State the definition of the *vector cross product*.
- What are two ways to compute the magnitude of a cross product in terms of components and magnitudes of the vectors?
- Explain why the value assigned to a particle's angular momentum depends on the choice of coordinate origin.
- If a particle is in uniform circular motion, how does its angular

momentum about the center of the circle depend on mass, radius, angular speed, time?
- State seven mathematical properties of the vector cross product.

§9.2 TORQUE

- Explain how experience in pushing a door open suggests defining torque as the cross product of position and force vectors.
- What is the relation between the torque acting on a particle and the particle's angular momentum?
- What is the relation among the power transmitted by a rotating system, the torque acting on the system, and its angular speed? Explain how this relation follows from the definition of work.

§9.3 THE CENTER OF MASS

- What is the *center of gravity* of an object? Explain how to use torque balance to locate an object's center of gravity.
- What is a *force-torque diagram*?
- How are an object's center of gravity and center of mass related? Is this relation completely independent of the object's location in the universe? Why or why not?
- What is the definition of the *center of mass* of a system of particles? In what way is the center of mass the *average* location of the system's mass?
- How is the motion of the center of mass related to the external forces acting on a system? Explain why this relation justifies the use of a particle model for everyday objects.
- What is the *center of mass reference frame*? Why is it useful to describe a system's behavior with respect to its CM reference frame?
- What do the terms *internal kinetic energy* and *internal angular momentum* mean?

§9.4 CONSERVATION OF ANGULAR MOMENTUM

- What is the reason *why* angular momentum is conserved, as illustrated by the example of astronauts grabbing a rope?
- State the law of conservation of angular momentum in its general form and in the form applicable to an isolated system.
- What is the *strong form* of Newton's third law? How does it enter into the proof of angular momentum conservation?

Basic Skill Drill

§9.1 ANGULAR MOMENTUM OF A PARTICLE

1. (a) A car of mass 1.0×10^3 kg is rounding a turn of radius 1.0 km at 55 km/h. What is the magnitude of the car's angular momentum about the center of the turn? (b) Compare the magnitude of the car's angular momentum with that of a 5.0×10^3 kg truck rounding the same turn at the same speed.

2. Can any of the vectors shown in ■ Figure 9.41 be the cross product of two of the others? Why or why not?

3. Find the cross product $\vec{a} \times \vec{b}$ of the vectors $\vec{a} = (5.0 \text{ m})\hat{i}$ and $\vec{b} = (4.3 \text{ m})\hat{k}$.

§9.2 TORQUE

4. A shopper pushes on the shop door at a point 0.80 m from the hinge with a force of 25 N perpendicular to the door. Find the torque about the hinge. What is the torque if the force is increased to 50 N? If the 25-N force is exerted 0.40 m from the hinge?

5. A monkey of mass $m = 10$ kg hangs from the end of a pole of length $\ell = 2$ m long, which extends from a wall (■ Figure 9.42). What torque does the monkey's weight exert about the point where the pole emerges?

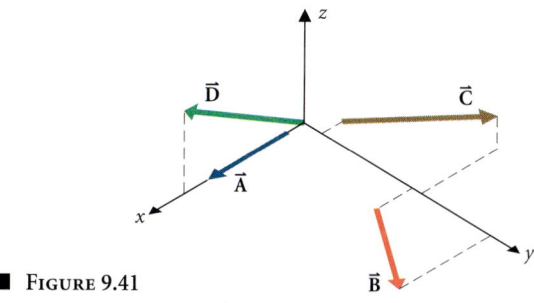

■ FIGURE 9.41

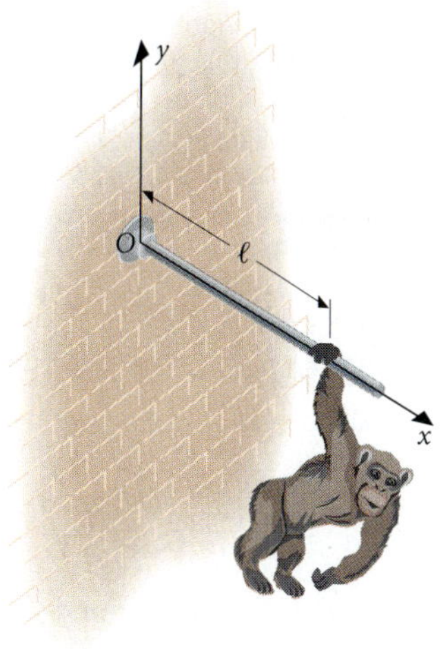

■ FIGURE 9.42    ■ FIGURE 9.43

6. What torque about the origin results when a force $\vec{F} = (670 \text{ N})(\hat{i}\sqrt{2} + \hat{j}\sqrt{3} - \hat{k}\sqrt{5})$ is applied at position $\vec{r} = (0.625 \text{ m})(-\hat{i}\sqrt{5} + \hat{k}\sqrt{3})$?

7. Two people are sitting on a bench (■ Figure 9.43). What happens if one of them gets up? Why?

8. The Nissan 300ZX Turbo automobile engine can exert a torque of up to 308 N·m at 3600 rpm. What is the maximum power output of the engine?

§9.3 THE CENTER OF MASS

9. Based on your experience, is it easier to balance an object by suspending it from above its center of gravity or by supporting it from below? In each case sketch a force-torque diagram for a slightly unbalanced object and explain the difference by discussing which way gravitational torque tends to rotate the object.

10. Four objects are located on a plane as shown in ■ Figure 9.44. Where is the center of mass of the system?

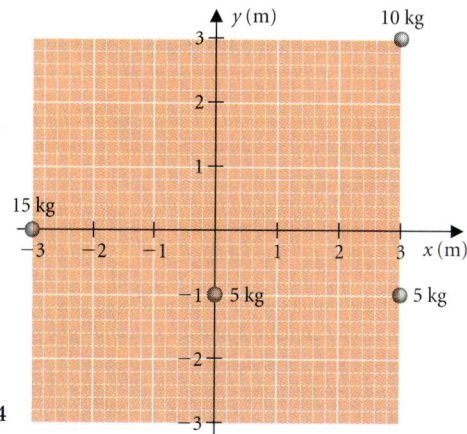

■ FIGURE 9.44

11. Four beads with masses of 10 g, 20 g, 30 g, and 40 g are placed on a straight wire of negligible mass at distances from its end of 10 cm, 20 cm, 30 cm, and 40 cm, respectively. Where is the center of mass of the system?

12. Two particles of equal mass 0.22 kg are located on the x-axis at $x = 1.50$ m and $x = -0.25$ m, moving with velocities $(2.5 \text{ m/s})\hat{i}$ and $(0.75 \text{ m/s})\hat{j}$, respectively. Find the position and velocity of their CM. What is the internal energy of the system?

§9.4 CONSERVATION OF ANGULAR MOMENTUM

13. Two balls, each of mass $M = 0.50$ kg, are held by two separate strings in uniform circular motion about point O (■ Figure 9.45). If each ball has speed $v = 2.0$ m/s, what is the total angular momentum of the system about O?

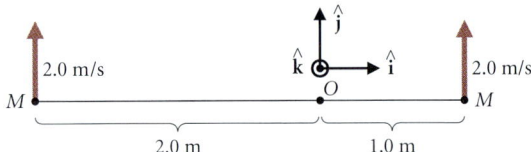

■ FIGURE 9.45

14. Two skaters, each of mass 70 kg, are spinning around and holding hands with a separation of 2 m and $\omega = 0.5$ rad/s. If they pull on their hands and reduce their separation to 1 m, what is their final angular speed?

Questions and Problems

§9.1 ANGULAR MOMENTUM OF A PARTICLE

15. ❖ Compare the angular momenta of two identical cars rounding the same turn, one at 75 km/h and the other at 55 km/h. In each case, take the origin at the center of the circle.

16. ❖ Compare the angular momenta of two identical cars, both traveling at 55 km/h, if one is rounding a turn of radius 1.0 km and the other is rounding a turn of radius 2.0 km.

17. ❖ Three particles of equal mass have very nearly the same position and have the velocities shown in ■ Figure 9.46. Which has the greatest magnitude of angular momentum with respect to O?

18. ❖ If two vectors $\vec{a}$ and $\vec{b}$ have the same cross product with a third vector $\vec{c}$—that is, $\vec{a} \times \vec{c} = \vec{b} \times \vec{c}$—are $\vec{a}$ and $\vec{b}$ equal? Explain why or why not.

19. ❖ A particle is located on the y-axis 2 m from the origin and has a velocity $\vec{v}$ in the x–y plane at 45° to the y-axis, $\vec{v} = (10 \text{ m/s})(\hat{i} + \hat{j})$. Locate the points about which (a) $\vec{L}$ is zero. (b) $\vec{L}$ is in the $+z$-direction. (c) $\vec{L}$ is in the $-z$-direction. (d) $\vec{L}$ is in the x–y plane. (e) If A is any point on the x-axis, describe at least one path the particle could follow while having constant angular momentum $\vec{L}_A$.

20. ❖ Show that for any constant C and any vectors $\vec{a}$ and $\vec{b}$, $(C\vec{a}) \times \vec{b} = C(\vec{a} \times \vec{b})$.

21. ◆ In a funfair ride, a car of mass 6.0×10^2 kg swings around once every 3.0 s on the end of an arm 5.0 m long. What is the magnitude of the car's angular momentum about the pivot point of the arm?

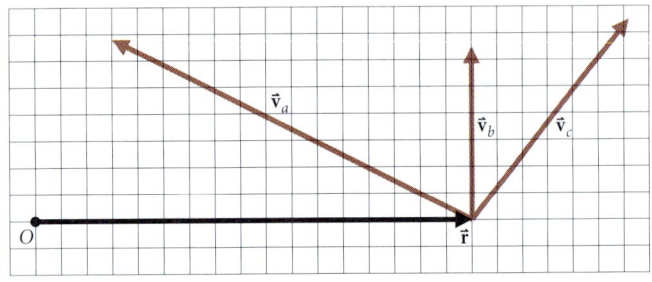

■ FIGURE 9.46

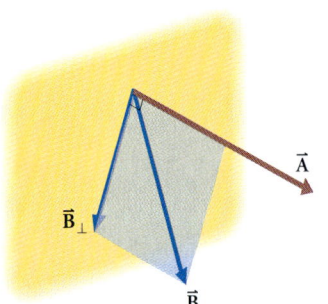

■ FIGURE 9.47

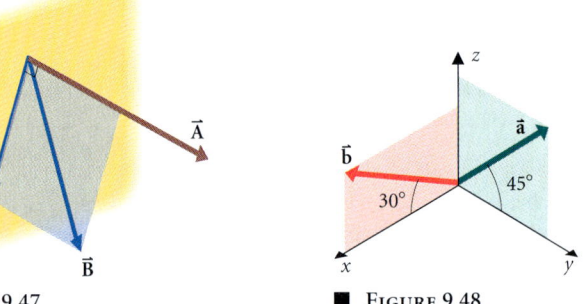

■ FIGURE 9.48

22. ◆ One skater (mass 85 kg) approaches another at a speed of 2.2 m/s, so as to pass 1.5 m from the second, stationary skater. Find the magnitude of the first skater's angular momentum about the second.

23. ◆ ■ Figure 9.47 shows the vector $\vec{A}$ and the plane perpendicular to it. Also shown are the vectors $\vec{B}$ and $\vec{B}_\perp$, its projection onto that plane. Show from the definition of the cross product that $\vec{A} \times \vec{B} = \vec{A} \times \vec{B}_\perp$.

24. ◆ A particle of mass m moves in the x–y plane, parallel to the x-axis, along the line $y = c$. Its velocity $\vec{v} = -v\,\hat{i}$ is constant. Find the particle's angular momentum about the origin.

25. ◆ Find the components of the vector $\vec{C} = \vec{A} \times \vec{B}$, where the components of $\vec{A}$ and $\vec{B}$ are: $\vec{A} = (0, 2.2 \text{ m}, 4.7 \text{ m})$ and $\vec{B} = (1.3 \text{ kg} \cdot \text{m/s}, 0, 0.71 \text{ kg} \cdot \text{m/s})$. What kind of physical property does vector $\vec{C}$ describe?

26. ◆ Find the cross product $\vec{a} \times \vec{b}$ of the two vectors shown in ■ Figure 9.48, if $|\vec{a}| = |\vec{b}| = 1$.

27. ◆ Compute the angular momentum about the origin of a particle of mass $m = 15$ kg at position $\vec{r} = (1.0 \text{ m})(\hat{i} + 2\hat{j})$ with velocity $\vec{v} = (5.0 \text{ m/s})(2\hat{i} + \hat{j})$.

28. ◆ Find the cross product $\vec{a} \times \vec{b}$ if $\vec{a} = 2\hat{i} - \hat{j}$ and $\vec{b} = \hat{j} - 2\hat{k}$. Show the two vectors and their cross product in a diagram.

29. ◆ Find the cross product $\vec{a} \times \vec{b}$ if $\vec{a} = \hat{i} - 2\hat{j}$ and $\vec{b} = \hat{j} - 2\hat{i}$. Show the two vectors and their cross product in a diagram.

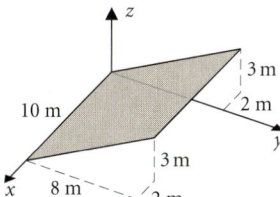

■ FIGURE 9.49

30. ♦ A particle has mass $m = 1.0$ kg, velocity $\vec{v} = (55 \text{ m/s})(\hat{i} - \hat{j})$, and position $\vec{r} = (32 \text{ m})(\hat{i} - \sqrt{2}\hat{j} + \sqrt{3}\hat{k})$. Compute its angular momentum with respect to the origin of the coordinate system.

31. ♦ Verify eqn. (9.9) for the cross product $\vec{a} \times \vec{b}$ in terms of components starting from expressions for $\vec{a}$ and $\vec{b}$ in components.

32. ♦♦ What is the magnitude of the Moon's angular momentum (a) about the Earth and (b) about the Sun? Use data from the inside front cover.

33. ♦♦ Use data from the inside front cover to find the magnitude of the solar system's angular momentum about the Sun. (Neglect any rotation of the Sun or planets about their own axes, and assume that all the planets orbit in the same plane.) Which planet contributes most of the angular momentum?

34. ♦♦ Find the magnitude of the Sun's angular momentum about Jupiter.

35. ♦♦ An aircraft of mass 10 000 kg is 100 km northwest of Pittsburgh at an altitude of 3000 m flying due east at 300 km/h. What is its angular momentum about Pittsburgh?

36. ♦♦ ■ Figure 9.49 shows a surface. Give the components of a vector that represents the area of the surface. What is the magnitude of the area?

37. ♦♦ Two sides of a parallelogram are defined by the vectors: $\vec{a} = (3.2 \text{ m})(\hat{i} + 2\hat{j} - \hat{k})$ and $\vec{b} = (2.6 \text{ m})(\hat{i} - \hat{j} + \hat{k})$. What is the area of the parallelogram and what angle does its normal make with the z-axis?

38. ♦♦ Show that in general $(\vec{a} \times \vec{b}) \times \vec{c} \neq \vec{a} \times (\vec{b} \times \vec{c})$. (*Hint:* Choose an easy special case and use geometrical arguments.)

39. ♦♦ Verify, using component notation, that:
$\vec{a} \times (\vec{b} \times \vec{c}) = \vec{b}(\vec{a} \cdot \vec{c}) - \vec{c}(\vec{a} \cdot \vec{b})$.
(*Hint:* Choose the coordinate system so that *one* of the three vectors is parallel to a coordinate axis. Can you simplify further still?)

40. ♦♦♦ A spiral coil is defined by the axis around which it coils, its radius R, and the distance it advances along the axis in each turn, called its pitch p (■ Figure 9.50). With the spiral axis along the z-coordinate axis, the equation for the position vector of points on the curve is:

$$\vec{r}(z) = z\hat{k} + R[\hat{i} \cos(2\pi z/p) + \hat{j} \sin(2\pi z/p)].$$

Compute the velocity and acceleration of a particle with mass m that moves upward along the spiral with constant vertical speed $dz/dt = v_o$. What is the angular momentum of the particle about the origin if $p = R$?

41. ♦♦♦ A particle's mass is m, and its position as a function of time is given by:

$$\vec{r}(t) = \vec{r}_o + \vec{v}_o t,$$

where the particle's constant velocity is $\vec{v}_o$. What is the particle's angular momentum $\vec{L}$ about the origin of the reference frame? Show explicitly that $\vec{L}$ does not change with the clock reading t. What is the magnitude of $\vec{L}$ when $\vec{r}_o$ and $\vec{v}_o$ are perpendicular? Show that for any given $\vec{r}_o$ and $\vec{v}_o$ there is a clock reading $t = t_\perp$ such that $\vec{r}(t_\perp)$ is perpendicular to $\vec{v}_o$. Express $|\vec{L}|$ in terms of $\vec{v}_o$ and $\vec{r}(t_\perp)$ in this general case. Sketch the general case in a diagram where $\vec{L}$ is perpendicular to

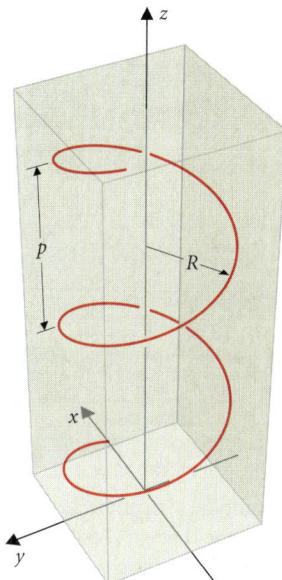

■ FIGURE 9.50

the plane of the diagram and discuss the geometrical significance of your results.

§9.2 TORQUE

42. ❖ A particle moves in a straight line with a constant acceleration $\vec{a}$. Point P is on the line and point Q is not. Discuss the value of the torque on the particle about P and about Q.

43. ❖ A particle moves in a straight line at constant speed. What net torque acts on the particle? Does the torque depend on the reference point chosen? Explain your reasoning.

44. ❖ A particle moves in uniform circular motion. What net torque acts on the particle? Does the torque depend on the reference point chosen? Explain your reasoning.

45. ♦ A diver of mass 80 kg stands on the end of a diving board, 6 m from its support. What torque about the support results from the diver's weight?

46. ♦ A man leaning against a fence exerts a force of 20 N normal to the fence at a point 1 m from the ground. What is the resulting torque about the base of the fence?

47. ♦ A child of mass 40 kg is sitting on one end of a bench, 20 cm from the leg (■ Figure 9.51). What is the torque about the foot of that leg due to the child's weight?

48. ♦ A Cessna 182Q aircraft is powered by a Continental O-470-U engine rated at 230 Hp. What torque does the engine exert at full power and 2400 rpm?

49. ♦♦ A food mixer draws a power of 270 W. The mixer element rotates at 2.0 cycles per second. If the mixer is 95% efficient, what torque is exerted on the food?

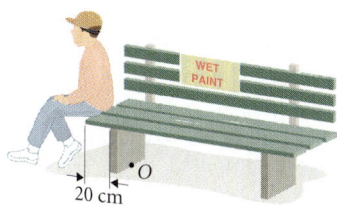

■ FIGURE 9.51

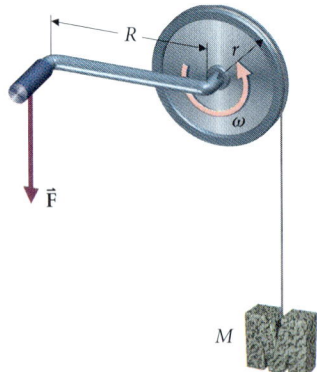

■ FIGURE 9.52

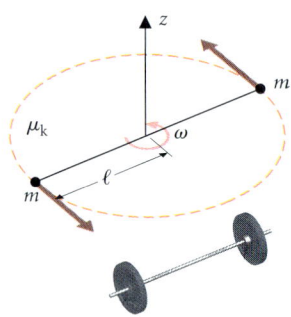

■ FIGURE 9.53

50. ◆◆ The motor of a model airplane (mass 1 kg) exerts a thrust along the axis of the airplane that exceeds the drag force on the plane by 1 N. The airplane is flown in a horizontal circle at the end of a string 5 m long. At what rate does the angular speed of the airplane increase? (Assume that lift on the wings is equal to and opposite from the weight of the airplane.)

51. ◆◆ The car in a funfair ride has mass 6.0×10^2 kg and turns on an arm 5.0 m long. Accelerating from rest, the car reaches its final angular speed of 0.33 rad/s in 10.0 s. What torque, assumed constant, acts on the car during this acceleration?

52. ◆◆ Water is raised from a well by a donkey who pulls a wooden arm attached to a screw. If the arm is 3.0 m long and the screw has pitch 3.00 cm, what is the mechanical advantage of the machine?

53. ◆◆ What force must you exert on the crank handle in ■ Figure 9.52 to turn the crank at constant angular frequency ω? What power is required? Compare the result with the power extracted from the block by gravity.

54. ◆◆ The torque acting on a particle as a function of time is:
$$\vec{\tau}(t) = (275 \text{ N·m})[(0.500)\hat{i} + (1.325 /s^2)t^2\hat{j} - (0.026 /s^3)t^3\hat{k}].$$
What can you say about the particle's angular momentum as a function of time?

55. ◆◆◆ A particle's angular momentum is given as a function of time by the formula:
$$\vec{L}(t) = L_o[(t/t_o)^2\hat{i} + \alpha(t/t_o)^4\hat{j} - \beta(t/t_o)^3\hat{k}].$$
(a) What is the torque $\vec{\tau}(t)$ acting on the particle? **(b)** Suppose the particle's mass is m and its position is given by:
$$\vec{r}(t) = \vec{v}_o t + \vec{a}_o(t^2/2) + \vec{J}_o(t^3/6).$$
Find the vectors $\vec{v}_o$, $\vec{a}_o$, and $\vec{J}_o$ in terms of m, L_o, t_o, α, and β.

56. ◆◆◆ A particle of mass M moves with constant angular speed ω counterclockwise around a circle of radius R centered on the origin. Compute the angular momentum of the particle about the point A, at position $\vec{r}_A = -R\hat{i}$, as a function of time. Compute the torque about A acting on the particle. Verify that: $\vec{\tau}_A = d\vec{L}_A/dt$.

57. ◆◆◆ A dumbbell consists of two identical particles of mass M connected by a stiff, massless rod of length 2ℓ. At time $t = 0$, the dumbbell is set spinning with angular speed ω_o on a surface with a small coefficient of friction μ_k, as in ■ Figure 9.53. **(a)** What is the angular momentum of the dumbbell? What is the torque exerted on the system by friction forces? At what time does the dumbbell stop? **(b)** What is the energy of the dumbbell at $t = 0$? What power is dissipated as heat by the friction forces? At what time does the dumbbell stop? Compare your two computed times.

§9.3 THE CENTER OF MASS

58. ❖ Four particles of equal mass M are placed at the four corners of the base of a cube. Four more particles, each of mass $2M$, are placed at the four corners of the top face of the cube. Where is the center of mass of the system?

59. ❖ An astronaut of mass M is sitting on a satellite of mass $2M$ while attempting to fix it. He is attached to the satellite by a safety cable of length L. A wrench lost by his partner is floating a distance $4L/5$ away, so he decides to leap off the satellite to fetch the wrench. Does he succeed? Why or why not?

60. ❖ ■ Figure 9.54 shows a group of skydivers. Describe the motion of the CM of the system. If the divers push apart horizontally, describe the motion of each after separation. Give as much detail as you can.

■ FIGURE 9.54

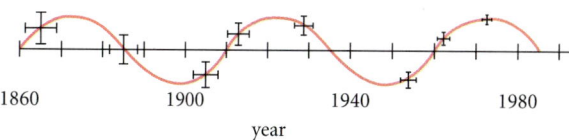

■ FIGURE 9.55

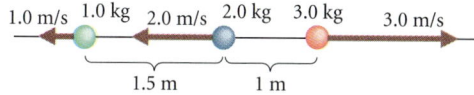

■ FIGURE 9.56

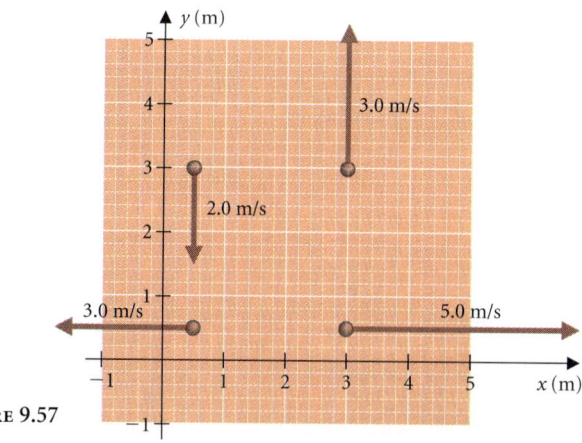

■ FIGURE 9.57

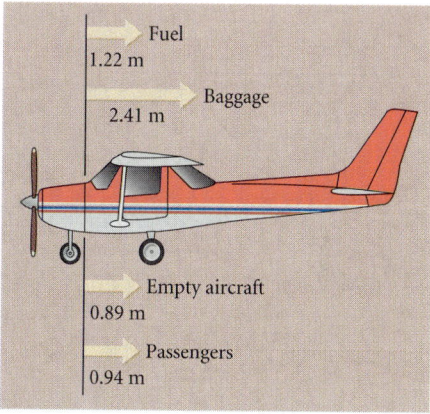

■ FIGURE 9.58

61. ❖ A distant star is observed to move along a wiggly path over a period of many tens of years (■ Figure 9.55). What would cause this kind of motion?
62. ❖ Two balls are tied together by a string of length ℓ. One ball is held fixed on a smooth table and the second is set moving in a circle with angular speed ω. Describe the motion of the system after the fixed ball is released.
63. ♦ At a certain instant, three particles are arranged in a line with the velocities shown in ■ Figure 9.56. Find the center of mass of the system and its velocity.
64. ♦ Two children, masses 50.0 kg and 30.0 kg, are sitting on opposite sides of a teeter-totter. If the 30-kg child sits at one end, 2.00 m from the support, where should the heavier child sit for the two to balance? If instead the two children sit on the ends of the teeter-totter, where should a third child of mass 40 kg sit to put the CM over the support?
65. ♦ Four equal masses m are suspended by strings at the 20.0-cm, 30.0-cm, 60.0-cm, and 90.0-cm markings on a meter stick whose mass is also m. Where should you support the meter stick so that it balances?
66. ♦ (a) Three particles of masses $m_1 = 5.0$ kg, $m_2 = 10.0$ kg, and $m_3 = 15.0$ kg are at positions $x_1 = -2.0$ m, $x_2 = +3.7$ m, and $x_3 = +4.0$ m along the x-axis. Where is their center of mass? (b) Obtain the same result by finding first the CM of particles 1 and 3, and then finding the CM of that combination and particle 2.
67. ♦ A massless ring of radius 0.50 m has six balls, each of mass 1.0 kg, attached to it at uniformly spaced intervals. The ring rotates at 6.0 revolutions per second. Considering the entire ring as a single *particle*, what is its kinetic energy? What is its internal kinetic energy?
68. ♦♦ The CM of a 1300-kg automobile is 0.5 m behind the front wheels. If 100 kg of luggage is placed in the trunk, 2 m behind the front wheels, where is the CM then?
69. ♦♦ Four particles, each of mass 1.0 kg, are moving in a plane with the velocities shown in ■ Figure 9.57. (a) Find the position and velocity of their center of mass. (b) What is the internal energy of the system? What is its internal angular momentum?
70. ♦♦ A space pirate approaches a stationary commercial ship at 0.10 km/s, passes it at a distance of 0.10 km, and attaches the ships together using a magnet on the end of a 0.10-km-long line. The pirate ship has a mass of 2.0×10^6 kg while the freighter has a mass of 1.0×10^6 kg. What is the final angular speed of the system and the velocity of its CM?

71. ♦♦ A uniform meter stick has a mass of 0.150 kg. The stick is balanced on a knife edge at the 30.0-cm mark, with an object suspended at the 15.0-cm mark. What is the object's mass?
72. ♦♦ When a uniform piece of wire of length L is bent at its center through an angle α, the wire's center of mass lies along the bisector of the angle. Find the distance of the CM from the center of the wire as a function of α.
73. ♦♦ A piece of wire of length L is bent so as to form three sides of a square. How far is the center of mass of the wire from the geometrical center of the square?
74. ♦♦ Three identical rods, each 1.00 m long, are welded perpendicularly to each other at a common end so as to represent the x-, y-, and z-axes of a Cartesian coordinate system. Where is the CM of this object?
75. ♦♦ (a) An empty aircraft has a mass of 817 kg and a center of mass 0.89 m behind the firewall (■ Figure 9.58). It is then loaded with 75 gallons of fuel (CM 1.22 m behind the firewall), two passengers (total mass 150 kg, CM 0.94 m behind the firewall), and 58 kg of baggage (CM 2.41 m behind the firewall). If the mass of 1 gallon of fuel is 2.7 kg, where is the center of mass of the loaded airplane? (b) An airplane becomes unstable if its center of mass is too far back toward the tail. If the aircraft must not have its CM more than 1.23 m behind the firewall, what is the maximum mass of baggage the plane may carry? Suppose one of the passengers (mass 72 kg) gets out of the plane. What maximum baggage may the plane now carry? Is your answer surprising?

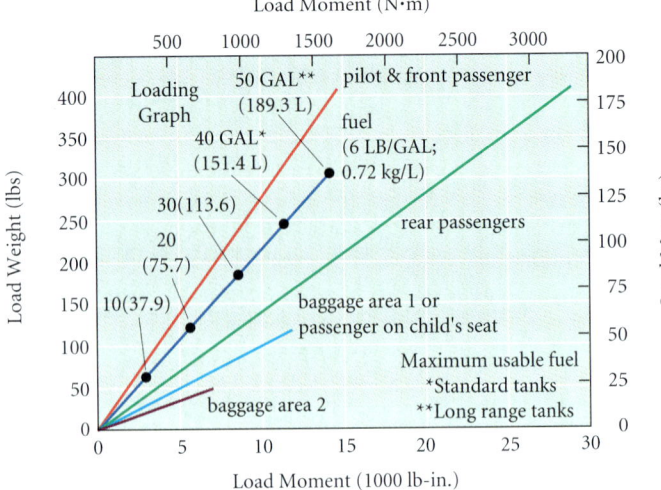

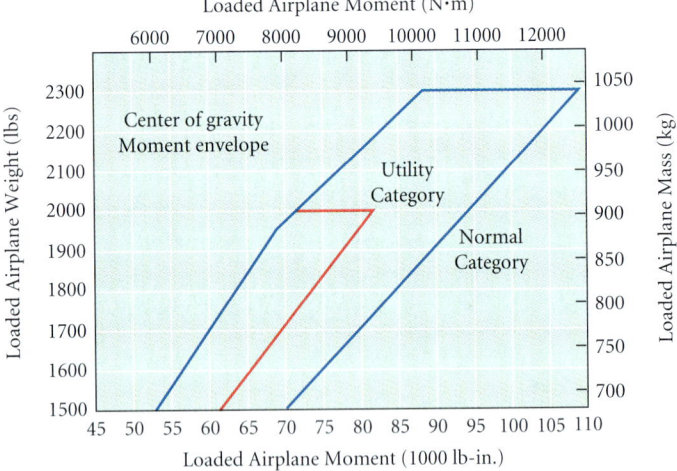

■ **FIGURE 9.59**
Data courtesy of Cessna Aircraft Corporation, *Cessna 172 pilot's operating handbook*, 1978: 6-10, 6-11.

76. ◆◆ ■ Figure 9.59 gives data for computing the center of mass of a Cessna 172 aircraft. The loading graph gives the product of weight and distance from the plane's firewall (moment) for different portions of the plane's load. The center of gravity moment envelope shows safe limits of weight and moment. If the airplane is loaded according to the following data, calculate its total weight and total moment and determine if it is within safe limits.

Load	Weight (lb)	Moment (1000 lb-in.)
Empty weight plus oil	1365	51.3
Pilot and front passenger	310	
Rear passenger	196	
Fuel (38 gallons)		

77. ◆◆ ■ Figure 9.60 shows the velocities of two aircraft: *A* has a mass of 3000 kg and *B* has a mass of 4000 kg. Find the velocity of the CM of the two aircraft.

78. ◆◆ In a highway safety test, an American car of mass 3.0×10^3 kg collides head-on with a Japanese car of mass 1800 kg. Both cars are moving at a speed of 35 m/s before the collision. If afterward

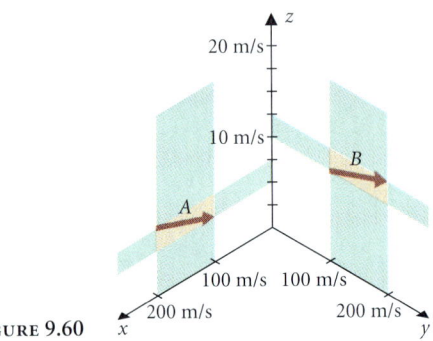

■ **FIGURE 9.60**

both cars are at rest in their CM reference frame, find the amount by which each car's speed changes in the collision.

79. ◆◆ ✉ If inhabitants of a planet orbiting Sirius were watching the Sun, they would observe a motion like the one shown in Figure 9.55. If the solar system is approximated by a two-body system consisting of the Sun and Jupiter, what is the amplitude of the Sun's motion; that is, how far does it move to either side of a straight line?

§9.4 CONSERVATION OF ANGULAR MOMENTUM

80. ❖ When two vehicles collide in a highway accident, often one vehicle ends up facing backwards—that is, opposite its original direction of motion. Explain why.

81. ❖ An astronaut, repairing a satellite in orbit, uses a wrench to exert a torque of 65 N·m on a fixture. Qualitatively describe the resulting motion of astronaut and satellite.

82. ◆ Two skaters approach each other from opposite directions on parallel paths 1.0 m apart. They have masses of 52 and 65 kg, respectively, and equal speeds of 3.0 m/s. They link hands as they reach each other, remaining 1.0 m apart. Describe the final motion of the system.

83. ◆◆ Two astronauts of equal mass $m = 80$ kg are attached to each other by a line 4 m long. When they are 4 m apart, one of the astronauts fires a maneuvering pistol that delivers an impulse of 160 N·s, perpendicular to the line. Describe the susbsequent motion of the two astronauts.

84. ◆◆ Due to an unfortunate accident, an astronaut of mass M is flung from the space station at speed v. Luckily, the astronaut's path passes a service and support module, mass $2M$, at distance $\ell/2$, where the astronaut grabs the end of a safety cable of length ℓ attached to the module. The cable jerks a few times as it comes taut, but the system settles down to a steady motion. Describe this motion of the astronaut/module combination in its CM reference frame. What is the velocity of the CM?

85. ◆◆ After pulling together, the skaters in Study Problem 8 let go of each other just as the 58-kg skater is moving parallel to the CM velocity. What speed does each skater then have over the ice?

86. ◆◆ We may model a small merry-go-round as six horses ("particles") each of mass 50 kg and spaced at regular intervals around a circle of radius 3 m. The merry-go-round is rotating clockwise at 1 rpm. What is its angular momentum? If a man of mass 100 kg jumps onto the edge of the merry-go-round, what is its final angular speed? (Ignore all frictional and other torques and assume that the man's initial velocity is toward the center of the merry-go-round.)

87. ◆◆◆ Begin with an expression for a system's angular momentum evaluated in the laboratory frame and transform it into an expression involving positions and velocities with respect to the CM. Show that the expression reduces to the form: $\vec{L} = \vec{R} \times \vec{P} + \vec{L}_{int}$, where $\vec{R}$ is the position of the CM with respect to the laboratory origin. Discuss why this decomposition works.

Additional Problems

88. ❖ What changes, if any, in the relation between torque and angular momentum would be necessary if a *left-hand rule* were used to assign directions to rotational quantities? What changes would there be in the properties of the cross product if its direction were defined by a left-hand rule?

89. ❖ Due to an unplanned explosion, a satellite disintegrates during its first orbit. One half is brought momentarily to rest and then falls radially inward and finally re-enters the Earth's atmosphere. What happens to the other half immediately after the explosion?

90. ◆◆ A communications satellite of mass 500 kg is in synchronous orbit (cf. Problem 0.14). Find the magnitude of its angular momentum about the center of the Earth.

91. ◆◆ An astronaut plans to grab a faulty satellite of mass 5.0×10^2 kg with a 3.0-m-long grappling arm so that it can be brought back for repair. As the end of the grappling arm contacts the satellite, the astronaut (mass 1.00×10^2 kg in her spacesuit) is moving toward it with a velocity of 0.50 m/s at an angle of 30° to the grappling arm. What is the final motion of satellite and astronaut?

92. ◆◆ ✉ Use dimensional analysis to estimate the angular momentum of a uniform sphere of mass M and radius R rotating at angular frequency ω. Do you expect your result to be an overestimate or underestimate? Explain your reasoning. Estimate the angular momentum of the Sun about its center using data from the inside front cover and compare with the result of Problem 33.

93. ◆◆ Two modules of a planetary exploration vehicle are connected by a long cable so that each half of the system experiences an acceleration of 1g to simulate Earth's gravity in the astronauts' quarters. **(a)** If the cable is 1.0 km long, and the mass of each module is $M = 3.6 \times 10^5$ kg, find the system's angular speed and angular momentum. **(b)** One of the modules experiences an emergency and has to jettison one-half the module, which is gently disconnected and released. Describe the subsequent motion of each piece of the system with respect to the CM of the intact spacecraft. Compute the angular momentum of each piece and verify that the total is unchanged.

94. ◆◆◆ A comet approaches the Sun on a hyperbolic orbit (■ Figure 9.61). If the comet has speed v_0 when a long way from the Sun, find its distance s from the Sun and its speed at the point of closest approach Q in terms of v_0, b and the Sun's mass.

95. ◆◆◆ A particle of mass m moves around a circular path of radius R in the x–y plane centered on the origin O. Its angular speed in-

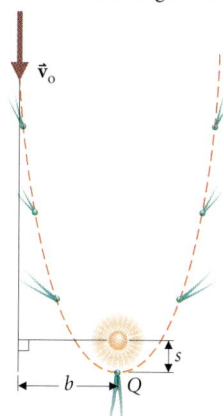

■ Figure 9.61

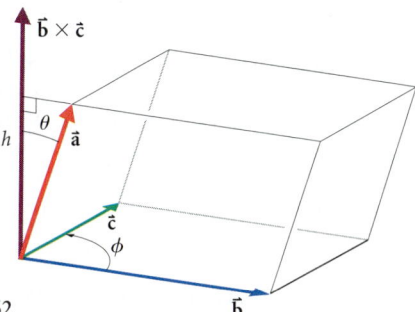

■ Figure 9.62

creases according to the formula $\omega = \alpha t$, and its angular position measured from the x-axis is:

$$\theta(t) = \int_0^t \omega \, dt = \alpha t^2/2.$$

(a) Find the position, velocity, and acceleration vectors of the particle as functions of time. **(b)** What is the magnitude of the force that must act on the particle as a function of time? **(c)** What is the angular momentum of the particle about the center of the circle as a function of time? Verify that $\vec{\tau}_O = d\vec{L}_O/dt$. **(d)** Show that for $t \ll \alpha^{-1/2}$ the required force is primarily needed to increase the particle's speed. **(e)** Show that for $t \gg \alpha^{-1/2}$ the primary requirement is for centripetal acceleration. **(f)** ✉ Through how much angle does the particle move in time $t \approx \alpha^{-1/2}$?

96. ◆◆◆ Prove that, in general, the CM of a system of particles may be found by dividing the system into several parts, finding the CM of each part separately, and then treating each part as a particle located at its individual CM. (*Hint:* Label the particles in the first part, 1, ..., n_1, in the second part $n_1 + 1, \ldots, n_2$, etc.)

97. ◆◆◆ Show that if a particle moves with constant speed and constant angular momentum about O, its path is either a straight line or a circle centered on O.

98. ◆◆◆ Using a combination of the dot and cross products we may define a *scalar triple product* of three vectors:

$$\vec{a} \cdot (\vec{b} \times \vec{c}).$$

Show that the scalar triple product $\vec{a} \cdot (\vec{b} \times \vec{c})$ is the volume of the solid figure with sides defined by the vectors $\vec{a}$, $\vec{b}$, and $\vec{c}$ (■ Figure 9.62). Use the result to argue that

$$\vec{a} \cdot (\vec{b} \times \vec{c}) = \vec{b} \cdot (\vec{c} \times \vec{a}) = \vec{c} \cdot (\vec{a} \times \vec{b}).$$

99. ◆◆◆ Show that $(\vec{a} \times \vec{b}) \cdot (\vec{c} \times \vec{d}) = (\vec{a} \cdot \vec{c})(\vec{b} \cdot \vec{d}) - (\vec{a} \cdot \vec{d})(\vec{b} \cdot \vec{c})$. (*Hint:* Use the results of Problems 39 and 98.)

Computer Problems

100. The masses, positions, and velocities of six objects are given in the file CH9P100 on the supplementary computer disk. Using a spreadsheet program, calculate the position of the CM and its velocity. Calculate the internal energy and angular momentum of this system.

101. Calculate the semimajor axis and eccentricity of an orbit that goes from Earth to Jupiter and then calculate the energy per unit mass E/m and angular momentum needed for this orbit. From the value of L/m, find the scaled value of the spacecraft's velocity when it leaves Earth: $v/\sqrt{GM/r_n}$. Using program *Kepler* on the supplementary computer disk, calculate the path of a spacecraft following this orbit, and find its period. Compare the computed result with Kepler's third law. (See Essay 3, p. 329.)

Challenge Problems

102. (a) Two astronauts, each of mass 75 kg, are tied together by a 100-m length of cable. They are undergoing uniform circular motion, each with a speed of 6 m/s. What is their speed after they let out a second 100 m of cable? What is the change in their kinetic energy during the process of lengthening their separation? Calculate the work done on each astronaut by cable tension and compare with the change in kinetic energy. **(b)** After an accident, a 100-kg astronaut is adrift at the end of a safety cable of length $\ell = 20$ m. The cable is attached to a service module 1.0 m in diameter, mass $2M = 200$ kg, and both are rotating about their CM with angular frequency $\omega = 0.15$ rad/s. How much work must the astronaut do to pull inward along the safety cable and reach the module? What force must the astronaut exert on the cable upon reaching the module? Is the feat possible? Why or why not?

103. Suppose that the variation of g with height above the Earth's surface is approximated by the formula $|\vec{g}| = g_o - \alpha(h - h_o)$ (cf. eqn. 5.3). Consider an ideal rod consisting of particles with masses m and M connected by a massless unbendable rod of length ℓ, as in the discussion in §9.3. With its center of mass held fixed at $h = h_o$, the rod is rotated to an angle θ with the horizontal. (For $\theta > 0$, M is above m.) Show that the rod's center of gravity is no longer the same as its center of mass. If $m = 1.0$ kg, $M = 2.0$ kg, $\ell = 0.50$ m, and $\theta = 45°$, what is the distance between the CG and the CM?

104. A particle of mass M slides without friction on the interior of a conical surface with apex angle α (■ Figure 9.63). Initially, the particle is at a height $h = h_o$ and has velocity $\vec{v} = v_o \hat{\imath}$. What value of $v_o \equiv v_c$ is required for the particle to remain at height h_o? If $v_o < v_c$, describe the motion of the particle qualitatively. What happens to angular momentum about the apex of the cone? Use conservation of angular momentum and/or energy to find the horizontal component of the particle's velocity as a function of its height h. Find the minimum height h_{min} to which the particle descends before beginning to rise again.

105. (a) Show that the rate at which work is done by external forces on a system of particles may be written:

$$\frac{dW}{dt} = \vec{F}_{ext} \cdot \vec{v}_{CM} + \sum_j \vec{F}_j \cdot \vec{v}_{j,CM} - \sum_j \sum_{i \neq j} \vec{F}_{ji} \cdot \vec{v}_{j,CM},$$

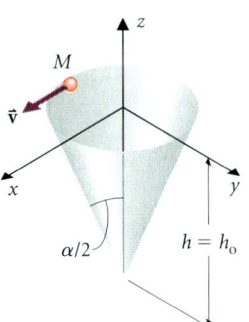

■ Figure 9.63

where $\vec{F}_{j,ext}$ is the external force acting on particle j, $\vec{F}_{ext} = \Sigma \vec{F}_{j,ext}$ is the net external force acting on the system, $\vec{v}_{CM}$ is the velocity of the system's CM, and $\vec{v}_{j,CM}$ is the velocity of particle j with respect to the CM. **(b)** Assuming that the internal forces $\vec{F}_{ji}$ are conservative, show that the third term is the rate of change of internal potential energy. **(c)** Recall the separation of the system's kinetic energy into internal and particle terms in §9.3.4. Show that the first term in dW/dt is the rate of change of the CM kinetic energy and that the second term in dW/dt is the rate of change of internal kinetic energy. **(d)** In the case where all of the particles of the system are at rest in the CM frame, show that this equation is equivalent to the work-energy theorem for a particle. **(e)** In the case where the only external forces acting are uniform gravity and a contact force acting on particle k, which remains stationary in the laboratory frame, show that the contact force does no net work on the system. Discuss the flow of energy in this case if the system's CM is initially at rest, rises a distance h, and again comes to rest.

106. Show that the external torque acting on a system may be expressed in the form:

$$\vec{\tau}_{ext} = \vec{r}_{CM} \times \vec{F}_{ext} + \Sigma \, \vec{r}_{j,CM} \times \vec{F}_{j,ext}.$$

Show also that the first term in $\vec{\tau}_{ext}$ is the rate of change of the system's CM angular momentum and that the second term gives the rate of change of the system's internal angular momentum.

ESSAY 3
Orbits

Maneuvering spacecraft in the solar system is one of the most important practical applications of Newton's law of gravitation. For a spacecraft to fire rockets continuously is extremely inefficient. It is more practical to *save fuel* and *spend time* coasting to the destination on an elliptical path. Thus spacecraft maneuvering consists largely of finding how to change one ellipse into another that goes to the destination. Since the rockets fire for short periods, we may think of them as making impulsive changes in the angular momentum and energy of the spacecraft. For this reason, it is useful for us to learn how to describe the elliptical orbit of a spacecraft in terms of its energy and angular momentum.

Study Hint: Review Kepler's laws (Chapter 0). Kepler's first law tells us the orbits are ellipses. See *Digging Deeper* in §0.2.3 for a description of ellipses.

If you burn enough fuel to escape the body you're orbiting, your path will be a hyperbola.

E3.1   Energy and the Semimajor Axis

We can find an expression for the semimajor axis a as a function of the spacecraft's energy by considering the craft's motion at the nearest (P_n) and farthest (P_f) points from the primary body (■ Figure E3.1). At these two points, the velocity of the spacecraft is perpendicular to its position vector, and its distance from the primary body is $r_n = a(1 - e)$ and $r_f = a(1 + e)$, where e is the eccentricity of the orbit. Since the gravitational force acting on the spacecraft causes no torque about the primary body, the spacecraft's angular momentum L is constant. Evaluating its magnitude at the near and far points, we find:

$$mr_n v_n = ma(1 - e)v_n \equiv |\vec{L}_n| \equiv L \equiv |\vec{L}_f| \equiv ma(1 + e)v_f = mr_f v_f.$$

This expression provides useful relations between v_n and v_f and between v_n and L:

$$v_f = \frac{1 - e}{1 + e} v_n \quad \text{and} \quad v_n = \frac{L}{ma(1 - e)}. \qquad \text{(E3.1)}$$

In addition, the gravitational force is conservative, so the spacecraft's total mechanical energy E is conserved. In particular, it is the same at the near and far points of the orbit:

Total mechanical energy	=	gravitational potential energy	+	kinetic energy
E	=	$-GMm/r$	+	$\tfrac{1}{2}mv^2$

Saturn's rings are a collection of orbiting particles.

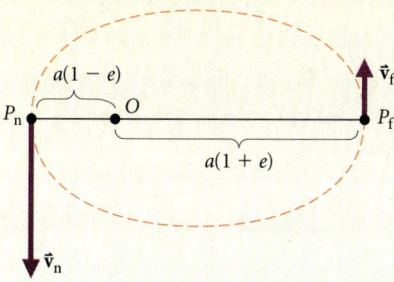

■ **Figure E3.1**
An elliptical orbit. The primary body is located at O. The orbiting body is closest to O at P_n and farthest from O at P_f. Note that $r_n = a(1 - e)$ and $r_f = a(1 + e)$, where e is the eccentricity of the orbit.

Remember: The gravitational force is conservative and causes no torque, so energy and angular momentum remain constant while the linear momentum varies.

Then:
$$-\frac{GMm}{a(1-e)} + \tfrac{1}{2}mv_n^2 = E_n = E = E_f = -\frac{GMm}{a(1+e)} + \tfrac{1}{2}mv_f^2.$$

We want to eliminate the two speeds from the expression for E, so we collect terms containing the speeds v_n and v_f on one side of this equation:

$$\frac{GM}{a}\left(\frac{1}{1+e} - \frac{1}{1-e}\right) = \tfrac{1}{2}(v_f^2 - v_n^2).$$

Next we use eqn. (E3.1) to eliminate the speed v_f at the far point:

$$\frac{GM}{a}\left[\frac{(1-e)-(1+e)}{(1-e^2)}\right] = \tfrac{1}{2}v_n^2\left[\left(\frac{1-e}{1+e}\right)^2 - 1\right].$$

Solve for v_n^2:
$$v_n^2 = \frac{2GM}{a}\left(\frac{-2e}{1-e^2}\right)\left(\frac{(1+e)^2}{(1-e)^2-(1+e)^2}\right)$$
$$= \frac{2GM}{a}\frac{(-2e)}{(1-e^2)}\frac{(1+e)^2}{(-4e)} = \frac{GM}{a}\frac{(1+e)^2}{(1-e^2)} = \frac{GM}{a}\frac{(1+e)}{(1-e)}. \quad (E3.2)$$

Now, use this result to eliminate v_n in the expression for the energy at P_n.

$$E_n = \frac{GMm}{a}\left[\frac{-1}{1-e} + \frac{1+e}{2(1-e)}\right] = \frac{GMm}{2a}\left(\frac{-2+1+e}{1-e}\right) = -\frac{GMm}{2a}.$$

The semimajor axis depends only on the energy of the spacecraft:

$$a = \frac{-GMm}{2E}. \quad (E3.3)$$

The minus sign deserves some thought, since the length a is definitely positive. The conventional choice of *infinity* as the reference state for potential energy means that the spacecraft is assigned a negative potential energy whenever it is closer than infinity—that is, always. A craft with enough kinetic energy to make E positive would be moving fast enough to escape the primary body. It would follow a hyperbolic orbit to which eqn. (E3.3) does not apply. For an elliptical orbit then, E is negative and a is positive, as it should be.

If we need to change the semimajor axis of a spacecraft's orbit, then we have to change E. During a short firing of its rocket motor, the spacecraft's displacement is negligible and its potential energy is unchanged. The pilot controls the craft's kinetic energy. Since E is negative, increasing kinetic energy makes it less negative, decreases $|E|$, and increases a. Conversely, decreasing kinetic energy increases $|E|$ and decreases a.

> The semimajor axis of a spacecraft's orbit changes in the same sense as the spacecraft's kinetic energy is changed.

Let's consider how the space shuttle, in a circular orbit 200 km above the Earth's surface, changes its orbit to land on Earth (■ Figure E3.2). To get the energy per unit mass in any orbit,

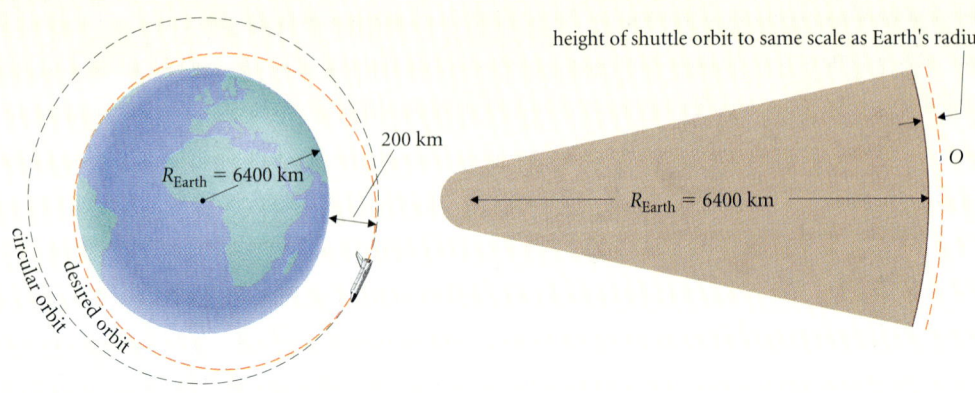

■ **Figure E3.2**
The space shuttle must reduce its energy to enter the landing orbit. The height of the orbit (and the size of the shuttle!) has been exaggerated for clarity. The actual height of the orbiter, drawn to scale, is indicated by point O.

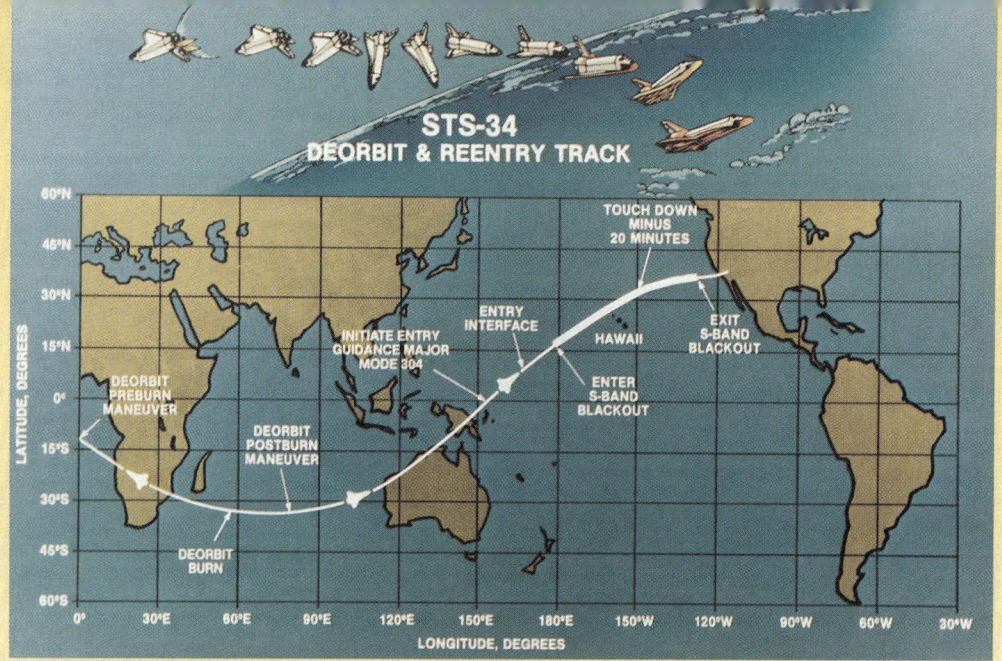

FIGURE E3.3
To deorbit, the space shuttle fires its engines in the opposite direction from its orbital velocity. The shuttle then rotates so as to land like an airplane half an orbit later.

we need only know the semimajor axis. For the shuttle's circular orbit, a is the radius: a_{circ} = radius of Earth + 200 km = 6.6×10^3 km. From eqn. (E3.3), the energy per unit mass is:

$$\frac{E}{m} = -\frac{GM}{2a_{\text{circ}}} = -\frac{(6.7 \times 10^{-11} \text{ m}^3/\text{kg} \cdot \text{s}^2)(6.1 \times 10^{24} \text{ kg})}{2(6.6 \times 10^6 \text{ m})} = -3.1 \times 10^7 \text{ J/kg}.$$

The orbit that descends to the surface has a semimajor axis:

$$a_{\text{descent}} = \tfrac{1}{2}(2 \times \text{radius of Earth} + 200 \text{ km}),$$

as illustrated in Figure E3.2. So, $a_{\text{descent}} = 6.5 \times 10^3$ km. Then:

$$\frac{\Delta E}{m} = -\frac{GM}{2a_{\text{circ}}} - \left(-\frac{GM}{2a_{\text{descent}}}\right) = \frac{GM(a_{\text{descent}} - a_{\text{circ}})}{2a_{\text{circ}}a_{\text{descent}}}$$

$$= -(-3.1 \times 10^7 \text{ J/kg})\left(\frac{-100 \text{ km}}{6.6 \times 10^3 \text{ km}}\right) = -4.7 \times 10^5 \text{ J/kg}.$$

In evaluating this expression we ignored the small difference between a_{circ} and a_{descent} except when they are subtracted. The minus sign indicates that, to descend, the shuttle's total energy must decrease. The pilot reduces the shuttle's kinetic energy by firing its rockets in the opposite direction from its orbital velocity (■ Figure E3.3).

E3.2 Angular Momentum and Eccentricity

To find an expression for the eccentricity of an orbit, we combine eqns. (E3.1) and (E3.2) for v_n and solve the resulting equation for e:

$$\left(\frac{L/m}{a(1-e)}\right)^2 = \frac{GM}{a}\left(\frac{1+e}{1-e}\right).$$

Rearranging gives:

$$\frac{(L/m)^2}{GMa} = \frac{(1-e)^2(1+e)}{1-e} = 1 - e^2.$$

Substituting for a from eqn. E3.4:

$$e^2 = 1 - \frac{(L/m)^2}{GMa} = 1 + \frac{2EL^2}{(GM)^2 m^3}.$$

Or:

$$e = \sqrt{1 + \frac{2EL^2}{(GM)^2 m^3}}. \quad\quad\quad (E3.4)$$

The eccentricity depends on both the energy and angular momentum of the spacecraft.

It was remarked to me by the late Mr. Charles Roupell . . . that to play billiards well was a sign of an ill-spent youth.

HERBERT SPENCER

CHAPTER 10
Collisions

CONCEPTS

Elastic collision
Inelastic collision

GOALS

Be able to:
Use conservation laws to analyze collisions.

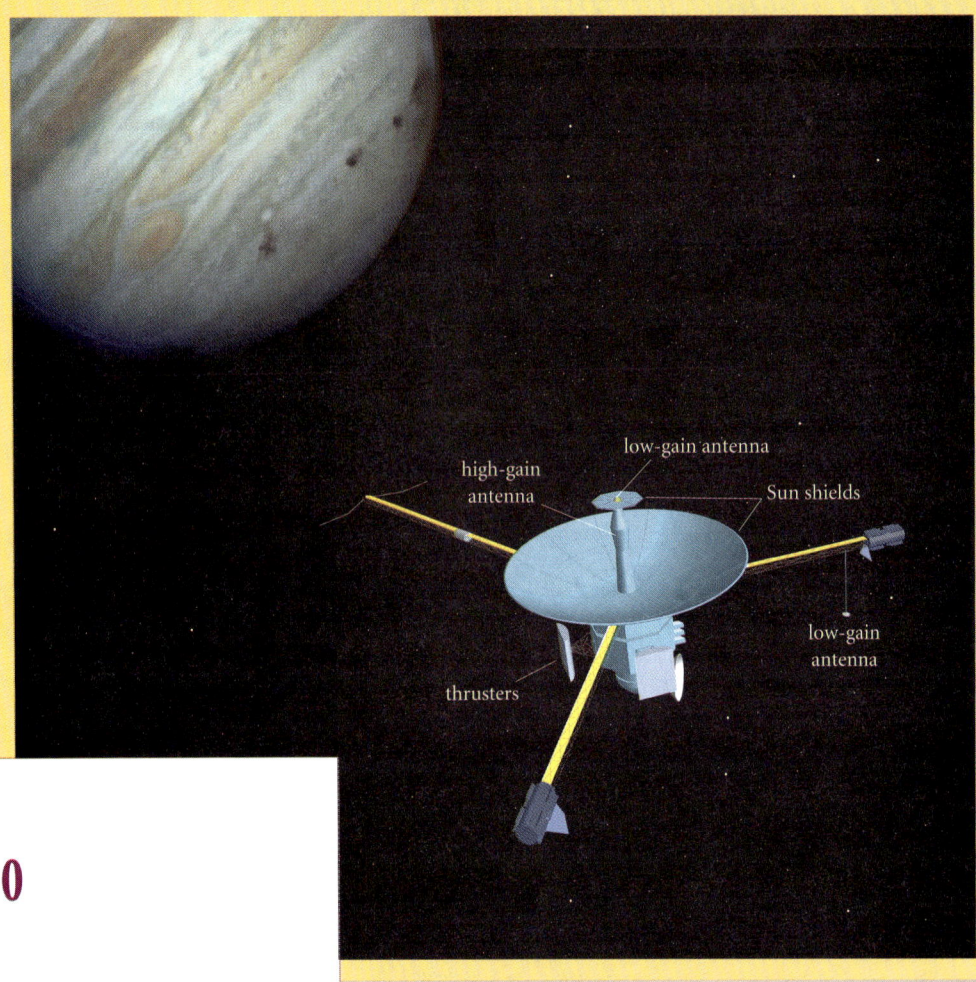

The Galileo spacecraft reached Jupiter in December, 1995. Once there, it entered an orbit around the planet and launched a probe into the Jovian atmosphere. Rockets sufficiently powerful to launch Galileo directly to Jupiter are unsafe as payload for the space shuttle. Thus Galileo used gravitational collisions with Venus and Earth to gain the energy it needed to reach Jupiter.

A booster rocket powerful enough to launch the *Galileo* spacecraft directly to Jupiter was too dangerous to carry on the space shuttle. So, NASA launched *Galileo* on a path first inward toward Venus, then twice around Earth (■ Figure 10.1), using gravitational collisions to provide the energy needed to get to Jupiter. Gravitational collisions? The spacecraft interacting with a planet has much in common with a baseball hitting a bat, a wreck on the freeway, or two high-energy protons colliding in an accelerator. Our goal in this chapter is to study the common features of these collision processes.

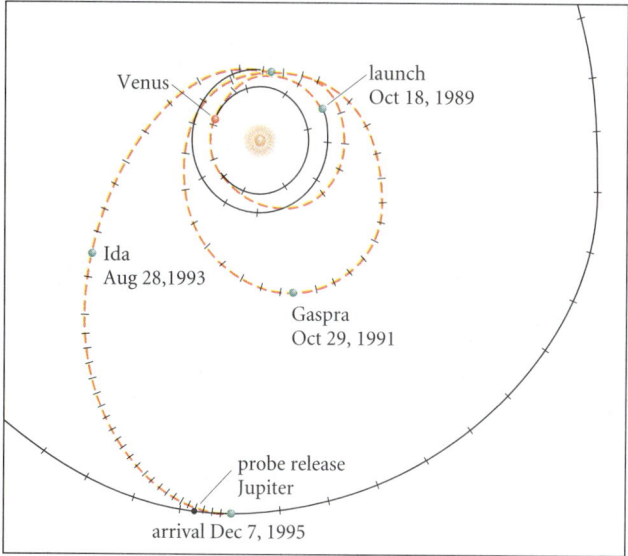

■ **Figure 10.1**
The planned track of *Galileo* en route to Jupiter weaves intricately among the orbits of the inner planets as it intersects the orbits of Venus, Earth, and the asteroids.

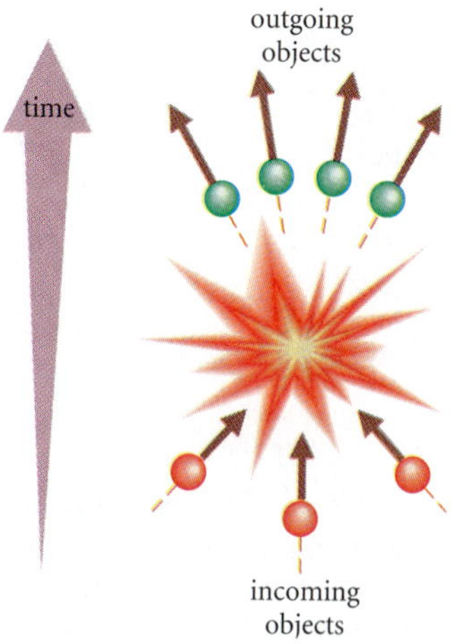

■ **Figure 10.2**
A *collision* occurs in three phases. A set of incoming particles approach each other freely, until they are close enough to interact. The encounter phase occurs in a relatively small region, takes place relatively quickly, and involves relatively large forces among the interacting particles. A set of particles (possibly different) emerges from the encounter, once again moving freely. The outgoing particles have the same total energy, linear momentum, and angular momentum as the incoming particles.

10.1 What Is a Collision?

A *collision* occurs when two or more objects, say a baseball and a window pane, initially far apart and not interacting with each other, come close together and exert large forces on each other, thus exchanging energy and momentum. Whatever objects exist after the interaction then move apart and no longer exert forces on each other. A collision thus involves an incoming set of objects, an encounter, and an outgoing set of objects (■ Figure 10.2). Sometimes the outgoing objects are not the same as the incoming objects. With an incoming baseball and window pane, the outgoing objects will likely be a baseball and a large number of dangerous glass fragments!

The baseball spends several seconds on its trajectory from the bat to the window and crashes through the window in about 10^{-2} s. The ball and glass spend another second or two scattering about the room. The encounter phase of a collision typically occurs much more rapidly than the incoming or outgoing phase. It involves forces that are very large but act only for a very short time, and in a complicated way that we may not know how to describe fully or accurately. As a result, collisions may be studied from two distinct points of view. Often the details of the encounter are uninteresting or unknown and we may study a collision using the conservation laws to relate a system's incoming and outgoing states. This is the view of the police officer who describes an accident as two incoming cars and an outgoing pile of wreckage. The opposite point of view is that of the safety engineer, who cares how the cars crumple and who wants to design them to protect the passengers.

The study of collisions traditionally begins by asking what we can learn from the conservation laws while ignoring the structure of the colliding particles and their interaction. In real collisions, one can rarely ignore the interaction completely; it is often the whole focus of a collision experiment. The conceptual models and examples we discuss toward the end of the chapter serve as an introduction to the art of working with both points of view.

Unless, of course, there is only one object after the collision.

Compare the discussion of impulse in §6.1.

10.2 COLLISIONS BETWEEN TWO PARTICLES

10.2.1 Elastic Collisions

SEE §9.3.4 FOR THE DEFINITION OF INTERNAL ENERGY.

In the encounter phase of any mechanical collision (Figure 10.2), the linear momentum, energy, and angular momentum of the system of colliding objects are all conserved. The individual objects may have complex structures, and substantial changes in their internal energy and angular momentum can occur. From the point of view that ignores internal changes, we only notice changes in the kinetic energy of the colliding *particles*. Since total energy is conserved, a change in one type, say thermal or elastic energy, produces a corresponding change in another kind, here kinetic energy. Thus it is useful to classify collisions according to whether or not internal energy changes. We shall first consider elastic (springy) collisions.

THE DETAILS OF THE INTERACTION DETERMINE WHICH COLLISIONS ARE ELASTIC. SEE §10.3. CONTRAST HARD BALLS (ELASTIC) WITH BLOBS OF PUTTY (INELASTIC).

> An *elastic* collision is one in which no change occurs in the internal energy of the colliding objects.

Each object's total energy is the sum of its internal energy and its kinetic energy, $\frac{1}{2}Mv_{CM}^2$, associated with the motion of its CM; thus:

> In an *elastic* collision, the total kinetic energy of the system of colliding objects, viewed as particles, is unchanged.

We begin by studying an elastic collision between two particles that are constrained to move along a single, straight line (■ Figure 10.3). Both the linear momentum and the kinetic energy of the system are conserved, so we can analyze the collision with the usual method for conservation law problems. We choose the x-axis to be along the direction of motion so that the velocities have only x-components. We use the symbol v for the x-components of the velocities before the collision, and u for the x-components of the velocities afterward.

	BEFORE	AFTER
x-component of linear momentum:	$P_x = m_1 v_1 + m_2 v_2.$	$P_x = m_1 u_1 + m_2 u_2.$
Kinetic energy:	$K = \frac{1}{2}m_1 v_1^2 + \frac{1}{2}m_2 v_2^2.$	$K = \frac{1}{2}m_1 u_1^2 + \frac{1}{2}m_2 u_2^2.$

Set quantities equal:

$$m_1 v_1 + m_2 v_2 = m_1 u_1 + m_2 u_2. \tag{i}$$

$$\tfrac{1}{2}m_1 v_1^2 + \tfrac{1}{2}m_2 v_2^2 = \tfrac{1}{2}m_1 u_1^2 + \tfrac{1}{2}m_2 u_2^2. \tag{ii}$$

We want to find u_1 and u_2 in terms of v_1 and v_2, so we solve eqn. (i) for u_2:

$$u_2 = v_2 + \left(\frac{m_1}{m_2}\right)(v_1 - u_1). \tag{iii}$$

Next multiply eqn. (ii) by 2 and rearrange:

$$m_1(v_1^2 - u_1^2) = m_2(u_2^2 - v_2^2),$$

$$m_1(v_1 - u_1)(v_1 + u_1) = m_2(u_2 - v_2)(u_2 + v_2). \tag{iv}$$

Substituting eqn. (iii) into eqn. (iv) gives:

$$m_1(v_1 - u_1)(v_1 + u_1) = m_2\left(\frac{m_1}{m_2}\right)(v_1 - u_1)(u_2 + v_2). \tag{v}$$

The masses cancel out, leaving a relation among the velocities.

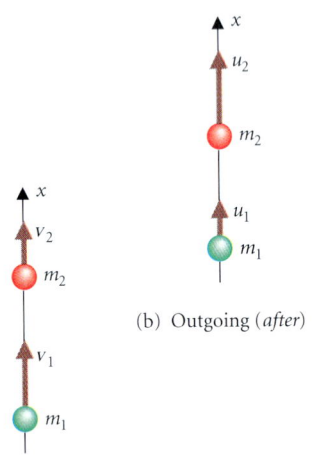

■ FIGURE 10.3
One-dimensional elastic collision.
(a) A particle of mass m_1 and velocity $v_1 \hat{\mathbf{i}}$ collides with another particle of mass m_2 with velocity of $v_2 \hat{\mathbf{i}}$. (b) The outgoing velocities of the two particles are $u_1 \hat{\mathbf{i}}$ and $u_2 \hat{\mathbf{i}}$. Linear momentum is conserved in all collisions; kinetic energy of the particles is unchanged in an elastic collision. The outgoing velocities in such a collision are determined by the ratio of the particles' masses.

One possible solution is: $v_1 - u_1 = 0 \Rightarrow v_1 = u_1$.

Then from eqn. (iii), it follows that $v_2 = u_2$. In this solution, neither particle's velocity changes; the particles do not interact at all. The solution we want has $v_1 \neq u_1$, so we may cancel the factor $(v_1 - u_1)$ from eqn. (v) to obtain:

$$v_1 + u_1 = v_2 + u_2. \qquad (vi)$$

Substituting eqn. (iii) into eqn. (vi), we have:

$$v_1 + u_1 = v_2 + v_2 + \left(\frac{m_1}{m_2}\right)(v_1 - u_1).$$

Solve for u_1: $u_1(1 + m_1/m_2) = 2v_2 + v_1(m_1/m_2 - 1).$

$$u_1 = \frac{2m_2 v_2 + v_1(m_1 - m_2)}{m_1 + m_2}. \qquad (10.1)$$

SPECIAL CASE: ONE-DIMENSIONAL.

Finally, we substitute eqn. (10.1) into eqn. (vi) and solve for u_2:

$$u_2 = v_1 - v_2 + \frac{2m_2 v_2 + v_1(m_1 - m_2)}{m_1 + m_2}$$

$$= \frac{2m_1 v_1 + v_2(m_2 - m_1)}{m_1 + m_2}. \qquad (10.2)$$

SPECIAL CASE: ONE-DIMENSIONAL.

Notice the symmetry between eqns. (10.1) and (10.2). It doesn't matter which object we label 1 and which we label 2, so such symmetry is required if the result is correct.

In the special case where the second object is initially at rest, $v_2 = 0$, we find:

$$u_1 = \frac{v_1(m_1 - m_2)}{m_1 + m_2} \quad \text{and} \quad u_2 = \frac{2m_1 v_1}{m_1 + m_2}. \qquad (10.3)$$

EXTRA SPECIAL CASE: ONE-DIMENSIONAL, $v_2 = 0$.

Thus v_1 and u_1 have opposite signs if $m_1 < m_2$, and the same sign if $m_1 > m_2$. Object 1 returns along its original path if it collides with a more massive object, but proceeds in the same direction if it collides with a less massive object. If $m_1 = m_2$, then $u_1 = 0$ and $u_2 = v_1$; the objects interchange roles. In all cases, u_2 has the same sign as v_1, and the second object is knocked forward in the direction of the original velocity v_1.

EXAMPLE 10.1 ♦ In bocci, a small target ball is thrown out and the players attempt to roll out larger balls so that they come to rest as close as possible to the target. In such a game a player's ball of mass $m_p = 0.36$ kg hits the target (mass $m_t = 0.12$ kg) with a speed of 2.5 m/s. If the collision is elastic and all incoming and outgoing velocities lie along the same line, find the final velocities of the two balls.

"INCOMING" REFERS TO THE SYSTEM BEFORE THE ENCOUNTER; AN INCOMING PARTICLE CAN BE AT REST.

MODEL The system of incoming particles is the two balls. The collision is one-dimensional and we are told it is elastic, so the total kinetic energy remains constant. We choose the x-axis to be along the direction of the initial velocity and solve for the x-components of the final velocities.

SETUP Since one ball is initially at rest, we call it ball 2 and use eqns. (10.3). Then $m_1 = m_p$ and $m_2 = m_t$.

SOLVE
$$u_p \equiv u_1 = \frac{v_1(m_p - m_t)}{m_p + m_t} = (2.5 \text{ m/s}) \left(\frac{0.36 \text{ kg} - 0.12 \text{ kg}}{0.36 \text{ kg} + 0.12 \text{ kg}}\right) = 1.2 \text{ m/s}.$$

$$u_t \equiv u_2 = \frac{2m_p v_1}{m_p + m_t} = (2.5 \text{ m/s}) \left(\frac{2(0.36 \text{ kg})}{0.36 \text{ kg} + 0.12 \text{ kg}}\right) = 3.8 \text{ m/s}.$$

ANALYZE After the collision, the target ball, with one-third the mass of the player's ball, moves three times faster. Both velocity components are positive, so both balls move in the same direction that ball 1 traveled.

There is an external force, friction, acting on the system. It is much smaller than the normal forces between the balls, so it can safely be ignored during the collision itself. After the collision, friction is essential if the balls are to roll and not slide. We stop our solution immediately after the collision and leave rolling to another day (Chapter 12). ∎

EXERCISE 10.1 ♦ What happens if instead the ball hits another player's ball of equal mass?

The same conservation principles apply to elastic collisions when the particles are free to move in two or three dimensions. We obtain a separate relation from each component of the momentum conservation equation.

EXAMPLE 10.2 ♦♦ Curling stones, all of the same mass, are used on the surface of a frozen lake to play a game of ice billiards. Sven needs to slide stone 1 so as to knock stone 2 into a net at the corner of the playing area (■ Figure 10.4). Sven projects stone 1 with initial speed $v_0 = 1.3$ m/s. Stone 2 is initially at rest. If he is successful, what is the speed of stone 2 as it enters the pocket? (Assume the stones collide elastically.)

MODEL The incoming particles (the two stones) are free to slide on the frictionless ice surface. Sven wants to knock stone 2 at an angle to the direction of stone 1's incident velocity, so the problem is two-dimensional.

SETUP We use the standard pattern for conservation of energy and momentum. Choose the x-axis to be along the direction of stone 1's initial velocity and the y-axis as shown in Figure 10.4. We are given stone 1's initial velocity and the direction of stone 2's outgoing velocity. ■ Figure 10.5 illustrates the notation for the outgoing velocities. We want to find u_2 in terms of the known speed v_0.

	BEFORE	**AFTER**
Kinetic energy:	$E = \tfrac{1}{2}mv_0^2 + 0.$	$E = \tfrac{1}{2}mu_1^2 + \tfrac{1}{2}mu_2^2.$
x-component of momentum:	$P_x = mv_0 + 0.$	$P_x = mu_1 \cos\theta + mu_2 \cos 45°.$
y-component of momentum:	$P_y = 0 + 0.$	$P_y = -mu_1 \sin\theta + mu_2 \sin 45°.$

Set quantities equal, and divide out the common factor m:

Energy: $\qquad v_0^2 = u_1^2 + u_2^2.$ (i)

x-component of momentum: $\qquad v_0 = u_1 \cos\theta + u_2/\sqrt{2}.$ (ii)

y-component of momentum: $\qquad 0 = -u_1 \sin\theta + u_2/\sqrt{2}.$ (iii)

SOLVE From eqn. (iii):

$$u_2 = (\sqrt{2})u_1 \sin\theta. \qquad (iv)$$

Then, substituting eqn. (iv) into eqn. (i),

$$v_0^2 = u_1^2(1 + 2\sin^2\theta), \qquad (v)$$

and eqn. (iv) into eqn. (ii),

$$v_0 = u_1(\cos\theta + \sin\theta). \qquad (vi)$$

Now, equating v_0^2 in eqns. (v) and (vi),

$$(1 + 2\sin^2\theta) = (\cos\theta + \sin\theta)^2 = 1 + 2\sin\theta\cos\theta.$$

(*Remember:* $\cos^2\theta + \sin^2\theta = 1$.) Thus $\sin\theta = \cos\theta$, so $\theta = 45°$. It follows from eqns. (iv) and (v) that:

$$u_1 = u_2 = v_0/\sqrt{2}.$$

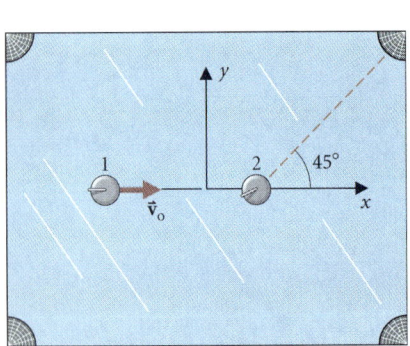

FIGURE 10.4
In a game of ice billiards Sven must launch curling stone 1 so that it strikes stone 2 and knocks it into a pocket at the corner of the playing area.

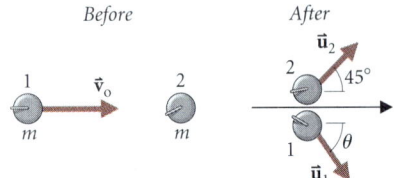

FIGURE 10.5
Velocities of the curling stones before and after the collision. We want to find the outgoing speed u_2 of stone 2.

ANALYZE In this example we needed to know that Sven succeeded; that is, we needed the direction of one outgoing velocity. That much information is in general both necessary and sufficient for a solution, as we show in the next section. In the one-dimensional case, we also knew the direction of one outgoing velocity. In that case, conservation of the y-component of momentum guarantees that if one outgoing object moves along the original line, so does the other.

The two stones' outgoing velocities are at right angles. The following example explores this result in more detail.

EXAMPLE 10.3 ♦♦ Show that if a particle collides elastically with an *identical* particle at rest, the two outgoing velocities are always perpendicular.

IT IS POSSIBLE TO FIND AN EXPRESSION FOR THE ANGLE BETWEEN OUTGOING VELOCITIES IN A GENERAL CASE. THE RESULT DEPENDS ON THE MASSES OF THE TWO PARTICLES. SEE PROBLEM 34.

MODEL To obtain a general proof, independent of special initial conditions, it is most efficient to express conservation of momentum in vector notation rather than with components. Otherwise, the algebraic approach is similar to that in Example 10.2.

SETUP Let $\vec{v}_o$ be the initial velocity of particle 1 and $\vec{u}_1$ and $\vec{u}_2$ be the two particles' outgoing velocities.

Conservation of energy: $\frac{1}{2}mv_o^2 = \frac{1}{2}mu_1^2 + \frac{1}{2}mu_2^2$.

Conservation of momentum: $m\vec{v}_o = m\vec{u}_1 + m\vec{u}_2$.

Each of these equations gives us an expression for v_o^2:

$$v_o^2 = u_1^2 + u_2^2,$$

and

$$v_o^2 = (\vec{u}_1 + \vec{u}_2) \cdot (\vec{u}_1 + \vec{u}_2) = u_1^2 + u_2^2 + 2\vec{u}_1 \cdot \vec{u}_2.$$

Subtracting the two equations, we find $\vec{u}_1 \cdot \vec{u}_2 = 0$. Thus $|\vec{u}_1||\vec{u}_2|\cos\theta = 0$, so the two velocities are perpendicular, unless one of them is zero.

ANALYZE One velocity is zero only if the particles miss each other and there is no collision at all or if they collide exactly head on as in Exercise 10.1. This example illustrates the power of vector algebra.

10.2.2 Elastic Collisions in the CM Reference Frame

The incoming velocities completely determine the outcome of a one-dimensional elastic collision. The same is not true for two-dimensional collisions, but the result of Example 10.3 is very suggestive: there the angle between the outgoing velocities is fixed, while the absolute direction of either velocity is not. Does any similar result hold in general?

We found in Chapter 9 that a system's behavior is most clearly described in the system's center of mass reference frame. This is particularly true for two-particle collisions (■ Figure 10.6). The particles' total momentum in the CM frame is zero, so their incoming momentum vectors are equal and opposite. The particles' outgoing momentum vectors are also equal and opposite. As we show in the next example, conservation of kinetic energy in an elastic collision ensures that the magnitude of *each* particle's momentum in the CM frame is unchanged by the collision.

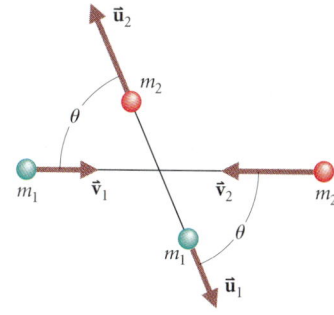

■ **FIGURE 10.6**
Two-particle elastic collision in the CM reference frame of the particles. Because the total momentum of a system in its CM frame is always zero and kinetic energy is conserved, each particle has the same speed after the collision as it did before: $|\vec{u}_2| = |\vec{v}_2|$; $|\vec{u}_1| = |\vec{v}_1|$. In a two-dimensional elastic collision, only the angle θ between incoming and outgoing velocities is not fixed by the incoming velocities.

EXAMPLE 10.4 ♦ Show that the magnitude of each particle's momentum in the CM reference frame is unchanged when the two particles collide elastically.

MODEL Once again we use the conservation laws for energy and momentum.

SETUP It is convenient to express the kinetic energy of each particle in terms of its momentum: $K = \frac{1}{2}mv^2 = \frac{1}{2}(mv)^2/m = p^2/(2m)$. The total momentum in the CM frame is zero, so the particles have equal and opposite momenta both before and after the collision;

$$|\vec{p}_{1,i}| = |\vec{p}_{2,i}| \equiv p_i \quad \text{and} \quad |\vec{p}_{1,f}| = |\vec{p}_{2,f}| \equiv p_f. \tag{i}$$

The total kinetic energy K of the system remains constant.

$$\text{Before: } K = \frac{p_{1,i}^2}{2m_1} + \frac{p_{2,i}^2}{2m_2} = p_i^2\left(\frac{1}{2m_1} + \frac{1}{2m_2}\right). \quad \text{(ii)}$$

$$\text{After: } K = \frac{p_{1,f}^2}{2m_1} + \frac{p_{2,f}^2}{2m_2} = p_f^2\left(\frac{1}{2m_1} + \frac{1}{2m_2}\right). \quad \text{(iii)}$$

SOLVE Equation (i) was used to simplify eqns. (ii) and (iii). Equating K in eqns. (ii) and (iii) gives $p_i = p_f$, as required.

ANALYZE Since the magnitude of each particle's momentum does not change, the only unknown is its direction. The direction is usually determined by some feature of the interaction and cannot be found using the conservation laws alone. The combined use of the CM reference frame and momentum in the expression for kinetic energy greatly streamlines this calculation.

SEE FIGURE 10.10.

EXERCISE 10.2 ♦♦ Derive eqns. (10.1) and (10.2) by analyzing a general one-dimensional collision in the two particles' CM reference frame.

10.2.3 Inelastic Collisions

An elastic collision is an ideal model. Even when things like billiard balls interact, a small amount of kinetic energy is converted to other forms. The opposite ideal case is a perfectly inelastic collision, in which the colliding objects stick together. The CM reference frame allows us to give a more precise definition:

WE DESCRIBED EXAMPLES OF PERFECTLY INELASTIC COLLISIONS WHEN DISCUSSING CONSERVATION OF MOMENTUM IN CHAPTER 6.

> In a *perfectly inelastic* collision, all the kinetic energy of the incoming objects in their CM reference frame is transformed to internal energy within the outgoing objects.

After a perfectly inelastic collision, the outgoing objects, viewed as particles, have no kinetic energy in the CM frame. They are all stationary in the CM frame—*stuck together*. In the lab frame the system still has its original, nonzero linear momentum and thus has some kinetic energy. The outgoing objects move together with their CM velocity.

NOTE: PERFECTLY INELASTIC DOESN'T MEAN THAT ALL THE KINETIC ENERGY MEASURED IN THE LAB FRAME IS TRANSFORMED. IT MEANS THAT AS MUCH AS POSSIBLE IS TRANSFORMED. ONLY IN THE CM FRAME IS IT EASY TO DETERMINE THAT MAXIMUM POSSIBLE AMOUNT—100%.

EXAMPLE 10.5 ♦♦ During an ice show, a bean bag of mass 1.4 kg is shot from a cannon with a horizontal velocity of 17.3 m/s. An amply padded clown of mass 62.0 kg catches the bag. If the clown is originally sitting at rest on the ice, what is her speed after she catches the bag?

MODEL The important concept is conservation of momentum. Some kinetic energy is converted to internal energy, so we cannot use conservation of energy. But we do know that both objects—bag and clown—have the same velocity after the collision. The clown remains on the ice, so we may ignore vertical motion.

SETUP We choose the x-axis to be along the direction of the bag's velocity as it leaves the cannon. The bag's horizontal velocity remains constant until the clown catches it.

BEFORE   AFTER

x-component of momentum: $m_b v_b + 0 = (m_b + m_c) v_f.$

SOLVE
$$v_f = v_b \frac{m_b}{m_b + m_c} = (17.3 \text{ m/s})\left(\frac{1.4 \text{ kg}}{1.4 \text{ kg} + 62.0 \text{ kg}}\right) = 0.38 \text{ m/s}.$$

ANALYZE The initial kinetic energy is $\frac{1}{2} m_b v_b^2 = 210$ J, while afterward it is $\frac{1}{2}(m_b + m_c) v_f^2 = 4.6$ J. Almost all the initial kinetic energy is converted to other forms in the collision.

The CM velocity initially (eqn. 9.17) is:

$$v_{CM} = \frac{m_b v_b + 0}{m_b + m_c} = v_f.$$

After the collision, both objects move at the CM velocity.

EXAMPLE 10.6 ♦♦♦ A limousine of mass $M = 2500$ kg runs into a compact car of mass $m = 1500$ kg parked illegally in a commute hours traffic lane. The sliding wreckage leaves marks on the pavement for a distance of $d = 5.2$ m. If the coefficient of friction between wreckage and pavement is $\mu_k = 0.34$, what was the limousine's initial speed (■ Figure 10.7)?

MODEL The incident occurs in two steps: first the cars collide and then the sliding wreckage is brought to rest by friction. We solve the problem by analyzing the two parts in reverse order. The marks on the pavement give the distance required for the frictional force of the pavement on the wreckage to bring it to rest. Together with the coefficient of friction, this information allows us to determine the speed u_1 of the wreckage immediately after the collision. We model the collision itself as perfectly inelastic and sufficiently rapid that we can ignore the effects of external forces during the collision. We also suppose that the distance the wreckage moves *during* the collision is negligible compared with the 5.2-m measured distance. From what we know about the masses of the two cars and the computed speed of the outgoing wreckage, we can find the initial speed v_1 of the limousine.

SETUP We choose the x-axis to be parallel to the direction of the limousine's incoming velocity. Figure 10.7d is the free-body diagram for the wreckage. We apply Newton's second law in the usual way:

$$\sum F_y = (m + M)a_y. \qquad \sum F_x = (m + M)a_x.$$
$$n - (m + M)g = 0. \qquad -f_k = (m + M)a_x.$$

The magnitude of the friction force is $f_k = \mu_k n$.
Combining these relations gives us $-\mu_k(m + M)g = (m + M)a_x$.

After the collision, the wreckage decelerates uniformly with $a_x = -\mu_k g$. We use eqn. (2.13) relating the acceleration a_x, distance $\Delta x = d$, and change in speed squared:

$$v_f^2 - v_i^2 = 2a_x \Delta x = -2\mu_k g d,$$

where v_f is zero and $v_i = u_1$ is the wreckage speed after the collision. Thus:

$$u_1 = \sqrt{2\mu_k g d}. \qquad (i)$$

Momentum is conserved in the perfectly inelastic collision, so:

$$P_x(\text{after}) = (M + m)u_1 = Mv_1 + 0 = P_x(\text{before}). \qquad (ii)$$

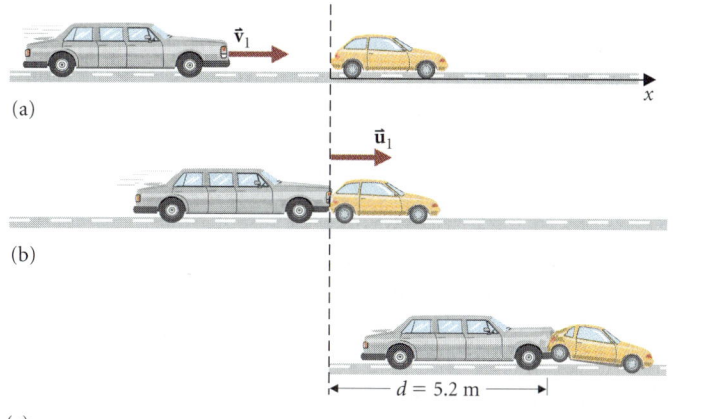

■ FIGURE 10.7
A limousine runs into a parked compact car. The wreckage slides for a distance $d = 5.2$ m. What was the initial speed of the limousine? (a) Limousine approaches with speed v_1. (b) Immediately after the collision, the wreckage has speed u_1. (c) The wreckage comes to rest after sliding a distance $d = 5.2$ m. (d) Free-body diagram for the wreckage.

SOLVE Substituting relation (i) for u_1 into (ii), the initial speed of the limousine is:

$$v_1 = \frac{(m+M)u_1}{M} = \frac{m+M}{M}\sqrt{2\mu_k g d}$$

$$= \frac{(4.0 \times 10^3 \text{ kg})\sqrt{2(0.34)(9.8 \text{ m/s}^2)(5.2 \text{ m})}}{2.5 \times 10^3 \text{ kg}}$$

$$= 9.4 \text{ m/s}.$$

ANALYZE Because of the assumptions in our model, the result is probably accurate only to a single significant figure. That's good enough for a traffic officer to know that the limousine driver was not speeding (9.4 m/s = 21 mph). One of the most practical uses of our results for elastic and inelastic collisions is as models that are *good enough* for rough calculations in sports or accident analysis.

Elastic and perfectly inelastic collisions are ideal limiting cases of how objects behave during the encounter phase. Better models require more information about how the colliding objects actually behave. Real balls provide an example only slightly more complicated than the ideal cases. A ball dropped on a hard surface rebounds with its energy reduced by a fixed fraction that is determined by measuring how high the ball bounces. A collision between two balls is therefore *inelastic*, but not perfectly inelastic. We can best make use of the experimental information by working in the CM reference frame.

EXAMPLE 10.7 ◆◆ A 0.100-kg rubber ball dropped onto a rubber mat from a height of 1.00 m bounces to a height of 0.70 m. If one such ball moving horizontally at 5.00 m/s collides with an identical, stationary ball, what are the velocities of the two balls after the collision? (Assume that the collision is one-dimensional.) How much thermal energy is produced?

MODEL Dropping the ball on the mat tells us how much of the ball's kinetic energy is transformed to internal energy when it is stopped in a head-on collision with another rubber surface. Before hitting the mat, its kinetic energy equals its earlier gravitational energy 1.00 m above the mat. After the bounce, it only has enough energy to rise 70 cm, where its potential energy is only 70% of the original value. Three-tenths of its energy is transformed when it is brought to rest and rebounds.

In the lab frame, the incoming ball is not stopped by the collision, so we do not know how to apply the given information. But in the CM frame (■ Figure 10.8), *both* balls stop and rebound; each dissipates 30% of its kinetic energy.

SETUP We choose the *x*-axis to be along the direction of the incident ball's velocity. The velocity of the CM before the collision has *x*-component (eqn. 9.17):

$$v_{\text{CM}} = \frac{m_1 v_1 + m_2 v_2}{m_1 + m_2} = \frac{(0.100 \text{ kg})(5.00 \text{ m/s}) + 0}{2(0.100 \text{ kg})} = 2.50 \text{ m/s}.$$

In the CM reference frame (Figure 10.8), the balls approach each other at $v = 2.50$ m/s along the *x*-axis and rebound with lesser but equal speeds u in opposite directions (cf.

■ **FIGURE 10.8**
Two balls undergo an inelastic collision. Experimental information is available about the energy dissipated by the balls when they are brought to rest and rebound. (a) In the lab frame, neither ball ever comes to rest, and it is unclear how to apply the experimental data. (b) In the balls' CM reference frame, they are both brought to rest, and the data are immediately applicable.

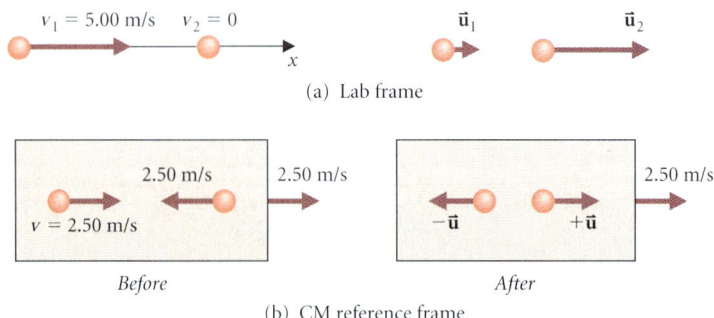

Example 10.4. The total momentum in the CM frame is zero.) The final kinetic energy of the system, $K_f = 2(\frac{1}{2}mu^2)$, is 70% of the initial energy:

$$2(\tfrac{1}{2}mu^2) = K_f = 0.70 K_i = (0.70)(2)(\tfrac{1}{2}mv^2).$$

SOLVE Thus each ball has speed $u = v\sqrt{0.70} = 2.1$ m/s in the CM frame after the collision. Since the balls move along the x-axis after the collision, the balls' velocities in the lab frame are:

$$\vec{u}_2 = [(2.1 + 2.50)\text{ m/s}]\hat{\imath} = (4.6 \text{ m/s})\hat{\imath}$$

and

$$\vec{u}_1 = [(-2.1 + 2.50)\text{ m/s}]\hat{\imath} = (0.4 \text{ m/s})\hat{\imath}.$$

The thermal energy produced equals the kinetic energy lost:

$$\Delta U_{\text{thermal}} = K_f - K_i = (0.30)(2)(\tfrac{1}{2}mv^2)$$
$$= (0.30)(0.10 \text{ kg})(2.50 \text{ m/s})^2 = 0.19 \text{ J}.$$

ANALYZE The thermal energy dissipated is only 15% of the initial kinetic energy $\frac{1}{2}(0.10 \text{ kg})(5.0 \text{ m/s})^2 = 1.25$ J measured in the *lab* frame. ∎

EXERCISE 10.3 ♦♦ A one-dimensional, partially inelastic collision is sometimes described by the coefficient of restitution ϵ:

$$\epsilon = \frac{\text{relative speed of separating objects}}{\text{relative speed of approaching objects}}$$
$$= \frac{|u_2 - u_1|}{|v_2 - v_1|}. \tag{10.4}$$

THE COEFFICIENT OF RESTITUTION DEPENDS ON THE CHARACTERISTICS OF THE PARTICULAR PAIR OF COLLIDING OBJECTS BUT NOT ON THE MAGNITUDE OF THEIR RELATIVE SPEED.

Values of ϵ range between 1 (elastic collision) and 0 (perfectly inelastic collision). Show that the coefficient of restitution for the ball hitting rubber in Example 10.7 equals $\sqrt{h_2/h_1}$, where h_1 is the 1.00-m height from which it is dropped and $h_2 = 0.70$ m is the height to which it rebounds. Verify that the incoming and outgoing velocities of the two colliding balls satisfy eqn. (10.4) in both the lab and CM frames. ∎

10.3 MODELS FOR ELASTIC AND INELASTIC COLLISIONS

The descriptions *elastic* and *inelastic* describe the outcome of a collision without providing any details of the interaction between the particles. When objects collide, they exert forces on each other that depend on the objects' detailed structure. The nature of those forces determine the changes in energy that occur. In this section we present examples that provide conceptual models for the different classes of collision. We begin with a one-dimensional elastic collision.

"ELASTIC" AND "INELASTIC" ARE PHENOMENOLOGICAL CLASSIFICATIONS— NAMES BUT NOT EXPLANATIONS—FOR PHENOMENA THAT OCCUR FREQUENTLY.

EXAMPLE 10.8 ♦♦ A block of mass m and initial speed v_0 collides with a spring attached to a stationary block of mass M. The spring has negligible mass (■ Figure 10.9). Describe the state of the system when the spring reaches maximum compression and when it has re-expanded.

■ **FIGURE 10.9**
A model for one-dimensional elastic collisions. In an elastic collision, the particles have to exert conservative forces on each other and thus store potential energy temporarily during their interaction. A spring provides a conceptual model of such an interaction. Here we show the state of the system in the three phases of the collision. (State a) A block approaches a stationary object. (State b) When the spring reaches maximum compression, both blocks move with the CM velocity; some of their kinetic energy has been transformed into elastic potential energy. (State c) After the spring re-expands, the system's energy is once again all in the form of kinetic energy. The collision is elastic.

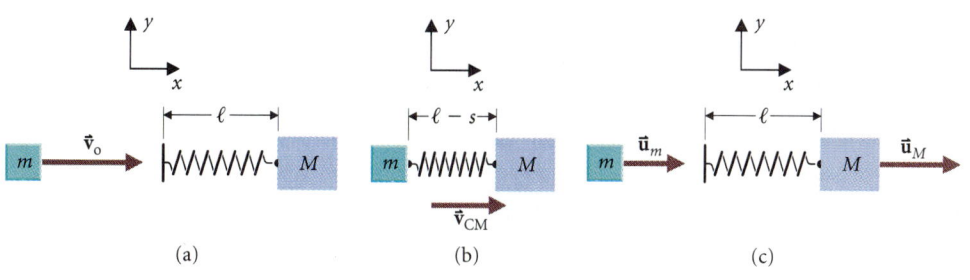

MODEL Only the conservative spring force acts on the blocks and so total mechanical energy is conserved. Thus this is a model of an elastic encounter. Linear momentum is also conserved. We ignore any distortion of the blocks themselves that might produce additional forces.

SETUP Three phases of the collision are shown in Figure 10.9.
1. The incoming block approaches the stationary object (state a).
2. As the block contacts the spring and compresses it, the resulting spring force slows the incoming block and accelerates the block-plus-spring object to the right. This diagram shows the system in state (b), when the spring has its maximum compression and both objects have the same velocity.
3. The spring re-expands, exerting a leftward force on one block and a rightward force on the other. Once the block on the left leaves the spring, forces are no longer exerted and the collision is over (state c).

We choose the x-axis to be along the direction of motion, as shown in the figures. Momentum is conserved throughout the encounter, and the CM velocity of the system remains constant. At the end of phase 2, state (b), both objects move at the CM velocity:

$$\begin{array}{ccc} & \textbf{STATE (A)} & \textbf{STATE (B)} \\ P_x = & mv_o & = (m+M)v_{CM}. \end{array}$$

SOLVE
$$v_{CM} = \frac{mv_o}{m+M}. \quad \text{(i)}$$

SETUP The kinetic energy of the two blocks moving together is less than the original energy of the smaller block. Some energy has been converted to elastic energy stored in the spring:

	STATE (A)	**STATE (B)**
Kinetic energy	$\frac{1}{2}mv_o^2$	$\frac{1}{2}(m+M)v_{CM}^2$
Elastic energy	0	$\frac{1}{2}ks^2$

Setting the totals equal gives:
$$\tfrac{1}{2}mv_o^2 = \tfrac{1}{2}(m+M)v_{CM}^2 + \tfrac{1}{2}ks^2.$$

Rearranging gives:
$$\tfrac{1}{2}ks^2 = \tfrac{1}{2}mv_o^2 - \tfrac{1}{2}(m+M)v_{CM}^2.$$

And, using eqn. (i):
$$\tfrac{1}{2}ks^2 = \tfrac{1}{2}mv_o^2\left[1 - \frac{m(m+M)}{(m+M)^2}\right]$$
$$= \tfrac{1}{2}\frac{mM}{m+M}v_o^2.$$

SOLVE So, the maximum spring compression is:
$$s = v_o\sqrt{\frac{mM}{k(m+M)}}.$$

SETUP After the spring re-expands, state (c), the two objects have velocities $u_m\hat{\imath}$ and $u_M\hat{\imath}$. We compare the linear momentum and energy in states (a) and (c). The spring is uncompressed in each state and there is no elastic energy.

	STATE (A)		**STATE (C)**	
Mass	m	M	m	M
x-component of momentum	mv_o	none	mu_m	Mu_M
Kinetic energy	$\tfrac{1}{2}mv_o^2$	none	$\tfrac{1}{2}mu_m^2$	$\tfrac{1}{2}Mu_M^2$

Set quantities equal:

Momentum $\quad mv_o = mu_m + Mu_M.$

Energy $\quad \tfrac{1}{2}mv_o^2 = \tfrac{1}{2}mu_m^2 + \tfrac{1}{2}Mu_M^2.$

SOLVE These equations are the same as those for the elastic collision of §10.2.1 and have the same solution. With $m_2 = M$ and $m_1 = m$ in eqn. (10.3), we have:

$$u_m = \frac{m - M}{M + m} v_o \quad \text{and} \quad u_M = \frac{2m}{m + M} v_o.$$

ANALYZE If $M > m$, then $u_m < 0$ and the incoming block returns to the left.

We do not pretend that a single-spring model describes the real structure of elastically colliding particles, but it does emphasize several points. For an elastic collision to occur, the particles have to exert conservative forces on each other—that is, have the ability to transform and store potential energy. Real systems also have limits, modeled here by the length of the spring. If the required compression s were greater than the spring length, the collision would instead be very inelastic. ∎

Billiard balls form a good *image* for the particles in an elastic collision, because their behavior is very nearly elastic. Billiard balls remain uniform spheres, undergoing no obvious change in their structure as they collide. The mass-and-spring system of the example is a better *model* for one-dimensional collisions, since it captures the idea that elastically colliding objects must have some way to exert conservative forces on each other.

We may model two-dimensional elastic collisions with billiard balls (■ Figure 10.10). Balls with finite radius can collide even though their centers follow paths that do not intersect. On contact, the balls exert impulsive normal forces on each other that give the direction of the momentum exchange in the collision. The direction of these normal forces is determined by the separation of the balls' incoming paths. This distance, called the *impact parameter*, determines the angle θ between the incoming and outgoing velocities in the CM frame.

NOTE THAT θ IS THE ONLY PARAMETER NOT DETERMINED BY THE CONSERVATION LAWS (CF. EXAMPLE 10.4).

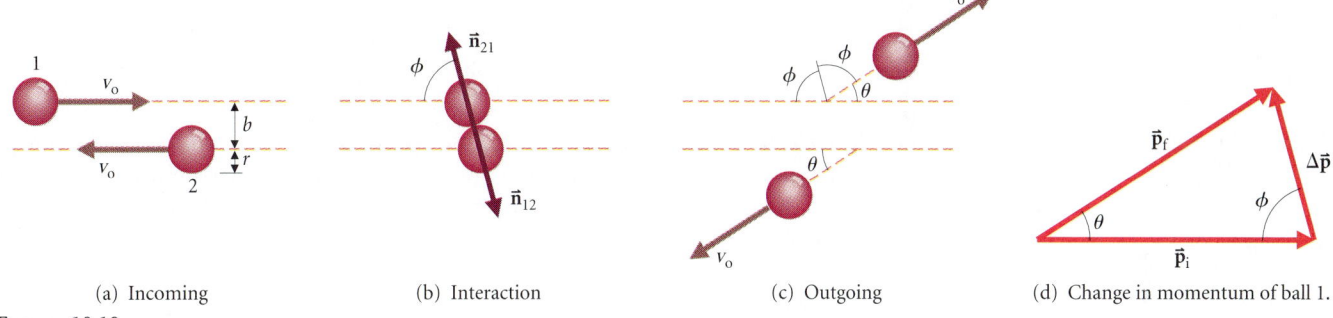

(a) Incoming   (b) Interaction   (c) Outgoing   (d) Change in momentum of ball 1.

■ **FIGURE 10.10**
Off-center collision of two billiard balls in the CM frame. (a) The centers of colliding billiard balls follow paths that are separated by a distance $b < 2r$, called the *impact parameter*. (b) The normal forces that the balls exert on each other deliver impulse at an angle ϕ to the balls' incoming velocities. (c) In an elastic collision, a ball's incoming velocity component along the line of the normal force is reversed, and its velocity component perpendicular to the normal force is unchanged. (d) Remember that $|p_f| = |p_i|$ in the CM frame, so the initial and final momenta of ball 1 form an isosceles triangle with apex angle $\theta = 180° - 2\phi$.

■ **EXERCISE 10.4** ♦ Show that $\sin \phi = b/(2r)$, where r is the radius of each ball. ∎

In the next example, a system of ideal billiard balls and springs provides a model for a class of inelastic collisions that excite a simple mode of energy storage. In the particle point of view, the ball-and-spring system is classed simply as a particle of mass $2m$. In those terms, this interaction is an inelastic collision in which half the incoming kinetic energy is transformed to internal energy.

■ **EXAMPLE 10.9** ♦♦ Two identical, ideal billiard balls A and B, initially at rest, are connected by a spring. A third billiard ball C makes a one-dimensional collision with this object (■ Figure 10.11). Each ball has mass m and the spring constant is k. Find the internal energy of the two-ball object after the collision and its kinetic energy when considered as a particle.

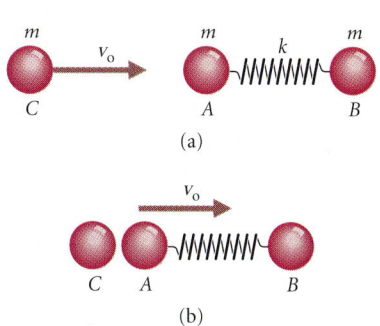

■ **FIGURE 10.11**
A model for one type of inelastic collision. The three balls A, B, and C each have the same mass. The collision of C with A is elastic: after their encounter C has delivered all of its energy and momentum to A. Part of A's kinetic energy is now internal energy of the A-plus-B system.

MODEL The incoming billiard ball C undergoes a rapid, elastic collision with ball A. This encounter is assumed to be so rapid that it is over before ball A is substantially displaced. Thus no compression of the spring occurs until after the collision is over, and ball B is not involved until later.

SETUP The collision of the two equal-mass balls A and C may be analyzed using the conservation laws, as in §10.1. In particular, we may use eqns. (10.3) with $m_1 = m_2$. Immediately after the collision (Figure 10.11b), ball C is at rest, and ball A moves to the right at speed v_o. The total energy of the two-ball object at that time is:

$$E_T = \tfrac{1}{2} m v_o^2.$$

Its CM moves in the x-direction with speed:

$$v_{CM} = \frac{m_A v_A + m_B v_B}{m_A + m_B} = \frac{v_o + 0}{2} = \frac{v_o}{2}.$$

The velocities of balls A and B in their CM frame have x-components (■ Figure 10.12):

$$v_{A,CM} = v_A - v_{CM} = v_o - v_o/2 = v_o/2,$$

and

$$v_{B,CM} = v_B - v_{CM} = 0 - v_o/2 = -v_o/2.$$

SOLVE The internal energy is the energy in the CM frame:

$$E_{int} = \tfrac{1}{2} m v_{A,CM}^2 + \tfrac{1}{2} m v_{B,CM}^2$$

$$= \tfrac{1}{2} m [(v_o/2)^2 + (v_o/2)^2] = m v_o^2/4.$$

Considered as a single particle AB, the combined object has kinetic energy:

$$K = \tfrac{1}{2}(m_A + m_B) v_{CM}^2 = m(v_o/2)^2 = m v_o^2/4.$$

ANALYZE After the collision, half of the energy is internal. As balls A and B approach each other, they compress the spring, and internal kinetic energy is transferred to internal elastic energy. The internal energy continues to oscillate between kinetic and elastic forms as the object vibrates. ∎

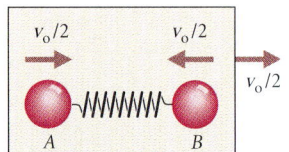

■ **FIGURE 10.12**
The combined object AB in its CM frame after the collision. The internal energy of the system oscillates between kinetic energy, as shown here, and potential energy when the two balls are at rest in this frame and the spring is compressed or extended.

EXERCISE 10.5 ♦♦ Calculate the coefficient of restitution for this collision.

Another model for an inelastic collision is illustrated in ■ Figure 10.13. The initial conditions are the same as in Example 10.8, except that clamps are designed to catch the incoming block and hold it to the spring. Thus the collision ends at state (b) in Figure 10.9, and the resulting object has internal energy stored in the compressed spring. Using the value for the maximum spring compression found in Example 10.8, the internal energy is:

$$U = \tfrac{1}{2} k s^2 = \tfrac{1}{2} \frac{m M v_o^2}{m + M}.$$

Since no external forces act, the total momentum, and hence the CM velocity, of the system is unchanged during the collision. The outgoing object moves with speed:

$$v_{CM} = \frac{m v_o + M(0)}{m + M} = \frac{m v_o}{m + M}.$$

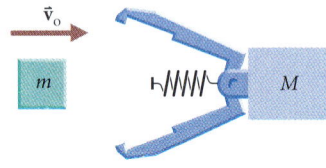

■ **FIGURE 10.13**
Model of a totally inelastic collision. The two particles are clamped together just as the spring reaches its maximum compression. The kinetic energy that "disappears" in the inelastic collision has been transformed to internal potential energy.

Seen from the perspective that ignores internal structure, this object models a collision in which two particles collide and stick together to form a single particle. The final particle has no kinetic energy in the CM frame. This is a model for a *perfectly inelastic* collision.

Neither of these models is, however, a good *image* for dissipation of energy as heat. The internal energy that results is stored in a very simple form. The essence of thermal energy is random distribution among many particles. So, the better a model represents thermal energy, the more complicated the model has to be, and the less useful it is for calculations. ∎ Figure 10.14 shows a model designed for lecture demonstrations.

The opposite image from the billiard ball is the (ideal!) putty ball, gooey and able to stick to anything. We shall exploit the putty ball as a good image for a perfectly inelastic particle. A more practical, though gruesome, example is an automobile (cf. Example 10.6).

EXERCISE 10.6 ♦♦ Joey throws a 0.050-kg pebble with a speed of 15 m/s at an empty 0.10-kg soda can resting on a tree limb and knocks it off. Later he discovers the pebble resting on the tree limb. How much energy was dissipated as heat in the collision between pebble and can? To which of the models in this section is this collision most similar? Explain your reasoning.

■ FIGURE 10.14
A device designed by Uri Ganiel for demonstrating totally inelastic collisions during a lecture. The frame of the cart supports several particles suspended on springs. When the cart undergoes a collision, the particles are set vibrating in a random manner, absorbing energy in a form that cannot be transformed back to kinetic energy of the cart as a whole. This randomness and irreversible behavior are the hallmarks of thermal energy. (See U. Ganiel, Elastic and inelastic collisions: A model, *The Physics Teacher*, **30**, Jan. 1992:18–19.)

✳ 10.4 SOME APPLICATIONS OF COLLISION THEORY
10.4.1 *Atomic and Subatomic Particles*

Collisions and disruption of particles occur frequently in the world of atomic and subatomic particles. The internal structure of such particles is very complex; purely elastic collisions occur, and so do inelastic encounters in which the emerging particles may have either less or *more* kinetic energy than the incident particles.

The first hint of this fascinating world of subatomic phenomena was the discovery in 1896 by Henri Becquerel that certain minerals spontaneously emit very energetic particles. We have since learned that Becquerel observed rapidly moving nuclei of helium atoms emitted when the nuclei of some larger atomic species break up. The classic example demonstrated by Marie Curie in 1898 is the decay of a nucleus of the element radium into a helium nucleus and the nucleus of a radon atom (∎ Figure 10.15):

$$\text{radium} \rightarrow \text{helium} + \text{radon}.$$

BECQUEREL CLASSIFIED THIS EFFECT AS "α-RADIATION." THE EMITTED He NUCLEI ARE CALLED α-PARTICLES.

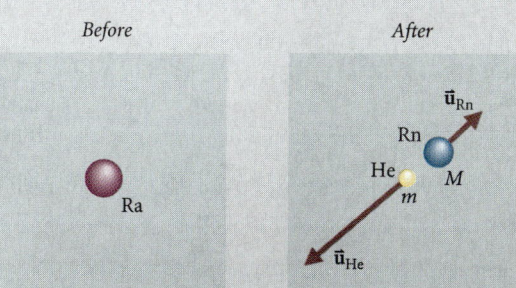

■ FIGURE 10.15
A radium nucleus decays into a radon nucleus and an α-particle (helium nucleus).

EXAMPLE 10.10 ♦♦ In a crude model of the radium nucleus, we imagine the helium and radon nuclei attached by a compressed spring. Decay occurs when the spring releases and the two pieces fly apart. Taking the mass of the helium nucleus as 4 units, the mass of a radon nucleus is 222 and that of radium is 226. What fraction of the outgoing kinetic energy does the helium nucleus receive?

MODEL To conserve momentum, initially zero, the helium and radon nuclei must move off in opposite directions. The initial potential energy of the particles in the nucleus, represented by the compressed spring, is transformed into kinetic energy in the outgoing state.

THE UNIT IS THE ATOMIC MASS UNIT, ABBREVIATED u. ITS NAME AND ITS VALUE ARE NOT IMPORTANT IN THIS PROBLEM.

SETUP The total momentum is:
$$0 = \vec{P}_i = \vec{P}_f = m\vec{u}_{He} + M\vec{u}_{Rn}.$$

Thus the outgoing speeds are related:
$$mu_{He} = Mu_{Rn}.$$

The outgoing kinetic energy is:
$$K = \tfrac{1}{2}mu_{He}^2 + \tfrac{1}{2}Mu_{Rn}^2.$$

SOLVE So, the fraction of the energy carried off by the helium is:
$$f = \frac{\tfrac{1}{2}mu_{He}^2}{\tfrac{1}{2}mu_{He}^2 + \tfrac{1}{2}Mu_{Rn}^2} = \frac{1}{1 + (M/m)(u_{Rn}/u_{He})^2} = \frac{1}{1 + (m/M)}$$
$$= \frac{1}{1 + 4/222} = \frac{222}{226} = 98.2\%.$$

ANALYZE The helium is the more easily observed of the products, because it has more energy. The decay is a *collision* with one incoming particle and two outgoing particles. The common feature of a particle decay and collisions is the rapid, complex transition between the initial and final states.

This example illustrates how we may use the familiar conservation laws to understand particle interactions, even when the objects that emerge are completely different from the original ones. Atomic nuclei are made up of numerous smaller particles known as neutrons and protons. A hydrogen nucleus consists of just one proton while a radium nucleus contains 88 protons and 138 neutrons. In the decay of radium, these particles are rearranged into a nucleus with 86 protons and 136 neutrons, plus another with 2 protons and 2 neutrons. The details of the process are governed by the rules of quantum mechanics.

Atomic nuclei may also fuse together with the release of energy. One of the most important such reactions occurs inside the Sun and in nuclear fusion experiments. A deuterium nucleus and a tritium nucleus may collide and fuse to form a helium nucleus and a neutron. In each reaction, 4×10^{-12} J of potential energy is converted to kinetic energy. The reaction requires very high temperatures. The nuclei exert repulsive electric forces on each other and so must approach each other with large initial speeds if they are to get close enough to react. High enough temperatures occur naturally in the Sun but are difficult to maintain in the laboratory. On the other hand, the energy released in the reaction is so much greater than the thermal energy required to make it go that controlled fusion could prove to be the primary energy source of the twenty-first century.

EXERCISE 10.7 ♦♦ Neglecting the initial kinetic energies of the deuterium and tritium nuclei in a fusion reaction, find the final energies of the neutron and the helium nucleus. Deuterium, tritium, and helium have masses of 2, 3, and 4 neutron-masses, respectively. If the neutron-mass is $m_n = 1.7 \times 10^{-27}$ kg, what are the final speeds of the neutron and the helium nucleus?

These reactions exhibit a common feature of atomic and subatomic collisions: they often rearrange the set of incoming particles into a very different outgoing set. In the quest for the fundamental stuff of which matter is made, physicists have discovered a host of different kinds of particles in addition to the neutrons, protons, and electrons that make up atoms. Neutrons and protons are themselves composed of smaller particles. To study these subparticles, experimenters use large accelerators that produce colliding protons with very high kinetic energies. ■ Figure 10.16 shows the rich detail that results from a single collision between a proton and another particle known as an antiproton.

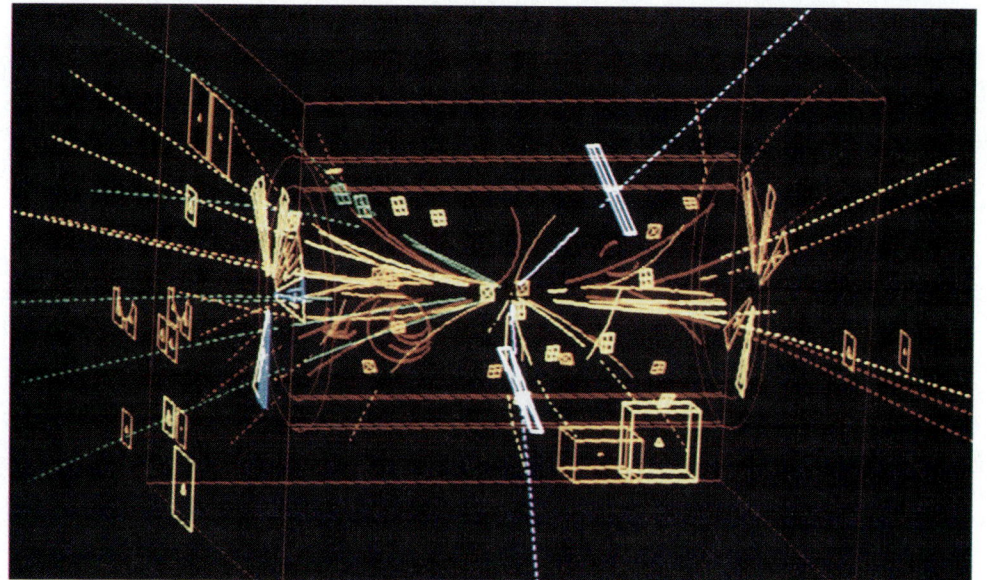

■ **Figure 10.16**
A computer reconstruction of particle tracks observed at the Super Proton Synchrotron Collider at CERN (European Laboratory for Particle Physics). In a collision between a particle and its antiparticle, an enormous number of particles may be created. The white tracks are those of an electron and a positron created in the decay of a Z_0 particle produced in the original collision. Detailed analysis of the tracks allows physicists to identify the momentum and charge of each particle. This experiment established the existence of the Z_0. (See Chapter 37.)

10.4.2 When Molecules Collide

The Sun emits large numbers of rapidly moving particles that bathe the Earth and are trapped by Earth's magnetic field. Collisions with these *solar wind* particles transfer energy to atoms and molecules in the upper atmosphere. The aurorae, or northern lights, are a beautiful and fascinating effect that arises when that energy is re-emitted as light. Light emission results from relatively small energy transfers. A more energetic collision can strip an electron from an atom, leaving an *ionized* atom behind. The ionization and recombination of atoms is an important process for the physics and chemistry of the upper atmosphere.

Once we consider the internal structure of colliding objects, we should also include the possibility of changing the internal angular momentum of the objects.

The aurora borealis. *These spectacular sheets of luminous gas result when energetic particles from the Sun are trapped by Earth's magnetic field and then collide with atoms in the upper atmosphere. These collisions transfer energy that is later released by the atoms as light.*

THE EXPRESSIONS WE USED FOR THE ANGULAR MOMENTA OF THE PARTICLES WERE FIRST CALCULATED IN EXAMPLE 9.2.

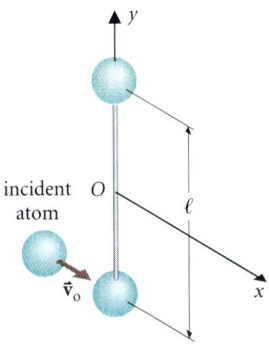

(a) *Before*

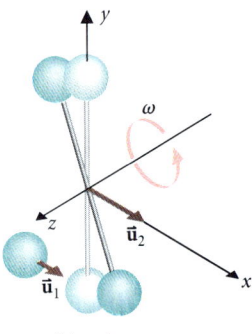

(b) *After*

■ **FIGURE 10.17**
A collision between an oxygen atom and an oxygen molecule. Some of the incident atom's kinetic energy is transformed to rotational kinetic energy, so the collision is inelastic. (a) Before the collision, the oxygen atom moves with speed v_o toward one of the molecule's atoms. (b) After the collision, the molecule moves to the right with speed u_2 and rotates about its center with angular speed ω. The incident atom has final velocity $\vec{u}_1$.

EXAMPLE 10.11 ♦♦♦ An oxygen atom moving with speed v_o collides with a stationary oxygen molecule as shown in ■ Figure 10.17. Describe the final state of the system and find the rotational kinetic energy of the molecule, assuming that the total kinetic energy is conserved.

MODEL We model the molecule as an ideal dumbbell: two particles (the oxygen atoms) joined by a massless rod. A convenient choice of origin is the center of the molecule. The impulsive force exerted by the incident oxygen atom exerts a torque about the center of the molecule, causing the molecule to rotate. The final state of the system is shown in Figure 10.17b.

SETUP We apply the conservation laws in the usual way, using the coordinate system shown in the figure. After the collision, the molecule has internal energy due to its rotation as well as kinetic energy associated with its linear motion. Each of its atoms has speed $v_{rot} = \omega r = \omega \ell /2$ with respect to the center of the molecule.

	BEFORE	**IMMEDIATELY AFTER**	
		Atom	Molecule
Kinetic energy	$\frac{1}{2}mv_o^2$	$\frac{1}{2}mu_1^2$	$+ \ \frac{1}{2}(2m)u_2^2 + 2[\frac{1}{2}m(\ell \omega/2)^2]$
x-component of linear momentum	mv_o	mu_1	$+ \ (2m)u_2$
Angular momentum $\vec{L}_O$ (z-component)	$m(\ell/2)v_o$	$m(\ell/2)u_1$	$+ \ 2m\omega(\ell/2)^2$

Set quantities equal: 2 × Energy $mv_o^2 = mu_1^2 + 2mu_2^2 + m\ell^2\omega^2/2.$ (i)

Linear momentum $mv_o = mu_1 + 2mu_2.$ (ii)

2 × (Angular momentum) $m\ell v_o = m\ell u_1 + m\omega \ell^2.$ (iii)

SOLVE Divide eqn. (ii) by m and eqn. (iii) by $m\ell$. Then eqn. (ii) becomes:

$$v_o - u_1 = 2u_2, \quad \text{(iv)}$$

and eqn. (iii) becomes: $v_o - u_1 = \omega \ell.$ (v)

Combining eqns. (iv) and (v) gives: $u_2 = \omega \ell /2.$ (vi)

Rearrange eqn. (i), and substitute for u_2 using eqn. (vi):

$$(v_o - u_1)(v_o + u_1) = v_o^2 - u_1^2 = 2u_2^2 + \omega^2 \ell^2/2 = \omega^2 \ell^2.$$

Using eqn. (v) again, we get:

$$(\omega \ell)(v_o + u_1) = (\omega \ell)^2 \ \Rightarrow \ v_o + u_1 = \omega \ell. \quad \text{(vii)}$$

Combining eqns. (v) and (vii) gives:

$$u_1 = 0 \quad \text{and} \quad \omega = v_o/\ell.$$

Finally, we use eqn. (iv) to find that $u_2 = v_o/2$.

The rotational kinetic energy of the molecule is:

$$K_{rot} = 2[\tfrac{1}{2}(m)(\omega \ell/2)^2] = mv_o^2/4.$$

This is equal to one-half the initial kinetic energy.

ANALYZE The incident atom remains stationary after the encounter. If the molecule were considered a particle, we would describe this as a partially inelastic collision.

The exchange of energy between linear and rotational motion in collisions is a significant feature of the thermal properties of gases (see Chapter 19). ■

EXERCISE 10.8 ❖ Show that the same results are obtained by assuming that the two oxygen atoms that collide undergo a rapid elastic collision before the third atom's motion can be altered (*Hint:* Look at the molecule and interpret the instantaneous velocities of its two atoms after the collision as a CM motion and a rotation of the molecule.)

10.4.3 Gravitational Collisions

A spacecraft approaching a planet from a very large distance experiences a conservative gravitational force, and so orbits the planet and returns again to large distances with the same kinetic energy (measured in the planet frame of reference) as it had on approaching. The encounter is an elastic collision of the spacecraft with the planet.

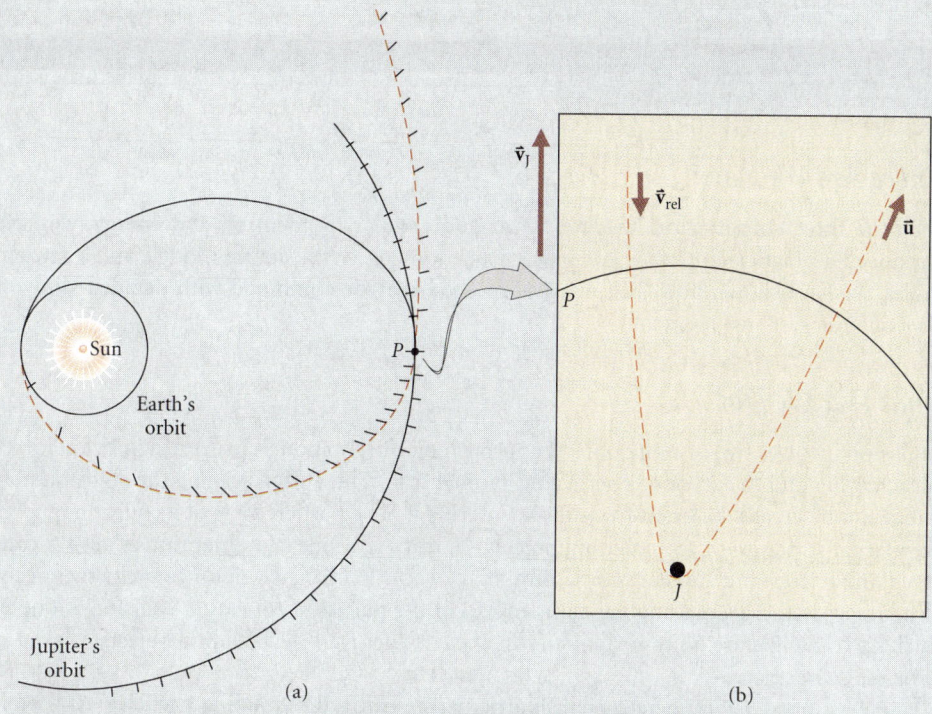

■ **FIGURE 10.18**
(a) A spacecraft from Earth approaches Jupiter on an elliptical orbit about the Sun. The tickmarks indicate its position at 30-day intervals. The spacecraft's initial orbit is tangent to Jupiter's orbit at point *P*, the launch having been timed so that the craft will arrive at that point at the same time Jupiter does. Jupiter's position every 30 days is also indicated by the tickmarks. Near *P*, Jupiter's speed is greater than the spacecraft's speed, as indicated by the greater distance between tickmarks. The "dot" at *P* indicates a region more than 50 times the size of Jupiter. The encounter with Jupiter puts the spacecraft on a path that proceeds outward rather than returning inward toward the Sun. (b) The interaction as seen in a reference frame moving with Jupiter. The circle represents the same region as the dot at *P* in (a). On the scale of this diagram, the spacecraft approaches from "infinity." In this reference frame, the spacecraft's energy is conserved in its encounter with Jupiter, so it leaves with the same speed as it arrives, but in a very different direction. Its speed *relative to the Sun* is greatly increased.

We may illustrate the idea by studying what happens when a spacecraft in an elliptical orbit about the Sun passes close to Jupiter (■ Figure 10.18). Jupiter's speed in its orbit is greater than the spacecraft's speed with respect to the Sun. In a reference frame moving with Jupiter (■ Figure 10.19), the spacecraft approaches from in front, swoops close to the planet's surface, and departs with its initial speed but almost in the opposite direction. Relative to the Sun, the spacecraft now has a speed greater than Jupiter's orbital speed and enough energy to escape the solar system.

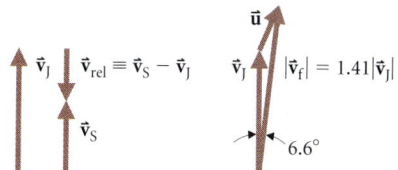

■ **FIGURE 10.19**
The velocity of the spacecraft with respect to the Sun is the vector sum of Jupiter's velocity with respect to the Sun and the spacecraft's velocity with respect to Jupiter. The spacecraft's initial speed relative to Jupiter is $0.43 v_J$, and its final speed relative to the Sun is $1.41 v_J$—sufficient to escape the solar system.

Such gravitational collisions made possible one of the loveliest achievements of space exploration, a "grand tour" of the outer solar system. Every few centuries, the planets are arranged so that a collision with one of the outer planets can be used to speed the craft on an orbit toward the next planet. The spacecraft carries instruments to study the planets on the way by. *Pioneers 10* and *11* passed by Saturn in 1979 and are now on their way out of the solar system carrying a message to any extraterrestrial civilization that might intercept them some millions of years hence. *Voyager 2*, launched in 1977, reached Uranus in 1986 and passed Neptune in 1989. Using gravitational collisions is now commonplace in solar system navigation. The *Galileo* spacecraft (Figure 10.1) visited Venus, made two passes by Earth, and reached Jupiter at the end of 1995. Truly, these are grand applications of collision theory.

Chapter Summary

Where Are We Now?

Using the three conservation laws for linear and angular momentum and energy, we have introduced a practical approach to collision phenomena. With simple models and a few examples, we have shown how this practical method may be combined with detailed study of how colliding systems interact.

What Did We Do?

A collision involves two or more objects approaching closely enough to exert significant forces on each other and to exchange energy and momentum. The objects leaving the collision may or may not be greatly different from those entering it. The particle view of a collision ignores the internal structure of the colliding objects. In this view, linear momentum is always conserved, but energy and angular momentum may *appear* to change if significant changes occur in the internal energy and angular momentum of the particles. An *elastic* collision is one in which such changes are unimportant and the *kinetic* energy of the colliding objects, viewed as particles, is conserved.

Analysis of collision phenomena is particularly clear in the colliding particles' CM reference frame. In an elastic collision of two particles, for example, the particles regain their original speeds in the CM frame and simply change direction. In a *perfectly inelastic* collision, the particles are at rest in the CM frame after the collision and move together in the lab frame. In a partially inelastic collision of rubber balls, information about how the balls dissipate kinetic energy is most easily applied in the CM frame.

Ideal mass–spring systems can model elastic and inelastic behavior and provide some insight into the behavior of actual objects. The models show that inelastic behavior in collisions can be represented by internal vibrations and rotations of the colliding objects.

Practical Applications

Collision theory has applications to such diverse problems as a billiard game and an automobile collision. It is particularly important in understanding the thermal properties of gases or investigating the structure and interactions of subatomic particles. Gravitational collisions are used to maneuver spacecraft. Particle encounters on the atomic and subatomic size scales often involve radical changes in the number and character of particles, and changes in internal energy can be much larger than the incoming kinetic energy. Nevertheless, these technologically important processes follow the same three conservation laws that apply in everyday life and in spacecraft encounters with planets.

Solutions to Exercises

10.1 A collision with another player's ball is one between equal-mass particles: $m_p = m_t$. So, from eqns. (10.3), we find $u_p = 0$ and $u_t = v_1$. The incoming ball is brought to rest and transfers all its energy and momentum to the target ball, whose outgoing speed is equal to the incoming speed of the first ball.

10.2 ■ Figure 10.20 shows the relations among the incoming and outgoing velocities in both lab and CM reference frames. All velocities are in the (positive or negative) x-direction, and so we look at their x-components. The CM velocity is:

$$v_{CM} = \frac{m_1 v_1 + m_2 v_2}{m_1 + m_2}. \quad (i)$$

The outgoing velocities in the CM frame are the incoming velocities reversed.

$$u_1 = u_{1,CM} + v_{CM} = -v_{1,CM} + v_{CM}$$
$$= -(v_1 - v_{CM}) + v_{CM} = 2v_{CM} - v_1.$$

Since the total momentum in the CM frame is zero:

$$m_2 v_{2,CM} + m_1 v_{1,CM} = 0.$$

Thus:

$$u_2 = u_{2,CM} + v_{CM} = -v_{2,CM} + v_{CM} = \frac{m_1}{m_2} v_{1,CM} + v_{CM}$$
$$= \frac{m_1}{m_2}(v_1 - v_{CM}) + v_{CM} = \frac{m_1}{m_2} v_1 + \frac{m_2 - m_1}{m_2} v_{CM}.$$

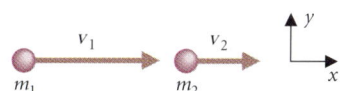

(a) Incoming (lab frame)

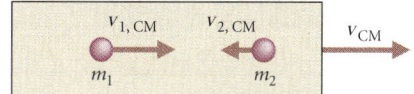

(b) Incoming (CM frame)

(c) Outgoing (CM frame)

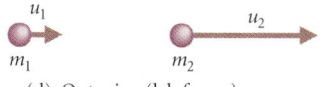

(d) Outgoing (lab frame)

■ **FIGURE 10.20**
The incoming phase of a one-dimensional collision viewed (a) in the lab frame and (b) in the CM frame. (c) The effect of the collision in the CM frame is to reverse the particles' velocities. (d) The outgoing phase viewed in the lab frame.

Substituting in eqn. (i) for v_{CM}:

$$u_1 = \frac{2(m_1 v_1 + m_2 v_2) - v_1(m_1 + m_2)}{m_1 + m_2}$$
$$= \frac{(m_1 - m_2)v_1 + 2m_2 v_2}{m_1 + m_2},$$

and

$$u_2 = \frac{m_1 v_1 (m_1 + m_2) + (m_2 - m_1)(m_1 v_1 + m_2 v_2)}{m_2(m_1 + m_2)}$$
$$= \frac{v_2(m_2 - m_1) + 2m_1 v_1}{m_1 + m_2}.$$

These are eqns. (10.1) and (10.2).

10.3 To compute the coefficient of restitution, we need the velocity of the ball just before and just after it collides with the ground. We use conservation of energy to relate the ball's speed to the height it rises:

$$\tfrac{1}{2} mv^2 = mgh \;\Rightarrow\; v = \sqrt{2gh}.$$

Thus: $\epsilon = \dfrac{|u_2 - u_1|}{|v_2 - v_1|} = \dfrac{|u_2 - 0|}{|v_2 - 0|} = \dfrac{\sqrt{2gh_2}}{\sqrt{2gh_1}} = \sqrt{\dfrac{h_2}{h_1}},$

as required. With $h_1 = 1.0$ m and $h_2 = 0.70$ m, $\epsilon = \sqrt{0.70} = 0.84$.

In Example 10.7, we found $u_2 = 4.6$ m/s and $u_1 = 0.4$ m/s; thus:

$$\epsilon = \frac{|4.6 \text{ m/s} - 0.4 \text{ m/s}|}{|5.00 \text{ m/s} - 0|} = 0.84,$$

as expected. Also, in the CM reference frame:

$$\epsilon = \frac{|2.1 \text{ m/s} - (-2.1 \text{ m/s})|}{|2.50 \text{ m/s} - (-2.50 \text{ m/s})|} = 0.84.$$

Because ϵ is independent of reference frame, the coefficient of restitution is a useful tool for comparing different collisions between a given pair of objects.

10.4 See ■ Figure 10.21. The hypotenuse of the right triangle is twice the ball's radius, so $\sin \phi = b/2r$, as required.

10.5 Initially, $v_1 = v_o$ and $v_2 = 0$. After the collision, $u_2 = 0$ and $u_1 = v_{CM} = v_o/2$. Thus:

$$\epsilon = \frac{|u_2 - u_1|}{|v_2 - v_1|} = \frac{v_o/2}{v_o} = 0.5.$$

10.6 This is an inelastic collision with the same outcome as in Example 10.9. The pebble has half the mass of the soda can and is left at

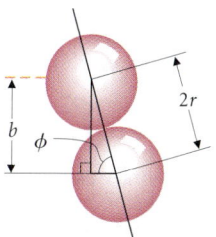

■ **FIGURE 10.21**

rest after the collision. As in the example, half the original kinetic energy of the pebble ends up as kinetic energy of the can and the other half is turned into thermal energy.

$$\Delta U = \tfrac{1}{2}(\tfrac{1}{2}mv_o^2) = \tfrac{1}{2}[\tfrac{1}{2}(0.050 \text{ kg})(15 \text{ m/s})^2] = 2.8 \text{ J}.$$

✽ **10.7** Since we neglect the particles' incoming kinetic energies, their CM reference frame is the lab frame. The emerging neutron and helium nucleus have equal and opposite momenta in the lab frame:

$$M_n \vec{v}_n = -M_{He}\vec{v}_{He}.$$

So, the speeds are related by:

$$v_n = (M_{He}/M_n)v_{He} = 4v_{He}.$$

The outgoing kinetic energy approximately equals the potential energy released in the reaction:

$$\Delta E \approx \tfrac{1}{2}M_n v_n^2 + \tfrac{1}{2}M_{He}v_{He}^2 = \tfrac{1}{2}M_n v_n^2[1 + 4(1/4)^2] = 5M_n v_n^2/8.$$

So:

$$v_n = \sqrt{8\,\Delta E/5M_n} = \sqrt{\frac{8(4 \times 10^{-12} \text{ J})}{5(1.7 \times 10^{-27} \text{ kg})}} = 6.1 \times 10^7 \text{ m/s}.$$

$v_{He} = v_n/4 = 1.5 \times 10^7$ m/s.

✽ **10.8** ■ Figure 10.22 shows the state of the molecules immediately after the encounter of the two equal-mass oxygen atoms. The incident atom is brought to rest, and the other travels with velocity $\vec{v}_o$ (cf. eqns. 10.3 with $m_1 = m_2$). The molecule has CM velocity:

$$\vec{v}_{CM} = \frac{mv_o + m(0)}{2m}\hat{\imath} = \frac{v_o}{2}\hat{\imath}.$$

In the CM reference frame, the two atoms have equal and opposite speeds $v_o/2$, corresponding to rotation about the center with angular speed:

$$\omega = \frac{v_o/2}{\ell/2} = \frac{v_o}{\ell}.$$

These are the results we found in Example 10.11.

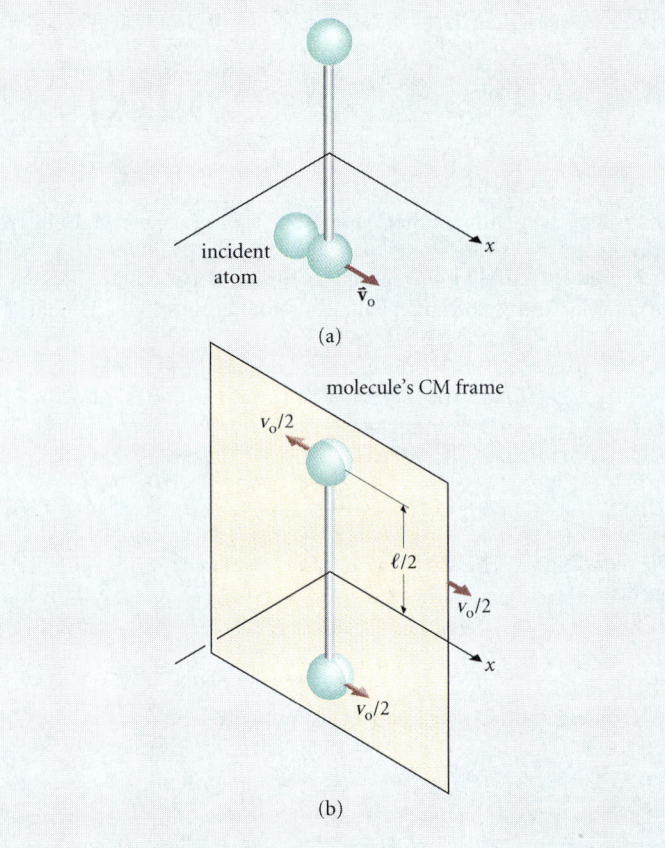

■ **Figure 10.22**
(a) When the two oxygen atoms collide, the incident atom is brought to rest, and the second atom moves off with speed v_o. The third atom remains momentarily at rest. (b) In the molecule's CM frame, each atom has speed $v_o/2$.

Basic Skills

Review Questions

§10.1 WHAT IS A COLLISION?

- What are the three phases of a collision? In what ways does the encounter phase differ from the other two?
- What are two points of view from which one might study a collision?

§10.2 COLLISIONS BETWEEN TWO PARTICLES

- Define an *elastic* collision. What quantities are conserved in an elastic encounter?
- Describe the general outcome of a two-particle elastic collision viewed in the particle's CM reference system.
- Why is the outcome of a two-dimensional elastic collision between two particles not determined by the incoming particle velocities alone?
- How much information about the outgoing state do you need to analyze a two-dimensional particle collision?
- What is a *perfectly inelastic* collision? Is all of a system's kinetic energy transformed in a perfectly inelastic collision? How does your answer depend on the reference frame in which the energy is measured?

§10.3 MODELS FOR ELASTIC AND INELASTIC COLLISIONS

- Compare a mass on a spring with a billiard ball. In what ways does a mass on a spring provide a better *model* for elastic collisions? In what ways is the billiard ball a better *image*?
- How do billiard balls allow us to model two-dimensional collisions?
- Explain how internal springs model an object's inelastic behavior. Why is thermal energy difficult to model well?

✸ §10.4 SOME APPLICATIONS OF COLLISION THEORY

- A nuclear decay, unlike the usual picture of a collision, has only one incoming particle; in what ways is it like a collision?
- What is a gravitational collision?

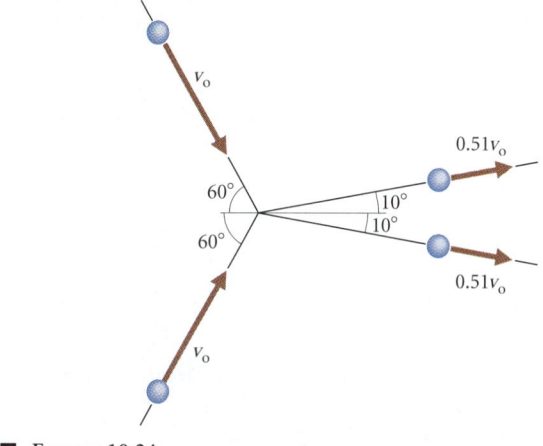

■ FIGURE 10.24

Basic Skill Drill

§10.1 WHAT IS A COLLISION?

1. The following sequence of events occurs in a baseball game: **(a)** The pitcher pitches the ball; **(b)** the batter hits the ball; **(c)** the left fielder leaps high and catches the ball. To what extent, if any, can each event be modeled as a collision? What features of a collision does each event exhibit, what features are absent, and what objects are colliding?

§10.2 COLLISIONS BETWEEN TWO PARTICLES

2. A bowling ball moving at $v = 3$ m/s strikes a pin one-third its mass in an elastic collision. Find the speeds of ball and pin after the collision. Assume they both move along the same line.
3. A ball, with initial speed $v_o = 1.0$ m/s, collides with a target ball of equal mass. After the collision, the first ball is moving with a speed of 0.1 m/s in the same direction as its initial velocity. Assuming the collision was elastic, what was the initial velocity of the target ball?
4. Two objects of equal mass and equal speed approach each other at right angles (■ Figure 10.23). It is claimed that the outgoing state resulting from their collision is also as shown in the figure. Do you believe this claim? Why or why not?
5. An astronaut of mass 200 kg moves directly toward a satellite of mass 1000 kg at a speed of 6 m/s. What is the CM velocity of the astronaut–satellite system? When the astronaut arrives at the satellite, she catches it and holds on. What are the velocities of satellite and astronaut in the CM frame before and after this collision?
6. The incoming and outgoing phases of a collision are shown in ■ Figure 10.24. Would you classify the collision as elastic, partially elastic, or totally inelastic? Explain your reasoning.

§10.3 MODELS FOR ELASTIC AND INELASTIC COLLISIONS

7. A molecule capable of vibrating is modeled as two point atoms of mass M connected by a spring (■ Figure 10.25). Initially, the molecule is not vibrating when it collides with a single atom of mass M, as shown in the figure. Decide whether the collision should be classed as elastic or inelastic, and explain your reasoning. If you decide it is inelastic, determine the amount of incoming kinetic energy that is transformed to internal energy.

✸ §10.4 SOME APPLICATIONS OF COLLISION THEORY

8. In a nuclear reaction a uranium nucleus (mass 238 units) decays into a thorium nucleus (mass 234 units) and an alpha particle (helium nucleus, mass 4.00 units). If the uranium nucleus is initially at rest, what fraction of the final kinetic energy does the α-particle have?

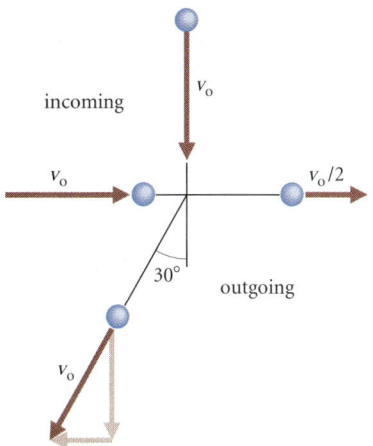

■ FIGURE 10.23

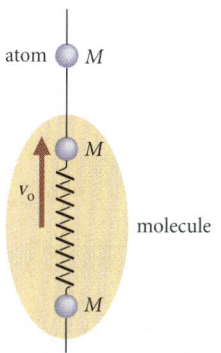

■ FIGURE 10.25

Questions and Problems

§10.2 COLLISIONS BETWEEN TWO PARTICLES

9. ❖ In a pool game the cue ball (white) and a blue ball are arranged as shown in ■ Figure 10.26. A player wants to shoot the cue ball so that after the collision the blue ball enters the center pocket on the right. Is this shot possible?

10. ❖ In 1978, a PSA jet and a Cessna 172 light aircraft collided over San Diego. The two planes were flying at the same altitude and in almost the same direction. The press reported that the Cessna (speed approximately 100 mph) had "hit the jet" (speed approximately 200 mph). Is this a convincing description of the accident? Why or why not?

11. ❖ Two objects of masses m_1 and m_2 collide. After the collision, m_2 has the same momentum as m_1 did before the collision. Can this happen in an elastic collision? Under what conditions?

12. ❖ Two identical balls, with equal and opposite velocities $v_o\hat{\mathbf{i}}$ and $-v_o\hat{\mathbf{i}}$, collide elastically. One ball's outgoing velocity is at angle θ with the x-axis; what can you say about the other ball's outgoing velocity?

13. ❖ Joe Heavyfoot hopes to improve his fate in a head-on collision by speeding up to have more momentum than the other car. Explain why Joe is dead wrong.

14. ◆ A rubber ball dropped from a height of 5.0 m rebounds to a height of 4.0 m. How far does it rise after the next bounce?

15. ◆ Engineers testing a new elastic bumper design for passenger cars collide a test car of mass 1100 kg moving at 6.5 m/s with a stationary car of mass 1200 kg. Find the speeds of both cars after the collision, assuming all the velocities lie along the same line.

16. ◆ Two identical balls of mass 0.52 kg are moving directly toward each other, each with speed $v_o = 2.5$ m/s. What is the velocity of each ball after they undergo an elastic one-dimensional collision?

17. ◆ A car of mass 1400 kg moving at 4.2 m/s bumps into another car of mass 1200 kg, initially moving at 2.1 m/s in the same direction. After the collision, the 1400-kg car is moving at 3.0 m/s. What is the final velocity of the other car? Was the collision elastic?

18. In shove-ha'penny (a game played in English pubs), coins are slid along a tabletop and collide with other coins at rest on the table. **(a)** ◆ If a small coin of mass m moving at a speed of 0.50 m/s collides elastically with a larger coin of three times the mass in a one-dimensional collision, find the speed of each coin after the collision. Ignore friction. **(b)** ◆◆ If instead the coin hits an identical coin and leaves the collision at an angle of 30° from its original direction, find its speed and the velocity of the other coin immediately after the elastic collision.

19. ◆◆ A mine car filled with coal and moving at 5.0 m/s collides elastically with a stationary, empty car at rest on the tracks. The mine cars each have a mass of 520 kg, and the outgoing speed of the empty car is 6.2 m/s. What is the mass of the coal in the loaded car?

20. ◆◆ In a pool game the cue ball, with incoming speed v_o, collides elastically with the eight ball, which is at rest. If the outgoing speed of the cue ball is $3v_o/5$, what is the angle between its incoming and outgoing velocities? What is the outgoing velocity of the eight ball?

21. ◆◆ A particle of mass m collides elastically with another of mass M. Find the condition under which m recoils in the opposite direction from its original velocity. What is the condition if M is initially at rest? (Such recoils in an experiment on the scattering of α-particles led Ernest Rutherford to the discovery of atomic nuclei. See Chapter 35.)

22. ◆◆ Dodgem cars at the funfair collide elastically. While moving with velocity $\vec{\mathbf{v}}$ ($|\vec{\mathbf{v}}| = 5.0$ m/s), your car hits another of equal mass moving in the same direction. The cars' outgoing velocities are each at a 22.5° angle from $\vec{\mathbf{v}}$. What was the velocity of the other car before you hit it?

23. ◆◆ A puck slides down an icy slope from a height h and collides elastically with a puck twice its mass at the bottom of the slope. Does the first puck move back up the slope? How far?

24. ◆◆ A particle of mass m moving at speed v_o collides with a stationary particle of mass $3m$. After the collision, the particle of mass m is moving in the same direction with speed $v_o/7$. What is the final velocity of the heavier particle? How much energy is dissipated in the collision?

25. ◆◆ A wooden ball dissipates 40% of its energy when it bounces from a wood floor. The ball collides at speed 12 m/s with an identical, stationary ball. After the collision, both balls move along the same line. What are the final velocities of the two balls?

26. ◆◆ Two identical rubber balls approach and collide with a coefficient of restitution of 0.78. If the speeds of the two balls as they approach are 1.5 m/s and 4.3 m/s, find their velocities after the collision.

27. ◆◆ Particles of mass m and $3m$, each with speed v_o, approach and collide as shown in ■ Figure 10.27. The outgoing particles are each of mass $2m$. One of them has a final velocity $3v_o\hat{\mathbf{i}}$. What is the final velocity of the other particle? How much internal energy is converted to kinetic energy in the collision? (Hint: Be careful with directions; your calculator may mislead you!)

28. ◆◆ A particle of mass m with velocity $\vec{\mathbf{v}}_o$ in the lab frame collides elastically with an equal-mass particle at rest in the lab frame. One particle's outgoing velocity in the CM reference frame is at an angle

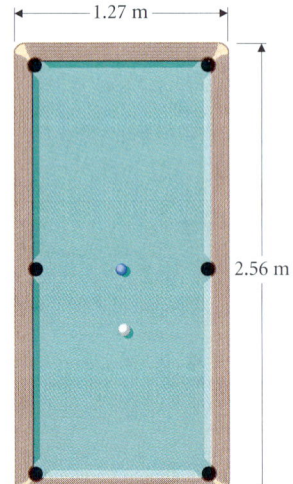

■ FIGURE 10.26

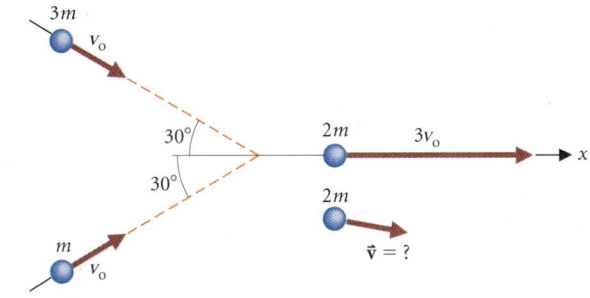

■ FIGURE 10.27

of 30° with the direction of $\vec{v}_o$. What are the two particles' outgoing velocities in the lab frame?

29. ◆◆ In a game of pool, the balls are set up as shown in Figure 10.26. The player wants to send the blue ball into the top left corner pocket. If the shot succeeds, what is the velocity of the cue ball after the collision? Give the cue ball's speed in terms of its speed before the collision.

30. ◆◆ Two hockey players of masses 89 kg and 94 kg both skate toward the puck. The first player skates at 15 m/s. The second player skates at a 60° angle to the first skater's path at 18 m/s. They collide, and slide on in a tangled mess over the ice. Find the direction and magnitude of their velocity after the collision.

31. ◆◆ Two identical hard rubber balls are each found to lose a fraction f of their energy when rebounding from a hard rubber surface. The balls undergo a one-dimensional collision, one at speed v_o, the other stationary. Analyze the collision in the CM reference frame following the method of Example 10.7. Then transform your results to the lab frame and compute the fraction f_{lab} of kinetic energy in the lab frame that is transformed into thermal energy. Give your result in terms of f. Check to see whether your formula is consistent with the results of Example 10.7.

32. Two identical rubber balls approach each other along parallel lines. Ball 1 has velocity $\vec{v}_1 = v_o \hat{i}$ and ball 2 has velocity $\vec{v}_2 = -(2.00)v_o \hat{i}$ in the lab frame. After the collision, ball 1's velocity is at an angle $\theta = 30.0°$ with the x-axis, and ball 2's speed is $(1.70)v_o$. **(a)** ◆◆ Analyze the collision in the lab frame to determine whether it is elastic. If not, find what fraction of the initial kinetic energy is transformed to thermal energy. **(b)** ◆◆ Check your analysis of the collision by transforming the outgoing velocities into the CM reference frame and verifying that they are equal and opposite. **(c)** ◆◆◆ Let ϕ be the angle between ball 1's incoming velocity in the CM frame and the balls' line of centers when they collide. Assuming each ball's momentum perpendicular to the line of centers is unchanged by the collision, determine ϕ. Find the change in magnitude of each ball's velocity component parallel to the line of centers. **(d)** ◆◆ How high would one of these balls bounce if dropped from a height of 1.0 m onto a hard rubber surface?

33. ◆◆◆ A very small *superball* of mass m is made of a rubber that is nearly perfectly elastic. It is placed on top of a very much larger superball of mass $M \gg m$, and the two are dropped together onto a very hard floor. Analyze the impact as an elastic collision of the large ball with the floor followed by an elastic collision between the large and small balls. Show that the small ball rises to nine times its original height.

34. ◆◆◆ A particle of mass m and initial speed v_o collides elastically with a stationary particle of mass M. Show that the angle ϕ between the final velocities of the two particles in the lab frame is given by:

$$\cos \phi = (m - M)\sqrt{\frac{1 - \cos \theta}{2(m^2 + M^2 + 2mM \cos \theta)}},$$

where θ is the angle between the final and initial velocities of particle m in the CM frame.

35. ◆◆◆ A wooden block of mass $M = 2.00$ kg slides without friction on a horizontal surface and is attached to a wall with a spring of constant $k = 1.00 \times 10^3$ N/m (■ Figure 10.28). A bullet of mass $m = 30.0$ g is fired into the wooden block, which then compresses the spring a distance $s = 30.0$ cm. What is the initial speed of the bullet?

§10.3 MODELS OF ELASTIC AND INELASTIC COLLISIONS

36. ❖ Two identical balls with incoming velocities $\vec{v}_A$ and $\vec{v}_B$ collide elastically. ■ Figure 10.29 shows the two balls at the instant of collision. Is it possible to determine which of the vectors 1–5 best represents the outgoing velocity of ball A? If so, which is it? Explain your reasoning.

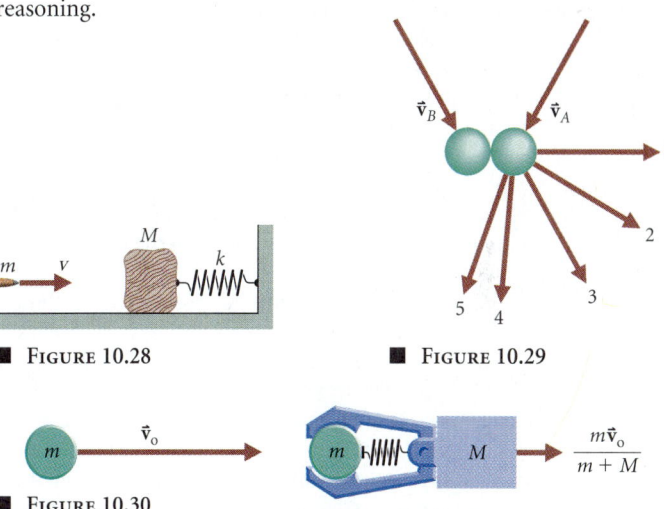

■ FIGURE 10.28   ■ FIGURE 10.29

■ FIGURE 10.30

37. ◆ A block of mass 0.35 kg moving at 1.7 m/s collides with a second block of mass 0.65 kg attached to a spring with spring constant $k = 2.5 \times 10^3$ N/m (cf. Figure 10.9). What is the maximum compression of the spring? Find the outgoing velocity of each block.

38. ◆ A block of mass 0.45 kg moving at 2.0 m/s makes an inelastic collision with a second block of mass 0.35 kg, at rest, equipped with a spring and a clamp (cf. Figure 10.13). How much energy is stored in the spring after the collision? If the spring compression is 0.95 cm, what is the spring constant?

39. ◆◆ Two pieces of a machine, with masses 17 kg and 35 kg, are mounted with a spring compressed between them. An astronaut accidentally releases the connection, but then catches the 35 kg piece. The 17-kg piece escapes and passes the space station at 2.5 m/s. What is the combined momentum of the astronaut and recaptured piece? How much energy was stored in the spring?

40. ◆◆◆ The object on the right in ■ Figure 10.30 was formed when a particle of mass m and speed v_o collided with a particle of mass M at rest. The two were clamped together just as the spring brought the two particles to rest in their CM reference frame. **(a)** Suppose now that a second particle of mass m and speed v_o strikes the system from the left, resulting in a one-dimensional elastic collision with the combined system of mass $m + M$. Shortly afterward, the effect of the collision causes the clamp to release and the two bound particles to separate. Find the outgoing velocities of the three resulting particles. **(b)** Suppose instead that the second particle of mass m impacts the system at speed v_o from the right. What then are the outgoing velocities of the three particles? **(c)** Finally, suppose that the second incoming particle, viewed in the combined object's CM frame, strikes the object with incoming speed v_o and perpendicular to the line between the two particles. Assume the initial collision is one-dimensional. What are the outgoing velocities of the three particles in the lab frame after the clamp releases?

✽ §10.4 SOME APPLICATIONS OF COLLISION THEORY

41. ◆◆ A hydrogen atom in its *first excited state* has $\Delta U = 1.63 \times 10^{-18}$ J greater potential energy than in its *ground state*. An electron with 1.8×10^{-18} J of kinetic energy collides with a stationary hydrogen atom in its ground state. If the collision is one-dimensional and the outgoing hydrogen atom is in its first excited state, estimate the outgoing velocities of the electron and the atom.

42. ◆◆ A platinum nucleus of mass 192 units decays by emitting an alpha particle (mass 4.00 units), leaving an osmium nucleus of mass 188 units. If 3.66×10^{-13} J of potential energy is converted to kinetic energy in the decay, find the speed of the α-particle.

43. ◆◆◆ A single oxygen atom with speed $|\vec{v}_i| = v_o$ collides elastically with one atom of an oxygen molecule as shown in ■ Figure 10.31. Its initial velocity makes an angle of 45° with the axis of the molecule, and its final velocity is at an angle of 30°, as in the figure. Treat all the atoms as particles, and find the final speed of the single atom, the CM velocity of the molecule, and the angular velocity of the molecule after the collision.

44. ◆◆◆ A model for a gas molecule consists of two spheres each of mass m connected by a spring. If another sphere of mass m collides with one of the spheres of the "molecule" with a velocity $\vec{v}$ making an angle $\theta = 30°$ with the molecule's axis, find the final velocity of the molecule, its rotational energy, and the energy stored in the vibrations of the spring. (Consider the spring rigid when describing the rotation of the molecule.) Model the original encounter as an elastic collision between equal-mass particles.

45. ◆◆◆ Two identical oxygen molecules, each moving with speed v_o in opposite directions, collide as shown in ■ Figure 10.32. Determine the final velocity of each molecule and its angular speed.

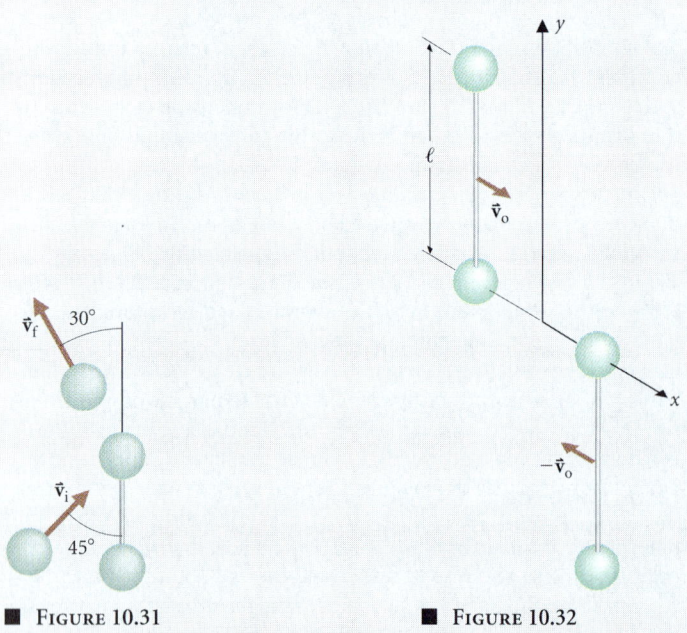

■ **FIGURE 10.31**   ■ **FIGURE 10.32**

Additional Problems

46. ❖ Five identical steel balls are suspended with threads from a frame so that they hang together (■ Figure 10.33). If one ball is pulled back and released, it strikes the line of four stationary balls and comes to rest. The single ball on the far end of the line is ejected with very nearly the same energy as the incoming ball had. **(a)** Discuss why this behavior is better explained as a series of collisions between equal-mass particles rather than as a single collision between a particle and one four times as massive. **(b)** Based on your reasoning in part (a), explain why two balls are ejected if two are originally pulled back. **(c)** When two identical, finite spheres collide, can they accelerate infinitely rapidly? In responding to this question, consider an individual atom somewhere inside one of the spheres; can it *know* it is supposed to accelerate before some change causes nearby atoms to *inform* it by exerting forces? In such a collision, one sphere decelerates and one accelerates. How do you suppose that the time intervals for these two processes compare? How does this line of reasoning support your conclusion in part (a)?

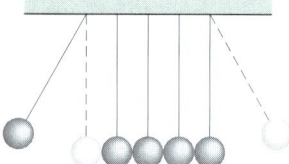

■ **FIGURE 10.33**

47. (a) ❖ An impulse $\Delta\vec{p}$ is delivered to one of the particles in a dumbbell. If the impulse is perpendicular to the rod connecting the two particles (■ Figure 10.34a), describe the resulting motion of the dumbbell. **(b)** ❖ If the impulse is delivered parallel to the rod (Figure 10.34b), what is the result? **(c)** ◆◆ What energy is given to the dumbbell in each of these two cases?

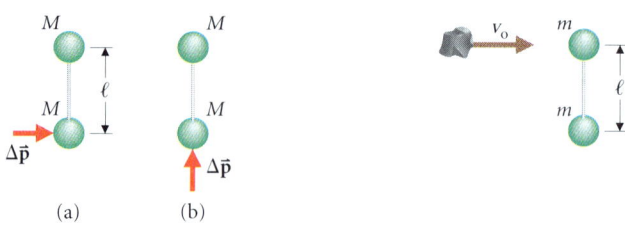

■ **FIGURE 10.34**   ■ **FIGURE 10.35**

48. ◆◆ A putty ball of mass m and initial speed v_o collides with a stationary dumbbell as shown in ■ Figure 10.35. The dumbbell consists of two particles of mass m connected by a rigid, massless rod of length ℓ. Find the amount of energy dissipated as heat.

49. ◆◆ In a target shooting competition, 35-g aluminum cans are placed on top of a post 1.5 m high. The target rifles fire 15-g bullets horizontally at 420 m/s. When a can is hit, it lands 2.3 m from the post. What is the velocity change of the bullet as it passes through a can? How much kinetic energy is transformed in the encounter?

50. ◆◆ A boxcar of mass $2m$ and a flatcar of mass m are each free to roll without friction along a railroad track. Initially, the boxcar has speed v_o (■ Figure 10.36) and the flatcar is at rest. As they collide, the

■ **FIGURE 10.36**

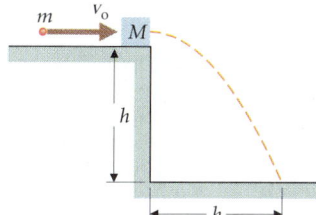

Figure 10.37

Figure 10.38

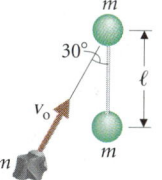

Figure 10.39

two cars link together and roll up the hill. At what height h do the cars come to rest?

51. ♦♦ A wooden block of mass $M = 2.0$ kg is at rest on a table. A bullet of mass 15 g is fired upward into the block through a small hole in the table and embeds itself in the block, which is projected upward from the table. An electronic timer measures the time interval between firing the bullet and the block falling back to the table as $\Delta t = 0.86$ s. What was the initial speed of the bullet, and how much thermal energy was transformed as it hit the block?

52. ♦♦ In a test of an advanced projectile system, a target block of mass M is at rest on the edge of a cliff of height h (■ Figure 10.37). A projectile of mass m strikes the block with an unknown speed v_o as shown. The block and projectile are then observed to land a distance h from the base of the cliff. If $M = 2$ kg, $m = 50$ g, and $h = 10$ m, what was the speed of the projectile?

53. ♦♦ Two small balls of equal mass are joined by a string and lie on a frictionless table with the string stretched out between them. Another ball of equal mass hits one of the first two with a velocity along the line of the string. **(a)** Describe the collision qualitatively. Find the velocities of all three balls a long time after the collision in terms of the original velocity of the third ball. Assume that any collision between individual balls is elastic. **(b)** What happens if the third ball's original velocity is directed at some angle θ to the string?

54. ♦♦ Two beads of mass m are sliding toward each other along a horizontal wire. Initially, they are a distance D apart and each moves with speed v_o. Their coefficients of kinetic friction with the wire are equal but otherwise unknown. They undergo a partially elastic collision and later come to rest a distance D apart. Find a relation among the initial speed, the unknown coefficient of friction, and the fraction of the beads' incoming kinetic energy dissipated as heat in the collision.

55. ♦♦ A star approaches another star of equal mass from a great distance with velocity $\vec{v}_o$. When the first star is again a great distance away, it is moving at 30° to its original direction. Find the velocities of both stars after the encounter.

56. ♦♦♦ A car of mass $m = 1.0 \times 10^3$ kg is moving in the x-direction with a speed of $v_o = 32$ m/s when it is struck by a car of unknown mass M and speed v moving in the y-direction. The combined wreckage of the two cars slides a distance of 30.0 m in a direction 30° from the x-axis. If the coefficient of friction between road and wreckage is $\mu = 0.32$, what are the mass and speed of the offending car?

57. ♦♦♦ Three hard rubber pucks are free to slide in a frictionless straight groove in the ice, which defines the x-axis. Initially, puck 1, of mass m and velocity $v_o \hat{\mathbf{i}}$, is incoming from the left. Puck 2, also of mass m, is stationary at the origin, and puck 3, of mass M, is stationary at $x = D$. Assuming the pucks collide elastically, describe the sequence of collisions that occur and the outgoing velocity of each puck. Find the final position of any puck whose outgoing velocity is zero. Treat the 3 cases $M < m$, $M = m$, and $M > m$.

58. ♦♦♦ ⊠ ■ Figure 10.38 shows a simplified model of an automobile for use in computing the effects of collisions. The mass of the car is evenly divided between the *putty* front end and the rear end. Each has mass 10^3 kg. The spring has constant $k = 4 \times 10^5$ N/m. If two such cars collide head-on with equal speeds of 30 m/s, find the maximum acceleration of the passengers and the force the seat belt must exert on a 70-kg passenger.

59. ♦♦♦ A putty ball of mass m and speed v_o strikes a dumbbell as shown in ■ Figure 10.39. What are the final linear and angular velocities of the dumbbell? How much thermal energy is produced?

60. ♦♦♦ In an English carnival game, a 120-g ball is launched at 10.0 m/s at an angle of 26.1° with the horizontal so as to hit a 1.5-kg coconut 4.0 m away and 1.0 m higher than the launch point of the ball. (See Problem 3.27.) **(a)** Verify that the ball hits the coconut and that its velocity is nearly horizontal just before the impact. **(b)** If the collision is elastic, what is the speed of the coconut after the collision? Where does it land? Assume that the velocities of ball and coconut are in the same direction immediately after the collision.

61. ♦♦♦ A particle collides elastically with an identical particle at rest. By analyzing the collision in the two particles' CM reference frame, show that their outgoing velocities in the lab frame are perpendicular.

62. ♦♦♦ In a pool game, the cue ball collides elastically with a red ball and a green ball that are initially touching (■ Figure 10.40). The red ball moves off at an angle θ with the cue ball's initial direction of motion, and the green ball moves off at an angle ϕ. Is this information sufficient to determine the outgoing velocity of the cue ball? If so, solve the problem. If not, state what additional information you might need. When the collision is symmetrical ($\theta = \phi$ and $u_2 = u_3$), determine the outgoing velocity of the cue ball. Check your answer against the results of §10.2 when $\theta = 0$. What happens when $\theta = \pi/2$?

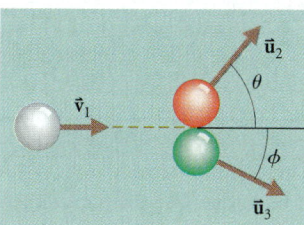

■ **Figure 10.40**

Computer Problem

63. A series of carts are arranged at rest along a straight track, separated by a distance d. Each succeeding cart has a mass $f = 0.90$ times the mass of the preceding one. The coefficient of friction between each cart and the track is 0.50. The first cart is set moving down the track at speed v_1, causing a sequence of elastic collisions. **(a)** If v_n is the speed with which cart n approaches cart $n+1$, show that the speeds of carts n and $n+1$ immediately after their collision are given by:

$$u_n = \frac{1-f}{1+f} v_n \quad \text{and} \quad u_{n+1} = \frac{2}{1+f} v_n.$$

Verify that these relations are consistent with the results in §10.2.

(b) Use results from Chapter 2 to find a formula for the speed v_{n+1} with which the $(n+1)$st cart collides with the $(n+2)$nd. (c) Set up a spreadsheet program to calculate u_{n+1} for $n = 1, \ldots, 20$. (d) Check your program by setting $\mu = 0$ and $f = 1$. Do you obtain the expected result? (e) If $d = 0.10$ m and $\mu = 0.50$, for what range of values of v_1 does u_{n+1} decrease as n increases? For what range of values of v_1 does u_{n+1} increase? If $v_1 = 2.0$ m/s, how many collisions occur before the system comes to rest?

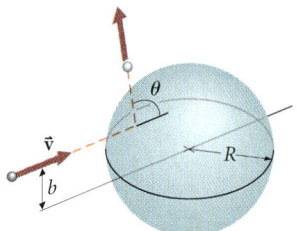

■ FIGURE 10.41

Challenge Problems

64. In a shove-ha'penny game (cf. Problem 18), a player slides a coin of mass m over a frictionless table so as to hit a stationary coin. The mass of the target coin is γm, and the incident coin leaves the collision at a 30° angle with its incoming direction. **(a)** Show that the angle ϕ between the incident coin's incoming velocity and the target coin's outgoing velocity satisfies:

$$\sqrt{3} \sin \phi \cos \phi = \sin^2 \phi - \tfrac{1}{2}(1 - 1/\gamma).$$

(b) Show that:

$$\sin^2 \phi = \frac{(4\gamma - 1) \pm \sqrt{3(4\gamma^2 - 1)}}{8\gamma}.$$

(c) What is the smallest value of γ that allows a solution for ϕ? What are the outgoing velocities for that value of γ? **(d)** For the case $\gamma = 1$, decide which of the two choices of sign for the square root in the formula for $\sin^2 \phi$ gives an acceptable result. Explain your reasoning. **(e)** What is the value of ϕ in the limit $\gamma \gg 1$?

65. (a) A small pellet of negligible mass collides elastically but off-center with a spherical ball of radius R and very large mass (■ Figure 10.41). Use the reasoning illustrated in Figure 10.10 to find the scattering angle $\theta(b)$, through which the pellet is deflected, as a function of b, the perpendicular distance between the pellet's line of approach and a parallel line through the ball's center. **(b)** If the pellet and target ball have comparable masses, the calculation of the function $\theta(b)$ has to be carried out in the particles' CM reference frame. Find $\theta(b)$ if the pellet has mass m and the ball has mass M. **(c)** Use the result of part (b) to find the deflection angle $\psi(b)$, measured in the lab frame. **(d)** Compute the function $\theta(b)$ for a very massive target ball [cf. part (a)], if the collision is inelastic. Assume the pellet is made of the same rubber as the balls in Example 10.7.

66. In an elastic linear collision among three objects of equal mass, only one has an initial velocity $v_{1,x} = v_o$ (■ Figure 10.42). Conservation of momentum and energy provide only two relations among the three outgoing final velocities. Use the requirement that the objects not pass through each other, $u_1 \leq u_2 \leq u_3$, to demonstrate the following limits:

$$-\frac{v_o}{3} \leq u_1 \leq 0, \quad 0 \leq u_2 \leq \frac{2v_o}{3}, \quad \text{and} \quad \frac{2v_o}{3} \leq u_3 \leq v_o.$$

(Hint: First show that $u_i = -u_j u_k/(u_j + u_k)$, where the index names i, j, and k stand for the subscripts 1, 2, and 3 in any order. Then, show that **both** u_1 and u_2 can be expressed in the form:

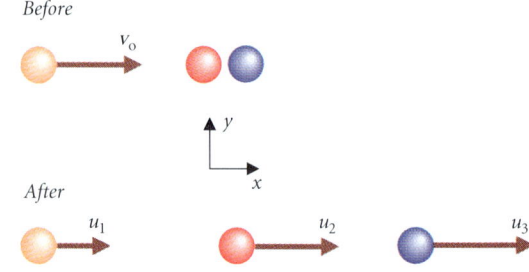

■ FIGURE 10.42

$$u_1, u_2 = \frac{(v_o - u_3) \pm \sqrt{(v_o - u_3)^2 - 4u_3(u_3 - v_o)}}{2}.$$

Finally, consider what conditions determine the maximum and minimum possible values of the velocities and derive the required inequalities.) See N. E. Brown, Impulsive thoughts on some elastic collisions, *The Physics Teacher*, Oct., 1985: 421–422.

67. Show that in the CM reference frame, the fraction of energy dissipated in an inelastic collision of two equal-mass objects is given by:

$$\frac{|\Delta E|}{K_i} = 1 - \epsilon^2,$$

where ϵ is the coefficient of restitution. Does this relation hold in the lab frame? Why or why not? Does the same relation hold if the masses are not equal?

68. If the incoming deuterium and tritium nuclei in a fusion reaction have equal energies $E = 3.2 \times 10^{-15}$ J before the reaction, find exact expressions for the final speeds of the outgoing particles. Assume a one-dimensional collision. What error is made in the final energies of the particles by neglecting E?

69. In an elastic collision between identical spheres, the probability that the particles' velocities will be deflected by an angle in the range θ to $\theta + d\theta$, measured in their CM reference frame, is:

$$dP = \tfrac{1}{2} \sin \theta \, d\theta.$$

If the target sphere is initially at rest in the laboratory, find the deflection angle of the incident sphere in the lab frame, ψ, as a function of θ. Find the probability that this deflection is in the range ψ to $\psi + d\psi$.

Part II Problems

1. ❖ Which of the following operations is permissible? In each case, give reasons for your answer: (a) Adding a displacement vector to a displacement vector. (b) Subtracting a displacement vector from a force vector. (c) Taking the dot product of a displacement vector with a force vector. (d) Taking the dot product of a force vector with a force vector.

2. ❖ A particle moves along a straight line with no forces acting on it. From the point of view of an observer at a point O not on the line, which of the following quantities remains constant? (a) The particle's linear momentum. (b) Its kinetic energy. (c) Its angular momentum about O. (d) Its angular momentum about any other point. Explain your reasoning. How do your answers change if the particle is subject to a force that is always directed toward O?

3. ♦♦ A uranium nucleus of mass 3.9×10^{-25} kg emits an α-particle (helium nucleus) with a mass of 6.7×10^{-27} kg and a speed of 2.2×10^7 m/s. What is the mass of the remnant nucleus? Find the magnitude of its momentum and its speed.

4. ♦♦ At Carisbrooke Castle on the Isle of Wight, water is raised from a well by a donkey who walks continuously inside a vertical wheel (■ Figure II.1). If the donkey has a mass of 100 kg and walks at a speed of 1 m/s, the angle θ is 15°, and the well is 10 m deep, what mass of water per second does the donkey raise from the well? Assume the mechanism is perfectly efficient.

5. ♦♦ A particle moves on a circle of radius R with angular speed ω. If the particle is at $\theta = 0$ at $t = 0$, find the force $\vec{F}(t)$ acting on it as a function of time. Integrate $\vec{F}(t)$ over time to find the impulse it delivers to the particle during a half-period of the motion. Compare your result with the change in the particle's momentum during the same time interval.

6. ♦♦ Jane, of mass 53 kg, is menaced by a large gorilla when Tarzan, of mass 85 kg, jumps from a branch 9.0 m above the ground, swings down hanging from a vine, and grabs her from in front of the startled gorilla. At what speed do they depart from the gorilla? To what maximum height do they rise?

7. ♦♦♦ A particle of mass M moving in the positive x-direction leaves the origin with a speed v_o and is acted on by a force described by the formula:

$$\vec{F} = F_o\left(\frac{x}{x_o}\right)^2 \hat{i}, \text{ for } 0 \leq x \leq x_o; \qquad \vec{F} \equiv 0, \text{ for } x > x_o.$$

Find the speed of the particle as a function of x. (Hint: Use the work-energy theorem.)

8. ♦♦♦ A sled with a mass of 75 kg slides without friction down a 2° slope. Down the slope 1.0×10^2 m from the point where it started, the sled connects with another, identical though stationary sled, and they slide on together to the bottom of the slope, a distance of another 1.0×10^2 m. How long does the whole procedure take?

9. ♦♦♦ Two identical passenger cars initially moving at right angles to each other collide at an intersection. The accident investigator notices that the wreckage slid 15 m diagonally across the intersection. The effective coefficient of friction between wreckage and road surface is $\mu_k = 1.0$. What was the speed of the wreckage immediately after the collision? Neglecting any effects of friction during the collision, use conservation of momentum first to argue that the cars had equal speeds before the collision and second to find the value of that speed. If each car's mass was 1200 kg, how much work was done by the forces that crumpled the cars into wreckage?

10. ♦♦♦ Two blocks of equal mass M are sliding on a frictionless surface at speeds v_1 and v_2 in the same direction. They are connected by a spring of constant k, which is initially relaxed (■ Figure II.2). If $v_2 = 2v_1$, find the maximum compression and the speed of the two blocks when the spring reaches maximum compression.

11. ♦♦♦ Two blocks of mass m are held against identical springs with spring constant k and compressed by the same amount s_o. The blocks are released and slide along frictionless tracks of different

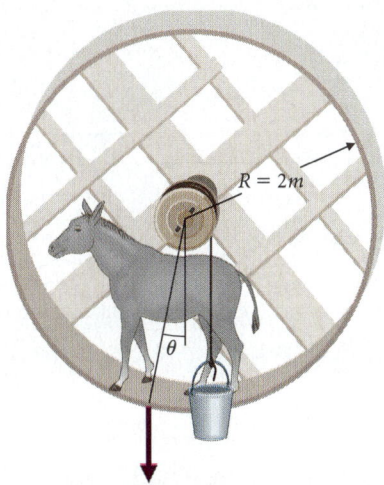

■ Figure II.1

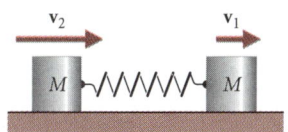

■ Figure II.2

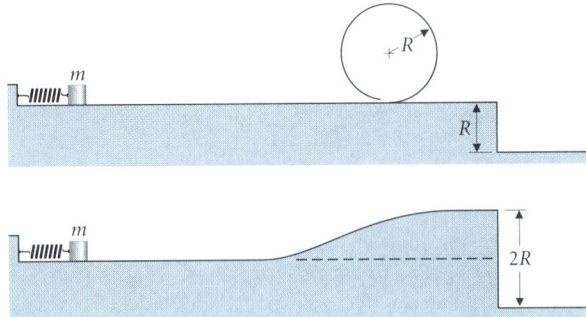

■ FIGURE II.3

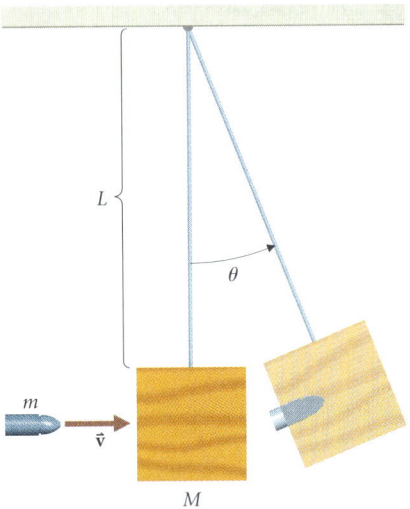

■ FIGURE II.4

shapes (■ Figure II.3). Which block lands a greater horizontal distance from the end of its track?

12. ♦♦♦ A ballistic pendulum is a device for measuring the speed of a bullet. The pendulum consists of a block of wood (mass M) suspended from a cord of length L (■ Figure II.4). A bullet of mass m is fired into the wood, and the pendulum then swings through an angle θ as in the figure. What was the original speed of the bullet?

Challenge Problems

13. ✉ In Example 6.4 we found the momentum change necessary to prevent a cometary nucleus from striking the Earth. Astronomers believe that old cometary nuclei may be made up of loose collections of small particles rather than being one solid mass. In that case, the nucleus would have to be held together by gravity as the impulse is delivered to it. **(a)** Compare the necessary change in the comet's speed with the escape speed from the comet's surface. Would a rapid delivery of the momentum work if the nucleus is a debris pile rather than a single solid object? **(b)** If the nuclear explosions used in the diversion process blow material off the comet with a speed of 3×10^4 m/s, what is the minimum energy required by the process? Express your result in megatons (1 megaton $\approx 10^{15}$ J).

14. ❖ As the Earth rotates, the Moon's tidal effect causes ocean water to flow over the ocean bottom and slosh against the continents. In effect, tidal motion produces a *friction* force that opposes the Earth's rotational motion. As tidal friction converts energy to thermal form, what happens to the Earth's angular momentum? What happens to the Moon's angular momentum? How does the Moon's orbit change? (The Apollo missions left optical reflectors on the Moon. The tidal effect on the Moon's orbit is easily measured by timing laser pulses reflected from these devices.)

Continuous Systems

PART THREE

Newton's laws are stated for *particles,* yet we have successfully applied them to everyday objects such as cars, which are not featureless points. We used the concept of center of mass to show why the particle model succeeds. As long as a system's internal angular momentum and energy are constant, the system behaves like a particle. However, when discussing the propulsion of automobiles (§8.5.3), inelastic collisions (Chapter 10), or a real object rotating about its center of mass, the particle model is inadequate. Clearly we need more powerful models.

RECALL THE DISCUSSION IN §9.3.

The next step is to model real objects as composed of very many particles, each subject to Newton's laws. For most purposes, we do not need to consider individual atoms; instead, we imagine *particles* to be arbitrary volumes very much smaller than a whole object but containing enough atoms that only their *average properties* are important. A dust mote with radius of approximately 10^{-2} mm contains about 10^{14} atoms and might be viewed, for example, as a collection of some 10^7 particles each containing 10^7 atoms.

For now, we do not have to consider how the forces between the particles hold them together as a system. Two rather different approximations give us useful models for solid objects and fluids. In *rigid* bodies like wheels, the internal forces maintain the body's shape independent of external influences: the particles maintain their positions relative to one another, although the object may move as a whole. In *fluids* such as water, the particles cannot separate but are able to slide past each other. Like the simple particle, these models capture important features of reality and allow us to increase our understanding of the physical world. Some properties of matter, such as the bending of a vaulter's pole or the stickiness of crude oil, remain out of reach until we build yet more complex models.

THE FACT THAT WE CAN DEFINE SUCH PARTICLES IS WHAT MAKES A SYSTEM *CONTINUOUS.*

CHAPTER 11
Rigid Bodies in Equilibrium

Concepts

Rigid body
Line of action of a force
Point of application of a force
Force-torque diagrams for rigid bodies
Stable and unstable equilibrium
Couples

Goals

Gain experience in computing torques

Be able to:
Apply the principles of force and torque balance to rigid bodies.
Find the CM of a continuous object.

Admittedly, it is hard to believe that anything can exist which is absolutely solid.

Lucretius

The road sign warns truck drivers about rolling over outwards if they round the turn too fast—yet the motorcycle leans inward! Why doesn't the truck lean inward too? Both trucker and cyclist want to accelerate around the turn without tipping over. To analyze these situations, we need to calculate both forces and torques acting on the vehicles. In this chapter we will learn techniques for solving such problems.

The first step is to understand the structure of a rigid body, exemplified by the bridge at Green River, Utah (■ Figure 11.1). Forces act within the bridge to maintain its shape. We can discover some of the properties of such forces by considering one of the triangles that compose the bridge. For a triangle's shape to remain constant, the lengths of its sides must be fixed. Any external forces tending to pull two joints apart are balanced by tension forces the strut exerts on the joints (■ Figure 11.2). Similarly, if external forces try to push the joints together, the strut is under *compression* and pushes outward to keep them apart. Any rigid body acts like a network of many such triangles: internal forces act along the lines between particles and keep the distances between them constant.

Rigidity has two important consequences:

1. Any motion of a system may be described as the motion of its center of mass (CM) combined with the motions of its parts with respect to the CM. If the system is a rigid body with fixed distances between its particles, the only possible motion with respect to the CM is rotation.
2. The internal forces have no effect on the internal energy or angular momentum of the body and serve only to transmit the effects of external forces. For most purposes, we may ignore the internal forces altogether and consider only the external forces acting on the body.

WE'LL SOLVE THIS PROBLEM IN §11.2.

WE DISCUSSED TENSION IN §5.3.

SEE §9.3.4.

IN §11.6 WE SHALL CALCULATE INTERNAL FORCES IN A TRUSS BRIDGE.

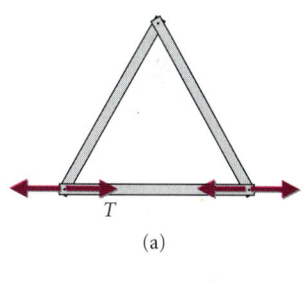
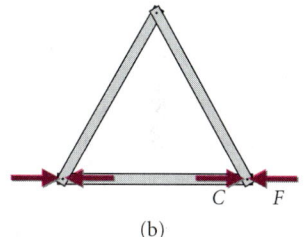

■ FIGURE 11.1
This truss bridge at Green River, Utah, illustrates the structure of a rigid body.

■ FIGURE 11.2
A triangle illustrates the forces that act within a rigid body. (a) If external forces act to stretch the base of the triangle, *tension* forces within the body oppose the external forces. (b) If external forces act to shrink the base, *compression* forces oppose the external forces. In both cases, the triangle maintains its shape until the external forces become so large that the material can no longer resist. The body ceases to be *rigid* and bends or breaks.

11.1 STATIC EQUILIBRIUM

11.1.1 Conditions for Equilibrium

A particle is in equilibrium only if it does not accelerate—that is, if the forces acting on the particle sum to zero. A solid body in static equilibrium has neither linear nor angular acceleration, so the total force acting on it and the total torque about any origin are zero:

$$\sum \vec{F} = 0 \quad \text{and} \quad \sum \vec{\tau} = 0.$$

To analyze torques, we must consider not only the magnitudes and directions of the forces, but also where they act. Usually, we consider a force to be exerted at a single point called its *point of application*. An important special case is weight, which may be considered as acting at the object's center of mass.

IN §9.3 WE USED TORQUE BALANCE TO DEFINE THE CENTER OF GRAVITY.

A force's line of action describes its direction:

> The *line of action* of a force is a line parallel to the force that passes through its point of application (■ Figure 11.3).

The line of action of an object's weight is a vertical line through its center of mass.

EXAMPLE 11.1 ♦♦ A painter has set up a plank of mass $M = 7.5$ kg and length 2.0 m on sawhorses set 1.0 m apart (■ Figure 11.4). If he sets his paint can (mass $m = 10.0$ kg) a distance of 0.25 m from the end of the plank, what normal force does each sawhorse exert on the plank?

MODEL Figure 11.4b is the force-torque diagram for the plank. The weights of plank and paint can are supported by the normal forces $\vec{n}_1$ and $\vec{n}_2$ that the sawhorses exert on the plank. Since the plank is in equilibrium, the total force and total torque on it are both zero.

SETUP Since all forces are vertical, they have only *y*-components:

$$\sum F_y = n_1 + n_2 - Mg - mg = 0.$$

We may choose *any* reference point for computing torques. Here we choose the reference point at O, the top of sawhorse 1. The weight of the plank acts at the midpoint (the CM) of the plank. All the torques are parallel to the *z*-axis, so we set the sum of the *z*-components to zero:

$$0 = \sum \tau_z = \sum (\vec{r} \times \vec{F})_z = \sum xF_y.$$

$$0 = (0.0 \text{ m})n_1 + (1.0 \text{ m})n_2 - (0.50 \text{ m})Mg - (1.25 \text{ m})mg.$$

SOLVE Then:

$$n_2 = (0.50M + 1.25m)g$$
$$= [0.50(7.5 \text{ kg}) + 1.25(10.0 \text{ kg})](9.8 \text{ m/s}^2) = 160 \text{ N}.$$

The value of n_1 follows from the force-balance equation:

$$n_1 = (7.5 \text{ kg} + 10.0 \text{ kg})(9.8 \text{ m/s}^2) - 160 \text{ N} = 12 \text{ N}.$$

ANALYZE The sawhorse nearest the paint can exerts the greater force.
It is usually best to choose a reference point on the line of action of one of the forces, here $\vec{n}_1$. That force then exerts zero torque about the reference point, and the resulting equations are simpler. ■

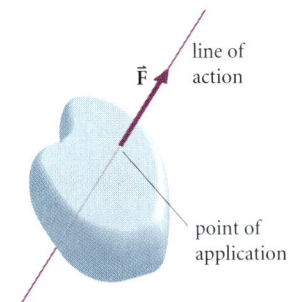

■ **FIGURE 11.3**
Line of action and point of application of a force.

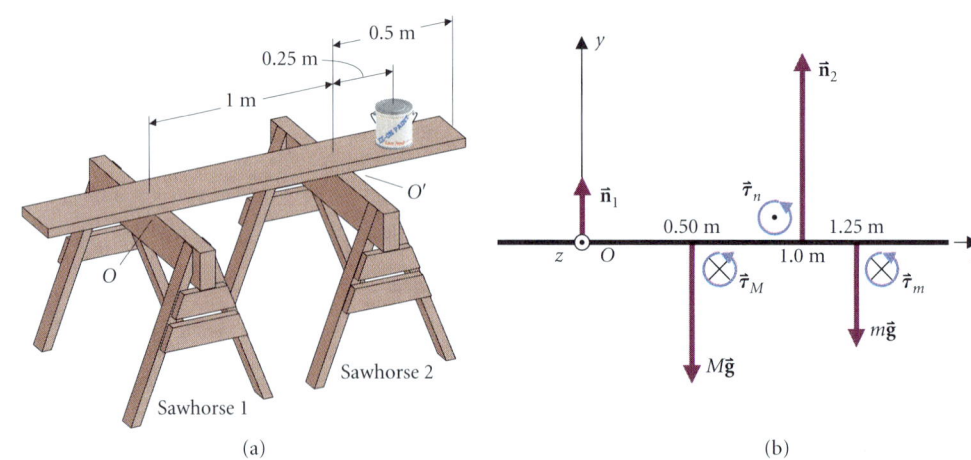

■ **FIGURE 11.4**
(a) Paint can on a plank. (b) Force-torque diagram for the plank. The reference point for torques is chosen to be point O, the top of sawhorse 1. All the torques are in the plus or minus *z*-direction.

EXERCISE 11.1 ♦♦ How close to the end of the plank may the painter put the paint can? (*Hint:* If the can is too close to the end, the plank rotates and topples over. What is the value of n_1 as the plank begins to topple?)

A force's line of action is not always obvious. For example, the forces acting on the shipping crate in ■ Figure 11.5 are its weight $\vec{W}$, friction $\vec{f}$, and the normal force exerted by the ramp, $\vec{n}$. Weight acts at the body's center of mass. Friction acts along the surface where the crate and the plane are in contact. The normal force doesn't act at a single point. Actually, a large number of small normal forces act all over the bottom surface of the crate. Their net torque about the CM balances the torque due to friction. The normal forces are not uniformly distributed across the base, but like weight, $\vec{n}$ has an effective point of application that we may find using torque balance.

EXAMPLE 11.2 ♦♦ The crate in Figure 11.5 has dimensions $\ell = 0.50$ m by $h = 0.75$ m and the ramp is inclined at $\theta = 30°$ to the horizontal. Find the effective point of application of the normal force $\vec{n}$.

MODEL The crate is in equilibrium. By applying force and torque balance, we may find where the normal force acts.

SETUP We choose the x-axis to be along the slope and the y-axis to be perpendicular to it. This choice of coordinates simplifies the calculation since $\vec{f}$ and $\vec{n}$ each have only one nonzero component. The sum of force components in each direction must be zero:

$$0 = \sum F_x = f - W \sin\theta$$

and

$$0 = \sum F_y = n - W \cos\theta.$$

We choose to compute the net torque about the CM of the crate. The weight acts at the CM, and so exerts no torque about it. The friction force acts along the surface between crate and ramp and exerts a torque $fh/2$, out of the paper (Figure 11.5b). The normal force acts at an unknown distance x from the downhill edge of the crate and exerts a torque $n[(\ell/2) - x]$ into the page. These torques sum to zero.

$$0 = \sum \vec{\tau} = \{f(h/2) - n[(\ell/2) - x]\}\hat{o} \text{ (out of the page).}$$

SOLVE From force balance, we have $f = W \sin\theta$ and $n = W \cos\theta$, so:

$$W(\sin 30°)(h/2) = W(\cos 30°)[(\ell/2) - x].$$

$$\frac{1}{2}\left(\frac{3}{8}\text{ m}\right) = \frac{\sqrt{3}}{2}\left(\frac{1}{4}\text{ m} - x\right),$$

or $1/4$ m $- x = \sqrt{3}/8$ m, and $x = (1/4 - \sqrt{3}/8)$ m $= 0.03$ m. The effective line of action of $\vec{n}$ is 0.03 m from the edge of the crate.

ANALYZE The point of application of $\vec{n}$ is important for the equilibrium of the crate. To show why, we discuss torques about point O on the downhill edge of the crate. Because the line of action of the frictional force $\vec{f}$ passes through O, $\vec{f}$ exerts no torque about O, and the torques due to $\vec{W}$ and $\vec{n}$ balance. Now imagine increasing the slope of the incline. As θ increases, the line of action of $\vec{W}$ comes closer to passing through O, so $\vec{W}$ exerts less torque about O. To keep the balance, the torque produced by $\vec{n}$ also decreases; $\vec{n}$'s line of action also comes closer to O. (The crate presses against the plane harder on the downhill side than on the uphill side.) A limit is reached when both $\vec{W}$ and $\vec{n}$ pass through O and the crate is balanced on its edge. Once $\vec{W}$ passes to the left of O (■ Figure 11.6), there is no possible location of $\vec{n}$ that allows torque balance: the crate tumbles over. For stability, the CM must be over the base of support OP.

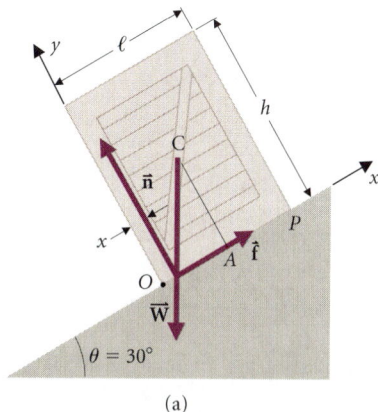
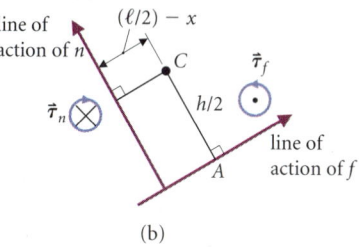

■ **FIGURE 11.5**
Shipping crate on a ramp. (a) The normal force is not uniformly distributed over the base but is concentrated on the downhill side. (b) Lines of action for the forces $\vec{f}$ and $\vec{n}$, together with their perpendicular distances from the center of mass at C. The two forces exert torques in opposite directions about the CM.

NOTE THAT THE SUBTRACTED NUMBERS ARE EACH SIGNIFICANT TO TWO DECIMAL PLACES, AND THEREFORE THE ANSWER IS ALSO.

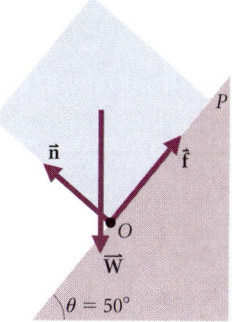

■ **FIGURE 11.6**
This crate cannot be in equilibrium. Its CM is not over its base OP, and $\vec{W}$ exerts an unbalanced torque about O. The crate will tumble over.

FIGURE 11.7
The tall, skinny box should fall over unless it is weighted at the bottom. If the CM is at the box's geometrical center, the weight's line of action falls to the left of the base of support.

We know intuitively that a stable object's CM must be over its base. For example, most people would conclude that the tall, skinny block in ■ Figure 11.7 is glued to the ramp, but are not surprised that the squat block is stationary. A reasonably uniform object has its center of gravity near its geometrical center. We expect the tall, skinny block to tumble because its geometrical center isn't over its base of support. However, it might be made of styrofoam and have a lead brick near its base. Architects can create surprising effects by designing structures that *look* as if they ought to fall over.

11.1.2 Stability

Try to stand a sharpened pencil on its point. You won't succeed, despite the fact that for an absolutely vertical pencil, both torques and forces balance. A pencil lying flat on its side is also in equilibrium and stays put. We say that the pencil on its side is in *stable* equilibrium, while the pencil standing on end is in *unstable* equilibrium.

The difference between the two equilibrium states is illustrated in ■ Figure 11.8, which shows pencils close to, but not quite at, equilibrium. In the unstable case, the torque about the point due to the pencil's weight tends to rotate the pencil away from vertical. In the stable case, the torque about O tends to rotate the pencil back toward equilibrium.

A second way to describe the stability of an equilibrium state is in terms of its energy. The CM of a nearly vertical pencil is lower than the CM of an exactly vertical pencil. Hence the potential energy decreases as the pencil falls away from the vertical equilibrium. In contrast, the pencil on its side increases its potential energy by moving slightly away from equilibrium.

> In *stable equilibrium* a system's potential energy is at a minimum, while in *unstable equilibrium* it is at a maximum.

Pencils with circular and rectangular cross sections (■ Figure 11.9) illustrate two other features of equilibrium. A circular pencil lying on its side experiences neither unbalanced forces or torques nor a change in potential energy if it is moved slightly; it is in *neutral* equilibrium.

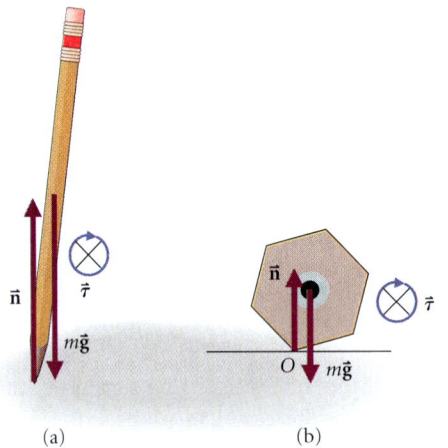

■ **FIGURE 11.8**
(a) Unstable equilibrium. When slightly displaced, the torque about the tip due to the pencil's weight causes the pencil to rotate further from equilibrium and fall down. (b) Stable equilibrium. When slightly displaced, the torque due to the weight rotates the pencil back to its original position on its side.

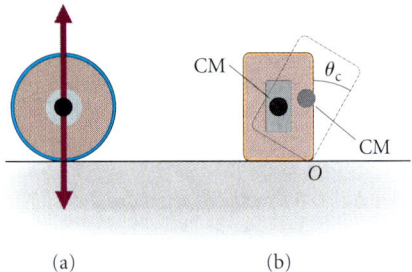

■ **FIGURE 11.9**
(a) The cylindrical pencil is in neutral equilibrium: when displaced, it neither returns to its original position nor moves further away from it. (b) A pencil with a rectangular cross section is stable for small displacements but will fall over onto its long side if displaced through an angle greater than θ_c, where the weight's line of action passes through the corner O of the block. (We use the subscript c for "critical" because the angle θ_c separates the two different kinds of behavior.)

The rectangular pencil is in stable equilibrium resting on its short side. Yet if it is displaced by a large enough angle (greater than θ_c), it will fall over onto its longer side and be in a new stable equilibrium with less potential energy than before. A stable equilibrium need only have less potential energy than other states that differ from it slightly. It need not have the least energy possible for the system.

EXERCISE 11.2 ❖ A ball is placed on the grooved rail shown in ■ Figure 11.10. Describe the stability of its equilibrium at points *A*, *B*, *C*, and *D*.

11.1.3 Couples

When two forces that are equal in magnitude and opposite in direction act on an object, equilibrium results only if the forces act along the same line. Equal and opposite forces that do not act along the same line cause a net torque (■ Figure 11.11). Choosing the reference point to be *O*, where $\vec{F}_2$ is applied, the total torque is due to $\vec{F}_1$ alone and has magnitude $|\vec{\tau}| = |\vec{r} \times \vec{F}_1| = \ell|\vec{F}_1|$, where ℓ is the distance between the parallel lines of action of the two forces. Such a pair of forces is called a *couple*. The torque exerted by a couple is independent of the origin about which it is computed, as illustrated by the following exercise.

EXERCISE 11.3 ♦ Compute the total torque exerted by $\vec{F}_1$ and $\vec{F}_2$ about a second reference point O', which is a distance *x* from *O* (Figure 11.11).

EXAMPLE 11.3 ♦♦ A framed photograph of mass 1.9 kg is hung on a wall using a nail 5.0 cm long. The nail is embedded 3.0 cm into the wall, and the wire supporting the photograph is at the far end of the nail, 2.0 cm from the wall (■ Figure 11.12). What forces does the nail exert on the wall?

MODEL The forces exerted by the nail on the wall are equal and opposite to the forces the wall exerts on the nail. To support the photograph, we might guess that the wall exerts a vertical contact force equal to the photograph's weight. But the weight of the photograph also exerts a clockwise torque about point *O*. To balance this torque, the wall exerts a *downward* force on the left end of the nail. The force exerted at *O* is correspondingly larger to maintain force balance. The forces the wall exerts are actually distributed across the 3-cm-long contact surface between nail and wall (Figure 11.12b). We model them with the force and couple shown in Figure 11.12a.

SETUP Force balance requires $\sum F_y = F_c + n - F_c - Mg = 0$.

SOLVE $n = Mg = (1.9 \text{ kg})(9.8 \text{ m/s}^2) = 18.6 \text{ N} = 19 \text{ N}$ (to 2 sig. fig.)

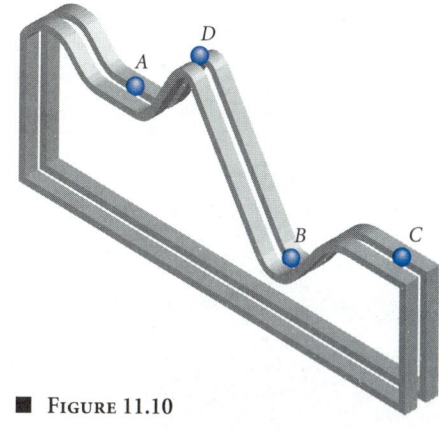

■ **FIGURE 11.10**

DO NOT CONFUSE THESE FORCES WITH NEWTON'S THIRD-LAW PAIRS. THOSE FORCES ACT ON DIFFERENT OBJECTS. THE TWO FORCES IN A COUPLE ACT ON THE SAME OBJECT.

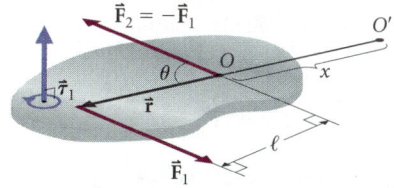

■ **FIGURE 11.11**
A couple consists of two equal and opposite forces that do not have the same line of action. A couple provides the same torque about any origin.

REMEMBER: ROUND TO TWO SIGNIFICANT FIGURES IN THE FINAL RESULT, BUT KEEP THREE IN INTERMEDIATE STEPS TO CONTROL ROUND-OFF ERROR.

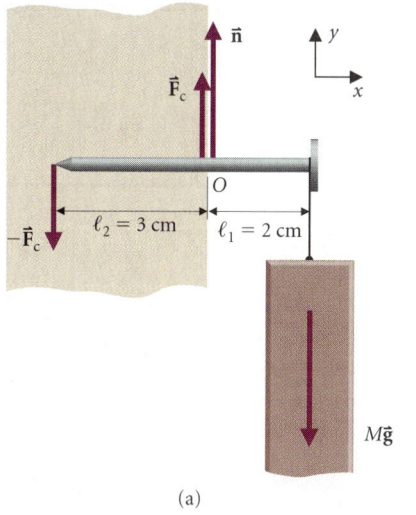
(a)

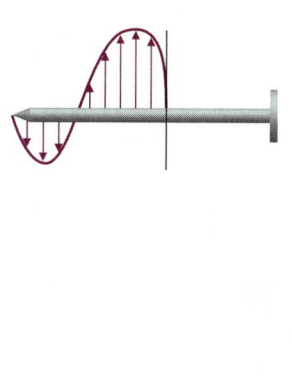
(b)

■ **FIGURE 11.12**
(a) A photograph hung on a wall using a nail. (b) The forces exerted by the wall are distributed along the length of the nail. We model these forces with the normal force $\vec{n}$ and the couple, $\vec{F}_c$ and $-\vec{F}_c$, shown in part (a).

SECTION 11.1 • STATIC EQUILIBRIUM

SETUP We compute torques about point O: $Mg\ell_1 = F_c\ell_2$.

SOLVE
$$F_c = Mg\frac{\ell_1}{\ell_2} = 18.6 \text{ N}\frac{2.0 \text{ cm}}{3.0 \text{ cm}} = 12 \text{ N}.$$

ANALYZE The forces exerted *on the wall* are $F_c + n = 31$ N downward at the edge of the wall and $F_c = 12$ N upward at the end of the nail. Thus the wall must be strong enough to withstand a force considerably greater than the photograph's weight.

EXERCISE 11.4 ♦ A flagpole extends horizontally through a 15-cm-thick plaster wall. The length of pole between the edge of the wall and the CM of the flag is 5.0 m. How heavy a flag can be hung from the pole if the plaster can withstand a force of 1.0×10^3 N exerted by the pole? (Neglect the weight of the pole and model the forces exerted as we did in Example 11.3.)

11.1.4 Three Forces in Equilibrium

There are only two possible situations in which three forces on a body are in equilibrium. In Example 11.1 three parallel forces act on the plank. In Example 11.2 the three forces on the crate are not parallel, but their lines of action intersect at a single point on the crate's base.
■ Figure 11.13 shows a different situation: three forces with lines of action that are not parallel and do not have a common intersection. If the intersection of two of the lines of action, O, is chosen as the reference point and ℓ is the distance from O to the line of action of $\vec{F}_3$, then the net torque about O is $|\vec{\tau}| = F_3\ell$. The object cannot be in equilibrium unless $\vec{F}_3$ or ℓ is zero—that is, unless all three lines of action intersect at O.

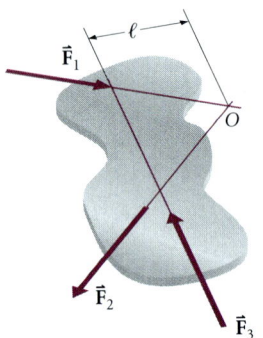

■ **FIGURE 11.13**
This object cannot be in equilibrium unless $\ell = 0$; $\vec{F}_3$ exerts an unbalanced torque about O.

WHEN ONLY THREE FORCES ACT, TORQUE BALANCE REDUCES TO THIS CONCISE RESULT.

> If three forces act on an object in equilibrium, their lines of action must be parallel or must intersect at a point.

EXAMPLE 11.4 ♦♦ While standing on rocky ground, Paul Bunyan, the legendary logger, holds a log against the edge of an ice cliff at a 45° angle (■ Figure 11.14). Point O is three-quarters of the way along the log. Find the magnitude and direction of the force Bunyan exerts on the lower end of the log.

MODEL Three forces act on the log, and they are not parallel. For the net torque on the log to be zero, the lines of action of all three forces must pass through a single point.

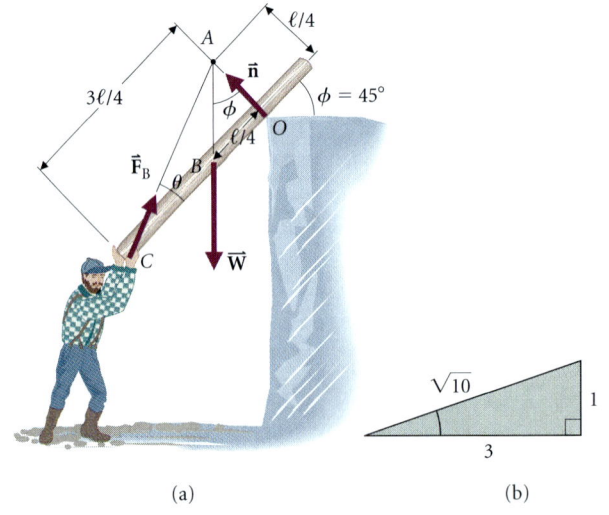

■ **FIGURE 11.14**
(a) What force must Paul exert? Point O, where the log rests on the ice cliff, is three-fourths of the way along the log. (b) Triangle for calculating the angle θ. The hypotenuse has length $\sqrt{1^2 + 3^2}$.

368 CHAPTER 11 • RIGID BODIES IN EQUILIBRIUM

So $\vec{F}_B$ must lie along a line through point A, the intersection of the lines of action of the log's weight $\vec{W}$ and the normal force $\vec{n}$.

SETUP The weight acts vertically downward, while $\vec{n}$ is at right angles to the log and at angle $\phi = 45°$ to $\vec{W}$. Then from the diagram, $OA = OB = \ell/4$, and $\vec{F}_B$ is at an angle θ to the log, where:

$$\tan\theta = \frac{OA}{OC} = \frac{\ell/4}{3\ell/4} = \frac{1}{3} \Rightarrow \theta = 18°.$$

To find the magnitude of $\vec{F}_B$, we balance the force components parallel to the log:

$$\sum F_\| = F_B \cos\theta - Mg \sin\phi = 0.$$

SOLVE So: $\qquad F_B = Mg(\sin\phi)/\cos\theta.$

Since $\tan\theta = \frac{1}{3}$, $\cos\theta = 3/\sqrt{10}$ (Figure 11.14b) and $\sin\phi = \sin 45° = 1/\sqrt{2}$. Then:

$$F_B = Mg\sqrt{5}/3 = 0.745Mg.$$

ANALYZE Not only does Bunyan support more than half the log's weight, he also has to push horizontally to balance the horizontal component of $\vec{n}$.

11.2 Dynamic Equilibrium

The same method of force and torque balance may also be applied to objects that are not at rest, provided that there is no linear or angular acceleration. Such systems are said to be in *dynamic equilibrium*.

EXAMPLE 11.5 ♦♦ Some archaeologists believe that the ancient Egyptians transported building stones up the pyramids by placing wheels around their ends and pulling them up ramps as shown in ■ Figure 11.15. How much force did the workers have to apply to the ropes?

MODEL Since the stone is rolled steadily up the ramp, it has neither linear nor angular acceleration: both force and torque balance apply. Figure 11.15b is the force-torque diagram. We choose coordinates with the x-axis parallel to the ramp, as shown.

SETUP The force balance equations are:

$$\sum F_x = T_1 + T_2 + f - Mg\sin\theta = 0. \quad \text{(i)}$$
$$\sum F_y = n - Mg\cos\theta = 0. \quad \text{(ii)}$$

TRIAL Equation (ii) gives an expression for n, but force balance alone does not allow us to find either the rope tension or the friction. What can we learn from torque balance?

SETUP Choosing the reference point at O, the point of contact of the wheel with the ramp, the total torque on the wheel is:

$$\sum\vec{\tau} = 2RT_1(-\hat{k}) + RW_x(-\hat{k}) = R(Mg\sin\theta - 2T_1)\hat{k} = 0. \quad \text{(iii)}$$

SOLVE From eqn. (iii), the required value of the rope tension is:

$$T_1 = Mg(\sin\theta)/2.$$

ANALYZE This example illustrates the importance of a carefully chosen reference point. By choosing a reference point that is on the line of action of both $\hat{f}$ and $\vec{T}_2$, those unknown forces are eliminated from the torque balance equation. Once T_1 is known, we can find the sum of T_2 and f from force balance. Finding the two forces separately requires a knowledge of the friction between the rope and the wheel.

We may extend the idea of dynamic equilibrium to include objects whose CM accelerates but that do not rotate about their CM. We may apply torque balance if we choose a reference point about which the angular momentum is constant.

■ **Figure 11.15**
(a) The rope is attached to a peg at the top of the ramp and passes around the wheel to the Egyptian workers. Because the wheel is rough, the tension in the rope is different on the two sides. The rough surface of the pyramid also exerts a frictional force on the wheel. (b) Force-torque diagram. Only the upper tension force $\vec{T}_1$ and weight exert torques about the contact point O. The other forces' lines of action pass through O.

Recall that an unbalanced torque causes a change in angular momentum: $\vec{\tau}_{net} = d\vec{L}/dt$.

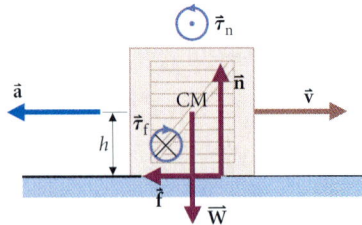

FIGURE 11.16
A sliding packing crate is abraded by friction. Because the normal force is concentrated toward the front of the crate, so are friction and abrasion.

EXAMPLE 11.6 ❖ A packing crate sliding along a horizontal surface is brought to rest by friction. Explain why the damage to the bottom surface is greater near the leading edge than toward the trailing edge.

MODEL The crate is not in equilibrium, but it is not rotating and its velocity has a constant direction. The crate's angular momentum remains zero about any reference point located on a line parallel to the velocity and *through the crate's CM*. Hence the torques about such a reference point must balance. In particular, we may choose a reference point that instantaneously coincides with the CM. The crate's weight exerts no torque about the CM, so the torques about this point due to friction and normal force balance each other (■ Figure 11.16). This can only occur if the normal force acts primarily on the front half of the block, as shown in the figure. Since the frictional force is proportional to n, it is also greater on the front half of the block, and it is friction that damages the crate. ∎

EXERCISE 11.5 ❖❖ Using the same reference point as in Example 11.6, find the point of application of the normal force on the block. Take $\mu_k = 0.5$. ∎

EXAMPLE 11.7 ❖❖❖ A truck with wheels 2.0 m apart and CM 1.1 m above the ground is rounding a turn of radius $r = 26$ m. Assuming that there is ample friction, find the fastest speed at which the truck can safely negotiate the turn.

MODEL The truck's CM is accelerating in uniform circular motion around the turn. However, if we choose our reference point at the center of the circle traced out by the CM, the truck's angular momentum is constant (■ Figure 11.17). (We obtained this result in §9.1.) "Ample friction" means that the friction forces f_1 and f_2 are large enough to provide the necessary centripetal acceleration. Friction also provides a torque about the center of the circle that tends to roll the truck outward. As the speed increases, so does the torque due to friction. The truck rolls slightly outward until the balancing torque due to the normal forces increases to compensate; that is, n_2 increases and n_1 decreases. Once n_1 becomes zero, the balancing torque can no longer increase, and any further increase in speed causes the truck to roll over. The condition $n_1 = 0$ gives the limiting speed.

SETUP We use the coordinate system shown in Figure 11.17. There is no vertical acceleration, so the y-components of the forces sum to zero.

$$\sum F_y = n_1 + n_2 - Mg = 0. \quad \text{(i)}$$

In the x-direction, the friction forces f_1 and f_2 cause the centripetal acceleration v^2/r (cf. Example 4.13).

$$\sum F_x = Ma_x \quad \Rightarrow \quad f_1 + f_2 = Mv^2/r. \quad \text{(ii)}$$

Next, we set the total torque about point O to zero. The z-component is:

$$\sum \tau_z = n_1 r_1 + n_2 r_2 - Mgr - (f_1 + f_2)h = 0. \quad \text{(iii)}$$

LOOK AT THE FIGURE CAREFULLY! THE CENTER OF THE CIRCULAR PATH IS NOT ON THE GROUND BUT AT THE HEIGHT OF THE CM.

ACCELERATION IN CIRCULAR MOTION WAS DISCUSSED IN §3.2.4.

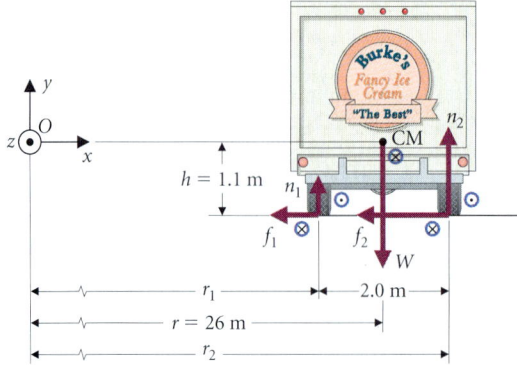

FIGURE 11.17
Force-torque diagram for the truck. Torques are taken about O, the center of the circle traced out by the truck's CM. Notice that O is a distance h above the ground. The truck leans outward slightly to make $n_2 > n_1$.

SOLVE Setting $n_1 = 0$ in eqn. (i), we find $n_2 = Mg$, and then combining eqns. (ii) and (iii) gives:

$$Mg(r_2 - r) = Mv^2h/r.$$
$$v^2 = g(r_2 - r)r/h$$
$$= (9.8 \text{ m/s}^2)(1.0 \text{ m})(26 \text{ m})/(1.1 \text{ m}) = 232 \text{ m}^2/\text{s}^2,$$

and
$$v = 15 \text{ m/s}.$$

ANALYZE The algebraic form of the answer confirms what experience tells us: that the limiting speed decreases as the height of the CM increases or the distance between the wheels $2(r_2 - r)$ decreases, or as the radius of turn r decreases.

EXERCISE 11.6 ❖ Using a similar reference point, explain why a motorcyclist negotiating the same turn as the truck must lean *inward* to balance torques.

11.3 EQUILIBRIUM OF SYSTEMS

In studying the equilibrium of a system of particles, we modeled the system as a number of separate free bodies, each in equilibrium and each experiencing zero net force. The connections, such as strings or contact forces that hold the system together, allow us to relate the forces acting on the different particles. The same techniques apply to a system of rigid bodies, with the additional requirement that torques as well as forces on each body must balance.

SEE §5.2 AND FIGURE 5.4.

EXAMPLE 11.8 ◆◆ In a factory, a pulley system is suspended from a light, horizontal beam supported from above by a cable (■ Figure 11.18). A worker is steadily lifting a gas pump of mass $M = 75$ kg. What forces do the beam and the cable exert on the wall?

MODEL We model the system as four separate objects: the pump, the pulley, the beam, and the wall. Figure 11.18b shows the necessary free-body diagrams and a force-torque diagram for the beam. Since the worker is "steadily lifting," the acceleration of the pump is zero. Neglecting the "light" beam's weight, there are three forces acting on the beam: forces due to the cable, the pulley system, and the wall. Since the cable and pulley forces both pass through the end of the pole, so does the force $\vec{F}_w$ exerted by the wall.

REMEMBER: FOR TORQUE BALANCE, THREE FORCES IN EQUILIBRIUM MUST PASS THROUGH A SINGLE POINT (CF. §11.1.4).

SETUP We analyze the free-body diagrams to find the forces, neglecting the mass of the pulley itself compared with that of the pump:

$$T_1 = Mg \quad \text{and} \quad T_2 = 2T_1 = 2Mg. \quad (i)$$

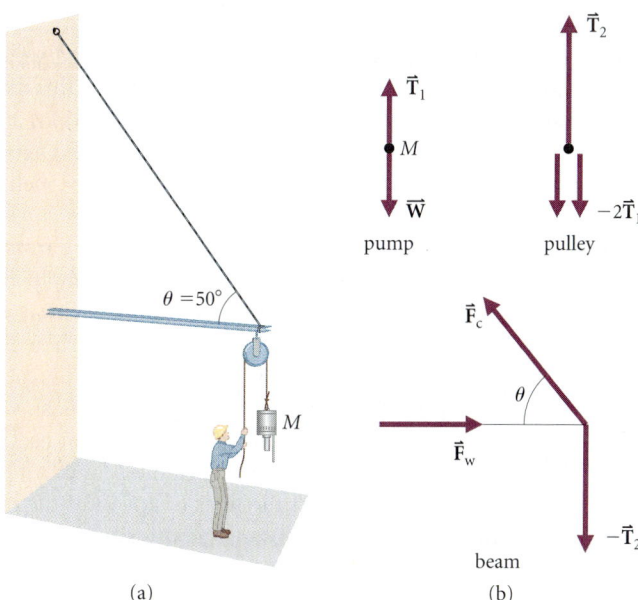

■ **FIGURE 11.18**
(a) A worker uses a pulley to lift a gas pump. (b) Free-body diagrams for pump and pulley, and the force-torque diagram for the beam. To achieve torque balance, the force exerted by the wall passes through the intersection point of $\vec{F}_c$ and $-\vec{T}_2$—that is, the end of the beam. The beam is in pure compression.

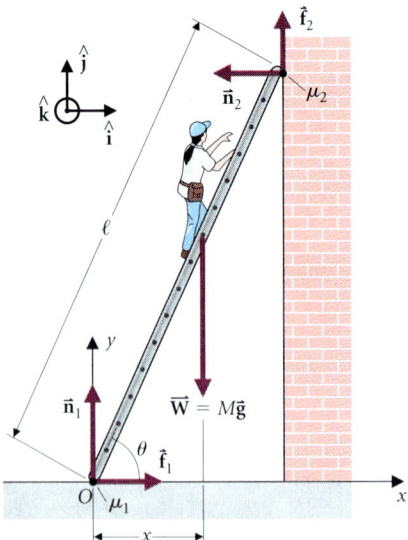

FIGURE 11.19
A carpenter climbs a ladder. With friction at both ends, analysis using a rigid-body model of the ladder fails. The forces acting on the ladder cause it to bend slightly.

Balancing horizontal and vertical force components acting on the beam, we find:

Horizontal components: $\quad F_c \cos\theta = F_w.$ (ii)

Vertical components: $\quad F_c \sin\theta = T_2.$ (iii)

SOLVE From eqns. (iii) and (i),

$$F_c = T_2/\sin\theta = 2Mg/\sin\theta$$
$$= 2(75 \text{ kg})(9.8 \text{ m/s}^2)/\sin 50° = 1.9 \times 10^3 \text{ N}.$$

Then using eqn. (ii):

$$F_w = F_c \cos\theta = (1.92 \times 10^3 \text{ N})\cos 50° = 1.2 \times 10^3 \text{ N}.$$

The force exerted on the wall by the beam is $-\vec{F}_w$, while the force exerted on the wall by the cable is $-\vec{F}_c$.

ANALYZE With both cable and pulley attached at its end, the beam is compressed but not bent—a clever design feature. Without the cable, the wall would have to exert a large couple on the end of the beam (cf. Example 11.3 and Exercise 11.4). The purely compressive load requires much less strength from the wall. In addition, a much thinner beam is needed to support a compressive load than to remain straight under a bending load. ∎

11.4 THE LADDER PROBLEM

A ladder leaning against a wall would seem to be a fairly good example of a rigid body to which the principles of force and torque balance apply. Let's see what we can learn about ladders—there's a surprise in store!

Suppose the ladder is made of light and strong material so that we may neglect its weight. A carpenter (mass M) places the ladder on a floor with a coefficient of friction μ_1 and against a wall with a coefficient of friction μ_2, then climbs partway up (■ Figure 11.19). We use the coordinate system shown, with the origin at the bottom of the ladder, and apply the conditions for equilibrium:

Force balance: $\quad \sum F_x = f_1 - n_2 = 0.$ (11.1)

$\quad\quad\quad\quad\quad\quad \sum F_y = n_1 + f_2 - Mg = 0.$ (11.2)

Torque balance about O:
$$\sum \vec{\tau}_O = -Mgx\hat{k} + f_2\ell\cos\theta\,\hat{k} + n_2\ell\sin\theta\,\hat{k} = 0$$
$$= (Mgx/\ell - f_2\cos\theta - n_2\sin\theta)\ell(-\hat{k}). \quad (11.3)$$

Here is the surprise: we have used all our physical principles, and we have only three equations for four unknowns (f_1, f_2, n_1, n_2). ■ Figure 11.20 illustrates this difficulty geometrically. The carpenter's weight and the contact forces (normal plus friction) at the ends of the ladder must have intersecting lines of action and must sum to zero. The figure shows two of many combinations of contact forces that satisfy these requirements. Modeling the ladder as rigid, we cannot determine which solution actually occurs.

This may be easier to understand if we imagine a gorilla hanging from a flexible pole (■ Figure 11.21). The same set of force and torque balance equations apply to the ladder and to the pole, but here it is clear that the pole's shape depends on the position of the gorilla. To solve the problem, we need to know how the pole changes its shape when a load is applied. The same is true of the ladder problem, even though the ladder flexes a great deal less. However, a ladder that is not rigid is beyond the scope of this book. So what can we do?

Example 11.9 shows how to simplify the ladder problem to obtain useful limits. A sensible carpenter would not attempt to use the ladder on a frictionless floor. After all, the normal force from a frictionless floor could not intersect the lines of action of the other two forces. Using the ladder against a frictionless wall is the most dangerous situation you might encounter. Let's look at it.

WE WEREN'T GIVEN THE LADDER'S MASS, AND IN A PRACTICAL CASE, IT IS LIKELY TO BE MUCH LESS THAN M.

WITH STATIC FRICTION, WE ONLY HAVE A LIMITING RELATION BETWEEN f AND n; $f \leq \mu n$. THERE IS NO GUARANTEE THAT THIS LIMIT WILL BE REACHED AT BOTH ENDS OF THE LADDER SIMULTANEOUSLY.

EVEN WHEN THE RIGID-BODY MODEL IS NOT ADEQUATE FOR A COMPLETE SOLUTION, IT CAN BE USED TO OBTAIN USEFUL LIMITS. INVESTIGATING LIMITING CASES LIKE THIS IS A FAVORITE TECHNIQUE OF PHYSICISTS.

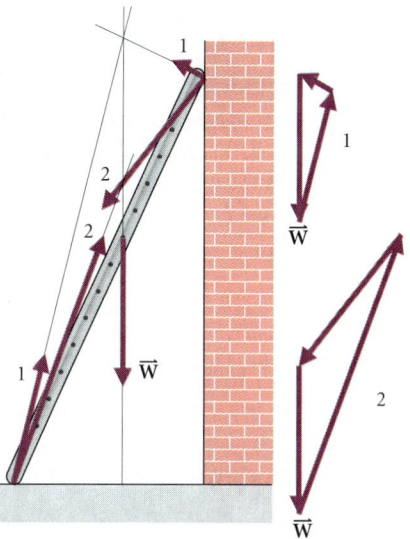

FIGURE 11.20
Either of the pairs of contact forces (normal and friction) labeled 1 or 2 would result in equilibrium of the ladder.

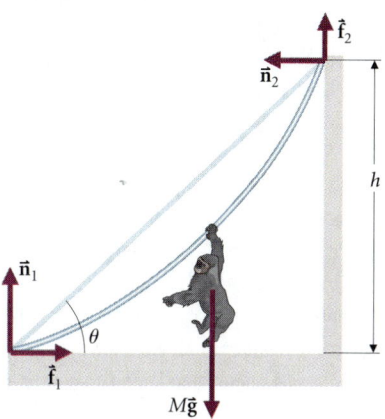

FIGURE 11.21
When a gorilla hangs from a pole, the flex is obvious.

EXAMPLE 11.9 ♦♦ In what range of angles θ may the carpenter safely use the ladder if there is no friction between the ladder and the wall ($\mu_2 = 0$)?

MODEL Safe use means that the ladder does not slide along the floor. The limit of safety occurs when the friction force on the floor reaches its maximum value.

SETUP With $\mu_2 = 0$, $f_2 = 0$. Substituting into eqns. (11.1) through (11.3), we may solve for the other forces. With no friction on the wall, there is only one solution, no matter how the ladder bends.

From eqn. (11.2): $\qquad\qquad n_1 = Mg.$ (i)

From eqn. (11.1): $\qquad\qquad n_2 = f_1.$ (ii)

Then from eqn. (11.3): $\qquad n_2 = \dfrac{Mgx}{\ell \sin \theta}.$ (iii)

For static friction: $\qquad\qquad f_1 \leq \mu_1 n_1.$ (iv)

SOLVE Combining these relations:

$$\frac{Mgx}{\ell \sin \theta} = n_2 = f_1 \leq \mu_1 n_1 = \mu_1 Mg.$$

So at the top of the ladder, where $x/\ell = \cos \theta$:

$$\tan \theta \geq 1/\mu_1.$$

We could also derive the result for f_1/n_1 geometrically using the results of §11.1.4 (■ Figure 11.22).

ANALYZE As the carpenter climbs up the ladder, her weight produces greater torque about the ladder's base. To balance this increased torque, the normal force $\vec{n}_2$ of the wall on the ladder has to increase. But force balance requires the frictional force $\vec{f}_1$ of the floor on the ladder to balance $\vec{n}_2$, and the frictional force is limited by $\mu_1 n_1$. For safety at the top of the ladder ($x = \ell \cos \theta$), the carpenter requires $\mu_1 \tan \theta \geq 1$. ■

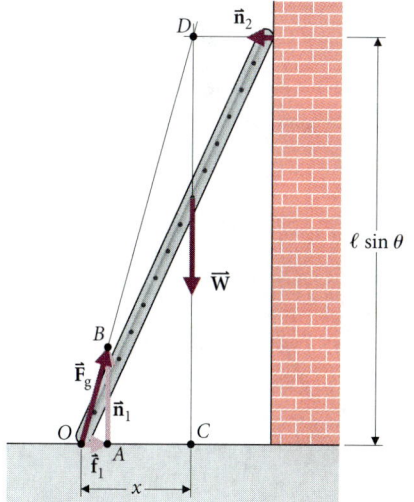

FIGURE 11.22
Ladder against a frictionless wall. In this case, the rigid-body model is adequate, and there is a well-defined solution. The contact force exerted by the ground is $\vec{F}_g = \vec{f}_1 + \vec{n}_1$. The force triangle OAB is similar to the geometrical triangle OCD, so $f_1/n_1 = OC/CD = x/(\ell \sin \theta)$.

11.5 THE CENTER OF MASS OF EXTENDED BODIES

In practical systems, weight is one of the most significant external forces. The weight $\vec{W} = M\vec{g}$ and its torque about any origin are the same as if the entire weight were exerted at the system's center of mass. In Chapter 9 we gave a general definition of the center of mass. Here we shall develop techniques for applying that definition to a continuous object.

For a system of individual particles with masses m_j and positions $\vec{r}_j$:

As we noted in Chapter 9, there is no need to distinguish center of gravity from center of mass for systems near the surface of the Earth.

$$\vec{r}_{CM} \equiv \frac{\sum m_j \vec{r}_j}{\sum m_j}. \qquad (11.4)$$

Equation (11.4) was also eqn. (9.16).

To apply the definition to a rigid body, we need to model the body as a collection of particles. In the simplest case, we know from experience that the CM of a symmetric object made of uniform material is at the object's geometrical center. If a body is composed of several such objects, they can serve as the *particles* in eqn. (11.4).

See Problem 9.96.

EXAMPLE 11.10 ♦ An L shape is cut from a thin, uniform sheet of metal (■ Figure 11.23). Find its center of mass.

MODEL The L shape is composed of two rectangles ABED (particle 1) and BFGC (particle 2). Since the metal sheet is uniform, the mass of each rectangle is proportional to its area, $M = \sigma A$, and its CM is at its geometrical center. (The constant of proportionality σ is called the mass per unit area, or surface density.) The problem then reduces to finding the center of mass for two particles located at the centers of the two rectangles (Figure 11.23b).

SETUP With the origin at A and the x-axis along AF:

$$\vec{r}_{CM} = \frac{M_1 \vec{r}_1 + M_2 \vec{r}_2}{M_1 + M_2} = \frac{\sigma A_1 \vec{r}_1 + \sigma A_2 \vec{r}_2}{\sigma A_1 + \sigma A_2} = \frac{A_1 \vec{r}_1 + A_2 \vec{r}_2}{A_1 + A_2}.$$

Here A_1 is the area of rectangle ABED: $A_1 = L(L/2) = L^2/2$. Similarly, A_2 is the area of rectangle BFGC: $A_2 = (L/2)^2 = L^2/4$.

SOLVE We find the x- and y-components separately:

$$x_{CM} = \frac{A_1 x_1 + A_2 x_2}{A_1 + A_2} = \frac{(L^2/2)L/4 + (L^2/4)3L/4}{L^2/2 + L^2/4}$$
$$= \frac{L^3/8 + 3L^3/16}{3L^2/4} = \frac{5L/4}{3} = \frac{5L}{12}.$$

$$y_{CM} = \frac{A_1 y_1 + A_2 y_2}{A_1 + A_2} = \frac{(L^2/2)L/2 + (L^2/4)L/4}{3L^2/4} = \frac{L(1 + 1/4)}{3} = \frac{5L}{12}.$$

ANALYZE It is not surprising that the CM is on a line of symmetry of the object.

We could model the plate as a uniform square of side L *minus* a square of side $L/2$. Then:

$$x_{CM} = \frac{L^2(L/2) - (L/2)^2(3L/4)}{L^2 - (L/2)^2} = \frac{L}{4}\left(\frac{8-3}{3}\right) = \frac{5L}{12},$$

as we obtained above. This model is equivalent to adding a small square of *negative* mass to the large square. Modeling a hole as equal amounts of positive and negative stuff is a very useful trick. ■

EXERCISE 11.7 ♦ Obtain the same result using rectangles AFGH and HCED as particles. ■

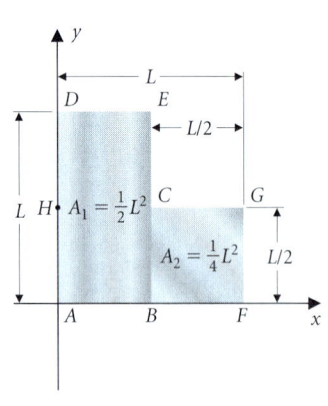

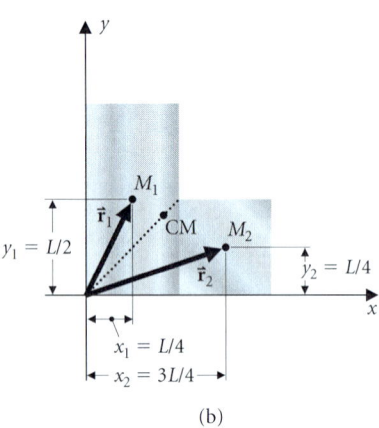

■ **FIGURE 11.23**
(a) The CM of the L shape is found by modeling it as two particles—the rectangles ADEB and BCGF. (b) The CM is on the symmetry line (shown dotted). The coordinates of the two *particles* are shown.

EXERCISE 11.8 ♦ A circular hole of radius $R/2$ is cut from a circular piece of uniform metal sheet of radius R (■ Figure 11.24). Locate the object's center of mass. (*Hint:* The easiest way is to model the object as a uniform circular sheet plus a smaller circular sheet of *negative* mass at the position of the hole.)

A general method for finding the CM of an arbitrary object is to dissect the object into differential pieces, each of which is regarded as a particle. Then we apply eqn. (11.4) by summing over the differential pieces. The sum is a physical integral of the kind we discussed in Interlude 2, and we may write eqn. (11.4) as:

$$\vec{r}_{CM} = \frac{\int \vec{r}\, dm}{\int dm} = \frac{\int \vec{r}\, dm}{M}. \tag{11.5}$$

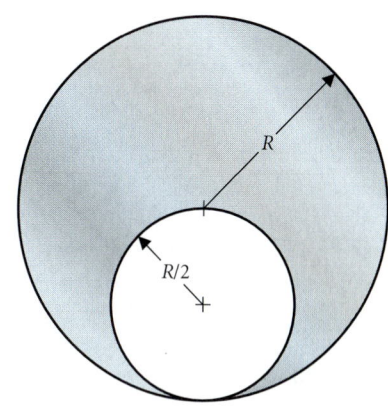

■ **FIGURE 11.24**

For example, we may slice a uniform rod (■ Figure 11.25) into particles with position described by the *x*-coordinate, and eqn. (11.5) becomes:

$$x_{CM} = \frac{\int x\, dm}{M}. \tag{11.6}$$

SINCE THE ROD IS ONE-DIMENSIONAL, WE ONLY NEED ONE COORDINATE TO DESCRIBE THE PARTICLES.

We evaluate this integral using the five-step procedure presented in Interlude 2.

EXAMPLE 11.11 ♦♦♦ Find the center of mass of a uniform rod of length L and total mass M.

MODEL
STEP I Model the system as a collection of differential elements and choose an appropriate coordinate system to describe it.
A convenient coordinate system has the *x*-axis along the rod (Figure 11.25), and we model the rod as a collection of very thin slices, each of thickness dx. We choose the origin O to be at one end of the rod.

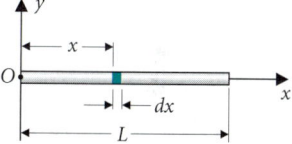

■ **FIGURE 11.25**
We choose the *x*-axis to be along the rod, with the origin at the left end, and slice the rod into differential pieces of length dx.

SETUP Identify a *typical* element and describe it using the chosen coordinates.
STEP II A *typical* slice is one that is not at a special position, such as at one end or at the exact center of the rod. The figure shows a typical slice with an arbitrary value of the *x*-coordinate.

STEP III Calculate the relevant physical quantity for the typical element in terms of the chosen coordinates. It is important that all dependence on the coordinates be shown explicitly.
Here the relevant physical quantities are position and mass. The position is given by the *x*-coordinate. Since the rod is uniform, the mass of any piece of the rod is proportional to its length. Our particle has length dx, so that its mass dm is:

$$\frac{dm}{M} = \frac{dx}{L} \quad \Rightarrow \quad dm = \frac{M}{L} dx.$$

STEP IV Determine the appropriate limits of integration.
We sum over all the particles in the rod with *x*-coordinates ranging from 0 at the one end to L at the other. We have represented the length of a slice by the coordinate change dx. Since length is positive, we make dx positive by proceeding along the rod *from $x = 0$ to $x = L$.*

SOLVE Integrate!
STEP V We have now completed the physical analysis and have a mathematics problem to solve:

$$\int x\, dm = \int_0^L x\frac{M}{L} dx = \frac{M}{L}\frac{x^2}{2}\bigg|_0^L = \frac{M}{L}\frac{L^2}{2} = \frac{ML}{2}.$$

Substituting this result into eqn. (11.6), we find:

$$x_{CM} = (ML/2)/M = L/2.$$

ANALYZE The center of mass is indeed at the center of the rod.

EXERCISE 11.9 ♦♦ It might be better to choose the coordinate origin at the center of the rod. Why? Repeat the calculation using that origin.

*11.6 BRIDGES

Bridge design is one of the most important practical applications of the principles we have been discussing. The simplest form of a bridge is a flat surface supported by rigid beams (■ Figure 11.26). The forces exerted by the bridge supports when a truck of mass M crosses the bridge may be found by direct application of force and torque balance to the beams. For simplicity, we ignore the weight of the beams themselves.

EXERCISE 11.10 ♦ Show that the support forces in Figure 11.26 are
$F_1 = Mg(\ell - x)/\ell$ and $F_2 = Mgx/\ell$.

The simple bridge works well for short spans or light loads. Since the forces on the beams cause bending, the necessary strength, and hence size, of the beams increases rapidly as load or span increases. More complex bridges use tension and compression members to help support the load and keep the bridge rigid. To illustrate how this can be done, we analyze a simple truss bridge with a load at midspan (■ Figure 11.27).

THE *LOAD* IS THE WEIGHT THE BRIDGE CARRIES. THE *SPAN* IS THE DISTANCE BETWEEN THE SUPPORTS.

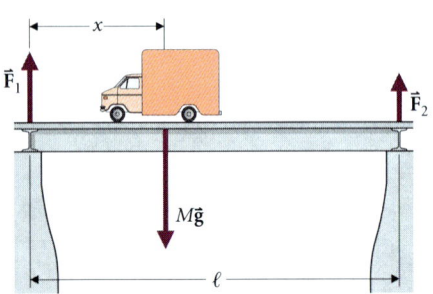

■ **FIGURE 11.26**
A truck crossing a simple bridge. Strictly speaking, the force the truck exerts on the bridge is a normal force. An equal and opposite normal force exerted by the bridge acts on the truck and supports its weight. Thus the magnitude of the normal force equals the truck's weight.

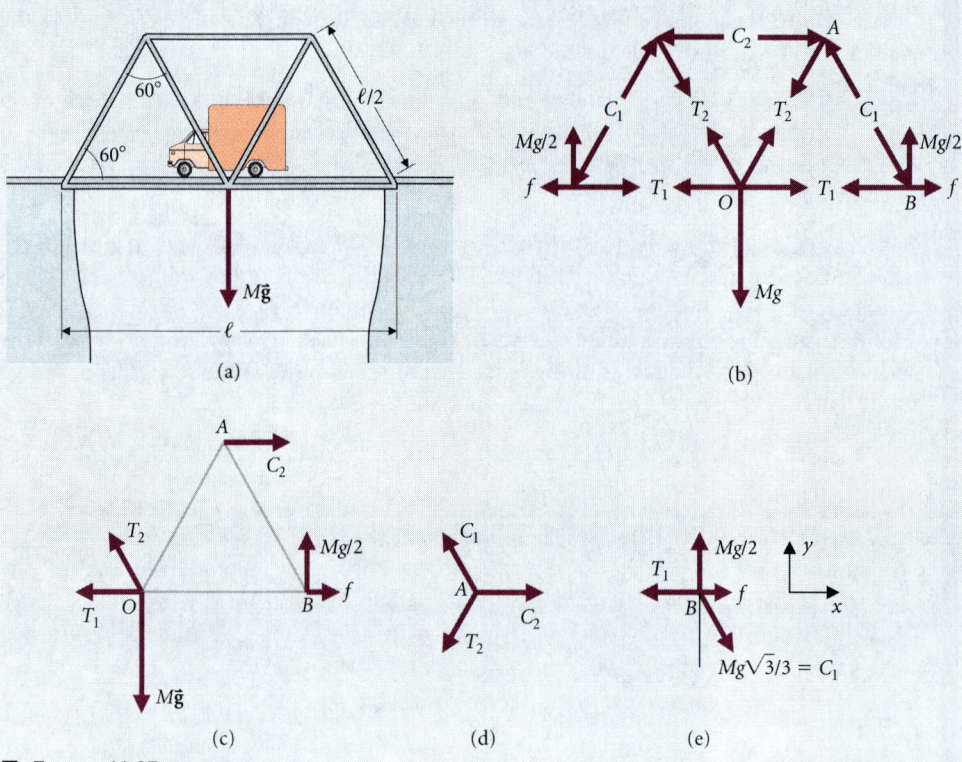

■ **FIGURE 11.27**
(a) A simple truss bridge. The truck is at the center of the span. (b) Tension and compression forces in the members of the truss. (c) External forces acting on the triangular piece OAB. (d) Free-body diagram for the junction at A. (e) Free-body diagram for the juncion at B.

Since the bridge is in equilibrium, the whole bridge, *or any part of it*, is in force and torque balance. In particular, we may model a junction where two or more members meet as a particle; in equilibrium, the net force on it is zero:

> The sum of all the forces exerted by the members and the external load at any junction is zero.

Each individual member is also in equilibrium. The forces exerted on it (forces exerted by the rest of the bridge on its ends, its weight, and any external load) are in force and torque balance. Again neglecting weight, the forces on a member without external load are equal and opposite and their line of action is along the member. Thus:

> If we neglect its weight, any member that does not carry external load is in either pure compression or pure tension.

In solving problems, we can apply force and torque balance to any combination of the bridge's members.

THE FORCES MUST BE EQUAL AND OPPOSITE FOR FORCE BALANCE AND ALONG THE SAME LINE FOR TORQUE BALANCE.

EXAMPLE 11.12 ♦♦♦ Neglecting the weight of the beams, find the tension or compression in each member of the bridge shown in Figure 11.27.

MODEL Three balanced external forces are exerted on the entire bridge: the weight of the truck and the two support forces. The vertical components of the support forces can be found in the same way as for the simple bridge (Exercise 11.10). Since the truck is at the midpoint of the bridge, each vertical component is $Mg/2$. The support forces may also include equal and opposite horizontal components, labeled f in the figures.

We model the bridge as a collection of members (Figure 11.27b). The directions shown for the forces represent educated guesses as to which members are in tension and which in compression.[1] If a guess is wrong, the computed value of the force will be negative. The magnitude will be correct, and the sign will tell us that the guessed direction should be reversed.

The next step is to apply force and torque balance to various parts of the truss to obtain a number of equations equal to the number of unknown forces. Here, fortunately, we obtain equations that determine the unknowns one at a time.

SETUP AND SOLVE Force balance at point O:

$$\sum F_y = 2T_2 \cos 30° - Mg = 0 \Rightarrow T_2 = Mg\sqrt{3}/3. \tag{i}$$

Next, we use torque balance on triangle OAB about reference point B (Figure 11.27c). The only forces acting on the triangle with lines of action not passing through B are C_2 acting at A, and Mg and T_2 acting at O. Thus:

$$\sum \vec{\tau} = (Mg - T_2 \cos 30°)\frac{\ell}{2}\hat{\mathbf{k}} + C_2 \frac{\ell}{2} \sin 60°(-\hat{\mathbf{k}}) = 0.$$

Dividing by $\ell/2$, and using the result of eqn. (i) for T_2:

$$\left[Mg - Mg\frac{\sqrt{3}}{3}\left(\frac{\sqrt{3}}{2}\right)\right] - C_2\frac{\sqrt{3}}{2} = \frac{Mg}{2} - C_2\frac{\sqrt{3}}{2} = 0.$$

---

[1] How did we guess? Since the purpose of the truss is to prevent bending of the horizontal beam, the center diagonals should be in tension. Bending of the bridge would compress the top member, so it ought to be in compression. The lower beam of a simple bridge would be in tension (we can think of it as a very stout cable that bends ever so slightly to support the weight of the truck), so we guess that the horizontal members of the truss are also in tension. The outside members must push up on the top member to oppose the downward forces of the center diagonals, and so should be in compression.

Thus:
$$C_2 = Mg\sqrt{3}/3. \quad \text{(ii)}$$

Force balance at A (Figure 11.27d) now gives the value of compression C_1. Notice that the three forces at A act at angles of 120° to each other and balance only if all three have the same magnitude. Thus $C_1 = C_2 = T_2 = Mg\sqrt{3}/3$.

Finally, we find $T_1 - f$ from force balance at B (Figure 11.27e). The horizontal components are:

$$\sum F_x = C_1 \sin 30° - T_1 + f = 0;$$

so

$$T_1 - f = C_1/2 = Mg\sqrt{3}/6.$$

ANALYZE We cannot determine how this total force is divided between T_1 and f. As in the ladder problem, the magnitude of f depends on small distortions of the structure.

Note that the tensions and compressions are of the same order of magnitude as the applied external load, Mg. Without the truss, the bridge would have to bend so that the tension forces at O have upward components that support the load (■ Figure 11.28): $2T \sin \theta = Mg$, so $T = Mg/(2 \sin \theta)$. If the bridge doesn't bend much, the angle is small, and the tension has to be much larger than the weight it supports. With the smaller forces in the truss design, the bridge members can be much lighter, more than compensating for their increased number.

WHEN ADDED, THE THREE FORCE VECTORS FORM AN EQUILATERAL TRIANGLE.

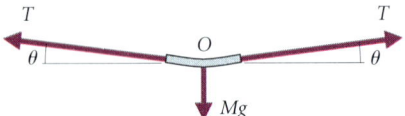

■ **FIGURE 11.28**
Forces on a piece of the roadway of a simple bridge. The bending is grossly exaggerated for clarity. The upward components of the tension have to support the weight of the road plus the load; and with a very small θ, the tension has to be much larger than the weight.

Chapter Summary

Where Are We Now?

In this chapter we have seen how to use the physical principles we have learned to understand the design of mechanical structures. We are now able to go beyond the particle model to discuss the equilibrium of structures that have size and shape but that are rigid.

What Did We Do?

A rigid body—one whose particles maintain constant separation from each other—has a constant shape. The internal forces that maintain that shape serve to transmit external forces and torques, and can be ignored in many problems.

A rigid body is in equilibrium if the net force and the net torque acting on it are both zero:

$$\sum \vec{F} = 0 \quad \text{and} \quad \sum \vec{\tau} = 0.$$

It may be assumed that a force acts at a single point called the point of application and along a line, called the line of action, that passes through the point of application. An object's weight acts at the center of mass, and its line of action is vertical. For a body to rest in equilibrium on a surface, the line of action of its weight must pass through the base of support. Equilibrium may be stable, unstable, or neutral, depending on the tendency of a body to return to or move further away from its equilibrium position following a small displacement.

Two equal and opposite forces whose parallel lines of action are separated by a distance ℓ form a couple. The torque exerted by a couple about any reference point is the magnitude of the force times ℓ. The need to balance torques as well as forces means that couples must often be exerted by supports and bearings. The magnitudes of the forces required from a support are often much greater than that needed for force balance alone.

If three forces act on a body in equilibrium, their lines of action either are parallel or intersect in a single point.

A system of objects may be split into component parts, each of which is separately in equilibrium, just as a system of particles can be split into separate particles, each in equilibrium. To find the center of mass of a rigid body, we imagine the body as composed of small elements (chosen for computational convenience) and sum over the elements using integration techniques:

$$\vec{r}_{CM} = \frac{\int \vec{r}\, dm}{M}.$$

Practical Applications

The principles of equilibrium are widely used in the design of mechanical structures. We discussed several examples, including factory equipment, ladders, and truss bridges.

A simple ladder is a system in which the rigid-body approximation is adequate only for special cases. If the ladder has friction at both ends, it is necessary to understand how the ladder bends to evaluate all the forces acting.

Truss bridges form an example of systems in which the use of suitable supports and struts can reduce the load on any one part of the system, leading to more economical designs. The same ideas are used in the design of buildings, factory equipment, cranes, and scaffolding, as well as more complicated bridges.

Principles of dynamic equilibrium are important in the design of vehicles such as automobiles, airplanes, and trains.

Solutions to Exercises

11.1 When the plank is just about to topple, it loses contact with sawhorse 1: $n_1 = 0$. Then n_2 supports the full weight of plank plus paint can: $n_2 = (M + m)g$. By taking torques about the second sawhorse, we can eliminate n_2. Let x be the distance of the paint can from the end of the plank. Then applying torque balance:

$$0 = \sum \vec{\tau}_{O'} = Mg(0.50 \text{ m})\hat{k} - mg(0.50 \text{ m} - x)\hat{k}.$$
$$(0.50 \text{ m} - x) = (7.5 \text{ kg})(0.50 \text{ m})/(10.0 \text{ kg}) = 0.375 \text{ m}.$$
$$x = 0.50 \text{ m} - 0.38 \text{ m} = 0.12 \text{ m}.$$

The paint can should be placed more than 0.12 m from the end of the plank.

11.2 At all the labeled points in Figure 11.10, the normal force exerted by the rail on the ball is equal and opposite to the weight, and the ball is in equilibrium. At A and C, a small displacement leaves the forces in balance: at these points the equilibrium is neutral. At B, a small displacement to the right results in the situation shown in ■ Figure 11.29a. The net force on the ball is to the left, tending to return the ball to point B. The equilibrium is stable. A small displacement from point D is shown in Figure 11.29b. The net force is to the right, tending to push the ball further away from D: the equilibrium is unstable.

We can also discuss the stability using energy. The ball's potential energy is entirely gravitational. It is a maximum at D, a minimum at B, and is unchanging near points A and C. Thus the equilibrium is unstable at D, stable at B, and neutral at A and C.

Finally, we should note that although point A is a point of neutral equilibrium, if the ball is displaced enough to get it over the hill at D, it will roll toward B. Similarly, if the ball is at C and is given a small velocity toward the left, it will roll all the way to B.

11.3 We need the perpendicular distances $O'B$ and $O'C$ from O' to the lines of action of $\vec{F}_1$ and $\vec{F}_2$, respectively (■ Figure 11.30). Since $OO' = x$, $O'C = x \sin \theta$, and $O'B = O'C + CB = O'C + OA = x \sin \theta + \ell$. Then the torque about O' has magnitude $F_1(\ell + x \sin \theta) - F_2(x \sin \theta) = F_1 \ell$, as before.

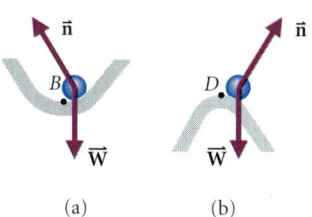

■ FIGURE 11.29
(a) When the ball is displaced from B, the net force tends to return it to B. (b) When the ball is displaced from D, the net force accelerates it away from D.

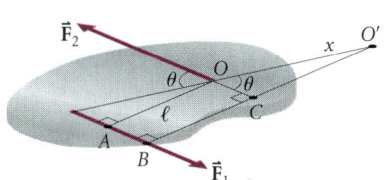

■ FIGURE 11.30

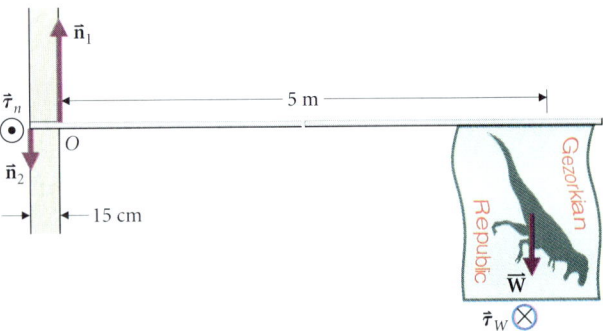

■ FIGURE 11.31
The forces acting on the flagpole are the weight of the flag and the normal force exerted by the wall, here modeled as two forces n_1 and n_2. The vectors are not drawn to scale: $n_1 = 34W$!

11.4 There is a torque about point O due to the weight of the flag (■ Figure 11.31). A couple exerted by the wall balances this torque. We model the forces exerted by the wall with the two normal forces $\vec{n}_1$ and $\vec{n}_2$. From force balance, the magnitude of the total upward force equals the weight of the flag:

$$n_1 - n_2 = W.$$

From torque balance, the magnitude of the total couple equals the torque due to the flag:

$$n_2(0.15 \text{ m}) = W(5.0 \text{ m}).$$

Thus: $\quad n_2 = W(5.0/0.15) = 33.3W.$
Then: $\quad n_1 = n_2 + W = 34.3W.$

The forces exerted on the wall by the flagpole are equal and opposite to the forces exerted by the wall on the flagpole. Thus the maximum force exerted on the wall is $34W$. This force must be less than 1.0×10^3 N, so:

$$Mg = W < (1.0 \times 10^3 \text{ N})/34.3 = 29.2 \text{ N}.$$

The maximum flag mass is $M = W/g = (29.2 \text{ N})/(9.8 \text{ m/s}^2) = 3.0$ kg.

11.5 Since the weight can be considered to act at the CM, we need only consider torques due to the frictional and normal forces. Let the normal force act a distance x ahead of the CM. Then:

$$0 = |\vec{\tau}_{CM}| = nx - fh = nx - \mu_k nh.$$

Thus $x = \mu_k h = 0.5h.$

11.6 As with the truck, friction of the road on the tire provides the force that accelerates the cycle's CM in circular motion (■ Figure 11.32). This force provides a torque tending to roll the cycle outward. Vertical force balance requires:

$$n = W.$$

If the cyclist leans inward, $\vec{W}$ and $\vec{n}$ form a couple whose torque balances the torque due to $\vec{f}$. If the cyclist leans outward, the couple and $\vec{f}$ both exert torques in the same direction, and the cycle falls over.

11.7 Rectangle $AFGH$ has area $L^2/2$ and is centered at the point $(L/2, L/4)$. Rectangle $HCED$ has area $L^2/4$ and is centered at $(L/4, 3L/4)$. Thus the center of mass is at:

$$x_{CM} = \frac{(L^2/2)L/2 + (L^2/4)L/4}{L^2/2 + L^2/4}$$

$$= \frac{L^3/4 + L^3/16}{3L^2/4}$$

$$= \frac{5L}{12}.$$

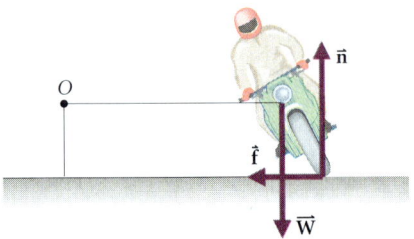

■ FIGURE 11.32
Forces acting on the motorcycle.

$$y_{CM} = \frac{(L^2/2)L/4 + (L^2/4)3L/4}{L^2/2 + L^2/4}$$

$$= \frac{L^3/8 + 3L^3/16}{3L^2/4}$$

$$= \frac{5L}{12}.$$

This is the same position that we found in Example 11.10.

11.8 We consider the sheet to be a complete circle of positive mass and a smaller circle of negative mass. Where they overlap, equal contributions of positive and negative sum to zero, forming the hole. Since the sheet is uniform, mass is proportional to area, and the center of mass of each circle is at that circle's center. We choose a coordinate system with the origin at the center of the big circle and the x-axis horizontal. Then the center of mass is on the y-axis with y-coordinate:

$$y_{CM} = \frac{\pi R^2(0) + [-\pi(R/2)^2](-R/2)}{\pi R^2 - \pi(R/2)^2} = \frac{R/8}{3/4} = \frac{R}{6}.$$

11.9 If the origin lies on a line of symmetry of the rod, one of the CM's coordinates is zero. With the origin at the center of the rod, we expect all the coordinates to be zero. Let's check it (■ Figure 11.33). Steps II, and III are the same as in Example 11.11. The limits of integration become $x = -L/2$ to $x = +L/2$. Thus:

$$x_{CM} = \frac{1}{M}\int_{-L/2}^{+L/2} x\frac{M}{L}\,dx = \frac{1}{L}\frac{x^2}{2}\Big|_{-L/2}^{+L/2}$$

$$= \frac{1}{2L}\left(\frac{L^2}{4} - \frac{L^2}{4}\right) = 0.$$

Once again, we find that the CM is at the center of the rod.

11.10 As usual, we balance forces and torques. With the reference point at the left-hand end of the bridge and the y-axis vertical:

$$\sum F_y = 0 = F_1 + F_2 - Mg$$

and

$$\sum \tau_z = 0 = F_2\ell - Mgx.$$

Then $F_2 = Mgx/\ell$ and $F_1 = Mg - Mgx/\ell = Mg(\ell - x)/\ell.$

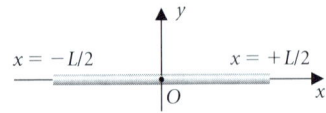

■ FIGURE 11.33

380 CHAPTER 11 • RIGID BODIES IN EQUILIBRIUM

Basic Skills

Review Questions

- Explain why the triangles in a truss bridge provide a model for the forces in a rigid body.
- What are two important consequences that follow from the property of rigidity?

§11.1 STATIC EQUILIBRIUM

- State two conditions for a rigid body to be in static equilibrium.
- What is the *point of application* of a force?
- What is the *line of action* of a force?
- Explain why the normal force acting on the base of a crate on an incline is not uniformly distributed over the bottom of the crate.
- What is meant by *stable equilibrium*? *Unstable equilibrium*? *Neutral equilibrium*? How can you distinguish the three using force or torque balance? Using energy considerations?
- What is a *couple*? Explain why couples are significant in analyzing the equilibrium of systems.
- What are the two possible configurations of three forces in equilibrium?

§11.2 DYNAMIC EQUILIBRIUM

- What is the condition for dynamic equilibrium? Explain why the analysis of dynamic equilibrium uses the same principles as for static equilibrium.
- Explain what is necessary to extend the idea of dynamic equilibrium to systems undergoing linear acceleration.

§11.3 EQUILIBRIUM OF SYSTEMS

- What are the similarities and differences between the methods of analysis for systems of particles and for systems of rigid bodies?

§11.4 THE LADDER PROBLEM

- What surprise emerges from the analysis of a ladder in equilibrium against a wall?
- What physical reason is responsible for this surprise, and why do we have to sidestep the problem? What sensible class of ladder problems can we solve with a rigid-body model of the ladder?

§11.5 THE CENTER OF MASS OF EXTENDED BODIES

- Describe how to find the center of mass of an object that can be modeled as a combination of symmetric objects.
- Describe the procedure for finding the CM of a continuous, non-symmetric object. In your discussion, review each of the five standard steps for integration problems.

* §11.6 BRIDGES

- State two principles for analyzing a bridge that follow from modeling the bridge as a collection of members and joints.

Basic Skill Drill

§11.1 STATIC EQUILIBRIUM

1. The structures shown in ■ Figure 11.34 are cut from uniform slabs of material. Which of them is in equilibrium? Justify your answers.
2. A wheelbarrow 0.91 m long is loaded with 25 kg of dirt so that its CM is 0.40 m from the front wheel (■ Figure 11.35). What force must the gardener apply to the handles to move the dirt, if the wheelbarrow makes an angle of 35° with the horizontal?

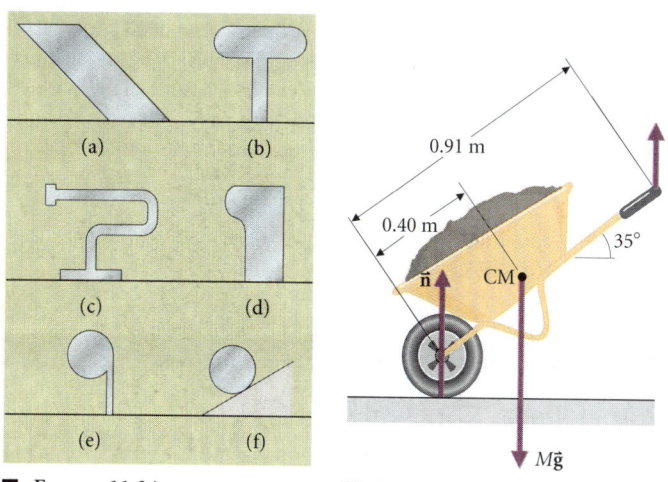

■ FIGURE 11.34   ■ FIGURE 11.35

3. A bench 1.0 m long has a total mass of 12 kg. Its legs are each 45 cm high, 0.25 m from an end, and of mass 3.0 kg. How close to one end can a 75-kg person sit?
4. What total force and couple are equivalent to the forces on the object shown in ■ Figure 11.36? Use point O as the reference point.

§11.2 DYNAMIC EQUILIBRIUM

5. A double-decker bus must turn corners more slowly than a sports car. Why?
6. A cyclist with CM 0.92 m above the ground (when riding straight) is riding around a turn with radius 20.0 m at 11 m/s. At what angle must the cyclist lean?

§11.3 EQUILIBRIUM OF SYSTEMS

7. Can the system in ■ Figure 11.37 be in equilibrium for any value of m, if $\mu = 0$? If $\mu = 1.0$, is equilibrium possible for $m = M$?

■ FIGURE 11.36   ■ FIGURE 11.37

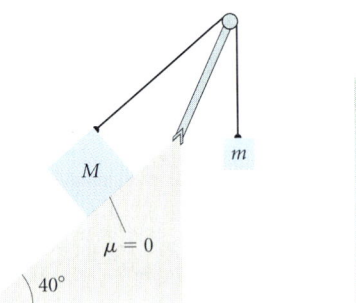

FIGURE 11.38

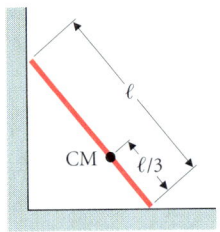

FIGURE 11.39

12. A rod 2.0 m long is made of two halves of different material, but of the same diameter. If the left half is of brass (density $\rho = 8.6 \times 10^3$ kg/m^3) and the right half is of iron ($\rho = 7.8 \times 10^3$ kg/m^3), where is the center of mass of the rod?

✱ §11.6 BRIDGES

13. A bench is designed with triangular braces as shown in ■ Figure 11.41. If a person of mass 85 kg sits in the middle of the bench, what is the compression in each brace?

8. If the ramp shown in ■ Figure 11.38 is frictionless and $M = 10$ kg, can the system be in equilibrium? If so, for what value of m?

§11.4 THE LADDER PROBLEM

9. The CM of an advertising sign of length ℓ is a distance $\ell/3$ from the bottom. The sign is set up against a wall as shown in ■ Figure 11.39. If both floor and wall are frictionless, is equilibrium possible? Discuss.

10. A ladder is set against a frictionless wall at an angle of 28° with the wall. Ignoring the ladder's mass, how high can a 72-kg electrician climb if the coefficient of friction between ladder and floor is 0.51?

§11.5 THE CENTER OF MASS OF EXTENDED BODIES

11. The T shape shown in ■ Figure 11.40 is cut from a uniform sheet of metal. Where is its center of mass?

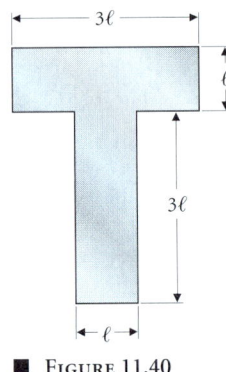

FIGURE 11.40

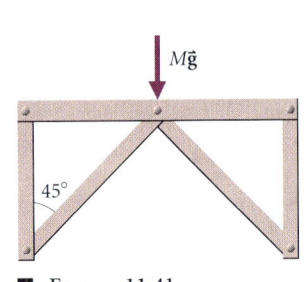

FIGURE 11.41

Questions and Problems

§11.2 STATIC EQUILIBRIUM

14. ❖ The Leaning Tower of Pisa (■ Figure 11.42) continues to lean further each year. It will eventually fall down. Why?

15. ❖ A cubical packing crate of side L contains an unknown object. The crate balances over the center of its base and is observed to tip over when placed on a ramp with an angle of 60° or more. Where is the CM of the crate and its contents?

16. ❖ Which of the following equilibrium states are stable, which unstable, and which neutral? (a) A cone standing on its point. (b) A cone on its base. (c) A cone lying on its side.

17. ❖ Why does a tall radio tower have guy wires around it?

18. ❖ Two friends on a canoe trip have to carry the canoe overhead on a portage between two lakes. How should they place themselves so that the stronger friend carries 50% more load than the weaker friend?

19. ◆ Mike, of mass $M = 53$ kg, sits on the end of a seesaw 2.0 m from the balance point. Jennie, of mass $m = 25$ kg, sits on the other side 1.0 m from the balance point. What force must their father exert on Jennie's end of the seesaw to hold it horizontal? Where should Mike and Jennie's father ask them to sit to make the required force zero?

20. ◆ What is the maximum mass the dredging crane in ■ Figure 11.43 can lift at one time without toppling if it has a mass $M = 2 \times 10^4$ kg?

FIGURE 11.42

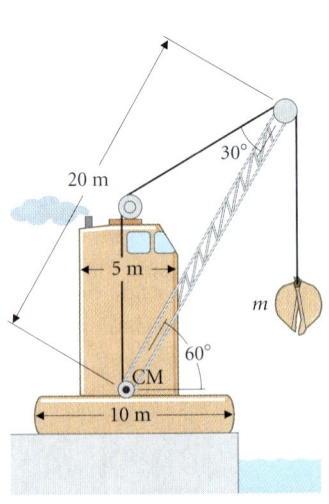

FIGURE 11.43

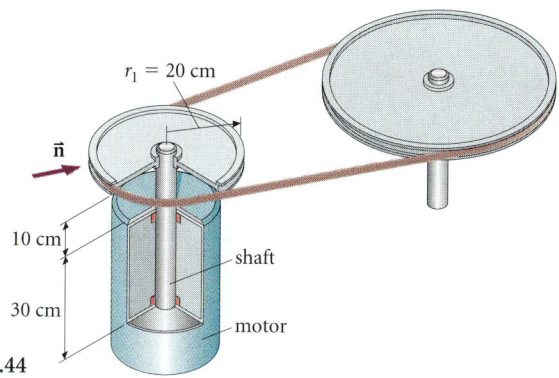

FIGURE 11.44

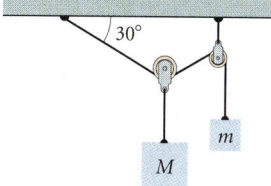

FIGURE 11.45

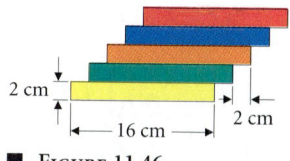

FIGURE 11.46

21. ♦ When both the front and back wheels of a truck are on the highway patrol's scales, the scales register a force of 5.30×10^4 N. When only the front wheels are on the scales, they register 2.45×10^4 N. What force will be registered when only the truck's back wheels are on the scales? If a truck's CM may be no more than 10% off-center between its wheels, is this truck legal?

22. ♦ A street sign is a metal rectangle 65 cm by 12 cm with a mass of 1.0 kg. Its short side is attached to a vertical post with a bracket. What force and couple must the bracket exert?

23. ♦ The belt of the machine shown in ■ Figure 11.44 exerts a normal force of 1.2×10^5 N on the small pulley. The shaft is supported on two bearings inside the motor, 10.0 cm and 40.0 cm from the pulley. What force and couple must the two bearings exert? What force is exerted by each bearing?

24. ♦♦ What mass m must be hung on the end of the rope in ■ Figure 11.45 for the system to be in equilibrium? Is the equilibrium stable?

25. (a) ❖ When climbing a steep mountainside, people often tend to lean forward, toward the hill. Why is this a poor practice?
(b) ♦ Compute the friction force on a climber's feet when standing on a mountainside with a 60° slope. What minimum coefficient of friction between shoes and rock is required? (c) ♦♦ If the climber leans forward (CM 20 cm horizontal distance in front of her feet) and puts her hands on the rock 60 cm in front of her feet, what is the maximum friction force that can act on her feet? Take μ at both hands and feet to have the value you found in part (b). What magnitude of force must her hands support?

26. (a) ♦ A vise of mass 30.0 kg sits 0.500 m from the end of a workbench 3.25 m long. The bench, of mass 25.0 kg, has two legs, one 0.750 m from each end. What are the normal forces exerted by the floor on each leg of the bench? (b) ♦♦ The factory designer can have more flexibility in placing tools if the workbenches are bolted to the floor. If the vise is moved to the end of the bench, what are the forces exerted by the bolts on each leg of the bench?

27. ♦♦ A child is stacking 6 cm × 2 cm × 16 cm blocks one on top of the other, as shown in ■ Figure 11.46. If each is displaced 2 cm from the next, how many blocks may be stacked before the stack falls down?

28. ♦♦ Find the maximum angle that the Leaning Tower of Pisa can lean before it falls down, assuming that it is a cylindrical shell of uniform material of inner radius $r = 5$ m and outer radius $R = 6$ m and height $h = 30$ m.

29. ♦♦ A diving board of mass 45 kg and 5.0 m long is supported by a post at one end and a second post 0.50 m away from the first. If a 75-kg diver stands on the pool end of the board, what forces must be exerted on the board by the posts?

30. ♦♦ A spherical boulder has fallen into a crack in a glacier and is wedged between two icy, plane surfaces (■ Figure 11.47). What force does the boulder exert on the vertical ice surface?

31. ♦♦ Prove that any set of forces acting on a rigid body is equivalent to a single force together with a couple. Does the representation depend on choice of reference point?

32. ♦♦ A bookshelf 25 cm wide and of mass 5.2 kg is to be attached to a wall by two right-angled brackets with 22-cm-long arms (■ Figure 11.48). Screw holes are located at 3.0, 10.0, and 17.0 cm from the end of each arm, but the carpenter has installed only the screws nearest the corners when lunchtime arrives. If the carpenter leaves 12 kg of tools on the shelf, with CM 11 cm from the wall, what total force must the screws exert during lunchtime?

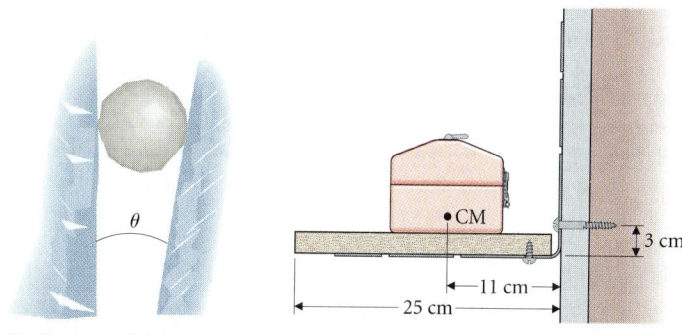

FIGURE 11.47　　**FIGURE 11.48**

33. ♦♦ An equipment rack holds electronic equipment by means of four screws at the corners of the front face of each piece of equipment (■ Figure 11.49). An amplifier of mass 9.7 kg is 18 cm deep and has a faceplate 48 cm (horizontal dimension) by 9.0 cm (vertical dimension) with the screws located 2.5 cm horizontally and 0.5 cm vertically from the edges. Find the total vertical force exerted by the screws and the horizontal force exerted by the upper pair of screws and by the lower pair. The equipment is installed by sliding it in and screwing the edges of the faceplate to the vertical bars of the rack behind the faceplate. Which screws should you put in first when installing the equipment in the rack?

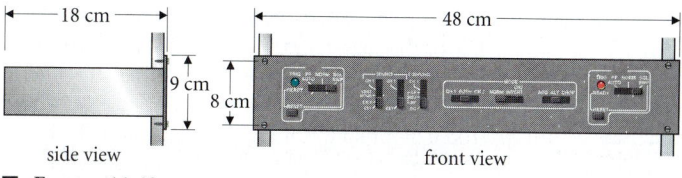

FIGURE 11.49

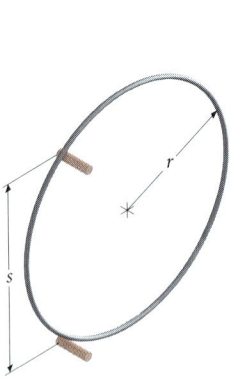
FIGURE 11.50

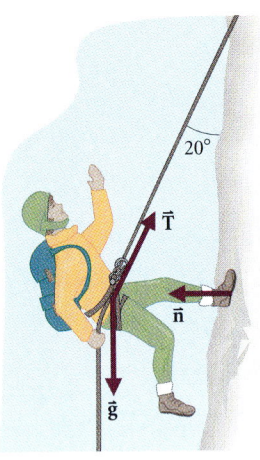
FIGURE 11.51

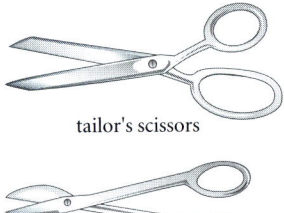
tailor's scissors

roofer's metal shears
FIGURE 11.52

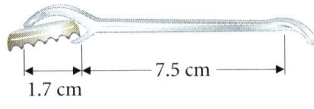
7.5 cm
1.7 cm
FIGURE 11.53

34. ♦♦ A circular steel hoop of radius R and mass M is supported in a vertical plane by two smooth, horizontal pegs (■ Figure 11.50). The points of contact of the hoop with the pegs are along a vertical line separated by a distance s, as shown. Find the forces exerted by the pegs on the hoop. Does your result change if one peg is smooth and the other rough? Why or why not?

35. ♦♦ Prove the following theorem: given a set of forces, acting in a plane, for which neither the net force nor the net torque is zero, there is a unique line of action along which a single force may be applied to produce equilibrium. Does the theorem hold for a set of forces acting in three dimensions? If not, under what conditions does it fail?

36. ♦♦ A door to the vault at Downtown Bank is 2.0 m high and 1.0 m wide. Its mass is 1.75×10^3 kg, and it is supported by two hinges located 25 cm from the top and the bottom of the door. What is the *horizontal* component of the force each hinge exerts on the door?

37. ♦♦♦ ■ Figure 11.51 shows an intrepid mountaineer performing a rappel, using rope friction to control the rate of slide down the rope. In a properly executed rappel, the climber's feet are flat against the wall. **(a)** If the wall is vertical and icy, and his body is held straight, where should the climbing rope be attached? **(b)** What difference, if any, does it make if his posture is as shown in the figure? **(c)** Suppose the climber's mass is M, his CM is a distance $d = 0.8$ m from his feet, and he wears a backpack of mass $M/4$, a distance $d/2$ above his CM (that is, closer to his head). What angle of bend at his ankles would maintain him in equilibrium with his body held straight? **(d)** What is the effect if he bends his body upward, as shown in the figure? If the bend is through an angle of $45°$ around a point $0.9d$ from his feet, what angle is required at his ankles? **(e)** If the face were rock, what minimum coefficient of friction would be necessary for the climber with his backpack on to hold his body straight and perpendicular to the wall?

38. ♦♦♦ A rectangular packing box 35.0 cm tall, sitting on a loading ramp, contains a cylindrical lamp of diameter 15.0 cm and mass 12.0 kg. The dimension of the box along the slope of the ramp is 25.0 cm, its mass when empty is 2.1 kg, and its CM is at its geometrical center. Due to a packing error, the lamp has rolled to the bottom corner of the box. Find all the forces acting on the box and on the lamp. What is the angle of the steepest ramp the box may sit on without toppling? Compare with the maximum angle possible if the lamp were still centered in the box.

39. ♦♦♦ A uniform rod 0.50 m long is hung from a hook by two strings of lengths 0.30 m and 0.35 m attached to its ends. At what angle to the horizontal will the rod hang?

§11.2 DYNAMIC EQUILIBRIUM

40. ❖ A cyclist who uses only the front brakes in a quick stop risks tumbling over the handlebars. Why?

41. ❖ Compare a tailor's scissors with a roofer's metal shears (■ Figure 11.52). Explain the difference in design.

42. ♦ A crowbar 2.00 m long is used to open a crate. The bar is inserted so that 10.0 cm of the bar is underneath the lid, and a force of 10.0 N is applied to the other end. Use torque balance to find the force exerted on the lid of the crate.

43. ♦ A pair of scissors measures 5.5 cm from the center of the handle to the pivot and 7.5 cm from the center of the pivot to the tip of the blade. If the user exerts a force of 0.50 N at the handle, what force is exerted at the tip of the blade?

44. ♦ What is the mechanical advantage of the bottle opener shown in ■ Figure 11.53?

45. ♦♦ Water is drawn from a well using a crank as shown in ■ Figure 11.54. Use torque balance to find the force on the crank needed to lift a 30-kg bucket of water.

46. ♦♦ An adjustable crescent wrench has a handle 24 cm long and a maximum jaw opening of 3 cm. What is the range of mechanical advantage available?

47. ♦♦ Compare the (normal) force that you must exert to open a screw-top bottle with a radius of 2 cm **(a)** with your bare hands and **(b)** with pliers 23 cm long. The coefficient of friction between your skin and the bottle top is 0.7, while that between the metal of the pliers and the bottle top is 0.5.

48. ♦♦ You have a crowbar 2.00 m long that has negligible mass. You want to lift a rock whose weight is 1.0×10^3 N. You intend to place a smaller stone underneath the crowbar to act as a pivot. Find the force you must exert to lift the rock as a function of the distance x of the pivot from the rock. What is the maximum distance x that allows *you* to lift the rock?

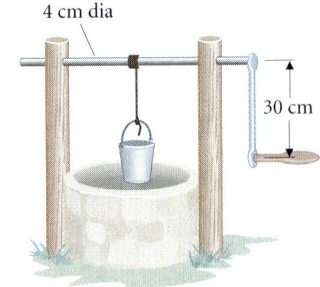
FIGURE 11.54

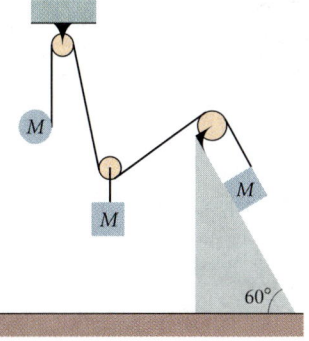

Figure 11.55

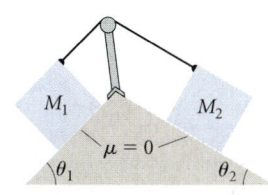

Figure 11.56

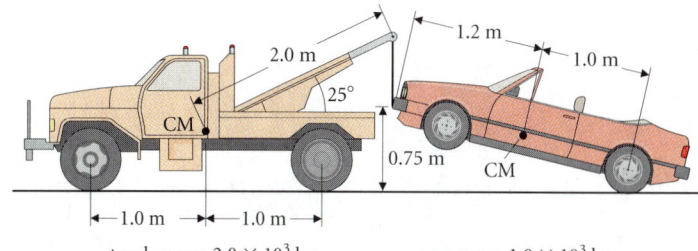

Figure 11.58

49. ♦♦ A rectangular box measuring 0.15 m × 0.20 m × 0.25 m is sitting on the bed of a pickup truck, with the 0.25-m side vertical and the 0.20-m side parallel to the velocity. How hard may the truck driver brake without causing the box to fall over? Assume friction is great enough that the box doesn't slide, and give the acceleration of the truck as your answer.

50. ♦♦ A cubical packing crate is to slide down a ramp at the warehouse at constant speed, and without toppling. Depending on the product inside the crate, the height of the CM varies from 0.20 to 0.60 of the side of the cube. What maximum coefficient of kinetic friction between the crate and the ramp is allowable? What constraint is there on the slope of the ramp?

51. ♦♦ A cyclist plus bicycle have a total mass of 115 kg with CM 0.95 m above the ground and halfway between the wheels (separation 1.20 m). What is the maximum braking force (due to friction of the road on the tires) that can be applied if only the front brake is used?

§11.3 EQUILIBRIUM OF SYSTEMS

52. ❖ List as many reasons as you can why the system shown in ■ Figure 11.55 cannot be in equilibrium.

53. ♦ Two cubical boxes of masses M_1 and M_2 sit respectively on ramps of angles θ_1 and θ_2 and are connected by a string that passes over a pulley at the apex of the structure (■ Figure 11.56). Is equilibrium possible? Under what circumstances?

54. ♦ A cylinder of radius R and mass M is on a ramp of angle θ. The coefficient of static friction between ramp and cylinder is μ_s. A string wrapped around the cylinder passes over a frictionless pulley and supports a particle of mass m (■ Figure 11.57). Find the mass m required to keep the system in equilibrium.

55. ♦♦ Suppose the cable breaks in the system of Figure 11.18 (Example 11.8); what forces must the wall exert on the beam? Assume that the beam is 1 m long and is embedded 5 cm into the wall.

56. ♦♦ A rod of length ℓ and negligible mass is attached to a wall by a frictionless hinge a distance ℓ below the ceiling. The other end of the rod is attached to the ceiling by a spring of constant k and relaxed length ℓ. The top end of the spring slides on a frictionless rail. A bead of mass M can be moved along the rod. Find the angle the rod makes with the horizontal as a function of the fractional distance f the bead has moved along the beam from the wall.

57. ♦♦ The tow truck in ■ Figure 11.58, mass 2.0×10^3 kg, tows a car of mass 1.0×10^3 kg at constant speed. What is the tension in the cable? What force does the road exert on each tire of the tow truck?

58. ♦♦ An object of mass M is suspended from the end of a horizontal strut of mass m (■ Figure 11.59). What horizontal and vertical forces does the wall exert on the strut? Is the vertical force up or down? Why? What coefficient of friction is needed between the wall and the strut to prevent slipping?

59. ♦♦ An object of mass $M = 5.0$ kg hangs from a beam of mass $m = 3.0$ kg (■ Figure 11.60). The maximum vertical force the hinge can exert is 55 N. What range of the coordinate x will not result in breaking the hinge?

60. ♦♦ A beam of mass m and length ℓ is connected to a wall by a hinge and a wire as shown in ■ Figure 11.61. What is the tension in the wire?

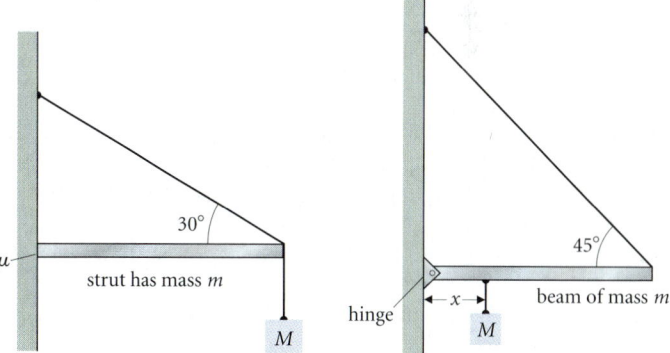

Figure 11.59   **Figure 11.60**

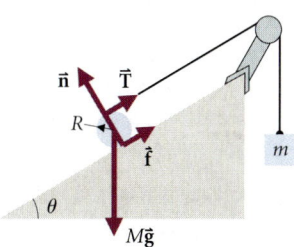

Figure 11.57

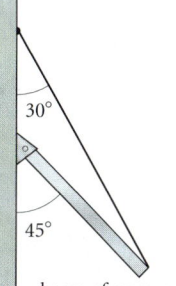

Figure 11.61   beam of mass m

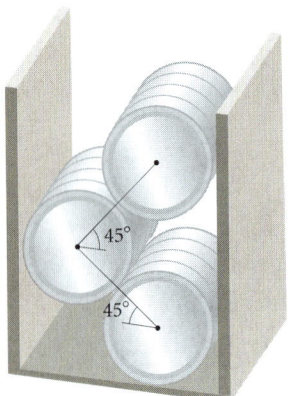

■ FIGURE 11.62

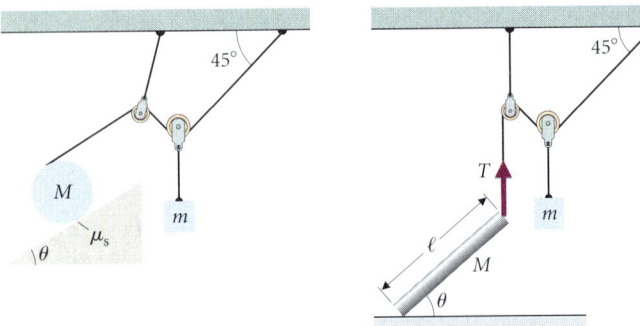

■ FIGURE 11.63   ■ FIGURE 11.64

61. ◆◆ Three identical cylindrical barrels, each of mass M, are stacked between frictionless walls as shown in ■ Figure 11.62. Find the forces that the barrels exert on each other and on the walls. Does it matter whether the coefficients of static friction between barrels or between the lower barrel and the floor are zero or not?

62. ◆◆ The pulleys in ■ Figure 11.63 are massless and frictionless, but the surface of the ramp has a coefficient of static friction μ_s and angle $\theta = 25°$. For what ratio m/M is the system in equilibrium? What is the minimum value of μ_s for which static equilibrium is possible?

63. ◆◆ One end of a rod of mass M and length ℓ rests on a frictionless surface. The other end is supported by the pulley system shown in ■ Figure 11.64. Find the mass m, the tension T, and the normal force n if the system is in equilibrium. Does the angle θ matter?

64. ◆◆◆ Two painters, one of mass 110 kg and one of mass 65 kg, are sitting on a uniform plank of mass 50.0 kg as shown in ■ Figure 11.65. The plank is 6.0 m long, each painter sits 2.0 m from an

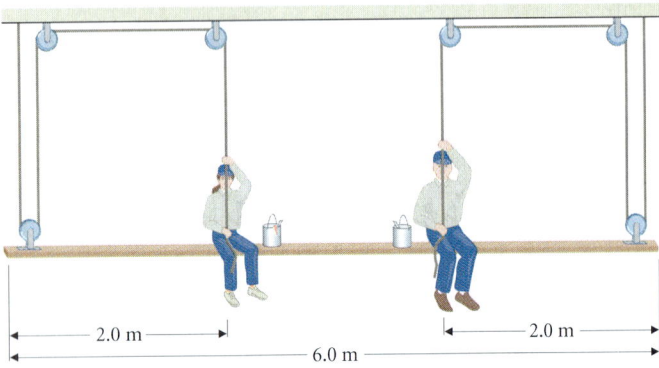

■ FIGURE 11.65

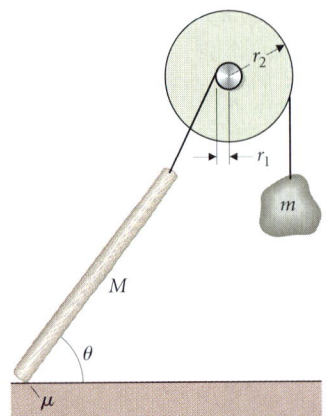

■ FIGURE 11.66

end of the plank, and they are lowering themselves with the rope and pulley system shown. The pulleys are attached 10 cm from each end of the plank. What force must each exert to keep the plank horizontal?

65. ◆◆◆ A wheel of radius r_2 is mounted on a shaft of radius r_1 (■ Figure 11.66). A rope wound around the shaft is attached to a log of mass M whose other end rests on a surface with coefficient of friction μ. A stone of mass m is suspended from a rope wound around the wheel. For what range of values of m is the system in equilibrium? Does the range of values depend on the angle θ of the log to the horizontal?

§11.4 THE LADDER PROBLEM

66. ◆ The Occupational Safety and Health Administration recommends that ladders be set up at an angle of 75° or greater with the horizontal. What minimum coefficient of friction between ladder and floor was assumed in producing this guideline?

67. ◆◆ A window cleaner of mass 110 kg has a 25-kg ladder that is 9.8 m long. If the ladder is placed at 75° to the horizontal on ground with a coefficient of friction $\mu_s = 0.7$, can he climb to the top? (Ignore friction at the top of the ladder.)

68. ◆◆ An advertising sign with length ℓ has its CM a distance $\ell/3$ from the bottom (Figure 11.39). What minimum coefficient of friction is needed between the advertising sign and the floor if the sign is in equilibrium at an angle of 40° with the frictionless wall?

69. ◆◆◆ In a carnival side show, a prize is placed at the end of a light but rigid pole mounted in a heavy log of mass $M = 5.0 \times 10^2$ kg. The log rests on a floor with coefficient of friction μ_s and lies at a 45° angle against a frictionless corner (■ Figure 11.67). If $\mu_s = 0.60$ and $\ell = 10.0$ m, what is the maximum mass of a person who could reach the prize? (*Hint:* The normal force at the corner is perpendicular to the log.)

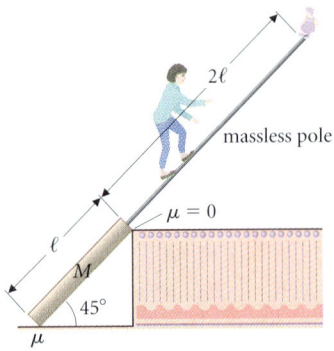

■ FIGURE 11.67

70. ♦♦♦ A painter places her ladder against a smooth wall that slopes away at an angle of 15° with the vertical. The coefficient of friction between the floor and the ladder is 0.50. If she places her ladder at a 60° angle to the floor, how high can she climb? (Ignore the mass of the ladder.)

§11.5 THE CENTER OF MASS OF EXTENDED BODIES

71. ❖ The shapes shown in ■ Figure 11.68 are each cut from a uniform sheet of metal. Which has the largest value of y_{CM}? Why?

72. ❖ Explain why the following statement is *false* and give a counterexample: the center of mass of any body must lie within the material of the body.

73. ❖ Where is the center of mass of a rod whose density decreases linearly with distance from the center of the rod?

74. ♦♦ A square piece of a uniform metal sheet of side L has a circular piece of radius $L/4$ removed from it. If the hole is tangent to one side of the square at its middle, where is the center of mass of the resulting object?

75. ♦♦ A rivet consists of a hemispherical top of radius R and a shaft of radius $R/2$ and length R (■ Figure 11.69). The center of mass of the rivet is found experimentally to be a distance $25R/22$ from the lower end of the shaft. Where is the center of mass of the hemispherical top?

76. ♦♦ A metal bracket of uniform material has three screw holes of radius 0.50 cm cut out of it, as shown in ■ Figure 11.70. Where is the center of mass of the bracket?

77. ♦♦♦ Find the center of mass of a semicircular object that has been cut from a uniform plane sheet of material.

78. ♦♦♦ A uniform right-circular cone has a base of radius a and a height h. Show that the center of mass of the cone is a distance $h/4$ above the base.

79. ♦♦♦ A triangle is removed from a uniform, square sheet so that the CM of the resulting object is at the apex P of the triangle (■ Figure 11.71). What is the height h of the triangle?

✱ §11.6 BRIDGES

80. ♦ A vertical mast stands at the edge of a canyon to support a telephone line. Coming from the canyon, the telephone cable meets the mast at an angle of 20° below horizontal, then runs vertically down the mast and continues underground. The tension in the telephone cable is $T_c = 1.0 \times 10^4$ N. A support cable runs from the ground to the top of the mast at an angle of 45°. What tension is needed in the support cable if the mast is to be in pure compression?

81. ♦♦ If the dredging crane in Figure 11.43 lifts a load of sand of mass $m = 5.0 \times 10^3$ kg, what is the compression in the boom of the crane?

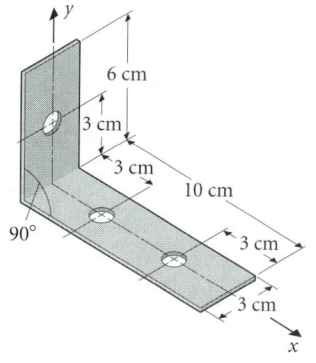

■ FIGURE 11.70

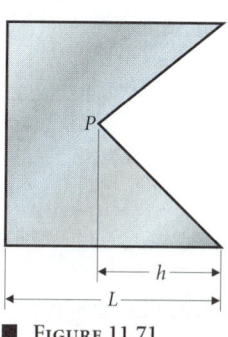

■ FIGURE 11.71

82. ♦♦♦ A sign of mass M is supported by the cantilever framework shown in ■ Figure 11.72. Find the compression or tension of each member and the forces exerted by the wall at the two support points.

83. ♦♦♦ A stepladder has sides of length 3.0 m and a crossbar located halfway along the sloping sides. The ladder forms a triangular shape with an apex angle of 30° when set up. An 85-kg person stands on a step one-third of the distance up one side of the ladder. What is the tension or compression in the crossbar? Ignore the mass of the ladder, and assume the floor is frictionless. (*Hint*: Consider the two sides as separate members, and take torques about the hinge.)

84. ♦♦♦ Find the forces acting in each member of the simple truss bridge in Figure 11.27 if the truck is a distance $\ell/4$ from the right end of the bridge. (*Hint*: Analyze the beam on which the truck rests and relate your results to Example 11.12. As in that case, the tensions in the roadbed cannot be determined.)

Additional Problems

85. ❖ You open the package containing your new decorative mobile and find three identical glass prisms, two sticks, one of length L and one of length $L/2$, and a length of string. Unfortunately, the instructions are missing. Explain how to construct the mobile so that it will hang in equilibrium with the sticks horizontal. You may cut the string.

86. ♦♦ A carpenter's square, cut from a uniform sheet of steel, has the dimensions shown in ■ Figure 11.73. If it is supported by a thin horizontal nail at point A, what angle does the long arm make with the horizontal?

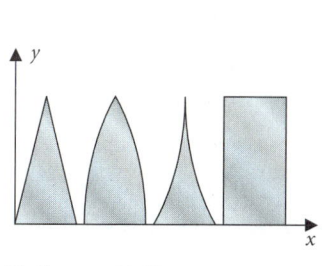

■ FIGURE 11.68

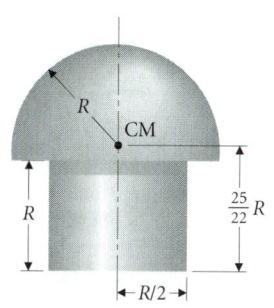

■ FIGURE 11.69

■ FIGURE 11.72

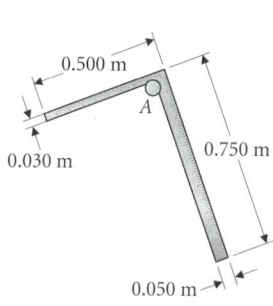

■ FIGURE 11.73

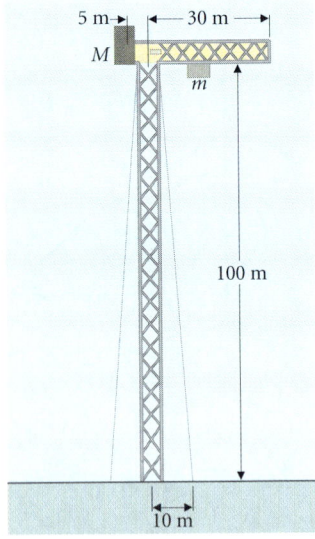
FIGURE 11.74

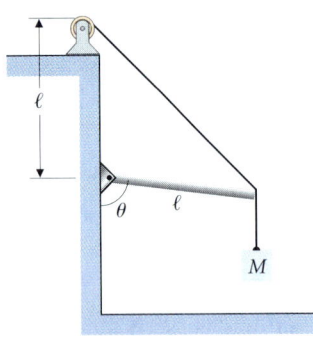
FIGURE 11.75

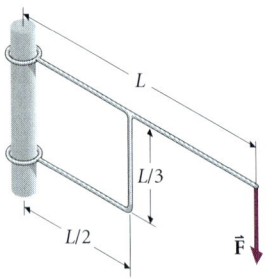
FIGURE 11.76

FIGURE 11.77

87. ♦♦ A tower crane (■ Figure 11.74) stands $L = 1.00 \times 10^2$ m above the ground and has a boom 30.0 m long. Guy wires leading from the ground are attached 10.0 m from the base of the crane. A counterweight $M = 2.50 \times 10^3$ kg remains fixed. A load $m = 5.0 \times 10^2$ kg is moved from the tower out to the end of the boom. Find the difference between the tensions in the guy wires when the load is at the tower and the difference between the tensions when the load is at the end of the boom. (Neglect the width of the crane tower.)

88. ♦♦ A sign is in the form of an isosceles right triangle with 0.30-m sides. The sign is attached to a post with two brackets, one at the top and one at the bottom. The right angle is against the post at the top. Is it possible to find the forces exerted by each of the brackets? If the top bracket supports 65% of the weight, find the magnitudes of the forces exerted by each bracket.

89. ♦♦ A crane system consists of a massless boom of length ℓ mounted on a frictionless pivot a distance ℓ below a crank (■ Figure 11.75). If the system is used to lift an object of mass M, find the tension in the cable as a function of angle θ.

90. ♦♦ An adjustable support (■ Figure 11.76) can slide along a vertical rod. What coefficient of static friction μ_s is necessary between the support and the rod if the support is not to slide when a force $\vec{F}$ is applied as shown?

91. ♦♦♦ A lawn roller of radius 25 cm is to be pulled over a terrace border 15 cm high (■ Figure 11.77). If the roller's mass is 75 kg, what minimum force on the handle is required? At what angle θ is the minimum force applied?

92. ♦♦♦ You wish to make a decoration using a thin, hemispherical, polished metal shell that is to hang with its circular edge in a vertical plane. Where would you insert the thread it is to hang from? If instead you suspended the shell from a point on its edge, at what angle would it hang?

93. ♦♦♦ An aircraft is supported by the lift force, which acts at the center of lift (CL). The weight acts at the CM, which is usually not directly above or below the center of lift. Another force is necessary for equilibrium. Why? This force is provided by the tail. If the CM is behind the CL, what is the direction of the force at the tail? What if the CM is ahead of the CL? If the plane noses up slightly, the lift force increases slightly. Examine the stability of each configuration to such a displacement. (Neglect any changes in the force exerted by the tail.) What do you conclude? If the plane has a mass of 1500 kg and the CM is 0.50 m forward of the CL, what is the force exerted at the tail (5.0 m behind the CL)? What is the magnitude of the lift force?

94. ♦♦♦ Og, the forgotten inventor of the wheel, found it much easier to construct regular polygons rather than circles. Thus Og's Mark IV experimental wagon (■ Figure 11.78) had square wheels. Assume the wheels are to "roll," not slide. **(a)** If the total mass of the wagon was M, what force F did Og's dinosaur have to exert to pull the wagon on level ground? **(b)** An advantage of Og's design is that the wagon will stand on a slope without brakes. On how steep a slope would the Mark IV wagon stand? **(c)** If Og experimented with a series of improved designs, the Mark N wagon having N-gonal wheels, how would the force needed from the dinosaur on level ground depend on N?

95. ♦♦♦ The maximum overhang in a stack of blocks (cf. Problem 27) can be obtained by placing the center of gravity of the top block over the edge of the block below, then putting the CG of the combined blocks above the edge of the third block down, and continuing until the CG of the entire stack above the first block is over the edge of the first. (See *Scientific American*, Nov. 1964.) What is the maximum overhang for a set of ten blocks each 8.00 cm long? What happens as the number of blocks increases?

96. ♦♦♦ If the coefficient of friction between the box and the ramp in Figure 11.38 is $\mu_s = 0.5$, for what values of m is the system in equilibrium?

97. ♦♦♦ One-third of a circular cylinder is removed to form the object resting against a wall in ■ Figure 11.79. The object's mass is M;

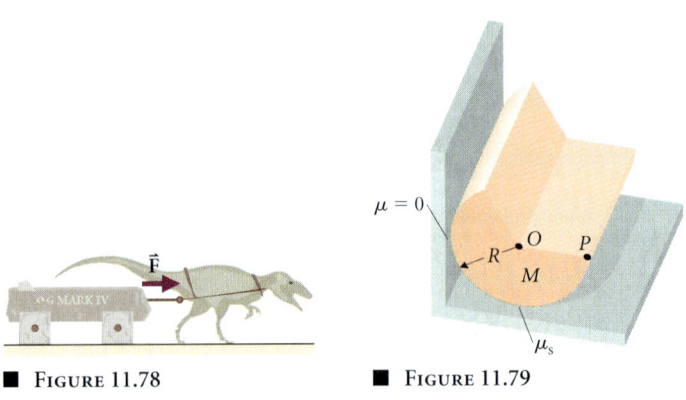

FIGURE 11.78

FIGURE 11.79

collection of rectangles for $N = 3$. **(c)** Set up a spreadsheet program to calculate h_N, the y-coordinate of the CM for a collection of N rectangles. Compute the result for $N = 4, \ldots, 10$. **(d)** Plot your results versus N, and show that h_N approaches $H/3 = 0.50$ m for large N. **(e)** Check your results against the exact formula

$$h_N = (H/3)[1 + 1/(2N^2)].$$

100. In an experimental support system for a structural beam, hydraulic actuators apply a continuously variable force per unit length along a horizontal segment of the beam between $x = -15.0$ cm and $x = +15.0$ cm. Sensors measure the force per unit length acting on the beam at intervals of 1.00 cm. (See file CH11P100 on the supplementary computer disk.) Use the trapezoidal rule (cf. Problem 6.91),

$$\int_0^{N\Delta x} f(x)\,dx \approx \frac{\Delta x}{2}\left[f(0) + 2\sum_{i=1}^{N-1} f(i\Delta x) + f(N\Delta x)\right],$$

to compute the total force acting on the beam and the total torque acting about $x = 0$. What forces acting at $x = -15.0$ cm and $x = +15.0$ cm model the net force and torque?

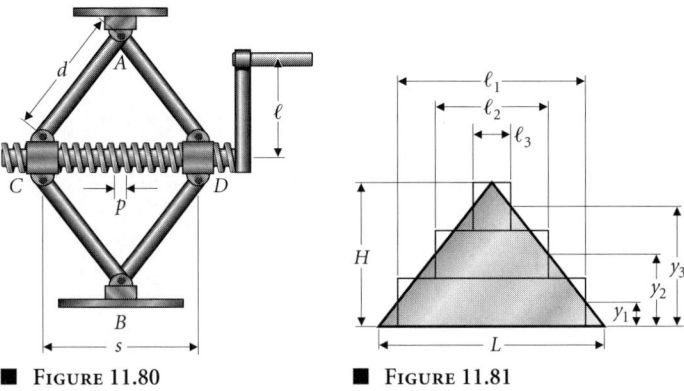

■ FIGURE 11.80   ■ FIGURE 11.81

the coefficient of static friction between it and the floor is μ_s; the wall is perfectly smooth; and the radius OP is horizontal. The CM is a distance $s = \sqrt{3}R/2\pi$ from the center of the cylinder (cf. Problem 104b). **(a)** What normal force does the object exert on the wall? What minimum value of μ_s is needed for equilibrium? **(b)** If the wall were removed, where along OP would you have to exert a downward force F to hold the object in equilibrium? Does the value of μ_s influence your result?

98. ♦♦♦ In the automobile jack shown in ■ Figure 11.80, a crank with arm length $\ell = 15$ cm turns a screw with pitch $p = 0.50$ cm. As the screw turns, points C and D of the frame move inward so that points A and B must move outward to lift the car. The sides of the frame have length $d = 20.0$ cm, and the separation s between points C and D varies from 30.0 cm to 10.0 cm. What is the mechanical advantage of the device as a function of s in this range? If the jack is used to raise a car of mass $M = 1.50 \times 10^3$ kg and wheelbase $L = 6.0$ m, estimate the force required on the crank when $s = 20.0$ cm.

Computer Problems

99. An isosceles triangle of base L and height H is cut from a uniform sheet of material (■ Figure 11.81). Approximate the triangle with N rectangles, each of height H/N and of length such that their ends intersect the triangle at their midpoints. (The case $N = 3$ is illustrated in Figure 11.81.) **(a)** Use a geometrical argument to show for any N that the total area of the rectangles equals the area of the triangle. **(b)** If $H = 1.50$ m and $L = 1.96$ m, locate the CM of the

Challenge Problems

101. A thin plank of mass M is laid against a cylinder, also of mass M, at an angle of 45°. The midpoint of the plank rests against the cylinder. All surfaces have the same coefficient of friction μ. What is the minimum value of μ such that the system can remain in equilibrium?

102. Find the center of mass of a uniform hemisphere of radius R. (*Hint:* Where is the CM of a disk?)

103. Most aircraft have brakes only on the main wheels, and not on the nosewheel. When a landing aircraft brakes, the pilot uses the controls to increase the downward force exerted by the tail surface. Why? For an aircraft of mass 2.0×10^3 kg landing at 2.0×10^2 km/h, by how much is the minimum landing roll decreased if the tail force is 1.0×10^3 N as compared with zero? Assume that the tail is 7.0 m behind the main wheels, the nosewheel is 2.0 m ahead of the mains, the CM is 0.70 m ahead of the mains, and the coefficient of friction is 1.0.

104. (a) A thin wire of length ℓ is bent into an arc of a circle of radius R. How far from the center of the circle is the center of mass of the wire? **(b)** A sector of radius R and central angle θ is cut from a uniform, circular plane sheet of metal. Use the result of part (a) to show that the distance of the sector's center of mass from the center of the circle is:

$$s = \frac{4R}{3\theta}\sin\frac{\theta}{2}.$$

105. Find the center of mass of a rod whose density varies with distance from one end: $\rho \propto x^2$, where x is distance from the end.

An acrobatic airplane performing at an air show. Notice the two-wing design that makes the plane more maneuverable. In this chapter we'll learn why the pilot might prefer this design.

CHAPTER 12
Dynamics of Rigid Bodies

Concepts

Body coordinates
Angular velocity
Angular acceleration
Rotational inertia
Rotational kinetic energy
Parallel axis theorem
Gyroscope
Precession

Goals

Be able to:

Apply the results for uniform angular acceleration.

Calculate the rotational inertia of a symmetrical rigid body about a given axis.

Apply the principles of dynamics to the rotation of a rigid body about a given axis.

The wheel is come full circle.
  WILLIAM SHAKESPEARE

Biplanes—with two wings stacked on top of each other—are inefficient for rapid transportation because their large air resistance limits them to relatively slow speeds. However, because they are more maneuverable, biplanes are often preferred by aerobatic pilots who perform complicated sequences of loops, rolls, and spins. In this chapter we shall study the principles of rigid-body dynamics that allow us to understand the biplane's advantage.

In particle mechanics we began with kinematics—describing motion. Then we discussed dynamics—relating accelerations to their causes. The study of rigid-body dynamics follows a similar program, and we have already completed much of the necessary work. We studied the kinematics of circular motion in Chapter 3. In Chapter 9 we developed the concept of torque and showed that the total torque acting on a system equals the rate of change of the system's angular momentum. In Chapter 11 we learned how to calculate the total torque acting on a rigid body. Now we continue this program. In §12.1 we apply the results of Chapter 3 to rotating rigid bodies. Then in §12.2 we show how to calculate the kinetic energy and angular momentum of a rotating body, and in §12.3 we use the relation $\vec{\tau} = d\vec{L}/dt$ to analyze how systems of rigid bodies behave.

Study Hint: This would be a good time to review §3.2 (circular motion) and §3.3 (relative motion).

12.1 Rotational Kinematics

12.1.1 Body Coordinates

Since the distances between particles in a rigid body are fixed, we do not have to describe the motion of each particle separately. Motion of the body is described by the velocity of its center of mass and its rotation about an axis through the center of mass. For example, the motion of a bicycle wheel is described by its rotation about the axle and the velocity of the axle along the road. We want to choose a coordinate system to match this description.

See §9.3 and the introduction to Chapter 11.

> A *body coordinate system* is defined by an origin and a set of coordinate axes that are fixed in a particular rigid body (■ Figure 12.1).

As always, we choose the origin and axes to simplify our calculations. For example, with the origin at the axle of the bicycle wheel, we might choose the x- and y-axes to be along two perpendicular spokes, and the z-axis to be along the axle.

Each particle of a rigid body has a position vector $\vec{r}_B$ that is fixed with respect to the body axes and that rotates together with the axes as the body rotates. The position vector $\vec{r}$ of the

The body frame is **not** inertial.

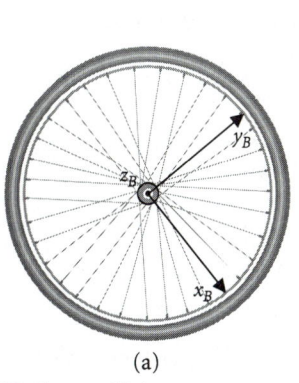
(a)

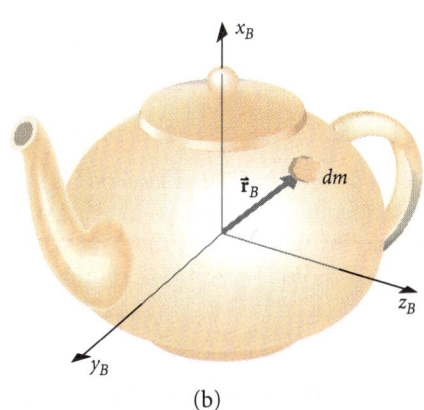
(b)

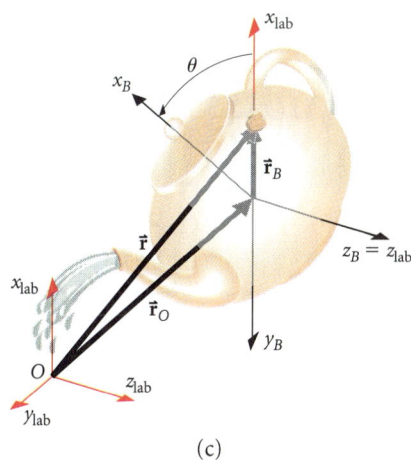
(c)

■ **Figure 12.1**
(a) A body frame of reference is a set of axes rigidly fixed to the body—here a bicycle wheel. (b) Body axes for an arbitrary object. (c) The position of a point at $\vec{r}_B$ in the body is known once the position of the body origin and the orientation of the body axes are known. Here the rotation axis is the z_{lab}-axis = z_B-axis and the orientation is described by the angle θ between the x_{lab}- and x_B-axes.

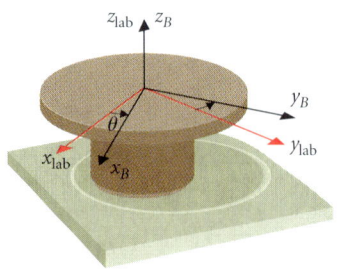

FIGURE 12.2
A potter's wheel rotates about the vertical (z) axis. Its orientation is determined by the angle θ between the x_{lab}- and x_B-axes. All particles in the wheel rotate with the same angular speed ω.

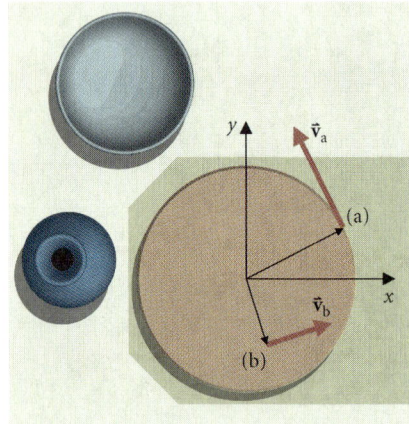

FIGURE 12.3
Each piece of a rotating wheel travels on a circle with the same angular speed. The linear speed of a particle is proportional to its distance from the wheel's center.

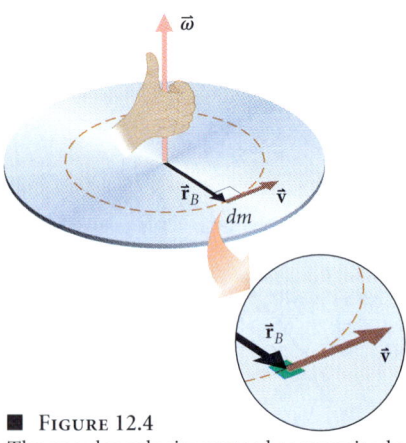

FIGURE 12.4
The angular velocity vector has magnitude equal to the angular speed and lies along the rotation axis, with its direction determined by a right-hand rule: curl your fingers in the direction of particle motion and your thumb gives the direction of $\vec{\omega}$.

NOTE: THIS IS THE SAME RIGHT-HAND RULE THAT IS USED TO DEFINE $\vec{\tau}$ AND $\vec{L}$.

particle in the inertial laboratory reference frame (Figure 12.1c) is the sum of its body position vector $\vec{r}_B$ and the position vector of the body origin in the inertial frame, $\vec{r}_O$.

$$\vec{r} = \vec{r}_B + \vec{r}_O.$$

The directions of the body coordinate axes completely specify the body's orientation.

For example, a potter's wheel (Figure 12.2) rotates about a vertical axis, which we take to be the z-axis in both the body and the laboratory frames of reference. The laboratory x- and y-axes are fixed in the horizontal plane of the floor. As the wheel rotates, the body axes rotate with it, and each individual particle in the wheel undergoes circular motion about the vertical (z) axis. The angle θ between the body and lab x-axes completely determines the orientation of the body coordinates in the lab frame. All the particles have the same angular speed $\omega = d\theta/dt$, since they are fixed rigidly in the wheel and rotate through the same angle $\Delta\theta$ in a given time interval Δt. This common value of ω is the angular speed of the wheel.

> The *angular speed* ω of a rigid body is the common angular speed of all its constituent particles about the body axis of rotation.

EXAMPLE 12.1 ♦ If the potter's wheel has a radius of 0.41 m and rotates with angular speed 3.0 rad/s, what is the linear speed of a particle (a) on its rim and (b) at a point 0.25 m from the axis?

MODEL Each particle in the wheel undergoes uniform circular motion in the x–y plane (Figure 12.3).

SETUP We use the relation $v = \omega r$ (eqn. 3.9), which relates the linear and angular speeds in circular motion.

SOLVE At the rim, $\omega r = (3.0 \text{ rad/s})(0.41 \text{ m}) = 1.2 \text{ m/s} = v$.
At point (b), $\omega r = (3.0 \text{ rad/s})(0.25 \text{ m}) = 0.75 \text{ m/s} = v$.

ANALYZE Points farther from the axis of rotation have greater linear speed.

12.1.2 Angular Velocity and Angular Acceleration

In particle kinematics, we found vectors to be a most useful mathematical tool. The same is true in rotational kinematics. Angular speed gives the magnitude of a rigid body's rotation rate but does not indicate either the direction of the rotation axis or the sense of rotation about that axis. A vector quantity $\vec{\omega}$, called *angular velocity*, includes this information.

> The *angular velocity* $\vec{\omega}$ of a rigid body is a vector along the body's rotation axis with magnitude equal to the body's angular speed. The direction of $\vec{\omega}$ is given by the right-hand rule illustrated in Figure 12.4: Curl the fingers of your right hand in the sense of the circular motion, and your thumb gives the direction of $\vec{\omega}$.

Using the coordinate system shown in Figure 12.2, the angular velocity of the potter's wheel in Example 12.1 is $\vec{\omega} = (3.0 \text{ rad/s})\hat{k}$.

Changes in a body's angular velocity are described by its *angular acceleration*:

$$\vec{\alpha} = \frac{d\vec{\omega}}{dt}.$$

TABLE 12.1 Definitions of Kinematic Quantities in Rotation About a Fixed Axis

Quantity	Rotational Motion	Linear Motion
Position	θ	y
Displacement	$\Delta\theta \equiv \theta_{final} - \theta_{initial}$	$\Delta y = y_{final} - y_{initial}$
Average velocity	$\omega_{av} = \Delta\theta/\Delta t$	$v_{av} = \Delta y/\Delta t$
Instantaneous velocity	$\omega = \lim\limits_{\Delta t \to 0} \omega_{av} = d\theta/dt$	$v = \lim\limits_{\Delta t \to 0} v_{av} = dy/dt$
Average acceleration	$\alpha_{av} = \dfrac{(\omega_{final} - \omega_{initial})}{\Delta t}$	$a_{av} = \dfrac{(v_{final} - v_{initial})}{\Delta t}$
Instantaneous acceleration	$\alpha = \lim\limits_{\Delta t \to 0} \alpha_{av} = d\omega/dt$	$a = \lim\limits_{\Delta t \to 0} a_{av} = dv/dt$

12.1.3 Rotation About a Fixed Axis

In general, the vectors $\vec{\alpha}$ and $\vec{\omega}$ may vary in both magnitude and direction. However, in many practical situations, rigid-body rotation occurs about a single axis, fixed in direction, with both $\vec{\alpha}$ and $\vec{\omega}$ lying along that axis. The rotation axis is then a natural choice for a coordinate axis and is usually called z. Then both $\vec{\alpha}$ and $\vec{\omega}$ have only z-components. The kinematic description of rotation about a fixed axis has the same mathematical structure as one-dimensional kinematics of a particle (● Table 12.1) because we need only one coordinate, an angle in the x–y plane, to represent the body's orientation. For example, if the body x- and y-axes of the potter's wheel are coincident with the lab axes at $t = 0$, the angle $\theta = \omega_{av}t$.

In the special case of uniform angular acceleration ($\vec{\alpha}$ = constant), the relations among angular quantities are similar to those for linear quantities in uniformly accelerated linear motion (● Table 12.2). The equations in the left-hand column are the same as those in the right-hand column if we replace y with θ, v with ω, and a with α, and they may be derived using the same steps that we used in §2.3 for linear motion. For example, let's derive eqn. 12.1. Angular accleration is constant:

$$\vec{\alpha} = \alpha_z \hat{k}.$$

The derivative of angular velocity is the angular acceleration:

$$\frac{d\omega_z}{dt} = \alpha_z.$$

REMEMBER: ANGLES ARE MEASURED IN RADIANS UNLESS CLEARLY STATED OTHERWISE. SEE §1.2.5 FOR THE DEFINITION OF THE RADIAN.

TABLE 12.2 Relations Among Kinematic Quantities with Uniform Acceleration

Rotational Motion		Linear Motion	
$\omega = \omega_i + \alpha t$	(12.1)	$v = v_i + at$	(2.10)
$\theta = \theta_i + \omega_i t + \tfrac{1}{2}\alpha t^2$	(12.2)	$y = y_i + v_i t + \tfrac{1}{2}at^2$	(2.11)
$\omega_{av} = (\omega_f + \omega_i)/2$	(12.3)	$v_{av} = (v_f + v_i)/2$	(2.12)
$\omega_f^2 - \omega_i^2 = 2\alpha(\theta_f - \theta_i)$	(12.4)	$v_f^2 - v_i^2 = 2a(y_f - y_i)$	(2.13)

Integrate:

$$\int_0^t \frac{d\omega_z}{dt} dt = \int_0^t \alpha_z \, dt.$$

$$\omega_z - \omega_i = \alpha_z t \Big|_0^t = \alpha_z t,$$

where ω_i is the angular velocity at $t = 0$. A rearrangement of this expression gives eqn. (12.1) in Table 12.2. Integrating again, we obtain eqn. (12.2). The other equations follow from the first two.

EXAMPLE 12.2 ♦ Friction on its axle causes a wheel with initial angular velocity 3.1 rad/s to decelerate at a uniform rate $\alpha = -0.050$ rad/s². After what time interval does the wheel come to rest? How many rotations does it make?

MODEL Uniform deceleration is consistent with our model of kinetic friction being independent of speed (§4.5.2).

SETUP We choose the z-axis to be along the axle of the wheel. Then ω has only a z-component that decreases from $\omega_i = 3.1$ rad/s to $\omega_f = 0$. We obtain the time from eqn. (12.1):

$$\alpha t = (\omega_f - \omega_i).$$

SOLVE So, $t = (\omega_f - \omega_i)/\alpha = (0 - 3.1 \text{ rad/s})/(-0.050 \text{ rad/s}^2) = 62$ s. The wheel stops after about 1 min.

SETUP The angular displacement $\Delta \theta$ (eqn. 12.4) is:

$$\Delta \theta = \frac{(\omega_f^2 - \omega_i^2)}{2\alpha} = \frac{-(3.1 \text{ rad/s})^2}{2(-0.050 \text{ rad/s}^2)}$$

$$= \frac{9.6}{0.10} = 96 \text{ rad}.$$

One complete rotation is an angular displacement of 2π rad, so:

SOLVE $\quad\quad N = \Delta\theta/2\pi = (96 \text{ rad})/(2\pi \text{ rad}) = 15.$

The wheel makes 15 rotations before it comes to a stop.

ANALYZE Compare the method of this example with that of Example 2.12. Once the time has been calculated, we could equally well use eqn. (12.2) to find $\Delta\theta$. ∎

EXERCISE 12.1 ♦ A flywheel is acclerated from rest with a uniform angular acceleration $\alpha = 0.010$ rad/s². What is the flywheel's angular speed after 1.0 h? How many revolutions has it made?

One of the most frequently encountered examples of rotational motion is that of a wheel on a vehicle. Wheels are designed to *roll without slipping;* that is, the part of the wheel that contacts the ground or rail does not move with respect to the ground. That automobile tires are made of rubber so they can grip the ground serves to emphasize how important the absence of slipping is for the safe control of the vehicle. The wheel's axle, being attached to the car, moves with the same linear velocity as the rest of the car. Points on the tire rotate about the axle with angular speed ω. Thus the point on the ground is actually moving backward with respect to the axle at speed $v = \omega R$. Its speed over the ground is $v_{\text{axle}} - \omega R = 0$. Thus v_{axle} and ω are related.

Rolling without slipping. At the bottom of the wheel where $\tilde{\mathbf{v}} = 0$ the spokes are in focus. At the top the wheel moves with speed $2v_{\text{axle}}$, and the spokes appear blurred.

Rolling without slipping: $$v_{axle} = \omega R. \tag{12.5}$$

Since this relation is true at every instant, the magnitudes of the accelerations are similarly related.

Rolling without slipping: $$a = \alpha R. \tag{12.6}$$

WE ALSO DERIVED THIS RELATION AS EQN. (3.14) IN §3.3.3.

REMEMBER: THESE ARE RELATIONS BETWEEN SPEEDS AND MAGNITUDES OF ACCELERATIONS, NOT BETWEEN VECTORS. THE VECTORS $\vec{v}$ AND $\vec{\omega}$ ARE PERPENDICULAR TO EACH OTHER.

EXAMPLE 12.3 ♦♦ A car accelerates from rest at 3.2 m/s². If the wheels have radii of 0.31 m and roll without slipping, how long does it take for the car to travel 0.35 km? How many revolutions have the wheels made in traveling that distance, and what is their final angular speed?

MODEL The wheel rolls without slipping, so we may apply eqns. (12.5) and (12.6). Also the car accelerates uniformly, so we may use the relations in Table 12.2. We want to find the time t, angular displacement $\Delta\theta$, and final angular speed ω_f.

SETUP Since the car accelerates from rest, $v_i = 0$, and, with the y-axis along the direction of motion (■ Figure 12.5), we may choose the origin at the car's initial position so that $y_i = 0$. We find the time t from the linear motion. Using eqn. (2.11):

$$y = \tfrac{1}{2}at^2 \Rightarrow t^2 = 2y/a. \tag{i}$$

With the x- and z-axes chosen as shown in Figure 12.5, the angular acceleration α is in the $+z$-direction, and its magnitude is given by eqn. (12.6). Then eqns. (i) and (12.2) relate the angular displacement of the wheel after time t to the given linear displacement:

$$\Delta\theta = \tfrac{1}{2}\alpha t^2 = \tfrac{1}{2}(a/R)t^2 = y/R. \tag{ii}$$

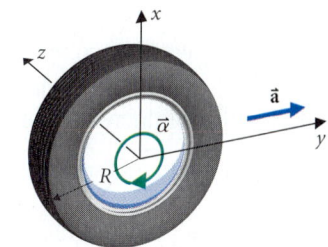

■ FIGURE 12.5
The car is accelerating to the right, in the $+y$-direction. With the coordinate axes chosen as shown here, α is in the $+z$-direction when the wheel rolls without slipping.

SOLVE From eqn. (i):

$$t = \sqrt{\frac{2y}{a}} = \sqrt{\frac{2(3.5 \times 10^2 \text{ m})}{3.2 \text{ m/s}^2}} = 15 \text{ s}.$$

Using eqn. (ii), we find that the number of revolutions is:

$$N = \frac{\Delta\theta}{2\pi} = \frac{y}{2\pi r} = \frac{350 \text{ m}}{2\pi(0.31 \text{ m})} = 180.$$

The final angular speed is (eqn. 12.1):

$$\omega = \alpha t = \frac{a}{R}t = \frac{3.2 \text{ m/s}^2}{0.31 \text{ m}}(15 \text{ s}) = 150 \text{ rad/s}.$$

Or, expressed in revolutions per minute (rpm), the frequency f (§3.2.2) is:

$$f = \frac{\omega}{2\pi} = \frac{(153 \text{ rad/s})(60 \text{ s})}{(2\pi \text{ rad/rev})(1 \text{ min})} = 1500 \text{ rpm}.$$

REMEMBER: KEEP AN EXTRA SIGNIFICANT FIGURE UNTIL THE END OF THE CALCULATION.

ANALYZE Notice that $\Delta\theta$, expressed as a total number of revolutions, is equal to (total distance traveled)/(circumference of wheel). Imagine that the wheel is covered with paint. As the wheel rolls over the ground without slipping, it deposits a stripe of paint one circumference long with each complete revolution. ■

EXERCISE 12.2 ♦ Find the car's speed after it has traveled 0.35 km and compare with the result for ω in Example 12.3.

Math Topic

VECTOR RELATIONS IN CIRCULAR MOTION

A particle moving uniformly in a circle of radius r has angular velocity vector $\vec{\omega}$ perpendicular to the plane of the circle (■ Figure 12.6). With origin at the center of the circle, we may find vector expressions for the linear velocity and acceleration of the particle in terms of $\vec{\omega}$. The magnitude of the particle's velocity is ωr, and its direction is perpendicular to both $\vec{\omega}$ and $\vec{r}$. The magnitude

$$|\vec{\omega} \times \vec{r}| = \omega r \sin 90° = \omega r = |\vec{v}|.$$

From the right-hand rule for cross products, $\vec{\omega} \times \vec{r}$ is in the direction of the velocity. Therefore, we conclude:

$$\vec{v} = \vec{\omega} \times \vec{r}.$$

Similarly, the acceleration of the particle has magnitude $\omega^2 r$. The cross product $\vec{\omega} \times \vec{v}$ is in the direction of $\vec{a}$ and has magnitude $\omega v \sin 90° = \omega(\omega r) = \omega^2 r$. So:

$$\vec{a} = \vec{\omega} \times \vec{v} = \vec{\omega} \times (\vec{\omega} \times \vec{r}).$$

We also know that $\vec{a} = -\omega^2 \vec{r}$ (eqn. 3.10) so that:

$$\vec{\omega} \times (\vec{\omega} \times \vec{r}) = -\omega^2 \vec{r}.$$

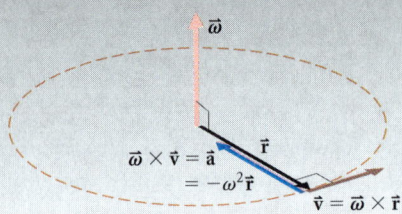

■ FIGURE 12.6
We express the velocity and acceleration of a particle in circular motion in terms of its angular velocity vector $\vec{\omega}$: $\vec{v} = \vec{\omega} \times \vec{r}$; $\vec{a} = \vec{\omega} \times \vec{v}$.

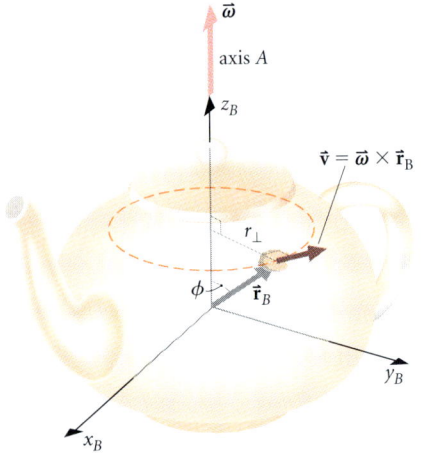

■ FIGURE 12.7
The speed of each particle in a body depends on its perpendicular distance $r_\perp$ from the rotation axis and on the angular speed ω: $|\vec{v}| = \omega r_\perp$.

REMEMBER: FOR A PARTICLE, $K = \frac{1}{2}mv^2$.

WE MADE THE SAME TRANSITION TO AN INTEGRAL IN DEFINING THE CENTER OF MASS (§11.5).

12.2 ROTATIONAL KINETIC ENERGY AND ANGULAR MOMENTUM

12.2.1 Energy

We begin our study of dynamics by expressing the energy and angular momentum of a rigid body in terms of its angular velocity. Since different parts of a rotating body have different linear velocities, we imagine each body as a sum of differential elements and sum the contributions from each element.

■ Figure 12.7 shows an object rotating about an axis labeled A, which coincides with the z-axis of the body coordinate system. Each particle in the body (with coordinates x_B, y_B, z_B) undergoes circular motion with angular speed ω and radius $r_\perp \equiv |\vec{r}_B| \sin \phi = \sqrt{x_B^2 + y_B^2}$, the distance of the particle from the rotation axis. The speed of the particle depends on its position: $v = \omega r_\perp$. So a particle of mass dm has kinetic energy:

$$dK = \tfrac{1}{2} dm |\vec{v}^2| = \tfrac{1}{2} dm \omega^2 r_\perp^2.$$

The total rotational kinetic energy is the sum over all such particles:

$$K = \sum_{\text{all particles}} \tfrac{1}{2} \omega^2 r_\perp^2 \, dm = \tfrac{1}{2} \omega^2 \left(\sum_{\text{all particles}} r_\perp^2 \, dm \right).$$

While v differs for each particle, the angular speed ω is the same for all the particles, and it can thus be factored from the sum. The sum over differential quantities is a physical integral:

$$K = \tfrac{1}{2} \omega^2 \int r_\perp^2 \, dm.$$

The resulting integral, $\int r_\perp^2 \, dm$, depends only on the mass distribution of the rigid body around the rotation axis and is called the *rotational inertia*, I. It relates the rotational kinetic energy of a body to its angular velocity in the same way that the mass of a particle relates its kinetic energy to its linear velocity:

$$K = \tfrac{1}{2}I\omega^2. \tag{12.7}$$

Rotational inertia's role in rotational dynamics is analogous to that of mass in particle dynamics. This is the first example of that analogy.

> The *rotational inertia* of a rigid body about an axis A is the sum over all the particles in the body of the particle's mass dm multiplied by the square of its perpendicular distance $r_\perp$ from the axis.
>
> $$I_A \equiv \int r_\perp^2 \, dm. \tag{12.8}$$

A rigid body does not have a single rotational inertia. In general, the body has a different value of rotational inertia for each possible rotation axis. • Table 12.3 (on page 398) lists values of rotational inertia for various objects and axes.

EXAMPLE 12.4 ♦ A weight lifter's barbell consists of two heavy disks of mass m and radius R connected by a light rod of length $\ell \gg R$ (■ Figure 12.8). Estimate its rotational inertia about an axis A through its CM and perpendicular to the rod. If $m = 75$ kg, $\ell = 2.0$ m, and the barbell has angular speed $\omega = 2.0$ rad/s, what is its rotational energy?

MODEL Since $\ell \gg R$, we model the heavy disks as particles for our estimate, and we ignore the mass of the light rod. With these approximations, we will make an error of only a few percent in the value of I.

SETUP Since there are only two particles in this system, each of mass m and perpendicular distance $\ell/2$ from the axis, the rotational inertia is:

$$I_A = \sum mr_\perp^2 = 2m(\ell/2)^2 = m\ell^2/2.$$

SOLVE The rotational kinetic energy is (eqn. 12.7):

$$K = \tfrac{1}{2}I_A\omega^2 = m\ell^2\omega^2/4 = (75 \text{ kg})(2.0 \text{ m})^2(2.0 \text{ rad/s})^2/4 = 3.0 \times 10^2 \text{ J}.$$

ANALYZE This is the same kinetic energy possessed by a 1-kg mass dropped from a height of 30 m. ∎

EXERCISE 12.3 ♦ Estimate the rotational inertia of the barbell about (a) an axis B through one of the disks and perpendicular to the rod and (b) an axis C through the CM and at 30° to the rod. ∎

12.2.2 Angular Momentum

The total angular momentum of a rotating rigid body is found by adding the angular momenta of its constituent particles. The method for summing the contributions is similar to that which we just used for finding the energy. A crucial difference is that we are adding vectors, so we must pay proper attention to their directions.

We choose the body's rotation axis to be the z-axis (■ Figure 12.9). The velocity of a particle of mass dm lies in the plane perpendicular to the rotation axis and has magnitude $\omega r_\perp$. The angular momentum of the particle is $d\vec{L} = dm(\vec{r}_B \times \vec{v})$ and is perpendicular to $\vec{r}_B$

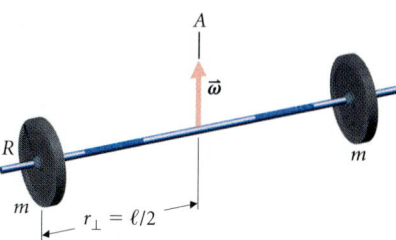

■ **FIGURE 12.8**
Since the disks have radius $r \ll \ell$, we estimate the rotational inertia of the barbell by modeling it as two particles at the ends of a rod with negligible mass. Each particle is at a distance $r_\perp = \ell/2$ from the rotation axis.

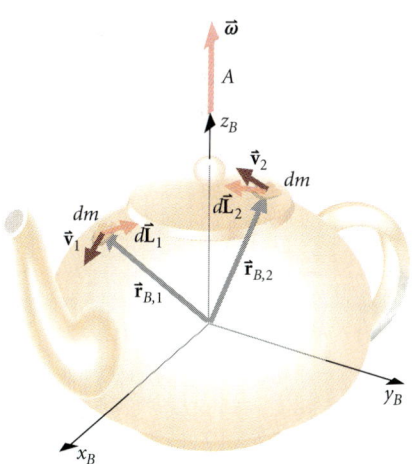

■ **FIGURE 12.9**
The angular momentum of a particle in the body is $d\vec{L} = dm(\vec{r}_B \times \vec{v})$. The contributions from different particles are in different directions.

REMEMBER: $\vec{L} = \vec{r} \times \vec{p} = \vec{r} \times (m\vec{v})$, AND WE MAY WRITE THIS AS $m(\vec{r} \times \vec{v})$ BECAUSE MASS IS A SCALAR.

ROTATIONAL INERTIA IS SOMETIMES CALLED MOMENT OF INERTIA.

TABLE 12.3 **Rotational Inertia of Various Objects**

Object	Axis	Rotational Inertia
Uniform rod	through center	$\dfrac{M\ell^2}{12}$
	about one end	$\dfrac{M\ell^2}{3}$
Uniform disk	through center, perpendicular to disk	$\dfrac{MR^2}{2}$
	a diameter	$\dfrac{MR^2}{4}$
Uniform rectangle	through center, perpendicular to plane	$\dfrac{M(\ell^2 + w^2)}{12}$
	through center, in the plane, perpendicular to edge	$\dfrac{M\ell^2}{12}$

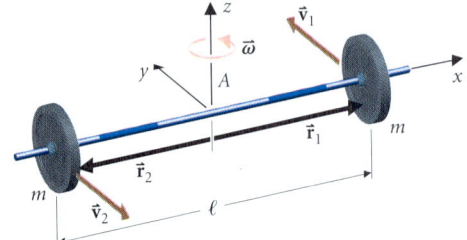

FIGURE 12.10
The barbell has two particles, each of which has angular momentum in the z-direction. The position vectors in the body coordinate system, $\vec{r}_1$ and $\vec{r}_2$, each have length $\ell/2$. Each particle has mass m and $r_\perp = \ell/2$.

and $\vec{v}$, as shown in the figure. The different contributions $d\vec{L}$ from different particles of the body are in different directions, and the direction of the sum is not immediately obvious. First we'll calculate the sum for a simple object.

EXAMPLE 12.5 ♦ In Example 12.4 the barbell is rotating with angular velocity $\vec{\omega}$ about its center of mass, where the vector $\vec{\omega}$ lies along axis A. Find its angular momentum.

MODEL As in Example 12.4, we model the barbell as two particles on the end of a massless rod. We define the z-coordinate axis to be along axis A and the x-axis to lie along the barbell axis at the instant shown (■ Figure 12.10).

398   CHAPTER 12 • DYNAMICS OF RIGID BODIES

TABLE 12.3 *(continued)*

WE'LL SHOW HOW TO CALCULATE THESE VALUES OF I IN §12.5.

Object	Axis	Rotational Inertia
Uniform cone	symmetry axis	$\dfrac{3MR^2}{10}$
Uniform cone	through the end, perpendicular to symmetry axis	$3M\left(\dfrac{h^2}{5} + \dfrac{R^2}{20}\right)$
Solid cylinder	symmetry axis	$\tfrac{1}{2}MR^2$
Solid cylinder	a diameter through the CM	$\dfrac{M}{4}\left(R^2 + \dfrac{h^2}{3}\right)$
Uniform sphere	a diameter	$\tfrac{2}{5}MR^2$
Uniform spherical shell	a diameter	$\tfrac{2}{3}MR^2$

SETUP Each particle is a distance $|\vec{r}_B| = \ell/2$ from axis A. The velocities are in the plus and minus y-directions, perpendicular to $\vec{r}_B$, and have magnitude $\omega\ell/2$.

SOLVE The angular momentum of the barbell is:

$$\vec{L}_A = m_1\vec{r}_1 \times \vec{v}_1 + m_2\vec{r}_2 \times \vec{v}_2$$
$$= m(\ell/2)(\omega\ell/2)(\hat{\mathbf{i}} \times \hat{\mathbf{j}}) + m(\ell/2)(\omega\ell/2)[(-\hat{\mathbf{i}}) \times (-\hat{\mathbf{j}})]$$
$$= (m\ell^2/2)\omega\hat{\mathbf{k}}.$$

In Example 12.4 we found the rotational inertia of this object: $I_A = m\ell^2/2$. So:

$$\vec{L}_A = I_A\vec{\omega}.$$

IN CHAPTER 9 YOU LEARNED THAT $\vec{L}$ FOR A PARTICLE DEPENDS ON THE ORIGIN O. SIMILARLY, HERE $\vec{L}$ DEPENDS ON THE AXIS OF ROTATION, SO WE LABEL $\vec{L}$ WITH THAT AXIS TO REMIND YOU.

NOTICE THE SIMILARITY BETWEEN $\vec{L} = I\vec{\omega}$ AND $\vec{p} = m\vec{v}$.

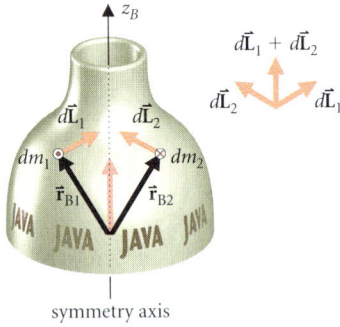

(a) Rotationally symmetric

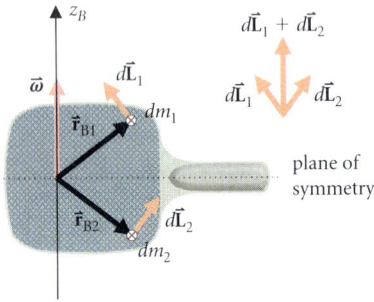

(b) Mirror symmetric

■ **FIGURE 12.11**
These symmetric objects have $\vec{L}$ parallel to $\vec{\omega}$. (a) Rotationally symmetric. Two elements at the same height z and equidistant from the rotation axis, with $dm_1 = dm_2$, have angular momenta of equal magnitude whose x- and y-components cancel while their z-components add. (b) Mirror-symmetric about the x–y plane. Here equal-mass elements located at the same distance above and below the symmetry plane have x- and y-components of angular momentum that cancel.

COMPARE THIS RESULT WITH NEWTON'S SECOND LAW FOR PARTICLES: FORCE = MASS × ACCELERATION.

ANALYZE The angular momentum $\vec{L}$ is parallel to $\vec{\omega}$ and proportional to the rotational inertia I_A. This pleasing result emerges because the barbell is a highly symmetric object in which the contribution to $\vec{L}$ from each particle is parallel to $\vec{\omega}$.

Fortunately, many rotating objects we may wish to study (like wheels, pulleys, or aircraft propellers) are symmetric, and their angular momenta are parallel to their angular velocities. There are two classes of such objects (■ Figure 12.11), and we shall restrict ourselves to studying them.

1. Rigid bodies that have rotational symmetry about an axis and rotate about that symmetry axis. Wheels are in this class of objects.
2. Rigid bodies that are mirror-symmetric about a plane and rotate about an axis perpendicular to that plane. Tennis rackets and baseball bats are in this category.

A rigid body that is rotationally symmetric about the z-axis or mirror-symmetric about the x–y plane has angular momentum about the z-axis equal to its rotational inertia about that axis multiplied by its angular velocity:

$$\vec{L}_A = I_A \vec{\omega}. \tag{12.9}$$

12.3 DYNAMIC BEHAVIOR OF RIGID BODIES

12.3.1 General Solution Method

We are now ready to analyze the behavior of rigid bodies when forces and torques act on them. The net force produces an acceleration of the body's CM that is described precisely as in the dynamic problems of Chapters 4 and 5. The net torque produces a change in the body's angular momentum and, consequently, an angular acceleration. Except in an optional section at the end of the chapter, we shall restrict our attention to situations in which both the net torque and the angular velocity lie along a symmetry axis of the rotating body. Then we can choose both laboratory and body z-axes to be along that symmetry axis. The body's angular momentum also lies along the z-axis and satisfies the relation $\vec{L} = I\vec{\omega}$. Furthermore, because the rotation occurs about a fixed body axis, the rotational inertia is constant; $dI/dt = 0$. Then:

$$\vec{\tau} = \frac{d\vec{L}}{dt} = \frac{d(I\vec{\omega})}{dt} = I\frac{d\vec{\omega}}{dt}$$

$$\vec{\tau} = I\vec{\alpha}. \tag{12.10}$$

torque = rotational inertia × angular acceleration.

This equation for rotational dynamics, like those for rotational kinematics, has the same form as the analogous equation for linear particle motion (● Table 12.4). Torque corresponds to force and rotational inertia corresponds to mass. Consequently, the method for analyzing a

TABLE 12.4 Dynamic Relations

	Rigid Body	Particle
Momentum	$\vec{L}_A = I_A\vec{\omega}$ (angular)	$\vec{p} = m\vec{v}$ (linear)
Kinetic energy	$K = \frac{1}{2}I_A\omega^2$	$K = \frac{1}{2}mv^2$
Governing relations	$\vec{\tau}_A = d\vec{L}_A/dt = I_A\vec{\alpha}$	$\vec{F} = d\vec{p}/dt = m\vec{a}$
Power	$P = \vec{\tau}\cdot\vec{\omega}$	$P = \vec{F}\cdot\vec{v}$

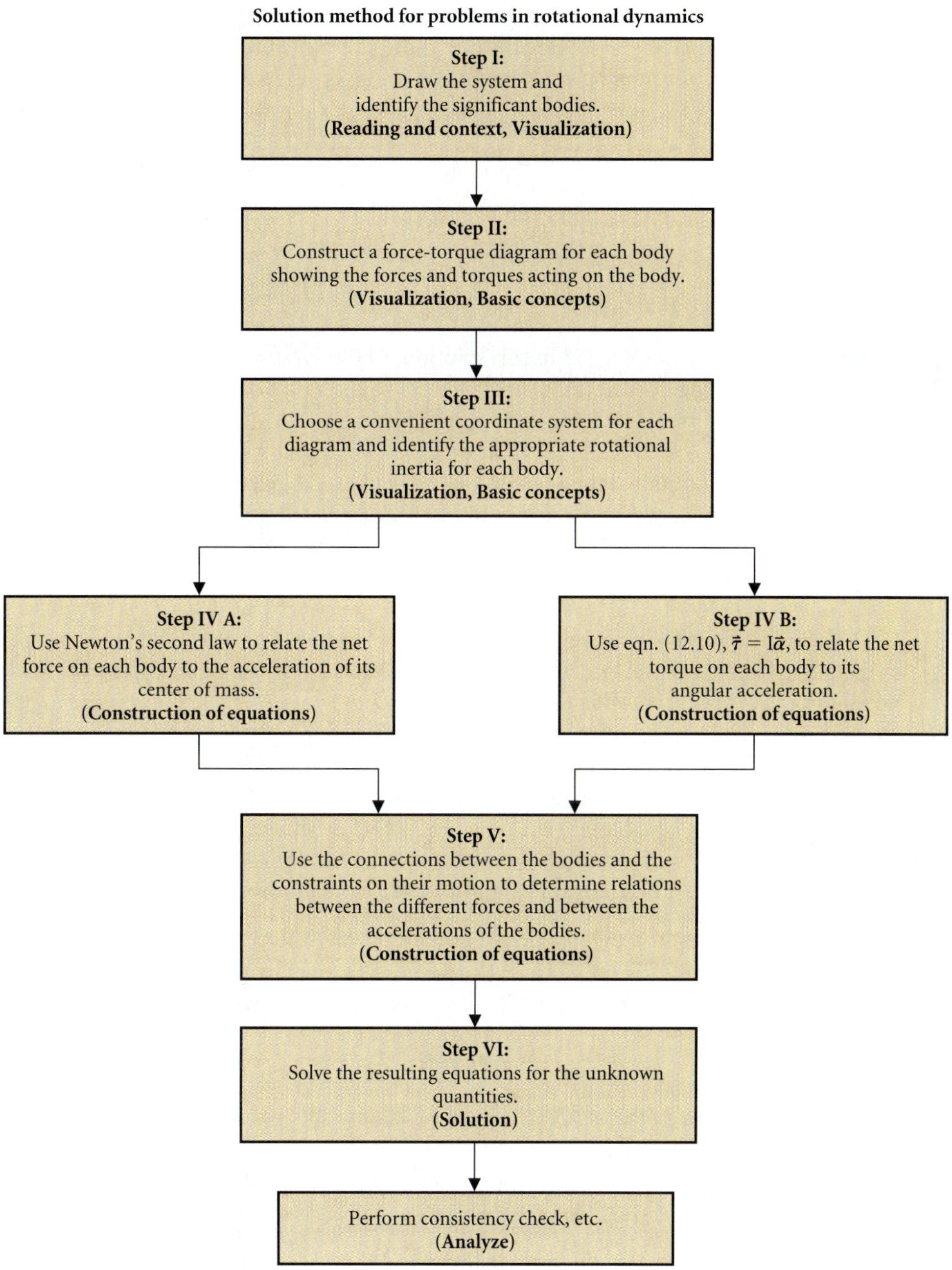

Solution method for problems in rotational dynamics

Step I:
Draw the system and identify the significant bodies.
(Reading and context, Visualization)

↓

Step II:
Construct a force-torque diagram for each body showing the forces and torques acting on the body.
(Visualization, Basic concepts)

↓

Step III:
Choose a convenient coordinate system for each diagram and identify the appropriate rotational inertia for each body.
(Visualization, Basic concepts)

Step IV A:
Use Newton's second law to relate the net force on each body to the acceleration of its center of mass.
(Construction of equations)

Step IV B:
Use eqn. (12.10), $\vec{\tau} = I\vec{\alpha}$, to relate the net torque on each body to its angular acceleration.
(Construction of equations)

Step V:
Use the connections between the bodies and the constraints on their motion to determine relations between the different forces and between the accelerations of the bodies.
(Construction of equations)

↓

Step VI:
Solve the resulting equations for the unknown quantities.
(Solution)

↓

Perform consistency check, etc.
(Analyze)

■ **FIGURE 12.12**
Compare this plan with Figure 5.4.

rigid body or system of bodies follows the same plan we developed for systems of particles (■ Figure 12.12).

EXAMPLE 12.6 ♦ A spool of string mounted on the wall has radius 0.12 m and mass 6.2 kg. It is free to rotate about an axle through its center. If someone pulls on the string's free end with a force of 5.2 N, what is the spool's angular acceleration?

MODEL We model the spool as a uniform cylinder.
STEP I

STEP II The torque about the spool's axle caused by string tension is $\vec{\tau} = \vec{r} \times \vec{F} = FR\hat{\otimes}$, (■ Figure 12.13). The axle exerts a normal force that opposes the tension and prevents the CM of the spool from accelerating, but it does not exert any torque about the axle.

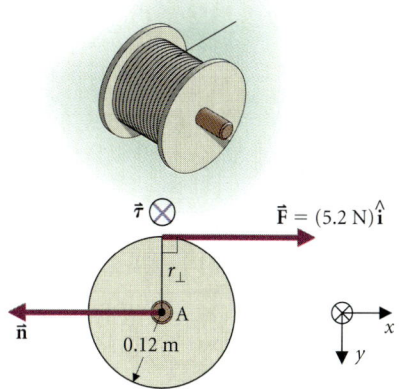

■ **FIGURE 12.13**
A spool of string rotates when you pull on the free end with force F. The string tension $T = F$ produces a torque about the axis of the spool. The normal force exerted by the axle on the spool prevents its CM from accelerating but exerts no torque about the axis A. *Remember:* The symbol $\otimes$ means "into the page" (§9.2.1).

STEP III The spool's rotational inertia about its axle is $\frac{1}{2}MR^2$ (Table 12.3). We choose the z-axis to be along the spool's axle, directed into the plane of the figure: $\hat{\mathbf{k}} = \hat{\otimes}$. Then both the torque and the angular acceleration are in the positive z-direction.

SETUP
STEP IV We use the z-component of eqn. (12.10):

$$\tau_z = I\alpha_z.$$

$$FR = \tfrac{1}{2}MR^2\alpha_z.$$

SOLVE
STEP VI

$$\alpha_z = \frac{2F}{MR} = \frac{2(5.2 \text{ N})}{(6.2 \text{ kg})(0.12 \text{ m})} = 14 \text{ rad/s}^2.$$

ANALYZE Step V is not needed in this solution. The corresponding linear acceleration of a point on the edge of the spool is $a = \alpha_z R = (14 \text{ rad/s}^2)(0.12 \text{ m}) = 1.7 \text{ m/s}^2$, a reasonable number for a gentle tug on a spool of string.

One of the most common examples of a constraint is the relation between the linear and angular motions of a rigid body. For example, wheels usually *roll without slipping*, and then the magnitudes of the linear and angular velocities and accelerations are related by eqns. (12.5) and (12.6).

EXAMPLE 12.7 ♦♦ A force $\vec{F}$ acts at the CM of a uniform wheel with mass M and radius R, rolling on a level surface. If the wheel rolls without slipping, what static frictional force acts on the wheel at the point of contact (■ Figure 12.14)? What is the maximum value of $|\vec{F}|$ for which the wheel can roll without slipping?

MODEL
STEPS I AND II To continue rolling without slipping as it accelerates, the wheel must have an angular acceleration. Friction, acting as shown in the figure, produces the necessary torque about the CM. As usual for static friction, the magnitude of f is less than or equal to $\mu_s n$. If this friction cannot provide enough torque, then the wheel slips.

STEP III We choose the origin to be at the center of the wheel, with the x-axis parallel to $\vec{F}$ and the z-axis into the page. The appropriate rotational inertia is that for a cylinder about its symmetry axis (see Table 12.3; a disk would give the same result):

$$I = \tfrac{1}{2}MR^2. \qquad \text{(i)}$$

SETUP
STEP IVa Normal force balances the wheel's weight and the net force in the x-direction accelerates it:

$$\sum F_x = F - f = Ma_x. \qquad \text{(ii)}$$

$$\sum F_y = Mg - n = 0. \qquad \text{(iii)}$$

STEP IVb Friction provides the torque about the origin, which is in the z-direction.

$$\sum \tau_z = fR = I\alpha_z. \qquad \text{(iv)}$$

STEP V If the wheel rolls without slipping, eqn. (12.6) relates the magnitudes of the linear and angular accelerations. We chose our axes so that if the wheel rolls without slipping, α_z is positive when a_x is positive (Figure 12.14). Thus:

$$a_x = \alpha_z R. \qquad \text{(v)}$$

SOLVE
STEP VI Combining eqns. (iv) and (v), we have:

$$fR = Ia_x/R \quad \text{or} \quad a_x = fR^2/I.$$

Then from eqn. (ii),

$$F - f = \frac{MfR^2}{I} \quad \text{or} \quad f = \frac{F}{(1 + MR^2/I)}.$$

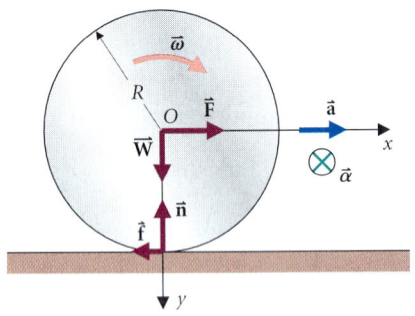

■ **FIGURE 12.14**
A force F acts at the CM of a uniform wheel of radius R. Torque due to friction causes the wheel to roll. Note the directions of $\vec{\alpha}$ and $\vec{a}$. When the wheel rolls without slipping, both a_x and α_z are positive.

Finally, using eqn. (i): $f = F/(1 + 2) = F/3$.   (vi)

SETUP This relation holds so long as $f \leq \mu_s n = \mu_s Mg$.

WE USED EQN. (iii).

SOLVE For $|\vec{F}| > F_{max} = 3f_{max} = 3\mu_s Mg$, the wheel slips.

ANALYZE The maximum acceleration for which the wheel continues to roll, from eqns. (ii) and (vi), is:

$$a_{max} = \tfrac{2}{3}F_{max}/M = 2f_{max}/M = 2\mu_s g.$$

Notice that the acceleration of the wheel is less than F/M, because the friction opposes $\vec{F}$.

Remember that eqn. (12.5) ($v = \omega R$) relates the magnitudes of $\vec{v}$ and $\vec{\omega}$. In solving problems we must also pay proper attention to the directions of these vector quantities. Careful choice of coordinates can simplify the process.

WE CHOSE THE DIRECTION OF THE z-AXIS SO THAT τ_z AND ω_z WOULD BE POSITIVE. WITH y UP AND z OUT OF THE PAGE, WE WOULD HAVE NEEDED A MINUS SIGN IN THE RELATION BETWEEN a_x AND α_z. REMEMBER: THE SIGN OF A COMPONENT INDICATES THE DIRECTION OF A VECTOR.

EXERCISE 12.4 ♦♦ What is the maximum acceleration of the wheel if the force $\vec{F}$ is applied at the top of the wheel?

EXAMPLE 12.8 ♦♦ Atwood's machine was used in early experimental tests of Newton's laws. In one version of this machine, two blocks with masses $m_1 = 0.50$ kg and $m_2 = 0.55$ kg are connected by a string. The string passes over a pulley with radius $R = 0.10$ m, rotational inertia $I = 1.3 \times 10^{-3}$ kg·m², and a frictionless bearing (■ Figure 12.15). If the system is released from rest, what is the acceleration of the larger block?

MODEL
STEPS I AND II We've always assumed that pulleys move with the string. If the pulley has significant mass, a torque is required, and the string tensions on either side of the pulley must differ. In this system the bearing is frictionless, but the groove of the pulley is not. Static friction between the string and the pulley causes the difference in tensions. Figure 12.15b shows the force-torque diagrams for the bodies in the system.

STEP III Since m_2 is larger than m_1, we expect block 2 to accelerate downward and the pulley to rotate clockwise. Thus in each diagram we choose coordinates with y down and z into the page. The appropriate rotational inertia is given in the problem statement.

SETUP
STEP IVa There are three bodies to consider in this problem. Only the two blocks have linear acceleration, and the forces on them are in the y-direction.

Forces on block 1: $\sum F_y = m_1 g - T_1 = m_1 a_{1y}$.   (i)

Forces on block 2: $\sum F_y = m_2 g - T_2 = m_2 a_{2y}$.   (ii)

We expect to find $a_{1y} < 0$, and $a_{2y} > 0$.

STEP IVb The only object that rotates is the pulley.

Torque about the center of the pulley: $\sum \tau_z = R(T_2 - T_1) = I\alpha_z$.   (iii)

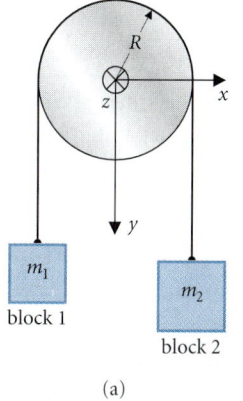

(a)

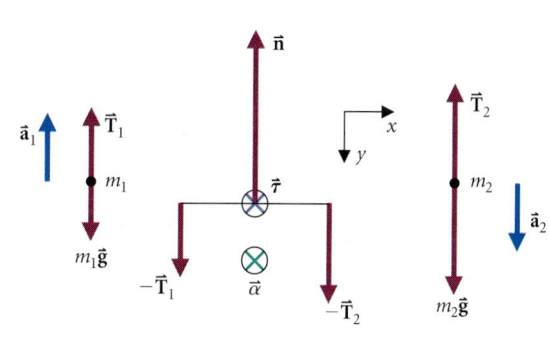

(b)

■ **FIGURE 12.15**
(a) Atwood's machine. The z-axis points into the page. (b) Free-body diagrams for the two blocks and the pulley. Because of friction between the string and the pulley, $T_1 \neq T_2$. In each diagram the x-axis is horizontal, y down, and z into the page.

THIS IS *CONSERVATION OF STRING*, CF. §5.2, ESPECIALLY EXAMPLE 5.2.

STEP V Block 1 moves up the same distance that block 2 moves down:

$$-a_{1y} = a_{2y} \equiv a. \qquad \text{(iv)}$$

There is also a relation between the linear acceleration of each block and the angular acceleration of the pulley. When the blocks move a distance $\Delta s = v\,\Delta t$, the pulley rotates through an angle $\omega\,\Delta t = \Delta\theta = \Delta s/R = v\,\Delta t/R$. So $\omega = v/R$, and:

$$\alpha_z = a/R. \qquad \text{(v)}$$

This relation is the same as that for rolling: the pulley rolls along the string.

SOLVE Combining eqns. (iii) and (v), $Ia/R = R(T_2 - T_1)$.

STEP VI Using eqns. (i), (ii) and (iv) to substitute for the tensions:

$$Ia/R^2 = m_2(g - a) - m_1(g + a).$$

Collecting terms: $a(m_1 + m_2 + I/R^2) = g(m_2 - m_1).$

Thus:
$$a = \frac{g(m_2 - m_1)}{m_1 + m_2 + I/R^2}$$

$$= \frac{(9.8 \text{ m/s}^2)(0.55 \text{ kg} - 0.50 \text{ kg})}{0.55 \text{ kg} + 0.50 \text{ kg} + (1.3 \times 10^{-3} \text{ kg} \cdot \text{m}^2)/(0.10 \text{ m})^2}$$

$$= 0.4 \text{ m/s}^2.$$

ANALYZE If $m_1 = m_2$, then $a = 0$ and the system is in equilibrium, as expected. The rotation of the pulley reduces the acceleration of the blocks compared with a system where the rope slides without friction over a smooth peg. The external gravitational force acting on the system accelerates the pulley as well as the two blocks. Notice that we lost a significant figure in subtracting the masses.

Study Problem 9 ♦♦♦ *Rockfall!*

A spherical boulder, sliding down an icy mountain slope, reaches a point where the ice has melted. Further down, the surface is rough, with a coefficient of kinetic friction $\mu_k = 0.50$ between boulder and slope. If the speed of the boulder when it reaches the rough surface is $v_o = 7.5$ m/s and the slope is at 45° with the horizontal, analyze the subsequent motion of the boulder and determine whether it ever rolls without slipping.

Modeling the System

Since the boulder initially slides on the ice, it does not roll until after it reaches the rough surface. Without friction, no torques act about its CM. Upon reaching the rough surface, it continues to slide, but kinetic friction produces a torque that causes angular acceleration. If the torque is great enough, then the boulder ends up rolling without slipping. Once that happens, static friction at the contact point adjusts itself to maintain the rolling motion.

Plan

We follow the standard method outlined in Figure 12.12. We also need the kinematic relations in Table 12.2. We shall solve for the linear motion and angular motion separately, and then compare them to determine whether rolling occurs.

Construction of Equations

STEPS I AND II. ■ Figure 12.16 is the force-torque diagram.

STEP III. We start the clock as the boulder leaves the ice with speed v_o. As it slides down the slope, the boulder's angular momentum about an arbitrary origin changes both because of the increasing speed of its CM and because of its changing angular velocity. However, if we choose

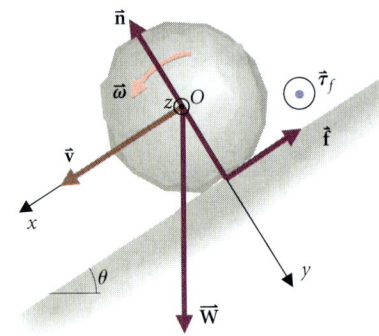

■ **FIGURE 12.16**
Rock falling down the mountainside. Torque about its CM due to friction starts it rolling. Does it stop slipping? The *z*-axis is out of the page.

the origin O to be the center of the boulder at $t = 0$, $\vec{v}_{CM}$ always lies along a line through O, $\vec{r} \times \vec{v}_{CM} = 0$, and $\vec{L}_O$ is due to the rolling alone. We also choose the x-axis to be down the slope and the z-axis to be out of the page. The appropriate rotational inertia is that for a sphere about an axis through its center (Table 12.3):

COMPARE EXAMPLE 11.6.

$$I = 2MR^2/5. \qquad (i)$$

STEP IV a.
$$\sum F_x = Mg \sin\theta - f = Ma_x. \qquad (ii)$$
$$\sum F_y = Mg \cos\theta - n = 0. \qquad (iii)$$

STEP IV b.
$$\sum \tau_z = fR = I\alpha_z. \qquad (iv)$$

STEP V. While the boulder is sliding, the magnitude of the friction force is:
$$f = \mu_k n. \qquad (v)$$

The boulder begins rolling without slipping if and when the constraint
$$v_x = \omega_z R \qquad (vi)$$
is satisfied. We also need the following kinematic relations.

Velocity of the boulder as a function of time: $\qquad v_x = v_o + a_x t. \qquad (vii)$

Angular velocity of the boulder as a function of time: $\qquad \omega_z = \alpha_z t. \qquad (viii)$

REMEMBER: THE BOULDER ALREADY HAS SPEED v_o WHEN IT REACHES THE EDGE OF THE ICE.

Solution

We find a_x by combining eqns. (ii), (iii), and (v):
$$a_x = g(\sin\theta - \mu_k \cos\theta). \qquad (ix)$$

We then find α_z by combining eqns. (iii), (iv), and (v):
$$\alpha_z = R\mu_k Mg (\cos\theta)/I.$$

Using eqn. (i), this becomes: $\qquad \alpha_z = 5\mu_k g (\cos\theta)/(2R).$

Putting these results into eqns. (vii) and (viii), we have:
$$v_x = v_o + g(\sin\theta - \mu_k \cos\theta)t$$

and
$$\omega_z = \frac{5\mu_k g \cos\theta}{2R} t.$$

The angular speed of the boulder increases as a result of the torque acting on it. If ω increases fast enough—that is, if μ_k is great enough—constraint (vi) is satisfied at some time t_r and then rolling without slipping begins (■ Figure 12.17a). If this happens, t_r is given by:

$$v_o + g(\sin\theta - \mu_k \cos\theta)t_r = v_x = \omega_z R = (5/2)t_r \mu_k g \cos\theta. \qquad (x)$$

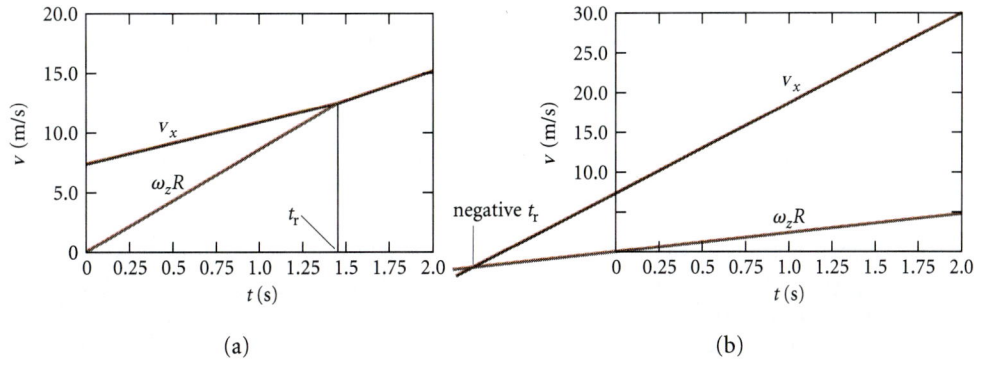

■ FIGURE 12.17
(a) Rock rolling down a 45° slope with coefficient of friction $\mu_k = 0.5$. The plot shows the linear speed of the rock's CM and the speed ωR of a point on its surface due to rotation about the CM. The angular speed increases more rapidly than the linear speed, and the condition $v = \omega R$ is satisfied at $t = t_r = 1.4$ s. Once the rock begins rolling without slipping, its linear speed increases more rapidly, and $\omega = v/R$ for $t > t_r$.
(b) $\mu_k = 0.1$. The angular speed increases more slowly than the linear speed, and the rolling motion never catches up. The two lines diverge, and their mathematical intersection, at $t_r < 0$, has no physical meaning.

That is:
$$t_r = \frac{v_0}{g[(7/2)\mu_k \cos\theta - \sin\theta]}$$
$$= \frac{7.5 \text{ m/s}}{(9.8 \text{ m/s}^2)[(7/2)(0.50)\cos 45° - \sin 45°]} = 1.4 \text{ s}.$$

Analysis

If there were insufficient frictional torque to spin the boulder up, its speed v would remain greater than ωR, and it would slide all the way down the slope (■ Figure 12.17b). In this case, there is no positive solution for t_r. From eqn. (x) this happens when

$$(7/2)\mu_k \cos\theta < \sin\theta,$$

or
$$\mu_k < \mu_{crit} \equiv (2/7)\tan\theta.$$

The same thing happens if μ is unchanged and the mountain is steeper: $\tan\theta > 7\mu/2$.

■ **EXERCISE 12.5** ❖ How would the results of Study Problem 9 be different if a cylindrical log were falling down the slope instead of a boulder? ■

✻ 12.3.2 Limitations of the Rigid-Body Model

LOOK AT THE CHIMNEY IN FIGURE 12.19!

The forces that hold a rigid body together are finite in strength, and so no body can be perfectly rigid. A rapidly rotating body may require very large forces to maintain its shape. If these forces exceed the maximum the material can exert, the body distorts or shatters. Thus analysis of the required forces is an important part of the design of rotating systems.

EXAMPLE 12.9 ♦♦ If a flywheel of radius 0.5 m rotates at $\omega = 1000$ rad/s, what internal force per kilogram is needed to hold particles at the rim on the wheel?

MODEL The internal forces must be sufficient to accelerate each particle of the flywheel on its circular path.

SETUP A particle at distance r from the rotation axis has acceleration $\omega^2 r$ (eqn. 3.10), and the force needed to produce this acceleration is:
$$F_c = m\omega^2 r.$$

SOLVE The force per unit mass for particles on the rim of the wheel is:
$$\frac{F_c}{m} = \omega^2 r = (1000 \text{ rad/s})^2 (0.5 \text{ m}) = 5 \times 10^5 \text{ N/kg}.$$

ANALYZE The required force on a particle at the rim of the flywheel is about 5×10^4 times the particle's weight, a considerable force indeed. ■

12.4 APPLICATION OF THE CONSERVATION LAWS

A system of rigid bodies, like any system, possesses energy and angular momentum that do not change if no external forces or torques act. Problems that ask for a comparison of two states of a system of bodies and are not concerned with the time required to move between the states will most probably yield to a conservation law approach.

EXAMPLE 12.10 ♦♦ A telephone pole of length $\ell = 21$ m and mass $M = 1.0 \times 10^3$ kg is accidentally released when standing nearly vertical on the ground. With what linear speed does the end of the pole hit the ground? Assume friction holds the end on the ground stationary.

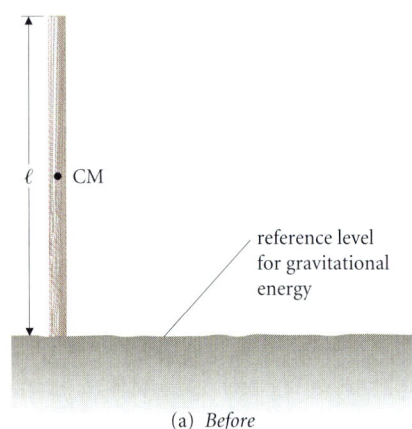

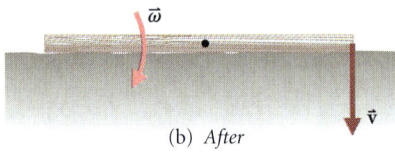

■ **FIGURE 12.18**
As a telephone pole standing on a rough surface falls, the end on the ground remains fixed and the pole rotates about that end. We use conservation of energy to find its angular speed as it hits the ground. The CM, initially at height $\ell/2$, is at the reference level as the pole hits the ground.

MODEL The nonconservative forces acting on the pole are the normal and static frictional forces exerted on its lower end. Since this end is stationary, these forces do no work and the total mechanical energy of the pole remains constant until the pole hits the ground. We use the standard approach for conservation law problems (Figure 6.10), with the reference level for gravitational energy at the ground. As the pole falls, it rotates about the stationary lower end. The CM of a uniform rod is $\ell/2$ from each end and its rotational inertia about its end is $M\ell^2/3$ (Table 12.3). The *before* and *after* states are shown in ■ Figure 12.18.

SETUP

	BEFORE	AFTER
	(pole standing vertically)	(pole about to hit the ground)
Kinetic energy	none	$\frac{1}{2}I\omega^2$
Gravitational energy	$Mg\ell/2$	none
Total energy	$Mg\ell/2$	$\frac{1}{2}I\omega^2$

Set them equal: $\quad Mg\ell/2 = \frac{1}{2}(M\ell^2/3)\omega^2.$

SOLVE
$$\omega = \sqrt{3g/\ell}.$$

The end of the pole hits the ground at speed: $\quad v = \omega\ell = \sqrt{3g\ell}.$

Plugging in the numbers, $v = [3(9.8 \text{ m/s}^2)(21 \text{ m})]^{1/2} = 25 \text{ m/s}.$

ANALYZE Note that the end of the pole is moving faster than a particle falling freely through a distance ℓ ($v = \sqrt{2g\ell}$, e.g., Example 8.5). Internal forces in the pole are needed to accelerate its end. ■ Figure 12.19 shows what happens when the internal forces in a falling chimney are insufficient to provide the necessary accelerations: it breaks.

■ **FIGURE 12.19**
This chimney began falling like the telephone pole in Example 12.10, but internal forces were insufficient to accelerate the top part, so it broke. Notice how the upper two-thirds has fallen behind the bottom third.

EXERCISE 12.6 ♦♦ Use conservation of energy to find the speed of the blocks in Atwood's machine (Example 12.8) when the large block has fallen a distance h. Is your result consistent with the result of the example? (*Hint:* Don't forget the rotational kinetic energy of the pulley.)

EXAMPLE 12.11 ♦♦ Satellites are often designed to rotate to maintain their stability. Sometimes it is necessary to de-spin them—that is, to reduce their angular speed. Two small objects, called yo-yos, each of mass $m = 1.0$ kg, are attached to the satellite by cords of length $\ell = 61$ m, which are wrapped around the satellite in a plane perpendicular to its angular velocity. To de-spin, the latches holding the yo-yos are opened and the cords unwind. The cords are designed to separate from the satellite when they make an angle $\phi = 80°$ with the satellite surface (■ Figure 12.20). If the initial spin rate is $\omega_i = 6.0$ rad/s,

(a) *Before*

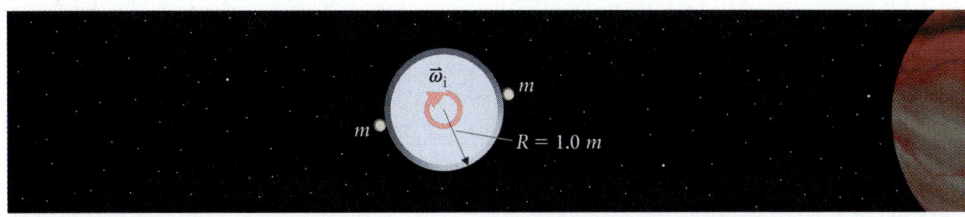

(b) *After*

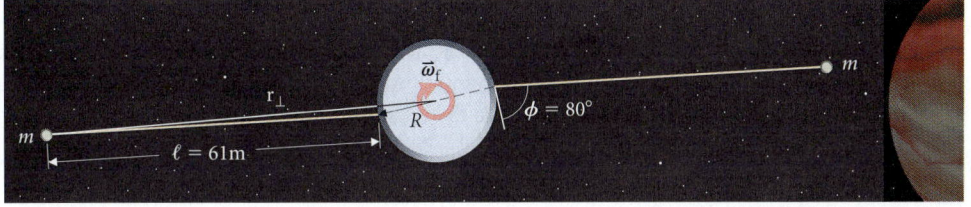

■ **FIGURE 12.20**
A satellite is de-spun by yo-yos. (a) Initially, the cords are wound around the satellite and the yo-yos are locked to its side. (b) After the latches are opened, the cords unwind. When the cords are fully unwound and make an angle $\phi = 80°$ with the satellite surface, as shown here, they separate from the satellite. The cords remain straight as they unwind. The full length ℓ is extended when they are released from the satellite.

the satellite's rotational inertia is $I = 1.5 \times 10^2$ kg·m^2, and its radius is 1.0 m, what is the satellite's final spin rate?

MODEL Since no external forces (other than gravity) act on the satellite, there are no external torques acting and so its angular momentum is constant. We choose the *before* state to be with the latches closed, and the *after* state to be with the cords fully extended and about to separate from the satellite. We find the perpendicular distance of the yo-yos from the rotation axis using the cosine rule.

SETUP
$$r_\perp^2 = R^2 + \ell^2 - 2R\ell \cos(\phi + 90°).$$

	BEFORE	AFTER
Angular momentum	$(I + 2mR^2)\omega_i$	$(I + 2mr_\perp^2)\omega_f$
Set totals equal:	$(I + 2mR^2)\omega_i = (I + 2mr_\perp^2)\omega_f.$	

SOLVE
$$\frac{\omega_f}{\omega_i} = \frac{I + 2mR^2}{I + 2m(R^2 + \ell^2 + 2R\ell \sin \phi)}.$$

The final angular velocity of the satellite is:
$$\omega_f = \frac{\omega_i(1 + 2mR^2/I)}{1 + (2mR^2/I)[1 + (\ell/R)^2 + 2(\ell/R)\sin \phi]}.$$

With $\ell/R = (61 \text{ m})/(1.0 \text{ m}) = 61$, and
$$2mR^2/I = 2(1.0 \text{ kg})(1.0 \text{ m})^2/(1.5 \times 10^2 \text{ kg·m}^2) = 0.013,$$
$$\omega_f = \frac{(6.0 \text{ rad/s})(1 + 0.013)}{1 + 0.013[1 + 61^2 + 2(61)\sin 80°]}$$
$$= 0.12 \text{ rad/s}.$$

ANALYZE It is expensive to carry things into space. This system is efficient because yo-yos of small mass can effectively de-spin the satellite. ∎

When a baseball player makes a hit, the ball imparts linear and angular impulses to the bat. In return, the bat exerts impulses on the batter's hands. Such situations, in which impulsive forces cause rapid changes in the energy and momentum of a rigid body, can often be analyzed with the conservation laws. To illustrate the method, we study the simpler (if old-fashioned) example of stickball, in which the bat may be modeled as a uniform rod.

EXAMPLE 12.12 ♦♦♦ The bat in a stickball game is a uniform rod of length $\ell = 0.80$ m. The player holds the bat at one end. Where should she hit the ball to minimize the impulse on her hands?

MODEL The ball hits the bat at a distance r_2 to the left of the CM (■ Figure 12.21). In the collision, the ball exchanges momentum with the bat and delivers an impulse of magnitude Δp to it.

SETUP The change in the CM velocity of the bat as a result of the collision is given by:
$$M \Delta v_{CM} = \Delta p.$$

The collision also changes the bat's angular momentum by an amount
$$\Delta L = r_2 \Delta p = I_{CM} \Delta \omega,$$

resulting in an angular velocity change of magnitude $\Delta \omega = \Delta L/I_{CM}$ about the CM. As the ball hits, the bat jerks backwards, but the rotational effect tends to move the handle end forward. The change in the bat's speed at a distance r_1 to the right of the CM immediately after the ball's impact is:
$$\Delta v = \Delta v_{CM} - r_1 \Delta \omega.$$

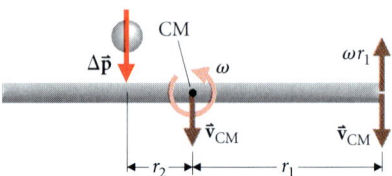

■ **FIGURE 12.21**
Stickball. The ball imparts an impulse $\Delta \vec{p}$ to the bat when it hits it a distance r_2 from the CM, and a corresponding angular impulse of magnitude $|\Delta \vec{L}| = r_2|\Delta \vec{p}|$. Where should the player hit the ball so that the resulting velocity at her hands (r_1 from the CM) is zero?

If this change is zero at the batter's hands, then no impulse is delivered to her hands. The value of r_2 that makes Δv zero at the batter's hands ($r_1 = \ell/2$) is given by:

$$\Delta v = \Delta v_{CM} - r_1 \Delta \omega = \frac{\Delta p}{M} - r_1 \frac{r_2 \Delta p}{I_{CM}} = 0.$$

The appropriate rotational inertia is that of a rod about its CM, $I = M\ell^2/12$.

SOLVE So: $\quad r_2 = \dfrac{I_{CM}}{Mr_1} = \dfrac{M\ell^2/12}{M\ell/2} = \dfrac{\ell}{6} = \dfrac{0.80 \text{ m}}{6} = 0.13 \text{ m}.$

The player should hit the ball 13 cm forward of the CM.

ANALYZE The result is independent of how hard the ball was hit (i.e., of the magnitude of the impulse Δp). The position we have calculated is not necessarily the best place to hit the ball. The player may be more interested in imparting a high speed to the outgoing ball than in protecting her hands. Vibrations in the bat (Chapter 14) may be more important in determining how sharp a sting the batter feels.

12.5 CALCULATION OF ROTATIONAL INERTIA

12.5.1 Integration

Rotational inertia (eqn. 12.8) is defined as an integral over a body. We developed the techniques for performing integrals in Interlude 2 and used them in Chapter 11 to locate a body's center of mass. Next we shall apply those techniques to calculate the rotational inertia of simple objects, for which only one coordinate is needed to describe differential elements of the body.

IN THIS SECTION THE STEP NUMBERS REFER TO INTERLUDE 2.

EXAMPLE 12.13 ♦♦♦ Find the rotational inertia of a uniform rod with length ℓ and mass M about an axis perpendicular to the rod and through its CM.

MODEL
STEP I We choose a coordinate system with the x-axis along the length of the rod (■ Figure 12.22) and the origin on the rotation axis—that is, at the rod's CM. The elements of the rod are differential pieces of length dx.

STEP II A typical element is located at an arbitrary coordinate x.

SETUP
STEP III The contribution of an element to the rotational inertia is $dI = r_\perp^2 \, dm$. With our choice of origin, the perpendicular distance from the axis of rotation is the magnitude of the coordinate: $r_\perp \equiv |x|$. Since the rod is uniform, the mass dm is proportional to the length of the element:

$$\frac{dm}{M} = \frac{dx}{\ell} \Rightarrow dm = \frac{M \, dx}{\ell}.$$

Then $dI = (M/\ell)x^2 \, dx$.

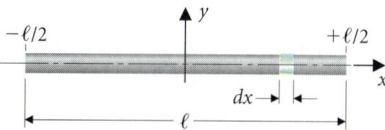

■ **FIGURE 12.22**
A uniform rod modeled as a collection of differential elements of length. We put the x-axis along the rod, with the origin at its center. An arbitrary differential element is shown. The perpendicular distance of this element from the rotation axis equals the magnitude of its x-coordinate.

STEP IV The rod extends from $x = -\ell/2$ to $+\ell/2$. Since we chose to use the coordinate increment dx to represent a length, dx must be positive; that is, we add the elements in order from left to right, from $-\ell/2$ to $+\ell/2$.

SOLVE
STEP V $\quad I = \displaystyle\int dI = \int_{-\ell/2}^{+\ell/2} \frac{M}{\ell} x^2 \, dx = \frac{M}{\ell} \frac{x^3}{3} \bigg|_{-\ell/2}^{+\ell/2} = \frac{M\ell^2}{12}.$

ANALYZE It is generally considered good form to express rotational inertia in terms of the object's total mass, as we did in Table 12.3. As Study Problem 9 illustrates, the object's total mass sometimes cancels from a problem solution.

EXERCISE 12.7 ❖ Explain why the rotational inertia of a thin circular ring of mass M and radius R about an axis through its CM and perpendicular to the plane of the ring is $I_{ring} = MR^2$.

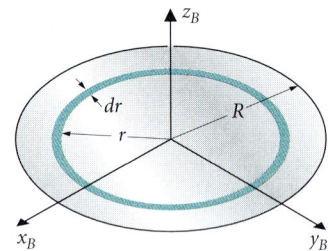

FIGURE 12.23
A disk is made up of differential rings of thickness dr.

THE TERM IN $(dr)^2$ IS MUCH SMALLER AND IS NEGLECTED.

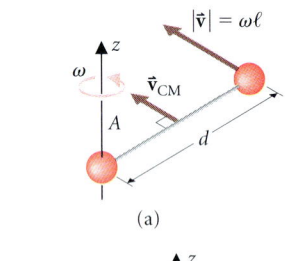

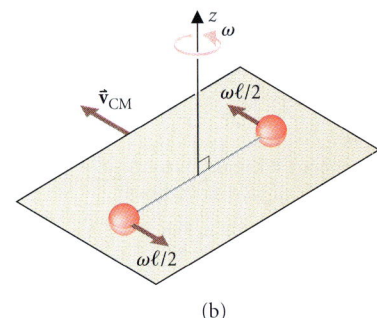

FIGURE 12.24
(a) A barbell rotating about one end. The CM moves with speed $v_{CM} = \omega\ell/2$. Its energy is due to the one moving particle. (b) In the CM frame, the barbell rotates about its CM. Its energy in the CM frame is due to both moving particles.

YOU SHOULD HAVE OBTAINED THIS AS THE RESULT OF EXERCISE 12.3A.

There is no new physical idea involved in computing rotational inertia for two- and three-dimensional objects. We may model symmetrical bodies as collections of simpler pieces whose rotational inertia is already known.

EXAMPLE 12.14 ♦♦♦ Find the rotational inertia of a uniform thin disk of mass M and radius R about an axis through its center and perpendicular to the disk.

MODEL STEP I We have already calculated the rotational inertia for a ring (Exercise 12.7), so we may model the disk as a collection of differential rings (■ Figure 12.23). We choose a coordinate system with the origin at the center of the disk and radius r measured outward.

STEP II The differential rings are defined by their radius r and width dr.

SETUP STEP III The relevant physical quantity is rotational inertia $dI = r^2\, dm$. Since the disk is uniform, the mass of the ring is proportional to its area:

$$\frac{dm}{M} = \frac{dA}{\pi R^2}.$$

The area dA is found by noting that the area of the ring is the difference in area between two circles whose radii differ by dr:

$$dA = \pi(r + dr)^2 - \pi r^2 = \pi[r^2 + 2r\, dr + (dr)^2 - r^2]$$
$$= 2\pi r\, dr.$$

So, $dm = (2M/R^2)r\, dr$ and $dI = (2M/R^2)r^3\, dr$.

STEP IV Since dr is the width of the ring, it must be positive. The limits of integration are from $r = 0$ to $r = R$.

SOLVE STEP V
$$I = \int dI = \int_0^R \frac{2M}{R^2}r^3\, dr = \frac{2M}{R^2}\frac{r^4}{4}\bigg|_0^R = \frac{MR^2}{2}.$$

ANALYZE Notice that each rotational inertia we compute is of the form

$$\text{mass} \times (\text{dimension of object})^2 \times \text{number},$$

where the number is less than (but of the order of) unity. Compare with Table 12.3. ■

12.5.2 The Parallel Axis Theorem

In Example 12.10 we used a tabulated value for the rotational inertia of a rod about an axis through one end. We could verify that value by repeating the calculation in Example 12.13 with the coordinate origin at one end of the rod, but there is a quicker alternative. The parallel axis theorem allows us to find an object's rotational inertia about any axis if we know its inertia about a parallel axis through the CM. This theorem is equivalent to the fact that any object's kinetic energy is a sum of its internal energy and its energy due to CM motion. Let's use a barbell (Example 12.4) to see how this works.

The barbell rotating with angular speed ω about an axis A through one of its ends (■ Figure 12.24) has total kinetic energy $K = \frac{1}{2}m(\omega\ell)^2$. If we interpret this as rotational energy:

$$K = \tfrac{1}{2}I_A\omega^2 \quad \text{and} \quad I_A = m\ell^2.$$

In the barbell's CM frame (Figure 12.24b), each end of the barbell has speed $\omega\ell/2$, so the internal energy is:

$$K_{\text{int}} = 2(\tfrac{1}{2}m)(\omega\ell/2)^2 = \tfrac{1}{2}(m\ell^2/2)\omega^2 = \tfrac{1}{2}I_{CM}\omega^2.$$

The kinetic energy due to the CM motion is:

$$K_{CM} = \tfrac{1}{2}(2m)v_{CM}^2 = m(\omega\ell/2)^2 = m(\ell/2)^2\omega^2.$$

Both calculations give the same result for the total kinetic energy:

$$\tfrac{1}{2}m(\omega\ell)^2 = K = K_{int} + K_{CM} = \tfrac{1}{2}(m\ell^2/2)\omega^2 + m(\ell/2)^2\omega^2.$$

In terms of rotational inertia and the barbell's total mass $M = 2m$:

$$\tfrac{1}{2}I_A\omega^2 = \tfrac{1}{2}I_{CM}\omega^2 + \tfrac{1}{2}M(\ell/2)^2\omega^2.$$

Thus:
$$I_A = I_{CM} + M(\ell/2)^2.$$

The barbell illustrates a result that is independent of an object's shape: rotation about an arbitrary axis is a combination of rotation about the CM and rotation of the CM about the arbitrary axis. The statement of this fact in terms of rotational inertia is the *parallel axis theorem*:

> The rotational inertia of a rigid body about any axis A equals the rotational inertia about a parallel axis through the CM plus the mass of the body multiplied by the square of the perpendicular distance h between the two axes:
>
> $$I_A = I_{CM} + Mh^2. \qquad (12.11)$$

You will find a general proof of this theorem in the Math Topic following.

Math Topic

General Proof of the Parallel Axis Theorem

Proof of eqn. (12.11) follows from the definition of rotational inertia, eqn. (12.8). Figure 12.25 shows a particle in the body with perpendicular distance $r_\perp$ from the axis A and $r'_\perp$ from a parallel axis through the CM. The perpendicular distance from the CM to axis A is h. Applying the law of cosines, we have:

$$r_\perp^2 = r'^2_\perp + h^2 - 2hr'_\perp \cos\theta.$$

So:
$$\int r_\perp^2 \, dm = \int (r'^2_\perp + h^2 - 2hr'_\perp \cos\theta) \, dm$$
$$= \int r'^2_\perp \, dm + h^2 \int dm - 2h \int r'_\perp \cos\theta \, dm.$$

But:
$$\int r'_\perp \cos\theta \, dm = \int x'_B \, dm$$
$$= \text{(total mass)} \times (x'_B \text{ coordinate of CM})$$
$$= 0,$$

since the CM is the origin of the prime coordinate system. (We used eqn. 11.5.)

Thus
$$I_A = \int r_\perp^2 \, dm = \int r'^2_\perp \, dm + Mh^2 = I_{CM} + Mh^2,$$

as required.

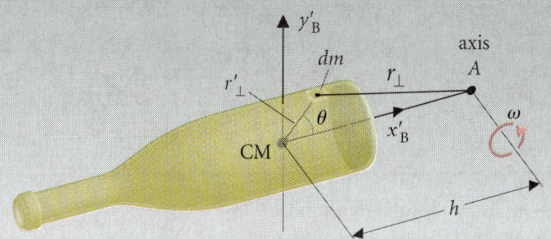

Figure 12.25
Proof of the parallel axis theorem. The bottle rotates about axis A. The element dm has perpendicular distance $r_\perp$ from axis A and $r'_\perp$ from an axis parallel to A through the CM. We use the cosine rule to relate these distances and hence to relate the two values of rotational inertia.

EXAMPLE 12.15 ♦ A circular hoop of radius 1.0 m and mass 1.0 kg is hanging from a peg pushed through its rim. If it rotates at 2.0 revolutions per second, find its energy and the magnitude of its angular momentum.

MODEL The hoop rotates about the fixed point on its circumference attached to the peg. We use the parallel axis theorem to find its rotational inertia about an axis through this point.

SETUP The rotational inertia about an axis through the center of the hoop is MR^2 (Exercise 12.7) and the distance between the two axes is R, so:

$$I_{\text{rim}} = I_{\text{CM}} + MR^2 = 2MR^2.$$

SOLVE The kinetic energy and angular momentum are:

$$E = \tfrac{1}{2}I_{\text{rim}}\omega^2 = \tfrac{1}{2}[2(1.0 \text{ kg})(1.0 \text{ m})^2][(2.0 \text{ rev/s})(2\pi \text{ rad/rev})]^2 = 160 \text{ J}.$$

$$|\vec{L}| = I_{\text{rim}}\omega = 2(1.0 \text{ kg}\cdot\text{m}^2)(2.0 \text{ rad/s})2\pi = 25 \text{ J}\cdot\text{s}.$$

ANALYZE In this case, half of the kinetic energy is internal energy, and half is due to motion of the hoop's CM. ∎

EXERCISE 12.8 ♦ Find the rotational inertia of a uniform rod with length ℓ about an axis through one of its ends. (Use the result of Example 12.13.) ∎

EXAMPLE 12.16 ♦♦♦ Find the rotational inertia of a thin, square plate of mass M and side ℓ about an axis perpendicular to the plate and through its center.

MODEL We model the object as a collection of thin rods of length ℓ and width dx (■ Figure 12.26), and use the parallel axis theorem to obtain the rotational inertia of an arbitrary rod about an axis not through its CM.

STEP I We choose a coordinate system with the origin at the center of the square and the z-axis perpendicular to the plate.

STEP II An arbitrary rod has coordinate x and width $dx > 0$.

SETUP The mass of a rod is $dm = M\, dA/A = M\ell\, dx/\ell^2 = (M/\ell)\, dx$. Its rotational
STEP III inertia about its CM (Example 12.13) is $dI_{\text{CM}} = dm\,\ell^2/12$, and so its rotational inertia about an axis through the origin is:

$$dI = dI_{\text{CM}} + dm\, x^2$$
$$= dm(\ell^2/12 + x^2) = (M/\ell)(\ell^2/12 + x^2)\, dx.$$

STEP IV The limits of integration are $x = -\ell/2$ to $x = \ell/2$, since we took $dx > 0$ in step II.

SOLVE Integrate.
STEP V

$$I = \int dI = \int_{-\ell/2}^{+\ell/2} \frac{M}{\ell}\left(\frac{\ell^2}{12} + x^2\right) dx = \frac{M}{\ell}\left(\frac{\ell^2}{12}x + \frac{x^3}{3}\right)\Bigg|_{-\ell/2}^{+\ell/2}$$

$$= 2M\ell^2(1/24 + 1/24) = M\ell^2/6.$$

ANALYZE Compare this result with the rotational inertia of a disk with radius $R = \ell/2$ about a similar axis: $\tfrac{1}{2}M(\ell/2)^2$ (Example 12.14) versus $\tfrac{2}{3}M(\ell/2)^2$ for the square. The square has additional material far from the rotation axis, hence the $\tfrac{2}{3}$ replaces the $\tfrac{1}{2}$. ∎

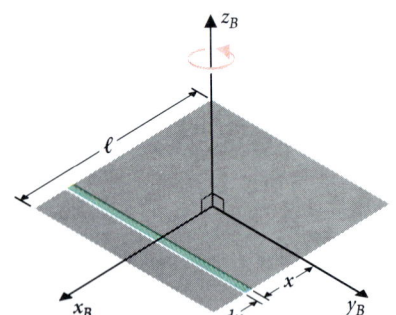

■ **FIGURE 12.26**
To find the rotational inertia of a square plate about an axis through its center, we model it as a collection of differential rods. An arbitrary rod rotates about an axis at distance x from its CM when the plate rotates about the z-axis.

We may now estimate the rotational inertia of a biplane about its longitudinal axis and compare it with that of a monoplane. The rotational inertia determines how much acceleration we get for a given torque, and thus indicates how maneuverable each aircraft is.

We model an aircraft as a uniform cylinder of mass M_c and radius R with plane rectangular wings measuring $\ell \times w$ (Figure 12.27). Both planes must have the same total wing area (and hence total wing mass M_w) to provide the same lift, so the biplane wings are half as long as the monoplane wings: $\ell_m = 2\ell_b$. We'll need the rotational inertia of a rectangle about an axis along its center line: $I = M\ell^2/12$ (Table 12.3). For the monoplane, with one wing of mass M_w attached at the top of the fuselage:

$$I_m = \tfrac{1}{2}M_c R^2 + M_w(\ell_m^2/12 + R^2).$$

For the biplane, with two wings each of mass $M_w/2$:

$$I_b = \tfrac{1}{2}M_c R^2 + 2\tfrac{1}{2}M_w(\ell_b^2/12 + R^2)$$
$$= \tfrac{1}{2}M_c R^2 + M_w(\ell_m^2/48 + R^2).$$

For a typical light aircraft, $M_c \sim M_w$ and $\ell_m \sim 14R$, so $I_m \sim 18M_w R^2$ and $I_b \sim 6M_w R^2$. Since ℓ_m is much larger than R, I_b is much smaller than I_m. With a smaller I, the biplane gets more angular acceleration for a given torque, resulting in an exciting aerobatic display.

NOTE THE USE OF THE PARALLEL AXIS THEOREM IN FINDING I FOR THE WINGS ABOUT THE FUSELAGE AXIS.

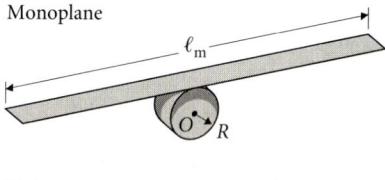
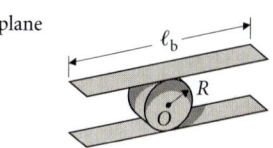

FIGURE 12.27
Models for a monoplane and a biplane. Each has rectangular wings and a cylindrical fuselage. The rotation axis is the axis of the cylinder. The total wing area is the same for each.

✻ 12.6 PRECESSION

One of the most intriguing examples of rigid-body mechanics is the motion of a top (Figure 12.28). The top is set spinning and then placed upon a small support. If the top were not spinning, it would be in unstable equilibrium, and even a tiny displacement would cause it to fall down. But the spinning top does not fall. Instead, its axis moves around on the surface of a cone with an apex at the support. This motion is called *precession*. A well-designed gyroscope mounted on a rod can even precess horizontally without falling (Figure 12.29).

The angular version of Newton's second law, $\vec{\tau} = d\vec{L}/dt$ (eqn. 12.10), describes precession. The forces acting on the gyroscope are its weight and a contact force at the support. The weight acts at the gyroscope's CM and contributes torque about the support, $\vec{\tau} = Wr_{CM}\hat{j}$. The angular momentum of the gyroscope is in the x-direction, $\vec{L} = I\vec{\omega} = I\omega\hat{i}$. The torque produces a change in the angular momentum vector, $d\vec{L} = \vec{\tau}\,dt$, at right angles to both $\vec{L}$ and $\vec{g}$. After a time dt, the new angular momentum vector, $\vec{L}(t + dt) = \vec{L}(t) + d\vec{L}$, has rotated through an angle $d\theta \sim dL/L$ in the x–y plane (Figure 12.29c). The gyroscope has precessed.

This relation between torque and angular momentum is analogous to the relation between force and velocity in uniform circular motion. In circular motion, the force is always perpendicular to the velocity, causing the velocity to change in direction while its magnitude remains constant. Similarly, torque perpendicular to the gyroscope's angular momentum causes $\vec{L}$ to rotate with constant magnitude. We may obtain the precession frequency Ω from the analogy with circular particle motion. In circular motion:

$$|\vec{F}| = \left|\frac{d\vec{p}}{dt}\right| = m|\vec{a}| = m|\vec{\omega}||\vec{v}| = |\vec{\omega}||\vec{p}| \;\Rightarrow\; \omega = \frac{|\vec{F}|}{|\vec{p}|}.$$

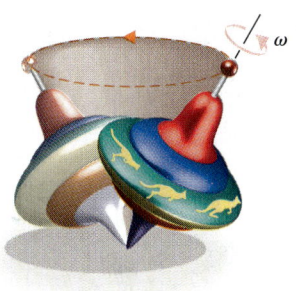

FIGURE 12.28
A spinning top demonstrates precession. It does not fall down, but its axis moves on the surface of a cone.

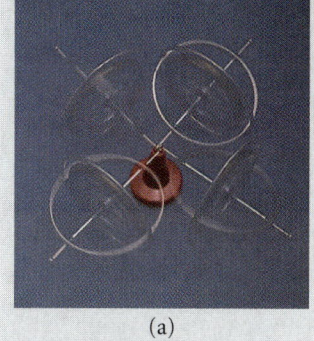
(a)

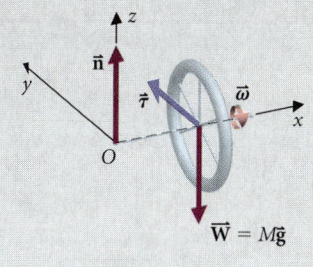
(b)

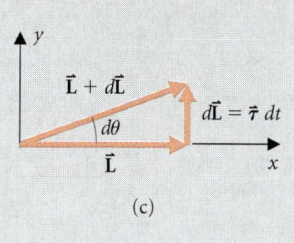
(c)

FIGURE 12.29
(a) A gyroscope can even precess with its rotation axis horizontal. (b) Its weight exerts a torque about the support that changes its angular momentum: $\vec{\tau} = d\vec{L}/dt$. (c) The angular momentum vector lies in the x–y plane. The change $d\vec{L} = \vec{\tau}\,dt$ is perpendicular to $\vec{L}$, so $\vec{L}$ rotates about the z-axis.

For the top:

$$|\vec{\tau}| = \left|\frac{d\vec{L}}{dt}\right| = \Omega|\vec{L}| \Rightarrow \Omega = \frac{|\vec{\tau}|}{|\vec{L}|}. \quad (12.12)$$

EXAMPLE 12.17 ♦ A gyroscope consists of a wheel of mass $M = 0.10$ kg and radius $R = 2.0$ cm mounted on a rod of negligible mass. The wheel's CM is $\ell = 5.0$ cm from the pivot point and its angular frequency $\omega = 1.0 \times 10^2$ rad/s. What is the angular frequency Ω of precession?

MODEL We model the gyroscope as a hoop with $I = MR^2$ (Exercise 12.7).

SETUP The magnitude of the torque acting about the pivot point is $\tau = \ell Mg$. The angular momentum of the wheel has magnitude $I\omega = MR^2\omega$.

SOLVE Using eqn. (12.12):

$\Omega = \tau/L = Mg\ell/(MR^2\omega) = g\ell/(R^2\omega)$
$= (9.8 \text{ m/s}^2)(5.0 \times 10^{-2} \text{ m})/[(2.0 \times 10^{-2} \text{ m})^2(1.0 \times 10^2 \text{ rad/s})] = 12$ rad/s.

ANALYZE The precession frequency is inversely proportional to ω. As friction slows the gyroscope's spin, it precesses faster.

Gyroscopes have many applications. ■ Figure 12.30 shows an aircraft heading indicator. The instrument contains a gyroscope with a mounting designed to exert no torque, no matter how the airplane turns. As the airplane turns, the gyro's spin axis maintains a fixed direction. A scale indicates the amount the airplane has turned with respect to this fixed direction. Gyros are also used in attitude indicators and rate-of-turn indicators. The rate-of-turn indicator uses precession of its gyro to determine turn rate.

The Earth is a spinning top subject to the gravitational forces of the Sun and the Moon. Since the Earth is not quite spherical, the gravitational forces acting on the Earth's bulges produce a torque that causes the Earth to precess with a period of roughly 26 000 y. This precession is observed as a change in location of the polestar with time. (■ Figure 12.31).

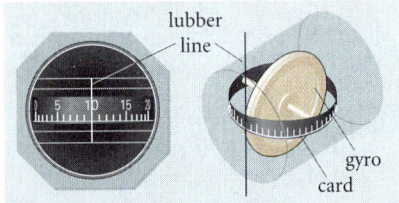

■ **FIGURE 12.30**
Aircraft heading indicator. The gyro maintains a fixed orientation and the aircraft turns around it.

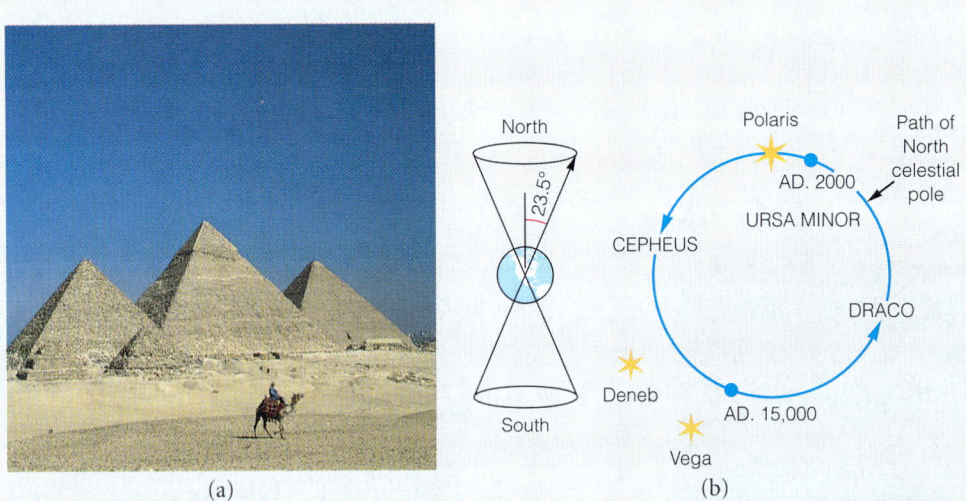

■ **FIGURE 12.31**
Precession of Earth's rotation axis. (a) The Great Pyramid at Gizeh has a shaft that allowed light from the polestar, Thuban, to shine on the tomb of the buried Pharoah. Now, 4000 y later, the pole is near Polaris but not near enough for starlight to shine down the shaft. (b) Over a 26 000-y period, the direction of the pole moves around a circle of radius $23\frac{1}{2}°$ in the northern sky.

Chapter Summary

Where Are We Now?

In this chapter we discussed the motion of objects with finite size. The size and shape of an object influence its motion through its rotational inertia. The examples in this chapter involve rigid, symmetrical objects with fairly simple motions. Further advance relies very little on new physical ideas but very heavily on mathematical technique. Real objects are not particles, nor are they rigid bodies. However, these two models provide a great deal of insight into the behavior of real things. While we are still limited in the kinds of motion and the kinds of object we can analyze, our ability to use the rigid-body model greatly increases the number of practical problems we can solve.

What Did We Do?

The position of a rigid body is completely determined by its orientation and the position of its CM. Orientation is most easily described with a coordinate system fixed to the body and rotating with it. The z-axis of such body coordinates is conventionally taken along the axis of rotation.

The kinematical relations for rotation of a rigid body about a fixed axis are very similar to those of one-dimensional particle kinematics (Tables 12.1 and 12.2). The equations of rotational dynamics are also similar to those of particle dynamics: rotational inertia plays the role of particle mass and torque replaces force. The rotational inertia of a body about an axis A,

$$I = \int r_\perp^2 \, dm,$$

includes the effect of the distribution of mass on the body's response to applied torques. The parallel axis theorem allows us to compute the rotational inertia about an axis not through the CM:

$$I_A = I_{CM} + Mh^2,$$

where h is the distance between axis A and a parallel axis through the CM. Values of rotational inertia for various bodies are listed in Table 12.3.

The primary equation of rotational dynamics is:

$$\vec{\tau} = I\vec{\alpha}.$$

Systems may be analyzed by including constraints that indicate how the different objects in the system influence each other's motions. Rolling without slipping is a common constraint that results in a relationship between a rolling object's CM speed and its angular speed: $v = \omega R$. A general plan for solving problems is given in Figure 12.12.

As in particle dynamics, conservation laws are useful in the solution of dynamical rigid-body problems, when initial and final states are compared. When no net external force or torque acts, the linear momentum, angular momentum, and energy of a system are all constant. The angular momentum and kinetic energy of a symmetrical, rotating rigid body with respect to its CM may be expressed in terms of the appropriate rotational inertia and the angular velocity:

$$\vec{L} = I\vec{\omega} \quad \text{and} \quad K = \tfrac{1}{2}I\omega^2.$$

✵ When the net torque acting on a rotating body is not parallel to its angular momentum vector, $\vec{L}$ rotates in a motion called precession. This is the principle behind the gyroscope.

Practical Applications

Dynamics of rigid bodies is important in the design of various kinds of machinery, from pulley systems to automobiles and aircraft instruments, from food mixers to Skil saws. The methods discussed in this chapter also help us to understand astronomical phenomena ranging from motions of the Earth to radiation from exotic rotating objects.

In the design of machinery, the rigid-body model is often good enough to serve as the starting point for calculating deviations from rigid behavior. An object rotating too rapidly will fly apart, as the internal forces necessary to accelerate the edge of the body become larger than the material can exert. We may use a rigid-body model to compute the necessary internal forces, but further knowledge of the properties of materials is required to determine whether a body can provide such forces.

Solutions to Exercises

12.1 The appropriate relation to use is eqn. (12.1): $\omega - \omega_i = \alpha t$. With $\omega_i = 0$, $\alpha = 0.010$ rad/s^2 and $t = 1.0$ h $= (60$ min$)(60$ s/min$) = 3600$ s, $\omega = (0.010$ rad/s$^2)(3600$ s$) = 36$ rad/s.

The number of revolutions follows from the relation for $\Delta\theta$, eqn. (12.2). $\Delta\theta = \omega_i t + \frac{1}{2}\alpha t^2 = 0 + \frac{1}{2}(0.010 \text{ rad/s}^2)(3600 \text{ s})^2 = 6.5 \times 10^4$ rad.

The number of revolutions is $\Delta\theta/(2\pi) = 1.0 \times 10^4$.

12.2 Using eqn. (2.13), we have: $v_f^2 = 2as = 2(3.2 \text{ m/s}^2)(350 \text{ m}) = 2240$ (m/s)^2. Then $v_f = 47$ m/s. We compare this result with $\omega r = (153 \text{ rad/s}) \times (0.31 \text{ m}) = 47$ m/s. The results agree.

12.3 (a) We continue to approximate the barbell as two particles. One particle is on the rotation axis and doesn't contribute. The rotational inertia is due to the second particle:

$$I_B = m\ell^2 = (75 \text{ kg})(2.0 \text{ m})^2 = 3.0 \times 10^2 \text{ kg·m}^2.$$

(b) (■ Figure 12.32) The perpendicular distances of the two particles from the rotation axis C are both $r_\perp = (\ell/2)\sin 30° = \ell/4$. The rotational inertia is:

$$I_C = 2m(\ell/4)^2 = m\ell^2/8 = 38 \text{ kg·m}^2.$$

12.4 Since the force is now applied at the top of the wheel, it produces a torque of magnitude $\tau = FR$ about the CM. The equation for α_z becomes:

$$\sum \tau_z = (f + F)R = I\alpha_z.$$

Equations (i), (ii), (iii), and (v) remain unchanged. The equation for a_x becomes: $a_x = (f + F)R^2/I$.

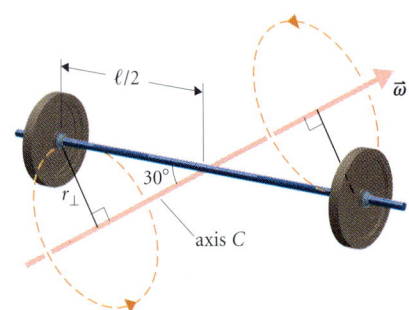

■ **FIGURE 12.32**
An axis through the CM of the barbell at a 30° angle with the rod. The two particles follow the circular paths shown, with radius $r_\perp = (\ell/2)\sin 30°$.

So, $f = F - Ma_x = F - (f + F)MR^2/I$,
or $f = F(1 - 2)/(1 + 2) = -F/3$.

The minus sign indicates that friction points in the opposite direction from that shown in Figure 12.14. The frictional force helps to accelerate the cylinder! We still require $|\vec{f}| \leq \mu Mg$, so that $F \leq 3\mu Mg$ remains the maximum force. The new maximum acceleration is then $a_{\max} = (F_{\max} - f_{\max})/M = 4F_{\max}/(3M) = 4\mu g$, twice the previous maximum.

12.5 The rotational inertia for the log, $MR^2/2$, is larger than for the spherical rock. The factor 2 replaces the 5/2 in the expression for ω_z, and 3 replaces the 7/2 in the expression for t_r. The log begins to roll after a longer time, because its angular acceleration is smaller.

12.6 We use the standard solution plan.

	BEFORE	AFTER
Kinetic energy	0	$\frac{1}{2}I\omega^2 + \frac{1}{2}(m_2 + m_1)v^2$
Gravitational potential energy	0	$-(m_2 - m_1)gh$

Set totals equal:

$$0 = \tfrac{1}{2}I\omega^2 + \tfrac{1}{2}(m_2 + m_1)v^2 - (m_2 - m_1)gh.$$

Solve for v, using $v = \omega R$:

$$v^2(m_2 + m_1 + I/R^2) = (m_2 - m_1)2gh.$$

$$v^2 = \frac{(m_2 - m_1)2gh}{m_2 + m_1 + I/R^2}.$$

This result is consistent with the acceleration we found in the example and eqn. (2.13).

12.7 We can consider the ring to be composed of a large number of particles, each of which is at the *same* perpendicular distance R from the axis, so $r_\perp^2 = R^2$ can be brought out of the integral:

$$I_{CM} = \int r_\perp^2 \, dm = R^2 \int dm = MR^2.$$

12.8 The axis through one end is a perpendicular distance $\ell/2$ from the axis through the CM. From the parallel axis theorem, we need only add $M(\ell/2)^2$ to I_{CM} to obtain the required rotational inertia:

$$I_{end} = (M\ell^2/4) + (M\ell^2/12) = M\ell^2/3.$$

This result is given in Table 12.3, and we used it in Example 12.10.

Basic Skills

Review Questions

§12.1 ROTATIONAL KINEMATICS

- What is a *body system of coordinates*? Explain why the position of the body origin and the orientation of the body axes determine the position of every particle in a rigid body.
- What is the *angular speed* of a rigid body; what is its *angular velocity*?
- What is *angular acceleration*?
- Explain why kinematics of rotation about a fixed axis is similar to that of one-dimensional particle motion. Compare the equations describing uniform angular acceleration with those describing uniform linear acceleration.
- Explain how the constraint of rolling without slipping relates the linear and angular accelerations of a rigid body.
- State the expressions that give velocity and acceleration of a particle in uniform circular motion as cross products of its angular velocity and position vectors.

§12.2 ROTATIONAL KINETIC ENERGY AND ANGULAR MOMENTUM

- Define the *rotational inertia* of a rigid body about a specified axis.
- How is rotational kinetic energy related to rotational inertia and angular speed?
- Explain why a rigid body's angular momentum may not be parallel to its angular velocity.
- Under what conditions is the body's angular momentum parallel to its angular velocity? In that case, what is the relation of $\vec{L}$ to $\vec{\omega}$?

§12.3 DYNAMIC BEHAVIOR OF RIGID BODIES

- Assuming a body's angular velocity and angular momentum are parallel at all times, what is the relation between the torque acting on the body and its angular acceleration?
- Discuss the similarities and differences between the solution plan (Figure 12.12) for problems in rotational dynamics and the plan for dynamics of systems of particles (Figure 5.4).
- What role does static friction between a pulley surface and the rope running over the pulley play in the pulley's dynamics?
- What is the analogy between pulley action and a wheel rolling without slipping?

§12.4 APPLICATION OF THE CONSERVATION LAWS

- What similarities are there between conservation law problems in rotational dynamics and in particle dynamics?
- Do you need to consider the effect of static friction between a pulley surface and the cord running over it when applying conservation of energy? Why or why not?

§12.5 CALCULATION OF ROTATIONAL INERTIA

- Review the five steps for evaluating a physical integral discussed in Interlude 2.
- Describe how to choose the differential elements in setting up an integral.
- Under what conditions is what we have called "the rotational kinetic energy of a rigid body" the same as the body's *internal* kinetic energy as defined in Chapter 9?
- Explain how the parallel axis theorem for rotational inertia values follows from the theorem that a system's kinetic energy is the sum of an internal term and a term due to the motion of the CM.

✶ §12.6 PRECESSION

- Explain how the precession of a gyroscope is analogous to the uniform circular motion of a particle.

Basic Skill Drill

§12.1 ROTATIONAL KINEMATICS

1. A space station is designed in the shape of a bicycle wheel with radial spokes. A cylindrical docking module is mounted perpendicular to the *wheel* at its center. If the station is rotating about the axis of the docking module, describe a useful set of body coordinates. Explain the advantages of your choice of axes.

2. A shaft of radius 0.010 m turns a pulley of radius 1.0 m at 3.0×10^3 rpm. Find the speed of a point (a) on the edge of the shaft, and (b) on the rim of the pulley.

3. A flywheel starting from rest is given an angular acceleration $\alpha = 1.0 \times 10^{-3}$ rad/s². How many rotations will the wheel make before its angular frequency reaches a value $\omega = 1.0 \times 10^2$ rad/s?

§12.2 ROTATIONAL KINETIC ENERGY AND ANGULAR MOMENTUM

4. Bruno the weight lifter has just won the world championship and celebrates by twirling a barbell at angular frequency $\omega = 12$ rad/s. The barbell has a 110-kg mass at each end of a 2.0-m rod. How much energy does Bruno use to set the barbell rotating?

5. An object with rotational inertia $I = 12$ kg·m² about axis A is rotating about A with angular frequency $\omega = 5.0$ rad/s. What is the magnitude of the object's angular momentum, assuming it is parallel to the angular velocity?

§12.3 DYNAMIC BEHAVIOR OF RIGID BODIES

6. A brake system applies a force of 27 N to a wheel with rotational inertia 5.2 kg·m² that is spinning at 5.0×10^2 rpm. If the force is tangential and is applied 15 cm from the axis of the wheel, how long does it take for the wheel to come to rest?

7. A cylindrical flywheel of mass $M = 220$ kg and radius $R = 2.0$ m has a cord wrapped around it in which a constant tension is maintained. The flywheel is supported on a frictionless axle. If the wheel starts from rest and turns through 250 rad in 120 s, find the tension T in the cord.

§12.4 APPLICATION OF THE CONSERVATION LAWS

8. A one-person space pod uses an internal gyroscope to control its orientation. The rotational inertia of the gyroscope alone is 2.5 kg·m², and it normally operates at angular speed $\omega = 1.20 \times 10^3$ rad/s. The rotational inertia of pod, occupant, and gyroscope together is 650 kg·m². All of a sudden, when the pod is not rotating, the gyroscope's bearing fails and it screeches to a halt. What is the final angular speed of the distressed space pod?

9. A diver leaps from the board with rotational inertia $I_i = 10.3$ kg·m² and angular speed $\omega_i = 3.14$ rad/s. If she then touches her toes to achieve a rotational inertia $I_f = 3.61$ kg·m², what is her angular speed in the tucked position? How much extra kinetic energy does she have in the tucked position?

§12.5 CALCULATION OF ROTATIONAL INERTIA

10. (a) A long cylindrical shell of mass M has inner radius R_1 and outer radius R_2 (■ Figure 12.33). Describe a suitable differential element for computing the rotational inertia of the object about its symmetry axis. (b) What is the difference in the shell's rotational inertia about its symmetry axis and about a parallel axis along its inner surface?

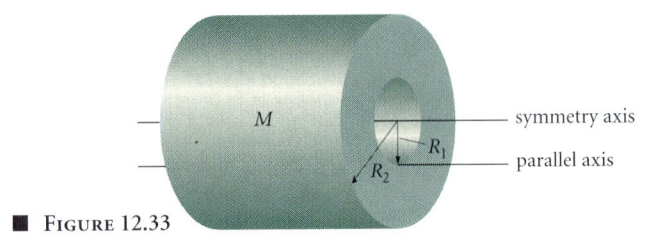

■ **Figure 12.33**

✱ §12.6 PRECESSION

11. A gyroscope used in an aircraft has a disk of radius 2.0 cm and mass 0.10 kg that spins at 3.0×10^3 rpm. If the CM of the disk is 2.0 mm from the support point, what is the precession rate of the gyro?

Questions and Problems

§12.1 ROTATIONAL KINEMATICS

12. ❖ What set of body axes might you choose to describe the rotational motion of an airplane about its center of mass?
13. ❖ Describe a useful set of body axes for studying the rotational motion of a clown on a unicycle about the point of contact of the tire with the ground.
14. ❖ Verify that the average angular velocity of an object undergoing uniform angular acceleration about a fixed axis is one-half the sum of its initial and final angular velocities. Use an argument from the discussion of uniform linear acceleration.
15. ❖ Adapt an argument from the discussion of uniform linear acceleration to verify the formula: $\omega_f^2 - \omega_i^2 = 2\alpha \, \Delta\theta$.
16. ❖ Draw a sketch showing how Earth's angular velocity ω_{Earth} is related to the horizon plane for a person at 45° north latitude. How would you answer a friend's question, Which way does Earth's angular velocity point?
17. ❖ One problem with the design of propeller-driven pursuit planes was that the speed of the propeller blade tips would exceed the speed of sound. Explain how this can occur when the plane's speed is less than that of sound.
18. ◆ A wheel of radius 0.5 m rotates at an angular frequency of 6 rad/s. Find the speed of a point on the rim.
19. ◆ In the gear system shown in ■ Figure 12.34, the 10-toothed gear rotates at angular speed $\omega = 0.4$ rad/s. What is the angular speed of the double gear and the linear speed of a tooth on its outer rim of radius 40 cm?
20. ◆◆ The wheel in a demonstration gyroscope has a radius of 5.0 cm. The string wrapped around the wheel has a total length of 0.90 m. To pull off the string, the demonstrator's hand accelerates uniformly during a time interval of 1.5 s. What are the angular acceleration and final angular speed of the gyroscope wheel?
21. ◆◆ The mean solar day is the average time between noon on two successive days. Draw a diagram of the solar system showing the plane of the Earth's orbit. Taking into account Earth's motion around the Sun, use your diagram to show that Earth's rotation period with respect to distant stars (the mean sidereal day) is not the same as the mean solar day. Which of the two *days* is the shorter and by how much? Explain your reasoning.
22. ◆◆ On a lark, a discus champion throws a circular metal plate with velocity 19 m/s at an angle of 45° to the horizontal. The plate is rotating in a vertical plane at a rate of $\omega = 22$ rad/s. How many rotations does the plate make before hitting the ground? Ignore air resistance.

§12.2 ROTATIONAL KINETIC ENERGY AND ANGULAR MOMENTUM

23. ❖ Each object in ■ Figure 12.35 has the same mass M. Which object has the greatest rotational inertia about an axis through its CM and perpendicular to the plane of the figure? Which has the least? Why?

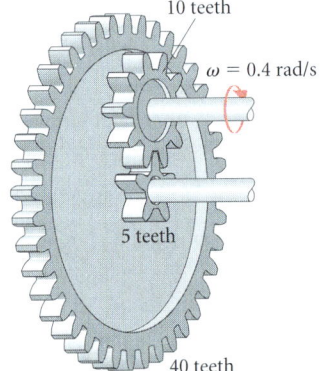

■ **Figure 12.34**

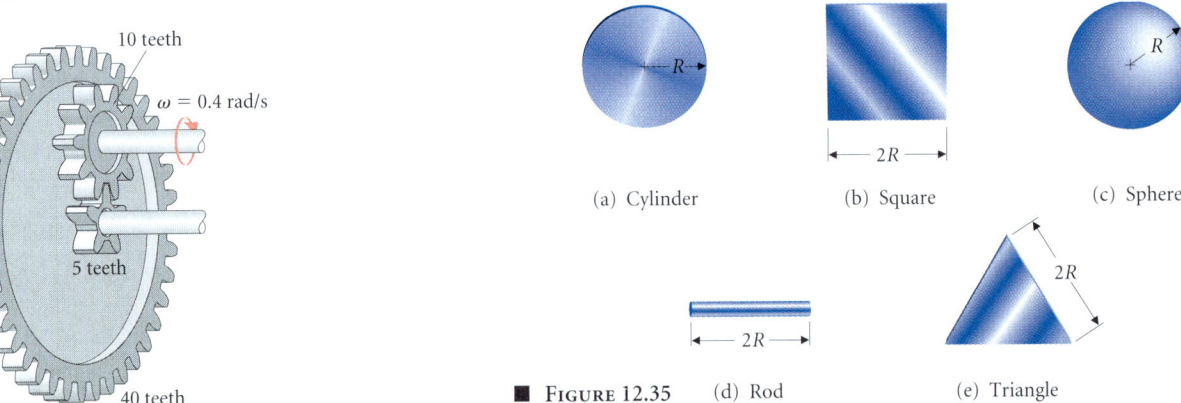

■ **Figure 12.35**

418 CHAPTER 12 • DYNAMICS OF RIGID BODIES

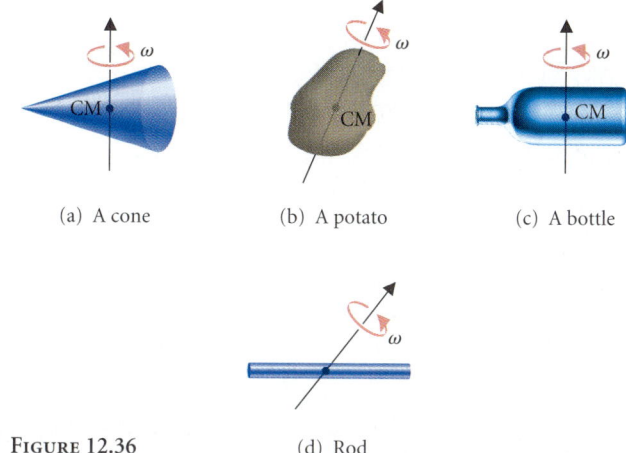

■ FIGURE 12.36

24. ❖ Each of the objects in ■ Figure 12.36 rotates as shown. Which of them has angular momentum parallel to its angular velocity? Give reasons for your answers.

25. ❖ An object consists of four identical particles at the corners of a tetrahedron. The object rotates about an axis through its CM and one of the particles. Show that its angular momentum is parallel to its angular velocity. What kind of rotational symmetry does the tetrahedron have about this axis? Can you describe a more general class of symmetry that implies $\vec{L} \parallel \vec{\omega}$?

26. ◆◆ Three small balls, each of mass m, are held together by rods of negligible mass in an equilateral triangle of side ℓ. What is the object's energy when it rotates with angular frequency ω about (a) an axis through its CM perpendicular to the plane of the triangle? (b) an axis through one ball and bisecting the opposite side of the triangle? (c) Find the object's angular momentum $\vec{L}$ for each of cases (a) and (b). Show that your results are consistent with the formula $\vec{L} = I\vec{\omega}$. (d) The object now rotates with angular speed ω about an axis through the midpoint of one side of the triangle and perpendicular to one of the other sides. With the origin where the axis intersects the midpoint of a side, compute the object's angular momentum. What is the angle between the object's angular velocity and its angular momentum?

27. ◆◆ A bicycle has a mass of 15 kg; the wheels each have a radius of 30.0 cm and mass 4.0 kg. If the bicycle is accelerating at 1.0 m/s², how much power must a 75-kg rider apply when the bike's speed is 5.0 m/s (ignore friction)?

28. ◆◆ A car of total mass 1.0×10^3 kg has four wheels, each modeled as a uniform disk of mass $m = 22$ kg, radius 31 cm, and thickness 15 cm. When the car is moving at 85 km/h, what is its total kinetic energy? What fraction of the total is rotational energy of the wheels?

§12.3 DYNAMIC BEHAVIOR OF RIGID BODIES

29. ❖ In the system shown in ■ Figure 12.37, a tension T is maintained in the string. Which way does the wheel roll?

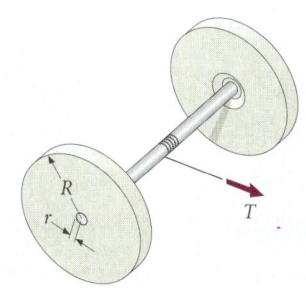

■ FIGURE 12.37

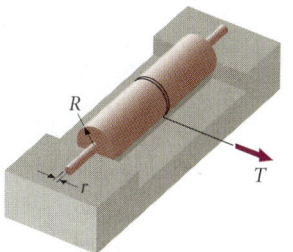

■ FIGURE 12.38

30. ❖ In the system shown in ■ Figure 12.38, a tension T is maintained in the string. Which way does the wheel roll?

31. ◆ A potter's wheel, initially spinning at 350 rpm, comes to rest in 45 s when the motor is turned off. Model the wheel as a uniform disk of mass 4.2 kg and radius 28 cm, with an axle of radius 2.5 cm. What frictional force is exerted on the axle? How much power does the motor draw when the wheel is spinning, if the frictional force does not change?

32. ◆◆ A solid cylinder of mass $M = 10.0$ kg and radius $R = 0.370$ m has a black stripe 0.15 cm wide painted on its surface parallel to its long axis. An electric motor exerts torque on the cylinder so that it undergoes uniform angular acceleration from rest. As the cylinder rotates, a photocell detects the number of times the black stripe passes by: 16 times in 0.874 s and 32 times in 1.236 s. What is the angular acceleration of the cylinder, and what torque is exerted by the motor? What are the cylinder's kinetic energy and angular momentum after 2.00 s? How many times has the stripe passed the photocell after 2.00 s?

33. ◆◆ The object in Figure 12.37 consists of two uniform disks, each of mass M, connected by an axle of negligible mass. What is the linear acceleration of the object when a tension T is maintained in the string? If the coefficient of static friction between the object and the surface is μ_s, what is the maximum value of T that allows rolling without slipping?

34. ◆◆ The cylinder in Figure 12.38 is uniform and has mass M; the axles have negligible mass. What is the linear acceleration of the cylinder when a tension T is maintained in the string? If the coefficient of static friction between the axles and the surface is μ_s, what is the maximum value of T that allows rolling without slipping?

35. ◆◆ A cylinder of mass M and radius R is held by two ideal strings attached to the ceiling (■ Figure 12.39). Each string has a length L wrapped around the cylinder. If the cylinder is released, how long does it take to reach the bottom of the strings? What are its CM and angular velocities at that time? What are its linear, rotational, and total kinetic energies?

36. ◆◆ A sphere of radius r is placed on a ramp at angle θ. What is the minimum coefficient of static friction that allows the sphere to roll without slipping?

37. ◆◆ A governor used to limit the angular speed of a steam turbine is shown in ■ Figure 12.40. Two balls of mass m are attached to the ver-

■ FIGURE 12.39

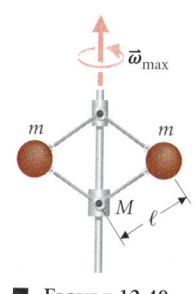

■ FIGURE 12.40

QUESTIONS AND PROBLEMS 419

Figure 12.41

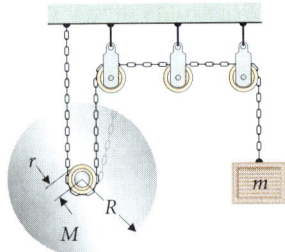

Figure 12.42

Figure 12.43

tical turbine shaft by massless pivot arms of length ℓ, as shown. The sleeve of mass M slides frictionlessly on the shaft. If the device shuts off power to the turbine when the pivot arms form a square, what is the turbine's maximum angular speed ω_{max}?

38. ◆◆ A particle of mass $m = 0.091$ kg is suspended by a string from a cylinder of radius $R = 0.35$ m, which rotates on frictionless bearings. If the mass is released from rest and falls 0.45 m in 1.36 s, what is the rotational inertia of the cylinder? If the cylinder is solid, what is its mass?

39. ◆◆◆ A mass of $M = 10.0$ kg is hung from the pulley system shown in ■ Figure 12.41. The pulleys each have mass $m = 0.550$ kg and radius $r = 0.100$ m. If the rope does not slip over the pulley, what force is required to accelerate the block upward at 1.00 m/s^2? Compare with the result for frictionless pulleys. (Treat the pulleys as uniform disks.)

40. ◆◆◆ A block of mass m is suspended from a frictionless chain that passes over frictionless rollers and around a gear of radius r on the shaft of a wheel, as shown in ■ Figure 12.42. The wheel is free to move vertically and has mass M, radius R, and rotational inertia I. Find the acceleration of the block, assuming $M < 2m$.

41. ◆◆◆ A uniform cylindrical flywheel of mass $M = 1.5 \times 10^3$ kg and radius $R = 0.50$ m rotates about a horizontal axis. It is supported on a shaft of radius $r = 0.10$ m whose bearings have a coefficient of kinetic friction $\mu_k = 0.20$ due to a lubrication system failure. Find the normal and frictional forces acting on the shaft. (*Hint:* The flywheel has no linear acceleration.) Find the frictional torque acting on the shaft and the time required for the wheel to come to rest from an initial angular speed $\omega = 120$ rad/s.

42. ◆◆◆ A wall-mounted can opener has a crank of length 8.5 cm that rotates a gear wheel with 10 teeth. This gear meshes with another one that has 17 teeth. The second gear is mounted on the same axle as the 1.0-cm-radius cutting wheel. What is the mechanical advantage of the machine?

43. ◆◆◆ A large steel barrel of inside radius R rotates with angular frequency Ω about its horizontal symmetry axis. A steel ball of radius r rolls without slipping on the inner surface of the barrel. If the barrel's angular speed is constant, where is the ball? Where is the ball if the barrel's angular acceleration is α? If the coefficient of static friction between ball and barrel is μ_s, what maximum value of α allows the ball to roll without slipping?

§12.4 APPLICATION OF THE CONSERVATION LAWS

44. ❖ A sphere and a cylinder of equal mass and radius are released from rest at the top of a ramp of angle θ and roll without slipping. Which one reaches the bottom first? Explain your reasoning. Would a toy truck rolling down the ramp on *small* wheels with frictionless axles win or lose against the sphere or against the cylinder?

45. ❖ You are standing on a small platform that is free to rotate about a vertical axis and are holding a spinning bicycle wheel with its axis horizontal (■ Figure 12.43). **(a)** Describe the motion of the system after you lift the bicycle wheel so that its axis points vertically. **(b)** Suppose instead that you are initially holding the bicycle wheel above your head so that it spins clockwise as you look up at it. Describe what happens as you move the wheel so its axle is horizontal. **(c)** Describe the forces you have to exert on the wheel's axis to carry out the changes in parts (a) and (b). **(d)** ◆◆ Estimate the magnitude of the effect you describe in part (b) if the wheel is rotating with a frequency $f = 3$ Hz.

46. ❖ A cylinder is attached to the ceiling by two ideal strings that are wrapped around the cylinder (Problem 35, Figure 12.39). If the bottom ends of the strings are firmly attached to the cylinder and the interaction of the strings and the cylinder is elastic, what will the cylinder do after reaching the end of the strings?

47. ❖ Describe an experiment you could use to determine the rotational inertia of an arbitrary body about a given axis.

48. ❖ A hoop and a solid sphere, each of radius R and mass M, roll at the same speed on level ground directly toward a ramp. Which has the greater kinetic energy? Which will roll further up the ramp before coming to rest?

49. ◆ A ball (uniform sphere) of mass $m = 1.0$ kg and radius $r = 0.10$ m is placed on a ramp of angle $\theta = 32°$ and rolls without slipping. What is its speed when it has rolled 1.0 m along the ramp?

50. ◆ An astronaut attempting to grab a satellite gives it an impulse $\Delta \vec{p} = (1.0 \times 10^2$ kg·m/s, along the line shown in ■ Figure 12.44 that passes 0.25 m from the center). If the satellite is a uniform cylinder of mass 120 kg, height 1.0 m, and radius 0.5 m, describe the final motion of the satellite.

51. ◆◆ An experimental automobile uses a flywheel as a power source. The mass of the car is 1.2×10^3 kg and the flywheel is a uniform disk with radius 0.50 m and mass 1.0×10^2 kg. What initial angular speed must the flywheel have if the car is to climb a hill 150 m high at constant speed? Neglect any friction losses.

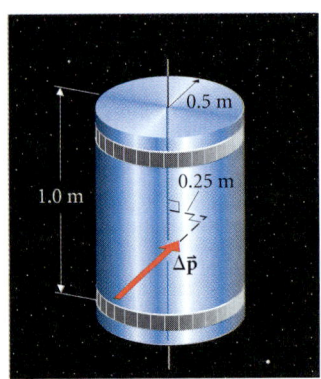

Figure 12.44

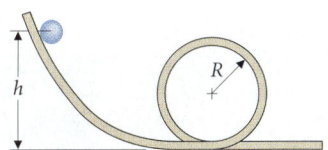

■ FIGURE 12.45

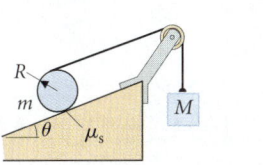

■ FIGURE 12.46

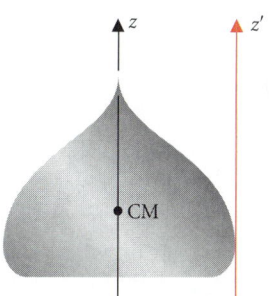

■ FIGURE 12.48

52. ◆◆ A playground merry-go-round is a disk of radius 4.0 m and mass 5.0×10^2 kg, initially stationary. If a child of mass 48 kg jumps tangentially and lands on the edge of the merry-go-round with a speed of 5.0 m/s, what is the resulting angular velocity?

53. ◆◆ A uniform spherical ball of radius r rolls without slipping on the looped track shown in ■ Figure 12.45. At what minimum height must it be released if it is to remain on the track at the top of the loop? What release height will suffice if the ball slides without friction on the track?

54. ◆◆ ✉ A student stands on a *small* platform that is free to rotate about a vertical axis. She is initially spinning with angular speed $\omega_i = 0.05$ rad/s while holding a 3.0-kg mass at arms length in each hand. If she draws the masses in close to her chest, estimate her angular speed.

55. ◆◆ A space station of mass M in the shape of a ring of radius R is equipped with spring launchers with spring constant k. Two probes, each of mass m, are launched simultaneously from springs at the opposite ends of a diameter, and in opposite directions tangent to the ring. Both springs have the same initial compression s. What are the final speeds of the probes, and what are the final speed and angular speed of the space station after launch?

56. ◆◆◆ A uniform cylinder of mass m is free to roll on a ramp with coefficient of static friction μ_s (■ Figure 12.46). A string wrapped around the cylinder passes over a massless, frictionless pulley to a block of mass M. What is the speed of the block when it has fallen a distance h, assuming the cylinder rolls without slipping?

57. ◆◆◆ You are on a long journey in a spherical space capsule with one window (■ Figure 12.47). You and the capsule are not rotating, and you are bored with the view. Explain how you can change the direction of your view out the window by rotating on your feet. If your rotational inertia is I_y and the rotational inertia of the capsule is I_c, by how much will your direction of view have rotated when you are next at rest looking through the window? Can you look (given enough effort) in any direction you desire? Explain how the answer to this question depends on the ratio I_y/I_c.

§12.5 CALCULATION OF ROTATIONAL INERTIA

58. ❖ Points 1 and 2 lie in a uniformly thick, plane metal sheet. Their position vectors from the sheet's center of mass are $\vec{r}_1$ and $\vec{r}_2$. If I_1 and I_2 are the rotational inertia values about axes perpendicular to the sheet through points 1 and 2, under what conditions does the following relation hold?

$$I_1 = I_2 + M|\vec{r}_1 - \vec{r}_2|^2.$$

59. ❖ A plane sheet of metal is to be set rotating about an axis perpendicular to the sheet. About which such perpendicular axis is the least energy required to rotate the sheet at an angular speed ω_o?

60. ❖ A pie-shaped wedge of mass M is cut from a uniform, circular sheet of material of radius R. Without performing an integration, determine the rotational inertia of the wedge about an axis through its point and perpendicular to its surface. (*Hint:* If you know the rotational inertia of several pieces of an object, how do you obtain that of the whole?)

61. ❖ If the object shown in ■ Figure 12.48 rotates about the axis z', will its angular momentum be parallel to its angular velocity? Why or why not? Is the answer the same or different for rotation about the z-axis through the CM? Do your answers support or violate the parallel axis theorem?

62. ◆ Use the parallel axis theorem to find the rotational inertia of a disk about an axis through its edge and perpendicular to the plane of the disk.

63. ◆ A particle of mass M is added to the end of a uniform rod of mass M and length ℓ. What is the rotational inertia of the system about axes perpendicular to the rod through its center and through the CM of the combined system?

64. ◆◆ A weight lifter's barbell consists of two uniform spheres of mass M and radius R connected by a massless, rigid rod, also of length R. What is the rotational inertia about axes z and z' in ■ Figure 12.49?

65. ◆◆ The rotational inertia of a uniform square sheet of material of mass M and side ℓ, about an axis through its CM and perpendicular to the sheet, is:

$$I = \alpha M \ell^2,$$

where α is a dimensionless number. Express I as the sum of contributions from the four quarters of the square, and so obtain a relation you may solve for α. Verify that the solution agrees with the result in Table 12.3.

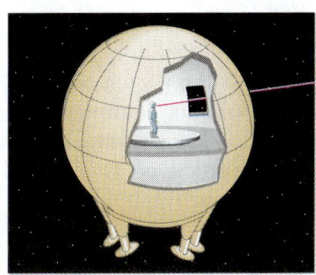

■ FIGURE 12.47

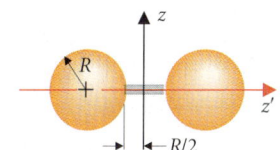

■ FIGURE 12.49

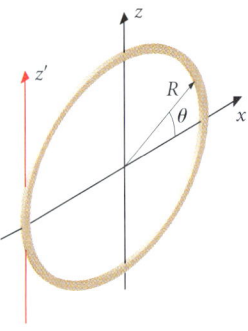

Figure 12.50

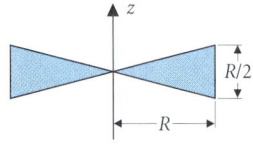

Figure 12.51

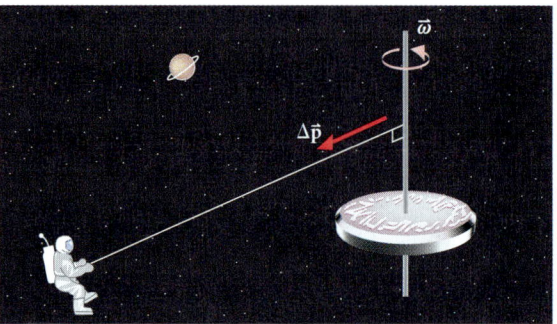

Figure 12.53

66. ♦♦♦ Express the rotational inertias about the z- and z'-axes of the thin hoop in ■ Figure 12.50 as integrals over angle θ. Evaluate the integrals.

67. ♦♦♦ Verify the formula in Table 12.3 for the rotational inertia of a plane, uniform square sheet about an axis through its CM, in its plane.

68. ♦♦♦ Find the rotational inertia of a plane, equilateral triangular object about a line bisecting one of the angles.

69. ♦♦♦ A flywheel of mass M and radius R has the cross section shown in ■ Figure 12.51. What is the rotational inertia of the flywheel about its z-axis?

70. ♦♦♦ Find the rotational inertia about its axis of a cylindrical shell of mass M with inner and outer radii r_1 and r_2.

71. ♦♦♦ A uniform sphere of mass M and radius R may be viewed as a stack of disks with z as the coordinate (■ Figure 12.52). Express the rotational inertia of the sphere about the z-axis as an integral over z. Evaluate the integral.

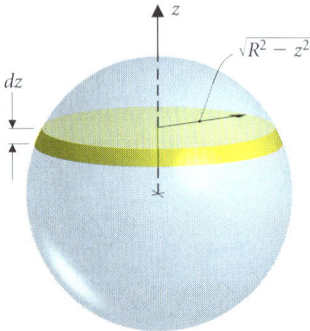

Figure 12.52

72. ♦♦♦ A uniform, thin sheet lies in the x–y plane of a body reference frame. Show that $I_z = I_x + I_y$. What relation holds among I_x, I_y, and I_z for the same sheet lying parallel to, but a distance h from, the x–y plane?

73. ♦♦♦ Three uniform rods, each of mass M, and length ℓ, are welded together to form an equilateral triangle. Find the rotational inertia of the object about an axis along one side and about a parallel axis through the object's CM.

* §12.6 PRECESSION

74. ❖ Satellites carrying astronomical instruments need to be able to point in a specified direction for long periods of time. Contrast the effect of a small torque on a rotating versus a nonrotating object, and explain why astronomical satellites are often set spinning with the instruments looking along the rotation axis.

75. ❖ An astronaut approaches an alien artifact in the shape of a disk spinning about a long axle through its center (■ Figure 12.53). Having slipped a small loop about the tip of the axle, the astronaut gives a sharp tug on the rope. Describe the resulting change in the artifact's angular momentum. In what direction does the tip of the axle move in response?

76. ♦♦ A top spinning at 500 rpm precesses with frequency $\Omega = 5$ rad/s. The mass of the top is 0.4 kg and its rotational inertia is 3×10^{-3} kg·m². Where is its center of mass?

77. ♦♦♦ ✉ Modeling the Earth as a sphere with a precession period of 26 000 y, estimate the torque exerted on it to cause this precession. Make a back-of-the envelope model of Earth's nonspherical shape by placing two particles, each of mass m, on the equator on opposite sides of the Earth. If the plane of the Moon's orbit, on average, makes an angle of 23° with the equator, what torque does the Moon exert on the Earth, as a function of m? Equate your two expressions for the torque to estimate m. What fraction of Earth's mass is it?

Additional Problems

78. ❖ Explain why the following statement is false: In a rigid body rotating about a fixed axis, each part of the body has the same linear acceleration.

79. ❖ In designing a flywheel-powered car, should the flywheel axis be mounted vertically or horizontally? Why?

80. ❖ A physics professor with a taste for practical jokes mounts a gyroscope in her suitcase. Describe what happens when the bell captain tries to carry the suitcase around a corner.

81. ❖ Describe how a child on a swing can "pump" the swing.

82. ❖ The directional controls of an aircraft are the ailerons, which allow the pilot to increase lift slightly on either of the wings while reducing lift on the opposite wing; the elevators, which allow the pilot to control lift on the tail surfaces of the aircraft; and the rudder, which allows control of sideways force on the tail surfaces. Explain how each control allows the pilot to rotate the aircraft about a specific body axis.

83. ♦ The upper stone grinding wheel in an old-fashioned flour mill (■ Figure 12.54) has a radius of 0.50 m. It is rotating on a shaft along its symmetry axis so that a point on its rim has a speed of $v = 2.6$ m/s. **(a)** Choose a set of body axes for the wheel and give the components of its angular velocity in your body coordinate system. **(b)** Friction acting on the shaft brings the wheel to rest with respect to the mill after 45 s. What are the components of the wheel's angular acceleration, assumed to be uniform? **(c)** If the mass of the upper wheel is 250 kg, use the rotational inertia formula from Table 12.3 to find the

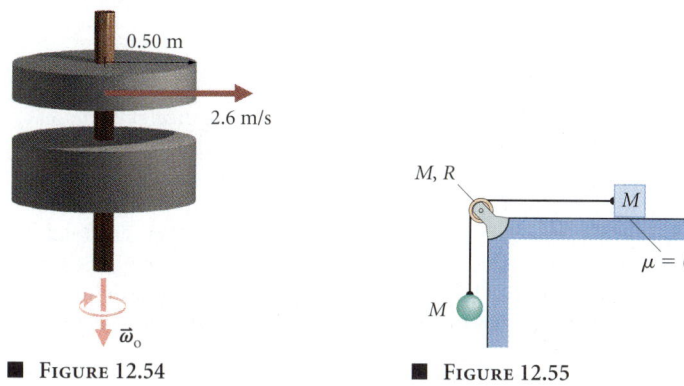

FIGURE 12.54

FIGURE 12.55

wheel's kinetic energy and angular momentum at $t = 0$. **(d)** Find the net torque that acted on the grinding wheel during the 45 s required to bring it to rest. Find the number of rotations of the wheel during this period, and use your result to find the net work done on the wheel by whatever exerts the torque. **(e)** The lower grinding wheel, mounted on the same shaft, is kept rotating at constant angular velocity $\vec{\omega}_o = -(1.3 \text{ rad/s})\hat{k}$, until at $t = 45$ s a mechanism drops the upper wheel onto the lower and friction between their surfaces quickly causes them to rotate together. Afterward, the combined system rotates on frictionless bearings. What is the system's final angular velocity? Assume the lower wheel has twice the rotational inertia of the upper wheel. Compute the kinetic energy of the system before and after the interaction. How much thermal energy is produced in the interaction?

84. ◆◆ A barbell (Figure 12.8) consists of a rod of length ℓ and mass m_r and two disks of mass m and radius r at the ends of the rod. Calculate the rotational inertia of the barbell about an axis perpendicular to the rod and through its center. If $r = \ell/8$ and $m_r = m/10$, how large is the correction to the estimate $I \approx m\ell^2/2$ that we found in Example 12.4?

85. ◆◆ A clock pendulum consists of a rod of mass 0.100 kg and length 0.509 m to which is attached a disk of radius 0.100 m and mass 0.500 kg. If the pendulum is pulled out to an angle of 30° with the vertical, what is its angular speed at the bottom of its swing?

86. ◆◆ A knife sharpener's grinding wheel has a radius of 0.10 m and rotates with an angular speed of 12 rad/s. If the craftsman presses a knife against the wheel with a normal force of 55 N and the motor driving the wheel draws 45 W of power, what is the coefficient of friction between grinding wheel and knife?

87. ◆◆ Two objects of the same mass M are connected by an ideal rope (■ Figure 12.55). The rope runs without slipping over a uniform cylindrical pulley of radius R and also of mass M, mounted on frictionless bearings. What are the accelerations of each object and the angular acceleration of the pulley?

88. ◆◆ A circular saw cuts through a board at a rate of 3.5 cm/s. The saw removes 0.60 cm of wood per rotation, and the normal force exerted on the board by the saw's teeth is 1.5×10^2 N. If the radius of the blade is 10.0 cm, how much power is needed to operate the saw?

89. ◆◆ Where must a billiard player's cue contact a ball if the ball is to roll without slipping immediately after the impact?

90. ◆◆ A uniform rod of mass M and length ℓ is initially horizontal, at rest, and supported by a frictionless pivot a distance $\ell/3$ from one end. If the rod is allowed to rotate about the pivot, what is its initial angular acceleration?

91. ◆◆ A gaucho's bola consists of three balls each of mass m and each attached to a string of length ℓ. The ends of the strings are tied together at the center of the device. When used to catch game, the bola rotates at angular frequency ω about an axis perpendicular to the plane of the three balls with equal angles between the strings. Show that the bola has (internal) angular momentum parallel to its angular velocity. Find its angular momentum and rotational energy.

92. ◆◆ A hoop and a solid cylinder are released from rest at the same time at the top of a hill that slopes at an angle of 30° with the horizontal. They roll down the hill on parallel paths without slipping. After 2.50 s, which is ahead and by how much?

93. ◆◆ Two cylinders are mounted on frictionless bearings, with their long axes parallel. A drive belt connecting the two cylinders is under tension so that static friction keeps it from slipping with respect to the surface of either cylinder (■ Figure 12.56). **(a)** If the linear speed of the belt is 5.30 m/s, what are the angular speeds of the two cylinders? **(b)** A ball of radius $r = 6.50$ cm and mass $m = 1.46$ kg is on top of the horizontal section of the belt. What is the angular speed of the ball if it remains in the same position? **(c)** ◆◆◆ If a motor now begins to increase the angular speed of the smaller cylinder at acceleration $\alpha = 0.048$ rad/s², what are the angular acceleration of the larger cylinder and the angular and linear accelerations of the ball, assuming the ball rolls without slipping?

94. ◆◆◆ ✉ A skater spinning around with his arms outstretched pulls them in. Model the skater as a uniform cylinder of height $h = 2$ m, radius $R = 20$ cm, and mass $M = 70$ kg. Model his arms as uniform rods of length $\ell = 1$ m, radius $r = 4$ cm, and mass $m = 3$ kg, which originally are extended horizontally from the edge of the cylinder and afterward are vertical on either side of the cylinder. If the skater's initial angular frequency is 1 rad/s, what is the final angular frequency? How much energy must he use to perform the maneuver?

95. ◆◆◆ A rod consists of two pieces, each of length 0.50 m, brazed together. One piece is of steel with mass 10.0 kg; the other is aluminum with mass 5.0 kg. Where is the CM of the rod, and what is its rotational inertia about axes through the CM and through each end, perpendicular to the rod?

96. ◆◆◆ In a factory, a heavy crate moves down a ramp at angle θ with the horizontal on rollers that eliminate friction. The crate is attached to a rope wrapped around a large drum with rotational inertia I, supported on frictionless bearings (■ Figure 12.57). After the crate

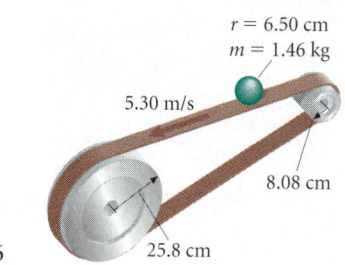

FIGURE 12.56

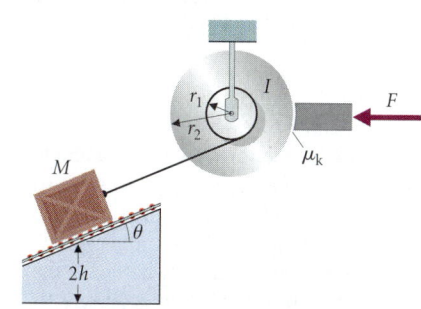

FIGURE 12.57

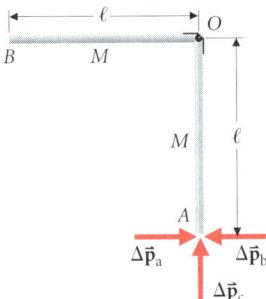

■ FIGURE 12.58

has descended a vertical distance h, a brake shoe with coefficient of friction μ_k exerts a normal force F against the drum, as shown. What is the angular acceleration of the drum after the brake is applied? If the crate is supposed to stop a vertical distance 2h below its initial position, what force F is required?

97. ♦♦♦ Two rods of mass M and length ℓ are connected at point O by a frictionless hinge (■ Figure 12.58). A particle collides with the system at point A and delivers a linear impulse of magnitude Δp. Describe the subsequent motion of the system if the direction of the impulse is (a) rightward parallel to OB, (b) leftward parallel to OB, and (c) parallel to AO. Hint: use the result of Problem 9.87.

98. ♦♦♦ A clown of mass $M = 85$ kg rides a unicycle whose wheel radius is 25 cm and whose mass is 5.0 kg. When riding at constant speed, the system of clown and cycle has its CM 1.3 m above the floor. In order to accelerate forward at 2.0 m/s^2, at what angle must the clown lean and what torque must he exert on the pedals?

99. ♦♦♦ A billiard ball of mass $m = 0.40$ kg and radius $r = 3.0$ cm is hit at a height equal to r with an impulse $\Delta p = 1.0$ kg·m/s. The coefficient of kinetic friction between ball and table is $\mu = 0.50$. How far does the ball slide before it begins to roll without slipping?

100. ♦♦♦ Find the acceleration of the hanging object of mass M in the system shown in Figure 12.46. If the coefficient of static friction between ramp and cylinder is μ_s, what is the maximum mass M for which the cylinder can roll without slipping?

Computer Problems

101. Compute the rotational inertia of a uniform rod of length $\ell = 1.00$ m about one end by dividing it up into N identical *particles*, each of length ℓ/N. Use a spreadsheet to compute the sum for $N = 5, 10,$ and 20. Compare the result with the exact answer from Table 12.3.

102. A stunt motorcyclist rides at speed $v = 8.5$ m/s on the interior surface of a right circular cylinder of radius $R = 5.00$ m. The total mass of cycle and rider is $M = 320$ kg, and, on a level road, the system CM is a distance $h = 0.80$ m above the road. Choose an origin as in Example 11.7. Ignoring the rotation of the wheels, show that the angle θ of the motorcyclist with the horizontal satisfies the relation:

$$\tan\theta = (gR/v^2)[1 - (h/R)\cos\theta].$$

Use a calculator or a computer to evaluate both sides of this equation for values of θ between 0 and 90°, and hence find the angle of lean.

103. A test system for determining friction in an experimental bearing for a video disk player consists of a disk mounted on the bearing. Air jets blowing on the disk exert a torque that would accelerate the disk with an angular acceleration $\alpha = 1.0783 \times 10^{-2}$ rad/s^2 in the absence of friction. The angular speed of the disk is measured by a photocell that detects the passage of uniformly spaced black stripes painted on the edge of the disk. (In all, 149 stripes are painted around the circumference.) In a trial run, the disk accelerates from rest and at intervals of 1.000×10^3 s, the photocell counts stripes during a time interval of 1.000 s. The counts for the first 20 measurements are given in file CH12P103 on the supplementary computer disk. (a) Determine whether the data are fit by uniform angular acceleration. (b) Any frictional torque is expected to follow the law $|\vec{\tau}| = bI\omega$, where I is the rotational inertia of the disk and b is an unknown parameter. An approximate prediction for $\omega(t)$ is then

$$\omega(t) \approx \alpha t[1 - bt/2 + (bt)^2/6],$$

and an exact prediction is:

$$\omega(t) = (\alpha/b)(1 - e^{-bt}).$$

Derive an expression for the number of counts predicted by each formula. (c) Devise a spreadsheet program that computes the predicted measurements, their differences from the measured data, and the sum of the squares of those differences for the first 5, the first 10, and all 20 measurements. Determine a value of b that minimizes the sums of squared differences and comment on the quality of the two predictions. (*Hint:* Make a first guess for b by fitting the fifth data point using only the first two terms in the approximate prediction.)

Challenge Problems

104. Earth's period of rotation increases at a rate of 7×10^{-4} s per century because of gravitational forces between the Moon and the bulges of water that are the ocean tides on Earth. Explain qualitatively how this works, and show that the Moon's orbital radius should increase with time as a result. Use conservation of angular momentum to find a relation between the rates of the two changes, neglecting any changes of the internal angular momentum of the Moon. Compute the rate of increase of the Moon's orbital radius.

105. ⌧ Estimate the change in length of Earth's day if all the polar icecaps were to melt. Model the ice as a pair of rings around the Earth at latitudes 80°N and 80°S (the average latitude of the ice in the caps). Assume that the melted ice would raise the sea level by 200 m. Model the ocean as covering the entire Earth with uniform thickness; that is, neglect the continents!

106. A *superball* is thrown against a vertical wall at angle θ with the perpendicular to the wall. A superball dissipates negligible energy during its impact with the wall, but because of static friction between its surface and the wall, substantial energy can be transferred between linear motion and rotational motion. Assume the component of the ball's velocity perpendicular to the wall is reversed during the collision without a change in magnitude. Find the relation between the component of linear impulse parallel to the wall and the angular impulse the ball receives during the collision, and determine the angle at which the superball rebounds.

107. Lenya and Hiroshi, UN Space Agency astronauts each of mass M, are assigned to investigate a gigantic alien space station discovered in orbit around Jupiter. The station is a disk of radius R and rotates at angular frequency ω about an axis perpendicular to its center. The dynamic duo enter a radial tube that leads directly from the edge toward the center. They crawl inward and reach a point $R/2$ from the center. Then, a defense system releases oil, and the helpless two slide without friction outward and emerge from the outer end of the tube. **(a)** Write expressions for the astronauts' velocity and acceleration as they crawl inward at constant speed v_o. Use an inertial Cartesian coordinate system whose x-axis is along the direction of the corridor at the instant they enter. **(b)** Calculate the internal energy each astronaut transforms during the crawl. (*Hint:* Decide what force(s) act on the astronauts and the work done by each. Discuss the energy flow during the crawling process.) **(c)** Discuss the reasons why the result of part (b) can be described in terms of an *effective radial* potential energy of each astronaut $U \equiv -\tfrac{1}{2}M\omega^2 r^2$, where r is the distance from the station's central axis. **(d)** Compute the astronauts' speed with respect to your inertial reference frame as they emerge from the tube. (*Hint:* The best approach is part conservation law and part kinematic.)

108. (a) A thin spherical shell of mass M and radius r may be viewed as a collection of hoops with the polar angle θ as coordinate (■ Figure 12.59). Express the rotational inertia of the shell about the z-axis as an integral over θ and evaluate the integral. **(b)** Use your result to express the rotational inertia of a uniform sphere of mass M and radius R as an integral over spherical shells. Evaluate the integral.

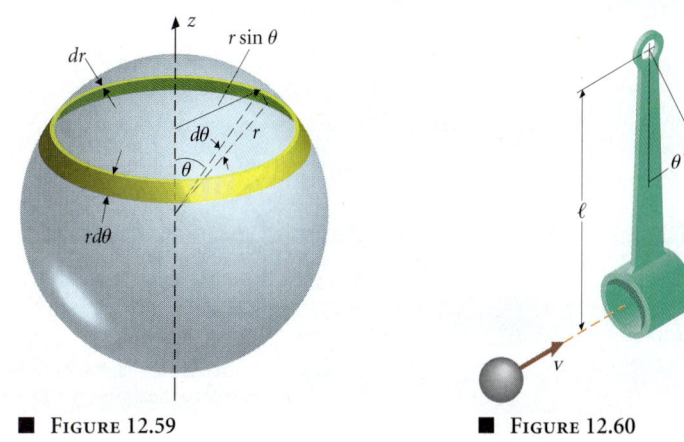

■ **Figure 12.59**

■ **Figure 12.60**

109. In a laboratory version of the ballistic pendulum, a steel ball of mass m and unknown speed v is caught at the end of a swinging arm of length ℓ (■ Figure 12.60). When empty, the arm has mass M and rotational inertia I about the pivot point. **(a)** Where should the CM of the catcher arm be located to avoid impact on the pivot when the ball is caught? **(b)** If the catcher arm rotates upward through an angle θ after the impact, what is the initial speed of the ball?

110. In the system of Figure 12.56 (but without the rolling ball) the smaller cylinder is initially rotating with angular speed $\omega_i = 5.00$ rad/s, the belt is slack, and the larger cylinder is stationary. A force $F = 5.00 \times 10^3$ N, to the left, is suddenly applied to the bearings of the larger cylinder so that the belt is drawn taut. If the coefficient of kinetic friction between belt and cylinder surfaces is $\mu_k = 0.425$, find the final angular speeds of the two cylinders and the time they require to reach those speeds. The mass of the larger cylinder is 16.43 kg, and that of the smaller cylinder is 4.17 kg.

Compare the final angular momentum of the system to its initial value, and explain how angular momentum is transferred to or from the system.

ESSAY 4
The Bicycle

The modern bicycle is a marvel of speed, stability, and maneuverability. Competition cyclists race downhill at speeds as high as 80 km/h.

A modern touring bicycle is designed for a thrilling downhill run as well as the more taxing uphill climb. It is also a practical tool, the most energy efficient of the common modes of transportation (■ Figure E4.1). Designing any machine involves matching a task with the power source available for performing the task. For a bicycle, the task is fast, safe transport of the person who serves as the power source.

The invention of the bicycle is credited to Karl von Drais, who discovered in 1817 that a steerable front wheel allows stable downhill coasting. Von Drais' vehicle was powered by the rider pushing on the ground, much the way a child propels a scooter. By 1870, pedals had become the standard propulsion mechanism, but without a gearing system that could withstand road grime, the only way to achieve speed was to drive a large wheel (■ Figure E4.2). The *ordinary* of the 1870s was thrilling to ride but incredibly unsafe. It gave way to the modern design with the invention of effective chain drives in the 1880s and the derailleur system of gearing in the 1920s. Today, gear ratios are quoted in terms of "effective wheel diameter"—a comparison to the ordinary's large wheel.

Experimental bicycles, designed to reduce wind resistance (■ Figure E4.3), are capable of achieving sustained operation at speeds up to 60 km/h. Bicycle-style power systems have even

EFFECTIVE WHEEL DIAMETER EQUALS ACTUAL DIAMETER DIVIDED BY GEAR RATIO.

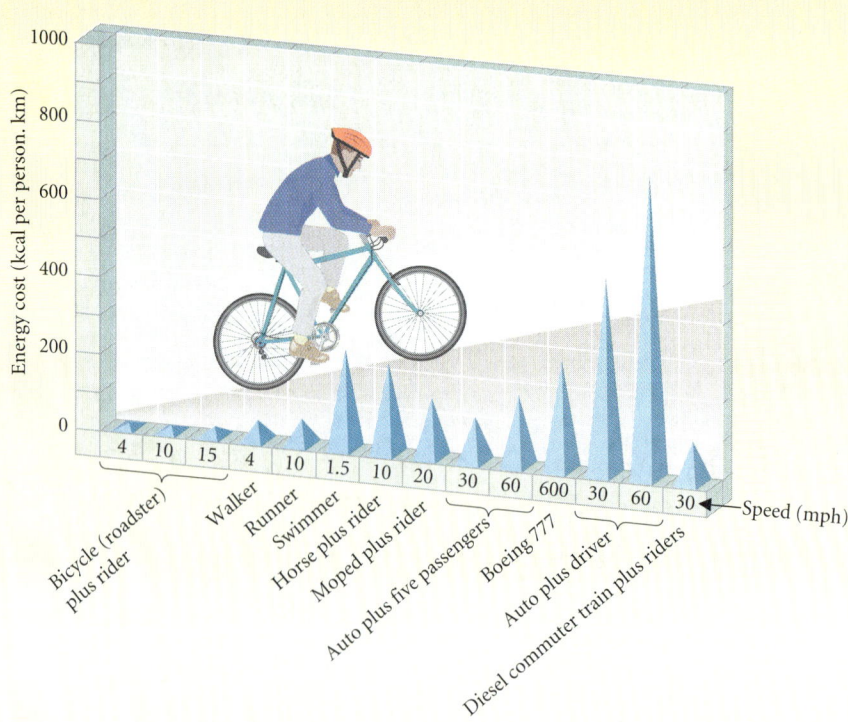

■ FIGURE E4.1
The energy use per passenger-kilometer is plotted for several common forms of transportation. On this basis, the modern touring bicycle is the most efficient.

FIGURE E4.2
The *ordinary* bicycle of the 1870s became wildly popular among the tall, athletic, single, daredevil men who could ride it and who could accept the risk of doing so. The recommended way to coast downhill was to ride with one's legs over the handlebars. That way the hapless rider could land on his feet if thrown off!

made human-powered flight possible (■ Figure E4.4). Such performance results from matching the human power source to the task of linear motion with an extremely simple machine closely tuned to human capabilities. The rider drives a bicycle's pedal-and-crank system in a sequence of rapid thrusts that are the optimum way for animal muscles to produce power. The chain transmitting power to the rear wheel is controlled by guides, called derailleurs, so that it passes around one gear wheel at the crank and a second gear at the hub of the rear wheel. The different choices of gear wheels ("speeds") allow the rider to pedal at a nearly optimum rate over a large range of linear speeds.

We look first at the design of the bicycle's chain drive. In such a *belt-and-pulley* system, the power input equals the torque acting on the drive gear multiplied by its angular speed: $P = \tau\omega$. The power output equals the corresponding product of angular speed and torque

FIGURE E4.3
Energy losses due to air resistance are the major power requirement for bicycles moving at more than a few meters per second. In bicycle-driven vehicles with good aerodynamic design, athletes can achieve sustained operation at speeds up to 60 km/h.

FIGURE E4.4
With modern materials such as plastics and carbon fiber struts, it is possible to build an aircraft with a bicycle-style drive system. The machine is light enough to fly, and its power requirements are within the range of a good athlete. The *Gossamer Albatross* flew across the English Channel in 1979, requiring 3 hours flight time and an average power of $\frac{1}{3}$ horsepower.

■ FIGURE E4.5
This touring bicycle has a choice of two gears attached to the pedal and crank and five attached to the rear wheel. The derailleurs, or chain guides, allow the rider to choose any of the ten combinations of front and back gears.

SEE THE DISCUSSION OF MACHINES IN §§7.2 AND 9.2.

acting on the output gear (■ Figure E4.5). The torque is due to the tension in the chain: $\tau = TR$. In an ideal system, the chain tension is the same on both wheels. Since the chain doesn't stretch, the linear speed $v = \omega R$ of each gear rim is the same. That is, the chain engages the same number of teeth per second on each gear. Together, these requirements guarantee that input and output powers are equal. Thus the torque ratio $r = \tau_O/\tau_I = \omega_I/\omega_O = R_O/R_I$, the ratio of gear radii, or equivalently, the ratio of tooth numbers on the gears:

$$r \equiv \frac{\text{\# of teeth on hub gear}}{\text{\# of teeth on crank gear}}.$$

• Table E4.1 gives the possible choices for r on one of the authors' Peugeot touring bicycle. The other dimensions we need are the radius of the rear wheel $R \approx \frac{1}{3}$ m and the length of the crank arm $\ell \approx 17$ cm.

Next, we need to know how much power is available. We discussed the general form of power curves in Chapter 7. Given the person's physical condition and how long the ride lasts, the only variable is the pedaling rate. The power a rider can produce rises from zero at low pedaling rate, passes through a maximum at the optimum rate, and decreases again toward zero at very rapid rates. The overall shape of the power curve holds for a wide variety of practical energy sources, though the details differ considerably. An automobile engine, for example, cannot deliver power at all near zero rotation speed and requires a clutch mechanism or fluid transmission to start the car moving. Here we can make reasonable estimates about a person's power curve and compare the results with measurements.

TABLE E4.1   Ideal Torque Ratios for a Touring Bicycle*

Number of crank teeth	Number of hub teeth				
	32	28	24	20	16
56	0.57	0.50	**0.43**	**0.36**	**0.29**
40	**0.80**	**0.70**	**0.60**	0.50	0.40

*Boldfaced values are the combinations recommended by the manufacturer.

Digging Deeper

Average Torque on the Bicycle Crank

The cyclist always pushes on the downward-moving pedal. Thrust directly downward on the pedal produces a torque of magnitude $F\ell \sin\theta$, where θ is the angle of the crank from vertical (Figure E4.6). As the cyclist's foot descends, the torque goes from zero at the top, through maximum when the crank is horizontal ($\theta = 90°$, $\sin\theta = 1$), and back to zero at the bottom. During the next half turn, the torque is exerted on the other pedal and varies through the same cycle. The overall average torque is thus equal to its average over any particular half turn of the pedals. $\langle|\vec{\tau}|\rangle = F\ell\langle\sin\theta\rangle$, where $\langle\rangle$ means "averaged over time," and $\theta = \omega t$. (For this estimate, we assume the pedals turn at a constant rate.)

$$\langle\sin\theta\rangle \equiv \frac{\int_0^{\pi/\omega} \sin\omega t\, dt}{\pi/\omega} = \frac{-\cos\omega t\Big|_0^{\pi/\omega}}{\pi} = \frac{2}{\pi}.$$

Expert cyclists, using toe-clips, can improve this ratio of average to maximum torque somewhat. A small further improvement is possible with elliptical crank gears.

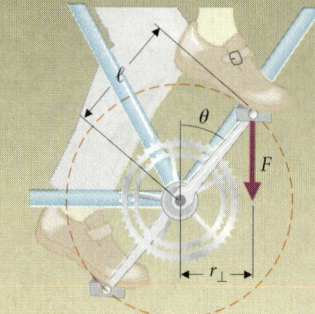

Figure E4.6
The torque acting on the crank varies as the gear turns:
$$|\vec{\tau}| = Fr_\perp = F\ell\sin\theta.$$

The upper limit on ω_1 is easiest to estimate. Around $\omega_1 \approx 20$ rad/s, each leg must go up and down 3 times per second. This is nearly everyone's upper limit.

At low values of ω_1, the power is determined by the maximum force the rider can exert on the pedals. By pulling upward on the handlebars, pushing downward on one pedal and tugging upward on the opposite pedal strap, a determined rider can exert a force on the pedals somewhat larger than the rider's weight. The resulting torque exerted on the crank varies as the crank arm rotates. The power input from the rider equals the average value of this torque multiplied by the angular speed ω_1. For this example, we assume a total mass of bicycle and rider of $m = 80$ kg and a maximum possible force equal to the weight of the whole system: $F_{max} \approx mg \approx 800$ N. The average torque is about $2/\pi$ times the maximum, or:

$$\langle\tau\rangle \approx (2/\pi)F_{max}\ell \approx 87\text{ N}\cdot\text{m}.$$

For small ω_1 (less than a few radians per second), the power output of the rider is directly proportional to angular speed: $P(\omega_1) = \langle\tau\rangle\omega_1$.

A rider can maintain peak power over a fairly broad range of ω_1 from about 5 rad/s to about 13 rad/s. Average people can maintain more than 1 horsepower for a short task such as running up stairs, while $\tfrac{2}{3}$ hp (500 W) maintained for an hour stretches the limits of world-class athletes. Here we assume a peak power of 500 W, representing either a short burst of power on a pleasure ride or about 20 minutes of world-class competition.

We now know about the power available to the bicycle; what are the power requirements? Overcoming air resistance (or drag) is the major power requirement when the bicycle's speed is greater than a few meters per second. In Figure E4.7, the black curve represents the rider's maximum sustained output. The green curves show the power remaining after subtracting the power used to overcome drag from the total power available. For a given gear setting, this remaining power is zero at the maximum speed of the bicycle. For $r = 0.8$, wind resistance hardly affects the maximum speed, while for $r = 0.29$, the rider's peak power output just overcomes wind resistance at a speed $v = \omega_0 R = \omega_1 R/r$ of about (12 rad/s)($\tfrac{1}{3}$ m)/(0.29) = 14 m/s. For comparison, the British speed record over 40 km is 13.6 m/s.

Recall that $\omega = 2\pi f$.

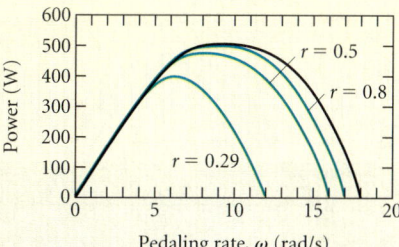

Figure E4.7
Power relations for a rider on a touring bicycle. The solid black curve shows the maximum power available from the cyclist as a function of pedaling rate. For this graph, the total mass of cycle and rider is 80 kg. The peak power of 500 W is realistic for a few minutes' sprint by an average rider or about 20 min of competition riding. Curves (in green) for each of three torque ratios, $r = 0.29, 0.5,$ and 0.8, show the power available after subtracting losses due to wind resistance.

Digging Deeper

Air Resistance

Drag force on the bicycle is proportional to the air density and the system's cross section perpendicular to the airflow, and nearly proportional to the square of the bicycle's speed:

$$F_D = \tfrac{1}{2}\rho C_D A v^2.$$

The power P_D required to counteract this force is the product of the drag force and the bicycle's speed, and so is proportional to the cube of the speed:

$$P_D = F_D v = \tfrac{1}{2}\rho C_D A v^3.$$

The drag coefficient C_D is determined experimentally; its value for racing bicycles is 0.9. Air density at sea level is 1.2 kg/m³, and the cross section of a rider fully crouched over the handlebars is about 0.9 m × 0.4 m = 0.4 m². With these numbers:

$$\tfrac{1}{2}\rho C_D A \approx 0.2 \text{ W}\cdot\text{s}^3/\text{m}^3.$$

The bicycle's speed over the ground is related to the angular speed of the wheel:

$$v = R\omega_O.$$

And, the angular speed of the wheel is related to that of the crank by the ideal torque ratio:

$$\omega_O = \omega_1/r.$$

So, the speed of the bicycle over the ground is $v = R\omega_1/r$, and, with $R = \tfrac{1}{3}$ m,

$$P_D = (0.2 \text{ W}\cdot\text{s}^3/\text{m}^3)\left(\frac{\omega_1 R}{r}\right)^3 \approx (7 \times 10^{-3} \text{ W}\cdot\text{s}^3)\left(\frac{\omega_1}{r}\right)^3.$$

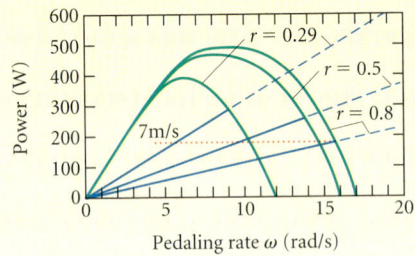

(a) Gear ratios and power requirements for hill climbing.

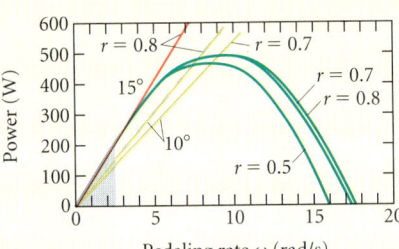

(b) Power requirements for various torque ratio.

■ Figure E4.8
(a) Hill climbing. The blue lines show the power required to climb a 2° hill with a given gear ratio. The dotted line is a line of power required to cycle at 7 m/s on this slope. (b) The linear curves show the power required to climb hills of slope 10° (yellow) and 15° (red) for several torque ratios. The shaded triangle at the bottom left represents a region where it is difficult to balance the bicycle. Our analysis does not apply in this region.

Going up a hill, power is required to increase the gravitational energy of bike and rider as they gain height h:

$$P = \frac{d}{dt}(mgh) = mg\frac{dh}{dt}.$$

The rate at which the bike gains height, dh/dt, depends on its speed over the ground and the angle θ of the slope,

$$dh/dt = v\sin\theta = \omega_1 R(\sin\theta)/r.$$

So, the power required is:

$$P = mgR\omega_1\frac{\sin\theta}{r} = (80 \text{ kg})(9.8 \text{ m/s}^2)(0.33 \text{ m})\omega_1\frac{\sin\theta}{r}$$

$$= (2.6 \times 10^2 \text{ N}\cdot\text{m})\frac{\sin\theta}{r}\omega_1.$$

Curves representing this power requirement are plotted in ■ Figure E4.8 for $\theta = 2°$ and 10° and for various values of r. For $\theta = 2°$, it is possible to ride the bicycle with any of the gear settings. For a given gear choice, the cyclist can achieve any speed up to the maximum, where the power available (maximum output minus loss to drag) equals that needed to go uphill. The maximum occurs where the green line of power available intersects the blue line of power required. For the lowest "speed" (highest torque ratio) $r = 0.8$, the intersection occurs at $\omega_1 = 16$ rad/s, corresponding to a maximum speed of 7 m/s. With $r = 0.29$ and $\omega_1 = 9$ rad/s, a speed of 10 m/s is possible. The dotted curve shows combinations of torque ratio and pedaling speed that achieve 7 m/s with power output less than the maximum available.

On the 10° slope, only the two highest torque ratios may be used. Pedaling at a modest rate $\omega_1 \approx 7$–8 rad/s, the cyclist can achieve a speed of approximately 3 m/s. The maximum slope the cyclist may ascend is $\theta \approx 15°$, corresponding to the maximum force limit with the highest torque ratio. At this limit, the assumption of a constant bicycle speed gets into trouble. The bike tends to accelerate and decelerate as the torque on the crank oscillates, and it becomes hard to balance the bike at low speed. Only an excellent athlete can ascend a 15° slope. Of course, automobiles have the same difficulties as cyclists, for the same reasons, and 15° slopes are rather rare. Over the range of slopes usually encountered, the bicycle is superbly adapted to its task.

*And sure the reverent eye must see
A Purpose in Liquidity*
RUPERT BROOKE

CHAPTER 13
Fluids

CONCEPTS

Fluid
Compressibility
Density
Pressure
Buoyant force
Streamline
Conservation of mass
Steady flow
Ram pressure

GOALS

Be able to:
Find the mechanical advantage of hydraulic machinery.

Compute pressure as a function of position in a static fluid.

Apply Archimedes' principle.

Apply Bernoulli's law and the equation of continuity in steady-flow problems.

Yachts in the America's Cup race. Fluid forces are responsible for floating each yacht (§13.4) and also for its propulsion (§13.5).

WE LISTED THE MAJOR TYPES OF FLUID FORCE IN TABLE 4.1 ALONG WITH THE OTHER FORCES OCCURRING IN MECHANICS.

WE SHOW HOW TO DO THIS IN EXAMPLE 13.7.

FOR WHAT WE MEAN BY *PARTICLE* IN THIS CONTEXT, SEE THE INTRODUCTION TO PART III.

The sailors in the America's Cup race rely on fluids—water and air—both to float their boats and to propel them across the finish line. The helmsman, with the wind in his face, has no problem understanding that air can exert powerful forces. It is more difficult for an airline passenger to realize that air is responsible for the success of the flight.

Why do the yachts float? What holds the airliner up? Why do your ears hurt when you dive to the bottom of the swimming pool? All three questions have the same answer—fluid pressure. Fluids exert force on their containers and on objects in contact with them. These forces are technologically important, their study has a long history, and it is time for us to think about them in some detail.

The earliest physical principle that we still accept in its original form was discovered by Archimedes (287–212 B.C.). The King of Syracuse asked him to determine whether a crown was made of pure gold, without damaging the crown. Archimedes solved this puzzle while taking a bath; as the story goes, he leapt out and ran down the street naked shouting, "Eureka!"—I have found it! He had realized that the fluid forces that made him float so comfortably in his bath would also allow him to measure the density of the crown.

During the Middle Ages, European miners developed an extensive practical technology for pumping water out of their mines. It was common knowledge that suction pumps could not lift water more than 10 m in a single stage. No one understood this annoying difficulty or connected it with floating bodies until Evangelista Torricelli (1608–1647), a student of Galileo, argued that the height of the water column in a suction pump is limited by air pressure on the water's surface. Torricelli tested his idea with a *barometer* (■ Figure 13.1) in which air supports a column of mercury in an inverted glass tube. In France, Blaise Pascal (1623–1662) conducted similar experiments. He compared a barometer set up on a mountaintop with one kept at low altitude on the same day. The thinner air on the mountain should support a shorter column of mercury. Returning from a rather hair-raising thunderstorm, the mountain team reported just that observation!

We shall model fluids, like solids, as collections of *particles* that are actually small volumes, each containing a large number of individual molecules. A *particle's* properties are defined by the average behavior of its molecules and are well defined if this average behavior changes slowly from one small volume to the next. Fluids obey Newton's laws, just as solids

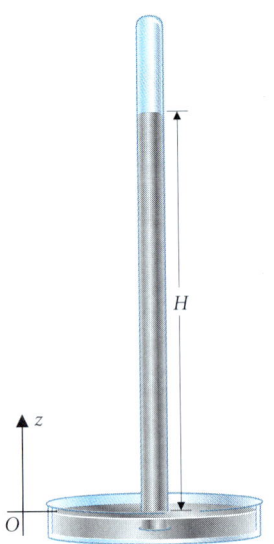

■ **FIGURE 13.1**
Torricelli's barometer. The column of mercury in the inverted glass tube is supported by atmospheric pressure. This barometer was used in early investigations of fluid pressure.

do, but the variation of properties within a single body of fluid is much more significant than in solids. We could ignore internal forces in the rigid-body model for solids. In contrast, we have to pay careful attention to the nature of internal forces acting within a fluid and how they arise from the interaction of individual molecules.

13.1 What is a Fluid?

Solid objects have a definite shape, and they resist efforts to change that shape. If we push on a rubber cube (■ Figure 13.2), the force components parallel to the surface of the cube—called *shear forces*—deform it slightly until the internal, elastic forces are able to balance the applied forces. We cannot do a similar experiment with water because water flows to take up the shape of its container; we can only have a cube of water if we put it in a cubical box. Water would not resist the distortion caused by shear forces on the cube, but would change its shape and continue to flow as long as any shear force remains. This ability to change shape is the defining characteristic of fluids.

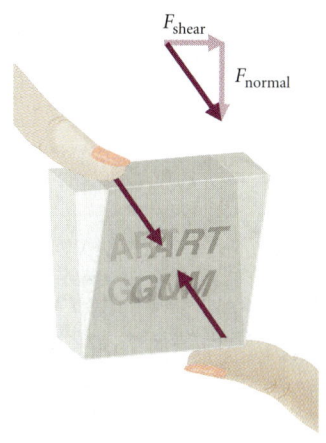

■ **Figure 13.2**
The tangential components of the forces applied to a solid cube's surface—called shear forces—distort the cube. Fluids cannot withstand shear forces.

> An *ideal fluid* is one that offers no resistance to changes in shape and cannot sustain shear forces.

Equivalently, the forces exerted by an ideal fluid on a surface have no tangential component but are perpendicular to the surface. Internal forces in the fluid act similarly: fluid elements on either side of an arbitrary plane somewhere inside a fluid (■ Figure 13.3) exert forces on each other that are perpendicular to the plane.

You might think of an ideal fluid as frictionless since it exerts only normal forces.

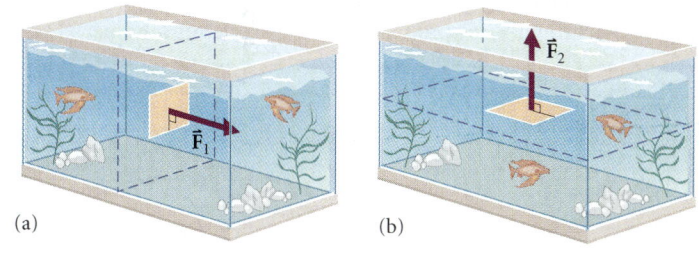

■ **Figure 13.3**
Force exerted by an ideal fluid is perpendicular to the surface on which it acts. (a) $\vec{F}_1$ is the force that fluid to the left of the surface exerts on fluid to the right. (b) $\vec{F}_2$ is the force that fluid below the surface exerts on fluid above. (c) $\vec{F}_3$ is the force that fluid below and to the left of the surface exerts on fluid above and to the right. All three forces have the same magnitude.

The mechanisms by which ideal fluids exert normal forces vary greatly. Water and air are good examples of this range of behavior. A cube of water resists compression nearly as strongly as a solid wooden cube. Large forces applied to the faces of either cube in an attempt to compress it would produce a very small change in the cube's volume (■ Figure 13.4). If, instead, we expand a container full of water, the water maintains its original volume and leaves part of the container empty. This behavior is described as *incompressible* and is typical of *liquids*. Air does not behave the same way at all. If we squeeze a container of air, the air gets smaller. If we expand the container, the air expands to fill the new volume. Such *compressible* behavior is typical of *gases*.

Whether a substance is a solid, liquid, or gas depends on the degree to which forces between the molecules of the substance determine its structure. In a solid the forces between the molecules are relatively strong and hold the molecules firmly in place. Molecular forces are weaker in liquids and negligible in dilute gases.

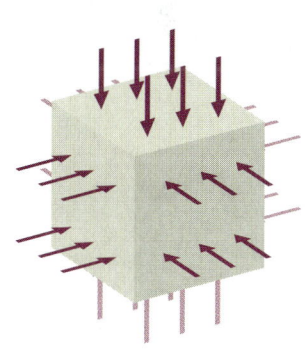

■ **Figure 13.4**
Normal forces applied to a cube's surfaces cause compression: if the cube were made of rubber, its volume would decrease slightly. If it were made of air, its volume would decrease substantially. If it were made of water, its volume wouldn't change.

In Chapter 20 we'll study the role of temperature in determining the structure of a substance.

13.2 BASIC PROPERTIES OF FLUIDS

13.2.1 Density

SEE §4.3.2 FOR A DISCUSSION OF MASS AND WHY IT IS FUNDAMENTAL.

Because fluids take on the shape of their containers, volume is a common measure of fluid quantity—grocers sell milk by the quart rather than by the kilogram (or pound). Mass is the fundamental measure of a substance, but we can accept volume as a practical measure for milk because mass and volume are directly related: we expect to get twice the mass when we buy twice the volume. These substances have a *uniform density*.

> The *density* of a uniform substance is its mass divided by the volume it occupies.
> $$\rho = M/V \tag{13.1}$$

WE CAN DEFINE DENSITIES OF QUANTITIES OTHER THAN MASS. "CHOCOLATE CHIP DENSITY" IS THE NUMBER OF CHOCOLATE CHIPS PER UNIT VOLUME IN A COOKIE; "FRECKLE DENSITY" IS THE NUMBER OF FRECKLES PER AREA OF SKIN. WHEN THE WORD "DENSITY" IS USED ALONE, IT MEANS MASS DENSITY.

Densities of fluids vary a great deal (● Table 13.1) and are often compared with the density of water ($1 \text{ g/cm}^3 = 10^3 \text{ kg/m}^3$).

THE DENSITY OF WATER EQUALS 1 g/cm^3 BECAUSE THIS WAS THE ORIGINAL DEFINITION OF THE GRAM.

> The *specific gravity* of a uniform substance is its density divided by the density of water.

Since specific gravity is the ratio of two quantities with the same physical dimensions, it has no units. The density of mercury is $1.36 \times 10^4 \text{ kg/m}^3$ and its specific gravity is 13.6. The specific gravity of air at atmospheric pressure is only about 10^{-3} (air has one-thousandth the density of water). Density (or specific gravity) can be a useful indicator of chemical composition: a winemaster determines the sugar content of wine by measuring its specific gravity. Similarly, an auto mechanic can determine the amount of acid in battery fluid by measuring its specific gravity.

WE HAVE TO SPECIFY THE PRESSURE OF THE AIR BECAUSE AIR IS COMPRESSIBLE. SEE §13.2.3.

TABLE 13.1 Densities

Solids	Density (kg/m³)	Fluids	Density (kg/m³)
Styrofoam	$\approx 1 \times 10^2$	Interstellar space	$\approx 10^{-27}$
Oak	$\approx 8 \times 10^2$	Sun (visible surface)	$\approx 2 \times 10^{-2}$
Ice	9.2×10^2	Hydrogen at 0°C, 1 atm	9.0×10^{-2}
Brick	$\approx 1.7 \times 10^3$	Steam at 100°C, 1 atm	0.6
Bone	$\approx 2.0 \times 10^3$	Air at 0°C, 1 atm	1.3
Glass	$\approx 2.5 \times 10^3$	Air at 20°C, 1 atm	1.2
Aluminum	2.7×10^3	Gasoline	6.8×10^2
Earth (average)	5.52×10^3	Alcohol	8.06×10^2
Iron	7.96×10^3	Water at 0°C, 1 atm	1.00×10^3
Copper	8.96×10^3	Seawater	1.025×10^3
Lead	1.13×10^4	Mercury	1.36×10^4
Gold	1.93×10^4	Sun (center)	$\approx 1.6 \times 10^5$
Platinum	2.14×10^4	Iron nucleus	$\approx 10^{18}$

EXAMPLE 13.1 ♦ A person walking along the road carries a cubical container 0.20 m on a side that is full and holds 5.4 kg of fluid. Compute the fluid's density and compare with Table 13.1. What is its specific gravity? Identify the fluid.

MODEL We assume the fluid has uniform density.

SETUP The density of the fluid is given by eqn. (13.1):

SOLVE
$$\rho = \frac{M}{V} = \frac{5.4 \text{ kg}}{(0.20 \text{ m})^3} = 675 \text{ kg/m}^3.$$

The specific gravity is:

$$\text{S.G.} \equiv \frac{\text{density of fluid}}{\text{density of water}} = \frac{675 \text{ kg/m}^3}{1.0 \times 10^3 \text{ kg/m}^3} = 0.68.$$

ANALYZE Based on its density (Table 13.1), the fluid is probably gasoline. ∎

For compressible fluids or nonuniform substances, eqn. (13.1) gives only an *average density*. We define density at a point as the limit of average density for ever-smaller volumes surrounding a point:

$$\rho(\vec{x}) = \lim_{V \to 0} \frac{M(\vec{x})}{V(\vec{x})}. \qquad (13.2)$$

We cannot actually let V go to zero. The density fluctuates wildly once the volume is so small that it contains only a few molecules. A well-defined limit is reached with a $V(\vec{x})$ that contains many molecules but is much smaller than the total volume of the fluid. The next example shows how small a volume we might need.

SEE ALSO THE DISCUSSION IN §2.1.4, WHICH DESCRIBES THE USE OF CALCULUS IN PHYSICS.

EXAMPLE 13.2 ♦♦ Test the validity of modeling a 1-cm cube of air as a particle. First compute the number of molecules in the cube. Given that the density in the Earth's atmosphere changes by an amount $\delta\rho = \rho \, \delta y/(10 \text{ km})$ in a height δy, what is the fractional change in density $\delta\rho/\rho$ from one side of the cube to the other?

REFER TO §13.3.4 FOR A JUSTIFICATION OF THIS EXPRESSION FOR $\delta\rho$.

MODEL Air is made up mostly of diatomic nitrogen molecules, each with an atomic mass of 28 (see the periodic table, back cover). Taking the value of the atomic mass unit from the inside front cover, $m \approx 28(1.7 \times 10^{-27} \text{ kg}) = 4.8 \times 10^{-26}$ kg.

SETUP From eqn. (13.1), the 1-cm cube has mass:

$$M = V\rho = (1 \times 10^{-2} \text{ m})^3 (1 \text{ kg/m}^3) = 1 \times 10^{-6} \text{ kg}.$$

SOLVE The number of air molecules in the small volume is:

$$N = M/m = (10^{-6} \text{ kg})/(4.8 \times 10^{-26} \text{ kg}) = 2 \times 10^{19} \text{ molecules}.$$

With $\delta y = 1$ cm, the fractional change in density across the cube is:

$$\delta\rho/\rho = (1 \text{ cm})/(10 \text{ km}) = 10^{-6}.$$

ANALYZE The density changes by only one part in a million across this small volume, yet it contains over a billion billion molecules: this cube is a suitable volume element for computing *density at a point* in the atmosphere. ∎

13.2.2 Pressure

A fluid resists normal forces exerted on it by its container, and normal forces also act between different parts of a fluid body. For example, gravity pulls down on the water at the top of a bucket, but the water does not fall because water underneath pushes upward. Unlike a rigid body, the water at the top of the bucket cannot be supported at a point. It must be supported over its entire bottom surface. Because fluid forces are spread out over extended surfaces, it is convenient to describe forces exerted by a fluid on a unit of surface area.

THIS DEFINITION IS IN TERMS OF THE MAGNITUDE OF THE FORCE. THAT MEANS THAT PRESSURE IS ALWAYS A POSITIVE QUANTITY.

> The *pressure* of a fluid on any surface is the magnitude of the normal force *per unit area* exerted by the fluid on the surface:
> $$P = |\vec{F}|/A. \qquad (13.3)$$

Pressure is the quantity that allows balloons to delight children and crushes submarines that descend too deep. The SI unit of pressure is the *pascal*.

> Fluid at a pressure of 1 *pascal* exerts a force of 1 newton on each square meter of surface:
> $$1 \text{ Pa} \equiv 1 \text{ N/m}^2.$$

Another common unit of pressure is the *atmosphere*. Actual atmospheric pressure varies from day to day and from place to place, so an average of actual atmospheric pressure at sea level is used to define an exact value for the *unit*:

$$1 \text{ atm} = 1.013250 \times 10^5 \text{ Pa}.$$

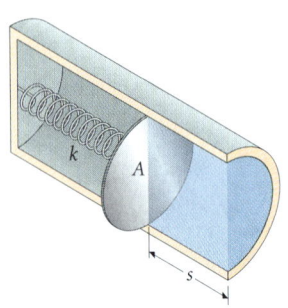

■ FIGURE 13.5
Concept for a pressure gauge. A force $F = PA$ acts on the disk and compresses the spring. In equilibrium, $ks = F$, so $s = PA/k$. The design is not practical because of difficulties with sealing the edges of the disk. The aneroid barometer (Figure 13.18) is a practical version of the same idea.

■ Figure 13.5 shows a conceptual device one might use to measure pressure and investigate its properties. A small disk has a known area A, and the force on the disk is measured by a spring of known spring constant k. We assume the volume behind the disk is completely empty and the gasket around the edge of the disk slides without friction.

SOME OF THE EXAMPLES IN THIS CHAPTER HAVE A SMALLER NUMBER OF SIGNIFICANT FIGURES THAN AVERAGE. WATER IS NOT AN IDEAL FLUID, ATMOSPHERIC PRESSURE VARIES FROM DAY TO DAY. WE DON'T WANT TO MISLEAD YOU BY CLAIMING MORE ACCURACY THAN IS REASONABLE.

EXAMPLE 13.3 ♦ In the pressure gauge shown in Figure 13.5, the area of the disk is $A = 10^{-4}$ m², the spring has constant $k = 10^4$ N/m, and the compression of the spring is $s = 1$ mm. What is the measured fluid pressure?

MODEL The force F acting on the outside of the disk is balanced by the spring force $F = ks$. (See §4.5.1.)

SETUP AND SOLVE We use the definition of pressure (eqn. 13.3):

$$P = F/A = ks/A = (10^4 \text{ N/m})(10^{-3} \text{ m})/(10^{-4} \text{ m}^2) = 10^5 \text{ N/m}^2 = 10^5 \text{ Pa}.$$

ANALYZE This is atmospheric pressure. To compress the probe by the same amount with your fingertip, you would have to exert a force of 10 N (2 lb). ∎

EXERCISE 13.1 ♦ Estimate the force due to atmospheric pressure on the top of your head. Why aren't you crushed?

To determine how pressure depends on direction, imagine rotating the pressure probe into different orientations, keeping the center of the disk at the same position. We find that the reading doesn't change. The spherical shape of a beach ball or of gas bubbles in a soda results from this property of pressure; they are spheres because the gas inside pushes out equally in all directions.

THE WORD COMES FROM THE GREEK ROOTS ISO = SAME AND TROPIC = TURNING.

> The magnitude of the force exerted by an ideal fluid on a small piece of surface does not depend on the orientation of the surface: fluid pressure is *isotropic*.

Digging Deeper

Von Guericke's Demonstration

The classic demonstration of atmospheric pressure was conducted by Otto von Guericke in 1660. He pumped the air from a large sphere ≈ 0.8 m in diameter that consisted of two hemispheres placed together. Teams of eight horses attached to each half failed to pull the pieces apart! (■ Figure 13.6a).

Von Guericke's result allows us to estimate the magnitude of atmospheric pressure. We need the result that the net force acting on one of the hemispheres is equal to atmospheric pressure times the cross section of the sphere (Figure 13.6b). Since the horses could not separate the hemispheres:

$$8 \text{ (force of horse)} < \text{(atmospheric pressure)} \times \text{(area)} = P_o \pi (d/2)^2.$$

A horse leaning forward at about 10° can exert a force on the ropes equal to about one-sixth its weight, or about 1.2 kN. Thus:

$$P_o > \frac{32(1.2 \text{ kN})}{\pi (0.8 \text{ m})^2} = 19 \text{ kN/m}^2.$$

In fact, air pressure greatly exceeds this value; about 40 horses would have been needed on each team to separate the two halves of the sphere.

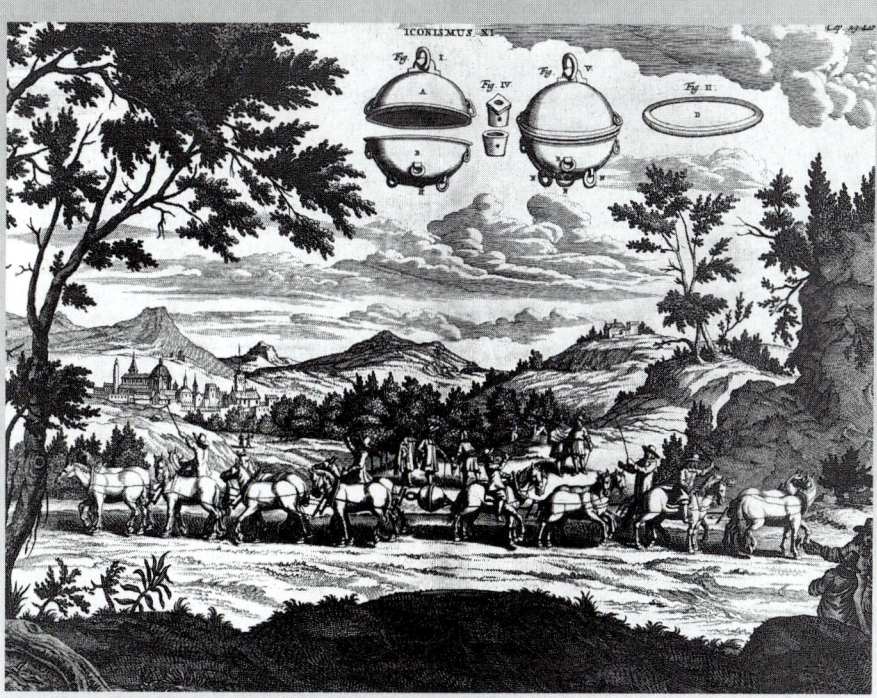
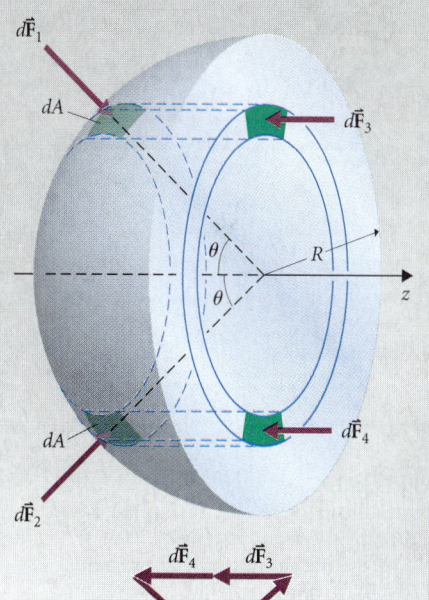

■ **Figure 13.6**
(a) Otto von Guericke's demonstration of the large forces caused by atmospheric pressure. Two hollow hemispheres were evacuated and hitched to teams of eight horses. The horses could not pull the spheres apart. (b) The force due to pressure on a hemisphere. For an object of this shape to be in equilibrium in the air, the net force on the curved side on the left is equal and opposite to that on the flat side: $F = \pi R^2 P$ and is perpendicular to the flat side.

Pressure does vary with position in a fluid. As with density, the definition of pressure at a point involves a limiting process:

$$P(\vec{x}) = \lim_{A \to 0} \frac{F(\vec{x})}{A(\vec{x})}. \tag{13.4}$$

When the pressure varies over a finite surface, the resulting force on the surface is found by adding the differential forces acting on each piece of it.

EXERCISE 13.2 ❖ A glass jar full of compressed air breaks. Do the glass fragments fly inward (implode) or outward (explode)? Why? What would happen if you smashed an old TV tube?

One of the most important applications of fluid pressure is in hydraulic machinery. Such machines operate at sufficiently large pressures that any variations of pressure within the fluid are negligible by comparison.

Fluid pressure is approximately constant throughout a high-pressure hydraulic system.

THIS IS A VERSION OF PASCAL'S LAW (EQN. 13.10), WHICH WE DERIVE IN THE NEXT SECTION. THIS VERSION IS VALID FOR SYSTEMS WITH PRESSURE MUCH GREATER THAN ATMOSPHERIC PRESSURE.

■ Figure 13.7 shows the fundamental idea. A force F_I is applied to the input piston of area A_I, producing a pressure $P = F_I/A_I$. The pressure is the same on the larger output piston and produces an output force of magnitude $F_O = PA_O$. The mechanical advantage (§7.2.2) of the machine is $F_O/F_I = (PA_O)/(PA_I) = A_O/A_I$.

The mechanical advantage of a hydraulic system is the ratio of areas of its output and input pistons.

EXAMPLE 13.4 ♦ The hydraulic lift in a mechanic's shop lifts a car of mass $M = 2.1 \times 10^3$ kg with a piston 22 cm in diameter. A pump forces fluid into the cylinder below the piston through an input line with inside diameter 1.0 cm. What pressure is required in the system? What force must the pump exert on the fluid to operate the system?

MODEL The output force of the system, the force resulting from pressure on the 22-cm piston, equals the weight of the car.

SETUP AND SOLVE The necessary pressure is:

$$P = \frac{F}{A} = \frac{Mg}{\pi d^2/4} = \frac{4(2.1 \times 10^3 \text{ kg})(9.8 \text{ m/s}^2)}{\pi (0.22 \text{ m})^2}$$
$$= 5.4 \times 10^5 \text{ Pa} = 5.4 \text{ atm}.$$

The mechanical advantage of the system is the ratio of areas:

$$\text{MA} = A_O/A_I = (22 \text{ cm}/1 \text{ cm})^2 = 484.$$

The required input force is then $F_I = F_O/\text{MA} = (2.1 \times 10^4 \text{ N})/484 = 43$ N.

ANALYZE To get this much mechanical advantage from a lever, with the load 1 m from the fulcrum, we'd need an arm a half kilometer long! ■

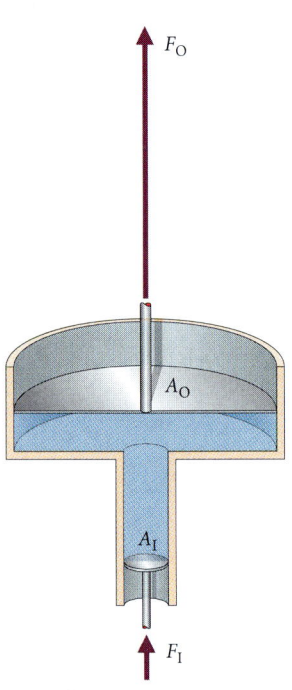

■ **FIGURE 13.7**
A hydraulic machine. Pressure applied to the fluid by the input piston results in a force on the output piston. The mechanical advantage of the machine is the ratio of the piston areas. Here a small input force F_I results in a large output force F_O.

13.2.3 What Causes Pressure?

Molecules in a fluid are always in motion; they collide with the walls of the container and with each other. Pressure results from momentum transfer in these collisions. We can illustrate this most clearly with a model of gases, because complications due to forces between individual molecules are unimportant for dilute gases. The simplest such model is called the *ideal gas*.

The *ideal gas* model rests on the following assumptions:

1. There are a very large number of molecules in any small volume.
   We demonstrated this for air in Example 13.2.
2. The molecules themselves occupy negligible volume.
   This seems reasonable, since the same number of molecules that make up 1 cm³ of water expand to form about 1800 cm³ of steam at atmospheric pressure. The molecules thus occupy less than $\frac{1}{1800}$ the volume of the steam.
3. The molecules are in continual, random motion.
   It is not obvious that the molecules are in motion, but it is this motion that causes pressure. Turning the argument around, pressure is experimental evidence for molecular motions!
4. Forces between molecules can be ignored, except during a collision.
5. All the collisions are elastic.
   The fourth and fifth assumptions are reasonable approximations for real gases in many circumstances.

WE'LL USE THIS MODEL EXTENSIVELY IN PART V.

THERE IS PLENTY OF OTHER EXPERIMENTAL EVIDENCE. WE DISCUSS SOME OF IT IN CHAPTER 19.

SEE CHAPTER 10 FOR A DISCUSSION OF ELASTIC COLLISIONS.

Let us use this model to calculate gas pressure on the walls of a small cubical box of side ℓ that contains N molecules of gas, each of mass m. For simplicity, we neglect the collisions of the molecules with each other. A molecule moving toward the left wall of the box with speed v_x (■ Figure 13.8) hits the wall, is reflected, and moves with speed v_x toward the right wall. The wall gives the molecule an impulse in the x-direction, $|\Delta \vec{p}| = |\vec{p}_f - \vec{p}_i| = mv_x - m(-v_x) = 2mv_x$, and receives an equal and opposite impulse. The molecule proceeds to the right wall in a time interval ℓ/v_x, is reflected, and returns to the left wall after a total elapsed time (since the first collision) of $2\ell/v_x$. The process then repeats. Momentum is transferred to the left wall at a rate:

BE CAREFUL TO DISTINGUISH $\vec{p}$, MEANING MOMENTUM, FROM P, MEANING PRESSURE.

IN A REAL GAS, MOLECULES SUFFER MANY COLLISIONS BEFORE THEY CAN CROSS A BOX OF FINITE SIZE. BUT BECAUSE THE COLLISIONS ARE RANDOM, ANY MOLECULE WITH INITIAL VELOCITY $\vec{v}$ IS REPLACED WITH ANOTHER THAT IS GIVEN THAT VELOCITY. OUR IDEALIZED CALCULATION MAKES NO SUBSTANTIAL ERROR.

$$\frac{|d\vec{p}|}{dt} = \frac{2mv_x}{2\ell/v_x} = \frac{mv_x^2}{\ell}.$$

Summing the contributions from all N molecules in the box, and recalling that force is the rate at which momentum is transferred (Newton's second law), we have:

$$F = \sum_N \frac{|d\vec{p}|}{dt} = \sum_N \frac{mv_x^2}{\ell} = Nm\langle v_x^2 \rangle/\ell,$$

REMEMBER: F IS THE MAGNITUDE OF THE FORCE.

where $\langle v_x^2 \rangle \equiv \sum v_x^2 / N$ is the average value of v_x^2 for all the molecules in the box. The pressure exerted by the gas on the wall is the force per unit area, F/A, where $A = \ell^2$ is the area of the wall:

$$P = \frac{F}{A} = \frac{(Nm/\ell)\langle v_x^2 \rangle}{\ell^2} = \frac{Nm}{\ell^3}\langle v_x^2 \rangle = \rho \langle v_x^2 \rangle,$$

where the density $Nm/\ell^3 = \rho$ is the total mass in the box divided by its volume. Since there is nothing special about the x-direction, similar calculations give $P = \rho \langle v_y^2 \rangle$ and $P = \rho \langle v_z^2 \rangle$ for the pressure on other sides of the box. Because the molecular motions are random (assumption 3), the average value of v_x^2 is the same as $\langle v_y^2 \rangle$ or $\langle v_z^2 \rangle$. This explains our earlier observation that pressure is isotropic. Since the total velocity squared is:

$$v^2 = v_x^2 + v_y^2 + v_z^2,$$

$\langle v_x^2 \rangle = \langle v^2 \rangle / 3$, and we have the following result:

> Pressure in an ideal gas: $P = \rho \langle v^2 \rangle / 3.$ (13.5)

The pressure is proportional to the density of the gas and to the mean square speed of the molecules.

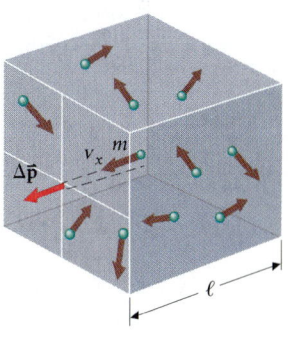

■ FIGURE 13.8
Molecules in a cubical box bounce back and forth off the walls. One molecule with speed v_x collides elastically with a wall and transfers momentum of magnitude $2mv_x$. Left to itself, the molecule would repeat the collision after a time $\Delta t = 2\ell/v_x$. Actually, random collisions among the molecules keep this particular molecule from returning but also replace it with another molecule moving the same way.

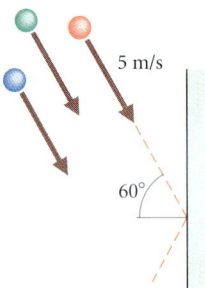

FIGURE 13.9
Hard balls collide elastically with a smooth wall. What is the pressure?

EXERCISE 13.3 ♦ A stream of ten balls per second, each of mass 200 g and moving at 5 m/s, impacts a smooth wall of area 1 m² at an angle of 60°, and is reflected (■ Figure 13.9). What is the pressure exerted on the wall by the balls?

EXERCISE 13.4 ❖ If the walls of the imaginary box we constructed above (Figure 13.8) are moving inwards, what happens to the speed of the molecule after it collides with the wall? After the box has been shrunk by a factor of 2 in volume, has the pressure inside increased, decreased, or stayed the same?

The ideal gas model illustrates how pressure is caused by molecular impacts and also explains how pressure of a compressible fluid increases if we put the same number of molecules in a smaller volume. Liquids behave very differently because the separation d of the molecules in a liquid is comparable to their size, and the molecules interact strongly. To get a visual image of how incompressible behavior might arise, we model the interactions by imagining the molecules in a liquid to be connected by springs with relaxed length about equal to d (■ Figure 13.10). Each molecule oscillates back and forth on the end of its spring. As a piston is lowered onto the liquid surface, the force is very small initially because very few of the oscillating molecules reach the piston. Those that do are moving slowly at the end of their oscillations and transfer very little momentum to the piston. When the piston is lowered an additional distance of the order of d, almost all the molecules within a distance d of the surface collide with it, and many are near the middle of their oscillation, where they move rapidly. Since d is of the order of 1 nm, a negligible change of volume produces an enormous change in pressure—the liquid is incompressible.

$$\text{Incompressible liquid:} \quad \rho = \text{constant, independent of } P. \tag{13.6}$$

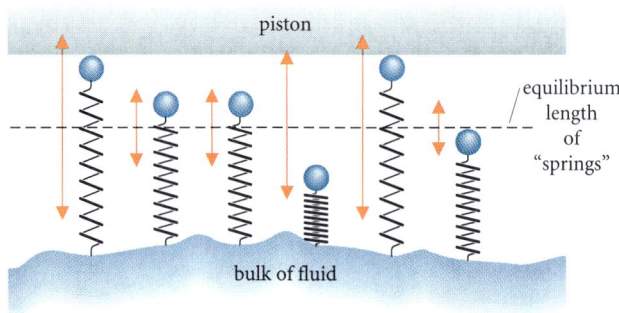

FIGURE 13.10
Rough model of liquid pressure. Beside each molecule is an arrow representing the range of that molecule's oscillation if the container wall were absent. In the low-pressure situation shown, very few molecules strike the piston. Moving the piston down a small distance would dramatically increase the rate of collisions and hence the pressure.

These models show how fluid pressure results from random motions of molecules. We have an explicit formula for the pressure of an ideal gas, but only a rough idea as to why liquids are difficult to compress. The ideal gas and incompressible liquid models illustrate the essential properties of fluids, and eqns. (13.5) and (13.6) are useful tools for understanding real fluids.

13.3 FLUIDS IN EQUILIBRIUM

13.3.1 Variation of Pressure in a Fluid at Rest

In a fluid at rest, each element of fluid is in equilibrium. On the Earth's surface, the forces in the fluid have to support the fluid against gravity. If the pressure were insufficient, fluid would start to fall, compress the bottom layers and thereby increase the pressure until the fall were stopped: we expect pressure to increase downwards in a fluid at rest. Since no external forces act horizontally, we would not expect pressure to vary in the horizontal direction. We may verify these conclusions using Newton's laws. For mathematical convenience, we work with a rectangular fluid element and choose the z-axis to be vertical.

ON AN ELEMENT *WITHIN* THE FLUID, THE ONLY EXTERNAL FORCE IS GRAVITY.

First we investigate how pressure varies in the x (horizontal) direction. The force on the vertical side of the element at position x due to fluid to the left is in the x-direction (■ Figure 13.11), and its magnitude equals the pressure times the area $dy\,dz$ of the face:

$$d\vec{F}(x) = \hat{i} P(x)\,dy\,dz.$$

The force on the opposite side at $x + dx$ is in the opposite direction:

$$d\vec{F}(x + dx) = (-\hat{i}) P(x + dx)\,dy\,dz.$$

The element is in equilibrium, so these forces balance:

$$\sum F_x = [P(x) - P(x + dx)]\,dy\,dz = 0.$$

Therefore $P(x) = P(x + dx)$. The pressure is the same on both sides of the element.

The same argument applies to the two faces at y and $y + dy$, since the total force in the y-direction is also zero. Any path between two arbitrary points A and B a finite distance apart but at the same level can be constructed from a sequence of differential elements (■ Figure 13.12). Since there is no pressure change across any of the elements, the total pressure change between A and B is a sum of zeros.

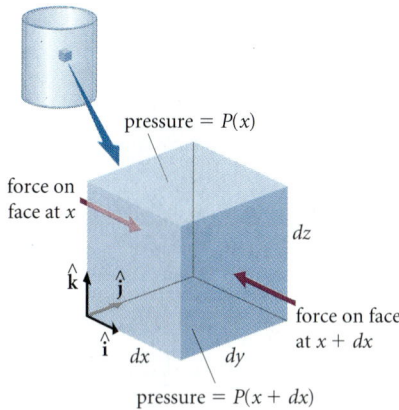

■ FIGURE 13.11
Horizontal forces on the sides of a rectangular fluid element are in balance.

> In a uniform fluid in equilibrium, the pressure has the same value at every point *with the same height.*

ALSO, SINCE PRESSURE IS ISOTROPIC, THE PRESSURE ACTING IN THE x-DIRECTION EQUALS THAT IN THE y-DIRECTION.

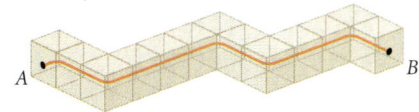

Pressure does vary in the vertical direction. The vertical forces exerted on a differential element of fluid are its weight and the forces due to pressure on its upper and lower faces (■ Figure 13.13). The mass of the element equals its volume multiplied by the density of the fluid, so its weight is:

$$d\vec{W} = dm\vec{g} = \rho\,dV\vec{g} = \rho(dx\,dy\,dz)g(-\hat{k}).$$

The forces on top and bottom are:

$$d\vec{F}(z + dz) = P(z + dz)\,dx\,dy(-\hat{k}) \quad \text{and} \quad d\vec{F}(z) = P(z)\,dx\,dy\,\hat{k}.$$

In equilibrium, these forces balance, so:

$$\sum F_z = P(z)\,dx\,dy - P(z + dz)\,dx\,dy - \rho g\,dx\,dy\,dz = 0.$$

Dividing by $dx\,dy\,dz$, we have:

$$\frac{P(z) - P(z + dz)}{dz} - \rho g = 0.$$

■ FIGURE 13.12
An arbitrary path between two points at the same height may be composed of rectangular elements. The pressure is the same on the sides of each rectangle and hence is the same at A and B.

REMEMBER: $\lim\limits_{dz \to 0} \dfrac{P(z + dz) - P(z)}{dz} \equiv \dfrac{dP}{dz}$

Rearranging gives:

$$\frac{P(z + dz) - P(z)}{dz} = -\rho g.$$

On the left-hand side, we may express the difference of pressures as dP to obtain the equation of *hydrostatic equilibrium*.

> In equilibrium, the pressure in a static fluid decreases with height at a rate given by the fluid density multiplied by the acceleration due to gravity:
>
> $$\frac{dP}{dz} = -\rho g. \qquad (13.7)$$

The pressure increases with depth (decreasing z) to support the weight of fluid layers above.

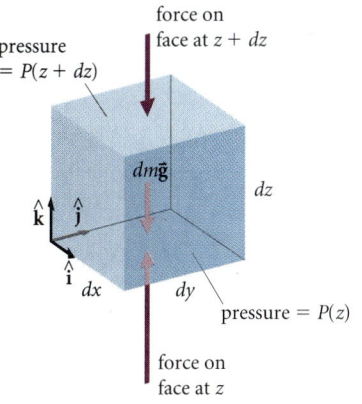

■ FIGURE 13.13
The vertical forces on the rectangular element are due to pressure and gravity.

SECTION 13.3 • FLUIDS IN EQUILIBRIUM

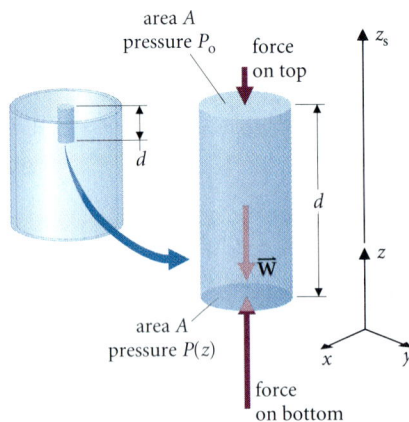

FIGURE 13.14
Forces on a finite cylinder of fluid are also in balance.

EXERCISE 13.5 ♦ A vertical cylinder of incompressible fluid has cross section A and extends from height z to the top of the fluid at z_s, where the pressure is P_o (■ Figure 13.14). Use the fact that the total force acting on the cylinder is zero, as we did in the derivation of eqn. (13.7), to find the pressure $P(z)$ at height z.

13.3.2 Hydrostatic Equilibrium

The law of hydrostatic equilibrium (eqn. 13.7) is used in problems as diverse as the design of dams and the computation of stellar structure, where the goal is to find the value of the pressure as a function of position. To use eqn. (13.7) in such problems, we need to know how density depends on pressure. In the simplest case, an incompressible fluid near the Earth's surface, both the density of the liquid ρ and the acceleration due to gravity $\vec{g}$ are constant. Then the derivative of pressure with height is a constant, and we may integrate eqn. (13.7) directly:

$$P_2 - P_1 = \int_{z_1}^{z_2} \frac{dP}{dz}\, dz = \int_{z_1}^{z_2} (-\rho g)\, dz = -\rho g(z_2 - z_1).$$

We usually know the value of the pressure at some position z_1 in the fluid. Coordinate z_2 can represent any point in the fluid, so we replace it with the variable coordinate z:

$$P = P_1 + \rho g(z_1 - z). \tag{13.8}$$

For example, when the top of the fluid is open to the atmosphere, as in Exercise 13.5, we set $P_1 = P_o$ (atmospheric pressure) at the free surface, $z_1 = z_s$. It is frequently convenient to measure distances downwards from the free surface so that if d is the depth, $d \equiv z_s - z$:

$$P = P_o + \rho g d. \tag{13.9}$$

When a piston exerts a force F on the fluid surface of area A, the pressure at the surface is $P_s = F/A$ (■ Figure 13.15). This value replaces atmospheric pressure in the previous solution:

$$P = P_s + \rho g d. \tag{13.10}$$

In these examples the pressure at the surface is transmitted throughout the fluid. Equation 13.8 expresses the general result, known as Pascal's principle, that is the basis for the operation of hydraulic machinery.

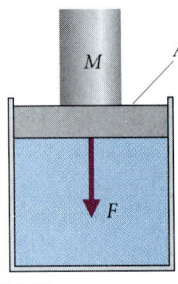

FIGURE 13.15
Pressure at the top of a body of fluid may differ from atmospheric pressure. Here the piston has area A and exerts a force of magnitude $F = P_o A + Mg$ on the fluid. The pressure at the top is $P_s = F/A = P_o + Mg/A$.

THIS IS A MORE CAREFUL STATEMENT THAN WE GAVE IN §13.2. THE TWO ARE CONSISTENT WHEN $P \gg \rho g d$.

Pascal's principle: In a uniform fluid in equilibrium, pressure increases linearly with depth. Pressure applied at any point is transmitted throughout the fluid.

EXAMPLE 13.5 ♦ A bathysphere (deep sphere) is a small spherical vessel used to transport scientists safely to the ocean floor. What pressure must the vessel withstand at the bottom of the Marianas trench, a depth of 10 915 m?

MODEL We may model ocean water as a constant-density fluid with a surface free to the atmosphere.

SETUP AND SOLVE We use eqn. (13.9):

$$P = P_o + \rho g d = 1.0 \times 10^5 \text{ Pa} + (1.03 \times 10^3 \text{ kg/m}^3)(9.8 \text{ N/kg})(1.1 \times 10^4 \text{ m})$$
$$= 1.1 \times 10^8 \text{ Pa} = 1100 \text{ atm}.$$

ANALYZE The air pressure at the top of the ocean is a small contribution to the pressure at the bottom. Similarly, the comfortable atmospheric pressure inside the bathysphere is little help to the engineer designing it.

Digging Deeper

THE HYDROSTATIC PARADOX

■ Figure 13.16 shows three flasks, each with the same area at the bottom and each containing the same height of liquid. According to eqn. (13.9), each flask has the same force acting on its bottom surface even though the three contain very different amounts of fluid. How can that be?

In flask A the forces due to fluid pressure on the side of the flask act horizontally outward, and the paired forces the flask exerts on the fluid are also horizontal; the sides make no contribution to supporting the weight of the fluid. In flask B the normal vectors to the side walls have vertical components, and so the forces exerted on the liquid have upward components, sufficient to support the fluid above the walls. The bottom only supports the column of fluid directly above it. In flask C, the forces exerted by the glass have a downward component. The bottom has to support the weight of fluid *and* balance the forces exerted by the walls of the flask on the fluid—a surprising result!

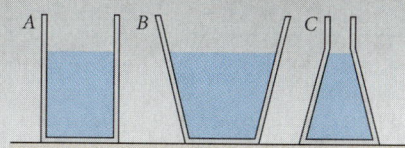

■ **FIGURE 13.16**
The hydrostatic paradox. The fluid pressure is the same on the bottoms of all three containers!

EXERCISE 13.6 ❖ How does pressure vary with depth in an astronaut's coffee cup on the space shuttle in orbit? ■

13.3.3 The Barometer

Torricelli's barometer (Figure 13.1) consists of a tube of liquid inverted in an open dish of the same liquid. When the tube is inverted, the liquid falls, leaving a vacuum above it and a column of liquid in the tube supported by the pressure at its bottom. Pressure everywhere in the fluid is determined by the pressure at a free surface and the fluid density. Thus, for a given fluid, the height of the column is determined by atmospheric pressure.

We choose the origin to be at the free surface in the dish and z to be directly upward. The pressure at the free surface is atmospheric, $P(0) = P_o$. The pressure in the fluid has the same value everywhere at $z_1 = 0$, including in the tube. We may apply eqn. (13.8) to the fluid in the tube, with $P = 0$ (vacuum) at $z = H$:

$$P = P_1 + \rho g(z_1 - z) \;\Rightarrow\; 0 = P_o + \rho g(0 - H).$$

Thus: $\quad P_o = \rho g H.$

If the liquid used were water with density 10^3 kg/m^3, H would be about

$$(10^5 \text{ N/m}^2)/[(10^3 \text{ kg/m}^3)(10 \text{ m/s}^2)] = 10 \text{ m}!$$

But using a dense liquid such as mercury, the column becomes a manageable length, about 0.8 m. Mercury barometers based on Torricelli's concept are in such frequent use that pressure is often measured in centimeters or inches of mercury.

EXERCISE 13.7 ◆ Compute the value of the pascal in centimeters of mercury (cm Hg). ■

13.3.4 The Atmosphere

So far, we have taken atmospheric pressure as a known quantity, but the behavior of the atmosphere is itself a hydrostatic problem with a *compressible* fluid. The ideal gas model applies quite well to air: atmospheric pressure is proportional to density and to the mean square

IN FACT, AIR TEMPERATURE DECREASES BY ABOUT 60°C IN THE FIRST 10 KM OF ALTITUDE, AND $\langle v^2 \rangle$ DECREASES BY ABOUT 20%. THIS CALCULATION, ALTHOUGH APPROXIMATE, SERVES AS A USEFUL INTRODUCTION.

RECALL §13.2.3. THE RELATION TO TEMPERATURE IS DISCUSSED IN §19.2.

RECALL THAT $(d/dx)(\ln x) = 1/x$, AND APPLY THE CHAIN RULE TO THE FUNCTION $f(P) = \ln(P)$.

RECALL THAT $\ln A - \ln B = \ln(A/B)$.

$e = 2.718\ldots, e^{-1} = 0.37, e^{-2} = 0.14, e^{-3} = 0.05$, ETC.

speed of the molecules. Air temperature, which we assume to be constant, determines the mean square speed. Then density and pressure are directly proportional:

$$\rho/\rho_o = P/P_o,$$

where P_o and ρ_o are the pressure and density of air at the surface of the Earth, which we take to be at $z = 0$. We also neglect variations in g; we will evaluate the validity of this assumption later. Equation (13.7) then becomes:

$$\frac{dP}{dz} = -\rho g = -\frac{\rho_o g}{P_o} P,$$

which may be rewritten:

$$-\frac{\rho_o g}{P_o} = \frac{1}{P}\left(\frac{dP}{dz}\right) = \frac{d}{dz}[\ln(P)].$$

In this case, the derivative of the *logarithm* of pressure is constant. Now we integrate:

$$\int_0^z \frac{d}{dz}[\ln(P)]\, dz = \int_0^z -\frac{\rho_o g}{P_o}\, dz,$$

$$\ln[P(z)] - \ln[P(0)] = -\frac{\rho_o g}{P_o} z,$$

and, since $P(0) = P_o$,

$$\ln(P/P_o) = -\frac{\rho_o g}{P_o} z.$$

The most convenient way to write the result is:

$$P = P_o \exp(-z/h_o), \quad \text{where } h_o \equiv P_o/(\rho_o g). \tag{13.11}$$

The length h_o is called the *scale height* of the atmosphere.

■ **EXERCISE 13.8** ♦ Show that h_o has dimensions of length.

The exponential function decreases rapidly when its argument gets less than -1. This means that atmospheric pressure decreases rapidly when the altitude is greater than one scale height (■ Figure 13.17). With typical values of $\rho_o = 1$ kg/m³, $P_o = 10^5$ Pa, and $g = 10$ m/s², **the scale height is about 10^4 m (30 000 ft). Thus changes of air pressure are unimportant on the scale of laboratory experiments or even buildings. But without a pressurized cabin, air would be too thin for people to breathe in a commercial airliner flying at 11 000 m. The aircraft's altitude is measured by an altimeter, which compares outside air pressure against the standard atmosphere** (■ Figure 13.18).

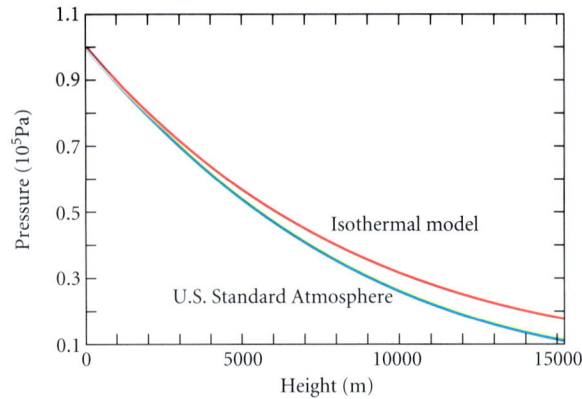

■ **FIGURE 13.17**
Pressure versus altitude in the Earth's atmosphere. The red curve represents the approximate calculation based on constant temperature. The green curve represents the U.S. Standard Atmosphere, a more precise model based on the average conditions in the atmosphere.

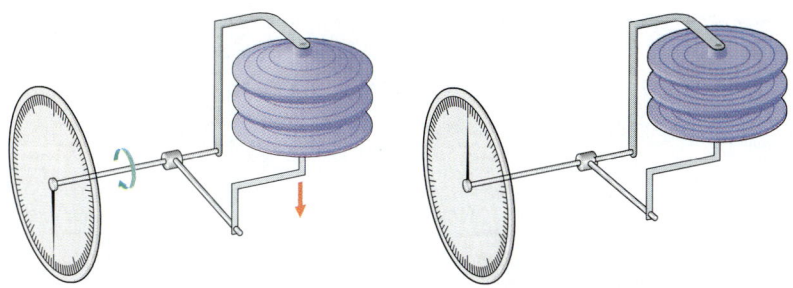

FIGURE 13.18
An aneroid barometer consists of an evacuated metal chamber, called a wafer. Atmospheric pressure compresses the wafer, which contracts when pressure increases and expands when pressure decreases. A mechanical linkage turns a dial on the instrument face as the wafer expands or contracts. One important application of this barometer is in aircraft altimeters. The device senses atmospheric pressure outside the aircraft. The dial is calibrated against the standard atmosphere calculation (Figure 13.17) and displays the aircraft's altitude.

To finish the discussion, we need to check our assumption of constant g. In §5.4 we showed that g changes by 1% in a height of about 30 km. Since the scale height in the atmosphere is only 10 km, assuming g to be constant gives an error that is negligible when compared with the assumption of constant temperature.

13.4 Archimedes' Principle
13.4.1 Buoyant Force

Boats, propelled by oars, wind, or water currents, were one of the first efficient modes of transportation. A ship floats because water exerts forces on it.

> The net upward force exerted by a fluid on an object is called *buoyant force*.

A thought experiment (■ Figure 13.19) shows us how to find the buoyant force exerted on a ship. Imagine removing the ship without allowing any water to move and then filling the resulting hole with water brought in from outside—shown green in the figure.

> The water necessary to fill the imaginary cavity left by the ship is called the water *displaced* by the ship.

In the imaginary situation, the displaced water is in equilibrium since it is part of a uniform body of fluid. The forces acting on it due to pressure exactly balance its weight. But the value of the pressure depends only on depth beneath the water surface, and the direction of the force depends only on the orientation of the surface on which it acts. The force exerted on the displaced water is determined by the shape of the cavity it fills. The same force acts on the ship. This is the idea that caused Archimedes to startle the good citizens of Syracuse!

WE DISCUSSED THOUGHT EXPERIMENTS IN §0.3.

SEE THE CHAPTER INTRODUCTION.

> *Archimedes' principle:* The buoyant force exerted by a fluid on an object would exactly balance the weight of the fluid displaced by the object.
>
> $$\vec{F}_{buoyant} = -\vec{g}(\rho_{fluid} V_{displaced}) \qquad (13.12)$$

THE BODY OF DISPLACED WATER IS ALSO IN TORQUE BALANCE. SO, THE BUOYANT FORCE ACTS THROUGH THE WATER'S CM. SEE §§9.3 AND 11.1, AND EXAMPLE 13.8.

EXERCISE 13.9 ♦ Verify Archimedes' principle for a submerged cube by finding the net force exerted on its surfaces by the surrounding fluid.

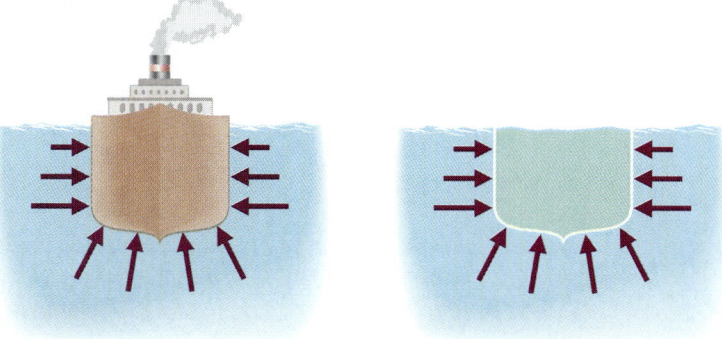

FIGURE 13.19
Archimedes' principle. The fluid forces exerted on the imaginary volume of water are the same as those actually exerted on the ship, since pressure depends only on depth beneath the surface. In the imaginary situation, the total force just balances the weight of the displaced water. Thus the actual buoyant force on the ship would just balance the weight of water displaced by the ship.

Digging Deeper

THE SHIP IN DRY DOCK

How much water is necessary to float a ship? The dry dock shown in ■ Figure 13.20 has dimensions only slightly larger than those of a large ship and contains a small amount of water. Yet the ship floats. The water in the dry dock rises to the water line of the ship, so the dock need only contain enough water to fill the blue volume in the figure. But Archimedes' principle says the ship displaces an equal weight of water!

The tricky word here is "displace." It conveys a common sense of "push out of position," but the correct meaning of displaced fluid in this context is "the amount of fluid that *would be required* to fill the submerged part of an object's volume." The ship in dry dock is "displacing" the green volume in the figure. The water that would fill this volume weighs as much as the ship. But the only water needed in the dry dock is that of the blue volume.

■ **FIGURE 13.20**
The ship's displacement is greater than the volume of water in the dock. *Displacement*, the volume shown in green, means the water that would be required to fill the submerged volume of the ship.

EXAMPLE 13.6 ♦♦ What fraction of an iceberg floats above water?

MODEL The floating iceberg displaces a weight of water equal to its own weight. If the ice had the same density as the water it floats in, the equal weight of water would also be an equal volume; the iceberg would have to be totally submerged to float! Actually, ice has a smaller density than seawater for two reasons: water is one of the rare substances that expand when they freeze, and seawater contains salt, which makes it denser than pure water. The two densities are $\rho_{\text{ice}} = 0.92 \times 10^3$ kg/m³ and $\rho_{\text{seawater}} = 1.03 \times 10^3$ kg/m³ (Table 13.1).

SETUP The volume of displaced water is equal to the volume V_1 of ice below the water surface (■ Figure 13.21). The total volume of the iceberg is $V = V_1 + V_2$. Applying Archimedes' principle:

Weight of iceberg = weight of displaced water.

$$\rho_{\text{ice}} V g = \rho_{\text{seawater}} V_1 g.$$

■ **FIGURE 13.21**
An iceberg floats in seawater. The water line divides the volume of the iceberg into two parts: $V = V_1 + V_2$, where V_1 is the submerged volume.

SOLVE So, the fraction of the iceberg below the surface is:

$$f_{below} \equiv \frac{V_1}{V} = \frac{\rho_{ice}}{\rho_{seawater}} = \frac{0.92 \times 10^3 \text{ kg/m}^3}{1.03 \times 10^3 \text{ kg/m}^3} = 0.89.$$

Then the fraction of the iceberg above water is $1 - f_{below} = 0.11$.

ANALYZE About 11% of the iceberg floats above water.

This example offers another way to think about Archimedes' principle. Objects float if their density is less than the density of water. The important quantity is the average density—a steel ship can float because it is full of air; a lump of solid steel sinks. ∎

EXAMPLE 13.7 ♦♦ Archimedes weighs the king's crown in air and again when it is immersed in water (■ Figure 13.22). If he finds the values $F_{air} = 27.44$ N and $F_{water} = 21.36$ N for the balancing forces, is the crown of pure gold?

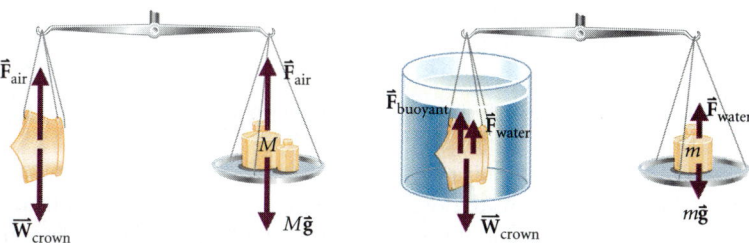

■ **FIGURE 13.22**
Archimedes' riddle. When the crown is submerged in water, the balancing mass is smaller because the buoyant force helps to support the crown's weight.

MODEL When the crown is immersed, part of its weight is supported by the buoyant force of the water so that less of it is supported by the scales. The difference between F_{air} and F_{water} is the magnitude of the buoyant force. According to Archimedes, the buoyant force also equals the weight of water displaced by the crown (eqn. 13.12). Since the crown is completely submerged, the volume of water displaced equals the volume of the crown. Knowing both the mass and volume of the crown, we can compute its density, and, comparing this with the known density of gold, we can determine whether it is genuine.

SETUP From the free-body diagram for the crown weighed in air, we have:

$$\sum F_y = F_{air} - W_{crown} = 0 \Rightarrow F_{air} = W_{crown}. \quad \text{(i)}$$

Similarly, when the crown is weighed while submerged, we have:

$$\sum F_y = F_{buoyant} + F_{water} - W_{crown} = 0 \Rightarrow F_{buoyant} = W_{crown} - F_{water}. \quad \text{(ii)}$$

From Archimedes' principle:

$$F_{buoyant} = \rho_{water} V_{crown} g. \quad \text{(iii)}$$

Using eqn. (13.1), the weight of the crown is:

$$W_{crown} = M_{crown} g = \rho_{crown} V_{crown} g. \quad \text{(iv)}$$

SOLVE Combining eqns. (i) and (ii):

$$F_{buoyant} = W_{crown} - F_{water} = F_{air} - F_{water}.$$

Then using eqns. (iii) and (iv):

$$\frac{\rho_{crown} V_{crown} g}{\rho_{water} V_{crown} g} = \frac{W_{crown}}{F_{buoyant}} = \frac{F_{air}}{F_{air} - F_{water}}.$$

Thus: $\rho_{crown} = \dfrac{\rho_{water} F_{air}}{F_{air} - F_{water}}$

$$= \frac{(1.00 \times 10^3 \text{ kg/m}^3)(27.44 \text{ N})}{27.44 \text{ N} - 21.36 \text{ N}} = 4.51 \times 10^3 \text{ kg/m}^3.$$

THE DENSITY OF WATER IS IN TABLE 13.1.

ANALYZE Since the density of gold is 19.3×10^3 kg/m³, the goldsmith is in big trouble! ∎

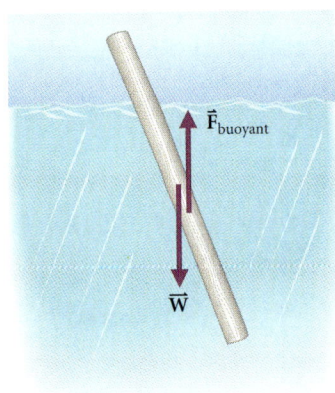

FIGURE 13.23
Why does a log float horizontally? The log's weight acts at its center of mass. The buoyant force, equal to the weight of water displaced, acts at the CM of the displaced water. The weight and buoyant force form a couple that rotates the almost vertical log to a horizontal position.

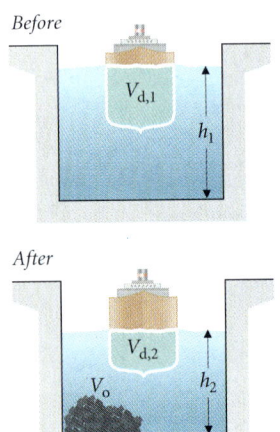

FIGURE 13.24
A ship accidentally unloads its dense cargo into a canal lock. As a result, the water level falls.

WE USED EQN. (13.1).

EXAMPLE 13.8 ❖ Why does a log float horizontally?

MODEL According to Archimedes' principle alone, the log could just as well float vertically, but the equilibrium is not stable. Consider the torques acting on a log floating not quite vertically (■ Figure 13.23). The weight of the log acts at its CM, a distance $L/2$ from the bottom. Since the buoyant force = weight of fluid displaced, it acts at the CM of the *submerged* portion of the log. The resulting torque tends to rotate the log farther, and it rises to a horizontal position. ∎

Study Problem 10 ❖❖❖ An Accident in the Lock

A cargo ship carrying iron ore floats in a canal lock. An accident in unloading the ship causes the ore to be dumped into the water (■ Figure 13.24). Does the water level in the lock rise or fall? If the density of iron ore is 5.0×10^3 kg/m^3, there are 1.0×10^5 kg of ore in the cargo, and the surface area of the lock is 1.0×10^3 m^2, what is the change in water level?

Modeling the System

This classic problem invites a leap to a faulty conclusion. It's tempting to say that the water rises because the iron ore is dumped into it. Let's see!

Since we are asked to compute a change in water level, we need to compare two states of the system. Before the accident, the ore is in the ship and the water depth is h_1. After the accident, the ore is at the bottom of the lock and the water depth is h_2. We want to find the difference $h_1 - h_2$.

In either state, the volume below the water level, $V = Ah$, is the sum of three parts: (1) the water itself, (2) the submerged volume of the ship, and (3) the volume of any ore on the bottom.

1. The volume of the water itself remains constant.
2. The cargo ship remains afloat during the incident, and Archimedes' principle applies to it throughout. Unloading the ship reduces its weight, so it requires less buoyant force to float. It rises and displaces less water. The submerged volume of the ship decreases.
3. When the ore is in the ship, it contributes to volume 2. So, volume 3 is zero until the ore falls out. Thus volume 3 increases.

Plan

Because we want to compare *before* and *after* states, the plan for this problem is like that for a conservation law problem. We'll calculate the total volume below the water line and compare the values before and after the ore is dumped.

Construction of Equations

1. Water volume: V_w is constant.
2. We find the displaced volume of the floating ship from Archimedes' principle. Let m_s and m_o be the masses of ship and ore, respectively.

Weight of displaced water $= V_d \rho_w g =$ weight of ship and cargo.

$$\text{BEFORE} \quad V_{d,1} = \frac{(m_s + m_o)g}{\rho_w g}. \qquad \text{AFTER} \quad V_{d,2} = \frac{m_s g}{\rho_w g}.$$

3. The volume of ore in the water:

$$\text{BEFORE} \quad V_{o,1} = 0. \qquad \text{AFTER} \quad V_{o,2} = m_o/\rho_o.$$

Total volume of ship, ore, and water:

$$\text{BEFORE} \quad Ah_1 = V_{d,1} + V_{o,1} + V_w. \qquad \text{AFTER} \quad Ah_2 = V_{d,2} + V_{o,2} + V_w.$$

Solution of Equations

Upon subtracting the two total volumes, the fixed volume of water cancels out.

$$A(h_1 - h_2) = V_{d,1} - V_{d,2} + V_{o,1} - V_{o,2}$$
$$= \frac{m_s + m_o}{\rho_w} - \frac{m_s}{\rho_w} + 0 - \frac{m_o}{\rho_o}.$$

So the difference in water levels is:

$$h_1 - h_2 = \left(\frac{m_o}{A}\right)\left(\frac{1}{\rho_w} - \frac{1}{\rho_o}\right)$$
$$= \left(\frac{1.0 \times 10^5 \text{ kg}}{1.0 \times 10^3 \text{ m}^2}\right)\left(\frac{1}{1.0 \times 10^3 \text{ kg/m}^3} - \frac{1}{5.0 \times 10^3 \text{ kg/m}^3}\right)$$
$$= 0.080 \text{ m} = 8.0 \text{ cm}.$$

The water level drops by 8.0 cm.

Analysis

The water level drops ($h_1 > h_2$) if the cargo sinks ($\rho_o > \rho_w$). Whether in the ship or out, the ore causes displacement of water. When the ore is inside the ship, the displacement has to be large enough to support the *weight* of the ore. Outside, the displacement equals the *volume* of the ore. Since the ore is denser than water, an amount of water with the same weight as the ore has a larger volume. Thus there is less displacement when the ore is on the bottom.

EXERCISE 13.10 ❖ Is the preceding solution valid for a cargo of wood? If not, what is the correct result for the change in water level?

13.5 Fluid Dynamics

Moving fluids exhibit some of the most interesting and complex phenomena in nature (■ Figure 13.25). Internal forces cause the swirls in this fluid to prepare the way for their own turbulent breakup. The study of such beautiful, chaotic motion is at the frontier of research, so we shall investigate simpler and more regular kinds of flow. Even in the simplest cases, internal forces are important. Water spurts from the side of a leaky bucket because of the interplay between gravity and pressure.

To describe a moving fluid completely, we need to know its velocity at each point as well as its pressure and density. This requires three relations among these variables. One, the relation between pressure and density, comes from our models for an incompressible liquid or an ideal gas. We obtain the other two relations from conservation of mass and energy.

13.5.1 Streamlines

The flow of air over the wing in ■ Figure 13.26 is highly organized spatially and is constant in time; it is *steady*. It is also *irrotational*: it does not show any of the swirling motion so pronounced in Figure 13.25. We shall restrict our study to flow that is both steady and irrotational.

> In a *steady* flow the measurable quantities of pressure, density, and velocity *at a given place* do not change with time.

These quantities do vary from place to place, and the velocity is not zero. Steady is not the same as static.

> *Irrotational* means that there are no swirling motions in the fluid.

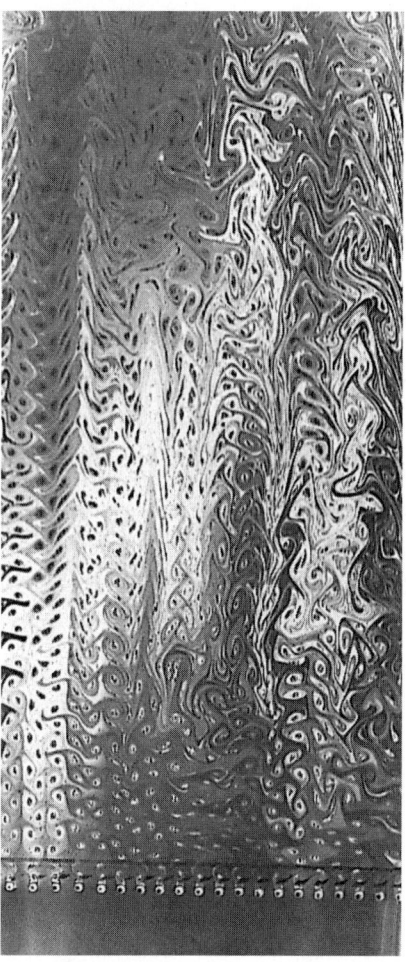

■ **Figure 13.25**
Vortices form in a soap solution as it flows past the teeth of a comb. Downstream, the vortices interact, and the flow becomes turbulent. Such complex flows, though beautiful, are difficult to understand. They are usually studied in more advanced courses.

Contrast this result with the behavior of rigid bodies. In Chapter 12 we could ignore the internal forces.

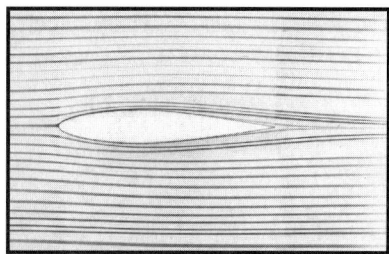

■ **Figure 13.26**
Water flows past an airfoil, from left to right. The flow is highly organized and steady. The streamlines are shown by colored fluid introduced upstream.

SEE §2.1.3.

In Figure 13.26 colored fluid makes the flow visible. Each colored particle is carried along with the fluid and traces out its path. These paths are called *streamlines*. Their technical definition follows from the fact that a fluid *particle's* velocity vector is tangent to its path.

| The velocity vector of the fluid at any point is tangent to the *streamline* at that point.

A bundle of streamlines passing through an element of area perpendicular to the fluid velocity (■ Figure 13.27) form a *stream tube*. The walls of the stream tube are imaginary and are defined by the motion of the fluid, but fluid remains inside the tube just as surely as if the walls were solid. We could imagine the entire flow as made of many pipes, each the size and shape of a stream tube. We shall obtain the laws of fluid motion in a convenient form by applying conservation laws to the fluid in a stream tube.

CONSERVATION OF MASS SEEMS INTUITIVELY OBVIOUS, AS MATTER NEITHER APPEARS NOR DISAPPEARS IN EVERYDAY EXPERIENCE. BUT WE MUST BE MORE CAREFUL WHEN DEALING WITH SUBATOMIC PARTICLE INTERACTIONS. SEE §34.4.2.

13.5.2 The Equation of Continuity

In steady flow, the amount of mass in any given volume doesn't change. When fluid flows steadily along the stream tube in ■ Figure 13.28, the same amount of fluid that enters the tube at Q in any time interval dt leaves at R during the same time interval. The distance traveled by the fluid at Q in time dt is $d\ell = v_Q\,dt$, so all the fluid within distance $d\ell$ of point Q crosses the surface of area A_Q at Q in time dt: the mass flowing across this surface in time dt is $dm = \rho_Q A_Q v_Q\,dt$. Similarly, the equal mass flowing out at R is $dm = \rho_R A_R v_R\,dt$. The mass flow rate at Q and R, or any other point along the stream tube, is:

$$dm/dt = \rho A v = \text{constant.} \qquad (13.13)$$

In steady flow, the product of fluid density, fluid speed, and stream tube area is constant along any stream tube.

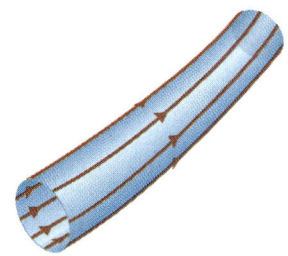

■ FIGURE 13.27
A bundle of streamlines defines a stream tube.

Equation (13.13) expresses the fact that fluid mass is neither created nor destroyed during its flow: it is known as the *equation of continuity*.

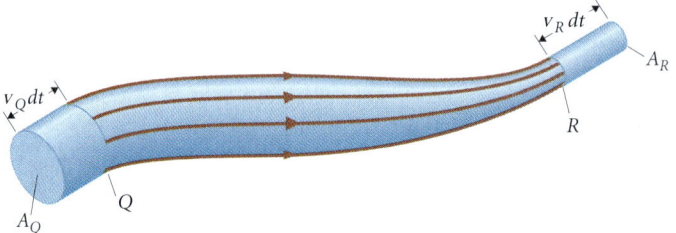

■ FIGURE 13.28
The same amount of material leaves the stream tube at R in time dt as enters at Q. Where the tube is wider, the flow is slower.

EXAMPLE 13.9 ♦ A pipe of circular cross section slowly widens from a diameter $d_1 = 10$ cm to a diameter $d_2 = 20$ cm. If water flows into the pipe at 1 m/s, what is the speed of the water at the exit?

MODEL The product of density, speed, and cross-sectional area is the same at both ends of the pipe. Water is an incompressible fluid; its density is the same throughout and so the density may be combined with the constant in eqn. (13.13).

REMEMBER: THE AREA A IS PERPENDICULAR TO THE FLUID VELOCITY.

SETUP Equation (13.13) becomes:

$$A_1 v_1 = \frac{\pi d_1^2 v_1}{4} = \frac{\pi d_2^2 v_2}{4} = A_2 v_2.$$

SOLVE
$$v_2 = \left(\frac{d_1}{d_2}\right)^2 v_1 = \left(\frac{10\text{ cm}}{20\text{ cm}}\right)^2 (1\text{ m/s}) = \tfrac{1}{4}\text{ m/s}.$$

ANALYZE In order to transport the same amount of fluid through a larger area, the flow has to slow down.

13.5.3 Bernoulli's Law

Next we shall apply the work-energy theorem to an incompressible fluid element of length $d\ell$ and mass $dm = \rho A\, d\ell$ moving along a stream tube (■ Figure 13.29). Work done on the element as it moves from O to Q changes its kinetic energy. The pressure, speed, and cross-sectional area all vary along the stream tube.

The forces acting on the element are due to pressure and gravity (Figure 13.29). To compute work, we need the force components parallel to the stream tube. We choose to call a component positive if the component vector is parallel to $\vec{v}$, so the work done is $W = \vec{F} \cdot d\vec{\ell} = F_\parallel\, d\ell$. Then the component of the gravitational force along the stream tube is $-(dm)g\cos\theta$, so:

$$dmg_\parallel = -\rho A\, d\ell\, g\cos\theta = -\rho A g\, dz.$$

WHENEVER THE SPEED VARIES, THE AREA MUST ALSO VARY, ACCORDING TO EQN. (13.13).

Forces due to pressure act on both ends and on the curved sides of the element. At O, the force is parallel to the stream tube:

$$F_{O,\parallel} = PA,$$

and similarly at Q,

$$\begin{aligned}F_{Q,\parallel} &= -(P + dP)(A + dA)\\ &= -PA - P\, dA - A\, dP.\end{aligned}$$

AS USUAL, WE HAVE IGNORED THE SECOND-ORDER TERM $dA\, dP$.

The forces acting perpendicular to the curved sides of area $2\pi r\, d\ell$ also have a component parallel to the displacement that exactly balances the $P\, dA$ term in $F_{Q,\parallel}$ (■ Figure 13.30):

$$\begin{aligned}\sum F_\parallel &= F_{O,\parallel} + F_{Q,\parallel} + F_{\text{sides},\parallel} + dmg_\parallel\\ &= PA - PA - P\, dA - A\, dP + P\, dA - \rho A g\, dz\\ &= -A\, dP - \rho A g\, dz.\end{aligned}$$

The change in the element's speed corresponds to a change in its kinetic energy:

$$\Delta K = \tfrac{1}{2}dm(v + dv)^2 - \tfrac{1}{2}dmv^2 = \tfrac{1}{2}dm\,2v\,dv = (\rho A\, d\ell)v\, dv.$$

WE IGNORE THE TERM $\tfrac{1}{2}dm\,(dv)^2$ SINCE IT IS MUCH SMALLER THAN THE OTHER TERMS.

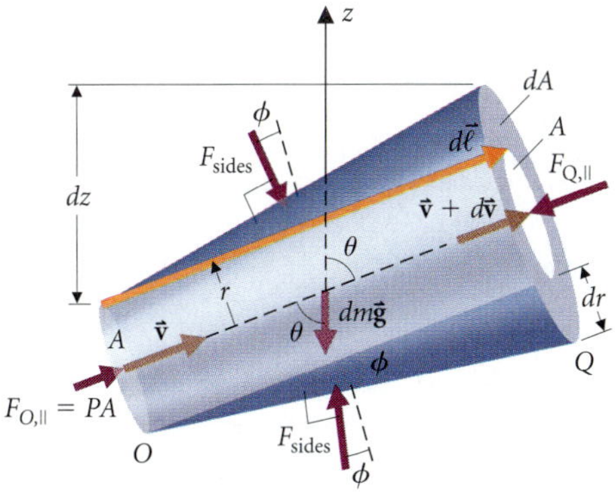

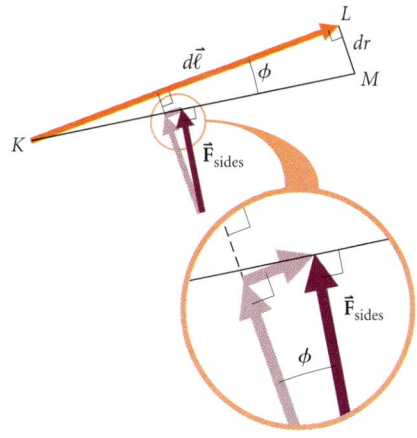

■ **Figure 13.29**
Bernoulli's law is derived by applying the work-energy theorem to a fluid element of mass $dm = \rho A\, d\ell$. The forces that do work are due to gravity and pressure on each surface of the element. Because the area of the tube changes, the sides flare with angle ϕ. Since $dr \ll d\ell$, $\phi \ll 1$.

■ **Figure 13.30**
As the radius of the stream tube changes by dr, the tube flares with angle ϕ, where $\tan\phi = dr/d\ell$ and ϕ is small. The change in area is $dA = d(\pi r^2)$. (See Example 12.14.) So, $dA = 2\pi r\, dr = 2\pi r(\tan\phi)\, d\ell \approx 2\pi r\phi\, d\ell$. The force on the curved sides of the stream tube (which is perpendicular to KM in the diagram) has a component parallel to $d\ell$: $F_{\text{sides}}\sin\phi \approx P(2\pi r\, d\ell)\phi = P\, dA$.

BERNOULLI'S LAW MAY ALSO BE DERIVED USING CONSERVATION OF MOMENTUM.

The work-energy theorem relates the work done by the net force to the change in kinetic energy:

$$\Delta K = W = \sum F_\parallel \, d\ell$$

$$(\rho A \, d\ell) v \, dv = -A \, dP \, d\ell - \rho A g \, dz \, d\ell.$$

Canceling a factor of $A \, d\ell$, we have:

$$-dP - \rho g \, dz = \rho v \, dv = \tfrac{1}{2} \rho \, dv^2.$$

Rearranging this expression, with g = constant and ρ = constant, we find:

$$d(P + \rho g z + \rho v^2/2) = 0.$$

IN THIS FORM, BERNOULLI'S LAW APPLIES ONLY TO INCOMPRESSIBLE FLUIDS. NOTE THAT THE CONSTANT MAY HAVE DIFFERENT VALUES ON DIFFERENT STREAMLINES.

Integrating, we get *Bernoulli's law*:

$$P + \rho g z + \rho v^2/2 = \text{constant along a streamline.} \quad (13.14)$$

The last two terms in eqn. (13.14) represent the gravitational potential energy and kinetic energy per unit volume of the fluid. The inclusion of the pressure term in this energy equation accounts for the work done by internal forces.

EXAMPLE 13.10 ♦♦ The pressure in the pipes of a city's water distribution system is 6.55×10^5 Pa, and the water main, 1.05 m below the street, has a diameter of 15.2 cm. A firefighter hooks up a hose with a 2.54-cm-diameter nozzle to a fire hydrant and stands on a ladder with the hose horizontal (■ Figure 13.31). If the firefighter needs a flow speed of at least 2.0 m/s, how high a ladder can she use? (Ignore any change in pressure in the water main due to the water flow.)

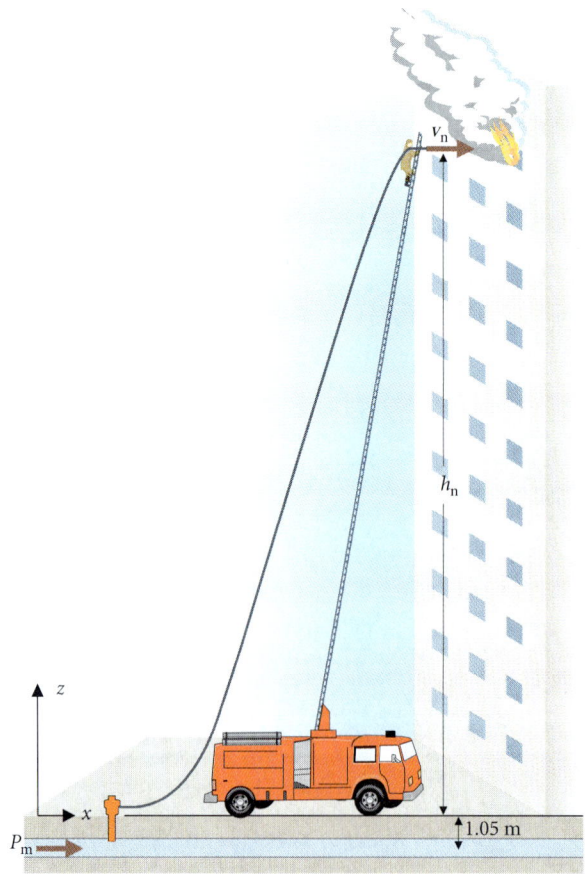

■ FIGURE 13.31
The firefighter needs a flow speed v_n of at least 2.0 m/s. How high may she climb on the ladder?

MODEL We assume steady flow and ignore transient effects that occur when the flow is turned on. We look at a streamline that runs through the center of the water main and the fire hose, and through the center of the stream of water after it leaves the hose. As the water leaves the hose, it is at atmospheric pressure.

SETUP The equation of continuity (13.13) and Bernoulli's law (13.14) are the fundamental relations we need. The area of the pipe is $A = \pi(d/2)^2$. Let "m" refer to the water main and "n" to the nozzle. Then $P_n = P_o$. We measure heights from street level.

Eqn. (13.13):
$$\pi(d_m/2)^2 v_m = \pi(d_n/2)^2 v_n. \quad \text{(i)}$$

Eqn. (13.14):
$$P_m + \tfrac{1}{2}\rho v_m^2 + \rho g h_m = P_n + \tfrac{1}{2}\rho v_n^2 + \rho g h_n. \quad \text{(ii)}$$

SOLVE From eqn. (i), $v_m = v_n(d_n/d_m)^2$. We solve eqn. (ii) for h_n and substitute for v_m in terms of v_n:

$$h_n = h_m + \frac{(P_m - P_o)}{\rho g} + \frac{v_n^2}{2g}\left[\left(\frac{d_n}{d_m}\right)^4 - 1\right].$$

The ratio of diameters, $d_n/d_m = (2.54\text{ cm})/(15.2\text{ cm}) = 0.167$. So, $(d_n/d_m)^4 = 7.8 \times 10^{-4}$ and is negligible compared with 1.00. Then:

$$h_n = -1.05\text{ m} + \frac{6.55 \times 10^5\text{ Pa} - 1.03 \times 10^5\text{ Pa}}{(1.00 \times 10^3\text{ kg/m}^3)(9.80\text{ m/s}^2)} - \frac{(2.0\text{ m/s})^2}{2(9.8\text{ m/s}^2)}$$

$$= -1.05\text{ m} + 56.3\text{ m} - 0.20\text{ m} = 55.0\text{ m}.$$

ANALYZE Check the units: $1\text{ Pa} = 1\text{ N/m}^2 = 1\text{ kg·m}/(s^2 \cdot m^2) = 1\text{ kg}/(m \cdot s^2)$. Thus

$$\frac{1\text{ Pa}}{(1\text{ kg/m}^3)(1\text{ m/s}^2)} = \frac{1\text{ kg}}{1\text{ m·s}^2} \cdot \frac{1\text{ m}^2\cdot s^2}{1\text{ kg}} = 1\text{ m},$$

as required. Since a building story averages about 5 m in height, the firefighter can train her hose onto an 11-story building. This calculation overestimates the height by neglecting effects such as viscosity and turbulence that dissipate energy. ∎

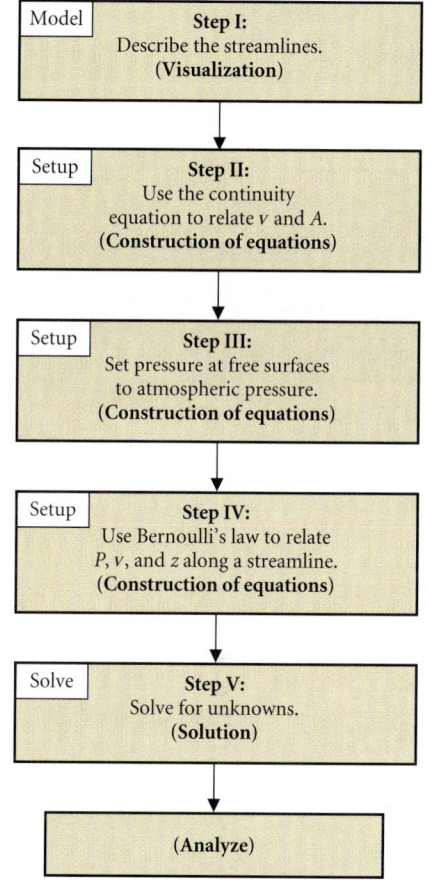

Solution plan for probems in steady, incompressible flow.

Step I: Describe the streamlines. (Visualization)

Step II: Use the continuity equation to relate v and A. (Construction of equations)

Step III: Set pressure at free surfaces to atmospheric pressure. (Construction of equations)

Step IV: Use Bernoulli's law to relate P, v, and z along a streamline. (Construction of equations)

Step V: Solve for unknowns. (Solution)

(Analyze)

■ **FIGURE 13.32**

13.5.4 Solving Problems in Fluid Dynamics

The continuity equation and Bernoulli's law allow us to solve some interesting problems in steady, incompressible fluid flow. A plan for solving such problems is outlined in ■ Figure 13.32. If the fluid is confined to a pipe, it is easy to see where the streamlines go. For unconfined flow, such as the flow of air over a wing, we must use other information to deduce the form of the streamlines. Then we use the continuity equation to relate fluid speed to the area of the stream tube. Bernoulli's law then determines how pressure varies along streamlines. The pressure satisfies two constraints:

1. The pressure of an ideal fluid can never be negative, and
2. A free surface (open to the atmosphere) is at atmospheric pressure.

EXAMPLE 13.11 ♦♦ Siphons were used in the middle ages, and it was known that a siphon could lift water over a height of no more than about 10 m. Explain this limit, and find the speed of water leaving the siphon shown in ■ Figure 13.33.

MODEL The solution plan follows the outline in Figure 13.32.

STEP I The streamlines go from the free surface of liquid in the big container (labeled O in the figure), to the bottom of the siphon tube (Q), then along the tube through the top (R) to the end (S). The pressure decreases along the stream tube to push the fluid up against gravity; $P(R) < P(Q)$. To start the siphon, the tube must be filled. One way to do this is to pump on the open end. Once started, the fluid continues to flow without further pumping.

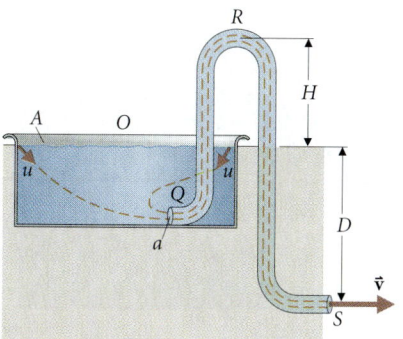

■ **FIGURE 13.33**
A siphon. Once the flow is started, it continues so long as the end S is below the water surface O, and the height $D + H$ is less than 10 m.

SETUP
STEP II The stream tube's area at point O is A, the area of the big container. At points Q, R, and S its area equals a, the cross-sectional area of the pipe. From the continuity equation, the fluid speed is the same at Q, R, and S. Let this speed be v, and the speed at O be u. Applying the continuity equation at O and Q:

$$u = (a/A)v.$$

In most siphons, a is much less than A, so u is much less than v (cf. Example 13.10).

STEP III The pressure at points O and S is atmospheric ($P = P_o$) because these places are open to the air.

STEP IV To see why the 10-m height limit exists, we apply Bernoulli's equation to points R and S of the streamlines shown in the figure. Measuring height with respect to the level of the liquid in the container:

$$P_R + \rho g H + \tfrac{1}{2}\rho v^2 = P_o + \rho g(-D) + \tfrac{1}{2}\rho v^2.$$

SOLVE This simplifies to:
$$P_R = P_o - \rho g(D + H).$$

Since P_R must be greater than zero,
$$D + H < P_o/(\rho g).$$

For water, $\rho = 10^3$ kg/m³, so:

$$D + H < (10^5 \text{ Pa})/[(10^3 \text{ kg/m}^3)(10 \text{ m/s}^2)] = 10 \text{ m},$$

or about 30 ft. To find the speed in the tube, we apply Bernoulli's law at points O and S of the streamline.

$$P_o + \rho g(0) + \tfrac{1}{2}\rho u^2 = P_o + \rho g(-D) + \tfrac{1}{2}\rho v^2.$$

Since $u \ll v$, we neglect the term in u^2. Then we have:

$$v = \sqrt{2gD}.$$

THIS RESULT IS CALLED TORRICELLI'S LAW.

ANALYZE This result is independent of the position of end Q of the tube, or of the height of R, provided that flow through the tube remains steady. To make a siphon flow faster, we should increase the vertical distance D between the end of the pipe and the water surface in the container. The end of the siphon at S must be below that surface ($D > 0$), otherwise v would be the square root of a negative number, which is impossible. Physically, we have to convert gravitational energy at O to kinetic energy at S, and this can only be done if O is above S. Since D must be greater than zero, $H < D + H < 10$ m is the maximum height over which water can be lifted. ■

13.6 Incompressible Flow of Air

In this section we shall investigate airflows. Air is a gas and thus is compressible. However, under some circumstances, it behaves as an incompressible fluid, and we can apply our two laws of incompressible fluid flow to air.

13.6.1 Airflow Over a Wing

Airflow over an airplane wing may be treated as though it were incompressible so long as the speed of the air over the wing is small compared with the speed of sound in air. This is true for light aircraft but is not a very good approximation for jets at high altitude.

As a tilted wing moves, it pushes downward on the air, which exerts an upward force on the wing according to Newton's third law. This is the lift force that supports the airplane. In the reference frame moving with the airplane, the airflow over the wing is steady (■ Figure 13.34). From the pilot's perspective, the air is approaching the wing at speed v. A long way from the wing, the streamlines are equally spaced horizontal lines. In Figure 13.34, the stream-

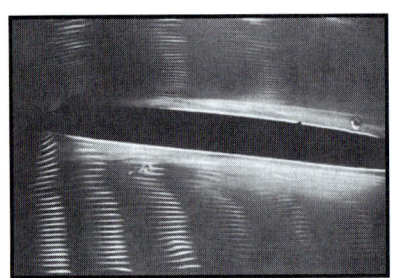

■ **FIGURE 13.34**
Airflow over a wing. The streamlines are illuminated during intervals of length 0.2 s. The greater spacing between pulses on the top of the wing indicates the higher speed there.

lines that pass over the wing are closer together than those that pass under. The area of a stream tube on top of the wing is less than the area of a stream tube below the wing. From the continuity equation, the speed of the air is greater over the top of the wing than under the wing. In the figure, the greater speed is illustrated by the greater separation of the tracer bubble pulses that mark the flow.

From Bernoulli's equation, the difference in speed is related to a difference in pressure; the pressure on top of the wing P_t is less than the pressure on the bottom P_b. The pressure difference causes a net upward force $= (P_b - P_t)A$, where A is the area of the wing. This is the lift produced by the wing. Its magnitude depends on the difference in flow speeds between the top and bottom of the wing, which in turn depends on the difference in size of the stream tubes, and that depends on the degree to which the wing is tipped up as well as on the wing's shape. The amount of tip is measured by the *angle of attack* α, which is the angle between the wing chord and the undisturbed airflow (■ Figure 13.35). Bernoulli's equation shows that the pressure difference depends on the density of the air times speed squared, so we obtain the relationship for the lift L:

$$L = \tfrac{1}{2} C_L \rho v^2 A, \qquad (13.15)$$

where C_L, the coefficient of lift, depends on the exact shape of the wing and is proportional to the angle of attack α. Lift increases as airspeed increases and as the angle of attack increases.

The theory discussed above breaks down if the angle of attack becomes too large because the air is no longer able to flow smoothly over the wing and becomes turbulent (Figure 13.35). Under these conditions, Bernoulli's equation is not applicable and the lift is reduced: the wing has *stalled*.

> **EXERCISE 13.11** ❖ Close to the surface of an object, air is forced to move with the object because of a kind of surface friction. Use this fact, and Bernoulli's equation, to explain why a spinning baseball (■ Figure 13.36) follows a curved path.

13.6.2 Ram Pressure

When a fluid is brought to rest abruptly, it exerts a force on the object that stops it. You can feel the force if you put your hand in a stream of water or out of a moving car's window. The force arises from ram pressure. When fluid flows around a blunt object, the streamlines pass around the object. The streamline at the exact center stops at the surface of the object. Air

WE MAY DESCRIBE THE GENERATION OF LIFT EITHER BY THE PRESSURE ON THE WING OR BY THE WING PUSHING THE AIR DOWN AND THE AIR PUSHING THE WING UP. THESE ARE NOT TWO DIFFERENT EFFECTS, BUT TWO ALTERNATIVE DESCRIPTIONS OF ONE EFFECT.

THE WING CHORD IS A STRAIGHT LINE FROM THE LEADING EDGE OF THE WING TO THE TRAILING EDGE. SEE FIGURE 13.35.

EXACT CALCULATION OF LIFT IS DIFFICULT, SO C_L IS USUALLY MEASURED EXPERIMENTALLY.

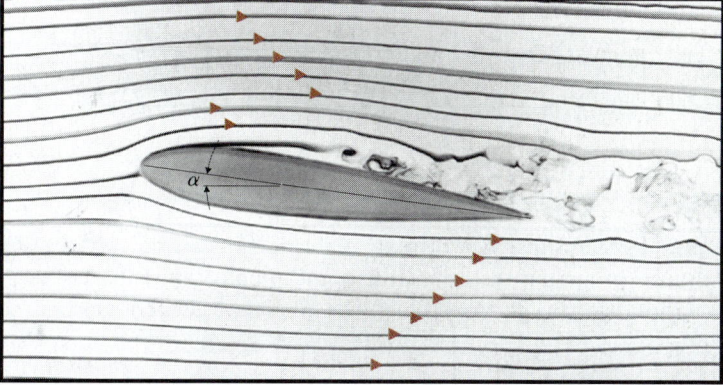

■ FIGURE 13.35
The angle of attack α is the angle between the wing chord and the undisturbed airflow (horizontal). If the angle of attack is too large, the streamlines no longer follow the shape of the wing, and turbulence results: the wing has stalled.

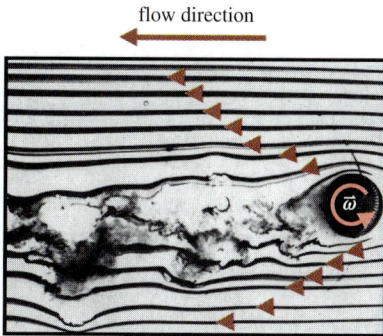

■ FIGURE 13.36
A spinning baseball also produces lift. The flow near the surface is dragged along, increasing the speed on one side and decreasing it on the other. Here the flow speed is about 25 m/s and the ball rotates at 630 rpm.

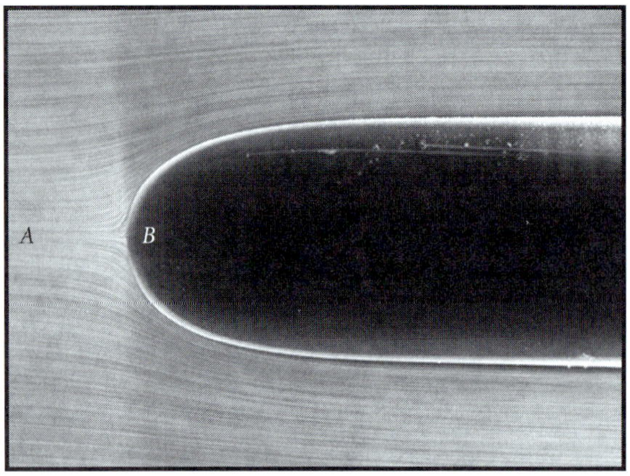

■ **FIGURE 13.37**
Flow past a blunt object. The central streamline stops at the surface of the object as the fluid is brought to rest.

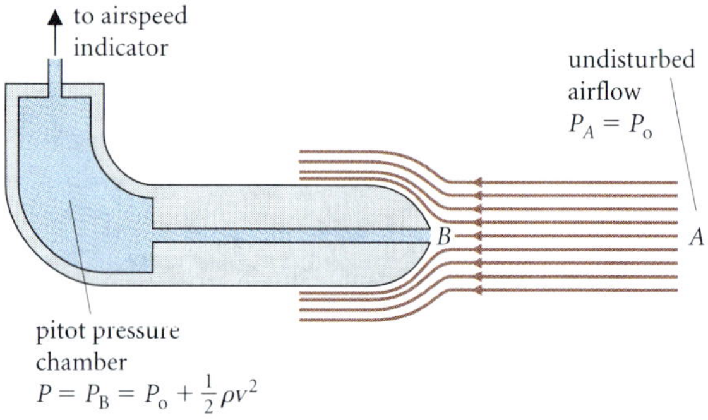

■ **FIGURE 13.38**
A Pitot tube measures the ram pressure at point B, where the fluid is brought to rest.

pitot pressure chamber
$P = P_B = P_o + \frac{1}{2}\rho v^2$

undisturbed airflow
$P_A = P_o$

flowing close to this streamline is almost brought to rest before it flows away along the surface of the object. Applying Bernoulli's law to the streamline AB in ■ Figure 13.37:

$$P_B = P_A + \tfrac{1}{2}\rho v^2. \tag{13.16}$$

The pressure at B is greater than at A, the difference being the *ram pressure* $\frac{1}{2}\rho v^2$.

A Pitot tube (■ Figure 13.38) uses ram pressure to measure an airplane's speed. Air flows around the tube just as for the blunt object. The instrument compares the pressure at B with the pressure of the undisturbed air P_o at the same altitude.

> **EXAMPLE 13.12** ♦ If the pressure in the Pitot tube is 1.01×10^5 Pa, the atmospheric pressure at the same altitude is 9.55×10^4 Pa, and the air density is 1.2 kg/m³, what is the speed of the airplane?
>
> **MODEL** We assume the airflow is incompressible.
>
> **SETUP** We apply eqn. (13.16): $\rho v^2/2 = P_B - P_A \Rightarrow v^2 = 2(P_B - P_A)/\rho$.
>
> **SOLVE**
> $$v^2 = 2\,\frac{(10.1 - 9.55) \times 10^4 \text{ Pa}}{1.2 \text{ kg/m}^3} = 9.17 \times 10^3 \text{ (m/s)}^2.$$
>
> $v = 96$ m/s.
>
> **ANALYZE** This is the speed (about 200 mph) of a light twin-engined aircraft. ■

Chapter Summary

Where Are We Now?

We have extended the range of systems that we can study to include fluids: liquids and gases. Unlike rigid bodies, fluids have properties that vary from point to point, and internal forces are important to their behavior. We have developed methods for computing those forces and studying their effects.

What Did We Do?

An important difference between a fluid and a solid is the ability of the fluid to change its shape; an ideal fluid can only exert or withstand normal forces. The basic properties of fluids are density (mass per unit volume) and pressure (force per unit area). The force exerted by an ideal fluid is perpendicular to the surface on which it acts. Ideal fluid pressure is isotropic; the magnitude of the force on an element of area does not depend on the orientation of the area.

Pressure is caused by collisions of fluid molecules. Ideal gases are *compressible*. The pressure exerted by a gas is proportional to its density. In an ideal liquid, both the rate and impulse of molecular collisions increase very rapidly as the liquid's volume is decreased. Liquids are approximately *incompressible*, with constant density.

Forces within a fluid support it against gravity, and so the pressure must increase with depth d in a fluid at rest.

$$P = P_s + \rho g d \tag{13.10}$$

Points at the same height in a uniform fluid are at the same pressure. A force applied to a confined fluid body increases the pressure by the same amount throughout the fluid.

The Earth's atmosphere is a compressible fluid; both pressure and density decrease with height. A rough approximation with constant air temperature (proportional to the mean square speed of the molecules) shows that pressure decreases exponentially with a scale height of about 10 km.

The total force due to pressure on an object floating or submerged in a fluid is called the buoyant force. It acts in the opposite direction from gravity, and, according to *Archimedes' principle*, its magnitude equals the weight of fluid the object displaces. An object floats in a fluid if its average density is less than the density of the fluid.

> CHECK THE DEFINITION OF "DISPLACED" IN §13.4.1. THIS IS A COMMON SOURCE OF CONFUSION.

In steady flow, the pressure, density, and velocity of the fluid at each point are constant in time. The fluid flows along streamlines. The fluid velocity is tangent to the streamline at each point. A bundle of streamlines defines a stream tube. Continuity (conservation of mass) requires that the product of density, fluid speed, and stream tube area be constant:

$$\rho v A = \text{constant along a streamline.} \tag{13.13}$$

The work-energy theorem applied to incompressible fluid gives Bernoulli's law:

$$P + \tfrac{1}{2}\rho v^2 + \rho g h = \text{constant along a streamline.} \tag{13.14}$$

Figure 13.32 shows a plan for solving problems in fluid dynamics.

Practical Applications

An understanding of fluid forces is essential in the design of ships, airplanes, hydraulic machinery, and dams as well as a host of other technological devices such as the brake system in your car. A static model of the Earth's atmosphere is the starting point for studies of daily weather and long-term climatic trends. The equation of hydrostatic equilibrium is used in modeling stars and planets. Fluid flow in pipes and around aircraft in subsonic flight are adequately represented as steady, incompressible flow.

Solutions to Exercises

13.1 If we take the top of a typical human head to have a radius of about 6 cm, the downward force due to atmospheric pressure is:

$$F = PA = (10^5 \text{ N/m}^2)\pi(0.06 \text{ m})^2 = 10^3 \text{ N},$$

equal to the weight of 220 lb. The force is not large enough to crush a person but would be exceedingly uncomfortable if you had to make a conscious effort to withstand it. Fluid pressure inside your body presses outward and balances inward pressure on your skin. One of the primary functions of a space suit is to protect an astronaut from the ill effects of unbalanced internal pressure.

13.2 The compressed air inside the jar exerts a greater pressure than the air outside the jar. The net force on the glass is outward, and is balanced by rigid-body forces in the glass. As the glass breaks, the rigid-body forces cease to act between the fragments, which are accelerated *outward* by the air pressure. The TV tube is evacuated (pressure inside ≈ 0), so the net force is due to the air outside the tube

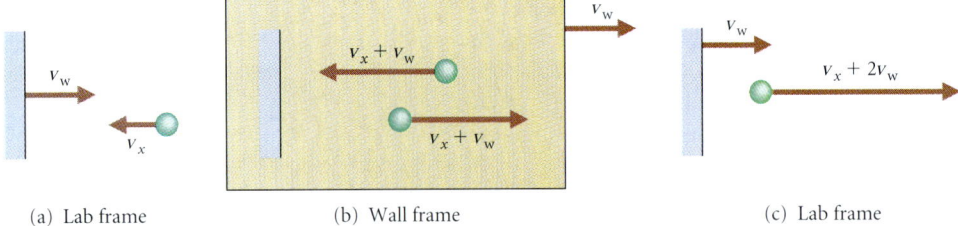

Figure 13.39
Collision of a molecule with a moving wall. (a) and (c) Lab frame. (b) Wall frame.

pushing inward on the glass. When the tube is smashed, the glass is accelerated inward. (Beware! Some fragments usually manage to pass through the wreckage and speed hazardously out the other side.)

13.3 When a ball collides with the smooth wall (no friction), the component of velocity parallel to the wall is not changed. Each ball transfers momentum

$$\Delta p = 2(0.2 \text{ kg})(5 \text{ m/s})\cos 60° = 1 \text{ kg·m/s}$$

to the wall, and ten balls hit a square meter of wall per second. The rate of momentum transfer per unit area is the pressure:

$$P = \frac{1}{A}\left(\frac{\Delta p}{\Delta t}\right) = \frac{(10 \text{ /s})(1 \text{ kg·m/s})}{1 \text{ m}^2} = 10 \text{ Pa}.$$

13.4 In an elastic collision with a wall, the molecule reverses its velocity *with respect to the wall*. If the wall is moving inward with speed v_w, a molecule with initial speed v_x in the lab frame has an x-component of velocity $-(v_x + v_w)$ with respect to the wall (■ Figure 13.39). After the collision, its velocity with respect to the wall has an x-component of $(v_x + v_w)$. In the lab frame, the component is $(v_x + 2v_w)$. Shrinking the volume of the box increases the pressure for two reasons: the molecules are moving faster and they are more densely packed.

13.5 The only forces acting on the cylinder in the vertical direction are its weight and the forces due to pressure on the bottom and the top. In equilibrium, these forces exactly balance:

$$\vec{F}_{bottom} + \vec{F}_{top} + \vec{W} = 0.$$
$$P(z)A - P_oA - (\rho Ad)g = 0.$$

Thus: $P(z) = P_o + \rho g d.$

This result, known as Pascal's principle, is also obtained in §13.3.2, eqn. (13.9).

13.6 Since the gravitational force on the coffee in the astronaut's cup is just sufficient to accelerate the fluid in the common orbit of the whole shuttle, none of the fluid's weight needs to be balanced by fluid forces. Pressure does *not* vary with depth in the astronaut's cup.

13.7 The pressure corresponding to 1 cm Hg is:

$$1 \text{ cm Hg} = \rho_{Hg}g(1 \text{ cm})$$
$$= (13.6 \times 10^3 \text{ kg/m}^3)(9.807 \text{ N/kg})(10^{-2} \text{ m})$$
$$= 1.334 \times 10^3 \text{ N/m}^2.$$

So $1 \text{ Pa} \equiv 1 \text{ N/m}^2 = 7.50 \times 10^{-4} \text{ cm Hg}.$

13.8
$$[h_o] = \frac{[P_o]}{[\rho_o][g]} = \frac{[\text{force/area}]}{[\text{mass/volume}][\text{length/time}^2]}$$
$$= \frac{(ML/T^2)(1/L^2)}{(M/L^3)(L/T^2)} = L, \text{ as required.}$$

13.9 (■ Figure 13.40) Let the side of the cube have length ℓ. The forces on the sides of the cube sum to zero, since at each depth the forces on a thin slice of the cube of height dz are equal and opposite. The pressure on the bottom is $P_b = P_o + \rho g d$, where d is the depth

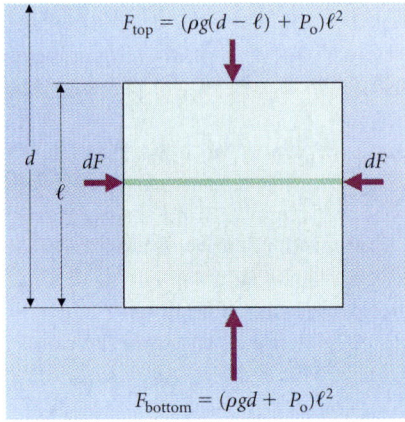

Figure 13.40
We may verify Archimedes' principle for a cube immersed in liquid.

of the bottom of the cube and ρ is the density of the fluid. The pressure on the top is $P_t = P_o + \rho g(d - \ell)$. The net upward fluid force exerted on the cube is:

$$\ell^2(P_b - P_t) = \ell^2[\rho g d - \rho g(d - \ell)] = \rho g \ell^3.$$

Since the submerged volume is ℓ^3:

$$F_{up} = V_{submerged}\rho g = M_{fluid \text{ displaced}} g,$$

and Archimedes' principle is verified.

13.10 No! Wood floats, so the same displacement is required to support the cargo whether it is in or out of the ship. In this case, the level doesn't change at all.

13.11 The airspeed is faster where the velocity of the baseball surface is parallel to the velocity of the air past it—that is, on the top in Figure 13.36. The situation is very similar to the airflow over a wing, with pressure less where speed is greater. There is a net fluid force on the baseball that causes its path to curve upward (■ Figure 13.41).

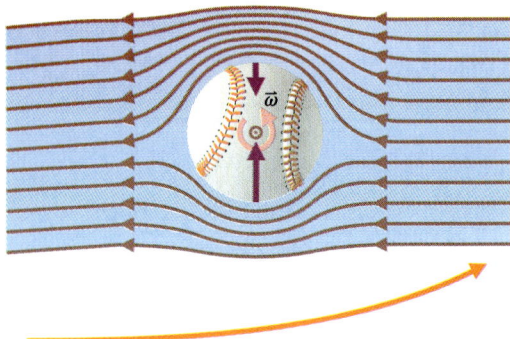

Figure 13.41
Forces due to pressure on the rotating baseball cause its path to curve. Compare this figure with Figure 13.36.

Basic Skills

Review Questions

§13.1 WHAT IS A FLUID?

- What is a *shear force*? What is an *ideal fluid*? Describe the direction of the force exerted by an ideal fluid on a surface.
- Contrast *incompressible* and *compressible* behavior. How do these types of behavior correspond to the properties of liquids and gases?

§13.2 BASIC PROPERTIES OF FLUIDS

- Define the *density* of a uniform substance. What is the *specific gravity* of a substance?
- How does the definition of density at a point differ from the definition of average density?
- Define the *pressure* of a fluid. What is the SI unit of pressure? Describe the *atmosphere* as a unit of pressure.
- What does it mean that the pressure of an ideal fluid is isotropic?
- Give an approximation for the behavior of pressure in a high-pressure hydraulic system. What is the mechanical advantage of a hydraulic system?
- What are the five assumptions that underlie the ideal gas model?
- Describe how pressure arises in the ideal gas model.
- Describe how a molecular model of a liquid explains *incompressible* behavior.

§13.3 FLUIDS IN EQUILIBRIUM

- Describe how the pressure in a fluid at rest in a gravitational field varies with horizontal position. How does the pressure vary with height?
- What is a *barometer*? What is the cm Hg as a unit of pressure?
- Describe how pressure varies with height in the Earth's atmosphere.

§13.4 ARCHIMEDES' PRINCIPLE

- Define *buoyant force*. How does it arise?
- What does the phrase "water displaced by a ship" mean?
- State Archimedes' principle.

§13.5 FLUID DYNAMICS

- What does *steady flow* mean? What is *irrotational* flow? What is a *streamline*?
- State the equation of continuity. What conservation law does it express?
- State Bernoulli's law for an incompressible fluid. What conservation law does it express?
- What two constraints must the pressure of an ideal fluid satisfy in steady flow?
- Describe the steps in the solution plan for steady-state fluid flow problems.
- What is a siphon? Explain why there is a maximum height over which a siphon can lift water.

§13.6 INCOMPRESSIBLE FLOW OF AIR

- Under what condition can we treat air as flowing incompressibly?
- Describe how Bernoulli's law relates the lift force on a wing to the difference between flow speeds over the top and bottom of the wing.
- What is *ram pressure*? Describe how it is used to measure the speed of an aircraft.

Basic Skill Drill

§13.1 WHAT IS A FLUID?

1. A horizontal rigid beam is supported at its two ends. Consider the forces that support a piece of the beam, say its middle half: are they shear or tension forces? How does your conclusion explain why a body of ideal fluid requires support at every point on its lower surface?

§13.2 BASIC PROPERTIES OF FLUIDS

2. A cubical block of wood 0.508 m on a side has a mass of 109 kg. What is its density? What is its specific gravity?

3. If the pressure in a fluid is 3.7×10^6 Pa, what is its pressure in atmospheres? What magnitude force does the fluid exert on a surface element of area 1.7×10^2 m²?

4. If an ideal fluid exerts a force of magnitude 7.9 N on a metal sheet of area 1.7×10^{-4} m² oriented parallel to the x–y plane, what magnitude force would the fluid exert on the sheet if it were rotated into the y–z plane?

5. A hydraulic press has two pistons, one of area 0.10 m² and one of area 1.0 m². If a load of 110 kg is to be raised, on which piston should it be placed? How much force must be exerted on the other piston to raise the load?

6. If a gas has density 1 kg/m³ and mean square speed $\langle v^2 \rangle = 10^6$ m²/s², what pressure does it exert on the walls of its container?

§13.3 FLUIDS IN EQUILIBRIUM

7. Determine the pressure at the bottom of a lake that is 20 m deep.

8. If today's atmospheric pressure is quoted as 30.0 inches of mercury, find the pressure in **(a)** pascals and **(b)** atmospheres.

§13.4 ARCHIMEDES' PRINCIPLE

9. What volume of water is displaced by a 750-metric-ton motor cruiser floating next to a dock?

10. A cubical metal box has side $L = 10.0$ cm and mass $m = 150$ g. How much mass must be placed inside the box if it is to float in water with its top at the surface?

§13.5 FLUID DYNAMICS

11. Water flows at speed 3.25 m/s into a device that divides it into three equal amounts flowing in different pipes. If the cross section of the pipe entering the device is 1.62×10^{-3} m² and the cross section of each of the exit pipes is 3.00×10^{-4} m², what is the water's speed in the exit pipes?

12. A pressure gauge on a natural gas pipeline reads gauge pressure, the difference between pressure in the pipeline and that in the surrounding atmosphere. Assuming the gauge functions properly, can it ever give a negative reading? What is the minimum possible reading of the gauge? Explain your reasoning.

13. On a day with an atmospheric pressure of 0.95×10^5 Pa, what is the maximum height over which a siphon can lift gasoline?

§13.6 INCOMPRESSIBLE FLOW OF AIR

14. In level flight, air flows over an airplane wing such that the speed of the air is 106 mph over the top of the wing and 88 mph over the bottom of the wing. Use Bernoulli's law to find the pressure difference between the top and bottom of the wing. Neglect any height difference. If the wing area is 170 ft², compute the lift force and determine the mass of the airplane.

Questions and Problems

§13.1 WHAT IS A FLUID?

15. ❖ Earthquakes cause rocks deep below the Earth's surface to vibrate rapidly and transmit shear forces. These same rocks flow in great circulating currents that cause continents to move over millions of years. What can you conclude about the solidness or fluidness of these rocks?

16. ❖ A large cardboard box is full of ping-pong balls. In what ways are the balls like an ideal fluid? In what ways are they nonideal? Which do they model more closely, a liquid or a gas? Why?

17. ❖ Can an ideal fluid stick to a vertical surface, even in small drops? Is water a completely ideal liquid?

18. ❖ If you apply a torque to a bucket that contains an ideal liquid, does the liquid begin to rotate? Why or why not? Explain why an uncooked egg is harder to set spinning than a hard-boiled egg. (Try it!)

§13.2 BASIC PROPERTIES OF FLUIDS

19. ❖ How do suction cups work?

20. ♦ A rubber ball of radius 5.0 cm has a mass of 620 g. What is the density of the rubber?

21. ♦ What is the mass of a brick measuring 10 cm × 5 cm × 2 cm if the density of brick is 2000 kg/m³?

22. ♦ Use the astronomical data on the inside front cover to find the density of Jupiter.

23. ♦ A neutron star has a mass equal to that of the Sun and a radius of 10 km. Find its density. Compare your result with the density of an iron nucleus (Table 13.1).

24. ♦ The air inside the space shuttle is at a pressure of 1 atm. What is the force per unit area on the walls of the space shuttle in orbit?

25. ♦ In a hydraulic elevator, a mass of 700 kg is supported by a piston 10 cm in radius. What fluid pressure is needed in the system?

26. ♦ What magnitude of force does atmospheric pressure exert on each side of the wall of a shed 5.0 m wide by 2.5 m high?

27. ♦ What force is necessary on the shaft of a bicycle pump with a piston diameter of $d = 2.0$ cm when the pressure in the tire is 5.0 atm?

28. ♦ ✉ In the movie *Goldfinger*, James Bond is saved when his foe, Auric Goldfinger, is "sucked" out the window of an airplane. What kind of force acts on Goldfinger? Is he actually sucked out the window, or is he pushed out? If the window is circular with radius 10 cm, make a rough estimate of the magnitude of the force by assuming that the pressure inside the airplane is 1 atm and the pressure outside is zero. (These assumptions overestimate the force.) Could such a thing really happen?

29. ♦ An artisan produces gold foil of thickness 25.0 μm. If the density of gold is 1.93×10^4 kg/m³, what area of foil can the artisan produce from 10.0 g of gold?

30. ♦♦ Assume that the area of contact between an automobile tire and the road is a rectangle with dimensions 15 cm × 10.0 cm. What tire pressure is needed to support a car of mass 1200 kg?

31. ♦♦ Equal masses of water and salad oil of specific gravity 0.80 are placed in a jar. What is the average density of the fluid in the jar? What is the average density if equal volumes are put in the jar?

32. ♦♦ Pumice is formed when gas bubbles are trapped inside basalt rock as it solidifies. If the specific gravity of basalt is 3.0 and a sphere of pumice 20 cm in radius has a mass of 10 kg, what fraction of the volume do the bubbles occupy?

33. ♦♦ The pressure in a bicycle tire is 5×10^5 Pa and the total mass of the bike and rider is about 100 kg. Estimate the area of contact between the bicycle tires and the road.

34. ♦♦ Two halves of a box are divided by a thin membrane. One-half contains oxygen molecules and the other half contains the same number of hydrogen molecules. What is the ratio of mean square speeds of the two kinds of molecules if the pressure is the same on both sides of the membrane?

§13.3 FLUIDS IN EQUILIBRIUM

35. ❖ Two teams of spelunkers (cave explorers) plan to explore two different, water-filled pools to see whether they are connected by an underground passage. If they find that the altitude of one pool is 720 m above sea level and the other is 750 m above sea level, can the two be connected? Explain your reasoning. Suppose the spelunkers test the altitudes of the pool surfaces by reading barometers placed at each surface. Can they determine whether the pools are connected? Explain your reasoning.

36. ❖ A dam is usually much thicker at the bottom than at the top. Why?

37. ❖ An engineer decides to use a successful dam design for a new lake of the same depth. If the new lake has ten times the surface area as the lake held by the original dam, what, if any, design changes are necessary for the new dam to hold back the increased amount of water? Explain your reasoning.

38. ❖ A person swings a bucket of water in a horizontal circle. Describe the variation of pressure of the water with "depth" in the bucket.

39. ❖ Two points at the same level in a uniform fluid have the same pressure. ■ Figure 13.42 shows a tube in which two fluids of different densities meet at a boundary. Use the reasoning of the text to show that points A and B have the same pressure while points C and D have unequal pressures.

40. ♦ Explain why a suction pump (■ Figure 13.43) can only lift water 10 m. How high can a suction pump lift mercury?

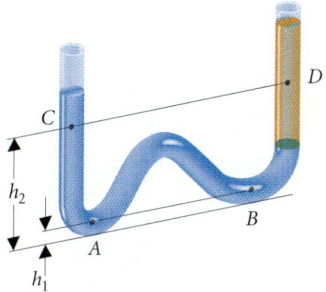

■ FIGURE 13.42

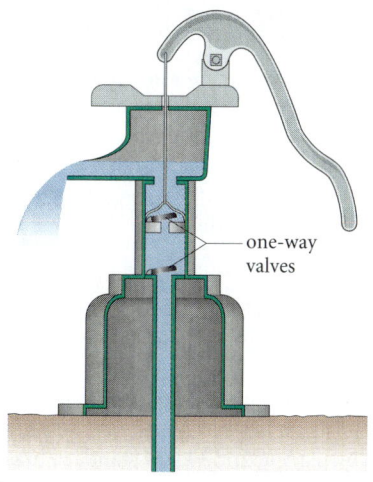

■ FIGURE 13.43

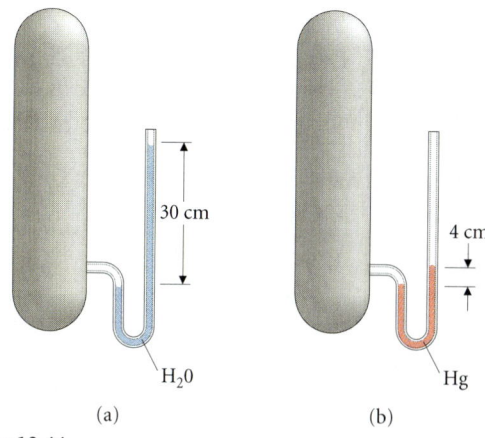

■ FIGURE 13.44

closed at the top. What is the hydrostatic force acting on the stopper's top surface?

50. ♦♦ A vessel of circular cross section contains water to a depth of 0.50 m and oil (specific gravity 0.60) to a depth of 0.50 m (■ Figure 13.47). What is the force acting on the stopper's top surface? (Take atmospheric pressure = 1.013×10^5 Pa.)

51. ♦♦♦ A quantity of mercury is at rest in a U-tube on a laboratory bench. Then Ali pours 10.0 cm of water into one side of the tube. Qualitatively describe what happens. Calculate the position of the liquid in each side of the tube at the end of the experiment, relative to the original mercury level.

41. ♦ If a barometer is to be made using a column of oil of density 8.0×10^2 kg/m³, how long should the vertical tube be?

42. ♦ What force per unit area does milk exert on the bottom of a milk carton of height 19.5 cm if the density of milk is 1.1×10^3 kg/m³? What is the maximum force per unit area exerted on the sides, and where does it occur?

43. ♦ What is the air pressure on top of Mount Whitney (height 4420 m), and on top of Mount Everest (height 8840 m)? (Assume pressure = 1 atm at sea level.)

44. ♦♦ An airplane is flying at 10 700 m. Inside the airplane, the cabin pressure is kept equal to the pressure in the atmosphere at an altitude of 2440 m. What are the magnitude and direction of the force per unit area on the walls of the aircraft cabin? Use the isothermal atmosphere model with $\rho_o = 1.3$ kg/m³.

45. ♦♦ An aircraft altimeter measures the pressure at the aircraft's altitude and compares it with the standard atmosphere (cf. Figure 13.17) to display the aircraft's altitude. If the pressure on the surface is 30 in. Hg and the altimeter senses a pressure of 25 in. Hg, what is the aircraft's altitude? (Use the isothermal model for the atmosphere to compute the answer with $\rho_o = 1.3$ kg/m³.)

46. ♦♦ Refer to ■ Figure 13.44. In which tank is the pressure greater? By how much?

47. ♦♦ A U-tube contains water and oil, as shown in ■ Figure 13.45. The water level is 10 cm below the oil level. Find the density of the oil.

48. ♦♦ A manometer is used to measure the pressure of gas in a cylinder, as shown in Figure 13.44b. The difference between the height of the mercury columns on either side of the U-tube is 4.00 cm. If a barometer reads 76.0 cm Hg, what is the pressure in the cylinder?

49. ♦♦ A container of water has the dimensions shown in ■ Figure 13.46. A cork stopper holds the water in the container, which is

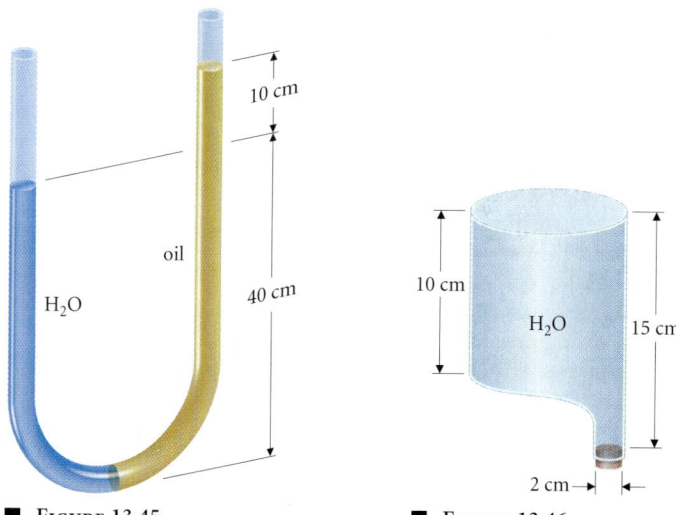

■ FIGURE 13.45

■ FIGURE 13.46

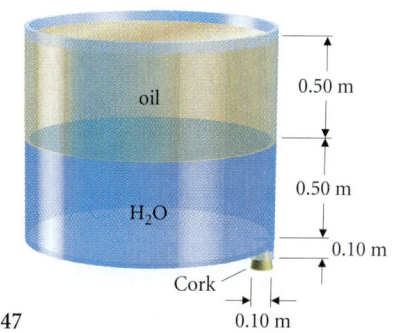

■ FIGURE 13.47

§13.4 ARCHIMEDES' PRINCIPLE

52. ❖ Which, if either, displaces more water when floating in a small pond: a single 1.5-ton mahogany log or a 1.5-ton bundle of balsa logs? Explain your reasoning.

53. ❖ Under which situation, if either, does your 1.0-m-long sailboat model displace more water: during a test float in your bathtub, or floating in a pond where you intend to race it against other boats?

54. ❖ Explain why it is easier to float in the Great Salt Lake than in a mountain lake.

55. ❖ Explain how a submarine can submerge and resurface.

56. ❖ A large Styrofoam cube and brass masses are on opposite pans of a laboratory balance inside an airtight glass bell jar. When the jar is filled with air at atmospheric pressure, the Styrofoam and brass masses are in balance. When air is pumped from the bell jar, do they remain in balance? If not, which side of the balance descends?

57. ❖ Large balloons are used to carry scientific instruments to high altitude. The balloon shell is very slack when the balloon is launched, to allow for expansion, but there is a maximum volume that can be achieved. The balloon rises to a fixed level in the atmosphere and remains at that altitude. Why?

58. ❖ You leave a glass of water full to the brim with an ice cube floating in it on the table as you run to answer the telephone. When you return, the ice has melted. Explain how to predict what has happened to the water level in the glass. Did it overflow?

59. ◆ What fraction of an iron lump's volume is above the surface when floating in mercury?

60. ◆◆ A polar bear of mass 500 kg is floating on an iceberg. As the ice melts, how small can the iceberg get before the bear gets wet feet? (Your answer should be the volume of the iceberg.)

61. ◆◆ A hydrometer is a device for measuring specific gravity (■ Figure 13.48). It consists of a weighted hollow tube that floats vertically in the liquid to be measured. When used to test the density of wine, a hydrometer of mass 1.00×10^2 g and cross-sectional area 5.00 cm^2 sinks to a depth of 19.0 cm. What is the density of the wine? Assume that the hydrometer is a perfect cylinder. (This measurement is the basis of the Brix scale for sugar content of wine.)

62. ◆◆ A piece of rock weighs 301 N in air and has an apparent weight of 202 N when suspended in water. What are the volume and density of the rock?

63. ◆◆ You (mass 80.0 kg) are the captain and sole crew of a barge with a rectangular cross section measuring 3.00 m × 10.5 m. Empty, it has a mass of 9020 kg. You wish to load it with sacks of coal, each

of which has a mass of 50.0 kg, and leave at low tide when the water depth is 1.00 m. How many sacks of coal may you load into the barge?

64. ◆◆ A hot-air balloon contains air of density 55% of the surrounding atmosphere. What volume is necessary to carry a 450-kg payload? What radius sphere is this? (Use the value 1.2 kg/m^3 for the density of air.)

65. ◆◆ The Goodyear blimp *Columbia* (■ Figure 13.49) is filled with 5740 m^3 of helium with density $\rho = 0.18$ kg/m^3. The ship has a mass of 5430 kg when empty. How much payload can the ship carry? (Take $\rho_{air} = 1.3$ kg/m^3.)

66. ◆◆◆ You plan to determine the mass of a steel cylinder ($\rho_{steel} = 8.5 \times 10^3$ kg/m^3) using a laboratory balance. You find that the cylinder is balanced by placing 2.153 kg of standard brass masses on the other pan of the balance. Estimate the error in your mass determination due to neglecting air buoyancy forces. ($\rho_{brass} = 8.9 \times 10^3$ kg/m^3; $\rho_{air} = 1.2$ kg/m^3.)

67. ◆◆◆ A glass beaker has a mass of 2.00×10^{-2} kg and holds 1.00×10^{-4} m^3 of water when completely full. If the beaker is set floating, 91.0 g of lead shot may be placed in its bottom before it sinks. What is the density of the glass?

68. ◆◆◆ A layer of oil (density 8.0×10^2 kg/m^3) floats on water in a container. A wooden cube 10.0 cm on a side with density 9.0×10^2 kg/m^3 floats so that its top surface is just barely submerged. How deep is the oil layer?

69. ◆◆◆ Verify Archimedes' principle by using integration to calculate the total force due to pressure acting on a cylinder floating with its axis horizontal.

■ FIGURE 13.48
A hydrometer.

■ FIGURE 13.49

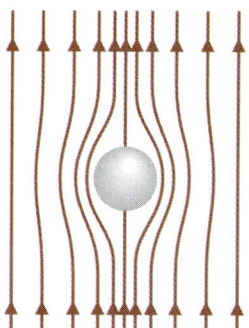

■ FIGURE 13.50

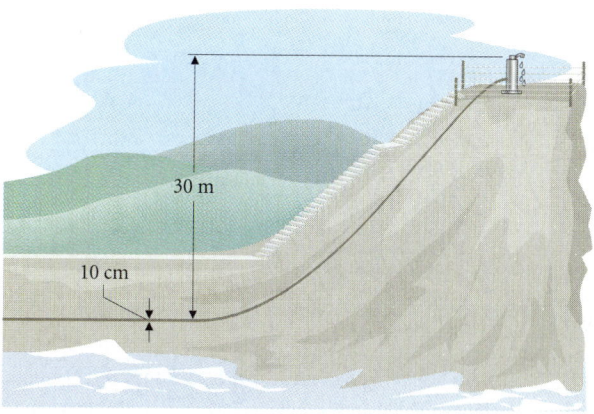

■ FIGURE 13.52

§13.5 FLUID DYNAMICS

70. ❖ Sketch the streamlines of airflow around the Goodyear blimp (Figure 13.49).

71. ❖ The Bay of Fundy in Nova Scotia has a broad opening to the sea that narrows to a thin channel. Tides in the channel have a vertical amplitude of nearly 20 m. Can you explain why?

72. ❖ In a popular physics demonstration, a ping-pong ball is suspended in a vertical jet of air that moves fastest in the center of the stream (■ Figure 13.50). If the ball is pushed slightly away from the center of the stream, it quickly returns. Why?

73. ❖ It is sometimes reported that hurricane winds suck windows outward on the side of a house parallel to the wind velocity. Can you explain why this might happen? Are the windows "sucked" outward or pushed outward?

74. ❖ Is the idea of a streamline valid in a rotational flow such as that shown in Figure 13.59? Why or why not?

75. ❖ Is the idea of a streamline valid in the turbulent flow shown in Figure 13.25? Why or why not?

76. ❖ A fluid flows through the pipe shown in ■ Figure 13.51. The tubes at points A, B, and C are open to the atmosphere. Are the fluid heights in the tubes consistent with Bernoulli's law for an incompressible, ideal fluid? If not, what would you conclude about the nature of the fluid?

77. ❖ Will a siphon used by a lunar colony lift water a greater or lesser height than on Earth? Assume that the colonists maintain a pressure of 1 atm in their dwellings.

78. ❖ Is the maximum height a siphon can lift gasoline greater or smaller than the maximum height it can lift water?

79. ❖ If the lower end of a siphon has a large amount of tube laid out on a horizontal surface, how does the flow rate of the siphon depend on the length of tube? (Assume the fluid is ideal.)

80. ◆ If a road narrows from three lanes to two, there is usually a sign saying, "Slow, road narrows." What should the sign say if having traffic flow like an incompressible fluid were the most important consideration? If traffic is moving at 50 mph on the three-lane road and the cars maintain their separation, at what speed should the traffic move on the two-lane road to avoid a traffic jam? If, instead, the traffic moves at 30 mph on the two-lane road, how much closer must the cars be to each other to maintain steady flow?

81. ◆ Assuming water to be an ideal fluid and ignoring turbulence, find how high water spurts from a broken water main if the pressure in the pipe is 6.9×10^5 Pa. (Assume the water travels directly upward and neglect the speed of the water inside the pipe.)

82. ◆ A pipe of constant diameter 10 cm carries water up a hill 30 m high. How much pressure is required at the bottom of the hill if the water is to reach the top? (■ Figure 13.52)

83. ◆ A siphon is used to empty a spa 0.95 m deep. The end of the hose is 0.25 m below the bottom of the spa. At what speed does the water run out of the hose (a) at the beginning and (b) at the end of the draining process?

84. ◆◆ Newly refined gasoline ($\rho = 0.70 \times 10^3$ kg/m^3) is being pumped from the refinery into storage tanks 15 m above the pump. A pressure gauge at the pump outlet reads $P_p = 2.7 \times 10^5$ Pa and a rate meter shows the gasoline flows from the pump at 2.5 m/s. If a pressure gauge at the storage tanks reads $P_s = 1.4 \times 10^5$ Pa, use Bernoulli's law to determine the gasoline's speed as it enters the tanks. Use the equation of continuity to determine the ratio of the pipe system's cross-sectional area at the pump to that at the tanks.

85. ◆◆ A pipe carries water over a hill 0.10 km high, as shown in ■ Figure 13.53. The water flows at 1.0 m^3/s, and the pipe is 12 cm in diameter. What is the minimum pressure that the water must have at point A if it is to flow steadily over the hill? How much power does the pump supply?

86. ◆◆ Water flows freely at a rate of 50.0 g/s out of the bottom end of a vertical pipe 2.0 cm in diameter. What is the diameter of the stream 2.0 m below the end of the pipe?

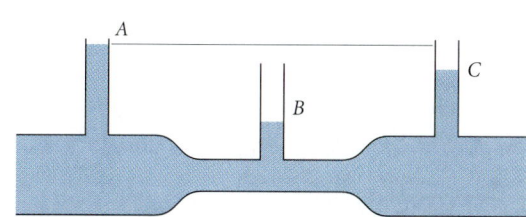

■ FIGURE 13.51

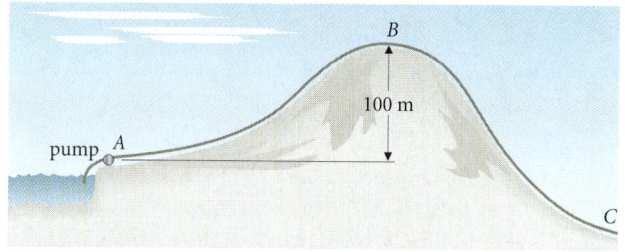

■ FIGURE 13.53

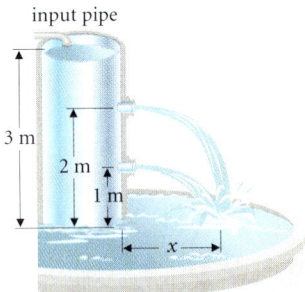

FIGURE 13.54

87. ◆◆ A fountain contains a cylinder with tubes placed at heights of 1.0 m and 2.0 m, as shown in ■ Figure 13.54. The water level in the cylinder is kept constant at a height of 3.0 m by an input pipe. As the water emerges from a tube, it strikes the fountain pool a distance x away from the cylinder. Find x for each of the two tubes. If both tubes are 2.5 cm in diameter, at what rate must the input pipe supply water?

88. ◆◆ A fluid of density ρ flows through the tube shown in ■ Figure 13.55. Find the pressure difference between points O and Q in terms of ρ, h, v_o, B, and b. If $B = 2b$, $h = 15$ m, and the fluid is water, for what values of v_o is $P_O > P_Q$?

89. ◆◆ Water flows smoothly down a 2° slope in a rectangular channel (■ Figure 13.56). At one point, the channel is 0.60 m wide and the stream is flowing 15 cm deep at 2.0 m/s. At another point 0.10 km farther down the slope, the channel is 1.0 m wide. What is the depth of the stream at the lower point?

90. ◆◆ Water flows into a nozzle at 3.00 m/s at a pressure of 1.31×10^5 Pa. What should the ratio of input to output diameter be if the flow is to remain steady? What is the flow speed at the exit?

91. ◆◆ An incompressible fluid of density ρ flows in a rectangular tube in which a tapered object of dimension $L \times W \times h/2$ is placed (■ Figure 13.57). Neglecting turbulence, and assuming that one-half of the fluid flows above the wing, compute the lift acting on the wing. (Assume $\rho g h \ll P_o$.)

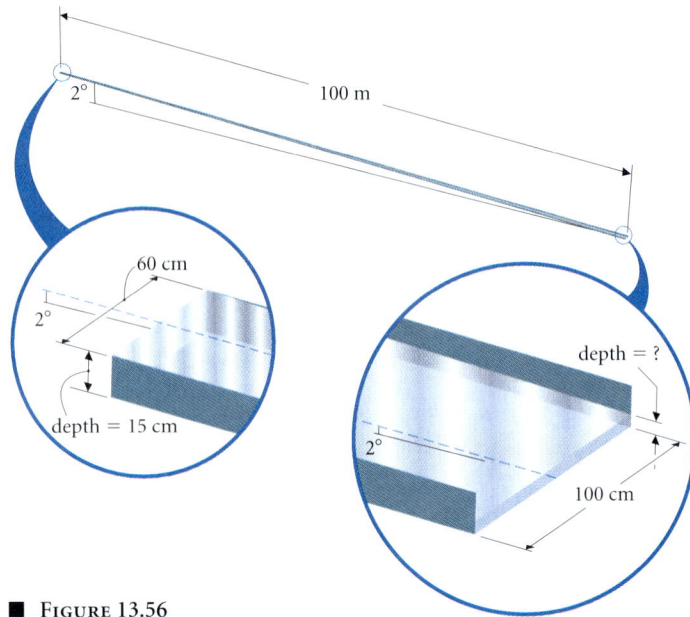

FIGURE 13.56

92. ◆◆ What is the force on the Vehicle Assembly Building at Kennedy Space Center in Florida when a 15-knot wind is blowing perpendicular to its side? The VAB measures 526 ft high by 495 ft wide, and 1 knot = 0.514 m/s.

§13.6 INCOMPRESSIBLE FLOW OF AIR

93. ❖ A sailboat can tack upwind with its sail arranged as shown in ■ Figure 13.58. Explain how the force on the sail is developed.

94. ❖ Airplanes and blimps both fly for reasons that you have learned in this chapter, but they are different reasons. Discuss.

95. ❖ Is the actual lift-off speed of an airplane at Leadville, Colorado (altitude 3023 m) the same as, greater than, or less than the speed of the same airplane at lift-off at Oakland, California (at sea level)? What about the airspeed indicator reading?

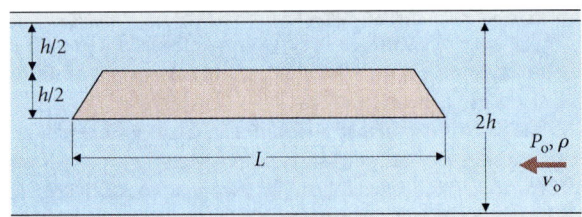

FIGURE 13.57

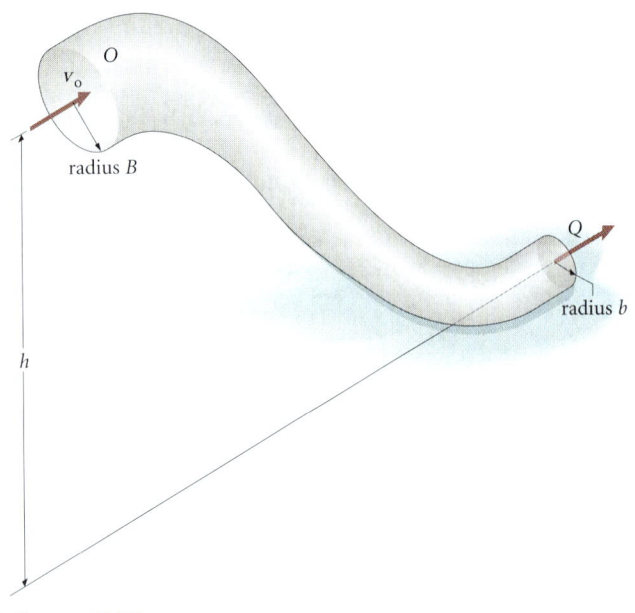

FIGURE 13.55

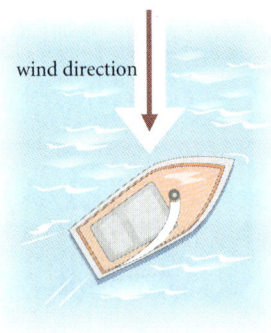

FIGURE 13.58

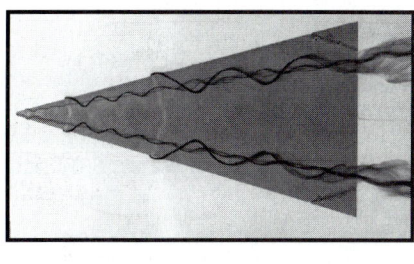

(a)

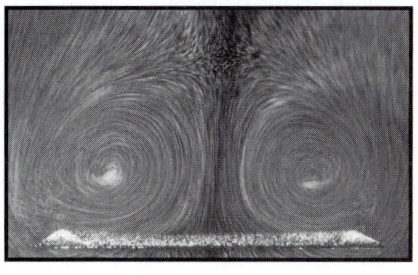

(b)

■ **Figure 13.59**
(a) Vortices above an inclined triangular wing. (b) Cross section of vortices at the trailing edge of the wing.

96. ❖ Large wingtip vortices form behind heavy aircraft on takeoff and landing (■ Figure 13.59). Smaller aircraft can be rolled over if they fly into the vortices. Explain why the vortices form. (*Hint:* Consider how lift is produced by the wing.)

97. ◆ An airplane is flying at 68.0 m/s at an altitude where the pressure is 0.99×10^5 Pa and the air density is 1.2 kg/m³. What pressure does the Pitot tube sense?

98. ◆◆ ■ Figure 13.60 shows the coefficient of lift for the NACA 63_1-012 airfoil as a function of angle of attack. If an airplane of mass 1500 kg using this airfoil has wings of total area 17 m², what is its speed when flying at a 3.0° angle of attack? (Take the air density to be 1.3 kg/m³.) (NACA—the National Advisory Committee for Aeronautics—was the predecessor of NASA.)

Additional Problems

99. ❖ What conservation principle ensures that the vortices in Figure 13.59 swirl in opposite directions?

100. ❖ ✉ Estimate the force required to propel a flat plate through air in a direction perpendicular to its surface. Assume the air behind the plate is turbulent and has zero average velocity with respect to the plate.

101. ❖ Why does the air inside a hot-air balloon need to be hotter to carry the same payload at Aspen, Colorado (8000 ft altitude) than in the Napa Valley in California (approximately 200 ft altitude)?

102. ◆◆ The system shown in ■ Figure 13.61 is in equilibrium. When the metal sphere of radius 2.00 cm is lowered into the beaker of water, does the system remain in equilibrium? If not, how much mass has to be suspended from which side to restore equilibrium?

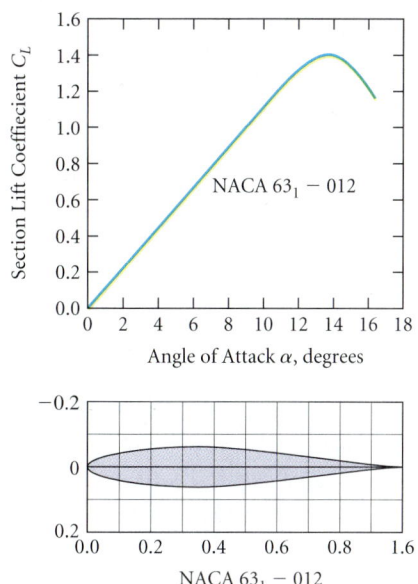

■ **Figure 13.60**

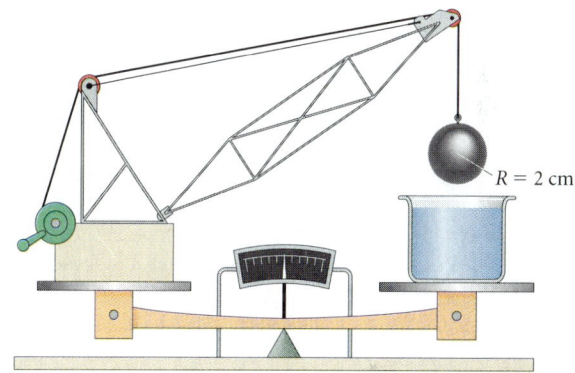

■ **Figure 13.61**

103. ◆◆ An airplane's altimeter measures atmospheric pressure of 2.96×10^4 Pa, and the Pitot tube measures a pressure of 3.33×10^4 Pa. Find the altitude and speed of the airplane. (Use data from Table 13.1 and Figure 13.17).

104. ◆◆◆ The lid of a pressure cooker (■ Figure 13.62) has a radius $r = 10.0$ cm and a handle of length $\ell = 20.0$ cm. If the coefficient of friction between the lid and the body of the cooker is $\mu = 0.300$, and the pressure inside the cooker is $P = 2.00$ atm, what force on the handle is required to turn the lid? (It is not smart to do this. Put it under the cold water tap first!)

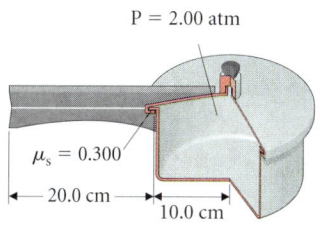

■ **Figure 13.62**

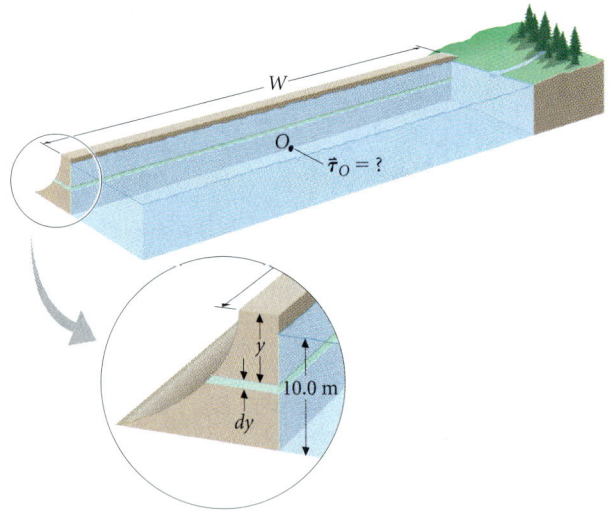

■ FIGURE 13.63

105. ♦♦♦ A dam of width $W = 103$ m is used to hold back water in a reservoir 10.0 m deep (■ Figure 13.63). What is the pressure of the water at a depth y below the surface? What is the net force on a strip of the dam of height dy at depth y? (Be careful here—what is on the other side of the dam?) Calculate the total force exerted on the dam.

What is the torque exerted on the dam about the midpoint of its base by the force at depth y that you calculated above? Find the total torque on the dam. Where is the "center of pressure"; that is, if the effects of pressure are replaced by a single force acting at depth D, what is D?

106. ♦♦♦ An incompressible fluid of density ρ flows in the pipe shown in ■ Figure 13.64. What is the speed of the fluid at X? Obtain a numerical value for $D = 3d$, $H = 5.0$ cm, and $h = 2.0$ cm.

107. ♦♦♦ Assume the fluid moving in the stream tube of Figure 13.29 is incompressible. Apply the impulse-momentum theorem to the fluid in the tube and show that it gives the same equation as Bernoulli's law. (Hint: $v = d\ell/dt$.)

108. ♦♦♦ An ideal fluid flows from the bottom of a container of circular cross section with inlet radius r_{in}, exit radius r_{out}, and height H. Find the radius of the container as a function of height if the fluid pressure equals atmospheric pressure throughout the fluid. At what rate must fluid be added to the top to maintain the level constant?

109. ♦♦♦ A propeller works in the same way as a wing; a force is produced by a difference in flow speed over the two sides of the propeller. A propeller for use in water has blades of length 1 m. What is the maximum angular speed of the propeller if it is used in water at a depth of 3 m and steady flow is to be maintained? (Hint: What is the constraint on the pressure?) What do you suppose happens when the pressure gets very low? (This effect is known as cavitation.)

Computer Problem

110. Use the ideal gas law to express pressure in the atmosphere in terms of density ρ and the mean square speed $\langle v^2 \rangle$ of the molecules. Let the subscript o indicate their values at sea level. Define the quantities $r = \rho/\rho_o$, $w = \langle v^2 \rangle$, and $\Delta \zeta = g\Delta z/w_o$. Substitute in eqn. (13.7) to obtain:

$$\Delta r = -r[\Delta w/w + (w_o/w)\Delta \zeta].$$

File CH13P110 on the supplementary computer disk contains the value of w/w_o for molecules in the Earth's atmosphere as a function of altitude. Taking $w_o = 85\,000$ (m/s)^2, use a spreadsheet program to calculate the change in density between z and $z + \Delta z$, where $\Delta z = 305$ m and z varies from 0 to 2745 m, and hence calculate the value of r at each height listed in the table. Also calculate the pressure divided by its value at sea level, $P/P_o = rw$. Use the spreadsheet to evaluate P/P_o using the isothermal atmosphere formula (13.11), and compare the two sets of values. (Take $\rho_o = 1.2$ kg/m^3 and $P_o = 1.01 \times 10^5$ Pa.)

Challenge Problems

111. A Cartesian diver (after René Descartes, 1596–1650, cf. Chapter 1) is a device that floats at the top of a fluid at atmospheric pressure but sinks when the surface pressure is increased. Often, these divers are designed to look like little devilish imps. We may study the effect with the less picturesque test tube in ■ Figure 13.65 that has a volume V and a mass m_o. It contains a mass $m_a \ll m_o$ of air. The mean square speed of the air molecules is $\langle v^2 \rangle$, and the fluid has density ρ. (a) Use Archimedes' principle to explain why the diver floats at low pressure and sinks at high pressure. (b) Show that the maximum pressure at which the diver floats is $P_{max} = \rho m_a \langle v^2 \rangle/m_o$. (Neglect the thickness of the glass in computing volumes and neglect the mass of air compared with the mass of glass.) (c) Beyond what depth will the diver be unable to rise to the surface, if the surface is at atmospheric pressure.

112. A centrifuge tube of length $\ell = 10$ cm is filled with water. It is then mounted in the centrifuge and spun about a vertical axis through one end with angular speed $\omega = 10^4$ rad/s. Find the pressure in the tube as a function of position.

113. (a) Explain why pressure varies with distance from the central axis of a spinning cylinder of fluid. A cylinder of radius R contains fluid of density ρ and spins with angular speed ω. Compute the pres-

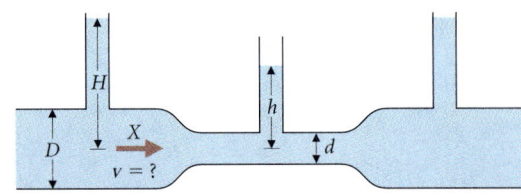

■ FIGURE 13.64

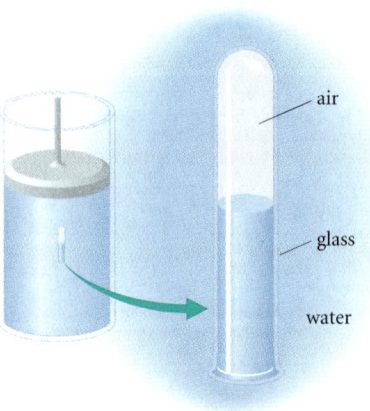

■ FIGURE 13.65

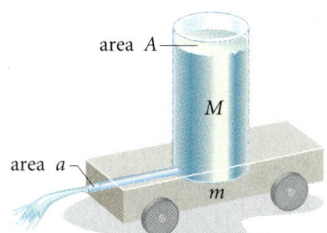

FIGURE 13.66

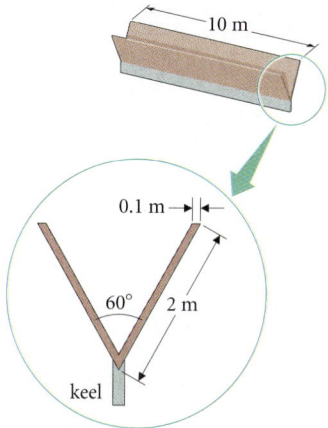

FIGURE 13.67

sure difference between the center and the circumference for water with $R = 0.175$ m and $\omega = 1550$ rpm. **(b)** How does pressure vary with depth below the surface in the rotating cylinder? Use your result from part (a) to find the shape of the fluid surface.

114. The physics demonstration shown in ■ Figure 13.66 consists of a reservoir of water mounted on a cart of mass m. The reservoir holds a mass M of water. Consider the flow of water from the reservoir during a time dt, and find the resulting change in the cart's velocity. What is the final velocity of the cart?

115. A wooden boat of triangular cross section has an iron keel (■ Figure 13.67). The wood has specific gravity 0.8 and the keel has a mass of 4000 kg. **(a)** To what depth is the boat submerged when floating empty? **(b)** Where is the center of mass? **(c)** Where is the center of pressure? (*Hint:* Use Archimedes' principle rather than a long calculation.) **(d)** Show that the boat is stable to small roll displacements. Neglect any change in position of the center of pressure during roll.

116. By considering forces on a small piece of a thin cylindrical shell, show that purely compressive forces in the shell can balance a uniform external pressure. If the shell has radius R and thickness t and is exposed to external pressure P, find the compressive force per unit area in the material.

Part III Problems

1. ❖ Initially, an ice skater is spinning on one toe with his arms fully extended. He then draws his arms in to place his hands on his hips. Determine whether each of the following increases, decreases, or stays the same: (a) the skater's angular speed, (b) the skater's angular momentum, (c) the skater's kinetic energy. Explain your reasoning. If you determine that the skater's kinetic energy or angular momentum change, describe the necessary transformation of energy or transfer of angular momentum.

2. ◆◆ In a carnival game, you place a ball on a ramp at your choice of position ℓ (Figure III.1). The ball then rolls without slipping along the ramp and leaves the ramp as a projectile. You win if the ball lands in the cup a distance D from the ramp. What value of ℓ should you choose?

3. ◆◆ Figure III.2 shows the water supply to a hydroelectric power plant. Water enters the generator turbines in the powerhouse at speed v_o and atmospheric pressure. Use Bernoulli's law to determine v_o. What is the mechanical power delivered to the turbines as a function of the pipe area? If the system produces electrical power with efficiency $e = 12\%$ and the desired power output of each turbine is $P_{elec} = 0.50$ MW, what is the diameter of the inlet pipe?

4. ◆◆◆ ✉ (a) A car may be modeled as a rectangular solid of length L, height h, width w, and mass M mounted on its wheels by four springs of constant k (two at each end). The driver jams on the brakes and locks the wheels in an emergency stop. If the coefficient of kinetic friction between tires and road is μ_k, find the compression of each spring. What is the maximum value of the coefficient of friction μ_{max} for which the car does not tumble forward? Take $M = 1.2 \times 10^3$ kg, $L = 5.0$ m, $h = 1.5$ m, $w = 2.1$ m, and $k = 2.2 \times 10^4$ N/m and evaluate μ_{max}. Is tumbling of the car a practical problem? (b) What is the minimum radius of a horizontal turn that the car can negotiate at $v = 32$ m/s without tumbling over?

5. ◆◆◆ An astronaut of mass M and height $h = 2.0$ m approaches a satellite of mass $5M$ and radius 0.50 m at speed $v = 0.50$ m/s, as shown in Figure III.3. Hoping to repair the satellite, the astronaut grabs onto it at the end of the 2.0-m-long radio mast. Describe the subsequent motion of the system and estimate the values of its final linear speed and angular speed.

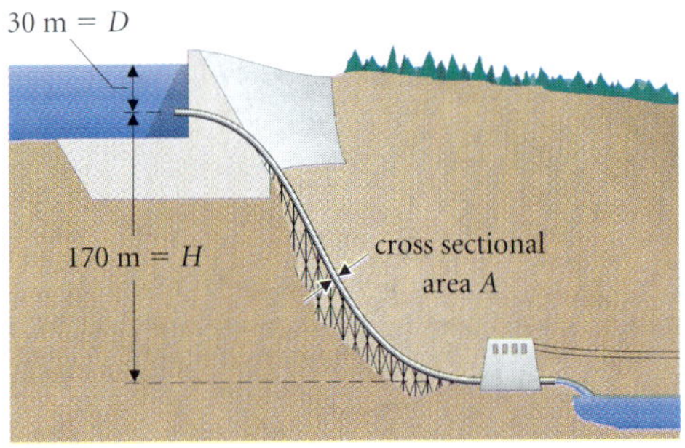

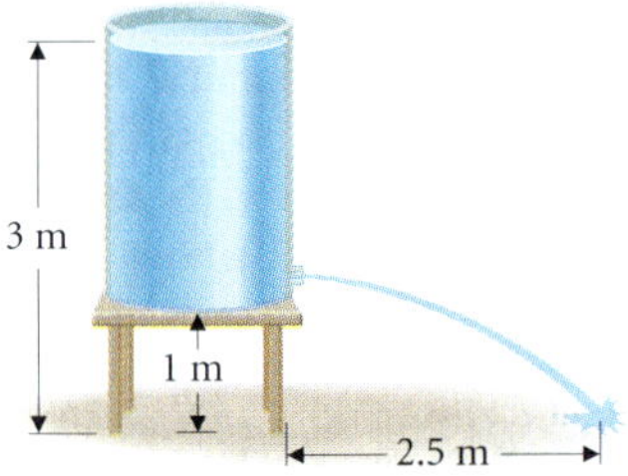

6. ◆◆◆ ✉ In 1054, a star in the constellation Taurus exploded, leaving behind a small, spherical remnant called a neutron star, which rotates with a period of 33 ms. The star has a mass of 3×10^{30} kg. (a) If the only force holding the star together is gravity, what is the maximum radius the star can have? What is its maximum possible rotational inertia? (b) The star radiates energy into space at a rate of 10^{31} W by changing its rotational energy into energy of radiation. Explain how to estimate the star's radius from observing the rate of change of its rotation period. (c) The observed rate of increase in period for the star is 35 ns/d. Estimate the radius of the star and comment on your result.

7. ◆◆◆ A rod of length ℓ and total mass M is tapered so that its mass per unit length λ varies linearly from $\lambda = 0$ at the pointed end to $\lambda = \lambda_o$ at the blunt end:

$$\lambda = \lambda_o r/\ell,$$

where r is the distance from the pointed end of the rod. Where is the rod's CM? What is its rotational inertia about an axis through the pointed end perpendicular to the rod?

8. ◆◆◆ A pointed float of density 500 kg/m³ is released at the bottom of a container of water 50 cm deep. How long does it take for the float to reach the surface? (Assume that the water offers no resistance to the motion of the float.)

9. ◆◆◆ Water flows from the container shown in Figure III.4 through a hole of area 1.0×10^{-4} m² and follows the parabolic path shown. How much power is dissipated as heat by turbulence at the exit of the container?

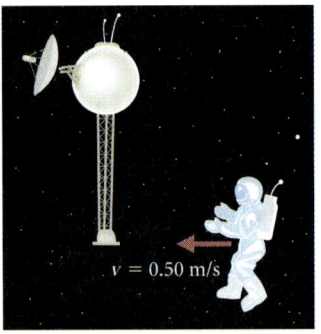

■ FIGURE III.3

■ FIGURE III.4

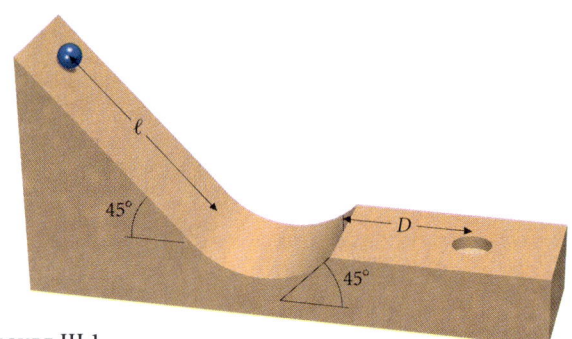

■ FIGURE III.1

■ FIGURE III.2

Oscillatory and Wave Motion

PART FOUR

The pendulum of a grandfather clock repeats its swing on a regular schedule that allows us to tell time: it is the prototype for oscillations. When you drop a pebble into a quiet pond, ripples spread outward over the surface. Each point on the surface moves upward as the crest of a ripple passes it and downward as a trough passes. Such oscillations, organized as a disturbance that moves away from an energy source, are characteristic of wave motion. Waves are extremely important for our perception of the world: sound is transmitted by waves in air that we detect when they displace sensitive fibers in the ear, and the light that brings us visual images exhibits all the properties of a wave disturbance. Oscillations and waves are the focus of this part.

Together, the ideas of waves and particles form the foundation for understanding all of physics. In 1905, Albert Einstein showed that a light wave also acts in many ways like a stream of particles. Then in 1924, Louis de Broglie presented a theory that material particles should also exhibit wave properties. This theory has been amply confirmed in experiments with atomic and subatomic particles, and we now believe that all fundamental natural processes are an inseparable combination of wave disturbance and particle motion.

In this part, we complete the toolkit of basic models that underlie all of physical theory. We begin by studying oscillations of simple systems (Chapter 14). In Chapter 15 we see how coordinated oscillations of the particles in a mechanical system form a wave that transmits energy through the system. At the same time, we develop the basic theory and vocabulary of waves. Chapter 16 is devoted to the common properties of sound and light. Sound waves are mechanical vibrations of fluids or solids. Though light waves are electromagnetic and not mechanical in nature, we can understand many interesting optical phenomena using only the wave properties of light. In Chapter 17 we describe how waves combine within a system, and we end Part IV with a discussion of optical instruments (Chapter 18).

THE ELECTRON MICROSCOPE IS BASED ON DE BROGLIE'S IDEA.

SEE CHAPTER 33 FOR THE ELECTROMAGNETIC PROPERTIES OF LIGHT.

*. . . time is come round,
and where I did begin,
there shall I end; . . .*
WILLIAM SHAKESPEARE

CHAPTER 14
Oscillatory Motion

CONCEPTS

Restoring force
Simple harmonic motion
Period and frequency
Amplitude
Phase
Resonance

GOALS

Be able to:
Identify systems that undergo simple harmonic motion and find their periods of oscillation.

Find the amplitude and phase of an oscillation.

Our modern society is very time conscious and we are dependent on clocks to order our lives. All modern clocks count oscillations of a regularly vibrating system. The pendulum of a grandfather clock is an obvious example. Many wristwatches count vibrations of a quartz crystal. In this chapter we shall look at some simple examples of oscillating systems. The pendulum is the subject of §14.2.

Accurate clocks are essential to our twentieth-century lives. You could not attend your 10 o'clock lecture if you did not know when it was 10 o'clock! All modern clocks measure time by counting oscillations: the rhythmic swing of a pendulum; the stretching and unstretching of a spring; the vibrations of a quartz crystal or a cesium atom. Galileo, watching a chandelier swinging in the cathedral at Pisa, was the first to recognize the regularity of oscillations. A half century later, his theories were used in constructing the first really accurate clock mechanisms.

In this chapter we shall study the least complicated form of oscillation: simple harmonic motion. This theory is the starting point for describing the vibration of systems and wave motion of all kinds.

THE VIBRATIONS OF CESIUM ATOMS ARE USED TO DEFINE THE SECOND, SEE §1.2.3.

WE DISCUSSED SOME OF GALILEO'S WORK IN §0.3.

14.1 SIMPLE HARMONIC MOTION
14.1.1 Restoring Forces

Oscillating systems—a clock pendulum, a boat bobbing up and down as waves pass beneath it, or a particle on the end of a spring—have a common feature: each system has a stable equilibrium state, and the oscillation involves motions about that state. In equilibrium, the net force and torque acting on each part of a system are zero. The equilibrium is stable if a small displacement results in a net force that tends to restore the system to the equilibrium state. Such *restoring forces* are a second characteristic of oscillating systems. In our principle example—an object on a spring—the restoring force is proportional to the object's displacement. In a system near equilibrium, the net force is usually approximately proportional to the displacement from the equilibrium position, and so an analysis of the particle-and-spring system is widely applicable.

SEE §11.1.2 FOR A DISCUSSION OF STABILITY.

EXERCISE 14.1 ❖ Describe the equilibrium state for each of the three systems mentioned in the preceding paragraph.

EXAMPLE 14.1 ❖ Describe qualitatively the motion of an ice cube sliding without friction near the bottom of a cylindrical pipe with its axis horizontal (■ Figure 14.1).

MODEL In equilibrium, the ice cube is at rest at the bottom of the pipe. If the ice cube is released a small distance from the lowest point of the pipe, the component of its weight acting along the pipe's surface accelerates the ice cube back toward the bottom. Gravity acts as the restoring force. When the ice cube reaches the bottom (Figure 14.1b), its momentum carries it through the equilibrium position and up the other side of the pipe. Now the component of weight parallel to the pipe's surface slows it to a stop (Figure 14.1c) and then accelerates it back toward the bottom again. In the absence of friction, this oscillatory motion would continue forever.

■ **FIGURE 14.1**
An ice cube near the bottom of a cylindrical pipe oscillates when slightly displaced. (a) The restoring force is the component of the cube's weight parallel to the surface of the pipe. Normal force balances the component perpendicular to the pipe. (b) As the ice cube passes through its equilibrium position, there is no horizontal force component ($\vec{n} + \vec{W}$ is vertical), so the acceleration parallel to the pipe's surface is zero. The cube's velocity is not zero, so it passes through the equilibrium point toward the left. (c) Once the cube passes the equilibrium point, the restoring force brings the block to rest and then accelerates it back toward the right.

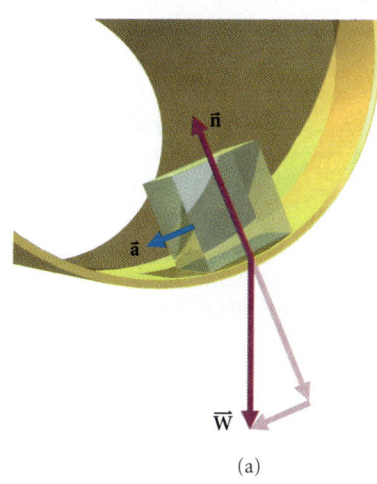

(a)

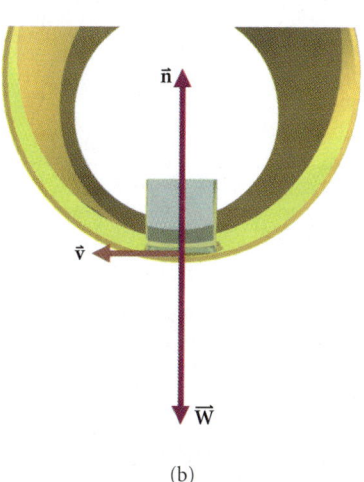

(b)

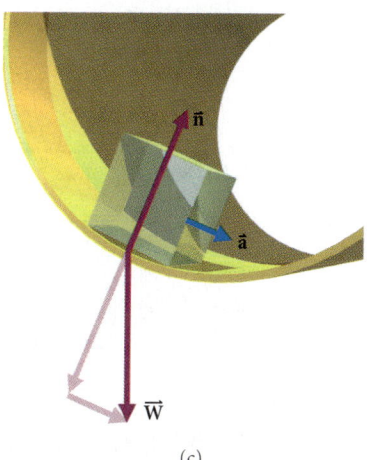

(c)

14.1.2 Equation of Motion for an Oscillating Particle

We describe the motion of an oscillating particle, or any particle, by expressing its position $\vec{r}(t)$ as a function of time. As is usual in dynamics problems, we begin with Newton's second law, which relates the restoring force to the particle's acceleration. In our basic example—one-dimensional motion of a particle on the end of a spring—both the restoring force and the displacement lie along a single line that we choose as the x-axis. If we put the origin at the particle's equilibrium position (■ Figure 14.2), then the stretch of the spring equals the particle's x-coordinate $x(t)$. The particle slides without friction on a horizontal surface, so the only force acting in the x-direction is the spring force. Then from Newton's second law:

$$ma_x = F_x = -kx(t).$$

SPRING FORCE WAS DISCUSSED IN §4.5.

Acceleration is the second derivative of position, so:

$$m\frac{d^2x}{dt^2} = -kx. \tag{14.1}$$

We want to find the function $x(t)$ that satisfies this relation. This is a new form of mathematical problem—it's more like a mathematical puzzle. As with any puzzle, guessing is a useful technique. Past experience is a guide for our guesses: eqn. (14.1) looks complicated, but we already know its solution!

14.1.3 Analogy with Circular Motion

Oscillations are closely related to another repetitive motion we have studied: uniform circular motion. Remember that a particle moving with constant angular speed ω on a circle of radius $r = A$ (■ Figure 14.3) accelerates toward the center of the circle:

$$\vec{a} = -\omega^2 \vec{r}.$$

WE DERIVED THIS RESULT IN §3.2.4.

The x-component of this equation is:

$$a_x = \frac{d^2x}{dt^2} = -\omega^2 x,$$

which has the same mathematical form as the equation for the particle on a spring.

THIS EQUATION IS EXTREMELY IMPORTANT. LEARN TO RECOGNIZE IT.

PARTICLE ON SPRING

$$\frac{d^2x}{dt^2} = -\frac{k}{m}x.$$

UNIFORM CIRCULAR MOTION x-COMPONENT:

$$\frac{d^2x}{dt^2} = -\omega^2 x.$$

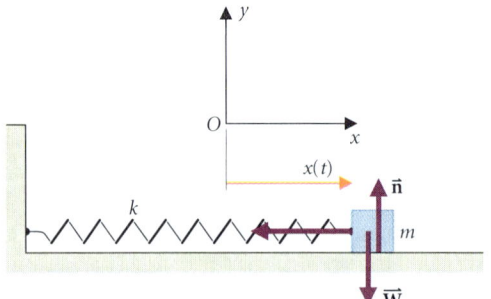

■ FIGURE 14.2
A particle on the end of a spring is an example of an oscillating system. When the block is displaced, the spring force acts to return it toward the equilibrium position. The origin is at the end of the unstretched spring, so x equals the stretch and the force on the block has x-component $F_x = -kx$. (The vertical forces—weight and normal force—balance each other.)

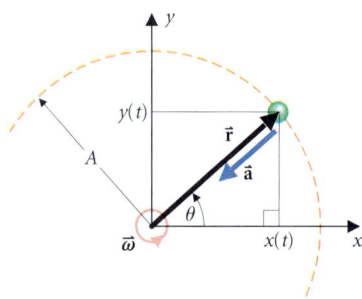

■ FIGURE 14.3
A particle moving in a circle of constant radius A. The particle's x-coordinate satisfies the same equation as the position of the particle on the spring:

$$x(t) = A\cos\theta(t) = A\cos(\omega t + \phi).$$

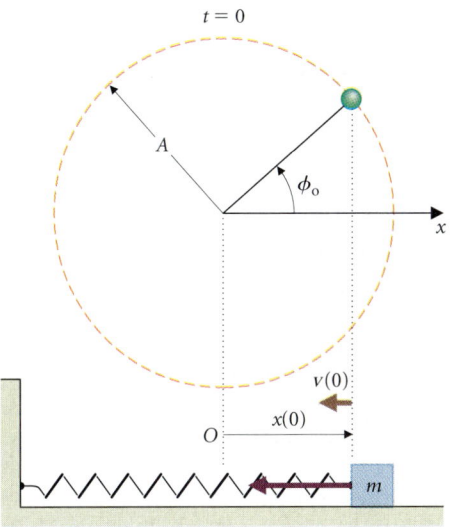
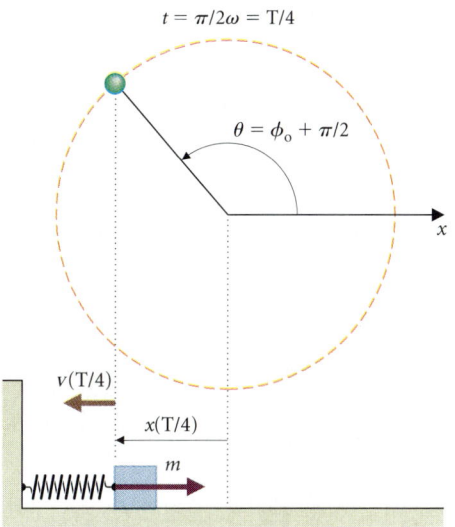
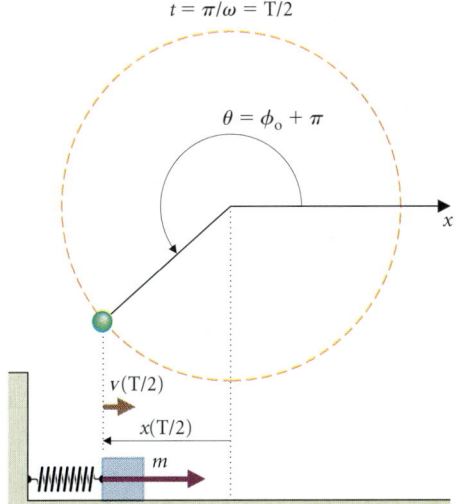
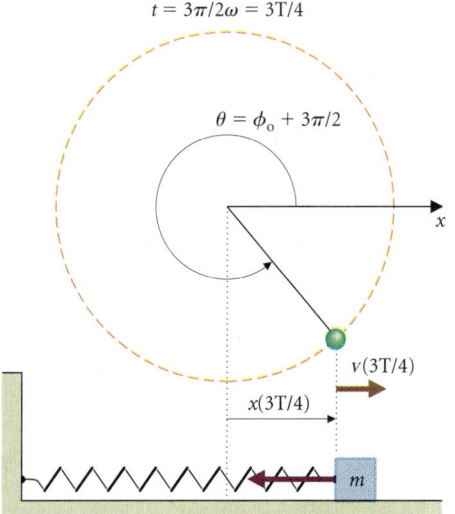

FIGURE 14.4
Comparison of the motion of the particle on the spring and the particle moving in a circle. With $\omega = \sqrt{k/m}$, the x-coordinates of the two particles are always the same, as are the x-components of their velocities.

If we define an *angular frequency*:
$$\omega = \sqrt{k/m} \qquad (14.2)$$

for the particle on the spring, then the equations for the particle on the spring and the x-component of the uniform circular motion become identical. They have identical solutions: $x = A \cos[\theta(t)]$. ■ Figure 14.4 illustrates this relation by showing both particles at several different times during one period of the circular motion. In both cases, the time for one complete cycle of the motion is called the period, T. Take care! Unlike a particle in circular motion, an oscillating particle passes through any position (except the maximum displacement) twice each period, but with velocities in opposite directions. A complete period is the time to reach the *same position* going in the *same direction*.

THIS DEFINITION IS MOTIVATED BY THE MATHEMATICS OF TRIGONOMETRIC FUNCTIONS.

IN DESCRIBING CIRCULAR MOTION, WE OFTEN USE "CYCLE" AND "REVOLUTION" INTERCHANGEABLY. FOR OSCILLATIONS, ONLY "CYCLE" IS APPROPRIATE.

> The *period* T of an oscillation is the time for a particle to return to the same position $\vec{r}$ with the same velocity $\vec{v}$. For any time t:
> $$\vec{r}(t + T) = \vec{r}(t) \quad \text{and} \quad \vec{v}(t + T) = \vec{v}(t).$$
> The *frequency* f of an oscillation is the number of cycles completed per unit time:
> $$f = 1/T.$$

THE RELATION BETWEEN PERIOD AND FREQUENCY FOR OSCILLATIONS IS THE SAME AS IN CIRCULAR MOTION.

SECTION 14.1 • SIMPLE HARMONIC MOTION

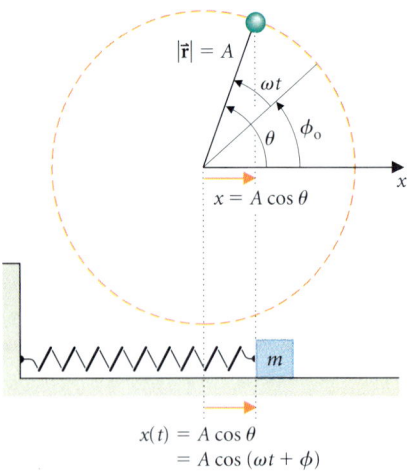

Figure 14.5
The amplitude of the oscillation equals the radius of the circle and the phase angle ϕ_o gives the location of the particle at $t = 0$.

The period $2\pi/\omega$ of the circular motion equals the period of the oscillation: $T = 2\pi\sqrt{m/k}$. The frequency of the oscillation is $f = \omega/2\pi$.

The position of the particle on a circle is specified by its angle θ from the x-axis, which changes uniformly in time:

$$\theta(t) = \omega t + \phi_o.$$

Then the value $x(t)$ for either the particle on the circle or the mass on the spring (Figure 14.5) is:

$$x(t) = A\cos(\omega t + \phi_o). \tag{14.3}$$

> The *amplitude A* of an oscillation is the magnitude of the particle's maximum displacement from the equilibrium position.

The oscillation amplitude is A, the maximum value of the spring stretch. It equals the radius of the analogous circle. The angle $\theta \equiv \omega t + \phi_o$ is called the *phase* of the oscillation, and ϕ_o, the value of the phase at $t = 0$, is the *phase constant*. This kind of oscillation is known as *simple harmonic motion* (SHM), and it is completely described by eqn. (14.3).

Math Toolbox

Harmonic Functions

The sine and cosine trigonometric functions are often called harmonic functions because they are important in the study of sound and music. The argument of the cosine function is called its phase. Because the function is periodic:

$$\cos(\theta + 2\pi) = \cos(\theta). \tag{14.4}$$

A particular phase is usually given as a number between 0 and 2π or between $-\pi$ and $+\pi$. For example, $\theta = 5\pi/2 = 2\pi + \pi/2$ corresponds to a phase of $\pi/2$.

The sine and cosine are closely related functions with the same periodic behavior. For example:

$$\cos(\theta - \pi/2) = \sin(\theta). \tag{14.5}$$

This relation between sine and cosine is described as a phase shift: the sine function is the cosine function shifted in phase by $\pi/2$. The graph of the sine function is the graph of the cosine function pushed to the right by $\pi/2$ (Figure 14.6). Since $\cos(\theta + \pi) = -\cos(\theta)$, inversion of the function corresponds to a phase shift of π. When a harmonic function is integrated or differentiated, the result is another harmonic function:

$$\frac{d}{d\theta}\cos(\theta) = -\sin(\theta); \quad \frac{d}{d\theta}\sin(\theta) = \cos(\theta).$$

After a second differentiation, the result is the negative of the original function:

$$\frac{d^2}{d\theta^2}\cos(\theta) = -\frac{d}{d\theta}\sin(\theta) = -\cos(\theta).$$

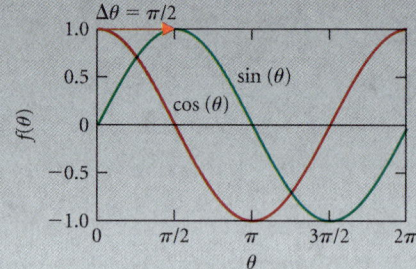

Figure 14.6
Harmonic functions are periodic with period 2π. The horizontal axis is the phase of the function. The sine function is the cosine function pushed to the right by phase $\pi/2$: $\cos(\theta - \pi/2) = \sin(\theta)$. Pushing another $\pi/2$ produces the function $f(\theta) = -\cos(\theta) = \cos(\theta - \pi)$: a phase shift of π inverts the cosine function.

Because of this property, both harmonic functions satisfy eqn. (14.1), the equation of simple harmonic motion.

In simple harmonic motion, the argument of the cosine function increases uniformly in time:

$$\theta = \omega t + \phi_o.$$

We are free to choose when to start the clock ($t = 0$). The phase constant ϕ_o is the value of the phase angle at that time. Two times t_1 and $t_2 = t_1 + T = t_1 + 2\pi/\omega$ are equivalent in the sense that both position and velocity are the same at each of these two times. The arguments $\theta_1 = \omega t_1 + \phi_o$ and $\theta_2 = \omega t_2 + \phi_o$ differ by $\omega T = 2\pi$, and so the phases are the same.

EXERCISE 14.2 ♦♦ Show that expression (14.3) for $x(t)$ satisfies eqn. (14.1) for any values of A and ϕ_o. [*Hint:* Differentiate $x(t)$ twice and substitute into eqn. (14.1).]

14.1.4 Using the Solution for Simple Harmonic Motion

In a mathematical sense, the problem of simple harmonic motion is solved by eqn. (14.3). The equation gives the position of the particle as a function of time, and its velocity is obtained by differentiation:

$$x(t) = A\cos(\omega t + \phi_o),$$

$$v(t) \equiv \frac{dx(t)}{dt} = \frac{d}{dt}[A\cos(\omega t + \phi_o)],$$

or

$$v(t) = -\omega A\sin(\omega t + \phi_o). \tag{14.6}$$

There are two (as yet) undetermined constants, A and ϕ_o, because the solution applies no matter where or when we start the oscillation. To apply the general solution (eqns. 14.3 and 14.6) to a particular situation, we use the given information to find these constants.

EXAMPLE 14.2 ♦♦ A particle of mass $m = 12$ kg is attached to the end of a spring with spring constant $k = 1.3 \times 10^4$ N/m (Figure 14.2). At $t = 0$, the spring is stretched 55 cm and the particle is released from rest. Describe the subsequent motion.

MODEL The particle on a spring system undergoes SHM, and we may apply eqns. (14.2), (14.3), and (14.6).

SETUP The angular frequency (eqn. 14.2) is:

$$\omega = \sqrt{\frac{k}{m}} = \sqrt{\frac{1.3 \times 10^4 \text{ N/m}}{12 \text{ kg}}} = \sqrt{1.1 \times 10^3 \text{ (rad/s)}^2} = 33 \text{ rad/s}.$$

The given position and velocity at $t = 0$ determine A and ϕ_o. Setting $t = 0$ in eqns. (14.3) and (14.6), we have:

$$x(t) = A\cos(\omega t + \phi_o) \Rightarrow 0.55 \text{ m} = x(0) = A\cos\phi_o, \quad \text{(i)}$$

and $$v(t) = -\omega A\sin(\omega t + \phi_o) \Rightarrow 0 = v(0) = -\omega A\sin\phi_o. \quad \text{(ii)}$$

Equation (ii) gives $\sin\phi_o = 0$, since neither the angular frequency nor the amplitude vanish. Thus $\phi_o = 0$, or π. Since x is positive at $t = 0$ and A is positive by definition, the solution we need is $\phi_o = 0$. Then from eqn. (i),

$$A\cos\phi_o = A = 0.55 \text{ m}.$$

REMEMBER: By convention, ϕ_o is taken to be between 0 and 2π or between $-\pi$ and $+\pi$.

SOLVE The position of the particle is given by:

$$x(t) = (0.55 \text{ m})\cos[(33 \text{ rad/s})t].$$

The maximum speed is $A\omega = (0.55 \text{ m})(33 \text{ rad/s}) = 18$ m/s. The velocity as a function of time is:

$$v(t) = -(18 \text{ m/s})\sin[(33 \text{ rad/s})t].$$

ANALYZE Notice that the product $A\omega$ has dimensions of L/T as required for a speed.
In this example, the given initial conditions generate two equations for the unknown constants A and ϕ_o. Two pieces of information are always needed. For example, position at two different times is sufficient.

We need to think carefully about the meaning of "different." If the times differ by a whole number of periods, then the phase is the same. These two times would not be considered "different." (See the *Math Toolbox*.)

EXERCISE 14.3 ♦♦ The particle-and-spring system described in Example 14.2 is set oscillating at $t = 0$. The particle is at $x = 0.0$ after 0.015 s and is at $x = -15$ cm after 0.030 s. Find $x(t)$.

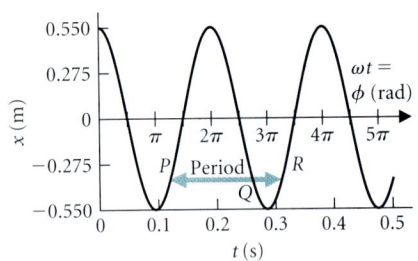
(a) Position graph.

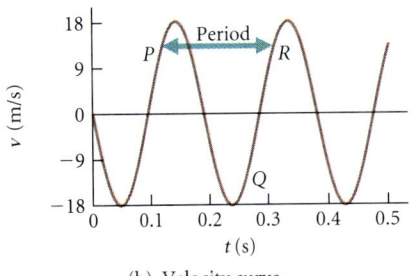
(b) Velocity curve.

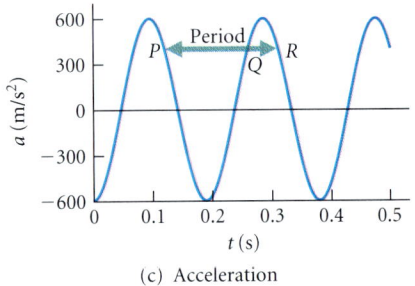
(c) Acceleration

■ **FIGURE 14.7**
Position, velocity, and acceleration for a particle in simple harmonic motion. (a) The position at $t = 0$ is at its greatest value, the amplitude $A = 0.55$ m. The time interval between points P and R represents one period. (b) The velocity curve is shifted to the left by $\Delta(\omega t) = \pi/2$ compared with the position graph. The velocity is maximum when the position is zero, and vice versa. The two curves are 90° *out of phase*. The particle's position is the same at P, Q, and R, but the velocity differs at Q. The period is PR, the time for the particle to return to the same position *with the same velocity*. (c) The acceleration is shifted by another $\Delta(\omega t) = \pi/2$ relative to the velocity curve. It is 180° out of phase with the position.

■ Figure 14.7 shows $x(t)$, $v(t)$, and $a(t)$ for the oscillation in Example 14.2. The velocity curve looks like the position curve shifted to the left by $\Delta(\omega t) = \pi/2$. The velocity is 90° *out of phase* with the position (see the *Math Toolbox*): when the displacement is maximum (phases 0, π, and 2π), velocity is zero; when the speed is maximum (phases $\pi/2$ and $3\pi/2$), the displacement is zero. Similarly, the acceleration curve is shifted by $\Delta(\omega t) = \pi$ from the position curve. The acceleration is 180° out of phase with the position: when position has its maximum positive value, acceleration has its maximum negative value. This happens because the restoring force is greatest when the displacement is greatest, and the restoring force is opposite the displacement. This is exactly the meaning of eqn. (14.1).

It is often useful to rewrite eqn. (14.3) as a sum of sine and cosine terms. Using a trigonometric identity from Appendix IB,

$$A \cos(\omega t + \phi_0) = A[\cos \phi_0 \cos \omega t - \sin \phi_0 \sin \omega t],$$

we may define two new constants that are functions of the constants A and ϕ_0: $B \equiv A \cos \phi_0$ and $C \equiv -A \sin \phi_0$. Then eqn. (14.3) becomes:

$$x(t) = B \cos \omega t + C \sin \omega t. \qquad (14.7)$$

The corresponding equation for the velocity is:

$$v(t) = -\omega B \sin \omega t + \omega C \cos \omega t. \qquad (14.8)$$

EXERCISE 14.4 ◆ Analyze the oscillation in Example 14.2 starting with eqns. (14.7) and (14.8). Use the initial conditions to find B and C.

EXAMPLE 14.3 ◆◆ A log of length $L = 3.00$ m and radius $r = 10.0$ cm is weighted at one end with a lump of lead so that it floats vertically (■ Figure 14.8). The total mass of the log and lead is $M = 60.0$ kg. If the log is pushed downwards a small amount $A = 10.0$ cm, show that simple harmonic motion results, and find the log's position as a function of time.

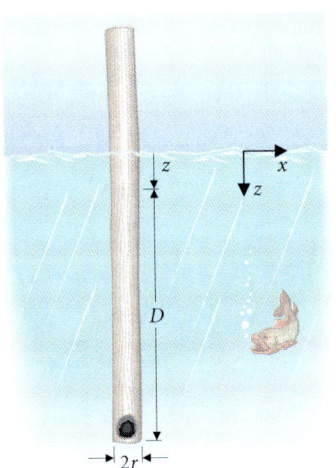

■ **FIGURE 14.8**
A floating log oscillates after being pushed downward. When displaced downward by an amount z, the excess buoyant force $\pi r^2 \rho_{\text{water}} gz$ is the restoring force. (The lead weight keeps the log vertical.)

MODEL The log floats because water pressure exerts an upward force at the bottom end. First we find the equilibrium position, then determine the restoring force.

SETUP According to Archimedes' principle (cf. §13.4), in the equilibrium state:

Total weight of log-plus-lead = weight of fluid displaced.

The log floats submerged to a depth D where:

$$Mg = \rho_{water} V_{submerged} g = \rho_{water}(\pi r^2 D)g. \quad (i)$$

Then: $\quad D = \dfrac{M}{\pi r^2 \rho_{water}} = \dfrac{60.0 \text{ kg}}{\pi(0.100 \text{ m})^2(1.00 \times 10^3 \text{ kg/m}^3)} = 1.91 \text{ m}.$

If the log is pushed down so that it is submerged to a depth $D + z$, a larger buoyant force acts on it because a greater weight of water is displaced. The buoyant force then exceeds the weight of log-plus-lead, and the log accelerates upward. Conversely, if the log were raised slightly above the equilibrium position, its weight would exceed the reduced buoyant force, and it would accelerate downward. The difference between weight and buoyant force acts as a restoring force, and the motion is oscillatory.

When the log is submerged to a depth $D + z$, the net downward force is:

$$\sum F_z = Mg - \pi r^2 (D + z)\rho_{water} g$$

But from eqn. (i), $\pi r^2 D \rho_{water} = M$, so:

$$\sum F_z = -\pi r^2 z \rho_{water} g.$$

Then, from Newton's second law:

$$M \dfrac{d^2 z}{dt^2} = \sum F_z = -(\pi r^2 \rho_{water} g) z.$$

SOLVE This equation has the same form as eqn. (14.1) for the mass on a spring. The displacement of the log is called z rather than x, and the effective spring constant $\pi r^2 \rho_{water} g \equiv k_{eff}$ replaces the constant k. The angular frequency is:

$$\omega = \sqrt{\dfrac{k_{eff}}{M}} = \sqrt{\dfrac{\pi r^2 \rho_{water} g}{M}} = \sqrt{\dfrac{g}{D}} = \sqrt{\dfrac{9.81 \text{ m/s}^2}{1.91 \text{ m}}} = 2.27 \text{ rad/s}. \quad \text{WE USED EQN. (i) AGAIN.}$$

The log is released from rest at $t = 0$. When initial conditions are given, it is usually easiest to use eqns. (14.7) and (14.8):

$$10.0 \text{ cm} = z(0) = B \quad \text{and} \quad 0 = v(0) = \omega C.$$

So: $\quad z(t) = (10.0 \text{ cm})\cos[(2.27 \text{ rad/s})t].$

ANALYZE Check the dimensions of ω:

$$[\omega] = \sqrt{\dfrac{L/T^2}{L}} = \dfrac{1}{T}, \quad \text{as required.}$$

Note that SHM occurs because the net z-component of the force on the log is proportional to $-z$.

14.2 THE PENDULUM

14.2.1 The Simple Pendulum

A simple pendulum is a particle of mass m suspended by a string of length ℓ. The particle—called the pendulum *bob*—swings without friction from a fixed suspension point. In equilibrium, the string hangs straight down, and its tension supports the particle against gravity (■ Figure 14.9). The lines of action of both forces pass through the suspension point, and there is no net torque about that point. When the string is displaced by an angle θ, the torque due to the particle's weight tends to restore the particle to its equilibrium position. We expect the system to oscillate.

With x to the right and y up, the net torque about the suspension point is:

$$\vec{\tau} = \vec{r} \times \vec{F} = -mg\ell\,(\sin\theta)\hat{k}.$$

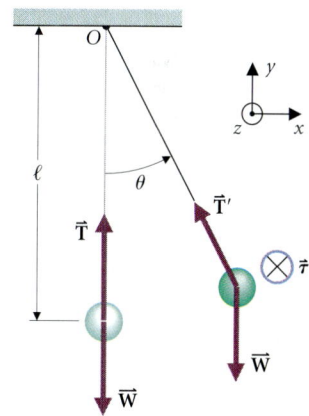

■ FIGURE 14.9
A simple pendulum is a particle on the end of a string or a massless rod of length ℓ. In equilibrium, tension in the string balances weight. When the bob is displaced, the weight exerts a torque about the support point O that accelerates the pendulum toward the equilibrium position.

WE CALCULATED THIS TORQUE IN EXAMPLE 9.5. REMEMBER TO USE THE RIGHT-HAND RULE.

It produces an angular acceleration α given by:

$$I\vec{\alpha} = \vec{\tau}.$$

REMEMBER: IDEAL STRING IS MASSLESS AND DOESN'T STRETCH.

Since the ideal string is massless, the rotational inertia of the pendulum is that of the particle alone: $I = m\ell^2$. Then,

$$m\ell^2 \frac{d^2\theta}{dt^2}\hat{k} = -mg\ell(\sin\theta)\hat{k},$$

or

$$\frac{d^2\theta}{dt^2} = -\frac{g\sin\theta}{\ell}.$$

SEE THE *MATH TOPIC* IN §3.2.3 FOR A DISCUSSION OF THIS APPROXIMATION TO THE SINE. REMEMBER THAT θ MUST BE IN RADIANS.

This equation is not of the standard form (14.1), and it has no simple solutions. However, when the amplitude is small, we may use the approximation $\sin\theta \approx \theta$. Then:

$$\frac{d^2\theta}{dt^2} \approx -\frac{g}{\ell}\theta,$$

COMPARE THIS WITH THE RESULT OF EXAMPLE 14.3. WHEN A SYSTEM MOVES UNDER THE INFLUENCE OF GRAVITY, THE EXPRESSION FOR ω^2 IS OF THE FORM $g/$(RELEVANT LENGTH SCALE). THE MASS OF THE BOB IS IRRELEVANT.

which is the equation for simple harmonic motion with θ as the displacement and with angular frequency $\omega = \sqrt{g/\ell}$. Thus:

$$\theta(t) \approx (\theta_{max})\cos\left(\sqrt{\frac{g}{\ell}}t + \phi_o\right).$$

Take care to distinguish $\omega = \sqrt{g/\ell}$ from $d\theta/dt$. The pendulum's angular speed, $d\theta/dt$, is not constant and does not equal the angular frequency ω of its oscillation.

WHEN NUMERICAL DATA ARE GIVEN, YOU CAN USE A COMPUTER TO INTEGRATE THE DIFFERENTIAL EQUATION. SEE PROBLEM 77.

Since $|\sin\theta| < |\theta|$ for any angle θ, the pendulum always accelerates less rapidly than it would in SHM. It takes more time to complete a swing than the SHM solution predicts. The exact solution is more complicated than we need here. The SHM solution is useful when the amplitude θ_{max} is less than about 10°.

The pendulum illustrates a very general result: large-amplitude oscillations of a system may be complex, but small oscillations are simple harmonic. The motion of a particle on a spring takes us a long way toward understanding oscillatory behavior in general.

EXERCISE 14.5 ♦ What fractional error is made in the angular acceleration using the SHM approximation when θ is exactly 1°, 10°, 20°, or 90°?

EXAMPLE 14.4 ♦ Each swing of a clock's pendulum from maximum displacement on one side to maximum displacement on the other is to take 1.00 s. What is the required length ℓ_o of a simple pendulum for this clock?

MODEL We model the clock's pendulum as a simple pendulum, assuming the amplitude is small.

SETUP The period is twice the 1.00 s required to go from maximum on one side to maximum on the other, $T = 2.00$ s. Using our result for the angular frequency ω:

$$T = \frac{2\pi}{\omega} = 2\pi\sqrt{\frac{\ell_o}{g}}.$$

SOLVE $\ell_o = g\left(\frac{T}{2\pi}\right)^2 = (9.81\text{ m/s}^2)\left(\frac{2.00\text{ s}}{2\pi}\right)^2 = 0.994$ m.

ANALYZE A pendulum this slow is appropriate for a grandfather clock.

14.2.2 The Physical Pendulum

A particle on a string is too simple a model for Galileo's chandelier or for a typical clock pendulum. A better model is the physical pendulum, a rigid body with its center of mass a distance ℓ below the support and a rotational inertia I about the support point (■ Figure 14.10). To see how the angular frequency depends on I and ℓ, we proceed as for the simple pendulum. The torque about the support point due to the pendulum's weight, $\vec{W} = m\vec{g}$, is:

$$\vec{\tau} = -mg\ell(\sin\theta)\hat{k}.$$

The torque produces an angular acceleration:

$$I\vec{\alpha} = \vec{\tau}.$$

$$I\frac{d^2\theta}{dt^2}\hat{k} = -mg\ell(\sin\theta)\hat{k}.$$

$$\frac{d^2\theta}{dt^2} = -\frac{mg\ell}{I}\sin\theta \approx -\frac{mg\ell}{I}\theta,$$

where we have used the small-angle approximation. The angular frequency of the pendulum is:

$$\omega = \sqrt{\frac{mg\ell}{I}}. \tag{14.9}$$

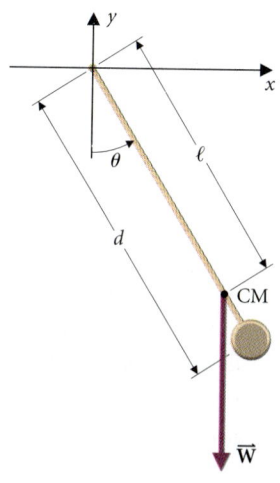

FIGURE 14.10
The physical pendulum. A real pendulum is a rigid body with rotational inertia I about the pivot point and center of mass a distance ℓ below the pivot. The pendulum shown here consists of a bob of mass m_2 attached to the end of a rod of length d and mass m_1: $m = m_1 + m_2$.

EXAMPLE 14.5 ♦♦ A clock pendulum is made out of a rod of length d and mass m_1 with a bob of mass m_2 at its end. What is the required length d of the rod for the clock in Example 14.4 ($T = 2.00$ s) if $m_1/m_2 \equiv \mu = 0.100$? Compare with the result for ℓ_o in Example 14.4.

MODEL To use eqn. (14.9) for the physical pendulum, we'll need to determine ℓ, the distance of the CM from the support, and the rotational inertia I for the rod-and-bob system. The tools we'll need are found in §11.5 and §12.5. We model the pendulum bob as a particle and the rod as uniform.

SETUP We use eqn. (11.4) [or equivalently, (9.16)] to locate the pendulum's CM:

$$\ell = \frac{m_1 d/2 + m_2 d}{m_1 + m_2} = d\left(\frac{1 + \mu/2}{1 + \mu}\right).$$

From Table 12.3, the rotational inertia of the rod about one end is $m_1 d^2/3$, so the rotational inertia of the bob-plus-rod about the support point is:

$$I = m_1 d^2/3 + m_2 d^2 = m_2 d^2(1 + \mu/3).$$

Then:
$$\omega^2 = (m_1 + m_2)g\ell/I.$$

$$\left(\frac{2\pi}{T}\right)^2 = \frac{(m_1 + m_2)gd(1 + \mu/2)/(1 + \mu)}{m_2 d^2(1 + \mu/3)}.$$

SOLVE Since $(m_1 + m_2)/m_2 = \mu + 1$, the expression on the right-hand side simplifies:

$$\left(\frac{2\pi}{T}\right)^2 = \frac{g}{d}\left(\frac{1 + \mu/2}{1 + \mu/3}\right).$$

Then:
$$d = g\left(\frac{T}{2\pi}\right)^2\left(\frac{1 + \mu/2}{1 + \mu/3}\right).$$

In Example 14.4 we found $\ell_o = g(T/2\pi)^2 = 0.994$ m. Thus with $\mu = 0.100$:

$$d = \ell_o\left(\frac{1 + \mu/2}{1 + \mu/3}\right) = \ell_o\left(\frac{1 + 1/20}{1 + 1/30}\right) = 1.02\ell_o = 1.01 \text{ m}.$$

ANALYZE The correction $d - \ell_o$ is only 2 cm, or 2%. It seems small, but a pendulum misadjusted by this much would lose 10 minutes per day! If greater accuracy is needed, we should include the finite size of the bob when calculating its rotational inertia. The correction reduces the calculated length slightly. At this level of accuracy, we should also worry about the local value of g. (See Problem 14.44.)

SECTION 14.2 • THE PENDULUM

14.3 Energy in Oscillatory Motion

Only conservative forces (gravitational or spring forces) do work on the oscillating systems we have discussed, so their total mechanical energy is conserved. As each system cycles back and forth, its energy continuously cycles between potential and kinetic forms (■ Figure 14.11). With its speed given by eqn. (14.6), the kinetic energy of a particle on a spring is:

$$K = \tfrac{1}{2}mv^2 = \tfrac{1}{2}m\omega^2 A^2 \sin^2(\omega t + \phi_o).$$

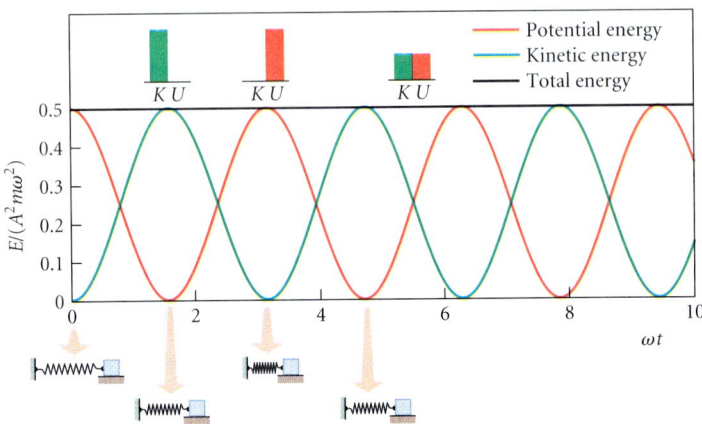

■ **Figure 14.11**
In simple harmonic motion, the energy oscillates between kinetic and potential forms. The total is constant. The dimensionless quantity $E_{\text{total}}/(A^2 m\omega^2) = 0.5$.

The spring stretch equals $x(t)$ (eqn. 14.3), so the system's potential energy is:

$$U = \tfrac{1}{2}kx^2 = \tfrac{1}{2}kA^2 \cos^2(\omega t + \phi_o).$$

Since $\omega = \sqrt{k/m}$ (eqn. 14.2), then $k = m\omega^2$, and the total energy is:

REMEMBER: $\sin^2\theta + \cos^2\theta = 1$ FOR ANY θ.

$$E = K + U = \tfrac{1}{2}kA^2[\sin^2(\omega t + \phi_o) + \cos^2(\omega t + \phi_o)] = \tfrac{1}{2}kA^2. \qquad (14.10)$$

The constant value of the energy equals the sum of $K + U$ at any time t. In particular, at maximum displacement, where $K = 0$, E equals the potential energy in the spring: $E = U_{\max} = \tfrac{1}{2}kA^2$. As the particle passes its equilibrium position where $U = 0$, E equals the kinetic energy $K_{\max} = \tfrac{1}{2}m\omega^2 A^2$. Thus the amplitude of an oscillation is directly related to its energy: $A = \sqrt{2E/k}$.

A similar result holds for the simple pendulum. Taking the reference level at the equilibrium position (■ Figure 14.12), the gravitational potential energy is:

$$U = mg\ell(1 - \cos\theta) \approx \tfrac{1}{2}mg\ell\theta^2 = \tfrac{1}{2}mg\ell\theta_{\max}^2 \cos^2(\omega t + \phi_o),$$

where we used the small-angle approximation $\cos\theta \approx 1 - \tfrac{1}{2}\theta^2$. The bob's speed is $v = \ell\, d\theta/dt$, so its kinetic energy is:

$$K = \tfrac{1}{2}mv^2 = \tfrac{1}{2}m\ell^2\omega^2\theta_{\max}^2 \sin^2(\omega t + \phi_o).$$

With $\omega^2 = g/\ell$, the total energy of the pendulum is:

$$E = K + U = \tfrac{1}{2}m\ell^2(g/\ell)\theta_{\max}^2 \sin^2(\omega t + \phi_o) + \tfrac{1}{2}mg\ell\theta_{\max}^2 \cos^2(\omega t + \phi_o)$$

$$E = \tfrac{1}{2}mg\ell\theta_{\max}^2. \qquad (14.11)$$

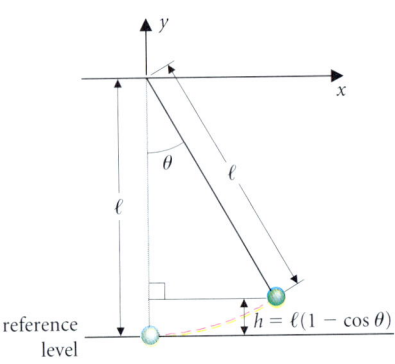

■ **Figure 14.12**
Potential energy of a simple pendulum. With the reference level chosen to be at the equilibrium position of the bob, $U = mgh = mg\ell(1 - \cos\theta)$.

For the pendulum, too, energy is proportional to the square of the amplitude. In fact, this result is true for any system undergoing simple harmonic motion.

EXAMPLE 14.6 ♦♦ A circus performer jumps for the trapeze as it swings 15 m below its support in the big top. He grabs it when the supporting cables make an angle of 10° with the vertical, his velocity is perpendicular to the cable, and his speed is 3.9 m/s (■ Figure 14.13). Neglecting the mass of the trapeze itself, find the amplitude of the resulting swing.

MODEL We model the performer-plus-trapeze as a simple pendulum, neglecting both the mass of the trapeze bar and the distance from the bar to the performer's CM.

SETUP We use the given speed and displacement to calculate the total energy of the system. The initial speed of the trapeze is not important since we are neglecting its mass. We assume that θ remains small throughout the oscillation and use eqn. (14.11) to find the amplitude:

$$E = K + U = \tfrac{1}{2}mv^2 + \tfrac{1}{2}mg\ell\theta^2 = \tfrac{1}{2}mg\ell\theta_{max}^2.$$

This equation relates the amplitude θ_{max} (in radians!) to the energy E. We were given θ in degrees, so we have to convert:

SOLVE
$$\theta_{max}^2 = \frac{2E}{mg\ell} = \frac{v^2}{g\ell} + \theta^2$$
$$= \frac{(3.9 \text{ m/s})^2}{(9.8 \text{ m/s}^2)(15 \text{ m})} + \left(\frac{10°}{180°}\pi\right)^2$$
$$= 0.10 + 0.03 = 0.13,$$

and $\theta_{max} = 0.37 \text{ rad} = 21°$.

ANALYZE There is about a 2% error in the solution of $\theta(t)$ obtained by setting $\sin\theta = \theta$ with $\theta = 20°$ (Exercise 14.5). The error in the cosine function, which is of interest here, is smaller. Our answer for θ_{max} was rounded from 0.366 rad. Then:

$$\cos(0.366) = 0.934; \quad 1 - \tfrac{1}{2}\theta^2 = 1.000 - 0.366^2/2 = 0.933.$$

The values differ by $\tfrac{1}{10}$%. Simple harmonic motion is an adequate approximation for the trapeze, especially given the uncertainties in our model. ■

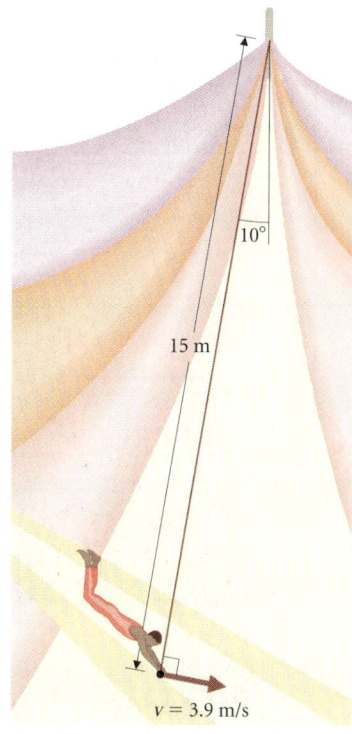

■ **FIGURE 14.13**
Through what angle does the trapeze performer swing?

Systems like the spring and the pendulum are described by a single position variable, and a graph of potential energy versus position is useful for analyzing their oscillations (■ Figure 14.14). As we showed in Chapter 11, the equilibrium state corresponds to a minimum of the potential energy. A restoring force tends to return the system to equilibrium, doing positive work ($\vec{F} \cdot d\vec{\ell} > 0$), and thus reducing the system's potential energy.

WHEN THE SYSTEM DOES WORK, ITS POTENTIAL ENERGY DECREASES (CHAPTER 8). WE DISCUSSED THE RELATION BETWEEN EQUILIBRIUM AND POTENTIAL ENERGY IN §11.1.2.

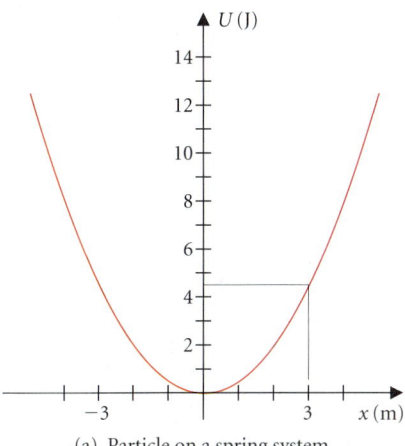
(a) Particle on a spring system.

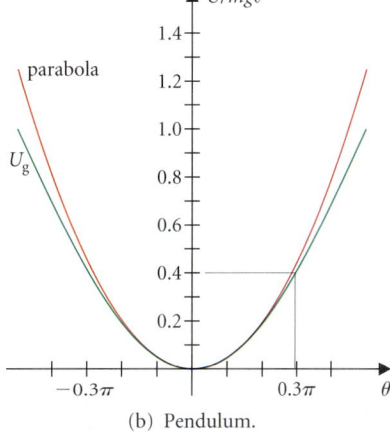
(b) Pendulum.

■ **FIGURE 14.14**
Potential energy as a function of position. At the end points of the oscillation, $E = U$, so these points may be read off the graph once E is known. (a) Particle on a spring system. The graph is a parabola: $U = \tfrac{1}{2}kx^2$. For example, if the system represented here has an energy of 4.5 J, its amplitude is 3 m. (b) Pendulum. The exact function is $U_g = mg\ell(1 - \cos\theta)$. The parabola is the small-angle approximation $U = \tfrac{1}{2}mg\ell\theta^2$. The two curves are indistinguishable at this scale for $\theta < 0.2\pi$ rad $= 0.63$ rad $= 36°$. The horizontal line at $U = 0.4mg\ell$ is the energy of the pendulum in Example 14.7.

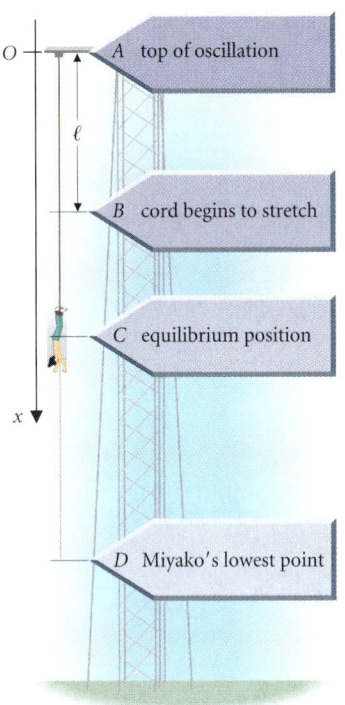

■ **FIGURE 14.15**
Bungee jumper. The bungee (= spring) has unstretched length ℓ. When Miyako steps off the top of the tower, she becomes an oscillating particle. During the first part of the oscillation—from A to B—she is in free fall. At B, the bungee begins to stretch, and Miyako undergoes SHM from B to D and back to B.

A bungee jumper leaps from a tower.

At the end points of an oscillation, the particle is momentarily at rest, so the kinetic energy is zero. If the total energy E in the oscillation is known, the end points where $U = E$ may be read off the potential energy graph.

The graph for the particle on a spring is a parabola, $U = \frac{1}{2}kx^2$, while the graph for the pendulum is only approximately a parabola: $U_g = mg\ell(1 - \cos\theta) \approx \frac{1}{2}mg\ell\theta^2$. The red curve in Figure 14.14b is a parabola that approximates the true curve near the equilibrium position—that is, for small-amplitude motion.

> **EXAMPLE 14.7** ♦♦ A pendulum has kinetic energy $0.40mg\ell$ at $\theta = 0$. Use Figure 14.14b to determine the amplitude of the oscillation.
>
> **MODEL** The total energy of the system remains constant. At $\theta = 0$, all the energy is kinetic. When the displacement is equal to the amplitude, all the energy is in the form of potential energy.
>
> **SETUP** We use the exact curve to determine the amplitude, since the approximate curve differs noticeably at $U = 0.40mg\ell$.
>
> **SOLVE** Reading off the graph, the potential energy is $0.40mg\ell$ when $\theta = 0.30\pi$ rad = $54°$.
>
> **ANALYZE** With an amplitude this large, the small-angle approximation is poor ($\sin 54° = 0.81$ versus $\theta = 0.3\pi = 0.94$). The error in the potential energy is 8%; the error in the position is 16%.

Study Problem 11 ♦♦♦ Bungee Jumping!

A bungee cord attached to a tower behaves like a spring of constant $k = 29.0$ N/m and unstretched length $\ell = 25.0$ m (■ Figure 14.15). Miyako, whose mass is $m = 55.0$ kg, steps off the top of the tower. Find how far Miyako falls, and the magnitude of the maximum acceleration she experiences.

Modeling the System

A bungee cord behaves like a spring when stretched, but it cannot be compressed. Until the cord begins to stretch, Miyako is in free fall! Once she reaches a distance ℓ below the top of the tower, the cord begins to stretch and exerts an upward force that slows her fall, stops her, and then accelerates her up again: she oscillates.

We choose the reference level for gravitational energy to be at Miyako's original position (A in Figure 14.15). When Miyako steps out, both her kinetic energy and her potential energy are zero. As she falls from the tower, her potential energy $U_g = -mgx$ decreases rapidly at first because the cord is slack and she gets closer to the Earth (■ Figure 14.16). But once the cord begins to stretch, its elastic energy $U_e = \frac{1}{2}k(x - \ell)^2$ increases. The total potential energy

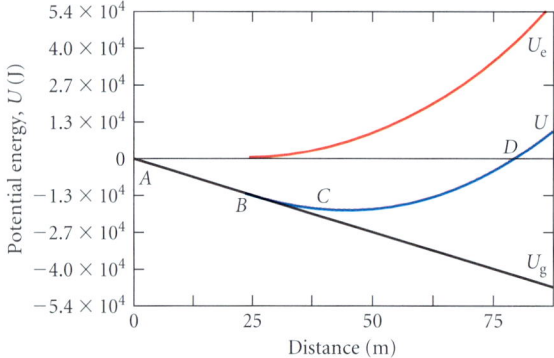

■ **FIGURE 14.16**
Miyako's potential energy plotted versus her distance below the top of the tower. The straight line represents her gravitational energy: $U_g = -mgx$. The elastic energy remains zero until point B ($x = \ell$), where the bungee cord begins to stretch. The total potential energy has a minimum at the equilibrium position, point C, where $x = 43.6$ m.

$U_e + U_g$ decreases to a minimum at point C and increases back to zero at point D. Miyako oscillates between her original position (point A) and point D. The motion is simple harmonic between B and D, where the cord is stretched, but she is in free fall between A and B. The maximum acceleration occurs where the cord's stretch is maximum—at the bottom.

Plan

Below point B, Miyako undergoes SHM about her equilibrium position, point C. With our chosen coordinates, the system's total energy $E = 0$ and Miyako's maximum distance below the tower may be read off the graph of potential energy, or found analytically by setting $U(x) = 0$. To find the acceleration, we need both the amplitude and the angular frequency of the simple harmonic motion.

Construction of Equations

We begin by finding the coordinate x_C of the equilibrium position. With the reference level for gravitational energy at the support point, the potential energy as a function of x is:

$$U(x) = U_g + U_e = \begin{cases} -mgx, & x < \ell; \\ -mgx + \tfrac{1}{2}k(x-\ell)^2, & x > \ell. \end{cases} \quad (i)$$

REMEMBER: $U(x)$ MEANS U AT x, NOT U TIMES x.

Point C is at the minimum of U, where $dU/dx = 0$. For $x < \ell$, there is no spring force to oppose the gravitational force. Thus x_C must be greater than ℓ.

At x_C: $\quad \dfrac{dU}{dx} = -mg + k(x - \ell) = 0.$

So: $\quad x_C = \ell + mg/k \quad (ii)$

$\quad\quad = 25.0 \text{ m} + \dfrac{(55.0 \text{ kg})(9.81 \text{ m/s}^2)}{29.0 \text{ N/m}}$

$\quad\quad = 25.0 \text{ m} + 18.6 \text{ m} = 43.6 \text{ m}.$

AT THIS POSITION, THE SPRING FORCE BALANCES THE WEIGHT. CHECK IT!

Next, we find how far Miyako falls. Her kinetic energy is again zero at her lowest position $x_D > \ell$, so the system's potential energy is also zero there.

$$0 = U(x_D) = -mgx_D + \tfrac{1}{2}k(\ell - x_D)^2.$$

$$x_D^2 - 2x_D(\ell + mg/k) + \ell^2 = 0.$$

Using the solution we found for x_C in eqn. (ii) gives:

$$x_D^2 - 2x_D x_C + \ell^2 = 0.$$

Solution

$$x_D = x_C \pm \sqrt{x_C^2 - \ell^2}.$$

One solution has $x_D > x_C$, and one has $x_D < x_C$. Since D is below C, $x_D > x_C$, and we need to take the + sign.

$$x_D = x_C + \sqrt{x_C^2 - \ell^2}. \quad (iii)$$

$\quad\quad = 43.6 \text{ m} + \sqrt{(43.6 \text{ m})^2 - (25.0 \text{ m})^2}$

$\quad\quad = 43.6 \text{ m} + 35.7 \text{ m} = 79.3 \text{ m}. \quad (iv)$

Construction of Equations

Below point B, the motion is simple harmonic with a frequency determined by k; $\omega = \sqrt{k/m}$. We may confirm this by writing the energy as a function of distance $z = x - x_C$ from the equilibrium point. Then $x = z + x_C = z + \ell + mg/k$, the spring stretch is $x - \ell = z + mg/k$, and eqn. (i) becomes:

$$U(z) = -mg(z + \ell + mg/k) + \tfrac{1}{2}k(z + mg/k)^2$$
$$= -mg\ell - (mg)^2/(2k) + \tfrac{1}{2}kz^2. \qquad \text{(v)}$$

The constant terms in $U(z)$ are without physical importance, since they do not change as Miyako moves. The term depending on z^2 is the same as for a spring in the absence of gravity. Miyako undergoes SHM about the equilibrium point with angular frequency:

$$\omega = \sqrt{\frac{k}{m}} = \sqrt{\frac{29.0 \text{ N/m}}{55.0 \text{ kg}}} = 0.726 \text{ rad/s},$$

and amplitude $A = x_D - x_C = 35.7$ m.

Solution

Miyako's acceleration in SHM is $-\omega^2 z$ (eqn. 14.1) and has its maximum value $-\omega^2 A$ when z has its maximum value $A = 35.7$ m:

$$|a_{\max}| = \omega^2 A = (0.726 \text{ rad/s})^2 (35.7 \text{ m}) = 18.8 \text{ m/s}^2.$$

Analysis

Miyako's total excursion from the starting point is 79 m! We hope she started at least that far off the ground. With a maximum acceleration of 18.8 m/s², or just under $2g$, her ride is exciting but not dangerous. By using a cord with a larger value of k, she could reduce the distance she falls, but at the expense of increasing the acceleration she must endure.

A real bungee jumper suffers significant air resistance during the oscillations and so never returns to the starting point. Miyako ultimately comes to rest at point C.

✱ 14.4 THE EFFECT OF EXTERNAL FORCES

14.4.1 Forced Oscillations and Resonance

When an external force is applied to an oscillating system, the resulting behavior varies greatly depending on the nature of the force. A constant force, such as the bungee jumper's weight, simply displaces the oscillator's equilibrium point without changing its oscillation frequency. In contrast, a periodic applied force alters the system's oscillations. At the right frequency, a relatively small applied force can produce oscillations large enough to disrupt a system. Let's see how this happens.

Roughly speaking, it is the variation of the external force over the system's natural oscillation period—its period when no external forces are applied—that determines how the oscillator responds to external influences. The system responds to a slowly varying external force almost as if the force were constant; the displacement from equilibrium is proportional to the force and varies slowly as the force varies. An external force that changes rapidly averages nearly to zero over a natural oscillation period and produces only a small effect. The greatest response occurs when the applied force operates at the system's natural frequency. Then the applied force remains parallel to the velocity of the system and continuously increases its energy. A child tugging on the ropes of a swing uses this principle to go higher.

Suppose a force $\vec{F}_{\text{ext}} = \hat{k} F_o \cos \omega t$ is applied to a system with natural oscillation frequency ω_o. Then the resulting oscillation also has angular frequency ω (see *Digging Deeper*):

$$z_{\text{ext}} = B \cos \omega t, \quad \text{with } B = \frac{F_o/m}{\omega_o^2 - \omega^2}. \qquad (14.12)$$

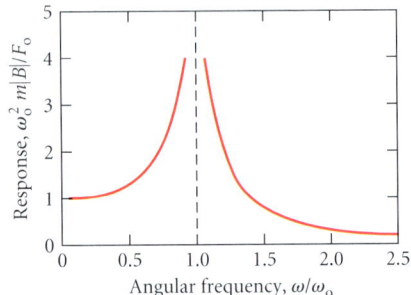

■ FIGURE 14.17
Response of an oscillator with natural frequency ω_o to a driving force at frequency ω. The horizontal axis shows the ratio of the driving frequency ω to the natural frequency ω_o. The y-axis displays the dimensionless amplitude $|B|m\omega_o^2/F_o$. The infinite response when the two frequencies are equal does not occur for real systems. In many cases, the model used to calculate the response fails as the amplitude grows: springs cease to obey Hooke's law; logs pop out of the water; a pendulum can go "over the top" (cf. Figure 14.18, which shows how friction can limit the amplitude).

At low frequencies, the amplitude tends to a constant value.

For $\omega \ll \omega_o$: $\qquad B \approx F_o/(m\omega_o^2)$.

Since B is positive, z_{ext} has the same sign as F_{ext}, so the displacement has the same direction as the applied force. At high frequencies, the amplitude tends to zero.

For $\omega \gg \omega_o$: $\qquad B \approx -F_o/(m\omega^2)$.

Here B is negative, which means the displacement is in the opposite direction from the applied force. The amplitude varies in frequency as we argued it should (■ Figure 14.17).

The amplitude shows its most interesting behavior near $\omega = \omega_o$, where the response becomes very large, a phenomenon known as *resonance*. The amplitude is formally infinite for ω exactly equal to ω_o. Our simplified model for the oscillator isn't valid near $\omega = \omega_o$. For example, a floating log (Example 14.3) would pop out of the water long before it could achieve infinite displacement! Real water exerts friction on the log that limits its amplitude when $\omega \approx \omega_o$ (■ Figure 14.18).

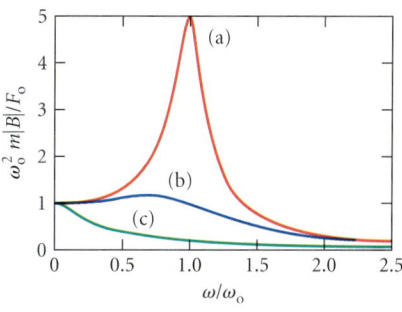

■ **FIGURE 14.18**
Response of the oscillator when friction is included. Curve (a) (small energy loss) is similar to Figure 14.17, but the amplitude remains finite at $\omega = \omega_o$. Greater energy loss (curves b and c) reduces or even eliminates the resonance. In terms of the damping time τ (§14.4.2), curves a, b and c correspond to $\omega_o \tau = 5$, 1, and 0.2, respectively.

Digging Deeper

THE AMPLITUDE OF A FORCED OSCILLATION

To verify these qualitative observations, we apply Newton's second law:

Net force = restoring force
    + external force
  = mass × acceleration.

The restoring force, like that for a spring, may be expressed in terms of the angular frequency ω_o of the oscillation it causes when acting alone. If $\vec{s}$ is the displacement of the system from equilibrium, the internal restoring force is:

$$\vec{F}_{int} = -k\vec{s} = -m\omega_o^2 \vec{s}.$$

We want to find the response to a periodic applied force with angular frequency ω:

$$\vec{F}_{ext} = \vec{F}_o \cos(\omega t),$$

which acts along the same line as the restoring force. We choose the z-axis to be along this line. Then $\vec{s} = z\hat{k}$ and a forced oscillation is described by:

$$m\vec{a} = \vec{F}_{int} + \vec{F}_{ext} = -m\omega_o^2 \vec{s} + \vec{F}_o \cos(\omega t),$$

or $\qquad m \dfrac{d^2 z}{dt^2}\hat{k} = (-m\omega_o^2 z + F_o \cos \omega t)\hat{k}.$ (i)

To solve this equation, we use the following two principles:

- The total displacement $z(t)$ is the sum of the response to the internal restoring force (a free oscillation) and a separate response to the external applied force:

$$z(t) = z_{int}(t) + z_{ext}(t), \qquad (ii)$$

where $\qquad z_{int}(t) = A \cos(\omega_o t + \phi_o).$ (iii)

- The response to an oscillating applied force is an oscillation at the same frequency:

$$z_{ext}(t) = B \cos(\omega t). \qquad (iv)$$

(Here again the technique for solving the puzzle is to make a guess and then test it. In math classes, z_{int} is described as the solution to the homogeneous equation, and z_{ext} is the solution to the inhomogeneous equation.)

Next we demonstrate that expressions (ii) through (iv) solve eqn. (i), and then we find the unknown amplitude B. First, we evaluate the acceleration:

$$\frac{d^2 z}{dt^2} = \frac{d^2 z_{int}(t)}{dt^2} + \frac{d^2 z_{ext}(t)}{dt^2}$$
$$= -\omega_o^2 z_{int}(t) - \omega^2 z_{ext}(t).$$

Then we substitute into eqn. (i):

$$m(-\omega_o^2 z_{int} - \omega^2 z_{ext}) = -m\omega_o^2 (z_{int} + z_{ext}) + F_o \cos \omega t.$$

The terms in z_{int} cancel. The remaining terms are:

$$-m\omega^2 z_{ext}(t) = -m\omega_o^2 z_{ext}(t) + F_o \cos \omega t.$$

Using expression (iv) for $z_{ext}(t)$:

$$0 = [(\omega^2 - \omega_o^2)B + F_o/m]\cos \omega t.$$

This relation must hold at all times t, when $\cos \omega t$ takes all values between -1 and $+1$. So the quantity in brackets vanishes. Thus:

$$B = \frac{F_o/m}{\omega_o^2 - \omega^2},$$

and $\qquad z_{ext}(t) = \dfrac{F_o \cos \omega t}{m(\omega_o^2 - \omega^2)}.$

The trial solution satisfies eqn. (i), and we obtain a specific expression for the response to the applied force.

Digging Deeper

A Driving Force

One possible origin of a driving force is a wave that raises and lowers the water level at the location of a floating log by an amount $h(t) = h_o \cos \omega t$ (Figure 14.19). Then the buoyant force on the log is:

$$\vec{F}_{\text{buoyant}} = g\rho_{\text{water}} \pi r^2 \{D + [z(t) + h(t)]\}(-\hat{k}).$$

The first term in this equation is the buoyant force that balances the weight of the log at equilibrium, so the net force on the log is:

$$\vec{F}_{\text{net}}(t) = \vec{W} + \vec{F}_{\text{buoyant}}$$
$$= (Mg/D)[z(t) + h(t)](-\hat{k})$$
$$= -M\omega_o^2 [\underbrace{z(t)}_{\text{restoring force}} + \underbrace{h(t)}_{\text{driving force}}]\hat{k},$$

where we used the result $\omega_o^2 = g/D$.

Usually, we see waves in a harbor or on a lake that have a low frequency compared with the oscillation frequency of a floating log. Consequently, the log appears simply to rise and fall with the water, and we don't perceive it as an oscillating system exposed to an oscillating force. That is precisely what we predict for the low-frequency response of the log.

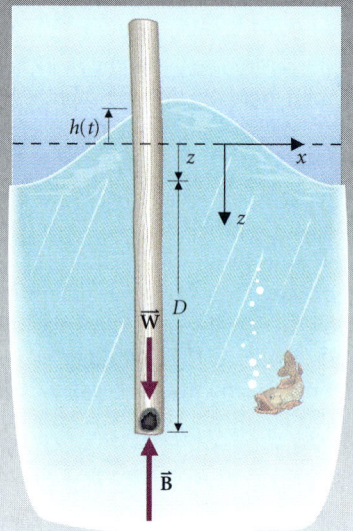

Figure 14.19
Model for a driving force. A wave passing a floating log changes the water depth by an amount $h(t) = h_o \cos \omega t$ and produces a harmonically varying buoyant force: $\vec{F} = F_o \cos \omega t (-\hat{k})$ with $F_o = -M\omega_o^2 h_o$.

This hotel balcony collapsed as a result of a resonant response to dancers' stomping in rhythm.

EXERCISE 14.6 ♦♦ In Study Problem 11, the external force is Miyako's weight, which does not vary in time. Thus the applied force has $\omega = 0$. Show that with $\omega = 0$ and $\omega_o^2 = k/m$, eqn. (14.12) gives the same solution that we obtained in the study problem: $z_{\text{res}} = x_C - \ell = mg/k$.

Resonant systems may be used to produce vibrations at a desired frequency, as in wristwatches and clocks, but sometimes the response is a nuisance. Resonant phenomena can produce surprisingly large motions and must be considered carefully in engineering design. The consequences of not being careful can be disastrous. In one instance, a hotel balcony collapsed when it resonated with the rhythmic stomping of dancing feet. The dancers provided an unanticipated driving force!

14.4.2 Damping

A car's shock absorbers are oscillation stoppers. With neither springs nor shocks, the body of the car would jolt impulsively every time the wheels went over a bump. The springs soften the rigid coupling between the wheels and the car's body. However, with springs alone, every bump would cause the car to oscillate, so shock absorbers are added to remove energy from these oscillations—that is, to *damp* the system.

We'll begin by analyzing one piece of the system—a shock absorber modeled as a piston in a cylinder of fluid (Figure 14.20). As the piston moves, fluid is forced through holes in the piston face and exerts a frictional force proportional to the piston's speed. When the block is given an impulse at $t = 0$, the shock absorber damps the resulting motion. According to Newton's second law:

Mass × acceleration of block = force on shock absorber piston.

$$m \frac{d\vec{v}}{dt} = -b\vec{v} = -bv\,\hat{i}. \qquad (14.13)$$

THIS IS THE SAME EQUATION WE FOUND IN THE ISOTHERMAL ATMOSPHERE CALCULATION (§13.3.4) AND ITS SOLUTION FOLLOWS THE SAME PATTERN.

The proportionality constant b is a measured property of the shock absorber. The minus sign appears because the force acts in the opposite direction from the velocity of the piston.

We want to solve eqn. (14.13) to find the piston's speed as a function of time: $v(t)$. First, we rearrange the equation so that all terms involving v appear on the left-hand side, and all terms involving t appear on the right-hand side. Then we integrate. The time integration goes from initial time $t = 0$ to final time t. The velocity integration goes from initial velocity v_i to final velocity v.

$$\int_{v_i}^{v} \frac{dv}{v} = -\int_{0}^{t} \frac{b}{m} dt.$$

$$\ln v \Big|_{v_i}^{v} = \ln v - \ln v_i = \ln(v/v_i) = -\frac{b}{m}t.$$

Taking the exponential of both sides:

$$v/v_i = e^{-(b/m)t} \Rightarrow v = v_i e^{-(b/m)t}. \tag{14.14}$$

Damping causes an exponential decay of the system's energy at a rate that is determined by the coefficient of t in the exponential.

> The *damping time* $\tau \equiv m/b$ is the time in which the initial speed of the particle is reduced by a factor e.

■ **EXERCISE 14.7** ♦ Show that m/b has dimensions of time.

The acceleration of the block, $-b\vec{v}/m$, has magnitude $(bv_i/m)e^{-t/\tau} = (v_i/\tau)e^{-t/\tau}$. Shock absorbers alone provide a hard suspension system, with maximum acceleration v_i/τ occurring at $t = 0$. Since they exert no force when $v = 0$, they cannot support the weight of the car. Springs that are added to the system to support the weight also soften the suspension and reduce the maximum vertical acceleration (■ Figure 14.21). The graphs in ■ Figure 14.22 illustrate this model's behavior when the car is given an initial speed v_i. For small values of the damping constant b, the damping time is long and the car oscillates with a slow exponential decrease of amplitude. As b is increased, the damping time τ becomes shorter while the oscillation period is altered only slightly. The character of the motion changes when the damping time becomes much shorter than the oscillation period: energy is removed so rapidly that the car does not oscillate at all.

> A system adjusted to be just at the transition between oscillation and exponential decay is said to be *critically damped*. The condition for critical damping is:

$$\omega_o \tau = \tfrac{1}{2}. \tag{14.15}$$

With critical damping, there is a compromise between the impact-absorbing properties of the spring and the energy-dissipating qualities of the shock absorber; the system removes the

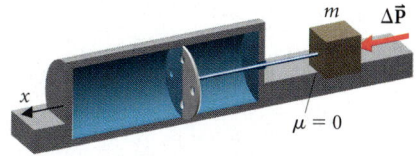

■ **FIGURE 14.20**
Model of a shock absorber. The block m represents the car. Sticky fluid in the cylinder passes through the holes in the piston, exerting a force that increases with the piston's speed and damps the motion.

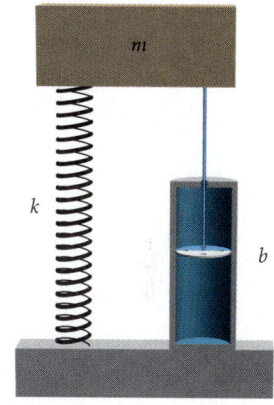

■ **FIGURE 14.21**
A car's suspension system includes both springs and shock absorbers. The absorber damps the spring oscillations. The simplified model shown here illustrates how the two components together influence the car's motion.

WE STATE THIS RESULT WITHOUT PROOF.

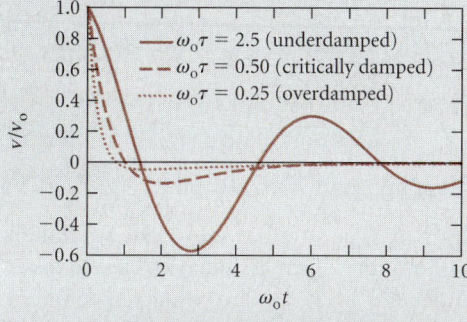

(a) Velocity of the block in Figure 14.21

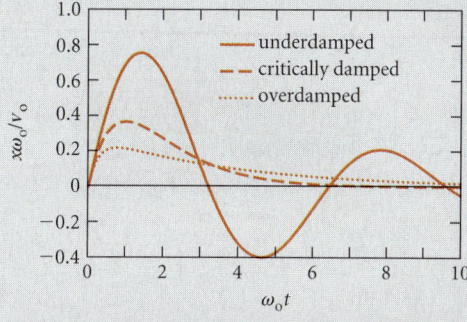

(b) Position of the block in Figure 14.21

■ **FIGURE 14.22**
(a) Velocity of the block in the model of Figure 14.21. With small damping, the system oscillates with a decreasing amplitude. Critically damped and overdamped systems do not oscillate at all. (b) A graph of the block's position shows the same general features, except that the position is zero at $t = 0$. Initially, the block moves away from equilibrium; then, as its velocity damps, the block moves back toward equilibrium.

This is why the factor on the right-hand side of eqn. (14.15) is $\frac{1}{2}$ and not 1.

undesirable oscillations of the undamped spring while minimizing the jolts. The automobile designer can choose the magnitude of damping to obtain a softer or harder ride: race cars have harder suspensions than luxury sedans.

In a damped system, dv/dt is proportional to v. In an oscillating system, the speed varies throughout the cycle as kinetic energy is converted to potential energy and back again. The average speed during the cycle is less than the maximum, so dv/dt, and hence dE/dt, is less than it would be without oscillations. In fact, the energy decays half as fast as in a damped system that does not oscillate.

EXAMPLE 14.8 ♦♦ In a real clock there is friction in the pendulum mount that damps the oscillations. Energy is fed into the system by descending "weights": the potential energy they lose increases the kinetic energy of the pendulum. A grandfather clock has a 1-m pendulum of mass $m = 0.5$ kg that oscillates with amplitude 0.1 rad. It has two brass cylinders of mass $M = 2$ kg that fall 1 m in 1 week. Estimate the damping time of the pendulum.

MODEL We may estimate the rate at which the system's energy changes in two ways: first, by calculating how much energy the descending cylinders feed into the pendulum; and second, by using the exponential decay caused by damping. We may use eqn. (14.14) to find the energy as a function of time. The pendulum's energy equals its kinetic energy at the bottom of the swing. If v is the speed of the pendulum at the bottom of its swing and v_i is the initial value of v, then:

$$E = \tfrac{1}{2}mv^2 = \tfrac{1}{2}m(v_i e^{-t/\tau})^2 = E_i e^{-2t/\tau}.$$

SETUP The potential energy difference $\Delta U = 2Mgh$ for the two cylinders (each of mass M and falling through a height h in $\Delta t = 1$ week) replaces the energy lost by the swinging pendulum at a rate:

$$\left|\frac{dE}{dt}\right| \sim \frac{\Delta U}{\Delta t} = \frac{2Mgh}{\Delta t}.$$

Damping decreases the pendulum's energy at a rate:

$$\frac{dE}{dt} = \frac{d}{dt}(E_i e^{-2t/\tau}) = -\frac{2}{\tau} E_i e^{-2t/\tau} = -\frac{2E}{\tau}.$$

The falling cylinders keep the energy nearly constant, $E \approx E_i$, so we may estimate the average energy loss rate for the 1-week period Δt to be:

$$\left|\frac{dE}{dt}\right| \sim \frac{2E_i}{\tau}.$$

The two rates have to be equal. The energy of the pendulum is obtained from its amplitude using eqn. (14.11): $E = \tfrac{1}{2}mg\ell\theta_{\max}^2$. Thus:

$$\frac{2Mgh}{\Delta t} = \frac{2(\tfrac{1}{2}mg\ell\theta_{\max}^2)}{\tau}.$$

So
$$\frac{\tau}{\Delta t} = \frac{mg\ell\theta_{\max}^2}{2Mgh}.$$

SOLVE
$$\frac{\tau}{1 \text{ wk}} = \frac{(0.5 \text{ kg})(1 \text{ m})(0.1 \text{ rad})^2}{2(2 \text{ kg})(1 \text{ m})} = 1.2 \times 10^{-3},$$

and $\tau = 1.2 \times 10^{-3}(7 \text{ d})(24 \text{ h/d})(60 \text{ min/h}) = 10$ min.

ANALYZE This estimate shows that the damping time is rather short. The pendulum would lose a substantial fraction of its energy in minutes if the cylinders did not descend steadily. The invention of a mechanism to put energy back into the swinging pendulum was essential before the first successful clocks could be constructed. ∎

This clock escapement transfers stored energy into the pendulum oscillations.

Chapter Summary

Where Are We Now?

We have discussed oscillations: motions of systems about a stable equilibrium position. These motions are important in their own right and are essential for understanding waves (Chapter 15).

What Did We Do?

If a system is displaced slightly from a stable equilibrium, it experiences a restoring force that returns it to the equilibrium position. For example, a particle on the end of a spring overshoots and oscillates about the equilibrium position: its motion is the same as a circular motion projected onto a straight line.

The particle on a spring executes *simple harmonic motion* (SHM) described by eqn. (14.1) and its solution (14.3): $x = A \cos(\omega t + \phi_o)$ with *angular frequency* (eqn. 14.2) $\omega = \sqrt{k/m}$. The maximum displacement of the particle from its equilibrium position is the *amplitude A* of the oscillation. The *period* of the oscillation T is the time to complete one full cycle: $T = 2\pi/\omega = 1/f$, and the *frequency f* is the number of cycles completed per unit time. The simple pendulum is an example of a system that executes simple harmonic motion so long as the displacement from equilibrium remains small. Its angular frequency $\omega = \sqrt{g/\ell}$. In oscillatory motions, the energy of the system changes from potential energy to kinetic energy and back during each cycle. The total energy of the system remains constant. The amplitude of the oscillation is proportional to the square root of the total energy.

✱ If an external force is imposed on the system, the response is very large when the frequency of the applied force equals the natural frequency of the system. This condition is known as resonance. In real systems with friction, or when frictionlike forces are applied, the oscillations damp. When the system is critically damped (damping time ≈ oscillation period), the oscillations disappear.

Practical Applications

We discussed two practical applications of oscillatory systems: a clock pendulum and an automobile suspension. Pendulum oscillations serve a useful purpose. In an automobile, the oscillations are a nuisance and the system is designed to suppress them. Oscillations are also of concern when designing buildings for earthquake safety and in studying aircraft stability. When a system such as a microscope must be isolated from external forces, it can be mounted on springs so that the system has a low resonant frequency. Higher-frequency external forces are averaged away and do not affect the system. A seismograph is vibrationally isolated in this way while the Earth and recording pen move.

Solutions to Exercises

14.1 The clock pendulum is in equilibrium when it hangs straight down, the boat when it is stationary on wave-free water, and the particle when it is at rest at the end of a vertical stretched spring, with the spring force balancing its weight. If the spring and particle lie on a horizontal surface, the particle is in equilibrium at the end of an unstretched spring.

14.2 $x(t) = A \cos(\omega t + \phi_o) \Rightarrow dx/dt = -\omega A \sin(\omega t + \phi_o)$,

and $\dfrac{d^2x}{dt^2} = -\omega^2 A \cos(\omega t + \phi_o) = -\omega^2 x$, for any values of A and ϕ_o.

Thus $m \dfrac{d^2x}{dt^2} = -m\omega^2 x = -m\left(\dfrac{k}{m}\right) x = -kx$, as required.

14.3 We begin with the standard form of the solution, $x = A\cos(\omega t + \phi_o)$ and evaluate it at the given times. From Example 14.2, $\omega = 33$ rad/s.

$x(0.015 \text{ s}) = A\cos[(0.015 \text{ s})(33 \text{ rad/s}) + \phi_o] = 0.0.$ (i)

$x(0.030 \text{ s}) = A\cos[(0.030 \text{ s})(33 \text{ rad/s}) + \phi_o] = -0.15$ m. (ii)

The cosine function has the value zero twice each period, when its argument is $\pi/2, 3\pi/2, \ldots$. We use the first solution: $\omega t + \phi_o = \pi/2$.
Thus:
$$\phi_o + 0.495 = \pi/2 = 1.571 \Rightarrow$$
$$\phi_o = 1.571 - 0.495 = 1.076 \text{ rad}.$$

Then eqn. (ii) becomes:

$A\cos(0.990 + 1.076) = A\cos(2.066) = -0.475A = -0.15$ m.

Or: $A = 0.32$ m.

So, the solution is $x = (0.32 \text{ m})\cos[(33 \text{ rad/s})t + 1.08 \text{ rad}]$.

14.4 We have:

$x(0) = 0.55 \text{ m} = B\cos(0) + C\sin(0) = B,$

and $v(0) = 0 = -\omega B\sin(0) + \omega C\cos(0) = \omega C.$

So $C = 0$, $B = 0.55$ m, and $x(t) = (0.55 \text{ m})\cos\omega t$, which is the same as the result of Example 14.2.

14.5 The angular acceleration is proportional to $\sin\theta$. In the approximation, $\sin\theta$ is replaced by θ (in radians!). The fractional error is given by:

$$\text{fractional error} = \frac{\theta - \sin\theta}{\sin\theta}.$$

Angle	$\sin\theta$	θ	Error (%)
1°	0.017452	0.017453	0.005
10°	0.1736	0.1745	0.5
20°	0.3420	0.3491	2
90°	1.000	$\pi/2 = 1.571$	57

14.6 The external force has x-component $F_{ext} = mg\cos(0) = mg$. The response also has zero frequency, and so is a constant displacement. Its magnitude is:

$$\frac{mg}{m\omega_o^2} = \frac{g}{k/m} = \frac{mg}{k}, \text{ as expected.}$$

14.7 The proportionality constant for the shock absorber is b, which has dimensions:

$$[b] = [F]/[v] = (ML/T^2)/(L/T) = M/T.$$

Thus $[m/b] = M/(M/T) = T$, as required.

Basic Skills

Review Questions

§14.1 SIMPLE HARMONIC MOTION

- What two features of a system are necessary for oscillations to occur?
- What is the differential equation satisfied by a particle undergoing simple harmonic motion?
- Explain the analogy between circular motion and simple harmonic motion.
- What is the *period* of an oscillation? What is the *amplitude*?
- Give a general expression for the displacement of a particle undergoing SHM. Explain how the constants in this expression are determined in a specific application.
- Sketch the displacement, velocity, and acceleration of a particle during one period of SHM.
- What does *out of phase* mean?
- What is the period of a particle of mass m oscillating on the end of a spring with constant k?

§14.2 THE PENDULUM

- What is a *simple pendulum*? Describe the restoring force that causes the pendulum to oscillate.
- Under what conditions does a pendulum undergo SHM?
- What is the pendulum's oscillation frequency?
- What is a *physical pendulum*? What is its period?

§14.3 ENERGY IN OSCILLATORY MOTION

- Describe how the energy of an oscillating particle varies between kinetic and potential forms during the oscillation.
- How does the energy depend on the amplitude of the oscillation?
- What is the shape of the potential energy curve for a particle undergoing SHM?
- If you know the total energy of an oscillation, explain how to find the range of its position during the oscillation from the potential energy curve.

✳ §14.4 THE EFFECT OF EXTERNAL FORCES

- What is *resonance*?
- Describe qualitatively how the amplitude of an oscillating system depends on the frequency of the applied force.
- If a system is damped by a force that is proportional to velocity, how does its velocity depend on time?
- What does it mean to say that an oscillating system is *critically damped*?

Basic Skill Drill

§14.1 SIMPLE HARMONIC MOTION

1. Identify the systems in the list below that have restoring forces and describe the kind of oscillation each makes. **(a)** A submarine submerged to a depth d. **(b)** A bead in the middle of a horizontal stretched string. **(c)** An air bubble in the bottom of a U-tube full of water. **(d)** A cylinder free to roll on an inclined plane and supported by a spring.

2. A particle is moving around a circle of radius 1.0 m at 2.0 m/s. At $t = 0$, it is at $\theta = \pi/4$. Give an expression for its x-coordinate as a function of time and find the period of its variation.

3. A block of mass 75 g is attached to the end of a spring of constant $k = 0.50$ N/m. The spring is compressed 7.5 cm and the block is released. Find the position and velocity of the block as a function of time.

§14.2 THE PENDULUM

4. A simple pendulum labeled A and another labeled B are identical except that the bob on the end of A has twice the mass of the bob on the end of B. Is the period of A longer than, shorter than, or the same as that of B? Explain your reasoning.
5. A simple pendulum labeled A and another labeled B are the same except that A's rod is twice as long as B's. Does A or B have the longer period? By how much?
6. Find the period of a simple pendulum consisting of a heavy, small ball on the end of a string of length 0.50 m.
7. A pendulum of length $\ell = 0.75$ m is pulled out to an angle of $3.0°$ and released. Find an expression for the angle θ from the vertical as a function of time.
8. A 75-kg metal sheet is suspended at a point on its edge $d = 0.63$ m from its center of mass. If the sheet oscillates at angular frequency $\omega = 0.20$ rad/s, what is its rotational inertia about an axis through the suspension point?

§14.3 ENERGY IN OSCILLATORY MOTION

9. What is the amplitude of a simple pendulum of length $\ell = 0.60$ m if the bob has a speed $v = 0.37$ m/s when passing through the equilibrium position?
10. A spring system with $m = 1.5$ kg and $k = 12$ N/m has an amplitude of 1.3 cm. What is the maximum speed of the mass?

✵ §14.4 THE EFFECT OF EXTERNAL FORCES

11. A driving force $F = (10\text{ N})\cos[(15\text{ rad/s})t]$ acts on a spring-and-mass system with $k = 10$ N/m and $m = 0.2$ kg. What is the amplitude of the resulting oscillation?
12. A damping system exerts a force $\vec{F} = -b\vec{v}$, $b = 27$ kg/s, on a mass $m = 6.9$ kg. How long does it take to reduce the speed by one-half?

Questions and Problems

§14.1 SIMPLE HARMONIC MOTION

13. ❖ A particle on the end of a spring oscillates with a period of 0.5 s. The particle is replaced with another of twice the mass. What is the period then? Explain your reasoning.
14. ❖ Two particle-on-spring systems have the same frequency, but the spring constant of one spring is twice that of the other. How do the particle masses compare?
15. ♦ A particle is moving around a circle of radius 0.14 m at 1.5 m/s. At $t = 0$, it is at $\theta = 30°$. Give an expression for its y-coordinate as a function of time and find the period of its variation.
16. ♦ At time $t = 0$, an object of mass 50 g on the end of a horizontal spring is moving to the right at 10 cm/s and is displaced 2 cm to the left from the equilibrium position. If the spring constant is 0.2 N/m, find an expression for the position of the particle as a function of time.
17. ♦ A block of mass 75 g attached to the end of a horizontal spring is given an impulse $\Delta \vec{p} = (0.22\text{ N·s})\hat{i}$, parallel to the spring. The spring constant is $k = 0.50$ N/m. Find the position of the block as a function of time.
18. ♦ A 2.0-kg ball is attached to the end of a spring of constant 1.0 N/m. At time $t = 0$, the ball is given an impulse $\Delta \vec{p} = (1.0\text{ N·s})\hat{i}$, where the x-direction is along the line of the spring. Find the position of the ball as a function of time.
19. ♦♦ In some Swiss cuckoo clocks, the "pendulum" is an object hanging on a spring that oscillates vertically. For an object of mass 20.0 g, what spring constant is required for a period of 0.50 s?
20. ♦♦ A cylindrical cork of radius 1.0 cm and mass 4.0 g is floating on wine of specific gravity 0.90. The cork is weighted at one end so that the axis of the cylinder is vertical. If the cork is pushed down slightly, find the period of its oscillations.
21. ♦♦ A hockey puck of mass $m = 0.30$ kg slides on a horizontal ice surface between two springs, each with spring constant $k = 1.2$ N/m

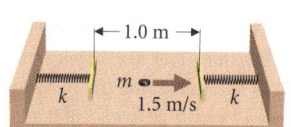

■ FIGURE 14.23

(■ Figure 14.23). When both springs are unstretched, the distance between their ends is 1.0 m. Show that the puck's motion is oscillatory. If its speed at the midpoint of the system is 1.5 m/s, find its period.
22. ♦♦ A block of mass m on a frictionless floor is attached to two springs, each of unstretched length ℓ. The springs are attached to two parallel walls and have spring constants k and $2k$, respectively. In equilibrium, the block is at rest and the two springs are unstretched. If the block is displaced slightly, show that it undergoes SHM, and find the angular frequency.
23. ♦♦ A particle of mass $m = 120$ g on a spring with constant $k = 12$ N/m has its equilibrium position at $x = 0$. If it is at $x = 0.15$ cm at $t = 0$ and $x = -0.045$ cm at $t = 0.22$ s, find its position as a function of time.
24. ♦♦ A particle of mass $m = 0.32$ kg is oscillating on the end of a spring with amplitude 0.25 m. It is at its equilibrium position ($x = 0$), moving in the positive x-direction, at $t = 0$, and 0.12 s later it is at $x = 0.20$ m. Find the spring constant.
25. ♦♦ A particle of mass 1.2 kg is oscillating on the end of a spring with spring constant 18 N/m. At $t = 0.50$ s, its position is $x = 0.33$ m and its velocity is $v = 1.7$ m/s. Find the amplitude of the oscillation.
26. ♦♦ A particle is oscillating on the end of a spring with spring constant $k = 15$ N/m. You observe that the particle is at rest at $x = 6.10$ cm at $t = 0.0$ s; it is at $x = 4.25$ cm at $t = 1.0$ s, and at $x = -0.18$ cm at $t = 2.0$ s. Find the mass of the particle.
27. ♦♦ A block of mass M is supported by a vertical spring, with spring constant k, attached to the floor. If the block is pushed down

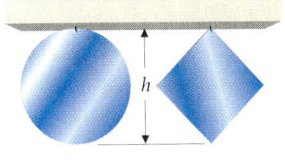

■ FIGURE 14.24

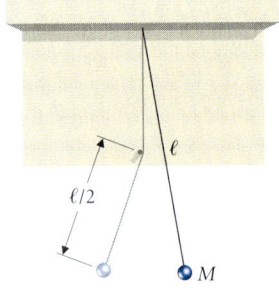

■ FIGURE 14.25

by a small amount and then released, describe the resulting motion and find its period. (*Hint:* First find the equilibrium position.)
28. ◆◆ A baby carriage is supported by four springs with spring constant $k = 22$ N/m, one at each wheel. The carriage has a mass of 15.0 kg. What is the oscillation period of the carriage when a baby of mass 5.2 kg is riding in it?

§14.2 THE PENDULUM

29. ❖ A pendulum has a movable disk attached to a rod. If the clock runs slow, should you move the disk toward the support or away from it? Why?
30. ❖ Each object in ■ Figure 14.24 is cut from a uniform metal sheet and suspended from a hole at one end, as shown. Which has the larger frequency of oscillation?
31. ◆ Find the period of a simple pendulum consisting of a heavy, small ball on the end of a string of length 0.75 m.
32. ◆ In the National Museum of American History there is a Foucault pendulum, which consists of a 115.4-kg brass ball suspended on a 22.25-m wire. What is the period of this pendulum?
33. ◆ A simple pendulum with period 2.000 s at the north pole ($g = 9.832$ m/s^2) is taken to the equator, where the value of g is 9.780 m/s^2. What is the period of small oscillations of this pendulum at the equator?
34. ◆ Find the length of a grandfather clock pendulum with period 2.00 s on the Moon ($g_{Moon} = 1.62$ m/s^2).
35. ◆ A pendulum of length $\ell = 0.75$ m is pulled out to an angle of 3.0° and released. Find an expression for the angle θ from the vertical as a function of time.
36. ◆ A child of mass 45 kg is swinging on a swing with ropes of length 3.0 m. How long does it take the child to swing from the maximum height on one side to the other?
37. ◆◆ A pendulum of length 2.50 m and mass 0.30 kg is set swinging by giving it an impulse of 0.12 N·s. Find an expression for θ as a function of time.
38. ◆◆ A small bead is constrained to slide on a circular hoop of radius $R = 0.13$ m standing in a vertical plane. Show that if the bead is displaced slightly from its equilibrium position, the resulting motion is approximately simple harmonic, and find its period.
39. ◆◆ A rod of length ℓ and mass M hangs from a peg inserted through a small hole at one end. Find the angular frequency of small oscillations of the rod about its equilibrium position.
40. ◆◆ A pendulum consists of a rod of length 0.500 m attached to a ball of radius 5.00 cm. The mass of the rod is exactly $\frac{1}{12}$ the mass of the ball. Find the period of the pendulum.
41. ◆◆ A hoop of radius 0.30 m is hung over a peg. Find the frequency of its oscillations.
42. ◆◆ A solid sphere of diameter 0.45 m is suspended from a small hook attached to its surface. With what period does it swing?

43. ◆◆ A pendulum is made by suspending a small object from the end of a string of length ℓ. Apply Newton's second law, $\vec{F} = m\vec{a}$, to derive the equation of SHM for small oscillations of this system. (*Hint:* Take components along and perpendicular to the string.)
44. ◆◆ A pendulum of period 2.000 s has a disk of radius 10.00 cm attached at its center to a rod of length d. Find the length d of the rod if the local value of g is 9.810 m/s^2 and the ratio of rod mass to disk mass is 0.1000. Compare with the result of Example 14.5.
45. ◆◆◆ A pendulum consists of a ball of mass M on the end of a string of length $\ell = 1.2$ m. There is a peg a distance $\ell/2$ below the suspension point (■ Figure 14.25). Find the period of small oscillations about equilibrium.
46. ◆◆◆ A rigid body is suspended by a rod that passes through the body a distance ℓ from its center of mass. At what other locations of the rod do small oscillations about equilibrium have the same period?
47. ◆◆◆ A small ball lies in the bottom of a hemispherical bowl of radius R. If the ball is slightly displaced, describe its resulting motion quantitatively. (Assume the ball rolls without slipping.)
48. ◆◆◆ A pendulum of length ℓ is given an impulse that causes the bob to move in a horizontal circle, with the string at an angle α to the vertical. Find the period of the pendulum, and compare with the period of the same pendulum swinging in a vertical plane.

§14.3 ENERGY IN OSCILLATORY MOTION

49. ❖ One pendulum oscillates with an amplitude of 1° and a second, identical pendulum has amplitude 2°. Which has more energy? By how much?
50. ❖ Two blocks of masses 1 kg and 2 kg, respectively, are oscillating on the ends of springs. The periods of the oscillations are the same, and they each have the same amplitude. Which has more energy? By how much? Explain your reasoning.
51. ◆ A particle-and-spring system with $k = 55$ N/m and $m = 17$ g is oscillating with an amplitude of 1.3 mm. What is the particle's speed when the displacement is 0.60 mm?
52. ◆◆ Find the speed of a pendulum bob at the bottom of its swing if its initial displacement is 5.7° and its period is 2.5 s.
53. ◆◆ A uniform cylinder of length 0.20 m and radius 9.0 cm is suspended from an axis through a diameter at one end of the cylinder (■ Figure 14.26). It is set oscillating with an amplitude of 2.0°. Find the angular frequency of the oscillation and the speed of the bottom of the cylinder as it passes through the equilibrium position.
54. ◆◆ A circus performer swinging on a trapeze reaches a maximum angle of 15°. If the trapeze is suspended 12 m below its support, find the performer's speed as a function of angle.
55. ◆◆◆ Refer to Study Problem 11. Find Miyako's maximum speed. Where does it occur?

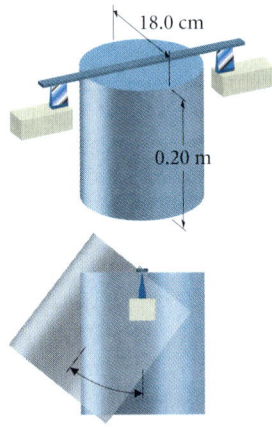

■ FIGURE 14.26

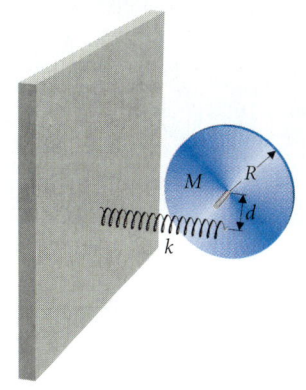

■ FIGURE 14.27

56. ♦♦♦ A large iron ball of mass 350 kg is attached to the end of a cable 35 m long. The system is used to demolish buildings. It is set up 1.0 m from the side of the building, pulled out 25°, and let go. What is its speed when it impacts the building? What angle is needed if the ball is to hit the building at 20.0 m/s?

✻ §14.4 THE EFFECT OF EXTERNAL FORCES

57. ♦ A particle of mass $m = 520$ kg is on the end of a spring. What spring constant k is required for the particle to resonate with an applied force having $\omega = 2.5$ rad/s?

58. ♦ A particle of mass $m = 320$ kg is mounted on a spring with $k = 19$ N/m. What damping constant b is required of a shock absorber if the system is to be critically damped?

59. ♦♦ A 0.55-m-long cylindrical buoy of plastic (specific gravity 0.50) is floating in a harbor. A wave of amplitude $h_o = 15$ cm and period 2.0 s passes by. What is the amplitude of the buoy's oscillations?

60. ♦♦♦ ✉ Refer to Study Problem 11. If, after one oscillation, Miyako rises just to the point where the bungee goes slack, what effective damping constant does the system have?

61. ♦♦♦ ✉ In Problem 28, a baby of mass 5.2 kg rides in a baby carriage of mass 15 kg supported on four springs, each having $k = 22$ N/m. Now the baby jumps up and down with angular frequency of 6 rad/s, just barely losing contact with the carriage at the top of each jump. What is the response of the carriage in the absence of damping?

Additional Problems

62. ♦ The piston in a Mazda Millenia S automobile has a diameter of 8.03 cm and travels a total distance of 7.42 cm per stroke. What is the amplitude of its oscillation? Find the maximum speed and maximum acceleration of the piston when it oscillates at 3700 rpm. Express your result for a_{max} in terms of the acceleration due to gravity, g. (Assume simple harmonic motion.)

63. ♦♦ A spring with spring constant $k = 92$ N/m is at the bottom of a frictionless ramp that makes an angle of 5.0° with the horizontal. A box of mass 11 kg is placed on the end of the spring. Show that the system undergoes SHM and find its period and amplitude.

64. ♦♦ A U-tube of radius 0.50 cm contains 0.50 kg of mercury. If pressure is applied to one side so that the mercury level in one side is 1.0 cm above the level in the other side, and then released, show that the system undergoes SHM and find the period. (*Hint:* If the two sides are at the same height, the system is in equilibrium. So, take the excess height as an additional mass of mercury whose weight is unbalanced. You also need to be careful in expressing acceleration in terms of the difference in heights!)

65. ♦♦ A torsion pendulum consists of a horizontal disk of radius 5.0 cm and mass 0.50 kg suspended by a fiber that passes through its center. When twisted, the fiber produces a torque of 1.0×10^{-3} N·m per radian of twist. If the disk is twisted through 5.0° and then released, find the period of the resulting oscillation, and the maximum angular speed achieved.

66. ♦♦ A bungee jumper of mass $m = 97$ kg is riding a bungee with spring constant 25 N/m. The bungee is suspended from a point 105 m above ground, and its unstretched length is 9.0 m. People on the tower observe the jumper to be at a height of 75 m at time $t = 0$ and at a height 52 m 5.0 s later. Find the jumper's position as a function of time. What is the amplitude of the oscillation?

67. ♦♦ A disk of mass M and radius R is mounted on a frictionless axle (■ Figure 14.27). A spring with spring constant k is attached to the disk a distance d below the axle, and the system is in equilibrium in the configuration shown. If the disk is rotated through a small angle θ_o, show that the system oscillates and find the relation between amplitude θ_o and energy in the oscillation.

68. ♦♦ A block of mass M is suspended between the floor and the ceiling by two springs with spring constants k_1 and k_2 and equal unstretched lengths ℓ (■ Figure 14.28). The separation of the floor and ceiling is 2ℓ. Find the frequency of small oscillations about equilibrium.

69. ♦♦ A block is attached to two springs with spring constants k_1 and k_2 and the same unstretched length (■ Figure 14.29). Find the period of small oscillations about equilibrium.

70. ♦♦♦ A bead of mass $m = 25$ g is at the middle of a wire of length $\ell = 0.75$ m. The tension in the wire is 45 N. What angle does the wire make with the horizontal? If the bead is pushed down slightly, show that it undergoes SHM and find the frequency.

71. ♦♦♦ A 2.0-cm-radius ball is released from a height of 0.50 m on a ramp that makes a 30° angle with the horizontal. It rolls down the ramp, across a 11-cm-long horizontal floor, and up a similar ramp on the other side. Find the period of the oscillation.

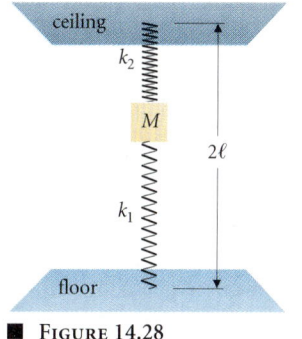

■ FIGURE 14.28

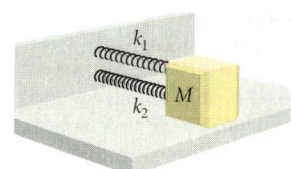

■ FIGURE 14.29

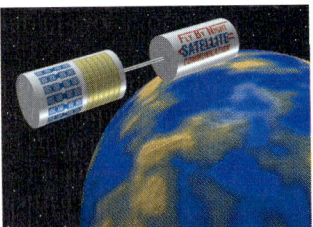

■ Figure 14.30

72. ♦♦♦ A satellite consists of two cylinders, each of mass 510 kg and radius 0.25 m. They are connected at their centers by a rod of length 0.50 m (■ Figure 14.30). The rod produces a torque of 1.0×10^5 N·m/rad when twisted. If one-half is twisted through $1.0°$, describe the subsequent motion of the satellite, and find its period.

73. ♦♦♦ Find the maximum speed of the log of Example 14.3 if it is initially pushed down to a depth $z_o = 1.0$ cm. What is its effective potential energy when displaced? Does this effective potential energy possess properties 1–4 of potential energy that were specified in Chapter 8? Explain.

74. ♦♦♦ A light plastic cylindrical container of gasoline floating in water is observed to oscillate at 1.2 Hz. Find the depth of the gasoline in the container. (Assume the container is flat enough to float stably with its axis vertical.)

75. ♦♦♦ A cart of mass M is attached to the end of a spring with spring constant k. A block of mass m sits on the cart. The coefficient of friction between the block and the cart is μ_s. What is the maximum amplitude of oscillation the cart can undergo without the block's sliding across the cart? (Neglect the rotational inertia of the cart's wheels.)

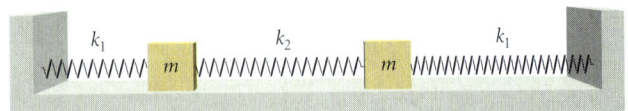

■ Figure 14.31

76. ♦♦♦ Two blocks of equal mass m are attached as shown in ■ Figure 14.31. The blocks are in equilibrium with all three springs relaxed, and there is no friction. Consider the following two motions of the system. **(a)** Both blocks move in the same direction and in phase—the two blocks have equal displacements from equilibrium at all times. **(b)** The blocks always move in opposite directions—the displacements of the blocks from equilibrium have the same magnitude but opposite directions. Find the angular frequency of the oscillation in each case.

Computer Problem

77. We may use a computer to solve the equation:

$$\frac{d^2\theta}{dt^2} = -\frac{g}{\ell}\sin\theta,$$

for the angular displacement of a pendulum whose amplitude is not small. First, using the Taylor series (Appendix IB), write an expression for $\theta(t + \Delta t)$ in terms of $\theta(t)$, $d\theta/dt$, $d^2\theta/dt^2$, and Δt. Similarly, write an expression for $d\theta/dt$ at $t + \Delta t$, keeping terms in $(\Delta t)^2$. Use the differential equation to express $d^2\theta/dt^2$ and $d^3\theta/dt^3$ in terms of θ and $d\theta/dt$. Calculate the period of a simple pendulum with length 1.00 m. Use one-hundredth of this time as the time step Δt. Use the following steps to calculate the actual period for the pendulum with an amplitude of $45°$ ($\pi/4$ rad).

In your spreadsheet, put the values of t in column 1 and θ in column 2. Start with $t = 0$ and $\theta = \pi/4$ in the first row. (Most spreadsheets expect the argument of a sine or cosine to be in radians.) Use an additional column for $d\theta/dt$. First, use your Taylor series to evaluate new values of θ and $d\theta/dt$ at $t + \Delta t$. In the second row, record the new values of θ and $d\theta/dt$, and update the time by adding the timestep Δt to the previous value of t. Continue until you have reached the value $\theta = \pi/4$ again. What is the final value of t? (This is the computed period.) Plot θ as a function of time, and compare with the solution for the simple pendulum (a cosine function). How do the periods compare?

Challenge Problems

78. A cylinder of mass M and radius R has a frictionless axle through its center that is attached at each end to a spring with spring constant k. The system is placed on an inclined plane as shown in ■ Figure 14.32. Find the frequency of small oscillations about equilibrium if the cylinder rolls on the plane. If the coefficient of friction between the plane and the cylinder is μ_s, what is the maximum amplitude oscillation for which the cylinder continues to roll without slipping?

79. Refer to Study Problem 11. Find the period of Miyako's oscillation.

80. A ball of mass 0.30 kg and radius $r = 15$ cm rolls without slipping between two springs, each with spring constant $k = 1.2$ N/m. When both springs are unstretched, the distance between their ends is 1.0 m. If the ball rolls at 1.5 m/s when not touching either spring, find the period of its oscillation. You may ignore the friction between the surface of the ball and the springs and assume that the springs touch the ball at height $h = r$ above the ground. Compare your answer with that of Problem 21.

81. A semicircular block of uniform wood stands on its curved edge on the floor. Find the CM of the block and its rotational inertia about its CM. What is the period of small oscillations about equilibrium?

82. A particle of mass m lying on a table is attached to a string of length ℓ, which passes through a hole in the table and is attached to a second particle of mass $M > m$ hanging underneath (■ Figure 14.33). Show that the system is in equilibrium if the particle on the table is set moving in a circle of radius r with an appropriate speed $v(r)$. Find $v(r)$. If the particle's speed is suddenly increased to $v(r) + dv$, show that the system undergoes oscillations and find their angular frequency and amplitude. (*Hint:* You will need to find the particle's velocity as a function of its position; conservation laws are the key.)

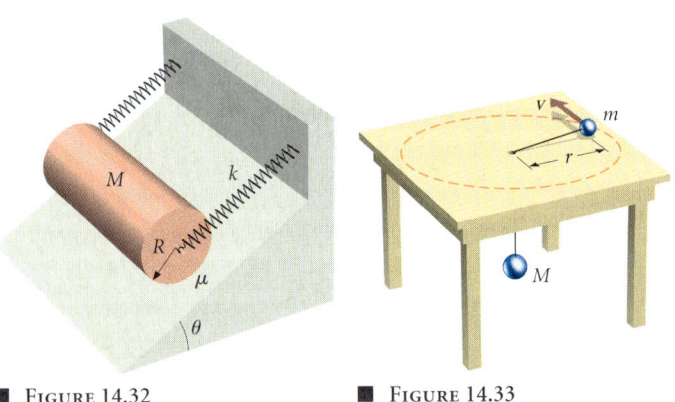

■ Figure 14.32   ■ Figure 14.33

Life is a wave, which in no two consecutive moments of its existence is composed of the same particles.
JOHN TYNDALL

CHAPTER 15
Introduction to Wave Motion

CONCEPTS
Pulse
Transverse wave
Longitudinal wave
Harmonic wave
Wave function
Wavelength
Superposition
Standing wave

GOALS
Know the terminology of waves.

Understand how waves propagate along a string.

Be able to determine the wavelength, frequency, and phase constant of a given harmonic wave.

Understand how superposition describes standing waves and reflection at boundaries.

Mechanical waves are the means by which the instruments in a symphony orchestra produce music. The musicians set up standing waves on their strings or drumheads, which in turn produce the sounds we hear. Waves on strings are the major focus of this chapter.

495

A symphony orchestra provides a wonderful physics demonstration: the violinists set up waves on their instruments' strings; the timpanist sets the drumhead vibrating; the pianist pounding the keys uses hammers to excite waves on the piano strings. All this activity is designed to produce sound—the subject of the next chapter—but the first step is the generation of mechanical waves in the instruments.

Ripples on a pond (■ Figure 15.1) don't produce a noticeable sound, but they too are mechanical waves. Spreading across the surface, the ripples transmit energy from the point where a stone strikes the water to the distant shores of the pond. Each element of water on the surface oscillates as ripples pass, both influencing and responding to the motion of nearby elements. As the ripples reach the shore, some energy is dissipated and some reflected as new ripples propagating back into the pond. This behavior exhibits all the characteristics of wave motion:

- Like a simple oscillation, wave motion occurs in a system with an equilibrium state. The wave is a disturbance of the system away from the equilibrium state.
- Unlike a simple oscillator, a wave involves motion at many different points in a distributed system. These motions are coupled, and energy can be transmitted throughout the system.
- By absorbing and reflecting waves, the system's boundaries have an important influence on its behavior.
- A system may support more than one wave disturbance at the same time. The original and reflected ripples on the pond form two simultaneous disturbances that pass through each other and retain their identity even though they may appear inextricably mixed.

Ripples spread out across a surface, but waves on a stretched string move along a single line. The ripples are easier to see, but the mathematics of waves on strings is much less complex. A description of waves on a string completes our toolbox of kinematic models. As we develop it, we'll learn why a musician adjusts the string tension to tune a cello.

15.1 Mechanical Waves

15.1.1 What Causes Mechanical Waves?

Let's see how oscillators can be coupled together. Take two identical objects attached to identical springs and hang one from each end of a rod pivoted at its center (■ Figure 15.2). Pull the object on the left downward while holding the rod fixed; then let go of both. The force exerted on the rod by the stretched spring is greater than that exerted by the undisturbed spring on the right. There is a net torque on the rod that accelerates it counterclockwise, stretching the spring on the right and setting the second object oscillating. The rod provides a connection that allows each spring to influence the other. The ability of one oscillator to transmit its energy to a similar, coupled oscillator is the basic mechanism of mechanical waves.

THE USE OF ELEMENTS TO DESCRIBE THE PROPERTIES OF A CONTINUOUS SYSTEM WAS FIRST DISCUSSED IN THE INTRODUCTION TO PART III.

WE GIVE AN EXAMPLE OF HOW THE MOTIONS ARE COUPLED IN §15.1.1.

SEE EXAMPLE 15.8.

■ **Figure 15.1**
Ripples on the water's surface illustrate the common features of mechanical waves. Elements of water oscillate as energy spreads outward from the initial disturbance.

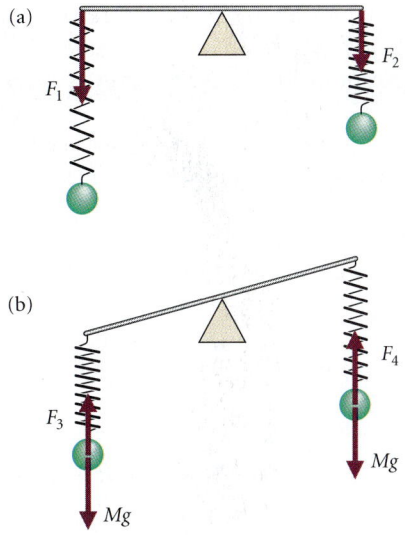

■ **FIGURE 15.2**
Two springs attached to a pivoted rod illustrate how oscillators can be coupled together. If the rod is released when the oscillating block at the left is below its equilibrium position (a), the excess force due to the stretched spring exerts a torque that rotates the bar, thus stretching the spring on the right (b) and starting it oscillating.

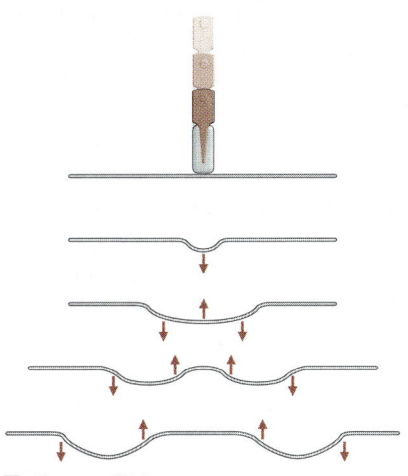

■ **FIGURE 15.3**
A blow to a string sets up wave pulses moving in both directions along the string. The tension in the string provides both the restoring force that acts on the originally displaced section as well as the coupling force that spreads the disturbance to neighboring elements.

The oscillating particles in a piano string are differential elements of the string. A sharp hammer blow (■ Figure 15.3) gives a small piece of the string a downward velocity. The displaced element under the hammer pulls downward on the string elements to either side, and those elements pull upward on the central piece. In turn, they pull downward on their outer neighboring elements. As a result, each element is first accelerated downward and then upward. Two outgoing wave pulses carry the effect of the hammer blow both right and left along the string.

> A *wave pulse* is a disturbance with a finite spatial size that propagates through the system.

Each element of the string acts as an oscillator strongly coupled to the other elements by the string tension. This coupling transmits energy from the original blow along the string, while each piece of the string moves down and up at a fixed horizontal position: a wave propagates along the string.

SEE §14.1.

The stretched string illustrates the two properties necessary for any mechanical system to support wave motion: inertia (mass), and elastic (restoring) forces that are necessary for individual pieces of the system to oscillate. In a system like a mass on a spring, the restoring force and inertia together determine the oscillation frequency. In a wave disturbance, they determine the speed. As we learned from Figure 15.3, the tension T provides the restoring force for waves on a string. The inertia is represented by the string's mass per unit length, μ. A greater tension produces greater accelerations, and the disturbance propagates more rapidly. With a greater mass per unit length, the string's acceleration is smaller and the disturbance propagates more slowly. As we shall show, the wave speed is:

$$v = \sqrt{T/\mu} \tag{15.1}$$

WE'LL DERIVE THIS RESULT IN §15.2.4.

WE CALL THE STRING TENSION F HERE TO AVOID CONFUSION WITH THE DIMENSION OF TIME, T. BE ALERT FOR THE MEANINGS OF THE SYMBOLS IN EACH EXPRESSION YOU USE.

THE USE OF DIMENSIONAL ANALYSIS IS DISCUSSED IN §1.3.3.

EXAMPLE 15.1 ❖ Using dimensional analysis, find how the speed v of waves on a stretched string is related to the tension F and mass per unit length μ. How do the wave speeds compare on two strings that are identical except that the tension in one is twice that in the other?

MODEL We need to find the appropriate combination of string tension and mass per unit length that has the dimensions of speed.

SETUP AND SOLVE $[v] = \dfrac{L}{T}$, $[F] = \dfrac{ML}{T^2}$, and $[\mu] = \dfrac{M}{L}$.

Since $[v]$ contains only L and T, we divide $[F]$ by $[\mu]$ to eliminate the dimension of mass.

$$\left[\dfrac{F}{\mu}\right] = \dfrac{L^2}{T^2} = [v]^2,$$

and we conclude $v \propto \sqrt{F/\mu}$. The string with doubled tension has the larger wave speed by a factor of $\sqrt{2}$.

ANALYZE We cannot find the constant of proportionality using dimensional analysis alone. ∎

15.1.2 The Language of Wave Theory

In this section we shall introduce a number of terms that are used to describe waves. Our example system is a very long stretched string. Imagine taking a photograph of the vibrating string (■ Figure 15.4); its shape, as captured on the film, is called the *waveform*.

■ **FIGURE 15.4**
Photograph of a wave on a long string. The shape of the string at a fixed time t is the waveform.

> A *waveform* is the shape of a wave disturbance on the entire system at a fixed time.

The pulses produced by a hammer blow to the string are examples of a finite waveform (Figures 15.3, ■ 15.5). Hitting the string continually at a fixed rate produces a series of propagating pulses (■ Figure 15.6). If the string is struck in exactly the same way each time, then all the pulses are identical, and the wave is *periodic*.

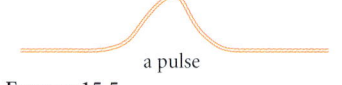

■ **FIGURE 15.5**
A pulse is a finite waveform. It is produced by a single brief excitation of the medium.

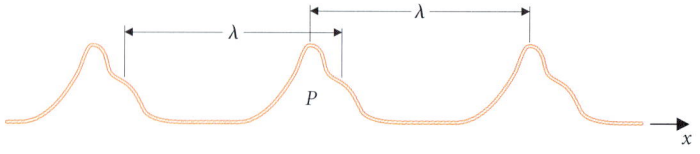

■ **FIGURE 15.6**
If we continue to hit the string, we produce a series of pulses. If the pulses occur at regular intervals, it is a periodic wave. The wavelength λ is the distance between peaks of the displacement in a photograph taken at a single time. Any other identifiable point on the waveform, such as the shoulder of this pulse, may also be used to define the wavelength.

NOTE THE COORDINATES CAREFULLY IN FIGURE 15.6, WHERE DISPLACEMENT IS PLOTTED AS A FUNCTION OF *POSITION*, AND IN FIGURE 15.8, WHERE IT IS PLOTTED AS A FUNCTION OF *TIME*.

> The *wavelength* λ of a periodic wave is the distance between successive pulses in the wave (Figure 15.6).

Most applications of wave behavior in physics involve such periodic waves.

EXERCISE 15.1 ❖ To what extent may the ripples on a pond caused by a falling pebble be described as a periodic wave or a wave pulse?

To record the wave motion, we could take a series of snapshots at successive times as a pulse passes a fixed point *P* (■ Figure 15.7). Equivalently, we could make a movie of the string at point *P*. In the movie, we see the string move upward and then downward as a pulse passes. ■ Figure 15.8 is a graph of the displacement of the string seen in the movie as a function of time.

> The *period T* of a periodic wave is the time interval between two pulses occurring at one place.

The period of the oscillations of the string element at *P* is also *T*.

> The *frequency f* of the wave is the number of pulses passing any fixed point in 1 second.
> The frequency is the reciprocal of the period: $f = 1/T$.

In Figure 15.8 the displacement is the same at times *A* and *B*, but the string is moving downward at *A* and upward at *B*. The cycle is not complete until time *C*, when the string is moving downward again. The period is the time between *A* and *C*.

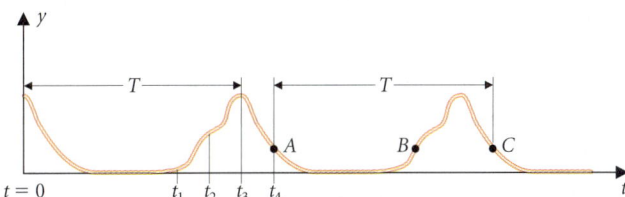

Figure 15.8 and the waveform (Figure 15.6) are closely related. As the wave moves by at speed *v*, say from left to right, each point on the waveform moves with it. Suppose maximum upward displacement occurs at point *P* at time $t = 0$, as in Figure 15.8. Then *P* next has maximum upward displacement when one wavelength of the wave has passed by (time t_3 in the figure). The wave travels a distance λ at speed *v* in time λ/*v*. Thus $t_3 - 0 = T = \lambda/v$, where *T* is the period of the wave. Since $T = 1/f$, we may write this relation as:

$$\lambda f = v. \qquad (15.2)$$

EXAMPLE 15.2 ◆ A wave on a guitar string has a frequency of 440 Hz and a wavelength of 1.2 m. What is the speed of the wave on the string?

MODEL We use the relation between wavelength and frequency, eqn. (15.2).

SETUP AND SOLVE $v = \lambda f = (1.2 \text{ m})(440 \text{ Hz}) = 530 \text{ m/s}$.

ANALYZE Notice how the units work out. The speed is large but typical of waves on strings. We'll see how the length of the string is related to the wavelength in §15.4. ■

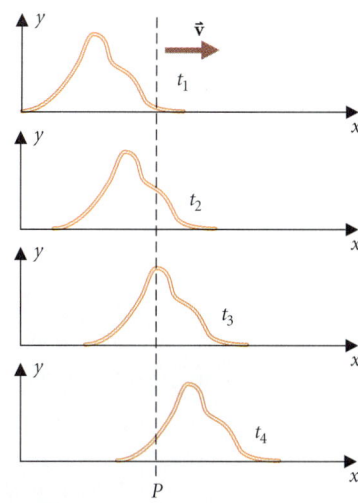

■ **FIGURE 15.7**
As the pulse moves, the displacement at point *P* changes. The snapshot in Figure 15.6 was taken at time t_3.

THE OSCILLATIONS ARE NOT NECESSARILY SHM UNLESS THE WAVES ARE HARMONIC. SEE §15.2.2.

■ **FIGURE 15.8**
If instead we take a movie of the string at one place, the time between the peaks is the wave period *T*; *T* is also the time between *A* and *C*. It is *not* the time between *A* and *B*. Although the displacement is the same at *A* and *B*, it is decreasing at *A* and increasing at *B*.

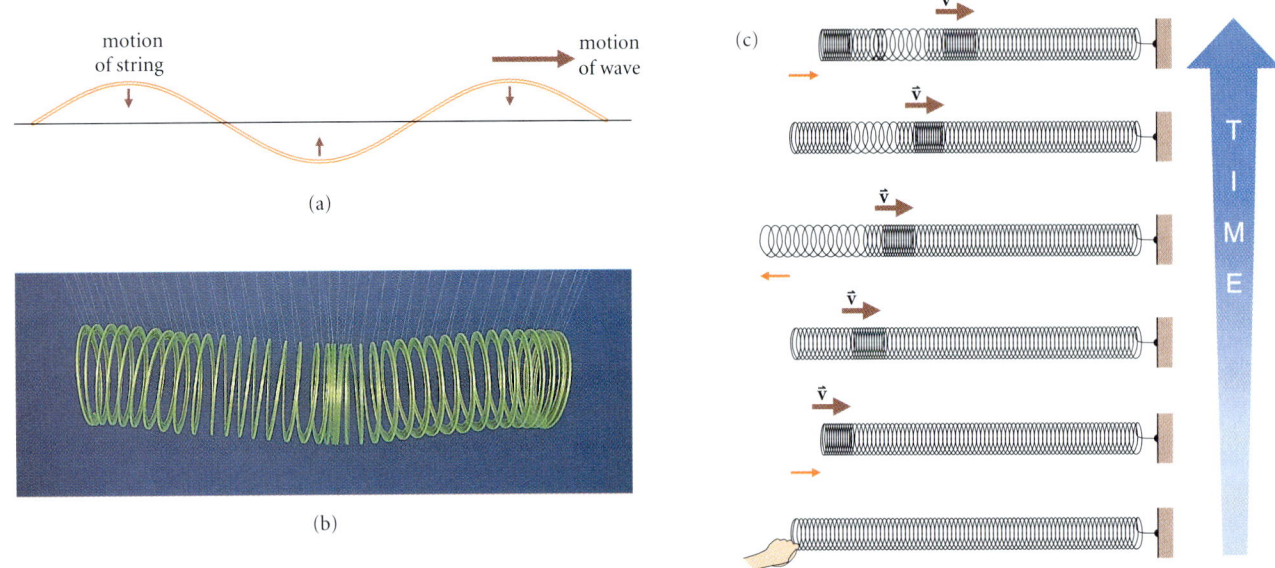

FIGURE 15.9
(a) In transverse waves, the displacement occurs perpendicular to the direction of wave motion. Waves on strings are of this type. (b) Here longitudinal waves are generated on a slinky. (c) In longitudinal waves, the displacement occurs along the same line as the wave motion.

LIGHT WAVES IN VACUUM ARE PECULIAR IN THAT THEY REQUIRE NO MEDIUM AND THERE ARE NO PARTICLE MOTIONS. HOWEVER, THEY ARE TRANSVERSE WAVES BECAUSE THE ELECTRIC AND MAGNETIC FIELDS IN THE WAVE ARE DESCRIBED BY VECTORS PERPENDICULAR TO THE DIRECTION OF WAVE MOTION. SEE CHAPTER 33.

Waves are classified according to the direction of motion of the individual elements as compared with the motion of the wave as a whole. Waves on a string are examples of transverse waves (■ Figure 15.9a). String particles move up and down while the wave moves along the string.

> In a *transverse wave,* the elements of the system move at right angles to the direction of wave motion.

If you move one end of a slinky back and forth along its length (Figure 15.9c), a wave travels along the slinky. The spring coils move along the spring, parallel to the direction of the wave motion. In sound waves, the air particles also move along the same line that the wave travels. These are examples of *longitudinal waves.*

> In *longitudinal waves,* the elements oscillate back and forth in the same direction as the wave motion.

Ripples on a water surface are neither pure transverse nor pure longitudinal: the particles move around small ellipses, with both transverse and longitudinal components of velocity (■ Figure 15.10).

EXERCISE 15.2 ❖ Use dimensional analysis to find how the speed of longitudinal waves in the spring depends on its mass m, spring constant k, and length ℓ.

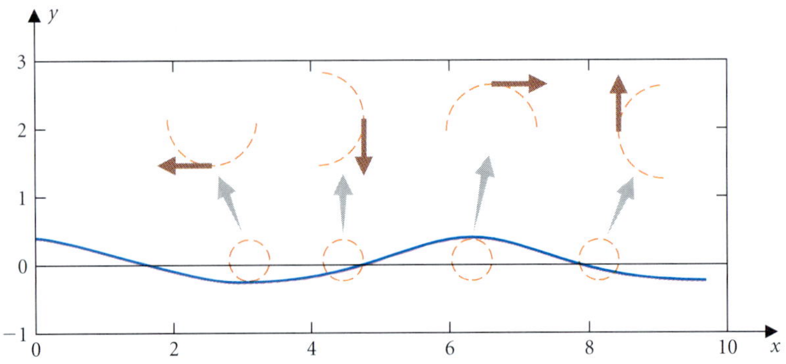

■ **FIGURE 15.10**
Particles of water move around almost circular paths as the ripples pass. In this wave moving to the right, water particles move to the right at the crests, to the left in the troughs, and vertically where the vertical surface displacement is zero. These waves are neither transverse nor longitudinal, but a mixture of both.

15.2 MATHEMATICAL DESCRIPTION OF A WAVE DISTURBANCE

15.2.1 The Wave Function

To describe the waveform of a disturbance at any time, we need a function of position—the displacement at each point in the system. But to describe how the waveform evolves, this mathematical function also depends on time. For example, if the x-axis of a coordinate system is taken along a stretched string with x increasing to the right, and the y-axis is in the direction of displacements caused by the wave, then the displacement $y(x, t)$ at each position x and time t gives a complete description of the disturbance. The function $y(x, t)$ is called the *wave function*.

So long as we don't strike a string too vigorously, the resulting pulses maintain their shape as they travel along the string. The wave function describes a disturbance with constant shape moving along the string at constant speed. As a result, the time and space dependences of the wave function are related. Our next task is to determine this relation.

At $t = 0$, the waveform is described by a function $f(x)$ that gives the displacement at any position x along the string:

$$f(x) \equiv y(x, 0). \tag{15.3}$$

■ Figure 15.11a is a snapshot of the wave at $t = 0$—in effect, a graph of $f(x)$. After a time interval $\Delta t = t_1 - 0$, the wave has moved a distance $v\,\Delta t$, where v is the wave speed (Figure 15.11b). Roughly speaking, the waveform f has moved to the right. If we can make this statement precise, we'll have a general description of possible wave functions.

At time $t = 0$ (Figure 15.11a), the piece of string at coordinate x_0 has displacement $A = f(x_0)$. At $t = t_1$ (Figure 15.11b), a different piece of string at coordinate $x_1 = x_0 + vt_1$ has the same displacement A:

(Displacement at $x = x_1$ and $t = t_1$) = (displacement at $x = x_0$ and $t = 0$).

$$y(x_1, t_1) = y(x_0, 0) \equiv f(x_0).$$

But $x_0 = x_1 - vt_1$, so:

$$y(x_1, t_1) = f(x_1 - vt_1). \tag{i}$$

The reasoning leading to eqn. (i) doesn't depend on the specific coordinate x_1 or the time interval $\Delta t = t_1 - 0$ so, in general:

$$y(x, t) = f(x - vt). \quad \text{(Wave moving to right)} \tag{15.4}$$

WE SHALL USE THIS CONVENTION (x INCREASES TO THE RIGHT) THROUGHOUT THE REST OF THIS CHAPTER BECAUSE IT AVOIDS CONTORTED LANGUAGE. WHEN USING THE RESULTS, REMEMBER THAT "TO THE RIGHT" MEANS "IN THE DIRECTION OF INCREASING x."

THE DISPLACEMENT y IS A FUNCTION OF THE TWO VARIABLES x AND t.

HERE THE FUNCTION $y(x, t)$ IS EVALUATED AT A FIXED VALUE, 0, OF THE VARIABLE t.

REMEMBER: f IS A FUNCTION OF ONE VARIABLE. THE WAVE FUNCTION $y(x, t)$ CAN DEPEND ONLY ON THE COMBINATIONS $x + vt$ OR $x - vt$. THE FUNCTION $f = 1/[(x - vt)^2 + a^2]$ IS OF THE CORRECT FORM, BUT $f = 3xvt$ IS NOT.

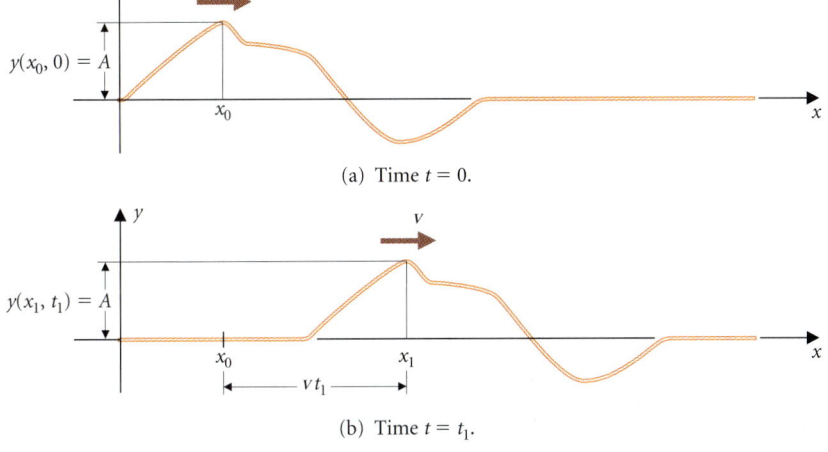

■ FIGURE 15.11
(a) At time $t = 0$, the wave peak $y = A$ is at $x = x_0$: $y(x_0, 0) = A$. (b) At time t_1, the wave peak has moved to position $x_1 = x_0 + vt_1$, so $y(x_1, t_1) = A$. Thus $y(x_1, t_1) = A = y(x_0, 0)$. Remember, the string itself moves up and down and does **not** move sideways.

In a wave moving to the left, the piece of string at coordinate $x_2 = x_0 - vt_1$ has the same displacement at time t_1 as the piece at x_0 had at $t = 0$.

$$y(x, t) = y(x + vt, 0) = f(x + vt). \quad \text{(Wave moving to left)} \quad (15.5)$$

Equations (15.4) and (15.5) describe precisely what it means for a waveform to move.

15.2.2 Harmonic Waves

We have not yet considered any specific form of the function f that appears in eqns. (15.3), (15.4), and (15.5). The shape of the string at $x = 0$ could be quite complex! The French mathematician Joseph Fourier (1768–1830) showed that any such waveform may be described as a superposition of simpler waveforms called harmonic waves. Thus, by studying harmonic waves, we can learn about the properties of all waves.

SEE THE *MATH TOOLBOX* ON HARMONIC FUNCTIONS IN CHAPTER 14. WE COULD ALSO USE A SINE WAVE.

| A *harmonic* wave is one whose waveform has the shape of a cosine function.

■ Figure 15.12 shows a harmonic wave propagating to the right on a stretched string, "photographed" at time $t = 0$. The function f that describes this waveform is:

$$f(x) = A \cos(kx + \phi_o),$$

where A is a constant called the *amplitude* of the wave. The argument $\phi = kx + \phi_o$ of the cosine is the *phase function* of the wave at $t = 0$.

THESE DEFINITIONS ARE EXACTLY THE SAME AS THOSE FOR OSCILLATIONS (CHAPTER 14).

| The *amplitude* of a harmonic wave is the magnitude of the maximum displacement.
| The *phase function* of a harmonic wave is the argument of the cosine function that describes the wave.

At a peak, the phase is zero or an integer multiple of 2π, and the displacement y equals the amplitude. The distance Δx between two consecutive peaks of the wave is its wavelength λ. Between the two peaks, the cosine function oscillates once and its phase changes by 2π:

$$\Delta \phi = k(x + \lambda) + \phi_o - (kx + \phi_o) = k\lambda = 2\pi.$$

Thus $k = 2\pi/\lambda$ and so the function $\phi(x)$ is:

$$\phi = \frac{2\pi}{\lambda}x + \phi_o.$$

Thus the wave function at $t = 0$ is:

$$y(x, 0) = f(x) = A \cos\left(\frac{2\pi}{\lambda}x + \phi_o\right).$$

BEWARE: DO NOT CONFUSE THIS k = WAVE NUMBER WITH k = SPRING CONSTANT. YOU MUST USE CONTEXT TO DISTINGUISH THEM.

The constant $k = 2\pi/\lambda$ is called the *wave number*. The dimensionless *phase constant* ϕ_o tells us where the first peak is. If $\phi_o = 0$, the first peak is at $x = 0$. The wave shown in our snapshot

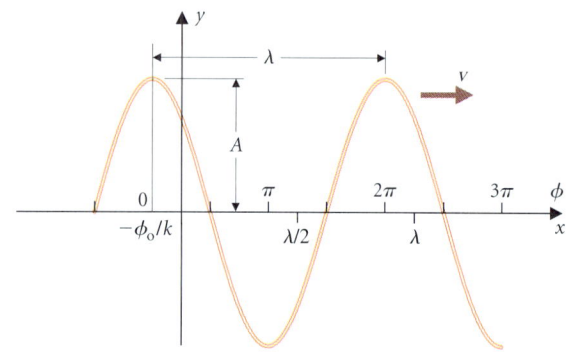

■ FIGURE 15.12
The waveform of a harmonic wave is a cosine. The maximum displacement is the amplitude A, $f(x) = A \cos[\phi(x)]$, where $\phi(x)$ is the wave phase. This waveform has phase constant $\phi_o = \pi/4$.

(Figure 15.12) has $\phi_o = \pi/4$. The complete wave function for a harmonic wave with speed v is $f(x \pm vt)$:

$$y(x, t) = A \cos\left[\frac{2\pi}{\lambda}(x \pm vt) + \phi_o\right]. \quad (15.6)$$

THE SYMBOL $\pm$ ALLOWS US TO WRITE ONE EXPRESSION ALLOWING FOR WAVE MOTION IN THE DIRECTION OF EITHER DECREASING OR INCREASING x.

An alternative way to evaluate how the phase depends on time is to observe its variation at a fixed point. The time for an element of string to go from its maximum displacement A to $-A$ and back to A again is the period T of the wave. So the argument of the cosine function changes by 2π as the time t increases by T. A term $2\pi t/T$ in the argument accomplishes this. Thus:

$$y(x, t) = A \cos\left(\frac{2\pi x}{\lambda} \pm \frac{2\pi t}{T} + \phi_o\right). \quad (15.7)$$

IN A HARMONIC WAVE, EACH PARTICLE UNDERGOES SHM WITH PERIOD T.

This result may also be obtained from eqns. (15.6) and (15.2). In terms of the wave number k and the *angular frequency* $\omega \equiv 2\pi f = 2\pi/T$, the wave function may be written:

$$y = A \cos(kx - \omega t + \phi_o), \quad \text{wave moving to the right,}$$
$$y = A \cos(kx + \omega t + \phi_o), \quad \text{wave moving to the left.} \quad (15.8)$$

and

Since $\lambda f = v$, then also $\omega/k = v$.

EXAMPLE 15.3 ♦ If the wave shown in ■ Figure 15.13a propagates to the right at $v = 100.0$ m/s, what are its wavelength λ, frequency f, and phase constant ϕ_o? Plot the displacement at $x = 0$ as a function of time.

MODEL The graph shows a snapshot of the wave taken at $t = 0$. It is a cosine waveform, so we may apply the results just developed.

SETUP AND SOLVE We may read the wavelength directly from the graph:

$$\lambda = \text{distance between peaks} = 10.0 \text{ m.}$$

At $t = 0$, the first peak is at $x = 5.0$ m. The phase at this peak is 0:

$$0 = \frac{2\pi}{10.0 \text{ m}}(5.0 \text{ m}) + \phi_o \Rightarrow \phi_o = -\pi.$$

Using eqn. (15.2), the frequency is:

$$f = \frac{v}{\lambda} = \frac{100.0 \text{ m/s}}{10.0 \text{ m}} = 10.0 \text{ Hz.}$$

We may summarize our results by writing the wave function. Reading off the graph, the amplitude is 3.0 cm, so:

$$y(x, t) = (3.0 \text{ cm})\cos[2\pi x/(10.0 \text{ m}) - 2\pi(10.0 \text{ Hz})t - \pi].$$

At $x = 0$, $y(0, t) = (3.0 \text{ cm})\cos[-2\pi(10.0 \text{ Hz})t - \pi]$. We have plotted this function in Figure 15.13b.

ANALYZE Even though we are looking at a snapshot of the wave and not a movie, we can determine the frequency if we know the wavelength and the speed. Notice that we can add an integer times 2π to the phase without changing the value of the function. So $\phi_o = -\pi + 2\pi = +\pi$ is equally valid. ■

We may think of the phase as defining a point on the waveform (a peak, a valley, or somewhere in between). The peak corresponds to phase 0, or equivalently, an integer multiple of 2π. The valley, where $y = -A$, corresponds to phase π, while zero displacement occurs at

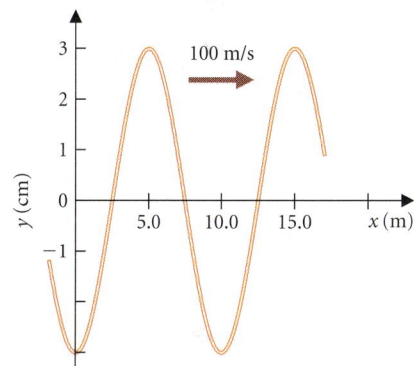

(a) A snapshot of a wave at $t = 0$.

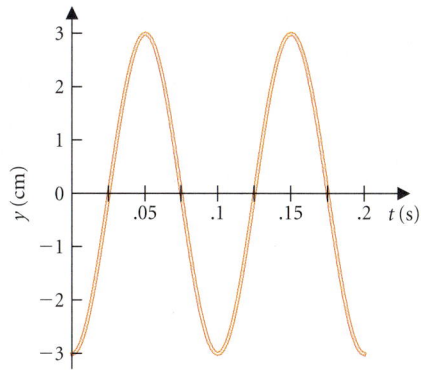

(b) Displacement of wave as a function of time.

■ **FIGURE 15.13**
(a) If we know the wave speed, this snapshot is sufficient to determine the wave function. (b) Displacement at $x = 0$ due to the same wave, plotted as a function of time. The period $T = 1/f = 0.1$ s.

SECTION 15.2 • MATHEMATICAL DESCRIPTION OF A WAVE DISTURBANCE phases $\pi/2$ and $3\pi/2$. Except at peaks and troughs, the phase depends on both the value of the displacement *and* whether it is increasing or decreasing. For example, at phase $\pi/2$ in Figure 15.12, the displacement is zero and increasing in time; at phase $3\pi/2$, the displacement is zero and decreasing in time.

EXAMPLE 15.4 ♦♦ A wave has the form:

$$y(x, t) = (10.0 \text{ cm})\cos[(1.5 \text{ rad/m})x - (0.25 \text{ rad/s})t + \pi/8].$$

At $t = 3.0$ s, find the points where the displacement equals 5.0 cm and is increasing in time.

MODEL Once we know the amplitude (10.0 cm here), the displacement is entirely determined by the wave phase.

SETUP The displacement $y = 5.0$ cm when the cosine function equals $\frac{1}{2}$—that is, at phase $\pm\pi/3$. With $\phi = (1.5 \text{ rad/m})x - (0.25 \text{ rad/s})t + \pi/8$, at any fixed place ($x = $ constant) the phase decreases as time increases. Looking at the graph of the cosine function (■ Figure 15.14), we see that the displacement increases as ϕ decreases at $\phi = +\pi/3$ but not at $\phi = -\pi/3$. So we choose $\phi = +\pi/3$.

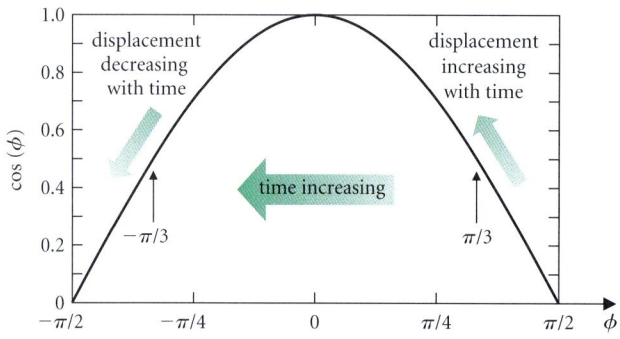

■ **FIGURE 15.14**
Graph of a cosine function for $-\pi/2 < \phi < +\pi/2$. At any fixed place, the phase of the given wave decreases as time increases. We are looking for a place where the value of the cosine is $\frac{1}{2}$ and increasing with time; that is, the phase is $+\pi/3$.

SOLVE Then:

$$\pi/3 = \phi = (1.5 \text{ rad/m})x - (0.25 \text{ rad/s})(3.0 \text{ s}) + \pi/8,$$

and

$$x = \frac{0.21\pi + 0.75}{1.5 \text{ rad/m}} = 0.94 \text{ m}.$$

ANALYZE This is not the only possible solution. The wavelength is $2\pi/(1.5 \text{ rad/m}) = 4.2$ m. The displacement is also 5.0 cm, and increasing, any whole number of wavelengths either side of the position we found, that is, at $x = -3.3$ m, 5.1 m, 9.3 m, ... These positions correspond to phases $\pi/3 \pm 2n\pi$, where n is any integer. ■

EXAMPLE 15.5 ♦♦ Find the wave function for a harmonic wave that propagates to the right along a string with tension $T = 1.2$ N and mass per unit length $\mu = 1.0 \times 10^{-2}$ kg/m. The maximum displacement anywhere on the string is 0.10 cm, and at $t = 0$, wave peaks occur at $x = 2.0$ cm and every 15 cm thereafter.

MODEL We may use eqn. (15.1) for the wave speed. Then the given information allows us to determine the appropriate form of the standard result (15.7) or (15.8).

SETUP The wave speed is:

$$v = \sqrt{T/\mu} = \sqrt{(1.2 \text{ N})/(1.0 \times 10^{-2} \text{ kg/m})} = 11 \text{ m/s}.$$

The amplitude is the maximum displacement, $A = 0.10$ cm. The wavelength is the distance between the maxima, $\lambda = 0.15$ m, and:

$$k = 2\pi/\lambda = 2\pi/(0.15 \text{ m}) = (13.3 \text{ rad/m})\pi = 42 \text{ rad/m}.$$

Then $\omega = kv = (13.3 \text{ rad/m})\pi(11 \text{ m/s}) = (146 \text{ rad/s})\pi = 460 \text{ rad/s}$.

The wave phase is $kx - \omega t + \phi_o$. The minus sign indicates that the wave moves to the right. To find ϕ_o, we use the initial conditions. A maximum (phase zero) occurs at $x = 2.0$ cm at $t = 0$:

$$0 = (13.3 \text{ rad/m})(0.020 \text{ m})\pi + \phi_o = 0.267\pi + \phi_o.$$

Hence $\phi_o = -0.267\pi = -0.84$.

SOLVE Thus the wave function is:

$$y(x, t) = (0.10 \text{ cm})\cos[(42 \text{ rad/m})x - (460 \text{ rad/s})t - 0.84 \text{ rad}].$$

ANALYZE Remember that the argument of the cosine function is in radians.

15.2.3 The Wave Equation

To understand the motion of any system—a racing car, a clock pendulum, or a rolling wheel—we apply Newton's second law to determine the accelerations. To show that the waves we have described are the correct result for the string's motion, we apply Newton's law to each element of the string. We'll outline the proof here and provide the details in the next section.

A string's wave function $y(x, t)$ depends on both position and time. The position variable labels an element of string, and the time dependence describes that element's motion. The displacement of a particular string element is given by $y(x, t)$ at a *fixed* value of the coordinate x. Its acceleration is the second time derivative of its displacement; that is, we hold x fixed and differentiate with respect to time. The resulting derivative is called a *partial derivative* and written $\partial^2 y/\partial t^2$:

$$a_y = \frac{\partial^2 y}{\partial t^2}.$$

A net force acts on an element because the string is curved, and so the tension forces on its ends don't quite balance. The net force depends on the string's curvature, the rate at which the string's slope changes (see §15.2.4). The slope (■ Figure 15.15a) is evaluated by calculating the x-derivative at a *fixed time*: the partial derivative with respective to x, $\partial y/\partial x$. Its rate of change with position is the second derivative $\partial^2 y/\partial x^2$. Then $F_y \propto T \partial^2 y/\partial x^2$, and Newton's law relates these derivatives:

Force on element = (mass of element)(acceleration of element).

$$CT \frac{\partial^2 y}{\partial x^2} = (\mu \, dx) \left(\frac{\partial^2 y}{\partial t^2} \right).$$

The derivation in the next section shows that $C = dx$, and so we obtain:

$$\boxed{\frac{\partial^2 y}{\partial t^2} = \left(\frac{T}{\mu}\right) \frac{\partial^2 y}{\partial x^2}.} \quad (15.9)$$

We asserted in eqn. (15.1) that T/μ is the square of the wave speed, so we may write:

$$\boxed{\frac{\partial^2 y}{\partial t^2} = v^2 \frac{\partial^2 y}{\partial x^2}, \text{ where } v^2 = T/\mu.} \quad (15.10)$$

Equation (15.9), or its equivalent (15.10), is the *wave equation* for the string.

Any real disturbance of the string is described by a function $y(x, t)$ that satisfies eqn. (15.10). The equation is linear, which means that if y_1 and y_2 both satisfy the equation, so does $y_1 + y_2$. Try it and see! As we show in the following *Math Topic*, the general solution is any function of the form $f(x + vt)$ or $f(x - vt)$, or a combination of the two. Equation (15.10) does not determine the wavelength of the wave, its amplitude, or its direction of motion.

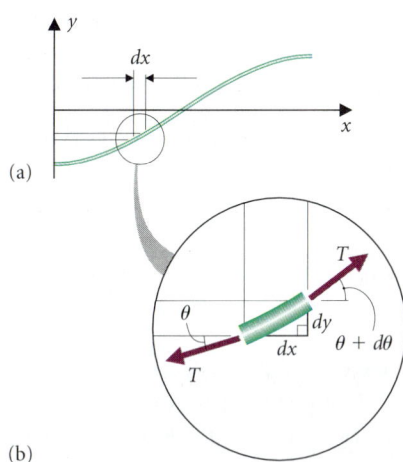

FIGURE 15.15
(a) A small element of displaced string.
(b) Because of the curvature of the element, the tension forces at the two ends do not quite balance: the string accelerates vertically. The angles have been exaggerated for clarity. In a real vibrating string, the angles are very small.

WE'RE LOOKING AT THE MOTION OF THE STRING IN THE MOVIE AT FIXED x.

IT'S THE STRING'S SLOPE IN THE SNAPSHOT OF THE WAVE TAKEN AT A FIXED TIME.

LEARN TO RECOGNIZE THIS EQUATION. IT HAS THE FORM: (2ND TIME DERIVATIVE) = SPEED² × (2ND SPACE DERIVATIVE).

THIS IS THE BASIS OF THE PRINCIPLE OF SUPERPOSITION. SEE §15.4.

Math Topic

DEMONSTRATION THAT $f(x \pm vt)$ SOLVES THE WAVE EQUATION

First we compute the two second derivatives that appear in eqn. (15.10). The necessary technique is the chain rule for derivatives:

$$\frac{d}{dx} f[g(x)] = \frac{df}{dg}\left(\frac{dg}{dx}\right).$$

The wave function f depends on the combination $u \equiv x \pm vt$ of the *two* variables x and t. So:

$$\frac{\partial f(u)}{\partial x} = \frac{df}{du}\left(\frac{\partial u}{\partial x}\right) \quad \text{and} \quad \frac{\partial f(u)}{\partial t} = \frac{df}{du}\left(\frac{\partial u}{\partial t}\right).$$

With

$$\frac{\partial u}{\partial x} = 1 \quad \text{and} \quad \frac{\partial u}{\partial t} = \pm v,$$

we have:

$$\frac{\partial f}{\partial x} = \frac{df}{du} \quad \text{and} \quad \frac{\partial f}{\partial t} = \pm v \frac{df}{du}.$$

The second derivatives are:

$$\frac{\partial^2 f(u)}{\partial x^2} \equiv \frac{\partial}{\partial x}\left(\frac{\partial f}{\partial x}\right) = \frac{\partial}{\partial x}\left(\frac{df}{du}\right)$$

$$= \frac{d}{du}\left(\frac{df}{du}\right)\frac{\partial u}{\partial x} = \frac{d^2 f}{du^2},$$

and

$$\frac{\partial^2 f(u)}{\partial t^2} = \frac{d}{du}\left(\pm v \frac{df}{du}\right)\frac{\partial u}{\partial t} = v^2 \frac{d^2 f}{du^2}.$$

Thus:

$$\frac{\partial^2 f}{\partial t^2} = v^2 \frac{d^2 f}{du^2} = v^2 \frac{\partial^2 f}{\partial x^2}.$$

That is, any function $f(x \pm vt)$ is a solution of eqn. (15.10).

EXERCISE 15.3 ♦♦ Taking f to be the harmonic function $A \cos(kx - \omega t)$, evaluate the partial derivatives and show that this function satisfies the wave equation if $\omega/k = v$.

✼ 15.2.4 Derivation of the Wave Equation for a String

In this section we shall apply Newton's second law to an element of stretched string and derive its equation of motion. The element has length dx, mass $dm = \mu\, dx$, and makes an angle θ with the horizontal (Figure 15.15). For systems like violin strings, the following two assumptions are justified:

1. The weight of a string element is negligible compared with the tension force.
2. The angle θ remains small for each piece of string.

Figure 15.15b shows a close-up view of the element and the tension forces acting on its ends. The element curves so the two tension forces act in slightly different directions. The net force in the horizontal direction is:

$$dF_x = T\cos(\theta + d\theta) - T\cos\theta.$$

Since θ is small, $\cos\theta \approx \cos(\theta + d\theta) \approx 1$, and $dF_x \approx 0$. The element does not accelerate in the horizontal direction provided that the tension is constant along the string, as we have assumed.

The vertical force component at the left end of the string element is:

$$F_y = -T\sin\theta.$$

REMEMBER THAT $\cos\theta \approx 1 - \frac{1}{2}\theta^2$, BUT $\sin\theta \approx \theta$. WE ARE WORKING TO FIRST ORDER IN THE SMALL QUANTITIES θ AND $d\theta$.

But $\sin\theta \approx \tan\theta$ for small angles, and $\tan\theta$ equals the slope of the curve $y(x, t)$—in a snapshot of the waveform taken at a *fixed* time t. It is given by the partial derivative with respect to x: $\tan\theta = \partial y/\partial x$. Then:

$$F_y \text{ (left end)} = -T\frac{\partial y}{\partial x}\bigg|_x \quad \text{and} \quad F_y \text{ (right end)} = T\frac{\partial y}{\partial x}\bigg|_{x+dx}$$

Thus the net force on the string segment is:

$$dF_y = T\left(\left.\frac{\partial y}{\partial x}\right|_{x+dx} - \left.\frac{\partial y}{\partial x}\right|_x\right) = T\frac{\partial}{\partial x}\left(\frac{\partial y}{\partial x}\right)dx = T\frac{\partial^2 y}{\partial x^2}dx.$$

WE USED THE DEFINITION OF DERIVATIVE:
$$\frac{df}{dx} = \lim_{dx \to 0} \frac{f(x+dx) - f(x)}{dx}.$$

The string element's acceleration a_y is the second partial time derivative of $y(x, t)$, and Newton's second law gives:

$$T\frac{\partial^2 y}{\partial x^2}dx = dF_y = (dm)a_y = (\mu\, dx)a_y = \mu\frac{\partial^2 y}{\partial t^2}dx.$$

So,
$$\frac{\partial^2 y}{\partial t^2} = \left(\frac{T}{\mu}\right)\frac{\partial^2 y}{\partial x^2},$$

which is eqn. 15.9. Since we assumed small angles θ in deriving this equation, it describes waves with amplitude much less than a wavelength.

USE A HARMONIC WAVE AS AN EXAMPLE:
$y = A\cos(kx - \omega t)$, $\theta \approx \partial y/\partial x \leq 2\pi A/\lambda$.
CHECK IT!

15.3 ENERGY TRANSMISSION BY HARMONIC WAVES

When a wave passes through a medium, the particles of the medium do not travel with the wave, but instead move only small distances from their original position. The quantity that moves through the medium is energy. In a wave on a string, each segment oscillates up and down at its original x-coordinate, while transmitting energy along the string. Here we calculate the power transmitted by the wave.

Imagine cutting a string into two pieces at point Q with coordinate x (■ Figure 15.16). At Q, the string to the left exerts a force $\vec{F}$ on the string to the right. Point Q has velocity $\vec{v}_Q = \hat{j}\, \partial y(x, t)/\partial t$. Thus the rate at which the string to the left does work on the string to the right is:

$$P = \vec{F} \cdot \vec{v}_Q = F_y v_{Q,y} = F_y \frac{\partial y}{\partial t}.$$

The horizontal component of the force does no work, as the string element doesn't move sideways. At Q, the string makes an angle θ with the horizontal, and the magnitude of $\vec{F}$ is the string tension T. So the y-component of the force is:

$$F_y = -T\sin\theta.$$

Since θ is small, $\sin\theta \approx \tan\theta$, and $\tan\theta = \partial y/\partial x$ is the slope of the string. Thus:

$$F_y \approx -T\tan\theta = -T(\partial y/\partial x).$$

So,
$$P = -T\frac{\partial y}{\partial x}\frac{\partial y}{\partial t}.$$

For a harmonic string function, $y = A\cos(kx \pm \omega t + \phi_o)$, the power transferred to the right at position x is:

$$P = -T[-kA\sin(kx \pm \omega t + \phi_o)][\mp\omega A\sin(kx \pm \omega t + \phi_o)]$$

$$P = \mp Tk\omega A^2 \sin^2(kx \pm \omega t + \phi_o). \qquad (15.11)$$

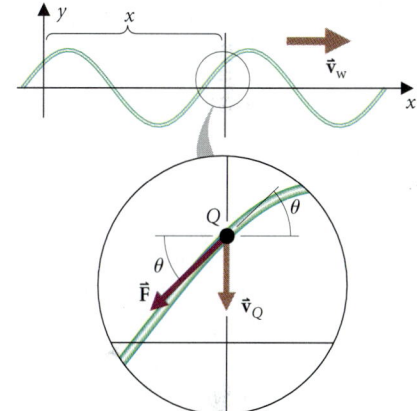

■ FIGURE 15.16
At point Q, the string to the left exerts a force of magnitude T on the string to the right. The rate at which work is done—the power—is $\vec{F} \cdot \vec{v}_Q$.

TAKE THE UPPER SIGN IN FRONT OF THE EXPRESSION WHEN TAKING THE UPPER SIGN INSIDE THE PARENTHESES.

There are several things to note about this result:
- Because the sine function is squared, the power has a constant sign. With the lower sign (rightward-moving wave), the power is positive: the string to the left does positive work on the string to the right. But with the upper sign (leftward-moving wave), the power is negative. The string on the *right* does positive work on the string to its *left*. Energy flows in the same direction as the wave motion.

THE FORCE THE STRING TO THE RIGHT EXERTS ON THE STRING TO THE LEFT IS $-\vec{F}$, WHILE THE VELOCITY OF THE RIGHT END OF THE LEFT-HAND STRING IS ALSO $\vec{v}_Q$. THUS THE POWER TRANSMITTED TO THE RIGHT EQUALS THAT EXTRACTED FROM THE LEFT.

- Power varies as the square of the wave amplitude. Alternatively, the amplitude of a wave is proportional to the square root of the power it transmits.
- Power is proportional to the square of the frequency. Since the wave speed $v_w = \omega/k$:

$$Tk\omega = T(\omega^2/v_w) = (\mu v_w^2)(\omega^2/v_w) = \mu v_w \omega^2.$$

- The power transmitted past one point of the string varies in time, from a minimum of zero to a maximum of $Tk\omega A^2$. The kinetic energy of a string segment $\frac{1}{2}v_y^2\,dm = \frac{1}{2}\mu\,dx(\partial y/\partial t)^2$. The elastic potential energy is proportional to the square of the string's stretch, which is proportional to $(\partial y/\partial x)^2$. Both are maximum at phases $\pi/2$ and $3\pi/2$. Maximum energy transmission occurs at these phases.
- The average power transmitted is obtained by averaging over many periods. The average value of $\sin^2\phi$ is one-half, so the average power is:

$$\langle |P| \rangle = \tfrac{1}{2}\mu v_w \omega^2 A^2. \tag{15.12}$$

> WE ALSO USED EQN. (15.1) TO ELIMINATE T.
>
> COMPARE THE POTENTIAL ENERGY STORED IN A SPRING (CHAPTER 8). SEE PROBLEM 85 FOR A FURTHER ANALYSIS OF THE POTENTIAL ENERGY OF THE STRING.
>
> THE QUICKEST WAY TO GET THE AVERAGE OF $\sin^2\phi$ IS TO NOTE THAT IT IS THE SAME AS THE AVERAGE OF $\cos^2\phi$. THEN $\sin^2\phi + \cos^2\phi = 1 \Rightarrow \langle \sin^2\phi \rangle = \langle \cos^2\phi \rangle = \tfrac{1}{2}$.

EXAMPLE 15.6 ♦ A wave traveling on a string of mass per unit length $\mu = 5.6$ g/m has the wave function:

$$y(x, t) = (0.75 \text{ mm})\cos[(4.8 \text{ rad/m})x - (1.5 \times 10^3 \text{ rad/s})t].$$

What average power is transmitted by the wave?

MODEL The wave function contains all the information about the wave. We'll use it to find v, A, and ω.

SETUP Comparing the wave function with the standard form $y = A\cos(kx - \omega t)$, we find $A = 0.75$ mm, $\omega = 1.5 \times 10^3$ rad/s, and $k = 4.8$ rad/m. The wave speed is:

$$v = \frac{\omega}{k} = \frac{1.5 \times 10^3 \text{ rad/s}}{4.8 \text{ rad/m}} = 310 \text{ m/s}.$$

SOLVE Using eqn. (15.12), the average power is:

$$\langle |P| \rangle = \tfrac{1}{2}(5.6 \times 10^{-3} \text{ kg/m})(310 \text{ m/s})(1.5 \times 10^3 \text{ rad/s})^2(0.75 \times 10^{-3} \text{ m})^2$$
$$= 1.1 \text{ W}.$$

ANALYZE Power is transmitted to the right, the same direction that the wave travels. ∎

15.4 SUPERPOSITION

If you throw two rocks into a pond together, two sets of ripples are formed. As the two sets start to overlap, the water surface shows the effects of both waves at once. As long as the total displacement remains small, it is the sum of the displacements due to the individual waves. This fact is called the *principle of superposition*:

> THE LATIN WORDS *POSARE*, TO PLACE, AND *SUPER*, ON TOP OF, GIVE A NICE IMAGE FOR THIS PRINCIPLE. INDIVIDUAL WAVE DISTURBANCES ARE PLACED ON TOP OF EACH OTHER.

> If a system supports two or more small-amplitude wave disturbances simultaneously, the overall wave function is the sum of the individual wave functions.

The principle of superposition is extremely important in understanding wave phenomena. Our first examples involve the superposition of a wave and its reflection.

15.4.1 Reflection of Waves at a Boundary

Strings in musical instruments have a finite length and are securely fixed at each end. When a pulse is sent along a piano string, it is reflected at the end and sent back along the string. The nature of the boundary determines the form of the reflected pulse.

A pulse $f(x + vt)$ propagating to the left along the string (the *incident* pulse) is reflected at a fixed end, where we set the coordinate $x = 0$. The *reflected* pulse propagates to the right and is described by another function, $g(x - vt)$. The displacement of the string is the sum of the incident pulse $f(x + vt)$ and the reflected pulse $g(x - vt)$:

$$y(x, t) = f(x + vt) + g(x - vt).$$

At the fixed end ($x = 0$), the displacement $y(0, t)$ is zero at all times:

$$f(+vt) + g(-vt) = 0 \Rightarrow g(-vt) = -f(+vt).$$

Thus g is the same shape as the incident pulse turned upside down and going the other way. ■ Figure 15.17 shows the two functions superposed at several times. The region $x < 0$ is not real, but imagining that the string extends into this region helps us to visualize the process of superposition.

Inversion of a harmonic wave corresponds to a phase change of π.

REMEMBER: $x + vt$ MEANS THE INCIDENT PULSE IS GOING LEFTWARD; $x - vt$ MEANS THE REFLECTED PULSE IS GOING TO THE RIGHT.

THIS IS OUR FIRST EXAMPLE OF A BOUNDARY CONDITION.

SEE, FOR EXAMPLE, FIGURE 14.6.

> At a *fixed* boundary, the reflected wave is inverted. For harmonic waves, the reflected wave's phase differs from that of the incident wave by π.

The opposite idealization is a free end: the end of the string at $x = 0$ is attached to a massless ring around a frictionless pole and is free to move vertically. Since the ring does not have infinite acceleration, it moves so that the vertical force acting on it is always zero (■ Figure 15.18):

$$T\frac{\partial y}{\partial x}\bigg|_{x=0} = 0.$$

RECALL THAT $F = ma$, SO F IS ZERO IF m IS ZERO, PROVIDED THAT a REMAINS FINITE.

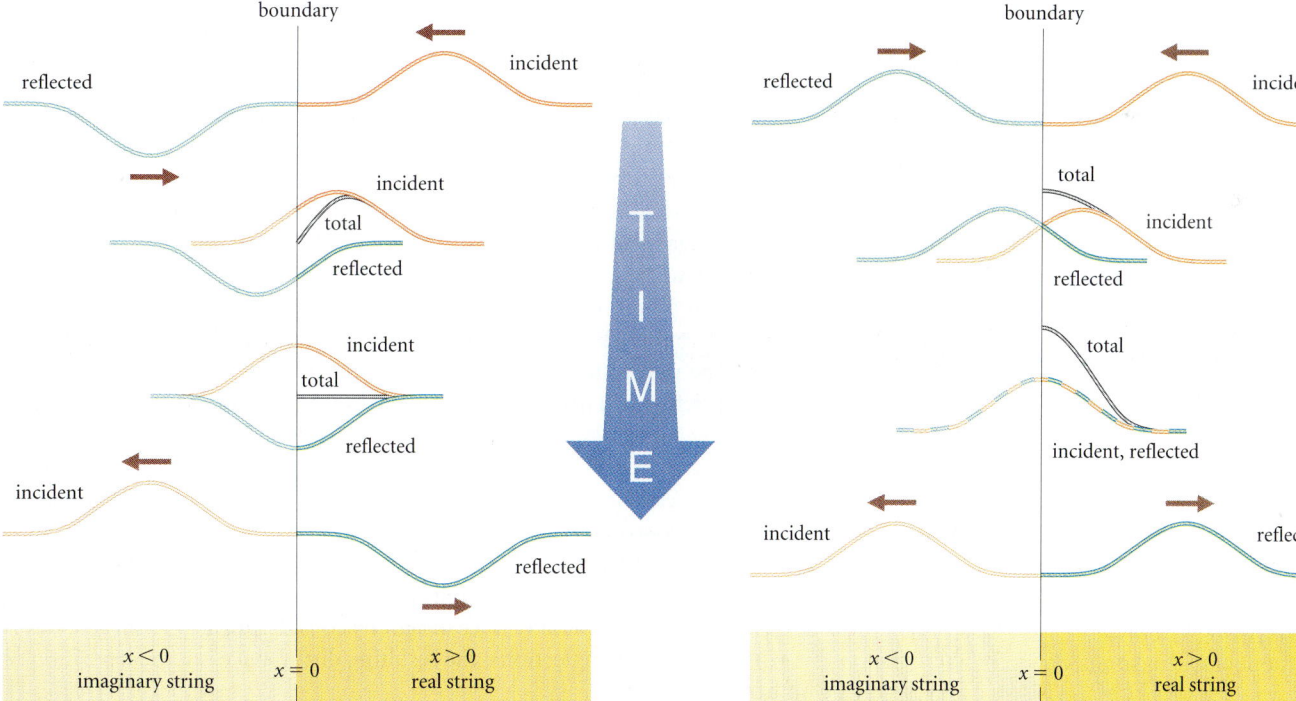

■ **FIGURE 15.17**
Reflection of a wave at a fixed boundary. Imagine that as the incident wave approaches the boundary, the reflected wave also approaches from the far (imaginary) side. The pulses merge at the boundary. At the fixed end, the displacement remains zero. This is possible only if the reflected pulse is inverted.

■ **FIGURE 15.18**
Reflection of a wave at a free boundary. Imagine the boundary as a mirror. The reflected wave behaves like the image of the incident wave, approaching the boundary from the "far side" and merging with the incident wave at the boundary. The slope of the disturbance remains zero at the boundary.

Since T is not zero, $\partial y/\partial x$ vanishes. The string remains perpendicular to the pole.

$$\left.\frac{\partial f(x+vt)}{\partial x}\right|_{x=0} + \left.\frac{\partial g(x-vt)}{\partial x}\right|_{x=0} = 0.$$

The two wave functions have equal and opposite slopes at $x = 0$. This is possible only if the reflected pulse is not inverted.

> At a *free* boundary, the reflected wave and incident wave functions have equal values and opposite slopes at any time. For harmonic waves, the reflected wave has the same phase as the incident wave.

EXAMPLE 15.7 ♦♦ A wave $y_i = A\cos(kx + \omega t)$ approaches a free boundary at $x = 0$. Assume that the reflected wave has the form $y_r = A\cos(kx - \omega t + \phi_o)$, and show that the phase constant ϕ_o is zero.

MODEL We use the principle of superposition.

SETUP The free-end condition that must be satisfied at all times t is:

$$\left.\frac{\partial y_i}{\partial x}\right|_{x=0} + \left.\frac{\partial y_r}{\partial x}\right|_{x=0} = 0.$$

$$-kA\sin(\omega t) - kA\sin(-\omega t + \phi_o) = 0.$$

SOLVE Since $\sin(-\theta) = -\sin(\theta)$, we may rewrite this equation as:

$$kA[\sin(\omega t) - \sin(\omega t - \phi_o)] = 0 \Rightarrow \phi_o = 0.$$

ANALYZE This calculation also demonstrates the equality of the two amplitudes. ■

15.4.2 Standing Waves

In musical instruments such as pianos, violins, or guitars, taut strings fixed at both ends produce specific musical notes. When the string is struck with a hammer, stroked with a bow, or plucked with a finger, waves propagate from the point of excitation toward each end. Upon reaching the end, the waves are reflected and return toward the middle, so there are waves propagating both to the right and to the left on the string. The resulting disturbance is the superposition of these waves.

SEE FIGURE 15.3.

A string of fixed length L cannot vibrate with any arbitrary frequency. To see why, imagine hitting the string with a hammer. Two pulses move away from the site of the blow at speed v, one in each direction. The rightward-moving pulse is reflected at the right end of the string and heads off toward the left end. There it is reflected again, and it returns to the right end a time $2L/v$ later. This pulse continues to return at intervals of $2L/v$. Meanwhile, the leftward-moving pulse is also reflected at each end and repeats its motion at intervals of $2L/v$. Hence the motion of the string is periodic with period $T = 2L/v$ and frequency $f = v/2L$. This is the *fundamental frequency* for the string.

The preceding qualitative argument shows why the string vibrates at the fundamental frequency, but complex pulses lead to more complicated behavior. In fact, there is an infinite set of allowed frequencies. We can obtain further insight by considering how harmonic waves combine on the string. After many reflections, the resulting disturbance is a superposition of waves moving both left and right.

REMEMBER: REFLECTION AT A FIXED BOUNDARY INTRODUCES A PHASE CHANGE OF π.

$$y(x, t) = \underbrace{A\cos(kx - \omega t + \phi_o)}_{\text{incident wave}} + \underbrace{A\cos(kx + \omega t + \phi_o + \pi)}_{\text{reflected wave}}.$$

THE MINUS SIGN IN THE SECOND TERM CAME FROM THE EXTRA π IN ITS PHASE: $\cos(\theta + \pi) = -\cos\theta$.

The total displacement is zero at the fixed ends of the string. With the origin at the left end of the string:

$$0 = y(0, t) = A\cos(-\omega t + \phi_o) - A\cos(\omega t + \phi_o).$$

Since $\cos(-\theta) = \cos(\theta)$, the boundary condition at $x = 0$ is satisfied at all times if the phase constant ϕ_o is zero. Then the displacement at any point on the string is:

$$y(x, t) = A\cos(kx - \omega t) - A\cos(kx + \omega t).$$

To simplify this result, we combine terms using the cosine addition rule (Appendix IB).

$$\cos\alpha - \cos\beta = 2\sin[(\alpha + \beta)/2]\sin[(\beta - \alpha)/2].$$

With $\alpha = kx - \omega t$ and $\beta = kx + \omega t$, we have:

$$y(x, t) = 2A\sin(kx)\sin(\omega t). \qquad (15.13)$$

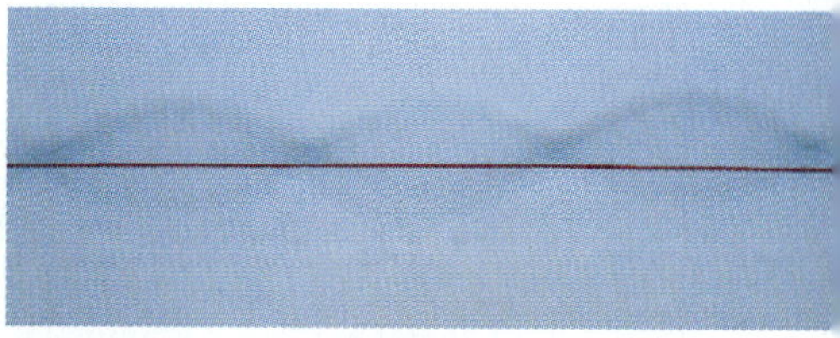

The disturbance on the string does not propagate as an ordinary wave does: it is a standing wave. At any one place, the string oscillates with amplitude $2A\sin kx$ and angular frequency ω. The curves $f(x) = \pm 2A\sin kx$, which pass through the end points of these oscillations, form the envelope of the wave.

> A *standing wave* is a disturbance that oscillates in time but does not propagate. The envelope of the oscillations has a fixed spatial pattern.

At certain places, called *nodes*, where $\sin(kx) = 0$, the string does not move at all!

| At a *node* of a standing wave, the displacement remains zero.

The sine function is zero whenever its argument is an integer m times π, so at the nodes: $kx = m\pi$, $m = 0, 1, 2, 3, \ldots$. At the points where $kx = (2m + 1)\pi/2$, $\sin(kx) = \pm 1$, and the oscillations have maximum amplitude $2A$; these points are called *antinodes*.

| At an *antinode* of a standing wave, the oscillations have maximum amplitude.

Since both ends of the string are fixed, both must be nodes. We have constructed the solution so that the displacement remains zero at the origin, but the displacement at the other boundary, $x = L$, is also zero at all times:

$$0 = y(L, t) = 2A\sin(kL)\sin(\omega t) \Rightarrow \sin(kL) = 0.$$

With $k = 0$, y is zero everywhere and there is no wave at all. The solution we need for kL is one of the other roots of the sine function:

$$kL = n\pi,$$

where n may be any nonzero integer. Since $k = 2\pi/\lambda$, the wavelength is:

$$\lambda_n = 2L/n, \quad n = 1, 2, 3, \ldots. \qquad (15.14)$$

Only certain wavelengths are allowed! (■ Figure 15.19)

When a guitarist plucks a string, standing waves are set up on it. Since $\omega/k = v = \sqrt{T/\mu}$, a string of a certain mass and length can support waves of frequency $f_n = v/\lambda_n = v(n/2L) = (n/2L)\sqrt{T/\mu}$. The *fundamental* frequency of the string occurs for $n = 1$, $f_1 = v/2L$, as indicated by our qualitative argument. The frequencies corresponding to $n > 1$ are the *harmonics*. The manner in which the string is plucked or struck determines the harmonics that are generated and the amplitude of each.

The fundamental frequency (and all the harmonics) of a string of fixed length can be increased by increasing the tension or decreasing the mass of the string. The manufacturer determines the length and mass of the string. Bass piano notes are produced by thick, heavy strings, while the high notes are produced by thin strings under high tension. The piano tuner adjusts the tensions to obtain exactly the desired frequency for each string.

Standing waves are important in a variety of situations. An organ pipe employs standing sound waves; some laser designs use standing light waves between two mirrors; and atoms are stable because the electrons form standing wave patterns around the atomic nuclei.

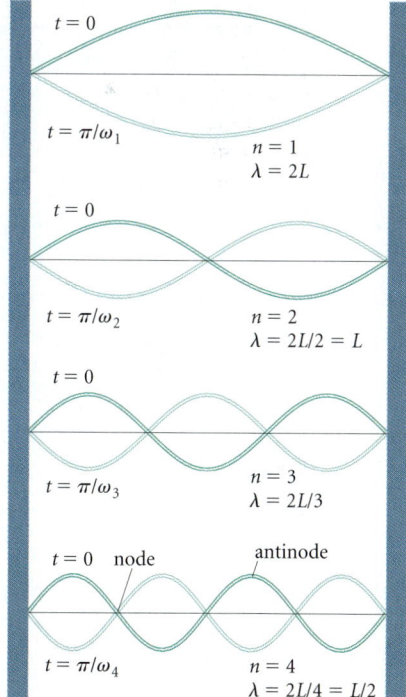

■ **Figure 15.19a**
Time-exposure photograph of a standing wave on a string. This photograph shows how oscillations of each element of the system form the envelope of the standing wave.

NODES OCCUR AT POINTS FOR A ONE-DIMENSIONAL WAVE, ALONG LINES FOR TWO-DIMENSIONAL WAVES, AND ALONG SURFACES FOR THREE-DIMENSIONAL WAVES.

(b) Standing waves on a string.

$t = 0$
$t = \pi/\omega_1$
$n = 1$
$\lambda = 2L$

$t = 0$
$t = \pi/\omega_2$
$n = 2$
$\lambda = 2L/2 = L$

$t = 0$
$t = \pi/\omega_3$
$n = 3$
$\lambda = 2L/3$

$t = 0$ node   antinode
$t = \pi/\omega_4$
$n = 4$
$\lambda = 2L/4 = L/2$

■ **Figure 15.19b**
Standing waves on a string. Nodes occur every half wavelength. A whole number of half wavelengths fit on the string. The nth mode has n antinodes.

Digging Deeper

Standing Waves and Musical Notes

Usually, a string on a musical instrument will not vibrate with a single standing wave mode. According to the principle of superposition, many wave disturbances can exist on the string at the same time. Because the ends of the string remain fixed, the motion of the string must be a superposition of standing waves. ■ Figure 15.20a shows a graph of the sound energy radiated by a violin string versus the frequency of the sound waves. The sharp, regularly spaced peaks occur at the frequencies of the string's possible standing wave modes. What musicians describe as the quality of the string's sound is determined by the relative amounts of energy radiated by the various modes. The musician selects the frequencies of the modes by controlling the length, mass, and tension of the string. Where and how vigorously the string is plucked determines the amplitude of each standing wave and thus the sound quality. Bowing the string produces a different set of harmonics and hence a different sound quality (Figure 15.20b). (The vibrations of an instrument's sounding board also affect its sound quality.)

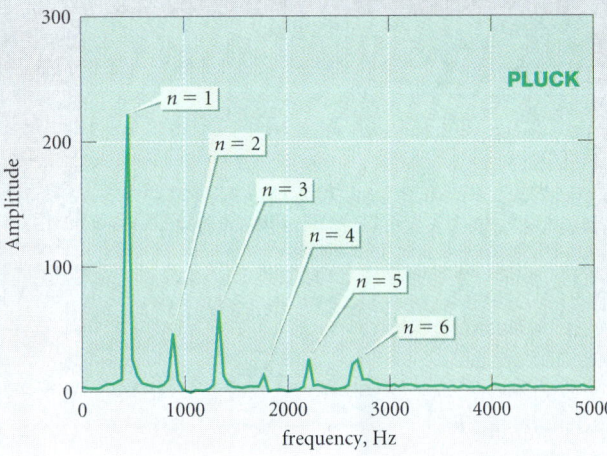

■ **Figure 15.20a**
Sound energy radiated by a plucked violin string as a function of frequency. The regularly spaced peaks indicate the standing wave modes. Harmonics up to $n = 6$ can be seen, but most of the energy is in the fundamental mode.

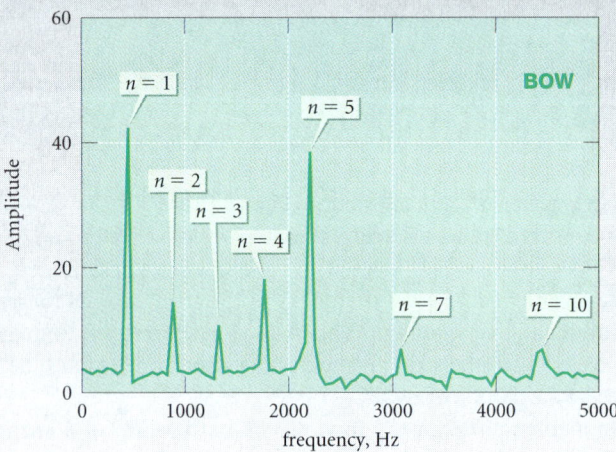

■ **Figure 15.20b**
The technique used to excite a string determines which modes are excited. Bowed strings have a different sound quality than plucked strings. Here we see that the $n = 5$ harmonic has a large amplitude when a violin string is bowed. Data courtesy of Alma Zook, Pomona College.

EXAMPLE 15.8 ♦♦ The A string on a cello has mass per unit length $\mu = 1.60$ g/m and length 0.70 m. What tension is required for the fundamental frequency to be 440 Hz?

MODEL Standing waves on the string have wavelength $2L/n$. The fundamental frequency has $n = 1$ and $\lambda = 2L$. The wavelength is related to the frequency by the wave speed.

SETUP The wave speed is: $\quad v = \sqrt{\dfrac{T}{\mu}} = \lambda f.$

SOLVE The required tension is:
$$T = \mu(\lambda f)^2 = \mu(2Lf)^2$$
$$= (1.60 \times 10^{-3} \text{ kg/m})[2(0.70 \text{ m})(440 \text{ Hz})]^2 = 610 \text{ N}.$$

ANALYZE At the beginning of the performance, the cellist tunes his instrument by adjusting the tension until the string sounds an A.

15.5 REFLECTION AND TRANSMISSION OF WAVES AT A JUNCTION OF TWO STRINGS

The results we obtained in §15.4.1 for waves reflected at fixed or free ends are special cases. We shall now see what happens in a more general case: reflection at a connection between two strings of different mass per unit length (■ Figure 15.21). The tension is the same in each string; otherwise the junction would accelerate rapidly to the right or left. Thus the wave speed is different on the two strings: $v_1/v_2 = \sqrt{\mu_2/\mu_1}$.

When a wave reaches the junction, at the origin $x = 0$, it is partially reflected and partially transmitted. Three separate waves are superposed on the system:

$$y_i = A_i \cos(k_i x - \omega_i t), \quad \text{incident wave moving to the right.}$$
$$y_r = A_r \cos(k_r x + \omega_r t), \quad \text{reflected wave moving to the left.}$$
$$y_t = A_t \cos(k_t x - \omega_t t), \quad \text{transmitted wave moving to the right.}$$

The amplitudes of the reflected and transmitted waves determine how much energy is reflected and how much transmitted.

The behavior of the string junction sets conditions the wave functions must satisfy at the boundary. The junction oscillates vertically; its position y_j is the same whether we consider it as the right end of the left-hand string **or** as the left end of the right-hand string. So, with $x = 0$ at the junction:

Right end of left-hand string: $\quad y_j = y_i(0, t) + y_r(0, t).$
Left end of right-hand string: $\quad y_j = y_t(0, t).$

$$A_i \cos(-\omega_i t) + A_r \cos \omega_r t = A_t \cos(-\omega_t t).$$

WE PREVIOUSLY FOUND A PHASE SHIFT OF ZERO OR π ON REFLECTION. ANTICIPATING A SIMILAR RESULT, WE CAN ACCOUNT FOR IT WITH A POSITIVE OR NEGATIVE VALUE FOR A_r.

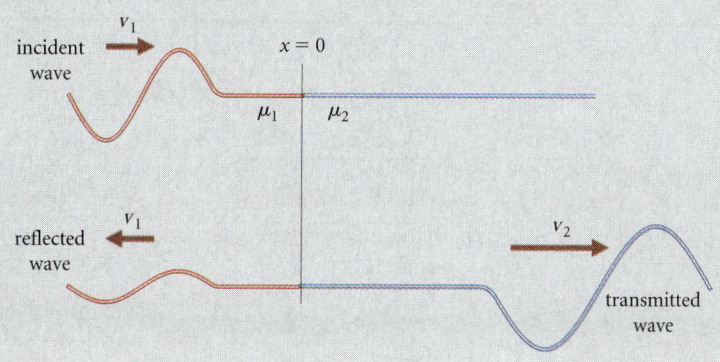

■ FIGURE 15.21
Two strings of different mass per unit length are joined together smoothly at $x = 0$. A wave incident from the left ($x < 0$) is partially reflected and partially transmitted.

This relation is true at all times only if all three frequencies are the same. Otherwise, by picking different values of t, we could generate an infinite number of equations that cannot all be satisfied with only two amplitudes to adjust:

$$\omega_r = \omega_t = \omega_i = \omega.$$

This is a general feature of matching waves at boundaries: all the waves involved have the same frequency. Then the cosine functions factor out and we have one relation among the constant amplitudes:

$$A_i + A_r = A_t. \tag{15.15}$$

With a common frequency, the incident and reflected waves have the same wave number: $k_r = k_i = \omega/v_1$. The transmitted wave has a different wave number: $k_t = \omega/v_2$.

There cannot be an abrupt kink in the string at the junction. Unless the tension forces act along the same line, the resulting net force would rapidly accelerate the massless junction to straighten the string. So the string's slope at the junction is the same whether evaluated on the left side or the right.

Right end of left-hand string: $\quad \dfrac{\partial y_j}{\partial x} = \dfrac{\partial y_i}{\partial x} + \dfrac{\partial y_r}{\partial x}.$

Left end of right-hand string: $\quad \dfrac{\partial y_j}{\partial x} = \dfrac{\partial y_t}{\partial x}.$

Equating the two expressions for the slope at the junction ($x = 0$), we have:

$$-k_i A_i \sin(-\omega t) - k_i A_r \sin(\omega t) = -k_t A_t \sin(-\omega t).$$

> RECALL THAT $k_i = k_r \neq k_t$.
>
> RECALL THAT $\sin(-\theta) = -\sin(\theta)$.

The sines factor out to give a second relation among the amplitudes:

$$k_i A_i - k_i A_r = k_t A_t. \tag{15.16}$$

Combining eqns. (15.15) and (15.16), we find:

$$A_t = \frac{2k_i}{k_i + k_t} A_i \quad \text{and} \quad A_r = \frac{k_i - k_t}{k_i + k_t} A_i. \tag{15.17}$$

EXAMPLE 15.9 ♦♦ Show that the power in the transmitted wave plus the power in the reflected wave equals the power in the incident wave.

MODEL Since the tension in the string is the same on both sides of the junction (as is the wave frequency), the time-averaged power transmitted by the wave is proportional to the amplitude squared times the wave number (eqns. 15.11 and 15.12).

SETUP We use eqns. (15.17) for the wave amplitudes.

Incident wave: $\quad P_i = \tfrac{1}{2} T\omega k_i A_i^2.$

Reflected wave: $\quad P_r = \tfrac{1}{2} T\omega k_r A_r^2 = \tfrac{1}{2} T\omega k_i \left(\dfrac{k_i - k_t}{k_i + k_t} \right)^2 A_i^2$

$\qquad = \tfrac{1}{2} T\omega \dfrac{k_i^2 + k_t^2 - 2k_i k_t}{(k_i + k_t)^2} k_i A_i^2.$

Transmitted wave: $\quad P_t = \tfrac{1}{2} T\omega k_t A_t^2 = \tfrac{1}{2} T\omega \dfrac{4 k_i^2 k_t}{(k_i + k_t)^2} A_i^2.$

SOLVE

$$\frac{P_r + P_t}{\tfrac{1}{2} T\omega} = A_i^2 k_i \frac{k_i^2 + k_t^2 - 2k_i k_t + 4k_i k_t}{(k_i + k_t)^2}$$

$$= \frac{k_i^2 + k_t^2 + 2k_i k_t}{(k_i + k_t)^2} k_i A_i^2 = k_i A_i^2 = \frac{P_i}{\tfrac{1}{2} T\omega}.$$

ANALYZE Energy is conserved in the reflection and transmission of the wave.

EXERCISE 15.4 ❖ Check that the amplitudes (eqn. 15.17) are correct for the case in which both strings have the same mass per unit length.

It is often useful to express the transmitted and reflected amplitudes in terms of the wave speeds. Using $v = \omega/k$, eqns. (15.17) become:

$$A_t = \frac{2v_2}{v_1 + v_2} A_i \quad \text{and} \quad A_r = \frac{v_2 - v_1}{v_1 + v_2} A_i. \tag{15.18}$$

If $v_2 > v_1$, then $k_t < k_i$ and A_r is positive, corresponding to no phase change. On the other hand, if $v_2 < v_1$, then $k_t > k_i$, and A_r is negative; the reflected wave is inverted, corresponding to a phase change of π.

WE COULD ALSO EXPRESS THIS RESULT IN TERMS OF THE MASS PER UNIT LENGTH OF THE STRING. SINCE $v = \sqrt{T/\mu}$, $v_2 < v_1$ IS EQUIVALENT TO $\mu_2 > \mu_1$. A PHASE CHANGE OCCURS WHEN THE WAVE REFLECTS OFF A HEAVIER STRING.

> A reflected wave will be inverted with respect to the incident wave (have a phase change of π) if the wave speed decreases across the boundary.

This result has important applications in optics (Chapter 17).

EXERCISE 15.5 ❖❖❖ Show that a fixed boundary is equivalent to the limit in which the second string is infinitely massive. Show that a free boundary corresponds to a second string with zero mass.

Chapter Summary

Where Are We Now?

By observing the behavior of waves on strings, we have exhibited the basic properties of wave behavior and developed terminology and mathematical descriptions for waves in general. We are ready to apply these basic results to sound and light waves.

What Did We Do?

A wave is a disturbance of a system away from a stable equilibrium state. A collection of coupled oscillators forms a good image for a system that supports waves. In reponse to the wave displacement, each part of the system exerts a restoring force on nearby parts that transfers energy and results in motion of the wave at a speed that is characteristic of the system. In similar systems, the wave speed is greater if the restoring force is greater or if the inertia of moving parts of the system is reduced. The speed of waves on a stretched string is

$$v = \sqrt{T/\mu}.$$

The shape of a wave disturbance at a fixed time is its *waveform*. A *pulse* is a waveform that is localized in a finite region of the system. The separation between pulses in a periodic wave is its *wavelength*, λ. A *harmonic* waveform has the shape of a cosine function. The *period* T of a periodic wave train is the time interval between two consecutive pulses. The *frequency* f is the reciprocal of the period. The *wave number*

$$k = 2\pi/\lambda,$$

and the *angular frequency*

$$\omega = 2\pi f.$$

Wave speed, frequency, and wavelength are related:

$$v = \lambda f = \omega/k.$$

In *transverse* waves, the displacement is perpendicular to the direction of wave motion. In *longitudinal* waves, displacement is along the direction of motion.

The *wave function* $y(x, t)$ gives wave displacement at each position and time. For any function f, $y(x, t) = f(x \pm vt)$ describes a wave moving at speed v ($+ \Rightarrow$ moving to the left, $- \Rightarrow$ moving to the right). Harmonic waves have the form:

$$y(x, t) = A\cos(kx \pm \omega t + \phi_o),$$

where ϕ_o is the *phase constant*.

Newton's second law applied to an element of string gives an equation for the wave function. The wave equation is:

$$\frac{\partial^2 y}{\partial t^2} = \frac{T}{\mu}\left(\frac{\partial^2 y}{\partial x^2}\right) = v^2 \frac{\partial^2 y}{\partial x^2}.$$

Its solutions are the general functions $y(x, t) = f(x \pm \sqrt{T/\mu}\, t)$.

Waves transmit energy in the direction of the wave motion at an average rate proportional to the square of the amplitude and the square of the frequency. Thus wave amplitude is proportional to the square root of the power transmitted.

Small-amplitude mechanical waves obey the *principle of superposition*:

The wave function for two or more wave disturbances propagating in the same system is the sum of the individual wave functions.

Reflection of waves at boundaries is described as the superposition of two waves moving in opposite directions. Waves on a string are inverted when reflected at a fixed end. Multiple reflections at the ends of a string of length L result in standing waves: disturbances that oscillate in time with a fixed spatial envelope and wavelength $\lambda_n = 2L/n$, where n is an integer. The corresponding set of frequencies is

$$f_n = v/\lambda_n = (n/2L)\sqrt{T/\mu}.$$

By adjusting its tension, the string of a musical instrument can be tuned to a specific fundamental frequency ($n = 1$). The string also vibrates with harmonics ($n > 1$) that influence its sound quality.

✱ At a boundary between two strings, an incident wave produces both reflected and transmitted waves at the same frequency. If the wave speed increases across the boundary, the reflected wave has no phase change; if the wave speed decreases across the boundary, the reflected wave has a phase shift of π.

Mathematical Results

A partial derivative of y with respect to x is a derivative with respect to x at a fixed time t, and is written $\partial y/\partial x$. Similarly, a partial derivative with respect to time, $\partial y/\partial t$, is taken at a fixed position, x.

Practical Applications

Standing waves are important in the operation of systems as diverse as violins and lasers. Civil engineers use wave reflection in the design of harbor breakwaters and ship canals. Seismologists use mechanical waves produced by earthquakes to study the interior of the Earth. Many aspects of communications technology rely on superposition of light or electrical waves. Finally, but very importantly, Fourier's theorems make harmonic wave functions a basic tool for advanced mathematical analysis of a variety of physical problems.

Solutions to Exercises

15.1 The surface at a point on the pond is displaced abruptly as the leading edge of the wave disturbance passes by. For a time, it oscillates periodically as ripples pass by, then it returns to equilibrium as the ripples die out. The disturbance is a pulse with structure described as a periodic wave.

15.2 By analogy with our results for the string, we expect the speed to depend on the restoring force, here determined by the spring constant k. Check its dimensions:
$$[k] = \frac{[F]}{L} = \frac{ML/T^2}{L} = \frac{M}{T^2}.$$
To eliminate the mass dimension, we divide by the spring's mass m:
$$[k/m] = 1/T^2. \quad T = \sqrt{\frac{m}{k}}$$
We need a length, and the only one we have is the length of the spring, ℓ.
$$\left[\ell\sqrt{\frac{k}{m}}\right] = \frac{L}{T} = [v]. \quad = \frac{\ell}{\sqrt{\frac{m}{k}}}$$
Thus: $v \propto \ell\sqrt{k/m} = \sqrt{S/\mu}$, where $S \equiv k\ell$ is called the stiffness of the spring and $\mu = m/\ell$ is its mass per unit length.

15.3 ✴ We are to mimic the calculation in the *Math Topic* with
$$f(\phi) = A\cos\phi \quad \text{and} \quad \phi = kx - \omega t.$$
So: $\dfrac{\partial f}{\partial x} = \dfrac{df}{d\phi}\left(\dfrac{\partial \phi}{\partial x}\right) = (-A\sin\phi)k,$

$\dfrac{\partial^2 f}{\partial x^2} = \dfrac{d}{d\phi}(-kA\sin\phi)\dfrac{\partial\phi}{\partial x} = -k^2 A\cos\phi,$

$\dfrac{\partial f}{\partial t} = \dfrac{df}{d\phi}\left(\dfrac{\partial \phi}{\partial t}\right) = (-A\sin\phi)(-\omega),$

$\dfrac{\partial^2 f}{\partial t^2} = \dfrac{d}{d\phi}(+\omega A\sin\phi)\dfrac{\partial\phi}{\partial t} = -\omega^2 A\cos\phi.$

Thus: $\dfrac{\partial^2 f}{\partial t^2} - \dfrac{T}{\mu}\left(\dfrac{\partial^2 f}{\partial x^2}\right) = \left(-\omega^2 + \dfrac{T}{\mu}k^2\right)A\cos\phi.$

So, for the right-hand side to be zero, $T/\mu = \omega^2/k^2 = v^2$.

15.4 ✴ If both strings are the same, the waves should act as if there were no junction at all. If $\mu_1 = \mu_2$, then $k_t = k_i$. From eqns. (15.17), $A_r = 0$ and $A_t = A_i$. There is no reflection and the transmitted wave is just the incident wave continued.

15.5 ✴ In the limit $\mu_2 \to \infty$, $v_2 = \sqrt{T/\mu_2} \to 0$ and $k_t = \omega/v_2 \to \infty$. Then:
$$A_t = \frac{2k_i}{k_i + k_t}A_i \to 0 \quad \text{and} \quad A_r = \frac{k_i - k_t}{k_i + k_t}A_i \to -A_i.$$
These are the same results as for reflection from a fixed boundary. If $\mu_2 \to 0$, then $v_2 \to \infty$ and $k_t \to 0$. In this limit:
$$A_t \to 2A_i \quad \text{and} \quad A_r \to A_i.$$
The result for A_r is what we expect for reflection from a free boundary. The "transmitted wave" has a large amplitude, but since $k_t = 0$, it carries no energy.

Basic Skills

Review Questions

§15.1 MECHANICAL WAVES

- In what ways is a wave like a simple oscillator? In what ways is it different?
- What two properties of a mechanical system are necessary for it to support waves?
- How does the speed of a wave on a string depend on the tension and mass per unit length of the string?
- Define *wave pulse* and *wavelength*. Describe a *periodic* wave.
- Define the *period* and *frequency* of a wave.
- How does a transverse wave differ from a longitudinal wave? Give an example of each type.

§15.2 MATHEMATICAL DESCRIPTION OF A WAVE DISTURBANCE

- What is the wave function?
- Define the *amplitude* and *phase* of a harmonic wave.
- What is the *wave number*? How is it related to angular frequency and speed?
- What is the general form for a wave equation?
- ✴ What approximations are made in deriving the wave equation for a string?

§15.3 ENERGY TRANSMISSION BY HARMONIC WAVES

- How does the average power transmitted by a wave depend on **(a)** its amplitude and **(b)** its frequency?
- Describe how the power transmitted past any point in a string varies with time.

§15.4 SUPERPOSITION

- State the principle of superposition.
- Describe how a wave pulse reflected from **(a)** a fixed boundary and **(b)** a free boundary is related to the incident pulse.

- What is a *standing wave*? How is the wavelength of a standing wave on a string related to the length of the string?
- What is the fundamental frequency for a string?

✷ §15.5 REFLECTION AND TRANSMISSION OF WAVES AT A JUNCTION OF TWO STRINGS

- Under what conditions is a pulse inverted when reflected from a junction of two strings?

Basic Skill Drill

§15.1 MECHANICAL WAVES

1. Two waves propagate on identical strings, but the tension in the first is three times that in the second. What is the ratio of the wave speeds, v_1/v_2?

2. A periodic wave has twice the wavelength of another wave propagating in the same system. How do the frequencies of the two waves compare?

3. A sound wave has a speed of 330 m/s and a wavelength of 0.75 m. What is its frequency?

4. A wave on a string has a wavelength of 27 cm and a frequency of 110 Hz. What is the speed of the wave?

§15.2 MATHEMATICAL DESCRIPTION OF A WAVE DISTURBANCE

5. A small flower is floating on the water as a surface wave comes by. It takes 0.12 s for the flower to rise to its maximum height from the equilibrium level. What is the period of the wave?

6. A wave is described by the function

$$y(x, t) = (10.0 \text{ cm})\cos[(5.0 \text{ rad/m})x + (220 \text{ rad/s})t].$$

Find the *velocity* of the wave, its wavelength, and its period.

7. The wave equation for a sound wave in air is:

$$\frac{\partial^2 y}{\partial t^2} = \frac{7}{5}\left(\frac{P}{\rho}\right)\left(\frac{\partial^2 y}{\partial x^2}\right),$$

where P is the air pressure and ρ its density. What is the speed of sound in air? Using typical values for P and ρ from Chapter 13, compute v.

§15.3 ENERGY TRANSMISSION BY HARMONIC WAVES

8. Two waves of the same frequency propagate along the same string, but one transmits twice as much power as the other. What is the ratio of their amplitudes?

9. Two strings are identical except that the tension in one is four times the tension in the other. If waves of equal frequency and amplitude propagate along each string, what is the ratio of the powers transmitted by the two waves?

10. Two waves of equal amplitude propagate along identical strings. If the frequency of one is three times the frequency of the other, what is the ratio of the transmitted powers?

11. Find the average power transmitted by the wave in Problem 6, if $\mu = 4.0$ g/m.

§15.4 SUPERPOSITION

12. Draw a graph showing the two wave disturbances:

$$y_1(x, t) = (5.0 \text{ cm})\cos[(3.0 \text{ rad/m})x + (5.0 \text{ rad/s})t],$$

and

$$y_2(x, t) = (3.0 \text{ cm})\cos[(6.0 \text{ rad/m})x + (10.0 \text{ rad/s})t],$$

and their superposition $y_1 + y_2$ at times $t = 0.0$ and $t = 0.40$ s.

13. If the wave y_1 in Problem 12 is incident on a fixed boundary at $x = 0$, write the wave function for the reflected wave. What differences, if any, would there be in the reflected wave function if the boundary were free?

14. A guitar string has length 0.670 m and mass per unit length 0.970×10^{-3} kg/m. What is the wavelength of the fundamental? If the fundamental frequency is to be G (392 Hz), what wave speed is necessary? What tension is required?

15. A standing wave of wavelength 1.0 m and frequency 110 Hz exists on a string with tension 120 N. What is the mass per unit length of the string if this is the fundamental frequency?

✷ §15.5 REFLECTION AND TRANSMISSION OF WAVES AT A JUNCTION OF TWO STRINGS

16. A wave of amplitude 0.50 cm is incident on a junction between two strings. The second string has mass per unit length twice the first. Find the amplitudes of the transmitted and reflected waves.

Questions and Problems

§15.1 MECHANICAL WAVES

17. ❖ Refer to Figure 15.13a. On a sketch of the string, draw vectors to show the velocity of string elements at the time illustrated.

18. ❖ Why do surface ripples not propagate on a pile of sand as they do on water?

19. ❖ Does the speed of a wave on a string depend on its direction of propagation? Why or why not?

20. ❖ A drummer sets up waves on the drumhead by hitting it. Are the waves transverse or longitudinal? Justify your answer.

21. ❖ ✉ A drumhead of radius R exerts a force F on the drum per unit length of its edge. The drumhead has mass μ per unit area. Use dimensional analysis to find how the speed of waves on the drumhead depends on F, μ, and R.

22. ❖ In what form is potential energy stored in a spring when it supports a longitudinal wave? When a transverse wave travels along a string, in what form is the potential energy stored?

23. ◆ ■ Figure 15.22 shows a siren. Puffs of air emerge through the holes as the disk turns, forming a periodic wave. If the disk has 11 holes and rotates 40 times per second, what is the frequency of the wave?

24. ◆ Breakers 9.0 m apart strike a beach every 5.0 s. Find the frequency, the wavelength, and the speed of the waves.

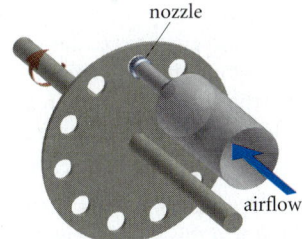

FIGURE 15.22

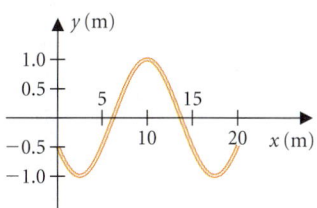

FIGURE 15.23

25. ◆◆ ✉ Dirt roads are subject to "washboarding"—the road surface develops ripples with their crests perpendicular to the road. The ripples form because most vehicles travel at about the same speed and their suspensions oscillate at roughly the same frequency. If the ripples are 3 m apart and the vehicles travel at 10 m/s, estimate the oscillation frequency of the suspension.

§15.2 MATHEMATICAL DESCRIPTION OF A WAVE DISTURBANCE

26. ◆ A harmonic wave has a period of 2.0 s and a speed of 11 m/s. Find the angular frequency ω and the wave number k of this wave.

27. ◆ The wave function for a harmonic wave on a string is:

$$y(x, t) = (1.00 \text{ mm})\sin[(62.8 \text{ rad/m})x + (314 \text{ rad/s})t].$$

In what direction does this wave travel, and what is its speed? Find the wavelength, frequency, and period of the wave, and the maximum displacement of any string segment.

28. ◆ At $t = 0$, a wave pulse on a string is described by the function:

$$f(x) = \frac{7.5 \times 10^{-6} \text{ m}^3}{x^2 + (0.050 \text{ m})^2}.$$

If the pulse moves to the left at 370 m/s, write an expression $y(x, t)$ for the displacement at $t > 0$. What is the maximum displacement?

29. ◆◆ A wave pulse on a string is described by the function:

$$y(x, t) = \frac{0.64 \text{ m}^3}{220 \text{ m}^2 + [x - (430 \text{ m/s})t]^2}.$$

Find the velocity of the wave. What is the maximum string displacement? Where does this maximum occur for $t = 0$ and for $t = 25$ ms?

30. ◆◆ A wave is described by the function:

$$y(x, t) = (5.0 \text{ cm})\cos[(3.0 \text{ rad/m})x - (5.0 \text{ rad/s})t + 3\pi/4].$$

Draw a graph of $y(x, 0)$ and a graph of $y(0.6 \text{ m}, t)$. On the first graph, mark two points one wavelength apart, and on the second, mark two points one period apart. (The number $3\pi/4$ is exact.)

31. ◆◆ A harmonic wave disturbance in a system is moving to the right at 10.0 m/s. The frequency is 50.0 Hz, the amplitude is 0.750 mm, and the displacement at time $t = 0$, coordinate $x = 3.25$ m, is zero and decreasing with time. Write the wave function for the wave.

32. ◆◆ The equation

$$y(x, t) = (2.00 \text{ cm})\cos[(6.28 \text{ rad/m})x - (3.14 \text{ rad/s})t + 1.047]$$

describes the displacement due to a wave on a string. Find the wavelength and the velocity of the wave. Sketch the wave at $t = 0$ s. Mark on your graph a place where the wave phase is $3\pi/4$.

33. ◆◆ A harmonic wave on a string has a speed of 5.0 m/s, a wavelength of 1.0 m, and is moving to the right. You notice that at $t = 0$ and $x = 0.33$ m, the string's displacement is zero but increasing with time. The maximum displacement anywhere on the string is 0.050 m. Write an equation for the wave function $y(x, t)$ and draw a graph showing the displacement as a function of x at $t = 0$ and at $t = 0.100$ s.

34. ◆◆ At $t = 0$, a wave disturbance has the waveform shown in ■ Figure 15.23. If the wave has a velocity of 510 m/s to the left, write the wave function in the form $y = A \cos(kx \pm \omega t + \phi_o)$, giving numerical values for A, k, ω, and ϕ_o.

35. ◆◆ ✉ A wave with amplitude 5.0 μm and wavelength 0.33 m propagates on a guitar string with mass per unit length 6.31 g/m and tension 290 N. Verify that the weight $\mu g \, dx$ of a string segment is much less than the maximum net y-component of tension $T(\partial^2 y/\partial x^2) \, dx$. For what amplitude wave would the two become comparable?

36. ◆◆◆ Quartz fiber exerts a torque when twisted that is proportional to the angle of twist per unit length: $\tau = \alpha \, d\theta/d\ell$. Find an equation for the displacement $\theta(x, t)$ along such a fiber in terms of α, the mass per unit length μ, and radius r of the fiber. What is the speed of torsional waves on the fiber?

§15.3 ENERGY TRANSMISSION BY HARMONIC WAVES

37. ❖ An oscillator is sending waves along a stretched string. If the power of the oscillator is held fixed, how does the amplitude of the waves depend on the oscillator frequency?

38. ❖ How does the average power transmitted by a wave compare with the peak power transmitted at a particular point?

39. ❖ Is the instantaneous power transmitted by a harmonic wave the same at every point at a given time? Is the average power the same at every point?

40. ◆ Find the average power transmitted by the wave in Problem 27 if $\mu = 6.0$ g/m.

41. ◆ Find the average power transmitted by the wave in Problem 30 if $T = 1500$ N.

42. ◆ Find the average power transmitted by a 256-Hz wave of amplitude 0.44 mm traveling along a string with tension 660 N and mass per unit length 8.0 g/m.

43. ◆◆ A 440-Hz oscillator with a power output of 120 W is sending waves along a string with tension 950 N and $\mu = 25$ g/m. What is the amplitude of the waves?

44. ◆◆ A wave of amplitude 0.35 mm and frequency 440 Hz is propagating along a string with mass per unit length $\mu = 5.0$ g/m. What is the tension in the string if the average power transmitted by the wave is 0.77 W?

45. ◆◆◆ ✉ An oscillating string loses energy by generating sound waves in the surrounding air. If the energy loss rate for a string with energy E oscillating with period T is given by $dE/dt = -0.005E/T$, by what fraction does the amplitude decrease per wavelength?

46. ◆◆◆ Show that the kinetic energy per unit length of a string is $u_k = \frac{1}{2}\mu(\partial y/\partial t)^2$. The potential energy per unit length is $u_p = \frac{1}{2}T(\partial y/\partial x)^2$. (Take this as given; it is not easy to prove!) Show that the power transmitted by a harmonic wave equals the wave speed multiplied by the total energy density.

$$P = v(u_k + u_p).$$

Using conservation of energy, make a qualitative argument supporting this result.

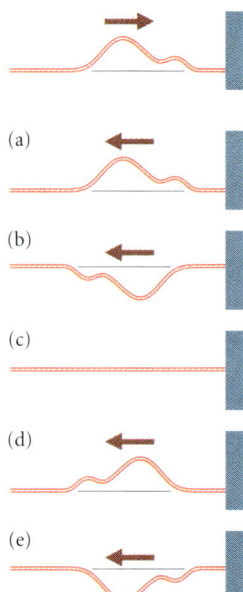

■ FIGURE 15.24

§15.4 SUPERPOSITION

47. ❖ Why don't waves of different wave number superpose to form standing waves?

48. ❖ What wave must be added to the following wave to form a standing wave?

$$y(x, t) = (0.30 \text{ cm})\cos[(15 \text{ rad/m})x - (120 \text{ rad/s})t].$$

49. ❖ A wave pulse with the shape shown in ■ Figure 15.24 is reflected from a fixed boundary. Which pulse most closely resembles the reflected pulse? Explain your reasoning.

50. ❖ A stretched string has a bead of mass M on it at its midpoint. Standing waves exist with the bead at a node. Why? What are their frequencies?

51. ❖ In a one-dimensional standing wave, what fraction of a wavelength is the distance between: **(a)** 2 nodes, **(b)** 2 antinodes, **(c)** a node and an antinode?

52. ◆ Find the fundamental frequency and the first three harmonics for a guitar string of length 0.650 m, mass per unit length 6.31×10^{-3} kg/m, and tension 289.9 N.

53. ◆◆ A guitar string, destined to be an upper E string, has length 99.06 cm and mass 0.430 g. What is its mass per unit length? What length of this string should be used on a guitar if the fundamental frequency is to be 330 Hz and its tension is to be 79.0 N?

54. ◆◆ What tension is required in a steel piano string tuned to 1760 Hz if its length is 0.11 m and its diameter is 0.80 mm? Compare with a string tuned to 110 Hz with length 0.80 m and diameter 2.0 mm ($\rho_{\text{steel}} = 7.86 \times 10^3$ kg/m^3).

55. ◆◆ Express the superposition of the following waves as a single harmonic wave:

$$y_1 = (5.0 \text{ cm})\cos[(2.0 \text{ rad/m})x + (4.0 \text{ rad/s})t],$$

and $\quad y_2 = (12 \text{ cm})\cos[(2.0 \text{ rad/m})x + (4.0 \text{ rad/s})t - \pi/2].$

What is the amplitude of the superposed wave disturbance? (*Hint:* Write the phase of y_1 as θ and express y_2 in terms of θ. Draw a diagram showing how y_1 and y_2 are each related to θ.)

56. ◆◆ The ends of a spring of mass m, constant k, and unstretched length ℓ are fixed rigidly to walls a distance ℓ apart. Use the result of Exercise 15.2 to find the frequencies of standing longitudinal waves in the spring.

57. ◆◆ A telephone wire ($\mu = 45$ g/m) is stretched between two poles 25 m apart. When wind gusts strike the wire 1.7 times per second, standing waves are set up on it. Assuming that the fundamental mode is excited, what is the tension in the wire?

58. ◆◆ A standing wave on a string has the form:

$$y(x, t) = (4.8 \text{ mm})\sin[x/(3.6 \text{ m})]\cos[(130 \text{ rad/s})t].$$

Determine the amplitude, wavelength, and speed of the traveling waves that combined to form this standing wave.

59. ◆◆ A string vibrates with a fundamental frequency of 435 Hz. By what percentage should the tension be changed to increase the frequency to 440 Hz?

✻ §15.5 REFLECTION AND TRANSMISSION OF WAVES AT A JUNCTION OF TWO STRINGS

60. ❖ Can you vary the fraction of power transmitted by a wave across a junction between two strings by adjusting the wavelength of the incident wave? Why or why not?

61. ◆ A wave of amplitude A and wave number k is incident on the junction of two strings. If the wave number in the second string is $2k$, find the amplitudes of the transmitted and reflected waves. What fraction of the incident power is transmitted?

62. ◆◆ A wave with wave function

$$y(x, t) = (0.20 \text{ cm})\cos[(11 \text{ rad/m})x - (150 \text{ rad/s})t]$$

is incident at a junction with a string of half the mass per unit length. How much power (on average) is transmitted to the second string if its mass per unit length is 6.50 g/m?

63. ◆◆ A wave of amplitude 0.35 mm and wavelength 0.45 m propagates along a string with $\mu = 6.5$ g/m, toward a junction with a second string. The amplitude of the transmitted wave is 0.29 mm. Find the mass per unit length of the second string. What is the amplitude of the reflected wave?

64. ◆◆ A wave propagates along a string toward a junction with a second string. The transmitted wave and the reflected wave each carry exactly one-half of the incident power. Find the ratio of the first string's mass per unit length to that of the second.

65. ◆◆◆ Three strings of equal tension and mass per unit length meet as shown in ■ Figure 15.25 with $\theta = 60°$. If a wave propagates toward the junction in one of the strings, find the reflected wave and the transmitted waves in each of the other two strings. (Assume the wave displacement is perpendicular to the plane of the figure.)

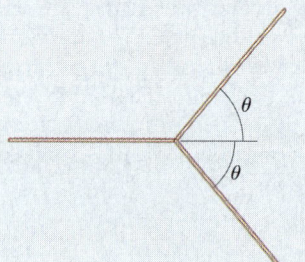

■ FIGURE 15.25

Additional Problems

66. ❖ Which of the following functions describe **(a)** a wave pulse, **(b)** a periodic wave, and which describe **(c)** neither?

(i) $y = \ln[a(x + vt)]$.   (iv) $y = A(x^2 - v^2t^2)$.
(ii) $y = \exp[-a(x - vt)^2]$.   (v) $y = \exp(-a|x + vt|)$.
(iii) $y = \tan[a(x + vt)]$.   (vi) $y = A\sin^2[a(x + vt)]$.

67. ❖ What forces act on a fluid element at the surface of a ripple? Sketch these forces at several points on the waveform. Can the fluid element move on a vertical line? What sort of path does the fluid element move along?

68. ❖ ✉ For surface waves on deep water, the relevant physical quantities determining the wave speed are gravitational acceleration g and wavelength λ. Use dimensional analysis to find the dependence of wave speed on g and λ. Compare the speed of surface waves propagating on mercury with waves on water. Would surface waves propagate more slowly or more rapidly on a swimming pool on Earth or on the Moon?

When the depth of the water becomes less than about one wavelength, the relevant length is the depth rather than λ. Use dimensional analysis to find the speed of surface water waves on shallow water.

69. ◆◆ A string of length ℓ and mass per unit length μ is attached to a second string of length 2ℓ and mass per unit length $\mu/4$. The other ends are fixed. Show that there exist standing waves on the string with nodes at the string junction, and find the frequencies.

70. ◆◆ Compute the time-averaged power transmitted past a point with coordinate x on a string of length ℓ when a standing wave exists on the string. Comment on your result.

71. ◆◆ Verify by direct substitution that the standing wave $y = A \sin kx \cos \omega t$ satisfies the wave equation (15.10) if $\omega/k = v$.

72. ◆◆◆ Compute the kinetic energy density $u_k = \frac{1}{2}\mu(\partial y/\partial t)^2$ and potential energy density $u_p = \frac{1}{2}T(\partial y/\partial x)^2$ for a standing wave $y(x, t) = A \sin(\pi x/L)\cos \omega t$ on a string of length L and mass per unit length μ. Show that the total energy of the string remains constant.

73. ◆◆◆ Find the tension in a spinning loop of chain of mass per unit length μ and angular speed ω. Show that the speed of waves on the chain is equal to its linear speed. If you strike the chain so as to excite two countermoving wave pulses, describe the resulting disturbance qualitatively.

74. ◆◆◆ The displacement of a string element when a transverse wave propagates is perpendicular to the string, but otherwise its direction is arbitrary. Set up a coordinate system with x along the string. **(a)** Find the superposition of the following two waves on the string:

$$y(x, t) = A \cos(kx - \omega t) \quad \text{and} \quad z(x, t) = A \sin(kx - \omega t).$$

Describe the motion of the string elements. **(b)** If the string has length $L = 2\pi/k$, standing waves can be set up on it with $y = 2A \sin kx \sin \omega t$ and $z = 2A \sin kx \cos \omega t$. Sketch the wave displacement at $t = 0$, $T/8$, and $T/4$. **(c)** To set up the standing wave mode described in (b) on a 1.5-kg rope 8.0 m long with a tension of 110 N, two circus performers turn the rope with frequency f. Find f. If two other performers jump rope at $x = L/4$ and $x = 3L/4$, what is the time interval between their jumps?

Computer Problems

75. Use a spreadsheet program to compute the following displacements at times $t = 0.0, 0.01, 0.02,$ and 0.03 s:

$$y_1 = (0.78 \text{ mm})\cos[(0.35 \text{ rad/m})x + (156 \text{ rad/s})t],$$

and $\quad y_2 = (0.65 \text{ mm})\cos[(0.35 \text{ rad/m})x - (156 \text{ rad/s})t].$

Make a plot showing y_1, y_2, and $y_1 + y_2$ between $x = 0$ and $x = 50$ m at each of these times. Is the superposition a standing wave?

76. Compute the superposition of the following two waves at $t = 0$ and $t = 0.04$ s, for values of x between 0 and 50 m:

$$y_1 = (0.50 \text{ mm})\cos[(1.1 \text{ rad/m})x + (363 \text{ rad/s})t],$$

and $\quad y_2 = (0.50 \text{ mm})\cos[(1.0 \text{ rad/m})x + (363 \text{ rad/s})t].$

How would you describe the results?

77. Compute the superposition of the following three waves at $t = 0$, 0.04 s, 0.08 s, and 1.12 s, for values of x between 0 and 50 m:

$$y_1 = (0.50 \text{ mm})\cos[(1.1 \text{ rad/m})x - (363 \text{ rad/s})t],$$

$$y_2 = (0.50 \text{ mm})\cos[(1.0 \text{ rad/m})x - (330 \text{ rad/s})t], \text{ and}$$

$$y_3 = (0.50 \text{ mm})\cos[(0.9 \text{ rad/m})x - (297 \text{ rad/s})t].$$

How would you describe the results?

78. Use a spreadsheet program to calculate the displacement due to an incoming pulse and its reflection from a boundary at $x = 0$. The shape of the incident pulse is given by $f(x - vt) = e^{-(x-vt)^2/a}$ with $v = 1$ m/s and $a = 2$ m². With a free boundary, the reflected pulse has the form $f(x + vt) = e^{-(x+vt)^2/a}$. Compute the total displacement between $x = -10$ m and 0 due to the two pulses at $t = -5$ s, -2 s, -1 s, 0 s, $+1$ s, $+2$ s, and $+5$ s. Plot your results at each time and comment on what they show. Is the slope of the string zero at $x = 0$ at all times? What is the wave function of the reflected pulse if the boundary is fixed? Compute and plot the total displacement at different times in this case also. Check your calculation by determining whether the displacement is zero at $x = 0$ at all times.

Challenge Problems

79. Can standing waves exist on the string described in Problem 69 if the junction is not a node? Discuss.

80. A rope of mass m and length L hangs vertically. How long does it take for a transverse wave to travel along the rope? First estimate the result using dimensional analysis; then perform an exact calculation.

81. Include the string's weight in the derivation of the wave equation, retaining the assumption that θ remains small. Show that the solutions to the resulting equation are of the form:

$$y(x, t) = \frac{\mu g x^2}{2T} + C_1 x + C_2 + f(x \pm \sqrt{T/\mu}\, t).$$

So long as θ remains small, the effects of weight and wave propagation are independent.

82. Two pendulums of mass M and length ℓ are coupled. When they have displacements θ_1 and θ_2, respectively, from equilibrium, the coupling torque exerted by one coupled pendulum on the second is:

$$\tau_{1 \text{ on } 2} = \epsilon(\theta_1 - \theta_2),$$

where $\epsilon \ll Mg\ell$. **(a)** Write differential equations for the motions of the two pendulums. **(b)** Add the two equations to obtain an equation for the sum $z_+ \equiv \theta_1 + \theta_2$ and subtract to get an equation for $z_- \equiv \theta_1 - \theta_2$. **(c)** Show that z_+ and z_- oscillate with frequencies $\omega_+ = \sqrt{g/\ell}$ and $\omega_- = \sqrt{g/\ell + 2\epsilon/M\ell^2}$, respectively. **(d)** If at $t = 0$, $\theta_1 = A$, and $\theta_2 = d\theta_1/dt = d\theta_2/dt = 0$, show that $\theta_1 = A \cos(\Omega_+ t)\cos(\Omega_- t)$ and $\theta_2 = -A \sin(\Omega_+ t)\sin(\Omega_- t)$, where $\Omega_+ = \frac{1}{2}(\omega_+ + \omega_-)$ and $\Omega_- = \frac{1}{2}(\omega_+ - \omega_-)$. Interpret the result as an exchange of energy between the pendulums.

83. A string 1.0 m long with tension 10.0 N has a standing wave on it with displacement:

$$y(x, t) = (1.2 \text{ mm})\sin[(\pi \text{ rad/m})x]\cos[(2.0 \times 10^2\pi \text{ rad/s})t].$$

Compute the power transmission along the string. Given that the potential energy density for the string is $u = \frac{1}{2}T(\partial y/\partial x)^2$, plot schematic graphs of the kinetic and potential energy densities and the power transmission at times $\omega t = 0$, $\pi/4$, $\pi/2$, $3\pi/4$, π. Argue qualitatively that your results are consistent with conservation of energy.

Figure 15.26

84. Discuss reflection of harmonic waves propagating to the left on the string shown in ■ Figure 15.26. *Hint:* There will be an incident wave and a reflected wave on the string to the right. The string to the left always has zero displacement at its fixed end. The mathematics is easiest if you take $x = 0$ at the junction between the two strings. Then we may write the displacements as:

$y_i = A \cos(k_2 x + \omega t + \phi_1)$, $\qquad y_r = A_r \cos(k_2 x - \omega t + \phi_2)$,

and $\qquad y_\ell = B \sin[k_1(x + \ell)]\cos \omega t$.

Use conservation of energy to express A_r in terms of A. Then apply the boundary conditions at the junction to determine ϕ_1, ϕ_2, and B. What is the maximum value of the amplitude $|B|$? For what values of the incident wavelength does it occur?

85. You are to compute the potential energy of a vibrating string. **(a)** Imagine holding a string in the shape $f(x) = hy(x, t_o)$ for $0 \leq h \leq 1$. Compute the force needed to hold each string element in place. **(b)** Compute the work done on each element of string to change from h to $h + dh$. **(c)** Integrate over h from 0 to 1 to find an expression for the total work done on each element. **(d)** Express the work done on the entire string as an integral, and integrate by parts to show that the string's potential energy is given by:

$$U = \int_0^L \frac{T}{2}\left(\frac{\partial y}{\partial x}\right)^2 dx.$$

86. Find the wave equation for longitudinal waves on a spring of mass m, constant k, and length L. **(a)** Assign as the coordinate of a point on the spring its position x when the spring is at rest. Let $y(x, t)$ be the displacement of the point from x at time t. Show that the amount an element of spring is stretched divided by its relaxed length dx is $\partial y/\partial x$. **(b)** From Hooke's law, the tension in any element of the spring is proportional to its stretch divided by its relaxed length: $T(x, t) \equiv S\, \partial y/\partial x$, where S is called the stiffness of the spring. By considering a uniformly stretched spring, show that the stiffness is given by $S = kL$. **(c)** Apply Newton's second law, as was done for a string in §15.2.4, to show that the wave equation for the spring is:

$$\frac{\partial^2 y}{\partial x^2} = \frac{m}{SL}\left(\frac{\partial^2 y}{\partial t^2}\right).$$

(d) Find the speed of the waves, and compare with the result of Exercise 15.2.

We all know what light is; but it is not easy to tell what it is.
SAMUEL JOHNSON

Rainbows are usually visible during or after a rainstorm, when water droplets are present in the air. The colored band is a result of dispersion (§16.5.5). The water droplets scatter the different wavelengths of light at slightly different angles.

CHAPTER 16
Sound and Light Waves

CONCEPTS

Spectrum
Intensity
Decibel
Wave front
Ray
Inverse square law
Doppler effect
Plane wave
Huygens' principle
Reflection
Refraction
Index of refraction
Dispersion

GOALS

Be able to:

Describe the relations among pressure, velocity, and displacement in a sound wave.

Describe the various regions of both the electromagnetic and the sound spectra.

Use the inverse square law to compare intensities of sound or light.

Compute the observed frequency of a wave using the Doppler effect.

Use the ray and wave front pictures to describe reflection and refraction.

Use Snell's law to follow rays through a sequence of plane boundaries.

523

SOUND AND LIGHT ARE VERY DIFFERENT WAVES, YET BECAUSE THEY ARE BOTH THREE-DIMENSIONAL WAVES, THEY HAVE MANY COMMON PROPERTIES. INCREASINGLY, SOUND (LIKE LIGHT) IS USED FOR IMAGING AND LIGHT (LIKE SOUND) IS USED FOR COMMUNICATION. IN THIS CHAPTER WE EMPHASIZE THE SIMILARITIES OF THE TWO IN PROPERTIES LIKE REFLECTION AND REFRACTION.

RADAR IS A FORM OF "LIGHT"; SEE §16.2.

A rainbow is a beautiful sight, even if there is no pot of gold at its end. Sunlight is composed of light waves with a range of frequencies. Water droplets in the air scatter each frequency at a different angle, separating the light into brilliantly colored arcs. Most sounds are also a mixture of frequencies. The nature of the mixture is what distinguishes music from noise.

There are both similarities and important differences between sound and light, and our bodies have evolved very elaborate and very different methods for detecting them. The human body uses light but not sound to form images of objects. However, sound wave imaging is used in medicine because the sound waves can penetrate the body without damaging living tissue (■ Figure 16.1), and bats track prey with reflected sound, much as radar is used to track aircraft. We begin our discussion with some basic facts about sound and light waves and then investigate some of their common properties as three-dimensional waves.

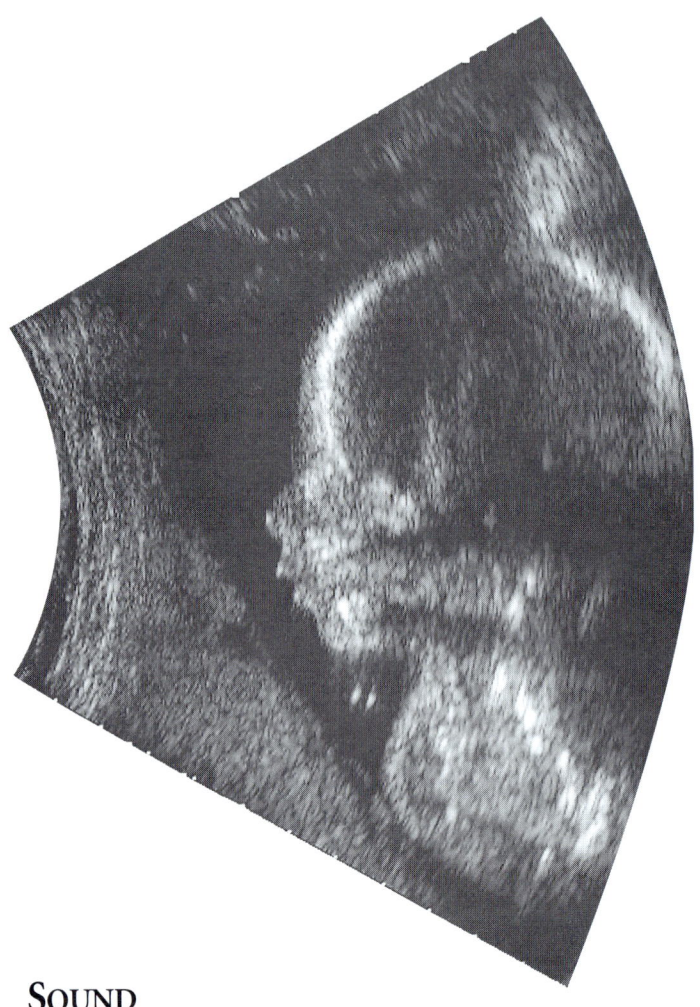

■ FIGURE 16.1
Image of a 5-month-old fetus taken with 3.5-MHz ultrasound. Ultrasound penetrates the body but is much less damaging than x rays, for example.

16.1 SOUND

Sound waves are longitudinal waves in which the restoring force results from pressure differences. Sound can propagate through any elastic medium—air, water, wood, iron—in fact, through any gas, liquid, or solid. We are most familiar with sound waves in air. A vibrating object—a string, a drum membrane, your vocal chords, or a loudspeaker cone—sets parcels of air in motion, oscillating back and forth in the direction that the wave is moving.

■ Figure 16.2 describes the acoustic (sound) spectrum. Healthy human ears can detect waves in the range roughly 20–20 000 Hz. Sound waves of higher frequency—called *ultrasonic*—are used for medical imaging and motion detection. Musical notes range from about 30 Hz to 4000 Hz. The present American musical standard for concert pitch designates 440 Hz as the note A.

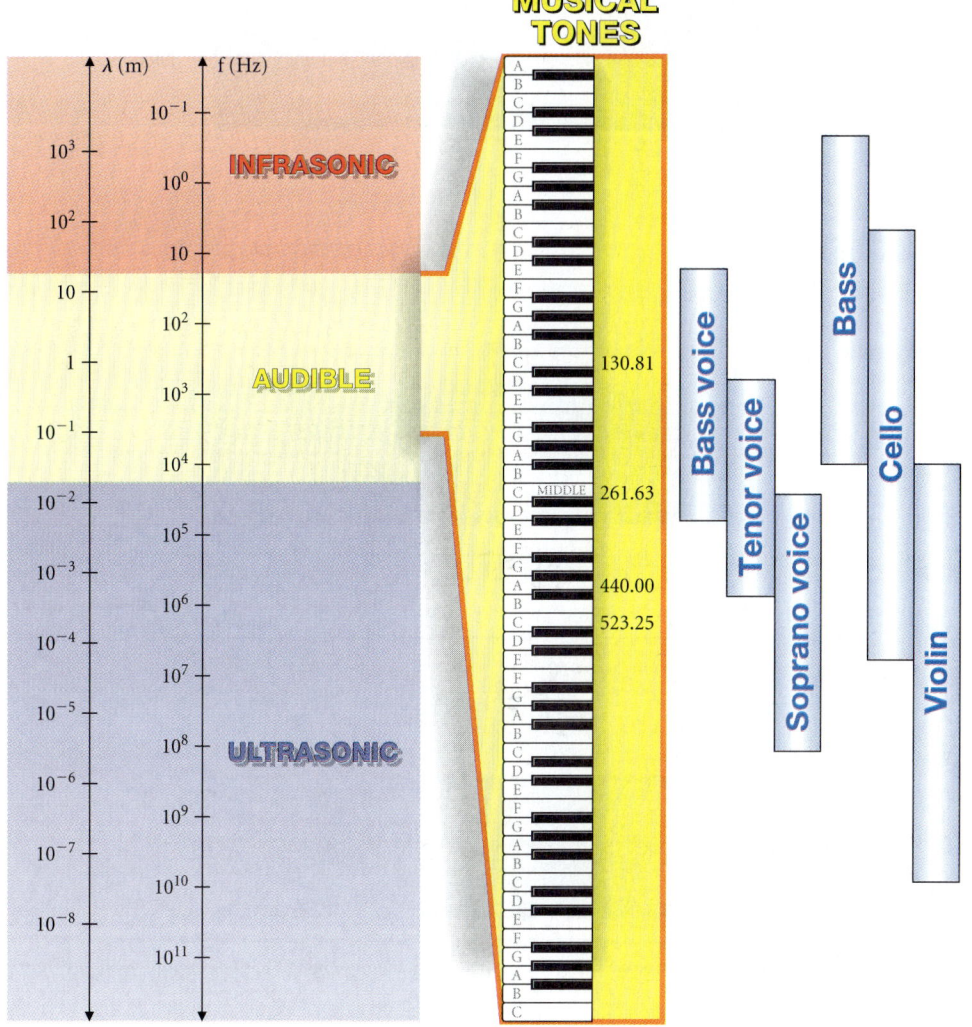

■ **Figure 16.2**
The acoustic spectrum. Infrasonic waves are not audible to the human ear, but they are detectable by elephants and whales. Dogs and bats can hear higher frequencies than humans. The musical scale shown is American concert pitch.

16.1.1 Sound Waves in a Tube

When a firecracker explodes (■ Figure 16.3), first you see the flash, then you hear the bang! For the sound to reach you, a mechanical wave must propagate from the firecracker to your ear. We learned in Chapter 15 that both restoring forces and inertia are required for a mechanical wave to propagate. The air provides them both. The explosion pushes air rapidly outward into the surrounding atmosphere, creating a spherical shell of compression. The higher pressure in this shell in turn pushes on air farther out, forming a pulse of high pressure that propagates outward. Pressure differences, pushing air from high-pressure regions toward

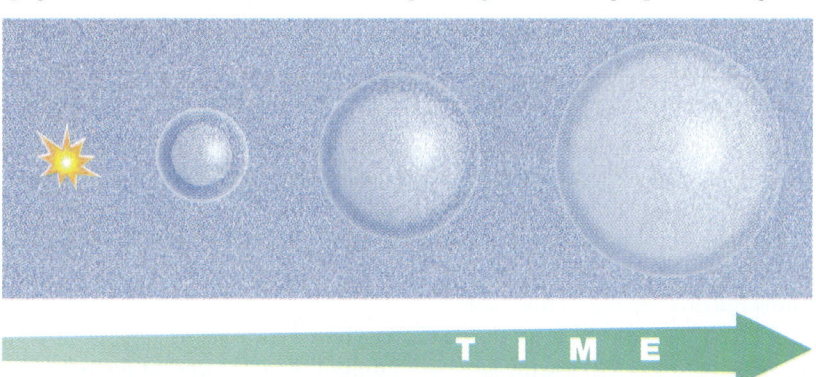

■ **Figure 16.3**
Sound wave generated by the explosion of a firecracker. A spherical shell of compressed air moves outward.

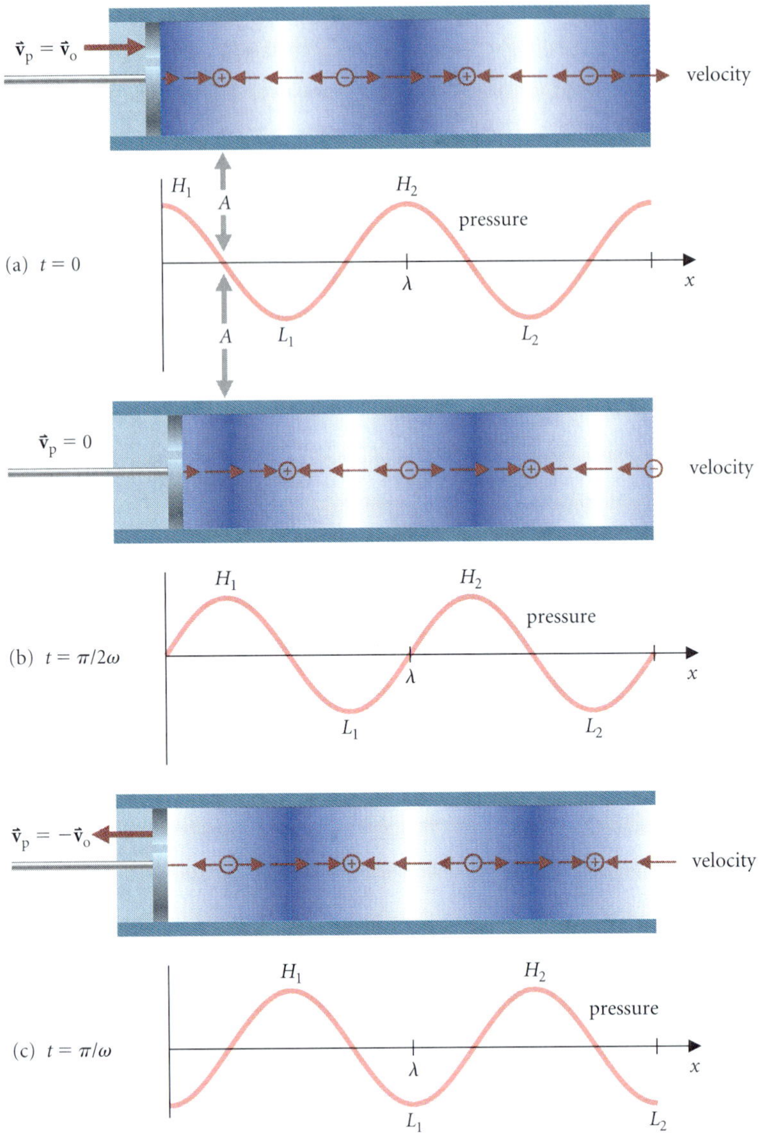

■ **FIGURE 16.4**
A piston generating sound waves in a tube. The piston has been oscillating long enough for a harmonic wave to be set up in the tube. (a) At $t = 0$, the piston has zero displacement and maximum velocity. Particles are converging to the points marked with a $+$ and diverging from points marked with a $-$. At point A, the net force is to the right. (b) At $t = \pi/2\omega$, the piston is at its maximum displacement; its velocity is zero. The high-pressure region [marked H_1 in part (a)] has moved to the right, to the place where the air was converging (the $+$) in part (a). At point A, the air is moving rapidly to the right. At the left-hand end of the tube, pressure increases toward the right, so air is pushed leftward. (c) At $t = \pi/\omega$, the piston has returned to its original position and has maximum velocity to the left. At the end of the tube, the pressure is a minimum and the air velocity is leftward. The high-pressure region H_1 has moved farther to the right.

WE CHOOSE THE x-AXIS TO BE ALONG THE TUBE, WITH THE ORIGIN AT THE EQUILIBRIUM POSITION OF THE PISTON.

low-pressure regions on either side, produce the restoring force and transmit the pulse outward. The air mass provides the necessary inertia.

A one-dimensional example we can examine in detail is the harmonic sound wave generated by a piston oscillating with amplitude s_o at the end of a long air-filled tube (■ Figure 16.4). The piston's position and velocity are:

$$x_p(t) = s_o \sin \omega t \quad \text{and} \quad v_p(t) = v_o \cos \omega t,$$

where $v_o \equiv \omega s_o$. At $t = 0$, the piston has zero displacement and maximum velocity. Particles are converging toward the points marked with a plus and diverging from points marked with a minus. Thus pressure is increasing at $+$ and decreasing at $-$. At point A, with pressure higher to the left and lower to the right, air is pushed to the right.

One-quarter period later, at time $t = \pi/2\omega$, the piston has reached maximum displacement; its velocity is zero. Where air was converging (the $+$) in Figure 16.4a, the pressure is a maximum in Figure 16.4b: the high-pressure region H_1 has moved to the right. At point A, the air is now moving rapidly to the right.

After one-half period, $t = \pi/\omega$, the piston has returned to its original position and has maximum velocity to the left, leaving a pressure minimum at the end of the tube. The high-pressure region H_1 has moved farther to the right.

EXERCISE 16.1 ❖ Sketch the variation of pressure and velocity in the tube at time $t = 3\pi/2\omega$. Give an argument supporting your sketch of the pressure distribution. Discuss how the system returns to the state shown in Figure 16.4a when $t = 2\pi/\omega$.

Air displacement, velocity, pressure, and density all vary in a sound wave. To see how these variations are related, we focus on an air slab of mass dm with its faces originally at coordinates x and $x + dx$ (■ Figure 16.5). Thus $dm = \rho_o A\, dx$, where ρ_o is the density of undisturbed air and A is the cross-sectional area of the tube. When a sound wave propagates in the tube, the air displacement is longitudinal—along the tube. Air originally at coordinate x is displaced to $x + s(x, t)$.

Connection Between Air Velocity and Air Displacement. The air velocity is the partial time derivative of its displacement—the time variation of s at a fixed place:

$$v(x, t) = \frac{\partial s}{\partial t}. \tag{16.1}$$

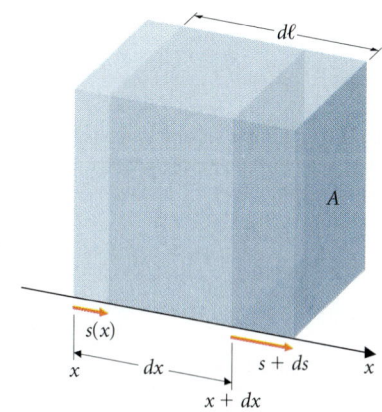

■ **FIGURE 16.5**
A slab of air with area A and sides at x and $x + dx$ is displaced in a sound wave so that its sides are at $x + s(x, t)$ and at $x + dx + s(x + dx, t)$. The length of the displaced slab is $d\ell = dx + ds$.

Connection Between Density Variations and Air Displacement. Different displacements of the slab's two faces change its density. For small displacements, the density variation is (see *Digging Deeper*):

$$\delta\rho = -\rho_o \frac{\partial s}{\partial x}. \tag{16.2}$$

Density variations in a sound wave are usually very small. At a rock concert, the maximum value of $\partial s/\partial x$ is about 10^{-4}.

SEE PROBLEM 90.

Connection Between Pressure Variation and Air Displacement. Since pressure is related to density, pressure variations in the wave, $\delta P \equiv P - P_o$ are proportional to the density variations:

$$\delta P \propto \delta\rho \propto \partial s/\partial x. \tag{16.3}$$

RECALL THE IDEAL GAS LAW, FOR EXAMPLE, WHERE P IS PROPORTIONAL TO ρ (§13.2).

Connection Between Air Displacement and Piston Displacement. At the end of the tube ($x = 0$), the air does not separate from the piston to leave a vacuum. So, at $x = 0$, the air displacement equals the piston displacement:

$$s(0, t) = s_o \sin \omega t = -s_o \sin(-\omega t).$$

Description of a One-Dimensional Sound Wave. The wave travels toward increasing x, so each quantity is described by a function of the single variable $kx - \omega t$. Using the connections described above, we find the following relations.

THIS WAVE IS CALLED A *TRAVELING* WAVE, IN CONTRAST WITH STANDING WAVES (§16.1.3).

Displacement: $\quad s(x, t) = -s_o \sin(kx - \omega t).$ (16.4)

Velocity: $\quad v(x, t) = \partial s/\partial t = \omega s_o \cos(kx - \omega t).$ (16.5)

Density: $\quad \rho(x, t) = \rho_o - \rho_o\, \partial s/\partial x = \rho_o - \rho_o s_o k \cos(kx - \omega t).$ (16.6)

Pressure: $\quad P(x, t) = P_o + P_* \cos(kx - \omega t),$ (16.7)

AS IN CHAPTER 13, THE SYMBOL P_o STANDS FOR ATMOSPHERIC PRESSURE.

where P_* is the amplitude of the pressure disturbance. In a sound wave traveling away from its source, like that shown in Figure 16.4, the air velocity is in phase with the pressure disturbance; velocity is maximum where pressure is maximum. The air displacement is out of phase with pressure by 90°; that is, displacement is zero at pressure maxima and minima. The density variation is 180° out of phase with the pressure. These phase relations are illustrated by Figure 16.4 and expressed by eqns. (16.4)–(16.7).

WE DO NOT CALL THE DISPLACEMENT y, AS WE DID FOR THE STRING, SINCE THAT NAME USUALLY IMPLIES A DISPLACEMENT PERPENDICULAR TO x.

Digging Deeper

DERIVATION OF EQN. (16.2)

The two faces of the air slab, originally at positions x and $x + dx$, are displaced to $x + s(x, t)$ and $x + dx + s(x + dx, t)$. The width of the displaced slab is:

$$d\ell = x + dx + s(x + dx, t) - [x + s(x, t)]$$
$$= dx + s(x + dx, t) - s(x, t)$$
$$= dx + \frac{\partial s}{\partial x} dx$$
$$= \left(1 + \frac{\partial s}{\partial x}\right) dx.$$

The slab's mass dm remains constant, so its density is inversely proportional to its volume $dV = A\, d\ell$:

$$\rho = \frac{dm}{dV} = \frac{dm}{A\, d\ell} = \frac{dm}{A\, dx(1 + \partial s/\partial x)} = \frac{\rho_\circ}{(1 + \partial s/\partial x)}. \quad (16.8)$$

Thus the change in density is:

$$\delta\rho = \rho - \rho_\circ = \rho_\circ\left(\frac{1}{1 + \partial s/\partial x} - 1\right) = -\frac{\rho_\circ\, \partial s/\partial x}{1 + \partial s/\partial x}.$$

We shall restrict our attention to small-amplitude waves in which $\partial s/\partial x$ is much less than unity and the denominator is nearly 1. For these waves:

$$\delta\rho = -\rho_\circ\, \partial s/\partial x. \quad (16.2)$$

16.1.2 The Speed of Sound

Expressions (16.4) through (16.7) hold for sound waves traveling through any medium. The speed of the waves and the relation of pressure amplitude P_* to displacement $s_\circ$ depend on the specific medium. From §15.1, we expect that a sound wave's speed should be related to the restoring force divided by inertia, and dimensional analysis confirms that $\sqrt{[P]/[\rho]} = L/T$. The correct expression for sound speed is:

$$v_s = \sqrt{dP/d\rho}.$$

IN CHAPTER 13 WE MODELED LIQUIDS AS INCOMPRESSIBLE FLUIDS (ρ = CONSTANT). IF A LIQUID WERE ABSOLUTELY (AND NOT APPROXIMATELY) INCOMPRESSIBLE, IT COULD NOT SUPPORT SOUND WAVES.

The density of liquids hardly changes as the pressure is increased, $dP/d\rho$ is relatively large and the sound speed is correspondingly high. The speed of sound in water is about 1500 m/s, compared with 340 m/s in air (● Table 16.1).

EXERCISE 16.2 ♦ Verify that $[P]/[\rho] = L^2/T^2 = [v]^2$.

Newton considered his ability to explain sound to be one of the strongest arguments for his theory of mechanics. He was aware that air pressure is proportional to density at constant temperature, so he set $dP/d\rho \equiv P/\rho$. But since air temperature rises with compression and air is a poor conductor of heat, the high-pressure region in the sound wave also has a higher temperature. This causes nearly a 20% increase in the value of v_s. Apparently, Newton became so frustrated at being unable to predict the correct speed that he developed and published a calculation using imaginative but erroneous corrections to get the measured result! In Chapter 19 we shall see that the correct relationship between pressure and density for air in sound waves is $P \propto \rho^\gamma$ with $\gamma = \frac{7}{5}$; so, $dP/d\rho = \gamma P/\rho$, and

$$v_s = \sqrt{\gamma P/\rho}. \tag{16.9}$$

FOR THE DERIVATION OF THE SOUND SPEED, DIG DEEPER.

Taking typical values of $P = 1$ atm $= 1.01 \times 10^5$ Pa and $\rho = 1.3$ kg/m^3 in eqn. (16.9), we find $v_s = 330$ m/s.

THE VALUES WERE TAKEN FROM CHAPTER 13.

EXAMPLE 16.1 ♦ On a chilly winter day, a hiker standing near a canyon wall yells loudly, and 0.75 s later she hears an echo. How far away is the canyon wall?

MODEL The sound travels to the canyon wall, is reflected, and travels back to the hiker. We take $v_s = 330$ m/s on a chilly day (Table 16.1, $T \approx 0°$C).

SETUP In the measured time, the sound travels twice the distance d to the wall:

$$2d = v_s t.$$

SOLVE So: $d = v_s t/2 = (330 \text{ m/s})(0.75 \text{ s})/2 = 120$ m.

ANALYZE We'll study the reflection process in §16.5. Since sound speed depends on temperature and humidity, our answer could be in error by up to 10%. ∎

EXERCISE 16.3 ♦♦ Use the relation $dP/d\rho = \gamma P/\rho$ to show that the amplitude of the pressure disturbance in a sound wave (eqn. 16.7) is:

$$P_* = \gamma k P_o s_o = \rho_o v_s^2 k s_o. \tag{16.10}$$

TABLE 16.1 Speed of Sound

Solids	Speed (m/s)	Liquids	Speed (m/s)
Beryllium	12890	Glycerol	1904
Diamond	12000	Ethylene glycol	1658
Aluminum	6420	Seawater	1531
Stainless steel	5790	Distilled water	1497
Pyrex glass	5640	Mercury	1450
Brass	4700	Kerosene	1324
Gold	3240	Turpentine	1255
Lucite	2680	Methanol	1103
Lead (annealed)	2160	Carbon tetrachloride	926
Polyethylene	1950		
Rubber	1600		

Speed of Sound in Dry Air

Temperature	$-25°$C	$0°$C	$25°$C	$50°$C
Speed (m/s)	316	331	340	361

At 25°C unless otherwise noted.

EXAMPLE 16.2 ♦♦ If the piston in the tube has a maximum displacement of $s_o = 1.0 \times 10^{-4}$ m and oscillates with angular frequency $\omega = 1.0 \times 10^2$ rad/s, find the amplitude P_* of the pressure variation.

MODEL The pressure amplitude of a sound wave is related to the displacement through eqn. (16.10). In any wave, the wave number k is related to the frequency and the wave speed: $v = \omega/k$.

SETUP Here we use eqn. (16.9) for the wave speed. The wave number is:
$$k = \omega/v_s = \omega/\sqrt{\gamma P_o/\rho_o}.$$

SOLVE From eqn. (16.10), the pressure amplitude is:
$$\begin{aligned} P_* &= \gamma k P_o s_o = \omega \sqrt{\gamma P_o \rho_o}\, s_o \\ &= (1.0 \times 10^2 \text{ rad/s})\sqrt{(1.4)(1.0 \times 10^5 \text{ Pa})(1.3 \text{ kg/m}^3)}\,(1.0 \times 10^{-4} \text{ m}) \\ &= 4.3 \text{ Pa} = 4.2 \times 10^{-5} \text{ atm}. \end{aligned}$$

ANALYZE Even though P_*/P_o is extremely small, this sound is very loud. See §16.3.2. ∎

16.1.3 Standing Sound Waves

REMEMBER: THE STANDING WAVE IS A SUPERPOSITION OF WAVES BOUNCING BACK AND FORTH BETWEEN THE ENDS OF THE ENCLOSURE.

Standing sound waves occur in a confined space, just as standing waves appear on a string of finite length. This phenomenon explains the operation of both organ pipes and wind instruments such as trumpets. Organ pipes are of two basic types: *open* pipes, which are open at both ends; and *closed* pipes, which are closed at one end. A trumpet acts like a pipe closed at the trumpeter's lips. Because air behaves differently at the two types of boundary, the two types of pipe support two different sets of standing waves.

At a closed end, the displacement of air particles is zero. Just as the displacement is zero at the fixed end of a string, air cannot move through a solid wall. A closed end is a node of displacement.

At an open end, the displacement of the particles is not constrained, but the pressure equals the ambient air pressure; that is, the pressure difference δP is zero. An open end is a pressure node. Since $\delta P \propto \partial s/\partial x$ (eqn. 16.3), a pressure node corresponds to an antinode of displacement; the displacement amplitude is a maximum there.

A closed pipe has a displacement node at one end and an antinode at the other. The distance between a node and the nearest antinode is one-quarter of a wavelength. Thus the pipe's length is an odd number of quarter wavelengths (■ Figure 16.7): $L = \lambda/4$, $L = 3\lambda/4$, $L = 5\lambda/4$, and so on.

Closed pipe: $\qquad \lambda_n = 4L/(2n - 1), \quad \text{where } n = 1, 2, 3, \ldots.$ \hfill (16.12)

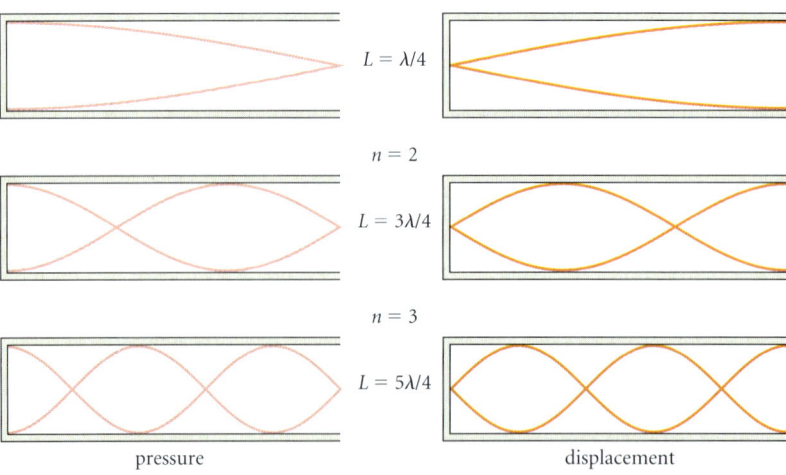

■ **FIGURE 16.7**
Standing sound waves in a pipe. This diagram shows the pressure perturbation δP and the air displacement s as a function of position in the pipe. A pipe closed at one end has a displacement node at the closed end and a pressure node (displacement antinode) at the open end. An odd number of quarter wavelengths fit into the pipe. *Remember:* Even though we have plotted both δP and s vertically, the displacement is longitudinal!

Digging Deeper

The Wave Equation for Sound

We find the wave equation by applying Newton's second law to a slab of air. The net force acting on the slab is due to the difference in pressure on its two faces (Figure 16.6):

$$dF_x = P(x)A - P(x + dx)A = -A\, dP.$$

The pressure difference $dP = P(x + dx) - P(x)$ across the element is related to the density difference. Using the definition of derivative and the chain rule:

$$dP = \frac{\partial P}{\partial x} dx = \frac{dP}{d\rho}\left(\frac{\partial \rho}{\partial x}\right) dx.$$

Air displacements change the density (eqn. 16.8):

$$\frac{\partial \rho}{\partial x} dx = \frac{dm}{A\, dx}\frac{\partial}{\partial x}\left(\frac{1}{1 + \partial s/\partial x}\right) dx$$

$$= -\frac{dm}{A}\left[\frac{\partial^2 s/\partial x^2}{(1 + \partial s/\partial x)^2}\right] \approx -\frac{dm}{A}\left(\frac{\partial^2 s}{\partial x^2}\right),$$

where the quantity $\partial s/\partial x$ in the denominator may be ignored for small-amplitude waves. The pressure difference causes the acceleration of the slab, so:

$$(dm)a_x = dF_x,$$

$$dm\frac{\partial^2 s}{\partial t^2} = -A(dP) = -A\frac{dP}{d\rho}\left[-\frac{dm}{A}\left(\frac{\partial^2 s}{\partial x^2}\right)\right],$$

and

$$\frac{\partial^2 s}{\partial t^2} = \frac{dP}{d\rho}\left(\frac{\partial^2 s}{\partial x^2}\right). \tag{16.11}$$

We have proved that the displacement of the particles satisfies a wave equation with wave speed $v_s = \sqrt{dP/d\rho}$. Waves of all frequencies satisfy eqn. (16.11).

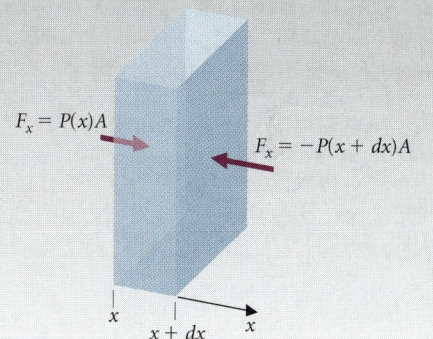

Figure 16.6
Forces due to pressure on a slab of air.

An open pipe has pressure nodes at each end, $\delta P = 0$ at $x = 0$ and at $x = L$ (Figure 16.8). Its length L is a whole number of half wavelengths.

The wavelengths of the standing waves are the same as for a string of length L (§15.4.2).

Open pipe: $\quad \lambda_m = 2L/m, \quad \text{where } m = 1, 2, 3, \ldots \quad (16.13)$

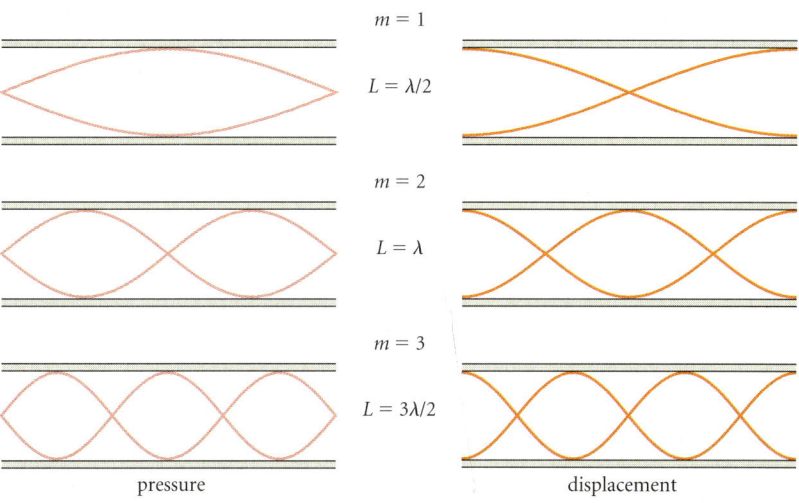

Figure 16.8
An open pipe has pressure nodes at both ends. Like waves on a string, a whole number of half wavelengths fit into the system.

As the air moves in and out of a pipe's open end, it excites outgoing waves in the surrounding air. Because of this coupling to outgoing waves, the open end is only approximately a pressure node. Experimentally, it is found that the effective length of a pipe exceeds its actual length by an amount roughly equal to its radius (■ Figure 16.9).

EXAMPLE 16.3 ♦♦ If you wish to build an organ pipe that will have a fundamental frequency of 262 Hz (C), how should you design it?

MODEL There are two ways to do this: use an open pipe or a closed pipe. Figures 16.7 and 16.8 show how standing waves fit into the two kinds of pipes.

SETUP Taking the speed of sound in air to be approximately 335 m/s, the wavelength of a 262-Hz sound wave is:

$$\lambda = v_s/f = (335 \text{ m/s})/(262 \text{ Hz}) = 1.28 \text{ m}.$$

So, we need a pipe that supports standing waves of length 1.28 m.

SOLVE An open pipe has a fundamental wavelength $\lambda_1 = 2L$, the longest wavelength that fits into the pipe with a node at both ends (eqn. 16.13). Then:

$$L_{\text{open}} = \lambda_1/2 = (1.28 \text{ m})/2 = 0.64 \text{ m}.$$

A closed pipe has a fundamental wavelength four times its length (eqn. 16.12 with $n = 1$): $\lambda_1 = 4L$. Thus:

$$L_{\text{closed}} = \lambda_1/4 = (1.28 \text{ m})/4 = 0.32 \text{ m},$$

half the length of the open pipe.

ANALYZE In music, the harmonics that occur are called overtones. The two pipes sound different because of the different harmonics of C that they also support. Your choice of design depends on which set of overtones you want. The overtones of the open pipe have wavelengths $\lambda_m = 2L/m = \lambda_1/m$. The frequencies, $f_m = v_s/\lambda_m$, are integer multiples of the fundamental frequency. These correspond to C (one octave above middle C), G, C yet another octave higher, and so on. (The first overtone is the second harmonic.) The closed pipe sounds overtones of wavelength $\lambda_n = 4L/(2n - 1) = \lambda_1/(2n - 1)$, where $n = 1, 2, 3, \ldots$. The frequencies are odd multiples of the fundamental frequency and correspond to G (1.5 octaves above middle C), E in the next octave up, and so on. (The first overtone is the third harmonic.) Figure 16.9 shows several pipes that display a variety of tone quality. ∎

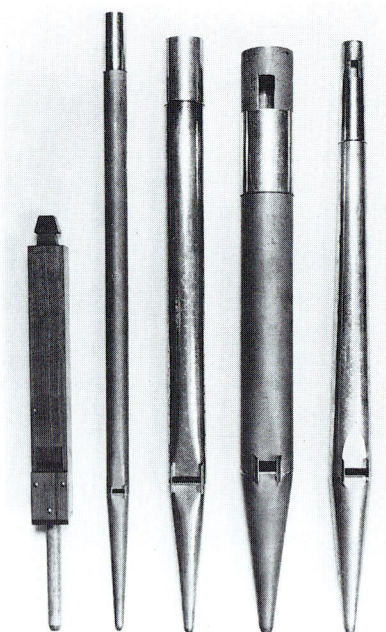

■ **FIGURE 16.9**
Pipes of different design yield notes of the same pitch but different tone—that is, different proportions of each overtone. The pipe at the far left is a closed pipe. It is half the length of the open pipes with the same fundamental frequency. The slightly different lengths are due to end corrections. Notice that the narrowest pipe is also the longest: its effective length is closest to its actual length.

16.2 LIGHT

In contrast with mechanical waves, light waves do not need a medium in which to move. They consist of oscillatory variations of electric and magnetic fields that can propagate in a vacuum as well as in various transparent materials like water and diamond. The fields are described by vectors perpendicular to the direction of wave motion, so light waves are transverse. We shall study the details of these waves after an introduction to electromagnetic theory. However, we can understand much of light's behavior simply by knowing that it is wavelike. It is in this spirit that we discuss light here.

SEE CHAPTER 33.

16.2.1 The Electromagnetic Spectrum

In addition to his work on mechanics, Newton also made fundamental contributions to the theory of light. He believed that light consists of particles that move in straight lines. Because of his immense prestige, the wave theory of light was not accepted until the beginning of the nineteenth century.

IN 1905, EINSTEIN SHOWED THAT LIGHT EXHIBITS BOTH WAVE AND PARTICLE BEHAVIOR. FOR THE PARTICLE THEORY OF LIGHT, SEE CHAPTER 35.

NEWTON PERFORMED THESE EXPERIMENTS IN 1666.

One of Newton's enduring discoveries in optics is the spectrum. In his experiment, sunlight entering a darkened room through a hole in the curtain passes through a glass prism and forms a continuous band of colors (■ Figure 16.10). By passing one end of the spectrum through a second hole to a second prism, he showed that the colors retain their identity: blue remains blue, red remains red, and no new colors are produced. Further, he combined the spectra produced by three prisms and recovered white light. Just as the frequency of a sound wave determines musical pitch, the frequency of light waves is associated with our sense of color.

• Table 16.2 illustrates the different regions of the electromagnetic spectrum. Visible light occupies the small range of frequencies between approximately 4×10^{14} Hz and 8×10^{14} Hz. The names given to the other regions of the spectrum reflect the history of their discovery and how they are produced, detected, and used. Infrared radiation, for example, was first detected by Sir William Herschel in 1800 when he put a thermometer next to the red portion of the spectrum of sunlight dispersed by a prism and noted a rise in temperature.

Many of our examples involve visible light, and it is useful to have a rough idea of the correspondence between wavelength and our sense of color. ■ Figure 16.11 shows the sensitivity of the human eye as a function of wavelength. The sensitivity of the eye is greatest at about 550 nm, close to the wavelength where the Sun's energy output is greatest. A beam of light with a single wavelength (*monochromatic* light) would cause one of the color sensations shown in the figure. The correspondence between wavelength and color is less precise when combinations of wavelengths are viewed. Other creatures have eyes with different wavelength sensitivities. Ants and bees, for example, see well in the ultraviolet, and the brilliant colors that we appreciate in wildflowers are often only secondary effects of the ultraviolet patterns visible to pollinating bees.

■ **FIGURE 16.10**
Newton's prism experiment. Sunlight enters the room through a small hole, passes through a glass prism in Newton's hand and forms the brilliantly colored band seen on the chair at right.

INFRARED MEANS, LITERALLY, "BELOW RED," BECAUSE THE FREQUENCIES OF THESE WAVES ARE LESS THAN THOSE OF VISIBLE RED LIGHT.

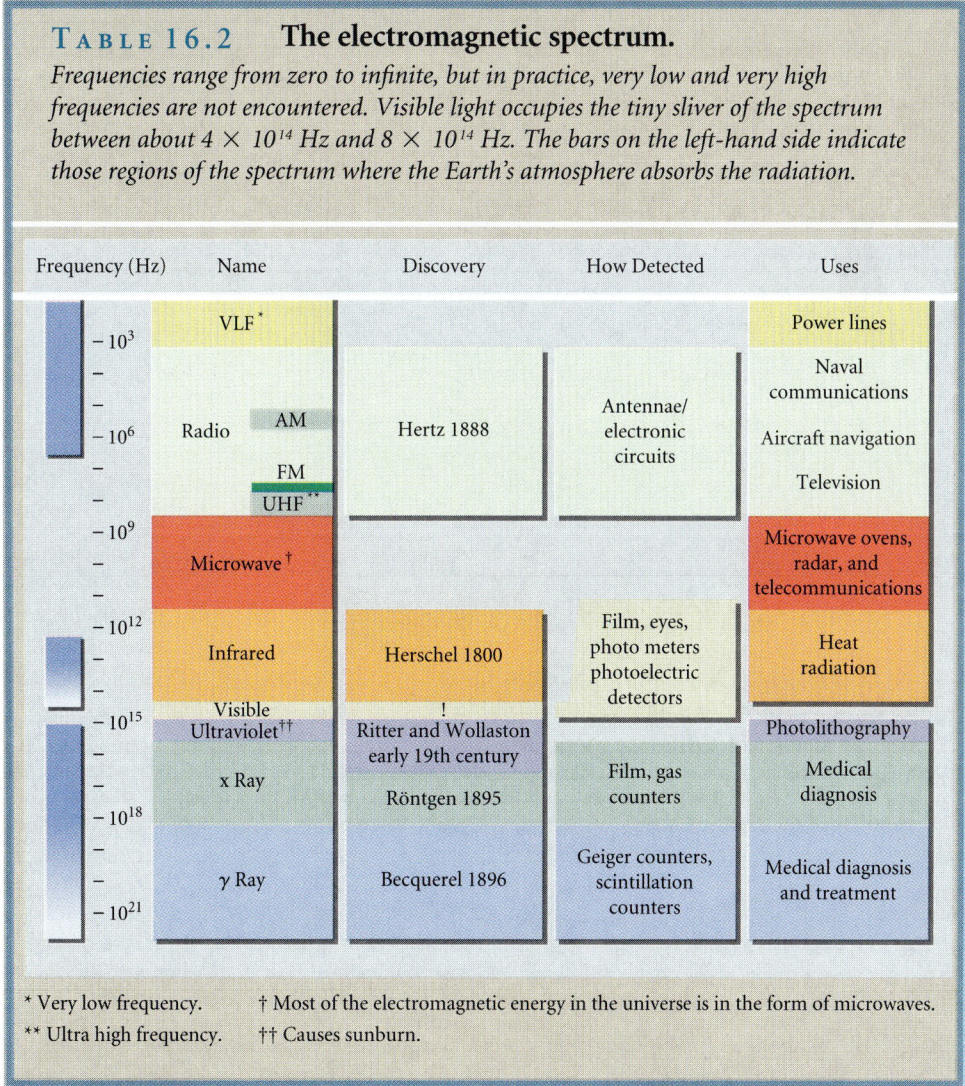

TABLE 16.2 The electromagnetic spectrum.
Frequencies range from zero to infinite, but in practice, very low and very high frequencies are not encountered. Visible light occupies the tiny sliver of the spectrum between about 4×10^{14} Hz and 8×10^{14} Hz. The bars on the left-hand side indicate those regions of the spectrum where the Earth's atmosphere absorbs the radiation.

* Very low frequency.   † Most of the electromagnetic energy in the universe is in the form of microwaves.
** Ultra high frequency.   †† Causes sunburn.

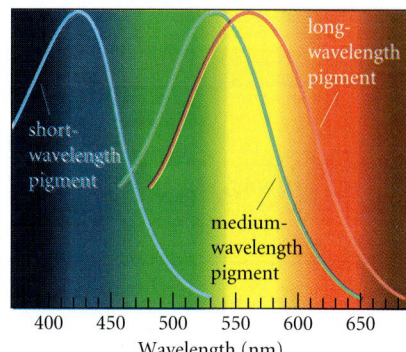

■ **FIGURE 16.11**
Wavelength sensitivity of the color pigments in the human eye. Each cone contains one of these pigments. The short-wavelength pigment absorbs light with wavelengths between 370 nm and 530 nm. The medium-wavelength pigment is most sensitive at 535 nm, and the long-wavelength pigment is most sensitive at 565 nm.

SECTION 16.2 • LIGHT

Digging Deeper

THE SPEED OF LIGHT

Until the Renaissance, most physicists and philosophers believed that the speed of light was infinite. Galileo made the first attempt to determine the speed of light experimentally (■ Figure 16.12). He could not determine whether light travels instantaneously, "but if not instantaneous, it is extraordinarily rapid."[1] In 1676, Olaus Roemer published a paper describing observations of the moons of Jupiter, which established that the speed of light must be finite (■ Figure 16.13). These conclusions were attacked by many prominent scientists, such as Hooke, although Newton and Huygens accepted the result. The controversy continued until 1729, when Bradley published his results on the aberration of starlight (■ Figure 16.14; see also Problem 34).

Modern methods for measuring c involve comparing the wavelength and frequency of a particular electromagnetic wave. Since 1983, $c \equiv 299\,792\,458$ m/s (exactly) has been used to define the meter (see §1.2.4).

---

[1] Galileo, *Discorsi*, 1638.

■ **FIGURE 16.12**
Galileo's experiment. Galileo sends a flash from a lantern to a colleague on a distant hill. The colleague immediately sends a flash back, and Galileo measures the time interval between the first flash and the return flash. The experiment fails because the light's travel time is much less than the reaction time of the experimenters.

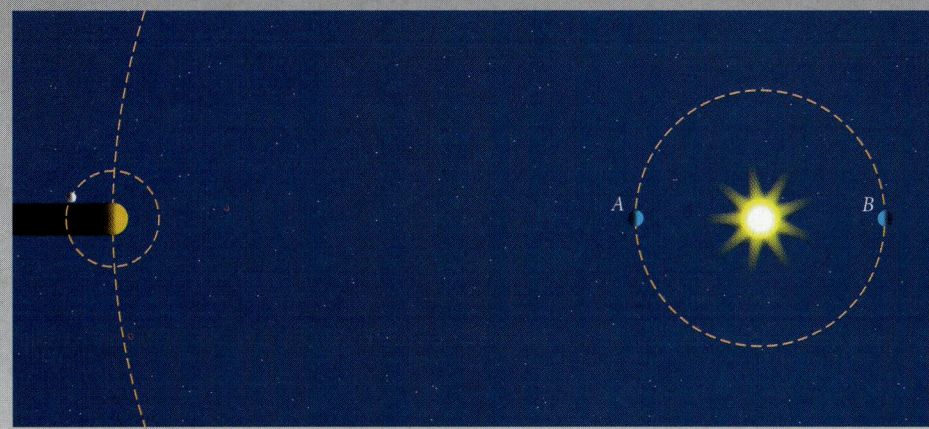

■ **Figure 16.13**
Roemer's method for detecting the finite speed of light. As the Earth moves around the Sun in its orbit, its distance from Jupiter varies, and the time for light to travel from Jupiter to Earth varies correspondingly. Roemer noticed that eclipses of the innermost moon of Jupiter did not occur at regular intervals but instead occurred earlier when the Earth was closest to Jupiter (point A) and later when the Earth was farthest from Jupiter (point B). He concluded that light takes about 22 min to travel the diameter of Earth's orbit. (This simplified discussion ignores the motion of Jupiter, which is slow compared with the Earth's motion.)

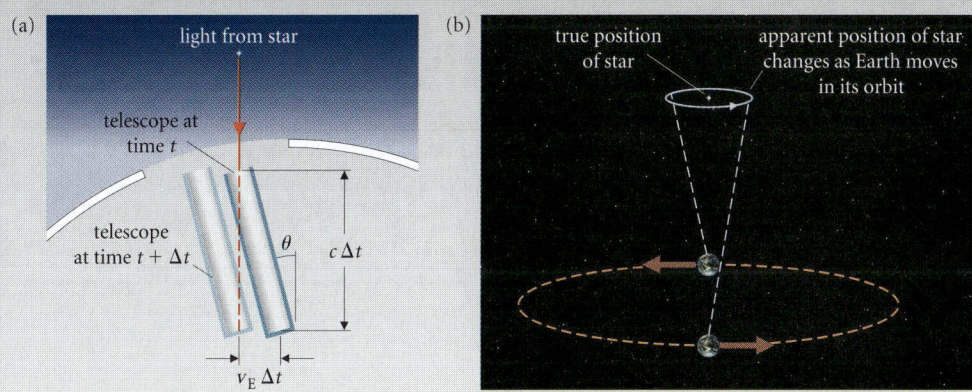

■ **Figure 16.14**
Aberration of starlight. (a) Because of the Earth's speed in its orbit, a telescope must be pointed slightly ahead of the star's position. Light from the star reaching the top of the telescope at time t reaches the bottom at time $t + \Delta t$ where Δt is the length of the telescope tube divided by the speed of light. At $t + \Delta t$, the Earth has moved, carrying the bottom of the telescope forward a distance $v_E \Delta t$. The telescope must be tilted at an angle to the star $\theta \approx v_E/c \approx 10^{-4}$ rad ≈ 21 seconds of arc. (b) The apparent position of a star depends on the direction of the Earth's motion, and so varies with a period of 1 y. James Bradley first noticed this effect, and correctly attributed it to aberration. He computed the light travel time from Sun to Earth to be 8 min 13 s and hence found $c = 3.04 \times 10^8$ m/s.

Digging Deeper

How Does the Eye Detect Light?

■ Figure 16.12 shows a human eye. Light enters through the pupil and is focused onto the retina by the cornea and the lens. Like our ears, our eyes can detect remarkably small quantities of energy. Under ideal conditions, an alert observer can detect as little as 4×10^{-17} J. Our eyes also respond well to very bright lights, sending high-quality images to the brain at light intensities a billion times brighter than the dimmest light we can see.

by Suzanne McKee

Several physiological mechanisms have evolved that enable our eyes to respond adequately over such a large range of brightness. The pupil operates like the aperture of a camera, opening to a larger size in the dark and closing down to a 2-mm diameter in very bright light. In the retina, there are two different kinds of light-sensitive cells that respond to different brightness ranges. One type, called rods because of their long cylindrical shape, responds to very low light levels. They are most important for night vision. The other type, called cones, is used for daylight vision. The cones are concentrated in the center of the retina in a region called the fovea centralis. Because the fovea can resolve fine detail, it is important for reading and other tasks requiring a fine-grain image.

The light reaching the cones and rods is absorbed by special photosensitive pigments. The absorbed energy is amplified and transmitted by other cells to the brain. To keep these neural cells operating within the optimal range, the retina changes its sensitivity as the average light intensity changes,

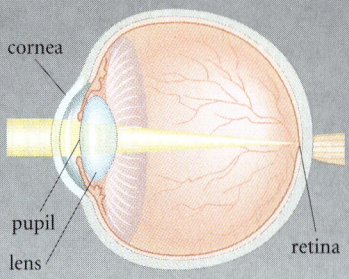

■ **Figure 16.15**
The human eye. Light enters through the pupil and is refracted at the cornea and the lens. Sensitive cells in the retina (rods and cones) respond to the light and send signals to the brain.

something like changing the volume control on a stereo receiver. These changes take time, and your eyes become more sensitive to weak lights after you sit in the dark for a few minutes.

Our eyes respond to the relatively narrow portion of the electromagnetic spectrum with wavelengths between 400 nm and 700 nm. The colors in a rainbow are produced by different wavelengths. The shorter wavelengths usually look bluish, and the longer wavelengths look reddish. It is difficult to predict exactly what color you will see from wavelength alone because the psychological experience of color is based on complex neural interactions that depend on the spatial distribution of wavelengths in adjacent areas.

Color information originates in the cones. There are three kinds of cones, each containing a different photosensitive pigment. Each pigment absorbs a different, but overlapping, range of wavelengths (Figure 16.11). You might imagine that each type of cone is looking at the world through a different colored filter. Because we have only three types of pigments, all the colors that we see can be produced by combining three primary wavelengths in different proportions.

16.2.2 The Speed of Light

All electromagnetic waves travel at the same speed in a vacuum, the *speed of light*: $c = 3 \times 10^8$ m/s. This speed plays a fundamental role in physics, as the maximum possible speed of anything that transports energy. In a material such as air or glass, visible light travels more slowly. Its speed v depends on both the properties of the material and the frequency of the wave, and is described by the *index of refraction*.

> The *index of refraction* n of a material is the ratio of the speed of light in vacuum to the speed of an electromagnetic wave in the material.
>
> $$n \equiv c/v. \tag{16.14}$$

TABLE 16.3	Index of Refraction for Some Common Substances (at 589.29 nm)			
Gases (at 0°C and 1 atm)			**Liquids (at 20°C)**	
Substance	n	Substance		n
Helium	1.000036	Water		1.3333
Air	1.000293	Ethyl alcohol		1.360
Carbon dioxide	1.00045	Carbon tetrachloride		1.461
Chlorine	1.00077	Benzene		1.501
Solids (at room temperature, except ice)				
Substance	n	Substance		n
Diamond	2.418	Glasses		
Sapphire	1.768	Fluor crown		1.464
Amber	1.546	Crown		1.523
Polystyrene	1.590	Dense flint		1.603
Sodium chloride (salt)	1.50	Dense baryte flint		1.702
Opal	1.4	Extra dense flint		1.785
Ice	1.31	Special baryte flint		1.865

The wavelength 589.29 nm is a standard defined by the radiation from sodium atoms.

For visible light in air, $n = 1.0003$; in glass, $n \approx 1.5$; and in water, $n \approx 1.3$. We give only approximate values here because n is a function of frequency. It is this variation that causes a glass prism to produce the spectrum in Newton's experiment. Blue light travels slightly more slowly in glass than does red light and is deflected more strongly by the prism. • Table 16.3 lists values of n for some common substances.

16.3 Energy in Sound and Light

16.3.1 Wave Fronts and Rays

A wave on a string moves in one dimension and transmits its energy along a single line until it encounters a boundary. The average power transmitted is the same at every point on the string. Waves moving at speed v_w in two or three dimensions spread out from a point source of energy in ever-expanding circles or spheres called wave fronts. With the origin at the wave source, the wave phase is:

$$\phi = kr - \omega t + \phi_o.$$

It has a fixed value on each wave front. As time passes, each wave front expands at the wave speed, but its phase remains constant:

$$\frac{d\phi}{dt} = k\frac{dr}{dt} - \omega$$
$$= kv_w - \omega = 0.$$

A *wave front* is a surface (or a line, for waves in two dimensions) on which the wave disturbance has constant phase.

Circular wavefronts spreading over the surface of a pond.

The wave speed $v_w = \omega/k$ is properly called the phase speed of the wave.

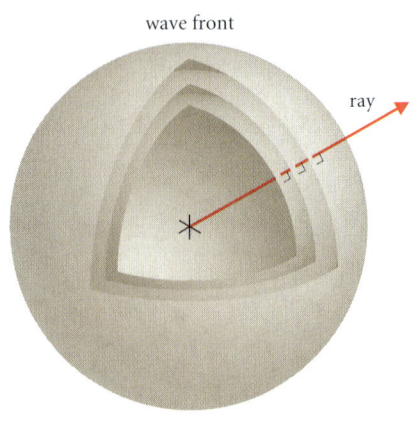

■ **Figure 16.16**
(a) Light rays provide a dramatic image for light propagation, as shown in this photograph of sunlight shining through trees. (b) Point sources emit spherical wave fronts that propagate away from the source. The corresponding rays are lines perpendicular to the wave fronts; that is, they are the radii of the spheres.

WAVE FRONTS CAN HAVE MANY DIFFERENT SHAPES DEPENDING ON THE NATURE OF THE WAVE SOURCE. A POINT SOURCE IS THE SIMPLEST.

Sound and light waves are inherently three-dimensional, and their wave fronts are two-dimensional surfaces. In the simplest case—a point source of energy emitting waves into a uniform environment—the resulting waves form spherical wave fronts centered on the source. We can also picture the energy source as emitting a large number of rays radially outward (■ Figure 16.16). Though intuitively appealing and very useful as tools for describing waves, rays are not real and are defined only by their relation to wave fronts.

> A *ray* is a line everywhere perpendicular to the wave fronts of a wave disturbance, pointing in the direction of the wave velocity.

16.3.2 Power in Sound Waves

To understand energy transport by sound, let's refer to the one-dimensional wave in a tube (Figure 16.4). As the piston oscillates, it does work on the air in contact with it. Energy is transferred to the air and transmitted down the tube by the sound wave. The amount of work done per second on an elementary slab (Figure 16.6) by the air to its left is:

WE USED EQNS. (16.7) AND (16.5).

$$\text{Power} = \vec{F} \cdot \vec{v} = AP(x)v(x) = A[P_o + P_* \cos(kx - \omega t)][\omega s_o \cos(kx - \omega t)]$$
$$= A[P_o \omega s_o \cos(kx - \omega t) + P_* \omega s_o \cos^2(kx - \omega t)].$$

The first term changes sign with time and averages to zero. It represents energy that flows back and forth within one wavelength. The second term is always positive, and represents energy the wave transmits over long distances. The power averaged over a whole number of periods (cf. §15.3) is:

$$\langle \text{Power} \rangle = A[P_o \omega s_o \langle \cos(kx - \omega t) \rangle + P_* \omega s_o \langle \cos^2(kx - \omega t) \rangle]$$
$$= A(0 + \tfrac{1}{2} P_* \omega s_o)$$

WE USED EQN. (16.10) TO WRITE s_o IN TERMS OF P_*. ALSO, $\omega/k = v_s = \sqrt{\gamma P_o/\rho}$.

$$= \tfrac{1}{2} A \omega P_* s_o = \tfrac{1}{2} \omega A \frac{P_*^2}{\gamma k P_o} = A \frac{v_s P_*^2}{2\gamma P_o} = A \frac{P_*^2}{2\rho_o v_s}.$$

As we discovered in Chapter 15, power is proportional to the square of the wave amplitude P_*. The average power is also proportional to A, the area of our slab. For this reason, *intensity* is a more useful quantity for sound and light waves.

The *intensity* of a wave is the average power transmitted per unit area of wave front.

The units of intensity are watts per meter squared (W/m²). For the sound wave:

$$I \equiv \frac{\langle \text{power} \rangle}{A} = \frac{P_*^2}{2\rho_o v_s}. \tag{16.15}$$

Using eqn. (16.10), we obtain the intensity in terms of displacement amplitude:

$$I = \tfrac{1}{2}\omega P_* s_o = \tfrac{1}{2}\omega\gamma k P_o s_o^2 = \tfrac{1}{2}\rho_o v_s \omega^2 s_o^2. \tag{16.16}$$

COMPARE THIS EXPRESSION WITH EQN. (15.12).

We use eqn. (16.5) to rewrite this result in terms of the velocity amplitude $v_o = \omega s_o$:

$$I = v_s(\tfrac{1}{2}\rho_o v_o^2).$$

Thus the intensity equals the maximum kinetic energy density times the sound speed.

In a sound wave, energy is transferred between kinetic energy and internal energy that is exhibited as an increase in temperature (See Chapter 19). The total energy density is constant and at any time equal to the maximum kinetic energy density. In general, wave intensity equals wave speed times energy density.

EXAMPLE 16.4 ♦♦ A sound wave has pressure amplitude $P_* = 1.5$ Pa and frequency $f = 1.0 \times 10^3$ Hz. What is its intensity? What is its displacement amplitude?

MODEL We assume the wave propagates in air at atmospheric pressure with a sound speed of 340 m/s. We may express the intensity in terms of either the pressure amplitude (eqn. 16.15) or the displacement amplitude (16.16).

SETUP AND SOLVE We obtain the air density from Table 13.1. Using eqn. (16.15):

$$I = \frac{P_*^2}{2\rho_o v_s} = \frac{(1.5 \text{ Pa})^2}{2(1.2 \text{ kg/m}^3)(340 \text{ m/s})} = 2.8 \times 10^{-3} \text{ W/m}^2.$$

The displacement amplitude is (eqn. 16.16):

$$s_o = \frac{1}{\omega}\sqrt{\frac{2I}{\rho_o v_s}} = \frac{1}{2\pi(1.0 \times 10^3 \text{ Hz})}\sqrt{\frac{2(2.8 \times 10^{-3} \text{ W/m}^2)}{(1.2 \text{ kg/m}^3)(340 \text{ m/s})}}$$

$$= 5.9 \times 10^{-7} \text{ m}.$$

ANALYZE The displacement of the fluid elements is remarkably small.

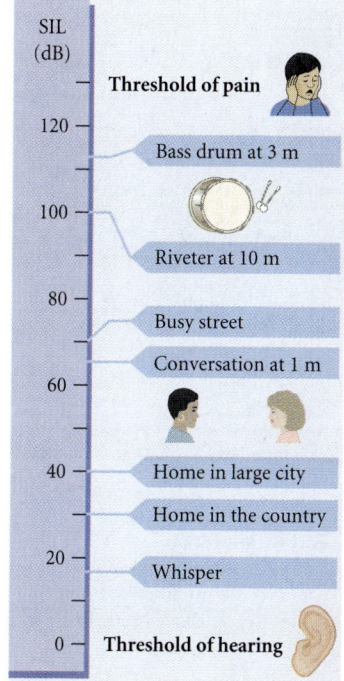

■ **FIGURE 16.17**
Sound intensity levels for some common sounds. The threshold of hearing at about 0 dB varies for different people and also with frequency.

How loud is the sound in Example 16.4? The intensity of sound waves is often compared with a reference intensity of 10^{-12} W/m², about equal to the smallest audible intensity:

The *sound intensity level* of a wave in *decibels* (dB) is:

$$\text{SIL} \equiv (10)\log_{10}[I/(10^{-12} \text{ W/m}^2)]. \tag{16.17}$$

THE DECIBEL IS, OF COURSE, ONE-TENTH OF A BEL, BUT THE BEL IS RARELY USED.

The decibel (dB) scale is named after Alexander Graham Bell (1847–1922), inventor of the telephone and an early student of hearing. Since it is a logarithmic scale, an increase of 10 dB implies an SIL 10 times as great.

The ear's sensitivity is a function of frequency, so the loudness of a sound depends on its frequency as well as its intensity. The threshold of hearing for an average person at 3 kHz is between 0 and 10 dB. People's perception of loudness correlates well with SIL. For moderately loud sounds (SIL > 40 dB), most people would describe an increase of 10 dB to be "twice as loud." ■ Figure 16.17 gives the decibel levels of some common sounds. The sound wave in Example 16.4 has an SIL of $10 \log(2.8 \times 10^{-3}/10^{-12}) = 94$ dB, and that's loud!

THIS RATHER SUBJECTIVE DESCRIPTION TURNS OUT TO GIVE A REMARKABLY CONSISTENT MEASURE OF LOUDNESS.

Digging Deeper

How Does the Ear Detect Sound?

The threshold of hearing at SIL ≈ 0 dB corresponds to displacements in the air of about 10^{-11} m, less than the diameter of an atom! The human ear achieves this amazing sensitivity with a remarkable sequence of structures (■ Figure 16.18).

The outer ear consists of the pinna (the part you can see), which channels sound waves to the auditory canal. In doing so, the pinna changes the response of the ear with the height of the source and enables us to determine the elevation of a high-pitched sound source. The ear canal resonates with sound waves at about 2700 Hz.

The eardrum and bones of the middle ear transmit air vibrations to the inner ear, where they are detected and result in nerve signals to the brain. The whole system has a peak response at about 3400 Hz.

■ Figure 16.19 is a schematic of the inner ear straightened out. Running the length of the inner ear is the basilar membrane, an elastic sheet crudely similar to a drumhead. Fluid motion excited by middle ear bones vibrating the oval window excites traveling waves on the basilar membrane. The point where these waves reach peak amplitude depends on their frequency, allowing the ear to discriminate the pitch of the incoming sound.

Nerve signals to the brain are produced by hair bundles (■ Figure 16.20) protruding from cells attached to the basilar membrane. The signals are caused by diffusion of ions into the tips of the hair bundles, and the rate of diffusion is strongly affected by the relative position of the hairs in a bundle. Studies of the diffusion chemistry show that air displacements of 10^{-11} m in the outer ear can in fact produce detectable differences in the diffusion rates at the tips of the hair bundles.

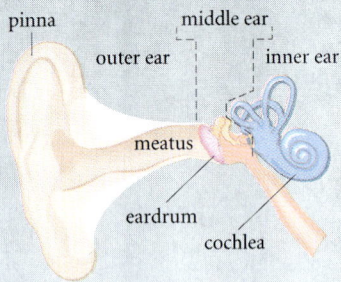

■ **Figure 16.18**
The ear comprises the outer ear (pinna and ear canal), the middle ear, and the inner ear. The middle ear uses a connecting set of bones to transfer vibrations from the outer ear to the inner ear.

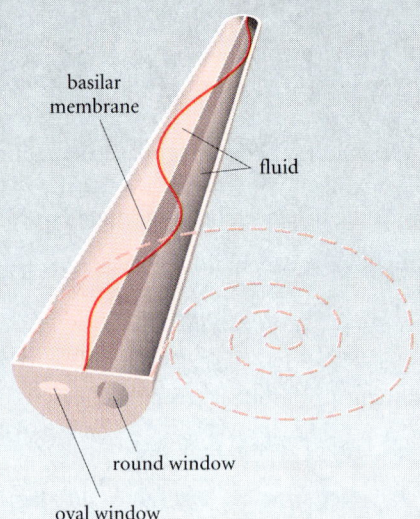

■ **Figure 16.19**
If the cochlea were unrolled, it would look like a fluid-filled tube of decreasing diameter with an elastic membrane (the basilar membrane) along its length. A wave's speed depends on its position along this membrane. Waves of different frequency deposit most of their energy at different places along the membrane, allowing the ear to discriminate different frequencies.

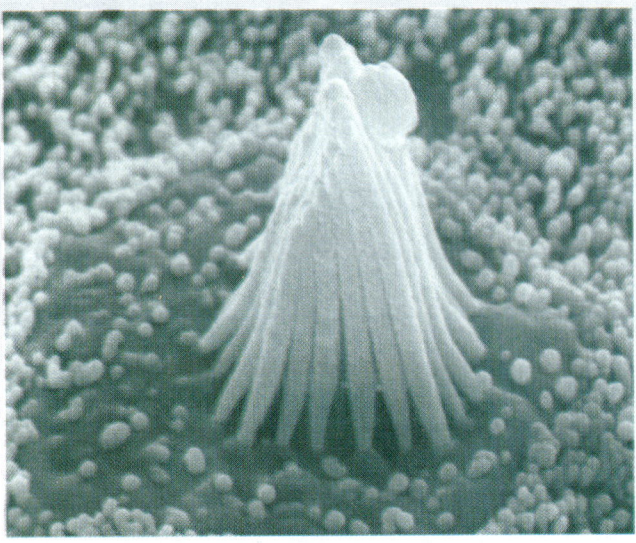

■ **Figure 16.20**
Hairs on the basilar membrane are sensitive to tiny displacements. The frog hair bundle in this picture is enlarged 14 000 times. This bundle is at rest. Bending it away from equilibrium causes electrical signals to be sent to the brain.

16.3.3 The Inverse Square Law

Waves from a point source of sound or light transmit power into ever-increasing volumes. A 50-W light bulb emits light waves with approximately spherical wave fronts. The same average power, $P_{av} = 50$ W, radiated by the bulb is transmitted at each wave front. The area of a wave front at a distance r from the source is $4\pi r^2$, so the intensity of the wave is:

$$I = \text{power/area} = P_{av}/(4\pi r^2).$$

The intensity of waves emitted by a point source decreases as $1/r^2$, a result known as the *inverse square law*. Since the intensity determines the loudness of a sound or the brightness of a light, sources of sound or light appear fainter with increased distance. The amplitude of the wave, which is proportional to the square root of the intensity, decreases as $1/r$.

THIS RESULT IS TRUE NO MATTER WHAT THE SHAPE OF THE BULB, PROVIDED THAT $r \gg$ DIAMETER OF BULB.

EXAMPLE 16.5 ♦ The intensity I_E of sunlight at the top of the Earth's atmosphere is 1.3 kW/m². What is the intensity of sunlight at Jupiter, which is 5.2 times as far from the Sun?

MODEL We use the inverse square law.

SETUP The intensity I_J at Jupiter is given by:

$$I_J/I_E = (r_E/r_J)^2.$$

SOLVE So:
$$I_J = I_E\left(\frac{r_E}{r_J}\right)^2 = \frac{1.3 \text{ kW/m}^2}{5.2^2} = 48 \text{ W/m}^2.$$

ANALYZE Because the intensity of sunlight decreases as the square of the distance from the Sun, only planets located at about the Earth's distance could support life as we know it. Much closer to the Sun, the intensity of light and heat is too great, and farther out, it is too small. ∎

THE ATMOSPHERE ALSO PLAYS AN IMPORTANT ROLE IN DETERMINING THE SURFACE TEMPERATURE OF A PLANET. FOR EXAMPLE, MARS COULD PERHAPS BE HABITABLE IF IT HAD A THICKER ATMOSPHERE.

EXAMPLE 16.6 ♦♦ A siren produces a wave at 5000 Hz with a pressure amplitude of 2.0 Pa at its surface, $r = 0.15$ m. If you measure a sound intensity level of 32 dB, how far are you from the siren?

MODEL The given information determines the intensity of the source at a distance of 0.15 m. We compare this result with the measured intensity, and use the inverse square law to find the distance.

SETUP We use eqn. (16.15) to determine the intensity I_1 at $r = 0.15$ m:

$$I_1 = \tfrac{1}{2}P_*^2/(\rho_o v_s).$$

The measured SIL (eqn. 16.17) gives the intensity I_2 at your position:

$$32 \text{ dB} = \text{SIL} = (10 \text{ dB})\log[I_2/(10^{-12} \text{ W/m}^2)].$$

So: $\quad 3.2 = \log[I_2/(10^{-12} \text{ W/m}^2)],$

and $\quad I_2 = (10^{-12} \text{ W/m}^2)10^{3.2} = 1.6 \times 10^{-9} \text{ W/m}^2.$

Using the inverse square law:

$$I_1/I_2 = (r_2/r_1)^2.$$

SOLVE
$$r_2 = r_1\sqrt{\frac{I_1}{I_2}} = \frac{P_*}{\sqrt{2\rho_o v_s I_2}}$$

$$= (0.15 \text{ m})\frac{2.0 \text{ Pa}}{\sqrt{2(1.2 \text{ kg/m}^3)(340 \text{ m/s})(1.6 \times 10^{-9} \text{ W/m}^2)}}$$

$$= 260 \text{ m}.$$

SINCE THE ANNOYANCE LEVEL OF NOISE DEPENDS ON THE FREQUENCY RESPONSE OF THE HUMAN EAR AND ON PSYCHOLOGICAL RESPONSE TO DIFFERENT FREQUENCIES, EXPERTS ON NOISE USE FREQUENCY-WEIGHTED DECIBEL SCALES (dBA) THAT DIFFER SLIGHTLY FROM SIL.

ANALYZE Check the units; 1 Pa = 1 N/m², 1 W = 1 N·m/s, and 1 N = 1 kg·m/s², so the fraction has units:

$$\frac{N/m^2}{\sqrt{(kg/m^3)(m/s)(N/m \cdot s)}} = \frac{(kg \cdot m/s^2)/m^2}{\sqrt{(kg \cdot m/s^2)kg/(m^3 \cdot s^2)}} = \frac{kg/(m \cdot s^2)}{kg/(m \cdot s^2)} = 1.$$

16.4 THE DOPPLER EFFECT

An approaching ambulance's siren is higher in pitch than it is after the vehicle passes you. The frequency difference between the police radar's transmitted pulse and the pulse reflected from your car reveals your speed. Light from a distant galaxy moving away from the Earth appears redder than the light from a similar galaxy that is not receding. These are examples of the Doppler effect. The Doppler effect for sound depends on whether the source or the observer is moving with respect to the air.

NAMED FOR JOHANN CHRISTIAN DOPPLER (1803–1853), WHO FIRST DESCRIBED IT.

16.4.1 Source Moving with Respect to the Air

When a stationary source emits waves of frequency f_o, each wave front is centered on the source (■ Figure 16.21a). But when the source moves, the center of each spherical wave front is at the position of the source *when that wave front was emitted*. In Figure 16.21b, the source moves at v_e, less than the wave speed v_w. At $t = 0$, the source is at the origin (point a) and emits wave front A. One period later, $t = T = 1/f_o$, the source is at point b ($x_b = v_e T$) and emits wave front B. After two periods, $t = 2T$, the source is at c, wave front A has radius $r_A = 2v_w T$, and wave front B has radius $r_B = v_w T$. The source moves directly toward an observer at O_1, who observes wave fronts separated by wavelength λ:

THE SUBSCRIPT e IN v_e STANDS FOR "EMITTER."

$$\lambda = x_A - x_B = r_A - (x_b + r_B) = 2v_w T - (v_e T + v_w T) = (v_w - v_e)T.$$

When the source is stationary, the wavelength is $\lambda_o = v_w T$, so:

$$\lambda = \left(1 - \frac{v_e}{v_w}\right)\lambda_o. \tag{16.18}$$

Because the wave fronts are closer together, more of them per second pass the ear of the observer at O_1, who hears a higher pitch. The observed time between wave front passages is $T_1 = \lambda/v_w$, and the observed frequency is:

$$f = \frac{1}{T_1} = \frac{v_w}{\lambda} = \frac{\lambda_o f_o}{\lambda} = \frac{f_o}{1 - v_e/v_w}. \tag{16.19}$$

■ **FIGURE 16.21**
(a) Wave fronts emitted by a stationary source. These wave fronts are each one wavelength apart. (b) When the source moves, each wave front is a sphere centered on the position of the source at the time that wave front was emitted. Wave fronts are pushed together on the side toward which the source moves and spread apart on the other side. The separation of the wave fronts along any direction still defines one wavelength. The figure shows the system at $t = 4T$.

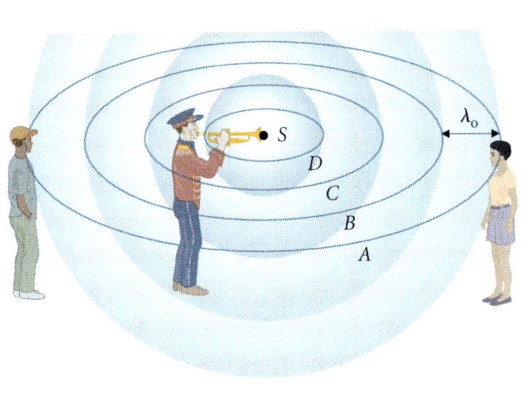
(a)

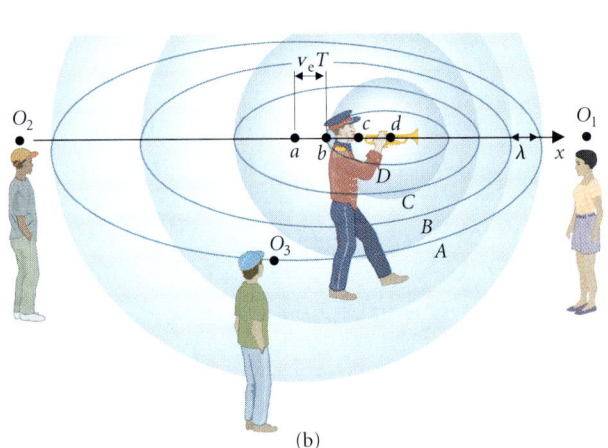
(b)

The source is moving *away* from the observer at O_2, who perceives an increased wavelength and a decreased frequency. Since the component of the source velocity toward O_2 is $-v_e$, we obtain the expressions for the observed wavelength and frequency at O_2 by changing the sign of v_e in eqns. (16.18) and (16.19).

EXERCISE 16.4 ♦ Repeat the derivation to find the observed wavelength when the source recedes from the observer. Show that the same result is obtained simply by changing the sign of v_e in eqns. (16.18) and (16.19).

At the time shown in Figure 16.21b, wave fronts passing an observer at O_3 move at right angles to the velocity of the source. That observer finds no change in the wavelength or frequency of the sound. The result for an observer at an arbitrary position is:

$$\lambda = \lambda_o \left(1 - \frac{\vec{v}_e \cdot \hat{v}_w}{v_w}\right) \quad \text{and} \quad f = \frac{f_o}{1 - (\vec{v}_e \cdot \hat{v}_w/v_w)}, \qquad (16.20)$$

SOURCE MOVES WITH VELOCITY $\vec{v}_e$.

where $\hat{v}_w$ is a unit vector in the direction of the wave velocity at the observer's position. The frequency shift represented by eqn. (16.20) occurs because the wavelength of the wave changes while its speed with respect to the observer stays the same.

EXAMPLE 16.7 ♦ A police car moving at 78 km/h sounds a siren with a frequency of 5.0 kHz. You are standing at the side of the road. What frequency do you hear (a) as the car approaches, (b) as the car passes you, and (c) as it recedes?

MODEL As the car approaches, you are in the position of observer O_1 in Figure 16.21b.

SETUP The frequency you hear is $f = f_o/(1 - v_e/v_w)$. The ratio of the police car's speed to the speed of sound is:

$$\frac{v_e}{v_w} = \frac{(78 \text{ km/h})(10^3 \text{ m/km})(1 \text{ h}/3600 \text{ s})}{330 \text{ m/s}} = 0.066.$$

SOLVE The observed frequency is:

$$f = \frac{5.0 \text{ kHz}}{1 - 0.066} = 5.4 \text{ kHz}.$$

MODEL When the car is passing, you are in the position of observer O_3; there is no difference in the frequency and you hear $f = 5.0$ kHz.
As the car recedes, you are in the position of observer O_2, and:

SOLVE
$$f = \frac{f_o}{1 + v_e/v_w} = \frac{5.0 \text{ kHz}}{1 + 0.066} = 4.7 \text{ kHz}.$$

ANALYZE The total change from approach to receding is about 15%, corresponding to about two and a half keys on the piano.

16.4.2 Observer Moving with Respect to the Air

When a source of sound moves through the transmitting medium, the wavelength of the waves is changed, producing a Doppler effect. A moving observer also experiences a change in the observed frequency, for a somewhat different physical reason. The wavelength of the sound waves does not vary, but the rate the wave fronts pass the observer's ear depends on the

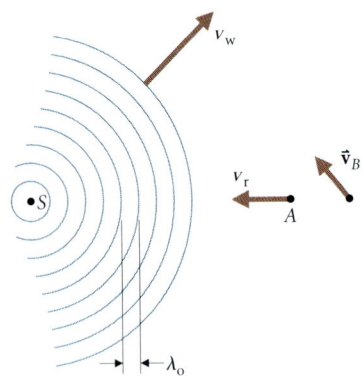
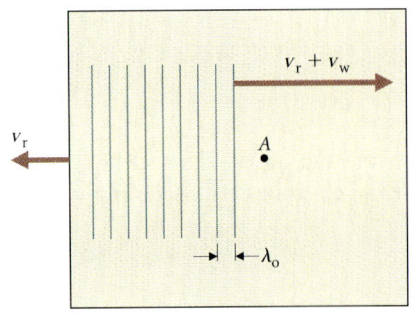
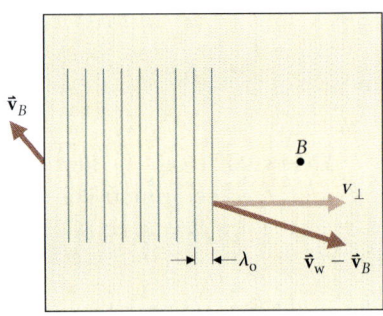

(a) Observers A and B move with respect to source S. (b) Frame moving with observer A. (c) Frame moving with observer B.

■ **FIGURE 16.22**
(a) When the observer moves, the wavelength is not changed, but the observed wave speed is: $f = v_{\text{rel}}/\lambda_o$. (b) Observer A's velocity is perpendicular to the wave fronts reaching him: $v_{\text{rel}} = v_r + v_w$. (c) In the frame moving with observer B, the relative velocity is not perpendicular to the wave fronts. Only the perpendicular component (shown in a lighter color) contributes to the Doppler effect.

THE r IN v_r STANDS FOR "RECEIVER."

OBSERVER MOVES WITH VELOCITY $\vec{v}_r$:

observer's velocity (■ Figure 16.22). The observer at A is approaching a stationary source at speed v_r and sees wave fronts, separated by wavelength λ_o, approach at a relative speed $v_r + v_w$. The observed frequency f is the rate these wave fronts pass the observer:

$$f = \frac{v_w + v_r}{\lambda_o} = \left(\frac{v_w + v_r}{v_w}\right)\frac{v_w}{\lambda_o} = \left(1 + \frac{v_r}{v_w}\right)f_o.$$

Figure 16.22c shows the wave fronts from the point of view of an observer with velocity $\vec{v}_r$ in an arbitrary direction. The rate at which wave fronts pass this observer is determined by the relative velocity perpendicular to the wave fronts:

$$\vec{v}_{\text{rel}} = \vec{v}_w - \vec{v}_r, \quad \text{and} \quad v_\perp = \vec{v}_{\text{rel}} \cdot \hat{\vec{v}}_w.$$

$$f = \frac{v_\perp}{\lambda_o} = \frac{\vec{v}_{\text{rel}} \cdot \hat{\vec{v}}_w}{\lambda_o} = \left[\frac{(\vec{v}_w - \vec{v}_r) \cdot \hat{\vec{v}}_w}{v_w}\right]\frac{v_w}{\lambda_o}.$$

$$f = \left(1 - \frac{\vec{v}_r \cdot \hat{\vec{v}}_w}{v_w}\right)f_o. \tag{16.21}$$

The Doppler effect for a moving observer occurs because the relative speed of the waves with respect to the observer is not the same as for an observer at rest, even though the wavelength stays the same.

16.4.3 Source and Observer Moving with Respect to the Air

When both source and observer move, both physical effects occur simultaneously. Let Δt be the time between the passage of wave fronts with a phase difference of 2π. With λ given by eqn. (16.20), the observed frequency is:

$$f = \frac{1}{\Delta t} = \frac{v_\perp}{\lambda} = \frac{v_w\left(1 - \dfrac{\vec{v}_r \cdot \hat{\vec{v}}_w}{v_w}\right)}{\lambda_o\left(1 - \dfrac{\vec{v}_e \cdot \hat{\vec{v}}_w}{v_w}\right)} = f_o\frac{\left(1 - \dfrac{\vec{v}_r \cdot \hat{\vec{v}}_w}{v_w}\right)}{\left(1 - \dfrac{\vec{v}_e \cdot \hat{\vec{v}}_w}{v_w}\right)}. \tag{16.22a}$$

When source and observer move with the same velocity, $\vec{v}_e = \vec{v}_r$, numerator and denominator are equal and there is no Doppler shift. When v_r and v_e are small compared with v_w, we may use the binomial expansion of the denominator to write:

RECALL $1/(1 - x) = 1 + x + x^2 + \cdots$. WHEN $x \ll 1$, WE DROP THE TERMS IN x^2 AND HIGHER POWERS.

$$f \approx f_o\left(1 - \frac{\vec{v}_r \cdot \hat{\vec{v}}_w}{v_w}\right)\left(1 + \frac{\vec{v}_e \cdot \hat{\vec{v}}_w}{v_w}\right)$$

$$f \approx f_o \left[1 - \frac{(\vec{v}_r - \vec{v}_e) \cdot \hat{v}_w}{v_w} \right], \qquad (16.22b)$$

SEE ALSO PROBLEM 54.

which emphasizes the dependence on the velocity difference $\vec{v}_r - \vec{v}_e$. The Doppler shift $|\Delta f|/f$ equals the speed at which source and observer separate divided by the sound speed. If a wind blows past a stationary source and observer, then both have the same velocity with respect to the air; there is no Doppler shift. If source, observer, and wind all have different velocities, expressions (16.22) apply in the reference frame of the air.

EXAMPLE 16.8 ♦♦ What is the observed frequency of the approaching police car siren (Example 16.7a) when a wind blows at 32 km/h parallel to the car's velocity?

MODEL We work in the reference frame of the air. The Doppler shift depends on the velocity of both source (the car) and observer.

SETUP With respect to the air, the car moves at 78 km/h − 32 km/h = 46 km/h and the observer moves at 32 km/h (■ Figure 16.23). We use v_w = 330 m/s to calculate the speed ratios.

Source: $\dfrac{v_e}{v_w} = \dfrac{(46\ \text{km/h})(10^3\ \text{m/km})(1\ \text{h}/3600\ \text{s})}{330\ \text{m/s}} = 0.039.$

Observer: $\dfrac{v_r}{v_w} = \dfrac{(32\ \text{km/h})(10^3\ \text{m/km})(1\ \text{h}/3600\ \text{s})}{330\ \text{m/s}} = 0.027.$

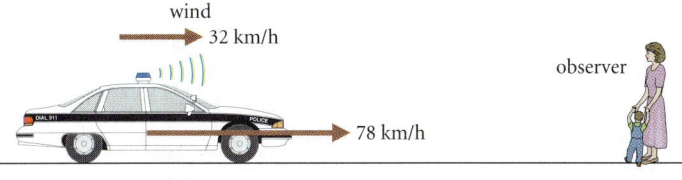
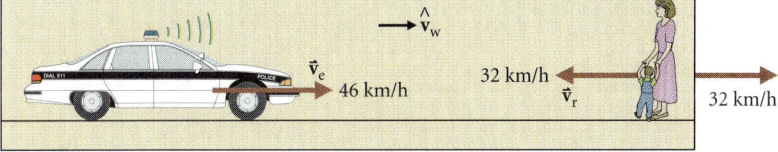

■ **FIGURE 16.23**
When a wind blows, we work in the reference frame of the air. Both source and observer have nonzero velocity in this frame.

SOLVE From eqn. (16.22a), the observed frequency is:

$$f = \frac{1 + v_r/v_w}{1 - v_e/v_w} f_o = \frac{1 + 0.027}{1 - 0.039}(5.0\ \text{kHz}) = 5.3\ \text{kHz}.$$

ANALYZE The effect of the wind is to decrease the observed frequency slightly. ■

16.4.4 The Doppler Effect for Light

According to Einstein's relativity theory, the speed of light is the same for all observers, so the velocity effect described in §16.4.2 cannot occur for light. The Doppler effect due to wavelength changes (§16.4.1) does occur, and it occurs whether the source *or* the observer moves! This difference in behavior arises because light waves are not vibrations in a medium, but are instead self-contained oscillations of electromagnetic fields. Only the relative velocity of source and observer can cause a Doppler effect. So long as this relative velocity is much less than the speed of light, the approximate formula (16.22b) applies with v_w equal to the speed of light c. But when the relative velocity is comparable to the speed of light, according to

THIS IS THE FIRST POSTULATE OF SPECIAL RELATIVITY. SEE §34.1.2.

Einstein's theory, time intervals measured at the source are not the same as time intervals measured by the observer. The result is an additional factor in the formula for Doppler effect. The relativistically correct formula is:

$$f = f_o \frac{\sqrt{1 - v_e^2/c^2}}{1 - (\vec{v}_e \cdot \hat{v}_w)/c}, \tag{16.23}$$

where $\vec{v}_e$ is the velocity of the source relative to the observer. For a receding source, $\vec{v}_e \cdot \hat{v}_w = -v_e$ and:

$$f = f_o \frac{\sqrt{1 - v_e^2/c^2}}{1 + v_e/c} = f_o \sqrt{\frac{1 - v_e/c}{1 + v_e/c}}. \tag{16.24}$$

THE TERM "QUASAR," SHORT FOR "QUASI-STELLAR RADIO SOURCE," MEANS AN ASTROPHYSICAL OBJECT THAT LOOKS LIKE A STAR ON A PHOTOGRAPHIC PLATE BUT IS ACTUALLY THE NUCLEUS OF A VERY DISTANT GALAXY.

EXAMPLE 16.9 ♦ The observed redshift of the distant quasar 3C9 (■ Figure 16.24) is $z \equiv (\lambda - \lambda_o)/\lambda_o = 2.012$. Find the speed at which 3C9 is receding from the Earth.

MODEL The large redshift shows that the wavelength has been changed by more than 100% and suggests 3C9 has a relativistic velocity. We should use eqn. (16.24).

SETUP Rearranging the given expression for z, we have:

$$z = \lambda/\lambda_o - 1 = f_o/f - 1 \Rightarrow f_o/f = 1 + z = 3.012.$$

SOLVE Using eqn. (16.24):

$$1 + z = \frac{f_o}{f} = \sqrt{\frac{1 + v_e/c}{1 - v_e/c}}.$$

Squaring gives:

$$(3.012)^2 = \frac{1 + v_e/c}{1 - v_e/c} \Rightarrow \frac{v_e}{c} = \frac{(3.012)^2 - 1}{(3.012)^2 + 1} = 0.8014.$$

So, $v_e = (0.8014)(2.998 \times 10^8 \text{ m/s}) = 2.403 \times 10^8$ m/s.

ANALYZE Large velocities of recession occur for many objects outside our own galaxy. They are interpreted as a general expansion of the universe.

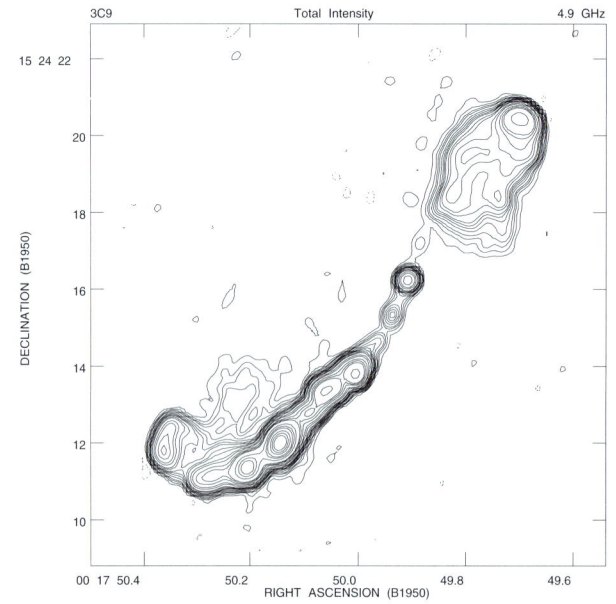

■ **FIGURE 16.24**
The quasar 3C9 is the 9th object listed in the Third Cambridge Catalogue of radio-emitting objects. Here we see the distribution of its 4.9-GHz radio waves. Its complex structure is typical of distant radio sources. The optical spectrum reveals spectral lines shifted by a factor of 2: $\Delta\lambda/\lambda = 2.012$.

EXAMPLE 16.10 ♦♦ The highway patrol uses the Doppler shift to determine whether cars are speeding. If the radar in a parked patrol car operates at 2450 MHz and the return signal has a frequency that is 550 Hz greater, what is the speed of the target vehicle?

MODEL We expect the speed of the vehicle to be much less than the speed of light, so we may use the approximate formula (16.22b). Since the return signal has a higher frequency, the vehicle is moving toward the transmitter at some speed v. Seen from the moving vehicle, the radar beam has a Doppler-shifted frequency. With $\vec{v}_e = 0$ and $\vec{v}_r \cdot \hat{v}_w = -v$:

$$f_1 = f_o(1 + v/c).$$

The car reflects a signal at this frequency and becomes a moving source of waves at frequency f_1 (■ Figure 16.25). The signal received by the police is Doppler shifted again.

SETUP With $\vec{v}_r = 0$ and $\vec{v}_e \cdot \hat{v}_w = +v$:

$$f_2 = f_1(1 + v/c) = f_o(1 + v/c)^2 \approx f_o(1 + 2v/c).$$

The frequency shift $\Delta f = f_2 - f_o = 550$ Hz corresponds to $2(v/c)f_o$, so:

SOLVE $\quad v = \dfrac{c}{2}\left(\dfrac{\Delta f}{f_o}\right) = (1.5 \times 10^8 \text{ m/s})\left(\dfrac{550 \text{ Hz}}{2450 \times 10^6 \text{ Hz}}\right) = 34$ m/s.

ANALYZE Check: $v/c = 10^{-7} \ll 1$, as expected. Although the frequency shift is small, it is easily detected using a phenomenon we shall describe in the next chapter. ■

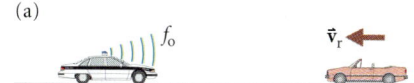
(a)

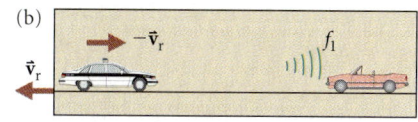
(b)

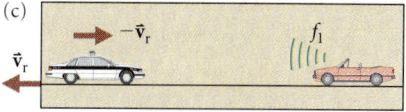
(c)

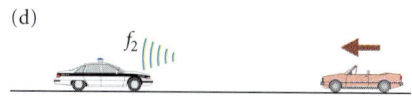
(d)

■ **FIGURE 16.25**
(a) A parked highway patrol car sends a radar signal at frequency $f_0 = 2450$ MHz toward a speeding vehicle. (b) In the frame in which the oncoming vehicle is at rest, the radar frequency has the Doppler-shifted value f_1. (c) The vehicle reflects a signal at frequency f_1. (d) The patrol car receives a signal at the Doppler-shifted frequency f_2.

16.5 REFLECTION AND REFRACTION OF SOUND AND LIGHT

16.5.1 Plane Waves

When a wave meets a boundary between two different materials—for example, when a light wave encounters a glass window—it is partially reflected and partially transmitted. The incident, reflected, and transmitted waves have the same frequency. The wave speeds in the two environments determine the wavelengths and the relative amplitudes. For two- and three-dimensional waves, the angle at which they meet the boundary is also important. This geometrical dependence is most easily studied with plane waves.

> The wave fronts of a *plane wave* are parallel planes.
> The rays of a plane wave are parallel straight lines at right angles to the wave fronts (■ Figure 16.26a).

SEE §15.5, WHERE WE PROVED THIS FOR WAVES ON STRINGS.

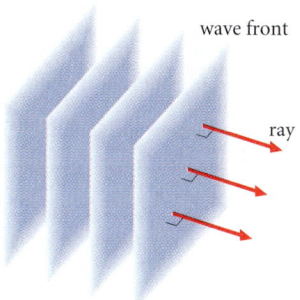

(a) Plane waves.

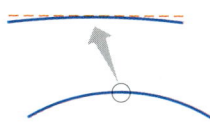

(b) Almost planar piece of spherical wave front.

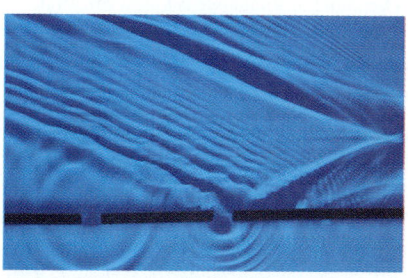

(c) Water ripples.

■ **FIGURE 16.26**
(a) Plane waves. The rays are parallel lines perpendicular to the wave fronts. (b) A small enough piece of any spherical wave front is almost plane. We can often approximate the wave fronts from distant sources as plane waves. (c) Water ripples with straight-line wave fronts reflect and refract like plane waves.

SEE §15.2.2.

The sound wave in a tube (§16.1.1) and the light emerging from a searchlight lens (Chapter 18) are examples of plane waves. A small enough piece of any wave front is approximately plane (Figure 16.26b). Furthermore, Fourier's theorem states that any disturbance is a superposition of harmonic plane waves. We can often find the response of a system by superposing its responses to individual plane waves. For these reasons, the plane wave is one of the most important conceptual tools in physics.

The water ripples in Figure 16.26c have straight-line wave fronts. Our results for plane waves apply without change to such two-dimensional waves.

16.5.2 Reflection of Plane Waves

When a plane wave is incident on a boundary surface, it produces a reflected plane wave (■ Figure 16.27). Experimentally, we find that the reflected wave fronts make the same angle with the surface as the incident wave fronts.

> The *angle of incidence* of a plane wave on a surface is the angle between the surface and the wave fronts. Equivalently, it is the angle between an incident ray and the normal to the surface.
>
> The incident ray and the normal define a plane called the *plane of incidence*.
>
> The *angle of reflection* is the angle between the surface and the reflected wave fronts. It is equal to the angle between a reflected ray and the normal to the surface.

The *law of reflection* describes the relation between these angles.

> The reflected ray lies in the plane of incidence. The angle of incidence equals the angle of reflection.

To understand the law of reflection, suppose that an incoming wave excites vibrations in the boundary surface, which then become the source for new spherical wavelets emitted from each point on the surface. This idea, known as *Huygens' principle*, is literally correct for light waves and helps us to visualize other kinds of waves. The reflected wave is the superposition of the individual spherical wavelets.

Figure 16.27b shows a plane wave incident on a flat surface at an angle θ. At $t = 0$, the wave front along the line AB intersects the surface at A, and the reflected wavelet from A begins to propagate outward. At time T, one period later, the wave front has moved forward one wavelength and meets the surface at C. Reflected wavelets are propagating outward from all the points between A and C. The spherical wavelets superpose so that the phase is constant on the tangent plane CD, which is a wave front of the overall reflected wave. The reflected wave is also a plane wave.

Now compare triangles ABC and CDA. Lines BC and AD, the incident and reflected rays, are at right angles to the incident and reflected wave fronts, AB and CD. Side AC is common

■ **FIGURE 16.27**
(a) The law of reflection. When plane waves reflect at a plane boundary, the angle of incidence equals the angle of reflection.
(b) Huygens' principle explains the law of reflection. The incident and reflected waves propagate at the same speed. The wave fronts shown are one wavelength apart. The angle of incidence is θ and the angle of reflection is ϕ.

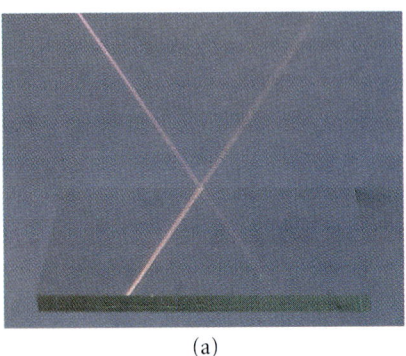

(a)

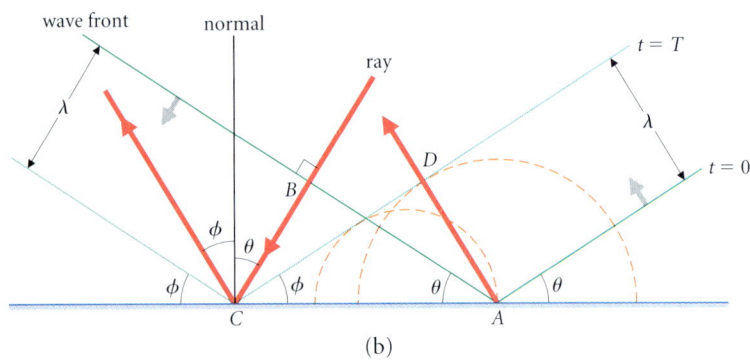

(b)

to the two triangles. Sides *BC* and *AD*, the distances traveled by the waves in time *T*, are of equal length because both waves travel at the same speed. Therefore, triangles *ABC* and *CDA* are congruent. The angle of incidence θ and the angle of reflection ϕ are equal.

EXAMPLE 16.11 ♦♦ A mechanic needs to locate a nut at point *B* on the underside of a vehicle, 24 cm above the ground (■ Figure 16.28). She may put a mirror on the ground and use a flashlight at point *A* to illuminate the nut, 72.0 cm away. At what angle θ should the flashlight be aimed?

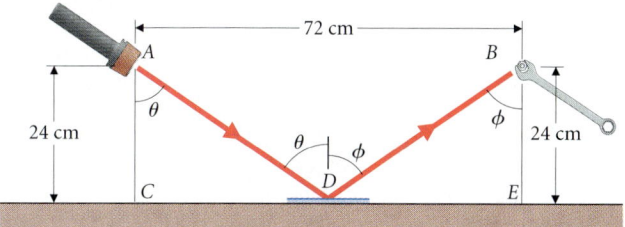

■ **FIGURE 16.28**
A flashlight and a mirror may be used to illuminate an awkwardly placed nut. According to the law of reflection, $\theta = \phi$.

MODEL The parallel rays in the flashlight beam obey the law of reflection.

SETUP Since the incident and reflected angles θ and ϕ in the figure are equal, the distances *CD* and *DE* are also equal: $CD = (72 \text{ cm})/2 = 36 \text{ cm}$.

SOLVE So: $\tan \theta = \dfrac{CD}{AC} = \dfrac{36 \text{ cm}}{24 \text{ cm}} = 1.5 \Rightarrow \theta = \tan^{-1}(1.5) = 56°$.

ANALYZE This type of illumination is useful when barriers prevent direct illumination and may also be used for special photographic effects. ■

16.5.3 Refraction of Plane Waves

If a wave is incident from air on a glass surface at an oblique angle of incidence θ, we observe experimentally that the transmitted wave propagates in a different direction from the incident wave (■ Figure 16.29). It is bent, or *refracted*.

> The *angle of refraction* is the angle between the transmitted wave fronts and the boundary surface. It is equal to the angle between the transmitted rays and the normal to the surface.

We may also use Huygens' principle to understand refraction. In Figure 16.29, a wave is incident at angle θ on a boundary that separates regions with different wave speeds v_1 and v_2.

THE REFRACTED RAY ALSO LIES IN THE PLANE OF INCIDENCE.

IN GENERAL, THERE IS ALSO A REFLECTED WAVE. WE HAVE NOT DRAWN IT, TO AVOID CLUTTERING THE FIGURE.

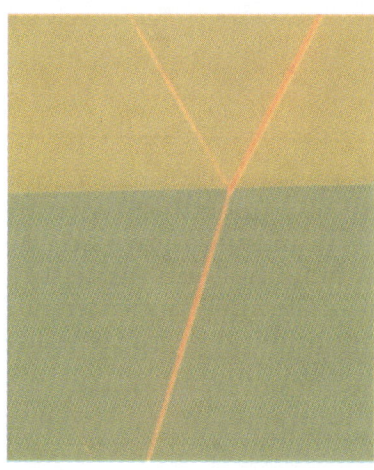

■ **FIGURE 16.29a**
Snell's law. Rays are refracted, or bent, as they pass from one medium to another.

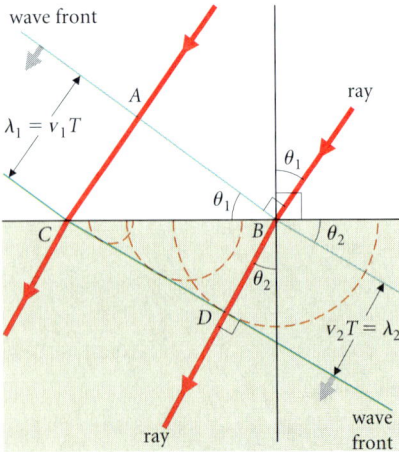

FIGURE 16.29b
The transmitted wave travels at a different speed in the second medium. Again, we show wave fronts one wavelength apart. The wavelength in the second medium is shorter when $v_2 < v_1$. The angle of incidence is θ_1, and the angle of refraction is θ_2.

SNELL'S LAW IS TRUE FOR PLANE WAVES IN GENERAL, SO IT APPLIES TO BOTH SOUND AND LIGHT.

In Figure 16.29b, we assume that $v_1 > v_2$. At time $t = 0$, an incident wave front meets the boundary at point B and generates a spherical wavelet moving outward from B into the second region. One period T later, the incident wave front intersects the surface at C, having generated wavelets at all the points between B and C. Plane CD is tangent to all the spherical wavelets and represents the overall refracted wave front. During the time T, the incident wave front moves the distance $AC = v_1 T = \lambda_1$ while the transmitted wave front moves the distance $BD = v_2 T = \lambda_2 < \lambda_1$. From triangles ABC and CBD, we find:

$$\sin \theta_1 = \frac{AC}{BC} = \frac{v_1 T}{BC} \quad \text{and} \quad \sin \theta_2 = \frac{BD}{BC} = \frac{v_2 T}{BC}.$$

Solving for T/BC, we obtain a relation among the angles and the wave speeds:

$$\frac{\sin \theta_1}{v_1} = \frac{\sin \theta_2}{v_2}, \tag{16.25}$$

which is known as Snell's law, after Willebrord Snell (1591–1626). For light waves, the law is usually written in terms of the refractive index $n = c/v$ (eqn. 16.14):

$$n_1 \sin \theta_1 = n_2 \sin \theta_2. \tag{16.26}$$

Further, since the frequency f is the same for each wave:

$$\frac{BD}{AC} = \frac{\lambda_2}{\lambda_1} = \frac{v_2/f}{v_1/f} = \frac{v_2}{v_1} = \frac{n_1}{n_2}. \tag{16.27}$$

The ratio of wavelengths in the two materials is equal to the ratio of wave speeds.

> **EXAMPLE 16.12** ♦ A beam of light is incident on the surface of a lake at an angle of 45° with the vertical. What is the direction of the refracted beam in the water?
>
> **MODEL** The light beam (a bundle of parallel rays) obeys Snell's law, and the incident medium is air, $n_1 = 1.00$.
>
> **SETUP** The angle of incidence is given as $\theta = 45°$, and from Table 16.3, the index of refraction for water is $n_2 = 1.33$.
>
> **SOLVE** So, using eqn. (16.26):
>
> $$\sin \theta_2 = \frac{n_1}{n_2} \sin \theta_1 = \frac{1.00}{1.33} \sin(45°) = 0.532 \quad \Rightarrow \quad \theta_2 = 32°.$$
>
> **ANALYZE** The light beam bends away from the surface (toward the normal) as it enters the water. ∎

16.5.4 Total Internal Reflection

As we saw in Example 16.12, light bends away from the surface (toward the normal) when it enters a medium with a lower wave speed. Conversely, when the wave speed increases across the boundary, the wavelength increases and the rays bend toward the surface (away from the normal). For a large enough angle of incidence, called the critical angle, the refracted wave fronts are at right angles to the boundary, and the rays run along the surface. The refracted wave does not propagate into the second region, and all the energy is reflected (■ Figure 16.30).

> The *critical angle* for a boundary is the angle of incidence for which the angle of refraction is 90°.

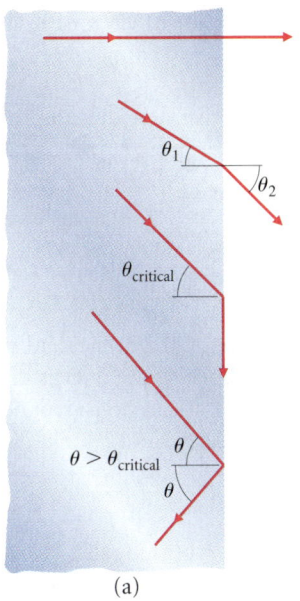 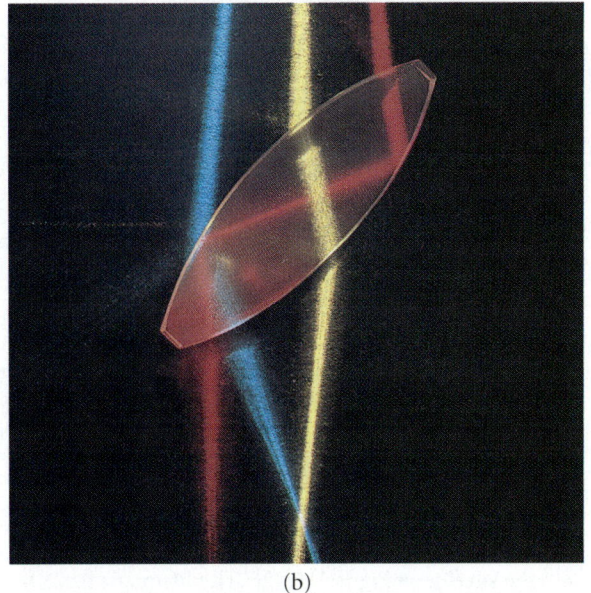

(a) (b)

■ **Figure 16.30**
(a) Definition of the critical angle. (b) The three light beams are refracted at the front surface of the lens and are incident on the back surface at differing angles of incidence. The yellow and blue beams are refracted out. The red beam's angle of incidence is greater than the critical angle, so it is internally reflected.

From Snell's law, the critical angle θ_c is given by:

$$\sin \theta_c = \frac{v_1}{v_2} \sin(90°) \quad \text{or} \quad \theta_c = \sin^{-1}\left(\frac{v_1}{v_2}\right) = \sin^{-1}\left(\frac{n_2}{n_1}\right).$$

If the angle of incidence is greater than θ_c, Snell's law gives $\sin \theta_2 > 1$! Of course, no such angle θ_2 exists; there is no transmitted wave and all the energy goes into the reflected wave. This phenomenon is called *total internal reflection*.

The property of total internal reflection is used in fiber-optic cables, which can transmit large amounts of information quickly (e.g., in the telephone system). Light is transmitted down a glass fiber, remaining trapped inside even when the fiber bends. In laboratory tests, signals have been transmitted for thousands of kilometers through carefully designed optical cables. ■ Figure 16.31 illustrates light trapped in a stream of water.

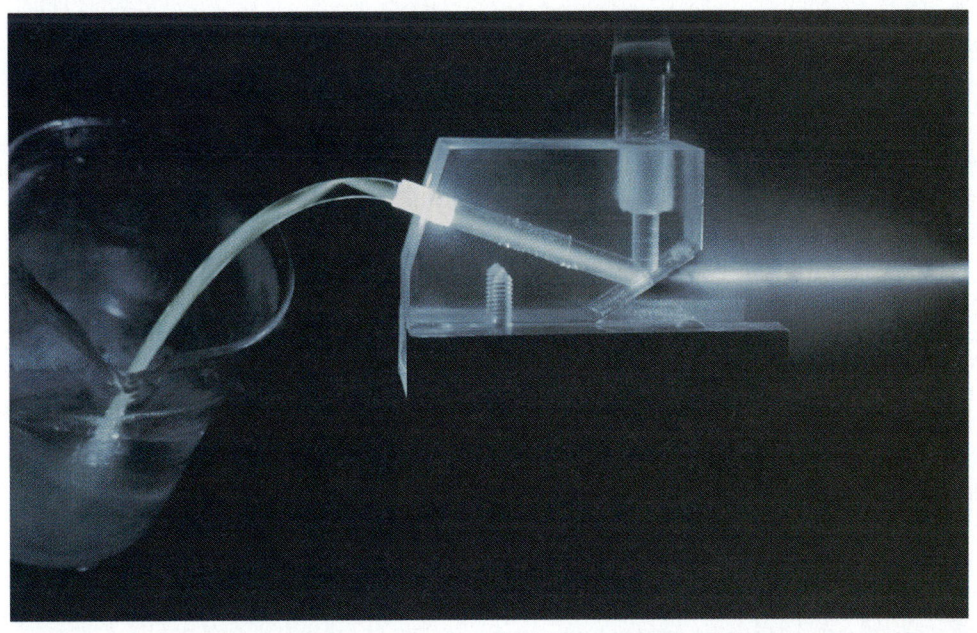

■ **Figure 16.31**
Total internal reflection traps light inside the water stream. This phenomenon has become very important in fiber optic technology for telecommunications.

EXAMPLE 16.13 ♦ Find the critical angle for light emerging from glass with an index of refraction $n = 1.52$.

MODEL We use Snell's law with a 90° angle of refraction. The second medium is air, $n_2 = 1.00$.

SETUP The angle of incidence is the critical angle, $\theta_1 = \theta_c$, and $n_1 = 1.52$.

SOLVE From eqn. (16.26): $n_1 \sin \theta_c = n_2 \sin \theta_2 = n_2 \sin 90° = n_2$. Thus:

$$\sin \theta_c = n_2/n_1 = 1.00/1.52 \quad \Rightarrow \quad \theta_c = \sin^{-1}(0.658) = 41°.$$

ANALYZE Light may be guided around gentle curves by a fiber optic. Light incident at greater than the critical angle (41°) is totally reflected inside the fiber. If the light is directed fairly accurately along the fiber at the start (well collimated), it can be contained within the fiber even at fairly large bends. Even at an abrupt (but less than 41°) kink, the beam bounces back and forth inside the fiber.

EXAMPLE 16.14 ♦♦ A lifeguard needs to warn a scuba diver swimming just below the surface (■ Figure 16.32). The lifeguard has a bullhorn that emits sound in a forward-directed beam. At what angle to the water should the lifeguard aim the horn?

MODEL Since the speed of sound in water is greater than in air, the lifeguard should ensure that the angle of incidence is less than or equal to the critical angle, if significant sound energy is to enter the water. Since the diver is just below the surface, sound aimed at slightly less than the critical angle is refracted almost parallel to the surface and directly at the diver.

■ **FIGURE 16.32**
A lifeguard sending a warning to a scuba diver needs to aim the bullhorn at about 12–13° to the normal, almost straight down, rather than pointing it at the diver.

SETUP We take the speed of sound in air and in water from Table 16.1, and use Snell's law in the form of eqn. (16.25):

SOLVE $\quad \sin \theta_c = \dfrac{V_{\text{air}}}{V_{\text{water}}} \sin(90°) = \dfrac{340 \text{ m/s}}{1500 \text{ m/s}} = 0.227 \quad \Rightarrow \quad \theta_c = 13°.$

The lifeguard should aim at slightly less than 13° to the normal or slightly more than $90° - 13° = 77°$ to the water.

See also Problem IV.5, which asks you to determine the reflected amplitude at normal incidence.

ANALYZE The small critical angle partly explains why the underwater world is so quiet: very little of the noise from the air is transmitted into the water.

16.5.5 Dispersion

The refractive index of most common substances is greater for light at the blue end of the spectrum than for light of lower frequency (greater wavelength). Thus different wavelengths of light are bent by different amounts when they pass from one medium to another, a phenomenon known as *dispersion* (■ Figure 16.33). When the incident medium is air ($n = 1.00$), shorter-wavelength light is bent more as it passes into the second medium. An incident beam that appears "white" includes many different wavelengths of light. After refraction, the different wavelengths are separated into a colored band, called a spectrum.

■ **FIGURE 16.33**
The prism disperses light into a spectrum: short-wavelength (blue) light is bent more at each boundary surface than long-wavelength (red) light.

EXAMPLE 16.15 ♦♦ Corning dense flint glass FDS D34-26 has an index of refraction $n = 1.7708$ at $\lambda = 435.8$ nm ("blue") and $n = 1.7273$ at $\lambda = 643.8$ nm ("red"). A beam of white light is incident from air on a slab of this glass. The angle of incidence is exactly 45°. Find the angles of refraction for the red and blue light.

MODEL Both wavelengths of light obey Snell's law. To five significant figures, the index of refraction of air is 1.0003 at both wavelengths (Table 16.3).

SETUP From eqn. (16.26), $n_1 \sin \theta_1 = n_2 \sin \theta_2$, so:

$$\sin \theta_2 = \frac{n_1}{n_2} \sin \theta_1 = \frac{1.0003}{n_2} \sin(45°).$$

SOLVE For the blue light:

$$\sin \theta_2 = \frac{1.0003}{1.7708}\left(\frac{\sqrt{2}}{2}\right) = 0.39943 \Rightarrow \theta_2 = 23.543°.$$

For the red light:

$$\sin \theta_2 = \frac{1.0003}{1.7273}\left(\frac{\sqrt{2}}{2}\right) = 0.40949 \Rightarrow \theta_2 = 24.173°.$$

ANALYZE The two wavelengths are separated by about half a degree. The separation is accentuated when the light reemerges into air.

Rainbows occur when sunlight is refracted by water droplets in the air. The primary rainbow is formed by rays that are refracted into the droplet, reflected once, and then refracted out (Figure 16.34). For most of the scattered light, the angle between the incident and scattered rays is about 138°.

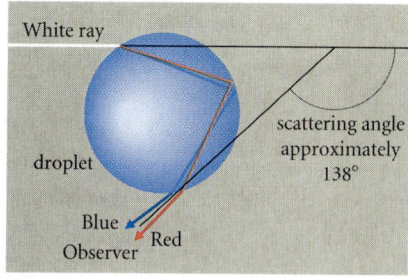

■ **FIGURE 16.34**
Formation of a rainbow. Light is refracted and dispersed by water droplets in the air. The scattering angle is about 138°.

SEE PROBLEM 91.

Chapter Summary

Where Are We Now?

Sound and light are both three-dimensional wave disturbances. We have studied some fundamental properties of these waves that highlight their similarities and their differences.

What Did We Do?

Sound is a mechanical vibration of molecules that travels in air with a speed $v_s = \sqrt{dP/d\rho} = \sqrt{\gamma P/\rho} \approx 330$ m/s. In a traveling sound wave, the velocity of an air element is in phase with the pressure disturbance, while displacement of the fluid elements is out of phase. Sound waves occur in a spectrum of frequencies ranging from nearly zero to more than 10^9 Hz. Audible sound lies in the range 20 Hz to 20 kHz. Ultrasonic waves ($f > 20$ kHz) are used by bats and also in medical imaging.

Standing sound waves occur in pipes, which may be closed or open. Different sets of allowed frequencies occur in the two kinds of pipe. In a pipe open at both ends, $\lambda_m = 2L/m$; in a pipe closed at one end, $\lambda_n = 4L/(2n-1)$, where n and m may each be any integer: $n, m = 1, 2, 3, \ldots$.

Visible light, with frequencies between 4×10^{14} and 8×10^{14} Hz, is one part of the spectrum of electromagnetic waves. This spectrum ranges from VLF (very low frequency) radio waves at $f \approx 1$ kHz to very energetic γ rays at $f \approx 10^{25}$ Hz. In vacuum, the speed of light waves is $c = 3 \times 10^8$ m/s. Light propagates at lesser speeds in ordinary materials. The index of refraction n of a material describes the speed of light in that material: $v_{\text{light}} = c/n$. The refractive index is a function of wavelength.

The strength of sound and light waves is measured by their *intensity:* the power transmitted per unit area of wave front. On the decibel scale, the intensity I of a sound wave is compared with the threshold of hearing at 10^{-12} W/m^2:

Sound intensity level in decibels = $(10)\log[I/(10^{-12} \text{ W/m}^2)]$.

Three-dimensional wave disturbances produced by a small energy source form nearly spherical wave fronts propagating outward from the source. Rays are defined as lines everywhere perpendicular to the wave fronts. The intensity of these waves decreases as the inverse square of distance from the source: $I \propto 1/r^2$. At great distances from the source, a small piece of a spherical wave front is very nearly a plane surface. The wave disturbance approximates a *plane* wave—that is, one whose wave fronts form a set of parallel planes.

In the Doppler effect, relative motion between source and observer alters the observed frequency of waves. The observed frequency increases if the two are approaching each other and decreases if they are receding. For sound, the exact relation depends on the motion of source and observer with respect to the air as well (eqn. 16.22). Since light requires no medium, its Doppler effect depends only on the relative velocity of source and observer. For light, however, we must include a correction for differences in the measurement of time intervals between the relatively moving source and observer (eqn. 16.23). In all cases, when the relative velocity of source and observer is much less than the wave speed, the frequency shift is $\Delta f = f_o v_\perp / v_w$. Only the component of relative velocity perpendicular to the wave fronts $v_\perp$ contributes to the Doppler shift when the motion is nonrelativistic ($v \ll c$).

When plane waves encounter a boundary, both reflected and transmitted waves may be produced. The angle of incidence θ_1 of a wave is the angle between its wave fronts and the boundary surface, or, equivalently, the angle between the incident rays and the normal to the boundary surface. The angle of reflection and angle of refraction θ_2, for the transmitted wave, are defined similarly. The incident ray, the reflected ray, and the normal to the surface lie in a single plane called the plane of incidence, and the angle of incidence equals the angle of reflection. The refracted ray also lies in the plane of incidence. The angle of refraction and the angle of incidence are related by Snell's law:

$$\frac{\sin \theta_1}{v_1} = \frac{\sin \theta_2}{v_2}.$$

For light, Snell's law may be written: $n_1 \sin \theta_1 = n_2 \sin \theta_2$.

Practical Applications

A description of sound waves provides the basic principles needed for the design of musical instruments and concert halls. Ultrasonic waves are used in medicine for diagnostics (e.g., viewing a fetus in the uterus) and treatment (kidney stones may be broken up by a well-directed ultrasonic beam), and in burglar alarms to detect moving objects. Ultrasonic waves are also used to weld gold wire to semiconductor chips, and in the micromachining of brittle materials.

Standing sound waves occur in organ pipes and other wind instruments; standing EM waves are used in lasers and microwave ovens.

Reflection and refraction of sound and light are important in forming images (Chapter 18). Fiber optics allow efficient information transfer using the phenomenon of total internal reflection.

The Doppler effect is used by astronomers to measure the velocities of distant objects and by police forces to determine which cars are exceeding the speed limit.

Solutions to Exercises

16.1 See ■ Figure 16.35. At $t = 3\pi/2\omega$, high-pressure regions occur at the points of converging motion shown in Figure 16.4c. The pressure at the piston itself is greater than it was at $t = \pi/\omega$, since the air was moving rapidly to the left at that time, while the piston itself has been slowing down. Air piles up at the piston.

As the piston moves back toward its original location, it compresses the air ahead of it, causing the pressure to increase still more. Noting the regions of converging and diverging air motion in Figure 16.35, we can determine the regions of high and low pressure. The system returns to the state shown in Figure 16.4a at time $t = 2\pi/\omega$.

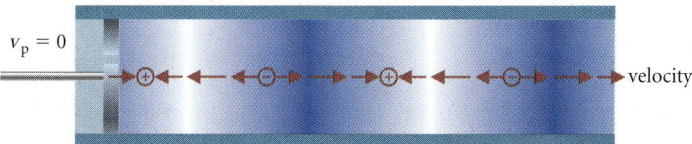
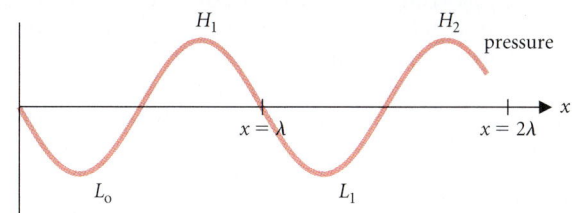

FIGURE 16.35
Wave in the tube at time $t = 3\pi/2\omega$.

16.2 Since pressure is force per unit area and density is mass per unit volume, we have:
$$\frac{[P]}{[\rho]} = \frac{[F/A]}{[M/V]} = \frac{MLT^{-2}\,L^{-2}}{ML^{-3}} = \frac{L^2}{T^2}, \quad \text{as required.}$$

16.3 Using the given expression for $dP/d\rho$, together with eqn. (16.2), we have:
$$\frac{dP}{P} = \gamma\frac{d\rho}{\rho} = -\gamma\frac{\partial s}{\partial x}.$$

Combining with eqns. (16.4) and (16.7):
$$\delta P \equiv P_* \cos(kx - \omega t) = -\gamma P_o \frac{\partial s}{\partial x}$$
$$= -\gamma P_o s_o [-k\cos(kx-\omega t)],$$
and
$$P_* = \gamma k P_o s_o.$$
Since
$$v_s^2 = \gamma P_o/\rho_o,$$
we may rewrite this as
$$P_* = \rho_o v_s^2 k s_o.$$

16.4 Refer to Figure 16.21b. The observer at O_2 observes a wavelength λ that we may calculate using wave fronts A and B. At time $2T$, wave front A is distance $2v_w T$ from the origin a, and wave front B, which was emitted at b, crosses the x-axis on the side closer to O_2 at distance $v_w T - v_e T$. Thus:
$$\lambda = 2v_w T - (v_w - v_e)T = (v_w + v_e)T$$
$$= Tv_w(1 + v_e/v_w) = \lambda_o(1 + v_e/v_w),$$
which is just eqn. (16.18) with the sign of v_e changed.

Basic Skills

Review Questions

§16.1 SOUND

- Which quantities are disturbed in a sound wave in air? Describe the phase relations between the disturbances in a traveling wave.
- Give an approximate value for the speed of sound in air. How is it related to the pressure and density of the air?
- Which quantities in a standing sound wave are zero at a pipe's open end? Which quantities are zero at a closed end?
- What are the wavelengths of standing waves in an open pipe and a closed pipe?

§16.2 LIGHT

- Which frequencies of EM waves contribute to visual light? Which frequencies are usually described as radio waves?
- State some ways in which light waves differ from sound waves.
- What feature of light waves is associated with our sense of color?
- What is the speed of light in vacuum?
- Define *refractive index*.

§16.3 ENERGY IN SOUND AND LIGHT

- Define *wave front* and *ray*.
- Define *intensity*.
- How does the intensity of a sound wave depend on its displacement amplitude?
- Define *sound intensity level*. What is its unit? What are its physical dimensions?
- What is the inverse square law? In what situations does it apply?

§16.4 THE DOPPLER EFFECT

- What is the Doppler effect?
- How does the frequency shift of sound due to the Doppler effect depend on the velocities of the source and of the observer?
- How can you compute the Doppler shift when a wind is blowing?
- What is the formula for Doppler shift that should be applied to light from distant astronomical objects? Why does it differ from the formula for sound waves from a receding source?

§16.5 REFLECTION AND REFRACTION OF SOUND AND LIGHT

- Define a *plane wave*. Why are plane waves important?
- Define the *angles of incidence, of reflection,* and *of refraction*, both in terms of wave fronts and in terms of rays. Explain why the two versions of each definition are equivalent.
- State the law of reflection.
- Describe Huygens' principle. How does it explain the law of reflection?
- State Snell's law.
- Describe how Huygens' principle accounts for Snell's law.
- What is *total internal reflection*? Under what conditions does it occur?
- What is *dispersion*? Why does it occur?

Basic Skill Drill

§16.1 SOUND

1. What is the wavelength corresponding to the note A (440 Hz) in air at 0° C? Two octaves higher, A has four times the frequency (1760 Hz). What is the corresponding wavelength?
2. An organ pipe 1.0 m long is closed at one end. What is the wavelength of its fundamental mode? What is its frequency?
3. A standing sound wave is excited in an open pipe 1.0 m long. What is the frequency of the fifth overtone ($n = 6$)? Sketch the pressure disturbance in the pipe for this wave. How many pressure nodes are there? How many displacement nodes?

§16.2 LIGHT

4. What is the wavelength of a 10^{19}-Hz x ray?
5. What is the speed of **(a)** a 10^{22}-Hz γ ray, **(b)** visible light, **(c)** 10-MHz radio waves traveling in vacuum?
6. What is the speed of light in plastic with an index of refraction of 1.6?
7. How long does it take light to travel **(a)** 10 m across a room, **(b)** around the world, **(c)** to the star α Centauri, a distance of 4×10^{16} m?

§16.3 ENERGY IN SOUND AND LIGHT

8. If one sound is 100 times more intense than another, how much greater is its sound intensity level?
9. A star radiates 10^{28} W of power. If the star is 5×10^{17} m away, what is the intensity of its light at Earth?
10. What is the SIL of a bass drum at 20 m?

§16.4 THE DOPPLER EFFECT

11. You are attending an outdoor concert on a day when the wind is blowing from the orchestra to the audience at 35 km/h. When the orchestra plays an A (440 Hz), what frequency do you hear?
12. A flutist on a moving train is playing the note of A (880 Hz). If the train is moving toward an observer at 41 m/s, and the speed of sound is 340 m/s, what note does the observer hear?
13. The star Canopus is moving away from us at 21 km/s. What is its redshift, $\Delta\lambda/\lambda_o$?
14. An aircraft travels at 220 m/s toward a stationary blimp. The blimp's engine whines at 21 kHz. What frequency does the pilot of the aircraft hear from the engine of the blimp? (Take $v_{sound} = 330$ m/s.)

§16.5 REFLECTION AND REFRACTION OF SOUND AND LIGHT

15. A light beam shining through a hole in a wall strikes a mirror at a 35° angle of incidence. If the wall and the mirror are parallel and 5.0 m apart, how far from the hole does the reflected beam strike the wall?
16. A sound wave is incident at a 5.0° angle on water. If the speed of sound is 330 m/s in the air and 1.5×10^3 m/s in the water, at what angle does the refracted wave leave the surface?
17. Find the critical angle for 590-nm light passing from diamond into air.

Questions and Problems

§16.1 SOUND

18. ❖ Why must loudspeakers designed to produce low notes (woofers) be bigger than loudspeakers for high notes (tweeters)?
19. ❖ An open organ pipe sounds a fundamental frequency of 262 Hz (C). If one end is closed with a stopper, what is the fundamental frequency then?
20. ♦ A 5.0-kHz sound wave has a pressure amplitude of 0.73 Pa. What are the corresponding density, velocity, and displacement amplitudes?
21. ♦ A jet aircraft is traveling at 0.82 of the speed of sound. If the temperature is 0° C, what is the jet's speed in km/h?
22. ♦ An organ pipe closed at one end has a fundamental frequency of 65.4 Hz. What is its length? What are the frequencies of the first three harmonics?
23. ♦ An organ pipe is to be built with a fundamental frequency of 32.7 Hz. How long must the pipe be **(a)** if it is open at both ends, and **(b)** if it is closed at one end?
24. ♦ The ear canal (Figure 16.18) behaves like a pipe and supports standing waves. Is it an open or closed pipe? The canal is about 2.7 cm long. Treating it as a straight pipe, what is the frequency of the fundamental standing wave in this pipe? (Take the speed of sound to be 340 m/s.)

25. ♦ What is the displacement amplitude for a sound wave with a pressure amplitude of 10^{-4} atm and a frequency of **(a)** 1 kHz, and **(b)** 50 Hz?
26. ♦♦ An organ pipe 25 cm long and open at one end emits sound at a frequency of 340 Hz. If the pipe is vibrating in its fundamental mode, what is the speed of sound? If $v_s = (330 \text{ m/s})(1 + T/546° \text{C})$, what is the temperature T (in °C)?
27. ♦♦ A vacuum chamber for testing satellite equipment is a 3.5-m-long cylinder. Find the frequencies of the fundamental mode and the first overtone of standing sound waves in this cylinder, when the doors at each end are closed but before the chamber is evacuated.

§16.2 LIGHT

28. ♦ What is the speed of light in diamond (index of refraction 2.4)?
29. ♦ When particles travel faster than the speed of light in a medium, radiation called Cerenkov radiation is emitted. What is the minimum speed of a particle in water if Cerenkov radiation is to be emitted? The detection of Cerenkov light in a large water tank was the basis for an experiment to detect the possible decay of protons.

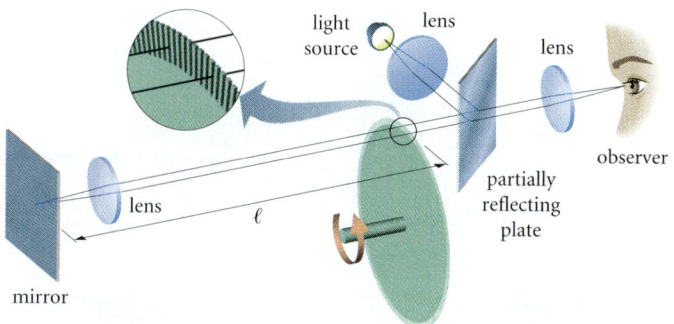

■ FIGURE 16.36
Fizeau's method for measuring the speed of light. The lenses and mirrors are used to shape and direct the light beam, and you may ignore them in Problem 34.

30. ♦ Distance to astronomical objects is often measured in light-years (ly). A light-year is the distance traveled by light in 1 year. How far is this?
31. ♦ How long does it take for light from the Sun to reach the Earth?
32. ♦ When the astronauts landed on the Moon, 4×10^8 m from Earth, what was the time delay between their communications to Earth and Mission Control's reply?
33. ♦♦ Some telephone signals are routed through orbiting geosynchronous satellites (see Example 5.12). What is the time delay caused by this routing? Is it noticeable?
34. ♦♦ Hippolyte Louis Fizeau (1819–1896) measured the speed of light by passing a beam through a spinning wheel with 720 teeth (■ Figure 16.36). The light passes through a gap between the teeth of the wheel, is reflected from a stationary mirror a distance ℓ away, and passes back through the wheel. When the wheel is rotating, the return beam can pass through only when the angular speed is adjusted so that a gap between the teeth is in the path of the beam. Find ω, the angular speed required to see the return beam, in terms of ℓ and c. In Fizeau's experiment, ℓ was equal to 8633 m. What value of ω was required?

§16.3 ENERGY IN SOUND AND LIGHT

35. ❖ Does the inverse square law hold exactly in practice? Why or why not? (*Hint:* Think about a streetlight on a foggy day.)
36. ❖ Ripples are formed on the water surface when a long, thin stick falls in, hitting the water with its long side. Describe the shape of the wave fronts formed (**a**) close to the stick and (**b**) a long way from the stick.
37. ♦ The power from the Sun reaching the top of the Earth's atmosphere is 1370 W/m². This number is called the solar constant. What is the intensity at the Sun's surface? Find the total power radiated by the Sun. (Solar data are given on the inside front cover.)
38. ♦ What is the intensity of a 25-W light bulb at a distance of 8.5 m?
39. ♦♦ If one 1-kHz sound has a sound intensity level 10 dB more than another, what is the ratio of the pressure amplitudes in the two waves?
40. ♦♦ A 75-W light bulb hangs from the ceiling. What is the visible light intensity on the floor, 3.0 m below the bulb, if 10% of its output is visible light?
41. ♦♦ A spherical loudspeaker of radius $R = 10.0$ cm oscillates in radius at frequency $f = 3.0 \times 10^2$ Hz with an amplitude of 0.050 mm. What is the SIL 3.0×10^2 m from the loudspeaker? Is it audible?

42. ♦♦ If conversation is barely audible when its SIL reaches 10 dB above a background noise level of 30 dB, use Figure 16.17 to determine the maximum distance at which normal conversation may be heard.
43. ♦♦ An explosion releases 1.0×10^7 J of energy in 1 s, of which 50.0% is converted to sound waves. Assuming spherical wave fronts, what is the sound intensity 110 m from the blast? How loud is this sound in decibels? (This corresponds to about one-thousandth of a ton of TNT.)
44. ♦♦ A politician is using a public address system mounted on a building to address a crowd in a square 10.0 m below. Assume that the wave fronts leaving the loudspeaker are spherical with a radius of 20.0 cm. If the desired sound level in the square is 99 dB, what is the SIL at the loudspeaker? What pressure and displacement amplitudes are required at 200 Hz? How much power is required if the system is 2% efficient?
45. ♦♦ An amplifier supplies 5.0 W of power to a perfectly efficient loudspeaker at one end of a 10.0-cm-diameter tube filled with air. If the applied signal has a frequency of 250 Hz, what is the intensity of the sound wave generated? What are the displacement amplitude and pressure amplitude of the wave?
46. ♦♦ The nearest star, α Centauri, is 4.2 light-years from Earth. It puts out almost the same power as the Sun. What is its intensity at the top of the Earth's atmosphere compared with that of the Sun (cf. Problem 37)?
47. ♦♦ Show that if the distance to a point source of sound is doubled, its SIL drops by 6 dB.
48. ♦♦ A circular water wave has an amplitude of 10.0 cm at $r = 5.0$ cm and an amplitude of 4.0 cm at $r = 25.0$ cm. What fraction of the wave energy is dissipated as heat between $r = 5.0$ cm and $r = 25.0$ cm?

§16.4 THE DOPPLER EFFECT

49. ❖ What is the speed of light emitted from a rocket moving at 10^8 m/s, as observed on Earth?
50. ❖ The motion of an astronomical object is determined using the Doppler effect. Light from Jupiter is observed to be shifted toward shorter wavelengths (*blueshifted*) on one side and redshifted on the other. What can you conclude from this observation?
51. ❖ Imagine a point source of sound waves moving in a straight line with speed v *faster* than the speed of sound v_w. Sketch wave fronts emitted by the source at intervals of one period. Find the apex angle of a cone tangent to these wave fronts as a function of v/v_w. The wave fronts emitted by the source superpose to form an abrupt disturbance along the tangent cone, called a shock wave or sonic boom.
52. ❖ Sound waves propagate around a triangle (■ Figure 16.37). Show that the wave returning to A has the same frequency as that leaving A regardless of wind speed.

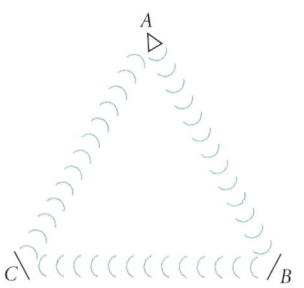

■ FIGURE 16.37

53. ♦ The Andromeda nebula is approaching us at 275 km/s. What is the Doppler shift $\Delta\lambda/\lambda_o$ for this galaxy? Is it a redshift or a blueshift?

54. ♦♦ A man standing on a hilltop signals to his friend by blowing a whistle (frequency 2.00 kHz). The wind is blowing from the signaler toward his friend at 7.00 m/s. What frequency does the friend hear? If the man bicycles down the hill toward the friend at 10.0 m/s, what frequency does the friend hear?

55. ♦♦ A stationary highway patrol vehicle is equipped with radar of frequency 2450 MHz. The beam is reflected from a car moving at 41 m/s. What is the frequency shift of the reflected signal received by the stationary highway patrol?

56. ♦♦ The quasi-stellar object 4C05.35 has a redshift $z = \Delta\lambda/\lambda_o = 2.877$. Use the relativistic Doppler shift formula to find the relative speed of this object and the Earth.

57. ♦♦ A siren sounding at 850 Hz is mounted on an ambulance traveling at 20.0 m/s. What frequency is heard by an observer standing on the sidewalk as the ambulance approaches, if $v_s = 340$ m/s?

58. ♦♦ A moving source of sound is at point A (■ Figure 16.38) at $t = 0$. An observer is at O. Find the arrival time at O of wave fronts emitted at point A and at point B, one period T later. From these arrival times, determine the frequency of the wave disturbance at O and so derive the Doppler shift for a moving source (eqn. 16.20). Assume $v_e T \ll r$.

59. ♦♦♦ A source of sound, emitting waves at frequency f_o, and an observer are separated by a distance D. (a) Assuming still air, how many wavefronts N, emitted at intervals of one period, are there between the source and the observer? (b) If the observer moves away from the source at speed v, how does the number N change with time? By comparing the rate at which wavefronts enter the region between the source and the observer with the rate at which they leave, calculate the observed frequency f. (c) If both observer and source are at rest, but a wind blows at constant velocity, use a similar argument to show that there is no Doppler shift.

60. ♦♦♦ A torpedo moving at 32 m/s is chasing a submarine moving in the same direction at 16 m/s. If the torpedo emits sound pulses at 12.0 kHz, what is the frequency of the signal reflected from the submarine that the torpedo detects?

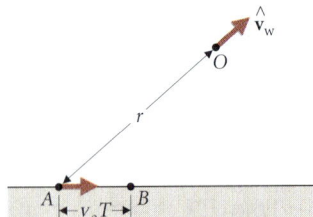

■ FIGURE 16.38

■ FIGURE 16.39
New York seen through a fish-eye lens.

§16.5 REFLECTION AND REFRACTION OF SOUND AND LIGHT

61. ❖ Why is a fish-eye lens so-called (■ Figure 16.39)?

62. ❖ Compare a fish's impressions of the world above water obtained (a) visually, and (b) aurally.

63. ❖ A light beam shines on a mirror with an angle of incidence θ. If the mirror is rotated through an angle α, show that the reflected beam is rotated through an angle 2α.

64. ❖ Sonar systems can detect submarines only within a distance of a few thousand meters. The systems are quite sensitive; lack of sound intensity is not the problem. What causes the limit? (Hint: Temperature, and hence sound speed, decreases with depth in the ocean.)

65. ❖ While standing on land, you want to illuminate an object underwater using a laser beam. Knowing that light is refracted as it enters the water, should you point the laser above, below, or directly at the position of the object as you see it? Explain your reasoning.

66. ♦ A light beam is incident on the mirror shown in ■ Figure 16.40 in the plane of the figure but at an arbitrary angle. By tracing the beam's path as it reflects off the mirror, show that it emerges parallel to its original direction. (Apollo astronauts placed reflectors shaped like the corner of a cube on the Moon to reflect laser signals from Earth and thus allow accurate measurement of the distance to the Moon. ♦♦♦ Can you extend your proof to such a *corner-reflector*?)

67. ♦ A light wave in air is incident on a boundary with ethyl alcohol (index of refraction 1.36) at a 30.0° angle. At what angle does the refracted ray leave the boundary?

68. ♦♦ What is the critical angle for light emerging from ice ($n = 1.31$) into air? Would fiber-optic cables made of ice be better or worse than glass? (Ignore problems of melting!)

69. ♦♦ If light is incident on a flint glass prism ($n = 1.65$) as shown in ■ Figure 16.41 where does it emerge?

70. ♦♦ Show that when light is incident at angle θ on a slab of glass with parallel sides distance d apart, it emerges parallel to the incident direction but displaced by an amount t, where $t/d = \theta(n - 1)/n$, provided that θ is small. Can such a slab be used to produce a spectrum?

71. ♦♦ (■ Figure 16.42) Light of wavelength 9.0×10^2 nm is incident at a 45° angle onto a quartz prism of apex angle exactly 30° and $n = 1.44$. At what angle with the normal does the light emerge? If the index of refraction at 4.0×10^2 nm is 1.47, at what angle does 400-nm light emerge?

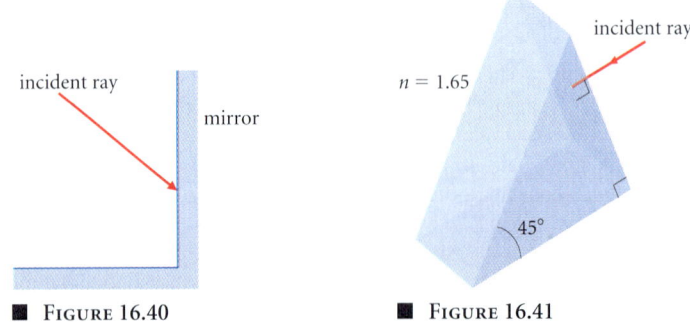

■ FIGURE 16.40

■ FIGURE 16.41

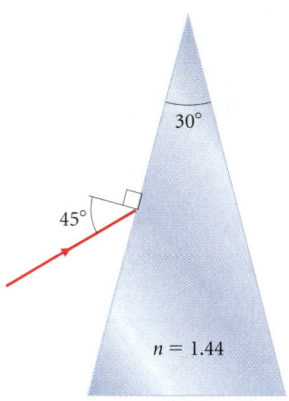
■ FIGURE 16.42

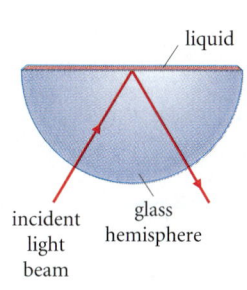
■ FIGURE 16.43

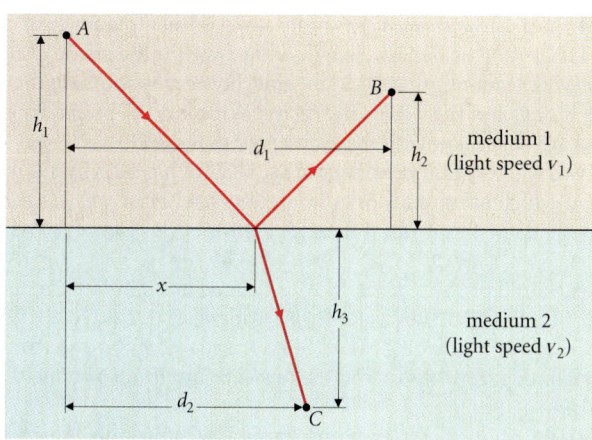
■ FIGURE 16.45
Fermat's principle.

72. ◆◆ ■ Figure 16.43 shows equipment used in a laboratory experiment to measure the refractive index of a liquid. With no liquid present, the beam is totally reflected for $\theta \geq 40.5°$. When carbon tetrachloride is poured onto the flat surface of the glass hemisphere, the beam is totally reflected for $\theta \geq 71.5°$. What is the refractive index of carbon tetrachloride? Are there any limits to the refractive indices that can be measured in this way?

73. ◆◆ A real mirror is a glass plate with a silvered, reflecting layer on the back side. Find where and at what angle the ray shown in ■ Figure 16.44 leaves the mirror if the refractive index of the glass is 1.5. Your answer will be a function of the mirror thickness t and the angle of incidence θ_i.

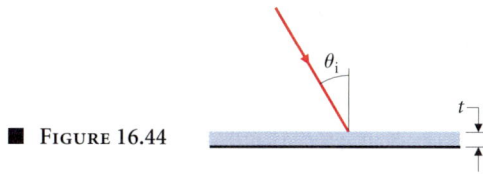
■ FIGURE 16.44

74. ◆◆ An equilateral glass prism is to be designed so that a light beam entering the prism perpendicular to a face will be totally reflected once and then emerge perpendicular to another face. Show that this is possible and find the minimum necessary index of refraction.

75. ◆◆◆ Pierre Fermat deduced the law of refraction from the assumption that light travels from one point to another along the path that takes the least time. Suppose that a light ray from point A could intersect the glass surface at any distance x and then pass through C (■ Figure 16.45). Compute the time required as a function of x and the light speeds v_1 and v_2 in the two media. Show that the path that requires the minimum time also satisfies Snell's law. Without further calculation, show that this path also contains the fewest number of wavelengths (it is the shortest *optical path length*) (cf. Problem 2.73).

76. ◆◆◆ Use Fermat's principle (Problem 75) to demonstrate the law of reflection. Suppose that a light ray from point A (Figure 16.45) could bounce off a mirror at any distance along the mirror and then pass through B. Compute the time required for the light to go from A to B, as a function of x. Show that the path that requires the minimum time also satisfies the law of reflection.

Additional Problems

77. ❖ In a one-way mirror, person A on one side sees his own reflection; an observer B on the other side sees person A. How do one-way mirrors work?

78. ◆ During a thunderstorm, you see a lightning flash and hear the thunderclap 3 s later. How far away is the storm?

79. ◆ Aircraft distance-measuring equipment (DME) sends a radio pulse to a ground station and uses the time interval until a reply is received to compute distance to the ground station. What is this time interval for a distance of (a) 5 km, and (b) 200 km? Does the speed of the airplane cause a significant error in the distance measurement? The DME can also compute aircraft speed. How?

80. ◆◆ A point source of sound emits spherical waves with wavelength $\lambda = 1$ m. On a wall distance $D = 100$ m away, the waves strike a circular hole, $R = 20$ cm in radius (■ Figure 16.46). Find the phase difference of the sound waves between the center of the hole and its circumference.

81. ◆◆ A room is lit by a single globe lamp in the ceiling with wavelength approximately 600 nm. What is the phase difference across a spot 0.50 m in diameter on the floor, 2.5 m below. Can the waves reaching the spot be regarded as plane?

82. ◆◆ A person standing in front of a set of stone steps hears a series of echoes in response to a hand clap. If each step is 30.5 cm long, what is the pitch of the reflected sound? (Take the sound speed to be 340 m/s.)

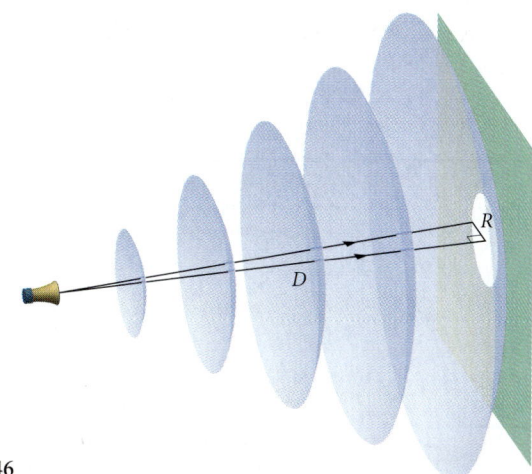
■ FIGURE 16.46

83. ♦♦ A microwave oven uses radio waves with a frequency of 2450 MHz. What is the wavelength of the wave? If the inside of the cavity is 36.8 cm long and 30.5 cm wide, how many wavelengths can fit (a) along, and (b) across the cavity? What kind of boundary conditions are consistent with these observed modes?

84. ♦♦ An argon gas laser operating at 488 nm has a cavity 1.0 m long. At which harmonic does the laser operate? How well can you tell? What is the frequency difference between this harmonic and the next ($n_{argon} = 1.00028$)?

85. ♦♦ Sound pressure level (SPL) is defined in a similar manner to SIL: SPL = $(20\ dB)\log_{10}(P_*/P_{ref})$. Taking $P_o = 1.0 \times 10^5$ Pa and $\rho_o = 1.2$ kg/m^3, show that SIL = SPL for $P_{ref} = 2.9 \times 10^{-5}$ Pa. Use the data in Figure 16.17 to find the pressure amplitude produced by a bass drum at 3 m.

86. (a) ♦♦ Show that in standing sound waves, pressure antinodes correspond to velocity nodes; that is, where the pressure is maximum, the air velocity is zero. (b) ❖ Sketch the pressure and velocity variations for $n = 3$ in a tube closed at one end. Show qualitatively how the air motions account for the changes in density and pressure.

87. ♦♦♦ A satellite is in a circular orbit about the Earth with an orbital period of 105 min. What is its orbital speed? The satellite transmits signals to Earth at 1600 MHz. Make a graph showing the observed frequency shift versus time as the satellite passes overhead. (*Hint:* Approximate the orbit as a straight line passing over the receiver and remember that only velocities along the line of sight produce Doppler shift.) If the receiver bandwidth is 50 Hz, does the Doppler shift need to be taken into account when listening for the satellite signal?

88. ♦♦♦ Acoustic signals from explosive charges placed underground are used to determine the depth of various rock layers. In one test, a charge is placed 60.0 m underground. Sound sensors (called *geophones*) are placed at intervals of 15 m along the surface, which slopes at 15° with the horizontal. For the single sound ray shown in ■ Figure 16.47, consider reflection and refraction at each boundary between layers. Determine which of the geophones detect sound energy from this ray. How long after the initial blast does each geophone record a signal? The sound speed in the various rock layers is given in ● Table 16.4.

89. ♦♦♦ Use the relationship between P and ρ for an ideal gas (§13.2.3) to show that the speed of sound depends only on the mean square speed of the molecules. Given that this quantity is proportional to temperature, how do the resonant frequencies of an organ pipe change when the temperature increases?

90. ♦♦♦ What is the maximum value of $\partial s/\partial x$ in a sound wave in terms of the displacement amplitude and wavelength? Find this maximum for the sound produced by a rock concert with an SIL of 120 dB.

TABLE 16.4 Speed of Sound in Rock

Rock Type	Speed (km/s)
Shale	2.2
Sandstone	4.2
Basalt	5.0
Granite	5.1
Limestone	6.0

Computer Problems

91. Trace the path of a light ray with impact parameter d (■ Figure 16.48) through a spherical water droplet. Assume the ray reflects at the back of the raindrop. Show that the scattering angle ϕ is given by:

$$\phi = 180° - 4\theta_2 + 2\theta_1,$$

where θ_1 is the angle of incidence of the ray as it first meets the raindrop, and θ_2 is the angle of refraction. Express θ_1 and θ_2 in terms of d. Use a spreadsheet program to calculate θ_1 and θ_2 for values of d/r from 0 to 1. Calculate the scattering angle in each case, and plot ϕ versus d/r. You should find a minimum at about 138°. Repeat the calculation for $n = 1.331$ ($\lambda = 656$ nm, "red"); $n = 1.333$ ($\lambda = 589$ nm, "yellow"); and $n = 1.337$ ($\lambda = 486$ nm, "blue"). By how much does the angle at the minimum change between red and blue?

92. An emergency vehicle traveling at 27 m/s carries a 925-Hz siren. As you stand on the sidewalk, the vehicle passes you at a distance of 5.0 m. Choose coordinates with $x = 0$ and $t = 0$ at the instant the vehicle passes you. Find the component of its velocity toward you in terms of its x-coordinate. Use the result to compute the Doppler-shifted frequency of the siren as a function of time. Use a spreadsheet program to compute the value of f as a function of t from $t = -5$ s to $+5$ s at intervals of 0.20 s. Draw a graph showing the observed frequency versus time.

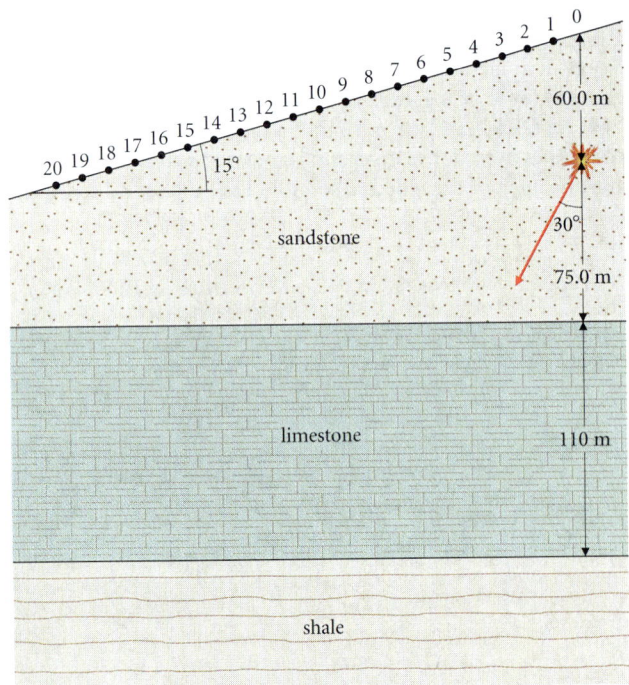

■ FIGURE 16.47

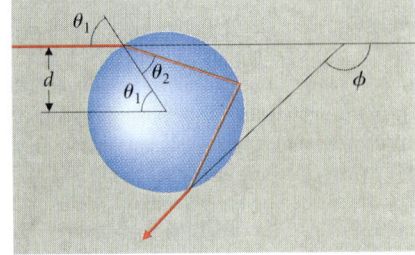

■ FIGURE 16.48

Challenge Problems

93. Olbers' paradox. This problem was originally recognized by Kepler, long before Olbers, for whom it is named. If the universe is infinite, then the sky should be bright at night. Why? (*Hint:* Consider spherical shells around the Earth and show that the intensity of light at Earth is the same for each shell). Since we now believe that the universe is infinite, why is the sky dark at night? (*Hints:* The universe is believed to be about 18 billion years old. It is expanding; distant galaxies are all receding from us.)

94. When light is directed onto an equilateral prism, the angle θ between the incident ray and the emergent ray is called the deviation (■ Figure 16.49). Find the conditions under which the deviation is minimum, and show that if θ_{min} is measured, the index of refraction of the prism may be found from:

$$n = 2\sin[\tfrac{1}{2}(\theta_{min} + 60°)].$$

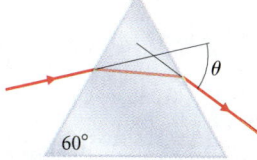

■ **FIGURE 16.49**
Light ray passing through a prism. The angle of deviation is θ.

95. (a) While Joel is hiking along a desert highway, he stands for a while on a small hill 15 m above the road. He notices that the pavement looks blurry at distances of more than 3 km. By how much does the index of refraction of the air just above the hot pavement differ from that of the air at Joel's location? (*Hint:* First argue that $n \sin \theta$ = constant describes the path of a light ray through air whose index of refraction varies continuously with height and then apply the formula to the rays from the pavement to Joel's eyes.) **(b)** Joel's goal is on a mountain peak 0.40 km above the desert floor and 60.0 km away. Could Joel see the peak from the road if there were no refraction effects? Can he actually see the peak from the road?

The bright colors in soap bubbles are due to interference: superposition of traveling waves. In this chapter we shall study several different types of interference. Soap films are discussed in §17.1.5.

CHAPTER 17
Interference and Diffraction

CONCEPTS
Constructive interference
Destructive interference
Beats
Coherence
Interferometer
Diffraction
Resolution
Phasor
Grating

GOALS
Be able to:

Compute the phase difference between two harmonic wave disturbances.

Relate phase differences to constructive and destructive interference.

Describe qualitatively the diffraction pattern of a circular aperture.

Compute the intensity of waves at different points in two-source interference and in a single-slit diffraction pattern.

*See the mountains kiss high Heaven
And the waves clasp one another.*

PERCY BYSSHE SHELLEY

Blowing bubbles is a fun pastime for kids of all ages. If you blow carefully, you can create a floating bubble that glows in the iridescent hues of the rainbow. The colors are due to *interference*—one of the many intriguing phenomena that result from the wave nature of light. Roughly speaking, interference occurs when two or more sources of waves are present at the same time, and *diffraction* occurs when an obstacle interrupts the passage of a wave.

In one dimension, the superposition of harmonic waves moving in opposite directions produces standing waves on a violin string or in an organ pipe. The superposition principle has even richer consequences for two- and three-dimensional waves. When such waves superpose (■ Figure 17.1), the resulting amplitude is zero along nodal lines or surfaces and maximum along antinodal lines or surfaces, where energy flow is concentrated. Concentration of wave energy by superposition explains such diverse phenomena as radio stations using several antennae to beam their signals and rogue waves that occasionally turn sailboats over or gouge chunks from the bows of supertankers. In this chapter we shall investigate several interference phenomena and develop the mathematical techniques for describing them.

■ **FIGURE 17.1**
Interference of water waves from two point sources. The nodal lines are clearly visible.

17.1 INTERFERENCE

17.1.1 Superposition of Two Harmonic Wave Functions

All interference phenomena are due to the superposition of waves with differing phase. The effect can be illustrated by the simplest such situation—the superposition of two one-dimensional, harmonic waves traveling in the same direction with the same amplitude, frequency, and wavelength, but with different phase. The sum of the two waves is also a harmonic wave with the same wavelength and frequency but with a different amplitude. In a snapshot of the waves, the phase difference causes one waveform to be shifted with respect to the other. The superposition of two waves with a very small phase difference has a larger amplitude than its components (■ Figure 17.2a), while a phase difference of π (a shift of one-half wavelength) between the original disturbances results in complete cancellation (Figure 17.2b).

In general, two such waves are represented by cosine functions whose phase differs by $\Delta\phi$:

$$y_1 = A\cos(kx - \omega t) \quad \text{and} \quad y_2 = A\cos(kx - \omega t + \Delta\phi).$$

REMEMBER: THE PHASE OF A WAVE IS THE ARGUMENT OF THE COSINE FUNCTION THAT DESCRIBES IT. ONE WAVELENGTH CHANGE IN DISTANCE CORRESPONDS TO A CHANGE OF 2π IN THE PHASE, AS DOES ONE PERIOD CHANGE IN TIME.

To obtain a useful expression for the sum of these waves, we use the identity:

$$\cos\alpha + \cos\beta = 2\cos\left(\frac{\alpha - \beta}{2}\right)\cos\left(\frac{\alpha + \beta}{2}\right). \quad (17.1)$$

THE PHASE DIFFERENCE $\Delta\phi$ NEED NOT BE A CONSTANT.

THIS TRIGONOMETRIC IDENTITY (APPENDIX IB) IS THE MATHEMATICAL BASIS OF MANY OF THE RESULTS IN THIS CHAPTER.

With $\alpha = kx - \omega t + \Delta\phi$ and $\beta = kx - \omega t$, the sum of these wave functions is:

$$y \equiv y_1 + y_2 = 2A\cos(\Delta\phi/2)\cos(kx - \omega t + \Delta\phi/2). \quad (17.2)$$

The resultant wave's phase is the average of the original wave phases, and its amplitude is:

$$A_r = 2A\cos(\Delta\phi/2). \quad (17.3)$$

The resultant amplitude A_r can take any value between 0 and $2A$, depending on the value of $\Delta\phi$. If the waves are exactly in phase, $\Delta\phi = 0$, the new amplitude is twice that of the two

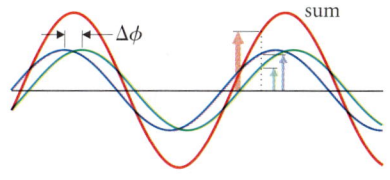
(a) Constructive interference.

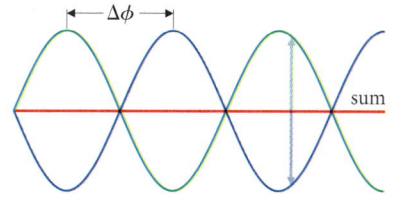
(b) Complete destructive interference.

■ **FIGURE 17.2**
Interference of equal-amplitude harmonic waves. When the phases of the two waves are almost equal (a), the amplitude of the superposition is almost twice the original amplitude. When the two waves are out of phase by π (b), the superposition has zero amplitude.

IF THE TWO WAVES HAVE DIFFERENT AM-
PLITUDES, ONLY PARTIAL CANCELLATION
RESULTS.

original waves. If the waves are exactly out of phase, $\Delta\phi = \pi$, the cosine factor reduces the amplitude to zero.

The general term *interference* is used to describe the superposition of traveling waves.

> Two wave disturbances *interfere constructively* if their sum has a greater amplitude than either component wave. In *complete constructive interference*, the amplitude of the sum equals the sum of the amplitudes of the interfering waves.
>
> The two waves *interfere destructively* if the resulting amplitude is less than that of at least one component wave. In *complete destructive interference*, the amplitude of the sum is zero.

Figure 17.2a illustrates constructive interference, while Figure 17.2b shows complete destructive interference.

EXERCISE 17.1 ♦ For what values of phase difference $\Delta\phi$ do equal-amplitude harmonic waves interfere constructively?

Since the rate a wave transmits energy is proportional to its amplitude squared, two waves with the same phase interfere constructively and transmit energy at four times the rate of one alone $[(2A)^2 = 4A^2]$. Two waves that suffer complete destructive interference transport no energy from their sources. Where does the energy go? It is impossible to create a perfect harmonic wave that doesn't change its form at all, in space or in time. Thus, while destructive interference may occur at one place or one time, interference is constructive somewhere else or at another time. Interference results in a redistribution of energy flow, rather than a change in the total amount. Two waves of slightly different frequency illustrate this redistribution, as we show next.

17.1.2 Beats

If two violinists both play an A but one violin is slightly out of tune, the combined sound pulses with varying loudness; the two notes *beat* with each other. The two sound waves have the same speed v_s, and we can analyze them most easily if the pressure disturbances in the two waves also have equal amplitude P_*:

BUT SEE ALSO PROBLEMS 53 AND 55.

$$\delta P_1 = P_* \cos(k_1 x - \omega_1 t) \quad \text{and} \quad \delta P_2 = P_* \cos(k_2 x - \omega_2 t),$$

with
$$v_s = \frac{\omega_1}{k_1} = \frac{\omega_2}{k_2}.$$

Using eqn. (17.1) with $\alpha = k_1 x - \omega_1 t$ and $\beta = k_2 x - \omega_2 t$, we may think of both waves as having the average angular frequency ω_o and a phase difference $\Delta\phi$ that depends on both position and time:

$$\frac{\alpha + \beta}{2} = \frac{k_1 + k_2}{2} x - \frac{\omega_1 + \omega_2}{2} t \equiv k_o x - \omega_o t,$$

and
$$\frac{\alpha - \beta}{2} = \frac{k_1 - k_2}{2} x - \frac{\omega_1 - \omega_2}{2} t \equiv \frac{\Delta k}{2} x - \frac{\Delta\omega}{2} t = \frac{\Delta\phi}{2}.$$

The pressure disturbance in the superposed waves is:

$$\delta P \equiv \delta P_1 + \delta P_2 = 2 P_* \cos[(\alpha - \beta)/2] \cos[(\alpha + \beta)/2].$$

$$\delta P = 2 P_* \cos(\Delta\phi/2) \cos(k_o x - \omega_o t) \equiv A_r(x, t) \cos(k_o x - \omega_o t). \tag{17.4}$$

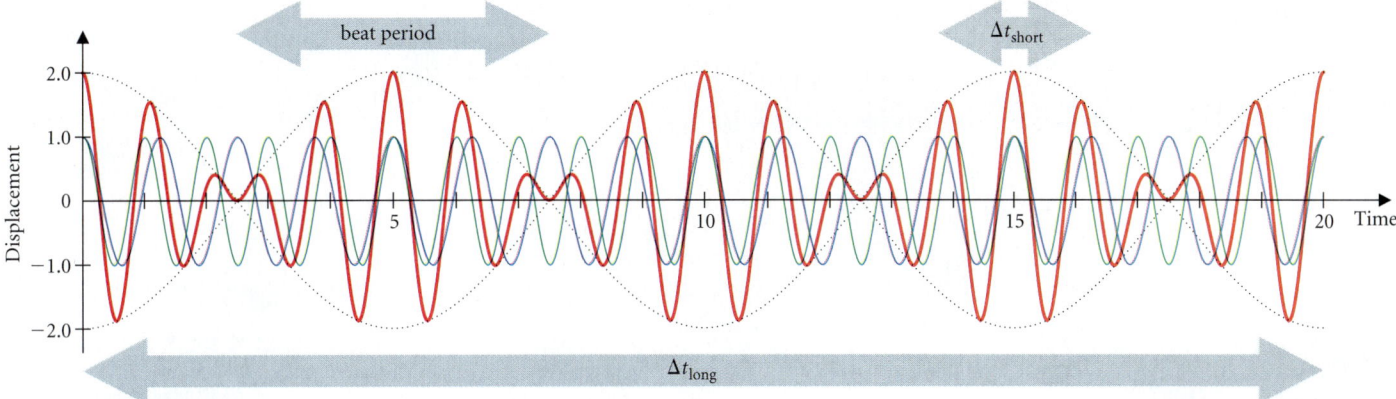

FIGURE 17.3
Beats occur when two waves of slightly different frequency are superposed. The superposition is a wave at the average frequency with an amplitude that varies in time (red line). Here we are looking at the amplitude of the superposition at a fixed place as a function of time. The variable amplitude A_r is represented by the dotted lines in this figure.

The superposition is a wave at the average frequency with a variable amplitude; see ■ Figure 17.3.

EXERCISE 17.2 ♦ Show that the amplitude $A_r(x, t)$ of δP is itself a harmonic wave that travels at speed $\Delta\omega/\Delta k = v_s$. What are the frequency and wavelength of this wave? ■

A listener at the origin ($x = 0$) hears a sound wave with a time-dependent amplitude:

$$A_r(0, t) \equiv 2P_* \cos\left(\frac{\Delta\phi}{2}\right)$$

$$= 2P_* \cos\left(\frac{\omega_1 - \omega_2}{2}t\right).$$

AN OBSERVER AT ANY FIXED PLACE HEARS A SIMILAR SOUND, BUT IT IS MATHEMATICALLY MOST CONVENIENT TO PUT THE OBSERVER AT THE ORIGIN.

Complete destructive interference (zero sound intensity) occurs whenever the cosine function is zero—that is, when $(\omega_1 - \omega_2)t/2 = \pi/2 \pm n\pi$, as in Figure 17.3. The interval between two successive zeros is $\Delta t = 2\pi/|\omega_1 - \omega_2|$. The listener perceives a sound at the average frequency $f_o = \omega_o/2\pi$, whose intensity varies at the *beat frequency* $f_b \equiv 1/\Delta t = |f_1 - f_2|$. The human ear can detect beats at frequencies f_b between approximately $\frac{1}{30}$ Hz and 10 Hz. When the concert-master plays an A, elimination of beats between their instruments and hers is one technique the musicians use to be sure they are in tune.

EXAMPLE 17.1 ♦ In tuning a piano string, a 262-Hz tuning fork is sounded at the same time that the string is struck. Beats are heard with a frequency of 6 per second. What is the frequency emitted by the string?

MODEL The two waves from string and tuning fork add to form a wave modulated at the beat frequency

$$f_b = |f_{string} - f_{fork}|,$$

which is observed to be 6 Hz.

SOLVE Thus f_{string} is either $\quad f_{fork} - f_b = 256$ Hz

or $\quad f_{fork} + f_b = 268$ Hz.

ANALYZE The piano tuner needs a good sense of pitch to determine whether the frequency of the piano string is lower or higher than the tuning fork. ■

Digging Deeper

Energy Redistribution in Beats

The intensity of the combined sound wave (cf. §16.3.2) is given by:
$$I = \frac{\text{Power}}{\text{Area}} = \frac{\vec{F} \cdot \vec{v}}{A} = Pv_x.$$

In §16.1 we found that v_x varies in phase with δP, with amplitude $v_o = \omega s_o = v_s P_*/\gamma P_o$. So, using expression (17.4) for δP:

$$I = \left[P_o + 2P_* \cos\left(\frac{\Delta\phi}{2}\right)\cos(k_o x - \omega_o t)\right]$$
$$\times \left[\frac{v_s}{\gamma P_o} 2P_* \cos\left(\frac{\Delta\phi}{2}\right)\cos(k_o x - \omega_o t)\right]$$
$$= \frac{2v_s P_*}{\gamma}\cos\left(\frac{\Delta\phi}{2}\right)\cos(k_o x - \omega_o t)$$
$$+ \frac{4P_*^2 v_s}{\gamma P_o}\cos^2\left(\frac{\Delta\phi}{2}\right)\cos^2(k_o x - \omega_o t).$$

Perceived loudness depends on the average transmitted intensity. Over a short time scale—$t_s \sim 2\pi/\omega_o$, equal to a few periods of the basic oscillation—the factor $\cos(\Delta\phi/2) \approx \cos(t\,\Delta\omega/2)$ is essentially constant because $\Delta\omega = \omega_1 - \omega_2 \ll \omega_o$ (Figure 17.3). With $\langle \cos\omega_o t\rangle = 0$, and $\langle\cos^2(\omega_o t)\rangle = \frac{1}{2}$, the short-term average is:

$$\langle I\rangle_{\text{short}} = \frac{2P_*^2 v_s}{\gamma P_o}\cos^2\left(\frac{t\,\Delta\omega}{2}\right).$$

At a fixed position, $\langle I\rangle_{\text{short}}$ varies in time at the beat frequency $f_b \equiv |f_1 - f_2|$ (■ Figure 17.4). The amplitude of $\langle I\rangle_{\text{short}}$ is four times the average intensity transmitted by a single wave. On the other hand, no power is transmitted past an observer at $x = 0$ at times when $\cos(t\,\Delta\omega/2) = 0$.

Averaging $\langle I\rangle_{\text{short}}$ again, now over a long time scale equal to several beats, we find:

$$\langle I\rangle_{\text{long}} = \frac{2P_*^2 v_s}{\gamma P_o}\langle\cos^2(t\,\Delta\omega/2)\rangle = \frac{P_*^2 v_s}{\gamma P_o}.$$

This is just twice the intensity transmitted by one of the component waves (cf. eqn. 16.15). Superposition hasn't changed the amount of energy transmitted by the two waves but has instead redistributed it into bursts.

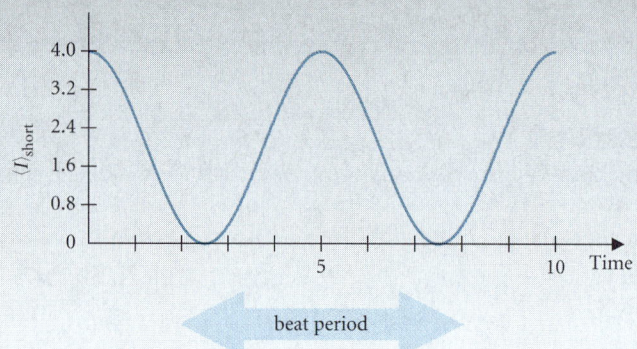

■ **Figure 17.4**
Averaged over a short time scale, equal to a few periods of the underlying waves, the intensity in the superposition at a fixed place oscillates in time at the beat frequency $f_b = |f_1 - f_2|$. We hear a loud-soft-loud-soft beat effect.

Digging Deeper

Phase Speed and Group Speed

Sound and light waves are used to transmit information. But a single harmonic wave looks the same at all times and all places: it carries no information. A modulated wave, such as that of beating musical notes, carries information via its varying amplitude. Each amplitude peak, and hence the information, moves at speed $\Delta\omega/\Delta k$ (Exercise 17.2). For the sound waves we have discussed, $\Delta\omega/\Delta k = v_s$ since $\omega/k = v_s$ for each frequency. More often, as for light waves traveling through glass (§16.5.5), the speed of harmonic waves depends on their frequency. In this case, the speed $\Delta\omega/\Delta k$, called the group speed v_g, differs from ω/k, which is called the phase speed v_ϕ. A careful analysis shows that the exact expression for group speed for an arbitrary superposition is $v_g = d\omega/dk$, the derivative of frequency with respect to wave number. Energy and information are transmitted at the group speed v_g. In some cases, such as x rays in glass, the phase speed can exceed the speed of light c, but v_g is *always* $\leq c$.

17.1.3 Interference Between Two Spatially Separated Wave Sources

Waves leaving two separated sources S_1 and S_2 with the same amplitude and phase arrive at a distant point with phases determined by the distances they travel. Their interference depends on position.

EXAMPLE 17.2 ❖ An observer at B sits in a boat exposed to waves originating at two gaps in a breakwater, S_1 and S_2. The point B is 50 wavelengths from S_1 and 52 wavelengths from S_2. Is the interference constructive or destructive at B? Later in the day, the wavelength of the ocean swell striking the breakwater has increased by 33%. What happens to the motion of the boat?

MODEL The nature of the interference, and the consequent motion of the boat, depends on the phase difference between the two waves arriving at B.

SETUP AND SOLVE Initially, the difference in distance the waves travel, their *path difference* from the two gaps to B, is exactly two wavelengths. Their phase difference is 4π and they interfere constructively. When the boat is on the 50th crest of the wave from S_1, it is on the 52nd crest from S_2 (■ Figure 17.5). The boat at B oscillates up and down at the same frequency as either single wave but with twice the amplitude.

REMEMBER: WE CAN ADD OR SUBTRACT INTEGER MULTIPLES OF 2π WITHOUT CHANGING THE WAVE PHASE; 4π IS EQUIVALENT TO ZERO.

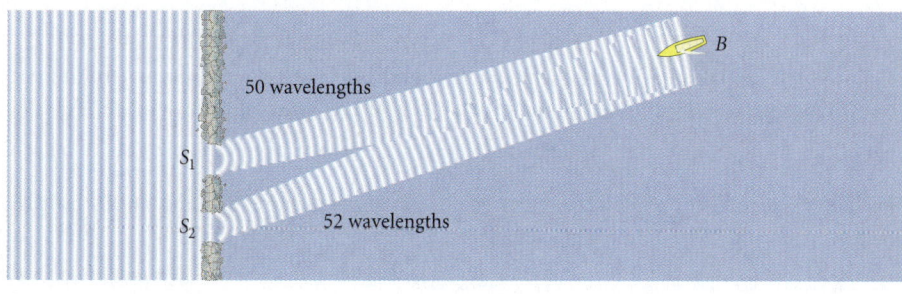

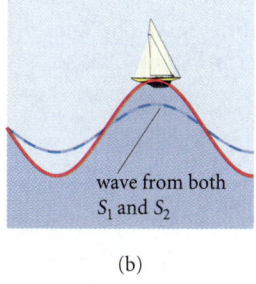

■ **FIGURE 17.5**
(a) Rays that reach the boat from the two openings. The boat is 50 wavelengths from S_1 and 52 wavelengths from S_2. (b) Thus it is on a crest of each wave at the same time, and bobs up twice as far as it would with one wave alone. (c) Later in the day, when the wavelength is 33% longer, the boat is on a crest from S_1 when it is on a trough from S_2. With equal-amplitude waves, the result is no displacement at all.

Later in the day, the wavelength is $\frac{4}{3}$ of its previous value. The boat is $(\frac{3}{4})52 = 39$ wavelengths from S_2 and $(\frac{3}{4})50 = 37.5$ wavelengths from S_1. The path difference is 1.5 wavelengths and the interference is destructive. When the wave from S_1 crests at B, the wave from S_2 has a trough there. The crest and trough cancel, leaving the boat at B motionless.

ANALYZE We do not always know the distances from the two sources to B, and it is only the difference in distances that matters. It is not important that the boat is on the 50th crest from S_1; what counts is that *some* crest from S_1 arrives at the same time as a crest from S_2.

We assumed that the waves from each gap have equal amplitude. If the amplitudes differ, complete destructive interference does not occur: the boat's motion is reduced but is not exactly zero.

The preceding example illustrates a basic principle: interference at any point P of waves from two sources depends on the difference in path length from each source to P. The wave originating at S_1 produces a disturbance at P with phase $\phi_1 = kr_1 - \omega t$ (■ Figure 17.6). Similarly, the wave originating at S_2 reaches P with phase $\phi_2 = kr_2 - \omega t$. The displacement y due to the superposition of these waves is given by:

$$y = A\cos(kr_1 - \omega t) + A\cos(kr_2 - \omega t)$$
$$= 2A\cos\left[k\left(\frac{r_1 - r_2}{2}\right)\right]\cos\left[k\left(\frac{r_1 + r_2}{2}\right) - \omega t\right].$$

WE USED EQN. (17.1) AGAIN, WITH $\alpha = kr_1 - \omega t$ AND $\beta = kr_2 - \omega t$.

This wave looks as if it originated at a point intermediate between S_1 and S_2 at distance $R = \frac{1}{2}(r_1 + r_2)$. Its amplitude is:

$$2A\cos\left[k\left(\frac{r_1 - r_2}{2}\right)\right] = 2A\cos(k\,\Delta r/2).$$

COMPARE THIS WITH BEATS, IN WHICH THE AMPLITUDE VARIES IN *TIME*. HERE THE AMPLITUDE VARIES IN *SPACE*. THE PHASE DIFFERENCE BETWEEN WAVES IS CAUSED BY THE DIFFERENT DISTANCES THAT THEY TRAVEL RATHER THAN BY DIFFERENT FREQUENCIES.

The two waves interfere constructively at some points and destructively at other points. Complete destructive interference occurs along the *nodal lines*, where $\cos(\Delta\phi/2) = 0$; that is, the phase difference is an odd multiple of π and the path difference is an odd number of half wavelengths:

FIGURE 17.1 ILLUSTRATES NODAL LINES FOR TWO SOURCES OF WATER WAVES.

$$\Delta\phi = k\,\Delta r = (2\pi/\lambda)\,\Delta r = (2m + 1)\pi, \quad m = 0, 1, 2, \ldots.$$

> Complete destructive interference occurs at points along nodal lines where the path difference from the sources is an odd number of half wavelengths:
>
> $$\Delta r_{min} = (2m + 1)\lambda/2.$$
>
> Complete constructive interference occurs at points along antinodal lines where the path difference is a whole number of wavelengths:
>
> $$\Delta r_{max} = 2m\pi/k = m\lambda.$$

When the observation point P is very distant from the sources ($R \approx r_1 \approx r_2 \gg d$), we may obtain a simple but excellent approximation for the path difference. The lines S_1P and S_2P from each source to the observation point are almost parallel (■ Figure 17.7) so:

$$\Delta r \equiv r_2 - r_1 \approx S_2Q = d\sin\theta,$$

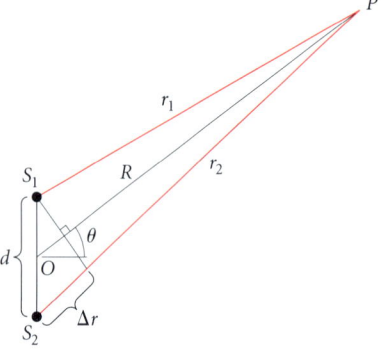

■ FIGURE 17.6
Interference between two spatially separated sources. The phase difference between waves arriving at P depends on the difference in path length $\Delta r = r_2 - r_1$ from the two sources.

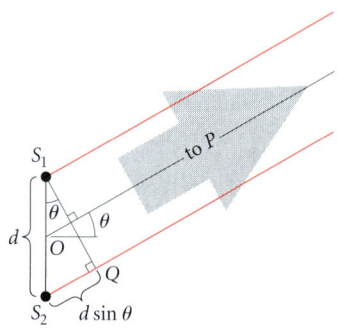

■ FIGURE 17.7
When the point P is very distant from the sources, $R \gg d$, the path difference $\Delta r \approx d\sin\theta$.

where θ is zero when the observation point is on the perpendicular bisector of the line joining the sources. Constructive interference occurs when

$$d \sin \theta = m\lambda, \qquad m = 0, 1, 2, \ldots, \qquad (17.5)$$

and destructive interference occurs when

$$d \sin \theta = (m + \tfrac{1}{2})\lambda, \qquad m = 0, 1, 2, \ldots. \qquad (17.6)$$

Far from the sources, the nodes and antinodes occur along straight lines pointing away from the sources (Figure 17.1). When the sources emit light waves, constructive interference produces bright light on a distant screen: bright and dark *fringes* appear at definite angles given by eqns. (17.5) and (17.6).

EXERCISE 17.3 ♦♦ Use the cosine rule to express the distances r_1 and r_2 in Figure 17.6 in terms of R, d, and θ. Show that when $R \gg d$, $\Delta r = d \sin \theta$. Taking $R = 10d$ and $\theta = 15°$, calculate r_1, r_2, $r_2 - r_1$, and $d \sin \theta$. How good is the approximation?

EXAMPLE 17.3 ♦ Two slits, separated by a distance of 10.0 μm, are illuminated by light of wavelength 6.00×10^2 nm (■ Figure 17.8). The light falls onto a screen 3.00 m from the slits. How far apart on the screen are the first few bright fringes?

Photograph of a 2-slit interference pattern.

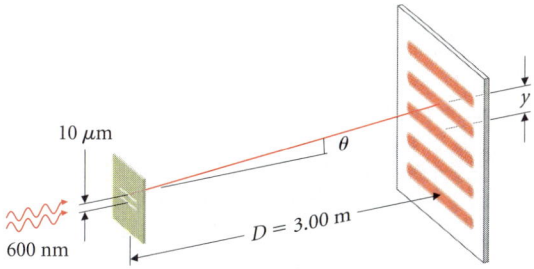

■ **FIGURE 17.8**
The screen is $D = 3.00$ m from the two slits, which are separated by $d = 10.0$ μm. Where are the interference maxima (bright fringes)? (The diagram is not drawn to scale.)

MODEL The bright fringes on the screen are places where constructive interference occurs—that is, where the path difference from the two slits is a whole number of wavelengths.

SETUP The bright fringes appear at angles θ where:

$$\sin \theta = m\lambda/d.$$

For small angles θ, we may use the approximation $\theta \approx \sin \theta \approx \tan \theta = y/D$, where D is the horizontal distance from source to screen and y is the vertical distance from the center of the screen ($\theta = 0$). Then:

$$y/D = m\lambda/d.$$

SOLVE
$$y = mD\frac{\lambda}{d} = m(3.00 \text{ m})\left(\frac{6.00 \times 10^{-7} \text{ m}}{10.0 \times 10^{-6} \text{ m}}\right) = m(0.180 \text{ m}).$$

Between two neighboring maxima, m increases by 1. The bright fringes are $\Delta y = 18.0$ cm apart on the screen.

ANALYZE This result is valid near the center of the screen, where the small-angle approximation is valid. Check: $y/D = (18.0 \text{ cm})/(3.00 \text{ m}) = 0.06 \ll 1$.

EXERCISE 17.4 ♦ The experiment of Example 17.3 is repeated with a green-colored laser beam. The screen is 3.00 m from the slits, and the first two dark fringes are located at distances of 8.30 cm and 24.8 cm from the center. What is the wavelength of the laser light?

17.1.4 Coherence

Light from an ordinary source such as a light bulb is made up of short pulses of varying length emitted at random intervals. The average pulse length is the *coherence length* of the source. Coherence lengths ℓ vary from about a micron for sunlight and several millimeters for gas discharge lamps to meters or more for lasers. With an *incoherent* source such as sunlight, the interference pattern changes roughly every 3 fs ($= \ell/c$) and is not visible (■ Figure 17.9).

Two-source interference for light waves was first demonstrated by Thomas Young in 1803. His device allows light to pass through a single pinhole before letting it fall on two pinholes, at equal distances from the first, that act as interfering sources (■ Figure 17.10). The

REMEMBER: fs = femtosecond = 10^{-15} s. (SEE TABLE 1.1.)

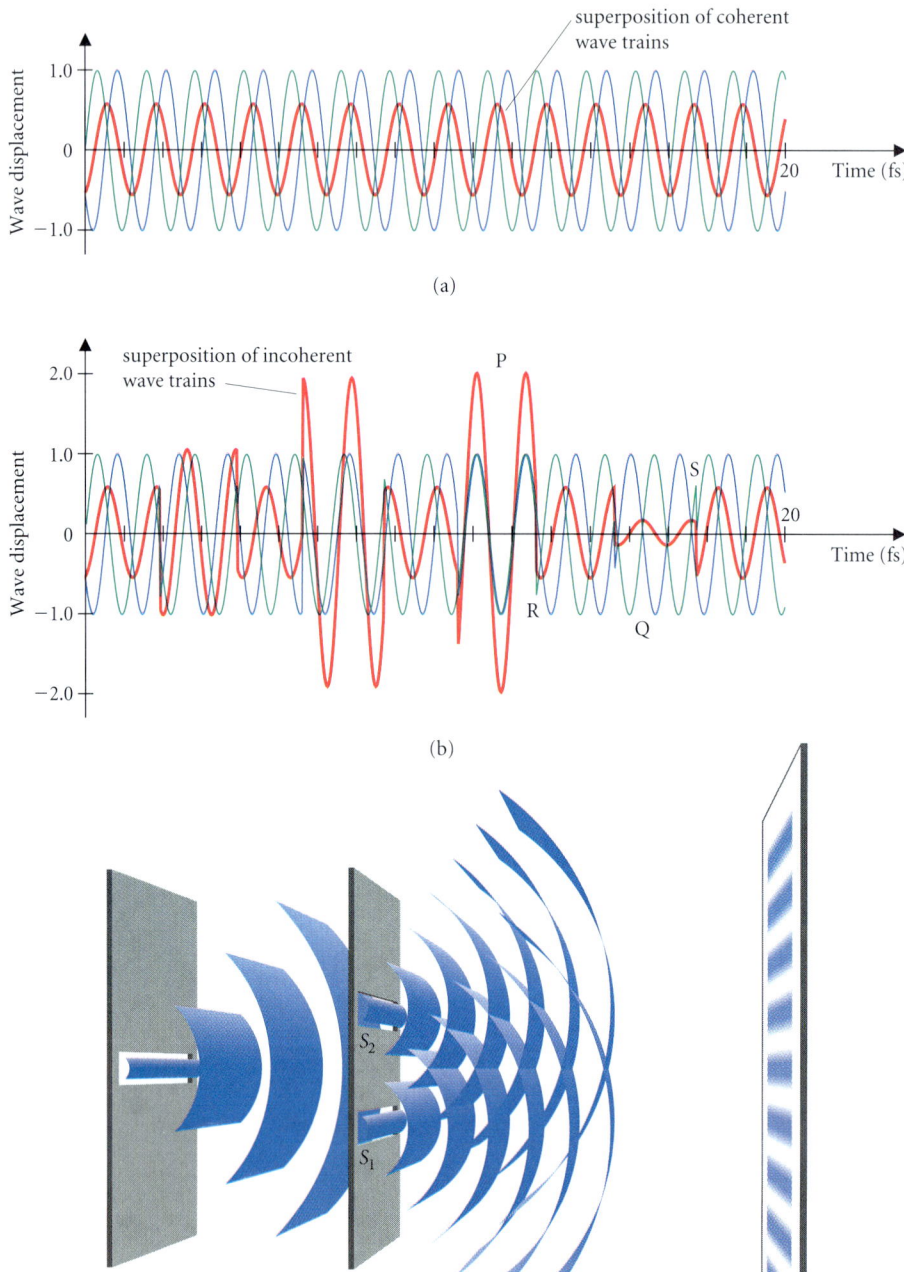

■ **FIGURE 17.9**
(a) When two coherent light sources of the same frequency are superposed, the amplitude of the superposition remains constant. (b) When incoherent wave trains superpose, the interference pattern changes so rapidly that it is not visible. The interference is constructive at time P and destructive at Q. Note the abrupt phase changes in the green curve at points R and S. The average time between such phase shifts is the coherence time of the wave.

■ **FIGURE 17.10**
Thomas Young created mutually coherent sources by using two equally spaced pinholes to split the wave fronts emerging from a single pinhole source. Here we illustrate the idea for waves passing through slits. The phase of the wave reaching the first hole changes randomly in time, but the phases of the waves emerging from the two holes always bear the same relation to each other.

first pinhole is small enough that phase variation across it is negligible. The phases of the light passing through each of the subsequent pinholes have the same relation to the phase at the first pinhole. These two phases change randomly with each new pulse but are the same at any particular time. The two sources are *mutually coherent,* provided that the path difference between the observation point and the two holes is less than the coherence length. Young's device is called a *wave-front–splitting* interferometer; it produces mutually coherent beams by splitting wave fronts from a single source into two separate sources with equal phase.

A laser is used in modern versions of Young's experiment. The coherence length of laser light is sufficiently long that the first pinhole is unnecessary.

17.1.5 Thin-film Interference

Interference also occurs when light passes through, or is reflected from, a thin film of transparent or partially transparent material. This effect is responsible for the colors of soap bubbles and of oil films on puddles after a rainstorm. When light of wavelength λ_o is incident at a small angle (normally incident) on a plane soap film of thickness d (■ Figure 17.11), the beam is partially reflected at the top surface of the film (beam 1) and partially transmitted into the film (beam 2). When the refracted beam (2) reaches the bottom side of the film, it is again partially reflected (beam 3) and partially refracted. Beam 3 then returns to the top side, where some of its energy leaves the film as beam 4. There is no need to follow the beams that reflect inside the film more than once, since they do not alter our conclusions. An observer at point P sees light that is essentially a combination of beams 1 and 4.

Two different effects contribute to the phase difference between the beams:

1. The light making up beam 4 travels twice across the film, a *path length* greater by $2d$ than beam 1.
2. The waves are reflected at the film boundaries, and a phase shift of π occurs whenever a wave reflects from a medium of higher refractive index (slower speed). (We proved this in §15.5.)

The soap film, being mostly water, has a refractive index $n = 1.33$, while air has a refractive index very close to 1. Thus the wave reflecting from the top surface of the film (beam 1) has a phase change of π. The wave reflecting from the bottom surface is reflecting from the air—a medium of lower refractive index than the soap solution—so there is no phase change at this reflection. Comparing the phase of each beam as it leaves the film with its phase as it reaches the film, we have:

$$\text{Beam 1} \quad \Delta\phi_1 = \pi \qquad \text{(due to reflection)},$$
$$\text{Beam 4} \quad \Delta\phi_4 = k_f(2d) = 4\pi d/\lambda_f$$
$$= 4\pi dn/\lambda_o \qquad \text{(due to path difference)}.$$

Here $\lambda_f = \lambda_o/n$ is the wavelength of light in the film.

The net phase shift between beams 1 and 4 is:

$$\Delta\phi_{net} = \Delta\phi_4 - \Delta\phi_1 = \pi(4nd/\lambda_o - 1).$$

Constructive interference occurs when this phase shift equals an integer times 2π:

$$\pi(4nd/\lambda_o - 1) = 2m\pi$$

or

$$2d = (2m+1)\lambda_o/2n = (m + \tfrac{1}{2})\lambda_f,$$

where $m = 0, 1, 2, \ldots$. The path difference $2d$ is an odd number of half wavelengths λ_f in the film.

Thin-film interference is limited by coherence. A film as thick as 1 mm does not show interference in daylight because the extra distance traveled by beam 4 exceeds the coherence length of the light and the beams have a random phase difference.

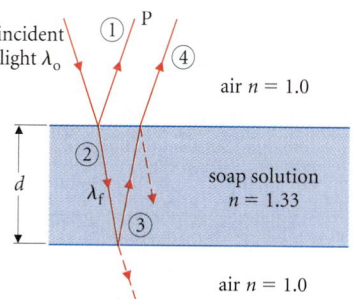

■ **Figure 17.11**
Thin-film viewed by reflection. Light incident on a boundary between two different materials is partly reflected and partly transmitted. Interference occurs between light that reflects at each side of the film (provided that the path difference $2d$ is less than the coherence length).

In Problem 85 you are asked to follow additional beams to prove this result.

See §16.2.2.

When a lens is placed on a glass plate and illuminated from above, a series of concentric light and dark bands are seen. These are known as Newton's rings. They are formed by interference between light reflected from the bottom of the lens and from the top of the plate.

Solution plan for thin film interference

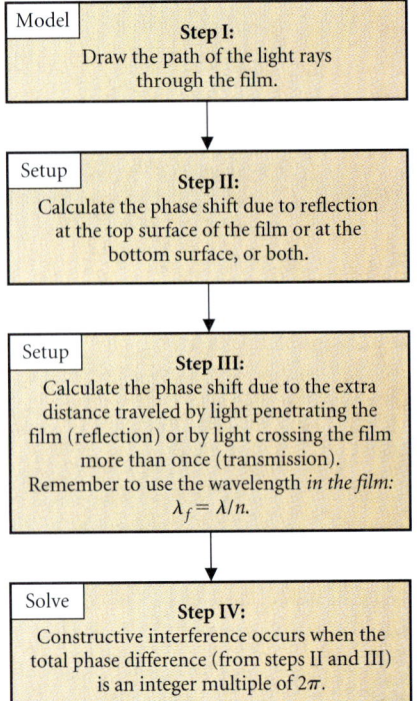

Model — **Step I:** Draw the path of the light rays through the film.

Setup — **Step II:** Calculate the phase shift due to reflection at the top surface of the film or at the bottom surface, or both.

Setup — **Step III:** Calculate the phase shift due to the extra distance traveled by light penetrating the film (reflection) or by light crossing the film more than once (transmission). Remember to use the wavelength *in the film:* $\lambda_f = \lambda/n$.

Solve — **Step IV:** Constructive interference occurs when the total phase difference (from steps II and III) is an integer multiple of 2π.

■ **FIGURE 17.12**

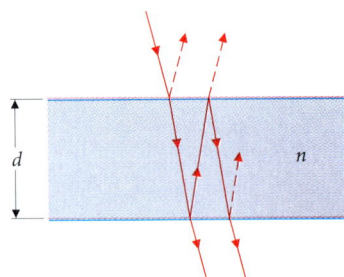

■ **FIGURE 17.13**
Thin-film viewed by transmission. Light that passes through the film is a superposition of light that passes directly through and light that bounces back and forth inside the film before exiting.

LIKE THE INPUT DATA, ANSWERS ARE GIVEN TO TWO SIGNIFICANT FIGURES.

In working thin-film interference problems, it is important to remember all the factors that contribute to the phase difference and to construct an appropriate formula for each problem from first principles (■ Figure 17.12).

EXAMPLE 17.4 ◆◆ White light is incident normally from air on a soap film of thickness $d = 4.4 \times 10^{-4}$ mm. What color does the film appear when reflected light is viewed?

MODEL When light is incident normally on the film, the extra distance traveled by the light reflected at the far side is $2d$. Remembering the phase change due to reflection at the near side, constructive interference occurs if this distance is an odd number of half wavelengths.

SETUP For constructive interference:
$$2d = (m + \tfrac{1}{2})\lambda_f = (m + \tfrac{1}{2})\lambda/n \quad \text{or} \quad \lambda = 2nd/(m + \tfrac{1}{2}).$$

The refractive index of the soap solution is 1.33.

SOLVE The wavelengths that appear bright (constructive interference) are:
$$\lambda = \frac{2(1.33)(4.4 \times 10^{-4} \text{ mm})}{m + \tfrac{1}{2}}$$
$$= \frac{1.17 \times 10^{-3} \text{ mm}}{m + \tfrac{1}{2}} = \frac{1170 \text{ nm}}{m + \tfrac{1}{2}}$$
$$= 2300 \text{ nm},\ 780 \text{ nm},\ 470 \text{ nm},\ 330 \text{ nm},\ \ldots.$$

Of these, only the light around 470 nm is visible, so the film appears blue-green.

ANALYZE Destructive interference occurs where $\lambda = 2dn/m$—that is, at 1200 nm, 590 nm, 390 nm, and so on. The absence of yellow light around 590 nm enhances the bluish appearance of the film. Because the intensity of the different beams differs, complete destructive interference does not occur. However, interference modulates the intensity of the superposition enough to give the film a distinct color. ■

EXAMPLE 17.5 ◆◆ What is the apparent color of the film in Example 17.4 when viewing light passing through it?

MODEL In this case, we consider interference between light passing directly through the film and light that is reflected once at each side before emerging (■ Figure 17.13).

1. Each reflection occurs at a boundary with air, a medium of lower refractive index (higher light speed) than soap solution. Thus there are no phase changes due to reflection.
2. The reflected beam travels an extra distance $2d$ and has a phase difference $\Delta\phi = k_f(2d) = 4\pi d/\lambda_f = 4\pi dn/\lambda$.

SETUP Constructive interference occurs for $\Delta\phi = m(2\pi)$:
$$2d = m(\lambda/n) \quad \text{or} \quad \lambda = 2nd/m.$$

SOLVE Putting in the numbers with $m = 1, 2, 3, \ldots$ gives:
$$\lambda = 2(1.33)\left(\frac{4.4 \times 10^{-4} \text{ mm}}{m}\right) = \frac{1170 \text{ nm}}{m},$$
$$= 1200 \text{ nm},\ 590 \text{ nm},\ 390 \text{ nm},\ 290 \text{ nm},\ \ldots.$$

Of these, the 590-nm light appears yellow-orange.

■ FIGURE 17.14
The same film viewed in reflected (left) and transmitted (right) light shows that the colored bands are in complementary colors.

ANALYZE The maximum at 390 nm could be wide enough to contribute some violet light as well. Compare these results with Example 17.4. The transmitted maxima coincide with the reflected minima. A film viewed by both reflected and transmitted light is shown in ■ Figure 17.14. ■

17.2 THE MICHELSON INTERFEROMETER AND THE MICHELSON–MORLEY EXPERIMENT

Equipment designed to exhibit interference between mutually coherent sources is called an *interferometer*. The Michelson interferometer is used in high-resolution spectrometers for both laboratory and astronomical work, and for accurate measurement of lengths. It achieves coherence through *amplitude splitting*: the light beam is divided by a beam-splitter or half-silvered mirror, which reflects half of the light and transmits half. The two half-beams are recombined after traversing two different paths. The difference in path length must be less than the coherence length of the light source.

■ Figure 17.15 is a schematic of the interferometer. Light is incident from the left onto the beam-splitter B. Half is transmitted, travels to mirror M_1, and is reflected back to the beam-splitter. There half is transmitted and half is reflected to the detector. Meanwhile, the other half of the incident light travels from B to mirror M_2 and is also reflected back to the beam-splitter, where half is reflected and half is transmitted through to the detector. The two light beams are combined and form an interference pattern at the detector.

Since the speed of light is less in glass than in air, the reflection of the beam from glass causes a phase change of π. Each beam is reflected twice (once at the beam-splitter and once at M_1 or M_2) between entering the device from the left and exiting to the detector, and so changes phase by 2π. The reflections produce no net phase *difference* between the beams.

The beam transmitted to the right through the beam-splitter passes through the glass plate, where its wavelength is shorter. This would contribute a phase difference were it not for the compensating plate placed in the other beam. Both beams pass through the same length of glass.

Since each beam travels the distance from the beam-splitter to the mirror twice, any phase difference between the two beams is due to the difference in path length $2(d_1 - d_2)$ they travel through the air. Constructive interference occurs when the path difference between the beams is equal to a whole number of wavelengths: $2(d_1 - d_2) = m\lambda$, or $d_1 - d_2 = m\lambda/2$. Similarly, for destructive interference $d_1 - d_2 = (2m + 1)\lambda/4$.

In 1887, Albert Michelson and his collaborator Edward Morley used his interferometer in an historically important experiment. Early theories of light waves were based on the prem-

THE MICHELSON INTERFEROMETER IS NAMED AFTER ITS INVENTOR, ALBERT MICHELSON (1852–1931).

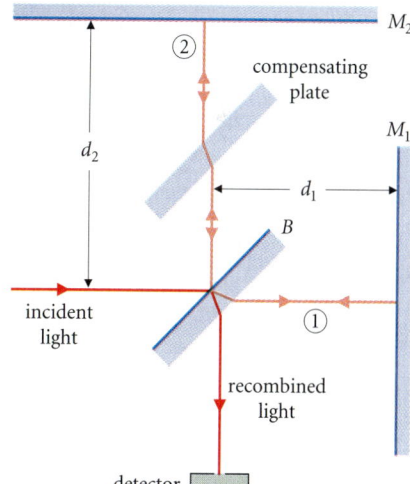

■ FIGURE 17.15
The Michelson interferometer creates coherent beams by splitting the amplitude at the beam-splitter B. The compensating plate ensures that each beam travels through the same length of glass. Each beam reflects once at the beam-splitter and once at either M_1 or M_2, so the phase difference is due solely to the small path difference $d_2 - d_1$.

■ FIGURE 17.16
The Michelson–Morley experiment. The mirrors were mounted on a sandstone slab floating in a trough of mercury.

ise that light, like other waves, moves through some medium. This medium, which was otherwise undetectable, was given the name "luminiferous ether." Michelson and Morley realized that their interferometer could detect motion through an ether and designed an experiment to look for it (■ Figure 17.16). To illustrate their experiment, we discuss how a similar instrument would work with sound waves.

EXAMPLE 17.6 ♦♦♦ A Michelson interferometer designed to use sound waves with frequency $f = 5.0$ kHz has arms of length $d = 1.0$ m (■ Figure 17.17). When a wind blows parallel to one of the arms, a difference $\Delta d = 0.50$ mm in the lengths of the arms is required to maintain complete constructive interference. What is the wind speed v?

MODEL In the reference frame moving with the wind, the interferometer moves with speed v. We need to calculate the distance traveled by each sound beam after it leaves the beam-splitter.

SETUP In Figure 17.17b the sound wave traveling parallel to the direction of wind flow leaves the beam-splitter at time $t = 0$ and reaches the reflector at time t_1. The reflector moves to the right during this interval by an amount vt_1, so the actual distance s_1 traveled by the sound is $s_1 = d + vt_1$. But $t_1 = s_1/v_s$, so $s_1 = d + (v/v_s)s_1$, and therefore:

$$s_1 = \frac{d}{1 - v/v_s}.$$

On the return trip, the beam-splitter is moving toward the sound, which travels the shorter distance:

$$s_2 = \frac{d}{1 + v/v_s}.$$

The sound wave that travels perpendicular to the wind flow must leave the beam-splitter aimed at the position the mirror will occupy at the time of the reflection (Figure 17.17c). The distance it travels to reach M_2 is:

$$s_3 = \sqrt{d^2 + v^2 t_3^2},$$

where $t_3 = s_3/v_s$ is the time required for sound to travel the distance s_3. Substituting for t_3, we obtain an equation for s_3, which has the solution:

$$s_3 = \frac{d}{\sqrt{1 - v^2/v_s^2}}.$$

The time and distance for the return trip are equal to those for the outgoing trip.

The distance traveled by the sound parallel to the wind flow differs from the distance traveled perpendicular to the wind flow by an amount Δs,

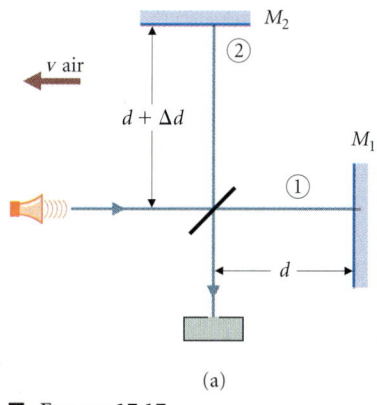
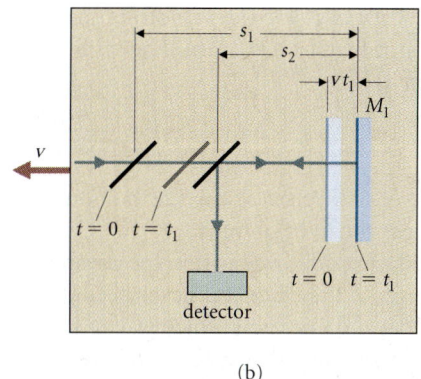
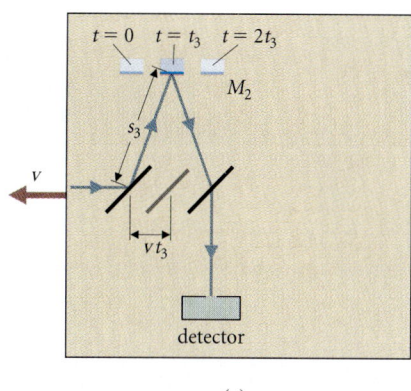

■ **FIGURE 17.17**
(a) A Michelson interferometer for sound illustrates the theory behind the ether-drift experiment. The wind is blowing to the left at speed v. A metal plate with holes in it makes a good beam-splitter for sound.

(b) In the frame moving with the wind, the whole interferometer moves to the right at speed v. Sound waves traveling back and forth parallel to the wind (beam 1) reflect from the mirror M_1 at time t_1 and from the beam-splitter at time t_2. They travel a distance $s_1 + s_2$.

(c) Waves moving perpendicular to the wind (beam 2) reflect from the beam-splitter at time $t = 0$ and from M_2 at time t_3. They travel a distance $2s_3$.

where
$$\Delta s = s_1 + s_2 - 2s_3$$
$$= \frac{d}{1 - v/v_s} + \frac{d}{1 + v/v_s} - \frac{2d}{\sqrt{1 - v^2/v_s^2}}$$
$$= 2d\left(\frac{1}{1 - v^2/v_s^2} - \frac{1}{\sqrt{1 - v^2/v_s^2}}\right).$$

Since v/v_s is small, we may expand these expressions, keeping terms up to the order of $(v/v_s)^2$.

$$\Delta s = 2d\left[\left(1 + \frac{v^2}{v_s^2}\right) - \left(1 + \frac{v^2}{2v_s^2}\right)\right] = d\frac{v^2}{v_s^2}.$$

REMEMBER: $1/(1 - x) = 1 + x + \cdots$ AND $1/\sqrt{1 - x} = (1 - x)^{-1/2} = 1 + \frac{1}{2}x + \cdots$. (APPENDIX IB)

We expect $v \ll v_s$, and so $\Delta s \ll d$.

A difference Δs in the distances traveled by the two sound beams produces a phase shift $k\,\Delta s$ between them.

$$\Delta\phi = 2\pi\,\Delta s/\lambda = 2\pi dv^2/(\lambda v_s^2). \tag{i}$$

SOLVE To maintain complete constructive interference, the path difference is eliminated by shortening one arm an amount $\Delta d = \Delta s/2$. With $\Delta d = 0.50$ mm, the inferred wind speed is:

$$v = v_s\sqrt{2\,\Delta d/d} = (340\text{ m/s})\sqrt{2(5.0 \times 10^{-4}\text{ m})/(1.0\text{ m})} = 11\text{ m/s}.$$

ANALYZE The phase shift is proportional to the square of the air speed (eqn. i). Notice that the air speed is obtained without detecting the air itself, but simply by detecting interference of the sound waves. ■

A device that measures the speed of a medium without detecting the medium itself is exactly what was needed to test the ether theory. Michelson and Morley reasoned that if an ether existed, it would have some velocity relative to the Earth. For example, if the ether were trapped in the solar system, then the Earth would move through it at approximately the Earth's orbital speed. That motion should have the same effect on light waves as the motion of wind on the sound waves in the preceding example. If there is no ether, the speed of light does not depend on its direction, and so there is no phase shift.

Let's calculate the size of the shift they might have expected to see. Their apparatus had an arm length $d = 11$ m. For visible light of wavelength 590 nm, the expected phase shift (eqn. i) is:

$$\Delta\phi = 2\pi\left(\frac{11\text{ m}}{5.9 \times 10^{-7}\text{ m}}\right)\left(\frac{v^2}{c^2}\right) = (1.2 \times 10^8)\left(\frac{v^2}{c^2}\right).$$

The orbital speed of the Earth about the Sun is approximately 3×10^4 m/s, so we might expect v/c to be about $(3 \times 10^4)/(3 \times 10^8) = 10^{-4}$. Then the phase shift would be about 1 rad, small but well within the capabilities of the experiment.

Since the ether had never been detected, its motion was not known, and Michelson and Morley did not know which way to set up the interferometer. They overcame this problem by slowly rotating their entire apparatus: the phase shift would be maximum with the two beams traveling along and perpendicular to the ether flow. They could see no phase shift for any orientation of the apparatus, even when they repeated the experiment at different times of day and in different months. The failure to find any phase shift led to the demise of the ether theory and helped people accept Einstein's later theory of a constant speed of light. Note that we used the ether theory to predict the outcome of the experiment, but since the prediction failed, we conclude that the theory is incorrect!

EINSTEIN PUBLISHED HIS THEORY IN 1905.

17.3 DIFFRACTION

When a pencil lead is placed in a light beam, it casts a shadow (■ Figure 17.18). But the center of the shadow is not completely dark, and dark stripes appear in the light on either side of the shadow. This *diffraction* pattern results from the interference of waves interrupted by the pencil lead.

Diffraction describes the behavior of wave fronts when they encounter and propagate past obstacles. The obstacle may be a sharp edge, an opening in an otherwise infinite barrier, or a large number of small objects. According to Huygens' principle, a wave emerging from an aperture may be considered as the sum of many interfering wavelets emanating from sources in the aperture. Thus diffraction is not a separate physical phenomenon, but a particular form of interference.

SEE §16.5.2 FOR HUYGENS' PRINCIPLE.

When light passes through a square opening of side a, an observer close to the aperture sees a rather complicated pattern—called the *Fresnel* diffraction pattern—due to the superposition of many waves of differing phase and amplitude (■ Figure 17.19). When the observer is at a large distance D from the aperture, the wavelets have equal amplitude but different phase and give rise to an orderly array of light and dark bands called the *Fraunhofer* diffraction pattern (■ Figure 17.20). Roughly speaking, the Fraunhofer limit requires the observation

AFTER AUGUSTINE FRESNEL (1788–1827).

AFTER JOSEPH FRAUNHOFER (1787–1826).

■ FIGURE 17.18
The shadow of a pencil lead shows a diffraction pattern. Light and dark bands extend outward from the edge of the shadow. Notice that the center of the shadow is not completely dark!

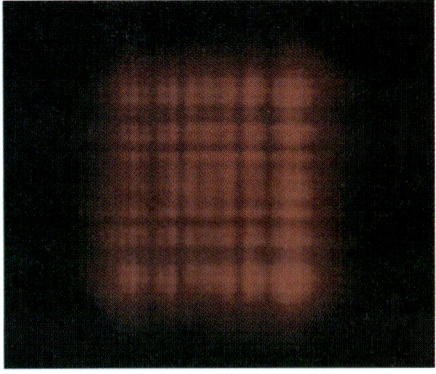

■ FIGURE 17.19
The Fresnel diffraction pattern for a square aperture is a complex mosaic of light and dark regions.

■ FIGURE 17.20
The Fraunhofer diffraction pattern visible at large distances from the aperture is a simpler sequence of light and dark spots.

point to be far enough away that rays reaching it from any part of the aperture are essentially parallel. The appropriate lengths for comparison are the wavelength of the light λ and the size of the aperture a. It turns out that Fraunhofer diffraction occurs when the distance D satisfies two constraints: $D \gg \lambda$ and $D \gg a^2/\lambda$.

17.3.1 The Rectangular Aperture

Because they involve interference among all the tiny wavelets envisioned by Huygens' principle, diffraction calculations tend to be complicated. For the moment, we shall consider only one analytic result that we can obtain from a qualitative argument—the locations of minimum intensity in the Fraunhofer diffraction pattern for a long, rectangular aperture with width a and length $\ell \gg \lambda$.

BUT SEE §17.4.

Light of wavelength λ diffracted by the aperture falls on a screen, a distance $D \gg a^2/\lambda$ from the slit. We imagine the aperture to be divided into a large number of elementary slits of width dw (■ Figure 17.21). We already know how to calculate interference between pairs of these slits (§17.1.3). If we can choose the pairs so that every pair interferes destructively, then we have found a diffraction minimum.

For every pair of slits to interfere destructively, they must all have the same separation. One way to achieve this is to divide the aperture in half and choose one slit from each half (Figure 17.21) so that each pair is separated by $a/2$. At a point P on the y-axis of the screen and at angle θ from the normal to the aperture, the path difference between waves from two such slits is $(a/2)\sin\theta$. The waves interfere destructively when the path difference is an odd number of half wavelengths. Thus a diffraction minimum occurs whenever:

COMPARE EQN. (17.6).

$$\frac{a}{2}\sin\theta = \frac{2m+1}{2}\lambda \quad \text{or} \quad \sin\theta = (2m+1)\frac{\lambda}{a}, \quad m = 0, 1, 2, \ldots. \quad \text{(i)}$$

NOTE: $\sin\theta$ EQUALS ANY ODD INTEGER TIMES (λ/a).

For any wave leaving the top half of the aperture at one of these angles, there is another wave from the bottom half, displaced by exactly one-half a slit width, which interferes with it destructively.

But we haven't yet found all the minima. If we divide the aperture into quarters, we can choose pairs of slits with separation $d = a/4$ (■ Figure 17.22). Waves from these pairs interfere destructively when $(a/4)\sin\theta = (2m+1)\lambda/2$, or:

$$\sin\theta = 2(2m+1)\left(\frac{\lambda}{a}\right). \quad \text{(ii)}$$

HERE $\sin\theta = 2, 6, 10, \ldots$ TIMES (λ/a).

Then any wave leaving the top quarter of the slit interferes destructively with a wave leaving the second quarter, and any wave leaving the third quarter interferes destructively with one

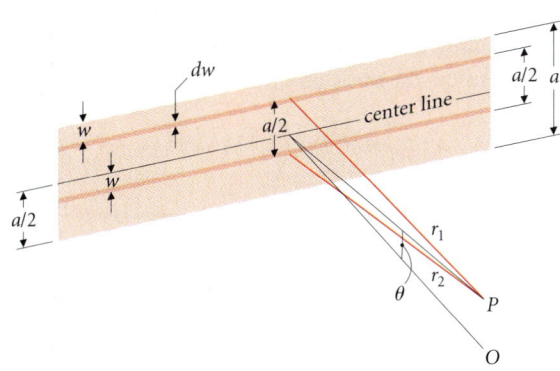

■ FIGURE 17.21
Diffraction through a long rectangular aperture of width a. We divide the aperture up into elementary slits of width dw. By superposing light from pairs of slits separated by distance $a/2$, located distance w below the top of the slit and the center line of the slit, respectively, we can find the positions of some of the diffraction minima. One such pair of slits is shown here.

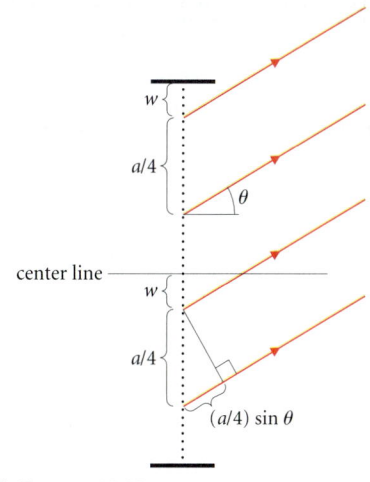

■ FIGURE 17.22
Dividing the slit into quarters and superposing light from elementary slits in each quarter allows us to find minima for which slit pairs separated by $a/4$ interfere destructively.

SECTION 17.3 • DIFFRACTION

HERE $\sin \theta = 4, 12, 20 \ldots$ TIMES (λ/a).

from the fourth quarter. Again, the net intensity is zero. If we divide the slit into eighths, we find still more minima when:

$$\sin \theta = 4(2m + 1)\lambda/a, \quad m = 1, 2, 3 \ldots \quad \text{(iii)}$$

Each time we divide the aperture more finely, we locate more diffraction minima, until every nonzero integer appears in the list:

$$\sin \theta = m\frac{\lambda}{a}, \quad m = 1, 2, 3, \ldots . \quad (17.7)$$

The observer sees a series of bright and dark bands. The brightest points are almost, but not exactly, halfway between the dark bands. The central spot O is bright.

EXAMPLE 17.7 ♦♦ Find the diffraction minima for light of wavelength $\lambda = 589$ nm shining through a slit 2.00 μm wide. What happens if infrared radiation with wavelength 2.00 μm is used?

MODEL The slit is assumed to be very long in its other dimension. We are to find the angles of the minima.

SETUP From eqn. (17.7), minima occur for:

$$\sin \theta = \frac{m\lambda}{a} = m\left(\frac{589 \text{ nm}}{2000 \text{ nm}}\right) = 0.294, 0.589, 0.884.$$

For $m > 3$, the equation gives a value greater than 1 for $\sin \theta$, which does not correspond to a real angle.

SOLVE Thus there are three minima at $\theta = 17.1°$, $36.1°$, and $62.1°$.
If $\lambda = 2$ μm, then $\sin \theta = 1$ for $m = 1$. The first minimum occurs at $\theta = 90°$, so there are no minima on the screen.

ANALYZE With $\lambda = 2$ μm, the wavelength equals the slit size. Whenever $a \leq \lambda$, there are no minima on the screen. ∎

17.3.2 The Circular Aperture

■ **FIGURE 17.23**
The diffraction pattern of a circular aperture shows many of the same features as a rectangular aperture: a broad, bright central maximum is surrounded by dark minima and secondary maxima.

NOTE THE SIMILARITY TO EQN. (17.7) WITH $m = 1$. THE ONLY DIFFERENCE IS THE NUMERICAL FACTOR 1.22.

Circular apertures are common in telescopes, microscopes, binoculars, and eyes! Diffraction through these apertures limits the performance of optical instruments. The intensity distribution for a circular aperture has a broad, intense central peak surrounded by less intense secondary maxima (■ Figure 17.23) and is qualitatively similar to the pattern for the rectangular aperture. The differences occur because wavelets that differ by π in phase do not have the same amplitude. The mathematics required for an exact calculation are beyond the scope of this book. One useful result is the position of the first minimum:

$$\sin \theta = 1.22\lambda/a, \quad (17.8)$$

where a is the diameter of the aperture.

17.3.3 Resolution

Diffraction limits our ability to distinguish two different sources of waves—stars, for example—which are observed through the same aperture. After the waves pass through the aperture, each source is represented by a diffraction pattern. The patterns of nearby sources merge (■ Figure 17.24), and it is difficult to distinguish them. As the separation of the sources increases, the central peaks of the diffraction patterns also separate. The minimum separation necessary to distinguish the sources depends on the techniques used—our eyes or a computer program, for example. We may obtain a useful estimate using a criterion proposed by Lord Rayleigh (1842–1919).

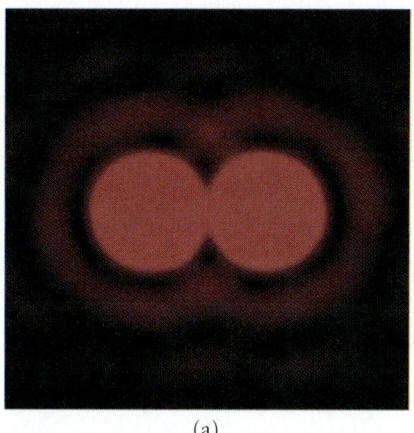
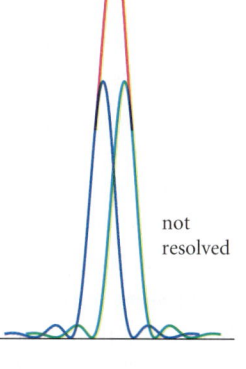
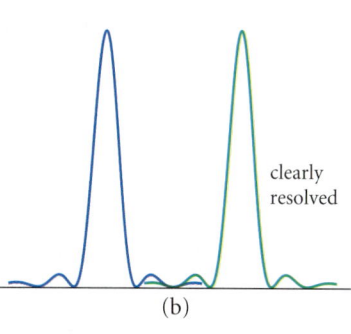
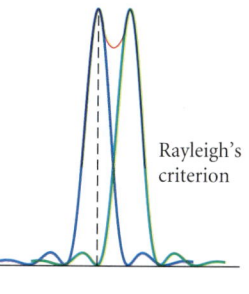

■ **FIGURE 17.24**
(a) Two closely spaced sources viewed through the same aperture have overlapping diffraction patterns that separate as the sources separate. (b) Two sources can be distinguished if they are separated by an angular distance greater than or equal to the distance from the center of the pattern to the first diffraction minimum (Rayleigh's criterion).

> *Rayleigh's criterion:* Sources can be *resolved* if they have an angular separation greater than or equal to that between the central diffraction peak and the first minimum of each source's diffraction pattern.

The angular resolution of a telescope is the angular separation of two sources it can just resolve. According to Rayleigh's criterion, it is given by eqn. (17.8).

EXAMPLE 17.8 ♦♦ At night, the human eye has an aperture with a diameter of about 8 mm. (It varies with the brightness of the environment.) If the headlights on a Honda are separated by 1.18 m, how far away can you distinguish the two lights?

MODEL According to Rayleigh's criterion, we can distinguish the lights if their angular separation is greater than about $1.22\lambda/a$. Here, a is the 8-mm diameter of the pupil. The wavelength λ is in the visible part of the spectrum, say about 600 nm.

SETUP Then:
$$\theta_{min} = \frac{(1.22)(600 \times 10^{-9} \text{ m})}{8 \times 10^{-3} \text{ m}} = 9 \times 10^{-5}.$$

The headlights are separated by $d = 1.18$ m, so their angular separation when they are a distance D away is $\theta = d/D$. Thus we require:

SOLVE $\quad \dfrac{1.18 \text{ m}}{D} > 9 \times 10^{-5} \quad$ or $\quad D < \dfrac{1.18 \text{ m}}{9 \times 10^{-5}} = 10$ km.

ANALYZE The eye is efficiently designed. Cones in the fovea are separated by a distance that corresponds to the diffraction limit. Of course, other flaws such as those in focusing (Chapter 18) may prove more limiting for any individual eye. ■

REMEMBER: WHEN θ IS SMALL, $\sin \theta \approx \theta$.

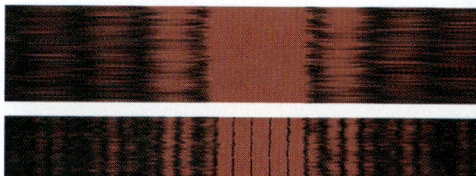

■ **FIGURE 17.25**
The influence of diffraction on double-slit interference. The top panel shows the diffraction pattern for a single slit with $a \approx d/3$. The lower panel shows the result when two such slits are illuminated together. The narrow lines are due to interference between the two slits. The broad bands show the effect of diffraction on the amplitude of waves from each slit.

WE SAY THAT THE DIFFRACTION PATTERN "MODULATES" THE INTERFERENCE PATTERN.

17.3.4 Combined Diffraction and Interference: The Double Slit

When we discussed double-slit interference, we ignored diffraction effects. We may do this only if the slit size a is much less than the wavelength of light. Otherwise diffraction does not spread the light uniformly in angle. If this condition is not met, diffraction strongly influences the wave amplitude from each slit, and the intensities of the interference maxima vary accordingly (■ Figure 17.25).

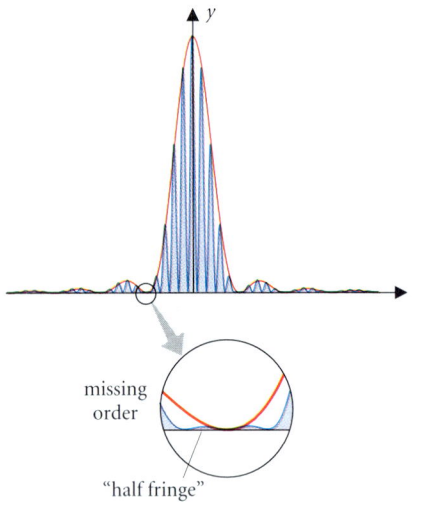

FIGURE 17.26
Intensity pattern in double-slit interference with $d = 5a$.

EXAMPLE 17.9 ♦♦ Two slits of width $a = 2.00$ μm and separation $d = 10.0$ μm are illuminated by light of wavelength $\lambda = 589$ nm. Describe the intensity pattern.

MODEL The light intensity is modulated by diffraction within a single slit: it is the diffraction patterns that superpose and interfere.

SETUP The diffraction minima for one such slit occur at $\sin\theta = m\lambda/a = m(0.294) = 0.294, 0.589$, and 0.884—that is, at $\theta = 17.1°, 36.1°$, and $62.1°$ (cf. Example 17.7). The double-slit interference maxima occur at $\sin\theta = n\lambda/d = (5.89 \times 10^{-2})n$: $\theta = 0, 3.38°, 6.77°, 10.2°, 13.6°, 17.1°$, and so on.

SOLVE The fifth interference maximum falls at the first diffraction minimum. There is no light in the diffracted waves in this direction, and so the interference maximum is missing from the pattern (■ Figure 17.26).

ANALYZE If the slit separation is an integer N times the slit width, then every Nth interference maximum is missing. Here $N = 5$. In Figure 17.25 every third fringe is missing, so $a = d/3$.

17.4 INTENSITY IN INTERFERENCE PATTERNS

17.4.1 Energy Redistribution in the Two-Slit Interference Pattern

So far, we have given only qualitative statements about intensity in interference patterns: "bright" or "dark." We begin a quantitative discussion by computing intensity in the interference pattern of two identical sources separated by distance d.

When two equal-amplitude waves with the same frequency superpose, the resulting amplitude is given by eqn. (17.3): $A_r = 2A\cos(\Delta\phi/2)$, where $\Delta\phi$ is the phase difference between the waves at the observation point. The intensity of the superposition is proportional to the square of its amplitude: $I_r \propto A_r^2$. Thus:

$$I_r = 4I_o \cos^2(\Delta\phi/2), \tag{17.9}$$

COMPARE EQNS. (15.12), (16.15), AND (16.16).

where $I_o \propto A^2$ is the intensity of either individual source. In two-slit interference, the phase shift is due to the difference in path length from each slit to a distant observation point P, $\Delta\phi = kd\sin\theta$, and the intensity (eqn. 17.9) is a function of the angle θ (■ Figure 17.27):

$$I(\theta) = 4I_o \cos^2\left(\frac{kd\sin\theta}{2}\right) = 4I_o \cos^2\left(\frac{\pi d\sin\theta}{\lambda}\right). \tag{17.10}$$

RECALL THAT $k = 2\pi/\lambda$.

We may also express the intensity as a function of the distance y of the observation point from the central maximum. When the angle θ is small ($y/D \ll 1$), $\sin\theta \approx \tan\theta = y/D$, and:

$$I(y) = 4I_o \cos^2\left(\frac{\pi y d}{\lambda D}\right).$$

COMPARE WITH *DIGGING DEEPER*: ENERGY REDISTRIBUTION, IN §17.1.2.

The total radiation from the slits is redistributed into a sequence of fringes rather than a uniform brightness. Intensity varies with distance as $\cos^2(\pi y d/\lambda D)$. So, the average intensity at the screen is one-half its maximum value: $I_{av} = \frac{1}{2}(4I_o) = 2I_o$, or, as expected, the sum of the intensities due to each source.

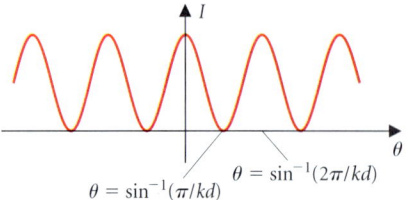

FIGURE 17.27
Intensity distribution in the two-slit interference pattern for very narrow slits. See eqn. (17.10). For small values of kd, the intensity peaks are evenly spaced in angle.

Digging Deeper

CLOSELY SPACED SOURCES

We have assumed so far that the sources of superposed waves continue to put out the same power while emitting waves of the same amplitude, with or without the superposition. But when the sources are very close together (less than about one wavelength), the waves from one source affect the other. When the separation $d \ll \lambda$, according to eqn. (17.10) the intensity is approximately constant, $I(\theta) \approx 4I_o$, and the total power transmitted is four times the power of an isolated source. To contribute a wave of the same amplitude, each source must double its power output.

For sound waves, it is fairly easy to see why this is the case. The power output from a speaker is the product of its velocity and the force acting on it (see §16.3.2). The speaker surface is exposed to its own wave *and* one of equal amplitude and phase from the other source; the pressure is doubled. In a sense, we cannot speak of the component waves as belonging to either source, as both sources do work on each wave! The same effect occurs for light waves for similar reasons, but to demonstrate this claim, we would have to delve deeper into the nature of light than we can do here.

■ Figure 17.28 shows how this effect depends on the separation of the sources when the amplitudes of the sources are held fixed. Notice that for some separations one source is out of phase with the wave from the other and the total power output is reduced rather than increased! This issue did not arise in the case of beats, since two sources of different frequency are alternately in phase and out of phase and their effects on each other average away.

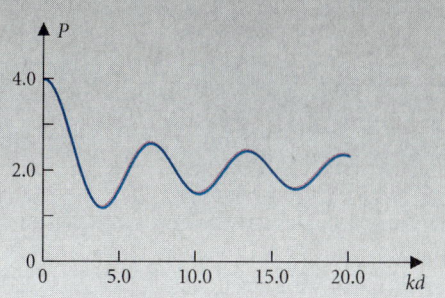

■ **FIGURE 17.28**
When the sources are very close together, they influence each other. The sources must put out more power than isolated sources to generate waves of the same amplitude. At slightly larger separations, less power is needed. As the separation increases, the power required approaches that for two isolated sources.

17.4.2 Phasors

To superpose more than two sources, we need an efficient way to add trigonometric functions with different phases and amplitudes. Here we develop a geometrical technique for understanding wave phase that also allows us to compute the intensity pattern resulting from the superposition of an arbitrary number of waves.

A vector of magnitude A that makes an angle

$$\phi = kx - \omega t + \phi_o$$

with the y-axis (■ Figure 17.29) has y-component:

$$y = A \cos(kx - \omega t + \phi_o),$$

which is equal to the wave displacement of a harmonic wave with amplitude A, propagating in the positive x-direction. The y- and z-directions in this diagram are not real spatial dimensions. The vector's y-component represents a wave displacement that could be the pressure disturbance at some point in a sound wave. The vector's z-component represents nothing and exists only for our convenience in using vectors to represent waves. The vector is called a *phasor*.

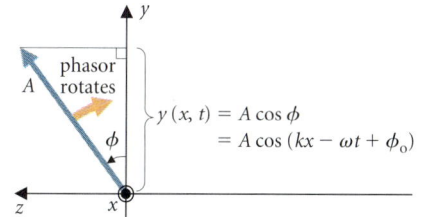

■ **FIGURE 17.29**
A phasor is a vector whose y-component represents the wave displacement. The phasor's length represents the amplitude and its angle from the y-axis, measured counterclockwise, is equal to the wave phase.

> A *phasor* is an abstract vector whose y-component represents the value of a wave disturbance. The vector's magnitude equals the amplitude of the wave, and the angle it makes with the y-axis, measured counterclockwise, is the wave phase.

As the wave phase decreases in time, the phasor rotates clockwise about the origin and its y-component varies between A and $-A$. As the spatial coordinate x increases, the phasor

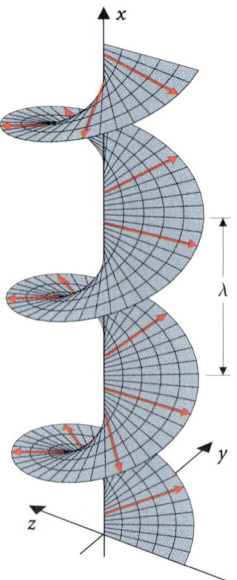

FIGURE 17.30
In a harmonic wave at a fixed time, the tips of the phasors lie along a spiral curve. As time progresses, the spiral moves along the x-axis at the wave speed.

rotates counterclockwise. At a fixed time, the tips of the phasors lie along a spiral curve (Figure 17.30). As time progresses, the spiral moves along the x-axis at the wave speed.

EXAMPLE 17.10 ♦ A plane sound wave is described by the following parameters: amplitude A = 4.0 Pa, wavelength λ = 0.100 m; frequency f = 3.30 kHz; phase constant ϕ_o = 0. Construct phasors to represent the wave disturbance at (a) x = 1.00 m, t = 1.00 ms; and (b) x = 2.00 m, t = 0.

MODEL Each phasor's length represents the amplitude of the wave (4.0 Pa) and the angle it makes with the y-axis is the wave phase.

SETUP The wave is described by the equation:

$$y \equiv \delta P = (4.0 \text{ Pa})\cos\left(\frac{2\pi}{0.100 \text{ m}}x - (2\pi)(3.30 \text{ kHz})t\right).$$

The wave phase at x = 1.00 m and t = 1.00 ms is:

$$\phi = 2\pi\left(\frac{1.00 \text{ m}}{0.100 \text{ m}} - 3.30\right) = 2\pi(6.70) \text{ rad}.$$

Every increment of 2π in phase means that the phasor has rotated around once and again lies along the y-axis. Thus we may ignore integer multiples of 2π: 6.70 × 2π is equivalent to 0.70 × 2π = 1.4π.

SOLVE Phasor (a) is a vector of magnitude 4.0 Pa, which makes an angle 1.4π with the y-axis (Figure 17.31).

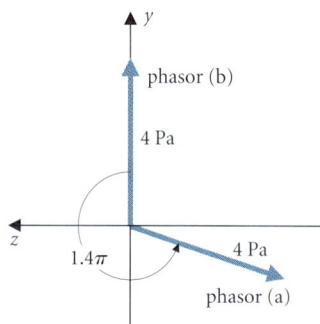

FIGURE 17.31
Phasor (a) represents the sound wave at x = 1.00 m, t = 1.00 ms and phasor (b) represents the sound wave at x = 2.00 m, t = 0.

At x = 2.00 m and t = 0, the wave phase is $\phi = 2\pi[(2.00 \text{ m})/(0.100 \text{ m}) - 0] = 2\pi(20.0)$. Phasor (b) lies along the y-axis as shown in the figure.

ANALYZE The y-components of the two phasors show how the pressure differs between the two positions and times: δP_1 = (4.0 Pa)cos(1.4π) = −1.2 Pa, while δP_2 = 4 Pa. ∎

SEE § 15.2.1 FOR THE DEFINITION OF WAVE FUNCTION.

To superpose two wave disturbances, we add their wave functions. The y-component of a sum of vectors equals the sum of the y-components. Thus superposition of waves corresponds to vector addition of their phasors.

EXAMPLE 17.11 ♦♦ The sound wave in Example 17.10 is combined with a second wave of the same frequency with amplitude 3.0 Pa and phase $\pi/2$ at x = 0.00 and t = 0.00. What sound pressure does the combined wave cause at x = 1.00 m and t = 1.00 ms? Find the intensity of the superposition in terms of the intensity I_o of the first wave.

MODEL The phasor representing the superposition is the vector sum of the phasors representing the two disturbances.

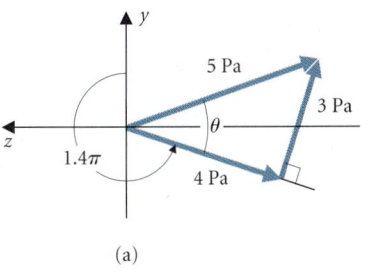
(a)

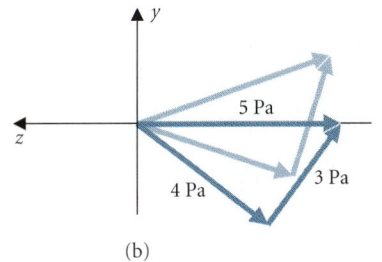
(b)

■ **Figure 17.32**
(a) The superposition of two waves is represented by their phasor sum, here shown at $t = 1.00$ ms, $x = 1.00$ m. (b) Phasor diagram for the same disturbance, one twentieth period later. Since both waves have the same frequency, the entire diagram rotates rigidly, clockwise in time.

SETUP We have already drawn the phasor representing the first wave at $x = 1.00$ m and $t = 1.00$ ms (Figure 17.31a). The phasor representing the second wave, with magnitude 3.0 Pa, has a phase $\pi/2$ greater and so is at right angles to the first.

SOLVE The vector sum of the two phasors (■ Figure 17.32) has magnitude 5.0 Pa. The angle θ in the figure is $\theta = \tan^{-1}(3.0/4.0) = 0.20\pi$. Thus the phase angle of the vector sum is: $\phi = 1.4\pi + 0.20\pi = 1.6\pi$. The amplitude of the sound disturbance at this point is $\frac{5}{4}A_1$, where $A_1 = 4.0$ Pa is the amplitude of the first wave, and so the intensity of the superposition is $I = 25I_o/16$.

ANALYZE Compare this analysis with the derivation of eqn. (17.3). ■

Since both of the phasors in Figure 17.32 have the same frequency, they rotate with time at the same rate. The phase difference between them, represented by the angle between them in the diagram, remains constant (Figure 17.32b). Phasors representing waves of the same frequency maintain a constant geometrical relationship as time passes. This principle applies to the superposition of any number of waves with the same frequency.

17.4.3 Interference of Multiple Sources

If we use three slits in a Young's experiment rather than two, the interference pattern is more complex (■ Figure 17.33). The brightest fringes in the three-slit pattern are at the same position as bright fringes from a two-slit pattern but are brighter and narrower. Secondary bright fringes occur between pairs of primary fringes. We may use phasor diagrams to illustrate this behavior.

Waves leave the three slits in phase but arrive at a distant point P with phase differences due to the different path lengths they traverse (■ Figure 17.34). The phase difference between the waves from S_1 and S_2 is $\Delta\phi = kd \sin\theta$, as in the case of two slits. The same phase difference occurs between the waves from slits S_2 and S_3. In a phasor diagram for the three waves, the

WE ARE ASSUMING HERE THAT EACH SLIT HAS WIDTH $a < \lambda$ SO THAT WE MAY IGNORE DIFFRACTION.

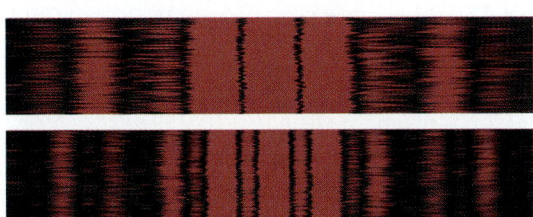

■ **Figure 17.33**
Three-slit interference (bottom panel). Bright maxima occur at the same locations as for two-slit interference (top panel), but they are narrower. In addition, secondary maxima occur between the principal maxima.

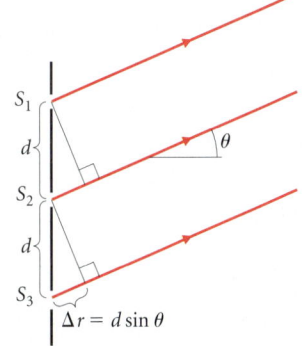

■ **Figure 17.34**
The path difference $d \sin\theta$, and hence the phase difference, between waves from slits S_1 and S_2 equals that between waves from slits S_2 and S_3.

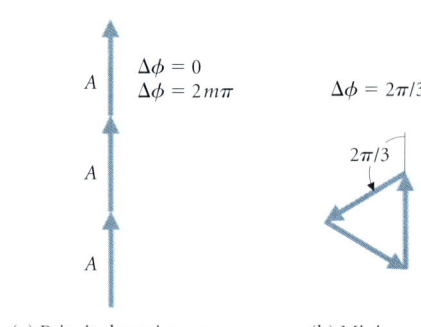 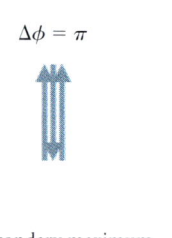 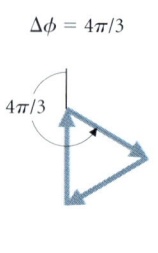 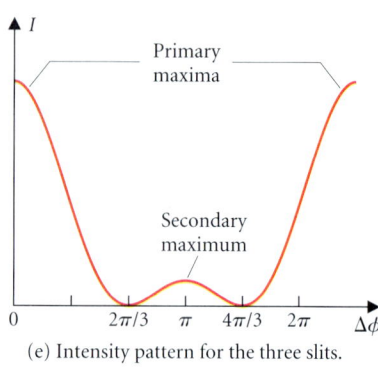

(a) Principal maximum. (b) Minimum. (c) Secondary maximum. (d) Minimum. (e) Intensity pattern for the three slits.

■ **FIGURE 17.35**
Phasor diagram for three-slit interference. (a) Principal maxima (amplitude = $3A$) occur where all three phasors line up: $\Delta\phi = 0$ or $2m\pi$. (b) Minima occur where the three phasors form a closed triangle: $\Delta\phi = 2\pi/3$ or $2m\pi + 2\pi/3$. (c) Secondary maxima occur where the three phasors lie on top of each other: $\Delta\phi = \pi$ or $= (2m + 1)\pi$. (d) Additional minima occur for $\Delta\phi = 4\pi/3$ or $4\pi/3 + 2m\pi$. Again, the phasors form an equilateral triangle. (e) The resulting intensity pattern for the three slits. There is one secondary maximum between the two minima.

REMEMBER: LIGHT INTENSITY IS PROPORTIONAL TO THE SQUARE OF THE WAVE AMPLITUDE, SO IT IS THE AMPLITUDE OF THE SUPERPOSITION THAT IS IMPORTANT IN DETERMINING THE INTERFERENCE PATTERN. AT A WAVE FREQUENCY $\sim 10^{15}$ Hz, OUR EYES AVERAGE OVER VERY MANY PERIODS.

COMPARE WITH EQN. (17.6) FOR TWO SLITS: $\sin\theta = (m + 1/2)(\lambda/d)$.

length of the individual arrows represents the amplitude A of the wave from any of the individual slits (■ Figure 17.35). The entire diagram rotates with the frequency of the wave, but the amplitude of the superposition remains constant. For convenience, we draw the diagram at the time when the wave from slit S_1 has phase zero.

The primary maxima occur when the phasors lie along a single line (Figure 17.35a) and the phasor sum has magnitude $3A$. This occurs when the phase difference between adjacent sources is $\Delta\phi = (2\pi d/\lambda)\sin\theta = 2m\pi$, or $\sin\theta = m\lambda/d$. Minima occur where the phasors sum to zero; that is, the three phasors form a closed triangle (Figure 17.35b). This happens for a phase difference $\Delta\phi = 2\pi/3 + 2m\pi$, or $\sin\theta = (m + 1/3)(\lambda/d)$. Since the minima are closer to the maxima, the bright fringes are narrower than in the two-slit pattern. For $\Delta\phi = \pi$, the phasors lie on top of one another (Figure 17.35c) and their sum equals the magnitude of a single phasor: this is a secondary maximum. Another minimum occurs for $\Delta\phi = 4\pi/3 + 2m\pi$ (Figure 17.35d). The resulting intensity pattern is shown in Figure 17.35e. Principal maxima occur with every 2π change in phase, and secondary maxima occur between the minima.

EXAMPLE 17.12 ❖ Use phasor diagrams to find the number of secondary maxima in the interference of five identical, equally spaced sources.

MODEL A secondary maximum is located between each pair of minima. At each minimum, the five phasors add to zero; that is, they form a closed figure.

SETUP To form a closed figure, the total phase shift $5\Delta\phi$ between the first phasor and the last equals an integer m times 2π. This occurs whenever the angle from one phasor to the next is $\Delta\phi = m(2\pi/5) = m(72°)$, as shown in ■ Figure 17.36. For $m = 5$, the phasors line up again: this is the next primary maximum.

SOLVE There are three secondary maxima between the primary maxima.

ANALYZE In general, with N sources, there are $N - 1$ minima and $N - 2$ secondary maxima between each pair of primary maxima. The primary maxima also become narrower as N increases. Roughly speaking, the effect of adding more sources is to distribute their energy into beams increasingly concentrated at the primary maxima. ■

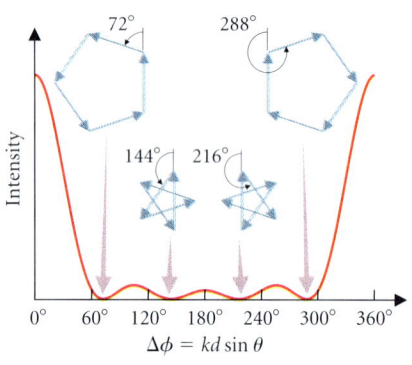

■ **FIGURE 17.36**
Five-slit interference. Phasor diagrams are shown for the four minima, corresponding to $\Delta\phi = m(2\pi/5)$, $m = 1, 2, 3, 4$; $m = 5$ corresponds to the next principal maximum.

EXERCISE 17.5 ❖ Find the positions of maxima and minima for four sources, and verify that there are two secondary maxima between the primary maxima. ■

17.4.4 Gratings

Diffraction gratings used for spectral analysis of light contain a very large number of closely spaced slits. Interference of light from the slits spreads different wavelengths through a wide range of angles and allows us to distinguish spectral features with small differences in wavelength.

For N slits, minima occur when the phase shift between waves from one slit and the next is $\Delta\phi = m(2\pi/N)$, where $m < N$. In the corresponding phasor diagram, N phasors, each with amplitude A, form a closed polygon of perimeter NA. At the first minimum ($m = 1$):

$$\Delta\phi = \frac{2\pi d \sin\theta}{\lambda} = \frac{2\pi}{N} \Rightarrow \sin\theta = \frac{\lambda}{Nd}. \quad (17.11)$$

For large N, θ is very small and the central maximum is very narrow. The intensities of the secondary maxima also decrease rapidly. The first secondary maximum occurs when the phasor chain wraps one and a half times around a "circle" of diameter D (■ Figure 17.37). The polygon approximates a smooth curve of length NA equal to 1.5 times the circumference πD. The amplitude of the phasor sum is $D \approx 2NA/3\pi$, and the intensity relative to the principal maximum (amplitude NA) $\approx 4/(9\pi^2) \approx 4.5\%$. Thus the grating concentrates light of a single wavelength into a few narrow, intense, widely separated peaks.

The mth principal maximum corresponding to wavelength λ is at $\sin\theta_\lambda = m\lambda/d$. The integer m is called the *order* of that maximum. A grating has high angular dispersion if it spreads different wavelengths out into a large range of angles. It has good *resolving power* if we can easily distinguish two closely spaced wavelengths.

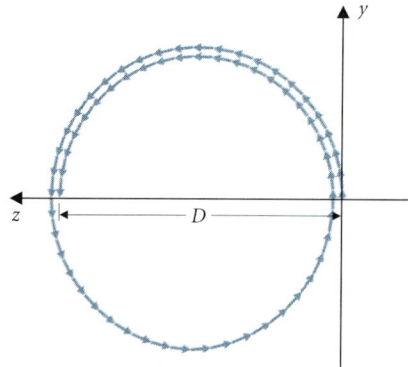

■ **FIGURE 17.37**
Gratings have a very large number of slits. So a chain of phasors representing the superposition of waves from the slits approximates a smooth curve of length NA. At the secondary maximum, that curve wraps one and a half times around a circle of diameter D where $\frac{3}{2}\pi D = NA$. Remember, intensity is proportional to amplitude squared. Since the diagram rotates rapidly, the orientation of the drawing is irrelevant.

EXAMPLE 17.13 ♦♦ A grating 1.00 cm across has 1.00×10^4 slits. It is used to observe the sodium "D lines" at wavelengths $\lambda_1 = 589.00$ nm and $\lambda_2 = 589.59$ nm. At what angle do the D lines appear in first order? If the telescope used to measure the dispersed spectrum can distinguish angles as small as $1.0'$, can the instrument resolve the D lines?

THE LINES WERE NAMED "D LINES" WHEN THEY WERE FIRST DISCOVERED IN THE SOLAR SPECTRUM AND THEIR ORIGIN WAS NOT YET KNOWN.

MODEL The important properties of the grating are the number of slits $N = 10^4$ and the slit separation $d = (1.00 \text{ cm})/(1.00 \times 10^4) = 1.00 \times 10^{-6}$ m. In first order ($m = 1$) the principal maxima occur where $\sin\theta = \lambda/d$. We are looking for small differences $\Delta\theta$ in angle, so we'll need a trigonometric relation from Appendix IB.

$$\sin(\theta + \Delta\theta) = \sin\theta\cos\Delta\theta + \cos\theta\sin\Delta\theta \approx \sin\theta + \Delta\theta\cos\theta.$$

SETUP At $\lambda = 589.00$ nm, the angle of the first principal maximum is given by:

$$\sin\theta = (589 \times 10^{-9} \text{ m})/(1.00 \times 10^{-6} \text{ m}) = 0.589, \quad \text{or} \quad \theta = 36.1°.$$

At $\lambda = 589.59$ nm:

$$\sin\theta + \Delta\theta_1 \cos\theta = \sin(\theta + \Delta\theta_1) = (\lambda + \Delta\lambda)/d.$$

Thus:

$$\Delta\theta_1 = \frac{\Delta\lambda}{d\cos\theta}$$

$$= \frac{0.59 \times 10^{-9} \text{ m}}{(1.00 \times 10^{-6} \text{ m})\cos(36.1°)}$$

$$= 7.3 \times 10^{-4} \text{ rad} = 0.042° = 2.5'.$$

The next minimum occurs where the phase shift has increased by $2\pi/N$, and thus the path length $d\sin\theta$ has increased by λ/N. Thus:

$$\sin\theta + \Delta\theta_2 \cos\theta = \sin(\theta + \Delta\theta_2) = (\lambda + \lambda/N)d,$$

and

$$\Delta\theta_2 = \lambda/(Nd\cos\theta).$$

A white light beam dispersed by a grating. The central maximum is white. The first and second order beams show the increase of θ with wavelength.

SOLVE The grating separates the two lines by 2.5 times the minimum angle the telescope can distinguish. We should also compare the separation of the lines with their width:

$$\frac{\Delta\theta_1}{\Delta\theta_2} = \frac{N\Delta\lambda}{\lambda} = (1.00 \times 10^4)\frac{0.59 \text{ nm}}{589 \text{ nm}} = 10.$$

ANALYZE Since the separation of the lines is ten times their width, the telescope is capable of distinguishing the two lines. Note that N is the number of slits *used*. If the grating is not fully illuminated, N is reduced.

EXERCISE 17.6 ♦ For the grating in Example 17.13, find the angle at which 8.00×10^2 nm occurs in first order, and show that $\theta_{400 \text{ nm}}$ (for $m = 2$) = $\theta_{800 \text{ nm}}$ (for $m = 1$).

For this grating, the 400-nm line in second order falls on top of the 800-nm line in first order. Because orders overlap, gratings must be designed and used carefully. Modern gratings are usually made by ruling grooves on a glass or metal surface. The grooves act as sources of scattered or reflected light. The behavior of the grating is similar for both transmission and reflection. With *blazing*—careful shaping of the rulings—light can be concentrated in the desired order.

17.4.5 Intensity in Diffraction Patterns

In this section we use phasors to find the intensity in the diffraction pattern for the rectangular slit. We divide the slit into a very large number N of pieces (■ Figure 17.38). Each wavelet travels a path $(a/N)\sin\theta$ longer than its neighbor above, giving rise to a phase shift $\Delta\phi = (2\pi/\lambda)(a/N)\sin\theta$ between each wavelet and its neighbor, and a total phase shift $\Delta\Phi_{\text{total}} = (2\pi/\lambda)a\sin\theta$ between the wavelets from the top and the bottom. At the central observation point O ($\theta = 0$), all the phasors line up to give the maximum amplitude $A_{\max}$ (■ Figure 17.39). At any other point P ($\theta > 0$), this same line of phasors is bent around to form a chain AB. Individual phasors have length $A_{\max}/N$, and each makes an angle $\Delta\phi$ with adjacent phasors. In the calculus limit $N \to \infty$, the chain of phasors AB is indistinguishable from a circular arc with radius $R = AD = DB$ and center at D. The line CB lies along the final phasor whose head is at B. Thus the total phase shift across the slit is the angle between the y-axis and line BC. From the quadrilateral $DBCA$, the angle ADB is $\Delta\Phi_{\text{total}}$, triangle ADB is isosceles, and hence the length of the phasor sum (line AB) is $2AE = 2R\sin(\Delta\Phi_{\text{total}}/2)$. The total arc length of the chain is $A_{\max} = R\Delta\Phi_{\text{total}}$, so $R = A_{\max}/\Delta\Phi_{\text{total}}$ and:

$$A(\theta) = 2R\sin\left(\frac{\Delta\Phi_{\text{total}}}{2}\right) = 2A_{\max}\left[\frac{\sin(\Delta\Phi_{\text{total}}/2)}{\Delta\Phi_{\text{total}}}\right].$$

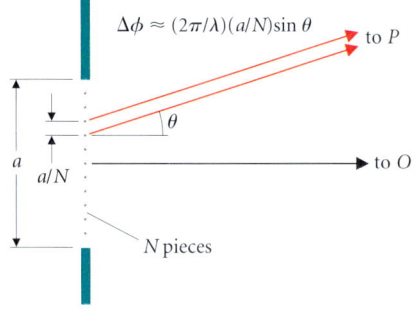

■ **FIGURE 17.38**
Calculation of intensity in diffraction from a rectangular aperture. The aperture is divided into elements of size a/N, each of which produces a wavelet of amplitude $A_{\max}/N$.

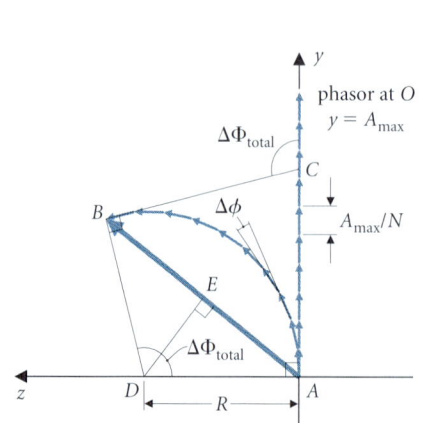

■ **FIGURE 17.39**
Phasor diagram for the wavelets. At the principal maximum O, all the phasors line up and the amplitude is $A_{\max}$. At an off-axis point P (Figure 17.38), the phasor chain is approximately an arc of a circle. Each phasor is displaced by an angle $\Delta\phi$ from its neighbor. Note that $\angle DAC$ and $\angle DBC$ are both right angles, so $\angle BDA = 180° - \angle BCA = \Delta\Phi_{\text{total}}$, the angle between the first and the last phasor. Chord AB represents the amplitude of the superposition.

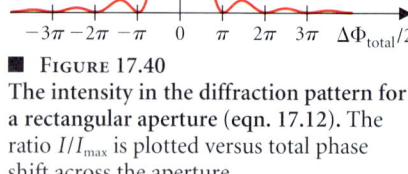

■ **FIGURE 17.40**
The intensity in the diffraction pattern for a rectangular aperture (eqn. 17.12). The ratio $I/I_{\max}$ is plotted versus total phase shift across the aperture.

Therefore:
$$I(\theta) = I_{max}\left[\frac{\sin(\Delta\Phi_{total}/2)}{\Delta\Phi_{total}/2}\right]^2 = I_{max}\left[\frac{\sin\left(\frac{\pi a}{\lambda}\sin\theta\right)}{\frac{\pi a}{\lambda}\sin\theta}\right]^2. \quad (17.12)$$

REMEMBER: $\Delta\phi$ IS IN RADIANS!

This function is illustrated in ■ Figure 17.40.

Near $\theta = 0$, the phase angle $\Delta\Phi_{total}/2 \equiv (\pi a/\lambda)\sin\theta$ is small. Then, since for a small angle $\alpha \ll 1$, $\sin\alpha \approx \alpha$, we have $I(0) = I_{max}$ as we expect from the qualitative discussion of the pattern. The intensity is zero whenever the phasor chain closes on itself—that is, when $\Delta\Phi_{total} = 2m\pi$. The numerator of eqn. (17.12) is zero in this case.

$$\Delta\Phi_{total} = \frac{2\pi a}{\lambda}\sin\theta = 2m\pi \Rightarrow \sin\theta = \frac{m\lambda}{a}, \quad m = 1, 2, 3, \ldots.$$

This is the same result we found in §17.3.

Most of the wave energy is concentrated in the central peak. Other maxima occur approximately where the numerator in eqn. (17.12) equals unity:

$$\frac{\pi a}{\lambda}\sin\theta \approx \frac{(2m+1)\pi}{2} \quad m = \text{integer} \geq 1.$$

(These positions are not exact, and $m = 0$ is not included, because of the effect of the denominator; see Problem 88.) At the second maximum, $\sin\theta \approx 3\lambda/2a$ and the intensity is:

$$I_2 \approx I_{max}\left[\frac{1}{(\pi a/\lambda)(3\lambda/2a)}\right]^2 = I_{max}\left(\frac{4}{9\pi^2}\right) = 0.045 I_{max}.$$

Subsequent peaks have even lower intensities, decreasing as m^{-2}.

The intensity in a two-slit interference pattern is obtained from eqn. (17.10) for the double slit with I_o given by eqn. (17.12) for a single slit (Figure 17.26).

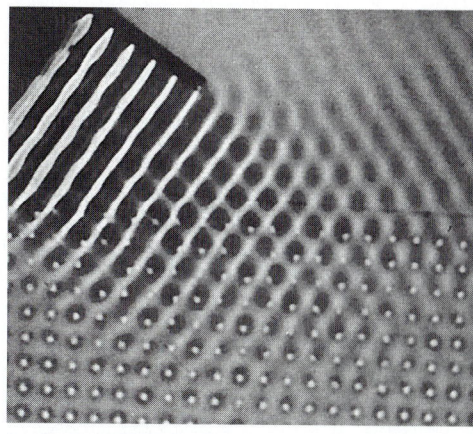

■ FIGURE 17.41
A symmetrical array of pegs scatters water waves at a specific angle. Atoms in crystals scatter light in the same way.

COMPARE THIS RESULT WITH THE INTENSITY OF SUBMAXIMA IN THE INTENSITY PATTERN OF A GRATING (JUST BELOW EQN. 17.11).

17.5 X RAY DIFFRACTION

A regular array of pegs scatters surface water waves at a specific angle (■ Figure 17.41). Wavelets scattered by the obstacles interfere constructively to produce a reflected wave in only one direction. Crystalline solids form such an array of scatterers and act as diffraction gratings for x rays with wavelengths of the order of a few crystal spacings. One of the simplest and historically most important such crystals is sodium chloride, common table salt. The sodium and chlorine atoms form a cubic array where the distance between atoms is about 0.28 nm (■ Figure 17.42).

■ FIGURE 17.42
(a) Crystals of common salt are formed from cubes of side 0.282 nm with sodium and chlorine atoms at the corners. (b) Eight similar cubes stack together to form the unit cell. (c) The unit cells stack together to form a crystal, which may also be viewed as a stack of parallel planes, one of which is shown. (d) A single plane of the crystal viewed from above.

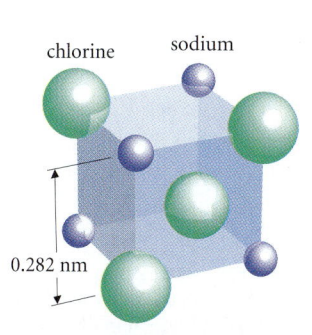

(a) Cubical element of a common salt crystal.

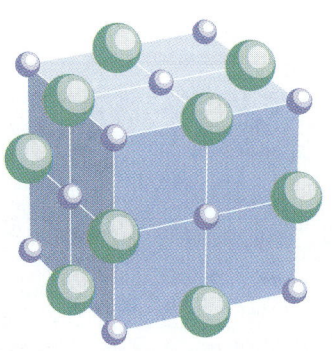

(b) The unit cell.

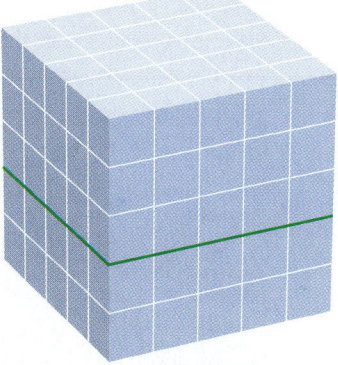

(c) A crystal.

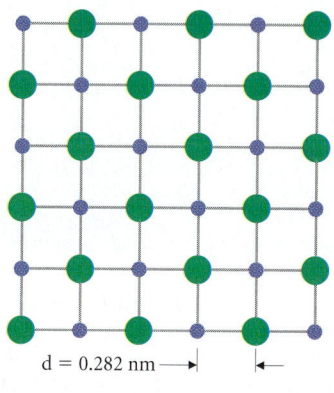

(d) Single plane of the crystal.

IN BRAGG DIFFRACTION, IT IS TRADITIONAL TO USE THE ANGLE BETWEEN THE RAY AND THE SURFACE. THIS IS NOT THE USUAL ANGLE OF INCIDENCE (I.E., ANGLE BETWEEN RAY AND NORMAL). BE CAREFUL!

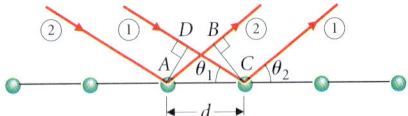

■ **FIGURE 17.43**
Bragg reflection. For constructive interference, the path difference $AB - DC$ between waves that scatter from neighboring atoms is a whole number of wavelengths. Each atom acts as a "slit" in a diffraction grating. The large number of scattering atoms results in tightly collimated beams.

Each atom in a crystal scatters incident x rays in all directions, but interference between x rays scattered from different atoms concentrates the radiation in certain preferred directions. ■ Figure 17.43 shows x rays scattered by neighboring atoms in the crystal lattice, which are separated by distance d. Both rays travel the same distance from the source at *infinity* to the line AD, and from BC to a distant detector. Between A and B, scattered ray 2 travels the distance $AB = d \cos \theta_2$, and incident ray 1 travels distance $DC = d \cos \theta_1$ between D and C. The path difference between the waves reaching a distant observer is:

$$\Delta r = d \cos \theta_1 - d \cos \theta_2.$$

For constructive interference, this path difference is a whole number of wavelengths:

$$d(\cos \theta_1 - \cos \theta_2) = m\lambda.$$

In particular, for $m = 0$, $\cos \theta_1 = \cos \theta_2$, $\theta_1 = \theta_2$. The x rays obey the usual law of reflection. This effect was observed by Sir William Henry Bragg (1862–1942) and is named Bragg scattering or Bragg reflection after him.

The intensity of the reflected wave depends on interference between waves reflected from successive layers in the crystal (■ Figure 17.44). For rays obeying the law of reflection, the path difference is $BCD = 2d \sin \theta$. For constructive interference: $2d \sin \theta = m\lambda$, or:

$$\sin \theta = m \frac{\lambda}{2d}. \tag{17.13}$$

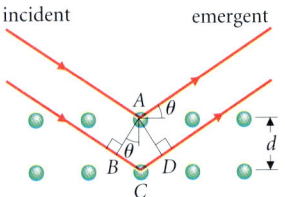

■ **FIGURE 17.44**
Waves scattering from neighboring crystal planes must also interfere constructively. Thus distance $BCD = 2d \sin \theta$ is also a whole number of wavelengths.

> **EXAMPLE 17.14** ♦ If x rays of wavelength 0.400 nm shine on a crystal of salt, at what angle to the surface is a strong reflected beam seen?
>
> **MODEL** The reflected beam and the incident beam make the same angle with the crystal surface. That angle is related to the wavelength and the crystal structure through eqn. (17.13). The relevant dimension d of the crystal is the distance between the top layer and the second layer of atoms, so $2d$ is the side of a unit cell (Figure 17.42b).
>
> **SETUP** $\qquad 2d = 0.564$ nm.
>
> A strong reflected beam occurs when: $\quad \sin \theta = \dfrac{m\lambda}{2d} = \dfrac{m(0.400 \text{ nm})}{0.564 \text{ nm}} = m(0.709).$
>
> **SOLVE** For $m = 1$, $\theta = 45.2°$.
>
> **ANALYZE** Higher values of m do not lead to real solutions for θ.

With a monochromatic x ray beam, Bragg diffraction is used to map out the structure of crystals (■ Figure 17.45). When a beam with a large spread of wavelengths is incident on a crystal, the diffraction process selectively scatters a particular wavelength; the crystal acts as an x ray spectrometer. The crystal may be rotated to select out different wavelengths in turn, or it may be made with a curved surface to disperse a range of wavelengths simultaneously (■ Figure 17.46). The type of crystal used depends on the range of wavelengths to be observed. Spectrometers used in the soft x ray band must have fairly large crystal spacings. Successful instruments have been built using stearates (soap)!

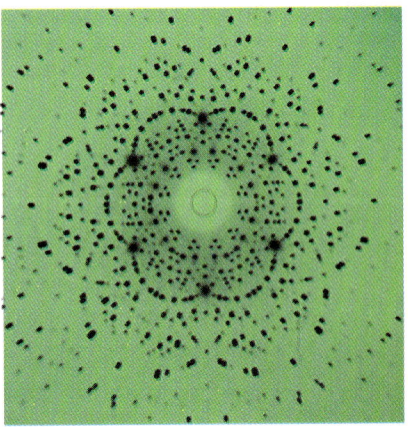

■ **FIGURE 17.45**
The x ray diffraction pattern for beryl. The various sharp diffraction maxima result from different sets of planes within the crystal. From the pattern and intensity of the maxima, it is possible to deduce the crystal structure.

■ **FIGURE 17.46**
Bragg spectrometer used on board the Einstein observatory. This photograph shows one of the curved crystals.

Chapter Summary

Where Are We Now?

We have developed the basic ideas of interference and diffraction, effects that arise from the superposition of traveling waves whose relative phase varies in space or time. These effects have important technological applications and are necessary for understanding the fundamental structure of matter (Chapter 35).

What Did We Do?

Two waves interfere constructively if their sum has an amplitude greater than that of any individual wave. They interfere destructively if the sum has a lesser amplitude than any of the waves. Two harmonic wave functions with equal amplitude but differing in phase by $\Delta\phi$ superpose to produce a harmonic wave with amplitude $2A \cos(\Delta\phi/2)$. Complete destructive interference occurs for $\Delta\phi = (2m + 1)\pi$.

Waves that differ slightly in frequency *beat* with each other. Their relative phase varies periodically with time, causing a variation in the amplitude of the resultant wave. The wave energy is concentrated into pulses that pass an observer at the beat frequency $f_b = |f_1 - f_2|$. The average power transmitted is the sum of the powers each wave alone would transmit.

Coherence describes the degree to which different sources maintain the same phase relation. If a light source emits wave trains of average length $\ell = c \, \Delta t$, then ℓ is the *coherence length* of the source and Δt is the coherence time. Ordinary light sources produce waves with very short coherence lengths, and two such sources are not mutually coherent. Lasers can have coherence lengths of meters or more.

Two mutually coherent point sources produce an interference pattern with destructive interference occurring along nodal surfaces and constructive interference along antinodal surfaces. At a point P very far from sources separated by distance d and at an angle θ from them, interference is constructive when the path difference $d \sin \theta = m\lambda$ and destructive when $d \sin \theta = (m + \frac{1}{2})\lambda$ (eqns. 17.5 and 17.6).

Light reflected from, or passing through, a thin film shows interference between waves that reflect within the film and those that do not. Phase changes on reflection as well as path length determine the phase difference between interfering waves (Figure 17.12).

A Michelson interferometer splits a single beam of light and superposes the two parts after they have traveled along different paths. Interference of the beams allows a sensitive comparison of the two path lengths. Michelson and Morley used such an instrument in an important experiment which showed that the speed of light is independent of the direction of propagation. Hence it was inferred that light needs no medium in which to propagate.

Diffraction occurs when a wave is intercepted by an obstacle. Light passing through an aperture forms a pattern of light and dark bands. The angular width θ of the central peak is approximately given by: $\sin \theta \approx$ wavelength/(size of aperture). Two sources may be resolved if they are separated by an angle greater than or equal to this θ. When light passes through two or more slits of width $\geq \lambda$, both interference and diffraction effects influence the intensity pattern.

Phasors are a convenient geometrical tool for superposing wave functions. A phasor is a vector with length proportional to the amplitude of a wave; its direction represents the wave phase. The y-component of the phasor equals the displacement due to the wave at a specific time and place. When waves superpose, their phasors add like vectors to give the amplitude and phase of the resultant wave.

A diffraction grating is a large number of closely spaced slits. The width of the primary maxima decreases as the number of slits N increases so that a grating produces narrow lines at each wavelength.

In x ray diffraction, x rays falling on a crystal with atomic separation d at the Bragg angle $\sin^{-1}(m\lambda/2d)$ are reflected and a bright beam is observed.

Practical Applications

Coatings that reduce the reflections of light from glass in display cases or camera lenses use the principles of thin-film interference. Through multiple-source interference, radio stations beam their broadcasts into selected directions, and steerable radar installations are made without moving parts. Diffraction limits the performance of both telescopes and the human eye. Gratings are used for spectroscopy in astronomy, industrial chemistry, and medicine, for example. We use x ray diffraction to determine the crystalline structure of substances and to make spectrometers for the x ray part of the spectrum. Wave front shaping through interference is used in the manufacture of microdevices. Destructive interference is the idea behind active noise control.

Solutions to Exercises

17.1 Interference is called constructive if the amplitude of the superposed waves exceeds that of one individual wave (eqn. 17.3); that is, if:

$$|\cos(\Delta\phi/2)| > \tfrac{1}{2}.$$

The corresponding ranges for $\Delta\phi$ are:

$$-2\pi/3 < \Delta\phi < 2\pi/3.$$

Two-thirds of the possible values for phase difference result in constructive interference, and one-third results in destructive interference.

17.2 The amplitude of δP is:

$$A_r(x, t) = 2P_* \cos\left(\frac{k_1 - k_2}{2} x - \frac{\omega_1 - \omega_2}{2} t\right).$$

This describes a harmonic wave traveling in the direction of increasing x with frequency $f = (\omega_1 - \omega_2)/4\pi$ and wavelength $\lambda = 4\pi/(k_1 - k_2)$. The wave speed is $v = f\lambda$:

$$v = \frac{\omega_1 - \omega_2}{k_1 - k_2} = \frac{k_1 v_s - k_2 v_s}{k_1 - k_2} = v_s.$$

The wave has frequency f equal to half the beat frequency $f_1 - f_2$ and travels at the sound speed. The sound is loud whether this amplitude wave is at maximum positive or negative pressure disturbance. The number of beats per second is twice the wave frequency.

17.3 Applying the cosine rule to triangle OS_1P,

$$r_1^2 = (d/2)^2 + R^2 - 2(d/2)R\cos(\pi/2 - \theta)$$
$$= R^2[1 - (d/R)\sin\theta + (d/2R)^2].$$

Applying the cosine rule to triangle OS_2P,

$$r_2^2 = (d/2)^2 + R^2 - 2(d/2)R\cos(\pi/2 + \theta)$$
$$= R^2[1 + (d/R)\sin\theta + (d/2R)^2].$$

Since $d/R \ll 1$, we take the square roots, and expand to first order in d/R:

$$r_1 \approx R[1 - (d/2R)\sin\theta] \quad \text{and} \quad r_2 \approx R[1 + (d/2R)\sin\theta].$$

So: $\qquad r_2 - r_1 \approx d\sin\theta,$

as required. With $R = 10d$ and $\theta = 15°$,
$r_1 = R\sqrt{1 - (0.1)\sin 15° + (0.05)^2} = 0.98824R$. Similarly, $r_2 = 1.01409R$, and $r_2 - r_1 = 0.02585R$. The approximate result is $d\sin\theta = 0.02588R$, which differs from the exact result by $3 \times 10^{-5}R$, or 0.1%. The approximation is valid to three significant figures.

17.4 According to eqn. (17.6), minima occur when $d\sin\theta = (m + \tfrac{1}{2})\lambda$. Since the screen is a long way from the slits $[D/d = (3.00\text{ m})/(10^{-5}\text{ m}) = 3 \times 10^5 \gg 1]$, we may approximate $\sin\theta \approx y/D$. Then:

$$\frac{yd}{D} = (m + \tfrac{1}{2})\lambda \quad \text{or} \quad \lambda = \frac{yd}{D(m + \tfrac{1}{2})}.$$

For $m = 0$:

$$\lambda = \frac{(8.30 \times 10^{-2}\text{ m})(10.0 \times 10^{-6}\text{ m})}{(3.0\text{ m})(\tfrac{1}{2})} = 553\text{ nm}.$$

The information about the second fringe gives another result. With $m = 1$:

$$\lambda = \frac{(24.8 \times 10^{-2}\text{ m})(10.0 \times 10^{-6}\text{ m})}{(3.0\text{ m})(\tfrac{3}{2})} = 551\text{ nm}.$$

Combining the results, $\lambda = (552 \pm 1)$ nm.

17.5 See ■ Figure 17.47. Minima occur when the four phasors combine to form a square ($\Delta\phi = 90°$ and $270°$), or when they lie on top of each other ($\Delta\phi = 180°$). There are $2 = 4 - 2$ secondary maxima between these $3 = 4 - 1$ minima.

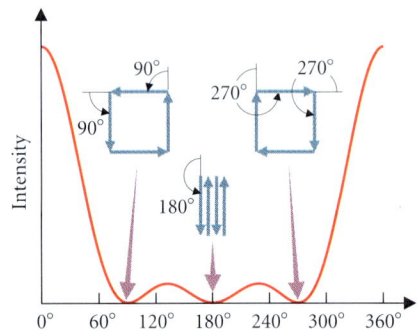

■ **FIGURE 17.47**
Phasor diagrams for four-slit interference. There are three minima and two secondary maxima.

17.6 At 800 nm, in first order ($m = 1$):

$$\sin\theta_{800\text{ nm}} = \frac{\lambda}{d} = \frac{800 \times 10^{-9}\text{ m}}{1.00 \times 10^{-6}\text{ m}} = 0.800.$$

So, $\theta_{800\text{ nm}} = 53.1°$.

In second order, the 400-nm line would appear at:

$$\sin\theta_{400\text{ nm}} = 2\left(\frac{400 \times 10^{-9}\text{ m}}{1.00 \times 10^{-6}\text{ m}}\right) = 0.800,$$

which gives the same angle as for 800 nm in first order.

Basic Skills

Review Questions

§17.1 INTERFERENCE

- Define *constructive* and *destructive interference*.
- Explain why interference results in a redistribution of energy flow.
- What are *beats*?
- When waves from two spatially separated sources interfere, how does the condition for constructive interference depend on the path lengths from each of the two sources to the observation point?
- What is the condition for destructive interference when the observation point is a long way from the two sources?
- What is *coherence*? Why is it important in interference experiments?
- What two effects must be considered in thin-film interference?

§17.2 THE MICHELSON INTERFEROMETER AND THE MICHELSON–MORLEY EXPERIMENT

- How is coherence achieved in the Michelson interferometer?
- Describe the Michelson–Morley experiment. What was the result?

§17.3 DIFFRACTION

- How does diffraction differ from interference? In what ways is it the same?
- What is the difference between Fresnel diffraction and Fraunhofer diffraction? Which is simpler to describe, and why?
- What is the angular width of the central diffraction peak due to light shining through a rectangular aperture? What is the angular width if the aperture is circular?
- What is Rayleigh's criterion for resolution of two closely spaced sources?
- If two slits are separated by four times their width, how does diffraction modify the interference pattern?

§17.4 INTENSITY IN INTERFERENCE PATTERNS

- How does the intensity of light in two-slit interference depend on the vertical distance y from the center of the screen?
- What is a *phasor*? What does it represent? How do phasors add?
- If all the waves that superpose have the same frequency, how does the phasor diagram evolve in time?
- How many interference minima occur between the principal maxima for seven equally spaced wave sources?
- What is a *grating*? How are the narrow principal maxima achieved?
- What is the intensity profile of the central diffraction peak for a rectangular slit? Roughly how bright is the second maximum compared with the first?

§17.5 X RAY DIFFRACTION

- Why does a salt crystal behave like a diffraction grating for x rays?
- What is the condition for Bragg reflection?

Basic Skill Drill

§17.1 INTERFERENCE

1. Light of wavelength $\lambda = 600.0$ nm falls on two narrow slits separated by a distance $d = 1.00$ mm. A screen is placed 2.00 m from the slits. Find the distances of the first three interference maxima from the center of the screen. What is the separation of the maxima? Is the separation the same for *all* maxima?

2. When tuning a guitar, a musician sounds a tuning fork of frequency 440 Hz and plucks the string. The resulting sound beats with a frequency of 5 Hz. What is the frequency of the string if it is flat (i.e., the frequency of the string is too low)? How can the musician correct the problem?

3. Nonreflecting glass is made by adding a thin film of a transparent substance such as magnesium fluoride ($n = 1.38$). What thickness of coating minimizes reflections of sunlight ($\lambda \approx 550$ nm)?

§17.2 THE MICHELSON INTERFEROMETER AND THE MICHELSON–MORLEY EXPERIMENT

4. A Michelson interferometer is set up so that a bright spot of monochromatic light ($\lambda = 550$ nm) is seen at the center of the interference pattern. If one mirror is moved until the spot becomes dark, how far is it moved?

§17.3 DIFFRACTION

5. Light from an argon laser ($\lambda = 488$ nm and $\lambda = 514.5$ nm) shines on a slit 3.00 μm wide. Describe the diffraction pattern. Is the green or the blue-green maximum wider? If the screen is 1.00 m away, what is the difference in the two widths?

6. A typical amateur's telescope has a 20-cm aperture. What is its angular resolution with 600-nm light?

7. Light of wavelength 600.0 nm is used to illuminate two slits with a width of 1.00 μm and a separation of 5.00 μm.
(a) Where are the interference maxima?
(b) Where is the diffraction minimum?
(c) Sketch the pattern observed on a distant screen.

§17.4 INTENSITY IN INTERFERENCE PATTERNS

8. Radio waves of frequency 100.0 MHz are transmitted in phase by two antennae 10.0 m apart. Find the ratio I/I_{max} 10.0 km away from the towers and 1.00 km off to the side of their center line.

9. Plot a phasor diagram to compute the sum of:

$$y_1 = (3 \text{ mm})\cos[(3 \text{ rad/m})x - (10^3 \text{ rad/s})t],$$

and $\quad y_2 = (1 \text{ mm})\cos[(3 \text{ rad/m})x - (10^3 \text{ rad/s})t + \pi/6].$

10. Light of wavelength $\lambda = 600$ nm emerges in phase from three parallel slits with separation $d = 1.0$ mm. Make a sketch of intensity versus position on a screen 10 m from the slits. Show the first two principal maxima on each side of the central spot.

11. Draw a phasor diagram for the first interference minimum produced by seven parallel slits. At what phase difference between individual phasors does this minimum occur?

12. A diffraction grating 5.0 cm × 5.0 cm has 360 grooves per mm. At what angle is the first-order maximum for $\lambda = 121.6$ nm? For $\lambda = 590.0$ nm?

§17.5 X RAY DIFFRACTION

13. X rays shine on a crystal of calcium chloride at an angle of 20°. The spacing between atomic planes is $d = 0.314$ nm. Which wavelength shows a strong diffracted beam?

Questions and Problems

§17.1 INTERFERENCE

14. ❖ If the double slit in Example 17.3 were placed under water, what would happen to the interference pattern?

15. ❖ When a soap film is held vertically, interference produces horizontal colored bands (Figure 17.14). Why?

16. ❖ The interference pattern produced by twin sources disappears when a piece of glass is placed in front of one of the sources (■ Figure 17.48). Discuss how this effect could be used to measure the coherence length of the light source.

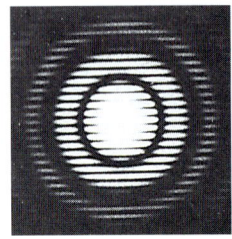
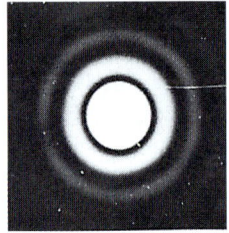

■ **Figure 17.48**

17. ❖ Show that for two point sources of waves, the nodal surfaces are hyperboloids. (The LORAN navigational system uses this fact.)

18. ◆ Laser light of wavelength $\lambda = 630$ nm shines on two slits, and interference fringes 1.0 mm apart are formed on a screen 1.0 m away. What is the separation of the slits?

19. ◆ White light is incident normally on a thin film of oil ($n = 1.20$) that lies on a puddle of water after a rainstorm. If the film is 250 nm thick, what color does it appear to be?

20. ◆ A soap film ($n = 1.33$) in air appears blue ($\lambda = 460$ nm) when viewed in reflected light and red ($\lambda = 690$ nm) when viewed in transmitted light. What is the thickness of the film? Assume normal incidence.

21. ◆ A mistuned cello string is sounded at the same time as a 256-Hz tuning fork. If 4 beats per second are heard, what are the possible frequencies of the cello string?

22. ◆◆ A thin film of oil ($n = 1.50$) floats on water ($n = 1.33$). If the film appears yellow-green ($\lambda = 550$ nm) when viewed at normal incidence, how thick is it?

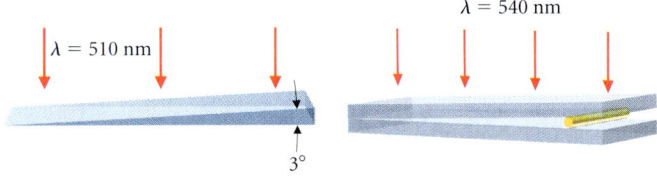

■ **Figure 17.49**   ■ **Figure 17.50**

23. ◆◆ A wedge of glass with a refractive index of 1.55 has an apex angle $\alpha = 3°$ (■ Figure 17.49). If light of wavelength 510 nm and coherence length 1.0×10^{-5} m is incident on the wedge as shown, how many bright fringes would be visible in a microscope viewing reflected light? Where does the last fringe occur?

24. ◆◆ Laser light of wavelength 488 nm shines on two slits with a separation of 1.00 mm and illuminates a screen 3.00 m away. Where are the bright fringes? A piece of plastic film ($n = 1.600$) 2.00 μm thick is placed in front of the top slit. What is the shift in the fringe pattern?

25. ◆◆ Two ribbon tweeters (loudspeakers) are separated by a distance of 20 cm. A person seated at an angle of 15° from the loudspeakers hears no sound when a pure tone is played. What is the frequency of the tone? (Take the speed of sound to be 340 m/s.)

26. ◆◆ Two glass plates are separated by a fine wire of diameter d (■ Figure 17.50). The plates are illuminated by light of wavelength 540 nm and six bright fringes are observed. What is the diameter of the wire?

27. ◆◆ Laser light of wavelength 630 nm shines on two narrow slits separated by 1.0 mm and illuminates a screen 2.0 m away. Where are the bright fringes? When a piece of glass 0.50 mm thick is placed in front of one of the slits, the central bright fringe moves up 15°. What is the refractive index of the glass at 630 nm?

28. ◆◆ Coherent plane waves from a laser are incident at an angle ϕ on a screen containing two thin slits separated by a distance d. Light passing through the slits produces an interference pattern on the other side of the screen. Find the condition for a maximum of this pattern.

29. ♦♦♦ (a) A glass lens is flat on one side. Its other side is spherical with radius $R = 1.00$ m. The lens is placed on a glass plate, and yellow light of wavelength $\lambda = 589$ nm is shone directly downward. The reflected light forms a series of concentric light and dark rings. Explain how the rings form and show that their radii are given by $r = \sqrt{\lambda R(m - \frac{1}{2})}$, $m = 1, 2, 3 \ldots$ Calculate the radii of the first three rings. (b) If the glass has index of refraction $n_g = 1.500$ and oil with $n_o = 1.52$ is placed in the space between lens and glass plate, what effect does this have on the rings?

30. ♦♦ A lens is tested by placing it on a plate made of the same glass and illuminating it with 436-nm light from a mercury lamp. Bright fringes are observed at radii of 0.43, 0.75, and 0.96 mm. What is the radius of curvature of the lens? What are the radii of the bright fringes if 589-nm sodium light is used? (Use the result of Problem 29.)

§17.2 THE MICHELSON INTERFEROMETER AND THE MICHELSON–MORLEY EXPERIMENT

31. ♦♦ To use a Michelson interferometer as a spectrometer, one of the mirrors may be tilted slightly so that the path length traveled by the light varies across the width of the beam. If the mirror is tilted 1 arc second, what is the difference in path length as a function of distance from the end of the mirror? How far from the edge of the mirror's image does 650-nm (red) light appear? Where does 430-nm (blue) light appear? (Hint: With such a small tilt, the beams are not appreciably deflected, but an extra wedge of air is introduced into the beam.)

32. ♦♦ A Michelson interferometer can be used to measure the refractive index of a gas that is close to 1. A 75-cm-long glass container with hydrogen in it is placed in one arm of the interferometer beam (■ Figure 17.51). A glass plate in the other beam ensures that both beams travel through the same thickness of glass. The entire apparatus is placed in a vacuum chamber. With $\lambda = 589$ nm, it is found that L_2 must exceed $L_1 = 1.00$ m by 99.0 μm for a bright spot to be seen at O. What is the index of refraction of the hydrogen sample?

33. ♦♦ A Michelson interferometer is used with a helium–neon laser of wavelength 632.8 nm. A glass-walled chamber of interior length 1.005 cm is placed in one of the beams, and the position of the mirror is adjusted until bright light is observed at the detector. The glass chamber is then evacuated. During the evacuation, the viewed light changes from bright to dark and back to bright nine times. What is the refractive index of air?

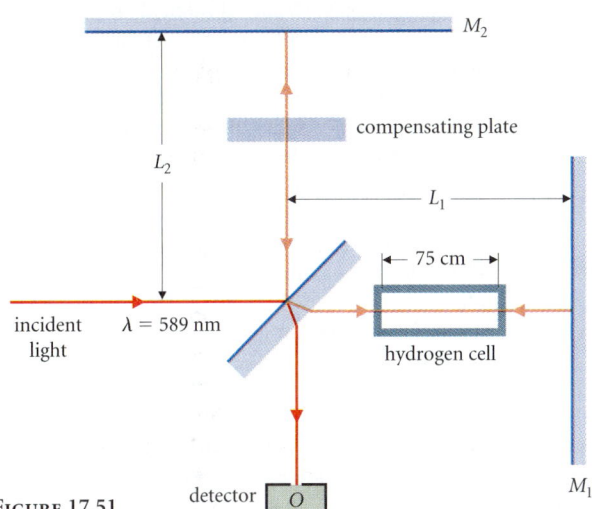

■ FIGURE 17.51

§17.3 DIFFRACTION

34. ❖ Explain why in everyday experience sound seems to propagate throughout a room, while light seems to cast sharp shadows.

35. ❖ Explain why you can hear the car radio for a longer time after you drive into a tunnel if you are playing an AM station ($f \approx 600$ kHz) rather than an FM station ($f \approx 100$ MHz).

36. ❖ Radio telescopes are much bigger than optical telescopes. One reason concerns the resolution obtainable with radio waves. Explain.

37. ❖ Describe the interference pattern of your ears for 3-kHz sound. Do you have a better sense of direction for lower- or higher-frequency sound?

38. ❖ Electrostatic tweeters produce high-frequency sound by moving a panel back and forth using electric forces. Three panels are used side by side, forming a speaker about 30 cm wide. Explain why electrostatic tweeters are directional (i.e., they sound good only if you sit in certain spots). (Hint: What is the wavelength of 2-kHz sound?)

39. ❖ Foghorns should deliver their message over a wide angle but also close to the water. Why does the shape shown in ■ Figure 17.52 do the job best?

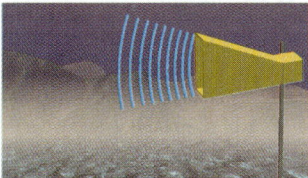

■ FIGURE 17.52

40. ❖ Why is there a bright line at the center of the shadow of the pencil lead (Figure 17.18)?

41. ♦ Light from a ruby laser at 694 nm shines through a slit of width 2.00 μm onto a screen 2.00 m away. Where is the first diffraction minimum?

42. ♦ Sketch the diffraction pattern observed on a screen 3.00 m from a 1.5-μm-wide slit illuminated by a He–Ne laser of wavelength 632.8 nm.

43. ♦ The red star UV Cet is actually a double-star system. The two stars are equally bright and have an apparent separation of 5.0″. What aperture telescope is needed to resolve the stars at 750 nm?

44. ♦ The star system Algol is a binary with a separation of 10^{10} m. The system is 5.1×10^{17} m from Earth. What is the angular separation of the stars? What size telescope would be needed to resolve them using visible light ($\lambda \approx 500$ nm)? What if 300-nm light could be used?

45. ♦ The Keck telescope in Hawaii has an aperture of 10 m. Considering only diffraction effects, what would be the smallest angular separation of two stars that can just be resolved with this telescope at 600 nm?

46. ♦ Atmospheric turbulence generally smears the image of a star over an angular size of 1″ of arc or more. What is the aperture of a telescope with a diffraction-limited resolution of 1″ (at 600 nm)? Why do astronomers use larger telescopes? The Hubble space telescope has an aperture of 2.4 m. What is its angular resolution?

47. ♦ Light of wavelength $\lambda = 633$ nm is incident on a two-slit system with separation $d = 0.50$ mm. What is the slit width a if every third interference maximum is missing?

48. ♦ Laser light shines on two slits of width 1.5 μm. In the interference pattern, every ninth fringe is missing. What is the separation of the slits?

49. ♦♦ How would you design a two-slit system with slits 2.0 μm wide so that, when used with laser light at λ = 488 nm, every other interference maximum is missing? If the laser is an argon laser emitting at λ' = 514.5 nm as well, sketch the diffraction/interference patterns at each of the two wavelengths.

50. ♦♦ In a Young's experiment using sodium light (λ = 589 nm), the slits are 1.2 μm wide. You notice that every fourth fringe is missing. What is the slit separation? How many fringes appear on the screen?

§17.4 INTENSITY IN INTERFERENCE PATTERNS

51. ♦ (a) Plot a phasor diagram to compute the sum of the two sound waves:

$$\delta P_1 = (3 \text{ Pa})\cos[(10 \text{ rad/m})x - (3.5 \times 10^3 \text{ rad/s})t + \pi/8],$$

and

$$\delta P_2 = (5 \text{ Pa})\cos[(10 \text{ rad/m})x - (3.5 \times 10^3 \text{ rad/s})t + \pi/4].$$

What is the sum of these two sound waves with the following third wave?

$$\delta P_3 = (2 \text{ Pa})\cos[(10 \text{ rad/m})x - (3.5 \times 10^3 \text{ rad/s})t - \pi/4].$$

52. ♦ Light from a He–Ne laser (wavelength 632.8 nm) shows a fringe at $\theta = 35°$ when shone through a diffraction grating. What is the spacing between rulings on the grating?

53. ♦♦ A piano tuner compares the frequency of a mistuned piano string (256 Hz) with that of a 262-Hz tuning fork (cf. Example 17.1). The string emits four times the sound intensity of the fork. Construct phasor diagrams that represent the combined sound disturbance at intervals of one-quarter of the beat period. What are the maximum and minimum intensities of the beats?

54. ♦♦ (a) Use a phasor diagram to find the sum of the waves:

$$y_1 = (3 \text{ cm})\cos[(3 \text{ rad/m})x - (900 \text{ rad/s})t],$$
$$y_2 = (4 \text{ cm})\cos[(3 \text{ rad/m})x - (900 \text{ rad/s})t + \pi/2],$$

and $y_3 = (4 \text{ cm})\cos[(3 \text{ rad/m})x - (900 \text{ rad/s})t - \pi/4].$

(b) Verify the result using trigonometric identities.

55. ♦♦ Two sound waves are described by the functions:

$$y_1 = (2.5 \text{ Pa})\cos[(0.100 \text{ rad/m})x + (34.0 \text{ rad/s})t],$$

and $y_2 = (3.5 \text{ Pa})\cos[(0.120 \text{ rad/m})x + (40.8 \text{ rad/s})t].$

They are emitted from two sources that are so close that they may be considered to be at the same place. What is the maximum pressure variation in the superposed sound? Plot a phasor diagram showing the sum $y_1 + y_2$ at position x = 10.0 m and times t = 0.895, 0.202, 2.281, and 3.436 s.

56. ♦♦ Consider the interference of six identical, equally spaced sources. Where are the first and second minima? What is your best guess for the phase difference at the first secondary maximum? Draw phasor diagrams at 5° intervals in phase around your best guess, and use them to find the phase shift at the secondary maximum. What is its intensity relative to the primary maximum?

57. ♦♦ Locate the angles of the minima in the interference pattern of six coherent wave sources equally spaced 3λ apart. Find the locations of any primary maxima that occur, and estimate the location of the secondary maxima.

58. ♦♦ Light from an argon laser (λ = 488 nm and λ = 514.5 nm) shines on a grating with 6.00 × 10² lines/mm. At what angles do the two wavelengths appear? How large a grating is required to resolve them?

59. ♦♦ A grating with 5.00 × 10² lines/mm is used to observe a spectral line at 447.6 nm. Where does this line appear in first order?

■ FIGURE 17.53

If the line is a blend of He (447.1 nm) and MgII (448.1 nm), what N is required to separate them?

60. ♦♦ What is the intensity of the second secondary maximum for a grating with $N \gg 1$?

61. ♦♦ (a) The *resolving power R* of a grating is the ratio of a wavelength to the minimum resolvable wavelength difference at that wavelength. Use Rayleigh's criterion to show that $R = Nm$.

(b) One of the gratings made at Mount Wilson in California was reported to have a resolving power of 676 000 in fifth order, 660 000 in sixth order, and 1 200 000 in eleventh order. The grating was 0.20 m wide. How many slits were on the grating? What was the slit separation? What is its theoretical angular dispersion $\Delta\theta/\Delta\lambda$ at 589 nm in second order?

62. ♦♦ A set of four gratings used at Mount Palomar were each 19 cm long with 400 grooves/mm and blazed for third order. For each grating, find the resolving power at 410 nm using the result of Problem 61a. The angular dispersion D of a grating is the rate at which the angular position of a principle maximum increases with wavelength. Show that $D = m/(d \cos \theta)$, and find the dispersion at 410 nm for each of the Mt. Palomar gratings. The four were used together to form a composite grating (■ Figure 17.53). How do the resolving power and angular dispersion of the composite compare with those of each individual grating?

63. ♦♦ The grooves on a compact disk act as a diffraction grating for reflected light. Laser light at 632 nm is incident on a small region of the disk where the grooves run vertically. If intensity maxima are observed at y = 0.44 m and 1.36 m on a screen 1.00 m away, what is the groove spacing on the disk?

64. ♦♦ In 1673, James Gregorie observed that a feather behaves like a diffraction grating. If sunlight admitted through a pinhole falls on a feather with a slit spacing of 5 μm, find the position of the first-order maximum at 450 and 600 nm.

65. ♦♦ A harp string tuned to 224 Hz beats with a 220-Hz tuning fork. The harp emits 3.0 times the intensity of the fork. What is the beat frequency? What are the maximum and minimum intensities of the beating sound?

66. ♦♦♦ Light of wavelength 589 nm is incident on twin slits 2.0 μm wide, separated by 10.0 μm (Example 17.9). Ignoring the variation of intensity due to diffraction over the width of one fringe, compare the total power in a bright fringe at $\theta = 45°$ with one at $\theta = 0$.

67. ♦♦♦ Laser light of wavelength 632 nm shines on a system of two 1.00-μm-wide slits, which are separated by 10.0 μm. The maximum intensity on the screen is I_o. Find the location and intensity of the fourth, eighth, and twelfth interference maxima.

■ FIGURE 17.54

■ FIGURE 17.57

§17.5 X RAY DIFFRACTION

68. ❖ ■ Figure 17.54 shows the diffraction pattern of x rays shining on an aggregate of very small, randomly oriented crystals (a "crystalline powder"). Explain why such a pattern consists of concentric rings.

69. ◆ The plane spacing in an x ray spectrometer made of lead stearate is 5.1 nm. Find the wavelengths of light reflected at Bragg angles of 15°, 30°, and 45°.

70. ◆ One of the materials used in the x ray crystal spectrometer on board the Einstein observatory was thallium acid phthalate (TAP) with a plane spacing d of 1.29 nm. At what angle should 1.24-nm x rays meet the crystal surface if they are to be Bragg reflected?

71. ◆ The plane spacing in a LiF crystal is 0.2014 nm. At what angles are x rays of frequency 2.5×10^{18} Hz reflected?

72. ◆ Crystal planes in silicon are separated by 0.313 nm. First-order Bragg reflection of a beam of x rays is observed at 14.2°. What is the wavelength of the x rays? At what angle is second-order Bragg reflection observed?

73. ◆◆ ■ Figure 17.55 shows two sets of planes formed by the atoms in a crystal. If a beam of x rays is incident on the crystal surface at angle θ, as shown in the figure, find the wavelengths that undergo Bragg reflection from each of the two sets of planes. At what angle to the surface does the reflected beam from the inclined set of planes emerge?

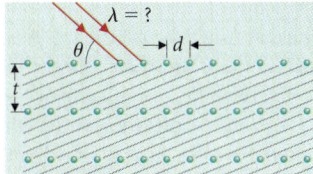
■ FIGURE 17.55

74. ◆◆◆ The crystal planes in a Bragg spectrometer follow a cylindrical surface with a radius of 20.0 cm, (■ Figure 17.56). The crystal is cesium chloride with a lattice spacing of 0.20 nm. Find the location of reflected x rays on the screen for $\lambda = $ 0.15, 0.20 and 0.25 nm.

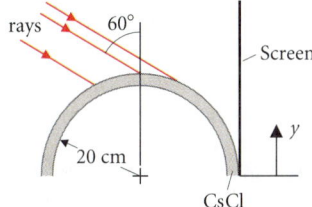
■ FIGURE 17.56

Additional Problems

75. ❖ Four radio antennae are at corners of a rectangle measuring $\lambda/2$ east-west by $\lambda/4$ north-south. The southward pair of antennae are driven with a phase lead of $\pi/2$ compared with the northward pair. Show that the array produces a beam directed northward.

76. ❖ Two radio antennae are separated by a distance $\lambda/2$. The signals transmitted have equal amplitude, but the left signal has a frequency that differs from the right by 60 Hz. Find how the resulting beam behaves as a function of time. [VHF Omnirange (VOR) aircraft navigation systems use this principle.]

77. ❖ Explain why the total power in a two-source interference pattern can be less than P_o for some values of kd (P_o is the power a single source would emit if it were alone). (See *Digging Deeper: Closely Spaced Sources*.)

78. ❖ ■ Figure 17.57 shows the Very Large Array of radio telescopes near Socorro, New Mexico. Explain why an array of many telescopes improves the resolution of an interferometer.

79. ◆◆ White light is incident normally on a film of polystyrene ($n = 1.595$) 0.50 μm thick. Find the wavelengths of the maxima for reflected visible light (400 nm − 650 nm) when the angle of incidence is **(a)** 30°, **(b)** 60°, and **(c)** 0. What color does the film appear?

80. ◆◆ Police radar of frequency 2450 MHz, reflected from an approaching car, beats at 550 Hz with the original signal. What is the speed of the car? (Assume the radar unit is at rest.)

81. ◆◆ Because radio telescopes have poor angular resolution (see Problem 36), interferometers are often used. An interferometer has two dishes separated by 5.0 km. Its resolution is the width of a fringe in the two-source interference pattern. Why? What is this resolution when the telescope is used to receive 21-cm wavelength radio waves?

82. ◆◆ A loudspeaker and a sound-reflecting wall are arranged so that the listener hears only sound directly from the speaker and reflected from the wall (■ Figure 17.58). If the distance d is 10.0 m and the sound frequency is 68.0 Hz, what is the smallest distance h such that destructive interference cancels the sound at the listener's position? (Take the speed of sound to be 340 m/s.)

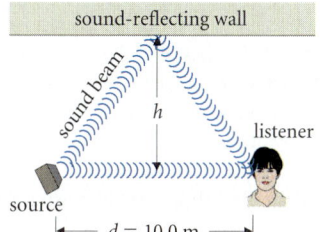
■ FIGURE 17.58

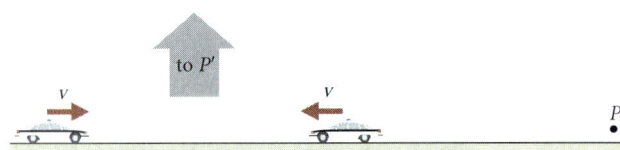

FIGURE 17.59

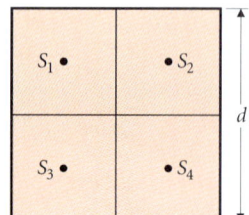

FIGURE 17.60

83. ♦♦♦ Find the energy density of two beating sound waves as a function of position and time. (*Hint:* the intensity of a wave equals the wave speed multiplied by the energy density.) Define short- and long-distance averages of the energy density, $\langle u \rangle_{\text{short}}$ and $\langle u \rangle_{\text{long}}$. What is the distance between two maxima of $\langle u \rangle_{\text{short}}$? What is the time interval between the passages of these maxima past an observer?

84. ♦♦♦ Two one-dimensional sound waves with amplitudes of P_1 and P_2 have frequencies of f_1 and f_2. Calculate the short time-scale intensity transmitted by the superposed waves, and show that it consists of a continuous intensity plus a pulsed intensity. What is the long time-scale average?

85. ♦♦♦ Light incident on a thin film undergoes multiple reflections within the film. At each reflection, a fraction f of the light intensity is transmitted and a fraction $1 - f$ is reflected. (A different fraction is reflected at the first encounter with the surface.) It is the superposition of all the beams that emerge from the top of the film, and whose path length is less than the coherence length of the light, that interferes with the directly reflected beam. Show that when beam 4 (cf. Figure 17.11) and beam 1 interfere constructively, subsequent beams are either completely in phase or completely out of phase with beam 1. What happens when beam 4 and beam 1 are out of phase by π? Use these results to argue that the apparent color of the film is not affected. Make a phasor diagram to find the correction to the intensity of beam 1 when $f = \frac{1}{2}$ at *each* interface and the coherence length is $10t$.

86. ♦♦♦ Two sources of sound waves whose frequency in still air is f_o are separated by a distance d and approach each other, each with speed v (■ Figure 17.59). The two sound waves are in phase as they are emitted from the sources. Observers at points P and P' are at distances much greater than the separation of the sources. Explain why the observer at P hears a beating sound, and compute the average frequency and the beat frequency. What does the observer at P' hear?

87. ♦♦♦ Draw the phasor diagram for two rectangular slits, including the effects of diffraction (cf. Figure 17.39), and use it to calculate the observed intensity at angle θ from the center of the slit system.

Computer Problems

88. (a) To find a more exact expression for the positions of the maxima in a single-slit diffraction pattern, write the expression for the intensity in terms of the phase shift variable $\alpha \equiv \Delta\Phi_{\text{total}}/2$, differentiate with respect to α, and set the derivative equal to zero. Show that the maxima occur for $\tan \alpha = \alpha$. (b) Plot a graph showing the two curves $y = \tan \alpha$ and $y = \alpha$, and find their intersections. These are the values of α corresponding to the maxima. (c) To obtain the values numerically using a spreadsheet, rewrite the equation as $\alpha = \tan^{-1}(\alpha)$. From the discussion in §17.3 and §17.4.5 we know that the maxima are close to $m\pi/2$ for $m = 3, 5, 7$, etc., so we choose a starting guess $x_1 = m\pi/2 - 0.05$. The second guess $x_2 = \tan^{-1}(x_1)$, etc. The spreadsheet function ATAN returns a value in radians between $-\pi/2$ and $+\pi/2$, so you will need to add the appropriate multiple of π. What are the values of α at the first three maxima, excluding the central maximum?

89. Explore the interference pattern due to four sources of sound with wavelength $\lambda = 20.0$ cm arranged in a square pattern. (This source pattern also approximates a square aperture; see ■ Figure 17.60.) Assume the sources are at the corners of a square with sides of 1.00 m, centered at the coordinate origin. Determine the distances from each source to a point P with coordinates (x, y, z), and hence determine the phase shifts $k \Delta r$ between the waves from each source arriving at P. Use a spreadsheet program to calculate the intensity at points P with $z = 10.0$ m and (a) x, y at 10-cm intervals between 0 and 1.0 m; and (b) x, y at 40-cm intervals between 0 and 4.0 m. Plot your results as graphs of intensity versus x at each value of y. Where are the interference maxima?

Challenge Problems

90. A light source with $\lambda = 650$ nm and unknown coherence length illuminates the first pinhole in a wave front–splitting interferometer. The two pinholes are separated by $d = 0.25$ mm. The interference pattern falls on a screen 1.0 m from the pinholes, and 15 fringes are visible. Explain how to determine the coherence length of the light source, and give an approximate value.

91. We can use x rays to determine molecular structure. To illustrate the principles, assume a beam of x rays of wavelength λ is incident on a diatomic molecule consisting of two atoms A and B separated by distance d. The molecule is oriented so that the line joining the atoms is perpendicular to the incident beam. The electric fields in the x rays cause the atoms to radiate at the same wavelength as the incident radiation so that the two atoms behave as point sources. The amplitude of the wave an atom emits is proportional to the number of electrons in that atom. The radiation emitted by the atoms is detected on a screen a large distance L from the molecule, and an interference pattern is seen. The separation of the maxima is $\ell \ll L$, and the ratio $I_{\max}/I_{\min} = 4$. (a) Find the atomic spacing d in terms of ℓ, L, and λ. (b) Determine the ratio of the number of electrons in the two atoms, N_A/N_B. (*Hint:* Phasors might be useful.)

92. Use the binomial theorem to expand the expressions for distance in the solution to Exercise 17.3 to third order—that is, include all terms in $(d/R)^3$. Use your results to find the correction to the approximate formula $r_2 - r_1 = d \sin \theta$. For $\theta \approx 0$, find the ratio d/R at which the approximate formula is in error by 1%.

93. A modification of the Michelson interferometer can be used to measure rotation with respect to an inertial reference frame. The beams emerging from the beam-splitter travel in opposite directions around a square that is rotating with angular velocity ω. Explain why the two beams travel around the square in different times, and show that the beam traveling in the direction of rotation takes a time longer by $\Delta t = 4\omega L^2/c^2$. If light of wavelength 630 nm is used in a device with side $L = 3.0 \times 10^2$ m to measure the rotation of the Earth, what phase shift would be observed? (Assume the experiment is carried out at 45° N latitude.) Michelson carried out this experiment in 1925. Some modern navigational systems measure rotation with three devices based on these principles and mounted on three perpendicular axes.

Astronaut Dr. Shannon Lucid is working in the Neutral Buoyancy Simulator, a tank full of water. Even though her helmet and suit are bulky, her head still looks too small. This effect is explained by the refraction of light at the curved air/water boundary (§18.4.2).

CHAPTER 18
Geometrical Optics

Concepts

Optical surface
Real image
Virtual image
Virtual object
Ray diagram
Focus
Lens
Microscope
Telescope
Aberration

Goals

Be able to:

Locate an image formed by a plane mirror, spherical mirror, or spherical refracting surface.

Determine the character of an image.

Use ray diagrams to construct the image of a finite object.

Compute the behavior of a compound optical system.

Give qualitative descriptions of the common aberrations in optical systems.

*The books are something like our books,
only the words go the wrong way.*

Lewis Carroll

THE MATERIAL IN THIS CHAPTER IS BASED ON THE LAWS OF REFLECTION AND REFRACTION (§16.5).

RAYS WERE DEFINED IN §16.3.1.

THE DIVERGENCE OF THE BEAM IS APPROXIMATELY DIAMETER DIVIDED BY WAVELENGTH (§17.3).

"SMALL ENOUGH" MUST STILL BE GREATER THAN A WAVELENGTH. THIS CONSTRAINT POSES NO PROBLEMS WHEN DEALING WITH VISIBLE LIGHT.

Why does Dr. Shannon Lucid's head appear so small as she works under water in simulated weightlessness? A similar shrinkage happens to objects seen in a vehicle's rearview mirror. Such illusions are often the basis of magicians' tricks or even of fraud. In this chapter our goal is to understand the formation of optical images—bizarre or realistic—using the techniques of geometrical optics.

In geometrical optics, we use the laws of reflection and refraction to follow light rays through an optical system. The rules for tracing rays follow naturally from the wave properties of light, which explain why the geometrical approach works and also what its limits are. If we try to isolate a single ray by passing light through a tiny hole (■ Figure 18.1), we obtain a pencil of light with a finite diameter, but only if the hole is much larger than the wavelength of light. With a smaller hole, diffraction produces a spherical wave instead. Modeling light as rays is a valid approximation whenever diffraction effects remain small. This means that the dimensions of the optical systems that we can study using rays are large compared with the wavelength of the light used.

The laws of reflection and refraction apply to waves incident on a plane surface. Since a small enough piece of any smooth surface is essentially plane, these laws apply at each piece of a curved surface as well. But all surfaces are bumpy at some level, and waves are sensitive to surface properties averaged over distances of roughly a wavelength. So the surfaces used in practical systems should be smooth on a scale of the wavelength used ($\sim 0.5\ \mu m$ for visible light) (■ Figure 18.2). This requirement meshes nicely with the concept of rays. Modeling light as rays and surfaces as smooth are useful approximations at size scales larger than a wavelength. Thus geometrical optics applies to a broad range of practical systems.

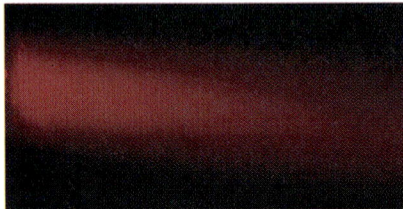

■ FIGURE 18.1
If we attempt to create a single ray by shining light through a hole, we always fail. A large hole produces a straight beam of light with a finite diameter. If we try to narrow the beam by using a smaller hole, diffraction causes the light to spread into a fan.

■ FIGURE 18.2
Optical surfaces must be smooth on the scale of the wavelength used. The surface of the 300-ft Greenbank radio telescope, which collapsed in November of 1988, was made of chicken wire. The holes had little effect on its ability to reflect radio waves with wavelengths of 0.2 m or more.

SEE §18.5.1 FOR A DISCUSSION OF THE EYE AS AN OPTICAL SYSTEM.

18.1 IMAGES FORMED BY PLANE SURFACES

When we view a scene, light reaches our eyes from a variety of sources: luminous objects such as stars, light bulbs, or computer screens; and the light scattered from the surfaces of tables, chairs, and other things. We may think of each piece of the scene as a point source of light that emits spherical wave fronts or, equivalently, of rays that pass radially outward. Your eyes are designed to focus a bundle of these diverging rays (■ Figure 18.3), which your brain interprets as a point source of light. Light rays passing through an optical system such as eyeglasses are bent so that the rays appear to come from a different point. The directions of the incoming

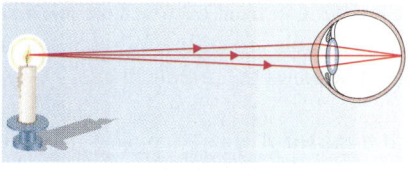
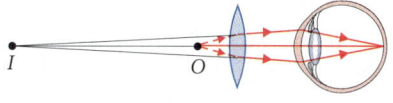
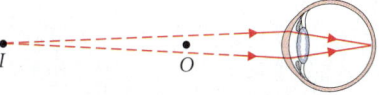

(a) (b) (c)

■ **Figure 18.3**

(a) Diverging light rays reaching the eye are focused onto the retina by the cornea and the lens. Retinal cells send signals to the brain, which interprets the light as a point source. (See §18.5.1 for further discussion of the eye.)

(b) This eyeglass lens bends light rays from the object at O so that they appear to come from a different point.

(c) The eye focuses the rays onto the retina as if they had come from the image point I rather than the object at O. The eye is unaware of the true origin of the rays. All the eye's information is obtained from the directions of the rays that enter it.

rays are all the information your eyes have about the location of the light source. Your eyes interpret the rays as if they came directly from an image of the original source. An ordinary plane mirror illustrates these ideas.

18.1.1 Images in a Plane Mirror

The image that stares back at you each morning from the bathroom mirror looks like a real person, about as far behind the mirror as you are in front of it (■ Figure 18.4a). To understand how this image is formed, look at the tip of your nose. Rays emerging from your nose reflect from the mirror into your eye (Figure 18.4b). It's usually easier to find the position of an image with rays chosen for that purpose, rather than with rays that actually enter the eye (Figure 18.4c). A ray perpendicular to the mirror is reflected directly backward at B. A ray striking the mirror at A reflects so that the angle of incidence $\angle OAN = \theta$ equals the angle of reflection $\angle NAE$. The two reflected rays (lines BO and AE) diverge from the image point I. The image is as far behind the mirror as the object is in front.

SEE §16.5.2.

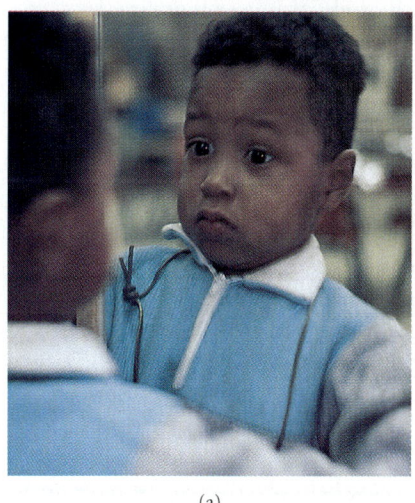

(a)

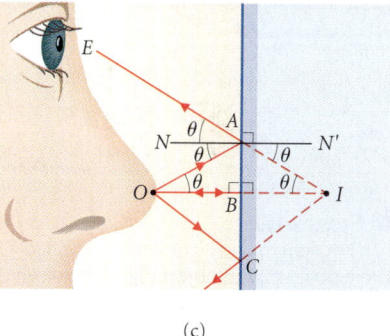

(b) (c)

■ **Figure 18.4**

(a) Mirrors have been used for centuries. The mirror image of a face is a familiar sight. (b) Rays from the tip of your nose reflect from the mirror to your eye. They appear to diverge from the image of your nose, behind the mirror.

(c) Since all the reflected rays diverge from the image, specially selected rays show the position of the image most clearly. Here a ray perpendicular to the mirror is chosen, as well as one that reflects at an arbitrary point A. The angle of incidence is $\angle OAN$, here called θ. It equals the angle of reflection NAE. Triangles OBA and IBA are congruent, so $OB = IB$. This conclusion does not depend on the actual value of $\angle OAN$, and so holds for *any* ray from O. All reflected rays from your nose appear to come from the image at I.

> The image of an object in a plane mirror lies on a line from the object perpendicular to the mirror and is the same distance behind the mirror as the object is in front.

EXERCISE 18.1 ♦ Standing 2.0 m from a mirror, you view the image of a candle 1.0 m from the mirror. How far away from you does the candle's image appear to be?

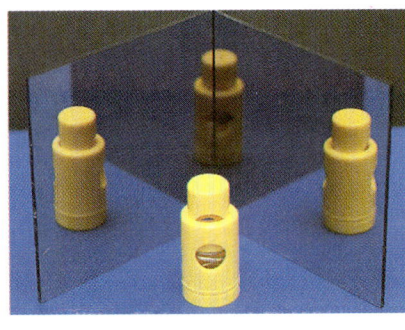

(a)

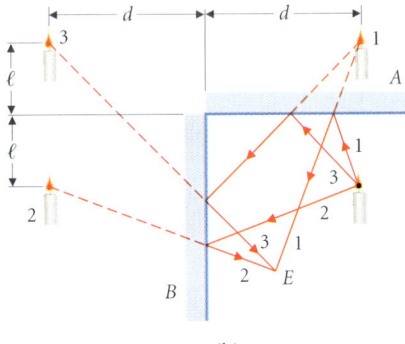

(b)

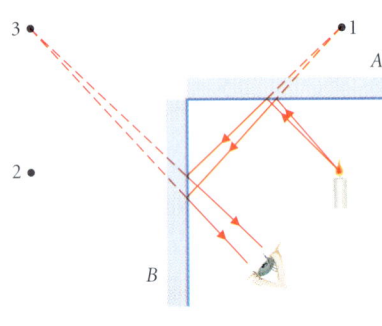

(c)

■ **FIGURE 18.5**
(a) Two mirrors at right angles produce three images of an object placed between them. (b) Three specially selected rays show how the images are formed. The third image is formed by rays like #3 that reflect twice before reaching the eye. (c) A set of diverging rays that reach the eye to form image 3.

HERE "REAL" HAS TAKEN ON A CONVENTIONAL TECHNICAL MEANING. WE SHALL USE "ACTUAL" TO SPEAK ABOUT MATERIAL OBJECTS (AS OPPOSED TO PHENOMENA) IN AN OPTICAL SYSTEM.

EXAMPLE 18.1 ❖ Brian holds a candle while standing in front of two plane mirrors set at right angles (■ Figure 18.5a). Three images of the candle are visible. Why? Sketch a pencil of rays from the candle that enter Brian's eye as he views the middle image.

MODEL Candlelight observed after reflecting from mirror A (ray 1) appears to come from image 1 behind mirror A. Similarly, light observed after reflecting from mirror B (ray 2) appears to come from image 2. Ray 3, however, reflects from mirror A and mirror B before observation. Such rays form the third image.

SETUP Rays like #3 approaching mirror B diverge from image 1 and so form a new image (#3) as if image 1 were an actual object. Brian sees light leaving mirror B as if it comes from image 3, which is as far behind mirror B as image 1 is in front.

SOLVE The pencil of rays in Figure 18.5c is constructed by drawing rays from image 3 to the eye, and then tracing them backward to the candle, using the law of reflection.

ANALYZE Image 3 may also be viewed with rays from image 2 in mirror B, as shown in the next exercise. It is the image of both primary images! ■

EXERCISE 18.2 ❖ Where can Brian put his eye to see image 3 with light that reflects first from mirror B?

18.1.2 Objects, Images, and Sign Conventions

When standing in front of a mirror, you know quite well that you are the object and the thing in the mirror is only an image. Yet we saw in Example 18.1 that an image can itself produce images in other mirrors; an image formed by one surface can act as an object for a second surface. So, we need to be careful with the words *image* and *object*. Their use is illustrated by the plane mirror, which is our first example of an optical surface.

| An *optical surface* is a surface that intercepts and changes the paths of light rays.

In Example 18.1 there are two optical surfaces, and each produces one or more images.

| An *image* is formed by light *leaving* an optical surface.

There are two kinds of image: the kind that appears in a mirror, and the kind that appears on a movie screen. The movie projector sends light onto the screen, which scatters the light to your eyes. The image on the screen is described as real:

| A *real image* is formed by light that reaches the image location.

A real image can also be observed directly with the eye (■ Figure 18.6a). Once the light passes through the image, it diverges again as if from an actual source.

In Figure 18.4 each ray from your nose is reflected from the mirror and then appears to emanate from the image point I behind the mirror. No light *actually passes* through I; it is a virtual image (Figure 18.6b).

| A *virtual image* is formed by light that does not pass through the image location.

The virtual image in your bathroom mirror is visible because light rays diverge from the image location even though nothing is actually there. A movie screen at the image location—behind the mirror, in your neighbor's bathroom—wouldn't reveal a thing.

The candle in Example 18.1 is an example of an optical object (Figure 18.6c).

| An *object* is the source of light rays moving toward an optical surface.

This definition includes our usual notion of objects, the light bulbs or candles that produce light energy, or such things as our bodies, which scatter light produced by something else. Images formed by other optical surfaces are also included.

Light rays spread out from all the objects we've discussed so far; these objects are real.

| A *real object* is one from which light diverges.

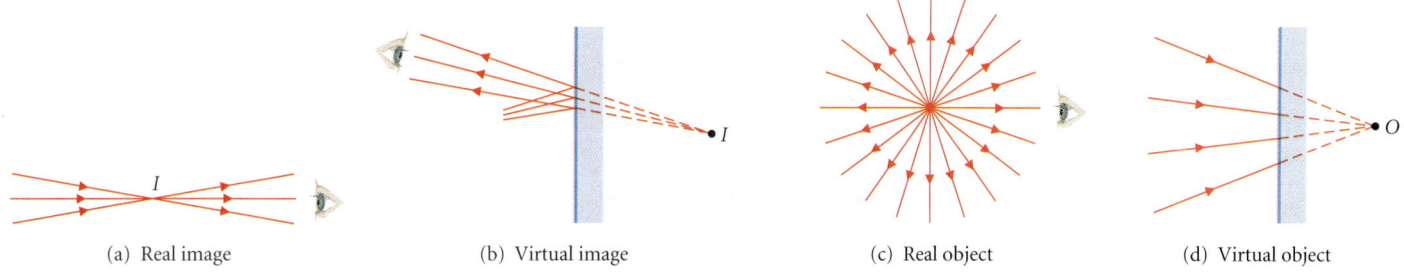

■ **FIGURE 18.6**
Definition of real and virtual images and objects. (a) Light rays converge to, pass through, and diverge from a real image. A screen may also be placed at *I*. (b) Rays appear to diverge from a virtual image. Virtual images may be produced by reflected or refracted rays. (c) Rays diverge from a real object. (d) Rays converge toward a virtual object. (The effect of the optical surface on the incident rays is not shown here.)

Image 1 in Example 18.1 is a real object for the second mirror, according to this definition. A real object does not have to be an actual tangible thing. The important point is that an optical object is associated with light approaching a surface, while an image is associated with light leaving an optical surface.

Suppose a mirror is placed in front of a movie screen. Light from the projector converges toward an image on the screen, but the mirror intercepts the light before it can form the real image. For such cases, we need the bizarre notion of a virtual object (Figure 18.6d):

> A *virtual object* is one toward which light converges.

Virtual objects are used for convenience in analyzing compound optical systems; they occur only when light does not actually arrive at the object location.

Plane light waves are a special case. The rays are parallel lines and the wave fronts are parallel planes. We may think of such waves as diverging from a real object at an infinite distance.

In geometrical optics, we are interested in relationships between objects and images, and we need a way to describe their positions. There are three rules (■ Figure 18.7):

1. Object and image coordinates are measured from the optical surface.
2. For *objects,* the positive direction of the coordinate x_o is backward along light rays approaching the surface.
3. For *images,* the positive direction of the coordinate x_i is forward along light rays leaving the surface.

These conventions guarantee that real objects and images have positive coordinates while virtual objects and images have negative coordinates (● Table 18.1). With this notation, our result for images in a plane mirror is written:

$$\text{Image in a plane mirror: } x_i = -x_o. \qquad (18.1)$$

If the nose in Figure 18.4 is 10 cm in front of the mirror, then $x_o = +10$ cm and $x_i = -10$ cm. The minus sign indicates that the image is virtual and behind the mirror.

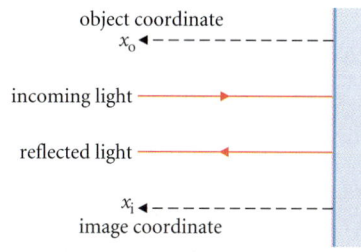

■ **FIGURE 18.7**
Sign conventions applied to reflection from a plane mirror. Coordinates are measured perpendicular to the optical surface independent of the angle of the rays.

TABLE 18.1   Types of Objects and Images

	Real (light actually present)	Virtual (light not present)
	COORDINATE POSITIVE	**COORDINATE NEGATIVE**
Object	Rays diverge from it.	Rays converge toward it.
Image	Rays converge toward it.	Rays appear to diverge from it.

SECTION 18.1 • IMAGES FORMED BY PLANE SURFACES

■ **FIGURE 18.8**
(a) When you look at a fish under the surface of the water, you actually see its virtual image formed by rays refracted at the water surface. (b) When observed from directly above, the image is formed by rays IB and IA. The angle θ_1 is exaggerated to make the geometry clear.

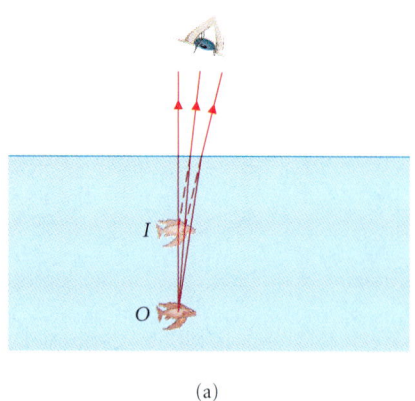

(a)

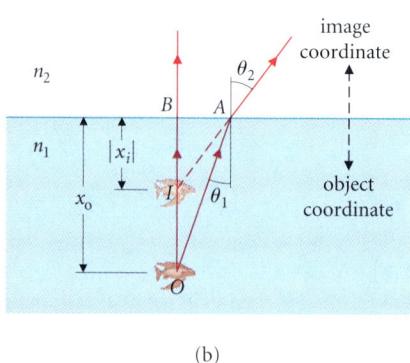

(b)

18.1.3 Images Formed by Plane Refracting Surfaces

Images are also formed by refraction. ■ Figure 18.8 shows a fish in a medium with refractive index n_1. Light from the fish passes into a second medium with refractive index n_2. A ray at right angles to the surface passes through undeflected at B. A second ray, making an angle θ_1 with the normal at A, is bent according to Snell's law. It leaves the surface at an angle θ_2 with the normal, where

$$n_1 \sin \theta_1 = n_2 \sin \theta_2.$$

A SINGLE RAY PERPENDICULAR TO THE SURFACE DOES PASS THROUGH I. REMEMBER THAT A BUNDLE OF RAYS IS NECESSARY TO FORM AN IMAGE.

These light rays in medium 2 diverge from point I. The image is virtual since light appears to come from I, but none actually passes through I.

When viewed from directly above, the depth of the image is determined by a small bundle of rays sufficiently close to the normal that $\sin \theta_1 \approx \theta_1 \approx \tan \theta_1$. The fish is a real object, so x_o is positive. The image is virtual, and its coordinate x_i is negative, so $|x_i| = -x_i > 0$. Thus,

$$x_o \sin \theta_1 \approx x_o \tan \theta_1 = AB = |x_i| \tan \theta_2$$
$$= -x_i \tan \theta_2 \approx -x_i \sin \theta_2$$
$$= -x_i (n_1/n_2) \sin \theta_1.$$

NOTICE THAT THE OBJECT IS IN THE MEDIUM WITH REFRACTIVE INDEX n_1 AND THE OBSERVER IS IN THE MEDIUM WITH AN INDEX n_2.

So:

$$x_o/n_1 = -x_i/n_2. \tag{18.2}$$

Light rays that strike the surface at larger angles appear to diverge from different image points. Since the eye of an observer intercepts only a narrow pencil of rays, an image is formed at a well-defined position for any angle of view. See Problem 32.

EXAMPLE 18.2 ♦ A tiny fish is swimming 1.0 m below the water surface. You observe it from directly above. Where does it appear to be?

MODEL The fish scatters sunlight that is refracted as it passes into the air, thus forming the image. We are viewing the fish from directly above, so we may use the small-angle approximation.

SETUP We use eqn. (18.2) with positive object distance $x_o = 1.0$ m, $n_1 = n_\text{water} \approx \frac{4}{3}$, and $n_2 = n_\text{air} = 1.0$.

SOLVE $x_i = -x_o n_2/n_1 = (1.0 \text{ m})(1.0)(\frac{3}{4}) = -0.75$ m.

ANALYZE The negative value of x_i reminds us that the image is virtual and on the opposite side of the surface from the emerging light. ■

EXERCISE 18.3 ♦ A bird is flying directly above the fish, 1.0 m above the water. Where is the bird's image, as seen by the fish? What kind of image is it? ■

18.2 Images Formed by Curved Surfaces

18.2.1 Spherical Mirrors

Curved mirrors have important applications in telescopes, automobile headlights, rearview mirrors, and the eyes of nocturnal animals. ■ Figure 18.9 illustrates how the behavior of a concave mirror depends on object distance. When the object is close to the surface the mirror produces a virtual image, as a plane mirror would. The mirror forms a real image of a distant object. To study these images, we apply the law of reflection to selected rays striking the mirror surface.

Directions and positions in an optical system are referred to a line called the *optical axis*. The optical axis of a simple system is its symmetry axis. Optical surfaces are often spherical because they are the easiest curved surfaces to manufacture. They are also the easiest to analyze mathematically. Any line through the center of curvature C of a spherical mirror is a line of symmetry, so we define the optical axis as the line passing through the object and C (■ Figure 18.10).

Spherical mirrors form a sharp image so long as the rays used to view it make very small angles with the optical axis.

> *Paraxial rays* are those that remain close to and make very small angles ($\ll 1$ rad) with the optical axis.

Rays at large angles to the axis produce a fuzzy blur around the paraxial image. The paraxial image itself is described by compact expressions that provide an excellent first approximation to the behavior of real systems. It is impractical to draw a diagram using only paraxial rays: it would not be legible! So we draw *construction rays*, which form large, clear diagrams and whose behavior is easy to analyze. Then we make the paraxial ray assumption during the algebraic calculation by assuming the angles are small.

First, we locate the image of an object placed at O, close to a concave spherical mirror of radius r (Figure 18.10). The object coordinate x_o is positive and less than $r/2$. The normal to the mirror at any point passes through the center of curvature C. A ray leaving O along the optical axis and striking the mirror at point A is normal to the mirror and is reflected back through C. A ray striking the mirror at P has angle of incidence $OPC = \theta$, so the reflected ray lies along PQ. The rays diverge from the virtual image I behind the mirror.

We determine the distance $AI = |x_i|$, using the paraxial ray approximation. At point P, $2\theta + \angle OPI = \pi$ rad. Also, from triangle OPI, $\alpha + \beta + \angle OPI = \pi$ rad. Thus: $2\theta = \alpha + \beta$. Similarly, in triangle CPI, angle θ equals the sum of the two internal angles γ and β:

$$\theta = \gamma + \beta.$$

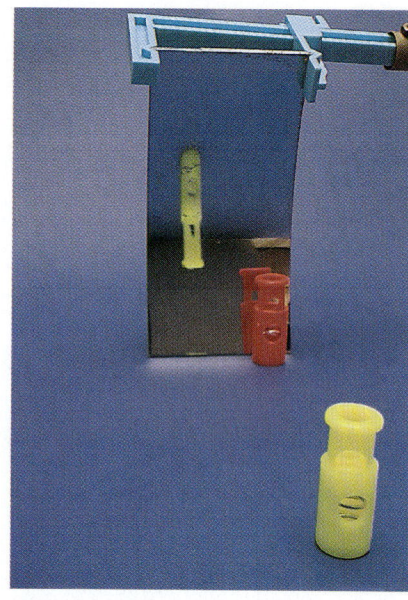

■ **FIGURE 18.9**
Images in a concave mirror. When the object is close to the mirror, the image is virtual. When the object is far away, the image is real.

NOTE: Paraxial means "close to the axis." Rays parallel to the axis but not close to it are NOT paraxial.

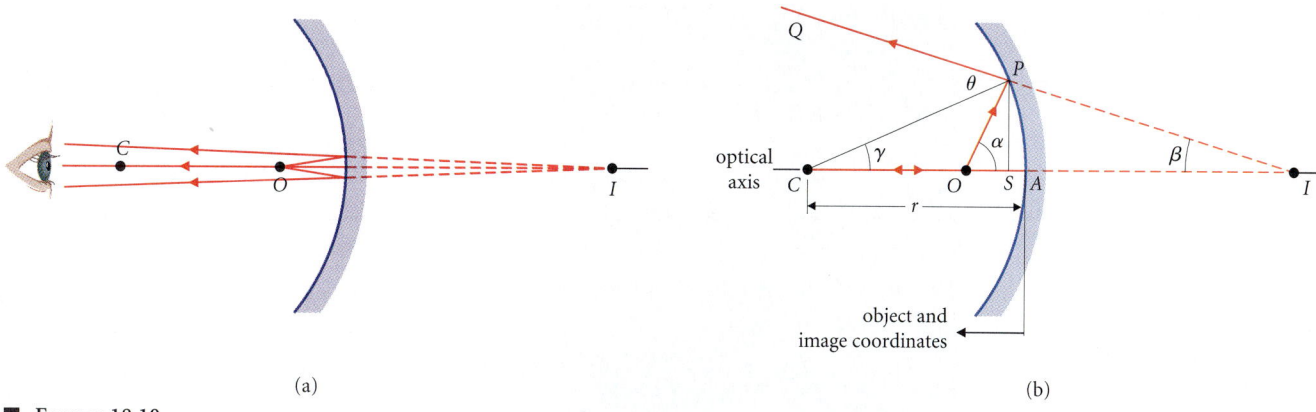

(a)   (b)

■ **FIGURE 18.10**
(a) Rays that remain close to the optical axis COI of a spherical mirror form a sharp image. These are paraxial rays. (b) Selected rays from an object O allow us to find the position of the image I. The optical axis is COA. Ray OA is perpendicular to the mirror and reflects back along the same line through O and C. Ray OPQ has an angle of incidence of θ. Both rays diverge from the image at I. True paraxial rays would not be distinguishable from the axis COA. The angles are drawn large for clarity.

REMEMBER: All angles are in radians. The angles in the figure are exaggerated.

Eliminating θ between the two expressions gives:

$$2\gamma = \alpha - \beta. \qquad (i)$$

Next we express these angles in terms of the distances AC, AO, and AI. The angle γ is the arc length PA divided by the radius $CA \equiv r$. For paraxial rays, the angle α is small, and we may make the approximations:

$$PS/OS = \tan \alpha \approx \alpha = PA/OA = PA/x_o.$$

The object is real, so its coordinate x_o is positive, and $x_o \equiv OA$. Similarly:

$$\beta \approx \tan \beta = PS/SI \approx PA/AI.$$

OUTGOING LIGHT *APPEARS* TO COME FROM I: IT IS A VIRTUAL IMAGE. SEE TABLE 18.1.

According to our sign convention, the image coordinate is negative and $AI \equiv -x_i$. Thus:

$$\beta \approx -PA/x_i.$$

Substituting these results into eqn. (i), we have:

$$\frac{2PA}{r} = \frac{PA}{x_o} + \frac{PA}{x_i},$$

or,

$$\frac{2}{r} = \frac{1}{x_o} + \frac{1}{x_i}, \qquad (18.3)$$

which is the relation we set out to find.

Now let's see what happens when the object is moved to a distance $x_o > r/2$ from the mirror (■ Figure 18.11). The light ray OP is reflected at P to the real image at I. The angle OPC between the incident ray and the normal to the mirror at P (CP) is equal to $\angle CPI$ between the reflected ray and the normal. To find the image distance $AI = x_i$, we note that $\beta = \alpha + 2\theta$, and $\gamma = \alpha + \theta$, so $\alpha + \beta = 2\gamma$. Once again, using the paraxial ray condition, we obtain eqn. 18.3, which applies for all values of the object coordinate.

When the object is at position $x_o = r/2$, the rays leaving the mirror are parallel. From eqn. (18.3), the image coordinate is:

$$\frac{1}{x_i} = \frac{2}{r} - \frac{1}{x_o} = \frac{2}{r} - \frac{2}{r} = 0 \Rightarrow x_i = \infty.$$

The image is located at *infinity*. Conversely, parallel incident light rays are all reflected through a single point—they are brought to a *focus* (■ Figure 18.12a). As $x_o \to \infty$, $x_i \to r/2$. This special image location is called the *focal point* of the mirror with focal coordinate $f = r/2$.

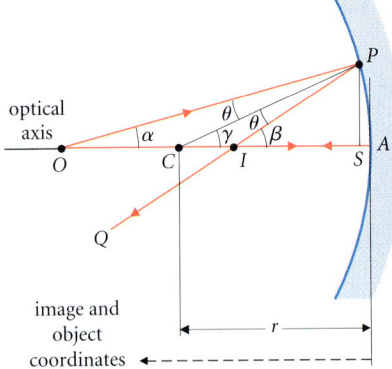

■ **Figure 18.11**
An object placed farther than $r/2$ from the mirror has a real image. The choice of rays to determine the image position is the same as in Figure 18.10b: $OCIA$ is normal to the mirror; OPI reflects at P. The angle α has been exaggerated for clarity.

THE EASIEST WAY TO REMEMBER THE MEANING OF CONCAVE IS TO NOTE THAT IT CONTAINS THE WORD "CAVE": THE MIRROR LOOKS LIKE A CAVE.

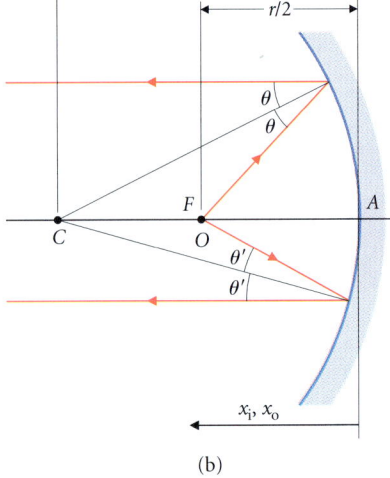

■ **Figure 18.12**
(a) When parallel light rays shine on the mirror, they are brought to a focus at $x_i = r/2$. If the object were at the focal point, the rays would emerge parallel to the axis to form an image at *infinity*. (b) The same ray diagram illustrates both situations if we reverse the arrows indicating the direction of propagation.

The *focal point F* of a mirror is the location of the image when parallel light falls on the mirror.

The *focal length* $|f|$ is the distance from the mirror to the focal point; for spherical mirrors, it equals one-half the radius of curvature.

When assigning coordinates to the focal point and the center of curvature, follow the rule for images.

THE IMPORTANCE OF THIS RULE WILL BECOME CLEAR AS WE CONTINUE OUR STUDY OF MIRRORS AND LENSES.

In Figure 18.12b, light leaves the mirror to the left. Both F and C also lie to the left of the mirror, so their coordinates are positive.

The focusing property of mirrors is widely used: automobile headlight bulbs are placed at the focus of a mirror; emergency mirrors are used to concentrate sunlight and start a fire.

We may express eqn. (18.3) in terms of f rather than r:

$$\frac{1}{x_o} + \frac{1}{x_i} = \frac{1}{f}. \qquad (18.4)$$

For spherical mirrors:

$$1/f = 2/r. \qquad (18.5)$$

A plane mirror may be regarded as a special case of a curved mirror with radius $r = \infty$. Then $2/r = 0$ and the mirror formula reduces to:

$$\frac{1}{x_o} + \frac{1}{x_i} = 0 \quad \Rightarrow \quad x_i = -x_o.$$

REMEMBER: THE IMAGE IS VIRTUAL.

The mirror eqns. (18.4) and (18.5) apply to all spherical mirrors, including the special case of the plane mirror. We shall find that eqn. (18.4) (but not 18.5) remains valid for thin lenses as well.

■ Figure 18.13 illustrates the behavior of a convex mirror. Parallel rays incident on the mirror from an object at infinity (Figure 18.13a) produce a virtual image at the focal point. In fact, the mirror produces a virtual image of a real object at any distance from the mirror (Figure 18.13b). To locate the image, we determine the path of paraxial rays like OPQ (■ Figure 18.14). The image is at I and the image coordinate $x_i = -AI$. The center of curvature is at C and, using the image rule, its coordinate is $r = -AC$.

■ **FIGURE 18.13**
(a) Parallel light reflected from a convex mirror appears to diverge from a focal point behind the mirror. According to our sign conventions, both f and r are negative. (b) The image formed by a convex mirror is virtual.

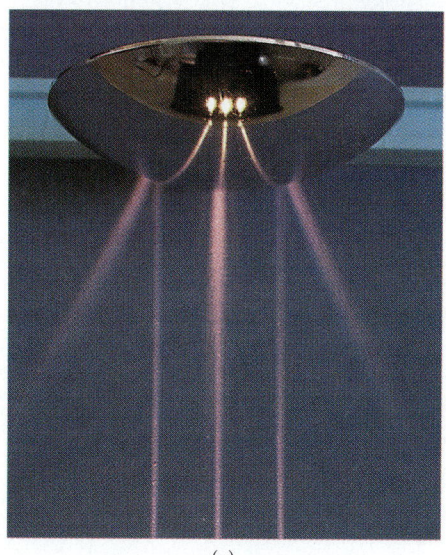

(a)   (b)

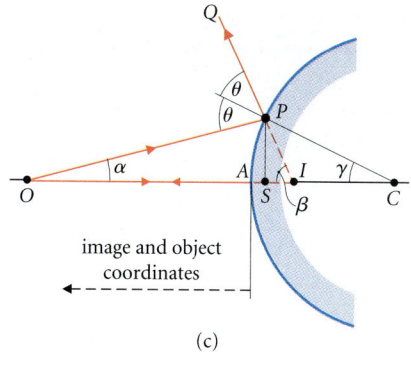

(c)

■ **FIGURE 18.14**
We select one ray normal to the mirror and another with an angle of incidence θ to find the image position. The angle α is exaggerated.

EXERCISE 18.4 ♦♦ Using a similar method to that for the concave mirror, show that eqns. (18.4) and (18.5) apply to the convex mirror.

EXAMPLE 18.3 ♦ A truck has a convex rearview mirror with a radius of curvature of 1.0 m. A light is 4.0 m from the mirror. What kind of image is formed, and where is it located?

MODEL The ray diagram (Figure 18.14) shows that the image is virtual.

SETUP The object coordinate is $x_o = +4.0$ m. The coordinates of the center of curvature and the focal point are negative: $r = -1.0$ m and $f = (-1.0 \text{ m})/2 = -0.50$ m.

SOLVE We use eqn. (18.4):

$$\frac{1}{x_o} + \frac{1}{x_i} = \frac{1}{f} \Rightarrow \frac{1}{x_i} = \frac{1}{f} - \frac{1}{x_o} = \frac{1}{-0.50 \text{ m}} - \frac{1}{4.0 \text{ m}} = -2.25 \text{ m}^{-1},$$

and $x_i = -0.44$ m.

ANALYZE The minus sign confirms that the image is behind the mirror and is virtual. The image is located 44 cm behind the mirror.

Digging Deeper

WAVE FRONTS

When light rays bend at an optical surface, the corresponding wave fronts change shape. Spherical wave fronts diverging from a point (real object or virtual image) or converging toward a point (virtual object or real image) are described by the curvature of their surfaces. The curvature of a spherical surface, equal to the reciprocal of its radius, is positive if the wave is converging and negative if the wave is diverging.

We can use eqn. (18.4) to compute the effect of a mirror on wave front curvature (■ Figure 18.15). Waves diverging from a real object at O have curvature $C_1 = -1/x_o$ when they encounter the mirror. After reflection, the wave fronts are converging on the image point and have curvature $C_2 = +1/x_i$. The change in curvature caused by reflection from the mirror is:

$$\Delta C \equiv C_2 - C_1 = \frac{1}{x_i} + \frac{1}{x_o} = \frac{1}{f}.$$

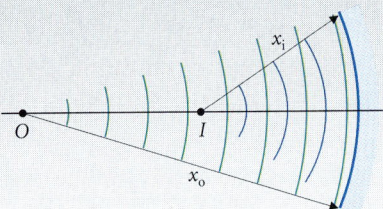

■ **FIGURE 18.15**
A mirror changes the curvature of wave fronts that reflect from it. In this illustration the incoming wave fronts are shown in green and the outgoing wave fronts in blue.

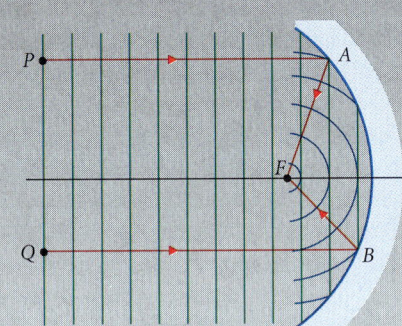

■ **FIGURE 18.16**
A parabolic surface brings all light rays parallel to its axis to a single focus.

A mirror adds curvature $1/f$ to the wave fronts. Opticians refer to the added curvature as the power of an optical element.

A spherical mirror focuses paraxial rays at a point, but rays displaced from the axis do not pass through the same point. This is called spherical aberration; see §18.6. We illustrate the use of the wave front picture by finding the shape of a surface that brings *all* rays parallel to the axis to a single focus.

As plane wave fronts approaching the surface are reflected (■ Figure 18.16), they continuously change shape into spheres centered on the focal point F. Any ray, such as PAF or QBF, crosses the same number of wave fronts between a specific incoming wave front and the focus, so wavelets from all parts of the wave front PQ interfere constructively at the focal point. (This is the general principle behind image formation in the wave front picture.) The sum of the distances from a point on the mirror surface (A or B) to a plane (the incoming wave front PQ) and a point (the focus F) is constant. The mirror is a parabola rotated about its symmetry axis. Parabolic mirrors are used in astronomical telescopes and automobile headlights.

18.2.2 Spherical Refracting Surfaces

Light entering our eyes is refracted at a series of curved surfaces so as to form a real image on the retina. Camera lenses and other precision optical instruments employ similarly intricate sequences of refracting surfaces. We begin the study of such systems by investigating the image produced by paraxial rays refracted at a spherical surface.

Light from a source at O travels through a material of refractive index n_1 and is refracted into a material with index $n_2 > n_1$ (■ Figure 18.17). A ray traveling along the optical axis is normal to the surface at A and passes undeflected through C. A different ray reaching the surface at P is refracted and intersects the first ray at the image I. The radius CP is normal to the surface at P, so θ_1 and θ_2 are the angles of incidence and refraction, respectively. To satisfy the paraxial condition, all angles are small, and Snell's law becomes:

$$n_1 \theta_1 = n_2 \theta_2. \quad \text{(i)}$$

REMEMBER: FOR SMALL ANGLES, $\sin \theta \approx \theta$.

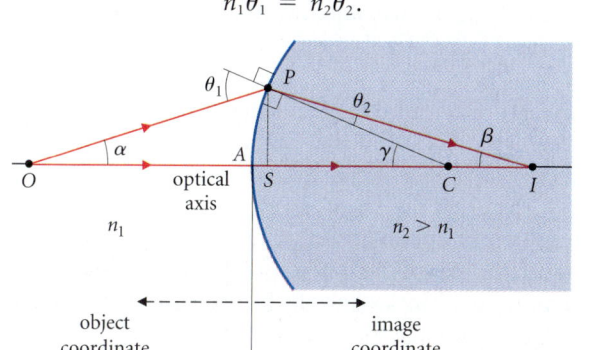

■ **FIGURE 18.17**
Light rays from an object O refracted at the curved surface AP form a real image at I. Angles α and β have been exaggerated for clarity.

We use the ray paths to determine the distance AI from the surface to the image. Since θ_1 is an external angle of triangle OPC:

$$\theta_1 = \alpha + \gamma. \quad \text{(ii)}$$

Similarly, the external angle of triangle OPI is $\theta_1 - \theta_2$:

$$\theta_1 - \theta_2 = \alpha + \beta. \quad \text{(iii)}$$

Using eqn. (i) to eliminate θ_2 and eqn. (ii) to eliminate θ_1, eqn. (iii) becomes:

$$\alpha n_1 + \beta n_2 = \gamma(n_2 - n_1). \quad \text{(iv)}$$

Since the angles are small, they are approximately equal to their tangents and the distance AS is negligible, so eqn. (iv) becomes:

$$n_1/OA + n_2/AI = (n_2 - n_1)/AC.$$

After meeting the surface, the light passes through and travels to the right. The image I and the center of curvature C are both to the right of the surface—that is, on the side of outgoing light—and so have positive coordinates, $x_i = AI$ and $r = AC$. The real object also has a positive coordinate, $x_o = OA$, so:

$$\frac{n_1}{x_o} + \frac{n_2}{x_i} = \frac{n_2 - n_1}{r}. \quad (18.6)$$

REMEMBER: WE USE THE IMAGE RULE TO ASSIGN THE COORDINATE r.

When using this formula, recall that the object lies in the medium whose refractive index is n_1 and the observer (but not necessarily the image) is in the medium with refractive index n_2.

EXAMPLE 18.4 ♦ The radius of curvature of the cornea is about 0.78 cm. The interior of the eye has an average index of refraction of about 1.38 (similar to water). Where would the image of a match flame 0.52 m from the eye be formed if the eye were just a single refracting surface?

SEE §18.5.1 AND PROBLEM 109 FOR FURTHER DISCUSSION OF THE EYE.

MODEL We model the cornea as a spherical refracting surface, and we use the paraxial ray approximation.

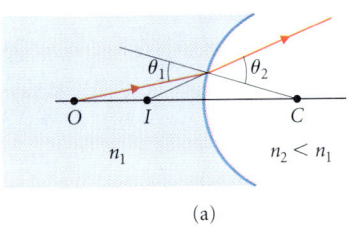

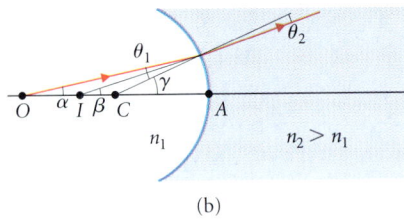

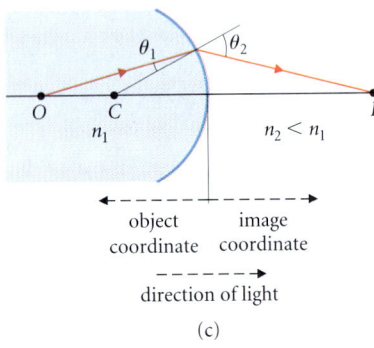

■ **FIGURE 18.18**
Both real and virtual images may be formed at curved refracting surfaces. The kind of image formed depends on the shape of the surface and the relative magnitudes of n_2 and n_1. (a) $n_2 < n_1$. Surface is convex toward the object, $r > 0$. The image is virtual. (b) $n_2 > n_1$. Surface is concave toward the object, $r < 0$. The image is virtual. (c) $n_2 < n_1$. Surface is concave toward the object, $r < 0$. The image is real, provided that the object is far enough from the surface, as shown here.

SETUP A real image is formed at coordinate x_i (Figure 18.17). Here $n_1 = 1.00$ (air), $n_2 = 1.38$, $x_o = +0.52$ m $= +52$ cm, and $r = +0.78$ cm.

SOLVE Applying eqn. (18.6):

$$\frac{n_1}{x_o} + \frac{n_2}{x_i} = \frac{n_2 - n_1}{r} \Rightarrow \frac{1}{x_i} = \frac{1}{n_2}\left(\frac{n_2 - n_1}{r} - \frac{n_1}{x_o}\right)$$

$$= \frac{1}{1.38}\left(\frac{0.38}{0.78 \text{ cm}} - \frac{1.0}{52 \text{ cm}}\right) = 0.339 \text{ cm}^{-1},$$

so
$$x_i = 2.9 \text{ cm}.$$

ANALYZE The positive value of x_i confirms that the image is real. Most refraction of light entering the eye does take place at the cornea, but the eye would not function very well if it were only a single surface, since the retina is fixed and image position would vary with object distance. ■

EXERCISE 18.5 ♦♦ Use eqn. (18.6) to show that a real image is formed by light refracted at a surface with $r > 0$ and $n_2 > n_1$ when the object is sufficiently far from the surface: $x_o > x_{o,min} = n_1 r/(n_2 - n_1)$. ■

Light propagating across a spherical surface from air into a material with a larger refractive index is bent toward the axis of the sphere. Different combinations of n_1, n_2, and surface shape result in different ray diagrams, some of which are drawn in ■ Figure 18.18. With proper use of the sign conventions, eqn. (18.6) applies to all of them.

We may define focal points for these refracting surfaces, as we did for mirrors.

> The *first focal point* is the location of the object whose image is at infinity. Assign its coordinate using the object rules.
>
> The *second focal point* is the image location for an object at infinity. Assign its coordinate using the image rules.

According to eqn. (18.6), with $x_i = \infty$, $f_1 = rn_1/(n_2 - n_1)$, and setting $x_o = \infty$, $f_2 = rn_2/(n_2 - n_1) = (n_2/n_1)f_1$. The two focal points are on opposite sides of the surface, and the two focal coordinates are different.

EXERCISE 18.6 ♦ Find the coordinates of the two focal points for the surface of the eye (cf. Example 18.4). ■

18.3 Lenses
18.3.1 Optical Surfaces in Series

In §18.1 and §18.2 we discussed images formed by simple surfaces. A more complex optical system is formed by combining two or more surfaces. To analyze any complicated optical system, we split it up into simple elements and calculate the effect of each element in sequence.

In Example 18.1 we saw that the image of one optical surface can act as the object for another surface. If the image is not formed before the light encounters the next surface, it becomes a virtual object. The effect of an optical surface is to change the direction of rays incident on it, and our expressions relating object and image distances are reliable tools for computing these effects.

Two curved refracting surfaces in series form a system called a *lens*. Its optical axis passes through both centers of curvature. ■ Figure 18.19 shows a lens consisting of two curved surfaces S_1 and S_2, which bound a material with refractive index n. Light approaching the lens from the object O in the material of index n_* encounters surface S_1 first. The bending of the rays at the surface is described by eqn. (18.6) with $n_1 = n_*$, $n_2 = n$. The coordinates of the

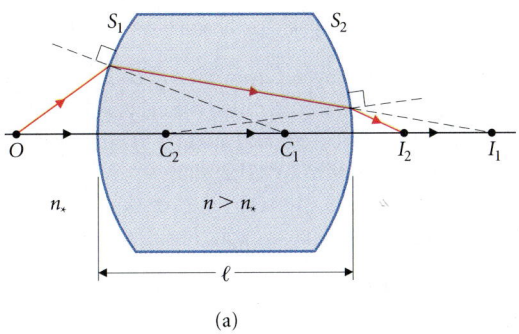
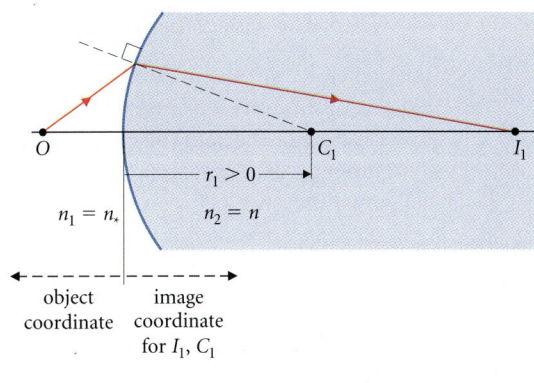

■ **FIGURE 18.19**
(a) A lens consists of two refracting surfaces in series. (b) When light from an object is incident on the lens, the rays are bent at the first surface toward a real image at I_1. (c) At the second surface, the rays are bent again to the real image at I_2. The image I_1 is a virtual object for S_2.

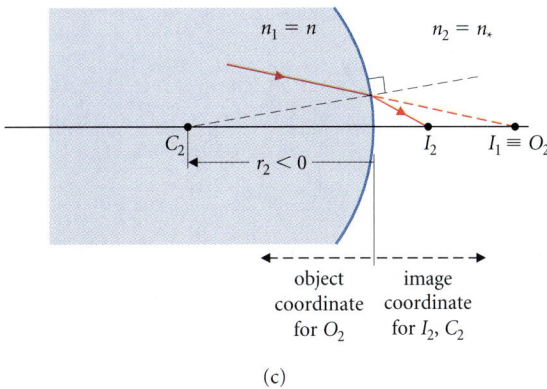

object and the center of curvature of the surface (x_o and r), are positive. The light rays converge toward the image I_1 (Figure 18.19b), and

$$\frac{n_*}{x_o} + \frac{n}{x_{i,1}} = \frac{n - n_*}{r_1} \quad \Rightarrow \quad \frac{n}{x_{i,1}} = \frac{(n - n_*)x_o - n_* r_1}{x_o r_1}.$$

Thus:
$$x_{i,1} = \frac{r_1 n x_o}{n x_o - (x_o + r_1)n_*}.$$

The image is real, as shown in the figure, if $x_{i,1}$ is positive—that is, if $n x_o > (x_o + r_1)n_*$. Otherwise, the image is virtual and to the left of the surface.

We have now completed our examination of the first surface and move on to the second. In Figure 18.19c, the light rays reach the second surface *before* converging at what would have been the image I_1. Here we have our first example of a virtual object: image I_1 forms a virtual object for surface S_2. To proceed with the calculation, we need its object coordinate $x_{o,2}$.

The *image* I_1 has become the *object* O_2, but the image coordinate $x_{i,1}$ is measured from the first surface, and the object coordinate $x_{o,2}$ is measured from the second surface. The distance of I_1 from S_2 is $x_{i,1} - \ell$, where ℓ is the distance separating the two surfaces, and so the *coordinate* of the virtual object is:

$$x_{o,2} = -(x_{i,1} - \ell) = \ell - x_{i,1}. \tag{18.7}$$

SEE FIGURE 18.6 FOR THE DEFINITION OF A VIRTUAL OBJECT.

EXERCISE 18.7 ♦♦ Show that the same expression for $x_{o,2}$ holds if I_1 is formed to the left of the second surface.

The next step is to use this result in eqn. (18.6) to locate the image formed by S_2. It's usually easier to do this numerically than to obtain an algebraic expression that works in all cases. (See Problems 53 and 54, for example.)

18.3.2 Thin Lenses

Thick-lens problems always require a sequential, surface-by-surface approach (§18.3.1). However, when the separation ℓ of the two surfaces is small compared with all other lengths in the problem—the radii of curvature and the distances of all objects and images from the lens surfaces—the lenses may be treated as *thin*. We may thus simplify the analysis by ignoring any difference between the distances of object or image from the front and back surfaces of the lens. So, we ignore ℓ in eqn. (18.7) and substitute $x_{o,2} \approx -x_{i,1}$ in eqn. (18.6).

WE NEED $\ell \ll r_1, r_2, x_o,$ AND $x_{i,1}$.

$$-\frac{n[nx_o - (x_o + r_1)n_*]}{r_1 n x_o} + \frac{n_*}{x_{i,2}} = \frac{n_* - n}{r_2}.$$

We simplify the first part of this expression by canceling a factor of nx_o in the first two terms and a factor nr_1 in the third:

$$\frac{n_* - n}{r_1} + \frac{n_*}{x_o} + \frac{n_*}{x_{i,2}} = \frac{n_* - n}{r_2}.$$

For thin lenses, we no longer need to distinguish the surfaces from which we measure object and image distances, so we replace $x_{i,2}$ with x_i. Then:

$$\frac{1}{x_i} + \frac{1}{x_o} = \frac{n - n_*}{n_*}\left(\frac{1}{r_1} - \frac{1}{r_2}\right). \qquad (18.8)$$

With eqn. (18.8) we may compute the coordinate of the final image formed by the lens, x_i, given the coordinate x_o of the object and the properties $n, n_*, r_1,$ and r_2 of the lens.

> **EXAMPLE 18.5** ♦♦ An old house has an ornate window made of glass 2.0 cm thick with a refractive index of $n = 1.50$. A portion of the window has curved surfaces of radii 1.0 and 0.50 m, respectively (■ Figure 18.20). A light bulb inside the house is 0.50 m from the window. Where is the image of the light bulb formed by the window?
>
> **MODEL** Since the window thickness (2.0 cm) is only 4% of the next smallest length in the problem (50 cm), the thin-lens approximation should serve fairly well.
>
> **SETUP** First, we evaluate the right-hand side of eqn. (18.8):
>
> $$\frac{n - n_*}{n_*}\left(\frac{1}{r_1} - \frac{1}{r_2}\right) = \frac{1.50 - 1.00}{1.00}\left(\frac{1}{-1.00\text{ m}} - \frac{1}{-0.50\text{ m}}\right)$$
> $$= 0.50\text{ /m}.$$
>
> **SOLVE** The object coordinate, measured from the center of the lens, is $x_o = 0.51$ m. Using eqn. (18.8):
>
> $$\frac{1}{0.51\text{ m}} + \frac{1}{x_i} = 0.50\text{ /m} \Rightarrow \frac{1}{x_i} = -1.46\text{ /m}$$
>
> or
>
> $$x_i = -0.68\text{ m}.$$
>
> **ANALYZE** The image is virtual and 68 cm from the center of the lens. This answer differs from a more exact calculation (Problem 53) by approximately 3%. ∎

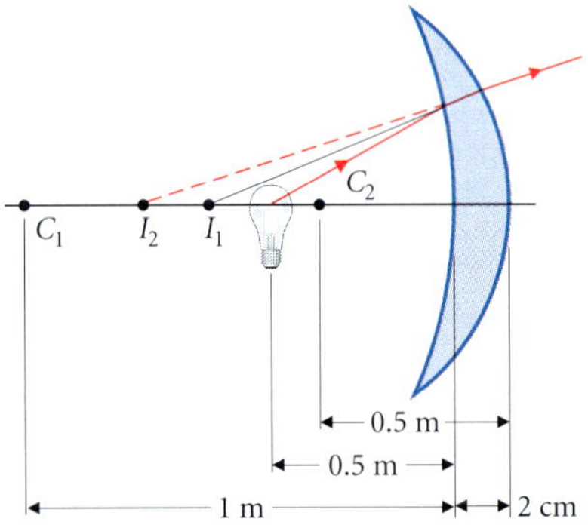

■ FIGURE 18.20
A light bulb is viewed through an ornate window. At the first surface, rays are bent so that they appear to come from the virtual image I_1. At the second surface the rays are bent again and appear to come from I_2.

The right-hand side of eqn. (18.8) gives the reciprocal of the lens' focal coordinates, a relation known as the *lens maker's equation*.

$$\frac{n - n_*}{n_*}\left(\frac{1}{r_1} - \frac{1}{r_2}\right) \equiv \frac{1}{f}. \qquad (18.9)$$

WITH THIS EXPRESSION FOR f, EQN. (18.4) APPLIES TO THIN LENSES AS WELL AS TO MIRRORS.

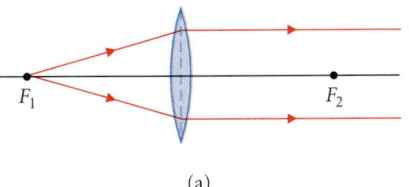
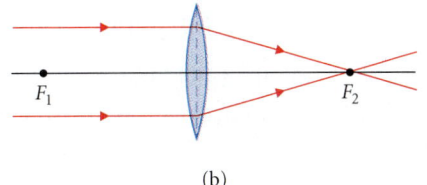

■ FIGURE 18.21
The focal points of a thin lens are at equal distances from the lens. (a) Light from an object at the first focal point emerges parallel to the optical axis. (b) Light rays parallel to the optical axis form a real image at the second focal point.

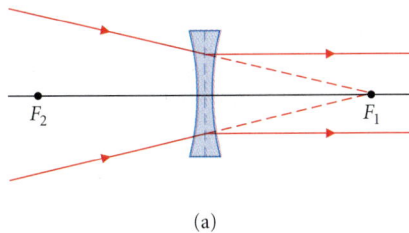
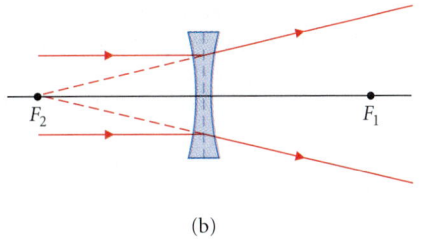

■ FIGURE 18.22
Focal points of a diverging lens ($f < 0$). (a) Light converging toward a virtual object at the first focal point emerges parallel to the optical axis. (b) Light rays parallel to the optical axis diverge from a virtual image at the second focal point.

To prove this claim, let $x_o = f$ in eqn. (18.8). Then $x_i \to \infty$ and f is the coordinate of the first focal point. Letting $x_o \to \infty$ gives $1/x_i = 1/f$, so that f is also the coordinate of the second focal point. The two focal coordinates of a thin lens are equal, but the focal points are on opposite sides of the lens (■ Figure 18.21). If the direction of light passing through the lens is reversed, then the focal points exchange names while remaining in the same positions. The focal coordinate f may be positive or negative. If f is positive, the lens is said to be *converging*, while if f is negative, the lens is *diverging* (■ Figure 18.22). The absolute value of f, $|f|$, is the *focal length* of the lens.

Usually a lens is used in air, so we may replace the refractive index n_* of the ambient medium by 1.

A TRICK FOR REMEMBERING WHICH FOCAL POINT IS WHICH IS "I BEFORE O": IMAGE AT ∞ GIVES F_1, OBJECT AT ∞ GIVES F_2.

$$\frac{1}{f} = (n - 1)\left(\frac{1}{r_1} - \frac{1}{r_2}\right) \quad \text{(lens used in air)}. \quad (18.10)$$

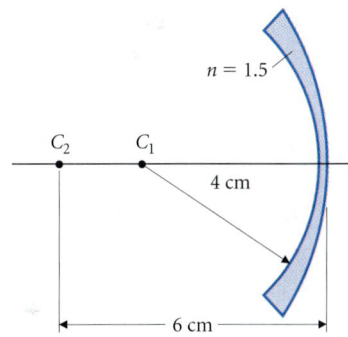

■ FIGURE 18.23
What is the focal length of this spectacle lens? This shape is called a meniscus lens.

EXAMPLE 18.6 ♦♦ A spectacle lens is made of glass with a refractive index of 1.50 and has radii of curvature of 4.0 cm and 6.0 cm (■ Figure 18.23). What is the focal length of the lens? Is it a converging or diverging lens?

MODEL We model the lens as thin and apply eqn. (18.10).

SETUP If we suppose that light is incident from the left, then using the image rules, the appropriate sign for each radius is negative: $r_1 = -4.0$ cm and $r_2 = -6.0$ cm.

$$\frac{1}{f} = (1.50 - 1.00)\left(\frac{1}{-4.0 \text{ cm}} - \frac{1}{-6.0 \text{ cm}}\right) = (0.50)\left(\frac{-3.0 + 2.0}{12.0 \text{ cm}}\right).$$

FOR LIGHT INCIDENT FROM THE RIGHT, BOTH SIGNS ARE POSITIVE AND THE SUBSCRIPTS 1 AND 2 ARE INTERCHANGED. THE SAME VALUE FOR f RESULTS.

SOLVE
$$f = -24 \text{ cm}.$$

The lens is diverging with a focal length of 24 cm.

ANALYZE An optician describes a lens in terms of its dioptric power, $1/f$, measured in units of diopters, where 1 diopter = $1/\text{m}$. This lens has a power of -4.2 diopters. ■

A lens maker designing a lens with a specified focal coordinate f has considerable freedom in choosing the radii r_1 and r_2. Only the combination $r_1^{-1} - r_2^{-1}$ determines f. Quite often $r_1 = -r_2$ and the lens is said to be *symmetric*. If one face is plane, $r_2 \to \infty$, the lens is called plano-convex ($r_1 > 0$) or plano-concave ($r_1 < 0$).

18.4 RAY TRACING AND IMAGES OF EXTENDED OBJECTS

So far, we have discussed ideal point objects. Actual objects have finite extent, and the size and shape of their images are as important as their location and nature (real or virtual). Here we develop a geometric method for finding images.

18.4.1 Images in Plane Mirrors

Each point of an image in a plane mirror is on a line perpendicular to the mirror from the corresponding object point and as far behind the mirror as the object is in front (■ Figure 18.24). Since triangles OQA and IJA are congruent, the image height h_i equals the object height h_o.

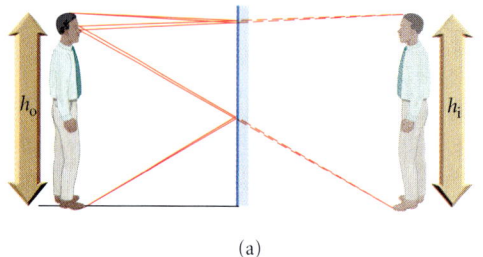

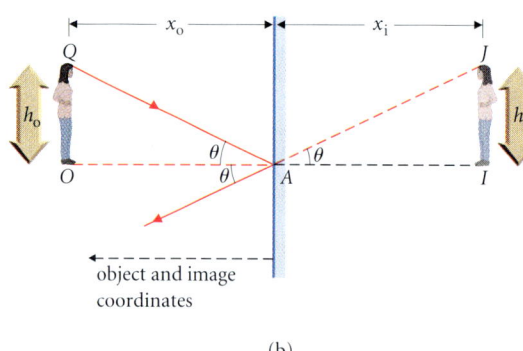

■ **FIGURE 18.24**
(a) The image in a plane mirror is virtual, unmagnified ($h_i = h_o$), and erect. Each point of the image is as far behind the mirror as the object is in front. Here we see rays from John's feet and from his head, reflecting from the mirror and entering his eye.

(b) Ray QA is one of the rays that forms image J. Similarly, ray OA is one of the rays that forms image I. Since $OA = IA$, these selected rays show that the image height equals the object height.

WARNING: WE SHALL NEED TO USE A NEGATIVE HEIGHT WHEN AN IMAGE IS UPSIDE DOWN.

The *magnification* of an image is the ratio of image height to the height of the corresponding object:

$$m \equiv h_i/h_o.$$

The magnification of the image in the plane mirror is $m = 1$.

The sizes of the object and image are geometrically related to the object and image coordinates:

$$\tan \theta = h_i/AI = h_o/OA \quad \text{so} \quad h_i/(-x_i) = h_o/x_o.$$

THIS FORMULA FOR THE MAGNIFICATION ALSO APPLIES TO CURVED MIRRORS AND LENSES (§18.4.2 AND §18.4.3).

Thus:

$$m \equiv h_i/h_o = -x_i/x_o. \tag{18.11}$$

REMEMBER: THE IMAGE IS VIRTUAL, SO x_i IS NEGATIVE AND $-x_i$ IS POSITIVE.

The image in a plane mirror has a negative coordinate, and its magnification is positive. The image has the same orientation as the object, and it is described as erect or upright.

| An *erect* image has the same orientation as the object.

The sign of the magnification describes the orientation of the image. Negative magnification corresponds to an inverted image.

| An *inverted* image has the opposite orientation from the object.

The image in a plane mirror is virtual, unmagnified, and erect.

Digging Deeper

Left, Right, and Parity

If a mirror doesn't turn an image upside down, why does it reverse right and left? The answer is easy: the mirror doesn't reverse right and left! The image of your right hand is directly in front of your right hand, just as the image of your forehead is directly in front of your forehead. While correct, this answer is not satisfying. The person in the mirror is definitely reversed, but front to back rather than left to right. Your image faces the opposite direction. In Lewis Carroll's *Through the Looking Glass,* Alice passes through a mirror into a world of fantasy. Could the world portrayed in a mirror be real? The question has important implications in physics, and the answer is no!

Suppose someone made a movie with the camera pointing at a mirror instead of directly at the actors. A sharp-eyed viewer might notice that most of the actors were left-handed. That's odd but not impossible. Only a few subtle experiments can tell the difference between the real world and the mirror world. None were known until, in 1957, T. D. Lee and C. N. Yang suggested several. For example, cobalt nuclei spontaneously decay into nickel nuclei, electrons, and weird particles called antineutrinos. C. S. Wu and her colleagues prepared cobalt so that all the nuclei had spin angular momentum in the same direction, and they found that the electrons were usually emitted in the opposite direction (■ Figure 18.25b). Now imagine viewing this experiment in a mirror; the spin would be reversed but the direction of the electrons would not. It turns out that the mirror result does not occur in the real world! The transformation of a system into its mirror image is an example of a symmetry transformation. Physicists call mirror-symmetry "parity." Other transformations include rotation and translation. These last two are believed to be exact symmetries of all physical laws. Parity is a symmetry for some laws (e.g., gravity and electromagnetism) but not all of them. The symmetry is said to be *broken* in weak interactions that govern the decay of the cobalt nuclei. Wu's result provided a major clue in the discovery of the fundamental laws that govern the behavior of matter at subatomic scales.

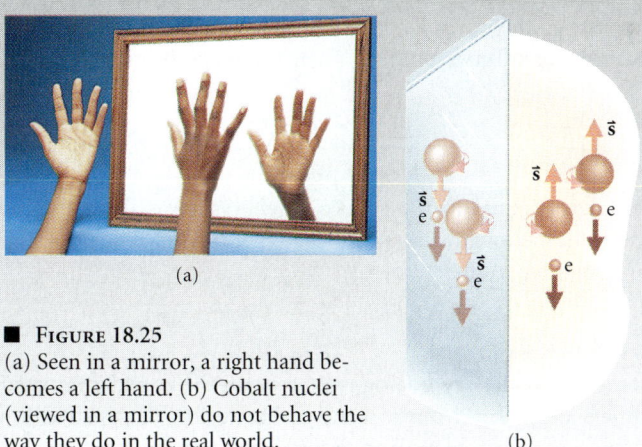

■ **Figure 18.25**
(a) Seen in a mirror, a right hand becomes a left hand. (b) Cobalt nuclei (viewed in a mirror) do not behave the way they do in the real world.

18.4.2 Images in Curved Mirrors

In §18.4.1 we found the image size using a derived rule (eqn. 18.1) for the image location. Such derived rules are nearly always easier to apply and consequently give more accurate drawings than direct use of the laws of reflection and refraction. An exact trace using a computer is better still. (See the essay on ray tracing.) Next we derive rules for curved mirrors and thin lenses that state the effect of the lens or mirror on a small number of selected rays.

An extended object perpendicular to the optical axis is represented by an arrow OQ (■ Figure 18.26). The image of the tail end O has coordinate x_i given by eqn. (18.4). The arrowhead Q is off the axis. We find its image by ray tracing.

A ray through the center of curvature C lies along a radius of the mirror, and so is incident normally. It is reflected back along the same path.

A ray parallel to the mirror axis looks like a ray from an object on the axis at infinity. (Recall Figures 18.12 and 18.13.) Such an object would have an image at the focal point, and so this ray passes through the focal point.

A ray that passes through the focal point looks like a ray from an object placed at the focal point. The image of that object is on the axis at infinity, and so the ray leaves the mirror parallel to the optical axis.

We may summarize these observations in three rules.

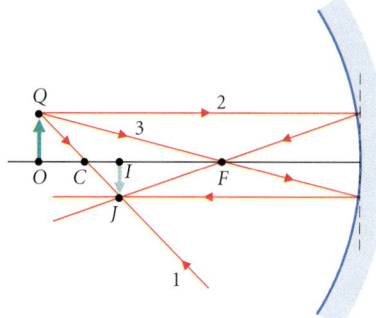

■ **Figure 18.26a**
Construction rays locate the image J of the object Q. The construction rays reflect at the mirror plane, shown as a dashed line, not at the drawn curve of the mirror. This construction is consistent with the paraxial ray assumption.

WE SHOW HOW TO APPLY THESE RULES TO CONVEX MIRRORS IN EXAMPLE 18.7.

RAY TRACING RULES FOR MIRRORS
1. A ray through the center of curvature is reflected back along the same path.
2. A ray parallel to the optical axis is reflected through the focal point.
3. A ray through the focal point is reflected parallel to the optical axis.

In Figure 18.26, these three rays intersect at the point J, which is the image of Q.

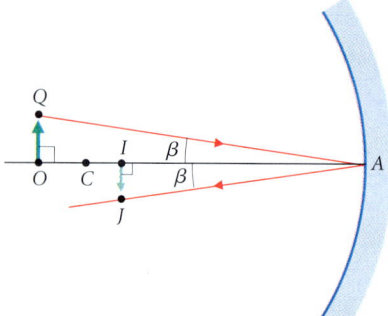

■ **FIGURE 18.26b** A ray that reflects at point A on the optical axis is useful for finding the height of the image.

We could use any two of the rules to locate the image of each point on the object. We find that the image IJ is perpendicular to the optical axis. (Equivalently, eqn. (18.4) correctly determines the image x-coordinate for off-axis points.) Then the magnification of the image (Figure 18.26b) is:

$$\left|\frac{h_i}{h_o}\right| = \frac{JI}{OQ} = \frac{IA}{OA} = \left|\frac{x_i}{x_o}\right|.$$

The ratio of image to object size is equal to the ratio of their distances from the mirror. According to our sign convention, the magnification is negative if the image is upside down, so, with x_i positive for a real image, we find:

$$m = -x_i/x_o.$$

WE MAY ALSO THINK OF THIS EQUATION AS GIVING THE y-COORDINATE OF AN IMAGE POINT. SEE PROBLEM 71.

Equation (18.11) also holds for curved mirrors.

REMEMBER: CONSTRUCTION RAYS ARE NOT REAL RAYS.
COMPUTER PROGRAMS READILY TRACE THE EXACT PATH OF RAYS THROUGH AN OPTICAL SYSTEM. ONE IS LISTED IN ESSAY 5.

It is important to remember that the ray tracing rules rely on the paraxial ray assumptions; that is, the mirror differs negligibly from a plane in the region of construction. To construct rays accurately, draw the mirror as a plane perpendicular to its axis; rays 2 and 3 reflect at this plane. In the figures, we draw this mirror plane as a dashed line.

EXAMPLE 18.7 ♦ You are standing 3.0 m from a convex mirror of focal length 1.0 m. Describe your image.

MODEL The image is perpendicular to the axis. To draw it, we could use any two of the three ray tracing rules. The focal point is behind the mirror, so we interpret the rules to mean that rays extended behind the mirror pass through C or F.

SETUP
1. A ray toward the center of curvature C is reflected back along the same line. The ray cannot actually pass through C, which is behind the mirror, but it is normal to the surface and is reflected back along the same line.
2. A ray parallel to the axis of the mirror diverges from the focal point F, which is behind the mirror (cf. Figure 18.13).
3. A ray that would pass through the focal point emerges parallel to the axis.

SOLVE In the ray diagram (■ Figure 18.27a), the virtual image of point Q is located where the backwards extensions of rays, 1, 2, and 3 intersect. By measuring the diagram, we find $AI = 3AF/4$ and $IJ = OQ/4$. Your image is erect and one-quarter your size.

ANALYZE We may confirm these results algebraically. The image is virtual and 0.75 m behind the mirror, so $x_i = -0.75$ m (see Exercise 18.8). The magnification is found from the similar triangles OQA and IJA (Figure 18.27b), giving the familiar formula:

$$m = -x_i/x_o = -(-0.75\text{ m})/(3.0\text{ m}) = +0.25.$$

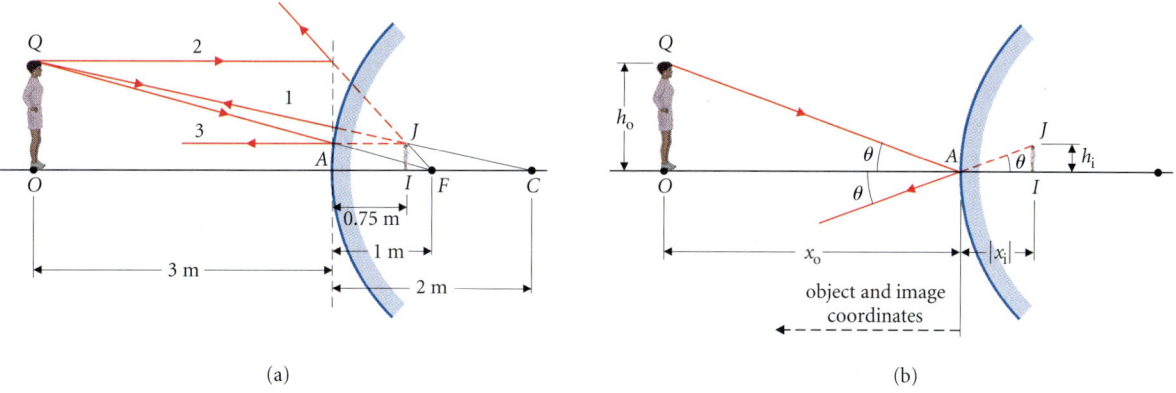

(a) (b)

■ **FIGURE 18.27**
(a) Construction rays for locating the image in a convex mirror. Note that the image lies on the backward extension of ray 3 parallel to the axis (shown in red) and not on the construction line to F (shown black). (b) A ray that reflects at A on the optical axis is useful for finding the magnification.

■ **EXERCISE 18.8** ♦ Verify the location of the image using the mirror eqn. (18.4). ■

Shannon Lucid's shrunken head is an image formed by a curved refracting surface. The object (her head) is in air, so $n_1 = 1.00$, and the material outside the helmet is water, $n_2 = 1.33$. We ignore the thickness of the helmet itself. Since her head occupies most of the space inside the helmet, we cannot assume paraxial rays, so we have traced rays accurately using a computer program. ■ Figure 18.28 shows rays from five points on a spherical object with radius equal to 60% of the helmet's radius. The refracted rays form a virtual image that is demagnified and distorted. The magnification is 0.88 at the front of the sphere and 0.74 at the top. Our subjective impression of "head shrinking" is greatly affected by the larger demagnification at off-axis points.

SEE THE FIRST PAGE OF THE CHAPTER.

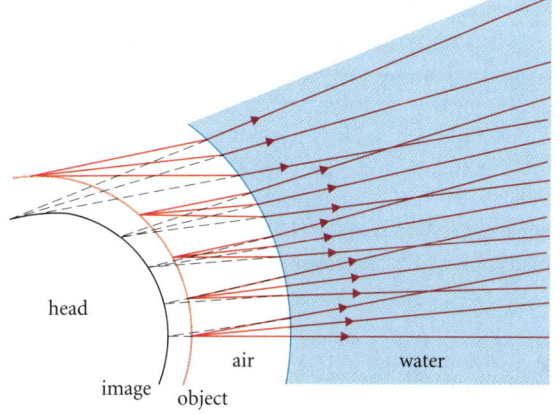

■ **FIGURE 18.28**
Results from a computer calculation of the paths of light rays leaving a spherical "head" with radius 60% of the helmet radius. The helmet has also been modeled as a sphere. The image is smaller than the object and also distorted. It looks too short. Compare this diagram with the photograph on the first page of this chapter.

18.4.3 Images Formed by Thin Lenses

There are also three rules for tracing rays through lenses. Near its center, the lens surfaces are very nearly parallel planes, and a ray passing through them emerges parallel to its original direction with a small sideways displacement proportional to the thickness of the lens (■ Figure 18.29). For a thin lens, we neglect this sideways displacement and draw the ray as a straight line through the center of the lens. The other two rules are the same as those for mirrors. Rays parallel to the lens axis, like those from an on-axis object at infinity, pass through the second focal point F_2. Rays through the first focal point F_1, like those from an object at F_1, emerge parallel to the axis.

> **RAY TRACING RULES FOR THIN LENSES**
> 1. A ray through the center of a thin lens continues in a straight line.
> 2. A ray parallel to the optical axis is refracted toward the second focal point F_2.
> 3. A ray passing through the first focal point F_1 emerges parallel to the optical axis.

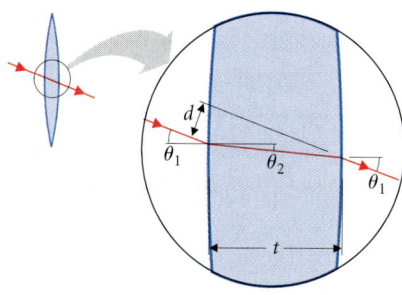

■ **FIGURE 18.29**
Selected ray 1 passes through the center of the lens and is undeflected. The small sideways displacement d may be neglected if the lens is thin and the ray is paraxial [$d = [(n-1)/n]\theta_1 t$; see Problem 16.70].

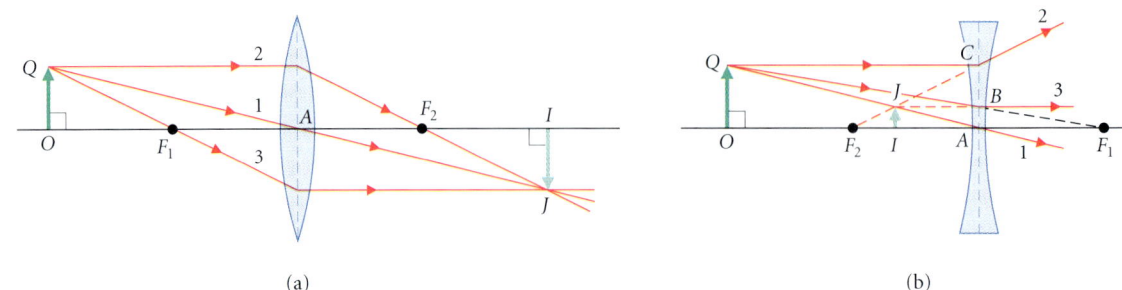

(a)   (b)

■ **FIGURE 18.30**
(a) Construction rays used to find the image formed by a converging lens. The image JI is perpendicular to the optical axis of the lens. The construction rays bend at the central plane of the lens (shown as a dashed line) so as to be consistent with the paraxial ray approximation. (b) Construction rays used to find the image formed by a diverging lens. Notice that F_2 is on the object side and F_1 is on the image side of the lens. Ray 2 is refracted as if from a virtual image at F_2, and ray 3 is aimed toward a virtual object at F_1 and refracted parallel to the axis. Note that the image point J does not lie on ray 3 but on the backward extension of the emerging ray (JB).

■ Figure 18.30 illustrates the use of these rules to construct images formed by converging and diverging lenses. The magnification of the image is found from the ratio:

$$\left|\frac{h_i}{h_o}\right| = \frac{IJ}{OQ} = \frac{AI}{AO} = \frac{|x_i|}{x_o}.$$

The image formed by the converging lens is inverted, so the magnification is negative. Again, $m = -x_i/x_o$.

These ray tracing rules are designed to agree with the algebraic results for paraxial rays passing through thin lenses. To be consistent, the rays must be drawn so that they bend at the central plane (shown as a dashed line in Figure 18.30). Do not draw construction rays bending at the lens surface!

When drawing a ray diagram for a diverging lens, remember why diverging lenses are so named—the lens spreads the rays apart. The second focal point is defined by the virtual image of an object at infinity: ray 2 *diverges from* the second focal point (Figure 18.30b). The first focal point is defined by a virtual object with image at infinity: ray 3 is *directed toward* the first focal point, beyond the lens, but is refracted away before it gets there. Both rays appear to diverge from the virtual image J. The image of Q does not lie on ray QB, but on JB, the backwards extension of the emerging ray. Similar care is needed when the object is inside the focal point of a converging lens.

EXAMPLE 18.8 ♦♦   A converging lens of focal length 5.0 cm is held 3.0 cm above a printed page, over a letter 2.0 mm long. Find the position and the magnification of the letter's image.

MODEL   The image is perpendicular to the axis. We use the ray tracing rules to find the image of the top of the letter, assuming the lens is thin.

SETUP   ■ Figure 18.31 shows three construction rays for this image. Ray 3 does not actually pass through F_1 because the object is in front of F_1. The ray appears to come from F_1.

SOLVE   The image is virtual and erect. Direct measurement shows that the image is 7.5 cm from the lens and that its magnification is 2.5.

ANALYZE   We may check our results using the lens formula (18.4):

$$\frac{1}{x_i} + \frac{1}{x_o} = \frac{1}{f} \Rightarrow x_i = \frac{x_o f}{x_o - f} = \frac{(3.0 \text{ cm})(5.0 \text{ cm})}{3.0 \text{ cm} - 5.0 \text{ cm}} = -7.5 \text{ cm}.$$

The magnification is: $m = -x_i/x_o = -(-7.5 \text{ cm})/(3.0 \text{ cm}) = 2.5$.

A magnifying glass is a converging lens used in this way.

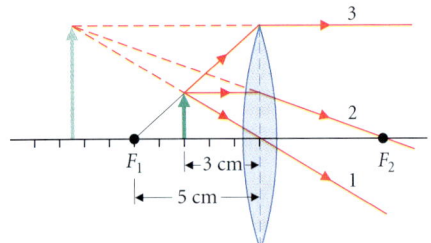

■ **FIGURE 18.31**
Construction rays for finding the image of an object viewed with a magnifying glass. The object is closer to the lens than its first focal point. The image is virtual. Note that ray 3 does not actually pass through F_1 but appears to come from F_1.

Ray diagrams are particularly useful for understanding image distortion. If an object lies entirely off the optical axis or makes a substantial angle with the axis, the image does not have the same shape as the object. As a consequence, there is no simple formula for the magnification. The image size may be found by careful measurement of the ray diagram. Off-axis distortion is an important factor in the design of optical instruments such as cameras and telescopes.

EXAMPLE 18.9 ♦♦ An object 1.0 cm long lies parallel to, and 1.0 cm from, the axis of a converging lens of focal length 3.0 cm, so that its head is 5.0 cm from the lens. Find the image formed.

MODEL Here the object is off axis, so we construct the image of both ends.

SETUP ■ Figure 18.32 shows construction rays 2 and 3 from each end of the object.

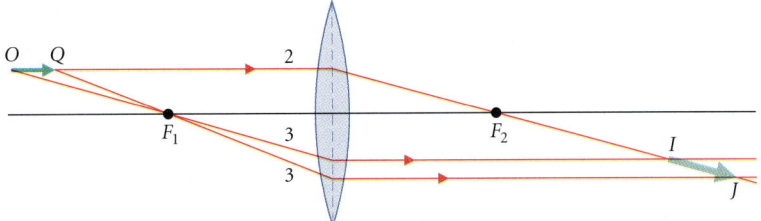

■ **FIGURE 18.32**
The image of an off-axis object is distorted. Construction ray 2 serves for both ends of the object and image. We also show ray 3 for both O and its image I, and Q and its image J. The image is not parallel to the axis.

SOLVE The real image IJ is not parallel to the lens axis. The size of the image is measured to be 1.6 times the size of the object.

THIS MAGNIFICATION CAN ALSO BE CALCULATED. SEE PROBLEM 71.

ANALYZE To avoid this distortion when using a magnifying glass, hold the lens perpendicular to the page, as in Example 18.8. ■

18.4.4 Visibility of Images

The rays used in constructing images are not necessarily real; they may miss the lens entirely (■ Figure 18.33). Such imaginary rays are still useful for finding images, but only light that passes through the lens forms an image. Real rays pass through the same image locations as construction rays, but along paths that are substantially more difficult to trace.

Once we have located an image, we can identify the set of rays that actually form it. In ■ Figure 18.34, we show rays from the foot O of the object, which form the image I in Example 18.9. An eye placed at point B in the cone of emerging rays can see the image point I but not J. Since J is farther from the lens, the cone of rays emerging from it has a smaller apex angle. To see the entire image, the eye must be placed inside this smaller cone—for example, at point A. The rays forming the sides of the smaller cone define the *limits of visibility* of the image IJ.

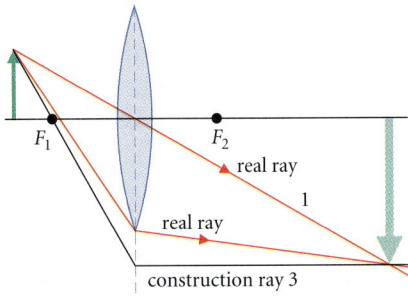

■ **FIGURE 18.33**
Construction rays are not necessarily real. Here construction ray 3 does not pass through the lens at all, so it cannot be a real ray. It still serves to locate the image, however. Once we know where the image is, it is straightforward to plot where the real rays go.

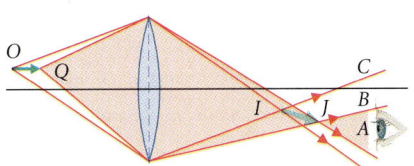

■ **FIGURE 18.34**
Rays that actually pass through the lens to form the image in Example 18.9. Since I is closer to the lens than J, the cone of rays that form image I is wider than the cone of rays forming image J. To view the whole image, the eye must be placed so that rays from both cones enter the eye (position A). An eye placed at B would see only part of the image; an eye placed at C would see no image at all.

18.5 COMPOUND OPTICAL SYSTEMS

Lenses and mirrors may be combined to form optical systems for specific uses. These instruments are coupled to photographic or electronic recording devices, or our own eyes! The human eye is itself an interesting optical system.

18.5.1 The Eye

SEE ALSO §16.2.1 AND FIGURE 16.15.

In the eye, a curved refracting surface (the cornea) is combined with a convex lens whose radii of curvature, and thus focal length, are controlled by muscles. Real images are formed on the retina (■ Figure 18.35). When the muscles attached to the lens of a normal eye are relaxed, objects at infinity are in focus. To focus on nearby objects, the eye muscles squeeze the lens to decrease its radii of curvature and hence its focal length. This *accommodation* of the lens maintains a constant image distance despite variations of object distance. The lens stiffens and thickens with age, and its ability to accommodate decreases. An eyeglass lens produces a virtual image that the wearer's eye can focus.

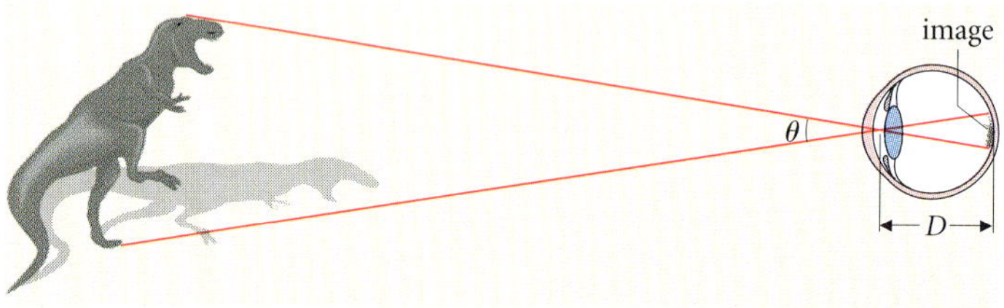

■ **FIGURE 18.35**
The size of an image on the retina is determined by the angular size of the object. The rays shown here are constructed according to rule 1 for the top and bottom of the object. The lens plus cornea form the lens system of the eye. The rays pass through the center of the lens system.

EXERCISE 18.9 ♦♦ The retina is 2.4 cm behind the cornea. Find the range of focal lengths of the lens-plus-cornea system necessary for the eye to accommodate to object distances from infinity to 25 cm.

The detail an eye can resolve depends primarily on the size of an image on the retina. Since the image distance from lens to retina is fixed, the image size is determined by the angular size of the object (Figure 18.35).

An object's *angular size* is the angle between rays from the top and bottom of the object that intersect at the eye.

Angular size may be increased by bringing an object closer to the eye. Sometimes, this is not possible—for example, if the object is the Moon. Furthermore, the eye's muscles can adjust its focal length over a limited range of values; objects too close to the eye cannot be brought into focus at all.

The *near point* of the eye is the closest position of an object that can be brought into focus.

The near point differs from individual to individual but is generally about 25 cm from the eye. The maximum angular size achievable with the unaided eye, θ_n, occurs with the object at the near point. Optical instruments can increase the apparent angular size of very small or very distant objects.

18.5.2 The Simple Magnifier

The simple magnifier, or magnifying glass, is a converging lens placed so that the object (height h_o) is closer to the lens than its focal point (cf. Example 18.8). The image formed is virtual and larger than the object.

The angular magnification produced by an optical instrument, m_θ, is given by:

$$m_\theta = \frac{\text{angular size of image}}{\text{maximum achievable angular size of object}}.$$

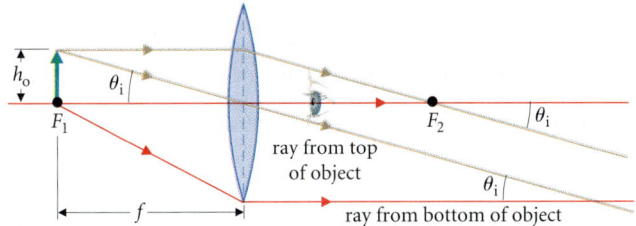

FIGURE 18.36
A simple magnifier is frequently used with the object at the focal point. The angular size of the image is h_o/f.

It is often recommended that the magnifier be used with the object at its focal point. Then the image is at infinity, and the eye is in its relaxed state as it views the image (■ Figure 18.36). The angular size of the image is then $\theta_i = h_o/f$, and, with $d_{np} \approx 25$ cm for an average eye, the angular magnification is:

$$m_\theta = \frac{\theta_i}{\theta_n} = \frac{h_o/f}{h_o/(25.0\ \text{cm})} = \frac{25.0\ \text{cm}}{f}. \qquad (18.12)$$

For the lens in Example 18.8, $f = 5.0$ cm and $m_\theta = 5.0$. This is often given as *the* angular magnification of the magnifier.

18.5.3 Microscopes

The simple magnifier is limited because it is difficult and expensive to make accurate lenses with very short focal length. A compound optical instrument is needed to obtain large magnification. The microscope achieves its magnification in two steps: the first lens, called the *objective*, produces a magnified real image that is viewed through the second or *eyepiece* lens used as a magnifier. The real image IJ acts as a real object at the focal point of the eyepiece lens (■ Figure 18.37).

We use the similar triangles OQA and IJA to relate the height of the real image, h_r, to the object height h_o: $h_r/h_o = -x_{i,1}/x_o$. The angular size θ_i of the virtual image is h_r/f_e, where f_e is the focal length of the eyepiece. The total angular magnification of the microscope is θ_i divided by the object's angular size at the near point of the unaided eye.

$$m_{\theta,\text{total}} = \frac{\theta_i}{\theta_n} = \frac{(-x_{i,1}h_o)/(x_o f_e)}{h_o/(25.0\ \text{cm})} = -\frac{x_{i,1}}{x_o}\left(\frac{25.0\ \text{cm}}{f_e}\right)$$

$$m_{\theta,\text{total}} \equiv m_o m_{\theta,e}. \qquad (18.13)$$

The total angular magnification is the linear magnification of the objective multiplied by the angular magnification of the eyepiece.

FOR EXAMPLE, A SIMPLE BICONVEX LENS WOULD HAVE SMALL r AND HENCE BE RATHER FAT. IT WOULD NOT PRODUCE CLEAR IMAGES, SINCE RAYS FORMING THE IMAGE WOULD NOT BE PARAXIAL.

REMEMBER: THE MINUS SIGN MEANS THE IMAGE IS INVERTED.

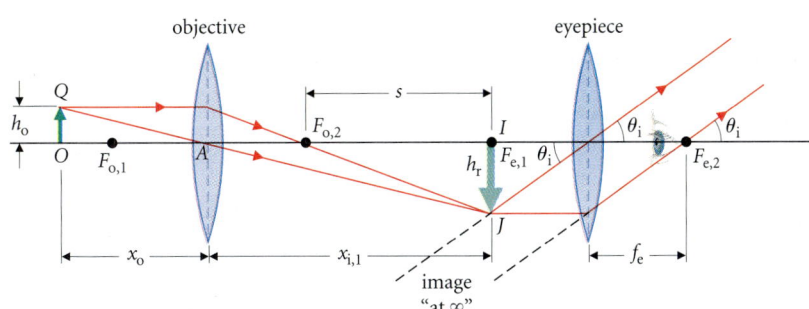

FIGURE 18.37
A simple microscope. Construction rays show the real image formed by the objective lens and the virtual image formed by the eyepiece lens, which acts as a magnifier. The trace is done in two sequential steps. Once the image IJ has been found, new construction rays originating at IJ are chosen to show the location of the final image.

EXAMPLE 18.10 ♦♦ In a physics laboratory exercise, an object is placed 8.0 cm from the objective lens (focal length 6.0 cm) of a microscope. The eyepiece lens has focal length 4.0 cm. The microscope is focused by adjusting the separation s of the two focal points, $F_{o,2}$ and $F_{e,1}$ (Figure 18.37). What magnification does this system achieve?

FIGURE 18.38
A modern research microscope works on the same basic principles as the simple microscope, but its lenses are carefully designed to give maximum magnification with minimum distortion. Design of the eyepiece is particularly important. This microscope has a magnification of 610.

AS USUAL, THE NEGATIVE VALUE MEANS THE IMAGE IS INVERTED.

FIGURE 18.39
A simple telescope. This design is called Keplerian. The eyepiece is placed so that its first focal point coincides with the second focal point of the objective. Hence the eyepiece forms a virtual image at *infinity*. Note we have positioned the object standing on the axis, as in all our other diagrams. (a) Construction rays used to find the image positions. (b) Rays that actually pass through the system to form the images. An on-axis object illuminates only part of the eyepiece. By placing the eye at the intersection of the emerging pencils of rays, the optimum field of view is obtained. The distance from the eyepiece to this point is called the *eye relief*.

MODEL Since the object is a fixed distance from the objective lens, its image I_1 is also at a fixed coordinate $x_{i,1}$. We focus the system by moving the eyepiece until the image I_1 is at the focal point $F_{e,1}$. Then eqn. (18.13) gives the magnification.

SETUP To determine $x_{i,1}$, we apply the lens equation to the objective lens, with $x_o = +8.0$ cm and focal coordinate $f_o = +6.0$ cm.

$$\frac{1}{x_{i,1}} + \frac{1}{x_o} = \frac{1}{f_o} \Rightarrow x_{i,1} = \frac{x_o f_o}{x_o - f_o} = \frac{(8.0 \text{ cm})(6.0 \text{ cm})}{8.0 \text{ cm} - 6.0 \text{ cm}} = 24 \text{ cm}.$$

The positive sign verifies that the image is real. We adjust the separation of the lenses until $s + f_e = x_{i,1}$—that is, until $s = 24$ cm $- 6.0$ cm $= 18$ cm. Comparing the similar triangles AIJ and AOQ, the image size h_r is then:

$$h_r = -\frac{x_{i,1}}{x_o} h_o = -\frac{24 \text{ cm}}{8.0 \text{ cm}} h_o = -3.0 h_o.$$

The angular size θ_i of the virtual image is:

$$\theta_i = \frac{h_r}{f_e} = \frac{-3.0 h_o}{4.0 \text{ cm}}.$$

We compare this with the maximum angular size of the object viewed at the near point, $\theta_n = h_o/(25 \text{ cm})$.

SOLVE The magnification $m_{\theta,\text{total}}$ of the microscope is:

$$m_{\theta,\text{total}} = \frac{\theta_i}{\theta_n} = -\frac{3.0 h_o/(4.0 \text{ cm})}{h_o/(25 \text{ cm})} = -19.$$

ANALYZE This value is much larger than one would expect for a simple magnifier with a reasonable value for f, but far less than is achieved by modern research microscopes (■ Figure 18.38).

18.5.4 Telescopes

Telescopes are used to view distant objects—usually so distant they are assumed to be at infinity. In the common Keplerian design, the principle of operation is similar to that of the microscope: form a real image and view it with a magnifying eyepiece. One measure of telescope performance is the angular size of the final image compared with the original angular size of the object.

A Keplerian telescope consists of two convex lenses (■ Figure 18.39). The first image is formed at the focal point $F_{o,2}$ of the objective, which coincides with the focal point $F_{e,1}$ of the eyepiece. The height h of this image is

$$h = -f_o \tan \theta_o \approx -f_o \theta_o,$$

assuming θ_o to be small. The final image is inverted and its angular size is:

$$\theta_i \approx h/f_e = -f_o \theta_o / f_e.$$

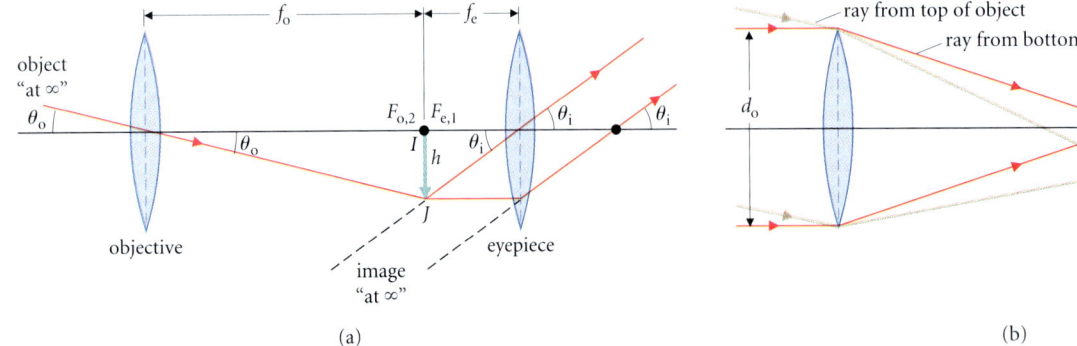

(a) (b)

The angular magnification is: $\quad m_\theta \equiv \theta_i/\theta_o = -f_o/f_e.\quad$ (18.14)

NOTE: THE OBJECT CANNOT BE MOVED CLOSER. ITS MAXIMUM ACHIEVABLE ANGULAR SIZE IS ITS ORIGINAL ANGULAR SIZE.

The rays that form the image are shown in Figure 18.39b. A point object on axis illuminates only a small fraction of the eyepiece. The emerging beam of light is described by the *exit pupil*. Rays from the top of the object emerge at the bottom edge of the eyepiece lens. The exit pupil's diameter is:

$$d_{\text{exit}} = d_o f_e/f_o = d_o/|m_\theta|.$$

The Keplerian design for refracting telescopes is traditional in astronomy, where inversion of the image is unimportant. Binoculars—one telescope for each eye mounted side by side—are designed to produce an erect image, usually using prisms (■ Figure 18.40). Binoculars designed for day use have an exit pupil of 3–5 mm, corresponding to the diameter of the pupil of the eye in average daylight conditions. Binoculars designed for night use have a large objective to gather more light and also a larger exit pupil, about 7 or 8 mm, since the pupil of the eye opens wider under low light levels.

EXAMPLE 18.11 ❖❖ You have been asked to design binoculars with $|m_\theta| = 10.0$, $d_{\text{exit}} = 3.0$ mm, and total length from objective to eyepiece of 25.0 cm. Find the required focal length of each lens and the diameter of the objective.

MODEL Each of the two identical telescopes in the binoculars is similar to the one shown in Figure 18.39. The parameters $d_{\text{exit}} = 3.0$ mm and $|m_\theta| \equiv f_o/f_e = 10.0$ are specified, as well as the total distance between the lenses, $f_o + f_e$. We are to find f_o, f_e, and d_o.

NOTE THAT f_o AND f_e ARE BOTH POSITIVE.

SETUP Since we have equations for f_o/f_e and $f_o + f_e$, we can solve for each focal length separately. We find the diameter d_o from the magnification and the size of the exit pupil:

$$f_o/f_e = |m_\theta| = 10.0, \quad f_o + f_e = 25.0 \text{ cm}, \quad \text{and} \quad d_o = |m_\theta| d_{\text{exit}}.$$

SOLVE $\quad f_o = 10.0 f_e \Rightarrow 11.0 f_e = 25.0$ cm $\quad$ and $\quad f_e = 2.27$ cm.

Then $\quad f_o = 22.7$ cm $\quad$ and $\quad d_o = (10.0)(3.0 \text{ mm}) = 3.0$ cm.

ANALYZE Binoculars are described both by their magnification and by the diameter of the objective lens: the specifications are given as $m_\theta \times d_o$. With $|m_\theta| = 10$ and $d_o = 30$ mm, these binoculars would be described as 10 × 30. ■

Study Problem 12 ❖❖❖❖ F-Stops

A camera is, in principle, a simple optical instrument that produces a real image on a strip of film. (The lenses used in modern cameras are far from simple, however! See ■ Figure 18.41.) Both the focal length f of the lens system and its diameter a (called the aperture) are adjustable. The image is well exposed when the total amount of light energy per unit area of film is near a specific value described by the film speed or ASA number. Light is admitted to the camera while the shutter is held open for a preset time called the exposure time. How should you choose the exposure time?

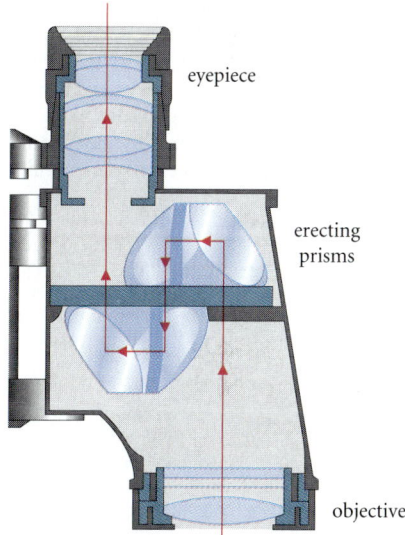

■ **FIGURE 18.40**
Binoculars are two telescopes mounted side by side. Since they are often used to view objects on Earth, two prisms are used to erect the image.

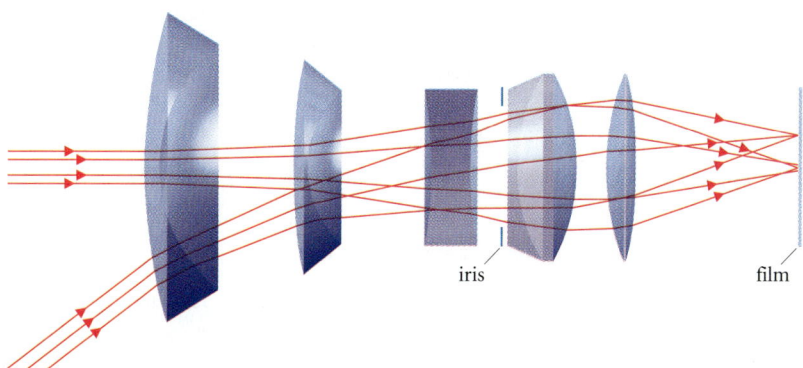

■ **FIGURE 18.41**
A fish-eye lens illustrates the complexity of modern camera lenses. The aperture is controlled by the iris between elements 3 and 4. Both on- and off-axis objects have images in focus at the film plane.

REMEMBER: THE DEFINITION OF INTENSITY = POWER/AREA. SEE, FOR EXAMPLE, §16.3.2.

Modeling the System

The exposure of the film is the product of the exposure time and the intensity of light falling on the film. The intensity depends on the area of the image and on the rate light from an object enters the camera lens. With a larger aperture, a greater amount of light enters the camera. With a larger focal length, the area of the image is greater. These two counteracting effects cause the exposure to depend on the ratio of focal length to aperture diameter, f/a, a quantity called the *f-ratio* or *f-stop*.

Setup

WE USED THE INVERSE SQUARE LAW.

An object with dimensions $h \times w$ at a distance D from the camera (■ Figure 18.42) emits light energy at a rate P, so the light intensity arriving at the lens is $I = kP/D^2$ (k is a constant that depends on the details of the emission). Light energy enters the camera at a rate $\pi a^2 I/4 = P(a/D)^2(k\pi/4)$ and spreads out over the area of the image on the film. The dimensions of the image are $(mh) \times (mw) = m^2 hw = m^2 A$, where m is the magnification and A is the area of the object. The image of a distant object is formed close to the focal point of the lens, so $m \approx f/D$. The film is properly exposed after a time t_{exp} when a fixed amount of light energy has been deposited on each unit area.

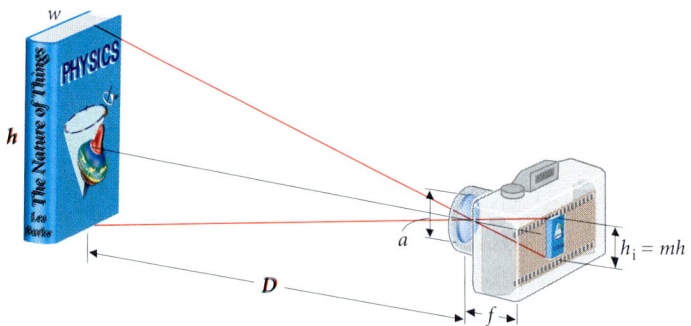

■ FIGURE 18.42
A camera contains a lens that forms a real image on the film. The exposure time depends on the light energy per unit area falling on the film. That depends on the light from the object that enters the camera and the area of the image. The lens demagnifies the image ($m < 1$).

Solution

$$\text{Energy per unit area on film} = \frac{k\pi P(a/D)^2 t_{\text{exp}}}{4A(f/D)^2} = t_{\text{exp}} \left(\frac{a}{f}\right)^2 \left(\frac{P}{A}\right) \left(\frac{k\pi}{4}\right).$$

To keep the energy per unit area constant, the exposure time should be adjusted as the square of the focal ratio.

$$t_{\text{exp}} \propto (f/a)^2.$$

Thus, when the *f*-stop is increased by a factor of $\sqrt{2}$, the proper exposure time increases by a factor of 2.

Analysis

The *f*-ratios of camera lenses are usually adjustable by factors of $\sqrt{2}$: f/1.4 (meaning $f/a = 1.4$), f/2.0, f/2.8, f/4.0, and so on, while exposure times vary by factors of 2: $\frac{1}{30}$ s, $\frac{1}{60}$ s, ... A lens with a small *f*-ratio f/a is often described as *fast*, since the image is relatively bright and the exposure time relatively short. Fast systems have some disadvantages, notably small depth of field.

DEPTH OF FIELD REFERS TO THE RANGE OF OBJECT POSITIONS THAT ARE FOCUSED TO AN ACCEPTABLY CRISP IMAGE.

Remarkably, for extended objects, the energy density on the film does not depend on the distance of the object at all.

GOOD RESOLUTION ALSO REQUIRES A LARGE OBJECTIVE (§17.3.3). WHILE THE ATMOSPHERE LIMITS RESOLUTION TO WELL BELOW THE DIFFRACTION LIMIT, MODERN TECHNIQUES USING INTERFEROMETRY CAN PRODUCE IMAGES CLOSE TO THE DIFFRACTION LIMIT.

18.5.5 Astronomical Telescopes—Reflectors

One major purpose of a good astronomical telescope is to gather as much light as possible, so the aperture of the objective should be as large as is practical. The telescopes we have described are refractors, a name that means they are constructed from lenses. Difficulties in producing,

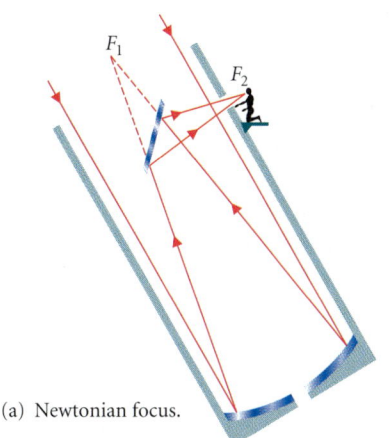

(a) Newtonian focus.

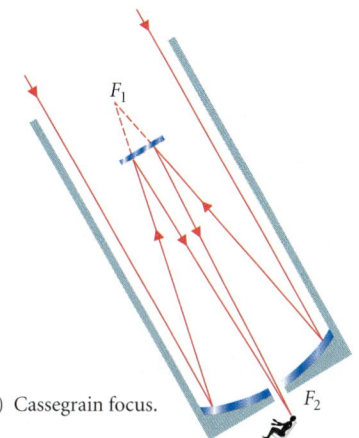

(b) Cassegrain focus.

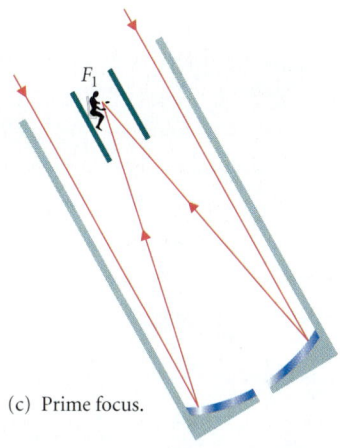

(c) Prime focus.

■ **FIGURE 18.43**
Several designs for reflecting telescopes.
(a) Newtonian focus. A plane mirror intercepts the light and reflects it through the support framework to a new focus F_2 outside the telescope tube.
(b) Cassegrain focus. A convex mirror intercepts light before it comes to a focus and directs it back to a new focus F_2 behind the primary mirror. The primary mirror has a hole in its center to allow the light to pass.
(c) Prime focus. The observer or automatic recording device is placed in a "cage," high up in the telescope structure at F_1.

supporting, and using large lenses limit refracting telescopes to objectives about 1 m in diameter. Mirrors can be made much larger, and they have other advantages too. They suffer much less from the kinds of image distortion or aberration that we discuss in §18.6. For example, parabolic mirrors bring parallel, off-axis rays to an accurate focus. For these reasons, the objectives of modern telescopes are concave mirrors. The Hale telescope on Mt. Palomar has a 5-m-diameter objective, and an 8.3-m mirror has been built for Japan's Subaru telescope.

Since a mirror focuses light in the middle of the telescope tube, reflecting telescopes present a novel design problem—getting at the focused light. ■ Figure 18.43 illustrates three solutions. In the Newtonian design, named after its inventor, a plane mirror sends the light out through the side of the telescope tube. The real image at F_1 becomes a virtual object for the plane mirror. A real image is formed at focus F_2. In the Cassegrain design, a convex secondary mirror placed in front of the primary mirror's focal point reflects the light through a hole in the primary. Large telescopes can carry a prime focus cage (■ Figure 18.44) in which an observer can ride. This arrangement has the shortest focal length and so is the *fastest*.

The different designs are suited to different kinds of astronomical observations, so telescopes can be changed from one configuration to another. Thanks to modern electronics, astronomers no longer have the romantic but chilly task of riding in the prime focus cage. Alas, professional astronomers rarely look through their telescopes any more.

■ **FIGURE 18.44**
Dr. Margaret Burbidge using the prime-focus cage at the 3-m telescope at the Lick Observatory near San Jose, California.

EXAMPLE 18.12 ♦♦ The primary mirror of a Cassegrain telescope has a radius of curvature $r_p = 30.00$ m. The secondary mirror is 10.00 m from the primary. What is its radius of curvature if the image of a distant star falls 1.00 m behind the primary?

MODEL The primary mirror forms a real image at its focal point, a distance $f = r_p/2 = 15.00$ m from the mirror. This image is a virtual object for the convex mirror, which in turn forms a real image behind the primary (Figure 18.43b).

SETUP With respect to the convex secondary mirror, the object coordinate is $x_o = -5.00$ m. The real image has coordinate $x_i = +11.00$ m.

THE MINUS SIGN INDICATES THAT THE OBJECT IS VIRTUAL.

SOLVE From eqn. (18.4), the required focal coordinate is given by:

$$f = \frac{x_i x_o}{x_i + x_o} = \frac{(11.00 \text{ m})(-5.00 \text{ m})}{11.00 \text{ m} + (-5.00 \text{ m})} = -\frac{55.00}{6.00} \text{ m} = -9.17 \text{ m}.$$

The required radius of curvature is twice the focal coordinate: $r_s = -18.3$ m.

■ FIGURE 18.45
The Keck telescope on Mauna Kea, Hawaii. Its mirror is a composite of many hexagonal segments that are controlled by computer.

SPHERICAL ABERRATION IN THE HUBBLE SPACE TELESCOPE WAS CAUSED BY ERRORS IN GRINDING THE MIRROR. SEE ESSAY 5.

| ANALYZE | The minus sign indicates that the center of curvature is behind the mirror; that is, the mirror is convex, as expected.

As larger and larger telescopes are designed, even mirrors become difficult and expensive to build. The twin Keck telescopes (10-m diameter) have a creative design in which many separate mirror segments form their images in the same place so that all the light can be combined in a single image (■ Figure 18.45). This technique involves continuous testing of image quality and adjusting of the component mirrors that has become possible in recent years because of advances in computer technology.

18.6 ABERRATIONS

In our discussion of optical systems, we have assumed that:

- Rays are paraxial.
- Lenses are thin.
- The focal lengths of the lenses do not depend on the wavelength of the light.

Each of these assumptions is an approximation for any real system: deviations from the ideal cause distortions of image quality known as aberrations. Here we comment on some of the most common aberrations.

Chromatic aberration occurs because the refractive index of a lens, and hence its focal length, depends on wavelength. The images of an object in red and in green light do not occur at the same place (■ Figure 18.46). The problem can be reduced by constructing composite lenses made of two or more materials with different refractive indices. Mirrors do not suffer chromatic aberration.

Spherical aberration occurs in both lens and mirror systems because nonparaxial rays are not focused at the same place as the paraxial rays. In effect, each annular region of the lens has a different focal length. Spherical aberration is primarily important in the design of "fast" optical systems that have both large aperture and short focal length. Both features require use of nonparaxial rays. Spherical aberration may be reduced or eliminated by using nonspherical surfaces, or by careful lens design. For example, a plano-convex lens with the plane side toward the image has less spherical aberration than a biconvex lens of the same focal length, when used to view a distant object.

Coma, or comatic aberration, is also caused by nonparaxial rays being focused at different positions. It occurs for off-axis objects and images, and may be thought of as a change in magnification across the lens. The central region of the lens/mirror forms a point image of a point source. Off-axis regions form a blurred image displaced from the point, giving rise to the comet shape that is responsible for the name. Coma occurs even when spherical aberration is absent (■ Figure 18.47).

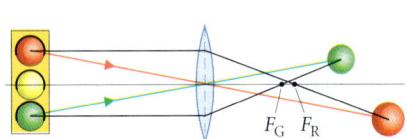

■ FIGURE 18.46
Chromatic (color) aberration occurs because the refractive index of glass depends on wavelength. The focal length of the lens varies with wavelength: $f_{\text{green}} < f_{\text{red}}$. The images of the red and green traffic lights are at different distances from the lens.

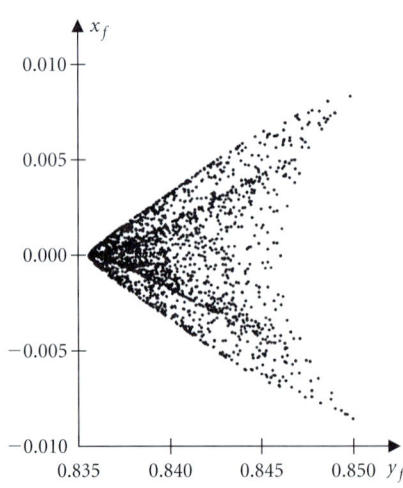

■ FIGURE 18.47
Coma. Computer ray trace of ray intersections in the focal plane. The model system has been corrected for spherical aberration. This image demonstrates almost pure coma. The image of an off-axis point produced by each annulus of the lens is a circle whose size increases with the size of the annulus. Note the circular edge of the "comet tail."

Where Are We Now?

Geometrical optics is the starting point for one of the most useful and widespread technologies: producing images. We have developed the basic theory of image formation by lenses and mirrors. We now leave the study of light until Chapters 33 and 35.

Chapter Summary

What Did We Do?

Images are of two types: virtual images (e.g., those formed by plane mirrors) and real images (like those formed by movie projectors). No light actually exists at a virtual image, though light propagates as if it came from the image. Objects, the sources of light in an optical system, may be tangible things or other images. A virtual object describes light converging toward a real image behind an optical surface.

We developed sign conventions for use in optical equations. Coordinates are measured from an optical surface. For the image coordinate x_i, the positive direction is forward along rays leaving the surface. For the object coordinate x_o, the positive direction is backward along rays approaching the surface. With these conventions, real objects and images have positive coordinates while virtual objects and images have negative coordinates (Table 18.1, Figure 18.7). The radius of curvature of an optical surface is given a sign using the image rules.

A single formula describes the location of images produced by spherical mirrors or thin lenses:

$$1/x_i + 1/x_o = 1/f. \tag{18.4}$$

The focal coordinate f is the image coordinate when the object is on axis at infinity. For a spherical mirror, $f = r/2$, where r is the coordinate of the mirror's center of curvature. Concave mirrors have positive r; convex mirrors have negative r. Convex mirrors form virtual images of real objects. A plane mirror is the limit of a spherical mirror as $r \to \infty$.

Light refracted at a spherical surface also forms images that can be located by eqn. (18.6). Two surfaces in series form a lens. Refracting surfaces and lenses have two focal points on the optical axis: F_1 is the location of an object whose image is at infinity; F_2 is the location of the image when the object is at infinity. For a thin lens, the two focal points are on opposite sides of the lens, but at the same distance $|f|$. If the lens is thin, eqn. (18.4) predicts the image position, with focal coordinate f given by the lens maker's equation (18.9 or 18.10):

$$\frac{1}{f} = (n-1)\left(\frac{1}{r_1} - \frac{1}{r_2}\right) \quad \text{(lens used in air)}.$$

Images of extended objects may be found by tracing rays. Three construction rays are particularly easy to locate.

1a. *Mirrors:* A ray through the center of curvature is reflected directly backward along the same line.
1b. *Thin lenses:* A ray through the center of the lens continues in a straight line.
2. A ray parallel to the optical axis passes through the second focal point.
3. A ray through the first focal point emerges parallel to the axis.

The two focal points of a mirror coincide.

The magnification m of an image is the ratio of image size to object size and is positive if the image is erect. For an object near and perpendicular to the optical axis, $m = -x_i/x_o$. For instruments like the eye, the angular size of the image is the most important factor in resolution of detail. Angular magnification is the angular size of the image divided by the maximum achievable angular size of the object.

Microscopes and telescopes use a combination of lenses to achieve greater angular magnification than can be achieved with a single lens. Modern astronomical telescopes use mirrors because they can be made larger than lenses. Some recent designs use combinations of mirrors in a computer-controlled array.

Chromatic and spherical aberrations and coma limit the performance of optical systems. These result from effects we ignore by assuming paraxial rays or constant index of refraction. Careful design can reduce or eliminate these defects. Professional optical engineers use computer ray tracing to design optical systems.

Practical Applications

Geometrical optics is widely used in civil engineering and surveying, camera and lens design, lithography of silicon chips, lightwave communications technology, medical and research microscopy, optometry, and astronomy. Some solar power plants use light concentrators designed with ray tracing.

Solutions to Exercises

18.1 ■ Figure 18.48 shows rays from the candle reaching the observer's eye. The rays appear to come from the image, which is 1.0 m behind the mirror and thus a total distance of 3.0 m from the observer. The candle appears to be 3.0 m away.

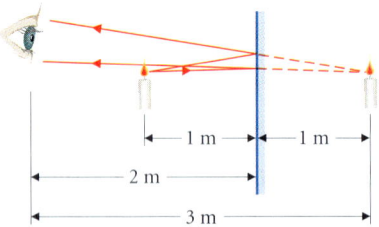

■ **FIGURE 18.48**
The image is 1.0 m behind the mirror, 3.0 m from the observer.

18.2 Brian should look toward image 3 along a line that intersects mirror A. ■ Figure 18.49 shows the region where this is possible, together with a ray bundle Brian might observe.

18.3 In this exercise, $n_1 = 1.00$ and $n_2 = 1.33$. The bird has a positive coordinate: $x_o = 1.0$ m. The image is located at coordinate:

$$x_i = -n_2 x_o / n_1 = -1.33(1.0 \text{ m})/1.00 = -1.33 \text{ m}.$$

The fish sees a virtual image 1.33 m above the water surface (a dangerous illusion)!

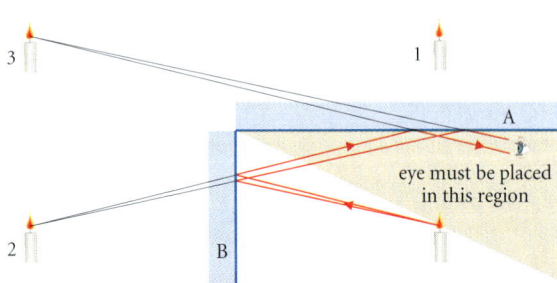

■ **FIGURE 18.49**
Using these rays to observe the third image, it appears as the image of image 2.

18.4 The image position for paraxial rays is found in a similar manner to that used for the concave mirror. Refer to Figure 18.14. As $2\theta = \alpha + \beta$ and $\theta = \alpha + \gamma$, we get $2\gamma = \beta - \alpha$. Using the paraxial ray assumption, this equation becomes:

$$2/AC = 1/AI - 1/OA.$$

The object coordinate is positive, $x_o = OA$, and the object is real. Similarly, $x_i = -AI$, and the image is virtual. The center of curvature C has coordinate $r = -AC$. Thus:

$$2/(-r) = -1/x_i - 1/x_o,$$

which is eqn. (18.3) with an overall sign change. Setting $f = r/2 = -AC/2$ for the convex mirror, we retrieve relation (18.4).

18.5 A real image is formed when $x_i > 0$. Rearranging eqn. (18.6):

$$\frac{n_2}{x_i} = \frac{n_2 - n_1}{r} - \frac{n_1}{x_o} = \frac{(n_2 - n_1)x_o - n_1 r}{x_o r}.$$

With $r > 0$, x_i is positive if $(n_2 - n_1)x_o > n_1 r$, or $x_o > x_{o,\min} = n_1 r/(n_2 - n_1)$.

18.6 The medium outside the eye is air, $n_1 = 1.0$, and $n_2 = 1.38$. Thus $f_1 = rn_1/(n_2 - n_1) = (0.78 \text{ cm})/0.38 = 2.1$ cm, and $f_2 = (n_2/n_1)f_1 = 1.38 f_1 = 2.8$ cm.

18.7 When a real image I_1 is formed to the left of the second surface, it becomes a real object for the second surface. Considered as an image, its coordinate $x_{i,1}$ is positive. Considered as an object, its coordinate $x_{o,2}$ is also positive and equals the distance $\ell - x_{i,1}$ to the second surface.

18.8 With object coordinate $x_o = +3.0$ m and focal coordinate $f = -1.0$ m, eqn. (18.4) becomes:

$$1/(3.0 \text{ m}) + 1/x_i = 1/(-1.0 \text{ m}).$$

Hence: $\dfrac{1}{x_i} = -\dfrac{1}{1.0 \text{ m}} - \dfrac{1}{3.0 \text{ m}} = \dfrac{-4.0}{3.0 \text{ m}} \Rightarrow x_i = -0.75$ m.

The image is 0.75 m behind the mirror.

18.9 The image coordinate is always $x_i = 2.4$ cm, so using eqn. 18.4, we have: $1/f = 1/x_o + 1/(2.4 \text{ cm})$. The focal coordinate f varies from 2.4 cm for objects at infinity to:

$$f = \frac{(25 \text{ cm})(2.4 \text{ cm})}{25 \text{ cm} + 2.4 \text{ cm}} = \frac{60 \text{ cm}^2}{27 \text{ cm}} = 2.2 \text{ cm}$$

for objects at the near point, 25 cm from the lens.

Basic Skills

Review Questions

- What happens if you try to isolate a single light ray?
- Explain why the ray approximation is valid whenever the laws of reflection and refraction apply.

§18.1 IMAGES FORMED BY PLANE SURFACES

- Where is an object's image in a plane mirror?
- Why do we often locate an image using rays that do not actually enter the eye?
- How many images are formed in a corner reflector (two mirrors at right angles)? Why?
- Define the terms: *optical surface, image,* and *object.*
- What is a *real image*? What is a *virtual image*?
- What is a *real object*? Is it always an actual, material thing?
- What is a *virtual object*?
- Define the object and image coordinates. Take particular care to define the positive sense of each.
- What is the equation relating the object and image coordinates for an image in a plane mirror? Explain the minus sign in this equation.
- Explain how images are formed by refraction at a plane surface.
- When you are watching a fish from directly above, does it appear closer to the water surface or farther away compared with its true distance?

§18.2 IMAGES FORMED BY CURVED SURFACES

- What kind of image is formed by a convex mirror? What kind or kinds of image are formed by a concave mirror? Explain carefully.
- What are *paraxial* rays? Why are they important?
- What is the *focal point* of a mirror? Where is it located?
- What is the equation relating the image and object coordinates when light reflects from a spherical mirror?
- What is the equation relating the image and object coordinates for an image formed by a spherical refracting surface?
- Define the *first focal point* and the *second focal point* of a curved refracting surface.

§18.3 LENSES

- Describe the procedure for finding the image formed by two refracting surfaces in sequence.
- When may a lens be described as *thin*?
- What is the *lens maker's equation*?
- What is the focal length of a lens?

§18.4 RAY TRACING AND IMAGES OF EXTENDED OBJECTS

- Define the magnification of an image. What is the magnification of an image in a plane mirror?
- What does it mean for magnification to be negative?
- How is the magnification of images of on-axis objects formed by mirrors or thin lenses related to the object and image coordinates?
- Describe three construction rays that allow you to find the image of an object in a mirror.
- Describe three construction rays for finding the image formed by a thin lens.

- How can you tell whether an image is visible, and if so, where it may be viewed?

§18.5 COMPOUND OPTICAL SYSTEMS

- How does the eye *accommodate* to focus objects at different distances on the retina?
- Define the *angular size* of an object. Why is it important?
- What is the *near point* of the eye?
- How does a simple magnifier work? How is its angular magnification defined?
- How does a microscope improve upon a simple magnifier? What is the total angular magnification of a simple microscope?
- How does a telescope differ from a microscope? In what ways are they similar?
- What is the angular magnification of a Keplerian telescope?
- What is the *f*-stop of a camera lens? Why is it important in photography?
- Why do modern astronomical telescopes use mirrors?
- Describe three designs of reflecting telescope.

§18.6 ABERRATIONS

- What is an optical *aberration*?
- Describe three common aberrations of lenses. Which of them also occur for mirrors? Why or why not?

Basic Skill Drill

§18.1 IMAGES FORMED BY PLANE SURFACES

1. Draw a diagram showing the rays that your eye would use to see a mirror image of your chin.

2. Refer to ■ Figure 18.50. For each diagram, describe the kind of object or image shown (real or virtual) and give the sign of its coordinate.

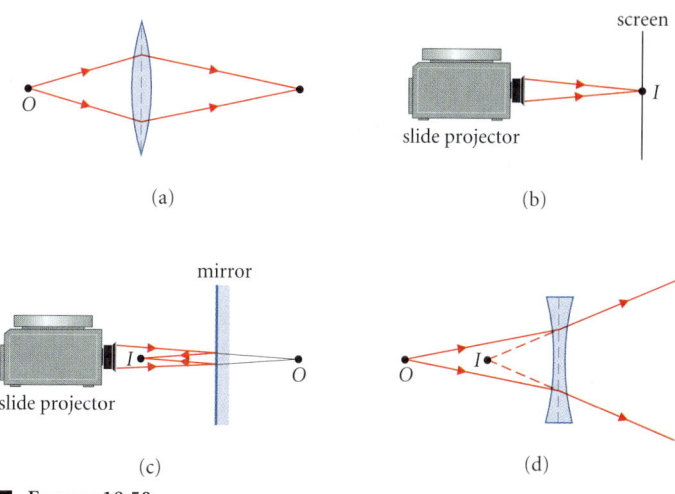

■ **FIGURE 18.50**

3. Looking in a plane mirror 1.0 m from you, you see an image of your physics teacher. The image appears to be 5.0 m in front of you. How far behind you is the teacher?

4. A coin lies on the bottom of a swimming pool 3.0 m deep. How far under the water does it appear to be, when viewed from directly above?

§18.2 IMAGES FORMED BY CURVED SURFACES

5. Two objects are placed in front of a concave spherical mirror with a radius of curvature of 8.0 cm. The objects are 10.0 cm and 3.0 cm from the mirror. Where are the images formed?

6. A star is reflected in a church dome, which is a hemisphere of radius 20.0 m. Where does the star's image appear to be?

7. A fish tank in a museum has a convex spherical front surface of radius 5.0 m. Find the image of a fish when it is **(a)** 3.0 m from the glass, and **(b)** 1.0 m from the glass. (You may ignore the thin layer of glass.)

8. Describe the image the fish in the previous problem sees when a person stands 1.0 m from the glass.

§18.3 LENSES

9. Two thin lenses have identical shapes. The index of refraction of lens A is 1.5 and the index of refraction of lens B is 2.0. What is the ratio of their focal lengths f_A/f_B?

10. An object is 15 cm from a converging lens with $f = +10.0$ cm. How far is the image from the object?

11. An object is 10.0 cm in front of a convex lens with a focal length of 25 cm. What kind of image is formed, and where is it?

§18.4 RAY TRACING AND IMAGES OF EXTENDED OBJECTS

12. An extended object is placed in front of a convex mirror at a distance equal to the focal length of the mirror. What is the magnification of the image?

13. A converging lens with a focal length of 57.0 cm is used to view a distant tree whose angular size is 10.0°. What is the size of the tree's image?

14. A lens with a focal length of 5.0 cm is held above a printed page so that the image of the printing is 25 cm below the lens. How far above the page is the lens held? What is the magnification?

15. An object is 10.0 cm in front of a concave mirror with a radius of curvature of 16 cm. Where is the image? What is the magnification? Is the image real or virtual, erect or inverted? Verify your answers by using construction rays to find the image.

16. An object 1.0 cm high is 3.5 cm from a diverging lens with a focal length of 2.0 cm. Use construction rays to find the position and size of the image. Use eqns. (18.4) and (18.11) to verify your result.

§18.5 COMPOUND OPTICAL SYSTEMS

17. A microscope has separation $s = 16$ cm between the two focal points. If the focal length of the objective is $f_o = 3.1$ cm and the eyepiece has focal length $f_e = 2.5$ cm, what is the magnification achieved by this microscope?

18. A Keplerian telescope has an objective with focal length $f_o = 15$ cm and an eyepiece with focal length $f_e = 2.5$ cm. What is the magnifying power of this telescope?

§18.6 ABERRATIONS

19. A convex lens is made of crown glass with index of refraction $n = 1.5296$ at 480.0 nm and $n = 1.5203$ at 656.3 nm. The radius of curvature of each face is 6.500 cm. Find the focal length of the lens at each wavelength.

Questions and Problems

§18.1 IMAGES FORMED BY PLANE SURFACES

20. ❖ Students practicing navigation measure the angle between a star and its image in a pool of water. How is the measured angle related to the angle of the star above the horizon?

21. ❖ Two plane mirrors are joined at a 60° angle. How many images are formed? Draw a ray diagram to show how each image could be viewed.

22. ❖ Two plane mirrors are joined at a 45° angle. How many images are formed? Draw a ray diagram to show how each image could be viewed.

23. ❖ Two mirrors are parallel to each other, distance d apart. An object is placed midway between them. How many images are there? What are their distances from the mirrors? What limits the number of images you can actually see?

24. ❖ Two plane mirrors join at an angle π/n, where n is an integer. Show that $n - 1$ images are formed and discuss their locations.

25. ❖ ✉ How can a system of two mirrors be used to see your own back? How far does the back of your head appear to be from your eye (in terms of the dimensions of your system)?

26. ❖ Explain how the rough surface of a movie screen makes the real image visible to the audience.

27. ♦ A sculpture is 2.0 m from a plane mirror. The photographer stands 3.0 m from the mirror and 5.0 m from the sculpture in the direction parallel to the mirror. For what distance should she focus the camera to photograph the sculpture's image in the mirror?

28. ♦ While working on a storage tank full of carbon tetrachloride, Jenny drops a wrench. The wrench appears to be 1.1 m below the surface. How deep is the liquid in the tank?

29. ♦ In a museum, a valuable document is protected by a sheet of crown glass 1.05 cm thick ($n = 1.53$). At what distance from the top of the glass sheet does the image of the writing appear when viewed from directly overhead?

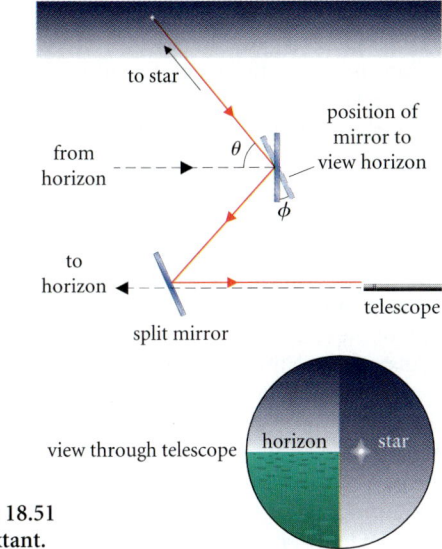

FIGURE 18.51
Marine sextant.

FIGURE 18.53

30. ♦♦ While swimming under water in a lake, you notice a branch that is 0.25 m above the water and try to grab it. Do you reach too high or too low? By how much?

31. ♦♦ ■ Figure 18.51 shows a marine sextant. The lower, split mirror is fixed. The central mirror rotates about a fixed axis. The observer, looking through the telescope, rotates the central mirror until both the horizon and a star are visible together. What is the relation between the altitude θ of the star and the angle ϕ that the mirror is rotated?

32. ♦♦♦ Find the position of a fish's image (■ Figure 18.52) when viewed at a large angle θ as shown. (*Hint:* Only a small bundle of rays enters the eye. Consider rays separated by a small angle $d\theta$, and work with the small triangle *IAB* shown in the figure.)

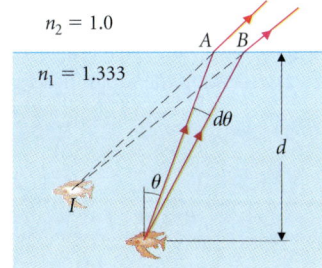

■ **FIGURE 18.52**

§18.2 IMAGES FORMED BY CURVED SURFACES

33. ❖ A concave spherical mirror with a 20-cm radius is placed in water ($n = 1.33$). What is the focal length of the mirror in the water?

34. ❖ Flashlights have silvered reflectors behind the bulb. What shape reflector is best? Why?

35. ❖ A spherical mirror has a radius of 5 cm. If an object is placed 2 cm in front of the mirror, and its image is virtual, can you tell whether the mirror is concave or convex?

36. ❖ Why can you see better under water if you wear goggles?

37. ❖ Why do large vehicles like buses have convex rearview mirrors?

38. ❖ (■ Figure 18.53) The photographer has made two self-portraits in a soap bubble. How did he do it?

39. ♦ A light bulb is 1.0 m from a concave mirror with a radius of curvature of 0.50 m. Where is the image? Describe it.

40. ♦ A candle is 10.0 cm inside a bay window with a radius of curvature of 3.0 m. What kind of image is formed by reflection off the glass, and where is it?

41. ♦ A truck's rearview mirror is convex with a radius of curvature of 22 cm. A car is 17 m from the mirror. Where is the car's image?

42. ♦♦ Show that a convex mirror produces a virtual image unless the object is virtual and closer than the focal point.

43. ♦♦ A lamp's oil reservoir is a thin-walled glass sphere of radius 8.5 cm. Along your line of sight, the wick is 6.0 cm from the surface. Where does the wick appear to be ($n_{oil} = 1.5$)?

§18.3 LENSES

44. ❖ Is a convex lens always a converging lens? Why or why not?

45. ❖ A fish-eye lens set into a door allows the occupant to view the porch. Explain how it operates. (See Figures 16.39 and 18.41.)

46. ♦ Find the focal length of the lens shown in ■ Figure 18.54.

47. ♦ A screen is set up 2.00 m in front of a slide projector lens ($f = +5.00$ cm). Where should the slide be placed if a real image is to be focused on the screen?

48. ♦ A plano-convex lens has a radius of curvature of 12.5 cm and an index of refraction of 1.57. What is the focal length of the lens?

49. ♦♦ Design a -4-diopter meniscus eyeglass lens, with $n = 1.5$. (There are many possibilities.)

50. ♦♦ Find the minimum separation between a real image and object for a convex lens of focal length f.

51. ♦♦ A thin lens with focal coordinate f produces an image with coordinate x_i of an object with coordinate x_o. Show that:

$$(x_o - f)(x_i - f) = f^2.$$

52. ♦♦ A luminous object and a screen are separated by a distance d. Show that there are two positions for a convex lens of focal length f

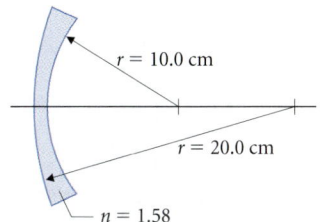

■ **FIGURE 18.54**

to be placed between object and screen so that a real image falls on the screen, provided that $f < d/4$. Find f in terms of d and the separation s of the two lens positions. (This is an accurate experimental method for determining f.)

53. ◆◆ Find the image of the light bulb in the ornate window (Example 18.5) *without* using the thin-lens approximation. Verify that the thin-lens approximation is good to 3%.

54. ◆◆ A fortune-teller's crystal ball has a radius of 10.0 cm and $n = 1.500$. Find the image of an object placed 1.00 m from its center formed **(a)** by reflection, and **(b)** by refraction. **(c)** Find the first and second focal points of the ball.

55. ◆◆ A sugar solution with $n = 1.42$ contains an air bubble of radius 1.00 cm. Where is the image of a sugar crystal 3.00 cm from the bubble's surface when viewed through the center of the bubble? Does the bubble behave like a converging or diverging lens?

56. ◆◆ An object lies at the bottom of a puddle 5.0 cm deep. On top of the water is a layer of oil ($n = 1.2$) 2.0 cm deep. Where does the object appear to be?

57. ◆◆◆ Show that for a symmetric biconvex lens to be considered thin, its diameter should be much less than the radius of curvature r of each surface.

58. ◆◆◆ A camera lens has focal length $f = 5.0$ cm. If the tolerable error in image position is $\delta x_i = 0.5$ mm, what range of object distances is properly focused near $x_o = 3.0$ m?

59. ◆◆◆ Find the focal points for a glass sphere of radius R and refractive index n. Show that the distances d_o and d_i of object and image from these two points satisfy the relation: $d_o d_i = f^2$, where f is the distance of each focal point from the center of the sphere.

60. ◆◆◆ **(a)** Two thin lenses of focal coordinates f_1 and f_2 are placed together. Find the focal coordinate f of the combination if it may still be considered thin. **(b)** A converging lens with a focal coordinate $f_1 = +10.0$ cm is placed in immediate contact with a diverging lens with $f_2 = -20.0$ cm. Find the focal coordinate of the combined lens. Is it converging or diverging?

§18.4 RAY TRACING AND IMAGES OF EXTENDED OBJECTS

61. ❖ Images of a very distant giraffe are produced by a diverging lens and by a converging lens of the same focal length f. Compare the sizes of the two images.

62. ❖ ■ Figure 18.55 shows a convex mirror used to allow cars to see around a blind intersection. Use ray diagrams to explain how the mirror works.

63. ❖ A convex mirror has a radius R. For which object distances is the magnification less than unity?

64. ◆ Describe the image of an extended object that is a distance $2f$ in front of a convex mirror of focal length f.

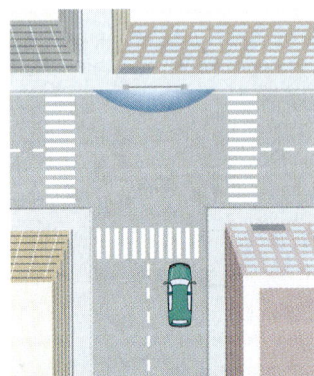

■ FIGURE 18.55

65. ◆ An object 0.50 cm high is placed 14 cm from a thin converging lens with a 4.5-cm focal length. Carefully draw construction rays to find the position and size of the image. Calculate these quantities, using eqns. (18.4) and (18.11). Do the results agree?

66. ◆◆ A symmetric concave lens is made of material of refractive index $n = 1.50$ and radii of curvature $|r| = 20.0$ cm. What is the focal length of the lens? An object 5.0 cm high is placed 30.0 cm from the lens. Compute the image location and describe the image completely. Construct the image using a ray diagram.

67. ◆◆ A biconcave lens is made of glass with a refractive index of 1.50 and has surfaces with radii of curvature 8.0 cm and 16 cm. What is the focal length of the lens? An object is placed 16 cm to the right of the lens. Calculate the position of the image formed. Draw an accurate diagram using construction rays to locate the image. What is the magnification? Is the image real or virtual?

68. ◆◆ A person 1.5 m tall stands 1.0 m away from a vertical plane mirror. What is the minimum length of the mirror that allows the person to view a complete image?

69. ◆◆◆ An object of length $\ell = 5.0$ mm lies along the optical axis of a concave spherical mirror whose focal length is 20.0 cm. If the middle of the object has coordinate $x_o = +50.0$ cm, what is the length of its image? The ratio of image length to object length is the longitudinal magnification of the mirror. How does the longitudinal magnification compare with the magnification of an object perpendicular to the optical axis?

70. ◆◆◆ Find the image of a square object of side $\ell = f/5$ that sits with one side on the optical axis of a converging lens of focal length f such that its corners are at distances $2f$ and $2.2f$ from the lens.

71. ◆◆◆ **(a)** Draw construction rays to find the image formed by the lens of the arrow shown in ■ Figure 18.56. Measure the relative sizes of object and image (the magnification). **(b)** Using the magnification formula (18.11), as well as eqn. (18.4), calculate the position of each end of the image and hence calculate the magnification. Do your answers to (a) and (b) agree? **(c)** Compare with the magnification for an arrow lying on the axis at the same distance from the lens.

72. ◆◆◆ Draw construction rays to find the image formed by the lens of the arrow shown in ■ Figure 18.57. Measure the relative sizes of object and image, and determine the magnification.

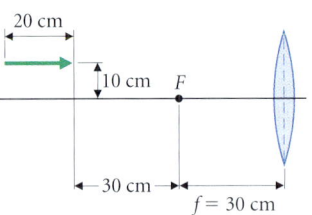

■ FIGURE 18.56

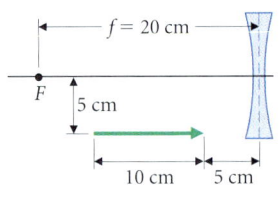

■ FIGURE 18.57

§18.5 COMPOUND OPTICAL SYSTEMS

73. ❖ Two identical plano-convex lenses are placed into contact in two different ways (■ Figure 18.58). Compare the focal lengths of the two combinations. Are they converging or diverging?

74. ❖ If a person wears diverging eyeglasses, will the image of a distant object be closer to, or farther from, the cornea than it is without eyeglasses?

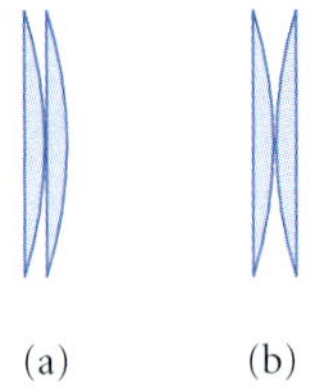

■ FIGURE 18.58 (a) (b)

75. ❖ A Galilean telescope uses a converging objective lens and a diverging eyepiece lens. Under what conditions does a Galilean telescope produce a real image?

76. ❖ For each of the following systems, discuss the importance of image character—real or virtual, inverted or erect, magnified or reduced: **(a)** astronomical telescope, **(b)** binoculars, **(c)** camera, **(d)** compound microscope, **(e)** movie projector, **(f)** eyeglasses, **(g)** searchlight mirror, **(h)** makeup mirror, **(i)** rearview mirror in a car.

77. ◆◆ A converging and a diverging lens are placed 5.0 cm apart, and an object O is placed 5.0 cm from the first lens. Each lens has a 5.0-cm focal length. What happens to the image position if the lenses are interchanged?

78. ◆◆ A small object is 20.0 cm from a thin converging lens of focal length 10.0 cm. To the right of the lens is a plane mirror that crosses the axis at the focal point of the lens and is tilted at 45° so that the reflected rays do not go back through the lens. Draw a ray diagram showing the final image. Is the image real or virtual?

79. ◆◆ Our eyes work by adjusting the focal length of the lens until the image falls on the retina. On planet Zorch, the inhabitants have eyes that act like cameras: the lens moves to keep the image on the retina. The lens of the Zorchian eye has a focal length of 2.5 cm. The relaxed eye focuses light from infinity on the retina. How much must the lens move, and in what direction, to see an object clearly 25 cm from the eye? (Fishes' eyes actually work this way.)

80. ◆◆ A converging lens of focal length f is placed $0.7f$ in front of a diverging lens of the same focal length and used to view a distant giraffe. Compare the resulting image with those obtained by the single lenses alone (cf. Problem 61.)

81. ◆◆ Each of three identical, thin converging lenses has a 5.0-cm focal length. The lenses are spaced 15 cm apart along a common optical axis. If an object OQ is placed 5.0 cm in front of the first lens, what is the magnification of the final image?

82. ◆◆ Two converging lenses, each of focal length 100 cm, are placed as shown in ■ Figure 18.59. The light source is located at the focal point of lens 1. Explain why the image formed by lens 2 doesn't depend on the distance between the lenses. Where is the image?

83. ◆◆ Two converging lenses and an object OQ are arranged as shown in ■ Figure 18.60. Compute the distance of the final image from the second lens and find its magnification. Use a ray diagram to construct the image.

84. ◆◆ The object in Figure 18.60 is moved inward until it is 1.0 m from the first lens. Where is the final image then? Construct the image using selected rays, and find its magnification.

85. ◆◆ Two lenses with equal focal lengths f are separated along their common optical axis by a distance $2f$. An object of height $f/5$ is on the axis a distance x_o to the left of the first lens. For each of the cases described below, use eqn. (18.4) to compute the location of the final image, and construct the image using a ray diagram: **(a)** Both lenses converging, $x_o = 3f/4$. **(b)** Both lenses converging, $x_o = f$. **(c)** Both lenses converging, $x_o = 1.5f$. **(d)** Lens 1 converging, lens 2 diverging, $x_o = 1.5f$. **(e)** Lens 1 converging, lens 2 diverging, $x_o = f$. **(f)** Lens 1 converging, lens 2 diverging, $x_o = 3f/4$. **(g)** Lens 1 diverging, lens 2 converging, $x_o = 1.5f$. **(h)** Lens 1 diverging, lens 2 converging, $x_o = f$. **(i)** Lens 1 diverging, lens 2 converging, $x_o = 3f/4$.

86. (a) ◆◆ A Galilean telescope has a converging objective lens that forms an image at the first focal point of the diverging secondary. Show that the final image is erect, find the angular magnification of such a telescope, and redesign the binoculars of Example 18.11 ($m_\theta = 10.0$, $d_{exit} = 3.0$ mm, objective-to-eyepiece distance = 25 cm) using Galilean telescopes. **(b)** ◆◆◆ What advantages or disadvantages would these binoculars have compared with a more conventional design using converging lenses and erecting prisms?

87. ◆◆◆ A telephoto lens consists of a converging front lens (focal coordinate f_1) and a diverging back lens (focal coordinate $-f_2$) (■ Figure 18.61). This has the advantage of giving a large image close to the back of the lens. **(a)** Compute the distance of the image from the diverging lens. This is called the back focal length. **(b)** Compute the effective focal length f_{eff} of the lens. (See figure.) **(c)** Compute the ratio of effective focal length to back focal length. **(d)** Show that the image size for an object with angular size θ is $f_{eff}\theta$.

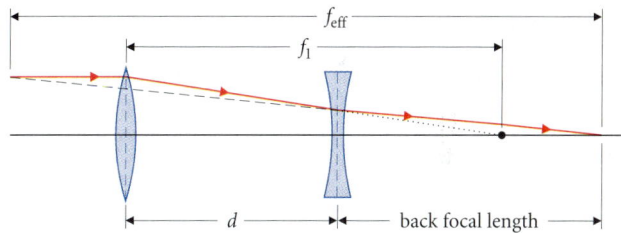

■ **FIGURE 18.61**
Telephoto lens.

88. ◆◆◆ ■ Figure 18.62 shows an optical system designed to observe radioactive samples at the tip of the object. Construct the final image of the object using a ray diagram. Show the limits of visibility of the sample.

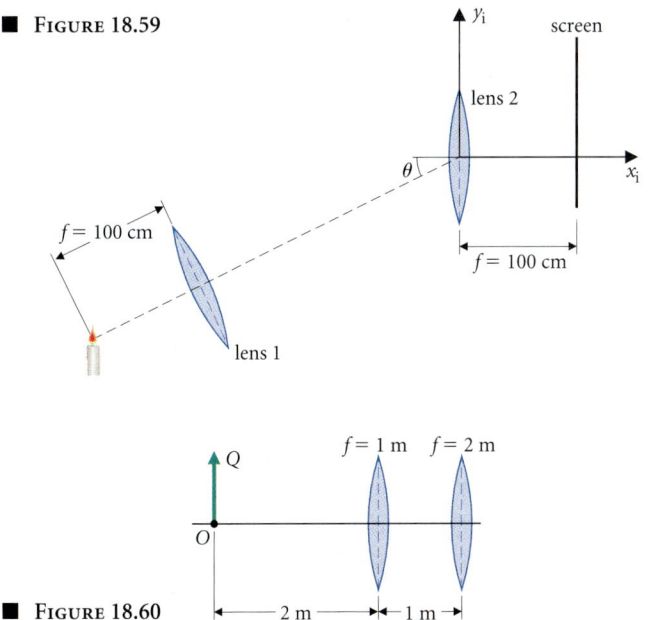

■ **FIGURE 18.59**

■ **FIGURE 18.60**

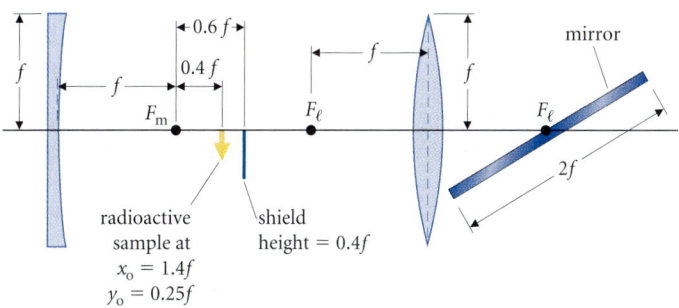

■ **FIGURE 18.62**

§18.6 ABERRATIONS

89. ♦♦ A thin, symmetric glass lens used in air has radii of curvature $r_1 = +10.0$ cm and $r_2 = -10.0$ cm. The index of refraction of the glass has the values $n = 1.500$ for $\lambda = 500$ nm and $n = 1.470$ for $\lambda = 700$ nm. Find the difference in image distance for the two colors for an object 3.0 m in front of the lens.

90. ♦♦ A ray of light parallel to the optical axis of a concave spherical mirror strikes the mirror at an angle of incidence of 30°. Find where the reflected ray crosses the optical axis. How far is this point from the focal point for paraxial rays?

91. ♦♦♦ The objective lens of a telescope consists of a symmetric converging lens made of crown glass ($n = 1.51$) and a diverging lens made of flint glass ($n = 1.66$) used to correct for chromatic aberration (■ Figure 18.63). The focal length of the crown glass lens alone is 6.0 m. What radius of curvature is required for the back surface of the flint glass element if the system is to focus light of wavelengths $\lambda = 486$ nm and $\lambda = 656$ nm at the same point? Assume the individual lenses and the combination are thin. The difference in the index of refraction at the two colors is n(at 486 nm) $-$ n(at 656 nm) $=$ 0.0085 for crown glass and $= 0.0206$ for extra dense flint glass. What is the focal length of the combined lens?

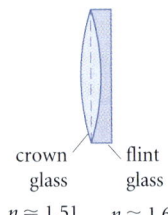

■ **FIGURE 18.63**
An achromatic lens.

Additional Problems

92. ❖ A symmetric convex lens focuses a real image onto a screen (cf. Figure 18.30). What would happen to the image if: **(a)** the lens were removed, **(b)** the screen were removed, or **(c)** the top half of the lens were covered with masking tape? In each case, justify your answer carefully using ray diagrams.

93. ❖ Use the wave front picture to show that every ray between the object and the image of a thin lens intersects the same number of wave fronts. How do the travel times of light between object and image compare for different rays?

94. ❖ Design a microscope that projects a real image on a screen.

95. ❖ Design a telescope to project an image of the Sun on a screen.

96. ❖ A paraboloidal surface has no spherical aberration if the object is at infinity. Does this remain true for objects at a finite distance? If not, what shape would be appropriate? (*Hint:* Look at Figure 0.16.)

97. ❖ ✉ In a "heads-up" display, the windshield of an airplane or automobile is used as a mirror to create a virtual image of certain instruments at infinity. To see how the system works, create a simple model using a plane mirror. Light from the instrument display is projected directly upward onto the windshield. At what angle to the horizontal should the mirror be placed for the image to be visible when you are looking straight ahead? What kind of light beam should be projected for the image to be at infinity? How can this be achieved?

98. ♦♦ Use the wave front method to show that a convex paraboloidal mirror has no spherical aberration for an object at infinity.

99. ♦♦ A glass lens ($n = 1.55$) has a focal length of 10.0 cm in air. What is its focal length when used in water? Does it make any difference whether the lens is converging or diverging?

100. ♦♦ Design a symmetric lens, made of water, with focal coordinate $f = +10.0$ cm.

101. ♦♦ Design a 10-diopter eyeglass lens, if it is to be made of glass with $n = 1.6$. (*Note:* There is not just one correct answer. You have many different options.)

102. ♦♦ A nearsighted person cannot focus objects more distant than 1.5 m onto the retina. (This is called the *far point*.) Should converging or diverging spectacle lenses be prescribed to correct the defect, and what power is required? (*Hint:* The image formed by the lens of a point at infinity should be at the person's far point.)

103. ♦♦ Determine whether it is possible for the surface in Figure 18.18b to produce a real image.

104. ♦♦♦ A telescope mirror has a diameter of 20.0 cm and a focal length of 1.0 m. How much does a parabolic mirror with these specifications differ from a spherical mirror?

105. ♦♦♦ Show, according to the following plan, that a paraboloidal mirror has no spherical aberration for objects at infinity. **(a)** Show that the formula $y = x^2/2R$ is the formula for a parabola that most closely matches a circle of radius R with center at $y = R$ and $x = 0$. **(b)** Find the angle of incidence for a ray parallel to the optical (y-) axis incident on the parabola at an arbitrary value of x. **(c)** Apply the law of reflection and show that the reflected ray intersects the y-axis at $y = R/2$.

Computer Problems

These problems make use of the computer program TRACE listed in Figure E5.1.

106. Run the ray tracing program to demonstrate spherical aberration in the Cassegrain telescope of Example 18.12 if it has spherical mirrors. An incoming ray, parallel to the optical axis, reflects from the mirrors and crosses the axis 1.0 mm from the paraxial focus. Determine the initial distance of the ray from the axis.

107. Use the ray tracing program to show how "bending" a lens can reduce its spherical aberration. Create a bi-convex lens of any focal length, and illuminate it with parallel light (corresponding to an object on axis at infinity). (A refractive index of 1.8 works well for this test.) Calculate the position of the paraxial image. Run a few rays through the system very close to the axis to verify this position. Next run a few rays farther from the axis and note how far from the paraxial image they cross the axis. Now reduce the radius of curvature of the first surface by a factor of 2 and set r_2 to infinity ($1/r_2 = 0$) to obtain a plano-convex lens of the same focal length. Run your paraxial rays again to find the paraxial image. It will have moved slightly due to a change in the effective center of the lens. Next run the nonparaxial rays. You should find that the aberration has been reduced by about a factor of 2. (This simple lens has aberrations other than spherical aberration so it is not possible to find a perfect image.)

108. Run the ray tracing program to investigate how to minimize spherical aberration in a Cassegrain telescope. Using the dimensions in Example 18.12, assume that the primary is paraboloidal ($s = 0$), and has an aperture of 3.0 m. Show that spherical aberration is minimized if the secondary has $s \approx 6.5$. This approximates the design of the Shane telescope at the Lick Observatory, University of California.

109. Put the following model of the eye into the computer program.

Surface	Radius of Curvature of Surface	Distance to Next Surface	Refractive Index
Cornea	7.8 mm	0.6 mm	1.376
Aqueous humor	6.4 mm	3.0 mm	1.336
Lens	10.1 mm	4.0 mm	1.404
Vitreous humor	−6.1 mm	16.9 mm	1.337
Retina			

Run rays parallel to the axis through the system and show that they are focused on the retina. Now change the lens radii to ±5.3 mm. Show that an object 50 cm away from the cornea forms an image on the retina. Be sure to use paraxial rays. Slopes less than about $u = 0.003$ work well.

Challenge Problems

110. In the microscope of Example 18.10, both the objective lens and the eyepiece have apertures of 1.0 cm. If the object is 1.0 cm long and centered on the optical axis, what fraction of its image is visible at one time?

111. The secondary mirror of a Cassegrain telescope ($f_{\text{primary}} = $ 15.0 m, distance between mirrors = 10.0 m; cf. Example 18.12) can

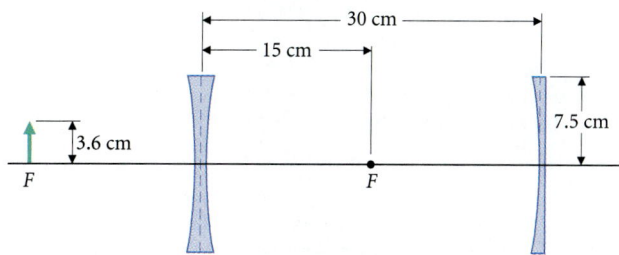

■ **Figure 18.64**

be moved back and forth a distance of 2 cm. What is the minimum distance of an object that can be focused at the image plane 1.00 m behind the primary mirror?

112. A concave lens and a concave mirror with equal focal lengths are arranged so that the focal point of the mirror and the first focal point of the lens coincide, ■ Figure 18.64. A small object is placed perpendicular to the optical axis of the system at the second focal point of the lens. Construct the final image using a ray diagram. Show the limits of visibility of the tip of the object if the diameter of each element is equal to $|f|$. (*Hint:* There are three images.)

113. Using construction rays, find the image of the object in Figure 18.64 if the concave lens is replaced by a convex lens with the same focal length $|f|$ and the same diameter. Show on your diagram the limits of visibility of the image's tip.

ESSAY 5
Ray Tracing Using a Computer

MICHAEL LAMPTON

In the preceding chapter you learned how an optical device produces an image: rays proceed from the object, encounter one or more optical surfaces that reflect or refract them, and finally arrive at a surface where they form an image or become visible. A professional optical engineer designing a camera lens or a microscope may have to examine the detailed paths of thousands of individual rays going through each trial system to evaluate its image quality. To handle this workload, rays are traced using a computer.

A ray tracing program has to deal with two entities: the optical system and the rays passing through it.

For a system producing a real image (e.g., a movie projector or a camera lens) the screen or film is included as the final optical surface. Information about image quality is obtained from the positions of the final rays, ignoring their arrival directions. However, for systems that produce a virtual image, such as a telescope or microscope, the final element is an "eye plane" where the viewer's eye will be located. Here, information about the image quality is obtained from the final ray directions rather than the positions of the final rays.

A ray tracing program needs the starting point and direction for each ray. The position and direction are customarily specified in Cartesian vector form: X, Y, and Z locate the point in space from which the ray is launched towards the optical system, and U, V, and W are the components of the unit vector along the ray. A ray aimed along the $+Z$-axis has a direction vector $U = 0$, $V = 0$, $W = 1$. The length of the ray direction vector is always unity.

The ray trace program called "TRACE" is listed in ■ Figure E5.1. The code is written in BASIC. Separate stages of the computation are marked by groups of line numbers: the setup of the optical train is done in the 2000's, the ray is set up in the 3000's, and the loop that manages the actual trace is in the 4000's. The remaining groups of statements are subroutines that do the numerical work. Like professional ray tracers, this program has messages that catch common error situations. If a surface intersection occurs behind the start of a ray segment, the message "backward" is displayed. If a ray is refracted into a lower-index medium at too large an angle, it may not be transmittable due to total internal reflection; a statement in the refraction subroutine handles this situation.

This program deals with optical systems that are coaxial: the surfaces are all lined up along the Z axis with their centers at $X = 0$, $Y = 0$, and none of them is tilted. The position of each surface is described by the Z-coordinate of the point where it crosses the axis. The surface figure is written as a mathematical function describing how deviation from a plane surface depends on the distance r from the axis of symmetry.

$$\text{Deviation from a plane} = \frac{Cr^2}{1 + \sqrt{1 - SC^2 r^2}}.$$

If $S = 0$, the surface is a paraboloid. It curves toward the $+Z$ or $-Z$-direction, depending on whether the curvature C is positive or negative. The shape factor S allows the same formula to describe other kinds of surfaces too. Negative values of S generate hyperboloids, while positive values of S describe ellipsoids (prolate for $S < 1$ and oblate for $S > 1$). The special value $S = 1$ represents a spherical surface, which is by far the most common in the design of lenses because spherical shapes are the most easily manufactured.

Figure E5.1
Ray tracing program

```
1000 REM   RAY TRACING PROGRAM
1010    DEFINT I-N: CLS : X0=1: Y0=0: Z0=0: U0=0: V0=0: W0=1
2000 REM SPECIFICATION OF THE OPTICS
2010    IF EN(1) < 1 THEN 2050
2020    PRINT : INPUT "CHANGE OPTICS?  Y/N/QUIT:", X$
2030    IF X$ = "q" THEN END
2040    IF X$ <> "y" THEN 3000
2050    INPUT "HOW MANY SURFACES?  ", N: IF N < 1 THEN END
2060    PRINT "IndexOfRefract, Z, Curv, 0=parab/1=sph, transmit/reflect"
2070    FOR I = 1 TO N
2080       INPUT EN(I), ZV(I), C(I), S(I), T$(I)
2090       IF EN(I) < 1 THEN EN(I) = 1
2100    NEXT I
3000 REM specification of the ray start
3010    PRINT : INPUT "New ray start? y/n/quit:", X$
3020    IF X$ = "q" THEN END
3030    IF X$ <> "y" THEN 4000
3040    PRINT "  x0      y0,     z0,    u0,   v0": INPUT X0, Y0, Z0, U0, V0
3050    IF U0*U0+V0*V0 > 1 THEN PRINT "Error: u0 v0 too big": GOTO 3040
3060    W0 = SQR(1-U0*U0-V0*V0)
4000 REM run the ray
4010    X=X0: Y=Y0: Z=Z0: U=U0: V=V0: W=W0
4020    PRINT "     x           y           z           u           v           w"
4030    PRINT USING "####.######"; X,Y,Z,U,V,W
4040    FOR I = 1 TO N                    : REM for each surface,
4050       Z = Z - ZV(I)                  : REM go to its frame
4060       GOSUB 5000                     : REM find ray intercept
4070       IF I = N THEN 4100             : REM all done
4080       IF T$(I) = "r" THEN GOSUB 6000: REM reflect it
4090       IF T$(I) = "t" THEN GOSUB 7000: REM refract it
4100       Z = Z + ZV(I)                  : REM return to lab frame
4110       PRINT USING "####.######"; X,Y,Z,U,V,W
4120    NEXT I
4130    GOTO 2000
5000 REM intercept subroutine finds new X,Y,Z
5010    XI=X: YI=Y: ZI=Z: OLDF=1000000   : REM save starting point
5020    IF ABS(W)>.7 THEN X=X-U*Z/W: Y=Y-V*Z/W: Z=0
5030    F = Z-.5*C(I)*(X*X+Y*Y+S(I)*Z*Z): REM f=0 at quadratic surface
5040    IF ABS(F) < .000001 THEN 5100    : REM converged
5050    IF ABS(F/OLDF) >= 1 THEN PRINT "ray missed surface": GOTO 2000
5060    OLDF=F: DFDS = -C(I)*X*U-C(I)*Y*V+(1-S(I)*C(I)*Z)*W
5070    DS=.1: IF ABS(DFDS) >.001 THEN DS=-F/DFDS
5080    X=X+U*DS: Y=Y+V*DS: Z=Z+W*DS     : REM take the step
5090    GOTO 5030
5100    DIR = (X-XI)*U+(Y-YI)*V+(Z-ZI)*W: IF DIR < 0 THEN PRINT "backwards"
5110    RETURN
6000 REM reflection subroutine finds new U,V,W
6010    GOSUB 8000
6020    U=U-2*DOT*P: V=V-2*DOT*Q: W=W-2*DOT*R: RETURN
7000 REM refraction subroutine finds new U,V,W
7010    GOSUB 8000
7020    D=EN(I+1)/EN(I): REM refraction ratio
7030    DOT2=1-(1-DOT*DOT)/(D*D)         : REM (emergent ray dot normal)^2
7040    IF DOT2 < 0 THEN PRINT "error: internal reflection": GOTO 2000
7050    GAMA = SQR(DOT2)*SGN(DOT)-DOT/D
7060    U=U/D+P*GAMA: V=V/D+Q*GAMA: W=W/D+R*GAMA: RETURN
8000 REM surface normal subroutine finds new P,Q,R,DOT
8010    P=-C(I)*X: Q=-C(I)*Y: R=1-S(I)*C(I)*Z: A=SQR(P*P+Q*Q+R*R)
8020    P=P/A: Q=Q/A: R=R/A: DOT=U*P+V*Q+W*R: RETURN
```

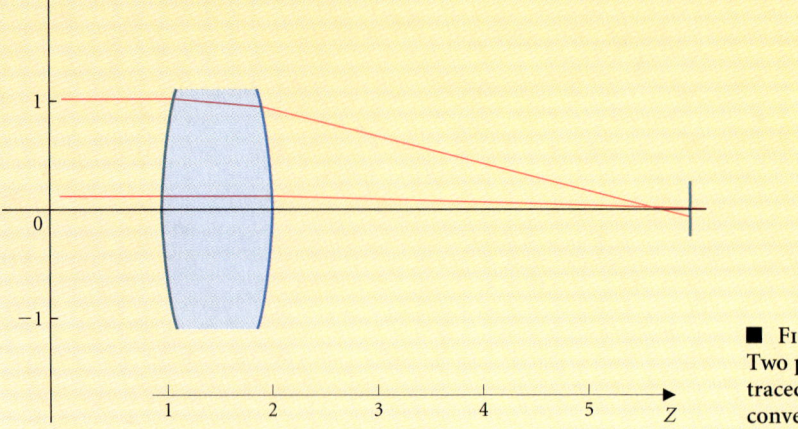

■ FIGURE E5.2
Two parallel rays traced through a biconvex glass lens.

To use this program, first draw a diagram of the optical system you want to study. Choose a $Z = 0$ point at the object if it is at a finite distance, or at some convenient place near the front of the instrument if the object is at infinity. For geometrical ray tracing, only the relative dimensions matter, so any convenient unit of length can be used. In optical work, lengths are often specified in inches or millimeters. Whatever unit you choose, be consistent. Surface curvatures are the reciprocals of the radii of curvature, with each radius in the same units you have used for the rest of the system. Number the surfaces in the sequence that light will arrive at them. Each surface needs a curvature and shape value. Flat surfaces have $C = 0$ and $S =$ anything. Mark the Z-axis values of the location of each surface—that is, the point where it crosses the Z-axis.

For our first example we shall trace two parallel rays of light through a thick biconvex glass lens (■ Figure E5.2). The rays come to an approximate focus on a flat screen near the focal plane. The optical media involved are air, whose refractive index is 1.0, and glass, assumed here to have a refractive index of 1.60. The media and surfaces are specified in the same sequence that light reaches them.

Run the program by typing GWBASIC TRACE or BASICA TRACE. The first thing the program asks is:

```
How many surfaces?
```

Our first example has three surfaces: the front of the lens, the back of the lens, and the screen. So, respond with a 3. The program then asks you to specify the three surfaces. Each surface is described by a list of five items separated by commas. The five things are the refractive index of the medium approaching the surface, the surface's location along the Z-axis, the curvature of the surface, the shape of the surface, and the letter "t" or "r" indicating whether the transmitted or reflected ray is to be followed (use "t" for lenses and "r" for mirrors). Press the Enter key after typing in each row of data. For our example, the three surfaces are specified as follows:

```
1,      1,      0.2,    1,   t
1.6,    2,     -0.2,    1,   t
1,      6,      0,      1,   t
```

The first row specifies a surface approached in air ($n = 1$), with the surface located 1 unit along the Z-axis. It is positively curved; that is, it curves towards positive Z. Since the medium on the left of the surface is air and the glass will be to the right, this makes the surface convex. Curvature is the reciprocal of the radius of curvature. With a curvature value of $+0.2$, its radius of curvature is 5 units, curved towards larger Z. The surface shape value of 1 indicates a sphere.

The second row starts off with a refractive index of 1.6, so the ray will be traveling through a dense glass medium as it travels from the first surface to the second. The second surface in our example is located at $Z = 2$ units. Unlike the first surface, this second surface is curved towards the negative Z-direction. Since the glass lies to the left of this surface and air is to the right, this second surface is also convex. Again it is spherical: shape = 1. Again, we follow the transmitted ray: "t".

The third row specifies that the ray will travel through air ($n = 1$) until it gets to the third surface at $Z = 6$. The third surface has zero curvature, which means it is flat. It is the image screen.

When you have entered all the surface information, the program then asks if you want to set up a ray to launch at the system:

New ray start? y/n /quit:

The program has a built-in default ray start that we can use. This default ray is one unit above the Z-axis and parallel to it. For this example we will use this default ray. Enter "n" to use this existing ray start.

Now the program has all the information needed to compute the ray trace. It works through the list of surfaces, displaying the ray position X, Y, and Z, and the ray direction components U, V, and W. The initial line of output shows the ray starting position and direction. The subsequent lines of output show where the ray intercepted the first, second, and third surfaces:

x	y	z	u	v	w
1.000000	0.000000	0.000000	0.000000	0.000000	1.000000
1.000000	0.000000	1.101021	−0.075957	0.000000	0.997111
0.938285	0.000000	1.911173	−0.240905	0.000000	0.970549
−0.076626	0.000000	6.000000	−0.240905	0.000000	0.970549

This completes an entire cycle of data input and output. The program then loops back to see whether you have any changes to make in the optical system and then asks if you want to change the ray start data. At either point, a "q" response will quit the program.

What can a ray trace tell us about our example lens? If the system were sharply focused on an object on the Z-axis at infinity, then this entering ray would have landed at the focal point on the axis. The final value of X would have been zero. But it is not zero! It is negative, showing that it crossed the axis on its way to the final image screen. For this ray, the screen is too far away from the lens. Should we trace again, moving the screen closer to the lens? Perhaps not. Let's trace again, this time using a ray that starts much closer to the axis, at $X = 0.1$, more closely imitating a paraxial ray:

x	y	z	u	v	w
0.100000	0.000000	0.000000	0.000000	0.000000	1.000000
0.100000	0.000000	1.001000	−0.007501	0.000000	0.999972
0.092513	0.000000	1.999144	−0.023109	0.000000	0.999733
0.000032	0.000000	6.000000	−0.023109	0.000000	0.999733

This time, the ray lands extremely close to the desired $X = 0$ point in the image, 2000 times closer, in fact. What is different? Real lenses, even when made from mathematically perfect spherical surfaces, have a variety of optical aberrations. We have just seen the effect of spherical aberration. The ray near the axis is very well focused, but the initial ray, starting ten times farther from the axis, landed badly out of focus.

Ray tracing is used in advanced work too. NASA's premier space observatory, Hubble Space Telescope, shown in ■ Figure E5.3, is a two-mirror instrument that was designed to

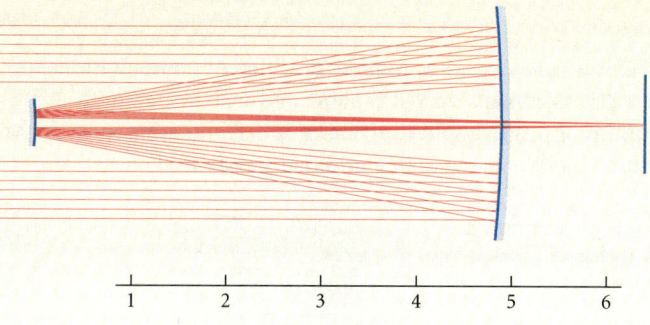

■ FIGURE E5.3
Hubble's two-mirror Optical Telescope Assembly.

ESSAY 5 • RAY TRACING USING A COMPUTER

produce images just a few microns in size, while employing a large 2.4-m aperture to gather useful amounts of light from extremely distant galaxies. The primary mirror is hyperbolic (but very nearly a paraboloid) and the secondary mirror is hyperbolic. Unfortunately, a manufacturing flaw caused the primary mirror to be made with a shape factor that is too negative. We can explore the consequences of this error with our ray tracer.

First, Hubble as it was designed (units are meters):

```
Number of surfaces: 3
    1,      4.906071,        -.09057971,         -.0023,         r
    1,      0,               -.736377,           -.49686,        r
    1,      6.4062,          0,                  0,              t
```

Using a typical ray 1 m from the axis (the default ray serves nicely), we get:

x	y	z	u	v	w
1.000000	0.000000	0.000000	0.000000	0.000000	1.000000
1.000000	0.000000	4.860781	-0.179684	0.000000	-0.983725
0.111315	0.000000	-0.004558	-0.017362	0.000000	0.999850
-0.000005	0.000000	6.406200	-0.017362	0.000000	0.999850

Notice the superb focus: for this ray, or any ray starting within 1.2 m of the axis, the final ray coordinate lies within a few microns of the axis. Moreover, rays making known small angles to the axis arrive at focal points that measure those incoming angles precisely, as you can verify by ray tracing. The image quality is superb.

Hubble's primary mirror, as built, had too negative a shape factor:

```
Number of surfaces: 3
    1,      4.906071,        -0.09057971,        -0.0120,        r
    1,      0,               -0.736377,          -0.49686,       r
    1,      6.4062,          0,                  0,              t
```

x	y	z	u	v	w
1.000000	0.000000	0.000000	0.000000	0.000000	1.000000
1.000000	0.000000	4.860783	-0.179676	0.000000	-0.983726
0.111350	0.000000	-0.004561	-0.017303	0.000000	0.999851
0.000409	0.000000	6.406200	-0.017303	0.000000	0.999851

Notice the final ray X-coordinate has increased enormously. This situation indicates a negative spherical aberration, the opposite of what we discovered in the simple glass lens example.

The Hubble Space Telescope can be focused in orbit by moving its secondary mirror slightly towards or away from the primary mirror. If the secondary mirror is shifted 200 microns farther away from the primary mirror, the edge rays are better focused while the rays nearer the axis are more poorly focused:

```
Number of surfaces: 3
    1,      4.906071,        -0.09057971,        -0.0120,        r
    1,     -0.000200,        -0.736377,          -0.49686,       r
    1,      6.4062,          0,                  0,              t
```

x	y	z	u	v	w
1.000000	0.000000	0.000000	0.000000	0.000000	1.000000
1.000000	0.000000	4.860783	-0.179676	0.000000	-0.983726
0.111314	0.000000	-0.004758	-0.017355	0.000000	0.999849
0.000034	0.000000	6.406200	-0.017355	0.000000	0.999849

Between 1990 and 1993, Hubble was operated in a number of different configurations, including this one. In December 1993, astronauts visited Hubble and installed an optical corrector that introduces a tiny amount of positive spherical aberration, eliminating most of the negative spherical aberration caused by the shape parameter S being excessively negative.

Part IV Problems

1. ❖ An observer 2.0 m from a mirror views the image of a candle 1.0 m from the mirror. Compare the intensity of light from the image at the observer's eye to that of light from the candle itself.
2. ❖❖ A pendulum consists of a copper ball (mass 1.9 kg) on the end of a steel wire of length 0.994 m and mass 6.2 g. What frequency of vibration would set up standing waves on the pendulum wire? Does it make a difference whether the pendulum is at rest or oscillating with an amplitude of 5°?
3. ❖❖ Spherical light waves from the Sun, of wavelength $\lambda = 5.5 \times 10^{-7}$ m, impinge on the Earth. Over how large a plane area does the phase of the wave differ by less than 0.1 rad? If the area is a circle, what is its radius?
4. ❖❖ Some notes on a piano are produced by two strings that are struck together. When a note is struck, it has an average frequency of 110 Hz with beats at 3.0 /s. If the tension in the good string is 3100 N, what adjustment is needed in the tension of the errant string?
5. ❖❖❖ Sound waves in air are incident normally on a water surface. What fraction of the sound energy is transmitted and what reflected? Use the method illustrated in §15.5.
6. ❖❖❖ Two loudspeakers are separated by a distance of 1.0 m. Two people are seated on a circle of radius 12 m, with the center between the two speakers. The person seated on the center line hears sounds of 30.0 Hz and 1.0 kHz with equal sound intensity level (SIL). The second listener hears the 30-Hz sound at 80.0 dB and the 1-kHz sound at 60.0 dB. How far is this listener from the center line of the speakers? How far must the person move to raise the 1-kHz sound intensity by 15 dB?
7. ❖❖❖ A diffraction grating with N lines of spacing d is placed between two converging lenses of focal length f (■ Figure IV.1). If the light source emits coherent, monochromatic light of wavelength λ, at what distances y along the screen do the principal maxima occur? What is the highest order principal maximum that can occur?

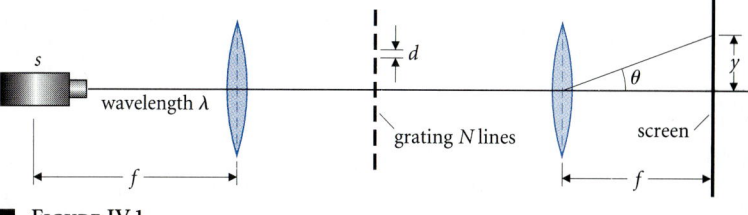

■ Figure IV.1

Challenge Problems

8. What happens to the frequency of standing waves in a tube as the length is slowly decreased at speed v? Think of the standing wave as a superposition of traveling waves reflecting back and forth in the tube, and find the Doppler shift of these waves as they reflect off the moving end. Use your result to obtain the standing wave frequency as a function of time. Do the two results agree?
9. Two coherent point sources of light, with wavelength $\lambda = 630$ nm, are separated by a distance of $a = 0.30$ mm and are located 2.0 m from a concave spherical mirror whose focal length and aperture are each 0.50 m, as shown in ■ Figure IV.2. Describe the interference pattern on a screen at the center of curvature of the mirror. Is diffraction due to the finite size of the mirror an important effect?

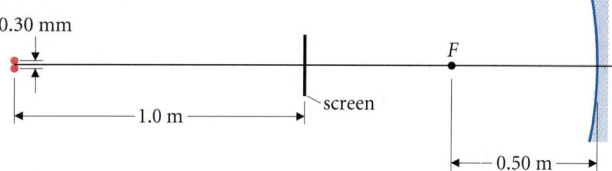

■ Figure IV.2

Thermodynamics

Conservation of energy is a fundamental principle of physics. When mechanical systems "dissipate energy as heat," the energy does not disappear. It is transformed into *thermal energy*—kinetic and potential energy randomly distributed among many molecules. In many cases, such as in an automobile's brake system, thermal energy is slowly transferred to the environment and lost from the system. In other instances, such as in the automobile's engine, thermal energy is used as a source of power. Thermodynamics is the study of the energy transformations that underlie all such processes.

The importance of heat to early human society is recorded world wide in myth and legend, where bringers of fire and forgers of metal are central figures in the construction of the universe. Machines that convert thermal energy to mechanical work are a much more recent development. The first practical steam engines were developed in eighteenth-century England when water-driven engines proved inadequate to pump dry the deepest coal mining shafts. In 1712, Newcomen invented a machine that used expanding steam to drive the pistons of the mine pumps, but his engine was rather inefficient and wasted the steam once it had been used. James Watt (1765) devised a method for condensing and recycling the spent steam (■ Figure V.1). His efficient, compact, and portable power source spurred the revolution in transportation and industrial production that occurred during the nineteenth and twentieth centuries. Because of their central role in technology, steam engines and their principles of operation became a focus of intellectual interest in the nineteenth century. In particular, Sadi Carnot (1824) laid the groundwork for the study of engine efficiency and reversibility of thermal processes.

Hero of Alexandria developed this engine in the third century A.D. The device spins on its axis, driven by steam jetting from the two nozzles. Hero regarded it as an intellectual curiosity, and it was never used as a practical machine for doing work.

PART FIVE

New theories of thermal energy also emerged toward the beginning of the nineteenth century. Previously, the mixing of hot and cold materials and the chemistry of combustion were explained as the exchange of a fluid substance, dubbed "phlogiston." An object was supposed to contain more phlogiston when hot than when cold. A piece of wood was supposed to possess more phlogiston than the resulting burned ash. This theory incorporated Aristotle's ancient idea of air as a fundamental element. New gases discovered in the mid-1700s were interpreted as the basic element—air—combined in various ways with phlogiston. The theory, though attractive, is quite wrong. Since the notion of thermal energy as a fluid remains a popular misconception, it is instructive to see why such theories failed.

In 1785, Antoine Laurent Lavoisier demonstrated conclusively that metals gain mass when they burn, whereas the phlogiston theory predicted no change. In light of this experimental evidence, the idea of heat as a special chemical fluid had to be abandoned. Chemistry came to be understood as reactions among numerous different chemical elements combining in various ways.

Even though the reality of phlogiston as a chemical fluid had been disproved, massless fluids remained an attractive explanation for physical phenomena. Benjamin Franklin had successfully explained the results of his electrical experiments as the transfer of a massless electrical fluid, and heat transfer was still described as the exchange of a massless fluid with a new name—"caloric"—but without chemical properties. In any such theory, the total amount of a real fluid has to be constant. While this remains a basic principle of electrical theory, it proved to be the downfall of the caloric explanation for heat flow.

■ **FIGURE V.1**
James Watt built the first practical steam engine that was able to recycle the steam used to drive the pistons. His invention helped stimulate the industrial revolution. This is his "new, improved" engine, patented in 1782.

In 1798, Count Rumford (■ Figure V.2), one of the more flamboyant characters of the eighteenth century, and also one of the more inventive physicists, was in charge of producing cannon for the Bavarian army. He observed that large amounts of water were required to cool the boring tools, particularly when they were dull and not cutting much metal. Heat production continued as long as the blunt tool was used. These observations contradict the caloric theory. If the heat were produced by the release of caloric stored in the metal, heat generation should have been greatest when the most metal was being cut, and the total amount of heat produced should have been limited to the stored caloric. Count Rumford concluded that heat is not stored in the metal but is produced from mechanical work done on the boring tool and thus is equivalent to mechanical energy.

During most of the nineteenth century, scientists understood chemical reactions to be various elements combining in different ways, and they believed heat to be a form of energy transfer. They left unstated what an element might be or what form thermal energy might take. This was a *macroscopic,* or large-scale, point of view. Toward the end of the century, a *microscopic* model gained acceptance. In this model, matter is composed of vast numbers of molecules, elements are different sorts of atoms, and thermal energy is the kinetic and potential energies of the molecules. In the microscopic view, thermal and chemical phenomena are the result of mechanical and electromagnetic principles operating on a very large number of atoms. We use both approaches in our discussion of thermodynamics and present the microscopic explanation for each macroscopic phenomenon we discuss.

Where should we begin? No matter where we start, we shall need to make some assumptions, which can be verified only when the theory is complete. As in a jigsaw puzzle, the picture appears in its full glory only when the last piece is inserted. Since we have to start somewhere, we begin with a familiar concept: temperature. You probably have a good intuitive idea of what it means. However, a precise definition involves equilibrium among systems, so in Chapter 19 our discussion begins with the idea of a thermodynamic system.

As we discussed in Chapter 13, the forces between molecules are least important in gases, for which we can develop a simple and surprisingly realistic model: the ideal gas. This model illustrates many of the important concepts of thermodynamics, while avoiding the complicated internal structure of real substances. In Chapter 19 we use this model to define the thermodynamic quantities of temperature and internal thermal energy. We then state the first law of thermodynamics, which expresses conservation of energy in all its forms.

In Chapter 20 we consider the thermodynamic behavior of real substances. First, we extend the ideal gas model to understand how gases condense to form liquids and solids. The expansion of objects with increasing temperature and calorimetry, the bookkeeping of thermal energy transfer, introduce the subject of practical temperature measurement.

In Chapter 21 we discuss heat flow, the transfer of thermal energy. We describe the three modes of heat transfer—conduction, convection, and radiation—and present several practical examples.

While energy conservation limits the processes that might occur, it does not determine which processes do occur. For that, we need a new physical property, entropy, which measures the randomness, or disorder, of a system (Chapter 22). It differs from the other physical quantities we have discussed through its statistical nature. Entropy is closely related to the availability of thermal energy to do mechanical work and to the practical inability to reverse processes involved in obtaining such work. The second law of thermodynamics states that entropy tends to increase and the availability of energy tends to decrease. Machines that obtain useful work from thermal energy underlie much of our society's technology, and their efficiency is limited by the second law of thermodynamics. There is plenty of energy on this planet; it is only available energy that is in short supply. Chapter 22 concludes with an essay on the philosophical significance of entropy and how the second law of thermodynamics bears on fundamental questions of biology, cosmology, and the nature of time.

■ **Figure V.2**
Count Rumford's observations of cannon boring convinced him that heat cannot be due to the release of caloric stored in an object but that it results instead from thermal energy generated when work is done.

It is rare to see Los Angeles and its environs with such clarity. Normally, smog from thousands of vehicles obscures the mountains. Occasionally, Santa Ana winds blow down from the mountains. The warm, oppressive wind clears out the smog. But California's worst fires have occurred during Santa Ana wind conditions. The hot, dry winds create optimum conditions for fires to start and then spread. In this chapter we begin our study of thermodynamics with an emphasis on the behavior of gases. We'll learn why the Santa Anas are hot, and we'll begin the study of processes that occur in automobile engines.

CHAPTER 19
Temperature and Thermal Energy

Concepts

Thermodynamic system
Temperature
Internal energy
Heat and work as thermodynamic processes
State variable
Specific heat
Isothermal process
Isobaric process
Isochoric process
Adiabatic process
Equipartition

Goals

Understand the definition and use of thermodynamic systems.

Be able to:
Convert among various temperature scales.

Apply the first law of thermodynamics to systems.

Represent and interpret processes with a *P-V* diagram.

> "It's a nice thing to think about," said Sheerin, "as a pretty abstraction, like a perfect gas, or absolute zero."
>
> Isaac Asimov

TO UNDERSTAND THE MATERIAL IN THIS CHAPTER, YOU SHOULD BE FAMILIAR WITH THE FLUID PROPERTIES OF DENSITY AND PRESSURE (§13.2), AND THE WORK-ENERGY THEOREM (CHAPTER 8).

Hot, dry winds descending from the mountains—the Chinook in Colorado, the Santa Ana in California, and the Mistral in the Mediterranean—can raise sandstorms or fan catastrophic fires. Most often, fortunately, they cause only oppressive weather and grumpy people. But why does hot air blow down from the mountains yet concentrate near the top of a heated room in winter? Atmospheric motion results from pressure forces that depend on the temperature, density, and composition of the gas. We emphasize the properties of gases as we begin the study of thermodynamics because they have a simpler structure than liquids and solids, and so are easier to understand. The first important property is temperature.

Temperature is the quantity we associate with sensations of hot and cold. Yet everyday experience is imprecise and depends on factors other than the thermodynamic quantity called temperature. For example, if you have had your hand in ice water for 10 minutes, cold tap water feels warm! A similar misjudgment in the temperature of photographic developer would cause disaster. To obtain a precise definition of temperature and methods for measuring it, we first need to define a thermodynamic system. So pour yourself a cup of coffee and we'll begin.

19.1 TEMPERATURE

19.1.1 Thermodynamic Systems

When freshly brewed coffee is poured into a cup sitting on the table (■ Figure 19.1), the coffee very quickly heats the cup. Both reach the same temperature, much hotter than room temperature. Then if no one drinks the coffee, both coffee and cup slowly cool to room temperature. This history illustrates a sequence of thermodynamic processes, each described as a system interacting with its environment. We define the *system* for convenience in our analysis. The *environment* is everything not in the system.

To study the freshly poured cup of coffee (■ Figure 19.2), we might choose system (a): the coffee in the cup. The corresponding environment is the cup, the room, and the outside world. At different stages in a process, it is often convenient to choose a different system. The cup rapidly reaches the same temperature as the coffee. Subsequently, system (b), the coffee and cup together, is most convenient. The environment of system (b) is the room and the outside world. Once the cup has cooled to room temperature, we might choose a third system (c), the coffee and the cup and the room it is in, with the outside world as its environment. Each possible choice of system is useful at a particular stage of the process.

Interactions between various systems and their environments typically occur on different time scales. The coffee stops sloshing about in the cup in a few seconds, brings the cup to equal temperature in less than a minute, then cools to room temperature in about 15 min. If we are interested in the coffee–cup interaction, we may ignore the slower interaction with the room. When a system interacts with its environment extremely slowly, we may model it as isolated:

| An *isolated system* has no interaction with its environment.

■ FIGURE 19.1
A cup of cooling coffee illustrates the idea of a thermodynamic system.

OUR FREEDOM TO CHOOSE A SYSTEM IS VERY SIMILAR TO OUR FREEDOM TO CHOOSE A COORDINATE SYSTEM.

■ FIGURE 19.2
For a given process, you should choose the system that allows the simplest description of the process. Here are three different ways to describe a cup of coffee. (a) The system is the coffee. Its environment is the cup, the room, and everything else. (b) The system is the coffee and the cup. Its environment is the room and the rest of the world. (c) The system is the coffee, the cup, and the room. The environment is everything outside the room.

(a)

(b)

(c)

The opposite of isolation is thermal contact:

> A system that can exchange thermal energy with its environment is in *thermal contact* with the environment.

SEE §8.5.1 FOR A DEFINITION OF THERMAL ENERGY.

19.1.2 Thermal Equilibrium and the Definition of Temperature

The thermal evolution of the coffee cup may be described by saying that the coffee and cup come into thermal equilibrium, then together approach equilibrium with the room.

> In thermodynamics, as in mechanics, an equilibrium state of an isolated system is one that has no tendency to change.
>
> Two systems are in thermal equilibrium with each other if there is no net transfer of thermal energy when they are put in thermal contact.

Two systems in thermal equilibrium need not actually be in thermal contact. However, if we were to test them by putting them in contact, there would be no net energy flow.

Now suppose that we have three systems and observe that systems 1 and 3 are in equilibrium when put in contact, as are systems 2 and 3. Experimentally, we always find that there is no net thermal energy transfer when systems 1 and 2 are put in contact (■ Figure 19.3).

> **THE ZEROTH LAW OF THERMODYNAMICS**
>
> Two systems each in thermal equilibrium with a third are in thermal equilibrium with each other.

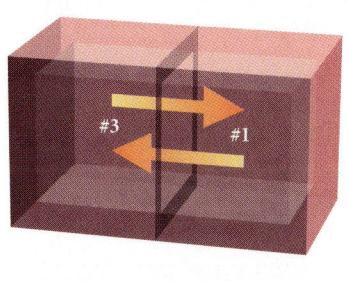
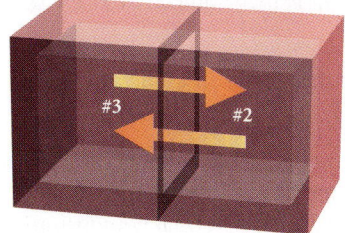
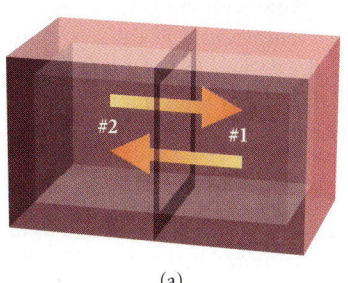

(a)

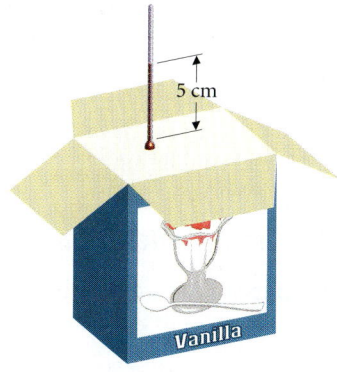
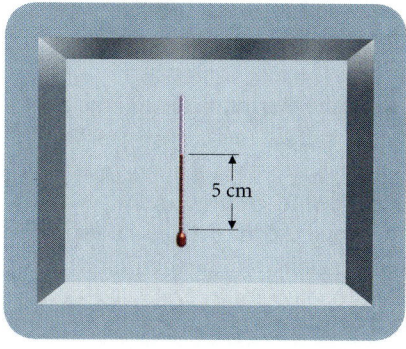

(b)

■ **FIGURE 19.3**
The zeroth law of thermodynamics.
(a) When systems 1 and 3 are put into thermal contact, we observe no net energy transfer between them. When systems 2 and 3 are put into thermal contact, we also observe no net energy transfer. Then we find there is no net energy transfer when systems 1 and 2 are put into thermal contact. (b) If system 3 is a thermometer, this law is the basis of temperature measurement. Here system 3 is a thermometer, system 1 is a box of ice cream, and system 2 is a freezer.

This law obtained its curious name because it logically precedes the first and second laws of thermodynamics, but its significance was appreciated only after those laws were already established. In the following example we apply the zeroth law to three familiar systems to see how it connects with our experience of temperature.

EXAMPLE 19.1 ❖ A thermometer (system 3) consists of a quantity of alcohol in a glass tube. When the thermometer is in thermal contact with soft ice cream in a box (system 1), the length of the alcohol column is 5 cm (Figure 19.3b). When placed in a freezer (system 2), the alcohol column remains 5 cm long. What happens to the ice cream when it is put into the freezer?

MODEL From experience, we know the length of the alcohol column measures temperature.

SETUP When they are in thermal contact, the ice cream and thermometer come to thermal equilibrium. Similarly, the thermometer comes to thermal equilibrium with the freezer when placed inside. Since the alcohol column has the same length in the freezer and in the ice cream, the thermometer is simultaneously in thermal equilibrium with both ice cream and freezer.

SOLVE By the zeroth law, the ice cream and freezer are in equilibrium with each other. There is no net exchange of energy, and the ice cream neither melts nor freezes hard.

ANALYZE We'll explore the behavior of the alcohol in the next chapter. ∎

In the previous example, the length of the alcohol column measures a scalar thermodynamic property that alone determines whether the ice cream and freezer are in thermal equilibrium. This scalar property is what we mean by *temperature*. We may restate the zeroth law as follows:

> Two systems are in thermal equilibrium with each other if and only if they have the same temperature.

In this form, the zeroth law *defines* temperature as the quantity that must be the same in two systems in thermal equilibrium.

19.1.3 Thermometers and Temperature Scales

To measure temperature, we need both a practical test system, or thermometer, and a way to assign numerical values to the measured result.

> A *thermometer* is a device with a measurable property that varies in a known way with temperature.

In a mercury thermometer, the property is the length of the mercury column. As temperature varies, the length changes in proportion to the change in temperature. To use a thermometer, we put it in thermal contact with the system to be measured. Once thermal equilibrium is established, the thermometer and the original system have the same temperature.

With a specific mercury thermometer, we could measure temperature in units of centimeters of mercury, but the numerical scale would depend on the design of that thermometer. Instead, we define temperatures for certain standard systems in precisely reproducible equilibrium states and use the defined values to calibrate the thermometer. The resulting *temperature scale* is independent of the thermometer used.

The two temperature scales in everyday use are those named after Anders Celsius (1701–1744) and Gabriel Fahrenheit (1686–1736). Fahrenheit assigned a temperature of zero to the coldest equilibrium state available to him, the freezing point of saturated brine, and 96° to the

NORMAL BODY TEMPERATURE IS ACTUALLY ABOUT 98.4°F. FAHRENHEIT MUST HAVE BEEN A COLD FELLA!

Digging Deeper

WHY A DEGREE?

According to one legend, Fahrenheit chose his unit size so that there would be 180 units between the freezing point and boiling point of pure water at sea level. Apparently, he wanted a scale analogous to the measure of angle. Consequently, the unit was given the name *degree*.

---

temperature of the human body. On this scale, the freezing point of pure water occurs at 32°F and the boiling point at 212°F. Celsius chose pure water as his standard. On the Celsius scale, the freezing point is 0°C and the boiling point is 100°C. There is no fundamental physical reason for preferring one scale over the other, but, as the Celsius scale is based on the more readily reproducible freezing point of water, it has become the standard in scientific work, as well as being the scale used in most countries other than the United States.

Both Celsius and Fahrenheit scales are in frequent use, so we need a way to convert between them. In both cases, the degree is obtained by dividing the temperature difference between freezing and boiling water into a number of equal parts. Thus the scales are linearly related to each other. Water boils at $100°C \equiv 212°F$; water freezes at $0°C \equiv 32°F$. The physical difference between the temperatures of the two states is independent of the scale we use. We express the temperature interval on both scales and equate the two values to find the conversion factor between Fahrenheit and Celsius degrees:

$$T_{\text{boiling}} - T_{\text{freezing}} = 180°F = 100°C,$$

so

$$1°F = \frac{100°C}{180} = \tfrac{5}{9}°C.$$

In other words, a particular temperature difference contains $\tfrac{5}{9}$ as many Celsius degrees as Fahrenheit degrees. To compare numerical values for temperature on the two scales, we must also account for the different zero points. Expressing the difference of any temperature from the freezing point of pure water, we find:

$$T(°C) - 0°C = T(°C) = \tfrac{5}{9}[T(°F) - 32°F]. \qquad (19.1a)$$

The converse relation is found by solving for $T(°F)$:

$$T(°F) = \tfrac{9}{5}T(°C) + 32°F. \qquad (19.1b)$$

THE CELSIUS SCALE WAS PREVIOUSLY KNOWN AS THE CENTIGRADE SCALE—CENTIGRADE = 100 GRADE. CELSIUS ACTUALLY ASSIGNED 100°C TO THE FREEZING POINT AND 0°C TO THE BOILING POINT. THE MODERN CELSIUS SCALE IS DUE TO HIS COLLEAGUE STROMER (1750).

EXAMPLE 19.2 ♦ A comfortable room temperature is 68°F. What is this temperature on the Celsius scale?

MODEL We need to convert to Celsius degrees and also change the zero point.

SETUP and SOLVE From eqn. (19.1a):

$$T(°C) = \tfrac{5}{9}(68 - 32) = [\tfrac{5}{9}(36)]°C = 20°C.$$

ANALYZE Remember this value. It will give you a feeling for Celsius temperatures. ∎

EXERCISE 19.1 ♦ Silver melts at 961.9°C. What is this temperature on the Fahrenheit scale?

WE HAVE MUCH MORE TO SAY ABOUT THE SIGNIFICANCE OF ABSOLUTE ZERO IN CHAPTER 22.

While there is no limit to high temperatures, there is a minimum temperature, called absolute zero.

> A system unable to transfer energy to any other system in thermal contact with it has the lowest possible temperature, called *absolute zero*.

Temperatures below about $-270°C$ do not occur naturally, although lower temperatures can be achieved in carefully designed experiments (● Table 19.1). Absolute zero has not been, and cannot be, reached. Two temperature scales recognize the significance of absolute zero by assigning it the numerical value of zero. The Kelvin scale, after Lord Kelvin (■ Figure 19.4), uses the official SI temperature unit, named the kelvin, which is the same size as the Celsius degree. The freezing point of water on the Kelvin scale is 273.15 K, and in general:

$$T = [T(°C) + 273.15 \text{ K}]. \qquad (19.2)$$

We reserve the symbol T by itself for temperatures expressed on the Kelvin scale. The symbol K stands for the unit regardless of whether the temperature is expressed on the Celsius or Kelvin scale. The symbols °C and °F mean that the quoted temperature is expressed on the Celsius or Fahrenheit scale. For example, we write a change in temperature as:

$$\Delta T = 363.15 \text{ K} - 293.15 \text{ K} = 90.00°C - 20.00°C = 70.00 \text{ K}.$$

The Rankine scale is often used in engineering practice. It also begins at absolute zero but uses the Fahrenheit degree as its unit. We shall not use the Rankine scale in this text. ■ Figure 19.5 illustrates the relations among the other three temperature scales.

■ FIGURE 19.4
William Thomson, Lord Kelvin (1824–1907) made numerous discoveries in thermodynamics. The Kelvin scale as well as the SI unit of temperature are named after him.

NOTE: THE DEGREE SYMBOL IS NOT USED WITH KELVIN. WRITE 70 K, NOT 70°K.

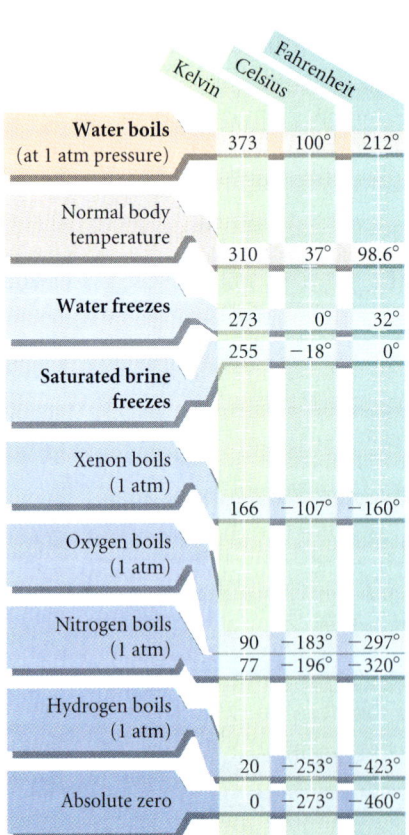

■ FIGURE 19.5
Comparison of the three common temperature scales. The Kelvin scale is used in theoretical work and begins at absolute zero. Several representative temperatures are shown, including the defining temperatures for the Celsius and Fahrenheit scales (bold face).

TABLE 19.1 Some Temperatures

Example	Temperature (K)
Gas between galaxies in clusters	$\approx 10^8$
Center of the Sun	1.5×10^7
Center of the Earth	6400
Iron melts	1808
Silver melts	1235
Average daytime temperature on Venus	720
Tin melts	505
Sodium melts	371
Average daytime temperature on Earth	295
Mercury melts	234
Dry ice sublimes	195
Average daytime temperature on Neptune	123
Oxygen boils at 1 atm	90
Nitrogen boils at 1 atm	77
Interstellar molecular clouds	≈ 70
Hydrogen boils at 1 atm	20
Helium boils at 1 atm	4.2
Temperature of the universal blackbody radiation	2.7
Lowest temperature achieved in a laboratory	a few $\times 10^{-9}$

19.2 Temperature in an Ideal Gas

19.2.1 The Ideal Gas Law

To further our understanding of temperature, we now look at the behavior of gases. A gas has a relatively simple structure, and we have already developed a theoretical model—the ideal gas model (cf. Chapter 13)—that works extremely well in situations as diverse as the center of the Sun and a helium balloon. At low density, real gases behave like ideal gases (■ Figure 19.6), and their common properties form the experimental description of an ideal gas. In this section we compare the theoretical model of the gas with its experimental behavior to see what temperature means on the microscopic scale.

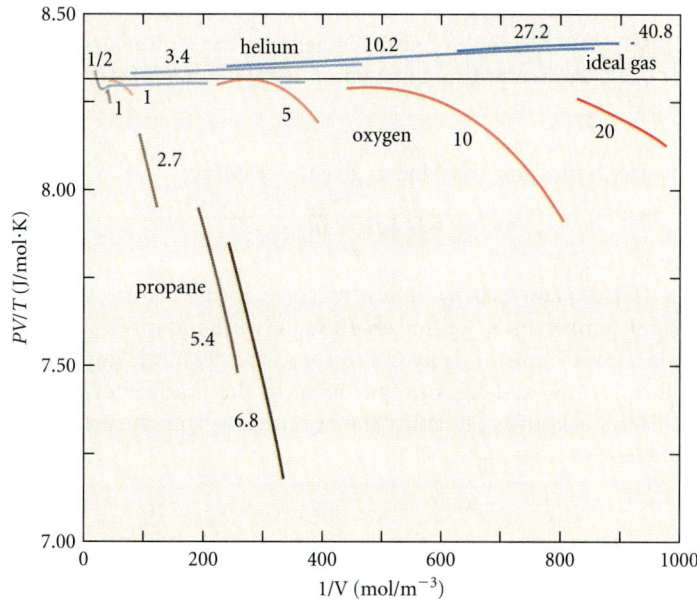

■ **Figure 19.6**
The quantity PV/T for 1 mol of gas (§19.2.2) should equal the gas constant $R = 8.31$ J/mol·K if the gas is ideal. Here we have plotted measured values of this quantity versus the gas density $1/V$ in mol/m³ for various different gases. Each curve is marked with the corresponding gas pressure in atmospheres. Temperature varies along each curve: $T = 33$–533 K for helium; 160–280 K for oxygen; 289–366 K for propane. Helium most closely approximates the behavior of an ideal gas. At 1 atm, the maximum deviation from ideal behavior is 0.9% for propane at 289 K. Real gases closely approximate the behavior of ideal gases under normal laboratory conditions.

The observed behavior of dilute gases is summarized in two laws:

> **BOYLE'S LAW**
>
> In a sample of gas at a constant temperature, the pressure and volume are inversely proportional:
>
> $$PV = \text{constant}. \tag{19.3}$$

ORDINARY AIR AT ATMOSPHERIC PRESSURE AND ROOM TEMPERATURE IS A PRETTY GOOD EXAMPLE OF AN IDEAL GAS. HELIUM BEHAVES LIKE AN IDEAL GAS AT TEMPERATURES ABOVE ABOUT 4 K. SEE ALSO §20.1.

NAMED AFTER ROBERT BOYLE, 1627–1691.

> **CHARLES' LAW**
>
> The volume of a sample of gas held at constant pressure is proportional to its temperature on the Kelvin scale:
>
> $$V = \text{constant} \times T. \tag{19.4}$$

NAMED AFTER JACQUES CHARLES, 1746–1823. (DISCOVERED INDEPENDENTLY BY JOSEPH LOUIS GAY-LUSSAC, 1778–1850.)

EXAMPLE 19.3 ♦ A sample of oxygen is held at a pressure of 1.3×10^4 Pa. Its volume is 0.750 m³ at 20.0°C. It expands to a volume of 0.900 m³ when heated. What is its final temperature?

MODEL We model the oxygen as an ideal gas and apply Charles' law (eqn. 19.4), expressing the given temperature on the Kelvin scale:

SETUP $\qquad 20.0°\text{C} = 273.15\text{ K} + 20.0\text{ K} = 293.2\text{ K}.$

Then: $$\text{Constant} = \frac{V}{T} = \frac{V_i}{T_i} = \frac{0.750 \text{ m}^3}{293 \text{ K}} = \frac{0.900 \text{ m}^3}{T_f} = \frac{V_f}{T_f}.$$

SOLVE The final temperature is:

$$T_f = \frac{0.900 \text{ m}^3}{0.750 \text{ m}^3}(293 \text{ K}) = 352 \text{ K}.$$

On the Celsius scale, the final temperature is:

$$T(°C) = 352 \text{ K} - 273 \text{ K} = 79°C.$$

ANALYZE Remember that T in eqn. (19.4) means temperature on the Kelvin scale.

EXERCISE 19.2 ♦ A sample of nitrogen held at a constant temperature of 300.0 K has a volume of 0.200 m³ at a pressure of 1.00×10^4 Pa. What is its volume at a pressure of 2.50×10^4 Pa?

Boyle's law and Charles' law combine to give the relation:

$$PV/T = \text{constant}, \tag{19.5}$$

which holds for a fixed quantity of gas at low pressure. If gas is added to a container of fixed volume at constant temperature, we find that the pressure increases. Hence the *constant* in eqn. (19.5) depends on the quantity of gas. Pressure arises from the impact of molecules on a container's walls (§13.2.3), and so is proportional to the number of molecules, N, in the sample. Thus N is the measure of quantity that we need. Setting the previous constant $= Nk$ gives *the ideal gas law*.

> For a sample of ideal gas, the product of pressure and volume is directly proportional to the product of temperature and number of molecules:
>
> $$PV = NkT. \tag{19.6}$$

The constant k is known as Boltzmann's constant. Its value is determined experimentally:

$$k = 1.381 \times 10^{-23} \text{ J/K}.$$

The ideal gas law is sometimes used as an experimental definition of temperature.

EXAMPLE 19.4 ♦♦ A sample of carbon dioxide (CO_2) gas occupies a volume of 0.500 m³ at a pressure of 1.00×10^4 Pa and temperature of 305 K. What is the mass of the sample?

MODEL An ideal gas is a reasonable model for carbon dioxide at this pressure.

SEE THE PERIODIC TABLE (INSIDE BACK COVER) FOR ATOMIC MASSES.

SETUP We find the number of molecules in the sample from the ideal gas law and then obtain the mass from the known molecular mass of CO_2.

$$N = \frac{PV}{kT} = \frac{(1.00 \times 10^4 \text{ Pa})(0.500 \text{ m}^3)}{(1.38 \times 10^{-23} \text{ J/K})(305 \text{ K})}$$

$$= 1.19 \times 10^{24} \text{ molecules}.$$

A molecule of CO_2 has one carbon atom (mass 12 u) and two oxygen atoms (each 16 u), so its mass is:

THE ATOMIC MASS UNIT
$$u = 1.66 \times 10^{-27} \text{ kg}.$$

$$m = (12 + 2 \times 16)(1.66 \times 10^{-27} \text{ kg})$$

$$= 7.3 \times 10^{-26} \text{ kg}.$$

SOLVE Therefore, the mass of the sample is:

$$M = mN = (1.19 \times 10^{24})(7.3 \times 10^{-26} \text{ kg}) = 87 \text{ g}.$$

ANALYZE If the molecules were packed tightly together, they would occupy a tiny fraction of the measured volume, which is mostly empty space.

19.2.2 Molecules and Moles

In experimental work it is often convenient, and in chemistry it is customary, to use a larger unit than a single molecule to describe the amount of gas in a sample. The SI unit for quantity of substance is the mole (abbreviation mol).

> One *mole* is the quantity of a substance whose mass in grams equals the chemical molecular mass of the substance.
>
> One mole equals Avogadro's number ($N_A = 6.02 \times 10^{23}$) of molecules.

NOTE: THE MASS IN THIS DEFINITION IS MEASURED IN GRAMS, NOT KILOGRAMS.

Thus, 1 mol of helium (mass 4 u) has a mass of 4.00 g and contains 6.02×10^{23} molecules. The number of moles of gas is $\mathcal{N} = N/N_A$, and the ideal gas law may be written:

$$PV = \mathcal{N}RT, \qquad (19.7)$$

where $R \;(\equiv kN_A) = 8.3145$ J/(mol·K) is called the *ideal gas constant*. Which form to use is primarily a matter of taste.

THE VALUE $R = 0.082$ L·atm/mol·K IS ALSO USEFUL.

EXERCISE 19.3 ♦ How many moles of CO_2 occupy a volume of 0.500 m³ at a pressure of 1.00×10^4 Pa and $T = 305$ K?

19.2.3 The Relationship Between Temperature and Internal Kinetic Energy in an Ideal Gas

In Chapter 13 we found that the pressure due to N ideal gas molecules, each of mass m, in a volume V is:

$$P = \frac{Nm\langle v^2 \rangle}{3V},$$

where $\langle v^2 \rangle$ is the average value of the square of the speed for all the molecules in the volume. The ideal gas law (eqn. 19.6) also relates pressure to the number of molecules and the volume:

$$P = \frac{N}{V}kT.$$

The expressions are consistent provided that:

$$kT = m\langle v^2 \rangle/3.$$

REMEMBER: THIS RELATIONSHIP HOLDS ONLY WHEN T IS EXPRESSED ON THE KELVIN SCALE.

Rearranging this expression, we find that in an ideal gas temperature measures the average translational kinetic energy per molecule:

$$K = \langle \tfrac{1}{2}mv^2 \rangle = \tfrac{3}{2}kT. \qquad (19.8)$$

A characteristic speed for the molecules, called the *root mean square* (rms) speed, is:

$$v_{\text{rms}} \equiv \sqrt{\langle v^2 \rangle} = \sqrt{\frac{3kT}{m}} \qquad (19.9)$$

Since each kind of molecule has the same average energy at a given temperature, lighter molecules move faster, on average, than heavy molecules. Hydrogen molecules in air move faster than nitrogen molecules.

TEMPERATURE ALSO DETERMINES HOW ENERGY IS DISTRIBUTED AMONG THE MOLECULES. WE'LL DISCUSS THE DISTRIBUTION OF ENERGY IN CHAPTER 22.

SEE EXAMPLE 19.2.

EXAMPLE 19.5 ♦ How fast does a typical molecule move in air at room temperature (68°F)?

MODEL Although air is a mixture of gases, it is about 80% nitrogen. So, we'll calculate the rms speed of a molecule in an ideal nitrogen gas.

SETUP Converting to the Kelvin scale, 68°F = 20°C = 293 K. Nitrogen has atoms of mass 14 u and is diatomic at room temperature.

SOLVE The root mean square speed $\sqrt{\langle v^2 \rangle}$ of an N_2 molecule at 293 K is (eqn. 19.9):

$$v_{\rm rms} = \sqrt{\frac{3kT}{m}} = \sqrt{\frac{3(1.38 \times 10^{-23} \text{ J/K})(293 \text{ K})}{2(14.0)(1.66 \times 10^{-27} \text{ kg})}} = 511 \text{ m/s} = 1140 \text{ mph!}$$

ANALYZE A typical air molecule moves more than twice as fast as a jet airliner.

EXERCISE 19.4 ♦ Find the rms speed of hydrogen molecules in Jupiter's atmosphere, where the temperature is 120 K.

19.2.4 Internal Energy of a Monatomic Ideal Gas

The molecular motions that cause the pressure of an ideal gas are random in direction and do not result in any net motion of the gas. The total energy associated with these random motions is called *internal thermal energy*. In general, internal energy may include other forms of kinetic and potential energy of the molecules, such as rotational energy. The essential feature of internal thermal energy is its random distribution among the molecules.

THERMAL ENERGY IS NOT THE ONLY FORM OF INTERNAL ENERGY (SEE §8.5). HOWEVER, IN OUR INTRODUCTION TO THERMODYNAMICS, WE SHALL RESTRICT OURSELVES TO SYSTEMS THAT DO NOT HAVE OTHER FORMS OF INTERNAL ENERGY. IN SUCH CASES, IT IS CUSTOMARY TO DROP THE WORD "THERMAL," AND WE SHALL DO SO.

> The *internal energy* U of a system is the total energy of its molecules in a reference frame in which the system's center of mass is at rest.

In a monatomic ideal gas, the structure of the atoms is not important. Only translational kinetic energy contributes:

K IS THE STANDARD SYMBOL FOR KINETIC ENERGY, AND WE USED IT IN CHAPTER 8. NOW WE ALSO HAVE BOLTZMAN'S CONSTANT, WHOSE SYMBOL IS k, AND THEY OFTEN APPEAR IN THE SAME EQUATION, AS IN (19.8). BE AWARE OF THE DISTINCTION BETWEEN UPPER- AND LOWERCASE LETTERS, USE CONTEXT, AND NOTE THAT k MOST OFTEN APPEARS BESIDE T! K MEANS KELVIN, SO BE ALERT.

$$U = NK = N\frac{m\langle v^2 \rangle}{2} \quad \text{or} \quad U = \tfrac{3}{2}NkT = \tfrac{3}{2}\mathcal{N}RT. \tag{19.10}$$

The internal energy is proportional to the number of atoms in the system. The temperature measures energy *per atom* and is independent of the total number of atoms. The internal energy of a system is distinct from any bulk kinetic energy the system may have if it moves as a whole (cf. §9.3).

EXAMPLE 19.6 ♦♦ A cylinder containing 0.10 kg of argon gas at temperature $T = 270$ K is carried on an airplane flying at 180 m/s. Find the internal energy of the argon gas. Show that the total kinetic energy of the gas may be divided into two parts—bulk kinetic energy and internal kinetic energy—and that the internal kinetic energy is the same as when the gas is at rest. Compare the magnitudes of the two parts.

MODEL We model the argon as an ideal gas with atoms of mass 39.95 u. To find the total kinetic energy of the gas, we sum up the energy of the individual atoms that compose it. A similar analysis was given in §9.3.

SETUP The internal energy is determined by the temperature of the gas (eqn. 19.10).

$$U = \frac{3}{2}NkT = \frac{3}{2}\left(\frac{M}{m}\right)(kT),$$

where M is the total mass of the sample and m the mass of a single atom.

SOLVE Thus:

$$U = \frac{3}{2}\left[\frac{0.10 \text{ kg}}{39.95(1.66 \times 10^{-27} \text{ kg})}\right](1.38 \times 10^{-23} \text{ J/K})(270 \text{ K}) = 8400 \text{ J}.$$

SETUP If the airplane's velocity is $\vec{v}_o$ and a gas atom has random velocity $\vec{v}_r$, the atom's total velocity is $\vec{v}_o + \vec{v}_r$. The mean square magnitude of the atoms' total velocity is:

$$\langle v^2 \rangle = \langle (\vec{v}_o + \vec{v}_r)^2 \rangle = \langle v_o^2 + 2\vec{v}_o \cdot \vec{v}_r + v_r^2 \rangle = v_o^2 + 2\vec{v}_o \cdot \langle \vec{v}_r \rangle + \langle v_r^2 \rangle.$$

Since the average value of the random velocity is $\langle \vec{v}_r \rangle = 0$,

$$\langle v^2 \rangle = v_o^2 + \langle v_r^2 \rangle.$$

This is just what is meant by "random."

SOLVE The total kinetic energy is:

$$K = \tfrac{1}{2}M(v_o^2 + \langle v_r^2 \rangle) = K_{CM} + U.$$

The first term is the kinetic energy due to motion of the center of mass (bulk motion). The second term is the thermal energy due to random motion. The internal energy is independent of the velocity $\vec{v}_o$ of the airplane. In particular, when the container of argon is at rest, $\vec{v}_o = 0$, and:

$$K = \tfrac{1}{2}M\langle v^2 \rangle = \tfrac{1}{2}M\langle v_r^2 \rangle = U.$$

SETUP The ratio of the two energies is:

$$\frac{K_{CM}}{U} = \frac{\tfrac{1}{2}Mv_o^2}{\tfrac{3}{2}NkT} = \frac{mv_o^2}{3kT},$$

where $m = M/N$ is the average mass of the gas molecules.

SOLVE With $v_o = 180$ m/s:

$$\frac{K_{CM}}{U} = \frac{(180 \text{ m/s})^2(39.95)(1.66 \times 10^{-27} \text{ kg})}{3(1.38 \times 10^{-23} \text{ J/K})(270 \text{ K})} = 0.19.$$

ANALYZE The bulk kinetic energy of the gas due to the airplane's velocity is only about 20% of the internal energy. ∎

19.3 THE FIRST LAW OF THERMODYNAMICS

Now that we have defined internal energy, we may extend the work-energy theorem to express conservation of energy in a form known as the first law of thermodynamics.

The work-energy theorem is described in Chapter 8.

19.3.1 Heat, Work, and the First Law of Thermodynamics

To increase the temperature of a system, we must add energy to it. We can increase a system's energy by doing work on it, as Rumford observed while boring cannon. When hot coffee is poured into a cup, however, the coffee increases the cup's energy by heating it. The coffee's molecules have more energy, on average, than the cup's molecules and transfer energy to the cup molecules via collisions.

Rumford's work is discussed in the introduction to this part.

> When energy is transferred from one system to another as a result of the temperature difference between them, the process is called *heat transfer*.
>
> *Heat* means the amount of energy that is transferred through a large number of random events.

As we mentioned in Chapter 1, common words take on precise meanings in physics. "Heat" is another example of this.

There are two important points to remember about these definitions. First, heat transfer is a process and does not involve any new form of energy. Heat transfer may increase the internal kinetic energy of a system, for example. Second, we are using "heat" as a description of the transfer process. Systems do not contain heat any more than they contain work. Systems *contain* internal energy that may be changed either by heat transfer *or* by doing work.

THIS WAS THE PROBLEM WITH THE THEORY OF CALORIC.

The *first law of thermodynamics* expresses conservation of energy during any process.

> The increase in a system's energy equals the heat transferred to the system plus the work done on the system.

BUT NOT EXCLUSIVELY! SOME BOOKS USE THE OPPOSITE CONVENTION, IN WHICH WORK DONE *ON* THE SYSTEM IS POSITIVE. WHEN USING ANOTHER REFERENCE, ALWAYS CHECK THE CONVENTION USED.

Because of the technological importance of work done by machines that convert thermal energy to more useful forms, the symbol W is frequently used for the work done *by* a system, while Q represents the heat added *to* the system. When work is done *on* the system, W is negative. In thermodynamics, it is customary to work in the CM frame, with all the system energy being internal thermal energy. Then the first law of thermodynamics (■ Figure 19.7) is written:

$$\Delta U = Q - W. \quad (19.11)$$

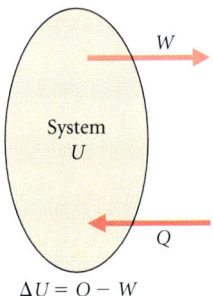

■ FIGURE 19.7
The first law of thermodynamics. Our sign convention is that W is positive when work is done by the system and Q is positive when heat is transferred to the system, as shown here.

EXAMPLE 19.7 ♦ In 1843, James Prescott Joule demonstrated the equivalence of mechanical and thermal energy (■ Figure 19.8). He used a falling mass to turn a set of paddles inside an insulated container of water, and measured the increase in temperature of the water due to the motion of the paddles. If the mass, 5.0 kg of lead, falls 10.0 m, what is the increase in internal energy of the paddles and the water?

MODEL We choose for our system the water and paddles inside the insulated container and neglect frictional losses in the external mechanism. Gravitational potential energy of the mass is converted to internal energy of the water.

SETUP Work is done on the system by the mass, so W is *negative*:

$$W = -mgh = -(5.0 \text{ kg})(9.8 \text{ m/s}^2)(10.0 \text{ m}) = -490 \text{ J}.$$

Since no heat transfer occurs, $Q = 0$ and the change in internal energy is:

SOLVE $\quad \Delta U = Q - W = 0 - (-490 \text{ J}) = 490 \text{ J}.$

ANALYZE Pay particular attention to the use of signs in this example. ■

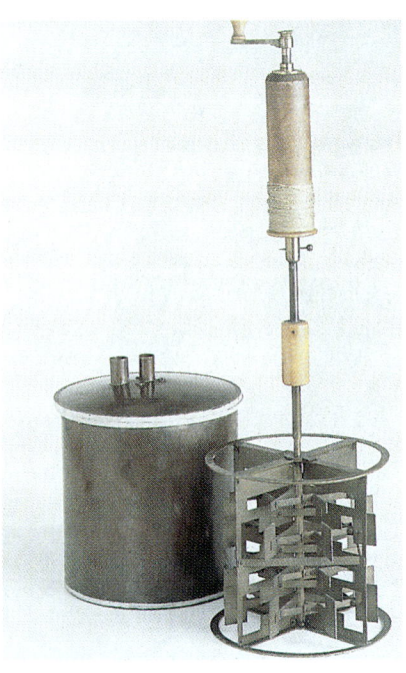

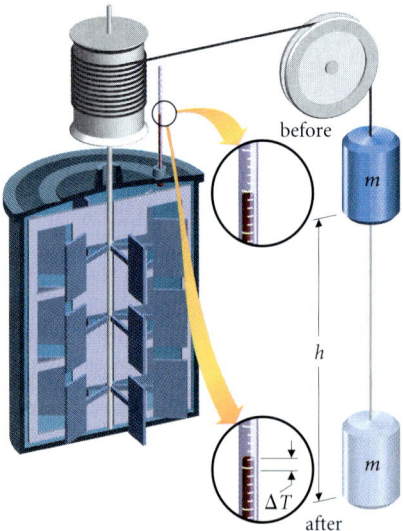

■ FIGURE 19.8
Joule's experiment to measure the relation between mechanical work and thermal energy, sometimes called "the mechanical equivalent of heat." The falling mass turns a set of paddles inside an insulated container of water. If the paddles are included in the system, $Q = 0$ and $\Delta U = -W$. The work done on the water increases its internal energy. Joule measured the temperature rise and the amount of work done to cause it. Thus he could relate the unit of work to the calorie, the traditional unit of heat transfer. (See Chapter 20 for the definition of the calorie.)

19.3.2 Work Done by an Ideal Gas

A common thermodynamic system consists of gas in a cylinder with a movable piston (■ Figure 19.9). (There are several in every automobile engine!) When the gas expands, it does work on the piston. The gas exerts a force normal to the piston's surface with a magnitude equal to the gas pressure times the area of the piston, $F = PA$. The piston pushes back with an equal but oppositely directed force. Since the pressure of the gas depends on its volume, $P = P(V)$, the force varies throughout the process. If the piston moves outward an infinitesimal amount ds, the gas volume increases by $dV = A\,ds$, and the work done by the gas on the piston is:

$$dW = F\,ds = PA\,ds = P\,dV.$$

SEE THE DEFINITION OF PRESSURE.

FOR EXAMPLE, IF TEMPERATURE IS CONSTANT, BOYLE'S LAW GIVES THE FUNCTION $P(V) = \text{CONSTANT}/V$.

REMEMBER: ds AND dV MAY BE NEGATIVE IF THE PISTON MOVES INWARD.

> The differential work done by a gas when its volume changes is the pressure times the differential change in volume.

The work done in a finite volume change from initial volume V_i to final volume V_f is:

$$W = \int_{V_i}^{V_f} P(V)\,dV. \qquad (19.12)$$

If we push the piston down against the gas pressure, we do work on the gas, V_f is less than V_i, and eqn. (19.12) gives a negative value for W. With no heat transfer, the internal energy of the system increases. If the gas expands, it does work on its environment; $V_f > V_i$, W is positive, and if $Q = 0$, the internal energy decreases.

EXAMPLE 19.8 ♦♦ Find the work done in an isothermal (constant temperature) expansion of an ideal gas from a volume $V_i = 0.80$ m^3 and pressure $P_i = 1.3 \times 10^5$ Pa to a final volume $V_f = 1.1$ m^3.

THE WORD *ISOTHERMAL* COMES FROM THE GREEK ROOT "ISO" (MEANING "SAME") COMBINED WITH THERMAL.

MODEL To find the work done, we evaluate the integral in eqn. (19.12) using the five-step method (see Interlude 2). Since T is constant, Boyle's law, $PV = $ constant, provides the relation between P and V during the process.

SETUP
STEPS I AND II We model the process as a series of differential volume changes dV and express all the physical quantities in terms of V (cf. Figure 19.9).

STEP III The pressure P_i and volume V_i at the start of the process determine the value of the constant in Boyle's law: constant $= PV = P_i V_i$. So:

$$P(V) = P_i V_i / V,$$

and $dW = P(V)\,dV$.

STEP IV The gas expands from an initial volume V_i to a final volume V_f, so the work done (eqn. 19.12) is:

$$W = \int_{V_i}^{V_f} dW = \int_{V_i}^{V_f} P(V)\,dV = \int_{V_i}^{V_f} \frac{P_i V_i}{V}\,dV.$$

SOLVE
STEP V Since it is constant, $P_i V_i$ may be brought out of the integral.

$$W = P_i V_i \int_{V_i}^{V_f} (1/V)\,dV = P_i V_i \ln V \Big|_{V_i}^{V_f}$$

$$W = P_i V_i \ln\left(\frac{V_f}{V_i}\right) \qquad (19.13)$$

$$= (1.3 \times 10^5 \text{ Pa})(0.80 \text{ m}^3)\ln\left(\frac{1.1 \text{ m}^3}{0.80 \text{ m}^3}\right) = 3.3 \times 10^4 \text{ J}.$$

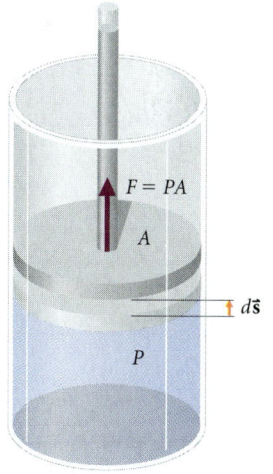

■ **FIGURE 19.9**
A common thermodynamic system. Gas is contained in a cylinder with fixed walls and cross-sectional area A. In this idealized version, the piston fits snugly but can move without friction. The gas has pressure P and exerts a force $F = PA$ on the piston. When the piston moves outward a distance ds, the work done by the gas is $dW = F\,ds = PA\,ds = P\,dV$. The total mass of gas remains fixed as the piston moves.

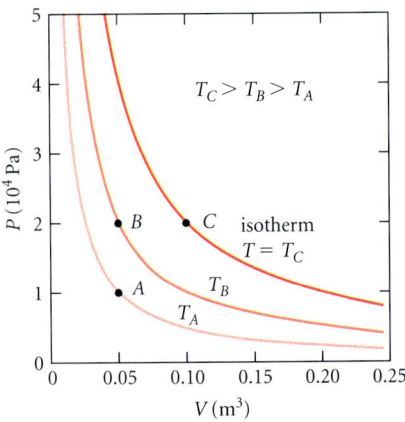

■ FIGURE 19.10
The P-V diagram. The diagram shows the state of a fixed mass of gas. The three curves shown are isotherms: the gas has a constant temperature along each line. The shape of each curve is a hyperbola. The isotherm through point C corresponds to a higher temperature than that through point A. The temperature at A is $T_A = P_A V_A / Nk = P_A V_A / \mathcal{N} R$, where N is the number of molecules of gas, $N = M/m$, and $\mathcal{N}$ is the number of moles. Thus $T_A = (60\text{ K}\cdot\text{mol})/\mathcal{N}$.

NOTE: ISOTHERMS CORRESPONDING TO DIFFERENT TEMPERATURES NEVER CROSS.

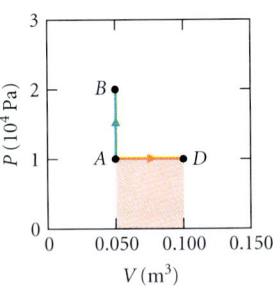

■ FIGURE 19.11
Gas processes are represented by lines in the P-V diagram. Line AB illustrates a constant-volume process in which the pressure increases by a factor of 2. Since B is on a higher isotherm, the gas temperature increases during the process. Line AD represents a constant-pressure process in which the volume increases by a factor of 2. The temperature also increases during this process. The work done during process AD is represented by the area under the line, shown shaded.

ANALYZE Since the gas does work as it expands, $W > 0$. The internal energy, and hence the temperature, would decrease unless heat were simultaneously added. For the moment, we won't worry about *how* this heat transfer occurs and will simply assume that the gas sample's environment is able to transfer heat at constant temperature. We'll consider this question more carefully in the following chapters.

19.4 THE P-V DIAGRAM

19.4.1 Representation of a Thermodynamic State in a P-V Diagram

In the discussion so far, we have referred loosely to the state of an ideal gas system, without specifying the information we need to describe it. At a certain time, the state of a system is precisely defined by the position and velocity of each of its molecules. Such information specifies a *microstate* of the system. Almost never can we measure, nor are we interested in, such detail. For example, there are many different microstates that all have the same average energy per molecule and thus the same temperature. For our purposes, these microstates are indistinguishable, and we shall describe a system by its macroscopic properties—quantities that are averages over the large number of molecules in the system. Technically, this specifies a *macrostate*, but we shall use the word "state" alone to mean macrostate. The macroscopic properties, such as temperature, that characterize a state are called *state variables*.

EXERCISE 19.5 ❖ List some state variables for an ideal gas system. How many of them are needed to specify a macrostate completely?

The state of an ideal gas system is specified by three state variables. For a fixed mass of gas, we may choose pressure P and volume V as independent variables and represent any state by a point on a graph of P versus V (■ Figure 19.10). States with the same temperature lie along a curve in the diagram called an *isotherm*. Along the isotherm through state A, the pressure corresponding to any volume V (cf. Example 19.8) is:

$$P(V) = P_A V_A / V.$$

The isotherm is a hyperbola. Isotherms corresponding to temperature $T > T_A$ have higher P at the same V, or higher V at the same P, and thus lie above and to the right of the isotherm through A.

19.4.2 Representation of Processes in the P-V Diagram

If a sample of gas slowly expands or is compressed, its pressure and volume change continuously and the point representing the system's state moves along a line in the P-V diagram. For example, if the gas expands at constant temperature, the point that describes the system moves along an isotherm. If instead pressure is increased at constant volume (■ Figure 19.11), the system point moves from state A to a new state B on a higher temperature isotherm. Whenever we know the state of a system throughout a process, that process is represented as a line in the P-V diagram, described by a function $P = P(V)$.

The work done by the gas during a process (eqn. 19.12) is:

$$W = \int_{V_i}^{V_f} P(V)\, dV,$$

which is represented by the area under the system path in the P-V diagram. In the constant-volume (*isochoric*) process A → B in Figure 19.11, no work is done. There is no change in volume ($dV \equiv 0$ throughout the process), and there is no area under the vertical line AB. In the constant-pressure (*isobaric*) process A → D, the work done is given by the area of the shaded rectangle:

$$W = P_A (V_D - V_A).$$

The internal energy of an ideal gas depends only on the temperature and the number N of molecules. Hence the isotherms in the P-V diagram are also lines of constant internal energy (■ Figure 19.12). Therefore, the change of an ideal gas system's internal energy during any process is determined from the isotherms passing through the initial and final states.

In an arbitrary process, both work *and* heat transfer occur. There is no direct way to evaluate the heat transfer during a process from the P-V diagram. We find the heat transfer using the first law of thermodynamics, once the work done by the gas and the change in internal energy are known.

EXAMPLE 19.9 ♦♦ Find the heat input during the constant-volume process AB shown in Figure 19.11.

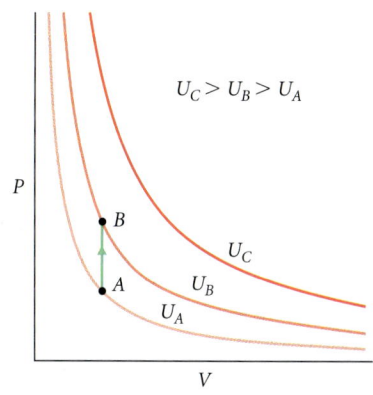

■ **FIGURE 19.12**
The isotherms are also lines along which the gas has constant internal energy. During process AB, no work is done, but the internal energy increases. Therefore, we conclude that heat transfer occurs during the process.

MODEL We model the gas as ideal. To find the heat transfer, we calculate the work done and the change in internal energy, then apply the first law of thermodynamics.

SETUP The constant-volume process takes the gas system to a final state B, with internal energy $U(B)$ greater than the internal energy $U(A)$ at A. The pressure increases from $P_A = 1.0 \times 10^4$ Pa to $P_B = 2P_A$ during the process. Since

$$U = \tfrac{3}{2}NkT = \tfrac{3}{2}PV,$$

the change in internal energy of the system is:

$$\Delta U = U_B - U_A = \tfrac{3}{2}P_B V_B - \tfrac{3}{2}P_A V_A$$

$$= 3P_A V_A - \tfrac{3}{2}P_A V_A = \tfrac{3}{2}P_A V_A = Q - W.$$

WE USED THE IDEAL GAS LAW.

SOLVE Since the volume does not change, no work is done in this process, $W = 0$, and the change in internal energy is due to heat transfer alone:

$$Q = \Delta U = \tfrac{3}{2}P_A V_A = 1.5(1.0 \times 10^4 \text{ Pa})(0.050 \text{ m}^3) = 750 \text{ J}.$$

ANALYZE It is important to note which of the quantities appearing in this discussion are state variables. The variables P, V, T, and U are all state variables, because they depend only on the state of the system (the position of the system point in the P-V diagram) and not on its past history (the path it took to get there). Work and heat transfer, on the other hand, depend *on the process*. They are properties of a line in the diagram, and not of a point. ■

REMEMBER: NOT ALL LINES ARE STRAIGHT.

EXERCISE 19.6 ♦♦ Evaluate the heat transfer during the isobaric (constant-pressure) process $A \rightarrow D$ in Figure 19.11. ■

EXAMPLE 19.10 ♦♦ Find the heat input to an ideal gas system during an isothermal expansion from volume 0.80 m³ and pressure 1.3×10^5 Pa to volume 1.1 m³.

MODEL The temperature and internal energy are constant throughout the process. So from the first law of thermodynamics, the heat transfer equals the work done by the gas:

SETUP $$0 = \Delta U = Q - W \Rightarrow Q = W.$$

We have already calculated the work done during the process (Example 19.8):

SOLVE $$Q = W = P_i V_i \ln(V_f/V_i) = 3.3 \times 10^4 \text{ J}.$$

ANALYZE In effect, the expansion converts the energy transferred as heat directly to work done by the gas. None of the energy input remains to increase the internal energy of the gas. ■

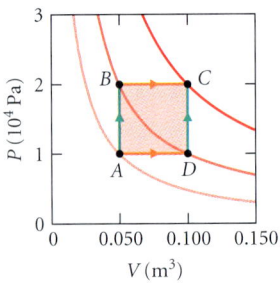

FIGURE 19.13
Two different processes that take a fixed mass of gas from state A to state C. More work is done during process ABC than during process ADC; the area under the line representing the process ABC is greater. Since the change in internal energy is the same for both processes, the heat transfer is also greater for process ABC.

EXAMPLE 19.11 ♦♦ ■ Figure 19.13 shows two processes ($A \to B \to C$ and $A \to D \to C$) that raise the temperature of an ideal gas sample. During which process does the gas do more work? By how much? By how much does the heat transfer differ in the two processes?

MODEL The work W_{ABC} done along path ABC is represented by the area under the line BC. It is larger than the area under the line AD, which represents the work done in process ADC: $W_{ADC} < W_{ABC}$. The difference is represented by the area of the square $ABCD$.

SETUP AND SOLVE
$$W_{ABC} - W_{ADC} = (P_D - P_A)(V_B - V_A)$$
$$= (1.0 \times 10^4 \text{ Pa})(0.050 \text{ m}^3) = 5.0 \times 10^2 \text{ J}.$$

During process ABC, the gas does 500 J more work than during process ADC.

MODEL Since both processes have the same initial and final states, they cause the same change in internal energy.

SETUP From the first law: $\Delta U = Q_{ABC} - W_{ABC} = Q_{ADC} - W_{ADC}$.

SOLVE Thus process ABC also requires a greater heat input:
$$Q_{ABC} - Q_{ADC} = W_{ABC} - W_{ADC} = 5.0 \times 10^2 \text{ J}.$$

ANALYZE In process ABC, the expansion occurs at a higher pressure, so more work is done. The energy required to do the extra work is provided by the greater heat transfer. ■

19.5 Specific Heats of an Ideal Gas

19.5.1 Heat Transfer and Specific Heat

The heat transfer required to produce a given temperature change is described by a quantity called specific heat.

> The *specific heat c* for a given process is the heat transfer required to change the temperature of a unit mass of substance by 1 degree using that process.
> $$c = \frac{Q}{M \, \Delta T}. \tag{19.14}$$

As we saw in the previous section, heat transfer depends on the process the system undergoes between its initial and final states, and it need not change the temperature at all. In fact, since temperature alone specifies a line in the P-V diagram, and not a point, a change in temperature does not by itself determine the final state of the system. The final state also depends on the process by which the temperature is changed. Specific heats are usually defined for two processes: constant volume and constant pressure. Tables of values usually list the specific heat at constant pressure (c_p) because it is easier to measure. However, the specific heat at constant volume (c_v) is easier to calculate.

19.5.2 Specific Heat of a Monatomic Ideal Gas at Constant Volume

FOR EXAMPLE, TO GO FROM POINT A TO POINT B IN FIGURE 19.13.

The heat transfer required to change the temperature of a gas at constant volume (eqn. 19.14) is:
$$Q = M c_v \, \Delta T,$$

where c_v is the *specific heat at constant volume* and ΔT is the change in temperature, $T_B - T_A$. Since $\Delta V = 0$, no work is done, so from the first law of thermodynamics, $\Delta U = Q$. For a monatomic ideal gas (eqn. 19.10),

$$U = \tfrac{3}{2}NkT$$

so

$$\Delta U = \tfrac{3}{2}Nk\,\Delta T,$$

and:

$$Mc_v\,\Delta T = Q = \Delta U = \tfrac{3}{2}Nk\,\Delta T.$$

Thus:

$$c_v = \frac{3Nk}{2M}.$$

The mass of gas is $M = Nm$, where m is the mass of each molecule, so

$$c_v = \frac{3k}{2m}. \tag{19.15}$$

c_v FOR AN IDEAL MONATOMIC GAS.

For a monatomic ideal gas, c_v is independent of temperature.

EXAMPLE 19.12 ♦ Using the ideal gas model, find the specific heat at constant volume for helium.

MODEL Helium is monatomic with molecules of mass $4.00(1.66 \times 10^{-27}\,\text{kg}) = 6.64 \times 10^{-27}$ kg. The ideal gas is a good model for helium at temperatures well above 4 K.

SETUP AND SOLVE Using eqn. (19.15),

$$c_v = \frac{3k}{2m} = \frac{3(1.38 \times 10^{-23}\,\text{J/K})}{2(6.64 \times 10^{-27}\,\text{kg})} = 3120\,\text{J/kg}\cdot\text{K}.$$

ANALYZE The experimental value is 3119 J/kg·K at 25°C. ∎

Because no work is done in a constant-volume process, the specific heat at constant volume is directly related to the internal energy change of the gas:

$$\Delta U = Q = Mc_v\,\Delta T. \tag{19.16}$$

So:

$$c_v = \frac{1}{M}\left(\frac{dU}{dT}\right). \tag{19.17}$$

THIS RELATION ALSO HOLDS FOR NONIDEAL SUBSTANCES, IF THE DERIVATIVE IS REPLACED BY A PARTIAL DERIVATIVE $\partial U/\partial T$. FOR MANY REAL SUBSTANCES, U DEPENDS ON OTHER PROPERTIES AS WELL AS T.

By measuring the specific heat at constant volume, we can learn how the internal energy of a substance depends on its temperature.

19.5.3 Specific Heat of a Monatomic Ideal Gas at Constant Pressure

How much heat transfer is needed to increase the gas temperature at constant pressure—for example, along path AD in Figure 19.13? Although the temperature change is the same as in the constant-volume process, the final state is not the same. The constant-pressure process takes the system to a different point on the T_B isotherm.

The *specific heat at constant pressure* c_p exceeds that at constant volume because the gas expands and does work during the process. Heat transfer must account for both the increase in internal energy (described by the temperature increase) *and* the energy used by the gas to do work.

From the definition (eqn. 19.14),

$$Mc_p\,\Delta T \equiv Q = \Delta U + W.$$

Since D is on the same isotherm as B, the internal energy of state D is the same as that of state B, and the change in internal energy ΔU_{AD} is equal to the change in internal energy for the constant-volume process, ΔU_{AB}, which we have already found in §19.5.2:

$$\Delta U_{AD} = \Delta U_{AB} = \tfrac{3}{2} Nk\, \Delta T.$$

The work done is given by the area under the line AD:

$$W = P_A(V_D - V_A).$$

Using the ideal gas law, we express the work in terms of the temperature change:

$$W = P_D V_D - P_A V_A = NkT_D - NkT_A = Nk\, \Delta T.$$

Finally, we have:

$$Mc_p\, \Delta T = Q = \Delta U + W = \tfrac{3}{2} Nk\, \Delta T + Nk\, \Delta T = \tfrac{5}{2} Nk\, \Delta T.$$

c_p FOR AN IDEAL MONATOMIC GAS.

And:
$$c_p = \frac{5k}{2m}. \tag{19.18}$$

As expected, the specific heat at constant pressure is larger than the specific heat at constant volume. The ratio of these two specific heats is important enough to deserve a special symbol: the Greek letter gamma.

$$\frac{c_p}{c_v} \equiv \gamma. \tag{19.19}$$

For an ideal monatomic gas, $\gamma = \tfrac{5}{3}$.

> **EXAMPLE 19.13** ♦♦ A loose plastic bag floating in a room contains 2×10^{-4} kg of helium. The room temperature rises from $20°C$ to $25°C$. What is the heat flow into the bag of helium?
>
> **MODEL** The room remains at atmospheric pressure, and the loose bag ensures that the helium also remains at atmospheric pressure. The heat transfer occurs at constant pressure.
>
> **SETUP** $T_i = 20°C$ and $T_f = 25°C$, so $\Delta T = 5°C = 5$ K.
>
> **SOLVE** Using eqns. (19.14) and (19.18), the heat flow into the helium is:
>
> $$Q = Mc_p\, \Delta T = M\left(\frac{5k}{2m}\right) \Delta T = (2 \times 10^{-4}\ \text{kg}) \frac{5(1.38 \times 10^{-23}\ \text{J/K})}{2(6.64 \times 10^{-27}\ \text{kg})} (5\ \text{K}) = 5\ \text{J}.$$
>
> **ANALYZE** Remember that ΔT in $°C$ is the same as ΔT in K.

19.5.4 The Relation Between c_p and c_v

WE DERIVED THIS EXPRESSION FOR W IMMEDIATELY PRECEDING EQN. (19.18).

For an ideal gas, the work done in a constant-pressure process is related to the temperature change: $W = Nk\, \Delta T$. Because of this, the two specific heats are related. The internal energy change also depends on ΔT and is the same for both processes (eqn. 19.16): $\Delta U = Mc_v\, \Delta T$. Referring to Figure 19.13, and using the first law:

$$Mc_v\, \Delta T = \Delta U_{AB} = \Delta U_{AD} = Q - W = Mc_p\, \Delta T - Nk\, \Delta T.$$

Dividing by $M\, \Delta T$ and using $m = M/N$, we have:

$$c_p = c_v + \frac{k}{m}. \tag{19.20}$$

Equation (19.20) holds for any ideal gas. To derive it, we assumed only that internal energy is constant along an isotherm—that is, determined by temperature alone. Most real gases (such as oxygen or water vapor) have diatomic or polyatomic molecules at room temperature. The internal energy of a polyatomic ideal gas also depends only on temperature, although the function $U(T)$ that describes its dependence on T differs from the expression for monatomic gases. The ratio of specific heats, γ, always exceeds unity since c_p always exceeds c_v by k/m. For air, $\gamma \approx 1.4$. We'll discuss these issues further in §19.7.

EXERCISE 19.7 ♦♦ Derive expressions for the molar specific heats c'_v and c'_p for a monatomic ideal gas—that is, the amount of heat required to raise the temperature of 1 mol of gas through 1 K. Show that $c'_p = c'_v + R$ for any ideal gas, where R is the ideal gas constant. [*Hint:* Describe the amount of gas by the number of moles $\mathcal{N}$ instead of the mass M and use the ideal gas law in the form of eqn. (19.7).]

19.6 ADIABATIC PROCESSES IN AN IDEAL GAS

In our discussion and examples, we have analyzed isothermal (constant-temperature), isochoric (constant-volume), and isobaric (constant-pressure) processes. All these processes require heat transfer (cf. Examples 19.9, 19.10, 19.11). However, heat transfer is relatively slow and is negligible in many processes that occur rapidly. Examples of such processes include pressure oscillations in a sound wave and the expansion phase in the operation of an automobile engine.

In an *adiabatic process*, a system does not exchange heat with its environment:

Adiabatic process ⇔ $Q = 0$.

An *adiabat* is the path followed by the system point in the *P-V* diagram during an adiabatic process.

In an adiabatic expansion, the system does work at the expense of its internal energy: $\Delta U = -W$. The adiabat drops to lower isotherms as the internal energy (and hence temperature) decreases. Adiabats are steeper than isotherms (see ■ Figures 19.14 and 19.15).

To calculate the work done in an adiabatic process, we need to know how P depends on V during the process—that is, the equation of an adiabat. An adiabatic compression takes the system from the point A (P_1, V_1, T_1) to a neighboring point E $(P_1 + dP, V_1 + dV, T_1 + dT)$ (Figure 19.14), where dP, dV, and dT represent differential changes in the thermodynamic state variables. The change in internal energy of an ideal gas system depends only on the temperature change dT, and so we may express it as $dU = Mc_v\, dT$ (eqn. 19.16), independent of the process actually performed. The work done by the system is $dW = P\, dV$. With no heat flow, $Q = 0$, the first law of thermodynamics requires:

$$Mc_v\, dT = dU = 0 - dW = -P\, dV. \quad (19.21)$$

To find an expression relating the variables P and V alone, we eliminate dT by differentiating the ideal gas law:

$$d(PV) = d(NkT),$$

so

$$(dP)V + P\, dV = Nk\, dT.$$

We substitute for dT in eqn. (19.21) to get:

$$Mc_v\, \frac{V\, dP + P\, dV}{Nk} = -P\, dV.$$

Collecting terms in dP and dV, and using $M = Nm$, we have:

$$V\, dP\left(c_v \frac{m}{k}\right) = -P\, dV\left(1 + c_v \frac{m}{k}\right).$$

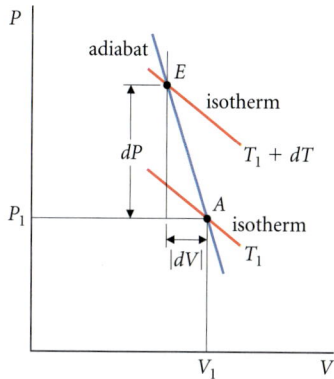

■ **FIGURE 19.14**
A differential piece of a *P-V* diagram showing two neighboring isotherms and an adiabat crossing them. When the gas is compressed along an adiabat, its pressure rises from P_1 to $P_1 + dP$. Work is done on the gas but no heat transfer occurs. According to the first law of thermodynamics, the internal energy and temperature of the gas increase. Thus point E is on a higher isotherm than A. Adiabats are steeper than isotherms.

NOTE THAT, IN FIGURE 19.14, dV IS NEGATIVE.

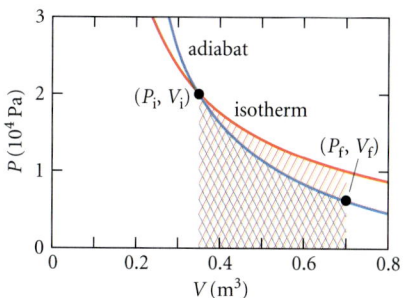

■ **FIGURE 19.15**
Comparison of an adiabatic expansion and an isothermal expansion from the same initial state. Each process doubles the gas volume. The final pressure is lower in the adiabatic expansion, and the work done during the process is less. The work done in each process is represented by the area under the corresponding curve.

So:
$$\frac{dP}{P} = -\frac{dV}{Vc_v}\left(c_v + \frac{k}{m}\right) = -\frac{dV}{V}\left(\frac{c_p}{c_v}\right) = -\gamma\frac{dV}{V}.$$

Integrating this equation, we have:
$$\int_{P_i}^{P_f} \frac{dP}{P} = \int_{V_i}^{V_f} -\gamma\frac{dV}{V} = -\gamma\int_{V_i}^{V_f} \frac{dV}{V},$$

Study Hint: Review the properties of logarithms and exponentials, Appendix I.

which becomes
$$\ln P\Big|_{P_i}^{P_f} = -\gamma \ln V\Big|_{V_i}^{V_f}.$$

Therefore:
$$\ln(P_f/P_i) = -\gamma \ln(V_f/V_i) = \ln[(V_i/V_f)^\gamma],$$

or
$$P_i V_i^\gamma = P_f V_f^\gamma = \alpha, \tag{19.22}$$

which is the required relationship between P and V. The parameter α serves to label adiabats in the same way that T labels isotherms. As expected, these lines are steeper than the isotherms, since γ is always greater than 1.

EXERCISE 19.8 ❖ A monatomic ideal gas expands adiabatically from a volume V_i to a volume $2V_i$. Is the work done less than or greater than the work done in an isothermal expansion from the same initial state to a volume of $2V_i$? Why?

EXAMPLE 19.14 ◆◆◆ Calculate the work done in the adiabatic process described in Exercise 19.8 with $V_i = 0.35$ m³ and $P_i = 2.0 \times 10^4$ Pa (Figure 19.15). Compare your answer with the work done in an isothermal expansion from V_i to $2V_i$.

MODEL The work done is represented by the area under the curve and equals $\int P\,dV$. We use eqn. (19.22) for $P(V)$ along the adiabat.

SETUP We model the process as a series of differential volume changes dV, with $dW = P(V)\,dV$ and $P(V) = P_i V_i^\gamma / V^\gamma$.

STEPS I, II, AND III

STEP IV Thus:
$$W = \int_{V_i}^{V_f} dW = \int_{V_i}^{V_f} P(V)\,dV = P_i V_i^\gamma \int_{V_i}^{V_f} \frac{dV}{V^\gamma}$$

STEP V
$$W = P_i V_i^\gamma \left[\frac{V^{-(\gamma-1)}}{-(\gamma-1)}\right]\Bigg|_{V_i}^{V_f}$$

$$= \frac{P_i V_i^\gamma}{\gamma - 1}\left[V_i^{-(\gamma-1)} - V_f^{-(\gamma-1)}\right].$$

Note: $V_i^\gamma = V_i V_i^{\gamma-1}$.

$$W = \frac{P_i V_i}{\gamma - 1}\left[1 - \left(\frac{V_i}{V_f}\right)^{\gamma-1}\right]. \tag{19.23}$$

SOLVE For the numbers given:
$$W = \frac{(2.0 \times 10^4 \text{ Pa})(0.35 \text{ m}^3)}{\frac{5}{3} - 1}\left[1 - \left(\frac{1}{2}\right)^{2/3}\right] = 3900 \text{ J}.$$

For an isothermal process, the work done (eqn. 19.13) is:
$$W = P_i V_i \ln\left(\frac{V_f}{V_i}\right) = (2.0 \times 10^4 \text{ Pa})(0.35 \text{ m}^3)\ln 2 = 4900 \text{ J}.$$

ANALYZE As expected, the work done in the isothermal process is greater. At the end of the adiabatic expansion, the gas has a lower temperature than it does during the isothermal expansion.

Adiabatic compression explains why Santa Ana winds are so hot and unpleasant. Pressure decreases with height in the Earth's atmosphere, so pressure increases in a wind blowing down off a mountain range. Because heat transfer through the air is slow, the compression is adiabatic and the temperature rises as the air is compressed. The temperature of about 70°F over inland high country can increase to over 100°F in air descending some 5000 ft as it flows to the coast.

WE DISCUSSED THE ATMOSPHERE IN §13.3.4.

Study Problem 13 ♦♦♦ *A Cycle in the P-V Diagram*

We have now completed our analysis of the most frequently encountered processes and their appearance in the *P-V* diagram. In this example, we show how to apply these results to problems involving gas systems.

Practical heat engines are systems that operate on a cycle; that is, the system undergoes a series of processes that ultimately return it to its initial state. The processes form a closed figure in the *P-V* diagram (■ Figure 19.16). One such cycle consists of an isobaric expansion from $V_1 = 0.15$ m³ to $2V_1$, an isochoric reduction of pressure from $P_1 = 6.0 \times 10^4$ Pa to $P_1/2$, and an isothermal compression back to the starting point. How much work is done by the system in going around the cycle, and what is the heat transfer?

Understanding the Physical System

The first step is to draw the cycle on the *P-V* diagram (■ Figure 19.17). The work done by the system is $\oint P\, dV$. The integral is taken around the whole cycle $A \to B \to C \to A$ and is represented by the area contained within the closed figure, *ABCA*. Since the system returns to the initial state, there is no net change of internal energy in a complete cycle. From the first law:

$$\Delta U = 0 = Q - W \quad \text{so} \quad Q = W.$$

The net heat transfer into the system equals the work done by the system, which is given by the area enclosed within *ABCA*. In general, a system following such a cycle performs work as a result of energy input in the form of heat transfer. We shall explore this aspect of cycles extensively in Chapter 22.

THIS IS TRUE FOR *ALL* CYCLIC PROCESSES.

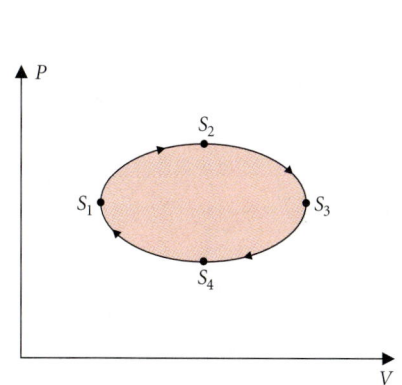

■ FIGURE 19.16
A cycle is a sequence of processes that return the gas to its initial state:
$S_1 \to S_2 \to S_3 \to S_4 \to S_1$.

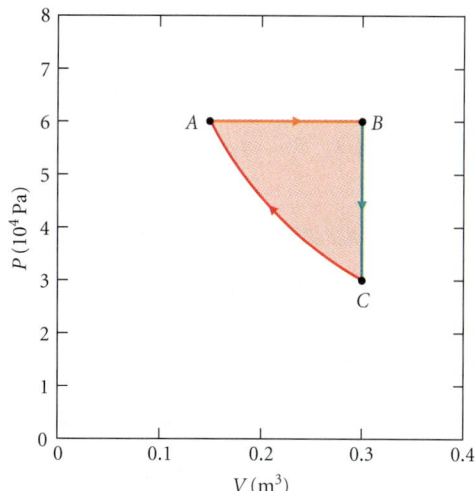

■ FIGURE 19.17
This cycle has three processes: a constant-pressure (isobaric) expansion from *A* to *B*, a constant-volume (isochoric) process from *B* to *C*, and an isothermal process from *C* to *A*. The work done is represented by the area enclosed within the curve representing the cycle. Convince yourself that $T_A = T_C$, and $T_B > T_A$. (*Hint:* What is *PV* at each point?)

SECTION 19.6 • ADIABATIC PROCESSES IN AN IDEAL GAS

Plan

We find the work done on each leg of the cycle and add the results to find W, then use the first law to find Q (■ Figure 19.18). We must take particular care with signs.

Setup of Solution

First, we calculate the work done on each leg. An isobaric process (P = constant) takes the system from A to B. The work done by the system is:

$$W_{AB} = \int_A^B P \, dV = P_1(2V_1 - V_1) = P_1 V_1.$$

No work is done during the isochoric process that takes the system from B to C: V = constant $\Rightarrow W_{BC} = 0$. An isothermal process takes the system from C back to A. Using eqn. (19.13):

$$W_{CA} = P_i V_i \ln(V_f/V_i) = P_1 V_1 \ln(\tfrac{1}{2}) = -P_1 V_1 \ln(2).$$

Solution

The net work done on the system is the sum of these terms:

$$W = W_{AB} + W_{BC} + W_{CA} = P_1 V_1 (1 - \ln 2) = +0.31 P_1 V_1.$$

The plus sign implies that the system does work on the environment, as expected. From the first law,

$$Q = W = 0.31 P_1 V_1 = 0.31 (6.0 \times 10^4 \text{ Pa})(0.15 \text{ m}^3) = 2800 \text{ J}.$$

THE MATHEMATICAL POINT THAT DESCRIBES THE SYSTEM MOVES IN A DIAGRAM. THE CYLINDER OF GAS DOES NOT MOVE PHYSICALLY!

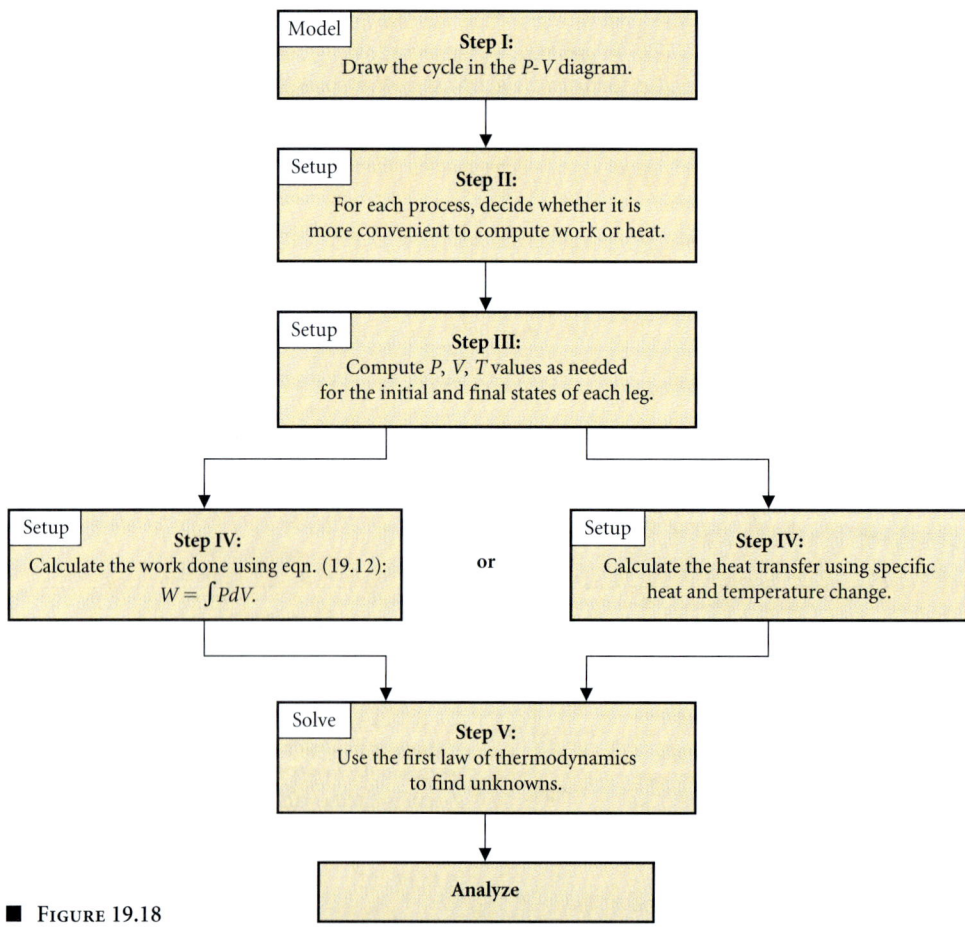

■ FIGURE 19.18

Analysis

From Figure 19.17, we notice that $T_B > T_A$. Thus the heat transfer to the system along AB is $Q = mc_p(T_B - T_A)$. Similarly, along BC, $Q = mc_v(T_C - T_B) = -mc_v(T_B - T_C)$. Along the isotherm CA the internal energy is constant and $Q = W$. We may summarize these results as shown in the table. Heat input ($Q > 0$) occurs along AB but heat output ($Q < 0$) occurs along BC and CA. Over the entire cycle, less energy is retrieved as work than is input as heat. Some energy is expelled as waste heat. This is a general property of all cyclic processes.

Process	T	W	Q
AB	increases	+	+
BC	decreases	0	−
CA	constant	−	−

✱ 19.7 Equipartition of Energy

According to the ideal gas model, the specific heat of any particular monatomic gas is constant, independent of the state variables P, V, and T. However, the specific heat of a real gas is a function of temperature. ■ Figure 19.19 shows the measured values of molar specific heat at constant volume for several gases. Only argon is consistent with the model discussed in §19.5. For the other gases shown, the specific heat increases in steps as the temperature increases. In this section we investigate why the specific heat increases with temperature and show how the steps depend on the structure of individual gas molecules. There are two important questions: how is the internal energy stored by the gas molecules, (what *modes* of storage does the gas have), and how is the energy distributed among these modes at a specific temperature?

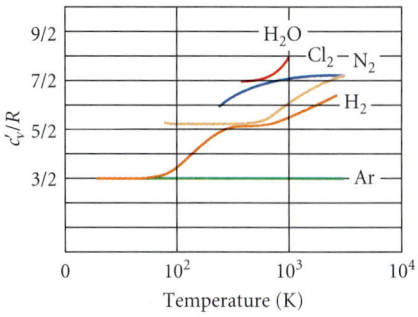

■ **Figure 19.19**
Measured values of molar specific heat at constant volume for several gases. Only the argon data are consistent with the theory in §19.5. To understand the behavior of the other gases, we look at the possible motions of polyatomic molecules.

19.7.1 Modes of Energy Storage

A good model for an atom in a monatomic ideal gas is a billiard ball: a hard, round, structureless object, capable only of linear motion. The atoms continually collide with each other, changing the directions of their velocities and exchanging energy. As a result, the average kinetic energy per atom,

$$K = \tfrac{1}{2}m\langle v^2 \rangle = \tfrac{1}{2}m(\langle v_x^2 \rangle + \langle v_y^2 \rangle + \langle v_z^2 \rangle),$$

consists of three equal parts:

$$\tfrac{1}{2}m\langle v_x^2 \rangle = \tfrac{1}{2}m\langle v_y^2 \rangle = \tfrac{1}{2}m\langle v_z^2 \rangle = K/3.$$

Since the total kinetic energy K of each atom has an average value of $\tfrac{3}{2}kT$ (eqn. 19.8), each of the three parts has an average value of $\tfrac{1}{2}kT$. Each part is a *mode of energy storage*, or *degree of freedom*, for the gas, and internal energy is distributed equally among the modes.

A billiard ball gas is a good model for helium or argon, but other common gases have more complex molecules. For example, oxygen and nitrogen molecules are diatomic and are more like dumbbells than like spheres. Ammonia molecules are pyramids (■ Figure 19.20). Rotation and vibration of such molecules offer additional ways to store energy.

A polyatomic molecule rotates about its center of mass and its rotational energy is also the sum of three terms:

$$E_{\text{rot}} = \tfrac{1}{2}(I_x \omega_x^2 + I_y \omega_y^2 + I_z \omega_z^2).$$

Just as for translational energy, each of the three terms corresponds to a possible mode of energy storage.

To illustrate how to count vibrational modes, we model a diatomic molecule as two billiard balls oscillating on the ends of a spring (■ Figure 19.21). As the balls oscillate, they have a relative displacement $\vec{s}$ and corresponding relative velocity $d\vec{s}/dt = \vec{v}_s$. The molecule's vibrational energy varies between kinetic and potential forms (cf. §14.3).

$$E_{\text{vib}} = \tfrac{1}{2}ks^2 + \tfrac{1}{2}m'v_s^2.$$

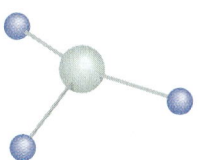

■ **Figure 19.20**
An ammonia molecule is pyramid shaped.

"Degree of freedom" is the correct technical term, but "mode of energy storage" describes the idea much better for our purposes here.

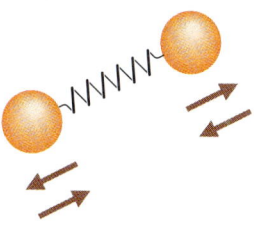

■ **Figure 19.21**
Atoms in a diatomic molecule may vibrate along the line joining them. We model the molecule as two particles joined by a spring. The energy of the molecule includes the kinetic energy of the vibrating atoms and the potential energy of the "spring" joining them.

The "reduced mass" m' is not quite equal to m. The difference is not important to the argument.

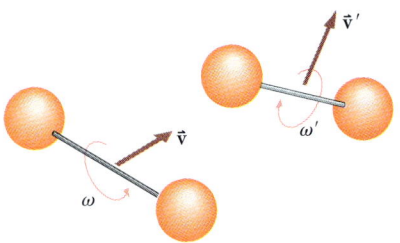

■ **Figure 19.22**
Polyatomic molecules also rotate as a result of off-center collisions among them. Diatomic molecules do not rotate about their longitudinal axis because it is almost impossible to hit them at points not on this axis. While these models help us to visualize the ideas, quantum mechanics governs the behavior of the molecules and determines those modes that are excited.

Again, each of the two terms corresponds to an energy storage mode. More complicated molecules vibrate with combinations of such simple oscillations. Since each oscillation provides two storage modes, the number of vibrational energy storage modes of a polyatomic molecule is twice the number of ways it can oscillate.

19.7.2 Energy Stored in Each Mode

Collisions transfer energy between modes and ensure that, on average, no mode stores more energy than any other. Off-center collisions, for example, cause the molecules to rotate (■ Figure 19.22; see also §10.4.2). Similarly, a rotating molecule causes another molecule to rebound like a baseball struck by a rotating bat. An individual molecule may have any amount of energy in translation and in rotation, but averaged over all the molecules, there is an equal amount of energy in each mode. A molecule has three modes of translational kinetic energy, accounting for an internal energy $U_{\text{trans}} = \frac{3}{2}kT$, or $\frac{1}{2}kT$ per mode for each molecule. At high enough temperatures, rotational and vibrational modes also store, on average, $\frac{1}{2}kT$ each per molecule.

A molecule does not always exhibit all possible modes of motion (we say that some modes are not "excited"). According to quantum theory (Chapter 35), there is a smallest amount of energy, called a quantum, that can exist in any mode. Each mode has a different size quantum. For vibrations, it is proportional to the frequency of the vibration, and for rotations, it is inversely proportional to the molecule's rotational inertia. The quantum of translational energy is negligible. If the size of the quantum for a particular mode exceeds $\frac{1}{2}kT$, only a very small fraction of collisions are sufficiently energetic to excite the mode, and very little energy is stored in that mode.

We may summarize these results in a statement known as the *principle of equipartition*.

> In thermal equilibrium, each gas molecule has, on average, $\frac{1}{2}kT$ of energy in each excited mode of energy storage.

19.7.3 Specific Heats of Polyatomic Ideal Gases

Because energy flow into a system is shared among all the modes, the temperature increase caused by a given heat input depends on the number of modes available. If a system has N_m excited modes of energy storage at temperature T, the average energy per molecule is:

$$E = \tfrac{1}{2}N_m kT.$$

Then the total internal energy of the system is:

$$U = \tfrac{1}{2}N_m NkT. \tag{19.24}$$

BE SURE TO DISTINGUISH N_m = NUMBER OF MODES FROM N = NUMBER OF MOLECULES.

The specific heat at constant volume for this system is (eqn. 19.17):

$$c_v = \frac{1}{M}\left(\frac{dU}{dT}\right) = \frac{N_m}{2}\left(\frac{k}{m}\right). \tag{19.25}$$

Similarly, the specific heat at constant pressure is:

$$c_p = c_v + \frac{k}{m} = \frac{N_m + 2}{2}\left(\frac{k}{m}\right), \tag{19.26}$$

and

$$\gamma = \frac{c_p}{c_v} = \frac{N_m + 2}{N_m}. \tag{19.27}$$

TABLE 19.2 Molar specific heats for several gases at 1 atm and 20°C

Gas	c'_v J/mol·K	c'_p J/mol·K	γ
helium	12.5	20.8	1.67
argon	12.5	20.8	1.67
hydrogen	20.4	28.8	1.41
oxygen	21.1	29.4	1.40
nitrogen	20.8	29.1	1.40
chlorine	25.7	34.7	1.35
carbon dioxide	28.5	37.0	1.30
sulphur dioxide	31.4	40.4	1.29
ammonia	27.8	36.8	1.31

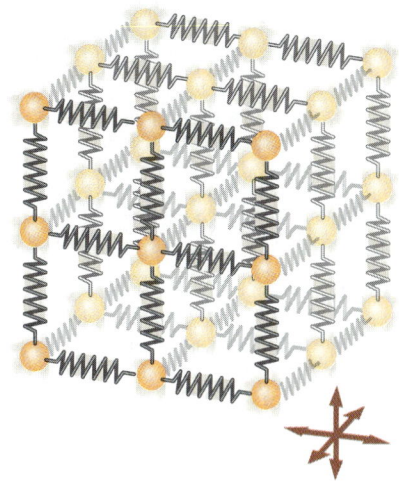

■ FIGURE 19.23
A simple model for a crystalline solid. The molecules are attached to each other with springs and can vibrate along three orthogonal directions. Each molecule therefore has six degrees of freedom.

COMPARE FIGURE 19.19.

The specific heats increase as N_m increases because, to increase the temperature by ΔT, we must add $\frac{1}{2} k \, \Delta T$ of energy to each available mode. Values of specific heats and γ for several gases are given in ● Table 19.2.

Hydrogen, a diatomic gas, has $c_v = \frac{3}{2}(k/m)$ at low temperature. The three translational modes are excited, but rotation and vibration are not. As temperature increases, c_v increases to $\frac{5}{2}(k/m)$, corresponding to five excited modes. The two additional modes correspond to rotation about the two axes perpendicular to the symmetry axis of the molecule. Because of the small rotational inertia about the symmetry axis, the energy quantum for rotation about that axis is very much larger than for the other two, and the mode is not excited. At higher temperatures, the specific heat increases again as molecules begin to vibrate. The molecules dissociate into atomic hydrogen before many are excited to higher vibrational states, and the value $c_v = \frac{7}{2}(k/m)$ is never reached. At high temperatures, nitrogen and chlorine have $c_v = \frac{7}{2}(k/m)$, corresponding to three translational, two rotational, and two vibrational modes.

We can gain some insight into the specific heat of solids from the following simple model. The molecules form a crystal lattice, considered to be cubical for simplicity (■ Figure 19.23). Each molecule is free to vibrate along each of the three axes defined by the lattice, and there is kinetic energy and potential energy associated with each direction. Overall translational motion is not possible. Thus there are six modes for each molecule to store energy. From eqn. (19.25):

$$c_v = \frac{N_m}{2}\left(\frac{k}{m}\right) = 3\frac{k}{m}. \quad (19.28)$$

SEE CHAPTER 20 FOR MORE ON SPECIFIC HEATS OF SOLIDS.

This result is known as the Dulong–Petit law and is valid for most solids at high enough temperatures, even if they do not have a cubic crystal structure. For each solid, there is a critical temperature called the Debye temperature below which the model fails.

19.7.4 Brownian Motion

Air molecules collide not only with each other, but with any other particles suspended in the air—dust, smoke, pollen, and so on. These particles contain many molecules and are much more massive than individual air molecules, but they are nevertheless in thermal equilibrium and have, on average, a kinetic energy of $\frac{1}{2} mv^2 = \frac{3}{2} kT$. In 1827, Robert Brown, a botanist, discovered the random motion of pollen particles suspended in water (■ Figure 19.24) and interpreted the motion to mean that the grains were alive! In 1905, Einstein's explanation of Brownian motion as random thermal motion provided a powerful method of measuring Avogadro's number and convinced the last skeptics that atoms exist.

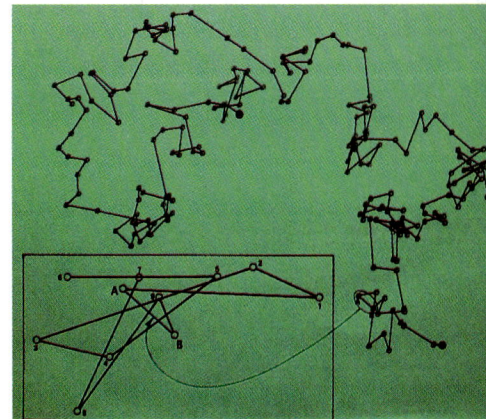

■ FIGURE 19.24
Brownian motion. Small particles suspended in a gas or liquid illustrate random motion due to collisions with fluid molecules. This sketch by the French physicist Jean Baptiste Perrin in 1912 shows the position of a suspended particle every 30 s. The path is an example of a *fractal*, appearing qualitatively similar on different scales. The bottom panel shows the position of the particle between points A and B at shorter time intervals.

Chapter Summary

Where Are We Now?

We have introduced the fundamental concepts we'll need in our study of thermodynamics: temperature, internal energy, heat transfer, and work. For an ideal gas, we have shown that internal energy is the random kinetic energy of the gas molecules and that temperature measures the average value of this energy per molecule. The first law of thermodynamics expresses the principle of energy conservation, including the internal thermal energy of a system. Now we are ready to extend these ideas to real substances.

What Did We Do?

A system is in thermal equilibrium if its thermodynamic properties show no tendency to change. Temperature is defined via the zeroth law of thermodynamics as the variable that determines whether systems are in thermal equilibrium. The temperature scales in common use are the Celsius and Fahrenheit scales. We may convert between them using eqns. (19.1). For scientific work, the Kelvin scale is used: water freezes at 273.15 K and boils at 373.15 K at atmospheric pressure.

We use the following definitions when discussing energy in thermodynamics:

- *Internal thermal energy:* Energy in any form stored within a system and distributed randomly among many molecules.
- *Heat transfer:* Thermal energy transferred during a process through many random events. Energy transferred as the result of a temperature difference is heat transfer.

In an ideal gas, internal energy is a function only of temperature: $U = U(T)$. The ideal gas law:

$$PV = NkT = \mathcal{N}RT$$

relates the pressure and volume of an ideal gas to the number of molecules N, the temperature T, and Boltzmann's constant $k = 1.38 \times 10^{-23}$ J/K. Equivalently, it may be expressed in terms of the number of moles $\mathcal{N} = N/N_A$ and the gas constant $R = kN_A = 8.3145$ J/mol·K. Avogadro's number $N_A = 6.02 \times 10^{23}$. Real gases behave like ideal gases at low pressure, provided the temperature is not very low.

The mean kinetic energy per molecule in a monatomic ideal gas is $\langle K \rangle = \frac{3}{2}kT$. The internal energy U of a monatomic ideal gas system is the total random kinetic energy of all its molecules: $U = \frac{3}{2}NkT$.

The first law of thermodynamics,

$$\Delta U = Q - W,$$

expresses conservation of energy. Internal energy may be increased by heat transfer Q or by doing work *on* the system. The work done *by* a gas as it expands is:

$$W = \int_{V_i}^{V_f} P\, dV.$$

The *P-V* diagram is an important tool for studying gas systems. The state of any ideal gas system is displayed as a point in the diagram. We can read off the state variables P and V immediately and then, given the amount of gas, deduce the variables T and U from the ideal gas model. A process is represented by a line in the diagram. The work done by the system during the process is the area under that line, and the heat input may then be deduced from the first law of thermodynamics. We described the following processes:

- Isothermal—constant temperature
- Isochoric—constant volume
- Isobaric—constant pressure
- Adiabatic—no heat transfer

The specific heat of a gas for a given process is the heat transfer required to raise the temperature of a unit mass of gas by 1 degree. Specific heats are defined for both constant-volume and constant-pressure processes. For a monatomic ideal gas, $c_v = \frac{3}{2}(k/m)$, where m is the mass of a gas molecule, and $c_p = c_v + k/m$. The ratio of specific heats $c_p/c_v \equiv \gamma$. During an adiabatic process, the pressure and volume of a gas are related by:

$$PV^\gamma = \text{constant}.$$

✱ Polyatomic gases can store energy in rotation and vibration of the molecules. According to the principle of equipartition, on average there is an equal amount of energy $\frac{1}{2}kT$ in each excited mode of storage. Translational kinetic energy provides three modes of storage corresponding to motion in three perpendicular directions.

Practical Applications

Any real gas is nearly ideal at low density. The ideal gas model is particularly useful in astrophysics, where it describes material in stars and between the stars. It also works quite well for air at room temperature. The equation we have derived for an adiabatic process describes the behavior both of air in sound waves and of the gases in a cylinder of an automobile engine during the expansion phase. The first law of thermodynamics and the ideas of specific heat are the starting point for understanding the behavior of real systems (Chapter 20).

THE AUTOMOBILE ENGINE IS DISCUSSED IN SOME DETAIL IN §22.2.

Solutions to Exercises

19.1 From eqn. (19.1),

$$T(°F) = [\tfrac{9}{5}(961.9) + 32]°F = 1763°F.$$

19.2 From eqn. (19.3), $PV = \text{constant} = (0.200 \text{ m}^3)(1.00 \times 10^4 \text{ Pa}) = V_f(2.50 \times 10^4 \text{ Pa})$. The volume at 2.50×10^4 Pa is:

$$V_f = (0.200 \text{ m}^3)\left(\frac{1.00 \times 10^4 \text{ Pa}}{2.50 \times 10^4 \text{ Pa}}\right) = 0.0800 \text{ m}^3.$$

19.3 The number of moles is:

$$\mathcal{N} = \frac{PV}{RT} = \frac{(1.00 \times 10^4 \text{ Pa})(0.500 \text{ m}^3)}{(8.31 \text{ J/mol·K})(305 \text{ K})} = 1.97 \text{ mol}.$$

The mass of the sample is the number of moles times the gram molecular mass of CO_2:

$$M = (1.97 \text{ mol})(44 \text{ g/mol}) = 87 \text{ g},$$

as we obtained in Example 19.4.

19.4 The mass of a hydrogen atom is 1.67×10^{-27} kg, so:

$$v_{\text{rms}} = \sqrt{\frac{3kT}{m}} = \sqrt{\frac{3(1.38 \times 10^{-23} \text{ J/K})(120 \text{ K})}{2(1.67 \times 10^{-27} \text{ kg})}} = 1200 \text{ m/s}.$$

19.5 The state variables for an ideal gas system include its pressure P, volume V, temperature T, internal energy U, and the number of molecules N (or the number of moles $\mathcal{N}$ or the mass M). They are related through the ideal gas law and eqn. (19.10), so with any three we may compute the other two.

19.6 Since $P_B V_B = (2.0 \times 10^4 \text{ Pa})(0.050 \text{ m}^3) = 1.0 \times 10^3 \text{ Pa·m}^3 = P_D V_D$, both B and D are on the same isotherm. Thus the change in internal energy between A and D is the same as that between A and B.

In Example 19.9, we found $\Delta U = 750$ J. The work done during the process is represented by the area under the line AD:

$$W_{AD} = P_A(V_D - V_A)$$
$$= (1.0 \times 10^4 \text{ Pa})(0.050 \text{ m}^3) = 5.0 \times 10^2 \text{ J}.$$

Using the first law of thermodynamics:

$$Q_{AD} = \Delta U + W_{AD} = 750 \text{ J} + 500 \text{ J} = 1250 \text{ J}.$$

It takes more heat transfer to effect the same temperature change along path AD than along path AB, since some of the energy transferred is used by the gas in doing work as it expands from V_A to V_D. We'll explore this further in the next section.

19.7 The derivation follows that for c_v and c_p. The heat required to raise 1 mol of gas through a temperature ΔT at constant volume is by definition $Q_v = c'_v \Delta T$. But we know $Q_v = \Delta U = \frac{3}{2}R \Delta T$, so $c'_v = 3R/2$. Similarly, the heat required to raise the temperature of 1 mol through ΔT at constant pressure is, by definition, $c'_p \Delta T$, and also equals $\Delta U + W$:

$$Q_p = \tfrac{3}{2}R \Delta T + R \Delta T = \tfrac{5}{2}R \Delta T \quad \text{so} \quad c'_p = \tfrac{5}{2}R.$$

Since $Q_p = Q_v + R \Delta T$, then:

$$c'_p = c'_v + R.$$

19.8 As the gas expands adiabatically, its temperature decreases, and so its pressure decreases faster than in an isothermal expansion (see Figure 19.15). The differential work $dW = P \, dV$ is less for each increment dV, so the total work done is less. Equivalently, the area between V_i and V_f under the adiabat passing through the initial state is less than the area between the same two volumes under the isotherm through the initial state.

Basic Skills

Review Questions

§19.1 TEMPERATURE

- What is a *thermodynamic system*? Give an example.
- When is a thermodynamic system *isolated*?
- What does it mean to say that two systems are in *thermal contact*?
- When is a thermodynamic system in thermal equilibrium?
- When are two systems in thermal equilibrium with each other?
- State the zeroth law of thermodynamics. How is it used to define temperature?
- What is a *thermometer*? What is a *temperature scale*?
- What are the freezing and boiling points of water on the Celsius and Fahrenheit scales? How are the two scales related?
- What is *absolute zero*?
- What is the *Kelvin scale*? What is the freezing point of water on this scale?

§19.2 TEMPERATURE IN AN IDEAL GAS

- State Boyle's law and Charles' law.
- State the ideal gas law. Under what conditions does it apply to real gases?
- What is a *mole*? How many molecules are in 1 mol of gas?
- What is the relation between temperature and the rms speed of ideal gas molecules?
- Define the *internal thermal energy* of an ideal gas system. How is it related to the gas temperature?

§19.3 THE FIRST LAW OF THERMODYNAMICS

- What is *heat transfer*?
- State the first law of thermodynamics. Which basic conservation law does it express?
- Write an expression for the work done by an ideal gas system during a process that changes its volume.

§19.4 THE P-V DIAGRAM

- List the state variables for an ideal gas system. How many of them are needed to specify a state? How is a state represented in a P-V diagram?
- What is an *isotherm*? What is its shape in a P-V diagram?
- How is a process represented in a P-V diagram?
- How may you evaluate the heat transfer during a process using a P-V diagram?
- Which of the following are properties of a point and which of a line in the P-V diagram? (a) heat transfer, (b) temperature, (c) volume, (d) work done, (e) pressure.

§19.5 SPECIFIC HEATS OF AN IDEAL GAS

- Define *specific heat at constant pressure* and *specific heat at constant volume*.
- What is the specific heat at constant volume for an ideal monatomic gas?
- What is the relation between c_p and c_v for an ideal gas?
- What does the symbol γ signify? What is its value for helium and for air?
- What is the molar specific heat at constant volume c'_v for a monatomic gas? How are c'_p and c'_v related?

§19.6 ADIABATIC PROCESSES IN AN IDEAL GAS

- Define an *adiabatic* process.
- What is an adiabat? What equation describes an adiabat in the P-V diagram?
- Is an adiabat steeper or shallower than an isotherm? Why?
- What is a thermodynamic *cycle*?
- How is the net heat transfer into a gas during a cycle related to the work done by the gas?

✱ §19.7 EQUIPARTITION OF ENERGY

- What is a *mode of energy storage*?
- How many modes of energy storage are there in a monatomic ideal gas?
- How many more modes of energy storage does a diatomic gas have? What are they?
- Are they all excited at room temperature? Why or why not?
- What is the Dulong–Petit law? When does it apply?
- What is *Brownian motion*?

Basic Skill Drill

§19.1 TEMPERATURE

1. Two separated systems A and B are each tested against a third system (a mercury-in-glass thermometer). In each case, the mercury column is 3 cm long when the thermometer is in contact with the system. Are systems A and B in equilibrium? Why or why not? Systems C and D are touching. The mercury column is 2.5 cm when in thermal contact with C and 3.5 cm when in contact with D. Are systems C and D in thermal equilibrium?

2. Iron melts at a temperature of 1808 K. Express this temperature on the Fahrenheit scale and on the Celsius scale.

3. Ethyl alcohol boils at 78.5°C. Express this temperature on the Kelvin scale and on the Fahrenheit scale.

§19.2 TEMPERATURE IN AN IDEAL GAS

4. Neon is a monatomic gas with a molecular mass of 20. What is the root mean square speed of neon molecules at 0°C? What is the internal energy of 1.0 kg of neon gas at 0°C?

5. A sample of gas in a 1-L flask has a mass of 2.5×10^{-5} kg. The gas exerts a pressure of 10^4 Pa at a temperature of $-80°C$. **(a)** How many moles of gas are in the flask? **(b)** What is the gram molecular mass of the gas? **(c)** What is the chemical composition of the gas?

6. A gas proportional counter for observing x rays is to be filled with xenon. The counter is a rectangular box measuring 0.50 m by 0.20 m by 5.0 cm. It operates at a temperature of 20°C and 1.0 atm pressure. How much xenon is required to fill the box?

§19.3 THE FIRST LAW OF THERMODYNAMICS

7. A nail is hammered into a heat-insulating material. You observe that the nail becomes hot. Take the system to be the nail plus the surrounding material. Has heat transfer into the system occurred?

8. A gas expands at a constant pressure $P = 875$ Pa from a volume of 1.0 m^3 to a volume of 2.5 m^3. How much work is done by the gas?

§19.4 THE P-V DIAGRAM

9. Show the positions of the following states of an ideal gas in a P-V diagram. **(a)** 3.0 mol at a temperature of 310 K and 1.0 atm pressure. **(b)** 2.0×10^{-3} m^3 at a pressure of 1.0 atm. **(c)** 2.0 mol at a volume of 1.0×10^{-3} m^3 and a temperature of 310 K.

10. Show the following processes in a P-V diagram. In each case, the system is a volume of ideal gas. **(a)** The system described in part (a) of Problem 9 expands at constant pressure to a volume of 0.11 m^3. **(b)** The system in part (b) of Problem 9 expands isothermally to a pressure of 0.50 atm. **(c)** The system in part (c) of Problem 9 is in a leaky box. The amount of gas decreases to 1.0 mol, at constant temperature.

§19.5 SPECIFIC HEATS OF AN IDEAL GAS

11. Using the ideal gas model, compute the specific heat at constant volume and at constant pressure for neon. Compare with the measured value $c_p = 1046$ J/kg·K at 25°C.

12. A cylinder of fixed volume contains 6.00 g of helium. The gas is heated from $-20.0°$C to $+10.0°$C. How much heat transfer must occur?

§19.6 ADIABATIC PROCESSES IN AN IDEAL GAS

13. In an adiabatic process, helium expands from a pressure of 2.0 atm and a volume of 0.25 m^3 to a volume of 0.35 m^3. What is the pressure after the expansion?

14. A cycle consisting of two constant-volume processes and two constant-pressure processes passes through the states shown in Figure 19.13: $A \rightarrow B \rightarrow C \rightarrow D \rightarrow A$. Find the total work done by the gas in one cycle and the net heat input.

✻ §19.7 EQUIPARTITION OF ENERGY

15. The measured value of γ for a gas at 75 K is $\frac{5}{3}$. The gas is known to be diatomic. What would you expect the value of γ to be at room temperature? Why?

16. Use the Dulong–Petit law to estimate the specific heat of copper. Compare your value with Table 20.4.

17. Use an ideal gas model to compute c_v for nitrogen gas at room temperature, assuming vibrations are not excited.

Questions and Problems

§19.1 TEMPERATURE

18. ❖ Define the system that would be most useful to consider in analyzing each of the following situations. **(a)** A ball is dropped to the ground from a height of 3 m. You wish to find the increase in temperature of the ball. **(b)** A bottle of wine is placed in an ice bucket. You wish to calculate how long it takes to chill. **(c)** Count Rumford needs to know how much water he should use to cool the cannon. **(d)** You wish to find out how hot the space shuttle gets during re-entry.

19. ❖ ✉ Are any of the following systems approximately isolated? If so, estimate the time scale for which the approximation holds. If not, describe a system in thermal contact with the given system. **(a)** An ice cube in a freezer. **(b)** Hot coffee in a thermos bottle. **(c)** Your body while lying on a beach. **(d)** The gas in a balloon rising through the air.

20. ❖ Imagine the following fanciful system: A room contains a number of elves and an equal number of balls. Each elf catches any approaching ball and throws it away in a random direction. The temperature is determined by the rate at which balls are thrown. This room is allowed to come into thermal contact with another identical room at a higher temperature by boring a number of holes in the adjoining wall, through which the balls can pass. Describe what happens. Do the elves in each room throw balls more or less often? Why?

21. ♦ Starting with the value on the Celsius scale, verify the numerical value of absolute zero on the Fahrenheit scale shown in Figure 19.5.

22. ♦ When people in the United States talk about "subzero" temperatures, they usually mean "less than 0°F." What is the corresponding range on the Celsius scale?

23. ♦ Mercury freezes at $-38.87°$C. What is this temperature on the Kelvin scale and on the Fahrenheit scale?

24. ♦♦ Find the temperature that has the same numerical value on both Fahrenheit and Celsius scales.

25. ♦♦ The length of a mercury column in a thermometer is 1.0 cm when the thermometer is immersed in freezing saturated brine and 10.0 cm in boiling water (at atmospheric pressure). What is the length of the mercury column in melting pure ice? What assumptions are you making about how the column length varies with temperature?

§19.2 TEMPERATURE IN AN IDEAL GAS

26. ❖ Air is a mixture of nitrogen and oxygen molecules. In a sample of air at 20°C, which molecules have the greater mean kinetic energy, the oxygen molecules or the less massive nitrogen molecules?

27. ❖ If two gases of different molecular masses, initially at the same temperature, are mixed together, which molecules will have the larger average kinetic energy? Which molecules will have the larger rms speed? Explain your reasoning.

28. ❖ In a certain process, the temperature of an ideal gas sample is held fixed. Does the rms speed of the molecules depend on the gas pressure? Why or why not?

29. ❖ Two systems are joined by a conducting wall that allows thermal contact. In equilibrium, the temperature of one system is T_1. What is the temperature of the other? Must the systems have the same internal energy? Justify your answer.

30. ❖ Two equal-volume containers of gas are in thermal contact. One contains 2 mol of gas and the other contains 3 mol. Which container has the greater internal energy, and why?

31. ❖ Two containers of gas at the same pressure and temperature are in thermal contact. One has a volume of 2 L, and the other has a volume of 1 L. Which container has the greater internal energy? Why?

32. ❖ Two different kinds of ideal gas (e.g., helium atoms and hydrogen molecules) are mixed in the same container. Model the gases as two separate thermodynamic systems in thermal contact. How does the pressure exerted by the mixture relate to the individual pressures each gas would exert alone? Explain your answer carefully. How, if at all, is the ideal gas law modified to describe the mixture?

33. ◆ What is the rms speed of hydrogen molecules at room temperature (20°C) and atmospheric pressure?

34. ◆ Use the ideal gas law to show that the density of an ideal gas is given by:
$$\rho = \frac{P}{T}\left(\frac{m}{k}\right).$$

35. ◆ The space between galaxies in rich clusters contains ionized gas at a temperature of 10^8 K. Find the rms speed of (a) the protons, and (b) the electrons in this gas.

36. ◆◆ ✉ A bicycle tire holds a gauge pressure (equals pressure minus atmospheric pressure) of 4.8 atm when parked in the shade at a temperature of 80°F. The tire explodes when ridden onto asphalt pavement at a temperature of 140°F. What was the gauge pressure in the tire as it exploded?

37. ◆◆ A soufflé is made from ingredients at 50°F and contains trapped air. When put into a 375°F oven, the soufflé rises by a factor of 1.4. What fraction of the soufflé's original volume was trapped air?

§19.3 THE FIRST LAW OF THERMODYNAMICS

38. ❖ Two heat-insulated cylinders each contain the same amount of an ideal gas (■ Figure 19.25). Both have initial volume V_1 and temperature T_1. System A has a very thin partition; system B has a frictionless piston. Compare the work done by system A when the partition is punctured with the work done by system B when the piston is allowed to rise slowly. Discuss the internal energy change in each cylinder. How does the temperature change in each cylinder?

39. ◆◆ Find the work done in the isothermal expansion of 2.5 g of helium at 290 K from a volume of 11 m³ to a volume of 18 m³.

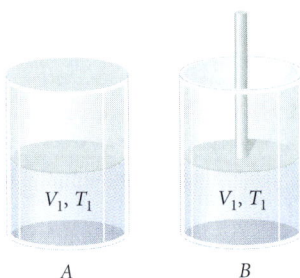

■ FIGURE 19.25

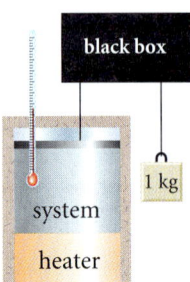

■ FIGURE 19.26

40. ◆◆ As morning sunlight falls on a tethered helium balloon, it expands from a volume of 1.05×10^3 m³ to a volume of 1.15×10^3 m³. How much work does the helium do on its environment (assuming the gas bag remains slack)? If the initial temperature of the helium is 270 K, how many moles of gas are in the balloon, how much does the helium's internal energy increase, and how many joules of solar energy are absorbed?

41. ◆◆ A system of 3.0 mol of a monatomic ideal gas is contained in an insulated box and is in contact with a heater, a thermometer, and a "black box" that raises or lowers a 1.0-kg block (■ Figure 19.26). (A black box is a machine with a variable mechanical advantage that does not produce or absorb any energy, but merely transmits it.) (a) The block is observed to rise through 2.4 m while the thermometer reads a constant temperature T. How much heat transfer occurs? (b) Now the block falls 1.2 m, and the thermometer indicates a 2.0 K temperature increase. How much heat transfer occurs? (c) The heater provides 15 J of heat transfer, but the thermometer reads a temperature drop of 1.0 K. What happens to the block?

42. ◆◆ A 10.0-kg block is attached through a *black box* to a piston in an insulated cylinder (cf. Figure 19.26) containing helium. The initial temperature of the gas in the cylinder is 305 K. The block falls 25 cm, and the final temperature is observed to be 295 K. How many moles of helium are in the cylinder?

§19.4 THE P-V DIAGRAM

43. ❖ On a pressure-*temperature* diagram locate the states A, B, C, D shown in Figure 19.13. Assume the sample contains 1.0 mol of ideal gas. How are the constant volume processes $A \rightarrow B$ and $D \rightarrow C$ represented in the P-T diagram?

44. ◆ Find the work done during each process shown in the P-V diagram in ■ Figure 19.27.

45. ◆◆ A sample of 0.60 m³ of monatomic ideal gas is held in a cylinder equipped with a piston. Its temperature is 2.0×10^2 K and its pressure is 0.050 atm. (a) How many moles of gas are in the cylinder? (b) The piston is slowly depressed, reducing the volume to 0.50 m³, and the temperature rises to 210 K. What is the pressure then? (c) Plot the initial and final states on a P-V diagram. Do you know enough to plot the process too? Why or why not?

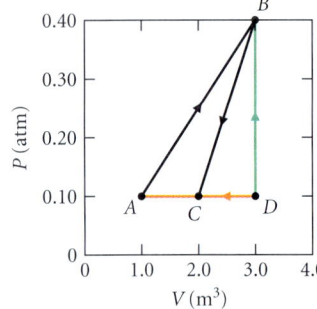

■ FIGURE 19.27

672 CHAPTER 19 • TEMPERATURE AND THERMAL ENERGY

46. ♦♦ A cylinder with a piston contains 2.00 mol of helium. The volume is 0.300 m³ and the temperature is 3.00 × 10² K. (a) What is the pressure? (b) The gas is to be compressed to a volume of 0.200 m³. Consider two processes.

1. A constant-pressure process.
2. Pressure increases to 0.210 atm. The process is represented by a straight line in the P-V diagram.

Plot both processes on a P-V diagram, and compute the work done on the gas in each case. (c) What is the final temperature in each of the processes in part (b)? (d) What is the heat transfer in each case?

47. ♦♦ A tank of argon gas has a volume of 0.0160 m³. The gas has a pressure of 5.00 atm and a temperature of 293 K. How much argon is in the tank? The tank is fitted with a piston and the gas is allowed to expand isothermally until its pressure is 3.00 atm. Subsequently, the pressure drops to 1.00 atm at constant volume. Find the final temperature and volume of the gas. How much work was done?

§19.5 SPECIFIC HEATS OF AN IDEAL GAS

48. ❖ Which has the greater specific heat at constant pressure, helium or krypton? Which has the greater ratio of specific heats γ?

49. ❖ If 10.0 J of energy are transferred to a monatomic ideal gas sample at constant volume and its temperature rises by 10.0 K, is it possible to determine the mass of the sample? The number of moles in the sample? If so, give the value.

50. ♦ The measured value of c_p for a monatomic gas is 515 J/kg·K. What is the gas?

51. ♦♦ It is found that 14.9 J of heat must be supplied to 1.00 g of a monatomic ideal gas to raise its temperature from 373 K to 473 K at constant volume. Identify the gas. What is its specific heat at constant pressure?

52. ♦♦ A cylinder of gas with a movable piston contains 1.0 × 10⁻³ m³ of helium at −100.0°C. The piston area is 4.8 cm² and it has a 5.0-kg mass. The entire assembly is in a vacuum chamber. What is the pressure of the helium? The gas is slowly heated to a temperature of 0.0°C. How much heat transfer occurs?

53. ♦♦ Use the ideal gas law to find the specific heat at constant volume for a mixture of 3.00 mol of neon and 2.00 mol of atomic hydrogen. What is the specific heat at constant pressure? Is the value of γ altered by mixing monatomic ideal gases?

§19.6 ADIABATIC PROCESSES IN AN IDEAL GAS

54. ♦ Air at atmospheric pressure occupies an insulated cylinder with a piston whose volume at the start of the experiment is 1.5 L. The piston is suddenly depressed until the pressure in the cylinder is 2.0 atm. What is the volume of the cylinder then?

55. ♦ An ideal monatomic gas, originally at atmospheric pressure, is subjected to intense sound waves in which a localized region of volume V is suddenly compressed adiabatically to a volume 0.980V. Find the instantaneous pressure in that localized volume.

56. ♦ An ideal gas system undergoes a cycle described by the path A → B → D → A in Figure 19.27. Find the total work done by the gas and the net heat transferred into the system.

57. ♦♦ In Chapter 16 we found that the speed of sound is given by: $v_s = \sqrt{dP/d\rho}$. Air in sound waves behaves adiabatically with $\gamma = 1.4$. Calculate the speed of sound in air at sea level at 0°C. (Air density is listed in Table 13.1.) Compare with v_{rms}.

58. ♦♦ One mole of a monatomic gas at atmospheric pressure undergoes an adiabatic compression that raises the temperature of the gas from 273 K to 373 K. How much work is done in the process?

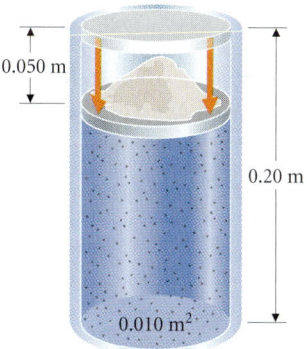

■ FIGURE 19.28

59. ♦♦ Oxygen gas at 280 K and atmospheric pressure is used to fill the oxygen tank in a Cessna 210 aircraft. The oxygen cylinder has a volume of 11.4 L, and the oxygen pressure when the tank is full is 122 atm. If the tank is filled rapidly, the oxygen is compressed adiabatically. How much oxygen fills the tank? What is the final temperature? How much oxygen fills the tank if it is filled slowly and isothermally? In practice, the process is never truly adiabatic, but it *is* important not to fill the tank too rapidly.

60. ♦♦ Prove that when an ideal gas expands adiabatically, the work done by the gas equals $Mc_v(T_1 - T_2)$, where T_1 is the initial temperature, T_2 is the final temperature, and M is the mass of the gas sample.

61. ♦♦ How much work is done in the power stroke of a four-cylinder Volkswagen Scirocco automobile engine? Model the power stroke as an adiabatic expansion with an initial pressure of 4.1 × 10⁶ Pa, a displacement ($V_f - V_i$) of 397 cm³ per cylinder, and a compression ratio (V_f/V_i) of 8.0.

62. ♦♦ In an experimental spacecraft landing system, an insulated cylinder of cross-sectional area 0.010 m² contains a monatomic ideal gas at T = 280 K. The system is being tested in a vacuum chamber (■ Figure 19.28). Initally the gas supports the piston at a height of 0.20 m. Sand is slowly placed onto the piston. When 10.0 kg of sand has been added, the piston has fallen 0.050 m. How much gas is in the cylinder? What is the mass of the piston?

63(a). ♦♦ Show that for an adiabatic process taking an ideal gas from P_1, V_1 to P_2, V_2, the work done by the gas is:

$$\frac{P_1 V_1 - P_2 V_2}{\gamma - 1},$$

where $\gamma = c_p/c_v$. (b) ❖ What is the internal energy change of the gas in the above process, expressed in terms of P_1, V_1, P_2, V_2, γ?

64. ♦♦ Two moles of an ideal monatomic gas expand adiabatically from a volume of 10.0 L and 1.00 atm pressure to a volume of 50.0 L. (a) Does the internal energy of the gas increase or decrease? (You should be able to answer this part of the question without doing any calculations!) (b) What is the change in internal energy? (c) If the gas is subsequently compressed at constant pressure to a volume of 10.0 L and its pressure is then increased at constant volume back to 1.00 atm, how much work is done by the gas in one whole cycle? What is the net heat transfer?

65. ♦♦♦ A bicycle tire has a volume of 5.0 × 10⁻³ m³ and is very flat ($P_{inside} = 1.0$ atm, T = 299 K). You want to pump the tire up to a pressure of 5.0 atm. The volume of the pump barrel is 7.0 × 10⁻⁴ m³ and n strokes of the pump are required. (a) How many moles of gas are originally in the tire? (b) How many moles fill the pump barrel at atmospheric pressure and 299 K? (c) What is the initial volume of the air that ends up in the tire, as a function of n?

✉ In parts (d) and (e), assume that the entire procedure is equivalent to a single adiabatic compression with $\gamma = \frac{7}{5}$ (cf. Problem 85). **(d)** Find n and the final number of moles of gas in the tire. **(e)** What is the final temperature? **(f)** If heat slowly leaks out of the tire, what is the pressure when the gas has cooled to 299 K?

This model does not apply well for ordinary humans, since we pump slowly enough that the gas cools considerably between pump strokes. Superman, maybe!

✱ §19.7 EQUIPARTITION OF ENERGY

66. ❖ Explain why $\gamma = 1.4$ for air.

67. ❖ The observed value of γ for a gas at room temperature is $\frac{9}{7}$. What can you say about the gas?

68. ❖ The temperature of a gas is raised by a factor of 4. By what factor would the rms rotation rate of its molecules change? Why?

69. ◆ Use an ideal gas model to compute the specific heat at constant pressure for steam, assuming no vibrations are excited. Compare your answer with the value in Table 20.4.

70. ◆ Compute c_p for methane (CH$_4$) assuming **(a)** no vibrations, and **(b)** six vibrational modes. Compare with the measured value 2200 J/kg·K at 25°C. What can you conclude about excitation of the vibrational modes? (Three rotational modes are excited.)

71. ◆◆ For a certain gas, the measured value of c_p is 1029 J/kg·K and the measured value of γ is 1.40. What is the gas?

72. ◆◆ In an experiment, 64.31 J of heat are required to raise the temperature of 10.00 g of a diatomic gas through 10.00°C at constant volume. What is the specific heat at constant volume of the gas? Assuming that the gas is ideal, identify it.

73. ◆◆ ✉ The quantum of rotational energy for an object is $E_{min} = (1 \times 10^{-68} \text{ J}^2 \cdot \text{s}^2)/I$, where I is the rotational inertia of the object. Model nitrogen molecules as two point atoms separated by 10^{-10} m, and estimate the rotational quantum for nitrogen. Compare your results with the data in Figure 19.19. Given that nitrogen liquifies at about 80 K, would you expect to encounter nitrogen gas with $c_v < \frac{5}{2}(k/m)$?

74. ◆◆◆ ✉ **(a)** According to one model, dust grains in the interstellar medium consist of 10^9 carbon atoms forming a cylinder with radius 10^{-7} m and height 4×10^{-7} m. In a gas cloud at temperature 100 K, what is the rms rotation rate of grains about the axis of the cylinder and about an axis along a diameter at the midpoint of the cylinder? **(b)** Use the formula for the quantum of rotational energy given in Problem 73 to find the corresponding quantum for rotation about each axis of a grain. Are rotational modes of grains excited at 100 K?

Additional Problems

75. ❖ Sailors escaping from capsized vessels are advised to blow out bubbles of air as they rise toward the surface. What do you suppose is the purpose of this advice?

76. ❖ Once we know the average energy of a molecule, it is possible to predict the rate of processes such as chemical reactions in a gas at temperature T. One reaction requires an energy input $\Delta E_1 = 5kT$, and a second requires $\Delta E_2 = kT/5$. Which proceeds more rapidly? Why?

77. ❖ A plasma consists of free electrons and singly ionized sodium atoms (one electron removed from each atom). Compare the rms speeds of electrons and sodium ions in equilibrium. Compare the contributions of electrons and ions to the pressure of the gas.

78. ◆ Find the rms speed of a dust mote of mass 10^{-14} g in air at a temperature of 300 K.

79. ◆◆ Starting with the ideal gas law and the definition of internal energy, show that for a monatomic ideal gas: $P = \frac{2}{3}u$, where P is the pressure and $u = U/V$ is the internal energy per unit volume (the energy density).

80. ◆◆◆ Show that in an ideal gas undergoing an adiabatic process the pressure P and absolute temperature T are related by:

$$P^{(\gamma-1)/\gamma} T^{-1} = \text{constant},$$

where γ is the ratio of the specific heat at constant pressure to the specific heat at constant volume. Find the relationship between a small pressure change ΔP and the corresponding temperature change ΔT for such an adiabatic process.

81. ◆◆◆ A bubble of air of radius 1.0 cm is formed at the bottom of a lake 20.0 m deep, where the temperature is 4°C, and then rises slowly to the top, where the temperature is 27°C. Calculate the radius of the bubble as it reaches the water surface. (Neglect surface tension.)

82. ◆◆◆ Helium balloons are used to carry scientific research instruments to high altitudes in the atmosphere. At launch, the balloons are very slack. Why? A balloon is designed to float at an altitude of 40 km. Its volume is 3×10^5 m^3 and the payload mass is 200 kg. **(a)** What is the atmospheric pressure at 40 km? Use the exponential model atmosphere of §13.3.4. **(b)** If the temperature at 40 km is 262 K, how much helium is required to fill the balloon? **(c)** If the temperature at sea level is 288 K, what fraction of the balloon is filled at sea level?

83. ◆◆◆ A primitive refrigerator consists of two 0.700-m^3 cylinders, each containing 30.0 mol of monatomic ideal gas at temperature 300.0 K and in thermal contact. One cylinder has a piston that is rapidly pulled out until the volume increases to 1.40 m^3. Find the temperature of the gas in each cylinder after equilibrium is established.

84. ◆◆◆ An insulated cylinder of cross section A and height $2h$ is divided into two equal volumes each containing $\mathcal{N}$ mol of monatomic ideal gas. The two volumes are separated by a fixed partition that allows the two gas samples to be in thermal contact (■ Figure 19.29). The upper gas sample supports a piston on which a constant force F acts. Initially, the gas samples have temperatures T_2 and $T_1 < T_2$. When the whole system reaches equilibrium, what displacement has the piston undergone?

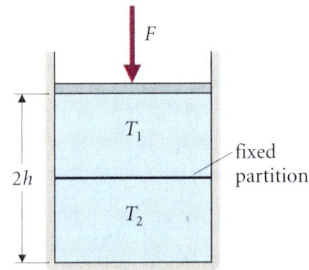

■ **FIGURE 19.29**

85. ◆◆◆ An ideal gas sample is confined in an insulated cylinder by two pistons A and B (■ Figure 19.30). The sample is initially divided in half by a thin, insulating partition and the two samples are initially in equilibrium. **(a)** Show that the following two processes lead to the same final state:

1. The partition is removed and the sample is compressed to $\frac{1}{10}$ its original volume by simultaneous use of both pistons.
2. Piston A compresses the left half of the sample to $\frac{1}{10}$ its original volume, piston B compresses the right half sample, then the partition is removed.

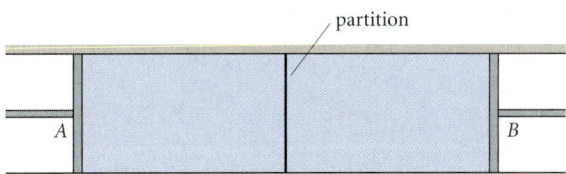

■ FIGURE 19.30

(b) Give an argument based on part (a) that shows that any sequence of adiabatic processes, involving any number of initial subvolumes, leading from a given initial volume of gas to a given final volume results in the same final state. Explain how this abstract theorem applies to using a bicycle pump to fill a tire (Problem 65).

86. ♦♦♦ **(a)** In a turbocharged airplane engine, turbine blades compress the ambient air before it enters the cylinders. At a fixed rpm, the engine power is approximately proportional to the air density at the cylinder intake valve. If the turbocharger achieves sea level pressure at 25 000 feet, what percentage of sea level power is achieved? (Use an isothermal atmosphere with a scale height of 30 000 feet, and assume the compression is adiabatic with $\gamma = 1.4$.) **(b)** An intercooler added to the engine increases its power output. Assume that the cooler causes an isobaric cooling of the compressed air as it enters the cylinder. How much cooling is required to obtain full power at 25 000 feet?

87. ❖ An electric oven consists of an electrical resistance inside a cooking volume surrounded by heat-insulating material. Consider the oven (including the resistance) as a thermodynamic system. When the oven is turned on—that is, the resistance is connected to an electrical outlet—does heat enter the oven? Does the temperature inside the oven go up? Why?

Computer Problem

88. A diatomic ideal gas with initial temperature 293 K is compressed from a volume of 1.000 m³ to 0.020 m³. Values of pressure as a function of volume during the process are listed in file CH19P88 on the supplementary computer disk. Set up a spreadsheet program to calculate the temperature at each volume, and plot it. Use a numerical algorithm such as the trapezoidal rule (see, e.g., Problem 6.91) to calculate the work done on the gas to compress it to a volume V. Also calculate the change in internal energy and the heat transfer. Plot each of these as a function of V, and comment.

Challenge Problems

89. In the lower layers of the Earth's atmosphere, the relation between density and pressure is approximately given by the adiabatic law (eqn. 19.22). Show that the pressure variation as a function of the height z is:

$$P = P_o \left[1 - \left(\frac{\gamma - 1}{\gamma}\right)\frac{gz\rho_o}{P_o} \right]^{\frac{\gamma}{\gamma - 1}}$$

where P_o and ρ_o are the pressure and density at $z = 0$ (the surface of the Earth).

90. A thin, flexible plastic bag contains 1.0 L of air at 15°C and atmospheric pressure. A small lead weight is attached to the bag so that the total mass is 0.50 kg. The bag is lowered into a lake at 15°C and then released. Show that there is a critical depth d_o such that the bag rises to the surface if released at a depth $d < d_o$ and sinks if released below d_o. Find d_o, neglecting the volume of the lead and plastic.

91. A neutron star has a mass of 2×10^{30} kg (the mass of the Sun), a radius of 10 km, and an atmosphere of hydrogen. Assume the hydrogen is an ideal monatomic gas, at a temperature of 10^4 K. Find the scale height of this atmosphere, where the pressure drops to $1/e$ of its value at the surface. (The answer may surprise you!) Comment on the validity of the constant g assumption.

92. A piston of mass M and area A is supported in a cylinder of volume V by $\mathcal{N}$ mol of an ideal gas with ratio of specific heats γ. If the piston is displaced slightly from its equilibrium position, show that it oscillates with frequency:

$$\omega = \sqrt{\frac{\gamma A}{V}\left(g + \frac{P_o A}{M}\right)},$$

where P_o is the external pressure.

CHAPTER 20
Thermodynamics of Real Substances

CONCEPTS

Van der Waals equation of state
Phase change
Latent heat
Coefficient of expansion
Heat capacity
Thermometry

GOALS

Understand:
Phase changes and latent heats.
Specific heats and heat capacities.
Calorimetry.

Be able to:
Compute the thermal expansion of bodies.

A thunderstorm shows dramatically how powerful thermodynamic effects can be. Driven by solar heating, the air becomes unstable and rises. Energy released by condensing water vapor accelerates the instability and the storm builds into a towering cloud with violent winds. In this chapter and the next, we'll learn about some of the processes that drive thunderstorms.

"The time has come," the Walrus said,
"To talk of many things:
Of shoes—and ships—and sealing wax—
Of cabbages—and kings—
And why the sea is boiling hot—
And whether pigs have wings."

LEWIS CARROLL

A thunderstorm is a dramatic demonstration of thermal processes in a real gas. The ideal gas model (Chapter 19) gives insight into the nature of thermal processes but does not describe the condensation of water from vapor that drives the thunderstorm's rapid growth. In this chapter we study thermal properties under a broader range of conditions, including liquids and solids, with an emphasis on the effects of temperature changes. We begin with an improved model of gases developed by Johannes van der Waals. This model shows how the forces between molecules, which we have ignored thus far, cause gases to condense into liquids at high pressure and low temperature.

20.1 THE BEHAVIOR OF REAL GASES

20.1.1 The van der Waals Equation of State

The major difference between gases and liquids is the degree to which intermolecular forces determine their behavior. The volume of a gas is determined by its container, whereas cohesion of molecules in a liquid causes it to have a fixed volume. To understand the condensation of gas into liquid, we modify the ideal gas model to include the effect of forces between the molecules. Van der Waals showed how to do this in an approximate way.

When molecules are very close together, the forces between them are repulsive. As a result, when molecules collide, they never touch but instead reach a minimum separation and then recede. To model this effect, we assume that each molecule has a finite volume v_m. Then, if N gas molecules occupy a container of volume V, the total volume available to an individual molecule—that is, volume not occupied by the other molecules—is $V - Nv_m$. The first improvement to the ideal gas law is to replace V by $V - Nv_m$.

At intermediate distances, the molecular forces are attractive and tend to pull the molecules together. These forces have the same effect as an additional external pressure, tending to increase the density (or equivalently, decrease the volume) of the gas. Molecular forces decline rapidly with distance, and so are more important at relatively high density N/V. The resulting correction to the pressure is proportional to $(N/V)^2$ (see *Digging Deeper*).

The ideal gas law $PV = NkT$ is modified by these two corrections:

$$\left[P + C\left(\frac{N}{V}\right)^2\right]\left[V - Nv_m\right] = NkT, \qquad (20.1)$$

> MOLECULAR FORCES ARE ELECTRICAL IN NATURE.

Digging Deeper

MOLECULAR FORCES

Molecular forces are electrical in nature, and we will understand them better after studying Part VI. Here we give an introduction.

An atom consists of a negatively charged electron cloud surrounding the positively charged nucleus. When these charges are symmetrically distributed, no net force is exerted on distant objects. As two molecules come closer, however, the two charge distributions exert electrical forces on each other that cause the electrons to be distributed asymmetrically about the nucleus. The molecular charge distribution takes a form called a dipole (cf. §24.5), whose effect on a charged object decreases as the cube of the distance. Thus, if the separation r of two molecules is doubled, the interaction produces $2^3 = 8$ times weaker dipoles and the force exerted over the doubled distance is another factor of 8 weaker. The attractive electric force between the two dipoles is proportional to r^{-6}. The average separation of the molecules is approximately the cube root of the volume per molecule: $r \sim (V/N)^{1/3}$, and $r^{-6} \sim (N/V)^2$. The pressure correction due to molecular forces is proportional to the average magnitude of the forces, and so is proportional to $r^{-6} \sim (N/V)^2$.

TABLE 20.1 Van der Waals Constants

$$\left(P + \frac{a}{V_m^2}\right)(V_m - b) = RT$$

Gas		a (L^2·atm/mol^2)	b (L/mol)
Ammonia	NH$_3$	4.17	0.037
Argon	A	1.345	0.032
Carbon dioxide	CO$_2$	3.59	0.043
Carbon monoxide	CO	1.485	0.03985
Carbon tetrachloride	CCl$_4$	20.39	0.138
Chlorine	Cl$_2$	6.49	0.056
Helium	He	0.034	0.024
Hydrogen	H$_2$	0.24	0.027
Hydrogen sulphide	H$_2$S	4.43	0.043
Krypton	Kr	2.32	0.040
Mercury	Hg	8.09	0.017
Neon	Ne	0.21	0.017
Nitrogen	N$_2$	1.39	0.039
Oxygen	O$_2$	1.36	0.032
Propane	C$_3$H$_8$	8.66	0.084
Sulphur dioxide	SO$_2$	6.71	0.056
Water vapor	H$_2$O	5.46	0.030
Xenon	Xe	4.19	0.051

REMEMBER: N_A = AVOGADRO'S NUMBER = 6×10^{23}.

where C is a constant. Expressed in terms of the volume per mole of gas, $V_m \equiv V/\mathcal{N} = VN_A/N$, the modified law is known as the *van der Waals equation of state*.

$$\left(P + \frac{a}{V_m^2}\right)(V_m - b) = RT. \tag{20.2}$$

The constants $a \equiv CN_A^2$ and $b \equiv N_A v_m$ are determined experimentally for each gas. Some values for these constants are given in ● Table 20.1.

EXAMPLE 20.1 ♦ A recreational vehicle stores 3.00 mol of propane cooking fuel in a 10.0-L cylinder. Find the pressure in the cylinder at 297 K.

MODEL We use the van der Waals gas model for propane, taking the values of a and b from Table 20.1.

SETUP The volume per mole is $V_m = (10.0 \text{ L})/(3.00 \text{ mol}) = 3.333$ L/mol.

SOLVE Rearranging eqn. (20.2):

$$P = \frac{RT}{V_m - b} - \frac{a}{V_m^2} = \frac{(8.315 \text{ J/mol·K})(297 \text{ K})}{(3.333 - 0.084) \text{ L/mol}} - \frac{8.66 \text{ L}^2\cdot\text{atm/mol}^2}{(3.333 \text{ L/mol})^2}$$

$$= \frac{(2470 \text{ J})(1 \text{ L})}{(3.249 \text{ L})(10^{-3} \text{ m}^3)} - 0.780 \text{ atm}$$

$$= (7.601 \times 10^5 \text{ N/m}^2)\left(\frac{1 \text{ atm}}{1.013 \times 10^5 \text{ N/m}^2}\right) - 0.780 \text{ atm}$$

$$= 7.503 \text{ atm} - 0.780 \text{ atm} = 6.72 \text{ atm}.$$

REMEMBER: THE CONVERSION FACTORS FOR THE UNITS ARE EXACT.

ANALYZE For comparison, the pressure of an ideal gas would be $P = NRT/V = 7.31$ atm. The difference is about 9%.

EXERCISE 20.1 ♦ If 0.650 mol of carbon tetrachloride in a 5.00-L cylinder exerts a pressure of 2.70 atm, what is the temperature?

20.1.2 Isotherms of the van der Waals Equation

Isotherms of the van der Waals equation drawn in a *P-V* diagram are a useful tool for studying the behavior of real gases (■ Figure 20.1). Curve (a) represents a gas at relatively high temperature and relatively large volume. Under these conditions, attractive intermolecular forces have little effect on the rapid motion of the molecules, and the volume occupied by the molecules is insignificant. The behavior of the gas is almost ideal. In the opposite case—low temperature or small volume—the isotherms differ substantially from those of an ideal gas. When the sample volume is nearly equal to the volume of the molecules, it is very difficult to compress the gas further. Large changes in pressure cause very small changes in volume, and the curves become nearly vertical as *V* approaches Nv_m. The van der Waals substance has become *incompressible*. In this region of the *P-V* diagram, the substance acts more like a liquid than a gas.

To understand the transition from gas to liquid, we imagine compressing the gas along one of the lower-temperature isotherms, curve (b) in Figure 20.1. We increase the pressure very slowly and precisely so that the gas remains extremely close to equilibrium. If we could compress the sample past point *A*, the pressure would have to decrease as the volume decreases (because of the increasing pull of the molecular forces). The sample would become unstable and collapse. In actual laboratory experiments, the substance forms a mixture of gas and liquid that contracts at constant pressure along the horizontal line in the figure. The proportion of liquid increases during the process, reaching 100% at the low-volume end of the horizontal line, at which point the pressure increases rapidly with further decrease in volume.

The size of the unstable region depends markedly on temperature. At a *critical temperature*, the isotherm (c) has an inflection point (point *C* in Figure 20.1), called the *critical point*.

WE INTRODUCED THE TERM "INCOMPRESSIBLE" IN CHAPTER 13.

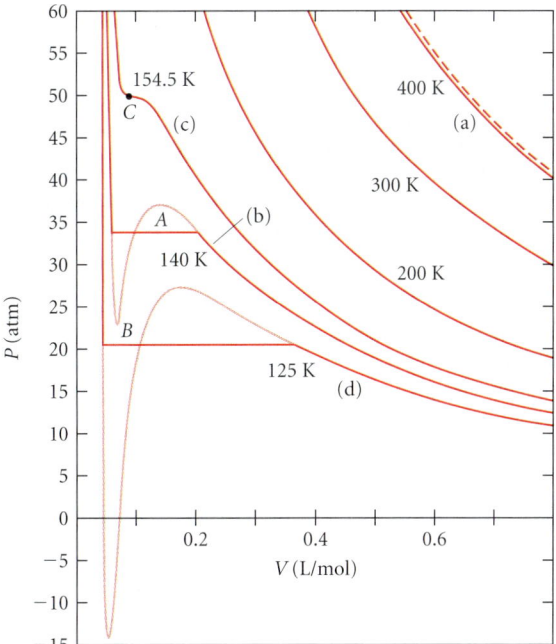

■ FIGURE 20.1
Isotherms of the van der Waals equation for oxygen. At high temperature (in the upper right corner of the diagram), the isotherms are very similar to the ideal gas isotherms. The dashed line at the upper right is the 400 K isotherm of the ideal gas law. Curve (a) is the 400 K isotherm of the van der Waals equation. As the temperature drops, the deviations from ideal behavior become apparent. The theoretical model described by eqn. (20.1) produces isotherms with a maximum (e.g., point *A*) and a minimum (point *B*). A gas being compressed along this isotherm would become unstable at point *A*. The 140 K and 125 K isotherms, labeled (b) and (d), show the transition to liquid as horizontal lines along which the substance is a mixture of gas and liquid. The steep curves to the left illustrate the incompressible behavior of liquids. Here volume changes very little even though pressure increases enormously. The 154.5 K isotherm has an inflection point at *C*. This is the critical point. At higher temperatures, there is no phase transition between gas and liquid.

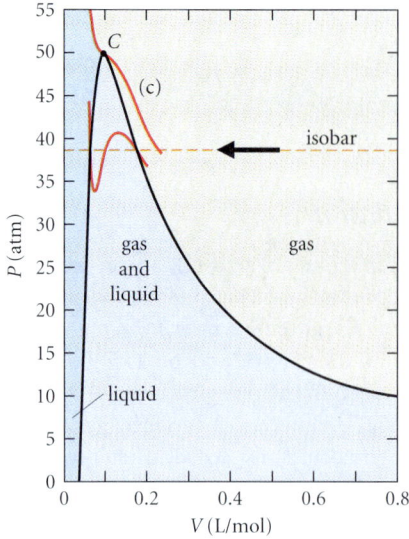

■ **FIGURE 20.2**
The isotherm through the critical point divides the P-V diagram into regions where oxygen is a gas (upper right), a liquid (far left), or a mixture of the two (center, below C). On isotherms just above the critical point (C), no transition between gas and liquid occurs, but the increasingly dense gas differs only subtly from nearby "liquid" states. We have included small pieces of two isotherms to show their relation to the phase boundaries.

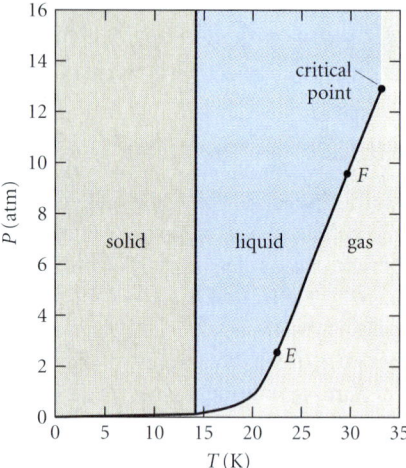

■ **FIGURE 20.3**
A phase diagram for hydrogen displays the phases as a function of pressure and temperature. At low temperatures, below about 15 K, hydrogen is a solid. At higher temperatures, the substance becomes liquid and then gas. The critical point is at 33 K and 13 atm. No liquid phase exists at any pressure once T exceeds 33 K. The liquid gas interface shows how the transition temperature increases with pressure between states E and F.

THE STATE OF THE SUBSTANCE—GAS, LIQUID, OR SOLID—IS CALLED A PHASE. WHEN CHANGING FROM ONE STATE TO ANOTHER, THE SUBSTANCE UNDERGOES A *PHASE TRANSITION*.

At temperatures above critical, there is no transition and the properties change continuously in an isothermal compression. Below the critical point, the P-V diagram divides into definite regions where the substance is a gas, a liquid, or a mixture of the two (■ Figure 20.2). Chemists often represent this behavior in a *phase diagram*, with pressure and temperature as the independent state variables. ■ Figure 20.3 is a phase diagram for hydrogen.

20.2 CHANGE OF PHASE

20.2.1 Phase Transitions at Constant Pressure

THE PRESSURE MUST BE BELOW THE CRITICAL POINT, OR THERE IS NO PHASE TRANSITION.

SEE *DIGGING DEEPER* FOR A DISCUSSION OF WHY THE TEMPERATURE REMAINS CONSTANT.

The easiest way to cause a phase transition is to heat or cool a substance at constant pressure—we do this every day in the kitchen. When a gas is cooled at constant pressure, the state point moves across isotherms from right to left on a horizontal line in the P-V diagram (Figure 20.2). The temperature drops at a rate determined by the rate of heat transfer, and the gas contracts. As the phase transition begins, molecules binding into liquid release potential energy that must be removed by heat transfer. We observe that the temperature remains constant during the transition even though internal energy is released. Once the transition is complete, the temperature of the resulting liquid decreases again.

The temperature at which a phase transition occurs depends on the pressure. Below the critical point, the phase transition occurs at a definite vaporization temperature—the boiling point—where a line of constant pressure (isobar) intersects the phase boundary. This temperature increases with pressure (compare points E and F in Figure 20.3). Water boils at a lower temperature at high altitude, and wise cooks boil the potatoes for 10 minutes longer in the mountains.

Digging Deeper

WHY IS TEMPERATURE CONSTANT DURING A PHASE TRANSITION?

During a phase transition, the system contains both liquid and gas. Molecules are continually being captured from the gas and escaping from the liquid at nearly equal rates. Energy flow into or out of the system controls the small difference between the two rates that is responsible for the transition. The capture rate is approximately the rate at which molecules collide with the liquid surface and varies only slightly with temperature. But only the most energetic molecules in the liquid can escape, so the escape rate is very sensitive to temperature. The two rates are nearly equal only at one well-defined temperature—the transition temperature. For example, as energy is added to boiling water, the escape rate increases but the temperature stays the same.

A similar phase transition occurs between liquid and solid phases. The molecules are tightly bound in a solid, and energy is released when molecules make the transition from the less tightly bound liquid state. Again, temperature remains constant during the phase transition. The van der Waals model is not sophisticated enough to describe liquid–solid transitions because the structure of solids depends on collective interactions of molecules rather than the simple two-molecule forces assumed in the model. The interactions among molecules in crystals can produce some fascinating behavior. Ordinary water provides an example. The melting temperature of most substances decreases as the pressure increases. The water–ice transition temperature increases slightly with pressure. There is also a peculiar tendency for a thin layer of water to remain on the ice surface down to temperatures of a few kelvin below the freezing point. These properties allow ice skates to work because liquid water under the blade lubricates the motion of the skate.

20.2.2 Latent Heat

When energy is added to a substance in one definite phase, its temperature increases since the average kinetic energy of its molecules steadily increases. During transitions from solid to liquid or liquid to gas, relatively large amounts of potential energy are stored without changing the average kinetic energy or the temperature.

■ Figure 20.4 illustrates how temperature changes as a sample of water of mass M is heated. The heat transfer along the flat portions of the curve—that is, the energy provided to

IN SOME KINDS OF PHASE TRANSITIONS, THERE IS NO CHANGE IN POTENTIAL ENERGY AND NO LATENT HEAT ASSOCIATED WITH THE TRANSITION.

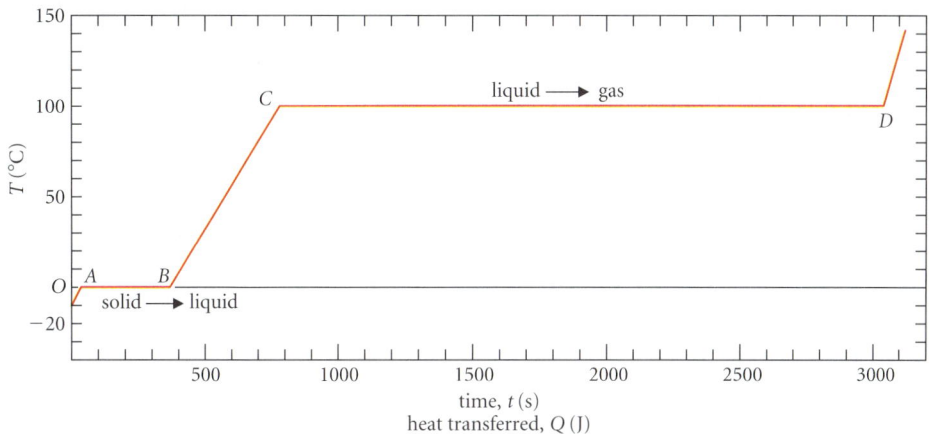

■ **FIGURE 20.4**
Graph of temperature versus time as a 1.0-g sample of ice is heated at a constant rate. Between O and A, the ice temperature rises until melting starts. Between A and B, the temperature stays constant at 0°C until the ice is all melted, then (interval BC) the water temperature rises until boiling begins. During interval CD, the temperature is constant at 100°C until the transition is complete. The heat transfer during each transition is a latent heat: it does not change the temperature but is used to increase the potential energy of the water molecules.

TABLE 20.2 **Latent Heats of Fusion and Vaporization**
(At atmospheric pressure, unless noted otherwise)

Substance	Melting Temperature (K)	L_f (kJ/kg)	Boiling Temperature (K)	L_v (kJ/kg)
Aluminum	932	399	2600	10530
Ammonia	198	452	240	1368
Copper	1356	205	2609	4796
Gold	1336	64	2873	1577
Helium	1 (at 26 atm)		4.6	25
Hydrogen	14	59	20	452
Iron	1808	289	3008	6320
Lead	601	23	2023	859
Mercury	234	11	630	295
Nitrogen			77	156
Oxygen	55	14	90	213
Silicon	1683	1654*	2628	10584
Silver	1235	111	2223	2356
Sodium	371	115	1187	1017
Tin	505	60	2543	1940
Water	273	334.7	373	2272
Zinc	693	102*	1180	1679
Freon 12			243	165
Freon 13			192	148

*Published values for these substances differ by up to 10%.

FOR A CHANGE OF PHASE AT CONSTANT PRESSURE, THE LATENT HEAT REQUIRED INCLUDES A CONTRIBUTION DUE TO THE WORK THAT MUST BE DONE IN EXPANSION. GENERALLY SPEAKING, THIS CONTRIBUTION IS SMALL COMPARED WITH THE INTERNAL ENERGY CHANGE.

accomplish a phase transition—is described by a *latent heat*: $(Q_D - Q_C)/M \equiv L_v$ is called the *latent heat of vaporization*, and $(Q_B - Q_A)/M \equiv L_f$ is the *latent heat of fusion*.

> The *latent heat* of a substance is the energy per unit mass provided to, or given up by, the substance during a phase change.

Energy must be provided to each molecule, so the heat needed for a transition depends on the number of molecules in the sample. Latent heat is measured in joules per kilogram (● Table 20.2). It is *latent,* or hidden, because it does not result in any temperature change. Phase transitions can go in either direction. The energy provided during melting (or boiling) is liberated again during solidification (or condensation). The latent *heat* changes the internal energy of the substance: according to the first law of thermodynamics, the energy could equally well be provided in the form of work.

EXAMPLE 20.2 ♦ How much energy is required to melt 0.106 kg of ice at 0°C and 1 atm?

MODEL The energy is required to increase the potential energy of the ice molecules; it is measured as latent heat (Table 20.2).

SETUP The latent heat of fusion for ice is 335 kJ/kg, at 1 atm pressure.

SOLVE To melt 0.106 kg of ice to water at 0°C, the energy required is:

$$Q = (3.35 \times 10^5 \text{ J/kg})(0.106 \text{ kg}) = 3.55 \times 10^4 \text{ J}.$$

ANALYZE Remember—this heat transfer does not change the temperature of the ice.

EXERCISE 20.2 ♦ How much steam must be condensed to release 3.55×10^4 J of energy? (Using a heat exchanger, this energy could be used to melt the ice in Example 20.2.)

A thunderstorm is powered by latent heat of vaporization; such storms begin when humid air is forced upward. The upward motion may be initiated by pressure differences in a frontal weather system or by solar heating that causes a region of air to expand and become less dense than its surroundings. As the air rises, its pressure drops, it cools, and the water vapor it contains condenses into droplets. The latent heat released during condensation keeps the rising air warmer and less dense than its surroundings. The air accelerates upward, generating towering cumulus clouds. We can estimate how fast the cloud grows by using a simplified model to find the acceleration of a small bubble of air.

Air with 50% humidity at 20°C contains about 10^{-2} kg/m³ of water vapor, which releases $\Delta E = (10^{-2} \text{ kg/m}^3)(2272 \text{ kJ/kg}) \approx 2 \times 10^4$ J/m³ of energy as it condenses. If all of this energy were used to heat the air at constant pressure, the temperature rise would be ΔT, where (eqn. 19.14):

$$\Delta E = \rho c_\text{p} \Delta T.$$

The change in temperature is related to the change in density. If the average mass of air molecules is m, the air density is $\rho = Nm/V$. Then the ideal gas law gives $P = NkT/V = \rho kT/m$, so at constant pressure ρT is constant, and:

$$\frac{\Delta \rho}{\rho} = -\frac{\Delta T}{T} = -\frac{\Delta E}{\rho c_\text{p} T} \approx \frac{-2 \times 10^4 \text{ J/m}^3}{(1 \text{ kg/m}^3)(1000 \text{ J/kg} \cdot \text{K})(293 \text{ K})} = -0.07.$$

The downward force acting on the bubble is its weight $W = (\rho + \Delta \rho)Vg$. According to Archimedes' principle, the upward buoyant force equals the weight of an equal volume of surrounding air: $F_\text{up} = \rho Vg$. The difference $F_\text{up} - W = (-\Delta\rho/\rho)W = Ma$, so the acceleration is $a = (-\Delta\rho/\rho)g \approx 0.07g$. The air accelerates upward at 7% of g. At this rate, the cloud could grow 1000 m in a minute. In the most severe storms, the water droplets freeze at high altitude, causing even more growth, and sometimes producing hailstones the size of golf balls. Thermal energy can produce powerful effects!

20.3 Thermal Expansion

20.3.1 Linear Expansion

A gas heated at constant pressure expands; according to the ideal gas law, $V \propto T$. Solids and liquids also change their volume when heated. For small changes in temperature, the increase in size is proportional to the temperature change ΔT. Since each molecule moves farther from its neighbors, the total increase $\Delta \ell$ in any linear dimension ℓ (■ Figure 20.5) is also proportional to the original dimension:

$$\Delta \ell = \alpha \ell \, \Delta T. \quad (20.3)$$

The *coefficient of linear thermal expansion*, α, for a substance is the fractional change in length per unit temperature change.

The coefficient of linear expansion [units K^{-1} or (°F)$^{-1}$] depends slightly on temperature and is usually measured experimentally. As a result, eqn. (20.3) is strictly true only for a differential temperature change dT. However, it is customary to use it for a finite ΔT with an average value of α over that temperature range. Values of α for some common substances are listed in ● Table 20.3. All the values are sufficiently small that $\Delta \ell / \ell \ll 1$ for typical values of ΔT.

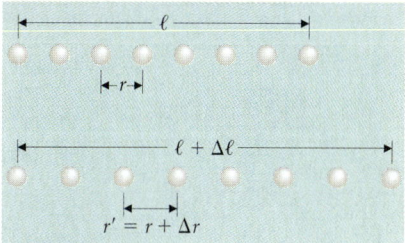

■ **FIGURE 20.5**
Thermal expansion. As temperature increases, each molecule separates from its neighbors. The linear dimension ℓ of a solid increases by an amount proportional to ℓ: $\Delta \ell = n \Delta r = (\ell/r)\Delta r$, where n is the number of molecules.

TABLE 20.3 Coefficient of Thermal Expansion for Some Common Substances

Substance	α $(10^{-6}$ K$^{-1})$ at 25°C
Aluminum	25
Copper	16.6
Glass	9
Glass (Pyrex)	3.2
Ice	51*
Invar 36†	1.6
Iron	12
Lead	29
Platinum	9
Rubber (hard)	80
Silicon	3
Silver	19
Sodium	70
Steel	11
Tin	20
Zinc	35

Substance	β (K^{-1}) at 25°C
Alcohol	1.1×10^{-3}
Mercury	1.81×10^{-4}
Water	2.6×10^{-4}

*This value of α is an average over the range -10°C to 0°C.
†Data on Invar 36 courtesy of the Carpenter Technology Corporation.

Digging Deeper

THERMAL EXPANSION OF SOLIDS

A solid expands because the average distance between its molecules increases with temperature. The potential energy of the molecules in a solid is a function of the distance r between them and has the same value at two different separations r_1 and r_2 (■ Figure 20.6). At temperature T_A, the internal energy of the molecules is divided between kinetic and potential energy. With, on average, $\frac{1}{2}kT_A$ per mode, the sum of kinetic plus potential energy is approximately kT_A. The molecules oscillate back and forth at height kT_A above the bottom of the potential well, between separations $r_{1,A}$ and $r_{2,A}$. (See §14.3 for an example of a particle oscillating in a potential well.) At a greater temperature T_B, the molecules oscillate higher in the potential well, between different separations $r_{1,B}$ and $r_{2,B}$. Because the potential is asymmetric, the average separation [roughly $r \approx \frac{1}{2}(r_1 + r_2)$] is greater at the higher temperature, and the solid expands.

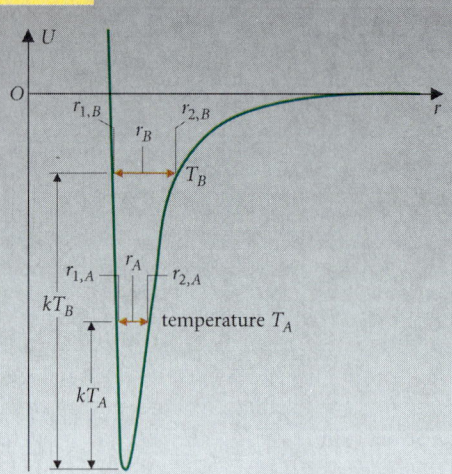

■ **FIGURE 20.6**
Potential energy of a molecule as a function of the separation r between molecules. The equipartition theorem (§19.7) applies to energy above the minimum possible, the lowest point on the curve. (*Remember:* Only changes in potential energy have physical significance.) At a temperature T, the total vibrational energy of a molecule is approximately kT above the minimum value, and the molecule oscillates between the two positions r_1 and r_2 where all its energy is potential energy. As temperature increases, the molecules oscillate higher up in the potential well and the average separation r increases. Compare r_B with r_A.

EXAMPLE 20.3 ♦ A steel bridge 100 m long is to be built in an area where annual temperature variations of 100°F are expected. How long should the expansion grid (■ Figure 20.7) for this bridge be?

MODEL Without the grid, expansion of the steel in hot weather could cause the bridge to buckle. The grid should be at least as long as the maximum expansion expected.

SETUP First, we convert the given temperature change to kelvin. Since we are concerned with temperature changes, not measured temperatures, the different zero points of the scales are irrelevant. Using eqn. (19.1a):

$$\Delta T = \frac{5 \text{ K}}{9°\text{F}} 100°\text{F} = 56 \text{ K}.$$

SOLVE Using eqn. (20.3), with α for steel taken from Table 20.3, the expected expansion is: $\Delta \ell = (11 \times 10^{-6} \text{ K}^{-1})(100 \text{ m})(56 \text{ K}) = 0.06 \text{ m} = 6 \text{ cm}.$

ANALYZE In this type of problem, safety requires that we err on the generous side. The grid must be *at least* 6 cm long. ∎

20.3.2 Volume Expansion

We may also describe the expansion of a solid by its change in volume.

The *coefficient of volume expansion*, β, is the fractional change in volume per unit temperature change:

$$\Delta V = \beta V \Delta T. \tag{20.4}$$

The coefficient of volume expansion is also measured in K^{-1}.

■ **FIGURE 20.7**
This expansion grid allows for changes in the bridge's length as the temperature changes. If this accommodation were not made, the resulting stresses could be large enough to damage the bridge.

EXAMPLE 20.4 ♦♦ Show that for an isotropic solid the coefficient of volume expansion is approximately $\beta = 3\alpha$, where α is the coefficient of linear expansion.

MODEL We model the solid as a collection of many small cubes. By making the cubes as small as necessary, we can model a solid of any shape. The volume of each cube is related to the length of its side, which expands as the temperature rises.

SETUP As the temperature increases from T to $T + \Delta T$, each side of a solid cube (■ Figure 20.8) expands from length L to $L + \Delta L$, in accordance with the law of linear expansion (eqn. 20.3). The new volume of the cube is then:

$$V + \Delta V = (L + \Delta L)^3 = L^3(1 + \alpha\,\Delta T)^3$$
$$\approx L^3(1 + 3\alpha\,\Delta T) = V(1 + 3\alpha\,\Delta T),$$

where we have neglected terms of order $(\alpha\,\Delta T)^2$ and higher powers, which are much less than unity. From the definition of β:

$$V + \Delta V = V(1 + \beta\,\Delta T).$$

SOLVE Comparing the two expressions, we find:

$$\boxed{\beta = 3\alpha.} \tag{20.5}$$

WE USE THE BINOMIAL EXPANSION (APPENDIX IB): $(1 + x)^n = 1 + nx + \cdots$.

ANALYZE The 3 is due to the three dimensions of the solid. See the following exercise.

EXERCISE 20.3 ♦♦ Show by a similar argument that the increase in the area of a plane object is $\Delta A = 2\alpha A\,\Delta T$. ■

The volume of a liquid increases in a similar way as the molecules move apart, and volume expansion is defined in the same way as for a solid.

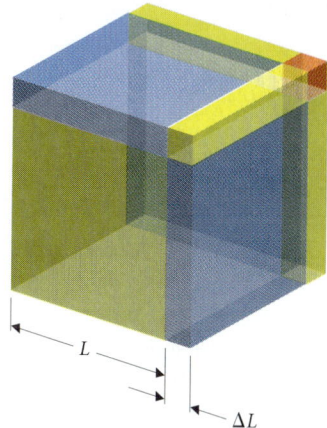

■ **FIGURE 20.8**
Thermal expansion of a cube. Each edge expands from L to $L + \Delta L$. The yellow and red volumes correspond to the terms ignored in the expansion of $(1 + \alpha\,\Delta T)^3 \approx (1 + 3\alpha\,\Delta T)$.

EXAMPLE 20.5 ♦♦ The mercury column in a thermometer has a diameter of 0.15 mm, and the reservoir holds 1.0 cm³ of mercury (■ Figure 20.9). Ignoring expansion of the glass, how far apart on the stem are marks indicating a 1.0 K temperature change?

MODEL Since the expected change in volume of the mercury is very small, we may use the 1.0-cm³ reservoir volume as the initial volume, ignoring the small amount of mercury initially in the stem. In any case, the stem is not usually immersed in the system to be measured.

SETUP The volume of mercury increases by $\Delta V = \beta V\,\Delta T$, with β from Table 20.3. This additional volume rises into the stem, forming a cylinder of cross-sectional area:

$$A = \frac{\pi d^2}{4} = \frac{\pi(1.5 \times 10^{-4}\ \text{m})^2}{4}$$
$$= 1.8 \times 10^{-8}\ \text{m}^2.$$

The height of the cylinder for $\Delta T = 1$ K is the required distance between marks.

SOLVE
$$\Delta V = A\,\Delta h.$$
$$\Delta h = \frac{\Delta V}{A} = \frac{\beta V\,\Delta T}{A}$$
$$= \frac{(1.8 \times 10^{-4}\ /\text{K})(1.0 \times 10^{-6}\ \text{m}^3)(1.0\ \text{K})}{1.8 \times 10^{-8}\ \text{m}^2}$$
$$= 1.0 \times 10^{-2}\ \text{m} = 1.0\ \text{cm}.$$

ANALYZE Since the glass also expands, the value we have found is a little too large. ■

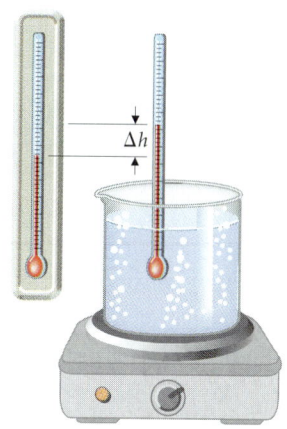

■ **FIGURE 20.9**
A mercury thermometer. When the thermometer is placed in a hotter system, the mercury in the bulb expands and moves up the stem.

SEE PROBLEM 45.

Digging Deeper

How Do Fish Survive the Winter?

Most substances contract when cooled; that is, they have positive values of α (or β), as indicated by the simple theory presented here. But between 0°C and 4°C, water *expands* when cooled (Figure 20.10). As water cools from a temperature above 4°C, it expands, becomes less dense, and so tends to float on the warmer water beneath. Hence a lake exposed to colder air freezes from the top down, and fish can live in the liquid water beneath the ice. (It's also important that the ice transmits heat poorly; see Chapter 21.)

The peculiar properties of water allow fish to live beneath the ice in a frozen mountain lake.

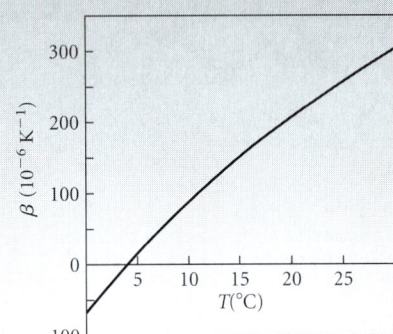

■ **FIGURE 20.10**
The coefficient of thermal expansion for water is negative between 0°C and 4°C. Water expands as it cools between 4°C and 0°C; ice is therefore less dense than water and floats.

Study Problem 14 ♦♦♦ A Thermal Switch

A bimetallic strip of length 2.0 cm is used as a switch. It consists of a strip of copper, of thickness 0.50 mm, bonded to a strip of invar, of thickness 0.50 mm (■ Figure 20.11). The device lies flat at a temperature of 293 K, and the copper side makes an electrical contact with a copper block beneath it. As its temperature rises, the device curves upward so that the contact is broken. If the strip is fixed at the end opposite the electrical contact, by how much does the strip clear the contact at 393 K?

Modeling the Physical System

As temperature rises, both metals expand. Since the coefficient of thermal expansion for copper is greater than that for invar, the copper expands more. To accommodate the resulting difference in length, the strip curls, breaking the contact. The thickness of each strip also expands, but since the thickness is small compared with the length, this change need not be considered (thickness/length = (0.50 mm)/(2.0 cm) = 0.025).

Plan

First, we calculate the expanded length of each strip. To accommodate the different expansions, the strip forms an arc of a circle with radius R, measured from the junction of the two metals. The invar strip lies along a circle of radius $R - \Delta R$, where ΔR is half the thickness of each strip, or 0.25 mm (■ Figure 20.12). The expanded length of the invar strip is the arc length along this center line. The arc length along the center line of the copper strip, radius $R + \Delta R$, is the new length of that strip. We determine the clearance d from R and θ.

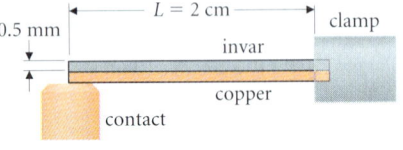

■ **FIGURE 20.11**
A thermal switch is made out of a bimetallic strip. The invar and copper strips, each of thickness 0.50 mm and length 2 cm, are welded together. The switch lies flat at 293 K, as shown here, and makes contact with the copper block.

Setup of Solution

With θ in radians, arc length $s = R\theta$, so the lengths of the two hot strips are:

$$(R + \Delta R)\theta = L + \Delta L_{Cu} = (1 + \alpha_{Cu} \Delta T)L, \quad \text{(i)}$$

and
$$(R - \Delta R)\theta = L + \Delta L_{invar} = (1 + \alpha_{invar} \Delta T)L. \quad \text{(ii)}$$

To solve for R, we eliminate the unknown angle θ by dividing these two equations:

$$\frac{R + \Delta R}{R - \Delta R} = \frac{1 + \alpha_{Cu} \Delta T}{1 + \alpha_{invar} \Delta T}.$$

Cross multiplying and gathering terms in R and ΔR gives:

$$R(\Delta T)(\alpha_{Cu} - \alpha_{invar}) = \Delta R[2 + (\alpha_{Cu} + \alpha_{invar})\Delta T].$$

Then:
$$R = \Delta R \frac{2 + (\alpha_{Cu} + \alpha_{invar})\Delta T}{(\alpha_{Cu} - \alpha_{invar})\Delta T}. \quad \text{(iii)}$$

Putting in the numbers, we have:

$$R = (0.25 \text{ mm})\left[\frac{2 + (1.66 + 0.16) \times (10^{-5} \text{ K}^{-1})(100 \text{ K})}{(1.66 - 0.16) \times (10^{-5} \text{ K}^{-1})(100 \text{ K})}\right]$$

$$= (0.25 \text{ mm})\left(\frac{2.0}{1.50 \times 10^{-3}}\right) = 333 \text{ mm}.$$

The strip clears the contact by an amount d (Figure 20.12), where:

$$d = R - R \cos \theta. \quad \text{(iv)}$$

From eqns. (i) and (ii), the angle $\theta \approx L/R \ll 1$. Thus $\cos \theta \approx 1 - \theta^2/2$, and we have:

$$1 - \cos \theta \approx \frac{\theta^2}{2} \approx \frac{1}{2}\left(\frac{L}{R}\right)^2.$$

Solution

The clearance d (eqn. iv) is: $d = \frac{R}{2}\left(\frac{L}{R}\right)^2 = \frac{L^2}{2R} = \frac{(2.0 \text{ cm})^2}{2(333 \text{ mm})} = 0.60 \text{ mm}.$

Analysis

The radius of curvature is inversely proportional to both ΔT and the difference between the expansion rates of the two metals. Thus the larger the temperature rise and the difference in expansion rates, the larger is the curvature and the smaller is the value of R.

We neglected the change in thickness t of the strips:

$$\Delta t = \alpha_{Cu} t \Delta T + \alpha_{invar} t \Delta T$$

$$= [(1.66 + 0.16) \times 10^{-5} \text{ K}^{-1}](0.50 \text{ mm})(100 \text{ K}) = 0.91 \text{ } \mu\text{m}.$$

As we anticipated, this change is much less than the calculated clearance of 600 μm.

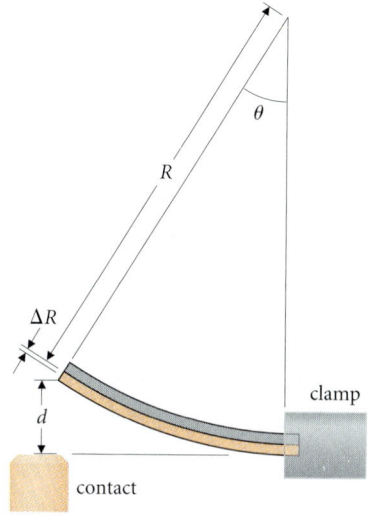

■ **FIGURE 20.12**
As the temperature rises, the copper expands more than the invar, and the strip curves away from the copper block, breaking the contact. The angle θ and the curvature of the strip are exaggerated in this drawing.

SEE APPENDIX IB FOR THE SERIES EXPANSION OF THE COSINE FUNCTION.

THE TERMS IN THE NUMERATOR OF EQN. (iii) THAT INVOLVE THE EXPANSION RATES ARE MUCH LESS THAN 2, AND HENCE HAVE LITTLE EFFECT.

20.4 CALORIMETRY

20.4.1 Specific Heat of Real Substances

In Chapter 19 we defined the specific heat of a substance as the heat transfer required to raise the temperature of a unit mass of the substance by 1 K. Thus:

$$c = \frac{Q}{M \Delta T}, \quad (20.6)$$

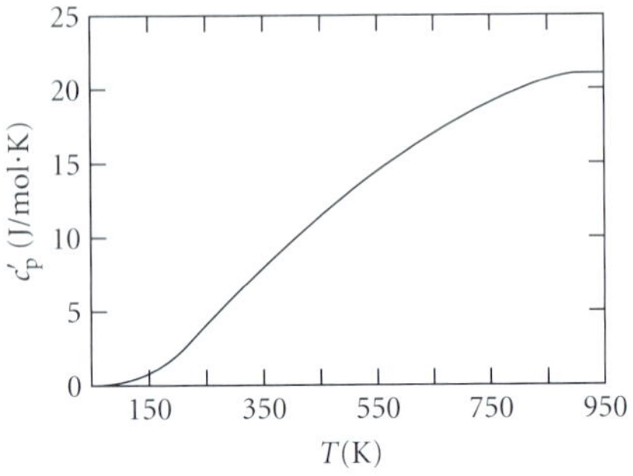

TABLE 20.4 Specific Heats at Constant Pressure
(At 25°C and 1 atm unless noted.)

Substance	c_p (J/kg·K)	Substance	c_p (J/kg·K)
Aluminum	900	Water	4186
Calcium	230	Ice (at 0°C)	2060
Chlorine	477	Common salt (NaCl)	864
Carbon (diamond)	507	Sugar	1147
Carbon (graphite)	712	Glass	490–830
Copper	385	Marble	863
Gold	130	Porcelain	1088
Helium	5188	Charcoal	670
Hydrogen	14300	Leather	1507
Invar 36	515		
Iron (cast)	502		
Iron (pure)	460		
Lead	130		
Magnesium	1017		
Mercury	138	Ammonia	2060
Neon	1030	Steam (at 110°C)	2009
Oxygen	920	Air	1005
Argon	519	Carbon dioxide	844
Sodium	1226		
Zinc	386		

■ **FIGURE 20.13**
Specific heat of diamond as a function of temperature. An asymptotic value is reached as the temperature approaches the Debye temperature, about 2000 K. Near room temperature, the specific heat varies strongly with temperature.

where M is the mass of the object, c is the specific heat of the material, and ΔT is the change in temperature. As we learned for gases, this definition is incomplete unless we specify the method of heat transfer. However, liquids and solids expand very little when heated, so the difference between their specific heats at constant pressure and at constant volume is negligible for most purposes. The specific heat is also a function of temperature, though for most substances under normal laboratory conditions, the temperature dependence is slight. Diamond is a notable exception (■ Figure 20.13). Values of specific heat for several common substances are listed in ● Table 20.4.

A substance's *molar specific heat* is the energy required to raise the temperature of 1 mole of the substance by 1 kelvin.

Digging Deeper

THE CALORIE

Since heat is nothing more than energy transferred because of a temperature difference, no special unit of heat is required. However, the calorie was defined early in the study of thermodynamics and has remained in common use.

A *calorie* (abbreviation cal) is the heat transfer required to raise the temperature of 1 g of water from 14.5°C to 15.5°C.

With this definition, the specific heat of water is 1 cal/(g·K) exactly (at 15°C). A British Thermal Unit (Btu) is defined similarly, using pounds and °F, so that the specific heat of water is also unity in British units (1 Btu/lb·°F). Joule, in his remarkable experiment of 1849, showed that 1 calorie is about 4 joules. There are at least four definitions of the calorie in modern use. The calorie defined here is the 15°C calorie, equal to 4.186 J. The Calorie (capital C), used to express the energy value of foods, is equal to 1 kcal (4186 J). The BIPM (§1.2.2) officially "deprecates" the use of the calorie.

At ordinary temperatures, the molar specific heat of many solids is close to 25 J/mol·K, a result known as the Dulong–Petit law. As for ideal gases, the constant value of the molar specific heat reflects the fact that the average energy per molecule is a simple function of temperature. The differing values of c_p in Table 20.4 reflect both the different molecular masses of the substances as well as discrepancies from the simple theory. Below a temperature called the *Debye temperature of the solid,* quantum effects become important and the Dulong–Petit law no longer applies. Debye temperatures vary widely: T_D is about 100 K for lead, and about 2000 K for diamond. Well below the Debye temperature specific heats typically decrease as T^3.

SEE §19.7.3, WHERE WE USED A SIMPLE MODEL OF A SOLID TO DERIVE THE DULONG–PETIT LAW.

AFTER PIETER DEBYE, 1884–1966.

20.4.2 Heat Capacity

Another useful concept related to heat transfer is heat capacity.

> The *heat capacity* of any object is defined as the heat transfer required to raise its temperature by 1 kelvin:
> $$C = \frac{Q}{\Delta T}. \tag{20.7}$$

BEWARE OF THIS TERMINOLOGY—IT IS NOT ACTUALLY A CAPACITY, IN THE USUAL SENSE OF A MAXIMUM AMOUNT THE SYSTEM CAN CONTAIN. HEAT IS ENERGY TRANSFERRED TO OR FROM, RATHER THAN CONTAINED IN, AN OBJECT. THIS IS ANOTHER RELIC FROM THE HISTORICAL MISUNDERSTANDING OF THE NATURE OF THERMAL ENERGY.

For an object made of a single material, heat capacity is the specific heat of the material times the mass of the object: $C = Mc$.

Heat capacity is a useful measure of how much thermal energy one body can transfer to another. It is possible to touch a very hot object with a small heat capacity and not be burned.

> **EXAMPLE 20.6** ◆ A 3.00-g fragment of charcoal briquet burning at 720 K comes into thermal equilibrium with a person at a final temperature of 310 K. How much thermal energy is transferred to the person? How much boiling water would transfer the same energy if T_f is the same?
>
> **MODEL** The heat transferred depends on the heat capacity of the briquet.
>
> **SETUP** Using the specific heat of charcoal (Table 20.4), the heat capacity of the briquet is:
> $$C = Mc = (3.00 \times 10^{-3} \text{ kg})(670 \text{ J/kg·K}) = 2.0 \text{ J/K}.$$
>
> **SOLVE** The heat transferred is:
> $$Q = C \Delta T = (2.0 \text{ J/K})(720 \text{ K} - 310 \text{ K}) = 0.82 \text{ kJ}.$$
>
> For boiling water at 100°C = 373 K, the heat transfer is:
> $$Q = Mc \Delta T = M(4186 \text{ J/kg·K})(373 \text{ K} - 310 \text{ K}) = M(2.64 \times 10^5 \text{ J/kg}).$$
>
> Setting the two heats equal, we find the mass of water required:
> $$M = \frac{0.824 \times 10^3 \text{ J}}{2.64 \times 10^5 \text{ J/kg}} = 3.1 \text{ g}.$$
>
> **ANALYZE** From the roughly equal masses, we'd judge the charcoal to be about as dangerous as boiling water in causing burns, even though the charcoal is at a higher temperature. In fact, it is much less dangerous because it transfers heat less efficiently (Chapter 21). ∎

20.4.3 Calorimetry

Specific heats are frequently measured using the experimental technique of calorimetry. A substance of known specific heat and known temperature is mixed with one of unknown specific heat at a different, known temperature in an insulated container called a calorimeter (■ Figure 20.14). The system is allowed to come into thermal equilibrium, and the final temperature is measured. In an ideal calorimeter, there is no interaction with the environment.

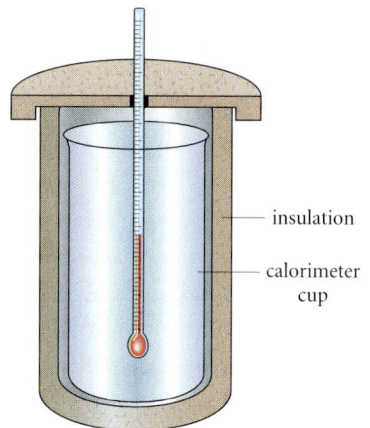

■ FIGURE 20.14
Calorimeter. The insulation isolates the system from its environment. When two systems at temperatures T_1 and T_2, respectively, are mixed in the calorimeter cup, they come to thermal equilibrium at a third temperature T_3, measured by the thermometer.

No thermal energy escapes, so the heat transferred from the hotter substance is gained by the cooler substance. Some heat is absorbed by the calorimeter itself. The heat capacity of the calorimeter cup can be measured in a separate experiment or calculated. We illustrate these techniques with a specific example.

> **EXAMPLE 20.7** ♦♦ A 50.0-g lump of metal at temperature 151°C is placed in a calorimeter cup containing 30.0 g of water at 23°C. After equilibrium is established, the temperature is 49°C. The heat capacity of the cup is 42 J/K. What is the specific heat c_m of the metal?
>
> **MODEL** There is no heat transfer to or from the combined system of cup, water, and metal. The system is initially out of equilibrium and reaches equilibrium by heat transfer from the water and calorimeter cup (part 1) to the metal lump (part 2).
>
> **SETUP** We calculate the heat transfer *to* each part of the system. We expect $Q_1 > 0$ and $Q_2 < 0$. The sum is the net heat transfer to the entire system—zero. We take the specific heat of water c_W from Table 20.4.
>
> 1. Cup + water, $T_1 = 23°C$:
>
> Heat capacity $\quad C_1 = C_{cup} + m_W c_W$
> $\quad\quad\quad\quad\quad\quad = 42$ J/K $+ (0.0300$ kg$)(4186$ J/kg·K$) = 168$ J/K.
>
> 2. Lump of metal, $T_2 = 151°C$:
>
> Heat capacity $\quad C_2 = m_m c_m = (0.050$ kg$) c_m$.
>
> 3. Final state, cup + water + lump of metal, $T_3 = 49°C$:
>
> Heat transfer to part 1 = heat capacity × change in temperature,
> $$Q_1 = C_1(T_3 - T_1).$$
>
> And: Heat transfer to part 2 = $Q_2 = C_2(T_3 - T_2)$.
>
> $0 = Q_{net} = Q_1 + Q_2 = C_1(T_3 - T_1) + C_2(T_3 - T_2)$ $\quad\quad$ (i)
> $\quad = (168$ J/K$)[(49 - 23)$ K$] + (0.050$ kg$)c_m[(49 - 151)$ K$]$.
>
> **SOLVE** $\quad c_m = \dfrac{(168 \text{ J/K})(26 \text{ K})}{(0.050 \text{ kg})(102 \text{ K})} = 8.6 \times 10^2$ J/kg·K.
>
> The specific heat of the metal is 860 J/kg·K.
>
> **ANALYZE** Water is used in the experiment because it is convenient and its specific heat is well known. ∎

REMEMBER: WE DEFINED Q TO BE POSITIVE WHEN ENERGY IS INPUT TO THE SYSTEM (§19.3.1).

THAT IS, $Q_1 = -Q_2$. HEAT LOST BY THE HOTTER PART IS GAINED BY THE COOLER PART.

> **EXERCISE 20.4** ❖ We have assumed here that no water is vaporized and lost from the system. If water were vaporized but recondensed inside the calorimeter, would the experimental results be affected? ∎

20.5 THERMOMETRY

20.5.1 Temperature Standards

Measurement of temperature consists of two parts: definition of standards for calibrating a thermometer, and use of the thermometer to compare an experimental system with the chosen standard. The constancy of temperature during a phase transition provides a useful standard. The Celsius scale is defined by the phase transition temperatures of water at atmospheric pressure. However, such standards are pressure dependent and can be determined only as accurately as pressure can be controlled during the transition. This difficulty is surmounted by using the *triple point* of water as a standard.

TABLE 20.5	Defining Temperatures on the International Practical Temperature Scale (1990)
State	**T (K)**
PRIMARY STANDARD	
Triple point of water	273.16
SECONDARY STANDARDS	
Triple point of	
Hydrogen	13.8033
Neon	24.5561
Oxygen	54.3584
Argon	83.8058
Mercury	234.3156
Melting point of Gallium	302.9146
Freezing point of	
Indium	429.7485
Tin	505.078
Zinc	692.677
Aluminum	933.473
Silver	1234.93
Gold	1337.33
Copper	1357.77

Note: Freezing and melting points are at 1 atm = 1.01325×10^5 Pa.

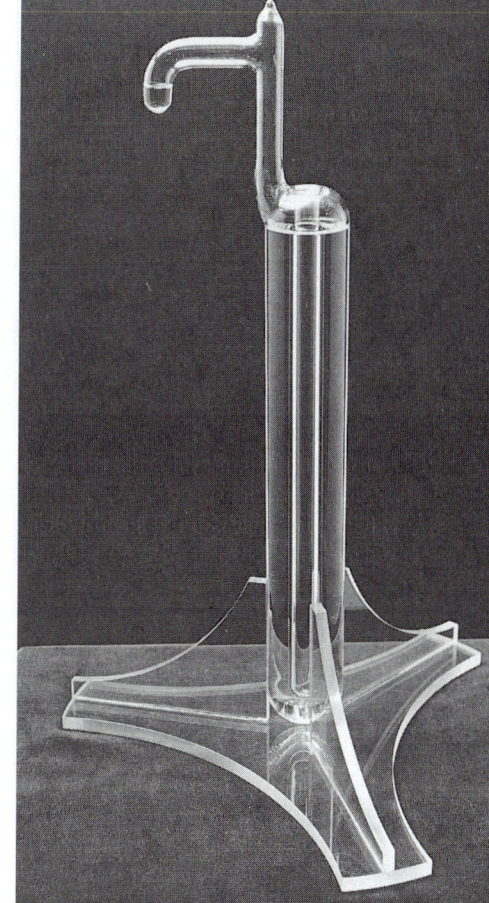

■ **FIGURE 20.15**
Apparatus for observing the triple point of water. This triple point cell at the National Institute of Standards and Technology (NIST) contains pure water at the triple point pressure. At the center is a hollow or *well*. The cell is prepared by cooling the inside of the thermometer well (e.g., by filling it with dry ice) so that water freezes outward, forming a mantle 4–10 mm thick around the well. Next the well is filled with chilled water, and a glass tube at room temperature is inserted just long enough to melt the ice mantle free of the well wall. The well may then be emptied, and the thermometer to be calibrated is inserted in the well. The temperature is reproducible to 0.0002 K.

The *triple point* of a substance is the state at which solid, liquid, and gas phases of the substance are observed to co-exist in equilibrium in the same sample. It occurs at a unique value of both temperature and pressure.

The triple point pressure for water, 4.58 mm Hg, is much less than atmospheric pressure, so apparatus must be specifically designed to achieve the triple point (■ Figure 20.15).

Two standards are needed to calibrate a temperature scale. Together with absolute zero, the triple point of water defines the magnitude of the kelvin. The triple point temperature for water is defined to be 273.16 K exactly. However, while triple point cells are practical laboratory systems, absolute zero cannot be realized at all (see Chapter 22). For practical use, a set of secondary standards has been established, forming the International Practical Temperature Scale (● Table 20.5).

A certain kind of thermometer measures temperatures over a limited range. There must be a standard system to define the temperature in each such range, and thermometers are designed with overlapping ranges so as to develop a consistent temperature scale. A common mercury-in-glass thermometer, for example, can measure temperatures in a range limited by the freezing temperature of mercury (234 K) and the temperature at which the glass softens (up to 600 K). In this range, the thermometer can be calibrated using the triple point of water and the melting point of gallium (303 K) or the freezing point of indium (430 K).

20.5.2 Practical Thermometers

There are several basic requirements for a useful thermometer. The first is an *easily measured property* that varies in a well-understood way with temperature. A linear variation is most convenient. One such property is the pressure of a fixed volume of ideal gas, which forms the basis for the constant-volume gas thermometer. In practice, a dilute real gas is used as an approximation to the ideal. This system works well down to a few kelvin, where even helium condenses. At high temperatures, the gas ionizes or its container melts, but other thermometers become more practical before that limit is reached. The gas thermometer is used when very precise values of temperature are needed, but it is not convenient for everyday use. Other common thermometric properties are the thermal expansion of metals, the electrical resistance of a coil of wire, and some other electrical properties of metals. A thermocouple uses the voltage developed across the junction between two different metals. The temperature sensor in your car probably is a thermocouple.

The second important requirement for a good thermometer is that it should *perturb the system of interest as little as possible.* In particular, it should have a small heat capacity, so that there is negligible heat transfer as it comes to equilibrium with the system.

Sometimes we want the thermometer to reach thermal equilibrium with the system and register the equilibrium temperature rapidly. This is important when measuring a constantly changing temperature as a function of time, for example. The mercury-in-glass thermometer is much slower than a thermocouple, though faster than thermometers relying on expansion of solids.

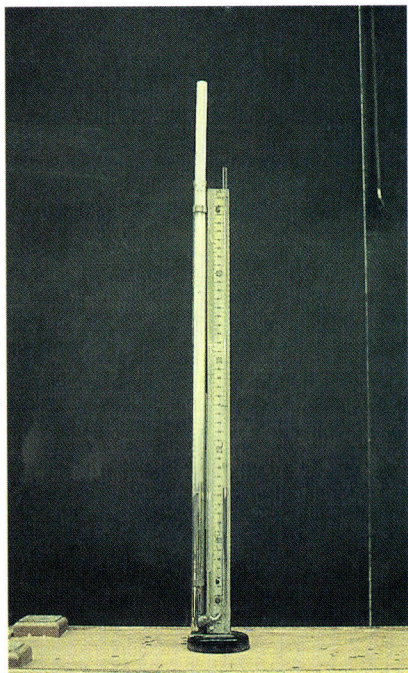

■ **Figure 20.16**
Student laboratory version of a constant-volume gas thermometer.

Remember: T is on the Kelvin scale in this expression!

EXAMPLE 20.8 ♦ A student uses a constant-volume gas thermometer in a physics experiment (■ Figure 20.16). A volume of nitrogen is trapped in a closed tube by a column of mercury. The pressure of the gas is 1.20×10^5 Pa at $0°C$. When immersed in hot liquid, the gas expands. The student adjusts the pressure to return the gas to its initial volume. The new pressure is found to be 1.44×10^5 Pa. What is the temperature of the liquid?

MODEL The thermometric property used here is gas pressure, which is related to temperature by the ideal gas law, $PV = NkT$.

SETUP Since V is kept constant, the ideal gas law reduces to:

$$\frac{P}{T} = \text{constant}.$$

SOLVE Thus: $T_2 = T_1 \dfrac{P_2}{P_1} = (273 \text{ K})\left(\dfrac{1.44 \times 10^5 \text{ Pa}}{1.20 \times 10^5 \text{ Pa}}\right) = 328 \text{ K}.$

The temperature of the hot liquid is $55°C$.

ANALYZE You may find it helpful to sketch these processes on a P-V diagram. ■

EXERCISE 20.5 ❖ When the bulb of a mercury thermometer is placed in a hot liquid, the mercury level falls slightly before rising. Why?

Chapter Summary

Where Are We Now?

We have used the ideas developed in Chapter 19 to understand thermal properties of real substances under a wide range of conditions. A model including forces between molecules explains qualitatively why gases condense into liquids. We have presented empirical relations for thermal expansion and specific heats, which may be understood in terms of the molecular model. These principles underlie the construction of thermometers and form the basis for practical thermodynamics.

What Did We Do?

We extended the ideal gas model to include some effects of intermolecular forces. Under conditions where these forces become dominant, a gas undergoes a phase transition and becomes liquid. During the phase transition, the temperature of the substance is constant and potential energy is released. Similar behavior occurs as the liquid condenses into solid form. To reverse the phase changes, energy must be returned to the substance. The energy per unit mass absorbed or released during a phase transition is called latent heat.

When a substance is heated or cooled, its volume changes. The process is described by the coefficient of linear (α) or volume (β) expansion for the substance. For example, the change in length of a rod when its temperature changes by ΔT is given by eqn. (20.3):

$$\Delta \ell = \alpha \ell \, \Delta T.$$

The coefficients α and β are functions of temperature; usually an average value for the temperature range in question is used.

Calorimetry is the bookkeeping of energy transfer. When two systems in thermal contact are otherwise isolated and no work is done, the heat transfer to one system equals the heat transfer from the other. The heat capacity C of a system measures the energy required to change the temperature of that system: $Q = C \Delta T$. Specific heat c is the heat capacity per unit mass: $c = C/M$.

Thermometry involves the comparison of a system's temperature against a standard. Standards are defined using phase transitions (see Table 20.5). A good thermometer has a well-defined and easily measured thermometric property and a small heat capacity.

Practical Applications

The behavior of systems under changing temperatures is important in many engineering problems, from the design of space vehicles and railways to stereo equipment and computers. Thermodynamic properties of the atmosphere are important in understanding weather. Thermometry is useful in medicine and cooking as well as in engineering and basic physics. The physics of phase changes is used in fabricating components out of metals and glasses. Calorimetry is used in measuring the metabolism of animals and in elementary particle experiments.

Solutions to Exercises

20.1 The volume per mole is $(5.00 \text{ L})/(0.650 \text{ mol}) = 7.69$ L/mol. Using eqn. (20.2), and values of a and b for CCl_4 from Table 20.1:

$$\left(1.01 \times 10^5 \frac{\text{Pa}}{\text{atm}}\right)\left[2.70 \text{ atm} + \frac{20.4 \text{ L}^2 \cdot \text{atm/mol}^2}{(7.69 \text{ L/mol})^2}\right]$$

$$\times \left[(7.69 - 0.14)\frac{\text{L}}{\text{mol}} \cdot \frac{10^{-3} \text{m}^3}{\text{L}}\right] = \left[8.31 \frac{\text{J}}{\text{mol} \cdot \text{K}}\right] T.$$

So: $T = (2320 \text{ Pa} \cdot \text{m}^3/\text{mol}) \div (8.31 \text{ J/mol} \cdot \text{K}) = 279$ K.

20.2 The latent heat of vaporization for water is 2272 kJ/kg (Table 20.2). The amount of steam to be condensed is:

$$M = \frac{E}{L_v} = \frac{35.5 \text{ kJ}}{2272 \text{ kJ/kg}} = 0.0156 \text{ kg} = 15.6 \text{ g}.$$

Because of its large latent heat of vaporization, steam can cause severe burns.

20.3 A square of material with side L expands to a new area:

$$A + \Delta A = (L + \Delta L)^2$$
$$= L^2(1 + \Delta L/L)^2$$
$$\approx L^2(1 + 2\,\Delta L/L)$$
$$= A(1 + 2\alpha\,\Delta T).$$

So $\Delta A = 2\alpha A \, \Delta T$, as required.

20.4 If water were vaporized, latent heat would have to be provided but would be released again when the vapor recondenses inside the calorimeter. There would be no net effect on the calculated specific heat.

20.5 The mercury is contained in glass that also expands, creating a reservoir of larger volume. The mercury level drops as mercury from the stem flows into the extra volume in the reservoir. As heat is conducted (Chapter 21) into the mercury, the mercury expands and the level in the stem rises.

Basic Skills

Review Questions

§20.1 THE BEHAVIOR OF REAL GASES

- Write the van der Waals equation for a gas. What physical effects are expressed by the terms containing *a* and *b* that appear in this equation?
- Sketch the van der Waals isotherms in a *P-V* diagram. In which regions of the diagram does the gas behavior approximate the ideal? Indicate on your diagram the region where the substance becomes liquid.
- What is the *critical point* for a gas?

§20.2 CHANGE OF PHASE

- How does the temperature of the liquid-vapor transition depend on pressure for common gases?
- What is *latent heat of fusion*?
- When water boils, its temperature remains constant. What happens to energy supplied as heat?

§20.3 THERMAL EXPANSION

- Define the *coefficient of linear thermal expansion* for a solid.
- Define the *coefficient of volume thermal expansion* for a substance.
- Describe two common situations in which thermal expansion is an important consideration.

§20.4 CALORIMETRY

- In Chapter 19 we discussed specific heat at constant volume and at constant pressure for gases. Table 20.4 lists specific heats at constant pressure for liquids and solids as well as a few gases. We do not list values of specific heat at constant volume for solids and liquids. Why not?
- What is the *heat capacity* of an object?
- What is *calorimetry*?
- What principle underlies the use of calorimetry for measuring specific heat?

§20.5 THERMOMETRY

- What is meant by a *temperature standard*?
- What is the *triple point* of a substance? Why is the triple point of water important?
- List two important properties of a good thermometer.

Basic Skill Drill

§20.1 THE BEHAVIOR OF REAL GASES

1. Find the pressure of 1.2 mol of N_2 in a 4.0-L container at 301 K.
2. If 1.70 mol of CO_2 in a 5.00-L container exerts a pressure of 7.2 atm, what is the temperature?

§20.2 CHANGE OF PHASE

3. How much energy is required to melt 10.0 kg of iron at 1808 K?
4. How much energy is removed from the inside of a refrigerator when 15 g of freon 12 is vaporized?

§20.3 THERMAL EXPANSION

5. A uniform brass rod 2.0 cm^2 in cross section and 55.6 cm long at 12°C is 0.42 cm longer at 412°C. Calculate the coefficient of linear expansion for brass.
6. How long should the expansion grid be at the end of an iron bridge 210 m long if the expected temperature variation is $-11°C$ to 35°C?

§20.4 CALORIMETRY

7. Why is it that a speck of hot ash from a fire does not burn you, but very hot tap water, which has a significantly lower temperature, causes severe burns? (*Hint:* Consider how much heat is transferred to your body in each case.)
8. A calorimeter has a sample cup with a heat capacity of 77 J/K. In an experiment to measure the specific heat of silver, a necklace of mass 185 g is placed in the cup, which is at room temperature (26°C). Then 92 g of hot water ($T = 89°C$) is poured into the cup. When the system comes to equilibrium, the temperature is 73°C. What is the specific heat of the silver?

§20.5 THERMOMETRY

9. A constant-volume gas thermometer contains neon. At lab temperature (19°C), the pressure is 1.00×10^3 Pa. What is the temperature of the bulb when the pressure is 1.10×10^3 Pa?
10. The length of the mercury column in a thermometer is 1.2 cm when placed in ice water and 15.4 cm when placed in boiling water. When left out on the lab bench, the length of the column is 4.3 cm. What is the temperature in the lab?

Questions and Problems

§20.1 THE BEHAVIOR OF REAL GASES

11. ❖ Discuss the effect of attractive forces between molecules on the work required to compress a van der Waals gas isothermally. Is the required work less than, the same as, or greater than that for a similar compression of an ideal gas? What would be the effect of the short-range repulsive forces? Which effect do you expect to be larger? Why?

12. ❖ Which gases in Table 20.1 have the smallest van der Waals constants? Which have the largest? Can you explain why?

13. ♦ A 5.50-L cylinder contains 2.50 mol of carbon dioxide at a temperature of 295 K. What is the pressure of gas in the cylinder?

14. ♦ A 6.50-L cylinder of compressed SO_2 contains 2.00 mol of gas at a pressure of 7.00 atm. What is the temperature in the cylinder? Compare with the result for an ideal gas under the same conditions.

15. ♦♦ ✉ Estimate the effective diameter of a chlorine molecule.

16. ♦♦ ✉ Estimate the effective diameter of a carbon tetrachloride molecule.

17. ♦♦♦ A 2.30-L cylinder of compressed oxygen contains gas at 6.10 atm and 288 K. How much oxygen is in the cylinder? [*Hint:* Try ignoring the smallest term on the left-hand side of eqn. (20.2). Is this a reasonable approximation?]

18. ♦♦♦ Two moles of ammonia exert a pressure of 1.5 atm at 300 K. Using the data in Table 20.1, make an accurate graph of the 300 K isotherm on a *P-V* diagram, and use the graph to find the volume of the gas.

§20.2 CHANGE OF PHASE

19. ❖ Traditionally, gold was extracted from ore by dissolving it in mercury, then boiling the mixture to evaporate the mercury. Use Table 20.2 to explain why this process works.

20. ❖ Why are steam burns more serious than burns inflicted by boiling water?

21. ♦ How much heat is required to melt a metric ton of iron?

22. ♦♦ Fresh snow is about one-tenth as dense as liquid water. How much latent heat is released in storm clouds per square kilometer of surface area by the freezing of water droplets if the storm deposits a 30-cm-deep layer of fresh snow?

23. ♦♦ A pot of water is boiling on a 100-W electric hotplate. Neglecting losses, how long does it take to vaporize 5×10^{-4} m^3 of water?

24. ♦♦ An advertisement for glass cooking pans shows an aluminum pan being melted inside a glass pan on a stovetop. How much heat must be provided to melt 0.30 kg of aluminum, once the melting point has been reached? How long would it take on a 2500-W burner, assuming no losses?

§20.3 THERMAL EXPANSION

25. ❖ An aluminum bar just fits inside an aluminum frame, as shown in ■ Figure 20.17. Initially, the bar is almost slipping. If the whole assembly is heated from 20°C to 100°C, does the bar fall?

26. ❖ A piece of glass contains an air bubble. If the temperature of the glass is raised, what happens to the diameter of the bubble?

27. ❖ If boiling water is poured into a fine crystal glass, it cracks. Why?

28. ❖ Why do frozen water pipes sometimes burst?

29. ♦ If the temperature of mercury is raised by 10°C, by how much does its density change?

30. ♦ When laying rails for a railway, one must allow for thermal expansion. What spacing should be allowed between 16-m steel rails being laid in Death Valley at 40°F, if a temperature of 130°F is expected on hot days?

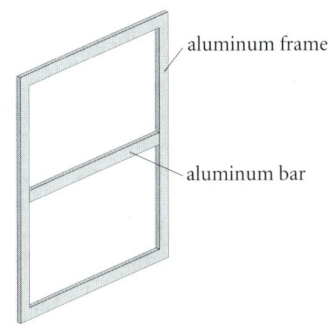

■ FIGURE 20.17

■ FIGURE 20.18

31. ♦♦ A household thermometer contains 2.5 cm^3 of alcohol. The diameter of the stem is 1.5 mm. How far apart are the marks for each °C and each °F?

32. ♦♦ A steel sleeve and a copper rod are shown in ■ Figure 20.18, where all the dimensions have been measured at 25°C. The steel sleeve alone is heated and placed onto the copper rod. As it cools, it is tightly held in place. Why? (This process is called "shrink-fitting.") The assembly is to be used in the Arctic winter. At what temperature does the steel sleeve separate from the copper rod?

33. ♦♦ A 1.0-L carafe is filled with white wine. Both wine and carafe are at a temperature of 5°C. The carafe is made of glass, with a volume coefficient of thermal expansion of 3.2×10^{-5} K^{-1}. When the carafe has risen to room temperature (25°C), 7 cm^3 of wine has overflowed. (What a waste!) What is the coefficient of thermal expansion for the wine?

34. ♦♦ Light aircraft fuel tanks are located in the wings. Each tank is equipped with a vent to allow air to enter as the fuel is drawn out. If the tanks are filled late in the evening when it is cold, fuel overflows out the vent tube the next day. If the tank holds 84 gal and is filled at 45°F, how much gas is lost when the temperature rises to 75°F the next day? (Take the volume coefficient of thermal expansion for gasoline to be 1.0×10^{-3} /K.) Estimate the error made by ignoring the expansion of the aluminum tank.

§20.4 CALORIMETRY

35. ❖ Water tubs are sometimes placed in greenhouses on cold nights. Why?

36. ♦ A Styrofoam coffee cup acts as a reasonably good calorimeter. If the cup contains 0.23 L of coffee at 90°C and 0.05 L of milk at 7°C is added, what is the final temperature of the coffee? (Assume that both coffee and milk have the same specific heat as water.)

37. ♦♦ Two solid bodies of heat capacity C_A and C_B, originally at temperatures T_A and T_B, respectively, are placed together inside a heat-insulating enclosure. Find the final temperature of each body. How would your answer differ if the enclosure into which the bodies were placed had imperfect insulation?

38. ♦♦ An insulated metal vessel of mass 5.0 kg contains 20.1 kg of water. A 2.5-kg piece of the same metal, initially at a temperature of 190°C is dropped into the water. If the water temperature rises from 16.0°C to 17.8°C, calculate the specific heat of the metal.

39. ♦♦ A heat sink used to dissipate heat in a stereo power amplifier is designed to operate at 60°C. If the heat sink for a 100-W amplifier is a 1.3-kg block of aluminum, how long would it take to reach 60°C after being turned on if room temperature is 20°C and losses are neglected? (Of course, losses cannot really be neglected in this case.)

40. ♦♦ In a chemical plant, water is used to cool a reactor chamber. Water at an initial temperature of 5°C flows at a rate of 16 gal/min over the cooling fins and emerges at a temperature of 30°C. At what rate is energy produced in the reaction chamber? (1 gal = 3.79 L.)

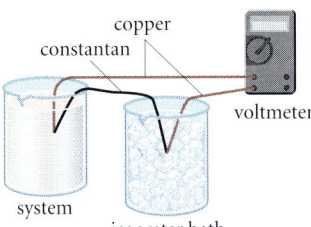

FIGURE 20.19

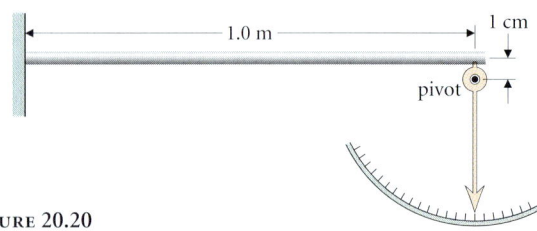

FIGURE 20.20

41. ◆◆ A mass $m_x = 0.300$ kg of an unknown substance at temperature $T_x = 401°C$ is put into a calorimeter containing 0.100 kg of water at $T_w = 20°C$. The calorimeter has a water equivalent (i.e., heat capacity divided by specific heat of water) of 0.050 kg. If the system comes to a temperature of $60°C$ after mixing, what is the specific heat of the unknown substance?

§20.5 THERMOMETRY

42. ◆◆ A thermometer consists of a platinum wire coil connected to a 2.0-V battery and an ammeter of resistance $R_a = 5.0\ \Omega$. The temperature coefficient of resistivity for platinum is $\alpha = 3.93 \times 10^{-3}/K$ between $0°C$ and $100°C$. The current in the ammeter is a function of temperature, and is given by:

$$I = \frac{\mathscr{E}}{R_a + R_p[1 + \alpha T(°C)]},$$

where $\mathscr{E}$ is the battery voltage and R_p is the resistance of the wire at $0°C$; I is in amps when $\mathscr{E}$ is in volts and R is in ohms (symbol Ω). At $22°C$, the current in the circuit is 0.190 A. When immersed in water, the current is 0.170 A. What is the temperature of the water?

43. ◆◆ A thermocouple consists of two wires of copper and constantan (an alloy of 45 percent Ni and 55 percent Cu) joined together. The free ends are joined to copper connecting wires (■ Figure 20.19). One junction between the two metals is placed in the system to be measured, and the other junction is placed in a reference system (often an ice bath at $0°C$). When the system to be measured is at a different temperature from the reference system, the needle of a voltmeter connected to the copper wires shows a deflection. A copper–constantan thermocouple produces a voltage $V = (0.0387\ mV/K)T_2$ when the reference junction is at $0°C$ and T_2 is the temperature of the second junction in $°C$. If the measured voltage is 0.523 mV, what is the temperature of the system? (The given formula is accurate to about $0.3°C$.)

44. ◆◆ A thermometer is designed as shown in ■ Figure 20.20. A metal rod 1.0 m long is mounted on a rigid base. Its coefficient of thermal expansion is $\alpha = 10^{-5}/K$. The free end of the rod is connected with a pointer, which pivots about a point 1 cm below the end of the rod. The pointer is vertical at $0°C$. Compute the angle the pointer makes with the vertical at $100°C$.

45. ◆◆ The bulb of a mercury thermometer is a sphere of radius R, and the stem has a bore of radius r. Show that the distance between two scratches, which would measure a temperature difference of $1°C$, is approximately given by $d = \frac{4}{3}R^3(\beta_m - \beta_g)/r^2$, where β_m and β_g are the coefficients of volume expansion of mercury and glass, respectively. (Assume that only the bulb is immersed in the system to be measured.)

46. ◆◆◆ The thermometer in the previous problem has a glass wall of thickness 1.0 mm and a specific heat of 555 J/kg·K. Estimate the heat capacity of the thermometer bulb if $R = 5.0$ mm and $r = 1.0$ mm. If the thermometer, initially at $20.0°C$, is used to measure the temperature of $210\ cm^3$ of water at $80.0°C$, what does the thermometer read?

47. ◆◆◆ A constant-volume gas thermometer contains 1.00×10^{-3} mol of helium in a glass container of volume $20.0\ cm^3$. If the mass of glass is 10.0 g and its specific heat is 5.00×10^2 J/kg·K, what is the heat capacity of the thermometer bulb? What is the pressure of the helium at room temperature (293 K)? When the thermometer is used to measure the temperature of a system with heat capacity 2.00×10^2 J/K, the pressure is 9.65×10^4 Pa. What is the temperature of the system? By how much has the thermometer perturbed the system temperature?

Additional Problems

48. ❖ How would the van der Waals equation be altered if molecular forces were repulsive instead of attractive at intermediate distances? What qualitative change in the isotherms would result? Would phase changes occur?

49. ◆◆ A calorimeter cup of heat capacity 10.0 J/K contains $30.0\ cm^3$ of water and 10.0 g of ice. Then $50.0\ cm^3$ of water at $75°C$ is added. What is the final temperature of the mixture?

50. ◆◆ A 2.0-g lead bullet moving with a speed of 220 m/s strikes and remains embedded in a 2.0-kg wooden block. Assuming that all the transformed energy raises the temperature of the bullet, calculate its change in temperature.

51. ◆◆ A 105-g sample of ice at $0.00°C$ is mixed with water at $20.0°C$. If the final temperature is to be $5.0°C$, how much water should you use?

52. ◆◆ A 50.0-g sample of ice at $0.0°C$ is dropped into 50.0 g of water at $60.0°C$, held in a beaker of negligible heat capacity. What is the final temperature of the system, and why?

53. ◆◆ Some heat transfer to the environment occurs during the experiment described in Example 20.7. Does the result of the example over- or underestimate the actual specific heat of the metal? Assuming that the metal obeys the Dulong–Petit law, use the result for c_m to find the metal's molecular mass. Identify the metal.

54. ◆◆ Compare the total energies required to heat 0.50 kg each of iron and gold from $20°C$ to their respective melting points, and then to melt them.

55. ◆◆ The interior of a refrigerator, including the food, has a heat capacity of 8.4×10^4 J/K. How much freon 12 must be vaporized to reduce the interior temperature by $15°C$, if all the heat of vaporization comes from the interior of the refrigerator?

56. ◆◆ A truck of mass 20.0 Mg descends a 7% slope at a constant speed of 25 m/s. If the driver (unwisely!) uses the brakes instead of the engine to maintain speed, at what rate is energy transformed to thermal energy by the brakes? If the brakes are made of cast iron and have a mass of 100.0 kg, how rapidly does their temperature rise, assuming no heat losses? If the brakes burn when they reach a tem-

perature of 500 K and the initial temperature is 300 K, how far may the truck go before burning its brakes? (It's the hydraulic fluid that burns!)

57. ♦♦ An object is placed on a block of ice, and its temperature falls from 523 K to 273 K while 10.0 g of ice melts. What is the heat capacity of the object?

58. ♦♦♦ A 5.0-kg sample of molten lead (specific heat = 143.4 J/kg·K) at 610 K is dropped into 1.0 L of water at 5.0°C. Assuming no losses, find the final temperature of the water.

59. ♦♦♦ A 4-oz (118.3-mL) drink of vodka (half ethyl alcohol and half water by volume), originally at 20°C, is cooled by adding 15 g of ice at 0°C. If all the ice melts, what is the final temperature of the drink? The specific heat of ethyl alcohol = 2.45 kJ/kg·K, and the density of ethyl alcohol = 789 kg/m³. Neglect the heat capacity of the glass and any inflow of heat from the environment.

60. ♦♦♦ A physics student is on a camping trip and runs out of fuel for his stove. He realizes that he can use his thermos bottle to heat water, by shaking it. If the bottle contains 0.5 L of water, the water falls 0.2 m per shake, and he can shake once per second, by how much does the water temperature change in 3 h of shaking?

Computer Problems

61. Use a spreadsheet or write a short program to plot the isotherms of carbon tetrachloride at 100 K, 200 K, 300 K, and 400 K in a diagram of P (atm) versus V_m (L/mol). Make two plots: one showing the region 0–15 L/mol and one showing 0–3 L/mol. Compare the 400 K isotherm with an ideal gas isotherm at 400 K. What is the minimum possible value of V_m?

62. Use the trapezoidal rule to calculate the work done when 11×10^{-3} m³ of propane at a pressure of 2.14×10^5 Pa expands by a factor of 2 at a constant temperature of 290 K. Compare with the result for an ideal gas of the same molecular weight.

Challenge Problems

63. A hollow aluminum ball is floating on water so that exactly one hemisphere is submerged. If the water is heated from 5°C to 80°C, find the position of the water level on the sphere when it is in the hot water. (Assume that the ball comes into thermal equilibrium with the water.)

64. In this problem you are asked to verify the behavior of a gas that obeys the van der Waals equation of state. **(a)** Find the derivative dP/dV for a van der Waals gas, using eqn. (20.2), and set this derivative equal to zero to find the positions of the maxima and minima of the isothermal curves. Show that these points occur at volumes V_m, which are given by:

$$\frac{RTb}{2a}\left(\frac{V_m}{b}\right)^3 = \left(\frac{V_m}{b} - 1\right)^2.$$

(b) Use a graphical method to study the solutions of this equation, by plotting both the left- and right-hand sides as functions of V_m/b, using $\alpha \equiv RTb/(2a) = 0.1$. Show that there are three solutions, one of which is unphysical. **(c)** How does the plot in part (b) change as α is increased? Specifically, show why there are no physical solutions when α is very large. **(d)** Calculate the value of α at which the two curves in your plot would be tangent to each other, and thus give an expression for the critical temperature of the van der Waals gas.

ESSAY 6
Low Temperatures and their Measurement

J. M. Lockhart

■ FIGURE E6.1
A superconducting receiver used at the Hat Creek Radio Observatory. The signal from a millimeter wave antenna is focused into the conical guide at upper right and fed to the detector in the cube below. In use, the entire assembly is placed in a refrigerator.

ELECTRICAL RESISTANCE IS DISCUSSED IN CHAPTER 26.

We saw in Chapter 19 that the random motion of atoms in a solid or liquid decreases with decreasing temperature. Below a temperature of 100 K or so, the increased order that prevails when random, thermally excited atomic motion is sufficiently reduced leads to the emergence of many important phenomena. The field of low-temperature physics, or *cryogenics,* deals with the study of solids and liquids in this region of temperature.

Many modern technical achievements rely on phenomena that occur only at low temperatures. For example, "high-temperature" superconductivity occurs typically at temperatures below 100 K, while "classical" superconductivity requires temperatures of less than about 20 K. Extremely stable magnets using superconductors are important for medical imaging, while detectors based on superconducting effects allow the direct measurement of the magnetic fields associated with human brain waves. Liquid helium, which has a boiling point of 4.2 K at atmospheric pressure, cools the components of telescopes used to observe infrared light so that radiation from the telescope itself does not interfere with the observation. Superconducting receivers are also important in radio astronomy (■ Figure E6.1). The world standards for the units of electrical voltage and resistance are based on low-temperature effects.

In addition to the technological uses of low temperatures, there is a great deal of research aimed at understanding some of the fascinating properties of materials at temperatures below a few kelvin. In particular, quantum mechanical effects can be observed and measured on a macroscopic scale when the "smearing" effects of random thermal motion are minimized. For example, liquid helium undergoes a peculiar phase transition at 2.2 K such that, at temperatures below this value, the liquid is a "superfluid" (■ Figure E6.2)—objects can move through it with no viscous drag, large volumes of liquid can move through tubes the diameter of a human hair with no pressure drop, boiling can no longer occur, and heat can propagate as a wave rather than by the diffusive process to be described in Chapter 21.

Low temperatures are generally attained through the use of liquefied gases (especially nitrogen and helium), which are obtained via adiabatic expansion processes. The temperatures of these liquids can be reduced below the atmospheric boiling point values by using a vacuum pump to reduce their vapor pressure, as discussed at the beginning of Chapter 20. Extremely low temperatures are obtained by a subtle technique using alignment of a sample's atoms in a magnetic field.

Thermometry from 100 K or so down to the lowest attainable temperatures (below a microkelvin) makes use of the temperature dependence of several different physical properties of materials, including pressure, electrical resistance, and magnetic susceptibility. It is useful to refer to two groups of thermometers: primary, for which the value of a physical property corresponding to a particular temperature can be precisely calculated (making the thermometer capable of absolute calibration); and secondary, for which relative temperature differences can be measured but calibration against a primary thermometer is required. Secondary thermometers generally offer significant advantages of operational simplicity and a convenient

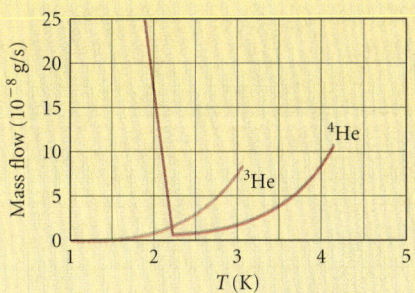

■ **Figure E6.2**
Liquid ^{3}He and ^{4}He flow under gravity through a fine hole. Notice how the flow rate for ^{4}He suddenly increases at about 2.2 K, indicating the onset of superfluid behavior. After D. W. Osborne, B. Weinstock, and B. M. Abraham, Phys Rev. 75, 988 (1949).

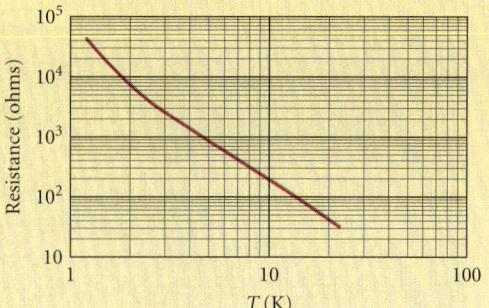

■ **Figure E6.3**
Electrical resistance of a germanium semiconductor thermometer as a function of temperature. Unlike metals near room temperature (Chapter 26), the resistance increases as temperature decreases.

form of temperature signal output. Constant-volume gas thermometers provide an important primary standard to a lower temperature of 2.6 K. Another (at least theoretically) primary standard is provided by liquid–vapor pressure thermometers, especially those using ^{4}He and ^{3}He, the latter being useful down to 0.3 K.

Electrical thermometers are quite convenient and popular secondary instruments in the range from room temperature to 0.1 K and can, in fact, be used to below 10 mK. From well above 300 K to a lower limit of around 10 K, platinum RTD (Resistive Temperature Detector) devices are common; these use the temperature dependence of the electrical resistance of a length of pure platinum wire. Below 20 K, the increase in the resistance of a semiconductor at low temperatures is often used, with single-crystal germanium a frequent choice. Graphite, in the form of carbon composition resistors, is also used. For both of these devices, the precise temperature dependence of the resistance is a complicated nonlinear function, and calibration against a primary thermometer at several temperatures is required. ■ Figure E6.3 shows the electrical resistance of a typical germanium semiconductor thermometer as a function of temperature in the range 1–20 K. It is important to minimize power dissipation (and the accompanying self-heating) during resistance measurement, so specialized low-power bridge circuits are employed.

Bridge circuits are discussed in §26.5.3.

Temperatures down to 1 mK can be measured by using the magnetic alignment (cf. §29.4) of atoms in a salt (often cerium magnesium nitrate, CMN) as the thermometric property. The sample's magnetic properties are measured with a bridge circuit or with a SQUID (Superconducting Quantum Interference Device) magnetometer. With SQUIDs, it is possible to measure the much smaller magnetic signals from nuclei of a metal such as copper; this method allows resolution of microkelvin temperatures. For these very lowest attainable temperatures, primary thermometers measure thermal electrical noise or the degree to which certain atomic nuclei are magnetically aligned.

What the hand dare seize the fire?
WILLIAM BLAKE

CHAPTER 21
Heat Transfer

CONCEPTS
Conduction
Convection
Radiation
Thermal resistance

GOALS
Be able to:
Compute heat transfer by conduction in symmetrical systems.

Find the temperature of a body using the Wien displacement law.

Compute the thermal radiation from a hot body.

Understand how and when convection occurs.

Skiers gathered around the fire illustrate all the methods of heat transfer. Heat from the fire is conducted through the metal of the stove and into the air. Convection currents rise from over the stove to heat the cabin, and the skiers feel the warm radiation from the hot metal. In this chapter we shall discuss all these effects.

Skiers gathered around the wood stove in their cabin heartily appreciate the processes of heat transfer. Energy from the burning wood is conducted through the metal of the stove. Heated air above the stove rises in convection currents that carry warm air throughout the cabin, while radiation from the hot metal warms the skiers directly. In the last two chapters we showed how the increase of a system's internal energy by heat transfer can change its temperature (warm the skiers) or its phase (melt the snow on their boots). Here we shall discuss *how* heat transfer occurs. There are three mechanisms: conduction, convection, and radiation.

21.1 CONDUCTION

Conduction is heat transfer by collisions. It is usually the most important mechanism for heat transfer through a solid body or between solid bodies in contact. Conduction occurs when temperature varies within the system, and heat flow occurs in an attempt to establish equilibrium.

Strictly speaking, temperature is defined only for a system in equilibrium. In analyzing continuous systems (Part III), we found that it is possible to divide a system into differential elements that contain huge numbers of molecules. Thus, for example, we can define density or pressure at a point in a continuous fluid. In the same way, we can define temperature at a point if the molecules within each differential element are in thermal equilibrium. When different elements are not in equilibrium with each other, the temperature is a function of position, and heat flows from regions of high temperature toward regions of low temperature.

THIS ASSUMPTION, CALLED *LOCAL* THERMODYNAMIC EQUILIBRIUM (LTE), IS VALID FOR THE SYSTEMS WE SHALL CONSIDER BECAUSE OF THE VERY LARGE NUMBER DENSITY OF MOLECULES IN EVERYDAY SYSTEMS (SEE EXAMPLE 13.2). IN SOME ASTROPHYSICAL SITUATIONS, THE NUMBER DENSITY OF MOLECULES IS SO LOW THAT LTE IS NOT A VALID APPROXIMATION.

21.1.1 Heat Transfer Along a Rod

When the ends of a uniform rod of length ℓ are maintained at different temperatures T_1 and T_2 (■ Figure 21.1), there is a *temperature gradient* along the rod. We shall restrict our attention to examples where the temperature changes at a constant rate along the rod. Then the temperature gradient is uniform, and:

$$\frac{dT}{dx} = \frac{T_2 - T_1}{\ell}.$$

Energy flows along the rod from the hotter end to the cooler end at a rate proportional to the temperature gradient. The heat flow is greater when the system is further from equilibrium.

Since heat transfer is by collisions, the transfer through a cross section of the rod depends on the number of interacting electrons and molecules and hence on the rod's cross-sectional area. The rate of transfer also depends on the detailed molecular or crystalline structure of the substance in a way that is difficult to compute and is usually measured experimentally. This dependence is expressed by a constant of proportionality k, called the *thermal conductivity* of the substance. The rate of heat transfer, or heat flux, H is:

$$H = kA \left| \frac{dT}{dx} \right|, \quad (21.1)$$

HERE IS YET ANOTHER k, TO ADD TO THE COLLECTION OF BOLTZMANN'S CONSTANT, KELVIN, KINETIC ENERGY, WAVE NUMBER, AND KILO- (AS IN KILOGRAM). THE SYMBOL k IS TRADITIONALLY USED FOR THERMAL CONDUCTIVITY. DO YOUR BEST TO KEEP THEM SORTED OUT.

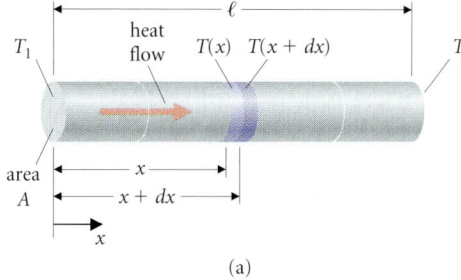
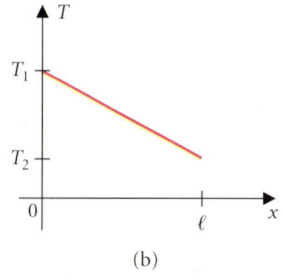

■ FIGURE 21.1
(a) When the ends of a rod are at different temperatures T_1 and T_2, there is a temperature gradient along it. With the x-axis chosen to be parallel to the rod, the temperature gradient is $dT/dx = (T_2 - T_1)/\ell$. Heat flows along the rod from the hot end to the cold end. The heat transfer rate H is proportional to the cross-sectional area A and $|dT/dx|$. (b) Temperature as a function of position along the rod.

THE RATE OF HEAT TRANSFER, LIKE THE RATE OF DOING WORK, IS A POWER.

ONCE AGAIN, DIAMOND IS THE EXCEPTION.

WE'LL DISCUSS THE COLLISIONS IN CHAPTER 22.

where A is the cross-sectional area normal to the direction of heat flow. The heat flux is measured in watts. Values of k for various substances are given in ● Table 21.1.

Generally speaking, thermal conductivities are greatest in metals, smaller for other solids and liquids, and lowest for gases. In metals, heat is transferred primarily by highly mobile electrons that travel relatively large distances between collisions (much greater than the separation of the atoms). In nonmetallic solids, interactions among nearby molecules result in less effective heat transfer. In gases, the molecules are far apart. They travel distances between interactions that are comparable to those of electrons in metals, but collisions occur much less often and conduction is inefficient. Heat transfer through liquids and gases is enhanced by large-scale motions in a process called convection (§21.2). When these motions are prevented, gases make good heat insulators. Goose down clothing is both light and warm. Heat flow is limited by the low conductivity of air trapped between the thin fibers of the feathers. In a more spectacular example, called the Leidenfrost effect, water droplets on a very hot plate bounce around and evaporate surprisingly slowly (■ Figure 21.2). Gas vaporizes at the bottom of the drops and forms an insulating layer that limits heat transfer into the liquid water.

The ceramic tiles that cover the bottom of the space shuttle were specially developed to have low thermal conductivity. Specifically, they are about 93% air. Seconds after emerging from an oven with an interior temperature of 1200°C, a cube of the ceramic can be safely held by its corners.

■ **FIGURE 21.2**
The Leidenfrost effect. Water droplets on a hot plate bounce around. Water vaporizes at the bottom of the drops. Because of its poor thermal conductivity, the water vapor forms an insulating layer that slows heat flow into the drop and prevents it from rapidly evaporating.

TABLE 21.1 Thermal Conductivity
(at 15°C and 1 atm unless noted)

Substance	Conductivity (W/m·K)	Substance	Conductivity (W/m·K)
SOLIDS		**GASES**	
Aluminum	240	Air	0.025
Copper	400	Ammonia	0.023
Diamond	990	Carbon dioxide	0.016
Gold	320	Chlorine	0.008
Iron	80	Helium	0.147
Lead	35	Hydrogen	0.182
Platinum	72	Methane	0.033
Silicon	150	Neon	0.047
Silver	430	Nitrogen	0.025
Stainless steel	30	Oxygen	0.026
Oak (across grain)	0.147	Water vapor	0.017
Ice (at 0°C)	1.7	**LIQUIDS**	
Corkboard	0.04		
Glass wool	0.042	Mercury	8.30
Felt	0.037	Water (15°C)	0.59
Sandstone	4	(60°C)	0.66
Marble	2–3	Ethyl alcohol (20°C)	0.17
Concrete	0.9–1.3	Carbon tetrachloride	0.11
Glass	0.7–0.9		

Digging Deeper

Why is Conductive Heat Flux Proportional to the Temperature Gradient?

To be specific, we'll discuss heat conduction in metals, although the basic ideas are similar in all substances. Energy is carried across a surface at coordinate x (■ Figure 21.3) primarily by electrons moving through the metal. The electrons travel an average distance λ between collisions. Particles crossing the surface to the right have energy corresponding to the temperature $T(x - \lambda)$ at the site of their last collision. [*Remember*: T is a function of x. The notation $T(x)$ means T measured at x.] Similarly, particles crossing to the left have temperature $T(x + \lambda)$. Equal numbers of particles pass in each direction, so the net energy transfer to the right is proportional to the temperature difference: $H \propto |T(x - \lambda) - T(x + \lambda)|$. From the definition of derivative:

$$\frac{dT}{dx} = \lim_{\lambda \to 0} \frac{T(x + \lambda) - T(x - \lambda)}{2\lambda}.$$

So:
$$H \propto 2\lambda |dT/dx|.$$

You may have noticed that there is a direction associated with both temperature gradient and heat flux: they are both vectors, and they have opposite directions. We shall be concerned primarily with the magnitudes of these quantities.

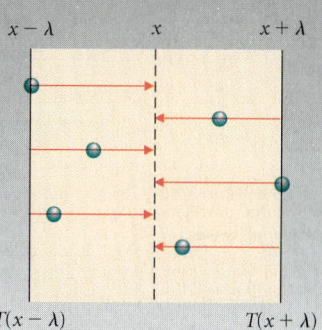

■ **FIGURE 21.3**
Heat conduction in a metal. Electrons crossing the imaginary plane at x have energy characteristic of the temperature at the site of their last collision, an average distance λ away: $T(x + \lambda) = T(x) + \lambda\, dT/dx$. Since equal numbers of particles travel in each direction, the net heat flux is proportional to the temperature difference $T(x - \lambda) - T(x + \lambda) = -2\lambda\, dT/dx$.

EXAMPLE 21.1 ♦ A copper rod of cross-sectional area 5.0 cm² is 0.85 m long. One end is at a temperature of 405 K and the other is at 310 K. What is the rate of heat flow along the rod?

MODEL The rate of heat flow is determined by the temperature gradient, which we assume is uniform.

SETUP The temperature gradient in the rod is:

$$\frac{dT}{dx} = \frac{T_2 - T_1}{\ell} = \frac{405 \text{ K} - 310 \text{ K}}{0.85 \text{ m}} = 110 \text{ K/m}.$$

SOLVE Using eqn. (21.1) and k for copper from Table 21.1, the heat flow is:

$$H = (400 \text{ W/K·m})(5.0 \times 10^{-4} \text{ m}^2)(110 \text{ K/m}) = 22 \text{ W}.$$

ANALYZE The result for H is correct to two significant figures because the value of k, like most of the others in the table, is accurate to two significant figures. ■

21.1.2 Steady State Heat Flow

Conduction through a copper rod with specified end temperatures (Example 21.1) is an example of *steady state* heat flow: the temperature varies with position but is constant in time. As a result, heat flows through each slice of the rod at the same rate. This principle is key to all steady state heat flow problems. Let's see why it's true.

We choose a coordinate system with the x-axis parallel to the direction of heat flow and focus on a small piece of an object, bounded by surfaces at x and $x + dx$ (■ Figure 21.4). Suppose more heat flows inward through the surface at x than leaves through the surface at $x + dx$.

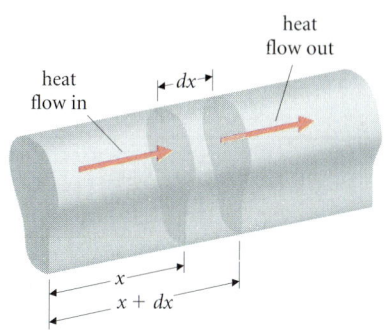

■ **FIGURE 21.4**
Steady state heat flow. The temperature in the region between x and $x + dx$ doesn't change, so the heat flow in at x equals the heat flow out at $x + dx$.

THIS ARGUMENT CLOSELY FOLLOWS THE DERIVATION OF THE CONTINUITY EQUATION FOR FLUIDS. SEE §13.5.2.

Then the internal energy between the surfaces would increase and the temperature there would rise. But if the temperature changes in time, the heat flow cannot remain steady. Similarly, heat outflow greater than inflow would decrease the temperature and disrupt steady flow.

> In a steady state, heat flow through any cross-sectional area normal to the temperature gradient is constant.

Take care in applying this principle when the system does not have a uniform cross-sectional area. The heat flow remains constant, but the temperature gradient varies with position.

EXAMPLE 21.2 ♦♦♦ A hot water pipe of radius 5.0 cm is surrounded by glass wool 1.0 cm thick (■ Figure 21.5). The pipe carries water at 82°C through a basement where the air temperature is 15°C. What is the heat loss per meter of pipe?

MODEL We assume the pipe itself is in thermal equilibrium with the water, at 82°. Heat flows radially outward through a length ℓ of the pipe, passing through concentric cylinders of increasing radius r and area $A(r) = 2\pi r\ell$. The temperature gradient dT/dr varies with radius in steady state heat flow because the constant heat flux H is proportional to $A(r)dT/dr$.

SETUP We use eqn. (21.1):

$$H = k2\pi r\ell \left|\frac{dT}{dr}\right| = \text{constant.} \quad (i)$$

The heat flux H is unknown but constant. We do know the temperature at the inner and outer radii, so we integrate the equation for H. Since the temperature decreases outward, dT/dr is negative:

$$\frac{dT}{dr} = \frac{-H}{k2\pi r\ell}.$$

Integrating,

$$T(r_2) - T(r_1) = \int_{r_1}^{r_2} \frac{dT}{dr}dr = \frac{-H}{k2\pi\ell}\int_{r_1}^{r_2}\frac{dr}{r} = \frac{-H}{k2\pi\ell}\ln\left(\frac{r_2}{r_1}\right).$$

SOLVE Using the value of k from Table 21.1, the heat loss per unit length is:

$$\frac{H}{\ell} = \frac{2\pi k[T(r_1) - T(r_2)]}{\ln(r_2/r_1)} = \frac{2\pi(0.042 \text{ W/K·m})(82 - 15) \text{ K}}{\ln[(6.0 \text{ cm})/(5.0 \text{ cm})]} = 97 \text{ W/m.}$$

ANALYZE The heat flux is proportional to the temperature difference between the inside and outside of the pipe, as expected. It would be a good idea to investigate whether steady state heat flow is a good approximation. See Exercise 21.1.

The back-of-the-envelope approach to this problem also gives a good result. In eqn. (i) for H, we use the average radius, $\langle r \rangle = 5.5$ cm, for r, and replace dT/dr with $\Delta T/\Delta r$. Comparing the resulting expression for H/ℓ with the exact solution, $\Delta r/\langle r\rangle = 1.0/5.5 = 0.18$ replaces $\ln(r_2/r_1) = \ln(1.2) = 0.18$! There is no difference to two significant figures.

EXERCISE 21.1 ♦♦ By how much would the rate of energy loss lower the temperature of the water in 1.0 m of the pipe in 1.0 s?

EXAMPLE 21.3 ♦♦♦ A bathysphere is a device used for deep sea exploration (■ Figure 21.6). We model one as an aluminum spherical shell with inner and outer radii $r_1 = 2.00$ m and $r_2 = 2.33$ m. If the bathysphere is used at a depth where the temperature is 3°C and the interior is to be kept at 15°C, how much power must the heater supply?

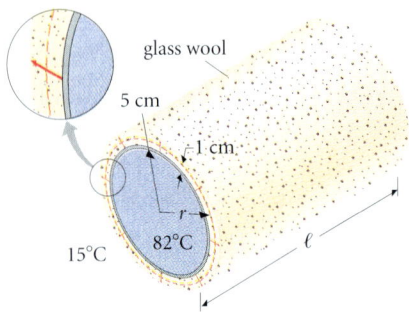

■ **FIGURE 21.5**
A hot water pipe carries water at 82°C through a basement at 15°C. Heat flows outward through the 1.0-cm-thick insulating cover. The area perpendicular to the heat flow increases with distance from the center of the pipe. The area of the dashed surface for the length ℓ of pipe shown is $2\pi r\ell$.

A BATHYSPHERE ALSO APPEARED IN EXAMPLE 13.5.

MODEL Heat is conducted out from the interior to the colder ocean through concentric spherical shells of decreasing temperature and increasing radius. The purpose of the heater is to maintain the internal temperature, so we may assume a steady state in which the constant heat flux H is equal to the power supplied by the heater.

SETUP At a radius r, the area normal to the direction of heat flow is $A(r) = 4\pi r^2$. In steady state heat flow, H is independent of r; that is,

$$H = A(r)k \left| \frac{dT}{dr} \right| = 4\pi k r^2 \left| \frac{dT}{dr} \right| = \text{constant}.$$

Since temperature decreases outward:

$$\frac{dT}{dr} = \frac{-H}{4\pi k r^2}.$$

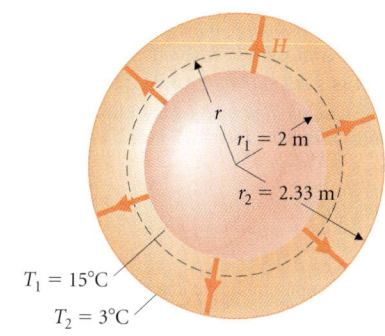

■ **FIGURE 21.6**
Heat flow through a spherical shell. The area of the dashed surface is $4\pi r^2$.

As in Example 21.2, we make use of the given temperatures at the boundaries by integrating:

$$T_2 - T_1 = \int_{r_1}^{r_2} \frac{dT}{dr} dr = \frac{-H}{4\pi k} \int_{r_1}^{r_2} \frac{1}{r^2} dr = \frac{-H}{4\pi k} \left(\frac{1}{r_1} - \frac{1}{r_2} \right).$$

Solving for H, we have:

$$H = 4\pi k (T_1 - T_2) \left(\frac{r_1 r_2}{r_2 - r_1} \right) \quad (21.2)$$

$$= 4\pi (240 \text{ W/K} \cdot \text{m})(12 \text{ K}) \frac{(2.00 \text{ m})(2.33 \text{ m})}{0.33 \text{ m}}$$

$$= 510 \text{ kW}.$$

ANALYZE This power level is quite impractical. It could be reduced considerably by insulating the bathysphere. (See Problem 26.) ■

Steady state heat flow is a good approximation for many practical applications, such as housing insulation, where external conditions change slowly compared with the time required to reach steady state. When conditions change rapidly, the conducting material gains or loses energy and heat flow is a function of position and time (see Problem 50).

21.1.3 Thermal Resistance

In applications to building insulation, eqn. (21.1) is used to give temperature difference in terms of heat flow:

$$\Delta T = \frac{\Delta x}{kA} H = RH, \quad (21.3)$$

where the *thermal resistance* of the insulation is

$$R \equiv \frac{\Delta x}{kA}. \quad (21.4)$$

In construction, thermal insulation is described by its R-factor (● Table 21.2), which is independent of the area:

$$R_f \equiv RA = \frac{\Delta x}{k}. \quad (21.5)$$

TABLE 21.2
Thermal Resistance of Building Materials

Material	R_f (m²·K/W)
Plaster board, 0.95 cm thick	0.056
Plywood, 1.27 cm thick	0.109
Tile	0.088
Asphalt roll roofing	0.026
Asphalt shingles	0.078
Window, single pane	0.159
Glass wool, 10 cm thick	2.5

The terminology is inspired by electrical circuits (Chapter 26). Heat flow is related to thermal resistance in the same way that electric current is related to electrical resistance.

EXAMPLE 21.4 ♦♦ Two insulators with R-factors $R_{f,1}$ and $R_{f,2}$ are used in *series*—that is, one behind the other, as shown in ■ Figure 21.7. Show that the R-factor of the combination is $R_{f,s} = R_{f,1} + R_{f,2}$; that is, thermal resistances in series add.

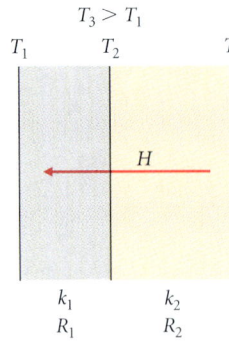

■ **Figure 21.7**
Two heat-insulating layers used in series. The heat flow is the same through each layer. The thermal resistance of the combination is the sum of the resistances of the individual layers.

MODEL The thermal resistance of an insulating layer relates heat flow to the temperature difference across that layer. The effective resistance relates heat flow to the temperature difference across the entire combination. In a steady state, the heat flow across each layer is the same.

SETUP Label the temperatures at the boundaries T_1, T_2, and T_3. Then the heat flux H across an area A of either insulator is:

$$H = k_1 A\left(\frac{T_2 - T_1}{\Delta x_1}\right) = k_2 A\left(\frac{T_3 - T_2}{\Delta x_2}\right). \quad \text{(i)}$$

We want to relate H to $T_3 - T_1$, by eliminating T_2. First, we find T_2 in terms of the other two temperatures and the R-factors. Using eqn. (21.5) in eqn. (i):

$$\frac{A}{R_{f,1}}(T_2 - T_1) = \frac{A}{R_{f,2}}(T_3 - T_2).$$

Rearranging, $\quad R_{f,2}(T_2 - T_1) = R_{f,1}(T_3 - T_2).$

Solving for T_2:
$$T_2 = \frac{R_{f,1} T_3 + R_{f,2} T_1}{R_{f,1} + R_{f,2}}.$$

Substituting back into eqn. (i), we obtain H in terms of T_1 and T_3:

$$H = \frac{A}{R_{f,1}}\left(\frac{R_{f,1} T_3 + R_{f,2} T_1}{R_{f,1} + R_{f,2}} - T_1\right)$$

$$= A\frac{T_3 - T_1}{R_{f,1} + R_{f,2}} = A\frac{\Delta T}{R_{f,s}}.$$

SOLVE Comparing with eqns. (21.3) and (21.5):

$$R_{f,s} = A\,\Delta T/H = R_{f,1} + R_{f,2}.$$

ANALYZE These thermal resistances add because the heat flows through each of them sequentially. When used side by side, more heat flows through the least resistant material, as shown in the next exercise. ■

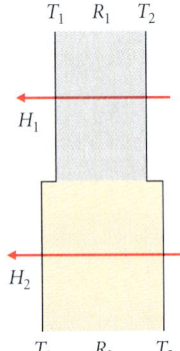

■ **Figure 21.8**
Heat-insulating layers used in parallel. The total heat flow is the sum of the heat flows through each layer, but the temperature difference across each layer is the same. The heat flow is greater through the material of lesser resistance.

EXERCISE 21.2 ♦♦ Show that for thermal resistances used side by side (in *parallel*, ■ Figure 21.8):

$$1/R_p = 1/R_1 + 1/R_2.$$

(*Hint*: The temperature difference across each layer is the same.)

21.2 Convection

Convection is heat transfer by large-scale motion of a fluid. One of the most common situations that leads to convection is a fluid heated at the bottom. Because most fluids expand when heated, the hot fluid is less dense than the cold fluid on top of it, and buoyant force causes it to flow upward toward the surface. Cold fluid flows in to take its place and is then heated so that the flow continues (■ Figure 21.9).

Convection is an important form of heat transfer both in the Sun (■ Figure 21.10) and in a kettle of boiling water. Convective flows in the Earth's interior cause Earth's magnetic field and drive continental drift. Convection is also responsible for many meteorological phenomena. On a warm, clear summer day over moderately flat terrain, as the Sun warms the ground, the temperature gradient in the atmosphere increases. Toward midday, the necessary temperature gradient reaches a critical value called the adiabatic value (see *Digging Deeper*), and the thermals so prized by the glider pilot begin to develop. Because atmospheric pressure decreases with altitude, the air expands further as it rises, and so it cools. If the air is sufficiently humid, water vapor condenses and clouds form at the top of the rising columns (■ Figure 21.11).

While the basic mechanism of convection is easy to describe, no simple theory of the resulting fluid motion has been found. Convective motions often have fascinating regularity. Fluid confined to a thin layer and heated uniformly from below develops a pronounced hexagonal pattern (■ Figure 21.12). Clouds frequently form in regular patterns known as cloud streets (■ Figure 21.13). Developing a complete understanding of such convective motions remains a fertile field of research.

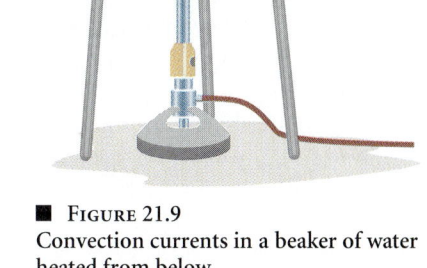

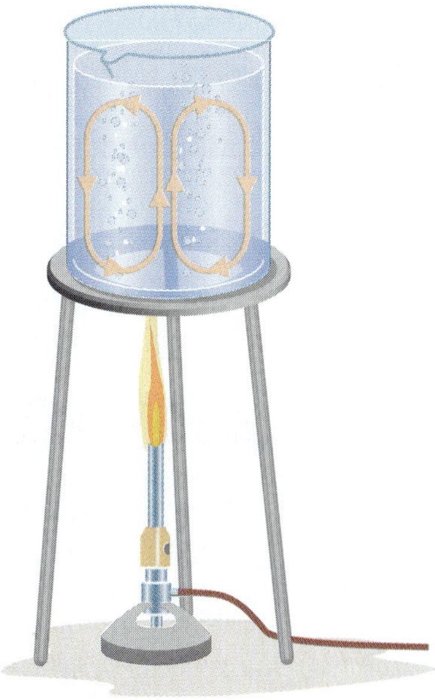

■ **Figure 21.9**
Convection currents in a beaker of water heated from below.

EVEN THE SIMPLEST THEORIES OF CONVECTION SUGGEST THAT CHAOTIC MOTION OCCURS.

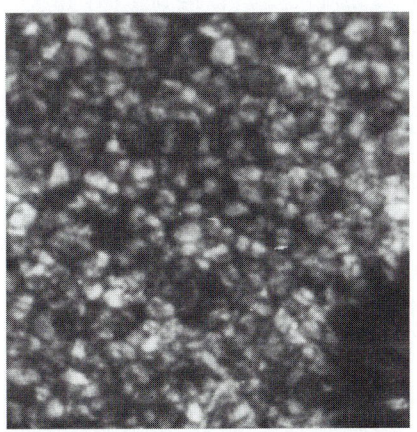

■ **Figure 21.10**
Convection is an important process for transferring heat outward in the Sun. Solar *granules* visible in this photograph are believed to be the tops of convective cells in the solar atmosphere. The dark boundary of each hexagonal cell is cool gas flowing back downward.

■ **Figure 21.11**
Cumulus clouds mark the site of convective currents in the Earth's atmosphere.

■ **Figure 21.12**
The hexagonal pattern of these convective cells is believed to be due to effects of surface tension.

■ **Figure 21.13**
Cloud streets indicate the regularity of the convective motions that produce them.

Digging Deeper

CONDITIONS FOR CONVECTION

In 1916, Karl Schwarzschild gave an argument that indicates when convection occurs. Imagine displacing a small blob of gas upward through the surrounding atmosphere (■ Figure 21.14). If the blob continues to rise on its own, the atmosphere is unstable and convection occurs. Conversely, if the blob falls back down, the atmosphere is stable against convection.

The forces acting on the blob are its weight and the buoyant forces due to the pressure of the surrounding air. According to Archimedes' principle (§13.4), the buoyant force exceeds the blob's weight if the blob is less dense than the surrounding air. Thus convection occurs if a displaced blob of air is less dense than its new surroundings.

Pressure decreases with height in the atmosphere (§13.3.4). Also, the temperature usually decreases with altitude in order to transfer heat upward from ground heated by the Sun. The pressure in a rising blob remains equal to the pressure in the surrounding atmosphere, but its temperature does not. The blob can respond to pressure changes at its surface by expanding or contracting at the speed of sound. Except in the unlikely case that the blob moves supersonically, it has time to expand and adjust its pressure. Heat transfer via conduction between the blob and the surrounding atmosphere is slow. So, to a good approximation, the blob expands adiabatically as its pressure decreases. With these models, let's compute the densities of blob and atmosphere and compare them.

First rewrite eqn. (19.22) in terms of the density

$$\rho = NM/V,$$

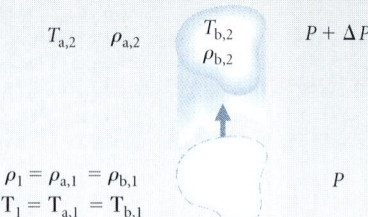

■ **FIGURE 21.14**
Conditions for convection. When a small blob of air is displaced upward, it continues to rise if it is less dense than its new surroundings. At the original position, both blob and atmosphere have density ρ_1 and temperature T_1. At the higher position, the atmosphere has density $\rho_{a,2} = \rho_1 + \Delta\rho_a$, while the blob has density $\rho_{b,2} = \rho_1 + \Delta\rho_b$.

where m is the molecular mass, N is the number of molecules in the blob, and V is its volume. We obtain a relation between pressure and density in an adiabatic expansion:

$$P\rho^{-\gamma} = \text{constant}.$$

Differentiating it, we obtain the change $\Delta\rho_b$ of the blob's density ρ_b as it rises from position 1 to position 2:

$$(\Delta P)\rho_b^{-\gamma} - \gamma\rho_b^{-\gamma-1}(\Delta\rho_b)P = 0,$$

So

$$\frac{\Delta\rho_b}{\rho_b} = \frac{\Delta P}{\gamma P}.$$

Similarly, we find the change $\Delta\rho_a$ of atmospheric density ρ_a by differentiating the ideal gas law in the form $\ln \rho = \ln P - \ln T - \ln(k/m)$:

$$\frac{\Delta\rho_a}{\rho_a} = \frac{\Delta P}{P} - \frac{\Delta T_a}{T_a}.$$

Notice that $\Delta\rho$, ΔP, and ΔT are all negative. The densities of the blob and the atmosphere at position 2 are:

$$\rho_{b,2} = \rho_1 + \Delta\rho_b = \rho_1 + \frac{\rho_1}{\gamma}\left(\frac{\Delta P}{P}\right),$$

and

$$\rho_{a,2} = \rho_1 + \Delta\rho_a$$
$$= \rho_1 + \rho_1\left(\frac{\Delta P}{P} - \frac{\Delta T_a}{T_a}\right).$$

Thus, the difference between blob and atmosphere is:

$$\frac{\rho_{b,2} - \rho_{a,2}}{\rho_1} = \frac{\Delta P}{P}\left(\frac{1}{\gamma} - 1\right) + \frac{\Delta T_a}{T_a}$$

$$= -\left(\frac{\Delta T}{T}\right)_{\text{adiabatic}} + \frac{\Delta T_a}{T_a}.$$

(We used the result of Problem 19.80 for the temperature change in an adiabatic process.)

Whether convection occurs ($\rho_{b,2} < \rho_{a,2}$) or the atmosphere is stable ($\rho_{b,2} > \rho_{a,2}$) depends on the temperature gradient in the atmosphere. We have seen that the heat flow by conduction increases as the temperature gradient increases, or to put it another way, a large heat flow *requires* a large temperature gradient. The same kind of relationship holds for heat transport by radiation (§21.3). But if the required temperature gradient is too large,

$$|\Delta T/T| > |(\Delta T/T)_{\text{adiabatic}}|,$$

then the fluid is unstable and heat is transported by convection instead.

21.3 RADIATION

21.3.1 The Nature of Thermal Radiation

The dull red glow of a poker pulled from the fire or of an electric stove's heating element, and the warmth we sense without touching them, result from electromagnetic radiation (Chapter 33) that carries energy away from the hot objects. Radiation is an important heat transport mechanism that operates in solid, liquid, and gaseous systems, and can transfer heat between systems that are not otherwise in physical contact, even across a vacuum. It is the mechanism by which the Sun heats the Earth. A full understanding of radiation processes requires a knowledge of quantum theory, so we present a simplified discussion here.

An individual molecule has a certain probability of changing its energy by emitting or absorbing electromagnetic radiation. For example, a molecule can change its vibrational state (§19.7) by radiation as well as by collisions. The frequency of the emitted radiation depends on the kind of molecules involved as well as the temperature. However, if the molecules are densely packed in the system, the radiation is reabsorbed and re-emitted many times before it escapes. This usually happens in solids, provided that T is not too large. After many interactions, the radiation escaping from the surface of the system takes on a characteristic form that depends *only* on the temperature. This form is called *blackbody radiation* because a body that absorbs all the radiation that falls on it, and so looks "black," emits radiation of this characteristic form. A dark, dull solid body (like a poker) is an everyday example that approximates this behavior.

Blackbody radiation has two major features: hot bodies radiate more than cold ones, and hot bodies radiate more of their energy at high frequencies (short wavelengths) than cold ones. Both characteristics are a result of the average energy of the molecules increasing with temperature. The distribution of the radiated power as a function of wavelength for three different temperatures is shown in ■ Figure 21.15.

More radiation emerges in a wavelength interval at the peak of each curve in Figure 21.15 than in other equal wavelength intervals. The wavelength λ_{max} of the peak is related to the temperature of the radiating body through the Wien displacement law:

$$\lambda_{max} T = 2.898 \times 10^{-3} \text{ K} \cdot \text{m}. \qquad (21.6)$$

The peak of the distribution occurs at shorter wavelength as T increases. That is why a lump of metal heated in a furnace changes color as the temperature increases, glowing first red, and later yellow.

EXAMPLE 21.5 ◆ The Sun's radiation peaks at a wavelength of 460 nm. What is the temperature of the region of the Sun from which the radiation emerges (conventionally called the *surface*)?

MODEL The temperature is related to the wavelength of peak emission by the Wien displacement law.

SETUP and SOLVE From eqn. (21.6), with $\lambda_{max} = 4.6 \times 10^{-7}$ m,

$$T = \frac{2.898 \times 10^{-3} \text{ K} \cdot \text{m}}{4.6 \times 10^{-7} \text{ m}} = 6300 \text{ K}.$$

ANALYZE This is called the *color temperature* of the Sun.

EXERCISE 21.3 ❖ The star Sirius appears to be bluish-white while Betelgeuse appears red. Compare the temperatures of Sirius and Betelgeuse.

EXERCISE 21.4 ❖ Some red dye molecules are thrown into a cavity at temperature T. What happens to the color of the emerging radiation?

UNFORTUNATELY, IN POPULAR USAGE THE TERM *RADIATION* HAS BECOME ALMOST SYNONYMOUS WITH "RADIOACTIVITY." RADIATION HERE MEANS THE EMISSION OF ELECTROMAGNETIC WAVES ("LIGHT") WITH ANY FREQUENCY FROM LOW-FREQUENCY RADIO TO γ RAYS.

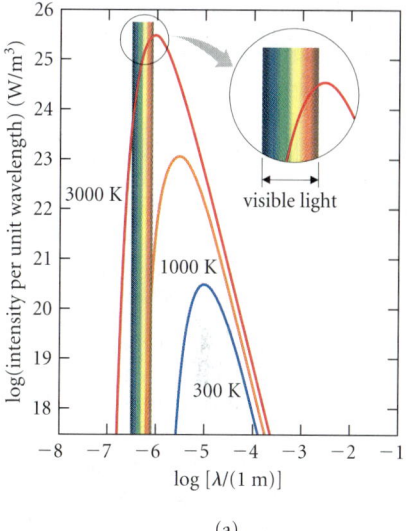

(a)

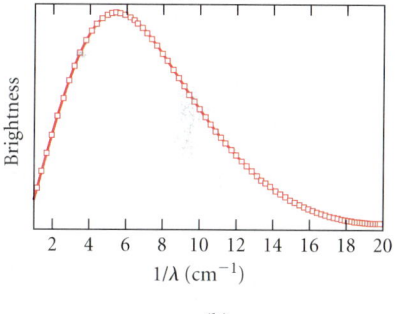

(b)

■ **FIGURE 21.15**
(a) Blackbody radiation. This graph shows how radiated power is distributed in wavelength at three different temperatures. The frequency range corresponding to visible light is indicated. At room temperature (about 300 K), most of the radiation occurs at longer wavelengths, called infrared. The position of the peak of each curve is related to the temperature through the Wien displacement law: $\lambda_{max} T = 3$ mm·K. (b) The universe is the best naturally occurring example of a blackbody. Observations from the Cosmic Background Explorer (squares) fit the theoretical curve almost perfectly. The observed temperature is 2.735 ± 0.06 K.

21.3.2 Radiation and Heat Transport

The rate at which radiation transfers heat from a hot blackbody is the total power radiated over the whole electromagnetic spectrum. The total power radiated per unit area, at all frequencies, is represented by the area under the curves in Figure 21.15. The resulting power is given by the Stefan–Boltzmann law:

POWER RADIATED BY A BLACKBODY.

$$\text{Power radiated} = \sigma A T^4, \qquad (21.7)$$

where the Stefan–Boltzmann constant $\sigma = 5.67 \times 10^{-8}$ W/m$^2 \cdot$K^4, A is the area of the body's radiating surface, and T is its temperature (in kelvin).

REMEMBER: INTENSITY = POWER PER UNIT AREA (§16.3.2).

EXAMPLE 21.6 ♦♦ The intensity of the Sun's radiation at the top of the Earth's atmosphere is 1400 W/m^2. The Earth is 1.5×10^{11} m from the Sun. What is the radius of the Sun?

THIS IS THE INVERSE SQUARE LAW; SEE §16.3.3.

MODEL The total power emitted by the Sun is the intensity at the Earth times the area of a sphere whose radius is the distance D between the Earth and the Sun. The Stefan–Boltzmann law relates this power to the Sun's surface area and temperature.

SETUP From eqn. (21.7), using the temperature found in Example 21.5, and expressing the Sun's area in terms of its radius, $A = 4\pi R^2$, we have:

$$\sigma 4\pi R^2 T^4 = \text{power} = 4\pi D^2 I.$$

THE TERM *CAVITY RADIATION* IS ALSO USED BECAUSE ONE WAY TO ENSURE THAT MANY ABSORPTIONS AND RE-EMISSIONS OCCUR IS TO HAVE A HOLLOW BODY WITH A TINY HOLE IN IT. RADIATION EMITTED BY MOLECULES AT THE INSIDE SURFACE OF THE CAVITY RATTLES AROUND A LOT INSIDE BEFORE FINDING ITS WAY OUT THE HOLE. THE RADIATION EMERGING FROM THE HOLE HAS THE CHARACTERISTICS OF BLACKBODY RADIATION.

SOLVE Then:

$$R = \frac{D}{T^2} \sqrt{\frac{I}{\sigma}} = \frac{1.5 \times 10^{11} \text{ m}}{(6300 \text{ K})^2} \sqrt{\frac{1400 \text{ W/m}^2}{5.67 \times 10^{-8} \text{ W/m}^2 \cdot \text{K}^4}} = 5.9 \times 10^8 \text{ m}.$$

ANALYZE This result is smaller than the value 7.0×10^8 m given on the inside front cover because the Sun is not a perfect blackbody (the emitted spectrum is not exactly as shown in Figure 21.15). The temperature computed from the total emitted radiation (called the *brightness temperature*) is slightly less than the color temperature.

A real body radiates less efficiently than a theoretical blackbody:

THE EMISSIVITY OF A REAL OBJECT IS ALSO A FUNCTION OF FREQUENCY, SO THE EMITTED SPECTRUM DIFFERS SLIGHTLY FROM FIGURE 21.15. WE SHALL IGNORE THIS COMPLICATION HERE.

$$\text{Power radiated} = \epsilon \sigma A T^4, \qquad (21.8)$$

where the *emissivity* ϵ is less than unity. For bodies that have dark, rough surfaces, $\epsilon \approx 1$, whereas for smooth, shiny objects, $\epsilon \ll 1$. In general, an object that reflects light well absorbs and emits weakly.

Since radiation can be transmitted over large distances, a system also absorbs radiation from its surroundings. For example, a coffee cup (cf. Chapter 19) at temperature T, receives radiation from the environment (here the table and the walls of the room) at temperature T_{env}. The net heat flow from a system at temperature T in an environment at T_{env} is:

$$\begin{aligned} H_{\text{net}} &= \text{radiation emitted} - \text{radiation absorbed} \\ &= dQ/dt = \epsilon \sigma A(T^4 - T_{\text{env}}^4), \end{aligned} \qquad (21.9)$$

where A is the surface area of the radiating object and ϵ is its emissivity.

EXAMPLE 21.7 ♦♦ Model yourself as a cylindrical grey body of temperature $T = 98°$F, $\epsilon = 0.5$. What is your temperature in kelvin? Ignoring production of body heat, how fast would your temperature drop due to radiation losses if you were in a room at 300 K?

MODEL Our bodies lose heat by radiation according to eqn. (21.9). We'll need to estimate the surface area of your body in order to calculate its heat loss. To calculate the temperature drop, we'll also need your heat capacity. We'll use the specific heat of water—the major constituent of your body.

SETUP Your body temperature is $T = [(98 - 32)°F][\frac{5}{9}°C/°F] = 37°C = 310$ K. The surface area of your body is $A \approx 2\pi rh$. Taking a geometric mean of the authors as typical dimensions for humans, $h \sim 2$ m, $2\pi r \sim 0.9$ m, and $M \sim 70$ kg. In a room at $T_{env} = 300$ K, the net heat loss by radiation is:

$$H_{net} = \epsilon A \sigma (T^4 - T_{env}^4)$$
$$\approx 0.5(2 \text{ m})(0.9 \text{ m})(5.7 \times 10^{-8} \text{ W/m}^2 \cdot \text{K}^4)[(310^4 - 300^4) \text{ K}^4]$$
$$= 60 \text{ W}.$$

SOLVE This heat loss reduces the body's energy at a rate:

$$H = Mc \left| \frac{dT}{dt} \right|,$$

or

$$60 \text{ W} = (70 \text{ kg})(4186 \text{ J/kg} \cdot \text{K}) \left| \frac{dT}{dt} \right|.$$

Therefore:

$$\left| \frac{dT}{dt} \right| = 2 \times 10^{-4} \text{ K/s}.$$

COMPARE EQN. (19.16), WHICH RELATES INTERNAL ENERGY TO TEMPERATURE AND SPECIFIC HEAT.

ANALYZE It would take about

$$(10 \text{ K})/(2 \times 10^{-4} \text{ K/s}) = 14 \text{ h}$$

for the body to reach room temperature due to radiation losses alone. Then radiation emitted by the body would be balanced by radiation absorbed from the room. But that doesn't happen—see the next exercise. ∎

EXERCISE 21.5 ♦♦ How many food Calories would it take to maintain your temperature at 98°F in spite of radiation loss? What important effects have you ignored in this calculation?

A FOOD CALORIE ACTUALLY PROVIDES 1 kcal OF HEAT BY CHEMICAL MEANS. SEE *DIGGING DEEPER* IN §20.4.1.

21.4 AN EMPIRICAL APPROACH TO HEAT TRANSFER: NEWTON'S LAW OF COOLING

In many circumstances, the temperature difference between a cooling system and its surroundings is not very large: $T = T_{env} + \Delta T$. The heat loss by radiation (eqn. 21.9) is proportional to $T^4 - T_{env}^4$, which is approximately:

$$(T^4 - T_{env}^4) = (T^2 + T_{env}^2)(T^2 - T_{env}^2)$$
$$\approx 2T^2(T + T_{env})(T - T_{env})$$
$$\approx 4T^3(T - T_{env}) = 4T^3 \Delta T.$$

EXAMPLE 21.7 ILLUSTRATES ONE SUCH SITUATION.

HERE WE ARE IGNORING THE DIFFERENCE BETWEEN T AND T_{env} EXCEPT WHEN THEY ARE SUBTRACTED.

The body cools in direct proportion to the temperature difference between it and its surroundings. A similar dependence holds for both conduction (e.g., eqn. 21.2) and convection, so the relationship

$$H \propto (T - T_{env})$$

holds regardless of the dominant cooling process. Newton discovered this relationship empirically, and so it is called Newton's law of cooling. It is a reasonable approximation so long as $T - T_{env} \ll T$ (on the Kelvin scale).

EXAMPLE 21.8 ♦♦♦ In a physics experiment conducted in a laboratory at 20°C, a beaker of water is observed to cool from 95°C to 85°C in 3.0 min. How long does it take to cool to 70°C?

MODEL The experimental data may be used to determine the unknown constant in Newton's law of cooling. We integrate the law to find the time to cool through a specific temperature difference.

SETUP Since $H = dQ/dt = C\, dT/dt$, we may write Newton's law of cooling as:

Compare eqn. (20.7).

$$\frac{dT}{dt} = -B(T - T_{\text{env}}),$$

where B is a constant that depends on the cooling rate and the heat capacity of the system. This is a differential equation for T (cf. §13.3.4) with solution:

$$T - T_{\text{env}} = (T_i - T_{\text{env}})\exp(-Bt),$$

Differentiate this expression for T and convince yourself that it solves the equation.

where $T_i = 95°C$ is the initial temperature. We find B from the given time to cool to 85°C:

$$(85 - 20)\, \text{K} = [(95 - 20)\, \text{K}]\exp[-B(3\text{ min})].$$

Thus: $$B = \frac{-1}{3.0\text{ min}}\ln\left(\frac{65}{75}\right) = 0.0477\,/\text{min}.$$

SOLVE Then the total time to cool to 70°C is:

$$t = \frac{-1}{0.0477\,/\text{min}}\ln\left(\frac{70 - 20}{75}\right) = 8.5\text{ min}.$$

The water takes 8.5 min to cool from 95°C to 70°C.

ANALYZE If it takes 3 min to cool through 10 K, it seems reasonable that it would take almost three times as long (8.5 min) to cool through two and a half times the temperature difference (25 K).

The constant B is likely to be a function of T, so a given empirical formula only works over a limited temperature range.

Chapter Summary

Where Are We Now?

We have studied the transfer of heat between systems and have presented useful expressions for heat transfer by conduction and radiation. Convection is more complicated and is not described by a simple formula. Empirical relations derived from measurements of heat transfer rates are often used when convection is important.

What Did We Do?

Heat is transferred by conduction, convection, and radiation.
- Conduction—transfer of heat by molecular and electron collisions.
- Convection—transfer of heat by bulk motion of fluid.
- Radiation—transfer of heat by emission and absorption of electromagnetic waves.

The rate of heat transfer by conduction is proportional to the temperature gradient dT/dx, and to the cross-sectional area A perpendicular to the direction of heat flow:

$$H_{\text{cond}} = kA\left|\frac{dT}{dx}\right|.$$

The proportionality constant k is the thermal conductivity (Table 21.1). In a steady state, the heat transfer is the same through each surface perpendicular to the direction of heat flow. Building materials are rated according to their R-factor $\Delta x/k$ and have thermal resistance $R = \Delta x/kA$. When used back to back (in series), resistances add: $R_s = R_1 + R_2$. Used side by side (in parallel), the inverses add to give the inverse of the total resistance: $1/R_p = 1/R_1 + 1/R_2$.

Hot, dense bodies radiate power with a well-defined spectrum that depends only on the temperature. The wavelength at which peak radiation occurs varies with temperature according to the Wien displacement law:

$$\lambda_{max} T = 3 \text{ K} \cdot \text{mm}.$$

Energy is radiated at a rate $P = \epsilon \sigma A T^4$, where the Stefan–Boltzmann constant $\sigma = 5.67 \times 10^{-8} \text{ W/m}^2 \cdot \text{K}^4$, A is the surface area of the body, and ϵ ($0 < \epsilon < 1$) is the emissivity of the body. Radiation is also absorbed from the surroundings with temperature T_{env}. The net heat transfer rate from the body is:

$$H_{rad} = \epsilon \sigma A (T^4 - T_{env}^4).$$

Convection occurs in fluids when the vertical temperature gradient exceeds a critical value. The fluid moves, carrying thermal energy with it.

Hot objects in colder surroundings cool at a rate proportional to the temperature difference between the object and the surroundings. This empirical law, due to Newton, is valid for all heat transfer mechanisms when the body is not very much hotter than its surroundings.

Practical Applications

An understanding of heat transfer is important in the design of energy efficient buildings and power plants. Proper cooling of aircraft engines can increase their power by 15%. Heat transfer is crucial in modeling stars and planets. The control of heat flow is also important in spacecraft, computers, and other electrical devices.

Solutions to Exercises

21.1 The heat capacity of the water in 1.0 m of pipe is:

$$C = Mc_{water} = \pi r^2 \ell \rho_{water} c_{water}.$$

The heat flow from the pipe changes the water temperature, $H = C\, dT/dt$; thus the temperature change in 1.0 s is:

$$\Delta T = \frac{dT}{dt} \Delta t = \frac{H}{C} \Delta t$$

$$= \frac{(97 \text{ W/m})(1.0 \text{ s})}{\pi (5.0 \times 10^{-2} \text{ m})^2 (1.0 \text{ m})(1.0 \times 10^3 \text{ kg/m}^3)(4186 \text{ J/kg} \cdot \text{K})}$$

$$= 3.0 \times 10^{-3} \text{ K}.$$

The water temperature changes extremely slowly, so steady state is indeed a reasonable approximation.

21.2 The total heat flow through the layer is the sum of the amounts flowing through the two resistances:

$$H = H_1 + H_2 = \frac{\Delta T}{R_1} + \frac{\Delta T}{R_2} = \Delta T \left(\frac{1}{R_1} + \frac{1}{R_2} \right) \equiv \frac{\Delta T}{R_p},$$

which gives the required result.

21.3 According to the Wien displacement law, Sirius, with a shorter λ_{max}, is hotter than Betelgeuse. Sirius is hotter (bluer) than the Sun, and Betelgeuse is colder (redder) than the Sun.

21.4 Nothing! The color of cavity radiation depends only on the temperature and not on the chemical nature of the system.

21.5 The number of food Calories required to maintain the body temperature is:

$$\frac{60 \text{ W}}{4186 \text{ J/Cal}} = 0.01 \text{ Cal/s} \quad \text{or} \quad \text{about } 1000 \text{ Cal/d}.$$

Medical experts say that a woman needs about 2000 Cal/d, a man about 2700 Cal/d, so that our composite person would need about 2300 Cal/d. Taken at face value, the calculation indicates that almost half our food Calories are used up just keeping us warm! The crucial fact we have ignored is the insulation provided by our clothes, which reduces the heat loss considerably.

Basic Skills

Review Questions

§21.1 CONDUCTION

- How is heat transferred by conduction?
- Do solids typically have greater or smaller thermal conductivity than gases? Why?
- What is *steady state* heat flow?
- If heat flows by conduction along a metal rod whose ends are kept at fixed temperatures T_1 and T_2, how does the heat flux depend on the area and length of the rod? How does the heat flux depend on the temperature difference between its ends?
- What is the *R-factor* of a building material? How is it related to thermal resistance of an insulating layer?
- If two insulating layers are placed back-to-back (in series), what is the thermal resistance of the combined layer?

§21.2 CONVECTION

- How is heat transferred by convection?
- Give three examples of systems in which convection is important.

§21.3 RADIATION

- How is heat transferred by radiation?
- If one body has temperature T_1 and another identical body has temperature $T_2 = 2T_1$, in what ways does the radiation emitted by the second body differ from that emitted by the first?
- How does the power radiated by a body depend on its temperature?
- What is the *emissivity* of a body? Is it possible for a body to have an emissivity of 2?

§21.4 AN EMPIRICAL APPROACH TO HEAT TRANSFER

- State Newton's law of cooling. When does it apply?

Basic Skill Drill

§21.1 CONDUCTION

1. Find the heat flow through a concrete wall of area 12 m² and thickness 7.0 cm when the outside temperature is 35°F and the inside temperature is 65°F.

2. Find the rate of heat flow along an iron rod of diameter 1.2 cm and length 2.0 m if one end is at 401 K and the other is at 280 K.

3. Compare the heat loss through a concrete wall of thickness 15 cm with an oak wall of thickness 3.0 cm. Which material makes the most energy-efficient building?

§21.2 CONVECTION

4. Why does convection occur in a pot of water set on a stove burner?

5. In which of the following systems is heat transferred by convection? Explain your answers. **(a)** A microwave oven **(b)** An electric oven **(c)** A room heated by a wood fire **(d)** A thermometer used to take a baby's temperature

§21.3 RADIATION

6. At what wavelength does radiated power peak for molten iron at 2000 K?

7. What is the wavelength of peak radiation from a human body at 98°F?

8. The giant star Antares has a luminosity of 1.9×10^{31} W and a temperature of 3300 K. What is its radius?

§21.4 AN EMPIRICAL APPROACH TO HEAT TRANSFER

9. ✉ An engine block cools from 95°C to 80°C in 10 min. Estimate how long it takes to cool from 80°C to 70°C.

Questions and Problems

§21.1 CONDUCTION

10. ❖ A pot of bubbling spaghetti sits on the stove. One handle of the pot retains its wooden covering, but the covering has broken off the other handle. Explain why you can hold the wooden handle, but not the metal one.

11. ❖ A goose down sleeping bag is designed to hold the down in small pockets so that it does not all clump up in one part of the bag. Why? In achieving this design, manufacturers avoid simply sewing the inner and outer surfaces together to form the pockets. Why?

12. ❖ Often people wet a finger before touching an object (quickly) to test if it's hot. Explain how the water protects against a burn.

13. ❖ When baking a potato, some people insert a metal skewer through it. Why?

14. ❖ A wire is made from a length of copper brazed at one end to an equal length of stainless steel. The wire is surrounded by thermal insulation and is stretched between two objects held at 100°C (at the copper end) and 0°C (at the steel end). Is the temperature at the

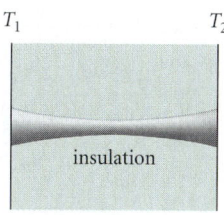

■ FIGURE 21.16

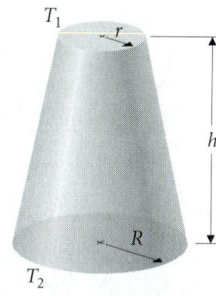

■ FIGURE 21.17

junction between the two metals equal to, greater than, or less than 50°C? Explain your reasoning.

15. ❖ Two metal plates held at different constant temperatures are separated by a tapered metal rod (■ Figure 21.16). Which of the following statements are correct?
(a) The heat flow is lowest where the rod is thinnest.
(b) The temperature gradient is greatest where the rod is thinnest.
(c) The heat flow through the rod is proportional to the temperature difference.
(d) The temperature varies linearly with position along the rod.
Explain why you chose your answers, and what is wrong with the statements you didn't choose.

16. ◆ Find the thermal resistance of a flat roof 11 m square made of 1.27-cm plywood covered with asphalt shingles.

17. ◆ A wall of plywood 10.0 m × 3.5 m has a 1.2 m × 2.3 m window in it. Compare the thermal resistance of the plywood wall and the window. (Use values from Table 21.2.) What is the thermal resistance of the complete wall, including the window?

18. ◆ Find the R-factor for a double-glazed window consisting of two single panes separated by a 0.50-cm air gap.

19. ◆ By how much does a 10-cm-thick layer of glass wool insulation decrease the heat loss through the roof of a house (area 100 m^2, $R_f = 0.15$ m$^2 \cdot$K/W) if the temperature difference between the interior and the exterior is 25 K?

20. ◆ A spa is made of 1.27-cm-thick plywood covered with tiles. Find the rate of heat loss per unit area through the walls of the spa when it contains water at 101°F and the outside temperature is 55°F.

21. ◆◆ A furnace wall is made of two different layers, the inner layer of thickness 0.10 m and thermal conductivity 0.17 W/K·m, the outer layer of thickness 0.20 m and thermal conductivity 0.84 W/K·m. The area of the wall is 1.0 m^2. If the inner surface is at a temperature of 601°C and the outer surface is at 260°C, calculate the temperature at the joining surface and the heat current flowing through the wall in a steady state.

22. ◆◆ A cylindrical steam pipe of outer radius 3.0 cm carries steam at a temperature of 112°C. The pipe is covered uniformly with a coating of a 6.0-cm-thick thermal insulator with a thermal conductivity of 1.7×10^{-3} W/K·m. The outer surface is at a temperature of 51°C. Calculate the heat loss per unit length of pipe.

23. ◆◆ A long tungsten wire with a diameter of 1.5 μm is encased in a porcelain sheath that has a diameter of 3.0 mm. The wire is heated electrically. At equilibrium, the temperature of the wire is 1200°C and the temperature of the outer surface of the porcelain is 150°C. Compute the power per unit length (in watts per meter) supplied to the tungsten wire by the electrical current in it. (Thermal conductivity of porcelain = 0.10 W/K·m.)

24. (a) ◆◆ An igloo is made of snow blocks 0.40 m thick and has an outer radius of 1.9 m. Two people, each generating 185 W of thermal power, are inside the igloo at night when the outside temperature is −40.0°C. If the conductivity of the snow is 0.20 W/K·m, what is the temperature at the inner surface of the igloo? (Neglect heat loss through the animal hides on the floor of the igloo.) **(b)** ❖ Can the igloo dwellers change the air temperature by taking off their heavy jackets? (Ignore initial transients and consider the temperature to be in steady state.)

25. ◆◆◆ The two end surfaces of a tapered rod with circular cross section are held at temperatures T_2 and T_1 (■ Figure 21.17). If the material of the rod has conductivity k, find the heat flow through the rod. How would the heat flow change if the rod were reversed?

26. ◆◆◆ You wish to insulate the bathysphere described in Example 21.3 by lining it with felt. If the occupant produces 100 W of thermal power, how much felt should be used so that no additional heating is needed?

§21.2 CONVECTION

27. ❖ The Sun heats the upper surface of a shallow pond. Is convection likely to occur? Why or why not?

28. ❖ On a cold windy day, conditions are described by the *wind-chill factor*, which gives the difference between the actual temperature and the temperature of still air that would carry heat away from a person at the same rate. Explain why this factor exists and why it increases rapidly with wind speed.

29. ❖ Explain why the flames in a fire rise. Would this occur equally well on a space station?

30. ❖ In spacecraft, fans are used to create air currents. If the air were still, the astronauts would suffocate. Why?

31. ❖ Explain why a convective flow must break up into many cells when its horizontal scale is much longer than its vertical scale. (*Hint:* If the flow doesn't break up, where does all the rising air go?)

§21.3 RADIATION

32. ❖ On a cold night, frost forms on dark, dirty cars but not on clean, light-colored ones. Why?

33. ❖ Some stars (like the binary companion of the bright star Sirius) are blue but very faint. What can you say about the size of the star?

34. ◆ The relic radiation from the hot, dense phase of the early universe has been measured to correspond to a blackbody at 2.7 K. At what wavelength does the radiation peak?

35. ◆ Glowing embers in a fire have a temperature of 900 K. At what wavelength does their emitted radiation peak?

36. ◆ Sirius B has a temperature of 25 000 K and a luminosity of 6×10^{25} W. Estimate its radius.

37. ◆◆ The temperature of the filament in an incandescent light bulb is about 2800 K. What is the surface area of the filament in a 100-W bulb if $\epsilon = 0.9$? Is the bulb efficient at producing light? Discuss.

38. ◆◆◆ An electric heating element produces 120 W of power in the interior of a highly conducting metal ball of radius $r = 1.0$ m. A thin, concentric shell of radius $R = 2.0$ m surrounds the ball, and the whole system is in vacuum and far from any outside source of radiation. All surfaces are perfectly absorbing. Find the temperature of each sphere when the system is in equilibrium.

§21.4 AN EMPIRICAL APPROACH TO HEAT TRANSFER

39. ♦♦ A vat of chemicals in a laboratory where the air temperature is 293 K cools from 405 K to 395 K in 7 min 10 s. What is the average value of dT/dt during this interval? From your result and Newton's law of cooling, estimate dT/dt at 350 K. Estimate the time required for the chemicals to cool from 355 K to 345 K.

40. ♦♦♦ If radiation is the dominant cooling process, the constant B in Newton's law of cooling is proportional to T^3. What is its dependence on T if conduction dominates? Write B as a linear combination of the two powers: $B = aT^3 + bT^n$. In an experiment, a system cools from 415 K to 402 K in 87 s and from 366 K to 356 K in 142 s. A thermometer in the lab reads 67°F. Find a and b. Which process dominates at 400 K? At room temperature?

Additional Problems

41. ❖ Double-glazed windows (■ Figure 21.18) are very effective in preventing heat loss from houses in cold climates. Discuss how this construction reduces heat loss by each of the mechanisms discussed in this chapter.

42. ❖ ■ Figure 21.19 shows the structure of a thermos bottle. Why is this bottle effective at keeping its contents at the temperature at which they were placed inside? Discuss the purpose of each design feature.

43. ❖ Fireplaces are usually built so as to be very shallow with high backs. Often they have structures to hold the burning logs apart. In a wood stove, the fire is inside a box that sits in the middle of the room. Describe the heating mechanisms used by each. Would the wood stove or the fireplace be more effective in a drafty medieval castle? Explain your answer.

44. ❖ As the Sun warms the surface of a frozen lake, the ice melts, leaving a layer of water at 0°C above the body of water that is at 4°C. (Why is the rest of the lake at 4°C?) Discuss what happens as the water is heated further. (*Hint:* What are the thermal expansion properties of water between 0°C and 4°C?)

45. (a) ❖ Heat sinks (■ Figure 21.20) are often used in electrical devices such as amplifiers to keep the temperature from rising too high. How do they work? **(b)** ♦♦ A heat sink made of 32 aluminum plates, each 4 cm × 10 cm × 1 cm, is used in a 200-W Hafler stereo amplifier. Considering only radiation, how hot would the heat sink get during operation in a room at 300 K? Take $\epsilon = 0.9$.

■ FIGURE 21.20

46. ♦♦ An insulated truck is rated at 40 Btu/(h·°F). (1 Btu = 1055 J.) The payload mass is 18 020 kg. The truck's carrying volume measures 3.0 m × 15 m × 5.0 m (external dimensions). If the inside temperature is −5.0°C and the outside is 20.0°C, find: **(a)** the total heat flux through the walls in watts, **(b)** the R-factor of the wall, **(c)** the temperature rise of the payload during an 8-h journey, assuming its specific heat is close to that of water.

47. ♦♦♦ A teapot may be modeled as a sphere of diameter 18 cm. The teapot contains tea at 95°C and is in a room at $T = 12°C$. **(a)** At what rate does the pot lose energy by radiation? (Take $\epsilon = 0.7$.) **(b)** The experimental result for the heat flow by convection from a sphere of diameter d at temperature T is:

$$H = (0.08 \text{ W/K·m})d(T - T_\infty)$$
$$\times \{1.7 + 0.3[1 + (5.6 \times 10^4 / \text{m}^3 \cdot \text{K})d^3(T - T_\infty)]^{1/4}\},$$

where T_∞ is the temperature at a great distance from the sphere. Compare the heat loss by convection from the teapot with heat loss by radiation.

48. ♦♦ A cooking pot has a copper bottom of radius 9.0 cm and thickness 3.0 mm. Water boils from the pot at 1.7×10^{-3} kg/s. What is the temperature of the outside surface of the pot bottom?

49. ♦♦♦ Copper rods link three heat reservoirs as shown in ■ Figure 21.21. Find the heat flux into or out of each reservoir in a steady state.

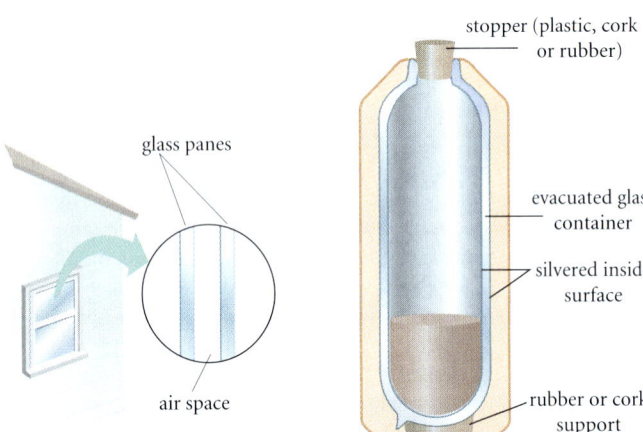

■ FIGURE 21.18
Structure of a double-glazed window.

■ FIGURE 21.19
A thermos bottle.

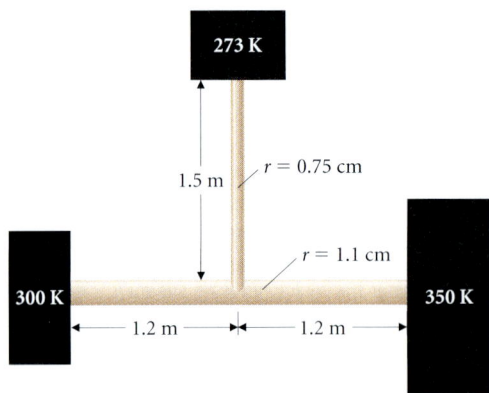

■ FIGURE 21.21

Computer Problem

50. A copper rod of length 1 m lies between two heat reservoirs at fixed temperatures of 500 K and 300 K, respectively. Initially, the whole rod is at 300 K. We may use a spreadsheet program to model heat flow through the rod and see how it reaches equilibrium. We divide the rod into six pieces, each represented by one column in your spreadsheet. Each reservoir also occupies one column, and each row represents a particular time. Place the fixed temperatures of the reservoirs in their respective columns, and place the 300 K initial temperature into each rod cell in the $t = 0$ row. Heat flows both into and out of each rod cell, according to eqn. (21.1), and the net heat flow changes the temperature: $H_{net} = dQ/dt = C\,dT/dt$, where C is the heat capacity of a cell. Show that $C = A\rho c_p\,\Delta\ell$, where A is the rod's cross-sectional area, ρ and c_p are respectively the density and specific heat of copper, and $\Delta\ell$ is the length of each cell. Argue that the time scale to reach equilibrium should be $\tau \sim (\rho c_p/k)L^2$, where k is the thermal conductivity and L is the length of the rod. The temperature in cell X_n is related to those around it by:

$$X_n = X_{n-1} + \left[\frac{(W_{n-1} - X_{n-1}) - (X_{n-1} - Y_{n-1})}{(\Delta\ell/L)^2}\right]\left(\frac{\Delta t}{\tau}\right).$$

In this expression W represents the column to the left of X, and Y represents the column to the right; $n - 1$ represents the row above n; and Δt is the time step. For a 1-m-long rod, $\Delta\ell = \frac{1}{6}$ m. Fill the rod cells with this formula, and copy down the page. Choose $\Delta t = 25$ s, and plot the temperature along the rod at times $t = 0$, 200 s, 400 s, 800 s, 1600 s, and 3200 s. Notice how the temperature distribution evolves toward a linear drop along the rod. Compare the computed time to reach equilibrium with the estimate τ.

Challenge Problems

51. An insulating material consists of cubical air pockets of side ℓ separated by solid walls of thickness $t = \alpha\ell$. Compute the thermal resistance of a single cell of the structure in terms of α, ℓ, and the thermal conductivities of air and the solid material. Use your results to compute the thermal conductivity of the bulk material. For what ratio α does this model reproduce the measured conductivity of glass wool?

52. When looking over a hot surface, you see that objects in the distance appear distorted in a way that varies in time. What causes this effect?

53. Paints can be formulated that have very different values of ϵ for visible and infrared light. ■ Figure 21.22 shows the value of ϵ as a function of wavelength for one such paint. Under what conditions would you use this paint? (*Hint:* The fraction of incident radiation absorbed by a surface at any frequency equals the value of ϵ at that frequency. What is the peak frequency of the Sun's radiation? What is the peak frequency of a building's radiation?)

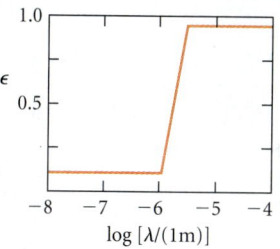

■ **FIGURE 21.22**
ϵ as a function of wavelength for a paint with special properties.

Steam trains were the first modern form of transportation. Fast and efficient, they linked the east coast of the United States to new settlements in the west. Steam engines of many types powered the industrial revolution in Europe and the United States.

CHAPTER 22
Entropy and the Second Law of Thermodynamics

Concepts
Availability of energy
Reversible versus irreversible processes
Disorder
Entropy
Heat engine
Efficiency
Carnot cycle
Refrigerator
Thermodynamic temperature scale
Absolute zero

Goals
Understand:
The second law of thermodynamics stated in terms of entropy, energy transfer, or engine efficiency.

The fundamental reasons why processes are irreversible.

Be able to:
Compute the efficiency of a heat engine cycle.

Compute the entropy change due to an arbitrary process.

*This is how the world ends
Not with a bang but a whimper.*

T. S. Eliot

Steam engines thundering across the prairie provided a practical means for crossing the vast expanses of the western United States. In such engines, burning fuel heats water to produce steam. Thermal energy stored in the steam is converted to mechanical energy of the train. You don't have to burn fuel to have a large amount of thermal energy. The air in a small room has about the same kinetic energy as a car going at 200 mph, but you can't use it to run the car. For that matter, less than half of the steam's thermal energy is converted to kinetic energy of the train. The engine's inefficiency is not a result of poor design: a joule of thermal energy in steam is fundamentally less available to do useful work than a joule of kinetic energy in a wheel. It is the random distribution of thermal energy among many particles that makes it different from the kinetic or potential energy of a large object, like the train, moving as a whole.

The reasons for this difference are described by the second law of thermodynamics, and involve the concepts of disorder in systems and reversibility of processes. In this chapter we investigate systems that convert thermal energy to useful work and develop a way to estimate the availability of thermal energy. We use a state variable called entropy that measures the degree of disorder in a system. Finally, we show how the second law and entropy give a deeper understanding of temperature and absolute zero.

22.1 WHY IS A SECOND LAW OF THERMODYNAMICS NECESSARY?

Simple mechanical processes that do work are reversible. For example, a mass attached to a pulley may descend and raise a second, equal mass. If the axle is frictionless, the process can be reversed by allowing the second mass to descend and raise the first. A slow adiabatic expansion of a gas is also theoretically reversible. To reverse it, simply compress the gas slowly and adiabatically. On the other hand, poking a hole in a tire is an irreversible process. The air in the tire rushes out through the hole and the tire deflates. You would be very surprised to see a punctured tire spontaneously re-inflate as air rushed back in!

> RECALL THAT IN AN ADIABATIC PROCESS, $Q = 0$.

A second example of an irreversible process is the melting of an ice cube. Fresh from the freezer, water molecules are firmly bound in a highly ordered crystal lattice. As the ice melts, the molecules are released from the lattice and flow into a puddle on the kitchen table; the system becomes more disordered. The reverse process does not happen. Water on the table does not spontaneously form ice cubes!

> A process is *reversible* if each step can be reversed exactly to move from the final state back to the initial state.

The first law of thermodynamics does not tell us why the adiabatic process is reversible and a tire's collapse is not. Energy is conserved in both processes no matter which way they go. We need a second law of thermodynamics to indicate which energetically possible processes actually occur.

It is not impossible for a punctured tire to re-inflate, just highly improbable: air molecules with disordered motions would have to join an ordered flow through the tiny hole back into the tire. Increasing disorder distinguishes possible processes from those so improbable they might as well be impossible. This observation provides a first statement of the *second law of thermodynamics*.

> **SECOND LAW, VERSION 1**
> The disorder of the universe tends to increase.

Increasing disorder in a system is the same as decreasing information about that system. Air molecules initially inside the tire are dispersed throughout the neighborhood in the final state. The water molecules initially confined in the ice cube eventually evaporate and end up

SEE § 19.3, ESPECIALLY EXAMPLE 19.8. WITH-OUT REMOVING HEAT, THE TEMPERATURE OF THE AIR IN THE TIRE WOULD INCREASE, AND DISORDER IN THE TIRE WOULD BE INCREASED.

similarly dispersed in the room. In both cases, we have less information about the location of the molecules in the final state than we did in the initial state.

The universe is important in the statement of the second law. We can pump air back into a flat tire, returning it to its initial state. However, we do work on the gas during the process. At the same time, heat transfer removes internal energy and increases the disorder of molecules outside the tire; it could melt ice, for example. Similarly, a freezer used to recreate the ice cube expels heat into the kitchen.

Adiabatic expansion of a gas neither disperses molecules like the collapsing tire nor absorbs heat like the melting ice. Ideally, the only interaction with the environment is an exchange of mechanical work that produces no disorder. Such reversible processes are important models for theoretical understanding of thermodynamics, but they can only approximate real processes. The second law tells us that any lapse from ideal behavior creates disorder and makes it impossible to return the universe to its initial state:

> **SECOND LAW, VERSION 2**
> No real process can be perfectly reversible.

Historically, the second law was first stated as a limit on the operation of heat engines. The connection between disorder and the availability of energy was not apparent to early workers, but they did recognize the limits of energy conversion. In 1851, Kelvin gave one of the earliest statements of the second law:

THESE STATEMENTS DO NOT TALK ABOUT WHAT CAN HAPPEN BUT INSTEAD TALK ABOUT WHAT CANNOT. THUS THE SECOND LAW IS DIFFERENT FROM OTHER LAWS WE HAVE DISCUSSED.

> **SECOND LAW, VERSION 3**
> No engine can be built that *completely* converts thermal energy to useful work in a cyclic process.

As we shall see, this is equivalent to the statement given by Rudolf Clausius in 1850 and by Max Planck (about 1895):

> **SECOND LAW, VERSION 4**
> It is not possible to build a machine whose task is to transfer heat from a cold place to a hot place without the input of energy as work.

Versions 3 and 4 are specific to engines. Our first task is to investigate their relation to the first two, more general statements of the second law.

22.2 HEAT ENGINES

22.2.1 Efficiency of Engines

An insulated cylinder of ideal gas with a piston is a simple system that extracts useful work from thermal energy. The gas expands adiabatically, doing work on the piston. Its temperature and internal energy decrease. From the first law of thermodynamics, the decrease in internal energy is equal to the work done by the system: $\Delta U = -W$. The system uses internal energy and converts it to work. However, once the pressure drops to atmospheric, the system can do no more work. Even though it still has internal energy, this energy cannot be extracted. For the system to continue operating, we could push the piston back in, which alone is useless since the net work done is then zero. Alternatively, we could replace, by heat transfer, the internal energy that has been expended.

REMEMBER: W IS POSITIVE IF WORK IS DONE BY THE SYSTEM, SO ΔU IS NEGATIVE.

A practical heat engine does both. By operating on a cycle (cf. §19.6), it can run repetitively and produce work as long as heat is supplied. Such devices produce a net positive amount W of useful work as a result of a heat input Q_{in}. The ratio of W to Q_{in} is called the efficiency of the engine:

$$\text{Efficiency} = \frac{\text{net work obtained}}{\text{heat input}}.$$

$$e = \frac{W}{Q_{in}}. \tag{22.1}$$

Both work done by the system (positive) and work done on the system (negative) are included in W, but only heat inflow to the system is counted in Q_{in}. This is an economic definition: to obtain work W is why you build the engine; Q_{in} is the energy cost to run it.

All cyclic heat engines have efficiency less than unity: not all the heat input can be extracted as work. As in the case of the one-step expansion we described above, any engine must be returned to its initial state so that it can continue to do work. This process always involves the rejection of waste heat to the environment; that is, energy is lost from the system *without* performing useful work.

22.2.2 The Otto Cycle

As an example, we analyze the Otto cycle, an idealization of the cycle commonly used to describe internal combustion engines (■ Figure 22.1). The cycle begins when the piston is pulled out with the intake valve open, and the cylinder fills with an air/fuel mixture (path AB in the figure). The valve closes at B, confining the mixture in the cylinder, where the piston rapidly reduces its volume from V_B to V_C. "Rapidly" means that there is no time for heat to flow, so the compression along curve BC is adiabatic. At C, the fuel/air mixture is ignited by the spark plug, and the thermal energy produced increases the pressure at constant volume V_C (path CD). The rapid expansion (adiabatic again) along path DE is the power stroke of the cycle; the expanding gas does work on the piston. Next, the valve is opened, resulting in a loss of heat to the environment and a drop in pressure (path EB). Finally, the piston is pushed in to expel the exhaust gases (path BF). Part of the work obtained during the power stroke DE is used to continue the cycle (BF, AB, and BC). The work and heat transfer during steps AB and BF, which are necessary to bring new fuel into the system, exactly cancel each other, and we may ignore them. The cycle of interest is thus $B \rightarrow C \rightarrow D \rightarrow E \rightarrow B$.

THE DETAILS OF THIS PROCESS ARE NOT IMPORTANT FOR OUR DISCUSSION HERE, BUT THEY ARE CRUCIAL TO THE DESIGN OF A GOOD AUTOMOBILE.

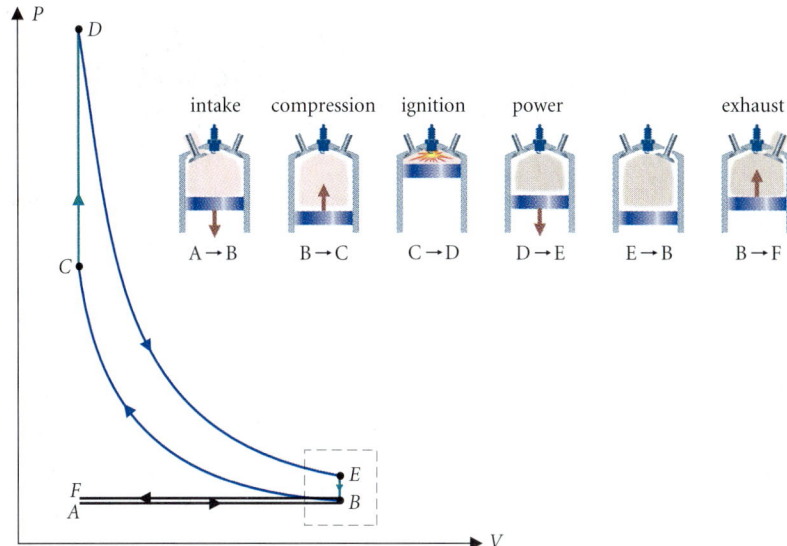

■ FIGURE 22.1
The Otto cycle is an idealization of the cycle used in a gasoline engine. The air intake and exhaust ($A \rightarrow B$ and $B \rightarrow F$) do not contribute to the cycle thermodynamically and may be ignored. The cycle has two constant-volume processes ($C \rightarrow D$ and $E \rightarrow B$) and two adiabatic processes. Heat transfer occurs during the constant-volume processes, and work is done during the adiabatic processes.

EXAMPLE 22.1 ♦♦♦ Find the efficiency of the Otto cycle.

MODEL There is no heat transfer along the adiabats BC and DE. Heat input occurs along CD and output along EB. No work is done during these constant-volume processes. Work is done on the engine along BC to prepare for the next ignition. Useful work is extracted only on the power stroke DE. To find the efficiency of the engine, we may analyze the cycle in terms of either work or heat transfer (Figure 19.18). By using heat transfer, we avoid doing integrals to find the work done along adiabats.

SETUP The heat input from burning fuel in the constant-volume process CD (eqn. 19.14) is:

$$Q_{CD} = Mc_v \, \Delta T_{CD}.$$

WE USED THE IDEAL GAS LAW.

The temperature at C is $T_C = P_C V_C/(Nk)$, and the temperature at D is $T_D = P_D V_C/(Nk)$, so the change in temperature ΔT_{CD} along CD is:

$$\Delta T_{CD} = \frac{(P_D - P_C)V_C}{Nk}.$$

Thus:
$$Q_{CD} = \frac{Mc_v(P_D - P_C)V_C}{Nk}.$$

Similarly, the heat input along EB is:

$$Q_{EB} = \frac{Mc_v(P_B - P_E)V_B}{Nk}.$$

Since the pressure P_B is less than P_E, Q_{EB} is negative, confirming that heat is rejected.

REMEMBER: U IS A STATE VARIABLE, SO U_B HAS ONE DEFINITE VALUE.

For the whole cycle $BCDEB$, the net change in internal energy ΔU is zero. So, from the first law of thermodynamics, the net heat input equals the work done:

$$Q_{CD} + Q_{EB} = Q = W.$$

SOLVE The efficiency is the ratio of the net work done by the engine (W) to the heat input (Q_{CD}). In terms of heat transfer alone:

$$\text{Efficiency} = e = \frac{Q_{CD} + Q_{EB}}{Q_{CD}} = 1 + \frac{Q_{EB}}{Q_{CD}}.$$

Since Q_{EB} is negative, the efficiency is less than 1, independent of any of the specific values $V_B, V_C, P_B, P_C, P_D, P_E$. Using our expressions for the heat transfers:

$$e = 1 + \frac{V_B(P_B - P_E)}{V_C(P_D - P_C)}.$$

Engines are frequently described by the compression ratio $r \equiv V_B/V_C$. To express e in terms of r, we recall that PV^γ remains constant along an adiabat (§19.6).
Along BC, $P_B V_B^\gamma = P_C V_C^\gamma.$
Along DE, $P_D V_C^\gamma = P_E V_B^\gamma.$
Subtracting these relations, $(P_B - P_E)V_B^\gamma = (P_C - P_D)V_C^\gamma$. Substituting in the expression for e, we find:

$$e = 1 + \frac{r}{-r^\gamma} = 1 - \frac{1}{r^{\gamma-1}}.$$

BUT THE RESULTING HIGHER TEMPERATURES CAUSE GREATER NITROGEN OXIDE POLLUTION!

ANALYZE An engine with a high compression ratio is thermodynamically more efficient than one with a low compression ratio. A typical modern automobile has a compression ratio of about 10. Since the gas is mostly air, $\gamma = \frac{7}{5}$, and the thermodynamic efficiency is $e = 1 - (0.1)^{2/5} = 60\%$ for the ideal Otto cycle. Dissipative processes such as friction and thermal conduction reduce a real engine's efficiency to less than one-half the thermodynamic efficiency.

22.3 THE CARNOT CYCLE

22.3.1 A Reversible Cycle

An Otto-cycle engine cannot be run in reverse, but its representation in the *P-V* diagram doesn't tell us why. To draw a process as a line in the *P-V* diagram, we must know the value of *P* and *V* at each step. We could then reverse the process by moving backwards along the path through the same values. In practice, whenever heat is transferred to a system whose temperature is changing, we do *not* know the system's pressure and volume in precise detail. To reverse the process, the heat transfer *at each temperature* should be exactly the opposite of its value in the initial process. This proves to be impossible.

In 1824, the French engineer Sadi Carnot (■ Figure 22.2) realized that one could obtain further insight into the performance of engines by studying a cycle that could theoretically be made reversible. He devised a cycle, named after him, in which heat is added or subtracted at a constant temperature—that is, along isotherms. Other portions of the cycle are adiabatic (■ Figure 22.3). A *heat reservoir* is used to transfer heat at a constant temperature.

> An *ideal heat reservoir* has an infinite heat capacity, so that heat transfer to or from it does not change its temperature, and an infinite thermal conductivity, so that temperature remains constant throughout the reservoir as thermal energy is transported.

A Carnot engine could be made reversible, in principle, by transferring heat with two such reservoirs and by eliminating all frictional and other losses. In practice, an object with a very large heat capacity and a high thermal conductivity approximates an ideal heat reservoir.

First, we analyze Carnot's cycle in terms of heat input, work done, and efficiency. The system expands from state *A* (Figure 22.3) along the isotherm at T_H to *B*. Work is done by the system ($W > 0$) and, since temperature is constant, the internal energy does not change. Heat transfer provides energy for the system to do work, so *Q* is positive and equals *W*.

The process *BC* is adiabatic, so the heat transfer is zero while the system is doing work: *Q* is zero and *W* is positive. The internal energy decreases, and *C* is at a lower temperature than *B*, $T_C < T_H$.

Process *CD* is isothermal at temperature T_C. An external agent does work on the system to compress the gas, so *W* and *Q* are negative.

The compression *DA* is adiabatic, with $Q = 0$ and *W* negative.

The work done along *AB* (eqn. 19.13) is:

$$W_{AB} = NkT_H \ln(V_B/V_A), \tag{22.2}$$

and the heat input is $Q_{AB} = W_{AB}$.

■ **FIGURE 22.2**
Sadi Carnot (1796–1832) was a French engineer who investigated the properties of heat engines. His most important work, "Reflections on the motive power of heat and on the machines adapted to develop this power," was published in 1824. It was translated into English in 1890.

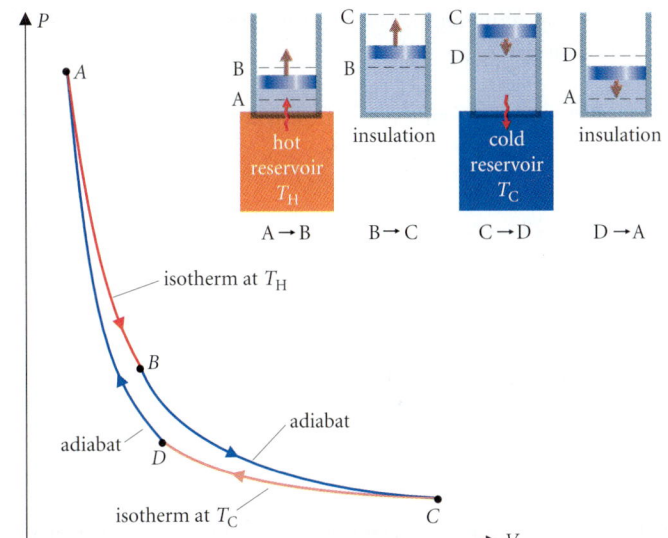

■ **FIGURE 22.3**
The Carnot cycle operates between two temperatures, T_H and T_C. Heat transfer occurs along the isotherms *CD* and *AB*.

STEP I (*AB*): The cylinder of gas is put into contact with the hot reservoir, and heat transfer into the cylinder occurs. The gas expands, and the system does work.

STEP II (*BC*): The cylinder is insulated, and the gas continues to expand and do work.

STEP III (*CD*): The cylinder is put into contact with the cold reservoir. Heat transfer into the cold reservoir occurs, and the piston is pushed inward so that work is done on the system.

STEP IV (*DA*): The cylinder is insulated and the piston is pushed farther inward until the system reaches its initial state.

NOTE THAT SINCE $V_D < V_C$, Q_{CD} IS NEGATIVE.

Similarly, along CD:
$$Q_{CD} = NkT_C \ln(V_D/V_C). \quad (22.3)$$

The efficiency of this cycle is:
$$e_c = W_{net}/Q_{in}.$$

From the first law of thermodynamics, $W_{net} = Q_{net}$ for a complete cycle, so:

$$e_c = \frac{Q_{net}}{Q_{in}} = \frac{Q_{AB} + Q_{CD}}{Q_{AB}} = 1 + \frac{Q_{CD}}{Q_{AB}}$$

$$= 1 - \frac{T_C \ln(V_C/V_D)}{T_H \ln(V_B/V_A)}.$$

WE HAVE WRITTEN $\ln(V_D/V_C) = -\ln(V_C/V_D)$ TO EMPHASIZE THAT $e < 1$.

As expected, e_c is less than unity.

We can simplify this expression by showing that the two volume ratios V_B/V_A and V_C/V_D are equal. Along an adiabat, PV^γ is constant, so

$$P_B V_B^\gamma = P_C V_C^\gamma$$

and
$$P_D V_D^\gamma = P_A V_A^\gamma.$$

Also, since T is constant along AB, Boyle's law gives:
$$P_A V_A = P_B V_B.$$

Similarly, along CD,
$$P_C V_C = P_D V_D.$$

Setting the product of the four left-hand sides of these equations equal to the product of the four right-hand sides and dividing by the product of the four pressures, we find:

$$V_B^\gamma V_D^\gamma V_A V_C = V_C^\gamma V_A^\gamma V_B V_D.$$

So:
$$\left(\frac{V_B}{V_A}\right)^{\gamma-1} = \left(\frac{V_C}{V_D}\right)^{\gamma-1}.$$

Hence
$$\frac{V_B}{V_A} = \frac{V_C}{V_D}, \quad (22.4)$$

and:
$$e_c = 1 - \frac{T_C}{T_H}. \quad (22.5)$$

ACTUALLY, THE RESULT APPLIES TO ANY REVERSIBLE CYCLE THAT OPERATES BETWEEN TWO TEMPERATURES. THE ONLY SUCH CYCLE THAT USES IDEAL GAS AS ITS WORKING FLUID IS THE CARNOT CYCLE.

This simple formula is appealing and easily memorized. Beware! It applies only to a Carnot cycle and NOT to an arbitrary cycle.

EXERCISE 22.1 ◆ Find the efficiency of a Carnot cycle operating between $T_H = 500$ K and $T_C = 300$ K.

22.3.2 Refrigerators

A Carnot heat engine requires the fewest (and most plausible) idealizations to produce a reversible cycle. Apart from its use in abstract arguments, the reversed Carnot cycle has a practical use. When the cycle is run forward, heat is extracted from a hot reservoir. The machine uses this energy to do work and rejects waste heat into a cold reservoir. Running in reverse, the cycle *uses* work to extract heat from the cold reservoir and reject heat to the hot reservoir: this machine is a refrigerator. Often, the cold reservoir is a box containing food, and the hot reservoir is your kitchen.

EXERCISE 22.2 ❖ On a hot day, you open the refrigerator door in an attempt to cool the kitchen. Do you succeed?

Reversing an ideal engine produces a poor refrigerator. The purpose of an engine is to produce as much work as possible with a minimum of waste heat. A refrigerator is supposed to extract as much heat as possible with a minimum of work.

> The *coefficient of performance K* of a refrigerator is defined as the heat removed from its cold reservoir divided by the work required to extract the heat.

$$K = \frac{Q_{\text{cold}}}{|W_{\text{net}}|}. \tag{22.6}$$

■ Figure 22.4 is a schematic of a refrigerator design common in home use. The central feature of the cycle is vaporization of the refrigerant from liquid to gas phase with absorption of latent heat from the interior. We investigate some features of refrigerators in the problems.

THE FREE EXPANSION IS LIKE AIR ESCAPING FROM A TIRE. SEE ALSO EXAMPLE 22.4.

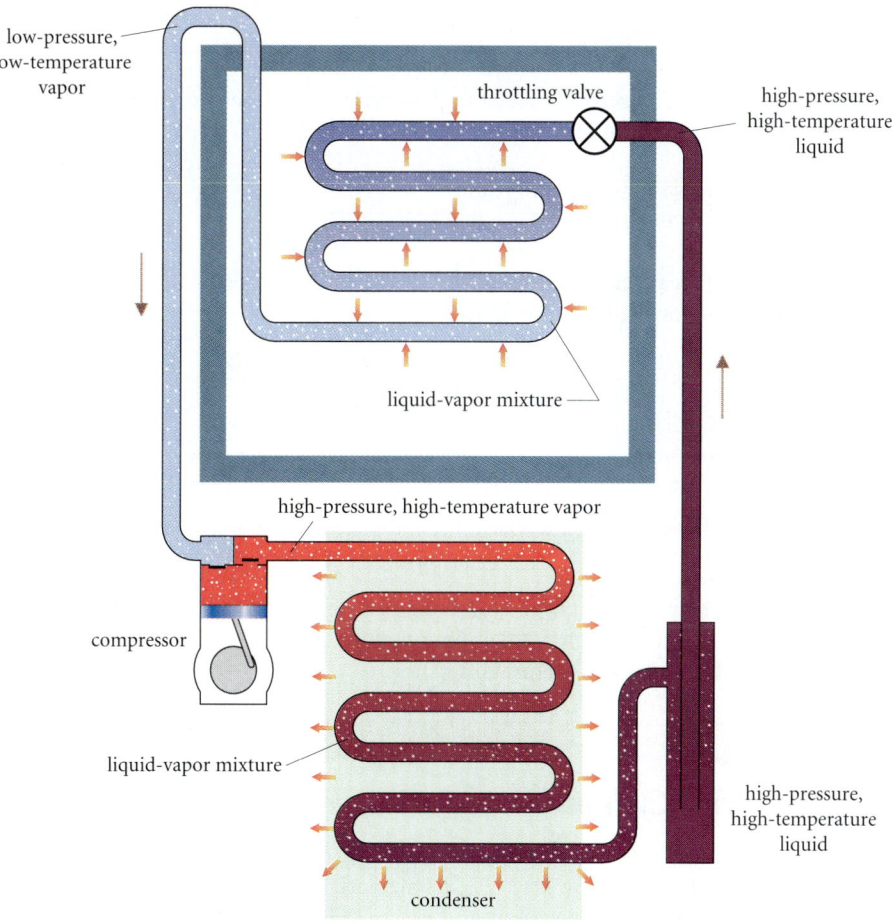

■ FIGURE 22.4
Schematic of a refrigerator for home use. Heat transfer from the cold reservoir occurs as the refrigerant vaporizes. Work is done by the compressor, and heat transfer to the hot reservoir occurs as the refrigerant liquefies. The hot reservoir is cooler than the hot vapor. The phase changes are irreversible processes.

> **EXAMPLE 22.2** ♦♦ What is the coefficient of performance K_c of a Carnot refrigerator operating between temperatures T_C and T_H?
>
> **MODEL** The refrigerator is a Carnot engine run in reverse. Heat is removed from the cold reservoir along the isotherm DC (■ Figure 22.5, overleaf).

SECTION 22.3 • THE CARNOT CYCLE   725

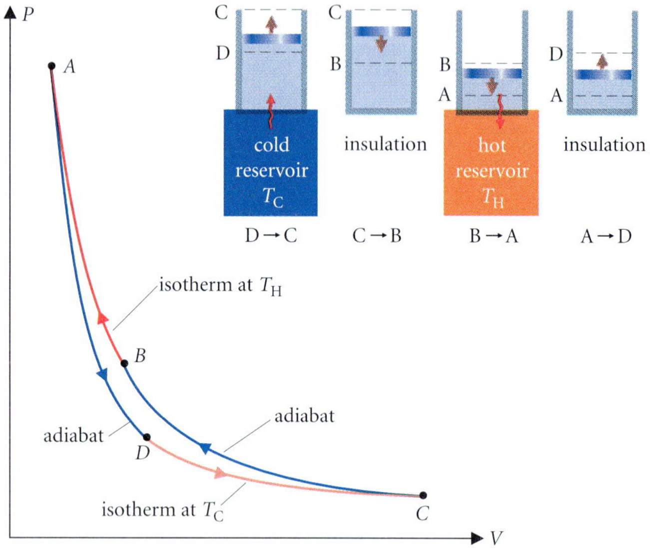

■ **FIGURE 22.5** Operation of a Carnot refrigerator. Compare with Figure 22.3.

NOTE THAT W IS NEGATIVE HERE.

SETUP As usual, ΔU is zero for the complete cycle, so the net work required to run the cycle equals the total heat input to the system, $W = Q_{DC} + Q_{BA}$, and the coefficient of performance is:

$$K_c = \frac{Q_{DC}}{|W|} = -\frac{Q_{DC}}{Q_{DC} + Q_{BA}}.$$

As the cycle is run in reverse, the values of Q are the negative of the corresponding quantities computed for the Carnot engine.

THIS SPECIFIC FORMULA APPLIES ONLY TO THE CARNOT CYCLE.

SOLVE
$$K_c = -\frac{Q_{CD}}{Q_{CD} + Q_{AB}} = -\frac{Q_{CD} + Q_{AB}}{Q_{CD} + Q_{AB}} + \frac{Q_{AB}}{Q_{CD} + Q_{AB}}$$
$$= -1 + \frac{1}{e_c} = \frac{1 - e_c}{e_c} = \frac{T_C}{T_H - T_C},$$

where e_c is the efficiency of the Carnot engine operating between the same two temperatures.

ANALYZE Performance as a refrigerator is inversely related to engine efficiency. A refrigerator performs best through a small temperature difference, while an engine performs best through a large temperature difference. ∎

22.4 ENTROPY

22.4.1 The Carnot Cycle and Entropy Change

We have argued that a process is not reversible if it increases disorder, but we do not yet know how to measure disorder. The tool we need is a state variable called *entropy,* symbol S. By accounting for entropy changes in a system and its environment, we can determine which processes are reversible.

> A process is reversible if and only if the entropy of the universe does not change during the process.

In terms of entropy, the second law of thermodynamics states:

> **SECOND LAW, VERSION 5**
> For any real process, the entropy change of the universe is positive.

These statements improve on the previous qualitative versions once we know how to calculate change of entropy in terms of the other thermodynamic variables.

Entropy is associated with disorder, so it increases as energy of random motion is added to a system. Therefore, entropy change is related to heat transfer. We use the reversibility of the Carnot cycle and a plausibility argument to investigate how they are related. The result does not improve our qualitative picture of entropy as disorder, nor does it allow us to compute a value of entropy for any given thermodynamic system. It does provide a method for computing the *change of entropy* between two states in terms of more familiar quantities and thus puts the second law of thermodynamics on a quantitative basis.

When a process is run backward, all the heat transfers change sign, so the sign of the net entropy change is reversed. But according to the second law, the entropy change of the universe can not be negative in either the original or the reversed process, so both changes are zero: $\Delta S_{\text{backward}} = -\Delta S_{\text{forward}} = 0$. The net entropy change around a reversible cycle is zero.

$$\Delta S_{\text{reversible cycle}} = 0. \tag{22.7}$$

CHANGE OF A QUANTITY BETWEEN TWO STATES IS USUALLY ALL WE NEED; COMPARE THE DISCUSSION OF POTENTIAL ENERGY (CHAPTER 8).

Thus we have two clues to help us find ΔS for a process: change in disorder is related to heat transfer (change in disordered thermal energy), and the total change in disorder vanishes for the *reversible* Carnot cycle. As no heat is transferred in the two adiabatic portions of the cycle, all the entropy changes occur during the isothermal processes. The heat transfers during these processes (eqns. 22.2 and 22.3) satisfy the following relation:

$$\frac{Q_{AB}}{T_H} = Nk \ln\left(\frac{V_B}{V_A}\right) = Nk \ln\left(\frac{V_C}{V_D}\right) = -\frac{Q_{CD}}{T_C}.$$

AN ADIABATIC PROCESS IS SOMETIMES CALLED AN *ISENTROPIC* PROCESS.

WE USED THE FACT THAT THE VOLUME RATIOS V_B/V_A AND V_C/V_D ARE EQUAL (EQN. 22.4).

So:

$$\frac{Q_{AB}}{T_H} + \frac{Q_{CD}}{T_C} = 0. \tag{22.8}$$

This equation has the same form as eqn. (22.7), which suggests that Q/T is the expression we need for the change in entropy during an isothermal process:

$$\Delta S_{\text{isothermal}} = \frac{Q}{T}. \tag{22.9}$$

With this result, we find that the entropy changes for the engine and its environment are *both* zero. The entropy of the Carnot engine increases during the process AB and decreases during the process CD. From the point of view of the heat reservoir at T_H, the heat flow is outward, as is the entropy flow:

$$\Delta S_{\text{reservoir}} = -\frac{Q}{T} = -\Delta S_{\text{engine}}.$$

In this ideal reversible situation, total entropy doesn't change, but flows from one place to another. The entropy gain of the engine along the isotherm at T_H is balanced by the entropy loss of the heat reservoir, and the entropy loss of the engine at T_C is gained by the cold reservoir. The Carnot cycle is unique in this respect.

22.4.2 Entropy Change in an Arbitrary Process

Now that we can calculate ΔS for adiabatic and isothermal processes, we'd like to extend the result to other processes also. For example, a sequence of n adiabats and isotherms (■ Figure 22.6) approximates the constant-volume portion of the Otto cycle, EB in Figure 22.1. As we increase n, each piece of this new line becomes shorter and we approximate the original process EB more closely. The total entropy change along the adiabat-isotherm approximation is:

$$\Delta S = \sum_{\substack{\text{all}\\\text{isotherms}}} \frac{Q_i}{T_i},$$

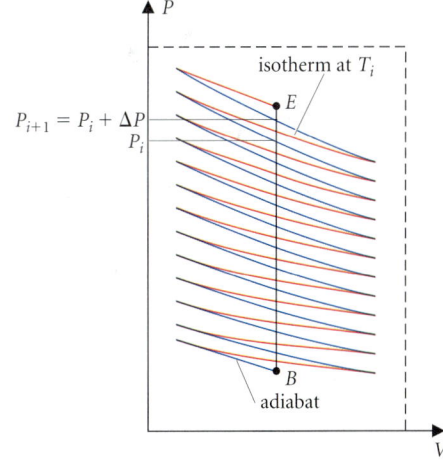

■ FIGURE 22.6
An adiabat-isotherm approximation to a constant-volume process. Compare with Figure 22.1. The region in the dashed box in that figure is plotted here. The vertical scale has been enlarged. By increasing the number of isotherms, the composite process approaches the original process more closely.

THE EXTENSION TO ARBITRARY PROCESSES IS NECESSARY TO PROVE THAT S IS A STATE VARIABLE.

where Q_i is the heat transfer during the ith isothermal process at temperature T_i. For short enough isotherms, Q_i/T_i equals dQ/T on the corresponding piece of the constant-volume process (see Problem 80). In the limit where we imagine the number of adiabats and isotherms increasing to infinity, the sum becomes:

$$\Delta S = \int_E^B \frac{dQ}{T}. \quad (22.10)$$

A similar construction works for any process represented by a path in the P-V diagram, and so eqn. (22.10) gives the entropy change for any *reversible* process.

> **EXAMPLE 22.3** ♦♦ One mole of monatomic ideal gas expands by a factor of 10. Compare the entropy changes of the gas when the expansion is carried out isobarically, isothermally, and adiabatically.
>
> **MODEL** Since no heat transfer occurs in the adiabatic process, the entropy change is zero. For the isothermal and isobaric processes, we evaluate the integral (22.10).
>
> **SETUP** For the isothermal process, there is no change in internal energy, and the heat transfer equals the work done by the gas.
>
> **STEPS I & II** We describe the process as a sequence of differential expansions dV, during each of which there is a heat transfer $dQ = dW = P\,dV$.
>
> **STEP III** We express the pressure in terms of volume using the ideal gas law for 1 mol of gas. Then $dS = dQ/T = P\,dV/T = R\,dV/V$.
>
> **STEP IV** The limits of integration are from V_i to $V_f = 10V_i$.
>
> **SOLVE**
> **STEP V**
> $$\Delta S_{\text{isothermal}} = \int \frac{dQ}{T} = \int_{V_i}^{V_f} \frac{R\,dV}{V} = R\ln\left(\frac{V_f}{V_i}\right) = R\ln(10).$$
>
> **SETUP**
> **STEP I** For the isobaric process, we calculate the heat transfer using the molar specific heat at constant pressure: $c_p' = \frac{5}{2}R$ for a monatomic gas. We describe the process as a sequence of differential temperature changes dT:
>
> **STEPS II & III** $\quad dS = dQ/T = c_p'\,dT/T = \frac{5}{2}R\,dT/T.$
>
> **SOLVE**
> **STEPS IV & V**
> $$\Delta S_{\text{isobaric}} = \int \frac{dQ}{T} = \int_{T_i}^{T_f} \frac{5}{2}\frac{R}{T}\,dT = \frac{5}{2}R\ln\left(\frac{T_f}{T_i}\right).$$
>
> **ANALYZE** From the ideal gas law, $T_f/T_i = V_f/V_i$, so: $\Delta S_{\text{isobaric}} = 2.5(\Delta S_{\text{isothermal}})$.
>
> Remember that the final states in the three processes are not the same. They each have the same volume, but the pressures differ: $P_{\text{isobaric}} > P_{\text{isothermal}} > P_{\text{adiabatic}}$.
>
> There are two kinds of information we may have about a molecule in a gas sample: its position and its velocity. We lose information about position when the volume occupied by the gas increases. Similarly, we lose information about velocity when the range of possible speeds increases, as happens when the temperature increases.
>
> In the isothermal expansion, only the volume changes, and $\Delta S_{\text{isothermal}} = R\ln(10)$ is the loss of information due to the volume increase. The temperature increase in the isobaric expansion causes an extra information loss of $1.5R\ln(10)$. Temperature decreases in the adiabatic expansion, causing an increase in velocity information that exactly balances the loss of position information. ∎

22.4.3 Entropy as a State Variable

If entropy is a state variable, as we have claimed, it describes the disorder of a particular thermodynamic state, independent of how the system got into that state. Thus the entropy change between two states is independent of the path chosen to evaluate the change. The change in S given by eqn. (22.10) is independent of the reversible path between two states

A and B, or, equivalently, $\int dQ/T$ around any closed path $A \to B \to A$ in the P-V diagram is zero.

To prove this assertion, we approximate each of two arbitrary paths between the states A and B by a sequence of adiabats and isotherms (■ Figure 22.7). The path $WXYZ$ between points W and Z may be replaced by the path $WXVZ$ for purposes of computing $\Sigma\ Q_i/T_i$, because $XVZYX$ is a Carnot cycle in which $Q_{VZ}/T_{VZ} = Q_{XY}/T_{XY}$ (eqn. 22.8, with $Q_{XY} = -Q_{YX}$). By a sequence of such alterations, we may change the two approximate paths to a single Carnot cycle, for which $\Sigma\ Q/T = 0$. Since $\Sigma\ Q/T = 0$ for all such approximations, the limit $\int dQ/T = 0$. The entropy difference between states A and B is independent of path, and S is a state variable.

Notice, however, that we have not derived (and shall not derive!) an expression for S itself, but only for the change in S due to any process. For practical purposes, we require only a method for computing changes, and that we have achieved.

SEE PROBLEM 33 FOR AN EXAMPLE OF HOW TO DO THIS.

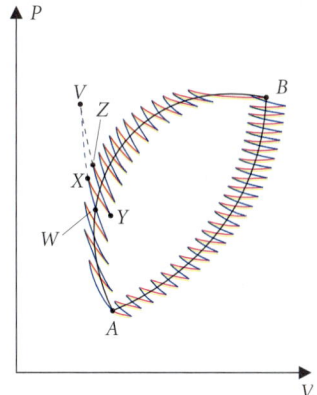

■ FIGURE 22.7
Adiabat-isotherm approximations to two arbitrary, reversible processes between states A and B. The entropy change along path XYZ is the same as that along XVZ. By a sequence of such alterations, we may replace the whole cycle ABA with a single Carnot cycle, for which the net entropy change ΔS is zero. Then ΔS is also zero for the cycle ABA. The entropy change between A and B is independent of path.

22.4.4 Entropy Change and Reversibility

We derived expression (22.10) for ΔS using reversible processes. We have no guarantee that it can be applied directly to a real, irreversible process (and indeed it cannot!). It is useful because we can always find an *imaginary* reversible process that links the initial and final states of a system. (The set of isotherms and adiabats is one such process.) Since entropy is a state variable, the change in entropy in going from one state to another is independent of the process, reversible or not. Once we have found the entropy change for a reversible process, we have found the entropy change for any process linking the same two states, regardless of how the system actually achieves its final state.

This argument has an odd feature. If the net entropy change of the universe for a reversible process is zero and we can always find an imaginary reversible process between beginning and end states of any process, how can any net entropy changes occur at all? To answer the question, let's look at the role of heat reservoirs in ideal processes.

The isothermal process in the Carnot cycle (AB in Figure 22.3) is reversible in an ideal world, so the entropy change for the universe is zero for this process. The entropy change for the system, however, is positive:

$$\Delta S_{\text{sys}} = \frac{Q_{AB}}{T_{\text{H}}} = Nk \ln\left(\frac{V_B}{V_A}\right).$$

The heat transfer into the system comes from the heat reservoir at temperature T_{H}, and the entropy change of that reservoir is negative, so:

$$\Delta S_{\text{universe}} = \Delta S_{\text{sys}} + \Delta S_{\text{res}} = 0.$$

The interaction of the system with the ideal heat reservoir is crucial. It makes the isothermal expansion reversible in principle. As we have seen, an ideally reversible process that is neither isothermal nor adiabatic is the limit of infinitely many adiabat-isotherm combinations, and so requires an infinite number of reservoirs to achieve the necessary heat transfer in a reversible manner. Even in a world where ideal heat reservoirs are available, this infinity requires an extra stretch of the imagination. It is the important difference between the Carnot cycle and other cycles.

In any real process, an infinite supply of heat reservoirs is not available. Heat transfer occurs by some practical means such as the combustion of fuel in the Otto cycle. This inability to provide reservoirs is the essential reason why such processes are truly irreversible.

To compute entropy change with a reversible process, we imagine heat reservoirs that interact with the system. The difference between the real process and the imaginary process lies in these interactions. The reservoirs have a negative net entropy change, which cancels the positive entropy change of the system:

$$0 = \Delta S = \underset{\text{(positive)}}{\Delta S_{\text{sys}}} + \underset{\text{(negative)}}{\Delta S_{\text{reservoirs}}}.$$

OF COURSE, REAL PROCESSES HAVE OTHER DIFFICULTIES AS WELL. THEY CANNOT PROCEED INFINITELY SLOWLY, FOR EXAMPLE. FOR THESE REASONS, NO REAL REVERSIBLE CARNOT CYCLE EXISTS.

EXAMPLE 22.4 ♦♦ The free expansion of a gas from a punctured tire (cf. §22.1) was our first example of an irreversible process. In a controlled version of this process, gas is originally confined to half a container by a partition, and the other half is evacuated. When a hole is punctured in the partition, the gas expands through the hole to fill the entire container (■ Figure 22.8). The container is perfectly insulated so that no heat transfer occurs during the expansion. Find the entropy change of the gas and its environment as a result of the free expansion. What is the actual entropy change of the universe?

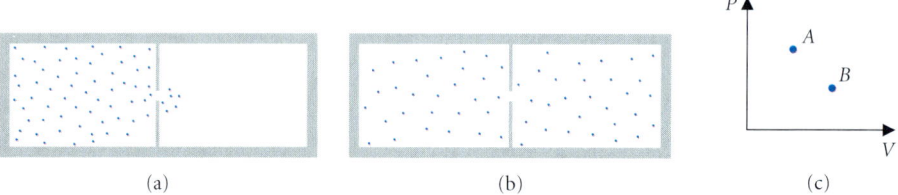

■ **Figure 22.8**
Free expansion of a gas is an irreversible process. (a) A hole is punctured in a partition that divides gas in one-half of an insulated container from the other, evacuated half. The gas expands to fill the container. (b) In the final state, the gas is uniformly distributed throughout the container. Its temperature has not changed. (c) We may plot the initial and final states on a P-V diagram. However, the gas is not in equilibrium during the process, so there is no way to plot the *process*.

MODEL Since the actual process is irreversible, we calculate the entropy change using an imaginary reversible process between the initial and final states of the system (■ Figure 22.9). Since no work is done in the free expansion (the gas has nothing to push against) and there is no heat transfer, $\Delta U = Q + W = 0$. Thus the temperature of the gas is the same after the free expansion as it was before. An isothermal expansion also links two states of differing volumes at the same temperature, and so it is a suitable imaginary reversible process. To perform this imaginary isothermal expansion, we introduce a heat reservoir that provides the necessary heat transfer at constant temperature and a movable piston that does work on the environment.

SETUP The heat transfer during the isothermal process (eqn. 19.13) is:

$$Q = NkT \ln(V_f/V_i) = NkT \ln 2.$$

SOLVE The entropy gained by the system in this process (eqn. 22.9) is:

$$\Delta S = Q/T = Nk \ln 2.$$

The imaginary reservoir has a negative entropy change:

$$\Delta S_{res} = -(Q/T) = -Nk \ln 2,$$

making the total entropy change of the universe for the imaginary reversible process equal to zero.

In the real process, the entropy change of the system (the gas in the box) is the same as in the reversible process, $\Delta S = Nk \ln 2$, positive, as required. However, there is no interaction with the environment, so the entropy change of the universe is just that of the system.

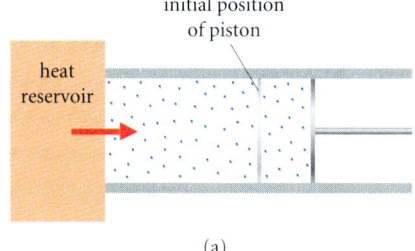

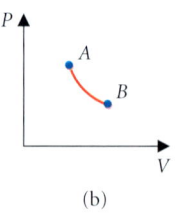

■ **Figure 22.9**
(a) An imaginary reversible process. An imaginary isothermal expansion links the initial and final states of the actual free expansion. Heat transfer from the reservoir is converted to work done on the piston. (b) The imaginary process can be plotted as a line in the P-V diagram: it is an isotherm linking states A and B.

ANALYZE In contrast with reversible heat transfer, where entropy flows from place to place like any other form of stuff, in the free expansion, entropy is created out of nothing.

If we were to use the same process (chosen to obtain the correct entropy change for the system) to calculate the entropy change of the environment, we'd get the wrong answer. We need a different imaginary process to get the correct answer for the environment: "do nothing." But this process doesn't give us the correct entropy change for the system. Construction of imaginary reversible processes is a useful computational tool, but no *one* process works for the whole universe. ∎

Heat conduction is a second process in which entropy is created. The irreversibility of this process explains why heat flows from hot to cold.

EXAMPLE 22.5 ♦♦ The ends of a uniform copper rod of length 2.0 m and cross-sectional area 0.80 cm² are maintained at different, constant temperatures $T_H = 405$ K and $T_C = 295$ K. Find the rate at which entropy is generated by heat conduction through the rod.

MODEL Heat conduction is an irreversible process. We need an imaginary reversible process to compute the entropy change. The rod links two heat reservoirs at temperatures T_H and T_C (■ Figure 22.10) and transfers heat between them. We obtain a suitable reversible process by replacing the rod with two more reservoirs at temperatures T_C and T_H and imagining heat flow in one direction between reservoirs at the same temperature. Then heat transfer from the original hot reservoir occurs at a constant temperature T_H, and heat transfer to the original cold reservoir occurs at T_C. No heat transfer occurs at other temperatures, so the process is reversible.

SETUP Heat flows along the rod (§21.1) at a rate:

$$H = kA \frac{(T_H - T_C)}{L}.$$

This changes the hot reservoir's entropy:

$$\frac{dS_H}{dt} = \frac{dQ_H/dt}{T_H} = \frac{-H}{T_H}.$$

Entropy is gained by the cold reservoir at a rate:

$$\frac{dS_C}{dt} = \frac{dQ_C/dt}{T_C} = \frac{H}{T_C}.$$

SOLVE The net rate of change in entropy for the two original reservoirs is thus:

$$\frac{dS_{net}}{dt} = \frac{dS_H}{dt} + \frac{dS_C}{dt} = -\frac{H}{T_H} + \frac{H}{T_C}$$

$$= H\left(\frac{1}{T_C} - \frac{1}{T_H}\right) = H\frac{(T_H - T_C)}{T_H T_C}$$

$$= kA\frac{(T_H - T_C)^2}{L T_C T_H},$$

which is positive. Thus entropy is created by heat conduction through the rod at a rate:

$$\frac{dS}{dt} = kA\frac{(T_H - T_C)^2}{L T_C T_H} = (400 \text{ W/m}\cdot\text{K})(0.80 \times 10^{-4} \text{ m}^2)\frac{(405 \text{ K} - 295 \text{ K})^2}{(2.0 \text{ m})(295 \text{ K})(405 \text{ K})}$$

$$= 1.6 \times 10^{-3} \text{ W/K}.$$

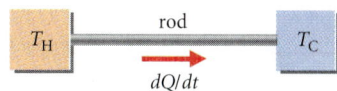

(a) Real process.

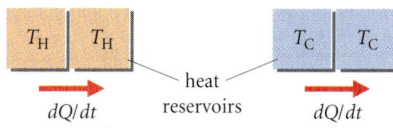

(b) Imaginary process.

■ **FIGURE 22.10**
Heat conduction through a rod. (a) The real process: heat flows through the rod from the hot reservoir to the cold reservoir. (b) Imaginary process: we replace the rod with two more reservoirs that allow heat transfer at constant temperature.

ANALYZE The two reservoirs we added to execute the imaginary process are not real, so the entropy they lose is also unreal. The whole universe gains entropy at the rate $dS/dt = 1.6 \times 10^{-3}$ W/K.

If heat were to flow from cold to hot, the signs of dS_H/dt and dS_C/dt would be reversed, and dS_{net}/dt would be negative, a result that is forbidden by the second law of thermodynamics.

From the point of view of randomness, heat conduction occurs because the system (rod) is trying to reach the more random state of uniform temperature.

Examples 22.4 and 22.5 exhibit the reasons why even the isothermal processes of the Carnot cycle are only ideally reversible. The material in a real reservoir has finite conductivity and so a temperature gradient is necessary for heat transfer. Entropy is produced as in Example 22.5. Furthermore, any expansion at a finite rate requires a pressure gradient. Flow of gas along this gradient is somewhat similar to the free expansion and also generates entropy. Attempting to reverse the process produces pressure and temperature gradients in the opposite direction, and so inevitably fails to repeat the previous states of the system.

22.5 A Limit on Efficiency

The Carnot cycle is the most efficient heat engine. The distinctive feature of the Carnot cycle is that it needs only two heat reservoirs for its operation. Even if we could invent another device—a superengine—operating on a cycle that needs only two reservoirs, it could not be more efficient than the Carnot cycle.

To demonstrate why, let's imagine a composite machine that uses the work output of the hypothetical superengine to run the Carnot cycle in reverse as a refrigerator (■ Figure 22.11). (In the discussion that follows, each symbol refers to the absolute value of the corresponding energy transfer.)

The superengine does work $W = Q_H - Q_C$ as a result of heat input Q_H from the hot reservoir and rejection of Q_C to the cold reservoir. This work drives the Carnot-cycle refrigerator, which removes heat Q'_C from the cold reservoir and dumps heat Q'_H into the hot reservoir. The net effect of the composite machine is to transfer heat between the cold and hot reservoirs. The total entropy change of the reservoirs per cycle is:

$$\Delta S = \frac{Q_C - Q'_C}{T_C} + \frac{Q'_H - Q_H}{T_H},$$

which must be positive.

Let us express the entropy change in terms of the efficiencies of the two engines. If the efficiency of the superengine is e_s, then $W = e_s Q_H$. For the Carnot cycle, $W = e_c Q'_H$. Since these amounts of work are equal:

$$Q_H = \frac{e_c}{e_s} Q'_H.$$

Now, from the first law of thermodynamics applied to the superengine,

$$Q_C = -W + Q_H = -W + \frac{e_c}{e_s} Q'_H,$$

and applied to the Carnot cycle,

$$Q'_C = -W + Q'_H.$$

We use these results to eliminate Q_C, Q_H, and Q'_C from the expression for ΔS:

$$\Delta S = \frac{-W + (e_c/e_s)Q'_H + W - Q'_H}{T_C} + \frac{Q'_H - (e_c/e_s)Q'_H}{T_H}$$

$$= Q'_H \left(\frac{e_c}{e_s} - 1\right)\left(\frac{1}{T_C} - \frac{1}{T_H}\right).$$

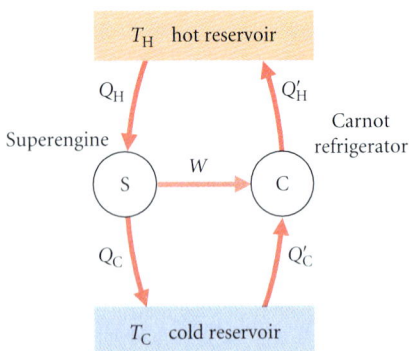

■ **FIGURE 22.11**
A hypothetical *superengine* operates between reservoirs at temperatures T_H and T_C. The superengine is used to run a Carnot refrigerator between the same two reservoirs. The composite machine transfers heat between the two reservoirs, but no work is done on the outside world. According to the second law, the heat transfer can go only one way: from hot to cold. This limits the efficiency of the superengine.

IN CONTRAST, TO BE REVERSIBLE THE OTTO CYCLE WOULD REQUIRE AN INFINITE NUMBER OF HEAT RESERVOIRS.

ANY CYCLE THAT OPERATES WITH AN IDEAL GAS AS ITS WORKING FLUID AND HAS ONLY TWO HEAT RESERVOIRS MUST BE A CARNOT CYCLE. OUR DEVICE WOULD HAVE TO USE A DIFFERENT MECHANISM, SUCH AS THERMAL EXPANSION OF A ROD, TO DO WORK.

Since $T_C < T_H$, the term in the second parentheses is always positive. Hence ΔS is positive if $e_c > e_s$ and zero if $e_c = e_s$, but negative if $e_s > e_c$. Since entropy always increases, the superengine cannot have an efficiency greater than that of the Carnot cycle.

If the superengine were reversible, then we could reverse the composite machine. By repeating the above argument, we could then prove that the Carnot cycle's efficiency cannot exceed that of the reversible superengine. Thus the two efficiencies are equal.

> Any reversible engine operating between two heat reservoirs at T_H and T_C has an efficiency:
> $$e_c = 1 - \frac{T_C}{T_H}.$$

An irreversible engine operating between these two reservoirs has a lesser efficiency.

Of course, most engines do not operate between only two reservoirs. In terms of temperatures, the efficiency of an Otto-cycle engine (cf. Figure 22.1) is:

$$e_{oc} = 1 - \frac{T_E}{T_D}.$$

■ **EXERCISE 22.3** ♦♦ Derive this expression from the results of §22.2. ■

The "cold" temperature for the Otto cycle is the temperature at the end of the expansion and the "hot" temperature is the temperature after the heat input. However, T_E is not the coldest temperature (T_B) reached during the cycle. Thus the Otto cycle, even if it could be made truly reversible, is inherently less efficient than a Carnot cycle.

Can any engine ever have perfect efficiency, $e = 1$? If we could build a Carnot cycle whose hot reservoir had $T_H = \infty$, or whose cold reservoir had $T_C = 0$ K, e would be unity. An infinite temperature is clearly impossible, since this would require infinite energy. A temperature of absolute zero is also impossible, as we discuss in §22.6. Thus all engines, even reversible ones, have an efficiency less than unity. There are no perfect heat engines.

> This is Kelvin's statement of the second law (see §22.1).

22.6 THE SIGNIFICANCE OF ABSOLUTE ZERO

22.6.1 The Thermodynamic Temperature Scale

We are now ready to investigate the nature of absolute zero temperature. In Chapter 19 we described temperature in terms of the average energy of ideal gas molecules. An ideal gas at absolute zero temperature would have zero gas pressure, and its molecules would have no kinetic energy. However, since no gases behave ideally at very low temperatures, an improved definition of temperature is needed. A thermodynamic definition based on heat transfer by reversible engines is appropriate at any temperature.

> Real gases would condense into liquid or solid form before reaching absolute zero.

> The ratio of the temperatures of two reservoirs is the ratio of the heat transfers between each reservoir and a reversible engine operating between them:
>
> $$\frac{T_H}{T_C} \equiv \frac{Q_H}{|Q_C|}. \qquad (22.11)$$

> This definition is based on relation (22.8).

Since an ideal gas can be used in a reversible Carnot engine, the thermodynamic and ideal gas temperature scales are identical. The thermodynamic definition, expressed for any reversible engine, is independent of any particular laboratory equipment; thus it is the basis for operational definitions that are especially useful at very low temperatures.

> See Essay 6: Low temperatures and their measurement.

■ **EXAMPLE 22.6** ♦ A Carnot-cycle engine using helium as a working fluid operates between a liquid nitrogen reservoir at 65 K and a system whose temperature is to be measured. In each cycle, 45 J of heat is removed from the nitrogen reservoir and 27 J of heat is dumped into the system. What is the system's temperature?

MODEL We use the thermodynamic temperature scale, in the form of eqn. (22.11).

SETUP Here $T_H = 65$ K, $Q_H = 45$ J, and $Q_C = -27$ J.

SOLVE Thus:

$$T_C = T_H \frac{|Q_C|}{Q_H} = (65 \text{ K})\left(\frac{27 \text{ J}}{45 \text{ J}}\right) = 39 \text{ K}.$$

ANALYZE Since $|Q_C| < Q_H$, $T_C < T_H$ as expected.

22.6.2 Absolute Zero

The efficiency of a Carnot cycle, $e_c = 1 - T_C/T_H$, would be unity if the cold reservoir were at absolute zero. The heat exhausted during the isothermal compression of the Carnot cycle, $Q = NkT_C \ln(V_f/V_i)$, is zero if $T_C = 0$ K. At absolute zero, isothermal processes involve no work and no heat transfer! Thus a cycle would have no waste heat transfer and a perfect efficiency. However, absolute zero is not achievable.

Suppose we try using a Carnot refrigerator to cool a system, with the aim of reaching absolute zero. We attach the system as the low-temperature reservoir of the Carnot refrigerator and adjust the refrigerator to remove an amount of heat $|Q_C|$ from the low-temperature reservoir (the system). The heat rejected to the high-temperature reservoir is then:

$$|Q_H| = \frac{T_H}{T_C}|Q_C|.$$

To operate the cycle, we must do work:

$$W = Q_{net} = \left(\frac{T_H}{T_C} - 1\right)|Q_C|.$$

As the temperature of the low-temperature reservoir drops during each cycle, the amount of work that we do increases and tends to infinity as $T_C \to 0$, no matter how small $|Q_C|$ is. We cannot reduce T_C to zero because it is impossible to do an infinite amount of work. It would be equally impossible to take a system at absolute zero and raise its temperature reversibly. This conclusion is often called the *third law of thermodynamics*:

> No system at finite temperature can be cooled to absolute zero.

AS FAR AS WE KNOW, THERE ARE NO SUCH SYSTEMS IN THE UNIVERSE. THE COLDEST TEMPERATURE THAT EXISTS ANYWHERE IN THE NATURAL COURSE OF THINGS IS PRESENTLY ABOUT 2.7 K, THE TEMPERATURE OF RADIATION THAT IS PRESENT THROUGHOUT THE UNIVERSE. CAREFUL LABORATORY EXPERIMENTS HAVE SUCCEEDED IN PRODUCING TEMPERATURES OF THE ORDER OF A FEW PICOKELVIN.

The significance of absolute zero lies in this inability to transfer heat into or out of a system at absolute zero. The internal energy of a system at absolute zero need not be (and in general is not) zero, but any internal energy it has cannot be extracted.

✷ 22.7 STATISTICAL MECHANICS

22.7.1 The Boltzmann Factor

To this point, we have discussed thermodynamic principles primarily from a macroscopic point of view, in which each state is described by a set of macroscopic variables such as P, V, T, U, and S. These variables derive from the properties of the individual molecules. For example, the gas pressure is related to the rms speed of the molecules. To understand entropy from a microscopic point of view, we need a better description of how molecules behave as temperature changes.

In Chapter 19, we discussed how a cup comes to equilibrium when hot coffee is poured into it. On the microscopic scale, molecules of the coffee continually collide with molecules in the cup that are also in thermal motion. Similar interactions occur between the coffee and the air. In thermal equilibrium, the net result of these interactions averages to zero: energy is transferred from coffee molecules to cup molecules at the same rate as from the cup to the coffee.

The molecules cannot all have the same energy. Even if we could carefully arrange initial conditions so that each molecule had exactly the average value $kT/2$ in each mode, the first collision would produce one molecule with more energy than average, and one with less. Thus the molecules have a distribution of energies. The principle of equipartition specifies only the average energy per molecule; it does not tell us the energy of any one molecule or how energy is distributed among the molecules. For that, we need a more powerful principle.

SEE § 19.7.1 FOR THE DEFINITION OF MODES.

It is remarkable, and certainly not obvious, that the balance of microscopic processes occurs not just on average, but in precise detail. That is, for each coffee molecule with an energy in the range E_1 to $E_1 + dE$ that gains energy ΔE from the cup, there is a corresponding molecule, initially with a higher energy, that loses an equal amount of energy ΔE and ends up with energy in the range E_1 to $E_1 + dE$. The effect of the interactions is merely to change which molecule does what. Once thermal equilibrium is established, the molecules have a well-defined distribution of energies such that the fraction of molecules in any given energy range is fixed.

DIG DEEPER FOR A MORE EXTENSIVE DISCUSSION OF DISTRIBUTIONS.

Most molecules have energy less than or about equal to $\frac{1}{2}kT$ in each mode. Those with an energy much greater than kT in any one mode are rare, since they would have to gain energy from each of several successive collisions. In equilibrium, the energy distribution decreases rapidly for $E > kT$. Ludwig Boltzmann (1844–1906) demonstrated that for two states that differ by energy ΔE, the ratio of the numbers of molecules in the two states is proportional to the *Boltzmann factor* $\exp(-\Delta E/kT)$. Equivalently, the probability of finding a molecule with energy E is proportional to $\exp(-E/kT)$.

Molecules can have the same energy even though they move in different ways. (For example, molecules moving at the same speed v in the x-direction or the y-direction have the same kinetic energy.) The number of molecules with a given energy is also proportional to the number of ways to achieve that energy.

> **THE BOLTZMANN RELATION**
>
> Let n_i be the number of molecules with energy E_i and g_i be the number of different ways for a molecule to achieve energy E_i. Then:
>
> $$\frac{n_2}{n_1} = \frac{g_2}{g_1} e^{-(E_2 - E_1)/kT} \qquad (22.12)$$

The Boltzmann relation explains the transitions between different values of specific heat (Figure 19.19). At low temperatures ($kT \ll$ the quantum of rotational energy E_R), almost no molecules can exchange enough energy in a collision to increase the rotational or vibrational energy of another molecule. As the temperature increases, the number of molecules with translational energy greater than E_R increases according to eqn. (22.12), and the fraction of collisions that cause rotation increases correspondingly. When $kT \gg E_R$, the Boltzmann factor $\exp(-E_R/kT)$ is nearly unity, and most molecules have enough energy to cause rotation; the rotational mode is fully excited. In hydrogen, for example, $E_R/k \approx 86$ K. The transition begins at about 70 K, where $\exp(-E_R/kT) = 0.3$, and is essentially complete at about 300 K, where $\exp(-E_R/kT) = 0.75$. Similar behavior occurs as kT approaches the necessary energy for vibrational transitions.

SEE §19.7.2, WHERE WE INTRODUCED THE QUANTA OF ROTATIONAL AND VIBRATIONAL ENERGY.

EXERCISE 22.4 ❖ Show that the expression for pressure in an isothermal atmosphere ($P = P_o e^{-\rho_o g y/P_o}$, §13.3.4) follows from the Boltzmann relation (22.12).

Digging Deeper

What Is a Distribution?

Just as gas molecules have a distribution of energies, physics students have a distribution of heights. The heights of students in two different physics classes are plotted in ■ Figure 22.12. The histogram has 5-cm bins; that is, all students with heights between 170 cm and 175 cm are counted together in that bin. (*Note:* The probability that a student selected at random has a specific height, say 165.00 cm, is zero. However, the probability that the height is between 162.5 cm and 167.5 cm is finite. We always need a range of values.) The histograms of the two classes are similar in overall shape (peak at about 175 cm, no students shorter than 150 cm or taller than 195 cm, etc.). The differences are due to the random selection of people in the two classes. By averaging over a large number of classes, we obtain a curve showing general features that depend on the entire population of students. This is a *distribution function* (black line) for heights of students. It does not represent any particular class but instead accurately describes the probability that a randomly chosen student's height lies in a specific range.

A gas system at one instant of time is like a single class of students. Collisions change the state so rapidly that any macroscopically interesting property is an average over many microstates and is best described by an average distribution function.

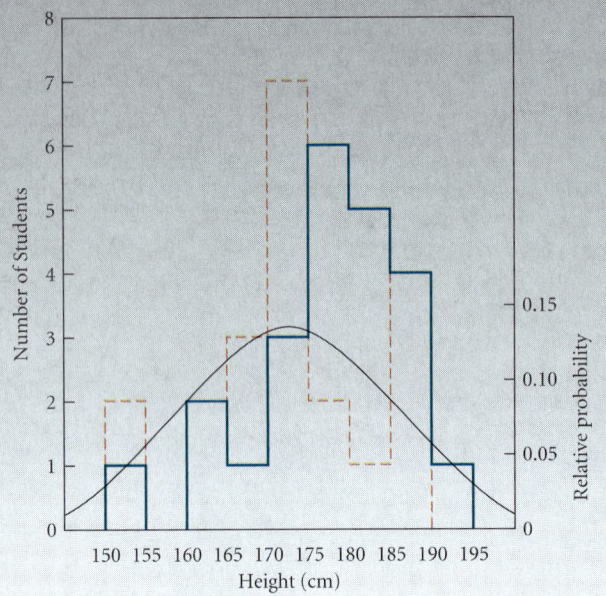

■ **Figure 22.12**
Distribution of heights for two classes of physics students. Solid green line: an upper division class. Orange line: a sophomore class. The bin size is 5 cm. The curve shows the average distribution function for this population of students. The value of the distribution function at a specific bin (the scale on the right) is the probability that a randomly chosen student has a height in that bin.

22.7.2 The Maxwell–Boltzmann Distribution

Using the Boltzmann relation, we can determine how the velocities of molecules are distributed. We begin by looking at one component of the velocity, say the x-component. Any one molecule can have any value of v_x from minus infinity to plus infinity. To define a state, we first define a bin size dv_x (see *Digging Deeper*). The energy of a molecule whose x-component of velocity is between a specified value v_x and $v_x + dv_x$ lies between the values $E = \frac{1}{2}mv_x^2$ and $E + dE = \frac{1}{2}m(v_x + dv_x)^2 \approx \frac{1}{2}mv_x^2 + mv_x\,dv_x$. The bigger we make dv_x, the larger the range of energies that constitute the state, and so the more molecules are in that state. Thus g in eqn. (22.12) is proportional to dv_x. Then according to Boltzmann's relation, the probability of finding a molecule with x-component of velocity between v_x and $v_x + dv_x$ is proportional to $\exp(-\frac{1}{2}mv_x^2/kT)dv_x$. The probability is maximum at $v_x = 0$, corresponding to the minimum possible energy. The average value of v_x is also zero; the gas as a whole does not move to right or left. This result is confirmed experimentally using apparatus similar to that shown in ■ Figure 22.13.

The term $(dv_x)^2$ is much smaller, so we drop it.

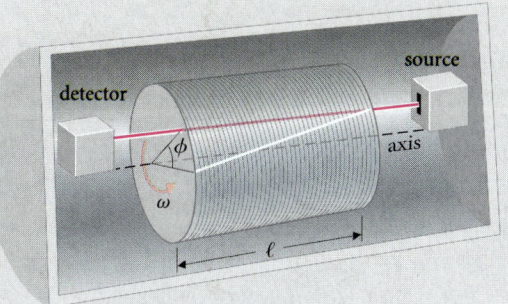

■ **Figure 22.13**
Apparatus for measuring the speed distribution of molecules. Hot gas in the box exits through a small hole and forms a beam that passes down a spiral groove in the rotating cylinder. Only those molecules with the specific speed $v = \omega \ell / \phi$ can pass down the groove. The molecular speed distribution is deduced from the variation of particle flux with the angular speed of the cylinder.

EXAMPLE 22.7 ❖ Why does the Moon have essentially no atmosphere?

MODEL The molecules that escape a planet are those with a vertical component of velocity greater than the escape speed (cf. Example 8.8), and with an altitude great enough that they have a small chance of colliding on the way out.

SETUP The gravitational acceleration at the surface and the corresponding escape speed are much less for the Moon than for the Earth. The temperature of the Earth's atmosphere is determined by the amount of solar radiation that falls on it. The Earth and Moon are, on average, the same distance from the Sun, and one might expect the temperature of any lunar atmosphere to be similar to that of the Earth.

SOLVE The distribution of vertical velocity components for nitrogen and hydrogen at 1600 K (the temperature of the Earth's upper atmosphere) is shown in ■ Figure 22.14. A much larger fraction of nitrogen molecules are able to escape from the Moon than from the Earth. As the fast molecules escape, leaving the slow molecules behind, they change the distribution. Collisions tend to restore equilibrium by producing more high-speed molecules, which then escape. In the 4.5 billion years since the Moon formed, all of the nitrogen has escaped.

THE ATMOSPHERE COOLS FROM THE SURFACE UPWARD, BUT ITS TEMPERATURE THEN RISES AGAIN AT VERY HIGH ALTITUDES, ABOVE ABOUT 20 KM.

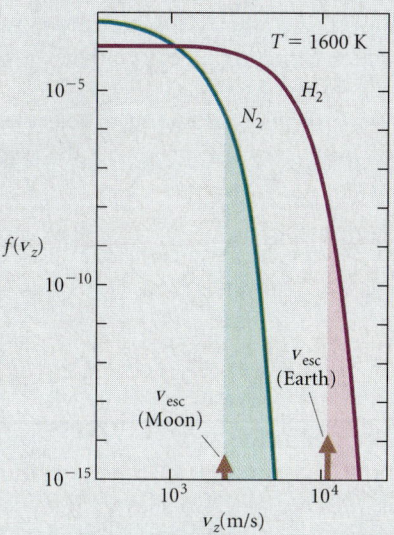

■ **Figure 22.14**
Distribution of vertical velocities for nitrogen and hydrogen at 1600 K, typical of the temperature at about 1000-km altitude in the Earth's atmosphere. The distribution has the characteristic exponential form. A fairly large fraction of nitrogen molecules, those with the highest speeds, can escape from the Moon. Far fewer molecules, too few to appear on this graph, could escape from the Earth. The values of v_{esc} for the Earth and the Moon are from Chapter 8.

ANALYZE Since more massive molecules move slower, the Moon has been able to retain an extremely tenuous atmosphere of argon atoms. The more massive Earth has retained its nitrogen, but has lost essentially all of its hydrogen gas. ■

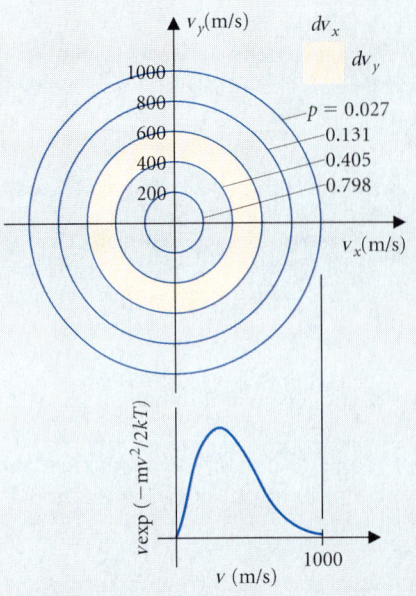

FIGURE 22.15
Two-dimensional speed distribution for nitrogen molecules at 300 K. The contours of constant Boltzmann factor are separated by a bin size of $\Delta v = 200$ m/s. A molecule with a speed v may lie anywhere on a circle of radius v centered at the origin. To be in the bin v to $v + \Delta v$, the molecule's velocity must lie in a narrow band of area $\Delta A = 2\pi v\, \Delta v$, and the probability that its speed is in the bin is the Boltzmann factor times that area. The Boltzmann factor decreases uniformly as v increases, but ΔA increases with v. The product shown here has a maximum at a finite value of $v > 0$.

THE RULES OF PROBABILITY ARE GIVEN IN APPENDIX IG.

Now let us consider molecules that move in two dimensions: x and y. The probability that a molecule has a certain x-component *and* a certain y-component of velocity is equal to the product of the probabilities:

$$\exp(-\tfrac{1}{2}mv_x^2/kT)\exp(-\tfrac{1}{2}mv_y^2/kT)dv_x\, dv_y = \exp[-\tfrac{1}{2}m(v_x^2 + v_y^2)/kT]dv_x\, dv_y$$
$$= \exp(-\tfrac{1}{2}mv^2/kT)dv_x\, dv_y,$$

where $v^2 = v_x^2 + v_y^2$.

■ Figure 22.15 illustrates how the distribution function $\exp(-\tfrac{1}{2}mv^2/kT)$ varies with v_x and v_y. The function is constant on circles $v = $ constant. The factor $dv_x\, dv_y$ is represented by an element of area in the figure. Probability is represented by the function value times the area. So the probability of finding a molecule with a speed between v and $v + dv$ is equal to the constant value of the distribution function times the area $2\pi v\, dv$ between two neighboring curves. This speed distribution function is zero at the origin because the area factor $2\pi v\, dv$ vanishes there. The highest probability occurs for a nonzero v.

Extending this argument to three dimensions, the probability of finding a molecule with speed v to $v + dv$ is proportional to $\exp(-\tfrac{1}{2}mv^2/kT)4\pi v^2\, dv$. We may write the result in terms of the number of molecules with speed in the given range:

$$n(v)\, dv = Nf_M(v)\, dv,$$

where N is the total number of molecules and $f_M(v)$ is the Maxwell–Boltzmann speed distribution:

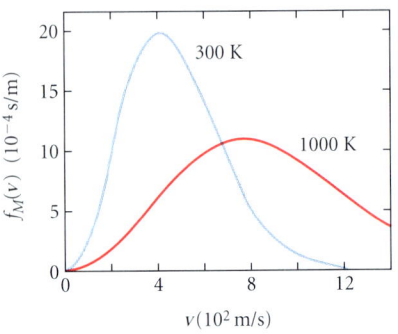

FIGURE 22.16
Speed distribution for nitrogen molecules at 300 K and 1000 K. As temperature increases, the distribution is pushed toward higher speeds.

$$f_M(v) = \left(\frac{m}{2\pi kT}\right)^{3/2} 4\pi v^2 \exp\left(-\frac{mv^2}{2kT}\right). \quad (22.13)$$

This proportionality constant ensures that $\int f_M(v)\, dv = 1$ and thus that the total number of molecules, summed over all possible speeds, is N. ■ Figure 22.16 shows the Maxwell–Boltzmann distribution for nitrogen molecules at two different temperatures.

THIS DISTRIBUTION IS OFTEN CALLED SIMPLY "THE MAXWELLIAN."

EXAMPLE 22.8 ♦♦ Compare the number of molecules having speed between v_{rms} and $v_{rms} + dv$ with the number whose speed is twice as great (between $2v_{rms}$ and $2v_{rms} + dv$).

MODEL We use the Maxwell–Boltzmann distribution to compare the numbers.

SETUP The rms speed is related to the temperature: $v_{rms}^2 = 3kT/m$ (eqn. 19.9).

SOLVE
$$\frac{n_2}{n_1} = \frac{v_2^2 \exp[-mv_2^2/(2kT)]}{v_1^2 \exp[-mv_1^2/(2kT)]} = \frac{4v_{rms}^2 e^{-6}}{v_{rms}^2 e^{-3/2}} = 4e^{-9/2} = 0.04.$$

ANALYZE The number of molecules with a speed of about $2v_{rms}$ is only 4% of those whose speed is about equal to v_{rms}.

Once we know the distribution of speeds, we can find the average value of any other quantity that depends on speed. Kinetic energy is one such quantity.

EXAMPLE 22.9 ♦♦♦ Use the Maxwellian distribution to show that the average kinetic energy of a molecule in a gas at temperature T is $3kT/2$ (eqn. 19.8).

MODEL The average energy is found by adding the energy of all the molecules and dividing by the total number of molecules.

SETUP
STEP I We use the integration scheme from Interlude 2. Each molecule is labeled with its speed v.

STEP II A typical element is an interval dv in speed.

STEP III The desired quantity is the total kinetic energy of the molecules with speeds between v and $v + dv$:
$$dE = \tfrac{1}{2} m v^2 n(v) \, dv = \tfrac{1}{2} m v^2 N f_M(v) \, dv.$$

STEP IV The limits of integration are $v = 0$ to $v = \infty$.
$$\langle E \rangle = \frac{E}{N} = \int_0^\infty \tfrac{1}{2} m v^2 f_M(v) \, dv.$$

REMEMBER: v IS SPEED, NOT VELOCITY, SO IT MUST BE POSITIVE.

SOLVE Substituting
STEP V
$$\frac{mv^2}{2kT} = x^2 \quad \text{and} \quad \sqrt{\frac{m}{2kT}} \, dv = dx,$$

we have: $\langle E \rangle = 4\pi \left(\dfrac{m}{2\pi kT}\right)^{3/2} \displaystyle\int_0^\infty \tfrac{1}{2} m v^4 \exp\left(-\dfrac{mv^2}{2kT}\right) dv = \dfrac{4kT}{\sqrt{\pi}} \displaystyle\int_0^\infty x^4 \exp(-x^2) \, dx.$

The value of the integral is $3\sqrt{\pi}/8$ (Appendix IF), so $\langle E \rangle = 3kT/2$, as expected.

ANALYZE Since we obtain the expected result for $\langle E \rangle$, this calculation verifies the proportionality constant in eqn. (22.13).

EXERCISE 22.5 ♦♦♦ Show that the average speed of the molecules is $\langle v \rangle = \sqrt{8kT/m\pi}$. (Hint: Use the substitution $y \equiv mv^2/2kT$.)

The average speed $\langle v \rangle$ is less than v_{rms}. In such averages, faster-moving particles are weighted more heavily than slower-moving particles; the higher the power of v, the greater is this bias. The most probable speed v_{mp}, at which the distribution function is maximum, is less than the average speed: $v_{mp} = \sqrt{2kT/m}$. The relations among these various speeds are shown in ■ Figure 22.17.

22.7.3 Mean Free Path and the Establishment of Equilibrium

We have often referred to the frequent collisions of molecules in our conceptual discussion of thermal processes. We are now ready to estimate the time interval and the distance a molecule travels between collisions. These estimates are important in understanding the approach of a system to equilibrium as well as processes like heat conduction that occur in systems far from equilibrium.

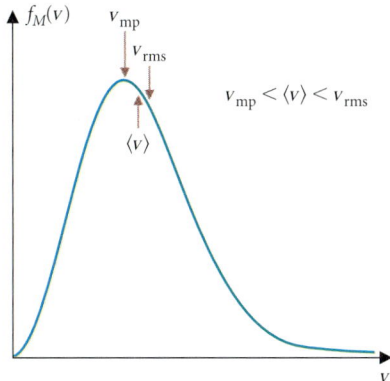

■ **FIGURE 22.17**
The most probable speed v_{mp} is less than the average speed $\langle v \rangle$, which is less than v_{rms}, but they are all close to the peak of the distribution.

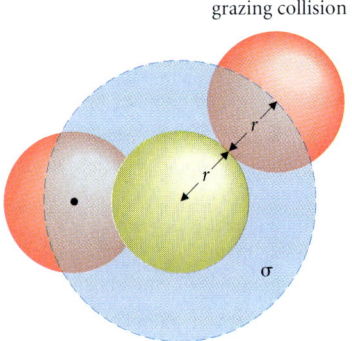

■ **FIGURE 22.18**
Molecular collisions. We model the molecules as spheres of radius r. Two molecules collide if the distance between their centers is less than the sum of their radii. The blue area is the molecule's collision cross section $\sigma = \pi d^2$.

STRICTLY SPEAKING, WE SHOULD INTEGRATE OVER THE DISTRIBUTION FUNCTION TO FIND THE COLLISION TIME FOR THE GAS SINCE EACH MOLECULE HAS A DIFFERENT SPEED. HOWEVER, WE HAVE IGNORED SEVERAL OTHER FACTORS, SUCH AS THE MOTION OF THE "TARGET" MOLECULES AND THE DETAILS OF THE COLLISION PROCESS. AT THIS LEVEL OF APPROXIMATION, WE MAY USE v_{rms} IN OUR EXPRESSION FOR Δt_c.

The size of molecules is an important factor in determining their collision rates. In this introduction, we ignore effects that depend on the structure of the molecules and model them as hard spheres of diameter d. Two such molecules collide if they are aimed so that their centers come closer than a distance d (■ Figure 22.18)—that is, if the second molecule's center is within an area $\sigma = \pi d^2$ centered on the first molecule. The area σ is called the *collision cross section* of the molecule.

To estimate how long it takes for one molecule to hit another, we treat the colliding molecule as an object with cross-sectional area σ and the target molecules as points. A molecule moving with speed v sweeps out a cylinder of volume $\sigma v \, \Delta t$ in time Δt. A collision occurs when this cylinder is large enough to contain at least one target molecule. In a gas of density n molecules per cubic meter, the average number of molecules in the cylinder is $n\sigma v \, \Delta t$, which equals unity when Δt equals the average time between collisions:

$$\Delta t_c = \frac{1}{n\sigma v}. \qquad (22.14)$$

EXAMPLE 22.10 ♦♦ What is the collision time for nitrogen molecules in air at room temperature and pressure? The molecular diameter is $d = 3 \times 10^{-10}$ m.

MODEL To use eqn. (22.14), we need values for n and v as well as σ. We find the density of air from the ideal gas law and use the rms speed for v.

SETUP
$$n = \frac{N}{V} = \frac{P}{kT} = \frac{1.0 \times 10^5 \text{ Pa}}{(1.4 \times 10^{-23} \text{ J/K})(300 \text{ K})} = 2 \times 10^{25} \text{ m}^{-3}.$$

The collision cross section is:
$$\sigma = \pi d^2 = \pi(3 \times 10^{-10} \text{ m})^2 = 3 \times 10^{-19} \text{ m}^2.$$

With $m = 28$ u for nitrogen molecules, the rms speed is:
$$v = \sqrt{\frac{3kT}{m}} = \sqrt{\frac{3(1.4 \times 10^{-23} \text{ J/K})(300 \text{ K})}{(28)(1.7 \times 10^{-27} \text{ kg})}} = 500 \text{ m/s}.$$

SOLVE So:
$$\Delta t_c = \frac{1}{n\sigma v} = \frac{1}{(2 \times 10^{25} \text{ m}^{-3})(500 \text{ m/s})(3 \times 10^{-19} \text{ m}^2)} = 3 \times 10^{-10} \text{ s}.$$

ANALYZE It is more difficult, and beyond the scope of our present discussion, to estimate the number of collisions per molecule required to establish thermal equilibrium. However, after each molecule has experienced several collisions, the speed distribution will be close to Maxwellian, except perhaps for the highest energies. Even allowing for a thousand collisions per molecule, which is plenty, atmospheric nitrogen would reach equilibrium at room temperature in less than a microsecond. The thermodynamic processes we plot on a P-V diagram are slow, quasi-equilibrium processes. "Slow" in this context means that a process occurs on a time scale much longer than a microsecond. ■

The average distance a molecule travels between collisions, the *mean free path* of the molecule, is:

$$\lambda \equiv v \, \Delta t_c = \frac{1}{n\sigma}. \qquad (22.15)$$

In the air around us (using numbers from Example 22.10):
$$\lambda = (500 \text{ m/s})(3 \times 10^{-10} \text{ s}) = 1.5 \times 10^{-7} \text{ m}.$$

Digging Deeper

THE DRUNKARD'S WALK

Diffusion is an example of a classic problem called the random (or drunkard's) walk. The displacement in a time t is the sum of many short, random displacements, each terminating in a collision (■ Figure 22.19). The directions of the displacements before and after a collision are unrelated. The total after N collisions is:

$$\vec{D} = \sum_{i=1}^{N} \vec{D}_i.$$

The magnitude of the displacement vector $|\vec{D}| = \sqrt{\vec{D} \cdot \vec{D}}$:

$$|\vec{D}|^2 = \left(\sum_{i=1}^{N} \vec{D}_i\right) \cdot \left(\sum_{j=1}^{N} \vec{D}_j\right) = \sum_{i=1}^{N} |\vec{D}_i|^2 + \sum_{i=1}^{N}\sum_{\substack{j=1 \\ j \neq i}}^{N} \vec{D}_i \cdot \vec{D}_j.$$

Since the steps are unrelated, $\vec{D}_i \cdot \vec{D}_j$ averages to zero. The average (rms) length of one step is $\sqrt{\langle|\vec{D}_i|^2\rangle} = \lambda$, the mean free path. So, the magnitude of the displacement after a large number of steps has the average value:

$$\langle|\vec{D}|^2\rangle = N\lambda^2.$$

The time for N collisions is $t = N\Delta t_c$, so $N = t/\Delta t_c = n\sigma v t$. The net distance traveled in time t is approximately:

$$s \approx \sqrt{N}\lambda = \frac{\sqrt{n\sigma v t}}{n\sigma} = \sqrt{\frac{vt}{n\sigma}}.$$

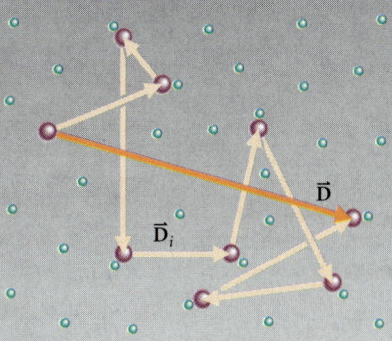

■ **FIGURE 22.19**
A random walk. The displacement $\vec{D}$ after a time t is the sum of the displacements $\vec{D}_i$ between successive collisions. After each collision, the molecule moves in a random direction, so the average value of $\vec{D}_i \cdot \vec{D}_j$ is zero.

The distance traveled increases as the square root of the time interval, rather than linearly with time as we would expect if there were a well-defined diffusion velocity.

Molecules that have a larger rms speed (light molecules) diffuse faster than slow (heavy) molecules. Diffusion of gas through a membrane may be used to increase the concentration of lighter molecules in a gas sample. This process is historically important as the first method for concentrating ^{235}U for use in nuclear fission reactors.

This distance is much greater than the molecule's diameter but very much less than the dimensions of a room. Although molecules move rapidly, they cannot move far before colliding. When the chemistry class produces hydrogen sulfide in the next room, the awful smell does not reach you at the speed of the molecules, several hundred meters per second, but much more slowly. The H_2S molecules move along an irregular path, making many collisions with air molecules on the way: the H_2S *diffuses* through the air.

22.7.4 A Microscopic View of Entropy

We may also describe entropy from a microscopic point of view. The entropy of a particular macrostate is determined by the number of possible microstates that correspond to the macrostate. Since we do not know which microstate the system is in, the more microstates that are possible, the less information is available and the greater is the entropy. The exact expression for the entropy of a macrostate is:

$$S = k \ln \Omega, \tag{22.16}$$

where k is Boltzmann's constant and Ω is the number of microstates.

ALL MICROSTATES ARE EQUALLY PROBABLE, SO THIS RESULT IS SOMETIMES STATED IN TERMS OF THE RELATIVE PROBABILITY OF THE MACROSTATE: $S \propto k \ln p$.

For example, the microstates of a gas with temperature T in a box of volume V are described by all the possible positions and velocities of the molecules. If the temperature of the gas remains fixed, the distribution of velocities does not change, and we need only consider the positions of the molecules to compare the number of microstates in each macrostate. The number of possible positions for each molecule is proportional to the volume of the box. For N molecules, the number of possible sets of positions of all N molecules is proportional to V^N: $\Omega = CV^N$. Then:

$$S = k \ln(CV^N) = Nk \ln V + k \ln C.$$

THE $k \ln C$ TERMS CANCEL IN THE SUBTRACTION.

For an isothermal or free expansion, the change in entropy is:

$$\Delta S = Nk \ln V_1 - Nk \ln V_2 = Nk \ln(V_1/V_2),$$

the same result we found in Examples 22.3 and 22.4 from the macroscopic description of entropy change.

EXAMPLE 22.11 ♦♦ A hypothetical system consists of three particles. Each particle can have an energy of 0, 1, or 2 units. How many macrostates have a total energy equal to 3 units? Compute the number of microstates in each macrostate and hence find the entropy of each.

WE ARE ASSUMING THAT THE THREE PARTICLES ARE DISTINGUISHABLE.

MODEL There are two ways to have a total energy of 3 units (■ Figure 22.20). With one particle having each allowable energy, the total energy is $0 + 1 + 2 = 3$. Alternatively, with all three particles having 1 unit of energy, the total is $1 + 1 + 1 = 3$.

SETUP To create the first macrostate, we choose one particle and give it 0 units of energy. There are three ways to make this choice. Then we take one of the remaining particles and give it 1 unit. There are two ways to do this. The remaining particle gets 2 units. The total number of microstates equals the number of different ways to arrange the particles, which is $3 \times 2 = 6$.

To create the second macrostate, we give each particle 1 unit of energy. There is only one way to do this, so the number of microstates is 1.

SOLVE The entropy of the second state is $k \ln 1 = 0$. The entropy of the first is $k \ln 6$.

ANALYZE The system is six times more likely to be in state 1, the state of greater entropy.

■ **FIGURE 22.20**
Each of the two states of a hypothetical three-particle system has a total energy of 3 units. State 1 can be created in six different ways, and so has the greater entropy.

22.7.5 Entropy and Equilibrium

Any real process results in an increase of entropy or disorder of the universe as a whole. *No process results in a decrease of total entropy.* Isolated systems, in time, come to equilibrium and thereafter undergo no further changes (§19.1). These two ideas together imply that equilibrium is a state of maximum entropy. For example, a system that reaches thermal equilibrium (i.e., constant temperature) by conduction of heat does so with an increase in entropy. For an isolated sample of ideal gas, we have a microscopic description of the equilibrium state, namely the Maxwell–Boltzmann distribution of velocities, which is a maximum entropy state. Boltzmann was also able to show that if the velocity distribution differs slightly from equilibrium, then collisions return the system to a Maxwell–Boltzmann distribution.

SEE EXAMPLE 22.5.

HOWEVER, THE ASSUMPTION OF A SMALL DEVIATION IS ESSENTIAL. EVEN FOR AN IDEAL GAS, IT HAS NOT BEEN SHOWN THAT GROSS DEVIATIONS FROM EQUILIBRIUM RESULT IN A RETURN TO EQUILIBRIUM THROUGH COLLISIONS.

A system can be perturbed from the equilibrium state only by reducing its entropy (and thus increasing the entropy of the environment), and it generally returns to the equilibrium state with an increase in system entropy. The equilibrium state may only be a local entropy maximum, in the sense that all nearby states have lower entropy. In such cases, however, the equilibrium may be unstable, and after a perturbation, the system moves to a state of higher entropy that is very different from the original state. Supercooled water is an example of this phenomenon. Clouds frequently contain liquid water droplets at a temperature less than 0°C.

> The entropy of this state is large because the water is dispersed throughout the cloud. States that differ only slightly, such as states in which the temperature is not uniform throughout the cloud, have lower entropy. So, the cloud is in an unstable equilibrium. A large disturbance, such as jostling of the water molecules by a passing airplane, results in an immediate freezing of the water with the accompanying release of latent heat. Ice is a more ordered state than water, but entropy is generated by conduction of the latent heat into the surrounding air and more than compensates for the entropy loss due to solidification. Thus the final state of ice and dry air has higher entropy than the original. It is a new equilibrium state that is far removed from the initial equilibrium.

Chapter Summary

Where Are We Now?

We have completed our development of the basic laws of thermodynamics. The versions in italics may aid your memory.

0. Temperature is the quantity that determines whether systems are in equilibrium. — *There's this game in town.*
1. Energy is conserved: $\Delta U = Q - W$. — *You can't win!*
2. The total entropy of the universe tends to increase. — *You can't even break even!*
3. Absolute zero is unattainable. — *You can't get out of the game!*

The second law of thermodynamics indicates which processes are probable and which are so improbable as to be effectively impossible. The direction of increasing entropy is also the direction of increasing time. Only in thermodynamics, of all the topics we have discussed, do we have a way to define the arrow of time.

What Did We Do?

We have studied the conversion of thermal energy to work in heat engine cycles. The efficiency of an engine is the net work produced divided by the heat input:

$$e = W_{net}/Q_{in}$$

and is always less than unity. In the Carnot cycle, heat transfer occurs at constant temperature, using two heat reservoirs at temperatures T_C and T_H. The efficiency of a Carnot engine has the maximum possible value: $e = 1 - T_C/T_H$.

Entropy is a state variable that measures the degree of disorder of a system. The entropy change of a system during a reversible process is:

$$\Delta S = \int \frac{dQ}{T},$$

where dQ is the heat input to the system at temperature T. To use this expression for a real process, we first imagine a reversible process that has the same initial and final states.

The second law of thermodynamics says that the entropy of the universe tends to increase.

Refrigerators are engines run backwards. The coefficient of performance of a refrigerator is the heat removed from the cold reservoir divided by the work required to run the cycle:

$$K = \frac{Q_{cold}}{|W_{net}|}.$$

By using a Carnot-cycle refrigerator in a thought experiment, we showed that it is impossible for any system at finite temperature to be cooled to absolute zero, and so proved the third law of thermodynamics.

✽ According to Boltzmann's law, the relative population of states whose energy differs by ΔE is proportional to $\exp(-\Delta E/kT)$. We used this result to derive the Maxwellian speed distribution: the fraction of molecules with speed in the range v to $v + dv$ in a gas at temperature T is:

$$f_M(v) = 4\pi \left(\frac{m}{2\pi kT}\right)^{3/2} v^2 \exp\left(-\frac{mv^2}{2kT}\right).$$

Molecules in a gas collide after traveling an average distance $\lambda = 1/n\sigma$. Because molecules cannot travel freely, a localized concentration of one kind of gas diffuses slowly through a uniform sample of another.

The entropy of a macrostate is $S = k \ln \Omega$, where Ω is the number of microstates that make up the macrostate.

Practical Applications

Heat engines are ubiquitous in modern society. There is one in every automobile, and every power plant is a huge heat engine. There is a refrigerator in almost every home in the United States. The same principles are used in air conditioners and heat pumps.

The fundamental concepts of statistical mechanics presented here are the starting point for a statistical theory of real substances. Collisions of particles are an important feature in thermal and electrical conduction in gases as well as in diffusion. The concept of mean free path is used in calculations of these transport properties and also in more exotic applications, such as calculations of nuclear explosions in stars.

Solutions to Exercises

22.1 $e_c = 1 - (300 \text{ K})/(500 \text{ K}) = 1 - 0.6 = 0.4$.

22.2 No! To remove heat from its interior, the refrigerator does work and dumps more heat into the kitchen. The kitchen will heat up, not cool down.

22.3 From §22.2.2, we have:

$$e = 1 + \frac{Q_{EB}}{Q_{CD}} = 1 + \frac{Mc_v(T_B - T_E)}{Mc_v(T_D - T_C)}.$$

But D and E, and C and B are each on an adiabat, so:

$$T_D V_2^{\gamma-1} = T_E V_1^{\gamma-1} \quad \text{and} \quad T_C V_2^{\gamma-1} = T_B V_1^{\gamma-1}.$$

So,
$$\frac{T_D}{T_C} = \frac{T_E}{T_B}.$$

Thus:
$$T_B - T_E = \left(\frac{T_C}{T_D} - 1\right) T_E = (T_C - T_D)\frac{T_E}{T_D}.$$

So:
$$e = 1 - T_E/T_D.$$

✽ **22.4** The coordinate y measures height above sea level. The pressure P_o is $NkT/V = n_o kT$, where n_o is the number of molecules per unit volume at sea level, and similarly $P = nkT$. Note also that

$$\rho_o gy = mn_o gy = n_o E,$$

where $E = mgy$ is the gravitational energy of one molecule at height y. Dividing through by kT, we may rewrite the expression for P as an expression for density of molecules:

$$n = n_o \exp(-E/kT)$$

or
$$n/n_o = \exp(-E/kT),$$

in agreement with eqn. (22.12) with equal gs for the two states.

22.5 The average speed is:

$$\langle v \rangle = \frac{\int vn(v)\, dv}{N} = \int_0^\infty v f_M(v)\, dv$$

$$= \int_0^\infty v 4\pi \left(\frac{m}{2\pi kT}\right)^{3/2} v^2 \exp\left(-\frac{mv^2}{2kT}\right) dv.$$

Using the suggested substitution, we find:

$$\langle v \rangle = \sqrt{\frac{8kT}{\pi m}} \int_0^\infty y e^{-y}\, dy.$$

Integrating by parts, we find the integral over y equals 1. Thus:

$$\langle v \rangle = \sqrt{8kT/\pi m}.$$

Basic Skills

Review Questions

§22.1 WHY IS A SECOND LAW OF THERMODYNAMICS NECESSARY?

- What is a *reversible process*? Give an example of a reversible process and an irreversible process.
- Can a real process be reversible? Why or why not?
- Give a statement of the second law of thermodynamics.

§22.2 HEAT ENGINES

- What is the *efficiency* of an engine? Are there any limits to its value?
- Describe the *Otto cycle*. What kinds of processes does it use? What is its efficiency?
- Give one example of a system that uses the Otto cycle.

§22.3 THE CARNOT CYCLE

- Define a *heat reservoir*.
- Describe the Carnot cycle. What processes does it use?
- What makes the Carnot cycle reversible, in principle?
- What is the efficiency of a Carnot cycle?
- What do you get when you reverse a Carnot cycle?
- What is the *coefficient of performance* of a refrigerator?

§22.4 ENTROPY

- What property of a system does entropy measure?
- How does the entropy of the universe change during a reversible process? An irreversible process?
- What is the entropy change of a system during an isothermal process?
- Give a formula for calculating the entropy change during an arbitrary reversible process.
- How should you calculate entropy changes during an irreversible process?

§22.5 A LIMIT ON EFFICIENCY

- Is it possible to build an engine that is more efficient than a Carnot engine? Why or why not?

§22.6 THE SIGNIFICANCE OF ABSOLUTE ZERO

- How is the *thermodynamic temperature scale* defined?
- State the third law of thermodynamics.
- Why is absolute zero *absolute*?

✷ §22.7 STATISTICAL MECHANICS

- What is the *Boltzmann factor*?
- What do the *g*-factors in the Boltzmann relation represent?
- What is the Maxwell–Boltzmann speed distribution? Where does the factor $4\pi v^2\, dv$ come from?
- How is the *collision time* for a molecule defined? On which properties of the molecule does it depend?
- What is the *mean free path*?
- How does the distance traveled by a diffusing molecule depend on the number of collisions that have occurred?
- How is the entropy of a macrostate expressed in terms of the microstates that make up that macrostate?
- How is the fact that a state is in equilibrium related to its entropy?

Basic Skill Drill

§22.1 WHY IS A SECOND LAW OF THERMODYNAMICS NECESSARY?

1. List three processes known from common experience to be irreversible. Describe how each process corresponds to an increase of disorder.

§22.2 HEAT ENGINES

2. What is the efficiency of an Otto air-cycle engine with a compression ratio of 5?

3. (a) One mole of a monatomic ideal gas is taken through the cycle shown in ■ Figure 22.21. Fill out a table giving the heat Q, the work W, and the internal energy change ΔU for each process and for the entire cycle. Express all quantities in joules with the appropriate signs (1 L·atm = 101 J). **(b)** Compute the efficiency of a reversible engine running on this cycle.

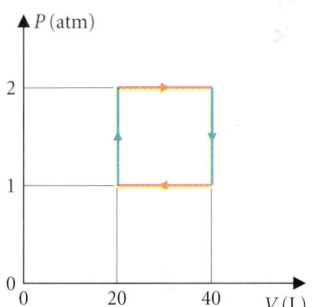

■ **Figure 22.21**

§22.3 THE CARNOT CYCLE

4. What is the efficiency of a Carnot engine operating between 1.0×10^3 K and 560 K?

5. A Carnot engine operates between two reservoirs at temperatures of 750 K and 410 K. In each cycle, 3.0×10^3 J of heat are supplied. How much work is done per cycle, and how much waste heat is exhausted?

6. An expensive home refrigerator with a coefficient of performance $K = 4$ operates between temperatures of 55.0°C and 0.0°C. What is the coefficient of performance of a Carnot-cycle refrigerator operating between the same temperatures?

§22.4 ENTROPY

7. When a system undergoes an irreversible process, is the change in entropy of the environment always positive?

8. Model the rush of air from a punctured tire as a free expansion of a diatomic ideal gas at 290 K from a volume of 0.016 m³ to 0.038 m³. The final pressure is atmospheric. How much entropy is created during the process?

§22.5 A LIMIT ON EFFICIENCY

9. An engine operates between a maximum temperature of 600 K and a minimum temperature of 300 K. What can you say about its efficiency?

§22.6 THE SIGNIFICANCE OF ABSOLUTE ZERO

10. A system whose temperature is to be measured is connected as the cold reservoir in a Carnot-cycle engine. During one cycle, 512 J of heat is removed from the hot reservoir at 2.00×10^2 K, and 3.50×10^2 J is dumped to the cold reservoir. What is the temperature of the system?

✱ §22.7 STATISTICAL MECHANICS

11. In a hydrogen atom, there are two possible states in which the electron has total energy -2.2×10^{-18} J and eight states with total energy -5.4×10^{-19} J. Find the ratio of numbers of atoms in the two states in a sample of atomic hydrogen gas at 10^4 K. (Despite the small value of the result, a star at this temperature has enough atoms in the more energetic state that their light is a characteristic feature of the star's spectrum.)

12. Find the collision time for CO_2 molecules in air at atmospheric pressure. Take $d = 3.8 \times 10^{-10}$ m and $T = 290$ K.

13. Find the mean free path of an electron in the solar corona, where $T \approx 10^6$ K and the density $\approx 10^{14}$ electrons/m^3. Take $\sigma = 6.7 \times 10^{-29}$ m^2.

14. The cards in a shuffled deck can be arranged in $52! = e^{156}$ ways. What is the entropy of a shuffled deck?

Questions and Problems

§22.1 WHY IS A SECOND LAW OF THERMODYNAMICS NECESSARY?

15. ✤ You turn a crank to lift a pail of water from a well. What features of the task are irreversible?

16. ✤ The elastic collision of two rigid spheres is reversible. Describe the irreversible features in a collision of two real billiard balls.

17. ✤ Your roommate demands that you decrease the disorder of your belongings. Describe the processes that increase the overall disorder of the universe while you carry out your friend's wishes.

§22.2 HEAT ENGINES

18. ✤ A real gas is put through a cycle of reversible processes and then through a cycle of irreversible processes. Compare the change in internal energy of the gas in the two cycles.

19. ✤ Which is more efficient, an Otto-cycle engine using diatomic or monatomic gas? Does the answer depend on the value of r?

20. ✦✦ An engine uses $\mathcal{N}$ mol of ideal monatomic gas in the cycle shown in ■ Figure 22.22. Find the efficiency of the engine.

21. ✦✦✦ The working fluid in an engine is $\mathcal{N}$ mol of an ideal monatomic gas. One cycle of the engine is represented in ■ Figure 22.23. **(a)** From A to B, the gas is compressed at constant pressure until its volume is $V_B = V_A/8$. Compute the gas temperatures at A and B in terms of P_A and V_A. **(b)** Compute the work done by the gas and the heat input to the gas on the leg AB. (Be careful with signs!) **(c)** On the leg BC, the gas is heated at constant volume. If the pressure at C is $P_C = 32P_B$, compute the temperature at C and the heat absorbed on this leg in terms of P_A and V_A. **(d)** The gas expands adiabatically from C to A. Compute the work done by the gas on this leg in terms of P_A and V_A. **(e)** What is the efficiency of this engine? **(f)** Prove that the given conditions are such that an adiabatic expansion from C actually returns to A.

22. ✦✦✦ The Brayton cycle (■ Figure 22.24) models a gas turbine engine. Process $A \to B$ is an adiabatic compression of air in the compressor. Fuel is mixed in and, after ignition, the gas expands at constant pressure ($B \to C$). The adiabatic expansion $C \to D$ drives the turbine. Finally, the cycle is closed with heat rejection at constant pressure. Find the efficiency of this engine in terms of the pressure ratio $r_p = P_B/P_A$.

23. ✦✦✦ ■ Figure 22.25 illustrates the sequence of processes in a diesel engine cycle. **(a)** Where does heat leave the engine and where does heat enter? Give a full discussion of your answer! **(b)** Where is work done? **(c)** Show that the efficiency of the Diesel cycle using ideal gas as a working fluid is:

$$e_d = 1 - \frac{T_B - T_C}{\gamma(T_A - T_D)},$$

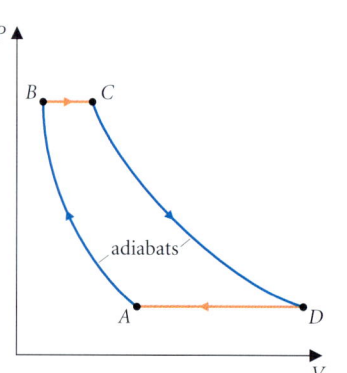

■ FIGURE 22.24
The Brayton cycle.

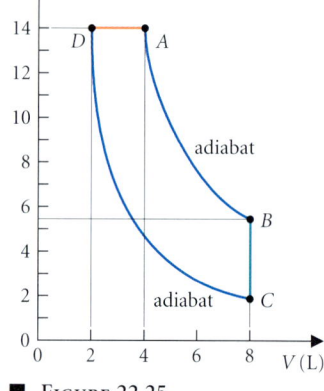

■ FIGURE 22.25
The Diesel cycle.

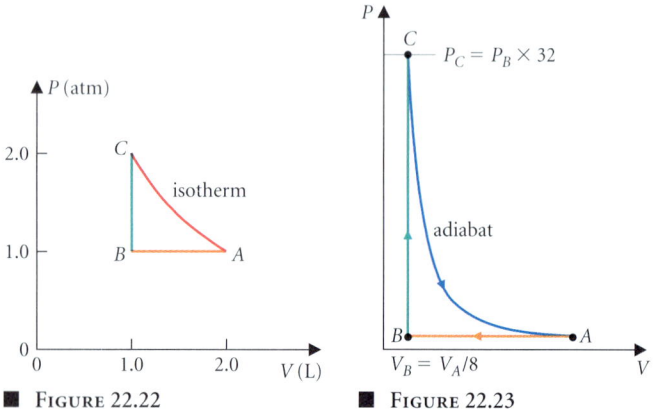

■ FIGURE 22.22

■ FIGURE 22.23

where γ is the ratio of specific heats of the gas. **(d)** Define the compression ratios $r_1 \equiv V_B/V_A = V_C/V_A$ and $r_2 \equiv V_C/V_D$. Then show that:

$$e_d = 1 - \frac{1}{\gamma}\left(\frac{r_1^{-\gamma} - r_2^{-\gamma}}{r_1^{-1} - r_2^{-1}}\right).$$

Compute the efficiency of a Diesel cycle with the compression ratios shown in Figure 22.25 and compare with an Otto cycle with a compression ratio of $r = r_2$. (*Hint:* The gas is diatomic.)

§22.3 THE CARNOT CYCLE

24. ❖ Why may a large volume of water be considered to be a good heat reservoir? In what way is it not ideal? In what way is a large block of copper a better heat reservoir? In what way is it worse?

25. ❖ The efficiency of a Carnot engine is to be improved by increasing the temperature of the hot reservoir. Describe qualitatively the corresponding changes in cylinder volume and operating pressure throughout the cycle, assuming the number of moles of gas and the minimum volume remain fixed. What changes occur if the efficiency is increased by decreasing the temperature of the cold reservoir?

26. ◆◆ A Carnot cycle uses a cold reservoir at 280 K. In each cycle, 2500 J of heat is supplied and 1200 J of work is done. What is the temperature of the hot reservoir?

27. ◆◆ A Carnot refrigerator is to keep its interior at 36°F when the outside temperature is 82°F. What is the coefficient of performance? If it uses 220 W of power, at what rate is heat rejected to the exterior?

28. ◆◆ A Carnot engine has an efficiency of 0.70. The hot reservoir is at 875 K. What is the temperature of the cold reservoir? If the designers are able to reduce the low temperature by 10 K, by how much does the efficiency change? If the power output is kept constant by increasing the time to complete each cycle, how does the rate of waste heat transfer change?

29. ◆◆ A Carnot freezer (Figure 22.5) keeps its interior at 25°F. When the outside temperature is 55°F, the freezer uses 75 W of power to remove heat leaking in from outside. Assuming the machine has a fixed volume ratio and uses the outside air as a high-temperature reservoir, how much does the power used by the freezer increase when the outside temperature increases to 70°F? (*Hint:* use Newton's law of cooling, §21.4.)

30. ◆◆◆ A refrigerator based on the Brayton cycle (Figure 22.24) is used in aircraft for cabin air conditioning. Find an expression for the coefficient of performance of this refrigerator in terms of $r_P = P_B/P_A$.

31. ◆◆◆ In a paper written in 1854, Clausius introduced the cycle shown in ■ Figure 22.26, which uses an ideal gas and three heat reservoirs. **(a)** Show that:

$$\frac{V_D}{V_C} = \frac{V_E}{V_F}\left(\frac{V_A}{V_B}\right).$$

(b) In the Clausius cycle, the heat input from the reservoir at T_2 equals the heat output to the reservoir at T_1. Using the result of (a), show that:

$$\frac{V_B}{V_A} = \left(\frac{V_E}{V_F}\right)^{1 - T_1/T_2}.$$

(c) Show that the efficiency of the cycle is:

$$e = \frac{T_3(T_2 - T_1)}{T_1 T_2 + T_3(T_2 - T_1)}.$$

§22.4 ENTROPY

32. ❖ Some warm water is mixed with ice and the ice melts. Assume that the entire process happens in an insulated box. Does the internal energy of the combined (ice + water) system change? Does the entropy of the combined system change? Now considering the ice and

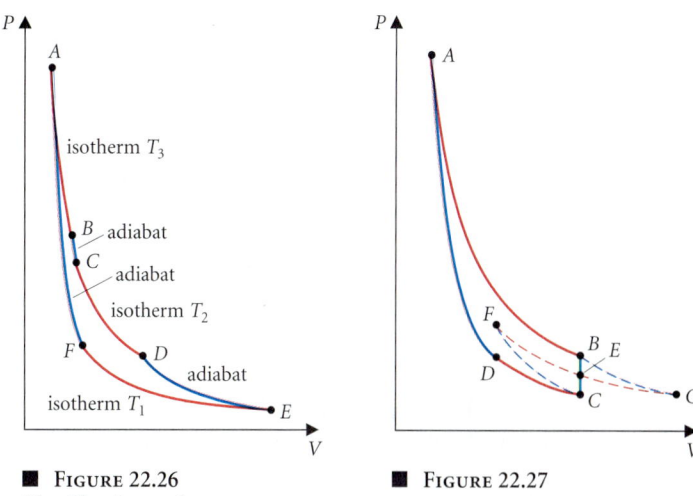

■ **FIGURE 22.26**
The Clausius cycle.

■ **FIGURE 22.27**

water as separate systems, discuss the change in internal energy and in entropy for each.

33. ❖ An adiabat-isotherm approximation for any cycle in the *P-V* plane has the same entropy change as a simple Carnot cycle—that is, zero. For example, ■ Figure 22.27 represents the process *ABCDA*, which consists of two isotherms (*AB* and *CD*), an adiabat (*DA*), and a constant-volume process (*BC*). Also shown in the figure is an adiabat-isotherm approximation to the constant-volume process. Find *two* Carnot cycles that are equivalent to the *approximate* cycle in terms of entropy change.

34. ❖ An inventor asks you to invest in a company that produces electric power from the thermal energy in seawater. Is the idea physically reasonable? If so, under what conditions? (See T. Penney and D. Bharathan, Power from the sea, *Scientific American,* Jan. 1987: 86.)

35. ❖ A heat pump extracts thermal energy from cold outside air and uses it to heat living quarters. Explain how this is possible.

36. ❖ A toy bird (■ Figure 22.28) absorbs water through a wick in its beak, becomes unbalanced, and lifts its beak from the water. Water then evaporates from the reservoir inside the bird, changing its balance so that it again dips its beak in the water. An enterprising inventor attaches a generator to the bird so as to produce electric power. Explain why this scheme does not violate the first law of thermodynamics. Why does it not violate the second law?

37. ◆ Discuss the entropy change of the Sun due to its radiation.

38. ◆◆ You put a kettle on to boil and get so engrossed in your physics problems that you forget about it. All of the 4.0×10^{-3} m^3 of water boils away. Estimate the entropy generated in the process.

39. ◆◆ Calculate the entropy change for 1.0 kg of water heated from 15°C to 80°C.

40. ◆◆ A roof of area 510 m^2 has $R_f = 0.20$ m$^2 \cdot$K/W. The inside temperature is 69°F and the outside temperature is 44°F. At what rate is entropy generated by heat loss through the roof?

■ **FIGURE 22.28**

41. ♦♦ Use the first law of thermodynamics to show that for an ideal gas system:
$$\Delta S = mc_v \ln(T_2/T_1) + Nk \ln(V_2/V_1).$$

§22.5 A LIMIT ON EFFICIENCY

42. ♦ What is the maximum efficiency of an engine that works between two reservoirs at $T = 185°F$ and $T = 45°F$?

43. ♦♦ An engine operating with two heat reservoirs has a measured efficiency of 0.17, known to be 55% of the maximum. The low-temperature reservoir is at 5°C. What is the temperature of the high-temperature reservoir?

§22.6 THE SIGNIFICANCE OF ABSOLUTE ZERO

44. ♦ A Carnot-cycle thermometer has a heat transfer of 25 J from a lab system and 74 J to its hot reservoir. If the hot reservoir is at 220 K, what is the temperature of the lab system?

45. ♦♦ A Carnot-cycle thermometer is used to measure the temperature of a system. Its high-temperature reservoir is at $-5°C$. Operating at 35 W, the rate of heat transfer to the hot reservoir is 82 W. What is the temperature of the system?

✻ §22.7 STATISTICAL MECHANICS

46. ❖ Use Boltzmann's law to explain why evaporation causes cooling.

47. ❖ If the temperature of the gas in a box is increased by a factor of 2, does the number of molecules in a speed range of v to $v + dv$ increase or decrease? Does your answer depend on the value of v (assuming $dv = $ constant)? How?

48. ❖ If the temperature of the Earth's atmosphere increased by 10%, would more or fewer nitrogen molecules escape? Why?

49. ❖ Helium in a tank on the space station leaks into space through a small meteorite puncture. Assuming the temperature T remains constant, how does the rate of gas outflow depend on the pressure in the tank? How does the pressure vary with time? Does the escape rate of molecules depend on their speed? Why or why not? Does the temperature remain constant?

50. ❖ Two identical gas samples occupy equal volumes, but one has twice the temperature of the other. How do the collision times in the two gases compare? What is the ratio of the mean free paths?

51. ❖ One sample of a gas has twice the density of another, but they have the same pressure. What are the ratios of the collision times and of the mean free paths for the two samples?

52. ♦ In the interstellar medium, one often finds regions filled primarily by hydrogen atoms with a density $n \approx 10$ cm^{-3} at a temperature $T \approx 100$ K. If the diameter of a hydrogen atom is $d \approx 10^{-10}$ m, compute its mean free path in such a region. How does this result compare with the size of the solar system?

53. ♦♦ For the triangular distribution of speeds shown in ■ Figure 22.29, compute v_{rms}, $\langle v \rangle$, and v_{mp}.

54. ♦♦ ● Table 22.1 shows the distribution of heights of physics students in a class. Plot a histogram showing the distribution of heights. What is the mean height? The rms height? The most probable height?

TABLE 22.1

Height range (cm)	Number of students
140–150	1
150–160	5
160–170	7
170–180	7
180–190	10
190–200	6
200–210	1

55. ♦♦ A sample of helium in a cylinder has a temperature of 20 K and a pressure of 0.5 atm. What is the mean free path of the He atoms in the cylinder? Take $d = 2 \times 10^{-10}$ m.

56. ♦♦ Assume that the air in a cylinder of an automobile engine has an average temperature of 1000°C and an average pressure of 5 atm. How many collisions does an average molecule make during the combustion and expansion phases (approximately a 0.05-s time interval)?

57. ♦♦ How many collisions does an H_2S molecule make in diffusing 5 m through air at 1 atm and 300 K? Take $\sigma(H_2S) = \sigma(\text{air})$.

58. ♦♦♦ At an altitude of 700 km, the temperature is 1600 K. Make an accurate graph of the one-dimensional Maxwellian distribution for helium at this temperature. Do He molecules escape the Earth rapidly? Discuss.

59. ♦♦♦ Knowing that the entropy change in an adiabatic expansion is zero (cf. Example 22.3), induce the dependence of entropy on temperature for an ideal monatomic gas. Write the result in terms of v_{rms}. Can you see the connection with eqn. (22.16)? (For another way to get this result, see Problem 41.)

60. ♦♦♦ By differentiating the Maxwell–Boltzmann distribution function, verify that the most probable speed of a molecule is $v_{mp} = \sqrt{2kT/m}$.

61. ♦♦♦ Write the velocity distribution for gas molecules in the form:
$$f(\vec{v}) = f(v_x)f(v_y)f(v_z).$$
Write the volume element $dv_x\, dv_y\, dv_z$ in spherical coordinates and integrate over the angles to retrieve the Maxwell–Boltzmann speed distribution.

62. ♦♦♦ As an illustration of the relationship between entropy and equilibrium, consider a cylindrical volume of cross section A and total length L. A movable partition separates two monatomic ideal gas samples with $\mathcal{N}_1$ and $\mathcal{N}_2$ mol of gas (■ Figure 22.30). The system is maintained at a constant temperature T. Then, a state of the system is determined by the position of the partition, L_1. **(a)** If the entropy of the state $L_1 = L/2$ is called S_o, compute the entropy of a state with an

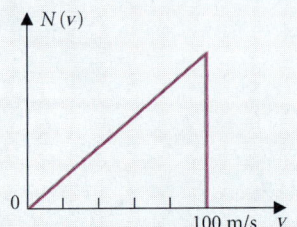
■ FIGURE 22.29

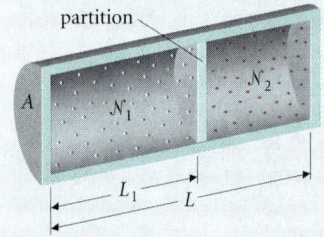
■ FIGURE 22.30

arbitrary value of L_1. **(b)** Show that the entropy of the system is a maximum when:
$$L_1/(L - L_1) = \mathcal{N}_1/\mathcal{N}_2.$$
(c) Show that in the maximum entropy state the pressures of the two gas samples are equal; that is, the system is in mechanical equilibrium with no net force acting on the partition.

Additional Problems

63. ❖ ✉ A gas sample contains 10^9 molecules, each of which undergoes a collision once every 10^{-10} s. After each collision, an individual molecule has a probability $p = \frac{1}{2}$ of being aimed toward the left half of the container. What is the probability that all molecules in the sample will be aimed toward the same half of the container? After N collisions by each molecule, what is the chance that all molecules will ever have been aimed simultaneously at the same half? Estimate how much time must pass before such an occurrence is reasonably likely.

64. ♦ A Carnot-cycle refrigerator operates between a cold reservoir at 2.0×10^2 K and a hot reservoir at 6.0×10^2 K. If 1.0×10^2 J of work are used per cycle to operate the device, how much heat per cycle does it remove from the cold reservoir, and how much heat does it reject to the hot reservoir? What is the entropy change per cycle for each reservoir?

65. ♦♦ A diesel engine operates through the cycle shown in Figure 22.25. If 0.70 mol of diatomic gas are used, what are the highest and lowest temperatures that occur during the cycle? Compute the efficiency of a Carnot engine running between these extreme temperatures, and compare with the result of Problem 23(c). Comment on the difference.

66. ♦♦ A lump of copper of mass 750 g at 450 K is added to 1.0 L of water at 280 K. What is the final temperature? How much entropy is created?

67. ♦♦ If 10.0 g of ice at 0.0°C is placed in an insulated container with 10.0 g of water at 80.0°C and the mixture is allowed to come to equilibrium, what is the final temperature of the system? Compute the entropy changes of ice and water and show that the net entropy change is positive.

68. ♦♦♦ Two moles of monatomic ideal gas are taken through the reversible cycle of processes shown in ■ Figure 22.31, where $T_1 = 280$ K and $T_2 = T_3 = 980$ K. **(a)** Compute P_2/P_1 and V_2/V_1. **(b)** Fill out a table listing the heat Q, work W, internal energy change ΔU, and entropy change ΔS for each process and for the whole cycle. **(c)** Compute the efficiency of an engine running on this cycle. **(d)** Compare with the efficiency of a Carnot cycle running between the same two temperatures.

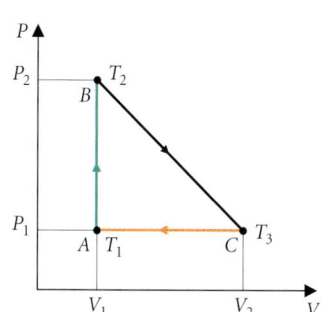

■ **Figure 22.31**

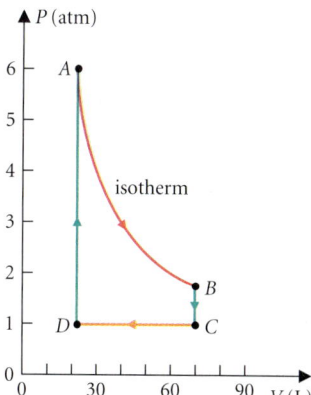

■ **Figure 22.32**

69. ♦♦♦ One mole of an ideal monatomic gas is taken through the complete cycle $ABCDA$ shown in ■ Figure 22.32. The cycle starts at state A ($P = 6.0$ atm, $V = 22.4$ L). **(a)** Find the temperature at each of the states A, B, C, and D. **(b)** Fill in a table, giving Q, W, ΔU, and ΔS for each process AB, BC, CD, and DA. Express all energies in joules with their appropriate signs. **(c)** Show that the sum of the entropy changes around the whole cycle is zero.

70. ♦♦♦ One mole of an ideal gas is heated from 200.0 K to 400.0 K isobarically. It is found that 5860 J of heat are needed to do this. **(a)** If the original volume of the gas was 22.4 L, what was the original pressure in atm? **(b)** Calculate the molar specific heats c'_p and c'_v. **(c)** Calculate ΔU, assuming U is independent of V. **(d)** What was the work done by the gas? **(e)** What is the final volume? **(f)** Assuming the process is reversible, what was the entropy change? **(g)** Is the gas monatomic, diatomic or polyatomic?

71. ♦♦♦ Find the efficiency of the Brayton cycle (Figure 22.24, Problem 22) in terms of the temperatures at A, B, C, and D. How does it compare with the efficiency of a Carnot cycle? Discuss.

72. ♦♦♦ Consider an attempt to produce absolute zero using a Carnot cycle with fixed parameters V_A, V_B, V_C, V_D. How much heat is removed from the cold reservoir during each cycle? If the heat capacity of the reservoir is C, what change in its temperature results? Writing
$$\frac{Nk}{C} \ln\left(\frac{V_B}{V_A}\right) \equiv \alpha,$$
show that after n cycles the temperature of the cold reservoir is: $T_n = (1 - \alpha)^n T_0$. Can the procedure succeed in achieving absolute zero?

✱ 73. ♦♦♦ We can estimate the possible uncertainty in our calculation of the mean collision time in an ideal gas by computing the average in a different way. Write the time until collision for an individual molecule as $t_{coll} = 1/n\sigma v$ and use the Maxwell–Boltzmann distribution to compute the average of $1/v$. Using this procedure, by how much does your value for the collision time differ from that derived in §22.7.3?

74. ♦♦♦ A beam of N_0 gas molecules is projected through a sample of similar gas molecules. The number of molecules in the beam will decrease with distance because of collisions that knock them out of the beam. Let $N(x)$ be the number of molecules that penetrate a distance x into the sample before colliding and let p be the probability per unit length of path that a beam molecule will suffer a collision. **(a)** Write an expression for the number of collisions that occur between x and $x + dx$. What is the change dN in the number of molecules in the beam? **(b)** Integrate the differential equation resulting from (a) to find $N(x)$. **(c)** Show that the mean free path is $\lambda = \langle x \rangle = 1/p$.

Computer Problems

75. What fraction of the molecules in a Maxwellian distribution have speeds greater than (a) v_{rms}, and (b) $3v_{\text{rms}}$? (Use a graphical or numerical method for the integral, or use integration by parts and a table of the error function.)

76. In file CH22P76 on the supplementary computer disk we list the specific heat of water as a function of temperature. By how much does the entropy of 1 kg of water increase when it is heated from 5°C to 25°C? How much error is made by using a constant value for the specific heat, taken from Table 20.4?

Challenge Problems

77. Boltzmann was able to calculate the Maxwellian speed distribution by considering an isothermal atmosphere. This problem follows his argument. (a) By how much does the density in the atmosphere decrease between heights z and $z + dz$? (Use the results of §13.3.4.) (b) Assume there is a height $z = z_o$ in the atmosphere above which collisions are not important. In what range v to $v + dv$ is a molecule's vertical speed when at z_o, if it rises to altitude z but does not go above $z + dz$? (c) The molecules in part (b) contribute to the observed density at all heights between z_o and $z + dz$ but not above $z + dz$. Those with a vertical speed of less than v contribute only below height z. Thus we can attribute the decrease in density between z and $z + dz$ to the number of molecules in the range v to $v + dv$. Make this identification, using the result of part (a), and hence find the fraction of molecules with vertical speed in the range v to $v + dv$.

78. (a) Home ice cream makers often use a brine solution to provide the low temperature necessary to freeze the ice cream. The brine is made by adding salt to a mixture of ice and water. Explain why adding salt lowers the temperature of the solution. (b) Compare two systems, one containing a lump of ice in equilibrium with pure water and the other containing ice in equilibrium with the same volume of salt solution. Compare entropy changes in the two systems when a small amount of heat ΔQ is transferred to each from the environment and show that the system containing salt must be at a lower temperature than the system without salt. Use the fact that ice contains no salt, and neglect any difference in the amount of ice that melts in the two systems. Show why the temperature difference is proportional to the concentration (moles per unit volume) of the salt.

79. Approximate a small piece CB of a constant-volume process by two adiabats CF and GB and an isotherm FG (cf. Figure 22.27). (a) Express the volume ratio V_G/V_F in terms of the pressure difference $\Delta P = P_B - P_C$. (b) Hence show that the heat transfer along the isotherm is:

$$Q_i = \frac{P_E V_E}{\gamma - 1} \ln\left(1 + \frac{\Delta P}{P_C}\right).$$

(c) Expand the logarithm, and show that, to first order in $\Delta T \equiv T_B - T_C$, $Q_i = Q_{cv}$, where Q_{cv} is the heat transfer in the constant-volume process CB. (d) Argue that the entropy change in a constant-volume process is $dS = dQ/T$.

80. The pressure of an ideal gas sample increases from P to $P + \Delta P$ along a small portion of a constant-volume process (cf. Figure 22.6). The volume of the sample is V. Construct an adiabat-isotherm approximation to the process using two isotherms, and one adiabat that passes through the point $(P + \Delta P/2, V)$. (a) Find the pressure and volume at each end of the adiabat. (b) Show that the total work done in the adiabat-isotherm process is:

$$W = \frac{PV}{\gamma - 1} \left\{ -\frac{\Delta P}{P}\left[1 + \ln\left(1 + \frac{\Delta P}{2P}\right)\right] + \left(1 + \frac{\Delta P}{P}\right)\ln\left(1 + \frac{\Delta P}{P}\right) \right\}.$$

(c) Expand the logarithms, keeping terms in $(\Delta P/P)^2$, and show that W is zero to that order. (d) Explain why W must be zero to at least the first order for the limit of infinitely many adiabats and isotherms to be equivalent to the constant-volume process.

81. A container of volume V contains $\mathcal{N}$ mol of a monatomic ideal gas and is initially in equilibrium at temperature T. There is a small hole in the container of cross-sectional area a. Assume that the collisions among the gas molecules are rapid enough to maintain the gas within the container at uniform density and temperature with an exact Maxwell–Boltzmann distribution of velocities. (a) Find the rate at which molecules exit from the hole. (b) Show that the average energy of the molecules leaving the hole is $2kT$. (c) Combine the results of parts (a) and (b) to find an expression for the rate of change of gas temperature in the container.

82. We wish to consider the entropy change per unit mass during the freezing of a cloud of supercooled water droplets. In such a cloud, the process may be assumed to occur uniformly so that a typical subvolume will experience no net heat flow and is effectively an isolated system. The initial state of such a subvolume is $M_1 = 1.00 \times 10^2$ g of air and $M_2 = 3.2$ g of water, all at $T_i = -10°C$. After a slight disturbance, the water freezes leaving air and ice particles in the final state. (a) Compute the final temperature T_f of the ice-air mixture. (b) Construct an imaginary reversible process that links the initial and final states, and compute the entropy change of the 103-g subsystem.

83. In the *Sadly Cannot* thermodynamic cycle, $\mathcal{N}$ mol of ideal gas undergo an adiabatic compression from state B to state A (■ Figure 22.33) followed by expansion along a straight line in the P-V diagram back to B. (a) Argue directly from the first law of thermodynamics that the net heat input Q_{AB} during the straight-line process equals the net work done by the gas during the entire cycle. Q_{AB} sadly cannot be input heat alone, since the efficiency would be unity. (b) The straight-line process is tangent to an adiabat at point D. Argue from the slope of adiabats intersecting the straight-line process at volumes $V < V_D$ that heat is entering the system between states A and D. Conversely, argue that heat leaves the system between states D and B. Show that:

$$V_D = \frac{\gamma}{\gamma + 1}\left(\frac{P_A V_B - P_B V_A}{P_A - P_B}\right).$$

(c) Compute the net work W for a complete cycle in terms of P_A, V_A, P_B, V_B, and γ. (d) Compute the heat input Q_{AD} in terms of P_A, V_A, P_D, V_D, and γ. (e) Taking $V_B = 3.00 V_A$ and $\gamma = \frac{5}{3}$, evaluate the actual efficiency of the cycle. (f) Argue that the maximum temperature in the cycle occurs where the straight-line portion is tangent to an isotherm. Find the volume at which the maximum temperature occurs and also the value of the maximum temperature in terms of P_A and V_A. (g) Find the efficiency of a Carnot cycle operating between the same maximum and minimum temperatures as the Sadly Cannot cycle.

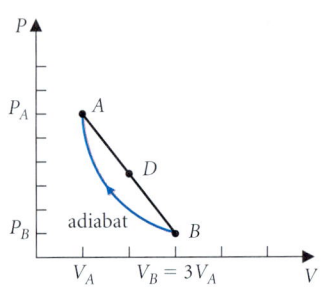

FIGURE 22.33
The Sadly Cannot cycle.

ESSAY 7
Entropy, Evolution, and the Arrow of Time

Physics provides us with several definitions of the direction of time. The second law of thermodynamics states that all processes cause a total entropy change of the universe with the same sign. Time is defined to increase in the same direction as entropy. Alternative definitions could be gleaned from the asymmetric behavior of some physical processes: the expansion of the universe or the decay of particles called K mesons. From a human perspective, we remember the past, but the future is still a closed book. Each of these definitions is internally consistent, but it is not necessary that they all define the same thing. In looking for a unity in the behavior of the universe, we would like to understand the apparent consistency among these definitions. This investigation is at the forefront of philosophy, and we would be presumptuous to claim that the question can be answered here. Thus we shall present but a brief commentary on the philosophy and controversy regarding entropy and the nature and direction of time.

Most of the laws of physics are time reversible. For example, a body acted upon by the gravitational force due to another body would still have acceleration GM/r^2 if the flow of time were reversed. Suppose that in some marvelous way NASA were to make a movie of planets orbiting the Sun under gravitational forces, but, in the Grand Premiere, they ran the movie backwards. Would it provoke laughter? No, because it would show planets moving quite naturally around the Sun. Time reversibility, then, means that a movie of some possible process run backwards shows another possible process.

A movie of a pool game run backwards would look amusing. But the impossibility of a reversed pool game is subtle. If you look only at a few frames (microscopic view), the reversed motion of individual balls is perfectly reasonable. Irreversibility of the whole game lies in the tremendous improbability of 16 rapidly moving balls coming together into a stationary triangle and ejecting a single, very rapidly moving white ball.

Until about 1960, it was thought that all physical laws are microscopically reversible, but in the domain of elementary particles this proves not to be so. Some of these particles, particularly one called the K meson, spontaneously change form in an irreversible manner. Grand Unified Field Theories, which we comment upon in the Epilogue, suggest a connection between this irreversibility and cosmology.

Controversy arises when we try to apply the second law of thermodynamics to the universe as a whole, or to biological systems. To do so, we must generalize the idea of entropy to apply to other kinds of systems from those we have considered so far. Individual organisms represent highly nonrandom systems, and the evolution of life represents a net trend toward increased organization over a time scale of many billions of years. How can this be if the second law of thermodynamics tells us that systems tend toward disorder?

Thermodynamics tells us no such thing! The disorder of the universe tends to increase, but we know that even simple gas systems can experience decreased disorder while the disorder of their environment increases. There is no prohibition against *sub*systems with decreasing entropy. The universe as a whole is very far from equilibrium and, in its trend toward equilibrium, produces very large flows of energy and entropy from place to place. Sunlight is an example of such a flow. In such a situation, it is quite possible for subsystems to exploit the entropy *flow* to maintain and increase their structure.

The universe as a whole also has an evolution that is apparently not time reversible. An amazing feature of this evolution is the appearance of order—in the form of galaxies, stars, and planets—out of the chaos of the early universe. How can both order and entropy increase as the universe expands? We must extend our ideas to include gravitational entropy, which increases as matter becomes condensed. To see why, consider the ultimate end point of gravitational collapse—a black hole (see Essay 8). When a particle falls into a black hole, we lose all information about it except its mass, electric charge, and angular momentum. This loss of information corresponds to an increase in entropy. The British mathematician Stephen Hawking has gone further and shown how a black hole can radiate its mass energy away, with a further increase in entropy. We may envision the end point of the universe as a dilute soup of radiation, which differs from the initial state through its larger volume and lower temperature. This dismal state of affairs need not worry us unduly, since it is many billions of years in the future!

Philosophers of the nineteenth and early twentieth centuries found the conclusions of thermodynamics depressing. The concept of a universal trend toward chaos and *heat death* is somewhat ugly. One need not be so pessimistic, however. The Sun is expected to maintain its energy flow for some 4 billion years, during which time life and its capabilities may well evolve as much as they have in the last 4 billion. The human race is in the springtime of its youth, and we may hope for great evolution of our knowledge and understanding in the aeons to come.

Part V Problems

1. ❖ Some of a madrona tree's surface is covered with a rough brown bark. On the rest of the surface, hard, red-colored wood is exposed. On a cool morning, the wood feels distinctly cold, while the bark feels only slightly cool. Both surfaces are in equilibrium with the surrounding air. What do you suppose so deceives your sense of temperature?

2. ◆◆ A rubber ball of mass 110 g is dropped from a height of 2.0 m and bounces to a height of 1.5 m. Describe what has happened to the energy of the ball. Has heat been added to the ball? If the specific heat of the rubber is 1.9×10^3 J/kg·K, find the increase in temperature of the ball.

3. ◆◆◆ ■ Figure V.4 illustrates a system before and after some process occurs inside. Nothing is known about the structure of the system. Apply the first law of thermodynamics to find the change in internal energy of the system during the process.

4. ◆◆◆ A mass M of ice at $-5°C$ surrounds an evacuated container of negligible heat capacity and of fixed volume V. The ice is in turn surrounded by a perfect insulator; no heat escapes. A sample of monatomic ideal gas at temperature $T_g > 0°C$ is suddenly introduced into the container. (a) If there are $\mathcal{N}$ moles of gas, what is the smallest temperature $T_g = T_{melt}$ for which the gas completely melts the ice? (b) In the case when the ice does not completely melt ($T_g < T_{melt}$), find the final ratio of the mass of melted water to the remaining mass of ice. (c) In the case when the ice completely melts ($T_g > T_{melt}$), find an expression for the final temperature of the system.

5. ◆◆◆ On a cold winter day, the uppermost layer of water in a lake is at $0.0°C$. The air temperature drops to $-10.0°C$ and the lake begins to freeze over. As ice forms, heat is lost by conduction through the ice already formed. Show that the rate of formation of the ice layer is:

$$\frac{dx}{dt} = \left(\frac{k_{ice}\,\Delta T}{L_f \rho_{water}}\right)\left(\frac{1}{x}\right),$$

where k_{ice} is the thermal conductivity of ice, L_f is the latent heat of fusion for water, ΔT is the temperature difference between the top and bottom of the ice layer, ρ_{water} is the density of water, and x is the thickness of the ice. How long does it take for a 5.0-cm-thick layer of ice to form?

6. ◆◆◆ The space shuttle has a flat bottom of area 100 m². The 2-mm-thick aluminum skin is covered with tiles with an average thickness of 4 cm and a thermal conductivity of 1.3×10^{-4} W/cm·K. The aluminum skin is attached to the interior by aluminum spars with an average length of 20 cm, which cover 1% of the total skin area. During re-entry, the skin is at a temperature of $1500°C$, and the interior of the cabin is at $15°C$. Assuming steady-state heat flow, find the rate at which heat flows into the cabin. If the shuttle interior may be regarded as a 7×10^4 kg mass of aluminum and re-entry takes 15 min, what is the temperature rise in the cabin during re-entry?

7. ◆◆ Sunlight, emitted by the Sun at a temperature of about 6000 K, is absorbed by the Earth at a rate of 1 kW/m² and is reradiated at approximately 300 K. Find the rate of entropy production per square meter due to this process. How thick a layer of ice must melt per hour to produce entropy at the same rate?

8. ◆◆ Iced tea is made by adding 150 g of ice cubes at $-2°C$ to a cup of tea at $80°C$. You may assume that the tea is made of water and the mass of tea is 250 g. Ignore heat loss to the surroundings. (a) What is the final temperature of the drink? (b) What is the entropy change of the universe due to this process?

9. ◆◆◆ Water at $90°C$ flows through a pipe of radius 3 cm insulated with glass wool 5 cm thick. The air surrounding the insulation is at $30°C$. What is the heat flux through the insulation? At what rate does entropy leave the pipe? At what rate does entropy leave the glass wool? How do you account for the difference?

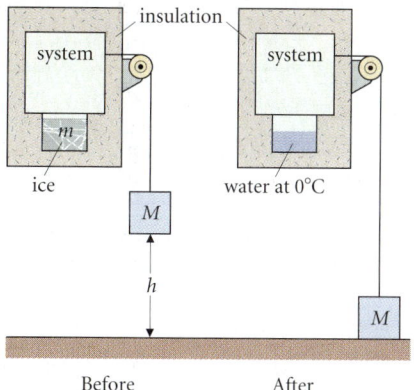

■ FIGURE V.4

Electromagnetic Fields

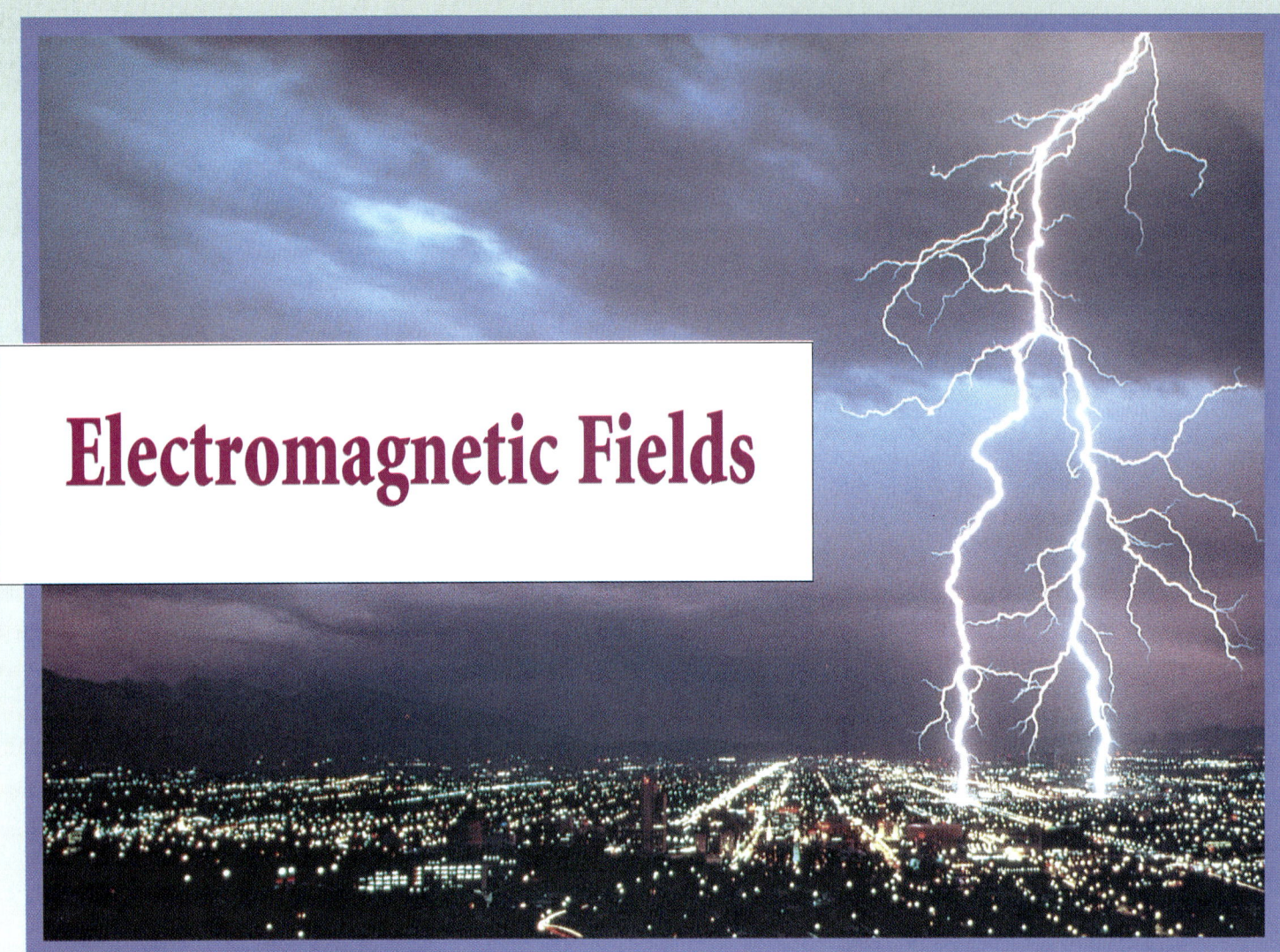

Fear no more the lightning flash,
Nor the all-dreaded thunder stone;
  WILLIAM SHAKESPEARE

The mechanical systems we studied in the first five parts are governed by Newton's laws of motion and the laws of thermodynamics. In Newtonian mechanics, the primary attribute of stuff that determines its behavior is *mass*, and the quantity that changes behavior is force. In Part I we discussed various different kinds of forces, such as friction and tension, but we presented a fundamental explanation only for gravity. Mechanical forces like tension in a rope result from electromagnetic interactions between the atoms that make up the rope. The laws that govern those interactions are the subject of the next two parts.

Until the last half of the eighteenth century, electricity was known as a curiosity, capable of producing entertaining effects. Magnetism, important because of the use of magnetic compasses in navigation, seemed to be quite distinct from electricity. But in the early nineteenth century, electric and magnetic phenomena were found to be closely related, and in the 1860s James Clerk Maxwell presented a unified theory of electricity, magnetism, and optics. Electricity and magnetism are different aspects of a single interaction that not only accounts for the nature of mechanical forces like friction but is also important for determining the laws of chemistry and understanding light. Applications of Maxwell's theory have caused profound revolutions in technology and communications and, through the work of Einstein, the theory has led to a fundamental revision of our concepts of space and time.

The unification of apparently diverse phenomena under a single theoretical umbrella is a major theme of modern physics. To study these phenomena, however, we have to break the theory into small pieces and learn them one at a time. Traditionally you first learn about static electricity, and weeks later you learn about magnetism. The coherence is lost. Our goal in this essay is to give a sense of the unity of electromagnetic phenomena.

Like mass in Newtonian mechanics, another fundamental property of matter—*electric charge*—is the basis of electromagnetism. Charge both produces electromagnetic (EM) fields and determines how objects respond to EM fields. Unlike particles, which are localized, the fields they produce extend throughout space. Fields play an essential role in electromagnetic phenomena, and their nature and description are major themes of this part.

SEE THE DISCUSSION OF THE GRAVITATIONAL FIELD IN ESSAY 2.

OVERVIEW OF ELECTROMAGNETISM
VI.1 Magnetic Field

Most people are familiar with the magnetic compass and its use in navigation. The word *magnetism* derives from the province of Magnesia, a region in Asia Minor with substantial deposits of the mineral magnetite. The ancient Greeks had noticed that pieces of this mineral, *lodestones*, would attract iron. Furthermore, a lodestone suspended from a string orients itself along a north-south line, providing a practical navigational device. The first documented use of a magnetic compass was in China in the eleventh century A.D., though both Chinese and Viking explorers probably used compasses as early as the sixth century. The magnetic compass remains a primary navigational tool (■ Figure VI.1).

By the thirteenth century it was known that magnets can exert both attractive and repulsive forces on each other (■ Figure VI.2). The end of a magnet that points north is called a *north-seeking*, *north*, or *positive* pole. The opposite end is a *south-seeking*, *south*, or *negative* pole. If we try to hold two magnets with their north poles or with

PART SIX

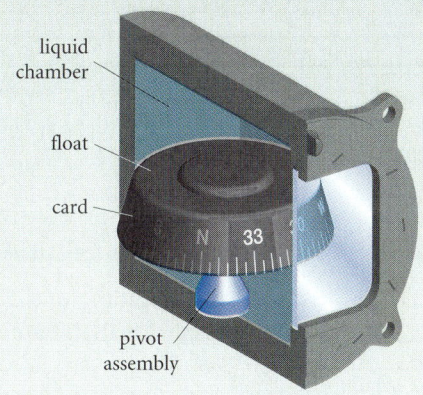

■ FIGURE VI.1
Navigational compass used in an aircraft. A small magnet floats in liquid and rotates so that its north-seeking pole points toward magnetic north.

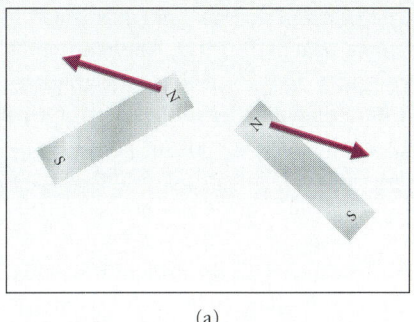
(a)

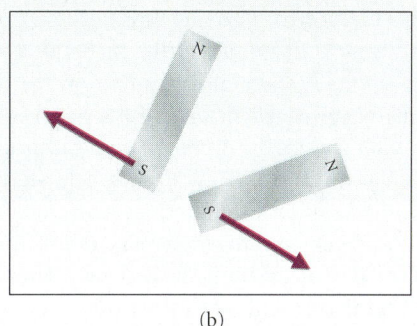
(b)

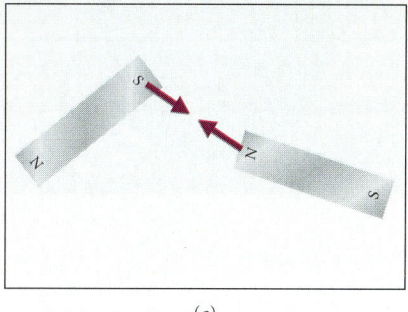
(c)

■ FIGURE VI.2
Magnetic forces exerted by bar magnets. Like poles repel (a and b); unlike poles attract (c). You can easily verify these results for yourself with magnets from your refrigerator door.

PART VI • ELECTROMAGNETIC FIELDS   755

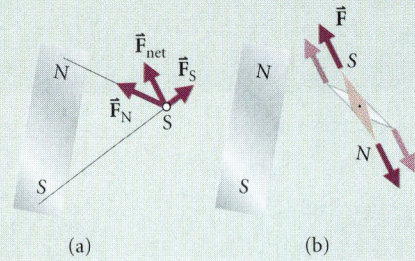

■ **Figure VI.3**
A compass needle placed near a magnet indicates the direction of the magnetic field at that point. (a) The needle's south(-seeking) pole is attracted to the magnet's north pole and repelled by the magnet's south pole. (b) Forces on the compass needle's north and south poles are equal and opposite. The resulting torque causes an angular acceleration of the needle. Once the resulting oscillations damp out, the compass needle is aligned with the magnetic forces.

their south poles together, they repel each other, while a north and a south pole attract each other.

| Like poles repel; unlike poles attract.

When a small compass needle is placed near a magnet, its south pole is attracted to the north pole of the magnet and repelled by the south pole (■ Figure VI.3). Opposite forces act on the north pole. The needle rotates until the net torque on it is zero, and it is parallel to the direction of the net force on each pole. Iron filings scattered on a sheet of paper near the magnet act like small compass needles that show the influence of the magnet on its surroundings—the magnetic field (■ Figure VI.4). The filings trace out lines from the north to the south pole of the magnet, forming a pattern called a *dipole field* pattern.

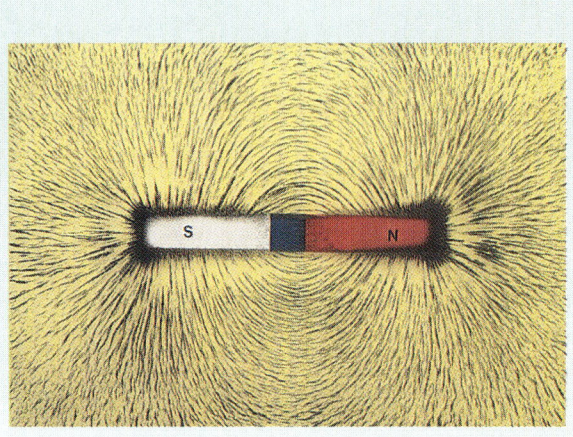

■ **Figure VI.4**
(a) Iron filings act like tiny compass needles. Each aligns with the magnetic forces acting on it. Together they trace out the pattern of the magnetic field.

(b) Computer-constructed drawing of the field lines of a bar magnet. Compare with the distribution of iron filings in Figure VI.4a. Each field line loop is closed by a line running from S to N within the magnet.

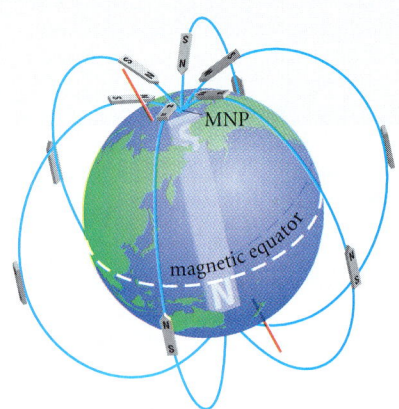

■ **Figure VI.5**
The magnetic field of the Earth is very nearly a dipole field. Magnetic rocks produce deviations from dipole behavior in local regions.

The first systematic investigations of electromagnetic phenomena were conducted by the Englishman William Gilbert (1540–1603), who distinguished electric and magnetic forces and described the behavior of a compass needle near another magnet. Gilbert realized that the Earth itself behaves like a great magnet, thus explaining how a compass senses north. A dipole pattern of field lines originates deep within the Earth (■ Figure VI.5), and a compass suspended freely aligns itself with those field lines. The magnetic field isn't aligned with the Earth's rotation axis, so a compass doesn't point exactly toward geographic north. Instead it points toward magnetic north, currently a point in Canada near Baffin Bay. The naming of magnetic poles arises historically from the use of the compass. The point that cartographers call the *North Magnetic Pole* of the Earth is the point where magnetic field lines converge as they enter the Earth. A compass placed parallel to the surface there would not point in any particular direction. If the Earth's field were exactly a dipole field, this point would actually mark the location of a south-seeking, or negative, magnetic pole, since a compass' north-seeking, or positive, magnetic pole points toward it.

The Earth's magnetic field arises from motion of the liquid rock in the core of the Earth. The field is constant over short time scales, but over decades it changes in both magnitude and direction. Over geologic time scales the poles wander about, and the field reverses direction (■ Figure VI.6). The last reversal occurred about 100 000 years ago.

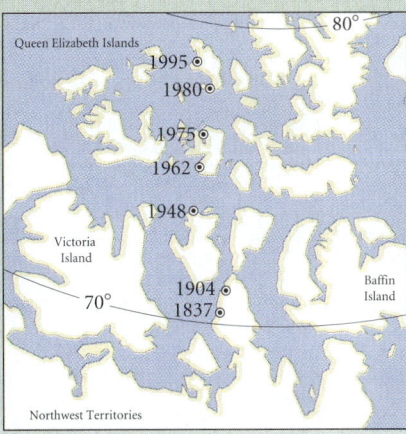

■ **Figure VI.6**
The Earth's magnetic poles wander about with time. Magnetic signatures in certain kinds of rock allow us to reconstruct the path of the poles over time.

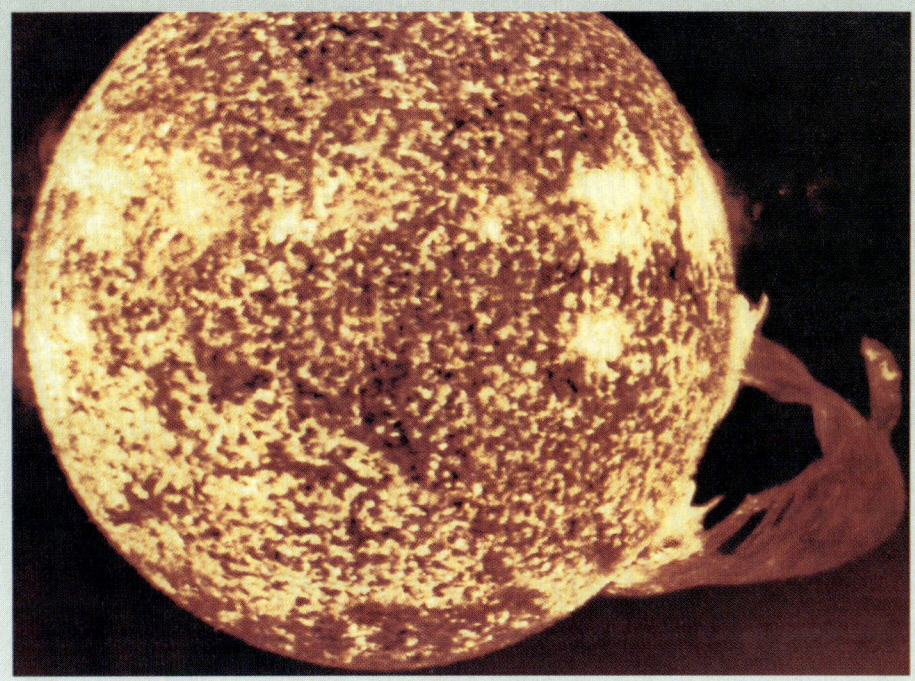

■ **Figure VI.7**
In a solar prominence we may observe the paths of charged particles that follow the magnetic field lines (see Chapter 29). The magnetic field emerges from the Sun in great loops.

The Sun displays some of the most visually dramatic magnetic phenomena. In a solar prominence (■ Figure VI.7) great loops of magnetic field erupt from the surface, and material flows along the field lines. Even more energetic outbursts on the Sun expel particles that would prove fatal if we were not shielded by the Earth's magnetic field. Most of the particles approaching the Earth are deflected away from the surface by the Earth's field, which is itself swept out into a long tail (■ Figure VI.8). Two bands of trapped particles, called the Van Allen belts, pose a radiation hazard to astronauts and leak particles into the atmosphere to produce the aurora.

What causes magnetic fields? Magnetic poles would have been William Gilbert's answer, but that answer cannot be correct. If you cut a magnet in half to isolate a magnetic pole, each piece ends up with both a north and south pole (■ Figure VI.9). It appears that a dipole is the simplest magnetic field source. So what produces a dipole? To answer this question, we must first understand some electric phenomena.

VI.2 Electric Charge

Lightning is a common and spectacular electric phenomenon that results from an electric field set up between the thundercloud and the ground. The electric spark that leaps from your finger to the doorknob after you shuffle over the rug on a dry day differs from lightning

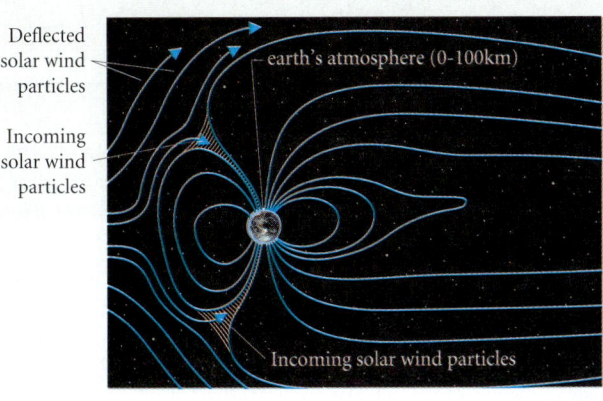

■ **Figure VI.8**
The Earth's magnetic field, which looks like a dipole near the surface, is drawn into a long tail by particles streaming from the Sun. Some of these particles are trapped in the field and contribute to phenomena such as the Van Allen belts and aurora.

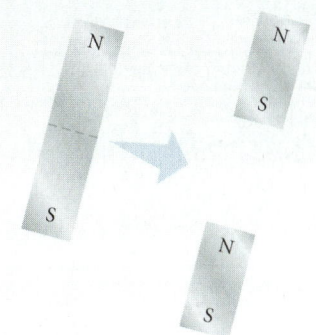

■ **Figure VI.9**
If we attempt to determine the source of magnetic fields by cutting a bar magnet in half, we obtain two smaller bar magnets. Magnetic poles always come in north-south pairs.

■ **Figure VI.10**
Benjamin Franklin (1706–1790). Born in Boston in 1706, Franklin's early years were spent as an assistant in his father's soap and candle factory, and as an apprentice in his brother's printing shop. By the age of 40 he was able to retire with a comfortable income from his own printing business in Philadelphia. Fascinated by a public lecture, he bought the lecturer's equipment and began his own research in electricity. His immense success as a scientist made him a natural choice as a representative of the American colonies in London and later as ambassador to France. His rustic American ways charmed the French court, but his suggestion for saving candles by getting up at dawn did not prove popular!

■ **Figure VI.11**
A lightning rod. The sharply pointed rod is connected to ground. As a thundercloud approaches, charge leaks from the ground into the air through the rod's point. The discharge is rapid enough to reduce the risk of a lightning strike. If the connection to ground is broken, the rod becomes a hazard. Instead of reducing the nearby concentration of electric charge, it becomes a relatively easy path for the lightning discharge.

primarily in the amount of energy released. Benjamin Franklin, ■ Figure VI.10, suggested the first experiments that showed the similarity between lightning and electric sparks. Franklin transformed the study of electricity from a display of freakish and amusing phenomena to a scientific discipline with great practical importance. His invention of the lightning rod, ■ Figure VI.11, was the first major practical application of the new discipline.

ABOUT 600 B.C., CF. CHAPTER 0.

The root *electr* derives from the Greek word *elektron,* meaning amber. Thales of Miletus observed that when amber is rubbed, it attracts other small bodies to itself. A few simple experiments illustrate the important features of this electric force. If we stroke a rubber rod with cat fur and hold it near a small ball suspended by a piece of thread, the ball is attracted to the rod (■ Figure VI.12). If the rod and ball are allowed to touch, the ball is then repelled. Afterwards, the ball is attracted by the cat fur. These experiments show that electric forces can be either attractive or repulsive and suggest that something is transferred between the rod and ball to cause the change from attraction to repulsion. That something is the quantity that allows us to understand electric phenomena: *electric charge.* There are two kinds of electric charge, the kind on the rod and the kind on the cat's fur.

Benjamin Franklin first named the two kinds of charge. As a result of his experiments, Franklin developed a theory of electric fluid that could flow from place to place: an object with excess fluid he called *electrised positively,* and one with a lack of fluid he called *electrised negatively.* The fluid was thought to attract ordinary matter but repel itself. His theory could not explain the repulsive forces between two negatively charged objects. Now we recognize two kinds of charge, still called positive and negative. Negative charge is associated with electrons that move more easily than the more massive, positively charged atomic nuclei. Usually objects are charged by transfer of negatively charged electrons, the opposite of Franklin's conjecture.

 (a)
 (b)
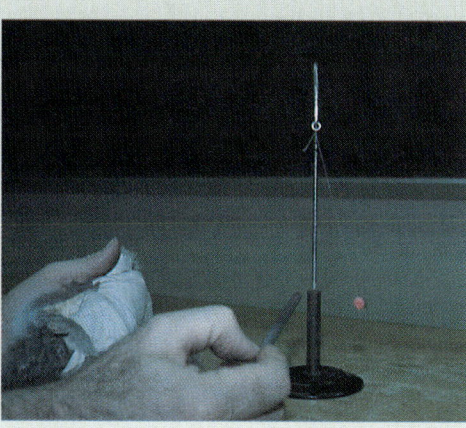 (c)
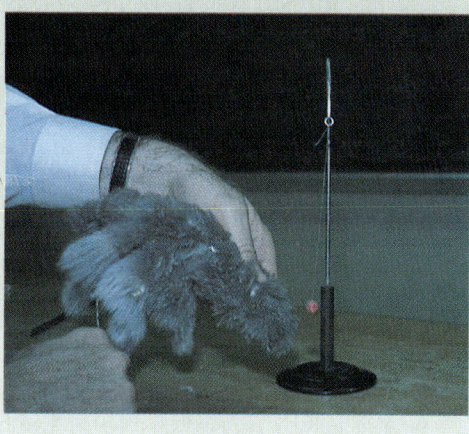 (d)

■ **FIGURE VI.12**
Experiment that demonstrates the existence and nature of electric charge. (a) A rubber rod is stroked with cat fur. (b) The rod attracts a small pith ball. (Pith is a light material found in the stems of certain plants.) (c) If the rod is allowed to touch the ball and is then removed, the ball is subsequently repelled by the rod. (d) The ball is also attracted to the fur. When the rod is rubbed on the fur, charge is transferred between them, leaving the fur and rod oppositely charged. When the charged rod touches the ball, some charge is transferred to the ball. Then the ball and rod have the same kind of charge and repel each other. The ball is attracted to the oppositely charged fur. (The reason why the ball is initially attracted to the rod is discussed in Chapter 27.)

VI.3 The Electrical Structure of Matter

To most of us, electric charge is an elusive concept. We never come across any sizable amount of it in the natural course of things, and so we have little intuition about it; yet it does fantastic things—light lamps, move trolleys, and shoot thunderbolts across the sky. It is astonishing to learn that electric interactions are 10^{40} times stronger than gravity! The elusive behavior of something so powerful arises from its occurrence in balancing positive and negative forms. We notice gravitational force because we live on an astronomical-sized chunk of mass. In stark contrast, it takes ingenuity to tease apart even a tiny amount of the Earth's electric charge. Ironically, the very strength of the electric force binding positive and negative charges together ensures that electric forces play a less obvious role than gravity in our everyday lives.

There is a close connection between charge and matter. Just how close only became clear early in the twentieth century with the triumph of the atomic theory. An atom consists of a nucleus surrounded by a comparatively large cloud of electrons (■ Figure VI.13). The nucleus is made up of particles called protons and neutrons. Each electron has a negative charge, given the traditional symbol e.

EVEN A SINGLE ELECTRON IN A HYDROGEN ATOM FORMS A *CLOUD*. WE'LL DISCUSS THIS BIZARRE NOTION IN CHAPTER 35.

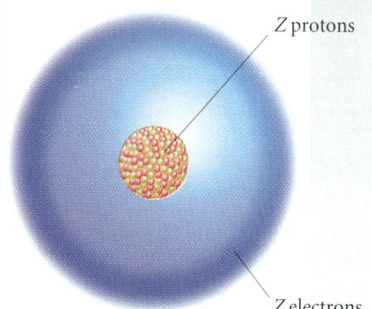

■ **FIGURE VI.13**
Schematic of an atom. Positively charged protons and uncharged neutrons crowd together in the nucleus. The less massive, negatively charged electrons form a cloud around the nucleus. The size of the electron cloud relative to the nucleus is much larger than we can show here: $r_{cloud}/r_{nucleus}$ is about 10^5. (This diagram does not attempt to show the electron distribution accurately. For more details, see Chapter 35.)

> Charge of an electron = $-e$.

Each proton carries a positive charge of the same magnitude.

> Charge of a proton = $+e$.

Neutrons are electrically *neutral*; they carry no net electric charge.

Z IS AN INTEGER.

A typical atom consists of Z protons and Z electrons, which determine the chemical nature of the atom (e.g., if $Z = 8$, it is an oxygen atom). The atom also contains a number of neutrons. Different isotopes of the same element have different numbers of neutrons (e.g., ^{16}O has 8 neutrons, while ^{18}O has 10). The patterns of the electrons around atomic nuclei, and the ways electrons can be added to, stripped from, or shared between atoms, determine the chemical properties of matter.

Each proton or neutron has roughly 2000 times the mass of an electron, so an atom's mass is concentrated in a tiny volume (its nucleus). All the *solid* objects that surround us are mostly vacuum. It is the electrical interactions that bind the structure together and produce its solid feeling.

VI.4   Electric Field

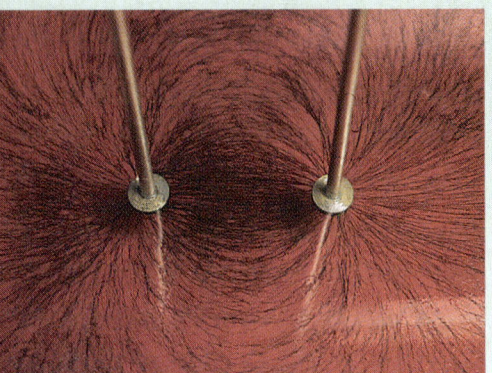

■ FIGURE VI.14
Small fibers suspended in an electric field behave like iron filings in a magnetic field: they align themselves with the field. The field line pattern produced by a positive charge and an equally large negative charge with a small separation looks like the pattern of a magnetic field produced by a bar magnet. This charge configuration is called an electric dipole. Charges in the individual fibers are slightly separated, so that each fiber itself behaves like a tiny electric dipole. A field line diagram for this system is also shown in Figure 23.14.

In ■ Figure VI.14 we see small fibers suspended in a fluid surrounding two concentrations of electric charge, one positive and the other negative. This charge configuration is called an electric dipole. The fibers lie along a pattern of lines strikingly similar to the pattern of magnetic field lines surrounding a magnet. The dipole influences its environment by producing an *electric field*. The electric field tugs on the electric charges in the atoms of the fibers, separating them slightly and forming small electric dipoles that respond to the electric field just as a compass needle responds to a magnetic field. Electric forces arise from a two-step process. One electrically charged object produces an electric field, while a second charged object, exposed to that field, experiences a force. The gravitational forces exerted by the Earth may be modeled similarly: the Earth produces a gravitational field described at each point by the acceleration due to gravity $\vec{g}$. The force exerted on an object is given by its mass multiplied by the gravitational field vector at the object's location:

$$\vec{F}_{grav} = m\vec{g}.$$

The electric field is also described by a vector at each point in space, called the electric field vector $\vec{E}$. The force exerted on an object with charge q is the charge multiplied by the electric field vector $\vec{E}$ at the object's location:

$$\vec{F}_{elec} = q\vec{E}.$$

Gravitational forces are always attractive, but electric forces may be attractive or repulsive because there are two kinds of charge.

Electric field lines indicate the direction of the electric field vector at each point in space. A diagram of the field lines is a useful tool for visualizing the electric field (■ Figure VI.15). The field near a single charge looks quite different from the pictures of magnetic field loops. Field lines emerge from isolated positive charges, which are the simplest sources of electric field. To produce an isolated electric charge, however, we have to separate equal amounts of positive and negative charge. When we look at a region large enough to include charges of both signs, the familiar looped field patterns reappear (Figure VI.14).

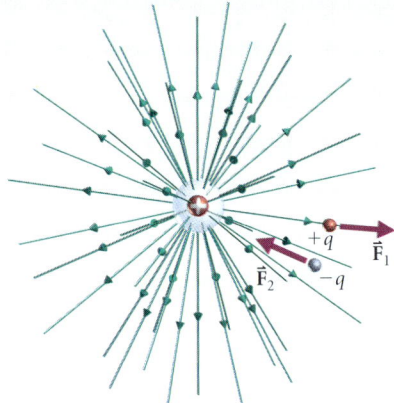

■ FIGURE VI.15
The electric field produced by a single positive charge. Positive charges are repelled by and negative charges are attracted toward the central positive charge.

VI.5   Moving Charge as the Source of Magnetic Field

When electric charges move, they produce magnetic field as well as electric field. Christian Oersted (1777–1851) demonstrated this relationship in 1820 during a public lecture. While performing a demonstration to illustrate the similarities between electricity and magnetism, he noticed that charges moving through a wire (*electric current*) influenced a compass needle. His experiment showed that the magnetic field lines form circles around the wire (■ Figure VI.16). The audience apparently was not impressed!

There is no obvious flow of charge in a bar magnet. Nonetheless, moving charge is the source of its magnetic field. In some atoms, the electrons' motions result in a net circulation around the nucleus, forming a miniature current loop, and each atom acts as a small magnetic dipole. In most materials, these atomic dipoles are oriented randomly and produce no net magnetic field. In iron, cobalt, nickel, and some alloys the atomic dipoles can be aligned in the same direction, producing large total magnetic field strengths.

THE LODESTONES DISCOVERED BY THE CHINESE AND THE ANCIENT GREEKS ARE COMPOSED OF AN IRON ORE NOW KNOWN AS MAGNETITE (Fe_2O_3).

■ **Figure VI.16**
Oersted discovered that magnetic field lines form circles around a wire carrying electric current. The direction of the field lines follows a right-hand rule: with your right thumb in the direction of the current, the field lines run around the wire in the direction that your fingers curl. Oersted demonstrated this rule using compass needles.

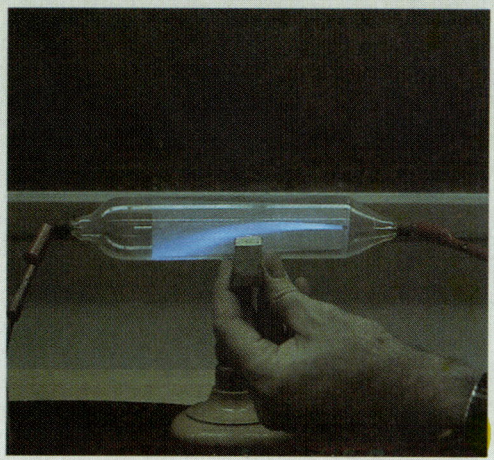

■ **Figure VI.17**
A phosphor screen indicates the acceleration of moving electrons in response to an applied magnetic field. The force is perpendicular both to the electrons' velocity and to the applied magnetic field.

VI.6 Magnetic Force on Moving Charges

Magnetic fields are produced by moving charges, and only moving charges experience magnetic force. ■ Figure VI.17 shows an electron beam moving from right to left in an evacuated tube. A phosphor screen (like a TV screen) intercepts a portion of the beam and makes its position visible. We may investigate the magnetic forces on the electrons by placing a bar magnet near the beam. The force is perpendicular both to the magnetic field $\vec{B}$ and to the velocity of the electrons.

Electrons moving through a wire form an electric current. If the wire passes through a region containing a magnetic field (■ Figure VI.18), the field exerts a force on each of the moving electrons. The total force on a piece of wire is the vector sum of the forces on all the individual electrons. The magnitude of the force is proportional to the magnetic field strength, the magnitude of the current, and the length of wire. Electric motors, ■ Figure VI.19, are a major technological application of magnetic forces on wires.

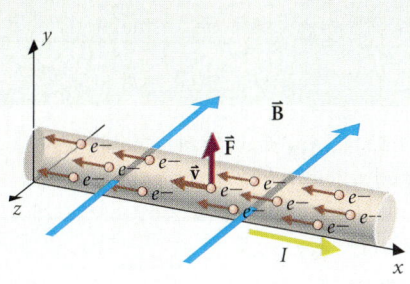

■ **Figure VI.18**
The electric current in a metal wire consists of moving electrons. When a magnetic field is applied, each electron feels a force. The total force on the wire is the vector sum of the forces on all the moving electrons. Thanks to Franklin's sign conventions, we describe negatively charged electrons moving to the left as an electric current I toward the right.

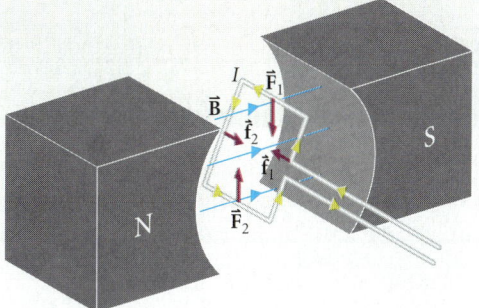

■ **Figure VI.19**
Electric motors make use of the forces on current-carrying wires. In this simple version, the current is in opposite directions on opposite sides of the square wire loop. The forces on opposite sides are equal and opposite. The resulting torque does mechanical work as the loop turns.

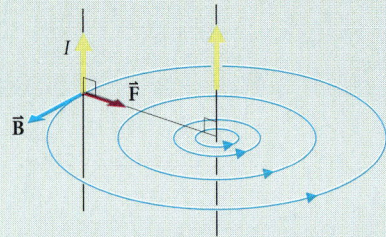

■ FIGURE VI.20
When each of two wires carries a current, the magnetic field produced by each wire causes the moving electrons in the other wire to experience a force. This observation forms the basis for the definition of the SI unit of current: the ampere.

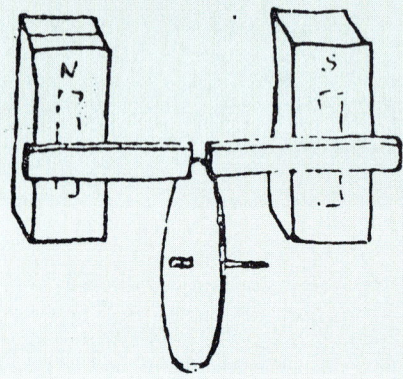

■ FIGURE VI.21
Diagram from Faraday's diary, August 29, 1831. Michael Faraday's experiment with a rotating metal disk showed some of the links between electric and magnetic fields (see Chapter 30) and paved the way for Maxwell's synthesis.

IN §3.3 WE DISCUSSED HOW THE REFERENCE FRAME AFFECTS THE DESCRIPTION OF MOTION.

VI.7 SI Units for Charge and Current

The magnetic field produced by the moving charges in a current-carrying wire influences the charges in a nearby wire (■ Figure VI.20). The force between two such wires is used to define the SI unit of electric current:

> If the same current is maintained in two long parallel wires 1 m apart that repel each other with a force of 2×10^{-7} N per meter of wire, then the current in each wire is 1 ampere (abbreviation A).

The definition of the charge unit follows from the ampere:

> When a current of 1 ampere exists in a wire, the charge flowing past any point of the wire in 1 second is 1 coulomb, (abbreviation C).
>
> $$1 \text{ C} \equiv 1 \text{ A·s}.$$

VI.8 Unity of the Electromagnetic Field

A compass seems unrelated to lightning, yet magnetic and electric phenomena are closely connected. Electric and magnetic fields are both produced by electric charge, and both exert forces on electric charge. In fact the two have an even deeper unity. In 1831 Michael Faraday discovered an electric current in a metal disk rotating between the poles of a magnet (■ Figure VI.21). Faraday described this effect as a production of electric field by the *cutting* of magnetic field lines. His image is illustrated in ■ Figure VI.22. When a wire *cuts* a field line, it retains a small loop of field corresponding to the current in the wire. Faraday's rotating disk continuously cuts through the field lines.

Magnetic field is also produced by changing the electric field. James Maxwell deduced this effect in 1864 from theoretical studies and predicted the existence of electromagnetic waves that move at the speed of light. In 1887 Heinrich Hertz produced and detected such waves, providing experimental verification of Maxwell's theory.

Albert Einstein showed in 1905 that there is a still closer relationship between electric and magnetic fields. His paper on special relativity established that the laws of electromagnetism are the same in every uniformly moving reference frame. Moving charge produces a magnetic field, but whether a charge is moving depends on the reference frame of the observer (■ Figure VI.23). A pure electric field in one reference frame is a mixture of electric and magnetic fields in another. Whether the field is electric, magnetic, or some combination of the two depends on whom you ask! Both kinds of field are aspects of a single electromagnetic field.

VI.9 Electromagnetic Waves

Maxwell's electromagnetic theory of 1865 was the first to incorporate the symmetry between $\vec{E}$ and $\vec{B}$. According to this theory, EM waves are emitted whenever electric charges are accelerated. ■ Figure VI.24 illustrates how a wave is generated when a charge is displaced rapidly

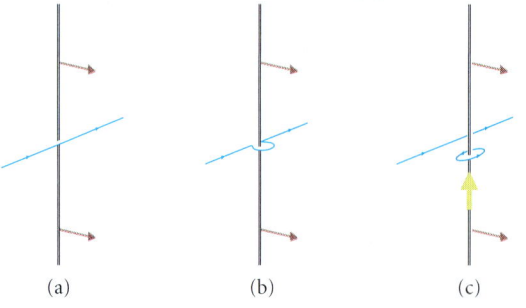

■ FIGURE VI.22
Michael Faraday used field lines as vivid images for his experimental results. In one such image, he imagined a moving metal wire passing through a magnetic field (a) and cutting off a loop (b). If the wire is part of a complete electrical circuit, a current arises as a result of the wire's passage through the field (c). The field line surrounding the wire indicates the presence of that electric current.

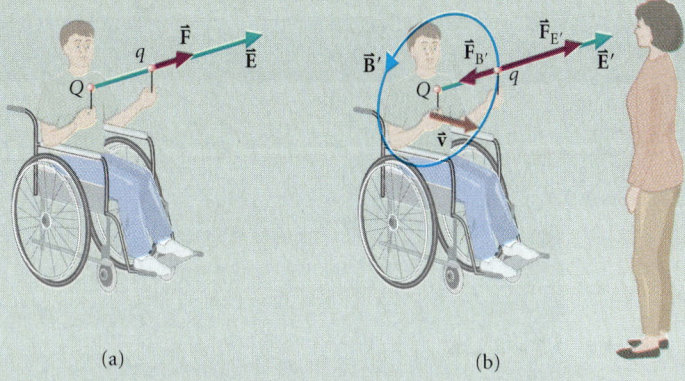

(a) (b)

and then left at rest. When the charge moves, its field lines cannot change everywhere at one instant. They respond somewhat like stretched rubber bands attached to the charge. By shaking the charge repeatedly, we can send a series of kinks along the field lines at the speed of light.

With Maxwell's identification of light as electromagnetic waves, electromagnetism grew from an obscure branch of physics into a fundamental theory of nature that describes a broad range of apparently disparate phenomena as the interaction of electric charge with fields. This unity inspired Einstein to search for a combined theory of electromagnetism and gravity. He didn't succeed, but electromagnetism has since been unified with the weak nuclear force, and grand unified theories are being constructed that link together all forces except gravity. Einstein's goal of unifying these forces with gravity has not yet been achieved.

As you study the chapters in Part VI, remember that these are the first steps along the road to understanding the unified structure of electromagnetism that does not become apparent until Part VII. So let's begin!

■ **Figure VI.23**
(a) Martin is holding two positive charges, one in each hand. He notices that the charge q experiences an electric field due to the charge Q and is repelled. According to Martin, the interaction between Q and q is purely electric. (b) If Martin is sitting in a wheelchair moving forward, Elizabeth notices that the charge Q moving with velocity $\vec{v}$ produces both electric and magnetic fields. According to Elizabeth, the repulsion of the moving charge q results from both electric and magnetic forces.

Martin and Elizabeth are both right. A pure electric field in Martin's reference frame is a combination of electric and magnetic fields in Elizabeth's frame. There is a single electromagnetic field. How we describe it is a matter of perspective—our reference frame.

SEE CHAPTER 33 FOR FURTHER DISCUSSION OF EM WAVES.

MAXWELL ALSO BRIEFLY CONSIDERED A POSSIBLE UNIFICATION OF EM THEORY AND GRAVITY.

THE WEAK NUCLEAR FORCE CONTROLS DECAY OF CERTAIN ATOMIC NUCLEI.

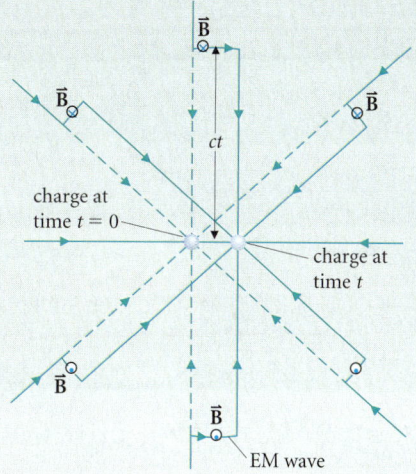

■ **Figure VI.24**
An idealized example shows how electromagnetic waves are generated. A charge initially at the origin is suddenly displaced to the right and abruptly brought to rest again. The field lines cannot move rigidly with the charge. It takes time for the lines to respond to the displacement. Dashed lines show the electric field before the displacement. The solid lines illustrate the field at time t after the displacement. The information that the charge has moved has traveled outward a distance ct. The field lines are centered at the charge's new position out to a distance ct but remain centered on the old position beyond that distance. At distance ct, the field lines are connected by a kink resulting from the charge's displacement.

During the displacement, the motion of the charge constituted an electric current in the negative x-direction that produced a magnetic field according to Oersted's rule (see Figure VI.16). This magnetic field, together with the kink in the electric field, is the wave disturbance that propagates away from the charge. The diagram shows that the electric and magnetic fields of the wave are perpendicular to each other and to the direction of travel of the disturbance. See Chapter 33 for a more detailed explanation.

CHAPTER 23
Charge and the Electric Field

Concepts

Electric charge
Point charge
Test charge
Coulomb's law
Electric field vector
Field line diagram
Superposition
Flux
Gauss' law

Goals

Understand:
The concept of field.

The relation between electromagnetic phenomena and the microscopic structure of matter.

Be able to:
Use Coulomb's law to find the force on a charged particle.

Find the electric field produced by a number of point charges.

Draw quantitative field line diagrams.

Use Gauss' law to relate flux to net charge for a set of point charges.

Intense electric fields above this demonstration device strip electrons from air molecules, producing a spectacular visual display.

> What objects are the fountains
> Of thy happy strain?
> What fields, or waves, or mountains?
> What shapes of sky or plain?
>
> PERCY BYSSHE SHELLEY

If you walk across a wool carpet on a dry winter day, you may get a nasty surprise when you reach for the door knob: an electric spark jumps out and your finger feels a sharp snap. While walking, you generate an electric field around your body that disrupts air molecules and causes the spark. The same effect produces an eerie purple glow hovering above demonstration equipment designed to maintain a large field continuously. Electric fields also hold the plastic wrap around your dish of leftovers, stick toner to paper in copy machines, and remove solid particles from exhaust gases in smokestacks. In this chapter we shall begin to study the electric fields that make these devices work.

As in mechanics, very small particles—*point charges*—obey the simplest rules. We use the particle model to introduce the ideas of electric charge and electric field and to develop field line diagrams as a tool for visualizing the electric field. Most practical electrical systems employ distributions of charge whose geometric shape is important. An electrostatic microphone, for example, uses a thin charged sheet of plastic to convert sound energy into an electric signal. In the next chapter we shall find that the rules and techniques for point charges provide a firm foundation for the study of such continuous systems.

SEE §23.1.4 FOR AN EXPLANATION OF THIS EFFECT.

BEFORE STARTING THIS CHAPTER, READ THE PART INTRODUCTION, WHERE WE SET THE STAGE FOR THE MATERIAL IN THIS PART AND THE NEXT.

ELEMENTARY PARTICLES COME CLOSEST TO THE THEORETICAL IDEAL OF A POINT CHARGE, BUT THEIR BEHAVIOR IS OFTEN COMPLICATED BY ADDITIONAL PHYSICAL PRINCIPLES.

23.1 ELECTRIC CHARGE

23.1.1 Charge and Matter

In mechanics, the most significant property of a particle is its mass, which determines its acceleration when forces act on it. In electromagnetic theory, there is another significant property: charge. There are two kinds of charge, called positive and negative. Ordinary matter is electrically neutral. Each atom contains a positively charged nucleus and enough negatively charged electrons to make its net charge zero. To study the properties of electric charge, we must first separate the positive from the negative. One way to do this is to stroke a rubber rod with cat fur. Before stroking, the fur and rod show no electric effects at all. In contact, they acquire electric charges of opposite kinds. Negative charge is removed from the fur and added to the rod, leaving the fur positively charged.

In all such experiments, we find that charge is neither created nor destroyed; it is simply transferred from one place to another. To produce a positively charged object means transferring some negative charge to one or more other objects. We may summarize this idea in a principle called the *law of conservation of charge*:

THE CHARGE ON THE NUCLEUS IS Ze, WHERE Z IS THE ATOMIC NUMBER. YOU CAN FIND THIS NUMBER IN THE PERIODIC TABLE OF THE ELEMENTS, INSIDE THE BACK COVER.

THIS EXPERIMENT IS DESCRIBED IN THE PART INTRODUCTION.

THUS WE SHOULD ADD CHARGE TO THE LIST OF CONSERVED QUANTITIES IN PART II.

> Electric charge may be transported, but it may not be created or destroyed. The total amount of charge in the universe is constant.

The SI unit of charge is the coulomb (C). The charge on an electron is $-e$: $e = 1.602 \times 10^{-19}$ C. Ordinary matter is composed of whole numbers of electrons and protons and is charged by addition or removal of particles (usually the electrons). We now believe that protons are composed of smaller particles called quarks, whose charges are multiples of $e/3$. Apparently, single, isolated quarks cannot exist outside the particles they compose, so for all practical purposes the smallest observable unit of charge is e. Thus charge is quantized: the charge on any object is an integer number N times e.

Because e is so small, N is usually very large in laboratory-scale phenomena. Just as air may be considered to be a continuous fluid even though it is a collection of individual molecules, a charge distribution may often be considered continuous even though it is made up of individual elementary charges.

A PROTON HAS A CHARGE $+e$. SEE THE PART INTRODUCTION.

FOR AN INTRODUCTION TO THE THEORY OF QUARKS, SEE CHAPTER 37.

EXAMPLE 23.1 ♦ Estimate the total amount of positive electric charge in your body.

MODEL We model a human body as approximately 100 kg of water.

SETUP Each water molecule contains two hydrogen nuclei, each with positive charge e, and one oxygen nucleus with positive charge $8e$. The total positive charge in each

REMEMBER: THE ATOMIC MASS UNIT u IS 1.66 × 10⁻²⁷ kg.

molecule is 10e. We find the number of molecules by dividing the mass of the body by the mass of one molecule, which is 18 u.

SOLVE The number of molecules is:

$$n = \frac{M_{body}}{M_{H_2O}} = \frac{100 \text{ kg}}{18(1.7 \times 10^{-27} \text{ kg})} = 3 \times 10^{27}.$$

Thus: $Q = Ne = n(10e) = (3 \times 10^{27})(10)(1.6 \times 10^{-19} \text{ C}) = 5 \times 10^9 \text{ C}$.

ANALYZE Of course, the body contains an equal amount of negative charge as well.

23.1.2 The Forces Between Charges

The law governing forces between charged objects is similar to the law of gravitation. Newton found that the expression for the gravitational force between two objects is simplest if they can be modeled as particles—objects whose size and structure are unimportant. A particle with a net electric charge also exerts and experiences electric force.

$F = Gm_1m_2/r^2$. SEE §5.4.

A *point charge* is a particle that carries a net electric charge.

The electric force law was discovered by Cavendish between 1771 and 1781, and independently by Coulomb between 1785 and 1789. Cavendish did not publish his results, so the law was named after Coulomb.

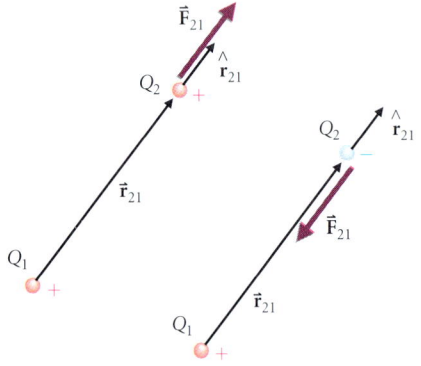

FIGURE 23.1
The force $\vec{F}_{21}$ on particle 2 due to particle 1 lies along the line joining them. The force is repulsive if the two particles have charges of the same sign and attractive if the particles have charges with opposite signs. The unit vector $\hat{r}_{21}$ points from particle 1 toward particle 2.

> **COULOMB'S LAW**
>
> Charges of like sign repel each other, while charges of unlike sign attract. The forces between point charges are proportional to the product of their charges and inversely proportional to the square of the distance between them. (■ Figure 23.1)
>
> $$\vec{F}_{21} = k\frac{Q_1 Q_2}{r^2} \hat{r}_{21}. \tag{23.1}$$

Here $\vec{F}_{21}$ is the force exerted on Q_2 by Q_1. The direction of the unit vector $\hat{r}_{21}$ is from Q_1 toward Q_2. Notice how the formula works: if Q_1 and Q_2 have the same sign, their product is positive, and the force is repulsive. Conversely, if they have opposite signs, their product is negative, so the force is opposite $\hat{r}_{21}$ and the force is attractive.

The constant of proportionality k, the Coulomb constant, is determined by measurement. Its value to three significant figures is:

$$k = 8.99 \times 10^9 \text{ N} \cdot \text{m}^2/\text{C}^2. \tag{23.2}$$

WE SHOULD SAY IT WAS *ORIGINALLY* DETERMINED BY MEASUREMENT. IN FACT k IS RELATED TO THE SPEED OF LIGHT (CHAPTER 33), WHOSE VALUE IS NOW CONSIDERED A FIXED CONSTANT. THE VALUE OF k IS THEREFORE ALSO FIXED.

The Coulomb constant is often expressed in terms of the quantity ϵ_0, which is most often used in conjunction with Gauss' law (§23.4) and capacitance (Chapter 27).

$$k \equiv \frac{1}{4\pi\epsilon_0}, \tag{23.3}$$

$$\epsilon_0 = 8.85 \times 10^{-12} \text{ C}^2/\text{N} \cdot \text{m}^2. \tag{23.4}$$

EXAMPLE 23.2 ♦ A small spherical ball with charge 5.6 C is 2.5 m from a second ball with charge −3.5 C. How large a force do the balls exert on each other?

MODEL We model the two balls as particles. The force between them is given by Coulomb's law.

SETUP Since the charges have opposite signs, the force between them is attractive and has magnitude:

$$|\vec{F}| = k\frac{|Q_1 Q_2|}{r^2}.$$

SOLVE
$$F = 9.0 \times 10^9 \frac{\text{N} \cdot \text{m}^2}{\text{C}^2} \frac{(5.6 \text{ C})(3.5 \text{ C})}{(2.5 \text{ m})^2} = 2.8 \times 10^{10} \text{ N}.$$

ANALYZE This is an enormous force! No known heavy machinery could hold the two "small balls" apart.

NOTICE THAT WE ROUNDED k TO TWO SIGNIFICANT FIGURES, SINCE THE OTHER NUMBERS ARE NOT KNOWN TO GREATER ACCURACY.

23.1.3 The Strength of the Electric Force

The coulomb is a very large amount of electric charge. The force between charges of a few coulombs separated by a few meters (Example 23.2) is about 10^{10} N, equal to the weight of a million metric tons. If you were within a meter of a 1-C point charge, it would exert forces on the electrons and protons in your body strong enough to pull you apart. In the following example, we investigate the amount of charge involved in a laboratory exercise.

EXAMPLE 23.3 ♦♦ Two charged pith balls, each of mass $m = 1.0$ g, are suspended from strings $\ell = 21$ cm long (■ Figure 23.2). The angle between the strings is $2\theta = 12°$, and the balls carry equal charges Q. What is Q?

MODEL We model each ball as a point charge in static equilibrium under three forces: weight, string tension, and electric repulsion.

SETUP We find the distance s between the balls using geometry, and the electric force by analyzing a free-body diagram (Figure 23.2b). Then we apply Coulomb's law to find the charge Q.

In the free-body diagram, choose the x-axis horizontal and y up. Then:

$$\sum F_x = T \sin \theta - F_e = 0$$

and
$$\sum F_y = T \cos \theta - mg = 0.$$

From Coulomb's law: $F_e = kQ^2/s^2$.

We eliminate the unknown tension by dividing the first two equations:

$$\tan \theta = \frac{F_e}{mg} = k\frac{Q^2}{mgs^2}.$$

The distance between the two charges is:

$$s = 2\ell \sin \theta,$$

so:
$$\tan \theta = \frac{kQ^2}{4mg\ell^2 \sin^2 \theta}.$$

SOLVE
$$Q = \pm \sqrt{\frac{4mg\ell^2}{k} \sin^2 \theta \tan \theta} = \pm 2\ell \sin \theta \sqrt{\frac{mg}{k} \tan \theta}$$

$$= \pm 2(0.21 \text{ m})\sin(6°) \sqrt{\frac{(1.0 \times 10^{-3} \text{ kg})(9.8 \text{ m/s}^2)}{9.0 \times 10^9 \text{ N} \cdot \text{m}^2/\text{C}^2} \tan(6°)}$$

$$= \pm 1.5 \times 10^{-8} \text{ C} = \pm 15 \text{ nC}.$$

ANALYZE Since $F \propto Q^2$, we cannot determine the sign from the given information. Either two positive or two negative charges would produce the same observed behavior. ■

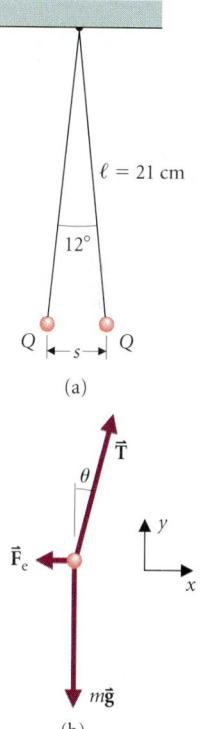

■ **FIGURE 23.2**
(a) Two charged pith balls suspended by strings from the same point are pushed apart by the Coulomb force between them. The angle between the strings is $2\theta = 12°$.
(b) Free-body diagram for the pith ball on the left. Three forces act on each ball: string tension, Coulomb force, and weight.

IN THE SPIRIT OF A BACK-OF-THE-ENVELOPE ESTIMATE, WE TAKE THE AVERAGE MASS OF A PITH BALL MOLECULE TO BE 10 u.

The net charge on the pith balls in Example 23.3 results from the addition (or removal) of about $(1.5 \times 10^{-8}\text{ C})/(1.6 \times 10^{-19}\text{ C}) = 100$ billion electrons. But each ball contains roughly

$$N \approx \frac{m_{\text{ball}}}{m_{\text{molecule}}} \approx \frac{m_{\text{ball}}}{10\text{ u}} = \frac{10^{-3}\text{ kg}}{10^{-26}\text{ kg}} = 10^{23} \text{ molecules},$$

and several electrons per molecule. So only about one part in 10^{13} of the ball's electrons are transferred. Even spectacular electric effects involve only a minuscule fraction of the charge present.

EXAMPLE 23.4 ♦ Compare the electrical and gravitational forces between a proton and an electron.

MODEL Both the electrical and gravitational forces vary in the same way with distance, so their ratio is independent of distance.

SETUP
$$\frac{|\vec{F}_{\text{elec}}|}{|\vec{F}_{\text{grav}}|} = \frac{ke^2/r^2}{Gm_p m_e/r^2} = \frac{ke^2}{Gm_p m_e}$$

SOLVE
$$= \frac{(9 \times 10^9 \text{ N·m}^2/\text{C}^2)(1.6 \times 10^{-19}\text{ C})^2}{(6.7 \times 10^{-11} \text{ m}^3/\text{kg·s}^2)(1.7 \times 10^{-27}\text{ kg})(9 \times 10^{-31}\text{ kg})}$$
$$= 2 \times 10^{39}.$$

ANALYZE Gravitational forces are completely unimportant to the structure of atoms. ■

THE POWER OF 10 IS THE IMPORTANT RESULT OF THIS CALCULATION, NOT THE 2. THAT IS WHY WE HAVE NOT USED MORE FIGURES.

23.1.4 Triboelectricity

We can understand many electrical phenomena by using Coulomb's force law qualitatively together with a crude model of a piece of solid matter (■ Figure 23.3). Heavy positive nuclei form a nearly rigid framework. Each nucleus is surrounded by a cloud of electrons held by the attractive electrical force between the electrons' negative charge and the positive charge on the nucleus. The outermost electrons are held by the weakest force for two reasons: They are farthest from the positive nucleus, and they are also repelled by the electrons closer in. The electrical properties of a solid depend on the relative freedom of outer electrons to move between atoms, so they are extremely sensitive to the strength of this binding force. The electrons are more mobile than the nuclei because of their much smaller mass.

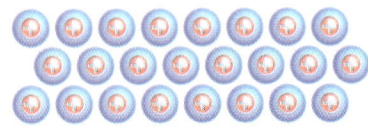

(a)

(b)

■ **FIGURE 23.3**
(a) A crude model of solid matter. The nuclei are surrounded by electron clouds. (b) A scanning tunneling microscope image of a gold surface shows the electron clouds as bumps separated by about 0.3 nm.

EXERCISE 23.1 ♦ If the same force F acts on a proton and an electron, what is the ratio of their accelerations?

The atoms in rubber attract their outer electrons more strongly than do atoms in cat fur. When cat fur and rubber are put in very close contact with each other (■ Figure 23.4), some electrons near the common surface are more strongly attracted to the rubber than to the cat fur, and so they move across the surface. When enough electrons have moved into the rubber, their repulsion overcomes the excess attraction of the rubber and stops further electrons from switching sides. When the two materials are separated, the rubber has taken a small fraction of the cat fur's electrons and so is negatively charged. The cat fur ends up positively charged. This phenomenon is called *tribo-* or contact electricity. In the triboelectric sequence (● Table 23.1) materials listed early in the sequence obtain positive charge from contact with those later in the list.

■ **FIGURE 23.4**
Triboelectricity. When the rubber and fur are put in contact, electrons move from the cat fur into the rubber, and the rubber becomes negatively charged. Migration stops when electrons in the fur are repelled as strongly by the excess electrons in the rubber as they are attracted by the rubber nuclei.

23.1.5 Conductors and Insulators

On a dry day, a rubber rod remains charged for several minutes, while a charged metal rod loses its charge immediately when you pick it up. The difference lies in the ability of charge to move through the bodies of the two rods. The metal is called a *conductor*, and the rubber is an *insulator*. Each molecule in the rubber holds its electrons strongly, while outer electrons in the metal are attracted as strongly by neighboring atoms as by their *own* atom and are thus free to wander throughout the material. Any charge placed on a conductor may move about freely or even pass through the body of the conductor. The most common conducting materials are metals. The same free electrons that sustain current in copper wire also transport heat efficiently through the bottom of a cooking pot and reflect light to produce its shiny appearance. (Values of conductivity are listed in Chapter 26.)

The degree to which electrons are free to move in a material is described by its *conductivity*. Conductivity varies enormously among materials. Metals allow charge to move some 10^{23} times more freely than does a typical insulator such as glass. Germanium and silicon have an intermediate number of mobile electrons per atom. Controlling conductivity of such materials via clever doping with impurities is the essence of the semiconductor industry.

TABLE 23.1 The Triboelectric Sequence

Positive
- Rabbit's fur
- Glass
- Mica
- Wool
- Cat's fur
- Silk
- Cellulose acetate
- Cotton
- Wood
- Amber
- Resins
- Natural rubber
- Metals (Cu, Ni, Co, Ag)
- Sulfur
- Metals (Pt, Au)
- Celluloid

Negative

Adapted from Gaylord P. Harnwell, *Principles of Electricity and Electromagnetism* (New York: McGraw-Hill, 1949), 3, and *Physics Today*, May 1986:51.

23.2 THE ELECTRIC FIELD OF A POINT CHARGE

The interaction between charges may be described as a two-step process. A charge creates an *electric field* in the region around it, while the electric field exerts force on any charge placed in it. The electric field exists at each point whether or not any charge is present to experience a force. To measure the field at a point, however, we place a charge there and measure the force on it. So as not to disturb the system significantly, we introduce as small a *test charge q* as possible. From the measured force on the test charge we find the value of the *electric field vector* at the location of the test charge:

$$\vec{E} \equiv \lim_{q \to 0} \frac{\vec{F}}{q}. \tag{23.5}$$

The magnitude of the electric field vector $|\vec{E}|$ is the *electric field strength*.

The vector $\vec{E}$ describes the electric field in the same way the vector $\vec{g}$ describes the gravitational field at a point.

$$\vec{g} = \lim_{m \to 0} \frac{\vec{F}_g}{m}.$$

Even objects as large as airplanes or supertankers experience the gravitational field of the Earth without themselves contributing significantly to it. We are surrounded by test masses!

■ Figure 23.5 illustrates electric field vectors produced by a point charge Q_1 at two nearby points P and S. A second charge Q_2 experiences a force exerted on it by the electric field (■ Figure 23.6).

SEE THE PART INTRODUCTION, §VI.4, FOR FURTHER DISCUSSION OF THE FIELD CONCEPT.

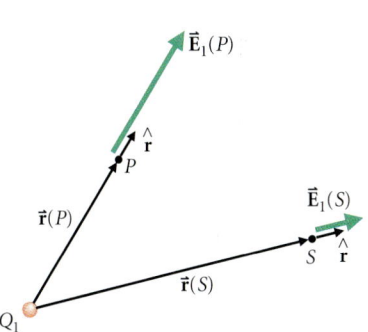

■ FIGURE 23.5
Electric field vectors due to the positive charge Q_1 at two points P and S. In each case the electric field points directly away from Q_1.

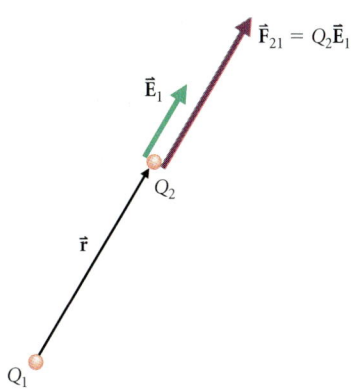

■ FIGURE 23.6
If another point charge Q_2 is placed at P, it experiences a force $\vec{F}_{21} = Q_2\vec{E}_1$.

Coulomb's experimental results (§23.1) for the field and force may be restated as follows:

> **COULOMB'S LAW FOR A POINT CHARGE**
>
> A point charge Q_1 produces an electric field vector $\vec{E}_1(\vec{r})$ at each point P. If P has position $\vec{r}$ with respect to Q_1, then:
>
> $$\vec{E}_1 = \frac{kQ_1 \hat{r}}{r^2}. \quad (23.6)$$
>
> The force exerted by the field on the point charge Q_2 at P is its charge multiplied by the electric field vector at P.
>
> $$\vec{F}_{21} \equiv Q_2 \vec{E}_1. \quad (23.7)$$

REMEMBER: THE ELECTRIC FIELD VECTOR IS A FUNCTION OF POSITION. IT MAY BE DIFFERENT AT EACH POINT OF SPACE.

IN CHAPTER 25 WE SHALL INTRODUCE ANOTHER UNIT FOR MEASURING FIELD: VOLT/METER.

In eqn. (23.6), the unit vector $\hat{r}$ points *from* the point charge *to* the field point P. Thus the electric field vector points away from a positive charge and toward a negative charge. The unit of electric field vector is the newton per coulomb (N/C) and has no special name.

Combining eqns. (23.6) and (23.7), with $\hat{r} = \hat{r}_{21}$, the unit vector from Q_1 to Q_2, we get back eqn. 23.1:

$$\vec{F}_{21} = Q_2 \frac{kQ_1}{r^2} \hat{r}_{21} = \frac{kQ_2 Q_1}{r^2} \hat{r}_{21}.$$

EXAMPLE 23.5 ♦ A point charge $Q = 10.0\ \mu C$ is at the origin. Find the electric field vector at the point P with coordinates $x = 2.0$ m, $y = 1.0$ m (■ Figure 23.7).

MODEL The electric field vector is given by Coulomb's law in the form of eqn. (23.6).

SETUP The distance from the origin to P may be found from Pythagoras' theorem. Then $\vec{E}(P)$ has magnitude:

$$|\vec{E}| = 9.0 \times 10^9\ \frac{\text{N} \cdot \text{m}^2}{\text{C}^2} \frac{10.0 \times 10^{-6}\ \text{C}}{[(2.0\ \text{m})^2 + (1.0\ \text{m})^2]} = 1.8 \times 10^4\ \text{N/C}.$$

The direction of $\vec{E}$ is directly away from the origin (Figure 23.7b), along:

$$\hat{r} = (\cos\theta)\hat{i} + (\sin\theta)\hat{j} = \frac{2\hat{i} + \hat{j}}{\sqrt{5}} = [(0.894)\hat{i} + (0.447)\hat{j}].$$

SOLVE $\vec{E} = (1.8 \times 10^4\ \text{N/C})\hat{r} = [1.6\hat{i} + 0.80\hat{j}] \times 10^4\ \text{N/C}.$

The field vector makes an angle $\theta = \tan^{-1}(0.50) = 27°$ with the positive x-axis, so we may also write:

$\vec{E} = (1.8 \times 10^4\ \text{N/C},$ in the x-y-plane at $\theta = 27°$ counterclockwise from the x-axis).

ANALYZE *Remember:* The electric field vector points away from the positive charge and lies in the plane containing both the charge Q and the point P. Thus in this example, the z-component is zero.

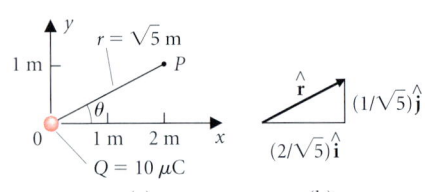

■ **FIGURE 23.7**
(a) A point charge $Q = 10.0\ \mu C$ is at the origin. The point P is $\sqrt{5}$ m away. (b) The unit vector $\hat{r}$ at P equals $(2/\sqrt{5})\hat{i} + (1/\sqrt{5})\hat{j}$. All the vectors, $\vec{r}$, $\hat{r}$, and $\vec{E}$, lie in the x-y-plane.

EXAMPLE 23.6 ♦ What is the force acting on a charge $Q_2 = -1.0\ \mu C$ at point P in Example 23.5?

MODEL The force on the charge is $\vec{F} = Q_2\vec{E}$, where $\vec{E}$ is the electric field vector at P, the position of Q_2.

SETUP The force is: $\vec{F}_{21} = Q_2\vec{E} = (-1.0 \times 10^{-6}\ \text{C})(10^4\ \text{N/C})[1.6\hat{i} + 0.80\hat{j}].$

SOLVE $\vec{F}_{21} = [-1.6\hat{i} - 0.80\hat{j}] \times 10^{-2}\ \text{N}.$

$1.8 = \sqrt{1.6^2 + 0.80^2}$; SEE ALSO EXAMPLE 23.5.

The force has magnitude 1.8×10^{-2} N and is directed toward the origin.

ANALYZE Since Q_2 is negative, it is attracted toward Q_1. Notice that this is automatically described by the mathematical expression for $\vec{F}_{21}$.

EXERCISE 23.2 ♦ In Figure 23.7, the 10.0 μC charge is moved to the point R with coordinates $x = -1.0$ m, $y = -1.0$ m. What is the electric field vector at P then?

If an electric force acts on the point charge Q_2, then an equal and opposite electric force acts on the charge Q_1. This force arises in the same way as its pair: Q_2 produces an electric field $\vec{E}_2$ (■ Figure 23.8) that exerts the force on Q_1. According to Coulomb's law, $\vec{E}_2$ at the location of Q_1 is:

$$\vec{E}_2 = \frac{kQ_2}{r^2}\hat{r}_{12},$$

and the force on Q_1 is: $\vec{F}_{12} \equiv Q_1\vec{E}_2 = Q_1\frac{kQ_2}{r^2}\hat{r}_{12} = -\frac{kQ_1Q_2}{r^2}\hat{r}_{21} = -\vec{F}_{21}.$

The Coulomb force satisfies the strong form of Newton's third law.

Taking Coulomb's law at face value, the field at a point charge's own location is both infinite in magnitude and undefined in direction. It wouldn't make much sense to compute with such a thing, and indeed, if we tried, we would find momentum not conserved. The field of a point charge does not act on the charge that produces it. The force acting on any point charge is computed from the field produced by *other* charges.

23.3 THE PRINCIPLE OF SUPERPOSITION

23.3.1 Superposition of Fields at a Point

Coulomb's law describes the electric field produced by a single charge. If several charges exist in a region of space, each contributes to the net electric field. Experimentally, we find that the total electric field is the vector sum of the individual contributions. The presence of one charge does not affect the contribution due to another. This rule is called *the principle of superposition*.

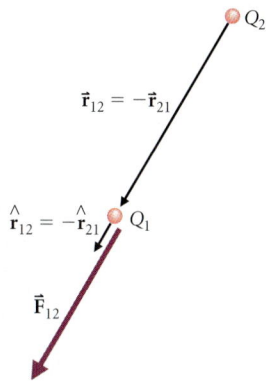

■ **FIGURE 23.8**
Coulomb forces on point charges obey Newton's third law. The charge Q_2 in Figure 23.6 exerts an equal and opposite force $\vec{F}_{12}$ on Q_1. The vector $\hat{r}_{12}$ points from 2 toward 1 and equals $-\hat{r}_{21}$.

WE HAVE TO BE CAREFUL APPLYING NEWTON'S THIRD LAW WHEN DEALING WITH SYSTEMS OF PARTICLES AND FIELDS. WE'LL HAVE MORE TO SAY ABOUT THIS IN CHAPTER 29.

> If several point charges Q_i exist in a region of space, the total electric field vector $\vec{E}$ at a point P is the sum of the electric field vectors produced by the individual charges.
>
> $$\vec{E} = \sum_i \vec{E}_i = \sum_i \frac{kQ_i}{r_i^2}\hat{r}_i. \quad (23.8)$$

EXAMPLE 23.7 ♦ A point charge $Q_1 = +10.0$ μC is at the origin, and a second charge $Q_2 = -5.0$ μC is placed on the y-axis at $y = 1.0$ m. What is the total electric field at the point P with coordinates $x = 2.0$ m, $y = 1.0$ m?

MODEL We use Coulomb's law to find the field due to each charge, and then we find the vector sum according to the principle of superposition (■ Figure 23.9).

SETUP We found the field $\vec{E}_1$ due to Q_1 in Example 23.5:

$$\vec{E}_1 = [1.6\hat{i} + 0.80\hat{j}] \times 10^4 \text{ N/C}.$$

The field at P due to the negative charge Q_2 is in the $-x$-direction:

$$\vec{E}_2 = k\frac{Q_2}{r_2^2}\hat{r}_2 = \left(9.0 \times 10^9 \frac{\text{N}\cdot\text{m}^2}{\text{C}^2}\right)\frac{(-5.0 \times 10^{-6} \text{ C})}{(2.0 \text{ m})^2}\hat{i} = (-1.1 \times 10^4 \text{ N/C})\hat{i}.$$

SOLVE $\vec{E} \equiv \vec{E}_1 + \vec{E}_2 = [1.6\hat{i} + 0.80\hat{j} - 1.1\hat{i}] \times 10^4$ N/C
$= [0.5\hat{i} + 0.80\hat{j}] \times 10^4$ N/C,

or $\vec{E} = (0.94 \times 10^4$ N/C, making an angle of 58° counterclockwise from the x-axis).

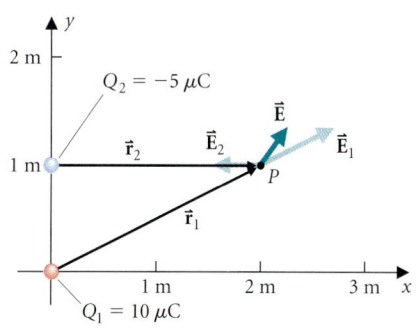

■ **FIGURE 23.9**
Two charges Q_1 and Q_2 lie on the y-axis. The electric field $\vec{E}$ at P is the sum of the electric fields $\vec{E}_1$ and $\vec{E}_2$ due to Q_1 and Q_2.

Digging Deeper

Measuring the Electron Charge

During the period 1910–1913, Robert A. Millikan developed the following experiment for measuring very small amounts of electric charge (■ Figure 23.9). An atomizer sprays a cloud of small oil droplets into the experimental chamber. As they are formed, the droplets acquire a small charge. Droplets occasionally fall through a small hole into a region between two metal plates. There the experimenter can observe the illuminated droplets against a dark background.

With no electric field between the plates, a drop falls at constant speed, its weight balanced by air resistance. When an electric field is applied by connecting a battery to the plates (see Chapters 24 and 25), the drop moves at a different speed, since air resistance now balances the vector sum of the drop's weight and the electric force $q\vec{E}$. (Air resistance is proportional to speed.) By observing the drop's speed in the two cases, the experimenter can deduce the electric charge q on the drop.

To 1% accuracy, Millikan found that all his measured charges were integer multiples of a basic unit that he identified as the electron charge e.

$$q = ne = \pm e, \pm 2e, \pm 3e, \ldots$$

Because the speed of an individual drop is very small, a drop may be followed for hours by varying the electric field. In such extended observations, Millikan could detect the transfer of charge between a drop and the surrounding air as the value of the drop's measured charge changed abruptly by one or two units. Millikan received the Nobel prize for physics in 1923 for this work and also for his measurements of the photoelectric effect (Chapter 35).

During the 1980s, in one of several attempts to detect free quarks, physicists at San Francisco State University used a modernized Millikan apparatus (■ Figure 23.10) to measure charges on droplets of a wide variety of substances, from mercury to mineral water. All the experiments confirmed Millikan's result:

> All objects larger than subatomic particles have electric charges that are integer multiples of e.

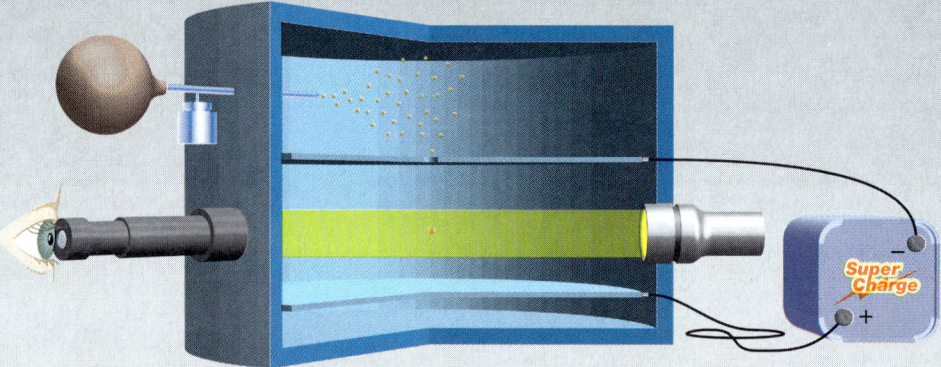

■ **Figure 23.10**
Millikan oil drop experiment. The speed of small oil drops moving through air varies as the applied electric field varies. By observing the drops' speeds, the experimenter can determine the charges on the drops.

■ **Figure 23.11**
The SFSU quark search experiment. This modernized Millikan apparatus was designed to observe charge on droplets from a wide variety of substances in a search for fractionally charged particles. The telescope for observing the drops is under the experimenter's hand. A laser beam entering the chamber through the small window (bottom left) provides a more accurate record of the drops' motion.

ANALYZE Remember that we obtain the total electric field at a point by adding the field vectors *at the same spatial position*. There is no physical meaning to the sum of field vectors at different positions.

EXAMPLE 23.8 ♦♦ Two positive and two negative point charges are at the corners of a square $\ell = 1.0$ m on a side (■ Figure 23.12). If $Q = 1.0$ μC, find the electric field at point P produced by the charges at the other three corners. What electric force acts on the charge at P?

MODEL Using the principle of superposition, the electric field is the sum of three contributions (Figure 23.12b). *Remember:* The charge at P does not contribute to the force acting on itself.

SETUP The contributions to the field at P due to the two negative charges have equal magnitude, $E_1 = E_2 = kQ/\ell^2$, and are at right angles. Their vector sum points toward the center of the square. The field due to the other positive charge points away from the center of the square and has magnitude $E_3 = kQ/(2\ell^2)$.

SOLVE The sum of all three vectors points toward the center and has magnitude:

$$|\vec{E}| = 2|\vec{E}_1|\cos 45° - |\vec{E}_3|$$

$$= 2\frac{kQ}{\ell^2}\frac{\sqrt{2}}{2} - \frac{kQ}{2\ell^2}$$

$$= \frac{kQ}{\ell^2}\left(\sqrt{2} - \frac{1}{2}\right)$$

$$= \frac{(9.0 \times 10^9 \text{ N}\cdot\text{m}^2/\text{C}^2)(1.0 \times 10^{-6}\text{ C})}{(1.0 \text{ m})^2}(1.41 - 0.50)$$

$$= 8.2 \times 10^3 \text{ N/C}.$$

The force is: $\vec{F} = Q\vec{E} = (8.2 \times 10^{-3}$ N, toward the center of the square).

ANALYZE We could have argued that the force is toward or away from the center of the square from the charges' symmetry with respect to the diagonal through P.

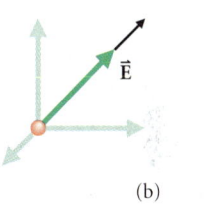

■ **FIGURE 23.12**
(a) Four point charges are at the corners of a square. Two are positive and two are negative. (b) The electric field at point P, the lower left-hand corner, is due to the three charges at the other three corners. The sum of the three vectors lies along the diagonal of the square.

23.3.2 Field Line Diagrams

A good way to visualize the electric field due to any charge distribution is to draw a field line diagram. The direction of the electric field vector at each point is tangent to the field line at that point. In other words, at any point the field line has the same direction as the electric field vector.

Electric field lines diverge from positive charges and converge into negative charges. We may imagine creating a positively charged object by pulling negative charge out. The electric field lines stretch between the two pieces of charge as we pull them apart. In ■ Figure 23.13 the field lines surrounding a single positive charge disappear at *infinity*, the final location of the negative charge that was removed.

In principle, a field line passes through every point. However, we obtain a useful picture by drawing some finite number N of field lines. These lines pass through ever-larger spheres centered on the object (Figure 23.13). The field lines that pass through the sphere at r_1 also pass through the sphere at r_2. Thus:

$$\frac{\text{number of lines per unit area at } r_1}{\text{number of lines per unit area at } r_2} = \frac{N/(4\pi r_1^2)}{N/(4\pi r_2^2)} = \left(\frac{r_2}{r_1}\right)^2.$$

According to Coulomb's law, this is also the ratio of the electric field strengths at the two positions.

$$\frac{|\vec{E}_1|}{|\vec{E}_2|} = \frac{kQ/r_1^2}{kQ/r_2^2} = \left(\frac{r_2}{r_1}\right)^2.$$

THE USE OF FIELD LINES IN STUDYING ELECTROMAGNETIC PHENOMENA WAS INSTITUTED BY MICHAEL FARADAY IN THE 1830S.

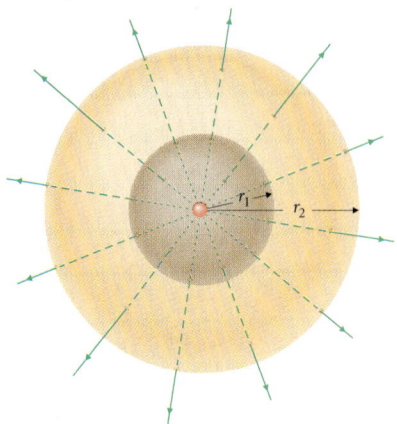

■ **FIGURE 23.13**
Field line diagram for a single positive point charge. If we draw a sequence of spheres centered on the charge, each field line passes through all of them. The number of lines per unit area of surface decreases with distance from the charge.

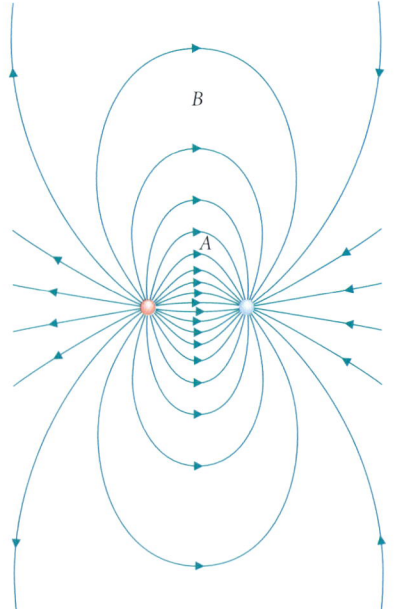

■ FIGURE 23.14
(a) Field line diagram for two equal and opposite point charges. (See also Figure VI.14.) The positive charge is at the left. All the lines emerging from the positive charge converge into the negative charge. Some of the lines form very large loops, much larger than the diagram. The lines that leave the left border of the diagram reenter at the right. The system of field lines is rotationally symmetric about the line joining the two charges; that is, the field line diagram is the same in any plane containing the two charges.

The distance between field lines at A, d_A, is about one-fourth that at B, $d_B \approx 4d_A$. The relative field strengths are determined by the separation of field lines.

The field line diagram describes the electric field according to the following rules:

> The direction of the electric field vector $\vec{E}$ at a point is tangent to the field line through that point.
>
> The strength of an electric field $|\vec{E}|$ at any location is proportional to the relative number of field lines passing through a unit area perpendicular to the lines.

The total number of field lines drawn in any diagram is arbitrary. However, the number of field lines drawn emerging from or converging into a point charge is proportional to the amount of charge, since electric field is proportional to the amount of charge that produces it. With two equal and opposite point charges (■ Figure 23.14), the same number of lines emerge from the positive charge as converge into the negative charge. None extend to infinity.

Field lines never cross each other. If two field lines were to cross at any point, the electric field vector would have two different directions. A test charge placed there would have to accelerate in both directions at the same time. That can't happen.

There are five rules for constructing field line diagrams for a collection of point charges:

1. Field lines begin at positive charge and end at negative charge.
2. The number of field lines shown diverging from or converging into a point charge is proportional to the magnitude of the charge.
3. Field lines are spherically symmetric near a point charge.
4. If the system has a net charge, the field lines are spherically symmetric at great distances.
5. Field lines never cross.

EXAMPLE 23.9 ♦♦ Estimate the relative strengths of the electric field at points A and B in Figure 23.14.

MODEL The strength of the electric field at a point is proportional to the number of field lines per unit area perpendicular to the lines.

SETUP The relative number of field lines per unit area (and thus the strength of the field) at a point is inversely proportional to the area surrounding each field line at that point. That area is roughly the square of the distance between two lines in a two-dimensional diagram (Figure 23.14).

SOLVE We measure the distance between lines near the two points and compute the ratio of their squares.

$$\frac{E_B}{E_A} \approx \left(\frac{d_A}{d_B}\right)^2 = \left(\frac{3 \text{ mm}}{13 \text{ mm}}\right)^2 \approx 0.05.$$

ANALYZE Field lines spread apart as the strength of the field decreases. ■

EXAMPLE 23.10 ♦ Two positive charges are at opposite corners of a square, and two negative charges occupy the other two corners. All four charges have the same magnitude. Sketch the field lines that lie in the plane of the square.

MODEL We choose to draw eight field lines for each charge.

SETUP Since the total charge of the system is zero, no field lines go to infinity (rule 4). Each line begins at a positive charge and ends at a negative charge (rule 1).

SOLVE We complete the diagram using rules 2 and 3 (■ Figure 23.15). Remember rule 5! The lines bend away from the center of the square; they do not cross there.

ANALYZE The field lines share the symmetry of the system about the diagonals of the square. ■

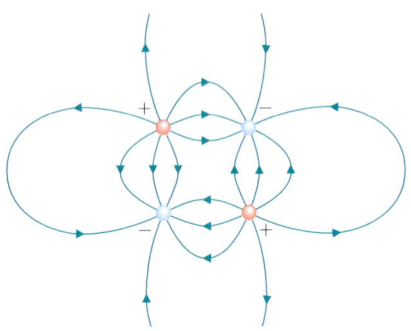

■ FIGURE 23.15
Field line diagram for a system of four charges, two positive and two negative, in the plane containing all four charges. (The diagram in other planes would look different.)

EXERCISE 23.3 ♦ A positive charge $2Q$ has two negative charges $-Q$ at equal distances on either side of it. Sketch a field line diagram for this system. ■

23.3.3 Calculation of Fields as a Function of Position

The electric field due to a system of charges exists at each point in space. Using Coulomb's law and the principle of superposition, we can express the electric field vector as a function of the point's coordinates. For an arbitrary point the calculation can be quite intricate. In this section we shall explore the methods and concepts involved in such calculations, but we shall minimize mathematical complexity by obtaining values of $\vec{E}$ for specially chosen points. Since the field lines are everywhere tangent to the electric field vector $\vec{E}$, it is usually helpful to begin by sketching the field line diagram.

EXAMPLE 23.11 ♦♦ Two equal positive charges Q are at positions $x = -a$ and $x = a$ on the x-axis of a coordinate system (■ Figure 23.16). Compute $\vec{E}$ at each point on the y-axis.

MODEL We begin with the field line diagram. Since the system is rotationally symmetric about the x-axis, we need to draw the field lines in only a single plane through the line of the charges. Showing six field lines per charge gives sufficient detail. We use the five rules given in §23.3.2.

1. There are no negative charges in the system, so the field lines end at *infinity*.
2. Six field lines emerge from each charge.
3. The field lines emerge symmetrically from the charges.
4. From a large distance, the separation of the charges is negligible, and the system appears as a point charge of magnitude $2Q$. The total of 12 lines diverge symmetrically to *infinity*.
5. The field lines do not cross between the charges, but bend away from each other.

The resulting diagram is shown in Figure 23.16a.

Now we calculate $\vec{E}$ at an arbitrary point P on the y-axis, as a function of the y-coordinate. According to the principle of superposition (Figure 23.16b), the total electric field is the sum of the two vectors $\vec{E}_1$ and $\vec{E}_2$ produced by the two charges. The two equal charges are at the same distance r from P.

SETUP We use Coulomb's law to find $\vec{E}_1$ and $\vec{E}_2$. The magnitudes are equal:

$$|\vec{E}_1| = |\vec{E}_2| = \frac{kQ}{r^2} = \frac{kQ}{a^2 + y^2}.$$

Since both vectors make equal angles with the y-axis, their vector sum lies along the y-axis. The x-components of $\vec{E}_1$ and $\vec{E}_2$ cancel, while their y-components add.

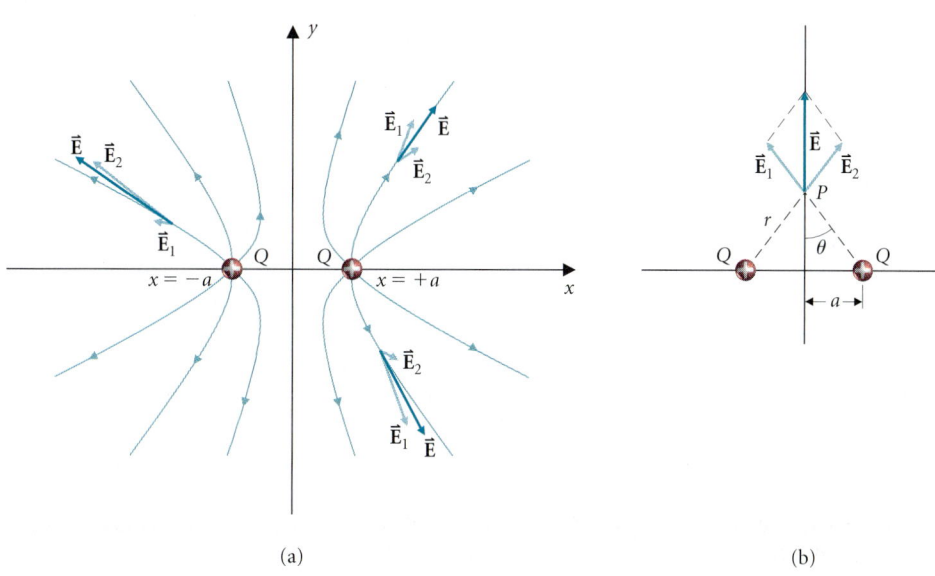

■ FIGURE 23.16
(a) Field line diagram for two equal positive point charges at $x = -a$ and $x = a$ on the x-axis. (b) The electric field at a point P on the y-axis is the sum of the fields due to each of the two charges. The vector sum lies along the y-axis.

SOLVE
$$\vec{E} = 2|\vec{E}_1| \cos\theta \, \hat{j} = 2 \frac{kQ}{r^2} \frac{y}{r} \hat{j} = \frac{2kQy}{(a^2 + y^2)^{3/2}} \hat{j}. \qquad (23.9)$$

ANALYZE The direction we computed for $\vec{E}$ is consistent with the field line diagram. At the origin, halfway between the charges, the electric field vectors produced by the two charges are equal and opposite, so $\vec{E} = 0$. The field lines bend away from the origin and none passes through the origin. Along the y-axis, the field lines crowd together before beginning to diverge, indicating that the electric field increases to a maximum strength before decreasing again at large distances. At points where $y \gg a$, the separation of the two charges becomes insignificant. In this limit we have:

$$\lim_{y \gg a} |\vec{E}| = \lim_{y \gg a} \frac{2kQy}{y^3(1 + a^2/y^2)^{3/2}} = \frac{2kQ}{y^2}.$$

This is just what we would expect for a point charge of magnitude $2Q$ at the origin.

The calculation actually gives the value of $\vec{E}$ at any point in the y-z-plane. Since the system is symmetric about the x-axis, the electric field at a point in the y-z-plane a distance $d = \sqrt{(y^2 + z^2)}$ from the origin points directly away from the x-axis and has magnitude $|\vec{E}| = 2kQd/(a^2 + d^2)^{3/2}$.

EXERCISE 23.4 ♦♦ Find the electric field at each point on the x-axis with $x > a$.

Study Problem 15 ♦♦♦ Two Unequal Charges

Two charged objects $Q_1 = +3Q$ and $Q_2 = -Q$ are separated by a distance D (■ Figure 23.17a). Find a point near the objects where $\vec{E} = 0$.

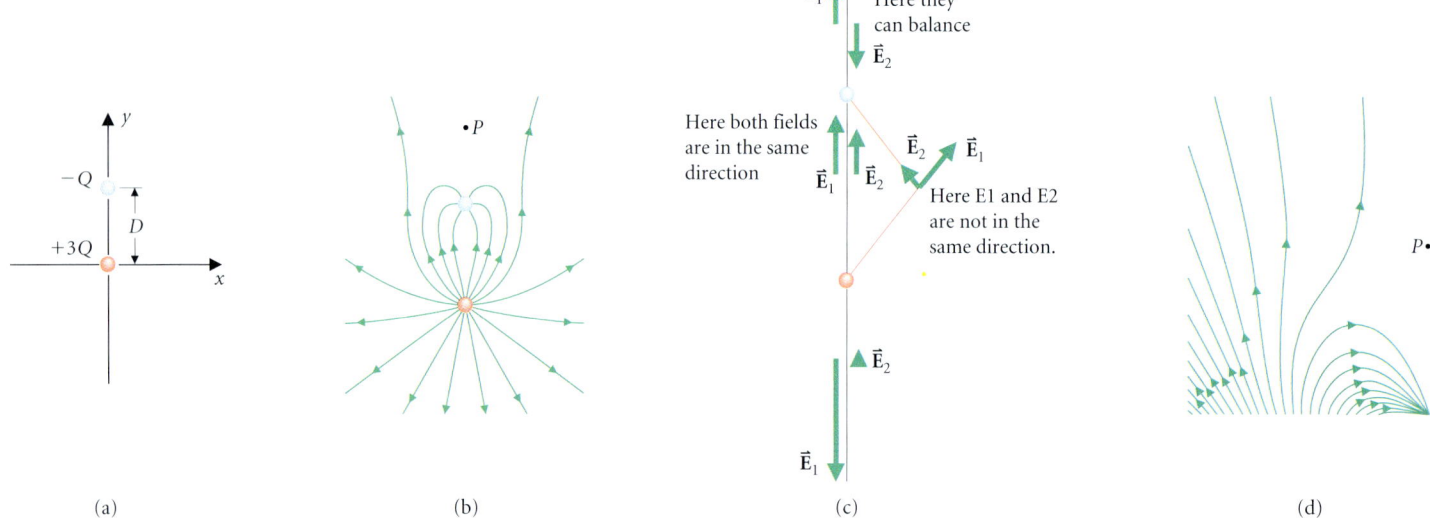

■ **FIGURE 23.17**
(a) Two unequal charges are on the y-axis: $+3Q$ at $y = 0$ and $-Q$ at $y = D$. (b) Field line diagram. One-third of the field lines emerging from the positive charge converge at the negative charge; the rest go off to infinity. (c) The net field can be zero only on the y-axis where the fields $\vec{E}_1$ and $\vec{E}_2$ are in opposite directions, that is, above the negative charge, at $y > D$. (d) An enlargement of the field line diagram in the region $y > D$. Many more field lines are shown. For $x > 0$, the diagram is the mirror image of that shown for $x < 0$. Notice how the lines bend away from P, where the field is zero.

Modeling the System

We begin by drawing the field line diagram (Figure 23.17b). The algebraic sum of the two charges is $+2Q$, so when viewed from a large distance (much greater than the separation D of the two charges), the system looks like a single point charge $+2Q$.

Applying the rules for field line diagrams from §23.3.2, we choose to represent a charge Q with 6 field lines; then 18 lines emerge from the positive charge $Q_1 = 3Q$, and 6 converge into the negative charge. The remaining 12 field lines are distorted by the presence of Q_2 but flare out symmetrically at large distances.

The total electric field $\vec{E}$ is the vector sum of contributions $\vec{E}_1$ and $\vec{E}_2$ produced by the charges $Q_1 = 3Q$ and $Q_2 = -Q$. The total field is zero at any point where $\vec{E}_1$ and $\vec{E}_2$ balance each other; that is, when $\vec{E}_1$ and $\vec{E}_2$ are (1) in opposite directions and (2) equal in magnitude (Figure 23.17c). The two vectors have opposite directions only at points on the line through the two charges, either below the positive charge or above the negative charge. The field of a point charge increases with the magnitude of the charge and decreases with distance from it. So the magnitudes $|\vec{E}_1|$ and $|\vec{E}_2|$ can be equal only at points closer to Q_2 than to Q_1, that is, above Q_2.

Setup

We put the origin at Q_1 and the y-axis along the line joining the two charges. We derive expressions for the electric field vectors at a point P on the y-axis due to Q_1 and Q_2, add them to find an expression for the total field vector $\vec{E}$, set this equal to zero, and solve for y.

From Coulomb's law:

$$\vec{E}_1 = k\frac{3Q}{y^2}\hat{j} \quad (y > 0),$$

$$\vec{E}_2 = -\frac{kQ}{(y-D)^2}\hat{j} \quad (y > D),$$

$$\vec{E} = \vec{E}_1 + \vec{E}_2 = kQ\left[\frac{3}{y^2} - \frac{1}{(y-D)^2}\right]\hat{j} \quad (y > D).$$

Solution of Equations

The field is zero where:
$$\frac{3}{y^2} - \frac{1}{(y-D)^2} = 0. \tag{i}$$

Setting the two terms equal and taking their square roots:

$$\frac{3}{y^2} = \frac{1}{(y-D)^2} \Rightarrow \sqrt{\frac{3}{y^2}} = \pm\sqrt{\frac{1}{(y-D)^2}}.$$

$$\frac{\sqrt{3}}{y} = \frac{\pm 1}{y-D} \Rightarrow (\sqrt{3} \mp 1)y = D\sqrt{3}.$$

$$y = \frac{\sqrt{3}}{\sqrt{3} \mp 1}D = \frac{3 \pm \sqrt{3}}{2}D.$$

DON'T FORGET THAT THERE ARE TWO SOLUTIONS, CORRESPONDING TO THE POSITIVE AND NEGATIVE SQUARE ROOT.

Are both of these solutions valid? No. In setting up the problem we established that $y > D$. For $y < D$, the direction of $\vec{E}_2$ is reversed, and eqn. (i) does not hold. The $+$ sign gives the correct result:
$$y = (3 + \sqrt{3})D/2 = 2.37D.$$

Analysis

The point P lies above the negative charge, as shown on the diagram. Notice that below this point, all field lines curve inward and terminate at the negative charge, while above this point the field lines terminate at infinity (Figure 23.17d). There is no well-defined direction for the field at P, corresponding to its zero magnitude.

23.4 GAUSS' LAW

23.4.1 The Relation Between Charge and Field Lines

Coulomb's law describes the electric field created by a charge distribution: given a point, you can calculate the field there. By drawing the field line diagram, we get a sense of how the charge distribution affects all space: it gives a more global perspective. The values of electric field vectors at different places are related because they are produced by the same charge distribution. In this section we introduce Gauss' law, which expresses this global relation of charge to field. It does not contain any new information, but it expresses the same information as Coulomb's law in a different way.

AFTER KARL FRIEDRICH GAUSS (1777–1855).

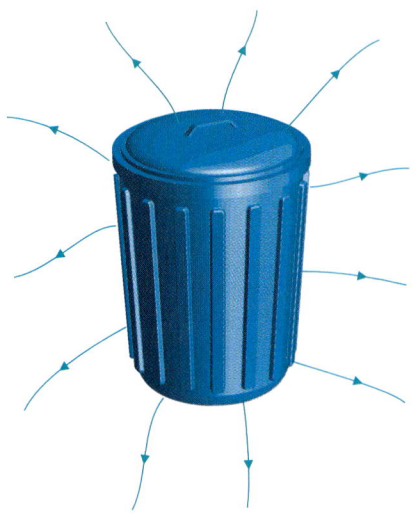

■ FIGURE 23.18
If field lines emerge from a plastic garbage pail, we may conclude that there is charge inside to produce the field. In this case the charge is positive, since the field lines point out of the pail.

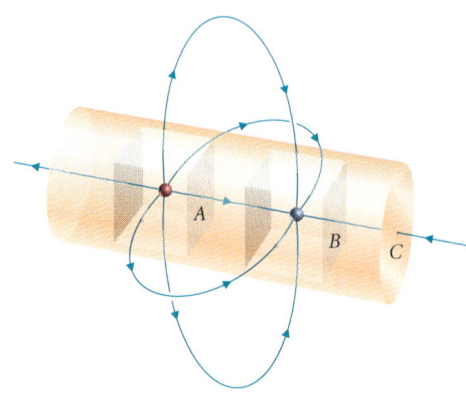

■ FIGURE 23.19
A system of two equal and opposite charges. The field line along the symmetry axis forms a very large loop seen leaving to the left and returning from the right.

Electric field lines begin and end on electric charges. Thus the field lines emerging from the pail in ■ Figure 23.18 indicate that there is charge inside to produce them. That is the main idea contained in Gauss' law. To begin developing a more quantitative statement, let's consider a specific system: two equal and opposite point charges (■ Figure 23.19). The six field lines we have drawn emerge from the positive charge and converge into the negative charge. The field lines that emerge from or pass into any volume indicate the charge contained *inside* that volume. For example, volume A contains a charge $+Q$, and six field lines emerge from its surface. Volume B contains charge $-Q$, and six field lines pass inward through its surface. If we describe a field line passing inward as a negative line coming out, then:

GAUSS' LAW: PRELIMINARY STATEMENT.

> The net number of field lines emerging from any volume is proportional to the net charge inside.

This result holds for a volume of arbitrary shape containing an arbitrary number of charged objects. Volume C contains both charges, with a total inside of $Q - Q = 0$. One of the field lines emerging from the positive charge doesn't intersect the surface of volume C at all. The rest of the lines emerge and then reenter the volume; they each count positive once and negative once, for a net contribution of zero.

EXAMPLE 23.12 ❖ It becomes increasingly difficult to draw accurate three-dimensional diagrams for more complicated charge distributions. So, imagine for the moment that we live in a two-dimensional world. Sketch the field line diagram for the system of three "point" charges shown in ■ Figure 23.20, and verify the relation between charge and field lines for volumes B and C.

MODEL If we draw six field lines per charge unit Q, 12 lines diverge from the positive charge and six lines converge into each of the negative charges. Four lines, labeled (a), (b), (c), and (d), form large loops: they leave and reenter the diagram.

SETUP Six field lines enter the elliptical volume B, which contains a total net charge $-Q$. Volume C contains a net charge $2Q$. Even though the volume wraps around the negative charge on the right, the charge is *not* inside the volume. All 12 field lines leaving the positive point charge leave volume C. At the lower right, field line (c) reenters the volume but then leaves again, so its net contribution is still one line leaving the volume.

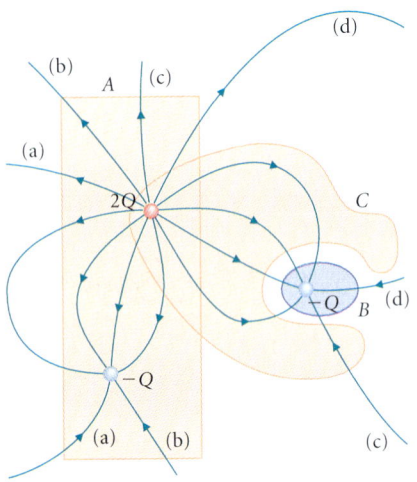

■ FIGURE 23.20
Gauss' law in a two-dimensional world. The number of field lines leaving each volume is proportional to the charge inside.

SOLVE The number of field lines entering each volume (-6 versus $+12$) is in the same ratio as the charge contained ($-Q$ versus $+2Q$).

ANALYZE A two-dimensional slice does not always give a true representation of a three-dimensional situation. However, a slice like this is an accurate representation of the field pattern if each of the "points" is actually an infinite *line* of charge lying perpendicular to the page.

SEE §24.2.

EXERCISE 23.5 ♦ Compare the number of field lines emerging from volume A in Figure 23.20 with the charge inside the volume.

23.4.2 Electric Flux

Field lines are a useful visual aid, but at best they offer a rough quantitative description of the electric field strength. Our next step is to find a precise definition of the quantity *electric flux* that we have just modeled as "number of field lines." The strength of the electric field is represented by the relative number of field lines passing through a unit area (§23.3.2):

$$|\vec{E}| \propto \frac{\text{number of field lines}}{\text{area}}.$$

■ Figure 23.21 shows a bundle of field lines passing through an area A_o perpendicular to the lines, and a second area A inclined at angle θ to the lines. The number of field lines in the bundle is represented by:

$$N \propto |\vec{E}|A_o = |\vec{E}|A \cos\theta = A\vec{E} \cdot \hat{n}.$$

The quantity on the right-hand side of this expression is the *electric flux* through the area A.

The *electric flux* $\Delta\Phi_E$ passing through an element of area ΔA with normal $\hat{n}$ is:

$$\Delta\Phi_E \equiv (\vec{E} \cdot \hat{n})\Delta A. \tag{23.10}$$

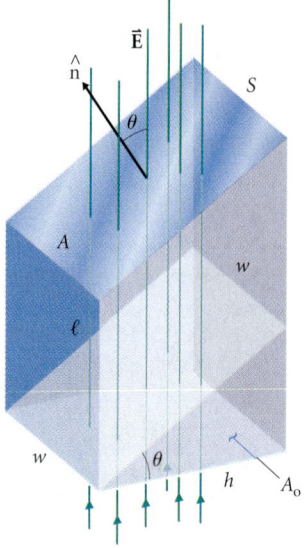

■ **FIGURE 23.21**
(a) The area of the tilted surface $A = w\ell = wh/\cos\theta = A_o/\cos\theta$. (b) The flux of electric field through the surface S is $\vec{E} \cdot \hat{n}A = |\vec{E}|A\cos\theta = |\vec{E}|A_o$.

Digging Deeper

FLUX

The word *flux* derives from the Latin for flow, and electric flux is mathematically very similar to the flow of liquid in a pipe. Imagine a pipe carrying water at constant velocity $\vec{v}$ (■ Figure 23.22). The volume of water flowing past any cross section A_o of the pipe per unit time is:

$$\frac{dV}{dt} = \frac{A_o|\vec{v}|\,dt}{dt} = A_o|\vec{v}|.$$

The same volume flows out of the diagonal surface at the end of the pipe. The *flux* out of the pipe is due to the velocity component $\vec{v} \cdot \hat{n}$ normal to the surface; motion parallel to the surface does not transport water across it.

$$\frac{dV}{dt} = A\vec{v} \cdot \hat{n} = A|\vec{v}|\cos\theta = A_o|\vec{v}|.$$

A similar expression (eqn. 23.10) describes electric flux.

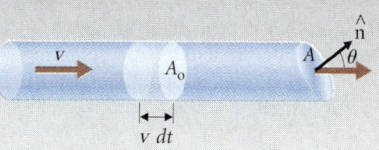

■ **FIGURE 23.22**
Flux of water through a pipe is
$$A\vec{v} \cdot \hat{n} = A_o|\vec{v}|.$$

A *CLOSED SURFACE* IS THE BOUNDARY OF A VOLUME. FOR EXAMPLE, A CLOSED CYLINDRICAL SURFACE SUCH AS A SOUP CAN INCLUDES THE TWO FLAT FACES AT THE ENDS. NO MATTER WHICH WAY YOU ORIENT THE SURFACE, THE SOUP WON'T RUN OUT.

EXERCISE 23.6 ♦ Find the electric flux through an element of area $\Delta A = 1.0 \text{ m}^2$ lying in the x-y-plane with normal in the positive z-direction (■ Figure 23.23) if the electric field is uniform:

$$\vec{E} = (1.0 \times 10^{-6} \text{ N/C}) \frac{(\hat{i} + \hat{k})}{\sqrt{2}}.$$

A surface that completely encloses a volume is called a *closed surface*. The total flux emerging from a volume is found by dividing the closed surface bounding the volume into elements of area and adding the flux through each of them (■ Figure 23.24). The normal for each element always points *outward* from the volume. (This is equivalent to counting emerging field lines as positive and entering field lines as negative.)

$$\Phi_E = \oint d\Phi_E = \oint \vec{E} \cdot \hat{n} \, dA. \tag{23.11}$$

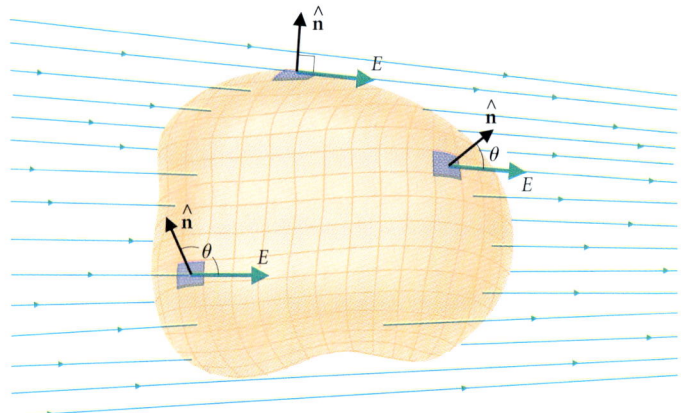

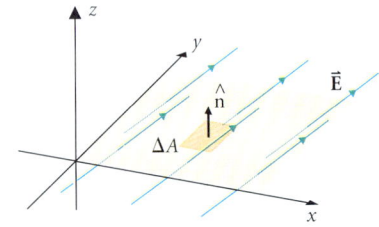

■ **FIGURE 23.23**
What is the electric flux through the area shown?

■ **FIGURE 23.24**
To find the electric flux through a closed, curved surface, divide the surface up into differential elements and sum the flux through all the elements. The normal to each element points outward from the volume enclosed by the surface.

EXAMPLE 23.13 ♦ Find the electric flux through a cubical box placed in a uniform field so that two of the faces are perpendicular to $\vec{E} = E\hat{i}$ (■ Figure 23.25).

MODEL Since $\vec{E}$ and $\hat{n}$ are each constant on any one side, we may calculate the flux through each surface of the box and add the results.

SETUP The electric field is parallel to the top and bottom of the box and also to two of the sides. On these faces, $\vec{E} \cdot \hat{n} = 0$, and the flux through them is zero. No field lines pierce these faces. Since normal vectors point outward from the box, $\hat{n} = \hat{i}$ on the side at $x = L$, and $\hat{n} = -\hat{i}$ on the side at $x = 0$. The flux through the side at $x = L$ is:

$$\Phi_L = (\vec{E} \cdot \hat{n})a^2 = \vec{E} \cdot \hat{i} a^2 = Ea^2.$$

The flux through the side at $x = 0$ is:

$$\Phi_0 = (\vec{E} \cdot \hat{n})a^2 = \vec{E} \cdot (-\hat{i})a^2 = -Ea^2.$$

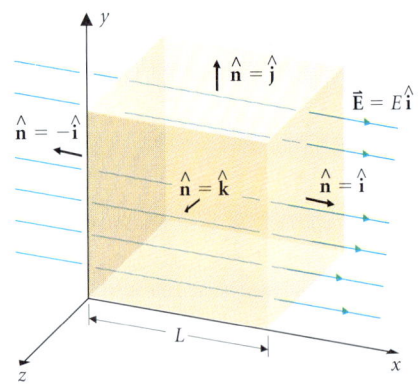

■ **FIGURE 23.25**
Flux of a uniform electric field through a cube. Notice that each field line that enters the cube at the left side leaves at the right. The net flux is zero.

SOLVE The net flux through the box is:

$$\oint \vec{E} \cdot \hat{n} \, dA = \Phi_L + \Phi_0 = Ea^2 + (-Ea^2) = 0.$$

ANALYZE There is no charge inside the box and the net flux through its surface is zero. All the field lines that pass in at $x = 0$ pass out again at $x = L$.

EXAMPLE 23.14 ♦♦ Find the electric flux through a sphere of radius $r = a$ surrounding a point charge Q (■ Figure 23.26).

MODEL Field lines emerge from the point charge and exit through the surface of the sphere. Since there is charge inside, the flux is not zero.

SETUP The normal to the surface at any point is along the outward radius, $\hat{n} = \hat{r}$. The electric field is also along the outward radius:

$$\vec{E} = \frac{kQ}{r^2} \hat{r}.$$

SOLVE The dot product $\vec{E} \cdot \hat{n} = kQ/r^2$ has the same value kQ/a^2 at each point on the surface $r = a$. Thus:

$$\Phi_E = \oint \frac{kQ}{a^2} dA = \frac{kQ}{a^2} \oint dA = \frac{kQ}{a^2} 4\pi a^2 = 4\pi kQ = \frac{Q}{\epsilon_0}.$$

ANALYZE The flux does not depend on the specific value $r = a$ of the sphere's radius. Once again the flux through the surface is proportional to the charge inside. More generally, outside any spherically symmetric distribution of charge:

$$\vec{E} \cdot \hat{n} = E_r \quad \text{and} \quad \Phi_E = \oint E_r \, dA = E_r \oint dA = 4\pi r^2 E_r.$$

The radial component of the field E_r may be factored from the integral because it is constant over the surface. This simplification occurs because the surface has the same symmetry—spherical symmetry—as the field line pattern.

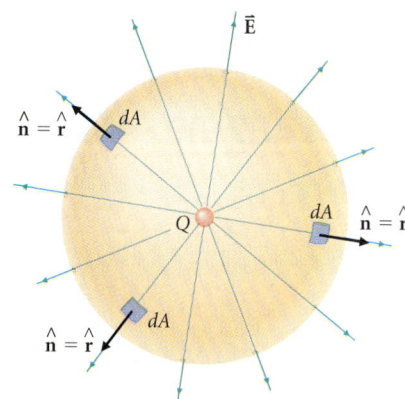

■ **FIGURE 23.26**
Electric flux through a sphere centered on a point charge Q. Here Q is positive. Everywhere on the surface of the sphere $\hat{n} = \hat{r}$ and is parallel to $\vec{E}$. The product $\vec{E} \cdot \hat{n} = |\vec{E}| = kQ/r^2$ has the same value everywhere on the surface of the sphere.

23.4.3 Gauss' Law for the Electric Field

A formal statement of the relation between charge and flux is known as Gauss' law.

> **GAUSS' LAW**
>
> The total electric flux emerging from an arbitrary volume equals the net charge enclosed within the volume divided by ϵ_0.
>
> $$\Phi_E \equiv \oint \vec{E} \cdot \hat{n} \, dA = \frac{Q_{\text{enclosed}}}{\epsilon_0}. \quad (23.12)$$
>
> The integral is taken over the closed surface surrounding the volume, and $\hat{n}$ is the outward normal to the surface element dA.

Examples 23.13 and 23.14 illustrate this statement. Notice that the total electric field, including that due to charges outside, is used to calculate the flux, but only the charge inside the surface contributes to Q_{enclosed}. Gauss' law is extremely useful for computing the electric field due to symmetric charge distributions.

EXAMPLE 23.15 ♦ A cubical box is known to contain a net charge of 6 μC. The measured flux through one face of the box is 8×10^5 N·m²/C. What is the total flux through the other five sides?

MODEL Gauss' law relates the total flux to the known charge inside. Since we are given the flux through one side, the flux through the other five sides equals the total flux minus the given flux.

SETUP From Gauss' law, the total flux through the walls of the cubical box is:

$$\Phi_E = \frac{Q}{\epsilon_0} = \frac{6 \, \mu\text{C}}{8.9 \times 10^{-12} \, \text{C}^2/\text{N}\cdot\text{m}^2} = 7 \times 10^5 \, \text{N}\cdot\text{m}^2/\text{C}.$$

SOLVE The flux through the remaining five sides is:

$$\Phi_E = 7 \times 10^5 \, \text{N}\cdot\text{m}^2/\text{C} - 8 \times 10^5 \, \text{N}\cdot\text{m}^2/\text{C}$$
$$= -1 \times 10^5 \, \text{N}\cdot\text{m}^2/\text{C}.$$

Karl Friedrich Gauss (1777–1855). A child prodigy, Gauss corrected his father's account books at age three. His family was not wealthy. Luckily his mathematical talents came to the attention of the Duke of Brunswick, who provided financial support until his career was established. Gauss received his Ph.D. in mathematics from the University of Helmstedt in 1799 and published his major work on arithmetic in 1801. He made contributions in number theory, non-Euclidean geometry, statistics, and the determination of cometary orbits. Most of his work in electricity and magnetism was accomplished between 1830 and 1840. He was one of the founders of the German Magnetic Union, which had the objective of making continuous observations of the Earth's magnetic field. He also developed an early version of the telegraph. The cgs-Gaussian unit system and its unit of magnetic induction both bear his name.

Digging Deeper

FORMAL PROOF OF GAUSS' LAW

The formal proof of Gauss' law uses the concept of *solid angle*.

The solid angle $d\Omega$ subtended by a surface of area dA at point P (Figure 23.28) is defined by:

$$d\Omega = \hat{\mathbf{r}} \cdot \hat{\mathbf{n}} \, \frac{dA}{r^2} = \frac{dA_\perp}{r^2}.$$

The unit of solid angle is the steradian; it is dimensionless. Qualitatively, the solid angle is the cone defined by the area dA and the point P. Any other area dA' perpendicular to $\vec{\mathbf{r}}$ and bounded by the cone also defines the solid angle: $d\Omega = dA'/r'^2$ (Figure 23.28b). In this way, a closed surface surrounding P can be replaced by a sphere centered on P. The sphere has area $4\pi r'^2$ and subtends solid angle $A/r'^2 = 4\pi r'^2/r'^2 = 4\pi$ at P. The solid angle subtended by any closed surface at a point within it is 4π steradians.

Consider a point charge q located at O inside the surface shown in Figure 23.29. The element of surface shown has normal $\hat{\mathbf{n}}$, and the flux through the surface element due to the single charge is:

$$d\Phi_E = \vec{\mathbf{E}} \cdot \hat{\mathbf{n}} \, dA = \frac{kq}{r^2} \hat{\mathbf{r}} \cdot \hat{\mathbf{n}} \, dA.$$

But $dA \, \hat{\mathbf{r}} \cdot \hat{\mathbf{n}}/r^2$ is the element of solid angle $d\Omega$ subtended at O by dA.

$$d\Phi_E = kq \, d\Omega.$$

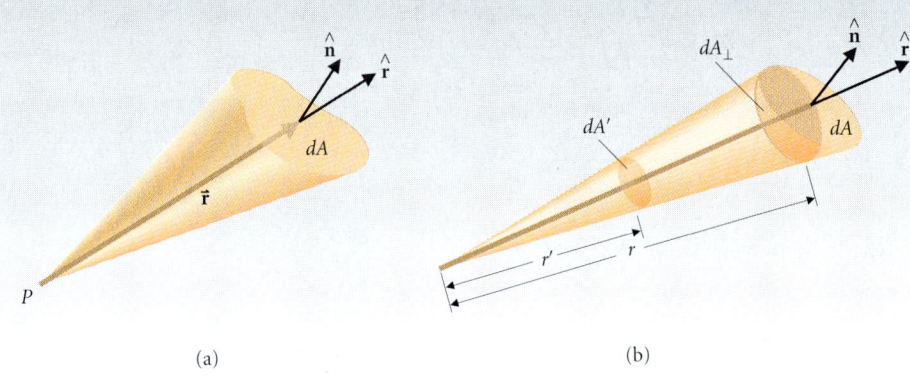

■ **FIGURE 23.28**
(a) The solid angle subtended at point P by the surface element of area dA is defined to be $d\Omega = \hat{\mathbf{r}} \cdot \hat{\mathbf{n}}/r^2$, where $\hat{\mathbf{r}}$ points from P to the surface element, and $\hat{\mathbf{n}}$ is the normal to the surface element. Qualitatively, $d\Omega$ is the angle of the cone formed by the area element with apex at P. (b) Any two areas bounded by the same cone define the same solid angle at P. Thus a closed surface and any sphere within it subtend the same solid angle at the center of the sphere. That solid angle is 4π steradians.

■ **FIGURE 23.27**
(a) The net charge in the box is $14 \, \mu C - 8 \, \mu C = 6 \, \mu C$. However, the flux through the top surface is positive, while the net flux through the rest of the box is negative. (b) Field lines from the positive charge outside the box pass inward, contributing negative flux through some of the faces. However, all these field lines also leave the box, contributing positive flux through other faces. The net flux through the entire surface corresponds to the 6-μC charge inside.

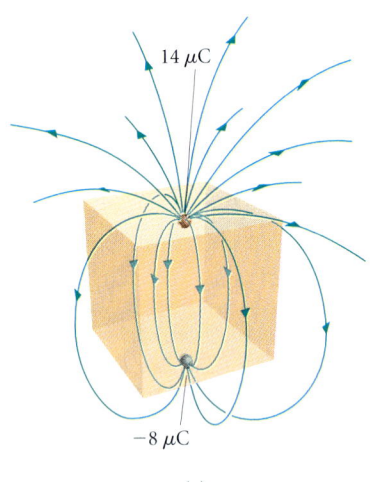

(a)

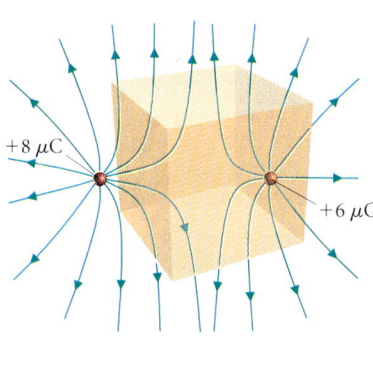

(b)

ANALYZE How could this situation arise? The box could contain both positive and negative charges, since the net flux through the other five sides is negative (■ Figure 23.27a). Alternatively, negative flux through one or more surfaces could be due to positive charge outside the box (Figure 23.27b).

EXERCISE 23.7 ♦ A spherical volume surrounds two point charges $Q_1 = 5 \, \mu C$ and $Q_2 = -3 \, \mu C$. What is the electric flux through the surface of the sphere?

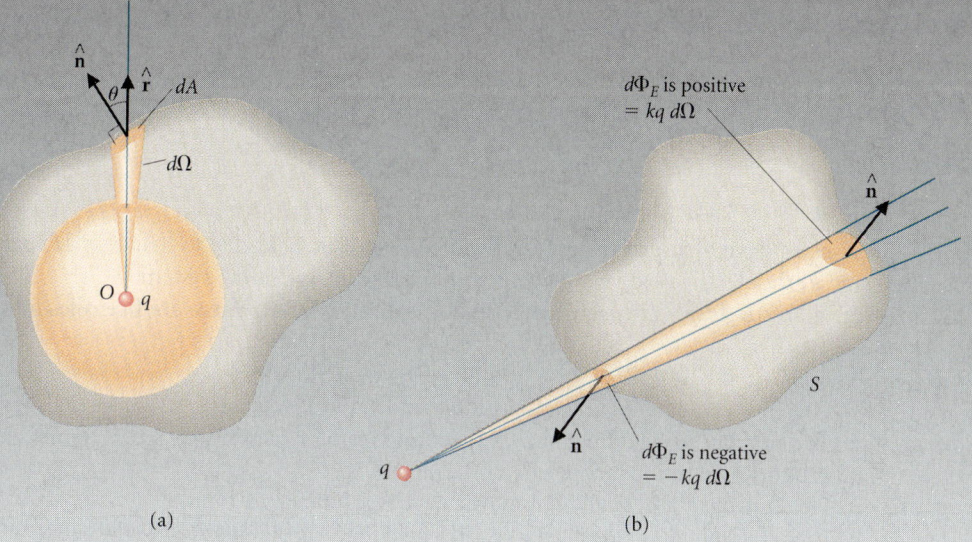

■ **FIGURE 23.29**
(a) We may express the electric flux due to a point charge through an element of surface in terms of the solid angle subtended by the surface at the position of the charge. (b) A cone of field lines from any charge outside a closed surface intersects the surface twice. The flux contributed by the element of area closest to the charge is negative, since $\hat{n}$ and $\hat{r}$ are almost opposite each other: $d\Phi_E = -kq\, d\Omega$. The element of area on the far side of the surface contributes a positive flux $kq\, d\Omega$, and the total is zero.

To find the total flux we sum over all the surface elements:

$$\Phi_E = \oint kq\, d\Omega = kq \oint d\Omega = 4\pi kq,$$

where $\oint d\Omega = 4\pi$ is the solid angle subtended by the whole surface at O. If there is more than one charge inside, we use the principle of superposition to find the flux they contribute:

$$\vec{E} = \sum_i \vec{E}_i.$$

Then: $\Phi_E = \oint \vec{E} \cdot \hat{n}\, dA = \oint \sum_i \vec{E}_i \cdot \hat{n}\, dA = \sum_i \oint \vec{E}_i \cdot \hat{n}\, dA$

$$= \sum_i 4\pi kq_i = 4\pi k \sum_i q_i.$$

If a charge is outside the surface (Figure 23.29b), the cone that defines the solid angle subtended at O by an element of the surface also cuts the surface at another place. The contributions to the flux through the surface contributed by the pair of surface elements are equal and opposite. So any charge outside the surface contributes zero net flux. Then the net flux due to all charges, both inside and outside, is:

$$\Phi_E = 4\pi k \sum_{\substack{\text{charges}\\ \text{inside}\\ \text{surface}}} q_i.$$

This is Gauss' law.

Chapter Summary

Where Are We Now?

We introduced electric charge as the fundamental property that produces electromagnetic fields and upon which those fields act. We learned how to use Coulomb's law to find the electric field due to a system of point charges, and we developed field line diagrams as a tool for visualizing electric fields. Gauss' law provides us a quantitative statement of global relations between charge and field. These ideas should form the basis for your thinking as we develop the concepts more fully in the chapters that follow.

What Did We Do?

There are two kinds of charge, labeled positive and negative. Charge can neither be created nor destroyed: electric phenomena occur when positive and negative charges are separated from each other. The SI unit of charge is the coulomb (C). Ordinary matter is composed of atoms, each with a positively charged nucleus surrounded by negatively charged electrons. A proton has charge $+e$ and an electron has charge $-e$: $e = 1.6 \times 10^{-19}$ C. The electrical behavior of everyday materials is determined by the forces the atoms exert on their outermost electrons.

A point charge Q produces an electric field vector at P according to Coulomb's law:

$$\vec{E} = k \frac{Q}{r^2} \hat{r},$$

where $\hat{r}$ is a unit vector that points from Q to P. The constant $k = 8.99 \times 10^9$ N·m²/C² $= 1/(4\pi\epsilon_0)$, where $\epsilon_0 = 8.85 \times 10^{-12}$ C²/N·m². The force exerted on a test charge q placed at P is

$$\vec{F} = q\vec{E}.$$

The principle of superposition states that the electric field due to a system of charges is the sum of the electric field vectors produced by each charge in the system.

Field line diagrams are useful for visualizing electric fields. There are five rules for drawing diagrams:

1. Field lines begin at positive charge and end at negative charge.
2. The number of field lines shown diverging from or converging into a point charge is proportional to the magnitude of the charge.
3. Field lines are spherically symmetric near a point charge.
4. If the system has a net charge, the field lines are spherically symmetric at great distances.
5. Field lines never cross.

The electric flux Φ_E through a surface S is a quantitative measure of field lines passing through that surface:

$$\Phi_E = \int \vec{E} \cdot \hat{n}\, dA.$$

Gauss' law states the relation between charge and flux:

> The electric flux Φ_E through a closed surface S enclosing a volume V is equal to the net charge inside the volume divided by ϵ_0.

Practical Applications

Charge buildup in clouds produces strong electric fields and lightning. The topics discussed in this chapter form the basis for a classical description of the structure of matter. Field line diagrams are a useful tool for estimating how different components of a semiconductor circuit interfere with each other.

Solutions to Exercises

23.1 From Newton's second law, the acceleration of a particle is $a = F/m$. If forces of equal magnitude act on two different particles, their accelerations are in the ratio:

$$\frac{a_1}{a_2} = \frac{F/m_1}{F/m_2} = \frac{m_2}{m_1}.$$

A proton is roughly 2000 times as massive as an electron, so

$$a_e/a_p \approx 2000.$$

23.2 See ■ Figure 23.30. The distance r of point P from the charge is given by:

$$r^2 = [2.0\text{ m} - (-1.0\text{ m})]^2 + [1.0\text{ m} - (-1.0\text{ m})]^2 = 13.0\text{ m}^2.$$

Then the electric field at P has magnitude:

$$E = \frac{kQ}{r^2} = \frac{(8.99 \times 10^9\text{ N·m}^2/\text{C}^2)(10.0 \times 10^{-6}\text{ C})}{13.0\text{ m}^2}$$

$$= 6.92 \times 10^3\text{ N/C}.$$

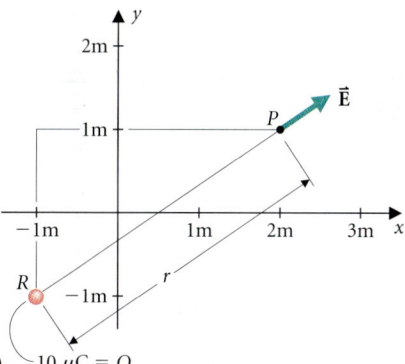

FIGURE 23.30 $10\,\mu C = Q$

The direction of the field is:

$$\hat{r} = \frac{3}{\sqrt{13}}\hat{i} + \frac{2}{\sqrt{13}}\hat{j} = 0.832\hat{i} + 0.555\hat{j}.$$

Finally we have: $\vec{E} = (5.8\hat{i} + 3.8\hat{j}) \times 10^3$ N/C.

23.3 Since the positive charge has twice the magnitude of each negative charge, we show twice as many field lines emerging from it as converging into each of the negative charges (■ Figure 23.31). The lines have spherical symmetry near each charge.

23.4 Each charge produces an electric field that points along the x-axis:

$$\vec{E}_1 = \hat{i}\frac{kQ}{(x-a)^2} \quad \text{and} \quad \vec{E}_2 = \hat{i}\frac{kQ}{(x+a)^2}.$$

Using the principle of superposition:

$$\vec{E} = \vec{E}_1 + \vec{E}_2 = \hat{i}kQ\left[\frac{1}{(x-a)^2} + \frac{1}{(x+a)^2}\right]$$

$$= \hat{i}kQ\frac{2(x^2+a^2)}{(x^2-a^2)^2}.$$

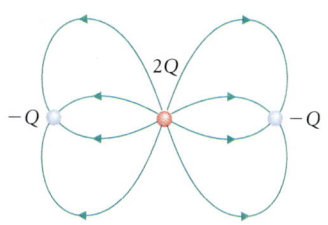

FIGURE 23.31

The field has magnitude $\approx 2kQ/x^2$ at large x, as expected from Coulomb's law.

23.5 Six field lines (the ones at the right) leave volume A. Three lines at the left (including those labeled (a) and (b)) leave and then reenter. Three lines pass directly to the negative charge inside the volume and do not leave at all. The net number of lines leaving is six, consistent with the net charge $2Q - Q = Q$ inside volume A.

23.6 We need the component of $\vec{E}$ normal to the surface:

$$\vec{E} \cdot \hat{n} = \frac{(1.0 \times 10^{-6}\text{ N/C})}{\sqrt{2}}(\hat{i} + \hat{k}) \cdot \hat{k} = 7.1 \times 10^{-7}\text{ N/C}.$$

Since the field is uniform across the area, the flux is:

$$\Phi_E = \vec{E} \cdot \hat{n}A = (7.1 \times 10^{-7}\text{ N/C})(1.0\text{ m}^2)$$
$$= 7.1 \times 10^{-7}\text{ N}\cdot\text{m}^2/\text{C}.$$

23.7 From Gauss' law, the charge inside the volume determines the flux through the surface. The net charge inside is $5\,\mu C - 3\,\mu C = 2\,\mu C$. Thus the flux through the surface of the sphere is:

$$\Phi_E = \frac{Q}{\epsilon_0} = \frac{2 \times 10^{-6}\text{ C}}{8.85 \times 10^{-12}\text{ C}^2/\text{N}\cdot\text{m}^2} = 2 \times 10^5\text{ N}\cdot\text{m}^2/\text{C}.$$

Basic Skills

Review Questions

§23.1 ELECTRIC CHARGE

- Describe an experiment that indicates there are two kinds of charge.
- State the *law of conservation of charge*.
- What is the charge on an electron?
- What is a *point charge*?
- State *Coulomb's law*.
- What is the *triboelectric sequence*?
- What is the difference between a *conductor* and an *insulator*?
- In what sense is the coulomb a large unit of charge?

§23.2 THE ELECTRIC FIELD OF A POINT CHARGE

- What is the operational definition of *electric field*?
- Describe the electric field due to a point charge Q. State how its magnitude and direction vary with position.

§23.3 THE PRINCIPLE OF SUPERPOSITION

- State the *principle of superposition*.
- What is a *field line diagram*? Explain how the field lines describe the magnitude and direction of $\vec{E}$.
- State the five rules for drawing field line diagrams for a collection of point charges.

§23.4 GAUSS' LAW

- Explain how a field line diagram illustrates Gauss' law.
- What is the electric flux through a plane surface of area A?
- What is a closed surface?
- State Gauss' law (a) in words and (b) as an equation. Carefully define all the symbols that appear in your equation.

Basic Skill Drill

§23.1 ELECTRIC CHARGE

1. A neutron is an unstable particle that decays into a proton, an electron, and an antineutrino. What is the charge on the antineutrino?

2. (a) A repulsive force of 1 N is exerted by a charge Q on a charge q. What force is exerted on Q by q? **(b)** If q is replaced by a charge $2q$ at the same location, what force is exerted by Q on the $2q$ charge? **(c)** If q is moved to twice its original distance from Q, what force is exerted on q by Q?

3. A person wearing a wool sweater and a cotton T-shirt steps onto an insulating platform and then removes the sweater. What sign of charge remains on the person? On the sweater?

4. Find the force on a charge $q = 1.0 \times 10^{-10}$ C placed 1.0 m from a charge $Q = -5.0 \times 10^{-3}$ C. (Give magnitude and direction.)

§23.2 THE ELECTRIC FIELD OF A POINT CHARGE

5. A point P is a distance $r = 1.0$ mm from a charge $Q = -1.0\ \mu$C. Compute the magnitude of the electric field vector at P. Describe its direction.

6. A particle with a charge of 1.0×10^{-9} C experiences a force of 3.0×10^{-5} N when placed at P. What is the magnitude of the electric field at P?

§23.3 THE PRINCIPLE OF SUPERPOSITION

7. Two identical positive charges, each of magnitude Q, are held fixed on the x-axis at $x = +1.0$ m and $x = -1.0$ m. A tiny test charge $+q$ is used to test the field of the two charges. Where, in addition to the points at infinite distance, is the electrostatic force on the test charge zero?

(a) Only at the exact midpoint between the two charges.
(b) At two points, $x = +2.0$ m and $x = -2.0$ m on the x-axis.
(c) At two points, $y = +1.0$ m and $y = -1.0$ m on the y-axis.
(d) At any point on the y-axis.
(e) Nowhere.
Explain your reasoning.

8. Two point charges of 15 μC each are on the x-axis at $x = 1.0$ m and $x = 1.5$ m. Find the electric field vector at the origin.

9. Three equal charges are at the corners of an equilateral triangle. Sketch the field line diagram in the plane of the triangle.

§23.4 GAUSS' LAW

10. An electric dipole (two equal and opposite point charges separated by a small distance) lies at the center of a spherical container. What is the electric flux through the sphere? Does your answer change if the dipole is not at the exact center? Explain your reasoning.

11. The point charges shown in ■ Figure 23.32 are held stationary. What are the values of electric flux through surfaces A, B, and C? If $Q = 4.425\ \mu$C, give numerical values for the three values of flux.

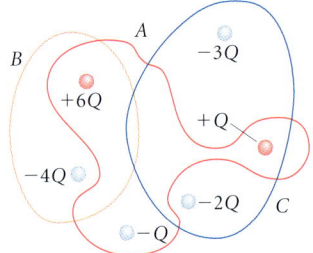

■ **Figure 23.32**

Questions and Problems

§23.1 ELECTRIC CHARGE

12. ❖ The distance between two charges is reduced by half. The force between them:
(a) becomes insignificant compared with gravity.
(b) becomes twice as large.
(c) becomes four times as large.
(d) becomes half as large.
(e) becomes one-fourth as large.
Explain how you chose your answer.

13. ❖ A rubber rod is stroked with wool and then allowed to come in contact with a pith ball. A glass rod is rubbed with silk and then placed in contact with a second pith ball. Is the force between the two charged pith balls attractive or repulsive? Explain your reasoning.

14. ❖ In the early universe and in stars, deuterium is produced from the combination of a proton and a neutron, with the release of a gamma ray. What is the charge on a deuterium nucleus?

15. ♦ ✉ Two electrons in a helium atom are separated by about 10^{-10} m. Estimate the magnitude of the repulsive force between them.

16. ♦ Two particles with equal mass have charges of 1.0 μC and 1.0 nC, respectively. What must be their mass if their mutual gravitational attraction is to overcome their Coulomb repulsion?

17. ♦ Two point charges of 1.0×10^{-6} C each are held together by a string 15 cm long. What is the tension in the string?

18. ♦♦ Two point charges 2.0 m apart exert repulsive forces of magnitude 2.7×10^{-2} N on each other. If one charge is known to be twice as large as the other, what are the two charges? Another pair of charges exert attractive forces of magnitude 2.7×10^{-2} N when 2.0 m apart. If the total charge of the pair is $+1.0\ \mu$C, what are the two charges?

19. ♦♦ Two objects, each carrying charge Q, are connected by a spring with force constant k_s and relaxed length zero. When the system is in equilibrium, what is the distance between the two charges?

20. ♦♦ In the cgs-Gaussian unit system, Coulomb's law takes the form $F = q_1 q_2 / r^2$, with no constant. Find the dimensions of charge in the Gaussian unit system.

21. ♦♦♦ Show that the force between two objects that share a total charge Q is largest when they share the charge equally, that is, when each has charge Q/2.

§23.2 THE ELECTRIC FIELD OF A POINT CHARGE

22. ♦ A point charge $Q = 16.5$ μC is at the origin. Find the electric field vector at the point P with coordinates $x = 15$ cm, $y = 25$ cm, $z = 0$.
23. ♦ The electric field strength 17 cm from a point charge is measured to be 56 N/C. What is the value of the charge?
24. ♦♦ A point charge $Q = -15$ nC is at the point $x = 1.0$ m, $y = 2.0$ m, $z = 0$. Find the electric field vector at the point $x = 2.0$ m, $y = -1.0$ m, $z = 1.0$ m.
25. ♦♦ A point charge $Q = 33$ nC is located at $x = 55$ mm, $y = 27$ mm, $z = 15$ mm. Find the electric field vector at the point P with coordinates $x = 15$ mm, $y = 35$ mm, $z = -6$ mm.
26. ♦♦ A pith ball with mass $m = 1.0$ g and charge 150 nC hangs from a thread in a region with uniform horizontal electric field of strength $E = 1.0 \times 10^4$ N/C. At what angle from the vertical does the ball hang?
27. ♦♦ What electric field vector is required to support an electron (mass 9×10^{-31} kg) against gravity?
28. ♦♦ A small oil drop contains exactly one excess electron. If the mass of the drop is 3×10^{-15} kg, what electric field vector is needed to support the drop against gravity?
29. ♦♦ On a windless day, when the electric field strength above the Earth's surface is 100 N/C, a fog droplet with a net charge of 250 electrons remains motionless. What is the mass of the droplet? What is its radius?

§23.3 THE PRINCIPLE OF SUPERPOSITION

30. ❖ A positive and a negative charge of equal magnitude are separated by a distance ℓ (■ Figure 23.33). What is the direction of the electric field vector at R and at P?

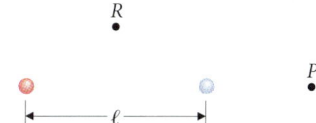

■ FIGURE 23.33

31. ❖ What is the direction of the electric field vector at R and at P in Figure 23.33 if the negative charge is replaced by a positive charge of equal magnitude?
32. ❖ A positive charge $2Q$ is at one vertex and two charges $-Q$ are at the other two vertices of an equilateral triangle. Sketch the field line diagram in the plane of the triangle.
33. ❖ Criticize the following statement: "An object cannot produce an electric field unless it has a net electric charge."
34. ❖ A charge of $+3Q$ is at the apex of a tetrahedron. Three negative charges, each $-Q$, are at the corners of the base. Sketch the field line diagram.
35. ❖ Positive charges $+Q$ and negative charges $-Q$ are spaced along a line with a separation d between neighboring charges, so that charges alternate in sign. Sketch a field line diagram for this system.
36. ❖ Six equal positive charges are equally spaced around a circle. Sketch a field line diagram in the plane of the circle and in a plane through the center and perpendicular to the circle.
37. ❖ Six equal charges Q are placed at the corners of a hexagon. What is the force on a test charge q placed at the center of the hexagon?

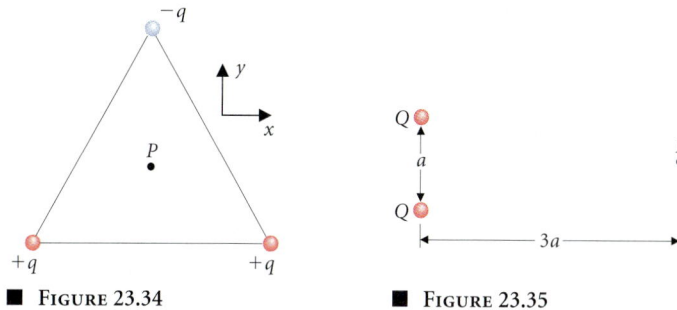

■ FIGURE 23.34

■ FIGURE 23.35

38. ❖ Three point charges (two $+q$, one $-q$) are arranged at the points of an equilateral triangle with sides of length L (■ Figure 23.34). The electric field at the center of the triangle:
(a) is zero.
(b) is infinite.
(c) is in the y-direction.
(d) is in the x-direction.
(e) is equal to q/L^2.
Explain what is wrong with the answers you don't choose.
39. ❖ Five equal negative charges are placed at the corners of a pentagon. A small positive test charge q is placed at the center of the pentagon. What is the force on the test charge? If one of the charges on the pentagon is removed, in which direction does the test charge accelerate?
40. ❖ Three charges, one $+Q$ and two with charge $-4Q$, are to be arranged so that the total electric force on each charge is zero. How can this be done? Is the arrangement stable?
41. ❖ If the distance between the two point charges Q in ■ Figure 23.35 is reduced by half, the magnitude of the electric field at P:
(a) decreases slightly.
(b) is reduced by a factor 2.
(c) is increased by a factor 2.
(d) is increased by a factor 4.
(e) increases slightly.
Explain what is wrong with the answers you don't choose.
42. ❖ Four equal positive charges are at the corners of a square. Sketch a field line diagram for this system in a plane perpendicular to the square and through a diagonal. Describe how the diagram is similar to that for two equal charges, and how it differs.
43. ♦ A point charge $Q = 3.0$ μC is located at the origin. (a) What is the magnitude of the electrostatic force on a positive charge $q = 1.0$ μC placed at point P with coordinates $x = 2.0$ m, $y = 2.0$ m, $z = 0$? (b) A second charge $-Q$ is placed at position R with coordinates $x = 0$, $y = 4.0$ m, $z = 0$. What is the total electrostatic force on q?
44. ♦ A point charge of 20.0 nC is at the origin. A second charge of -30.0 nC is on the z-axis at $z = 1.00$ m. Find the electric field vector at the point $x = 0$, $y = 2.00$ m, $z = 1.00$ m.
45. ♦ Three charges $+Q$ and one charge $-Q$ are at the corners of a square of side a. Find the magnitude of the electric field at the center of the square produced by each charge. Find the magnitude of the total electric field at the center. What direction is the force on a small negative test charge at the center?
46. ♦♦ Two point charges, each of 15 nC, are on the x-axis at $x = 1.0$ m and $x = 1.5$ m. Find the electric field as a function of position on the x-axis.
47. ♦♦ Four equal charges $+Q$ are placed at the corners of a square of side a. Find the electric field strength (a) at the center of the square, (b) at one corner, and (c) at the midpoint of one side. In each case, state which charges contribute to the field you compute, and why.
48. ♦♦ A charge $2Q$ and two equal charges $-Q$ lie on a line with each negative charge a distance d from the positive charge (see Exercise 23.3). What is the electric field vector: (a) at the position of the

positive charge? **(b)** at the position of one of the negative charges? **(c)** at an arbitrary point on the positive *y*-axis? **(d)** on the *x*-axis, at a point with $x > d$? In each case, state which charges contribute to the field you compute, and why.

49. ♦♦ Two point charges are held fixed on the *x*-axis. One with charge $Q_1 = -2.0\ \mu C$ is at $x = -3.0$ m, and the other with charge $Q_2 = +4.0\ \mu C$ is at $x = +1.0$ m. Find a point where a third charge may be placed and experience no net electric force.

50. ♦♦ Three positive charges of 1.0×10^{-5} C are each at the corner of an equilateral triangle of side 10.0 cm. Find the force on a fourth charge of 1.0×10^{-8} C placed at the midpoint of one side of the triangle.

51. ♦♦ Find the electric field at the origin due to the system of charges in ■ Figure 23.36.

52. ♦♦♦ Find the electric field produced by the system of charges in ■ Figure 23.36 as a function of position on the *z*-axis.

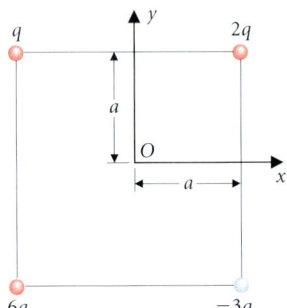

■ **FIGURE 23.36**

§23.4 GAUSS' LAW

53. ❖ If the electric flux through a closed surface is zero, is the field identically zero everywhere on the surface? Discuss.

54. ❖ An opaque cubical box is observed to have a net number of electric field lines emerging from it. What can you say about the net charge in the box?
(a) Nothing.   **(b)** Its sign but not its magnitude.
(c) Its magnitude but not its sign.
(d) Both its sign and its magnitude.
Explain how you chose your answer, and if you chose (b), (c), or (d), give the sign or magnitude, as appropriate.

55. ❖ A point charge Q is in the exact center of a cubical surface of side a. What is the electric flux emerging through one face of the cube? Can you answer the question easily if the charge is within the cube but not at the exact center? Why or why not?

56. ❖ You observe that field lines are emerging from a closed box, but every field line that leaves the box reenters. Which of the following statements are true? Explain your answers.
(a) There is no charge in the box.
(b) The net charge in the box is positive.
(c) The net charge in the box is negative.
(d) There are both positive and negative charges in the box, but the net charge is zero.
(e) None of the statements (a)–(d) is true: it is impossible to say anything about the charge in the box.

57. ♦ A tetrahedron contains a point charge $Q = 65$ nC. Find the electric flux through the surface of the tetrahedron. Can you find the flux through each face separately? Under what conditions?

58. ♦ A cube of side $\ell = 2.0$ m is centered at the origin, with the coordinate axes perpendicular to its faces. Find the flux of the electric field $\vec{E} = (15\ \text{N/C})\hat{i} + (27\ \text{N/C})\hat{j} + (39\ \text{N/C})\hat{k}$ through each face of the cube.

59. ♦♦ The electric field in a certain region of space is described by the function:

$$\vec{E} = (6.0\ \text{N/C})\hat{i} + (7.0\ \text{N/C})\hat{j}.$$

Find the electric flux through the surface $x = 6y$, $0 < x < 6.0$ m, $0 < z < 1.0$ m.

60. ♦♦ A tetrahedron of side ℓ has one face in the *x*-*y*-plane. The electric field in the region has magnitude E_o and is in the positive *z*-direction. What is the electric flux through the surface of the tetrahedron? Find the flux through each surface of the tetrahedron.

61. ♦♦♦ Find the flux of the electric field $\vec{E} = (250\ \text{N/C})\hat{j}$ through the surface $x^2 + y^2 + z^2 = 1.0\ \text{m}^2$, $x > 0$, $y > 0$, $z > 0$. (*Hint:* The easiest way is to use Gauss' law.)

Additional Problems

62. ❖ How can you use a field line diagram to estimate the electric field strength at a point? Use this result to explain why the electric flux through a surface is proportional to the number of field lines crossing that surface.

63. ❖ Three equal charges are in a line with equal separations between them. The outer two are fixed; the center one is free. Show that the force on the central charge is zero. Is the equilibrium stable or unstable? What if the central charge is negative? Is the magnitude of the central charge important for its stability?

64. ❖ A positive test charge q is in equilibrium at the center of a square that has four equal charges $+Q$ at its corners. Is the equilibrium stable against small displacements: **(a)** perpendicular to the plane of the square? **(b)** in the plane of the square and perpendicular to one side? **(c)** in the plane of the square and toward one corner?

65. ♦♦ What is the total amount of negative charge contained in a raindrop 3.0 mm in diameter? If by magic we could suddenly remove all this negative charge, leaving the positive charge behind, what would be the acceleration of two raindrops 1 m apart?

66. ♦♦ ✉ Suppose an electron is removed from this book and placed 1 mm above the page. Calculate its acceleration, and *estimate* how long it would take to return to its proper place.

67. ♦♦♦ In the system of Study Problem 15, describe how the strength of the electric field varies along the *y*-axis with $y > D$. At what point is the *y*-component of the field maximum? What feature of the field line diagram shows the location of this point?

68. ❖ While standing vertically, brushing dust from your formal suit, you notice a piece of lint moving at constant speed toward your coat at an angle of 45° to the vertical. Why does the lint not fall vertically? Compare the weight of the lint, the electrical force on the lint, and the air resistance to its motion.

69. ♦♦♦ Two equal point charges Q are on the *x*-axis at $x = \pm a$ (cf. Example 23.11). Where on the *y*-axis is the electric field magnitude a maximum?

Computer Problems

70. Set up a spreadsheet program to calculate the *x*-, *y*-, and *z*-components of the electric field due to the collection of point charges described in the file CH23P70 on the supplementary computer disk. Calculate the field components at points P_1 and P_2 with coordinates $(-3.0$ m, -2.0 m, 2.3 m) and $(1.5$ m, 2.0 m, 1.3 m).

71. A point charge $+Q$ is at the origin, and two charges $-Q$ are on the x-axis at $x = +L$ and $x = -L$. Locate qualitatively two points where $\vec{E} = 0$. Using 12 lines per charge, sketch the electric field line diagram for the system in the x-y-plane. Discuss the relation of your diagram to the $\vec{E} = 0$ points. Use Coulomb's law to find the exact location of the two points.

72. Two point charges of $1.00\ \mu C$ and $-5.00\ \mu C$ are at the ends of a spring of unstretched length $\ell = 5.00$ cm and $k = 5.00 \times 10^4$ N/m. What is the separation of the charges in equilibrium?

73. Use the program "FIELDLINE" on your supplementary computer disk to draw the field line diagram for the following sets of charges: **(a)** A negative charge of -1 unit at $x = 200$, $y = 250$; a charge of -1 unit at $x = 300$, $y = 175$; and a charge of $+3$ units at $x = 200$, $y = 200$. Use a step size of 1.0, 10 lines per charge unit, and starting angle $\pi/4$. The program works best if you enter the largest charge first. **(b)** Four charges of ± 1 unit at the corners of a square, arranged so that charges of the same sign are at opposite ends of each diagonal. Comment on the diagrams. In particular note how many lines go off to infinity, where the field is weak, and where it is strong.

Challenge Problems

74. Show that a displacement along a field line lying in the x-y-plane satisfies the equation:
$$\frac{dy}{dx} = \frac{E_y}{E_x}.$$

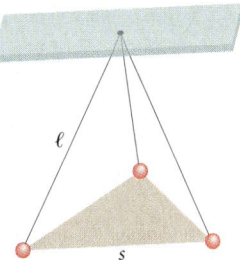

■ FIGURE 23.37

75. Find an analytic expression for the x- and y-components of the electric field due to two positive charges on the x-axis at $x = a$ and $x = -a$. Use the result of Problem 74 to find a differential equation for the field lines. (There is no need to solve the equation.)

76. Three identical charged objects of mass m and charge Q are suspended from a common point by massless strings as shown in ■ Figure 23.37. Show that the dimension s satisfies the equation:
$$s^3 = \frac{3kQ^2}{mg}\sqrt{\ell^2 - s^2/3}.$$

Solve this equation approximately in the two limits **(a)** $s \ll \ell$ and **(b)** $s \approx \sqrt{3}\ell$. Under what conditions are the approximate solutions valid?

Computer problem: For the cases $m = 1.0$ g, $\ell = 0.25$ m, **(a)** $Q = 55$ nC and **(b)** $Q = 450$ nC, use a numerical method to solve this equation exactly for $x = s/\ell$, and compare each result with the appropriate approximate solution.

In a television set, electric fields accelerate electrons in the television tube. Magnetic forces (Chapter 29) move the electrons across the end of the tube to form the picture. To understand how the television works, we must be able to compute the electric field in the tube and to determine how the electron responds. In this chapter we begin to develop the necessary techniques. The electron motion is discussed in Example 24.8.

CHAPTER 24
Static Electric Fields

CONCEPTS

Field line symmetry
Charge density
Dipole

GOALS

Be able to:

Find the electric field produced by a continuous distribution of charge.

Use symmetry arguments to simplify electric field calculations.

Compute the motion of a test charge in a given electric field.

And I have tried to apprehend the Pythagorean power by which number holds sway above the flux. A little of this, but not much, I have achieved.

BERTRAND RUSSELL

The picture on your television screen is produced by a beam of electrons accelerated down the tube by an electric field. To design and build a good TV set, the manufacturer needs to control the electric field in the tube precisely and to understand the electron's response to that field. Photocopiers and ink-jet printers require a similarly well-tuned electric field if they are to produce high-quality documents and graphics. In this chapter we'll study how to calculate such fields and the behavior of charged particles exposed to them.

The electric field in a television set or a photocopier is due to a distribution of charge spread over surfaces rather than to a few isolated point charges. Of course, the charge is a property of individual electrons and atomic nuclei, but their number is so immense that we may consider their distribution as continuous. We may model any distribution as a collection of differential elements, each containing many electrons and nuclei but small enough to be treated as a point charge. Then the principle of superposition gives $\vec{E}$ as a physical integral of the kind described in Interlude 2. If the charge distribution has appropriate symmetry, we may use Gauss' law to calculate the field.

BEFORE BEGINNING THIS CHAPTER, REVIEW THE STEPS IN THE INTEGRATION METHOD FROM INTERLUDE 2.

REMEMBER: MODELING THE SYSTEM AS A COLLECTION OF DIFFERENTIAL ELEMENTS IS THE FIRST STEP IN THE INTEGRATION METHOD. AN EXAMPLE IS ILLUSTRATED IN FIGURE 24.8.

24.1 USING GAUSS' LAW TO CALCULATE ELECTRIC FIELD

Before beginning to calculate the electric field of any charge distribution, it is wise to sketch a field line diagram that shows the field's overall behavior. When the charge distribution has appropriate symmetry, the resulting field pattern is also symmetric, and Gauss' law provides the simplest method for computing the field. The general method is outlined in ■ Figure 24.1.

THERE ARE THREE IMPORTANT CASES: SPHERICAL, CYLINDRICAL, AND PLANE SYMMETRY. WE'LL ENCOUNTER EXAMPLES OF EACH.

Method for finding symmetric electric fields using Gauss' law

Model — **Step I:** Sketch a field line diagram to illustrate the direction of the field. Use the system's symmetry and Coulomb's law to help you. Then choose a coordinate system so that the electric field vector at each point has a single component.

Setup — **Step II:** For a typical point P at which $\vec{E}$ is to be found, choose a suitable closed surface for applying Gauss' law. The point P must lie on the chosen surface, and each part of the surface should be either parallel to or perpendicular to the field at that point.

Setup — **Step III:** Express the flux emerging from the surface as a single component of the electric field vector multiplied by an area.

Setup — **Step IV:** Determine the net charge inside the volume enclosed by the chosen surface.

Solve — **Step V:** Apply Gauss' law to find the electric field. The point P is usually typical of an entire region and your calculation will determine $\vec{E}$ as a function of position throughout that region.

If the charge distribution divides space into different regions, such as inside or outside a sphere, you will need to repeat steps II – V for each of those regions.

■ FIGURE 24.1

REMEMBER: A GAUSSIAN SURFACE COMPLETELY ENCLOSES A VOLUME.

IF THERE IS NOT ENOUGH SYMMETRY TO REDUCE THE FLUX INTEGRAL IN THIS WAY, GAUSS' LAW WON'T HELP YOU.

To find the electric field vector at a typical point P, construct an imaginary closed surface that passes through the point. This surface, which is called a *Gaussian surface,* should share the symmetry of the charge distribution and at each point be either parallel or perpendicular to $\vec{E}$. Then, with *properly chosen coordinates,* $\vec{E}$ will have only one component. This component will be constant on all parts of the surface where $\vec{E} \cdot \hat{n}$ is not zero. The flux integral $\oint \vec{E} \cdot \hat{n} \, dA$ for the Gaussian surface then reduces to an area multiplied by this constant component. We use the known charge distribution to calculate the total charge enclosed by the surface, apply Gauss' law to find the one component of $\vec{E}$, and so determine $\vec{E}$ at each point on the surface.

EXAMPLE 24.1 ♦♦ Find the value of $\vec{E}$ at any point either inside or outside a uniformly charged, thin, spherical shell with radius R and total charge Q.

MODEL The charge distribution has spherical symmetry, so we may use Gauss' law to find the field. We'll need to do two calculations: one for the region outside the shell and one for the region inside.

STEP I The electric field lines share the spherical symmetry of the charge distribution (■ Figure 24.2). Outside the shell the field lines extend radially outward, making the same pattern that would be produced by a point charge located at the center of the shell.

Inside the shell, any field lines leaving the positive charge on the shell would have to point radially inward, converging on the center. Field lines can converge only on electric charge, and there is none at the center, so field lines cannot exist inside the shell.

SETUP We choose a spherical surface of radius r, concentric with the charged shell.
STEP II The normal to the surface is $\hat{n} = \hat{r}$ and is parallel to the electric field vector everywhere on the surface.

STEP III Because of the spherical symmetry, the radial component of the field has the same value at each point on the shell. We calculate the flux through the surface as follows:

STUDY HINT: A SIMILAR LOGIC CHAIN APPLIES WHENEVER GAUSS' LAW IS USED.

Definition of flux: $\quad \Phi_E = \oint \vec{E} \cdot \hat{n} \, dA$,

normal $\hat{n} = \hat{r}$: $\quad\quad\quad\quad = \oint \vec{E} \cdot \hat{r} \, dA$,

$\vec{E} \cdot \hat{r} = r$ component of $\vec{E}$: $\quad = \oint E_r \, dA$,

E_r is constant over the surface: $\quad = E_r \oint dA$,

Surface area of a sphere is $4\pi r^2$: $\quad = 4\pi r^2 E_r$.

STEP IV The charge inside the Gaussian sphere is the total charge Q on the shell.

SOLVE Applying Gauss' law:
STEP V
$$\Phi_E = 4\pi r^2 E_r = Q/\epsilon_0. \quad\quad (24.1)$$

Thus anywhere outside the shell:

$$\vec{E} = E_r \hat{r} = \frac{Q}{4\pi r^2 \epsilon_0} \hat{r} \quad (r > R).$$

SETUP AND SOLVE The field inside the shell is found in exactly the same way, except that there is no charge inside any Gaussian sphere within the shell. Setting $Q = 0$ in eqn. (24.1), we find that the electric field is zero anywhere inside the shell.

$$\vec{E} = 0 \quad (r < R).$$

ANALYZE The electric field outside the shell is identical to the field produced by a point charge Q at the center of the shell, but the field inside the shell is zero. The need to make separate arguments for separate spatial regions is typical of calculations with Gauss' law. ■

EXERCISE 24.1 ❖ What is the electric field outside a ball with total charge Q uniformly distributed throughout its volume?

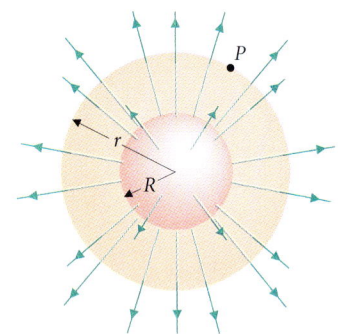

■ **FIGURE 24.2**
Electric field due to a uniformly charged spherical shell. Outside the shell, the field lines have spherical symmetry: they diverge from the origin. The field line pattern is the same as that due to a point charge at the origin, and the mathematical expression for the electric field is the same, too. Inside the shell, there are no field lines at all, and $\vec{E} = 0$. We choose a sphere as the volume to which we apply Gauss' law.

24.2 THE ELECTRIC FIELD DUE TO A LINEAR CHARGE DISTRIBUTION

Next we examine the electric field produced by charge uniformly distributed on a long straight filament such as the central electrode of a Geiger counter (■ Figure 24.3). This example illustrates the interplay between approximation and exact calculation in obtaining useful answers.

At large distances from any finite charge distribution, no matter how complicated, the system's structure is not noticeable. If the distribution has nonzero net charge, its field line diagram looks like that of a point charge. The electric field is approximately given by Coulomb's law with Q equal to the total net charge in the distribution.

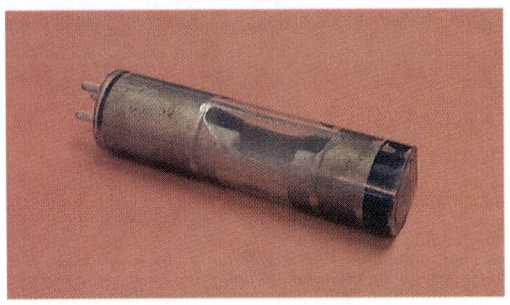

■ FIGURE 24.3
The central electrode of a Geiger counter is an example of a charged filament.

From nearby points, the distribution appears large. Often we may model it as infinite and find enough symmetry to apply Gauss' law. Even though no real distribution is truly infinite, this method usually provides an excellent approximation in a region close to the charge distribution. At intermediate distances, the electric field vector is found using Coulomb's law and integration. Because $\vec{E}$ and each differential contribution $d\vec{E}$ are vectors, we must take particular care with direction when evaluating the integral. It is usually best to calculate the x-, y-, and z-components of $\vec{E}$ separately. These calculations are often intricate. The approximations at near and far distances provide a valuable check on the results.

24.2.1 An Infinitely Long, Uniformly Charged Filament

We model the filament as having negligible diameter. The amount of charge on the filament is described by a linear charge density:

> The *linear charge density* λ at a point P of a filament is the amount of charge per unit length along the filament. The charge dQ on a length $d\ell$ of the filament is given by:
>
> $$dQ = \lambda \, d\ell. \quad (24.2)$$

A *uniformly* charged filament with length ℓ and total charge Q has the same value of λ at each point, and $\lambda = Q/\ell$.

Before calculating, we draw the field line diagram for the filament to see what we should expect. If the filament is very long, or if we consider points very close to it, it appears *infinitely* long. Such an infinitely long, uniformly charged filament is a highly symmetric object with a symmetric electric field. Its properties—charge density and diameter—don't depend on position along its length, and so the field lines do not bend along the direction of the filament but lie in planes perpendicular to it (■ Figure 24.4).

> **EXAMPLE 24.2** ♦♦ Find the electric field due to an infinite filament with uniform linear charge density λ.
>
> **MODEL** The charge density has cylindrical symmetry, and we may use Gauss' law to compute the field it produces.

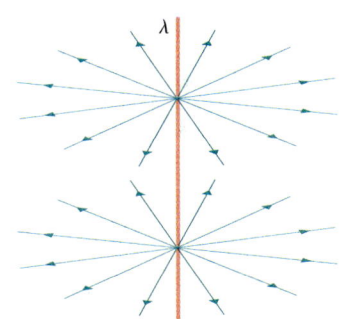

■ FIGURE 24.4
Field line diagram for an infinitely long, uniformly charged filament. Since the properties of the filament are the same everywhere along it, the field lines do not bend but emerge perpendicular to the filament, like the spokes of a wheel.

COMPARE WITH THE REASONING IN EXAMPLE 24.1.

BE CAREFUL TO DISTINGUISH THE UNIT VECTOR $\hat{r}$ IN CYLINDRICAL COORDINATES FROM THAT IN SPHERICAL COORDINATES. SOME BOOKS USE THE GREEK LETTER $\hat{\rho}$ FOR THE UNIT VECTOR IN CYLINDRICAL COORDINATES.

STEP I The field lines radiate from the filament, perpendicular to it and uniformly distributed around it (Figure 24.4). In cylindrical coordinates, the electric field has a single component E_r.

SETUP
STEP II For the Gaussian surface we choose a cylinder of height h and radius r centered on the filament. The normal to this surface is parallel to $\vec{E}$ at points on the curved surface and perpendicular to $\vec{E}$ on the flat ends (■ Figure 24.5).

STEP III The flux through the flat ends is zero (no field lines pass through the ends; $\vec{E} \cdot \hat{n} = 0$). The flux through the curved surface ($\hat{n} = \hat{r}$) is:

$$\Phi_E = \int d\Phi_E = \oint \vec{E} \cdot \hat{n}\, dA = \oint \vec{E} \cdot \hat{r}\, dA = \oint E_r\, dA.$$

Because of the rotational symmetry, the electric field component E_r is constant along the curved surface, and so it may be pulled out of the integral:

$$\Phi_E = E_r \oint dA = E_r 2\pi rh. \qquad (24.3)$$

This is the expression for flux through a cylindrical Gaussian surface in any example where the charge distribution has cylindrical symmetry.

STEP IV The charge inside the cylinder is $Q = \lambda h$.

SOLVE Applying Gauss' law:
STEP V

$$\Phi_E = \oint \vec{E} \cdot \hat{n}\, dA = 2\pi rhE_r = \lambda h/\epsilon_0$$

$$\Rightarrow \vec{E} = \left(\frac{\lambda}{2\pi r\epsilon_0}, \text{ radially outward from the filament}\right). \qquad (24.4)$$

ANALYZE Notice that the arbitrarily chosen height h of the Gaussian cylinder cancels.
The electric field due to an infinite line decreases as 1/(distance from the line). Compare with the $1/(\text{distance})^2$ behavior of the field due to a point charge. ■

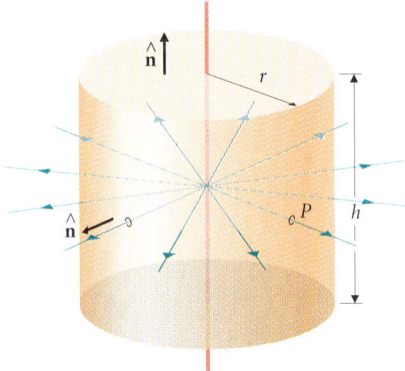

■ **FIGURE 24.5**
To find the field due to an infinite filament, we choose a cylindrical surface of height h and radius r. The electric field is parallel to $\hat{n}$ on the curved surface and perpendicular to $\hat{n}$ on the flat ends. Since a length h of the filament is inside the cylinder, charge $Q = \lambda h$ is enclosed inside.

24.2.2 A Finite, Uniformly Charged Filament

If the filament has a finite length ℓ, it has a finite charge $Q = \lambda \ell$. At very large distances from the filament, the field line diagram looks like that for a point charge Q (■ Figure 24.6) and we expect $|\vec{E}| \approx kQ/r^2$. Very close to the filament, the field lines emerge at right angles, and the field is given by Gauss' law (§24.2.1). At intermediate distances, we evaluate the field using Coulomb's law and the principle of superposition.

EXAMPLE 24.3 ♦♦♦ A filament of length ℓ has total charge Q distributed uniformly along it. Find the electric field at any point P in a plane bisecting the filament.

MODEL ■ Figure 24.7 is the field line diagram. We model the filament as a collection of differential pieces and use Coulomb's law for the field due to each piece. We sum the contributions $d\vec{E}$ from each piece using the principle of superposition.

SETUP We choose the plane containing the filament and point P to be the x-y-plane,
STEP I with the x-axis along the filament and the y-axis along its bisector, through P (■ Figure 24.8). Each differential piece of the filament has length dx.

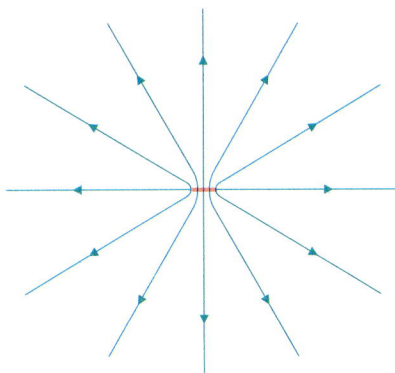

■ **FIGURE 24.6**
Field line diagram for a real charged filament viewed from a large distance away. The field lines diverge as from a point charge.

HERE THE STEPS REFER TO THE INTEGRATION METHOD IN INTERLUDE 2.

STEP II A typical element is a piece of the filament with coordinate x, length dx, and charge $dQ = \lambda\, dx$. Since the filament is uniformly charged, λ is independent of x and equals Q/ℓ.

STEP III We wish to find the electric field at the arbitrary point P as a function of its y-coordinate. The contribution of the element dQ to this field is:

$$d\vec{E}_1 = \frac{k\, dQ}{r^2}\hat{r} = \frac{k\lambda\, dx}{r^2}\hat{r} \qquad (\text{where } r^2 = x^2 + y^2)\ (\text{Figure 24.8}).$$

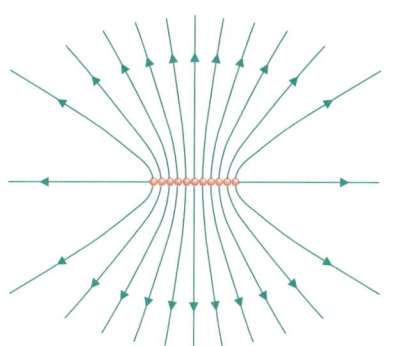

FIGURE 24.7
Field line diagram for a finite, uniformly charged filament at intermediate distances. The field lines emerge perpendicular to the filament and diverge at large distances.

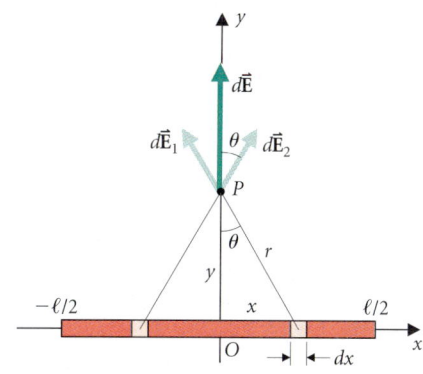

FIGURE 24.8
To calculate $\vec{E}$ at a point on the y-axis, we look at the field $d\vec{E}_1$ produced by a differential element of the filament at x. By summing the contributions from elements located symmetrically at x and $-x$, we determine that the electric field at P is in the y-direction. The element $d\vec{E}_1$ makes an angle θ with the y-axis, with $\tan\theta = x/y$.

We may simplify the calculation by using the symmetry about the y-axis. A second element of charge at coordinate $-x$ has an equal charge dQ, since λ is the same at both positions. The contribution $d\vec{E}_2$ from this element has the same magnitude as that due to the first element and makes an equal angle with the y-axis, on the opposite side. Thus the x-components of the two contributions sum to zero. For the y-component of the sum we have (cf. eqn. 23.9):

$$d\vec{E} = 2\hat{j}\, dE_{1,y} = 2\hat{j}|d\vec{E}_1|\cos\theta = 2\hat{j}\,\frac{k\,dQ}{r^2}\frac{y}{r} = 2\hat{j}\,\frac{ky\lambda\,dx}{r^3}.$$

THIS IS THE FIELD ON THE PERPENDICULAR BISECTOR OF TWO EQUAL POINT CHARGES (EXAMPLE 23.11).

We obtain the final result by adding the contributions of all such pairs of charge elements. Remember that r is a function of x and y.

STEP IV We add contributions from charge elements with coordinates ranging from $x = -\ell/2$ to $x = +\ell/2$. However, we add them as symmetrical pairs labeled by the coordinate of their right-hand element, which varies from $x = 0$ to $x = +\ell/2$.

SOLVE
STEP V
$$\vec{E}(y) \equiv \int d\vec{E} = \int 2\hat{j}\,\frac{ky\lambda\,dx}{r^3} = 2k\lambda y\hat{j}\int_0^{\ell/2}\frac{dx}{(x^2+y^2)^{3/2}}.$$

IF YOU PREFER, YOU MAY CALCULATE EACH COMPONENT OF $\vec{E}$ SEPARATELY AND INTEGRATE OVER INDIVIDUAL ELEMENTS BETWEEN $-\ell/2$ AND $+\ell/2$. THE CALCULATION IS LONGER, BUT THE ANSWER IS THE SAME, OF COURSE.

Since $x = y\tan\theta$, we may convert the calculation to an integral over θ as the independent variable:

$$dx \equiv d\theta\left(\frac{dx}{d\theta}\right) = y\sec^2\theta\,d\theta,$$

THIS IS A USEFUL TRICK THAT WE'LL WANT TO USE OFTEN.

and $x^2 + y^2 = y^2(1+\tan^2\theta) = y^2\sec^2\theta$. So:

$$\int\frac{dx}{(x^2+y^2)^{3/2}} = \int\frac{y\sec^2\theta}{y^3\sec^3\theta}\,d\theta = \frac{1}{y^2}\int\cos\theta\,d\theta = \frac{\sin\theta}{y^2}.$$

REMEMBER: WE ARE INTEGRATING OVER x, HOLDING y CONSTANT.

From Figure 24.8, $\sin\theta = x/r = x/(x^2+y^2)^{1/2}$, so:

$$\vec{E}(y) = \frac{2k\lambda\hat{j}}{y}\,\frac{x}{(x^2+y^2)^{1/2}}\bigg|_0^{\ell/2} = \frac{k\lambda\ell\hat{j}}{y(y^2+\ell^2/4)^{1/2}} = \frac{kQ}{y\sqrt{y^2+\ell^2/4}}\,\hat{j}.$$

ANALYZE It is the symmetry of the *charge* distribution that we used here. If the left half of the filament were negatively charged, its contributions $d\vec{E}_2$ would reverse direction. The y-components would then sum to zero, and the resulting field would be in the negative x-direction. Always remember to sketch a field line diagram first.

24.2.3 Comparison of Exact and Approximate Calculations

In the near limit $y \ll \ell$, the exact expression for the finite filament becomes:

$$\lim_{y \ll \ell} \vec{E}(y) = \frac{k\lambda\ell}{y\sqrt{\ell^2/4}} \hat{\jmath} = \frac{2k\lambda}{y} \hat{\jmath},$$

which verifies our conclusions from Gauss' law and the field line diagram for the infinite filament.

In the far limit $y \gg \ell$, the exact expression should agree with Coulomb's law for a point charge. Indeed, if we neglect $\ell^2/4$ compared with y^2:

$$\lim_{y \gg \ell} |\vec{E}(y)| = \frac{kQ}{y\sqrt{y^2}} = \frac{kQ}{y^2}.$$

Comparing the graphs of these two approximations with the exact result (■ Figure 24.9), we obtain rough rules for estimating $\vec{E}$: if we want answers accurate to 10%, then *near* means closer than about 0.2ℓ and *far* means more distant than 0.8ℓ.

NOTICE HOW THE SYMBOLS $\hat{\jmath}$ AND y ARE READ AS "VECTOR RADIALLY OUTWARD FROM THE FILAMENT" AND "DISTANCE FROM THE FILAMENT" WHEN COMPARING WITH EQN. (24.4).

THESE ESTIMATES APPLY TO THE ELECTRIC FIELD OF THE FILAMENT. THEY ARE NOT GENERAL STATEMENTS ABOUT ANY CALCULATION.

24.2.4 More Complicated Linear Charge Distributions

Once we know the field due to a uniformly charged filament, such filaments can be used as elements in more complicated problems.

EXAMPLE 24.4 ♦♦ Two parallel, infinitely long filaments lying in the *x-y*-plane are separated by a distance $d = 1.0$ mm. Each has uniform charge density $\lambda = 1.0$ nC/m (■ Figure 24.10). What is the electric field vector at point *P* on the *z*-axis?

MODEL The electric field at *P* is the sum of the two field vectors produced by the individual filaments, each found using Gauss' law, and given by eqn. (24.4).

SETUP Figure 24.10b shows a view along the filaments, with the field vectors due to each and their sum. The *x*-components cancel, and the *z*-components add.

SOLVE We substitute $2k = 1/(2\pi\epsilon_0)$ into eqn. (24.4):

$$\vec{E} = \vec{E}_1 + \vec{E}_2 = 2|\vec{E}_1|(\cos\theta)\hat{k} = 2\frac{2k\lambda}{r}(\cos\theta)\hat{k}$$

$$= 4k\lambda\frac{d}{r^2}\hat{k} = \frac{4k\lambda d}{d^2 + d^2/4}\hat{k} = \frac{16k\lambda}{5d}\hat{k}.$$

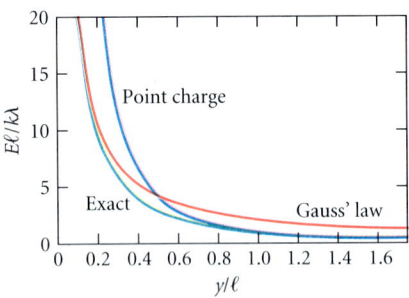

■ FIGURE 24.9
Comparison of exact and approximate solutions for the electric field due to a finite filament. The near solution, found using Gauss' law (red line) is close to the exact solution (green line) when $y \lesssim 0.2\ell$. The far (point charge) solution (blue line) is close to the exact solution for $y \gtrsim 0.8\ell$.

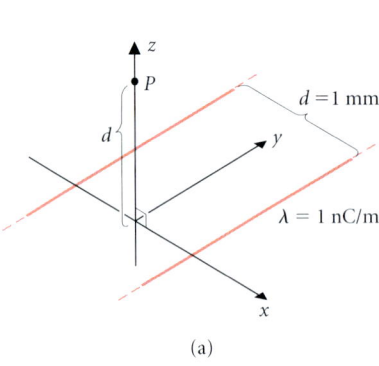
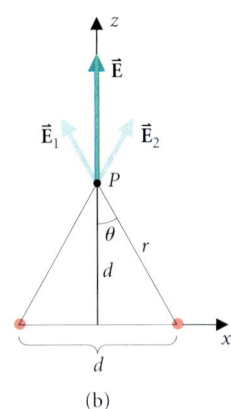

■ FIGURE 24.10
(a) Electric field due to two parallel, infinite filaments separated by a distance *d*. The filaments lie in the *x-y*-plane, and we find the field at *P*, on the *z*-axis at $z = d$.

(b) View in the *x-z*-plane of the two contributions to $\vec{E}$ at *P*. The filaments are perpendicular to the page, and $\cos\theta = d/r$.

$$|\vec{E}| = \frac{16(9.0 \times 10^9 \text{ N} \cdot \text{m}^2/\text{C}^2)(1.0 \times 10^{-9} \text{ C/m})}{5.0 \times 10^{-3} \text{ m}} = 2.9 \times 10^4 \text{ N/C}.$$

Thus: $\vec{E} = (2.9 \times 10^4 \text{ N/C})\hat{\mathbf{k}}.$

ANALYZE Since λ has dimensions [charge]/[length], λ/d has dimensions of [charge]/[length]^2, and $k\lambda/d$ has the correct dimensions for electric field. ■

EXERCISE 24.2 ♦♦ Point P in ■ Figure 24.11 is a distance z above the center of a uniformly charged circular filament with radius a and total charge Q. Show that the electric field vector at P is given by:

$$\vec{E} = \frac{kQz}{(z^2 + a^2)^{3/2}} \hat{\mathbf{k}}. \tag{24.5}$$

(*Hint:* Use a symmetry argument to show that the field vector $\vec{E}$ is in the z-direction.) ■

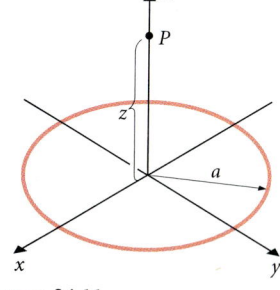

■ **FIGURE 24.11**
The circular filament lying in the x-y-plane has a uniform charge density.

24.3 ELECTRIC FIELD DUE TO SURFACE AND VOLUME CHARGE DISTRIBUTIONS

24.3.1 Surface Charge Distributions

When charge is distributed over a surface, we describe it with a surface charge density:

> The *surface charge density* σ at a point is the amount of charge per unit area on a surface. The charge dQ on an area dA surrounding the point is given by:
>
> $$dQ = \sigma \, dA. \tag{24.6}$$

In a *uniform* distribution, σ has the same value at every point on the surface. Our first example is a uniformly charged infinite plane sheet. It has sufficient symmetry that we may use Gauss' law to find the field.

EXAMPLE 24.5 ♦♦ Find the electric field due to an infinite plane sheet that carries a uniform surface charge density σ (■ Figure 24.12).

MODEL
STEP I Since the sheet has the same charge density everywhere in the plane, the field lines emerging from it do not bend at all. They are perpendicular to the surface and parallel to each other. The separation of field lines, and hence $|\vec{E}|$, doesn't change. The magnitude of the field is the same on both sides and remains finite at the sheet. We choose the plane of the sheet as the x-y-plane; then only the z-component of $\vec{E}$ is nonzero, and E_z changes sign at $z = 0$. We use Gauss' law to find E_z.

SETUP
STEP II We choose as our Gaussian surface a box with vertical sides (parallel to $\vec{E}$) and with top and bottom parallel to the sheet (perpendicular to $\vec{E}$). The shape and area of the horizontal faces are unimportant. The box extends as far below the sheet as above, to make use of the symmetry.

Pulling plastic wrap from the roll produces a charged sheet.

IN CONTRAST, FIELD LINES CONVERGE AT POINT CHARGES AND FILAMENTS, WHERE THE FIELD MAGNITUDE BECOMES INFINITELY LARGE.

ACTUALLY, THE VALUE OF E_z TURNS OUT TO BE INDEPENDENT OF DISTANCE FROM THE SHEET, AS WE ARGUED IN OUR MODELING SECTION. LOCATING THE GAUSSIAN SURFACE SYMMETRICALLY IS ALWAYS A SAFE APPROACH. THE AREA A CANCELS FROM THE FINAL RESULT.

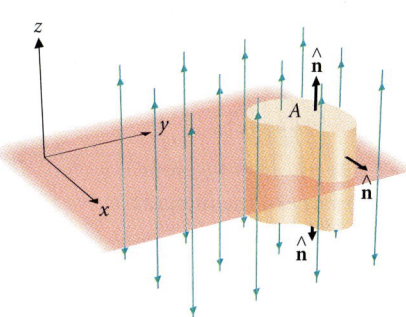

■ **FIGURE 24.12**
An infinite plane sheet has a uniform surface charge density σ. We choose the z-axis perpendicular to the sheet. The field lines emerge perpendicular to the sheet and remain parallel as they go off to infinity. Since the separation of fields lines indicates the field strength, the field due to the sheet is uniform.

STEP III Since the field is perpendicular to $\hat{\mathbf{n}}$ at points on the vertical sides of the box, there is no flux through the sides. On the top and bottom $\vec{\mathbf{E}}$ is parallel to $\hat{\mathbf{n}} = \pm\hat{\mathbf{k}}$, and $|E_z|$ is constant over each surface, so the net flux is twice that emerging from the top surface:

$$\Phi_E = 2AE_{z,\text{top}}.$$

STEP IV The box contains a charge σA.

SOLVE
STEP V Applying Gauss' law:

$$2AE_{z,\text{top}} = \frac{\sigma A}{\epsilon_0}$$

$$\Rightarrow \vec{\mathbf{E}}(z) = \frac{\sigma}{2\epsilon_0}\hat{\mathbf{k}} \qquad (z > 0). \tag{24.7}$$

Below the sheet the electric field changes direction: $\vec{\mathbf{E}} = (\sigma/2\epsilon_0)(-\hat{\mathbf{k}})$ for $z < 0$.

ANALYZE The surface charge density has dimensions [charge]/[length]^2, so σ/ϵ_0 has the correct dimensions for electric field. The uniform field lines can't go on forever. In a real system, the lines terminate on distant charges of the opposite sign that were removed to create the charged sheet. ∎

EXAMPLE 24.6 ♦♦♦ An element of an electrostatic microphone is a circular plastic disk of radius a that carries a total charge Q uniformly distributed over its surface. Find the electric field $\vec{\mathbf{E}}$ at point P a distance z above the center of the disk.

MODEL As always, we begin by sketching a field line diagram for the system (■ Figure 24.13a). Far away from the disk, the field lines diverge as from a point charge. Since the disk is uniformly charged, the same number of field lines emerge from each unit of area. The lines do not diverge from a single point or line. Near the surface they resemble the field of an infinite sheet, and $|\vec{\mathbf{E}}|$ has a finite limit at the surface of the disk. When $z \ll a$, $\vec{\mathbf{E}}$ is perpendicular to the disk and approaches the value found in Example 24.5 using Gauss' law. To calculate the field at intermediate distances, we use the principle of superposition.

SETUP
STEP I Since we know the electric field produced by a circular filament (Exercise 24.2), such filaments are appropriate differential elements to use in analyzing the disk. We use polar coordinates r and θ in the plane of the disk and z perpendicular to the disk (Figure 24.13b).

STEP II A typical element is a ring of radius r and thickness dr, which has area $dA = 2\pi r\, dr$. The charge density on the disk is $\sigma = Q/(\pi a^2)$, and the charge on the element is:

$$dQ = \sigma\, dA = \frac{Q}{\pi a^2}\, 2\pi r\, dr = \frac{2Qr\, dr}{a^2}.$$

STEP III The electric field at P produced by this ring is given by eqn. 24.5:

$$d\vec{\mathbf{E}} = \frac{k(dQ)z}{(z^2 + r^2)^{3/2}}\hat{\mathbf{k}} = \frac{kQz}{a^2}\frac{2r\, dr}{(z^2 + r^2)^{3/2}}\hat{\mathbf{k}}.$$

STEP IV Since we expressed a positive ring thickness as dr, we sum the rings from the center toward the edge. That is, the radius r ranges from 0 to a.

SOLVE
STEP V Each element of field $d\vec{\mathbf{E}}$ is in the z-direction, so the sum $\vec{\mathbf{E}}$ is, too. The unit vector $\hat{\mathbf{k}}$ is constant and may be brought out of the integral:

$$\vec{\mathbf{E}} = \hat{\mathbf{k}}\frac{kQz}{a^2}\int_0^a \frac{2r\, dr}{(z^2 + r^2)^{3/2}}.$$

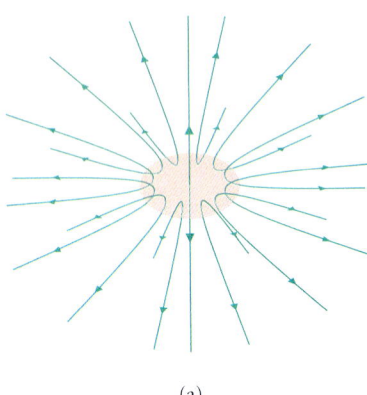

(a)

■ **FIGURE 24.13** (a) Electric field due to a uniformly charged disk. The field lines emerge perpendicular to the disk (cf. Figure 24.12), and at large distances they diverge as from a point charge.

HERE THE STEP NUMBERS REFER TO THE INTEGRATION METHOD IN INTERLUDE 2.

THAT IS, WE USE CYLINDRICAL COORDINATES.

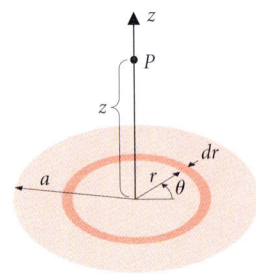

(b)

■ **FIGURE 24.13** (b) We model the disk as a collection of rings, each with thickness dr and charge $dQ = \sigma 2\pi r\, dr$. The field at P due to each ring is in the z-direction, perpendicular to the disk.

We change variables to $u = z^2 + r^2$. Then $du = 2r\, dr$, and we have:

$$\int_0^a \frac{2r\, dr}{(z^2 + r^2)^{3/2}} = \int_{z^2}^{z^2+a^2} \frac{du}{u^{3/2}} = \left.\frac{u^{-1/2}}{-1/2}\right|_{z^2}^{z^2+a^2}$$

$$= -2\left(\frac{1}{\sqrt{z^2+a^2}} - \frac{1}{z}\right).$$

Thus:
$$\vec{E} = \frac{2kQ\hat{k}}{a^2}\left(1 - \frac{z}{\sqrt{z^2+a^2}}\right).$$

ANALYZE Checking the limits, we find for $z \ll a$:

$$|\vec{E}| \approx \frac{2kQ}{a^2} = 2\pi k\sigma, \tag{24.8}$$

which agrees with eqn. (24.7) from Gauss' law. For $z \gg a$, we expand the square root using the binomial expansion:

$$|\vec{E}| \approx \frac{2kQ}{a^2}\left[1 - \left(1 - \frac{a^2}{2z^2}\right)\right] = \frac{kQ}{z^2},$$

which is the expected result for a point charge located at the origin.

NOTICE THAT WE CANNOT IGNORE THE a^2 TERM ENTIRELY BECAUSE WE WOULD GET A LIMIT OF ZERO. ALTHOUGH CORRECT, THIS IS NOT VERY INFORMATIVE.

24.3.2 Volume Charge Distributions

Charge distributed throughout a volume is described by a volume charge density:

> The volume *charge density* ρ at a point is the amount of charge per unit volume. The charge dQ in a volume element dV surrounding the point is given by:
>
> $$dQ = \rho\, dV. \tag{24.9}$$

WHEREAS σ IS CALLED "SURFACE CHARGE DENSITY," ρ IS USUALLY CALLED JUST "CHARGE DENSITY."

The techniques for finding the resulting electric field are the same ones that we used for linear and surface distributions: Gauss' law, or Coulomb's law and the principle of superposition.

EXAMPLE 24.7 ♦♦ Find the electric field due to an infinite plane slab of thickness w that carries a uniform charge density ρ (■ Figure 24.14).

MODEL The system is sufficiently symmetric that we may use Gauss' law to calculate the electric field. We choose coordinates with the x-y-plane as the midplane of the slab.

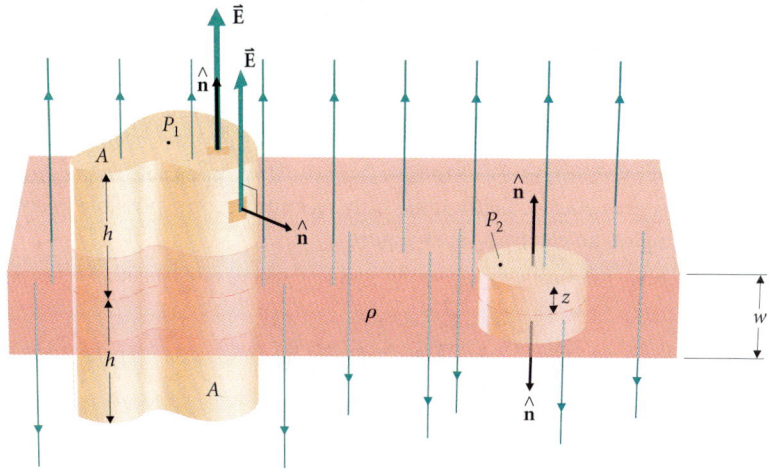

■ FIGURE 24.14
Electric field due to a uniformly charged slab of thickness w. We choose the z-axis perpendicular to the slab. The field lines are everywhere perpendicular to the sides of the slab; however, inside the slab, new field lines originate at the charge within each layer. The field strength increases with distance from the central plane. To apply Gauss' law, we choose a box with sides parallel and perpendicular to the slab.

STEP I The field lines outside the slab have the same symmetry as the lines from the plane sheet (Example 24.5). At points outside the slab ($|z| > w/2$), it looks like a charged sheet with surface charge density $\sigma = \rho w$. The calculation of the electric field proceeds exactly as in Example 24.5, and the result is the same:

$$\vec{E} = \frac{\sigma}{2\epsilon_0}\hat{k} = \frac{\rho w}{2\epsilon_0}\hat{k} \qquad (z > w/2).$$

Since the charge distribution within the slab also has plane symmetry, the field lines there remain parallel to the z-axis. However, in each successive layer field lines begin on the charges in that layer. Thus the magnitude of the field increases with distance from the center of the slab, and E_z is a function of z. Because the slab is symmetric about its midplane, we expect $E_z(-z) = -E_z(z)$.

REMEMBER THAT $E(z)$ MEANS "E AT COORDINATE z."

SETUP
STEP II We choose the Gaussian surface to be a box with vertical sides and with top and bottom faces inside and parallel to the slab. The box extends as far below the midplane of the slab as above, to make use of the symmetry, and its top is at $z < w/2$. The volume of the box is $V = 2zA$.

STEP III Since the field is perpendicular to $\hat{n}$ on the vertical sides of the box, there is no flux through the sides. On both the top and bottom $\vec{E}(z) = E_z(z)\hat{k}$ is parallel to $\hat{n} = \pm\hat{k}$. Since $E_z(-z) = -E_z(z)$, the total flux through the box is:

$$\Phi_E = 2E_z(z)A.$$

STEP IV The total charge inside the box is $Q = \rho V = \rho 2Az$.

SOLVE
STEP V Applying Gauss' law:

$$\Phi_E = 2E_z(z)A = \frac{2\rho Az}{\epsilon_0}$$

$$\Rightarrow \vec{E}(z) = \frac{\rho z}{\epsilon_0}\hat{k} \qquad (|z| < w/2).$$

ANALYZE The results for the fields inside and outside give the same expression at the edge of the slab, where $z = w/2$. The electric field at the midpoint of the slab is zero, as expected from the symmetry. ■

EXERCISE 24.3 ❖ Why is it necessary for the ends of the box to be equidistant from the midplane when computing $\vec{E}$ for $|z| < w/2$? Is this requirement necessary for computing the field outside the slab? Can you find another box that works for the field inside?

EXERCISE 24.4 ❖❖ Find the electric field within a uniformly charged sphere with charge Q and radius R.

24.4 MOTION OF CHARGES IN AN ELECTRIC FIELD

If a particle with charge q is in a region with electric field $\vec{E}$, it experiences a force $\vec{F} = q\vec{E}$. Frequently, this electric force is much stronger than any other force acting on the particle and we may neglect the other forces. For example, because it is impossible to produce an environment completely free of electric fields, it has only recently become possible to detect the gravitational force acting on an electron (see ■ Figure 24.15).

Neglecting other forces, the acceleration of a particle in an electric field $\vec{E}$ is:

$$\vec{a} = \frac{\vec{F}}{m} = \frac{q}{m}\vec{E}. \qquad (24.10)$$

Measuring the acceleration of a particle in a known electric field is one method of determining its *charge-to-mass ratio* q/m.

FIGURE 24.15
Apparatus for measuring the gravitational force on an electron. Electrons are contained in the cylinder just left of center, which extends into the room below. Great pains must be taken to minimize electric and magnetic fields inside the cylinder and to correct for the effects of any fields that do remain.

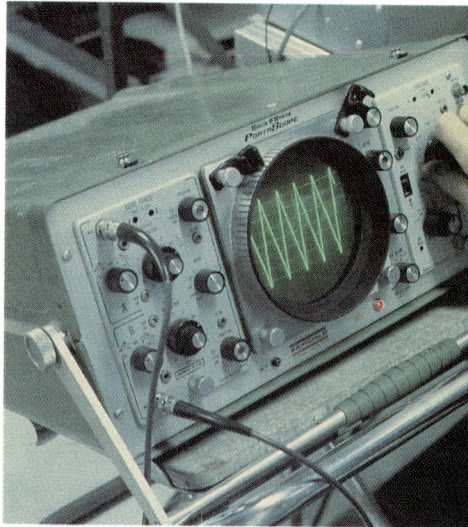

Electric fields along the oscilloscope tube accelerate the electrons in the beam. Electric fields across the tube control their sideways deflection to produce the image on the screen.

EXAMPLE 24.8 ♦ In the *electron gun* of a color TV set the electric field $|\vec{E}| = 2.5 \times 10^6$ N/C in a region 1.0 cm long (■ Figure 24.16). Electrons enter the region at low speed and are accelerated into a beam that produces light when it strikes phosphors on the tube face. What is the acceleration of each electron? What is the speed of an electron as it emerges from the gun?

MODEL We model the electric field as uniform, so the electron (charge $-e$) has a constant acceleration while it is in the 1.0-cm-long region where $|\vec{E}|$ is nonzero. We may find the electron's speed using the kinematic equations from Chapter 2.

SETUP The electron's acceleration has magnitude:

$$a = \frac{F}{m} = \frac{e}{m} E.$$

SOLVE The electron's speed after moving $s = 1.0$ cm from rest at this acceleration is (eqn. 2.13):

$$v = \sqrt{2as} = \sqrt{2 \frac{e}{m} Es}$$

$$= \sqrt{(2) \frac{1.60 \times 10^{-19} \text{ C}}{9.11 \times 10^{-31} \text{ kg}} (2.5 \times 10^6 \text{ N/C})(1.0 \times 10^{-2} \text{ m})}$$

$$= 9.4 \times 10^7 \text{ m/s}.$$

ANALYZE These very energetic electrons are moving at about a third the speed of light. In addition to exciting the phosphors on the TV screen, they can produce x rays. TV manufacturers have to put x-ray-absorbent coatings on the front of the TV screen to avoid exposing viewers to hazardous radiation.

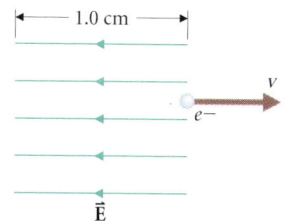

■ **FIGURE 24.16**
In a television set, electrons are accelerated by an electric field, modeled here as uniform. Remember that the force on a negatively charged electron is opposite $\vec{E}$.

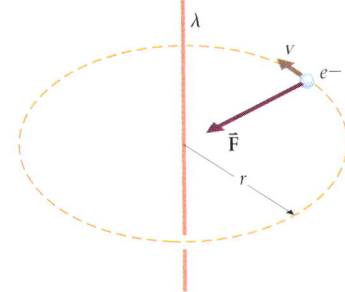

FIGURE 24.17
An electron orbits a charged filament. To produce an inwardly directed force, the filament must have a positive charge.

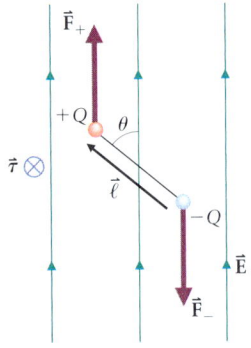

FIGURE 24.18
A dipole placed in a uniform field. The two charges in the dipole experience equal and opposite forces, forming a couple whose torque accelerates the dipole toward a configuration with $\vec{p}$ parallel to $\vec{E}$.

MOLECULAR FORCES AND THEIR CONSEQUENCES WERE DISCUSSED IN CHAPTER 20.

SEE CHAPTER 27 FOR MORE ON THIS TOPIC.

THE TWO FORCES FORM A *COUPLE* (CHAPTER 11). REMEMBER: THE TORQUE EXERTED BY A COUPLE ABOUT ANY POINT IS THE SAME.

When a calculation using Newtonian mechanics leads to an answer for v that is a large fraction of the speed of light, we should check the result using relativistic mechanics (see Chapter 34). In this example, the Newtonian calculation overestimates the electron speed by about 4%.

EXAMPLE 24.9 ♦♦♦ An electron is observed to move in a circle around a long, uniformly charged filament. If the speed of the electron is 6.0×10^6 m/s, what is the charge density on the filament?

MODEL The electric field of a positively charged filament is radially outward. The electron moves in a circle around the filament (■ Figure 24.17) because the force $\vec{F} = q\vec{E} = -e\vec{E}$ acting on it is always directed toward the center of the circle.

SETUP The electric field due to the filament is given by eqn. (24.4). Letting r be the radius of the observed circular path and λ be the charge density on the filament, we calculate the magnitude of the electric force on the electron:

$$F = eE = e(2k\lambda/r).$$

This force causes the observed centripetal acceleration:

$$F = ma.$$
$$2k\lambda e/r = mv^2/r.$$

SOLVE The distance of the electron from the filament cancels:

$$\lambda = mv^2/(2ke)$$
$$= \frac{(9.11 \times 10^{-31} \text{ kg})(6.0 \times 10^6 \text{ m/s})^2}{2(9.0 \times 10^9 \text{ N} \cdot \text{m}^2/\text{C}^2)(1.60 \times 10^{-19} \text{ C})}$$
$$= 1.1 \times 10^{-8} \text{ C/m}.$$

ANALYZE Remember that the electron has a negative charge, so the electric force on it is opposite the electric field vector.

24.5 THE DIPOLE

An electric dipole—two equal but opposite charges separated by a small distance—is one of the most important and widely applicable physical models. For example, the forces between uncharged molecules of vapor condensing into a liquid occur because the molecules act like electric dipoles. Since it has no net charge, the dipole is a useful model for predicting the behavior of neutral matter exposed to electric fields. The dipole field pattern describes the simplest magnetic field that can occur and approximates the magnetic field produced by a complex system like the Earth. In this section, we investigate the behavior of electric dipoles.

EXAMPLE 24.10 ❖ Two point charges $+Q$ and $-Q$, each with mass m, are at the ends of a massless rod of length ℓ. The object is placed in a region where there is a uniform electric field $\vec{E}$ (■ Figure 24.18). Describe the motion of the dipole.

MODEL Each of the charges in the dipole experiences an electric force. The net force and net torque on the dipole determine its linear and angular accelerations.

SETUP The positive charge experiences a force $\vec{F}_+ = Q\vec{E}$, and the negative charge experiences a force $\vec{F}_- = -Q\vec{E} = -\vec{F}_+$. The total force:

$$\vec{F}_t \equiv \vec{F}_+ + \vec{F}_- = Q\vec{E} - Q\vec{E},$$

is zero, and the dipole's center of mass does not accelerate. With the origin at the negative charge, the torque on the dipole is:

$$\vec{\tau} = \vec{r} \times \vec{F} = \vec{\ell} \times Q\vec{E} = QE\ell \sin\theta \, \hat{\otimes} \quad \text{(into the plane of the figure)}.$$

This torque causes the dipole to begin rotating toward alignment with the electric field. When the dipole is parallel to $\vec{E}$, the torque is zero, but the dipole has gained angular momentum. It continues to rotate beyond alignment, and the torque reverses direction. The dipole oscillates about an equilibrium position parallel to the field direction.

ANALYZE The example gives no mechanism for the dipole to lose energy, so it will continue to oscillate indefinitely. Real dipoles, such as the particles in Figure VI.14 in the overview, experience friction. Consequently their oscillations die out, and they align with the field. A compass needle is a small *magnetic* dipole that points north as a result of a similar alignment process.

THERE ARE OTHER DAMPING MECHANISMS TOO. FOR EXAMPLE, AN OSCILLATING ELECTRIC DIPOLE RADIATES EM WAVES (CHAPTER 33).

The torque exerted on the dipole depends on the magnitude of the point charges Q, their separation ℓ, and their orientation with respect to the field. Together these define a vector called the dipole moment.

> The *dipole moment* $\vec{p}$ of a pair of equal but opposite charges has magnitude equal to the product of the charge magnitude and the separation of the pair.
>
> $$|\vec{p}| = Q\ell. \quad (24.11)$$
>
> The direction of $\vec{p}$ is from the negative charge toward the positive charge.

In terms of the dipole moment, the torque on the dipole is

$$\vec{\tau} = \vec{p} \times \vec{E}. \quad (24.12)$$

EXERCISE 24.5 ♦♦♦ Find the force on a dipole placed at a distance r from a point charge and oriented with $\vec{p}$ along a radial line from the charge. Assume $\ell \ll r$.

EXAMPLE 24.11 ♦♦♦ Find an expression for the electric field at a point P on the x-axis due to a dipole at the origin if the distance to the point is much greater than the size of the dipole and $\vec{p} = p\hat{\imath}$.

MODEL The electric field is the sum of the field vectors $\vec{E}_+$ and $\vec{E}_-$ due to the two charges in the dipole. When the separation of the two charges is small, their field vectors nearly cancel, and the sum is relatively small.

SETUP We choose the origin to be at the negatively charged end of the dipole (■ Figure 24.19). From the principle of superposition and Coulomb's law:

USING AN ORIGIN SOMEWHERE ELSE ON THE DIPOLE DOESN'T CHANGE THE EXPRESSION WE OBTAIN FOR $\vec{E}$.

$$\vec{E} = \vec{E}_+ + \vec{E}_- = \frac{kQ}{(x-\ell)^2}\hat{\imath} + \frac{k(-Q)}{x^2}\hat{\imath} = kQ\hat{\imath}\left[\frac{x^2 - (x-\ell)^2}{x^2(x-\ell)^2}\right]$$

$$= kQ\hat{\imath}\left[\frac{2\ell x - \ell^2}{x^2(x-\ell)^2}\right] = kQ\ell\hat{\imath}\left[\frac{2x - \ell}{x^2(x-\ell)^2}\right].$$

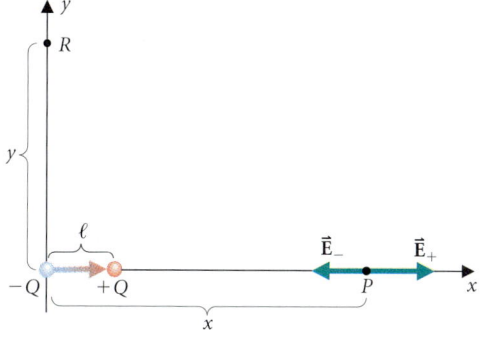

■ FIGURE 24.19
A dipole lies along the x-axis. The electric field at P is the superposition of the fields due to each of the charges that make up the dipole. Their magnitudes differ slightly, because of the different distances of the two charges from P. (The diagram is not drawn to scale: $x \gg \ell$.) In Exercise 24.6, you are asked to find the electric field at point R, with $y \gg \ell$.

THIS IS AN EXAMPLE OF THE RULE IN INTERLUDE 1.

SOLVE Having simplified as much as possible, we may now use the fact that $\ell \ll x$. We neglect ℓ with respect to x whenever they are added or subtracted but not when they are multiplied. Thus:

$$\vec{E} \approx kQ\ell \hat{i}\, \frac{2}{x^3}.$$

$$\vec{E} = \frac{2kp}{x^3}\, \hat{i}. \qquad (24.13)$$

ANALYZE *Remember:* This result is correct for points on the dipole axis, with $x \gg \ell$. However, the decrease of $|\vec{E}|$ with distance cubed is a general feature of the dipole field. ∎

EXERCISE 24.6 ♦♦♦ Show that the electric field at a point on the y-axis in Figure 24.19 is: $\vec{E} = -k\vec{p}/|y|^3$. ∎

EXAMPLE 24.12 ♦♦♦ Water molecules have a dipole moment p of approximately 6×10^{-30} C·m. Two water molecules are separated by 10^{-7} m, and their dipole moments are in the same direction along their line of separation (■ Figure 24.20a). What electric force acts on each molecule?

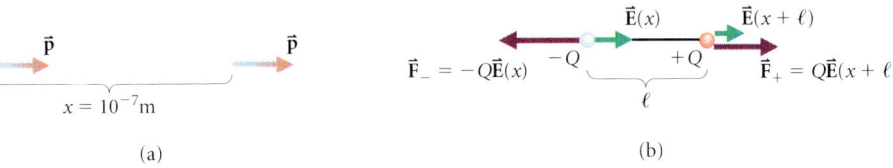

■ **FIGURE 24.20**
(a) The force that one dipole exerts on another depends on their relative orientation. Here the two dipoles are parallel and colinear, separated by a distance x.

(b) Enlargement of the dipole on the right, showing the electric field due to the other dipole and the forces exerted on the two charges.

MODEL We model each dipole as two point charges $\pm Q$ separated by a distance ℓ, with $p = Q\ell$. One dipole produces an electric field (eqn. 24.13) at the location of the other. As a result, each charge in the second dipole experiences a force, and the net force on the second dipole is the sum of the forces acting on its two charges (Figure 24.20b).

SETUP With the origin at the first dipole, the net force on the second is:

$$\vec{F} = Q\vec{E}(x + \ell) - Q\vec{E}(x)$$

$$= Q\left[\frac{2kp\hat{i}}{(x+\ell)^3} - \frac{2kp\hat{i}}{x^3}\right]$$

$$= 2kpQ\hat{i}\left[\frac{x^3 - (x+\ell)^3}{(x+\ell)^3 x^3}\right]$$

$$= 2kpQ\hat{i}\left[\frac{-3x^2\ell - 3x\ell^2 - \ell^3}{(x+\ell)^3 x^3}\right].$$

NOTE ALSO: $\ell^3/(3x^2\ell) = \frac{1}{3}(\ell/x)^2 \ll 1$.

SOLVE Since the dimension ℓ of a molecule is approximately 10^{-9} m $\ll 10^{-7}$ m $= x$, we ignore ℓ compared with x whenever they are added or subtracted:

$$\vec{F} = \hat{i}\, 2kQp\ell\left[\frac{-3x(x+\ell) - \ell^2}{(x+\ell)^3 x^3}\right] \approx \hat{i}\, 2kQp\ell\, \frac{-3x^2}{x^6} = -\hat{i}\, \frac{6kp^2}{x^4}.$$

The minus sign indicates that the force is attractive. Its magnitude is:

$$|\vec{F}| = \frac{6(9 \times 10^9 \text{ N·m}^2/\text{C}^2)(6 \times 10^{-30} \text{ C·m})^2}{(10^{-7} \text{ m})^4}$$

$$= 2 \times 10^{-20} \text{ N}.$$

The force on the other dipole is equal and opposite.

ANALYZE The mass of each molecule (two hydrogen atoms plus one oxygen atom) is roughly 18 u, or about 3×10^{-26} kg, so each molecule has a huge acceleration:

$$|\vec{a}| = |\vec{F}|/m \approx 10^6 \text{ m/s}^2,$$

or about 10^5 g! This acceleration occurs in a gas when two molecules come very close together and the influence of other molecules can be ignored: that is, during a *collision*. In liquid water the molecules are influenced by the fields of many other nearby molecules as well.

Chapter Summary

Where Are We Now?

We have shown how to use Coulomb's law and Gauss' law to compute the electric field due to a distribution of charge. We have discussed the motion of charges under the influence of electric fields. We described a configuration of charge called a dipole.

What Did We Do?

The principle of superposition states that the electric field vector due to a charge distribution is the sum of the electric field vectors produced by each charge element in the distribution. The vector sum is evaluated by integration taking proper account of the directions of the vectors. When the system has appropriate symmetry, Gauss' law provides an efficient way to compute the field. A solution plan for using Gauss' law is given in Figure 24.1. We used these two principles to find the electric field due to a continuous charge distribution. Two important results are the field due to a long filament ($E = 2k\lambda/r$, eqn. 24.4) and the field due to an infinite plane sheet ($E = \sigma/2\epsilon_0$, eqn. 24.7).

A particle's acceleration when placed in an electric field depends on its charge-to-mass ratio, $\vec{a} = (q/m)\vec{E}$. A dipole comprises a positive charge Q and negative charge $-Q$ separated by a distance ℓ. The dipole moment $\vec{p}$ has magnitude $Q\ell$ and points from the negative charge toward the positive charge. A dipole placed in an external electric field feels a torque $\vec{\tau} = \vec{p} \times \vec{E}$ that rotates it toward alignment with that field. The electric field of a dipole decreases as the inverse cube of distance from the dipole.

Practical Applications

Static electric fields accelerate particles inside the cathode ray tubes used in oscilloscopes and television sets, and in linear accelerators for high-energy particle research. Electrostatic fields are also used in some designs of high-frequency loudspeakers and microphones, and in photocopiers. Dipoles are a good model for molecules exposed to electric fields (Chapter 27) and form the basis for the $1/r^6$ behavior of molecular forces in the van der Waals model of gases (Chapter 20).

Solutions to Exercises

24.1 The field line diagram outside the ball looks just the same as the field line diagram outside the spherical shell. The flux through any sphere surrounding and centered on the ball is the same as we calculated in Example 24.1: $\Phi_E = 4\pi r^2 E_r$, and the charge enclosed is Q. From Gauss' law, the electric field outside the ball is the same as for a point charge Q at the center of the ball.

24.2 Consider the ring to be made up of differential elements with coordinate ϕ and length $a\, d\phi$ (■ Figure 24.21). Each element has charge $dQ = \lambda a\, d\phi = (Q/2\pi a)a\, d\phi$. The electric field vectors contributed by two such elements on opposite ends of a diameter are shown in Figure 24.21. Once again we have two equal charges, and the electric field vector produced by the pair is along the z-axis and has magnitude (eqn. 23.9):

$$dE = \frac{2k(dQ)z}{(a^2 + z^2)^{3/2}} = \frac{kQz(d\phi)}{\pi(a^2 + z^2)^{3/2}}.$$

Now we add up the contributions of all such pairs. We need to consider the coordinate of only one charge in each pair, which ranges from $\phi = 0$ to $\phi = \pi$.

$$\vec{E} = \int d\vec{E} = \hat{k}\, \frac{kQz}{\pi(a^2 + z^2)^{3/2}} \int_0^\pi d\phi = \hat{k}\, \frac{kQz}{(a^2 + z^2)^{3/2}},$$

as required. The integrand is independent of ϕ because all the differential elements are the same distance from P.

24.3 Since the electric field outside the slab is independent of z, we could put the box ends anywhere. So long as one end is on each side of the slab, the net flux is the same. The field inside the slab does depend on z; here we need to make full use of the symmetry and put

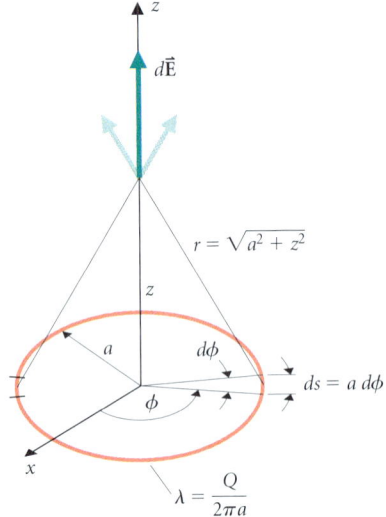

■ **Figure 24.21**
Two charge elements at opposite ends of a diameter form a pair of equal charges, as in Example 23.11.

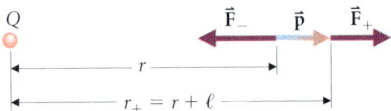

■ **Figure 24.22**
The electric forces acting on each charge of the dipole are not quite equal in magnitude because of the different distances from the point charge Q.

the ends of the box at the same distance from the midplane so as to have the same magnitude of $\vec{E}$ on each end.

Another method would be to put one end of the box outside the slab and one end inside. Then the known value of the field outside could be used to find the flux through that end of the box. Alternatively, we could put one end of the box at the midplane itself, where $\vec{E}$ is zero due to the symmetry.

24.4 We choose a spherical Gaussian surface of radius r, concentric with the sphere of charge. The flux through the surface is $4\pi r^2 E_r$, and the charge enclosed in the volume within the surface is $\tfrac{4}{3}\pi r^3 \rho$. The charge density $\rho = Q/(\tfrac{4}{3}\pi R^3)$. Applying Gauss' law:

$$4\pi r^2 E_r = \left(\frac{r}{R}\right)^3 \frac{Q}{\epsilon_0} \;\Rightarrow\; \vec{E} = \frac{Qr}{4\pi R^3 \epsilon_0}\, \hat{r}.$$

At the surface of the charged sphere, where $r = R$, this expression gives the same result that we found in Exercise 24.1 for the electric field vector outside the sphere.

24.5 In ■ Figure 24.22 the point charge is Q, and the dipole moment is $p = q\ell$. The electric field produced at the negative end of the dipole is:

$$\vec{E}_- = \frac{kQ}{r^2}\, \hat{r},$$

so the force on the dipole's negative charge is:

$$\vec{F}_- = -\frac{kQq}{r^2}\, \hat{r}.$$

At the positive end:

$$\vec{E}_+ = \frac{kQ}{r_+^2}\, \hat{r} \quad \text{and} \quad \vec{F}_+ = \frac{kQq}{r_+^2}\, \hat{r} = \frac{kQq}{(r+\ell)^2}\, \hat{r}.$$

Therefore:

$$\vec{F} = \vec{F}_+ + \vec{F}_- = kQq\left[\frac{1}{(r+\ell)^2} - \frac{1}{r^2}\right]\hat{r}.$$

The term in brackets is:

$$\frac{-2r\ell - \ell^2}{r^2(r+\ell)^2} \approx \frac{-2\ell}{r^3},$$

where, since $\ell \ll r$, we have dropped ℓ whenever it is added to r. Thus:

$$\vec{F} = -\frac{2kQq\ell}{r^3}\, \hat{r} = -\frac{2kQ\vec{p}}{r^3}.$$

You can check the result by using eqn. 24.13 to compute the force exerted by the dipole on the point charge, then applying Newton's third law.
Since

$$-2kQ/r^3 = (d/dr)(kQ/r^2),$$

the force may also be written:

$$\vec{F} = p\frac{d\vec{E}}{dr}.$$

Only in a nonuniform field does a dipole experience a net force.

24.6 We compute the fields produced by each charge at point R:

$$\vec{E}_- = \frac{kQ}{y^2}(-\hat{j})$$

and

$$\vec{E}_+ = \frac{kQ}{(y^2 + \ell^2)^{3/2}}\left(y\hat{j} - \ell\hat{i}\right).$$

If we ignore ℓ with respect to y when they are added, the y-components cancel. The net electric field is:

$$\vec{E} = \vec{E}_- + \vec{E}_+ = \frac{-kQ\ell}{(y^2)^{3/2}}\hat{i} = \frac{-k}{|y|^3}\vec{p}.$$

Basic Skills

Review Questions

§24.1 USING GAUSS' LAW TO CALCULATE ELECTRIC FIELDS

- Under what conditions may Gauss' law be used to find the electric field?
- How should the Gaussian surface be matched to the symmetry of the field line diagram?
- What is the electric field inside a uniformly charged spherical shell?

§24.2 THE ELECTRIC FIELD DUE TO A LINEAR CHARGE DISTRIBUTION

- When evaluating $\vec{E}$ using integration, what feature of the calculation requires particular care?
- Define *linear charge density*.
- How does the electric field of a uniformly charged infinite filament depend on distance from the filament?

§24.3 ELECTRIC FIELDS DUE TO SURFACE AND VOLUME CHARGE DISTRIBUTIONS

- Define *surface charge density*.
- What is the electric field due to an infinite, uniformly charged plane sheet?
- Define *volume charge density*.
- Describe a volume that you could use to find the electric field inside a uniformly charged slab using Gauss' law.

§24.4 MOTION OF CHARGES IN AN ELECTRIC FIELD

- What feature of a particle describes its response to an applied electric field?

§24.5 THE DIPOLE

- What is an *electric dipole*? Define its electric dipole moment $\vec{p}$.
- How does a dipole behave when placed in a uniform electric field?
- How does the electric field at a point P due to a dipole depend on the distance of P from the dipole?

Basic Skill Drill

§24.1 USING GAUSS' LAW TO CALCULATE ELECTRIC FIELDS

1. Find the electric field 16 cm from the surface of a uniformly charged sphere of radius 5.0 cm carrying a total charge of 16 nC.

§24.2 THE ELECTRIC FIELD DUE TO A LINEAR CHARGE DISTRIBUTION

2. A uniformly charged rod 12 cm long carries a charge of 16 μC. Find the linear charge density on the rod.

3. An infinitely long filament carries a uniform charge density $\lambda = 175\ \mu$C/m. What is the electric field 0.50 m from the filament?

§24.3 ELECTRIC FIELDS DUE TO SURFACE AND VOLUME CHARGE DISTRIBUTIONS

4. A uniformly charged square plate 11.6 cm on a side carries a total charge of 98.7 μC. What is the surface charge density on the plate?

5. What is the electric field near the center of and 5.5 mm above a large flat plate carrying a uniform charge density 4.6 mC/m^2?

6. A uniformly charged rectangular solid measuring 15 cm $\times$ 25 cm $\times$ 35 cm carries a total charge of 750 μC. What is the charge density?

§24.4 MOTION OF CHARGES IN AN ELECTRIC FIELD

7. A proton is released from rest 10.0 cm above an infinite plane with a charge density of $-175\ \mu$C/m^2. What is its acceleration? How long does it take to reach the plane?

8. A helium nucleus with charge $2e$ and mass 4.0 u is placed 3.0 μm from an infinite uniformly charged filament with $\lambda = 5.0$ nC/m. Find the magnitude and direction of its initial acceleration.

§24.5 THE DIPOLE

9. An object 1.5 μm long has a dipole moment of 25×10^{-14} C$\cdot$m. What is the magnitude of each charge in the dipole?

10. A dipole of moment 0.85×10^{-15} C$\cdot$m is placed in a uniform field $E = 4.3$ N/C, with $\vec{p}$ perpendicular to $\vec{E}$. What torque acts on the dipole?

Questions and Problems

§24.1 USING GAUSS' LAW TO CALCULATE ELECTRIC FIELDS

11. ❖ One half of a sphere carries a charge $+Q$ uniformly distributed on its surface. The other half of the sphere has a charge $-Q$. Can you use Gauss' law to find the electric field produced by the sphere? Why or why not?

12. ❖ A cube carries a charge Q uniformly distributed throughout its volume. Can you use Gauss' law to find the electric field produced by the cube? Why or why not?

13. ◆ ✉ In good weather, the electric field near the surface of the Earth is about 100 N/C and points inward. Estimate the net charge on the Earth's surface. If the charge is uniformly distributed, how much is there on each square meter of surface? Assume the Earth is a perfect sphere.

14. ◆ Two concentric spherical surfaces with radii R_1 and R_2 each carry a total charge Q. What is the electric field between the two shells?

15. ◆◆ Two concentric plastic spherical shells carry uniformly distributed charges, Q on the inner shell and $-Q$ on the outer shell. Find the electric field (a) inside the smaller shell, (b) between the shells, and (c) outside the larger shell.

16. ◆◆ Three concentric glass spherical shells each carry a charge Q, uniformly distributed over the shell. The radii of the shells are $a < b < c$. Find the electric field everywhere. (*Hint:* There are four distinct regions.)

§24.2 THE ELECTRIC FIELD DUE TO A LINEAR CHARGE DISTRIBUTION

17. ❖ What is the electric field at the center of an equilateral triangle made out of three uniformly charged rods, each carrying the same charge density λ?

18. ❖ An infinite, uniformly charged filament is perpendicular to the plane of a uniformly charged circular filament and through its center. What is the total force each filament exerts on the other? Does the answer depend on the signs or magnitudes of their charge densities?

19. ◆ A long glass filament carries a charge density $\lambda = +1.34\ \mu C/m$. If two 1.0-mm-long pieces are cut from the filament, how much charge does each piece have? What force would the segments exert on each other if they were held 1.0 m apart?

20. ◆ A long glass filament carries a charge density $\lambda = -2.7$ nC/m. What is the magnitude of the electric field 0.67 mm from the filament?

21. ◆ A circular filament of radius $R = 0.65$ m has a total charge of 6.3 μC. What electric field does it produce on its axis a distance R from its center? Compare with the field produced on its bisector at a distance $\sqrt{2}\ R$ by a straight filament of length $2\pi R$ with the same total charge.

22. ◆ The tube of a Geiger counter is 15 cm long and has a radius of 1.2 cm. The electric field near the outer wall is 1.5×10^4 N/C. If electrons are attracted toward the central filament, what is the linear charge density on the filament? What is the total charge on the filament?

23. ◆◆ Two uniformly charged infinite filaments are parallel and separated by a distance d. If each has charge density λ, find the force per unit length each exerts on the other.

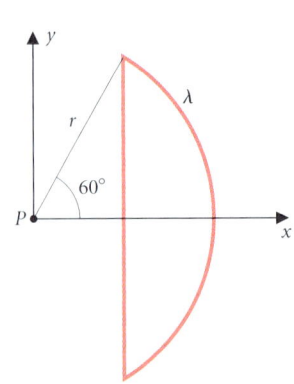

■ FIGURE 24.23

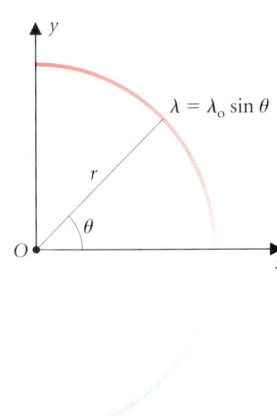

■ FIGURE 24.24

24. ◆◆ A point charge Q is on the axis of a uniformly charged circular filament with charge density λ and radius a. If the point charge is a distance z from the center of the filament, show that the force it exerts on the filament is:

$$\vec{F} = \left(\frac{2\pi k a \lambda Q z}{[z^2 + a^2]^{3/2}},\ \text{along the axis of the filament}\right).$$

25. ◆◆ A uniformly charged rod of length ℓ with charge density λ lies along the x-axis, with its midpoint at the origin. Find the electric field at a point on the x-axis, with $x > \ell/2$.

26. ◆◆ A semi-infinite filament with uniform charge density λ lies along the positive x-axis. Find the electric field at points on the y-axis.

27. ◆◆◆ A uniformly charged filament is in the shape of a circular arc and a chord (■ Figure 24.23). Find the electric field at the center of the circle (point P).

28. ◆◆◆ Two semicircular filaments have a common diameter. One lies in the x-y-plane and one in the x-z-plane. The curved portion of each carries a uniform line charge density λ. Find the electric field at the common center of the semicircles.

29. ◆◆◆ A semicircular piece of wire carries a charge density $\lambda = \lambda_o \sin \theta$, where θ is zero at the midpoint of the wire (■ Figure 24.24). Find the electric field at the center of the circle (point O).

30. ◆◆◆ Two semi-infinite filaments lie along the positive x- and y-axes, respectively. Each carries the same charge density λ. Find the electric field at a point on the diagonal (■ Figure 24.25).

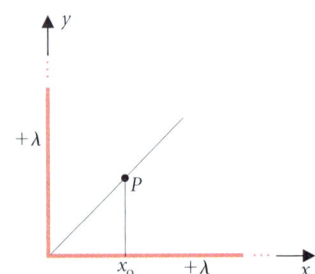

■ FIGURE 24.25

808 CHAPTER 24 • STATIC ELECTRIC FIELDS

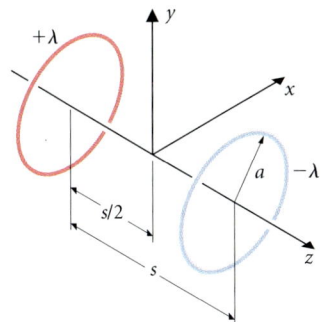

■ FIGURE 24.26

31. ◆◆◆ Two uniformly charged circular filaments with charge densities $+\lambda$ and $-\lambda$ are parallel, have a common axis, and are separated by a distance s (■ Figure 24.26). Find the electric field vector at points on the axis between the filaments as a function of z. Show that $d\vec{E}/dz = 0$ at $z = 0$.

§24.3 ELECTRIC FIELDS DUE TO SURFACE AND VOLUME CHARGE DISTRIBUTIONS

32. ❖ A regular tetrahedron of side L carries a total positive charge Q uniformly distributed over its surface. At point P, at very large distance from the object ($D \gg L$), the electric field due to the object is:
(a) kQ/L^2. **(b)** kQ/D^2. **(c)** kQD/L^2. **(d)** kQL/D. **(e)** kQD/L^3. Explain what is wrong with the answers you don't choose.

33. ❖ A regular tetrahedron carrying a total positive charge Q uniformly distributed over its surface has one of its faces in the x-z-plane (■ Figure 24.27). At point R just below the center of this face, the direction of the electric field is:
(a) undetermined, since the field is zero.
(b) in the negative y-direction.
(c) in the positive x-direction.
(d) in the negative z-direction.
(e) impossible to determine without a detailed calculation.
Explain why you chose your answer.

34. ❖ Find the electric field on the axis near the center of a very long, uniformly charged cylindrical tube.

35. ❖ Two uniformly charged spherical shells have radii R_1 and R_2, with $R_1 - R_2 \ll R_1$. At a point P between the shells they look like infinite planes, but the electric field is not the same as that between two infinite charged planes. Why? Find a model for the sphere, using infinite planes, that correctly describes the field between the shells near point P.

36. ◆ A uniformly charged glass plate carries surface charge density $\sigma = 4.3$ nC/m^2. If a circular disk of radius $R = 3.2$ cm is cut from the plate, what charge does the disk carry?

37. ◆ What is the electric field 0.35 cm above the surface of the glass plate described in Problem 36?

38. ◆ A glass sphere of radius 0.025 cm is bombarded with ions so as to have a total charge of 7.6 pC uniformly distributed throughout its volume. What is the sphere's charge density? What is the electric field 0.05 mm above the surface of the sphere?

39. ◆◆ The cylindrical drum of a copy machine is 40 cm long and has a radius of 5 cm. If charge is uniformly distributed over the curved surface of the drum, and the electric field near the surface is 2×10^5 N/C, estimate the amount of charge on the drum.

40. ◆◆ Two parallel, infinite planes are separated by a distance d. Find the electric field everywhere **(a)** if both planes carry a surface charge density σ and **(b)** if one plane has charge density σ and the other $-\sigma$. (This problem has application in studying capacitors, Chapter 27.)

41. ◆◆ Two uniformly charged infinite planes with equal charge densities σ intersect at right angles. Find the electric field everywhere.

42. ◆◆ A thick spherical shell with inner radius R_1 and outer radius R_2 has a uniform charge density ρ. What is the total charge on the shell? Use Gauss' law to find the electric field everywhere. Sketch the magnitude of the field as a function of distance r from the center of the spheres.

43. ◆◆ A very long cylindrical shell with inner radius a and outer radius b carries a uniform charge density ρ. Use Gauss' law to find the electric field produced by the shell in the regions $r < a$, $a < r < b$, and $r > b$.

44. ◆◆◆ Find the electric field due to an infinite plane by treating the plane as a collection of infinite, uniformly charged filaments, and using superposition.

45. ◆◆◆ Find the electric field on the z-axis due to a disc of radius a lying in the x-y-plane with center at the origin and carrying a charge density $\sigma = \sigma_0(r/a)^2$.

46. ◆◆◆ An infinite plane slab parallel to the x-y-plane occupies the region $-w < z < w$. The charge density in the slab varies with z according to the formula $\rho = \rho_0|z|/w$, for $|z| < w$. Find the electric field produced by the slab for $|z| > w$ and for $|z| < w$.

47. ◆◆◆ Find the electric field everywhere due to the slab of charge in Problem 46 if the charge density is given by $\rho = \rho_0(z/w)$, for $|z| < w$.

48. ◆◆◆ A sphere of radius a has a variable charge density $\rho(r) = \rho_0(r/a)^2$. Find the electric field everywhere inside and outside the sphere. Plot $|\vec{E}(r)|$ versus r. Make a sketch of the field line diagram, showing where field lines begin.

49. ◆◆◆ A sphere of radius R has uniform charge density ρ except within a spherical hole of radius $R/2$, where $\rho = 0$ (■ Figure 24.28). Use the result of Exercise 24.4 to show that $\vec{E}$ is uniform inside the hole, and find its magnitude.

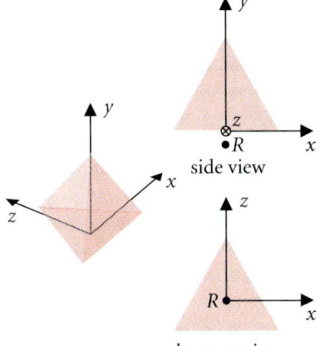

■ FIGURE 24.27

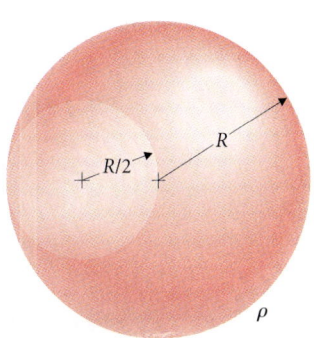

■ FIGURE 24.28

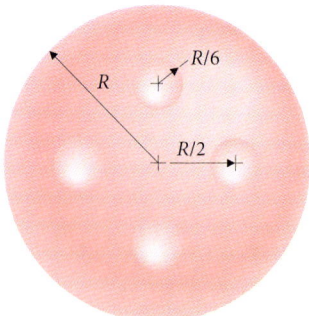

■ FIGURE 24.29

50. ◆◆◆ A sphere of radius R has uniform charge density ρ except within four spherical holes of radius $R/6$. The centers of all five spheres lie in the same plane (■ Figure 24.29). What is the electric field at the center of the large sphere? What is the electric field at the center of one of the holes? (*Hint:* Use the result of Exercise 24.4.)

51. ◆◆◆ An infinitely long cylindrical object with radius R has a charge distribution that depends on distance from the axis according to the formula:

$$\rho = Ar + Br^2 \qquad (r \le R).$$

Apply Gauss' law to find the electric field produced by the object. What ratio A/B results in zero field outside the object? For this ratio, sketch plots of ρ and $|\vec{E}|$ as functions of r.

§24.4 MOTION OF CHARGES IN AN ELECTRIC FIELD

52. ❖ If a charged particle is released in a region with an electric field $\vec{E}$, will the particle move along a field line? Why or why not?
53. ❖ An electron is set in motion in the x-direction at a distance d above an infinite sheet with surface charge density $+\sigma$ (■ Figure 24.30). The electron will:
(a) continue to move in the x-direction with a uniform speed.
(b) be accelerated in the x-direction.
(c) be decelerated and finally move off in the $-x$-direction.
(d) be accelerated in the $+y$-direction.
(e) be accelerated in the $-y$-direction.
Explain what is wrong with the answers you don't choose.
54. ❖ An electron is originally at point P, 1.0 m away from a very large uniformly charged plane surface. The acceleration of the electron is (a, away from the plane). Later, the electron is at point R, 2.0 m away from the surface. Its acceleration has magnitude:
(a) $2a$.   **(b)** $a/2$.   **(c)** $4a$.   **(d)** a.   **(e)** $a/4$.
Explain how you chose your answer.
55. ❖ Three charges of equal magnitude are held at the corners of a triangle as shown in ■ Figure 24.31. If the negative charge is released, describe its subsequent motion qualitatively.
56. ◆ A charged particle of mass 5.0 g is placed in an electric field of 10.0 N/C. It is observed to accelerate antiparallel to the field at 8.0 m/s². What is the charge on the particle?
57. ◆ Two identical charged particles of mass 5.0 g and charge 17 nC are held 0.10 m apart and released. What is the initial acceleration of each particle?
58. ◆ If the mass of an electron is 9.11×10^{-31} kg, find the electric field necessary to accelerate the electron at $a = 1.76 \times 10^{10}$ m/s². The mass of a proton is 1.67×10^{-27} kg. What is its acceleration in the same electric field?
59. ◆ At a certain time an electron is observed to accelerate to the right at $a_e = 1.6 \times 10^5$ m/s². At the same time a charge is known to be 1.0 m to the left of the electron. If the unknown charge produces the only force on the electron, what is the value of the charge?
60. ◆◆ What is the acceleration of an electron in an electric field of 10^3 N/C? What distance does the electron travel in such a field before its speed equals one-tenth the speed of light? How far would a proton have to travel to reach the same speed? [At faster speeds the laws of special relativity (Chapter 34) are necessary for accurate results.]
61. ◆◆ An object with unknown charge q and mass $m = 1.0$ g moves in a circle about a very much more massive object with charge $Q = 1.0\ \mu$C. The distance between the two is $r = 2.0$ cm, and the smaller object moves at speed $v = 5.0$ m/s. What is q?
62. ◆◆ An infinite plane surface carries a uniform charge density $\sigma = +20.0$ nC/m². An electron emerges from a small hole in the surface with a speed of 2.7×10^7 m/s at an angle of 60° from the normal. Describe the motion of the electron. What time interval passes before the electron returns to the surface? At what distance from the hole does it strike the surface?
63. ◆◆ If a proton emerges from the hole in Problem 62 with the same initial velocity, find the position of the proton as a function of time.
64. ◆◆ An electron is released 1.0 cm above a uniformly charged infinite plate with charge density 1.0×10^{-9} C/m². What is the speed of the electron when it hits the plate?
65. ◆◆ A particle of mass 1.5 g is moving in a circular path around a filament carrying a charge density of 1.0 μC/m. The charge on the particle is -15 nC. Find the speed of the particle.
66. ◆◆ Three parallel, infinite, uniformly charged planes are arranged as shown in ■ Figure 24.32. A small hole passes through the middle plane. At $t = 0$ an electron emerges from the hole moving perpendicular to the planes with speed v_o. Assuming v_o is small enough that the electron does not collide with the negative plate, show that its motion is periodic, and find the period.
67. ◆◆ In an ink-jet printer, drops of ink are given a controlled amount of charge before they pass through a uniform electric field generated by two charged plates (■ Figure 24.33). If the electric field between the plates is 1.5×10^6 N/C over a region 1.7 cm long, and drops of mass 0.12 ng enter the electric field region moving horizon-

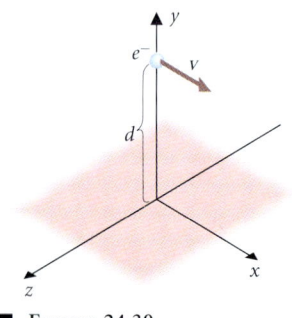

■ FIGURE 24.30

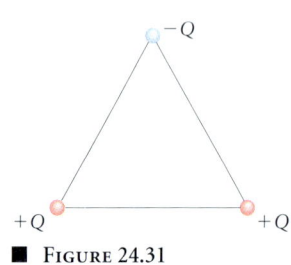

■ FIGURE 24.31

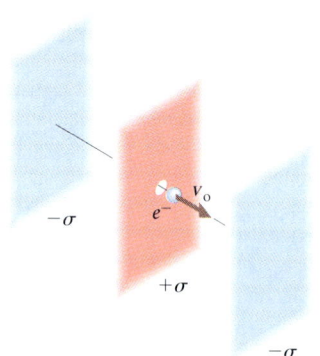

■ FIGURE 24.32

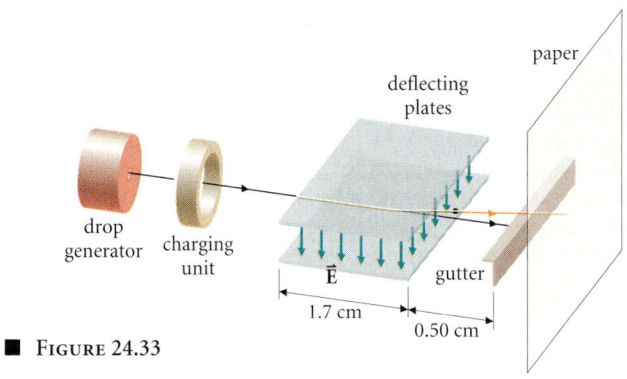

Figure 24.33

tally at 17 m/s, how much charge on a drop is necessary to produce a vertical deflection of **(a)** 0.50 mm and **(b)** 0.75 mm at the end of the electric field region? What is the deflection at the position of the paper, 0.50 cm from the end of the charged plates, in each case? Ignore other forces such as gravity and air resistance, and assume the field drops abruptly to zero at the edge of the plates.

68. ♦♦♦ In a cathode-ray tube used in an oscilloscope, an electric field $\vec{E}_o$ parallel to the tube accelerates the electrons to a speed v (cf. Example 24.8). A second electric field $\vec{E}$ perpendicular to the tube deflects them sideways (■ Figure 24.34). Find the deflection angle θ in terms of v, E, and the length of the electric field region, ℓ. For a tube with $\ell = 10$ cm, and $v = 9 \times 10^7$ m/s, how large an electric field is required to give a deflection of 40°?

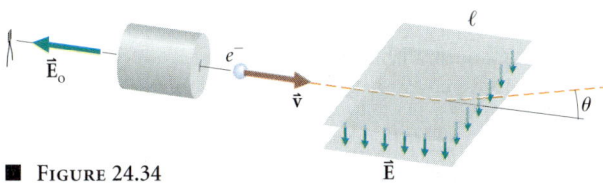

Figure 24.34

§24.5 THE DIPOLE

69. ❖ A dipole is placed near a long, uniformly charged filament. Describe the orientation the dipole achieves if it can rotate about its center, and the resulting oscillations damp.

70. ❖ A charge configuration consists of three equally spaced charges on a line. The central charge is $+2Q$; the other two have charge $-Q$ and are at a distance a on either side of the center. Draw a diagram showing the forces acting on the charge configuration in a uniform external electric field. Show that the total force and total torque on the configuration are both zero.

71. ❖ A circular filament lying in the x-y-plane has charge density $\lambda = \lambda_o \sin \theta$ (■ Figure 24.35). What is the direction of the electric field at each of the points A, B, C, and D?

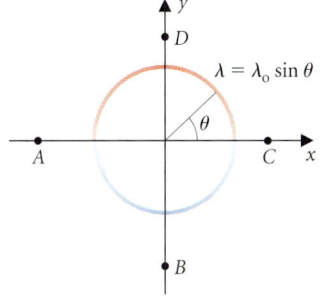

Figure 24.35

72. ♦ A needle 2.0 mm long has a charge of 1.0 nC at one end and -1.0 nC at the other. Find the dipole moment of the needle.

73. ♦ The needle described in Problem 72 lies along the x-axis with its center at the origin. Find the electric field on the x-axis at $x = 33$ cm.

74. ♦ A needle suspended from a string hangs horizontally. The electric field at the needle's location is horizontal with magnitude 3.7×10^3 N/C and is at an angle of 30° with the needle. There is no net electrical force acting on the needle, but the string exerts a torque of 3.7×10^{-4} N·m to hold the needle in equilibrium. What is the needle's dipole moment?

75. ♦♦ A dipole of moment 9.0 nC·m is placed in a region where the electric field is 1.5 N/C. If the dipole may be modeled as a rod 5.0 mm long with a mass of 1.0 g, find the period of small oscillations about the equilibrium position.

76. ♦♦♦ A dipole of moment $\vec{p}$ is placed near a point charge Q. Find the force and torque on the dipole as a function of distance r from the charge and of the orientation of the dipole.

Additional Problems

77. ❖ Two uniformly charged spherical plastic shells of diameter D each carry charge Q. They are separated by a distance $3D$ center to center. What is the electric field halfway between the shells? At the center of each shell?

78. ♦♦ A uniformly charged filament with charge density λ is parallel to a uniformly charged plane, with density σ, a distance s away. At what positions would an electron be in static equilibrium? Is the equilibrium stable? Would a proton be in stable equilibrium?

79. ♦♦ ✉ A thin, square sheet of side ℓ carries a total charge Q uniformly spread along its surface. Use Gauss' law to find the electric field *very near* the center of the square. From Coulomb's law obtain the magnitude of the electric field at *very large* distances from the square. By comparing these two approximations, estimate the distance from the square at which the infinite plane approximation becomes invalid.

80. ♦♦ In an experiment to measure charge, a small water droplet 8.41 μm in diameter falls through an evacuated region in which there is an electric field 1.10×10^6 N/C downward. A laser beam system measures the times when the drop passes successive slits 2.00 cm apart to be 0.0000 s, 0.0179 s, 0.0330 s, and 0.0463 s. What is the acceleration of the drop? What is the charge on the drop in units of e? What would the acceleration of the drop be if the field were reversed?

81. ♦♦♦ A classical model of an atom has electrons in orbit about a positively charged nucleus. Consider an electron in a circular orbit about a proton. Find the radius of the orbit in terms of the angular momentum of the electron.

82. ♦♦♦ ✉ The machine shown in ■ Figure 24.36 uses two oppositely charged sheets to support the block against gravity. Using the

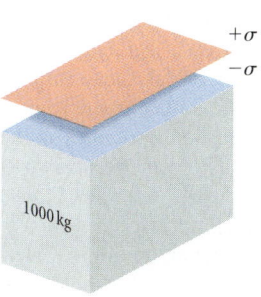

Figure 24.36

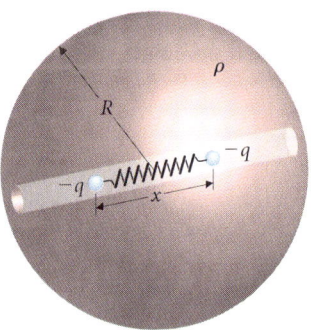

Figure 24.37

result for an electric field due to an infinite plane to approximate the electric field due to each sheet, what charge density is required if the sheet is 1 m on a side and the block has a mass of 1000 kg?

83. ◆◆◆ A small hole is drilled through a sphere with uniform charge density ρ. Two charges $-q$ are attached to a spring of constant k and zero relaxed length (■ Figure 24.37). Find the equilibrium separation x of the charges. Consider both $x < 2R$ and $x > 2R$.

84. ◆◆◆ An infinite plane slab of material of thickness $2s$ contains a variable charge density $\rho = \rho_o(x^2/s^2)$ (■ Figure 24.38). Apply Gauss' law to find the electric field due to the slab.

85. ◆◆◆ A uniform rod with mass $M = 0.75$ kg, charge density $\lambda = 1.5\ \mu$C/m, and length $L = 1.2$ m is attached with a frictionless hinge to an infinite, vertical plane surface with uniform charge density $\sigma = 85\ \mu$C/m². At what angle with the vertical does the rod hang? What force does the hinge exert?

86. ◆◆◆ A very thin hole is cut along a diameter of an otherwise uniformly charged sphere with radius $R = 0.20$ m and total charge $Q = 1.0\ \mu$C. If an electron is released from rest at one end of the hole, show that its motion is simple harmonic, and find the period.

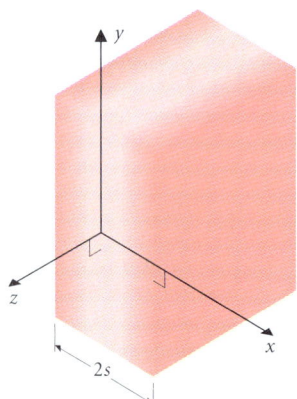

Figure 24.38

Computer Problems

87. A line charge of length 1.00 m with $\lambda = 1.00$ nC/m lies along the x-axis with its center at the origin. Divide the line into 20 segments each of length 0.05 m. Compute the x- and y-components of the electric field produced by each segment at a point P_1 on the y-axis at $y_1 = 0.50$ m. Use the trapezoidal rule to sum the contributions to find the electric field at P_1, and compare your result with the result of Example 24.3. Now compute the x- and y-components of the electric field at points P_2 and P_3 with coordinates $x_2 = 0.25$ m, $x_3 = 0.50$ m, $y_2 = y_3 = 0.50$ m.

88. A circular filament of radius 1.0 m with $\lambda = 1.0$ nC/m lies in the x-y-plane with its center at the origin. Calculate the components of the electric field at the point P with coordinates $x = -1.0$ m, $y = 0.0$ m, $z = +1.0$ m.

89. Two identical dipoles with $\vec{p} = (1.0\ \mu\text{C}\cdot\text{m})\hat{j}$ are located at the origin and at a distance $r = 5.0$ m from the origin. Calculate the force exerted on the second dipole by the first, when the line from the first to the second makes an angle of **(a)** 30°, **(b)** 60°, and **(c)** 90° with the x-axis. *Hint:* Model each dipole as two point charges of magnitude 10 μC separated by a distance 0.10 m. Compare your result for (c) with the result of Example 24.12.

Challenge Problems

90. A rod of length 2ℓ lies along the x-axis with its center at the origin. The rod carries a charge density $\lambda = \lambda_o(x/\ell)$. Find the electric field at a point on the y-axis.

91. Show that the electric field at an arbitrary point P with polar coordinates r and θ due to a dipole at the origin is:

$$\vec{E} = \frac{kp}{r^3}[\sin\theta\ \hat{\theta} + 2\cos\theta\ \hat{r}].$$

92. Show that a displacement along a field line satisfies the relation:

$$\frac{dr}{d\theta} = \frac{rE_r}{E_\theta},$$

where r and θ are polar coordinates in the plane containing the field line. Using the result of Problem 91, obtain a differential equation for the field lines due to a dipole at distances $r \gg$ the length ℓ of the dipole. Show that each field line satisfies the equation:

$$r = r_m \sin^2\theta,$$

where r_m is the maximum distance of the field line from the origin.

93. A hemispherical shell has a uniform charge density σ spread across its surface. What is the value of the electric field at point P (■ Figure 24.39)? If the electric force on a point charge q placed at P just balances gravity, what is the mass of the particle? Is the equilibrium stable?

94. A quadrupole consists of four charges at the corners of a small square of side ℓ. Two positive charges are at the ends of one diagonal, and two negative charges are at the ends of the other diagonal. Find the electric field due to this charge distribution **(a)** at a point on the perpendicular through the center of the square and **(b)** at a point in the plane of the square on the bisector of one side. Can you find a reasonable definition of *quadrupole moment*?

95. Find the electric field at a point P at distance y above the center line of a uniformly charged, infinitely long strip of width L (■ Figure 24.40). The strip has charge density σ.

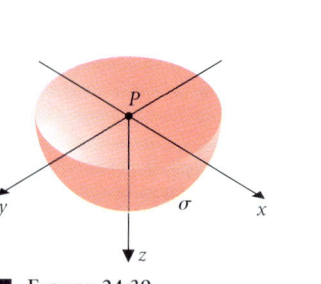

Figure 24.39

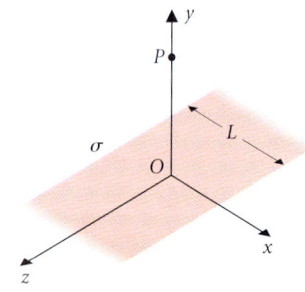

Figure 24.40

CHAPTER 25
Electric Potential Energy

CONCEPTS
Electric potential energy
Electric potential
Equipotential surface
Conductor
Electron volt

GOALS
Be able to:
Compute the potential energy of a system of point charges and use the results in energy conservation problems.

Calculate the potential due to a discrete or continuous charge distribution.

Sketch the equipotential surfaces for a given charge distribution.

Understand:
The relation between electric potential and the electric field vector.

The use of potential in explaining the behavior of conductors.

In the Boston Museum of Science, operator Don Salvatore sits in a metal cage. The giant towers behind him are Van de Graaff generators. The large spheres at their tops are charged up to 2.5 million volts. The electric fields that surround them are great enough to ionize the air, providing current paths that allow sparks to jump to nearby wires and smaller metal spheres. Inside his cage, Mr. Salvatore is protected from the artificial lightning storm that surrounds him. In this chapter we'll investigate properties of conductors that explain how the cage can protect its occupant (§25.6). We introduce electric potential—measured in volts—and see how it helps us to understand situations like this. (Robert J. Van de Graaff built this model in 1931.)

I sometimes fancy that such nearness to nature as I have described keeps the spirit sensitive to impressions not commonly felt, and in touch with unseen powers.

OHIYESA

Study hint: Review the four properties of potential energy in §8.1.

$\mathcal{E}$lectric energy provides spectacular entertainment in museum displays that snap and crackle like miniature lightning storms. Huge sparks look and sound scary, but the operator is quite safe in his cage. The display relies on the ability of electrons to flow within the bars of the metal cage and on the fact that the sparks transmit very little total charge though impressive amounts of energy per electron.

We have already discovered energy in several different forms: kinetic energy, gravitational and elastic potential energy, and thermal energy. Electric potential energy—our next addition to this list—is the key to understanding such diverse phenomena as chemical interactions and high-voltage power transmission as well as sparks in museums. In this chapter, we shall introduce electric potential energy and explore its relation to the electric field vector $\vec{E}$.

25.1 Potential Energy of a Pair of Point Charges

25.1.1 Work Done by the Coulomb Force

In Chapters 23 and 24 we discussed the electric field, described by the vector $\vec{E}$, and the force $\vec{F} = Q\vec{E}$ it exerts on an electrically charged particle. The force is a function of position only and is similar in form to the gravitational force. Also like the gravitational force, the Coulomb force is conservative (cf. §8.3), and the system of charged particles and electric field possesses electric potential energy. If a particle is released, the electric force causes it to accelerate and gain kinetic energy at the expense of the system's potential energy.

These are the decay products of radon.

EXAMPLE 25.1 ♦♦ A polonium nucleus (charge $Q_1 = 1.3 \times 10^{-17}$ C) and an alpha particle (charge $Q_2 = 3.2 \times 10^{-19}$ C) are a distance $d_1 = 9.1 \times 10^{-15}$ m apart. Assume the polonium is fixed and the alpha particle is free to move. Find the work done on the alpha particle when it moves to a new position at distance $d_2 = 2d_1$ from the polonium nucleus (■ Figure 25.1). If the alpha particle is initially at rest, find its speed at its new position.

MODEL The alpha particle accelerates away because the electric field $\vec{E}_1$ of the polonium nucleus exerts a repulsive force on it. Since the force depends on position, we find the work done as an integral of the differential amounts of work done during each differential displacement of Q_2. The speed of the alpha particle can be found by setting the work done on it equal to the change in its kinetic energy.

See Interlude 2 and §8.2.2.

SETUP The force exerted by $\vec{E}_1$ when Q_2 is a distance r from Q_1 is:

$$\vec{F}(r) = \frac{kQ_2Q_1}{r^2}\,\hat{r}.$$

STEPS I AND II The path of Q_2 between its initial and final positions is along a radial line. The appropriate coordinate is the distance r, and a differential displacement is described by: $d\vec{r} = dr\,\hat{r}$.

STEP III The work done on Q_2 during a differential displacement is:

$$dW = \vec{F} \cdot d\vec{r} = \frac{kQ_2Q_1}{r^2}\,\hat{r} \cdot (dr\,\hat{r}) = \frac{kQ_2Q_1\,dr}{r^2}.$$

STEPS IV AND V The particle moves from $r = d_1$ to $r = d_2$, so:

$$W \equiv \int dW = \int_{d_1}^{d_2} \frac{kQ_2Q_1\,dr}{r^2} = kQ_2Q_1\left(-\frac{1}{r}\right)\Big|_{d_1}^{d_2} = kQ_2Q_1\left(-\frac{1}{d_2} + \frac{1}{d_1}\right). \quad (25.1)$$

SOLVE
$$W = \frac{(9.0 \times 10^9 \text{ N·m}^2/\text{C}^2)(1.3 \times 10^{-17}\text{ C})(3.2 \times 10^{-19}\text{ C})}{(9.1 \times 10^{-15}\text{ m})}\left(-\frac{1}{2} + \frac{1}{1}\right)$$
$$= 2.1 \times 10^{-12} \text{ J}.$$

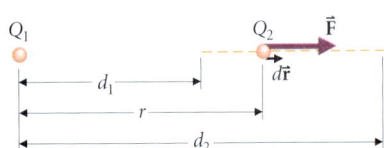

■ **FIGURE 25.1**
The electric field due to a fixed polonium nucleus exerts a force on the nearby alpha particle. The polonium nucleus does work on the alpha particle, which accelerates away.

SETUP Using the work-energy theorem:

$$\tfrac{1}{2}mv_f^2 - \tfrac{1}{2}mv_i^2 = W.$$

With $v_i = 0$, $\quad v_f = \sqrt{2W/m}$.

SOLVE The mass m of the alpha particle $= 4.0$ u $= 6.6 \times 10^{-27}$ kg, so:

$$v_f = \sqrt{\frac{2(2.1 \times 10^{-12} \text{ J})}{6.6 \times 10^{-27} \text{ kg}}} = 2.5 \times 10^7 \text{ m/s}.$$

ANALYZE The final velocity is 8% of the speed of light. This is another indication of the strength of the electric force. This example provides an approximate model for the nuclear decay of radon.

25.1.2 The Electric Force as a Conservative Force

The Coulomb force is conservative. That is, the work done moving a test charge between two points is independent of the path. ■ Figure 25.2a shows an arbitrary path between two points at distances d_1 and d_2 from a point charge Q. The work done by the electric field on a test charge q when it undergoes a displacement $d\vec{\ell}$ is:

$$dW = \vec{F}_e \cdot d\vec{\ell} = q\frac{kQ}{r^2}\hat{r} \cdot d\vec{\ell}.$$

From Figure 25.2b, $\hat{r} \cdot d\vec{\ell} = dr$, and $dW = q\dfrac{kQ\, dr}{r^2}$.

This is the same expression for dW that we obtained in Example 25.1, where we calculated the work done along a radial path. Thus eqn. (25.1) also gives the work done in separating two charges from d_1 to d_2 along an *arbitrary* path.

$$W = kqQ\left(\frac{1}{d_1} - \frac{1}{d_2}\right). \tag{25.2}$$

25.1.3 Potential Energy of a Pair of Charges

The work done by a conservative force decreases the *potential* energy of the system:

$$W = U(d_1) - U(d_2).$$

REMEMBER: ONLY CHANGES OF POTENTIAL ENERGY ARE MEASURABLE.

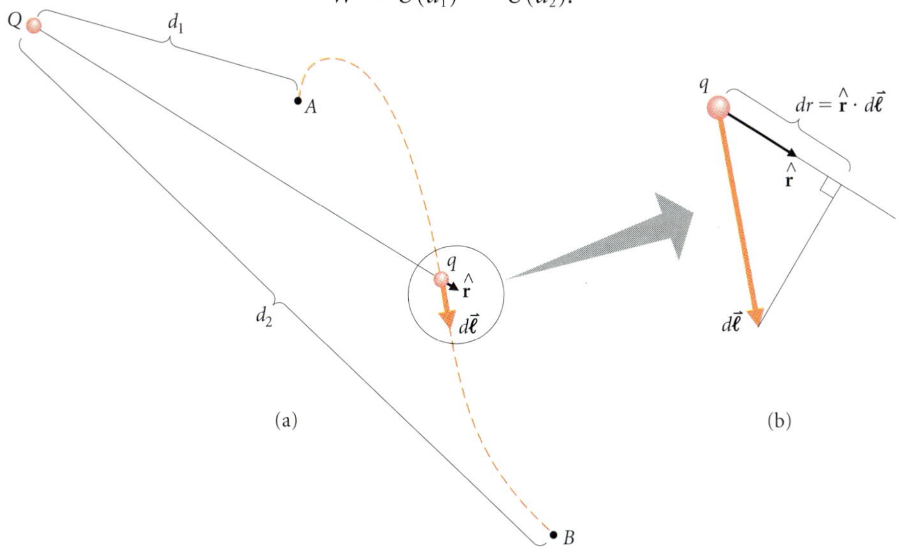

■ **FIGURE 25.2**
The Coulomb force is conservative; that is, the work done by the Coulomb force on a charge that moves from A to B is independent of the path taken between the two points. Since the field points radially outward, the work done in a differential displacement $d\vec{\ell}$ depends only on the dot product $\hat{r} \cdot d\vec{\ell} = dr$, the change in distance from Q_1.

Comparing this equation with eqn. (25.2), we find the potential energy as a function of the distance between a pair of charges:

$$U(d) \equiv \frac{kqQ}{d} + \text{arbitrary constant}.$$

As usual, the potential energy function contains an arbitrary constant because we may choose where to set the function equal to zero for *our convenience* (cf. §8.2). It is customary to set the energy function to zero when the two particles are infinitely far apart. In that case the arbitrary constant is zero, and the potential energy function is:

THE SAME CONVENTION IS USED FOR GRAVITATIONAL POTENTIAL ENERGY. SEE §8.2.

$$U(d) = \frac{kqQ}{d}. \tag{25.3}$$

Recall that a *system* possesses potential energy, not its individual particles. The system here is the pair of charges, and the potential energy we have calculated is that of a pair of charges separated by distance d. If instead of only one charge moving, both are free to move, they both accelerate and share the energy of the system.

EXAMPLE 25.2 ◆◆ In the first stage of a controlled nuclear fusion reaction, nuclei of deuterium (D, mass 2.0 u) and tritium (T, mass 3.0 u) react to form a helium nucleus and a neutron. The D and T nuclei each have charge $+e$, so the Coulomb force between them is repulsive. Assume the center of mass of the two particles is at rest in the lab frame. How fast must each particle be moving as they approach each other from a great distance if they are to reach a minimum separation of 2.0×10^{-15} m?

MODEL We are given an initial state of the system and asked about a final state, without being asked how the system arrives at the final state or how much time it requires. These are the features that identify a conservation law problem (cf. §6.2). Both particles come to rest in their CM frame as they reach minimum separation. The initial kinetic energy of the system is converted to potential energy. The linear momentum of the system is zero throughout. The Coulomb force lies along the line between the particles, so the angular momentum of each particle about the CM remains zero and we need not consider it further. The final state of the system is illustrated in ■ Figure 25.3.

SETUP

	BEFORE		AFTER	
Energy:	Potential	Kinetic	Potential	Kinetic
	zero	$\tfrac{1}{2}M_D v_D^2 + \tfrac{1}{2}M_T v_T^2$	$\dfrac{kQ_D Q_T}{d}$	zero
Momentum (x-component):		$M_D v_D - M_T v_T$	zero	

Set values *before* equal to values *after*:

$$\tfrac{1}{2}M_D v_D^2 + \tfrac{1}{2}M_T v_T^2 = \frac{kQ_D Q_T}{d}. \tag{i}$$

$$M_D v_D - M_T v_T = 0. \tag{ii}$$

SOLVE From eqn. (ii):

$$v_D = (M_T/M_D)v_T.$$

Substituting in eqn. (i):

$$\frac{kQ_D Q_T}{d} = \tfrac{1}{2}M_T v_T^2 + \tfrac{1}{2}M_D\left(\frac{M_T v_T}{M_D}\right)^2 = \tfrac{1}{2}M_T v_T^2\left(1 + \frac{M_T}{M_D}\right).$$

Then:

$$v_T^2 = \frac{2kQ_D Q_T}{M_T d}\frac{M_D}{(M_T + M_D)}.$$

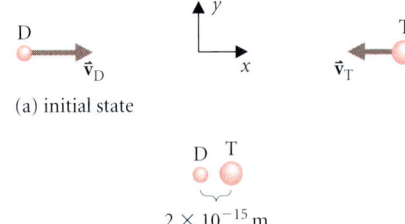

(a) initial state

(b) final state:

■ FIGURE 25.3
In nuclear fusion, a deuterium nucleus and a tritium nucleus combine to form helium. Since both reacting nuclei are positively charged, they repel each other. They must move toward each other at high speed to get close enough to react. We choose the x-axis along the line joining the two nuclei. (a) Nuclei separated by a large distance. (b) Nuclei at rest with a separation of 2 fm.

So: $v_T = \sqrt{\dfrac{2kQ_D Q_T}{M_T d} \dfrac{M_D}{(M_T + M_D)}}$ and $v_D = \sqrt{\dfrac{2kQ_D Q_T}{M_D d} \dfrac{M_T}{(M_D + M_T)}}$.

$$v_T = \sqrt{\dfrac{2(9.0 \times 10^9 \text{ N}\cdot\text{m}^2/\text{C}^2)(1.6 \times 10^{-19} \text{ C})^2}{(2.0 \times 10^{-15} \text{ m})(3.0)(1.66 \times 10^{-27} \text{ kg})} \dfrac{(2.0 \text{ u})}{(2.0 \text{ u} + 3.0 \text{ u})}}$$

$$= 4.3 \times 10^6 \text{ m/s}.$$

$$v_D = \dfrac{M_T}{M_D} v_T = \dfrac{3.0 \text{ u}}{2.0 \text{ u}} \, 4.3 \times 10^6 \text{ m/s} = 6.5 \times 10^6 \text{ m/s}.$$

ANALYZE Notice that we can use any unit for the mass in a ratio such as M_T/M_D, provided that we use the same unit in the numerator and the denominator. Since the speed of the particles in a gas increases as the square root of the temperature (Chapter 19), the deuterium/tritium gas must be very hot (at least 10^7 K) for the nuclei to achieve the necessary speeds.

25.2 ELECTRIC POTENTIAL

25.2.1 Potential Energy of a Charge in an Arbitrary Electric Field

In an oscilloscope, charges on the electron gun and the deflection plates produce electric fields that accelerate and guide the electron beam. The forces between different electrons in the beam are comparatively small, and we may think of the individual electrons as test charges moving in the fields of the tube. As each electron accelerates, it gains kinetic energy, and the system of electron plus electric field loses potential energy.

To find the potential energy of a charge in an arbitrary electric field, we begin by calculating the work done on the charge by the field. As the charge moves through a displacement $d\vec{\ell}$ the work done is:

$$dW = \vec{F} \cdot d\vec{\ell} = q\vec{E} \cdot d\vec{\ell}.$$

If the charge moves from P_1 to a new position P_2 (■ Figure 25.4), the work done on it by electric forces is:

$$W = q \int_{P_1}^{P_2} \vec{E} \cdot d\vec{\ell}. \tag{25.4}$$

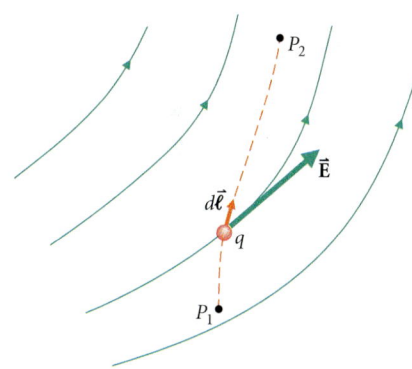

■ **FIGURE 25.4**
A charged particle moves from point P_1 to P_2 in a region where the electric field is $\vec{E}$, illustrated here by the field line diagram. The field does work on the particle, corresponding to a change in its potential energy.

The electric force is conservative, so W is independent of the path the electron travels between P_1 and P_2. This work is done at the expense of potential energy belonging to the system of the electron and the charges that produce the field:

$$-\Delta U \equiv -(U_2 - U_1) = W = q \int_{P_1}^{P_2} \vec{E} \cdot d\vec{\ell}.$$

Thus:

$$\Delta U = U_2 - U_1 = -q \int_{P_1}^{P_2} \vec{E} \cdot d\vec{\ell} = q \int_{P_2}^{P_1} \vec{E} \cdot d\vec{\ell}. \tag{25.5}$$

THIS DEVELOPMENT FOLLOWS THAT IN CHAPTER 8. COMPARE EQN. (25.5) WITH EQN. (8.3).

In the second form of this integral, we move along the path from P_2 to P_1 rather than from P_1 to P_2. The direction of each differential displacement $d\vec{\ell}$ is reversed, and therefore so is the sign of the integral. You may use whichever form is more convenient. The two expressions have slightly different physical interpretations: doing work *on* the system increases its potential energy; when the *system* does work, its potential energy decreases.

25.2.2 Electric Potential

We defined the electric field vector at any point as the force per unit charge exerted on a test charge placed at that point. Similarly, we may measure the change in energy of the system as a test charge moves between two points and interpret it in terms of a field quantity, called

IN CHAPTER 23, EQN. (23.5).

electric potential V, produced by the other charges. Like the field vector $\vec{E}$, potential describes the electric field itself and is independent of the test charge that moves through the field.

> The *difference in electric potential* ΔV between two points equals the change in potential energy of the system when a test charge is moved between the same two points, divided by the value of the charge.
>
> $$\Delta V = \lim_{q \to 0} \frac{\Delta U}{q}. \tag{25.6}$$

Using eqn. (25.5) to express ΔU in terms of $\vec{E}$, we have:

$$\Delta V \equiv V_{21} = V(P_2) - V(P_1) = \frac{U_2 - U_1}{q}.$$

$$\Delta V = -\int_{P_1}^{P_2} \vec{E} \cdot d\vec{\ell}. \tag{25.7}$$

A useful rule for remembering the signs in eqn. (25.7) is:

> The electric field vector points toward decreasing potential.

THE DIRECTION OF $\vec{E}$ NEED NOT BE ALONG A STRAIGHT LINE BETWEEN THE TWO POINTS. THE RULE GIVES THE GENERAL TREND.

Just as we have the freedom to choose an arbitrary constant in defining potential energy, we can also choose an arbitrary constant in defining potential. Relative to a reference point R where we choose to define the potential as zero, the electric potential is given by:

$$V(P) = -\int_R^P \vec{E} \cdot d\vec{\ell}. \tag{25.8}$$

The value of the potential at P is independent of the path taken between R and P. In deriving the expression for the potential energy of a pair of point charges, we chose to assign zero potential energy to a pair separated by an infinite distance. Similarly, a common choice for the reference point R is at *infinity*, particularly when dealing with a system of point charges. In laboratory practice, electrical *ground* is taken as the reference point. Connecting a conducting wire to the water pipes is a satisfactory way to construct a ground reference for simple electric circuits. The wiring in modern buildings includes a ground wire that is connected to a distant buried metal plate.

AFTER ALESSANDRO VOLTA (1745–1827), THE INVENTOR OF THE BATTERY.

Since potential is energy per unit charge, its SI unit is the joule per coulomb. This unit is given the name volt (symbol V):

$$1 \text{ V} \equiv 1 \text{ J/C}.$$

Combining the definition (eqn. 25.6) with eqn. (25.3) for the potential energy of a pair of point charges with R at infinity, we find the electric potential at a point P a distance d from a single point charge Q:

SPECIAL CASE: POTENTIAL AT DISTANCE d FROM A SINGLE POINT CHARGE, WITH REFERENCE POINT AT INFINITY.

$$V = k\frac{Q}{d}. \tag{25.9}$$

EXERCISE 25.1 ♦♦ Derive the formula $V = kQ/d$ for the potential produced by a single point charge from eqn. (25.8).

EXAMPLE 25.3 ♦ Find the potential V at a point P a distance $d = 30.0$ cm from a point charge $Q = -1.00 \, \mu\text{C}$.

MODEL We choose the reference point at infinity and use eqn. (25.9).

SETUP AND SOLVE

$$V = k\frac{Q}{d} = (8.99 \times 10^9 \text{ N·m}^2/\text{C}^2)\frac{(-1.00 \times 10^{-6} \text{ C})}{(0.300 \text{ m})}$$
$$= -3.00 \times 10^4 \text{ N·m/C} = -3.00 \times 10^4 \text{ V}.$$

ANALYZE It would require $qV = 3 \times 10^{-5}$ J to bring a -1-nC charge to this position from infinity. ∎

EXAMPLE 25.4 ◆◆ An electron moves from rest at P_1 with potential $V_1 = 9.0$ V to a point P_2 with potential $V_2 = 10.0$ V. What is the electron's speed when it reaches P_2?

MODEL Since the electron has negative charge, its motion results in a decrease of the system's potential energy and a corresponding increase in the electron's kinetic energy.

SETUP

	BEFORE		**AFTER**	
Energy:	Kinetic	Potential	Kinetic	Potential
	zero	$-eV_1$	$\frac{1}{2}mv^2$	$-eV_2$

Set the total energy *before* equal to its value *after*:

$$-eV_1 = \tfrac{1}{2}mv^2 - eV_2.$$

SOLVE

$$\tfrac{1}{2}mv^2 = e(V_2 - V_1).$$

$$v = \sqrt{2\frac{e(V_2 - V_1)}{m}} = \sqrt{\frac{2(1.6 \times 10^{-19} \text{ C})(10.0 \text{ V} - 9.0 \text{ V})}{9.1 \times 10^{-31} \text{ kg}}}$$
$$= 5.9 \times 10^5 \text{ m/s}.$$

ANALYZE Since the potential decreases between P_2 and P_1, according to the rule following eqn. (25.7), the electric field points generally from P_2 toward P_1. The negatively charged electron experiences a force opposite $\vec{\mathbf{E}}$, that is, toward P_2. ∎

The kinetic energy gained by the electron in Example 25.4 is a very convenient energy unit in electronics and atomic physics.

An *electron volt* (eV) is the work done on an electron that moves through a potential difference of one volt. Equivalently, 1 eV is the change in an electron's potential energy when it moves through a potential difference of one volt.

BECAUSE THEY ARE RELATED BY A FACTOR OF 1 V, THE NUMERICAL VALUE OF THE ELECTRON VOLT IN JOULES EQUALS THE NUMERICAL VALUE OF THE ELECTRON CHARGE IN COULOMBS.

$$1.00 \text{ eV} = 1.60 \times 10^{-19} \text{ J}. \qquad (25.10)$$

EXAMPLE 25.5 ◆◆◆ A battery maintains a potential difference of 10.0 V between two large parallel metal plates separated by a distance of 1.0 mm (■ Figure 25.5). Electrons emerge in all directions from a small hole in the positive (high-potential) plate. If the electrons all have the same initial speed $v_o = 2.0 \times 10^6$ m/s, but they emerge at different angles θ with the normal to the plates, for what range of angles θ can electrons reach the negative plate?

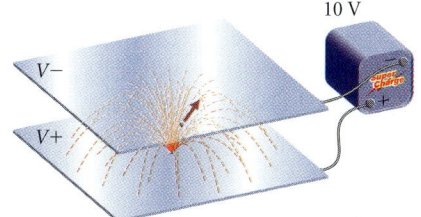

■ **FIGURE 25.5** (a)
A battery hooked up to two metal plates maintains a potential difference of 10.0 V between them.

MODEL The battery is designed to maintain a constant potential difference between the two plates. In this case, the negative plate is held at 10.0 V lower potential than the positive plate. If we choose the reference point at the positive plate, $V_+ \equiv 0$, then the potential at the negative plate $V_- = -10.0$ V.

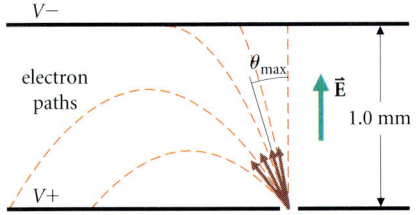

■ **FIGURE 25.5** (b) Electrons passing through a hole in the plate at higher potential follow parabolic paths as the electric field accelerates them back toward that same plate. Electrons that emerge at angle θ_{max} to the normal just make it to the other plate.

The electrons move on parabolic paths similar to that of a baseball accelerated by gravity (Figure 25.5b). The electric field between the plates decelerates the electrons, reducing the component of velocity perpendicular to the plates while leaving the velocity parallel to the plates unchanged. We choose the x-axis parallel to the plates and y perpendicular. Electrons leave the hole with $v_y = v_o \cos \theta$. An electron that just reaches the second plate arrives there with v_y equal to zero. This electron has the maximum possible exit angle, θ_{max}. Electrons with smaller exit angles leave the hole with a larger v_y and are able to reach the negative plate; those with larger exit angles do not. We may find θ_{max} with the conservation law technique.

SETUP The x-component of velocity, $v_o \sin \theta$, remains constant during the motion.

	BEFORE (AT POSITIVE PLATE)		**AFTER** (AT NEGATIVE PLATE) $v_y = 0$	
Energy:	Kinetic	Potential	Kinetic	Potential
	$\frac{1}{2}mv_o^2$	$-eV_+ = 0$	$\frac{1}{2}m(v_o \sin \theta_{max})^2$	$-eV_-$

Set value *before* equal to value *after*:

$$\tfrac{1}{2}mv_o^2 = \tfrac{1}{2}mv_o^2 \sin^2 \theta_{max} - eV_-.$$

SOLVE
$$\tfrac{1}{2}mv_o^2(1 - \sin^2 \theta_{max}) = -eV_-.$$

$$\cos^2 \theta_{max} = \frac{-2eV_-}{mv_o^2} = \frac{-2(1.6 \times 10^{-19} \text{ C})(-10.0 \text{ V})}{(9.1 \times 10^{-31} \text{ kg})(2.0 \times 10^6 \text{ m/s})^2} = 0.88.$$

$$\theta_{max} = 20°.$$

ANALYZE A 14% increase in the potential difference stops electrons from reaching the negative plate altogether ($\cos^2 \theta_{max}$ increases to 1.00, and θ_{max} becomes zero). A similar reduction in the potential difference more than doubles the number of electrons reaching the plate. This principle is used to control the intensity of electron beams. The brightness control of a TV tube is one example.

Notice how we used the freedom to choose a reference point to simplify the calculation. Here we chose $V = 0$ at the positive plate; the choice $R = \infty$ would not make much sense in this example.

25.2.3 Calculation of Field from Potential

ACTUALLY WE CAN ALWAYS CALCULATE THE FIELD IF WE KNOW ENOUGH MATH. SEE THE MATH TOPIC ON GRADIENTS.

Equation 25.8 allows us to calculate the potential difference between two points from a known electric field. Conversely, we can calculate the field from the potential. To use eqn. (25.8), we need to know the direction of the field and how it varies along the path.

EXAMPLE 25.6 ♦♦ Given that the electric field between the plates in Example 25.5 is uniform and perpendicular to the plates, find its magnitude.

COMPARE WITH EXAMPLE 24.5. THE ELECTRIC FIELD HERE IS THE SUPERPOSITION OF THE FIELD DUE TO TWO UNIFORM SHEETS.

MODEL We begin with the relation between the given potential difference and the electric field (eqn. 25.7). The integral may be carried out along any path between the plates, but it is simplest along a straight line perpendicular to the plates (■ Figure 25.6).

SETUP The electric field vector is antiparallel to each displacement $d\vec{\ell}$ along the chosen path and has constant magnitude. The integral simplifies accordingly:

RECALL HOW WE SIMPLIFY FLUX INTEGRALS WHEN APPLYING GAUSS' LAW. THE SAME KIND OF REASONING IS USED HERE.

$$\Delta V = -\int_-^+ \vec{E} \cdot d\vec{\ell} = -\int_-^+ -|\vec{E}| \, d\ell = |\vec{E}| \int_-^+ d\ell = |\vec{E}|d,$$

where $d = 1.0$ mm is the distance between the plates.

SOLVE
$$|\vec{E}| = \frac{\Delta V}{d} = \frac{V_+ - V_-}{d} = \frac{10.0 \text{ V}}{1.0 \times 10^{-3} \text{ m}} = 1.0 \times 10^4 \text{ V/m}.$$

ANALYZE These units look unfamiliar, but notice that:
$$\frac{\text{V}}{\text{m}} = \frac{\text{J/C}}{\text{m}} = \frac{\text{N} \cdot \text{m}}{\text{C} \cdot \text{m}} = \frac{\text{N}}{\text{C}}.$$

Volts per meter is a more common unit for electric field strength than newtons per coulomb.

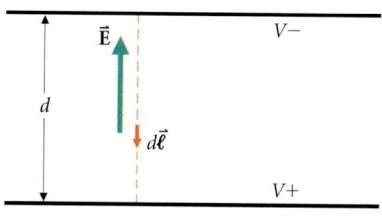

■ **FIGURE 25.6**
The electric field between the plates is known to be uniform and perpendicular to the plates. An integral of $-\vec{E} \cdot d\vec{\ell}$ along any path between the plates gives the potential difference. If we choose a path with $d\vec{\ell}$ antiparallel to $\vec{E}$, that is, perpendicular to the plates, the integral simplifies to $|\vec{E}|d = \Delta V$.

25.3 POTENTIAL ENERGY OF A SYSTEM OF CHARGES
25.3.1 The Principle of Superposition

Using the definition of potential in terms of the electric field vector $\vec{E}$, we may show that the principle of superposition (§23.3) may be restated as an addition law for potential. Suppose $\vec{E}$ is the sum of a finite number N of different contributions $\vec{E}_i$, $i = 1 \ldots N$. (For example, each contribution $\vec{E}_i$ could be the field due to a point charge, a line of charge, or a sheet of charge.)

$$\vec{E} = \sum_{i=1}^{N} \vec{E}_i.$$

Then:
$$V(P) = -\int_R^P \vec{E} \cdot d\vec{\ell} = -\int_R^P \sum_{i=1}^{N} \vec{E}_i \cdot d\vec{\ell} = \sum_{i=1}^{N} \left(-\int_R^P \vec{E}_i \cdot d\vec{\ell} \right).$$

The ith term in this sum is a separate contribution V_i to the potential at P due to the electric field $\vec{E}_i$ alone (eqn. 25.8). Thus:

$$V(P) = \sum_{i=1}^{N} V_i(P). \quad (25.11)$$

For a set of point charges, this result becomes:

$$V(P) = \sum_{i=1}^{N} V_i(P) = \sum_{i=1}^{N} \frac{kQ_i}{r_i}, \quad (25.12)$$

where r_i is the distance of P from Q_i.

The potential at a point due to an arbitrary, finite set of point charges is the sum of the potentials at that point due to the individual charges.

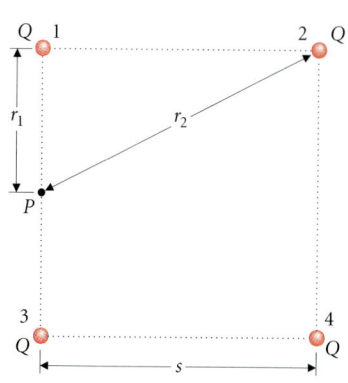

EXAMPLE 25.7 ♦♦ Find the potential at P, the midpoint of one side of a square of side s with a charge Q at each corner (■ Figure 25.7).

MODEL We use the principle of superposition. The potential at P is the sum of potentials produced by each of the charges. With the charges labeled as shown in the figure:
$$V(P) = V_1(P) + V_2(P) + V_3(P) + V_4(P).$$

SETUP To find the individual potential contributions, we need only the distances of the charges from P. For Q_1 and Q_3, the distances are $r_1 = r_3 = s/2$. For the other two charges:
$$r_2 = r_4 = \sqrt{s^2 + (s/2)^2} = \sqrt{5}s/2.$$

■ **FIGURE 25.7**
To find the potential at point P, we use the principle of superposition to sum the potentials due to the four charges at the corners of the square.

SOLVE $$V(P) = kQ\left(\frac{2}{r_1} + \frac{2}{r_2}\right) = \frac{kQ}{s}\left[2(2) + 2\frac{2}{\sqrt{5}}\right] = \frac{4kQ}{s}\left(\frac{5+\sqrt{5}}{5}\right) = 5.8\frac{kQ}{s}.$$

ANALYZE The principle of superposition for potential is easier to use than that for the electric field vector because potential is a scalar quantity.

EXERCISE 25.2 ♦♦ A dipole (§24.5) with moment $\vec{p} = p\hat{i}$ is at the origin. Find the potential it produces at an arbitrary point on the *x*-axis.

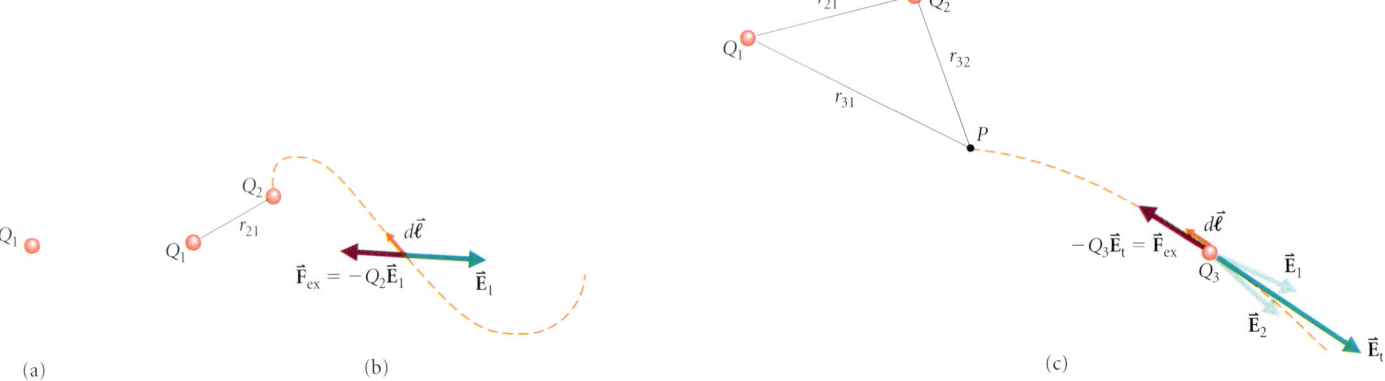

FIGURE 25.8
To find the potential energy of this system of three charges, we calculate the work done to assemble the system. (a) No work is needed to put the first charge in place. (b) Next we bring up the second charge, doing work kQ_1Q_2/r_{21}. (c) Finally, we bring up the third charge and have to do work against the electric force exerted by both the other charges. The work done may be expressed as the sum of the potential energies of the two new charge pairs formed. The two new pairs are Q_1Q_3 and Q_2Q_3.

25.3.2 Potential Energy of Systems of Charges

The potential energy of any physical system equals the work done to assemble the system (§8.4) and is independent of the procedure we choose. By convention the potential energy of a system of point charges is taken to be zero when the charges are very far apart. First, let us look at an arbitrary system of three charges (■ Figure 25.8).

We imagine assembling the system one charge at a time, beginning with them all very far apart. No work is required to put the first charge Q_1 in place, since no electrical forces act on it: $W_1 = 0$. To place the second charge in position (Figure 25.8b), we do work against the Coulomb force exerted by the first charge. This work equals the final potential energy of the system of two charges (cf. eqn. 25.3 and Example 25.1):

$$W_2 = Q_2 V_1(r_{21}) = \frac{kQ_2 Q_1}{r_{21}}.$$

ACTUALLY IN EXAMPLE 25.1 WE CALCULATED THE WORK DONE *BY THE FIELD* TO EXPEL THE CHARGE TO INFINITY. THAT EQUALS THE WORK DONE *BY US* TO ASSEMBLE THE PAIR.

The force $\vec{F}_{ex}$ we exert on the third particle balances the electric force exerted by the first two charges. From the principle of superposition, that electric force is due to the sum of the electric field vectors produced by the two individual charges:

$$\vec{F}_{ex} = -Q_3 \vec{E} = -Q_3(\vec{E}_1 + \vec{E}_2).$$

Now we find the work needed to put Q_3 in place at P:

$$W_3 = \int_\infty^P \vec{F}_{ex} \cdot d\vec{\ell} = \int_\infty^P [-Q_3(\vec{E}_1 + \vec{E}_2)] \cdot d\vec{\ell}$$

$$= \int_\infty^P (-Q_3 \vec{E}_1) \cdot d\vec{\ell} + \int_\infty^P (-Q_3 \vec{E}_2) \cdot d\vec{\ell}.$$

REMEMBER: EACH OF THE WORK INTEGRALS IS INDEPENDENT OF PATH.

Each integral is the work required to bring Q_3 into place near a single other charge:

$$W_3 = Q_3 V_1(r_{31}) + Q_3 V_2(r_{32}) = \frac{kQ_3 Q_1}{r_{31}} + \frac{kQ_3 Q_2}{r_{32}}.$$

Notice that $W_3 = Q_3 \Delta V = Q_3[V(P) - 0]$, where $V(P)$ is the potential at P due to the other two charges.

As Q_3 is put in place it forms two new charge pairs, one with each charge already in position, and two new terms are added to our expression for the potential energy of the system. This argument extends readily to a system containing any finite number of point charges. Bringing in a fourth charge creates three new charge pairs and adds three new terms to the system energy, and so on. The potential energy of an arbitrary system of charges is (■ Figure 25.9):

$$U = W = \sum_{\substack{\text{all} \\ \text{distinct} \\ \text{pairs}}} \frac{kQ_i Q_j}{r_{ij}}. \qquad (25.13)$$

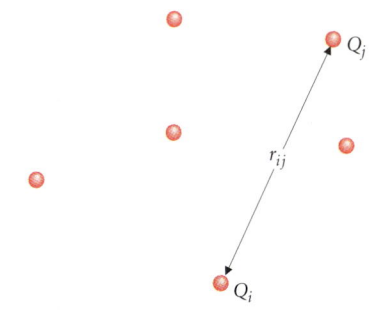

■ FIGURE 25.9
The energy of an arbitrary system of charges is stored in charge pairs. One such pair, Q_i and Q_j separated by distance r_{ij}, is shown here.

EXAMPLE 25.8 ♦♦ Find the potential energy of a system of three identical charges Q placed at the corners of an equilateral triangle of side s.

MODEL We imagine assembling the system one charge at a time.

SETUP No work is required to place the first charge at one corner of the triangle, $W_1 = 0$. To bring up the second charge, we do work against the Coulomb force exerted by the first charge:

$$W_2 = QV_1(s) = kQ^2/s.$$

Finally we bring up the third charge, doing work against the Coulomb force exerted by the first two (■ Figure 25.10):

$$W_3 = QV_1(s) + QV_2(s) = 2kQ^2/s.$$

SOLVE The potential energy of the system equals the total work done to assemble it:

$$U = W_1 + W_2 + W_3 = 0 + \frac{kQ^2}{s} + \frac{2kQ^2}{s} = \frac{3kQ^2}{s}.$$

ANALYZE To compare with eqn. (25.13), count up the number of distinct pairs in the system and add the potential energies of the pairs. Here we have three pairs (1-2, 2-3, and 3-1) and the potential energy of each pair is kQ^2/s, so the potential energy of the system is $3kQ^2/s$. ■

IN PROBLEM 93 WE ASK YOU TO VERIFY THIS RESULT BY DIRECT INTEGRATION OF THE WORK DONE, USING EQN. (23.9) FOR THE ELECTRIC FIELD.

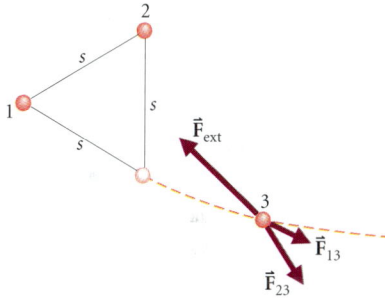

■ FIGURE 25.10
Bringing up the third charge to form an equilateral triangle.

EXERCISE 25.3 ♦ Use eqn. (25.13) to show that the potential energy of a system of four identical charges at the corners of a square of side s is: $U = (4 + \sqrt{2})kQ^2/s$. ■

If we bring a fifth charge q up from infinity and place it at P in Figure 25.7, the potential energy of the system of five charges increases by an amount $qV(P)$. It is tempting to think of this quantity as energy *possessed by* the charge q and to calculate the energy of the system by adding up energies *possessed by* each charge in the system. That procedure is faulty because it implicitly assumes that each charge is brought up with all the others already in place. Potential energy is unavoidably a property of an *entire system,* not of its individual particles.

The sum $\sum Q_i V_i$ overestimates the system energy by a factor of 2:

$$\sum_i Q_i V_i = \sum_i Q_i \sum_{j \neq i} \frac{kQ_j}{r_{ij}} = \sum_{j \neq i} \sum_i \frac{kQ_i Q_j}{r_{ij}}.$$

Let's compare this expression with eqn. (25.13). Each term in the double sum is the potential energy of a pair of charges i and j. However, each pair appears twice in the sum. For example, the pair of charges labeled 2 and 3 appears once with $i = 2$ and $j = 3$, and once with $i = 3$ and $j = 2$. Thus the sum is exactly twice the total potential energy of the system. For a system of point charges then:

$$U_{\text{system}} = \tfrac{1}{2} \sum_i Q_i V_i. \qquad (25.14)$$

IN THIS EXPRESSION V_i IS THE POTENTIAL AT THE POSITION OF CHARGE Q_i DUE TO ALL THE OTHER CHARGES.

EXAMPLE 25.9 ♦ Compute the potential energy of the system of four charges Q at the corners of a square of side s using eqn. (25.14).

MODEL To use eqn. 25.14 we first calculate the potential at the position of each of the four charges due to the other three. Because the system is symmetric, the result is the same at all four corners.

SETUP The potential at one corner of the square produced by charges at the other three corners is:

$$V_{corner} = \frac{2kQ}{s} + \frac{kQ}{s\sqrt{2}} = \frac{kQ}{s}\left(2 + \frac{1}{\sqrt{2}}\right) = \frac{kQ}{s}\left(\frac{4+\sqrt{2}}{2}\right).$$

SOLVE Equation (25.14) becomes the sum of four equal terms:

$$U_{system} = \tfrac{1}{2}\sum_{i=1}^{4} Q_i V_i = \tfrac{1}{2}(4QV_{corner}) = \frac{kQ^2}{s}(4+\sqrt{2}).$$

ANALYZE You should have obtained the same result in Exercise 25.3. ∎

Study Problem 16 ♦♦♦ The Collapsing Square

Four particles, two with charge $+Q$ and two with charge $-Q$, are placed at the corners of a square of side a (■ Figure 25.11a). Each has mass m, and they are released from rest. What is the speed of each charge when the side of the square has decreased to ℓ?

Modeling the System

Each particle is attracted to its two oppositely charged neighbors and repelled by the identical charge across the diagonal of the square. The two attractive forces have the same magnitude and sum to give an attractive force directly opposite the repulsive force. The net force on each charge is inward along the diagonal (cf. Example 23.8), so the square shrinks while maintaining its shape (Figure 25.11b). At any one time, the speed of each charge is the same. No *external* forces act, so the total energy of the system is conserved. The charges gain kinetic energy as the electric potential energy of the system decreases. We compute the system's potential energy as a function of the square's size. Conservation of energy gives the final speed of each charge.

Setup

The potential energy of the system is the potential energy of the six pairs of charges AB, AC, AD, BC, BD, and CD. For any given pair, the energy $U = kQ_1Q_2/d$, where d is the separation of the pair. When the square has sides of length s, the energy of each pair is:

Pair	Q_1	Q_2	d	U
AB	Q	$-Q$	s	$-kQ^2/s$
AC	Q	Q	$s\sqrt{2}$	$kQ^2/(s\sqrt{2})$
AD	Q	$-Q$	s	$-kQ^2/s$
BC	$-Q$	Q	s	$-kQ^2/s$
BD	$-Q$	$-Q$	$s\sqrt{2}$	$kQ^2/(s\sqrt{2})$
CD	$-Q$	Q	s	$-kQ^2/s$

The total energy of the system is thus:

$$U_{total} = \frac{kQ^2}{s}\left(-1 + \frac{1}{\sqrt{2}} - 1 - 1 + \frac{1}{\sqrt{2}} - 1\right)$$

$$= \frac{kQ^2}{s}\left(-4 + \frac{2}{\sqrt{2}}\right) = -\frac{kQ^2}{s}(4-\sqrt{2}) = -2.6\frac{kQ^2}{s}.$$

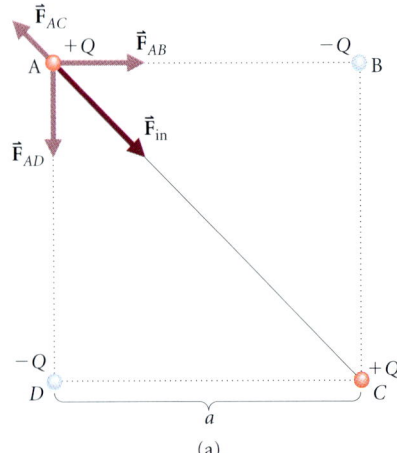

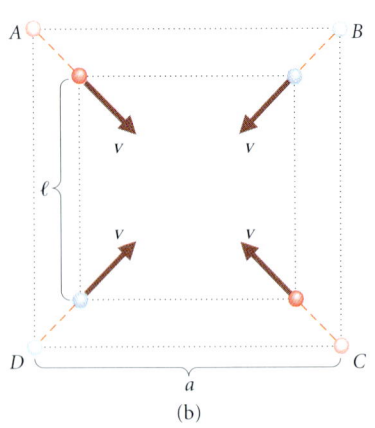

■ **FIGURE 25.11**
(a) Four charges of alternating sign are placed at the corners of a square. Each charge feels forces from the other three. The two attractive forces $\vec{F}_{AB} + \vec{F}_{AD}$ sum to give a force $\vec{F}_{in}$ inward across the diagonal. The net force $\vec{F}_{in} + \vec{F}_{AC}$ is toward the center of the square. (b) As the charges accelerate inward, each has the same speed at any time. The square shrinks while maintaining its shape.

Once again, we apply the standard method for using conservation laws.

	Before		**After**	
	(side of square $s = a$)		(side of square $s = \ell$)	
	Kinetic	Potential	Kinetic	Potential
Energy:	zero	$-\dfrac{kQ^2}{a}(4 - \sqrt{2})$	$4(\tfrac{1}{2}mv^2)$	$-\dfrac{kQ^2}{\ell}(4 - \sqrt{2})$

In Problem 104 we ask you to calculate the time it takes for the square to collapse.

Solution of Equations

Set total energy *before* equal to total energy *after*:

$$-\frac{kQ^2}{a}(4-\sqrt{2}) = 4(\tfrac{1}{2}mv^2) - \frac{kQ^2}{\ell}(4-\sqrt{2}).$$

$$2mv^2 = kQ^2(4-\sqrt{2})\left(\frac{1}{\ell} - \frac{1}{a}\right).$$

So:
$$v = Q\sqrt{\frac{k(4-\sqrt{2})}{2m}\left(\frac{1}{\ell} - \frac{1}{a}\right)}.$$

Analysis

The particles give up potential energy as they move. Since we chose $U = 0$ for $s \to \infty$, the particles move toward smaller values of s, where the potential energy is less (U is a larger negative number). With $U = 0$ at infinity, a negative potential energy indicates a bound system, that is, one that tends to collapse rather than fly apart. With all the charges positive as in Example 25.9, the potential energy of the system is positive: the system is unbound and tends to expand to infinity.

25.4 Equipotential Surfaces

25.4.1 The Relation Between Equipotential Surfaces and Field Lines

Field line diagrams give one method of visualizing the electric field. An alternate picture of the field is provided by equipotential surfaces.

> An *equipotential surface* is a surface on which the electric potential has a constant value.

The potential due to a point charge depends only on distance from the charge; the equipotential surfaces are spheres centered on the charge (■ Figure 25.12). If the charge is positive, the value of the potential decreases outward. The field lines are radial, cross the spherical equipotential surfaces at right angles, and point in the direction of lower potential. In Figure 25.12, the potential difference between successive surfaces is the same, so they concentrate near the charge. The field lines are also closest together near the charge, indicating that the field strength is greatest where the potential is changing most rapidly.

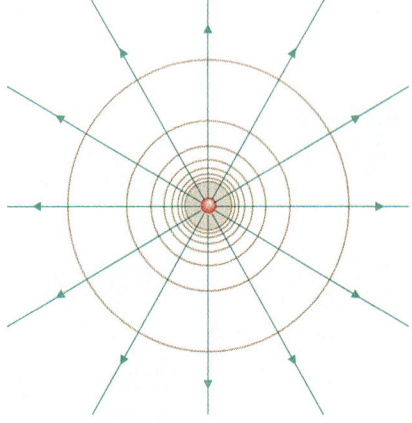

■ **Figure 25.12**
A positive point charge illustrates the relation between field lines and equipotential surfaces. Since $V = kQ/r$, the equipotential surfaces are spheres with $r = $ constant. The field lines are radial, so they are perpendicular to the equipotential surfaces. Note that if we draw surfaces with a constant potential difference, they are closer together near the charge. (In the central shaded region the surfaces are too close to be distinguished in this diagram.) The field lines are also closer together at the center. With a central charge of 3 nC, the potential difference between the surfaces shown is 1 kV.

EXAMPLE 25.10 ♦♦ A constant electric field $\vec{E} = E_o\hat{\imath}$ exists in a region of space. Describe the equipotential surfaces.

MODEL The relation between potential and field is given by eqn. (25.8). We may use this relation to find the potential V at an arbitrary point as a function of the coordinates. Equipotential surfaces are defined by the relation $V(x, y, z) = $ constant.

SETUP We choose the x-axis to be parallel to $\vec{E}$ and put the reference point R on the x-axis with coordinate $x = r$ (■ Figure 25.13). Then the potential at a point P with coordinates (x, y, z) is:

$$V(P) = -\int_R^P \vec{E}\cdot d\vec{\ell} = -\int_R^P E_o\hat{\imath}\cdot d\vec{\ell}.$$

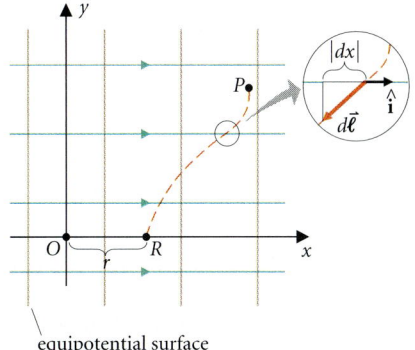

■ **Figure 25.13**
When the electric field is uniform, the equipotential surfaces are planes perpendicular to $\vec{E}$. Here $\vec{E} = E_o\hat{\imath}$.

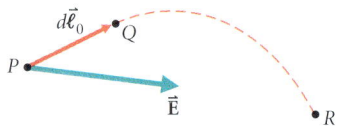

FIGURE 25.14
The potential difference between P and Q is $dV = -\vec{E} \cdot d\vec{\ell}_o$. If P and Q are on the same equipotential surface, $dV = 0$ and so $d\vec{\ell}_o$ is perpendicular to $\vec{E}$. Note also that the electric field points toward lower potential.

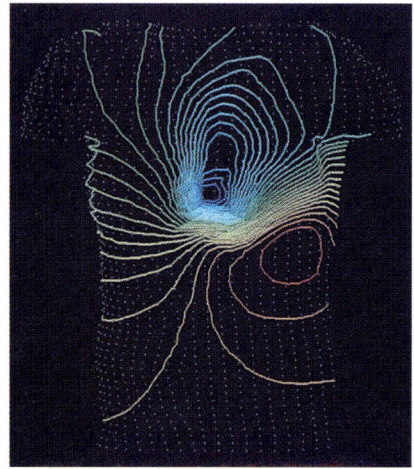

The contour lines in this picture show where equipotential surfaces cross the skin on a person's body. The lines are color coded from blue = most negative to red = most positive. They indicate an electric dipole. Compare with the field line diagram for a dipole (Figure 23.14).

For every path element, $\hat{i} \cdot d\vec{\ell} = dx$, so:

$$V(P) = V(x, y, z) = -\int_r^x E_o \, dx = E_o(r - x).$$

SOLVE The function $V(x, y, z)$ depends only on x. The equipotential surfaces are surfaces of constant x, that is, planes perpendicular to the x-axis.

ANALYZE The field lines are again perpendicular to the equipotential surfaces and point toward smaller values of V.

In the case of a point charge or a uniform field, the field lines are perpendicular to the equipotential surfaces. This is always true (■ Figure 25.14). The potential difference between two points P and Q separated by a differential displacement $d\vec{\ell}_o$ is (eqn. 25.7):

$$dV_{QP} \equiv V(Q) - V(P) = -\int_P^Q \vec{E} \cdot d\vec{\ell} = -\vec{E} \cdot d\vec{\ell}_o. \quad (25.15)$$

If P and Q are on the same equipotential surface, $dV_{QP} = 0$, so $\vec{E} \cdot d\vec{\ell}_o = 0$ and $\vec{E}$ is perpendicular to $d\vec{\ell}_o$.

25.4.2 Equipotential Surfaces for a System of Point Charges

As we did for field line diagrams, we may determine a set of rules for drawing equipotential surfaces. Near a point charge, the equipotential surfaces are spheres with the point charge at the center (Figure 25.12). At a large distance from a system of point charges, the equipotential surfaces are spheres surrounding the whole system. In between, some rather odd shapes can result. There is usually a transition surface that may be double-lobed or may touch itself. At such connections there is no definite direction perpendicular to the surface, and the direction of $\vec{E}$ is undefined. Thus these surface connections can occur only at points where $\vec{E} = 0$.

Avoid confusing $\vec{E} = 0$ and $V = 0$! That $\vec{E} = 0$ means only that V is not changing and says nothing about its value. (The point is a maximum, minimum, or saddle point.) Conversely, $V = 0$ by itself says nothing about the value of $\vec{E}$. For example, between two point charges of opposite sign, the potential varies continuously from a positive value near the positive charge to a negative value near the negative charge. It is zero somewhere between them, where $\vec{E}$ points toward the negative charge. The following example illustrates these ideas.

Math Topic

ELECTRIC FIELD AS THE GRADIENT OF POTENTIAL

A vector field like $\vec{E}$ assigns a vector to each point. A scalar field like V assigns a number to each point. In mathematics the *gradient* of a scalar field, such as electric potential, is a vector that points in the direction in which the scalar field varies most rapidly. Its magnitude is the maximum rate of change of the scalar field. Perpendicular to the gradient, the scalar field does not change at all. The relation between the electric field vector and the electric potential is summarized by the statement:

$$\vec{E} = -\text{gradient } V.$$

In Cartesian coordinates:

$$\text{gradient } V = \frac{\partial V}{\partial x}\hat{i} + \frac{\partial V}{\partial y}\hat{j} + \frac{\partial V}{\partial z}\hat{k} = -\vec{E}$$

The uniform field in Example 25.10 has potential $V(x, y, z) = E_o(r - x)$, so $\partial V/\partial x = -E_o$, $\partial V/\partial y = \partial V/\partial z = 0$, and

$$\vec{E} = -\text{gradient } V = E_o \hat{i},$$

as expected.

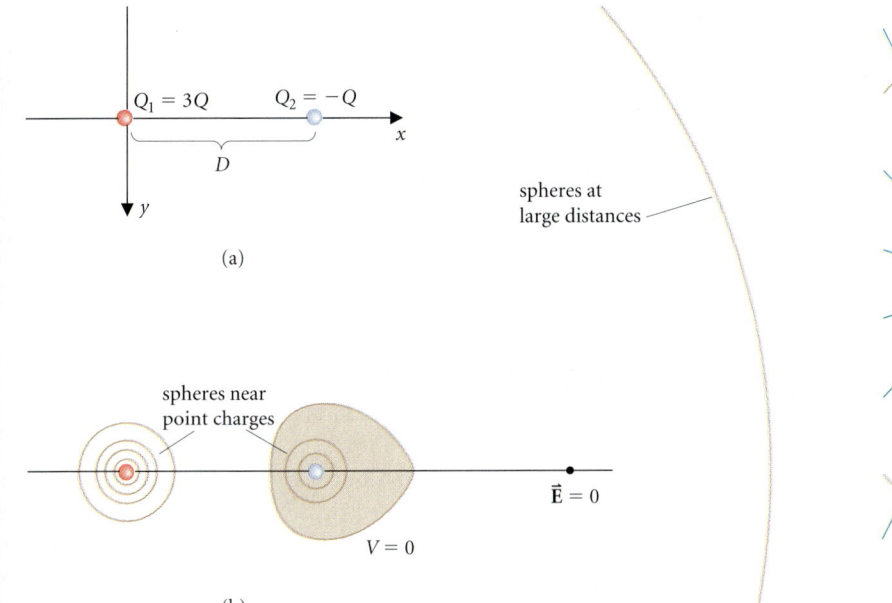

EXAMPLE 25.11 ♦♦♦ Two particles with charges $Q_1 = 3Q$ and $Q_2 = -Q$ are on the x-axis separated by a distance D (■ Figure 25.15a). Sketch the equipotential surfaces in the x-y-plane, and find two points on the x-axis, other than $\pm\infty$, where $V = 0$.

MODEL Near each charge the equipotential surfaces are spheres surrounding the charge. We adopt the usual convention that $V \to 0$ as $r \to \infty$. The potential is $+\infty$ at the positive charge and is $-\infty$ at the negative charge. Far from the system, at a distance much greater than D but less than infinity, the potential is positive, since the system's net charge $(+2Q)$ is positive. Separating the regions of positive and negative potential is an equipotential surface on which $V = 0$ (Figure 25.15b).

SETUP Since the $V = 0$ surface surrounds the negative charge, one of the two required points lies between the two charges $(x < D)$, and one is to the right of the negative charge $(x > D)$.

For $x > D$: $\quad V(x) = \dfrac{k(3Q)}{x} + \dfrac{k(-Q)}{x - D} = kQ\left(\dfrac{3}{x} - \dfrac{1}{x - D}\right).$

SOLVE The potential is zero where:

$$\dfrac{3}{x} = \dfrac{1}{x - D} \quad \text{or} \quad 3x - 3D = x \;\Rightarrow\; x = \dfrac{3D}{2}.$$

SETUP For $0 < x < D$: $\quad V(x) = \dfrac{k(3Q)}{x} + \dfrac{k(-Q)}{D - x} = kQ\left(\dfrac{3}{x} - \dfrac{1}{D - x}\right).$

SOLVE This function is zero where:

$$\dfrac{3}{x} = \dfrac{1}{D - x} \quad \text{or} \quad 3D - 3x = x \;\Rightarrow\; x = \dfrac{3D}{4}.$$

ANALYZE In order for the potential to be zero at a point, the point must be closer to the smaller, negative charge so that the closeness can overcome the factor of 3 in the sizes of the charges. Check the ratio of distances in the two cases:

$$\dfrac{\text{Distance to + charge}}{\text{Distance to - charge}} = \begin{cases} \dfrac{3D/2}{(3D/2) - D} = 3 & \text{for } x = 3D/2. \\[2ex] \dfrac{3D/4}{D - (3D/4)} = 3 & \text{for } x = 3D/4. \end{cases}$$

■ **FIGURE 25.15**
(a) Two point charges, $3Q$ and $-Q$, are on the x-axis separated by distance D.
(b) Since $V \to +\infty$ at the positive charge and $V \to -\infty$ at the negative charge, it is zero somewhere in between. Also, since the net charge is $+2Q$, the potential is positive at large distances from the system. Thus the potential is also zero somewhere to the right of the negative charge. The surface of zero potential bounds the shaded region, where the potential is negative. (c) The equipotential surfaces are perpendicular to the field lines. Since the equipotential surface through P touches itself there, $\vec{E}$ is zero at P. The potential decreases away from P along the x-axis and increases away from P perpendicular to the x-axis.

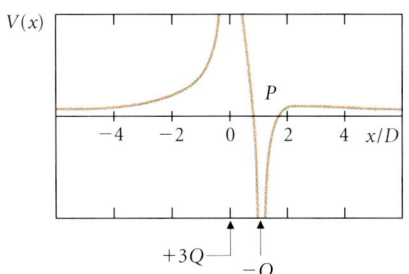

FIGURE 25.16
Graph of $V(x)$ along the x-axis through the two charges. The potential has a local maximum at point P, where the electric field vector is zero.

Notice that we need two different expressions for $x > D$ and $x < D$ because the potential at a point depends on the distance of that point from each charge, which is by definition positive. The denominator $x - D$ becomes negative for $x < D$, so it is replaced with $D - x$. We need yet a third expression for $x < 0$.

The system of charges in this problem is the same as that in Study Problem 15. The equipotential surfaces are everywhere normal to the field lines (Figure 25.15c). In Study Problem 15 we found that $\vec{E} = 0$ on the x-axis at point P where $x = (3 + \sqrt{3})D/2$. Note that one of the equipotential surfaces in Figure 25.15c touches itself on the x-axis at this point, which corresponds to a maximum in the function $V(x)$ (■ Figure 25.16). ■

25.5 Potential Due to a Continuous Distribution of Charge

There are two ways to find the electrical potential produced by charge distributed continuously on an object. The first way is to compute the electric field vector and then integrate it along an appropriate path, using eqn. (25.8). The second way is to model the charge distribution as a collection of differential elements and use the principle of superposition to express the potential function as an integral. The second method fails for infinite objects, because the interchange of sum and integral in the derivation of eqn. (25.12) is valid only if both sum and integral converge. However, for symmetric infinite objects we may use Gauss' law to find the electric field, and the first method usually works well.

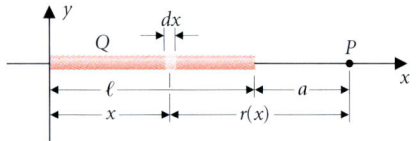

FIGURE 25.17
A finite, uniformly charged rod. A differential element with coordinate x is a distance $r = a + \ell - x$ from P.

EXAMPLE 25.12 ◆◆◆ Find the potential at distance a from one end of a uniformly charged rod of length ℓ and total charge Q, as shown in ■ Figure 25.17.

MODEL We model the rod as a collection of differential elements, and integrate.

SETUP We choose the x-axis along the rod, with the origin at its far end. A typical
STEPS I AND II element is a differential piece of rod with coordinate x, length dx, and charge:

$$dQ = \frac{Q}{\ell}\, dx.$$

STEP III The distance of the element from P is $r = a + \ell - x$, and it produces potential dV at P:

$$dV = k\frac{dQ}{r} = \frac{(kQ/\ell)\,dx}{a + \ell - x}.$$

STEP IV The rod extends from $x = 0$ to $x = \ell$, and we have chosen dx to represent a positive length, so we integrate from 0 to ℓ.

SOLVE
STEP V
$$V = \int_0^\ell \frac{kQ\, dx}{\ell(a + \ell - x)} = -\frac{kQ}{\ell}\,[\ln(a + \ell - x)]\,\Big|_0^\ell.$$

$$V = \frac{kQ}{\ell}\ln\!\left(\frac{a + \ell}{a}\right).$$

ANALYZE If $a \gg \ell$, we may use the approximation $\ln(1 + \epsilon) \approx \epsilon$ (Appendix IB) to obtain:

$$V \approx \frac{kQ}{\ell}\frac{\ell}{a} = \frac{kQ}{a} \quad (a \gg \ell).$$

As expected, this is the result for a point charge. ■

EXERCISE 25.4 ◆◆ Find the potential at points on the axis of a uniformly charged ring of radius a and total charge Q (■ Figure 25.18).

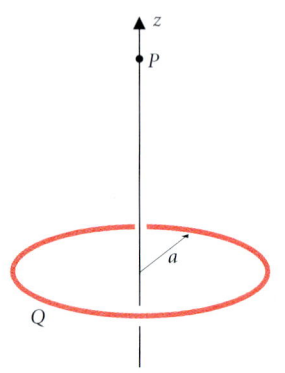

FIGURE 25.18

EXERCISE 25.5 ♦♦♦ Use the result of Exercise 25.4 to find the potential on the axis of a uniformly charged disk of radius R.

EXAMPLE 25.13 ♦♦♦ Find the potential at a perpendicular distance a from an infinite, uniformly charged filament.

MODEL The method used in Example 25.12 may not be used here. The filament is infinitely long: its total charge is infinite, and the integral is again a logarithm that diverges as $\ell \to \infty$. Instead, we obtain the potential as an integral of the field vector (eqn. 25.8).

SETUP The electric field vector due to the infinite filament is (eqn. 24.4):

$$\vec{E} = \frac{2k\lambda}{r} \hat{r}.$$

> Using Gauss' law is the easiest way to get this result; λ is the linear charge density on the filament, cf. §24.2.

SOLVE The potential difference between P and R is:

$$V(P) - V(R) = -\int_R^P \vec{E} \cdot d\vec{\ell} = -\int_R^P \frac{2k\lambda}{r} \hat{r} \cdot d\vec{\ell}.$$

For any path we choose between P and R, $\hat{r} \cdot d\vec{\ell} = dr$, the change in radial distance from the line. Let $r = r_0$ at the reference point R, where we choose $V(R) \equiv 0$. Then:

$$V(P) = -\int_{r_0}^{a} \frac{2k\lambda}{r} dr = -2kr\lambda \ln(r) \Big|_{r_0}^{a}$$

$$V(P) = 2k\lambda \ln(r_0/a). \tag{25.16}$$

ANALYZE There is no obvious place for the reference point. We cannot choose the reference point R at $r_0 = 0$ or $r_0 = \infty$, since the logarithm is infinite at both places. We must choose a position at some intermediate distance r_0 from the filament. Of course, no charge distribution is truly infinite. Using such a model in a practical situation, you would choose the reference point by considering the relation of the filament to other objects. The potential *difference* between any two points at finite distance is well defined:

$$V(P_1) - V(P_2) = 2k\lambda \ln(a_2/a_1).$$

> See Problem 66.

25.6 THE BEHAVIOR OF CONDUCTORS

25.6.1 Response of a Conductor to an Electric Field

Conductors are materials in which some of the electrons move relatively freely. The copper wire inside an electrical cable is a typical, efficient conductor. Any net electric field within the body of such a material exerts a force on electrons that causes them to move. For example, if we try to produce an electric field within a piece of conducting material by bringing a positively charged object near it, electrons accelerate toward the surface of the conductor closest to the object (■ Figure 25.19). The resulting deficit of electrons on the opposite side of the

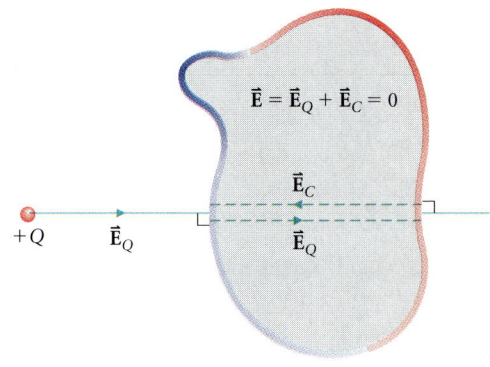

■ **FIGURE 25.19**
When a charged object is brought near a conductor, equilibrium is established rapidly. At each point within the conductor the electric field $\vec{E}_c$ produced by the surface charge distribution is equal to and opposite the field $\vec{E}_Q$ produced by the point charge. The surface charges also distort the field outside the conductor (see Figure 25.22).

THE TIME REQUIRED IS APPROXIMATELY THE LIGHT TRAVEL TIME ACROSS THE CONDUCTOR (SEE PROBLEM 82).

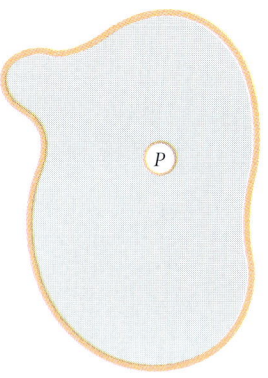

■ FIGURE 25.20
There is no flux through any Gaussian surface that is entirely within the conductor. Here we see two such surfaces. A small surface surrounding the arbitrary point P shows that $\rho = 0$ at P. A large surface reaching to the boundary of the conductor shows that all the charge on the conductor must be on its surface.

CHARGES ON THE SURFACE ARE ACTUALLY CONCENTRATED WITHIN ROUGHLY AN ATOMIC DIAMETER OF THE SURFACE.

IN REALITY THE CHARGE IS DISTRIBUTED OVER THE ENTIRE SURFACE AS WELL AS THE ENDS OF THE NEEDLE.

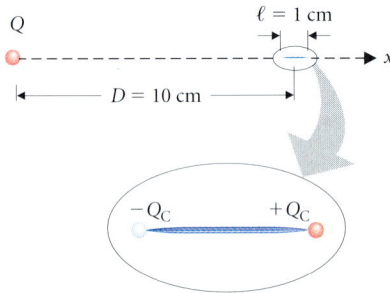

■ FIGURE 25.21
An external point charge induces surface charge on a metal needle, here modeled with two point charges $\pm Q_c$ on the ends. The external point charge exerts a net attractive force on the needle because the negative charge $-Q_c$ is closer to Q than the positive charge $+Q_c$.

conductor leaves a positive charge there. These surface charges produce an electric field within the conductor that opposes the electric field due to the charged object. As long as any electric field remains inside the conductor, electrons continue to respond until the distribution of charge on the conductor surface exactly neutralizes the external field within the conductor and leaves it in equilibrium. This process happens extremely rapidly. Since all the electrons move, each electron travels only a very short distance in creating the necessary charge distribution.

> In static equilibrium, the electric field $\vec{E}$ is zero at every point within the body of a conducting material.

Any net charge within the conductor would be the source of electric field. If the field is zero, then the charge must be also. To see why, we apply Gauss' law to a closed surface within the metal. Since $\vec{E} \equiv 0$, the flux through *any* such surface is zero, and thus the surface encloses zero net charge. We may choose a very small Gaussian surface, surrounding an arbitrary point P, or one so large that it coincides with the surface of the conductor (■ Figure 25.20). Thus the charge density is zero at every point inside the conductor, and any net charge on the conductor must be on its surface.

> In static equilibrium, net electric charge can exist only on the surfaces of a conductor, not within its body.

A charged object exerts a substantial force on a neutral piece of conductor, because the surface charge is not all at the same distance from the object.

EXAMPLE 25.14 ♦♦ A point charge $Q = +1\ \mu C$ is 10 cm from the middle of a thin metal needle 1 cm long (■ Figure 25.21). Estimate the charges on the ends of the needle required to maintain $\vec{E} \equiv 0$ within the metal. Estimate the net attractive force on the needle.

MODEL We model the charges on the needle as point charges $\pm Q_c$ at its ends and estimate their magnitude by requiring $\vec{E} = 0$ at the middle of the needle. Since negative charge is attracted toward the positive charge Q, and positive charge is repelled, the end of the needle closest to Q is negatively charged. This model is only approximate, since it does not make $\vec{E}$ zero throughout the needle.

SETUP The field at the center of the needle is the superposition of the fields due to the charges at the two ends, plus the field due to the external point charge Q:

$$\vec{E}_Q = \frac{kQ}{D^2}\hat{i}, \qquad \vec{E}_{c+} = \frac{kQ_c}{(\ell/2)^2}(-\hat{i}), \qquad \text{and} \qquad \vec{E}_{c-} = \frac{kQ_c}{(\ell/2)^2}(-\hat{i}).$$

The net electric field is zero:

$$0 = \vec{E}_Q + \vec{E}_{c+} + \vec{E}_{c-} = \frac{kQ}{D^2}\hat{i} + \left[-\frac{2kQ_c}{(\ell/2)^2}\hat{i}\right].$$

SOLVE
$$Q_c = \frac{Q\ell^2}{8D^2} = \frac{10^{-6}\ \text{C}}{8}\left(\frac{1\ \text{cm}}{10\ \text{cm}}\right)^2 = 1\ \text{nC}.$$

SETUP The force exerted by Q on the needle is the sum of the forces on the two charges $\pm Q_c$. Using Coulomb's law:

SOLVE
$$\vec{F} = \frac{k(-Q_c)Q}{(D-\ell/2)^2}\hat{i} + \frac{kQ_cQ}{(D+\ell/2)^2}\hat{i}.$$

We simplify this expression by factoring the common quantities:

$$\vec{F} = -\frac{kQ_cQ\hat{i}}{D^2}\left[\frac{1}{(1-\ell/2D)^2} - \frac{1}{(1+\ell/2D)^2}\right].$$

Then we combine terms over a common denominator:

$$\vec{F} = -\frac{kQ_cQ\hat{i}}{D^2}\left[\frac{(1+\ell/2D)^2 - (1-\ell/2D)^2}{(1-\ell^2/4D^2)^2}\right].$$

Now, since $\ell^2/4D^2 = 0.002$, we neglect it compared with 1:

$$\vec{F} \approx -\frac{kQ_cQ\hat{i}}{D^2}\left(\frac{2\ell/D}{1}\right) = \frac{-2kQ_cQ\ell\hat{i}}{D^3}.$$

Then substituting the value we found for Q_c gives:

$$\vec{F} \approx \frac{-kQ^2\ell^3\hat{i}}{4D^5} = \frac{-\hat{i}(9\times 10^9\text{ N}\cdot\text{m}^2/\text{C}^2)(10^{-6}\text{ C})^2(10^{-2}\text{ m})^3}{4(10^{-1}\text{ m})^5}$$
$$= -\hat{i}(2\times 10^{-4}\text{ N}).$$

ANALYZE The direction $(-\hat{i})$ confirms that the force is attractive. A steel needle 1 mm in diameter has a mass m of approximately 6×10^{-5} kg and would accelerate at $|\vec{a}| = |\vec{F}|/m \approx 3$ m/s², or at about $\frac{1}{3}g$.

25.6.2 Conductors and Electric Potential

Since the electric field vanishes within a conductor, there is no potential difference between any two points within the conductor (eqn. 25.7). The entire volume is at a constant potential, and its surface is an equipotential surface. Electric field lines are always perpendicular to equipotential surfaces (§25.4.1), so the field lines that emanate from or converge into charges on the conductor are perpendicular to the surface.

■ Figure 25.22 shows a conducting sphere placed in a region where the electric field is initially uniform and parallel to the x-axis. The external electric field points to the right, pushing electrons within the sphere to its left surface. A net positive charge remains at the right surface. The field produced by these charges neutralizes the applied field within the sphere and also modifies the field outside. External field lines connect to these charges and meet the sphere at right angles to its surface. There is no electric field within the sphere.

Now we can see why Don Salvatore is safe within his metal cage at the Boston Museum of Science. Imagine trying to make an electric field within a cavity inside a conductor (■ Figure 25.23a). Surface charges keep $\vec{E}$ out of the conducting material, and since the surface is an equipotential, field lines may not run from one point on the surface to another across the cavity. There is no field at all inside the cavity. The metal cage isn't a perfect shield because of the holes between bars (Figure 25.23b), but it is good enough.

EXERCISE 25.6 ❖ Is it safe to sit in your car during a lightning storm? What happens when you get out?

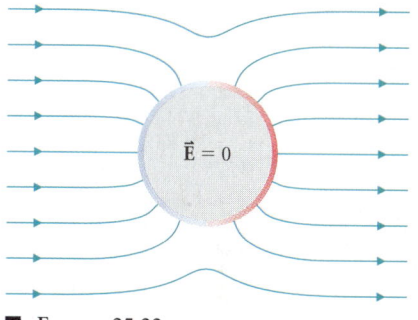

■ **FIGURE 25.22**
A conducting sphere is placed in a uniform field. The applied electric field creates a surface charge distribution on the sphere that distorts the field outside. The field lines meet the sphere perpendicular to the surface. The electric field within the sphere is zero.

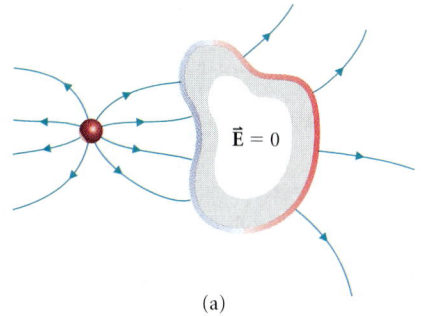
(a)

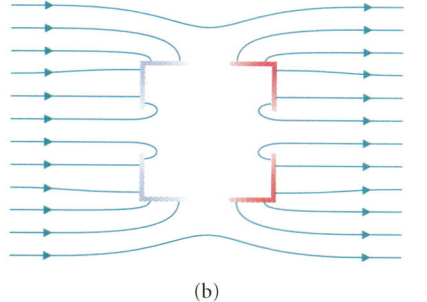
(b)

■ **FIGURE 25.23**
(a) Since the whole conductor is at the same potential, no field lines pass from one point of the conductor to another. That includes field lines inside the cavity. Since there is no charge inside the cavity, it is not possible to have any static electric field in the cavity. The surface charge induced by the external charge is all on the outer surface of the conductor. (b) Even if there are holes in the conducting surface, the electric field does not penetrate very far into the cavity.

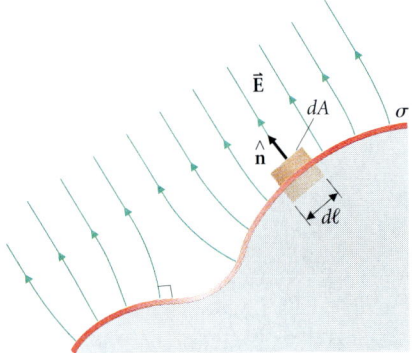

■ FIGURE 25.24
To find the relation between electric field and the surface charge density on a conductor, we apply Gauss' law to the small pillbox shown here. It is constructed with its curved sides perpendicular to the conductor surface and thus parallel to the field just outside the surface. The flat faces are parallel to the conductor surface and perpendicular to $\vec{E}$.

IF THE FIELD LINES POINT TOWARD THE SURFACE, σ IS NEGATIVE; IF THEY POINT AWAY, σ IS POSITIVE.

25.6.3 The Relation Between Field and Surface Charge Density on a Conductor

We may use Gauss' law to find the relation between the surface charge density on a conductor and the electric field vector just outside. We choose a small volume (a *pillbox*) with one face just outside the conductor and the pillbox side walls perpendicular to the conductor surface (■ Figure 25.24). If the conductor surface is not flat, we choose the pillbox area dA small enough that the curvature is negligible. That is, we take $d\ell \sim \sqrt{dA} \ll R$, where R is the radius of curvature of the surface at that point. Since there is no electric field inside the conductor, there is no electric flux through the pillbox walls inside the conductor. With $\vec{E}$ perpendicular to the surface, there is no flux through the side walls outside the conductor either. The remaining face is parallel to the conductor surface, $\vec{E}$ is parallel to $\hat{n}$ on this face, and the flux through it is $E\,dA$. The charge inside the pillbox is all on the surface of the conductor: $Q = \sigma\,dA$. Applying Gauss' law, we find:

$$\Phi_E = \oint \vec{E} \cdot \hat{n}\,dA = E\,dA = \frac{Q}{\epsilon_0} = \frac{\sigma\,dA}{\epsilon_0}.$$

So:

$$\vec{E} = \frac{\sigma}{\epsilon_0}\hat{n} = 4\pi k\sigma\hat{n}. \qquad (25.17)$$

Equivalently, the charge density is: $\quad |\sigma| = \epsilon_0|\vec{E}|. \qquad (25.18)$

The electric field near a large, uniformly charged plane with charge density σ has magnitude $\sigma/(2\epsilon_0)$ (eqn. 24.7). Electric field lines emerge from both sides perpendicular to the surface (■ Figure 25.25a). If a conducting surface has the same charge density, the same number of field lines come out, but all on one side (Figure 25.25b). The field outside is double the value for a single charged plane (Figure 25.25c).

EXERCISE 25.7 ♦ The charge density at a point on the surface of a conductor is $1.2\ \mu C/m^2$. What is the electric field just outside the conductor at that point?

25.6.4 The Relation Between the Shape of a Conductor and the Electric Field at its Surface

Benjamin Franklin was the first to note that sharply pointed objects leak charge into the air much more rapidly than do blunt objects. Electric fields are strongest near points, and sufficiently intense fields can strip electrons from molecules in the air, increasing its conductivity. We can determine the relationship between electric field and the shape of a conductor by considering how conductors in contact share charge.

■ FIGURE 25.25
(a) A plane sheet of charge with surface charge density σ. The electric field is normal to the sheet and has magnitude $\sigma/(2\epsilon_0)$. (b) The same charge density on the surface of a plane conductor. All the field lines emerge outward from the surface, and the field is twice as large: $|\vec{E}| = \sigma/\epsilon_0$. (c) The field near a conducting surface at P is the superposition of the field due to the nearby surface charge density $[|\vec{E}_s| = \sigma/(2\epsilon_0)]$ and the field $\vec{E}_e$ due to charges elsewhere on the surface of the conductor. In order that the field inside the conductor be zero, the electric field $\vec{E}_e$ must be equal to and opposite the field *within* the conductor produced by the surface charge at P, and thus its magnitude is also $\sigma/(2\epsilon_0)$. Outside the conductor, $\vec{E}_s$ reverses direction, but $\vec{E}_e$ does not. The total field outside has magnitude σ/ϵ_0.

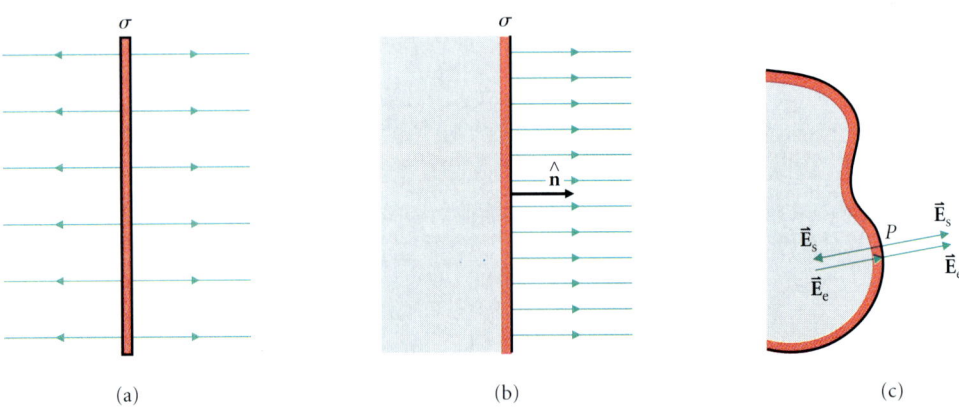

EXAMPLE 25.15 ♦♦ Two conducting spheres of radii R_1 and R_2 are far apart but connected by a thin conducting wire (■ Figure 25.26). They share a total charge Q. Find the charge density on each sphere and the electric field strength at each sphere's surface.

MODEL The two spheres connected by a wire form a single conducting object. No electric field can exist within this object, so the electric potential is constant everywhere on it. Since the two spheres are far apart, they act like isolated spherical charges with electric fields and potentials given by Coulomb's law (cf. Example 24.1 and Exercise 25.1). The charge on the wire is negligible so long as it is thin.

SETUP The potential at distance R from the center of a sphere carrying charge Q is kQ/R. This formula holds even at the surface of the sphere. If the two spheres have charges Q_1 and Q_2, they have potentials $V_1 = kQ_1/R_1$ and $V_2 = kQ_2/R_2$. Since the spheres are connected by a wire, they are at the *same* potential. Thus:

$$Q_1/R_1 = Q_2/R_2.$$

Since the total charge $Q = Q_1 + Q_2 = Q_1 + (R_2/R_1)Q_1$, we have:

$$Q_1 = \frac{QR_1}{R_1 + R_2} \quad \text{and} \quad Q_2 = \frac{QR_2}{R_1 + R_2}.$$

Since the spheres are far apart, they do not much influence each other's charge distributions, which we may assume uniform. Then:

$$\sigma_1 = \frac{Q_1}{4\pi R_1^2} \quad \text{and} \quad \sigma_2 = \frac{Q_2}{4\pi R_2^2}.$$

SOLVE Substituting the values of Q_1 and Q_2 into these equations, we find:

$$\sigma_1 = \frac{QR_1}{4\pi R_1^2(R_1 + R_2)} = \frac{Q}{4\pi R_1(R_1 + R_2)},$$

and similarly:

$$\sigma_2 = \frac{Q}{4\pi R_2(R_1 + R_2)}.$$

The electric field strength at the surface of one of the spheres is $|\vec{E}| = 4\pi k\sigma$ (eqn. 25.17). So:

$$|\vec{E}_1| = \frac{kQ}{R_1(R_1 + R_2)},$$

and:

$$|\vec{E}_2| = \frac{kQ}{R_2(R_1 + R_2)}.$$

ANALYZE $|\vec{E}| \propto 1/R$, that is:

$$\frac{|\vec{E}_1|}{|\vec{E}_2|} = \frac{R_2}{R_1}.$$

This is a general relation; the electric field is large near a sharp point because the radius of curvature is small (■ Figures 25.27, 25.28).

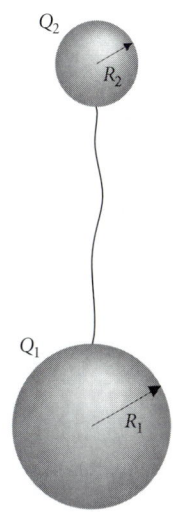

■ **FIGURE 25.26**
Two spheres sharing a total charge Q are connected by a long, thin wire. The spheres and wire form a single conductor with a uniform potential. The charge on each sphere is proportional to its radius.

THE WIRE CHANGES THE CHARGE DISTRIBUTION SLIGHTLY WHERE IT TOUCHES THE SPHERES. WE ALSO IGNORE THIS EFFECT.

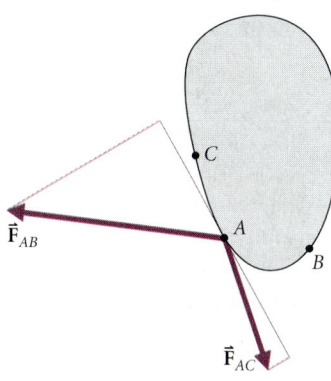

■ **FIGURE 25.27**
Charge is in equilibrium on a conducting surface when the net force parallel to the surface acting on any element of charge is zero. Two charge elements at points B and C exert very different forces on the element at A, though they are the same distance away. Because of the greater curvature between B and A, the force $\vec{F}_{AB}$ is more nearly perpendicular to the surface than is $\vec{F}_{AC}$. Since the tangential components balance, $|\vec{F}_{AB}| > |\vec{F}_{AC}|$, and the charge element at B is greater than the charge element at C. The charge density, and hence the electric field, is greater at B than at C.

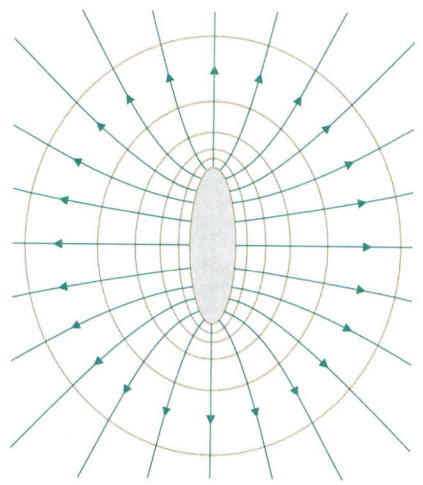

■ **FIGURE 25.28**
Field lines and equipotential surfaces for an object with varying curvature. The surfaces evolve toward spherical sphape and are more closely spaced where their initial curvature is greatest. Thus electric field is greatest where curvature is greatest.

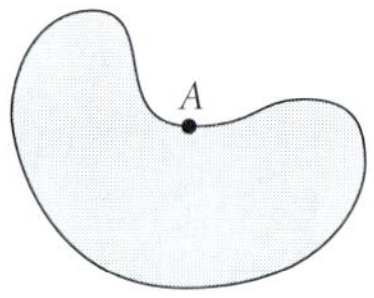

EXERCISE 25.8 ❖ The object in ■ Figure 25.29 is a charged conductor. Sketch a field line diagram for this object, and convince yourself that the surface charge density is small at point A, where the surface has a dimple, that is, a region of negative curvature.

■ FIGURE 25.29

Chapter Summary

Where Are We Now?

We have discussed the electric potential energy of systems of charges. This is the third form of potential energy we have studied, the others being elastic and gravitational energy. We used potential energy to introduce the idea of electric potential. Electric potential is a scalar quantity that describes the effect of a charge on its surroundings and is closely related to the electric field vector. We studied the behavior of conductors. We now have enough tools to begin a serious study of electric systems.

What Did We Do?

Electric potential describes the work done to move a charge between two points in an electric field: $\Delta U = q \Delta V$. The potential difference between two points P and R is:

$$V(P) - V(R) = -\int_R^P \vec{E} \cdot d\vec{\ell}.$$

If R is the reference point where we choose to assign the potential a value of zero, this equation defines the potential $V(P)$. Potential is measured in volts. One volt is 1 joule per coulomb.

When the charge distribution is a collection of point charges, the reference point R is usually chosen at infinity. Then the potential due to a point charge is kQ/r, where r is the distance from the charge (r is always positive). The potential due to a finite, continuous distribution of charge may be found by integrating the potentials due to each differential charge element.

We found the potential energy of a system of charges by computing the work done to assemble the system. The energy of such a system is the sum of the energies kQ_iQ_j/r_{ij} of each distinct pair of charges Q_i and Q_j in the system. The energy may also be computed from the potential at the position of each charge in the system:

$$U = \tfrac{1}{2}\Sigma Q_i V_i.$$

When a charge q is moved from the reference point R to P, the potential energy of the system changes by

$$\Delta U = qV(P).$$

Surfaces of constant potential are always perpendicular to field lines, and the value of the potential decreases along field lines. In static equilibrium, the electric field is zero within a conductor, and the surface of a conductor is an equipotential surface. The charge density on the surface of the conductor is related to the electric field at the surface:

$$|\sigma| = \epsilon_0 |\vec{E}|.$$

Practical Applications

Potential is important for understanding electric circuits (Chapter 26). Many electronic devices require carefully designed patterns of electric field to operate properly. Since conductors are equipotential surfaces, the field pattern can be controlled by properly shaping the metal electrodes used in a device. Electric potential is an essential concept in the design of television sets, photocopiers, x-ray machines for medical diagnosis, and particle accelerators for physics research.

Solutions to Exercises

25.1 We put the origin at the point charge and integrate along a radial line, which we'll call the x-axis. The reference point R is at ∞ and P is at $x = d$.

$$V(P) = -\int_\infty^P \vec{E} \cdot d\vec{\ell} = -\int_\infty^d \frac{kQ}{x^2}\hat{i} \cdot (dx\,\hat{i})$$

$$= -\int_\infty^d \frac{kQ\,dx}{x^2} = \frac{kQ}{x}\bigg|_\infty^d = \frac{kQ}{d}.$$

25.2 Model the dipole as two point charges, $-Q$ at the origin and $+Q$ at $x = a$ (Figure 25.30). The potential at $x > a$ is:

$$V_a(x) = kQ\left(\frac{1}{x-a} - \frac{1}{x}\right) = kQ\frac{a}{x(x-a)}.$$

The potential produced by an ideal dipole of moment $p = Qa$ is:

$$V(x) = \lim_{a \to 0} V_a(x) = \frac{kp}{x^2}.$$

FIGURE 25.30

25.3 (Figure 25.31) There are six charge pairs, AB, AC, BD, CD, AD, and BC. The first four have separation s, and the other two have separation $s\sqrt{2}$. Thus:

$$U_{\text{system}} = 4\frac{kQ^2}{s} + 2\frac{kQ^2}{s\sqrt{2}} = \frac{kQ^2}{s}(4 + \sqrt{2}).$$

25.4 STEPS I AND II Take the z-axis as the axis of the ring, with the origin at its center (Figure 25.32). A typical element of the ring at angle θ has length $a\,d\theta$ and charge $dQ = Q\,d\theta/(2\pi)$.

STEP III The potential at P due to the element is:

$$dV = k\,dQ/s.$$

The distance $s = \sqrt{z^2 + a^2}$ is the same for each point on the ring.

STEPS IV AND V The angle θ runs from 0 to 2π, so:

$$V = \int_0^{2\pi} \frac{kQ}{2\pi} \frac{d\theta}{\sqrt{z^2 + a^2}} = \frac{kQ}{\sqrt{z^2 + a^2}}\int_0^{2\pi}\frac{d\theta}{2\pi} = \frac{kQ}{\sqrt{z^2 + a^2}}.$$

Note: Since all the charge is at the same distance:

$$V = k(\text{total charge})/(\text{distance}).$$

25.5 The disk is a collection of differential rings of radius r and width dr (cf. Example 24.6). The charge on such a ring is:

$$dQ_{\text{ring}} = \sigma\,dA = \frac{Q}{\pi R^2}(2\pi r\,dr) = \frac{2Q}{R^2}r\,dr.$$

From Exercise 25.4 the potential due to the ring is:

$$dV = \frac{k\,dQ_{\text{ring}}}{\sqrt{z^2 + r^2}} = \frac{2kQr\,dr}{R^2\sqrt{z^2 + r^2}}$$

Thus:

$$V = \int dV = \int_0^R \frac{kQ}{R^2}\frac{2r\,dr}{\sqrt{z^2 + r^2}}$$

$$= \frac{kQ}{R^2}\int_{z^2}^{z^2+R^2}\frac{du}{\sqrt{u}}, \quad \text{where } u = z^2 + r^2.$$

So:

$$V = \frac{kQ}{R^2}\left(\frac{\sqrt{u}}{\frac{1}{2}}\right)\bigg|_{z^2}^{z^2+R^2} = \frac{2kQ}{R^2}\left(\sqrt{z^2 + R^2} - z\right).$$

25.6 Your car acts like the metal cage in the Boston Museum of Science, so you'll be relatively safe. If a lightning bolt does strike the car, wait several minutes before getting out so any charge can leak away, equalizing potentials of car and ground. This happens faster if you connect the car frame to ground with something metallic. Use gloves! You don't want your body to be the path the current takes!

25.7 From eqn. (25.17), the field is normal to the surface and has magnitude:

$$E = 4\pi k\sigma = 4\pi(9.0 \times 10^9\,\text{N}\cdot\text{m}^2/\text{C}^2)(1.2 \times 10^{-6}\,\text{C/m}^2)$$
$$= 1.4 \times 10^5\,\text{N/C} = 1.4 \times 10^5\,\text{V/m}.$$

25.8 See Figure 25.33. The field lines are perpendicular to the surface and as distance from the object increases, they bend toward the radial pattern expected for a point charge. To accomplish this, they spread apart near the surface at A. The increased distance between the lines indicates a reduced field magnitude and a reduced surface charge density.

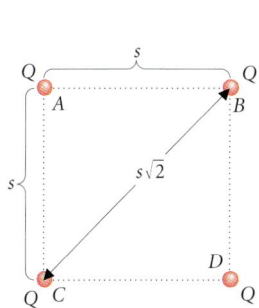

FIGURE 25.31

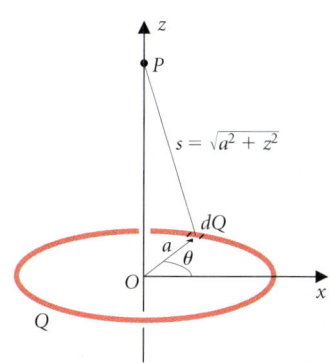

FIGURE 25.32

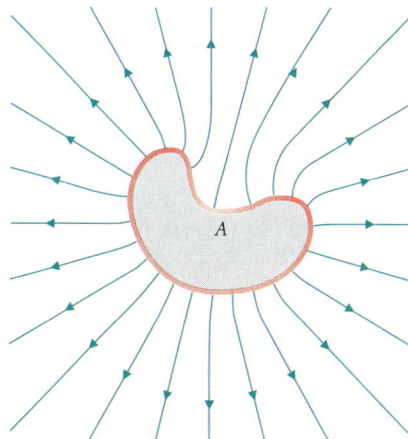

FIGURE 25.33

Basic Skills

Review Questions

§25.1 POTENTIAL ENERGY OF A PAIR OF POINT CHARGES

- What feature of the Coulomb force allows us to define electric potential energy?
- How does the potential energy of a pair of charges depend on the distance between them?

§25.2 ELECTRIC POTENTIAL

- What is the operational definition of electric potential difference?
- State a useful rule for remembering how potential difference is related to electric field.
- When the electric field is due to a system of point charges, where do we usually choose the reference point R such that $V(R)$ is zero?
- What is the unit of electric potential?
- Write two versions of the SI units for electric field.
- What is the electric potential at distance d from a point charge Q if the reference point is at infinity?
- What is an *electron volt*? What physical quantity is measured in electron volts?

§25.3 POTENTIAL ENERGY OF A SYSTEM OF CHARGES

- State the principle of superposition for electric potential.
- Describe a procedure for calculating the potential energy of a system of point charges.
- How is the potential energy of a system related to the energies of its charge pairs?
- How is the potential energy of a system of charges expressed in terms of the potential at the location of each charge?

§25.4 EQUIPOTENTIAL SURFACES

- How are equipotential surfaces related to field lines?
- What can you say about the electric field vector at a position where an equipotential surface crosses itself?
- Is the electric field necessarily zero where the potential is zero? Why or why not?

§25.5 POTENTIAL DUE TO A CONTINUOUS DISTRIBUTION OF CHARGE

- Can the principle of superposition be used to find the potential due to an infinite plane sheet of charge? Why or why not?

§25.6 THE BEHAVIOR OF CONDUCTORS

- What is the static electric field within a conductor?
- How does electric potential vary within a conducting object? Does it make a difference whether the conductor is solid or hollow?
- What is the direction and magnitude of the electric field vector at the surface of a conductor?
- How does the surface charge density on a conductor's surface depend on the shape of the conductor?

Basic Skill Drill

§25.1 POTENTIAL ENERGY OF A PAIR OF POINT CHARGES

1. Two protons are separated by a distance of 5.0 nm. What is the potential energy of the system?
2. Two protons are initially separated by 5.0 nm and are at rest. What is the speed of each proton after they have moved far apart?

§25.2 ELECTRIC POTENTIAL

3. How much work is required to transfer a charge of 6 C against a potential difference of 100 V?
4. What is the potential difference between two points $x = 0$, $y = 0$ and $x = 1.30$ m, $y = 0$ in a region where the field is uniform and equal to $(1.50 \times 10^4 \text{ V/m})\hat{\imath}$?

§25.3 POTENTIAL ENERGY OF A SYSTEM OF CHARGES

5. Three identical charges are fixed at the corners of an equilateral triangle. At the center of the triangle:
(a) only the electric field is zero.
(b) only the electric potential is zero.
(c) both the electric field and the potential are zero.
(d) neither the electric field nor the potential is zero.
Explain why you chose your answer.
6. Three equal charges Q lie in a line, with the distance between two neighboring charges equal to s. What is the potential energy of the system?

§25.4 EQUIPOTENTIAL SURFACES

7. Sketch the equipotential surfaces due to a system of two equal point charges Q separated by a distance d.
8. ■ Figure 25.34 shows several equipotential surfaces in a certain region of space. Each surface is labeled with the value of the potential on the surface. In which region is the magnitude of the electric field greatest? In which direction does $\vec{E}$ point? Explain your reasoning.

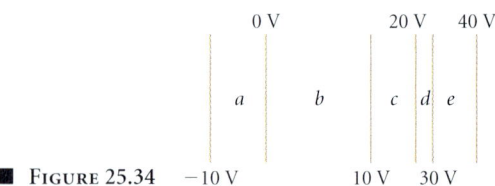

■ **Figure 25.34**

§25.5 POTENTIAL DUE TO A CONTINUOUS DISTRIBUTION OF CHARGE

9. Find the potential at the center of a hemispherical shell of radius a carrying a total charge Q uniformly distributed over it.
10. A filament of length $L = 3.0$ cm carries a uniform charge density $\lambda = 6.7$ μC/m. Estimate the potential at a point on a perpendicular bisector of the filament and 1.8 m away.

§25.6 THE BEHAVIOR OF CONDUCTORS

11. A large point charge Q is held near a metal shell and causes the charge distribution shown in ■ Figure 25.35. Which choice best describes the magnitude of the electric field at point A within the shell?
(a) kQ/D^2 (b) $2\pi k\sigma$ (c) zero (d) $kQ\sigma/(Dd)$ (e) $k\sigma/d^2$
Explain why you chose your answer.

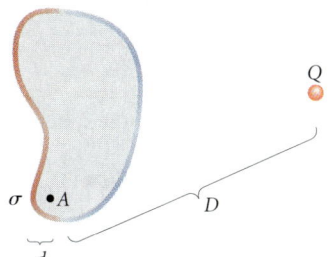

■ FIGURE 25.35

12. A point charge Q is initially a distance R from the center of a *solid* metal sphere of radius $a \ll R$. If the charge is moved to a distance $R/2$, the magnitude of the electric field at the center of the sphere:

(a) increases by a factor 2. (b) decreases by a factor 2.
(c) stays the same.
Explain why you chose your answer.

13. A conducting sphere of radius 10 cm carries a charge Q. A second conducting sphere of radius 20 cm is held at the same potential as the first sphere. The charge on the second sphere is:
(a) Q. (b) $2Q$. (c) $Q/2$. (d) $4Q$. (e) $Q/4$.
Explain your reasoning.

14. A solid conducting object is placed between two infinite, uniformly charged planes (■ Figure 25.36). Which arrow best describes the direction of the electric field at point A? What is wrong with the answers you don't choose?

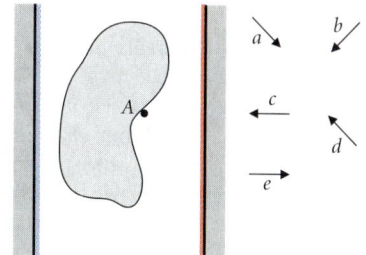

■ FIGURE 25.36

Questions and Problems

§25.1 POTENTIAL ENERGY OF A PAIR OF CHARGES

15. ❖ A point charge Q is at the center of a square. How much work is required to move a second charge q slowly along one side of the square?

16. ◆ (a) Find the potential energy of two point charges of $+3.0\ \mu C$ and $-0.20\ \mu C$ separated by 5.0 cm. (b) How much work would be required to increase their separation to 10.0 cm?

17. ◆ A system of two equal charges separated by 7.5 μm has a potential energy of 25 eV. What is the value of the charge?

18. ◆◆ Two identical point charges, each with charge $Q = 1.00 \times 10^{-6}$ C and mass $m = 1.00 \times 10^{-3}$ g, are released from rest when separated by $d = 0.100$ m. What is their speed when they are very far apart?

19. ◆◆ A point charge Q is fixed in position, and a second object with charge q and mass m moves directly toward it from a great distance. If the initial speed of the object is v, compute the minimum distance between the two objects. If $Q = 1.0\ \mu C$, $q = 1.0$ nC, $m = 1.0 \times 10^{-5}$ kg, and $v = 3.0 \times 10^5$ m/s, what is the minimum distance of approach?

20. ◆◆ Two α-particles (diameter 1.9×10^{-15} m) are headed directly toward each other with equal speeds. Compute the minimum energy in electron volts each particle must have if they are to collide. What initial speed must each particle have?

21. ◆◆ A helium nucleus has radius $r_{He} = 1.9$ fm, mass $m = 6.6 \times 10^{-27}$ kg, and charge $+2e$. A gold nucleus has charge $+87e$ and radius 7.0 fm. What initial speed must a helium nucleus have if it is to come into contact with a fixed gold nucleus in a head-on collision?

(Ernest Rutherford studied this kind of collision in 1910 to establish the small size of atomic nuclei. An accurate calculation requires the use of special relativity, but this answer is good to about 10%.)

22. ◆◆ A uranium nucleus in a reactor captures a slow neutron and divides, or fissions, into two smaller *daughter* nuclei. Assuming the nucleus divides into two equal daughters with charge $Q = 46e$ and diameter $d = 2 \times 10^{-14}$ m, calculate their electric potential energy. (The electrostatic energy is only about 1% of the particles' total energy.)

§25.2 ELECTRIC POTENTIAL

23. ❖ At point C in ■ Figure 25.37, is the potential greater than, less than, or the same as the potential at B? Explain your reasoning.

24. ❖ Is there any physical significance to a uniform change in the value of the electric potential everywhere in space? Why or why not?

25. ❖ What is the potential difference between two points on the x-axis at $x = 1$ cm and $x = 10$ cm if the electric field in the region is $(1.0 \times 10^3\ V/m)\hat{j}$?

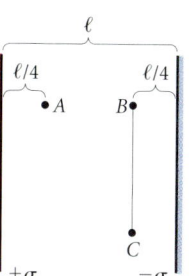

■ FIGURE 25.37

26. ❖ Points A, B, and C are distinct points in space. If it is known that $V_A - V_B = 10.0$ V, and $V_C - V_B = 2.0$ V, calculate:

$$\int_A^C \vec{E} \cdot d\vec{\ell}.$$

27. ◆ Show from the formula for potential due to a point charge that the units of the Coulomb constant must be $[k] = $ V·m/C. Derive this same expression from the known units $[k] = $ N·m²/C².

28. ◆ A proton is accelerated from rest through a potential difference of 0.100 MV. What is the final speed of the proton? Why don't we need to know the electric field strength? ($m_p = 1.67 \times 10^{-27}$ kg)

29. ◆ If $\vec{E} = (16$ V/m$)\,\hat{\imath} + (8.5$ V/m$)\,\hat{\jmath}$, and the potential is zero at the origin, find the potential at point P with coordinates $x = 1.5$ m, $y = 3.5$ m.

30. ◆ Compute the potential difference between points $O(x = 0, y = 0)$ and $P(x = 3.00$ cm, $y = 2.00$ cm$)$ if the electric field in the region is $\vec{E} = (2.50$ kV/m$)(0.300\hat{\imath} + 0.500\hat{\jmath})$.

31. ◆ Point P is 6.6 cm from a uniformly charged concrete wall. Point Q is 27.0 cm from the wall. The potential difference between P and Q is $V_{PQ} = -75.0$ V. Find the electric field vector in the region between P and Q. What is the surface charge density on the wall?

32. ◆◆ An infinite plane sheet of charge has uniform surface density $\sigma = 1.00$ nC/m². What is the separation between equipotential surfaces differing in potential by 10.0 V?

33. ◆◆ Two concentric metal spherical shells have radii $R_1 = 0.200$ m and $R_2 = 0.300$ m. The shells carry equal and opposite charges $Q_1 = -Q_2 = 3.27$ μC. What is the potential difference $\Delta V = V_2 - V_1$ between the shells?

34. ◆◆ An electric field $\vec{E}(x) = E_o(x/a)\hat{\imath}$ exists in a certain region of space. If $E_o = 5.0 \times 10^3$ V/m and $a = 1.0$ cm, find the potential difference between the planes $x = 1.5$ cm and $x = 2.8$ cm.

35. ◆◆ A Geiger counter (Figure 24.3) is operated at a potential difference of 11 kV. If the inner wire has a radius of 25 μm and the chamber radius is 1.05 cm, find the electric field at the edge of the wire and at the outer radius of the chamber.

36. ◆◆◆ Two equal point charges $Q = 1.5$ nC are on the x-axis at $x = -2.5$ cm and $x = +2.5$ cm. Using eqn. (23.9) for the electric field on the y-axis, compute the potential difference between the points $x = 0, y = 1.0$ cm and $x = 0, y = 10.0$ cm.

§25.3 POTENTIAL ENERGY OF A SYSTEM OF CHARGES

37. ❖ A system consists of five point charges. How many distinct charge pairs are there in the system?

38. ❖ If the charge Q in the upper right-hand corner of the square in ■ Figure 25.38 is moved to the center of the square, the energy of the system:
(a) decreases. (b) increases. (c) remains the same.
(d) depends on the path the charge travels on its trip to the center.
Explain how you arrive at your answer.

39. ◆ Compute the electrostatic energy of the system in ■ Figure 25.38.

40. ◆ A positive charge Q and two negative charges $-Q$ are at the corners of an equilateral triangle (■ Figure 25.39). What is the potential at point P?

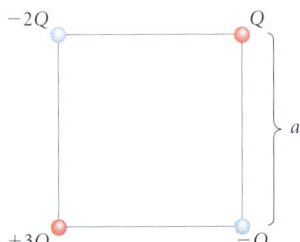

■ FIGURE 25.38

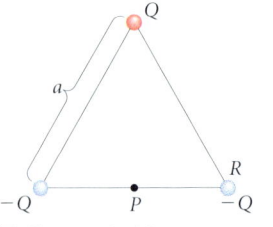

■ FIGURE 25.39

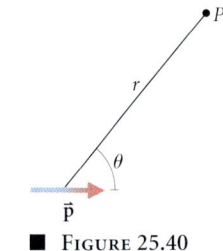

■ FIGURE 25.40

41. ◆ What is the potential energy of the system of charges in ■ Figure 25.39?

42. ◆ What is the potential at point R in ■ Figure 25.39? Which charges do you include in your calculation of potential? Why?

43. ◆◆ Four identical particles, each with mass m and charge Q, are initially at the four corners of a square with side a. If all four particles are simultaneously released from rest, what is their speed when they are a large distance apart? Take $m = 1.0 \times 10^{-3}$ g, $Q = 1.0$ μC, and $a = 1.0$ cm.

44. ◆◆ A point charge $Q = +3.0$ μC is fixed at the origin, and two particles with mass $m = 0.010$ kg and charge $q = +1.0$ μC are on the x-axis at $a = +1.0$ mm and $x = -1.0$ mm. What is the electrical potential energy of the system? If the two particles are simultaneously released from rest, what is their speed when they are a very large distance apart?

45. ◆◆ Two identical charges $+Q$ are fixed at two corners of an equilateral triangle of side ℓ. A third charge q is placed at the third corner and then released. Find the kinetic energy of the third charge when it has reached a great distance from the other two.

46. ◆◆ Two equal positive charges with $Q = 120$ nC are separated by a distance $D = 15$ cm. A particle of mass $m = 1.5$ g and charge $q = 1.0$ nC is placed midway between them. Is the particle in equilibrium? If it is displaced slightly, determine the direction of its motion and find its speed when it is a large distance from the system of $2Q$.

47. ◆◆◆ Find the potential due to a dipole at an arbitrary point with polar coordinates r and θ with respect to the dipole (■ Figure 25.40).

§25.4 EQUIPOTENTIAL SURFACES

48. ❖ Two charges $+Q$ and $-2Q$ are placed as shown in ■ Figure 25.41. In which of the three regions A, B, or C is it possible for the potential to be zero? Explain your reasoning.

49. ❖ Sketch the equipotential surfaces for two equal and opposite charges a distance s apart.

50. ❖ The equipotential surfaces in ■ Figure 25.42 are spaced at equal intervals of 10^{-3} V. Using the distance scale at the bottom of the figure, estimate the magnitude and show the direction of the electric field at points A and B.

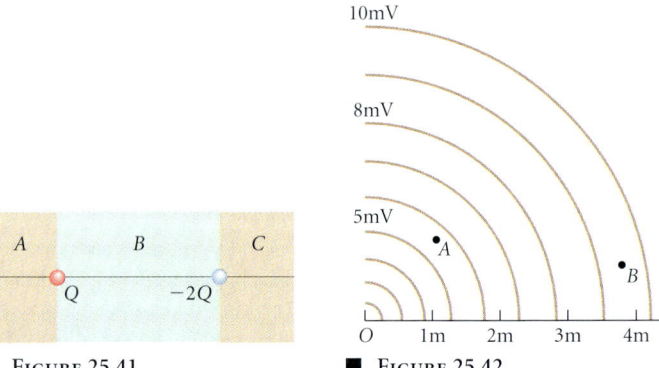

■ FIGURE 25.41

■ FIGURE 25.42

CHAPTER 25 • ELECTRIC POTENTIAL ENERGY

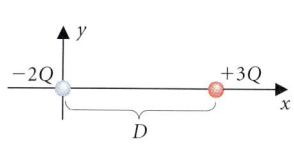

■ FIGURE 25.43

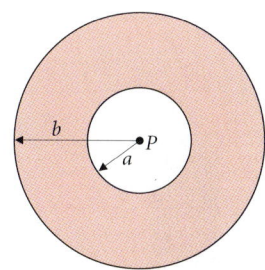

■ FIGURE 25.44

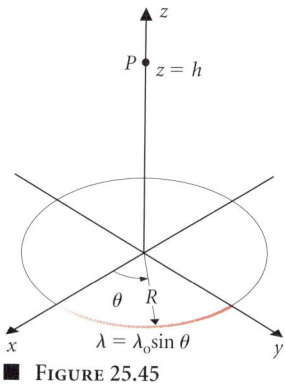

■ FIGURE 25.45

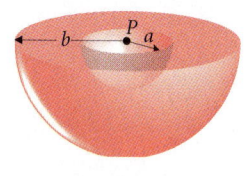

■ FIGURE 25.46

51. ❖ Three equal positive charges are fixed at the corners of an equilateral triangle. Sketch the pattern of equipotential surfaces for the system where they intersect the plane of the triangle. Using eight lines per charge, sketch the electric field line diagram in the plane of the triangle and compare with the equipotential surfaces.

52. ❖ A point charge $+Q$ is at the origin, and two negative charges $-Q$ are on the x-axis at $x = -a$ and $x = +a$. Sketch the equipotential surfaces for the system where they intersect the x-y-plane. Compare with the electric field line diagram in the x-y-plane.

53. ❖ A point charge Q is a distance D above a uniformly charged plane sheet with surface charge density σ. Both Q and σ are positive. Sketch the field line diagram and the equipotential surfaces for this system.

54. ◆◆ Two point charges $-2Q$ and $+3Q$ are on the x-axis at the origin and at $x = D$ (■ Figure 25.43). Find two points on the x-axis (not at infinity) where the electric potential is zero.

55. ◆◆ Two point charges of $+1.5\ \mu C$ and $-3.0\ \mu C$ are on the y-axis at $y = 0$ and $y = 2.0$ cm. Find all the points on the y-axis (not at infinity) where **(a)** $V = 0$ and **(b)** $\vec{E} = 0$. What is the value of the potential where $\vec{E} = 0$?

56. ◆◆ A point charge $3Q/4$ is at the origin, and two charges $-Q/2$ are on the x-axis at $x = -L$ and $x = +L$. Find points on the x-axis and on the y-axis where $V = 0$.

§25.5 POTENTIAL DUE TO A CONTINUOUS DISTRIBUTION OF CHARGE

57. ❖ A filament of length L carries a uniform charge density λ. Is the potential at P larger **(a)** when the filament is formed into a circle with its center at P or **(b)** if the filament is straightened out and lies along a tangent to the original circle?

58. ❖ Sketch the equipotential surfaces produced by a uniformly charged ring. Make the sketch in a plane perpendicular to the plane of the ring and through its center.

59. ◆ ✉ Model yourself as a sphere 30 cm in diameter receiving charge from the rug as you walk over it. How much charge do you need to reach a potential of 10^4 V?

60. ◆ A charge of $1.0\ \mu C$ is uniformly distributed on a thin spherical shell of radius 0.20 m. Find the potential at the center of the shell. What is the potential at points outside the shell? Draw a graph showing the value of the electric potential as a function of distance from the center of the shell.

61. ◆◆ Points A and B lie between two infinite, uniformly charged planes with surface charge densities $\pm\sigma$ (Figure 25.37). The potential difference $\Delta V = V_A - V_B$ is:
(a) 0. **(b)** $+4\pi k\sigma\ell$. **(c)** $-4\pi k\sigma\ell$. **(d)** $+2\pi k\sigma\ell$. **(e)** $-2\pi k\sigma\ell$.
What is incorrect about the answers you didn't choose?

62. ◆◆ A test particle with charge q and mass m is projected from a large distance along the axis of a uniformly charged ring of radius a and charge Q. What initial speed is required if the particle is to pass through the center of the ring?

63. ◆◆◆ A rod lies between $x = -L/2$ and $+L/2$. It carries a linear charge density λ C/m. Find expressions for the potential at a point on the y-axis by three methods. **(a)** For $y \ll L$, treat L as infinite. **(b)** For $y \gg L$, use Coulomb's law. **(c)** Do an exact calculation using integration. Compare the results of (a), (b), and (c) graphically as functions of y and determine the ranges of y in which (a) and (b) are reasonable approximations. Compare your results with those from §24.2.3.

64. ◆◆◆ Find the potential at the center of a plane annular ring of inner radius a and outer radius b carrying a surface charge density σ (■ Figure 25.44).

65. ◆◆◆ An object in the shape of a quarter circle with radius R lies in the x-y-plane (■ Figure 25.45). If the object carries a charge per unit length given by $\lambda = \lambda_o \sin\theta$, find the electric potential at a point P at $z = h$ on the z-axis.

66. ◆◆◆ **(a)** Find the potential at an arbitrary point due to two parallel, oppositely charged infinite filaments (linear charge densities λ and $-\lambda$) separated by a distance d. **(b)** Explain why the mathematical difficulties that occur for a single infinite filament do not occur for this pair.

67. ◆◆◆ A uniformly charged hemisphere of radius b and charge density ρ has a hemispherical depression of radius a cut from its center (■ Figure 25.46). Find the potential at point P, the center of the depression.

§25.6 THE BEHAVIOR OF CONDUCTORS

68. ❖ Each of the infinitely long, cylindrical, conducting objects shown in ■ Figure 25.47 carries the same total charge. Sketch a field line diagram in the x-y-plane for each.

69. ❖ Two lumps of metal, one uncharged and the other carrying a net charge $-Q$, are placed near each other. Make a sketch of the situation showing qualitatively the charge densities on the surfaces of the conductors. Is there any net electric force acting between the conductors? Explain your reasoning.

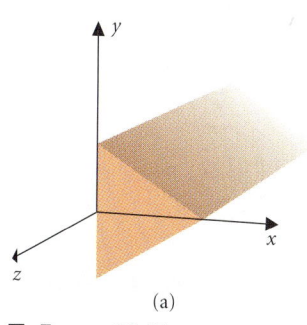
(a)

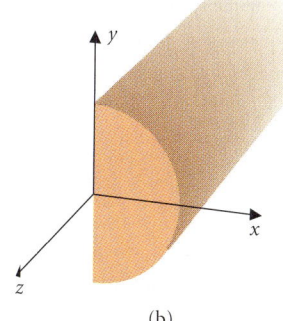
(b)

■ FIGURE 25.47

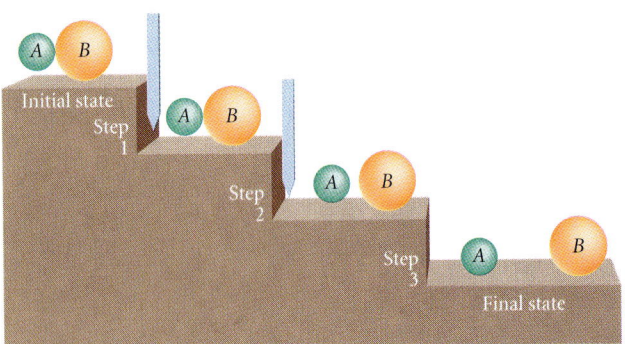

■ FIGURE 25.48

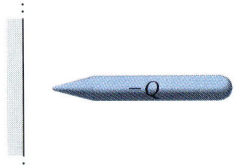

■ FIGURE 25.50   ■ FIGURE 25.51

70. ❖ Two conducting spheres are initially in contact and uncharged. They undergo the process depicted in ■ Figure 25.48. Step 1: A negatively charged object is brought up and held close to the smaller sphere. Step 2: With the charged object held fixed, the two spheres are separated using insulated gloves. Step 3: The charged object is removed. Let Q_A and Q_B be the charges on the spheres in the final state. Then Q_A is:
(a) greater in magnitude than Q_B and of the same sign.
(b) greater in magnitude than Q_B and of the opposite sign.
(c) lesser in magnitude than Q_B and of the same sign.
(d) lesser in magnitude than Q_B and of the opposite sign.
(e) equal in magnitude to Q_B and of the opposite sign.
Explain your reasoning. If the two spheres are then separated by a great distance, which will be at the greater potential? Why? Which potential will be of greater magnitude? Why?

71. ❖ (a) A spherical conducting shell and a solid conducting ball are a great distance apart (■ Figure 25.49a). The shell has a positive charge Q_s, and the ball has a negative charge Q_b. $|Q_b| > |Q_s|$. With $V = 0$ at infinity, which of the diagrams in Figure 25.49b correctly represents the relation between the potential of the shell V_s and of the ball V_b? Explain your reasoning. (b) The ball is placed inside the shell without touching it (Figure 25.49c). Which of the diagrams in Figure 25.49b now represents the potentials of the two objects?

72. ❖ Explain why the electric field between the plates in Example 25.5 should be very nearly uniform.

73. ❖ A coaxial cable consists of a long, straight metal wire at the center of a thin cylindrical sheath of woven metal fibers (■ Figure 25.50). If the wire has a uniform charge density λ, describe the charge density (if any) that resides on the interior and exterior surfaces of the sheath.

74. ❖ A lump of conducting material with no net charge has an irregular hole within it, inside of which is a point charge Q. What charges reside on the inner and outer surfaces of the lump? *Approximately* what is the electric field at a great distance r from the lump?

75. ❖ Sketch a field line and equipotential surface diagram for the configuration of conductors shown in ■ Figure 25.51, if the object carries a charge $-Q$ and the plane is grounded.

76. ❖ A hollow conducting bowl, ■ Figure 25.52, carries a large amount of charge. The location of the charge is tested by touching the bowl with a proof plane—a small metal disk mounted on an insulating rod. Compare the charges received by a plane that touches the bowl at B and one that touches at A. Explain your reasoning.

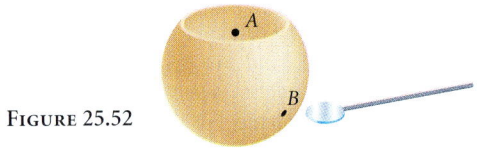

■ FIGURE 25.52

77. ❖ An electroscope (■ Figure 25.53) consists of a metal ball mounted on a rod that connects to a metal plate and a thin metal foil. When there is a net charge on the plate and foil, repulsive electric force causes the foil to hang away from the plate at an angle. When a charged object is brought near the electroscope ball, the foil rises. Why? If you touch the ball of the electroscope with your finger while the charged object is held near, the foil falls. Why? If you remove first your finger and then the charged object, the foil lifts once again. Explain this behavior. If the charge on the object is positive, what is the sign of charge remaining on the electroscope?

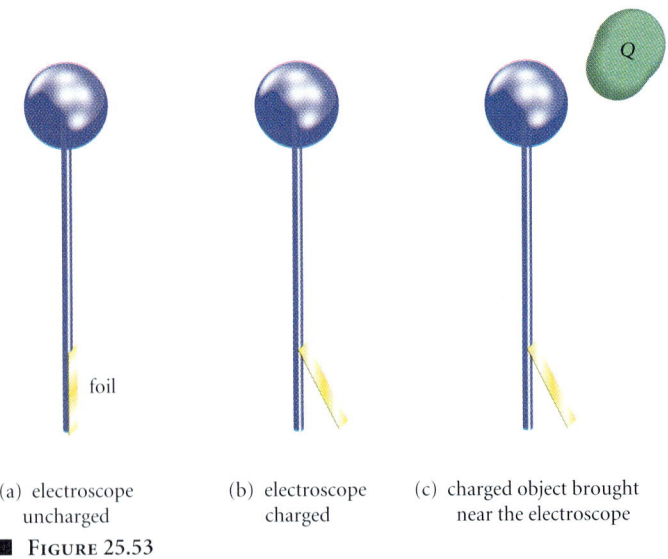

(a) electroscope uncharged   (b) electroscope charged   (c) charged object brought near the electroscope

■ FIGURE 25.53

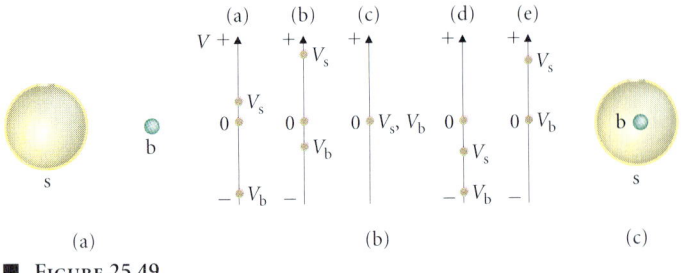

(a)

(b)

(c)

■ FIGURE 25.49

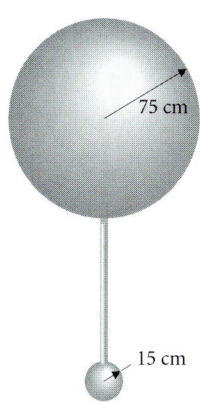

■ FIGURE 25.54

78. ◆ The electric field at point P on the surface of a conductor is measured to be 157 kV/m. What is the surface charge density on the conductor at P?

79. ◆ A copper cube has a charge density of 0.78 $\mu C/m^2$ at the center of each face. What is the electric field outside the cube near the center of the face? Discuss *qualitatively* how the field strength changes near the edges and corners of the cube.

80. ◆◆ ✉ A dumbbell-shaped conductor has spherical ends of radius 15 cm and 75 cm, respectively (■ Figure 25.54). The electric field at the surface of the small end is measured to be 98 kV/m. What is the electric field at the other end? What is the potential at the surface? Estimate the total charge on the object.

81. ◆◆ ✉ A steel needle 0.7 cm long hangs from a string in front of a plastic sheet that has a uniform surface charge density of 300 $\mu C/m^2$. Estimate how large a point charge at each end of the needle models the induced charge distribution. What is the net force on the needle?

82. ◆◆◆ ✉ Assume that the 1-cm-long needle in Example 25.14 is made of copper (density 9×10^3 kg/m^3). A copper atom has a mass of approximately 10^{-25} kg and contributes two free electrons to the metal. What fraction f of the needle's electrons must be displaced to balance the external charge? Rather than some electrons moving the length of the needle, all the electrons are displaced a small distance $\Delta \ell$. Show that $\Delta \ell / \ell \approx f$, where ℓ is the length of the needle. Assuming constant acceleration, calculate the time required for an electron to move the same fraction of the needle's length if it is exposed to the unbalanced electric field of the point charge. (Use the value of the electric field at the center of the needle.) A change in the electric field propagates at roughly the speed of light. At that speed how long would it take for an abrupt change in the electric field to propagate through the needle? Which calculation provides the better estimate of the time required for the needle to reach equilibrium?

Additional Problems

83. ◆ The chemical energy stored in a gallon of gasoline is about 130 MJ. How many coulombs would have to be stored at a potential of 12 V to provide the same energy as a 15-gallon tank of gas? If one battery can deliver 5 C/s for 10 h, how many would be needed to match the energy in the gas tank?

84. ◆ Use the fact that charge multiplied by potential difference gives potential energy to express the volt in terms of the meter, the coulomb, the kilogram, and the second.

85. ◆◆ Two concentric spherical metal shells of radii 0.350 m and 0.150 m are maintained at a potential difference of 175 kV. What is the electric field at a radius of 0.250 m?

86. ◆◆ In the Bohr model of the hydrogen atom, a single electron orbits a single proton in a circle of radius $a_0 = 5.29 \times 10^{-11}$ m. Find the energy needed to ionize the atom, that is, to remove the electron to infinity. Express your result in electron volts.

87. ◆◆ Two parallel plastic sheets are separated by a distance $d = 1.3$ mm. If they carry equal and opposite surface charge densities $\sigma = 1.3\ \mu C/m^2$, what is the potential difference between them? Describe what happens if a metal slab of thickness $t = 1.0$ mm is inserted between the plastic sheets. What is the new potential difference between the sheets?

88. ◆◆ A proton with initial kinetic energy of 5 MeV is headed directly toward a distant positively charged sphere of total charge $Q = 5 \times 10^{-5}$ C and radius $R = 0.1$ m. Does the proton hit the sphere? If so, what is the kinetic energy of the proton as it hits? With what speed does the proton hit the sphere?

89. ◆◆ The probe tip for a scanning tunneling microscope has radius of curvature $r \approx 10^{-9}$ m. If the electrode is held at a potential of 100 mV with respect to the surface to be scanned, estimate the electric field at the tip of the electrode.

90. ◆◆ The breakdown field for dry air is about 10^6 V/m. What charge on an aluminum sphere 20 cm in radius will result in a spark discharge from the sphere?

91. ◆◆ A Van de Graaff machine has a dome of radius 2.0 m (■ Figure 25.55). A motor causes a charged insulating belt to travel continuously between grounded spikes at the bottom of the device and another set of spikes connected to the dome at the top. Charge is transferred through the upper spikes to the outer surface of the dome at a rate of 17 μC/s. How long does it take to charge the dome to 3.0 MV? How much power is required to operate the machine when the dome potential is 3.0 MV?

92. ◆◆ At what temperature do two protons have enough energy on average to reach a separation of 2 fm in a head-on collision? How can the Sun (central temperature about 1.3×10^7 K) succeed in generating energy by nuclear fusion? (*Hint:* Do all the protons have the same energy in thermal equilibrium?)

93. ◆◆◆ Verify the result of Example 25.8 by calculating the work done to bring up the third charge using eqn. (23.9) for the electric field of the first two charges.

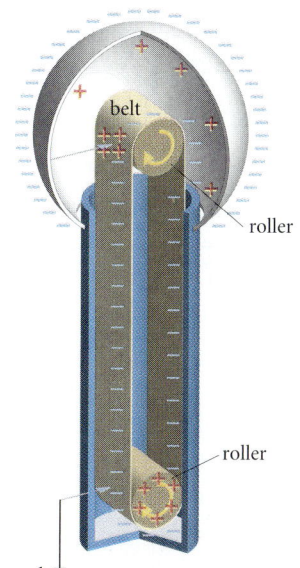

■ FIGURE 25.55 to ground

Computer Problems

94. Three point charges are placed on the x-axis. Two, each 1.0 μC, are at $x = -1.0$ m and $x = 0.0$. The third, -1.0 μC, is at $x = +1.0$ m. Find all the points on the x-axis, not at infinity, where (a) $V = 0$ and (b) $\vec{E} = 0$. What is the value of the potential where $\vec{E} = 0$?

95. Set up a spreadsheet program to calculate the potential due to a set of point charges. Find the potential at points P_1 (coordinates $x = 1.5$ m, $y = 2.5$ m, $z = 0.3$ m) and P_2 ($x = -0.45$ m, $y = 1.7$ m, $z = -2.2$ m) due to the set of charges listed in file CH25P95 on the supplementary computer disk.

96. Set up a spreadsheet program to calculate the potential energy of a system of charges. Find the potential energy of the system in file CH25P95 (cf. Problem 95).

Challenge Problems

97. A filament extends from $x = -a$ to $x = +a$. Its charge density is given by: $\lambda(x) = \lambda_o(x/a)$. Find the potential produced by the filament at a point on the x-axis with $|x| > a$. Compare your result for $|x| \gg a$ with the result for a dipole (cf. Exercise 25.2), and determine the dipole moment of the filament.

98. A point charge $+3Q$ is at the origin, and two negative charges $-Q$ are on the x-axis at $x = d$ and $x = 2d$. Sketch the equipotential surfaces for the system where they intersect the x-y-plane. Compare your sketch with the field line diagram in the x-y-plane.

99. (a) Find the potential due to a uniformly charged sphere of radius R and charge density ρ, at points both outside and inside the sphere. (b) Use the result of part (a) to find the electric potential energy stored in the sphere. Compare your result with the work necessary to assemble the sphere by adding successive thin spherical shells of charge.

100. Find the potential on the axis of a disk of radius a, a distance d from it, if the disk has a charge density $\sigma = \sigma_o(r/a)^2$.

101. A hollow cylindrical object of radius a and length $2L$ (■ Figure 25.56) is uniformly charged with density σ. Find the electric potential at an arbitrary point P on the z-axis.

102. Three point charges, one $+Q$ and two $-Q$, lie at the vertices of an equilateral triangle. Draw two diagrams, one showing electric field

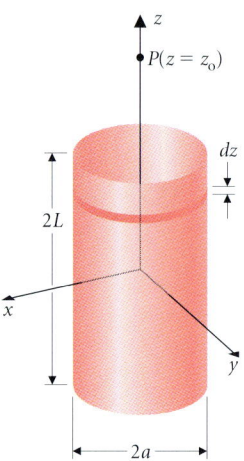

■ **FIGURE 25.56**

lines and the other showing equipotential surfaces in the plane of the triangle. *State* how you use symmetry, behavior of the fields at large distances, and behavior of the fields near the charges. Show qualitatively the locations of any points where $\vec{E} = 0$ and the $V = 0$ equipotential surfaces.

Hint for problems 103 and 104: Show in each case that the problem reduces to solving the differential equation:

$$\frac{d\ell}{\sqrt{\dfrac{1}{\ell} - \dfrac{1}{\ell_o}}} = -(\text{constant})\ dt,$$

where ℓ is a characteristic dimension of the system, ℓ_o is its value when the system is at rest, and the constant is of the form $\sqrt{2kq^2/m}$ times a factor that depends on the specific system. Then show that the equation can be solved using the substitution $\ell/\ell_o = \cos^2 \theta$.

103. Two particles each with mass m but with equal and opposite charges q and $-q$ are released from rest when separated by distance d. Find the time required for them to collide.

104. Find the time taken for the square of charges in Study Problem 16 to collapse to zero size. Take $a = 1.0$ m, $Q = 1.0$ μC, and $m = 1.0$ mg. *Hint:* When the square side shrinks by ds, how far does each charge move?

This person is using a simple electric circuit. The car battery supplies the power to operate the light. Wires from the battery form a closed current path. In this chapter we'll discuss some basic circuit designs and the principles that allow power transfer from electrical sources to practical devices.

CHAPTER 26
Introduction to Electric Circuits

Concepts
Current
Electromotive force
Resistance
Conductivity/resistivity
Ohm's law
Ohmic losses
Series and parallel circuits
Equivalent circuit
Kirchhoff's rules

Goals
Be able to:
Compute the resistance of a simple object given its resistivity.

Analyze simple circuits using series/parallel relations or Kirchhoff's rules.

Compute the power requirements for simple circuits.

*A smell of burning fills the startled air—
The electrician is no longer there!*

Hilaire Belloc

A person reading a road map in a car late at night is using one of the simplest and most common applications of electric current. Charge flows through wires to the roof lamp and back to the battery. Energy is supplied by the car's battery and is converted to heat and light in the lamp. This system illustrates the three basic elements in all circuits: an energy source, a closed current path, and a practical device that uses energy (■ Figure 26.1). In this chapter we shall study the principles on which these elements operate and then discuss how to apply these principles in more complex situations. For the moment we consider only circuits in which current is constant in time.

26.1 Basic Circuit Behavior

26.1.1 Electric Current

Some electrons in a conductor can move in response to an applied electric field. Electrons quickly build up at the surface of the conductor and neutralize the applied field (§25.6). But if we remove electrons from the conductor as they arrive at the surface and put them back in at the other end, the result is a continuous flow of electrons—an electric current. You can make a device that does this by sticking an iron nail and a copper spike into a lemon. Chemical reactions at the spike remove electrons from the copper, while reactions at the nail push electrons into the iron. The device drives a small current through a wire connected between its iron and copper *electrodes* (■ Figure 26.2).

Benjamin Franklin envisioned electric fluid as positively charged. This historical fact has left us with the sign convention illustrated in ■ Figure 26.3. Negatively charged electrons flowing through a conducting wire toward the left constitute a current that is described as an equivalent flow of positive charge to the right. (The actual positive charges—atomic nuclei—oscillate about their equilibrium positions and do not contribute to the current.)

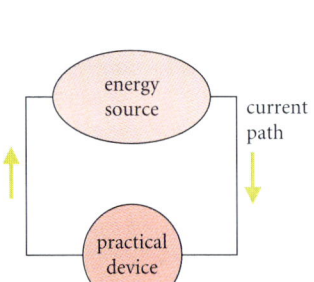

■ Figure 26.1
Schematic of an electric circuit. The purpose of any circuit is to transfer power from the electric source to the practical device.

This definition refers to the magnitude of the current. Sign conventions are important in analyzing circuits. See §26.4.

> The current in a wire is the rate at which charge flows through any cross section of the wire:
> $$I \equiv \pm dQ/dt. \tag{26.1}$$

The SI unit of current is the ampere: $\quad 1 \text{ A} = 1 \text{ C/s}.$

Remember: The ampere is the SI base unit: $1 \text{ C} = 1 \text{ A} \cdot \text{s}$. See Part VI introduction and Chapter 29.

26.1.2 Batteries and Electromotive Force

We can make an analogy between the flow of electrons in a wire and fluid flow. Imagine a large pile of rocks on a hillside that is porous enough for water to flow through it, over and around the rocks (■ Figure 26.4). Water starts at the top of the hill and flows down through

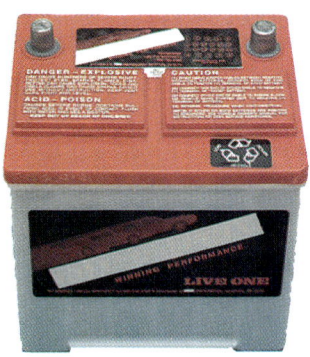

■ Figure 26.2
A simple battery is easily constructed from a lemon, a nail, and a piece of copper wire. The voltmeter indicates the potential difference between the terminals.

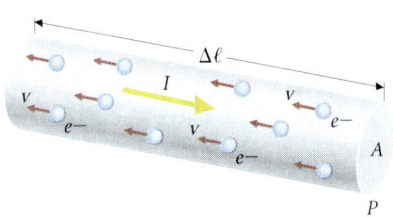

■ Figure 26.3
Definition of electric current. Negatively charged electrons moving to the left constitute a rightward electric current I. The current is the rate at which charge flows through any cross section of the wire.

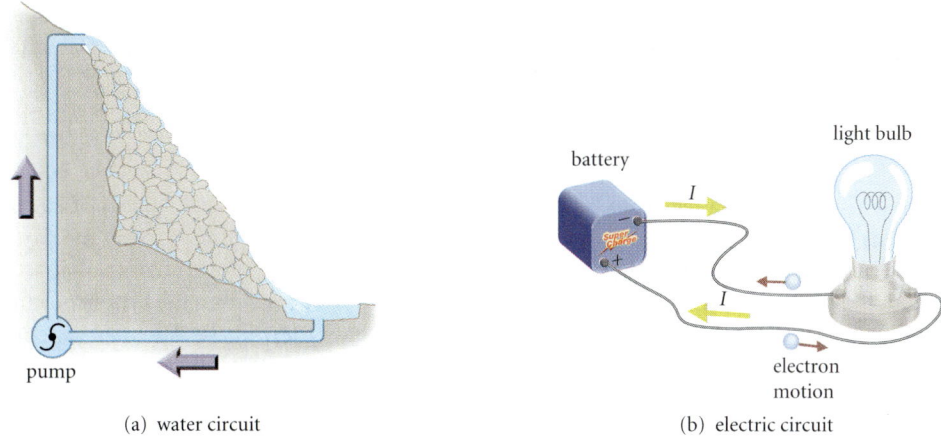

■ **FIGURE 26.4**
(a) A water circuit consists of a pump and a continuous stream of water that flows downhill through a pile of rocks. (b) In an electric circuit, the battery raises the electric potential of charge in the same way that the pump raises the gravitational potential of water. Here the arrow shows the direction of electric current I. The electrons move in the opposite direction, indicated by the brown arrows.

the pile. The gravitational potential energy of the water is less at the bottom than at the top, and the difference is used up by *friction* as the water flows through the pile. If the water is to circulate continuously, a pump is needed to lift it from the bottom to the top of the hill. In an electrical system an energy source such as a battery acts as a pump, and a device such as a lamp corresponds to the pile of rocks. The electrical "pump" is called a source of *electromotive force,* or emf.

Batteries use stored chemical energy to replace energy lost by the electrons moving through the conducting circuit. Chemical energy is another form of electric potential energy. Electrons arrange themselves around atomic nuclei so as to have the minimum possible potential energy. If two atoms come close together and can share their electrons so as to reduce their combined potential energy, they bind together to form a chemical compound. In a car battery (■ Figure 26.5) reactions occur at each terminal. At the positive terminal, lead oxide reacts with positive ions in the sulfuric acid solution, and at the negative terminal metallic lead reacts with negative sulfate ions. As a result of these reactions, charges of opposite sign build up on the terminals. The electric field arising from this charge separation creates a potential difference between the two terminals. A single lead-acid battery cell develops a 2-V potential difference. Six such cells are connected together in a standard 12-V battery. The potential difference developed by the battery is called its electromotive force. In spite of its name, emf is *not* a force but is the energy transferred to each unit of charge that moves between the terminals.

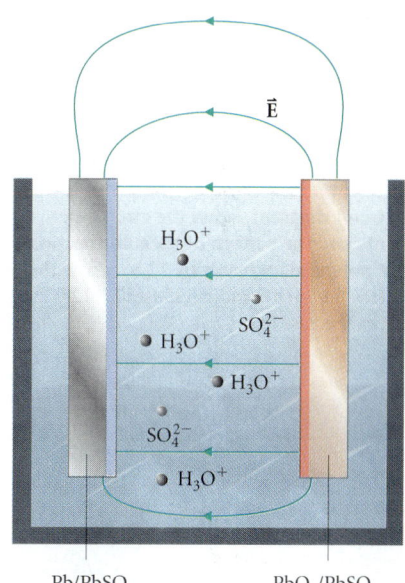

■ **FIGURE 26.5**
At the negative terminal of the battery, lead atoms react with sulfate ions (SO_4^{2-}) in the solution. Each reaction delivers a charge $-2e$ to the negative terminal. At the positive terminal, lead oxide PbO_2 reacts with hydronium ($H_3O_2^+$ = water plus an extra proton) and sulfate ions to produce lead sulfate and water. Each reaction delivers a charge $+2e$ to the positive terminal. The chemical reactions release approximately 2 eV for each electron removed from the positive plate and added to the negative plate and deplete the acid in the battery. The reactions reach equilibrium when the potential difference between the plates is about 2 V. When the battery is used, electrons flow from the negative plate to the positive plate through the external circuit, disturbing the chemical equilibrium. Inside the battery, more electrons flow to the negative terminal to compensate. To charge the battery, current is forced through it in the opposite direction, reversing the chemical reactions and replenishing the battery acid.

> The *electromotive force* (emf) $\mathcal{E}$ of a battery or other electric power source is the value of the potential difference it maintains between its terminals in the absence of current.

Generators, solar power cells, and the like are also sources of emf that convert energy in other forms to electric potential energy. ■ Figure 26.6 shows the symbol used to represent an emf in a circuit. The longer bar represents the positive terminal.

Electrons flowing through a battery collide with electrons and molecules in the battery fluid. Some of the electrical energy they gain is dissipated as heat, and the potential difference

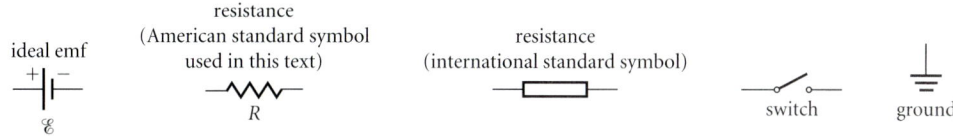

■ **FIGURE 26.6**
Standard circuit symbols for an ideal emf or battery, a resistance, a switch, and connection to ground. The positive battery terminal is indicated by the longer line. Batteries are sometimes indicated by a series of these symbols. In this text we use the American standard symbol for resistance. The international standard symbol is also shown here. Potential is taken to be zero at ground.

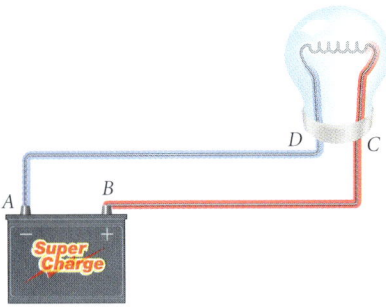

■ FIGURE 26.7
The circuit connecting a car dome light to the battery. The charge distribution in the circuit is set up immediately after the switch is closed (Problem 25.82) to produce the necessary electric fields. The field in the wires is small compared with that in the light bulb filament. The potential difference across the lamp equals that across the battery.

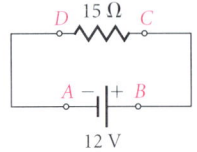

■ FIGURE 26.8
A circuit diagram shows the circuit elements and the pattern of their connections. The potential increases by 12 V across the battery ($A \rightarrow B$), and decreases by 12 V across the resistor ($C \rightarrow D$).

■ FIGURE 26.9
Georg Simon Ohm (1787–1854). Based on his experiments with batteries, Ohm derived eqn. (26.2) in 1826. He was motivated by the connection between temperature difference and heat flow (see Chapter 21). Ohm's theory was confirmed experimentally by Fechner (1801–1887) and Wheatstone (1802–1875) (see §26.5). However, Ohm did not receive official recognition for his discovery until 1841, when he was awarded the Copley medal of the Royal Society. He was finally promoted to a university professorship 22 years after his discovery, 5 years before his death.

across the battery terminals is reduced. When a battery provides current to an external circuit, the potential difference across its terminals is less than its emf. The measured potential difference is called the battery's *terminal voltage.*

26.1.3 Resistance

We may use an ideal battery with no internal resistance to illustrate how the ideas of electric potential that we developed in Chapter 25 are applied to electric circuits. Closing the switch on the car's dome light completes a conducting path from the battery through the light and back to the battery (■ Figure 26.7). The complex chemical reactions within the battery produce a simple result: a fixed potential difference across the battery terminals. An equal potential difference exists across the light bulb. Immediately after the switch is closed, a very small amount of charge flows through the battery and is distributed in the circuit to produce the necessary electric field. A negligible electric field is sufficient to maintain current through the thick connecting wires; a much larger field is required in the light bulb's thin filament. The bulb is an example of an electrical *resistance.* ■ Figure 26.8 is an abstract diagram of the circuit using the standard symbol for resistance.

To see why the potential difference across the light equals that across the battery, recall that the potential difference between any two points is independent of the path used to compute it:

$$\Delta V = V_B - V_A = V_{BA} = -\int_A^B \vec{E} \cdot d\vec{\ell}.$$

If the two points A and B are the terminals of the ideal battery (Figure 26.8), the potential difference V_{BA} is the electromotive force of the battery. Along a path from A to B through the connecting wires and the light bulb:

$$V_{BA} = V_{BC} + V_{CD} + V_{DA} = (V_{BC} + V_{DA}) + V_{CD}.$$

$$V_{BA} = -\underbrace{\left(\int_C^B + \int_A^D\right)\vec{E} \cdot d\vec{\ell}}_{\text{connecting wires}} - \underbrace{\int_D^C \vec{E} \cdot d\vec{\ell}}_{\text{light bulb}}.$$

The first term may usually be ignored because it is much smaller than the second. The second term is the potential difference across the light bulb.

The potential difference V_{CD} and the resulting current I through the light bulb are the two quantities that describe the bulb's behavior in the circuit. Their ratio gives an operational definition of the bulb's resistance to charge flow:

> The *electrical resistance* of a circuit component is the potential difference between its terminals divided by the current that is established in response to that potential difference:
>
> $$R \equiv \Delta V / I. \qquad (26.2)$$

The unit of resistance is the ohm (symbol Ω), after Georg Simon Ohm (■ Figure 26.9). One ohm equals 1 volt per ampere:

$$1\ \Omega \equiv \frac{1\ \text{V}}{1\ \text{A}} = \frac{1\ \text{J/C}}{1\ \text{C/s}} = 1\ \text{J} \cdot \text{s}/\text{C}^2.$$

EXAMPLE 26.1 ♦ The potential difference between a light bulb's terminals is 12 V when it carries a current of 0.80 A. What is the resistance of the bulb?

MODEL We apply the definition of resistance, eqn. (26.2):

SETUP AND SOLVE
$$R = \frac{\Delta V}{I} = \frac{12 \text{ V}}{0.80 \text{ A}} = 15 \text{ }\Omega.$$

THE SYMBOL Ω IS THE GREEK CAPITAL OMEGA (APPENDIX IIA).

ANALYZE Another light bulb carrying a current of 1.60 A with the same potential difference across it would have half the resistance, or 7.5 Ω.

A useful model for a real battery is an ideal, lossless emf together with an *internal resistance*. The resistance accounts for the difference between emf and terminal voltage as well as for the energy required to move charge through the battery (■ Figure 26.10). Efficient batteries have very small internal resistance. A *dead* battery has a large internal resistance.

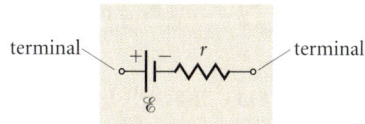

■ **FIGURE 26.10**
A model for a real battery. Power loss within the battery is represented by the internal resistance r connected to the ideal emf. This end-to-end connection is called a series connection (see §26.3.1).

EXAMPLE 26.2 ♦♦ A 12.0-V car battery has a terminal voltage of 11.7 V when supplying a current of 4.8 A to the automobile circuit. What is the battery's internal resistance?

MODEL The terminal voltage is the emf minus the potential drop across the internal resistance.

SETUP We use eqn. (26.2) to express the potential difference ΔV_{int} across the internal resistance r in terms of the current I:

$$\Delta V_{\text{term}} = \mathcal{E} - \Delta V_{\text{int}} = \mathcal{E} - Ir.$$

SOLVE
$$r = \frac{\mathcal{E} - \Delta V_{\text{term}}}{I} = \frac{12.0 \text{ V} - 11.7 \text{ V}}{4.8 \text{ A}} = 0.06 \text{ }\Omega.$$

ANALYZE A good car battery has an internal resistance of less than 0.02 Ω. This internal resistance is much larger than normal; the battery is in bad shape.

26.1.4 Energy Relations in a Simple Circuit

As electrons flow through a circuit, chemical reactions in the battery raise their potential energy. In the rest of the circuit, this potential energy is converted to thermal energy. In ordinary conductors, the electrons are accelerated by the electric field but collide frequently in the material. After a collision, an electron's velocity is in a random direction, unrelated to the direction of the applied force. Its kinetic energy is converted to energy of random motion (thermal energy), and a rise in temperature is observed. Electric current nearly always dissipates energy as heat. This is a nuisance in a computer but is the purpose of an electric toaster.

The potential difference between the ends of the resistor in Figure 26.8 is $\Delta V = V_{CD} = V_C - V_D$. An amount of charge δq passing through this component loses potential energy $\delta U = V_{CD} \, \delta q$. If the charge δq passes through the resistor in time δt, the energy is dissipated as heat at a rate:

$$P_\Omega \equiv \frac{\delta U}{\delta t} = \frac{V_{CD} \, \delta q}{\delta t} = I \, \Delta V. \tag{26.3a}$$

ELECTRIC CURRENT IN SUPERCONDUCTORS IS AN IMPORTANT EXCEPTION. SEE DIGGING DEEPER.

WE'LL DISCUSS THE COLLISIONS FURTHER IN §26.2.

WE USED EQN. (26.1) FOR I. SEE §7.2 FOR THE DEFINITION OF POWER.

This power is called rate of ohmic loss or Joule heating. The former term emphasizes loss of electrical energy, and the latter emphasizes conversion to thermal energy.

Using eqn. (26.2), we may express the power in terms of resistance:

$$P_\Omega = I(IR) = I^2 R, \tag{26.3b}$$

and

$$P_\Omega = \frac{\Delta V}{R} \Delta V = \frac{(\Delta V)^2}{R}. \tag{26.3c}$$

An equal amount of power is generated by the battery to maintain the current in the resistor. Charge δq passing through the battery gains potential energy $\delta U = \mathcal{E}\delta q$, so the power output of the battery is:

$$P_{\mathcal{E}} \equiv \frac{\delta U}{\delta t} = \frac{\mathcal{E}\delta q}{\delta t} = I\mathcal{E}.$$

THIS RESULT IS REQUIRED BY CONSERVATION OF ENERGY.

Since the current is the same in each case, and $\Delta V = \mathcal{E}$, $P_{\mathcal{E}} = P_{\Omega}$.

EXAMPLE 26.3 ♦ A toaster draws a current $I = 6.7$ A and has resistance $R = 18\,\Omega$. How much power does the toaster require?

MODEL Since we are given I and R, we use eqn. (26.3b) to compute the power dissipated.

SETUP AND SOLVE $P_{\Omega} = I^2 R = (6.7\text{ A})^2\,(18\,\Omega) = 0.81$ kW.

ANALYZE Check the units:

$$\text{A}^2 \cdot \Omega = \text{A}^2\,\frac{\text{V}}{\text{A}} = \frac{\text{C}}{\text{s}}\,\text{V} = \frac{\text{C}\cdot\text{J}}{\text{s}\cdot\text{C}} = \text{W}.$$

REMEMBER: EMF IS A SOURCE OF ENERGY. IT IS NOT ACTUALLY A FORCE AT ALL.

EXERCISE 26.1 ❖ A simple circuit contains an emf connected to a resistance. If the resistance is increased while the emf remains the same, does the dissipated power increase, decrease, or stay the same?

26.1.5 Safety Considerations

Despite the warning sign (■ Figure 26.11a), high voltage alone is not necessarily dangerous. The student in Figure 26.11b is at a potential of 100 000 V and is smiling! She is not in danger because she is standing on an insulating mat, and her whole body is at the same potential. It *is* very dangerous to send large electric currents through the body. The human body is a conductor, so a current is established in response to a potential *difference* maintained across different parts of the body. Currents in the human body can cause burns and can disrupt the function of muscles such as those in the heart and around the lungs. Even small currents, 5 mA or less, cause the sensation of shock; 100 mA can be fatal. Thus it is important to avoid any situation in which the human body forms part of a conducting circuit.

Old electrical equipment becomes dangerous if the insulation wears thin and *live* parts (those at high potential) can be touched. Electrical appliances in kitchens and bathrooms can be particularly dangerous because water is a reasonably good conductor, and charge flows more easily into a wet hand than a dry one. Proper ground connections greatly improve the safety of any appliance. Current takes the path of least resistance, through the ground wire rather than your body.

WE'LL DEMONSTRATE THIS IN §26.3.

■ FIGURE 26.11
High voltage itself is not dangerous. It is a large potential *difference* that can be dangerous if it is applied across part of a human body. Always exercise care when handling electrical equipment.

26.2 A Model for Current and Resistance

To demonstrate how current depends on potential difference, we use a simple model of electron flow in a conducting substance. An accurate analysis requires quantum theory, but the model we present allows us to understand important qualitative features of resistance and obtain rough quantitative estimates.

In a conductor, a large number of electrons, usually one or two per atom, can move through the material with speeds of the order of 10^6 m/s. As they travel, the electrons collide with the surrounding material. The electrons accelerate in response to an electric field, and the collisions amount to a resistive force akin to friction. A reasonable model is that the electrons move with a constant velocity, with the resistive force balancing the applied force.

The atoms in a solid are arrayed in a regular pattern called a lattice. In a cold, perfectly regular lattice, electrons move almost unimpeded. But in a real substance the lattice has imperfections (missing atoms, atoms slightly out of place, even atoms of a different kind), and as temperature rises, the atoms oscillate with increasing frequency. An electron's motion is disturbed by these imperfections. For convenience we shall refer to these interactions as collisions with atoms, but you should remember that the electron actually interacts with the lattice as a whole.

The instantaneous velocity of an electron is redirected in each collision, so the electrons move on random paths that on average go nowhere (■ Figure 26.12). When an electric field is applied, each electron is accelerated between collisions and gains a small additional velocity, always in the direction opposite the electric field (■ Figure 26.13). An electron that collided at time t_o and obtained a velocity $\vec{v}_o$ as a result of that collision, has a velocity $\vec{v}$ at time t, where:

$$\vec{v} = \vec{v}_o + \vec{a}(t - t_o).$$

The acceleration $\vec{a}$ is $\vec{F}/m = -e\vec{E}/m$, so:

$$\vec{v} = \vec{v}_o - \frac{e}{m}\vec{E}(t - t_o).$$

The average velocity of the electrons at any time is:

$$\langle \vec{v} \rangle = \left\langle \vec{v}_o - \frac{e}{m}\vec{E}(t - t_o) \right\rangle.$$

The average velocity after a collision is random, so $\langle \vec{v}_o \rangle = 0$. Thus:

$$\langle \vec{v} \rangle = -\frac{e}{m}\vec{E}\left\langle (t - t_o) \right\rangle.$$

The average time since the last collision, $\langle (t - t_o) \rangle$, equals the average time between collisions, τ. Thus the free electrons drift with an average velocity $\vec{v}_d$ given by:

$$\vec{v}_d = -\frac{e}{m}\vec{E}\tau. \tag{26.4}$$

According to quantum mechanics, the electron speed is determined by the properties of the material.

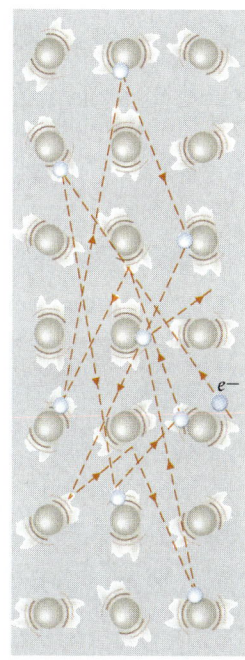

■ Figure 26.12
Classical model of electron motion in a solid. A rapidly moving electron suffers frequent collisions with the lattice of atoms in the solid that redirect its motion. On average, it goes nowhere.

The argument proving this result is quite subtle. See, for example, *The Feynman Lectures on Physics*, v. 1, Ch. 43.

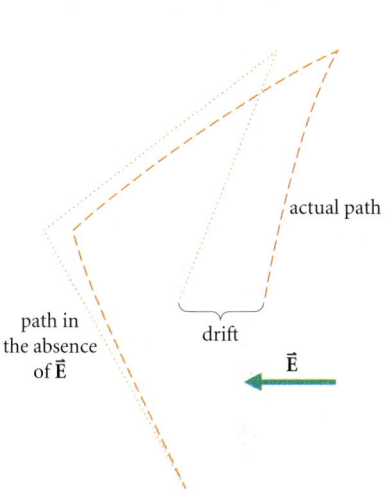

■ Figure 26.13
When an electric field is applied, the electron accelerates in the direction opposite $\vec{E}$ between collisions and travels on a parabolic path. The result is a net drift in the direction opposite $\vec{E}$. The size of the drift is vastly exaggerated in this drawing, to make it visible. ($v_d/v = \frac{1}{10}$ instead of the actual 10^{-10}!)

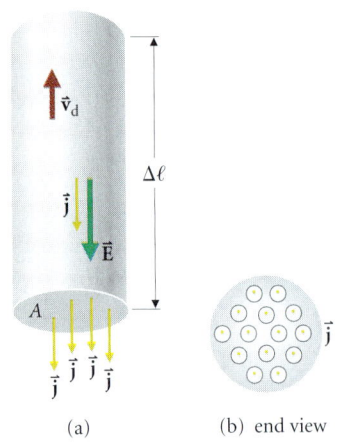

FIGURE 26.14
(a) Electrons, n_e per unit volume, drifting with velocity $\vec{v}_d$, produce a current density $\vec{j} = -en_e\vec{v}_d$ parallel to $\vec{E}$. (b) End view of the wire. The total current $I = jA$.

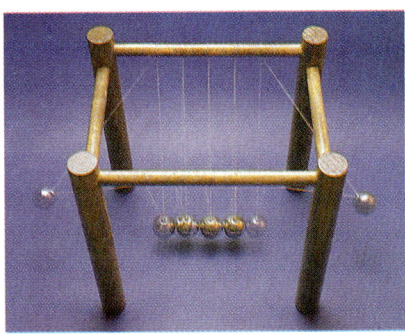

before

after

FIGURE 26.15
When a ball collides with a set of identical balls, initially in contact, one ball is rapidly ejected from the end of the row. The signal is transmitted by compression waves in the middle three balls, which hardly move at all. This system forms a mechanical analogue of signal transmission in electric circuits. No one electron moves very far; the signal is transmitted by the electric field.

THE CURRENT $I = \int \vec{j} \cdot \hat{n}\, dA$ IS THE FLUX OF THE CURRENT DENSITY.

REMEMBER THE SIGN CONVENTION: CURRENT IS IN THE DIRECTION OPPOSITE ELECTRON VELOCITY.

A piece of wire of length $\Delta\ell$ and cross-sectional area A with n_e free electrons per unit volume contains $N_e = n_e A\, \Delta\ell$ free electrons (■ Figure 26.14). These electrons move out of the piece of wire in a time $\Delta t = \Delta\ell/v_d$, so the resulting current is:

$$I = \frac{|dQ|}{dt} = \frac{en_e A\, \Delta\ell}{\Delta t} = n_e e A v_d. \tag{26.5}$$

Electron drift speeds are remarkably small compared with the speed of the electrons between collisions, or the speed of electrical signals along a wire. Signals are transmitted by the electric field, not by the electrons themselves (■ Figure 26.15).

EXAMPLE 26.4 ♦♦ A 12-gauge copper wire has a diameter of 2.0 mm. If the wire carries a 15 A current, find the drift velocity of the electrons in the copper. (The number of atoms per cubic meter of copper is 8.5×10^{28} and there is one free electron per atom.) If the electric field in the wire is 77 mV/m, and the electrons have a speed of 2×10^6 m/s, estimate the average distance an electron travels between collisions.

MODEL The drift velocity is related to the current in the wire through eqn. (26.5).

SETUP The cross-sectional area of the wire is πr^2, with $r = d/2 = 1.0$ mm.

SOLVE
$$v_d = \frac{I}{n_e e A}$$
$$= \frac{15\text{ A}}{(8.5 \times 10^{28}\text{ m}^{-3})(1.6 \times 10^{-19}\text{ C})\pi(1.0 \times 10^{-3}\text{ m})^2}$$
$$= 3.5 \times 10^{-4}\text{ m/s}.$$

SETUP The distance traveled between collisions, λ_c, is the collision time multiplied by the speed of the electrons. Using eqn. (26.4), we find:

$$\tau = \frac{mv_d}{eE}$$
$$= \frac{(9.1 \times 10^{-31}\text{ kg})(3.5 \times 10^{-4}\text{ m/s})}{(1.6 \times 10^{-19}\text{ C})(77 \times 10^{-3}\text{ V/m})} = 2.6 \times 10^{-14}\text{ s}.$$

SOLVE $\lambda_c = v\tau = (2 \times 10^6\text{ m/s})(2.6 \times 10^{-14}\text{ s}) = 5 \times 10^{-8}$ m.

ANALYZE The mean free path λ_c is roughly 150 atomic spacings.
Check the units of τ:

$$\frac{1\text{ kg}\cdot\text{m/s}}{1\text{ C}\cdot\text{V/m}} = \frac{1\text{ N}\cdot\text{s}}{1\text{ C}\cdot\text{V/m}} = \frac{1\text{ (N/C)}\cdot\text{s}}{1\text{ V/m}} = 1\text{ s}.$$

The total current depends on the wire's cross section. The quantity that relates charge flow directly to electric field and the properties of the material is the current per unit area, or *current density*:

$$\vec{j} = -n_e e \vec{v}_d. \tag{26.6}$$

Using eqn. (26.4) for $\vec{v}_d$, we have:

$$\vec{j} = \frac{n_e e^2 \tau}{m}\vec{E} = \sigma\vec{E}, \tag{26.7}$$

where the *conductivity*,

$$\sigma = \frac{n_e e^2 \tau}{m}, \tag{26.8}$$

TABLE 26.1 Resistivities of Common Materials (at 20°C)

Conductors

Substance	ρ ($\Omega \cdot$m)	Temperature Coefficient α (per °C)
Silver	1.6×10^{-8}	3.8×10^{-3}
Copper	1.7×10^{-8}	3.9×10^{-3}
Gold	2.4×10^{-8}	3.4×10^{-3}
Aluminum	2.8×10^{-8}	3.9×10^{-3}
Magnesium	4.6×10^{-8}	4×10^{-3}
Tungsten	5.6×10^{-8}	4.5×10^{-3}
Platinum	1.0×10^{-7}	3×10^{-3}
Iron	1.0×10^{-7}	5.0×10^{-3}
Steel (piano wire)	1.2×10^{-7}	4×10^{-3}
Lead	2.2×10^{-7}	3.9×10^{-3}
Constantan	4.9×10^{-7}	1×10^{-5}
Invar 36	8.2×10^{-7}	1.1×10^{-3}
Mercury	9.6×10^{-7}	8.9×10^{-4}
Nichrome	1.15×10^{-6}	4.0×10^{-4}

Semiconductors

Substance	ρ ($\Omega \cdot$m)
Carbon (graphite)	1.375×10^{-5}
Germanium	0.46
Silicon	0.03

Insulators

Substance	ρ ($\Omega \cdot$m)
Wood	$10^8 - 10^{11}$
Glass	$10^{10} - 10^{14}$
Sulfur	2×10^{17}
Quartz	7.5×10^{17}

Data from *CRC Handbook*, 64th ed. Data on Invar courtesy of Carpenter Technology Corporation.

is a property of the conducting material. The relation is often written as:

$$\vec{E} = \rho \vec{j}, \quad (26.9)$$

where

$$\rho \equiv \frac{1}{\sigma} = \frac{m}{n_e e^2 \tau} \quad (26.10)$$

is the *resistivity* of the material. Its unit is the ohm-meter ($\Omega \cdot$m). Resistivity is small if there are many electrons per unit volume and they collide infrequently. Resistivities of several common materials are listed in ● Table 26.1.

Do not confuse σ = conductivity with σ = surface charge density or ρ = resistivity with ρ = charge density.

EXERCISE 26.2 ♦ Use eqn. (26.10) to show that the units of ρ are $\Omega \cdot$m.

EXERCISE 26.3 ♦ Use the data in Example 26.4 to estimate the resistivity of copper. Compare your answer with the value listed in Table 26.1.

The resistance of a wire depends on its length ℓ, its area A, and the resistivity of the material. For a uniform wire, eqns. (26.5) and (26.9) give a relation between electric field and total current:

$$I = jA = (A/\rho)E.$$

FOR CALCULATING FIELD FROM POTENTIAL SEE, FOR INSTANCE, EXAMPLE 25.6.

Thus when I, A, and ρ are constant along a wire, E is also constant. Since $\vec{E}$ is parallel to $\vec{j}$, and hence to $d\vec{\ell}$, $E = \Delta V/\ell$, where ΔV is the potential difference between the ends of the wire. So:

$$I = \left(\frac{A}{\rho \ell}\right) \Delta V.$$

Comparing with the definition of resistance (eqn. 26.2), we obtain:

$$R = \frac{\rho \ell}{A}. \tag{26.11}$$

Ohm discovered that in many conductors current is directly proportional to applied voltage; that is, resistance is independent of current.

> A material obeys *Ohm's law* if its resistivity is constant, independent of the applied potential difference and the resulting current.

Materials that obey Ohm's law are called *ohmic materials.* If its temperature is held fixed, metal wire is an excellent example of an ohmic material.

In our model, resistivity depends on a number of well-defined physical quantities all of which, except the collision time τ, are constants for any given material. In most substances, τ decreases as temperature increases, and resistance is observed to increase linearly with temperature. If ρ_0 is the resistivity at temperature T_0, then:

$$\rho = \rho_0[1 + (T - T_0)\alpha]. \tag{26.12}$$

Table 26.1 also lists the temperature coefficients α for several substances.

EXAMPLE 26.5 ♦ An automobile dome light is connected to the positive terminal of the battery with copper wire of length 1.0 m, radius 0.50 mm, and resistivity $1.6 \times 10^{-8}\ \Omega \cdot$m. What is the resistance of the wire?

MODEL The resistance is given by eqn. (26.11).

SETUP The area of the wire is $\pi r^2 = \pi(5.0 \times 10^{-4}\ \text{m})^2$.

SOLVE From eqn. (26.11), the resistance is:

$$R = \frac{\rho \ell}{A} = \frac{(1.6 \times 10^{-8}\ \Omega \cdot \text{m})(1.0\ \text{m})}{\pi(25 \times 10^{-8}\ \text{m}^2)} = 0.020\ \Omega.$$

ANALYZE The resistance of ordinary wires is much less than that of most circuit components and is usually ignored in analyzing circuits. When a circuit requires a measured amount of resistance, a *resistor* is used. Resistors are usually made of graphite in a clay substrate and have resistances ranging from about $1\ \Omega$ to $22\ \text{M}\Omega$ (■ Figure 26.16).

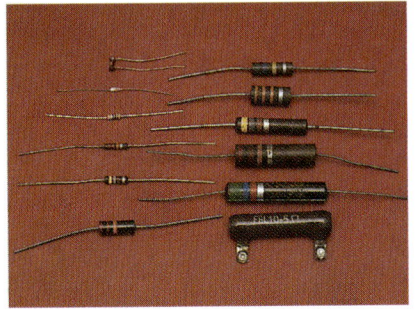

■ **FIGURE 26.16**
Commercial resistors are usually made of graphite in a clay substrate. The colored bands identify the magnitude of the resistance and the accuracy of that number as specified by the manufacturer.

EXAMPLE 26.6 ♦♦ Use Table 26.1 to find the resistivity of tungsten at $1.00 \times 10^3\,°$C.

MODEL We assume the resistivity increases linearly with temperature up to 1000°C.

SETUP Table 26.1 lists values of resistivity and temperature coefficient at 20°C. Using eqn. (26.12), we find the resistivity $\rho(T)$ at temperature T:

$$\rho(T) = \rho(20°\text{C})[1 + (T - 20°\text{C})\alpha].$$

SOLVE $\rho(1000°\text{C}) = (5.6 \times 10^{-8}\ \Omega \cdot \text{m})[1 + (980\ \text{K})(4.5 \times 10^{-3}\ /\text{K})]$
$= 3.0 \times 10^{-7}\ \Omega \cdot \text{m}.$

REMEMBER: A TEMPERATURE *DIFFERENCE* ON THE CELSIUS SCALE IS EXPRESSED IN KELVIN.

ANALYZE Ordinary metals have small values of α and obey Ohm's law quite well over a moderate range of temperatures.

EXERCISE 26.4 ♦♦ The filament of a light bulb is to have a resistance of 16 Ω and is to be made from 10.0 cm of tungsten wire wound into a helical coil. If its operating temperature is 1000°C, what diameter wire should be used?

26.3 SERIES AND PARALLEL COMBINATIONS OF RESISTORS

Nearly all interesting electric circuits consist of more than a single energy source and a single energy-consuming device. When several elements are combined in a circuit, each influences the behavior of the others. Electric potential and the conservation of charge are the key physical ideas that allow us to understand and predict these influences. In this section, we apply these ideas to the simplest combinations of resistive elements—series and parallel connections.

26.3.1 Resistors in Series

The indicator (mounted in the panel) and the sensor (in the tank) of an automobile fuel gauge are connected in *series* (end to end) (■ Figure 26.17a). Once a steady state is established, charge does not build up anywhere in the circuit, since the changing electric fields would alter the electric current. Thus charge passes point A in the circuit at the same rate as it passes points B and C. The current is the same through both components.

THIS IS THE SAME ARGUMENT THAT WE USED IN DISCUSSING STEADY STATE FLOW OF FLUIDS (CHAPTER 13) AND HEAT (CHAPTER 21).

> Two or more resistors are in *series* if there is a single current path through all of them; that is, the current in each element is the same.

Potential differences are the key to analyzing the circuit. The potential difference between points A and C equals the sum of potential differences across the two resistors (Figure 26.17b):

$$V_{AC} = V_A - V_C$$
$$= (V_A - V_B) + (V_B - V_C)$$
$$\equiv V_{AB} + V_{BC}.$$

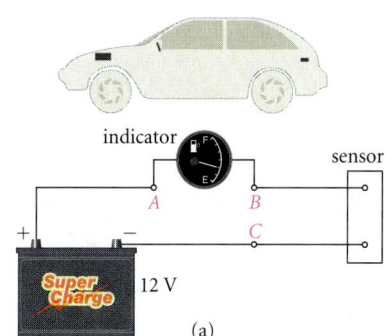

The potential differences are related to the current and the individual resistances (eqn. 26.2):

$$V_{AB} = IR_1$$

and
$$V_{BC} = IR_2.$$

Thus
$$V_{AC} = IR_1 + IR_2 = I(R_1 + R_2),$$

and the *equivalent resistance* for the series combination is:

$$R_s \equiv \frac{V_{AC}}{I} = R_1 + R_2.$$

Here, *equivalent* means that the actual circuit and the *equivalent circuit*, Figure 26.17c, draw the same current from the battery and have the same potential difference between points A and C. The only differences between the two circuits are within the series combination itself. The method for finding R_s applies to any number of components connected in series.

> The equivalent resistance for any number of components connected in series is the sum of their individual resistances:

$$R_s = \sum_i R_i. \tag{26.13}$$

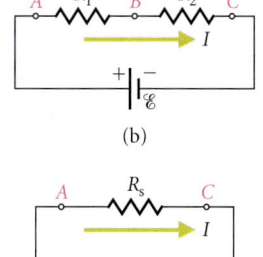

■ **FIGURE 26.17**
(a) An automobile fuel gauge. There are two resistances in series: the indicator in the panel and the sensor in the tank. The resistance of the tank sensor depends on the quantity of fuel in the tank. (b) Circuit diagram for the fuel gauge. (c) Equivalent circuit. The current I is the same as in the actual circuit, and the potential difference between points A and C is also the same.

Digging Deeper

Nonohmic Devices

I: The Light Bulb

The resistance of a metal wire is independent of the current flowing through it only so long as the metal's temperature is held constant. Increase of temperature increases the thermal speed of the electrons and the vibration of the atoms, decreasing the collision time and increasing the resistivity. ■ Figure 26.18 is a graph of current in a tungsten filament versus potential difference across it. When the filament's temperature is held constant, the graph is a straight line whose slope $1/R$ measures the constant resistance of the filament. When the filament is used in a light bulb, its resistance increases with temperature, and the corresponding graph curves downward. This contrast emphasizes that Ohm's law is a practical relation rather than a fundamental one. It works well for some systems in specific circumstances. The graph of voltage versus current, or *characteristic curve*, for a device provides a useful tool for describing its behavior.

II: Superconductivity

In some materials resistivity decreases abruptly to zero at very low temperatures. Electrons form pairs that avoid collisions altogether, allowing current to flow with no resistance at all. For mercury, this transition occurs at 4 K (■ Figure 26.19). Since 1986 it has been found that certain metallic oxides become superconducting at temperatures as high as 125 K = −148°C. Since this transition temperature is above the temperature of liquid nitrogen, it is relatively easy to achieve. Such materials are potentially very important for making practical superconducting devices. However, the materials tend to be brittle, so it is not easy to manufacture flexible wires from them. Understanding the mechanism by which superconductivity occurs in these materials, and learning how to build practical devices with them, remain fertile fields of research.

III: Semiconductors

■ Figure 26.19 shows the characteristic curve of a semiconductor diode, whose resistance depends strongly on the direction of current through it. In silicon, the major material of such a device, four valence electrons are shared among neighboring

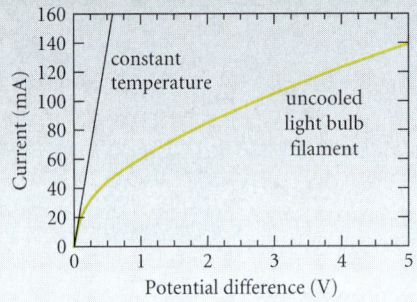

Figure 26.18
Current in a tungsten light bulb filament versus applied voltage. As the current increases, the power dissipated increases and the light bulb heats up. Its resistance increases, and so the rate at which current changes with ΔV decreases.

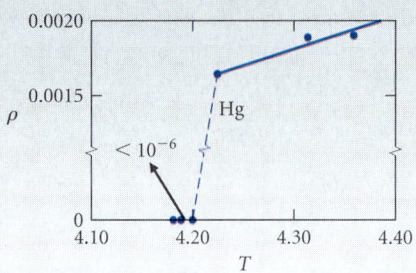

Figure 26.19
Superconducting transition for mercury, as observed by H. Kamerlingh Onnes in 1911. The transition begins at between 4.2 K and 4.3 K (about −269°C). When the transition is complete, the resistivity is truly zero. Current in such material persists forever without any electric field to drive it. Experiments have demonstrated current persistence over periods of years.

EXAMPLE 26.7 ♦ In the fuel gauge shown in Figure 26.17, the resistance of the indicator is 11 Ω. The resistance of the sensor is 75 Ω when the tank is half-full. What is the current in the circuit when the indicator reads half-full? How much power does the fuel gauge use then?

MODEL The two resistances are connected in series.

SETUP The equivalent resistance in the circuit is given by eqn. (26.13).

$$R_s = R_1 + R_2 = 11\ \Omega + 75\ \Omega = 86\ \Omega.$$

SOLVE The current in the circuit is given by eqn. (26.2):

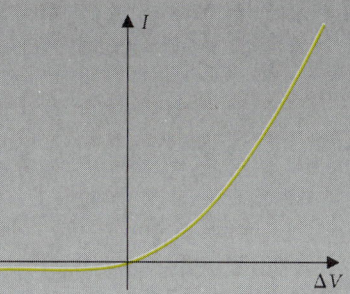

■ **FIGURE 26.20**
Characteristic curve for a semiconductor diode. The device permits current in only one direction and so acts like an electric "valve."

atoms but usually are not free to conduct current through the material. Silicon is a *semi*conductor because at normal temperature only a small fraction of its electrons have enough energy to become conduction electrons. Each electron that does so enhances conductivity, both by its own motion and by the curious ability of the "hole" (or lack of an electron) it leaves behind to move from atom to atom.

The conductivity of semiconductors is enhanced by *doping*. Replacement of a silicon atom with gallium, which has only three valence electrons, immediately produces a hole due to the lack of a valence electron. Thus silicon doped with gallium has a supply of positive charge carriers and so has a larger conductivity than pure silicon. It is called a *p-type* semiconductor. Conversely, doping silicon with arsenic, which has five valence electrons, gives a conduction electron for each arsenic atom, producing an *n-type* semiconductor. When p-type and n-type semiconductors are put in contact (■ Figure 26.21), the excess of free electrons on one side of the junction and holes on the other side causes a net flow of charge, until equilibrium is established by a buildup of negative charge in the p-type side and positive charge on the n-type side. In equilibrium a small number of thermal electrons are produced in the p-type material and flow toward the n-type region. These are balanced by electrons from the n-type region with enough kinetic energy to cross the potential barrier. A similar balance occurs between the much smaller flows of holes. An external electric field in the "forward-bias" direction reduces the internal potential barrier and allows electrons to flow from the n-type region. In contrast, an electric field in the "reverse-bias" direction suppresses the n-type flow. The resulting asymmetry of current can be very dramatic (Figure 26.21). Diodes are often used in circuits to ensure that current occurs in only one direction.

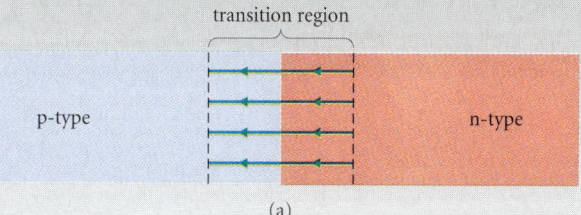
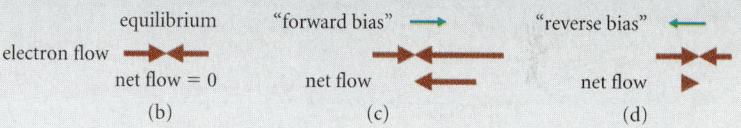

■ **FIGURE 26.21**
(a) Semiconductor junction diode. Silicon doped with gallium has a shortage of valence electrons in its atomic lattice—it is called p-type. Silicon doped with arsenic is n-type; it has an excess of valence electrons. Electrons migrate from the n-type material to the p-type until the electric field produced by the resulting charge separation stops the migration. (b) A small number of electrons in the p-type region escape from silicon atoms and flow toward the right. Meanwhile a few electrons in the n-type region have sufficient kinetic energy to cross the transition region. In equilibrium, the potential difference between the two regions is just right to keep these flows equal. (c) An applied field in the *forward bias* direction greatly increases the number of electrons able to cross the transition region. (d) Reverse bias decreases the number of electrons able to cross the transition region. The current is almost zero.

$$I = \frac{\Delta V}{R_s} = \frac{12 \text{ V}}{86 \text{ }\Omega} = 0.14 \text{ A.}$$

The power used is (eqn. 26.3c):

$$P = \frac{(\Delta V)^2}{R_s} = \frac{(12 \text{ V})^2}{86 \text{ }\Omega} = 1.7 \text{ W.}$$

ANALYZE The additional resistance of the indicator actually reduces the power used by the sensor. Connecting additional resistance in series reduces the current in the circuit and hence reduces the power $I^2 R$ used by each resistance. ■

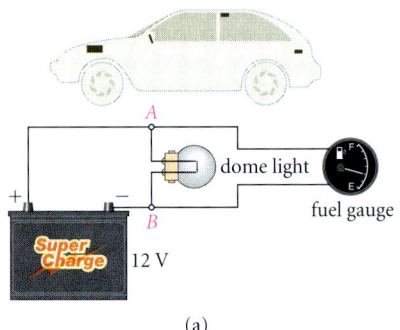

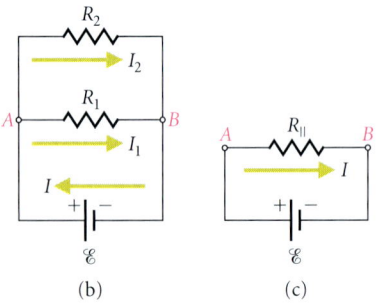

FIGURE 26.22
(a) A car dome light and fuel gauge are connected in parallel across the car battery. This configuration allows each to be turned on independently of the other and provides the full battery emf across each device.
(b) Circuit diagram for the two devices connected to the battery. The current divides at point A and recombines at point B. (c) Equivalent circuit.

26.3.2 Resistors in Parallel

If two components are connected to a single power source, but you want the option of turning on the power to one at a time, a series connection is not appropriate. In addition, neither component operates at full power when they are connected in series. The appliances in a car are actually connected so that each has the full emf of the battery across its terminals; they are connected in *parallel*. As you stop to read a map at night, the dome light and the fuel gauge are connected in parallel across the battery (■ Figure 26.22). The current I from the battery divides at point A into a current I_1 through the dome light and a current I_2 through the fuel gauge. Since charge is neither created nor destroyed, in a steady state the current arriving at A from the battery equals the total current leaving A and passing through the appliances:

$$I = I_1 + I_2.$$

> Two or more resistors are in *parallel* when they are all connected between the same two points in a circuit, and the potential difference across each is the same.
>
> The currents in parallel resistors derive from a common source and recombine into a common outgoing current.

The equivalent resistance for a parallel combination is defined by the potential difference and the total current:

$$R_\| \equiv \frac{V_{AB}}{I}.$$

The individual currents are: $I_1 = \dfrac{V_{AB}}{R_1}$ and $I_2 = \dfrac{V_{AB}}{R_2}$.

To use the relation between the currents, we invert the expression for $R_\|$:

$$\frac{1}{R_\|} = \frac{I}{V_{AB}} = \frac{I_1 + I_2}{V_{AB}} = \frac{1}{R_1} + \frac{1}{R_2}.$$

A similar result holds for any number of resistors in parallel.

> For a set of parallel resistors, the reciprocal of their equivalent resistance equals the sum of reciprocals of their individual resistances:
>
> $$\frac{1}{R_\|} = \sum_i \frac{1}{R_i}. \tag{26.14}$$

EXAMPLE 26.8 ♦ Find the equivalent resistance of an automobile fuel gauge ($R = 86\ \Omega$) and dome light ($R = 18\ \Omega$) connected in parallel across a 12-V battery (Figure 26.22). Determine the current in each device and the total current through the battery. What is the ratio of the currents in the two devices?

MODEL The devices are connected in parallel.

SETUP From eqn. (26.14):

$$\frac{1}{R_\|} = \frac{1}{R_1} + \frac{1}{R_2} = \frac{R_2 + R_1}{R_1 R_2}.$$

$$R_\| = \frac{R_1 R_2}{R_1 + R_2} = \frac{(86.0\ \Omega)(18.0\ \Omega)}{86.0\ \Omega + 18.0\ \Omega} = 14.9\ \Omega.$$

SOLVE The currents are:

$$I_1 = \frac{V_{AB}}{R_1} = \frac{12\ \text{V}}{86\ \Omega} = 0.14\ \text{A}.$$

$$I_2 = \frac{V_{AB}}{R_2} = \frac{12\ \text{V}}{18\ \Omega} = 0.67\ \text{A}.$$

The total current is: $I = \dfrac{V_{AB}}{R_{\|}} = \dfrac{12 \text{ V}}{14.9 \text{ }\Omega} = 0.81 \text{ A} = I_1 + I_2.$

Since $V_{AB} = I_1 R_1 = I_2 R_2$: $\quad \dfrac{0.14 \text{ A}}{0.67 \text{ A}} = \dfrac{I_1}{I_2} = \dfrac{R_2}{R_1} = \dfrac{18 \text{ }\Omega}{86 \text{ }\Omega} = 0.21.$

ANALYZE The currents are inversely proportional to the resistances; the dome light draws five times as much current as the fuel gauge.

EXERCISE 26.5 ◆◆ Find the power used by each of the devices in Example 26.8.

EXERCISE 26.6 ◆◆ The resistance of the two devices in parallel is 15 Ω, less than the resistance of either alone. Show that for any parallel combination of resistors the equivalent resistance is less than any of the individual resistances.

26.3.3 Combined Series and Parallel Circuits

Because a battery has internal resistance, a parallel connection of two components to a car battery is actually a series-parallel connection. The circuit diagram is shown in ■ Figure 26.23a. Such circuits can be simplified by replacing each parallel or series combination of resistors with its equivalent resistance.

EXAMPLE 26.9 ◆◆ A 12.0-V auto battery has an internal resistance of $r = 0.02$ Ω. The 0.200-Ω starter motor and a 1.20-Ω headlight are connected in parallel across it. What is the current in the battery? What is the total power required from the emf? At what rate is energy dissipated as heat within the battery itself?

MODEL Two resistors R_1 and R_2 in parallel can be replaced with their equivalent resistance, as shown in Figure 26.23b. The equivalent resistance forms a series combination with the internal resistance of the battery.

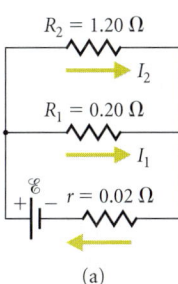

SETUP
$$\dfrac{1}{R_{\|}} = \dfrac{1}{R_1} + \dfrac{1}{R_2}$$
$$= \dfrac{1}{0.200 \text{ }\Omega} + \dfrac{1}{1.20 \text{ }\Omega} = (5.00 + 0.833) \text{ }\Omega^{-1} = 5.83 \text{ }\Omega^{-1}.$$

So: $R_{\|} = 0.171$ Ω.

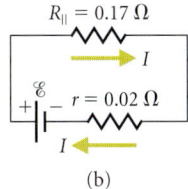

The equivalent resistance $R_{\|}$ is in series with the internal resistance r. Together they are equivalent to a single resistor $R_s = r + R_{\|} = 0.02$ $\Omega + 0.171$ $\Omega = 0.19$ Ω (Figure 26.23c).

SOLVE The total current in circuit (c) is:
$$I = \dfrac{\mathcal{E}}{R_s} = \dfrac{12.0 \text{ V}}{0.19 \text{ }\Omega} = 63 \text{ A}.$$

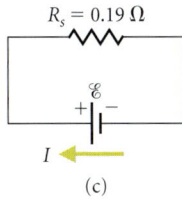

This is the current through the battery in the actual circuit. The battery supplies power:
$$P_{\mathcal{E}} = \mathcal{E}I = (12.0 \text{ V})(63 \text{ A}) = 760 \text{ W}.$$

The current is the same through the internal resistance of the battery, resulting in an ohmic loss:
$$P_r = I^2 r = (63 \text{ A})^2 (0.02 \text{ }\Omega) = 80 \text{ W},$$
which is dissipated as heat in the battery.

ANALYZE The remaining 680 W is delivered to the starter and the headlight. Since the power dissipated in each resistor is $(\Delta V)^2/R$, the amounts of power delivered to each component are inversely proportional to their resistances. The starter uses six times as much as the headlight, or $\tfrac{6}{7} = 86\%$ of the total.

■ **FIGURE 26.23**
(a) Circuit diagram for a real battery, with internal resistance, connected to a starter motor and a headlight. (b) The two parallel resistances are reduced to one equivalent resistance. (c) The two series resistances combine to form this equivalent circuit.

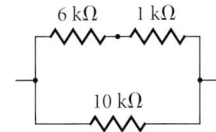

FIGURE 26.24

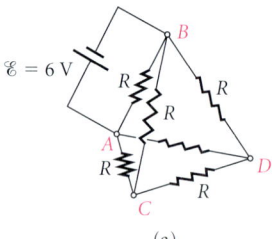

(a)

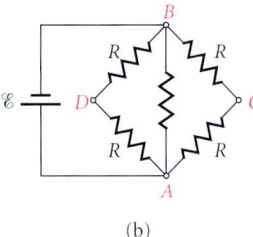

(b)

FIGURE 26.25
(a) This tetrahedron is not a simple series/parallel connection but may be reduced to one once we realize that there is no current between C and D. (b) The equivalent series/parallel circuit diagram.

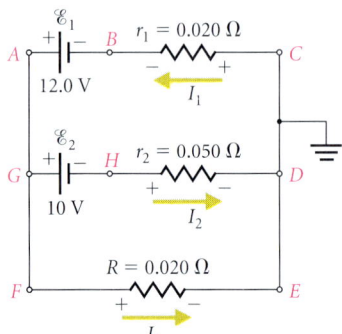

FIGURE 26.26
This circuit, used to jump-start a car, is not a series/parallel connection. The two different emfs make the potentials at B and H different. The negative terminal of the battery is connected to the car chassis, which serves as *ground* ($V = 0$), indicated by the multiple bar symbol. (*Remember:* Resistances r_1 and r_2 are *inside* the batteries.)

EXERCISE 26.7 ♦ Find the equivalent resistance of the three resistors in the circuit shown in ■ Figure 26.24.

Circuits containing any number of resistors with series and parallel connections may be analyzed by this technique of sequentially replacing parts of the circuit with equivalent single resistors to obtain an equivalent circuit that contains a single resistor and a single emf. Often, complex resistor combinations may be reduced to series/parallel circuits using symmetry arguments.

EXAMPLE 26.10 ♦♦♦ Six equal 15-kΩ resistors are connected to form a tetrahedron (■ Figure 26.25a). A 6.0-V battery is connected to points A and B. How much power is drawn from the battery?

MODEL To find the power, we need the equivalent resistance of the object between points A and B. At point A, current divides into three currents along paths AB, ACB, and ADB. The system of resistors is symmetric. If we imagine standing the tetrahedron on edge AB, we have mirror symmetry about AB. Thus the currents along paths ACB and ADB are the same, and so are the potential differences $V_{AC} = V_{AD}$. Thus C and D are at the same potential. With no potential difference between C and D, there is no current along CD, and eliminating it leaves an equivalent circuit. The circuit simplifies to the series/parallel combination in Figure 26.25b.

SETUP Each of the series combinations R_{ACB} and R_{ADB} equals $R + R = 2R$. The parallel combination between A and B is then:

$$\frac{1}{R_{\parallel}} = \frac{1}{R_{AB}} + \frac{1}{R_{ACB}} + \frac{1}{R_{ADB}} = \frac{1}{R} + 2\left(\frac{1}{2R}\right) = \frac{2}{R}.$$

Thus: $R_{\parallel} = R/2.$

SOLVE The power used by the tetrahedron is:

$$P = \frac{\mathcal{E}^2}{R_{\parallel}} = \frac{2\mathcal{E}^2}{R} = \frac{2(6.0 \text{ V})^2}{15 \times 10^3 \ \Omega} = 4.8 \text{ mW}.$$

ANALYZE Current is zero along CD provided that the resistances are *exactly* equal. For this to be true, the connecting wires should be exactly the same length on each side, too, to make the small resistances of the wires themselves equal along AC, AD, CB, and DB. ■

26.4 KIRCHHOFF'S RULES

Circuits containing only series and parallel combinations of resistors may be analyzed with a sequence of equivalent circuits using the methods of §26.3. However, these rules are not sufficient for a circuit with two batteries in parallel connected to a load (■ Figure 26.26). The internal resistances of the two batteries are *not* a parallel combination, because the emfs can make the potentials differ at points B and H, with the result that $V_{BC} \neq V_{HD}$. In this section, we shall outline a procedure that works for any circuit. The two basic principles are those we have been using all along: charge conservation and path independence of potential difference (or equivalently, energy conservation). Their formal statements are called Kirchhoff's rules, after Gustav Kirchhoff (1824–1887). We begin with a few definitions and then state the rules.

> An *arm* of a circuit is a portion in which there is a single current. Examples are the arms $GABCD$, GHD, and $GFED$ in Figure 26.26.
>
> A *junction* is a point where three or more arms join. Examples are points G and D in the figure.
>
> A *loop* is a path through the circuit that returns to its starting point. Examples are paths $ABCDHGA$ and $ABCDEFGA$ in the figure.

The fact that charge does not continuously build up at any junction requires

> *Kirchhoff's junction rule:* The sum of currents entering a junction equals the sum of currents leaving the junction.

$$\sum I_{in} = \sum I_{out}.$$

THE JUNCTION RULE IS A STATEMENT OF *CHARGE CONSERVATION.*

The fact that the potential has a definite value at any point in the circuit requires

> *Kirchhoff's loop rule:* The algebraic sum of the potential changes around any loop in a circuit is zero.

$$\sum_{\text{closed loop}} \Delta V = 0.$$

THE LOOP RULE IS A STATEMENT OF *ENERGY CONSERVATION.*

In the following examples and study problem we'll apply these rules, using the plan shown in ■ Figure 26.27.

EXAMPLE 26.11 ♦ In a flashlight two 1.5-V batteries are connected in series with a light bulb ($R = 1.0 \, \Omega$). How much power does the light bulb draw from the batteries?

MODEL We can use Kirchhoff's loop rule to determine the potential difference across the light bulb. ■ Figure 26.28 is the circuit diagram.

SETUP The circuit has only one loop, and we traverse it clockwise:

$$\Delta V = -\mathcal{E}_1 - \mathcal{E}_2 + \Delta V_{bulb} = 0 \Rightarrow \Delta V_{bulb} = \mathcal{E}_1 + \mathcal{E}_2 = 3.0 \text{ V}.$$

SOLVE The power used by the bulb is (eqn. 26.3c):

$$P = (\Delta V)^2 / R = (3.0 \text{ V})^2 / (1.0 \, \Omega) = 9.0 \text{ W}.$$

ANALYZE When batteries are connected in series, their emfs add. ■

EXAMPLE 26.12 ♦♦ In an automobile starter circuit, a battery is connected to a starter with resistance $R = 0.20 \, \Omega$. When your battery is low, your friend Jane can help out by attaching jumper cables from her battery to yours. If Jane's battery has an emf of 12.0 V and internal resistance of 0.020 Ω, and your battery has an internal resistance of 0.050 Ω and emf of 10.0 V, how much current is delivered to the starter?

MODEL Figure 26.26 is the circuit diagram. The current variables for the three arms are I_1, I_2, and I_3; the arrows define their positive directions. Here we can guess the actual directions of the currents and define the directions so as to get positive values for the current variables. Notice that current I_2 through your battery (#2) is in the direction opposite that expected in normal operation. Jane's battery is not only powering the starter but is charging your battery too.

SETUP We apply the junction rule at G:

$$I_1 = I_2 + I_3. \quad \text{(i)}$$

There are three currents and they all appear in this one equation, so we do not need to look at any more junctions. (The junction rule at D generates the same equation.)

We may apply the loop rule to any loop and count potential changes in either direction around the loop. There are three possible choices of loop: *ABCDHGA, ABCDEFGA,* or *GHDEFG.* Any two include all the arms of the circuit. We choose to use loops *ABCDHGA* and *GHDEFG,* both traversed clockwise as indicated by the sequence of letters. Potential drops across a resistor to drive current through it. The + and − signs drawn on the resistors are a convenient shorthand for keeping track of the signs in loop equations. The plus sign is always where the current arrow "enters" the resistor.

Loop *ABCDHGA* $\quad \Delta V = V_{AB} + V_{BC} + V_{DH} + V_{HG}.$

$$0 = -\mathcal{E}_1 + I_1 r_1 + I_2 r_2 + \mathcal{E}_2. \quad \text{(ii)}$$

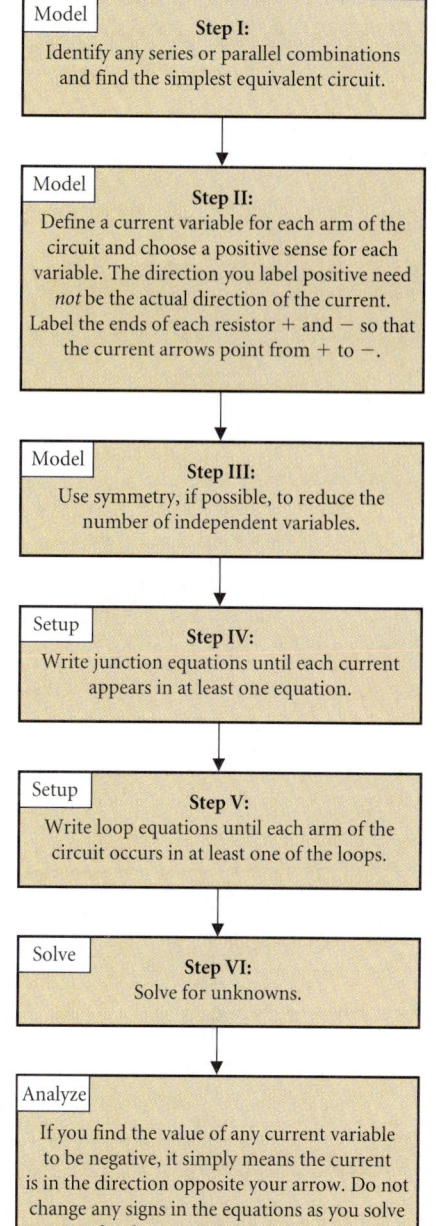

Solution plan for circuit problems using Kirchhoff's rules

Model — **Step I:** Identify any series or parallel combinations and find the simplest equivalent circuit.

Model — **Step II:** Define a current variable for each arm of the circuit and choose a positive sense for each variable. The direction you label positive need *not* be the actual direction of the current. Label the ends of each resistor + and − so that the current arrows point from + to −.

Model — **Step III:** Use symmetry, if possible, to reduce the number of independent variables.

Setup — **Step IV:** Write junction equations until each current appears in at least one equation.

Setup — **Step V:** Write loop equations until each arm of the circuit occurs in at least one of the loops.

Solve — **Step VI:** Solve for unknowns.

Analyze — If you find the value of any current variable to be negative, it simply means the current is in the direction opposite your arrow. Do not change any signs in the equations as you solve for the remaining unknowns!

■ FIGURE 26.27

IT IS NOT NECESSARY TO GUESS THE ACTUAL DIRECTIONS. WE SHALL EXPLORE THIS POINT FURTHER IN THE STUDY PROBLEM.

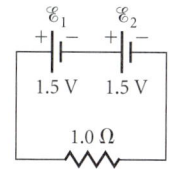

■ FIGURE 26.28
The circuit diagram for a flashlight with two batteries.

IF WE TRAVERSED THE LOOPS COUNTER-CLOCKWISE, EACH TERM WOULD CHANGE SIGN, BUT THE RESULTING EQUATIONS WOULD BE THE SAME.

Loop $GHDEFG$ $\quad \Delta V = V_{GH} + V_{HD} + V_{EF}.$

$$0 = -\mathcal{E}_2 - I_2 r_2 + I_3 R. \quad \text{(iii)}$$

We now have three equations in three unknowns (I_1, I_2, and I_3). ■ Figures 26.29a and b are graphs of potential versus position around these two loops.

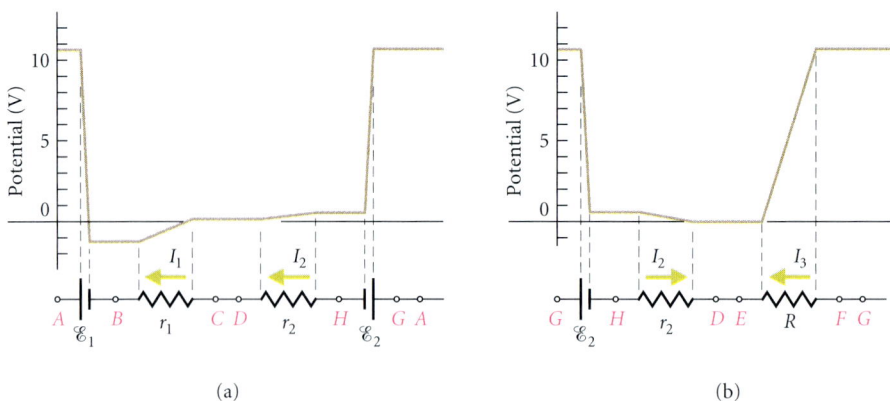

■ **FIGURE 26.29**
Potential changes around the loops in the circuit diagramed in Figure 26.26. We chose the zero level at ground, points C, D, and E. The potential is almost constant in the connecting wires and increases or decreases across batteries and resistors according to the polarity and direction of current. You could sketch a curve like this before solving the problem. The numerical values shown here are taken from the solution. (a) Loop $ABCDHGA$. (b) Loop $GHDEFG$.

SOLVE Using eqns. (i) and (ii), we eliminate I_1:

$$-\mathcal{E}_1 + r_1(I_2 + I_3) + I_2 r_2 + \mathcal{E}_2 = 0,$$

or:
$$I_2 = \frac{\mathcal{E}_1 - \mathcal{E}_2 - r_1 I_3}{r_1 + r_2}.$$

Substituting this result into eqn. (iii) gives:

$$-\mathcal{E}_2 - \frac{(\mathcal{E}_1 - \mathcal{E}_2 - r_1 I_3)}{(r_1 + r_2)} r_2 + I_3 R = 0,$$

from which we obtain:

$$I_3 = \frac{(\mathcal{E}_1 - \mathcal{E}_2)r_2 + \mathcal{E}_2(r_1 + r_2)}{R(r_1 + r_2) + r_1 r_2} = \frac{r_2 \mathcal{E}_1 + r_1 \mathcal{E}_2}{R(r_1 + r_2) + r_1 r_2}$$

$$= \frac{(0.050\ \Omega)(12.0\ \text{V}) + (0.020\ \Omega)(10.0\ \text{V})}{(0.20\ \Omega)(0.020\ \Omega + 0.050\ \Omega) + (0.020\ \Omega)(0.050\ \Omega)}$$

$$= \frac{0.80\ \text{V}}{0.015\ \Omega} = 53\ \text{A}.$$

ANALYZE The results for the other currents are $I_2 = [2.0\ \text{V} - (0.020\ \Omega)(53\ \text{A})]/(0.070)$ = 13 A, and $I_1 = I_2 + I_3 = 66$ A. All the values are positive, indicating that the currents are in the directions we expected. Note the nice symmetry in the expression for I_3. ■

EXERCISE 26.8 ◆◆ Compare the current delivered to the starter by the parallel batteries in Example 26.12 with that delivered by the low battery alone.

Study Problem 17 ◆◆ A Compound Circuit
Find the current in each arm of the circuit diagramed in ■ Figure 26.30a. What total power is required of $\mathcal{E}_2$ if $\mathcal{E}_1 = 3\mathcal{E}_2 = 30.0$ V, and $R = 10.0\ \Omega$?

Modeling the System

First, we notice that the two resistors between points A and B are in parallel, so the circuit reduces to the equivalent circuit shown in Figure 26.30b. The equivalent circuit has five arms, and we define five current variables. Here it is not easy to guess the actual current directions, so some of our current variables may have negative values. We still label the + and − ends of the resistors *as if* all the current values were positive. These conventions ensure that we will obtain correct equations from Kirchhoff's rules regardless of the choice of directions for the current variables.

FOR A DIFFERENT CHOICE, SEE EXERCISE 26.10.

Setup of Solution

Applying the junction rule, we find:

$$\text{At } A \quad I_1 + I_3 = I_2. \tag{i}$$
$$\text{At } B \quad I_4 = I_3 + I_5. \tag{ii}$$

All five current variables appear in these two equations, so an equation for junction C is unnecessary.

Applying the loop rule, we find:

$$\text{Loop } ACDA \quad \Delta V = V_{AC} + V_{DA} = -I_2 R + \mathcal{E}_1 = 0. \tag{iii}$$

$$\text{Loop } ABCA \quad \Delta V = V_{AB} + V_{BC} + V_{CA}.$$

$$0 = I_3 \frac{R}{2} - \mathcal{E}_2 + I_2 R. \tag{iv}$$

$$\text{Loop } BEFCB \quad \Delta V = V_{EF} + V_{CB} = -I_5 R + \mathcal{E}_2 = 0. \tag{v}$$

AS USUAL, THE SEQUENCE OF LETTERS INDICATES THE DIRECTION AROUND THE LOOP, CLOCKWISE HERE.

We do not need any more loop equations because each arm of the circuit has appeared in at least one of the loops. We have five equations for the five unknowns.

Solution

Equations (iii) and (v) give I_2 and I_5 immediately in terms of the known emfs:

$$I_2 = \frac{\mathcal{E}_1}{R} \quad \text{and} \quad I_5 = \frac{\mathcal{E}_2}{R}.$$

Substituting for I_2 in eqn. (iv) gives I_3:

$$I_3 \frac{R}{2} - \mathcal{E}_2 + \mathcal{E}_1 = 0 \;\Rightarrow\; I_3 = \frac{2(\mathcal{E}_2 - \mathcal{E}_1)}{R}.$$

Then from eqn. (ii):

$$I_4 = I_3 + I_5 = \frac{3\mathcal{E}_2 - 2\mathcal{E}_1}{R},$$

and from eqn. (i):

$$I_1 = I_2 - I_3 = \frac{3\mathcal{E}_1 - 2\mathcal{E}_2}{R}.$$

Substituting the known values into the equations gives:

$$I_1 = \frac{3(30.0 \text{ V}) - 2(10.0 \text{ V})}{10.0 \; \Omega} = 7.00 \text{ A}, \quad I_2 = \frac{30.0 \text{ V}}{10.0 \; \Omega} = 3.00 \text{ A},$$

$$I_3 = \frac{2(10.0 \text{ V} - 30.0 \text{ V})}{10.0 \; \Omega} = -4.00 \text{ A},$$

$$I_4 = \frac{3(10.0 \text{ V}) - 2(30.0 \text{ V})}{10.0 \; \Omega} = -3.00 \text{ A}, \quad \text{and} \quad I_5 = \frac{10.0 \text{ V}}{10.0 \; \Omega} = 1.00 \text{ A}.$$

The power output from battery #2 is:

$$P_2 = \mathcal{E}_2 I_4 = (10.0 \text{ V})(-3.00 \text{ A}) = -30.0 \text{ W}.$$

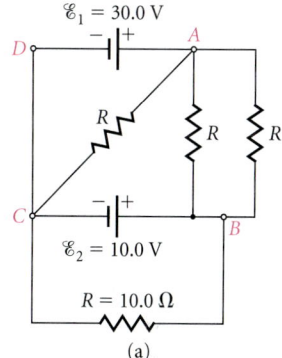

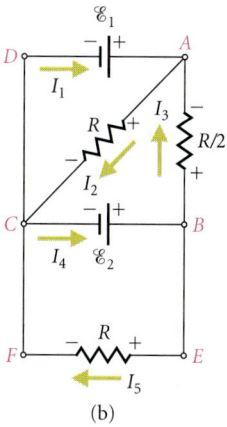

■ **FIGURE 26.30**
(a) Circuit diagram for a compound circuit. (b) The two parallel resistors may be replaced with a single equivalent resistance to form this circuit. We define the current variables as shown. *Remember:* The arrows are not necessarily in the actual directions of the currents.

BATTERIES ARE NOT CHARGED REALLY. THEY ARE ENERGIZED. ORDINARY SPEECH CAN BE MISLEADING HERE.

MANY COMPUTER MATH PACKAGES OFFER ROUTINES FOR SOLVING SUCH SETS OF EQUATIONS. MATRIX METHODS MAY ALSO BE USED. SEE PROBLEM 85.

Analysis

As a quick check on the arithmetic, note that the answers do indeed satisfy eqns. (i) and (ii). Two of the variables, I_3 and I_4, have negative values; those currents are in the direction opposite the arrows in the diagram. The negative value of P_2 means that battery #2 is absorbing energy from the circuit: it is being charged.

Do not be discouraged by the number of equations that result from applying Kirchhoff's rules. They are linear equations and always succumb to a systematic process of elimination.

EXERCISE 26.9 ♦♦ The junction equation for point C in Figure 26.30 would give redundant information. Show that it can be derived from eqns. (i) and (ii). Similarly, derive the loop equation for loop *ABCDA* from eqns. (iii) and (iv).

EXERCISE 26.10 ♦♦ Do the problem again, choosing the current variable in arm *AB* of Figure 26.30 pointing from *A* to *B*, and the current variable in arm *CB* pointing from *B* to *C*. Show that the physical results are the same.

✻ 26.5 ELECTRICAL MEASUREMENT

The art of electrical measurement relies on clever circuit design. Precision measurements now use sophisticated solid-state circuitry with designs beyond the scope of an introductory text. However, many of these designs rely on principles we have studied, and we can get a sense of electronic design by studying a few simple systems.

In a *galvanometer* (■ Figure 26.31) magnetic force on current in a wire coil results in a deflection of the meter needle proportional to the current. Carefully constructed meters of this type are sensitive to currents as small as 10^{-8} A. Through careful design, such current meters may be used to measure potential difference or resistance.

MAGNETIC FORCE ON A CURRENT-CARRYING WIRE IS DISCUSSED IN §29.2.

Modern solid-state devices can be constructed to measure potential differences of a few volts with great sensitivity. These devices are the basis of practical multimeters for use in analyzing electric circuits. In this section we'll explore several aspects of their design.

Whenever a meter is used to measure voltage or current in a circuit, the circuit that is being measured is the one containing the meter, not the original circuit. Thus an important aspect of meter design is that it disturb the circuit as little as possible.

26.5.1 An Ammeter

To measure the current in an arm of a circuit, an ammeter is inserted into the arm, in series with the circuit components (■ Figure 26.32a). The resistance of the ammeter increases the total resistance in that arm and thus decreases the current. Thus an ammeter should have as small a resistance as possible. To make an ammeter, we may connect a small shunt resistance r_s in parallel with a galvanometer (Figure 26.32b). Most of the current in the circuit passes through the shunt, while a small amount passes through the galvanometer. This design has the additional benefit of protecting the galvanometer, since any other than a very small current would damage it.

■ FIGURE 26.31
A galvanometer is an accurate instrument for measuring current. Magnetic force on current in a wire coil results in a deflection of the meter needle that is proportional to current. Since even moderate currents can damage the mechanism, meters using a galvanometer are designed so that only a small current passes through the galvanometer itself.

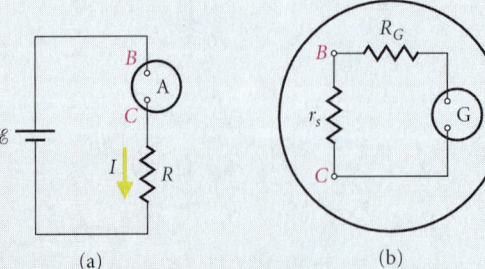

■ FIGURE 26.32
(a) To measure current through a resistor, connect an ammeter in series with it. (b) In an ammeter, the galvanometer (resistance R_G) is in parallel with a shunt resistance $r_s \ll R_G$ that carries most of the current. The small shunt resistance disturbs the current in the circuit only slightly.

EXAMPLE 26.13 ♦♦ In the circuit whose diagram is shown in Figure 26.32, $\mathcal{E} = 110$ V, $R = 1.0$ kΩ, $R_G = 35$ Ω, and $r_s = 1.0$ Ω. Find the current in the galvanometer.

MODEL Since the resistance of the ammeter is much less than the 1.0-kΩ resistance in the circuit, we may ignore any change of current caused by the ammeter.

SETUP The resistance of the ammeter is R_A, where:

$$R_A = \left(\frac{1}{R_G} + \frac{1}{r_s}\right)^{-1} = \left(\frac{1}{35\ \Omega} + \frac{1}{1.0\ \Omega}\right)^{-1} = 0.97\ \Omega.$$

The current through the ammeter is $I \approx \mathcal{E}/R = (110\text{ V})/(1.0\text{ k}\Omega) = 0.11$ A, and the potential difference across the ammeter is $V_{BC} = IR_A = 0.11$ V.

SOLVE The current through the galvanometer is then:

$$I_G = \frac{V_{BC}}{R_G} = \frac{0.11\text{ V}}{35\ \Omega} = 3.1\text{ mA}.$$

ANALYZE Connecting the ammeter into the circuit increases the total resistance (and decreases the current through R) by less than 0.1%.

THE AMMETER IS CALIBRATED TO READ THE TOTAL CURRENT THROUGH IT, NOT THE CURRENT IN THE GALVANOMETER ITSELF.

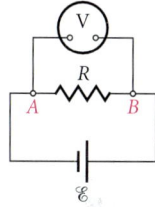

■ **FIGURE 26.33**
A voltmeter is connected in parallel with a circuit component to measure potential difference across it.

26.5.2 Voltmeters

To measure the potential difference between two points in a circuit, a voltmeter is connected between the points, in parallel with the circuit components (■ Figure 26.33). To disturb the circuit as little as possible, the meter should have a large resistance. One voltmeter design (■ Figure 26.34) has a sensitive galvanometer and a large resistance R_m in series with it.

EXAMPLE 26.14 ♦♦ A voltmeter has a resistance $R_m = 1.0 \times 10^5$ Ω, and the galvanometer can measure current with an uncertainty of $\delta I = \pm 5.0 \times 10^{-7}$ A. What is the minimum potential difference the meter can measure to 10% accuracy?

MODEL If the uncertainty in the measurement is to be 10%, the current in the galvanometer should be at least $I_G = 10\ \delta I = 5.0 \times 10^{-6}$ A, so that $\delta I/I \leq \frac{1}{10}$.

SETUP AND SOLVE The required potential difference is

$$\Delta V = R_m I_G = (1.0 \times 10^5\ \Omega)(5.0 \times 10^{-6}\text{ A}) = 0.50\text{ V}.$$

ANALYZE We ignored the galvanometer's resistance, which is much less than $10^5\ \Omega$. ■

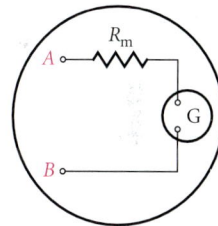

■ **FIGURE 26.34**
In one design of voltmeter, a galvanometer is in series with a large resistance. The meter draws only a small amount of current from the original circuit.

Modern solid-state voltmeters use semiconductor devices that have extremely large resistance ($\geq 10^{12}$ Ω). However, they are limited to potential differences of at most a volt or two. To reduce the potential difference across the device, it is connected in parallel with a set of resistors in series that are used as a *voltage divider* (■ Figure 26.35). A typical voltmeter

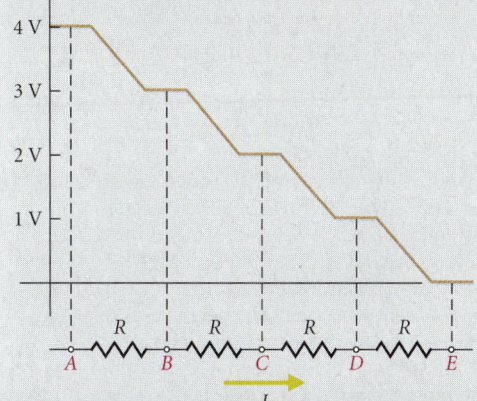

■ **FIGURE 26.35**
Resistors in series act as a voltage divider. The sum of potential differences across each resistor equals the total potential difference across the entire series. Thus the resistors *divide* the total potential difference. With the four equal resistors shown here, $V_{AB} = \frac{1}{2}V_{AC} = \frac{1}{3}V_{AD} = \frac{1}{4}V_{AE}$.

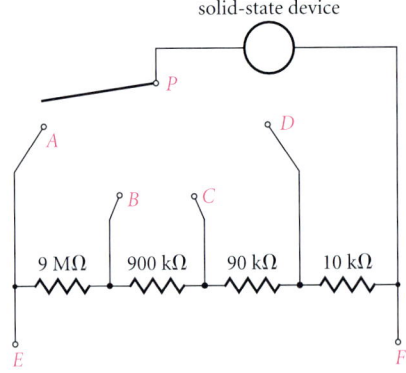

FIGURE 26.36
A modern voltmeter uses a solid-state device to measure the potential difference. The series of resistors allows the voltmeter to operate over four different ranges, depending on the connection of P to one of the four points A, B, C, or D. The solid-state device itself has an extremely high resistance. For all practical purposes we may assume it draws no current at all. The 10-MΩ resistance of this meter is typical.

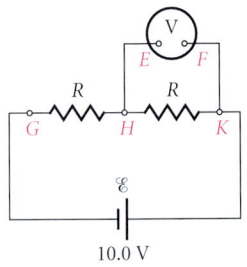

FIGURE 26.37
The voltmeter in Figure 26.36 is connected across one of two identical resistors R in a circuit.

SUCH ACCURACY IS NEEDED WHEN MEASURING EXTREMELY LOW TEMPERATURES, FOR EXAMPLE. (SEE ESSAY 6.)

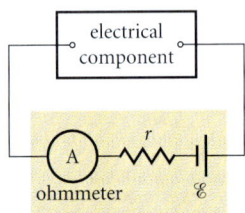

FIGURE 26.38
An ohmmeter contains an emf and an ammeter. The component to be measured is disconnected from its circuit and then connected across the ohmmeter.

(Figure 26.36) has a 10 MΩ resistance, divided into 9 MΩ, 900 kΩ, 90 kΩ, and 10 kΩ. A switch moves connection P to points A, B, C, or D depending on the magnitude of the voltage to be measured. Connected at D, the solid-state device measures the potential across the 10-kΩ resistor, which is $\frac{1}{1000}$ of the total potential drop across the series combination. Similarly, when connected at C, the device measures the potential drop across 100 kΩ of resistance, or $\frac{1}{100}$ of the total. Connected at B, the device measures $\frac{1}{10}$ of the total potential difference.

EXAMPLE 26.15 ♦♦ A voltmeter is used to measure the potential difference across one of two identical resistors connected in series with a 10.0-V emf (Figure 26.37). If the solid-state device is limited to 0.2 V, which connection, A, B, C, or D (Figure 26.36), should be used?

MODEL The voltage across the solid-state device cannot exceed 0.2 V, so we use the voltage divider to reduce the measured voltage to less than 0.2 V.

SETUP Since we are measuring the potential difference across one of two identical resistors in series, the result should be half of the battery emf, or 5 V.

SOLVE We need to reduce the measured voltage by at least a factor of 5/0.2, or 25. Ten won't do, so we choose to reduce by a factor of 100 and use the connection at point C. ($R_{CF} = 100$ k$\Omega = \frac{1}{100} R_{EF}$, so $V_{CF} = \frac{1}{100} V_{EF}$.)

ANALYZE Always err on the side of caution. You don't want to blow out the meter. ∎

26.5.3 The Wheatstone Bridge

A simple ohmmeter for measuring resistance is a battery in series with an ammeter and a resistor r (Figure 26.38). A component to be measured is disconnected from any circuit, and the ohmmeter is placed across it. Then the equivalent resistance R_{eq} is the battery emf divided by the ammeter current, and the component's resistance is $R_{eq} - r$. An ohmmeter is very convenient to use, but the accuracy of its measurements is limited by the internal resistance of the battery and the accuracy of the galvanometer. When extremely accurate measurements are needed, a Wheatstone bridge may be used (Figure 26.39, after Charles Wheatstone, 1802–1875).

The Wheatstone bridge uses four resistances. Resistor R_1 and the ratio R_4/R_3 have known values, and the ratio may be varied. R_x is the resistance to be determined. In use, the bridge is balanced by adjusting the known ratio R_4/R_3 until the current through the galvanometer is zero. Thus the potential difference between points A and B is zero.

The potential at A is related to the potentials at C and D and to the potential differences across the resistors:

$$V_A = V_C - V_{CA} \quad \text{or} \quad V_A = V_D + V_{AD}.$$

Similarly:

$$V_B = V_C - V_{CB} \quad \text{or} \quad V_B = V_D + V_{BD}.$$

Since $V_A = V_B$, we may subtract these equations to find:

$$V_{CA} = V_{CB} \quad \text{and} \quad V_{AD} = V_{BD}.$$

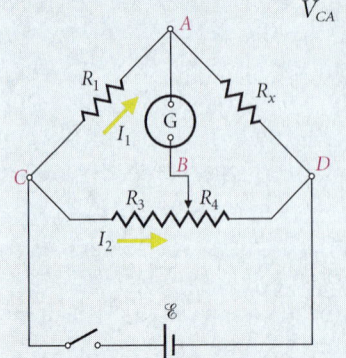

FIGURE 26.39
A Wheatstone bridge uses a *null measurement* to determine the unknown resistance R_x. One of the other resistors, say R_1, is precisely known. The ratio of the other two is variable. Here we show a single coil that is divided by the tap B. The ratio of the resistances R_3 and R_4 equals the ratio of the corresponding lengths of coil. This device is called a potentiometer. With no current in the galvanometer, the potential at A equals the potential at B. When the bridge is balanced, the galvanometer reading does not change when the switch is opened and closed. Such null measurements are the basis of the most accurate instruments.

Since there is no current between A and B, both R_1 and R_x have the same current I_1, while R_3 and R_4 have the same current I_2. So:

$$I_1 R_1 = I_2 R_3 \quad \text{and} \quad I_1 R_x = I_2 R_4.$$

Dividing these equations, we obtain:
$$\frac{R_x}{R_1} = \frac{R_4}{R_3}. \tag{26.15}$$

The unknown resistance is determined from the values of R_1, R_3, and R_4.

Since the measurement uses a zero reading from the galvanometer, most uncertainties are eliminated. You can easily find the galvanometer reading for zero current by disconnecting it from the circuit. Since no current or potential difference is measured, battery stability is unimportant. Resistance R_1 should be manufactured with great precision, but only the ratio R_4/R_3 needs to be accurately known. If both are made of the same material, temperature variations do not affect the ratio, for example. Similar use of *null* results is essential for many precision measurement techniques.

Chapter Summary

Where Are We Now?

We have developed methods for analyzing direct current circuits. These methods are the starting point for understanding and designing electrical devices.

What Did We Do?

In an ideal battery, chemical reactions at each terminal maintain a constant potential difference between the terminals. This potential difference is the electromotive force of the battery. Charge flowing through a circuit gains energy from a source of emf and delivers that energy to practical devices. The resistance of a circuit component is the ratio of the potential difference between its ends to the current through it: $R = \Delta V/I$. Metal wire at constant temperature is a material whose resistance does not vary with the current. Such a material follows Ohm's law: voltage is proportional to current. Ohm's law is widely used to describe circuits, but a large variety of interesting and useful devices show deviations from this simple behavior.

The power output of a battery equals the current in the battery multiplied by its emf. Similarly, the power dissipated by a resistor equals the current multiplied by the potential difference across the resistor:

$$P_{\mathcal{E}} = I\mathcal{E} \quad \text{and} \quad P_\Omega = I\,\Delta V = I^2 R = (\Delta V)^2/R.$$

Two or more resistors in *series* are connected end to end and have the same current in each of them. Replacing resistors in series with an equivalent resistance, equal to the sum of the individual resistances, produces an equivalent circuit, one that behaves in the same way as the original circuit:

$$R_s = \sum_i R_i.$$

Two or more resistors in *parallel* are connected across the same two points in a circuit and have the same potential difference across each of them. Current in parallel resistors divides and then recombines after passing through the resistors. The reciprocals of the individual parallel resistances add to give the reciprocal of the equivalent resistance:

$$\frac{1}{R_\parallel} = \sum_i \frac{1}{R_i}.$$

The equivalent resistance $R_\|$ is less than each individual resistance R_i. Combined series and parallel networks of resistors may be analyzed by using these rules to construct a sequence of simplified equivalent circuits.

A network of batteries and resistances may be analyzed by a systematic procedure based on Kirchhoff's rules. Kirchhoff's junction rule—the sum of currents entering a junction equals the sum of currents leaving—embodies conservation of charge in circuits. Kirchhoff's loop rule—the algebraic sum of potential changes around any loop in a circuit equals zero—embodies conservation of energy. Each point in the circuit has a definite value of potential.

Electrical measurement involves careful design of measuring devices. The Wheatstone bridge illustrates the use of null techniques to achieve precision in measurement. Ammeters and voltmeters necessarily disturb the circuit they measure. Simple models of such meters allow estimation of and correction for such disturbance.

Practical Applications

The circuit theory developed in this chapter applies directly to power supply systems in aircraft and the space shuttle as well as in cars, and to most battery-powered devices, such as flashlights. These are direct current (DC) circuits. The current in your house oscillates in time; nevertheless, at any instant the currents and potential differences have the same relations they would have in a direct current system. DC circuit analysis with a proper averaging over time gives an adequate description of your household circuitry. (We'll investigate this topic further in Chapter 32.) Because the electric signals can travel across the chip much faster than the input signal changes, direct current techniques also apply to micron-sized solid-state circuit elements used in microwave technology, where signals oscillate at frequencies of about 10^{11} Hz!

Solutions to Exercises

26.1. From eqn. (26.3c), $P = (\Delta V)^2/R$, so if R is increased with ΔV held constant, P decreases.

26.2. Replacing each algebraic quantity on the right-hand side of eqn. (26.10) with its appropriate unit gives us the units of ρ:

$$\frac{\text{kg} \cdot \text{m}^3}{\text{C}^2 \cdot \text{s}} = \frac{(\text{kg} \cdot \text{m}^2/\text{s}^2) \cdot \text{m}}{\text{C} \cdot (\text{C}/\text{s})}$$

$$= \frac{\text{J} \cdot \text{m}/\text{C}}{\text{A}} = \frac{\text{V} \cdot \text{m}}{\text{A}} = \Omega \cdot \text{m}.$$

26.3. From eqn. (26.10):

$$\rho = \frac{1}{\sigma} = \frac{m}{ne^2\tau}$$

$$= \frac{9.11 \times 10^{-31} \text{ kg}}{(8.5 \times 10^{28} \text{ m}^{-3})(1.6 \times 10^{-19} \text{ C})^2(2.6 \times 10^{-14} \text{ s})}$$

$$= 1.6 \times 10^{-8} \ \Omega \cdot \text{m}.$$

Table 26.1 gives $\rho = 1.7 \times 10^{-8} \ \Omega \cdot \text{m}$ for copper at 20°C.

26.4. From eqn. (26.11):

$$A = \frac{\pi d^2}{4} = \frac{\rho \ell}{R}$$

Inserting the value of ρ from Example 26.6, we obtain:

$$d = \sqrt{\frac{4\rho\ell}{\pi R}} = \sqrt{\frac{4(3.0 \times 10^{-7} \ \Omega \cdot \text{m})(0.10 \text{ m})}{\pi(16 \ \Omega)}}$$

$$= 4.9 \times 10^{-5} \text{ m}.$$

26.5. The potential difference across each device is 12 V. Using eqn. (26.3c), we find the power used by each:

$$P_1 = \frac{\mathcal{E}^2}{R_1} = \frac{(12 \text{ V})^2}{86 \ \Omega} = 1.7 \text{ W}$$

and

$$P_2 = \frac{(12 \text{ V})^2}{18 \ \Omega} = 8.0 \text{ W}.$$

26.6. Suppose a set of resistors are connected in parallel across an emf $\mathcal{E}$. Adding another resistor doesn't change the current through the resistors already in place but does increase the total current by the amount that now exists in the new resistor. The equivalent resistance, $R_\| \equiv \mathcal{E}/I_{\text{total}}$, is reduced by the increase in current.

We may also give a mathematical proof:

$$R_\| = \frac{1}{\sum_i (1/R_i)}.$$

For any individual resistor, labeled j,
$$\frac{R_\parallel}{R_j} = \frac{1/R_j}{\sum_i (1/R_i)} < 1.$$

26.7. The two resistors in series are equivalent to a 7-kΩ resistor. This combination is in parallel with the 10-kΩ resistor. The equivalent of the parallel combination is:
$$R_\parallel = \frac{(7\text{ k}\Omega)(10\text{ k}\Omega)}{7\text{ k}\Omega + 10\text{ k}\Omega} = 4\text{ k}\Omega.$$

26.8. The current delivered by the single battery is:
$$I' = \frac{\mathscr{E}_2}{R + r_2} = \frac{10.0\text{ V}}{0.25\text{ }\Omega} = 40\text{ A} \quad \text{(2 significant figures).}$$
This current is only 75% of the current delivered when Jane helps.

26.9. The junction equation for point C is:
$$I_2 + I_5 = I_1 + I_4.$$
This is the sum of eqns. (i) and (ii). The loop equation for loop $ABCDA$ is:
$$\Delta V = +I_3 R/2 - \mathscr{E}_2 + \mathscr{E}_1 = 0.$$
This is the sum of eqns. (iii) and (iv).

26.10. ■ Figure 26.40 shows the circuit with the revised sign conventions. The resulting changes in the Kirchhoff's equations are underlined:

At A $\quad\quad I_1 = I_2 + \underline{I'_3}.$

At B $\quad\quad \underline{I'_3} = \underline{I'_4} + I_5.$

Loop $ACDA$ $\quad \Delta V = -I_2 R + \mathscr{E}_1 = 0.$

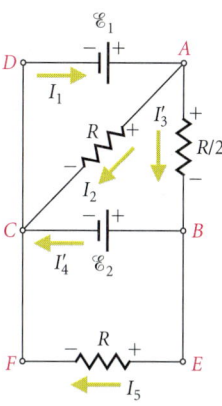

■ **Figure 26.40**

Loop $ABCA$ $\quad \Delta V = -I'_3 R/2 - \mathscr{E}_2 + I_2 R = 0.$
Loop $BEFCB$ $\quad \Delta V = -I_5 R + \mathscr{E}_2 = 0.$

Following the same sequence of steps as in the study problem, we obtain:
$$I'_3 = \frac{2(\mathscr{E}_1 - \mathscr{E}_2)}{R} = +4.00\text{ A},$$

and
$$I'_4 = \frac{2\mathscr{E}_1 - 3\mathscr{E}_2}{R} = +3.00\text{ A}.$$

Both sets of results describe the same currents. The signs of the results for the current variables are reversed ($I_3 = -4.00$ A, $I'_3 = +4.00$ A), but so are the meanings of "positive" and "negative." This is why you don't have to guess actual directions of currents before defining your current variables.

Basic Skills

Review Questions

§26.1 BASIC CIRCUIT BEHAVIOR

- Define the *electric current* in a wire.
- How is current related to electron motion in the wire?
- What is the *electromotive force* of a battery? How does it differ from *terminal voltage*?
- Define the *electrical resistance* of a circuit component.
- Why is the resistance of connecting wires in a circuit usually ignored?
- Give three expressions for the rate of energy dissipation in a resistor.
- What situations should you avoid in order to use electrical equipment safely?

§26.2 A MODEL FOR CURRENT AND RESISTANCE

- What is the electron drift velocity in a conductor? How is it related to the average time between collisions?
- Define *current density*. How is it related to electric field?
- What is *resistivity*? Which properties of a material determine its resistivity?
- Which properties of a wire determine its resistance?
- State *Ohm's law*.
- How does resistance depend on temperature?
- What is the order-of-magnitude value of resistivity for a conductor? A semiconductor? An insulator?

§26.3 SERIES AND PARALLEL COMBINATIONS OF RESISTORS

- What does it mean for resistors to be connected in *series*?
- What is an *equivalent circuit*?
- What is the equivalent resistance of a set of resistors connected in series? Is it greater than or less than the value of each individual resistance?
- What does it mean for resistors to be connected in *parallel*?
- What is the equivalent resistance of a set of resistors connected in parallel? Is it greater than or less than the value of each individual resistance?

§26.4 KIRCHHOFF'S RULES

- Define an *arm*, a *junction*, and a *loop* in a circuit.
- State Kirchhoff's two rules of circuit behavior. Which physical principles do they express?
- When should you stop writing junction equations? Loop equations?
- What does it mean if one of the current variables in your solution has a negative value?

✳ §26.5 ELECTRICAL MEASUREMENT

- How is an ammeter connected into a circuit?
- What are the components of a simple ammeter? How are they connected?
- Is a small or a large resistance needed in an ammeter? Why?
- How is a voltmeter connected into a circuit?
- What are the components of a simple voltmeter? Draw a circuit diagram indicating how they are connected.
- How does a modern solid-state voltmeter differ from this simple design?
- How would you use an ohmmeter to measure the resistance of a circuit component?
- How does a Wheatstone bridge allow accurate measurement of resistance? Why is it more accurate than an ohmmeter?

■ FIGURE 26.41

■ FIGURE 26.42

§26.3 SERIES AND PARALLEL COMBINATIONS OF RESISTORS

6. A 36-kΩ resistor and a 5.0-kΩ resistor are connected in series. What is the equivalent resistance?
7. A 36-kΩ resistor and a 5.0-kΩ resistor are connected in parallel. What is the equivalent resistance?
8. What is the equivalent resistance between points *A* and *B* in ■ Figure 26.41?

§26.4 KIRCHHOFF'S RULES

9. What terminal voltage is required of an emf used to charge a battery of emf 10.0 V and internal resistance 0.60 Ω with a current of 2.0 A?
10. Refer to the circuit diagram in ■ Figure 26.42. What is the current in the circuit?

✳ §26.5 ELECTRICAL MEASUREMENT

11. A Wheatstone bridge is used to measure an unknown resistance. The bridge balances when configured as shown in ■ Figure 26.43. What is the value of the unknown resistance?
12. An ammeter contains a galvanometer of resistance 15.00 Ω and a shunt resistance $r = 0.25\ \Omega$. It is connected into a circuit with a 6.00-V power supply and a 25.00-Ω resistor (■ Figure 26.44). What current does it measure? What was the current before the ammeter was inserted in the circuit?

Basic Skill Drill

§26.1 BASIC CIRCUIT BEHAVIOR

1. A current of 5.0 A exists in a 10-Ω resistor for 4.0 min. How much charge passes through each cross section of the wire in this time?
2. A 12-V battery is connected to a 7.5-Ω light bulb. What is the current in the circuit?
3. A simple circuit has an ideal emf of 5.0 V and contains a 15-W lamp. What is the current in the circuit? What is the resistance of the lamp?

§26.2 A MODEL FOR CURRENT AND RESISTANCE

4. A copper wire 2.0 mm in diameter extends 4800 km from New York to Bristol, England, as part of an undersea cable. What is the resistance of the wire?
5. One wire is 1.0 m long and 2.0 mm in diameter. A second wire of the same material is 0.75 m long and 1.0 mm in diameter. Find the ratio R_1/R_2 of the resistances of the two wires.

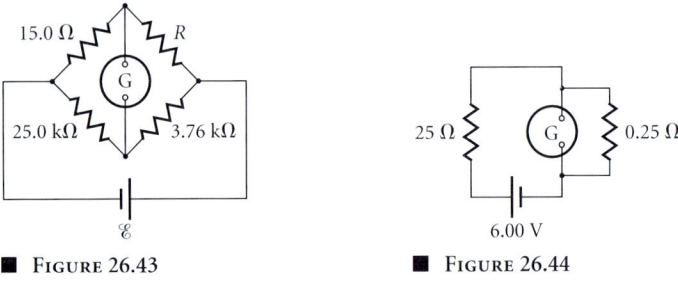

■ FIGURE 26.43

■ FIGURE 26.44

Questions and Problems

§26.1 BASIC CIRCUIT BEHAVIOR

13. ❖ What current is produced by a ring of radius *r* and uniform charge density λ rotating about its axis at angular frequency ω?
14. ♦ An average lightning flash lasts 0.19 s and transfers 25 C of charge. What is the average electric current during the flash?
15. ♦ A 12-V car battery is rated at 120 A·h. Assuming that the terminal voltage remains 12 V, for how long can this battery supply power at 29 W? How much charge flows through the battery in this time? How much energy is initially stored in the battery?
16. ♦ The electron beam in a cathode-ray tube carries 5.0×10^{15} electrons per second to a phosphor screen. What is the electric current in the tube?
17. ♦ A 1.50-V battery has an internal resistance of 0.025 Ω. What is the battery's terminal voltage when it is delivering 2.33 A current?

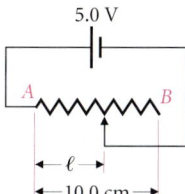

■ FIGURE 26.45

18. ♦♦ A single electron in uniform circular motion constitutes an average current around the circle. If the electron moves clockwise with angular speed ω in a circle of radius r, find the current. Evaluate I if $\omega = 2 \times 10^{16}$ rad/s and $r = 5$ nm.

19. ♦♦ ■ Figure 26.45 shows a circuit containing a potentiometer (variable resistance). The resistance of the potentiometer is proportional to the length connected into the circuit. If the total resistance of the potentiometer when the slide wire is at end B is 1.0 kΩ, and the length AB is 10.0 cm, find the current in the circuit as a function of the position of the slidewire from end A.

20. ♦♦ A simple circuit consists of an ideal emf of 12 V and a resistance of 20 Ω. The negative pole of the battery is grounded (this is usually the case in an automobile). Plot the potential as a function of position around the circuit.

21. ♦♦ What size (voltage) battery is required to operate a 25-W bulb with a resistance of 0.15 Ω?

§26.2 A MODEL FOR CURRENT AND RESISTANCE

22. ❖ Which has the larger resistance, a copper kettle with electrical connections at spout and handle, or a thin copper wire just long enough to stretch from spout to handle along the surface of the kettle?

23. ❖ Which has the larger resistance, a gold wire or a copper wire of the same length and diameter?

24. ❖ Which causes the greatest change in the resistance of a wire: (a) changing its diameter by a factor of 2, (b) changing its length by a factor of 2, or (c) raising its temperature from 300 K to 600 K. Which causes the least change? Explain your reasoning.

25. ♦ An aluminum transmission line extends 150 km from a power plant to a nearby city. What diameter should the line have if its resistance is to be less than 0.50 kΩ?

26. ♦ Eighteen-gauge copper wire (diameter 1.024 mm) is rated for a maximum current of 5 A. What is the maximum current density? What electric field in the copper is required to maintain the maximum current?

27. ♦♦ A potential difference of 10.0 V is applied to a wire 1.0 m long and 3.0 mm in diameter. The resistance of the wire is found to be 0.017 Ω. Compute the current density in the material and the resistivity of the wire. Using Table 26.1, can you identify the material?

28. ♦♦ The toaster in Example 26.3 ($R = 18 \Omega$) uses a nichrome strip of 1.0 mm $\times$ 0.25 mm rectangular cross section for its heating elements. If the operating temperature is 310°C, what is the length of the strip?

29. ♦♦ A copper wire has a diameter of 0.80 mm. What diameter tungsten wire has the same resistance per unit length? Find the electric field in each wire when it is carrying a current of 5.0 mA.

30. ♦♦ A mercury-filled rubber tube is 1.00 m long and has 1.00 mm inside diameter. The tube is wrapped around a person's arm and is stretched 0.04 cm by expansion of the arm each time the person's heart beats. What is the fractional change in resistance of the mercury during a heartbeat? (Assume the cross section remains circular, and the volume of mercury remains constant.)

31. ♦♦ Aluminum has three free electrons per atom. Using data from Tables 26.1, 13.1, and the periodic table, determine the drift

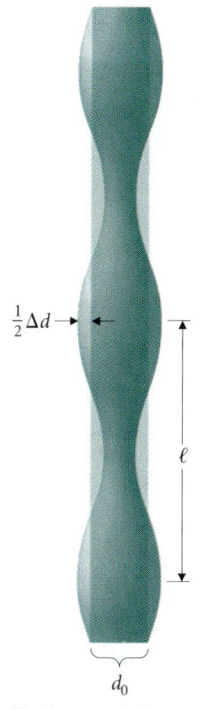

■ FIGURE 26.46

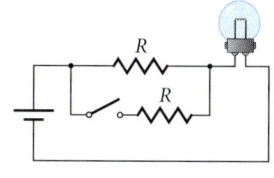

■ FIGURE 26.47

speed of conduction electrons in an aluminum wire 1.0 mm in diameter at 20°C when the current in the wire is 1.0 mA. What is the average time between collisions?

32. ♦♦♦ Because of a manufacturing error the diameter of a strand of wire varies periodically along its length (■ Figure 26.46).

$$d(z) = 1.00 \text{ mm} + (0.01 \text{ mm})\{\sin[2\pi z/(1 \text{ cm})]\}.$$

Calculate *approximately* the difference between the average resistance per unit length of this wire and that of a wire of the same material with a constant diameter of 1.0 mm.

§26.3 SERIES AND PARALLEL COMBINATIONS OF RESISTORS

33. ❖ Refer to the circuit diagram in ■ Figure 26.47. Does the lamp burn more brightly with the switch open or closed?

34. ❖ A power transmission line has a resistance r. Is it better to transmit power with high emf and low current or low emf and high current? Why?

35. ❖ A string of decorative lights could be connected in series or in parallel. What simple experiment, requiring no equipment, allows you to determine how they are connected? Explain your reasoning.

36. ♦ A wire of length L and resistance R is cut in half, and the two halves are connected in parallel. What is the resistance of the parallel combination?

37. ♦ Find the equivalent resistance of the circuit diagramed in ■ Figure 26.48.

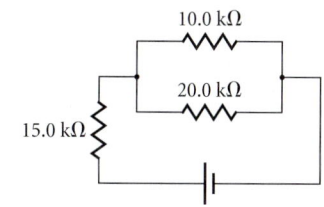

■ FIGURE 26.48

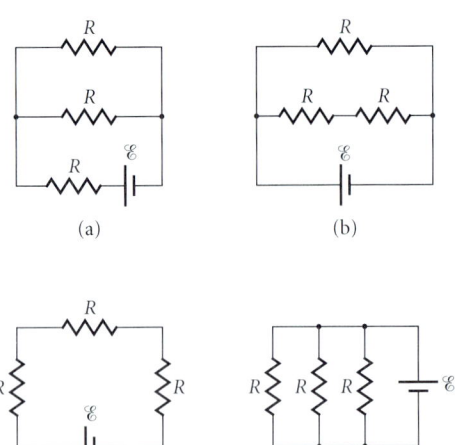

■ FIGURE 26.49

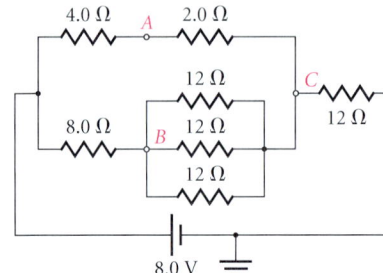

■ FIGURE 26.51

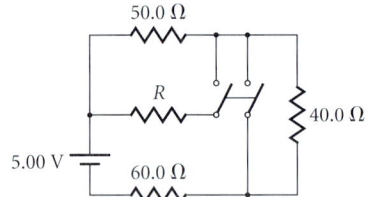

■ FIGURE 26.52

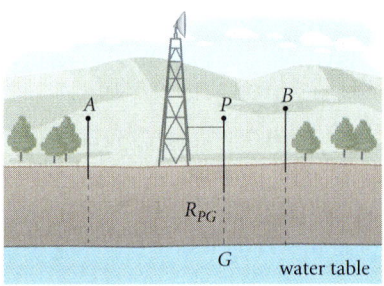

■ FIGURE 26.53

38. ◆◆ If a total current I passes through resistors R_1 and R_2 connected in parallel, show that the currents in the individual resistors are:

$$I_1 = \frac{R_2}{R_1 + R_2} I \quad \text{and} \quad I_2 = \frac{R_1}{R_1 + R_2} I.$$

39. ◆◆ In ■ Figure 26.49 are diagrams of four different circuits that may be constructed with an ideal battery and three equal resistors. Which has the greatest current through the battery? Which has the least? If $\mathcal{E} = 10.0$ V and $R = 5.00\ \Omega$, what are the greatest and least currents?

40. ◆◆ Find the power used by each resistor in the circuit diagramed in ■ Figure 26.49a. How much less power does each resistor use than it would if it were the only resistor in the circuit?

41. ◆◆ A real battery has emf $\mathcal{E} = 50.0$ V. When it is connected to a 20.0-Ω resistor, the current is 2.45 A. What are the internal resistance of the battery and the potential difference between its terminals?

42. ◆◆ A 5.0-Ω and a 20.0-Ω resistor are connected in parallel across an ideal 10.0-V emf. What is the current through each circuit element? Find the power output of the battery and the power dissipated by each resistor.

43. ◆◆ In the circuit shown in ■ Figure 26.50, find the current in each resistor and the values of the potential at points A, B, and C. (Take $V = 0$ at ground.)

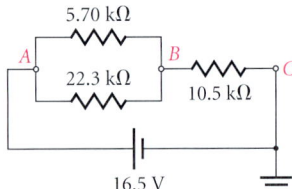

■ FIGURE 26.50

44. ◆◆ There is a total current I through resistors R_1, R_2, and R_3 connected in parallel. Derive expressions for the currents in the individual resistors.

45. ◆◆◆ Find the equivalent resistance of the combination shown in ■ Figure 26.51. Find the current through the battery and through each resistor. Find the potential V at points A, B, and C.

46. ◆◆◆ In the circuit diagramed in ■ Figure 26.52, the current through the 50-Ω resistor is the same whether the switch is open or closed. What is R?

47. ◆◆◆ To protect a radio transmitter against lightning, it is connected to a metal post in the ground (■ Figure 26.53). Proper protection requires that the resistance between the post and the water table be less than 10 Ω. To test for this, the station engineer drives two spikes into the ground at A and B. The resistance between the two spikes is measured to be $R_{AB} = 40\ \Omega$. The resistance between A and the post (R_{AP}) is measured as 28 Ω, and the resistance between B and the post (R_{BP}) is 30 Ω. Is the radio tower adequately grounded?

§26.4 KIRCHHOFF'S RULES

48. ❖ In a compound circuit with more than one battery, does current always circulate around a loop from the positive battery terminal to the negative terminal? Why or why not?

49. ❖ If the 100-Ω resistor in ■ Figure 26.54 is increased to 101 Ω, does the current through the resistor increase, decrease, or stay the same? Does the potential difference across it increase, decrease, or stay the same?

50. ◆ A flashlight uses two 1.5-V batteries to light a bulb. The current in the circuit is 0.53 A. What is the resistance of the bulb? How much power does it consume?

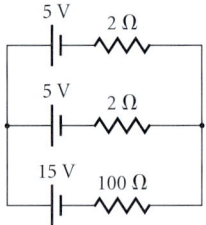

■ FIGURE 26.54

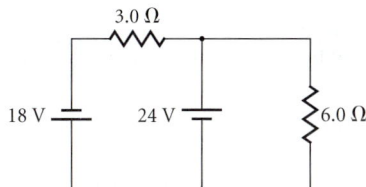
■ FIGURE 26.55

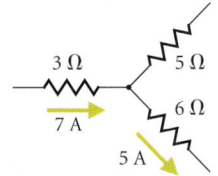
■ FIGURE 26.56

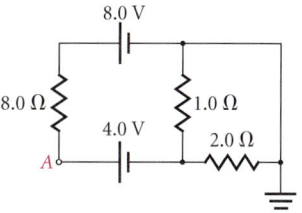
■ FIGURE 26.58

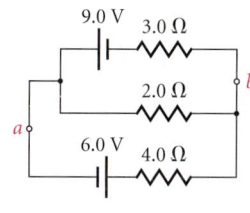
■ FIGURE 26.59

51. ◆ A 12.0-V battery with internal resistance of 0.02 Ω is used to charge a battery with an emf of 10.5 V and an internal resistance of 0.15 Ω. What is the charging current?

52. ◆ A flashlight contains two batteries in series, each with an emf of 1.50 V and internal resistance of 0.17 Ω. The bulb has a resistance of 25.0 Ω. Draw a circuit diagram, and find the current in the circuit. How much power does the bulb draw?

53. ◆◆ Find the current through each resistor in ■ Figure 26.55.

54. ◆◆ What value of the emf $\mathscr{E}_2$ in the circuit of Figure 26.30 is necessary if that battery is to deliver power to the circuit rather than absorbing power from the circuit?

55. ◆◆ What is the potential difference across the 100-Ω resistor in the circuit of ■ Figure 26.54?

56. ◆◆ Two automobile batteries are connected in parallel to power a wheelchair. If each of the batteries has an emf $\mathscr{E} = 12.0$ V and internal resistance $r = 0.020$ Ω, and the wheelchair motor has resistance $R = 1.00$ Ω, find the current provided to the motor. What would be the current delivered to the motor if the batteries were connected in series? What are the relative advantages of series and parallel connections?

57. ◆◆ Show that the currents I_1 and I_2 in Figure 26.26 (Example 26.12) are given by:

$$I_1 = \frac{(\mathscr{E}_1 - \mathscr{E}_2)R + \mathscr{E}_1 r_2}{R(r_1 + r_2) + r_1 r_2}$$

and

$$I_2 = -\frac{(\mathscr{E}_2 - \mathscr{E}_1)R + \mathscr{E}_2 r_1}{R(r_1 + r_2) + r_1 r_2}.$$

58. ◆◆ In the circuit segment diagramed in ■ Figure 26.56, find the potential difference across each resistor.

59. ◆◆ Three ideal batteries and three identical resistors are connected in the circuit diagramed in ■ Figure 26.57. Find the current in each arm of the circuit.

60. ◆◆ A 3-Ω resistor, a 1-Ω resistor, a 3-V emf, and a 9-V emf are all connected in series. Draw diagrams of two circuits that fit this description but that carry different currents. Apply Kirchhoff's loop rule to each circuit. Find the current and draw a graph showing the variation of potential around the circuit.

61. ◆◆ What would be the current if the good battery in Figure 26.26 (Example 26.12) were connected in series with the dead battery? What would you have to do to make the connection? Would it be worth it?

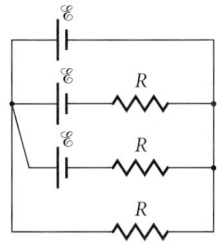
■ FIGURE 26.57

62. ◆◆ Find all the currents in the circuit diagramed in ■ Figure 26.58. What is the potential at point A?

63. ◆◆ Find all the currents in the circuit shown in ■ Figure 26.59 and the potential difference between points a and b.

64. ◆◆ Find the resistance of the combination in ■ Figure 26.60.

65. ◆◆ Twelve identical resistors are connected together to form a cube (■ Figure 26.61). What is the cube's equivalent resistance when connected into a circuit at points A and D? Hint: Use symmetry to equate the currents in some of the arms before applying Kirchhoff's rules.

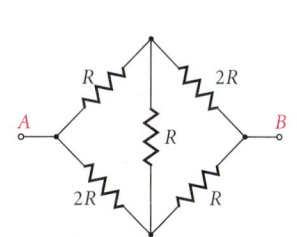
■ FIGURE 26.60

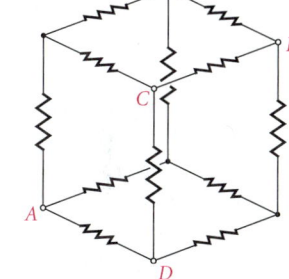
■ FIGURE 26.61

*§26.5 ELECTRICAL MEASUREMENT

66. ◆ A Wheatstone bridge is used to determine the resistance of a piece of steel wire. With $R_1 = 4.00$ Ω, the ratio of R_4 to R_3 is found to be 0.092. What is R_x? If the wire is 45.0 cm long and 0.50 mm in diameter, what is its resistivity?

67. ◆ A voltmeter with a resistance $R_m = 5.00 \times 10^5$ Ω is used to measure the potential difference across one of two 3.00-kΩ resistors connected in series to a 5.00-V battery (■ Figure 26.62). What potential difference does the voltmeter measure? What is the true potential difference before the voltmeter is connected?

68. ◆◆ A multirange ammeter uses a series of shunt resistances and a solid-state voltage detector (■ Figure 26.63). The solid-state device is sensitive to potential differences of 0 to 0.2 V. The ammeter is con-

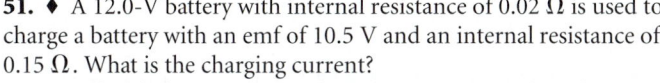

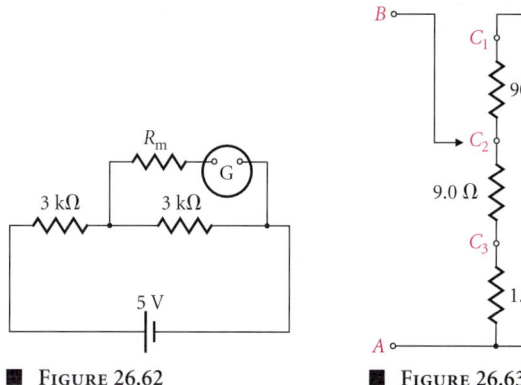
■ FIGURE 26.62   ■ FIGURE 26.63

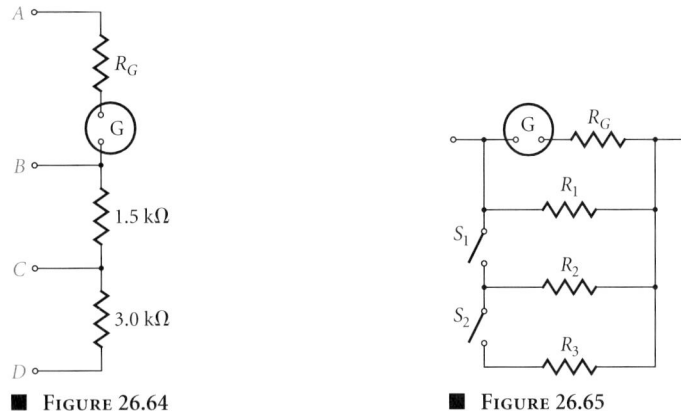

Figure 26.64

Figure 26.65

nected into the circuit between points A and B, with B connected to one of points C_1, C_2, or C_3. If the current to be measured is 3.00 mA, which connection should be used? If the current is produced in a circuit with an emf of 1.50 V and a resistance of 0.500 kΩ, what is the measured current? By what percentage does the ammeter disturb the circuit?

69. ♦♦ A voltmeter is made using a galvanometer of internal resistance 1.5 kΩ and two resistances of 1.5 kΩ and 3.0 kΩ (■ Figure 26.64). The galvanometer gives a full-scale deflection when there is a current of 5 μA in it. What voltages across terminals AB, AC, and AD produce a full-scale deflection?

70. ♦♦♦ An ohmmeter is constructed by connecting a known emf in series with a galvanometer of resistance 15 Ω. The circuit is completed by connecting the unknown resistance in series. If the galvanometer sensitivity is 50.0 μA at full scale and a 2.0-V cell is used as the known emf, what is the minimum resistance that can be measured? If the uncertainty in meter reading is $\frac{1}{100}$ of full scale, what is the largest resistance that can be measured to 10% accuracy? What known resistance, in series with the galvanometer, will protect it against damage when the ohmmeter leads are touched together?

71. ♦♦♦ An ammeter is constructed from a galvanometer with resistance $R_G = 18.0$ Ω and three resistances connected in parallel (■ Figure 26.65). $R_1 = 2.0$ Ω. The galvanometer shows a full-scale deflection when a 50.0-μA current passes through it. What is the maximum current the ammeter can measure when the switches are both open? If switch S_1 is closed, does the ammeter become more or less sensitive? (A more sensitive ammeter can measure a smaller current.) What resistances R_2 and R_3 are required to change the sensitivity by a factor of 10 when switch S_1 is closed, and another factor of 10 when both switches are closed?

72. ♦♦♦ A potentiometer may be used to measure potential differences without requiring any current (■ Figure 26.66). The emf $\mathcal{E}_s$ is accurately known. To measure a potential difference V the switch is first set at position a and the sliding contact is moved until $\ell = \ell_s$, and the galvanometer detects no current. The switch is then set at position b, and the contact is moved again until $\ell = \ell_x$, and the galvanometer detects no current. Show that $V = \mathcal{E}_s(\ell_x/\ell_s)$.

Additional Problems

73. ❖ A disk of radius R and uniform charge density ρ rotates with angular speed ω about an axis perpendicular to the disk at its center. Describe the resulting distribution of current.

74. ♦ Electric companies charge for power in units of kW·h. How many joules equal 1 kilowatt-hour? If the power company charges 12¢/kW·h, and supplies power at 120 V, what is the cost per coulomb?

75. ♦♦ A 1.50-V battery delivers a current of 0.235 A when connected to a 6.35-Ω load. What is the internal resistance of the battery?

76. ♦♦ A battery delivers a current of 1.98 A when connected to a 6.00-Ω resistance and 2.96 A when connected to a 4.00-Ω resistance. Find the emf and internal resistance of the battery.

77. ♦♦ A 12.00-V battery is connected to two identical resistors, each of resistance R. With the resistors connected in series, the measured current is 0.5985 A. With the resistors connected in parallel, the measured current is 2.376 A. Find the internal resistance of the battery and the resistance R.

78. ♦♦ A whimsical physics instructor connects 12 identical 10.0-Ω resistors together to make the cubical resistor shown in Figure 26.61. An ideal battery of emf 4.0 V is connected between points A and B (across a body diagonal). Find the current through the battery. Use symmetry arguments to simplify the problem as much as possible before applying Kirchhoff's rules. What is the current through a real battery with an internal resistance of 0.30 Ω?

79. ♦♦♦ The resistor network in ■ Figure 26.67 extends infinitely to the right. What is the resistance between points A and B? *Hint:* If you remove the two leftmost resistors, the network is unchanged.

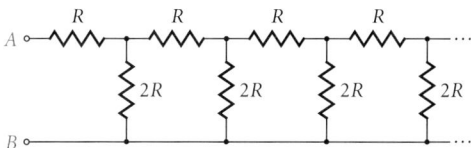

Figure 26.67

80. ♦♦♦ A 55-kg woman pedaling a bicycle wheel powers an electric generator (source of emf). The generator maintains a current of 5.0 A through a 20.0-Ω resistance. Using this power more enjoyably, how rapidly may she gain altitude while riding a 15-kg bicycle?

81. ♦♦♦ Show that maximum power is delivered to a load when the load resistance is equal to the internal resistance of the power supply.

82. ♦♦♦ The circuit in ■ Figure 26.68 has geometric symmetry. Make use of the symmetry to find the currents. Solve the problem using

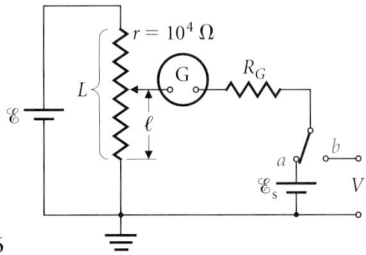

Figure 26.66

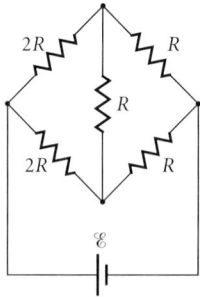

Figure 26.68

Kirchhoff's rules with five current variables and show that the first method gives correct results.

83. ♦♦♦ A battery with $\mathcal{E} = 1.0$ V is connected between points A and C (across a face diagonal) of the cubical resistor in Figure 26.61. Each resistance is 10.0 Ω. Find the current through the battery. (*Hint:* Use geometric symmetry to reduce the circuit to a series/parallel combination.)

Computer Problems

84. File CH26P84 on the supplementary computer disk contains measured values of current when a potential difference ΔV is applied across a light-emitting diode. Use a spreadsheet program to plot the characteristic curve (I vs. ΔV). The slope of the characteristic curve is called the dynamic resistance. Use the spreadsheet to estimate the dynamic resistance of the diode at 1.50 V and 1.70 V.

85. Applying Kirchhoff's rules to a circuit, we obtain a set of linear equations for the currents that may be described by a matrix equation:

$$MI = J,$$

where I is the vector of current values with components $I_1, I_2, \ldots, I_n$, and the matrix M contains the coefficients of the currents in each of the equations. The equations in Example 26.12 are:

(i) $\quad I_1 - I_2 - I_3 = 0.$
(ii) $\quad (0.02 \ \Omega)I_1 + (0.05 \ \Omega)I_2 = 2.0$ V.
(iii) $\quad -(0.05 \ \Omega)I_2 + (0.20 \ \Omega)I_3 = 10.0$ V.

The matrix M contains the coefficients of the Is:

$$M = \begin{pmatrix} 1 & -1 & -1 \\ 0.02 & 0.05 & 0 \\ 0 & -0.05 & 0.20 \end{pmatrix} \Omega,$$

and the vector $J = (0, 2.0, 10.0)$ V. The problem is solved by inverting the matrix:

$$I = M^{-1}J.$$

Use the matrix inversion program on the supplementary computer disk to show that the inverse of the matrix is:

$$M^{-1} = \begin{pmatrix} 2/3 & 50/3 & 10/3 \\ -4/15 & 40/3 & -4/3 \\ -1/15 & 10/3 & 14/3 \end{pmatrix} \Omega^{-1}.$$

So then: $\begin{pmatrix} I_1 \\ I_2 \\ I_3 \end{pmatrix} = \begin{pmatrix} 2/3 & 50/3 & 10/3 \\ -4/15 & 40/3 & -4/3 \\ -1/15 & 10/3 & 14/3 \end{pmatrix} \begin{pmatrix} 0 \\ 2.0 \\ 10.0 \end{pmatrix}$ A.

Each element in the product matrix is the sum of the products of the numbers in a row of the matrix times the numbers in the vector J:

$$\begin{pmatrix} (2/3) \times 0 + (50/3) \times (2.0) + (10/3) \times (10.0) \\ (-4/15) \times 0 + (40/3) \times (2.0) + (-4/3) \times (10.0) \\ (-1/15) \times 0 + (10/3) \times (2.0) + (14/3) \times (10.0) \end{pmatrix} A = \begin{pmatrix} 66.7 \\ 13.3 \\ 53.3 \end{pmatrix} A.$$

Use this method to find the currents in the circuit in ■ Figure 26.69. (Spreadsheet programs and calculators often have the ability to do matrix inversion and matrix multiplication.)

Challenge Problems

86. A superposition theorem states that one may solve a circuit by first solving a number of simplified circuits, each containing only a single emf but otherwise the same as the original circuit. The current in any arm of the actual circuit is the sum of the currents in the same arm of each simplified circuit. Use this method to verify the results of Example 26.12. Find currents I_1, I_2, and I_3 in Figure 26.26 if $\mathcal{E}_1 = 12.0$ V, $\mathcal{E}_2 = 10.0$ V, $r_1 = 1.5 \ \Omega$, $r_2 = 1.2 \ \Omega$, and $R = 36.0 \ \Omega$.

87. Loop currents: An alternative method for using Kirchhoff's rules defines one loop current for each loop used in the solution. The actual current in any arm of the circuit is the algebraic sum of the loop currents associated with that arm. **(a)** Using the circuit in ■ Figure 26.70 as an example, show that the loop current method automatically satisfies Kirchhoff's junction rule. **(b)** Use Kirchhoff's loop rule to write a set of equations for the loop currents, and solve for i_1 and i_2. **Warning:** Never use a combination of loop currents and arm currents in one circuit. Pick one method and stick to it.

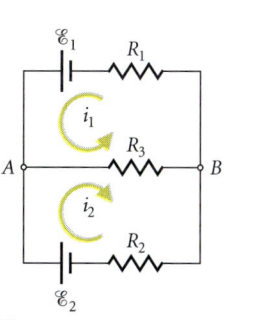

■ **Figure 26.70**

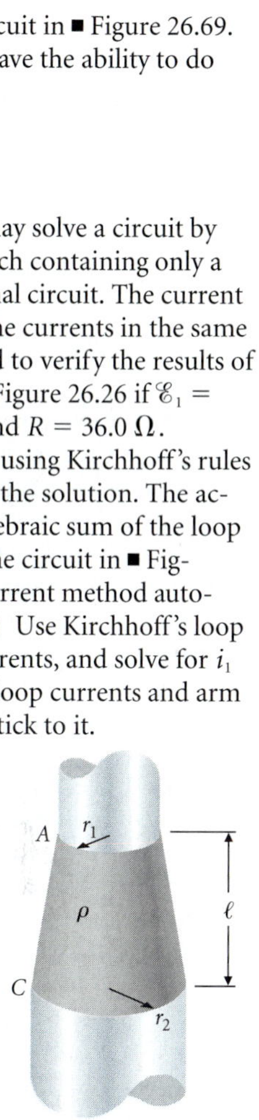

88. If the galvanometer in a Wheatstone bridge circuit (Figure 26.39) has 20.0 Ω of resistance and an uncertainty of ±1 μA in its reading, what is the uncertainty in the measured value of R_x? Take $R_1 = 5.00 \ \Omega$, $R_3 = 10.0 \ \Omega$, $R_4 = 20.0 \ \Omega$, and $\mathcal{E} = 5.00$ V. Is the uncertainty significant?

89. A device designed to connect two wires of different radii has the shape of a conical frustum of height ℓ, smaller radius r_1, and larger radius r_2 (■ Figure 26.71). If the device is made of material with resistivity ρ, compute its resistance approximately. Assume $r_2 - r_1 \ll \ell$.

90. Thevenin's theorem states that any portion of a circuit containing only emfs and resistors and with two junctions that connect to the rest of the circuit is equivalent to a single battery with emf $\mathcal{E}_{TH}$ and internal resistance R_{TH} (■ Figure 26.72). **(a)** Show that E_{TH} and R_{TH}

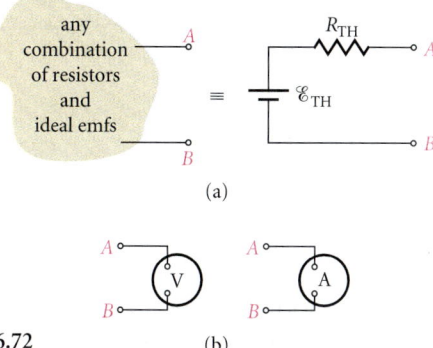

■ **Figure 26.72**

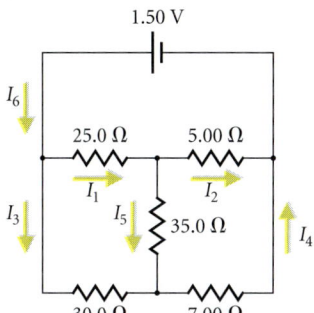

■ **Figure 26.69**

■ **Figure 26.71**

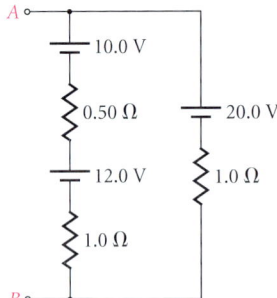

FIGURE 26.73

FIGURE 26.76

can be found by measuring two quantities: (i) the open-circuit potential difference between points B and A and (ii) the short-circuit current between A and B (Figure 26.72b). **(b)** Find the Thevenin equivalent of the power supply in ■ Figure 26.73. What is the current through a 1.00-kΩ load? **(c)** Use Thevenin's theorem to show that if any single resistance in a circuit is increased, the current through it decreases and the potential difference across it increases.

91. Norton's theorem states that any combination of emfs and resistors with two junctions that connect to the rest of the circuit is equivalent to a current source I_N in parallel with a resistance R_N (■ Figure 26.74). **(a)** Show that I_N and R_N may be found by measuring (i) the open-circuit potential difference between points A and B and (ii) the short-circuit current from A to B. (A current source is a device that adjusts its emf to maintain a constant current.) **(b)** Find the Norton equivalent of the circuit in ■ Figure 26.75.

92. Refer to Problems 90 and 91. Prove that the Norton and Thevenin resistances for any combination of emfs and resistors are the same.

93. *Star-Delta Transformation* ■ Figure 26.76 shows a *star* and a *delta* resistor combination. Each connects to the rest of a circuit at three points. Show that replacing the star with the delta (or vice-versa) produces an equivalent circuit if and only if:

$$r_1 = \frac{R_2 R_3}{R_1 + R_2 + R_3}$$

and

$$r_2 = \frac{R_1 R_3}{R_1 + R_2 + R_3}$$

and

$$r_3 = \frac{R_1 R_2}{R_1 + R_2 + R_3},$$

or:

$$R_1 = \frac{r_1 r_2 + r_2 r_3 + r_3 r_1}{r_1},$$

with similar expressions for R_2 and R_3. A battery with $\mathcal{E} = 10.0$ V is connected between points A and D of the cubical resistor (Figure 26.61) with $R = 10.0 \, \Omega$. Use star-delta transformations to reduce the problem to a series-parallel circuit, and find the total current through the battery.

94. In a lead-acid battery, the reactions at each terminal are reversible. When a charged battery is idle, the reactions in fact proceed in both directions at equal rates with no net energy use or charge transfer. Explain how connecting the battery in a circuit and allowing charge to flow from the terminals causes the reactions to replace the charge on the terminals.

95. An unknown resistance is connected to an unknown ideal battery in a simple circuit. A voltmeter and ammeter are to be used to measure current and potential difference simultaneously. Both the voltmeter and ammeter use galvanometers with 30.0-Ω resistance. The voltmeter has a 1.00-kΩ resistor in series with the galvanometer, and the ammeter has a 0.500-Ω shunt. When the voltmeter is connected in parallel with the resistor, the meters read $\Delta V = 4.76$ V and $I = 0.481$ A. What are the emf and resistance in the circuit? What would the meter readings be if the voltmeter were connected in parallel with the battery? If you take the voltmeter reading as the emf and the voltmeter reading divided by the ammeter reading as the resistance, which position gives the more accurate measurements?

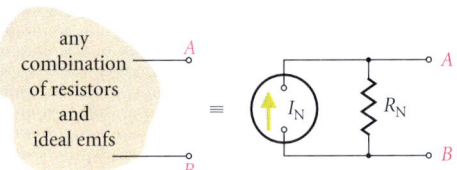

FIGURE 26.74

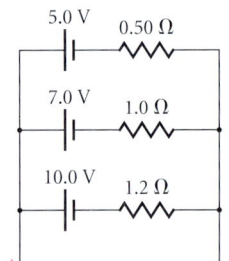

FIGURE 26.75

Bank of capacitors designed at the Ernest O. Lawrence Berkeley National Laboratory and used to power a neutral beam injector. Each bank consists of ten modules of 272 capacitors. (See also Figure 27.11).

CHAPTER 27
Capacitance and Electrostatic Energy

CONCEPTS

Capacitor
Capacitance
Dielectric
Bound charge
Free charge
Electrostatic energy density

GOALS

Be able to:

Calculate the capacitance of symmetric objects.

Compute the equivalent capacitance of series and parallel combinations.

Compute the energy stored in a capacitor.

Compute the electrostatic energy density at a point.

They were hoarding and dispersing energy while the inanimate universe was running down around us.

LOREN EISELEY

A MORE EXTENSIVE DISCUSSION OF CAPACITOR CIRCUITS WILL BE GIVEN IN CHAPTERS 31 AND 32.

Plentiful energy for a billion years from the hydrogen in seawater—this is the hope of nuclear fusion research. The energy is available; the problem is how to extract it. All the experiments striving to achieve fusion need large pulses of energy either to inject the fuel into the experiment chamber or to compress and heat it. For example, a neutral beam injector designed at the University of California, Berkeley, produces pulses lasting 50 ms that deliver nearly 1.6 MJ of energy. Such pulses greatly exceed the capability of the local power company and require a method of storing electrical energy for quick release. A bank of 8160 individual *capacitors* stores the energy. In this chapter we shall study capacitors and give an introduction to their use in circuits. We shall also gain a deeper understanding of electrostatic energy and its relation to the electric field.

27.1 CAPACITANCE

27.1.1 The Parallel Plate Capacitor

Any object that can store charge is a capacitor. The simplest capacitor to analyze is made from two parallel conducting plates, each of area A (■ Figure 27.1). In use, one plate holds a positive charge density σ while the other has an equal negative charge density. We model the system as two infinite uniformly charged sheets. Each sheet produces an electric field of magnitude $E = \sigma/(2\epsilon_0)$ (eqn. 24.7) perpendicular to the plates. The two fields reinforce each other to give $E = \sigma/\epsilon_0$ between the plates and cancel to give zero outside (Figure 27.1b). The potential difference between the plates is $|\Delta V| = Ed = \sigma d/\epsilon_0$. The total charge on one plate is $Q = \sigma A$, so $\sigma = Q/A$. We find that the potential difference ΔV between the plates is proportional to the charge Q:

RECALL: σ = CHARGE/AREA.

WE USED EQN. (25.7) WITH E = CONSTANT.

$$|\Delta V| = \frac{d}{\epsilon_0 A} Q. \tag{27.1}$$

■ FIGURE 27.1
(a) A charged parallel plate capacitor. The two conducting plates are separated by a distance d. Each plate has area A; one has a uniform charge density σ, and the other, $-\sigma$. (b) The electric field vector is the sum of the vectors produced by the two plates. The electric field vector produced by a uniform sheet of charge has magnitude $\sigma/(2\epsilon_0)$ and is constant. Between the plates, the vectors are in the same direction, and $E = \sigma/\epsilon_0$. On either side, the vectors are in opposite directions and sum to zero.

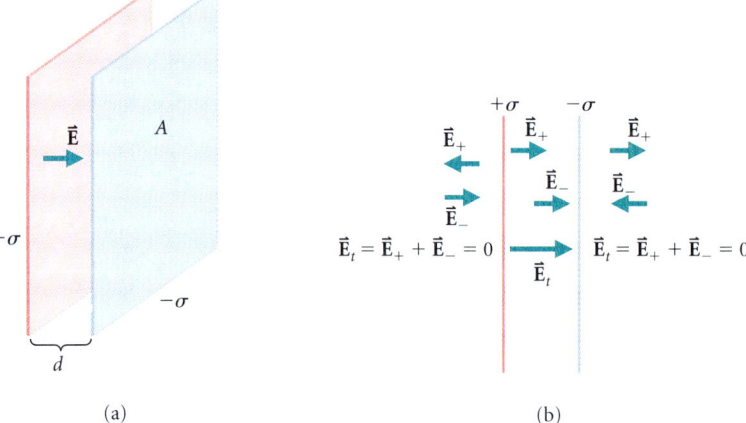

In fact, for *any* pair of conductors, charge and potential difference are proportional (■ Figure 27.2). Their *ratio* depends only on the shapes, sizes, and relative positions of the conductors and gives an operational definition of capacitance:

The *capacitance* of two conductors is the magnitude of the ratio of the charge on one of them to the potential difference between them, when they are oppositely charged:

$$C \equiv \left|\frac{Q}{\Delta V}\right|. \tag{27.2}$$

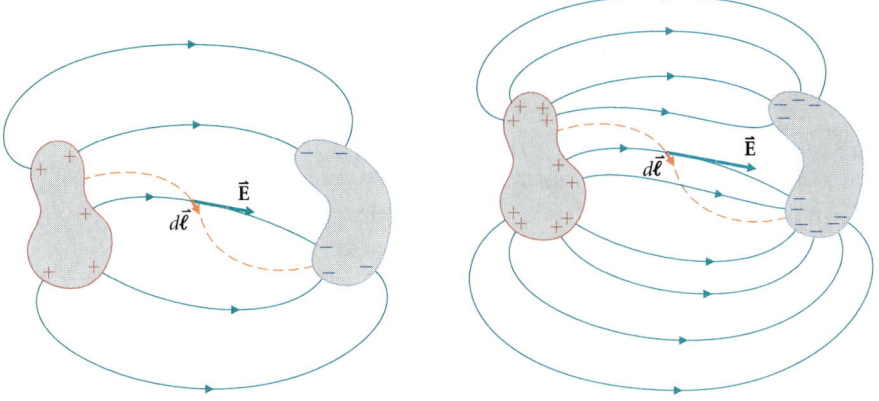

FIGURE 27.2
The potential difference between any two conductors with equal and opposite charge is proportional to that charge. Charge is distributed on the surface of each conductor so that $\vec{E}$ is zero within the conductor, and just outside the conductor, $\vec{E}$ is perpendicular to the surface. Only one pattern of charge density accomplishes this. If the total charge is doubled, the charge density at each point doubles, leaving the pattern unchanged. As a result, the electric field strength doubles, whereas the direction at each point is unchanged. Thus the potential difference $\int \vec{E} \cdot d\vec{\ell}$ between the conductors also doubles.

We find the capacitance of the two parallel plates by combining eqns. (27.1) and (27.2):

$$C = \frac{\epsilon_0 A}{d}. \tag{27.3}$$

STRICTLY SPEAKING THIS EXPRESSION IS CORRECT WITH VACUUM BETWEEN THE PLATES. AIR MAKES LITTLE DIFFERENCE (BUT SEE §27.3).

Once the capacitor is made, the plate area A and separation d are set, and the capacitance does not change. The amount of charge on the plates does change as the potential difference between the plates is varied.

The unit of capacitance is the farad (F). Conductors with a capacitance of 1 farad at a potential difference of 1 volt store 1 coulomb of charge:

$$1\ \text{F} \equiv \frac{1\ \text{C}}{1\ \text{V}}.$$

Just as a coulomb is a large unit of charge, a farad is a very large unit of capacitance.

NAMED AFTER MICHAEL FARADAY (1791–1867). FARADAY MADE FUNDAMENTAL CONTRIBUTIONS TO A WIDE RANGE OF FIELDS. HIS FULL NAME IS USED FOR A UNIT OF CHARGE IN ELECTROCHEMISTRY. THE PASSAGE OF 1 FARADAY OF CHARGE THROUGH A SOLUTION DEPOSITS 1 MOLE OF A UNIVALENT ION ON THE ELECTRODES.

EXAMPLE 27.1 ♦ Two parallel metal plates are separated by 2.0 mm. What area should they have to form a 1.0-F capacitor?

MODEL We use eqn. (27.3) for a parallel plate capacitor.

SETUP AND SOLVE
$$A = \frac{dC}{\epsilon_0} = \frac{(2.0 \times 10^{-3}\ \text{m})(1.0\ \text{F})}{8.85 \times 10^{-12}\ \text{C}^2/\text{N} \cdot \text{m}^2} = 2.3 \times 10^8\ \text{m}^2.$$

ANALYZE Square plates 15 km on a side would do the job! Check the units:

$$\frac{\text{N} \cdot \text{m}^3 \cdot \text{F}}{\text{C}^2} = \frac{\text{J} \cdot \text{F} \cdot \text{m}^2}{\text{C}^2} = \frac{(\text{C} \cdot \text{V}) \cdot (\text{C}/\text{V}) \cdot \text{m}^2}{\text{C}^2} = \text{m}^2.$$

The units of ϵ_0, $\text{C}^2/\text{N} \cdot \text{m}^2$, are equivalent to F/m. ∎

EXERCISE 27.1 ♦ What area plates, separated by 2.0 mm, would form a 1.0-nF capacitor? How much charge would this capacitor store with a potential difference of 2000 V? ∎

27.1.2 Calculating Capacitance

The definition of capacitance reflects a capacitor's practical use: we apply a known potential difference ΔV to the capacitor, which then stores charge. To calculate the capacitance of a given pair of conductors we work in the opposite direction and compute the potential difference between the conductors when a charge Q is stored.

FOR EXAMPLE, A BATTERY IS CONNECTED ACROSS THE PLATES AND HOLDS CHARGE ON THEM.

OUTSIDE BOTH SHELLS, THE NET FIELD IS ZERO.

SINCE WE ARE INTEGRATING FROM R_2 TO R_1, EACH dr IS NEGATIVE. DON'T BE TEMPTED TO ADD EXTRA MINUS SIGNS!

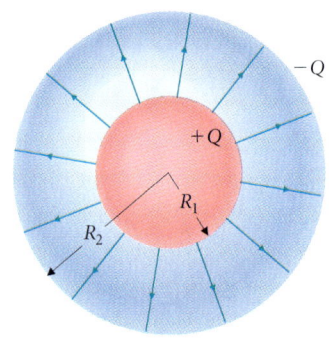

■ FIGURE 27.3
A spherical capacitor consists of two concentric spherical shells. The electric field due to the outer sphere is zero between the shells.

EXAMPLE 27.2 ♦♦ A spherical capacitor consists of two concentric metal spheres of radii R_1 and R_2 (■ Figure 27.3). The inner sphere carries a charge Q and the outer sphere, $-Q$. What is the potential difference between the spheres? What is their capacitance?

MODEL A uniformly charged spherical shell produces zero net field in its interior. Outside, it produces the same field strength as a point charge at its center (cf. Example 24.1). Thus the electric field between the two shells of the capacitor is due only to the inner shell, and $|\vec{E}| = kQ/r^2$.

SETUP Field lines run from the inner shell to the outer, so the inner shell is at the higher potential. The potential difference between the shells is (eqn. 25.7):

$$\Delta V = V_1 - V_2 = -\int_{R_2}^{R_1} \vec{E} \cdot d\vec{\ell} = -\int_{R_2}^{R_1} \frac{kQ}{r^2} \hat{r} \cdot dr\, \hat{r} = -\int_{R_2}^{R_1} \frac{kQ}{r^2}\, dr = \left.\frac{kQ}{r}\right|_{R_2}^{R_1}$$

$$= kQ\left(\frac{1}{R_1} - \frac{1}{R_2}\right) = kQ\,\frac{R_2 - R_1}{R_1 R_2}.$$

SOLVE The capacitance of the object is:

$$C = \frac{Q}{\Delta V} = \frac{R_1 R_2}{k(R_2 - R_1)} = \frac{4\pi\epsilon_0 R_1 R_2}{R_2 - R_1}. \tag{27.4}$$

ANALYZE Again, the capacitance depends only on the physical dimensions of the capacitor. ■

EXERCISE 27.2 ♦ What is the capacitance of a pair of spherical shells with $R_1 = 2.0$ cm and $R_2 = 3.0$ cm?

EXERCISE 27.3 ♦♦ Show that when $R_2 - R_1 \ll R_1$, the capacitance of two nested spheres is the same as that of parallel plates with the same surface area and separation. ■

An isolated conductor may be viewed as one of a pair of conductors, the other of which is at infinity; then we may speak of the capacitance of an isolated object.

EXAMPLE 27.3 ♦♦ Find the capacitance of an aluminum ball with a radius of 10.0 cm.

MODEL This is a special case of the capacitor in Example 27.2, in which $R_2 \to \infty$.

SETUP When taking the limit, it is easiest to use the intermediate result in Example 27.2:

$$C = \frac{Q}{\Delta V} = \lim_{R_2 \to \infty} \frac{1}{k(1/R_1 - 1/R_2)} = \frac{R_1}{k} = 4\pi\epsilon_0 R_1.$$

SOLVE With $R_1 = 10.0$ cm:

$$C = \frac{1.00 \times 10^{-1}\text{ m}}{8.99 \times 10^9\text{ N}\cdot\text{m}^2/\text{C}^2} = 11.1\text{ pF}.$$

ANALYZE The ball would have to be huge to make a 1-F capacitor. The Earth is good for about 0.7 mF. ■

Our result for a sphere gives a useful rule for estimating the capacitance of any isolated object:

Capacitance ~ (linear dimension)/k.

For the sphere, the linear dimension is the radius of the sphere, and the rule gives an exact result. With more complicated systems, the result is cruder. For parallel plates, the only combination of A and d with dimensions of length is A/d. The rule gives $C \sim A/kd$, which is too large by a factor 4π.

EXAMPLE 27.4 ♦♦ Find the capacitance per unit length of coaxial cable if its inner wire has a radius of 1.5 mm and its outer conductor has an inside radius of 4.5 mm. (■ Figure 27.4).

MODEL We calculate the potential difference between the two conductors when they carry charges per unit length $+\lambda$ and $-\lambda$; then we apply the definition of capacitance.

SETUP A charged cylindrical shell, like a spherical shell, produces zero net field inside itself. The electric field the shell produces outside itself is (eqn. 24.4):

$$\vec{E} = \left(\frac{\lambda}{2\pi\epsilon_0 r}, \quad \text{radially outward}\right).$$

Between the conductors in the cable, the net field is that due to the charge on the inner cylinder. The potential difference is:

$$\Delta V = V_1 - V_2 = -\int_{r_2}^{r_1} \vec{E} \cdot d\vec{\ell} = -\int_{r_2}^{r_1} \frac{\lambda}{2\pi\epsilon_0 r} dr = \frac{\lambda}{2\pi\epsilon_0} \ln(r_2/r_1).$$

SOLVE The capacitance per unit length is:

$$\frac{dC}{d\ell} = \frac{dQ/d\ell}{\Delta V} = \frac{\lambda}{\Delta V} = \frac{2\pi\epsilon_0}{\ln(r_2/r_1)} = \frac{1}{2k \ln(r_2/r_1)}. \tag{27.5}$$

Putting in the numbers, we get $r_2/r_1 = (4.5 \text{ mm})/(1.5 \text{ mm}) = 3.0$. Then:

$$\frac{dC}{d\ell} = \frac{1}{2(9.0 \times 10^9 \text{ N}\cdot\text{m}^2/\text{C}^2)\ln(3.0)} = 0.051 \text{ nF/m}.$$

ANALYZE The capacitance of a finite length of cable is approximately its length times the capacitance per unit length. The appropriate dimension of a cylinder to use in our rule for estimating capacitance is the length. ■

EXERCISE 27.4 ♦ Estimate the capacitance of a person.

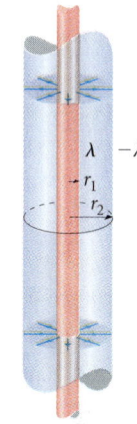

■ **FIGURE 27.4**
A coaxial cable has an inner conducting cylinder surrounded by a conducting cylindrical shell. Each carries charge per unit length $\pm\lambda$. Here we assume the inner conductor is positively charged. As in the spherical capacitor, the electric field due to the outer conductor is zero inside the cable.

Digging Deeper

GROUNDING A DC CIRCUIT

When a DC circuit is grounded, why isn't the circuit disrupted by a current through the ground connection? The circuit in ■ Figure 27.5 operates disconnected from ground when the switch is open, and carries a current of 7 mA. The circuit can be grounded at either terminal of the battery by moving the switch to A or to B. When the switch is moved from A to B the potential of every point in the circuit increases by $\Delta V = 14$ V. To produce this change, a charge ΔQ flows through the ground wire, where $\Delta Q = C \Delta V$, and C is the capacitance of the circuit. (The other conductor is at infinity, cf. Example 27.3). Using the estimation rule with a linear dimension of 10 cm for the circuit, we obtain $C \approx (10^{-1} \text{ m})/k \approx 10$ pF, and $\Delta Q = C \Delta V \approx 0.14$ nC.

The current of 7 mA in the circuit can distribute this charge in a time $\tau \approx \Delta Q/I \approx 20$ ns. Charge does flow from ground—for this very short time—but once equilibrium is established, no further charge transfer occurs.

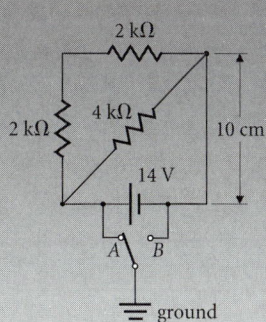

■ **FIGURE 27.5**
This DC circuit may be grounded at either terminal of the battery by placing the switch at point A or at point B. When the switch setting is changed, the circuit requires about 20 ns to adjust.

27.2 Energy Storage in Capacitors

27.2.1 Charging a Capacitor

WE'LL INVESTIGATE THIS CIRCUIT MORE CAREFULLY IN CHAPTER 31.

To charge a capacitor, we connect it to an emf that supplies energy to separate charge into equal positive and negative amounts on the two conductors. Unavoidably the circuit contains some resistance r (■ Figure 27.6a). At $t = 0$, the switch is closed in position A, and charge begins to flow. As time passes and charge accumulates on the capacitor, the potential difference across the capacitor becomes equal to the emf of the battery, and the current decreases to zero. Each of the two conductors then has charge of magnitude:

$$Q = C\,\Delta V = C\mathcal{E}.$$

The emf does work both to charge the capacitor and to drive current through the resistor. The work done on the resistor is dissipated as heat. What happens to the work done on the capacitor? If the switch is subsequently moved to position B (Figure 27.6b), the potential difference across resistor R drives current in the right-hand loop as the capacitor discharges. The energy dissipated in resistor R must have been stored in the capacitor.

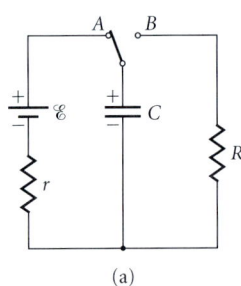
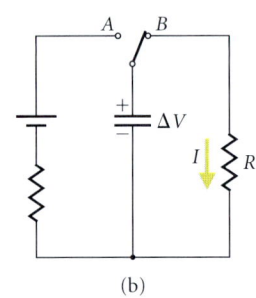

■ **FIGURE 27.6**
Circuit for charging and discharging a capacitor. (a) With the switch in position A, as shown, the capacitor stores charge as the potential difference across it approaches the emf of the battery. (b) When the switch is moved to position B, the capacitor discharges through the resistor. Energy is stored in the capacitor when it is charged and is subsequently dissipated as heat by the resistor.

The electrostatic energy stored in the capacitor equals the work done to charge it. The energy depends only on the amount of charge, not how it was put in place. So we may compute the energy by analyzing any imaginary process that would charge the capacitor (cf. §25.3). So we imagine a process that is easy to compute: carrying charge across the gap from one plate of a parallel plate capacitor to the other (■ Figure 27.7). When the capacitor is partially charged, the electric field strength between the plates is $E = 4\pi k\sigma$. The surface charge density σ is related to the amount of charge q already on the plate, $\sigma = q/A$, and the potential difference between the plates is:

REMEMBER: THIS IS AN *IMAGINARY* PROCESS!

WE USE A CAPITAL D FOR THE PLATE SEPARATION HERE TO AVOID CONFUSION WITH THE DIFFERENTIAL QUANTITIES dq AND dW.

$$\Delta V = ED = 4\pi k\sigma D = 4\pi kqD/A.$$

To increase the charge on the plate by dq we do work:

$$dW = \Delta V\,dq = \frac{4\pi kqD}{A}\,dq.$$

Since the work done depends on the charge already transferred, we compute the total work done by integrating. The physical process is a sum of differential charge transfers dq, each involving work dW. The coordinate q ranges from 0 (uncharged) to Q (fully charged). Thus the total work done to charge the capacitor is:

$$W = \int_0^Q \Delta V\,dq = \frac{4\pi kD}{A}\int_0^Q q\,dq = \frac{4\pi kD}{A}\frac{Q^2}{2} = \frac{Q^2}{2C}.$$

We set the work done equal to the stored energy:

$$U = Q^2/2C. \qquad (27.6a)$$

Since $Q = C\,\Delta V$, we may also write this as:

$$U = \tfrac{1}{2}C(\Delta V)^2, \qquad (27.6b)$$

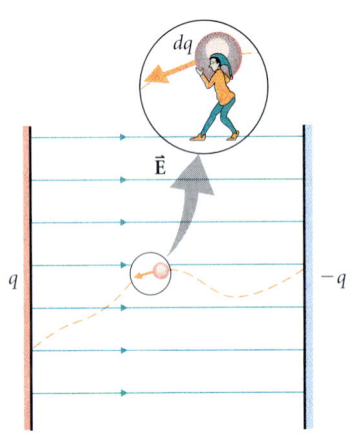

■ **FIGURE 27.7**
An imaginary process for charging a capacitor. An elf carries positive charge from the negative plate to the positive plate. The elf has to do work against the electric field produced by the charge already in place.

or: $$U = \tfrac{1}{2} Q \Delta V. \qquad (27.6c)$$

Compare the expression $U = \tfrac{1}{2} Q \Delta V$ with eqn. (25.14).

These expressions apply to any capacitor. If we replace the capacitance $A/4\pi kD$ with the arbitrary capacitance C, the argument goes through unchanged.

EXERCISE 27.5 ♦ A parallel plate capacitor has plates of area 10.0 m² separated by a gap 0.10 mm wide. How much energy does it store when connected to a 5.0-V battery?

According to eqn. (27.6b), to store more energy in a given capacitor we should increase the potential difference across it. But the potential difference is limited: too large an electric field destroys the capacitor. For example, an electric field greater than a few million volts per meter ionizes air, and charge can flow from one plate to another across the air gap. The largest individual capacitors used in the Berkeley neutral beam injector have a capacitance of 6.0 mF and a limiting voltage of 450 V. Large numbers of these capacitors are connected together to increase both the total potential difference and the total energy stored in the system. Next we'll investigate how these connections work.

27.2.2 Capacitors in Parallel

Stored energy is increased with no change in voltage by using a *parallel* combination.

Two or more capacitors are in *parallel* when they are connected between the same two points in a circuit, and the potential difference across each is the same.

The two parallel capacitors in ■ Figure 27.8 are connected across the same potential difference $\Delta V = \mathcal{E}$. Thus:

$$\Delta V = \frac{Q_1}{C_1} = \frac{Q_2}{C_2}.$$

The total amount of charge stored is $Q \equiv Q_1 + Q_2$. Thus the equivalent capacitance of the combination is:

$$C_\| \equiv \frac{Q}{\Delta V} = \frac{Q_1 + Q_2}{\Delta V} = C_1 + C_2.$$

The same argument extends to any number of capacitors in parallel, so the equivalent capacitance of any parallel combination of capacitors is the sum of the individual capacitances:

$$C_\| = \sum_i C_i. \qquad (27.7)$$

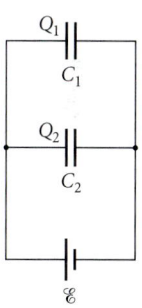

■ **FIGURE 27.8**
Two capacitors in parallel. The potential difference across each capacitor is the same. The total charge is the sum of the amounts of charge on each capacitor: $Q = Q_1 + Q_2$.

EXAMPLE 27.5 ♦ A 1.0-μF and a 6.0-μF capacitor are connected in parallel. What is the capacitance of the combination? How much energy is stored when they are connected to a 10.0-V power supply?

MODEL The capacitance of the parallel combination is the sum of the individual capacitances.

SETUP AND SOLVE $C_\| = C_1 + C_2 = 7.0 \; \mu\text{F}.$

The stored energy is (eqn. 27.6b):

$$W = \tfrac{1}{2} C_\| (\Delta V)^2 = \tfrac{1}{2}(7.0 \times 10^{-6} \text{ F})(10.0 \text{ V})^2 = 3.5 \times 10^{-4} \text{ J}.$$

ANALYZE Check the units:

$$\text{F} \cdot \text{V}^2 = \frac{\text{C}}{\text{V}} \text{V}^2 = \text{C} \cdot \text{V} = \text{C} \, \frac{\text{N} \cdot \text{m}}{\text{C}} = \text{J}.$$

The total energy stored is the sum of the energies stored in the individual capacitors. ■

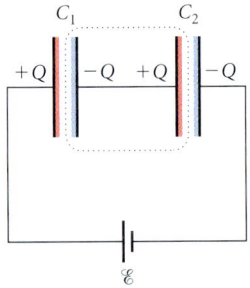

■ **FIGURE 27.9**
Two capacitors in series. The dotted curve encloses an electrically isolated piece of the circuit whose total charge must remain constant and equal to zero. The charge on the negative plate of C_1 and that on the positive plate of C_2 have the same magnitude. The capacitors store equal amounts of charge.

A QUANTITY CALLED *REACTANCE* MEASURES VOLTS PER AMPERE, AND REACTANCES COMBINE LIKE RESISTANCES (SEE CHAPTER 32).

27.2.3 Capacitors in Series

A series connection produces a compound capacitor able to withstand a greater total potential difference.

> Two or more capacitors are in *series* if they are connected in sequence in one arm of a circuit.

The capacitors in ■ Figure 27.9 are in series. The dotted curve in the figure includes one conducting surface from each capacitor and the wire used to connect them. Together these elements form one unified conducting body that is electrically isolated from the rest of the circuit and so contains zero net charge. Any electrons removed from the positive plate of C_2 are deposited on the negative plate of C_1. The two capacitors store equal charges: $Q_1 = Q_2 = Q$. The total potential difference across the two capacitors is the sum of the potential differences across each one:

$$\mathcal{E} = \Delta V = \Delta V_1 + \Delta V_2 = \frac{Q_1}{C_1} + \frac{Q_2}{C_2} = Q\left(\frac{1}{C_1} + \frac{1}{C_2}\right).$$

In terms of the total stored charge Q and the equivalent capacitance of the combination, C_s, the potential difference is: $\Delta V = Q/C_s$. So:

$$\frac{1}{C_s} = \frac{1}{C_1} + \frac{1}{C_2} \quad \text{or} \quad C_s = \frac{C_1 C_2}{C_1 + C_2}.$$

Once again, the argument extends to an arbitrary number of capacitors.

> For any number of capacitors in series, the reciprocal of the equivalent capacitance equals the sum of the reciprocals of the individual capacitances:

$$\frac{1}{C_s} = \sum_i \frac{1}{C_i}. \tag{27.8}$$

> **EXERCISE 27.6** ♦ What is the equivalent capacitance when the 1.0-μF and 6.0-μF capacitors in Example 27.5 are connected in series? What is the stored energy when they are connected across the 10.0-V power supply?

The rules for combining capacitors in series are similar to those for combining resistors in parallel; capacitors in parallel add in the same way as resistors in series. This reversal arises from the role of potential difference in the definitions of capacitance and resistance: capacitance is coulombs per volt, but resistance is volts per ampere.

27.2.4 Series and Parallel Combinations

A capacitor network containing both series and parallel combinations of capacitors may be analyzed in the same way as a resistor network. We find a sequence of equivalent circuits that reduces to a single equivalent capacitor.

> **EXAMPLE 27.6** ♦♦ What is the equivalent capacitance of the network shown in ■ Figure 27.10a?
>
> **MODEL** We proceed to simplify each series or parallel combination, as we did in Chapter 26 with resistor combinations.
>
> **SETUP** The two series combinations have the following equivalents:
>
> $$C_1 = \frac{(12\ \mu\text{F})(12\ \mu\text{F})}{12\ \mu\text{F} + 12\ \mu\text{F}} = 6.0\ \mu\text{F} \quad \text{and} \quad C_2 = \frac{(24\ \mu\text{F})(6.0\ \mu\text{F})}{24\ \mu\text{F} + 6.0\ \mu\text{F}} = 4.8\ \mu\text{F}.$$

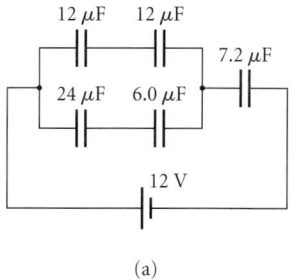 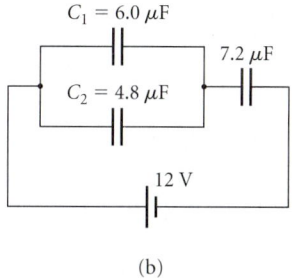 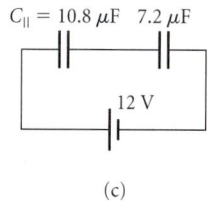 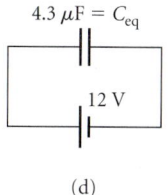

(a)   (b)   (c)   (d)

■ **FIGURE 27.10**
(a) A series-parallel capacitor combination is simplified in steps. (b) The series combinations are replaced with their series equivalents. (c) The parallel combination is replaced with its equivalent. (d) The final series combination is replaced with its equivalent.

Figure 27.10b is the equivalent circuit with these substitutions. Next, the parallel combination of C_1 and C_2 is equivalent to:

$$C_{\parallel} = 6.0~\mu\text{F} + 4.8~\mu\text{F} = 10.8~\mu\text{F}.$$

SOLVE The 7.2-μF capacitor is in series with $C_{\parallel}$ (Figure 27.10c). The equivalent capacitance of the whole combination is:

$$C_{eq} = \frac{(7.2~\mu\text{F})(10.8~\mu\text{F})}{7.2~\mu\text{F} + 10.8~\mu\text{F}} = 4.3~\mu\text{F}.$$

ANALYZE The capacitance of a series combination is less than any of the individual capacitances. The capacitance of a parallel combination is greater than each individual capacitance. Again these are the opposite of the rules for resistors. ■

EXAMPLE 27.7 ♦♦ What is the charge on the 24-μF capacitor in Figure 27.10a?

MODEL We proceed by using the sequence of equivalent circuits found in the previous example, working backward toward the individual capacitors.

SETUP The total charge stored by the combination (Figure 27.10d) is:

$$Q = C_{eq}\mathcal{E} = (4.3~\mu\text{F})(12~\text{V}) = 52~\mu\text{C}.$$

Since series capacitors have the same charge, both the 7.2-μF capacitor and the 10.8-μF parallel combination of C_1 and C_2 hold equal charges Q (Figure 27.10c). Capacitors C_1 and C_2 share this charge so as to have the same potential difference: $Q_1 + Q_2 = Q$ (Figure 27.10b). Also:

$$\Delta V_{\parallel} = \frac{Q_1}{C_1} = \frac{Q_2}{C_2} = \frac{Q}{C_{\parallel}} = \frac{Q}{C_1 + C_2}.$$

Thus the parallel capacitors share the charge in proportion to their capacitances. So:

$$Q_2 = \frac{C_2}{C_1 + C_2} Q = \frac{4.8}{10.8}(52~\mu\text{C}) = 23~\mu\text{C}.$$

SOLVE Capacitor C_2 is actually the series combination of the 6.0-μF and 24-μF capacitors (Figure 27.10a), each of which holds $Q_2 = 23~\mu\text{C}$.

ANALYZE If we replaced the 7.2-μF capacitor with a larger one, the charge on the 24-μF capacitor would increase. Do you see why? ■

Individual capacitors in the Lawrence Berkeley facility have capacitance $C_0 = 2.0$ mF. Each side of the module shown in ■ Figure 27.11 holds a series string of 136 capacitors (17 rows of 8). The two strings are connected in parallel to make the module, and ten modules are connected in parallel to form a *bank*.

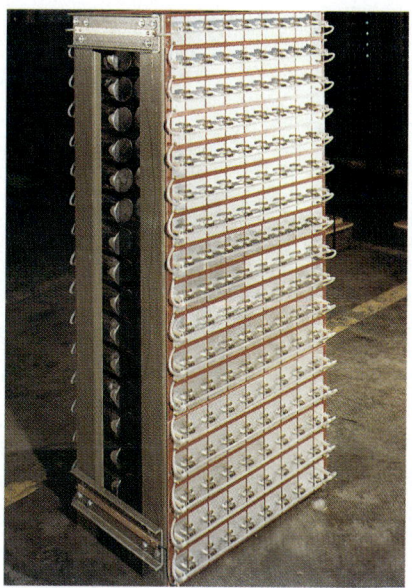

■ **FIGURE 27.11**
Module of capacitors used at Lawrence Berkeley Laboratory. Each side of the module holds a string of 136 2-mF capacitors (17 rows of 8). The two strings on each side are connected in parallel.

RECALL: EACH CAPACITOR IN A PARALLEL COMBINATION HAS THE SAME POTENTIAL DIFFERENCE ACROSS IT.

RECALL: CAPACITORS IN SERIES HAVE THE SAME Q.

SECTION 27.2 • ENERGY STORAGE IN CAPACITORS   **883**

EXERCISE 27.7 ♦♦ The capacitor banks are charged to a potential difference of 54.4 kV. Find the potential difference across an individual capacitor and the total energy stored in the bank.

To achieve a greater potential difference, the three banks of capacitors are charged while connected in parallel and then switched to a series connection. This increases the potential difference across the entire array of capacitors to three times the emf of the charging supply. Since the three banks act as identical capacitors, there is no current during the switching process, and no energy is needed to achieve this increase of potential difference!

IN PROBLEM 33 WE ASK YOU TO PROVE THIS RESULT.

27.3 Dielectrics and Practical Capacitors

Design of a practical capacitor to store large amounts of energy presents several problems. The device needs conducting surfaces with large surface area and small separation that fit into a reasonably small volume. A design with parallel plates separated by air meets none of these needs. Charged capacitor plates attract each other and so they must be held apart mechanically (■ Figure 27.12). The electric field between the plates can produce electrical breakdown (ionization) of the air, which leaks the charge across the gap, or, at high enough voltage, can produce sparks that ruin the device. All these problems are solved by placing a layer of insulating material or *dielectric* between the conducting plates (■ Figure 27.13). In this section we shall study the electrical properties of dielectrics.

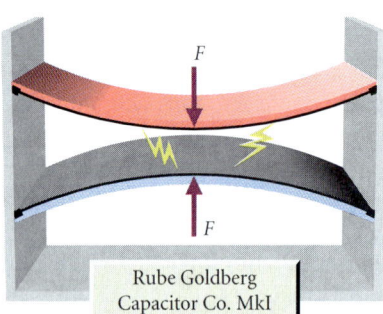

■ **Figure 27.12**
Faulty design for a large capacitor. The conductors in a capacitor are oppositely charged, so the Coulomb force between them is attractive. The device must be mechanically strong enough to resist this force. If the electric field between the plates is large enough, charge flow between the plates discharges the capacitor. For air, the limiting field is a few million volts per meter, and for dielectrics it is typically five to ten times that for air.

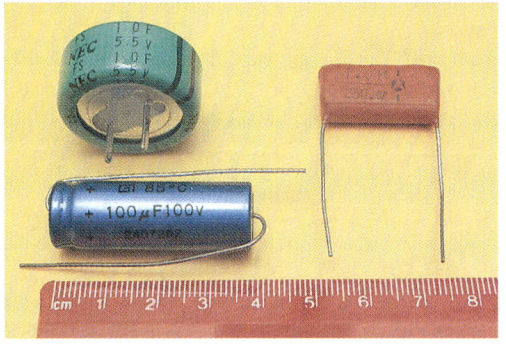

■ **Figure 27.13**
Commercially available capacitors. In the red capacitor, Mylar plastic and aluminum foil conductors form layers that are rolled into a compact cylinder. The blue capacitor was made by electrodepositing a thin dielectric layer on one conductor. This electrolytic capacitor is *polar* (one-way!)—charging the wrong conductor positive removes the dielectric layer. Compact capacitors as large as 1.0 F are now available.

27.3.1 The Dielectric Constant

In an insulating substance, electrons are not free to move from atom to atom when an electric field is applied. Instead, the electric force produces *polarization* (■ Figure 27.14). A small displacement of the electrons in each atom or molecule of the substance forms a small dipole, oriented in the direction of the electric field. Any volume in the interior of the dielectric remains electrically neutral, but on the dielectric's surface the ends of these atomic dipoles produce a net charge density, σ_B, called *bound* charge. Like the free charge density on the surface of conductors, this bound charge occurs within the outermost atomic layer of the

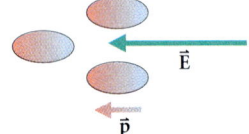

■ **Figure 27.14**
Electric field acting on individual molecules of an insulating substance polarizes them. Each molecule becomes a small dipole $\vec{p}$ oriented in the same direction as the electric field vector $\vec{E}$.

884 CHAPTER 27 • CAPACITANCE AND ELECTROSTATIC ENERGY

dielectric surface. Unlike the free charge, however, bound charge is attached to its parent atoms. It cannot flow throughout the volume of the dielectric or into the external circuit.

Bound charge on the surface of a dielectric material, like surface charge on a conductor, produces a net electric field that opposes the applied field (■ Figure 27.15). But unlike surface charge on a conductor, bound charge is not large enough to eliminate the net electric field within the dielectric material entirely. The polarization of individual atoms is proportional to the field acting on them, so the amount of bound charge and the internal field it produces are both proportional to the applied field. Consequently, the actual net field $\vec{E}$ that exists within the material is proportional to the external applied field. The ratio between the two is a property of the particular material.

IN SOME MATERIALS, PARTICULARLY CRYSTALS, THE DIPOLE ORIENTATION DEPENDS ON THE STRUCTURE OF THE MATERIAL AS WELL AS THE DIRECTION OF $\vec{E}$. THIS EFFECT IS NOT IMPORTANT FOR OUR DISCUSSION OF CAPACITORS.

SEE §27.3.2 FOR DETAILS.

> The *dielectric constant* κ of a material is the ratio of the applied electric field to the net electric field within the material:
>
> $$\kappa = \frac{|\vec{E}_a|}{|\vec{E}|}. \qquad (27.9)$$

κ IS THE GREEK LETTER KAPPA (APPENDIX IIA).

● Table 27.1 lists dielectric constants of some materials commonly used in capacitors. The dielectric constant of air is very close to unity, and for most purposes we may ignore the difference.

When the applied electric field is very large, the force on individual molecules becomes large enough to strip off electrons, and the material begins to conduct. The maximum electric field that a material can withstand before it loses its insulating properties is called its *dielectric strength*. The value of dielectric strength is quite variable, because it depends on temperature, presence of impurities, and so forth. Values vary from about 3 MV/m for air to a few tens of MV/m for ceramics and plastics.

TABLE 27.1 Dielectric Constants of Some Common Materials

Material	Dielectric Constant
Air (1 atm)	1.0006
Water	80
Ruby mica	5–7
Glass	4–10
Class I ceramic	5–450
Class II ceramic	200–12000
Oil-impregnated paper	4–6
Mylar	3
Polytetrafluoroethylene (Teflon)	2.1
Polyethylene	2.4
Polystyrene	2.5
Polycarbonate	3
Plexiglas	3.1
Epoxy	3.6
Polyvinyl chloride	4.6
Aluminum oxide	11
Tantalum oxide	28
Diamond	5.7

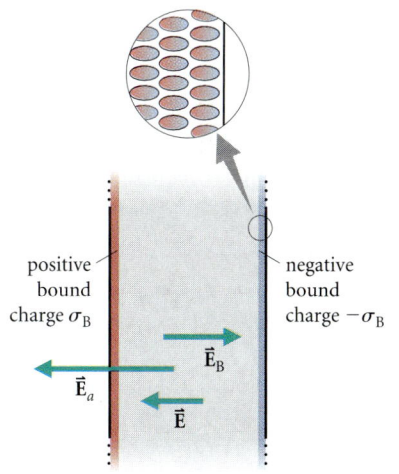

$\vec{E}_a$ applied external field

$\vec{E}_B$ field produced by bound charge

$\vec{E} = \vec{E}_a + \vec{E}_B$ total field within the material

■ FIGURE 27.15
At the surface of a dielectric material the molecular dipoles form an unbalanced layer of charge known as *bound charge*. The bound charge produces an electric field that opposes the externally applied field within the material.

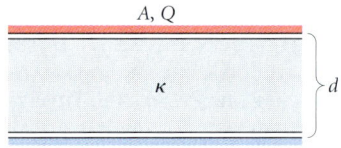

FIGURE 27.16
A parallel plate capacitor filled with dielectric. Because the potential difference between the plates for a given charge Q is reduced, the dielectric increases the capacitance by a factor κ.

EXAMPLE 27.8 ♦♦ A material with dielectric constant κ fills a parallel plate capacitor with plate area A and separation d (■ Figure 27.16). When the magnitude of charge on each plate is Q, what is the electric field between the plates? What is the potential difference between the plates? What is the capacitance of the device?

MODEL The applied field is produced by the charged plates:

$$|\vec{E}_a| = \frac{\sigma}{\epsilon_0} = \frac{Q}{\epsilon_0 A}.$$

The field within the dielectric is reduced by the induced bound charge, so the potential difference is reduced, and the capacitance (eqn. 27.2) is increased.

SETUP From eqn. (27.9):

$$|\vec{E}| = \frac{|\vec{E}_a|}{\kappa} = \frac{Q}{\kappa \epsilon_0 A}.$$

SOLVE The potential difference is:

$$\Delta V = |\vec{E}|d = \frac{Qd}{\kappa \epsilon_0 A},$$

and the capacitance is: $\quad C \equiv \dfrac{Q}{\Delta V} = \dfrac{Q \kappa \epsilon_0 A}{Qd} = \kappa \dfrac{\epsilon_0 A}{d}.$

ANALYZE The capacitance of the plates is increased by a factor of κ over its value with vacuum between the plates (eqn. 27.1). An alternative definition of the dielectric constant is the factor by which a material increases the capacitance of a system. ■

EXERCISE 27.8 ♦ The red device shown in Figure 27.13 has a capacitance $C = 1.0\ \mu$F. If the dielectric material is a 2.5×10^{-5}-m-thick sheet of mylar, what is the surface area of each aluminum foil conductor?

EXAMPLE 27.9 ♦♦ A parallel plate capacitor with plates of area $A = 16$ cm^2 and separation $d = 0.50$ cm is connected to a 12.0-V battery. How much energy is stored in the capacitor? A slab of polyvinyl chloride (PVC) that just fits between the plates is inserted into the capacitor. What is the energy stored then? What difference does it make if the battery is disconnected before the PVC slab is inserted?

MODEL The battery charges the capacitor, and the stored energy is given by eqn. (27.6). Inserting the PVC slab increases the capacitance. With the battery connected, the potential difference remains fixed, and the stored energy increases. If the battery is disconnected first, the *charge* remains fixed. Using eqn. (27.6a), we see that the stored energy decreases as the capacitance is increased.

SETUP With air between the plates, the capacitance is (eqn. 27.3):

$$C_{air} = \frac{\epsilon_0 A}{d} = \frac{(8.85 \times 10^{-12}\ \text{F/m})(16 \times 10^{-4}\ \text{m}^2)}{5.0 \times 10^{-3}\ \text{m}} = 2.83\ \text{pF}.$$

Connected to the 12.0-V battery, the stored charge is:

$$Q = C\,\Delta V = (2.83\ \text{pF})(12.0\ \text{V}) = 34.0\ \text{pC}.$$

SOLVE The stored energy at 12.0 V is:

$$U_{air} = \tfrac{1}{2} C_{air}(\Delta V)^2 = \tfrac{1}{2}(2.83\ \text{pF})(12.0\ \text{V})^2 = 0.20\ \text{nJ}.$$

SETUP After insertion of the PVC slab ($\kappa = 4.6$ from Table 27.1), the capacitance becomes:

$$C_{PVC} = \kappa C_{air} = (4.6)(2.83\ \text{pF}) = 13.0\ \text{pF}.$$

SOLVE The stored energy at 12.0 V is then:

$$U_1 = \tfrac{1}{2}C_{\text{PVC}}(\Delta V)^2 = \kappa U_{\text{air}} = 4.6(0.204 \text{ nJ}) = 0.94 \text{ nJ}.$$

If instead the battery is disconnected before the slab is inserted, the new energy is:

$$U_2 = \frac{Q^2}{2C_{\text{PVC}}} = \frac{(3.40 \times 10^{-11} \text{ C})^2}{2(13.0 \times 10^{-12} \text{ F})} = 0.044 \text{ nJ}.$$

ANALYZE When it remains connected to the battery, the capacitor's stored energy increases as the slab enters. That means work is *done on* the system of capacitor and slab. The charge on the capacitor after the slab is inserted is (eqn. 27.6c) $2U_1/\Delta V = 0.16$ nC. The extra charge is drawn from the battery, which also supplies the required energy.

If the battery is disconnected first, the energy decreases as the slab is inserted. That means the *system does work*; it sucks the slab in! The final potential difference is $Q/C_{\text{PVC}} = 2.6$ V.

REMEMBER: KEEP AN EXTRA FIGURE IN INTERMEDIATE RESULTS.

✱ 27.3.2 Polarization and Susceptibility

■ Figure 27.17 illustrates the connection between the bound charge at the surface of a dielectric and the polarization of the molecules. The bound charge per unit area equals the number of molecules per unit area at the surface times the bound charge at the end of each molecule:

$$\begin{aligned}\text{Bound charge per unit area} &= \text{molecules per unit area} \times \text{charge per molecule}\\ &= \text{molecules per unit area} \times \frac{\text{dipole moment}}{\text{length of molecule}}\\ &= \text{molecules per unit volume} \times \text{dipole moment}.\end{aligned}$$

$$\sigma_B = np,$$

where n is the number of molecules per unit volume and p is the dipole moment per molecule. The product np is the dipole moment per unit volume in the material. Together with the direction of the dipoles this quantity forms a vector called the *polarization vector* $\vec{P} = n\vec{p}$. The electric field produced by the bound charge has magnitude $E_B = \sigma_B/\epsilon_0$ and is in the direction opposite the polarization of the material:

$$\vec{E}_B = -\frac{n\vec{p}}{\epsilon_0} = \frac{-\vec{P}}{\epsilon_0}.$$

The net electric field $\vec{E}$ is the sum of the applied field $\vec{E}_a$ and the field $\vec{E}_B$ produced by the bound charge:

$$\vec{E} = \vec{E}_a + \vec{E}_B = \vec{E}_a - \vec{P}/\epsilon_0.$$

The polarization vector $\vec{P}$ results from the dipole moments of individual molecules, each proportional to the net field:

$$\vec{P} \equiv n\vec{p} \equiv \epsilon_0 \chi \vec{E}.$$

The proportionality constant χ is called the *dielectric susceptibility* of the material. Combining these relations, we find:

$$\vec{E}_a = \vec{E} + \frac{\vec{P}}{\epsilon_0} = \vec{E} + \frac{\epsilon_0 \chi \vec{E}}{\epsilon_0} = \vec{E}(1 + \chi). \qquad (27.10)$$

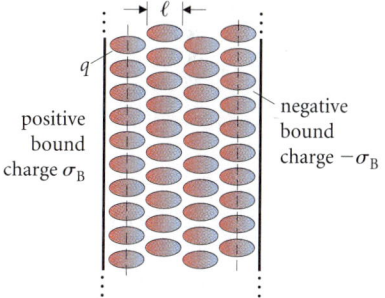

■ **FIGURE 27.17**
The bound charge arises because the end of each molecule at the surface has the same sign of charge. The charge on its other end is balanced by molecules one layer in from the surface. The amount of bound charge contributed by each molecule equals its dipole moment $p = q\ell$ divided by its length ℓ.

A GREATER FIELD PRODUCES A GREATER CHARGE SEPARATION AND THUS A GREATER DIPOLE MOMENT.

See §24.5, especially Example 24.10.

Comparing eqn. (27.10) with eqn. (27.9), we find that the dielectric constant is given by:

$$\kappa = 1 + \chi. \quad (27.11)$$

In some materials, such as water, each molecule has a dipole moment even without an applied field. An applied field acts to align the otherwise randomly oriented molecular dipoles. Because the degree of alignment is proportional to $\vec{E}_a$, eqns. (27.10) and (27.11) still apply.

✱ 27.3.3 Electric Displacement

Electric field within a dielectric material is produced by both free and bound charge. It is sometimes more convenient to work with a vector called electric *displacement* $\vec{D}$, which is related to the free charge alone. The displacement is defined by its relation to the electric field vector and the polarization:

$$\vec{D} \equiv \epsilon_0 \vec{E} + \vec{P} = \epsilon_0 \vec{E} - \epsilon_0 \vec{E}_B = \epsilon_0 \vec{E}_a.$$

$$\vec{D} = \epsilon_0 \kappa \vec{E}. \quad (27.12)$$

EXAMPLE 27.10 ♦♦ A plastic slab with dielectric constant $\kappa = 2.50$ fills half the volume between the plates of a capacitor (■ Figure 27.18). The plates have area $A = 25$ cm^2 and separation 1.50 mm. The electric field in the air is $|\vec{E}_a| = 1.00$ kV/m. Find $|\vec{E}_{in}|$ in the dielectric and $|\vec{D}|$ both in the air and in the dielectric. What is the capacitance?

MODEL Outside the dielectric, the two bound charge layers produce canceling fields, so the electric field in the air gap is just the applied field due to the free charges on the capacitor plates. Within the dielectric, the applied field is reduced by $\vec{E}_B$.

SETUP In the air gap: $|\vec{E}| \equiv |\vec{E}_a| = \sigma/\epsilon_0 = 1.00$ kV/m.

SOLVE The electric field in the dielectric is reduced by a factor κ, so:

$$|\vec{E}_{in}| = |\vec{E}_a|/\kappa = (1.00 \text{ kV/m})/(2.50) = 0.400 \text{ kV/m}.$$

In air, $\kappa \approx 1$, so that $|\vec{D}| = \epsilon_0 |\vec{E}| = \epsilon_0 \sigma/\epsilon_0 = \sigma$.

This expression shows directly the relation between D and the free charge. Numerically:

$$D = (8.85 \times 10^{-12} \text{ F/m})(1.00 \text{ kV/m}) = 8.85 \times 10^{-9} \text{ C/m}^2.$$

In the dielectric the magnitude of $\vec{D}$ is:

$$|\vec{D}_{in}| = \epsilon_0 \kappa |\vec{E}_{in}| = \epsilon_0 \kappa \frac{|\vec{E}_a|}{\kappa} = \epsilon_0 \frac{\sigma}{\epsilon_0} = \sigma,$$

the same as in the air.

ANALYZE The displacement vector has the same value both inside and outside the dielectric and depends only on σ. Note that its units are C/m^2—the units of surface charge density.

SETUP To find the capacitance we need the potential difference:

$$|\Delta V| = \int \vec{E} \cdot d\vec{\ell} = E_a \frac{d}{2} + E_{in} \frac{d}{2} = E_a \frac{d}{2}\left(1 + \frac{1}{\kappa}\right).$$

SOLVE Then: $\dfrac{1}{C} = \dfrac{\Delta V}{Q} = \dfrac{\sigma}{\epsilon_0}\dfrac{d}{2}\left(1 + \dfrac{1}{\kappa}\right)\dfrac{1}{\sigma A} = \dfrac{d}{2\epsilon_0 A}\left(1 + \dfrac{1}{\kappa}\right).$

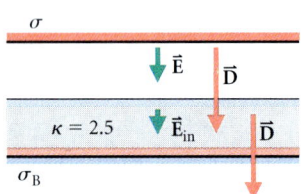

■ **Figure 27.18**
A capacitor half-filled with dielectric. The electric field $\vec{E}$ is reduced within the dielectric, whereas the displacement $\vec{D}$ is uniform throughout the capacitor's interior.

So: $C = \dfrac{2(8.85 \times 10^{-12} \text{ F/m})(25 \times 10^{-4} \text{ m}^2)}{(1.5 \times 10^{-3} \text{ m})(1 + 1/2.50)} = 21 \text{ pF}.$

ANALYZE This system behaves like two capacitors in series, with capacitances

$$C_1 = 2\epsilon_0 A/d \quad \text{and} \quad C_2 = 2\kappa\epsilon_0 A/d$$

combining according to eqn. (27.8).

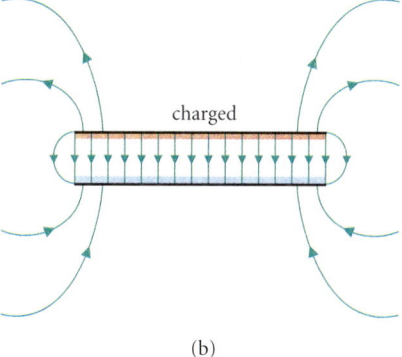

FIGURE 27.19
A charged capacitor differs from an uncharged one because of the separated charge on the plates and the electric field between the plates. Thinking of the field lines as stretched rubber bands gives an image for the electrostatic energy stored in the field.

27.4 ENERGY IN THE ELECTRIC FIELD

27.4.1 Electrostatic Energy Density in Vacuum

A capacitor stores electrostatic potential energy, but we don't yet have a picture of how or where it is stored. ■ Figure 27.19 shows a parallel plate capacitor before and after being charged. The important difference between the two states is the presence of the charges and their electric field. The electric field exerts forces on the charged plates, much as if the field lines were a collection of stretched rubber bands. Thinking of the field lines in this way gives us an image of the energy stored *in the field*. We shall use this capacitor to derive an expression for the amount of energy stored in a unit volume containing electric field.

The work done to charge the capacitor is (eqn. 27.6b) $W = \tfrac{1}{2} C(\Delta V)^2$. The capacitance is given by eqn. (27.3) ($C = \epsilon_0 A/d$), and the potential difference between the plates is $|\Delta V| = Ed$. So:

$$W = \dfrac{1}{2}\left(\dfrac{\epsilon_0 A}{d}\right) E^2 d^2 = \left(\dfrac{\epsilon_0 E^2}{2}\right)(Ad).$$

The stored energy is the product of the volume Ad between the capacitor plates and the uniform density u_E of electrostatic energy in vacuum:

$$u_E \equiv \dfrac{E^2}{8\pi k} = \dfrac{1}{2}\epsilon_0 E^2. \tag{27.13}$$

ENERGY DENSITY IS ENERGY PER UNIT VOLUME.

Although we obtained this expression using a special model—a parallel plate capacitor—it is a completely general result for electric energy density in vacuum.

SEE PROBLEM 75 AND §27.4.2 FOR FURTHER DISCUSSION OF THIS POINT.

If dielectric materials are present, energy is stored both in the average electric field within the material and in the increased potential energy of the polarized molecules. For practical purposes we need an expression for the total energy density in terms of the electric field and the material's dielectric constant.

We derived eqn. (27.6a) for stored energy in terms of charge and capacitance by calculating the work required to charge the capacitor. A similar derivation works for capacitors with dielectrics. For a parallel plate capacitor, the stored energy is:

$$U = \dfrac{1}{2}\left(\dfrac{Q^2}{C}\right) = \dfrac{1}{2}\left(\dfrac{Q^2}{\kappa\epsilon_0 A/d}\right).$$

WE USED EQN. (27.6a) AND THE RESULT OF EXAMPLE 27.8.

The electric field magnitude is $E = Q/(\kappa\epsilon_0 A)$ (cf. Example 27.8), so:

$$Q = \kappa\epsilon_0 AE,$$

and:

$$U = \dfrac{1}{2}\dfrac{(\kappa\epsilon_0 AE)^2 d}{\kappa\epsilon_0 A} = \tfrac{1}{2}\kappa\epsilon_0 E^2(Ad).$$

Thus the energy density in a dielectric medium is the energy divided by the volume:

$$u_E = \tfrac{1}{2}\kappa\epsilon_0 E^2. \tag{27.14}$$

COMPARE WITH EQN. (27.13); $\kappa = 1$ IN VACUUM.

EXAMPLE 27.11 ♦♦ A spherical high-voltage terminal with a radius of 2.5 mm is insulated with a layer of epoxy. If the terminal potential is 10.0 kV, what is the electric field energy density in the epoxy 1.0 mm from the surface of the terminal?

MODEL Using the known potential, we may find the electric field in vacuum at any distance from the terminal. The field in the epoxy is reduced by the dielectric constant. Once we know the value of E, eqn. (27.14) gives the energy density.

SETUP The potential due to a charged spherical conductor is $V = kQ/r$. At the surface, $r = r_s$, we set V equal to the given value of $V_s = 10.0$ kV. The applied electric field has magnitude:

$$E_a = \frac{kQ}{r^2} = \frac{kQ}{r_s}\frac{r_s}{r^2} = \frac{V_s r_s}{r^2}.$$

Within the epoxy, the electric field strength is:

$$E = \frac{E_a}{\kappa} = \frac{V_s r_s}{\kappa r^2}.$$

SOLVE From Table 27.1, $\kappa = 3.6$ for epoxy. The energy density at distance d from the center of the terminal, $r = r_s + d = 2.5$ mm $+ 1.0$ mm $= 3.5$ mm, is:

$$u = \tfrac{1}{2}\kappa\epsilon_0 E^2 = \kappa\epsilon_0 \frac{(V_s r_s)^2}{2(\kappa r^2)^2} = \frac{\epsilon_0(V_s r_s)^2}{2\kappa(r_s + d)^4}$$

$$= \frac{(8.85 \times 10^{-12} \text{ F/m})(1.0 \times 10^4 \text{ V})^2(2.5 \times 10^{-3} \text{ m})^2}{2(3.6)(3.5 \times 10^{-3} \text{ m})^4}$$

$$= 5.1 \text{ J/m}^3.$$

ANALYZE Check the units:

$$\frac{(\text{F/m})\cdot\text{V}^2\cdot\text{m}^2}{\text{m}^4} = \frac{\text{F}\cdot\text{V}^2}{\text{m}^3} = \frac{\text{C}\cdot\text{V}^2}{\text{V}\cdot\text{m}^3} = \frac{\text{C}\cdot\text{V}}{\text{m}^3} = \frac{\text{J}}{\text{m}^3}.$$

The net effect of the dielectric is to reduce the energy density. The increase of potential energy due to formation of dipoles is outweighed by the reduction of $|\vec{E}|$. ∎

✻ 27.4.2 *Electrostatic Energy of Two Point Charges*

In Chapter 25 we computed the potential energy of a pair of point charges Q_1 and Q_2: $U = kQ_1Q_2/r$ (eqn. 25.3). Now we have another way of computing electrostatic energy, using the energy density u_E. These two descriptions ought to agree.

The electric field vector at a point P due to the two charges is the sum of their individual fields:

$$\vec{E} = \vec{E}_1 + \vec{E}_2.$$

Thus at P:
$$u_E = \tfrac{1}{2}\epsilon_0 E^2 = \tfrac{1}{2}\epsilon_0(\vec{E}_1 + \vec{E}_2)\cdot(\vec{E}_1 + \vec{E}_2)$$
$$= \tfrac{1}{2}\epsilon_0(E_1^2 + E_2^2 + 2\vec{E}_1\cdot\vec{E}_2).$$

The total electrostatic energy of the two charges is expressed as the sum of three terms:

$$U = \int_{\text{all space}} \tfrac{1}{2}\epsilon_0 E_1^2 \, dV + \int_{\text{all space}} \tfrac{1}{2}\epsilon_0 E_2^2 \, dV + \int_{\text{all space}} \epsilon_0 \vec{E}_1\cdot\vec{E}_2 \, dV \equiv U_1 + U_2 + U_{\text{int}}.$$

> THE INTEGRAL IS OVER "ALL SPACE" BECAUSE THE ELECTRIC FIELD EXTENDS TO INFINITY.

The first two terms, U_1 and U_2, are the same energies we would calculate for the individual charges if they were isolated point charges. They are called the *self-energies* of the two charges.

The *interaction energy* U_{int} results from the combined fields and is the only part of the energy that changes as the particles move. By evaluating the integral, it can be shown that:

$$U_{int} = \frac{kQ_1Q_2}{r}.$$

This interaction energy is the potential energy we described in Chapter 25.

The self-energy term for an individual charge is the energy stored throughout all space outside the particle. The electrostatic energy outside a spherical distribution of charge may be found by integrating the energy density (■ Figure 27.20). The answer is intriguing.

STEPS I AND II We model the region outside a sphere of radius R as a set of spherical shells with radius r and thickness dr.

STEP III The electric field at radius $r > R$ has magnitude $|\vec{E}| = kQ/r^2$. The corresponding energy density is:

$$u_E = \tfrac{1}{2}\epsilon_0 E^2 = \frac{1}{2}\epsilon_0 \left(\frac{kQ}{r^2}\right)^2 = \frac{kQ^2}{8\pi r^4}.$$

The volume dV of a spherical shell of radius r and thickness dr is:

$$dV = 4\pi r^2\, dr,$$

and the electrostatic energy stored in the shell is the energy density times the volume:

$$dU = u_E\, dV = \frac{kQ^2}{8\pi r^4} 4\pi r^2\, dr = \frac{kQ^2\, dr}{2r^2}.$$

STEPS IV AND V The total energy stored outside radius R is then:

$$U = \int_R^\infty dU = \int_R^\infty \frac{kQ^2}{2r^2}\, dr = -\left.\frac{kQ^2}{2r}\right|_R^\infty = \frac{kQ^2}{2R}.$$

What inner radius R should we use? With $R = 0$ the self-energy is infinite! Clearly, point charges cannot exist—or can they?

■ **FIGURE 27.20**
The self-energy of a charged shell is the total electrostatic energy stored in the region $r > R$.

REMEMBER: $k = 1/(4\pi\epsilon_0)$.

✱ 27.4.3 The Classical Electron Radius and Renormalization

In his special theory of relativity Einstein argued that mass is a form of energy, according to the famous relation $E = mc^2$. By extending this idea, he supposed that the observed mass of an electron might be none other than its electrical self-energy!

EXAMPLE 27.12 ◆◆ Model an electron as a thin spherical shell of charge. If its mass-energy equals its electrical self-energy, what is the radius r_s of the shell?

MODEL We choose to model the electron as a spherical shell, since then all its charge is at the same potential, $V = ke/r_s$.

SETUP The potential energy of the shell is (eqn. 25.14) $U = \tfrac{1}{2}eV$. Setting the mass-energy and the electrical self-energy equal, we have:

$$mc^2 = \frac{1}{2}\left(\frac{ke^2}{r_s}\right).$$

SOLVE
$$r_s = \frac{1}{2}\left(\frac{ke^2}{mc^2}\right)$$
$$= \frac{(9.0 \times 10^9\ \text{N·m}^2/\text{C}^2)(1.6 \times 10^{-19}\ \text{C})^2}{2(9.1 \times 10^{-31}\ \text{kg})(3.0 \times 10^8\ \text{m/s}^2)^2} = 1.4 \times 10^{-15}\ \text{m}.$$

ANALYZE The quantity $r_e \equiv ke^2/mc^2 = 3 \times 10^{-15}$ m is known as the *classical electron radius*. Unfortunately, this beautiful idea is not correct. Modern experiments show that the electron radius is much less than 10^{-15} m. ■

IF WE MODEL THE ELECTRON AS A SOLID SPHERE, ONLY THE NUMERICAL FACTOR $\tfrac{1}{2}$ DIFFERS IN THE EXPRESSION FOR U. SEE PROBLEM 73.

The self-energy of elementary particles, even if not truly infinite, is enormous. Since the self-energy is constant, we may simply agree to ignore it and work with finite interaction energies only! The subtraction of infinite constants from physical properties of particles is a common feature of current theories and is called *renormalization*. The process produces consistent theories that predict phenomena with remarkable accuracy. Some recent, highly speculative, and as yet untested ideas—the superstring theories—do not require renormalization. It remains to be seen if infinite constants are a property of nature or an inadequacy of our ideas.

Chapter Summary

Where Are We Now?

We have studied capacitors as devices for storage of charge and energy. Electrostatic energy is stored throughout space wherever there is electric field.

What Did We Do?

Two conducting objects form a capacitor. The potential difference between the conductors when they carry equal and opposite charges is proportional to the amount of charge. Their capacitance is the ratio of charge to potential difference:

$$C \equiv \left| \frac{Q}{\Delta V} \right|.$$

The unit of capacitance is the farad (F). The capacitance of any conductor is roughly ϵ_0 times an appropriate linear dimension. For a parallel plate capacitor,

$$C = \epsilon_0 A/d.$$

Work must be done to store charge on a capacitor and increase its electric potential energy. The amount of stored energy is proportional to the square of the stored charge: $U = \frac{1}{2} Q^2/C = \frac{1}{2} C(\Delta V)^2$ (eqns. 27.6a, b, c).

To store large amounts of energy, capacitors are connected into banks. Capacitors connected across the same potential difference (in parallel) have an equivalent capacitance equal to the sum of their individual capacitances.

$$C_\| = \sum_i C_i. \tag{27.7}$$

Each capacitor in a series connection stores the same amount of charge. The reciprocal of the equivalent capacitance is the sum of the reciprocals of the individual capacitances.

$$\frac{1}{C_s} = \sum_i \frac{1}{C_i}. \tag{27.8}$$

Dielectric insulators provide mechanical support and improve the electrical properties of practical capacitors. Polarization of molecules in the dielectric results in bound charge at the surface of the dielectric that reduces the electric field within the dielectric material. The dielectric constant κ is the ratio of the applied electric field to the field in the dielectric medium (eqn. 27.9, Table 27.1). Dielectrics increase the capacitance of a device by a factor κ, since decreasing the electric field results in a lower potential difference for the same stored charge.

Electric potential energy is actually stored throughout space wherever there is electric field. The energy density is proportional to the square of the field vector:

$$u_E = \tfrac{1}{2}\epsilon_0 E^2.$$

> ✻ The electric displacement vector $\vec{D} = \kappa\epsilon_0\vec{E}$ is useful for analyzing systems with dielectrics, because it is related directly to the free charge. The polarization vector $\vec{P}$, or dipole moment per unit volume, is related to the bound charge.

Practical Applications

Capacitors are used in electrical circuits whenever temporary storage of small amounts of charge is necessary. Flash units on cameras are a familiar example. We'll explore these applications further in Chapter 31. Large banks of capacitors can store and quickly release energy in amounts that would be very inconvenient or very expensive to draw directly from commercial power systems. Such systems are used to power lasers, electromagnets, and neutral beam accelerators used in nuclear fusion research.

Solutions to Exercises

27.1. From eqn. (27.3), the area is:
$$A = \frac{dC}{\epsilon_0} = \frac{(2.0 \times 10^{-3}\ \text{m})(1.0 \times 10^{-9}\ \text{F})}{(8.85 \times 10^{-12}\ \text{F/m})} = 0.23\ \text{m}^2.$$
The charge stored is:
$$Q = CV = (10^{-9}\ \text{F})(2 \times 10^3\ \text{V}) = 2\ \mu\text{C}.$$

27.2. We use eqn. (27.4):
$$C = \frac{(2.0 \times 10^{-2}\ \text{m})(3.0 \times 10^{-2}\ \text{m})}{(9.0 \times 10^9\ \text{N}\cdot\text{m}^2/\text{C}^2)(3.0 \times 10^{-2}\ \text{m} - 2.0 \times 10^{-2}\ \text{m})}$$
$$= 6.7\ \text{pF}.$$

27.3. When $R_1 \approx R_2$, both inner and outer spheres have area:
$$A \approx 4\pi R_1^2 \approx 4\pi R_2^2 \approx 4\pi R_1 R_2.$$
Their separation is $d = R_2 - R_1$, so from eqn. (27.4):
$$C = \frac{4\pi\epsilon_0 R_1 R_2}{(R_2 - R_1)} \approx \frac{\epsilon_0 A}{d},$$
which is the same expression as for parallel plates.

27.4. A typical person has dimension of the order of 1 m and a capacitance:
$$C \sim (1\ \text{m})/(9 \times 10^9\ \text{N}\cdot\text{m}^2/\text{C}^2) = 0.1\ \text{nF}.$$

27.5. Using eqn. (27.6b) relating stored energy to potential difference, we have:
$$U = \tfrac{1}{2}C(\Delta V)^2 = \frac{\epsilon_0 A}{2d}(\Delta V)^2$$
$$= \frac{(8.85 \times 10^{-12}\ \text{F/m})(10.0\ \text{m}^2)(5.0\ \text{V})^2}{2(1.0 \times 10^{-4}\ \text{m})} = 11\ \mu\text{J}.$$

27.6. The series capacitance is:
$$C_s = \frac{C_1 C_2}{C_1 + C_2} = \frac{(1\ \mu\text{F})(6\ \mu\text{F})}{(1\ \mu\text{F} + 6\ \mu\text{F})} = \frac{6}{7}\ \mu\text{F}.$$
Then the energy stored is:
$$U = \tfrac{1}{2}C_s(\Delta V)^2$$
$$= \frac{1}{2}\left(\frac{6}{7}\ \mu\text{F}\right)(10.0\ \text{V})^2 = 4.3 \times 10^{-5}\ \text{J}.$$

27.7. The capacitor bank consists of 20 parallel strings, each of 136 capacitors in series. Each string is connected across the full 54.4-kV potential difference. The potential difference across each individual capacitor is:
$$V_1 = (54.4\ \text{kV})/136 = 0.400\ \text{kV}.$$
Each individual capacitor in the bank contains the same energy, so the total stored is:
$$U = (20)(136)\tfrac{1}{2}(2.0 \times 10^{-3}\ \text{F})(400\ \text{V})^2 = 0.44\ \text{MJ}.$$

27.8. The dielectric constant for Mylar is 3 (Table 27.1). The capacitance is $C = \kappa\epsilon_0 A/d$, so the area is:
$$A = \frac{dC}{\kappa\epsilon_0}$$
$$= \frac{(2.5 \times 10^{-5}\ \text{m})(1.0 \times 10^{-6}\ \text{F})}{3(8.85 \times 10^{-12}\ \text{F/m})} = 0.94\ \text{m}^2.$$

Basic Skills

Review Questions

§27.1 CAPACITANCE

- Define *capacitance*.
- How does the capacitance of a parallel plate capacitor depend on the plate area? On the plate separation?
- What procedure is used to calculate the capacitance of any pair of conductors?
- Describe a *back-of-the-envelope* method for estimating capacitance.

§27.2 ENERGY STORAGE IN CAPACITORS

- How does the energy stored in a capacitor depend on its charge? On the potential difference?
- Describe a parallel connection of capacitors.
- What is the equivalent capacitance of a set of capacitors connected in parallel?
- What is the advantage of a parallel connection?
- Describe a series connection of capacitors.
- What is the equivalent capacitance of a set of capacitors connected in series?
- Why might you use a series connection?

§27.3 DIELECTRICS AND PRACTICAL CAPACITORS

- What is *polarization*?
- What is the *dielectric constant* of a material?
- What is *bound charge*? Describe how it results from polarization of a material.
- If a capacitor is filled with material whose dielectric constant is $\kappa > 1$, how does its capacitance change?
- How does inserting a dielectric into a capacitor change its stored energy **(a)** if it is kept connected to a battery and **(b)** if it is charged but insulated from its surroundings? Describe the energy flow in each of these processes.
- ✸ What are the *polarization vector* $\vec{P}$ and the *displacement vector* $\vec{D}$?

§27.4 ENERGY IN THE ELECTRIC FIELD

- How is the *energy density* in the electric field in vacuum related to the field strength?
- How is the expression for energy density modified to account for energy of polarization?
- ✸ What is *interaction energy*? How does it differ from *self-energy*?

Basic Skill Drill

§27.1 CAPACITANCE

1. A 10.0-pF capacitor is connected to a 12-V battery. What is the charge on the capacitor?

2. A capacitor connected to a 1.5-V battery carries a 27-μC charge. What is its capacitance?

3. You want to make a parallel plate capacitor from two soup-can ends of radius 4.25 cm. If the capacitance is to be 16 pF, what should be the plate separation?

4. What is the capacitance of a steel ball bearing of radius 0.75 cm?

§27.2 ENERGY STORAGE IN CAPACITORS

5. A 3.9-nF capacitor is connected to a 6.0-V battery. How much energy does it store?

6. How large a capacitor would you need to store 16 mJ at 120 V?

7. A 1.8-pF and a 4.7-pF capacitor are connected in parallel. What is the equivalent capacitance?

8. A 1.8-μF capacitor and a 0.56-μF capacitor are connected in series. What is the equivalent capacitance of the combination?

§27.3 DIELECTRICS AND PRACTICAL CAPACITORS

9. Find the capacitance of a parallel plate capacitor filled with polystyrene, with $A = 11$ cm^2 and $d = 0.74$ mm.

10. A parallel plate capacitor has square plates of side ℓ with separation d. If the capacitor is filled with plexiglas and d is unchanged, how much smaller is the value of ℓ for the same capacitance? If ℓ were unchanged, how much larger would d have to be?

§27.4 ENERGY IN THE ELECTRIC FIELD

11. What is the electrostatic energy density inside a parallel plate capacitor with a plate separation of 2.2 mm and connected to a 1.5-V battery?

12. What is the electrostatic energy density 6.0 cm from a point charge of 3.0 μC?

Questions and Problems

§27.1 CAPACITANCE

13. ✸ ✉ Estimate the capacitance of a car.

14. ✸ Two parallel plate capacitors have the same separation d, but one has twice the area of the other. If both have the same potential difference ΔV between their plates, what is the ratio of **(a)** their stored charges and **(b)** their charge densities?

15. ✸ A parallel plate capacitor is charged and then disconnected from the battery. The plates are then moved apart using insulated handles. Which of the following results?
(a) The charge on the capacitor increases.
(b) The charge on the capacitor decreases.
(c) The capacitance of the plates increases.

(d) The voltage between the plates remains the same.
(e) The voltage between the plates increases.
Explain what is wrong with the answers you didn't choose.

16. ◆ What is the radius of an isolated conducting sphere with a capacitance of 1 F? Compare your result with the radius of the Moon's orbit about the Earth.

17. ◆ A parallel plate capacitor with $A = 125$ cm^2 and $d = 1.25$ mm is charged to a potential difference of 10.0 V. What is the charge density on each plate?

18. ◆ A coaxial cable consists of a cylindrical metal sheath of inner radius $b = 0.3$ cm surrounding an inner wire of outer radius $a = 0.1$ cm. Find the capacitance per unit length of the cable.

19. ◆◆ A 28.0-pF parallel plate capacitor has plates of area 420 cm^2. What is the electric field between its plates when it is connected across a potential difference of 12 V?

20. ◆◆ A parallel plate capacitor has plate area A and separation d. A metal slab of thickness $d/3$ and area A is inserted between the plates (■ Figure 27.21). The capacitor is given a charge Q_0. Explaining your reasoning at each step, find: **(a)** the charge density on each plate, **(b)** the electric field between the top plate and the slab, within the slab, and between the slab and the lower plate, **(c)** the potential difference between the plates, and **(d)** the capacitance of the system.

21. ◆◆◆ Two long parallel wires, each of radius a, are separated by a distance d (■ Figure 27.22). One wire carries a uniform charge density $+\lambda$, and the other carries $-\lambda$. Calculate the electric field due to the wires by superposing contributions from the two filaments. Use the result to estimate the capacitance per unit length of the wires.

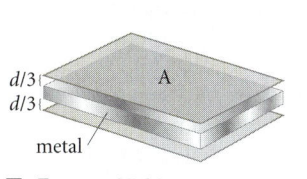
■ FIGURE 27.21

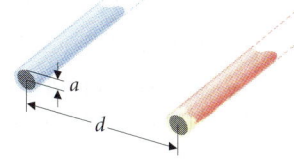
■ FIGURE 27.22

§27.2 ENERGY STORAGE IN CAPACITORS

22. ❖ Two capacitors are charged to the same potential difference. The second has twice the capacitance of the first. What is the ratio of the energies stored in each?

23. ❖ Two capacitors carry the same charge Q. If $C_1 = 3C_2$, what is V_1/V_2? What is the ratio of the stored energies?

24. ❖ If the charge stored in a capacitor is doubled, by what factor is the stored energy increased?

25. ❖ You have six identical 5-μF capacitors. How can you connect them together to form a 6-μF module?

26. ◆ What is the equivalent capacitance of the combination in ■ Figure 27.23 if it is connected into a circuit at points a and b?

27. ◆ What is the equivalent capacitance of the combination in ■ Figure 27.24 if it is connected into a circuit at points a and b?

28. ◆ What is the equivalent capacitance of the combination in ■ Figure 27.25 if it is connected into a circuit at points a and b?

29. ◆◆ Three capacitors have capacitances of 2.0 μF, 4.0 μF, and 8.0 μF. How would you connect them across a battery in order to

■ FIGURE 27.23 ■ FIGURE 27.24

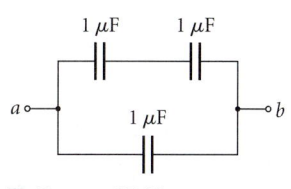
■ FIGURE 27.25

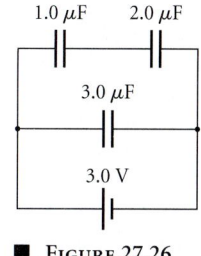
■ FIGURE 27.26

minimize the potential difference across each capacitance? What is the equivalent capacitance in this configuration?

30. ◆◆ How would you connect three capacitors (2.0 μF, 4.0 μF, and 8.0 μF) across a battery to store maximum charge? What is their equivalent capacitance in this configuration?

31. ◆◆ What values of capacitance can be produced by various combinations of a 5.0-mF, a 10.0-mF, and a 30.0-mF capacitor? Draw a circuit diagram for each case.

32. ◆◆ Find the charge on each capacitor in the system in ■ Figure 27.26. What is the total energy stored?

33. ◆◆ Three identical capacitors with capacitance C are connected in parallel with an emf $\mathcal{E}$. Find the charge on each capacitor and the total stored energy. A person using insulating gloves then disconnects the emf and connects the capacitors in series. Find the final values of the charge on each capacitor, the total stored energy, and the potential difference across the combination.

34. ◆◆◆ Capacitors C_1 and C_2 are initially connected in parallel and carry a combined charge Q_0 (■ Figure 27.27). Find the charge on each capacitor and the energy stored in the system. The switches are opened, C_2's terminals are reversed, and then the switches are closed again. Find the stored charge and the energy in the final state.

35. ◆◆◆ What is the capacitance of the semi-infinite network in ■ Figure 27.28? (*Hint:* Removing the leftmost two capacitors leaves an identical semi-infinite network!)

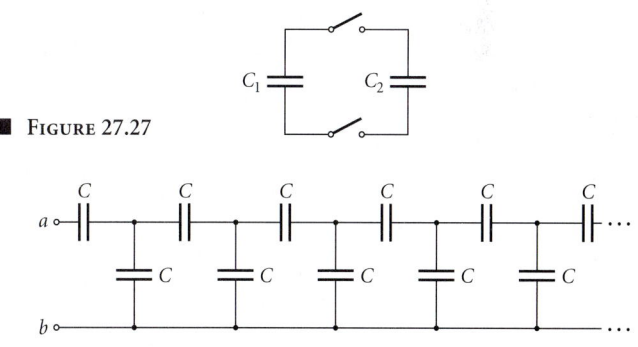
■ FIGURE 27.27

■ FIGURE 27.28

§27.3 DIELECTRICS AND PRACTICAL CAPACITORS

36. ❖ Each plate of a parallel plate capacitor has a charge density of magnitude σ_{free}, and is insulated from its surroundings. If a thin conducting plate the same shape as the capacitor plate is inserted into the capacitor, the induced charge density on the conductor:
(a) is greater than σ_{free}. **(b)** is less than σ_{free}. **(c)** equals σ_{free}.
(d) can be either greater than or less than σ_{free} depending on the conductivity of the conductor.
Explain why you chose your answer.

37. ❖ Each of the plates of a parallel plate capacitor has a charge density of magnitude σ_{free} and is insulated from its surroundings. If a thin

dielectric slab the same shape as the capacitor plate is inserted into the capacitor, the induced bound charge density on the dielectric:
(a) is greater than σ_{free}. (b) is less than σ_{free}. (c) equals σ_{free}.
(d) can be either greater or less than σ_{free} depending on the dielectric constant of the slab.
Explain why you chose your answer.

38. ❖ A parallel plate capacitor consists of two conducting plates of area A separated by an air space of thickness d. If a plastic slab of dielectric constant $\kappa > 1$ is placed between the plates:
(a) the electric field between the plates increases.
(b) the potential difference between the plates increases.
(c) the capacitance of the capacitor increases.
(d) the charge density on the plates decreases.
Explain what is wrong with the answers you didn't choose. Does your answer change if the capacitor is connected to a battery?

39. ◆ Two capacitors have identical metal plates and carry the same charge. One has no dielectric; the other is filled with epoxy. If the electric field in the empty capacitor has magnitude 7.5×10^4 V/m, find the electric field in the epoxy.

40. ◆ A demonstration capacitor has circular plates 15 cm in radius that are separated by $d = 0.30$ cm. A dielectric with constant $\kappa = 2.0$ is inserted between the plates. What is the capacitance of the device?

41. ◆ Find the capacitance of a Pyrex shell ($\kappa = 4.7$) of 1.0-cm inner radius and 1.3-cm outer radius with copper plated on its inner and outer surfaces.

42. ◆◆ A rectangular block of Teflon with sides 3 cm × 2 cm × 1 cm is to be plated with metal on two opposite sides to produce a capacitor. Use a parallel plate capacitor model to decide which choice of coated sides results in the greatest capacitance. What is that capacitance? For each pair of sides, comment on whether parallel plates or parallel wires are a better model for the conductors.

43. ◆◆ Find the capacitance per unit length of a coaxial cable filled with polystyrene. The inner conductor has a radius of 0.75 mm, and the outer conductor has a radius of 0.75 cm.

44. ◆◆ You need a 6.0-pF capacitor. What length of coaxial cable should you use if its inner wire has a radius of 2.3 mm, its outer conductor has an inner radius of 1.1 cm, and it is filled with epoxy?

45. ◆◆ Find the capacitance of the parallel plate capacitor in ■ Figure 27.29 if the plate area is $A = 25$ cm^2, $d = 2.0$ mm, $\kappa_1 = 1.5$, and $\kappa_2 = 2.5$.

46. ◆◆ A spherical capacitor with inner radius of 2.5 cm and outer radius of 3.5 cm is filled with epoxy. It is connected to a 1.5-V battery. What is the charge on the capacitor? Find the surface charge density on the inner and outer conductors. What is the bound charge density on the inner and outer surfaces of the epoxy?

47. ◆◆◆ Two capacitors are connected in parallel with a 12-V battery. Each has plates of area 16 cm^2. The plate separations are $d_1 = 2.0$ mm and $d_2 = 3.0$ mm. The first capacitor has a 2.0-mm slab of plexiglas between the plates. Find the two capacitances, the stored charge, and the stored energy. The capacitors are now disconnected from the battery but remain connected in parallel with each other. The plexiglas slab is moved from the first to the second. What are the two capacitances then? Calculate the charge on each and the total energy stored. How much work is done to move the slab?

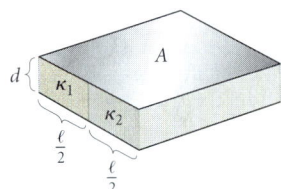

■ FIGURE 27.29

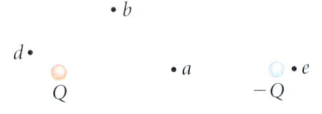

■ FIGURE 27.30

§27.4 ENERGY IN THE ELECTRIC FIELD

48. ❖ You have measured the energy density at a distance 1 m from a certain point charge. If the charge is increased by a factor of 2, by what factor does the energy density change? If the charge is replaced with an equal and opposite charge, does the energy density change?

49. ❖ Two point charges are located as shown in ■ Figure 27.30. At which of points a, b, c, d, or e is the electric field energy density greatest? Why?

50. ❖ Which charge configuration has the greater electrostatic energy, a charge Q on the surface of a metallic conductor with radius R or the same charge Q distributed uniformly throughout a sphere of radius R?

51. ◆ What is the electrostatic energy density 10 cm from a 6-μC point charge?

52. ◆ Find the energy density: (a) midway between two point charges $+Q$ a distance D apart. (b) midway between two point charges $+Q$ and $-Q$ a distance D apart. (c) 1.0 cm from a filament carrying 1.0 μC/m.

53. ◆◆ What is the energy density in vacuum: (a) 5 cm from a point charge of 3 μC? (b) 5 cm from a charge of -3 μC? (c) midway between a 3-μC and a -3-μC charge separated by 10 cm? (d) midway between two $+3$-μC charges or two -3-μC charges separated by 10 cm? Are the third and fourth answers consistent with the first two? Discuss.

54. ◆◆ A total charge of 5.0 μC is spread over (a) a sphere of radius $R = 2.0$ cm and (b) a square plate with the same area (side $4\sqrt{\pi} = 7.1$ cm). Find the electrostatic energy density at a point 1.0 mm from the surface in each case. How do the values compare? (For the square, use the infinite plane result for $\vec{E}$.)

55. ◆◆ A charged metal ball has radius R. Find the radius r of a spherical surface that contains half the electrostatic energy.

56. ◆◆ Three equal 5.3-nC charges are at the corners of an equilateral triangle of side 2.5 cm. Find the electric energy density (a) at the center of the triangle and (b) at the center of one side.

57. ◆◆ Two point charges $\pm Q$ are on the x-axis at $x_Q = \pm \ell$. Find the electrostatic energy density at a point P on the x-axis with (a) $|x_P| < \ell$ and (b) $|x_P| > \ell$. In particular, find the energy density at $x = 2\ell$.

58. ✳ ❖ Show that the interaction energy density produced by two equal but opposite point charges is positive within a sphere with the two charges at opposite ends of a diameter, and negative outside that sphere.

Additional Problems

59. ❖ ■ Figure 27.31 shows a combination of capacitors that is neither series nor parallel. Kirchhoff's loop rule applies without change to such a circuit. How would you restate Kirchhoff's junction rule to relate the charges on the capacitors? (Remember to state carefully the meaning of the algebraic symbol Q_i describing the charge on a capacitor.)

60. ❖ How would you expect the dielectric constant of a dilute gas to vary with the density of the gas?

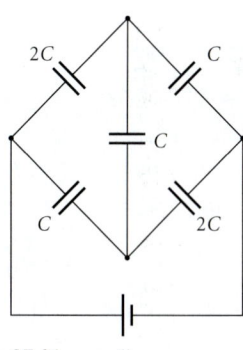

■ FIGURE 27.31

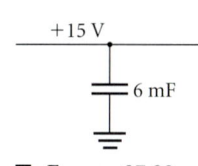

■ FIGURE 27.32

61. ♦♦ Two parallel plate capacitors have the same dimensions, but one is filled with forsterite ($\kappa = 6.2$, dielectric strength = 9.4 MV/m), and the other is filled with paraffin ($\kappa = 2.2$, dielectric strength = 9.8 MV/m). Which can hold more charge? How much more? Explain your reasoning carefully.

62. ♦♦ ■ Figure 27.32 shows part of the circuit diagram for the LZ-1 stereo preamplifier. What is the charge on the capacitor when the amplifier is in use, with +15 V on the positive terminal of the capacitor? How much energy is stored?

63. ♦♦ After the switch is moved to position A in Figure 27.6, some energy from the battery is stored in the capacitor, and some is dissipated as heat by the resistor. Show the two amounts of energy are equal. (The easiest method is to calculate the work done by the battery.)

64. ♦♦ You want to construct a parallel plate capacitor with capacitance C and volume V. Find the required values of A and d in terms of V and C.

65. ♦♦ ✉ If the maximum dielectric strength of any material is of order 10 MV/m, estimate the minimum volume of a capacitor that could store 1 MJ of energy.

66. ♦♦♦ (■ Figure 27.33) With the switch at position f, a total charge $Q_0 = C\mathcal{E}$ resides on the isolated circuit segment between the series capacitors. Find the charge on each capacitor, the energy stored in the capacitors, and the potential at point h. The switch is moved to position g. Find the new values of these quantities, a long time after the switch is moved. How much energy is dissipated as heat in the resistor during this process?

67. ♦♦♦ A tunable capacitor for use in a radio has two sets of semicircular conducting plates (■ Figure 27.34). One set can be rotated with respect to the other set. If there are n positive and n negative plates, how many capacitors are there? How are they connected? Find the capacitance of the set of plates when the negative ones make an angle θ with the positive ones.

68. ♦♦♦ Starting from the electrostatic energy density, find the energy stored in a spherical capacitor with inner radius r_1 and outer radius r_2. Use the expression $U = \frac{1}{2}Q^2/C$ to obtain a value for C. Compare with eqn. (27.4).

Computer Problems

69. A -1.0-μC charge is at the origin, and a 2.0-μC charge is at $x = 1.0$ cm. Either analytically or numerically, compute the energy density at points on the x-axis for both $x < 1.0$ cm and $x > 1.0$ cm. Plot $u(x)$ for $x > 1.0$ cm. How large must x be for $u(x)$ to differ by less than 20% from the value due to a single point charge of 1.0 μC? (Think carefully about the correct position for the 1-μC charge!)

70. Three 1.0-μC point charges are at the corners of an equilateral triangle of side 1.0 m centered at the origin. Use a spreadsheet program to calculate the electric field components and the electric energy density at points along the x-axis and along parallel lines at $y = 0.5$ m and $y = 1.5$ m. Choose points with $x = 0, 0.2, 0.4, \ldots, 2$ m. Plot the energy density as a function of x for each y, and comment.

71. File CH27P71 on the supplementary computer disk lists values of dielectric constant as a function of position between the plates of a parallel plate capacitor with plate area 1.0 cm². Compute the capacitance. Compare with the value you would obtain by using an average value of the dielectric constant. (The coordinate x varies from $x = 0$ at one of the plates to $x = 1.0$ μm at the other.)

Challenge Problems

72. A capacitor is made from a sheet of material consisting of two layers of aluminum foil and two layers of Mylar (■ Figure 27.35a). The foil thickness is 5×10^{-6} m, and the thickness of the Mylar is to be determined. The foil sheet is $L = 3.0$ cm wide and is rolled into a cylinder 1.0 cm in radius (Figure 27.35b). If the capacitance of the device is to be 0.10 μF, what thickness of Mylar should be used?

73. Model an electron as a uniform sphere of total charge $-e$. What radius must it have for its electrostatic energy to equal its mass-energy?

74. Choose sign conventions for the five unknown capacitors in ■ Figure 27.31. Use Kirchhoff's rules to obtain five equations for the unknown charges. Solve for the charges and find the effective capacitance of the combination. (*Hint:* Make use of the symmetry of the configuration to simplify the calculations.)

75. An arbitrary capacitor has two conductors carrying opposite charges $\pm Q$. Use the following plan to show that the electrostatic energy density inside the capacitor is $u = \frac{1}{2}\epsilon_0 E^2$. Consider a tube of field lines between an element of charge dQ on one plate and an element $-dQ$ on the other. Express the contribution dU of these two elements to the energy stored in the capacitor in terms of dQ and an integral along the field lines between the elements. Use Gauss' law to relate the field E at any point along the tube to the electric field at one of the conductors, and so express dQ in terms of E. Hence, show that:

$$dU = \int_{\substack{\text{field}\\ \text{tube}}} \tfrac{1}{2}\epsilon_0 E^2 \, dV.$$

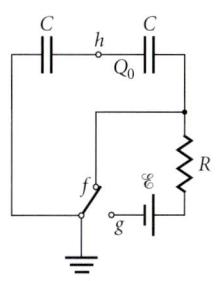

■ FIGURE 27.33

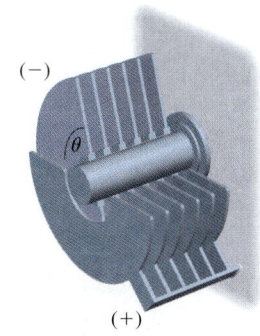

■ FIGURE 27.34

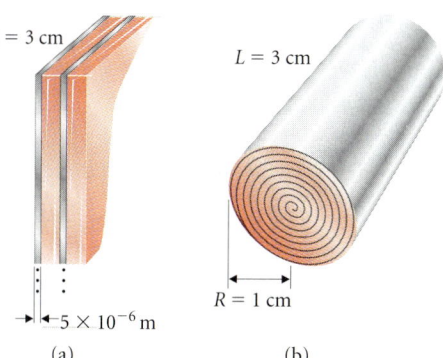

■ FIGURE 27.35 (a) (b)

CHAPTER 28
Static Magnetic Fields

CONCEPTS

Magnetic force
Magnetic field
Magnetic moment
Ampère's law
Circulation of magnetic field

GOALS

Be able to:

Apply the Biot-Savart law to find the magnetic field due to a current distribution.

Compute the magnetic moment of a loop.

Apply Ampère's law to calculate magnetic field produced by symmetric systems.

Interior of the Princeton Tokamak. The carefully constructed magnetic field of this device has been able to confine a hot plasma long enough for nuclear fusion reactions to occur. In 1994 these reactions lasted long enough to liberate as much energy as it took to operate the machine. This is a major step on the long road toward building a fully functioning nuclear fusion power plant.

The nation that controls magnetism will control the universe!
CHESTER GOULD, "DICK TRACY"

Magnetic forces are used to control the motion of charged particles in numerous devices, from television sets to controlled nuclear fusion experiments. The fuel in a successful nuclear fusion reactor is so hot that no ordinary container can be used for the experiment chamber. Instead, an elaborately constructed *magnetic bottle* confines the particles. Containing them long enough for the desired nuclear reactions to occur depends on understanding both the particle motion and the design of magnetic fields. In this chapter, we'll investigate how magnetic fields are shaped by their current sources. In the next chapter, we'll study the motion of particles in response to magnetic forces.

VECTOR CROSS PRODUCTS ARE USED EXTENSIVELY IN THIS CHAPTER. YOU MAY WANT TO REVIEW §9.1, ESPECIALLY THE MATH TOOLBOX.

28.1 MAGNETIC FORCE

In the overview of electromagnetism we appealed to your experience with magnets and the forces they exert on each other. But forces between magnets do not provide an operational definition of magnetic field because magnetic poles always occur in pairs, and the force on an isolated pole cannot be measured. Instead, as we did for electric field, we investigate the force exerted on a point *charge*.

NOW WOULD BE A GOOD TIME TO READ THE OVERVIEW AGAIN.

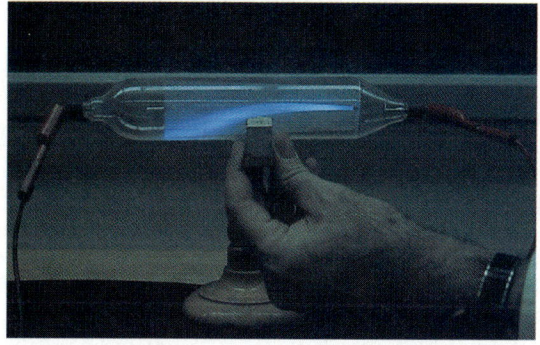

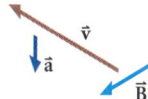

■ FIGURE 28.1

(a) An electron beam. The tube contains a phosphor screen that makes the beam visible. By bringing a bar magnet near the beam, we can investigate the behavior of magnetic force.

(b) The beam bends as the magnet is brought near. The magnetic force is perpendicular to both $\vec{v}$ and $\vec{B}$, so the beam moves downward.

We observe that no magnetic force acts on a stationary charge, whereas a moving charge experiences a force perpendicular to its velocity. ■ Figure 28.1a shows an electron beam moving from right to left in an evacuated tube. When we bring a horizontal bar magnet close to the tube, the beam bends downward. The magnetic force on electrons in the beam is perpendicular to both their velocity and the magnetic field. Further experiments show that the force on a charged particle is proportional to its speed, its charge, and the magnetic field strength:

WE ASSUME THAT WE KNOW THE DIRECTION OF THE MAGNETIC FIELD PRODUCED BY THE BAR MAGNET; SEE, FOR EXAMPLE, FIGURE VI.4 IN THE OVERVIEW.

> The magnetic force on a moving point charge equals its charge multiplied by the cross product of its velocity with the magnetic field:
>
> $$\vec{F}_{mag} = q\,\vec{v} \times \vec{B}. \qquad (28.1)$$

Figure 28.1b shows how this rule accounts for the bending of the beam in the evacuated tube.

We can determine the magnitude of the magnetic field by measuring the forces acting on equal point charges moving in different directions. The maximum force acts on a point charge with velocity perpendicular to $\vec{B}$, and we define the magnetic field strength by:

$$B = \frac{F_{max}}{qv}. \qquad (28.2)$$

The direction of $\vec{B}$ is parallel to the velocity of a point charge that is not deflected.

■ **FIGURE 28.2**
The SI unit of magnetic induction, the tesla, is named after Nikola Tesla (1856–1943), who worked out the basic design of our power transmission system, developed numerous improvements in electrical equipment, and investigated the possibility of wireless power transmission.

IT IS INFORMALLY CALLED MAGNETIC FIELD, AND WE SHALL CONTINUE TO USE THIS NAME.

TABLE 28.1 Magnetic Fields

	Magnetic Field Strength (T)
At the surface of a neutron star	10^8
Near a superconducting magnet	up to 20
In the T-3 Tokamak	3.5
In a high-quality moving-coil loudspeaker	1.5
In the Advanced Light Source at Lawrence Berkeley Laboratory	1.04
In the center of the largest sunspot	0.39
In MRI instruments for medical imaging	0.35
Inside a color TV set	2×10^{-2}
On the Earth's surface, at the equator	3.2×10^{-5}
Due to a power line, on the ground directly below it (This contribution should be added to the Earth's magnetic field.)	10^{-6}
In interstellar space	10^{-10}
In a magnetically shielded experimental environment	10^{-14}

The formal name for the field vector $\vec{B}$ is *magnetic induction*. Its SI unit is the tesla (symbol T) after Nikola Tesla (■ Figure 28.2).

$$1\,\text{T} = 1\,\frac{\text{N}\cdot\text{s}}{\text{C}\cdot\text{m}} = 1\,\frac{\text{N}}{\text{A}\cdot\text{m}}.$$

The cgs unit of magnetic induction, the gauss, is also commonly used: $1\,\text{G} \equiv 10^{-4}\,\text{T}$. The largest magnetic field you are likely to encounter in everyday experience is about 1 T (● Table 28.1).

EXAMPLE 28.1 ◆◆ A particle with a charge of 6.0 μC is used in an experiment to measure magnetic field. The particle moves at 1500 m/s. No force is observed when the particle moves along the *x*-axis. The maximum force of 7.3 mN is observed when the particle moves along the *y*-axis. When the particle moves in the positive *y*-direction, the force it experiences is in the positive *z*-direction. Find $\vec{B}$.

MODEL Since the particle experiences no force when moving along the *x*-axis, the magnetic field is in either the positive or negative *x*-direction. Since the force is in the $+z$-direction when $\vec{v}$ is in the $+y$-direction, we conclude that $\vec{B}$ is in the *negative* *x*-direction:

$$\hat{j} \times (-\hat{i}) = \hat{k}.$$

SETUP We use eqn. (28.2) to find the field strength:

$$B = \frac{F_{max}}{qv} = \frac{7.3 \times 10^{-3}\,\text{N}}{(6.0 \times 10^{-6}\,\text{C})(1500\,\text{m/s})} = 0.81\,\text{T}.$$

SOLVE $\quad \vec{B} = -(0.81\,\text{T})\,\hat{i}.$

ANALYZE This is a strong magnetic field, such as might be produced near an electromagnet. ■

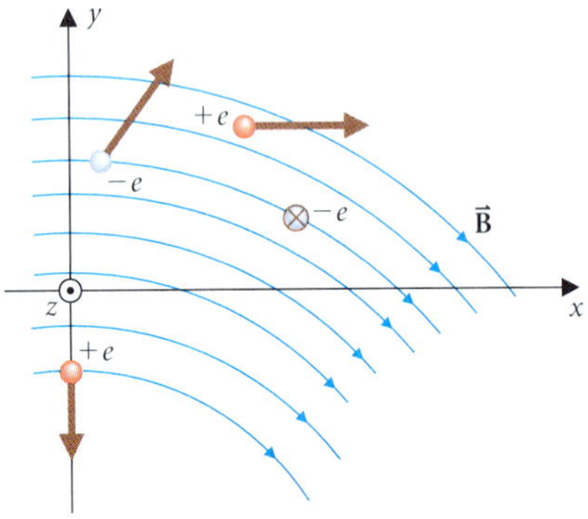

EXERCISE 28.1 ❖ ■ Figure 28.3 shows the velocity vectors of several charged particles in a magnetic field represented by the field lines in the figure. What is the direction of the force on each particle? ■

■ **FIGURE 28.3**
What is the direction of the magnetic force acting on each of these particles?

28.2 CURRENT AS THE SOURCE OF MAGNETIC FIELDS

28.2.1 The Biot-Savart Law

In 1820 Oersted demonstrated that the magnetic field lines produced by a long straight wire form circles centered on the wire. Later the same year, Jean Baptiste Biot (1774–1862) and Felix Savart (1791–1841) showed that the magnetic field strength decreases inversely with distance from the wire. These results provide excellent qualitative images for describing magnetic fields. They were the starting point for finding an exact relation for the magnetic field produced by an arbitrary distribution of current.

The simplest static electric field is that of a point charge. But a point charge in motion does not produce a static magnetic field. As the charge approaches a given point the magnetic field increases to a maximum and then dies away as the particle recedes. Static magnetic field is produced only by steady electric current. The magnetic field is a superposition of contributions from small elements of the current path. Each element $d\vec{\ell}$ is in the direction of the conventional current with magnitude I. A description of the field $d\vec{B}$ produced by such an element is named after Biot and Savart.

WE SHALL REFER TO THIS FACT AS OERSTED'S RULE. FOR A DISCUSSION OF OERSTED'S DISCOVERY, SEE THE OVERVIEW, ESPECIALLY FIGURE VI.16.

THE ELECTRIC FIELD OF A POINT CHARGE IS GIVEN BY COULOMB'S LAW (EQN. 23.6).

> **THE BIOT-SAVART LAW (■ FIGURE 28.4):**
>
> $$d\vec{B} = \frac{\mu_0}{4\pi} I \frac{d\vec{\ell} \times \hat{r}}{r^2}. \qquad (28.3)$$

THE BIOT-SAVART LAW IN THIS FORM WAS ANNOUNCED ON OCTOBER 30, 1820, AT A MEETING OF THE FRENCH ACADEMY OF SCIENCES.

In the Biot-Savart law, as in Coulomb's law, the field strength is directly proportional to the source (here $I\,d\vec{\ell}$) and inversely proportional to distance squared. The rule for magnetic field direction, however, is very different from that for electric field. The source $I\,d\vec{\ell}$ is a *vector* quantity, and the magnetic field vector it produces is perpendicular to both the direction of the source and the direction of the position vector $\hat{r}$ from the source to the field point. Notice that the direction of the vector $d\vec{\ell}$ is determined by the direction of the current.

The constant μ_0 is *defined* to have the exact value:

$$\mu_0 = 4\pi \times 10^{-7} \text{ N/A}^2.$$

COMPARE EQN. (28.3) WITH EQN. (23.6). THE DEFINITION OF THE UNIT VECTOR $\hat{r}$ IS THE SAME IN BOTH CASES.

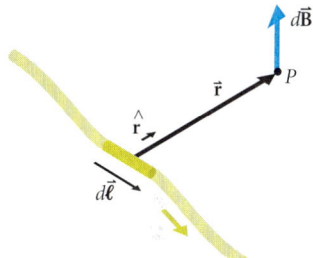

■ **FIGURE 28.4**
The Biot-Savart law describes the magnetic field produced by a small segment of current-carrying wire. The vector $d\vec{\ell}$ describes the direction of the wire and the sense of the current in it. The vector $\hat{r}$ points *from* the source point *toward* the field point. [The same definition is used in Coulomb's law, eqns. (23.1) and (23.6).] The direction of the magnetic field at P is perpendicular to both $d\vec{\ell}$ and $\hat{r}$, in the direction of their cross product.

EXERCISE 28.2 ♦ Show that the source term $I\,d\vec{\ell}$ in the Biot-Savart law equals the number of electrons in the element of wire multiplied by their drift velocity $\vec{v}_d$ and their charge $(-e)$.

EXERCISE 28.3 ❖ Show that the magnetic field vector $d\vec{B}$ vanishes at points along the line of the wire element and has maximum magnitude in the plane perpendicular to the element.

EXAMPLE 28.2 ♦♦ A thin wire carrying current $I = 5.0$ A passes through the origin parallel to the x-axis (■ Figure 28.5a). Find the magnetic field at point P on the y-axis at $y = 3.0$ m produced by a 1.0-mm piece of the wire centered at the origin.

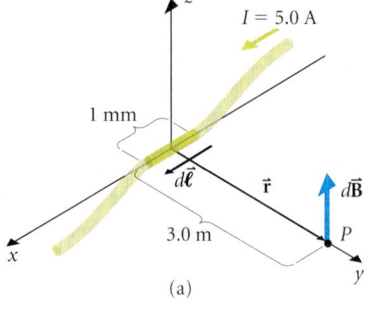
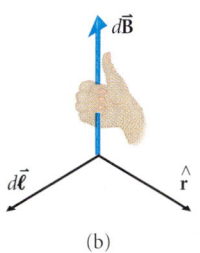

■ **FIGURE 28.5**
(a) Magnetic field at P due to a segment of wire at the origin. Here $d\vec{\ell}$ is in the x-direction, $\hat{r}$ is in the y-direction, and $d\vec{B}$ is in the z-direction. (b) Using the right-hand rule to find the direction of $d\vec{B}$.

MODEL The piece of wire is very small compared with its distance to P, so we may treat it as a single element and find the field it produces directly from eqn. (28.3).

SETUP The element of wire carries current in the +x-direction, so $\vec{\ell} = (1.0 \text{ mm})\hat{i}$. The position vector $\vec{r} = (3.0 \text{ m})\hat{j}$, and the unit vector $\hat{r} = \hat{j}$.

SOLVE
$$\vec{B} = \frac{\mu_0}{4\pi} I \frac{\vec{\ell} \times \hat{r}}{r^2}$$
$$= (10^{-7} \text{ N/A}^2)(5.0 \text{ A}) \frac{(1.0 \times 10^{-3} \text{ m})(\hat{i} \times \hat{j})}{(3.0 \text{ m})^2}$$
$$= \frac{5.0 \times 10^{-10}}{9.0} \frac{\text{N}}{\text{A} \cdot \text{m}} \hat{k}$$
$$= (5.6 \times 10^{-11} \text{ T})\hat{k}.$$

ANALYZE The direction is found equally well from formal rules for cross products of unit vectors or from the right-hand rule illustrated in Figure 28.5b. ∎

28.2.2 The Magnetic Field Produced by a Straight Wire Segment

The magnetic field due to a longer piece of wire is found from the principle of superposition:

STRICTLY SPEAKING, THE INTEGRAL IS ALWAYS OVER A CLOSED PATH (HENCE $\oint$), SINCE STEADY CURRENT ALWAYS FOLLOWS A CLOSED PATH. IF IT DID NOT, CHARGE WOULD BUILD UP AT THE END OF THE PATH.

PRINCIPLE OF SUPERPOSITION

The magnetic field produced by an extended current distribution is the vector sum of contributions due to its individual elements:

$$\vec{B} = \oint d\vec{B} = \oint \frac{\mu_0}{4\pi} I \frac{d\vec{\ell} \times \hat{r}}{r^2}. \quad (28.4)$$

We first apply this principle to obtain the field due to current in a straight segment of wire.

EXAMPLE 28.3 ♦♦♦ Find the magnetic field produced by a straight segment of wire of length ℓ carrying current I. Compare your result with the observations of Oersted, Biot, and Savart for a long straight wire ($\ell \to \infty$).

MODEL We are asked for the value of $\vec{B}$ at an arbitrary point P (■ Figure 28.6). We use the principle of superposition and use the integration procedure from Interlude 2.

SETUP STEPS I, II, AND III We take the y-axis along the wire segment and the x-axis through the field point P. From the Biot-Savart law, the contribution of a typical element of the wire, $d\vec{\ell} = dy\,\hat{j}$, to the field at P is:

$$d\vec{B} = \frac{\mu_0}{4\pi} I \frac{d\vec{\ell} \times \hat{r}}{r^2}.$$

The cross product has magnitude:

$$|d\vec{\ell} \times \hat{r}| = |d\vec{\ell}||\hat{r}| \sin \phi = |d\vec{\ell}| \sin(180° - \phi) = |d\vec{\ell}| \frac{x}{r}.$$

The direction of $d\vec{B}$ is perpendicular to both $\hat{r}$ and $d\vec{\ell}$. According to the right-hand rule, the field $d\vec{B}$ at P is in the minus z-direction for every element $d\vec{\ell}$. Since the direction is perpendicular to $\hat{r}$, it is tangent to the circle through P centered on the wire, in agreement with Oersted's observations. The most convenient coordinate for describing $d\vec{\ell}$ is

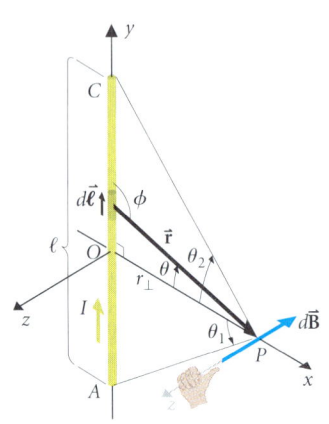

■ **FIGURE 28.6**
Magnetic field due to a finite segment of wire is the superposition of the contributions due to the individual wire segments. For a straight segment like this, the perpendicular distance $r_\perp = r \sin \phi = x$ is the same for each element. The angles $\theta_1 = \angle OPA$ and $\theta_2 = \angle OPC$ have signs determined by the position of the end of the wire segment with respect to the foot of the perpendicular at O. Going from O to C, we travel in the same direction as the current, so $\angle OPC$ is positive. Similarly, going from O to A, we travel opposite I, so $\angle OPA$ is negative.

the angle θ measured from the x-axis. The angle is positive for elements with positive y-coordinate. Then:

$$y = x \tan \theta \quad \text{and} \quad r = x \sec \theta.$$

The length of the wire element is:

$$|d\vec{\ell}| = dy = d(x \tan \theta) = x \sec^2 \theta \, d\theta.$$

So:

$$|d\vec{B}| = \frac{\mu_0 I x^2 \sec^2 \theta \, d\theta}{4\pi (x \sec \theta)^3} = \frac{\mu_0 I}{4\pi x} \cos \theta \, d\theta.$$

WE USED THE SAME ANGLE IN EXAMPLE 24.3 TO FIND THE ELECTRIC FIELD DUE TO A LINE CHARGE.

STEP IV The limits of integration run from θ_1 (negative for the case shown in the figure) to θ_2.

SOLVE
STEP V Since the contributions $d\vec{B}$ from all elements of the wire are in the same direction, we may pull the constant unit vector $-\hat{k}$ out of the integral:

$$\vec{B} = \int d\vec{B} = \int dB_z \, \hat{k} = -\hat{k} \int_{\theta_1}^{\theta_2} \frac{\mu_0 I}{4\pi x} \cos \theta \, d\theta = -\hat{k} \frac{\mu_0 I}{4\pi x} \sin \theta \bigg|_{\theta_1}^{\theta_2}.$$

$$\vec{B} = -\hat{k} \frac{\mu_0 I}{4\pi x} (\sin \theta_2 - \sin \theta_1). \tag{28.5}$$

For a very long wire, ($\ell \gg x$), $\theta_1 \approx -\pi/2$, and $\theta_2 \approx +\pi/2$. Then $\sin \theta_2 = 1$, and $\sin \theta_1 = -1$. The distance x is the perpendicular distance $r_\perp$ of P from the wire segment, so:

$$|\vec{B}| = \frac{\mu_0 I}{2\pi r_\perp}. \tag{28.6}$$

ANALYZE Equation (28.6) reproduces Biot and Savart's observations that $|\vec{B}| \propto 1/r_\perp$ and gives the constant of proportionality $\mu_0/(2\pi)$. The direction of $\vec{B}$ also follows the subsidiary right-hand rule we observed in the Overview of electromagnetism: with the thumb of your right hand parallel to the current, your fingers curl about the wire in the direction of the magnetic field. ∎

EXERCISE 28.4 ❖ Two long straight parallel wires carry equal currents in opposite directions. What magnetic field do they produce at a point halfway between them?

28.2.3 The Magnetic Field of Loops and Coils

In Chapter 24 we found that simple models describe the electric field at a large distance from a system of charges. A system with net charge looks like a point charge, while a system with no net charge looks like a dipole. The magnetic field produced by a current distribution also simplifies at large distances. Experiments show that the simplest magnetic field that occurs is a dipole.

REMEMBER: AN ELECTRIC DIPOLE CONSISTS OF TWO EQUAL AND OPPOSITE POINT CHARGES SEPARATED BY A SMALL DISTANCE (CF. §24.5).

> At large distances from any distribution of current, the magnetic field it produces resembles an ideal dipole field.

As an example we use the Biot-Savart law to compute the field a circular current loop produces at points on its axis. An exact calculation of the field at other points is mathematically intricate and goes beyond our needs. By comparing the field on the axis with the expression computed in Chapter 24 for an electric dipole, we can identify the magnetic dipole moment of the current loop and obtain an excellent approximation to the magnetic field away from the loop axis.

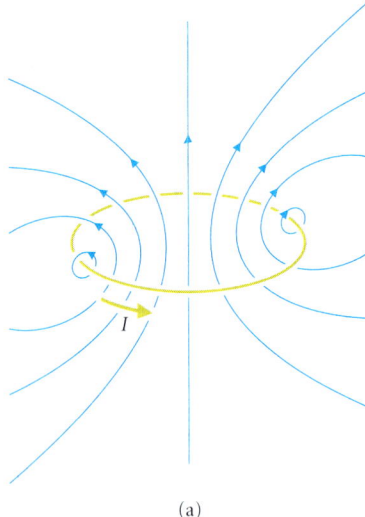

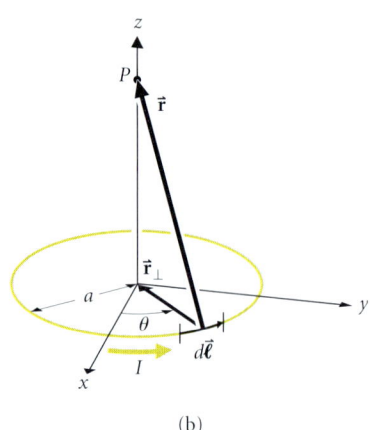

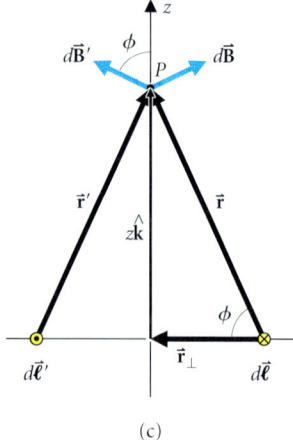

(a) (b) (c)

■ **Figure 28.7**
(a) The magnetic field on the axis of a circular current loop. Magnetic field lines circle the wire. On the axis, the sum of the contributions from each element of the loop produces a field along the axis. At large distances from the loop, the field pattern resembles that of a dipole.

(b) We describe each element of the loop by its position at angle θ from the x-axis and its length $d\ell = a\, d\theta$. The z-axis is perpendicular to the loop through its center.

(c) View in a plane perpendicular to the loop. Two line elements at opposite ends of a diameter contribute magnetic field elements of equal magnitude. Their z-components add, and their perpendicular components cancel:
$d\vec{B} + d\vec{B}' = 2\, dB \cos\phi\, \hat{k} = 2\, dB_z\, \hat{k}.$
The vector $\vec{r} = \vec{r}_\perp + z\hat{k}.$

Like electric field lines, magnetic field lines cannot cross at the center of the loop.

EXAMPLE 28.4 ♦♦ Find the magnetic field on the axis of a thin circular wire of radius a carrying current I.

MODEL According to Oersted's rule, the magnetic field lines close to the wire circle it, as shown in ■ Figure 28.7a. Along the axis, the magnetic field is perpendicular to the loop.

The magnetic field vector at any point is the sum of the contributions $d\vec{B}$ from the individual line segments. We use the principle of superposition and evaluate the integral (eqn. 28.4) using the integration process from Interlude 2.

SETUP
STEP I To make use of the current loop's symmetry, we choose cylindrical coordinates with the origin at the loop's center, z-axis perpendicular to its plane, and angle θ measured in the sense of the current I (Figure 28.7b). Then the field point has coordinates $(0, 0, z)$.

STEP II A typical current element $d\vec{\ell}$ at position θ measured from the x-axis has length $|d\vec{\ell}| = a\, d\theta$.

STEP III The magnetic field $d\vec{B}$ produced by $d\vec{\ell}$ at P has a component along the z-axis and a component radially outward, parallel to the x-y-plane (Figure 28.7c). The element $d\vec{\ell}'$ at the opposite end of a diameter makes a contribution $d\vec{B}'$ with an opposite radial component. Their z-components add. The total magnetic field vector $\vec{B}$ at P is the sum of contributions from such pairs and is in the z-direction.

For each element $d\vec{\ell}$:

$$d\vec{\ell} \times \hat{r} = \frac{d\vec{\ell} \times \vec{r}}{r} = \frac{d\vec{\ell} \times (\vec{r}_\perp + z\hat{k})}{r}.$$

Both vectors $d\vec{\ell}$ and $\vec{r}_\perp$ are in the x-y-plane, so $d\vec{\ell} \times \vec{r}_\perp$ is in the z-direction. The term $(z/r)d\vec{\ell} \times \hat{k}$ is the radial component. Since $|\vec{r}_\perp| = a$, and $d\vec{\ell} \perp \vec{r}_\perp$:

$$[d\vec{\ell} \times \hat{r}]_z = \frac{a\, d\ell}{r},$$

where $r = \sqrt{z^2 + a^2}$ is the same for each element. Thus:

$$dB_z = \frac{\mu_0}{4\pi} \frac{a\,d\ell}{r^3}$$

$$= \frac{\mu_0}{4\pi} \frac{a\,d\ell}{(z^2 + a^2)^{3/2}}.$$

SOLVE Both z and a are independent of position on the loop, so:

STEPS IV AND V

$$\vec{B} = \hat{k} \oint \frac{\mu_0 a\,d\ell}{4\pi(z^2 + a^2)^{3/2}}$$

$$= \hat{k} \frac{\mu_0 a}{4\pi(z^2 + a^2)^{3/2}} \oint d\ell.$$

The integral $\oint d\ell$ means the total length around the loop—its circumference, $2\pi a$. So:

$$\vec{B} = \frac{\mu_0 I a^2}{2(z^2 + a^2)^{3/2}} \hat{k}. \tag{28.7}$$

ANALYZE Notice how similar this expression is to eqn. (24.5) for the electric field on the axis of a ring of charge. An important difference between the expressions is that $\vec{E}$ reverses direction on opposite sides of a ring; $\vec{B}$ does not.

At large distances from the loop, details of its size and shape are no longer important, and its field line diagram looks like that of a dipole (compare Figure 28.7a with Figure VI.4 in the Overview). For $z \gg a$, eqn. (28.7) becomes:

$$B_z \approx \frac{\mu_0 I a^2}{2z^3} = \frac{\mu_0}{4\pi} \frac{2\pi a^2 I}{z^3}.$$

For comparison, the field due to an electric dipole on its axis is (eqn. 24.13):

$$E_z = \frac{2kp}{z^3}.$$

By analogy:

$$B_z = \frac{2k_m m}{z^3},$$

where $k_m = \mu_0/4\pi$, and $\vec{m} = \pi a^2 I \hat{k}$ is the *magnetic dipole moment* of the loop.

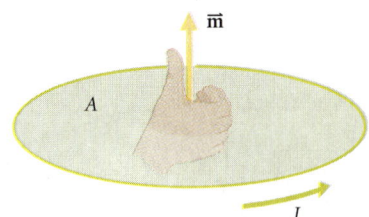

■ **FIGURE 28.8**
Definition of magnetic moment of a loop. The direction of $\vec{m}$ is determined by a right-hand rule. Curl the fingers of your right hand around the loop in the direction of I; your thumb gives the direction of $\vec{m}$. The magnitude $m = IA$.

> A current loop with area A and carrying current I has a *magnetic dipole moment* $\vec{m}$ of magnitude:
>
> $$m = IA. \tag{28.8}$$
>
> The direction of $\vec{m}$ is perpendicular to the loop according to a right-hand rule: when the fingers curl in the direction of the current, the thumb points along $\vec{m}$ (■ Figure 28.8).

USUALLY $\vec{m}$ IS CALLED SIMPLY THE MAGNETIC MOMENT, WITH DIPOLE BEING UNDERSTOOD.

Modeling any current loop as a dipole gives an accurate value for the magnetic field at large distances from the loop.

A simple coil is made by wrapping several turns of wire around the same loop. If there are N turns of wire in the coil, it is equivalent to N separate loops producing equal contributions to the field at any point. The coil has a dipole moment with magnitude $m = NIA$ and direction determined as shown in Figure 28.8.

Digging Deeper

MAGNETIC MOMENT OF AN ARBITRARY PLANAR LOOP

We use the Biot-Savart law to calculate the magnetic field at a large distance from an arbitrary current loop that lies in a single plane. We choose the origin at a point inside the loop (■ Figure 28.9a), and the z-axis perpendicular to the plane of the loop. Then the magnetic field at a point on the z-axis due to the element $d\vec{\ell}$ is:

$$d\vec{B} = k_m I \frac{d\vec{\ell} \times \hat{r}}{r^2} = k_m I \frac{d\vec{\ell} \times \vec{r}}{r^3}.$$

Writing $\vec{r} = \vec{r}_\perp + z\hat{k}$ and assuming $z \gg |\vec{r}_\perp|$, we have:

$$d\vec{B} \approx \frac{k_m I}{z^3} d\vec{\ell} \times (\vec{r}_\perp + z\hat{k}).$$

The product $d\vec{\ell} \times \vec{r}_\perp$ is in the z-direction, while $d\vec{\ell} \times \hat{k}$ lies in the x-y-plane. To find the loop's dipole moment, we use the z-component of the field. (At this large distance from the loop, B_x and B_y are zero; see Problem 71.)

$$dB_z \approx \frac{k_m I}{z^3} |d\vec{\ell} \times \vec{r}_\perp|.$$

From Figure 28.9b, $\frac{1}{2}|d\vec{\ell} \times \vec{r}_\perp|$ is the differential area dA (cf. Chapter 9). Thus:

$$B_z = \int dB_z = \int \frac{k_m I}{z^3} 2\, dA = \frac{2k_m IA}{z^3}.$$

So $m = IA$ is the correct expression for the magnetic dipole moment of an arbitrary current loop.

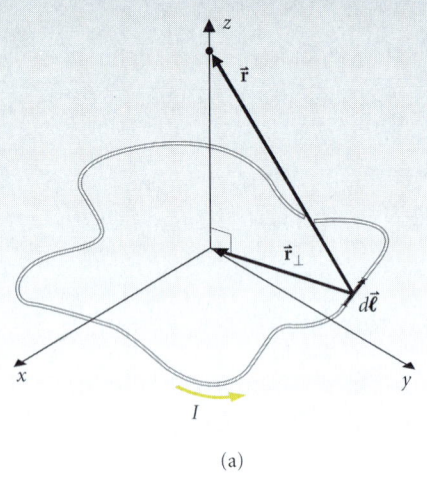
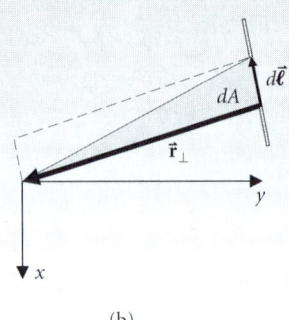

■ **FIGURE 28.9**
(a) Geometry for computing the magnetic field produced by a plane loop of arbitrary shape. (b) Remember that the cross product of two vectors represents the area of the parallelogram formed by the two vectors. Thus $\frac{1}{2}|d\vec{\ell} \times \vec{r}_\perp| = dA$, the triangular area element formed by the line element and the center of the loop.

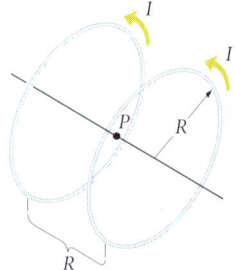

■ **FIGURE 28.10**
Helmholtz coils (after Hermann von Helmholtz, 1821–1894). Two parallel coils carrying current in the same direction produce a nearly uniform field in the region around point P, halfway between them. (For proof of the uniformity, see Problem 39.)

EXAMPLE 28.5 ♦♦ Helmholtz coils, ■ Figure 28.10, produce a nearly uniform field near the point halfway between their centers. To create a region in the laboratory almost free of magnetic field, a pair of coils with radius $R = 30$ cm and $N = 10^2$ turns of wire each is used to produce a magnetic field equal to and opposite the Earth's magnetic field. If the Earth's field in the laboratory has magnitude $B_e = 0.3$ G, what current is required in the coils?

MODEL The magnetic field on the axis of a coil is the same on either side of the coil. Thus the magnetic field produced by the two Helmholtz coils at their midpoint is just twice that of a single coil.

SETUP The magnetic field of one loop is given by eqn. (28.7) with $z = R/2$ and $a = R$. We have N loops per coil and two coils, so:

$$|\vec{B}_e| = 2N \frac{\mu_0 I R^2}{2(5R^2/4)^{3/2}} = \frac{8\mu_0 NI}{(5)^{3/2} R}.$$

The field produced by the coils must be opposite the Earth's field and have the same magnitude, so:

$$I = \frac{(5)^{3/2}RB_e}{8\mu_0 N} = \frac{11(0.3 \text{ m})(0.3 \text{ G})(10^{-4} \text{ T/G})}{8(4\pi \times 10^{-7} \text{ N/A}^2)(10^2)} = 0.1 \text{ A}.$$

ANALYZE The current is not particularly large and is easily produced in laboratory equipment. Properties of Helmholtz coils are explored further in the problems.

28.3 Integral Laws for Static Magnetic Fields

Gauss' law for the electric field gives a global relation between the electric field and charge and a useful technique for computing the fields produced by symmetric sources. A relation that plays a similar role for magnetic field is known as Ampère's law. It is based on a different geometric property: magnetic field lines circulate around their sources rather than diverging from them. In this section we shall describe Ampère's law and a version of Gauss' law that applies to magnetic field and give some examples of their use.

28.3.1 Gauss' Law for the Magnetic Field

Just as electric field lines begin and end on electric charge, we would expect magnetic field lines to begin and end on single magnetic poles. But experiments have repeatedly failed to detect any such poles. Thus magnetic field lines form loops that neither begin nor end. Any field line that enters a volume also leaves it (■ Figure 28.11), and no net magnetic flux emerges. Gauss' law for the magnetic field is quite simple:

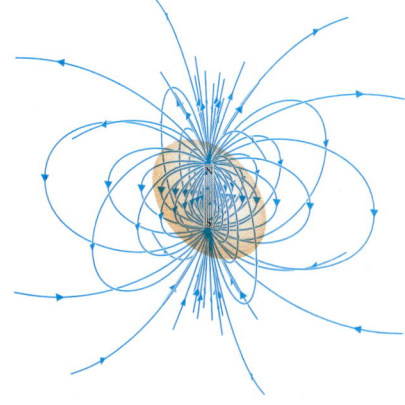

■ FIGURE 28.11
Gauss' law for the magnetic field. The flux through any closed surface is zero: every field line entering the volume through surface S leaves again.

> The net magnetic flux Φ_B emerging from any volume is zero:
>
> $$\Phi_B \equiv \oint \vec{B} \cdot \hat{n} \, dA = 0. \tag{28.9}$$

The integral is taken over the closed surface surrounding the volume.

The currents that are the sources of $\vec{B}$ do not appear in eqn. (28.9), so Gauss' law is not helpful for calculating magnetic field. But it is very useful for understanding the different behavior of electric and magnetic fields at a boundary. The electric field changes abruptly at the surface of a conductor because surface charge is present. In contrast, since there can be no magnetic charge, the normal component of $\vec{B}$ is continuous across any boundary.

See §25.6.3 for the relation between σ and E.

28.3.2 Circulation and Ampère's Law

Flux measures the divergence of electric field lines from their sources. Magnetic field lines form loops around their current sources, and the mathematical quantity that measures this behavior is named *circulation*. Like flux, the term "circulation" is derived from fluid theory, where it describes how the fluid swirls as it flows. To get a rough intuitive idea of circulation, imagine paddling a canoe around a closed curve (■ Figure 28.12). At all points along curve A water flows almost parallel to the boat's path. The water has positive circulation. As you paddle

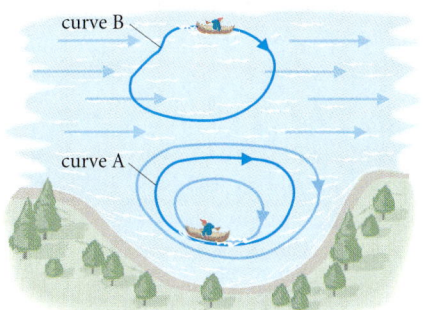

■ FIGURE 28.12
Paddling a canoe around a closed curve, we determine whether the water flow has *circulation* around the curve by noticing whether we mostly paddle downstream ($\mathscr{C} > 0$), paddle upstream ($\mathscr{C} < 0$), or paddle downstream as often as we paddle upstream ($\mathscr{C} = 0$).

STRUGGLING THE OTHER WAY AROUND CURVE A, YOU WOULD GET A CLEAR IMPRESSION OF NEGATIVE CIRCULATION.

around curve B, the water moves in the direction opposite the boat about as often as it moves in the same direction. There is no net circulation.

The circulation $\mathcal{C}$ of a vector field $\vec{v}$ around a closed curve C is:

$$\mathcal{C} \equiv \oint_C \vec{v} \cdot d\vec{\ell}. \tag{28.10}$$

The circulation of $\vec{B}$ around C describes the extent to which field lines follow the direction of the curve.

■ Figure 28.13 shows an application of magnetic circulation. The ammeter measures current when it is clamped around a wire. It does not need to be connected into the circuit because it senses the magnetic field produced by current in the wire. The clamp defines a closed path surrounding the wire, and the circulation of magnetic field around this path provides enough information to determine the current.

■ FIGURE 28.13
This ammeter is *not* connected into the circuit. Placed *around* a current-carrying wire, it measures the magnetic field.

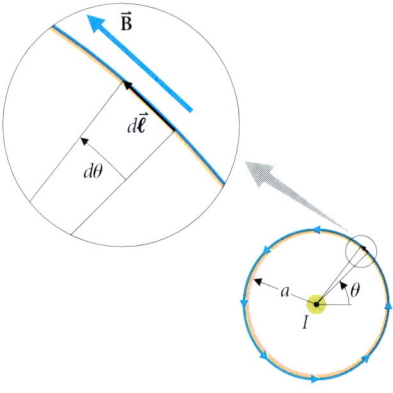

■ FIGURE 28.14
The circulation around the circle $r = a$ equals μ_0 times the current I in the central wire. On the circle, $\vec{B} \cdot d\vec{\ell} = B\, d\ell$ for each line element.

EXAMPLE 28.6 ♦♦ Find the circulation of $\vec{B}$ around a circle of radius a centered on a wire carrying current I (■ Figure 28.14).

MODEL The magnetic field lines form circles around the wire, everywhere parallel to the curve C. We choose to go around C counterclockwise, so we expect the circulation to be positive.

SETUP The magnetic field has constant magnitude around the circle and is given by
STEPS I, II, III eqn. (28.6) with $r_\perp = a$:

$$\vec{B} = \frac{\mu_0 I}{2\pi a}\hat{\theta}.$$

We use the angle θ as the coordinate to describe each element of the circle:

$$d\vec{\ell} = a\, d\theta\, \hat{\theta}.$$

Thus:
$$d\mathcal{C} = \vec{B} \cdot d\vec{\ell} = \frac{\mu_0 I}{2\pi a} a\, d\theta = \frac{\mu_0 I}{2\pi} d\theta.$$

SOLVE The circulation is:
STEPS IV, V
$$\mathcal{C} = \oint \vec{B} \cdot d\vec{\ell} = \frac{\mu_0 I}{2\pi} \oint_0^{2\pi} d\theta = \mu_0 I.$$

ANALYZE Note the relation between circulation and current.

REMEMBER: THIS IS OERSTED'S RULE.

EXERCISE 28.5 ❖ What is the circulation of a static electric field around any closed curve?

Example 28.6 illustrates the general result known as Ampère's law.

> **AMPÈRE'S LAW**
>
> The circulation $\mathscr{C}$ of the magnetic field around any closed curve C equals the current through C multiplied by μ_0.
>
> $$\mathscr{C} = \oint_C \vec{B} \cdot d\vec{\ell} = \mu_0 I. \qquad (28.11)$$

The sign convention for current is determined by a right-hand rule: curl the fingers of your right hand around the curve C in the sense of the vectors $d\vec{\ell}$. The current is positive if it is in the direction of your thumb (■ Figure 28.15). Ampère's law contains the same information as the Biot-Savart law but expressed in a different form.

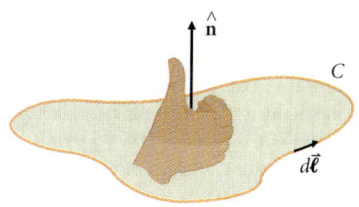

■ **FIGURE 28.15**
A right-hand rule defines the positive sense of current in Ampère's law. First choose a direction to travel around the curve C. This defines the direction of each element $d\vec{\ell}$. Curl the fingers of your right hand around the curve in the chosen direction; your thumb defines the positive sense for I, or for $\hat{\mathbf{n}}$ in eqn. (28.12).

BY ANALOGY WITH "GAUSSIAN SURFACE," WE SHALL CALL CURVES USED IN AMPÈRE'S LAW "AMPÈRIAN CURVES."

Digging Deeper

DEMONSTRATION OF AMPÈRE'S LAW

A general proof that Ampère's law follows from the Biot-Savart law is mathematically complex. Here we demonstrate the result for a combination of long straight wires. ■ Figure 28.16 shows two arbitrary closed curves near a single long straight wire. Curve (a) surrounds the wire, whereas curve (b) does not. The z-axis is taken along the wire, and polar coordinates r and θ are used in the x-y-plane. A typical element $d\vec{\ell}$ of the curve has components:

$$d\vec{\ell} = dr\,\hat{\mathbf{r}} + r\,d\theta\,\hat{\boldsymbol{\theta}} + dz\,\hat{\mathbf{z}},$$

and the magnetic field produced by the wire is:

$$\vec{B} = \frac{\mu_0 I}{2\pi r}\,\hat{\boldsymbol{\theta}}.$$

So, since $\hat{\boldsymbol{\theta}} \cdot \hat{\mathbf{r}}$ and $\hat{\boldsymbol{\theta}} \cdot \hat{\mathbf{z}}$ are both zero,

$$\vec{B} \cdot d\vec{\ell} = \frac{\mu_0 I}{2\pi}\,d\theta,$$

and the circulation is

$$\mathscr{C} \equiv \oint \vec{B} \cdot d\vec{\ell} = \frac{\mu_0 I}{2\pi} \oint d\theta.$$

Around curve (a) the angle θ increases uniformly from 0 to 2π:

$$\oint_a d\theta = 2\pi,$$

and

$$\oint_a \vec{B} \cdot d\vec{\ell} = \mu_0 I.$$

Around curve (b) the angle θ increases from a minimum value to a maximum and back again. The integral is zero.

The circulation around a curve from any combination of long straight wires follows from the principle of superposition. A wire passing through the curve contributes $\mu_0 I$; a wire that does not pass through the curve makes no net contribution.

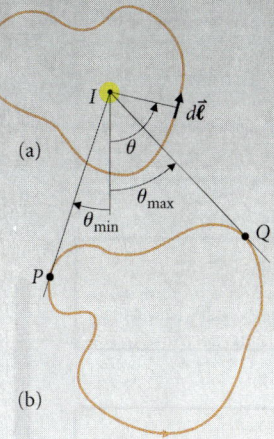

■ **FIGURE 28.16**
Demonstration of Ampère's law. Around curve a, θ increases from 0 to 2π. Around curve b, θ increases from $\theta_{\min}$ (which is negative in this case) to $\theta_{\max}$ and back again as we go from P to Q to P:

$$\oint_b d\theta = 0.$$

The surface is like a soap film attached to a wire loop. As you blow the bubble, the film changes shape but remains attached to the loop. This surface can also have any shape provided that its edge is the chosen curve.

Method for finding symmetric magnetic fields using Ampere's law

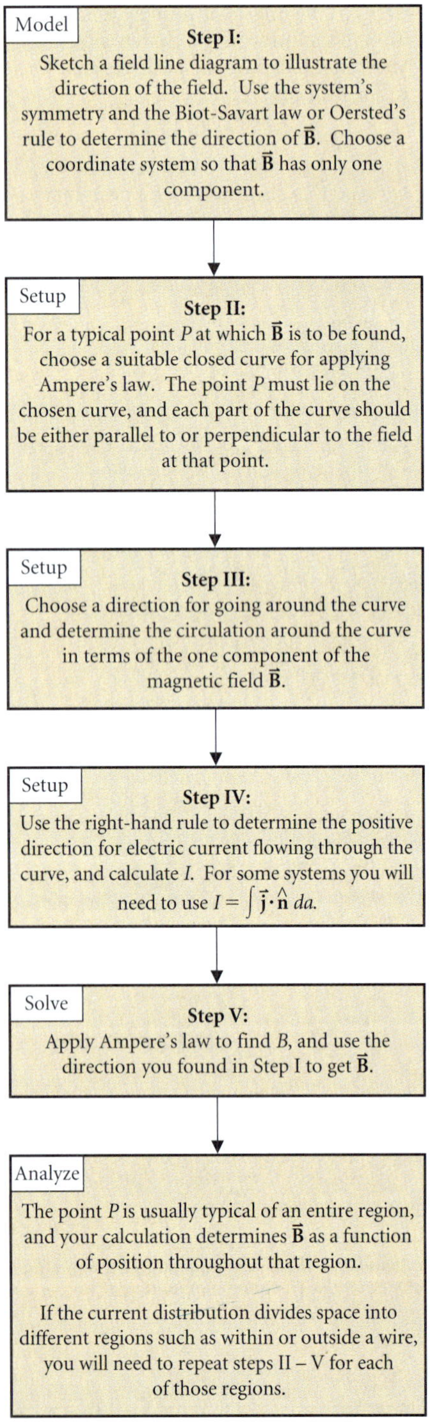

Model — **Step I:** Sketch a field line diagram to illustrate the direction of the field. Use the system's symmetry and the Biot-Savart law or Oersted's rule to determine the direction of $\vec{B}$. Choose a coordinate system so that $\vec{B}$ has only one component.

Setup — **Step II:** For a typical point P at which $\vec{B}$ is to be found, choose a suitable closed curve for applying Ampere's law. The point P must lie on the chosen curve, and each part of the curve should be either parallel to or perpendicular to the field at that point.

Setup — **Step III:** Choose a direction for going around the curve and determine the circulation around the curve in terms of the one component of the magnetic field $\vec{B}$.

Setup — **Step IV:** Use the right-hand rule to determine the positive direction for electric current flowing through the curve, and calculate I. For some systems you will need to use $I = \int \vec{j} \cdot \hat{n} \, da$.

Solve — **Step V:** Apply Ampere's law to find B, and use the direction you found in Step I to get $\vec{B}$.

Analyze — The point P is usually typical of an entire region, and your calculation determines $\vec{B}$ as a function of position throughout that region.

If the current distribution divides space into different regions such as within or outside a wire, you will need to repeat steps II – V for each of those regions.

■ **Figure 28.19**

EXERCISE 28.6 ♦ ■ Figure 28.17 shows several long straight wires carrying current perpendicular to the plane of the page. Use Ampère's law to find the circulation around each curve when traversed in the direction indicated.

Sometimes it is useful to express Ampère's law in terms of current density, $\vec{j}$ = current per unit area (§26.2). Imagine a surface S whose edge is the curve C (■ Figure 28.18). The direction of the normal to the surface at any point is chosen according to the right-hand rule, as above. Then we may write Ampère's law as:

$$\oint_C \vec{B} \cdot d\vec{\ell} = \mu_0 I = \mu_0 \int_S \vec{j} \cdot \hat{n} \, dA. \qquad (28.12)$$

The circulation of $\vec{B}$ around C is equal to μ_0 multiplied by the flux of $\vec{j}$ through any surface S spanning the curve C.

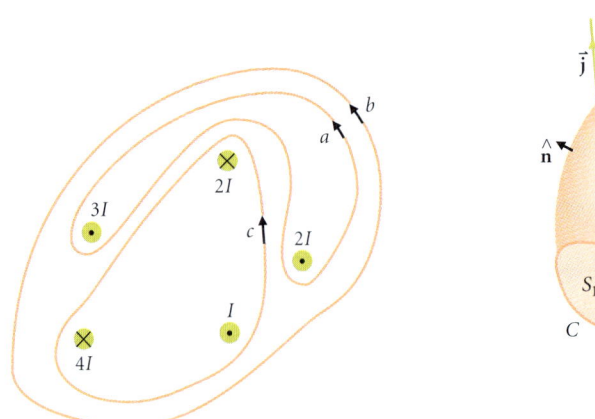

■ **Figure 28.17**

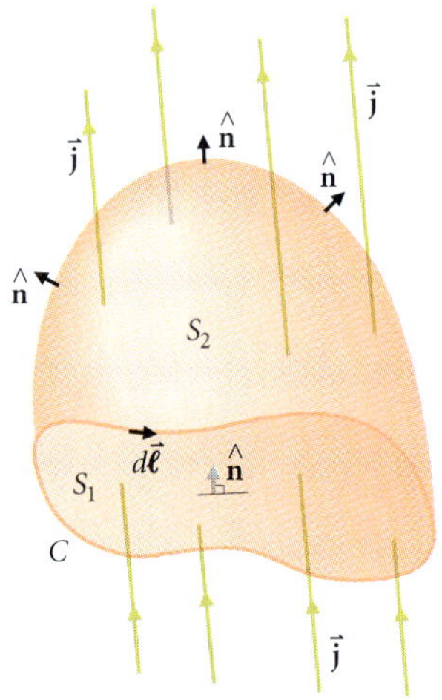

■ **Figure 28.18**
With steady current, the flux of $\vec{j}$ is the same through any surface whose edge is the curve C. Two such surfaces are shown here. This result is akin to Kirchhoff's junction rule: if the fluxes were not the same, charge inside the volume would change with time.

28.3.3 Finding Magnetic Fields with Ampère's Law

When the current distribution is symmetric, Ampère's law often provides the most efficient method for calculating the magnetic field. The method is very similar to that for finding electric field from Gauss' law (■ Figure 28.19).

EXAMPLE 28.7 ♦♦ Use Ampère's law to find the magnetic field produced by the two oppositely directed currents I in a coaxial cable (■ Figure 28.20a).

MODEL — STEP I The cable has cylindrical symmetry. As in the case of the long straight wire, we expect the magnetic field lines to form circles centered on the symmetry axis.

SETUP — STEP II There are two regions to consider: outside the entire system and between the two conductors. For each region we choose as an Ampèrian curve a circle of arbitrary radius r centered on the symmetry axis, so that $\vec{B}$ is parallel to $d\vec{\ell}$ everywhere on the curve.

STEP III The magnitude of $\vec{B}$ is constant everywhere on the curve. Going around the curve counterclockwise as shown in Figure 28.20b, the circulation is:

$$\mathcal{C} \equiv \oint \vec{B} \cdot d\vec{\ell} = \oint B_\theta \, d\ell = B_\theta \oint d\ell = 2\pi r B_\theta. \qquad (28.13)$$

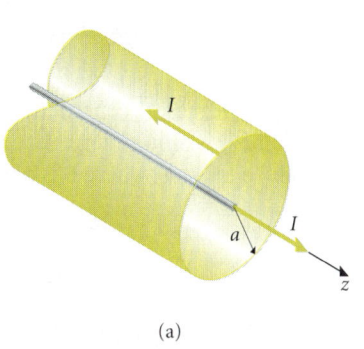
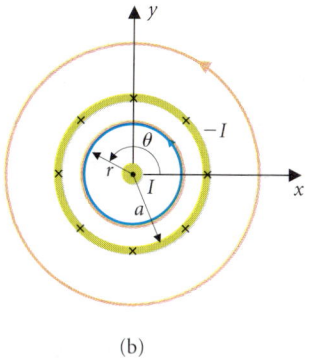

(a) (b)

■ FIGURE 28.20
A coaxial cable carries current in opposite directions in the inner wire and the outer cylindrical conducting sheath. From the symmetry of the current distribution, we determine that $\vec{B}$ has only a θ component. Using Ampère's law, we find that magnetic field is confined to the space between the conductors.

Equation (28.13) gives the circulation in terms of B_θ for a circular curve in any cylindrically symmetric problem.

STEP IV According to the right-hand rule, the positive direction for current is out of the page. The total current through the curve between the conductors is $+I$, while the current through a curve outside the system is $+I - I = 0$.

SOLVE
STEP V Thus for $r_\text{wire} < r < a$:

$$2\pi r B_\theta = \mu_0 I \quad \text{and} \quad \vec{B} = \frac{\mu_0 I}{2\pi r}\hat{\theta},$$

and for $r > a$, $\vec{B} \equiv 0$.

ANALYZE The current in the outer cylinder does not contribute to the magnetic field inside. Outside the cable, the two currents produce magnetic field in opposite directions, and the net field is zero. ∎

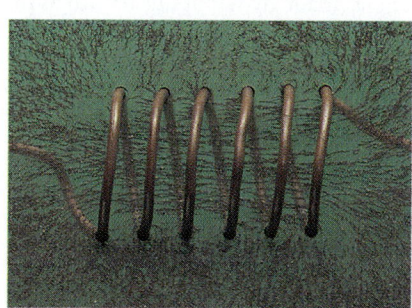
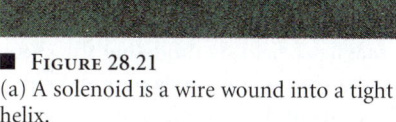
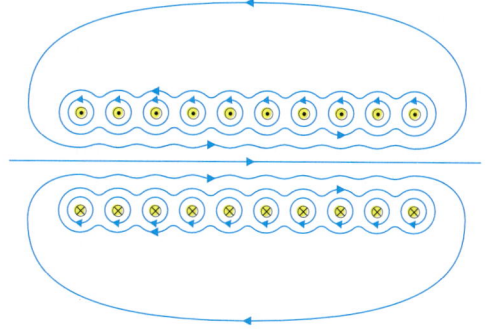
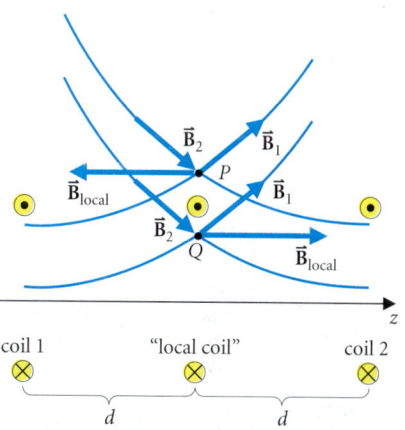

■ FIGURE 28.21
(a) A solenoid is a wire wound into a tight helix.
(b) The field inside the solenoid is almost uniform and parallel to the axis; outside, the field is almost zero.

■ FIGURE 28.22
The magnetic field produced by a solenoid at points P and Q just inside and just outside the coil is the sum of contributions from nearby coils ($\vec{B}_\text{local}$) and from pairs of coils at equal distances on either side of P and Q. One such pair is shown here, along with its contributions $\vec{B}_1$ and $\vec{B}_2$ whose sum is a vector in the $+z$-direction. The field $\vec{B}_\text{local}$ is in the $+z$-direction inside the solenoid at Q but in the $-z$-direction at P outside. Thus at P the local field opposes the field due to distant pairs of coils, and the sum is almost zero. Inside, at Q, the contributions reinforce each other. (The lengths of vectors $\vec{B}_1$ and $\vec{B}_2$ are shown exaggerated for clarity. Many such pairs sum to give $\vec{B}$ approximately equal to $-\vec{B}_\text{local}$ at P.)

A solenoid is a coil in which the wire is wound in a helix (■ Figure 28.21). The field near a loosely wound helix is quite intricate. Very close to the wire, the field lines form circles around the wire in accordance with Oersted's rule, but at distances comparable to the separation between loops, the lines merge into curves surrounding the entire coil. At great distances, the field lines evolve into the familiar dipole pattern. For long, tightly wound coils, the pattern simplifies: within the coil the field is nearly uniform, while just outside the coil it is nearly zero (■ Figure 28.22). Calculation of the field for a tightly wound coil is correspondingly simplified.

EXAMPLE 28.8 ◆◆ An automobile starter uses a solenoid with radius $a = 2.0$ cm, length $L = 10$ cm, and $N = 500$ turns. Estimate the magnetic field inside the solenoid when it carries a current of 50 A. Assume the solenoid is very long and tightly wound.

MODEL
STEP I We model the solenoid as infinite and assume that the magnetic field inside is parallel to the solenoid's axis, which we choose to be the z-axis. The field outside is nearly zero. The number of turns per unit length is $n = N/L$.

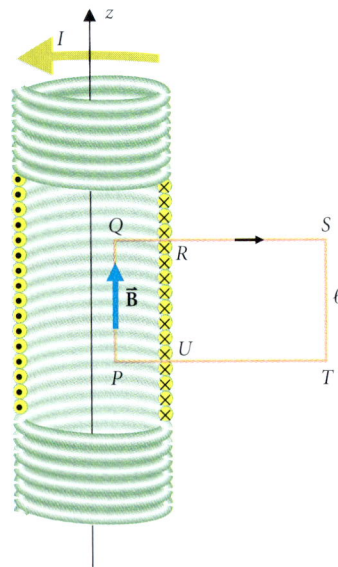

FIGURE 28.23
Inside a long, tightly wound solenoid, the magnetic field is parallel to the axis. Outside the solenoid, the field is nearly zero. To apply Ampère's law, we choose a rectangular curve of side ℓ with PQ parallel to the solenoid axis.

SETUP
STEP II
A rectangle is a suitable curve (■ Figure 28.23). The magnetic field lies along the side parallel to the solenoid's axis, is perpendicular to the two sides that cut the solenoid, and is zero outside.

STEP III We go around the rectangle clockwise. Along QR and UP, $\vec{B}$ is perpendicular to $d\vec{\ell}$, so $\vec{B} \cdot d\vec{\ell} = 0$. Along RS, ST, and TU the magnetic field is almost zero. The only contribution to the circulation comes from side PQ parallel to the axis, where $\vec{B}$ is parallel to $d\vec{\ell}$ and has constant magnitude:

$$\mathcal{C} = \oint \vec{B} \cdot d\vec{\ell} = \int_P^Q \vec{B} \cdot d\vec{\ell} = B_z \ell.$$

STEP IV The positive direction for current is into the page. The number of loops of wire passing through the rectangle is $n\ell$, and each carries current I, so the total current through the rectangle is $In\ell$.

SOLVE Applying Ampère's law gives:
STEP V
$$B_z \ell = \mu_0 n \ell I.$$

Thus:
$$\vec{B} = \mu_0 n I \hat{k}. \qquad (28.14)$$

For the numbers given:
$$B = (4\pi \times 10^{-7} \text{ N/A}^2) \frac{500}{(0.1 \text{ m})} (50 \text{ A}) = 0.3 \text{ T}.$$

ANALYZE The arbitrarily chosen length ℓ of the rectangle does not appear in the final expression for $\vec{B}$, nor does the position of side PQ. Inside the long solenoid, the field has uniform magnitude as well as constant direction. ∎

EXERCISE 28.7 ♦♦ A coil (called a Rowland ring) is made by wrapping 10^4 turns of wire around a torus (■ Figure 28.24). If the wire carries a current $I = 0.10$ A, apply Ampère's law to find the magnetic field along the center line of the torus. ∎

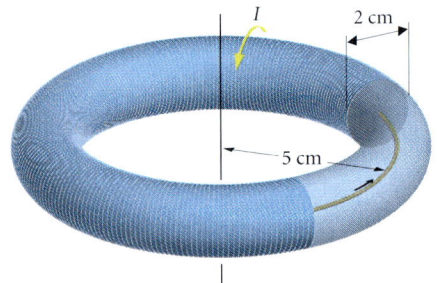

FIGURE 28.24

EXAMPLE 28.9 ♦♦♦ Current is not always confined in cylindrical wires. On semiconductor chips, flat conducting strips, much wider than their thickness, carry current between circuit elements. Current sheets also occur where the solar wind meets the Earth's magnetic field. Find the magnetic field produced by an infinite plane current sheet of thickness s that carries a uniform current density $\vec{j} = j_0 \hat{i}$ (■ Figure 28.25).

MODEL
STEP I
We may determine the direction of the magnetic field by modeling the sheet as a set of parallel wires (■ Figure 28.26). We choose the z-axis perpendicular to the sheet. The field directly above a single "wire" points in the $-y$-direction. "Wires" to the right produce magnetic field vectors with negative y- and z-components, while

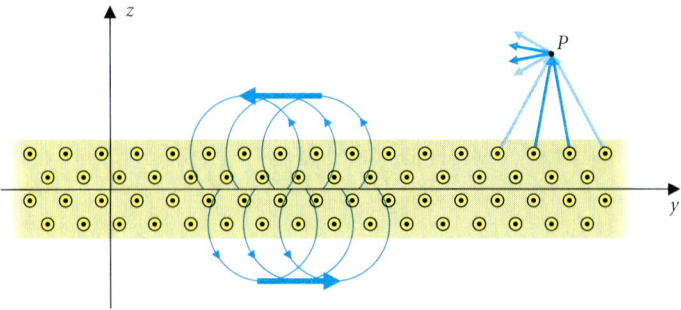

FIGURE 28.26
We model the sheet as an infinite set of parallel wires. Each wire produces circular field lines. The superposition of fields due to different wires is parallel to the sheet and perpendicular to I. To be more precise, at any point P, we choose pairs of wires equidistant from P. The magnetic field due to each pair is in the $-y$-direction. The direction of $\vec{B}$ reverses across the sheet.

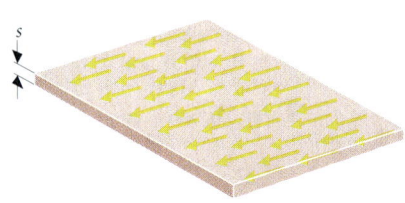

FIGURE 28.25
Current in a conducting sheet.

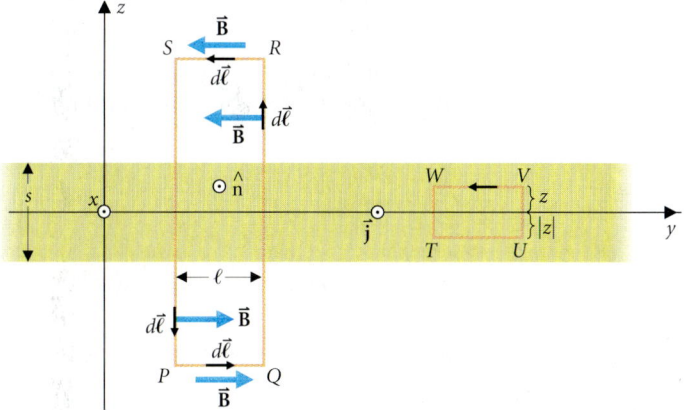

■ FIGURE 28.27
To find the magnetic field due to a uniform current sheet, we choose rectangular curves with two sides parallel to the sheet and the same distance away. The magnetic field is parallel to $d\vec{\ell}$ along PQ, RS, TU, and VW. It is perpendicular to $d\vec{\ell}$ along QR, SP, UV, and WT.

"wires" to the left produce magnetic field vectors with negative y- and positive z-components. In an infinite sheet, every wire on the right has a corresponding wire an equal distance away on the left. The sum of the vectors due to the two wires is in the $-y$-direction. Thus the total magnetic field is in the $-y$-direction. The same argument shows that the magnetic field is in the $+y$-direction below the sheet. At equal distances from the sheet, the field strength is the same both above and below the sheet: B_y (above) $= -B_y$ (below). The same symmetry exists within the sheet.

SETUP
STEP II There are two regions: inside the sheet and outside. In each region we choose as Ampèrian curve a rectangle with two sides parallel to the sheet and at equal distances from the midplane (■ Figure 28.27). These two sides are also parallel to the magnetic field. The other two sides are perpendicular to the sheet and perpendicular to the magnetic field.

STEP III First we find the field outside the sheet. We proceed around the curve PQRS counterclockwise. The term $\vec{B} \cdot d\vec{\ell}$ is zero on the vertical sides. The magnitude of $\vec{B}$ is constant along the horizontal sides, and:

$$\mathscr{C} = \oint \vec{B} \cdot d\vec{\ell} = \int_P^Q \vec{B} \cdot dy\, \hat{y} + \int_R^S \vec{B} \cdot dy\, \hat{y} = B_y(P) \int_P^Q dy + B_y(R) \int_R^S dy$$
$$= \ell[B_y(P) - B_y(R)] = -2\ell B_y(R).$$

HERE WE USE $\hat{y}$ FOR THE UNIT VECTOR IN THE y-DIRECTION, TO AVOID CONFUSION WITH THE CURRENT DENSITY VECTOR $\vec{j}$.

A similar expression gives the circulation around rectangle TUVW in terms of $B_y(V)$.

NOTE: $d\vec{\ell} = dy\, \hat{y}$ IS THE CORRECT EXPRESSION ON BOTH ENDS OF THE LOOP. GOING FROM R TO S, $dy < 0$ AND $d\vec{\ell}$ POINTS LEFT.

STEP IV The direction of $\hat{n}$ is out of the page. The current through rectangle PQRS is:

$$I = \int \vec{j} \cdot \hat{n}\, da = j_0 \int da = j_0 s\ell.$$

SOLVE
STEP V Applying Ampère's law gives:

$$-2\ell B_y(R) = \mu_0 j_0 s\ell \;\Rightarrow\; B_y(R) = -\tfrac{1}{2}\mu_0 j_0 s.$$

$\vec{B} = -\tfrac{1}{2}\mu_0 j_0 s\, \hat{y}$ above the sheet and $\vec{B} = \tfrac{1}{2}\mu_0 j_0 s\, \hat{y}$ below the sheet.

STEP IV Next we find the magnetic field inside the sheet. The current through rectangle TUVW is:

$$I = 2|z|\ell j_0.$$

SOLVE
STEP V Applying Ampère's law gives:

$$-2\ell B_y(V) = 2z\ell j_0.$$
$$\vec{B} = -\mu_0 j_0 z\, \hat{y}.$$

ANALYZE Since z changes sign below the y-axis, one formula works for $\vec{B}$ everywhere inside the sheet. The solutions for the magnetic field inside and outside the sheet give the same answer at the edge of the sheet, where $z = \pm s/2$. The field outside the sheet is constant, independent of the distance from the sheet.

28.3.4 Summary of the Integral Laws for Static Fields

We have now derived flux laws and circulation laws for both electric and magnetic fields. The circulation law for static electric field arises because the Coulomb force is conservative (see Exercise 28.5).

> Flux laws:
> $$\Phi_E = \oint \vec{E} \cdot \hat{n}\, dA = \frac{Q}{\epsilon_0} \quad (23.12) \quad \text{and} \quad \Phi_B = \oint \vec{B} \cdot \hat{n}\, dA = 0 \quad (28.9)$$
> (Gauss)
>
> Circulation laws:
> $$\oint \vec{E} \cdot d\vec{\ell} = 0 \quad (28.15) \quad \text{and} \quad \mathscr{C} = \oint \vec{B} \cdot d\vec{\ell} = \mu_0 I \quad (28.11)$$
> (Ampère)

These relations would be symmetric were it not for the lack of isolated magnetic poles to provide source terms in eqns. (28.9) and (28.15). These equations are equivalent to Coulomb's law and the Biot-Savart law and as such may seem little more than a convenience in a limited class of calculations. However, when fields vary in time (Chapter 30), the circulation laws are profoundly modified, and the four integral laws become our most powerful tool for understanding electromagnetic fields.

Study Problem 18 ♦♦♦ An Electron Beam

A beam of electrons has radius R and contains n electrons per cubic meter moving with velocity $\vec{v}$ along the beam (■ Figure 28.28a). Find the electric and magnetic fields produced by the beam and the resulting force on an electron at the edge of the beam. Does the radius of the beam expand or contract as a result of this force? Assuming the force remains constant, estimate the change in the beam radius during the time the electrons move a distance s along the beam. Is this an important effect in the beam of a TV tube with $s = 15$ cm, $v = 9.4 \times 10^7$ m/s, $R = 0.1$ mm, and $n = 10^{10}$ /m^3?

Modeling the System

A beam that is much longer than its diameter forms a cylindrically symmetric distribution of charge and current. Repulsive electric forces between the electrons tend to disrupt the beam. Electron motion to the right constitutes a current to the left, with magnetic field lines circulating about it (Figure 28.28b). The resulting magnetic force on the electrons is inward. The beam expands if the electric force exceeds the magnetic force. So long as the expansion is slow, the approximations of cylindrical symmetry and static fields remain valid.

Setup

We use Gauss' law and Ampère's law to calculate the fields and then evaluate the force on each electron produced by these fields. We may estimate the beam's expansion by finding the radial displacement of an electron on the edge of the beam in the time s/v that it takes to travel a distance s along the beam.

STEP I The charge distribution has cylindrical symmetry, so the electric field points radially inward (cf. Example 24.2).

STEP II We apply Gauss' law to a cylinder of length ℓ lying along the beam with its outer surface at radius r (■ Figure 28.29).

STEP III The flux emerging from this cylinder is (cf. eqn. 24.3):
$$\Phi_E = 2\pi r \ell E_r.$$

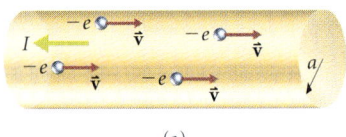
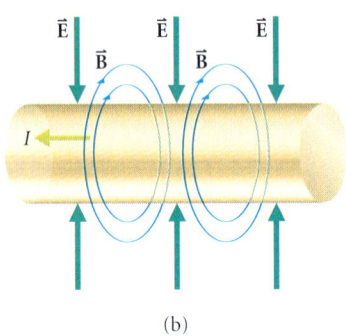

■ **FIGURE 28.28**
(a) An electron beam has n electrons per cubic meter, each traveling with velocity $\vec{v}$ toward the right. The electrons form a current $I = nev$ toward the left. (b) The electric field vector produced by the negatively charged electrons is radially inward. The magnetic field lines form rings surrounding the beam axis, with direction determined by Oersted's right-hand rule.

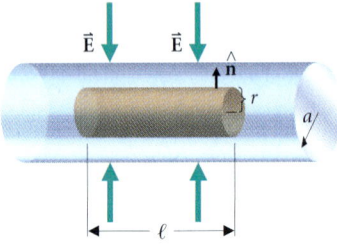

■ **FIGURE 28.29**
We apply Gauss' law to a cylinder of radius $r < a$ and length ℓ whose axis coincides with the beam axis. The normal to the curved surface is opposite $\vec{E}$, so the flux through this Gaussian surface is negative, as is the charge enclosed.

STEP IV The total charge inside the cylinder is:

$$Q = -ne\pi r^2 \ell.$$

Solution

STEP V Applying Gauss' law gives: $\Phi_E = Q/\epsilon_0.$

$$2\pi r \ell E_r = \frac{-ne\pi r^2 \ell}{\epsilon_0} \;\Rightarrow\; \vec{E} = -\frac{ner}{2\epsilon_0}\hat{r}.$$

Setup

STEP I The magnetic field lines are circles around the beam axis.

STEP II We apply Ampère's law to a circle of radius r centered on the beam axis, traversed as shown in ■ Figure 28.30.

STEP III The circulation of $\vec{B}$ around the circle is (cf. eqn. 28.13):

$$\mathscr{C} = 2\pi r B_\theta.$$

STEP IV The positive direction for current is along the axis parallel to $\vec{v}$. The current passing through the circle is:

$$I = -\pi r^2 j = -\pi r^2 nev.$$

Solution

STEP V Applying Ampère's law we obtain: $\mathscr{C} = \mu_0 I.$

$$2\pi r B_\theta = -\mu_0 \pi r^2 nev \;\Rightarrow\; \vec{B} = \tfrac{1}{2}\mu_0 nevr(-\hat{\theta}).$$

The total force acting on an electron is the sum of the electric and magnetic forces (■ Figure 28.31):

$$\vec{F} = -e(\vec{E} + \vec{v} \times \vec{B}) = e\left(\frac{ner}{2\epsilon_0} - \mu_0 \frac{nerv^2}{2}\right)\hat{r}$$

$$= \frac{ne^2 r}{2\epsilon_0}(1 - \mu_0 \epsilon_0 v^2)\hat{r}.$$

The product $\mu_0 \epsilon_0$ has dimensions of (speed)$^{-2}$. The corresponding speed is:

$$\frac{1}{\sqrt{\mu_0 \epsilon_0}} = \frac{1}{\sqrt{(4\pi \times 10^{-7}\text{ N/A}^2)(8.85 \times 10^{-12}\text{ C}^2/\text{N}\cdot\text{m}^2)}}$$

$$= 3 \times 10^8 \text{ m/s},$$

the speed of light! Only if the electrons were moving at the speed of light could the magnetic force equal the electric repulsion. Since $v < c$, the beam expands.

To find the displacement of an electron at the edge of the beam, we evaluate the fields at $r = R$. The displacement we want is:

$$s_\perp \sim \tfrac{1}{2}a(\Delta t)^2 = \frac{F}{2m}\left(\frac{s}{v}\right)^2 = \frac{ne^2 R s^2}{4m\epsilon_0 v^2}(1 - \mu_0 \epsilon_0 v^2).$$

For the TV beam:

$$s_\perp = \frac{(10^{10}\text{ /m}^3)(1.6 \times 10^{-19}\text{ C})^2(10^{-4}\text{ m})(0.15\text{ m})^2}{4(9.11 \times 10^{-31}\text{ kg})(8.85 \times 10^{-12}\text{ C}^2/\text{N}\cdot\text{m}^2)(9.4 \times 10^7\text{ m/s})^2}\left[1 - \left(\frac{9.4}{30}\right)^2\right]$$

$$= 2 \times 10^{-9}\text{ m}.$$

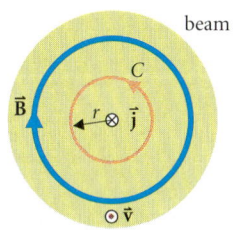

■ FIGURE 28.30
We apply Ampère's law to a circle of radius $r < a$ centered on the beam axis. Going around the circle opposite $\vec{B}$ (counter-clockwise in this end view), $\hat{n}$ is out of the page, while $\vec{j}$ is into the page. The current through the curve is negative.

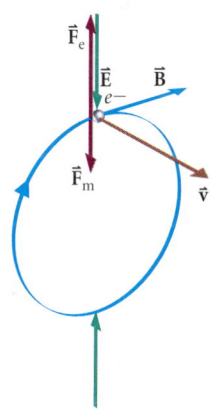

■ FIGURE 28.31
The electric and magnetic forces on an electron in the beam are both radial: $\vec{F}_e$ is outward, and $\vec{F}_m$ is inward.

THIS IS NO ACCIDENT! WE'LL SEE WHY IN CHAPTER 33.

Analysis

The expansion of the beam is 2 parts in 10^5 of the beam radius, and is unimportant. (This justifies our use of constant acceleration in calculating the expansion.) In accelerators where beams travel long distances the expansion is significant, and efforts must be made to refocus the beam. When current is carried in an ionized but neutral beam, only the magnetic force acts, and it can *pinch*, or confine, the beam.

In our everyday experience magnetic forces occur in motors, loudspeakers, or between two bar magnets. These forces seem strong compared with the electric forces we encounter, yet the magnetic force is inherently weaker by the ratio v/c. Electric forces only seem to be weaker because of the very small amounts of unbalanced charge, and correspondingly weak fields, that commonly occur.

Chapter Summary

Where Are We Now?

We have discussed methods for calculating the magnetic field produced by an arbitrary distribution of current. Ampère's law gives a global relationship between the magnetic field and its sources that complements the Biot-Savart law. We have now established four integral laws for static electric and magnetic fields.

What Did We Do?

The magnetic force exerted on a moving charge is $\vec{F} = q\vec{v} \times \vec{B}$, perpendicular to both $\vec{v}$ and $\vec{B}$. The force law gives an operational definition of $\vec{B}$.

The Biot-Savart law describes the magnetic field produced by an element $d\vec{\ell}$ of wire carrying current I:

$$d\vec{B} = I \frac{d\vec{\ell} \times \hat{r}}{r^2}.$$

Using the principle of superposition, the magnetic field due to a finite current-carrying wire can be found by integration. The magnetic field due to a long straight wire is:

$$\vec{B} = \left(\frac{\mu_0 I}{2\pi r}, \quad \text{tangent to a circle centered on the wire} \right). \quad (28.6)$$

At large distances from a current loop, its magnetic field is well approximated by the field of an ideal dipole with magnetic moment $|\vec{m}| = IA$, where A is the area of the loop. The direction of $\vec{m}$ is found from a right-hand rule: curl the fingers of your right hand around the loop in the direction of the current, and $\vec{m}$ lies along your thumb.

The *circulation* of $\vec{B}$ around a curve C,

$$\mathscr{C} = \oint_C \vec{B} \cdot d\vec{\ell},$$

describes the extent to which the field lines follow the direction of the curve. Ampère's law relates the circulation of $\vec{B}$ around a closed curve to the current passing through that curve:

$$\mathscr{C} = \mu_0 I.$$

The positive direction for I is chosen according to a right-hand rule (Figure 28.15). Ampère's law can be used to calculate fields in some important symmetric cases. A solution plan is given in Figure 28.19.

The magnetic field inside a long solenoid, $B = \mu_0 n I$ (eqn. 28.14), is nearly uniform, while the field just outside the solenoid is nearly zero.

Practical Applications

Magnetic fields have numerous practical applications, many of which will be discussed in the next chapter. To design magnetic field structures such as those used for plasma confinement, engineers use both carefully constructed iron pole pieces and electric current in loops and coils of various shapes. In this chapter we have seen how the current path shapes the magnetic field. The current meter mentioned in §28.3.2 is a direct practical application of Ampère's law. Most applications are more subtle, and many require the dynamic relations we shall develop in Chapter 30. Ampère's law is necessary for understanding the behavior of magnetic fields at the boundaries between two different media.

Solutions to Exercises

28.1. See ■ Figure 28.32. Remember that when the charge is negative, the magnetic force is *opposite* $\vec{v} \times \vec{B}$.

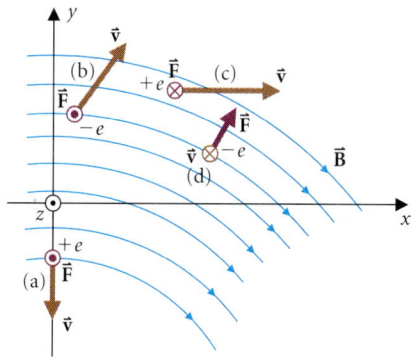

■ FIGURE 28.32
The direction of each magnetic force $\vec{F}_m = q\vec{v} \times \vec{B}$ is determined by the right-hand rule for cross products. The resulting vectors are shown here.

28.2. The current is the charge flowing through a cross section (area A) of the wire per second: $I = nev_d A$. Thus

$$I\,d\vec{\ell} = nev_d A\,d\vec{\ell} = e\vec{v}_d(n\,dV),$$

where $n\,dV$ is the total number of electrons in the element of wire. Electrons moving with velocity $\vec{v}_d$ produce current in the opposite direction:

$$I\,d\vec{\ell} = -e\vec{v}_d(n\,dV).$$

28.3. The magnitude of the field produced by an element of wire $d\vec{\ell}$ is proportional to the magnitude of the cross product $|d\vec{\ell} \times \hat{r}| = d\ell \sin\theta$, where θ is the angle between $d\vec{\ell}$ and $\hat{r}$. Along the line of $d\vec{\ell}$, $\theta = 0$ and $|d\vec{B}| = 0$. The maximum value of $|d\vec{B}|$ occurs for $\theta = 90°$.

28.4. The magnetic field at P (■ Figure 28.33) is the sum of two contributions, one from each wire. According to Oersted's rule, each contribution is into the page. Thus from eqn. (28.6), the net field is:

$$\vec{B} = 2\left(\frac{\mu_0 I}{2\pi(\ell/2)}\right)\hat{\otimes} = \frac{2\mu_0 I}{\pi\ell}\hat{\otimes}.$$

28.5. The electric potential $\int \vec{E} \cdot d\vec{\ell}$ between any two points A and B is independent of the path used to evaluate the integral. Thus the change in potential around any closed curve (which equals the circulation of $\vec{E}$ around the curve) may be written:

$$\mathscr{C}_{SE} = \oint \vec{E}\cdot d\vec{\ell} = \int_A^B \vec{E}\cdot d\vec{\ell} + \int_B^A \vec{E}\cdot d\vec{\ell}$$

$$= \int_A^B \vec{E}\cdot d\vec{\ell} - \int_A^B \vec{E}\cdot d\vec{\ell} = 0.$$

The circulation of a static electric field around any closed curve is zero.

28.6. All three curves are traversed counterclockwise, so the positive direction for current is out of the page. All we need do to find the circulations is to compute the current through each curve.

Curve	Net Current	Circulation
a	$3I + 2I = 5I$	$5\mu_0 I$
b	$3I + 2I + I - 2I - 4I = 0$	zero
c	$I - 2I - 4I = -5I$	$-5\mu_0 I$

28.7. The magnetic field inside a solenoid is parallel to its axis. The Rowland ring is a solenoid bent into a circle, and we expect the magnetic field lines to take the form of circles threading the ring. Choosing the circle along the center line as our Ampèrian curve, and going around it counterclockwise, we find that the circulation is $2\pi R B_\theta$, where $R = 5$ cm. The current through the curve is due to all 10^4 turns carrying 0.1 A each. Its direction is downward, so the net current is -10^3 A. Applying Ampère's law, $\oint \vec{B}\cdot d\vec{\ell} = \mu_0 I$:

$$2\pi(0.05\text{ m})B_\theta = -\mu_0(10^3\text{ A}).$$

$$B_\theta = -\frac{(4\pi \times 10^{-7}\text{ N/A}^2)(10^3\text{ A})}{2\pi(0.05\text{ m})} = -4 \times 10^{-3}\text{ T}.$$

The direction of $\vec{B}$ is around the circle clockwise.

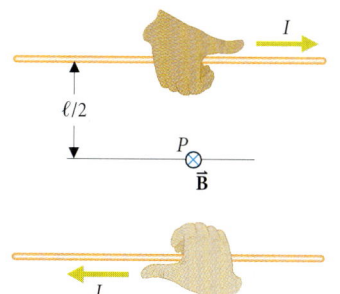

■ FIGURE 28.33

Basic Skills

Review Questions

§28.1 MAGNETIC FORCE

- What magnetic force is exerted on a point charge at rest?
- Describe how the magnetic force exerted on a point charge depends on its velocity.
- If the magnetic field strength is doubled, by how much does the force exerted on a point charge change?
- How could you determine the direction of $\vec{B}$ by measuring the force exerted on a point charge?
- What is the formal name for $\vec{B}$? What is its SI unit?

§28.2 CURRENT AS THE SOURCE OF MAGNETIC FIELD

- What did Oersted discover about the magnetic field produced by a long straight wire carrying current?
- State the Biot-Savart law.
- When calculating static magnetic field, why do we not begin with the field due to a point charge?
- How is the Biot-Savart law similar to Coulomb's law? How does it differ?
- What is the principle of superposition for magnetic fields? Why is the integral in eqn. (28.4) written as $\oint$?
- What magnetic field is produced by a straight wire segment of length ℓ carrying current I? How do you determine the angles that appear in this formula?
- What is the *magnetic moment* of a current loop? Under what circumstances is it useful in determining the field produced by the loop?
- What are *Helmholtz coils*?

§28.3 INTEGRAL LAWS FOR STATIC MAGNETIC FIELDS

- State Gauss' law for the magnetic field.
- Define the *circulation* of magnetic field around a curve C.
- State Ampère's law for the static magnetic field.
- Explain how to assign a positive direction for current when using Ampère's law.
- What is the magnetic field inside a solenoid with n turns per unit length?
- What two forces act on charged particles moving in a beam with velocity $\vec{v}$? Which force is greater? Why?

Basic Skill Drill

§28.1 MAGNETIC FORCE

1. (■ Figure 28.34) What are the directions of the forces acting on the electrons at C and at D?

2. A particle with charge $q = 15$ nC is moving along the x-axis at a speed of 250 m/s. The magnetic field in the region is 3.5×10^{-3} T in the y-direction. What is the force on the particle?

3. The magnetic field in a region of space is in the x-direction and has magnitude 0.48 T. A particle with a charge of 0.65 μC is moving

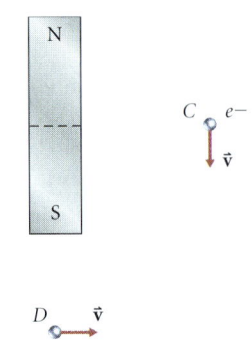

■ FIGURE 28.34

at 375 m/s. For which direction(s) of the particle's velocity will the magnetic force on it be greatest? What is the maximum force?

§28.2 CURRENT AS THE SOURCE OF MAGNETIC FIELD

4. A long straight wire carries a current of 15 A. What is the magnetic field 35 cm from the wire?

5. Point P is a perpendicular distance 5.0 cm from the center of a 17-cm-long straight wire segment carrying a current of 1.0 mA. What is the magnetic field at P produced by the wire?

6. You wish to design a simple coil of radius 0.100 m that produces a 6.00×10^{-3}-T magnetic field at its center. If the coil is to carry a current of 1.00 A, how many turns of wire are necessary?

7. Find the magnetic moment of a triangular wire loop with three equal sides 3.5 cm long carrying a 7.7-A current.

§28.3 INTEGRAL LAWS FOR STATIC MAGNETIC FIELDS

8. What is the magnetic flux through the surface of the Earth?

9. Three wires carry currents as shown in ■ Figure 28.35. Find the circulation of $\vec{B}$ around each of the curves C_1, C_2, and C_3.

10. What is the magnetic field on the axis of a solenoid with 25 turns per centimeter carrying a 15-A current?

11. Find the magnetic field at $r = 0.75$ cm between the conductors of a coaxial cable carrying a current of 5.5 mA. What is the magnetic field outside the cable?

12. The electron beam at the Stanford Linear Accelerator Center (SLAC) has a diameter of 1.9 cm and carries an average current of 25 μA. What is the magnetic field at the edge of the beam? Assume the electrons move at speed $c = 3 \times 10^8$ m/s, and estimate the electron density in the beam. (Ignore relativistic effects.)

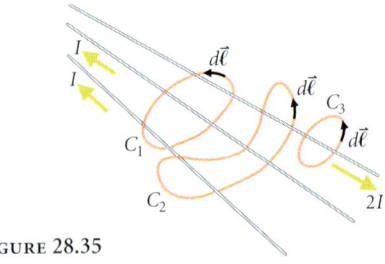

■ FIGURE 28.35

Questions and Problems

§28.1 MAGNETIC FORCE

13. ❖ A charged particle is moving in the x-direction. A bar magnet is brought up with its north pole pointing in the y-direction. The particle accelerates in the −z-direction. What is the sign of the charge on the particle?

14. ❖ An electron moving in the −x-direction experiences no magnetic force. A second electron moving in the +y-direction accelerates in the +z-direction. What is the direction of $\vec{B}$?

15. ❖ A charged particle moving to the right is deflected upward. How could you determine whether the force on the particle is electric or magnetic?

16. ♦ From the operational definition of magnetic field strength, prove the following units relations:

$$1\ T = 1\ V \cdot s/m^2$$
$$= 1\ \Omega \cdot C/m^2$$
$$= 1\ kg/C \cdot s.$$

17. ♦ The space shuttle develops a net charge Q of approximately 2 μC during an experiment on plasma beams. The shuttle's speed is 10 km/s, and the Earth's magnetic field B_E at the orbital radius is approximately 3×10^{-5} T. Estimate the magnetic force on the shuttle. Does this effect produce a problem for NASA?

18. ♦ The magnetic field near the surface of a neutron star is about 10^8 T. Estimate the magnetic force on an electron moving at 10^7 m/s perpendicular to $\vec{B}$.

19. ♦ A particle with a charge of −27 nC moves in a region where the magnetic field is 0.80 T$\hat{k}$. Find the force acting on the particle when its velocity is:
(a) 3500 m/s $\hat{i}$.
(b) 2.7×10^4 m/s $\hat{j}$.
(c) $(1.0 \times 10^5\ m/s)(\hat{i} + \hat{k})$.

20. ♦ The magnetic field $\vec{B}$ is a vector pointing into the page with magnitude 0.050 T. Find the magnitude and direction of the magnetic force acting on:
(a) an electron traveling to the right at $v = 3 \times 10^7$ m/s.
(b) a proton traveling down at $v = 6 \times 10^7$ m/s.

21. ♦ A magnetic field $\vec{B}$ has magnitude 0.057 T and direction 30.0° upward from the horizontal. An iron nucleus with charge 26e is moving horizontally to the right at 1800 m/s. What is the magnetic force exerted on the nucleus?

22. ♦♦ A charged particle with q = 16 μC moving at 175 m/s experiences a maximum force of 17×10^{-3} N in the +z-direction when it moves at 45° to the +x- and +y-axes. What is the magnitude of the magnetic field? Can you determine its direction? What force does the particle experience when moving along the y-axis?

§28.2 CURRENT AS THE SOURCE OF MAGNETIC FIELD

23. ❖ Which produces greater magnetic field at the center of a loop with radius R: the loop itself carrying current I or a straight wire segment of length $2\pi R$, tangent to the loop and carrying the same current I? Explain your reasoning, arguing directly from the Biot-Savart law.

24. ❖ The points a, b, and c in ■ Figure 28.36 lie in the same plane as the rectangular current loop. At which point is the magnetic field the greatest? Explain your reasoning.

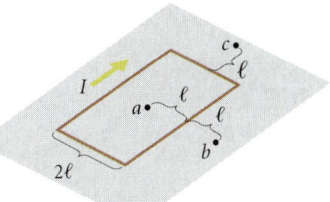

■ FIGURE 28.36

25. ❖ You have a given length of wire and wish to make a device with the largest possible magnetic moment.
(a) What shape loop would you use? Explain why.
(b) Would you wrap the wire in a coil or use a single turn? Explain why.

26. ❖ (a) Two parallel long straight wires carry current in the same direction. Draw a diagram showing the pattern of magnetic field lines in a plane perpendicular to the two wires.
(b) Draw a diagram of the field lines if the wires carry current in opposite directions.

27. ♦ An electric power transmission line is 5 m above ground and carries a current of 50 A. What magnetic field does it produce on the ground 10 m from a point directly below the line?

28. ♦ You are using jumper cables to help a friend start his car. The current in the cable is 66 A. What is the magnetic field strength 15 cm from a cable, if it is approximately straight?

29. ♦ If the two jumper cables in Problem 28 form a circle of radius 0.45 m, what is the magnetic field 15 cm above the center of the circle?

30. ♦♦ Two loops of wire with radii R_2 = 12 cm and R_1 = 9.0 cm are placed concentrically in the x-y-plane (■ Figure 28.37). The outer loop carries current I_2 = 15 A, and the inner loop carries current I_1 = 12 A. What is the magnetic field at the common center of the loops?

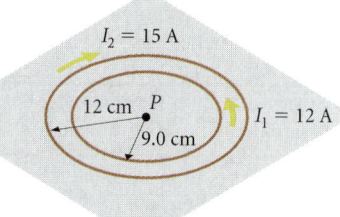

■ FIGURE 28.37

31. ♦♦ A straight wire segment 17 cm long lies along the positive x-axis with one end at the origin. It carries a current I = 1.0 mA in the positive x-direction. Find the magnetic field contributed by this segment at the point P with coordinates x = 8.5 cm, y = −5.0 cm, z = 0.

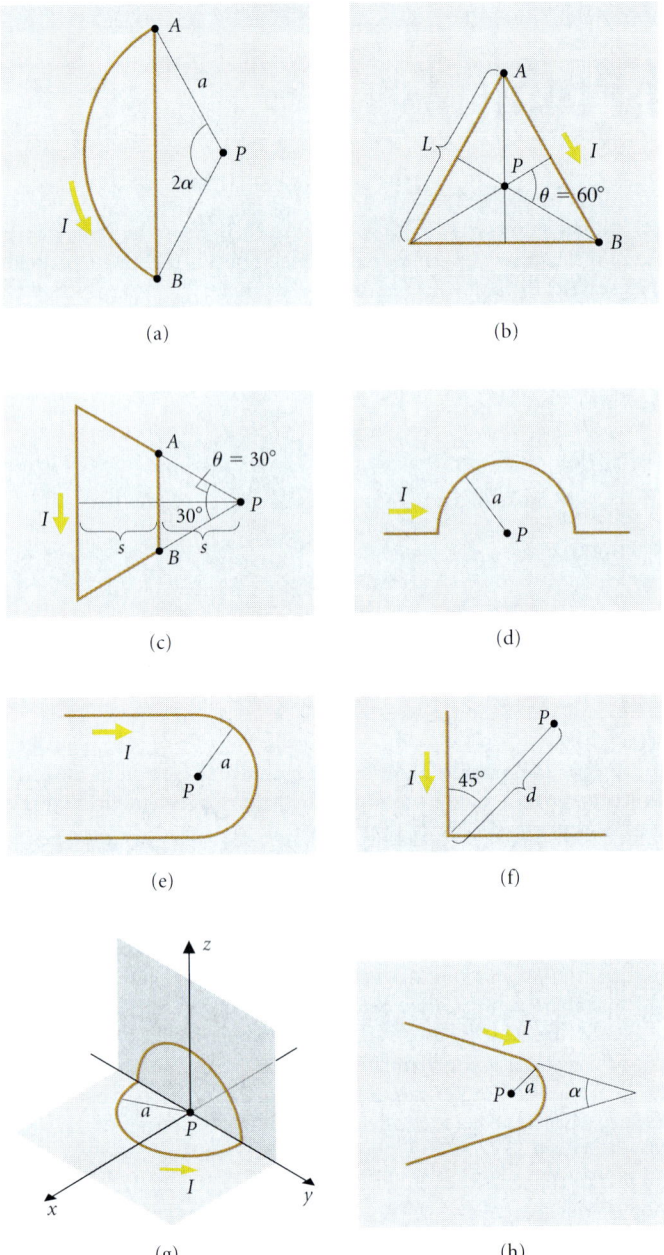

FIGURE 28.38

32. ♦♦ ■ Figure 28.38 shows several wires, each carrying current I. In each case, find the magnetic field at point P in the diagram.

33. ♦♦ Two parallel wires are 12 cm apart. Each carries a 25-A current in the same direction. Find the magnetic field midway between the wires and outside the wires at a point in the same plane 6.0 cm from one wire.

34. ♦♦ Assume that the magnetic moment of a finite, tightly wound solenoid equals the number of loops in the solenoid multiplied by the magnetic moment of each loop. Show that the magnetic moment is also given by the formula:

$$m \approx \frac{BV}{\mu_0},$$

where B is the magnetic field and V is the volume inside the solenoid.

35. ♦♦ A thin, uniformly charged ring of radius $a = 0.10$ m carries a total electric charge $Q = 3.3$ μC and rotates with angular speed $\omega = 120$ rad/s about its symmetry axis. Find the magnetic field it produces at its center.

36. ♦♦ A wire loop in the shape of a semicircle of radius 7.5 cm carries a 15-A current. Find: **(a)** the magnetic moment of the loop. **(b)** the magnetic field strength 1.0 m directly above the loop. **(c)** the magnetic field strength in the plane of the loop 1.0 m from the center of the straight side.

37. ♦♦ A square loop of side 3.50 cm carries a 4.77-A current. Find the magnetic field 7.00 cm above the center of the square. Compare your result with the approximate result using the magnetic moment of the loop and the dipole field. How large is the error in the approximate result?

38. ♦♦ Two solenoids are used to produce magnetic field across a cathode-ray tube (■ Figure 28.39). Each coil has 1.0×10^3 turns, is 10.0 cm long, has a radius of 5.0 cm, and carries a 1.1-A current. The CRT is 8.0 cm in diameter. **(a)** Use eqn. (28.14) for the magnetic field inside a solenoid to calculate the magnetic field applied to the CRT. **(b)** Use the result of Problem 41 to calculate the magnetic field at the center of the CRT and at one edge. By how much does each value differ from the result of (a)? By how much do the values differ from each other?

39. ♦♦♦ Helmholtz coils are often used to produce a uniform field in the laboratory. Each of two parallel circular coils of radius R and separated by a distance R carries a current I in the same sense (Figure 28.10). Find the magnetic field at point P and show that $\partial B/\partial z$, $\partial^2 B/\partial z^2$, and $\partial^3 B/\partial z^3$ are all zero at P.

40. ♦♦♦ Find the magnetic field on the axis of a disk of radius $R = 5.0$ cm carrying a uniform charge density $\sigma = 1.0$ μC/m² and spinning at $\omega = 12$ rad/s, at distance $d = 3.0$ cm from the disk.

41. ♦♦♦ Show that the magnetic field on the axis of a finite, tightly wound solenoid (■ Figure 28.40) is given by:

$$\vec{B} = 2\pi k_m nI(\cos \theta_1 + \cos \theta_2)\hat{k}.$$

(*Hint:* Consider each element dz of the solenoid to be a circular coil with $n\,dz$ turns, and use eqn. (28.7). Be careful with the limits of integration.) What error is made in Example 28.8 by assuming the solenoid's length to be infinite?

42. ♦♦♦ Use the result of Problem 41 to find how the magnetic field of a finite, tightly wound solenoid varies at large distances along the axis of the solenoid. From the result deduce the magnetic moment of the solenoid. Compare your result with the number of loops in the solenoid multiplied by the magnetic moment of each loop. (*Hint:* Use the binomial theorem to expand the square roots in your expressions for $\cos \theta_1$ and $\cos \theta_2$.)

43. ♦♦♦ A charged ring with $Q = 3.3$ μC and $a = 0.10$ m (cf. Problem 35) is rotating with $\omega = 120$ rad/s about an axis in the plane of

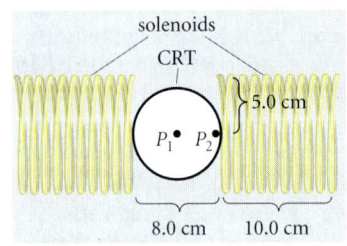

■ **FIGURE 28.39**

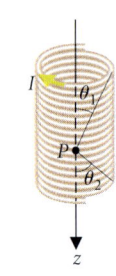

■ **FIGURE 28.40**

the ring. Find the magnetic field it produces at points on the rotation axis a large distance from the ring.

44. ♦♦♦ A thin, uniformly charged spherical shell of radius $a = 0.10$ m carries a total charge $Q = 3.3$ μC and rotates with angular speed $\omega = 120$ rad/s. Find the magnetic field it produces at points on its rotation axis a large distance from the shell.

§28.3 INTEGRAL LAWS FOR STATIC MAGNETIC FIELDS

45. ❖ A wire carrying current I runs along the axis of a cylindrical box of radius r. What is the magnetic flux through the box?

46. ❖ ■ Figure 28.41 shows a uniform magnetic field between the pole faces of an electromagnet, with zero field elsewhere. Apply Ampère's law to the orange rectangle to prove that the field cannot go abruptly to zero as in the figure.

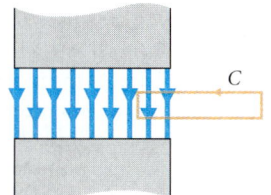

■ **FIGURE 28.41**

47. ♦♦ A cylinder of radius r has 100 wires closely spaced around its circumference, each running the length of the cylinder and carrying current I. Find the magnetic field: **(a)** inside the cylinder. **(b)** outside the cylinder. In each case consider points at a distance greater than $r/100$ from the surface of the cylinder. What is the purpose of this restriction?

48. ♦♦ A wire of radius 1.5 mm carries a current of 0.22 A. The current density within the wire is uniform. Find the magnetic field inside and outside the wire, and plot $|\vec{B}|$ as a function of radius.

49. ♦♦ A conducting cylinder of radius b with a hole of radius a bored along its length (■ Figure 28.42) carries a uniform current density $\vec{j}$. First use Ampère's law to find an expression for the magnetic field within a wire with uniform current density, then use superposition to show that $\vec{B}$ is constant within the hole.

50. ♦♦♦ A long conducting cylindrical shell carries a uniform current density around its circumference (■ Figure 28.43). Apply Ampère's law to find the resulting magnetic field. (*Hint*: The shell is similar to a solenoid.)

51. ♦♦♦ The current density within a long wire of radius a varies as $j(r) = j_0(r/a)$. Find the magnetic field inside and outside the wire. Plot $B(r)$ as a function of r.

52. ♦♦♦ Uniform surface current in an infinite plane sheet divides evenly between the two halves of a long cylindrical shell then reunites in a second plane sheet diametrically opposite the first (■ Figure 28.44).

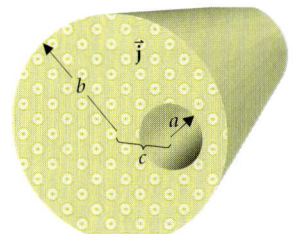

■ **FIGURE 28.42**

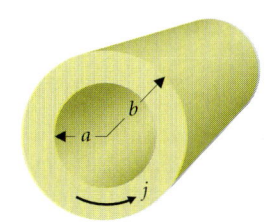

■ **FIGURE 28.43**

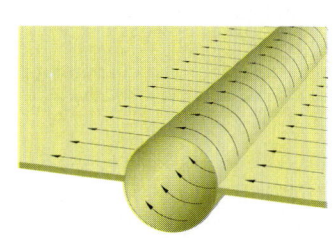

■ **FIGURE 28.44**

■ **FIGURE 28.45**

The surface current density in the sheets is K. (K is the product of the current density in each sheet times the very small thickness of the sheet.) **(a)** Describe the symmetry properties of this system and the conclusions you may draw about the symmetry of the magnetic field. **(b)** Determine Ampèrian curves you might use to check each of your conclusions about the field symmetry. **(c)** Show that the magnetic field is zero at any point inside the cylinder. **(d)** Find the magnetic field outside the cylinder.

53. ♦♦♦ **(a)** Each of two long straight wires carries current $I/2$ to an electrode at the center of an infinite conducting sheet (■ Figure 28.45). The current spreads out uniformly along radial lines in the sheet. **(a)** Describe the symmetry of the system. What conclusions can you draw about the symmetry of the magnetic field? **(b)** Describe a curve that allows you to use Ampère's law to find the difference in magnetic induction on opposite sides of the sheet. Does your result support your description of the field in part (a)? **(c)** Further verify your result using curves passing around the wires. **(d)** If the current is delivered to the electrode by a single wire carrying current I, what is the magnetic field on the side of the sheet without the wire? On the side with the wire?

54. ♦♦ A model for a jet that might power an extragalactic radio source is a beam 200 light years in radius carrying a current of 10^{15} A. (A light year is approximately 10^{16} m). **(a)** What is the magnetic field at the edge of the jet? **(b)** If the current were provided by electrons moving at one-tenth the speed of light, what would be the electron density in the jet? **(c)** What is the radial acceleration of an electron at the edge of the beam? If the beam travels 3×10^5 light years, is this a reasonable model for the jet? (*Hint*: Assume that the acceleration is constant, and compare the radial displacement of the electron with the beam radius.)

Additional Problems

55. ❖ A bar magnet is supported at its center of mass by a string so that it may rotate freely. Will it lie horizontally in equilibrium? If not, what might you do to make it lie horizontally?

56. ❖ Two bar magnets are suspended on strings. When the suspension points are close together, the ends of the magnets nearest each other are painted red. When the suspension points are moved far apart, the red end of one magnet points north. In what direction does the red end of the other magnet point?

57. ❖ Is there anywhere on Earth where a compass points toward geographical south?

58. ❖ Ship's compasses are *compensated* by placing magnets near the compass. For what are these magnets compensating?

59. ❖ An electron is moving radially away from a current-carrying wire. Describe the force on the electron.

60. ❖ A positively charged particle is moving parallel to a long straight wire, in the direction opposite the current. Describe the force on the particle.

61. ❖ A charged particle passes through the center of a current loop with velocity perpendicular to the plane of the loop. What force acts on the particle?

62. ❖ In what direction must an electron move near a long straight current-carrying wire to experience no magnetic force? Is this possible?

63. ❖ An electron passing through the plane of a current loop is momentarily unaccelerated. What may you conclude about the direction of the electron's velocity?

64. ❖ Use the fact that the circulation of a static electric field around any closed curve is zero (cf. Exercise 28.5) to show that the electric field in a parallel plate capacitor cannot cut off abruptly at the edge of the plates.

65. ❖ While it is turning, an airplane is banked at an angle to the horizontal. Explain why the airplane's compass reads incorrectly during the turn. Does the error depend on the direction of the turn, on the direction of flight, or on the hemisphere of the Earth where the plane is flying?

66. ❖❖ In a Van de Graaff electric generator, a belt carries a charge density σ and moves with speed v. Find $\vec{E}$ and $\vec{B}$ near the surface of the belt.

67. ❖❖ Two plane current sheets are separated by a distance d. Each carries a surface current density K. Find the magnetic field everywhere if the two current densities are (a) parallel, (b) antiparallel, and (c) at right angles.

68. ❖❖ Three wires in a circuit form a Y-shaped junction (■ Figure 28.46). How are the currents in the wires related? If $I_1 = I/3$, find the magnetic field at points A, B, and C, each distance d from the junction.

69. ❖❖ ✉ The magnetic field at the Earth's equator ($R = 6400$ km) is 3×10^{-5} T. What is the Earth's magnetic moment? Modeling the current source as a circular loop of radius $R/2$, calculate how large a current would be required to produce the Earth's magnetic field.

70. ❖❖❖ A wire loop is in the form of an ellipse with semimajor axis a and semiminor axis b. Find the magnetic field at a focus, if a current I flows in the loop.

71. ❖❖❖ An arbitrarily shaped current loop lies in the x-y-plane. First, explain why the term $d\vec{\ell} \times (z\hat{k})$ that we ignored while Digging Deeper integrates to zero. Second, show that for points on the z-axis, $r = \sqrt{z^2 + r_\perp^2}$, so that the approximation $r \approx z$ neglects terms of order $r_\perp^2/z^2$ compared with 1.

72. ❖❖❖ A conducting slab on a silicon chip is 50 μm wide and $a =$ 1.5 μm thick. It carries a current density

$$\vec{j} = (1.0 \times 10^3 \text{ A/m}^2)(z/a)\hat{i}$$

in the region $0 < z < a$. (a) Modeling the slab as infinitely long and wide, use Ampère's law to find the magnetic field inside and outside the slab. (b) Model the slab as a set of long straight wires each having width dy and carrying current $dI = dy \int j\, dz = dy(0.75 \times 10^{-3} \text{ A/m})$ in the x-direction. Superpose the contributions of these wires to find the field at a point on the z-axis a distance d above the center of the slab. Estimate the error made by using Ampère's law to compute the field.

Computer Problems

73. Two Helmholtz coils are parallel to the y-z-plane (Figure 28.10) and each has a 12.00-cm radius. One has its center at the origin, and the other is centered on the x-axis at $x = 12.00$ cm. Each carries a 7.750-A current. Calculate the magnetic field on the x-axis at $x = 0$, $x = 2.00$ cm, $x = 4.00$ cm, and $x = 6.00$ cm. Use your results to plot the field along the axis as a function of position. (See also Problem 39).

74. The magnetic field on the axis of a finite solenoid with a radius a and $n = N/L$ turns per unit length is $B = \mu_0 nI(\cos\theta_1 + \cos\theta_2)/2$. (See Problem 41, Figure 28.40) Express the magnetic field as $B = B_\infty f(z/L)$, where B_∞ is the result for an infinite solenoid. Taking $a = 2.00$ cm and $L = 10.00$ cm, compute values of $B/B_\infty = f(z/L)$ for values of z/L between 0 and 10.00, and plot the results. Compare your answer with the result for an infinite solenoid. Outside the solenoid, its field may be approximated by a dipole field with magnetic moment $m = NIA$, where $A = \pi a^2$ (cf. Problems 34 and 42). Compare your result for B with the dipole field. At what values of z/L is the dipole approximation better than 5%?

Challenge Problems

75. A circular wire loop of radius a and carrying current I lies in the x-y-plane with its center at the origin. Find the magnetic field at a point slightly off axis, with coordinates $(x, 0, z)$, $x \ll z$. (a) Use the Biot-Savart law to express $\vec{B}$ as an integral. (b) Use the binomial theorem to expand the $-\frac{3}{2}$ power in the integrand. Assume that x/z is small and neglect terms with powers higher than $(x/z)^1$. (c) Perform the integration, and show that your result is consistent with the dipole formula $\vec{B} = (k_m m/r^3)(\hat{\theta}\sin\theta + 2\hat{r}\cos\theta)$ for $\theta \ll 1$ and $z \gg a$.

76. Use the Biot-Savart law to write an integral that gives the magnetic field due to a circular current loop of radius a at a point in the plane of the loop a distance d from the axis. (a) Assume $d/a \ll 1$, and expand the integrand in powers of d/a, keeping terms through d^2/a^2. Perform the integration to find $\vec{B}$. (b) Repeat part (a) assuming $d/a \gg 1$.

77. An infinitely long wire in the shape of a spiral is described by the formula:

$$\vec{r}(\theta) = \frac{p\theta}{2\pi}\hat{k} + R(\cos\theta\,\hat{i} + \sin\theta\,\hat{j}).$$

If the wire carries a current I flowing toward increasing θ, show that the z-component of the magnetic field at the origin is:

$$B_z(\text{origin}) = \frac{\mu_0 I}{p}.$$

Compare your result with eqn. (28.14) for a tightly wound solenoid.

78. A current I flows in a wire along a diameter of a thin conducting spherical shell of radius R (■ Figure 28.47). The current then flows back to the beginning of the wire through the shell. The current density in the shell is uniformly distributed in angle. Describe the magnetic field produced by the system and find $\vec{B}$ everywhere.

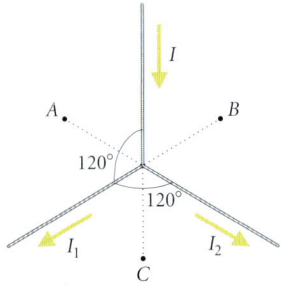

■ FIGURE 28.46

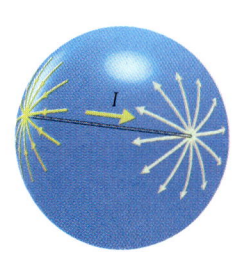

■ FIGURE 28.47

Charged solar wind particles are deflected around the Earth by its magnetic field. Some particles are captured and move in spiral paths along the magnetic field lines. Near the magnetic poles, the particles reach the Earth's upper atmosphere. Particles colliding with air molecules are responsible for the brilliantly colored displays of light called the aurora australis, or southern lights. A similar phenomenon occurs at the North pole.

CHAPTER 29
Static Magnetic Fields: Applications

Concepts
Larmor radius
Cyclotron frequency
Magnetic mirror
Lorentz force
Hall effect
Paramagnetism
Diamagnetism
Ferromagnetism

Goals
Be able to:
Determine the path of a particle moving in a magnetic field.

Compute force and torque exerted by magnetic fields on current-carrying wires.

Understand how atomic behavior results in macroscopic magnetic properties of materials.

*She that had all Magnetique force alone,
To draw, and fasten sundred parts in one;*

JOHN DONNE, THE FIRST ANNIVERSARY

Our lives are more pleasant because our planet is magnetic. The Sun continually spews out streams of charged particles that would endanger life on the surface of an unprotected planet. The Earth's magnetic field shields us by channeling the flow of particles away from the Earth's surface. When the Sun is particularly active, solar wind particles pose a radiation hazard for astronauts in the shuttle or space station. The aurora is a dramatic result of energetic particles that descend to the upper atmosphere.

Magnetic fields are used extensively in modern society. Magnetic forces help generate the sound from your stereo and start your car. Should you fall seriously ill, magnetic fields help doctors to see inside your body without surgery. A machine in which electrons orbit in a magnetic field can produce x-ray beams for fabricating microcomputer circuits. By accelerating carbon atoms in a similar machine, the different isotopes can be separated to allow dating of historical samples. The high-energy plasmas used in fusion research are confined by carefully designed magnetic fields.

Magnetic field acts on and is produced by electric charge in motion. In this chapter we'll study magnetic forces and look at some of their technological applications.

29.1 Motion of Charged Particles in a Magnetic Field

In Chapter 28 we observed that the force acting on a particle with charge q moving with velocity $\vec{v}$ in a magnetic field $\vec{B}$ is (eqn. 28.1):

$$\vec{F} = q\,\vec{v} \times \vec{B}.$$

The force is always perpendicular to the particle's velocity and so does no work on the particle:

$$dW = \vec{F} \cdot d\vec{s} = \vec{F} \cdot \vec{v}\, dt = 0, \quad \text{since } \vec{F} \text{ is perpendicular to } \vec{v}.$$

Magnetic forces can alter the direction of a particle's motion but cannot alter its energy. Since the force is also perpendicular to $\vec{B}$, it cannot change a particle's velocity component parallel to the field. The character of a particle's motion depends on the direction of its velocity with respect to the field.

29.1.1 Motion Perpendicular to a Uniform Magnetic Field

If a particle's velocity is initially in the plane perpendicular to $\vec{B}$ (■ Figure 29.1), it remains in that plane, since its velocity component parallel to $\vec{B}$ remains zero. The particle's speed is constant, and in a uniform field the magnitude of the force acting on the particle is constant. With $\vec{v}$ perpendicular to $\vec{B}$:

$$|\vec{F}| = qvB.$$

A force of constant magnitude always perpendicular to a particle's velocity causes the particle to move in a circle (Figure 29.1; cf. §§3.2 and 4.7). The radius of the circle, called the *Larmor*

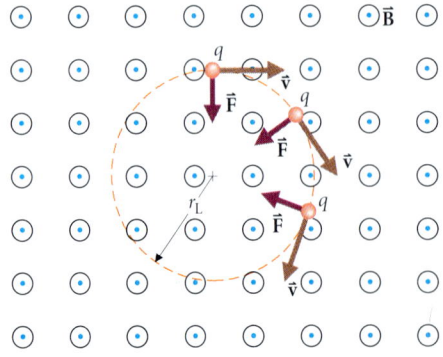

■ Figure 29.1
A positively charged particle moves in the plane perpendicular to $\vec{B}$. The magnetic force is perpendicular to both $\vec{v}$ and $\vec{B}$. The particle moves in a circle with radius $r_L = mv/(qB)$.

radius or gyro-radius r_L, is found by recognizing that magnetic force causes the centripetal acceleration v^2/r. Applying Newton's second law, we obtain:

$$\frac{mv^2}{r} = qvB.$$

SEE EQN. 3.11.

Thus the radius of the orbit is:
$$r = \frac{mv}{qB} \equiv r_L. \tag{29.1}$$

The angular speed of a particle circling in a magnetic field is:

$$\omega = \frac{v}{r} = v\frac{qB}{mv} = \frac{qB}{m} \equiv \omega_c. \tag{29.2}$$

SEE EXAMPLE 29.2 AND *DIGGING DEEPER: MORE ON CYCLOTRONS*.

The frequency is *independent of the particle's speed*. A faster particle moves in a larger circle but requires the same time per orbit. Because this fact is crucial for the design of cyclotron particle accelerators, ω_c is called the *cyclotron frequency*.

STRICTLY, ω_c SHOULD BE CALLED THE CYCLOTRON ANGULAR FREQUENCY, BUT THAT IS AWKWARD AND NOT TRADITIONAL. THE SAME NAME IS USED FOR $f_c = \omega_c/2\pi$: WATCH THE CONTEXT!

EXAMPLE 29.1 ◆ Find the Larmor radius and cyclotron frequency for an electron with kinetic energy 5.0 eV in the Van Allen belts (cf. Figure VI.8), where the Earth's magnetic field is 1.5×10^{-7} T. Assume $\vec{v}$ is perpendicular to $\vec{B}$.

MODEL The electron moves on a circle in the plane perpendicular to $\vec{B}$.

SETUP The electron's kinetic energy is $E = \frac{1}{2}mv^2$, and 1 eV = 1.6×10^{-19} J.

SOLVE The Larmor radius is then (eqn. 29.1):

$$r_L = \frac{mv}{eB} = \frac{m}{eB}\sqrt{\frac{2E}{m}} = \frac{\sqrt{2Em}}{eB}$$

$$= \frac{\sqrt{2(5.0 \text{ eV})(1.6 \times 10^{-19} \text{ J/eV})(9.1 \times 10^{-31} \text{ kg})}}{(1.6 \times 10^{-19} \text{ C})(1.5 \times 10^{-7} \text{ T})} = 50 \text{ m}.$$

The cyclotron frequency is:

$$\omega_c = \frac{eB}{m} = \frac{(1.6 \times 10^{-19} \text{ C})(1.5 \times 10^{-7} \text{ T})}{9.1 \times 10^{-31} \text{ kg}} = 2.6 \times 10^4 \text{ rad/s}.$$

ANALYZE Compared with the scale of the system (cf. $R_E \approx 6000$ km), the radius of the electron's orbit is tiny. Thus charged particles are constrained to remain close to the magnetic field lines; they are unable to flow freely from the Sun to Earth's surface. ■

EXERCISE 29.1 ❖ Describe how the orbit of a positron with the same energy would differ from that of the electron. (A positron has the same mass as an electron but the opposite electric charge.)

EXERCISE 29.2 ◆ Find the Larmor radius and cyclotron frequency for a 5.0-eV proton in the Van Allen belts. ■

29.1.2 Practical Applications of Circular Particle Motion

■ Figure 29.2 illustrates the operation of a mass spectrometer used for analysis of chemical composition. In the ion source the sample is vaporized and ionized (one or more electrons are removed from each atom). The ions are then accelerated by an electric field and pass through a grid into a region of uniform magnetic field. There the ions travel along circles determined by their Larmor radii and arrive at various points on a position-sensitive detector. In the original mass spectrometer design, photographic film was used as the detector.

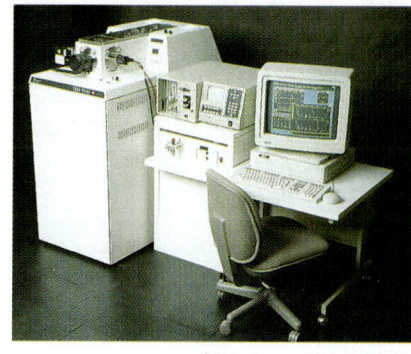

(a)

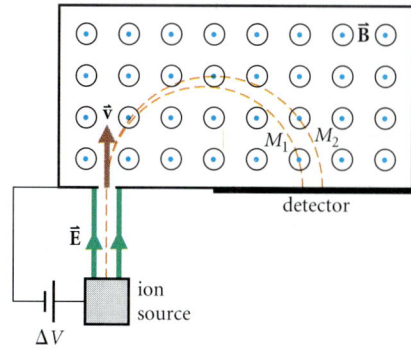

(b)

■ **FIGURE 29.2**
(a) A mass spectrometer. (b) Atoms are stripped of one or more electrons in the ion source and accelerated by the electric field so that they enter the spectrometer chamber with velocity $\vec{v}$. The radius of an ion's circular path in the chamber depends on its charge-to-mass ratio q/m. Particles with the same charge and differing mass reach different spots on the detector.

IN THIS REGION THE ION IS ACCELERATED BY AN ELECTRIC FORCE.

THE CIRCULAR MOTION IS DUE TO A MAGNETIC FORCE.

An ion with mass m and charge q enters the magnetic field with kinetic energy gained by crossing a potential difference ΔV. Its kinetic energy is $\frac{1}{2}mv^2 = q\,\Delta V$, so its speed is:

$$v = \sqrt{\frac{2q\,\Delta V}{m}}.$$

Its Larmor radius in the magnetic field is:

$$r_L = \frac{mv}{qB} = \frac{(2m\,\Delta V/q)^{1/2}}{B}.$$

For a given ΔV and $\vec{B}$, an ion's arrival point at the detector depends only on the ion's charge-to-mass ratio, q/m: $\quad r_L \propto \sqrt{m/q}$.

EXERCISE 29.3 ♦ The magnetic field in a mass spectrometer is $B = 0.20$ T, and the accelerating voltage is $\Delta V = 1.0 \times 10^4$ V. At what distance from the beam entry point are singly ionized oxygen ions collected? Where are nitrogen ions collected? ∎

Acceleration of particles to high energies has found many uses in medicine and industry as well as physics research. Electrostatic accelerators are severely limited by the difficulty of maintaining large potential differences without breakdown. In 1930 Ernest O. Lawrence (■ Figure 29.3) invented the cyclotron, which uses magnetic fields in conjunction with much smaller potential differences.

In a cyclotron (■ Figure 29.4) particles are accelerated by an electric field between two hollow, D-shaped copper electrodes called *dees*. There is no electric field inside the dees, so the particles follow circular orbits in the magnetic field. Regardless of their energy, they require a time $\Delta t = \pi/\omega_c$ between passages across the gap between the dees. During this half-cycle the potential difference between the dees is reversed, so that the particles are accelerated each time they cross the gap. The particles are injected near the center of the dees and spiral outward as they speed up [*Remember:* r increases as v increases (eqn. 29.1)]. The maximum speed is limited by the radius of the dees. Since the particles cross the gap many times, a relatively small potential difference can produce a large energy gain.

■ **FIGURE 29.3**
Ernest O. Lawrence (1901–1958) with his original cyclotron.

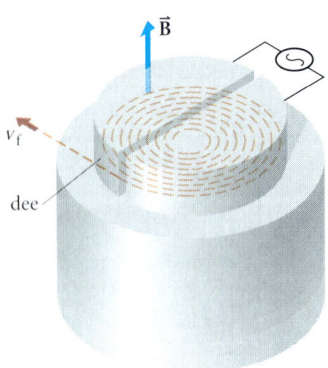

■ **FIGURE 29.4**
A cyclotron. A potential difference that changes sign at intervals $\Delta t = \pi/\omega_c$ is applied across the dees to accelerate particles spiraling inside. (An electric field exists between the dees.) The diagram shows the acceleration of positively charged particles.

EXAMPLE 29.2 ♦♦ In Lawrence's 1939 cyclotron, the dees had a diameter of 60.0 in. The device used a potential difference of 0.20 MV between the dees to produce a beam of singly ionized 14.5-MeV deuterium nuclei. What magnetic field did the device require? How much time did each particle spend in the machine?

MODEL The magnetic field controls the size of the particle's circular path. When the particle's speed and energy are greatest, its path barely fits inside the machine. This constraint determines the necessary field. The particle completes one revolution each cyclotron period, crosses the gap twice, and gains $2e\,\Delta V = 400$ keV of energy each revolution. Comparing this value with the total energy gain, we can find the number of revolutions, and hence the time.

SETUP The maximum energy E_{max} occurs when the particle's Larmor radius equals the radius of the dees: $r_D = r_L = mv/qB$. So for $E_{max} = 14.5$ MeV, the required field is:

$$B = \frac{mv}{qr_D} = \frac{m\sqrt{2E_{max}/m}}{qr_D} = \frac{\sqrt{2mE_{max}}}{qr_D}.$$

A deuterium nucleus contains one neutron and one proton, and so its mass is 2.0 u. "Singly ionized" means the charge is $+e$. So:

SOLVE
$$B = \frac{[2(2.0)(1.66 \times 10^{-27}\text{ kg})(1.45 \times 10^7\text{ eV})(1.6 \times 10^{-19}\text{ J/eV})]^{1/2}}{(1.6 \times 10^{-19}\text{ C})(30.0\text{ in.})(2.54 \times 10^{-2}\text{ m/in.})}$$

$$= 1.0 \text{ T.}$$

Every time a particle passes from one dee to the other it gains $\Delta E = 200$ keV. To gain $E_{max} = 14.5$ MeV of kinetic energy in 200-keV steps the particle must pass between the dees $N = E_{max}/\Delta E = (1.45 \times 10^7 \text{ eV})/(2.0 \times 10^5 \text{ eV}) = 73$ times. A particle spends half the cyclotron period in each dee, so the total time for acceleration is:

$$T = 73 \left(\frac{1}{2} \frac{2\pi}{\omega_c} \right)$$

$$= 73\pi \frac{m}{qB}$$

$$= \frac{73\pi (2.0)(1.66 \times 10^{-27} \text{ kg})}{(1.6 \times 10^{-19} \text{ C})(1.02 \text{ T})}$$

$$= 4.7 \ \mu\text{s}.$$

WE KEPT AN ADDITIONAL FIGURE IN THE INTERMEDIATE RESULT FOR B.

ANALYZE Since $E_{max} \propto (Br_D)^2$, to obtain more energy per particle requires a larger device or a larger magnetic field.

29.1.3 Motion in Combined Electric and Magnetic Fields

If a charged particle moves through a region that contains both electric and magnetic fields, the two fields act independently on the particle, and the total force is the sum of the separate electric and magnetic forces:

$$\vec{F} = q(\vec{E} + \vec{v} \times \vec{B}). \quad (29.3)$$

This equation is known as the *Lorentz force law*.

AFTER HENDRIK A. LORENTZ (1853–1928).

To gain intuition about the effects of combined fields, we shall look at some particular examples in which both fields are constant in time and uniform in space. In the simplest case, the electric field is parallel to $\vec{B}$. It accelerates the particle along the field lines, changing the component of velocity parallel to $\vec{B}$. Any velocity component perpendicular to $\vec{B}$ would be influenced only by the magnetic force, resulting in an independent circular motion around the field lines.

Next we'll look at the motion of a particle when the electric and magnetic fields are perpendicular. The magnetic force $q\vec{v} \times \vec{B}$ is perpendicular to $\vec{B}$, so if a charged particle moves with a particular velocity $\vec{v}_o$, the net Lorentz force on it is zero:

$$0 = \vec{F} = q(\vec{E} + \vec{v}_o \times \vec{B})$$

$$\Rightarrow \vec{v}_o \times \vec{B} = -\vec{E}.$$

To find $\vec{v}_o$, we choose the x-direction along $\vec{E}$ and the y-direction along $\vec{B}$. Then:

$$\vec{v}_o \times \vec{B} = -v_z B \hat{i} + v_x B \hat{k}.$$

For this to equal $-\vec{E} = -E\hat{i}$, $v_x = 0$ and $v_z = E/B$. Thus:

$$\vec{v}_o = \frac{E}{B} \hat{k} = \frac{\vec{E} \times \vec{B}}{B^2}. \quad (29.4)$$

A device called a velocity selector uses this effect to produce a beam of particles with nearly identical velocities. The beam must pass through two small holes along a line parallel to $\vec{E} \times \vec{B}$. Only particles with velocity very nearly equal to $\vec{v}_o$ can pass through both holes.

EXERCISE 29.4 ❖ Describe the motion of a particle with initial velocity $\vec{v} = \vec{v}_o + v_y \hat{j}$, where the y-direction is parallel to the magnetic field.

Digging Deeper

More on Cyclotrons

Cyclotrons as archaeological tools

Since accelerating particles cross the dees many times, the potential difference across the dees must vary at precisely the correct cyclotron frequency, otherwise the particles enter the gap when the potential difference acts to decelerate them, and they lose rather than gain energy. With a fixed magnetic field, only particles of a predetermined charge-to-mass ratio are accelerated. Richard Muller of the University of California at Berkeley has taken advantage of this fact to use a cyclotron as a very sensitive mass spectrometer.

Most natural carbon is in the form of ^{12}C, but ^{14}C is present in trace amounts, and it decays at a known rate to ^{14}N (Chapter 36). While alive, organisms take in carbon from the atmosphere or from food and maintain a relatively constant ratio of ^{14}C to ^{12}C. After the organism's death, the proportion of ^{14}C to ^{12}C decreases as the ^{14}C decays. In the technique of carbon dating, the ratio of ^{14}C to ^{12}C is used as a measure of a sample's age. To date very old archaeological samples, a small number of ^{14}C atoms must be detected from among much larger numbers of other atoms. The cyclotron can be tuned to accelerate only singly ionized ^{14}C and ^{14}N, which have almost the same charge-to-mass ratio q/m. The beam exiting the cyclotron then passes through a dilute gas that strips electrons from atoms in the beam. Because of their greater electric charge, the stripped nitrogen nuclei lose energy more rapidly than the carbon nuclei, which are the only species that remains in the beam. Because a very small number of atoms can be counted, this method allows dating of tiny samples of material.

Synchrotrons

According to Einstein's theory of special relativity, the momentum of a particle (§34.4.1) is given by the formula:

$$p = \frac{mv}{(1 - v^2/c^2)^{1/2}} \equiv \gamma m v,$$

where c is the speed of light. The factor γ is not present in Newton's theory, which we have used to analyze the cyclotron (γ is very close to 1 unless v is almost equal to c). Use of the correct relativistic mechanics gives more accurate expressions for the frequency and radius of the particle's motion:

$$\omega = \frac{qB}{\gamma m} \quad \text{and} \quad r = \frac{\gamma m v}{qB}.$$

If a particle in a cyclotron gains enough energy to make γ substantially greater than unity, the particle gets out of phase with the accelerating potential and ceases to gain energy. Most modern accelerators are *synchrotrons*, which use a variable magnetic field to maintain a constant Larmor radius and a variable-frequency accelerating potential that is synchronized

Figure 29.5
Dr. Michel Ter-Pogossian with the first medical cyclotron used in a U.S. hospital, at Washington University, St. Louis, MO.

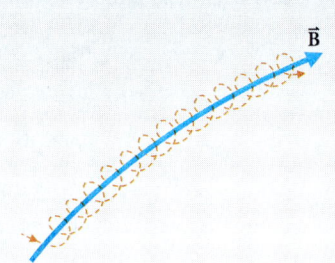

Figure 29.6
When a particle has a velocity component along $\vec{B}$, its path is a helix along the field line. This picture shows the path of a positively charged particle.

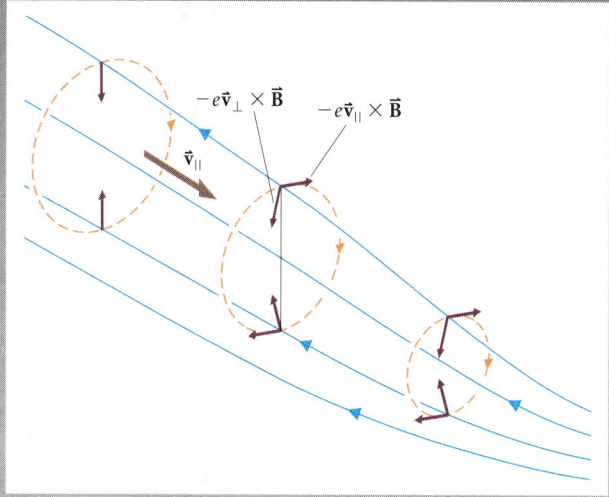

■ **Figure 29.7**
Particle spiraling in a nonuniform field. Here we show a negatively charged electron. The magnetic force has two components due to the two components of $\vec{v}$: $v_\perp$ (the circular motion) and $v_\parallel$ (along the field line). The force component $-e\vec{v}_\perp \times \vec{B}$ does not point exactly at the center of the circle but has a component along the field line, which reduces $v_\parallel$. The force component $-e\vec{v}_\parallel \times \vec{B}$ points along the circumference of the circle and increases $v_\perp$. At the mirror point, $v_\parallel = 0$ and all the particle's energy is in the circular motion.

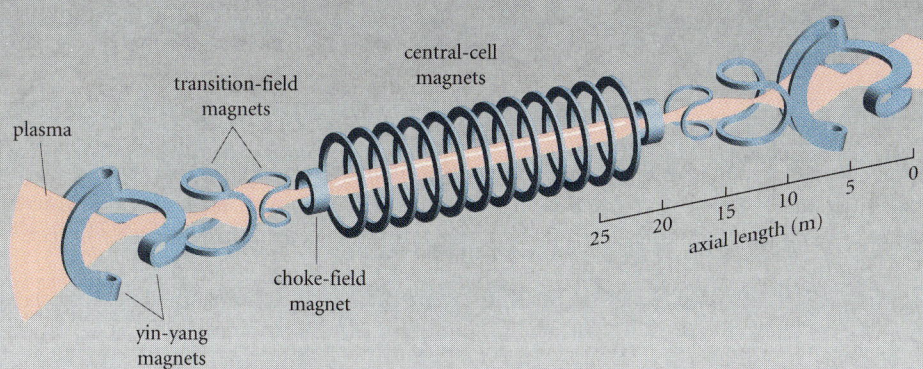

■ **Figure 29.8**
Overall configuration of the tandem mirror fusion reactor developed at Lawrence Livermore Laboratory. Plasma confinement is achieved by the double mirror at each end of the chamber and by electrostatic forces exerted by plasma clouds trapped in the mirror cells.

with the particles as they gain energy. The synchrotron at Fermilab achieves a proton beam with energy of 10^{12} eV ($\gamma \approx 10^3$). Whenever a charged particle is accelerated, it radiates (cf. the Overview §VI.9). Radiation emitted by electrons in synchrotrons is tightly beamed in the direction of the particles' motion. Synchrotrons produce an intense and controllable x-ray beam that can be used in medical and industrial applications (■ Figure 29.5).

The Magnetic Mirror

Particle motion exactly perpendicular to a magnetic field is an ideal case. A particle in a uniform field usually has some velocity component parallel to $\vec{B}$. This constant velocity along the field combines with the particle's circular motion perpendicular to the field to form a helix (■ Figure 29.6).

In a nonuniform magnetic field the field lines converge or diverge. A particle spiraling along converging field lines toward a region of greater $|\vec{B}|$ encounters a force component that reduces the particle's parallel velocity (■ Figure 29.7). In a strongly converging field, this force reverses the particle's velocity parallel to the field, reflecting it from the region of intense field. The converging field lines form a *magnetic mirror*. Only particles originally moving nearly parallel to the field can penetrate the mirror. Particles trapped in the Earth's magnetic field travel along spirals between reflections by magnetic mirrors at the North and South poles. Of those particles that travel nearly parallel to the magnetic field at the equator, a small fraction escapes through the mirror at each pole and enters the Earth's atmosphere to provide power for the polar aurora. Collisions continually replenish the supply of escaping particles and maintain equilibrium between particles captured from the Sun and those escaping to the atmosphere.

Magnetic mirrors have also been used in controlled nuclear fusion experiments such as the Tandem Mirror Fusion Experiment at Lawrence Livermore National Laboratory (■ Figure 29.8). This experiment has now been terminated in favor of the Tokamak design.

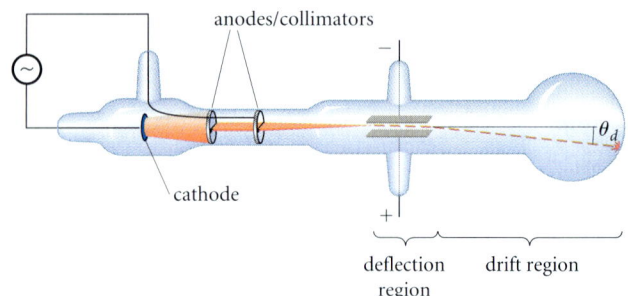

■ **FIGURE 29.9**
J. J. Thompson used this cathode-ray tube in his 1897 experiment to determine the charge-to-mass ratio e/m for an electron.

In 1897 J. J. Thompson (1856–1940) used the result in eqn. (29.4) to determine the ratio of charge to mass for an electron. He used a cathode-ray tube (■ Figure 29.9) in which a parallel electric field accelerates a beam of electrons along the tube. An electric field $\vec{E}_d$ between the deflection plates and perpendicular to the tube bends the beam sideways. If the electrons enter the deflection region with speed v, they remain between the plates for a time $t = L/v$ and have an acceleration across the tube $a = F/m = eE_d/m$. They attain a sideways velocity,

$$v_s = at = \frac{eE_d t}{m} = \frac{eE_d L}{mv},$$

and are deflected through a small angle:

REMEMBER: FOR SMALL ANGLES, $\theta \approx \tan\theta$.

$$\theta_d \approx \tan\theta_d = \frac{v_s}{v} = \frac{eE_d L}{mv^2},$$

which is measured to give a relation between the unknowns v and e/m. Helmholtz coils placed on either side of the tube are then used to create a perpendicular magnetic field. Current in the coils is adjusted until the electron beam remains undeflected. Then the electron speed $v = E_d/B$, and:

$$\frac{e}{m} = \frac{E_d \theta_d}{B^2 L}.$$

Thompson measured $v = 1.5 \times 10^7$ m/s in his tube, and his value for e/m was within 4% of the modern value of 1.76×10^{11} C/kg.

A particle moving perpendicular to $\vec{B}$ with a velocity other than $\vec{v}_o$ undergoes circular motion under the influence of the magnetic force, but the electric field causes the center of the particle's circular orbit to drift in the direction of $\vec{E} \times \vec{B}$. ■ Figure 29.10a illustrates this effect for a particle with positive charge q and mass m. Starting at point a, the particle moves in the direction of $\vec{E}$ and gains speed. After passing point b, the particle moves opposite $\vec{E}$ and slows down. Since the particle's Larmor radius is proportional to its speed (eqn. 29.1), the radius increases between a and b and decreases between b and d. The speed increases again as the particle moves to point e, and the cycle repeats. The particle's path is not a closed circle, because the curvature $1/r$ is greater where the speed is smaller. Point e is to the left of point a: the particle has drifted to the left, in the direction of $\vec{E} \times \vec{B}$. Figure 29.10b shows that the same result is obtained for a negative particle. The drift velocity $\vec{v}_E$ of the center turns out to equal $\vec{v}_o$ independent of any of the particle's properties!

■ **FIGURE 29.10**
Motion of a particle in a region with $\vec{E}$ perpendicular to $\vec{B}$. (a) Positively charged particle. The particle's speed, and hence its Larmor radius, increases as it moves parallel to $\vec{E}$ (from a to b). The Larmor radius decreases again as the particle moves from b to d and slows down. As a result, the orbit does not close: e is to the left of a. (b) A negatively charged particle orbits in the opposite direction but gains energy when it moves opposite $\vec{E}$. The resulting drift is the same as for a positively charged particle.

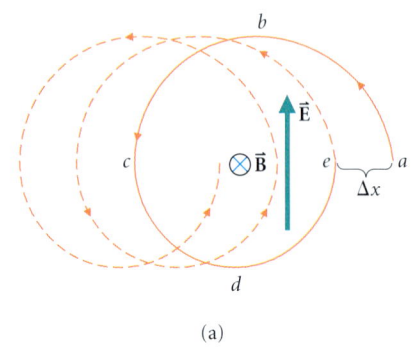
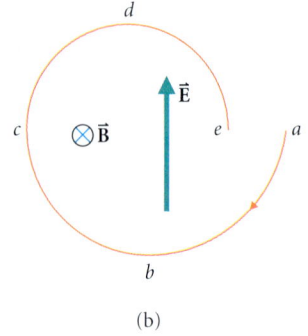

EXAMPLE 29.3 ♦ Near the surface of the Earth at the equator typical values for $\vec{E}$ and $\vec{B}$ are (100 V/m, down) and (3×10^{-5} T, north), respectively. If it were not slowed by the atmosphere, with what velocity would a charged particle drift?

MODEL The velocity is in the direction of $\vec{E} \times \vec{B}$, or (down) $\times$ (north) = east (■ Figure 29.11).

SETUP AND SOLVE We use eqn. (29.4) to find the magnitude $|\vec{v}_E| = |\vec{v}_o|$.

$$v_E = \frac{E}{B} = \frac{100 \text{ V/m}}{3 \times 10^{-5} \text{ T}} = 3 \times 10^6 \text{ m/s}.$$

ANALYZE This speed is surprisingly fast. Neglecting the atmosphere, it would take $2\pi R/v_E$, or approximately 10 s, to circle the Earth! Check the units:

$$\frac{V}{m \cdot T} = \frac{N}{C \cdot T} = \frac{N \cdot C \cdot m}{C \cdot N \cdot s} = \frac{m}{s}.$$

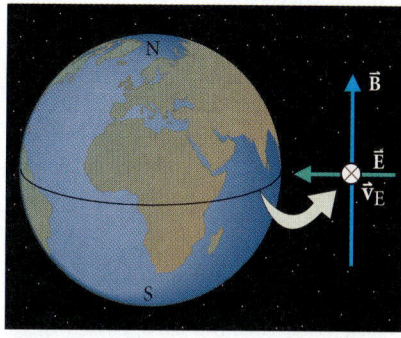

■ **FIGURE 29.11**
A charged particle near the Earth's surface would circle the Earth in about 10 s due to $\vec{E} \times \vec{B}$ drift if there were no atmosphere. In this diagram $\vec{v}_E$ is into the page.

29.2 FORCES ON CURRENT-CARRYING WIRES

29.2.1 Force on a Wire Segment

Current in a wire arises from the motion of many individual electrons. When a magnetic field is present, the forces acting on the individual electrons add to produce a net force on the wire. We begin by calculating the force on a length $d\ell$ of wire with cross-sectional area A carrying a current I in a magnetic field $\vec{B}$ (■ Figure 29.12). If a material contains n mobile (*conduction*) electrons per unit volume, the number in a length $d\ell$ of the wire is $nA\,d\ell$, and each drifts with an average velocity $\vec{v}_d$ (§26.2). The total force on the wire segment is the sum of the forces on each of these electrons:

$$d\vec{F} = \sum_{\substack{\text{all} \\ \text{moving} \\ \text{electrons}}} -e\vec{v}_d \times \vec{B} = -nA\,d\ell\,e\vec{v}_d \times \vec{B}. \quad \text{We used eqn. (28.1).}$$

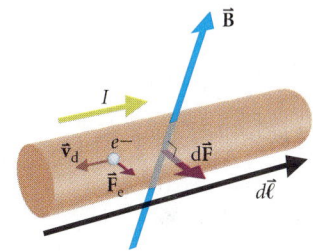

■ **FIGURE 29.12**
The force on a wire segment is the sum of the forces on all the moving electrons inside. The ions do not move and so do not contribute to the force.

From eqn. (26.5), the current is $I = nAev_d$. Since the vectors $d\vec{\ell}$ and $\vec{v}_d$ are antiparallel, we find:

$$d\vec{F} = I\,d\vec{\ell} \times \vec{B}. \tag{29.5}$$

REMEMBER: BY CONVENTION, THE DIRECTION OF $d\vec{\ell}$ IS THE DIRECTION OF CONVENTIONAL CURRENT, I.

In §29.3 we'll show how this force is transferred to the wire as a whole.

EXERCISE 29.5 ♦ Find the magnetic force on a 1.0-mm segment of wire carrying a current $I = 5.0$ A in the x-direction if the magnetic field is 0.60 T in the y-direction. ■

EXAMPLE 29.4 ❖ Show that the net torque about the center of a straight wire segment carrying current I in a uniform field is zero.

MODEL Two symmetrically located elements, equidistant from the center, experience equal forces $d\vec{F}_1 = d\vec{F}_2$ (■ Figure 29.13). These forces produce equal and opposite torques $d\vec{\tau}_1 = -d\vec{\tau}_2$ about the center. The total torque is the sum of such balancing pairs and so is zero.

ANALYZE The total force acts as if it were all applied at the center of the segment. ■

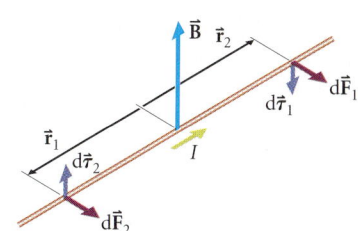

■ **FIGURE 29.13**
The net torque about the center of a straight wire segment in a uniform magnetic field is zero: the line of action of the net magnetic force passes through the center of the wire.

COMPARE WITH OUR DISCUSSION OF CENTER OF MASS IN §9.3 AND §11.1.

In deriving eqn. (29.5) we assumed that all the electrons have the same drift velocity and experience the same magnetic field. Finite lengths of wire usually are not straight, or they pass through regions of variable field. To calculate forces and torques acting on a finite wire, we model it as a collection of elementary segments and add the forces acting on the individual elements.

■ **FIGURE 29.14**
Moving-coil loudspeaker. (a) Top view. The radial magnetic field is produced by pole pieces. A coil attached to the loudspeaker cone carries current I, driven by the amplifier. The force is downward (into the page here) when the current direction is clockwise viewed from above. Reversing the current direction changes the direction of the force. When the current oscillates, so does the speaker cone, and sound waves are produced (cf. Chapter 16). (b) Side view. The gap is shown enlarged for clarity.

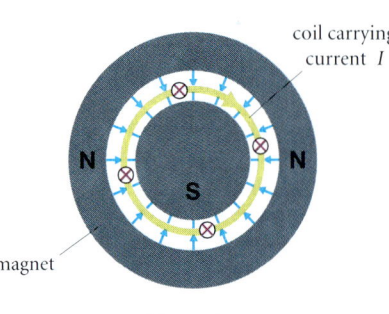
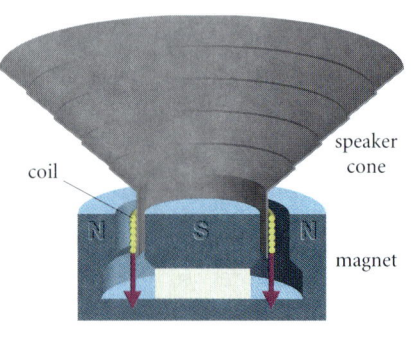

(a) top view

(b) side view

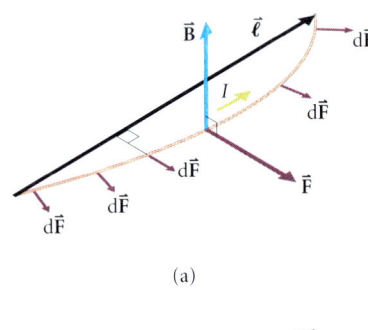

(a)

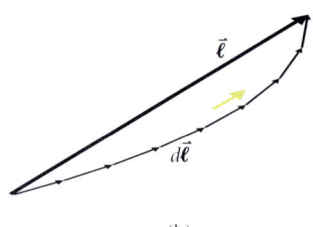

(b)

■ **FIGURE 29.15**
(a) Force on a wire segment in a uniform magnetic field is the sum of the forces on the individual elements of wire: $\vec{F} = I\vec{\ell} \times \vec{B}$. (b) The sum of the vectors $d\vec{\ell}$ is the vector $\vec{\ell}$ from one end of the wire to the other.

WE BRING $\vec{B}$ OUT OF THE INTEGRAL BECAUSE IT IS CONSTANT, BUT WE DO *NOT* CHANGE THE ORDER OF VECTORS IN THE CROSS PRODUCT. THAT WOULD CHANGE THE SIGN OF THE ANSWER.

EXAMPLE 29.5 ♦♦ A cone-type loudspeaker contains a permanent magnet surrounded by a coil of wire (■ Figure 29.14). Magnetic force on a sinusoidally varying current in the coil oscillates the coil and the attached speaker cone to produce sound waves. If the magnetic field strength at the position of the coil is 1.0 T, the coil has 100 turns of radius 4.0 cm, and the current is 1.0 A, what is the force on the cone?

MODEL The loudspeaker is designed so that the magnetic field is perpendicular to the wire everywhere. With current clockwise around the coil (Figure 29.14a), the direction of the force on each element of the coil, $d\vec{F} = I\, d\vec{\ell} \times \vec{B}$, is into the page (downward in Figure 29.14b). We add the contributions due to each element of wire.

SETUP Each element $d\vec{\ell}$ is perpendicular to the field, so:

$$|d\vec{F}| = I\, d\ell\, |\vec{B}|.$$

Finally, $|\vec{B}|$ is the same at each element, and the total length of the wire is the number N of turns in the coil times the circumference $2\pi r$ of each coil.

SOLVE The total force has magnitude:

$$F = I\ell B = (1.0 \text{ A})(100)(2\pi)(4.0 \times 10^{-2} \text{ m})(1.0 \text{ T}) = 25 \text{ N}.$$
$$\vec{F} = (25 \text{ N, downward in Figure 29.14b}).$$

ANALYZE When the direction of current is reversed, the direction of the force reverses too. To increase the force, we should increase I, ℓ, or B. The cheapest alternative is to increase the length of wire. ■

Next we shall derive a convenient expression for the force acting on a wire segment of arbitrary shape carrying current I in a uniform magnetic field $\vec{B}$ (■ Figure 29.15a). The force on any differential piece of the wire is given by eqn. (29.5):

$$d\vec{F} = I\, d\vec{\ell} \times \vec{B}.$$

The current I and the magnetic field $\vec{B}$ are the same for each element, so the total force is:

$$\vec{F} = \int d\vec{F} = \int I\, d\vec{\ell} \times \vec{B} = I\left(\int d\vec{\ell}\right) \times \vec{B}.$$

The integral is the vector sum of the line element vectors and equals the vector $\vec{\ell}$ joining the end points of the segment (Figure 29.15b). Thus:

$$\vec{F} = I\vec{\ell} \times \vec{B}. \tag{29.6}$$

EXAMPLE 29.6 ♦ A semicircular wire segment of radius $a = 2.54$ cm lies in the x-y-plane with its two ends on the x-axis (■ Figure 29.16a). The current in the loop is 1.35 A, and the magnetic field in the region is $\vec{B} = (2.95 \times 10^{-5} \text{ T})\hat{j} - (4.76 \times 10^{-5} \text{ T})\hat{k}$. Find the force on the wire segment.

MODEL The vector $\vec{\ell}$ from one end of the segment to the other lies along the x-axis. The magnetic field is uniform, so we may use eqn. (29.6).

SETUP
$$\vec{F} = I\vec{\ell} \times \vec{B} = I\ell\hat{i} \times (B_y\hat{j} + B_z\hat{k})$$
$$= I\ell[B_y\hat{k} + B_z(-\hat{j})].$$

SOLVE Notice that the given B_z is negative:
$$\vec{F} = (1.35 \text{ A})(2)(2.54 \text{ cm})[(2.95 \times 10^{-5} \text{ T})\hat{k} + (4.76 \times 10^{-5} \text{ T})\hat{j}]$$
$$= (2.02 \text{ }\mu\text{N})\hat{k} + (3.26 \text{ }\mu\text{N})\hat{j}.$$

ANALYZE Check the units:
$$\text{A}\cdot\text{m}\cdot\text{T} = \text{A}\cdot\text{m}\cdot\frac{\text{N}}{\text{A}\cdot\text{m}} = \text{N}.$$

Check the direction using the right-hand rule (Figure 29.16b).

Since electric current in wires is a source of magnetic field, one such wire exerts a magnetic force on another. The force between two long parallel wires is used in the SI definition of the ampere (■ Figure 29.17). The wire on the left produces a magnetic field (eqn. 28.6),
$$\vec{B}_1 = \frac{\mu_0 I_1}{2\pi d}\hat{j},$$
at the position of the wire on the right. A length $d\ell$ of the second wire experiences a force:
$$d\vec{F} = I_2\, d\vec{\ell} \times \vec{B}_1$$
$$= I_2\, d\ell\,\hat{k} \times \hat{j}\,\frac{\mu_0 I_1}{2\pi d}$$
$$= \frac{\mu_0 I_1 I_2}{2\pi d}\, d\ell(-\hat{i}).$$

Thus the force per unit length acting on the second wire has magnitude:
$$\boxed{\frac{dF}{d\ell} = \frac{\mu_0 I_1 I_2}{2\pi d}.} \quad (29.7)$$

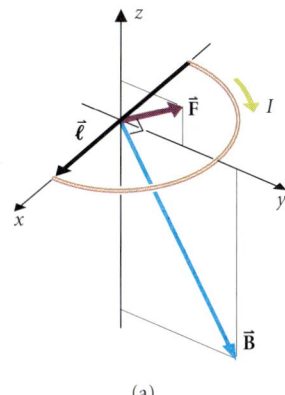
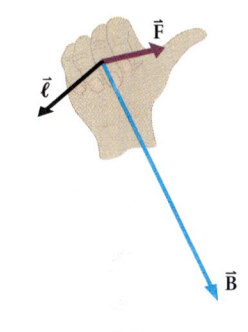

■ **FIGURE 29.16**
(a) Force on a semicircular piece of wire. The wire lies in the x-y-plane. The magnetic field is uniform with $\vec{B}$ in the y-z-plane. (b) The right-hand rule shows that the force is also in the y-z-plane, perpendicular to $\vec{B}$.

The force is attractive if the two wires carry current in the same sense, and repulsive if the currents are in opposite directions. If both currents equal 1 A exactly, and $d = 1$ m exactly, then the force per unit length is:
$$\frac{dF}{d\ell} = \frac{2(10^{-7}\text{ N/A}^2)(1\text{ A})^2}{1\text{ m}} = 2 \times 10^{-7}\text{ N/m}.$$

The definition of μ_0 (exactly $4\pi \times 10^{-7}$ N/A^2) is effectively the definition of the ampere.

> If the same current is maintained in two long parallel wires that attract each other with a force of 2×10^{-7} N per meter of wire when the wires are 1 meter apart, then the current in each wire is 1 *ampere* (adopted 1948).

EXERCISE 29.6 ♦ Show that the force per unit length on the first wire is equal to and opposite that on the second.

When calculating forces, a simple model of the current distribution often provides remarkably good approximate results.

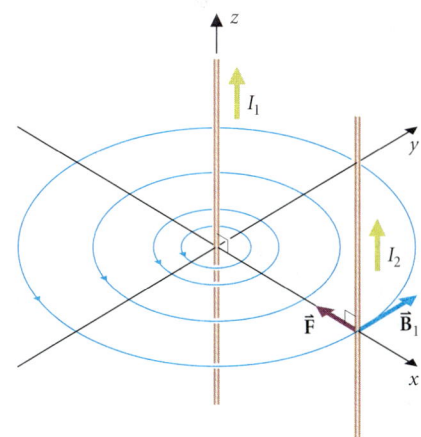

■ **FIGURE 29.17**
Two parallel wires carrying equal currents exert attractive forces on each other. The wire on the left produces a magnetic field at the position of the wire on the right. The moving charges in the second wire experience a magnetic force toward the first wire.

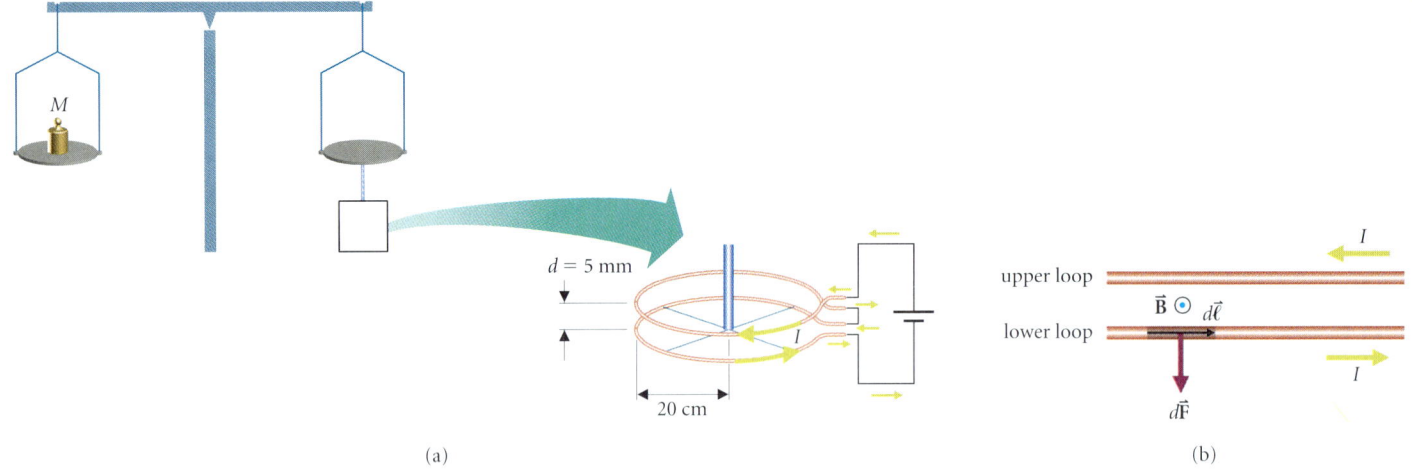

(a) (b)

FIGURE 29.18
(a) A current balance. Currents are in opposite directions in the two coils. The repulsive force between them is balanced by the gravitational force on the other arm of the balance. (b) The magnetic field at a position on the lower loop may be estimated by modeling the upper loop as a long straight wire: $\vec{B}$ is out of the page, and $d\vec{F}$ is downward.

AN EXACT CALCULATION SHOWS THAT THIS METHOD OVERESTIMATES THE FORCE BY 0.3%.

WE MAY ADD THE MAGNITUDES BECAUSE EACH ELEMENT OF FORCE $d\vec{F}$ HAS THE SAME DIRECTION.

Galvanometer movement. Torque acting on a current-carrying loop causes the pointer movement in this galvanometer.

EXAMPLE 29.7 ♦♦♦ ■ Figure 29.18a shows a schematic of a current balance for measuring the forces between parallel current-carrying wires. Two parallel coils of radius $a = 0.20$ m and 150 turns each are connected so that the currents in the coils have the same magnitude but opposite directions. The upper coil is fixed, and the lower coil moves with the balance. In equilibrium, the coils are separated by a distance $d = 5.0$ mm. Estimate the mass M required to balance the system when the current in the coils is 1.0 A.

MODEL The system balances when the weight Mg equals the repulsive force between the two coils. Because the two coils are parallel and close together ($d \ll a$), a reasonable estimate of the field at any position on the lower coil is obtained by modeling the upper coil as a long straight wire tangent to the circle at that point (Figure 29.18b). (This works because the separation of the wires is small compared with their radius of curvature.) The magnetic field lines form circles around the upper wire and so are at right angles to the lower wire.

SETUP The magnetic field strength produced by the upper coil is the same for each position on the lower coil (eqn. 28.6): $B = \mu_0 NI/(2\pi d)$. Figure 29.18b shows the magnetic field and the resulting force on a typical element of the lower coil:

$$d\vec{F} = (NI\, d\ell\, B, \quad \text{downward}).$$

Since the force per unit length is the same for each element, the total force has magnitude:

$$F = \int dF = NIB \int_{\text{lower coil}} d\ell = 2\pi a NIB = \frac{a\mu_0(NI)^2}{d}.$$

The weight required to balance this force is $Mg = F$.

SOLVE Hence the mass required on the balance is:

$$M = \frac{a\mu_0(NI)^2}{gd} = \frac{(0.20 \text{ m})(4\pi \times 10^{-7} \text{ N/A}^2)(150)^2(1.0 \text{ A})^2}{(9.8 \text{ N/kg})(5.0 \times 10^{-3} \text{ m})} = 0.12 \text{ kg}.$$

ANALYZE A similar current balance is used in laboratory equipment that provides the standard for the ampere. ■

29.2.2 Force and Torque on Current Loops

Many applications of magnetic forces involve rotating machinery in which torques exerted by the field are important. In this section we shall investigate the torque exerted on a current loop. The result is conveniently expressed in terms of magnetic moment.

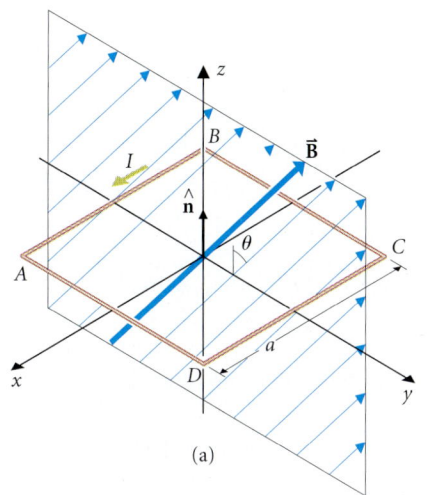
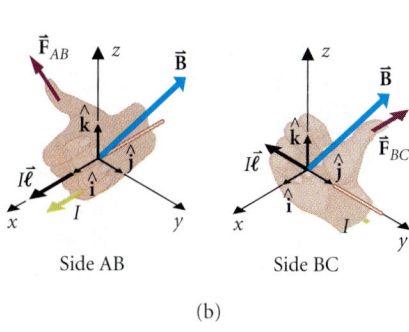
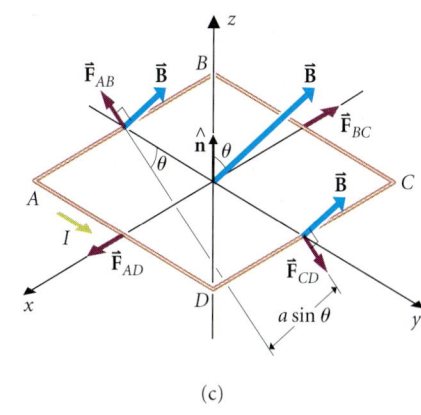

(a) (b) (c)

EXAMPLE 29.8 ♦♦ A square current loop is placed in a uniform magnetic field that makes an angle θ with the normal to the loop (■ Figure 29.19a). Find the net force acting on the loop and the net torque about its center.

MODEL We may apply eqn. (29.6) to each side of the loop or to the loop as a whole. If we consider the whole loop as a single wire segment, both its ends are at the same point. The vector $\vec{\ell}$ in eqn. (29.6) is zero, so the net force on the loop is zero. We look at the forces on each side separately to calculate the torque. The line of action of each force is through the center of the corresponding side (Example 29.4).

SETUP We choose the z-axis perpendicular to the loop, with the x- and y-axes parallel to sides AB and BC. Equation (29.6) applies to each side of the square, and $\vec{B}$ has y- and z-components, but no x-component. Vector $\vec{\ell}$ for side AB is in the x-direction (Figure 29.19b), and thus the force acting on side AB is:

$$\vec{F}_{AB} = Ia\hat{i} \times \vec{B} = Ia\hat{i} \times (B \sin\theta \, \hat{j} + B \cos\theta \, \hat{k})$$
$$= IaB (\sin\theta \, \hat{k} - \cos\theta \, \hat{j}).$$

The force on side CD is: $\vec{F}_{CD} = -Ia\hat{i} \times \vec{B} = -IaB(\sin\theta \, \hat{k} - \cos\theta \, \hat{j})$.

These forces are equal and opposite. The same is true of the forces on sides BC and AD:

$$\vec{F}_{AD} = Ia\hat{j} \times \vec{B} = Ia\hat{j} \times (B \sin\theta \, \hat{j} + B \cos\theta \, \hat{k}) = IaB (\cos\theta \, \hat{i}) = -\vec{F}_{BC}.$$

The net force on the loop is zero.

SOLVE The lines of action of forces $\vec{F}_{AD}$ and $\vec{F}_{BC}$ both pass through the center of the loop and so produce no torque about the center of the loop. However, $\vec{F}_{AB}$ and $\vec{F}_{CD}$ exert torques of equal magnitude and direction that act to rotate AB up and CD down. The net torque is:

$$\vec{\tau} = \Sigma \vec{r} \times \vec{F} = -\left(\frac{a}{2}\right)\hat{j} \times IaB \sin\theta \, \hat{k} + \left(\frac{a}{2}\right)\hat{j} \times (-IaB \sin\theta \, \hat{k})$$
$$= (Ia^2)B \sin\theta(-\hat{i}).$$

Recognizing $Ia^2\hat{k} = IA\hat{k}$ as the magnetic moment of the loop (eqn. 28.8), we have:

$$\vec{\tau} = mB \sin\theta(-\hat{i}) = \vec{m} \times \vec{B}. \tag{29.8}$$

ANALYZE The total force on any closed loop in a uniform $\vec{B}$ field is zero. According to eqn. (29.6), the total force on any piece of wire is $\vec{F}_T = I\vec{\ell} \times \vec{B}$, where $\vec{\ell}$ is the vector between the end points of the wire. For any closed loop the end points are the same, and $\vec{\ell} = 0$. Equation (29.8) describes the torque on any loop in a *uniform* magnetic field, but the proof is left to the problems. The torque tends to align the dipole moment of the loop with the field and the plane of the loop perpendicular to the field (cf. §24.5). ■

■ **FIGURE 29.19**
(a) The magnetic field makes an angle θ with the normal to the plane of a current loop. The loop lies in the x-y-plane, and $\hat{n} = \hat{k}$. (b) Vector diagrams showing the forces on sides AB and BC. (c) All four forces on the loop. The forces on sides BC and AD are equal and opposite and lie along the same line. $\vec{F}_{AB}$ and $\vec{F}_{CD}$ are equal and opposite, but their lines of action are separated by a distance $a \sin\theta$. There is a net torque on the loop given by $\vec{\tau} = \vec{m} \times \vec{B}$.

$\vec{F}_{AB}$ AND $\vec{F}_{CD}$ FORM A COUPLE WITH LINES OF ACTION SEPARATED BY $a \sin\theta$ (FIGURE 29.19C) AND A TORQUE OF MAGNITUDE $\tau = F_{AB}(a \sin\theta) = (IaB)(a \sin\theta) = (Ia^2)B \sin\theta$ (see §11.1.3).

COMPARE WITH EQN. (24.12).

Digging Deeper

Magnetic Forces and Newton's Third Law

■ Figure 29.20a shows two separate elements of wire and the magnetic forces they exert on each other. The forces are not equal and opposite; the magnetic field of element 1 exerts an upward force on element 2, but element 2 exerts no force at all on element 1! The magnetic field due to element 2 is zero at the position of element 1. The mutual magnetic forces on two point charges in motion display the same behavior (Figure 29.20b). What has happened to Newton's third law? In Part II we showed that Newton's third law expresses conservation of linear and angular momentum in a system of particles. The conservation laws are fundamental principles, required by the symmetry of space and time. They must still apply when magnetic forces occur. A proper analysis of momentum conservation for a system of moving charges must include the electric and magnetic fields, which extend over all space. We have already discovered that electric field stores energy (§27.4). The electromagnetic field can also store momentum. The apparent failure of Newton's third law tells us that momentum is being exchanged between particles and fields in the system. Momentum of the entire system of particles and fields *is* conserved. The forces between two finite, closed current loops do obey Newton's third law, since the net momentum exchanged with the fields is zero in that case. Nevertheless, you should always be wary about applying Newton's third law when magnetic forces occur.

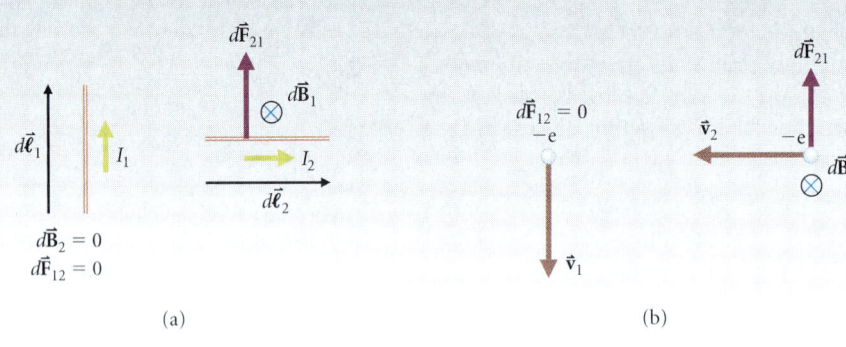

■ **Figure 29.20**
(a) The forces exerted by two elements of current-carrying wire on each other are not equal and opposite! Here, the force exerted by element 2 on element 1 is zero, while the force exerted by 1 on 2 is not. (b) The same effect occurs for two electrons moving at right angles to each other. These apparent violations of Newton's third law result from exchange of momentum between particles and fields.

EXAMPLE 29.9 ♦ Many satellites use current loops interacting with the Earth's magnetic field to control their orientation in space. A system designed for the *Extreme Ultraviolet Explorer* (EUVE) satellite (■ Figure 29.21) has a coil of area $A = 1$ m^2 and $N = 10^3$ turns. What current is required to produce a torque of 0.06 N·m if the Earth's magnetic field is $B = 3 \times 10^{-5}$ T in the plane of the coil?

MODEL We may apply eqn. (29.8) directly.

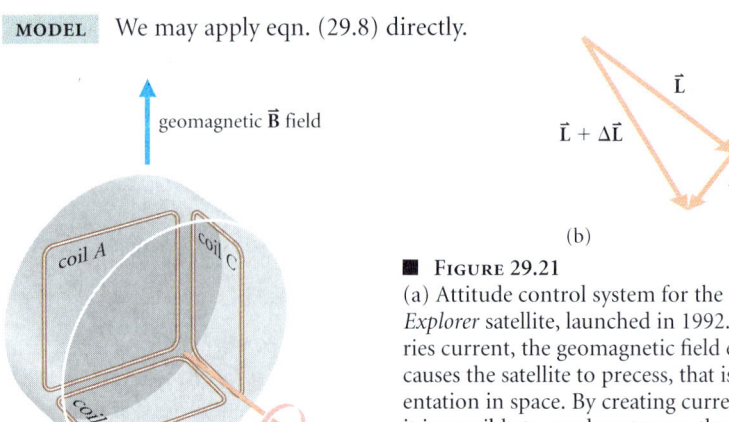

■ **Figure 29.21**
(a) Attitude control system for the *Extreme Ultraviolet Explorer* satellite, launched in 1992. When coil A carries current, the geomagnetic field exerts a torque that causes the satellite to precess, that is, to change its orientation in space. By creating current in coils B and C it is possible to produce torque that changes the satellite's spin rate. (b) Vector diagram showing the satellite's angular momentum and the change $\Delta \vec{L} = \vec{\tau} \Delta t$ caused by the magnetic torque.

SETUP The magnetic moment of the coil has magnitude $m = NIA$ and is perpendicular to the plane of the coil. The torque exerted by the Earth's field is $\vec{\tau} = \vec{m} \times \vec{B}$. If the field is in the plane of the coil, then $\vec{m}$ is perpendicular to $\vec{B}$ and:

$$\tau = mB = NIAB.$$

SOLVE
$$I = \frac{\tau}{NAB} = \frac{0.06 \text{ N} \cdot \text{m}}{1000(1 \text{ m}^2)(3 \times 10^{-5} \text{ T})} = 2 \text{ A}.$$

ANALYZE As designed for *EUVE*, this system requires 24 kg of wire and uses 100 W of power.

EXAMPLE 29.10 ♦♦ Each of two current loops has N turns, radius a, and area $A = \pi a^2$. One loop lies in the x-y-plane with its center at the origin. The other is parallel to the x-z-plane with its center at $y = D \gg a$. Estimate the torque the first loop exerts on the second if each carries current I.

MODEL ■ Figure 29.22 shows the loops, together with the magnetic field produced by the first loop. We calculate the field at the center of one loop due to the other and then compute the torque using eqn. (29.8). We may model the first loop as a dipole (§28.2.3) in calculating the field it produces.

SETUP We find the magnetic field in the plane of a loop by adapting the result of Exercise 24.6 to a magnetic dipole (i.e., replacing $k\vec{p}$ with $\mu_0 \vec{m}/4\pi$). The magnetic moment of the first loop is in the z-direction and has magnitude IA. At the center of the second loop $y = D$:

$$\vec{B}_1 = \frac{\mu_0}{4\pi y^3}(-\vec{m}_1) = \frac{\mu_0 m_1}{4\pi D^3}(-\hat{k}) = \frac{\mu_0 IA}{4\pi D^3}(-\hat{k}).$$

The magnetic moment of the second loop is $\vec{m}_2 = IA(-\hat{j})$, and since $D \gg a$, the magnetic field $\vec{B}_1$ is almost constant over the area of the second loop.

SOLVE Thus the torque is given by eqn. (29.8):

$$\vec{\tau}_{21} = \vec{m}_2 \times \vec{B} = -\frac{\mu_0}{4\pi} \frac{\vec{m}_2 \times \vec{m}_1}{D^3} \tag{29.9}$$

$$= -\frac{\mu_0}{4\pi} \frac{(NIA)^2}{D^3}(-\hat{j} \times \hat{k}) = \frac{\mu_0}{4\pi} \frac{(NIA)^2}{D^3} \hat{i}.$$

ANALYZE The torque rotates the second loop toward an orientation where the two magnetic moments are antiparallel. Although we derived eqn. (29.9) for a specific orientation of the loops, it is true for any two loops separated by a distance much greater than either's size. ■

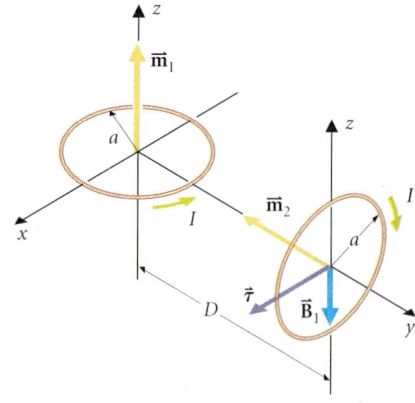

■ **FIGURE 29.22**
Two current loops lie in orthogonal planes. To find the torque one exerts on the other, we calculate the magnetic field the first loop produces. If the loops are far apart ($D \gg a$), we may use the dipole approximation.

29.3 THE HALL EFFECT

In 1879 E. H. Hall discovered that a conductor carrying current in a magnetic field develops a small electric field perpendicular both to the magnetic field and to the direction of the current. Hall used this observation to prove that current in metals is carried by negative charge.

Current in a metal strip is driven by an applied electric field $\vec{E}$ (■ Figure 29.23). Each conduction electron in a stationary strip has an average drift velocity $\vec{v}_d$. In a region of uniform magnetic field, the magnetic force $-e\vec{v}_d \times \vec{B}$ causes electrons to drift toward one side of the strip (side a in the figure), leaving excess positive charge on side b. These charge concentrations produce an electric field $\vec{E}_H$. The sideways motion ceases when the electric force $-e\vec{E}_H$ on each electron balances the magnetic force $-e\vec{v}_d \times \vec{B}$. The resulting potential difference across the strip,

$$|\Delta V_H| = E_H L = v_d B_\perp L, \tag{29.10}$$

is known as the *Hall potential*.

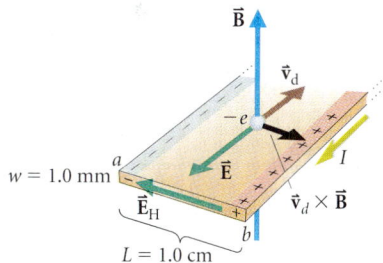

■ **FIGURE 29.23**
The Hall effect. A strip of conductor carries current I perpendicular to a magnetic field $\vec{B}$. The moving electrons feel a magnetic force that causes them to drift to one edge of the strip. This concentration of charge produces an electric field across the strip. In equilibrium, the electric force component $-e\vec{E}_H$ on each electron balances the magnetic force component $ev_d B$, and so $\vec{E}_H = -\vec{v} \times \vec{B}$.

$B_\perp$ IS THE COMPONENT OF $\vec{B}$ PERPENDICULAR TO THE DIMENSION L OF THE STRIP.

EXERCISE 29.7 ❖ Show that the sign of the Hall potential would be reversed if current were carried by positive instead of negative charges.

EXAMPLE 29.11 ❖❖ A copper strip has a cross section of 1.0 mm × 1.0 cm (Figure 29.23). The strip carries a current of 5.0 A, and the magnetic field is $|\vec{B}| = 1.0$ T. What is the Hall potential?

MODEL The number of conduction electrons per unit volume in copper is $n = 8.5 \times 10^{28}$ /m^3 (cf. Example 26.4). As these electrons move, forming the electric current, magnetic force acts on them.

SETUP The drift speed of the electrons is $v_d = j/ne = I/neA$. The Hall potential difference is (eqn. 29.10):

$$|\Delta V_H| = v_d BL = \frac{IBL}{neLw} = \frac{IB}{new}.$$

Thus:

$$|\Delta V_H| = \frac{(5.0 \text{ A})(1.0 \text{ T})}{(8.5 \times 10^{28} \text{ /m}^3)(1.6 \times 10^{-19} \text{ C})(1.0 \times 10^{-3} \text{ m})} = 0.37 \ \mu\text{V}.$$

ANALYZE Hall fields are very small because drift speeds are very small (cf. Example 26.4). Nevertheless, they are very important.[1] Since $\Delta V_H \propto (1/ne)$, measurement of the Hall potential with a known magnetic field may be used to determine the sign and number of the charge carriers in a conductor. Alternatively, using a conductor with known properties, the Hall effect may be used to measure $\vec{B}$.

In equilibrium, the Hall electric force balances the magnetic force on moving electrons, so the component across the strip of the net force on each electron is zero. However, the force on each stationary, positively charged ion is due only to the Hall field and is not balanced by any magnetic force. The Hall electric field transmits the magnetic force to the ions and thus to the strip as a whole.

29.4 MAGNETIC MATERIALS

29.4.1 Atomic Model of Magnetization

SEE THE OVERVIEW OF ELECTROMAGNETISM §VI.5.

Electron motion in individual atoms constitutes a current that can produce magnetic field. The behavior of atoms follows the principles of quantum mechanics, so classical models of magnetism are limited in accuracy. Nevertheless, we can obtain a reasonable, qualitative understanding from a classical approach.

Most materials show only a weak response to an applied external field. They are classified as *paramagnetic* if the field within the material is greater than the applied field and as *diamagnetic* if the internal field is less than the applied field. Paramagnetic materials show a weak attraction toward regions of strong magnetic field, whereas diamagnetic materials are repelled. *Ferromagnetic* materials—primarily iron, cobalt, nickel, and their alloys—show a strong response to applied fields and can form permanent magnets.

Paramagnetism. Each atom in a paramagnetic material has a permanent magnetic moment that is associated with the net angular momentum of its electrons. Normally these atomic moments are randomly oriented and produce no net magnetic field. An external field tends to align the atomic dipoles, which then produce a net field that adds to the applied field. The atoms do not align completely, since thermal motion tends to randomize their directions.

---

[1] In 1985 the Nobel prize in physics was awarded to Klaus von Klitzing for the discovery of the quantum Hall effect—a related phenomenon that occurs in semiconductors.

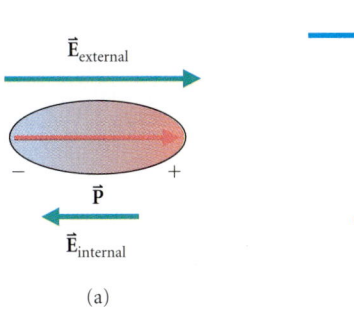
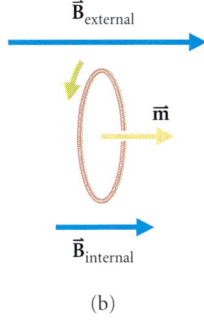

■ **FIGURE 29.24**
(a) Electric dipoles aligned with an externally applied electric field produce an electric field opposite $\vec{E}_{external}$ within the material (cf. Figure 27.15). (b) Magnetic dipoles aligned with an externally applied magnetic field produce a magnetic field within the material that is in the same direction as $\vec{B}_{external}$.

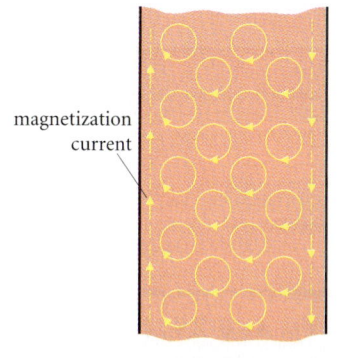

■ **FIGURE 29.25**
Magnetization current. Each atom behaves like a miniature current loop. Between atoms the currents are in opposite directions, and cancel. At the edge of the material, there is no cancellation. The net current produces magnetic field in the material with direction into the page.

A useful analogy for this effect is a box containing ping-pong balls, each with a small bead of lead attached to its inside surface. The balls tend to settle with the lead bead on the bottom, but if the box is shaken violently, the balls bounce about, and the lead beads appear in random positions. Roughly speaking, greater temperature corresponds to greater shaking of the atoms, and so paramagnetism decreases as temperature increases.

Electric dipoles in a dielectric align with an applied electric field and reduce the applied field. Magnetic dipoles in a paramagnetic material align with an applied magnetic field and increase the applied field (■ Figure 29.24). This reversal of behavior occurs because electric field is produced by charge, and magnetic field is produced by current. ■ Figure 29.25 illustrates what happens when the atoms, each modeled as a small current loop, are aligned. The currents between atoms are in opposite directions and cancel. On the surfaces of the material the atoms produce a net current called *magnetization current* that enhances the external field.

Diamagnetism. In a diamagnetic material with no applied magnetic field the individual atoms have no net magnetic moment; the magnetic effects of the electrons in each individual atom cancel. An external magnetic field changes the motion of the electrons and induces a net magnetic moment in each atom.

THE PATTERN OF ELECTRONS IN AN ATOM DETERMINES ITS MAGNETIC MOMENT. SEE *DIGGING DEEPER*.

An electron in orbit about a nucleus (■ Figure 29.26) forms a current loop with magnetic moment $m = I(\pi r^2)$, where r is the radius of the electron orbit. The current is $I = (e/T) = (ev/2\pi r)$, where T is the period of the orbit and v is the electron's speed. Thus:

REMEMBER: THIS IS A CLASSICAL MODEL.

$$m = \frac{ev}{2\pi r} \pi r^2 = \frac{evr}{2}.$$

WE USED EQN. (28.8) FOR m.

The electric force F_e exerted on the electron of mass M by the atomic nucleus causes the centripetal acceleration:

$$F_e = Mv^2/r.$$

An external magnetic field applied parallel to $\vec{m}$ causes a magnetic force $\vec{F}_m = -e\vec{v} \times \vec{B}$, which is radially outward. Because of the reduced inward force, the centripetal acceleration of the electron decreases. Thus either the orbit radius r increases or the speed decreases. In fact it is the speed that changes. With the new speed $v + dv$, Newton's second law becomes:

A FORCE PARALLEL TO VELOCITY CHANGES THE SPEED. WE'LL SEE IN CHAPTER 30 HOW THIS FORCE ARISES.

$$\frac{M(v + dv)^2}{r} = F_e - evB.$$

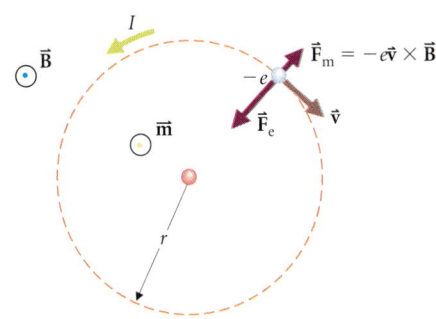

■ **FIGURE 29.26**
Diamagnetism. The applied field $\vec{B}$ causes a reduction of the net force acting on the electron. The electron's orbital radius remains the same, but its speed is reduced (we'll see why in Chapter 30), and the magnetic moment of the atom changes by $\Delta \vec{m} \propto -\vec{B}$.

Expanding the square and ignoring the term in $(dv)^2$, we have:

$$\frac{Mv^2}{r} + \frac{2Mv\,dv}{r} = F_c - evB.$$

The first term on the left-hand side equals the first term on the right, so:

$$dv = -\frac{eBr}{2M}.$$

Remember: Speed decreases, so dv is negative.

The change in speed is proportional to the applied field B. The magnetic moment changes by an amount:

$$dm = d\left(\frac{evr}{2}\right) = \frac{er}{2}\,dv = \frac{er}{2}\left(-\frac{eBr}{2M}\right) = -\frac{e^2r^2B}{4M}.$$

The magnetic moment of the electron orbit is reduced by an amount proportional to the applied field B.

EXERCISE 29.8 ♦♦ Show that if $\vec{B}$ is applied antiparallel to $\vec{m}$, then m is increased, so that $d\vec{m}$ is always antiparallel to $\vec{B}$.

The magnetic moments of the electrons in each atom sum to zero when there is no applied field. But since $d\vec{m}$ is in the same direction for each electron, the net result of an applied $\vec{B}$ is to create an induced moment opposite $\vec{B}$.

The diamagnetic effect occurs in all substances, but the induced magnetic moments are much smaller than the permanent moments of paramagnetic atoms. However, the diamagnetic effect is independent of temperature and becomes the dominant effect at high temperature as the alignment of permanent moments decreases.

Ferromagnetism. In ferromagnetic materials a quantum mechanical interaction between electrons in neighboring atoms causes the atomic dipole moments to align with each other even in the absence of an external field. Below a particular temperature, called the *Curie temperature,* the tendency to align overcomes thermal randomization and the material spontaneously *freezes* into *domains* in which all atomic dipoles have the same direction (■ Figure 29.27, ● Table 29.1). Domain formation is similar to the growth of ice crystals as water freezes.

TABLE 29.1
Curie Temperatures

Material	Curie Temperature (°C)
Iron	770
Cobalt	1127
Nickel	358
Gadolinium	16
Dysprosium	−168
Chromium dioxide	117
Ferrites:	
NiFe$_2$O$_4$	585
CuFe$_2$O$_4$	455
NiAlFeO$_4$	198

After Pierre Curie, 1859–1906.

Note: The degree of magnetization is vastly greater in ferromagnetic materials than in paramagnetic materials.

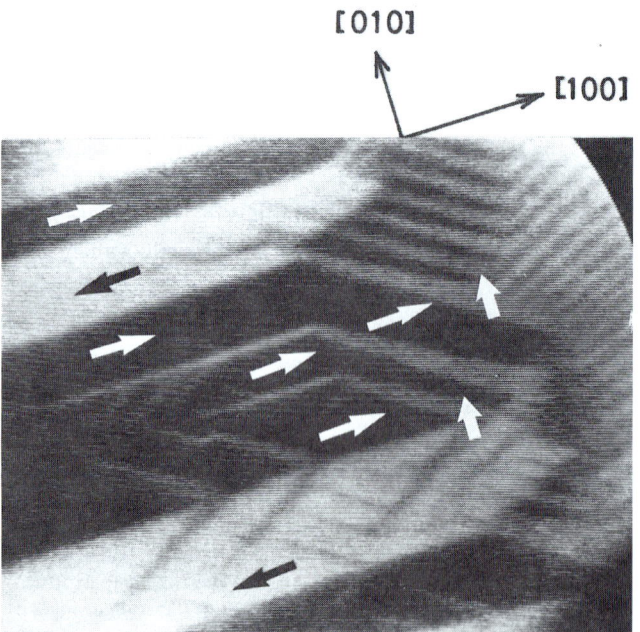

■ **Figure 29.27**
Ferromagnetism. The magnetic domain structure of silicon iron, as revealed by a scanning electron microscope. The arrows indicate the magnetic axis of the domains.

Digging Deeper

Magnetic Moment and Angular Momentum

The magnetic moment of a particle with mass M and charge e in a circular orbit is:

$$m = \frac{evr}{2} = \frac{e}{2M} L,$$

where $L = Mvr$ is the angular momentum of the particle about the center of the circle. Magnetic moment is also proportional to angular momentum for electrons and nuclei in atoms. Each particle has an intrinsic angular momentum—called spin—as well as its orbital angular momentum, and its magnetic moment is related to L by an empirical factor called g:

$$m = g \frac{e}{2M} L.$$

For an electron, the value of g that relates its intrinsic magnetic moment to its spin is 2.002319304386.

Using the principles of quantum mechanics it is possible to compute the total angular momentum of an atom and hence to compute its magnetic moment m. The same principles determine the chemical behavior of atoms, and so magnetic properties are related to chemical properties. For example, helium is chemically inert, and its magnetic moment is zero. In general, the relations are quite complicated and beyond the scope of this text.

When an external field is applied to a ferromagnetic material, it becomes strongly magnetized. In a weak applied field, domains with magnetization parallel to the applied field grow at the expense of neighboring domains. A strong applied field causes the direction of magnetization within each domain to rotate toward the applied field direction (■ Figure 29.28). In some ferromagnetic materials, especially alloys such as neodymium–iron and Alnico (an alloy of nickel, cobalt, aluminum, and copper), the domains remain aligned when the applied field is removed, forming a permanent magnet. These materials are described as magnetically *hard*. In *soft* materials such as the ferrites (iron oxides combined with iron, nickel, or cobalt), the domains do not remain aligned when the applied field is removed. Pure iron is magnetically soft.

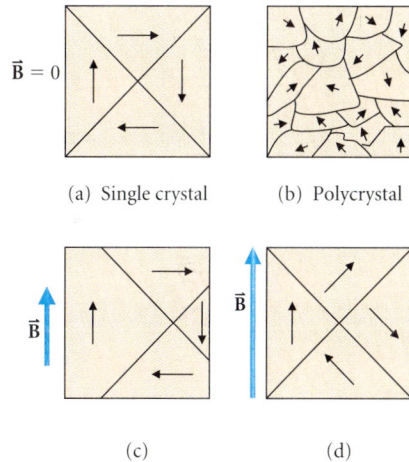

■ **Figure 29.28**
In an unmagnetized ferromagnetic material (a and b), the magnetic axes of different domains point in random directions, and there is no net magnetic field. A weak applied field (c) causes domains magnetized parallel to $\vec{B}$ to grow. A strong applied field (d) rotates the domains.

✱ 29.4.2 The Magnetic Field Intensity $\vec{H}$

The magnetic induction $\vec{B}$ at any point within a magnetic material arises both from distant sources and from nearby atomic magnetization. In advanced work, it is useful to separate these two contributions by defining a vector field $\vec{H}$ known as the *magnetic field intensity*. The *magnetization* of the material is described by $\vec{M}$, the magnetic dipole moment per unit volume, and the field intensity is defined by:

$$\vec{H} \equiv \frac{\vec{B}}{\mu_0} - \vec{M}.$$

Frequently, as we saw in the classical model of diamagnetism, the magnetization $\vec{M}$ is proportional to the applied field:

$$\vec{M} = \chi_m \vec{H}.$$

The constant χ_m is called the *magnetic susceptibility*. Then $\vec{B}$ and $\vec{H}$ are also proportional:

$$\vec{B} = \mu_0(1 + \chi_m)\vec{H} \equiv \mu \vec{H}. \qquad (29.11)$$

TABLE 29.2 Magnetic Susceptibilities of Some Common Materials

Paramagnetic Materials (at room temperature)		Diamagnetic Materials	
Substance	Susceptibility	Substance	Susceptibility
Aluminum	2.1×10^{-5}	Ammonia	-2.6×10^{-6}
Cesium	5.1×10^{-5}	Carbon (graphite)	-1.6×10^{-5}
Iron oxide (FeO)	7.2×10^{-3}	Copper	-9.7×10^{-6}
Lithium	1.4×10^{-5}	Hydrogen (H_2 gas)	-2.1×10^{-9}
Magnesium	1.2×10^{-5}	Lead	-1.6×10^{-5}
Oxygen (O_2 gas)	1.8×10^{-6}	Mercury	-2.8×10^{-5}
Platinum	2.8×10^{-4}	Nitrogen (N_2 gas)	-5.0×10^{-9}
Tungsten	6.8×10^{-5}	Sodium chloride	-1.4×10^{-5}
		Water	-9.1×10^{-6}

SOURCE: *Handbook of Chemistry and Physics*, 64th ed. (Cleveland: Chemical Rubber Co.)

The constant of proportionality μ is called the *magnetic permeability*. (The constant μ_0 is called the permeability of the vacuum, since in vacuum $\mu = \mu_0$.) • Table 29.2 lists magnetic susceptibilities for some common diamagnetic ($\chi_m < 0$) and paramagnetic ($\chi_m > 0$) materials. Note that $|\chi_m|$ is much less than unity for all the materials in the table.

Practical ferromagnetic materials are not well described by eqn. (29.11). For iron, quite reasonable values of H (about 80 A/m) result in complete alignment of all domains at $B = 2.15$ T. The field is said to be *saturated*. In addition, the domains do not completely return to a random orientation as the external field is removed. The value of B is not uniquely determined by the value of H (■ Figure 29.29). As H is increased and then decreased, B depends on earlier values of H as well as on its present value. This phenomenon is known as *hysteresis*, and the closed curve in a graph of B versus H is called the *hysteresis loop*. The hysteresis loop for soft materials is long and narrow (Figure 29.29a). For hard materials the loop is wider, and a substantial B remains when $H = 0$; such materials are used to make *permanent* magnets (Figure 29.29b).

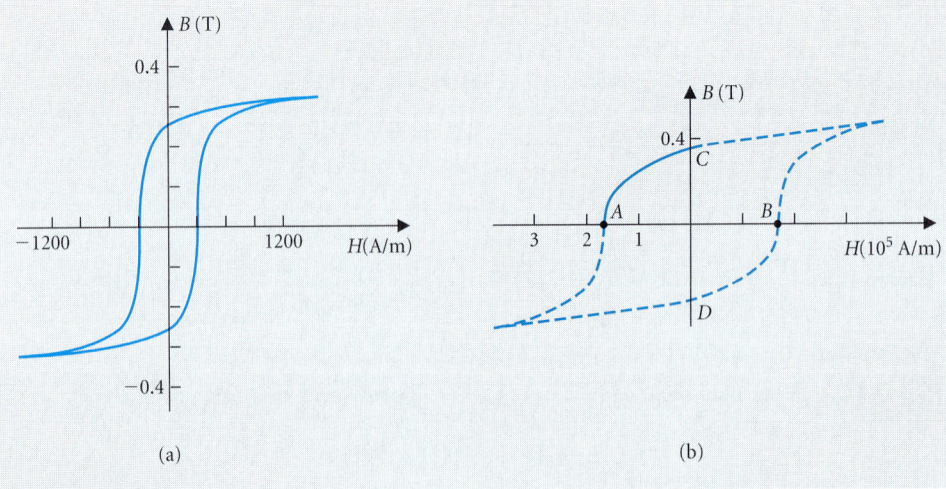

■ **FIGURE 29.29**
(a) Hysteresis curve for 4C4 Ferrite at 25° C. Ferrite is a *soft* material used in transformer cores. (b) Demagnetization curve for Ferroxdur ($BaFe_{12}O_{18}$), a *hard* magnetic material used for permanent magnets. *Solid line:* data; *dashed line:* schematic. Note how the scale differs from Figure 29.29a. $H = 10^5$ A/m is required to reduce B to zero in Ferroxdur, whereas only 300 A/m is required for Ferrite.

Digging Deeper

Magnetic Resonance Imaging

■ Figure 29.30 shows a view of the interior of a person's head made with the technique of nuclear magnetic resonance. This powerful imaging method uses magnetic field and radio waves to replace the damaging x and γ radiation used by competing techniques. Many atomic nuclei have a net angular momentum and a net magnetic moment. (The magnetic moments of nuclei are several orders of magnitude smaller than the typical moment of a paramagnetic atom.) An external magnetic field $\vec{B}_o$ causes alignment of the nuclear dipoles and a small bulk magnetization. Individual nuclei act like spinning tops. The external magnetic torque causes the nuclear top to precess like an actual top in a gravitational field (cf. §12.6 and Problem 59). The precession frequency, or *Larmor* frequency, is determined by the external magnetic field and the particular kind of nucleus; it typically falls in the radio frequency range.

If the magnetic field of a radio wave oscillating at the precession frequency is applied at right angles to $\vec{B}_o$, the nuclei absorb energy from the oscillating field, becoming aligned opposite $\vec{B}_o$ (cf. Problem 76). This resonant absorption of energy gives the technique its name.

When the applied radio wave is turned off, the nuclei realign with $\vec{B}_o$ and emit radio waves at the Larmor frequency. Interactions between the magnetic moments of neighboring nuclei (and hence the local chemical properties of the medium) determine the time the nuclei take to realign. Introducing small spatial variations in B_o causes the frequency of the emitted radiation to depend on position. Observations of decay time versus frequency can then be translated into a picture of chemical properties as a function of position—hence the image in Figure 29.30.

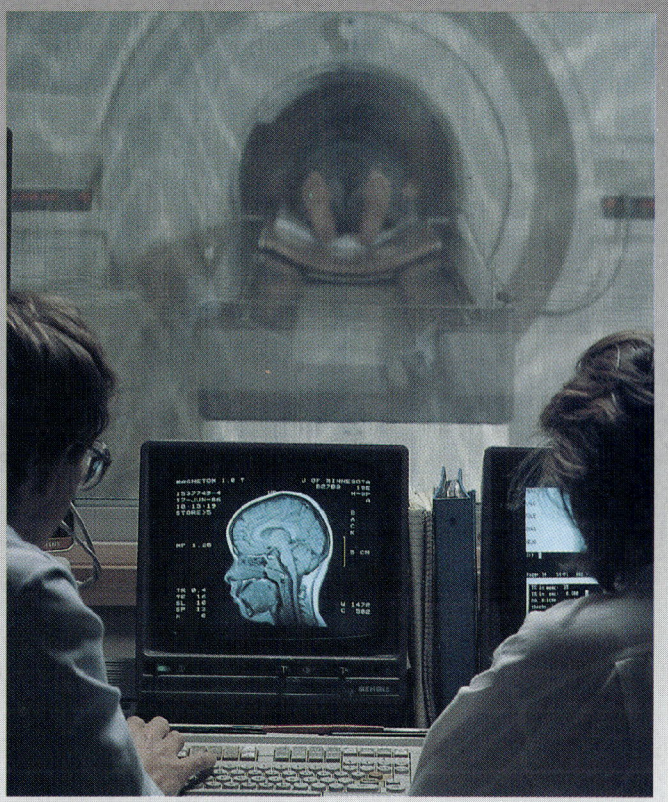

■ **Figure 29.30**
MRI image of a brain. Combinations of proton density and rates at which nuclei realign with $\vec{B}_0$ are interpreted as different kinds of tissue.

Chapter Summary

Where Are We Now?

We have studied the motion of particles in response to magnetic forces, and the forces and torques on current-carrying wires. These ideas allow us to understand many important technical applications of magnetic fields.

What Did We Do?

The magnetic force exerted on a moving charge is $\vec{F} = q\vec{v} \times \vec{B}$, perpendicular to both $\vec{v}$ and $\vec{B}$. A charged particle moving perpendicular to a uniform magnetic field $\vec{B}$ travels in a circle with an angular frequency $\omega_c = qB/m$ that is independent of its speed. The radius of the circle is the Larmor radius $r_L = mv/qB$. The particle's velocity parallel to $\vec{B}$ is unaffected by the magnetic field. In a uniform field particles move in regular spirals around the field lines.

Converging field lines form magnetic mirrors that can reverse particle motion along the field lines. This effect causes particles to be trapped in the Earth's Van Allen belts.

The magnetic force on a segment $d\vec{\ell}$ of a current-carrying wire is:

$$d\vec{F} = I\, d\vec{\ell} \times \vec{B}.$$

The magnetic force between two wires is used to define the ampere:

> If the same current is maintained in two long parallel wires that attract each other with a force of 2×10^{-7} N per meter of wire when the wires are 1 meter apart, then the current in each wire is 1 ampere.

The net force on a current loop in a uniform magnetic field is zero. The torque on a current loop is:

$$\vec{\tau} = \vec{m} \times \vec{B},$$

where $\vec{m}$ is the magnetic moment of the loop. The force and torque on a wire in a nonuniform field must be found using integration.

A conductor carrying current in a magnetic field develops an electric field perpendicular to both the field and the current. The sign of the resulting Hall potential shows that current in many metals, such as copper, is carried by negative charges. The magnetic force exerted on electrons in the wire is transmitted to the atoms by the Hall field.

When a magnetic field is applied to a medium, it causes two opposing effects: (a) lining up of the individual atomic magnetic moments and (b) induction in each atom of a magnetic moment that opposes the field. These effects give rise to paramagnetic and diamagnetic properties of materials. Ferromagnetism arises from a quantum mechanical interaction between atoms that causes the individual magnetic moments to line up in large regions called domains.

✶ When computing fields in magnetic materials it is useful to define a new field vector $\vec{H}$, the magnetic field intensity. For linear materials, $\vec{H}$ is related to $\vec{B}$ by the magnetic permeability μ:

$$\vec{B} = \mu\vec{H} = \mu_0(1 + \chi_m)\vec{H}.$$

Practical Applications

The spiral motion of particles in magnetic fields is used to advantage in the design of cyclotrons and synchrotrons for accelerating particles to high energy, in mass spectrometers, and in magnetic mirror machines for confining high-temperature plasma. Particles produced in high-energy physics experiments can be identified by the tracks they leave in bubble chambers. A magnetic field causes the tracks to curve (Figure 29.31). Magnetic force is also used to control the electron beam in a TV tube. The magnetic torque on a current loop is used to detect current in some electrical meters. Magnetic forces on current-carrying wires are the basis of electrical motors and moving-coil loudspeakers. Magnetic torques on current loops are used to control the orientation of spacecraft. Nuclear magnetic resonance is used in medical imaging. The direction of magnetization in ferromagnetic materials is used to store information on tapes used in tape recorders, VCRs, and credit cards.

Solutions to Exercises

29.1. The only difference between an electron and a positron is the sign of the charge. The force on the positron is equal in magnitude but opposite in direction. The positron orbits in the opposite sense.
■ Figure 29.31 shows tracks of an electron and a positron in a bubble chamber.

29.2. The proton has a positive charge e and mass 1837 times that of the electron. The cyclotron frequency is thus $\frac{1}{1837}$ that of the electron, or $(2.6 \times 10^4 \text{ rad/s})/1837 = 14$ rad/s. The Larmor radius is proportional to $\sqrt{m}$ and so is $\sqrt{1837} = 43$ times larger than that of the electron and equals 43(50 m), or 2 km.

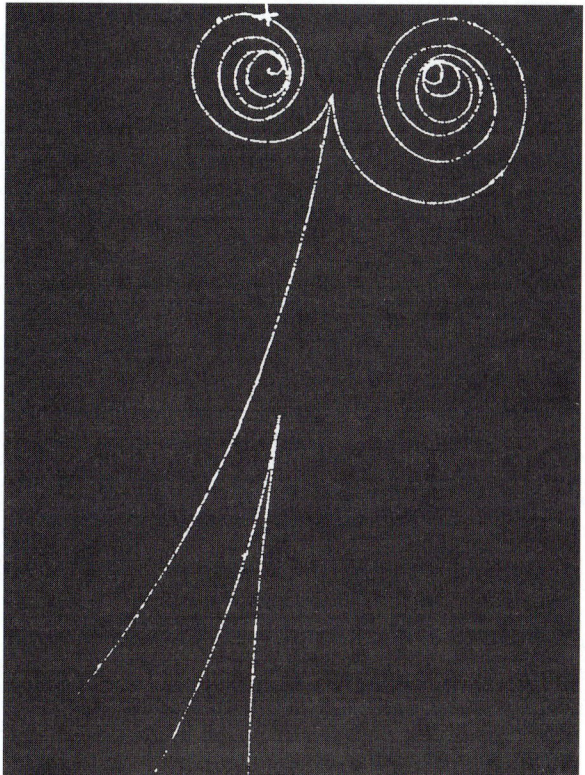

■ FIGURE 29.31
Tracks of an electron and positron in a bubble chamber. The particles are created when a high-energy γ ray scatters off an atomic electron. The direction of the curvature is determined by the sign of each particle's charge. As they lose energy, the particles follow a spiral of decreasing Larmor radius.

29.3. The mass of an oxygen ion is 16 u. Each singly ionized atom has charge $+e$. The required distance is twice the Larmor radius, so:

$$D = 2r_L = 2\frac{mv}{eB} = 2\frac{(2m\,\Delta V/e)^{1/2}}{B}.$$

For oxygen ions:

$$D_O = 2\frac{[2(16)(1.66 \times 10^{-27}\text{ kg})(1.0 \times 10^4\text{ V})/(1.6 \times 10^{-19}\text{ C})]^{1/2}}{0.20\text{ T}}$$

$$= 0.58\text{ m}.$$

For a nitrogen atom (mass 14 u):

$$D_N = \sqrt{M_N/M_O}\,D_O = \sqrt{14/16}\,(0.58\text{ m}) = 0.54\text{ m}.$$

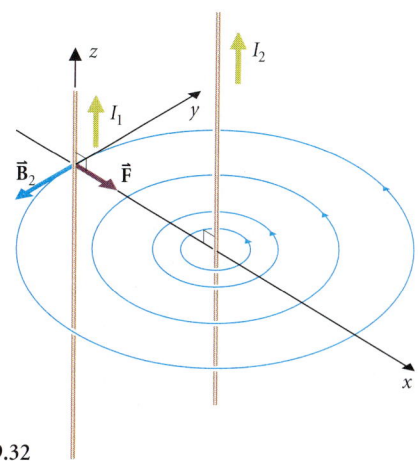

■ FIGURE 29.32

29.4. The Lorentz force on the particle is:
$$\vec{F} = q(\vec{E} + \vec{v} \times \vec{B}) = q(\vec{E} + \vec{v}_o \times \vec{B} + v_y\hat{j} \times \vec{B}).$$
From the definition of $\vec{v}_o$, $\vec{E} + \vec{v}_o \times \vec{B} = 0$, and $\hat{j} \times \vec{B}$ is also zero. The particle's velocity remains constant.

29.5. From eqn. (29.5):
$$\vec{F} = I\,d\vec{\ell} \times \vec{B} = (5.0\text{ A})(1.0 \times 10^{-3}\text{ m})(0.60\text{ T})(\hat{i} \times \hat{j})$$
$$= 3.0 \times 10^{-3}\text{ N }\hat{k}.$$

29.6. (■ Figure 29.32) The wire on the right produces magnetic field:
$$\vec{B}_2 = \frac{\mu_0 I_2}{2\pi d}(-\hat{j})$$
at the position of the left-hand wire. So a length $d\ell$ of that wire experiences a force:
$$d\vec{F} = I_1\,d\ell\,\hat{k} \times \frac{\mu_0 I_2}{2\pi d}(-\hat{j}) = \frac{\mu_0 I_1 I_2}{2\pi d}\,d\ell\,\hat{i},$$
which has the same magnitude as but direction opposite the force per unit length on the right-hand wire.

29.7. To produce the same current, the positive charges would have to drift in the opposite direction. With both charge and velocity changing sign, the magnetic force would be in the same direction as with negative charge carriers. Thus positive charge carriers would drift to the same edge as negative carriers do and give rise to Hall electric field in the opposite direction. The sign of the potential difference would be reversed. Certain materials (e.g., beryllium) do have positive charge carriers.

29.8. If $\vec{B}$ is antiparallel to $\vec{m}$, then the magnetic force is radially inward, increasing the centripetal acceleration. The speed increases and so does $|\vec{m}|$.

Basic Skills

Review Questions

§29.1 MOTION OF CHARGED PARTICLES IN A MAGNETIC FIELD

- When a charged particle moves perpendicular to a uniform magnetic field, what is the shape of its path?
- Give a formula for the *Larmor radius*. If a particle's speed doubles, how does the Larmor radius change?
- What is the *cyclotron frequency*? How does it depend on $\vec{B}$? Which properties of the particle determine ω_c?
- How is a cyclotron able to accelerate particles to high energy with a small potential difference?

- If $\vec{E}$ is perpendicular to $\vec{B}$, particles with a certain velocity feel no force. What is that velocity? On which properties of the particle does it depend?
- Describe the motion of a particle moving in the plane perpendicular to $\vec{B}$ when there is a uniform electric field in that plane.

§29.2 FORCES ON CURRENT-CARRYING WIRES

- What force is exerted on a wire segment $d\vec{\ell}$ carrying current I in a region of space with a magnetic field $\vec{B}$?
- Describe the force on a finite wire segment carrying current I in a region where $\vec{B}$ is uniform. Explain why the line of action of the net force acts at the center of the segment.
- What is the official SI definition of the ampere?
- What is the net force acting on a closed current loop in a uniform magnetic field?
- What is the net torque on a closed current loop in a uniform magnetic field?

§29.3 THE HALL EFFECT

- Describe the Hall effect. Explain how it allows us to determine the sign of the moving charges in a conductor.

§29.4 MAGNETIC MATERIALS

- Describe the differences between *diamagnetic* and *paramagnetic* materials.
- What is *ferromagnetism*? Name at least one ferromagnetic material.
- What is the Curie temperature?
- ✱ What are the magnetic field intensity $\vec{H}$ and the magnetization $\vec{M}$? How are they related?
- What is *hysteresis*?

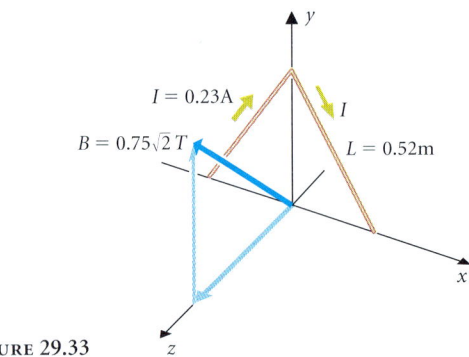

FIGURE 29.33

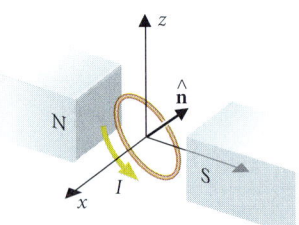

FIGURE 29.34

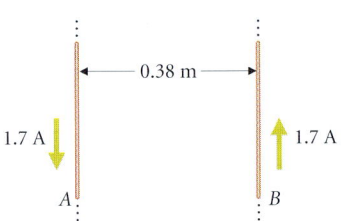

FIGURE 29.35

Basic Skill Drill

§29.1 MOTION OF CHARGED PARTICLES IN A MAGNETIC FIELD

1. The magnetic field strength in interstellar space is about 0.3 nT. What are the Larmor radius and cyclotron frequency of an interstellar electron with an energy of 10^{-19} J?

2. Find the Larmor radius and cyclotron frequency for a proton in the solar corona with $v = 1.0 \times 10^5$ m/s and $B = 0.10$ T.

3. The original cyclotron constructed by Lawrence in 1931 was 4 in. in diameter and accelerated H_2 ions to 80 keV. What was the value of the magnetic field?

§29.2 FORCES ON CURRENT-CARRYING WIRES

4. A wire segment of length $d\ell = 1.0 \times 10^{-3}$ m lies along the y-axis and carries current $I = 0.67$ A toward the positive y-direction. The magnetic field at the location of the segment is $\vec{B} = (1.2 \text{ T})(\hat{i} - \hat{k})$. What magnetic force acts on the segment?

5. A segment of wire carrying current $I = 0.23$ A lies along two sides of an equilateral triangle in the x-y-plane (■ Figure 29.33). The third side of the triangle lies along the x-axis. The magnetic field has the uniform value $\vec{B} = (0.75 \text{ T})(\hat{j} + \hat{k})$. What is the net magnetic force acting on the segment?

6. ■ Figure 29.34 shows a current loop that lies between two magnetic pole pieces. The normal $\hat{n}$ to the loop lies in the y-z-plane. Which of the following statements are correct?
(a) The loop does not move, since the net force acting on it is zero.
(b) The loop rotates so that $\hat{n}$ points toward the south pole S.
(c) The loop rotates so that $\hat{n}$ points toward the north pole N.
(d) The loop rotates so that $\hat{n}$ lies along the x-axis.
(e) The loop rotates clockwise about the x-axis and accelerates along the y-axis.
Explain what is wrong with the answers you didn't choose.

7. Two long straight wires carry a current $I = 1.7$ A each, in opposite directions (■ Figure 29.35). If the separation of the wires is 0.38 m, what is the force per unit length on wire B due to wire A?

8. A compass needle has dipole moment $m = 0.1$ A·m². What torque acts on the needle if it is oriented east-west on the Earth's surface at the equator, where the Earth's magnetic field is $B = 3 \times 10^{-5}$ T?

§29.3 THE HALL EFFECT

9. Find the Hall potential across a copper wire 1.0 mm in diameter carrying a 2.5-mA current perpendicular to a magnetic field $B = 1.3$ T.

§29.4 MAGNETIC MATERIALS

10. The magnetic susceptibility of a material is:
(a) necessarily positive.
(b) always small and negative.
(c) always proportional to temperature.
(d) generally $\gg 1$.
(e) none of the above.
Explain what is wrong with the answers you didn't choose.

11. ✱ The magnetic field intensity H in a platinum bar is 1.00000×10^2 A/m. What are the magnetization M and the induction B?

Questions and Problems

§29.1 MOTION OF CHARGED PARTICLES IN A MAGNETIC FIELD

12. ❖ A negatively charged electron moves in vacuum parallel to a long straight wire carrying current as shown in ■ Figure 29.36. The electron is deflected:
(a) left. (b) up. (c) right. (d) down. (e) not at all.
Explain why you chose your answer.

13. ❖ An electron is injected with speed v into a region where the magnetic field points into the page (■ Figure 29.37). Describe its subsequent motion.

■ FIGURE 29.36   ■ FIGURE 29.37

14. ❖ ■ Figure 29.38 shows several possible paths of particles in a uniform magnetic field. Electrons and protons with equal energy enter the system from the right. Which pair of paths represent their motion?
(a) electrons (2) protons (4)   (b) electrons (4) protons (2)
(c) electrons (1) protons (3)   (d) electrons (3) protons (1)

15. ❖ Can a cyclotron accelerate particles with either sign of charge? If so, what changes are necessary?

16. ❖ A long straight filament carries a uniform charge density λ, and an electron is moving in a circular orbit about the filament. What happens to the electron's orbit if a current I is established in the filament?

17. ❖ Low-energy cosmic rays reach the Earth's surface in much greater numbers in northern latitudes than near the equator. Assuming these charged particles approach the Earth uniformly from all directions, explain the observed distribution at the Earth's surface.

18. ❖ If an electric field exists parallel to $\vec{B}$, describe the motion of a charged particle with initial velocity (a) parallel to $\vec{B}$ and (b) perpendicular to $\vec{B}$.

19. ◆ (a) Some playful thinkers have imagined the ultimate particle accelerator: a ring around the Earth's equator, using the Earth's magnetic field to hold the particles in orbit. The Earth's field at the equator $B_E \approx 3 \times 10^{-5}$ T. What is the momentum of a proton circling in such an accelerator? Express your results in the physicists' unit GeV/c. (b) What proton momentum could such an accelerator achieve if 20-T superconducting magnets were used instead of the Earth's field?

20. ◆ Radiation from the magnetron in your microwave oven is emitted by electrons orbiting in a magnetic field. If the radiation is emitted at the cyclotron frequency f_c, what field strength is necessary to produce microwaves of frequency 3 GHz?

21. ◆◆ Find the Larmor radius and the cyclotron frequency f_c for a proton and an electron in the outer Van Allen belt at 3.4 Earth radii above the equator. Assume each particle has a speed of 3.0×10^5 m/s perpendicular to the magnetic field, and the Earth's magnetic field is a dipole with $B = 3.12 \times 10^{-5}$ T at the equator.

22. ◆◆ A 10-in. diameter cyclotron was built by Lawrence and Livingstone in 1932. It produced 1-MeV protons. What magnetic field was necessary?

23. ◆◆ With magnets that produce a field of 0.13 T, what radius cyclotron would accelerate a beam of protons to an energy of 350 keV? At what frequency should the potential difference across the dees be changed?

24. ◆◆ The synchrotron at Fermi National Laboratory produces a proton beam with a kinetic energy of 10^{12} eV per proton. The radius of the accelerator is 1 km. What magnetic field strength is required? (At this energy, a proton moves with very nearly the speed of light, and its effective mass γm is 1.8×10^{-24} kg.)

25. ◆◆ A mass spectrometer is designed to operate with a magnetic field of 1.3 T and a fixed distance of 0.30 m between the beam entrance and the detector. What potential difference is necessary for the device to detect singly ionized benzene molecules? (The mass of a benzene molecule is 78 u.)

26. ◆◆ A beam of carbon nuclei emerges from an accelerator with kinetic energy $K = 10$ GeV. A "bending magnet" is used to aim the beam toward an experimental chamber. If the beam is to be bent in a circle of 1.0-m radius, what magnetic field is required? Use the relativistic formula for the Larmor radius with $\gamma = 1 + K/mc^2$ (see *Digging Deeper: More on Cyclotrons*), and assume the nuclei have speed $v \approx c$.

27. ◆◆ Deflection of the electron beam in a modern TV set is accomplished by magnetic forces. The electrons are accelerated until they have kinetic energy of 25 keV (cf. Example 24.8). What magnetic field is required to produce a deflection of 3.3 cm if the magnetic field region is 8.0 cm long? (Newtonian mechanics is adequate for this problem. Relativistic effects make about a 3% correction.) If

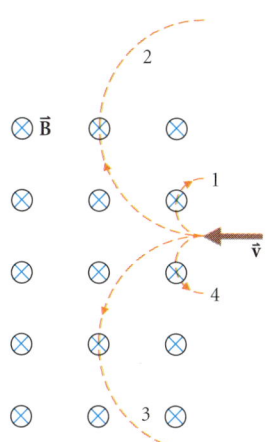

■ FIGURE 29.38
Relative sizes of small and large circles are not to scale.

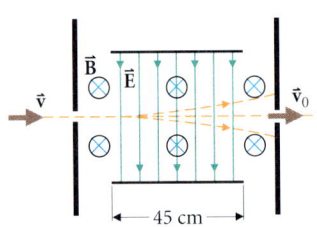

FIGURE 29.39

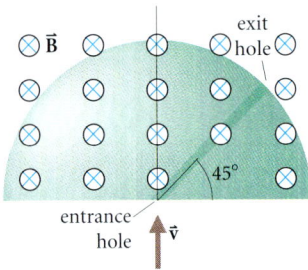

FIGURE 29.40

there is a 22-cm gap between the end of the magnetic field region and the end of the tube, how far from the center of the screen does the beam strike?

28. ♦♦ **(a)** A velocity selector for electrons (■ Figure 29.39) has electric and magnetic fields at right angles. If the magnitudes of the fields are $E = 1.5$ kV/m and $B = 0.75$ T, at what speed v_0 can particles pass through the selector? **(b)** If a particle's speed is $v_0 + \delta v$ and its charge-to-mass ratio is $q/m = \alpha$, what are its acceleration and its displacement perpendicular to its velocity after a time interval Δt? **(c)** If the length of the field region is 45 cm, the exit aperture has radius $r = 0.50$ cm, and all particles enter exactly perpendicular to both $\vec{E}$ and $\vec{B}$, what range of initial speeds exit the selector?

29. ♦♦ A velocity selector for electrons is an evacuated hemisphere within which the magnetic field is parallel to the flat base. Electrons are admitted through a hole at the center of the base and exit through a second hole halfway to the top of the hemisphere (■ Figure 29.40). Find the speed of electrons that exit the hole if the radius of the hemisphere is 0.15 m and $B = 5.4 \times 10^{-4}$ T.

30. ♦♦ Find the separation between the last two orbits in the MIT cyclotron that accelerates deuterons ($m = 2.0$ u, $q = +e$) to an energy of 16 MeV. The potential difference between the dees is 80.0 kV, and the radius of each dee is 18.74 in.

§29.2 FORCES ON CURRENT-CARRYING WIRES

31. ❖ A length of wire carrying current I lies between two magnet pole pieces (■ Figure 29.41). What is the direction of the force on the wire?

32. ❖ A long straight wire runs along the axis of a circular wire loop. If both wires carry current I, what is the net force acting **(a)** on the straight wire and **(b)** on the loop? Justify your answer.

33. ❖ A long straight wire lies along the z-axis and carries current I_1. A small element of wire $d\vec{\ell}$ carrying current I_2 is placed at P, distance d from the long wire, with $d\vec{\ell}$ lying in the x-z-plane. What orientation of the wire segment maximizes the y-component of the force acting on the segment? Explain your reasoning.

34. ❖ A loop carrying a current I has a magnet of dipole moment m at its center (■ Figure 29.42). Both loop and dipole lie in the x-y-

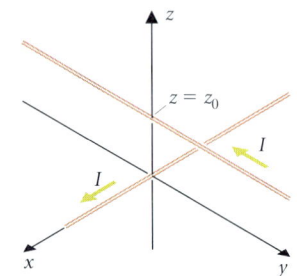

FIGURE 29.43

plane, and the magnet is fixed. Which of the following statements is correct?
(a) The loop is in equilibrium.
(b) The loop rotates around line BD with C moving out of the page.
(c) The loop rotates around line BD with A moving out of the page.
(d) The loop rotates around line AC with B moving out of the page.
(e) The loop rotates around line AC with D moving out of the page.
Explain how you chose your answer.

35. ❖ Two long straight wires carry currents of equal magnitude I in the directions shown in ■ Figure 29.43. One lies along the x-axis, and the other is parallel to the y-axis and passes through the point $x = 0$, $y = 0$, $z = z_0$. What is the net force on the second wire due to the first?

36. ❖ Two coils A and B both lie with their axes along the y-axis, and each carries a current I (■ Figure 29.44). What is the direction of the net force on coil B?

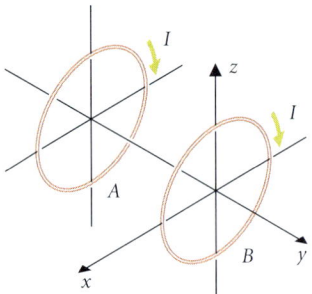

FIGURE 29.44

37. ❖ A small current loop lies in the x-y-plane centered at the origin. A long straight wire is parallel to the x-axis and passes through the point $(0, 0, z_0)$. Currents are in the directions shown in ■ Figure 29.45. The loop experiences:
(a) neither a net force nor a net torque.
(b) no torque but a net force in the $+z$-direction.
(c) no net force but a net torque in the $-y$-direction.
(d) no net force but a net torque in the $+x$-direction.
(e) a net force in the $-y$-direction and a net torque in the $+x$-direction.
Explain how you chose your answer.

FIGURE 29.41

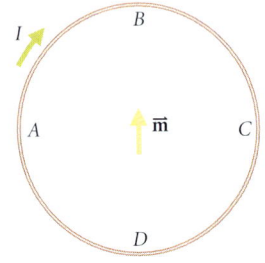

FIGURE 29.42

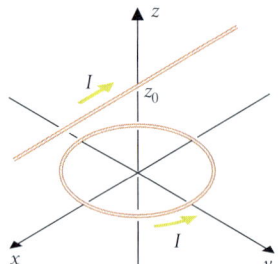

FIGURE 29.45

38. ❖ Two concentric wire loops lying in the same plane carry currents in the same direction. Sketch the magnetic field lines produced by the outer loop. What net force and net torque are exerted on the inner loop? Repeat the analysis when the inner loop has been rotated through 90° so that its plane is perpendicular to the plane of the outer loop. Discuss the stability of the first configuration. Is the system stable when the two loops lie in the same plane but have currents in opposite directions?

39. ❖ Two long wires carry opposite currents I parallel to the x-axis at $z = 0$, $y = \pm d$. Describe qualitatively how the net force and net torque on a small square current loop vary as it is moved along the z-axis, along the x-axis, and around a circle centered on the wire at $z = d$. Assume the loop is always oriented parallel to the x-y-plane.

40. ◆ A wire segment 1.2 cm long lies along the x-axis and carries a current of 4.3 mA. The magnetic field is uniform and equals $(0.045\text{ T})(\hat{\imath} + \hat{\jmath})$. Calculate the force on the wire segment.

41. ◆ A rectangular coil 6.0 cm by 12.0 cm has six turns and carries a 1.1-A current. It lies between the poles of a permanent magnet with its plane parallel to the 0.25-T magnetic field. What is the torque on the loop?

42. (a) ◆ A small current loop of area $A = 1.5\text{ cm}^2$ carries current $I = 3.6$ mA. What torque acts on it in a lab where its magnetic moment is perpendicular to the Earth's magnetic field of 4.2×10^{-5} T? (b) ◆◆ Find its angular acceleration if its rotational inertia about a diameter is 6.2 g·cm^2.

43. ◆◆ Two parallel, semi-infinite wires are connected by a semicircular wire of radius R (■ Figure 29.46). The magnetic field is uniform, parallel to the straight wires, and perpendicular to the plane of the semicircle. Find the total magnetic force on the object.

44. ◆◆ Find the force and torque on the rectangular loop shown in ■ Figure 29.47.

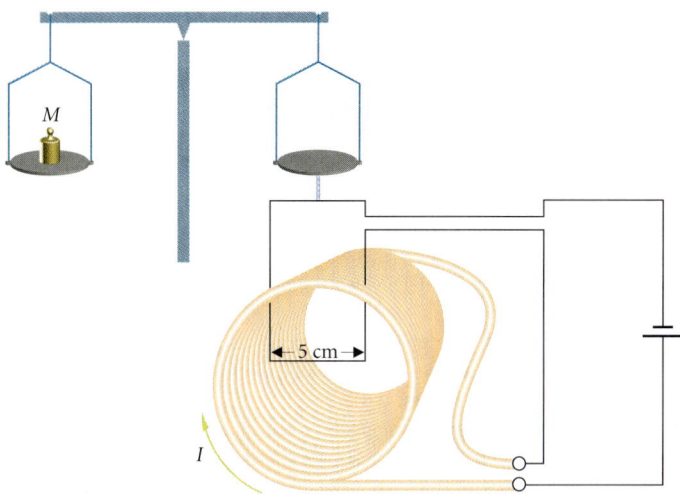

■ FIGURE 29.48

wire segment of length $\ell = 5.0$ cm (■ Figure 29.48). What mass M is necessary for equilibrium when the current in the system is 10.0 A?

47. ◆◆ A galvanometer (cf. §26.5) consists of a square coil of side 2.0 cm with 1.0×10^3 turns lying in the magnetic field produced by a permanent magnet (■ Figure 29.49). The coil experiences a torque when it carries current. Why? Rotation of the coil is opposed by a spring that exerts a torque $\tau = \Gamma\theta$ when the coil is deflected through angle θ (in radians). The coil is attached to a pointer. If the magnetic field at the position of the coil is 0.70 T, find the relation between I and θ. What value of Γ is required if a full-scale deflection of 45° is to correspond to 1.0 mA?

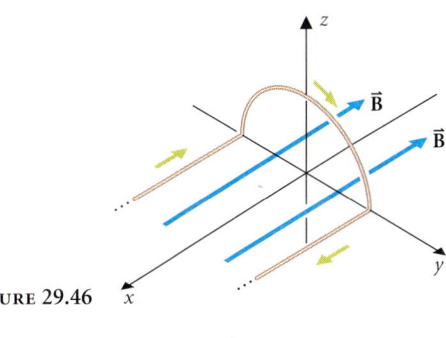

■ FIGURE 29.46

■ FIGURE 29.49

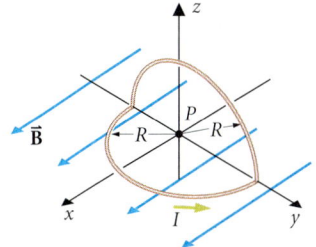

■ FIGURE 29.47

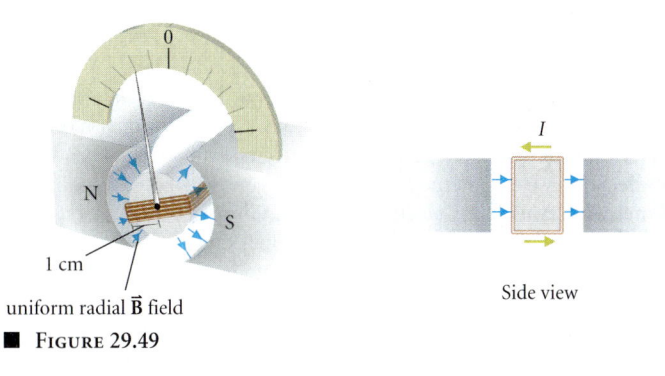

■ FIGURE 29.50

45. ◆◆ A wire loop is in the form of an equilateral triangle. The magnetic field lies in the plane of the triangle, perpendicular to one side. Find the force acting on each side of the triangle, and verify that the net force is zero. Find the torque about the center of the loop, and verify that it equals $\vec{m} \times \vec{B}$.

46. ◆◆ A current balance is constructed to measure the force between a solenoid with $n = 1.0 \times 10^4$ turns per meter and a straight

48. ◆◆◆ The wire loop shown in ■ Figure 29.50 carries a current I. If the magnetic field is uniform, $\vec{B} = B_0\hat{\imath}$, compute the total force and torque about the origin acting on the loop.

§29.3 THE HALL EFFECT

49. ❖ A conducting strip has length ℓ and width w perpendicular to a magnetic field B, and thickness d parallel to the field. The material

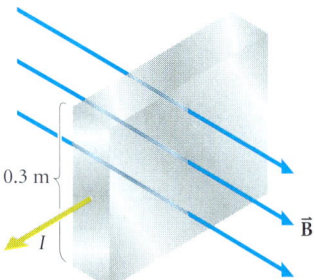

■ FIGURE 29.51

has resistivity ρ and a potential difference V is maintained between the ends of the strip. How does the Hall potential difference depend on each of these parameters?

50. ♦ There is a current along a silver bus bar 0.3 m wide (■ Figure 29.51), immersed in a magnetic field $B = 1$ T. If the drift speed of the electrons required to carry the current is 10^{-5} m/s, compute the potential difference between the top and the bottom of the bus bar required to counteract the magnetic force on conduction electrons. Is the top at positive or negative potential with respect to the bottom?

51. ♦♦ The Hall potential measured across a copper wire 1.17 cm in diameter carrying a 175-A current is 1.05 μV. What is the component of B perpendicular to the current? (Copper has 8.5×10^{28} /m^3 conduction electrons.)

52. ♦♦ Find the current in a copper strip 1.2 cm $\times$ 0.30 cm with a magnetic field of 0.96 T perpendicular to the 1.2-cm side if the Hall potential difference is 0.74 μV. (Density of conduction electrons is 8.5×10^{28} /m^3.)

53. ♦♦ The electric *third rail* in a subway system is made of steel, measures 10.0 cm high by 2.0 cm wide, and carries a 150-A current. Calculate the Hall potential difference across the rail due to the Earth's field with components $B_{\text{horiz}} = 3.0 \times 10^{-5}$ T and $B_{\text{vert}} = 2.7 \times 10^{-5}$ T. (Assume steel has 8×10^{28} /m^3 conduction electrons.)

54. ♦♦♦ An aluminum strip carries a current $I = 30$ A in a magnetic field $B = 1.2$ T. The strip has a rectangular cross section with dimensions 1.0 mm parallel to the magnetic field and 3.00 cm perpendicular to the field. A Hall potential $V_H = 1.2$ μV is observed across the 3-cm dimension. How many electrons per aluminum atom are free to carry current? (data for Al: density = 2700 kg/m^3; mass of Al atom = 27 u)

55. (a) ♦♦ To extract energy from the system of Example 29.11, the Hall potential is allowed to drive current I' through each of $N = 100$ shunt resistors (■ Figure 29.52). If each shunt resistance is $r = 1.0$ kΩ, find the power extracted ($L = 1.0$ cm, $w = 1.0$ mm, $B = 1.0$ T, $I = 5.0$ A). **(b)** ♦♦♦ Show that the power output of the battery equals the power dissipated in the bar plus the power dissipated in the shunts.

§29.4 MAGNETIC MATERIALS

56. ❖ What would happen to an iron bar magnet if it were heated to 800° C?

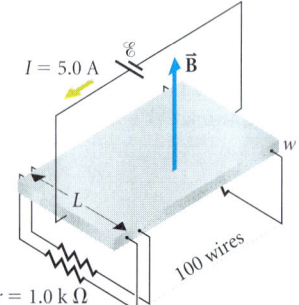

■ FIGURE 29.52

57. ❖ A sphere of magnetic material is centered on the axis of a current loop (but not at its center). Model the current loop and the sphere as bar magnets and explain why the loop attracts a paramagnetic sphere and repels a diamagnetic sphere.

58. ♦ The magnetic moment of a finite, tightly wound solenoid may be expressed as (cf. Problem 28.34) $m \approx BV/\mu_0$, where B is the magnetic field and V is the volume inside the solenoid. A cylindrical bar magnet is 10 cm long and has a 5-cm radius. The magnetic induction within the material is $B = 1.3$ T. Use the preceding relation to estimate its magnetic moment.

59. ✳ A magnetic dipole of moment $\vec{m}$ is spinning about its axis with angular momentum $\vec{L}$. If the dipole is at an angle θ to a uniform magnetic field $\vec{B}$, find the precession frequency in terms of L/m. (*Hint:* Refer to Example 12.17.)

Additional Problems

60. ❖ A sphere with both mass and charge distributed uniformly rotates with angular velocity ω. Show that the angular momentum $\vec{L}$ and the magnetic moment $\vec{m}$ of the sphere are both parallel to ω.

61. ❖ **(a)** One could design a cyclotron to correct for relativistic variation of the cyclotron frequency by allowing the magnetic field to increase with radius. Explain why. **(b)** This design has an inherent flaw. In a cylindrically symmetric field, the field lines bend toward regions of lesser field strength (■ Figure 29.53). Show that this makes the beam unstable against any deviation from the central plane of the machine.

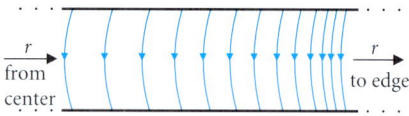

■ FIGURE 29.53

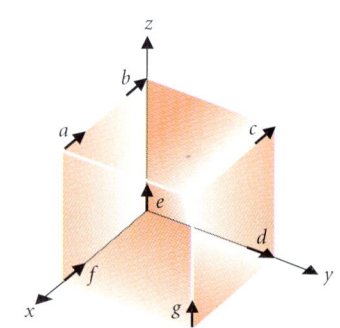

■ FIGURE 29.54

62. ❖ Refer to ■ Figure 29.54. Which pairs of wire elements, if any, exert forces that obey Newton's third law applied to the elements alone?

63. ❖ A long straight wire carries current along the axis of a long, loosely wound solenoid. Is there a net force on either object? What is the effect of the wire on the solenoid?

64. ❖ Two identical coils, each carrying current I, are suspended near each other by threads. When the two coils are in equilibrium, are their currents in the same or in opposite senses? Discuss.

65. ♦ Electrons in Jupiter's magnetosphere radiate cyclotron radiation in the dekameter band. Assuming that the electrons radiate at the cyclotron frequency, estimate Jupiter's magnetic field.

66. ♦ An electron with speed $v = 3 \times 10^5$ m/s is moving parallel to the current in a long straight wire 15 cm away. If the wire carries $I = 1.3$ mA, what is the acceleration of the electron?

67. ♦ A whimsical engineer has designed a train to use the vertical component of the Earth's magnetic field ($\approx 10^{-5}$ T) for propulsion. What current through the 3-m-long axle is necessary to provide a force of 10^5 N? If the axle has a 1-Ω resistance, what power is necessary? Comment on your result.

68. ♦♦ A small magnetic dipole, $m = 1.0 \times 10^{-4}$ A·m², lies in the x-y-plane at the origin. A current loop of radius $a = 12$ cm carrying a 1.0-mA current also lies in the x-y-plane with its center at the origin. How large is the torque acting on the magnetic dipole?

69. ♦♦ A particle moves in a spiral whose axis is a long straight wire carrying current I. The particle's velocity component parallel to the wire $v_\parallel$ is in the direction of I. What is the sign of the particle's charge Q? What relation must hold between Q, I, $v_\parallel$, $v_\perp$, and r?

70. ♦♦ What is the Larmor radius of a cosmic ray proton with energy $E = 10^9$ GeV and momentum $p = E/c$ in the intergalactic magnetic field of 0.1 μG? If the size of the galaxy is about 3×10^{20} m, what is the maximum energy proton the galaxy can retain?

71. ♦♦ A tangent galvanometer used to measure the strength of the Earth's magnetic field consists of a circular coil with N turns and a small compass needle suspended at its center (■ Figure 29.55). The device is set up with the coil in a north-south plane, so that the compass needle is in the plane of the coil. With a current I in the coil, the needle is deflected by an angle θ. Find the relation between θ, I, the coil's radius a, and the horizontal component of the Earth's field.

72. ♦♦ A charged particle moving in a circle constitutes a current loop. Find the magnetic moment m of the loop in terms of the radius of the circle and the angular frequency. If the circular motion is caused by magnetic force, the appropriate radius is the Larmor radius. Express the magnetic moment in terms of B. When a particle moves in a field of variable strength, its magnetic moment remains constant. (We'll see why in Chapter 30.) This is another way to understand the magnetic mirror effect. The particle's kinetic energy K also remains constant. (Why?) Express the energy in terms of the components of velocity perpendicular ($v_\perp$) and parallel ($v_\parallel$) to $\vec{B}$, and use the result to express $v_\parallel$ in terms of m and K. At what value of B does the mirror reflection occur?

73. ♦♦♦ A nonuniform field in a region of space has $\vec{B} = (10^{-4}$ T$)\hat{\imath}$, $dB/dx = 10^{-5}$ T/m. Use Gauss' law for the magnetic field applied to the volume shown in ■ Figure 29.56 to find the slope of a field line that is 0.2 m from a line that lies along the x-axis. What is the x-component of acceleration for a proton with Larmor radius 0.2 m moving along the central field line in the +x-direction?

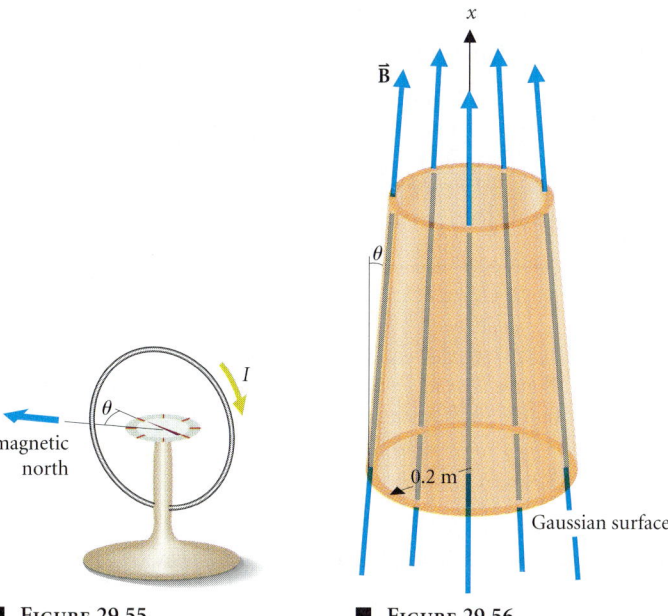

■ FIGURE 29.55

■ FIGURE 29.56

Computer Problem

74. (a) Program MAGMOT on your supplementary computer disk computes the motion of a charged particle in a uniform magnetic field. The field is in the z-direction, and the particle starts at $x = -1$, $y = z = 0$, moving in the y-direction. The program asks you to input the initial y-component of v; 1 works well. You also need to input a time step (try 0.01) and how often you want the program to output the particle's position. The program then plots the particle's path in the x-y-plane on the screen. Try changing the initial speed and seeing how the orbit varies. The particle has unit charge and unit mass, and the field strength is also 1 unit. How does the computed orbit radius compare with the expected value? How does the computed period compare with the expected value? (b) Program MAGTWO computes the motion of the same particle in a variable magnetic field with components $B_z = 1 + z/5$, $B_x = -(x+1)/10$, and $B_y = -y/10$. The particle starts with $v_z = 0.5$. Choose $v_y = 1$. Where does the particle *mirror*? What happens if you increase v_y? What if you decrease v_y?

Challenge Problems

75. The conclusion that $\vec{E}_H = -\vec{v}_d \times \vec{B}$ is valid in a fixed wire. What correction is necessary to find the Hall electric field in a segment of wire that is accelerated by the magnetic force?

76. (a) Find the work required to rotate a magnetic dipole from a direction parallel to the magnetic field to a direction at angle θ to the field. Show that your result can be interpreted in terms of a potential energy of the dipole given by $U = -\vec{m} \cdot \vec{B}$. (b) In a nonuniform magnetic field, the potential energy computed in part (a) varies if the dipole is displaced without changing its orientation. This change in energy results from a net force acting on the dipole. Describe how to find the force acting on a dipole in a nonuniform field. Check your idea by repeating Example 29.8 with magnetic field given by $B_0(1 + \alpha x)\hat{k}$. (c) Compute the torque exerted by the second loop in Example 29.10 on the first. Note that it is not equal to and opposite the torque exerted by loop 1 on loop 2. (d) What couple must act on the two loops if the net torque exerted by the system on itself is zero? (e) Use the result of part (b) (force = $\Delta(\vec{m} \cdot \vec{B})$/displacement) to calculate the force acting on each loop, and show that the required couple does act.

77. A coil with $N = 1.0 \times 10^4$ turns and radius $r = 1.0$ cm carries current $I = 10.0$ A. The coil lies in the x-y-plane with its axis along the z-axis. A long straight wire parallel to the x-axis passes through the z-axis at $z = 10.0$ cm. (See also Figure 29.45 and Problem 37). The wire carries current $I' = 5.0$ A toward $-\hat{\imath}$. (a) Modeling the coil as a dipole, compute the force and torque it exerts on the wire. (b) Compare each with the force and torque the wire exerts on the dipole. (Use the method of Problem 76e.)

78. Use the vector identity

$$\vec{a} \times (\vec{b} \times \vec{c}) = \vec{b}(\vec{a} \cdot \vec{c}) - \vec{c}(\vec{a} \cdot \vec{b})$$

to show that the torque on an arbitrary plane current loop in a uniform field is given by $\vec{\tau} = I \oint (\vec{r} \cdot \vec{B}) d\vec{\ell}$. (*Hint:* Show that $\oint \vec{r} \cdot d\vec{\ell} \equiv 0$ for a closed loop.) Choose coordinates with the z-axis perpendicular to the loop and the y-axis perpendicular to $\vec{B}$. Let θ be the angle between $\vec{B}$ and the z-axis. Show that the torque is:

$$\vec{\tau} = IB \sin \theta \oint x \, d\vec{\ell}.$$

Show by a geometric argument that:

$$\oint x \, d\vec{\ell} = \hat{\jmath}(\text{area of loop}).$$

Thus show that $\vec{\tau} = \vec{m} \times \vec{B}$ for the arbitrary loop.

Part VI Problems

1. ❖ It is possible to increase the magnetic field of a wire loop either (a) by increasing the current by a factor N or (b) by using N identical loops in a coil. Show that the ratio of power used by the coil to power used by the loop with the larger current is $1/N$. Power consumption is one of the most important considerations in magnet design.

2. ◆◆ A beam of ions is removed from a cyclotron using a deflector (■ Figure VI.25). The deflector is at a high negative potential with respect to the dee so that an outward electric force is applied to the ions. The MIT cyclotron with $R = 18.75$ in. and deflector spacing of 0.3 in. accelerates deuterons (^{2}H ions) to 16 MeV. What potential difference across the deflector is required to increase the orbit radius by 20%?

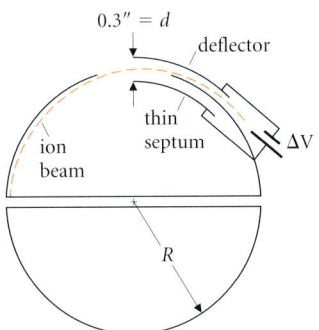

■ FIGURE VI.25

3. ◆◆◆ A crude but effective model of a hydrogen atom consists of a proton attached to an electron by a spring with spring constant $k = 1.6 \times 10^3$ N/m. Find the polarizability of a hydrogen atom, that is, the ratio of its electric dipole moment to the electric field that causes the moment. Estimate the force exerted by a water molecule on a hydrogen atom 10^{-7} m away (cf. Example 24.12 and Exercise 24.6).

4. ◆◆◆ Suppose Kirchhoff's junction rule were not quite obeyed. Specifically, in ■ Figure VI.26 suppose that $I_3 = 0.999(I_1 + I_2)$ and that the remaining 0.1% of the current deposited charge at junction A, 10 cm from each of the two 12-V batteries. How long could this situation persist before the potential at each battery due to the point charge at A exceeded the emf of the battery?

5. ◆◆◆ A parallel plate capacitor has plates of area A and separation d. It is initially connected to a battery whose emf is $\mathcal{E}$ and then disconnected when fully charged. Compute the energy stored in the capacitor. Find the force acting on one plate by computing the change in stored energy as the plate separation is increased by an amount dx. Compute the electric field between the plates. Compare the force on the plate with its charge times the electric field. If the two results are different, explain why. (*Hint:* See §25.6.)

6. (a) ❖ Suppose in Problem 5 that the battery were left connected. Would there be any difference in the forces acting on the capacitor plates? Why or why not? Discuss the significance of the battery in computing the force from a change of potential energy. (b) ◆◆◆ Compute the change in stored energy of the system as the plate separation is increased by dx and derive from your result the force acting on one plate.

7. ❖ ■ Figure VI.27 shows a system that we might model as two capacitors in series. Sketch field lines for the distribution of charge shown, and then use the circulation law for static electric field to show that this charge distribution cannot be correct. Show that a consistent field pattern results if the magnitude of charge on the inner plate of the system is slightly less than that on the outer plate. Argue that the capacitance of the series combination of capacitors is slightly greater than the value given by the usual series formula.

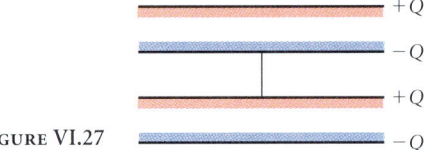

■ FIGURE VI.27

Computer Problem

8. Two charges $+3Q$ and $-Q$ are on the x-axis separated by a distance D with the origin at $+3Q$ (cf. Example 25.11). Compute the potential V_0 at the point on the x-axis at $x = (3 + \sqrt{3})D/2$ where $\vec{E} = 0$. Find two other points on the x-axis where the potential is V_0. Show that points with potential V_0 on the line through $-Q$ parallel to the y-axis may be found from the equation:

$$-\frac{D}{|y|} + \frac{3}{[1 + (y/D)^2]^{1/2}} = 4 - 2\sqrt{3}.$$

Find the solutions to this equation numerically. Compare your answers with Figure 25.15c, and use them to understand the shape of the equipotential surface through the $E = 0$ point.

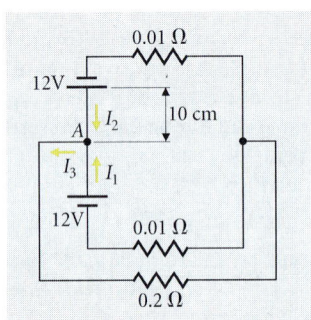

■ FIGURE VI.26

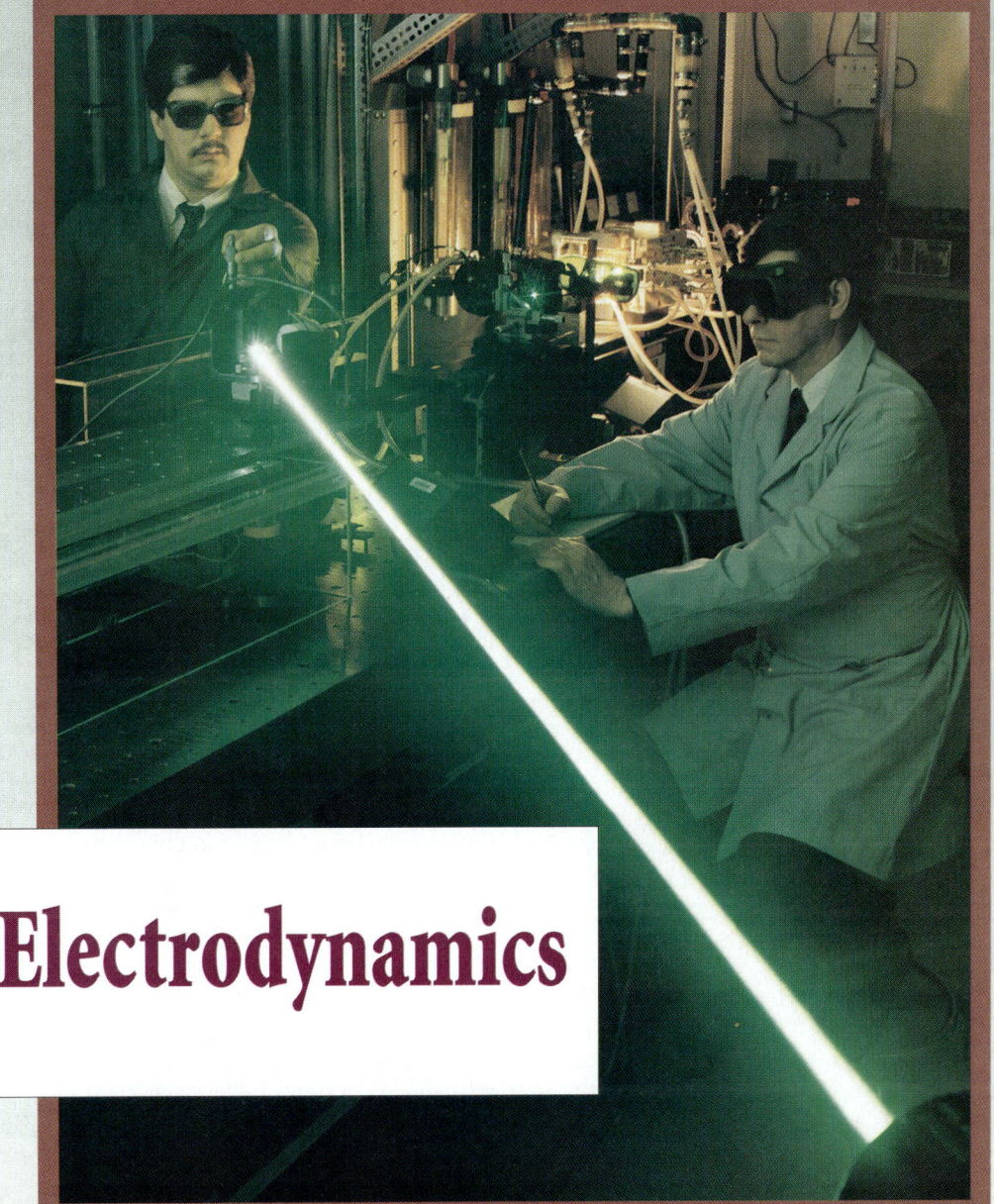

A laser exemplifies the power and beauty of electromagnetic waves. In this part we develop the full set of Maxwell equations and show how they describe the electromagnetic nature of light.

Electrodynamics

PART SEVEN

Static electric fields power your flashlight, and static magnetic fields turn your compass needle toward North. These seem to be unrelated physical effects, but electricity and magnetism appear distinct only when we consider static fields in a single reference frame. Time-varying fields *are* closely related, and the distinction between electric and magnetic depends on the reference frame in which we measure. The subject of this Part is electrodynamics. We'll look at electromagnetic fields that vary in time and learn how to describe electric and magnetic effects in reference frames that move with respect to one another. This study establishes a unity between electric and magnetic phenomena and shows how light waves arise as oscillating fields.

■ FIGURE VII.1
James Clerk Maxwell (1831–1879). Maxwell, a Scot, probably picked up his interest in science from his father. He was educated by his mother until her death, after which time he was sent to school in Edinburgh. He began his undergraduate studies at Edinburgh University. When his mathematical talents became apparent, he moved in 1850 to the university that could offer the best mathematical training, Cambridge, where he graduated second in his class. He began his important work in electromagnetic theory immediately and published his first paper on the topic in 1855. This work culminated in his *Treatise on Electricity and Magnetism*, published in 1873. Maxwell also made important contributions in statistical physics (see §22.7).

■ FIGURE VII.2
Albert Einstein (1879–1955). Einstein was born in Ulm, Germany, on March 14, 1879. He first came to the United States in 1932, and in 1933, he took up a position at the Institute for Advanced Study in Princeton, where he remained until his death.

Einstein made contributions to almost every branch of physics, but he is probably best known for his work on relativity. His first paper on special relativity, published in 1905, was titled *On the electrodynamics of moving bodies*. His first paper on the general theory was published in 1916. In 1921, he was awarded the Nobel Prize for his work on the photoelectric effect (cf. Chapter 35), also published in 1905. Einstein is also remembered for his humanitarian works. In his later years, he campaigned vigorously for a ban on nuclear weapons.

The theory of electrodynamics was put into its modern form by James Clerk Maxwell (■ Figure VII.1) and Albert Einstein (■ Figure VII.2). In this part, we shall study Maxwell's theory and touch on its applications to electric power technology, electronic circuits, optics, and microwaves. Each of the theory's basic equations bears the name of the physicist who is credited with its discovery: Gauss, Ampère, or Faraday. As a set, the equations are named after Maxwell because he was the first to develop them into a complete and consistent theory.

Maxwell's equations relate electromagnetic fields to their sources. We have already studied the two equations that relate the electric and magnetic fields to charge and current, respectively: Gauss' law (Chapters 23 and 24) and the time-independent form of Ampère's law (Chapter 28). Next, we shall see that charge and current are not the only source of these fields. A *changing* magnetic field creates an electric field and, similarly, a changing $\vec{E}$ creates $\vec{B}$. In Chapter 30 we study the laws that describe these effects: Faraday's law and Maxwell's extension of Ampère's law. In Chapters 31 and 32 we discuss circuits with time-varying currents, and in Chapter 33 we show how Maxwell's equations describe electromagnetic waves.

Grand Coulee Dam. In a hydroelectric power plant, gravitational potential energy of the water is converted first to kinetic energy as the water falls to the powerhouse at the bottom of the dam, and then to electric energy as the water is made to turn the generators in the powerhouse. The basic principle behind all generators is Faraday's law, the major topic of this chapter. A model of a simple generator is discussed in §30.2.2.

CHAPTER 30
Dynamic Fields

Concepts
Induced electromotive force
Lenz's law
Motional electromotive force
Motor/generator
Induced electric field
Eddy currents
Displacement current

Goals
Be able to:

Recognize systems in which emf is induced and apply Faraday's law to compute the emf.

Compute induced electric field in symmetric systems.

Recognize systems in which displacement current occurs and compute the current.

. . . creating a new theory . . . is rather like climbing a mountain, gaining new and wider views, discovering unexpected connections between our starting point and its rich environment. But the point from which we started out still exists and can be seen, although it appears smaller and forms a tiny part of our broad view gained by the mastery of the obstacles on our adventurous way up.

Albert Einstein and Leopold Infeld

WE'LL DISCUSS THESE GENERATORS IN §30.2.2.

Hydroelectric power plants are common in the Pacific Northwest. Gravitational potential energy of the water in the reservoir is converted to kinetic energy as the water falls to the powerhouse at the base of the dam, where the water turns generators that produce electric power. On a smaller scale, similar generators provide power to motor homes in remote campsites and to hospitals when the main power source is interrupted. In each of these generators, electric current is created when a coil is forced to rotate in a magnetic field.

In 1820, Oersted's discovery of magnetic effects caused by electric current created a great interest in a search for electric effects produced by magnetic fields—electromagnetic induction. Ten years passed before Michael Faraday (■ Figure 30.1) was able to demonstrate such effects. Ampère had misinterpreted several experiments because he was looking for electric phenomena caused by *static* magnetic fields. Faraday's crucial discovery was the need for *change* in the magnetic field.

■ **FIGURE 30.1**
Michael Faraday (1791–1867). The historian and physicist Sir Edmund Whittaker said of Faraday: "Among experimental philosophers Faraday holds by universal consent the foremost place." Faraday made numerous discoveries in physics and chemistry, and is the only scientist to have two units named after him. He also promoted science education for the public.

Faraday's discovery is the basis of the electric power system. Power is transmitted across country at high voltage and low current to minimize ohmic losses in the power lines, yet domestic power is supplied at 110 V. On a pole near your home is a transformer that uses dynamic (i.e., changing) magnetic fields to link the distribution circuit to your household circuitry. Generators at the power plant and electric motors that use the power are equally dependent on dynamic fields.

POWER LINES WITHIN A CITY OPERATE AT 12 KV.

Changing electric fields also produce magnetic fields. This fact was not discovered experimentally, since the effect would have been negligible in the laboratory experiments of the early nineteenth century. It was predicted theoretically by Maxwell during the years 1857–1865 in studies designed to develop a firm mathematical and conceptual basis for electromagnetic theory. He suggested that a changing electric field acts like an additional *displacement current* in Ampère's law. Maxwell's theoretical discovery is the key to understanding the electromagnetic nature of light.

DISPLACEMENT CURRENT IS DISCUSSED IN §30.6.

In this chapter we shall study the two dynamic relations between electric and magnetic fields and arrive at the complete set of Maxwell equations, which form the basis for all of classical electromagnetic theory.

30.1 INDUCED EMF
30.1.1 Faraday's Law

In a classroom demonstration of electromagnetic induction (■ Figure 30.2), an ammeter is connected across a coil, and a magnet is moved back and forth along the axis of the coil. So long as the magnet is held fixed, nothing happens, but while the magnet is in motion, the ammeter deflects, indicating the existence of current and an electromotive force in the coil-ammeter circuit:

REMEMBER: EMF MEANS ELECTROMOTIVE FORCE. SEE CHAPTER 26.

- A static magnetic field produces no electrical effect.
- A changing magnetic field induces an emf in the coil.

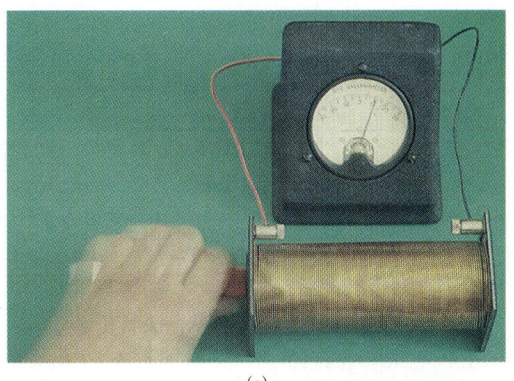
(a)

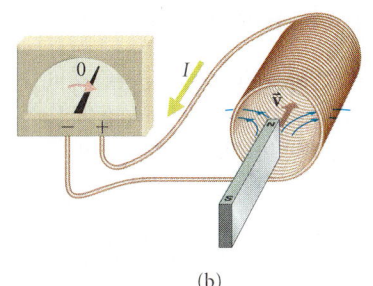
(b)

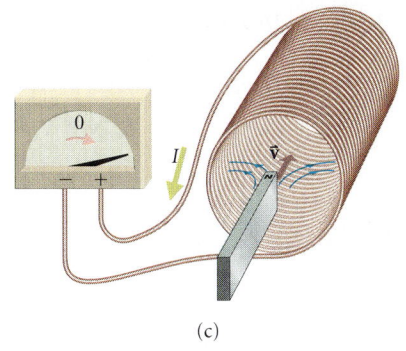
(c)

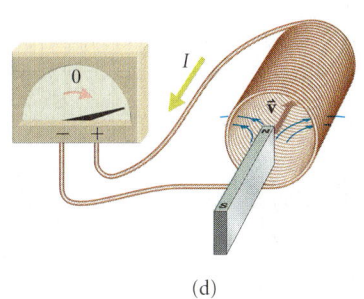
(d)

■ **Figure 30.2**
(a) An experiment that demonstrates Faraday's law. As the magnet is moved toward the coil, the ammeter indicates the presence of current in the circuit. (b) Schematic of the experiment shown in the photograph. (c) When the same magnet is used with a coil of greater radius, the ammeter reading indicates that the current is greater. (d) When the magnet is moved more rapidly, the current is again increased.

Several other tests reveal important features of induction. Repeating the experiment with the same magnet but a larger coil results in a larger electromotive force: the emf induced in the coil is proportional to its *area*. It is not change in $\vec{B}$ itself that is important, but the change in its flux $\Phi_B = \int \vec{B} \cdot \hat{n} \, dA$ through the area of the coil.

- Induced emf results from change of magnetic flux.

Finally, experiment shows that the ammeter reading is proportional to the number of turns in the coil and to the speed with which the magnet moves.

- The total magnetic flux through a coil with N turns is the sum of the fluxes passing through each of its turns.

$$\Phi_B = \sum_{i=1}^{N} \Phi_{B,i} = N\Phi_{B,\text{each loop}}.$$

- The induced emf $\mathcal{E}$ is proportional to the rate at which flux, Φ_B, changes.

In his experiments, Faraday demonstrated that $\mathcal{E}$ is always *proportional* to $d\Phi_B/dt$: in SI they are equal. Thus we may state *Faraday's law*.

MAGNETIC FLUX IS DEFINED IN THE SAME WAY AS ELECTRIC FLUX. SEE §23.4.2 AND §28.3.1.

> The magnitude of the emf $\mathcal{E}$ induced in a circuit equals the rate at which magnetic flux through the circuit changes:
>
> $$|\mathcal{E}| = \left| \frac{d\Phi_B}{dt} \right|. \qquad (30.1)$$

The SI unit of magnetic flux is the weber, after Wilhelm Weber (1804–1891):

$$1 \text{ Wb} = 1 \text{ T} \cdot \text{m}^2.$$

■ **EXERCISE 30.1** ◆ Show that $1 \text{ Wb/s} \equiv 1 \text{ T} \cdot \text{m}^2/\text{s} = 1 \text{ V}$.

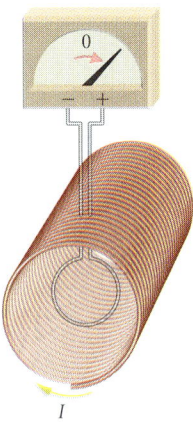

■ FIGURE 30.3
A small circular loop is placed around the axis of a large solenoid. As the current in the solenoid rises from zero, the magnetic field inside the solenoid increases. The increasing flux through the loop generates an emf in the loop.

EXAMPLE 30.1 ♦♦ A small loop of resistance 1 mΩ and radius $a = 1$ cm lies on the axis of a solenoid with $n = 10\,000$ turns per meter (■ Figure 30.3). The current in the solenoid is increased uniformly from zero to 0.1 A in 10 ms. What is the current through an ammeter connected to the small loop?

MODEL As the current in the solenoid increases, so does the magnetic field it produces. Consequently, magnetic flux through the loop increases, inducing an emf in the loop and producing current in it. We assume the magnetic field inside the solenoid is uniform at any one time (cf. Example 28.8).

SETUP At any instant, the magnetic field inside the solenoid has magnitude $|\vec{B}| = \mu_0 n I$ (eqn. 28.14). Since $\vec{B}$ is perpendicular to the loop, the flux through the loop at any time is:

$$\Phi_B = \int \vec{B} \cdot \hat{n}\, dA = \pi a^2 |\vec{B}| = \pi a^2 \mu_0 n I.$$

From Faraday's law, the emf induced in the loop equals the rate of change of this flux. The current I is the only time-dependent variable in Φ_B:

$$|\mathcal{E}| = \left|\frac{d\Phi_B}{dt}\right| = \pi a^2 \mu_0 n \left|\frac{dI}{dt}\right|.$$

SOLVE Finally, the current in the loop is found from the definition of resistance (eqn. 26.2):

$$I_\ell = \frac{|\mathcal{E}|}{R} = \frac{\pi a^2 \mu_0 n}{R}\left|\frac{dI}{dt}\right|$$

$$= \left[\frac{\pi(10^{-2}\text{ m})^2(4\pi \times 10^{-7}\text{ N/A}^2)(10^4\text{ /m})}{10^{-3}\ \Omega}\right]\left(\frac{0.1\text{ A}}{10^{-2}\text{ s}}\right) = 0.04\text{ A}.$$

ANALYZE The current in the loop exists only while the solenoid current is changing. It is all right to use the result of static field calculations to find $\vec{B}$ when the current and field are varying as slowly as in this example. In rapidly varying systems, however, the fields propagate through the system as waves at the speed of light. We discuss such systems in Chapter 33. To decide whether the field is slowly varying, ask whether the field changes substantially during the time it takes for light to cross the system. In this example, that time is $(1\text{ cm})/(3 \times 10^8\text{ m/s}) = 3 \times 10^{-11}$ s, which is vastly less than the 10-ms time scale for changes in B and I. ■

30.1.2 Lenz's Law

We can learn more from experiments with the magnet, coil, and ammeter (■ Figure 30.4). When the magnet is pulled away from the coil, rather than pushed toward it, the direction of current through the ammeter reverses.

- When the magnetic flux is changed at a fixed rate, the magnitude of induced emf is the same whether flux increases or decreases.
- The sense of induced emf depends on the sense of flux change.

The rule that gives the sense of induced emf is credited to Heinrich Lenz (1804–1865).

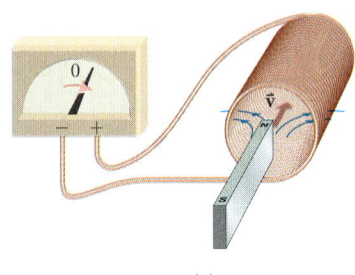

(a)

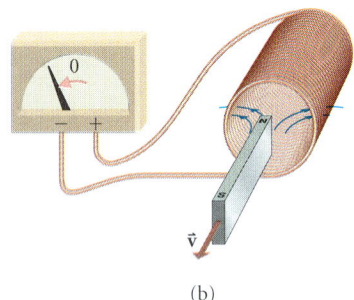

(b)

■ FIGURE 30.4
Lenz's law. Moving the magnet at the same rate in opposite directions generates currents with the same magnitude but opposite directions.

> **LENZ'S LAW**
> Induced emf acts so as to oppose the change in flux that creates it.

EXAMPLE 30.2 ❖ A magnet lies along the axis of a conducting loop with its N pole pointing toward the loop (■ Figure 30.5). If the magnet is pushed toward the loop, what is the direction of current around the loop?

MODEL According to Lenz's law, the current in the loop opposes any change in magnetic flux caused by the movement of the magnet. The magnetic field strength decreases with distance from the magnet, and its direction is away from the N pole. As the magnet approaches the loop, the magnetic field at the position of the loop increases, and hence the magnetic flux through the loop increases. According to Lenz's law, the current in the loop produces an induced magnetic flux *through the loop* that opposes this change.

The magnetic field vector at the center of the loop is the sum of the contributions from the magnet and from the loop itself: $\vec{B}_{net} = \vec{B}_{magnet} + \vec{B}_{loop}$. As $\vec{B}_{magnet}$ increases, the field produced by the loop in response must be in the opposite direction in order to reduce the increase of $\vec{B}_{net}$: $|\vec{B}_{net}| = |\vec{B}_{magnet}| - |\vec{B}_{loop}|$. Thus the field lines produced by the loop point toward the magnet and, according to the right-hand rule, the current is counterclockwise around the loop in ■ Figure 30.6.

ANALYZE We do not have to refer to magnetic flux when using Lenz's law. We may view the motion of the magnet toward the loop as the change that induces emf in the loop. Then the current in the loop must cause a force that decelerates the incoming magnet, opposing that change. That is, the magnetic dipole of the current loop has its N pole toward the N pole of the moving magnet. Again, we find that the current is counterclockwise. ■

EXERCISE 30.2 ❖ Suppose the magnet in Example 30.2 passes through the loop. Show that the current in the loop is clockwise as the magnet recedes. ■

Lenz's law is necessary for conservation of energy. If the current in Example 30.2 were in the opposite sense, the magnet would be attracted toward the loop and would gain kinetic energy! We could use the increased kinetic energy of the magnet to do work while using the induced emf to operate electrical machinery. Repeating the process would produce endless free energy, but this is, of course, impossible.

Put another way, work must be done on the system to produce energy. If the loop has resistance R, thermal energy is produced in it by the current at a rate I^2R. Thus we have to push the magnet toward the loop *against the opposing force*, doing work at a rate $\vec{F} \cdot \vec{v} = I^2R$.

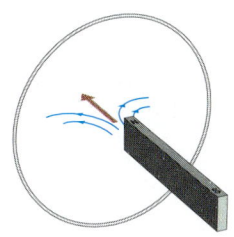

■ **FIGURE 30.5**
A magnet is pushed, N pole first, toward a conducting loop. If we choose the normal to the plane of the loop to be into the page, magnetic flux through the loop increases because the field at the loop increases as the magnet approaches.

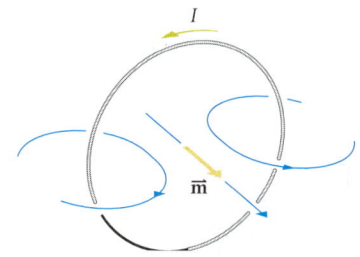

■ **FIGURE 30.6**
According to Lenz's law, the current in the loop opposes the increase in flux by making magnetic field in the opposite direction. The loop's magnetic moment is opposite that of the incoming magnet, so the loop and magnet repel each other.

30.1.3 Induced Electric Field

Faraday's law in the form of eqn. (30.1) and Lenz's law together describe the emf arising from a change in flux but do not tell us what produces the emf. Current exists in an ordinary conducting circuit only if work is done on the individual electrons to replace the energy they lose in collisions. There must be an induced electric field in the coil that exerts the necessary force on the electrons (■ Figure 30.7). Induced electric field does not arise directly from electric charge; its source is the changing magnetic field, and its field lines circulate around the source rather than emerging from it. In this way, induced electric field is like magnetic field and unlike static electric field.

Around a fixed circuit C, the emf may be expressed in terms of the induced electric field:

$$\mathcal{E} = \oint_C \vec{E}_{ind} \cdot d\vec{\ell}. \qquad (30.2)$$

WE'LL SAY MORE ABOUT INDUCED ELECTRIC FIELD IN §30.4.

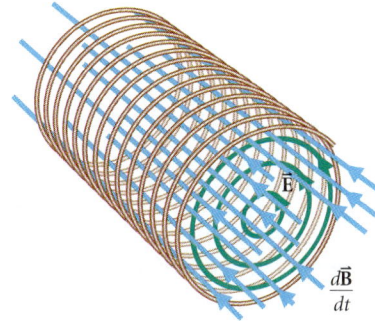

■ **FIGURE 30.7**
Induced electric field circulates around its source (changing magnetic field). Magnetic field circulates around its source (electric current) in a similar way. There is an important difference between the two cases: $\vec{E}_{ind}$ and $\vec{B}$ circulate in opposite directions around their sources.

THIS IS THE SAME CONVENTION USED IN AMPÈRE'S LAW, §28.3, ESPECIALLY FIGURE 28.15.

30.1.4 Sign Conventions

Faraday's law may be written in a compact form by defining a set of sign conventions. First, we choose the positive direction for current (or equivalently, induced electric field). Then the positive direction for flux is defined from the positive direction for current using a right-hand rule. For example, if we choose current to be positive when it is counterclockwise around the loop in ■ Figure 30.8, then the normal to the area of the loop is out of the page. We use this normal to calculate the flux, which is positive if $\vec{B}$ is in the same direction as $\hat{n}$. Once this sign convention has been defined, we can write Faraday's law in an equation that encompasses Lenz's law as well:

$$\mathcal{E} = -\frac{d\Phi_B}{dt}. \tag{30.3}$$

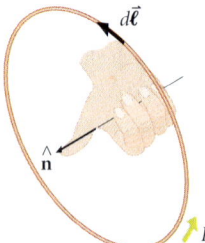

■ FIGURE 30.8
Sign conventions for emf and flux. Choosing the direction for $d\vec{\ell}$ determines the positive direction of $\hat{n}$, and hence the positive sense of flux. The emf $\mathcal{E}$ is positive if it drives current in the direction of $d\vec{\ell}$, as shown here.

Using this sign convention in Example 30.2, we might define current to be positive if it is counterclockwise in Figure 30.5. Then the normal to the loop is out of the page. The magnet produces magnetic field into the page and hence negative flux through the loop. As the magnet approaches the loop, the field strength increases and the negative flux increases in *magnitude*: $d\Phi_B/dt$ is negative. Using Faraday's law in the form of eqn. (30.3), the emf is positive (counterclockwise), so the current is also counterclockwise as we concluded in the example.

EXAMPLE 30.3 ♦♦ A square loop of wire has sides 3.5 cm long parallel to the x- and y-axes, and a resistance of 0.10 Ω. The magnetic field in the region is increasing with time: $\vec{B} = B_o(\hat{i} + \hat{k})$, where $B_o = (0.050 \text{ T/s}^2)t^2$. Find the current induced in the loop at $t = 1.0$ ms and $t = 0.50$ s.

MODEL Since the magnetic field increases as the square of the time, the rate of change of flux, and hence the current, are also functions of time. We begin by choosing the positive direction around the loop, then we apply eqn. (30.3).

SETUP Choose the counterclockwise direction around the loop to be positive so that $\hat{n}$ is in the z-direction (■ Figure 30.9). The flux through the loop is:

$$\Phi_B = \int \vec{B} \cdot \hat{n} \, dA = \int B_o(\hat{i} + \hat{k}) \cdot \hat{k} \, dA$$

$$= B_o(0 + 1) \int dA = B_o \ell^2.$$

We pulled B_o out of the integral since it does not depend on position. Next we apply Faraday's law:

$$\mathcal{E} = -\frac{d\Phi_B}{dt} = -\frac{dB_o}{dt}\ell^2 = -\frac{d}{dt}[(0.050 \text{ T/s}^2)t^2](3.5 \text{ cm})^2$$

$$= -(0.050 \text{ T/s}^2)2t(3.5 \text{ cm})^2 = -(122 \; \mu\text{V/s})t.$$

SOLVE The current in the loop is:

$$I = \frac{\mathcal{E}}{R} = \frac{(-122 \; \mu\text{V/s})t}{0.10 \; \Omega} = (-1.22 \text{ mA/s})t.$$

At $t = 1.0$ ms, the current is:

$$I = (-1.22 \text{ mA/s})(1.0 \times 10^{-3} \text{ s}) = -1.2 \; \mu\text{A}.$$

And, at $t = 0.50$ s, it is:

$$I = (-1.22 \text{ mA/s})(0.50 \text{ s}) = -0.61 \text{ mA}.$$

ANALYZE The current is negative, which means it is clockwise. It produces a magnetic field in the negative z-direction within the loop, opposing the increase of applied magnetic field.

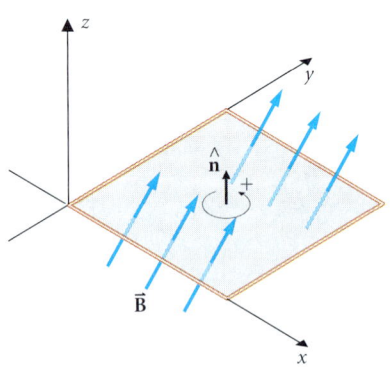

■ FIGURE 30.9
A square loop in the x-y-plane has resistance $R = 0.10 \; \Omega$. The magnetic field is at 45° to the x-y-plane and is increasing at a rate that itself increases with time. We choose the positive direction around the loop to be counterclockwise, as shown, so that $\hat{n} = \hat{k}$. Then both flux and $d\Phi_B/dt$ are positive for $t > 0$.

30.2 MOTIONAL EMF

30.2.1 EMF in Circuits with Moving Boundaries

Faraday's law applies whenever the magnetic flux through a circuit changes for any reason. Varying the magnetic field is only one way of altering flux.

> *Motional emf* occurs when flux through a circuit varies because the circuit moves or changes its size or shape.

The following example shows how motional emf arises.

EXAMPLE 30.4 ♦♦ Two horizontal conducting rails 1.0 m apart lie in a region where the magnetic field is directly downward with magnitude $B = 0.10$ T (■ Figure 30.10). At one end, the rails are connected by a conducting cross tie. The resistance of rails and cross tie is negligible. A rod is pulled along the rails at constant speed $v = 0.50$ m/s. If the resistance between the two contact points of the rod on the rails is $R = 1.2$ mΩ, what is the current in the circuit? How much power is needed to keep the rod moving?

MODEL The rod, rails, and cross tie form an electric circuit. As the rod moves, the area of the circuit and the magnetic flux through it both increase. Current is established by the resulting emf. From Lenz's law:

1. Magnetic force on the induced current in the rod opposes the rod's motion.
2. Magnetic field lines produced by the current pass upward through the circuit to oppose the increase in downward flux due to the rod's motion.

From either argument, we conclude that the current is counterclockwise.

SETUP Figure 30.10 shows the rod at a distance x from the cross tie. Since $\vec{B}$ is uniform and perpendicular to the plane containing the rails, the flux through the circuit at this instant is $\Phi_B = BA = B\ell x$. The flux changes at a rate:

$$\left|\frac{d\Phi_B}{dt}\right| = \frac{d}{dt}(B\ell x) = B\ell \frac{dx}{dt} = B\ell v = (0.10 \text{ T})(1.0 \text{ m})(0.50 \text{ m/s}) = 5.0 \times 10^{-2} \text{ V}.$$

SOLVE We use this result in Faraday's law to find the emf in the circuit. Using eqn. (30.1), the resulting current is:

$$I = \frac{|\mathcal{E}|}{R} = \frac{1}{R}\left|\frac{d\Phi_B}{dt}\right| = \frac{5.0 \times 10^{-2} \text{ V}}{1.2 \times 10^{-3} \ \Omega} = 42 \text{ A}.$$

Since the current is perpendicular to the magnetic field, the force on the rod has magnitude:

$$|\vec{F}| = I\ell B = (42 \text{ A})(1.0 \text{ m})(0.10 \text{ T}) = 4.2 \text{ N}.$$

To maintain the rod's speed, an equal and opposite mechanical force must be applied. The required power is:

$$P = Fv = (4.2 \text{ N})(0.5 \text{ m/s}) = 2.1 \text{ W}.$$

ANALYZE The electric power used by the circuit is:

$$P = \mathcal{E}I = (5.0 \times 10^{-2} \text{ V})(42 \text{ A}) = 2.1 \text{ W}.$$

The mechanical power used to move the rod is dissipated as heat by the electrical resistance. ■

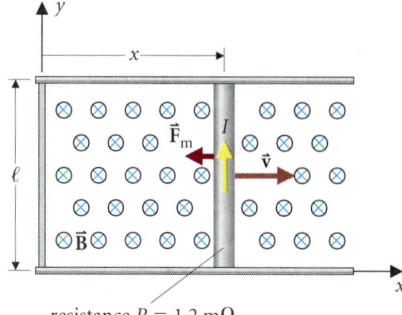

■ **FIGURE 30.10**
A conducting rod slides in the x-direction on conducting rails. When the rod is pulled at speed v, a current is induced in the circuit. The resulting magnetic force $\vec{F}_m = I\vec{\ell} \times \vec{B}$ on the rod opposes its motion.

In Example 30.4 it is a magnetic force that drives current through the resistive rod. The electrons in the moving rod have a velocity component $\vec{v}$, and thus each experiences a magnetic force component along the rod: $\vec{F} = -e\vec{v} \times \vec{B}$. Motional emf emphasizes two features of Faraday's law:

- The circuits to which the law applies may have moving boundaries.
- Magnetic forces contribute to emf.

As the electrons respond to the magnetic force, they gain a drift velocity $\vec{v}_d$. We'll investigate the consequences of this velocity component in §30.3.

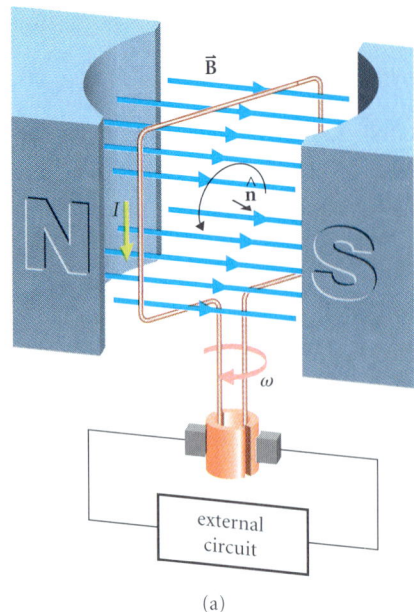
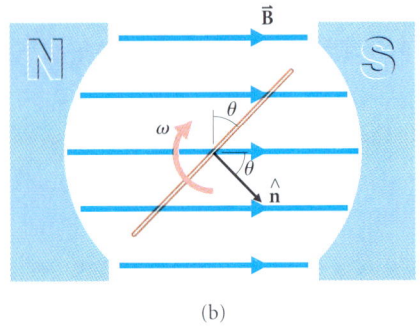

■ **FIGURE 30.11**
(a) A simple generator. A wire loop rotates between the poles of a magnet. We define the positive direction around the loop as shown. (b) The loop viewed from above. We assume the magnetic field between the poles is uniform. As the loop rotates, the angle θ between its normal $\hat{n}$ and the magnetic field vector $\vec{B}$ increases: $\theta = \omega t$.

30.2.2 Generators and Motors

One of the most important applications of motional emf is in the design of electric generators and motors. A generator converts mechanical work to electrical energy. A motor works the other way, converting electrical energy to mechanical work. Here we look at a simple device that can act as either a generator or a motor.

When a wire loop rotates in a magnetic field as shown in ■ Figure 30.11, there is a changing flux through the loop and hence an induced emf around it. If we define current as positive when it has the direction shown in the figure, then the flux through the loop at the instant shown is:

$$\Phi_B = \int \vec{B} \cdot \hat{n}\, dA = BA \cos(\theta) = BA \cos(\omega t),$$

where ω is the angular speed of the loop and we have chosen to start the clock when the loop is perpendicular to the field. As the loop rotates, the flux changes at a rate:

$$\frac{d\Phi_B}{dt} = \frac{d}{dt}[BA \cos(\omega t)] = -BA\omega \sin(\omega t). \qquad (30.4)$$

The induced emf $\mathcal{E} = BA\omega \sin \omega t$ varies sinusoidally between 0 and $\pm \omega BA$, and the current changes direction twice every period. The device is a simple *generator*. In generators designed for commercial power plants (■ Figure 30.12), $\omega = 2\pi f$ is set by the standard 60-Hz frequency of the power grid. The magnetic field is limited to about 1 T by saturation of the iron magnet cores. Thus the output emf of the generator is determined by the area of the coil and the number of turns of wire.

■ **FIGURE 30.12**
A generator used in a commercial power plant.

EXAMPLE 30.5 ◆ A simple generator has a rectangular loop of dimensions 10.0 cm × 5.0 cm with ten turns rotating in a field of 1.5 T. Find the maximum emf generated when the loop rotates at 60.0 Hz.

MODEL The motional emf is produced by rotation of the loop and varies sinusoidally.

SETUP The flux through the generator coil is ten times the flux through one loop.

SOLVE From eqn. (30.4) and Faraday's law, the maximum emf occurs when $\sin(\omega t) = 1$.

$$\mathcal{E} = NBA\omega = (10)(1.5\text{ T})(0.10\text{ m})(0.050\text{ m})(2\pi)(60.0\text{ Hz}) = 28\text{ V}.$$

REMEMBER: $\omega = 2\pi f$.

ANALYZE Slip rings and brushes connect the rotating loop into a stationary circuit. Their design determines whether the current in the circuit is unidirectional (DC) (■ Figure 30.13) or changes direction with the emf (AC).

By turning the generator loop, we obtain an electric current. Conversely, if we pass a current through the loop, magnetic forces act on the wire, creating a torque. This device is a simple electric motor.

EXAMPLE 30.6 ◆◆ If a current of 1.0 A is passed through the loop described in Example 30.5 (■ Figure 30.14), find the torque on the loop as a function of the angle θ between the magnetic field and the normal to the loop.

MODEL Since the magnetic field is uniform, we may use the magnetic moment of the loop to compute the torque (see §29.2.2).

SETUP The magnetic moment $\vec{m} = NIA\hat{n}$, where $\hat{n}$ is the normal to the loop and A is its area. Using eqn. (29.8) with the current direction shown in the figure, the torque on the current loop is:

$$\vec{\tau} = \vec{m} \times \vec{B} = (NIAB \sin\theta, \text{counterclockwise as viewed in Figure 30.14b})$$
$$= [(1.5\text{ T})(5.0 \times 10^{-3}\text{ m}^2)(10)(1.0\text{ A})\sin\theta, \text{counterclockwise}]$$
$$= [(7.5 \times 10^{-2}\text{ N}\cdot\text{m})\sin\theta, \text{counterclockwise}].$$

ANALYZE This torque also varies sinusoidally (Figure 30.14c). To keep the motor rotating in the same direction, the direction of the current must be reversed every half turn. This may be accomplished using brushes and a slip ring. ■

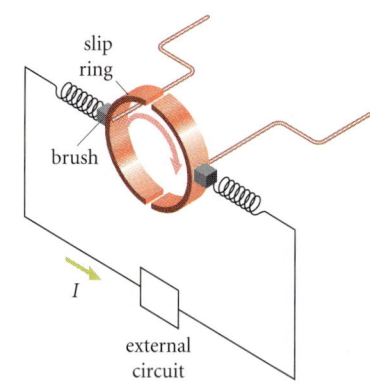

■ **FIGURE 30.13**
Brushes and a rotating slip ring convert a generator's AC output to DC. Each end of the generator loop is attached to the slip ring. *Brushes* are held against the slip ring and slide along them as they rotate. The brushes transfer current to the stationary wires in the external circuit. In the arrangement shown here, the connections between generator wires and the external circuit are reversed each half rotation as the brushes cross the gaps in the slip ring. The resulting current in the external circuit is always in the same direction. Brushes are typically made of graphite, which both conducts electricity and lubricates the slip rings.

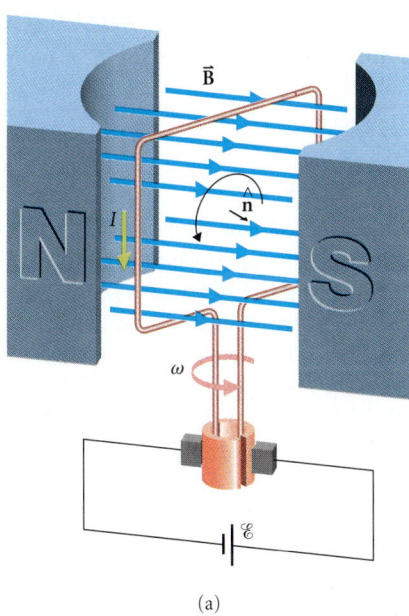

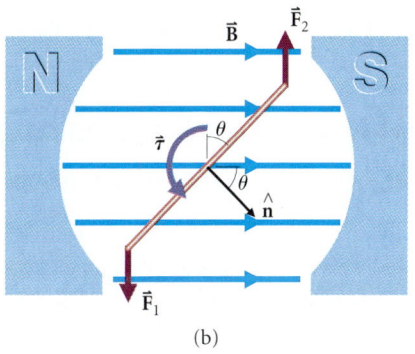

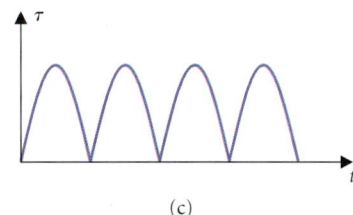

■ **FIGURE 30.14**
(a) The DC emf drives current through the wire loop. (b) Magnetic force acting on this current develops a net torque on the loop. (c) The net torque varies sinusoidally with the position of the loop. The emf is connected via brushes and slip rings (cf. Figure 30.13) so that the torque has constant direction. Its magnitude is $\tau \propto |\sin(\omega t)|$.

If the coil has a resistance r, then an emf of magnitude Ir is required to drive the current. When the motor is operating, there is also an induced emf of magnitude $\mathcal{E}_{ind} = BA\omega|\sin\omega t|$ opposing the current, so an emf $\mathcal{E} = Ir + \mathcal{E}_{ind}$ is needed to maintain the current. Similarly, mechanical force is needed to rotate the loop when it is used as a generator.

EXAMPLE 30.7 ◆◆ Find the power produced by the generator of Example 30.5 when connected in series with a resistance R. Show that it is equal to the mechanical power required to operate the generator.

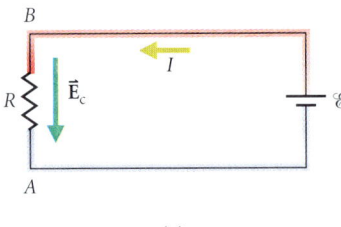

(a)

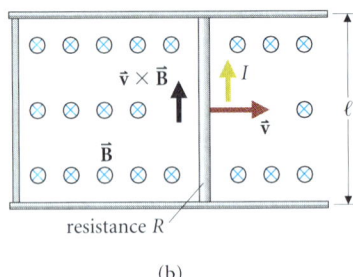

(b)

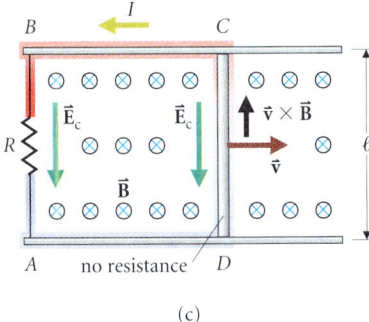

(c)

■ FIGURE 30.15
EMF and potential difference in circuits. (a) In a battery-powered circuit, energy is provided by chemical reactions in the battery and is used at the resistor. Charge is distributed throughout the circuit, as indicated schematically by the red and blue colors, and there is a potential difference $V_{BA} = \mathcal{E}$ across the resistor. (b) Conducting rod on rails. The rod has a resistance R. The magnetic force drives electrons through the rod. There are no charge distributions, Coulomb fields, or potential differences. (c) Here the rod is perfectly conducting, but the rails are connected by a *fixed* resistor R. A Coulomb electric force opposes the magnetic force in the perfectly conducting rod. Energy gained by the electrons in the moving rod is transferred to the resistor, where it is used. The potential difference $V_{BA} = V_{CD} = vB\ell$ indicates that the transfer is occurring.

THE ENERGY IS TRANSPORTED BY THE FIELDS, AS WE'LL SEE IN CHAPTER 33.

WE'LL CALL FIELDS DUE TO CHARGE "COULOMB FIELDS" (WITH SUBSCRIPT C) TO DISTINGUISH THEM FROM INDUCED FIELDS.

MODEL The electrical power equals the emf times the current in the circuit (eqn. 26.3). To use this relation, we need to determine the current. The current is the induced emf divided by the total resistance. The mechanical power equals the angular frequency times the torque, $P_m = \omega\tau$ (eqn. 9.14).

SETUP The total resistance in the circuit equals the resistance of the generator coil plus the external resistor in series with it: $R_{tot} = R + r$. The current is:

$$I = \frac{\mathcal{E}}{R+r} = \frac{NBA\omega}{R+r} \sin \omega t.$$

SOLVE The electrical power produced by the generator and used by the circuit is:

$$P_e = I\mathcal{E} = \frac{\mathcal{E}^2}{R+r} = \frac{(NBA\omega)^2 \sin^2 \omega t}{R+r}.$$

From Example 30.6, the magnetic torque exerted on the loop is $\tau = NIAB \sin \omega t$. Then, using our result for I:

$$\tau = NAB \sin \omega t \frac{NBA\omega}{R+r} \sin \omega t = \frac{(NBA)^2 \omega}{R+r} \sin^2 \omega t.$$

An equal mechanical torque must be applied to keep the generator turning at constant angular speed. The mechanical power required to exert this torque is:

$$P_m = \omega\tau = \frac{(NBA\omega)^2}{R+r} \sin^2 \omega t.$$

ANALYZE As expected, the required mechanical power input P_m equals the electrical power output of the generator P_e. ■

EXERCISE 30.3 ♦♦ When the loop is used as a motor, show that the electrical power input equals the mechanical power output plus the joule heat dissipated in the motor loop.

30.3 THE NATURE OF EMF

30.3.1 Potential Difference and EMF

Any circuit has both an energy source and an energy sink. In our study of DC circuits, we used the chemical storage battery as a model for the energy source. The emf of an ideal battery is the energy it provides to the circuit per unit of charge that passes through it, and also equals the potential difference the battery maintains between its terminals. Induced emf is another kind of energy source for circuits, but it is not related so simply to potential difference. We must carefully distinguish the roles of potential difference and energy input.

■ Figure 30.15 shows three circuits, each with the same emf, resistance, and current. Their differences are in the nature of the emf and in the role of static fields.

Circuit (a) is a DC circuit like those we studied in Chapter 26. The battery provides energy that is used *in a different place*—at the resistor. Electric potential provides the bookkeeping system that describes where energy is provided and where it is used. Potential rises across the battery and decreases again across the resistor.

Circuit (b) is the rod-on-rails system we studied in Example 30.4. Here the energy source is the external agent pulling the rod. The motional emf arising from magnetic force on electrons in the rod makes the energy available to the circuit. The energy is dissipated as heat by resistance *in the rod*. In the rest of the circuit, no energy is needed and no fields are required to drive current. Consequently, there is no electric field, or potential difference.

In circuit (c) the energy source and emf are the same as in circuit (b), but here the rod has negligible resistance and the rails are connected by a *stationary* resistor. As in circuit (a), the energy is used in the resistor, distant from the source. To understand this circuit, let's look at what happens as the rod begins to slide. The magnetic force $\vec{F}_m = -e\vec{v} \times \vec{B}$ on conduction electrons in the rod accelerates them rapidly toward end D, but there is no force on the electrons in the stationary resistor to oppose the frictionlike force due to the electrical resistance. The electrons stop and negative charge builds up between A and D, leaving positive charge between B and C. The electric field $\vec{E}_c$ produced by the charge distribution creates a potential difference of magnitude $E_c\ell$ between BC and AD. Once the electric and magnetic forces on the electrons in the rod balance ($\vec{E}_c = -\vec{v} \times \vec{B}$), the charge buildup ceases. The Coulomb electric force drives current through the resistor, and electrons flow at constant speed through the perfectly conducting rod, where the force on them is now zero. The potential rises by $V_{BA} = vB\ell$ across the rod, indicating where energy enters the circuit, and decreases again across the resistor, indicating where the energy is used.

THIS WHOLE PROCESS HAPPENS VERY RAPIDLY ($\lesssim 1$ NS).

EXAMPLE 30.8 ♦♦ An automobile axle is 1.5 m long. If the car is moving at 15 m/s and the vertical component of the Earth's magnetic field is $B_\perp = 2.5 \times 10^{-5}$ T, what is the potential difference between the ends of the axle?

MODEL Initially, electrons move toward end Y of the axle in response to the magnetic force resulting from its motion (■ Figure 30.16). Once the magnetic force is balanced by the electric force due to the charge distribution in the axle, no more electrons move. In equilibrium, the net force on electrons in the axle is zero. The Coulomb electric field produces a potential difference between the ends of the axle.

SETUP In equilibrium, $\vec{F} = -e(\vec{E}_c + \vec{v} \times \vec{B}) = 0$, so $\vec{E}_c = -\vec{v} \times \vec{B}$. The potential difference between the ends of the axle (eqn. 25.7) is:

$$\Delta V \equiv V_X - V_Y = -\int_Y^X \vec{E}_c \cdot d\vec{\ell} = |\vec{E}_c|\ell = vB_{vert}\ell.$$

SOLVE With the numbers given:

$$\Delta V = (15 \text{ m/s})(2.5 \times 10^{-5} \text{ T})(1.5 \text{ m}) = 0.56 \text{ mV}.$$

ANALYZE We can estimate how much charge is required at each end of the axle to produce the required Coulomb electric field. We approximate the charge distribution with two point charges $\pm q$ at the ends, which produce an electric field:

$$2\frac{kq}{(\ell/2)^2} \sim vB \quad \Rightarrow \quad q \sim \frac{vB\ell^2}{8k} = \frac{\ell \Delta V}{8k} = \frac{(1.5 \text{ m})(0.56 \times 10^{-3} \text{ V})}{8(9 \times 10^9 \text{ N} \cdot \text{m}^2/\text{C}^2)} = 10^{-14} \text{ C}.$$ ■

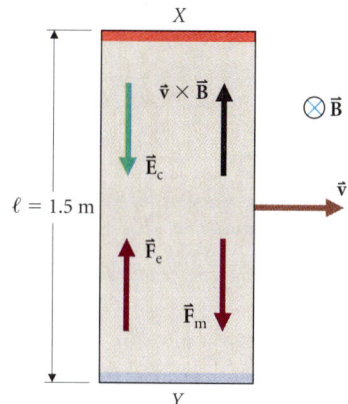

■ **FIGURE 30.16**
An autombile axle moving through the Earth's magnetic field. The magnetic force due to the vertical component of the Earth's field drives electrons toward end Y of the axle. The resulting Coulomb electric field creates a potential difference between the ends of the axle. In equilibrium, the electric force balances the magnetic force, $\vec{E}_c = -\vec{v} \times \vec{B}$.

SEE EXAMPLE 25.14 FOR A SIMILAR APPROXIMATION.

EXERCISE 30.4 ♦ What potential difference exists between the wing tips of a Boeing 747 aircraft (60-m wingspan) flying at 1000 km/h through a magnetic field 2.5×10^{-5} T perpendicular to the wing? ■

✶30.3.2 Mathematical Properties of EMF and Potential Difference

The distinction between emf and potential difference is apparent in their mathematical properties. The potential difference between any two points is given by the integral of the *Coulomb* electric field along any path between the two points:

$$\Delta V \equiv V_B - V_A = -\int_A^B \vec{E}_c \cdot d\vec{\ell}.$$

THIS IS EQN. (25.7).

If the integral is taken around a closed path starting and ending at A, the potential difference is zero:

$$V_A - V_A = \oint \vec{E}_c \cdot d\vec{\ell} = 0.$$

This is Kirchhoff's rule—the change of potential around any closed path is zero.
  In contrast, Faraday's law tells us that the emf around a closed path is not zero. As we have seen, the emf describes energy sources and results from magnetic effects as well as induced electric fields. The *total* electromagnetic force acting on a charge is $\vec{F} = q(\vec{E} + \vec{v} \times \vec{B})$.

FOR KIRCHHOFF'S RULES, SEE §26.4.

So:

> The *induced emf* acting around a closed path is the work done per unit charge as the charge is moved around the path:
>
> $$\mathcal{E} \equiv \frac{W}{q} \equiv \oint (\vec{E} + \vec{v} \times \vec{B}) \cdot d\vec{\ell}. \tag{30.5}$$

THE FORCE EXERTED BY THE INDUCED FIELD IS NOT CONSERVATIVE. SEE §8.3.1.

Although the total electric field appears in eqn. (30.5), only the induced electric field contributes to the emf, since the integral of the Coulomb field around a closed curve is zero. On the other hand, potential is defined by the Coulomb field alone; the induced field does not contribute.

✻30.3.3 EMF and Choice of Reference Frame

If we push a magnet toward a wire loop at speed v (■ Figure 30.18a), the changing magnetic field induces electric field at the loop, and current flows in the loop. If instead, we push the loop toward the magnet at speed v (Figure 30.18b), magnetic forces produce a motional emf that drives current in the loop. The current is the same in both cases. Einstein, in his famous 1905 paper on special relativity, observed that these two experiments *must* give the same result because they are really the same experiment! Viewed from a reference frame fixed to the magnet (Figure 30.18c), the first experiment looks just like the second experiment. Induced electric field and motional emf are not separate physical effects, but two ways of describing the same effect in different reference frames. A similar correspondence exists between the electric and magnetic fields themselves. The quantity $\vec{E} + \vec{v} \times \vec{B}$ that gives the contribution to emf at a particular element $d\vec{\ell}$ of a moving circuit *is* the electric field $\vec{E}'$ measured in the reference frame moving with that element.

Digging Deeper

MAGNETIC FORCE AND EMF

Since magnetic force is always perpendicular to the velocity of the charge it acts on, magnetic fields cannot themselves do work on charges. Yet we have used $dW = q(\vec{v} \times \vec{B}) \cdot d\vec{\ell}$ in eqn. (30.5). This term is important in some circuits (e.g., the rod on rails). In fact, it is the Hall electric field $\vec{E}_H = -\vec{v}_d \times \vec{B}$ (§29.3) that does the work. The power delivered to a single electron (■ Figure 30.17) is:

$$P_e = -e\vec{E}_H \cdot \vec{v}$$
$$= eE_H v = evBv_d.$$

The power delivered to all the electrons in a length $d\ell$ of the rod is:

$$dP = n\, dV\, P_e = nA\, d\ell\, evBv_d$$
$$= neAv_d Bv\, d\ell = I\, d\mathcal{E},$$

with $d\mathcal{E} = Bv\, d\ell$, in agreement with eqn. (30.5).

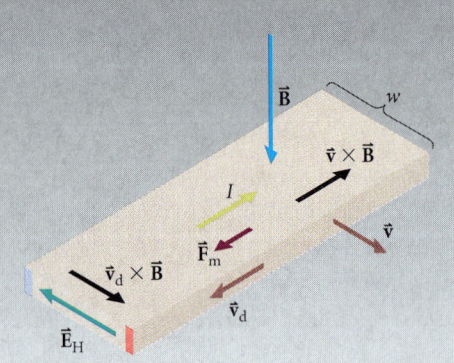

■ FIGURE 30.17
Enlargement of a conducting rod in which a motional emf is developed. When current exists in the rod, electrons move along it with a drift velocity $\vec{v}_d$. A Hall electric field $E_H = v_d Bw$ develops across the rod. Though $\vec{v} \times \vec{B}$ appears in the formula for emf, it is the Hall field that does work on charge.

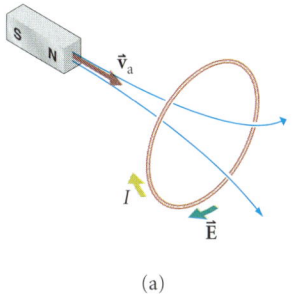 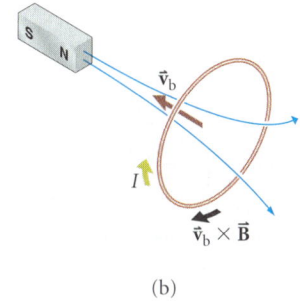 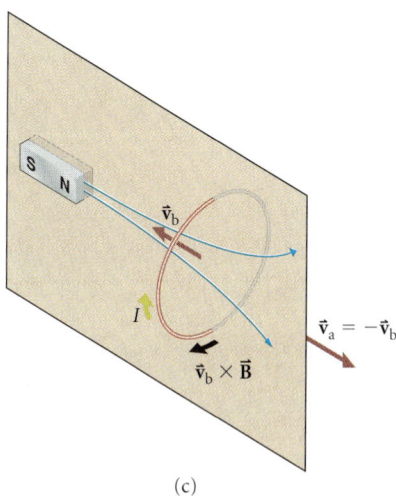

(a)   (b)   (c)

■ **FIGURE 30.18**
(a) In the lab frame, as we push a magnet toward a wire loop, the increasing flux through the loop creates an induced electric field that drives current. (b) If instead we push the loop toward the magnet, magnetic force $-e\vec{v} \times \vec{B}$ drives the current in the loop. (c) Experiment (a) viewed in a reference frame moving with the magnet is indistinguishable from experiment (b). Thus the electric field $\vec{E}$ in experiment (a) equals $\vec{v} \times \vec{B}$ in experiment (b).

30.4   CALCULATION OF INDUCED ELECTRIC FIELD

In symmetric systems, Faraday's law may be used to calculate the induced electric field. The method, outlined on the left-hand side of ■ Figure 30.19, is similar to that used with Gauss' law and Ampère's law. Faraday's law applies to any curve, whether or not it follows an electric circuit. You should choose a stationary curve that makes full use of the symmetry of the system.

Digging Deeper

THE COMPLETE MATHEMATICAL STATEMENT OF FARADAY'S LAW

A mathematical statement that includes all the features of induction we have discussed is:

$$\mathcal{E} = \oint (\vec{E} + \vec{v} \times \vec{B}) \cdot d\vec{\ell}$$
$$= -\frac{d}{dt} \oint \vec{B} \cdot \hat{n}\, dA \quad (30.6)$$
$$= -\frac{d\Phi_B}{dt}.$$

The emf around any closed curve, regardless of how it is moving, equals the rate of change of flux through the curve. The emf is the work done per unit charge as the charge moves around the curve. The work done as the unit of charge moves along an element $d\vec{\ell}$ of the circuit has an electrical contribution $\vec{E} \cdot d\vec{\ell}$, and there may also be a motional contribution $(\vec{v} \times \vec{B}) \cdot d\vec{\ell}$.

The minus sign expresses Lenz's law. By convention, the choices of direction for the elements $d\vec{\ell}$ and the normal vectors $\hat{n}$ follow the right-hand rule shown in Figures 30.8 and 28.15.

The time derivative outside the integral sign symbolizes the fact that flux change may be caused by change in shape of the circuit (the rod on rails), change in the value of $\vec{B}$ (pushing the magnet through the loop), or change in the orientation of $\hat{n}$ (the generator). If the only time variable quantity is $\vec{B}$ itself, then there is induced electric field but no motional ($\vec{v} \times \vec{B}$) contribution.

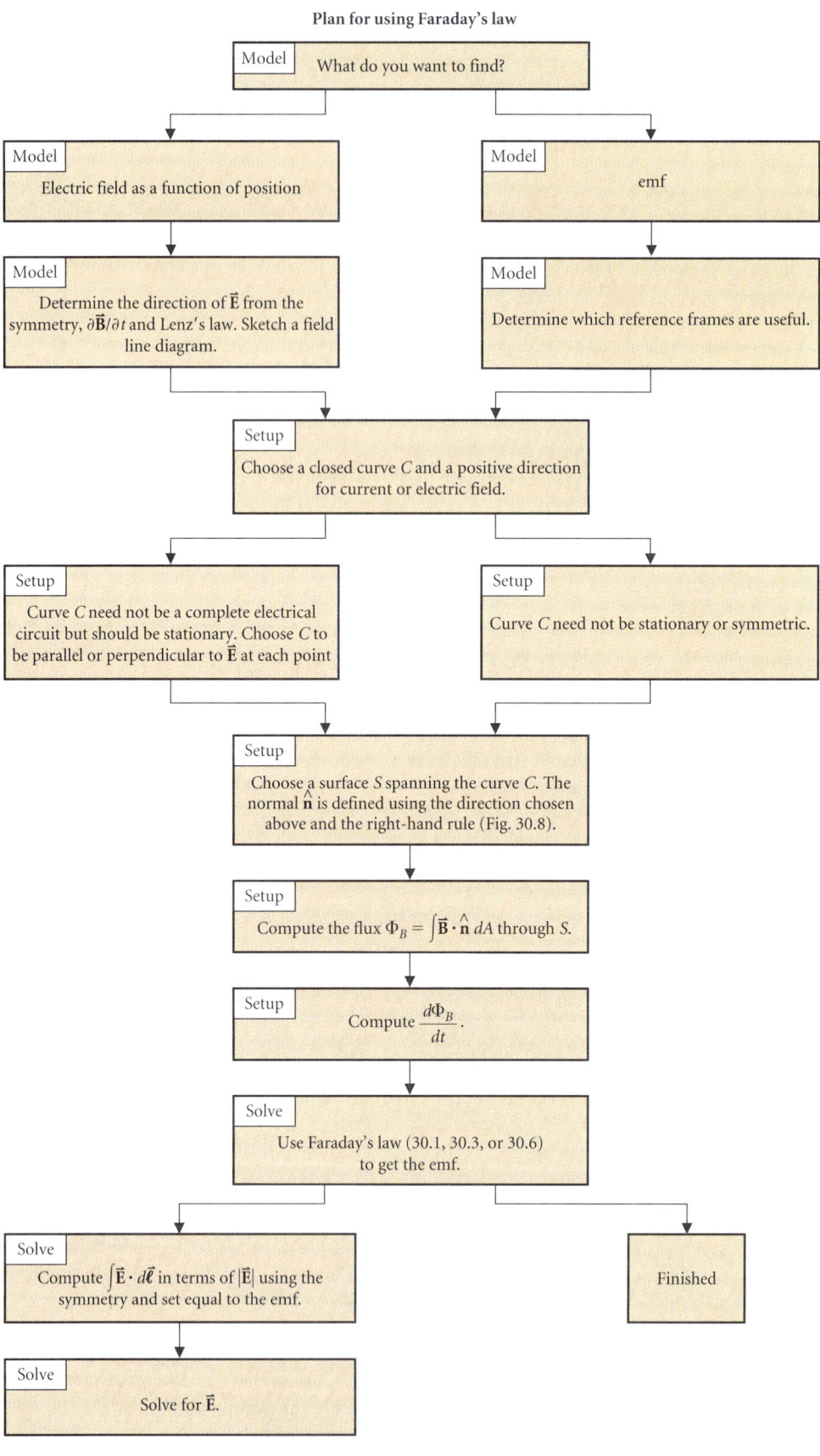

■ FIGURE 30.19

EXAMPLE 30.9 ♦♦♦ The magnetic field inside a solenoid is increasing at a constant rate $|d\vec{B}/dt| = \alpha$. Find the induced electric field inside the solenoid.

MODEL The changing magnetic field has cylindrical symmetry, and the electric field lines circulate around this source like the magnetic field lines around a cylindrical current distribution. The electric field lines form circles around the axis of the solenoid. To find the magnitude of the field at any point, we may apply Faraday's law to a circular path passing through that point (■ Figure 30.20). The direction of the field lines in the figure follows from Lenz's law.

IF THERE COULD BE A CURRENT AROUND THE CURVE, IT WOULD OPPOSE THE CHANGE OF FLUX.

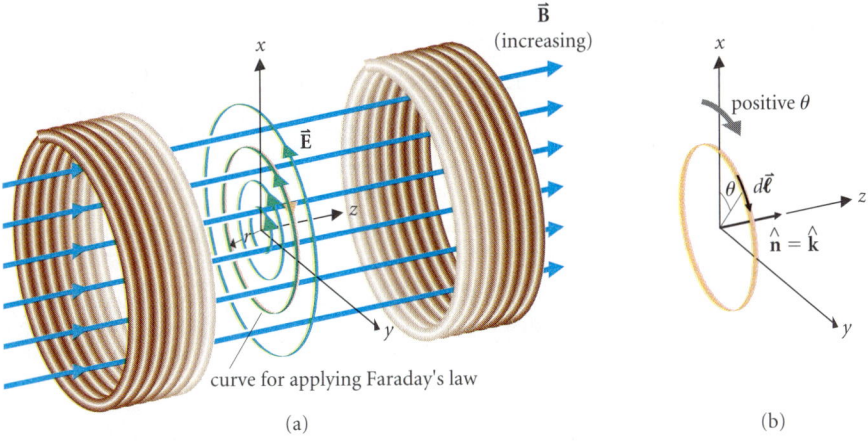

■ **FIGURE 30.20**
(a) Finding the induced electric field inside a solenoid. We choose the z-axis to be along the axis of the solenoid, parallel to $\vec{B}$, and apply Faraday's law to a circle of radius r centered on the axis and parallel to the x-y-plane. (b) We choose to go around the circle in the same direction that the angle θ increases, as shown, so that according to the right-hand rule, $\hat{n} = \hat{k}$.

SETUP We choose the z-axis to be along the axis of the solenoid, with the circle in the x-y-plane. Using the right-hand rule and choosing $\hat{n} = \hat{k}$, we go around the circle as shown in the figure. Since B_z is spatially uniform, the flux passing through the area of the circle is $\Phi_B = \pi r^2 B_z$. Then:

$$\frac{d\Phi_B}{dt} = \pi r^2 \frac{dB_z}{dt} = \pi r^2 \alpha.$$

Here the curve is stationary, so there is no motional contribution to the emf. The electric field has only a θ-component, and its magnitude is the same at each point on the chosen circle. As a result, the integral reduces to a multiplication:

$$\oint \vec{E} \cdot d\vec{\ell} = \oint E_\theta \, d\ell = E_\theta \oint d\ell = 2\pi r E_\theta.$$

COMPARE EQN. (28.13) FOR THE CIRCULATION OF $\vec{B}$ AROUND A CIRCLE WHEN THERE IS CYLINDRICAL SYMMETRY.

SOLVE Applying Faraday's law (eqns. 30.2 and 30.3), we find:

$$2\pi r E_\theta = \oint \vec{E} \cdot d\vec{\ell} = \mathcal{E} = -\frac{d\Phi_B}{dt} = -\pi r^2 \alpha,$$

or

$$E_\theta = -\alpha r/2.$$

ANALYZE The electric field exists whether or not there is a material circuit within the solenoid. If a conducting loop were placed there, the induced field would cause a current. The negative value of E_θ confirms the direction of $\vec{E}$ shown in the figure.   ■

EXERCISE 30.5 ♦♦ Find the induced electric field outside the solenoid in Example 30.9 if the magnetic field is changing at the same rate.   ■

EXAMPLE 30.10 ♦♦♦ A circular wire loop is perpendicular to the axis of the solenoid (■ Figure 30.21). A small gap of width w is cut in the loop. Assume $w \ll d$, the diameter of the wire, which is in turn much less than the radius R of the loop. What is the electric field within the gap?

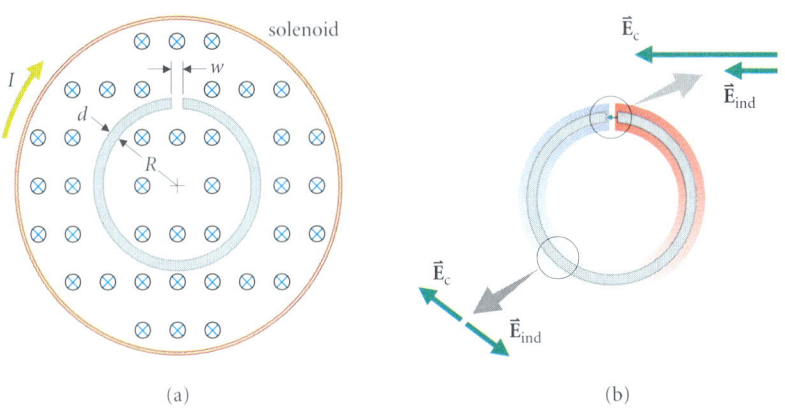

■ **FIGURE 30.21**
(a) A wire loop inside a solenoid has a small gap of width w. (b) No current can exist because of the gap in the loop. Within the metal, the induced electric field $\vec{E}_{ind}$ is opposed by a Coulomb field $\vec{E}_c$ due to the charge distribution on the loop. Within the gap, $\vec{E}_c$ and $\vec{E}_{ind}$ reinforce each other, creating a large total field.

MODEL The solenoid produces the same induced electric field whether the loop is present or not. Because of the gap, steady current cannot exist, so the net force on any electron in the loop has to be zero. Charge densities develop on the loop, as shown schematically in Figure 30.21b, and produce a Coulomb electric field $\vec{E}_c$ that opposes the induced field within the loop:

$$\vec{E} = \vec{E}_c + \vec{E}_{ind} = 0 \text{ within the metal of the loop.}$$

SETUP From Faraday's law, the circulation of the electric field around the loop is given by:

$$\oint \vec{E} \cdot d\vec{\ell} = -\frac{d\Phi_B}{dt} = -\pi R^2 \alpha.$$

SEE EXAMPLE 30.9. WE USED THE SAME SIGN CONVENTIONS.

The electric field $\vec{E}$ is the total electric field, here a superposition of induced and Coulomb field. The only contribution to the circulation integral comes from the portion of the circle in the gap, which is small enough to be modeled as a single element of the circle:

$$\oint_{\text{circle}} \vec{E} \cdot d\vec{\ell} = \int_{\text{gap}} \vec{E} \cdot d\vec{\ell} \approx w E_{\theta,\text{gap}}.$$

SOLVE Thus the electric field in the gap is:

$$E_{\theta,\text{gap}} \approx -\pi R^2 \alpha / w.$$

SEE ALSO PROBLEM 82.

ANALYZE Because the gap width can be made very small, the electric field in the gap can be very large even when the flux changes slowly. This field is mostly Coulomb field, produced by the charges on the ring. Heinrich Hertz used such a device as an antenna in his experimental discovery of electromagnetic waves. ■

Study Problem 19 ♦♦♦ *The Betatron*

The betatron is designed to accelerate electrons using induced electric field. ■ Figure 30.22b is a schematic drawing of the mechanism. The electrons orbit on a circle of fixed radius within the ceramic *doughnut*. The current in the electromagnet windings oscillates sinusoidally at 60 Hz. Electrons are injected just as the magnetic field is zero, but increasing downward. Show

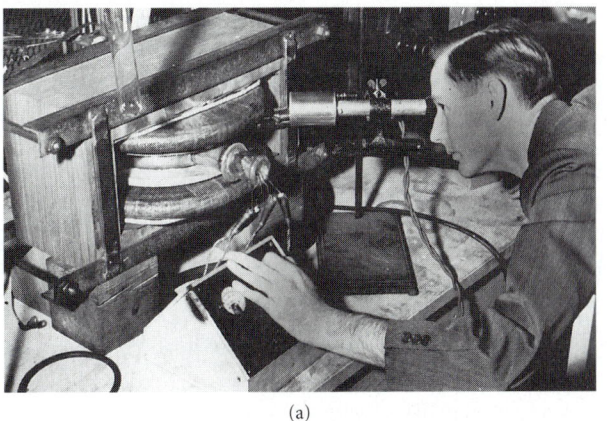
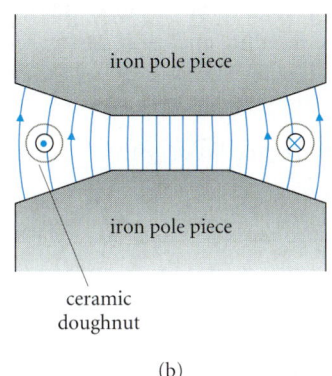

■ **Figure 30.22**
(a) D. W. Kerst with the first betatron in 1941. This betatron could accelerate electrons to an energy of 2.3 MeV. (b) Schematic of a betatron. The pole pieces are shaped so that the magnetic field at the location of the particles is one-half the average value of the field within the orbit.

that (a) the electron energy increases during the first quarter of an oscillation cycle but would decrease during the second quarter; and (b) for the device to operate, the magnetic field at the radius R of the electron orbit must always be precisely half the average field inside the orbit.

Modeling the System

The magnetic field plays a double role in the betatron. Magnetic force is required to accelerate the electrons along their circular path, and the change of magnetic field in time creates the electric field that increases the electrons' energy. We are to demonstrate that a simple constraint on the field allows it to serve both functions.

Since the magnetic field remains cylindrically symmetric, the electric field lines are circles about the symmetry axis, tangent to the orbit of the electrons. ■ Figure 30.23a shows the field pattern during the first quarter cycle. The electric field accelerates the electrons in the clockwise direction, and the magnetic force points inward. During the second quarter cycle (Figure 30.23b), the direction of $d\vec{B}/dt$ reverses as the magnetic field begins to decrease. The electric field also reverses and the electrons would lose energy. During the third quarter of the cycle (Figure 30.23c), the electric field would accelerate the electrons in the reverse direction, and in the fourth quarter, $\vec{E}$ would again produce deceleration.

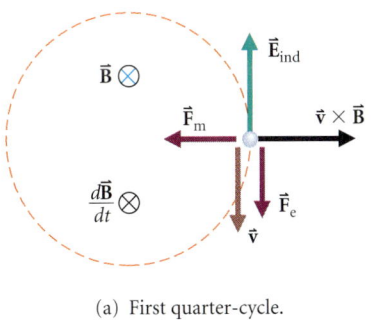

(a) First quarter-cycle.

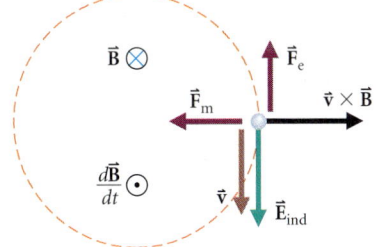

(b) Second quarter-cycle.

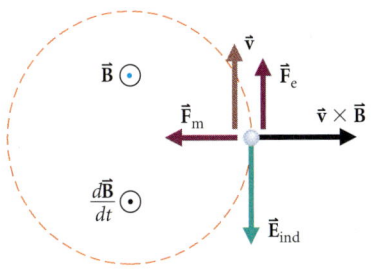

(c) Third quarter-cycle.

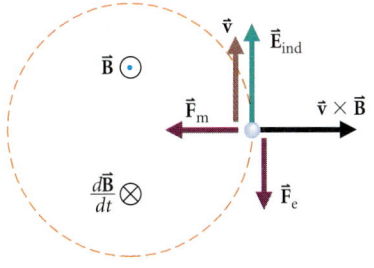

(d) Fourth quarter-cycle.

■ **Figure 30.23**
Forces on an electron in a betatron. In an operating betatron, electrons remain in the machine during the first quarter cycle only. (a) During the first quarter cycle, the magnetic field points into the page and is increasing. The induced electric field is counterclockwise. The electric force increases the electron's speed, while the magnetic force causes the centripetal acceleration v^2/r. (b) During the second quarter cycle, the magnetic field strength is decreasing. The induced electric field is clockwise, and the electric force would decelerate the electrons if they were allowed to remain in the machine. (c) During the third quarter cycle, the magnetic field points out of the page and is increasing. The electrons would be accelerated counterclockwise. (d) During the fourth quarter cycle, the magnetic field strength is decreasing, and the electron energy would be reduced by the electric force opposite $\vec{v}$.

SEE DIGGING DEEPER: MORE ON CYCLOTRONS IN CHAPTER 29. THERE WE GIVE THE RELATIVISTIC EXPRESSION FOR MOMENTUM.

Because the electrons reach speeds approaching the speed of light, a classical description of their motion should be treated skeptically. In this case, results obtained using classical mechanics and expressed in terms of momentum are the same as the exact relativistic results. Only the relation between the electron velocity and momentum differs.

Setup of Solution

Since magnetic force produces the centripetal acceleration of the electrons, the magnetic field at the orbital radius R is related to the electron speed v:

$$-e\vec{v} \times \vec{B} = -evB\hat{r} = -\frac{mv^2}{R}\hat{r} \Rightarrow p \equiv mv = eBR. \quad (i)$$

The electric force acts parallel to the electron velocity during the first quarter cycle and increases the speed:

$$\frac{d|\vec{p}|}{dt} = |\vec{F}_e| = eE. \quad (ii)$$

REMEMBER: FORCE PARALLEL TO VELOCITY CHANGES SPEED; FORCE PERPENDICULAR TO VELOCITY CHANGES DIRECTION. SEE §2.2 AND §3.2.6.

We find the electric field by applying Faraday's law to a circle of radius R. We go around the circle shown in Figure 30.23a counterclockwise.

COMPARE THE CALCULATION OF $\vec{E}$ IN EXAMPLE 30.9.

$$\oint \vec{E} \cdot d\vec{\ell} = 2\pi R E_\theta = -\frac{d\Phi_B}{dt} \Rightarrow E_\theta = -\frac{1}{2\pi R}\left(\frac{d\Phi_B}{dt}\right). \quad (iii)$$

Solution of Equations

Substituting the result (iii) for E into eqn. (ii) gives:

$$\frac{dp}{dt} = eE = \frac{e}{2\pi R}\left|\frac{d\Phi_B}{dt}\right|.$$

Since both the momentum and the flux are approximately zero when the electrons are injected, the solution for p is:

$$p = \frac{e|\Phi_B|}{2\pi R}.$$

Thus, using relation (i) between p and B,

$$eBR = \frac{e|\Phi_B|}{2\pi R}$$

or

$$B = \frac{|\Phi_B|}{2\pi R^2}.$$

The flux within radius R divided by the area πR^2 is just the average magnetic field within radius R. Thus the magnetic field at radius R is half the average value within the orbit, as we were asked to show.

Analysis

With our choice of directions for $d\vec{\ell}$ and $\hat{n}$, Φ_B is negative and E_θ is positive.

Betatrons built for medical research have produced beams with energies as high as 100 MeV.

EXERCISE 30.6 ♦ The average magnetic field in a betatron reaches a maximum of 1.2 T. If the radius of the electron orbits is 0.60 m, what is the final momentum of electrons accelerated by the device?

30.5 Eddy Currents

When a conductor moves through a magnetic field, Lenz's law predicts a force that opposes the relative motion (see Examples 30.2, 30.4, and 30.7). In each case, the force is due to current, driven by an induced emf, interacting with the magnetic field. A clever design can exploit such forces to drive motors or to build a propulsion system. ■ Figure 30.24 shows a metal plate between two sets of magnet coils. If current is turned on and off in each pair of coils, progressing in sequence from left to right, the coils simulate a magnet moving to the right past the plate. According to Lenz's law, the resulting forces accelerate the plate to keep pace with the moving magnet. This is the principle of the linear induction motor and of mass drivers envisioned as the primary means of launching cargoes from future space colonies. The experimental vehicle shown in ■ Figure 30.25 uses a similar effect to advantage as it "floats" on a cushion of magnetic field using linear synchronous motors in an electrodynamic levitation system.

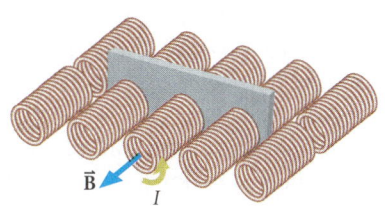

■ **Figure 30.24**
(a) Professor Gerard O'Neill, late of Princeton University, incorporated mass drivers like this in his dream of colonizing space. Large space colonies in orbit about the Sun would need an economic source of raw materials. A mass driver capable of launching projectiles at speeds of 3 km/s could deliver payloads directly from the Moon's surface to a colony. In this simplified design for either a linear induction motor or a mass driver, a sequence of electromagnets is arranged along a track so that a metal blade passes between their poles in sequence. The current in each magnet coil is increased from zero as the leading edge of the blade approaches, reaches a maximum as the middle of the blade passes, and decreases back to zero as the receding edge of the blade passes. (b) The magnetic field acts like a sequence of magnets always moving faster than the blade and dragging it along. Experimental mass drivers currently achieve an acceleration of 1800 g!

■ **Figure 30.25**
An experimental MAGLEV train. Magnetic forces are used both to support the vehicle and to accelerate it.

In a demonstration of this effect (■ Figure 30.26), a metal pendulum swinging between the poles of a permanent magnet is rapidly brought to a halt. If slots are cut in the metal, interrupting the current path, the pendulum swings more freely. To investigate the force that stops the pendulum, we model the pendulum as a large, thin metal sheet moving with velocity

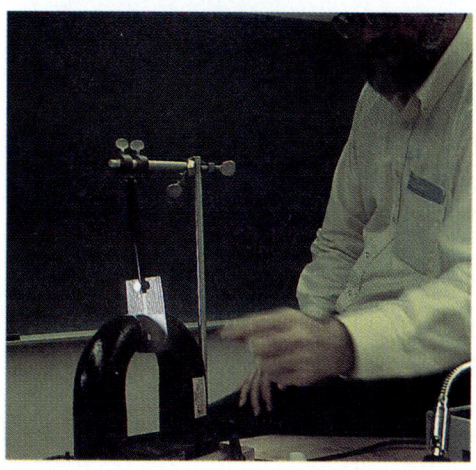

■ **Figure 30.26**
(a) When a metal plate is set swinging between the poles of a magnet, strong damping rapidly brings it to rest on the first swing. (b) The slotted plate swings for a much longer time before coming to rest.

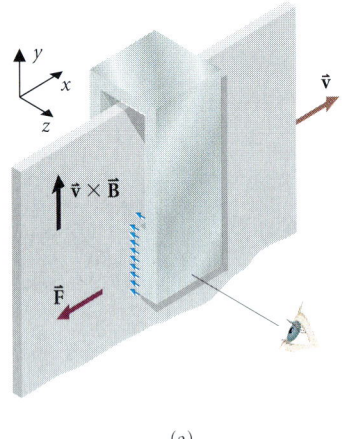
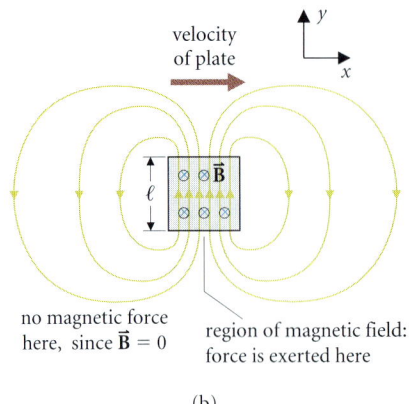
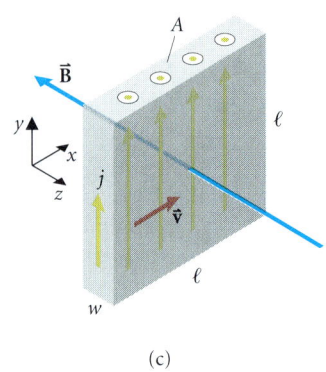

no magnetic force here, since $\vec{B} = 0$ region of magnetic field: force is exerted here

(a)   (b)   (c)

■ **FIGURE 30.27**
(a) A very large, thin, flat plate moves between square magnet poles. (b) The plate moves in the x-direction, with $\vec{B}$ in the $-z$-direction. Thus the magnetic force drives electrons in the $-y$-direction, and the current is in the $+y$-direction between the magnet poles. The current loops close outside the pole region, driven by a charge distribution on the plate. (c) The plate has thickness w, so the current passes through area $A = \ell w$.

$\vec{v}$ between the square poles of a magnet (■ Figure 30.27a). Electrons in the metal between the magnet poles experience a magnetic force that drives current in the direction of $\vec{v} \times \vec{B}$. The current circulates in the characteristic pattern of loops, or *eddies*, illustrated in Figure 30.27b. The magnetic force on this current is the decelerating force required by Lenz's law.

Eddy currents convert energy to thermal form via joule heating. This dissipation accounts for the kinetic energy lost by the swinging pendulum. It can be a substantial nuisance in electrical machinery. Like the slotted plate, devices such as transformers are built from many separate conducting pieces rather than one continuous piece so as to minimize eddy current effects.

Eddy currents also produce a drag force on magnetically levitated (MAGLEV) vehicles (Figure 30.25). Fortunately, the ratio of drag force to support force decreases as the vehicle's speed increases so that practical designs for high-speed systems may be possible.

Digging Deeper

FORCES DUE TO EDDY CURRENTS

To estimate the magnitude of the force on the swinging pendulum, we approximate the magnetic field as uniform within the square region between the magnet poles, and zero elsewhere. Within that region of the plate, magnetic force acts on the electrons and causes them to drift in the direction of the force. We choose the x-axis to be parallel to $\vec{v}$, as shown in Figure 30.27b. The current density in the field region is related to the force that drives it:

$$\vec{j} \sim \frac{\vec{v} \times \vec{B}}{\rho} = \frac{v\hat{i} \times (-B\hat{k})}{\rho} = \frac{vB}{\rho}\hat{j},$$

where ρ is the resistivity of the metal in the plate. [This is eqn. (26.9) with $\vec{E}$ replaced by $\vec{v} \times \vec{B}$.] The current emerges uniformly from one face of a rectangular volume (Figure 30.27c), and the total current is:

$$I = jA = \frac{vB}{\rho}\ell w.$$

The magnetic field acts on this current over a length ℓ and exerts a force opposite $\vec{v}$:

$$\vec{F} \approx I\vec{\ell} \times \vec{B} \sim I\ell\hat{j} \times B(-\hat{k})$$
$$= \frac{vB^2\ell^2 w}{\rho}(-\hat{i}).$$

(In fact, charge densities on the plate are necessary to drive current outside the magnetic field region. The resulting electric field opposes the current in the B field region. An exact calculation for circular magnet poles shows that this effect reduces the net force on the plate by a factor of 2.)

In this result, the dependence of eddy current force on the relative speed v, the resistivity ρ, and the square of the magnetic field B is exact. To use the result in other situations, interpret $B^2\ell w$ as an average of B^2 multiplied by the volume V of metal exposed to the field:

$$F \sim \frac{vB_{av}^2 V}{\rho}.$$

30.6 The Ampère–Maxwell Law

Faraday's law and Ampère's law in its static form are oddly asymmetric. A changing magnetic field creates an electric field. Faraday's law relates the circulation of electric field to change in magnetic flux. We might expect that a changing electric field would create a magnetic field, but Ampère's law in its static form relates the circulation of magnetic field only to current. (There is no "current" term in Faraday's law because there are no magnetic currents.) For symmetry, Ampère's law should have a term involving change in electric flux. Maxwell showed that the additional term is not just aesthetic but is necessary to make Ampère's law self-consistent.

The circulation of $\vec{B}$ around a closed curve is related to the current through a surface bounded by the curve. Mathematically, one may prove that the same relation must hold for *any* surface bounded by the curve, not just the obvious plane surfaces we used in the examples of Chapter 28. In a steady state situation, this mathematical requirement is automatically satisfied. Current through one surface bounded by a curve also passes through any other surface spanning that curve (■ Figure 30.28). However, in a dynamic system this need not be so. Consider charging a capacitor using very long wires (■ Figure 30.29). As the capacitor charges, the current I in the wires produces a magnetic field $\vec{B}$. Applying Ampère's law to the curve C and the surface S_1, we obtain the usual result: $B = \mu_0 I/(2\pi r)$. But no current passes through the surface S_2. The static form of Ampère's law applied to S_2 predicts $B = 0$—a contradiction. Note, however, that the electric flux between the plates changes as the capacitor charges. If A is the area of the plates and σ is the charge density, then:

> Sir George Stokes first published this result as a problem on a Cambridge University prize exam taken by Maxwell!

> This is eqn. (28.6).

> We derived the expression for E_z in §27.1.1.

$$\frac{d\Phi_E}{dt} = A\frac{dE_z}{dt} = \frac{A}{\epsilon_0}\left(\frac{d\sigma}{dt}\right) = \frac{1}{\epsilon_0}\left(\frac{dQ}{dt}\right) = \frac{I}{\epsilon_0}.$$

This thought experiment shows both the problem with Ampère's law and how to fix it. The current may differ through two surfaces with a common boundary, but only if the charge inside changes. The change of electric flux through the surfaces makes up for the difference in current. The quantity

$$I + \epsilon_0 \frac{d\Phi_E}{dt}$$

is the same for every surface. The second term is called *displacement current*. When charging the capacitor, the conduction current due to the motion of charges through the wires equals the displacement current through the capacitor due to changing electric flux.

> Up to this point "conduction current" has been called simply "current." We use this new name to distinguish it from displacement current.

The *displacement current* through a fixed surface is ϵ_0 times the rate of change of electric flux through the surface:

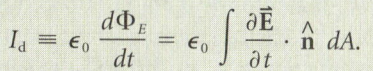

$$I_d \equiv \epsilon_0 \frac{d\Phi_E}{dt} = \epsilon_0 \int \frac{\partial \vec{E}}{\partial t} \cdot \hat{n}\, dA. \qquad (30.7)$$

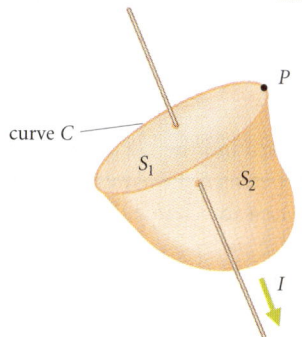

■ **Figure 30.28**
Steady current in a wire. The current is the same through any surface whose edge is the circle through P. For example, I passes through both the plane surface S_1 and the curved surface S_2.

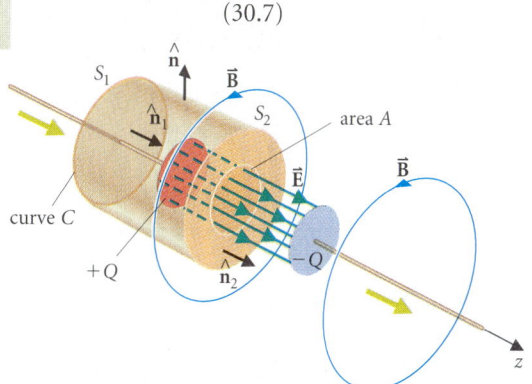

■ **Figure 30.29**
The process of charging a capacitor illustrates the problem with the static form of Ampère's law. We apply Ampère's law to the circle C centered on the wire, and choose the z-axis to be parallel to the wire as shown. Surface S_1 is a plane surface with $\hat{n}_1 = \hat{k}$, bounded by curve C. Surface S_2 consists of the curved wall of a cylinder also bounded by C, together with a flat surface parallel to S_1 but between the capacitor plates. On the flat part of S_2, the normal is also equal to $\hat{k}$. Current I flows through S_1, but there is no conduction current through S_2.

Digging Deeper

CONTINUITY OF TOTAL CURRENT

■ Figure 30.30 shows two surfaces S_1 and S_2 with a common boundary curve C. Together they enclose a volume V. If the current I_2 leaving V through S_2 is not equal to the current I_1 entering through S_1, then the charge Q inside the volume changes:

$$I_1 - I_2 = \frac{dQ}{dt}.$$

But, from Gauss' law, Q/ϵ_0 equals the electric flux Φ_E emerging from V. When applying Gauss' law, the normal outward from the volume is always used, but in Ampère's law, the right-hand rule determines the direction of the normal. Thus:

$$\Phi_E = \oint_S \vec{E} \cdot \hat{n} \, dA$$

$$= \int_{S_1} \vec{E} \cdot (-\hat{n}_1) \, dA + \int_{S_2} \vec{E} \cdot \hat{n}_2 \, dA$$

$$= \frac{Q}{\epsilon_0} = -\Phi_1 + \Phi_2,$$

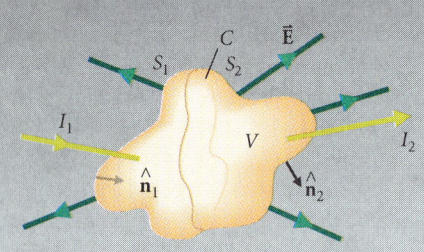

■ FIGURE 30.30
A closed surface S is divided into two pieces S_1 and S_2 by a plane surface. If the current I_1 entering the volume through S_1 does not equal the current I_2 leaving through S_2, then the charge inside the volume changes. Gauss' law relates the electric flux through S to the charge inside. When applying Gauss' law, the normal outward from the volume is always used. Thus $\hat{n} = -\hat{n}_1$ on S_1 and $\hat{n} = +\hat{n}_2$ on S_2.

where Φ_2 is the flux leaving through S_2, and Φ_1 is the flux entering through S_1. Thus:

$$I_1 - I_2 = \frac{dQ}{dt} = \epsilon_0 \frac{d\Phi_E}{dt} = \epsilon_0 \frac{d\Phi_2}{dt} - \epsilon_0 \frac{d\Phi_1}{dt}$$

$$= I_{d,2} - I_{d,1}.$$

Thus:
$$I_1 + I_{d,1} = I_2 + I_{d,2}.$$

The sum of regular current and displacement current is the same for each surface.

The complete form of the Ampère–Maxwell law includes displacement current on an equal footing with regular current. For a fixed curve C:

TO APPLY THIS FORM OF AMPÈRE'S LAW, USE A FIXED CURVE C.

$$\oint_C \vec{B} \cdot d\vec{\ell} = \mu_0(I + I_d) = \mu_0 \int_S \left(\vec{j} + \epsilon_0 \frac{\partial \vec{E}}{\partial t} \right) \cdot \hat{n} \, dA. \tag{30.8}$$

To apply Ampère's law, we may use the method outlined in Figure 28.19, if we include the displacement current. With this change, the plan looks like the one for Faraday's law (Figure 30.19), except that the conduction current through the surface must also be computed.

EXAMPLE 30.11 ♦♦ Find the magnetic field between the circular capacitor plates of radius R in Figure 30.29.

MODEL We expect the magnetic field lines to form circles around the z-axis between the capacitor plates as well as outside them. The electric field between the plates is due to the surface charge already placed there: $\vec{E} = (\sigma/\epsilon_0)\hat{k}$. Modeling the charge distribution as uniform, $\sigma = Q/\pi R^2$.

SETUP We choose a circle of radius r centered on the z-axis as our Ampèrian curve. Going counterclockwise in ■ Figure 30.31, the circulation of $\vec{B}$ around this curve is $2\pi r B_\theta$. The normal to the surface S is in the z-direction, and the electric field is constant over the surface, so the electric flux through the surface is:

$$\Phi_E = \int \vec{E} \cdot \hat{n} \, dA = \int (\sigma/\epsilon_0)\hat{k} \cdot \hat{k} \, dA = (\sigma/\epsilon_0)(\pi r^2) = (Q/\epsilon_0)(r/R)^2.$$

There is no conduction current through the surface.

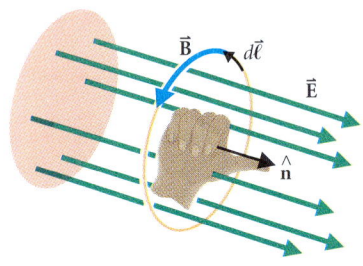

■ FIGURE 30.31
To calculate the magnetic field between the capacitor plates, we choose a circular Ampèrian curve and traverse it counterclockwise, as shown here. Then the electric flux through the curve is positive and increasing.

SOLVE Applying Ampère's law:

$$2\pi r B_\theta = \mathscr{C} = \mu_0(I + I_d) = \mu_0\left(0 + \epsilon_0 \frac{d\Phi_E}{dt}\right)$$

$$= \mu_0\epsilon_0\left(\frac{1}{\epsilon_0}\right)\left(\frac{dQ}{dt}\right)\left(\frac{r^2}{R^2}\right) = \mu_0 I \left(\frac{r}{R}\right)^2.$$

Thus:
$$B_\theta = \frac{\mu_0 I}{2\pi R}\left(\frac{r}{R}\right).$$

ANALYZE At the edge of the plates, the magnetic field reaches the value $B_\theta = \mu_0 I/(2\pi R)$ expected for the current I in the long wire. The capacitor acts like a fat wire with displacement current distributed uniformly across it. The magnetic field increases outward correspondingly.

Chapter Summary

Where Are We Now?

Faraday's law and the Ampère–Maxwell law complete the set of Maxwell's equations. The inclusion of dynamic (i.e., changing) fields allows the discussion of a wide variety of practical systems. Maxwell's equations, together with the Lorentz force law and Newton's laws, are sufficient to solve any problem in classical electrodynamics, although many problems require advanced mathematical techniques.

Maxwell's Equations

(Gauss' law) The electric flux emerging from a volume equals the total charge within the volume divided by ϵ_0:

$$\Phi_E \equiv \oint \vec{E} \cdot \hat{n}\, dA = \frac{Q}{\epsilon_0} = \frac{1}{\epsilon_0}\int \rho\, dV. \qquad \text{EQN. 23.12, §23.4.3}$$

The magnetic flux emerging from any volume is zero:

$$\Phi_B \equiv \oint \vec{B} \cdot \hat{n}\, dA = 0. \qquad \text{EQN. 28.9, §28.3.1}$$

(Faraday's law) The electromotive force induced around any closed curve equals the rate of change of the magnetic flux through any surface bounded by the curve. If $\vec{v}$ is the velocity of the curve element $d\vec{\ell}$, then:

$$\mathscr{E} = \oint (\vec{E} + \vec{v}\times\vec{B})\cdot d\vec{\ell} = -\frac{d\Phi_B}{dt} = -\frac{d}{dt}\int \vec{B}\cdot\hat{n}\, dA. \qquad \text{EQN. 30.6, §30.1–30.3}$$

(Ampère–Maxwell law) The circulation of the magnetic field around any fixed closed curve is μ_0 times the total (conduction plus displacement) current through any surface bounded by the curve:

$$\oint \vec{B}\cdot d\vec{\ell} = \mu_0(I + I_d) = \mu_0\left(I + \epsilon_0\frac{d\Phi_E}{dt}\right) = \mu_0\int\left(\vec{j} + \epsilon_0\frac{\partial\vec{E}}{\partial t}\right)\cdot\hat{n}\, dA. \qquad \text{EQN. 30.8, §30.6}$$

This particular statement of Ampère's law is valid for fixed, but not moving, curves.

What Did We Do?

Faraday's experiments show that a change in magnetic flux through a circuit, for any reason, induces an electromotive force around the circuit. Change in the magnetic field in time induces an electric field that produces the emf.

According to Lenz's law, the response of a system to an induced emf opposes the change in flux that causes the emf. Lenz's law is a consequence of energy conservation; it gives the direction of an induced electric field and predicts the direction of forces acting on a system.

Magnetic forces on electrons in a moving conductor drive current. To find the motional emf, apply Faraday's law to a curve attached to the moving conductor.

Electromotive force describes the energy input to a circuit. Electric potential indicates how energy gained in one part of a circuit is transmitted for use elsewhere. Thus potential differences are intimately related to emf but are not the same thing.

✱ Electric and magnetic fields are different features of a unified electromagnetic field. In the reference frame moving with a circuit element, the emf arises from the action of the electric field $\vec{E}'$. In another reference frame, the emf appears to have both electric and motional (magnetic) contributions:

$$\vec{E}' = \vec{E} + \vec{v} \times \vec{B}.$$

In symmetric systems, induced electric field may be computed from Faraday's law in the same way that magnetic field is found from Ampère's law. (See Figure 30.19.)

Relative motion between a conductor and a magnetic field causes eddies of current within the conductor. Magnetic forces on these eddy currents strongly damp the relative motion.

The static version of Ampère's law is incomplete when applied to nonsteady currents. Maxwell discovered that displacement current must be included with conduction current to obtain a law valid for dynamic fields. The displacement current through a surface is ϵ_0 times the rate of change in the electric flux through the surface.

Practical Applications

One of the most important applications of Faraday's law is the production of electrical energy from mechanical energy. We discussed a simple generator in Examples 30.5 and 30.7. Devices working on the same principle produce electrical energy in commercial power plants. On a smaller scale, magnetos used in piston aircraft engines and lawn mowers use induced emf to produce ignition sparks: the engines run without battery power.

Another important application is the electric motor. The simplest motor is a generator run backwards (Example 30.6). A few more complex designs are explored in the problem set.

Eddy currents resulting from induced emf generate unwanted heat in many electrical devices, especially transformers, which must be designed to suppress the currents. However, mass drivers use eddy current forces to accelerate projectiles. Similar designs have been proposed for transportation systems.

Displacement current is fundamental to the understanding of electromagnetic waves (Chapter 33).

Solutions to Exercises

30.1. From the force law $\vec{F} = q\vec{v} \times \vec{B}$,

$$1\text{ T} = \frac{1\text{ N}}{(1\text{ C})\cdot(1\text{ m/s})},$$

so $1\,\dfrac{\text{T}\cdot\text{m}^2}{\text{s}} = 1\,\dfrac{\text{N}\cdot\text{s}\cdot\text{m}^2}{\text{C}\cdot\text{m}\cdot\text{s}} = 1\,\dfrac{\text{N}\cdot\text{m}}{\text{C}} = 1\text{ J/C} = 1\text{ V}.$

30.2. Argument 1: The force exerted by the loop should decelerate the receding magnet. Thus the N pole of the loop's dipole moment points toward the S pole of the magnet, and so the current is clockwise.

Argument 2: The magnetic flux produced by the magnet is now decreasing, so the response of the loop is to produce field lines in the same direction as those of the magnet. Again, we conclude the current is clockwise.

30.3. The mechanical power output of the motor is its torque multiplied by its angular speed: $P_m = \omega\tau = \omega NBAI \sin\theta$. As the motor rotates, changing flux results in an emf $\mathscr{E}_{\text{ind}} = d\Phi_B/dt = \omega NBA \sin\theta$, and the electrical power supplied is:

$$P_e = \mathscr{E}I = (Ir + \mathscr{E}_{\text{ind}})I = I^2r + \omega NBAI \sin\theta = P_{\text{joule}} + P_m,$$

as required.

30.4. We are given the component of $\vec{B}$ perpendicular to the (horizontal) airplane wing. We convert the speed to m/s (1 h = 3600 s); then the required potential difference is:

$$\Delta V = B_{\text{vert}} \ell v$$
$$= (2.5 \times 10^{-5} \text{ T})(60 \text{ m})[(10^6 \text{ m})/(3.6 \times 10^3 \text{ s})]$$
$$= 0.4 \text{ V}.$$

30.5. We apply Faraday's law to a circle of radius $r > R$ centered on the axis of the solenoid (■ Figure 30.32). Here we choose to go around the curve counterclockwise (in the expected direction of $\vec{E}$), with $\hat{n}$ out of the page, antiparallel to $\vec{B}$. All the flux in the solenoid passes through the surface S, and the magnetic field outside the solenoid is approximately 0, so the flux through S is $\Phi_B = -\pi R^2 B$. The emf is:

$$\mathcal{E} = -\frac{d\Phi_B}{dt} = +\pi R^2 \alpha = 2\pi r E_{\text{ind}}.$$

So, $E_{\text{ind}} = R^2 \alpha/(2r)$. The direction of $\vec{E}_{\text{ind}}$ is shown in the figure.

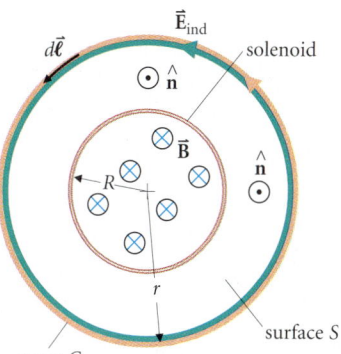

■ **FIGURE 30.32** curve C

30.6. The field at the electron orbit is $B = \frac{1}{2}B_{\text{av}} = 0.60$ T. So:
$$p = eBR = (1.6 \times 10^{-19} \text{ C})(0.60 \text{ T})(0.60 \text{ m})$$
$$= 5.8 \times 10^{-20} \text{ N·s}.$$

The energy of such an electron is 110 MeV.

Basic Skills

Review Questions

§30.1 INDUCED EMF

- What happens when a bar magnet is pushed N pole first toward a coil connected to an ammeter? How does the effect depend on the size of the coil and the speed of the magnet?
- State Faraday's law.
- Under what conditions is it possible to use static field calculations for time-variable systems?
- State Lenz's law. To which conservation law is it related?
- Describe sign conventions that allow you to combine Faraday's law and Lenz's law into a single mathematical statement. What is that statement?

§30.2 MOTIONAL EMF

- When does motional emf occur?
- Describe the operation of a simple generator. How does the magnetic flux through the loop change? How does the emf generated depend on the angular speed of the loop, its area, and the magnetic field strength?
- How much power is required to operate a simple motor?
- What are *brushes*? In which devices are they used, and why?

§30.3 THE NATURE OF EMF

- Does induced electric field contribute to potential difference? Why or why not?
- Define emf in terms of work.
- How is emf related to the electric and magnetic fields in an arbitrary circuit?
- ✶ If the electric and magnetic fields in a certain reference frame are $\vec{E}$ and $\vec{B}$, what is the electric field in a second reference frame moving with velocity $\vec{v}$ with respect to the first?
- If the magnetic force does no work (cf. §29.1), how does motional emf transform energy?

§30.4 CALCULATION OF INDUCED ELECTRIC FIELD

- Describe how induced electric field is related to its source.

§30.5 EDDY CURRENTS

- What are *eddy currents* and when do they occur?
- Give one example in which eddy currents are useful and one example in which they are a nuisance.

§30.6 THE AMPÈRE–MAXWELL LAW

- Why is the static form of Ampère's law incomplete?
- Define *displacement current*.
- State the complete Ampère–Maxwell law.

Basic Skill Drill

§30.1 INDUCED EMF

1. The magnetic flux through each loop of a coil with 150 turns increases from 1.60 Wb to 2.40 Wb in a time interval of 0.15 s. What is the average emf developed in the coil during that time interval?

2. A loop of wire lies flat on a table with an ammeter connected in it. Also lying on the table is a bar magnet oriented as shown in ■ Figure 30.33. Without being raised from the table, the magnet is turned 180° to reverse the positions of the N and S poles. While this is done, conventional current through the ammeter is:
(a) momentarily to the right.
(b) momentarily to the left.
(c) first to the right and then to the left.
(d) first to the left and then to the right.
(e) essentially zero.
Explain why you chose your answer.

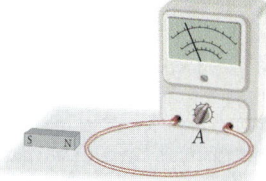

■ **FIGURE 30.33**

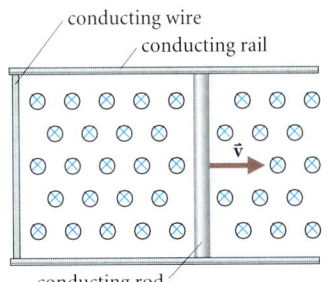

FIGURE 30.34

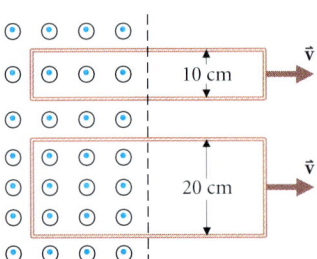

FIGURE 30.35

3. The magnetic flux through a loop is given by:
$$\Phi_B(t) = (1\text{ mWb})(At^4 + \cos\omega t),$$
where t is the time in seconds, $A = 10^{-4}/\text{s}^4$, and $\omega = 400$ rad/s. What is the induced emf as a function of time?

§30.2 MOTIONAL EMF

4. A rod moves on metal rails with a velocity $\vec{v}$ as shown in ■ Figure 30.34, in a region where the magnetic field is into the page. The force on the rod is:
(a) zero.   (b) infinite.   (c) in the direction of $\vec{v}$.
(d) in the direction opposite $\vec{v}$.   (e) perpendicular to $\vec{v}$.
Explain why you chose your answer.

5. Two rectangular wire loops each have a resistance of 10 Ω, but one is twice as wide as the other (■ Figure 30.35). They are both pulled out of a strong magnetic field $B = 0.5$ T at the same constant speed $v = 10$ m/s. Use Faraday's law to compare the emfs developed around the two loops and the heat losses in each. On which loop must the greater force act?

6. A simple generator has a rectangular wire loop with dimensions 2.0 cm by 4.0 cm and 150 turns. If it is rotated at $f = 60.0$ Hz in a field of 0.80 T, what is the peak emf produced?

7. A motor has a square coil with 1.0-cm sides and 50 turns that rotates in a magnetic field of 0.75 T. If a current of 2.5 A is supplied to the motor, what is the maximum torque it exerts?

§30.3 THE NATURE OF EMF

8. A railroad supervisor decides to check on his engineers to see whether they are driving the trains at the correct speed. He places his voltmeter leads on either side of the 2.0-m-wide tracks and measures a potential difference 1.3 mV. If the vertical component of the Earth's field at that location is 2.3×10^{-5} T, what is the train's speed?

§30.4 CALCULATION OF INDUCED ELECTRIC FIELD

9. What induced electric field is produced inside a solenoid, 1.0 cm from its axis, if it has 25 turns/cm and the current is increased at 15 A/s?

10. Kerst's 1942 betatron accelerated electrons to 20 MeV at an orbit radius of 19 cm. The momentum of a 20-MeV electron, allowing for relativistic effects, is $39mc$, where m is the electron mass and c is the speed of light. What maximum magnetic field was required?

§30.5 EDDY CURRENTS

11. ✖ A copper plate of thickness 0.2 cm moves at 2 m/s between the poles of a magnet. The magnet poles are circular with radius 3 cm, and the magnetic field strength between them is 0.3 T. Estimate the force required to keep the plate moving.

§30.6 THE AMPÈRE–MAXWELL LAW

12. A uniform sheet lying in the x-y-plane has a uniform charge density σ that is increasing at a uniform rate $d\sigma/dt = 16$ mC/m²·s. Find the displacement current through a circle of radius $r = 2.5$ cm lying parallel to the x-y-plane at $z = z_0 = 50$ cm.

13. A parallel plate capacitor is being charged. At one instant, the surface charge density on the plates is increasing at 35 mC/m²·s. Find the magnetic field 1.0 cm from the center of the plates at this instant.

Questions and Problems

§30.1 INDUCED EMF

14. ❖ An ammeter is attached to a circular loop of wire. You grab a bar magnet and push it through the loop N pole end first. Sketch the ammeter reading as a function of time.

15. ❖ What is the direction of current through the ammeter in Example 30.1?

16. ❖ Two loops of wire lie side by side on the table (■ Figure 30.36). One loop includes a battery and a switch. The other includes an ammeter. The switch, initially open, is closed. At the time of closure, current through the ammeter is:
(a) momentarily to the right (The ammeter reading is positive.).
(b) momentarily to the left (The ammeter reading is negative.).

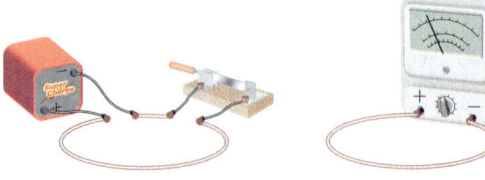

■ **FIGURE 30.36**

(c) steadily to the right.
(d) steadily to the left.
(e) first to the right and then to the left.
Explain what is wrong with the answers you didn't choose.

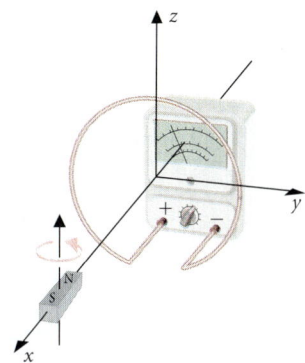

■ FIGURE 30.37

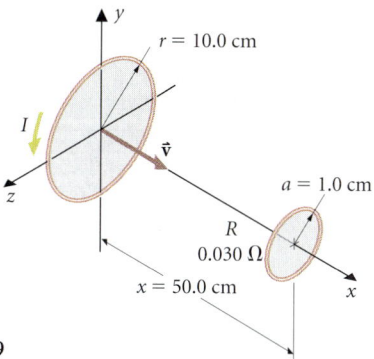

■ FIGURE 30.39

17. ❖ A circular loop of wire lies in the *y-z*-plane with its center at the origin and is connected to an ammeter that reads positive when the current enters the + terminal. A magnet lying along the *x*-axis is rotated 180° about an axis parallel to *z* (■ Figure 30.37), until its N pole points toward the loop. The ammeter reads
(a) first plus then minus.
(b) essentially not at all.
(c) plus while the magnet is rotating.
(d) minus while the magnet is rotating.
(e) first minus then plus.
Explain what is wrong with the answers you didn't choose.

18. ❖ Two conducting loops lie perpendicular to a magnetic field that is increasing at 1.0 mT/s. Loop *a* is circular with radius 1.5 cm. Loop *b* is square with 2.5-cm sides. Both have the same resistance. In which loop is the current greater? Justify your answer.

19. ❖ Two identical conducting loops lie perpendicular to a magnetic field. At the position of loop A, the field is 1.5 T and it is increasing at 0.2 T/s. At the position of loop B, the field is 1.2 T and it is decreasing at 0.3 T/s. In which loop is the emf greater? Why?

20. ◆ A coil with 150 turns is connected in series with a voltmeter. If the average voltmeter reading during a 2.7-s interval is 71 V, what change of magnetic flux through the coil occurs during the interval?

21. ◆ A circular loop of radius 2.50 cm lies perpendicular to a magnetic field that is decreasing at 0.305 T/s. What emf is induced in the loop?

22. ◆ A circular wire loop of radius 3.33 cm lies in the *y-z*-plane with its center at the origin. The magnetic field in the region is given by:
$$\vec{B} = [0.350 \text{ T} + (0.125 \text{ T/s})t](\hat{i} - \hat{j} - \hat{k}).$$
What is the emf induced in the loop at $t = 0$?

23. ◆◆ The magnetic field in a region of space is:
$$\vec{B} = (15 \text{ T})\hat{k}\{[t/(1.0 \text{ s})]^2 - [t/(3.0 \text{ s})]^3\}.$$
A circular loop of radius $r = 3.0$ cm and resistance $0.050 \, \Omega$ lies in the *x-y*-plane. (a) Find the current in the loop at $t = 0.50$ s, 1.0 s, 2.0 s, and 5.0 s. (b) When does the maximum current occur, and what is its value?

24. ◆◆ Two long, tightly wound solenoids have a common axis (■ Figure 30.38). The inner solenoid has radius $a = 2.0$ cm, $N = 10^3$ turns,

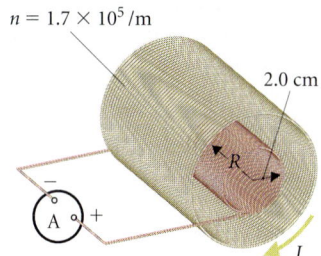

■ FIGURE 30.38

and a resistance $R = 10.0 \, \Omega$. The outer solenoid has $n = 1.7 \times 10^5$ turns per meter and carries a current that is increasing at a constant rate $dI/dt = 0.60$ A/s. What is the current through an ammeter connected to the inner solenoid? What is its direction?

25. ◆◆ (■ Figure 30.39) Two circular loops are parallel to the *y-z*-plane with their centers on the *x*-axis. The loop on the right, with resistance $R = 0.030 \, \Omega$ and radius $a = 1.0$ cm, is fixed in position. The larger loop, a distance $x = 50.0$ cm to the left, carries current $I = 0.15$ A and has radius $r = 10.0$ cm. It moves directly toward the smaller loop with a velocity of $(2.2 \text{ m/s})\hat{i}$. Find the magnitude and sense of the current induced in the smaller loop. [*Hint:* Use the result for the magnetic field on the axis of a current loop, eqn. (28.7), assuming $a \ll r \ll x$.]

26. ◆◆ A circuit contains a 10.0-Ω resistor, a 12-V battery, and a 15-cm-radius loop immersed in a 1.0-T magnetic field (■ Figure 30.40). How fast must the magnetic field be changed to prevent current in the circuit? Should it be increased or decreased?

27. ◆◆ A loop of wire is in the form of an equilateral triangle of side $\ell = 3.0$ cm. The wire has resistivity
$$\rho = 1.7 \times 10^{-8} \, \Omega \cdot \text{m}$$
and a diameter of 1.0 mm. What is the resistance of the loop? The loop lies in the *x-y*-plane and the magnetic field is:
$$\vec{B} = B_o\hat{k}[(t/t_o)^2 - (t/t_o)^4],$$
where $B_o = 0.33$ T and $t_o = 1.0$ s. Find the current in the loop as a function of time and plot it. Show that the current reaches a maximum at 0.41 s. What is the power being dissipated in the loop at this time? When, other than at $t = 0$, is the power zero? What happens after that?

§30.2 MOTIONAL EMF

28. ❖ A circular loop is moving at speed *v* parallel to a diameter (■ Figure 30.41). A uniform magnetic field $\vec{B}$ is perpendicular to the plane of the loop. Is there a current around the loop? Why or why not?

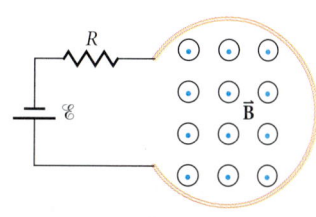

■ FIGURE 30.40

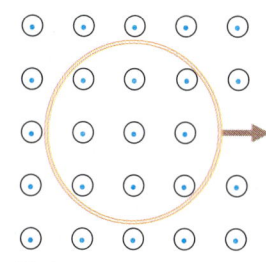

■ FIGURE 30.41

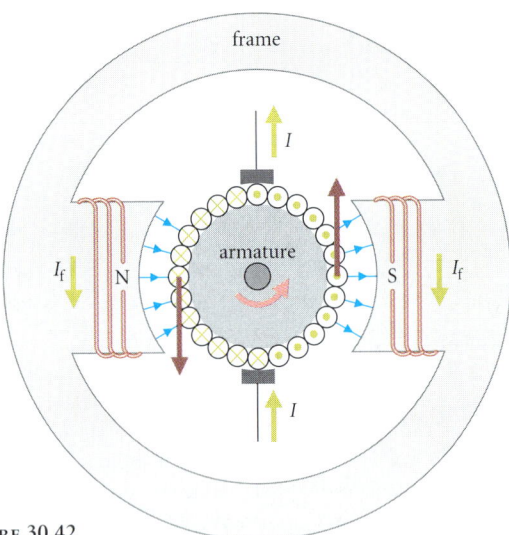

FIGURE 30.42

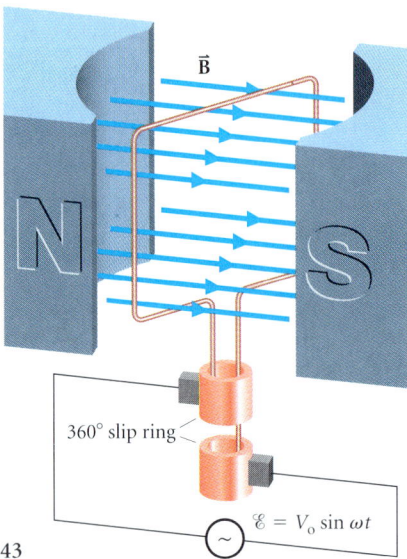

FIGURE 30.43

29. ❖ ■ Figure 30.42 shows a schematic of a practical DC motor. Discuss how the output torque varies. Describe the advantages, both mechanical and electrical, of this design over the simple motor discussed in the text.

30. ◆ A simple generator has a coil of area 15.0 cm² and 125 turns. It rotates at 60.0 Hz. What magnetic field is required if the peak voltage generated is to be 75.0 V?

31. ◆ A simple generator has a rectangular coil measuring 15.0 cm × 35.0 cm. It rotates at 60.0 Hz in a 1.25-T magnetic field. How many turns are required in the coil for the generator to produce a peak emf of 8.50 kV?

32. ◆◆ A simple generator (cf. Figure 30.11) has a coil with dimensions 0.10 m × 0.20 m, which is rotated at 60.0 Hz in a 0.50-T magnetic field. How many windings are needed if the generator is to develop an average power of 0.10 kW when the total resistance in the circuit is 0.10 kΩ?

33. ◆◆ A simple motor has 250 turns of wire on a coil measuring 12 cm × 4.0 cm. Its resistance is 6.0 Ω. The magnetic field between the pole pieces is 0.95 T. How much current in the coil is needed to produce a maximum torque of 30.0 N·m? What maximum and minimum emf is needed to drive the motor at 60 Hz if the current has constant magnitude?

34. ◆◆ A circular wire loop with resistance $R = 1$ mΩ and radius $a = 5$ cm is placed in a constant magnetic field $B = 1$ T with $\vec{B}$ perpendicular to the plane of the loop. You pull on opposite ends of a diameter and draw the loop into a very narrow rectangle. (a) Explain why there is a current in the loop. (b) Explain why the total charge that moves past any point on the loop is independent of the time you take to pull the wire taut. (c) ✉ If you pull the loop closed in $\frac{1}{2}$ s, estimate how much thermal energy is produced in the loop and the average force you have to exert on the wire.

35. ◆◆ A uniform magnetic field $\vec{B} = B_o \hat{k}$ is present in the region $y > 0$. A semicircular wire loop lying in the x-y-plane has radius a, resistance R, and is centered at the origin. It rotates about the $+z$-axis with a constant angular speed ω. The straight diameter of the loop lies along the x-axis at $t = 0$, with the curved side at positive y. Find the emf induced in the loop as a function of time and sketch a graph of the emf versus time for one complete rotation of the loop.

36. ◆◆ A generator is made from a metal rod of length 1.00 m. At one end, a hole in the rod fits smoothly over a metal peg. At the other end, the rod slides smoothly on a circular metal track. The rod rotates about the peg at 50.0 rpm, and a magnetic field of 0.350 T is perpendicular to the plane of rotation. What is the current through a 1.25-kΩ resistor connected between the peg and the track?

37. ◆◆ A rectangular loop of resistance $R = 1.0$ mΩ measuring 12 cm × 3.5 cm is projected with velocity $\vec{v} = (7.7$ m/s, parallel to its long side) into a region where $\vec{B} = (0.65$ T, perpendicular to the loop). If the mass of the loop is 38 g, compute the emf around the loop, the current in it, the force on the loop, and its acceleration.

38. ◆◆◆ ■ Figure 30.43 shows a simple design for a synchronous motor. An oscillating potential difference $V(t) = V_o \sin \omega t$ is applied to the terminals of the rotating loop. (a) When it is not driving a mechanical load, the motor rotates at angular speed ω, and there is no current in the loop. Explain why. (b) When the motor is running without load, what value of V_o is required? What is the position of the loop as a function of time? (c) If the motor is to produce an average mechanical power P_o and its windings have resistance r, by how much must the potential difference V_o be increased?

§30.3 THE NATURE OF EMF

39. ❖ A wooden boat is sailing on a conducting (saltwater) sea at speed $v = 12$ m/s relative to the water. The vertical component of the Earth's field is $B_\perp \approx 10^{-5}$ T. A metal rod 1 m long, lying perpendicular to $\vec{v}$, is on board the boat. An engineer on the boat connects a voltmeter to the two ends of the rod. What does the voltmeter read? Another engineer sitting on a buoy reaches up and touches his voltmeter leads to the ends of the rod as the boat passes. What does his voltmeter read? The engineer on board, believing his voltmeter to be broken, tosses the leads into the water where they land 1 m apart. What does the voltmeter read then? If the engineer on the buoy tosses his leads into the water, what does his meter read?

40. ❖ Suppose, in the system of Example 30.4, that both the sliding rod and the cross tie have equal resistance $R/2 = 0.5$ mΩ. Is the current in the circuit changed? Describe the charge densities that arise on the conducting surfaces (i.e., compare them with the case in which the cross tie has the entire resistance). Describe the fields acting within the rod and the cross tie.

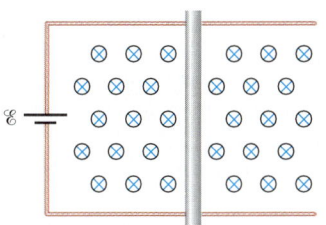

FIGURE 30.44

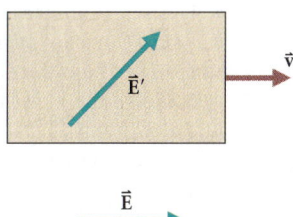

FIGURE 30.45

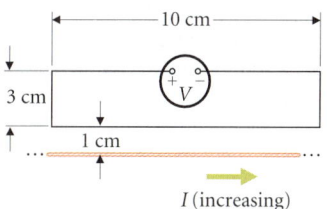

FIGURE 30.46

41. ❖ A thin conducting rod rotates about one end with angular speed ω in a region of space with a uniform magnetic field. Looking along the field lines, the rod rotates clockwise. The potential difference between the stationary end A of the rod and its moving end C, $\Delta V = V_A - V_C$, is:
(a) positive. (b) negative. (c) proportional to ω^2.
(d) inversely proportional to ω. (e) zero.
Explain what is wrong with the answers you didn't choose.

42. ❖ At what speed and in what direction must the rod in ■ Figure 30.44 move if there is to be no current in the circuit? Describe qualitatively the surface charges on the system and the fields within the moving rod.

43. ❖ A conducting rod rotates about its center in a region of uniform magnetic field $\vec{B}$ inclined at an angle θ to the rotation axis. Explain why the potential difference between the rod's center and one of its ends is proportional to $\cos\theta$.

44. ❖ ❖ ■ Figure 30.45 shows the direction of the electric field $\vec{E}'$ measured by an observer on a moving cart, and the field $\vec{E}$ measured by an experimenter in the lab. What is the direction of the magnetic field $\vec{B}$ measured in the lab?

45. ♦ A 1-km-long conducting cable tethers a satellite to the space shuttle. The cable moves through the Earth's magnetic field ($B \sim 2 \times 10^{-5}$ T) at a speed of 7 km/s. Estimate the potential difference between the two ends of the tether. The Earth's field near the equator is northward and the space shuttle moves eastward in its orbit. The cable points radially outward from the shuttle to the satellite. Which end of the cable is at higher potential?

46. ❖ ♦ The electric and magnetic fields in a particular region are $\vec{E} = (1.0 \times 10^{-3} \text{ V/m})\hat{i}$ and $\vec{B} = (1.0 \text{ T})\hat{j}$. A particle moves so that the electric field in its rest frame is zero. What is the particle's velocity?

47. ♦♦ A horizontal copper rod of length 0.20 m rotates about one end at angular speed $\omega = 520$ rad/s in a vertical magnetic field $B = 0.25$ T. What is the potential difference between its ends? The rod rotates clockwise when viewed along the field lines, and $\vec{B}$ is parallel to the rotation axis. Which end is at greater potential?

48. ❖ ♦♦ The electric field in a region of space is 125 V/m in the $+x$-direction. The magnetic field is 5×10^{-8} T in the $+z$-direction. Aliens on a spacecraft moving at 2×10^7 m/s in the $+y$-direction measure the electric field in the surrounding space. What is the result of their measurement?

49. ❖ ♦♦ A particle is moving at $(3.00 \times 10^4 \text{ m/s})(\hat{i} + \hat{j})$ in an accelerator experiment where the magnetic field in the lab is $(5.50 \times 10^{-3} \text{ T})(\hat{j} + \hat{k})$. An electric field is applied so that the electric field in a frame moving with the particle is $(0.550 \text{ kV/m})\hat{i}$. What is the electric field applied in the lab?

50. ❖ ♦♦♦ A long straight wire near a rectangular circuit carries current that is increasing at a rate $dI/dt = 10^{-3}$ A/s in the sense shown (■ Figure 30.46). The only resistance in the circuit is in the voltmeter. The voltmeter V reads positive when its + terminal is at a higher potential than its − terminal. Compute the reading on the voltmeter. Pay particular attention to the sign of the reading.

§30.4 CALCULATION OF INDUCED ELECTRIC FIELD

51. ❖ Suppose we wanted to use a betatron to slow down electrons. Then when should we inject them into the betatron?

52. ♦ The current in a solenoid with 10^3 turns per meter and radius 1.5 cm is increasing at 6 mA/s. Find the electric field at radii 0.5 cm and 2.5 cm (i.e., inside and outside the solenoid).

53. ♦♦ The poles of an electromagnet have circular cross-sectional area $A = 0.30$ m². If the field between the poles increases at a rate 0.020 T/s, find the magnitude of the induced electric field at points 0.20 m and 0.50 m from the symmetry axis of the magnet.

54. ♦♦ In Kerst's original betatron, the electrons (mass m) had a final momentum equal to $4.5mc$, where c is the speed of light. Find the maximum induced electric field that acts on the electron in this betatron.

55. ♦♦ The magnetic induction $\vec{B}$ within a long cylindrical shell of magnetic material is increasing at the rate of 10^{-2} T/s (■ Figure 30.47). The magnetic field outside the shell remains negligible. Find the induced electric field as a function of radius from the cylinder axis.

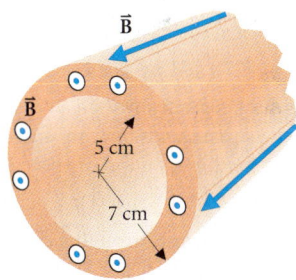

FIGURE 30.47

56. ♦♦♦ Two infinite, plane current sheets, each parallel to the x-z-plane, are located at $y = \pm d/2$. The sheets each carry current of magnitude K A/m in the $\pm z$-directions in the top and bottom sheets, respectively. Find the magnetic field produced by the sheets. If K increases with time at a constant rate, find the induced electric field everywhere.

§30.5 EDDY CURRENTS

57. ❖ A uniform magnetic field $\vec{B} = B_0\hat{i}$ exists in the region $y < 0$. A thin, rectangular metal plate lying in the y-z-plane measures L in the z-direction and w in the y-direction. It is pulled at constant velocity $\vec{v} = v_0\hat{j}$ out of the magnetic field. Describe how the force on the plate varies with position as it is pulled. Sketch the pattern of current in the plate. Explain why current density is not parallel to $\vec{v} \times \vec{B}$ everywhere.

58. ❖ A thin metal plate is pulled past the poles of a horseshoe magnet so that the field lines penetrate the plate in two patches with opposite polarity. Sketch the resulting pattern of eddy currents. Do the eddy currents caused by the two patches cancel or reinforce each other? Why?

59. ❖ Since the braking force due to eddy currents is proportional to the volume of conductor exposed to magnetic field, wouldn't the largest braking force be achieved by immersing the entire conductor in a uniform magnetic field? Justify your answer.

60. ♦♦ ✉ Magnet poles create a rectangular region of magnetic field 3.5 cm × 2.5 cm. The magnetic field strength between the poles is 0.85 T. A sheet of copper 2.5 mm thick moves between the poles at 6.5 m/s. Estimate the power required to keep the plate moving.

61. ♦♦ ✉ An aluminum disk with radius $R = 0.50$ m and thickness $t = 1.0$ cm rotates at 1500 rpm about an axle through its center. Magnet poles produce a magnetic field 0.25 T perpendicular to the disk in a circular region 3.0 cm in diameter centered 38 cm from the axle. **(a)** Sketch the eddy current pattern in the disk. **(b)** Calculate the torque acting on the disk. **(c)** Estimate the time for the disk to come to rest.

62. ♦♦ ✉ A metal plate swings between the square poles of a permanent magnet as in §30.5. Obtain an approximate expression for the acceleration a of the plate. Estimate the time required for the plate to come to rest by computing v_o/a_o. Estimate the distance traveled as v_o^2/a_o. Obtain numerical values if the mass of the plate is $M = 1$ kg, the magnetic field is $B = 1$ T, the plate thickness is $w = 0.03$ m, the resistivity is $\rho = 10^{-7}\ \Omega\cdot$m, the size of the magnetic field region is $\ell = 0.1$ m, and the initial speed of the plate is $v_o = 10$ m/s.

§30.6 THE AMPÈRE–MAXWELL LAW

63. ♦ A constant current $I = 15$ A charges a parallel plate capacitor with circular plates of radius $R = 3.0$ cm. Use the Ampère–Maxwell law to find the magnetic field between the capacitor plates as a function of r, for $r < R$.

64. ♦♦ The electric field in a region of space is:
$$\vec{E} = (270\ \text{kV/m}\cdot\text{s})t(\hat{i} - 2\hat{j} + 3\hat{k}).$$
Find the displacement current through a circle of radius 0.65 m in the x-y-plane, and a square with 0.50-m sides in the y-z-plane.

65. ♦♦ The electric field in a region of space is:
$$\vec{E} = (69.5\ \text{kV/m}\cdot\text{s}^2)t^2(3\hat{i} + 4\hat{j}).$$
Find the displacement current through a circle of radius 15 cm in the x-z-plane, centered at the origin, at $t = 0.5$ s and $t = 7.7$ s.

66. **(a)** ♦♦ Two equal and opposite point charges lie on the x-axis: -0.1 C at $x = 0$ and $+0.1$ C at $x = 10$ cm. Each charge is increasing in magnitude at 10^{-3} C/s. Find the displacement current through a circle centered at the point $x = 5$ cm, $y = 0$, $z = 0$, with radius 4 cm. **(b)** ♦♦♦ Use the result of part (a) to find the magnetic field at the point $x = 5$ cm, $y = 4$ cm, $z = 0$. **(c)** ♦♦ The charges are increasing because there is a constant current I along the x-axis from $+\infty$ to $x = 10$ cm and from $x = 0$ to $-\infty$. Find the magnetic field using the Biot-Savart law (§28.2.1) and show that the result agrees with your answer to part (b).

67. ♦♦♦ A point charge is created by passing a constant current I along a semi-infinite wire as shown in ■ Figure 30.48. Find the magnetic field at point P.

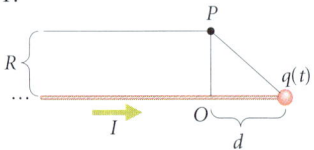

■ FIGURE 30.48

68. ♦♦♦ Two semi-infinite straight wires lie along the positive and negative x-axes. Constant currents I in each wire form a point charge $Q(t)$ at the origin. Find the displacement current through a circle of radius r parallel to the y-z-plane with center on the x-axis at an arbitrary value of x. Apply the Ampère–Maxwell law to find the magnetic field at a point on the circle. Is your result consistent with the Biot-Savart law?

69. ♦♦♦ A coaxial cable lying on the negative x-axis carries current I along its inner conductor of radius r and a return current $I/2$ along the outer sheath at radius $R \gg r$. The cable ends on a conducting sphere of radius R, centered at the origin, which accumulates charge at a rate $dQ/dt = I/2$. Assuming that the charge distribution on the sphere is uniform, calculate the magnetic field at $x = 3R$, $y = 3R$; $x = -3R$, $y = 3R$; and $x = -3R$, $y = R/2$.

Additional Problems

70. ❖ A bar magnet is dropped along the axis of a long, vertical conducting cylinder. The magnetic moment of the magnet lies along the axis of the cylinder. What produces the upward force predicted by Lenz's law? How does the force depend on the speed of the magnet? Describe the resulting motion of the magnet.

71. ❖ The iron cores of transformers are not made of solid iron, but of many thin sheets separated by insulating layers. Why?

72. ♦♦ A wire loop made of copper (resistivity $\rho = 1.7 \times 10^{-8}\ \Omega\cdot$m) has a radius $r = 0.1$ m and is made of wire with cross-sectional area $A = 10^{-5}$ m². The loop lies in a horizontal plane and is immersed in a uniform, horizontal magnetic field $B = 0.5$ T. The vertical component of the field is zero but is increasing at a rate $dB/dt = 0.01$ T/s. **(a)** Find the resistance of the wire loop. **(b)** Use Lenz's law to determine the sense of current in the loop. **(c)** Use Faraday's law to find the emf around the loop. **(d)** What is the current in the loop? **(e)** What is the magnetic moment of the loop? **(f)** What is the torque acting on the loop?

73. ♦♦ A long, rectangular wire loop of width 15.0 cm is pulled parallel to its long side at 0.505 m/s into a region where the magnetic field is perpendicular to the loop. The magnetic field strength is decreasing at 0.250 T/s. The loop begins to enter the magnetic field at $t = 0$, when the field strength is 1.500 T. Calculate the rate at which flux through the loop is changing. What is the induced emf at $t = 1.00$ s and $t = 5.00$ s?

74. ♦♦ A small coil with 35 turns of wire and area $A = 6.2 \times 10^{-4}$ m² is pulled into a position between the poles of a magnet with its plane perpendicular to the magnetic field. If it takes 2.7 ms from the time that the coil begins to enter the field region until it is in place, and the average emf developed by the coil, as measured by a voltmeter, is $+9.6$ V, what is the magnetic induction between the poles of the magnet? If the coil is then rotated 180° in 3.4 ms, what average emf is measured by the voltmeter? Explain your choice of sign carefully. If the coil is then removed from the magnetic field region in 4.2 ms, what average emf is measured?

75. ♦♦ The magnetic field in a region of space is described by the function:
$$\vec{B} = B_o(x\hat{i}/a - y\hat{j}/a).$$
A square wire loop of side b and resistance R lies parallel to the y-z-plane and moves along the x-axis at constant speed v. It is at the origin at $t = 0$. Find the current in the loop, and evaluate it for $B_o = 0.81$ T, $v = 15$ m/s, $a = 10.0b$, $b = 1.2$ cm, and $R = 0.18\ \Omega$. What force is needed to move the loop? Show that the mechanical power used to move the loop equals the electrical power dissipated in the resistance of the loop.

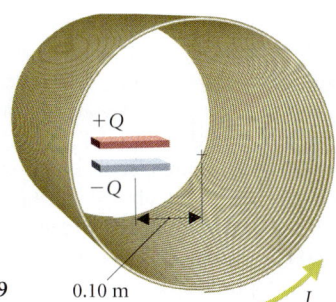

■ **Figure 30.49**

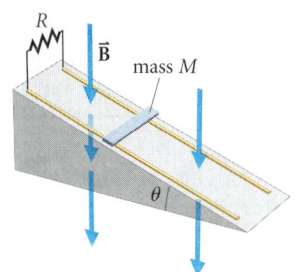

■ **Figure 30.51**

76. ♦♦ ✉ A small cylindrical magnet 1.0 cm in diameter, with a mass of 4 g, slides at a constant speed of 3 cm/s down a smooth ramp making a 30° angle with the horizontal. The ramp is made from an aluminum sheet 2.5 mm thick. Estimate the magnetic field produced by the magnet at its surface.

77. ♦♦ A parallel plate capacitor with plate area $A = 2.0 \times 10^{-2}$ m^2 and separation $d = 1.0 \times 10^{-4}$ m carries charge $Q = 1.3$ pC. The capacitor is inside a solenoid with $n = 4.0 \times 10^4$ turns/m and current increasing at a rate $dI/dt = 1.0 \times 10^4$ A/s (■ Figure 30.49). The center of the capacitor is at $r = 0.10$ m from the solenoid axis. (a) What is the static electric field between the plates? (b) What is the potential difference between the plates? (c) What is the induced electric field near the center of the capacitor plates? (d) What is the total electric field between the plates? (e) For what charge Q would the induced field just balance the Coulomb field? Would there still be a potential difference in this case?

78. ♦♦ A perfectly conducting rod slides on frictionless, perfectly conducting rails as shown in ■ Figure 30.50. The magnetic field strength increases at a constant rate $dB/dt = \alpha$. Describe how the rod moves. Find its position as a function of time, if it is initially at rest at $x = x_0$.

79. (a) ♦♦ Two metal rods, each of mass $m = 1.75$ kg and length $\ell = 0.65$ m, are joined at the ends by two metal springs. Each spring has spring constant $k = 9.0 \times 10^{-5}$ N/m and equilibrium length ℓ. The total resistance R in the circuit is 1.5 Ω. A uniform magnetic field $B = 0.37$ T is perpendicular to the plane containing the rods and springs. One rod is fixed and the other is free to move, remaining parallel to the fixed rod. At time $t = 0$, the movable rod is pulled out to a distance 2ℓ from the fixed rod and released. Describe the subsequent motion qualitatively. Calculate the net force on the rod when it is a distance x from the fixed rod and has speed v. (b) ♦♦♦ Assume that the position of the rod may be described by the function $x = \ell + A_1 e^{-b_1 t} + A_2 e^{-b_2 t}$. Find A_1, A_2, b_1, and b_2.

80. ♦♦♦ If the rod in Figure 30.44 (Problem 42) has mass M and resistance R, show that it approaches its final speed exponentially: $v = v_f(1 - e^{-\alpha t})$. What is the final speed v_f if there is a coefficient of friction μ between the rod and rails? In that case, show that, when $v = v_f$, the power expended by the battery is equal to the mechanical power needed to overcome friction plus the power dissipated as joule heating in the resistance.

81. ♦♦♦ A perfectly conducting rod of mass M slides on frictionless rails on an inclined plane (■ Figure 30.51) in a region where the magnetic field $\vec{B}$ is vertical. The rails are joined at the top of the plane by a wire of resistance R. At what speed does the rod slide down the plane? Show that the gravitational energy lost by the rod is dissipated as heat in the resistor.

82. ♦♦♦ Apply Gauss' law to find the charge density on the gap faces of the wire loop in Example 30.10. Estimate the total charge on each gap face and show that the two charges form a dipole with moment p independent of the gap width. Compute the static electric field produced by this dipole at a point P in the loop diametrically opposite the gap. Show that the dipole is inadequate to produce the entire static field known to exist at P. Discuss, qualitatively, how the static field is produced.

83. ♦♦♦ In Example 30.4 the cross tie is replaced by a second rod with resistance $R = 1.0$ mΩ and mass $M = 1.0$ kg. The second rod is free to slide on the rails without friction. At $t = 0$, the two rods are together at $x = 0$, and the one rod is pulled with a constant speed $v = 10.0$ m/s as in the example. Find the velocity and position of the second rod as functions of time. [Hint: v is of the form $v = v_f(1 - e^{-\alpha t})$.]

84. ♦♦♦ Between August and November of 1831, Faraday conducted a series of experiments demonstrating electromagnetic induction. In one of them, he rotated a copper disk between the poles of a magnet (■ Figure 30.52) and obtained a steady current in a circuit between the center and edge of the disk. Today, this device is called a *homopolar generator* because it produces DC current. Mercury contacts at the center and the edge of the disk connect it in series with a resistor $R = 15$ Ω and an ammeter. The radius of the disk is $a = 25$ cm, its angular speed is $\omega = 2\pi$ rad/s, and the magnetic field is $B = 1.0$ T. Find the emf developed in the circuit and the reading of the ammeter.

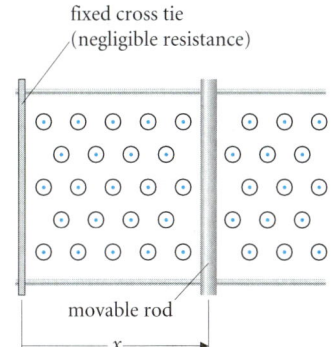

■ **Figure 30.50**

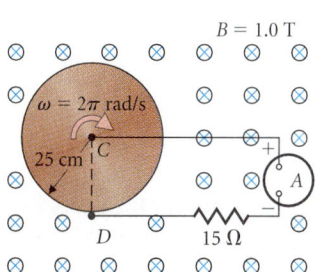

■ **Figure 30.52**

Computer Problems

85. File CH30P85 on the supplementary computer disk lists the emf recorded in a circuit as a function of time. If the flux through the circuit is 0 at $t = 0$, what is the flux through the circuit at $t = 2.0$ s?

86. A metal bar 1.00 m long with mass 0.500 kg slides on rails as shown in Figure 30.44. The emf of the battery is 10.0 V and the resistance of the bar is 1.00 Ω. Find an expression for the acceleration of the bar as a function of its velocity. If the bar starts at the battery at $t = 0$ s, use a numerical method of your choice to compute its velocity and position for $t > 0$. Plot the results and comment. (An analytic solution also exists; compare Problem 80. If you can find it, plot it also and compare with your computed results.)

Challenge Problems

87. A charge Q is at the origin and an equal and opposite charge $-Q$ is at $x = \ell$. There is a current I in a wire joining them along the x-axis. (a) Find the electric field at the point $x = \ell/2$, y. The current causes Q to decrease: $I = -dQ/dt$. Use this result to find the displacement current through a circle of radius $\ell/2$ in terms of I. Use the Ampère–Maxwell law to calculate the magnetic field at the point $x = \ell/2$, $y = \ell/2$. Verify your result using the Biot-Savart law. (b) Calculate the potential difference between the two charges if each lies on a sphere of radius $a \ll \ell$. If the connecting wire has a resistance R, how is the current I related to the charge Q? How does Q vary in time? The fields you found in part (a) vary in time as Q does. If fields propagate at the speed of light, under what circumstances do the fields you calculated above give sensible results for the instantaneous fields? (You should be able to state the condition as a minimum value for the resistance R.)

88. A rectangular loop of wire with dimensions $\ell \times w$ has resistance R and mass m. It falls under the influence of gravity into a region of vertical extent w in which the magnetic field $\vec{B}$ is horizontal and uniform (■ Figure 30.53). At time $t = 0$, the loop is just about to enter the region and its speed is zero. Find the position of the loop as a function of time.

89. (a) A magnetic dipole lies along the axis of a circular wire loop with its magnetic moment pointing toward the center of the loop. The magnetic moment of the dipole is increasing. Draw a diagram showing the resulting current and how the force predicted by Lenz's law is produced. What changes occur if the magnetic dipole moment decreases? (b) The loop has radius a and is held fixed at a distance z_o from the dipole. The dipole moment varies sinusoidally as $\vec{m} = \vec{m}_o \sin \omega t$. If the loop has radius R, find the force acting on it as a function of time. Assume that the magnetic field is described by the static dipole formula and that the resistance of the loop obeys Ohm's law. What is the average force on the loop?

90. The magnetic field in a region of space remains uniform, but its strength increases in time. Consider a charged particle moving in the plane perpendicular to the field and assume that its orbit remains circular, though its orbital radius may vary. Show that the following relation holds between the magnetic field and the particle's speed v:

$$\frac{1}{v}\left(\frac{dv}{dt}\right) = \frac{1}{2B}\left(\frac{dB}{dt}\right).$$

Integrate this equation and use the result to show:

$$r^2 B = \text{constant}.$$

The magnetic flux through the particle's orbit remains constant. This is one example of a phenomenon called "flux freezing": conducting matter and magnetic field are strongly coupled. For example, when an ordinary star collapses to form a neutron star 10 km in diameter, the magnetic field lines are dragged inward, resulting in field strengths as high as 10^8 T.

91. ■ Figure 30.54 illustrates the synchronous induction motor. An oscillating current passes through the coils of each electromagnet. The three currents are out of phase with each other by 120°. (Commercial power is usually supplied this way.) Describe the magnetic field produced by the coils. Show that as the currents in the coils vary, the magnetic field pattern they produce in the central region rotates. Hence explain how the field produces a torque on the shaft. Induction motors are particularly useful in applications where sparks are dangerous, since they do not require brushes or slip rings.

92. (a) Surface current K spreads charge across the capacitor plates in Problem 63. Show that the surface current at radial distance r from the center of each plate has magnitude:

$$|K(r)| = \frac{I}{2\pi r}\left(1 - \frac{r^2}{R^2}\right).$$

(Hint: Show that this distribution of current delivers charge at the required rate to each annular ring of radius r and width dr on the surface of the capacitor plates.) (b) Use an Ampèrian curve consisting of circular arcs of radius r on each side of one capacitor plate, connected by lines through and normal to the plate, to show that the surface current is just sufficient to account for the discontinuity of $\vec{B}$ across the capacitor plate.

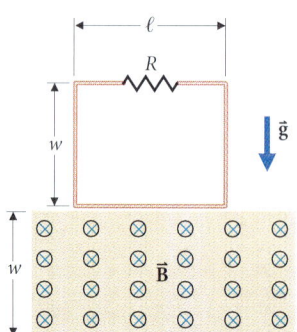

■ Figure 30.53

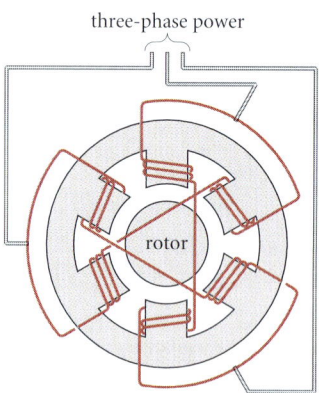

■ Figure 30.54

A flashing lamp on a road sign is designed to get your attention. The lamp is part of a circuit that uses a battery and a capacitor. Energy from the battery is stored in the capacitor and periodically released in the bulb's flash. In this chapter we'll investigate several circuits that exhibit similar time-dependent behavior. This circuit is analyzed in Example 31.3.

CHAPTER 31
Introduction to Time-Dependent Circuits

Concepts

Exponential decay
Time constant
Inductance
Magnetic energy density
Transformer
Damped oscillation
Coupled circuits

Goals

Be able to:

Describe qualitatively the behavior of time-dependent circuits.

Use the basic solutions to the exponential and harmonic equations.

Compute the magnetic energy stored in a system.

Recognize systems that exhibit self-induction or mutual induction and estimate their inductance.

"Mine is a long and a sad tale!" said the mouse, turning to Alice, and sighing. "It is a long tail, certainly," said Alice, looking down with wonder at the mouse's tail; "but why do you call it sad?"

Lewis Carroll

WE'LL DISCUSS MAGNETIC ENERGY IN §31.2.2.

As you drive along the highway at night, the bright flash of a hazard warning catches your eye. Your engine hums and the spark plugs fire in steady rhythm, as you carefully negotiate the danger. Specially designed electric circuits determine the performance of the emergency warning light and the engine's ignition system. Each circuit draws energy slowly from its source, stores it as electric or magnetic field, and then releases it in a rapid burst. Flashguns for photography use a similar circuit. We already have the tools we need to understand the behavior of these circuits: they are the conservation laws for energy and charge, expressed as Kirchhoff's rules (§26.4). In this chapter we'll use these rules to investigate the behavior of circuits in which current changes in time.

31.1 RESISTOR–CAPACITOR CIRCUITS

31.1.1 Discharging a Capacitor

The time evolution of charge and current in any circuit follows two characteristic patterns: oscillation and/or exponential decay. We'll learn to recognize the pattern of each circuit that we study before we develop the mathematics. In our first example a resistor and a capacitor are connected in series.

■ Figure 31.1 is the circuit diagram for a charged capacitor in series with a resistor and a switch. When the switch is closed, the potential difference across the capacitor drives current through the resistor, and the capacitor discharges. As the charge decreases, so do the potential difference and the current. Ultimately, both charge and current decrease to zero. The energy originally stored in the capacitor is converted to thermal energy in the resistor.

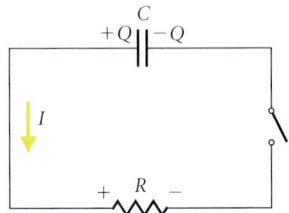

■ FIGURE 31.1
Discharge of a capacitor. When the switch is closed, current through the resistor reduces the charge Q. The arrow defines the positive sense for the current variable $I(t)$.

The charge on the positive plate of the capacitor is represented by the function $Q(t)$. The current is represented by the function $I(t)$. A positive value for this function represents current in the sense shown by the arrow in the figure. In this case, we expect both functions to be positive at all times and to approach zero as time passes.

To analyze the circuit behavior quantitatively, we apply Kirchhoff's rules. On a counterclockwise path around the circuit, the change in potential across the capacitor is Q/C and across the resistor is $-IR$. From Kirchhoff's loop rule:

REMEMBER: THE DEFINITION OF CAPACITANCE, EQN. (27.2), IS $C = Q/\Delta V$.

$$\frac{Q}{C} - IR = 0.$$

When the function $I(t)$ is positive, current in the circuit reduces the charge $Q(t)$ on the capacitor:

NOTE THE MINUS SIGN: IT'S IMPORTANT! WE'LL DISCUSS SIGNS FURTHER IN THE MATH TOOLBOX ON SIGN CONVENTIONS.

$$I = -\frac{dQ}{dt}.$$

Combining these two equations: $\dfrac{Q}{C} + \dfrac{dQ}{dt} R = 0,$

or

$$\frac{dQ}{dt} + \frac{Q}{RC} = 0. \tag{31.1}$$

Thus the loop rule gives a differential equation for the charge as a function of time. To solve it, we collect all the terms in Q on one side of the equation, leaving t on the other side:

$$\frac{dQ}{Q} = \frac{-1}{RC} dt.$$

Next we integrate both sides between $t = 0$, when the switch is closed, and a later time $t > 0$. During this time, Q decreases from its initial value $Q(0) = Q_0$.

WE ARE USING THE DUMMY VARIABLES Q' AND t' IN THE INTEGRANDS TO DISTINGUISH THEM FROM THE LIMITS Q AND t.

$$\int_{Q_0}^{Q} \frac{dQ'}{Q'} = \int_0^t \left(\frac{-1}{RC}\right) dt'.$$

Thus $\ln Q' \Big|_{Q_0}^{Q} = \ln Q - \ln Q_0 = \ln\left(\dfrac{Q}{Q_0}\right) = \dfrac{-t}{RC},$

or
$$Q(t) = Q_0 e^{-t/RC}. \tag{31.2}$$

Because its solution involves the exponential function, eqn. (31.1) is often called the *exponential equation*. Its characteristic feature is that the derivative of a quantity, here Q, is proportional to the quantity itself.

The current is given by:
$$I(t) = -\frac{dQ}{dt} = \frac{Q_0}{RC} e^{-t/RC}, \tag{31.3}$$

and the potential difference across the capacitor is:
$$\Delta V(t) = \frac{Q(t)}{C} = \frac{Q_0}{C} e^{-t/RC} = \Delta V_0 e^{-t/RC}. \tag{31.4}$$

■ Figure 31.2 is a graph of the potential difference as a function of time. The constant $\Delta V_0 = Q_0/C$ is the potential difference at $t = 0$, when the switch is closed. In a formal mathematical sense, the charge requires infinite time to decay to zero. In practice, once it has decreased to a few percent of its initial value, deviations of the real circuit from the ideal RC model become important. A practical measure of the time for the circuit to discharge is the *time constant*

$$\tau_C \equiv RC. \tag{31.5}$$

At $t = \tau_C$, the charge is reduced by a factor e, and at $t = 3\tau_C$, it is $1/e^3$, or about 5% of its initial value.

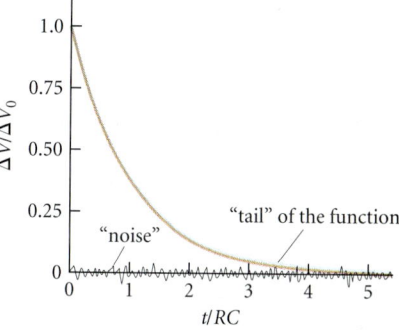

■ **FIGURE 31.2**
Potential difference across the discharging capacitor decays exponentially as a function of time. Random fluctuations (noise) in the circuit exceed the potential difference predicted by the model after a few time constants (here $t \approx 4.5\tau = 4.5RC$), and the potential difference is effectively zero. The part of the function that asymptotically approaches zero ($t \gtrsim 3\tau$) is called the "exponential tail." Current displays the same type of exponential decay.

REMEMBER $e = 2.718\ldots$, OR ABOUT 3 FOR ORDER OF MAGNITUDE ESTIMATES. SIMILARLY, $e^2 \sim 10$.

THE ARGUMENT OF AN EXPONENTIAL FUNCTION, LIKE THAT OF A LOGARITHM, IS ALWAYS DIMENSIONLESS.

EXERCISE 31.1 ♦ Show that the product RC has dimensions of time. That is, show $1\,\Omega\cdot F = 1\,s$.

EXAMPLE 31.1 ♦♦ A 10.0-μF capacitor is in series with a 2.5-kΩ resistor. The capacitor is initially charged to a potential difference of 120 V. What is the time constant of the circuit? What are the initial current through the resistor and the current after one time constant?

MODEL The capacitor has an initial charge $Q_0 = C\,\Delta V_0$. The results we have just developed describe its discharge through the resistor.

SETUP The time constant (eqn. 31.5) is:
$$\tau_C \equiv RC = (2.5\text{ k}\Omega)(10.0\ \mu F) = 2.5 \times 10^{-2}\text{ s}.$$

The current as a function of time (eqn. 31.3) is:
$$I(t) = \frac{Q_0}{RC} e^{-t/RC}.$$

SOLVE The initial current is:
$$I(0) \equiv I_0 = \frac{Q_0}{RC} = \frac{C\,\Delta V_0}{RC} = \frac{\Delta V_0}{R}$$
$$= \frac{120\text{ V}}{2.5 \times 10^3\ \Omega} = 0.048\text{ A}.$$

This is the maximum value of the current. After one time constant,
$$I(\tau_C) = I_0/e = 0.018\text{ A}.$$

ANALYZE To increase the time constant, we may increase either C or R. At a fixed voltage ΔV_0, increasing R also decreases the initial current in the circuit, whereas increasing C increases the initial charge on the capacitor.

It is the resistor that is responsible for the exponential decay of charge and current in the RC circuit because resistors are the only circuit elements that dissipate energy. In its initial state, the charged capacitor contains electrostatic energy that is converted into thermal energy in the resistor as the capacitor discharges. No energy remains stored in the circuit in its final state.

> **EXAMPLE 31.2** ♦♦♦ Show that when a capacitor is discharged the energy dissipated as heat in the resistor equals the initial energy stored in the capacitor.
>
> **MODEL** The initial stored energy is $U = Q_0^2/(2C)$ (eqn. 27.6a), and the energy is dissipated in the resistor at a rate $P(t) = I^2R$ (eqn. 26.3b). Since the current decreases with time, the power dissipated by the resistor also decreases, so we integrate the power over time to obtain the total energy dissipated.
>
> **SETUP** We use eqn. (31.3) for $I(t)$.
>
> **SOLVE**
> $$U = \int_0^\infty P\,dt = \int_0^\infty I^2 R\,dt = \int_0^\infty R\left(\frac{Q_0}{RC}e^{-t/RC}\right)^2 dt$$
> $$= \int_0^\infty \frac{Q_0^2}{RC^2}e^{-2t/RC}\,dt = \frac{Q_0^2}{RC^2}\left(-\frac{RC}{2}e^{-2t/RC}\right)\bigg|_0^\infty = \frac{Q_0^2}{2C}.$$
>
> **ANALYZE** As expected, the total energy dissipated in the resistor equals the initial stored energy.

Math Toolbox

Sign Conventions

To analyze a circuit, we describe the physical quantities, charge and current, with mathematical functions, $Q(t)$ and $I(t)$. Both the magnitude and sign of these functions give important information. The function $Q(t)$ means the amount of charge on a specific capacitor plate that we choose. The sign of Q determines which plate is positive. Similarly, the function $I(t)$ means current in a specific direction that we choose: a negative value means current is in the opposite direction. This freedom of definition is similar to our freedom to choose the positive coordinate directions in a mechanics problem. It is an essential freedom, since we cannot always guess in advance the actual behavior of a complicated circuit.

For example, consider two different conventions in the capacitor discharge problem. In our previous discussion, current away from the positive plate of the capacitor was called positive. This choice leads to the relation:

$$I = -\frac{dQ}{dt},$$

and the solution has $I(t) > 0$; charge flows off the capacitor.

If we reverse the definition of I (■ Figure 31.3) so that current *toward* the positive plate of the capacitor is called positive, then:

$$I = +\frac{dQ}{dt}.$$

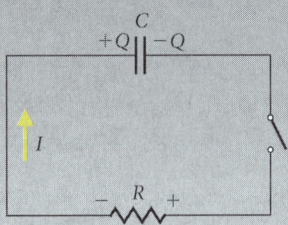

■ **FIGURE 31.3**
You may define the current variable to have either sense. With I defined as shown here, $I = +dQ/dt$. The solution has the opposite sign from that in eqn. (31.3), but the same physical meaning.

On a counterclockwise path around the circuit, the potential difference across the resistor is $+IR$, and Kirchhoff's rule becomes:

$$\frac{Q}{C} + IR = 0 \Rightarrow \frac{Q}{RC} + \frac{dQ}{dt} = 0.$$

The differential equation for Q and its solution are unchanged. But, in this case,

$$I = +\frac{dQ}{dt} = -\frac{Q_0}{RC}e^{-t/RC},$$

is negative, again indicating that charge flows off the capacitor. The mathematical function has a different sign, but the physical meaning is the same. Some people like to choose I so that $I = +dQ/dt$; other people prefer $I > 0$. You can't have both, so make a decision and be aware of what you have done.

31.1.2 The Solution Method

Our analysis of the discharging capacitor illustrates how the behavior of time-dependent circuits depends on relations among charges, currents, and their rates of change; how Kirchhoff's rules applied to these circuits generate differential equations; and how qualitative physical reasoning acts as a guide to their solution. Each kind of system has a typical behavior, which you should learn to recognize. From this first example we have learned that:

> Resistor–capacitor systems evolve exponentially from an initial state to a final state with a characteristic time constant.

In the series RC circuit we discussed, the initial state is a charged capacitor, the final state is a discharged capacitor, and the time constant is

$$\tau_C = RC.$$

The method we have to used to solve for $I(t)$ in §31.1.1 is summarized in ■ Figure 31.4.

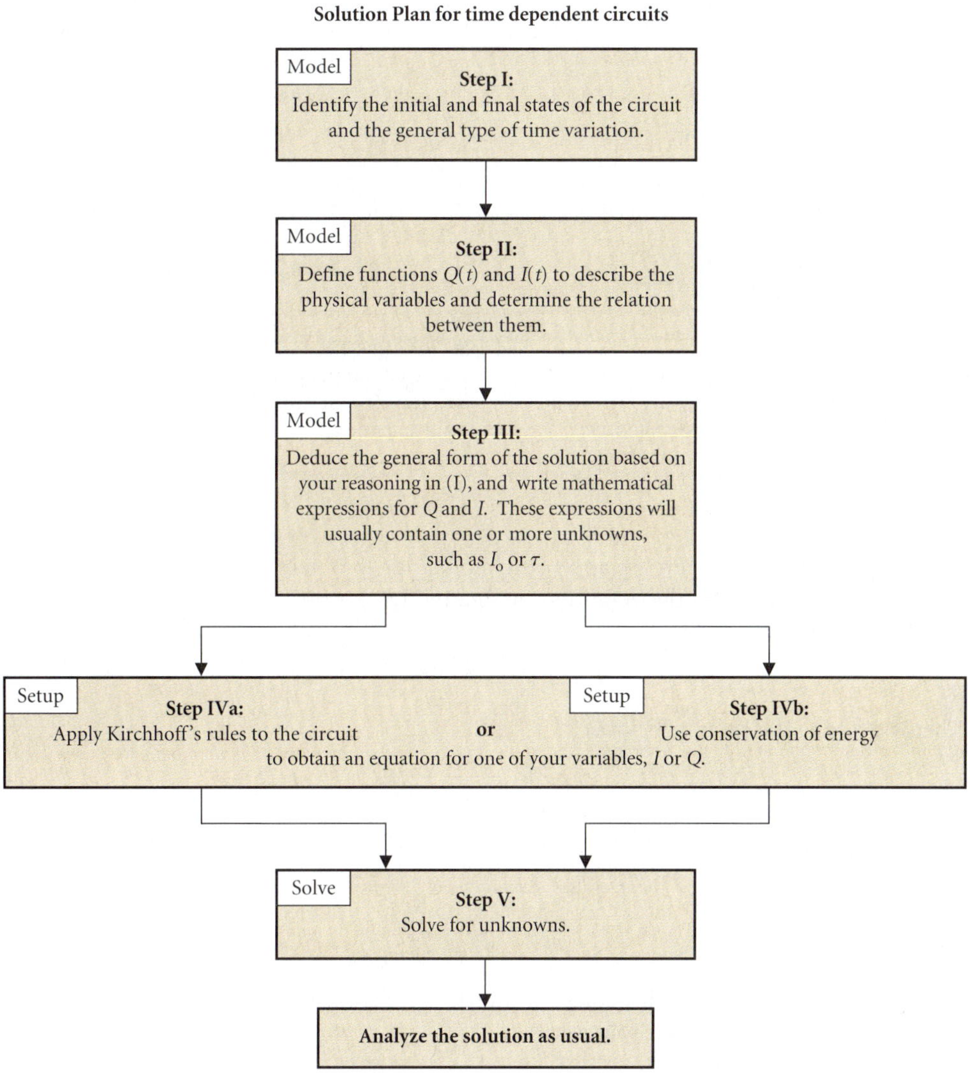

Solution Plan for time dependent circuits

Model — **Step I:** Identify the initial and final states of the circuit and the general type of time variation.

Model — **Step II:** Define functions $Q(t)$ and $I(t)$ to describe the physical variables and determine the relation between them.

Model — **Step III:** Deduce the general form of the solution based on your reasoning in (I), and write mathematical expressions for Q and I. These expressions will usually contain one or more unknowns, such as I_o or τ.

Setup — **Step IVa:** Apply Kirchhoff's rules to the circuit
or
Setup — **Step IVb:** Use conservation of energy

to obtain an equation for one of your variables, I or Q.

Solve — **Step V:** Solve for unknowns.

Analyze the solution as usual.

■ FIGURE 31.4
Solution method for solving time-dependent circuit problems. Compare this with the plan for DC circuits (Figure 26.27).

31.1.3 Charging a Capacitor

THIS IS STEP I OF OUR SOLUTION PLAN, FIGURE 31.4.

■ Figure 31.5 is the diagram of a circuit used to charge a capacitor. Initially, the capacitor is uncharged. At $t = 0$, the switch is closed. The battery drives current through the resistor, building up charge on the capacitor, until the potential difference across the capacitor equals the emf of the battery. Then the potential difference across the resistor and the current in it are zero. The capacitor has a final charge:

$$Q_f = C\mathcal{E}.$$

THE VARIABLES Q AND I ARE DEFINED IN THE DIAGRAM (STEP II).

Following the pattern for RC circuits, the charge evolves exponentially from its initial to its final value. That is, the *difference* $\Delta Q \equiv Q_f - Q(t)$ between the final value of the capacitor charge and its value at time t decays exponentially to zero. ■ Figure 31.6 is a graph of the corresponding potential difference $\Delta V = Q/C$ across the capacitor as a function of time. We expect the time constant to be $\tau_C = RC$. That is, we guess:

$$\Delta Q \equiv Q_f - Q(t) = Q_f e^{-t/RC},$$

THIS IS STEP III.

or

$$Q(t) = C\mathcal{E}(1 - e^{-t/RC}). \quad (31.6)$$

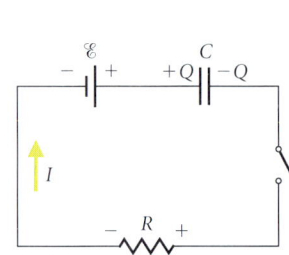

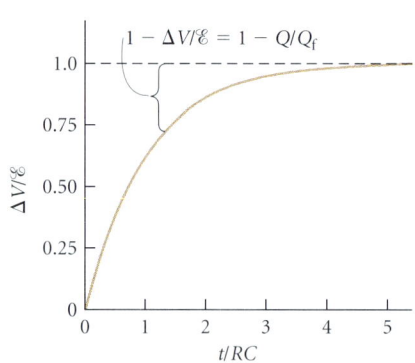

■ **FIGURE 31.5**
Charging a capacitor. Current increases the capacitor charge until it reaches its final value: with $\Delta V = \mathcal{E}$, $Q_f = \mathcal{E}C$. As the potential difference across the capacitor increases, the current decreases. (Marking Q and I on this figure is Step II of the solution plan.)

■ **FIGURE 31.6**
Potential difference across the charging capacitor as a function of time. $\Delta V/\mathcal{E} = Q/Q_f$ approaches unity exponentially. Notice that

$$1 - Q/Q_f = (Q_f - Q)/Q_f = \Delta Q/Q_f \to 0$$

as t increases.

To verify this conclusion, we apply Kirchhoff's rule to obtain a differential equation for the capacitor charge. With the definitions in Figure 31.5, positive current corresponds to increasing capacitor charge:

THIS IS STEP IVA.

$$I(t) = +\frac{dQ}{dt}.$$

Then, traversing the circuit clockwise,

$$0 = \mathcal{E} - \frac{Q}{C} - IR.$$

Combining these two equations, we find:

$$\frac{dQ}{dt} + \frac{Q}{RC} = \frac{\mathcal{E}}{R}. \quad (31.7)$$

BUT SEE THE *MATH TOOLBOX* IN §31.3.3.

Although we shall not solve this equation directly, it is straightforward to demonstrate that expression (31.6) is the correct solution.

EXERCISE 31.2 ♦♦ Differentiate eqn. (31.6) and substitute the result into eqn. (31.7) to show that we guessed the correct solution.

We find the solution for $I(t)$ by differentiating:

$$I(t) = \frac{dQ}{dt} = \frac{d}{dt}C\mathcal{E}(1 - e^{-t/RC}) = \frac{C\mathcal{E}}{RC}e^{-t/RC} = \frac{\mathcal{E}}{R}e^{-t/RC}.$$

Thus $I(0) = \mathcal{E}/R$, and $I \to 0$ as $t \to \infty$, as we predicted.

EXAMPLE 31.3 ♦♦ A highway emergency flasher uses a 120.0-V battery, a 1.0-MΩ resistor, a 1.0-μF capacitor, and a neon flash lamp in the circuit shown in ■ Figure 31.7. The flash lamp has a resistance of more than 10^7 Ω when the voltage across it is less than 110.0 V. Above 110.0 V, the neon gas ionizes, the lamp's resistance drops to 10 Ω, and the capacitor discharges completely. Find the time between flashes, estimate the duration of each flash, and find the energy released in each.

MODEL Until the capacitor voltage reaches the breakdown voltage $V_b = 110.0$ V, the large resistance of the flash lamp ensures that it draws a negligible current. The capacitor charges as if the lamp were absent. At V_b, however, the lamp resistance quickly becomes negligible, and the capacitor discharges through the lamp as if the battery and series resistor were absent. The time between flashes is the time for the capacitor to charge to V_b. The flash duration is roughly the time for the capacitor to discharge through the lamp, or about three time constants of the capacitor–lamp circuit. The flash energy is the stored energy in the capacitor at 110.0 V.

SETUP Equation (31.6) describes the charging of the capacitor. The flash interval is found by solving for the time when the capacitor voltage is $V_b = 110.0$ V:

$$V_b = Q(t)/C = \mathcal{E}(1 - e^{-t/RC}).$$

So
$$e^{-t/RC} = 1 - \frac{V_b}{\mathcal{E}} = 1 - \frac{110.0 \text{ V}}{120.0 \text{ V}} = 1 - 0.9167 = 0.0833.$$

SOLVE The time between flashes is:

$$t = -RC \ln(0.0833) = 2.5RC = 2.5(1.0 \times 10^6 \text{ }\Omega)(1.0 \times 10^{-6} \text{ F}) = 2.5 \text{ s}.$$

The flash duration is:

$$t_f \approx 3\tau_C = 3CR_{\text{lamp}} = 3(1.0 \times 10^{-6} \text{ F})(10 \text{ }\Omega) = 30 \text{ }\mu\text{s}.$$

The energy in the flash is:

$$U_f = \tfrac{1}{2}CV_b^2 = \tfrac{1}{2}(1.0 \times 10^{-6} \text{ F})(110 \text{ V})^2 = 6.1 \text{ mJ}.$$

ANALYZE During the flash, the light is as bright as a 200-W light bulb (power = U_f/t_f). ■

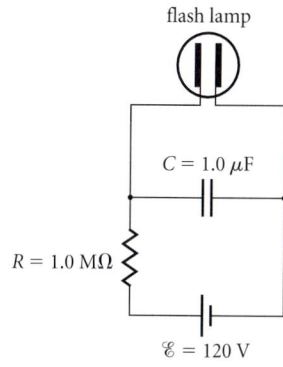

■ **FIGURE 31.7**
Circuit diagram for a highway emergency flasher. The lamp itself has an extremely large resistance until the gas inside ionizes, when its resistance drops to 10 Ω.

SEE THE DISCUSSION IN §31.1.1, ESPECIALLY EQN. (31.5).

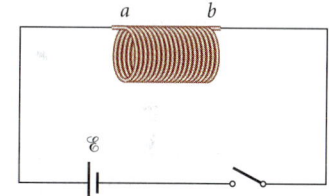

■ **FIGURE 31.8**
The coil is made out of perfectly conducting wire. When the switch is closed, current in the circuit increases from zero.

31.2 INDUCTANCE

31.2.1 Self-Inductance

The simple circuit diagrammed in ■ Figure 31.8 contains a battery (emf $\mathcal{E}$) and a solenoid connected in series with a switch. When the switch is closed, we might expect that a large current $\mathcal{E}/R$ would pass through the low resistance of the wire. Yet even for a perfect conductor ($R = 0$), the current is zero immediately after the switch is closed and increases at a finite rate. As the current in the solenoid increases, the magnetic field inside it also increases, inducing an electric field that opposes the increase in current according to Lenz's law.

Let us first assume the solenoid has zero resistance. We'll look at a coil with finite resistance in the next section. The net electric field within the *perfectly conducting* wire, equal to the sum of the Coulomb field produced by the battery and the induced field, must be zero. The current increases at just the right rate to make the net electric field in the wire remain zero:

$$0 = \vec{E}_{\text{net}} = \vec{E}_{\text{ind}} + \vec{E}_c \Rightarrow \vec{E}_{\text{ind}} = -\vec{E}_c.$$

REMEMBER: ACCORDING TO FARADAY'S LAW, CHANGING B CREATES AN E.

THE ELECTRONS IN THE *PERFECT* CONDUCTOR WOULD HAVE INFINITE ACCELERATION IF THE NET FIELD WERE NOT ZERO.

REMEMBER: POTENTIAL DIFFERENCE IS DUE TO THE COULOMB FIELD.

Consequently, the induced emf developed by the inductor equals the potential difference maintained by the battery:

$$\mathcal{E} = \int_b^a \vec{E}_{ind} \cdot d\vec{\ell} \bigg|_{\text{along wire}} = -\int_b^a \vec{E}_c \cdot d\vec{\ell} \bigg|_{\text{along wire}} = V_{ab}. \qquad (31.8)$$

Because the induced field is opposite the Coulomb field, the induced emf is often called a *back emf*.

EXERCISE 31.3 ❖ What happens if the switch is suddenly opened when there is current in the coil? Can you explain why switches used to control such devices usually have a capacitor in parallel with the switch?

THIS IS THE MAGNETIC FIELD INSIDE AN INFINITE SOLENOID. WE MAY USE THIS RESULT IF THE SOLENOID IS MUCH LONGER THAN ITS DIAMETER.

We may compute the induced emf of the solenoid from Faraday's law. When there is a current I in the solenoid, the magnetic field inside (eqn. 28.14) is:

$$|\vec{B}| = \mu_0 n I = \mu_0 N I / \ell,$$

where N is the number of turns in the coil and ℓ is its length. As the current increases, B increases and the induced electric field circulates around the increasing flux (cf. Examples 30.1 and 30.9). The flux through one turn of the coil (area A) is BA, and the flux through the whole solenoid is:

$$\Phi_B = NBA = \frac{\mu_0 N^2 A}{\ell} I.$$

The induced emf equals the rate of change of flux:

$$|\mathcal{E}| = \left|\frac{d\Phi_B}{dt}\right| = \frac{d}{dt}\left|\frac{\mu_0 N^2 A I}{\ell}\right| = \frac{\mu_0 N^2 A}{\ell}\left|\frac{dI}{dt}\right|.$$

The induced emf is proportional to the rate of change of current. The constant of proportionality is called the *inductance* and is given the standard symbol L.

The *inductance* of a circuit element is the ratio of the induced emf in it to the rate of change of current through it:

$$L \equiv \frac{|\mathcal{E}_{ind}|}{|dI/dt|}. \qquad (31.9)$$

In an ideal solenoid with no resistance, the induced emf equals the potential difference $|V_{ab}|$ across the solenoid (eqn. 31.8):

WE'LL NEED THIS RELATION IN THE NEXT SECTION.

$$|V_{ab}| = \frac{\mu_0 A N^2}{\ell}\left|\frac{dI}{dt}\right| \equiv L\left|\frac{dI}{dt}\right|. \qquad (31.10)$$

Since

$$|\mathcal{E}| = \left|\frac{d\Phi_B}{dt}\right| = L\left|\frac{dI}{dt}\right|,$$

an equivalent definition of inductance is the ratio of flux to current:

$$L = \Phi_B / I. \qquad (31.11)$$

For the solenoid, with $n = N/\ell$ turns per unit length:

WE USED EQN. (28.14) AGAIN.

$$L = \mu_0 A N^2 / \ell = \mu_0 n^2 A \ell. \qquad (31.12)$$

Like capacitance, inductance depends on the geometrical structure of a circuit element and is independent of variables such as current or dI/dt.

A circuit element designed to have a particular inductance is called an *inductor*. ■ Figure 31.9 shows the circuit symbol and how the potential changes across an inductor when the current is changing in the direction of the arrow. The sign convention reflects the fact that $\vec{E}_c$ points in the direction of current change.

The unit of inductance is the henry (symbol H):

$$1 \text{ H} = 1 \frac{\text{V} \cdot \text{s}}{\text{A}}.$$

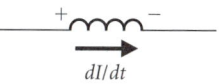

■ **FIGURE 31.9**
Circuit diagram symbol for an inductor. The sign convention indicates that potential drops across the inductor in the direction that current increases.

AFTER JOSEPH HENRY (1797–1878).

EXAMPLE 31.4 ♦ Find the inductance of a solenoid with 1.0×10^3 turns, length 3.0 cm, and radius 0.50 cm.

MODEL We model the solenoid as *very long* and apply eqn. (31.12).

SETUP The area of the solenoid is $A = \pi r^2$.

SOLVE
$$L = \mu_0 \pi r^2 \frac{N^2}{\ell}$$
$$= (4\pi \times 10^{-7} \text{ N/A}^2) \pi (0.5 \times 10^{-2} \text{ m})^2 \left[\frac{(1.0 \times 10^3)^2}{3 \times 10^{-2} \text{ m}}\right]$$
$$= 3.3 \text{ mH}.$$

ANALYZE Check the units: $\dfrac{\text{N} \cdot \text{m}}{\text{A}^2} = \dfrac{\text{J}}{\text{A}^2} = \dfrac{\text{J} \cdot \text{s}}{\text{A} \cdot \text{C}} = \dfrac{\text{V} \cdot \text{s}}{\text{A}} = \text{H}.$

EXERCISE 31.4 ♦ Verify that $1 \text{ H} = 1 \text{ Wb/A}$.

31.2.2 Energy Storage in an Inductor

Once the switch is closed in Figure 31.8, the constant potential difference maintained by the battery causes current to increase with time. The battery is doing work on charge flowing through it, but no heat is dissipated by resistance. Energy delivered by the battery is stored in the inductor.

The power output of the battery is:

$$P(t) = \mathcal{E}I(t) = \frac{\mu_0 N^2 A}{\ell}\left(\frac{dI}{dt}\right)I = LI\frac{dI}{dt}.$$

REMEMBER: WE ARE STILL ASSUMING THE SOLENOID HAS ZERO RESISTANCE. FOR A REAL INDUCTOR WITH RESISTANCE, THE CURRENT INCREASES LINEARLY FOR $t \ll L/R$. WE'LL DISCUSS THE COMPLETE TIME DEPENDENCE IN THE NEXT SECTION. SEE ALSO EXERCISE 31.10.

The stored energy U at time t equals the total work done to establish the current I in the inductor:

$$U = \int_0^t P \, dt = \int_0^t LI\frac{dI}{dt} \, dt = \int_0^I LI \, dI = \tfrac{1}{2}LI^2 \Big|_0^I$$

$$U = \tfrac{1}{2}LI^2. \qquad (31.13)$$

EXERCISE 31.5 ♦ Find the energy stored in a 3.3-mH solenoid when it carries a current of 1.1 A.

Equation (31.13) is similar to the result for energy stored in a capacitor (§27.2): $U = \tfrac{1}{2}Q^2/C$. In that case, electrostatic energy is stored in the electric field itself throughout the volume of the capacitor. The inductor also stores energy throughout its volume, in the form of magnetic field. For the long solenoid, we have:

$$U = \tfrac{1}{2}LI^2 = \tfrac{1}{2}\mu_0\frac{N^2 A}{\ell}I^2 = \tfrac{1}{2}\left(\mu_0\frac{N}{\ell}I\right)^2\left(\frac{\ell A}{\mu_0}\right) = \tfrac{1}{2}\ell A B^2/\mu_0.$$

The stored energy equals the volume ℓA inside the solenoid multiplied by an energy density:

$$u_B = B^2/(2\mu_0). \tag{31.14}$$

Although we derived this expression for the energy density inside the solenoid, it is true generally.

Compare eqn. (31.14) with the corresponding energy density due to electric field ($u_E = \epsilon_0 E^2/2$, eqn. 27.13). When both electric and magnetic fields are present, both store energy and the total energy density is:

$$u \equiv u_E + u_B = \tfrac{1}{2}\left(\epsilon_0 E^2 + \frac{B^2}{\mu_0}\right). \tag{31.15}$$

EXAMPLE 31.5 ♦♦ Find the magnetic field energy density 6.0 mm from a long straight wire carrying a steady current of 10.0 mA.

MODEL We calculate the magnetic field, and then use eqn. (31.14).

SETUP The magnetic field is $B = \mu_0 I/(2\pi r)$ (eqn. 28.6).

SOLVE So the field energy density is:

$$u_B = \frac{B^2}{2\mu_0} = \frac{(\mu_0 I)^2}{2\mu_0(4\pi^2 r^2)} = \frac{\mu_0 I^2}{8\pi^2 r^2} = (4\pi \times 10^{-7}\ \text{N/A}^2)\left[\frac{(1.0 \times 10^{-2}\ \text{A})^2}{8\pi^2(6.0 \times 10^{-3}\ \text{m})^2}\right]$$

$$= \frac{1}{2\pi(0.36)} 10^{-7}\ \text{N/m}^2 = 4.4 \times 10^{-8}\ \text{J/m}^3.$$

ANALYZE Twice as far from the wire, the energy density is one-quarter as great. ∎

EXERCISE 31.6 ♦♦ How would the answer to Example 31.5 change if the wire also carries a charge density of 5.0 pC/m?

EXAMPLE 31.6 ♦♦♦ The inner conductor of a coaxial cable has radius $a = 0.50$ mm. The radius of the outer sheath is $b = 0.50$ cm (■ Figure 31.10). What is the inductance per unit length of the cable? (Assume the current in the inner conductor is on its outer surface.)

THIS ASSUMPTION IS STRICTLY TRUE ONLY FOR AC CURRENT (CHAPTER 32). WE USE IT HERE TO SIMPLIFY THE CALCULATION.

MODEL The inner conductor and the outer sheath carry equal and opposite currents. Thus, from Ampère's law, the magnetic field is zero everywhere except between the two conductors (cf. Example 28.7). We may compute the inductance of the cable from the stored energy using eqn. (31.13): $L = 2U/I^2$.

SETUP With $\vec{B}$ given by eqn. (28.6), $B = \mu_0 I/(2\pi r)$, the energy density at radius $r > a$ inside the cable is:

$$u_B(r) = \frac{B^2}{2\mu_0} = \frac{[\mu_0 I/(2\pi r)]^2}{2\mu_0} = \frac{\mu_0 I^2}{8\pi^2 r^2}.$$

The energy density is a function of radius. The energy stored in a cylindrical shell of length ℓ and width dr is:

$$dU = u_B(r)\ dV = u_B(r)(2\pi r\ell\ dr).$$

Thus the energy stored between radii a and b in length ℓ of the cable is:

$$U = \int u_B\ dV = \int_a^b \frac{\mu_0 I^2}{8\pi^2 r^2}(2\pi r\ell)\ dr = \frac{\mu_0 I^2 \ell}{4\pi}\int_a^b \frac{dr}{r} = \frac{\mu_0 I^2 \ell}{4\pi}\ln\left(\frac{b}{a}\right).$$

SOLVE The inductance per unit length is:

$$\frac{L}{\ell} = \frac{2U}{I^2 \ell} = \frac{\mu_0}{2\pi}\ln\left(\frac{b}{a}\right) = (2 \times 10^{-7}\ \text{N/A}^2)(\ln 10) = 4.6 \times 10^{-7}\ \text{H/m}.$$

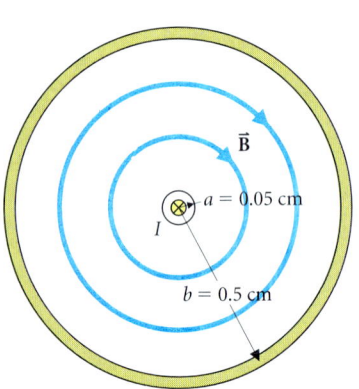

■ **FIGURE 31.10**
A coaxial cable carrying current I stores magnetic energy between the conductors. The current in the outer sheath is equal and opposite to the current in the inner wire so the net current is zero.

ANALYZE The units calculation is an intermediate result in Exercise 31.4. The result does not depend on the actual values of b and a, but only on their ratio. Does that surprise you? The inductance could also be calculated from eqn. (31.9).

SEE PROBLEM 37.

31.2.3 Mutual Inductance

Two circuits may influence each other even if there are no wires connecting them. When a changing magnetic field produced by one circuit passes through another, the resulting induced field produces an emf in the second circuit (cf. Example 30.1).

> The *mutual inductance* between two circuit elements is the ratio of emf developed in one element to the rate of change of current in the other element:
>
> $$M \equiv \frac{|\mathcal{E}_2|}{|dI_1/dt|}. \qquad (31.16)$$

Conversely, changing current in circuit 2 induces an emf in circuit 1. The corresponding mutual inductance has the same value:

$$M \equiv \frac{|\mathcal{E}_1|}{|dI_2/dt|}.$$

(We omit the proof of this assertion.) The mutual inductance of two circuit elements is usually calculated as the flux through one element divided by the current in the other:

$$M = \Phi_1/I_2 = \Phi_2/I_1. \qquad (31.17)$$

SEE PROBLEM 32 FOR AN EXAMPLE THAT DEMONSTRATES THIS EQUALITY.

EXAMPLE 31.7 ♦♦ Find the mutual inductance between a solenoid with 1.0×10^4 turns/m and a single wire loop of radius $r = 1.0$ cm placed inside the solenoid with its plane perpendicular to the axis (■ Figure 31.11).

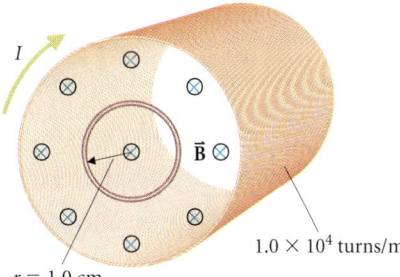

■ FIGURE 31.11
A wire loop is placed inside a long solenoid with its plane perpendicular to the solenoid axis.

MODEL We calculate the magnetic flux Φ_2 through the loop due to a current I_1 in the solenoid and use eqn. (31.17).

SETUP The field inside the solenoid is $B = \mu_0 n I_1$, so the flux through the loop is:

$$\Phi_2 = BA = \mu_0 n I_1 (\pi r^2).$$

SOLVE The mutual inductance is:

$$M = \Phi_2/I_1 = \mu_0 n \pi r^2$$
$$= (4\pi \times 10^{-7}\ \text{H/m})(1.0 \times 10^4\ \text{/m})\pi(0.010\ \text{m})^2 = 3.9\ \mu\text{H}.$$

ANALYZE Note that the mutual inductance is independent of I_1. Calculating the flux through the solenoid due to current in the loop is much more difficult.

SEE PROBLEM 90.

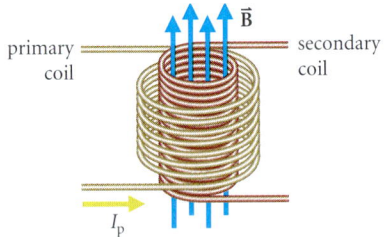

(a) An air core transformer.

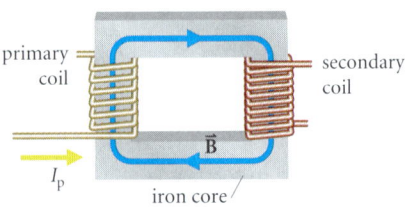

(b) An iron core transformer.

■ **FIGURE 31.12**
(a) In an air core transformer, magnetic flux produced by the primary coil links the secondary. When current changes in the primary, induced electric field drives current in the secondary. (b) An iron core transformer operates on the same principle, but the magnetic flux is trapped inside the core and threads both coils even though they are spatially separated.

■ **FIGURE 31.13**
Circuit diagram symbol for a transformer.

One of the most important practical applications of mutual inductance is the electrical transformer. A transformer consists of two wire coils arranged so that flux produced by one links the other. In an air core transformer (■ Figure 31.12a), one coil lies inside the other. Power transformers use an iron core (Figure 31.12b). Nearly all of the magnetic flux remains within the iron and thus links both coils. ■ Figure 31.13 is the circuit symbol for a transformer.

A transformer is frequently used to match the voltage of a power source to that needed by an appliance. A sinusoidally varying potential difference is applied to the *primary* (input) coil, causing a time-varying current in the coil. The resulting variable flux produces an induced electric field that both opposes the applied potential difference and produces an emf in the *secondary* (output) coil.

In a well-designed transformer, resistance in the coils is a relatively small effect that we may neglect here. Then (eqn. 31.8) the induced emf in the primary coil exactly balances the applied potential difference $\mathcal{E}_p$:

$$\mathcal{E}_p = -\mathcal{E}_{induced} = N_p \frac{d\Phi_B}{dt}.$$

Here N_p is the number of turns in the primary coil and Φ_B is the magnetic flux passing through each turn. The same flux also passes through each turn of the secondary coil, so the induced emf in the secondary has magnitude:

$$|\mathcal{E}_s| = N_s \frac{d\Phi_B}{dt}.$$

Since the flux through each turn Φ_B is the same in each expression, the ratio of output emf to input emf equals the ratio of the numbers of turns in the coils:

$$\left|\frac{\mathcal{E}_s}{\mathcal{E}_p}\right| = \frac{N_s}{N_p}. \qquad (31.18)$$

A *step-down* transformer has $N_s < N_p$, and the voltage in the secondary is less than that in the primary. A *step-up* transformer has $N_s > N_p$ and $|\mathcal{E}_s| > |\mathcal{E}_p|$.

The transformer equation (31.18) arises from the fact that the two coils are mutual inductors but does not depend on the value of their mutual inductance. As long as the coil resistances are truly negligible, the result is independent of the current or the power transmitted by the device. The magnitude of the mutual inductance is important when we consider the power transmitted to the secondary circuit, including the effect of resistance.

EXAMPLE 31.8 ◆◆ The line voltage in a city power supply is 12 kV. The voltage supplied to each house is 120 V. What turns ratio is required in the transformer that converts the voltage? What is the ratio of current drawn by the household to current in the power line?

MODEL The turns ratio equals the voltage ratio (eqn. 31.18). The ratio of currents in each coil equals the ratio of fluxes through them (eqn. 31.17).

SETUP AND SOLVE Using eqn. (31.18):

$$\frac{12\,000\text{ V}}{120\text{ V}} = \left|\frac{\mathcal{E}_p}{\mathcal{E}_s}\right| = \frac{N_p}{N_s} = 1.0 \times 10^2.$$

Since the flux through each coil is the flux through one turn times the number of turns and the flux through one turn is the same for each coil, then from eqn. (31.17):

$$\frac{I_s}{I_p} = \frac{\Phi_p}{\Phi_s} = \frac{N_p}{N_s} = 1.0 \times 10^2.$$

ANALYZE In an ideal transformer, the power provided by the primary equals the power drawn by the secondary: $I_s\mathcal{E}_s = I_p\mathcal{E}_p$. Transformers used by the Pacific Gas & Electric Company are rated at a few tens of kV·A power, corresponding to a secondary current of up to a few hundred amps.

EXERCISE 31.7 ♦♦ In an ideal transformer, the same flux passes through each turn of the primary and secondary coils. If L_p and L_s are the self-inductances of the two coils, show that their mutual inductance is $M = \sqrt{L_p L_s}$.

31.3 INDUCTOR CIRCUITS

31.3.1 The LR Circuit

A real solenoid has some resistance as well as inductance, and we can model it as a pure inductance in series with a resistance (■ Figure 31.14). The behavior of this circuit is similar to the series RC circuit we discussed in §31.1. When the switch has been in position A for a long time, there is a steady current $I_0 = \mathcal{E}/R$ in the circuit, magnetic energy $\frac{1}{2}LI_0^2$ is stored in the inductor, and the potential difference across the resistor equals the battery emf. (Since $dI/dt = 0$, there is no potential difference across the inductor.) When the switch is moved to position B at time $t = 0$, the current decreases exponentially to zero as the energy stored in the inductor is dissipated by the resistor:

$$I(t) = I_0 e^{-t/\tau}. \tag{31.19}$$

COMPARE WITH THE BEHAVIOR OF THE RC CIRCUIT, ESPECIALLY EQN. (31.3).

The time constant τ for the decay is determined by the inductance L and resistance R. The only combination of these with dimensions of time is L/R, so we guess that the time constant for the circuit is $\tau = \tau_L = L/R$.

The potential difference across the inductor due to the changing current (eqn. 31.10) is equal and opposite the potential difference IR across the resistor. As I decreases, so does its rate of change, and both potential differences decrease to zero.

THIS ARGUMENT DOES NOT RULE OUT A DIMENSIONLESS NUMBER LIKE 2 OR π MULTIPLYING L/R. IF SUCH A NUMBER EXISTS, IT WILL SHOW UP IN THE KIRCHHOFF'S RULE ANALYSIS.

EXERCISE 31.8 ♦ Show that the ratio L/R has dimensions of time—that is:

$$(1 \text{ H})/(1 \text{ }\Omega) = 1 \text{ s}.$$

We may verify our conclusions about the circuit's behavior by applying Kirchhoff's loop rule. The sign conventions for potential drops across the resistor and inductor are fixed by the positive sense of $I(t)$ chosen in Figure 31.14. (See Figure 31.9). Here we have $I > 0$, and we expect $dI/dt < 0$. Then $\Delta V_R = IR > 0$, and $\Delta V_I = L \, dI/dt < 0$ (eqn. 31.10). Traversing the circuit clockwise with the switch in position B, Kirchhoff's rule gives:

$$0 = \Delta V_I + \Delta V_R = L\frac{dI}{dt} + IR.$$

As expected, this is the exponential differential equation. Substituting expression (31.19), we find:

$$L\frac{dI}{dt} + IR = L\frac{d}{dt}(I_0 e^{-t/\tau}) + RI_0 e^{-t/\tau}$$
$$= LI_0\left(-\frac{1}{\tau}\right)e^{-t/\tau} + RI_0 e^{-t/\tau}$$
$$= I_0 e^{-t/\tau}\left(-\frac{L}{\tau} + R\right) = 0.$$

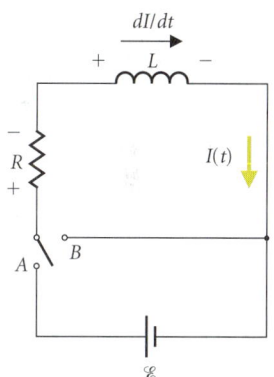

■ FIGURE 31.14
An LR circuit. When the switch has been in position A for a long time, the current in the circuit is $\mathcal{E}/R$. If the switch is then moved to position B, the current decays to zero exponentially.

The differential equation is satisfied for any value of I_0 if:

$$\tau = \tau_L = L/R, \tag{31.20}$$

as we guessed.

The inductor also regulates the rate at which current increases in the circuit when the switch is returned to position A. When the switch has been left in position B for a time interval $\Delta t \gg \tau_L$, the current is zero. Just after the switch is moved to position A, $I = 0$, all the potential difference of the battery is developed across the inductor and the current increases rapidly. As the current builds, more of the potential drop occurs across the resistor and less

THE CURRENT DECREASES TO ABOUT 5% OF ITS INITIAL VALUE IN A TIME $t \approx 3\tau = 3L/R$. WITH A LARGER R, THE CURRENT DECREASES MORE RAPIDLY. COMPARE WITH THE TIME TO DISCHARGE A CAPACITOR (§31.1).

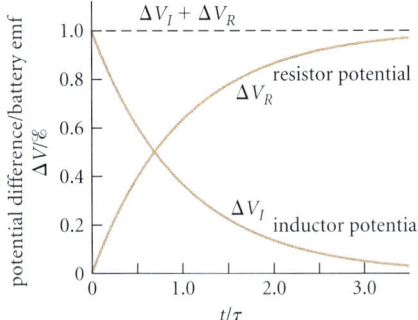

FIGURE 31.15
Potential differences across the inductor and the resistor in an *LR* circuit as functions of time after the switch is closed. As the current builds, the potential difference across the resistor increases and across the inductor decreases. The sum always equals the battery emf.

across the inductor; the current increases less rapidly. After a long time, the current approaches the final value $\mathcal{E}/R$ and $dI/dt = 0$. All the battery's potential difference is developed across the resistor (■ Figure 31.15). Again, we expect an exponential evolution:

$$I(t) = \frac{\mathcal{E}}{R}(1 - e^{-Rt/L}). \qquad (31.21)$$

EXERCISE 31.9 ♦ Apply Kirchhoff's rule to the *LR* circuit (Figure 31.14) to obtain a differential equation for the current after the switch is moved from *B* to *A*. Show that eqn. (31.21) is the correct solution.

EXERCISE 31.10 ♦♦♦ Show that the solution (31.21) predicts $I(t) \propto t$ for $t \ll \tau_L = L/R$. Plot a graph or use the series expansion for the exponential function (Appendix IB). ■

The spark plug in an automobile engine requires a potential difference of 20 kV for a period of a few microseconds during each cycle of the engine (about 0.02 s). The circuit that provides the spark uses an inductor as the energy source and operates somewhat like the emergency flasher in Example 31.3. ■ Figure 31.16 shows a simplified version of the circuit. A rotor synchronized mechanically with the engine closes the switch (the points) and allows current to build through the inductor. When the switch opens, the current decreases rapidly, and a large enough potential difference develops across the inductor for charge to jump the spark plug gap. At any time, there is current in a single loop of the spark coil circuit. In each loop, the inductor and a resistor are in series.

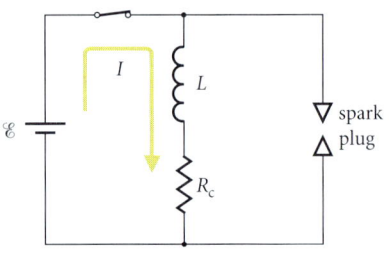

(a) Switch closed.

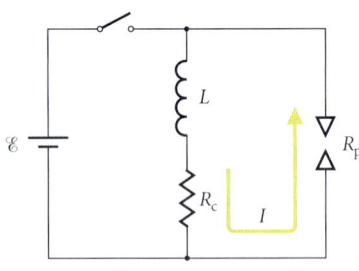

(b) Switch open.

FIGURE 31.16
Simplified diagram of a spark coil circuit. (a) Switch closed. The battery drives current through the coil, represented by a resistor and inductor in series. Energy is stored as magnetic field in the coil. (b) Switch open. Current in the battery loop is interrupted. The inductor prohibits the current from falling immediately to zero, so a spark jumps the gap across the plug, igniting the fuel.

EXAMPLE 31.9 ♦♦ An automobile spark coil circuit (Figure 31.16) has $\mathcal{E} = 12$ V, $L = 10.0$ mH, $R_c = 10\ \Omega$, and $R_p = 7.0$ kΩ (when sparking). Estimate: (a) the time the points must remain in contact to build the current, (b) the duration of the spark, and (c) the energy released in the spark.

MODEL With the switch closed, the battery supplies current to the coil in the single loop circuit on the left-hand side of Figure 31.16a. The current builds exponentially according to eqn. (31.21). Once the switch is opened, the inductor prevents the current from dropping immediately to zero, so a spark jumps the gap on the right.

SETUP (a) The time constant for the single loop circuit on the left when the points are in contact is:

$$\tau_L = L/R_c = (1.0 \times 10^{-2}\ \text{H})/(10\ \Omega) = 1\ \text{ms}.$$

SOLVE The points must be closed for $\approx 3\tau_L = 3$ ms to build current in the inductor to within 5% of its maximum value.

ANALYZE This time is much less than the 20-ms rotation period of the crank shaft at 3000 rpm.

SETUP (b) With the points open, current through the spark plug in the right loop decays according to eqn. (31.19). The spark duration is roughly the time constant of a circuit containing the inductor and spark plug in series. The resistance of the coil is negligible compared with the 7-kΩ resistance of the plug: $R = R_c + R_p \approx R_p$.

SOLVE $\tau_s \approx L/R_p = (1.00 \times 10^{-2}\ \text{H})/(7.0 \times 10^3\ \Omega) = 1.4\ \mu$s.

SETUP (c) The energy in the spark comes from the magnetic energy stored in the inductor (eqn. 31.13). The current in the circuit just before the spark occurs is $I \approx \mathcal{E}/R$.

SOLVE
$$U_s = \tfrac{1}{2}LI^2 \approx \tfrac{1}{2}(10^{-2}\ \text{H})\left(\frac{12\ \text{V}}{10\ \Omega}\right)^2 = 7\ \text{mJ}.$$

ANALYZE The spark duration is much less than the time to build current in the inductor. During the spark, the inductor delivers about (7 mJ)/(1.4 μs) = 5 kW of power. ■

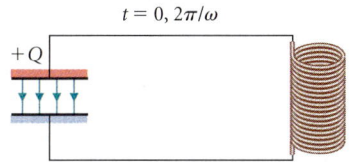

(a) Energy stored in the capacitor.

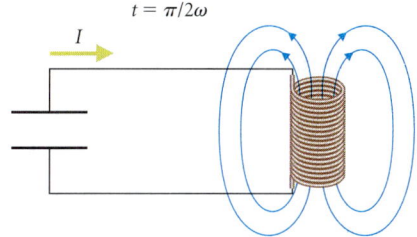

(b) Energy stored in the inductor.

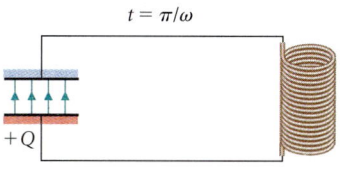

(c) Energy stored in the capacitor again.

■ **FIGURE 31.17**
Energy in the LC circuit. (a) Initially, all the energy is stored in the capacitor as electric field energy. When the switch is closed, the capacitor begins to discharge. The positive charge is on the upper plate.

(b) As current builds in the inductor, the magnetic field energy increases, and energy is transferred from the capacitor to the inductor. When the charge is zero, the current is maximum and all the energy is in the inductor. In this idealized circuit, there are no ohmic losses.

(c) The current begins to charge the capacitor in the opposite sense, transferring energy back to electric form. The current through the inductor decreases until the capacitor is fully charged and the current is zero. The positive charge is then on the lower plate.

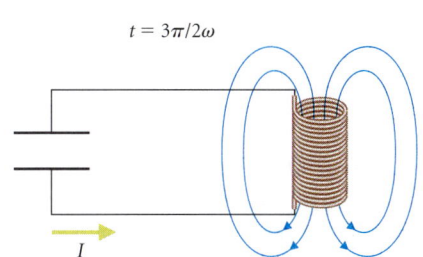

(d) Energy stored in the inductor.

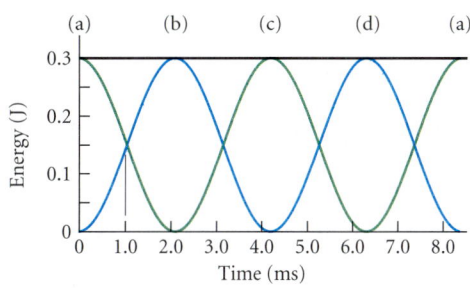

(e) Energy in an LC circuit as a function of time.

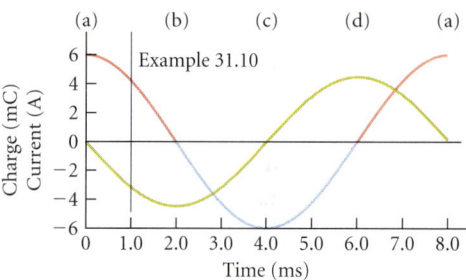

(f) Charge and current in the circuit as a function of time.

(d) The capacitor discharges again, the current increases, and energy is transferred back to the inductor. The current and magnetic field directions are the reverse of their directions in (b). The current then returns the system to state (a), and the cycle repeats indefinitely.

(e) Energy in an LC circuit as a function of time. The values were taken from Example 31.10. The corresponding picture, (a) through (d), is marked at the top of the graph. At 1.0 ms (cf. Example 31.10), the energy is almost equally divided between electric and magnetic forms.

(f) Charge and current in the circuit as functions of time. Compare the values at 1.0 ms as computed in Example 31.10 with the values shown here.

31.3.2 The LC Circuit

An inductor and capacitor in series (■ Figure 31.17) form a circuit with two ways to store energy: as electric energy in the capacitor and as magnetic energy in the inductor. When the capacitor is charged and the switch is closed, energy flows back and forth between the inductor and capacitor. When the inductor stores energy, a current exists that changes the capacitor charge and hence its stored energy. Charge on the capacitor produces a potential difference that causes the inductor current to change. With no resistance to dissipate energy, an ideal LC circuit oscillates forever. Such oscillations are the second basic pattern of circuit behavior. The frequency of the oscillation can depend only on L and C, and the combination that has the dimensions of $[T^{-1}]$ is $1/\sqrt{LC}$.

COMPARE WITH THE ENERGY FLOW BETWEEN KINETIC AND POTENTIAL FORMS IN A PENDULUM, §14.3.

■ **EXERCISE 31.11** ♦ Show that $1/\sqrt{LC}$ has the dimensions of frequency—that is, $1\ H\cdot F = 1\ s^2$.

As usual, Kirchhoff's rules give a differential equation that governs the circuit's behavior. We define Q and I as shown in ■ Figure 31.18, and proceed counterclockwise. Kirchhoff's loop rule requires:

$$0 = +L\frac{dI}{dt} + \frac{Q}{C}.$$

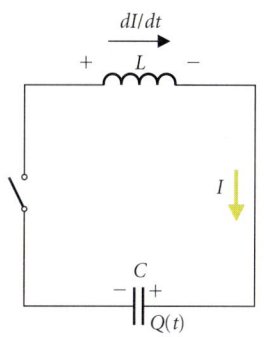

■ **FIGURE 31.18**
Circuit diagram for an LC circuit. In this idealized version, there is no resistance. The current variable has been chosen so that $I = +dQ/dt$.

Positive current $I(t)$ increases the charge $Q(t)$, so:

$$I = +\frac{dQ}{dt}.$$

Substituting:

$$0 = L\frac{d^2Q}{dt^2} + \frac{Q}{C}. \tag{31.22}$$

Rearranging:

$$\frac{d^2Q}{dt^2} = -\frac{1}{LC}Q.$$

THE GENERAL FORM OF THE EQUATION FOR SHM IS $d^2x/dt^2 = -\omega^2 x$.

This is the equation for simple harmonic motion (SHM) (eqn. 14.1). Its general solution (cf. eqn. 14.7) is:

$$Q(t) = Q_0 \cos\omega t + Q_1 \sin\omega t, \tag{31.23}$$

or, equivalently (cf. eqn. 14.3):

$$Q(t) = Q_2 \cos(\omega t + \phi_0). \tag{31.24}$$

The angular frequency is:

$$\omega \equiv \omega_0 = 1/\sqrt{LC}, \tag{31.25}$$

consistent with our argument on dimensional grounds. Equation (31.23) describes any oscillation of the circuit. The constant Q_0 is the charge on the capacitor at $t = 0$ and Q_1 is related to the current at $t = 0$, as we show in the next example. Their values may be found from known values of charge and current at any specific time, or from two values of either charge or current at two different times.

REMEMBER: THE TWO TIMES MUST BE AT DIFFERENT PHASES OF THE OSCILLATION.

> **EXAMPLE 31.10** ♦♦ In Figure 31.18, $L = 30.0$ mH and $C = 60.0$ μF. At $t = 0$, the capacitor carries a charge $Q_0 = +6.0$ mC, when the switch is closed. (a) Find the charge on the capacitor and the current through the inductor as functions of time. (b) Show that the total energy of the system is constant and find its value. (c) Give numerical values for the capacitor charge and inductor current 1.0 ms after the switch is closed.
>
> **MODEL** Initially, the capacitor is fully charged but the current is zero. All the stored energy is in the capacitor. The potential difference across the capacitor (and hence across the inductor) causes current through the inductor to increase, discharging the capacitor and storing magnetic energy in the inductor. We may use the results that we have just developed [eqns. (31.23) through (31.25)].
>
> **SETUP** (a) Just after the switch closes at $t = 0$, the current through the inductor remains zero although its derivative is nonzero.
>
> $$I(0) = \left.\frac{dQ}{dt}\right|_{t=0} = -\omega Q_0 \sin(0) + \omega Q_1 \cos(0) = \omega Q_1.$$
>
> Since $I(0) = 0$, then $Q_1 = 0$.
> The frequency of the oscillation is:
>
> $$\omega_0 = \frac{1}{\sqrt{LC}} = \frac{1}{\sqrt{(3.00 \times 10^{-2}\text{ H})(6.00 \times 10^{-5}\text{ F})}} = 745 \text{ rad/s}.$$
>
> **SOLVE** Thus the charge as a function of time is:
>
> $$Q(t) = Q_0 \cos(\omega_0 t) = (6.0 \text{ mC})\cos[(745 \text{ rad/s})t].$$
>
> The current is:
>
> $$I(t) = -Q_0\omega_0 \sin(\omega_0 t) = -(6.0 \text{ mC})(745 \text{ rad/s})\sin[(745 \text{ rad/s})t]$$
> $$= -(4.5 \text{ A})\sin[(745 \text{ rad/s})t].$$

IN COMPUTING NUMERICAL VALUES [PART (C)], WE USE THE UNROUNDED AMPLITUDE, 4.47 A.

SETUP (b) The total energy at any time is the sum of the energies stored in the inductor and capacitor.

SOLVE Thus: $U = \frac{1}{2}LI^2 + \frac{1}{2}\frac{Q^2}{C} = \frac{1}{2}L\left[\frac{Q_0^2}{LC}\sin^2(\omega_0 t)\right] + \frac{1}{2C}[Q_0^2 \cos^2(\omega_0 t)]$

$= \frac{Q_0^2}{2C}[\sin^2(\omega_0 t) + \cos^2(\omega_0 t)] = \frac{Q_0^2}{2C},$

and is constant. Its value is: $U = \dfrac{(6.0 \times 10^{-3}\,\text{C})^2}{2(6.00 \times 10^{-5}\,\text{F})} = 0.30\,\text{J}.$

SETUP (c) We obtain values of Q and I from the expressions obtained in (a), with $t = 1.0 \times 10^{-3}$ s.

SOLVE $Q(1.0\,\text{ms}) = (6.0\,\text{mC})\cos(0.745\,\text{rad}) = 4.4\,\text{mC},$

and $I(1.0\,\text{ms}) = -(4.5\,\text{A})\sin(0.745\,\text{rad}) = -3.0\,\text{A}.$

ANALYZE The minus sign in the answer for I means that the current at $t = 1.0$ ms is opposite the arrow in Figure 31.18. The capacitor is discharging. With no resistance in the circuit, energy is conserved and the oscillation goes on forever. Of course, no real circuit behaves exactly this way. An electrical circuit without resistance of some kind is like a mechanical system without friction—impossible! Even a superconducting LC circuit loses energy by radiating electromagnetic waves (Chapter 33). In the next section we'll see how resistance damps oscillations.

31.3.3 The LRC Circuit

A pure LC circuit is useful as an idealized illustration of current oscillations, but any real inductor also has a resistance that eventually dissipates the energy. In series with a capacitor, it forms a circuit that exhibits both LC and LR circuit behavior: the current oscillates, but the oscillation damps exponentially.

We can use Kirchhoff's rule to verify this behavior for the LRC circuit. There is a charge Q_0 on the capacitor in ■ Figure 31.19 when the switch is closed. Defining the capacitor charge and current as shown in the figure,

$$I(t) = -\frac{dQ}{dt},$$

and going counterclockwise around the circuit, Kirchhoff's loop rule gives:

$$-\frac{Q}{C} + L\frac{dI}{dt} + IR = 0.$$

Combining these two equations and rearranging, we find:

$$\boxed{\frac{Q}{LC} + \frac{R}{L}\left(\frac{dQ}{dt}\right) + \frac{d^2Q}{dt^2} = 0.} \qquad (31.26)$$

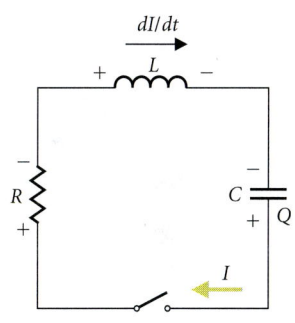

■ **FIGURE 31.19**
A series LRC circuit. With charge and current variables defined as shown, $I = -dQ/dt$.

This equation combines aspects of the exponential equation (first derivative proportional to Q) and the equation for SHM (second derivative proportional to Q). Its solution is a damped oscillation. As a trial solution, we assume Q has the form:

$$Q(t) = Q_0\cos(\omega t - \phi)e^{-\alpha t},$$

WE DISCUSSED DAMPED OSCILLATIONS IN §14.4.2.

with decay rate $\alpha \equiv 1/\tau$. When the switch is closed, the inductor prevents any instantaneous change of current, so $I(0) = 0$.

$\left.\dfrac{dQ}{dt}\right|_{t=0} = \left.-\omega Q_0 \sin(\omega t - \phi)e^{-\alpha t} - \alpha Q_0 \cos(\omega t - \phi)e^{-\alpha t}\right|_{t=0}$

$= \omega Q_0 \sin\phi - \alpha Q_0 \cos\phi = 0.$

Thus $\tan\phi = \alpha/\omega$. Differentiating again, we find:

$$\frac{d^2Q}{dt^2} = Q_0[-\omega^2\cos(\omega t - \phi) + 2\alpha\omega\sin(\omega t - \phi) + \alpha^2\cos(\omega t - \phi)]e^{-\alpha t}$$

$$= Q_0[(\alpha^2 - \omega^2)\cos(\omega t - \phi) + 2\alpha\omega\sin(\omega t - \phi)]e^{-\alpha t}.$$

Substituting the expressions for Q, dQ/dt, and d^2Q/dt^2 into eqn. (31.26), we have:

$$\frac{Q_0}{LC}\cos(\omega t - \phi)e^{-\alpha t} + \frac{R}{L}Q_0[-\omega\sin(\omega t - \phi) - \alpha\cos(\omega t - \phi)]e^{-\alpha t}$$
$$+ Q_0[(\alpha^2 - \omega^2)\cos(\omega t - \phi) + 2\alpha\omega\sin(\omega t - \phi)]e^{-\alpha t} = 0.$$

The expression $Q_0 e^{-\alpha t}$ is a factor in each term, and may be divided out. Collecting terms, we have:

$$\left(\frac{1}{LC} - \frac{R}{L}\alpha + \alpha^2 - \omega^2\right)\cos(\omega t - \phi) + \left(-\frac{R}{L}\omega + 2\alpha\omega\right)\sin(\omega t - \phi) = 0. \quad \text{(i)}$$

There are two unknowns: α and ω. We can get two different equations for them by evaluating eqn. (i) at two different times. At $t = \phi/\omega$, $\sin(\omega t - \phi) = 0$, $\cos(\omega t - \phi) = 1$, and

$$1/LC - \alpha(R/L) + \alpha^2 - \omega^2 = 0. \quad \text{(ii)}$$

At $t = (\phi + \pi/2)/\omega$, $\cos(\omega t - \phi) = 0$, $\sin(\omega t - \phi) = 1$, so

$$\omega(R/L - 2\alpha) = 0.$$

Thus:
$$\alpha = \frac{R}{2L} \equiv \frac{1}{\tau_{LRC}}. \quad (31.27)$$

Then from eqn. (ii):

$$\omega^2 = \frac{1}{LC} - \alpha\frac{R}{L} + \alpha^2 = \omega_0^2 - \alpha^2 = \frac{1}{LC} - \frac{R^2}{4L^2} \equiv \omega_1^2. \quad (31.28)$$

So the solution for Q is:

$$Q(t) = Q_0\cos(\omega_1 t - \phi)e^{-Rt/2L}, \quad (31.29)$$

with ω_1 given by eqn. (31.28) and $\phi = \tan^{-1}(\alpha/\omega)$.

The time constant $\tau_{LRC} = 2L/R$ is twice that for the LR circuit. Since the current oscillates (Figure 31.20), the average power dissipated (which is proportional to current squared) is one-half what it would be for a steady current, and so the amplitude of the oscillating current decays only half as rapidly as a steady current. The resistance also reduces the frequency of the oscillations compared with an ideal LC circuit.

We determined the constants α and ω by evaluating eqn. (i) at two specific times, $t = \phi/\omega$ and $t = (\phi + \pi/2)/\omega$. Sine and cosine functions vary differently with time, so the only way to satisfy the differential equation at *all* times is to set the coefficients of the sine and cosine terms equal to zero separately. Choosing two specific times at which either $\sin(\omega t - \phi)$ or $\cos(\omega t - \phi)$ is zero emphasizes this point. We'll use this technique often.

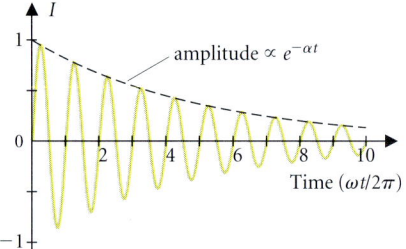

■ **FIGURE 31.20**
Current in the LRC circuit as a function of time. The values of L, R, and C are those in Example 31.11: $L = 4.7$ mH, $R = 1.2$ kΩ, $C = 18$ pF. The current oscillates, but the amplitude decreases (dashed line) as energy is dissipated in the resistor.

COMPARE EQNS. (31.25) AND (31.28).

EXAMPLE 31.11 ♦ A radio tuning circuit is a series LRC circuit with $C = 18$ pF, $R = 1.2$ kΩ, and $L = 4.7$ mH. Find the natural frequency ω_1 and damping rate α for this circuit.

MODEL We may use eqns. (31.27) and (31.28).

SETUP AND SOLVE The damping rate is:

$$\alpha = \frac{R}{2L} = \frac{1.2 \times 10^3 \, \Omega}{2(4.7 \times 10^{-3} \, \text{H})} = 1.3 \times 10^5 \text{ /s}.$$

The natural frequency is:

$$\omega_1^2 = \frac{1}{LC} - \alpha^2 = \frac{1}{(4.7 \times 10^{-3}\text{ H})(1.8 \times 10^{-11}\text{ F})} - (1.3 \times 10^5\text{ /s})^2$$

$$= 1.2 \times 10^{13}\ (\text{rad/s})^2,$$

so: $\omega_1 = 3.4 \times 10^6$ rad/s.

ANALYZE Equivalently, $f_1 = \omega_1/2\pi = 550$ kHz, in the AM band.

Since the square of the angular frequency is the difference of two terms, it is possible for ω_1^2 to be zero or even negative. This happens when $\alpha \geq \omega_0$ (eqn. 31.28), or equivalently, $R \geq 2\sqrt{L/C}$. With a resistance this large, the energy in the circuit is dissipated sufficiently fast that no oscillations appear. If $\omega_1 = 0$, the circuit is said to be *critically damped*; with $\omega_1^2 < 0$, the circuit is *overdamped*.

An *LRC* circuit is analogous to a mechanical system of a block pushed and pulled through a fluid by a spring. The inductance plays the roll of mass, the capacitance of spring constant, the charge is like displacement, and the current is like velocity. The resistance in the circuit is like drag on the block as it moves through the fluid. If the resistance is small, it doesn't affect the oscillations much, but a large resistance, like strong drag, eliminates the oscillations altogether.

Math Toolbox

How to Solve a Linear Differential Equation

The procedure we have used to solve the differential equations for time-dependent circuits is as follows:

1. Guess the mathematical form of the solution on physical grounds.
2. Plug into the differential equation to find time constants and/or frequencies.
3. Choose values for the adjustable constants in the solution to fit the known initial conditions or other given information.

Using qualitative reasoning in this way is a valuable skill, but what if your guesses don't work? Try this method.

1. Write the equation in standard form with the coefficient of the highest derivative equal to unity, all terms involving Q on the left-hand side (the homogeneous part) and terms not involving Q on the right-hand side (the inhomogeneous part).

    Example: *RC* circuit with battery (eqn. 31.7)

    $$\underbrace{\frac{dQ}{dt} + \frac{1}{RC}Q}_{\text{homogeneous part (H)}} = \underbrace{\frac{\mathcal{E}}{R}}_{\text{inhomogeneous part (I)}}.$$

2. Replace the right-hand side with zero and solve the resulting equation. The text gives a solution to any equation you will encounter in the problems. The solution contains as many adjustable constants as the order of the highest derivative—one for the exponential equation and two for simple harmonic motion.

    Example: $Q_H(t) = \underset{\underset{\text{adjustable constant}}{\uparrow}}{A} e^{-t/RC}$.

3. Find any function that satisfies the original equation. (This is the hard part.) If the right-hand side is not a function of time, you can get a result by setting all the derivatives on the left-hand side to zero.

    Example: With $dQ/dt = 0$, eqn. (31.7) becomes:

    $$\frac{Q}{RC} = \frac{\mathcal{E}}{R} \Rightarrow Q_I = \mathcal{E}C.$$

    If the right-hand side is a function of time, you must rely on physical intuition, guesswork, experience, or your differential equations text!

4. The complete solution is the sum of the two solutions you have found:

    $$Q(t) = Q_H + Q_I.$$

    Example: $Q(t) = Ae^{-t/RC} + \mathcal{E}C.$

5. Solve for the constants from initial conditions or other given information in the usual way.

    Example: $Q = 0$ at $t = 0 \Rightarrow A + \mathcal{E}C = 0$
    $\Rightarrow A = -\mathcal{E}C.$

    The solution is $Q(t) = \mathcal{E}C(1 - e^{-t/RC})$—eqn. (31.6).

EXAMPLE 31.12 ♦♦ If the values of L and R are fixed in the circuit described in Example 31.11, what value of C results in critical damping?

MODEL Critical damping occurs when $\omega_1^2 = 0$.

SETUP AND SOLVE From eqn. (31.28), $\omega_1^2 = 0$ when:

$$0 = \frac{1}{LC} - \frac{R^2}{4L^2} \quad \text{or} \quad C = \frac{4L}{R^2} = \frac{4(4.7 \times 10^{-3}\ \text{H})}{(1.2 \times 10^3\ \Omega)^2} = 1.3 \times 10^{-8}\ \text{F}.$$

ANALYZE With $\omega_1 = 0$ in eqn. (31.29), the charge decays exponentially. ∎

EXERCISE 31.12 ♦♦ For the overdamped case, $R > 2\sqrt{L/C}$, assume $Q(t) = Q_0 e^{-\beta t}$ and show that:

$$\beta = \frac{R}{2L} \pm \sqrt{\frac{R^2}{4L^2} - \frac{1}{LC}}.$$

✱ 31.4 Multiloop Circuits

If a circuit has more than one loop, we need to apply Kirchhoff's rules according to the plan in Figure 31.3, and we obtain a set of coupled differential equations. The next example illustrates this technique.

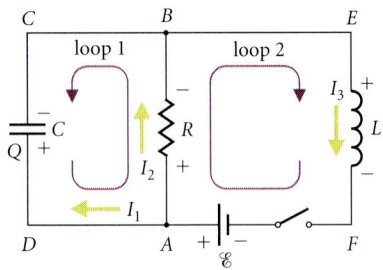

FIGURE 31.21
We have defined the current variables so that $I_1 = dQ/dt$, and $I_3 = I_1 + I_2$.

EXAMPLE 31.13 ♦♦♦ The capacitor in ■ Figure 31.21 is uncharged at $t = 0$ when the switch is closed. Find the potential difference across the capacitor for $t > 0$.

MODEL What is the expected behavior? A long time after the switch is closed, the capacitor is fully charged and there is a constant current through the inductor. Because the circuit contains all three kinds of element, we expect a decaying oscillation toward the final equilibrium.

STEP I

STEP II Defining currents I_1, I_2, and I_3 as shown in the diagram, the expected final state has:

$$I_{1,f} = 0, \quad I_{2,f} = I_{3,f} = \mathcal{E}/R, \quad Q_f/C = I_{3,f}R \Rightarrow Q_f = \mathcal{E}C.$$

Initially, $Q = 0$, and since the inductor does not allow instantaneous change of current, $I_3(0) = 0$ also.

SETUP Next we apply Kirchhoff's rules (cf. Figure 26.27):
STEP IV

$$\text{Junction rule at } A: \quad I_3 = I_1 + I_2. \tag{i}$$

$$\text{Loop 1 } (ABCD): \quad -I_2R + Q/C = 0. \tag{ii}$$

$$\text{Loop 2 } (ABEF): \quad -I_2R - L\frac{dI_3}{dt} + \mathcal{E} = 0. \tag{iii}$$

The relation between I_1 and Q is:

$$I_1 = \frac{dQ}{dt}, \tag{iv}$$

and from eqn. (ii), $I_2 = Q/(RC)$. Then from eqn. (i),

$$I_3 = \frac{Q}{RC} + \frac{dQ}{dt}. \tag{v}$$

Substituting into eqn. (iii), we get:

$$\frac{d^2Q}{dt^2} + \frac{1}{RC}\left(\frac{dQ}{dt}\right) + \frac{Q}{LC} = \frac{\mathcal{E}}{L}. \tag{vi}$$

SOLVE
STEPS III AND V This equation can be solved using the method given in the *Math Toolbox*. The homogeneous equation is like the *LRC* circuit eqn. (31.26), with different coefficients [the *LR* circuit decay rate R/L is replaced by the *RC* circuit decay rate $1/(RC)$]. We may adopt that solution (eqn. 31.29) with the appropriate change:

$$Q_H = Q_0 \cos(\omega t - \phi) e^{-\alpha t} \tag{vii}$$

with
$$\alpha = \frac{1}{2RC} \quad \text{and} \quad \omega^2 = \frac{1}{LC} - \frac{1}{(2RC)^2}. \tag{viii}$$

The solution to the inhomogeneous equation is $Q_I = \mathcal{E}C$. Thus:

$$\begin{aligned} Q(t) &= Q_I + Q_H \\ &= \mathcal{E}C + Q_0 \cos(\omega t - \phi) e^{-\alpha t}. \end{aligned} \tag{ix}$$

We complete the solution by finding the constants Q_0 and ϕ. Applying the initial conditions:

$$Q(0) = \mathcal{E}C + Q_0 \cos \phi = 0 \;\Rightarrow\; Q_0 = -\mathcal{E}C/\cos \phi.$$

Both I_3 and Q are zero at $t = 0$, so according to eqn. (v), dQ/dt is also zero.

$$\left. \frac{dQ}{dt} \right|_{t=0} = \left. -\omega Q_0 \sin(\omega t - \phi) e^{-\alpha t} - \alpha Q_0 \cos(\omega t - \phi) e^{-\alpha t} \right|_{t=0}$$

$$= \omega Q_0 \sin \phi - \alpha Q_0 \cos \phi = 0 \;\Rightarrow\; \tan \phi = \alpha/\omega,$$

as before. Putting these results into the general solution (eqn. ix), with α from eqn. (viii), we find the potential difference across the capacitor, $\Delta V = Q/C$:

$$\Delta V(t) = \mathcal{E}\left[1 - \frac{\cos(\omega t - \phi)}{\cos \phi} e^{-t/2RC}\right].$$

ANALYZE Check that this solution has the expected behavior. For example, $\Delta V \to \mathcal{E}$ as $t \to \infty$.

Chapter Summary

Where Are We Now?

Using Kirchhoff's rules, we have analyzed the behavior of circuits in which current varies with time. These circuits have many practical applications and bring together all of the electromagnetic effects we have studied so far.

What Did We Do?

Coils resist change in the current they carry. The rate of change of current in such an inductor is proportional to the potential difference applied to it:

$$\Delta V = L\, dI/dt,$$

where L is the self-inductance of the coil. The static electric field in the inductor points in the direction of dI/dt. A similar effect occurs when flux produced by one circuit element threads another, giving rise to mutual inductance (eqn. 31.16). Electrical transformers are an application of mutual inductance.

The energy stored in an inductor carrying current I is $\frac{1}{2}LI^2$. The energy is stored in the magnetic field itself, with an energy density $u_B = \frac{1}{2}B^2/\mu_0$. Combining this result with eqn. (27.13) for electric energy, the total energy stored in (electric plus magnetic) fields has density:

$$u = u_E + u_B = \tfrac{1}{2}(\epsilon_0 E^2 + B^2/\mu_0).$$

We studied four types of single-loop circuit: *RC*, *LR*, *LC*, and *LRC* circuits. These simple circuits demonstrate two important types of behavior: exponential decay and oscillation.

Decay results from energy dissipation in a resistor. The exponential time constants for the *RC* and *LR* circuits are $\tau_C = RC$ and $\tau_L = L/R$.

A system can oscillate only if it has more than one way to store energy. In the *LC* circuit, energy oscillates between electric field energy in the capacitor and magnetic field energy in the inductor. The angular frequency of the oscillation is $\omega_0 = 1/\sqrt{LC}$. The *LRC* circuit exhibits damped oscillations as the resistor dissipates the stored energy. The time constant $\tau_{LRC} = \tfrac{1}{2}L/R = 1/\alpha$ is one-half the time constant for the *LR* circuit and the oscillation frequency is $\omega_1 = \sqrt{\omega_0^2 - \alpha^2}$. With a large enough resistance, the damping rate $\alpha > \omega_0$ and no oscillations occur. The circuit is overdamped.

Kirchhoff's rules are the tools used to analyze all circuits. They produce linear differential equations for the charge or current. In the *Math Toolbox* we have outlined a method for solving such equations. The first step in any solution is to determine the kind of behavior that is expected and the final equilibrium that is achieved. A solution plan is outlined in Figure 31.4.

✶ The same principles apply to the solution of multiloop circuits. Kirchhoff's rules produce coupled differential equations. The kinds of behavior—oscillation and decay—are the same.

Practical Applications

Capacitors are frequently used in switching circuits to prevent sparks. Shunt resistors are connected in parallel with capacitor banks to ensure that the stored energy is dissipated safely when the banks are not in use, and *RC* circuits are also used to produce light flashes, either at regular intervals (warning signals) or on demand (camera flash attachments). Inductors find numerous uses in electric circuits, such as smoothing variations in current and storing energy. Inductor circuits provide the spark to automobile engines. Transformers use mutual inductance to couple circuits that transmit power at different voltages. Applications in AC circuits are the subject of the next chapter.

Solutions to Exercises

31.1.
$$(1\ \Omega)\cdot(1\ \text{F}) = \frac{1\ \text{V}}{1\ \text{A}} \cdot \frac{1\ \text{C}}{1\ \text{V}} = \frac{1\ \text{C}}{1\ \text{C/s}} = 1\ \text{s}.$$

31.2. Differentiating eqn. (31.6),
$$\frac{dQ}{dt} = \frac{d}{dt}[C\mathcal{E}(1 - e^{-t/RC})] = -C\mathcal{E}\left(-\frac{1}{RC}\right)e^{-t/RC} = \frac{\mathcal{E}}{R}e^{-t/RC}.$$

Substituting into eqn. (31.7):
$$\frac{\mathcal{E}}{R}e^{-t/RC} + \frac{C\mathcal{E}(1 - e^{-t/RC})}{RC} = \frac{\mathcal{E}}{R}.$$

The two exponential terms cancel and the equation is satisfied.

31.3. The inductor resists change in current. With the circuit broken, it does so by forcing a spark to jump between the blades of the switch as it is opened. Such sparks damage switches; a capacitor in parallel with the switch allows the inductor to charge the capacitor rather than form a spark.

31.4.
$$1\ \text{H} = 1\ \frac{\text{V}\cdot\text{s}}{\text{A}} = 1\ \frac{\text{J}\cdot\text{s}}{\text{C}\cdot\text{A}} = 1\ \frac{\text{N}\cdot\text{m}}{\text{A}^2} = 1\ \frac{\text{N}}{\text{A}\cdot\text{m}} \cdot \frac{\text{m}^2}{\text{A}}$$
$$= 1\ \frac{\text{T}\cdot\text{m}^2}{\text{A}} = 1\ \frac{\text{Wb}}{\text{A}}.$$

31.5. $U = \tfrac{1}{2}LI^2 = \tfrac{1}{2}(3.3 \times 10^{-3}\ \text{H})(1.1\ \text{A})^2 = 2.0\ \text{mJ}.$
(*Note:* From Exercise 31.4, 1 H = 1 J/A².)

31.6. The charge density on the wire produces an electric field at a radius r from the wire (eqn. 24.4) $E = \lambda/(2\pi\epsilon_0 r)$, so we must include the electric field energy density:
$$u_E = \tfrac{1}{2}\epsilon_0 E^2 = \tfrac{1}{2}\epsilon_0[\lambda/(2\pi\epsilon_0 r)]^2.$$

The electric energy density 6.0 mm from the wire is:
$$u_E = \left(\frac{0.5}{8.85 \times 10^{-12}\ \text{F/m}}\right)\left[\frac{5.0 \times 10^{-12}\ \text{C/m}}{2\pi(6.0 \times 10^{-3}\ \text{m})}\right]^2$$
$$= 9.9 \times 10^{-10}\ \text{J/m}^3.$$

The electric field increases the energy density by 2%. Since both kinds of field decrease as $1/r$, the ratio of their energy densities is independent of distance from the wire.

31.7. The flux through each turn of the primary coil is: $\Phi_p = I_p(L_p/N_p)$. The same flux passes through each turn of the secondary coil, so the total flux through the secondary is:

$$\Phi_s = N_s\Phi_p = I_p L_p N_s/N_p.$$

The mutual inductance is: $M = \Phi_s/I_p = L_p N_s/N_p$. The mutual inductance may also be found from the flux through the primary: $M = \Phi_p/I_s = L_s N_p/N_s$. Multiplying these together:

$$M^2 = \left(L_p \frac{N_s}{N_p}\right)\left(L_s \frac{N_p}{N_s}\right) = L_p L_s.$$

31.8. $\dfrac{1\text{ H}}{1\ \Omega} = \dfrac{1\text{ V·s/A}}{1\text{ V/A}} = 1$ s, as required.

31.9. Traversing the circuit clockwise,

$$0 = \mathcal{E} - IR - L\frac{dI}{dt}.$$

Differentiating eqn. (31.21):

$$\frac{dI}{dt} = \frac{\mathcal{E}}{R}\left(\frac{R}{L}\right)e^{-Rt/L} = \frac{\mathcal{E}}{L}e^{-Rt/L}.$$

Substituting into Kirchhoff's rule:

$$0 \stackrel{?}{=} \mathcal{E} - R\frac{\mathcal{E}}{R}(1 - e^{-Rt/L}) - L\frac{\mathcal{E}}{L}e^{-Rt/L} = 0.$$

The assumed solution satisfies the equation.

31.10. In ■ Figure 31.22 we have plotted the ratio $I/I_f = IR/\mathcal{E}$ of the current I to its final value as a function of the dimensionless time variable t/τ. The straight line is extremely close to the exact solution for $t/\tau < 0.1$. The difference is 4% at $t/\tau = 0.1$.

Alternatively, using the exponential series:

$$e^x = 1 + x + \tfrac{1}{2}x^2 + \cdots,$$

we have: $1 - e^{-t/\tau} = 1 - [1 - t/\tau + \tfrac{1}{2}(t/\tau)^2 + \cdots]$
$$= t/\tau - \tfrac{1}{2}(t/\tau)^2 + \cdots.$$

For $t \ll \tau$, we ignore the squared term and all the higher powers. Then, using eqn. (31.21),

$$I(t) \approx (\mathcal{E}/R)(t/\tau) = (\mathcal{E}/L)t,$$

proportional to time. Compare this result with eqn. (31.9).

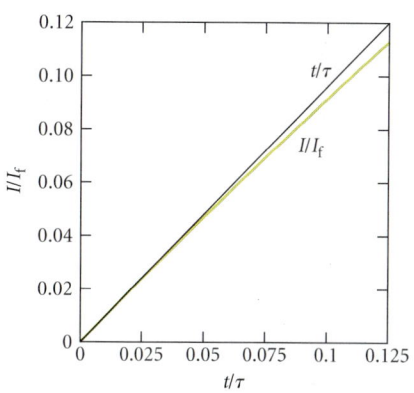

■ **Figure 31.22**
Graph of the current in an inductor as a function of time for $t/\tau < 0.12$. The linear function $I/I_f = t/\tau$ is shown for comparison. The two curves are indistinguishable for $t/\tau < 0.05$ and differ by 4% at $t/\tau = 0.1$.

31.11. $(1\text{ H})(1\text{ F}) = (1\text{ V·s/A})(1\text{ C/V})$
$$= \frac{(1\text{ C})\cdot(1\text{ s})}{1\text{ C/s}} = 1\text{ s}^2.$$

31.12. With the assumed form for Q,

$$-I(t) = \frac{dQ}{dt} = -\beta Q_0 e^{-\beta t} \quad\text{and}\quad -\frac{dI}{dt} = \frac{d^2Q}{dt^2} = \beta^2 Q_0 e^{-\beta t}.$$

Substituting into eqn. (31.26):

$$\beta^2 - \frac{R}{L}\beta + \frac{1}{LC} = 0.$$

Solving the quadratic:

$$\beta_\pm = \frac{R}{2L} \pm \sqrt{\frac{R^2}{4L^2} - \frac{1}{LC}},$$

as required. There are two time constants, corresponding to the plus and minus signs. The complete solution has two terms, one with each time constant:

$$Q = Q_1\exp(-\beta_+ t) + Q_2\exp(-\beta_- t).$$

Both β_+ and β_- are positive so both terms decay with time.

Basic Skills

Review Questions

§31.1 RESISTOR–CAPACITOR CIRCUITS

- Which mathematical function describes the potential difference across a discharging capacitor as a function of time?
- What is the time constant for an RC circuit?
- Formally, how long does it take for the capacitor to discharge? What is a *practical* measure of the discharge time?
- Is the current variable $I(t)$ always related to the charge variable $Q(t)$ by $I = +dQ/dt$? Why or why not?
- When a capacitor discharges in an RC circuit, what happens to the energy that it stored?
- Describe the characteristic behavior of an RC circuit as it evolves from its initial state to its final state.

§31.2 INDUCTANCE

- In a circuit containing a superconducting coil, a battery, and a switch, what is the current immediately after the switch is closed? Why isn't it infinite?
- Give two different expressions for the *inductance* of a circuit element.

- What is the SI unit of inductance?
- How does the magnetic energy stored by an inductor depend on the current in it?
- How does magnetic energy density depend on magnetic field strength?
- Define the *mutual inductance* of two circuit elements.
- What is a transformer? Give an example of how one is used.

§31.3 INDUCTOR CIRCUITS

- How does current vary with time in an *LR* circuit? What is the time constant?
- How does current vary with time in an *LC* circuit?
- In mechanical oscillations, energy varies between potential and kinetic forms. What forms of energy occur in an electrical circuit oscillation?
- What happens if resistance is added to an *LC* circuit? How does the oscillation frequency change?
- Does an *LRC* circuit always oscillate? Why or why not?

✳ §31.4 MULTILOOP CIRCUITS

- Which rules are used to analyze multiloop circuits?
- Do any new forms of behavior appear?

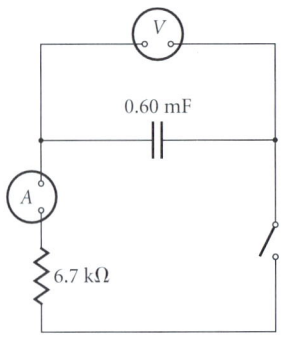

■ FIGURE 31.23

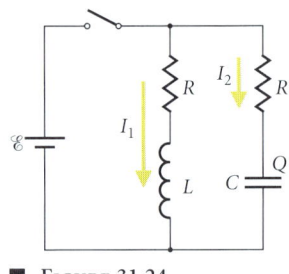

■ FIGURE 31.24

§31.2 INDUCTANCE

4. Find the inductance of a solenoid with 750 turns, length 2.5 cm, and radius 0.45 cm.

5. Find the magnetic energy density at a point 15 cm from a long, straight wire carrying a current of 1.7 A.

6. A transformer with 100 turns of wire on the primary coil and 500 turns on the secondary coil is connected to a potential difference $\Delta V = (100 \text{ V})\cos \omega t$. Find the potential difference across the secondary terminals.

§31.3 INDUCTOR CIRCUITS

7. An *LR* circuit contains a 3.0-Ω resistor and a 4.0-H inductor in series with an ammeter. At $t = 4.0$ s, the ammeter reads 0.050 A. What was the current at $t = 0$?

8. When a 0.16-μF capacitor is connected in series with a 12-mH inductor, the circuit is found to be critically damped. What is the resistance of the inductor?

9. A 56-μF capacitor is connected in series with a 3.0-mH inductor. What is the oscillation frequency of the resulting circuit? At $t = 0$, the capacitor is uncharged and the current in the circuit is 17 mA. What is the energy in the inductor? Describe how the current and the total energy in the circuit vary with time.

✳ §31.4 MULTILOOP CIRCUITS

10. The switch in ■ Figure 31.24 is closed at $t = 0$. What are the values of I_1, I_2, and Q immediately after the switch is closed? What are the values a long time later? Do the inductor and capacitor arms of the circuit affect each other's behavior? Why or why not?

Basic Skill Drill

§31.1 RESISTOR–CAPACITOR CIRCUITS

1. Refer to ■ Figure 31.23. If the resistance in the circuit is 6.7 kΩ, and the initial charge on the capacitor is $Q_0 = 0.27$ C, find the meter readings 6.0 s after the switch is closed.

2. A lightning protection system contains a 3.0-mF capacitor connected to ground through a 1.0-Ω resistance. A lightning bolt delivers a charge of 0.70 C to the system. What energy is absorbed by the system, and how long does it take to discharge the energy to ground?

3. A 10.0-V battery, a 56-μF capacitor, a 10.0-kΩ resistor, and a switch are connected in series. What is the final charge on the capacitor a long time after the switch is closed? What is the final current? How much time is required for the charge to reach 95% of its final value? What is the initial current in the circuit?

Questions and Problems

§31.1 RESISTOR–CAPACITOR CIRCUITS

11. ❖ Explain why the charge on a capacitor cannot change abruptly when the switch is closed in an *RC* circuit.

12. ❖ If the resistance in an *RC* circuit is doubled, how does the time constant change? What if the capacitance is doubled? In each case, give a physical argument to support your answer.

13. ♦ What time is required for the charge on a capacitor in an *RC* circuit to decrease to one-half its original value?

14. ♦ **(a)** What is the time constant of a circuit containing a 32-nF capacitor in series with a 14-kΩ resistor? **(b)** If the charge on the capacitor is initially $Q_0 = 32$ μC, what time interval is required for it to decrease to 1.6 μC? **(c)** What is the initial current in the circuit?

15. ♦ A 16-μF and a 20-μF capacitor are connected in parallel. What resistance must be connected across them for the circuit to have a time constant of 1.0 ms?

16. ♦♦ What is the initial current in an *RC* circuit (Figure 31.1) if the initial charge on the capacitor is Q_0? Show that if the current were constant, the capacitor would discharge completely in one time constant.

17. ♦♦ Refer to the circuit diagram in Figure 31.5. At $t = 0$, when the switch is closed, the capacitor carries a charge $Q_0 = 2C\mathcal{E}$. Find the charge and current as functions of time after the switch is closed.

18. ♦♦ A 65.0-nF capacitor is connected in series with a resistor. A 12.0-V battery is connected across the capacitor. Immediately after the battery is disconnected, there is a 0.240-A current in the resistor. What is the resistance? Find the current 3.0 μs later.

19. ♦♦ Refer to the circuit diagram in Figure 32.23. At 4.0 s after the switch is closed, the voltmeter reads 150 V and the ammeter reads 30.0 mA. What is the value of the resistance *R*? What will the meters read 8.2 s later? What was the original charge on the capacitor?

20. ♦♦ Find the charge on each capacitor in ■ Figure 31.25 as a function of time after the switch is closed.

21. ♦♦ You have two capacitors, 56 μF and 22 μF, and two resistors, 18 kΩ and 33 kΩ. You need to build an *RC* circuit with a time constant of 0.91 s. Can you do it? If so, how?

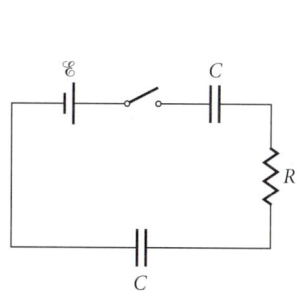

■ FIGURE 31.25

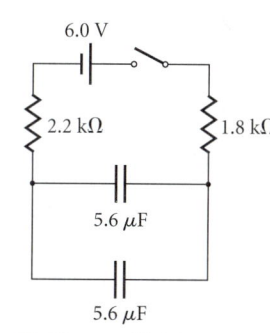

■ FIGURE 31.26

22. ♦♦ A circuit contains two resistors and two capacitors connected as shown in ■ Figure 31.26. What is the charge on each capacitor a long time after the switch is closed? How long does it take to reach that state? How much energy is dissipated in the resistors?

23. ♦♦♦ Two unequal capacitors C_1 and C_2 are connected in series with a resistor *R* and a switch. Initially, one capacitor has a charge Q_1 and the second is uncharged. Then the switch is closed. Show that the capacitors share the final charge like parallel capacitors, but in the time constant they combine like series capacitors.

§31.2 INDUCTANCE

24. ❖ Two long wires, each carrying current *I*, are perpendicular to the page, as shown in ■ Figure 31.27. At which point is the magnetic energy density greatest? Explain why you chose your answer.

25. ❖ Compare the total magnetic energy in two solenoids, each with *N* turns, area *A*, and current *I*, but lengths ℓ and 2ℓ.

26. ❖ An inductor is to be made by wrapping a fixed length of wire around a cardboard tube. You have a choice of tubes with the same length but different radii. Does the inductance depend on which tube you choose? If so, how?

27. ♦ Find the magnetic energy density at a point 5.0 m from a power cable carrying a current of 0.80 kA.

28. ♦ Find the magnetic energy per kilometer of depth inside a large sunspot with area 6×10^{15} m^2 and $B = 4 \times 10^{-3}$ T.

29. ♦ Find the total electromagnetic energy density near the surface of the Earth, where $B = 4 \times 10^{-5}$ T and $E = 100$ V/m. What fraction of the total energy density is electric?

30. ♦♦ Find the magnetic energy density on the axis of a current loop with radius 2.0 cm carrying a current of 6.0 mA, at a distance of 3.0 cm from the center of the loop.

31. ♦♦ A constant current *I* in a semi-infinite wire lying along the negative *z*-axis forms a point charge *Q* at the origin. Find the total electromagnetic energy density at point *P* with coordinates $z = -d$, $x = R$, with $R = 2.0$ cm, $I = 6.0$ mA, and $d = 1.5$ cm at a time when $Q = 3.0$ pC. Compare Problem 30.67 for the calculation of $B = (\mu_0 I/4\pi R)(1 + d/\sqrt{d^2 + R^2})$.

32. ♦♦ ✉ Two solenoids lie along the same axis. The larger has a radius $a = 5.0$ cm, length $\ell = 15$ cm, and $N = 1.3 \times 10^4$ turns. The smaller solenoid has radius $b = 3.0$ cm, length $\ell' = 10.0$ cm, and $N = 5 \times 10^3$ turns. Estimate the mutual inductance of the two solenoids: first by assuming a current I_1 in the large solenoid and computing the flux Φ_2 through the smaller; and second by assuming a current I_2 in the smaller solenoid and calculating Φ_1.

33. ♦♦ A coil consists of 1.5×10^4 turns of wire wrapped around a toroidal (doughnut-shaped) tube. The center of the tube has radius 10.0 cm and the tube's cross section has radius 2.0 cm (cf. Figure 28.24). Estimate the inductance of the ring.

34. ♦♦ (a) Estimate the mutual inductance of two concentric, coplanar circular wire loops, one of radius 5 cm and one of radius 0.5 cm. (b) Estimate the mutual inductance of the two loops when they are separated by 50 cm (center to center) in their common plane. (*Hint:* Use magnetic moment.)

35. ♦♦♦ Two solenoids of radii a_1 and $a_2 < a_1$ are mounted one inside the other and connected in series so that each carries the same current (■ Figure 31.28). Each has length ℓ and *N* turns. (a) Is the inductance greater if the coils are wound in the same sense (as shown in the figure) or in the opposite sense? Why? (b) Compute the inductance of the combination. (You may use the results for long solenoids in your calculations.)

36. ♦♦♦ Use the result $B = \mu_0 nI(\cos\theta_1 + \cos\theta_2)/2$ (Problem 28.41, Figure 28.40) for the magnetic field inside a finite solenoid to improve the estimate of a solenoid's inductance. Assume that the magnetic field is uniform over any cross section of the solenoid though it varies with position along the axis.

37. ♦♦♦ In a coaxial cable, apply Faraday's law to find the emf $\mathcal{E}$ around a rectangular curve, with length ℓ parallel to the cable and whose two ends are radial lines between the two conductors. Find the inductance per unit length of the cable from the relation $L/\ell = (\mathcal{E}/\ell)/(dI/dt)$, (cf. Example 31.6).

38. ♦♦♦ Two parallel wires, each with radius 0.23 mm, carry current in opposite directions. Their centers are separated by 3.0 cm. Calculate the inductance per unit length of the two wires. (a) Assume the current is on the outside of the two wires. (b) Does it make a difference if the current is uniform within the two wires? If so, how much?

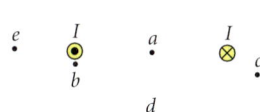

■ FIGURE 31.27

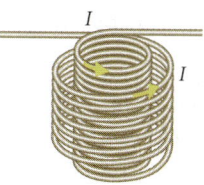

■ FIGURE 31.28

§31.3 INDUCTOR CIRCUITS

39. ❖ Explain why the current through an inductor cannot change abruptly when the switch is closed in an *LR* circuit in series with a battery.

40. ❖ If the resistance in an *LR* circuit is doubled, how does the time constant change? State the physical reason for the change.

41. ❖ If the resistance in an *LRC* circuit is doubled, how does the time constant change? What happens to the frequency ω_1?

42. ❖ Make an analogy between an *LC* circuit and a particle on a spring (cf. §14.1), and discuss the roles of inductance and capacitance in the circuit.

43. ❖ If the inductance in an *LRC* circuit is doubled, how does this affect the time constant and the frequency ω_1? Give a physical explanation for your answer.

44. ❖ The switch in ■ Figure 31.29 remains in position ② for 0.2 s, then snaps quickly to position ①, remains there for 0.2 s, then snaps back to ②, and so on. Sketch a graph showing the current through the resistor as a function of time. How would the graph change if the oscillation frequency of the switch were increased? If it were decreased?

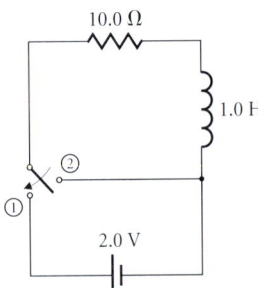

■ FIGURE 31.29

45. ❖ (a) Show that the combined inductance of two inductors connected in series is $L_s = L_1 + L_2$. (Assume the inductors are separated widely enough that flux from one does not thread the other.) (b) Show that the combined inductance of two inductors connected in parallel is given by $1/L = 1/L_1 + 1/L_2$. (Assume that no flux from one threads the other.)

46. ◆ A circuit contains a 3.0-mH inductor, a 6.0-μF capacitor, and an ammeter. The ammeter has negligible effect on the circuit. If the ammeter reads 0.50 A at $t = 0$ and 0.20 A at $t = 0.21$ ms, find an expression for the current at an arbitrary time.

47. ◆ For a long time prior to $t = 0$, the switch in the circuit of ■ Figure 31.29 has been in position ①. At $t = 0$, the switch is put into position ②. What are the magnitude and direction of current at $t = 0.30$ s?

48. ◆ What turns ratio is needed in a transformer to convert 33-V generator output voltage to 110 V to run a household appliance? How could you build a transformer to do the job? (*Remember:* Turns come in whole numbers!)

49. ◆ A 10.0-V battery, a 10.0-kΩ resistor, and a 0.30-H inductor are connected in series with a switch. The switch is closed at $t = 0$ after remaining open for several seconds. What is the current in the inductor immediately after the switch is closed? What is the final current in the inductor? How much time is required for the current to reach 95% of its final value?

50. ◆ You have a coil with a resistance of 25 Ω and an inductance of 75 mH. What size capacitor should you connect in series with it to make a critically damped circuit?

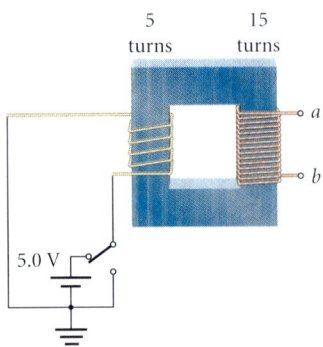

■ FIGURE 31.30

51. ◆◆ The primary coil of the transformer shown in ■ Figure 31.30 is connected to a switching circuit that switches the coil between ground and the positive terminal of a 5.0-V battery at a frequency of 4.0 Hz. Make a sketch showing how the potential difference V_{ab} across the ends of the secondary coil varies in time. How is the behavior affected by the self-inductance of the transformer?

52. ◆◆ An *LRC* circuit contains a 15-μF capacitor and a 23-mH inductor. What resistance is necessary for the circuit to be critically damped ($\omega_1 = 0$)? If the resistance is half this value, find the frequency and time constant for the circuit.

53. ◆◆ You have two capacitors, $C = 5.6\ \mu$F and $C = 2.2\ \mu$F, and two inductors, $L = 22$ mH and $L = 56$ mH. How should you connect them to obtain a circuit that oscillates with a 2.2-ms period?

54. ◆◆ A current I_0 flows initially in a series *LR* circuit. Show that the energy dissipated by the resistor equals the energy initially stored in the inductor.

55. ◆◆ The switch is closed at $t = 0$ in a circuit containing a battery, resistor, and inductor in series. Find the current in the circuit as a function of time. Show that at any time $t > 0$, the energy dissipated by the resistor plus the energy stored in the inductor equals the work done by the battery since $t = 0$.

56. ◆◆ A 56-μF capacitor is connected in series with a 0.30-H inductor. The system oscillates with a total energy of 27 μJ. Compute the maximum values of the capacitor voltage and of inductor current. If the volume within the inductor is 1.9×10^{-4} m^3, estimate the maximum value of the magnetic field in the inductor. If the maximum value of the electric field is 1.0×10^6 V/m and the capacitor is air-filled, find its dimensions.

57. ◆◆ Resistivity of the wire ensures that a real solenoid has both inductance and resistance. If the solenoid is tightly wound with a single layer of wire of length ℓ, show that its *LR* time constant depends only on the resistivity and diameter of the wire and the radius of the coil.

58. ◆◆ A circuit contains a known resistance of 2.7 kΩ, and an unknown inductance. When a 12-V battery is connected, what is the current in the circuit a long time after the switch is closed? If it takes 3.7 ms for the current to reach 95% of this value, what is the inductance in the circuit?

59. ◆◆◆ ■ Figure 31.31 represents the spark coil circuit in an automobile engine. Compare with Figure 31.16. The additional capacitor prevents unwanted sparking across the points. Assume that the switch has been closed for a long time. What are the capacitor charge and the current through the coil? Find the potential difference across the coil as a function of time after the switch is opened but before the plug sparks. What is the maximum voltage achieved? (See also Problem 67.)

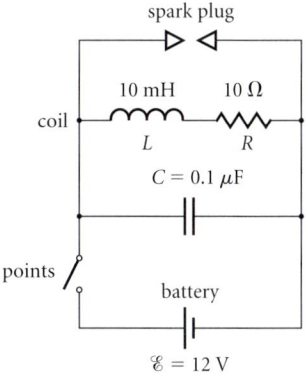

FIGURE 31.31

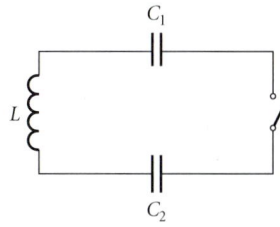

FIGURE 31.32

60. ♦♦♦ There is a charge Q_0 on capacitor C_1 in ■ Figure 31.32, and capacitor C_2 is uncharged. At $t = 0$, the switch is closed. What is the current in the circuit immediately after the switch is closed, and a long time after the switch is closed? Find the current in the circuit and the two charges as functions of time for $t > 0$.

61. ♦♦♦ In ■ Figure 31.33, both capacitors are initially uncharged, and all the switches are open. First S_2 is closed, then S_1 is closed. Find the charge on each capacitor. At time $t = 0$, S_1 is opened and S_4 is closed. What is the maximum current through the inductor and when does it occur? At the instant of maximum current, S_2 is opened and S_3 is closed. What is the maximum charge on C_2 and when does it occur?

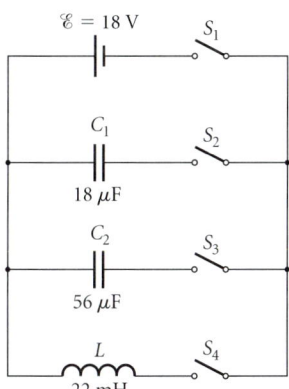

FIGURE 31.33

62. ♦♦♦ A 1.2-kΩ resistor, a 3.3-mH inductor, and a 2.5-nF capacitor are connected in series with a 12-V battery. Find the current in the circuit as a function of time after the switch is closed.

63. ♦♦♦ Assume the resistor in a series LRC circuit is sufficiently small that the solution for $Q(t)$ is of the form $Q(t) = Q_0 a(t)\sin \omega t$, where $a(t)$ changes negligibly during one cycle of the oscillation. Compute the heat dissipated by the resistor during one cycle, treating $a(t)$ as a constant. Then set the energy dissipated equal to the change in the stored energy ($\frac{1}{2}Q_0^2 a^2/C$) to obtain an approximate differential equation for $[a(t)]^2$. Solve the equation, and compare the resulting expression for $a(t)$ with the exact solution found in the text.

64. ♦♦♦ An LRC circuit contains a 5.0-mH inductor, a 0.10-kΩ resistor, and a 3.0-μF capacitor in series. For a long time prior to $t = 0$, a 6.0-V battery has been connected across the capacitor. Find the charge on the capacitor and the current in the circuit. At $t = 0$, the battery is disconnected from the capacitor. Find the charge and current as functions of time for $t > 0$, and give numerical values at $t = 0.10$ ms.

✻ §31.4 MULTILOOP CIRCUITS

65. ❖ Describe the behavior of the circuit in ■ Figure 31.34 a long time after the switch is closed.

66. ❖ If the resistance in ■ Figure 31.34 is doubled, how is the circuit behavior affected?

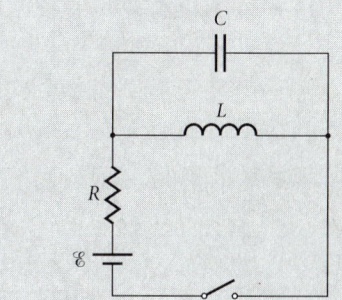

FIGURE 31.34

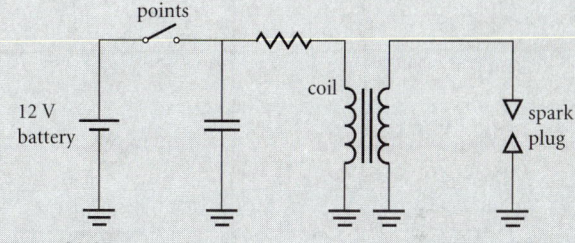

FIGURE 31.35

67. ❖ ■ Figure 31.35 is a yet more accurate version of the automotive spark coil circuit. Explain the purpose of the capacitor and of the transformer. (See also Problem 59.)

68. ♦♦ Two identical capacitors, each in series with a resistor R, are connected in parallel (■ Figure 31.36). The combination is connected across a third resistor R. There is a switch between the capacitors, and a charge Q_0 is on one of the capacitors. At $t = 0$, the switch is closed. What is the final state of the system? Find the charge on each capacitor and the current through the third resistor as functions of time.

69. ♦♦ What is the charge on each of the capacitors in ■ Figure 31.37 when the switch has been closed for a long time? Apply Kirchhoff's rules or an energy method to determine the current supplied by the battery as a function of time after the switch is closed. What happens when the switch is opened again?

70. ♦♦♦ The circuit in Example 31.3 should really be treated as a multiloop circuit. Show that when this is done, the time constant becomes $\tau = \frac{10}{11}RC$. How big is the error in the time between flashes computed in the example?

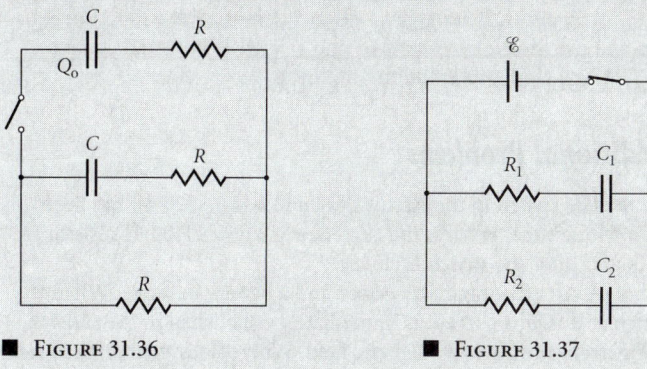

FIGURE 31.36   **FIGURE 31.37**

QUESTIONS AND PROBLEMS   **1013**

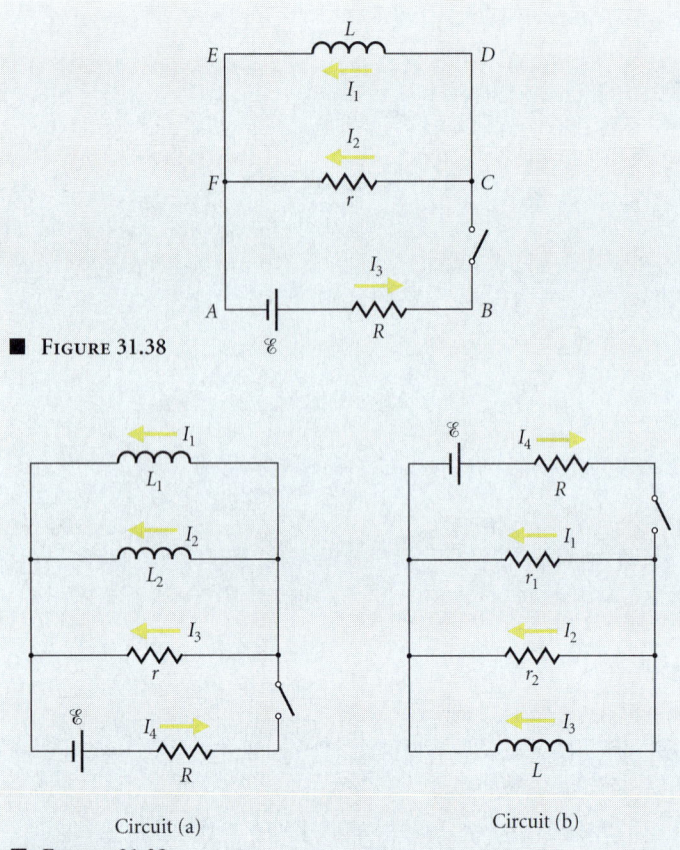

■ FIGURE 31.38

Circuit (a)   Circuit (b)

■ FIGURE 31.39

71. ◆◆◆ Apply Kirchhoff's rules to the junction at C and to loops $ABCF$ and $CDEF$ in ■ Figure 31.38 to obtain equations relating I_1, I_2, I_3, and their derivatives. Manipulate these equations to obtain a single equation of the form:

$$\mathcal{E} - L_T \frac{dI_3}{dt} - R_T I_3 = 0.$$

Comment on the value of the time constant. Solve the equation and find I_3, I_1, and I_2 if the switch is closed at $t = 0$.

72. ◆◆◆ Find the charge on the capacitor in the circuit shown in ■ Figure 31.34 as a function of time after the switch is closed. Show that the time constant is $2RC$.

73. ◆◆◆ For each of the circuits in ■ Figure 31.39, find the currents I_1, I_2, I_3, and I_4 as functions of time after the switch is closed. (*Hint:* For circuit (a), use Kirchhoff's loop rule to express dI_2/dt, dI_1/dt, and dI_4/dt in terms of dI_3/dt and I_3. Then differentiate the junction equation and use these expressions to find a single differential equation for I_3. Use a similar method for circuit (b).)

Additional Problems

74. ❖ The switch in the circuit shown in ■ Figure 31.40 has been open a long time. What is the current through the 100-Ω resistor immediately after the switch is closed?

75. ❖ A circuit contains a resistor and a battery in series. When the switch is closed, a current is immediately established in the resistor. The current produces a magnetic field. Where does the magnetic field energy come from? What does "immediately" really mean in this case?

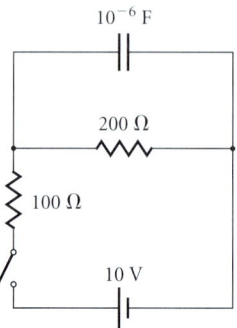

■ FIGURE 31.40

76. ❖ If the resistance in an LR circuit is increased, the time constant decreases. If the resistance in an RC circuit is increased, the time constant increases. Explain the difference.

77. ◆◆ Find the combined inductance of two coils of inductance L_1 and L_2 connected in series if they are close together and have mutual inductance M. Does it make a difference if the two coils wind in the same or opposite senses? Find their combined inductance when they are connected in parallel, again assuming the mutual inductance is M. (Compare with Problem 45.)

78. ◆◆ You have three 220-Ω resistors and a coil with an inductance of 2.5 mH and a resistance of 15 Ω. You wish to construct an LR circuit with a time constant of 15 μs. How may this be done?

79. ◆◆ A circular loop of radius 3.5 cm carries a charge of 2.7 mC/m and spins about an axis through its center at 1500 rpm. What is the total electromagnetic energy density at the center of the loop?

80. ◆◆ A battery of emf $\mathcal{E}$ has been connected for a long time in series with a resistor R and a parallel plate capacitor of capacitance C. What is the charge on the capacitor and the current in the circuit? At time $t = 0$, a slab of dielectric with dielectric constant κ is quickly inserted into the capacitor, so that it fills the space between the plates. What is the charge on the capacitor a long time later? Find the current in the circuit and the charge on the capacitor as functions of time for $t > 0$.

81. ✱ ◆◆ Find the current through the battery in ■ Figure 31.41 as a function of time after the switch is closed.

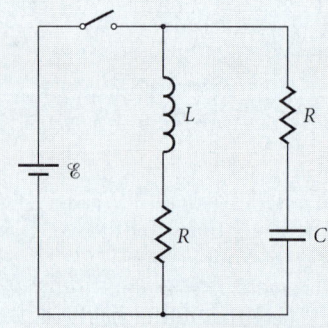

■ FIGURE 31.41

82. ◆◆◆ ✉ A loop of radius R is made of wire with radius r. Make two estimates of the loop's self-inductance. (a) First assume the flux through the loop equals the area of the loop multiplied by the magnetic field strength at the center. Argue that this is a lower limit for the flux. (b) Second, assume the magnetic field in the loop's interior varies with distance from the wire in the same way as for a long straight wire. Do you think this is a better or worse approximation than that in part (a)? Why? (c) Evaluate your results for $r = 0.5$ mm and $R = 3.40$ cm. By what factor do they differ?

83. ✳ ♦♦♦ A variable inductor is made by inserting a ferrite plunger partway into a coil. Find the inductance of the coil in terms of the number of turns N, the radius R, the length ℓ, and the distance d that the plunger is inserted. Assume that the ferrite may be described by an effective permeability μ. (Variable inductors of this type are frequently used in tuning circuits for automobile radios.)

Computer Problems

84. The current in an inductor increases linearly for $t \ll \tau_L$. (See Exercise 31.10.) If $L = 1.0$ H and $R = 1.0\ \Omega$, after how long does the actual current differ from the linear increase by 1%? By 10%?

85. A 56-nF capacitor is connected in series with a 12-V battery and a 1.8-kΩ resistor. At $t = 0$, the switch is closed. Plot the energy stored in the capacitor, the total energy dissipated in the resistor, and the total energy drawn from the battery as functions of time. Verify that energy is conserved throughout the charging process.

86. An LRC circuit has a 2.2-nF capacitor, a 2.2-mH inductor, and a 780-Ω resistor connected in series across a 12-V battery. The switch is closed at $t = 0$. Modifying eqn. (31.29) to account for the different initial conditions, write down the solution for Q as a function of time, and use it to calculate $I(t)$. Using a spreadsheet or another computer program, calculate the potential difference across the capacitor, the inductor, and the resistor every microsecond from $t = 0$ to $t = 10\ \mu s$. Verify that the sum of the three potential differences equals 12 V at each time.

Challenge Problems

87. Any multiloop circuit containing a resistor has a final state in which currents and capacitor charges are constant in time. True or false? Justify your answer.

88. Two LR circuits are coupled by mutual inductance. In one loop, an inductor with $L = 10$ mH is in series with a 100-Ω resistor, a 12-V battery, and a switch. In the other loop, a 20-mH inductor is in series with a 200-Ω resistor. The mutual inductance is 10 mH. Find the current in each loop as a function of time after the switch is closed. What are the values of the two currents at $t = 0.1$ ms?

89. Two LR circuits are coupled by their mutual inductance (■ Figure 31.42). If $\mathcal{E} = 12$ V, $L = 2M = 3.0$ mH, and $R = 1.6$ kΩ, find the

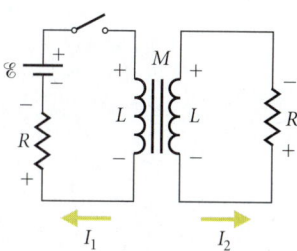

■ FIGURE 31.42

currents in the circuits as functions of time after the switch is closed. Give numerical values for the currents at $t = 1.0\ \mu s$.

90. (a) Calculate the mutual inductance of a small coil carrying current I, with dipole moment $m\hat{k}$, and a wire loop of radius a that lies in the x-y-plane a distance d from the dipole, with its center on the z-axis. [Hint: At distance r from the z-axis, B_z due to the dipole is $B_z = k_m m[-1 + 3z^2/(z^2 + r^2)]/(z^2 + r^2)^{3/2}$.]
(b) Use the result of part (a) to compute the mutual inductance of the dipole and an infinite solenoid with radius a and n turns per meter when the dipole is aligned along the solenoid axis.
(c) Compare your result in part (b) with a calculation of the mutual inductance from the flux produced by the solenoid linking the cross section of the small coil.

91. Analyze the charging and discharging of a capacitor in series with a nonohmic device in which current varies with applied voltage as:

$$I = AV^\alpha.$$

Show that for $\alpha < 1$ the process requires a finite time, while for $\alpha \geq 1$ it requires infinite time. In each case, define a time constant and give its value in terms of α, A, and C.

92. The terminals of a superconducting solenoid are connected by a superconducting wire and, initially, there is a current I in the circuit. Imagine an external force squeezing on the solenoid so as to change its radius an amount dr. No emf can develop between the solenoid's terminals without causing infinite current in the superconducting circuit. Use this fact and Faraday's law to find the change in magnetic field within the solenoid. Compute the change in stored energy and the force per unit area needed to compress the solenoid. Your result gives the pressure exerted by a magnetic field. Also find the force per unit area directly from the formula for magnetic force on a current-carrying wire. Compare with the similar result for electric force on a charged, conducting surface (cf. Part VI, Problem 5).

CHAPTER 32
Introduction to Alternating Current Circuits

Transformers at an electric substation. In the nation's power grid, power is transmitted at high voltage and low current. In our houses, we use power at much lower voltage but a higher current. This arrangement minimizes ohmic losses in the power grid. A transformer is used to effect the transition to lower voltage. Additional transformers on the power pole outside your home complete the transition to 120-V power.

CONCEPTS
Phase shift
Reactance
Impedance
Phasor
Power factor
Resonance
Quality

GOALS
Be able to:
Compute the response of a simple circuit to an AC power source.
Determine the power used in a circuit driven by an oscillating emf.
Use phasors to analyze series and parallel AC circuits.

Such harmony is in immortal souls;
But, whilst this muddy vesture of decay
Doth grossly close it in, we cannot hear it.
WILLIAM SHAKESPEARE

In 1882, Nikola Tesla (cf. Chapter 28) demonstrated the great practical advantages of oscillating electrical voltage for the generation, transmission, and use of electrical energy. For example, the transformer in a power company substation reduces the voltage from several hundred thousand volts used in cross-country transmission to the tens of thousands used for transmission within a city. A time-varying potential is essential to the transformer's operation (cf. §31.2.3). Following Tesla's ideas, electric power is now supplied to domestic and industrial users as alternating current (AC). When you plug an appliance into a wall socket in your home, the current in the circuit oscillates sinusoidally in time. Such diverse appliances as industrial motors, vacuum cleaners, and FM radios use AC power. In this chapter we shall study the response of resistors, capacitors, and inductors to oscillating voltage, and develop the fundamentals of AC circuits.

> ONCE AGAIN WE SHALL MAKE MAJOR USE OF KIRCHHOFF'S RULES (§26.4) IN ANALYZING CIRCUITS. YOU MAY ALSO WISH TO REVIEW THE MATERIAL ON OSCILLATIONS (CHAPTER 14).

32.1 SINGLE-ELEMENT CIRCUITS

32.1.1 Voltage and Current

EMF. The simplest DC circuit contains a battery to provide energy and a resistor to use it. To build the simplest AC circuit, we replace the battery with an oscillating emf. A single-loop generator (§30.2.2) produces a sinusoidally varying potential difference across its terminals as it rotates. It provides a good mental image for practical generators that operate on the same principles and produce an output of similar form:

> DC MEANS DIRECT CURRENT. THE CURRENT IS ALWAYS IN THE SAME DIRECTION.

$$\mathcal{E}(t) = V_0 \cos \omega t. \tag{32.1}$$

■ Figure 32.1 shows the circuit symbol for an ideal AC generator with output given by eqn. (32.1) and with no internal resistance.

> THE USE OF A COSINE FUNCTION RATHER THAN A SINE IN EQN. (32.1) IS RELATED TO OUR CHOICE OF ORIGIN FOR TIME. BECAUSE OF THE REPETITIVE NATURE OF THE OSCILLATIONS, SUCH CHOICES DO NOT AFFECT ANY OF THE RESULTS. OUR CHOICES ARE CONSISTENT WITH THOSE IN CHAPTER 14 (SEE, E.G., EQN. 14.3).

Resistance. The currents in a DC network reach a steady state extremely rapidly, in the time required for the electric field to propagate across the circuit at the speed of light. So long as the field and charge distribution that drive current respond much more rapidly than the applied potential varies, resistors behave the same way in an AC circuit as in a DC circuit. This requirement is satisfied for AC power if light can cross the circuit in less than an oscillation period. With 60-Hz commercial power, the circuit could be the size of Texas! Any network made up only of resistors may be analyzed as if it were a DC circuit but with all values of current and potential difference oscillating sinusoidally.

> SEE THE ANALYSIS IN EXAMPLE 30.1, WHERE WE ALSO DISCUSSED TIME-VARYING FIELDS.

In the circuit diagrammed in ■ Figure 32.2, the generator develops a sinusoidally varying potential difference that drives current in the resistor:

$$I(t) = \frac{\Delta V(t)}{R} = \frac{V_0}{R} \cos \omega t. \tag{32.2}$$

The current varies synchronously with the potential difference: it is *in phase*. Its amplitude is

$$I_0 = V_0/R.$$

■ **FIGURE 32.1**
Circuit diagram symbol for an AC power supply. The symbol illustrates the oscillating form of the potential: $\mathcal{E} = V_0 \cos \omega t$.

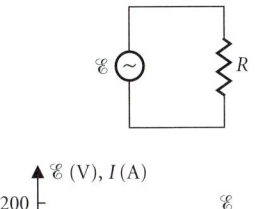

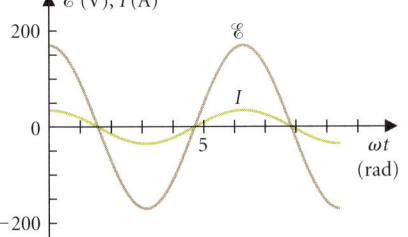

■ **FIGURE 32.2**
Potential difference across and current in a resistor in series with an AC power supply. The current oscillates in phase with the emf. Its amplitude is $I_0 = V_0/R$. The voltage and current are those for the resistance in Exercise 32.1.

Capacitors. Similarly, when an AC generator is connected to a capacitor, the potential difference across the capacitor, and hence the charge on it, varies in phase with the emf:

$$Q(t) = C\,\Delta V(t) = CV_0 \cos \omega t. \tag{32.3}$$

Current in the circuit changes the stored charge (Figure 32.3):

$$I(t) = \frac{dQ}{dt} = CV_0(-\omega)\sin \omega t = \omega CV_0 \cos\left(\omega t + \frac{\pi}{2}\right). \tag{32.4}$$

$\cos(\theta + \pi/2) = -\sin\theta$. See the Math Toolbox on harmonic functions, Chapter 14.

Charge cannot move across a capacitor, so in DC circuits no steady current can exist in a circuit arm containing a capacitor. In AC circuits, however, a capacitor arm can carry oscillating current with constant amplitude $I_0 = \omega CV_0$. The current charges and discharges the capacitor every cycle.

Don't confuse constant amplitude with constant magnitude! The magnitude of the current varies sinusoidally.

Capacitor current is not in phase with the emf. The current is zero when the charge is maximum or minimum, and the current has its maximum value when the charge is zero and is increasing most rapidly. Maximum current occurs in the circuit one-quarter period earlier than the maximum applied voltage (Figure 32.3). We say that the current *leads* the voltage. Equivalently, the voltage follows, or *lags*, the current.

Inductors. When the generator is connected across an inductor (Figure 32.4), the potential difference is related to the rate at which current changes (eqn. 31.10):

$$\Delta V(t) = V_0 \cos \omega t = L\frac{dI}{dt}.$$

To obtain the relation between $I(t)$ and $\Delta V(t)$, we integrate:

$$I(t) = \int \frac{\Delta V(t)}{L}\,dt = \frac{V_0}{L}\int \cos \omega t\, dt = \frac{V_0}{\omega L}\sin \omega t.$$

Since there is no DC current, the integration constant is zero.

$$I(t) = \frac{V_0}{\omega L}\cos\left(\omega t - \frac{\pi}{2}\right). \tag{32.5}$$

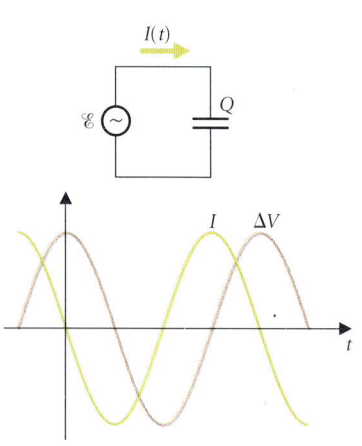

■ **Figure 32.3**
Voltage and current in a capacitor circuit. The current variable is defined as shown in the circuit diagram, so that $I = +dQ/dt$. Current leads the voltage by $\pi/2$ rad (one-quarter period).

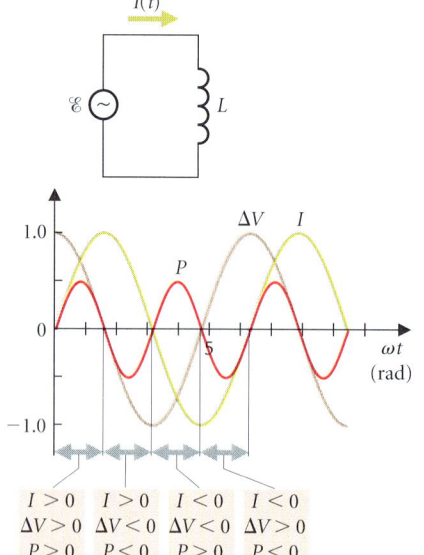

■ **Figure 32.4**
Current and voltage in an inductor circuit. The current lags the voltage by one-quarter period. The power $P(t) = \Delta V(t)I(t)$ oscillates twice as fast. Whenever voltage and current have the *same* sign, positive or negative, the generator provides energy to the circuit. When the two signs are opposite, the circuit returns energy to the generator. With a pure inductance, there is no net energy transfer.

The potential difference is zero when the current is maximum or minimum and has its maximum value when the current is zero and changing most rapidly. Again, there is a phase shift between current and emf, but now the current lags the voltage by $\pi/2$—that is, the voltage leads the current.

32.1.2 Power

Ideal capacitors and inductors store energy in electric or magnetic fields but have no way to convert it to thermal energy. On average, they draw no power from an AC generator. Let's see how the phase shifts between emf and current bring this about.

The instantaneous power input to a circuit element carrying current $I(t)$ is:

$$P(t) = \Delta V(t) I(t).$$

WE USED EQN. (26.3A).

For a resistor in an AC circuit,

$$P(t) = \Delta V(t) \frac{\Delta V(t)}{R} = \frac{V_0^2}{R} \cos^2 \omega t.$$

The average power used by the resistor over a cycle is:

$$P_{av} \equiv \langle P(t) \rangle = \frac{\langle [\Delta V(t)]^2 \rangle}{R} = \frac{V_0^2}{R} \langle \cos^2 \omega t \rangle.$$

Power relations in circuits are usually stated in terms of the *root mean square* (rms) voltage:

$$V_{rms} \equiv \langle [\Delta V(t)]^2 \rangle^{1/2}. \quad (32.6)$$

Then the power used by the resistor circuit is:

$$\langle P(t) \rangle = V_{rms}^2 / R. \quad (32.7)$$

Equations (32.6) and (32.7) hold for an arbitrary variation of emf with time.

To evaluate V_{rms} for sinusoidal variations, we need the average value of $\cos^2 \omega t$:

WE FIRST ENCOUNTERED THIS AVERAGE IN §15.3. SEE ALSO PROBLEM 26.

$$\langle \cos^2 \omega t \rangle = \frac{1}{T} \int_0^T \cos^2 \omega t \, dt = \tfrac{1}{2}.$$

Then:

$$V_{rms} = V_0 / \sqrt{2}, \quad (32.8)$$

and:

$$\langle P(t) \rangle = \tfrac{1}{2} V_0^2 / R.$$

It is the rms voltage, not the voltage amplitude, that is normally used to rate an AC power source. For standard U.S. household power, $V_{rms} = 120$ V. The corresponding voltage amplitude is $\sqrt{2} V_{rms} = 170$ V.

EXERCISE 32.1 ♦ Find the average power input to a 5-Ω room heater coil using commercial power at 60 Hz and $V_{rms} = 120$ V.

The power output of the generator in an inductor circuit is:

$$P(t) = \mathcal{E}(t) I(t) = (V_0 \cos \omega t) \left(\frac{V_0}{\omega L} \sin \omega t \right) = \frac{V_0^2}{2\omega L} \sin(2\omega t).$$

WE USED THE RESULT $\sin(2\theta) = 2 \sin \theta \cos \theta$; SEE APPENDIX IB.

The generator provides energy to the circuit when it increases the potential energy of charge flowing through it. With our sign conventions, this occurs when $\mathcal{E}$ and I have the same sign, either both positive or both negative (Figure 32.4). When the signs are opposite, energy returns to the generator. Thus the power output of the generator oscillates at twice the frequency of the potential difference. Energy is drawn from the generator and is stored in the inductor, but the energy flows back into the generator a quarter cycle later. In a capacitor circuit, current leads the voltage rather than lags, but the power it draws from the generator varies in a similar way. In both cases, the generator's power output averages to zero.

RECALL OUR DISCUSSION OF BATTERY CHARGING IN §26.4, ESPECIALLY EXAMPLE 26.12.

EXERCISE 32.2 ♦♦ Show that the power output of the generator in a capacitor circuit is also proportional to $\sin(2\omega t)$.

32.1.3 Reactance and Phase Shift

The current in an AC circuit has a *phase shift* ϕ relative to the emf driving it:

$$\mathcal{E}(t) = V_0 \cos \omega t \quad \text{and} \quad I(t) = I_0 \cos(\omega t - \phi). \tag{32.9}$$

With this notation, the phase shift is positive when voltage leads current, as in an inductor, and negative when voltage lags current, as in a capacitor.

Compare eqns. (32.2), (32.4), and (32.5). Remember: With a single inductor or capacitor, the peak current occurs a quarter cycle earlier or later than the peak voltage.

Like resistors, where $V_0 = I_0 R$, capacitors and inductors draw AC current with an amplitude proportional to the applied voltage amplitude. Because of the phase shift between voltage and current in capacitors or inductors and the corresponding absence of power dissipation associated with resistance, the constant of proportionality X in the relation $V_0 = I_0 X$ is given a different name: *reactance*.

> The magnitude of the *reactance* $|X|$ of a capacitor or inductor in an AC circuit is the ratio of the applied voltage amplitude to the resulting current amplitude.
>
> $$|X| \equiv V_0/I_0. \tag{32.10}$$

For an inductor, the reactance (eqn. 32.5) is:

$$X_L = \omega L. \tag{32.11}$$

In advanced work, the reactance is often given a sign that describes the phase shift between the potential difference and the current. We shall not use this convention here, but we'll use absolute-value signs on X_C. Then all our expressions are correct, no matter which convention is used.

For a capacitor, the reactance (eqn. 32.4) is:

$$|X_C| = 1/(\omega C). \tag{32.12}$$

Since reactance has the same physical dimensions as resistance, it is also measured in ohms. Values of X and ϕ for single-element circuits are summarized in ● Table 32.1.

Unlike resistance, reactance depends on the frequency of the applied AC potential. At very low frequencies, flux in an inductor changes slowly, induced electric fields are unimportant, and $X_L \to 0$. At high frequencies, rapidly changing flux produces large induced fields and $X_L \to \infty$. The current required to charge and discharge a capacitor approaches zero at low frequency and $|X_C| \to \infty$. At high frequency, a large current is required to change the charge rapidly, and $X_C \to 0$.

EXERCISE 32.3 ♦ Show that $1 \text{ s/F} = 1 \, \Omega$ and $1 \text{ H/s} = 1 \, \Omega$.

TABLE 32.1 **Summary of Phase Relations in a Single-Element Circuit**

Generator voltage $\mathcal{E} = V_0 \cos \omega t$

Element	Current	Resistance/Reactance	Phase Shift
Resistor	$I = \mathcal{E}/R$	R	0
Capacitor	$I = \omega C V_0 \cos(\omega t + \pi/2)$	$1/\omega C$	$-\pi/2$
Inductor	$I = (V_0/\omega L) \cos(\omega t - \pi/2)$	ωL	$+\pi/2$

EXAMPLE 32.1 ♦ What is the reactance of a 3.5-μF capacitor used in a 60.0-Hz AC circuit? What current amplitude results when the capacitor is connected to a 120-V (rms) power supply?

MODEL We may use relations (32.10) and (32.12).

SETUP The angular frequency ω corresponding to $f = 60$ Hz is $\omega = 2\pi f$.

SOLVE The reactance of the capacitor is:

$$|X_C| = \frac{1}{\omega C} = \frac{1}{2\pi(60 \text{ Hz})(3.5 \times 10^{-6} \text{ F})} = 760 \text{ }\Omega.$$

The voltage amplitude is $V_0 = \sqrt{2}\,(120 \text{ V}) = 170$ V. The current amplitude is:

$$I_0 = V_0/|X_C| = (170 \text{ V})/(760 \text{ }\Omega) = 0.22 \text{ A}.$$

ANALYZE Remember that the current charges and discharges the capacitor. The maximum rate of charge is 0.22 C/s. ∎

32.2 Two-Component Circuits

Analysis of an AC circuit network follows essentially the same methods we used for DC circuits. In each case, we expect a current oscillating at the same frequency as the generator and phase shifted with respect to the emf. In this section we'll look at series connections of two circuit elements.

32.2.1 Steady State Response

An RC Circuit. ∎ Figure 32.5 is a circuit diagram for a capacitor and resistor in series with an AC generator. To analyze the circuit, we use Kirchhoff's loop rule. With I and Q defined as shown in the figure, and proceeding clockwise, we have:

$$V_0 \cos \omega t - I(t)R - \frac{Q(t)}{C} = 0.$$

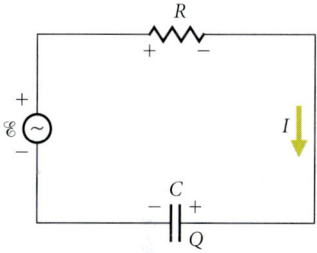

∎ **FIGURE 32.5**
An *RC* circuit. The current that charges and discharges the capacitor leads the generator voltage by a phase less than 90°.

The results of §32.1 guide us in finding a solution to this equation. The current oscillates with the same frequency as the applied AC voltage but may be out of phase. The resistor alone would give a current in phase with the generator; the capacitor alone would give a current out of phase by $\pi/2$. We expect a combination of both behaviors. Thus we assume the current has the form:

$$I(t) = I_0 \cos(\omega t - \phi).$$

With current and charge defined as in the figure, current increases stored charge, $I = +dQ/dt$, and a similar expression holds for $Q(t)$:

$$Q(t) = Q_0 \sin(\omega t - \phi),$$

with

$$I_0 = \omega Q_0.$$

Substituting this assumed form into the equation relating I, Q, and $\mathcal{E}$:

$$V_0 \cos \omega t - RI_0 \cos(\omega t - \phi) - \frac{I_0}{\omega C} \sin(\omega t - \phi) = 0.$$

This equation must be satisfied at *all* times. In particular, at $t = \phi/\omega$, $\cos(\omega t - \phi) = 1$, $\sin(\omega t - \phi) = 0$, and the equation reduces to:

$$V_0 \cos \phi - RI_0 = 0. \qquad \text{(i)}$$

SEE §31.3.3, WHERE WE ARGUED THAT THE SINE AND COSINE TERMS VANISH SEPARATELY.

Similarly, at $t = (\phi + \pi/2)/\omega$, $\cos(\omega t - \phi) = 0$, $\sin(\omega t - \phi) = 1$, and the equation becomes:

$$-V_0 \sin \phi - I_0/(\omega C) = 0. \qquad \text{(ii)}$$

Dividing eqns. (i) and (ii), we obtain:

$$\tan \phi = -1/(\omega RC) = -|X_C|/R. \quad (32.13)$$

COMPARE THE DISCUSSION IN §32.1.3, FOLLOWING EQN. (32.9).

The phase shift is negative and has a magnitude less than 90°. A negative phase shift means that the voltage lags the current. Similarly, squaring and adding, we find I_0:

$$V_0^2 = V_0^2 \cos^2 \phi + V_0^2 \sin^2 \phi = (RI_0)^2 + I_0^2/(\omega C)^2.$$

Thus:

$$I_0 = \frac{V_0}{\sqrt{R^2 + 1/(\omega C)^2}} = \frac{V_0}{\sqrt{R^2 + X_C^2}}.$$

The quantity

$$Z = \sqrt{R^2 + X_C^2} \quad (32.14)$$

is called the *impedance* of the circuit. It expresses the combined effect of resistance and reactance on current amplitude. In terms of impedance, the current is:

$$I(t) = \frac{V_0}{Z} \cos(\omega t - \phi), \quad (32.15)$$

and from eqn. (i), we also have:

$$\cos \phi = R/Z. \quad (32.16)$$

The effect of adding a resistor to the circuit is to reduce both the current amplitude and the magnitude of the phase shift we'd get with a capacitor alone. The sign of the phase shift is not changed.

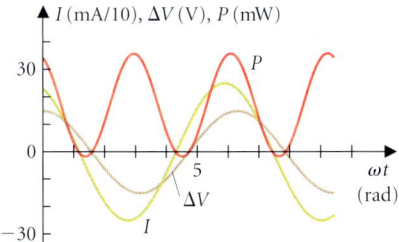

■ FIGURE 32.6
In this *RC* circuit, voltage lags current by 24°. The power $P(t) = \Delta V(t)I(t)$ oscillates through mostly positive values. On average, the generator transfers energy to the circuit to become thermal energy in the resistor.

REMEMBER: DON'T ROUND INTERMEDIATE ANSWERS UNTIL THE END OF THE CALCULATION.

EXAMPLE 32.2 ♦ An *RC* circuit contains a 0.82-μF capacitor in series with a 5.6-kΩ resistor and an AC generator with angular frequency $\omega = 5.0 \times 10^2$ rad/s and a peak voltage of 15 V. Find the amplitude of the current in the circuit and the phase shift with respect to the applied voltage.

MODEL We apply relations (32.13) through (32.15).

SETUP The reactance of the capacitor is:

$$|X_C| = 1/\omega C = 1/[(500 \text{ rad/s})(0.82 \times 10^{-6} \text{ F})] = 2.44 \text{ k}\Omega.$$

The impedance in the circuit is:

$$Z = \sqrt{R^2 + X_C^2} = \sqrt{(5.6 \text{ k}\Omega)^2 + (2.44 \text{ k}\Omega)^2} = 6.11 \text{ k}\Omega.$$

SOLVE The current amplitude and phase shift are:

$$I_0 = \frac{V_0}{Z} = \frac{15 \text{ V}}{6.11 \text{ k}\Omega} = 2.5 \text{ mA}.$$

and $$\phi = \tan^{-1}\left(\frac{-|X_C|}{R}\right) = \tan^{-1}\left(\frac{-2.44 \text{ k}\Omega}{5.6 \text{ k}\Omega}\right) = \tan^{-1}(-0.44) = -24°.$$

ANALYZE The voltage lags the current by 24° (see ■ Figure 32.6). ■

An LR Circuit. Next we apply Kirchhoff's loop rule to an *LR* circuit with an AC generator to obtain similar relations for current amplitude and phase. Traversing the loop clockwise in ■ Figure 32.7, we find:

$$V_0 \cos \omega t - L\frac{dI}{dt} - IR = 0.$$

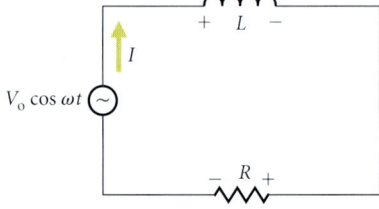

■ FIGURE 32.7
An *LR* circuit. The current lags the voltage by less than 90°.

Again, we expect the current to be of the form:

$$I(t) = I_0 \cos(\omega t - \phi).$$

Substituting this assumed form into the equation above, we have:

$$V_0 \cos \omega t - L[-\omega I_0 \sin(\omega t - \phi)] - I_0 R \cos(\omega t - \phi) = 0.$$

Setting $t = \phi/\omega$:

$$V_0 \cos \phi - I_0 R = 0. \tag{i}$$

And, setting $t = (\phi + \pi/2)/\omega$:

$$-V_0 \sin \phi + \omega L I_0 = 0. \tag{ii}$$

Dividing eqns. (i) and (ii) gives:

$$\boxed{\tan \phi = \omega L/R = X_L/R.} \tag{32.17}$$

Squaring and adding:

$$V_0^2 = I_0^2[R^2 + (\omega L)^2] \Rightarrow I_0 = V_0/\sqrt{R^2 + (\omega L)^2}.$$

The complete expression for the current is

$$\boxed{I(t) = \frac{V_0}{Z} \cos(\omega t - \phi),} \tag{32.18}$$

where the *impedance* of the circuit is:

$$\boxed{Z = \sqrt{R^2 + X_L^2}.} \tag{32.19}$$

The voltage leads the current ($\phi > 0$) by less than 90°. Equations (32.17) through (32.19) differ from the results for the *RC* circuit (eqns. 32.13–32.15) only in the sign of the reactance terms. Once again, with resistance in the circuit, the magnitude of the phase shift is reduced, but its sign is not changed.

The impedance of an *LR* circuit is dominated by resistance at low frequencies, $\omega \ll R/L$. At high frequencies, $\omega \gg R/L$, the impedance is dominated by inductance and grows approximately linearly with frequency (■ Figure 32.8). This variation is useful for filtering electrical signals that have a mixture of frequencies. For example, stereo speaker systems usually combine low-frequency "woofers" with high-frequency "tweeters." An *LR* combination in series with the woofer selectively passes the portion of the signal the woofer is designed to reproduce well.

THEY ARE IDENTICAL IF CAPACITIVE REACTANCE IS DEFINED TO BE NEGATIVE.

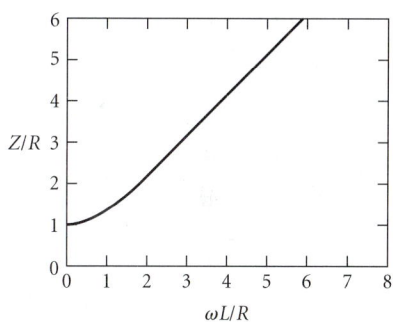

■ **FIGURE 32.8**
Impedance $Z = R\sqrt{1 + (\omega L/R)^2}$ of an *LR* circuit as a function of angular frequency. The impedance is dominated by the resistance when $\omega \ll R/L$, and by the inductance when $\omega \gg R/L$. Thus Z grows linearly with frequency when $\omega \gg R/L$.

EXAMPLE 32.3 ♦♦ The filter for a woofer is designed to attenuate (i.e., reduce the amplitude of) signals by a factor of 5.0 at 5.0 kHz. If the resistance in the circuit is 8.0 Ω, what inductance is required?

MODEL To attenuate by a factor of 5, the impedance at 5 kHz should be five times the value at zero frequency, where only the resistor contributes.

SETUP Thus the impedance at 5.0 kHz should be $5R = 5.0(8.0\ \Omega) = 40.0\ \Omega$. Using eqn. (32.19), and converting f to $\omega = 2\pi f$:

$$Z^2 = R^2 + X_L^2 = R^2 + (\omega L)^2$$
$$= (8.0\ \Omega)^2 + [(2\pi)(5.0 \times 10^3\ \text{Hz})L]^2.$$

SOLVE We set $Z = 40.0\ \Omega$ and solve for L:

$$L^2 = \frac{(40.0\ \Omega)^2 - (8.0\ \Omega)^2}{[2\pi(5.0 \times 10^3\ \text{Hz})]^2} = 1.6 \times 10^{-6}\ \text{H}^2,$$

and $L = 1.2$ mH.

ANALYZE This value of inductance is readily available at an electrical supply store. Because the ear is insensitive to the relative phase of sounds at different frequencies, the phase shift of the woofer signal is unimportant. ∎

32.2.2 Power

The power drawn from the generator in an *LR* or *RC* circuit is:

$$P(t) = \mathcal{E}(t)I(t) = V_0 \cos \omega t \left[\frac{V_0}{Z} \cos(\omega t - \phi) \right].$$

WE USED EQN. (32.18) FOR $I(t)$.

The average power output is:

$$\langle P(t) \rangle = \frac{V_0^2}{Z} \langle \cos \omega t \cos(\omega t - \phi) \rangle.$$

REMEMBER: The notation $\langle x \rangle$ means the average value of x; see Appendix IIC.

Expanding $\cos(\omega t - \phi)$ (see Appendix IB) gives:

$$\langle \cos \omega t \cos(\omega t - \phi) \rangle = \langle \cos \omega t (\cos \omega t \cos \phi + \sin \omega t \sin \phi) \rangle$$
$$= \cos \phi \langle \cos^2 \omega t \rangle + \sin \phi \langle \cos \omega t \sin \omega t \rangle.$$

WE HAVE USED THIS TRIGONOMETRIC IDENTITY BEFORE, IN CHAPTER 14, FOR EXAMPLE.

The average value of $\cos^2 \omega t = \frac{1}{2}$, but $\cos \omega t \sin \omega t$ averages to zero, so:

$$\langle P(t) \rangle = \frac{V_0^2}{Z} [\tfrac{1}{2} \cos \phi + (0) \sin \phi].$$

$$\langle P(t) \rangle = \frac{V_0^2}{2Z} \cos \phi = \frac{V_{rms}^2}{Z} \cos \phi. \tag{32.20}$$

In an AC circuit, impedance plays the same role as resistance in a DC circuit. The *power factor*, $\cos \phi$, arises because current out of phase with the emf does not contribute to the average power.

$$\text{DC circuit: } P = \frac{V^2}{R}; \qquad \text{AC circuit: } \langle P \rangle = \frac{V_{rms}^2}{Z} \cos \phi.$$

The power factor is $\cos \phi = R/Z$ (eqn. 32.16). No net energy is used unless the circuit contains resistance.

EXAMPLE 32.4 ◆ Find the power used by an *LR* circuit with $L = 10.0$ mH and $R = 22\ \Omega$ when a 60.0-Hz, 120-V power supply is applied.

MODEL We apply eqn. (32.20). Don't forget to convert f to ω.

SETUP The impedance in the circuit is given by:

$$Z^2 = R^2 + (\omega L)^2 = (22\ \Omega)^2 + [2\pi(60.0\ \text{Hz})(0.010\ \text{H})]^2 = 498\ \Omega^2.$$

(Notice that we don't need to take the square root. The next step involves Z^2, not Z.)

SOLVE The rms voltage is 120 V, so:

$$\langle P \rangle = \frac{V_{rms}^2}{Z} \left(\frac{R}{Z} \right) = \frac{(120\ \text{V})^2}{498\ \Omega^2} (22\ \Omega) = 640\ \text{W}.$$

ANALYZE For comparison, the power dissipated in a DC circuit with 120 V applied to a 22-Ω resistor would be $\mathcal{E}^2/R = (120\ \text{V})^2/(22\ \Omega) = 650$ W. ∎

Reactance can be a significant nuisance in an electrically powered facility. Current out of phase with the applied voltage delivers no energy to the facility but results in costly resistive losses in the power company's distribution system. The remedy is to reduce the reactance.

An inductor and a capacitor each draw current proportional to $\sin \omega t$ (eqns. 32.4 and 32.5). Thus the current through a series combination of *L* and *C* is $I = I_0 \sin \omega t$, and the potential difference across the combination is:

$$\Delta V = \Delta V_L + \Delta V_C = I_0 \omega L \cos(\omega t) - \frac{I_0}{\omega C} \cos(\omega t)$$

SOME REFERENCES DEFINE X_C TO BE NEGATIVE: $X_C = -|X_C|$. THEN $X_{series} = X_L + X_C$ AND REACTANCES COMBINE LIKE RESISTANCES.

$$= I_0 \left(\omega L - \frac{1}{\omega C} \right) \cos(\omega t) = I_0 (X_L - |X_C|) \cos(\omega t) = V_0 \cos \omega t.$$

Comparing this with eqn. (32.10), we conclude that the reactance of a series combination equals the inductive reactance minus the capacitive reactance. Since the motors in a factory usually have inductive reactance, a capacitor bank installed in series reduces the total reactance and the corresponding phase shift.

32.2.3 Transient Response

The exponential behavior of the *LR* and *RC* circuits that we studied in Chapter 31 is also present in AC circuits. The general solution for the current is:

$$I(t) = (V_0/Z)\cos(\omega t - \phi) + A e^{-t/\tau}. \tag{32.21}$$

The exponential term is called the *transient* solution because its magnitude decreases with time, ultimately becoming insignificant compared with the first term (the *steady state* response). In nearly all practical applications, the steady state response is the important result. In what follows we shall neglect transient behavior (but see Problem 38).

32.3 CIRCUIT ANALYSIS USING PHASORS
32.3.1 Phasors

In Chapter 14 we discovered that uniform circular motion and simple harmonic motion are closely related. We may describe an oscillating object's position with one component of a position vector that rotates uniformly around a circle. We can also use this analogy in AC circuits; all currents and potentials vary sinusoidally with the same frequency ω, and each may be represented by a rotating vector, or *phasor*.

The phasor diagram shown in ■ Figure 32.9 represents the relation between current and voltage for a capacitor. The voltage phasor $\mathcal{V}$ has magnitude V_0 and makes an angle ωt with the x-axis. The actual potential difference across the capacitor is $V(t) = V_0 \cos \omega t = |\mathcal{V}|\cos \omega t$, the x-component of the phasor. The y-component of the phasor has no physical meaning. Similarly, the current phasor $\mathcal{I}$ makes an angle $\chi = \omega t + \pi/2$ with the x-axis, and the actual current is the x-component: $I(t) = I_0 \cos(\omega t + \pi/2) = |\mathcal{I}|\cos \chi$. The magnitudes of the current and voltage phasors are related by the capacitive reactance:

$$|\mathcal{V}| = |X_C||\mathcal{I}| = |\mathcal{I}|/(\omega C). \tag{32.22}$$

As time proceeds, both $\mathcal{I}$ and $\mathcal{V}$ rotate counterclockwise with angular frequency ω.

EXERCISE 32.4 ♦♦ Draw a diagram similar to Figure 32.9 showing phasors representing the voltage and current for an inductor. (*Hint:* The current lags the voltage, eqn. 32.5.)

32.3.2 Phasor Representation of a Series Circuit

We may illustrate the phasor method using a series *RC* circuit. In a series circuit, the current is the same in all elements, so the phasor diagram (■ Figure 32.10) has a single current phasor. Since the diagram rotates with time, we may draw the current phasor at any orientation. It is convenient to put it along the x-axis.

There are three voltage phasors, corresponding to the potential differences across the resistor, the capacitor, and the generator. The voltage across the resistor has the same phase as the current and is represented by a phasor parallel to the current phasor. The capacitor voltage lags the current by 90° and is represented by a phasor at right angles to the current phasor. The capacitor voltage phasor lies behind (lags) the current phasor as they both rotate counterclockwise in the diagram.

According to Kirchhoff's loop rule, the sum of the potential differences around the circuit is zero:

$$V_0 \cos \omega t - IR - Q/C = 0 \quad \Rightarrow \quad V_0 \cos \omega t = IR + Q/C.$$

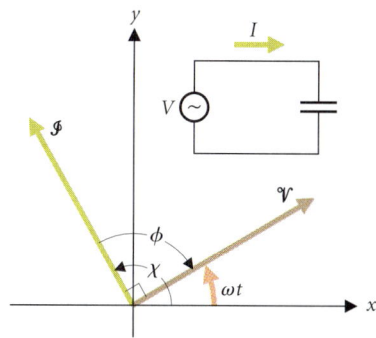

■ FIGURE 32.9
Voltage and current phasors for a capacitor. Both phasors rotate counterclockwise at angular frequency ω with a fixed angle between them. The current phasor precedes the voltage phasor as they rotate: the current *leads* the voltage by 90°. The measured value of current at any time *t* equals the x-component of the current phasor. The angle ϕ is drawn from the current phasor to the voltage phasor, with the counterclockwise direction taken as positive. In this diagram, ϕ is negative. (Compare with eqn. 32.4 and Table 32.1.)

WE USED A SIMILAR DEVICE WHEN STUDYING INTERFERENCE; SEE §17.4.2.

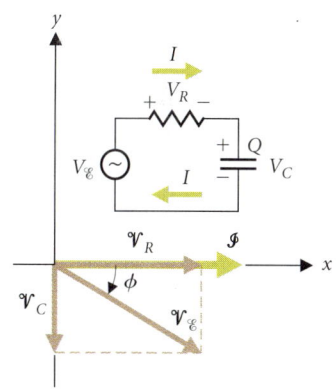

■ FIGURE 32.10
Phasor diagram for a series *RC* circuit. We may draw the phasor diagram at any orientation. It is convenient to put the single current phasor along the x-axis. The voltage across the resistor is in phase with the current, so it also lies along the x-axis. The voltage across the capacitor lags the current by 90° (see Figure 32.9), so it lies along the negative y-axis. By Kirchhoff's loop rule, the generator voltage is the phasor sum of the resistor and capacitor voltages. It lags the current by angle $|\phi|$, or, equivalently, the current leads the voltage by $|\phi|$. (*Remember:* ϕ is drawn from the current phasor to the voltage phasor. In this diagram, ϕ is negative.)

REMEMBER: If $\vec{a} + \vec{b} = \vec{c}$, then $a_x + b_x = c_x$.

Since each potential difference is represented by the x-component of the corresponding phasor, the *vector* sum of the resistor and capacitor voltage phasors equals the generator voltage phasor.

> The total potential difference across a series combination is the phasor sum of the individual voltage phasors.

For the resistor and capacitor, the magnitude of the phasor sum is:

$$|\mathcal{V}_\mathcal{E}| = \sqrt{|\mathcal{V}_R|^2 + |\mathcal{V}_C|^2} = \sqrt{R^2|\mathcal{I}|^2 + X_C^2|\mathcal{I}|^2} = |\mathcal{I}|\sqrt{R^2 + X_C^2} = |\mathcal{I}|Z.$$

THIS IS THE SAME CONVENTION DESCRIBED ALGEBRAICALLY IN §32.1.3.

By convention, the phase shift is represented by the angle *from* the current phasor *to* the voltage phasor, with the counterclockwise direction taken as positive. When the voltage lags the current, as in this case, the phase shift is negative. Its magnitude is given by:

$$\tan|\phi| = \frac{|\mathcal{V}_C|}{|\mathcal{V}_R|} = \frac{|X_C|}{R}.$$

The current leads the voltage. We have retrieved eqns. (32.13) through (32.15).

> **EXAMPLE 32.5** ◆◆ A circuit driven harmonically at 60.0 Hz contains a 10.0-μF capacitor and a 0.500-kΩ resistor. What is the current in the circuit if $V_0 = 156$ V?
>
> **MODEL** The phasor diagram (Figure 32.10) shows the relations we need. The voltage lags the current.
>
> **SETUP** The angular frequency $\omega = 2\pi f$. The capacitive reactance is:
>
> $$|X_C| = \frac{1}{\omega C} = \frac{1}{2\pi(60.0 \text{ Hz})(10.0 \ \mu\text{F})} = 265 \ \Omega.$$
>
> The total impedance is:
>
> $$Z = \sqrt{R^2 + X_C^2} = \sqrt{(500 \ \Omega)^2 + (265 \ \Omega)^2} = 566 \ \Omega.$$
>
> **SOLVE** The current amplitude equals the magnitude of the current phasor:
>
> $$I_0 = |\mathcal{I}| = \frac{|\mathcal{V}|}{Z} = \frac{156 \text{ V}}{566 \ \Omega} = 0.276 \text{ A}.$$
>
> The phase angle ϕ is negative. Its magnitude is given by:
>
> $$\tan|\phi| = \frac{|X_C|}{R} = \frac{265 \ \Omega}{500 \ \Omega} = 0.530 \Rightarrow |\phi| = 27.9°.$$
>
> **ANALYZE** We might write our answer as $I = (0.276 \text{ A}) \cos(120\pi t + 0.155\pi)$, where we have converted ϕ to radians. ■ Figure 32.11 is a graph of this function. The voltage lags the current by $0.155\pi/2\pi = 0.08$ of a cycle. The voltage phasors for the resistor and capacitor have magnitude $|\mathcal{V}_R| = |\mathcal{I}|R = (0.276 \text{ A})(500 \ \Omega) = 138$ V and $|\mathcal{V}_C| = |\mathcal{I}||X_C| = (0.276 \text{ A})(265 \ \Omega) = 73$ V. These values were used in Figure 32.10.

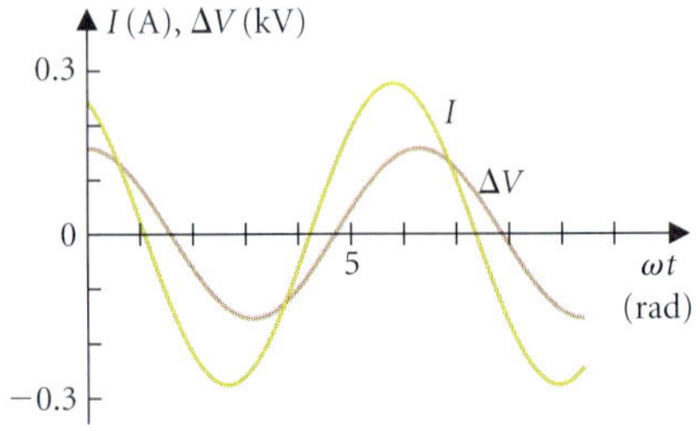

■ **FIGURE 32.11**
Voltage and current as a function of time. The current leads the voltage by 28° = 0.155π rad = 0.08 cycle.

> **EXERCISE 32.5** ◆◆ Draw a phasor diagram for a series LR circuit and use it to obtain eqns. (32.17) through (32.19) for the current amplitude and phase.

The expression for average power used by an AC circuit (eqn. 32.20),

$$\langle P(t) \rangle = \tfrac{1}{2} V_0 I_0 \cos \phi,$$

has an elegant form as a dot product of phasors:

$$\langle P(t) \rangle = \tfrac{1}{2} \mathcal{V} \cdot \mathcal{I}. \qquad (32.23)$$

> **EXERCISE 32.6** ◆ What is the time-averaged power used in the *RC* circuit of Example 32.5?

32.3.3 Phasor Representation of a Parallel Circuit

When circuit elements are connected in parallel, the potential difference across each element is the same. For example, in a parallel combination of a resistor and an inductor, there is one voltage phasor, $\mathcal{V}_R = \mathcal{V}_L = \mathcal{V}_\mathcal{E}$. In ■ Figure 32.12, we have drawn the phasor diagram, with the voltage phasor along the x-axis for convenience. The resistor current is in phase with the voltage, while the inductor current lags the inductor voltage by 90°.

> According to Kirchhoff's junction rule, the total current through a parallel combination is the phasor sum of the individual current phasors.

The total current is the vector sum of the individual current phasors $\mathcal{I}_R$ and $\mathcal{I}_L$. The magnitude of $\mathcal{I}$ is:

$$|\mathcal{I}| = \sqrt{|\mathcal{I}_R|^2 + |\mathcal{I}_L|^2} = |\mathcal{V}_\mathcal{E}|\sqrt{(1/R)^2 + (1/X_L)^2}.$$

The voltage leads the current by a phase shift ϕ, where: $\tan\phi = \dfrac{|\mathcal{I}_L|}{|\mathcal{I}_R|} = \dfrac{|\mathcal{V}_\mathcal{E}|/X_L}{|\mathcal{V}_\mathcal{E}|/R} = \dfrac{R}{X_L}$.

As with any circuit, Kirchhoff's rules form the basis of our solution method (■ Figure 32.13). When analyzing an AC circuit, the voltages and currents are represented by phasors and we add them like vectors. In the following example we show how to use the method to analyze a circuit that has both a series and a parallel combination of elements.

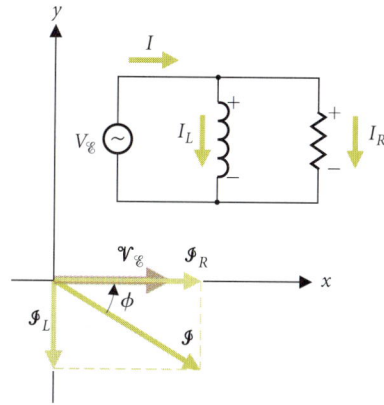

■ **Figure 32.12**
Phasor diagram for a parallel combination of inductor and resistor. The generator and the two circuit elements in parallel have a common voltage phasor, here drawn along the x-axis. According to Kirchhoff's junction rule, the current phasor $\mathcal{I}$ equals the sum of the phasors $\mathcal{I}_L$ and $\mathcal{I}_R$. The resistor current is in phase with the voltage, so we draw it along the x-axis. The inductor current lags the voltage by 90° (cf. Exercise 32.4). The total current lags the voltage by angle ϕ.

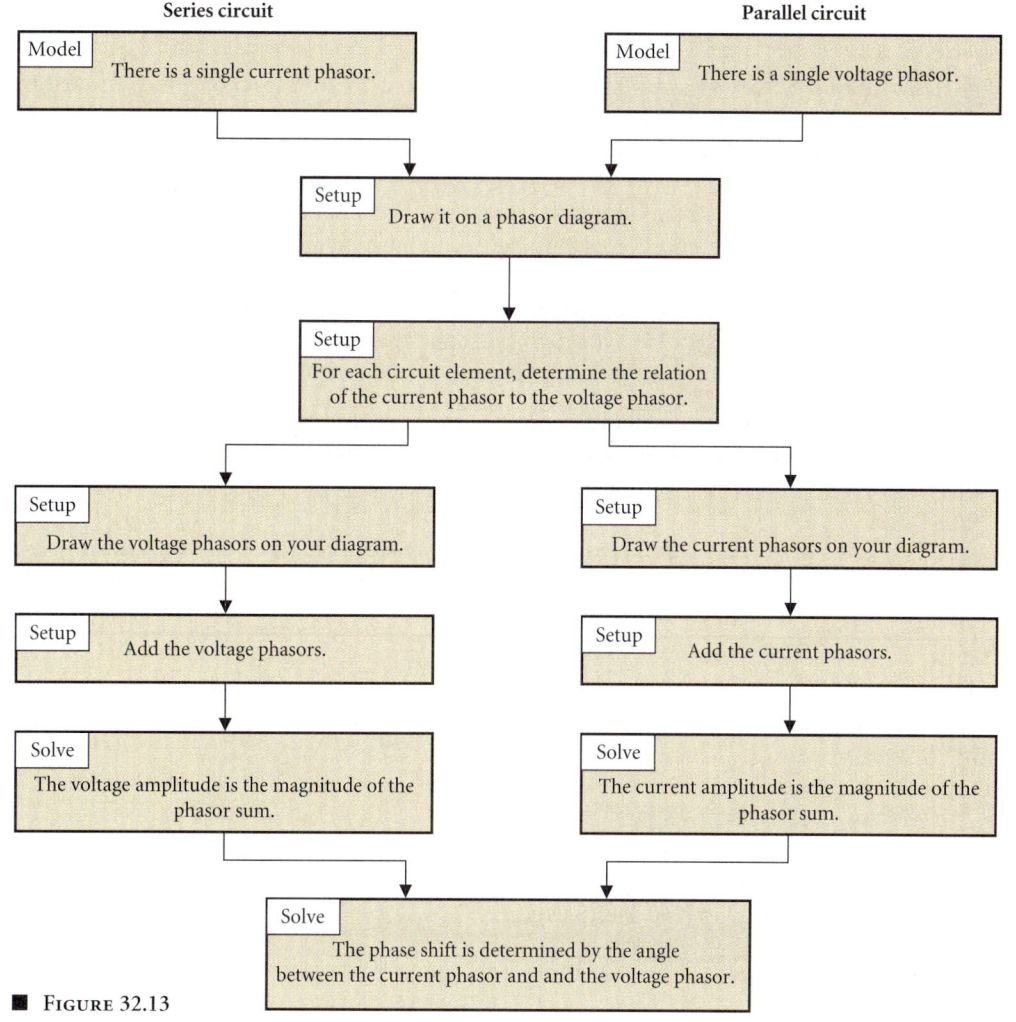

But don't try to add a voltage to a current!

■ **Figure 32.13**

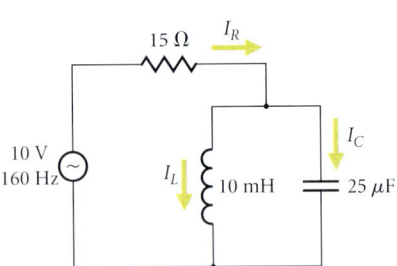

FIGURE 32.14
The resistor current divides at the junction with the parallel combination of inductor and capacitor: $I_R = I_L + I_C$.

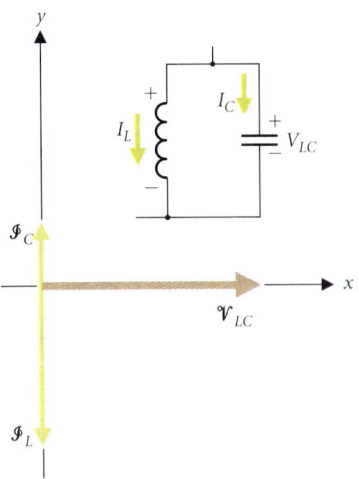

FIGURE 32.15
Phasor diagram for the parallel combination of inductor and capacitor. We have chosen to draw the voltage phasor along the x-axis. The capacitor current leads the voltage, while the inductor current lags. These phasors lie along the same line but in opposite directions. The phasor sum has magnitude $|\mathcal{I}| = ||\mathcal{I}_L| - |\mathcal{I}_C||$.

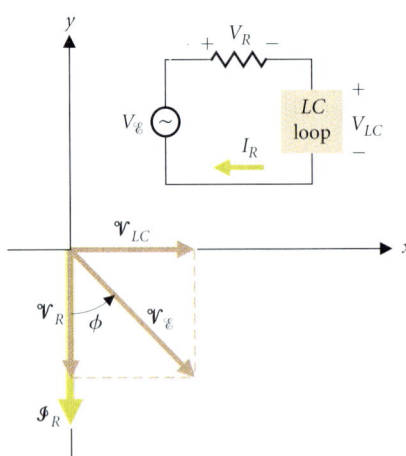

FIGURE 32.16
Phasor diagram for the resistor in series with the LC combination. The LC voltage lies along the x-axis, and the current lags this voltage by 90°, as in Figure 32.15. The resistor voltage is in phase with the current, so it also lies along the negative y-axis. The generator voltage is the phasor sum of $\mathcal{V}_{LC}$ and $\mathcal{V}_R$. The current lags the voltage by $\phi = 42°$.

EXAMPLE 32.6 ♦♦ A 10.0-mH inductor and a 25-μF capacitor are connected in parallel. The combination is in series with a 15-Ω resistor and a 10.0-V, 160-Hz emf (■ Figure 32.14). What is the current through the resistor?

MODEL The generator voltage is the sum of the potential difference across the resistor and the potential difference across the parallel combination of inductor and capacitor. The current through the resistor divides between the inductor and the capacitor. The inductor current lags the voltage and the capacitor current leads. We begin our analysis with the LC combination (■ Figure 32.15). Once we have found the voltage phasor for the parallel combination, we may add it to the voltage phasor for the resistor in series (■ Figure 32.16).

SETUP The same potential difference V_{LC} exists across the inductor and the capacitor. The current through each element is related to the potential difference by the reactance:

$$|\mathcal{I}_L| = |\mathcal{V}_{LC}|/X_L \quad \text{and} \quad |\mathcal{I}_C| = |\mathcal{V}_{LC}|/|X_C|.$$

The inductor current lags the voltage by $\pi/2$, while the capacitor current leads by $\pi/2$ (eqns. 32.5 and 32.4). The two current phasors are opposite each other (Figure 32.15). The current through the resistor is the phasor sum:

$$\mathcal{I}_R = \mathcal{I}_L + \mathcal{I}_C,$$

with magnitude
$$|\mathcal{I}_R| = |\mathcal{V}_{LC}|\left(\frac{1}{X_L} - \frac{1}{|X_C|}\right) = \frac{|\mathcal{V}_{LC}|}{X_{LC}}, \tag{i}$$

where
$$\frac{1}{X_{LC}} = \frac{1}{X_L} - \frac{1}{|X_C|} = \frac{1}{\omega L} - \omega C$$

$$= \frac{1}{2\pi(160\text{ Hz})(0.0100\text{ H})} - 2\pi(160\text{ Hz})(25 \times 10^{-6}\text{ F})$$

$$= 0.0743\ /\Omega.$$

So: $X_{LC} = 13.5\ \Omega.$

THE IMPEDANCE OF A PARALLEL COMBINATION SHOULD REMIND YOU OF THE RULES FOR COMBINING RESISTORS IN PARALLEL.

Figure 32.16 shows the diagram for the series combination of resistor and *LC* loop. The generator phasor is the vector sum of $\mathcal{V}_R$ and $\mathcal{V}_{LC}$. Each of these phasors is related to the current phasor $\mathcal{I}_R$: $|\mathcal{V}_R| = R|\mathcal{I}_R|$ and, by eqn. (i), $|\mathcal{V}_{LC}| = X_{LC}|\mathcal{I}_R|$. The resistor voltage $\mathcal{V}_R$ is in phase with $\mathcal{I}_R$, but $\mathcal{V}_{LC}$ leads by $\pi/2$ (Figure 32.16). Thus:

$$|\mathcal{V}_\mathcal{E}| = |\mathcal{I}_R|\sqrt{R^2 + X_{LC}^2}. \tag{ii}$$

SOLVE Relation (ii) is what we need to solve for $\mathcal{I}_R$:

$$|\mathcal{I}_R| = \frac{|\mathcal{V}_\mathcal{E}|}{\sqrt{R^2 + X_{LC}^2}} = \frac{10.0\ \text{V}}{\sqrt{(15\ \Omega)^2 + (13.5\ \Omega)^2}} = 0.50\ \text{A},$$

and the phase angle is given by

$$\tan\phi = \frac{|\mathcal{V}_{LC}|}{|\mathcal{V}_R|} = \frac{X_{LC}}{R} = \frac{13.5\ \Omega}{15\ \Omega} = 0.90 \Rightarrow \phi = 42°.$$

The current lags the voltage by 42°.

ANALYZE Our final result may be written: $I = (0.50\ \text{A})\cos[(2\pi \times 160)t - 0.23\pi]$, see ■ Figure 32.17.

If capacitive reactance is taken to be negative, then eqn. (i) shows that reactances in parallel combine like resistances in parallel. ■

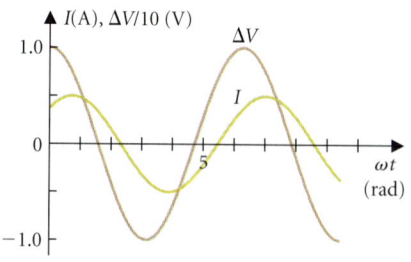

■ **FIGURE 32.17**
Current and generator voltage as functions of time. The current lags the voltage by 42°.

32.4 THE *LRC* CIRCUIT

In Chapter 31 we found that the transient behavior of a series *LRC* circuit is a damped oscillation. When the resistance is small, $R \ll 2\sqrt{L/C}$, the circuit oscillates with very nearly the natural frequency $\omega_0 = 1/\sqrt{LC}$. When an AC generator is connected to the circuit, the current undergoes a steady state, forced oscillation at the applied frequency (cf. §14.4). When the applied frequency equals the natural frequency ω_0, the current is in phase with the applied voltage, the circuit efficiently absorbs energy from the generator, and the current amplitude is at a maximum. This phenomenon is known as *resonance*. An *LRC* circuit can be used to detect weak signals near its resonant frequency or to act as a filter that emphasizes a narrow range of frequencies near ω_0.

WE ARE ASSUMING THAT THE CIRCUIT DAMPING IS LESS THAN CRITICAL; SEE EXAMPLE 31.12.

As in any series circuit, the *LRC* circuit has a single current phasor $\mathcal{I}$, which we place along the *x*-axis in ■ Figure 32.18. As time passes, this phasor rotates counterclockwise about the *z*-axis. The voltage across the resistor is in phase with the current, while the voltages across the inductor and capacitor lead and lag by 90°, respectively. The generator voltage $\mathcal{E} = V_0\cos\omega t$ equals the sum of the potential differences across the three elements. The phasor sum is:

$$\mathcal{V}_\mathcal{E} = \mathcal{V}_L + \mathcal{V}_C + \mathcal{V}_R.$$

The inductor and capacitor voltage phasors lie along opposite directions, so the magnitude of their sum is the difference of their magnitudes:

$$|\mathcal{V}_L + \mathcal{V}_C| = ||\mathcal{V}_L| - |\mathcal{V}_C|| = |X_L|\mathcal{I}| - |X_C||\mathcal{I}|| = |X_L - |X_C|||\mathcal{I}|.$$

The resistor voltage phasor is at right angles, so we find the magnitude of the sum using the Pythagorean theorem:

$$|\mathcal{V}| = |\mathcal{I}|\sqrt{R^2 + (X_L - |X_C|)^2} = |\mathcal{I}|Z.$$

Thus the impedance of a series *LRC* circuit is:

$$Z = \sqrt{R^2 + [\omega L - 1/(\omega C)]^2}. \tag{32.24}$$

The voltage leads the current. The phase shift ϕ is given by:

$$\tan\phi = \frac{X_L - |X_C|}{R} = \frac{\omega L - 1/(\omega C)}{R} = \frac{\omega^2 LC - 1}{\omega RC}. \tag{32.25}$$

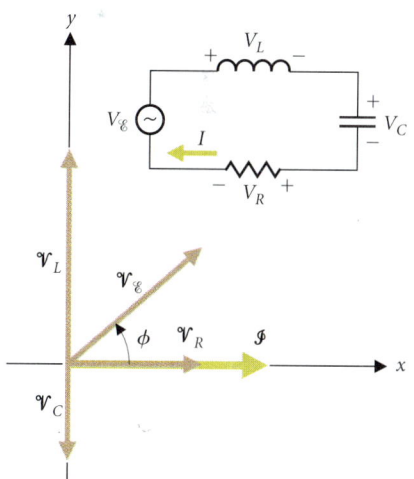

■ **FIGURE 32.18**
Phasor diagram for a series *LRC* circuit. We have chosen to put the single current phasor along the *x*-axis. The resistor voltage is in phase with the current (along the *x*-axis), the inductor voltage leads the current, and the capacitor voltage lags. The generator voltage is the phasor sum of the potential differences across the resistor, capacitor, and inductor. In this example it leads the current.

So the current is:
$$I(t) = \frac{V_0}{\sqrt{R^2 + [\omega L - 1/(\omega C)]^2}} \cos(\omega t - \phi). \quad (32.26)$$

WE ALSO FOUND THIS RESULT FOR X_{LC} IN §32.2.2. COMPARE WITH THE RESULT FOR A PARALLEL LC COMBINATION IN EXAMPLE 32.6.

Results (32.24) through (32.26) are consistent with the relations we developed for the RC circuit (eqns. 32.13–32.15) and the LR circuit (eqns. 32.17–32.19), and the result that the reactance of a series inductor/capacitor combination is $X_{LC} = \omega L - 1/(\omega C)$.

The current amplitude is a maximum when the impedance Z is a minimum—that is, when the combination of capacitive and inductive reactances is zero. This happens at angular frequency ω_0, where:

$$X_{LC} = \omega_0 L - 1/(\omega_0 C) = 0,$$

FOR $R = 0$, ω_0 IS THE NATURAL OSCILLATION FREQUENCY OF THE CIRCUIT. COMPARE EQNS. (31.25) AND (31.28).

or
$$\omega_0^2 = 1/(LC).$$

If there were no resistance, the current amplitude would become infinite at this *resonant frequency*. Resistance damps the oscillations, keeping the amplitude, $I_0 = V_0/Z$, finite. At the resonant frequency, the phase shift vanishes. The amplitudes of the voltage across the inductor and the capacitor may be each much larger than the generator voltage, but they are equal and opposite.

LOOK AT EQN. (32.25). THE NUMERATOR IS ZERO AT THE RESONANT FREQUENCY.

We can investigate the behavior of the circuit at other frequencies by looking at the capacitor voltage amplitude as a function of frequency:

$$V_C(\omega) = I_0/(\omega C) = V_0/(\omega C Z).$$

REMEMBER THAT Z IS ITSELF A FUNCTION OF ω.

At resonance, the impedance Z reduces to the resistance R, and $\omega = \omega_0$:

$$V_C(\omega_0) = \frac{V_0}{\omega_0 C R} = \frac{V_0 \sqrt{LC}}{CR} = \frac{V_0}{R}\sqrt{\frac{L}{C}}.$$

THE INDUCTOR VOLTAGE MAY ALSO BE USED. IT EXHIBITS THE SAME KIND OF RESONANT BEHAVIOR.

The voltage across the capacitor is increased (*amplified*) by the factor:

$$\frac{V_C(\omega_0)}{V_0} \equiv Q \equiv \frac{1}{R}\sqrt{\frac{L}{C}}. \quad (32.27)$$

The *quality factor* Q of the series LRC circuit describes its effectiveness as an amplifier of weak signals. It is much larger than unity in most practical circuits. The quality factor also describes how well an LRC circuit can discriminate between different applied frequencies. The capacitor voltage amplitude in terms of Q is:

$$V_C(\omega) = \frac{V_0 Q}{\sqrt{(\omega/\omega_0)^2 + Q^2[(\omega/\omega_0)^2 - 1]^2}}. \quad (32.28)$$

■ Figure 32.19 shows this function for four different values of Q. As Q increases, the curve reaches a higher maximum value and peaks more sharply. These are desirable characteristics in a circuit designed to respond to a narrow range of frequencies around one selected value.

EXERCISE 32.7 ◆◆ Show that the quality factor may be written $Q = \frac{1}{2}\omega_0 \tau$, where τ is the damping time constant of the free LRC circuit (eqn. 31.27).

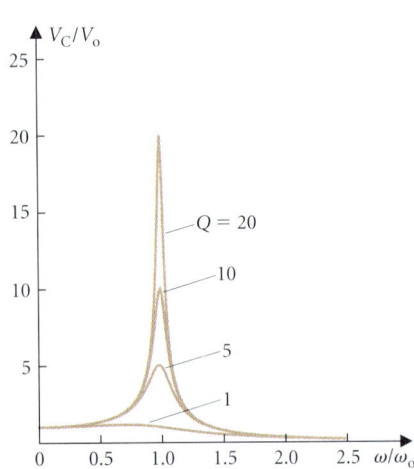

■ **FIGURE 32.19**
Capacitor voltage as a function of frequency for a series LRC circuit. Each curve is labeled with the value of its quality factor Q. As $Q = (1/R)\sqrt{L/C}$ increases, the amplitude peak at resonance becomes taller and narrower. This is a useful feature in circuits designed for frequency selectivity (see Example 32.8).

In a radio tuning circuit, a variable capacitor or inductor is used to adjust the resonant frequency. The AC driving voltage is provided by the electric field of radio waves arriving at the antenna.

EXAMPLE 32.7 ◆◆ An AM radio antenna has a resistance of 51.0 Ω. A radio enthusiast wants to build a series LRC circuit to tune over a range of frequencies from 0.50 to 2.0 MHz with a Q of 50.0 at 1.00 MHz. If the inductor resistance is 1.0 Ω, what value of L is required? Over what range must the capacitor be adjustable?

MODEL At resonance, the capacitive and inductive reactances are equal.

$$X_L = |X_C| = \sqrt{L/C} \equiv X.$$

From eqn. (32.27), the quality factor is $Q = X/R$ so, with a known resistance in the circuit, we can find the value of X that produces a desired Q value. Once we have X at the given resonant frequency (1.00 MHz), we can find the required values of L and C.

Tuning a desired frequency means adjusting the circuit's resonant frequency to that value. The adjustment is made in this circuit by varying the capacitance, with L and R fixed.

Do not confuse this Q with charge.

SETUP The total resistance in the circuit is the series combination of the antenna and the inductor: $R = 51.0 \,\Omega + 1.0 \,\Omega = 52.0 \,\Omega$. Then the reactance X needed for a Q value of 50.0 is:

$$X = QR = 50.0(52.0 \,\Omega) = 2.60 \text{ k}\Omega.$$

With known values for $X = \sqrt{L/C}$ and the resonant frequency $\omega_0 = 2\pi f_0 = 1/\sqrt{LC}$, we can find both L and C.

SOLVE With $f_0 = 1.00$ MHz, the required inductance is:

$$L = \sqrt{\frac{L}{C}}\sqrt{LC} = \frac{X}{\omega_0} = \frac{2.60 \times 10^3 \,\Omega}{2\pi(1.00 \times 10^6 \text{ Hz})} = 0.414 \text{ mH}.$$

Since $\omega_0^2 = 1/(LC)$, the capacitance needed to tune a given frequency is:

$$C = \frac{1}{\omega_0^2 L} = \frac{1}{[2\pi(5.0 \times 10^5 \text{ Hz})]^2(0.414 \text{ mH})} = 0.245 \text{ nF at 500 kHz},$$

and $$C = \frac{1}{[2\pi(2.0 \times 10^6 \text{ Hz})]^2(0.414 \text{ mH})} = 15.3 \text{ pF at 2 MHz}.$$

The required value of C ranges from 15 pF for 2 MHz to 245 pF for 500 kHz.

ANALYZE The required capacitance ranges over a factor of 16 to produce a factor of 4 in the frequency range. An alternative design would have a variable inductance. We could have found the capacitance required for a Q of 50 at 1 MHz, and the corresponding range of L.

See Problem 27.67, which describes a tunable capacitor. Problem 31.83 describes a variable inductor.

EXAMPLE 32.8 ♦♦ Two radio stations broadcast at 0.810 MHz and 0.850 MHz. When the antenna circuit in Example 32.7 is tuned to 0.810 MHz, it also responds to signals oscillating at 0.850 MHz. If the amplitude of the input signals reaching the antenna is the same at the two frequencies, how large is the unwanted output signal at 0.850 MHz?

MODEL A modern radio uses a second circuit to isolate the desired signal further. The input to the second circuit is the antenna circuit capacitor voltage. In this problem, we want to see how well the antenna circuit itself discriminates, so we compare the peak capacitor voltages at the two frequencies.

SETUP Using the result of Example 32.7, the capacitance required to tune the radio to 0.810 MHz is:

$$C = \frac{1}{\omega_0^2 L} = \frac{1}{[2\pi(8.10 \times 10^5 \text{ Hz})]^2(0.414 \text{ mH})} = 93.25 \text{ pF}.$$

The value of Q at this frequency is:

$$Q = \frac{1}{R}\sqrt{\frac{L}{C}} = \left(\frac{1}{52.0 \,\Omega}\right)\sqrt{\frac{4.14 \times 10^{-4} \text{ H}}{93.25 \times 10^{-12} \text{ F}}} = 40.52.$$

The capacitor voltage amplitude at the resonant frequency (810 kHz) is related to the input voltage amplitude through eqn. (32.27): $V_C(\omega_0) = V_0 Q$. We obtain the capacitor voltage amplitude at a nearby frequency from eqn. (32.28).

$$\frac{V_C(\omega)}{V_0 Q} = \frac{1}{\sqrt{(\omega/\omega_0)^2 + Q^2[(\omega/\omega_0)^2 - 1]^2}}.$$

For the 850-kHz signal,

$$\frac{\omega}{\omega_0} = \frac{2\pi f}{2\pi f_0} = \frac{f}{f_0} = \frac{0.850 \text{ MHz}}{0.810 \text{ MHz}} = 1.049.$$

SOLVE The ratio of the capacitor voltage at 850 kHz to that at 810 kHz is:

$$\frac{V_C}{V_0 Q} = \frac{1}{\sqrt{1.049^2 + 40.52^2(1.049^2 - 1)^2}} = 0.24.$$

ANALYZE The circuit response to the unwanted signal at 0.850 MHz is 24% of its response to the 0.810-MHz signal. Better selectivity requires a larger Q value. In practice, the second circuit improves the discrimination by as much as a factor of 10.

NOTE: THE SUBTRACTION REDUCES THE NUMBER OF SIGNIFICANT FIGURES IN THE RESULT FROM 3 TO 2.

Chapter Summary

Where Are We Now?

We have extended our analysis of electric circuits to include sinusoidally varying emfs. Such emfs produce alternating current (AC). These circuits are the basis of most household appliances as well as many industrial devices.

What Did We Do?

Circuits, AC as well as DC, are analyzed using Kirchhoff's laws. In an AC circuit the current oscillates at the same frequency as the emf but may be out of phase with the voltage. Thus the power used by the circuit is generally less than the average emf times the average current. The circuit behavior is determined by its resistance R, reactance X, and impedance, $Z = \sqrt{R^2 + X^2}$. The *reactance* is contributed by inductors and capacitors in the circuit. If ω is the angular frequency of the applied emf, the inductor reactance is $X_L = \omega L$ and the capacitor reactance is $|X_C| = 1/(\omega C)$. The reactance of a series combination of inductors and capacitors is the total inductive reactance minus the total capacitive reactance:

$$X_s = X_L - |X_C|.$$

Then: $\quad I = \dfrac{V_0}{Z} \cos(\omega t - \phi), \quad \tan\phi = \dfrac{X_s}{R}, \quad \text{and} \quad \langle P \rangle = V_{\text{rms}}^2 \dfrac{R}{Z^2}.$

The current through, or voltage across, each component may be described using *phasors*—rotating vectors that describe the magnitudes and phase relations of these quantities. In a series circuit, the voltage phasors add vectorially to give the applied potential; in a parallel circuit, the current phasors add vectorially to give the total current.

A series LRC circuit resonates—has peak current amplitude—at frequency $\omega_0 = 1/\sqrt{LC}$. At this frequency, the total reactance is zero and $Z = R$. The breadth of the resonance is determined by the quality factor:

$$Q = \tfrac{1}{2}\omega_0 \tau = \frac{1}{R}\sqrt{\frac{L}{C}}.$$

Practical Applications

The methods we have developed allow us to analyze a large number of practical circuits that do not include solid state devices. Familiar examples we have discussed include radio tuner circuits, loudspeaker crossover circuits, and circuits coupled by transformers.

Solutions to Exercises

32.1 The average power input is:

$$\langle P \rangle = \frac{V_{\text{rms}}^2}{R} = \frac{(120 \text{ V})^2}{5.0 \text{ }\Omega} = 2.9 \text{ kW}.$$

32.2 Using eqn. (32.4) for $I(t)$, the power output is:

$$P(t) = \mathcal{E}(t)I(t) = V_0 \cos \omega t \, (-\omega C V_0) \sin \omega t$$

$$= -\frac{\omega C V_0^2}{2} \sin(2\omega t).$$

32.3 $\dfrac{1 \text{ s}}{1 \text{ F}} = \dfrac{1 \text{ s}}{1 \text{ C/V}} = \dfrac{1 \text{ V}}{1 \text{ C/s}} = \dfrac{1 \text{ V}}{1 \text{ A}} = 1 \text{ }\Omega.$

Similarly, $\dfrac{1 \text{ H}}{1 \text{ s}} = \dfrac{1 \text{ V}}{(1 \text{ A/s})(1 \text{ s})} = \dfrac{1 \text{ V}}{1 \text{ A}} = 1 \text{ }\Omega.$

32.4 See ■ Figure 32.20. The current lags the voltage by 90°.

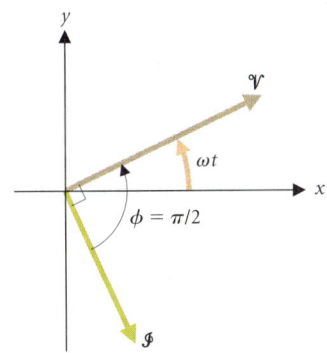

■ **Figure 32.20**
Phasor diagram for an inductor. The phasors rotate counterclockwise with angular speed ω.

32.5 (■ Figure 32.21) The inductor voltage leads the current. The total potential difference is the vector sum of the resistor and inductor voltages:

$$|\mathcal{V}| = \sqrt{|\mathcal{V}_R|^2 + |\mathcal{V}_L|^2} = |\mathcal{I}|\sqrt{R^2 + (\omega L)^2}$$

(cf. eqns. 32.18 and 32.19), and

$$\tan \phi = \frac{|\mathcal{V}_L|}{|\mathcal{V}_R|} = \frac{|\mathcal{I}|\omega L}{|\mathcal{I}|R} = \frac{X_L}{R}, \text{ which is eqn. (32.17).}$$

32.6 $\langle P \rangle = \frac{1}{2}\mathcal{V} \cdot \mathcal{I} = \frac{1}{2}(156 \text{ V})(0.276 \text{ A})\cos(27.9°) = 19.0 \text{ W}.$ The power as a function of time is shown in ■ Figure 32.22.

32.7 The damping time constant is $\tau = 2L/R$, so:

$$\tfrac{1}{2}\omega_0 \tau = \frac{1}{2}\left(\frac{1}{\sqrt{LC}}\right)\left(\frac{2L}{R}\right)$$

$$= \frac{1}{R}\sqrt{\frac{L}{C}} = Q,$$

as required. Qualitatively, Q is π times the number of oscillations the current undergoes in one damping time. Using an acoustic analogy, a bell with high Q rings, but a drum with low Q thuds.

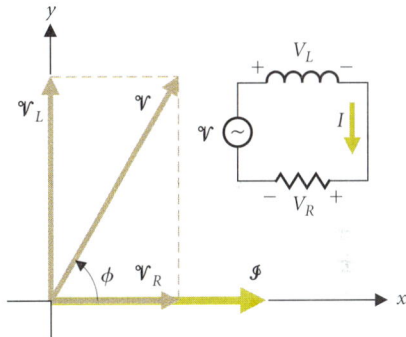

■ **Figure 32.21**
Phasor diagram for a series LR circuit. We have chosen to place the current phasor along the x-axis.

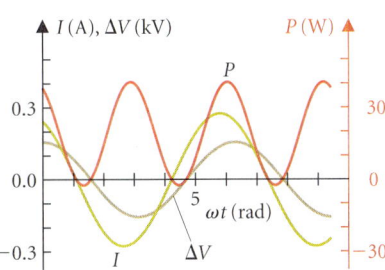

■ **Figure 32.22**
Voltage, current, and power for the RC circuit. The power is positive for almost all of the cycle because the phase shift between voltage and current is small.

Basic Skills

Review Questions

§32.1 SINGLE-ELEMENT CIRCUITS

- What constraint exists on the size of a circuit to which you can apply the techniques of this chapter? What is the reason for the constraint?
- What does it mean to say that current *leads* or *lags* voltage?
- Does current in a circuit arm containing a capacitor lead or lag the potential difference across the capacitor?
- Does current in an inductor lead or lag the potential difference across it?
- When AC current passes through a resistor, how does the power dissipated depend on the applied potential difference? What is the time-averaged power?
- Define V_{rms}.
- What is the time-averaged power dissipated when an AC voltage is applied to a capacitor? What about an inductor?
- Define the *reactance* of a capacitor or inductor.
- How does the reactance of an inductor depend on the frequency of the AC power supply?
- How does the reactance of a capacitor depend on the frequency of the AC power supply?

§32.2 TWO-COMPONENT CIRCUITS

- What is *impedance*?
- In a series connection of two circuit elements, how does the impedance depend on resistance and reactance?
- When an AC power supply is connected to a resistor and an inductor in series, how is the current related to the applied voltage? How does the phase shift depend on the reactance? On the resistance?
- What is the *power factor*? How does it depend on resistance?
- Why might you want to reduce the total reactance in a circuit?

§32.3 CIRCUIT ANALYSIS USING PHASORS

- What is a *phasor*?
- In a phasor diagram, how should you measure the phase shift ϕ?
- In a series circuit, how many current phasors are there? How many voltage phasors?
- How do the voltage phasors combine?
- Does it matter at what angle you draw the current phasor? Why or why not?
- Express the power used by a circuit in terms of phasors.
- In a parallel combination of two circuit elements, how many voltage phasors are there? How many current phasors? How do the current phasors combine?

§32.4 THE *LRC* CIRCUIT

- What is the impedance of a series combination of resistor, inductor, and capacitor?

- What is the resonant frequency of a series *LRC* circuit?
- What happens to the current at *resonance*? How are the capacitor and inductor voltages related at resonance?
- What is the quality factor Q for a series *LRC* circuit? What *quality* does it measure?

Basic Skill Drill

§32.1 SINGLE-ELEMENT CIRCUITS

1. What is the reactance of a 5.6-μF capacitor used with a 60.0-Hz power supply?

2. What is the reactance of a 1.2-mH inductor used with a 150-Hz power supply?

3. Find the resistance of a 60-W light bulb designed to operate off the power network (60 Hz, 120 V rms).

4. Find the current amplitude when a 32-μF capacitor is connected to an AC generator providing 25 V rms at 1.0 kHz.

§32.2 TWO-COMPONENT CIRCUITS

5. A coil with an inductance of 5.0 mH and a resistance of 10.0 Ω is connected to an AC power supply providing 10.0-V voltage amplitude at 0.50 kHz. What is the amplitude of the current in the circuit? How much power is used?

6. A 1.0-kΩ resistor and a 2.5-nF capacitor are connected in series to an AC power supply providing 12 V rms at 15 kHz. What is the current amplitude in the circuit? How much power is used?

7. Find the current as a function of time in a circuit with a 2.20-mH inductor and a 5.60-Ω resistor connected to a 60.0-Hz, 120-V power supply.

§32.3 CIRCUIT ANALYSIS USING PHASORS

8. A 120-V, 60-Hz AC generator is connected to a coil with a resistance of 3.0 Ω and an inductance of 2.40 mH. Draw a phasor diagram at $t = 0$, 5 ms, and 10 ms.

9. A 4.8-μF capacitor and a 120-Ω resistor are connected in series with a 120-V, 60.0-Hz power supply. Draw a phasor diagram and use it to find the average power used by the circuit.

§32.4 THE *LRC* CIRCUIT

10. Find the resonant frequency f_0 and Q value for a series *LRC* circuit with $L = 32$ μH, $C = 4.7$ μF, and $R = 6.0$ Ω.

11. You have a 56-nF capacitor. You want to make an *LRC* circuit with a resonant frequency at 610 kHz and a Q of 60.0. What values of L and R should you choose? If the input voltage is $V_0 = 0.68$ mV, what is the current amplitude at resonance?

Questions and Problems

§32.1 SINGLE-ELEMENT CIRCUITS

12. ❖ Two capacitors have capacitance C and $2C$. Which has the larger reactance? By what factor? Explain the physical reason for the difference.

13. ❖ Two inductors have inductances L and $2L$. Which has the larger reactance? By what factor? Why?

14. ❖ The frequency of an AC power source is doubled. By how much does the reactance of an inductor in the circuit change? Explain the physical reason for the change.

15. ❖ The frequency of an AC power source is halved. By how much does the reactance of a capacitor in the circuit change? Explain the reason for the change.

16. ❖ What is the rms voltage of a rectified cosine wave
$$V(t) = V_0|\cos \omega t|?$$

17. ❖ Estimate the maximum frequency at which each of the following systems could operate with designs that ignore the response time of electric and magnetic field.
(a) A power grid for the entire United States.
(b) A power grid for a medium-sized city like Denver.
(c) The circuits in a 2-m-diameter supercomputer.
(d) The circuit on a single microprocessor chip 1 cm across in a laptop computer.

18. ♦ A 60.0-mH inductor has a resistance of 5.0 Ω. At what frequency does its resistance equal its reactance?

19. ♦ What is the current amplitude in a circuit with a 15-V rms, 750-Hz power supply connected to a 1.2-nF capacitor?

20. ♦ Commercial power (120 V, 60.0 Hz) is connected to an inductor, and the resulting current amplitude is 1.75 A. What is the inductance?

21. ♦ A hair dryer manufactured for use in the United States is rated at 1200 W. What is its resistance? (Assume it has no inductance or capacitance.)

22. ♦♦ When a power supply is connected to a 5.6-μF capacitor, the current amplitude is 17 A. Connected to a 2.8-mH inductor, the current amplitude is 25 A. What is the frequency of the power supply? How much power is used when the same source is connected to a 5.6-kΩ resistor?

23. ♦♦ What amplitude sine wave produces the same average power in a resistor as a square wave generator with maximum voltage V_0 (■ Figure 32.23)?

24. ♦♦♦ What is the rms voltage of a sawtooth wave (■ Figure 32.24)?

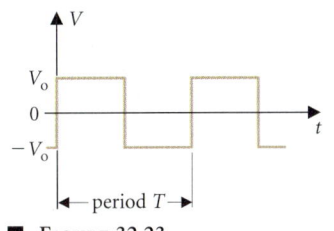
■ FIGURE 32.23

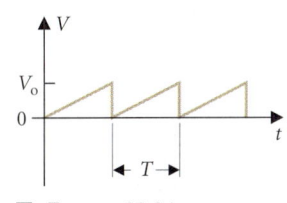
■ FIGURE 32.24

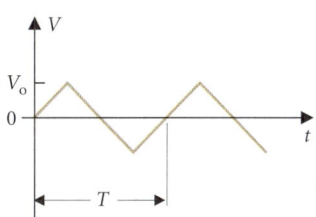
■ FIGURE 32.25

25. ♦♦♦ What is the rms voltage of a triangle wave (■ Figure 32.25)? Based on your results, what would you guess is the rms voltage for a rectified triangle wave?

26. ♦♦♦ Show by direct integration that the average value of $\sin^2 \omega t$ or $\cos^2 \omega t$ over one period is $\frac{1}{2}$.

§32.2 TWO-COMPONENT CIRCUITS

27. ❖ Explain why a resistor uses less power when connected in series with a capacitor or an inductor than when connected alone.

28. ❖ An inductor and a capacitor each have the same reactance X. Each is connected into a circuit with a resistor R and a power supply. Compare the power used by the two circuits.

29. ❖ A resistor is connected in series with an unknown circuit element. If the power consumption of the circuit increases as the frequency of the applied voltage increases, is the unknown element a capacitor, a resistor, or an inductor?

30. ♦ An inductor and a capacitor are connected in series with an AC generator. What is the current in the circuit? Do you believe your solution at $\omega = 1/\sqrt{LC}$? Qualitatively discuss what would happen in a real circuit.

31. ♦ An electric motor has a 75-mH inductance and a 24-Ω resistance. What is the phase shift between current and voltage when it is connected to the commercial power network? What is the average power used?

32. ♦ A coil used in a physics experiment has a 33-mH inductance and an 85-Ω resistance. It is connected to an AC generator with a frequency of $\omega = 1.0 \times 10^3$ rad/s and a peak voltage of 15 V. Find the amplitude of the current through the coil and the phase shift with respect to the applied voltage.

33. ♦ A factory has a total inductance of 8.0 H. What capacitance connected in series is required to make the total reactance at 60.0 Hz zero?

34. ♦♦ Show that the average power used by a series AC circuit is the square of the current amplitude times half the resistance.

35. ♦♦ A coil has resistance $R = 10.0$ Ω and inductance $L = 5.0$ mH. At what frequency is the power used by the coil a maximum? Give a numerical value for the maximum power used when $V_0 = 10.0$ V. Compare with the power used at 60.0 Hz.

36. ♦♦ A 1.2-kΩ resistor and a 2.2-μF capacitor are connected in series across a 120-V, 60-Hz power supply. Find the current in the circuit as a function of time. What is the average power drawn from the power supply?

37. ♦♦ In a stereo crossover network, a 1.9-mH inductor is in series with a 53-μF capacitor. Plot a graph showing the ratio of potential difference across the inductor to the potential difference across the capacitor as a function of frequency. At what frequency is there an equal potential difference across each element? (This is called the crossover frequency.) Across which element should the tweeter be connected, and which the woofer?

38. ♦♦♦ A circuit contains a 33-mH inductor and an 80.0-Ω resistor in series with a 60.0-Hz AC power supply and a switch. The switch is closed at $t = 0$. Find the current in the circuit as a function of time. How many periods of oscillation occur before the transient response damps to 10% of the amplitude of the steady state current?

§32.3 CIRCUIT ANALYSIS USING PHASORS

39. ❖ Does the phasor diagram for a series circuit depend on the order in which circuit elements occur? Why or why not?
40. ❖ Are the voltage phasors due to capacitive and inductive reactance always directly opposite each other in a series circuit? In a parallel circuit?
41. ♦ A 15-V, 25-Hz power supply is connected in series to a 15-kΩ resistor and a 0.56-μF capacitor. Draw the phasor diagram at $t = 0$, 0.01 s, and 0.02 s. What is the current amplitude?
42. ♦ A 35-mH inductor and a 5.6-Ω resistor are connected in parallel across a 44-V, 30.0-Hz power supply. Draw the phasor diagram at $t = 0$, 0.01 s, and 0.02 s. What is the current amplitude?
43. ♦♦ A food mixer has a resistance of 35 Ω and an inductance of 22 mH in series. Draw a phasor diagram for the voltage and current when the mixer is used with home power (120 V, 60.0 Hz). How much power does the mixer draw? How much power would it draw if the inductance were zero?
44. ♦♦ Use a phasor diagram to find the current through the power supply shown in ■ Figure 32.26. Compare and contrast your results with the behavior of a series RC circuit (e.g., Example 32.5).

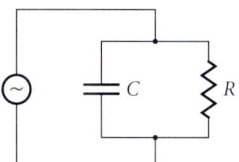

■ FIGURE 32.26

45. ♦♦ Household wiring is arranged so that when different appliances are plugged in, they are in parallel across the power supply. A vacuum cleaner with $R = 55$ Ω and $L = 33$ mH and a television set with $R = 85$ Ω are operating at the same time. Draw a circuit diagram and a phasor diagram. How much power is used? What fraction is used by the vacuum and what fraction by the television?
46. ♦♦ Use the phasor method to find the current through the power supply in the circuit diagramed in ■ Figure 32.27.
47. ♦♦♦ Two coils are connected in parallel across a 75-V, 55-Hz power supply. One coil has an inductance of 15 mH and a resistance

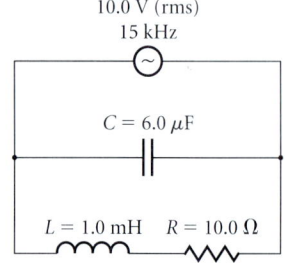

■ FIGURE 32.27

of 3.5 Ω; the second has an inductance of 25 mH and a resistance of 6.7 Ω. Draw a phasor diagram and find the currents in the circuit. What is the power absorbed?

§32.4 THE LRC CIRCUIT

48. ❖ If the inductance in a series LRC circuit is doubled, how do the resonant frequency, time constant, and quality factor change?
49. ❖ If the capacitor in a series LRC circuit is changed for one with four times the capacitance, how do the resonant frequency, time constant, and quality factor change?
50. ❖ If the resistance in a series LRC circuit is reduced by a factor of 2, how do the resonant frequency, time constant, and quality factor change?
51. ♦ Find the resonant frequency and Q value for a circuit with $R = 18$ Ω, $C = 5.6$ μF, and $L = 33$ mH.
52. ♦ You have a 5.6-mH inductor. You want to build a circuit with a resonant frequency of 2.5 kHz and a Q of 45. What values of C and R should you choose?
53. ♦ You have two inductors, 2.2 mH and 0.75 mH, and two capacitors, 56 pF and 1.0 nF. Each inductor also has 0.95-Ω resistance. Using one inductor and one capacitor, you build a series LRC circuit. (a) What is the greatest resonant frequency you can achieve? What is the Q of that circuit? (b) What is the greatest Q you can achieve? What is the resonant frequency of that circuit?
54. ♦ Show that the power input to an LRC circuit is maximum at resonance.
55. ♦♦ An LRC circuit using a 15-μH inductor is tuned to 0.10 MHz. What maximum resistance can the circuit have if an unwanted signal at 90 kHz is amplified by less than 10% of the amplification at 100 kHz?
56. ♦♦ A tuning circuit is adjusted with a variable inductance. If $R = 5.6$ Ω and $C = 1.2$ pF, what range of inductance is required to tune the resonant frequency over the range 75 MHz to 250 MHz? How does the Q value for the circuit change over this range? At what values of ω/ω_0 does the capacitor amplitude equal 10% of the peak value at $f_0 = 75$ MHz and 250 MHz?
57. ♦♦ Use a phasor diagram to find the currents in each arm of the parallel LRC circuit in ■ Figure 32.28. Find the current amplitude and the average power drawn from the generator. Compare with the behavior of the series LRC circuit.

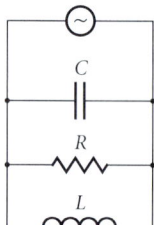

■ FIGURE 32.28

58. ♦♦ Find the potential difference across the inductor in a series LRC circuit in terms of ω/ω_0 and Q. Plot this function and compare your graph with Figure 32.19. Give a physical reason for each difference.
59. ♦♦ A series LRC radio tuning circuit has $L = 0.56$ mH, $R = 2.2$ Ω, and a variable capacitance. What value of C is required to tune the circuit to 660 kHz? What is the Q value? If another station at 600 kHz has the same amplitude, how large is the unwanted output signal compared with the selected signal?
60. ♦♦ In the FM band, 88–108 MHz, stations may be separated by as little as 200 kHz. In a modern FM tuner, the input signal is converted to an intermediate frequency of 10.7 MHz. Then a secondary

circuit discriminates between signals from different stations. What value of Q is required in the secondary circuit if the output signal at 10.5 MHz is 25% of that at 10.7 MHz? Assume equal input amplitudes at the two frequencies. What value of L is required in the circuit if the total resistance is 1.2 Ω? What capacitance is required?

61. ♦♦♦ Show that the maximum amplitude of the capacitor voltage in a series LRC circuit occurs at a frequency $\omega = \omega_0 \sqrt{1 - 1/(2Q^2)}$ and has the value:

$$V_{C,max} = \frac{V_0 Q}{\sqrt{1 - 1/(4Q^2)}}.$$

62. ♦♦♦ Show that the maximum amplitude of the inductor voltage in a series LRC circuit occurs at a frequency:

$$\omega = \frac{\omega_0}{\sqrt{1 - 1/(2Q^2)}}$$

and has the same value as the maximum capacitor voltage (Problem 61).

63. ♦♦♦ In a series LRC circuit driven at frequency ω, show that the ratio of maximum energy stored in the capacitor during a cycle to maximum energy stored in the inductor is:

$$\frac{U_C}{U_L} = \frac{\omega_0^2}{\omega^2}.$$

Show that the difference in the peak energies equals the work done by the AC generator during the quarter cycle in which the capacitor discharges and the inductor reaches peak current, less the thermal energy transformed in the resistor.

64. ♦♦♦ One measure of the width of the resonance curve is the full-width half-maximum, the difference between two angular frequencies at which the circuit response is half its maximum value. Show that, for large Q, the fwhm is approximately $\sqrt{3}\omega_0/Q$.

65. ♦♦♦ In a series LRC circuit, show that Q is 2π times the maximum energy stored in the capacitor at resonance divided by the energy dissipated per cycle (at resonance).

Additional Problems

66. ✦ Describe the low- and high-frequency behavior of the circuit shown in ■ Figure 32.29.

67. ✦ Discuss the behavior of the circuit in ■ Figure 32.30. What is the effect, if any, of the capacitor?

68. ✦ A circuit contains a resistor connected to a "black box" and a power supply. As the power supply frequency is increased from zero, you notice that the power absorbed by the circuit decreases. Is the component in the black box a resistor, a capacitor, or an inductor?

69. ✦ How does the rms power depend on the frequency of the AC power source if: **(a)** the circuit contains only resistance? **(b)** the circuit contains a resistor and a capacitor in series?

70. ♦♦ Two capacitors are connected in series to an AC power supply. Draw a phasor diagram showing the current and the potential differences across each capacitor. What is the current in the circuit? Do the capacitors combine according to the rules you learned in Chapter 27?

71. ♦♦ A coil with a 20-Ω resistor and a 5-mH inductor is connected in parallel with a 30-μF capacitor. What is the impedance of the combination at 330 Hz?

72. ♦♦♦ Find the current I in the circuit shown in ■ Figure 32.31.

73. ♦♦♦ Use phasor diagrams to analyze the circuit shown in ■ Figure 32.32 if the frequency of each generator is 350 Hz. What is the current in the resistor? How much power is drawn from each emf?

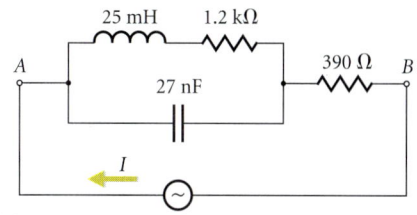

■ **FIGURE 32.31** $V_0 = 15$ V, $f = 10.0$ kHz

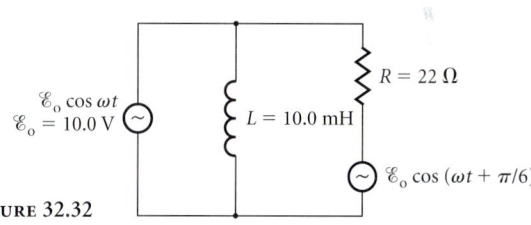

■ **FIGURE 32.32**

74. ♦♦♦ The quality factor for a parallel antenna circuit (■ Figure 32.33) is the inverse of Q for a series circuit with the same components: $Q_\| = R\sqrt{C/L}$. Derive this expression using the results $Q = \frac{1}{2}\omega_0\tau$ (Exercise 7) and $\tau = 2RC$ (Problem 31.72). Choosing values of L, R, and C, design an antenna circuit with a Q of 5 at 102 MHz. Find the capacitor voltage in terms of Q, and plot $V_C/\mathcal{E}$ versus ω/ω_0 for the circuit you have designed.

75. ♦♦♦ **(a)** A series LRC circuit is used in an instrument for measuring the level of liquid in a tank (■ Figure 32.34). The capacitor has parallel plates of dimension $\ell = 20.0$ cm by $w = 5.0$ cm and separation $d = 1.0$ mm. The liquid has dielectric constant $\kappa = 2.3$. The level of liquid is determined by measuring the resonant frequency of the circuit. Show that the liquid level is given by:

$$x = \frac{7.0 \times 10^{14} \text{ m/s}^2}{\omega_0^2} - 0.15 \text{ m}.$$

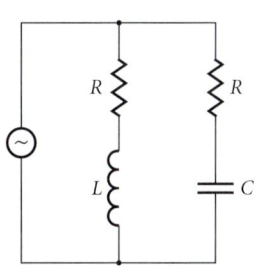

■ **FIGURE 32.29**

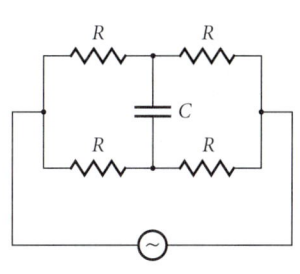

■ **FIGURE 32.30**

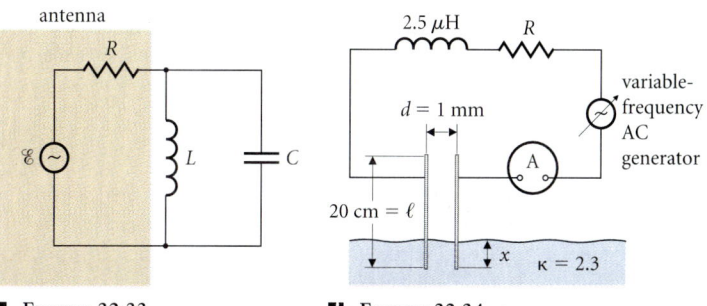

■ **FIGURE 32.33**   ■ **FIGURE 32.34**

(b) If the ammeter uncertainty in measuring peak current amplitude is 1%, what Q is needed to measure the liquid level to 1% accuracy? What maximum resistance in the circuit will allow this accuracy?

Computer Problems

76. A circuit with $L = 1.2$ mH and $R = 17\,\Omega$ is connected to a 35-V, 2.0-kHz power supply. Use a spreadsheet program to calculate and plot $V(t)$, $I(t)$, and $P(t)$ over two cycles. What is the time-averaged power? (You may calculate the time average analytically or numerically.)

77. A circuit with $R = 56\,\Omega$ and $C = 1.2\,\mu$F is connected to a 5.0-V, 130-kHz power supply. Use a spreadsheet program to plot $V(t)$, $I(t)$, and $P(t)$ over two cycles. What is the time-averaged power?

78. Express the power used by a series LRC circuit as a function of generator frequency in dimensionless form. That is, express the ratio $P \div (V_0^2/2R)$ as a function of the ratio f/f_0. Plot graphs of the function for $Q = 1, 10, 20$. Comment.

Challenge Problems

79. A practical radio antenna uses a parallel LRC circuit (cf. Figure 32.33) in which the incoming signal acts as the emf. Use phasors to find the ratio of capacitor voltage to antenna voltage. The coil has inductance $L = 2.5\,\mu$H and resistance $r = 65\,\Omega$. Take $C = 150$ pF, antenna resistance $R = 325\,\Omega$, and the operating frequency of the circuit $f = 0.82$ MHz.

80. Use phasors to find the currents in each arm of the circuit shown in ■ Figure 32.35 and the power drawn from the generator.

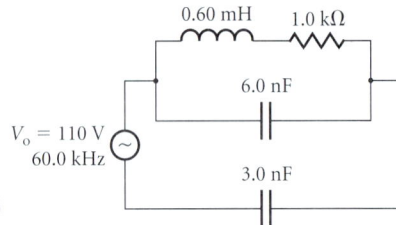

■ **FIGURE 32.35**

81. (a) Show that if the voltage applied to a series LRC circuit is the sum of two oscillating voltages, the voltage across the capacitor is the sum of the two voltages that would result from each of the two inputs applied separately. **(b)** A series LRC circuit with resonant frequency ω_0 and quality factor $Q = 10$ is connected to a signal generator that produces a voltage $V(t) = V_0 \cos(\omega_0 t) + (V_0/3) \cos(3\omega_0 t)$. What is the resulting capacitor voltage as a function of time? **(c)** Fourier's theorem states that any periodic function can be represented as a sum of trigonometric functions. ■ Figure 32.36 shows a square wave produced by alternately switching a battery into and out of a circuit. The square wave is a sum of cosine functions, of which the first two are shown in the figure:

$$S(t) = \frac{4V_0}{\pi}[\cos(\omega_s t) - \tfrac{1}{3}\cos(3\omega_s t) + \tfrac{1}{5}\cos(5\omega_s t) + \cdots].$$

What capacitor voltage results if the LRC circuit in part (b) is connected to the square wave voltage with $\omega_s = \omega_0$? How good a sine wave generator does this circuit make? **(d)** Describe the capacitor voltage when the frequency of the square wave is (i) $\omega_s = \omega_0/3$, and (ii) $\omega_s = \omega_0/5$. **(e)** When the frequency of the square wave is much

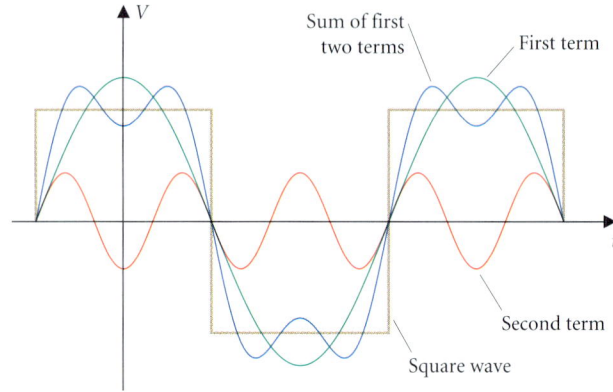

■ **FIGURE 32.36**

less than the resonant frequency of the LRC circuit, describe the behavior of the capacitor voltage qualitatively, using the methods of Chapter 31. How is this result consistent with parts (a) through (d)?

82. A coil is wound with twin-lead wire (two separate wires bound together so as to remain parallel to each other, but not electrically connected) to form two inductors with the same length ℓ, cross-sectional area A, and number of turns N (■ Figure 32.37). One of the wires is connected to a capacitor C. The other wire is connected in series to a second, identical capacitor C and a signal generator that puts out a voltage $V_0 \cos \omega t$. **(a)** Compute the self-inductance L of one of the wires wrapped around the coil. This will be the self-inductance in each of the two circuits. Also compute the mutual inductance M of the two circuits. How is M related to L? **(b)** With the sign conventions shown in the figure, use Kirchhoff's laws to write differential equations for the two charges $Q_1(t)$ and $Q_2(t)$. Solve for the steady state response of the system, and find its resonant frequency. **(c)** Plot a graph showing the voltage across each capacitor as a function of frequency squared. (Use $\omega^2 LC$ as the frequency variable.) **(d)** Discuss the behavior of the system at $\omega = 1/\sqrt{LC}$.

83. In the jumping rings demonstration (■ Figure 32.38), a metal ring is placed over an iron core that penetrates a solenoid. The ring has resistance R_r and inductance L_r in series. The coils around the central magnet form another LR circuit with inductance L_c and resistance R_c, driven by an AC generator. The mutual inductance is M. Find the current in the ring. Use the result to show that there is a net upward force on the ring.

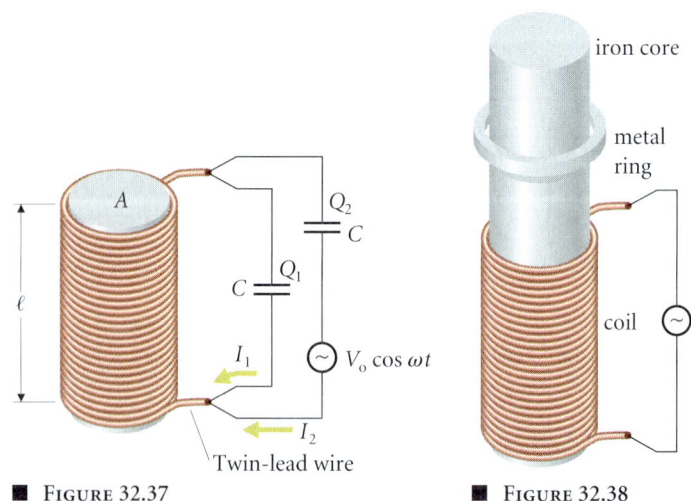

■ **FIGURE 32.37**   ■ **FIGURE 32.38**

CHAPTER 33
Electromagnetic Waves

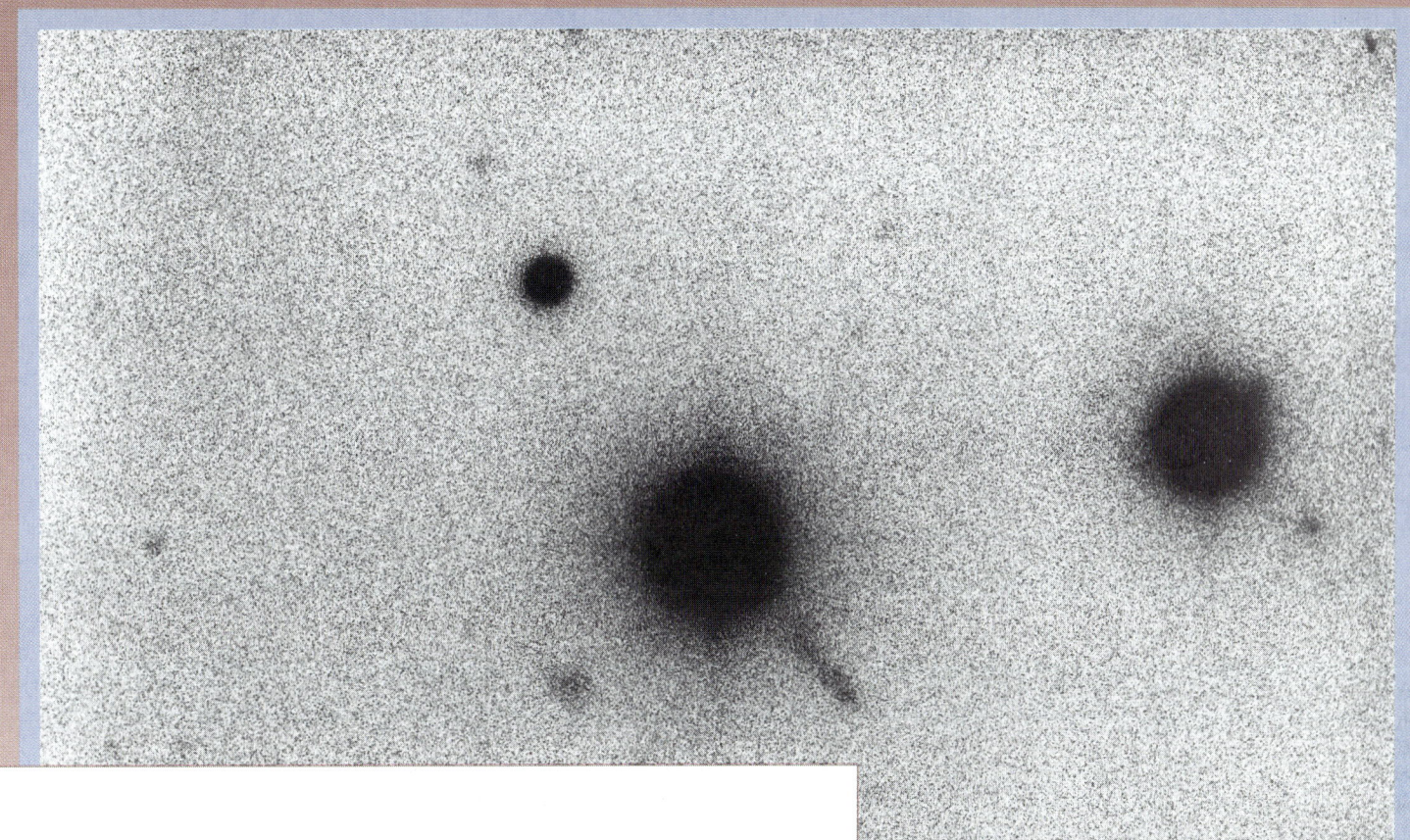

CONCEPTS

Electromagnetic wave
Poynting vector
Radiation pressure
Linear polarization
Circular polarization
Cavity oscillator
Waveguide

GOALS

Understand:
The configuration of fields in a plane EM wave.
How polarized light is produced.
Why a waveguide has a cutoff frequency.

Be able to:
Compute the radiation pressure exerted by an EM wave.
Compute the wave intensity transmitted by a sequence of polarizing filters.

The quasi-stellar object 3C273. (The name means it is the 273rd object in the third Cambridge catalogue of radio-emitting objects.) Quasi-stellar objects (QSOs), as their name suggests, look like stars on optical photographs, but they are believed to be the bright nuclei of distant galaxies. They are the most distant known objects in the universe. To outshine far closer objects, they must emit enormous amounts of radiation throughout the EM spectrum, from radio waves to γ rays. This radiation is the only information we have about them. By carefully studying these EM waves (their distribution in frequency, time variability, total power, etc.), we can build models of the enigmatic QSOs to learn more about them and their role in the evolution of the universe.

I have also a paper afloat, containing an electromagnetic theory of light, which, till I am convinced to the contrary, I hold to be great guns.

JAMES CLERK MAXWELL

According to special relativity theory (Chapter 34), nothing can travel faster than light does in vacuum.

Light from a distant quasar has been traveling toward Earth for billions of years. This electromagnetic radiation carries information about the composition and structure of the quasar and also about gas clouds that lie between us and the quasar. Almost everything we know about the universe beyond our solar system we have learned by analysis of electromagnetic (EM) radiation. Maxwell's identification of EM waves with light was one of the most exciting and beautiful discoveries in physics and has been one of the most significant technologically as well. Electromagnetic waves provide the fastest and most efficient way to transmit information, not only across the universe but on Earth as well, through fiber optic cables and satellite links. This chapter is a brief introduction to the theory of EM waves. We hope also to share an appreciation of the elegance and power of Maxwell's theory.

33.1 Plane Electromagnetic Waves

33.1.1 Origin and Structure of a Plane EM Wave

Reread the section on EM waves in the Overview of electromagnetism.

Accelerating charges generate EM waves. In Figure VI.24 we illustrated how an electron that is rapidly displaced from rest and then abruptly stopped again forms a wave pulse. Displacement of the electron produces outward-moving kinks in the electric field lines, and the electron's motion constitutes a current that produces a closely coupled magnetic field pulse. If the electron oscillates, a train of oscillations propagates outward along each electric field line, together with an oscillating magnetic field pattern (■ Figure 33.1). The electric and magnetic fields in the wave are perpendicular to each other and oscillate in phase. The direction of wave propagation, labeled by the wave vector $\vec{k}$, is perpendicular to both $\vec{E}$ and $\vec{B}$ (■ Figure 33.2):

Study Hint: Review the material on waves, especially Chapter 15. We'll encounter the wave equation again in this chapter.

$$\vec{k} \parallel \vec{E} \times \vec{B}. \tag{33.1}$$

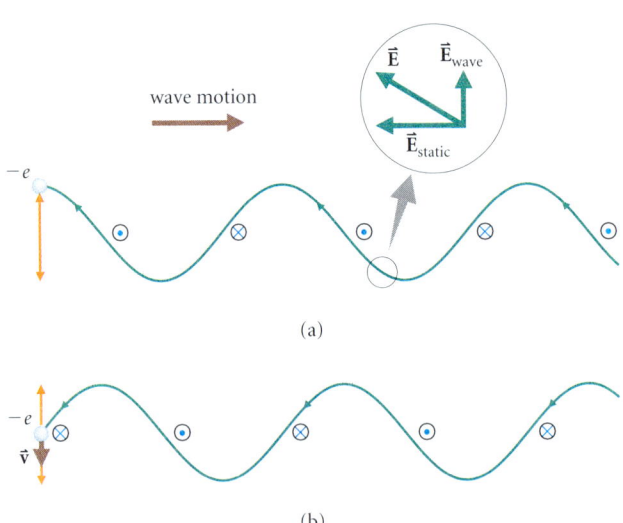

(a)

(b)

■ Figure 33.1
An oscillating electron produces a train of pulses along each electric field line. Notice that since the electron charge is negative, the electric field direction is toward the electron. The wave itself moves outward. As shown in the enlarged view, the electric field near the electron is a mixture of *static* field, the Coulomb field produced by the electron at rest, and *wave* field, due to the electron's oscillations. The electron in motion constitutes electric current that generates the magnetic field in the wave. Here we show the fields in the direction perpendicular to the electron's motion. The wave amplitude varies with direction, but the field pattern is similar. No radiation is emitted in the direction of the electron's acceleration. (a) At the end of an oscillation, the electron is instantaneously at rest. At this time, both $\vec{E}_{wave}$ and $\vec{B}_{wave}$ are zero at the electron's position (left end of the diagram). (b) One quarter-cycle later, the electron has its maximum speed, and the field line (due to $\vec{E}_{static} + \vec{E}_{wave}$) has its maximum slope. Both E_{wave} and B_{wave} are maximum.

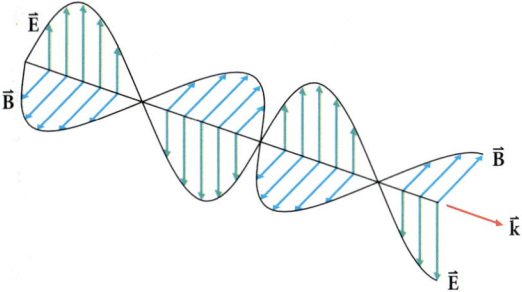

■ **Figure 33.2**
A long way from its source, the static fields have become much smaller than the wave fields, and the wave is approximately plane. This diagram is a snapshot of the wave at a fixed time at points along a single ray. The diagram would be identical along any other ray. The wave vector $\vec{k}$ describes the direction of propagation: it is parallel to $\vec{E} \times \vec{B}$.

The magnitudes of static electric and magnetic fields decrease as one over the distance squared, while, as we shall see in §33.2, the wave fields decrease as one over distance. Thus, at a large distance from the source, the static fields are unimportant and the wave dominates. In this chapter we focus attention on the oscillating (wave) fields. Let's begin by describing the structure of a plane EM wave. In §33.1.2 we shall demonstrate that this structure satisfies Maxwell's equations.

The oscillating electron emits spherical wave fronts. Close to any point P, a great distance d from the accelerating charge, the wave fronts form a set of very nearly parallel planes. We may describe the fields near P as a harmonic plane wave (cf. §16.5.1 and eqns. 15.6–15.8):

$$\vec{E} = \vec{E}_0 \cos(kx - \omega t) \quad \text{and} \quad \vec{B} = \vec{B}_0 \cos(kx - \omega t). \qquad (33.2)$$

The wave has angular frequency ω (equal to $2\pi f$, where f is the frequency of the electron oscillations), wave number $k = |\vec{k}| = 2\pi/\lambda$, and wavelength λ. Its speed is:

$$v = \omega/k = f\lambda. \qquad (33.3)$$

The minus sign in the expression for the wave phase $\phi = kx - \omega t$ indicates that the wave is traveling in the direction of increasing x.

SEE COULOMB'S LAW, §23.2, EQN. (23.6), AND THE BIOT–SAVART LAW, §28.2, EQN. (28.3).

HERE "CLOSE TO" MEANS "WITHIN A DISTANCE MUCH LESS THAN d." SEE §16.5.1 FOR FURTHER DISCUSSION OF THE PLANE WAVE APPROXIMATION.

THIS IS ALSO EQN. (15.2).

AT A FIXED VALUE OF $\phi = \phi_1$, $x = \phi_1/k + vt$ INCREASES AS t INCREASES. SEE ALSO THE DISCUSSION IN CHAPTER 15.

33.1.2 The Wave Equation for $\vec{E}$ and $\vec{B}$

In this section we show that the structure we described in §33.1.1 for an EM wave in vacuum is consistent with Maxwell's equations. We shall see that, according to Maxwell's theory, the waves travel at the speed of light, and we shall find the relation between $\vec{E}_0$ and $\vec{B}_0$.

We have four equations to satisfy (Chapter 30 summary). Gauss' laws for both the electric and magnetic fields are automatically satisfied by the plane wave in vacuum. The field lines for $\vec{E}$ and $\vec{B}$ are parallel sets of straight lines that neither begin nor end. The flux of either through any closed surface is zero (cf. Example 23.13), regardless of the wave's amplitude or speed.

Faraday's law and Ampère's law describe how changes of each field in time are related to the variation of the other field in space. In order to satisfy these equations, we shall find specific requirements on the wave speed and field amplitudes. We begin by applying Faraday's law (eqn. 30.3):

$$\oint \vec{E} \cdot d\vec{\ell} = -\frac{d}{dt} \int \vec{B} \cdot \hat{n} \, dA.$$

■ Figure 33.3 shows how the wave fields vary in a typical small region at a particular time. Following the plan in Figure 30.19, we choose a curve that is parallel or perpendicular to $\vec{E}$ at each point—in this case, the differential rectangle $GHIJ$ in the x-y-plane with height h and

$\oint \vec{E} \cdot \hat{n} \, dA = Q/\epsilon_0$ AND $\oint \vec{B} \cdot \hat{n} \, dA = 0.$

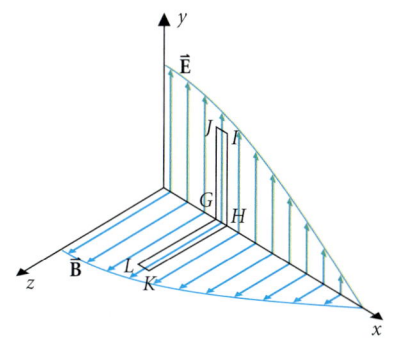

■ **Figure 33.3** (a)

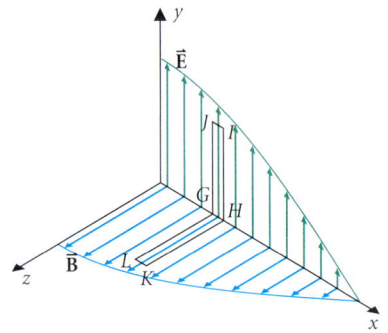
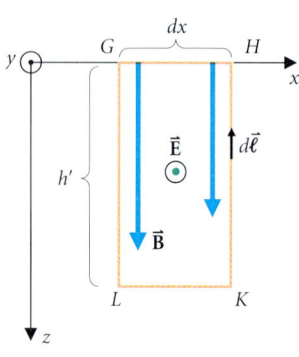

(a) An enlargement of the wave snapshot.　(b) Curve for applying Faraday's law.　(c) Curve for applying Ampère's law.

■ FIGURE 33.3
(a) An enlargement of the wave snapshot showing the $\vec{E}$ and $\vec{B}$ vectors at points along the x-axis. In a plane wave, the field vectors are independent of y and z: the picture would look the same along any line parallel to the x-axis. The wave moves in the $+x$-direction, with $\vec{E}$ in the y-direction and $\vec{B}$ in the z-direction. Notice that $\vec{E} \times \vec{B} = EB\hat{x}$. We apply Faraday's law to the rectangle $GHIJ$, traversed counterclockwise. The normal is then in the $+z$-direction. We apply Ampère's law to the rectangle $GLKH$, traversed counterclockwise, with normal in the $+y$-direction. (b) Curve for applying Faraday's law. The fields are shown at a fixed time. (c) Curve for applying Ampère's law.

THROUGHOUT THIS CHAPTER WE USE $\hat{x}$, $\hat{y}$, AND $\hat{z}$ FOR THE UNIT VECTORS, INSTEAD OF $\hat{i}$, $\hat{j}$, AND $\hat{k}$, BECAUSE OF POSSIBLE CONFUSION WITH THE WAVE VECTOR $\vec{k}$ AND CURRENT DENSITY $\vec{j}$.

REMEMBER: E IS A FUNCTION OF x; $E(x)$ MEANS "E AT x," NOT "E TIMES x."

width dx. The vectors $d\vec{\ell}$ circulate counterclockwise around the rectangle and the normal vector $\hat{n} = +\hat{z}$. The electric field $\vec{E}$ is parallel to $d\vec{\ell}$ on HI, perpendicular to $d\vec{\ell}$ on GH and IJ, and antiparallel to $d\vec{\ell}$ on JG. Thus:

$$\oint_{GHIJ} \vec{E} \cdot d\vec{\ell} = hE(x + dx) - hE(x).$$

The magnetic field is parallel to the normal $\hat{n} = +\hat{z}$ so:

$$\int \vec{B} \cdot \hat{n} \, dA = B(x)h \, dx.$$

We have neglected the small change in B between x and $x + dx$ compared with $B(x)$. The change of flux is caused entirely by change of $\vec{B}$ with time. So from Faraday's law:

REMEMBER: THE PARTIAL DERIVATIVE $\partial B / \partial t$ MEANS DERIVATIVE WITH RESPECT TO TIME AT A FIXED PLACE.

$$h[E(x + dx) - E(x)] = -\frac{d}{dt}[B(x)h \, dx] = -h \, dx \, \frac{\partial B(x)}{\partial t}.$$

Dividing by $h \, dx$, we find:
$$\frac{E(x + dx) - E(x)}{dx} = -\frac{\partial}{\partial t} B(x).$$

In the limit $dx \to 0$, the left-hand side becomes the derivative of E with respect to x at a fixed time (i.e., in the snapshot shown in Figure 33.3):

$$\frac{\partial E}{\partial x} = -\frac{\partial B}{\partial t}. \tag{33.4}$$

Next we apply the Ampère–Maxwell law (eqn. 30.8) using a similar plan (Figure 28.19). In a region free of moving charge, only displacement current appears:

$$\oint \vec{B} \cdot d\vec{\ell} = \epsilon_0 \mu_0 \frac{d}{dt} \int \vec{E} \cdot \hat{n} \, dA.$$

We choose the differential rectangle $GLKH$ in the x-z-plane with height h' and width dx (Figure 33.3c). The vectors $d\vec{\ell}$ circulate counterclockwise, and the normal vector $\hat{n} = +\hat{y}$. The circulation of $\vec{B}$ around this curve is:

$$\oint_{GLKH} \vec{B} \cdot d\vec{\ell} = [B(x) - B(x + dx)]h' = -[B(x + dx) - B(x)]h'.$$

1042　CHAPTER 33 • ELECTROMAGNETIC WAVES

The electric field $\vec{E}$ is parallel to the normal $\hat{n} = \hat{y}$, and the electric flux through the rectangle is:
$$\int \vec{E} \cdot \hat{n} \, dA = E(x) \, dx \, h'.$$

Therefore, applying Ampère's law:
$$-[B(x+dx) - B(x)]h' = \epsilon_0 \mu_0 \frac{\partial E}{\partial t} h' \, dx.$$

$$-\frac{\partial B}{\partial x} = \epsilon_0 \mu_0 \frac{\partial E}{\partial t}. \tag{33.5}$$

Maxwell's equations provide relations between the first derivatives of $\vec{E}$ and $\vec{B}$, while the wave equation (15.10) involves second derivatives. This suggests that we should differentiate eqns. (33.4) and (33.5) again. In particular, differentiating eqn. (33.5) with respect to time at a fixed x, we obtain:

$$-\frac{\partial^2 B}{\partial t \, \partial x} = \epsilon_0 \mu_0 \frac{\partial^2 E}{\partial t^2}.$$

And, differentiating eqn. (33.4) with respect to x at a fixed time gives us:

$$-\frac{\partial^2 B}{\partial x \, \partial t} = \frac{\partial^2 E}{\partial x^2}.$$

Since the two mixed derivatives of B are equal (Exercise 33.1), we may combine these two equations to obtain:

$$\epsilon_0 \mu_0 \frac{\partial^2 E}{\partial t^2} = \frac{\partial^2 E}{\partial x^2}, \tag{33.6}$$

which is a wave equation for E. The coefficient of the second time derivative determines the wave speed:

$$v^2 = \frac{1}{\epsilon_0 \mu_0} = \frac{1}{(8.85 \times 10^{-12} \text{ C}^2/\text{N} \cdot \text{m}^2)(4\pi \times 10^{-7} \text{ H/m})}$$

$$= 8.99 \times 10^{16} \text{ m}^2/\text{s}^2.$$

So the speed v, equal to the speed of light c, is:

$$c \equiv \sqrt{1/(\epsilon_0 \mu_0)} = 3.00 \times 10^8 \text{ m/s}. \tag{33.7}$$

SEE §15.2.3 FOR FURTHER DISCUSSION OF THE WAVE EQUATION.

This remarkable result shows that the speed of light is determined entirely by local electrodynamic properties of space. It not only confirms the electromagnetic nature of light, but it led Einstein to develop the special theory of relativity that profoundly changed our ideas of space and time (see Chapter 34). In accord with that theory, c is now taken to have a fixed value (cf. §1.2.4). Since the value of μ_0 is fixed by the definition of the ampere (§29.2.1), eqn. (33.7) determines the value of ϵ_0.

EXERCISE 33.1 ♦♦ The partial derivatives $\partial^2/\partial x \, \partial t$ and $\partial^2/\partial t \, \partial x$ are equal for any function. Demonstrate this for the harmonic wave function (eqn. 33.2).

EXERCISE 33.2 ♦♦ Differentiate eqn. (33.4) with respect to time and eqn. (33.5) with respect to x to show that B satisfies the same wave equation as E.

The harmonic functions (33.2) are solutions of the wave eqn. (33.6), provided that $\omega/k = c$. The amplitudes E_0 and B_0 are not independent. Using eqn. (33.2):

CHECK THIS BY DIFFERENTIATING EQNS. (33.2) AND SUBSTITUTING INTO EQN. (33.6).

$$\frac{\partial E}{\partial x} = -kE_0 \sin(kx - \omega t) \quad \text{and} \quad \frac{\partial B}{\partial t} = \omega B_0 \sin(kx - \omega t).$$

These relations are consistent with eqn. (33.4), provided that $kE_0 = \omega B_0$, so:

$$B_0 = E_0/c. \tag{33.8}$$

Furthermore, eqns. (33.4) and (33.5) each require that the phases of E and B are the same, as we have assumed. This completes the demonstration that plane waves traveling at the speed of light are a solution of Maxwell's equations in vacuum. Plane waves are not the only solutions, but as we mentioned in Chapter 15, any wave disturbance may be expressed as a sum of harmonic plane waves.

Equations (33.2) are solutions of Maxwell's equations for any value of amplitude and frequency. Thus Maxwell's theory not only explains visible light, but predicts EM waves at all frequencies. Maxwell's theory was triumphantly confirmed in 1887 when Hertz produced and then detected radio waves in the laboratory. By measuring both wavelength and frequency independently, Hertz showed that the radio waves move at the speed of light. We discussed the EM spectrum and the properties that EM waves share with other waves in Chapter 16. In the rest of this chapter, we shall investigate their specific electromagnetic properties.

EXAMPLE 33.1 ♦ An EM wave in the visible part of the spectrum has a wavelength of 550 nm and an electric field amplitude of 670 V/m. Determine the frequency of the wave and the magnetic field amplitude. If the wave travels in the $+x$-direction and its phase is zero when x and t are both zero, write expressions for $E(x, t)$ and $B(x, t)$.

MODEL We use the given information in eqns. (33.2), (33.3), and (33.8).

SETUP AND SOLVE From eqns. (33.3) and (33.7):

$$c = \omega/k = \lambda f = (550 \text{ nm})f = 3.00 \times 10^8 \text{ m/s}.$$

Thus:
$$f = \frac{3.00 \times 10^8 \text{ m/s}}{550 \times 10^{-9} \text{ m}} = 5.5 \times 10^{14} \text{ Hz},$$

and
$$\omega = 2\pi f = 3.4 \times 10^{15} \text{ rad/s}.$$

We use eqn. (33.8) for the magnetic field amplitude:

$$B_0 = \frac{E_0}{c} = \frac{670 \text{ V/m}}{3.0 \times 10^8 \text{ m/s}} = 2.2 \text{ } \mu\text{T}.$$

Finally, $k = 2\pi/\lambda = 1.1 \times 10^7$ rad/m. Thus we may apply eqns. (33.2) to this wave.

$$E(x, t) = (670 \text{ V/m})\cos[(1.1 \times 10^7 \text{ rad/m})x - (3.4 \times 10^{15} \text{ rad/s})t],$$
and $\quad B(x, t) = (2.2 \text{ } \mu\text{T})\cos[(1.1 \times 10^7 \text{ rad/m})x - (3.4 \times 10^{15} \text{ rad/s})t].$

ANALYZE The directions of the electric and magnetic fields are at right angles. We were not given enough information in the problem statement to write vector equations for $\vec{E}$ and $\vec{B}$. The magnitude of B is about one-tenth of the Earth's field (cf. Table 28.1). ∎

33.2 ENERGY AND MOMENTUM TRANSPORT BY EM WAVES

33.1.2 Energy Density and the Poynting Vector

The warmth of sunshine on a cold winter day and your summer sunburn are both the result of electromagnetic energy transport. Energy is stored in a wave by the electric and magnetic fields, and is transported as the field pattern moves. The EM energy density in the wave (eqn. 31.15) is:

$$u = u_E + u_B = \epsilon_0 \frac{E^2}{2} + \frac{B^2}{2\mu_0}.$$

In an EM wave, the magnitudes of E and B are related ($B = E/c$, eqn. 33.8), so:

$$u_B = \frac{B^2}{2\mu_0} = \frac{\epsilon_0 B^2}{2\epsilon_0\mu_0} = \tfrac{1}{2}\epsilon_0 c^2 \left(\frac{E}{c}\right)^2 = \tfrac{1}{2}\epsilon_0 E^2 = u_E.$$

The electric and magnetic fields have equal energy density, and their total is:

$$u = \epsilon_0 E^2 = \epsilon_0 E_0^2 \cos^2(kx - \omega t).$$

The distribution of energy in space at a time t is shown in ■ Figure 33.4. At a later time $t + dt$, the distribution has moved to the right a distance $c\,dt$ as the wave carries energy with it. The energy crossing a plane surface of area A perpendicular to the wave in time dt is the energy contained in a volume of thickness $c\,dt$ along the wave:

$$dU = u\,dV = uAc\,dt.$$

The wave *intensity* is the power it transfers per unit area:

$$\frac{1}{A}\left(\frac{dU}{dt}\right) = cu = c\epsilon_0 E^2 = c^2 \epsilon_0 EB = \frac{EB}{\mu_0}.$$

WAVE INTENSITY WAS DEFINED IN §16.3.

REMEMBER: $c^2 = 1/(\epsilon_0\mu_0)$.

The energy moves in the direction of $\vec{k}$—that is, in the direction of $\vec{E} \times \vec{B}$. Thus the rate and direction of energy transport per unit area are both described by the *Poynting vector*:

NAMED AFTER JOHN HENRY POYNTING (1852–1914).

$$\boxed{\vec{S} \equiv \frac{\vec{E} \times \vec{B}}{\mu_0}.} \qquad (33.9)$$

EXAMPLE 33.2 ◆ Find the average rate at which energy is transported per unit area by a plane EM wave with amplitude $E_0 = 17$ V/m.

MODEL We use eqn. (33.9) together with eqn. (33.8) relating the electric and magnetic field amplitudes.

SETUP The magnitude of the Poynting vector is:

$$|\vec{S}| = \frac{|\vec{E} \times \vec{B}|}{\mu_0} = \frac{|\vec{E}||\vec{B}|}{\mu_0} = \frac{|\vec{E}|^2}{\mu_0 c} = \frac{E_0^2}{\mu_0 c}\cos^2(kx-\omega t).$$

SOLVE The average value of the cosine squared is $\tfrac{1}{2}$, so:

$$\langle|\vec{S}|\rangle = \frac{E_0^2}{2\mu_0 c} = \frac{(17\text{ V/m})^2}{2(4\pi \times 10^{-7}\text{ H/m})(3.0 \times 10^8\text{ m/s})} = 0.38\text{ W/m}^2.$$

ANALYZE Compare this with the intensity of a 60-W light bulb: $\langle S \rangle = (60\text{ W})/(4\pi r^2) = 0.38$ W/m² at $r = 3.5$ m. ■

EXAMPLE 33.3 ◆◆ The Sun is 1.5×10^{11} m from the Earth (on average), and its electromagnetic power (its *luminosity*) is $L = 3.9 \times 10^{26}$ W. What is the mean amplitude of the electric field in radiation from the Sun at the top of the Earth's atmosphere?

MODEL The mean solar energy transported across unit area per unit time is the average magnitude of the Poynting vector at Earth.

SETUP The energy from the Sun passes through a sphere of radius $4\pi R^2$, where R is the Earth–Sun distance (cf. §16.3.3). So:

$$\langle S \rangle = \frac{L}{4\pi R^2} = \frac{3.9 \times 10^{26}\text{ W}}{4\pi(1.5 \times 10^{11}\text{ m})^2} = 1.4 \times 10^3\text{ W/m}^2.$$

But $\langle S \rangle$ is also related to the electric field:

$$\langle S \rangle = E_0^2/(2\mu_0 c).$$

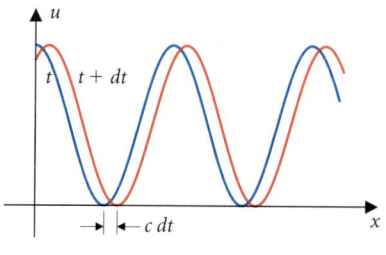

(a)

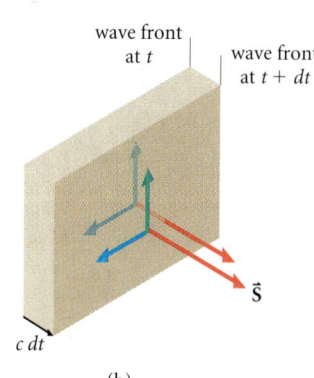

(b)

■ **FIGURE 33.4**
(a) Distribution of EM energy in an EM wave propagating in the x-direction. *Blue line*: distribution at time t. *Red line*: distribution at time $t + dt$. As the wave moves, the energy moves with it at speed c in the direction of the wave vector $\vec{k}$. (b) The energy that passes through a surface of area A in time dt is the energy stored in a volume of thickness $c\,dt$.

$\langle S \rangle = 1400$ W/m² IS CALLED THE *SOLAR CONSTANT*.

SOLVE Solving for E_0 and using the calculated value of $\langle S \rangle$:

$$E_0 = \sqrt{2\mu_0 c \langle S \rangle}$$

$$= \sqrt{2(4\pi \times 10^{-7} \text{ H/m})(3.0 \times 10^8 \text{ m/s})(1.4 \times 10^3 \text{ W/m}^2)}$$

$$= 1.0 \times 10^3 \text{ V/m}.$$

ANALYZE This electric field magnitude is about 10^{-3} of the maximum electric field in a capacitor. Note that this is an average electric field amplitude due to all waves at all frequencies. Since $\langle S \rangle \propto 1/R^2$ (the inverse square law, §16.3.3), $E_0 \propto 1/R$.

33.2.2 Momentum Density and Radiation Pressure

Material particles in motion have linear and angular momentum as well as energy. The same is true of EM waves. Light, absorbed or reflected by a surface, transfers momentum to the surface producing a radiation pressure.

COMPARE WITH THE DEFINITION OF FLUID PRESSURE, §13.2.2, EQN. (13.3).

| *Radiation pressure* is the force per unit area exerted normal to a surface by EM waves.

While basking in the summer sunshine, you absorb nearly a kilowatt of power (cf. Example 33.3), yet you do not feel pushed toward your shadow! Radiation pressure is usually smaller than fluid pressure because radiation has a much smaller ratio of momentum to energy than does an ordinary fluid particle. For a particle with speed v, the ratio is $p/K = mv/(\frac{1}{2}mv^2) = 2/v$. For EM radiation, $p/U = 1/c$.

> A wave incident normally on a surface of area A carries momentum across the surface at a rate:
>
> $$\frac{d\vec{p}}{dt} = A\frac{\vec{S}}{c}. \qquad (33.10)$$

While radiation pressure is negligible in everyday situations, it is crucial in the atmospheres of very luminous stars and in nuclear fusion research experiments where the pressure of laser light compresses target pellets to a very high density.

EXAMPLE 33.4 ♦ Find the radiation pressure exerted on a mirror by sunlight incident normally.

MODEL The pressure is the force per unit area exerted by the light, and the force is the rate of momentum transfer. If we assume that the mirror is a perfect reflector, then the momentum of the wave is reversed, and $\Delta\vec{p} = \vec{p}_{\text{out}} - \vec{p}_{\text{in}} = -2\vec{p}_{\text{in}}$. An equal and opposite momentum is transferred to the mirror.

SETUP Thus momentum is transferred to a unit area of the mirror at a rate (eqn. 33.10):

$$\frac{|\Delta\vec{p}|}{A\,\Delta t} = \frac{|\vec{S}_{\text{in}} - \vec{S}_{\text{out}}|}{c} = \frac{2|\vec{S}_{\text{in}}|}{c}.$$

SOLVE Using the value of S from Example 33.3, the pressure is:

$$P = \frac{2(1.4 \times 10^3 \text{ W/m}^2)}{3.0 \times 10^8 \text{ m/s}}$$

$$= 9.3 \times 10^{-6} \text{ N/m}^2.$$

ANALYZE Compare this value with typical fluid pressures: 1 atm $\approx 10^5$ N/m².

Digging Deeper

THE MOMENTUM OF LIGHT

The ratio of momentum to energy in an EM wave is $1/c$. A rigorous proof of this fact requires mathematics beyond the scope of this text, but we can see that it makes sense by looking at a specific example. We calculate the force an EM wave exerts on a weakly conducting medium that contains a low density of charged particles. The wave causes these particles to move and hence to absorb some of the wave's energy and momentum. (A negligible amount of energy and momentum is reflected.) Radiation from the Sun traveling through interplanetary space toward Earth is an example of this situation.

Energy and momentum are both conserved in the interaction of the wave with the medium. Thus the energy gained by the medium is equal to the energy lost by the wave. The same is true of the momentum. By finding the ratio of energy to momentum absorbed, we can deduce the ratio of energy to momentum in the wave.

At a fixed position, $x = 0$, the wave's electric field oscillates in time:

$$\vec{E}(t) = \vec{E}_0 \cos(\omega t).$$

If the conductivity of the medium is σ, the electric field drives a current density:

$$\vec{j}(t) = \sigma \vec{E}(t) = \sigma \vec{E}_0 \cos(\omega t),$$

consisting of n electrons per unit volume, each moving with an average drift velocity (eqn. 26.6, ■ Figure 33.5):

$$\vec{v}_d = \frac{\vec{j}}{-en} = -\frac{\sigma \vec{E}_0}{en} \cos(\omega t).$$

Each electron feels an electric force $\vec{F}_E = -e\vec{E}$. So the power per unit volume delivered to the electrons by the electric field is:

$$\frac{dW}{dt} = n\vec{F}_E \cdot \vec{v}_d = n(-e\vec{E}) \cdot \vec{v}_d$$

$$= -ne\vec{E}_0 \cos(\omega t) \cdot \left[-\frac{\sigma \vec{E}_0}{ne} \cos(\omega t) \right]$$

$$= \sigma E_0^2 \cos^2(\omega t).$$

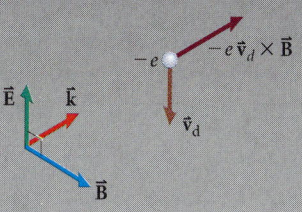

■ FIGURE 33.5
When an electric field encounters an electron, the electron is accelerated opposite $\vec{E}$ and gains a drift velocity $\vec{v}_d$ through the surrounding material. The magnetic force $-e\vec{v}_d \times \vec{B}$ is in the direction of wave propagation.

The magnetic force perpendicular to $\vec{v}$ does no work, and so delivers no power to the electrons.

Because the electric field oscillates, the net momentum it delivers to the electrons averages to zero. The magnetic field of the waves also exerts force on the drifting electrons. The momentum absorbed by the electrons per unit volume is:

$$\frac{d\vec{p}}{dt} = \vec{F}_B = n(-e\vec{v}_d \times \vec{B})$$

$$= -ne\left[-\frac{\sigma \vec{E}_0}{ne} \cos(\omega t) \right] \times \vec{B}_0 \cos(\omega t)$$

$$= \sigma \vec{E}_0 \times \vec{B}_0 \cos^2(\omega t).$$

The rate of momentum transfer divided by the power absorption is:

$$\frac{|d\vec{p}/dt|}{dW/dt} = \frac{E_0 B_0}{E_0^2} = \frac{B_0}{E_0} = \frac{1}{c}.$$

This ratio is independent of the wave amplitude, the conductivity, or any other feature of the specific material. The ratio of momentum to energy gained by the medium from the wave is constant because it is a property of the *wave*:

$$\frac{|\vec{p}_{\text{wave}}|}{U_{\text{wave}}} = \frac{1}{c}.$$

EXAMPLE 33.5 ♦♦ A laser beam with $S = 1.0 \times 10^6$ W/m² is incident normally on a sheet of plastic; 70% is reflected and 30% is absorbed. Find the radiation pressure on the plastic.

MODEL   When light is absorbed ($\vec{p}_{\text{out}} = 0$), the change in momentum is:

$$\Delta \vec{p} = \vec{p}_{\text{out}} - \vec{p}_{\text{in}} = 0 - \vec{p}_{\text{in}} = -\vec{p}_{\text{in}},$$

half the result for reflected light. In this example we must treat the reflected light and the absorbed light separately.

■ **FIGURE 33.6**
A technician inside the target chamber of the NOVA laser at the Lawrence Livermore Laboratory. A pellet of nuclear fuel is compressed by the radiation pressure due to the laser beams. Ten laser beams with a total power of 10^{14} W converge on the 0.05-mm-diameter pellet.

SETUP For the reflected fraction, $P_r = f_r(2S/c)$.

For the absorbed fraction, $P_a = (1 - f_r)S/c$.

SOLVE The total pressure exerted on the plastic is:

$$P = P_r + P_a = (1 + f_r)\frac{S}{c} = (1.7)\frac{1.0 \times 10^6 \text{ W/m}^2}{3.0 \times 10^8 \text{ m/s}} = 5.7 \times 10^{-3} \text{ N/m}^2.$$

ANALYZE Even lasers don't push very hard.

EXERCISE 33.3 ♦ Find the radiation pressure exerted by the NOVA laser with a power of 10^{14} W focused onto and absorbed by a deuterium/tritium pellet 0.05 mm in diameter (■ Figure 33.6).

33.2.3 Energy Transport in Circuits

Energy provided by the battery in a DC circuit is dissipated in resistance, but how does it get from one place to another? The metal wires provide a path for current around the circuit, and one might expect the energy to flow through the metal as well, but that's not the case. The Poynting vector shows where energy flows.

■ Figure 33.7 shows a constant current in a long straight wire with nonzero resistance. A uniform electric field $\vec{E}$ in the wire drives the current, and the current produces a magnetic field that circulates around the wire. The Poynting vector $\vec{S} = (\vec{E} \times \vec{B})/\mu_0$ is perpendicular to the wire's surface and shows that energy flows into the wire from the surrounding space.

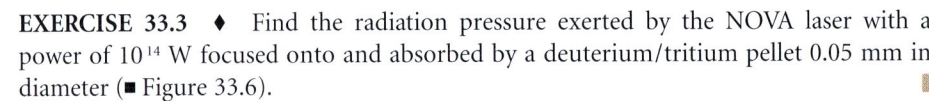

■ **FIGURE 33.7**
The electric field parallel to a long straight wire drives current in the wire. The magnetic field produced by the current circulates around the wire. The Poynting vector $\vec{S} = \vec{E} \times \vec{B}/\mu_0$ points inward and delivers the energy that heats the wire.

The Poynting vector gives the energy flow per unit area per unit time, so we can use it to compute the amount of energy that flows into a length ℓ of the wire with resistance R. Because the fields are cylindrically symmetric around the wire, $|\vec{S}|$ is constant over the surface of the wire, and the energy inflow is $|\vec{S}|$ times the surface area:

$$\frac{dU}{dt} = P = 2\pi a \ell |\vec{S}| = 2\pi a \ell \frac{EB}{\mu_0}.$$

NOTE: *P* HERE MEANS POWER NOT PRESSURE.

This energy is used as joule heating in the wire. To see this, we should express the result in terms of current and resistance. At the surface of the wire,

$$B = \mu_0 I/(2\pi a).$$

Digging Deeper

OBLIQUE INCIDENCE

When light is incident at an oblique angle, the energy reaching an area dA of the surface is the amount crossing area $dA_\perp = dA \cos\theta$ of the wave front (■ Figure 33.8). Thus the rate at which energy reaches the surface is reduced by a factor $\cos\theta$:

$$\frac{dU}{dt}\text{(per unit area)} = |\vec{S} \cdot \hat{n}| = S\cos\theta.$$

Only the component of momentum perpendicular to the surface, $p_\perp = p\cos\theta$, contributes to radiation pressure. Including both effects, the radiation pressure is reduced by a factor $\cos^2\theta$.

EXAMPLE 33.6 ♦♦ A microwave beam with an intensity of 98 kW/m² is incident on a plane metal surface at a 30° angle and is reflected. What is the radiation pressure exerted on the surface?

MODEL Momentum is transferred to the surface at a rate

$$dp_\perp/dt = 2S\cos^2\theta/c.$$

The factor of 2 appears because the beam is reflected, and the $\cos^2\theta$ accounts for the oblique incidence.

SETUP $P_r = 2S\cos^2\theta/c$

SOLVE
$= 2(98 \times 10^3 \text{ W/m}^2)\cos^2(30°)/(3.0 \times 10^8 \text{ m/s})$
$= 4.9 \times 10^{-4} \text{ N/m}^2.$

ANALYZE The 30° angle of incidence reduces the pressure by 25% compared with normal incidence. ■

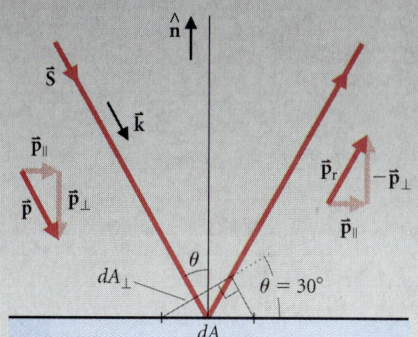

■ **FIGURE 33.8**
When light is incident at an angle θ, the radiation pressure is reduced by a factor of $\cos^2\theta$:
(i) the energy reaching area A of the surface is the energy crossing area $dA_\perp = dA\cos\theta$ of the wave front; and
(ii) only the normal component $p\cos\theta$ of the momentum contributes to the pressure.

The product $E\ell$ is the potential difference ΔV between the ends of the wire segment, which is related to the segment's resistance R:

$$E\ell = \Delta V = IR.$$

REMEMBER: $\vec{E}$ IS UNIFORM.

Thus the power flowing into the wire is:

$$P = 2\pi a(IR)\frac{\mu_0 I}{\mu_0 2\pi a} = I^2 R.$$

All of the EM energy converted to thermal form in the wire is brought in from the surrounding space by the fields. The current and the charge distribution are necessary for the fields to exist, but energy is not carried by the charge itself. Remember, the electron drift velocity is so small ($v_d \sim 10^{-3}$ m/s, §26.2) that electrons would take hours to reach a light bulb from the switch across the room, yet the light comes on almost instantaneously.

SOME OF THE THERMAL ENERGY PRODUCED LEAVES THE WIRE AS INFRARED RADIATION. THUS $\vec{S}$ ACTUALLY CONSISTS OF A DC INWARD COMPONENT AND AN OUTWARD COMPONENT THAT OSCILLATES AT INFRARED FREQUENCIES.

EXERCISE 33.4 ❖ For any steady current, show that the Poynting vector is everywhere parallel to the equipotential surfaces (cf. §25.4). ■

EXAMPLE 33.7 ❖ A battery and resistor are connected at opposite ends of a diameter of a circular wire loop. Assuming the wire is perfectly conducting, sketch the Poynting vectors in the plane of the figure.

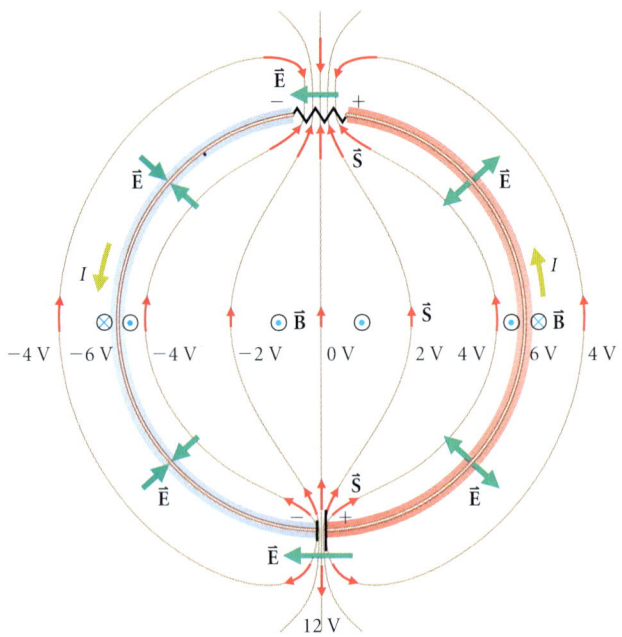

■ **Figure 33.9**
Poynting flux in a DC circuit. The connecting wires are assumed to be perfectly conducting. Electric field lines run from right to left across the circuit. Magnetic field lines circle the wire and come up out of the page inside the current loop (cf. Figure 28.7). The Poynting vector runs across the circuit from the battery to the resistor. The energy flow is greatest where E and B are greatest: near but outside the perfectly conducting wire.

MODEL See ■ Figure 33.9. The Poynting vector is parallel to the equipotential surfaces (see Exercise 33.4), so we begin by sketching them. There is no electric field within the perfectly conducting wires: it's not necessary because there is no resistance. Thus each semicircle of wire is at a constant potential. The potential increases across the battery and decreases across the resistor. The equipotential surfaces join points at equal potential and so run across the circuit between the battery and the resistor. The electric field, perpendicular to the equipotentials, crosses the circuit from one side to the other. The electric field is perpendicular to the perfectly conducting wires and is largest close to the wires (cf. §24.2.1, §25.6). Near the wire, the magnetic field forms loops around the wire, and at large distances it forms a dipole pattern. The Poynting flux is along the equipotential surfaces and is strongest near the wire, where the electric and magnetic fields are most intense.

ANALYZE Essentially, the fields transport EM energy over the surface of the wires and not through them. The circuit sets up exactly the right field pattern to deliver the energy where it is needed.

33.3 POLARIZATION

33.3.1 Linear Polarization

Sunlight or the light from a desk lamp comes from many atoms, each producing a short wave train. The light is the sum of many waves like the one shown in Figure 33.2. The electric fields in the various wave trains bear no special relation to each other, and the light is said to be *unpolarized* (■ Figure 33.10). By contrast, current in a dipole radio antenna flows back and

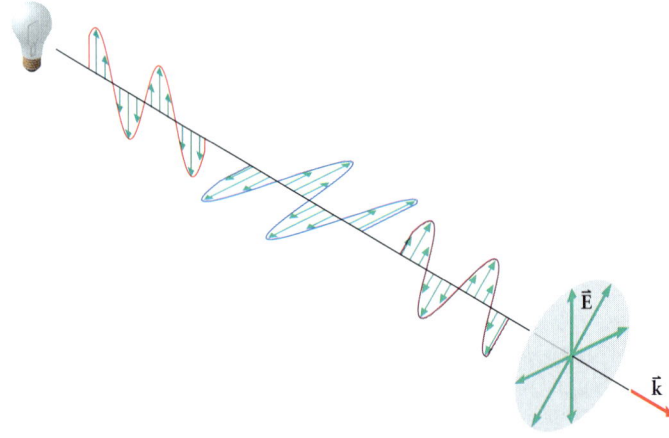

■ **Figure 33.10**
Unpolarized light. The electric field vectors in the various wave pulses bear no special relation to each other, except that they are all perpendicular to $\vec{k}$. At any fixed place the direction of $\vec{E}$ varies randomly in time.

forth along a single direction and produces EM waves with electric fields that are all along the same line: the radiation is *linearly polarized*. Certain light sources, such as lasers, also produce linearly polarized light.

Polarized light can be produced from an unpolarized light source by using a *polarizing filter* that transmits only radiation with its electric field along one direction. That direction defines the *transmission axis* of the filter. Unpolarized light falling on the filter emerges linearly polarized but with a reduced intensity. Half the wave energy falling on the filter is absorbed.

To understand how such a filter works, consider microwaves falling on a set of parallel wires (■ Figure 33.11). Waves with electric field parallel to the wires cause current in the wires. Energy is drained from the waves and dissipated as heat. Electric field perpendicular to the wires causes no current, so waves polarized perpendicular to the wires are not absorbed. In an effective filter the separation of the wires is comparable to the wavelength of the radiation, and the wire thickness is much less than the wavelength. A filter for visible light requires spacings of atomic dimensions. A commercial polarizing filter contains molecules of polyvinyl alcohol that are pulled into long parallel lines by stretching a thin sheet. The sheet is then treated with iodine that is absorbed and provides conduction electrons. Each iodized polyvinyl alcohol molecule acts like one of the wires in the microwave filter.

Polarizing filters are also used to analyze the polarization of incident radiation. A wave polarized at an angle to the transmission axis is partially absorbed and partially transmitted. The incident wave's electric field $\vec{E}_i$ has components along, and perpendicular to, the axis (■ Figure 33.12). In effect, it is the sum of two waves, with two perpendicular polarizations whose amplitudes are the two components of $\vec{E}_i$. Only the wave polarized parallel to the filter's axis is transmitted.

If the incident wave is polarized at angle θ to the transmission axis, then the component of E along the axis is $E_\parallel = E_i \cos \theta$. The intensity of the waves (given by the Poynting vector) is proportional to the square of the electric field amplitude:

$$I = |\vec{S}| = |\vec{E} \times \vec{B}|/\mu_0 = E^2/\mu_0 c.$$

So, the transmitted intensity I_t is:

$$I_t = I_i \cos^2 \theta, \tag{33.11}$$

where I_i is the intensity incident on the filter. This relation was discovered by Etienne Malus in 1809; it is known as the law of Malus.

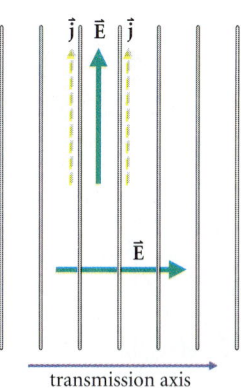

■ **FIGURE 33.11**
Microwaves incident on a set of parallel wires. Electric field $\vec{E}$ parallel to the wires drives current, and energy is drained from the wave. If $\vec{E}$ is perpendicular to the wires, there is no current and the wave is not absorbed. The transmission axis lies perpendicular to the wires, as shown.

SEE ALSO §16.3.

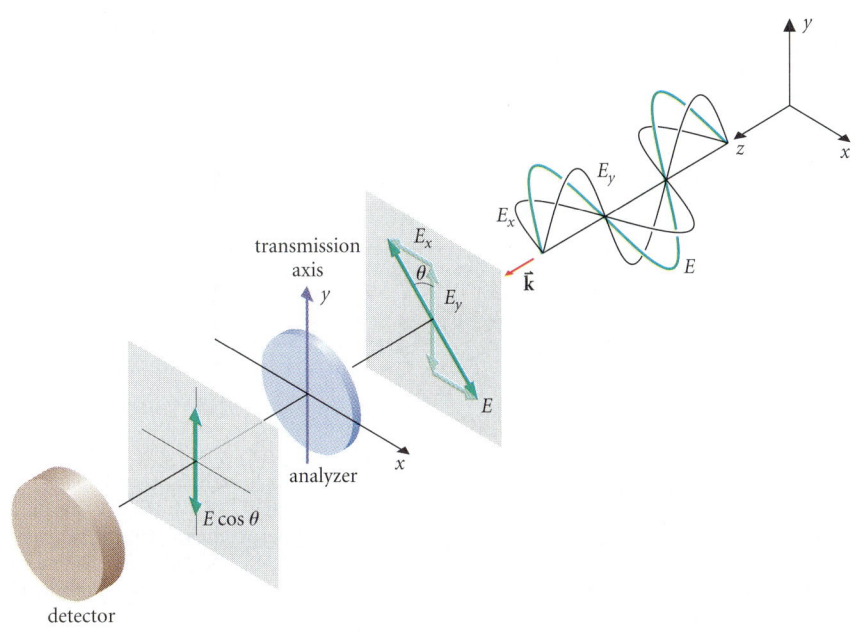

■ **FIGURE 33.12**
The law of Malus. A polarized wave incident on the analyzer is composed of two waves polarized along, and perpendicular to, the transmission axis. Only the component along the transmission axis is transmitted. Its electric field amplitude is $E \cos \theta$, and the corresponding intensity is $I_i \cos^2 \theta$.

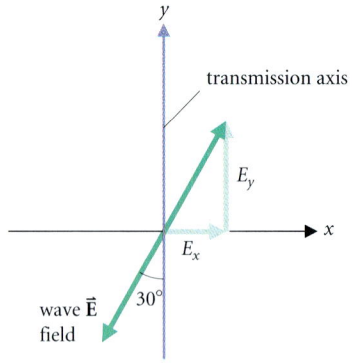

FIGURE 33.13
Polarized light is incident on a polarizing filter lying in the plane of the page. Only the component of $\vec{E}$ along the transmission axis is transmitted. We have chosen coordinates with y along the transmission axis.

EXAMPLE 33.8 ♦ Light polarized with $\vec{E}_i$ at 30° to the vertical is incident on a polarizing filter with a vertical transmission axis (■ Figure 33.13). Describe the transmitted wave. What percentage of the incident intensity is transmitted?

MODEL The incident electric field may be resolved into components along (y) and perpendicular (x) to the transmission axis. Only the y-component is transmitted.

SETUP Since $\cos 30° = \sqrt{3}/2$, and $\sin 30° = \frac{1}{2}$: $\vec{E}_i = \dfrac{E_i}{2}(\hat{x} + \sqrt{3}\hat{y})$.

Only the vertical (y) component is transmitted: $\vec{E}_t = \dfrac{\sqrt{3}}{2}E_i\hat{y}$.

SOLVE The transmitted wave is vertically polarized with an amplitude $\sqrt{3}/2$ of the incident wave. Its intensity is given by:

$$\frac{I_t}{I_i} = \left(\frac{E_t}{E_i}\right)^2 = \frac{3}{4}.$$

Thus 75% of the light is transmitted.

ANALYZE When analyzing the effects of polarizers, always choose components of $\vec{E}$ along and perpendicular to the transmission axis. ■

EXAMPLE 33.9 ♦♦ The transmitted wave in Example 33.8 is incident on a second polarizing filter with its transmission axis at 60° to the first—that is, perpendicular to the wave's original polarization (■ Figure 33.14). What fraction of the original intensity is transmitted by both filters?

MODEL The light emerging from the first filter is vertically polarized. To analyze how it interacts with the second filter, we need to resolve it into the sum of waves polarized perpendicular and parallel to that filter's transmission axis (Figure 33.14b). Only the parallel component is transmitted.

SETUP From Example 33.8, the electric field in the wave incident on the second filter is $\vec{E}_2 = (\sqrt{3}/2)E_i\hat{y}$. Its component parallel to the transmission axis of the second filter is:

$$E_{\parallel,2} = E_2 \cos(60°) = \left(\frac{\sqrt{3}}{2}E_i\right)\frac{1}{2} = \frac{\sqrt{3}}{4}E_i.$$

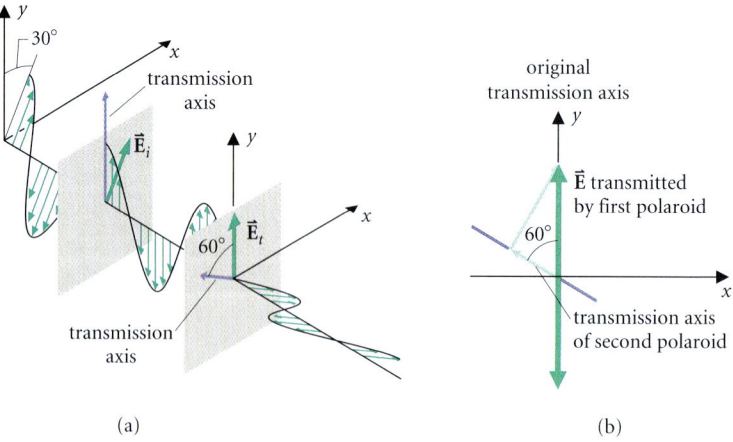

FIGURE 33.14
(a) A second filter is placed with its transmission axis at a 60° angle with the first—that is, perpendicular to the polarization of the light incident on the entire system. (b) We resolve the electric field transmitted by the first filter into components along and perpendicular to the transmission axis of the *second* filter. Only the parallel component is transmitted. The beam intensity decreases upon passing through the first filter. Nevertheless, more light passes through the combined system than if the first filter were absent.

SOLVE The transmitted intensity $I_{t,2}$ is proportional to the square of this field:

$$I_{t,2} = \tfrac{3}{16} I_i.$$

ANALYZE The second filter alone would transmit no light, since its transmission axis is perpendicular to the electric field in the wave incident on the entire system. The intermediate filter rotates the polarization of the wave and so allows $\tfrac{3}{16}$ of the original intensity to pass both filters!

EXERCISE 33.5 ♦♦♦ Unpolarized light is incident on a polarizing filter. Use Malus' law to show that 50% of the intensity is transmitted.

33.3.2 Polarization by Reflection

Reflected light is often highly polarized (■ Figure 33.15). To understand why, we look at what causes reflection and refraction—electrons set oscillating by the wave. These electrons themselves radiate waves polarized parallel to their acceleration. Waves emitted perpendicular to the acceleration have the greatest amplitude; no energy is emitted parallel to the acceleration (cf. Figure 33.1). Thus the direction of the electron motion determines the amplitude and polarization of the reflected wave.

Electrons in the second medium oscillate parallel to the electric field in the transmitted wave. This direction is in turn determined both by the direction of transmitted rays and by the polarization. When the incident wave is polarized parallel to the surface, electrons oscillate parallel to the surface and produce reflected and refracted rays with the same polarization (■ Figure 33.16a). In contrast, when the incident magnetic field is parallel to the surface, the incident electric field makes an angle θ_1 with the surface and the electric field in the transmitted wave makes an angle θ_2 with the surface. The electrons in the second medium oscillate parallel to $\vec{E}_2$, at an angle θ_2 to the surface (Figure 33.16b). The reflected ray makes a small angle ϕ with the direction of electron acceleration, so its amplitude is small.

■ **FIGURE 33.15**
(a) A photograph taken through a window shows unwanted reflections in the glass. (b) A polarizer placed in front of the camera lens reduces the reflections that appear in the photograph because the reflected light is partially polarized.

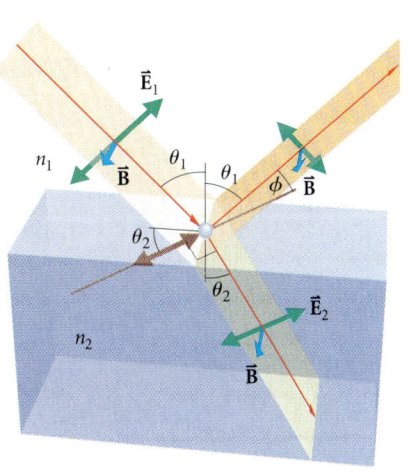

(a) Incident light polarized with $\vec{E}$ parallel to the surface.

(b) Incident light polarized with $\vec{B}$ parallel to the surface.

(c) Reflected ray at right angles to the refracted ray.

■ **FIGURE 33.16**
Polarization by reflection. (a) Incident light polarized with $\vec{E}$ parallel to the surface. Electron oscillations in the medium are also parallel to the surface, as is the polarization of the reflected and transmitted light. (b) Incident light polarized with $\vec{B}$ parallel to the surface. The electric field lies in the plane of incidence, perpendicular to the ray. Oscillations of electrons in the medium, parallel to $\vec{E}$ in the medium, produced a reflected wave with reduced amplitude. (c) Reflected ray at right angles to the refracted ray. Light polarized with $\vec{B}$ parallel to the surface [cf. (b)] would have to be emitted along the line of the electron acceleration. No radiation is emitted along this line, and there is no reflected ray. Light polarized with $\vec{E}$ parallel to the surface [cf. (a)] is reflected. The incident angle that produces this configuration is called Brewster's angle.

Digging Deeper

Polarization in Nature

Polarization of Skylight

The whole sky looks bright because sunlight is scattered by molecules in the Earth's atmosphere. Because the incident sunlight is unpolarized, electrons in the molecules oscillate in all possible directions in the plane perpendicular to the incident light (■ Figure 33.17). An observer looking at right angles to the Sun is looking along this plane and sees radiation only from electrons oscillating in the single direction at right angles to the line of sight. The observed radiation is polarized (■ Figure 33.18). At other angles, the scattered light is partially polarized. Photographers sometimes use Polaroid filters to reduce the apparent brightness of the sky in their photographs.

Polarization of Starlight

Dust grains in the region between stars scatter and absorb starlight (■ Figure 33.19). Elongated grains tend to align per-

■ **FIGURE 33.18**
A pair of polarizers with their transmission axes at right angles. The right Polaroid is noticeably darker than the left one, indicating the partial polarization of skylight.

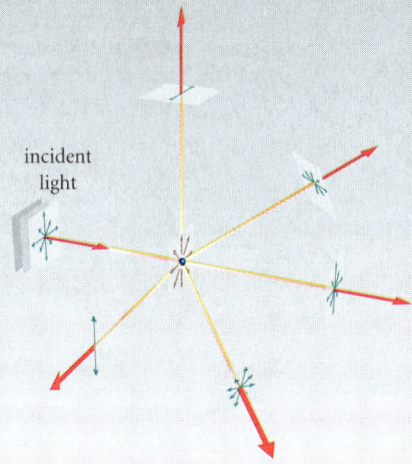

■ **FIGURE 33.17**
Light from the Sun is scattered by air molecules to illuminate the whole sky. The molecules oscillate along the electric field direction, perpendicular to the rays, and light emitted in the plane of oscillation is polarized. This is the primary reason why a Polaroid filter on a camera can emphasize clouds, or "cut through" haze.

pendicular to the interstellar magnetic field, and scattered light is polarized with $\vec{E}$ parallel to the interstellar field direction. By observing the polarization, it is possible to map out the interstellar magnetic field. The observations also allow us to deduce valuable information about the nature of the grains themselves.

Interstellar Masers

Water molecules in a giant gas cloud where stars are forming emit very intense, highly polarized microwaves (■ Figure 33.20). The molecules are excited to a high-energy state by an energy source in the cloud, perhaps shock waves, and emit coherent radiation, forming a natural *maser*.

Masers and lasers operate on the same principle though at different frequencies. An electromagnetic wave stimulates the excited molecules to emit radiation in phase with the incident wave and also with the same polarization. Hence the acronym: Microwave Amplification by Stimulated Emission of Radiation.

For one particular incident angle θ_B, known as Brewster's angle, the refracted and reflected rays at are right angles (Figure 33.16c). If the incident light is polarized with $\vec{B}$ parallel to the surface, the reflected wave would have to be parallel to the oscillating electrons' acceleration. The electrons cannot emit in this direction, so there is no reflected wave. Incident light polarized with $\vec{E}$ parallel to the surface is reflected as usual. Thus when unpolarized light is reflected from a surface at Brewster's angle, the reflected light is completely polarized parallel to the surface. Light reflected at other angles is partially polarized.

We may obtain an expression for Brewster's angle from Snell's law (eqn. 16.26):

$$n_1 \sin \theta_1 = n_2 \sin \theta_2.$$

FIGURE 33.19
(a) The Horsehead nebula in the constellation of Orion. The dark shape is caused by dust clouds that absorb the light coming from behind them. Elongated dust particles are aligned perpendicular to the local magnetic field.

(b) Light scattered by dust grains, like microwaves scattered by wires (Figure 33.11), is polarized with $\vec{E}$ perpendicular to the grains and thus parallel to the interstellar magnetic field. Each mark on this diagram shows the measured polarization of light from a single star. Longer lines indicate greater polarization. The diagram maps the direction of the galactic magnetic field. Notice that $\vec{B}$ tends to lie along the center of the diagram—parallel to the plane of our galaxy (the Milky Way).

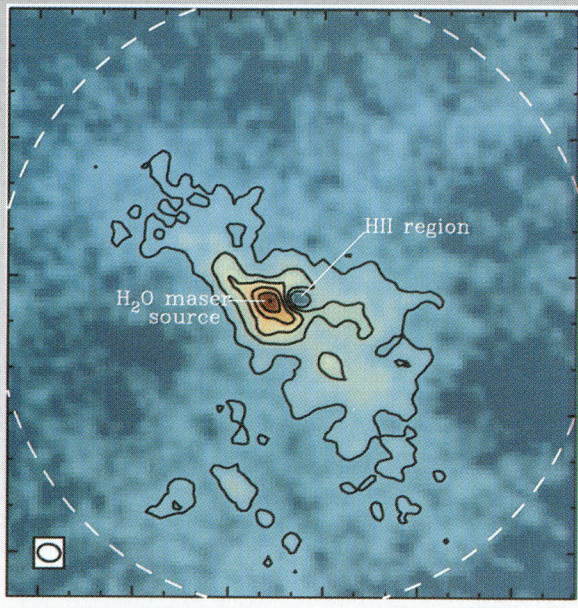

FIGURE 33.20
This region of the sky—called W3—is believed to be a birthplace for stars. In the central region, shown in red in this false-color radio map, there is a very bright, highly polarized source of EM waves. Both the intensity of the source and the strong polarization help the identification of this source as a maser.

When the refracted and reflected rays are at right angles (Figure 33.16c), $\theta_2 = 90° - \theta_1$, and Snell's law becomes:
$$n_1 \sin \theta_1 = n_2 \sin(90° - \theta_1) = n_2 \cos \theta_1.$$

Thus, with $\theta_1 = \theta_B$:
$$\tan \theta_B = \frac{n_2}{n_1}. \tag{33.12}$$

Polaroid sunglasses take advantage of the polarization of reflected sunlight. The sunglasses are polarizing filters with a vertical transmission axis. They are particularly effective in reducing the glare from water and highways.

FIGURE 33.21
A *smart* antenna array uses 24 helical antennae to receive rocket telemetry data from any direction in any polarization.

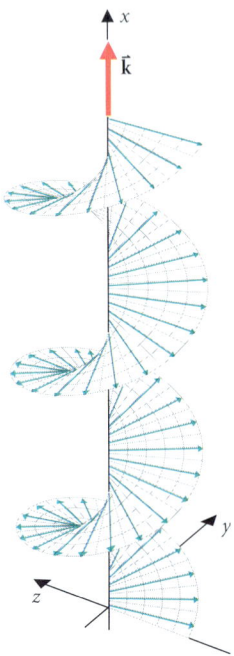

FIGURE 33.22
Circular polarization. This picture is a snapshot of the wave at a fixed time. The tips of electric field vectors lie on a helix. In this example the polarization is right circular—with the thumb of your right hand pointing along $\vec{k}$, your fingers indicate the direction of the helix.

EXAMPLE 33.10 ◆ For what angle of incidence is light reflected from water completely polarized?

MODEL The reflected light is completely polarized when the light is incident at Brewster's angle.

SETUP The refractive index of water (Table 16.3) is $n_2 = 1.33$, and $n_1 = 1.00$ is the refractive index of air.

SOLVE Brewster's angle for water is given by eqn. (33.12):

$$\tan \theta_B = n_2/n_1 = 1.33 \quad \text{and} \quad \theta_B = 53°.$$

ANALYZE Light incident at other angles is partially polarized, since the reflected light polarized with $\vec{B}$ parallel to the surface has a smaller amplitude than the reflected light polarized with $\vec{E}$ parallel to the surface.

33.3.3 Circular Polarization

Radio signals transmitted to a satellite from Earth are quite weak by the time they reach its orbit. If the waves were linearly polarized, the receiver antenna would have to be carefully aligned with the wave electric field, a prohibitive requirement for a satellite. Circular polarized waves have a rotating electric field that can be transmitted and received efficiently by a helical antenna (■ Figure 33.21), which is not sensitive to orientation.

Circular polarization can be produced by combining two linearly polarized waves that are 90° out of phase. The electric fields in the two waves are at right angles:

$$\vec{E}_1 = E_0 \hat{y} \cos(kx - \omega t) \quad \text{and} \quad \vec{E}_2 = E_0 \hat{z} \cos(kx - \omega t - \pi/2).$$

The electric field in the superposed wave is:

$$\vec{E} = E_0[\hat{y} \cos(kx - \omega t) + \hat{z} \sin(kx - \omega t)].$$

■ Figure 33.22 shows how this electric field vector varies in space at one time. The tips of the vectors form a right-handed helix. This polarization is called *right circular*. ■ Figure 33.23 shows how the electric field varies in time at a fixed place. An observer looking toward the source sees the electric field vector rotate clockwise with constant angular speed ω.

If the wave polarized along $\hat{z}$ has a phase lead of 90° instead of a lag, the superposed wave rotates counterclockwise, and the $\vec{E}$ vectors lie on a left-handed helix. Such a wave has *left circular* polarization.

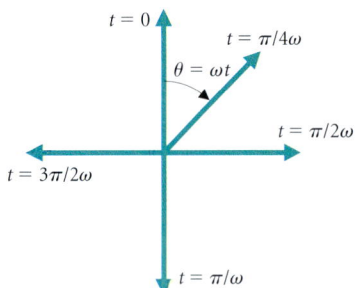

FIGURE 33.23
A movie of the wave at a fixed place shows the electric vector rotating. In this right circular wave propagating out of the page toward you, the vector rotates clockwise with angular speed ω. Imagine the helix in Figure 33.22 moving along the axis past you. The electric field vector in the plane of the page changes as the helix moves.

One way to produce circularly polarized light is to pass unpolarized light through a filter in which two linear polarizations travel at different speeds. Calcite is one of a class of substances, called *birefringent*, that have this property. If the calcite crystal has the right thickness, one linear polarization emerges retarded in phase by 90° compared with the other polarization. This filter is called a *quarter-wave plate*.

EXAMPLE 33.11 ❖ Show that a linearly polarized wave is a superposition of right and left circular polarized waves with equal amplitude.

MODEL At a fixed position, the electric field of the right circular wave is rotating clockwise, and $\vec{E}$ in the left circular wave is rotating counterclockwise (■ Figure 33.24). At some time t, both lie along the y-axis. At a later time $t + \Delta t$, they have rotated through equal angles in opposite directions. The sum still lies along the y-axis, although its magnitude is reduced. This is what we expect for a linearly polarized wave.

SETUP AND SOLVE At time $t + \Delta t$, the electric field has magnitude:

$$E = 2E_0 \cos \theta = 2E_0 \cos(\omega \, \Delta t).$$

This is the electric field of a linearly polarized wave with amplitude $2E_0$.

ANALYZE To understand the propagation of a linearly polarized wave through a polarizing filter, we needed to model it as a sum of two waves, each polarized along or perpendicular to the transmission axis (e.g., Example 33.8). Similarly, to understand the propagation of such a wave through a solution of spiral molecules or a magnetized plasma in interstellar space, it is essential to model the wave as a sum of two circular polarizations. (See Problem 51.) This example shows how to do that. Unpolarized light may be modeled as a superposition of linearly or circularly polarized waves.

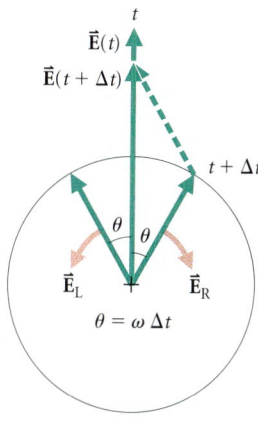

■ **FIGURE 33.24**
The superposition of a right-circular wave and a left-circular wave, each with the same amplitude E_0 and angular frequency ω, is a linearly polarized wave. In this example the sum of the two electric field vectors always lies along the y-axis and varies between $2E_0 \hat{y}$ and $-2E_0 \hat{y}$.

✻ 33.4 ELECTROMAGNETIC OSCILLATIONS AND MICROWAVES

33.4.1 Cavity Oscillators

You need a resonant circuit to detect oscillating radio signals. The *LRC* circuits we discussed in Chapter 32 are effective detectors of FM signals because wavelengths in the FM band, roughly 3 m, are much larger than the size of the circuit. Light can cross the circuit in much less than one oscillation period of the wave and the circuit responds coherently. However, at higher frequencies (shorter wavelengths), this is no longer true. An effective detector or generator for microwaves should be a compact resonant device.

A cylindrical cavity oscillator (■ Figure 33.25) illustrates how such devices operate. Electromagnetic oscillations can occur in the hollow space surrounded by conducting walls. Like the different standing wave patterns on a string, many different modes of oscillation can occur. Their wavelengths are comparable to the dimensions of the cavity, a few centimeters in the microwave band.

SEE THE DISCUSSION IN §32.1.1.

SEE §15.4.2 FOR STANDING WAVE MODES ON A STRING.

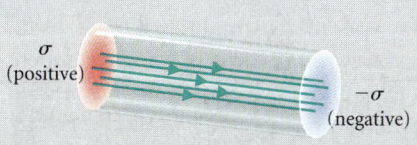

■ **FIGURE 33.25**
Electromagnetic oscillations in a cylindrical cavity. (a) $t = 0$. Electric field runs along the length of the cavity. Charge density exists on the two ends.

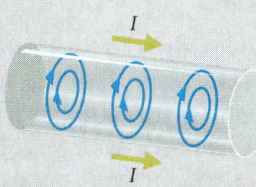

(b) $t = \pi/2\omega = T/4$. As the electric field changes, magnetic field circulates around the displacement current. Current in the conducting walls changes the surface charge density σ.

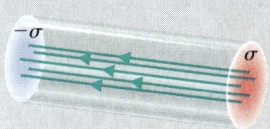

(c) $t = \pi/\omega = T/2$. Electric field and charge density have reversed relative to (a).

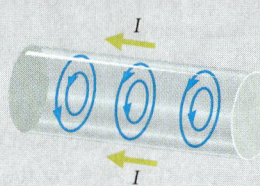

(d) $t = 3\pi/\omega = 3T/4$. Magnetic field circulates around the displacement current: $\vec{B}$ and I are opposite from those in (b).

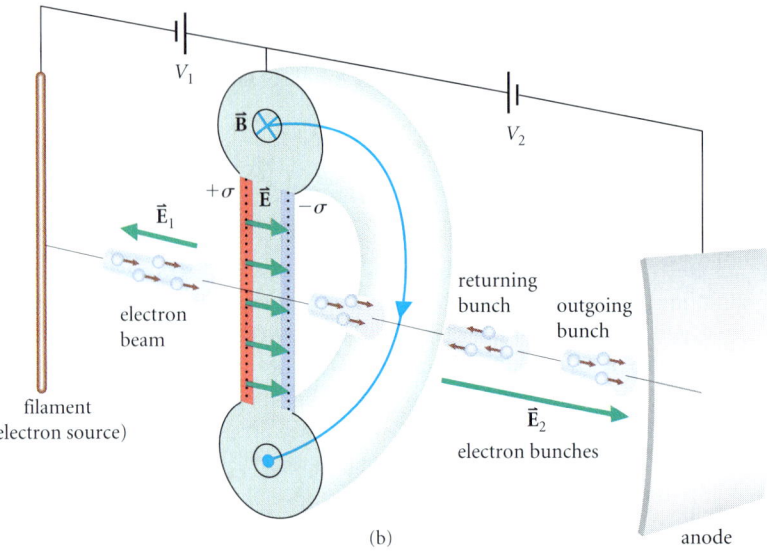

■ **FIGURE 33.26**
(a) A Klystron tube. (b) Magnetic field runs around the doughnut-shaped tube, while electric field runs across the doughnut hole. An electron beam passes through a metal grid at the doughnut hole, is reflected by a charged plate, and passes back through the hole. With proper timing of the beam, it transfers its kinetic energy into electromagnetic energy in the cavity.

The mode illustrated in Figure 33.25 has electric field lines parallel to the long axis of the cylinder. The electric field is maximum along the axis and zero at the conducting walls. Magnetic field circulates around the displacement current that is produced as the electric field changes. Current flows back and forth along the inner surface of the cavity as the charges on the end faces vary with the electric field. The cavity stores energy in the form of electric and magnetic field, and behaves like an *LC* circuit in which both inductor and capacitor occupy the same volume. Energy loss by the surface current damps the oscillator, just as the resistance in an *LRC* circuit damps its oscillations. An important practical application of resonant cavities is the Klystron tube (■ Figure 33.26), developed as a generator for high-frequency radar signals.

EACH POSSIBLE MODE OF THE CAVITY HAS A DIFFERENT FREQUENCY, DIFFERENT PATTERN OF FIELDS, AND DIFFERENT EFFECTIVE INDUCTANCE AND CAPACITANCE.

33.4.2 Waveguides

Antennae and transmission lines for microwaves employ waveguides, conducting tubes inside which the microwaves propagate. The structure of EM waves propagating in a guide is different from that of unconfined waves. If the tube's walls were perfectly conducting, the electric field inside any wall would have to be zero. The electric field in the cavity would be exactly normal to the walls (cf. §25.6). Inside ordinary conductors, the field is extremely small and decreases to zero within a small distance from the surface. We shall make very little error if we assume $E = 0$ within the metal. A static magnetic field inside a conductor could exist, but a changing magnetic field would generate an electric field and so cannot occur. Since the normal component of $\vec{B}$ is continuous across any boundary (cf. §28.3.1) and $\vec{B}$ is zero within the conductor, $\vec{B}$ inside the guide is tangential at the surface. Thus the boundary conditions at the surface of the wave guide are:

THE MAGNITUDE OF THE ELECTRIC FIELD INSIDE THE CONDUCTOR COMPARED WITH THAT OUTSIDE IS APPROXIMATELY $\sqrt{\omega \epsilon_0 / \sigma}$, OR ABOUT $\sqrt{(10^{-19}\text{ s})\omega}$ FOR COPPER.

- $\vec{E}$ is normal to the surface, or zero.
- $\vec{B}$ is tangential to the surface, or zero.

We can understand the main features of waveguides by looking at a simple system: parallel metal plates with separation a (■ Figure 33.27). A wave propagating along the guide in the x-direction and polarized with $\vec{E}$ normal to the plates (in the z-direction) and $\vec{B}$ parallel to the plates (the y-direction) satisfies the boundary conditions and so is a possible mode. It is called a *transverse electromagnetic* or TEM mode. A wave polarized with $\vec{E}$ parallel to the

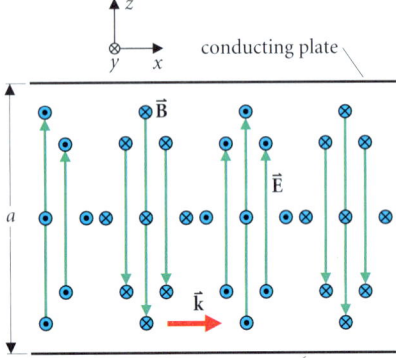

■ **FIGURE 33.27**
An EM wave propagating between two parallel plates. In the TEM mode, both $\vec{E}$ and $\vec{B}$ are perpendicular to the wave vector, just as in a free-space wave.

plates cannot propagate directly along the guide because $\vec{E}$ is tangential to the surface. Such a *transverse electric* (TE) mode can propagate as a superposition of waves reflecting off the plates as shown in ■ Figure 33.28. Rays (in red) indicate the reflecting waves. Wave fronts, perpendicular to the rays, are indicated by the electric and magnetic field vectors. There is a phase change of π (the electric field reverses direction) at each reflection, which keeps $\vec{E} = 0$ on the guide surface, as required. The components of $\vec{B}$ perpendicular to the plates also sum to zero at the surface, so $\vec{B}$ is tangential. The electric and magnetic fields have maximum strength in the center of the guide (■ Figure 33.29), where the superposition of the magnetic field vectors is perpendicular to the plates.

We discussed such phase changes in §15.4 and §17.1.5.

Only certain reflection angles are possible for a wave with a given frequency because of the way in which waves have to superpose in the guide. Each allowed angle corresponds to a particular wavelength of the disturbance in the guide. The allowed wavelengths are given by:

$$\lambda_g = \frac{\lambda_f}{\sqrt{1 - (\lambda_f/2a)^2(2m-1)^2}}, \tag{33.13}$$

Dig Deeper to see how this works.

where a is the plate separation and m is an integer. The wavelength of any disturbance in the waveguide (λ_g) is greater than the free-space wavelength (λ_f). When $\lambda_f = 2a/(2m-1)$, the guide wavelength becomes infinite. (This happens when the reflection angle θ is 0.) In our model, this corresponds to the two waves bouncing back and forth in the z-direction, and not traveling in the x-direction at all. If λ_f is greater than $2a$, no mode number m gives a real solution for the wavelength in the guide, and no wave can propagate. The wavelength $\lambda_c = 2a$ is called the *cutoff wavelength* for the guide.

EXAMPLE 33.12 ♦♦ Two parallel conducting slabs are separated by 2.50 cm. What is the cutoff wavelength for this waveguide? What is the wavelength in the guide of radiation with a free-space wavelength of 3.00 cm, propagating in the TE mode?

MODEL The cutoff wavelength is twice the slab separation, or 5.00 cm. Since 3 cm is less than this cutoff wavelength, the 3.00-cm radiation can propagate between the slabs. Allowed wavelengths in the TE mode are given by eqn. (33.13) with integer values of m. However, for $m = 2$, $(\lambda_f/2a)^2(2m-1)^2 = (\frac{3}{5})^2(3^2) = 3.24 > 1$. For $m \geq 2$, λ_g is not defined, so only the lowest mode ($m = 1$) can propagate.

SETUP We use eqn. (33.13). With $m = 1$, $a = 2.50$ cm, and $\lambda_f = 3.00$ cm:

SOLVE
$$\lambda_g = \frac{3.00 \text{ cm}}{\sqrt{1 - [(3.00 \text{ cm})/(5.00 \text{ cm})]^2}} = 3.75 \text{ cm}.$$

ANALYZE The TEM mode could also propagate. In practice, engineers usually design guides for a single mode.

EXERCISE 33.6 ❖ Describe the radiation that emerges when unpolarized radiation with a 6.0-cm wavelength enters the waveguide in Example 33.12.

It is also possible for the reflecting waves in a guide to have $\vec{B}$ parallel to the walls, $\vec{B} = B\hat{y}$. These are called transverse magnetic (TM) modes. In the parallel plate waveguide, the TM modes happen to have the same cutoff frequencies as the TE modes, although in many waveguides there are two different sets of frequencies.

A practical waveguide has two connected sets of plates forming a rectangular cross section. Wave propagation in such a guide shows the same general features that we have discussed for the parallel plates, except that the TEM mode does not propagate. In particular, the TE and TM modes each exhibit cutoffs. Some aspects of these guides are explored in the problems.

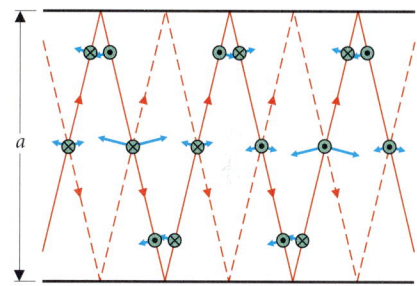

■ **Figure 33.28**
In the TE mode, the electric field is parallel to the plates. Here we model the mode with waves reflecting back and forth off the walls. The electric field vectors sum to zero at each plate, while the magnetic field vectors sum to give $\vec{B}$ parallel to the plates at the surface.

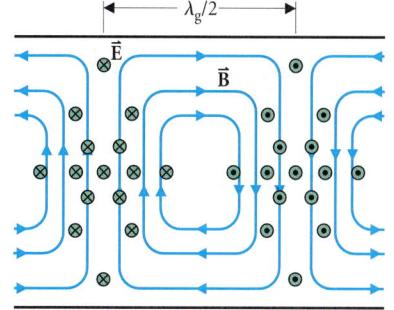

■ **Figure 33.29**
Pattern of electric and magnetic fields in a waveguide. This is the mode described in Figure 33.28. The magnetic field circulates around the displacement current. The magnetic field is tangential at the plates, and the electric field is zero there. The wave propagates to the right. (Check the direction of the Poynting vector.)

Digging Deeper

SUPERPOSITION OF REFLECTING WAVES

Figure 33.30 shows a wave front and a ray of the reflecting wave for an angle of incidence θ that is possible. The electric field must have the same phase everywhere on the wave front PQ. Thus along the ray $APCQ$, the phase difference between P and Q is an integer multiple of 2π. The calculation is easiest if we compute the phase changes from A to P and from A to Q along the ray and then subtract.

The straight-line distance s from A to Q is given by the plate separation a and the angle of incidence θ:

$$s = a \tan \theta.$$

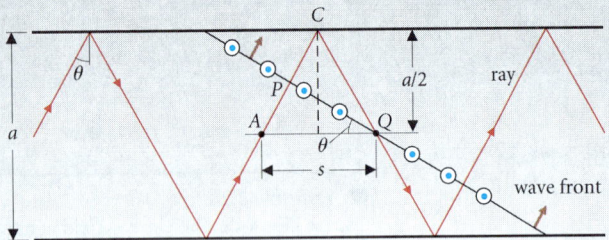

FIGURE 33.30
The wave phase is constant along each wave front, such as the wave front PQ shown here. Only for certain angles of incidence θ with $\cos \theta = (2m - 1)\lambda_f/2a$ do the reflecting waves superpose to produce allowed modes in the guide. (Remember that a large number of such wave fronts are superposed in the guide.)

Then the distance AP is $s(\sin \theta) = a(\tan \theta)(\sin \theta)$. The phase difference from A to P is:

$$\phi_P - \phi_A = \frac{2\pi s(\sin \theta)}{\lambda_f} = \frac{2\pi}{\lambda_f}\left(\frac{a \sin^2 \theta}{\cos \theta}\right), \quad \text{(i)}$$

where $\lambda_f = c/f$ is the wavelength of the wave in free space. Now the path length along the ray from A to Q is $a/\cos \theta$, and the phase difference includes a phase shift of π due to reflection at C:

$$\phi_Q - \phi_A = \frac{2\pi}{\lambda_f}\left(\frac{a}{\cos \theta}\right) + \pi. \quad \text{(ii)}$$

Since P and Q are on the same wave front, separate calculations of the wave phase at the two points can only differ by an integer multiple of 2π:

$$\phi_Q - \phi_P = \frac{2\pi}{\lambda_f}\left(\frac{a}{\cos \theta}\right) + \pi - \frac{2\pi}{\lambda_f}\left(\frac{a \sin^2 \theta}{\cos \theta}\right) = 2m\pi,$$

where m is an integer. Then:

$$\lambda_f(2m - 1) = \frac{2a}{\cos \theta}(1 - \sin^2 \theta) = 2a \cos \theta,$$

and $\quad \cos \theta = (2m - 1)\lambda_f/(2a). \quad \text{(iii)}$

The corresponding wavelength λ_g of a given mode is the length s along the guide between two points Q and A for which $\phi_Q - \phi_A = 2\pi$. Using eqn. (i) with $s = \lambda_g$:

$$2\pi = 2\pi(\lambda_g/\lambda_f) \sin \theta.$$

So, using eqn. (iii):

$$\lambda_g = \frac{\lambda_f}{\sin \theta} = \frac{\lambda_f}{\sqrt{1 - \cos^2 \theta}} = \frac{\lambda_f}{\sqrt{1 - (\lambda_f/2a)^2(2m - 1)^2}}.$$

Chapter Summary

Where Are We Now?

We have completed our exploration of electromagnetism and demonstrated its unity with optics. Throughout this part we have used simple models of the atomic structure of matter to show how electromagnetic theory can explain phenomena. To proceed further, we need the quantum theory of atoms and molecules (Chapter 35) and more advanced mathematical methods.

What Did We Do?

An oscillating electron produces EM waves. At large distances from the electron, the waves are plane, with electric and magnetic fields perpendicular and $\vec{E} \times \vec{B}$ in the direction of wave motion. Maxwell's equations show that the wave speed in vacuum is the speed of light,

$$c = 1/\sqrt{\mu_0 \epsilon_0} = 3 \times 10^8 \text{ m/s}.$$

The magnetic field amplitude is related to the electric field amplitude by $E_0 = cB_0$.

The energy flux per unit area in an EM wave or any electromagnetic field is described by the Poynting vector:

$$\vec{S} = \vec{E} \times \vec{B}/\mu_0.$$

When a wave is reflected or absorbed, it exerts a radiation pressure because it transfers momentum. The momentum flux in the wave is $\vec{S}/c$, and the radiation pressure exerted on a surface is:

$$P = (K/c)|\vec{S}|\cos^2\theta,$$

where θ is the angle of incidence at the surface and K is a constant that equals 2 for total reflection or 1 for total absorption.

The direction of the electric field vector describes the *polarization* of a wave. When all the $\vec{E}$ vectors oscillate along a single direction, the wave is linearly polarized. When the $\vec{E}$ vector rotates, forming a circular helix, the wave is circularly polarized. Most naturally occurring light is unpolarized. Linearly polarized light is produced by passing light through a polarizing filter, or by scattering or reflection. When light is incident on a dielectric surface at Brewster's angle, given by

$$\tan\theta_B = \frac{n_2}{n_1},$$

the reflected light is completely polarized with $\vec{E}$ parallel to the surface.

✱ Cavity oscillators and waveguides are used in microwave generators and receivers. The oscillator is a closed cavity containing EM waves. A set of plane waves bouncing back and forth off the walls model wave propagation in a guide. Wavelengths longer than the *cutoff wavelength* cannot propagate in the guide.

Practical Applications

Light and radio waves are used extensively for communication. Microwaves are increasingly used for the transmission of television signals from satellites. Radiation pressure forms comet tails and prevents gravitational collapse of bright stars. Radiation pressure from laser beams is used in nuclear fusion experiments and in traps where atomic properties of cold neutral atoms can be tested. Polarization of reflected and scattered light is used to advantage by people who wear Polaroid sunglasses and by photographers who use polarizing filters. Polarization of transmitted light due to birefringence is used in engineering stress analysis and in minerology.

Solutions to Exercises

33.1. The first derivatives with respect to x and t are calculated in the text. The second derivatives are:

$$\frac{\partial^2}{\partial t\,\partial x}[\cos(kx - \omega t)] = \frac{\partial}{\partial t}[-k\sin(kx - \omega t)]$$
$$= \omega k \cos(kx - \omega t),$$

and:

$$\frac{\partial^2}{\partial x\,\partial t}[\cos(kx - \omega t)] = \frac{\partial}{\partial x}[\omega\sin(kx - \omega t)]$$
$$= k\omega \cos(kx - \omega t).$$

These are equal, as required.

33.2. The derivative of eqn. (33.4) with respect to time is:

$$\frac{\partial^2 E}{\partial t\,\partial x} = -\frac{\partial^2 B}{\partial t^2}.$$

And, the derivative of eqn. (33.5) with respect to x is:

$$-\frac{\partial^2 B}{\partial x^2} = \epsilon_0\mu_0 \frac{\partial^2 E}{\partial x\,\partial t}.$$

Combining these gives:

$$\frac{\partial^2 B}{\partial t^2} = \frac{1}{\epsilon_0\mu_0}\frac{\partial^2 B}{\partial x^2}, \quad \text{as required.}$$

33.3. The Poynting flux in the laser beam is the total power divided by the area onto which it is focused, $A = \pi d^2$:

$$S = \frac{10^{14}\,\text{W}}{\pi(0.05 \times 10^{-3}\,\text{m})^2} = 1.3 \times 10^{22}\,\text{W/m}^2.$$

Since the radiation is totally absorbed, the radiation pressure is:

$$P = \frac{S}{c} = \frac{1.3 \times 10^{22}\,\text{W/m}^2}{3 \times 10^8\,\text{m/s}} = 4 \times 10^{13}\,\text{Pa}.$$

A big enough laser focused finely enough can be pushy.

33.4. The equipotential surfaces are perpendicular to the Coulomb field lines. (There is no induced field if the current is steady.) Since $\vec{S} \propto \vec{E} \times \vec{B}$, $\vec{S}$ is also perpendicular to $\vec{E}$, so $\vec{S}$ runs along equipotential surfaces.

33.5. Unpolarized light is the superposition of individual wave pulses, each polarized at a random angle θ ($0 < \theta < \pi$) to the transmission axis. At a particular time, any value of this angle is equally likely. Averaged over time, the incident intensity with polarization in a range of angles $d\theta$ depends only on the range $d\theta$, and not on the value of the angle.

$$\frac{dI}{d\theta} = \frac{I_0}{\pi}.$$

According to Malus' law, a fraction $\cos^2\theta$ of this intensity passes the filter. The total transmitted intensity is found by summing up the contributions from all the incident waves:

$$I = \int dI_{\text{transmitted}}$$
$$= \frac{I_0}{\pi} \int_0^\pi \cos^2\theta\, d\theta$$
$$= \frac{I_0}{2}.$$

33.6. Since $2a = 5.0$ cm, the free-space wavelength is greater than the cutoff. Waves polarized parallel to the plates cannot propagate. The TEM mode can propagate, so the emerging radiation is linearly polarized perpendicular to the plates.

Basic Skills

Review Questions

§33.1 PLANE ELECTROMAGNETIC WAVES

- How are electromagnetic waves generated?
- Describe the configuration of electric and magnetic field vectors in a plane EM wave.
- Write the wave equation for the electric field.
- What is the relation between the amplitudes of the electric and magnetic fields in an EM wave?
- Describe the phase relation between $\vec{E}$ and $\vec{B}$ in an EM wave.
- Why do we believe that light is an EM wave?

§33.2 ENERGY AND MOMENTUM TRANSPORT BY EM WAVES

- What is the *Poynting vector*? How is it related to $\vec{E}$ and $\vec{B}$?
- Which physical property of an EM wave is described by the Poynting vector?
- What is the relation between energy and momentum in an EM wave?
- Define *radiation pressure*.
- How could you calculate the radiation pressure on a surface when a light wave is incident on it?
- How is energy transported from the battery to the resistor in a simple circuit?

§33.3 POLARIZATION

- Describe the electric field vectors in *linearly polarized* light.
- How does a polarizing filter work?
- State the law of Malus.
- How is light polarized by reflection?
- What is Brewster's angle?
- Describe a left circularly polarized wave.
- Discuss how a circularly polarized wave can result from the superposition of two linearly polarized waves.
- What kind of wave results from the superposition of a right and a left circular wave, each with the same amplitude and phase?

* §33.4 ELECTROMAGNETIC OSCILLATIONS AND MICROWAVES

- Describe the configuration of fields in one mode of oscillation for a cylindrical cavity oscillator.

- What boundary conditions apply to the electric and magnetic fields in a waveguide?
- What is the TEM mode?
- Sketch the configuration of fields in the TE mode between two parallel conducting plates.
- Why is there a cutoff wavelength below which the TE mode cannot propagate?

Basic Skill Drill

§33.1 PLANE ELECTROMAGNETIC WAVES

1. An electromagnetic wave has a magnetic field amplitude of 1.56×10^{-8} T. What is the electric field amplitude?

2. The electric field amplitude in an EM wave is $(0.45$ V/m$)\hat{z}$. The magnetic field is in the $-x$-direction. What is the magnetic field amplitude, and what is the direction of propagation?

3. The electric field in a plane EM wave propagating in the z-direction at a certain place and time is $\vec{E} = (6.0$ V/m$)(\hat{x} + \hat{y})$. Find $\vec{B}$.

4. The magnetic field in an EM wave is given by:
$$\vec{B} = (4.3 \times 10^{-10}\text{ T})\hat{z}\,\cos\{[\pi/(1.0\text{ mm})](x - ct)\}.$$
Find $\vec{E}$.

§33.2 ENERGY AND MOMENTUM TRANSPORT BY EM WAVES

5. The power reaching the Earth from a distant star is 2×10^{-8} W/m^2. Find E_0 in the radiation from the star.

6. A 1.5-mW laser beam produces light with an electric field amplitude of 0.50 kV/m. What is the diameter of the beam?

7. A plane EM wave has $E_0 = 3.0$ V/m. What is the time-averaged magnitude of $\vec{S}$? At what rate is momentum carried over a 1.5-m^2 surface perpendicular to $\vec{k}$?

8. A 15-W light beam 3.0 mm in diameter, incident normally on a human hand, is absorbed. What is the radiation pressure on the hand?

§33.3 POLARIZATION

9. Light with intensity $I_0 = 7.75$ W/m^2 polarized along the y-axis falls on a polarizing filter with its transmission axis at 20° to $\hat{y}$. What is the intensity of the transmitted light?

10. Linearly polarized light traveling in the x-direction is incident on a polarizing filter with its transmission axis in the y-direction. The transmitted intensity is 45% of the incident intensity. What is the angle of polarization in the incident beam?
11. Find Brewster's angle for diamond in air (cf. Table 16.3).
12. A laser beam is incident from air on a plane surface. You test the reflected ray with a polarizing filter and determine that the reflected beam is completely polarized when the angle of incidence is 55°. What is the refractive index of the material?

* §33.4 ELECTROMAGNETIC OSCILLATIONS AND MICROWAVES

13. Two parallel conducting plates are 3.56 cm apart. What is the cut-off frequency for waves propagating in the TE mode in this waveguide?
14. Radio waves are passed between two parallel plates. The emerging radiation is unpolarized when $\lambda < 1.25$ cm but is polarized with $\vec{E}$ perpendicular to the plates when $\lambda > 1.25$ cm. What is the separation of the plates?

Questions and Problems

§33.1 PLANE ELECTROMAGNETIC WAVES

15. ❖ An EM wave propagating in the x-direction has $\vec{B}$ in the y-direction at a certain place and time. What is the corresponding direction of $\vec{E}$?
16. ❖ Show that a plane wave with $\vec{E}$ parallel to $\vec{B}$ cannot satisfy Maxwell's equations.
17. ❖ Show that a plane wave with $\vec{E}$ parallel to $\hat{k}$ and $\vec{B}$ perpendicular to $\hat{k}$ cannot satisfy Maxwell's equations.
18. ♦ In a wave propagating in the y-direction, the magnetic field amplitude is: $\vec{B}_0 = (1.2 \times 10^{-7}\,\text{T})(\hat{x} + \sqrt{3}\,\hat{z})$. Find $\vec{E}_0$.
19. ♦ A 16.5-MHz radio wave is propagating in the x-direction. The electric field amplitude is 0.792 V/m. Determine the wave vector $\vec{k}$ and the magnetic field amplitude.
20. ♦♦ An EM wave with a frequency of 1.80 GHz has a magnetic field amplitude of $(4.75\,\text{nT})\,\hat{z}$ and propagates in the y-direction. The magnetic field is maximum in the positive z-direction at $y = 0.250$ m, $t = 0.650$ ns. Write expressions for $\vec{E}$ and $\vec{B}$ in the wave.
21. ♦♦ Two EM waves with equal electric field amplitudes of 0.44 V/m are propagating in the x-direction. One wave has $\vec{E}$ in the y-direction, and the other has $\vec{E}$ in the z-direction. Find the magnetic field vector amplitude in the superposed wave.
22. ♦♦ The direction of the electric field vector in an EM wave is $(1/\sqrt{6})(\hat{x} + \sqrt{3}\,\hat{y} + \sqrt{2}\,\hat{z})$, and the direction of the magnetic field is $(0.29)(2\hat{x} + \hat{y} - 2.6\hat{z})$. What is the direction of propagation of the wave?
23. ♦♦ An EM wave has wave vector $\vec{k} = (\hat{x} + 2\hat{y})$ rad/m. The magnetic field amplitude is $\vec{B}_0 = (1.43 \times 10^{-10}\,\text{T})(2\hat{x} - \hat{y} + \hat{z})$. Determine the electric field vector amplitude and the frequency of the wave.

§33.2 ENERGY AND MOMENTUM TRANSPORT BY EM WAVES

24. ❖ (a) Two plane EM waves propagating in the same direction have angular frequencies of ω and 2ω. They have the same electric field amplitude. Compare the energy transported by each wave.
(b) Two plane EM waves have the same frequency and direction, but the electric field amplitude of one is twice the amplitude of the other. Compare the energy transported by each wave.
25. ❖ (a) A power supply drives current through the central wire of a perfectly conducting coaxial cable, through a resistor at the far end, and back through the cable's outer cylinder. Sketch the electric and magnetic fields in the region between the conductors, and show that the Poynting flux flows along the cable in the direction of the inner current. (b) Describe the pattern of the Poynting vector in the coaxial cable if the conductors have a small resistivity.
26. ❖ Discuss how the flow of energy from a battery to a resistor changes as the circuit is made larger in diameter.
27. ♦♦ A solar collector on a satellite is proposed to provide 10^9 W of power for the city of Chicago. Assuming perfect absorption and 1% efficiency in conversion to useful power, what area should the solar collector have? The power is beamed down to a perfectly efficient collector on the surface of the Earth. If the collector's area is 2.5×10^4 m^2, what microwave intensity is required? What is the electric field amplitude in the microwave beam at the Earth's surface?
28. ♦♦ Find the average electric and magnetic field amplitudes in EM waves 1 m from a 100-W light bulb.
29. ♦♦ A laser beam reflecting normally off a surface with an area of 3.0 cm^2 exerts a force of 2.4 N on it. What is the electric field amplitude in the laser beam?
30. ♦♦ A possible spacecraft design uses a solar sail made of aluminized plastic. As sunlight reflects off the sail, radiation pressure drives the spacecraft outward. If the sail material has a density of 700 kg/m^3, what is the maximum thickness of the sail for which the force due to radiation pressure exceeds the gravitational force on the sail? If the sail area is 10^6 m^2, its thickness is 1 μm, and the craft carries a 100-kg payload, what is its acceleration at the radius of Earth's orbit?
31. ♦♦ A scheme for powering an interstellar starship envisions using a laser on the Moon for propulsion. What laser power is needed to propel a 30 000-kg starship with an acceleration of $0.01g$ if the laser beam is completely absorbed by the starship?
32. ♦♦ An aircraft tracking radar radiates 1 kW of power into a fan-shaped beam that is thin in the horizontal direction and spreads through an angle of 45° in the vertical direction. At a distance of 10 km, the beam's horizontal dimension is 200 m. Assuming perfect reflection, what radiation pressure does the beam exert on an aircraft 10 km away?
33. ♦♦ A satellite has an average power consumption of 350 W. Assuming 5.5% efficiency for conversion of solar energy to electrical energy, what area is required for the solar panels?
34. ♦♦ A 1.0-kW laser with a beam area of 1.0 cm^2 is incident on a mirror with an angle of incidence equal to 15°. Assuming perfect reflection, what is the force due to radiation pressure on the mirror?

35. ♦♦♦ Perfectly absorbing interplanetary dust grains have density $\rho = 3 \times 10^3$ kg/m^3. Show that spherical grains with radii less than a critical value R_c are blown out of the solar system. Calculate R_c.

36. ♦♦♦ A plane EM wave with $\vec{E} = E_0 \hat{y} \cos(kx - \omega t)$ is traveling along the x-axis. How much EM energy U exists at time t in a rectangular volume of dimensions dy by dz between x and $x + dx$? Compute the time rate of change of energy in this volume as the wave passes. Show that dU/dt equals the energy transported into the volume at x by the Poynting vector minus the energy transported out at $x + dx$. Show that:

$$\frac{\partial u}{\partial t} + \frac{\partial S}{\partial x} = 0,$$

where u is the energy density in the wave. This is the differential form of the energy conservation law for EM field.

37. ♦♦♦ Show that the total power transported down a coaxial cable is $I \Delta V$, where ΔV is the potential difference between the core and the sheath (cf. Problem 25).

§33.3 POLARIZATION

38. ❖ A beam of light passes through two polarizing filters with vertical transmission axes. A third filter is placed between the other two with its transmission axis at an angle θ to the vertical. For what angle θ does the transmitted beam have maximum intensity? For what angle is it minimum? How do the answers change if the two outer filters have perpendicular transmission axes? Explain your reasoning.

39. ❖ You wish to photograph a reflection of trees in a lake. Can you use a polarizing filter to make the reflection appear brighter? How?

40. ❖ A beam of light is a superposition of unpolarized light with intensity I_u and a plane-polarized wave with intensity I_p. Describe how the intensity transmitted by a polarizing filter varies as the filter rotates in the plane perpendicular to the light beam.

41. ♦ Laser light is linearly polarized with its electric field vector along the x-axis. What is the intensity of the beam, as a fraction of its initial intensity, after passing through: (a) A polarizing filter with its axis at 37.0° to the x-axis? (b) A polarizing filter with its axis at 77.0° to the x-axis?

42. ♦ Find Brewster's angle for a water–glass interface if the glass has a refractive index of 1.6.

43. ♦ Light shines on a glass surface at an angle of incidence equal to 58°. The reflected light is 100% polarized. What is the refractive index of the glass?

44. ♦♦ An EM wave with amplitude $\vec{E}_0 = (340 \text{ V/m})(2\hat{x} - \hat{y})$ passes through a polarizing filter with its axis parallel to the vector $\hat{x} + 3\hat{y}$. What is the intensity of the transmitted light?

45. ♦♦ Laser light of intensity 1.6 W/m^2, polarized so that its electric field vector makes an angle 45° with the x-axis, passes through two polarizing filters. The transmission axis of the first is along the x-axis, and the second filter's axis is parallel to the incident polarization. What is the intensity of the transmitted light? How would the result differ if the second transmission axis were perpendicular to the initial polarization?

46. ♦♦ An unpolarized light beam of intensity I_0 passes along the z-axis through three polarizing filters. The first filter's transmission axis is parallel to $\hat{x}$. The second filter's axis makes an angle of 25° with the y-axis, and the third's axis is parallel to $\hat{y}$. What is the intensity of the transmitted light?

47. ♦♦ A beam of unpolarized light of intensity I_0 is incident at Brewster's angle on a slab of glass with a refractive index of 1.5. The reflected beam intensity is 9.5% of I_0. Assume that there is no absorption. (a) Calculate Brewster's angle for the glass in air. (b) The percent polarization of a light beam is:

$$p_\% = 100 \left| \frac{I_1 - I_2}{I_1 + I_2} \right|,$$

where I_1 and I_2 are the intensities in two orthogonal polarizations. Find the percent polarization of the reflected ray. (c) Find the percent polarization of the transmitted ray.

48. ♦♦ Compute the y- and z-components of the electric field vector in the superposition of two waves with electric field vectors given by:

$$\vec{E}_1 = (25.0 \text{ V/m})\hat{y} \cos[(6.0 \text{ rad/m})x - (1.8 \times 10^9 \text{ rad/s})t],$$

and

$$\vec{E}_2 = (30.0 \text{ V/m})\hat{z} \sin[(6.0 \text{ rad/m})x - (1.8 \times 10^9 \text{ rad/s})t],$$

at $x = 0$ and $t = 0$, $T/4$, and $T/2$. Draw the vectors. How would you describe the polarization of the superposition?

49. ♦♦♦ A quarter-wave plate produces a phase shift of 90° between waves polarized at angles of $\pm 45°$ with the vertical. Unpolarized light of intensity I_0 passes through a polarizing filter with its transmission axis vertical, then through the quarter-wave plate, and finally through another filter with a horizontal transmission axis. What is the intensity of the light emerging from the system?

50. ♦♦♦ The transmission axes of two polarizing filters make an angle θ_0. A third filter may be rotated between the other two. Prove that the three filters transmit maximum intensity when the middle filter's axis makes an angle $\theta_0/2$ with each of the other two.

51. ♦♦♦ *Faraday rotation.* A linearly polarized EM wave with its electric field vector along the z-axis propagates into an ionized medium. In this medium, right-circular waves propagate faster than left-circular waves. The difference in speed is $\Delta v/c = 7 \times 10^{-5}$, and the wave frequency is 15 GHz. What is the direction of polarization after the wave has traveled 25 m? (*Hint:* Describe the linearly polarized wave as the superposition of two circularly polarized waves, as in Example 33.11. Write the y- and z-components of $\vec{E}$ after the wave has propagated a distance x, remembering that k_R and k_L are different, and calculate the angle E makes with the z-axis from $\tan \theta = E_y/E_z$.)

✶ §33.4 ELECTROMAGNETIC OSCILLATIONS AND MICROWAVES

52. ❖ Use the Poynting vector to describe the energy flow in the cylindrical cavity oscillator discussed in §33.4.1.

53. ❖ At high frequencies, a coaxial cable acts as a waveguide. Show that a TEM mode can propagate along the cable, and sketch the configuration of the fields. Is there a cutoff?

54. ❖ Use the Poynting vector to describe the energy flow in the TE mode shown in Figure 33.29.

55. ❖ Sketch the field lines for the $m = 1$ TM mode in a parallel plate waveguide.

56. ❖ The electric field of the $m = 2$ TE mode in a parallel plate waveguide is proportional to $\sin(2\pi z/a)$. Sketch the field configuration in the guide.

57. ❖ A microwave oven is a rectangular resonant cavity. Describe the configuration of fields inside in the lowest mode. What disadvantages would the lowest mode have for cooking?

58. ♦ You wish to use two parallel conducting plates to make a waveguide for 2.7-cm wavelength TE microwaves, but 3.0-cm wavelength waves are not to propagate. What should the separation of the plates be?

59. ♦ Calculate the cutoff frequency for the TE mode between two parallel plates with a 15-cm separation.

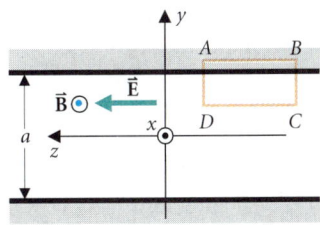
(a) Circuit ABCD.

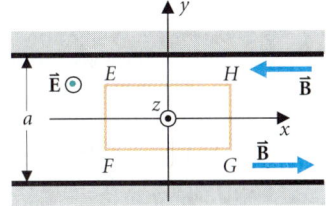
(b) Circuit EFGH.

■ FIGURE 33.31

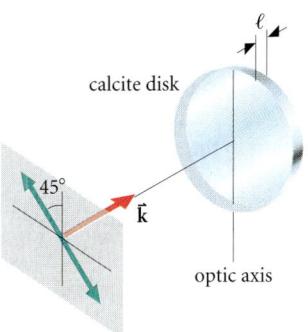
■ FIGURE 33.32

60. ◆ What separation of conducting plates is required to make a waveguide with 165-MHz cutoff frequency in the TE mode?

61. ◆◆◆ A waveguide consists of two conducting plates with a 2.0-cm separation. It is excited so that the $m = 1$ TE mode and the TEM mode have equal amplitude. The TM mode is not excited. The radiation has free-space wavelength $\lambda_f = 3.0$ cm. What length of guide is necessary if the emerging wave is to be circularly polarized?

62. ◆◆◆ A cavity oscillator is formed by two infinite, parallel, and perfectly conducting metal plates with separation a. The magnetic field inside the cavity is given by:
$$\vec{B} = -B_0 \sin(\pi y/a) \cos(\omega t)\hat{x}.$$
(a) Assuming that $\vec{E}$ is in the z-direction, apply Faraday's law to circuit ABCD (■ Figure 33.31a) to find the electric field inside the cavity.
(b) Compute the displacement current everywhere inside the cavity.
(c) Using the result of (b), apply Ampère's law to circuit EFGH (Figure 33.31b) to obtain another expression for $\vec{B}$ in the cavity. (Hint: Use the symmetry of $\vec{B}$). (d) What value of ω makes the result of (c) the same as the given expression for $\vec{B}$?

63. ◆◆◆ Two infinite, flat, parallel, and perfectly conducting metal plates lie in the planes $y = \pm a/2$. The space between the plates acts as a cavity oscillator in which the electric field is given by:
$$\vec{E} = E_0 \cos(\pi y/a) \sin(\omega t)\hat{z}.$$
(a) Assuming that $\vec{B}$ is in the x-direction, compute the displacement current density and apply Ampère's law to a rectangular circuit inside the cavity to obtain B everywhere inside. (b) What is $\vec{B}$ within the conducting plates? Use Ampère's law to compute the current per meter on the metal surfaces, and compare with the displacement current per meter between the plates.

Additional Problems

64. ❖ Assuming that the amplitude of the ripple in a field line of an oscillating electron is proportional to the amplitude of its acceleration component perpendicular to the field line, argue that the intensity radiated by the electron is proportional to $\sin^2 \theta$, where θ is the angle between the field line and the direction of oscillation.

65. ◆ What charge on the Sun would make the static electric field at the Earth's orbit equal the electric field amplitude (about 1 kV/m, Example 33.3) in sunlight?

66. ◆◆ What current around the Sun's equator would produce a magnetic field at the Earth's orbit equal to the average magnetic field amplitude in radiation reaching the Earth? (Hint: Use magnetic moment.)

67. ◆◆ ✉ Lasers are increasingly being used for surgery. The laser cauterizes and seals the wound as it cuts. To get a rough idea of how the process works, use the following model. Model flesh as water at 300 K and imagine that flesh is removed when the water boils. Assuming total absorption, what power laser is required to remove 1 mm of flesh per second over an area 1 mm in diameter? How much force is exerted on the wound as a result of this process?

68. ◆◆◆ Calcite is a mineral in which the speed of an EM wave depends on its polarization. Electrons in calcite oscillate more freely along certain crystal planes than perpendicular to them. The direction perpendicular to the planes is called the optic axis. Waves polarized parallel to the optic axis travel more rapidly than waves polarized perpendicular to it. Light with wavelength λ_0 and polarized at 45° to the optic axis falls on the calcite disk shown in ■ Figure 33.32. If $\Delta n = n_\perp - n_\parallel$ is the difference in the refractive index for the two polarizations, describe the polarization of the emerging wave if the thickness ℓ of the disk is (a) $\lambda_0/(4\Delta n)$, (b) $\lambda_0/(2\Delta n)$, and (c) $\lambda_0/(3\Delta n)$.

Computer Problems

69. Use a spreadsheet program to compute the electric field vectors
$$\vec{E}_1 = (3 \text{ V/m})\cos[(5 \text{ rad/s})t]\hat{x},$$
and
$$\vec{E}_2 = (3 \text{ V/m})\sin[(5 \text{ rad/s})t]\hat{y}$$
every 0.04 s for one complete cycle. Compute the components of the superposition $\vec{E}_1 + \vec{E}_2$, and plot the vector. How would you describe the polarization? What happens if the magnitude of $\vec{E}_2$ is 2 V/m?

70. In Example 33.9 we showed how a polarizer placed between two others with their axes perpendicular allows some light to be transmitted. Take the y-axis along the first polarizer's transmission axis, and the x-axis along the final polarizer's axis. Set up a spreadsheet program to calculate the amplitude of the transmitted wave when the intermediate polarizer's axis makes an angle θ with the y-axis. For what value of θ is the most light transmitted? Now add another polarizer between the second and the last, with its axis at 2θ to the y-axis. What value of θ maximizes the transmitted light? How does the transmitted intensity compare with the previous case? Now, add another polarizer and repeat the calculations. Can you draw any general conclusions?

Challenge Problems

71. An electron moves uniformly in a circle. In the spirit of §33.1.1, describe the field line emerging from the electron perpendicular to the plane of its motion. Show that the corresponding wave is circularly polarized and that its Poynting vector is directed away from the electron. Describe how the polarization of waves emitted by the electron depends on direction.

72. A nonreflective coating for TV screens consists of a polarizing filter on top of a quarter-wave plate. Explain how it works.

73. Consider a rectangular guide measuring $a = 3$ cm by $b = 5$ cm. Argue that the $m = 1$ TE mode can propagate in this guide as well as one with $b = \infty$. What is the cutoff? What cutoff results from the finite value of b? If a 4-GHz wave is introduced into the guide, what is the polarization of the emerging wave?

74. A uniformly, positively charged filament lies along the y-axis with its center at the origin. A small coil with dipole moment $\vec{m} = m\hat{z}$ is also at the origin. Sketch a diagram showing field lines for the electric and magnetic fields in the x-z plane. From your diagram, determine the direction of the Poynting vector at points in the x-z plane. From your results, argue that the EM field carries angular momentum in the z-direction. Assuming the filament is free to rotate about the z-axis, describe what happens when the current in the coil is reduced to zero. Where does the angular momentum go? Use Faraday's law to show how the necessary force is exerted on the filament.

75. A plane EM wave propagating into a material with resistivity ρ has the form:

$$\vec{E}(z, t) = E_0 \hat{y} e^{-\alpha z} \cos(kz - \omega t),$$

and

$$\vec{B}(z, t) = -B_0 \hat{x} e^{-\alpha z} \cos(kz - \omega t + \phi_0).$$

Apply Maxwell's equations and the relation $\vec{E} = \rho \vec{j}$ to find α, k, ϕ_0 and the ratio B_0/E_0 as functions of ω.

Part VII Problems

1. ♦♦ A coil is made by wrapping 24 m of 18 gauge copper wire (157 feet per ohm at 68° F, diameter 4.1 mm) around a cardboard tube 2 in. in diameter. Calculate the impedance of the coil when connected to 240-Hz, 2.2-V AC power. What is the current in the coil?

2. ♦♦♦ An electron is held in a circular orbit about an atomic nucleus by the attractive Coulomb force between them. The electron forms a current loop with magnetic moment $\vec{m}$. A magnetic field $\vec{B}$ is applied to the atom, parallel to $\vec{m}$. Assuming cylindrical symmetry about the center of the atom, find the induced electric field in terms of dB/dt, as the field is changed from 0 to $\vec{B}$. Show that the change in the electron's speed as a result of the applied $\vec{B}$ is $\Delta v = -(er/2m)B$ (cf. §29.4.1).

3. ♦♦♦ A current $I = 1.0$ A is charging a circular capacitor with plates of radius 4.0 cm. Compute the electric and magnetic energy densities at a point 3.0 cm from the axis of symmetry when the charge density on the plates is 10.0 nC/m². In considering the energy used to charge a capacitor, we neglected magnetic energy density. Is this justified?

4. ♦♦♦ A parallel plate capacitor filled with a dielectric is connected to a resistor and a battery. If you suddenly pull the dielectric slab out of the capacitor, what happens to the charge on the capacitor? What happens to the energy? Taking into account the battery, the resistor, and the work you do, calculate the energy input and output from the circuit. Verify that energy is conserved.

5. ♦♦♦ (a) A star has mass M, luminosity L, and radius R. Find the average value of the Poynting flux at the surface of the star.
(b) If an electron acts like an absorbing surface with effective area $\sigma = 6.7 \times 10^{-29}$ m², find the force exerted on an electron at the surface by absorption of radiation from the star.
(c) If the star's luminosity is large enough, the force due to radiation pressure blows off the star's outer layers. Because of the electric force between them, each electron drags a proton along with it. Find the luminosity—called the Eddington luminosity—for which the force on an electron at the star's surface due to radiation pressure balances the gravitational force on a proton. What is the Eddington luminosity for a star with the mass of the Sun?
(d) If the luminosity of a star is proportional to the 3.45 power of its mass, what is the largest mass a star can have and still have a luminosity less than its Eddington limit?

Twentieth-Century Physics

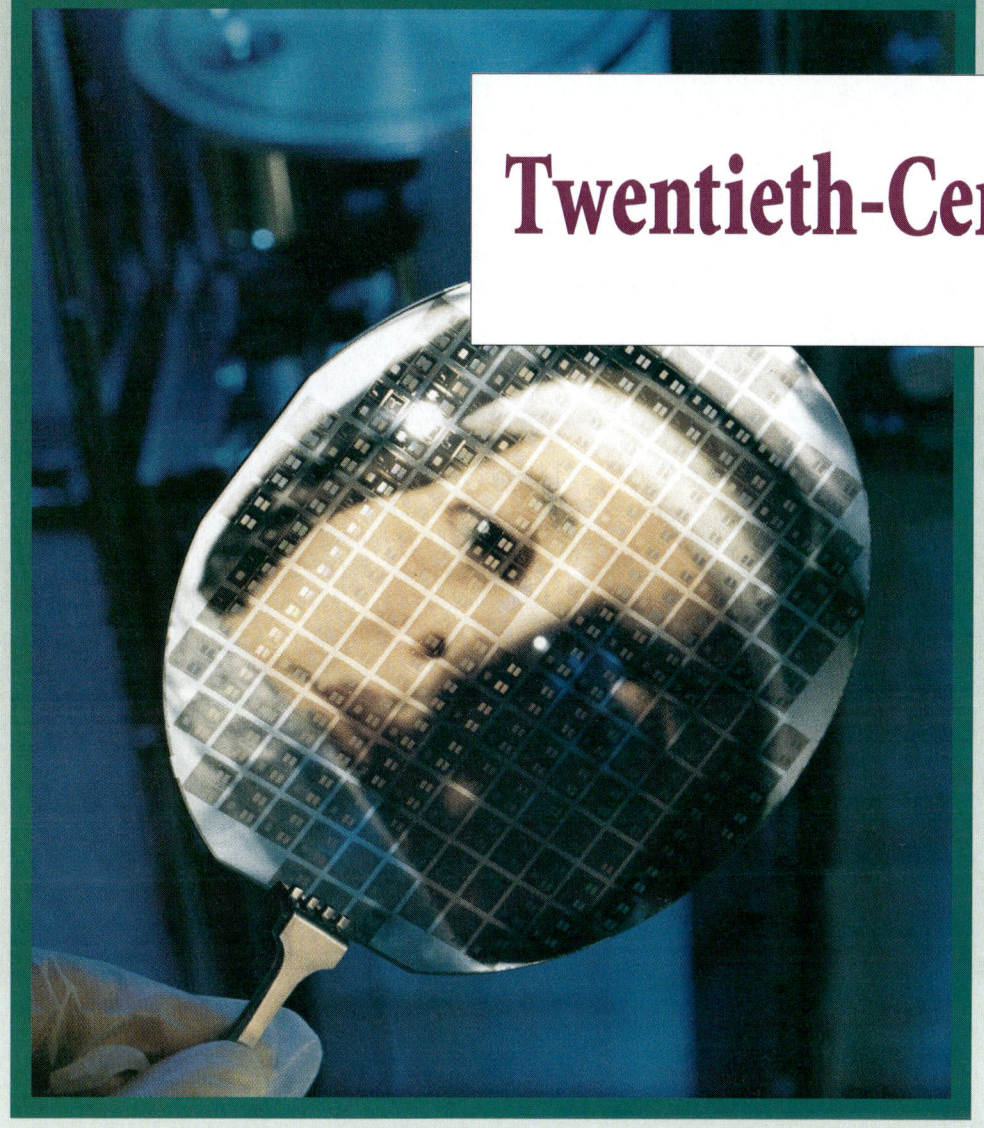

PART EIGHT

Physicists could perhaps be excused for feeling smug in the years following Maxwell's work. Newton's theory of mechanics and Maxwell's electromagnetic theory seemed to be completely accurate and capable, in principle, of describing all known phenomena. Astronomers understood the universe as a roughly spherical collection of stars with the Sun near its center. The same laws of mechanics, light, and chemistry could be observed at work in the stars and on Earth. With the basic rules governing the universe known, all that remained was to work out the details.

But it's never wise to be smug! In the years following 1890, this neat picture began to come apart, and our view of the universe underwent a complete revolution. As a result of several careful experiments dealing with the propagation of light, the theory of electromagnetic waves had to be modified and several ad hoc assumptions were required—the theory became ugly. Einstein instead suggested a complete revision of Newtonian ideas about space and time. Max Planck's explanation of cavity radiation and Einstein's explanation of how electrons are ejected from illuminated metal surfaces both showed that light acts like a stream of particles as well as like an electromagnetic wave. The discovery of radioactivity introduced a new kind of process, neither gravitational nor electromagnetic, and provided the particles Ernest Rutherford used to prove that atoms are mostly vacuum—very small heavy nuclei, each surrounded by a cloud of electrons.

While physicists were revolutionizing our understanding of atomic phenomena, astronomers were making equally profound discoveries about the nature of the universe. The visible sphere of stars is just a small region near the edge of a much larger system that is itself but one of millions. Edwin Hubble showed that the entire universe is expanding. Aleksandr Friedmann and others used Einstein's general theory of relativity to explain the observations in terms of a universe that abruptly came into existence 10–20 billion years ago in an extremely hot, dense state.

The modern conception of the physical world is a four-dimensional composite of space and time in which a sea of elementary particles forms and dissolves. The behavior of such particles can only be predicted on average. Because the physics of everyday events involves very large numbers of particles, these averages are well defined, and the Newtonian picture we have used thus far is an extremely good approximation to this average behavior. Modern physics does not replace Newtonian physics, but extends it.

In this part we introduce the exciting ideas of twentieth-century physics. In Chapter 34 we describe Einstein's ideas of relativity and space-time. In Chapter 35 we explore how the wave/particle nature of both light and matter leads to an understanding of the structure of atoms and their chemical behavior. In Chapters 36 and 37 we describe the structure of atomic nuclei and the principles that underlie the behavior of subnuclear particles. We conclude with an epilogue on current attempts to unify all of physics and how these ideas relate the smallest scale structure of matter with the structure and evolution of the universe.

The solar eclipse of 1991. Bright streamers of the solar corona become visible when the Moon blocks the disk of the Sun. The star Delta Geminorum is visible between the streamers to the right (west) of the Sun. By comparing a star's position measured during an eclipse with its position when not near the Sun, astronomers can observe the effect of the Sun's gravity on light from the star. The first such measurements were made by Arthur Eddington and Andrew Crommelin. Their results were announced at a meeting of the Royal Astronomical Society on November 6, 1919. The next day The Times carried the headline "Revolution in Science." These observations provided the first confirmation of the general theory of relativity.

CHAPTER 34
Relativity and Space-Time

Concepts

Relative quantity
Invariant
Simultaneity
Event
Space-time
World line
Time dilation
Length contraction
Lorentz transformation
Rest mass energy
Energy-momentum invariant

Goals

Be able to:

Compute the effects of time dilation and length contraction.

Describe the world lines of particles and light rays in a space-time diagram.

Relate the energy and momentum of a particle in different reference frames.

Henceforth space by itself and time by itself are doomed to fade away into mere shadows, and only a kind of union of the two will preserve an independent reality.

HERMANN MINKOWSKI

BEFORE STARTING THIS CHAPTER, REVIEW THE MATERIAL ON RELATIVELY MOVING REFERENCE FRAMES IN §3.3.

EINSTEIN FIRST PREDICTED THE BENDING OF LIGHT IN 1907. IT WASN'T UNTIL 1915 THAT HE WAS ABLE TO PREDICT ITS MAGNITUDE CORRECTLY.

*I*n May 1919, two groups of British scientists, led by Arthur Eddington and Andrew Crommelin, traveled halfway around the world to observe a solar eclipse. Their goal was to test the prediction of a young German physicist, Albert Einstein, that the Sun's gravity bends light rays. Beginning in 1905, Einstein had published a series of papers that forever changed the way we look at the world. Eddington and Crommelin verified his predictions, and his name became a household word.

As a young man, Einstein asked himself how a light beam would appear to someone traveling with it. To such an observer, the sinusoidal pattern of electric and magnetic fields in the wave would appear fixed in space. That is a mystery, for such a stationary pattern of fields violates Maxwell's equations. The mystery is resolved by Einstein's special theory of relativity, published in 1905. Convinced that fields are a more fundamental concept than particles, Einstein assumed that Maxwell's equations are exactly correct for all observers, and he showed that the Newtonian concepts of space and time needed fundamental revision. A consequence of Einstein's theory is that no observer can travel at the speed of light!

Since 1905, the predictions of special relativity have been verified in many experiments. While most often used for exotic applications in particle physics and astronomy, the theory also touches our lives through nuclear power plants, aircraft navigation, and synchrotron x-ray sources that are used in the production of computer chips or in medical applications such as the study of cells in the human body (■ Figure 34.1) and the diagnosis of heart disease. In §34.1–34.3 we shall study the relativistic view of space and time, and in §34.4 we'll look at the implications of Einstein's theory for particle dynamics.

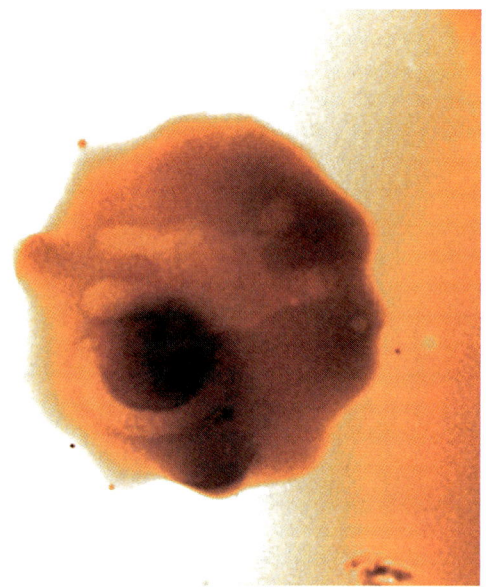

■ FIGURE 34.1
A human red blood cell infected with the malaria parasite. The parasite is in the ring-structured phase of its development. The image was taken with an x-ray microscope using 2.4 nm x rays from the Advanced Light Source at the Ernest O. Lawrence Berkeley National Laboratory. The short wavelength x rays allow higher resolution than would be possible with visible light. The ALS, completed in 1993, is a synchrotron light source that produces the brightest beams of soft-x-ray and ultraviolet light.

34.1 SPECIAL RELATIVITY

34.1.1 What Is a Relativity Theory?

WE DISCUSSED GALILEO'S THOUGHT EXPERIMENT IN §0.3.

In the seventeenth century, Galileo reasoned that a passenger in a ship's cabin cannot tell from the behavior of things on board whether the ship is anchored in harbor or moving at constant velocity. We verify Galileo's conclusion continuously as we revolve around the Sun with a speed of 30 km/s and fall toward the Virgo cluster of galaxies at 600 km/s, without any sense of our motion. All physical phenomena proceed as they would in the absence of such motions. A theory of relativity describes *how* the laws of physics agree with this observation. We shall illustrate this idea using *Galilean* relativity, which is valid only for relative speeds much less than the speed of light. In §34.1.2 we shall see how Einstein extended the idea to develop a theory valid for all possible speeds.

BUT WE ACCELERATE AROUND THE SUN, YOU MIGHT SAY. THE EARTH IS AN ACCELERATED REFERENCE FRAME! HOWEVER, THE EARTH FALLS FREELY UNDER THE INFLUENCE OF GRAVITY. ACCORDING TO GENERAL RELATIVITY, IT IS THEREFORE AN INERTIAL FRAME. (SEE THE ESSAY FOLLOWING THIS CHAPTER.)

Newton's second law exhibits the important features of any relativity theory. Suppose a person on a train moving at constant velocity $\vec{v}_{t,E}$ throws a baseball upward so that it falls back into the same car it started from (■ Figure 34.2). A person standing beside the tracks (i.e., on the Earth) describes the ball's velocity at a particular time with a vector $\vec{u}_E$. A passenger on the train describes the ball's velocity at the same time with a different vector $\vec{u}_t$. The ball's velocity is an example of a relative quantity.

> A *relative quantity* is one whose value depends on the reference frame in which it is measured.

Relative quantities are said to *transform* between reference frames. The Galilean transformation law for the baseball's velocity is:

$$\vec{u}_t = \vec{u}_E - \vec{v}_{t,E}. \qquad (34.1)$$

The mass of the baseball and the gravitational force on it have the same value in both reference frames; they are *invariant*.

> An *invariant* is a quantity that is independent of reference frame.

The most important invariant in this example is Newton's second law itself. Its form, $\vec{F} = m\vec{a}$, is the same in both frames. That is, in the Earth reference frame $\vec{F}_E = m\vec{a}_E$, and in the train reference frame $\vec{F}_t = m\vec{a}_t$. Let's see why this is true.

In the Earth frame,

$$\vec{F}_E = m\vec{a}_E = m\frac{d\vec{u}_E}{dt}. \qquad (i)$$

Since the force is invariant, the force measured in the train frame equals the force measured in the Earth frame: $\vec{F}_t = \vec{F}_E$. The mass is also invariant. Using the transformation law (eqn. 34.1), we may write $\vec{u}_E$ in terms of $\vec{u}_t$ to find the force in the train frame in terms of $\vec{a}_t$.

$$\vec{F}_t = \vec{F}_E = m\frac{d}{dt}(\vec{u}_E)$$
$$= m\frac{d}{dt}(\vec{u}_t + \vec{v}_{t,E}) = m\left(\frac{d\vec{u}_t}{dt} + \frac{d\vec{v}_{t,E}}{dt}\right).$$

Since $\vec{v}_{t,E}$ is constant, $d\vec{v}_{t,E}/dt = 0$, and so:

$$\vec{F}_t = m\frac{d\vec{u}_t}{dt} = m\vec{a}_t. \qquad (ii)$$

Equation (ii) relates the force to the acceleration in the train frame in the same way that eqn. (i) relates those vectors in the Earth frame. Newton's second law thus illustrates the *special principle of relativity*:

> The laws of physics have the same form in any two reference frames in uniform relative motion.

The laws that govern the behavior of physical systems and the fundamental properties of objects (like charge) are invariant.[1] Practical computations are usually carried out in one particular reference frame and involve relative quantities. Thus a complete theory of relativity

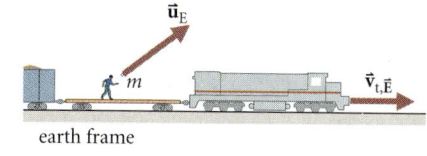

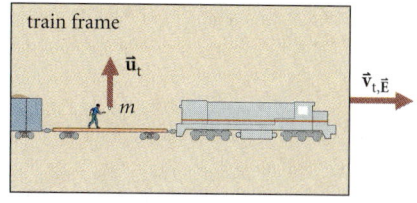

■ FIGURE 34.2
Galilean relativity. The mass of the ball is the same in the Earth frame (a) and in the train frame (b). It is an invariant. The ball's velocity at any instant differs in the two frames: $\vec{u}_t = \vec{u}_E - \vec{v}_{t,E}$. It is a relative quantity.

TO CONVINCE YOURSELF THAT FORCE IS AN INVARIANT, LOOK AT THE EXPRESSIONS FOR, SAY, GRAVITATIONAL FORCE OR COULOMB FORCE. THE FORCES DEPEND ON FUNDAMENTAL PROPERTIES OF PARTICLES, LIKE MASS, AND ON PARTICLE SEPARATION, WHICH IS ALSO A GALILEAN INVARIANT.

Note: THIS DERIVATION SHOWS THAT THE ACCELERATION IS ALSO INVARIANT.

---

[1] Many popular expositions of relativity miss this point entirely, leading to misguided claims that science justifies relativism in philosophy.

TABLE 34.1 Some Invariant and Relative Quantities in Galilean Relativity

(Velocity of prime frame as measured by an observer in the unprime frame is $\vec{v}$.)

Invariants

Mass	m
Time	t
Distance	s
Acceleration	$\vec{a}$
Force	$\vec{F}$
Charge	Q

Relative Quantities		Transformation Law	Text Reference
Position	$\vec{r}$	$\vec{r}\,' = \vec{r} - \vec{v}t$	§3.3
Velocity	$\vec{u}$	$\vec{u}\,' = \vec{u} - \vec{v}$	§3.3
Momentum	$\vec{p} = m\vec{u}$	$\vec{p}\,' = \vec{p} - m\vec{v}$	§9.3.4
Kinetic energy	$K = \tfrac{1}{2}mu^2$	$K' = K + \tfrac{1}{2}mv^2 - m\vec{u}\cdot\vec{v}$	§9.3.4
Electric field	$\vec{E}$	$\vec{E}\,' = \vec{E} + \vec{v} \times \vec{B}$	§30.3.3

states which physical quantities are relative and which are invariant, gives the transformation laws for relative quantities, and describes physical laws in an invariant manner. In our discussion of classical physics we have developed the set of rules for Galilean relativity, some of which are collected in • Table 34.1.

34.1.2 Einstein's Postulates

Maxwell's equations defy Galilean relativity: they predict a speed $c = 3 \times 10^8$ m/s for electromagnetic waves. According to the Galilean velocity transformation law (eqn. 34.1), this prediction can hold in only one reference frame. Thus, either the Galilean transformation laws are incorrect or, as physicists believed in the nineteenth century, the special principle of relativity does not apply to electromagnetism. At that time, physicists thought of electromagnetic waves as disturbances in a medium known as the ether, and Maxwell's equations were considered valid only in the frame in which the ether is at rest. Toward the end of the nineteenth century, measurements of the refractive index of a moving liquid, stellar aberration (see Figure 16.14), and Michelson's attempts to detect motion with respect to the ether (§17.2) had forced physicists, led by H. A. Lorentz, to modify the ether theory and to introduce theoretically unmotivated assumptions such as ether drag.

> IN HIS LATER YEARS, EINSTEIN COULD NOT RECALL THAT MICHELSON AND MORLEY'S RESULT HAD INFLUENCED HIS THINKING.

Einstein's approach was quite different. He assumed the special principle of relativity to be valid for all physical laws, discarded the ether concept, and took Maxwell's equations as the proper, invariant description of electromagnetism. With these assumptions, Galilean relativity can be only an approximation.

In his 1905 paper on relativity, Einstein stated his premises in the form of two postulates. The first of these is the special principle of relativity, and the second is a consequence of Maxwell's equations.

POSTULATES OF SPECIAL RELATIVITY

1. The laws of physics take the same form in all inertial reference frames.
2. In any given inertial frame, the speed of light c is observed to be the same no matter what the velocity of observer or emitter.

> WE DEFINED AN INERTIAL REFERENCE FRAME FOR NEWTONIAN PHYSICS IN §4.8.1. EINSTEIN MODIFIED THE DEFINITION.

Invariance of the speed of light seems at odds with common sense. If you run toward a baseball, it approaches your glove faster than if you run away. A flash of light approaches you at the same speed whether you run toward the source or away from it! Since we have no personal experience with things that move almost as fast as light, we must build intuition by applying Einstein's postulates in special situations to derive relations that we can check against experiments.

34.1.3 Time Dilation

Our first situation involves a clock that uses light pulses to define a time interval. Using it, we find that time is a relative quantity.

This clock uses the reflection of light between two parallel mirrors to measure equal intervals of time (■ Figure 34.3). In one cycle of the clock, a light pulse is emitted at A, reflects from the upper mirror at B, and returns to A, where it is detected. In one frame of reference (#1), the clock is at rest (Figure 34.3a). The pulse travels a distance $2L$ straight up and down and takes a time $\Delta t_1 = 2L/c$ to complete one cycle.

A second frame moves to the left, parallel to the mirrors, at speed v with respect to the first. According to observers in frame 2, the clock moves to the right at a constant speed v, and the light pulse follows the triangular path $A'B'D'$ shown in Figure 34.3b. According to Einstein's second postulate, the speed of the pulse is c, the same as in frame 1, so the time interval Δt_2 between emission and reception of the light pulse is $\Delta t_2 = 2\ell/c$. Since the light has to keep pace with the moving clock, $\ell > L$, and $\Delta t_2 > \Delta t_1$. The clock cycle takes more time in a reference frame where the clock is in motion. To find how much more, we note that ℓ depends on Δt_2. Applying the Pythagorean theorem to triangle $A'B'C'$:

$$\left(\frac{c\,\Delta t_2}{2}\right)^2 = \ell^2 = L^2 + \left(\frac{v\,\Delta t_2}{2}\right)^2.$$

NOTE: A' MEANS EVENT A (LIGHT PULSE IS EMITTED) AS OBSERVED IN FRAME 2. THE LIGHT PULSE IS RECEIVED AT LOCATION D', DIFFERENT FROM A', BECAUSE THE CLOCK IS MOVING.

So,
$$\Delta t_2 = \frac{2L}{c}\left(\frac{1}{\sqrt{1 - v^2/c^2}}\right) = \frac{\Delta t_1}{\sqrt{1 - \beta^2}} \equiv \gamma\,\Delta t_1,$$

where we have introduced the standard abbreviations:

THE EFFECT IS INDEPENDENT OF THE CONSTRUCTION OF THE CLOCK OR ITS DIRECTION OF MOTION. FOR ANOTHER EXAMPLE, SEE PROBLEM 27.

$$\beta \equiv \frac{v}{c} \quad \text{and} \quad \gamma \equiv \frac{1}{\sqrt{1 - v^2/c^2}} = \frac{1}{\sqrt{1 - \beta^2}}. \tag{34.2}$$

The time interval between the emission and detection of the light pulse is a relative quantity, different for different observers. This effect is an example of *time dilation*. The concept of time interval is inseparably linked to specific pairs of events and to the observers measuring the interval.

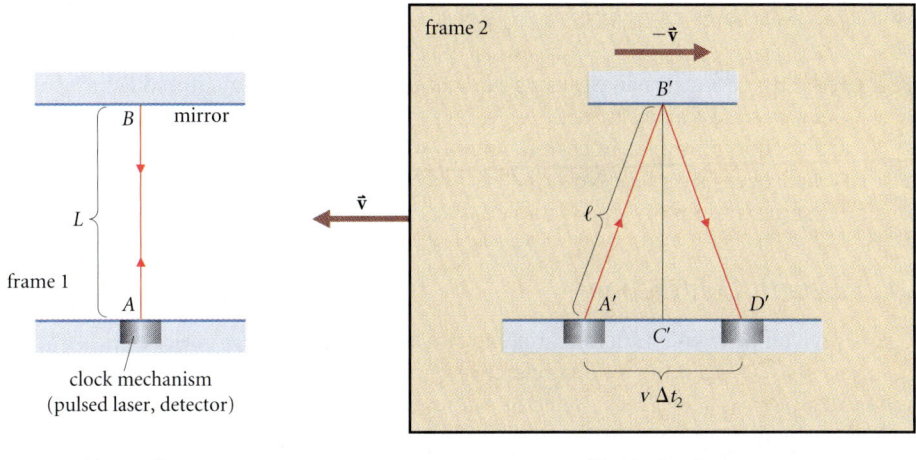

■ FIGURE 34.3
A clock measures the time interval between the emission and reception of a light pulse. (a) Rest frame. In one cycle the pulse travels from A to B and back to A, taking time $\Delta t = 2L/c$. (b) Moving frame. The clock moves at speed v and the light pulse follows a triangular path $A'B'D'$, still at speed c. The time for the light pulse to travel the longer path is greater than the interval Δt measured in the rest frame.

(a) Rest frame
(b) Moving frame

> The smallest time interval Δt_1 between two events is measured in the reference frame in which both events occur at the same place. In another frame moving at speed v with respect to the first, the measured time interval is:
>
> $$\Delta t_2 = \gamma \, \Delta t_1. \qquad (34.3)$$

Cesium clocks are accurate enough to measure the time dilation effect between two similar clocks, one on the Earth's surface and one carried in a jet aircraft. Substantial time dilations occur for objects moving close to the speed of light and are observed in the interactions of subatomic particles.

EXAMPLE 34.1 ♦♦ Laboratory experiments produce low-speed μ^- mesons (muons), which then decay spontaneously with an average lifetime of $\tau = 2.197$ μs. Cosmic rays striking the Earth's atmosphere at an altitude $h = 25.00$ km also produce muons that are later detected at the surface. What speed is necessary for an average muon to reach the surface before decaying?

MODEL The muon's average lifetime is determined by its physical structure. Crudely speaking, in the reference frame in which it is at rest, each muon carries a clock with a definite period. The events of muon production and detection occur within 2.197 μs (on average) measured in this muon rest frame, where they occur at the same place—the position of the muon. The production and detection events occur at different places in the Earth reference frame, and the time interval between them is longer: $\Delta t_E = \gamma \tau$.

SETUP The required speed of the muons is:

$$v = h/\Delta t_E = h/(\gamma \tau).$$

Using eqn. (34.2) for γ in terms of β:

$$\beta \equiv \frac{v}{c} = \frac{h}{\gamma \tau c} = \frac{h}{\tau c} \sqrt{1 - \beta^2}.$$

SOLVE
$$\beta^2 = \left(\frac{h}{\tau c}\right)^2 (1 - \beta^2) \Rightarrow \beta^2 \left[1 + \left(\frac{h}{\tau c}\right)^2\right] = \left(\frac{h}{\tau c}\right)^2.$$

Putting in the numbers:

$$\frac{h}{\tau c} = \frac{25.00 \times 10^3 \text{ m}}{(2.197 \times 10^{-6} \text{ s})(2.998 \times 10^8 \text{ m/s})} = 37.96.$$

Thus:
$$\beta = \frac{37.96}{\sqrt{37.96^2 + 1}} = 0.9997.$$

ANALYZE By contrast, naively using Newtonian theory without time dilation would give a speed $v = (25 \text{ km})/\tau = 38c$!

The detection of muons in cosmic ray showers (in 1937) occurred long after special relativity had been accepted. Nevertheless, it constitutes one of the earliest and most forceful demonstrations of time dilation. ∎

34.1.4 Length Contraction

In the Earth reference frame, time dilation explains how muons can reach detectors on the Earth's surface. However, in the muon's reference frame, the detector has only 2.2 μs to reach the muon. At a speed of $0.9997c$, the detector cannot move 25 km in that time. The resolution of this dilemma is that the detector doesn't have to travel 25 km; its distance from the muon is reduced by a factor $1/\gamma$!

EXERCISE 34.1 ♦ Calculate γ for $\beta = 0.9997$ and show that the detector must travel 660 m in the muon's frame. Show that it can just do this in 2.2 μs.

Like time intervals, distances depend on the measuring procedure. Let's look at what happens when each observer computes distance as speed multiplied by time interval. An observer on the Earth sees the muon travel a distance ℓ_E in a time $t_E = \gamma\tau$ and computes the muon's speed as $v = \ell_E/t_E = \ell_E/\gamma\tau$. Conversely, the muon "sees" the detector travel a distance ℓ_μ in a time $t_\mu = \tau$ and computes the detector's speed as $v = \ell_\mu/t_\mu = \ell_\mu/\tau$. But the relative speed of Earth and muon is the same in each of their reference frames. Thus $\ell_\mu/\tau = \ell_E/\gamma\tau$, and so $\ell_\mu = \ell_E/\gamma$.

NOTICE THAT WHEN WE COMPUTE THE SPEED OF SOMETHING THIS WAY WE MUST MEASURE BOTH ℓ AND t IN THE *SAME* FRAME.

EXAMPLE 34.2 ♦♦ A rocket 1.0×10^2 m long (measured by engineers on board) flies past a space station at $v = 0.80c$ (■ Figure 34.4). How long does the rocket take to pass an astronaut on the space station? What is the length of the rocket as measured by the astronaut?

MODEL The astronaut measures the time interval between two events: E_1—nose of rocket passes astronaut, and E_2—tail of rocket passes astronaut. The astronaut determines the rocket's length from its speed times the time interval. We compare these measurements with those of the rocket engineers to see how the rocket lengths in the two frames compare.

SETUP First, we calculate the time as observed by the rocket engineers:

$$\Delta t_R = \frac{\ell_R}{v} = \frac{100 \text{ m}}{0.80(3.0 \times 10^8 \text{ m/s})} = 0.42 \ \mu\text{s}.$$

The two events E_1 and E_2 occur at the same place in the astronaut's frame, so the time interval in that frame is the shorter. From eqn. (34.3):

$$\Delta t_A = \Delta t_R/\gamma = \ell_R/(v\gamma) = (0.42 \ \mu\text{s})\sqrt{1 - 0.80^2}$$
$$= (0.42 \ \mu\text{s})(0.60) = 0.25 \ \mu\text{s}.$$

SOLVE The length of the rocket in the astronaut's frame is:

$$\ell_A = v \ \Delta t_A = v \ \Delta t_R/\gamma = \ell_R/\gamma = (100 \text{ m})(0.60) = 60 \text{ m}.$$

ANALYZE It is important to identify clearly the *pair of events* that defines the measurement of any length.

■ **FIGURE 34.4**
A rocket flies past an astronaut A in the space station, who measures the rocket's length as $\ell = v \ \Delta t_A$.

The length of the rocket and the distance traveled by the muon are relative quantities. This effect is called *length contraction*.

> The length ℓ of an object is longest in the reference frame in which it is at rest. In another frame moving *parallel* to the object, its length is:
>
> $$\ell' = \ell/\gamma. \qquad (34.4)$$

Like time dilation, length contraction depends on the value of γ, which is greater than or equal to 1 and becomes infinite as v approaches c. This is our first indication that physical objects cannot exceed the speed of light.

34.1.5 Simultaneity

When events happen before our eyes, we can usually tell whether they happen at the same time or not. But how do we know about the timing of separated events? In action thrillers, the team meets to synchronize watches, or waits for a radio message and then follows a set

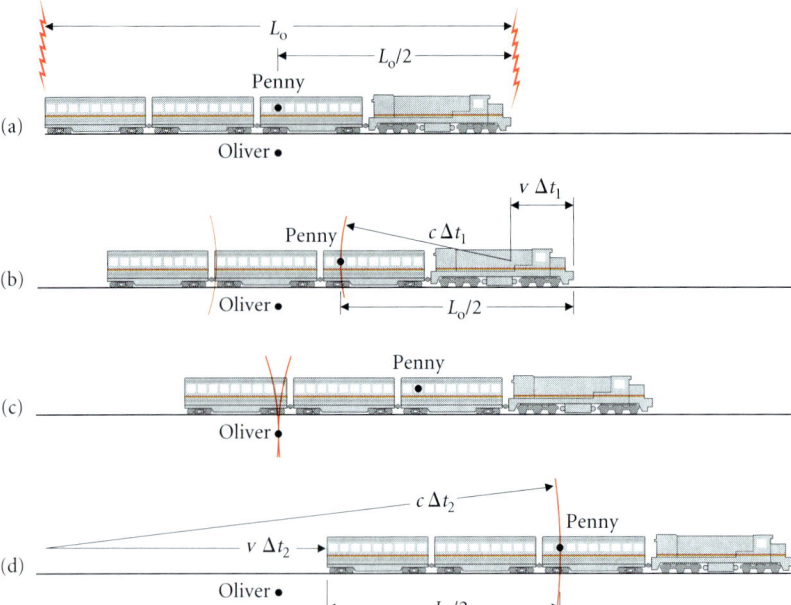

■ **FIGURE 34.5**
Simultaneity. Lightning bolts strike each end of the train at the same time, as measured by the observer Oliver. The sequence of events in Oliver's frame is: (a) Lightning bolts strike the train as Penny passes Oliver. (b) Light signal from front lightning bolt (bolt 1) reaches Penny after a time Δt_1. (c) Light signals from both bolts reach Oliver. (d) Light signal from rear bolt (bolt 2) reaches Penny a time Δt_2 after (a).

OLIVER CONCLUDES THAT BOTH FLASHES OCCURRED AT THE SAME TIME BECAUSE THEY REACH HIM FROM EQUAL DISTANCES AT THE SAME TIME.

schedule. We do the same in everyday life. One of the most intriguing consequences of Einstein's postulates is that this method of synchronizing events depends on the reference frame. To see why, let's study one of Einstein's thought experiments.

A train, the Relativistic Flyer, moves to the right at a large speed v. A passenger P (Penny) sits in the middle of the train, and an observer O (Oliver) stands beside the tracks. Two lightning bolts strike the ends of the train. ■ Figure 34.5 shows the sequence of events in Oliver's reference frame.

Oliver's interpretation: The two bolts strike at the same time, just as Penny passes him. Penny is moving to the right and encounters the light flash from bolt 1 before it reaches him. Next, both flashes arrive together at Oliver's position. Finally, the light flash from the second bolt reaches Penny.

Penny's interpretation of these events is different. The two flashes start at equal distances from her, travel at the same speed, and thus require the same time interval to reach her. The only way the flashes can arrive one after the other is for them to have occurred at different times. According to Penny, bolt 1 strikes before Oliver passes by and bolt 2 strikes later.

EXERCISE 34.2 ❖ Illustrate the sequence of events in Penny's reference frame, beginning with bolt 1 striking. (Assume the relative velocity of Oliver and Penny is such that bolt 2 strikes before Penny sees the flash from bolt 1.)

EXAMPLE 34.3 ◆◆◆ If the train's length is $L_P = 5.0 \times 10^2$ m (measured by Penny) and it is moving at $4c/5$, what time difference does Penny observe between the occurrence of the two bolts?

MODEL Since the light from each bolt travels at the same speed c in Penny's frame, the difference in the arrival time of the flashes equals the difference between the times the bolts struck. That difference is most readily calculated in Oliver's reference frame and then transformed to Penny's reference frame using the time dilation formula.

SETUP Since the train is at rest in Penny's frame, her measurement of its length is greater than Oliver's. So its length in his reference frame is $L_O = L_P/\gamma$ (eqn. 34.4). In Oliver's frame, a time interval Δt_1 elapses between the bolts striking and Penny seeing the first flash. During this interval Penny travels a distance $v\,\Delta t_1$ and the flash travels $c\,\Delta t_1$. The sum of these distances is $L_O/2$ (Figure 34.5):

$$(c + v)\Delta t_1 = L_O/2 = L_P/(2\gamma).$$

Similarly, Penny sees the second flash after an interval Δt_2, in Oliver's frame, where:

$$c\,\Delta t_2 = L_O/2 + v\,\Delta t_2,$$

or

$$(c - v)\Delta t_2 = L_P/(2\gamma).$$

The time interval in Oliver's frame between the two flashes reaching Penny is:

$$\delta t_O = \Delta t_2 - \Delta t_1 = \frac{L_P}{2\gamma}\left(\frac{1}{c-v} - \frac{1}{c+v}\right) = \frac{L_P v}{\gamma(c^2 - v^2)} = \frac{\gamma L_P v}{c^2}.$$

SOLVE The two flash receptions occur at different positions in Oliver's frame but at the same position in Penny's frame. Thus the time interval measured by Penny is the shorter:

$$\delta t_P = \frac{\delta t_O}{\gamma} = \frac{L_P v}{c^2} = \left(\frac{500\text{ m}}{3.0\times 10^8\text{ m/s}}\right)\left(\frac{4}{5}\right) = 1.3\times 10^{-6}\text{ s}.$$

ANALYZE Like time dilation and length contraction, the disagreement among observers over simultaneity of events is appreciable only when the relative speed of the observers is close to that of light.

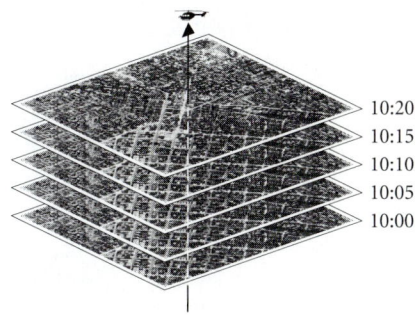

■ **FIGURE 34.6**
The helicopter parable. Photographs taken from a helicopter hovering over 42nd and Broadway are stacked to form a record of the history of New York City. The intersection's image is always at the center of the photographs.

34.2 SPACE-TIME

Time dilation, length contraction, and relativity of simultaneity are closely related phenomena. For example, length measurements involve the distance between two positions *at the same time* so that length contraction is just another form of disagreement over simultaneity. Hermann Minkowski realized in 1905 that these phenomena have a natural interpretation if we abandon space and time as separate concepts and think of them as unified in a single, four-dimensional *space-time*. To visualize this concept, let us imagine that an artist decides to record a history of New York City by making a sequence of aerial photographs from a helicopter hovering high above the city. Each photograph is labeled with the time it is taken, and the history is constructed by stacking the photographic plates vertically (■ Figure 34.6). This marvelous work of art shows all details of the city's life, including the surging crowds at 42nd and Broadway and poor Mr. Smith, who trips while running to catch the bus. The artist has constructed a *space-time diagram*.

BECAUSE THE SPEED OF LIGHT IS FINITE, A PHOTOGRAPH USING LIGHT CAN ONLY BE A METAPHOR FOR A RECORD OF SIMULTANEOUS EVENTS. IN §34.2.1 WE SHALL DESCRIBE A METHOD THAT ELIMINATES THE INACCURACIES OF A PHOTOGRAPH.

34.2.1 Representation of Space-time in a Single Reference Frame

Every occurrence (e.g., Mr. Smith trips) has a location in space and time. An ideal *event* occurs at a point in space and at an instant in time. It is described by the time when it happens and the three Cartesian coordinates of its position (t, x, y, z). Space-time is the four-dimensional collection of all events. Space evolves through time. A point, such as 42nd and Broadway, has a history, a sequence of events, represented by a line in space-time called a *world line*. An alternative view of space-time is the collection of world lines corresponding to each spatial point.

The most common convention for drawing space-time is to place the time axis vertically. All three spatial dimensions are perpendicular to the time axis, so at least one has to be left out of the diagram! Then simultaneous events occur in parallel planes perpendicular to the time-axis (■ Figure 34.7). It is convenient to use ct as the coordinate along the time axis so that the units on all axes are meters. Equivalently, we could put (distance/c) on each space-axis and use units of seconds.

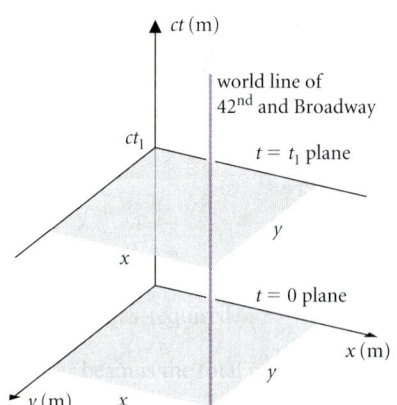

■ **FIGURE 34.7**
Space-time diagram with two spatial coordinates, and time (coordinate ct) vertical. Simultaneous events occur in parallel planes. Two such planes are shown here. A spatial point has a history, represented by a line in the diagram called a *world line*. The world line of 42nd and Broadway is vertical because the street intersection does not move in this reference frame.

Digging Deeper

DEFINING COORDINATES IN A REFERENCE FRAME

To establish a reference frame, we need a way to compare the times of events on different world lines. We can use standard clocks to measure time but, because of time dilation, they may not move with respect to each other. Thus we need many clocks and many observers to read them. In real space-time, there are usually only a few spacecraft or pieces of laboratory equipment available to receive and respond to signals. So for thought experiments, we use an ideal space-time that is filled with observers, one at each spatial point. One particular observer's world line is the time-axis of the frame, and an event along this world line is chosen as the origin. The method for synchronizing clocks and assigning spatial coordinates to the observers closely mimics measurement procedures in a real experiment. Each observer must be equipped with a device for sending EM waves (e.g., radar).

An observer, A, first measures her distance from the origin by sending a radar signal as shown in ■ Figure 34.8. The observer at the origin sends a reply immediately upon receiving the signal. According to Einstein's postulate, the signals travel at speed c, so if the first signal is emitted at time t_e and the reply is received at time t_r, the signal travels a total distance $c(t_r - t_e) = 2d$, and so observer A's measured distance from the origin is:

$$d = \tfrac{1}{2}c(t_r - t_e).$$

The observers check that they are mutually stationary by repeating the measurements and ensuring that d = constant.

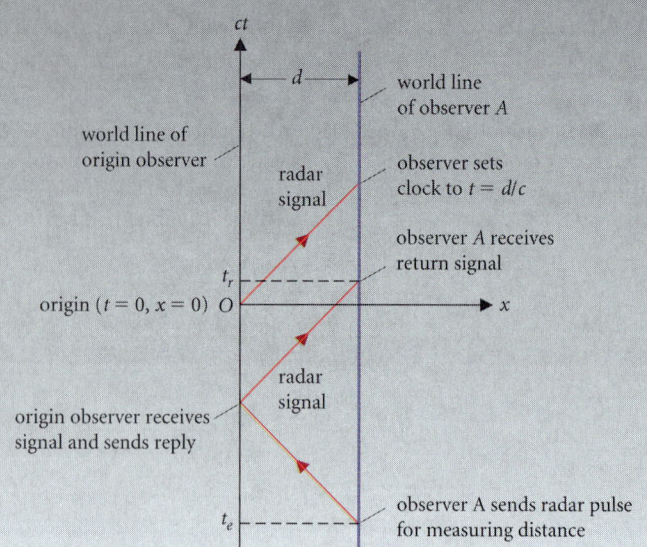

■ **FIGURE 34.8**
Defining coordinates. A radar signal is sent at time t_e from observer A to the observer at the origin, and back. The reply is received at time t_r. The distance d is determined to be $\tfrac{1}{2}c(t_r - t_e)$. Once A knows the distance from O, A's clock is synchronized using a signal sent from the origin.

Next, they synchronize their clocks using a light flash emitted at the origin. (*Remember:* The origin means $t = 0$ as well as $x = y = z = 0$.) Knowing the signal travels at speed c, observer A, distance d away, sets her clock to $t = d/c$ as the light flash passes. Now the observers are ready to conduct experiments.

EXAMPLE 34.4 ❖ Draw a space-time diagram illustrating the events in the helicopter parable.

MODEL As the helicopter hovers over the city, it remains for an extended time in one spatial position that we choose to call $x = 0$, $y = 0$, $z = 0$. Thus there is a sequence of events with increasing time coordinates but fixed space coordinates that describe the helicopter. This is the helicopter's *world line*. Since its spatial position is the spatial origin, this line is also the time-axis. We pick one event on this line as the origin ($t = 0$, $x = 0$, $y = 0$, $z = 0$).

Mr. Smith, running for the bus at speed v, occupies a sequence of positions

$$x = x_o + vt = x_o + (v/c)ct = x_o + \beta(ct).$$

His world line is the set of points with coordinates $(t, x, y, z) = (t, x_o + \beta ct, 0, 0)$—that is, a line with slope $1/\beta$ (■ Figure 34.9). When he trips, his speed becomes zero. His world line changes direction and moves vertically (x = constant), indicating that he is at rest.

The helicopter pilot sees Mr. Smith by observing a light wave that propagates from Mr. Smith to the helicopter. It travels at speed c ($\beta = 1$), so its world line has slope 1 and is therefore at 45° to each axis.

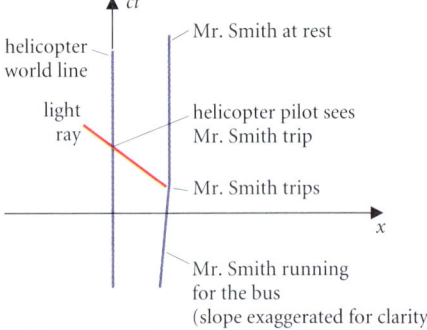

■ **FIGURE 34.9**
Events in the helicopter parable shown in a space-time diagram. Only one space dimension is shown. The part of Mr. Smith's world line where he is running has a barely discernible slope ($v \ll c$) even though we have exaggerated it enormously. After he trips, his world line is vertical, indicating that he is at rest in this frame.

> **ANALYZE** We have to exaggerate the slope of Mr. Smith's world line in the diagram to notice it at all. For any reasonable estimate of the speed at which he can run, β would be tiny.

34.2.2 Space-time Interval

Since space and time coordinates are not invariant, familiar concepts like past and future need careful definition. There is an invariant quantity, called the space-time interval, that allows us to order events in space-time and provide those definitions.

Suppose a flash of light is emitted in all directions at an event A. The expanding wave front comprises the events labeled *light cone* in ■ Figure 34.10. For example, light moving in the positive x-direction defines events with coordinates t, $x = x_A + c(t - t_A)$. These coordinates describe the straight line shown in the diagram. Light moving in all possible directions in the x-y-plane gives a set of straight lines forming the surface of a cone, hence the name.

The light cone from A divides space-time into several regions. Suppose event A is you at this moment reading this page. Your future is all the events you can influence. For example, by shining a flashlight on a piece of photographic film, you can expose it. The flashlight beam travels at the speed of light, and the event *film is exposed* would lie on the surface of the light cone. If you throw a baseball through a window, the event *window breaks* would be well inside your future light cone since the ball travels at a speed much less than c. Anything leaving A carrying energy or information moves at a speed $\leq c$ and has a world line on or within the light cone: this defines the future of A.

WE CANNOT EASILY DRAW THE RESULT OF LIGHT MOVING IN ALL THREE SPATIAL DIRECTIONS.

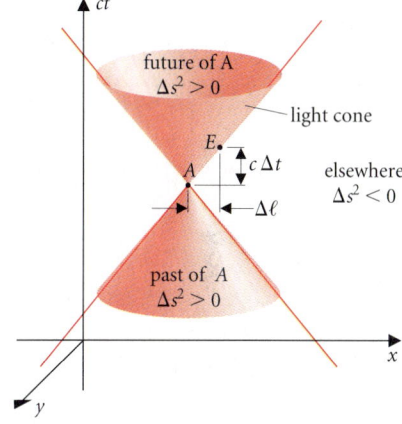

■ FIGURE 34.10
The past and future. Two spatial dimensions are shown here. The past and future of event A are defined by the *light cone:* the set of events corresponding to a flash of light emitted in all directions at A. At E, $\Delta \ell / \Delta t = c$. The space-time interval between events on the light cone is zero. Events not in the past or future of A are *elsewhere*.

> The *future* of an event A is the set of events that could receive a signal from A—that is, the events within the light cone that begins at A.

Event A also has a past light cone. For example, your sunburned nose is a result of light from the Sun reaching your nose. Your strained shoulder muscle is a result of your walk to the library carrying your books. All these events lie within the light cone that reaches event A from the past.

> The *past* of an event A is the region within the past light cone—that is, the set of events from which a signal could reach A.

Events that are in the past or future of A are said to have a *timelike* separation from A. Events that can neither send a signal to, nor receive a signal from, A are *elsewhere* and are said to have a *spacelike* separation from A.

If one event causes another, we would expect that relation to be independent of reference frame. A quantity called the *space-time interval* describes such relations between events. The correct expression for interval is most easily discovered by looking at events on the same light cone. The time interval Δt and the spatial distance $\Delta \ell$ between two such events A and E are related: $\Delta \ell = c \Delta t$ (Figure 34.10). Thus:

$$\Delta s^2 \equiv c^2 (\Delta t)^2 - (\Delta \ell)^2$$

is zero for every event E on the light cone. According to Einstein's postulate, the speed of light is the same in every reference frame so that $\Delta s^2 = 0$ is an invariant description of the light cone.

Next, we notice that for events in the past or future of A, $\Delta \ell < c \Delta t$, so:

$$\Delta s^2 = c^2 (\Delta t)^2 - (\Delta \ell)^2 > 0,$$

and for events elsewhere, $\Delta s^2 < 0$. These distinctions are also invariant. Combining these results, we may deduce the basic geometric property of space-time.

COMPARE WITH THE PYTHAGOREAN THEOREM: $\Delta\ell^2 = \Delta x^2 + \Delta y^2 + \Delta z^2$. BUT NOTE THAT THE MINUS SIGN IN EQN. (34.5) IS NOT A TYPOGRAPHICAL ERROR.

> The square of the space-time interval,
> $$\Delta s^2 \equiv c^2(\Delta t)^2 - (\Delta\ell)^2, \tag{34.5}$$
> between any two events in space-time is an invariant.

This interval may be used to define invariant quantities that replace the Newtonian concepts of time interval and length.

THE PROPER TIME BETWEEN TWO EVENTS IS ONLY DEFINED FOR EVENTS WITH A TIMELIKE SEPARATION.

> The *proper time* interval $\Delta\tau$ between two events with a timelike separation is the space-time interval between them divided by c:
> $$\Delta\tau = \Delta s/c = \sqrt{(\Delta t)^2 - (\Delta\ell)^2/c^2}. \tag{34.6}$$

The proper time interval $\Delta\tau$ equals the coordinate time interval Δt in a reference frame where the events occur at the same place ($\Delta\ell = 0$). If the two events are on the world line of a particle, this is the reference frame in which the particle is at rest. Proper time measures the rate of all physical processes—the decay of muons or the rate at which your heart beats. Two events in your life occur at the same place in your reference frame. Observers in any other reference frame measure a spatial separation between those events and a different time interval, but the proper time they calculate is the interval shown by your watch.

THERE IS NO SUCH THING AS A PROPER DISTANCE BETWEEN EVENTS WITH A TIMELIKE SEPARATION.

> The *proper distance* between two events with a spacelike separation is the magnitude of the space-time interval between them:
> $$L = \sqrt{-\Delta s^2} = \sqrt{(\Delta\ell)^2 - c^2(\Delta t)^2}. \tag{34.7}$$

In the reference frame where an object is at rest, the locations of its ends at some time t define two events with $\Delta t = 0$. The proper distance between these events is called the proper length or *rest length* of the object.

EXAMPLE 34.5 ♦ A sigma particle produced in an accelerator experiment is captured in a detector 0.150 m away 0.520 ns later. Find the speed of the particle and the proper time interval between its production and detection.

MODEL We are given the laboratory coordinate time interval. Ordinary kinematics applies within a given reference frame to determine the speed. To obtain the proper time, we apply eqn. (34.6).

SETUP AND SOLVE The particle's speed is:

$$v = \Delta\ell/\Delta t = (0.150 \text{ m})/(0.520 \times 10^{-9} \text{ s}) = 2.88 \times 10^8 \text{ m/s}.$$

The proper time interval is:

$$\Delta\tau = \sqrt{(\Delta t)^2 - (\Delta\ell)^2/c^2} \tag{i}$$
$$= \sqrt{(0.520 \times 10^{-9} \text{ s})^2 - (0.150 \text{ m})^2/(3.00 \times 10^8 \text{ m/s})^2} = 0.143 \text{ ns}.$$

ANALYZE As we saw in the muon experiment (Example 34.1), it is time in the particle's rest frame—the proper time—that determines whether a particle is likely to reach a detector before it decays. The mean lifetime of a sigma particle—0.15 ns—is longer than the proper time we calculated, consistent with its capture in the detector. Einstein's theory is verified daily by such events in accelerators worldwide.

WE'LL HAVE MORE TO SAY ABOUT PARTICLE DECAY IN CHAPTERS 36 AND 37.

Since $\Delta\ell = v\,\Delta t$, eqn. (i) may be written $\Delta\tau = \Delta t/\gamma$, so it is just the time dilation formula, eqn. (34.3). ∎

34.3 The Lorentz Transformation

34.3.1 Coordinate Transformation

Frequently, we need to compare representations of space-time in different reference frames. For this we require the transformation equations relating the coordinates of an arbitrary event in one frame to its coordinates in the other frame. Relative motion is the important difference between the two frames, so we avoid irrelevant complications by choosing both x-axes to be along the direction of the relative velocity and placing the origin at the same event in each frame. The coordinates in the second, prime, frame are labeled t', x', y', and z'. Thus the *prime frame* moves with velocity $\vec{v} = v\hat{i}$ with respect to the *unprime frame*. Dig Deeper (p. 1082) to see that the transformation is given by:

$$ct' = \gamma(ct - \beta x), \quad x' = \gamma(x - \beta ct),$$
$$y' = y, \quad z' = z. \quad (34.8)$$

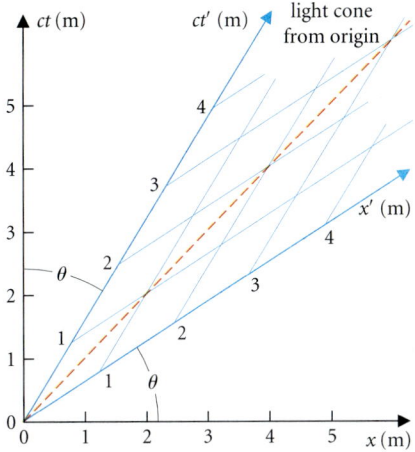

FIGURE 34.11
The Lorentz transformation. The prime axes are shown in the unprime frame. Both axes are rotated toward the light cone from the origin, with $\tan\theta = \beta$. Keep in mind when using this diagram that the scales on the two sets of axes are not the same!

Using the Lorentz transformation (eqns. 34.8), we can determine how the axes of the prime frame appear in a space-time diagram drawn by unprime observers (■ Figure 34.11). Along the t'-axis, x', y', and z' equal zero. At points on this line, $x = \beta ct$. (Use eqns. 34.8 with $x' = 0$). This is the equation of a particle (the observer at the prime origin) moving to the right at speed βc. Similarly, along the x'-axis, t', y', and z' equal zero. At points along this line, $t' = 0 \Rightarrow ct = \beta x$, which describes a line making angle $\theta = \tan^{-1}\beta$ with the x-axis. Both the time- and space-axes are rotated toward the light cone. In the limit as $v \to c$, both axes coincide along a 45° line. This is a second indication that a relative speed greater than c cannot occur since, if it did, the space- and time-axes would trade places.

NOTE THAT $\beta c \equiv v$. AS $\beta \to 0$, $\gamma \to 1$, AND EQNS. (34.8) REDUCE TO THE GALILEAN TRANSFORMATION IN TABLE 34.1.

EXERCISE 34.3 ♦♦ In the prime frame, the unprime frame moves with velocity $\vec{v} = -v\hat{i}$. Use this fact to write down the Lorentz transformation from prime to unprime coordinates. Locate the unprime axes in a space-time diagram drawn by prime observers. ■

EXAMPLE 34.6 ♦♦ A rocket leaves Earth traveling at $v = 4c/5$. After 1 day, a radio signal is sent to the rocket from Earth. The rocket immediately sends a reply. Find the following: (a) The time at which the signal is received by the rocket, as perceived by the rocket crew. (b) The distance of the rocket from Earth at that time. (c) The time at which the return signal is received on Earth, as measured by both Earth and rocket observers.

MODEL We begin by drawing the space-time diagram in the Earth frame (■ Figure 34.12) with the events E_0 (signal sent to rocket), E_1 (signal received at rocket), and E_2 (return signal received at Earth). We use the given information to find the coordinates of these events in the Earth frame and then use the Lorentz transformation to find the coordinates in the rocket frame. We choose the x-axis to be parallel to the rocket's velocity, with the origin at the event *rocket leaves Earth*.

SETUP E_0: $x_0 = 0$, $t_0 = 1$ day.

E_1: E_1 occurs at the intersection of the rocket's world line and the light cone from E_0. The rocket moves at speed v and thus its x-coordinate at time t is $x = vt$. The radio signal travels at speed c. At time t, it has been traveling for $\Delta t = t - t_0$ and its x-coordinate is $x = c(t - t_0)$. The radio signal meets the rocket when their x-coordinates are the same—that is, at t_1 when:

$$vt_1 = c(t_1 - t_0),$$

or
$$t_1 = \frac{ct_0}{c - v} = \frac{t_0}{1 - \beta} = \frac{1 \text{ day}}{1 - 4/5} = 5 \text{ days}.$$

Then $x_1 = vt_1 = 4$ light-days. [This is the answer to (b).]

E_2: The light signal takes 4 days to reach the rocket and an equal time to come back.

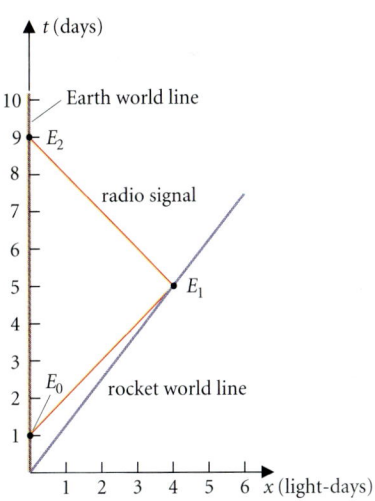

FIGURE 34.12
Space-time diagram for a rocket leaving Earth. The rocket launch is taken as the origin. Events E_0 (radio signal sent from Earth), E_1 (signal received at rocket), and E_2 (reply received at Earth) are shown. The distance unit is a light-day: the distance traveled by light in 1 day.

Digging Deeper

DERIVATION OF THE LORENTZ TRANSFORMATION

Since both the prime and unprime frames are inertial, the transformation equations should be linear in the coordinates so that uniform motion in one frame is uniform in the other. When the relative speed of the frames is much less than c, the transformation reduces to the Galilean result (Table 34.1). These requirements are satisfied by a transformation of the form:

$$ct' = Act + Bx, \quad x' = Dct + Ex,$$
$$y' = Fy, \quad z' = Fz, \tag{i}$$

where the coefficients A, B, D, E, and F depend only on the relative velocity. The same coefficient F occurs for y' and z' because with both directions perpendicular to the relative velocity, relative motion should affect them in the same way.

To determine the coefficients, we use the invariance of the space-time interval: Δs^2 between any two events is the same in both frames. We choose the origin and another event with $y = z = 0$ but arbitrary values of x and t. Then also $y' = z' = 0$, and the space-time interval between the two events is:

$$\Delta s^2 = c^2 t^2 - x^2 = c^2 (t')^2 - (x')^2.$$

Applying the transformation (i) to the prime coordinates, we find:

$$c^2 t^2 - x^2 = (Act + Bx)^2 - (Dct + Ex)^2.$$

Gathering up like terms, we have:

$$c^2 t^2 (A^2 - D^2 - 1) + 2xct(AB - DE) + x^2(B^2 - E^2 + 1) = 0.$$

Since this equation holds for arbitrary values of x and t, the three coefficients each vanish separately:

$$A^2 - D^2 - 1 = 0, \tag{ii}$$
$$AB - DE = 0, \tag{iii}$$

and

$$B^2 - E^2 + 1 = 0. \tag{iv}$$

From eqn. (iii), $AB = DE$, so we can eliminate B from eqn. (iv):

$$0 = \frac{D^2 E^2}{A^2} - E^2 + 1 = E^2 \left(\frac{D^2 - A^2}{A^2} \right) + 1.$$

Then using eqn. (ii), we find: $E^2 \left(\frac{-1}{A^2} \right) + 1 = 0$ or $A^2 = E^2$.

Both coefficients A and E are positive, since t and t' increase together, as do x and x'. So $A = E$, and it follows that $D = B$.

Next we consider a particular event whose coordinates we know. The spatial origin of the prime frame ($x' = 0$, $y' = 0$, $z' = 0$) moves to the right with speed v in the unprime frame. Thus at any time T in the unprime frame, its coordinates are: $x = vT$, $t = T$ and $x' = 0$, $t' = T'$.

Using the second of eqns. (i) to relate these coordinates, we have:

$$x' = 0 = Dct + Ex = DcT + EvT,$$

or

$$D = -(v/c)E = -\beta E = -\beta A.$$

Then from eqn. (ii),

$$(1 - \beta^2)A^2 - 1 = 0 \;\Rightarrow\; A = \frac{1}{\sqrt{1 - \beta^2}} \equiv \gamma.$$

With $E = A$ and $D = -\beta A$, the resulting Lorentz transformation is:

$$ct' = \gamma(ct - \beta x) \quad \text{and} \quad x' = \gamma(x - \beta ct).$$

We complete the derivation by noting that F equals 1 (Figure 34.13 and Exercise 34.4).

EXERCISE 34.4 ♦♦ By considering an event with $y, z \neq 0$, show that $F = 1$.

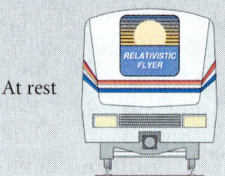
At rest

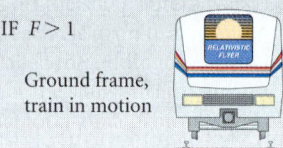
IF $F > 1$
Ground frame, train in motion

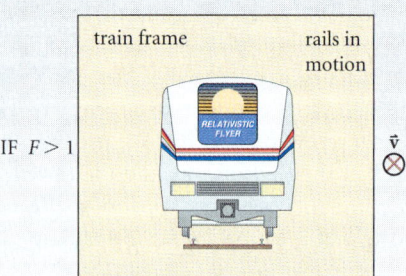
IF $F > 1$
train frame — rails in motion

■ **FIGURE 34.13**
What if rapidly moving objects were to contract in the direction perpendicular to their motion—that is, if the coefficient F in the Lorentz transformation were not unity? The Relativistic Flyer is designed to fit exactly on the rails when at rest. For $F > 1$, an engineer on the ground would observe the train falling between the tracks as it sped up. An engineer on the train would observe the moving rails contracting between the wheels of the train. These two opposite disasters cannot both happen. A similar argument rules out width expansion ($F < 1$). This argument relies on the fact that both engineers agree on the simultaneity of the events that they use to measure widths (see Problem 16). They can agree because the separation of the events is perpendicular to the velocity.

SOLVE $x_2 = 0$, $t_2 = 5$ days $+ 4$ days $= 9$ days.

SETUP It remains to find t_1' and t_2'. From eqn. (34.8):

$$t_1' = \gamma(t_1 - \beta x_1/c),$$

with

$$\gamma = \frac{1}{\sqrt{1-\beta^2}} = \frac{1}{\sqrt{1-16/25}} = \frac{5}{3}.$$

SOLVE $t_1' = \frac{5}{3}[5 \text{ days} - \frac{4}{5}(4 \text{ days})] = (\frac{5}{3})(\frac{9}{5})$ days $= 3$ days.

Similarly: $\qquad t_2' = \frac{5}{3}(9 \text{ days}) = 15$ days.

ANALYZE Notice that for two events occurring at the same place in the rocket frame (*rocket leaves Earth* and *rocket receives signal*), the rocket interval of 3 days is shorter than the Earth interval of 5 days. For events that occur at the same place in the Earth frame, the reverse is true.

EXERCISE 34.5 ❖ Draw these events in the rocket frame.

Study Problem 20 ❖❖❖ *The Student's Revenge*

A chemistry professor is a passenger on the Relativistic Flyer, a train that travels at $0.80c$. A disgruntled physics student who failed the chemistry course notices that the train, length $\ell = 150$ m, must pass through a tunnel $L = 0.10$ km long. The student knows the train is contracted to a length $\ell/\gamma = \frac{3}{5}(150 \text{ m}) = 90$ m and arranges to close gates at both ends of the tunnel when the train is inside, thus trapping the professor (■ Figure 34.14). The professor is not worried. Knowing the tunnel is contracted to a length of $\frac{3}{5}(100 \text{ m}) = 60$ m, the professor concludes that the train will never be completely inside the tunnel and the student's scheme will fail. Who is right?

Modeling the System

The professor is certainly correct in that, from his point of view, the tunnel is shorter than the train. However, he has forgotten that events that are simultaneous in the student's frame, such as closing gates at the two ends of the tunnel, are not simultaneous in his frame. If the tunnel exit is closed first, the train can continue to enter the shortened tunnel (crumpling as it goes) until it is completely inside. Then the entrance closes and the professor is trapped.

To demonstrate the truth of this conjecture, we identify the four relevant events and look at their coordinates in both reference frames.

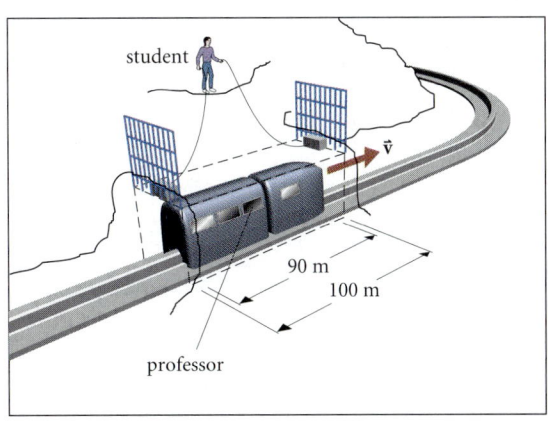

(a) Student's picture.

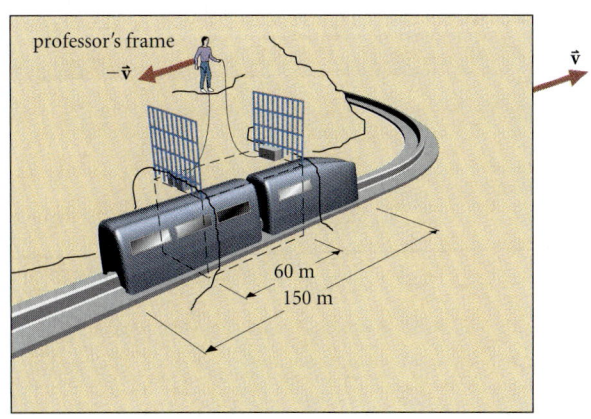

(b) Professor's belief.

■ **FIGURE 34.14**
(a) The disgruntled student is poised to close the gates and trap the Lorentz-contracted train inside the tunnel.
(b) The chemistry professor believes that, in his frame, the train will never be completely inside the Lorentz-contracted tunnel and thinks that he is safe.

■ **FIGURE 34.15**
Space-time diagram for the events in the student's frame. The rest length of the tunnel is $L = 100$ m and the length of the train is $\ell/\gamma = 3(150 \text{ m})/5 = 90$ m. The world lines of both ends of the train and the tunnel are shown. Events E_1 and E_2, gates closing at the tunnel entrance and exit, are simultaneous in this frame.

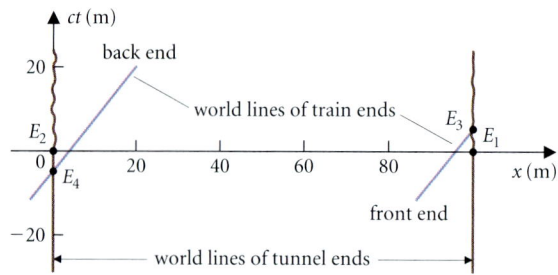

Setup of Solution

The four events are as follows.

E_1: Tunnel exit closes.

E_2: Tunnel entrance closes.

E_3: Front end of train reaches tunnel exit.

E_4: Back end of train reaches tunnel entrance.

We choose the origin to be at the tunnel entrance, at the instant the student closes the gate. We assume the train is centered in the tunnel at $t = 0$ in the student's (unprime) frame. The tunnel has length L (■ Figure 34.15), so the first two events have coordinates:

$$x_1 = L, \; ct_1 = 0 \quad \text{and} \quad x_2 = 0, \; ct_2 = 0.$$

In the student's frame, the train has length ℓ/γ. At $t = 0$, its front end is a distance $(L - \ell/\gamma)/2$ from the tunnel exit and its back end is an equal distance from the tunnel entrance. Thus the front reaches the exit at time $ct_3 = (L - \ell/\gamma)/(2\beta)$ and the back end was at the entrance at time $ct_4 = -ct_3$. Thus events 3 and 4 have coordinates:

$$x_3 = L, \; ct_3 = \frac{L - \ell/\gamma}{2\beta} \quad \text{and} \quad x_4 = 0, \; ct_4 = \frac{-(L - \ell/\gamma)}{2\beta}.$$

Next, we use eqns. (34.8) with $\beta = \frac{4}{5}$, $\gamma = \frac{5}{3}$ to find the time coordinates in the professor's (prime) frame:

$$ct_1' = \gamma(ct_1 - \beta x_1) = \gamma(0 - \beta L) = -\gamma\beta L.$$

$$ct_2' = \gamma(ct_2 - \beta x_2) = \gamma(0 - 0) = 0.$$

$$ct_3' = \gamma(ct_3 - \beta x_3) = \gamma\left(\frac{L - \ell/\gamma}{2\beta} - \beta L\right) = \frac{\gamma L(1 - 2\beta^2) - \ell}{2\beta}.$$

$$ct_4' = \gamma(ct_4 - \beta x_4) = \gamma\left[\frac{-(L - \ell/\gamma)}{2\beta} - 0\right] = \frac{-\gamma L + \ell}{2\beta}.$$

NOTE: E_2 IS THE ORIGIN IN BOTH FRAMES.

Solution

We evaluate the times of the four events in the professor's frame:

$$ct_1' = -\left(\tfrac{5}{3}\right)\left(\tfrac{4}{5}\right)(100 \text{ m}) = -133 \text{ m}.$$

$$ct_2' = 0.$$

$$ct_3' = \frac{(5/3)(100 \text{ m})(1 - 32/25) - 150 \text{ m}}{8/5} = -123 \text{ m}.$$

$$ct_4' = -\frac{(5/3)(100 \text{ m}) - (150 \text{ m})}{8/5} = -10.4 \text{ m}.$$

In this frame the sequence of events is as follows. First, the tunnel exit is blocked [E_1 at $t' = -(133 \text{ m})/c = -0.44 \text{ } \mu\text{s}$]. Next, the front of the train smashes into the closed tunnel exit [E_3

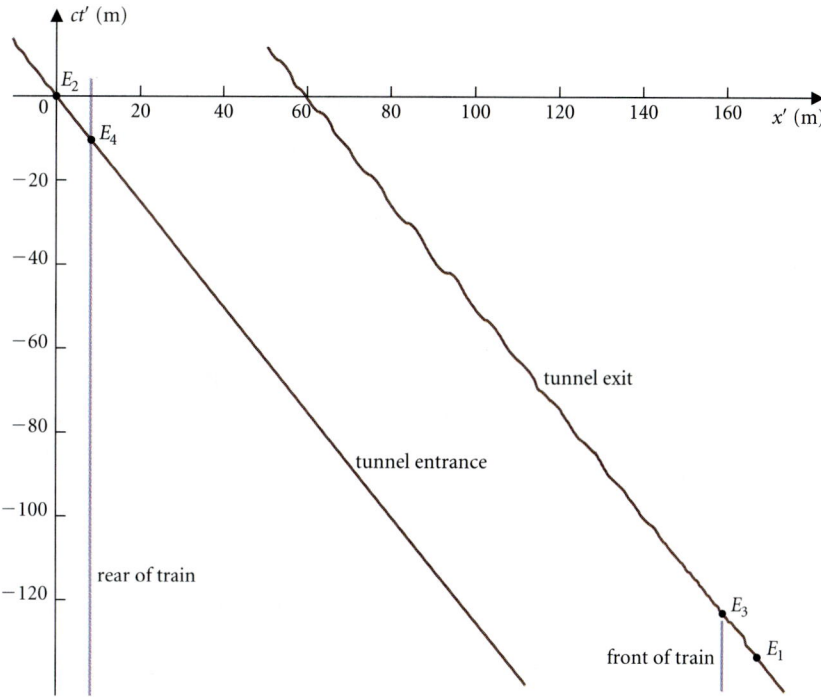

■ **FIGURE 34.16**
The tunnel viewed in the professor's frame. The tunnel has length $L/\gamma = 3(100 \text{ m})/5 = 60 \text{ m}$ and the train has length 150 m. Events E_1 and E_2 are not simultaneous in this frame: E_1 occurs first. However, E_1 still precedes E_3, and E_4 precedes E_2. Compare with Figure 34.15.

at $t' = (-123 \text{ m})/c = -0.41 \text{ }\mu\text{s}$]. Then the back end of the train enters the tunnel [E_4 at $t' = (-10.4 \text{ m})/c = -0.035 \text{ }\mu\text{s}$]. Finally, the tunnel entrance is closed and the professor is trapped (E_2 at $t' = 0$). The events in the professor's frame are shown in ■ Figure 34.16.

Analysis

The sequence of cause and effect is preserved by the Lorentz transformation, even though events that are simultaneous in one frame are not simultaneous in another. Notice that in both frames E_1 precedes E_3 (the tunnel exit is blocked before the train runs into it) and E_4 precedes E_2 (the back end of the train enters the tunnel before it is blocked). However, the sequence of events that do not affect each other (E_1 and E_2, or E_3 and E_4) can change.

The moral of this story is never to jump to conclusions when solving problems in special relativity! And, of course, the student could (and should) have found a better way to handle her frustrations!

34.3.2 Velocity Transformation

From the Lorentz transformation for coordinates, we may derive rules for transforming an object's velocity components between frames. The definition of instantaneous velocity involves events separated by a small time interval along the world line of an object. Suppose two such events are separated by a displacement $d\vec{r}'$ and a time dt' in the prime frame. The velocity in that frame is then:

$$\vec{u}' = \frac{d\vec{r}'}{dt'} = \frac{dx'\hat{i} + d\vec{r}'_\perp}{dt'},$$

where $d\vec{r}_\perp = d\vec{r}'_\perp$ is the component of $d\vec{r}$ in the y-z-plane (■ Figure 34.17). The prime frame has velocity $\vec{v} = \beta c \hat{i}$ relative to the unprime frame. By differentiating eqns. (34.8), we can write both $d\vec{r}'$ and dt' in terms of unprime coordinates:

$$\vec{u}' = \frac{\gamma(dx - \beta c\, dt)\hat{i} + d\vec{r}_\perp}{\gamma(dt - \beta\, dx/c)},$$

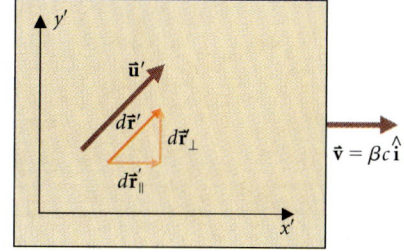

(a) Prime frame.

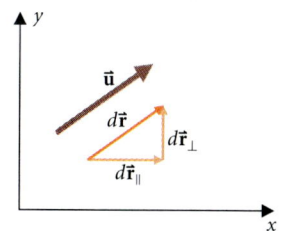

(b) Unprime frame.

■ **FIGURE 34.17**
Lorentz transformation for velocities. Here we are looking at the velocity vectors in the x-y-plane. In the prime frame (a), a particle undergoes a displacement $d\vec{r}'$ in time dt'. In the unprime frame (b), its displacement is $d\vec{r}$ in time dt.

IN THIS SECTION WE USE $\vec{v}$ FOR THE VELOCITY OF THE PRIME FRAME MEASURED BY OBSERVERS IN THE UNPRIME FRAME, AND $\vec{u}$ AND $\vec{u}'$ FOR THE VELOCITY OF AN OBJECT MEASURED WITHIN EACH FRAME. DISTINGUISH THEM CAREFULLY!

Dividing top and bottom by dt, we find:

$$\vec{u}' = \frac{(dx/dt) - \beta c}{1 - (\beta/c)(dx/dt)}\hat{i} + \frac{d\vec{r}_\perp/dt}{\gamma[1 - (\beta/c)(dx/dt)]}.$$

In components:

WHEN USING EQNS. (34.9), BE SURE TO SPECIFY CLEARLY WHAT DEFINES EACH FRAME AND WHICH OBJECT HAS THE VELOCITY $\vec{u}$.

$$u'_x = \frac{u_x - \beta c}{1 - \beta u_x/c} \quad \text{and} \quad \vec{u}'_\perp = \frac{\vec{u}_\perp}{\gamma(1 - \beta u_x/c)} \tag{34.9}$$

In the limit $\beta \ll 1$, the denominators in each expression tend to 1, and we recover the Galilean results (Table 34.1). In the special case $\vec{u} = 0$, we find $\vec{u}' = -\beta c\hat{i}$, as expected. In general, the velocity components parallel and perpendicular to $\vec{v}$ transform differently. The γ factors cancel in u'_x because both time dilation and length contraction influence the ratio dx'/dt'. Length contraction has no effect on the ratio $d\vec{r}'_\perp/dt'$ as lengths perpendicular to $\vec{v}$ do not change. The factor $1 - \beta u_x/c$ that appears in both expressions arises from the disagreement about simultaneity between the frames.

EXERCISE 34.6 ♦ A rocket moving with speed v directly away from a planet emits a flash of light. Use eqns. (34.9) to find the speed of the flash as measured by an observer on the planet.

EXAMPLE 34.7 ♦ As a rocket passes a space station, the inhabitants of the station measure the rocket's speed to be $v = 0.65c$. At the instant the rocket passes, they launch a supply package at speed $0.78c$ in the same direction. How fast does the supply package approach the rocket, as measured by the rocket crew?

MODEL The space station defines the unprime frame and the rocket defines the prime frame. The package has velocity $\vec{u}$ in the unprime (space station) frame and $\vec{u}'$ in the prime (rocket) frame. The velocities of rocket and package are parallel.

SETUP To convert to the rocket frame, we choose the x-axis to be along the rocket's path and apply eqns. (34.9).

SOLVE We have $u_x = 0.78c$ and $\beta = 0.65$. Thus:

$$u'_x = \frac{0.78c - 0.65c}{1 - (0.65)(0.78)} = \frac{0.13c}{0.49} = 0.26c.$$

ANALYZE Compare with the Galilean result of $0.78c - 0.65c = 0.13c$. The supply package approaches the rocket twice as fast as the Galilean result would predict.

EXAMPLE 34.8 ♦♦ A second rocket is approaching the space station at right angles to the track of the first, as seen from the space station, with speed $0.45c$. Describe its velocity as seen by the first rocket's crew (■ Figure 34.18).

MODEL Again, we choose the space station to define the unprime frame and the first rocket to define the prime frame, with the x- and x'-axes along the first rocket's path. In the space station frame, the second rocket moves in the y-direction. In the prime (first rocket's) frame, the second rocket's velocity has both x'- and y'-components.

SETUP We use eqns. (34.9) with $\beta = 0.65$, $u_x = 0$, and $u_y = 0.45c$. Then:

$$\gamma = \frac{1}{\sqrt{1 - 0.65^2}} = 1.32.$$

SOLVE

$$u'_x = \frac{u_x - \beta c}{1 - \beta u_x/c} = \frac{0 - 0.65c}{1 - 0} = -0.65c,$$

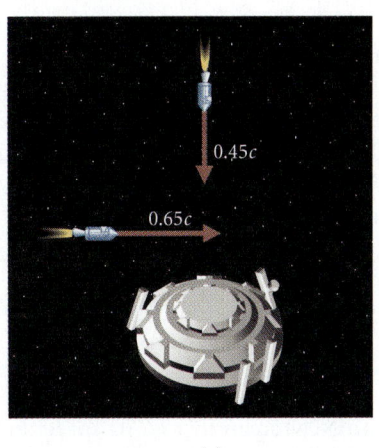
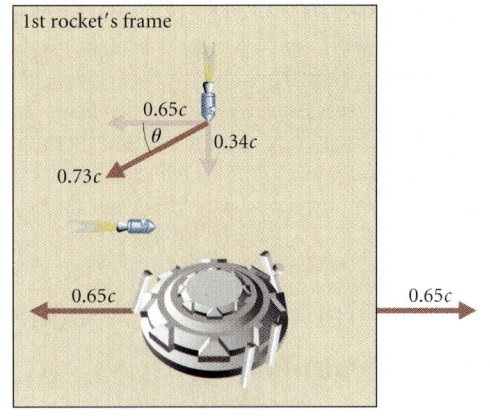

(a) (b)

■ **Figure 34.18**
(a) View in the space station frame as the second rocket approaches. (b) The situation in the first rocket's frame. The second rocket is approaching from in front.

and
$$u'_y = \frac{u_y}{\gamma(1 - \beta u_x/c)} = \frac{0.45c}{1.32(1 - 0)} = 0.34c.$$

Equivalently, the second rocket's speed is:
$$u' = c\sqrt{(0.65)^2 + (0.34)^2} = 0.73c,$$

and it is moving at an angle
$$\theta = \tan^{-1}(0.34/0.65) = 28°$$ to the first rocket's path.

ANALYZE Compare with the Galilean result: $\theta_G = \tan^{-1}(0.45/0.65) = 35°$. The relativistic expression for $\tan \theta$ is smaller by a factor γ. As $u \to c$, everything appears to approach the rocket from almost directly in front, and the geometry of space appears distorted. ■

For more on this kind of distortion, see Problem 56.

34.3.3 Acceleration in Special Relativity

Special relativity is *special* because it describes the relations between *uniformly moving* reference frames. Acceleration involves a change of reference frame that we must describe carefully. A famous thought experiment considers the dilemma of the traveling twin.

One twin (mobile Melia) volunteers for the first trip to Sirius (8.6 light-years away) and her sister (domestic Delia) stays home. The expedition accelerates rapidly, cruises at $\beta = \frac{4}{5}$, turns around quickly at Sirius after encountering its hostile inhabitants, and returns at $\beta = \frac{4}{5}$. On Earth, 21.5 years have passed when the expedition returns, but the expedition members have aged only $\Delta t' = \Delta t/\gamma = 12.9$ years. Melia, the traveling twin, is 8.6 years younger than her sister.

Melia is perplexed. After all, according to rocket observations, the Earth sped away and later returned. Sister Delia moves with the Earth, so why doesn't she end up younger?

Acceleration at Sirius provides the answer (■ Figure 34.19). Mobile Melia occupies one inertial reference frame G while going to Sirius and a second frame R while returning. With special relativity, she correctly calculates the time intervals LB and CR. The time interval BC is left out. Event B is simultaneous with the rocket firing (event O) in Melia's outbound frame, and event C is simultaneous with O in her inbound frame. Events B and C are certainly not simultaneous from domestic Delia's point of view! Because Melia accelerates and so changes from one inertial frame to another, she cannot successfully apply results from special relativity that are valid only when the observer remains in a single, inertial reference frame.[2]

In 1960, Chalmers Sherwin pointed out that iron nuclei could be used as the twins in an experimental test of round-trip time dilation, and R. V. Pound and G. A. Rebka were able to

This thought experiment is known as the twin paradox.

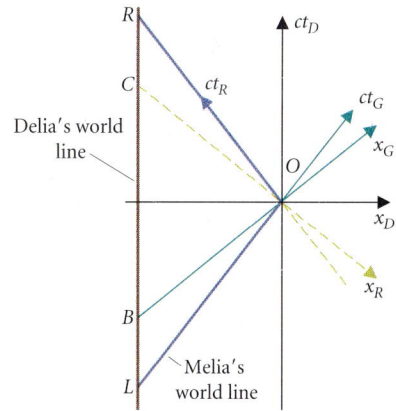

■ **Figure 34.19**
Space-time diagram for the twins Melia and Delia. Origin is chosen at the event *rocket engines fire at Sirius*. Melia's reference frame changes from G to R as a result of the acceleration caused by the rocket engines. Using special relativity to compute the time intervals in her rest frame, she misses the time interval BC due to the change of frame: O and B are simultaneous in frame G; O and C are simultaneous in frame R.

---
[2] In spite of the name, there is no real paradox here. Note carefully that Melia's youth is not a *dynamic* effect due to forces acting on her as she turns around. It is a purely *kinematic* effect due to the change of reference frame, as experiment confirms.

IN PROBLEMS 60–62 WE ASK YOU TO INVESTIGATE FURTHER ASPECTS OF THE TWIN PARADOX.

FOR THERMAL MOTION, SEE CHAPTER 19, ESPECIALLY §19.2. FOR A DETAILED COMPARISON OF THE EXPERIMENT WITH THEORY, SEE PROBLEM 83.

measure the effect. Because of its thermal motion, an iron nucleus undergoes periods of constant velocity with short periods of abrupt acceleration that change its velocity, rather like Melia on her rocket trip. The average speed of the traveling atom increases as temperature increases. The iron nucleus also carries an internal clock whose period can be measured by looking at the frequency of light emitted by the nucleus. Pound and Rebka found that an iron sample 1 K hotter than another had a lower effective frequency, $\Delta f/f = -2 \times 10^{-15}$. The lower frequency corresponds to the lesser aging of the traveling twin.

34.4 RELATIVISTIC DYNAMICS

34.4.1 Momentum

IN THIS SECTION WE SHALL CONTINUE THE CONVENTION OF USING $\vec{v}$ FOR THE VELOCITY OF A FRAME AND $\vec{u}$ FOR THE VELOCITY OF A PARTICLE IN A FRAME.

When a constant force acts on a particle, according to Newtonian theory, its speed and its momentum $|\vec{p}_N| \equiv mu$ increase uniformly with time. According to special relativity, the speed is limited by the finite value $u = c$, and so the Newtonian momentum also approaches a limit mc, apparently contradicting Newton's second law. According to Einstein's first postulate, if Newton's second law in the form $\vec{F} = d\vec{p}/dt$ is invariant, the Newtonian expression for momentum cannot be exact.

Although $\vec{p}$ cannot remain proportional to speed when $u \sim c$, we have no reason to doubt its dependence on mass or that its direction is parallel to $\vec{u}$. So we suppose that the correct relativistic expression is of the form:

$$\vec{p} = mf(u)\vec{u},$$

IN EQN. (34.10), $\gamma = (1 - u^2/c^2)^{-1/2}$. PREVIOUSLY WE USED $\gamma = (1 - v^2/c^2)^{-1/2}$, WHERE v IS THE RELATIVE SPEED OF TWO REFERENCE FRAMES. YOU MUST USE CONTEXT TO DISTINGUISH THE TWO γS. WHEN BOTH APPEAR IN THE SAME EXPRESSION, WE USE γ_u FOR THE VALUE OF γ THAT PERTAINS TO A PARTICLE WITH SPEED u.

where $f(u)$ corrects the dependence on speed. Since Newton's expression is correct for $u \ll c$, we know that $f(u) \to 1$ as $u \to 0$. *Digging Deeper,* we show that $f(u) = \gamma$. The correct relativistic expression for momentum is:

$$\vec{p} = \gamma m\vec{u}. \qquad (34.10)$$

As the particle's speed approaches the speed of light, a large increase in momentum corresponds to a negligible increase in speed.

Sometimes eqn. (34.10) is interpreted by saying that a particle's mass increases with speed according to the formula $m = \gamma m_o$, where m_o is the *rest mass.* We prefer to reserve the word "mass" and the symbol m for rest mass, which is an invariant property of a particle or system. This is the convention used in eqn. (34.10) and throughout this book.

EXAMPLE 34.9 ♦ A cosmic ray electron of mass $m = 9.11 \times 10^{-31}$ kg is traveling with speed $u = 0.995c$. What is its momentum?

MODEL With a speed almost equal to the speed of light, we should use the relativistic formula for momentum, eqn. (34.10).

SETUP First we calculate γ using the electron's speed:

$$\gamma = (1 - u^2/c^2)^{-1/2} = (1 - 0.995^2)^{-1/2} = 10.0.$$

SOLVE The momentum has magnitude:

$$p = \gamma mu$$
$$= (10.0)(9.11 \times 10^{-31} \text{ kg})(0.995)(3.00 \times 10^8 \text{ m/s})$$
$$= 2.72 \times 10^{-21} \text{ kg} \cdot \text{m/s}.$$

ANALYZE The momentum is ten times the Newtonian value. *Remember:* When using the expression γmu, use the particle's speed u to compute γ. ∎

EXERCISE 34.7 ♦ At what speed does an electron have twice the momentum predicted by the Newtonian formula?

Digging Deeper

Relativistic Momentum

We can determine the function $f(u)$ from a thought experiment about colliding particles, analyzed in two different reference frames. Two identical particles each of mass m undergo the elastic collision shown in ■ Figure 34.20a. The first particle moves with speed u at a small angle α to the x-axis, while the second moves with speed $w \ll c$ along the y-axis. The particles collide, and the y-component of each particle's velocity is reversed.

Let's look at the momentum of the two particles before and after the collision:

	Before	After
p_x	$muf(u)\cos\alpha + 0$	$muf(u)\cos\alpha + 0$
p_y	$-muf(u)\sin\alpha + mwf(w)$	$muf(u)\sin\alpha - mwf(w)$

Next we apply conservation of momentum. The x-components are already equal. For the y-components to be equal, we need:

$$wf(w) = uf(u)\sin\alpha. \quad \text{(i)}$$

Now let's look at the same collision in a frame where particle 1 moves parallel to the y-axis. This (prime) frame moves to the right with speed $v = u\cos\alpha$. We use eqns. 34.9 with $\beta = (u/c)\cos\alpha$ and $\gamma = [1 - (u/c)^2\cos^2\alpha]^{-1/2}$ to Lorentz-transform the velocities.

For the first particle:

$$u'_{1,y} = \gamma u \sin\alpha \equiv u' \quad \text{and} \quad u'_{1,x} = 0.$$

Its speed is u'.

For the second particle:

$$u'_{2,x} = w'_x = -u\cos\alpha \quad \text{and} \quad u'_{2,y} = w'_y = w/\gamma.$$

Its speed is given by $(w')^2 = (w'_x)^2 + (w'_y)^2$:

$$w' = \left[u^2\cos^2\alpha + w^2\left(1 - \frac{u^2\cos^2\alpha}{c^2}\right)\right]^{1/2}$$

$$= u\left[\cos^2\alpha\left(1 - \frac{w^2}{c^2}\right) + \frac{w^2}{u^2}\right]^{1/2}.$$

As in the unprime frame, we apply conservation of momentum in the y-direction:

	Before	After
p'_y	$-mu'f(u') + mw'_y f(w')$	$mu'f(u') - mw'_y f(w')$

Set before = after:

$$u'f(u') = w'_y f(w'). \quad \text{(ii)}$$

We find $f(u)$ by considering the limit $w/c \to 0$. We may neglect quantities of order w^2/c^2 (or equivalently α^2) in eqns. (i) and (ii). As $w \to 0$, $f(w) \to 1$, and:

$$\cos\alpha = \sqrt{1 - \sin^2\alpha} \to 1.$$

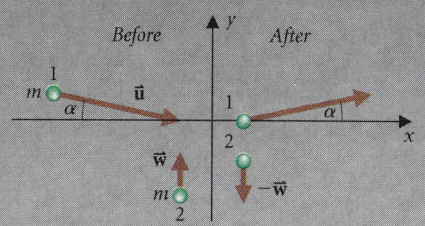

(a) Unprime frame.

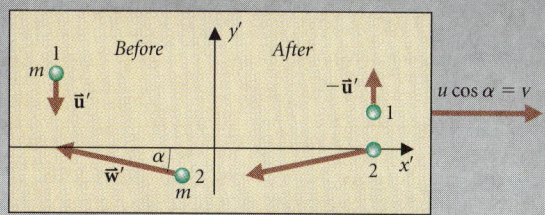

(b) Prime frame.

■ **Figure 34.20**
(a) Two identical particles collide. One particle has a large speed u but makes a small angle α with the x-axis. The other particle has a small speed $w \ll c$ and travels parallel to the y-axis. During the collision, each particle's y-component of momentum is reversed. (b) The same collision in the prime frame. This frame moves at speed $v = u\cos\alpha$ in the x-direction with respect to the unprime frame. The first particle moves parallel to the y-axis with a small speed u', and the second particle moves at a large speed w', making a small angle with the x'-axis. When $w \ll c$, this diagram is the mirror image of (a).

So:
$$\gamma \to \frac{1}{\sqrt{1 - u^2/c^2}} \equiv \gamma_u,$$

$$w' \to u,$$

$$w'_y \to w/\gamma_u$$

$$u' \to u\gamma_u \sin\alpha.$$

Then eqn. (i) becomes:

$$w = u\sin\alpha f(u), \quad \text{(iii)}$$

and eqn. (ii) becomes:

$$u\gamma_u \sin\alpha f(u\gamma_u \sin\alpha) = (w/\gamma_u)f(u). \quad \text{(iv)}$$

We know that $f \to 1$ as its argument approaches 0. On the left-hand side of eqn. (iv), f multiplies $\sin\alpha$, which is already small, so any correction to $f = 1$ gives a term that is even smaller and may be neglected. Thus eqn. (iv) becomes:

$$u\gamma_u \sin\alpha = (w/\gamma_u)f(u). \quad \text{(v)}$$

Combining eqns. (iii) and (v) to eliminate $\sin\alpha$, we have:

$$f(u) = \gamma_u.$$

So, the correct relativistic formula for momentum is:

$$\vec{\mathbf{p}} = \gamma m\vec{\mathbf{u}}.$$

In Chapter 6 we expressed Newton's second law in its most fundamental form: $\vec{F} = d\vec{p}/dt$. In this form, Newton's law remains valid for particles whose speed approaches c, provided that we use the proper relativistic expression for $\vec{p}$ and work in a single reference frame. To find the corresponding relation between force and acceleration, we differentiate expression (34.10).

$$\vec{F} = \frac{d\vec{p}}{dt} = \frac{d}{dt}(\gamma m \vec{u}) = m\left(\gamma \frac{d\vec{u}}{dt} + \frac{d\gamma}{dt}\vec{u}\right).$$

Since
$$\frac{d\gamma}{dt} = \frac{d}{dt}\left(\frac{1}{\sqrt{1-u^2/c^2}}\right) = \frac{(-\frac{1}{2})(-2u/c^2)}{(1-u^2/c^2)^{3/2}}\left(\frac{du}{dt}\right) = \gamma^3\left(\frac{u}{c^2}\right)\left(\frac{du}{dt}\right),$$

$$\vec{F} = \gamma m\left[\frac{d\vec{u}}{dt} + \gamma^2\left(\frac{u}{c^2}\right)\left(\frac{du}{dt}\right)\vec{u}\right] = \gamma m\left[\vec{a} + \gamma^2\left(\frac{u}{c^2}\right)\left(\frac{du}{dt}\right)\vec{u}\right].$$

Now we resolve $\vec{a}$ into components parallel and perpendicular to $\vec{u}$. The parallel component changes the particle's speed: $a_\| = du/dt$. Choose the x-axis to be along $\vec{u}$. Then $\vec{a}_\| = a_\| \hat{i}$, and:

$$\vec{F} = \gamma m\left(\vec{a}_\perp + a_\|\hat{i} + \gamma^2 \frac{u}{c^2} a_\| u \hat{i}\right) = \gamma m\left[\vec{a}_\perp + \vec{a}_\|\left(1 + \gamma^2 \frac{u^2}{c^2}\right)\right].$$

The quantity multiplying $\vec{a}_\|$ simplifies:

$$1 + \gamma^2\beta^2 = 1 + \beta^2/(1-\beta^2) = 1/(1-\beta^2) = \gamma^2.$$

So:
$$\vec{F} = \gamma m \vec{a}_\perp + \gamma^3 m \vec{a}_\|. \tag{34.11}$$

Since the coefficients of $\vec{a}_\perp$ and $\vec{a}_\|$ are different, the acceleration is in general not parallel to the force. It is more difficult to increase the particle's speed than to change its direction of motion. The Newtonian result $\vec{F} = m\vec{a}$ is the limit of eqn. (34.11) as $u \to 0$ and $\gamma \to 1$.

34.4.2 Mass and Energy

One of Einstein's most famous discoveries is that mass is a form of energy. As a consequence, mass may be converted into other forms of energy, and particles may be annihilated and created as these changes occur. To show how Einstein's relation arises, we calculate the work done on a particle and, as in Newtonian physics, interpret it as an increase in energy. Suppose, for example, that an accelerator applies a constant electric force $\vec{F}$ to an electron. Using eqn. (34.11) for the component of $\vec{F}$ parallel to $\vec{u}$, the force does work at a rate:

ALTHOUGH WE HAVE NOT PROVED IT, THE EXPRESSION $P = \vec{F} \cdot \vec{u}$ REMAINS VALID AS $u \to c$.

$$P = \vec{F} \cdot \vec{u} = \gamma^3 m u \frac{du}{dt} = mc^2\left[\frac{(1/2)(d/dt)(u^2/c^2)}{(1-u^2/c^2)^{3/2}}\right] = \frac{d}{dt}(\gamma m c^2).$$

According to the work-energy theorem, the work done on the electron increases the quantity γmc^2, which is the electron's energy:

FOR THE WORK-ENERGY THEOREM, SEE §7.1.

$$E = \gamma m c^2. \tag{34.12}$$

For a particle at rest, $\gamma = 1$ and its energy is $E_0 = mc^2$, the *rest energy* discovered by Einstein. The kinetic energy—that is, the energy due to motion—is given by:

$$K = E - E_0 = (\gamma - 1)mc^2. \tag{34.13}$$

EXERCISE 34.8 ♦♦ Expand γ in powers of u^2/c^2 and show that the first two terms in eqn. (34.13) are the Newtonian kinetic energy and a relativistic correction:

$$K \approx \tfrac{1}{2}mu^2\left[1 + \frac{3}{4}\left(\frac{u^2}{c^2}\right)\right] \quad \text{(for } u \ll c\text{)}.$$

EXAMPLE 34.10 ♦♦ An electron in a television tube has a kinetic energy of about 35 keV. Cosmic ray electrons reach energies of many GeV. What is the rest energy of an electron? Find the speed of an electron in the television tube and of a 1-GeV cosmic ray electron. How accurate is the Newtonian kinetic energy in each case?

MODEL In each case, we compare the kinetic energy (eqn. 34.13) with the Newtonian value $K = \frac{1}{2}mu^2$.

SETUP The mass of an electron is 9.110×10^{-31} kg.

SOLVE Its rest energy is:

$$E_0 = mc^2 = \frac{(9.11 \times 10^{-31}\text{ kg})(2.998 \times 10^8 \text{ m/s})^2}{1.602 \times 10^{-19} \text{ J/eV}} = 511 \text{ keV}.$$

SETUP The electron's kinetic energy is $(\gamma - 1)mc^2$. We find its speed from γ:

$$\gamma^2 = \frac{1}{1-\beta^2} \Rightarrow \beta = \sqrt{1 - \frac{1}{\gamma^2}}.$$

SOLVE In the television tube:

$$\gamma - 1 = K/mc^2 = (35 \text{ keV})/(511 \text{ keV}) = 0.0685.$$

With $\gamma = 1.0685$, we find β:

$$\beta = \sqrt{1 - \frac{1}{1.0685^2}} = 0.352.$$

For small β, the correction term to the Newtonian K (Exercise 34.8) is:

$$\tfrac{3}{4}\beta^2 = 0.093.$$

For the cosmic ray, we have:

$$\gamma = E/mc^2 = (10^6 \text{ keV})/(511 \text{ keV}) = 2 \times 10^3.$$

The corresponding speed is:

$$\beta = \sqrt{1 - (2 \times 10^3)^{-2}} = 0.99999987,$$

which is indistinguishable from 1, to the significance we are working.

ANALYZE For the electron in the television set, the Newtonian kinetic energy is accurate to 9%, but the Newtonian formula underestimates the cosmic ray kinetic energy by a factor of almost 4000. ∎

Unlike the arbitrary constants we encountered with potential energies, the rest energy has real physical significance. It is unchanged when a particle is accelerated, but it can be released when the particle combines with another and changes form. This is precisely what happens in nuclear reactions—for example, in power plants and the Sun. Because of this, particle physicists often write masses in energy units: the mass of a proton is written

$$m_p = 938 \text{ MeV}/c^2.$$

EXAMPLE 34.11 ♦♦ The principal energy-producing reaction in the Sun fuses protons into a helium nucleus. Four protons, each with mass 1.0079 u, form a helium nucleus with mass 4.00260 u (1 u = 1.66×10^{-27} kg). How many reactions occur in the Sun each second? Estimate how long the Sun can continue to shine.

MODEL First we find the energy liberated in each reaction. Then the number of reactions per second equals the total energy radiated per second divided by the energy per reaction. We can obtain an upper limit to the Sun's life by calculating the time for it to convert all its hydrogen to helium.

THE SUN'S MASS AND LUMINOSITY ARE GIVEN INSIDE THE FRONT COVER.

SETUP In each reaction the total rest mass decreases by

$$\Delta m = 4m_H - m_{He} = (4 \times 1.0079 - 4.0026) \text{ u} = 0.029 \text{ u},$$

releasing an energy:

$$\Delta E = \Delta mc^2 = 0.029(1.66 \times 10^{-27} \text{ kg})(3.0 \times 10^8 \text{ m/s})^2 = 4.3 \times 10^{-12} \text{ J}.$$

The Sun's luminosity is 3.9×10^{26} W, so the number of reactions per second is:

SOLVE
$$\frac{dn}{dt} = \frac{L}{\Delta E} = \frac{3.9 \times 10^{26} \text{ W}}{4.3 \times 10^{-12} \text{ J}} = 9.0 \times 10^{37} \text{ /s}.$$

SETUP The mass of helium produced per second is:

$$\frac{dM}{dt} = \frac{dn}{dt} m_{He}.$$

And, the time to convert the entire Sun to helium is:

$$T \sim \frac{M}{dM/dt} = \left(\frac{M}{m_{He}}\right)\left(\frac{1}{dn/dt}\right).$$

SOLVE The Sun's mass is 2.0×10^{30} kg, so:

$$T \sim \left(\frac{2.0 \times 10^{30} \text{ kg}}{4(1.7 \times 10^{-27} \text{ kg})}\right)\left(\frac{1}{9.0 \times 10^{37} \text{ /s}}\right) = 3.3 \times 10^{18} \text{ s} \approx 10^{11} \text{ y}.$$

ANALYZE This crude estimate is about ten times too large. When its inner 10% is converted to helium, the Sun will undergo structural changes and destroy the Earth. Of its 10-billion-year lifetime, about 4.5 billion years have passed since the Sun's formation. ∎

As a particle is accelerated towards the speed of light, its energy and momentum increase dramatically. To reach light speed would require infinite energy. Again, we see that no particle with $m > 0$ can travel at (or above) the speed of light.[3]

34.4.3 The Energy-Momentum Invariant

Neither energy nor momentum is an invariant quantity since they both depend on a relative quantity: the particle's speed. However, the combination $E^2 - p^2c^2$ is invariant:

$$E^2 - p^2c^2 = (\gamma mc^2)^2 - (\gamma mu)^2c^2 = \gamma^2 m^2 c^4\left(1 - \frac{u^2}{c^2}\right).$$

$$\boxed{E^2 - p^2c^2 = m^2c^4.} \tag{34.14}$$

This energy-momentum invariant has important similarities to the interval invariant (§34.2.2). The location of an event in space-time has four components: time and the three components of the position vector. The invariant is:

$$\text{Invariant} = (\text{time component})^2 - (\text{length of spatial vector})^2.$$

Equation (34.14) has the same form if we identify energy as the time component that goes with the momentum vector. Thus energy and momentum form a four-dimensional vector whose invariant length is determined by the mass. Three Newtonian concepts—mass, energy, and momentum—are unified into a single quantity that obeys a four-dimensional vector conservation law.

---

[3] It has been suggested that tachyons, particles with imaginary mass, travel with speed $> c$. Such particles could never go slower than light! There is no evidence that such particles actually exist.

■ FIGURE 34.21
The Bevatron at the University of California, Berkeley, shortly after its construction. This machine was designed to accelerate protons to an energy sufficient to generate antiprotons (see Example 34.12).

In 1954, the Bevatron accelerator was constructed at the University of California, Berkeley (■ Figure 34.21). A major purpose of the machine was to create antiprotons—particles with the same mass as a proton but with opposite values of charge and certain other physical properties. Accelerated protons would collide with stationary protons to create pairs of protons and antiprotons out of kinetic energy:

2 protons + kinetic energy → 3 protons + 1 antiproton.

To design the Bevatron, it was important to determine how much kinetic energy would be needed. To see how this was done, we apply eqn. (34.14) to the total energy and momentum of the system. This step is valid, although we omit the proof here.

THE TOTAL MOMENTUM OF A SYSTEM WAS DISCUSSED IN CHAPTER 6.

EXAMPLE 34.12 ♦♦♦ Compute the minimum energy necessary in the Bevatron beam to produce antiprotons.

MODEL If all the particles were at rest in the laboratory after the collision, $2mc^2$ of kinetic energy would suffice. However, the momentum of the protons in the beam is conserved: the reaction products cannot be at rest in the lab frame (■ Figure 34.22). As in nonrelativistic mechanics, the description is simplest in a frame where the system has zero momentum. [The name is changed from the center of mass frame to the center of energy (CE) frame in relativistic discussions.] In the CE frame, both the beam and the target protons have kinetic energy that is converted into the particle pair. At the minimum beam energy, there is no energy left in kinetic form: the final state has four stationary particles in the CE frame.

To find the minimum energy, we note that the energy-momentum invariant has the same value in the CE frame after the collision as in the lab frame before the collision. In any one frame, energy and momentum are separately *conserved*, and therefore so is their invariant combination. The invariant has the same value before and after the collision. Because it is *invariant*, its value is the same in both the lab and CE reference frames.

(a) Lab frame.

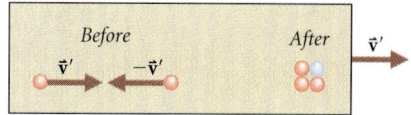

(b) CE frame.

■ FIGURE 34.22
(a) Proton–proton collision in the lab frame results in three protons and one antiproton. All four particles have the same mass. (b) Here the four particles are at rest in the CE frame after the collision. The initial energy of the proton in the lab frame is then the minimum necessary for production of an antiproton.

SETUP We use the standard conservation law plan. In the lab frame before the collision, there is one moving proton ($E = \gamma mc^2$, $p_x = \gamma mu$) and one proton at rest ($E = mc^2$, $p_x = 0$). In the CE frame after the collision, there are four particles, each with equal mass and energy ($E = mc^2$), all at rest ($p_x = 0$).

$$\begin{aligned}
\text{CE After} &\qquad\qquad \text{Lab Before}\\
E^2 - p^2c^2 = (4mc^2)^2 - 0 &= (mc^2 + \gamma mc^2)^2 - (\gamma mu)^2 c^2.\\
16 m^2 c^4 &= m^2 c^4 [(\gamma + 1)^2 - \gamma^2 u^2/c^2]\\
&= m^2 c^4 [\gamma^2 (1 - u^2/c^2) + 2\gamma + 1]\\
&= 2 m^2 c^4 (\gamma + 1).
\end{aligned}$$

SOLVE Thus $\gamma + 1 = 8$. To produce antiprotons requires a minimum γ factor of 7 and a minimum kinetic energy:

$$K_{\min} = (\gamma - 1) mc^2 = 6(0.938 \text{ GeV}) = 5.6 \text{ GeV}.$$

ANALYZE The Bevatron was designed to provide 6.4 GeV per proton. In general, the probability that a reaction will occur is exceedingly small at the minimum energy. The extra 0.8 GeV provided in the design ensures a reasonable production rate. The Bevatron was a proton synchrotron (see *Digging Deeper*, §29.1).

A BeV (for <u>b</u>illion <u>e</u>lecron <u>v</u>olts) is an old name for GeV—hence the Bevatron's name.

Chapter Summary

Where Are We Now?

According to the special principle of relativity, physical laws have the same form in any inertial reference frame. Einstein developed his special theory of relativity so that Maxwell's electromagnetic theory should be consistent with this principle. Together with his later revision of gravitational theory (general relativity), these theories require that we recognize a four-dimensional space-time as the arena in which physical events occur. We have developed a basic description of space-time and studied the kinematics of special relativity. We have learned how expressions for a particle's momentum and energy are modified when it moves at speeds approaching that of light, and how the structure of space-time leads us to define a single energy-momentum invariant that is independent of reference frame.

What Did We Do?

A relativity theory classifies physical quantities as relative or invariant, describes how to express relative quantities in different reference frames and how to express physical laws in invariant form.

Physics is described by events occurring in a four-dimensional space-time. A particle follows a world line through space-time that consists of the sequence of events in the particle's history. Space-time diagrams provide a convenient geometrical representation of the relations among events and world lines. The Lorentz transformation shows how to convert relative quantities from one frame to another. When the prime frame moves with velocity $\vec{v} = v\hat{i}$ as measured in the unprime frame, the coordinates are related by:

$$ct' = \gamma(ct - \beta x), \qquad x' = \gamma(x - \beta ct), \qquad y' = y, \quad \text{and} \quad z' = z.$$

The space-time interval $\Delta s^2 = c^2 \Delta t^2 - \Delta \ell^2$ is invariant.

Special relativity is valid in inertial reference frames. The inertial frame in which an object's velocity is zero is called its rest frame. While an object accelerates, its rest frame changes continuously. Calculations involving acceleration may require the use of several reference frames.

TABLE 34.2 Invariant and Relative Quantities in Special Relativity (Relative velocity of frames is $\vec{v} = \beta c \hat{1}$.)

Invariants

Rest mass of a particle	m
Charge of a particle	Q
Space-time interval	$\Delta s^2 = c^2 \Delta t^2 - |\Delta \vec{x}|^2$
Energy-momentum invariant	$m^2 c^4 = E^2 - p^2 c^2$
Electromagnetic field invariants*	$\vec{E} \cdot \vec{B}$
	$|\vec{E}|^2 - c^2 |\vec{B}|^2$

Relative Quantities		Transformation Law	Text Reference
Time coordinate	t	$ct' = \gamma(ct - \beta x)$	34.8
Spatial coordinate	x	$x' = \gamma(x - \beta ct)$	34.8
Velocity of a particle	$\vec{u}$	$u'_x = \dfrac{u_x - \beta c}{1 - \beta u_x / c}$	34.9
		$\vec{u}'_\perp = \dfrac{\vec{u}_\perp}{\gamma(1 - \beta u_x/c)}$	34.9
Energy	$E = \gamma_u mc^2$	$E' = \gamma(E - \beta p_x c)$	Problem 77
Momentum	$\vec{p} = \gamma_u m \vec{u}$	$cp'_x = \gamma(cp_x - \beta E)$	
		$\vec{p}'_\perp = \vec{p}_\perp$	
Electric field*	$\vec{E}$	$\vec{E}'_\perp = \gamma(\vec{E}_\perp + c\vec{\beta} \times \vec{B})$, $\quad E'_x = E_x$	
Magnetic field*	$\vec{B}$	$\vec{B}'_\perp = \gamma(\vec{B}_\perp - \dfrac{\vec{\beta}}{c} \times \vec{E})$, $\quad B'_x = B_x$	

*Not derived in the text. Note that the Lorentz transformation for $\vec{E}$ differs from the transformation in Table 34.1 only by the factor γ in the expression for $\vec{E}'_\perp$.

As a particle approaches the speed of light, its momentum, $\vec{p} = \gamma m \vec{u}$, increases without bound. In any reference frame, no material object can have a speed equal to or greater than c. The relativistic expression for energy, $E = \gamma mc^2$, shows that each particle has a rest energy mc^2. When particles interact, energy may be transformed between rest energy and other forms.

We have listed some important invariant and relative quantities in • Table 34.2.

Practical Applications

Newtonian physics is an excellent approximation in everyday situations, but relativity theory must be used in calculations involving objects moving near the speed of light. Examples include analyzing the results of accelerator experiments or studying cosmic rays. A recent application is the free-electron laser. Relativistic effects are important in the production of synchrotron radiation by accelerator beams. Synchrotron radiation is finding increasing use in manufacturing, diagnostic, and healthcare applications.

Einstein's relation $E = \gamma mc^2$ is the basis for the operation of nuclear reactors and also allows us to understand energy generation in stars. We'll explore these topics further in Chapter 36.

The relativistic expression for momentum is essential in understanding the behavior of particles in accelerators (cf. §29.1.2, §30.4) and in cosmic rays.

Solutions to Exercises

34.1 From Example 34.1,
$$\gamma = h/(\beta\tau c) = 38/\beta = 38/0.9997 \approx 38.$$
At the muon's birth, the detector's distance is:
$$d = (25 \text{ km})/\gamma = (25 \text{ km})/38 = 0.66 \text{ km}.$$
The detector is just able to travel this distance in the 2.2-μs lifetime of the muon: $v\tau \approx c\tau = (3.0 \times 10^8 \text{ m/s})(2.2 \times 10^{-6} \text{ s}) = 660 \text{ m}$.

34.2 See ■ Figure 34.23. **(a)** Bolt 1 strikes. **(b)** Bolt 2 strikes. **(c)** Flash from bolt 1 reaches Penny. **(d)** Flashes from both bolts reach Oliver. **(e)** Flash from bolt 2 reaches Penny.

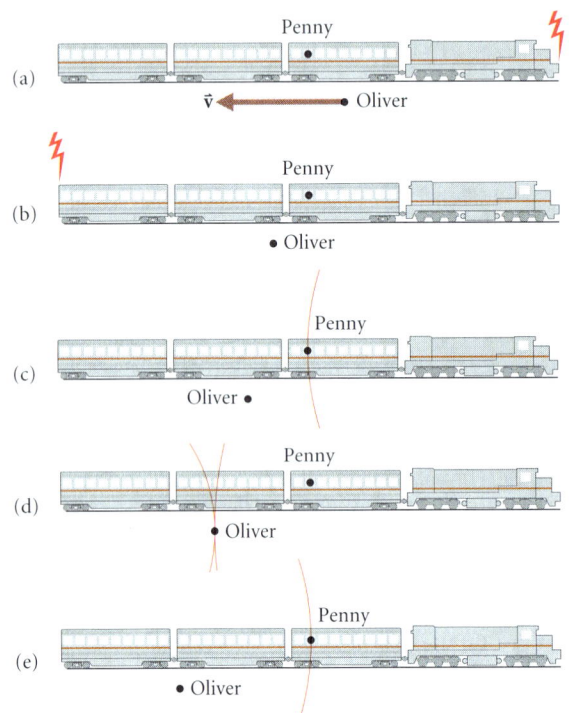

■ **FIGURE 34.23**
The lightning bolts in Penny's reference frame. Remember, the light from the bolts moves at speed c, and Oliver moves with speed v in this frame.

34.3 The reverse Lorentz transformation is given by eqns. (34.8), with the prime and unprime switched, and $\beta \to -\beta$:
$$ct = \gamma(ct' + \beta x') \quad \text{and} \quad x = \gamma(x' + \beta ct').$$
(You could also derive these results by solving eqns. (34.8) for x and ct in terms of x' and ct'.)

Along the x-axis, $t = 0$. Thus $ct' = -\beta x'$. This line has a negative slope and makes an angle θ with the x'-axis (■ Figure 34.24), where:
$$\tan\theta = |ct'|/x' = \beta.$$
Along the t-axis, $x = 0$ so that $x' = -\beta ct'$; again, the angle between

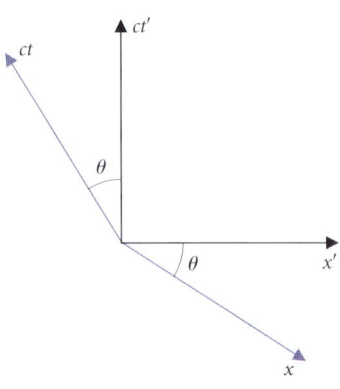

■ **FIGURE 34.24**
The unprime frame has velocity $-\beta c\hat{\mathbf{i}}$ with respect to the prime frame. If the prime axes are drawn perpendicular, then the unprime axes are at an angle $90° + 2\tan^{-1}\beta$. The relative positions and the angle between the prime and unprime axes are the same no matter which axes we choose to draw perpendicular.

the axes is $\theta = \tan^{-1}\beta$. The angles between the axes are the same as in Figure 34.11.

34.4 The interval between the origin and the new event is:
$$s^2 = c^2t^2 - x^2 - y^2 - z^2 = c^2(t')^2 - (x')^2 - (y')^2 - (z')^2.$$
We have already shown that $c^2t^2 - x^2 = c^2(t')^2 - (x')^2$, so we must have:
$$y^2 + z^2 = (y')^2 + (z')^2 = F^2(y^2 + z^2) \implies F^2 = 1.$$
The argument for choosing the positive root is the same as that given for A.

34.5 See ■ Figure 34.25. In the rocket frame the Earth moves with $\beta = 0.8$. The light signal travels a much smaller distance to the rocket than it does returning to earth.

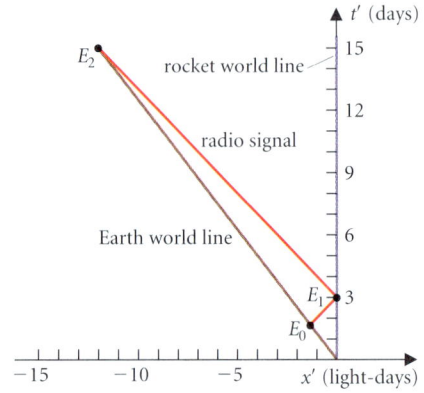

■ **FIGURE 34.25**
The events of Example 34.6 in the rocket frame. The signal emitted from Earth to the rocket travels a short distance, but the signal from the rocket has to chase the receding Earth: it travels a much larger distance.

34.6 The planet moves at speed $v = \beta c$ with respect to the rocket. The speed of the flash is c in the rocket frame. In the planet frame, it is:

$$c' = \frac{c - \beta c}{1 - \beta c/c} = c, \text{ as required by Einstein's postulates.}$$

34.7 With $\gamma = 2$, the momentum is twice that given by the Newtonian formula. Then $1/\gamma^2 = 1 - u^2/c^2 = 1/4$, and so $u = (\sqrt{3}/2)c = 0.87c$.

34.8

$$\gamma = \left(1 - \frac{u^2}{c^2}\right)^{-1/2} = 1 + \frac{1}{2}\left(\frac{u^2}{c^2}\right) + \left(-\frac{1}{2}\right)\left(-\frac{3}{2}\right)\frac{1}{2}\left(\frac{u^4}{c^4}\right) + \cdots$$

Therefore: $(\gamma - 1)mc^2 = \frac{1}{2}mc^2\frac{u^2}{c^2} + \frac{3}{8}mc^2\frac{u^4}{c^4} + \cdots$

$$\approx \frac{1}{2}mu^2\left[1 + \frac{3}{4}\left(\frac{u}{c}\right)^2\right].$$

Basic Skills

Review Questions

§34.1 SPECIAL RELATIVITY

- What is a *relative quantity*? Give an example of a relative quantity in Galilean relativity and in special relativity.
- What is an *invariant*? Give an example of an invariant in Galilean relativity and in special relativity.
- What are the features of a complete theory of relativity?
- List Einstein's postulates for special relativity. Which of them is known as the *special principle of relativity*?
- What is *time dilation*?
- How can you tell which observer measures the shortest time interval between two events?
- Describe *length contraction*. Are all dimensions contracted? If not, which one or ones?
- If you determine that two events occur at the same time, do all observers agree that they are simultaneous? Why or why not?

§34.2 SPACE-TIME

- What is *space-time*? How might you represent it on paper?
- Describe a procedure for setting up a coordinate system in space-time.
- How is the *future* of an event defined? How is the *past* defined?
- What is the space-time *interval* between two events?
- What is a *spacelike* interval? A *timelike* interval?
- Define *proper time* and *proper distance*.

§34.3 THE LORENTZ TRANSFORMATION

- What assumptions were made in deriving the Lorentz transformation for the coordinates in the form (34.8)?
- Describe how to locate the coordinate axes of a relatively moving reference frame in a space-time diagram.
- What three effects are important in the Lorentz transformation for velocities? Do they all play a role in transforming each component of $\bar{u}$?

§34.4 RELATIVISTIC DYNAMICS

- How do we know that the Newtonian expression for momentum cannot be correct as $u \to c$?
- What is the correct relativistic expression for a particle's momentum?
- What is a particle's *rest energy*? Can it ever be extracted?
- What is the *energy-momentum invariant*? What relation among mass, energy, and momentum does it suggest?
- Using energy and momentum as examples, carefully state the difference between conserved and invariant quantities.

Basic Skill Drill

§34.1 SPECIAL RELATIVITY

1. A rocket ship with length 0.10 km in its own rest frame is moving at $\beta = 0.70$ toward an asteroid. What is the length of the ship as measured by an observer on the asteroid?

2. Neutral pions (a kind of subatomic particle) have an average lifetime of 8.4×10^{-17} s in their own rest frame. In a lab experiment pions are produced with a speed $\beta = 0.995$. What is the average pion lifetime in the lab? How far do these pions travel, on average?

3. The captain of a spaceship en route from Aldebaran to Sirius sends signals to Earth every 12 h as she passes the solar system. Earth Central receives signals every 12 h 3 min. What is the speed of the spaceship?

§34.2 SPACE-TIME

4. Draw the following events in a space-time diagram: a rocket leaves Earth at $0.3c$; after departure, the rocket sends a signal to Earth; Earth receives the signal 17 h after the rocket departure. Use your diagram to determine when and where the signal was sent (as measured in the Earth frame). Calculate the space-time interval between the events *rocket leaves Earth* and *rocket sends signal*.

5. Event E_1 occurs at $t = 0$, $x = 3.50$ m. Event E_2 occurs at $t = 1.3$ ns, $x = -0.75$ m. Calculate the space-time interval between the events. Is the interval spacelike or timelike?

§34.3 THE LORENTZ TRANSFORMATION

6. An event occurs at $t = 0.27$ s, $x = 7900$ km. Find the coordinates of this event in a reference frame moving at speed $v = 1.4 \times 10^8$ m/s along the x-axis with respect to the first frame.

7. Two rockets leave a space station along parallel paths. The space station personnel record their speeds as $v_1 = 0.85c$ and $v_2 = 0.73c$. What is the speed of the first rocket, as measured by the crew on the second?

§34.4 RELATIVISTIC DYNAMICS

8. A positron is the antiparticle of an electron; it has the same mass but opposite charge. When a positron and an electron collide, they annihilate, producing two photons each with the same energy. If the particles approach with speed $v \ll c$, what is the photon energy?

9. A proton has kinetic energy $K = 450$ MeV. What is the value of γ for the proton? What is its momentum? What is its energy-momentum invariant? (*Note:* γ rays with this energy have been observed arriving from the center of the galaxy.)

Questions and Problems

34.1 SPECIAL RELATIVITY

10. ❖ Decide whether each of the following statements is true or false, and explain your answers. **(a)** The wave fronts propagating outward from a flash of light are spherical according to any observer in an inertial reference frame. **(b)** The wave fronts remain centered on their source according to every inertial observer.

11. ❖ For each transformation listed, decide whether each quantity is relative or invariant. Justify your answers.

Quantity	Transformation
Speed	Galilean transformation
Velocity	translation of coordinates
Length	reflection of *y*-coordinate axis in *x-z*-plane
Mass	rotation of coordinates
Position	Lorentz transformation

12. ❖ If a perfectly rigid rod *AB* is struck at one end, the whole rod accelerates simultaneously (■ Figure 34.26). Show that such a rigid rod cannot exist because, for some observers, the far end *B* would accelerate before the rod is struck.

13. ❖ A space navy cruiser of rest length *L* passes directly by a disabled space freighter at speed β (■ Figure 34.27). Two astronauts leap from the front and the back of the cruiser so as to pass simultaneously (according to the freighter crew) through two entry ports distance L/γ apart on the freighter. How is their procedure described by the crew of the cruiser?

14. ❖ A rod of rest length *L* slides at speed $\beta \approx 1$ parallel to its length on a frictionless surface. A groove of width *L* is cut in the surface perpendicular to the rod's path. While crossing the groove, the rod accelerates downward and strikes the far edge of the groove in a fearful collision (■ Figure 34.28). How does the system appear in the rest frame of the rod? What must happen to the rod if events in this reference frame are to be consistent with those described in the groove rest frame?

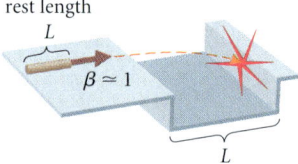

■ FIGURE 34.28

15. ❖ A champion runner who runs at nearly the speed of light holds a mirror in front of herself as she runs. Can she see her reflection in the mirror? Use the special principle of relativity in your answer. Would the answer be different in Galilean relativity? Why or why not?

16. ❖ Consider the following variation on the tale of Oliver, Penny, and the lightning bolts. Oliver is standing in the middle of a bridge that crosses over the railroad tracks at right angles. Two lightning bolts strike the ends of the (fortunately well-grounded) bridge, simultaneously in Oliver's frame, just as Penny, on the train, passes beneath. Show that Penny and Oliver agree that the bolts strike simultaneously.

17. ❖ What is the minimum number of observers needed to test time dilation if each observer can make measurements only at her own location? Specify the number of observers needed in each reference frame, and explain your reasoning.

18. ❖ There once was a young fencer named Fisk
Whose swordplay was exceedingly brisk.
So fast was his action
The Fitzgerald contraction
Reduced his épée to a disk.

Is this possible? If so, describe the motion of the épée relative to the observer.

19. ◆ A space tug is towing a square space station panel past a space colony at one-quarter the speed of light. What is the shape of the panel in the reference frame of a colonist?

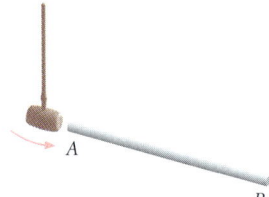

■ FIGURE 34.26

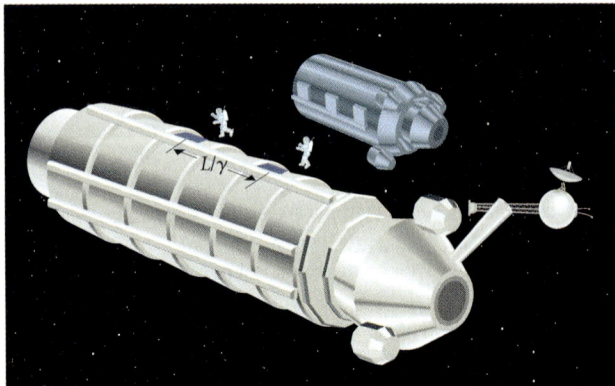

■ FIGURE 34.27

20. ♦ Which has the greater length according to lab frame observers, a rod of rest length 1.25 m moving at $\beta = 0.60$ or a rod of rest length 1.333 m moving at $\beta = 0.80$? Assume that each rod's velocity is parallel to its length.

21. ♦ Lambda particles have an average lifetime $\tau = 2.9 \times 10^{-10}$ s. If in a collision, lambda particles are produced with speed $v = 0.95c$ in the lab frame, how far do they travel, on average, before decaying?

22. ♦ A transmitter that emits radar pulses at intervals of 7.600 μs is mounted on an alien spacecraft that passes Earth at a speed $\beta = 0.9975$. At what frequency are the pulses received on Earth just as the spacecraft passes?

23. ♦ An omega particle moving at $\beta = 0.9990$ traveled 0.8386 m in the lab before decaying. How long did it exist in its own rest frame before it decayed?

24. ♦♦ Two particles emerge from a collision experiment. The first has speed $\beta = 0.975$ and decays after 26.39 ns in its own rest frame. The second particle has speed $\beta = 0.880$ and decays after 34.79 ns in its rest frame. Which particle survives longer in the lab frame? By how much?

25. ♦♦ While mining an asteroid, you observe a rocket pass by. You measure its length to be 417 m, and it takes 3.0 μs to pass you. What is the speed of the rocket and its length as measured by its crew?

26. ♦♦ How regularly would your heart have to beat to detect the effects of time dilation when you ride on a jet aircraft traveling at 250 m/s?

27. ♦♦ Suppose the light clock in §34.1.3 is in motion parallel to the direction of the light beams. Show that the same time dilation formula holds.

28. ♦♦ *The transverse Doppler effect.* An atom radiates light at frequency f_o as measured by an observer at rest with respect to the atom. What is the frequency measured by an observer moving at speed v perpendicular to the line from observer to atom?

29. ♦♦ A centrifuge with rotor arms 10 cm long spins with an angular speed $\omega = 10^5$ rad/s. By what fraction $\Delta\ell/\ell$ is the length of a molecule at the rim of the centrifuge contracted with respect to a similar molecule at rest in the laboratory?

30. ♦♦♦ *The Doppler effect for light.* Consider an atom radiating light of frequency f_o and moving directly away from the observer with speed $v = \beta c$. By considering the time interval between the arrival of successive wave fronts, derive eqn. (16.24):

$$f/f_o = \sqrt{(1-\beta)/(1+\beta)}.$$

§34.2 SPACE-TIME

31. ❖ Is an interval between two events on a particle's world line spacelike or timelike? Explain.

32. ❖ Two rocket ships depart Earth 20 min apart heading in the same direction, with $\beta = 0.8$. The trailing rocket emits light flashes at regular intervals $\Delta t'$ in its frame. Show that the interval between flashes as observed by the leading rocket is also $\Delta t'$.

33. ❖ Two events P and Q have a spacelike separation. Show that both the future and past light cones of P and Q overlap. This result shows that such events may have a common cause in the past and may mutually influence events in their common future. Can the past of one intersect the future of the other?

34. ❖ The light cone is the set of events with zero interval ($\Delta s = 0$) from the origin. Describe the set of all events that have the same spacelike separation ($\Delta s^2 = \Delta s_o^2 < 0$) from the origin. Also describe the set of events with the same timelike separation ($\Delta s^2 = \Delta s_1^2 > 0$).

35. ♦ (■ Figure 34.29) Calculate the space-time interval between events A and B, and between events C and D. Assume $y = z = 0$ for all the events. Decide whether each interval is spacelike or timelike.

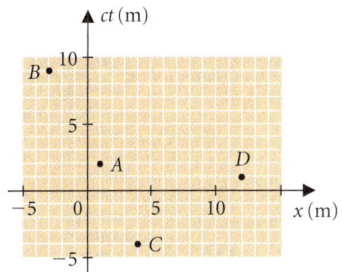

■ FIGURE 34.29

36. ♦ The coordinates of two events are $ct_1 = 0.70$ m, $x_1 = 2.0$ m, and $ct_2 = 10.37$ m, $x_2 = 35.6$ m. Calculate the proper distance between the events. Does proper time between the two events have any meaning? Why or why not?

37. ♦ The coordinates of two events are $ct_1 = 0.45$ m, $x_1 = 16.29$ m, and $ct_2 = 37.56$ m, $x_2 = 23.49$ m. Calculate the proper time between the two events. Does proper length between the events have any meaning? Why or why not?

38. ♦♦ Draw the following events on a space-time diagram. At $t = 0$, a rocket leaves a space station at $0.67c$ enroute to Earth, 7.5 light-hours away. After 0.5 hours the space station sends a radio message to Earth announcing the departure; 0.5 hours after receiving the signal, Earth sends a message to the rocket clearing it to land. Assume the space station is at rest with respect to Earth. Use your diagram to determine when (in Earth-time) the rocket receives clearance to land. How far is it from Earth then? How much proper time has elapsed on board the rocket since it departed the space station?

§34.3 THE LORENTZ TRANSFORMATION

39. ❖ Construct a space-time diagram that represents two reference frames moving at speed β with respect to each other and using different events as origin. How are the equations of the Lorentz transformation modified to describe this case?

40. ❖ Using a space-time diagram, show that for any two events separated by a timelike interval, there is a reference frame with both events on its time-axis.

41. ❖ Using a space-time diagram, show that for any two events separated by a spacelike interval, there is a reference frame with both events on its space-axis.

42. ❖ Suppose that (just in case the student is correct) the chemistry professor (Study Problem 20) plans to look for light from the crash as the front end of the train hits the tunnel's blocked exit, and escape by rocket if a crash occurs. Can the professor escape the tunnel by this method? Use a space-time diagram to justify your answer.

43. ♦ What error is made by using the Newtonian velocity addition formula to find the relative speed of two jet aircraft, both with a speed of 1800 km/h, approaching each other head on.

44. ♦ Consider two events on the time-axis of the prime reference frame, separated by a time interval $\Delta t'$. Use the Lorentz transformation to find the time interval between the events in the unprime frame and so derive the time dilation formula.

45. ♦ Two signaling lasers are pointed directly at each other. Show that the relative speed of two light pulses, one from each laser, is c.

46. ♦ A space navy cruiser moving parallel to the x-axis of a space station at $\beta = 0.80$ launches a rescue pod in the y-direction with speed $\beta = 0.90$ in the cruiser reference frame. What is the pod's speed relative to observers on the space station?

47. ♦♦ Refer to Figure 34.29. Determine the velocity of a reference frame with origin O in which event A occurs on the time-axis. Find the coordinates of events A, B, C, and D in this frame. Calculate the

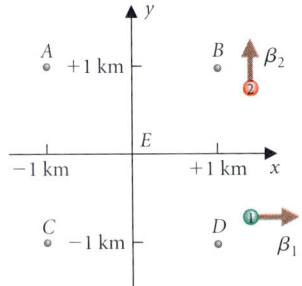

FIGURE 34.30

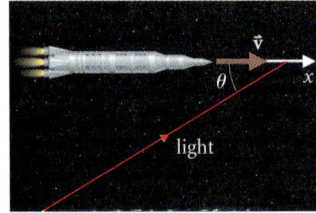

FIGURE 34.31

space-time intervals between event pairs AB and CD, and verify that they are the same as those calculated in the original frame. (See also Problem 35.)

48. ♦♦ At $t = 0$, a rocket departs Earth at $0.90c$ on a journey to a colony 12 light-years away. After 1 year, a radio signal is sent from the Earth to the rocket and after 2 years a signal is sent from the colony to the rocket. Draw these events on a space-time diagram. Which signal arrives first? Find the arrival times of the signals in both the Earth and rocket frames.

49. ♦♦ Events A, B, C, D, and E occur simultaneously according to observers at rest in the reference frame shown in ■ Figure 34.30. Observers 1 and 2 each move at speed $\beta = 0.800$ with respect to that frame, observer 1 along the x-axis and observer 2 along the y-axis. Determine the sequence of events in each observer's frame. Assuming both observers choose event E as the origin, compute the coordinates of events A–D in each frame.

50. ♦♦ In the Earth reference frame, flashes occur simultaneously at evenly spaced points 1.0 km apart along the x-axis. A rocket ship with velocity $(0.80c)\hat{\mathbf{i}}$ is at $x = 0$ when the flash occurs there. Find the time interval between the rocket's reception of flashes from $x = +N$ km and $x = -N$ km. What is this time interval in the rocket's frame? What time of occurrence does the rocket record for each flash?

51. ♦♦ A hedgehog-class automated interstellar exploration module approaches the Betelgeuse system at speed $\beta = 0.80$ along the x-axis. The hedgehog splits into six submodules that depart with speed $\beta' = 0.60$, one each along the directions $\pm x$, $\pm y$, and $\pm z$. **(a)** What is the speed of each submodule with respect to the Betelgeuse system? **(b)** What is the relative speed of each pair of submodules with opposite directions? **(c)** With perpendicular directions?

52. ♦♦ Use the velocity transformation law to show that a light pulse traveling along the y-axis (i.e., perpendicular to $\vec{\mathbf{v}}$) in the unprime frame also has speed c in the prime frame. (As usual, the prime frame is moving at $\vec{\mathbf{v}} = v\hat{\mathbf{i}}$ with respect to the unprime frame.)

53. ♦♦ In Figure 34.11, suppose that 1 cm represents a time $t = 1\ \mu s$ ($ct = 300$ m) along the time-axis and a length 300 m on the space-axis. What length represents $1\ \mu s$ on the t'-axis? Show that the same length represents 300 m on the x'-axis.

54. ♦♦ Two light pulses are moving in the positive x-direction. In the unprime frame, they are separated by distance d. Show that in a prime frame moving at velocity $u\hat{\mathbf{i}}$ with respect to the unprime frame, they are separated by distance $d' = d\sqrt{(c + u)/(c - u)}$. From this result, derive the Doppler effect for light (eqn. 16.24). (See also Problem 30).

55. ♦♦ Suppose the prime frame could have speed c with respect to the unprime frame. Show that all objects moving with speed less than c in the unprime frame would move at speed c in the prime frame.

56. ♦♦ A rocket is moving at speed v in a certain reference frame (■ Figure 34.31), and a beam of light is traveling at an angle θ to the x-axis in the same frame. Use the velocity transformation law to show that in the rocket frame the speed of light is still c but that its angle of approach to the rocket is changed. Find an expression for $\tan \theta'$. Discuss qualitatively the appearance of the sky as seen by observers on a rocket traveling directly away from Earth toward a star-filled universe.

57. ♦♦ Using the Lorentz transformation, derive the following law for the transformation of a position vector $\vec{\mathbf{r}}$ into a reference frame (prime) moving with velocity βc with respect to the unprime frame:

$$\vec{\mathbf{r}}' = \vec{\mathbf{r}} + (\gamma - 1)(\vec{\mathbf{r}} \cdot \hat{\boldsymbol{\beta}})\hat{\boldsymbol{\beta}} - \gamma \vec{\boldsymbol{\beta}} ct.$$

From this, show that the velocity transformation law is:

$$\vec{\mathbf{w}}' = \frac{\vec{\mathbf{w}} + (\gamma - 1)(\vec{\mathbf{w}} \cdot \hat{\boldsymbol{\beta}})\hat{\boldsymbol{\beta}} - \gamma \vec{\boldsymbol{\beta}} c}{\gamma(1 - \vec{\boldsymbol{\beta}} \cdot \vec{\mathbf{w}}/c)}.$$

Show that this expression reduces to eqns. (34.9) when $\vec{\boldsymbol{\beta}} = \beta \hat{\mathbf{i}}$.

58. ♦♦♦ A quantity called rapidity, $\xi \equiv \tanh^{-1} \beta$, is sometimes used as a measure of an object's speed. If an object moves parallel to the x'-axis and has rapidity ξ_1 measured in the prime frame, and if the prime frame moves parallel to the x-axis and has rapidity ξ_2 with respect to the unprime frame, show that the object's rapidity in the unprime frame is $\xi = \xi_1 + \xi_2$.

59. ♦♦♦ Derive the Lorentz transformation for acceleration by differentiating the velocity transformation.

60. ❖ We have supposed that acceleration at Sirius has no direct effect on proper time as experienced by mobile Melia. With the cooperation of a Sirian scientist, Melia could test the hypothesis. How?

61. ♦♦♦ To investigate the Case of the Youthful Twin, each twin transmits a radio pulse every 2 y, as measured in her own frame. Draw a space-time diagram in domestic Delia's reference frame showing mobile Melia's world line (constant speed out, instantaneous acceleration, same speed back), and the world lines of the radio pulses sent by each. Show that by counting received pulses, each twin may conclude that Delia is the older upon reunion.

62. ♦♦♦ Reversing her velocity at the destination, mobile Melia switches from the outgoing inertial frame G to the returning inertial frame R. In these two reference frames, different events E_1 and E_2 on Delia's world line are simultaneous with the turnaround event. Find the time interval in Delia's frame between E_1 and E_2. Show that when this extra time is added to Delia's 7.7 y of aging computed in Melia's rest frame, the result is consistent with the computation in Delia's frame. (This calculation resolves the paradox by correctly computing proper time along Delia's world line. Notice that we must use Delia's frame during the time that Melia is accelerating. We can add the results because proper time is invariant.)

§34.4 RELATIVISTIC DYNAMICS

63. ❖ Modern accelerators such as LEP (Linear Electron Positron Collider) use two beams of particles that approach each other with equal speeds before colliding. What is the advantage of this design?

64. ❖ The energy-momentum invariant of a system of particles is observed to be M^2c^4 in the lab frame. What is its value in a frame moving at speed $\beta = 0.977$ in the laboratory? If the system remains isolated but undergoes a reaction that reduces the number of particles in the system, how does the invariant change? Describe how the invariant would change if the reaction also produces energetic γ rays that are lost from the system.

65. ♦ In nuclear fusion experiments, deuterium (mass 2.01410 u) and tritium (3.01605 u) combine to form helium (4.0026 u) and a neutron (1.00867 u). How many reactions per second are needed to provide 1.0 kW of power?

66. ♦ In nuclear power plants, a nucleus of ^{235}U (mass 235.044 u) combines with a neutron (mass 1.0087 u) to produce a krypton nucleus (mass 91.926 u), a barium nucleus (141.916 u), and two more neutrons. How much uranium is processed per second in a 1.00-GW power plant?

67. ♦ When a proton and an electron combine to form neutral hydrogen, 13.6 eV of energy is released. By what fraction is the mass of the system decreased because of this energy release? (The 13.6 eV is called the binding energy of the system.)

68. ♦ What value of γ would be required for a single proton to have the same kinetic energy as a 10^3-kg car traveling at 20 m/s? How would the momentum of the proton compare with that of the car?

69. ♦ What is the energy-momentum invariant for an electron with a kinetic energy of 75.0 eV? Of 6.9 GeV? What magnitude of momentum does each electron have?

70. ♦ A proton has a total energy of 2.75 TeV. Find its energy-momentum invariant and its momentum.

71. ♦♦ In the later stages of stellar evolution, the primary energy-producing reaction is the triple-α process in which three helium nuclei (mass 4.0026 u) combine to form a carbon nucleus (mass 12.000 u). How much energy is produced per reaction? If a star in this phase has 400 times the luminosity of the Sun and 6.3 times the Sun's mass, how long would it take to change 1% of its mass from helium to carbon?

72. ♦♦ How many years' worth of the Earth's current power usage (roughly 10 TW) would be required to accelerate a 70-kg person to a speed of $0.997c$ (assuming perfect efficiency)?

73. ♦♦ A cosmic ray proton with an energy of 10^8 GeV is absorbed by a stray pea (mass 5.0 g) floating in the space shuttle galley. What is the resulting speed of the pea?

74. ♦♦ A particle moving at $0.85c$ along the x-axis collides inelastically with another identical particle at rest. After the collision, one particle moves off along the x-axis in the same direction as the incoming particle at $\beta = 0.74$. Find the velocity of the other particle after the collision. How much of the initial kinetic energy was converted to other forms in the collision?

75. ♦♦♦ Two particles of mass m_1 and m_2 and speeds u_1 and u_2 approach each other at an angle α and collide. During the collision, the particles coalesce, forming a new particle of mass m. Show that:
$$m^2 = m_1^2 + m_2^2 + 2m_1m_2\gamma_1\gamma_2(1 - \beta_1\beta_2 \cos \alpha).$$
Show that $m > m_1 + m_2$ for any values of u_1, u_2, and α. Explain this result. (*Hint:* Compare expressions for the system's energy-momentum invariant before and after the collision.)

76. ♦♦♦ Cosmic rays are often described by their rigidity, $R = pc/(Ze)$, where p is the momentum and Ze the charge of the nucleus. Show that the Larmor radius (§29.1, eqn. 29.1) of the particle is $r_L = R/(Bc)$, so particles of high rigidity deviate little from straight-line paths.

77. ♦♦♦ (a) Use the Lorentz velocity transformation (eqns. 34.9) to show that the particle's γ factor transforms according to the rule: $\gamma_{u'} = \gamma_u\gamma(1 - \beta u/c)$, where γ (without a subscript) = $(1 - v^2/c^2)^{-1/2}$, $\gamma_u = (1 - u^2/c^2)^{-1/2}$, and similarly for $\gamma_{u'}$. **(b)** Use this result to verify the transformation laws for energy and momentum given in Table 34.2.

Additional Problems

78. ❖ We have tacitly assumed that relatively moving observers agree on their relative speed. Describe an experiment that would test this hypothesis.

79. ❖ When studying the Newtonian dynamics of collisions, we discovered that when two identical particles collide elastically, their speeds in the CM frame are unchanged, although the direction of the velocity may change. Is the same true if the particle speed is relativistic? Why or why not?

80. ❖ In one reference frame, a plane rectangular wire loop is at rest. A uniform magnetic field is perpendicular to the plane of the loop and increasing in time. A second reference frame moves with speed v parallel to one side of the loop. Show that the magnetic field in the second frame is nonuniform as well as time variable. Use the motional emf in the second frame to argue that a Faraday-type law must exist.

81. ❖ Evaluate the electromagnetic field invariants (Table 34.2) for an EM wave propagating in vacuum. Use the results to show that an EM wave has the same structure (as given in §33.1) in any two inertial frames.

82. ♦♦ Two lamps that flash once every 50 ns (each in its rest frame) are mounted one on the tunnel entrance and one on the back end of the train in Study Problem 20. Each lamp emits a flash just as they pass each other. Locate the next flash of each light in a space-time diagram in both the tunnel and train frames. Show that observers in each reference frame observe time dilation of the other's light flashes. Explain why there is no paradox in the results.

83. ♦♦ How would you expect the emitted frequency of a particular spectral line to vary with temperature because of the speed of the emitting particles? Find the fractional change per kelvin, $(1/f)\,df/dT$, for iron nuclei. [*Hint:* Use the relation between temperature and rms speed for an ideal gas of nuclei (§19.2), and use time dilation to relate the time intervals (a) between the emission of two wave fronts, and (b) between the reception of two wave fronts. Compare with the results of the Pound–Rebka experiment (§34.3.3).]

84. ♦♦ Draw the following events on a space-time diagram. A rocket pirate ship leaves its base for a mineral-rich asteroid, 1.375 light-years away, traveling at $0.800c$. Light from the pirates' rocket thrust at departure is observed by the space patrol at its port, 0.170 light-years from the asteroid and 1.545 light-years from the pirate base. At what speed must the space patrol travel to reach the asteroid before the pirates? How much does the patrol age during its journey?

85. ♦♦♦ As the space patrol approaches the asteroid (Problem 84) at $\beta = 0.978$, what is its speed as observed by the pirates who approach the asteroid from the opposite direction at $\beta = 0.800$? If the pirates observe the space patrol's rocket burn as it leaves port, how much time do they have to plan a retreat between receiving the signal and the arrival of the patrol?

86. ♦♦♦ A cylinder lying along the x-axis is rotating about its axis with angular velocity ω. Show that if the cylinder is viewed from a frame moving in the x-direction with speed βc, the cylinder is twisted by an amount $\Delta\phi = \omega\beta/c$ per unit length.

87. ♦♦♦ (a) The helicopter that we imagined photographing Manhattan in §34.2.1 is 1 km above the ground and records an area 1 km on a side. Explain why the events shown in a photograph are not exactly simultaneous, and find the maximum time difference between events shown in a single photograph. **(b)** As the photographs were taken, an airplane was flying past the helicopter at a speed of 600 km/h. In the airplane's frame, what is the maximum time difference between events that are simultaneous for the helicopter pilot? **(c)** Is the travel time of light or the aircraft's motion the more important source of error in constructing a space-time diagram from photographs?

88. ♦♦♦ *The symmetric space-time diagram.* The representation of prime and unprime frames in Figure 34.11 is asymmetric and disguises the equivalence of the two frames for describing physics. For example, representation of length contraction is unconvincing because the scales on the two sets of axes are different. We can construct a symmetric representation by using a third reference frame (labeled *) in which the original two frames move with equal speed in opposite directions (■ Figure 34.32). In the asterisk frame, the unprime frame moves with speed $c\beta_*$. **(a)** Explain why the prime and unprime axes make equal angles ϕ with the * axes, as shown in Figure 34.32. Explain why the x- and ct'-axes are perpendicular, as are the x'- and ct-axes. **(b)** Consider the prime origin as a particle with velocity $\vec{u}_* = \beta_* c \hat{i}$ in the asterisk frame and $\vec{u} = \beta c \hat{i}$ in the unprime frame. Use the velocity transformation law to prove that

$$\beta_* = (\beta - \beta_*)/(1 - \beta\beta_*).$$

(c) Show that $\beta_* = (1 - \sqrt{1 - \beta^2})/\beta$. Explain why you may reject the + sign when solving the quadratic. **(d)** Use the result of (b) to express β in terms of β_*. Then substitute $\tan\phi = \beta_*$ (cf. Figure 34.11) to show that $\sin 2\phi = \beta$.

89. ✦ Explain why the length and time scales are identical on all four axes in the symmetric space-time diagram (Problem 88, Figure 34.32). Describe how to locate **(a)** all events that are simultaneous with a given event in either reference frame, and **(b)** all events that occur at the same place in either reference frame.

90. ♦♦ In a symmetric space-time diagram (Figure 34.32, Problem 88), construct the world lines of the ends of a meter stick that moves with speed $\beta = \frac{4}{5}$ along the x-axis. Identify events corresponding to length measurements in the prime and unprime frames to show that, in this diagram, the rod is clearly shortened in the frame in which it moves.

91. ♦♦♦ Physical quantities in special relativity are expressed most simply in four dimensions as functions of proper time (§34.2). For example, the four-dimensional position vector $r = (ct, \vec{r})$ has components equal to the coordinates of an event. The four-dimensional velocity vector of a particle is the rate of change of its four-dimensional position vector with respect to proper time: $u = dr/d\tau$. Find the time and space components of u in terms of the three-dimensional position vector $\vec{u}$. Show that the invariant magnitude [(time component)^2 − (spatial vector)^2] of u equals c^2, and also show that the relativistic momentum-energy vector (§34.4) is $p = mu$.

92. ♦♦♦ An observer living on a rotating disk decides to measure the radii and circumferences of circles centered on the rotation axis, using a standard meter stick. Show that the observer finds the circumference of a circle of radius r to be $2\pi r/\sqrt{1 - \omega^2 r^2/c^2}$.

93. ♦♦♦ Use the transformation laws for $\vec{E}$ and $\vec{B}$ given in Table 34.2 to compute $\vec{E}' \cdot \vec{B}'$ and $(E')^2 - c^2(B')^2$ in terms of $\vec{E}$ and $\vec{B}$. Show that each quantity is invariant.

Computer Problems

94. At $t = 0$, a rocket, the *Interstellar Ramjet*, leaves Earth with a constant acceleration a' as measured by accelerometers on board. At time t, the rocket has speed $v(t)$. In a "prime" frame moving at speed v, the rocket has speed $dv' = a' \, dt'$ after a time dt' measured in the prime frame. Show that the corresponding time interval in the Earth frame is $dt = dt'/\gamma$. Use the velocity transformation to show that the speed in the Earth frame at time $t + dt$ is:

$$v(t + dt) = \frac{v(t) + a' \, dt'}{1 + v(t)a' \, dt'/c^2}.$$

Use a spreadsheet or other computer program to compute dt' and $v(t + dt)$, given $v(t)$ and dt. You may find it convenient to use dimensionless variables $\beta = v/c$ and time variable t/t_o, where $t_o \equiv c/a'$. If the acceleration a' has magnitude $g = 9.8$ m/s^2, after how long is the rocket's speed = 75% of c? What is the rocket's speed after 2 y? Plot the world line of the *Interstellar Ramjet* on a space-time diagram. Show that a light signal emitted in the x-direction from an event at $ct = 0$ and $x < -c^2/a'$ never reaches the *Ramjet*.

95. Compute the magnitude of the momentum and the kinetic energy of a particle with speed v and compare with the Newtonian values. Make plots of p/p_N and K/K_N as functions of β. For what values of β is each ratio equal to 2? To 10?

Challenge Problems

96. A distant observer takes a photograph of a cube with sides of proper length ℓ moving at speed v perpendicular to the line of sight as shown in ■ Figure 34.33. The photograph is made with light that crosses line AB in the observer's frame simultaneously. Show that light from face PS contributes to the image, which looks like a rotated cube. What is the apparent angle of rotation?

97. In this problem we demonstrate that the existence of electric field and two signs of charge implies the existence of magnetic field. A long, straight wire is electrically neutral and carries a current I produced by electrons moving at speed v. A single electron outside the wire also moves parallel to the wire at speed v. **(a)** Show that the wire is electrically charged in a reference frame moving with the electrons. (*Hint:* Consider two lengths of wire, one containing N electrons and one containing N positive charges. The wire is electrically neutral because these two lengths are the same in the lab frame. How do they compare in the electron frame?) **(b)** Compare the electric force on the isolated electron computed in the electron frame with the magnetic force computed in the laboratory.

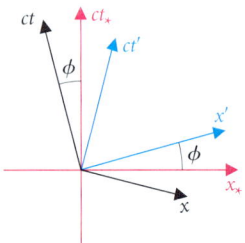

■ FIGURE 34.32

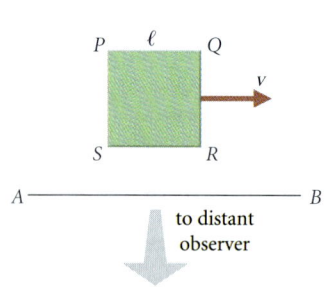

■ FIGURE 34.33

> *This is news distinctly shocking and apprehensions for the safety of confidence even in the multiplication table will arise...*
>
> NEW YORK TIMES EDITORIAL, NOVEMBER 11, 1919

ESSAY 8
General Relativity: A Geometric Theory of Gravity

E8.1 GRAVITY AS AN INERTIAL FORCE

When a bus decelerates rapidly, loose packages slide, briefcases fall over, and passengers stand at an angle as if, for a moment, gravity had been tilted toward the front of the bus. In the accelerated reference frame of the bus, each object is subject to an apparent force that overturns the briefcases and throws the passengers forward.

A physics student on the sidewalk interprets events differently; for example, the sliding packages move at constant velocity while the bus around them decelerates. In this reference frame, the apparent force doesn't exist at all; it is an example of an *inertial force*.

> An *inertial force* is an apparent force needed to account for the acceleration of free bodies with respect to a noninertial reference frame.

Inertial forces mimic gravitational force. The acceleration resulting from the noninertial nature of the reference frame is the same for all free objects. The inertial force on an object equals its mass multiplied by this acceleration. For example, the inertial force on an object of mass m in the bus is:

$$\vec{F}_{\text{inertial}} = -m\vec{a},$$

where $\vec{a}$ is the acceleration of the bus with respect to the sidewalk.

Prototype of the Laser Interferometer Gravitational-Wave Observatory (LIGO). General relativity relates the structure of space to the matter within it, and predicts the existence of gravitational waves: ripples in space-time. Interference between laser beams reflected from mirrors at the ends of the two arms measures the path difference to within a fraction of a wavelength. Passage of a gravitational wave (§E8.4) alters these path lengths and is detectable as a change in the interference pattern.

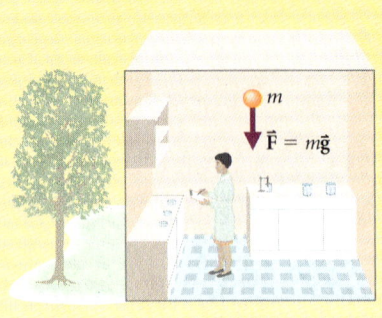
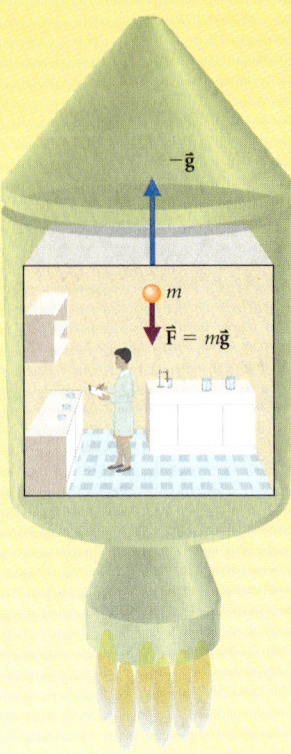

■ **Figure E8.1**
The equivalence principle. The experimenter cannot determine whether the lab is in a uniform gravitational field or in an accelerating rocket.

Einstein realized that gravity and inertial forces are more than just similar; gravity *is* an inertial force. Consider two identical laboratories (■ Figure E8.1), one on Earth in which there is a gravitational acceleration $\vec{g}$, and one far from any astronomical object but accelerating at $-\vec{g}$ due to rocket thrust. Each piece of equipment in the rocket lab is subject to an inertial force equal to the gravitational force on the corresponding equipment in the Earth lab. The result of any mechanical experiment is the same in each lab. Einstein generalized this conclusion in the *equivalence principle*.

> No experiment of any kind can distinguish a uniform gravitational field from a uniform acceleration of the reference frame.

Gravitational force appears in frames that are *not* inertial. Conversely, the local inertial frame is one in which gravity disappears—that is, a freely falling frame. Practical examples are the interior of the orbiting space shuttle, or NASA's *Vomit Comet* flown on a parabolic path.

Einstein's concept of an inertial frame differs from the definition we have used in Newtonian physics. Previously, we described gravity as a force in the same class as spring or electric forces. An inertial frame meant one in which every acceleration is produced by an identifiable force. In Einstein's picture, the gravitational force is not on the force list at all: it is a fictitious force due to acceleration of the reference frame. Inertial frames are defined to be ones in which there is no gravitational force and no gravitational acceleration, and all *other* accelerations are produced by identifiable forces.

Physics in these inertial frames is described by special relativity. Using special relativity with the equivalence principle allows us to deduce some remarkable effects of gravity.

Gravity causes time dilation. To demonstrate this effect, imagine dropping a clock to Earth from a great distance (■ Figure E8.2). Since it remains in an inertial frame, the clock measures the rate proper time passes in the absence of gravitational effects. We compare it with another identical clock at rest with respect to the Earth and at a distance R from the center. The freely falling clock passes the fixed clock at speed $v = \sqrt{2GM/R} = \sqrt{-2\Phi}$, where $\Phi = -GM/R$ is the gravitational potential at radius R. We use the special relativistic time dilation formula to compare the time interval between ticks of the earthbound clock (in the prime frame) and the freely falling clock (in the unprime frame). The time interval in the falling frame is the shorter, since the earthbound clock ticks do not occur at the same place in *any* inertial reference frame. Thus the period of the earthbound clock $\Delta t'$ is:

$$\Delta t' = \gamma \, \Delta t = \frac{\Delta t}{\sqrt{1 - v^2/c^2}}$$

$$= \frac{\Delta t}{\sqrt{1 - 2GM/Rc^2}},$$

where Δt is the period of the freely falling standard clock. In general, if $\Delta t'$ is the time interval between clock ticks where the gravitational potential is Φ,

$$\Delta t' \sqrt{1 + 2\Phi/c^2} = \text{constant}. \tag{E8.1}$$

Figure E8.2 also compares the readings on three stationary clocks at one instant. All the clock readings were the same when the moving clock was released. Stationary clocks near the Earth read earlier times than the falling one. Clocks closer to the Earth show a greater difference.

To see how large the effect is, let's compare the rate of aging of twin brothers, Joe, who stays at his home at sea level, and Stan, who climbs to the top of Mount Everest (8848 m above sea level). Because the gravitational potential is different at the two locations, the brothers' internal clocks tick at different rates.

Stan is 8848 m farther from the center of the Earth than Joe. The Earth's mass is 6.0×10^{24} kg and its radius is 6.4×10^3 km. The quantity

$$\frac{GM}{Rc^2} = \frac{(6.7 \times 10^{-11} \text{ m}^3/\text{kg} \cdot \text{s}^2)(6.0 \times 10^{24} \text{ kg})}{(6.4 \times 10^6 \text{ m})(3.0 \times 10^8 \text{ m/s})^2}$$

$$= 7.0 \times 10^{-10}$$

is so small that we can approximate the square root using the binomial expansion (Appendix IB): $(1 + x)^{1/2} \approx 1 + \frac{1}{2}x$.

The intervals between the twins' heartbeats are in the ratio (eqn. E8.1):

$$\frac{\Delta t_J}{\Delta t_S} = \frac{\sqrt{1 - 2GM/(R_S c^2)}}{\sqrt{1 - 2GM/(R_J c^2)}} \approx \frac{1 - GM/(R_S c^2)}{1 - GM/(R_J c^2)}.$$

We use the binomial expansion again, $(1 + x)^{-1} \approx 1 - x$:

$$\frac{\Delta t_J}{\Delta t_S} = 1 - \frac{GM}{c^2}\left(\frac{1}{R_S} - \frac{1}{R_J}\right)$$

$$= 1 - \frac{GM}{c^2}\left(\frac{R_J - R_S}{R_S R_J}\right)$$

$$= 1 + \frac{GM}{Rc^2}\left(\frac{R_S - R_J}{R}\right),$$

where we ignore the difference between the two radii except when they are subtracted.

$$\frac{\Delta t_J}{\Delta t_S} = 1 + (7.0 \times 10^{-10})\left(\frac{8848 \text{ m}}{6.4 \times 10^6 \text{ m}}\right)$$

$$= 1 + 9.6 \times 10^{-13}.$$

The difference between their rates of aging is 1 part in 10^{12}. To make the total difference equal to 1 min, Stan would have to stay on Mt. Everest for a time T, where $10^{-12} \, T = 1$ min, or $T = 2$ million years!

Since Δt_J is greater than Δt_S, Joe's heart beats slower than Stan's, and Joe ages less. The difference should not dissuade Stan from making the trip. Only very carefully designed experiments can measure relativistic effects on the Earth.

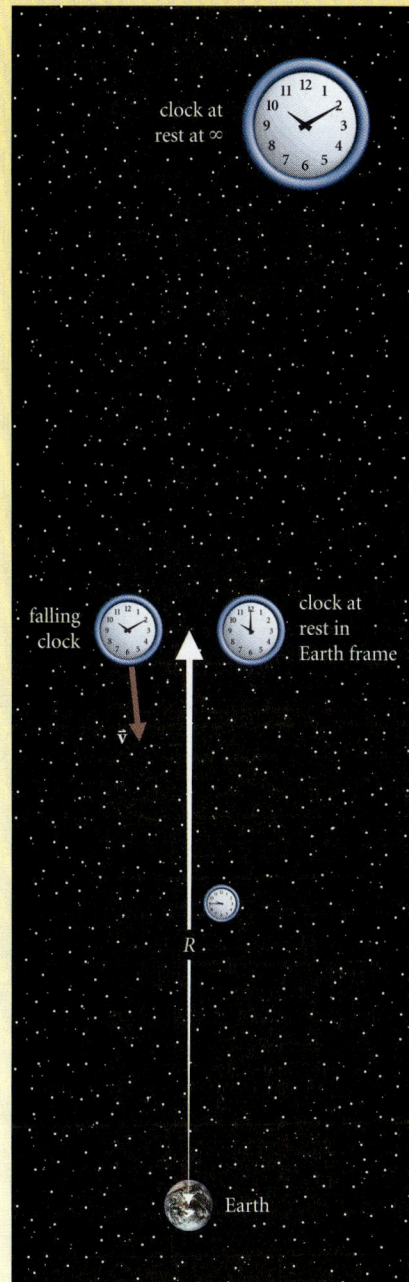

Figure E8.2
Gravitational time dilation. A freely falling clock measures proper time between events. A clock at rest in the Earth's frame is initially synchronized with an identical clock that is dropped from a great height. The falling clock registers a greater elapsed time than the stationary clock it is passing.

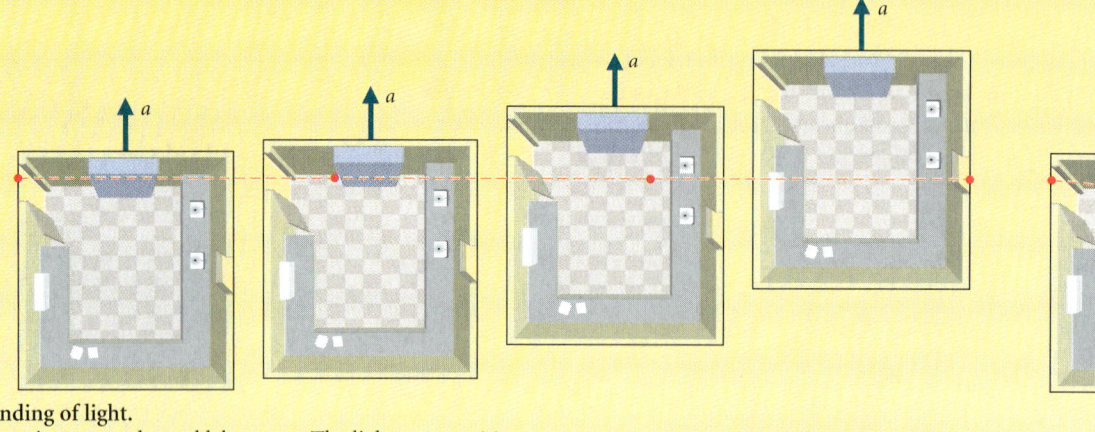

■ **FIGURE E8.3**
Gravitational bending of light.
(a) Light pulse crossing an accelerated laboratory. The light follows a straight line in an unaccelerated (inertial) frame.
(b) Relative to the laboratory, the light follows a curved path.

Digging Deeper

WHY SHOULD WE BELIEVE THE EQUIVALENCE PRINCIPLE?

The equivalence principle has been extensively tested. If it is correct, then everything should fall at the same rate, even though nuclear and electromagnetic energy contribute in different proportions to the masses of different chemical substances. ■ Figure E8.4 shows an experiment that demonstrates that gravity and inertial forces have identical effects on these forms of energy. The experiment was performed by Roland von Eötvös in 1889.[1] Two objects of equal mass but different chemical composition are at the ends of a bar suspended by a thin fiber. Both gravity and inertial force due to the Earth's rotation act on the objects. Any difference in the gravitational accelerations would cause a torque on the fiber. Eötvös found no difference to within 1 part in 10^9. More recent experiments have improved the accuracy to 1 part in 10^{13}.

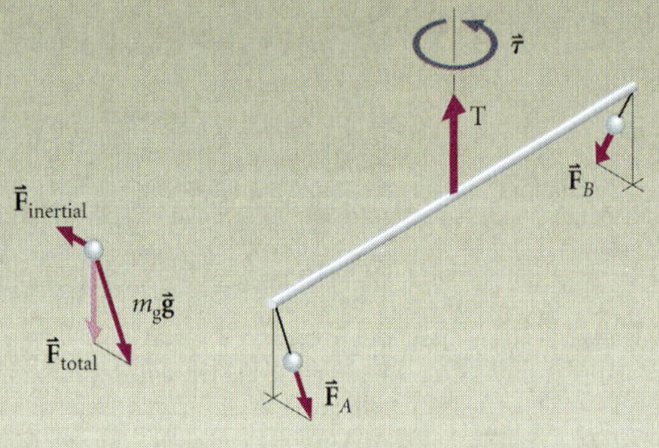

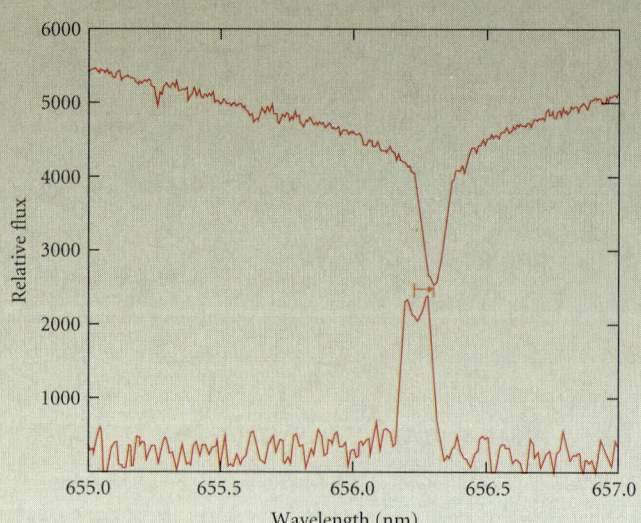

(a) (b)

■ **FIGURE E8.4**
The Eötvös experiment. Two objects of equal inertial mass are suspended from a bar. (a) The objects hang at the local vertical, defined by the gravitational force plus the inertial force in the rotating frame of the Earth. (b) If the two gravitational masses are not quite equal, the objects hang at different angles and a torque is exerted on the bar. (The angles are shown exaggerated for clarity.)

■ **FIGURE E8.5**
The spectrum of the white dwarf star 40 Eridani B shows a gravitational redshift. The upper spectrum shows the Balmer α absorption line of hydrogen (§35.2.2) in the white dwarf spectrum. The lower spectrum shows the same line in emission in the red dwarf binary companion 40 Eridani C. There is an obvious offset between the centers of the two lines of about 0.06 nm, of which 7% is a Doppler shift (§16.4) due to the motion of the two stars, and the rest is a gravitational redshift. The mass of the white dwarf is about 0.5 times the sun's mass, but its radius is only about 8000 km so GM/Rc^2 is about 10^{-4}.

---

[1] Eötvös' aim was to show, in the context of Newtonian theory, that the mass that produces gravitational force is the same as the mass that responds to force. This must be true if the equivalence principle holds.

A second consequence of the equivalence principle is that gravity accelerates light as well as matter. Since gravity is an effect of the reference frame, everything in the frame is accelerated equally. ■ Figure E8.3 shows a light pulse moving through an accelerated laboratory. The light moves in a straight line as seen by an observer in an inertial frame but traces a curved path across the accelerated frame. In a gravitational field, light travels along straight lines in a freely falling frame. Relative to the Earth's surface, light falls downward, pulled by gravity.

E8.2 Curvature of Space-Time

A real gravitational field is uniform only in small regions, and a freely falling inertial frame can only be defined *locally*. We need different inertial frames to discuss physics on board the space shuttle and in New Orleans. To construct a reference frame that covers the entire solar system, we must imagine a patchwork of many local reference frames. The extended reference frame so constructed describes a curved space-time. Navigation on the Earth's surface provides a good example of what this means.

When light is emitted by atoms close to a massive body, gravitational time dilation causes a distant observer to record the passage of wave fronts at a rate slower than that for waves emitted by identical atoms near the observer. This effect, called the gravitational red shift, has been observed in the spectra of certain stars (■ Figure E8.5). It has also been tested on Earth. Iron nuclei emit γ rays with an energy of 14.4 keV. In 1960, Pound and Rebka measured the difference in frequency between γ rays emitted at the bottom of a 22.5-m-high tower and the γ rays absorbed at the top. In 1965, Pound and Snider improved the experiment, measuring a frequency shift $\Delta f/f = (2.56 \pm 0.03) \times 10^{-15}$, in agreement with the predictions of general relativity.

Gravitational bending of light by the Sun was first detected during the solar eclipse of 1919 (■ Figure E8.6). More accurate results are obtained using radio waves emitted by distant quasars. The results confirm Einstein's predictions to within 1%. ■ Figure E8.7 shows multiple images formed as the light from a single quasar is deflected by a massive cluster of galaxies between us and the quasar. There are several known examples of such gravitational lenses.

■ Figure E8.6
Photograph of the May, 1919 solar eclipse taken by Eddington and Cottingham at Principe Island. Measurement of star positions on this photograph gave the first evidence of gravitational bending of light rays. Unfortunately, because of the age of the photograph, the star images are no longer distinguishable.

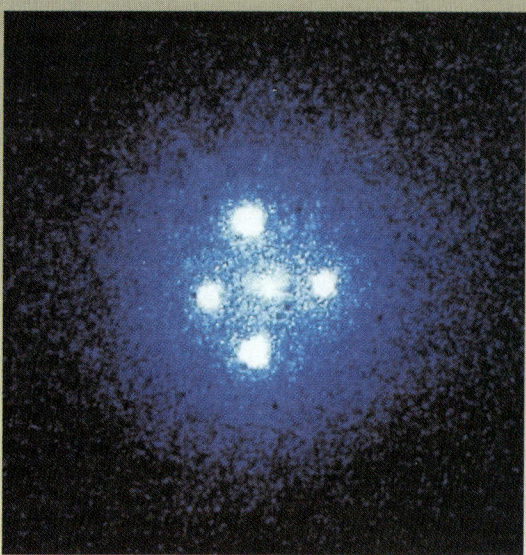

■ Figure E8.7
The *Einstein cross*: a gravitational lens. Light from a distant quasar (approximately 8 billion light-years away) is bent by the gravitational field of an intervening galaxy (the diffuse central object, approximately 400 million light-years away) to form four distinct images of the quasar. The number and position of the images formed depends on the precise alignment of background and foreground objects.

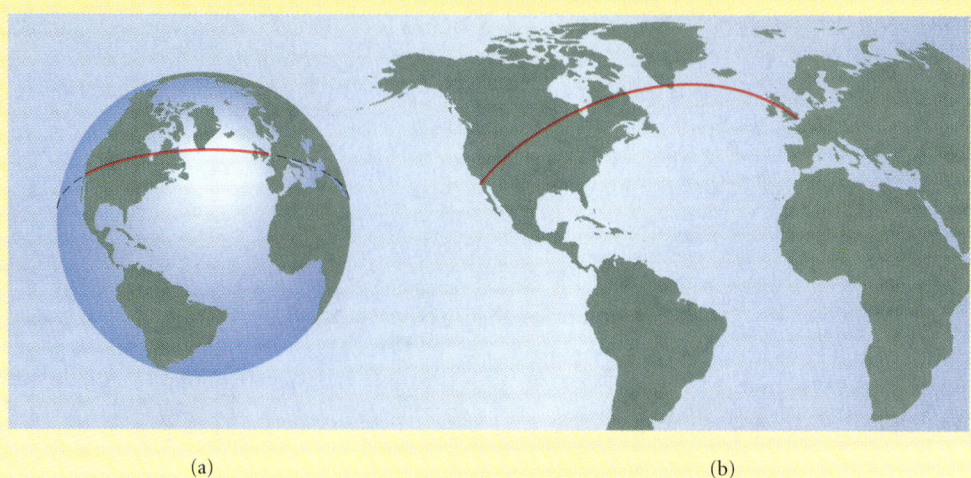

(a)   (b)

■ **Figure E8.8**
(a) The airline route from Los Angeles to London is part of a great circle around the globe. (b) When drawn on a flat piece of paper, the route looks curved.

An airplane traveling between Los Angeles and London flies over northern Canada and Greenland, heading first north and then south (■ Figure E8.8). The plane's path is the shortest distance from Los Angeles to London, a straight line on the curved surface of the Earth. It *seems* to change direction because the rectangular grid of north-south-east-west is laid down *locally* on small patches of the Earth's surface and does not describe the whole Earth well. On a map showing the whole Earth on a flat piece of paper, the *straight line* route from Los Angeles to London *looks* curved.

In an inertial frame, a particle moves in a straight line unless acted upon by a force. Its world line is the path with minimum proper time between any two events. This principle also holds in a global reference frame. The nearly circular orbit of the Earth and the highly elliptical orbit of a comet around the Sun are both straight world lines in curved space-time. They appear curved both because we are looking at projections into three-dimensional space, and because the Newtonian description of space distorts its geometry, like the flat map of the Earth's surface. Like an ant crawling in a straight line over a plum, the Earth's path curves right around and comes back to the same place in space (although not to the same event in space-time)!

As in the Newtonian model, mass causes gravity. The mass of the Sun distorts space-time, stretching and curving it (■ Figure E8.9). In the solar system, differences between Einstein's geometric theory of gravity (general relativity) and Newtonian theory are small. Among the successful predictions of Einstein's theory are a rotation of Mercury's elliptical orbit by 43 arc seconds per century and corrections of the order of 100 μs in timing of radar signals reflected from Mercury and Venus. Near compact, massive stars or in the dynamics of the universe as a whole, space-time curvature produces dramatic effects.

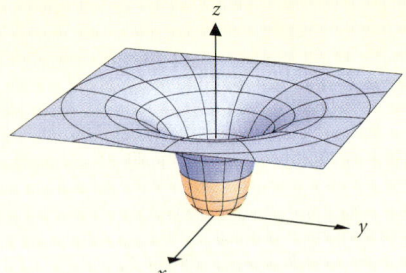

■ **Figure E8.9**
In the curved space around a star, circles of radius r have a circumference less than $2\pi r$. This diagram helps us visualize the effect. Distances in a plane through the equator of a star are represented by distances on a sheet that curves in an imaginary third dimension. This diagram corresponds to a very compact star: a neutron star. The orange region is the interior of the star. For a black hole, the central region would extend infinitely *downward*, forming a hole in the sheet.

E8.3   Black Holes

In 1798, Pierre Simon de Laplace imagined a star so massive that the escape velocity from its surface exceeds the speed of light. No light could escape from such a star: it would be a *black hole*, swallowing things up through its gravitational pull but invisible to an outside observer.

A particle leaving the surface of a sphere just escapes to infinity if its total (kinetic plus potential) energy $E = \frac{1}{2}mv^2 - GMm/R$ is zero (cf. Example 8.8). It is said to have escape speed v_{esc}, where $v_{\text{esc}} = 2GM/R$. The escape speed equals the speed of light if the sphere's radius is:

$$R = 2GM/c^2 \equiv R_S. \tag{E8.2}$$

This radius is called the Schwarzschild radius of the star, after Karl Schwarzschild, who first calculated the results of Einstein's theory for a point mass.

The mass of the Sun is 2.0×10^{30} kg (front cover), so its Schwarzschild radius is:

$$R_S = \frac{2GM}{c^2} = \frac{2(6.7 \times 10^{-11} \text{ m}^3/\text{kg}\cdot\text{s}^2)(2.0 \times 10^{30} \text{ kg})}{(3.0 \times 10^8 \text{ m/s})^2} = 3.0 \text{ km}.$$

The Sun is much larger than this ($R = 7 \times 10^{10}$ m), but stars called neutron stars exist with radii equal to a few Schwarzschild radii.

Although we used Newtonian theory to calculate the Schwarzschild radius, eqn. (E8.2) is also the correct result according to Einstein's theory. The entire future of an event within R_S lies within the black hole, and there can be no communication with the outside. Any world line that passes within R_S remains inside.

From the outside, the object is invisible.[2] Nothing escapes from within R_S, and gravitational time dilation, which is infinite at R_S, ensures that any signal from just outside takes infinite time to reach an outside observer. We would never actually see anything fall in. Far from the hole, its gravitational field is the same as for any other star of the same mass, so it might be observed by its influence on the motion of nearby visible objects.

Normal astronomical objects are supported by internal pressure at radii much greater than their Schwarzschild radii. However, some exploding stars may compress their central regions enough to form a black hole. A few astronomical x-ray sources have been interpreted as black holes accreting gas from binary companions in orbit about the holes. The x rays are emitted by hot infalling gas swirling around the hole before being swallowed (■ Figure E8.10). The central regions of a star cluster in a galactic nucleus could collapse to form a black hole with a billion times the mass of the Sun. Observations of our own galactic center show the violent motions that we would expect near such a large point mass. Hubble telescope observations of the centers of some distant galaxies show dense concentrations of rapidly moving gas and stars (■ Figure E8.11): exactly what would be expected in the vicinity of massive black holes.

■ **FIGURE E8.10**
Artist's conception of a galactic x-ray source. Gas pulled from a companion star forms a disk of matter spiraling onto a black hole at the disk's center. The x rays are emitted by hot gas at the inner edge of the disk.

E8.4 GRAVITATIONAL WAVES

Mass causes gravity by curving space-time. Accelerating mass can produce oscillating distortions—that is, gravitational waves—that propagate at the same speed as light. Such waves are extremely difficult to detect directly, and experimental attempts have so far proved unsuccessful. However, astronomers have discovered a remarkable system in which two very condensed stars orbit each other. Observed changes in the system are consistent with the predicted loss of energy and momentum to gravitational waves. In 1993, Richard Hulse and Joe Taylor received the Nobel prize for this discovery.

E8.5 THE HISTORY OF THE UNIVERSE

Observations of distant galaxies show a uniform distribution of matter on the largest scales (> 300 million light-years). According to general relativity, such a universe expands from an initial state of infinite density. Whether the expansion continues, or stops and is followed by contraction, depends on the specific values of the expansion velocity and the density of the universe. Distant galaxies do have relative speed proportional to separation, as predicted by the theory. The expansion rate is somewhat uncertain, but probably in the range 15–30 km/s per million light-years, corresponding to an age of the universe between 10 and 20 billion years. The observed mass density of the universe is more uncertain, since it is likely that only a small fraction of the total material emits light. According to the best current estimates, the universe will expand forever.

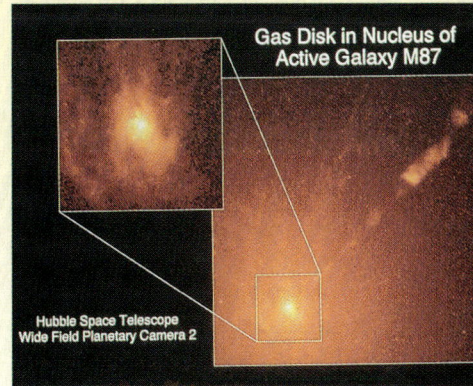

■ **FIGURE E8.11**
A disk of gas in the core of the giant galaxy called M87 (the 87th object in Messier's catalogue). Doppler-shifted emission lines (cf. §16.4.4 and Chapter 35) show that the gas is moving rapidly (about 550 km/s) in opposite directions on opposite sides of the disk. This rapid rotation is expected if there is a large mass in a small volume at the galaxy's center—a black hole.

---

[2] Quantum physics modifies this result. The hole emits blackbody radiation at a temperature inversely proportional to R_S^2. This effect is negligible for a hole with the mass of a star.

Scanning tunneling microscope image of a copper surface. Researchers at IBM's Almaden Research Center built a corral of iron atoms, carefully positioned on the surface. The electrons on the surface within the corral form a wave with circular "ripples." The system is cooled to 4 K to prevent the iron atoms from moving over the surface.

CHAPTER 35
Light and Atoms

Concepts

Photon
Quantization
Correspondence principle
De Broglie waves
Pauli principle
Uncertainty principle

Goals

Be able to:
Relate energy and momentum of particles to wavelength and frequency of the corresponding waves.

Compute the frequencies of photons emitted during transitions in hydrogenic atoms.

Understand:
Why the photoelectric effect requires a particle description of light.

How a wave description of matter explains atomic stability.

How wave and particle descriptions complement each other.

I don't think anyone understands quantum mechanics.

RICHARD P. FEYNMAN

With a scanning tunneling microscope, we can directly observe how electrons behave on the surface of a metal. Confined within a *corral* of iron atoms arranged on a copper surface, the electrons form a standing wave that looks like ripples on a pond. This is an odd picture that seems to conflict with our previous model of electrons as particles—pointlike concentrations of mass and charge.

In classical physics, the particle model provides a valuable description of objects, even atoms. For example, we used it to study the thermal properties of gases (Chapter 19). In Part IV we found that waves transport energy and momentum through a system. They describe the collective behavior of the particles or fields that comprise the system. At the beginning of the twentieth century, physicists discovered that the best way to understand the structure of matter was to combine elements of both the particle and wave pictures. Light is a wave in the electromagnetic field, but light interacts with atoms in particlelike bundles called photons. Similarly, an electron beam shows wave diffraction when reflected from a crystal.

The picture that emerges is bizarre by classical standards. Its philosophical interpretation is still hotly debated, but its predictions are verified with incredible precision in many experiments. The electronic switches that operate your pocket calculator or route your telephone call are but one development based on the principles of atomic physics. The great chemical diversity of the world, important not only for technology but for our very existence, is a consequence of these principles, which determine the patterns of electrons in atoms. In this

Digging Deeper

THE ORIGINS OF THE QUANTUM IDEA

Quantum theory is based on the concept that certain physical properties of systems—energy, momentum, and angular momentum—are not truly continuous variables, but occur in small chunks called quanta. In Newtonian theory, we ignore this granularity because the systems we study contain enormous numbers of quanta. A spinning ice skater, for example, has about 10^{35} quanta of angular momentum. Using Newtonian physics is like modeling charge distributions as continuous while knowing that charge comes in integer multiples of e.

Max Planck first suggested the idea of energy quanta in December 1900, as "an act of desperation." Planck had worked for decades attempting to understand thermal radiation from hot objects (§§21.3 and 35.1.3). In September 1900, fresh laboratory data led him to the precise empirical formula we know as Planck's law, but he could not yet explain why the law should hold. He knew that radiation was emitted and absorbed by atoms and molecules (oscillators) in the object, and he had obtained a relation between the electromagnetic energy at frequency f and the energy of the oscillators at the same frequency. Planck made two further, astonishing assumptions.

The first was the *quantum postulate*:

> The energy of each oscillator in the body is an integer times a basic energy unit E:
>
> $$U = NE.$$

The second step was a novel procedure for determining the number of possible oscillator energy states (§22.7), which implied that the basic energy unit is proportional to frequency:

$$E = hf.$$

Both steps were bizarre by the standards of classical physics, and Planck himself considered them tentative.

As in many other problems of the time, it was Einstein who saw the significance of the quantum hypothesis. In 1905, he introduced the quantum of energy in radiation. He was motivated by his derivation of the entropy of radiation, which he based on the observed frequency dependence of thermal radiation at high frequencies. He immediately used his new idea to explain the photoelectric effect (§35.1.1), and his theories were fully confirmed in 1916 by Millikan.

In 1906, Einstein realized that "Planck's theory makes implicit use of the light-quantum hypothesis." In 1909, Einstein first used the word "pointlike" in connection with the light quanta. The idea that the light quanta carry momentum was first stated by Johannes Stark in the same year. In 1916, Einstein established the connection between Bohr's theory of atomic spectra (§35.2) and Planck's formula, and explicitly added momentum to his quantum theory. The photon was born.

chapter we study how a particle description of light and a wave description of electrons fuse into a model of atomic structure. We then look briefly at the meaning of this wave/particle description and the fundamental changes it requires in relations of cause and effect.

35.1 Photons

35.1.1 The Photoelectric Effect

Since its discovery, the photoelectric effect (■ Figure 35.1) has become an important part of modern technology. In a television camera, light falls on a semiconductor surface and liberates electrons, converting an image to an electronic signal. In a photomultiplier tube (■ Figure 35.2), used to detect extremely faint light sources, light falls on the metal cathode surface and ejects electrons. Each electron ejected from the cathode creates a current that is amplified within the tube and detected by an electronic circuit connected to it.

THE EJECTED ELECTRONS ARE CALLED *PHOTOELECTRONS*.

Heinrich Hertz was the first to observe the photoelectric effect. In 1887, while performing an experiment to confirm Maxwell's theory of electromagnetic waves, he noticed that ultraviolet light from sparks generated by his transmitting coil influenced the sensitivity of his receiving coil. Careful experiments by Hertz and others over the next 18 years revealed several important characteristics of the photoelectric effect:

- The rate of electron emission is proportional to light intensity.
- The energy of the ejected electrons is independent of light intensity.
- For any given metal, there is a minimum frequency f_0 of light that can eject electrons.
- The energy of electrons ejected by light of frequency f is proportional to the frequency difference $f - f_0$.
- Electrons are emitted immediately (within 10^{-9} s) after light exposure begins, regardless of intensity.

Of these features, only the first is easily understood using classical electromagnetic theory: the rate electrons absorb energy from the light is proportional to the rate energy is delivered to the metal surface. The next three features are puzzling, since most classical models predict a much more complicated frequency dependence. Classical theory cannot account for the rapid start of electron flow, as we illustrate in the following example.

■ **FIGURE 35.1**
The photoelectric effect. Light waves incident on a surface eject electrons.

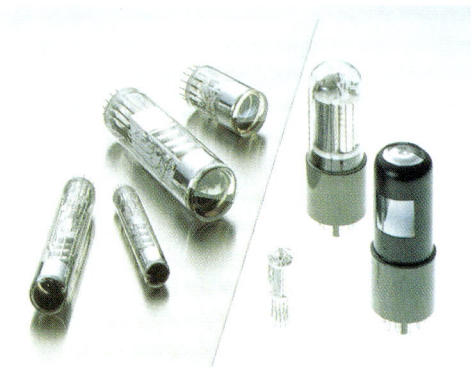

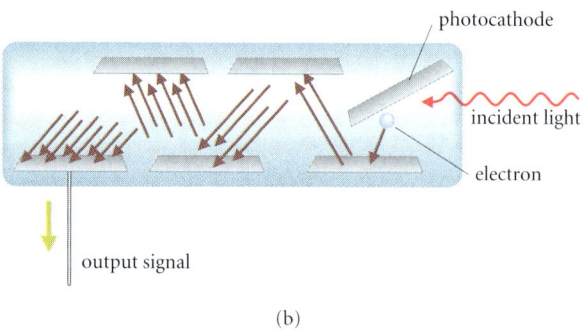

(b)

■ **FIGURE 35.2**
(a) Commercial photomultiplier tubes make use of the photoelectric effect.

(b) Incident light releases electrons at the photocathode. Cathodes inside the tube, called dynodes, are held at increasingly positive potentials so that the electrons are accelerated from dynode to dynode. At each dynode, the electrons eject more electrons, increasing the total current in the tube. A measurable current results from a tiny amount of incident light.

EXAMPLE 35.1 ♦♦ Light of intensity $I = 10$ W/m² falling on a metal surface ejects electrons, each with 3 eV of energy. According to classical theory, how much time is required for an atom of radius 10^{-10} m to receive the necessary 3 eV of energy?

MODEL In a metal, there are at most a few free electrons per atom, so the cross section of an atom is a reasonable guess for the absorbing area per electron. To estimate the shortest time before electrons can be ejected, we also suppose that the energy is absorbed with perfect efficiency:

SETUP
$$\Delta t_{classical} = \frac{\text{observed energy}}{\text{absorption rate}} = \frac{\text{observed energy}}{\text{intensity} \times \text{area}}.$$

SOLVE
$$\Delta t_{classical} = \frac{(3 \text{ eV})(1.6 \times 10^{-19} \text{ J/eV})}{(10 \text{ W/m}^2)(\pi)(10^{-10} \text{ m})^2} = 2 \text{ s}.$$

ANALYZE The classical prediction is vastly greater than the observed time required for electron ejection.

In 1905, Einstein gave the correct explanation of the photoelectric effect: though propagating as a wave according to Maxwell's theory, the electromagnetic field delivers energy to electrons in the metal in chunks, or *quanta*. The magnitude of an energy quantum is proportional to the wave frequency:

$$E = hf. \tag{35.1}$$

The constant of proportionality h is Planck's constant:

$$h = 6.626 \times 10^{-34} \text{ J} \cdot \text{s}.$$

Planck first obtained this value in 1901 by fitting a theoretical function to the experimentally determined blackbody radiation spectrum. Millikan obtained it independently from his photoelectric data (■ Figure 35.3).

The incident light is a stream of energy quanta in the form of particles, or *photons*, each with energy hf and each with an equal probability of ejecting an electron. Thus the number of electrons ejected per second is proportional to the number of photons per second reaching the surface, and so is proportional to the light intensity. Electrons can be ejected as soon as the first photon strikes the metal.

Einstein's model directly explains the minimum frequency f_0 and the proportionality of the observed electron energy and frequency. Electrons are free to move within the metal but are bound by the attractive force of the metal atoms with a potential energy called the *work function*.

The *work function* ϕ of a metal is the minimum energy required to remove a single electron from the metal.

PLANCK'S 1901 VALUE WAS 6.55×10^{-34} J·s.

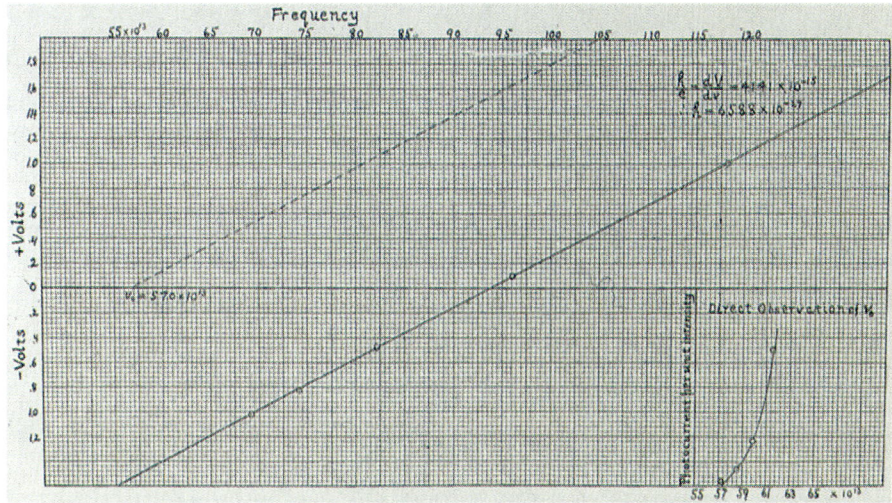

■ FIGURE 35.3
Millikan's original data on photoemission from lithium. The potential of the anode that is just sufficient to suppress the photocurrent is plotted against the wavelength of light. Millikan calculated h from the slope of the line, obtaining the value 6.588×10^{-34} J·s. The value written on his graph is in cgs units of erg·s.

■ FIGURE 35.4
The work function. An incident photon is absorbed by an electron beneath the surface of a metal, where the electron's potential energy is $-\phi$. The electron escapes from the metal. Outside the metal, the electron has zero potential energy, and its kinetic energy is the amount hf absorbed from the photon less the amount ϕ of potential energy gained.

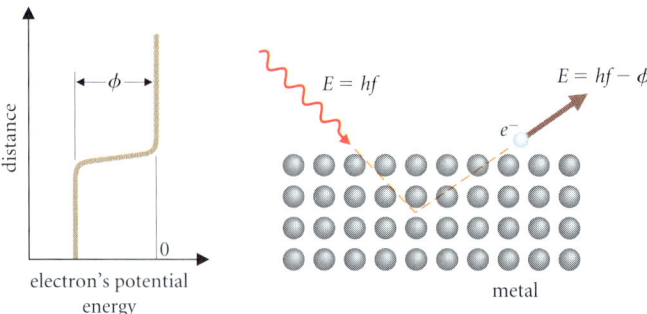

When the incident light wave delivers a quantum of energy to an electron, a minimum amount ϕ is used to escape from the metal. So electrons emerge with a range of kinetic energy up to a maximum (■ Figure 35.4):

$$\tfrac{1}{2}mv^2 = hf - \phi. \tag{35.2}$$

The minimum frequency f_0 is that for which the energy of a quantum just equals the work function: $hf_0 = \phi$. Not every photon ejects an electron.

> The fraction of photons reaching a given metal surface with sufficient energy that actually eject an electron is the *quantum efficiency* of the metal.

Typical metals (like aluminum, copper, and gold) have work functions of 4–5 eV and peak quantum efficiency of 10–20% at about $\lambda = 100$ nm. ■ Figure 35.5 is a plot of quantum efficiency for a gold photodiode as a function of photon wavelength. Few materials are efficient photoemitters in the visible. Cesium antimonide (CsSb) is a notable exception, with a peak quantum efficiency of 15% at 450 nm.

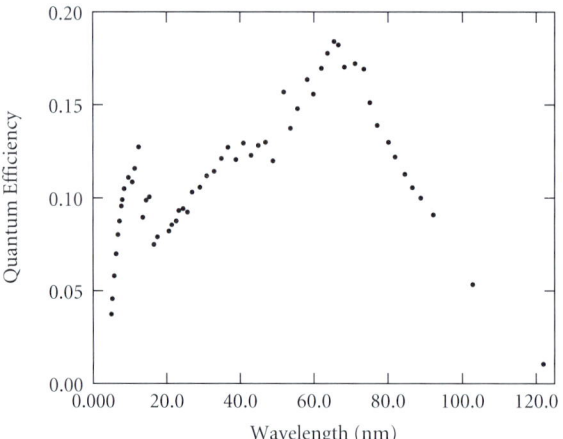

■ FIGURE 35.5
Quantum efficiency of a gold photodiode as a function of frequency. The systematic uncertainty in the measurements is about 10%.

MONOCHROMATIC (SINGLE COLOR) MEANS HAVING A NARROW RANGE OF FREQUENCIES.

EXAMPLE 35.2 ◆ A monochromatic light beam falls on a potassium surface. If the photoelectrons emerge with a maximum kinetic energy of 2.0 eV, what is the frequency of the light?

MODEL The energy of an individual light quantum equals the work function plus the kinetic energy of the photoelectrons. The frequency of the light is determined directly from the energy of a quantum using eqn. (35.1).

SETUP We take the work function for potassium from ● Table 35.1.

SOLVE From eqn. (35.2),

$$f = \frac{\tfrac{1}{2}mv^2 + \phi}{h} = \frac{(2.0 \text{ eV} + 2.3 \text{ eV})(1.6 \times 10^{-19} \text{ J/eV})}{6.6 \times 10^{-34} \text{ J} \cdot \text{s}} = 1.0 \times 10^{15} \text{ Hz}.$$

TABLE 35.1   Photoelectric Properties of Selected Metals

Metal		Work Function (eV)	Spread in Measured Values
Aluminum	Al	4.25	±0.13
Barium	Ba	3	*
Calcium	Ca	2.9	*
Cerium	Ce	2.9	*
Copper	Cu	4.6	±0.16
Gold	Au	5.3	±0.13
Iron	Fe	4.7	±0.11
Lithium	Li	3	*
Mercury	Hg	4.5	*
Nickel	Ni	5.15	±0.12
Platinum	Pt	5.85	±0.25
Potassium	K	2.3	*
Silver	Ag	4.6	±0.09
Sodium	Na	2.75	*
Tungsten	W	4.6	±0.32

*Only one source available.

ANALYZE This corresponds to a wavelength $\lambda = c/f = 290$ nm, which is in the ultraviolet.

EXAMPLE 35.3 ♦♦ A star known to be as bright as the Sun (luminosity ≈ 4×10^{26} W) is 3×10^{18} m from Earth. A photoelectric detector with a quantum efficiency of $e_q = 0.1$ and a 1-cm² area is used to observe that star. Estimate the photon count rate.

MODEL Without much more detailed information both about the star and about the detector, we can only obtain a rough estimate. So we assume that the given quantum efficiency is an appropriate average over the star's spectrum and that there is no absorption of starlight along its path to Earth. Then, if the star is at distance R, the power per unit area reaching the detector is the star's luminosity L divided by the area of a sphere of radius R. The power incident on the detector of area A is:

$$P = AL/(4\pi R^2).$$

SETUP This power (= energy/time) equals the number of photons per second reaching the detector times the energy per photon:

$$P = \frac{dn}{dt}hf \quad \Rightarrow \quad \frac{dn}{dt} = \frac{P}{hf}.$$

A fraction e_q of these photons actually eject an electron, so the count rate is:

$$r = \frac{e_q P}{hf} = \frac{e_q AL}{4\pi R^2 hf}.$$

For a star like the Sun, an appropriate average wavelength is 500 nm, and $f = c/\lambda$.

SOLVE $r = \dfrac{(0.1)(10^{-4} \text{ m}^2)(4 \times 10^{26} \text{ W})(5 \times 10^{-7} \text{ m})}{4\pi(3 \times 10^{18} \text{ m})^2(6.6 \times 10^{-34} \text{ J}\cdot\text{s})(3 \times 10^8 \text{ m/s})} = 90$ /s.

ANALYZE This is quite a large count rate by astronomical standards. Multiplication devices can increase the electron yield considerably. An astronomer could easily observe the star with this detector.

ODDLY, WITH DIFFERENT MATERIALS YOU GET THE WORK FUNCTION OF THE ANODE. SEE PROBLEM 98.

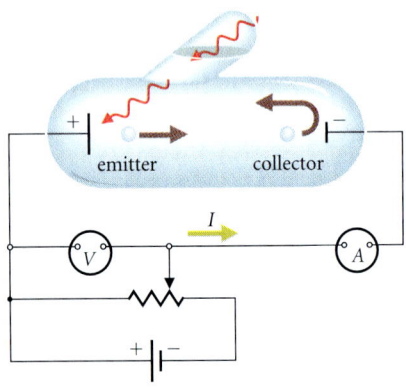

■ FIGURE 35.6
Apparatus for measuring the work function. Incident photons eject electrons from the emitter, or cathode. The collector, or anode, is maintained at a negative potential with respect to the emitter so that the resulting electric field decelerates the emitted electrons. When V is set to the stopping potential, no electrons reach the collector, and the measured current in the circuit drops to zero.

IN PROBLEMS LIKE THIS, THE COMBINATION $hc = 1.24 \times 10^3$ eV·nm IS OFTEN USEFUL.

IN THE EXAMPLE WE FOUND $\Delta V = -(h/e)f + \phi/e$, A LINE WITH SLOPE $-h/e$.

EXERCISE 35.1 ♦♦ A light beam of intensity 1.0 W/m² and wavelength 100 nm falls on a gold photodiode (Figure 35.5). Find the photoelectron current emitted per unit area of the metal surface.

To determine the work function for a metal, we could use an experimental photocell (■ Figure 35.6). Incident light with a fixed, known frequency liberates electrons at a surface called the *cathode*. Electrons collected on another surface called the *anode* produce current in the external circuit. If the anode is held at a negative potential with respect to the cathode, the resulting electric force on the ejected electrons decelerates them. With a great enough negative potential, no electrons reach the anode at all. The potential difference that is just great enough to stop the electrons is called the *stopping potential*. If both cathode and anode are made of the same material, the stopping potential measures the work function of that material.

EXAMPLE 35.4 ♦♦ In a photoelectric experiment, light of wavelength 253.5 nm falls on a lithium cathode. The electron curent is suppressed when the lithium anode is held at a potential of −1.85 V with respect to the cathode. Find the work function for lithium.

MODEL Electrons leave the cathode with kinetic energy that is converted to electric potential energy at the anode. With a stopping potential of about 2 V, the electrons are nonrelativistic. Einstein's relation (35.2) determines the work function.

SETUP The electron kinetic energy may be found using the usual conservation law methods. As the potential difference between anode and cathode is made more negative, the potential energy of electrons arriving at the anode is increased and their kinetic energy is decreased. With the anode at the stopping potential, the electrons have zero kinetic energy there.

	BEFORE (at cathode)	AFTER (at anode)
Kinetic energy	$\frac{1}{2}mv^2$	0
Potential energy	$-eV_{cathode}$	$-eV_{anode}$
Set totals equal:	$\frac{1}{2}mv^2 - eV_{cathode}$ =	$-eV_{anode}.$

Use eqn. (35.2) to find ϕ:

$$-e(V_{anode} - V_{cathode}) = \tfrac{1}{2}mv^2 = hf - \phi = h\frac{c}{\lambda} - \phi.$$

SOLVE $\phi = h\dfrac{c}{\lambda} + e\,\Delta V = \dfrac{(6.63 \times 10^{-34} \text{ J·s})(3.00 \times 10^8 \text{ m/s})}{(1.60 \times 10^{-19} \text{ J/eV})(253.5 \times 10^{-9} \text{ m})} - 1.85 \text{ eV}$

$= (4.90 - 1.85) \text{ eV} = 3.05 \text{ eV}.$

ANALYZE From Table 35.1, the work function for lithium is 3 eV, in good agreement with this result.

If stopping potential is measured as a function of wavelength, the value of Planck's constant may be found from the slope of a graph of ΔV versus $f = c/\lambda$ and the work function from its intercept. Millikan measured the slope (Figure 35.3) and showed it to be the same for several different metals.

35.1.2 The Compton Effect

In 1922, Arthur H. Compton obtained experimental evidence that photons behave like particles. Compton shone a beam of x rays with wavelength $\lambda = 0.07$ nm onto a sample of graphite and observed that the wavelength of x rays emerging at 90° from their original direction was increased by 0.002 nm (■ Figure 35.7). The energy of the x-ray photons,

$$E = hc/\lambda \approx 18 \text{ keV},$$

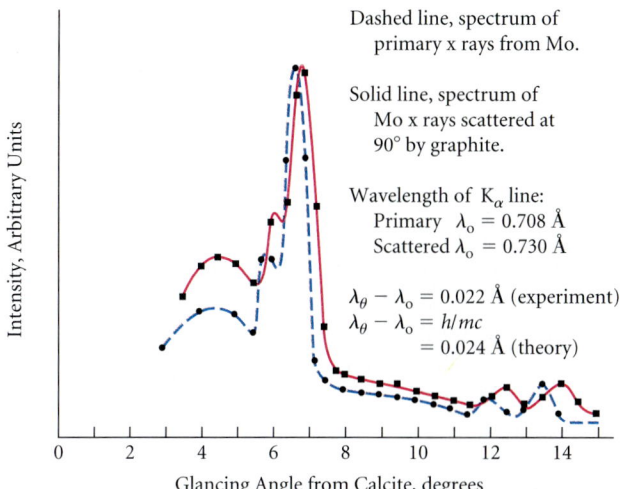

■ FIGURE 35.7
Compton's spectrum of x rays from molybdenum. Dashed line—primary spectrum. Solid line—spectrum of x rays scattered through 90° by graphite. The x-ray wavelength was determined by diffraction using a calcite crystal (§17.5). Large glancing angles correspond to longer wavelengths. The uniform increase in wavelength is evident. Redrawn from Figure 4, Physical Review second series, vol. 21, #5, May 1923.

is much larger than the few eV binding energy of electrons to the carbon atoms in the graphite block. In effect, the electrons act like free particles exposed to the x-ray beam. Classically, such free electrons would scatter an electromagnetic wave with no change in wavelength. Compton explained the observed wavelength change as the result of elastic collisions between x-ray photons and stationary free electrons. The effect is now named after him.

Before a Compton collision, the incident photon has frequency f_1, and the electron is assumed to be at rest (■ Figure 35.8). After the collision, the photon has frequency f_2 and travels at angle θ to the incident beam. The electron recoils at angle ϕ. Electromagnetic radiation carries momentum given by its energy divided by the speed of light. Thus, the initial momentum of the photon is:

$$p = E_1/c = hf_1/c. \tag{35.3}$$

COMPARE THE DISCUSSION IN §33.2.2.

The electron recoils at high speed and must be described using special relativity. Treating the encounter as a particle collision, we apply conservation of energy and momentum:

THUS COMPTON SCATTERING ALSO TESTS THE RELATIVISTIC EXPRESSIONS FOR ENERGY AND MOMENTUM (§34.4).

	BEFORE	AFTER	
Energy	$hf_1 + mc^2$	$= hf_2 + \gamma mc^2.$	(i)
x-component of momentum	hf_1/c	$= (hf_2/c)\cos\theta + \gamma mv \cos\phi.$	(ii)
y-component of momentum	0	$= (hf_2/c)\sin\theta - \gamma mv \sin\phi.$	(iii)

For comparison with Compton's data, we need a relation among the two photon wavelengths and the scattering angle θ. First, we use the two momentum equations, (ii) and (iii), to eliminate ϕ. (Recall $\beta \equiv v/c$ and $\gamma \equiv 1/\sqrt{1-\beta^2}$.)

$$(\gamma mv)^2(\cos^2\phi + \sin^2\phi) = (h/c)^2[(f_1 - f_2\cos\theta)^2 + (f_2\sin\theta)^2],$$

or,

$$\gamma^2\beta^2 = \gamma^2 - 1 = (h/mc^2)^2(f_1^2 + f_2^2 - 2f_1f_2\cos\theta). \tag{iv}$$

From the energy equation (i):

$$(\gamma - 1)^2 = (h/mc^2)^2(f_1 - f_2)^2. \tag{v}$$

Subtracting (v) from (iv) and using the energy equation again:

$$(h/mc^2)^2[2f_1f_2(1 - \cos\theta)] = 2(\gamma - 1) = 2(h/mc^2)(f_1 - f_2).$$

Dividing by $2hf_1f_2/(mc^3)$:

$$\frac{h}{mc}(1 - \cos\theta) = \frac{c(f_1 - f_2)}{f_1 f_2}.$$

The factor,

$$h/(mc) \equiv \lambda_C = 2.43 \times 10^{-3} \text{ nm}, \tag{35.4}$$

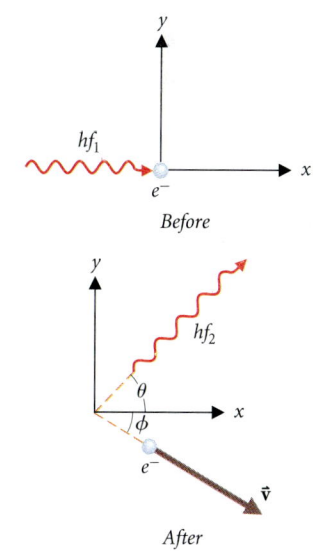

■ FIGURE 35.8
Compton scattering. We have chosen the x-axis to be along the direction of the incoming photon. The electron is initially at rest. After the collision, the photon has frequency f_2 and travels at an angle θ to its incoming direction. The electron recoils with speed v at angle ϕ to the x-axis.

is known as the *Compton wavelength* of the electron. Converting from frequency to wavelength, $f = c/\lambda$, we find:

$$\lambda_2 - \lambda_1 = \lambda_C(1 - \cos\theta). \quad (35.5)$$

For photons scattered at 90° ($\cos\theta = 0$), eqn. (35.5) predicts a wavelength increase $\Delta\lambda = \lambda_C$, in agreement with Compton's results. Compton performed a second experiment in which 565-keV γ rays from a radium source were scattered at various angles from iron, aluminum, and paraffin, and he measured the wavelength of the scattered photons as a function of angle. Equation (35.5) provided an excellent fit to his measured values (■ Figure 35.9).

Compton's experiment also exhibits the simultaneous validity, or *duality*, of wave and particle pictures in the description of light. While the effect being measured is understood to result from collisions of x-ray particles with electrons, it is observed as the difference in two wavelengths that are measured from wave interference effects in a crystal spectrometer (cf. §17.5).

EXAMPLE 35.5 ♦ If x rays of wavelength 0.12 nm are scattered from free electrons in a plasma sample, what change in wavelength is observed for x rays scattered at 45° to the incident beam? (A plasma is a gas in which the particles are charged.)

MODEL The assumption of zero initial electron velocity is reasonable so long as the electrons are nonrelativistic.

SETUP We use eqn. (35.5) with $\cos 45° = \sqrt{2}/2$.

SOLVE $\lambda_2 - \lambda_1 = (2.43 \times 10^{-3}\text{ nm})(1 - \sqrt{2}/2) = 7.12 \times 10^{-4}\text{ nm}$.

ANALYZE The wavelength increases by 7.1×10^{-4} nm, or 0.59%. ■

Using the relation between photon wavelength and energy, $\lambda = c/f = hc/E$, eqn. (35.5) becomes:

$$\frac{\lambda_2 - \lambda_1}{\lambda_1} \equiv \frac{\Delta\lambda}{\lambda} = \frac{E_1}{mc^2}(1 - \cos\theta). \quad (35.6)$$

The fractional change in λ is determined primarily by the ratio of photon energy to the electron's rest mass energy, $mc^2 = 0.511$ MeV. Substantial effects occur for very high energy photons (γ rays).

EXERCISE 35.2 ♦♦ Find the fractional change in energy when 1.0-MeV γ rays are scattered through 90°. ■

✵ 35.1.3 The Planck Radiation Law

Max Planck introduced the idea of energy quanta in 1900 to explain blackbody radiation. Planck's formula for the electromagnetic energy per unit volume du between the frequencies f and $f + df$ in a cavity at temperature T is:

$$du = \frac{8\pi h f^3}{c^3} \frac{df}{\exp(hf/kT) - 1}. \quad (35.7)$$

We can see why photons are necessary to explain blackbody radiation by considering a special case: the radiation inside a perfectly reflecting, cubical box. In equilibrium, according to classical thermodynamics, there is, on average, $\frac{1}{2}kT$ of energy in each possible mode of energy storage. An oscillator has two modes of energy storage: kinetic and potential energy. So each of a system's possible oscillation modes has an energy kT (cf. §19.7). Thus, for a classical

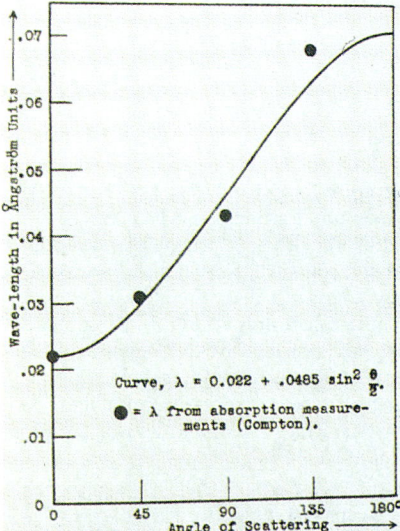

■ **FIGURE 35.9**
Compton's original data, published in 1923, show the variation in wavelength of 0.22-nm γ rays scattered by iron, aluminum, and paraffin. The incident γ rays were from a radium source (see Chapter 36). Compton determined the scattered frequency by absorbing the γ rays in lead, whose absorption properties were well known. The measurements are in good agreement with the theoretical curve (eqn. 35.5).

SEE § 21.3 FOR A DISCUSSION OF BLACK-BODY RADIATION.

calculation, we need only determine the number of modes of oscillation within the cube. The result was derived independently by Rayleigh, Jeans, and Einstein, and is called the *Rayleigh–Jeans law* for the energy density in blackbody radiation:

$$du_{\text{R-J}} = \frac{8\pi f^2 df}{c^3} kT. \tag{35.8}$$

The Rayleigh–Jeans law gives the correct energy density in the low-frequency limit but fails disastrously at high frequencies. Because the number of possible oscillation modes increases without limit as frequency increases, classical theory predicts an infinite energy density and correspondingly infinite energy flow from any source of radiation. Sitting before your fireplace, you should be vaporized by an infinite flux of γ radiation!

THIS NONSENSICAL RESULT IS CALLED THE "ULTRAVIOLET CATASTROPHE."

The flaw in classical reasoning lies not in the counting of modes, but in the assumption of energy equipartition among them. Because the energy is bundled into photons, a mode of frequency f cannot have an energy less than hf. When this quantum is much larger than kT, according to Boltzmann's law (§22.7), it is improbable that a mode contains any energy at all, much less an equal share with all the other modes. At high frequencies, the number of modes at frequency f should be multiplied not by the equipartition energy kT, but by a product of the photon energy hf and a Boltzmann factor $\exp(-hf/kT)$, which gives the probability of the mode being excited.

35.2 BOHR'S ATOMIC MODEL
35.2.1 The Structure of Atoms

At the beginning of the twentieth century, three experimental facts about atoms were known:
- The volume occupied by a single atom in solids is approximately a sphere with a radius of a few times 10^{-10} m.
- Atoms contain negatively charged electrons that can be separated from and recombined with the main, positively charged body of the atom.
- An atom of a specific element emits and absorbs specific frequencies of light (■ Figure 35.10).

The plum pudding model of the atom (■ Figure 35.11) had been proposed based on these properties. The plum pudding atom was a uniform sphere of positively charged material with a fixed pattern of electrons embedded in it. In 1910, Ernest Rutherford and his students Hans Geiger and Ernest Marsden devised an experiment to test this model. They bombarded a thin gold foil with a beam of α-particles emitted by a radioactive substance. The α-particles, which

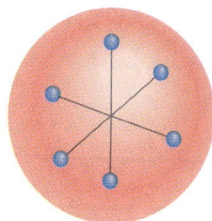

■ **FIGURE 35.11**
Plum pudding model of the atom. In this model, positively charged material forms a uniform sphere of radius $\approx 10^{-10}$ m, the observed size of the atom. Pointlike electrons are embedded in the sphere in a regular lattice. While this model can explain the simplest properties of atoms, Rutherford's experiment showed that it can't be correct.

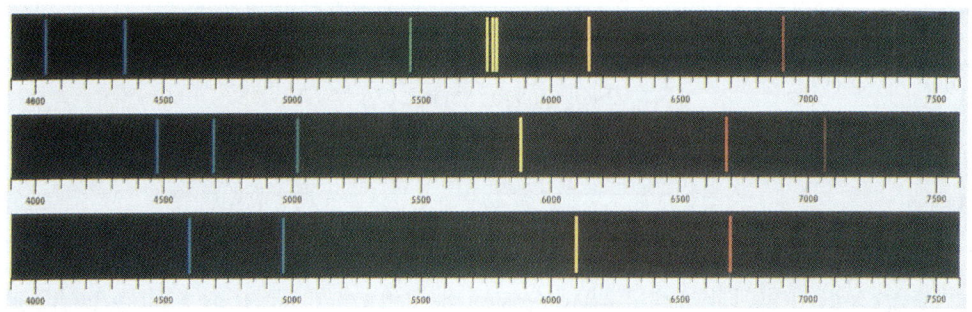

■ **FIGURE 35.10**
Spectra of light from atoms is emitted at a set of discrete frequencies. This photograph shows the light emitted by mercury (top), helium, and lithium (bottom).

we now know are bare nuclei of helium atoms, are positively charged and are deflected by electrical forces encountered while passing near, or through, atoms in the foil. Rutherford measured the directions at which the α-particles emerged (■ Figure 35.12).

EXAMPLE 35.6 ♦♦♦ Use the plum pudding model to predict the results of Rutherford's experiment. Estimate the maximum deflection of a 1-MeV α-particle (assumed pointlike) caused by a uniform sphere of radius 4×10^{-10} m with the total positive charge of a gold nucleus ($+79e$).

MODEL The force exerted on the α-particle at any distance r from the center of a sphere with uniform charge density ρ is:

$$F \propto \frac{\text{charge within radius } r}{r^2} = \frac{q(r)}{r^2}.$$

For paths that pass within the sphere, $r < R$, $q(r) = \rho V = \frac{4}{3}\pi r^3 \rho$, and $F \propto r$. For paths that pass outside the sphere, $r > R$, $q(r)$ equals the total charge of the particle, and $F \propto 1/r^2$. Thus the force is maximum at the surface of the sphere. A particle that just skims the surface is deflected the most (■ Figure 35.13). The particle gains momentum Δp perpendicular to its original velocity and is deflected through a small angle $\theta \approx \tan \theta = \Delta p / p$. If the deflection is small, we may ignore any deviation of the path from a straight line *during* the interaction.

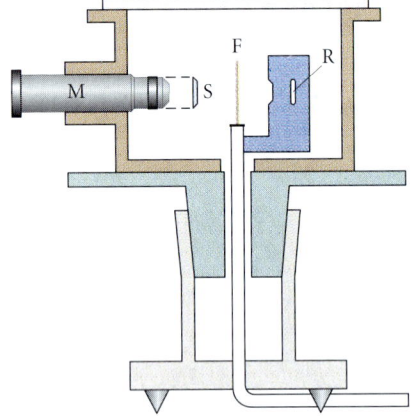

■ **FIGURE 35.12**
Rutherford's apparatus. Helium nuclei (α-particles) from a radioactive source R pass through a thin gold foil F. Particles striking screen S produce flashes that are observed through telescope M. The telescope is rotated and the rate α-particles strike the screen is observed as a function of telescope position.

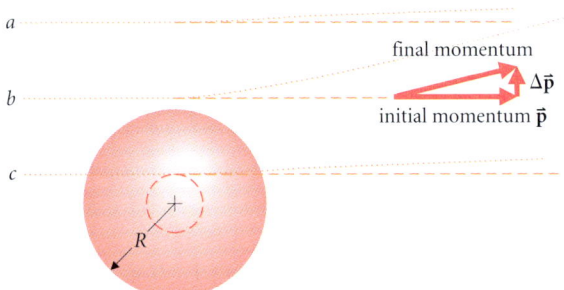

■ **FIGURE 35.13**
Deflection of charged particles by a uniform sphere of charge (the *plum pudding* model). The deflection is greatest when the particle just skims the surface (track b). Particles that pass farther (track a) from the sphere experience a lesser force because of the inverse square Coulomb force law. Particles that pass through the sphere (track c) experience a lesser force because a smaller amount of charge (that within the dashed sphere) produces force on the particle.

SETUP The particle travels at speed v and is near the sphere for a time $\Delta t \approx 2R/v$. Assuming for simplicity that the maximum force acts for a time Δt, it produces a change in momentum:

$$\Delta p \approx F_{\text{max}} \Delta t = \frac{kQ_\alpha Q_{\text{Au}}}{R^2}\left(\frac{2R}{v}\right).$$

In terms of the kinetic energy $E_\alpha = \frac{1}{2}mv^2$ of the α-particle, the deflection angle is:

$$\theta \approx \frac{\Delta p}{p} = \frac{2kQ_\alpha Q_{\text{Au}}/Rv}{mv} = \frac{kQ_\alpha Q_{\text{Au}}}{RE_\alpha}$$

$$= \frac{(9 \times 10^9 \text{ N}\cdot\text{m}^2/\text{C}^2)(2)(79)(1.6 \times 10^{-19} \text{ C})^2}{(4 \times 10^{-10} \text{ m})(10^6 \text{ eV})(1.6 \times 10^{-19} \text{ J/eV})}$$

$$= 6 \times 10^{-4} \text{ rad}.$$

ANALYZE This calculation overestimates the deflection caused by a plum pudding atom because it neglects the effects of the negatively charged electrons. Thus Rutherford expected to see his α-particles undergo small deflections. ■

To Rutherford's amazement, some of the particles were scattered to large angles, and some even returned in the direction they had come, contradicting the predictions of the plum pudding model. Rutherford showed that his data are explained exactly by a model in which the positive charge of the atom and almost all its mass are concentrated in a point at the center (■ Figure 35.14). Modern scattering experiments show that a typical atomic nucleus is a few times 10^{-15} m in radius.

> **EXERCISE 35.3** ♦♦ How close can a 1-MeV α-particle get to a pointlike gold nucleus if initially aimed directly at it? What happens to the α-particle subsequently?

35.2.2 Balmer's Spectrum and Bohr's Atom

Rutherford's experiment leaves no doubt that solid material is made up of small particles surrounded by empty space. Yet this discovery posed impossible problems for classical physics. Electrons could orbit an atomic nucleus to form an atom but, because they are accelerating, the electrons should radiate electromagnetic waves, lose energy, and plunge into the nucleus within 10^{-8} s. Material objects should immediately collapse by a factor of 10^{15} in volume while emitting a tremendous burst of radiation!

In 1913, Niels Bohr showed how the photon model of light could explain the stability of atoms. He assumed that an electron orbiting an atomic nucleus cannot radiate energy continuously; it emits or absorbs radiation one photon at a time, making abrupt transitions between orbits with different energies. Furthermore, for a given atom, only specific photon frequencies are possible.

The frequencies of light emitted or absorbed by a hydrogen atom can be described using a formula discovered by Johann Balmer in 1885:

$$f_{nn'} = f_o \left(\frac{1}{n^2} - \frac{1}{(n')^2} \right), \quad (35.9)$$

where n and $n' > n$ are integers and $f_o = 3.288 \times 10^{15}$ Hz. The spectral lines corresponding to $n = 2$, called the Balmer series, lie in the visible (■ Figure 35.15). The Lyman series, $n = 1$, is in the ultraviolet. The Paschen series, $n = 3$, and the Brackett series, $n = 4$, lie in the infrared.

In Bohr's model, a photon with energy $hf_{nn'}$ is emitted when an electron makes a transition from orbit n' to orbit n (■ Figure 35.16). Since only certain photon energies are emitted,

SEE CHAPTER 36 FOR MORE DETAILS OF NUCLEAR STRUCTURE.

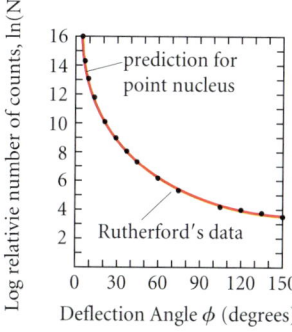

■ FIGURE 35.14
Rutherford-scattering data. The dots represent scattering rates of α-particles passing through gold foil, as measured by Rutherford's students, Geiger and Marsden. The solid line is the theoretical expectation for scattering by a point nucleus.

ACTUALLY, BALMER'S FORMULA HAD $n = 2$. LATER WORKERS GENERALIZED HIS RESULT TO ARBITRARY VALUES OF n.

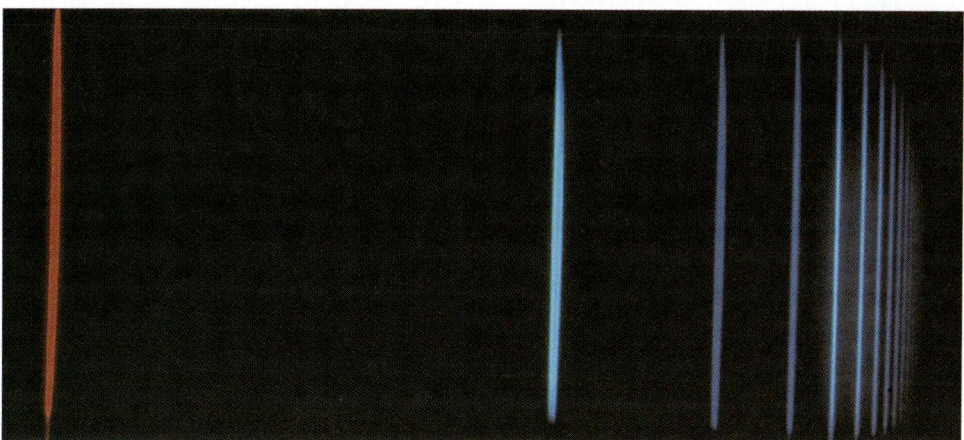

■ FIGURE 35.15
The hydrogen spectrum. The lines that appear in the visible are called the Balmer sequence: the brightest line, at the red end of the spectrum, is Balmer alpha. It corresponds to $n = 2$, $n' = 3$ in eqn. (35.9). The rest of the lines are in the blue part of the spectrum.

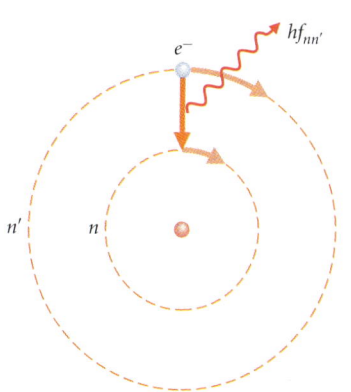

■ FIGURE 35.16
Photon emission in Bohr's model of the hydrogen atom corresponds to an electron jumping from one allowed orbit, labeled by integer n', down to another, labeled by integer n.

the electron orbits in hydrogen are also restricted. Balmer's formula suggests that the possible electron orbits in a hydrogen atom can also be labeled by an integer n. Since each orbit corresponds to a specific value of the electron's angular momentum, Bohr assumed that the angular momentum is quantized.

THE SYMBOL $\hbar$ IS READ "H BAR." ITS VALUE IS $\hbar = h/(2\pi) = 1.0546 \times 10^{-34}$ J·s.

> The angular momentum of an electron in its orbital motion is an integer multiple of Planck's constant divided by 2π.
>
> $$L = nh/(2\pi) \equiv n\hbar. \qquad (35.10)$$

Working from this bold assumption, Bohr used *classical* (Newtonian) physics to investigate the properties of the atom. Each electron moves in a circular orbit of radius r_n with speed v_n. Bohr's principle (35.10) imposes a relation between the radius of the nth orbit and the electron's speed v_n in that orbit (■ Figure 35.17):

$$L = mv_n r_n = nh/(2\pi) \equiv n\hbar.$$

WE USE NONRELATIVISTIC EXPRESSIONS FOR L AND K. WE'LL CHECK THE VALIDITY OF THIS LATER.

The electric (Coulomb) force due to the positively charged nucleus causes the electron's acceleration as it moves in a circle:

$$ke^2/r_n^2 = mv_n^2/r_n. \qquad (i)$$

Combining these two relations, we find:

$$ke^2 = \frac{(mv_n r_n)^2}{mr_n} = \frac{(n\hbar)^2}{mr_n}.$$

Solving for the orbit radius:

$$r_n = n^2 \frac{\hbar^2}{mke^2}. \qquad (35.11)$$

Using eqn. (i), the kinetic and potential energies of the electron are:

$$K_n \equiv \tfrac{1}{2}mv_n^2 = \tfrac{1}{2}ke^2/r_n \quad \text{and} \quad U_n = -ke^2/r_n.$$

So, the total energy of the electron is:

$$E_n \equiv K_n + U_n = -\tfrac{1}{2}ke^2/r_n. \qquad (35.12)$$

Using eqn. (35.11) for r_n, the total energy of the electron is:

$$E_n = -\frac{m(ke^2)^2}{2\hbar^2}\left(\frac{1}{n^2}\right). \qquad (35.13)$$

Thus using the assumption that angular momentum is quantized together with classical ideas about motion, Bohr obtained expressions for both the orbital radius r_n and the energy in that orbit E_n that depend solely on fundamental constants (e, m, k, $\hbar$) and an integer n.

When an electron drops from orbit n' to orbit $n < n'$, its energy decreases. A photon is emitted and carries away the energy lost by the electron:

$$hf_{nn'} = E_{n'} - E_n = \frac{m(ke^2)^2}{2\hbar^2}\left[\frac{1}{n^2} - \frac{1}{(n')^2}\right]. \qquad (35.14)$$

Thus Balmer's empirical formula for the hydrogen spectrum arises from the principle that angular momentum exists in multiples of a fundamental quantum $\hbar$.

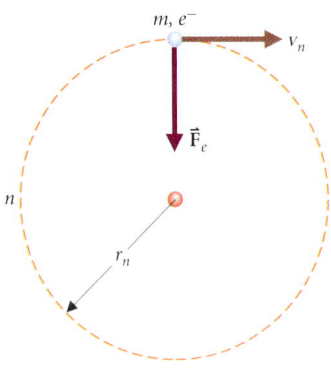

■ FIGURE 35.17
Calculation of allowed orbits in Bohr's model for the hydrogen atom. The electron follows a circular orbit under the influence of the Coulomb force due to the nucleus. Bohr assumed that angular momentum $L_n = mv_n r_n$ is quantized, $L_n = n\hbar$, and thus there is a specific relation between the radius r_n of the nth orbit and the electron's speed v_n in that orbit.

We may also write eqn. (35.14) as:

$$E_{n'} - E_n = E_o \left[\frac{1}{n^2} - \frac{1}{(n')^2} \right],$$

with
$$E_o = \frac{m(ke^2)^2}{2\hbar^2} = \frac{m(ke^2)^2(2\pi)^2}{2h^2} = \frac{me^4}{8\epsilon_0^2 h^2} \qquad (35.15)$$

$$= \frac{(9.109 \times 10^{-31} \text{ kg})(1.602 \times 10^{-19} \text{ C})^4}{8(8.854 \times 10^{-12} \text{ F/m})^2(6.626 \times 10^{-34} \text{ J·s})^2}$$

$$= 2.180 \times 10^{-18} \text{ J} = 13.6 \text{ eV}.$$

Each possible state of a hydrogen atom is labeled by the integer n of the electron orbit and is represented in an *energy-level diagram* (■ Figure 35.18). The vertical axis represents the energy of the atom, while the horizontal axis is used only for convenience in representing processes with the diagram. The lowest possible energy ($n = 1$) occurs in the *ground state* of the atom. States with $n > 1$ are called *excited* states. When an electron makes a transition between two states, $n_i \to n_f$, it is accompanied by emission ($n_f < n_i$) or absorption ($n_f > n_i$) of a photon. States with finite n have negative energy; the electron is bound to the atom. A state with positive energy represents a free electron passing by an isolated proton with positive kinetic energy and zero potential energy. The energy required to raise an electron from the ground state ($n = 1$) to $n' = \infty$ is called the *ionization potential* $\chi = E_o$. When a photon with energy $hf > \chi$ is absorbed by an atom, an electron is ejected, and the atom is said to be *ionized*.

COMPARE WITH OUR DISCUSSIONS OF GRAVITATIONAL (§8.2.2) AND ELECTRICAL (CHAPTER 25) POTENTIAL ENERGY.

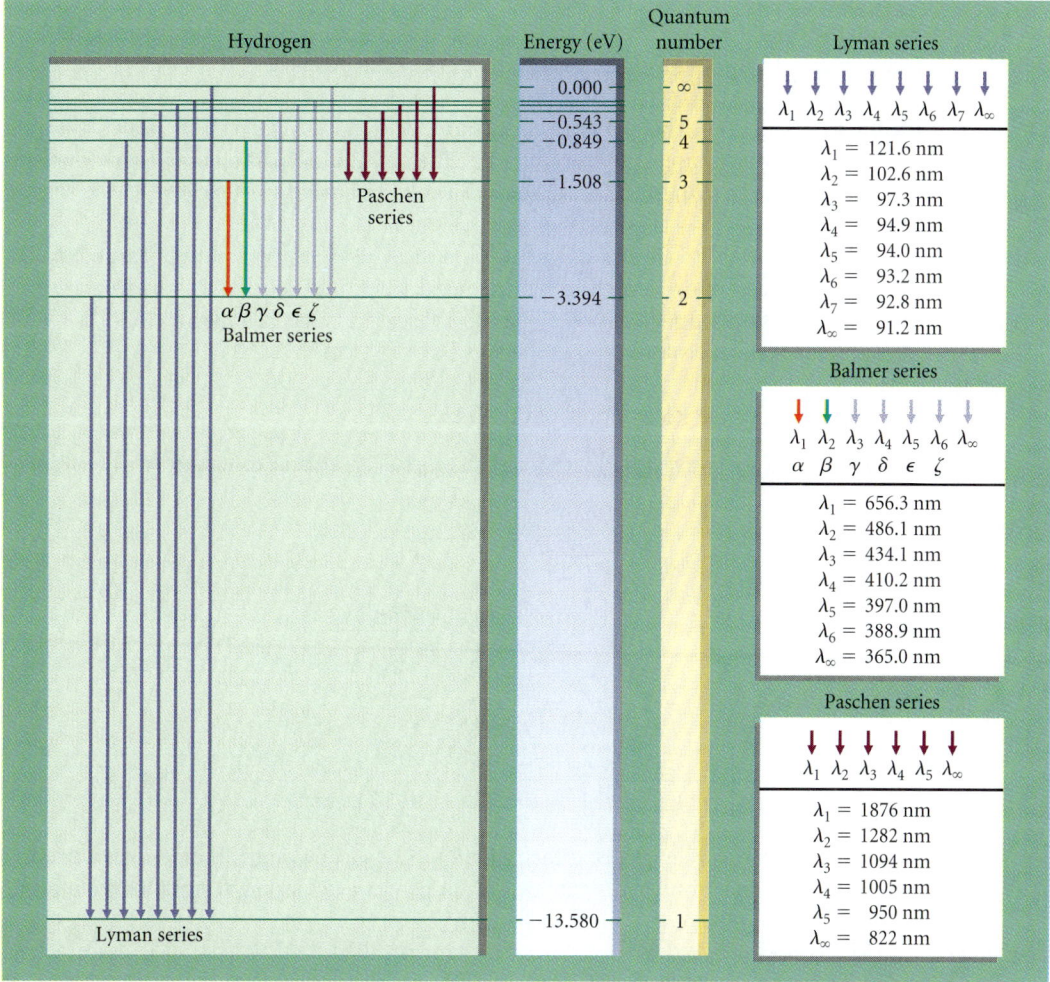

■ FIGURE 35.18
Energy-level diagram for hydrogen, according to the Bohr model, showing the transitions responsible for some of the observed lines. The levels corresponding to large n are closer together than those corresponding to small n. Transitions between closely spaced levels with large n values produce lines in the radio range.

NOTE: r_1 IS OFTEN WRITTEN a_o AND IS CALLED THE *BOHR RADIUS*.

EXAMPLE 35.7 ♦♦ Find the radius of the first Bohr orbit ($n = 1$) for hydrogen and the speed of an electron in that orbit.

MODEL Equation (35.11) gives the radius. We expect the answer to be of the order of 10^{-10} m. Knowing the radius, we apply Bohr's condition on angular momentum to get the speed.

SETUP Setting $n = 1$ in eqn. (35.11), we have:

$$r_1 = \frac{\hbar^2}{mke^2}$$

SOLVE
$$r_1 = \frac{(1.0546 \times 10^{-34} \text{ J·s})^2}{(9.1094 \times 10^{-31} \text{ kg})(8.9876 \times 10^9 \text{ N·m}^2/\text{C}^2)(1.6022 \times 10^{-19} \text{ C})^2}$$

$$= 0.5292 \times 10^{-10} \text{ m}.$$

The speed is:

$$v_1 = \frac{\hbar}{mr_1} = \frac{1.05 \times 10^{-34} \text{ J·s}}{(9.11 \times 10^{-31} \text{ kg})(0.529 \times 10^{-10} \text{ m})}$$

$$= 2.18 \times 10^6 \text{ m/s}.$$

ANALYZE Since this speed is less than 1% that of light, we are justified in ignoring special relativity in the Bohr theory for hydrogen. ∎

EXERCISE 35.4 ♦ Verify that to three significant figures Bohr's theory gives the correct numerical value for the constant f_o in Balmer's formula.

EXERCISE 35.5 ♦ The dimensionless number $\alpha \equiv ke^2/(\hbar c) = \frac{1}{137}$, called the *fine structure constant*, is useful in comparing values of atomic parameters. Show that the energy E_o of an electron in the first Bohr orbit is $\frac{1}{2}\alpha^2 mc^2$, and $v_1 = \alpha c$. ∎

EXAMPLE 35.8 ♦ Clouds of hydrogen in the Milky Way typically appear red (■ Figure 35.19). Find the wavelength of light emitted when the electron in a hydrogen atom makes a transition from the $n = 3$ state to the $n = 2$ state (*Balmer* α), and use your answer to explain the color.

MODEL We use the Bohr model of the atom.

SETUP First, we convert Balmer's formula (eqn. 35.9) to an expression for wavelength:

$$f_{nn'} = f_o \left[\frac{(n')^2 - n^2}{(nn')^2} \right].$$

So:

$$\lambda_{nn'} = \frac{c}{f_{nn'}} = \frac{c}{f_o} \left[\frac{(nn')^2}{(n')^2 - n^2} \right].$$

SOLVE With $n = n_f = 2$ and $n' = n_i = 3$,

$$\lambda_{23} = \frac{c}{f_o} \left[\frac{(nn')^2}{(n')^2 - n^2} \right] = \frac{(2.998 \times 10^8 \text{ m/s})(2^2)(3^2)}{(3.288 \times 10^{15} \text{ Hz})(3^2 - 2^2)} = 656 \text{ nm}.$$

ANALYZE This wavelength is at the red end of the visible range. Balmer α is a prominent emission line from hydrogen gas at temperatures of a few thousand kelvin, and accounts for the red color.

The constant $f_o/c = R = 1.097 \times 10^7$ /m is called the *Rydberg*. It is useful when using the Balmer formula to compute wavelengths. ∎

Niels Bohr, 1885–1962

EXAMPLE 35.9 ♦ What is the minimum frequency of a photon that can ionize a hydrogen atom from the $n = 3$ state?

MODEL The photon would just raise the atom to the $n' = \infty$ state; we find its frequency from Balmer's formula.

SETUP AND SOLVE
$$f_{\min} = f_o\left[\frac{1}{3^2} - \frac{1}{\infty^2}\right] = \frac{f_o}{9} = 3.65 \times 10^{14} \text{ Hz}.$$

ANALYZE The light would be described as infrared (cf. Table 16.2). The corresponding wavelength is $c/f = 0.82\ \mu\text{m}$.

An atom from which all electrons but one have been stripped is called a *hydrogenic atom*. Bohr's model is easily modified to apply to any such atom. If the charge on the atomic nucleus is Ze, then the force between the nucleus and electron is given by $F = kZe^2/r^2$. All other results are also modified by replacing the factor ke^2 by kZe^2.

UPON BEING TOLD THAT BOHR'S MODEL CORRECTLY PREDICTS E_o FOR HELIUM, EINSTEIN REMARKED, "THEN IT IS ONE OF THE GREATEST DISCOVERIES."

EXERCISE 35.6 ♦♦ Repeat the calculations of the Bohr model for a hydrogenic atom with nuclear charge Ze. Show that the radius of the nth orbit is reduced by a factor Z and the binding energy of the atom is multiplied by Z^2.

EXAMPLE 35.10 ♦ Hot gas in distant clusters of galaxies emits an x-ray spectral line with a photon energy of 6.7 keV. If the emission is from the $n' = 2$ to $n = 1$ state in a hydrogenic atom, what is the nuclear charge of the atom?

MODEL We use the expression for photon energy from Bohr's model, multiplied by the square of the nuclear charge (see Exercise 35.6).

SETUP Thus:
$$Z^2 E_o\left(\frac{1}{1^2} - \frac{1}{2^2}\right) = 6.7 \text{ keV},$$
$$Z^2 E_o(\tfrac{3}{4}) = 6.7 \text{ keV}.$$

SOLVE So:
$$Z = \left[\frac{4(6.7 \times 10^3 \text{ eV})}{3(13.6 \text{ eV})}\right]^{1/2} = 26.$$

ANALYZE This is the charge of an iron nucleus. Astronomers use observations of the 6.7-keV line to estimate the amount of iron in the gas. That information provides important clues about the origin of the gas in the cluster.

FIGURE 35.19
The central portion of the Eta Carinae Nebula (also known as the Keyhole Nebula). The red color is due to emission from the Balmer α line of hydrogen.

EXERCISE 35.7 ♦ What is the second ionization potential of helium? That is, once the first electron is removed, how much energy is required to remove the second?

35.2.3 The Correspondence Principle

According to quantum theory, atoms undergo abrupt transitions between states as they absorb or emit photons, while classical theory considers continuous fields and continuous change in macroscopic systems. Both theories are very successful, and both describe the behavior of charged objects responding to electromagnetic forces. Bohr realized that classical theory must be a limiting case of a complete quantum theory; the two should give corresponding results for systems that are large on the atomic scale, though small on the macroscopic scale, or for systems in which the energy of a single photon is negligible.

We have seen an example of the correspondence principle in the spectrum of thermal radiation. In the low-frequency limit, $hf \ll kT$, a single photon's energy is much less than the typical thermal energy, and the classical Rayleigh–Jeans formula gives the correct spectrum. A second example is provided by radiation from electrons in Bohr orbits with very large quantum numbers n. The frequency of a photon emitted in a transition from $n' = n + 1$ to n is given by:

$$f_{n+1,n} = f_o \left[\frac{1}{n^2} - \frac{1}{(n+1)^2} \right] = f_o \frac{2n+1}{n^2(n+1)^2} \approx \frac{2f_o}{n^3} \quad \text{for } n \gg 1.$$

Classically, the radiation frequency should equal the frequency of the electron in its orbit: $f_c = v/(2\pi r_n)$. Using Bohr's principle to express v in terms of n and r_n, and eqn. (35.11) for r_n:

$$f_c = \frac{n\hbar}{2\pi m r_n^2} = \frac{n\hbar}{2\pi m} \left(\frac{mke^2}{n^2 \hbar^2} \right)^2.$$

Next, we substitute eqn. (35.15) for $E_o = hf_o = 2\pi\hbar f_o$:

$$f_c = \frac{(ke^2)^2 m}{2\pi \hbar^3 n^3} = \frac{E_o}{\pi n^3 \hbar} = \frac{2f_o}{n^3}.$$

The classical and quantum results correspond when n is large.

Atoms having electrons in large n states (called *highly excited* atoms) are big ($r \gg r_1$) and fragile. They do not occur naturally on Earth. In 1965, astronomers detected radiation from hydrogen atoms in interstellar space undergoing transitions between levels with n near 100. Values of n as large as 350 have since been discovered. Large values of n (10–100) can be obtained in the laboratory using a laser to excite the atoms. These atoms provide an excellent vehicle for testing the correspondence principle.

35.3 ELECTRON WAVES
35.3.1 De Broglie's Hypothesis

Bohr's atomic model can explain the stability of atoms and connect atomic structure with the photon description of light. But what constrains the angular momentum of an electron? The answer, discovered by Louis de Broglie in 1923, is that electrons, like photons, have both a wave and a particle nature. When a wave is confined in a system with finite dimensions, a standing wave pattern is formed. Certain wavelengths are reinforced by reflections at the boundaries of the system, while other wavelengths interfere destructively. Each state of a hydrogen atom corresponds to a different standing-wave pattern (■ Figure 35.20).

The energy and momentum of a photon are related to the frequency and wavelength of the corresponding electromagnetic wave (eqn. 35.1):

$$E = hf$$

and

$$p = E/c = hf/c = h/\lambda.$$

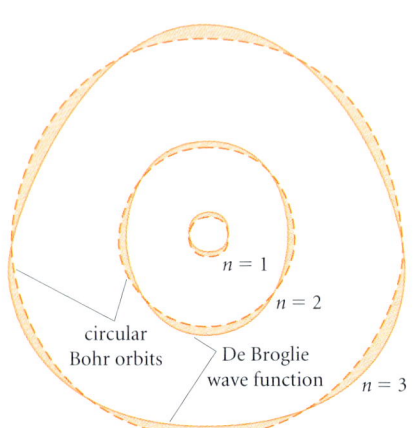

■ **FIGURE 35.20**
De Broglie waves and the Bohr theory. Exactly n de Broglie waves fit into the circumference of the nth Bohr orbit. If the wavelength were slightly longer or shorter, the wave would suffer destructive interference as it wrapped around the orbit. This picture serves as a useful image of how quantization occurs: it is not an accurate model. Historically, it served as a transition between the Bohr and Schrödinger descriptions.

COMPARE §15.4.2 AND §33.4.

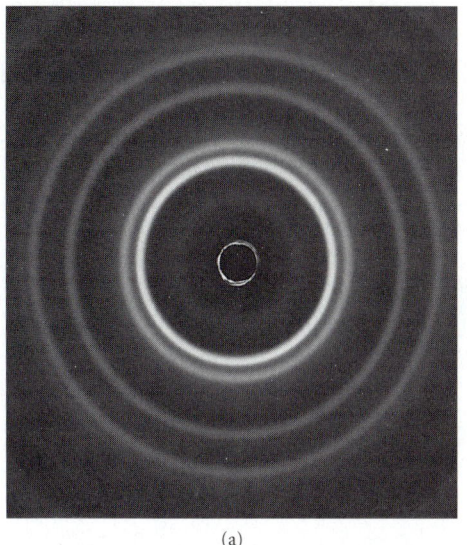
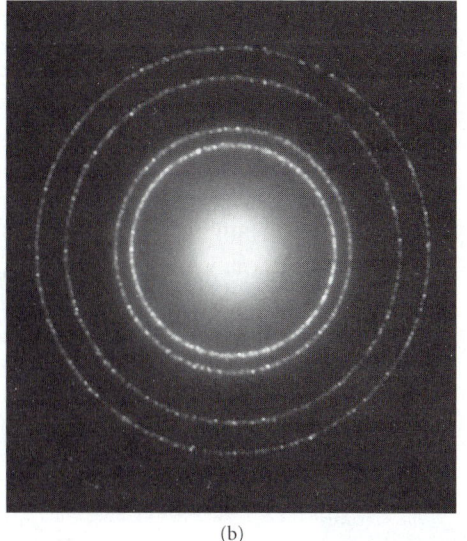

(a) (b)

■ **FIGURE 35.21**
The wave nature of electrons is confirmed by the diffraction pattern produced when electrons scatter off crystals. (a) The x-ray diffraction pattern of aluminum. A lead disk was placed over the center of the film for most of the exposure to avoid overexposing the film in these bright regions. (b) Diffraction pattern of aluminum using electrons. Both patterns show the characteristic diffraction rings.

The frequency and wavelength of a particle's associated de Broglie wave are related to the particle's total energy $E = \gamma mc^2$ and momentum $p = \gamma mv$ in the same way:

$$f = E/h \quad \text{and} \quad \lambda = h/p. \tag{35.16}$$

When an electron beam scatters from a crystal surface, a diffraction pattern is observed (■ Figure 35.21). Since diffraction is purely a wave phenomenon and cannot be explained with particles, this observation provides direct experimental confirmation of de Broglie's hypothesis.

EXERCISE 35.8 ◆ Show that the wave speed $v_d = \lambda f$ of a de Broglie wave $= c^2/v$.

We may use relations (35.16) to show that exactly n de Broglie wavelengths fit into the nth Bohr orbit. With $p = mv_n$, and L from eqn. (35.10):

$$\frac{2\pi r_n}{\lambda_n} = \frac{2\pi r_n}{h/mv_n} = \frac{2\pi}{h} mv_n r_n = \frac{L}{\hbar} = \frac{1}{\hbar}\hbar n = n.$$

EXAMPLE 35.11 ◆ What is the wavelength of electrons in a television set that have a kinetic energy of 35 keV?

MODEL Since the kinetic energy is much less than the 511-keV rest energy of an electron, we may use the Newtonian formula for the electron momentum in eqn. (35.16).

SETUP $K = \frac{1}{2}mv^2,$ and $p = mv = m\sqrt{2K/m} = \sqrt{2mK}.$

SOLVE Then:

$$\lambda = \frac{h}{p} = \frac{h}{\sqrt{2mK}}$$

$$= \frac{6.63 \times 10^{-34} \text{ J} \cdot \text{s}}{[2(9.11 \times 10^{-31} \text{ kg})(3.5 \times 10^4 \text{ eV})(1.6 \times 10^{-19} \text{ J/eV})]^{1/2}}$$

$$= 6.6 \times 10^{-12} \text{ m}.$$

ANALYZE This is some 10^5 times smaller than the wavelengths of visible light. Electron microscopes can achieve much greater resolution than is available with light microscopes, since wave diffraction is no longer a limiting factor.

DIFFRACTION AND ITS EFFECTS ON RESOLUTION ARE DISCUSSED IN §17.3.

35.3.2 Schrödinger's Picture of the Hydrogen Atom

The work of Bohr and de Broglie led to a qualitative understanding of atomic structure and provided quantitative results for simple atoms. Bohr's model of the atom explained both Rutherford's data and the hydrogen spectrum, and de Broglie's waves offered a believable mechanism for quantization. However, careful examination of the spectra of certain elements such as sodium showed that some of the spectral lines are split—the line is actually two lines with a very small separation in frequency (■ Figure 35.22). Bohr's theory could not account for this

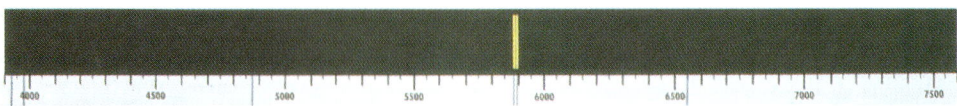

■ **Figure 35.22**
The bright sodium line in the yellow part of the spectrum is actually two very closely spaced lines. Such line *splitting* is not explained by Bohr's model.

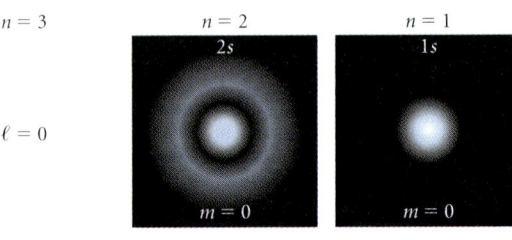

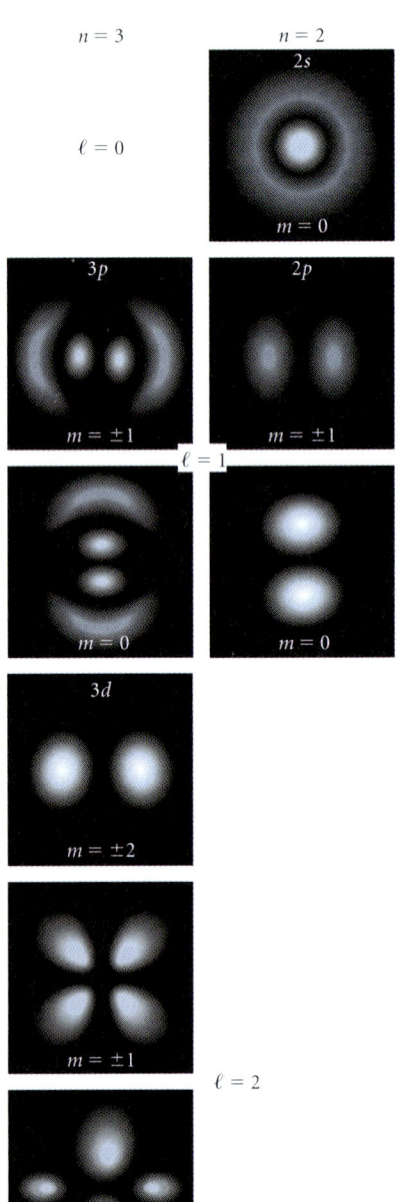

■ **Figure 35.23**
Electron wave functions in hydrogen. The $n = 1$ state has only one possible value of ℓ—0. This is called the 1s state. The wave function is spherically symmetric and $|\psi|^2$ is greatest at the origin. When $n = 2$, there are two possible values of ℓ. The 2s state has $\ell = 0$ and the 2p state has $\ell = 1$. With $\ell = 1$, there are three possible values of m: 0 and ± 1. All $\ell = 0$ states are spherically symmetric. Note that as ℓ increases, the distribution becomes less centrally concentrated. In general, ψ is a complex-valued function; its magnitude is real-valued.

phenomenon, nor could it explain the structure of atoms with many electrons. In 1926, Erwin Schrödinger produced a theory capable of explaining the distribution of electrons in atoms with $Z > 1$ and many details of atomic spectra. His theory is valid where relativistic effects are unimportant, and it is used for fundamental calculations in chemistry and atomic physics. We shall not perform any calculations with Schrödinger's theory, but we can discuss its qualitative consequences.

Schrödinger's equations describe the behavior of the electron wave function ψ. A pulse of light can be modeled with photons or EM waves and, similarly, a particle such as an electron can be described by a matter (de Broglie) wave. Just as the energy density in electromagnetic fields is proportional to the square of the field amplitudes, it is the squared magnitude of the wave function that describes electron density. One may view the density $|\psi|^2$ at a point as the probability of an electron being in a unit volume surrounding the point, or as the fraction of time the electron spends in such a unit volume, or as a description of the charge density in an electron *cloud* surrounding the atomic nucleus. ■ Figure 35.23 illustrates this cloud for a few of the lowest energy states in hydrogen.

In a hydrogen atom, the electron is modeled with a three-dimensional standing wave, described by three integers that give the number of nodes of the pattern in each of the coordinate directions. In spherical coordinates, Bohr's n, the *principal* quantum number, corresponds to the radial coordinate. Nodes in the polar (θ) direction are described by ℓ and nodes in the azimuthal (ϕ) direction are described by m. The energy of each state depends on n and, except for small corrections we'll discuss later, is the same as that predicted by Bohr (eqn. 35.13). In Schrödinger's theory, ℓ and m are related to angular momentum. The total angular momentum of the electron is:

$$L = \sqrt{\ell(\ell + 1)}\hbar, \tag{35.17}$$

and the z-component of the angular momentum is $m\hbar$. The numbers ℓ and m satisfy the restrictions:

$$0 \leq \ell \leq n - 1 \quad \text{and} \quad -\ell \leq m \leq \ell. \tag{35.18}$$

Thus, for each value of ℓ, there are $2\ell + 1$ possible values of m (ℓ negative values, ℓ positive values, and zero). For $n = 1$ (the ground state), both ℓ and m are zero. Thus $L = 0$ in the ground state, whereas in Bohr's approximate model $L = \hbar$. The set of possible states in Schrödinger's theory is much richer than in the Bohr model it replaced, but the basic concept—quantization of energy and angular momentum—remains fundamental. Schrödinger found that quantization is not an ad hoc assumption, but a direct consequence of the electron's wave nature.

■ Figure 35.24 is an energy-level diagram for hydrogen, including all the levels found in the Schrödinger theory. The diagram is very similar to that for the Bohr theory, except that for $n > 1$, each level is a composite of states with different values of ℓ and m. Careful measurements of the hydrogen spectrum show that states with different ℓ actually have slightly different energies.

This *fine structure* in the spectrum arises from relativistic motion and electron spin. Modeling the electron as a small, rotating, charged ball is a useful visualization so long as you do not take it too literally. Such a ball has a magnetic moment and behaves like a small magnetic dipole. The electron's magnetic moment is parallel or antiparallel to its orbital angular momentum. Thus an individual electron has a spin angular momentum described by the quantum number $s = \frac{1}{2}$ and with z-component of spin described by the quantum number m_s, which takes only the values $\pm\frac{1}{2}$. In a reference frame where the electron is at rest at any instant, the positively charged nucleus is in motion. The moving nucleus acts as an electric

SEE § 29.4.1 FOR THE RELATION BETWEEN MAGNETIC MOMENT AND ANGULAR MOMENTUM.

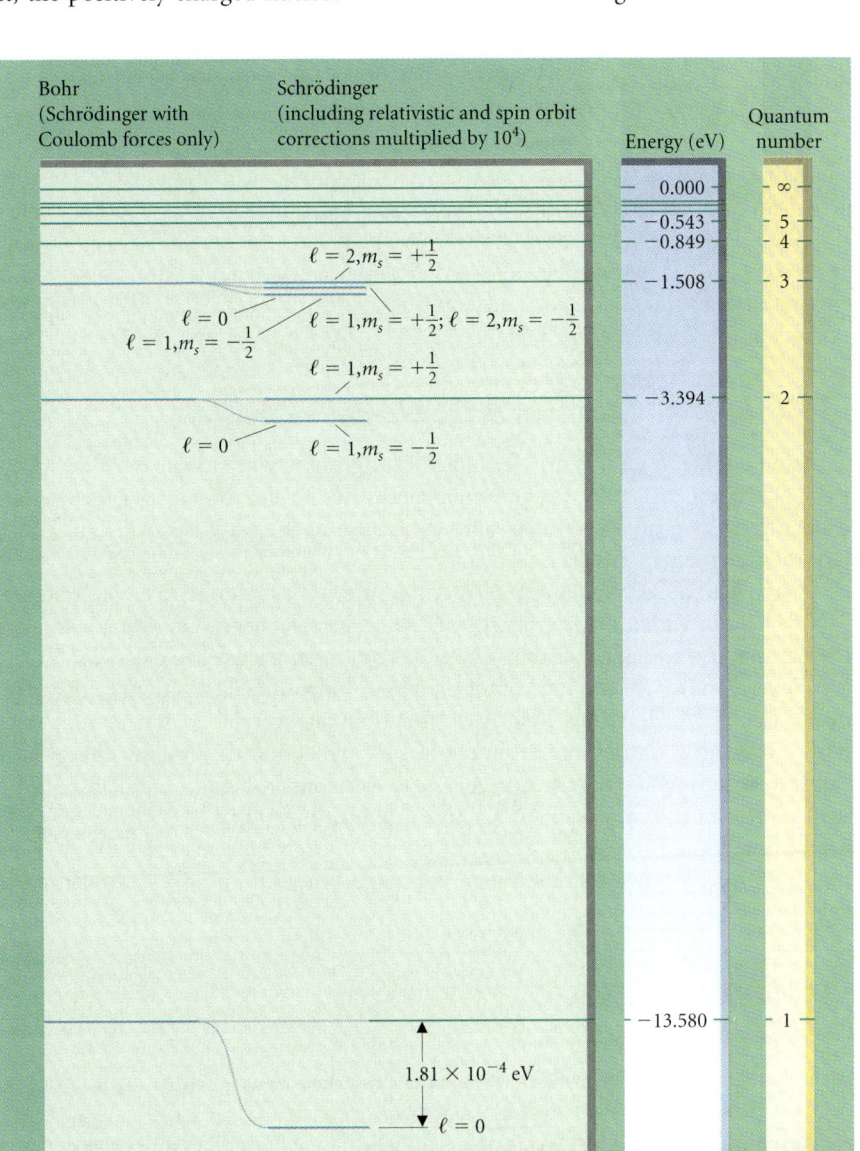

■ **FIGURE 35.24**
Energy-level diagram for hydrogen. The Bohr and Schrödinger theories make identical predictions when relativity is ignored and only Coulomb forces are considered (left column). Using the correct relativistic expression for the electron's energy and momentum, and including the electron's magnetic properties in Schrödinger's theory, we find small shifts and splittings of the levels, as illustrated in the right column of the diagram for $n \leq 3$. These effects are about 10^4 times smaller than the energy differences between Bohr levels and are shown greatly exaggerated for clarity. These predictions of Schrödinger's theory are in excellent agreement with measurements.

■ FIGURE 35.25
Zeeman splitting of hydrogen absorption lines in the spectrum of a white dwarf star. Compare the Balmer ($n = 2$) absorption features at 435 and 485 nm in the two spectra shown. The strong magnetic field (about 300 T) of the white dwarf PG 1658+441 causes different m states to have slightly different energies and results in the observed triple line structure. *Note:* These stars are known only by their numbers in specialized astronomical catalogues.

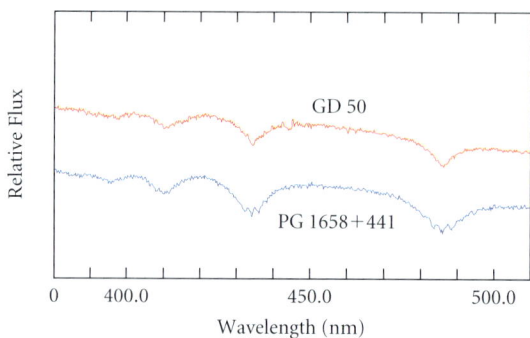

AFTER PIETER ZEEMANN (1865–1943).

THE PERIODIC TABLE IS INSIDE THE BACK COVER.

AFTER WOLFGANG PAULI (1900–1958).

BECAUSE OF ENRICO FERMI'S BASIC WORK ON THE RELATION BETWEEN A PARTICLE'S SPIN AND ITS THERMODYNAMIC BEHAVIOR, PARTICLES THAT SHARE THIS PROPERTY OF ELECTRONS ARE CALLED *FERMIONS*.

current and produces a magnetic field. As a result, the spinning electron has a small amount of magnetic potential energy. Energy differences caused by this spin–orbit interaction are of the order of 10^{-4} eV. In a spectroscope, the result is an apparent splitting of a spectral line into several closely spaced components.

Hyperfine structure due to the spin of the nucleus causes line splittings roughly 100 times smaller still. The electron energy changes when its spin flips between states parallel and antiparallel to the nuclear spin. Low-energy photons emitted as a result of such spin flips in the hydrogen ground state are very important in astronomy. The resulting radio waves with wavelength $\lambda = 21$ cm can traverse thousands of light-years of interstellar gas, allowing radio astronomers to observe distant regions of our galaxy that cannot be seen in visible light.

States with different quantum numbers respond differently to outside influences. Different values of m can be distinguished if an external magnetic field is applied to the atom. The magnetic field distinguishes one direction—parallel to the field—from all others. The orbital motion of the electron gives the atom as a whole a magnetic moment and a magnetic potential energy ($\approx 10^{-3}$ eV/T), which depends on the component of angular momentum parallel to the external field and hence on m. The resulting energy differences cause spectral line splittings known as the *Zeeman effect* (■ Figure 35.25).

35.3.3 The Pauli Exclusion Principle and Chemistry

In the periodic table, elements are organized into groups with similar chemical properties. Using the Schrödinger theory, we can understand why the elements form these groups. Carbon differs from silver and resembles silicon because of the ways electrons are arranged in atoms of these elements.

Hydrogen has the simplest atomic structure, since it has but one electron. The next simplest case is helium with a nuclear charge of $2e$ and two electrons. On average, the two electrons are farther from each other than either is from the nucleus. The electrical forces of the electrons on each other are less important than their individual interactions with the nucleus. Then each electron is described by a wave function very similar to the ground state wave function of hydrogen, though the orbital radius is smaller and the binding energy larger because of the greater charge on the nucleus. *Both* electrons in helium are held tightly by the nucleus and helium is a chemically unreactive (noble or inert) gas.

Even in helium, however, the electrons do not entirely behave as if each were alone. A new physical principle with no classical counterpart comes into play. It is the *Pauli exclusion principle*.

> No more than one electron in a system can occupy any quantum state.

The quantum state of an electron in an atom is described by its quantum numbers n, ℓ, m, and m_s. According to the Pauli principle, no two electrons in an atom can have the same values of all four numbers. The two electrons in helium may have the same principal quantum number $n = 1$ and the same angular momentum quantum numbers $\ell = m = 0$, but if they do, their spins *must* be oriented in opposite directions: $m_s = +\frac{1}{2}$ and $m_s = -\frac{1}{2}$. These quantum numbers describe the ground state of helium.

In atoms with three or more electrons, the exclusion principle forces electrons to be in higher energy states. In lithium, as in helium, the first two electrons fill the $n = 1$ state with opposite spins. The third electron must enter a much less tightly bound state with $n = 2$, $\ell = m = 0$. Lithium is a highly reactive metal that readily shares the one loosely bound electron in chemical bonds. Beryllium's four electrons fill the $n = 2$, $\ell = m = 0$ state. It is less reactive than lithium, but much more so than helium. Six electrons can be put in the three states with $n = 2$, $\ell = 1$, and $m = -1, 0,$ or $+1$. These correspond to the elements from boron through neon (● Table 35.2). Neon, with all the $n = 2$ states filled is, like helium, an inert gas. When all the quantum states in a certain n level are filled in this way, we say the atom has a *closed shell*. Fluorine, lacking one electron from such a complete shell, is a highly corrosive gas that, in reactions, takes electrons from atoms such as lithium.

The entire periodic table can be understood in terms of the quantum numbers of the electrons. After the third period, however, the mutual electrical forces among the electrons are important in determining the order in which the states are filled. We use the distributions shown in Figure 35.23 to see how this happens.

In states with $\ell = 0$, the wave function is spherically symmetric and has large values close to the origin. When ℓ is not zero, the wave function vanishes at the origin and is concentrated at a distance r, which increases with ℓ. In a multielectron atom, the presence of other electrons

TABLE 35.2

Element	Z	$n=1$	$n=2$		$n=3$			$n=4$
		$\ell=0$	$\ell=0$	$\ell=1$	$\ell=0$	$\ell=1$	$\ell=2$	$\ell=0$
H	1	1						
He	2	2	$n=1$ shell filled					
Li	3	2	1					
Be	4	2	2					
B	5	2	2	1				
C	6	2	2	2				
N	7	2	2	3				
O	8	2	2	4				
F	9	2	2	5				
Ne	10	2	2	6	$n=1$ and $n=2$ shells filled			
Na	11	2	2	6	1			
Mg	12	2	2	6	2			
Al	13	2	2	6	2	1		
Si	14	2	2	6	2	2		
P	15	2	2	6	2	3		
S	16	2	2	6	2	4		
Cl	17	2	2	6	2	5		
Ar	18	2	2	6	2	6		
K	19	2	2	6	2	6		1
Ca	20	2	2	6	2	6		2
Sc	21	2	2	6	2	6	1	2

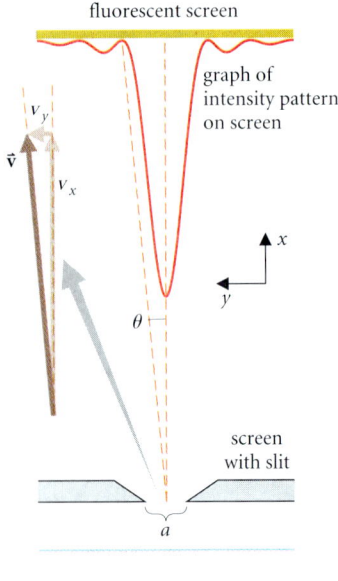

■ FIGURE 35.26
Measuring the position of an electron beam with a horizontal slit of width a. The uncertainty in the position is $\delta y = a$. A diffraction pattern forms on the screen. In order to reach a point on the screen at angle θ from the slit, the y-component of the electron's velocity must satisfy $v_y/v = \sin \theta$. The uncertainty in the electron's position within the diffraction maximum corresponds to an uncertainty $\delta p_y = mv_y = mv \sin \theta = p\lambda/a = h/a$ in the electron's momentum.

has the effect of shielding the nucleus, reducing its effective charge. This shielding effect is greater for states with higher ℓ since they are concentrated farther from the center of the distribution. Electron states with greater ℓ see a smaller effective nuclear charge and have a correspondingly smaller binding energy. In the fourth period of the periodic table, this effect becomes sufficiently pronounced that the $n = 4$, $\ell = 0$ states fill before the $n = 3$, $\ell = 2$ states.

Since 1926, Schrödinger's picture has been developed into a complete theory of chemical bonding. Computational complexity is now the main limit to understanding chemical reactions from first principles.

35.4 QUANTUM MECHANICS

The success of the photon and electron wave models in explaining atomic phenomena leaves no doubt that dual wave/particle behavior is a fundamental aspect of nature. That duality at the atomic level does not contradict our experience of distinct particles and waves on the macroscopic scale, but the duality introduces new principles that we don't encounter in the everyday world. We conclude our discussion of atomic physics by showing that wave mechanics imposes fundamental limits on measurement accuracy and on our ability to predict the behavior of physical systems.

35.4.1 The Heisenberg Uncertainty Principle

Suppose we attempt to measure the y-coordinate of electrons by passing a beam through a long, horizontal slit (■ Figure 35.26). The experiment determines the position of each electron in the beam with an uncertainty equal to the width a of the slit: $\delta y = a$. At the same time, each electron's wave function is diffracted by the slit and forms the usual single-slit diffraction pattern (cf. §17.3.1). Almost all the electrons strike the screen within the central diffraction maximum. This means that each individual electron leaves the slit at some unknown angle between 0 and θ, where, for electrons of wavelength λ, θ is given by $\sin \theta = \lambda/a$. Uncertainty in the y-direction of the electron's motion corresponds to uncertainty in the y-component of its momentum, given by:

$$\delta p_y \sim p \sin \theta = p\lambda/a.$$

Using de Broglie's formula for the electron wavelength, $\lambda = h/p$, we find:

$$\delta p_y \sim h/a,$$

or

$$\delta y \, \delta p_y \sim h.$$

In principle, the accuracy of the position measurement can be improved indefinitely by decreasing the width of the slit. However, there is a penalty for increased accuracy; the uncertainty in momentum increases in proportion as the central diffraction maximum expands. Improving our knowledge of the electron's position inevitably reduces the information available about its momentum.

Werner Heisenberg showed in 1927 that this limit is not a peculiar feature of any particular experiment but a basic feature of wave/particle duality. To see why, let's investigate how waves superpose to form a localized, particlelike disturbance. A single plane wave has a well-defined wavelength λ, but spreads uniformly throughout space (■ Figure 35.27a). Superposing a second wave of slightly different wavelength results in constructive interference where the two waves are in phase and destructive interference where they are out of phase (Figure 35.27b). Because the probability of finding an electron at a point is proportional to the wave amplitude squared, such a wave superposition describes a somewhat localized particle. By superposing a large number of individual waves with a range of wavelengths, we can arrange to have constructive interference in one small region with destructive interference everywhere else (Figure 35.27c). This superposition represents a particle located within a range of positions δx and represented by waves with a range of momenta δp. The localization in space can be achieved only by superposing waves with a spread in wavelength. According to Fourier's theorem of wave superpositions, the range of wavelengths is related to the spatial localization by:

$$\delta x\,\delta(1/\lambda) \geq 1/(4\pi).$$

Since $p = h/\lambda$, $\delta(1/\lambda) = \delta p/h$, and so:

$$\delta x\,\delta p \geq \hbar/2. \tag{35.19}$$

This analysis shows that eqn. (35.19) is more than just a limit on the accuracy of measurement; it is a fundamental relation between actual physical properties of a system.

A similar general relation holds between uncertainties in energy and time measurements. If we observe the wave representation of a localized particle passing a fixed observation point, the time we would record for its passage is uncertain by:

$$\delta t \sim \delta x/v,$$

where $v = p/m$ is the speed of the particle. The particle's energy is uncertain by:

$$\delta E = \delta(p^2/2m) = (p\,\delta p)/m.$$

So, the product of uncertainties in energy and time is:

$$\delta E\,\delta t = \frac{p\,\delta p}{m}\left(\frac{\delta x}{v}\right) = \delta p\,\delta x \geq \frac{\hbar}{2}. \tag{35.20}$$

EXAMPLE 35.12 ♦♦ Spectral lines from atoms are not infinitely sharp. Atoms typically remain in one state for about 10^{-8} s before emitting a photon. Use Heisenberg's uncertainty relations to estimate the uncertainty in the energy of an emitted photon. Compare this uncertainty with the fine structure splitting.

MODEL The uncertainty in the energy of the photon is related to the uncertainty in the time of its emission, represented here by the typical time the atom remains in one state before emitting a photon (the *lifetime* of the state).

SETUP The uncertainty in the energy is given by eqn. (35.20):

$$\delta E_{\text{photon}} \sim \frac{\hbar}{2\,\delta t} = \frac{1.06 \times 10^{-34}\ \text{J}\cdot\text{s}}{2(10^{-8}\ \text{s})(1.6 \times 10^{-19}\ \text{J/eV})} = 3 \times 10^{-8}\ \text{eV}.$$

ANALYZE This is a few tenths of a percent of the fine structure splitting (§35.3.2) and is roughly comparable with hyperfine effects. ∎

EXERCISE 35.9 ♦ Excited iron nuclei emit the 14.4-keV x rays used in the Pound–Rebka experiment (cf. Essay 8: General Relativity). The excited state has a mean lifetime of 10^{-7} s. What is the uncertainty of the photon energy? Compare with the energy shift $\Delta E/E \sim 2 \times 10^{-15}$ that Pound and Rebka were trying to measure. ∎

35.4.2 The Meaning of the Wave Function

Atomic phenomena are statistical: experiments with a single atom suspended by laser radiation show transitions between states occurring at randomly distributed times. Quantum theory accurately predicts the average time interval but cannot predict when an individual transition will occur. This strange situation is inconsistent with classical physics. If we knew the position and velocity of every particle, we could imagine using Newton's laws to compute the complete future of the universe. In quantum mechanics, the uncertainty principle denies us accurate knowledge of both position and velocity, and only allows us to find the probability that any future state of the universe will occur.

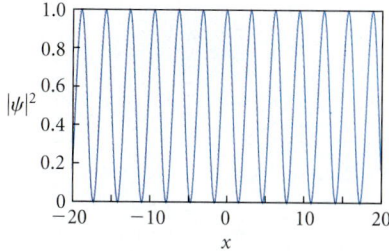

(a) Single wave

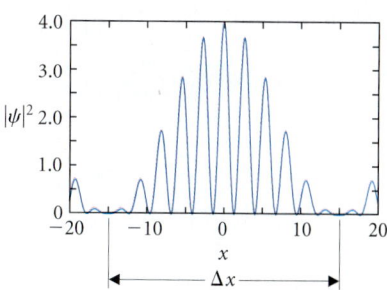

(b) Two waves

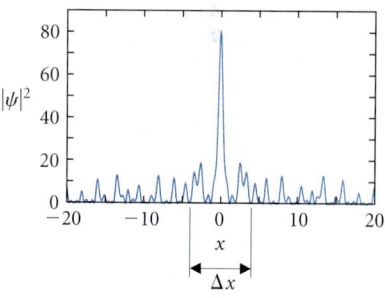

(c) Nine waves

■ **FIGURE 35.27**
Wave superposition. Each plot shows $|\psi|^2$ as a function of x. (a) A wave with a single precisely known wavelength is not localized at all in space. (b) The superposition of two waves with slightly different wavelengths produces a modulated wave: here $\Delta\lambda$ is one unit and the wave function is concentrated in the region $-15 \leq x \leq +15$. The pattern repeats periodically in both positive and negative x-directions. (c) Superposition of nine waves ($\Delta\lambda \approx 8$) produces a wave packet that is localized to a fairly small region about the origin. Again, the pattern repeats to left and right. An infinite number of waves must be superposed to eliminate this repetition and localize the particle at a single spot.

To see how this interpretation follows from wave/particle duality, consider the interference of electrons passing through a screen with two small holes (■ Figure 35.28) and striking a television screen. An intense electron beam produces a two-source interference pattern on the screen. If the intensity is reduced so that electrons arrive one at a time, isolated flashes occur randomly on the screen but never at minima of the diffraction pattern. How can we interpret such behavior? The experiment tells us that electric charge leaves the electron gun in a bundle and that sometime later a flash occurs on the screen, but only our theoretical description tells us what it is that propagates from the gun to the screen. It can't be a classical particle, which would have to pass through one hole or the other and couldn't show interference at all. What propagates is Schrödinger's wave function ψ, which undergoes two-source interference in the standard way. But the wave strikes the entire screen and the flash occurs at an isolated spot! The wave represents not a substance, in any sense we are used to, but the probability of an event occurring. The event—*electron leaves gun*—will be followed by a flash; a total probability of unity leaves the gun, propagates as a wave, and spreads out in an interference pattern to strike the screen. Once the event—*flash*—occurs somewhere, it has no further chance of occurring elsewhere. According to the standard interpretation of quantum mechanics, the wave function abruptly ceases to exist everywhere else.

The standard interpretation of quantum mechanics provides extremely accurate methods of calculation and has satisfied every experimental test yet devised, but it has provoked fierce philosophical debate. Einstein was never able to accept it as final, because it seems to deny that any objective real thing connects events such as *charge leaves gun* and *flash*. Measurement—locating the flash on the screen—occurs as a discontinuous event involving collapse and regeneration of the wave function describing the electron. The role of conscious observers in such measurements has sparked particularly lively debate. As an amusing and provocative introduction to such topics, we close the chapter with a parable devised by Schrödinger, which assumes the quantum theory applies to macroscopic systems as well as microscopic.

A cat is placed in an airtight box with an oxygen supply and with a glass vial containing cyanide gas to be released if a radiation detector is struck by a particle from a radioactive sample. The radioactive sample is a quantum system for which we can accurately compute the *probability* a particle will be released in a given time interval. We propose to place the cat in the box just long enough for the probability to be one-half, then open the box, and see what has happened.

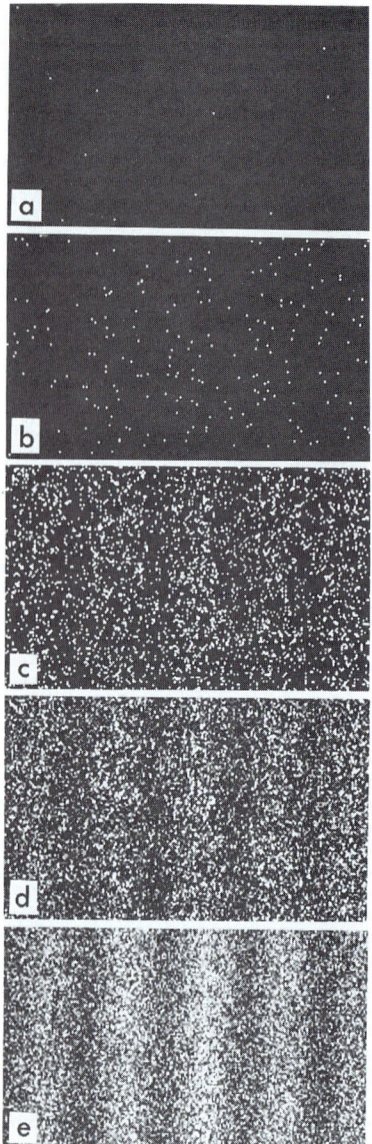

■ **Figure 35.28**
Electron interference. In this two-slit experiment, the electrons are separated by electric forces. Electrons leave their source at so slow a rate that two electrons are not present between the source and the detector at the same time. Thus each electron interferes with itself. Individual electrons striking the screen produce flashes of light. All the flashes, taken together, produce a two-source interference pattern. Most electrons strike the screen near the interference maxima. (a) 10 electrons have reached the screen; their positions appear completely random. (b) 200 electrons; an overall pattern is detectable if you know what to look for. (c) 3000 electrons; a striped pattern is easily visible in the distribution of impact points. The contrast between interference maxima and minima becomes more pronounced as the number of electrons increases. (d) 20 000 electrons. (e) 70 000 electrons.

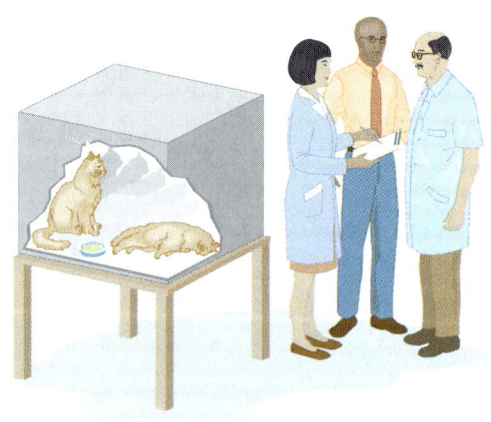

Schrödinger's cat. The cat has a 50% chance of surviving the experiment. According to quantum mechanics, the cage contains neither a live nor a dead cat, but an even mixture of live-cat and dead-cat states—whatever that means!

Erwin Schrödinger, 1887–1961

According to quantum mechanics, what is in the box just before you open it is neither a dead cat nor a live cat but a wave function describing equal probabilities for finding a live cat or a dead cat. When you open the box, the wave function collapses and you observe one or the other state. While you are pondering how strange this description is, your research supervisor approaches the lab room knowing that it contains a wave function describing equal probabilities of finding a researcher who will report a live cat and one who will report a dead cat . . .

Chapter Summary

Where Are We Now?

We have seen that a dual wave/particle model is needed to explain atomic phenomena. By applying the model in simple cases, we have been able to extract some fundamental results such as the uncertainty principle. In the next two chapters we shall use this model to discuss atomic nuclei and the many kinds of subatomic particle discovered in the quest for the ultimate building blocks of nature.

What Did We Do?

Light incident on certain metal surfaces ejects electrons with energy that depends on the light frequency. Einstein explained this *photoelectric effect* by assuming electromagnetic energy is absorbed as photons, particles whose energy is related to their wave frequency by:

$$E = hf.$$

When the energy of a photon is completely absorbed, an electron may be ejected with kinetic energy:

$$K = hf - \phi.$$

The work function ϕ is the energy required to remove an electron from the metal's surface. The probability that an electron is ejected is the quantum efficiency of the metal at that frequency.

The scattering of x rays by stationary electrons—the *Compton effect*—is best understood as elastic particle collisions between photons and free electrons. The momentum of a photon is:

$$p = E/c = hf/c = h/\lambda.$$

In collisions between photons and stationary electrons, the photon wavelength is increased as its energy is decreased:

$$\lambda_2 - \lambda_1 = \lambda_C(1 - \cos \theta),$$

where the Compton wavelength $\lambda_C = h/mc$. For an electron, $\lambda_C = 2.43 \times 10^{-3}$ nm.

✻ Classical electromagnetism and thermodynamics predict a thermal radiation spectrum that becomes ever more intense with increasing frequency. The photon picture of radiation leads to the correct spectrum, which decreases exponentially at high frequency.

Rutherford's α-particle-scattering experiment shows that the positive charge in an atom is concentrated in a nucleus vastly smaller than the atom as a whole. According to classical theory, such atoms should collapse as orbiting electrons radiate electromagnetic waves. In Bohr's atomic model, electrons orbit the nucleus and can only radiate energy in the form of photons as they make abrupt transitions between orbits. The frequency of the radiation is given by the Balmer formula:

$$f_{nn'} = f_o \left(\frac{1}{n^2} - \frac{1}{(n')^2} \right), \quad \text{where } f_o = 3.288 \times 10^{15} \text{ Hz}.$$

The angular momentum of an electron in its orbit is quantized:

$$L = nh/2\pi \equiv n\hbar.$$

The correspondence principle states that classical and quantum predictions must be the same in situations involving many quanta (e.g., $n \gg 1$ in the expressions above).

Electrons, like photons, have wave as well as particle properties. The frequency and wavelength of a particle are related to its energy and momentum in the same way as for photons:

$$E = hf \quad \text{and} \quad p = h/\lambda.$$

An integral number of electron wavelengths fit into a Bohr orbit, indicating that atomic states result from standing-wave patterns. A proper three-dimensional calculation of standing-wave patterns in hydrogen follows from Schrödinger's theory of wave mechanics. The sequential filling of these standing-wave states, which occurs as the number of electrons in an atom is increased, explains the observed regularities in the chemical behavior of the elements.

Wave/particle duality places fundamental limits on the accuracy of measurement. According to Heisenberg's uncertainty principle, the product of uncertainties in position and momentum measurements or of energy and time measurements has a minimum value:

$$\delta x \, \delta p_x \geq \tfrac{1}{2}\hbar \quad \text{and} \quad \delta E \, \delta t \geq \tfrac{1}{2}\hbar.$$

Atomic phenomena appear to be inherently statistical; current theory is limited to predicting the probability of future events. Schrödinger's wave function represents the probability of an event occurring. The wave function collapses abruptly when the event occurs. This description is a subject of lively philosophical debate.

Practical Applications

Photoelectric phenomena are used to detect light in applications ranging from TV cameras to astronomical research. Wave mechanics of electrons is essential for the design of microwave generators and any of the semiconductors used in electronics. Chemical research relies on quantum theory to understand chemical bonding and to predict the molecular structure and properties of new chemicals.

Solutions to Exercises

35.1 The rate electrons (charge e) are emitted per unit area equals the quantum efficiency e_q multiplied by the rate photons strike the surface (intensity I divided by the energy per photon). We take the quantum efficiency for gold at 100 nm from Figure 35.5. The photocurrent per unit area j is:

$$j = e \frac{I}{hf} e_q = e \frac{I\lambda}{hc} e_q$$

$$= \frac{(1.6 \times 10^{-19} \text{ C})(1.0 \text{ W/m}^2)(100 \times 10^{-9} \text{ m})}{(6.6 \times 10^{-34} \text{ J·s})(3.0 \times 10^8 \text{ m/s})}(0.06)$$

$$= 5 \times 10^{-3} \text{ A/m}^2.$$

35.2 With $\theta = 90°$, $\cos \theta = 0$, and eqn. (35.6) gives:

$$\frac{\lambda_2 - \lambda_1}{\lambda_1} = \frac{E_1}{mc^2} = \frac{1.0 \text{ MeV}}{0.511 \text{ MeV}} = 2.0.$$

Using $\lambda = hc/E$, we find:

$$\frac{(1/E_2) - (1/E_1)}{1/E_1} = \frac{E_1 - E_2}{E_2}$$

$$= 2.0$$

Thus $E_1 = 3.0 E_2$

and so $(E_1 - E_2)/E_1 = 2/3.$

35.3 This is a conservation law problem. At the minimum distance d, the gold nucleus and the α-particle are at rest with respect to each other. Since the gold nucleus is much more massive than the α, the two are nearly stationary in the lab frame as well, and we may ignore their kinetic energy. Then, their electrical potential energy equals the initial kinetic energy K_α of the α:

$$kQ_\alpha Q_{Au}/d = K_\alpha.$$

Then:
$$d = \frac{kQ_\alpha Q_{Au}}{K_\alpha} = \frac{(9 \times 10^9 \text{ N·m}^2/\text{C}^2)(2)(79)(1.6 \times 10^{-19} \text{ C})^2}{(1 \times 10^6 \text{ eV})(1.6 \times 10^{-19} \text{ J/eV})}$$
$$= 2 \times 10^{-13} \text{ m}.$$

The directly aimed α-particle penetrates to about 0.002 of the atomic diameter, which is still approximately 50 nuclear diameters from the nucleus. Subsequently, the α-particle reaccelerates back along its original path and appears as a particle deflected by 180°.

35.4 Comparing Balmer's empirical formula with Bohr's theoretical result gives:
$$f_o = \frac{E_o}{h} = \frac{2.180 \times 10^{-18} \text{ J}}{6.626 \times 10^{-34} \text{ J·s}} = 3.290 \times 10^{15} \text{ Hz},$$
as expected. (The difference in the fourth significant figure is due to the assumption that the proton remains stationary.)

35.5 The quantity $ke^2/\hbar$ that appears in the expression for E_o equals αc. Thus:
$$E_o = \frac{m(ke^2)^2}{2\hbar^2} = \frac{m}{2}(\alpha c)^2 = \tfrac{1}{2}\alpha^2 mc^2.$$
Similarly:
$$v_1 = \frac{\hbar}{mr_1} = \frac{\hbar}{m}\left(\frac{mke^2}{\hbar^2}\right) = \frac{ke^2}{\hbar} = \alpha c.$$

35.6 Equating electric force Zke^2/r^2 to mass times acceleration and using quantization of angular momentum gives:
$$Zke^2 = mv_n^2 r_n = (n\hbar)^2/(mr_n).$$

Then,
$$r_n = \frac{n^2}{Z}\left(\frac{\hbar^2}{mke^2}\right) = \frac{n^2}{Z}a_o,$$
as required. The kinetic, potential, and total energies are:
$$K_n = \tfrac{1}{2}mv_n^2 = \tfrac{1}{2}kZe^2/r_n, \quad U_n = -kZe^2/r_n,$$
and
$$E_n \equiv K_n + U_n = -\tfrac{1}{2}kZe^2/r_n.$$
Using the result above for r_n, we obtain:
$$E_n = -\frac{m(ke^2)^2}{2\hbar^2}\left(\frac{Z^2}{n^2}\right) = -E_o\frac{Z^2}{n^2}, \quad \text{as required.}$$

35.7 Since $Z = 2$ for helium, the energy of the electron in the ground state of singly ionized helium is $Z^2 = 4$ times the corresponding energy in hydrogen. Thus the second ionization potential of helium is:
$$4hf_o = 4(13.6 \text{ eV}) = 54.4 \text{ eV}.$$

35.8 $v_d = \lambda f = (h/p)(E/h) = E/p = \gamma mc^2/(\gamma mv) = c^2/v.$

35.9 Heisenberg's uncertainty principle (eqn. 35.20) gives the energy spread directly:
$$\delta E = \frac{\hbar}{2\,\delta t} = \frac{1.05 \times 10^{-34} \text{ J·s}}{2(1.6 \times 10^{-19} \text{ J/eV})(10^{-7} \text{ s})} = 3 \times 10^{-9} \text{ eV},$$
and
$$\frac{\delta E}{E} = \frac{3 \times 10^{-9} \text{ eV}}{14.4 \times 10^3 \text{ eV}} = 2 \times 10^{-13}.$$
The line width is more than 100 times the expected shift.

Basic Skills

Review Questions

§35.1 PHOTONS

- Describe the *photoelectric effect*, and state five experimental facts about it. Which fact is most devastating for classical theory?
- How is the energy of a light quantum related to its frequency?
- What is the *work function* of a metal?
- Define *quantum efficiency*. Is it a constant for a given metal?
- What is *stopping potential*?
- Describe the *Compton effect*. How does it differ from the classical description of light scattered from electrons?
- Which length determines the wavelength change for scattered light at any specific angle of scattering?
- ✹ What explains the exponential decrease toward high frequency in Planck's radiation law?

§35.2 BOHR'S ATOMIC MODEL

- What experimental evidence suggests that atomic nuclei are very small?
- How did Balmer describe the frequencies of light emitted by a hydrogen atom?
- What hypotheses did Bohr make in his model of the hydrogen atom?
- Roughly how large is the first Bohr orbit for hydrogen?
- How much larger is the second Bohr orbit as compared with the first?
- What is an *energy-level diagram*? Sketch the diagram for hydrogen.
- What is the *ionization potential* of an atom?
- What is the *correspondence principle*?

§35.3 ELECTRON WAVES

- How does the idea of electron waves help clarify the Bohr atomic model?
- What physical property of a particle determines its de Broglie wavelength?
- Bohr's n is the *principal* quantum number in Schrödinger's model of the atom. What are the others? What properties do they represent? How are they related to fine structure and hyperfine structure in the energy-level diagram?
- What is the Pauli *exclusion principle*? Explain how it determines the structure of the periodic table.
- What is a *closed shell*? What is its significance for an atom's chemical properties? Describe the chemical properties of atoms with nearly closed shells.

§35.4 QUANTUM MECHANICS

- State two forms of the Heisenberg uncertainty principle.
- What is the standard interpretation of a particle's *wave function*?

Basic Skill Drill

§35.1 PHOTONS

1. What are the wavelength and frequency of a 1.0-keV photon?
2. An FM station broadcasts a power of 250 kW at 102.2 MHz. At what rate does the station emit photons?
3. What is the energy of electrons emitted from a silver surface when UV photons with an energy of 15.0 eV shine on the surface?
4. A photon of wavelength 0.135 nm scatters from an electron and returns along its original path. What is the wavelength of the scattered photon?

§35.2 BOHR'S ATOMIC MODEL

5. What wavelength photon is emitted when a hydrogen atom undergoes a transition from $n' = 5$ to $n = 2$?
6. Compare the radius of the second Bohr orbit to the wavelength of a photon emitted in an $n' = 2$ to $n = 1$ transition.
7. Find the energy required to remove the eighth electron from an oxygen atom, leaving a totally ionized atom.

§35.3 ELECTRON WAVES

8. At what speed would an electron have a de Broglie wavelength of 1.0 m?
9. How many different angular momentum states are there in the third shell ($n = 3$)? What is the maximum number of electrons an atom can have in the third shell?
10. Would you expect oxygen and fluorine to form a stable chemical compound? Why or why not? What about sodium and fluorine?

§35.4 QUANTUM MECHANICS

11. A certain apparatus can measure the momentum of an electron to 1 part in 10^3. If the speed of the electron is 1.5×10^8 m/s, what is the minimum uncertainty in its position after the measurement?
12. An electron is known to be between $x = 0$ and $x = 1.5$ nm. What is the minimum uncertainty in the x-component of its velocity?
13. An electron remains in the $n = 4$ state of a nitrogen atom for 6×10^4 s, on average, before it transitions to $n = 2$. What is the uncertainty in the energy of the photon emitted? (Lines with such long lifetimes are observed in space but not on Earth.)

Questions and Problems

§35.1 PHOTONS

14. ♦ Compare the momentum of a 10-keV photon with that of an electron that has 10 keV of kinetic energy.
15. ♦ A certain organic molecule dissociates into two smaller molecules when it absorbs 7.4 eV. What is the maximum-wavelength photon that can dissociate the molecule?
16. ♦ What is the binding energy of a molecule that can be dissociated by photons with wavelengths less than 555 nm?
17. ♦ What is the energy of a photon whose wavelength is 425 nm?
18. ♦ Show that the Compton wavelength of the electron equals the wavelength of a photon with energy $E = mc^2$.
19. ♦ What is the energy of photoelectrons emitted from a sodium surface when the incident light has wavelength $\lambda = 415$ nm?
20. ♦ What is the longest wavelength of light that can eject photoelectrons from a platinum surface?
21. ♦ The Rayleigh–Jeans law is valid when $hf < kT$. Determine whether it is adequate to describe the following situations: **(a)** Cosmic radio waves with temperature $T = 2.7$ K at wavelength $\lambda = 3$ cm. **(b)** The flux of "green" light ($\lambda = 550$ nm) from the Sun ($T \approx 6000$ K). **(c)** Infrared radiation from the Sun at wavelength $\lambda = 10$ μm.
22. ♦ Light of wavelength 550 nm has intensity $|\vec{S}| = 1.5$ W/m². What is the area of a surface that receives 1 photon/s, on average?
23. ♦♦ A photon with wavelength $\lambda = 4.00$ nm is Compton scattered through a 90° angle. What is the fractional change in wavelength?
24. ♦♦ A radio station broadcasts 150 kW at a frequency of 95.0 MHz. Assuming the station radiates uniformly in all directions, how many photons per unit area and per unit time strike a surface 1.5 km from the station? A detector with area 10 cm² and 1.5 km from the station can count individual photons at rates up to 10^{12} /s. Can it distinguish whether the station emits photons rather than a continuous wave?
25. ♦♦ One hundred days after light from the explosion first reached Earth, Supernova 1987a in the Large Magellanic Cloud was observed to emit visible radiation at a rate of 10^{35} W. How many photons per second would be collected by a telescope 5 m in diameter? (The Large Magellanic Cloud is 1.7×10^5 light-years from Earth.)
26. ♦♦ The intensity of light from a barely visible object is 1.5×10^{-11} W/m². At what minimum rate must photons enter the eye for an object to be visible? Take the diameter of the pupil to be 0.70 cm and the photon wavelength to be 560 nm.
27. ♦♦ A celestial x-ray source emits 10^{26} W of 6-keV x rays. If the source is 3×10^{19} m away, at what rate do photons arrive at a 900-cm² detector in low Earth orbit?
28. ♦♦ A 1.3-Mev x ray makes a head-on collision with a stationary electron and is scattered through an angle of 180°. What are the final velocity of the electron and the frequency of the scattered photon?
29. ♦♦ What is the final speed of an electron that has just scattered a 10.0-keV photon through an angle of 45°? A 1.0-MeV photon?
30. ♦♦ An x ray of wavelength 3.00 nm is Compton scattered through an angle of 60.0°. What is the recoil energy of the electron?
31. ♦♦ In a Compton-scattering experiment, both the scattered photon and the recoil electron are detected. If the electron has 61 keV of kinetic energy and the scattered photon has an energy of 150 keV, what is the initial wavelength of the x rays?
32. ♦♦ What is the maximum energy that can be transferred to electrons by Compton scattering of x rays with an initial wavelength of 5.0 nm?
33. ♦♦ The maximum kinetic energy transferred to electrons as they scatter x rays is observed to be 6.0 keV. What is the wavelength of the incident x rays?
34. ♦♦ What is the maximum energy a photon of UV light ($\lambda = 400$ nm) can transfer to an electron in a Compton collision? Could such a collision eject the electron from a metal with a work function of 1.0 eV? (*Note:* Your result proves that Compton collisions are *not* the mechanism of the photoelectric effect.)

35. ♦♦ Photoelectrons ejected from a surface by light of wavelength 4.0×10^2 nm are stopped by a potential difference of 0.80 V. What potential difference is required to stop electrons ejected by light of wavelength 3.0×10^2 nm?

36. ♦♦ A telescope on a satellite receives light over a 0.78-m² area and focuses it onto a gold photodiode. The system is used to observe a star's ultraviolet light. When light with wavelength in the range λ = 85 nm–95 nm falls on the detector, 7 electrons per second are liberated. What is the observed intensity of the star's radiation in this wavelength range?

37. ♦♦ You want to build a photodetector for blue light (∼ 420 nm). Which of the materials in Table 35.1 would make a suitable detector? Justify your choice.

38. ❋ ♦♦ In a cavity with volume 1.0 m³ at temperature 300 K, find the number of photons in the frequency ranges f = (3.00 to 3.03) × 10^{13} Hz, f = (2.00 to 2.02) × 10^{14} Hz, and f = (3.00 to 3.03) × 10^{14} Hz. What does a photon number less than unity mean?

39. ♦♦ Both the emitting and collecting surfaces of a photocell (Figure 35.6) are made of cerium. What electrical potential difference V_s (stopping potential) is required to eliminate the flow of photoelectrons between the surfaces if the incident light has wavelength λ = 320 nm? What is the change in stopping potential if the wavelength is decreased to 300 nm?

40. ♦♦♦ Show that the directions of a Compton-scattered photon and the recoil electron are related by $\cot\phi = (1 + hf/mc^2)\tan(\theta/2)$.

41. ♦♦♦ Use conservation of energy and momentum to show that a free electron cannot completely absorb a photon. How can an electron in metal absorb a photon in the photoelectric effect?

42. ♦♦♦ Energy transfers to or from an oscillating system are quantized according to the same rule as photons: $\Delta E = nhf$, where n is an integer. What is the smallest energy that can be transferred to an object of mass 5 g suspended by a spring with a spring constant of 20 N/m? How many energy quanta are needed for this system to oscillate with an amplitude of 1.0 cm?

§35.2 BOHR'S ATOMIC MODEL

43. ❖ On the basis of Bohr's theory, which has more lines in the optical portion of its spectrum, hydrogen or doubly ionized lithium?

44. ❖ Museums often exhibit mineral samples that are drab when seen by ordinary room lighting but glow with brilliant colors when bathed in ultraviolet (*black*) light. This phenomenon is called fluorescence. The electrical discharge in a fluorescent light produces ultraviolet radiation, which causes a coating material on the inside surface of the tube to fluoresce. Use the concept of atomic energy levels to explain how fluorescence works.

45. ❖ Chemical isotopes are atoms with the same nuclear charge but different mass. For example, deuterium (*heavy hydrogen*) has an extra neutron, giving its nucleus a mass about twice that of a proton. Do all isotopes of a given element have the same spectrum; that is, are their energy levels the same? Why or why not?

46. ♦ Use Balmer's formula to determine the wavelengths of visible lines in the hydrogen spectrum.

47. ♦ What transition in hydrogen produces higher-frequency photons, n' = 5 to n = 3, or n' = 20 to n = 4?

48. ♦ For what initial Bohr states is it possible for a 950-nm photon to ionize hydrogen?

49. ♦ What is the binding energy of an electron in the n = 3 state of hydrogen?

50. ♦♦ Find the short-wavelength limits (n' = ∞) of the Lyman, Balmer, Paschen, and Brackett series for hydrogen. In what part of the spectrum does each limit lie?

51. ♦♦ Is it possible for the short-wavelength limit of a spectral series with n = N + 1 to have a shorter wavelength than the α line (n' = N + 1 to n = N) of the spectral series with n = N? If so, for which series is it possible?

52. ♦♦ Through what minimum potential difference must an electron be accelerated so that it can transfer energy to a hydrogen atom in its ground state via an inelastic collision?

53. ♦♦ An ultraviolet photon of wavelength 85 nm is absorbed by a hydrogen atom. What is the kinetic energy of the emerging electron?

54. ♦♦ A photon emitted when an electron makes a transition from the n' = 2 state to n = 1 state in hydrogenic helium is later absorbed by a hydrogen atom in its ground state. What is the kinetic energy of the resulting electron?

55. ♦♦ An electron typically remains in the n = 2 state of hydrogen for 10^{-8} s. How many revolutions does a typical electron make according to the Bohr theory before a transition occurs?

56. ♦♦ According to classical physics, an electron in circular motion at frequency f radiates EM waves at frequencies f, $2f$, $3f$, and so on. Consider the allowed transitions between energy levels at high n (cf. the calculation in §35.2.3) to show that Bohr's model makes this same prediction in the limit of large n.

57. ♦♦ Compute the first four energy levels for hydrogenic oxygen. Determine the energy and wavelength of spectral lines corresponding to n' = 2, n = 1 and n' = 3, n = 2. In which part of the spectrum do these lines lie?

58. ♦♦ Find the angular momentum quantum number for the orbit of the Moon around the Earth. If the angular momentum of the Moon were quantized, could you tell? Why or why not? What about a bicycle wheel?

59. ♦♦ What is the radius of the electron in the n = 350 orbit of hydrogen? Compare the atom's diameter with the average separation between atoms in a gas with p = 1 atm, T = 300 K and in the interstellar medium, where the density is about 10^3 m^{-3} and T = 100 K.

60. ♦♦♦ Show that the frequency of the photon emitted in a transition from the nth to the $(n - 1)$th Bohr orbit lies between the two revolution frequencies of the electron in those two orbits.

61. ♦♦♦ ✉ The classical formula for electromagnetic power radiated by an electron is $P = e^2 a^2 / (6\pi\epsilon_0 c^3)$, where a is the magnitude of the electron's acceleration. Evaluate the classical power radiated by an electron in the nth Bohr orbit, and use the result to estimate the time for transition to the $(n - 1)$th orbit. For what values of n do you expect your value to be correct? How does your time for n = 2 compare with the observed time of about 10^{-8} s?

62. ♦♦♦ Neutral hydrogen atoms in a gas discharge tube, originally in the ground state, are excited by collisions with free electrons with 12.2 eV of kinetic energy. What possible wavelengths can subsequently be emitted by the hydrogen atoms?

§35.3 ELECTRON WAVES

63. ❖ A particle is confined between two rigid plane surfaces, one at x = 0 and one at x = L. Schrödinger's wave function ψ for the lowest energy state of the particle is proportional to $\sin(\pi x/L)$. If you observe at a random time, where is the particle most likely to be found?— Least likely to be found? Compare with the corresponding predictions for a classical particle bouncing (elastically) back and forth between the surfaces. Higher energy states have wave functions $\psi \propto \sin(n\pi x/L)$. For large n, describe where the particle is most likely to be found. Do the quantum and classical descriptions correspond in the limit of very large n? Why or why not?

64. ❖ Can an electron and a photon have both the same wavelength and the same frequency? Why or why not?

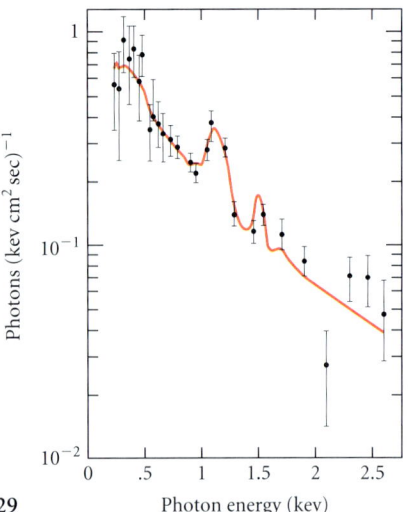

■ FIGURE 35.29

65. ❖ An absorption line at 6.4 keV is observed in the spectra of galactic x-ray sources. It is identified as the transition from $n = 1$ to $n' = 2$ in iron. Why does the value not agree with that from the Bohr formula (Example 35.10)? What is the effective charge internal to the transitioning electron?
66. ◆ Find the de Broglie wavelength of a 0.22 rifle bullet with a mass of 0.010 kg and a speed of 150 m/s.
67. ◆ Compare the de Broglie wavelengths of an electron, a proton, and a 3.0-g pea, each with speed 100 km/s.
68. ◆◆ Find the de Broglie wavelength of a cosmic ray proton with a kinetic energy of 3.0 GeV.
69. ◆◆ Compute the first four Bohr energy levels for hydrogenic iron (i.e., an iron nucleus with one electron around it). A spectral line at 1.15 keV is observed in the spectrum of the Virgo cluster of galaxies (■ Figure 35.29). Can you identify which transition this is? What is the effective charge on the iron nucleus?
70. ◆◆ What accelerating voltage is necessary to produce an electron beam with a de Broglie wavelength equal to the 0.15-nm spacing between atoms in a crystal target?
71. ◆◆ A radium atom ejects an α-particle (helium nucleus) with 5.87 MeV of kinetic energy. Compare the de Broglie wavelength of the α-particle with the radius of the radium nucleus (2×10^{-14} m).
72. ◆◆ What energy electrons are required in an electron microscope with a 1-nm resolution? (Assume the electron wavelength should be less than one-tenth the desired linear resolution.)
73. ◆◆ ✉ The diameter of the ring of iron atoms shown on the chapter title page is 144 nm. Estimate the wavelength of the electron wave and hence the electron's kinetic energy.
74. ◆◆ What is the de Broglie wavelength of an electron in the $n = 2$ orbit of helium? Compare with the orbital radius.
75. ◆◆ At what energy does an electron's de Broglie wavelength equal its Compton wavelength? (*Hint:* The electron is relativistic.)
76. ◆◆◆ An electron is confined between two infinite walls at $x = 0$ and $x = 1.0$ μm, as in Problem 63. What is the de Broglie wavelength of the electron in its lowest energy state? What is the electron's energy? How many electrons can have this energy? If there are six electrons, what is the total energy of the system in its lowest energy state?

§35.4 QUANTUM MECHANICS

77. ❖ Electrons in highly excited states of a hydrogen atom ($n \gg 1$) can remain in those states for rather long times (a millisecond or longer). Use the uncertainty principle to explain why.

78. ◆ If the speed of an electron in the x-direction is measured with an accuracy of 10^{-6} m/s, to what accuracy can its position along the x-axis be located?
79. ◆◆ Transitions from a certain atomic energy level to the ground state occur on average in 10^{-8} s. If the energy released in the transition is 2.48 eV, what is the minimum uncertainty in the wavelength of the resulting photon?
80. ◆◆ An electron is confined between two plane surfaces a distance L apart. From Heisenberg's uncertainty principle, estimate the minimum kinetic energy the electron can have. Evaluate the result when L is the size of an atom, about 10^{-10} m, and when L is the size of an atomic nucleus, about 10^{-15} m.
81. ◆◆ Estimate from Heisenberg's uncertainty principle the minimum momentum of an electron confined in a region equal in size to the diameter of the first Bohr orbit. Compare the result with the momentum deduced from Bohr's theory.
82. ◆◆ An electron is confined in a cubical box with 1.0-mm sides. What is the uncertainty in the electron's energy?

Additional Problems

83. ◆◆ What error is made in computing the de Broglie wavelength of a 35-keV electron using the Newtonian formula for its momentum? Of a 1-MeV electron?
84. ◆◆ Show that the ratio of a particle's Compton wavelength to its de Broglie wavelength is $\lambda_C/\lambda_d = \sqrt{\gamma^2 - 1}$.
85. ◆◆ Calculate the de Broglie wavelength λ_d of the recoiling electron in Compton scattering with $\theta = 90°$. Examine the three cases: (a) $\lambda_1 \ll \lambda_C$, (b) $\lambda_1 \approx \lambda_C$, and (c) $\lambda_1 \gg \lambda_C$. In each case, show that $\lambda_d \approx \lambda_1$.
86. ◆◆ When a hydrogen atom emits a photon, the atom recoils to conserve momentum. Show that this effect causes a change in wavelength of an emitted photon that is essentially independent of photon energy. Compute the value of the wavelength change.
87. ◆◆ Show that the classical electron radius (Example 27.12), the Compton wavelength, and the radius of the first Bohr orbit are approximately in the ratio $\alpha^2 : \alpha : 1$.
88. ◆◆◆ A photoemitting metal surface with work function $\phi = 2.30$ eV is in a region of uniform magnetic field strength $B = 0.635$ T. Find the maximum radius of the paths traveled by photoelectrons as a function of the wavelength of incident light.
89. ◆◆◆ The first step in forming an image on an ordinary photographic plate is the dissociation of silver bromide (AgBr) molecules by incident light. The dissociation energy of AgBr is measured to be 1.00×10^5 J/mol. Find the energy in electron volts required to dissociate a single molecule of AgBr and estimate the maximum wavelength to which the photographic plate is sensitive.
90. ◆◆◆ A silver sphere (work function 4.6 eV) is suspended in a vacuum chamber by an insulating thread. Ultraviolet light of wavelength 0.20 μm shines on the sphere. What is the resulting electric potential of the sphere?
91. ◆◆◆ In the Bohr model for hydrogen, we assumed that the proton remains at rest. In a better model, both electron and proton orbit about their common center of mass. Taking this into account, assume that the total angular momentum of the system is quantized and calculate E_o. Show that the result is given by eqn. (35.15) if we replace m with the *reduced mass* $\mu = m_e m_p/(m_e + m_p)$. How large is the correction to the wavelength of Balmer α?
92. ◆◆◆ Use the result of Problem 91 to calculate the frequency difference between Balmer α lines produced by hydrogen and deute-

rium. What resolution is necessary for a spectrograph to resolve lines from the two isotopes? (Compare §17.4.4 and Problem 17.61.)

93. ♦♦♦ Calculate a Bohr model for positronium—an electron in orbit about a positron—assuming that the total angular momentum of the system is quantized. Be careful to include the motion of both particles (cf. Problem 91). What is the separation of the particles in the ground state? What is the frequency of light emitted in a transition from $n' = 2$ to $n = 1$?

94. ♦♦♦ A beam of electrons with an energy of 5 keV passes through a powder of silver crystals. If the interatomic spacing of silver atoms in a crystal is 0.408 nm, find the angular radius of the first-order diffraction ring from the principal Bragg planes (cf. §17.5).

Computer Problem

95. In 1916, Millikan published measurements of the potential difference required to suppress the photocurrent when a freshly cut sodium surface is illuminated by each of six different spectral lines from a mercury vapor lamp. (See file CH35P95 on the supplementary computer disk.) Graph these data and determine Planck's constant h from the slope of the graph.

Challenge Problems

96. A sample of hydrogen gas with density $\rho = 0.065$ kg/m^3 is heated by an electric discharge to a temperature of several thousand kelvin. Explain why only a few Balmer lines are visible in the light emission from the sample, even though from the temperature alone one would expect a sufficient fraction of atoms to be in levels with very high n. Use Bohr's theory to estimate the number of lines that can be observed. (*Hint:* How big can an atom get without interfering with nearby atoms?)

97. ✉ Compute the ratio of magnetic moment to angular momentum for a ring of mass M and charge Q rotating about an axis perpendicular to the ring. Assuming this same ratio holds for electron orbital angular momentum and spin, find the orbital and spin magnetic moments of an electron in the $n = 1$ Bohr orbit of hydrogen. (This assumption makes an error of about a factor of 2 in the spin magnetic moment.) To estimate the energy of interaction between the electron's orbital and spin magnetic moments (spin-orbit coupling), treat the orbital magnetic moment as at the center of the atom and parallel to the spin magnetic moment, which is located at the electron. Estimate the magnitude of the hyperfine energy assuming the same ratio to find the nuclear magnetic moment from its spin of $\frac{1}{2}\hbar$.

98. Draw a schematic graph of the potential energy of electrons versus position in the circuit shown in Figure 35.6 if emitter and collector are made of different metals. From your graph, argue that the stopping potential is given by:

$$V = hf/e - \phi_c,$$

where ϕ_c is the work function of the collector.

99. Model a particle as the superposition of two waves with wavelength λ_1 and λ_2. Show that the superposition moves with speed:

$$v_s = \frac{f_1 - f_2}{(1/\lambda_1) - (1/\lambda_2)}.$$

Express v_1 and v_2 as $v \pm \Delta v/2$ and show that $v_s = v$, ignoring terms of order $(\Delta v/c)^2$. (See Exercises 17.2 and 35.8.)

100. (a) A μ meson is an elementary particle with the same charge as an electron but with 207 times the mass. Its average lifetime before decay into lighter particles is 2.22 μs. Rework the Bohr theory for a μ meson in orbit about a nucleus of charge $+Ze$ to find the radii and energies of the stationary states of such a μ-mesic atom. What is the energy of a photon emitted in the $n' = 2$ to $n = 1$ state for a μ-mesic atom ($Z = 82$). (b) The measured value of the photon energy in an $n' = 2$ to $n = 1$ transition in μ-mesic lead is 6.0 MeV, which disagrees with the result of part (a). The radius of a lead nucleus is 7.1×10^{-15} m. Show that the first Bohr orbit lies inside the nucleus. Find a corrected radius and energy of the first Bohr orbit, taking this fact into account, and recompute the photon energy. (*Hint:* How do the electric force on the μ meson and its potential energy depend on distance from the center of the nucleus, when the meson is inside the nucleus?) How much better is this approximation?

ESSAY 9
The Scanning Tunneling Microscope

Shirley Chiang
Department of Physics
University of California at Davis

The scanning tunneling microscope (STM) is a remarkable instrument which permits the measurement of the topography of a surface down to the atomic level. It was invented by Gerd Binnig and Heinrich Rohrer in 1982, and they shared half of the 1986 Nobel Prize in Physics for this achievement. The operation of this instrument is based on the phenomenon of quantum mechanical tunneling of electrons from a sharp tip to a solid surface through a vacuum gap when a voltage is applied between the tip and the surface. Classically, a potential barrier exists between the tip and the surface, and an electron whose energy is smaller than the barrier height could not exist within the barrier, nor would it have enough energy to surmount the barrier. But because electrons also have wave properties, for a barrier of finite width and height, a quantum mechanical calculation shows that the electron has a finite probability of being within the barrier, and a finite probability of "tunneling" through the barrier and appearing on the other side. (■ Figure E9.1). The probability that the electron can tunnel through the barrier increases as the barrier becomes narrower.

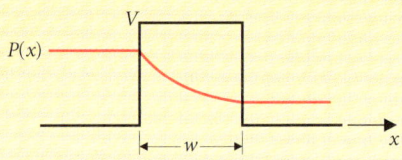

■ **Figure E9.1**
$P(x)$ indicates the probability of an electron incident on the square barrier with height V being at position x. Note that the probability does not include the reflected electron wave at the barrier but gives a qualitative idea of the exponential decrease of the probability with the width w of the barrier.

In the STM, the distance between the tip and sample is on the order of a few atomic diameters, and the quantum mechanical tunneling current I has the following dependence: $I \propto \exp(-Az)$, where A is a variable which can depend on the materials used for tip and sample, and z is the distance between the tip and sample. For a particular set of tip and sample materials, A can usually be considered to be constant. When z is in Ångstroms (1 Å = 10^{-8} cm, and a typical lattice spacing in a solid is 2.5 – 3 Å), $I \propto 10^{-z}$, i.e. the current changes by one order of magnitude when the distance changes 1 Å. This tunneling current is measured and kept constant by a feedback circuit. The tip in the STM is moved over the surface in three dimensions (■ Figure E9.2) using piezoelectric crystals, which are materials which expand or contract by an amount proportional to the voltage which is applied to their electrodes. As the tip is scanned in a raster pattern in the x-y plane over the surface, the feedback circuit moves the tip to and from the surface so as to keep the current constant. Thus, the tip traces a path which is a constant distance away from the surface. A computer is normally used to make a plot of the voltages applied to the x, y, and z piezoelectric elements, which can be calibrated to yield a three-dimensional (3D) topographic map of the surface. Because of the exponential dependence of the current on the distance, only the atoms closest to the end of the tip contribute significantly to the tunneling current. When the tip is sufficiently sharp, perhaps with a single atom at its apex, one can obtain atomically resolved images of the surface for some materials.

The STM has been used to measure many types of materials, including metals, semiconductors, and molecules. ■ Figure E9.3 shows a 3D view of individual atoms of a gold (Au(111)) surface, which are arranged in a close-packed hexagonal array. For many metals, the STM can resolve individual atoms which are adjacent to one another. Many surfaces have reconstructed top layers, in which the equilibrium structure of the top layer of the surface is different from the bulk crystal below it. One such case is the top layer of a clean silicon (Si(111)) surface, in which the top layer of the surface has a rhombus-shaped unit cell which is 7 times as long along each axis as the unit cell in the rest of the sample. Therefore, this structure is called a 7 × 7 reconstruction. Although the structure of the surface layer is now understood, it is quite complicated, and ■ Figure E9.4 shows only the 12 Si "adatoms," which sit above the

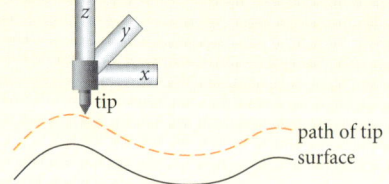

■ **Figure E9.2**
Schematic diagram showing operation of Scanning Tunneling Microscope. The tunneling tip is shown moving at a constant distance above the surface, as its height is regulated by a constant current feedback loop. x, y, and z indicate the piezoelectric elements which allow the movement of the tip in three dimensional space with respect to the sample.

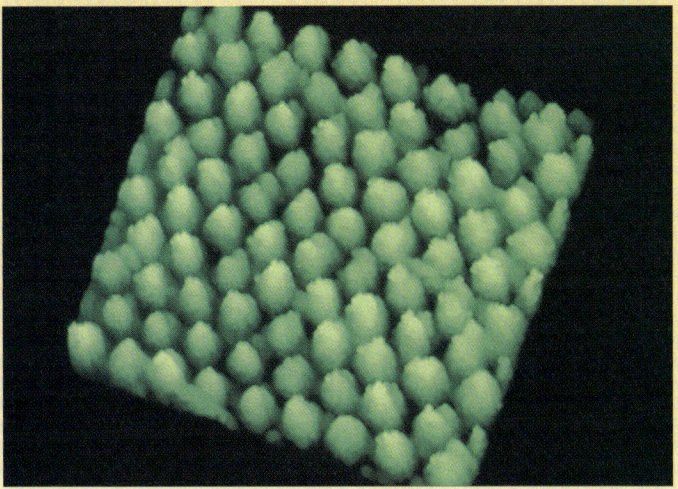

■ FIGURE E9.3
Three-dimensional view of STM image of gold (Au(111)) surface, in which atoms are arranged in a close-packed hexagonal array.

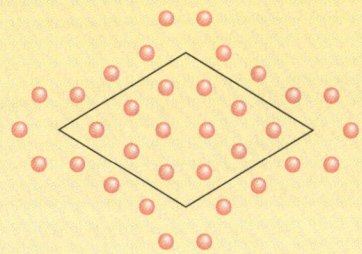

■ FIGURE E9.4
Schematic diagram showing the 12 Si "adatoms" that sit above the surface in the Si(111) 7 × 7 reconstruction. These are the atoms which are observed in the STM image in Figure E9.5. The surface unit cell is indicated by the rectangle. Note that the structure has 6 adatoms around holes in the corners of the unit cell.

surface, in each unit cell. As these atoms are actually higher than other structures in the surface layer, the STM images only these atoms. ■ Figure E9.5 shows a 3D view of a Si(111) 7 × 7 surface as measured in an STM image. Distinctive features in the structure are not only the 12 adatoms in a unit cell, but also the arrangement of 6 adatoms around what appear to be holes in the corners of the unit cell. This particular image also shows an atomic step on the surface, so that one sees the heights of two different layers on the surface simultaneously. Most STM images of semiconductors, however, cannot be interpreted so simply as pictures of individual atoms, as the STM often measures the electronic orbitals of atoms of the surface rather than the positions of individual atoms.

The STM is an instrument which has developed very quickly over the last decade and is now used actively in many fields of research, ranging from physics and chemistry to materials science and biology. It can be used in air, in liquids, and in vacuum environments. It has often been used to study metal and semiconductor substrates, some of which have atoms and molecules adsorbed on top. In addition, it has been used to study such widely varying samples as electrodes in electrochemical reactions, superconducting and layered samples, molecular and metallic thin films, as well as some biological materials.

THE PICTURE ON PAGE VI IN THE PROLOGUE SHOWS BENZENE MOLECULES.

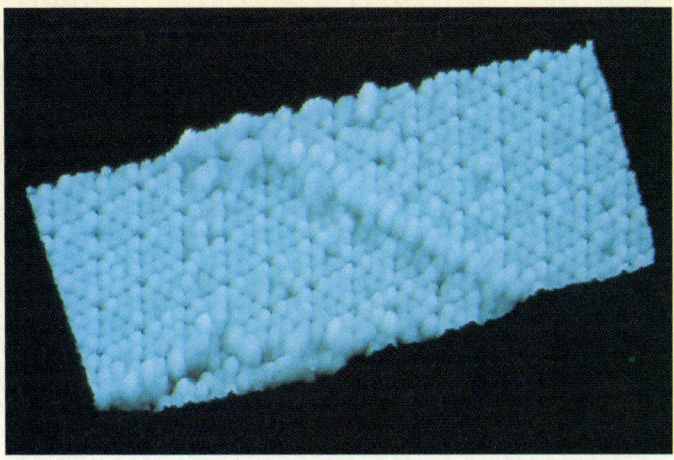

■ FIGURE E9.5
Three-dimensional view of an STM image of Si(111) 7 × 7, showing the 12 adatoms per unit cell which are sketched in Figure E9.4. This image also shows clearly an atomic step on the surface.

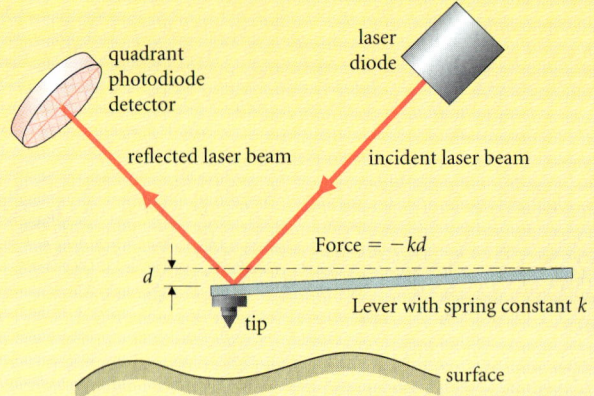

■ **Figure E9.6**
Schematic diagram showing operation of an atomic force microscope.

In addition, many other studies on non-conducting samples are now performed using an atomic force microscope (AFM), also called a scanning force microscope, a close cousin of the STM which was an invention of Gerd Binnig, together with Calvin Quate and Christoph Gerber in 1985. This instrument senses the amount of deflection of a small lever near the surface to measure the interaction force between a tip mounted on the end of the lever and the surface (see ■ Figure E9.6). The most common method of sensing this deflection is to reflect a laser diode beam from the end of the lever. The deflection of the beam is measured using a position sensitive detector—a photodiode divided into 2 or 4 parts so that signals can be compared among the parts. Such difference signals indicate the movement of the beam across the face of the detector and therefore the deflection of the lever. Using this technique, many types of forces have been measured, including atomic repulsion, Van der Waals, electric, magnetic, and frictional forces. Surface topography can often be measured in the contact mode, in which the tip is pushed directly against the surface, while the deflections of the lever are measured. In addition, images can be obtained at a constant force or a constant force derivative, permitting surface topographic imaging without actually having contact between the tip and surface. Such contact can damage soft samples. AFM images are less likely to have atomic resolution than STM images and are more dependent on the actual tip shape. The AFM has been used to measure samples ranging from magnetic and compact disks to soft biological materials. In addition to research applications, AFM's are now being used in development labs and for characterization of manufacturing problems.

This nuclear power plant is operated by Pacific Gas and Electric Company at Diablo Canyon, California. Unlike fossil fuels, nuclear fuel is sufficient to last for centuries. Yet the use of nuclear power has become a politically charged issue. The safety of such power plants has been questioned since the accident at Chernobyl, Ukraine, even though most plants used worldwide are of a safer design. Radioactive wastes must be safely stored for millennia.

CHAPTER 36
Atomic Nuclei

Concepts

Atomic number
Atomic mass number
Atomic mass unit
Strong nuclear force
Weak nuclear force
Isotope
Radioactive series
Antiparticles
Quantum numbers

Goals

Be able to:

Write reaction equations for α, β, and γ decays.

Use mass tables to determine whether a particular nucleus is unstable against α or β decay.

Relate observed radiation rates to the number of radioactive atoms in a sample.

A new thing had just been born; a new control; a new understanding of man, which man had acquired over nature.

I. I. Rabi

36.1 BASIC NUCLEAR STRUCTURE

36.1.1 Charge and Mass

An ordinary hydrogen atom has the simplest nucleus: a particle called a *proton* (■ Figure 36.2). Each proton has a positive electric charge equal in magnitude to the electron charge; its mass is about 2000 times larger than the electron mass:

$$q_\text{p} = +e = 1.60218 \times 10^{-19} \text{ C} \quad \text{and} \quad M_\text{p} = 1.67265 \times 10^{-27} \text{ kg} \approx 1836 m_\text{e}.$$

Because the proton and electron have opposite charges, a hydrogen atom is electrically neutral. All neutral atoms have a similar structure: the electric charge on any nucleus is a whole multiple Z of the proton charge ($q = +Ze$) and is exactly balanced by the charge on the surrounding electrons.

> The *atomic number* Z of an element equals the number of protons in an atomic nucleus of that element.

As we saw in Chapter 35, the chemical properties of an element are determined by the number of electrons surrounding each atom.

IF ATOMS WERE NOT *EXACTLY* NEUTRAL, BECAUSE OF A SMALL DIFFERENCE BETWEEN THE SIZES OF THE PROTON AND ELECTRON CHARGES, THE RESULTING ELECTRICAL FORCES WOULD CAUSE ALL PARTS OF THE UNIVERSE TO ACCELERATE AWAY FROM EACH OTHER. ASTRONOMICAL OBSERVATION SHOWS NO SUCH ACCELERATIONS, AND LIMITS ANY CHARGE IMBALANCE TO 1 PART IN 10^{30}.

Nuclear power plants operating normally have excellent safety records and emit much less pollution (including radiation!) than coal-fired plants. Smaller reactors contribute to public health through life-saving medical therapies (■ Figure 36.1) and effective long-term sterilization of food products. But they also raise the specters of radioactive waste products that must be stored for centuries, reactor accidents such as occurred at Chernobyl, Ukraine, in 1986, causing worldwide nuclear pollution, and the diversion of plutonium for weapons use. Nuclear energy has become one of the major technical and political issues of our time. France, for example, has embarked on an ambitious nuclear power program, while in the United States, the nuclear power industry is in decline.

Both the benefits and risks of the nuclear age stem from fundamental research accomplished between 1890 and 1940. In this chapter we describe the model of nuclear structure that emerged from that research and discuss some applications of the model in technology, medicine, and astrophysics.

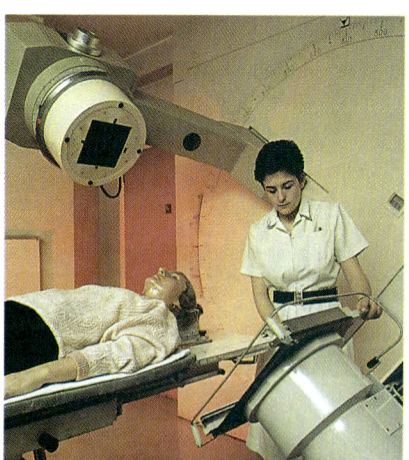

■ FIGURE 36.1
A patient being treated with a beam of radiation from radioactive cobalt nuclei prepared in a nuclear reactor. In modern medicine, radioactive nuclei are used in both diagnosis and treatment of disease.

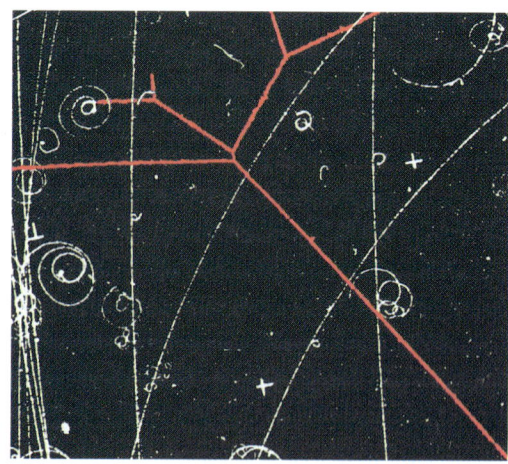

■ FIGURE 36.2
Track formed by a proton passing through a bubble chamber. Liquid hydrogen is maintained at its boiling point so that bubbles form around atoms ionized by the passing proton. The proton undergoes a collision with a stationary proton—the nucleus of a hydrogen atom in the liquid. The two protons subsequently undergo further collisions. Notice that the outgoing tracks are at 90° (cf. Example 10.3).

The *atomic number Z* of an element is also the number of electrons in a neutral atom of the substance.

The masses of atomic nuclei may be measured directly with a mass spectrometer (§29.1.2). Such measurements show that individual nuclear masses are nearly integral multiples of the proton mass:

$$M \approx AM_p.$$

The integer A is the *atomic mass number* of the nucleus. Since the chemical properties of an atom depend only on atomic number, nuclei with the same Z form atoms of the same element regardless of the value of A.

Nuclei with the same atomic number but different atomic mass form different *isotopes* of the same element.

A naturally occurring sample of an element is often a mixture of several isotopes. Atomic masses determined by chemical methods give the average atomic mass number for the mixture, which is usually not an integer. Chemical atomic masses are the values most often found in tables.

CHEMICAL MASSES ARE LISTED IN THE PERIODIC TABLE INSIDE THE BACK COVER.

Nuclei with particular values of A and Z belong to a *nuclear species*. The symbol for a species denotes the atomic number by the name of the corresponding chemical element. The atomic mass is written as a superscript preceding the name:

$$^A(\text{Name}).$$

Examples are ^{4}He for the helium ($Z = 2$) isotope with mass number $A = 4$, and ^{56}Fe for the iron ($Z = 26$) isotope with $A = 56$. The atomic mass number A and atomic number Z of a nucleus are generally different. The only exception is hydrogen, for which both equal unity. For light nuclei, A is roughly twice Z (■ Figure 36.3).

SOME AUTHORS PREFER TO INCLUDE THE ATOMIC NUMBER AS A SUBSCRIPT: $^4_2$He AND $^{56}_{26}$Fe.

The atomic mass number only gives an approximate value for the mass of a nucleus. Precise mass values are tabulated in terms of the atomic mass unit:

One *atomic mass unit* (u) is one-twelfth the mass of a neutral ^{12}C atom.

$$1 \text{ u} = 1.66054 \times 10^{-27} \text{ kg}.$$

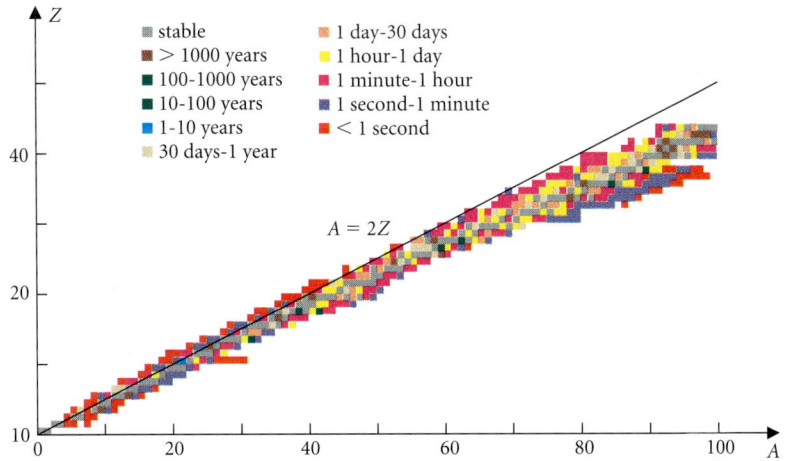

■ **FIGURE 36.3**
Light, stable nuclei occur along a line in a Z/A diagram with $A \sim 2Z$. As Z increases, A tends to be slightly larger than $2Z$. In this diagram the gray boxes indicate stable nuclei. The colors indicate the lifetimes of unstable nuclei.

TABLE 36.1 Masses of Selected Atoms (in u)

Z\A	2Z−3	2Z−2	2Z−1	2Z	2Z+1	2Z+2	2Z+3	2Z+4	2Z+5	2Z+6	2Z+7	2Z+8
1 H			1.007825	2.01410	3.01605							
2 He			3.01603	4.00260	5.0122							
3 Li			5.0125	6.01512	7.016005							
4 Be		6.01973	7.01693	8.005305	9.01218							
5 B			9.01333	10.01294	11.00931	12.01435						
6 C				12.000000	13.00336	14.00324	15.01060					
7 N			13.00574	14.00307	15.00011	16.00610						
8 O	13.02481	14.00860	15.00307	15.99492	16.99913	17.99916						
9 F		16.01148			18.998405							
10 Ne				19.99244	20.99385	21.99139						
11 Na					22.98977							
12 Mg			22.99413	23.98504	24.98584	25.98259						
13 Al					26.98154	27.98191						
14 Si				27.97693	28.97650	29.97377						
15 P					30.97376							
16 S				31.97207	32.97146	33.96787		35.96708				
17 Cl					34.96885		36.96590					
18 Ar				35.96755	36.96677	37.96273		39.96238				
19 K					38.96371	39.96400	40.96183					
20 Ca				39.96259		41.95863	42.95878	43.95549		45.95369		47.95253
21 Sc							44.95592					
22 Ti						45.95263	46.95177	47.94795	48.94787	49.94478		
23 V								49.94716	50.94396			
24 Cr						49.94605		51.94051	52.94065	53.93888		
25 Mn									54.93805			
26 Fe						53.93961		55.93493	56.93639	57.93328		
27 Co									58.93319	59.93388		
28 Ni						57.93534		59.93078	60.93105	61.92834		63.92796
29 Cu									62.92959		64.92779	

Masses are often given in energy units using Einstein's relation; $m = E/c^2$:

$$1 \text{ u} = 931.49 \text{ MeV}/c^2.$$

Tabulated masses are those of neutral atoms. To obtain the mass of an isolated nucleus, you should subtract the mass of Z electrons. The binding energy of the electrons to the atom also affects the mass of the atom but is typically a few keV/c^2 or less and is negligible in nearly all practical calculations. We list the masses of some selected nuclei in • Table 36.1.

EXAMPLE 36.1 ♦♦ Assuming natural chlorine (chemical atomic mass 35.453) consists entirely of the isotopes ^{35}Cl and ^{37}Cl, estimate the proportion of ^{35}Cl in the natural mixture.

MODEL The two isotopes contribute to the measured chemical mass in direct proportion to their fraction in the natural mixture.

SETUP Letting f be the fraction of ^{35}Cl,

(Chemical atomic mass) = f(mass of ^{35}Cl) + $(1 - f)$(mass of ^{37}Cl).

Table 36.1 (Continued)

Z		A	mass	A	mass	A	mass	A	mass	A	mass	A	mass	A	mass	A	mass
30	Zn	64	63.92914	66	65.92604	67	66.927132	68	67.924848	70	69.925325						
35	Br	79	78.918332	81	80.91629	87	86.9199										
36	Kr	78	77.92040	80	79.91638	82	81.91348	83	82.91413	84	83.91151	86	85.91062	92	91.9262		
56	Ba	130	129.90628	132	131.90449	134	133.90449	135	134.90567	136	135.90456	137	136.9058	138	137.90524	141	140.9141
57	La	138	137.90716	139	138.9064	143	142.916										
58	Ce	136	135.90718	138	137.90603	140	139.90548	142	141.9093	144	143.91359						
59	Pr	141	140.9077														
60	Nd	141	141.90777	143	142.90986	144	143.91013	145	144.9126	146	145.91315	148	147.9169	150	149.92092		
62	Sm	144	143.91207	146	145.9131	147	146.9149	148	147.91485	149	148.91721	150	149.9173	152	151.91976	154	153.92222
80	Hg	196	195.96582	198	197.9668	199	198.96828	200	199.96832	201	200.9703	202	201.97064	204	203.9735		
81	Tl	203	202.97235	205	204.97444												
82	Pb	204	203.97305	206	205.97448	207	206.9759	208	207.97666								
83	Bi	209	208.9804	211	210.98729	212	211.99129	213	212.99439	214	213.99873						
84	Po	203	202.9813	204	203.9802	205	204.9810	206	205.9804	207	206.98161	208	207.98125	209	208.98243	210	209.98288
		211	210.98666	212	211.98887	213	212.99286	214	213.99521	215	214.99945	216	216.00192	217	217.0064	218	218.0090
85	At	205	204.9860	206	205.9865	207	206.9857	208	207.9865	209	208.98617	210	209.9870	211	210.98751	212	211.99074
		213	212.99294	214	213.99634	215	214.99865	216	216.00243	217	217.00472	218	218.00871	219	219.0113		
86	Rn	207	206.9906	208	207.9895	209	208.9902	210	209.98956	211	210.99062	212	211.99072	213	212.99389	214	213.99537
		215	214.99875	216	216.00028	217	217.00394	218	218.00562	219	219.00951	220	220.0114	221	221.0155	222	222.01761
87	Fr	209	208.9959	210	209.9962	211	210.9955	212	211.9962	213	212.99619	214	213.99887	215	215.00036	216	216.0032
		217	217.00464														
88	Ra	211	211.0008	212	211.9997	213	213.0002	214	213.99997	215	215.00274	216	216.0035	217	217.00632	218	218.00715
		219	219.01008	220	220.0110	221	221.01393	222	222.01539	223	223.01853	224	224.02021	225	225.0236	226	226.02544
92	U	233	233.03965	234	234.0410	235	235.04394	236	236.04558	238	238.05082						
93	Np	231	231.0383	233	233.0408	235	235.04408	237	237.04819								
94	Pu	233	233.0430	234	234.0433	235	235.0453	236	236.04605	237	237.04843	238	238.04958	239	239.05218	240	240.05383
		241	241.05687	242	242.05877	243	243.062	244	244.06423								

boldface figures = stable

Taking the masses of ^{35}Cl and ^{37}Cl from Table 36.1, we have:

$$35.453 = f(34.969) + (1 - f)(36.966).$$

SOLVE
$$f = \frac{36.966 - 35.453}{36.966 - 34.969} = 0.7576.$$

ANALYZE Natural chlorine is about 76% ^{35}Cl and 24% ^{37}Cl. ∎

36.1.2 The Size of Nuclei

Although they contain almost all the mass, nuclei occupy a tiny fraction of the volume of matter. Ernest Rutherford and his students first carried out experiments that established the very small size of atomic nuclei (cf. §35.2.1). Using the Coulomb force law to explain the results of their scattering experiments, they determined that the radii of gold, silver, and copper nuclei were less than 32, 20, and 12 fm, respectively. Scattering α-particles (helium nuclei) from aluminum nuclei, they discovered deviations from the Coulomb model when the α-particles came within 8 fm of the center of the atom. This was the first indication of a distinct *nuclear* force.

THE SI SYMBOL FOR 10^{-15} m, 1 fm = 1 FEMTOMETER, IS ALSO AN ABBREVIATION FOR ENRICO FERMI'S LAST NAME. FERMI MADE NUMEROUS EARLY DISCOVERIES IN NUCLEAR PHYSICS, AND THIS UNIT IS OFTEN CALLED A FERMI IN HIS HONOR.

REMEMBER: THE ELECTRON WAVE HAS WAVELENGTH $\lambda = h/p$, §35.3. DIFFRACTION IS DISCUSSED IN §17.3 AND §17.4.5.

When relativistic electrons are used as the scattering particles, they are diffracted by the nucleus. The diffraction pattern provides an accurate measurement of the nuclear size (■ Figure 36.4). The results of these experiments show that the radii of nuclei are described by the formula:

$$R \approx (1.2 \text{ fm}) A^{1/3}. \tag{36.1}$$

The volume of a nucleus is roughly proportional to the mass number A so that nuclei all have about the same mass density.

36.1.3 Nucleons

The word "atom" was introduced by ancient Greek philosophers to mean a particle that cannot be cut into smaller pieces. We now think of atoms as composites of nuclei and electrons. The electrons may be removed by photon absorption or by collisions, and they may be shared among atoms in chemical reactions. The concept of a chemical element is useful because these electron reactions do not alter the atomic nuclei. However, much more energetic reactions can change the structure of nuclei (■ Figure 36.5), indicating that nuclei are themselves composite structures of more fundamental particles.

The electric charges of nuclei are easily explained by supposing that a nucleus of atomic number Z contains Z protons, but these protons account for only a part of the nuclear mass. The rest is contributed by a number $N = A - Z$ of electrically neutral particles called *neutrons* (■ Figure 36.6), each with nearly the same mass as a proton. Neutrons are unstable and cannot exist outside a nucleus for more than about 10 min. Their role in explaining nuclear properties was the first clue to their existence. The reality of neutrons was established in 1932 by James Chadwick, who was able to produce a beam of free neutrons and observe their collisions with protons in hydrogen-containing materials (■ Figure 36.7). The properties of neutrons are summarized, along with those of protons and electrons, in ● Table 36.2. (We discuss the decay of neutrons in §36.2.3.)

Atomic structure is remarkably simple: all of the ordinary matter we observe in the universe is made up of protons, neutrons, and electrons. Protons and neutrons are both referred to as *nucleons*. Next, we describe how the nucleons interact to form atomic nuclei.

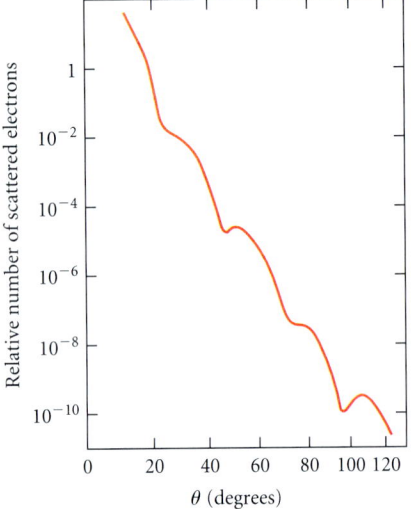

■ FIGURE 36.4
In an experiment to measure the size of a nucleus, 450-MeV electrons were scattered by ^{58}Ni nuclei. The electron wavelength ($\lambda \approx 3$ fm) is small enough to probe the nuclear size. Diffraction minima should occur at $\sin \theta = m\lambda/d$ (§17.3). Here the experimenters have plotted the number of electrons scattered in each direction. The positions of the minima indicate a nuclear radius equal to 4.1 fm. Adapted from a figure by Ingo Sick, Institut für Physik der Basel. Used with permission.

■ FIGURE 36.5
This cloud chamber photograph, taken in 1925 by Blackett and Lees, shows one of the first nuclear reactions observed visually. An α-particle, moving upward in the photograph, collides with a nitrogen atom in the chamber, producing an oxygen nucleus and a proton. The proton recoils backward and to the left. The cloud chamber contains supersaturated gas. Moisture condenses on charged ions produced by the passage of the fast particles. Cloud chambers were important in the early days of nuclear physics but have been replaced by more modern technologies.

■ FIGURE 36.6
A nucleus is composed of Z protons (colored red in the diagram) and $N = A - Z$ neutrons (colored tan) bound together in a region with a diameter of the order of 10^{-15} m ≡ 1 fm. This diagram represents a nucleus with mass number $A = 19$, atomic number $Z = 9$, and $N = A - Z = 10$ neutrons—that is, ^{19}F.

TABLE 36.2	Properties of Atomic Constituents				
	Mass			Electric Charge (e)	Spin ($\hbar$)
	(10^{-27} kg)	(u)	(MeV)		
Electron	9.109×10^{-4}	5.49×10^{-4}	0.5110	-1	$\frac{1}{2}$
Nucleons					
Proton	1.67262	1.00727	938.272	$+1$	$\frac{1}{2}$
Neutron	1.67493	1.00866	939.566	0	$\frac{1}{2}$

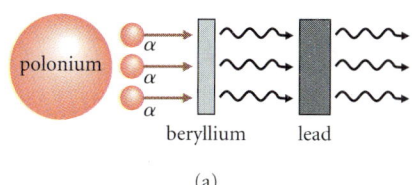

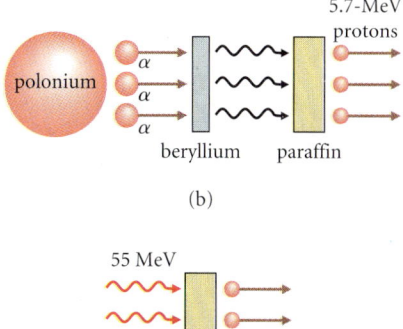

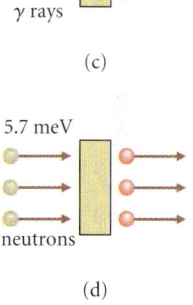

36.1.4 Nuclear Forces

Electrical repulsion between its positively charged protons would rapidly destroy a nucleus unless some other force acts to hold it together. This force is called the *strong nuclear force*. Collision experiments show that the strong force between protons overwhelms their electrical repulsion when the protons are less than about 10^{-15} m apart but decreases much more rapidly than electrical forces for distances greater than 10^{-15} m. The strong force doesn't distinguish between the two kinds of nucleon; it acts in the same way between a pair of neutrons or between a neutron and a proton as it does between a pair of protons. There is no

■ FIGURE 36.7
James Chadwick, a protégé of Rutherford, conducted experiments that established the existence of neutrons. Chadwick's colleagues described him as "the personification of the ideal experimentalist." (a) Radioactive polonium emits α-particles that strike a beryllium target and produce radiation that penetrates a thick lead shield. (b) The particles from the interaction in beryllium eject 5.7-MeV protons from a paraffin target. (c) To eject 5.7-MeV protons from paraffin using γ rays, the γ rays would have to have at least 55 MeV of energy, much more than is available from the nuclear reaction in beryllium. (d) Chadwick showed that neutral particles with about the same mass as the proton need only about 5.7 MeV of energy to eject the protons via elastic collisions.

Digging Deeper

WHY THE NEUTRON IS NECESSARY

The proton is named after Proteus, a mythical Greek figure who could take on any form at will. The name was given when an atomic nucleus was thought to consist of A protons and $N \equiv A - Z$ electrons bound together to give the proper charge and mass. Such an economical description of all matter is very attractive. However, it has some major problems that can be solved by including neutrons in the theory.

- According to the uncertainty principle, an electron confined within a nucleus must have a large momentum. The resulting kinetic energy is ten times greater than the observed energies of electrons emitted by radioactive nuclei (§36.2).
- The spin angular momentum of nuclei is observed to depend only on the mass number A. According to quantum mechanics, spin is determined by the total number of particles in a system, which depends on both A and Z in the proton + electron model.
- Electrons have magnetic moments 600 times larger than do protons, while nuclear magnetic moments are comparable to that of a proton.

To address these observations, we might suppose an improbable cancellation of magnetic moments and abandon the rules of angular momentum so crucial to understanding chemistry and atomic spectra. Such ideas are not logically faulty but have an ugliness that in physics is often a symptom of error. By contrast, neutrons explain all three observations in a straightforward way. Chadwick believed in the existence of neutrons long before he discovered them experimentally.

Chadwick's discovery illustrates an important theme of modern physics. Given the choice between accepting an ugly theory or a new kind of particle, physicists tend to bet on the new particle. Experiment has often confirmed this approach.

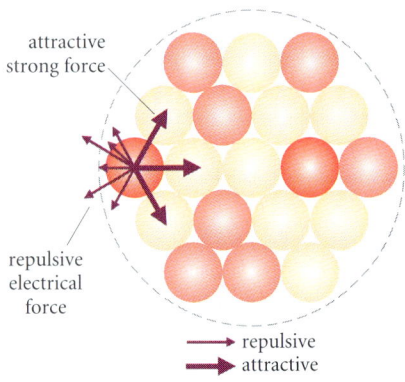

■ FIGURE 36.8
A nucleus is bound together by the attractive *strong* force its nucleons exert on each other. Electric forces between the protons are repulsive and tend to disrupt the nucleus. In a large nucleus, the short-range strong force acts primarily between a nucleon and its nearest neighbors. A proton, however, feels the long-range electrical repulsion of every other proton in the nucleus. For this reason, nuclei with $A \gtrsim 240$ are unstable.

THE BINDING ENERGY OF THE NUCLEUS IS THE *NEGATIVE* OF ITS POTENTIAL ENERGY. THUS WHEN BINDING ENERGY INCREASES, SOME OTHER FORM OF ENERGY ALSO INCREASES CORRESPONDINGLY.

NOTICE THAT THE ELECTRON MASS CANCELS OUT IN EQN. (36.2). THAT ISN'T TRUE IN ALL THE EQUATIONS WE SHALL WRITE, SO CHECK CAREFULLY!

IN BOTH CASES, THE TOTAL BINDING ENERGY OF THE NUCLEI PRODUCED IS GREATER THAN THAT OF THE ORIGINAL PARTICLES.

simple formula, like Coulomb's law, that completely describes the strong force exerted by two nucleons on each other. In this chapter we shall rely on approximate models of the strong force.

The short range of the strong force limits the size of nuclei. In a nucleus very much larger than 1 fm, protons near the surface would feel the strong attraction of relatively few nearby nucleons while feeling the electrical repulsion of every other proton (■ Figure 36.8). In large nuclei, the number of neutrons N exceeds the number of protons Z. The additional neutrons increase the effects of the strong force without increasing electrical repulsion, thus tending to hold a nucleus together. The largest known stable nuclei have $A \approx 240$. Larger nuclei can be produced in carefully designed experiments, but they quickly reduce their size by expelling nucleons in the form of smaller nuclei.

A second kind of nuclear interaction due to the *weak nuclear force* causes a neutron to disrupt, or β-decay, into a proton and an electron (see §36.2.3). Inside a nucleus, β decay is inhibited by the strong force. In a nucleus with too many neutrons, the weak decay process overwhelms the strong force, and the nucleus reduces its neutron number by emitting electrons. Either too few or too many neutrons cause a nucleus to be unstable. That is why stable nuclei cluster near a line in a plot of Z versus A (Figure 36.3).

36.1.5 Binding Energy

The strong force pulls nucleons together and binds them into a nucleus. As with the electrons in an atom, this binding can be described either by the force itself or by the corresponding potential energy. The potential energies of nuclei are large enough to be observed in measurements of nuclear masses; a stable nucleus has less mass than its nucleons would if they could be separated as free particles. According to Einstein's mass-energy relation (§34.4.2), this mass difference corresponds to the energy needed to separate the nucleons—the binding energy B of the nucleus.

> The *mass defect* of a nucleus equals the difference between the total mass of its individual nucleons and the mass of the assembled nucleus:
>
> $\Delta M(A, Z) = Z \times$ (proton mass) $+ N \times$ (neutron mass) $-$ (nuclear mass).

Data are usually tabulated for neutral atoms. For example, in Table 36.1,

$$M_\text{H} = 1.0078 \text{ u} = 1.0073 \text{ u} + 0.0005 \text{ u} = M_\text{p} + m_\text{e}.$$

Thus a formula for the mass defect in terms of tabulated data must include the mass of the electrons:

$$\Delta M(A, Z) = ZM_\text{H} + NM_\text{n} - M(A, Z). \tag{36.2}$$

> The *binding energy* of a nucleus equals its mass defect expressed in energy units:
>
> $$B = c^2 \, \Delta M. \tag{36.3}$$

The binding energy per nucleon (B/A) in a nucleus determines how easy it is to break up the nucleus. When B/A is large, the nucleus is tightly bound; if B/A is small, the nucleus is more easily split.

The short range of the strong force is evident in a plot of B/A versus atomic mass number for stable nuclei (■ Figure 36.9). Adding a nucleon to an element less massive than iron causes each nucleon to be more tightly bound to the larger nucleus. Adding a nucleon to atoms more massive than iron increases the size of the already large nucleus and decreases the attraction between nucleons already present. The shape of the binding energy curve has great importance for nuclear technology. Energy may be released either by breaking apart large nuclei, as in fission reactors using uranium, or by fusing together lighter nuclei. Fusion reactors under development would operate by fusing deuterium and tritium into helium.

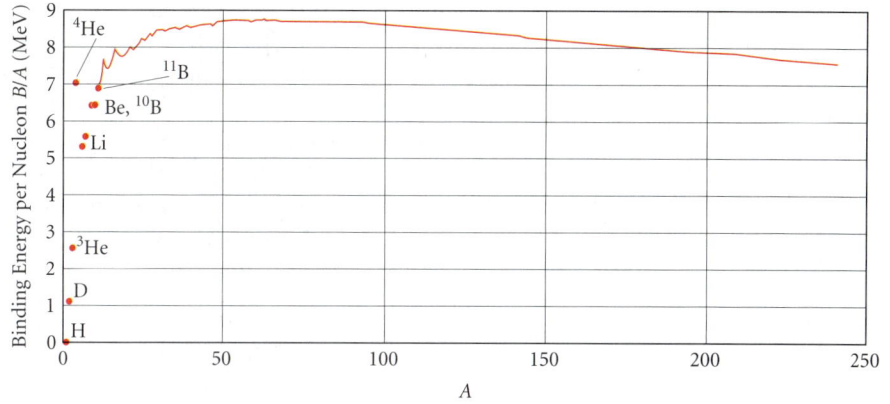

■ **Figure 36.9**
Binding energy per nucleon versus mass number *A* for stable nuclei. The curve has a maximum at elements near iron. Notice also that helium is more tightly bound than its neighbors.

EXAMPLE 36.2 ♦ Find the binding energy of a ⁴He nucleus. Compare the binding energy per nucleon with the value in Figure 36.9.

MODEL A helium nucleus contains two neutrons and two protons. We use eqns. (36.2) and (36.3).

SETUP From Table 36.1, the mass of a ⁴He atom is: $M(^4\text{He}) = 4.00260$ u. So, using values for M_p, m_e, and M_n from Table 36.2, the mass defect of ⁴He is:

$$\Delta M(4, 2) = 2(M_p + m_e) + 2M_n - M(^4\text{He})$$
$$= [2(1.00727 + 0.00055 + 1.00866) - 4.00260]\ \text{u} = 0.03036\ \text{u}.$$

SOLVE This corresponds to a binding energy of:

$$B = c^2\ \Delta M = c^2(0.03036\ \text{u})(931.49\ \text{MeV}/c^2\cdot\text{u}) = 28.28\ \text{MeV}.$$

ANALYZE The binding energy per nucleon is $B/A = (28.28\ \text{MeV})/4 = 7.07$ MeV/nucleon, as compared with a value of 7.1 MeV/nucleon read directly from the figure. ∎

EXERCISE 36.1 ♦♦ Through a sequence of reactions in stars, six helium nuclei fuse into one magnesium nucleus. Use Figure 36.9 to estimate the energy released.

36.2 NATURAL RADIOACTIVITY

36.2.1 Conservation Laws and Quantum Numbers

The nuclear age began in 1896, when Becquerel discovered the spontaneous emission of energetic particles by samples of uranium. In 1898, Marie and Pierre Curie (■ Figure 36.10) discovered similar radiation from thorium and radium, and gave the phenomenon its modern name: radioactivity. Three kinds of radiation are common in nature. They were originally named according to their *range*—the distance they can penetrate into surrounding material:

- α rays (bare helium nuclei), which typically penetrate a few centimeters of air;
- β rays (electrons) capable of penetrating several meters of air;
- γ rays (photons) whose range depends strongly on energy but is typically much greater than for α or β rays.

In each case, the emission of a particle changes a *parent* nucleus into a different, *daughter* nucleus. We typically have accurate descriptions of the parent nucleus and of the daughter products. The intermediate processes by which decay occurs are only now beginning to be understood in detail. Whenever we wish to compare initial and final states without reference

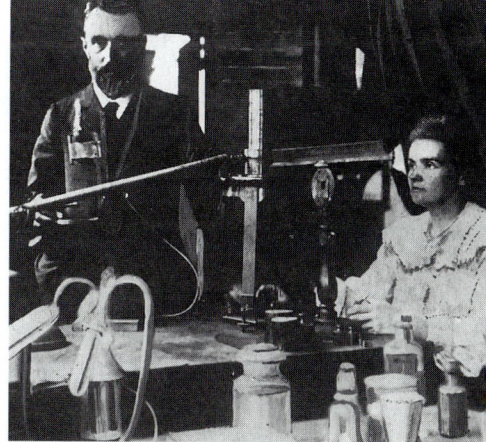

■ **Figure 36.10**
Marie and Pierre Curie in their laboratory. With Henri Becquerel, they were awarded the Nobel prize in 1903 for their discovery of radioactivity. Marie Curie won a second Nobel prize, in chemistry, for her later work with radium. Their daughter, Irene Joliot-Curie, also won a Nobel prize for her work on the production of radioactive substances by bombardment with α-particles.

to intermediate states, conservation laws provide the most powerful tools. The familiar conservation laws for energy, linear and angular momentum, and electric charge remain valid in nuclear processes. In addition, nuclear decays obey two conservation laws that are not apparent on a larger scale.

Technical names for the conserved quantities are *lepton* (light particle) *number* and *baryon* (heavy particle) *number*. We have already discussed one kind of lepton, the electron, and two kinds of baryon, the nucleons. As such new and unfamiliar particle properties enter our tale, it is useful to think of each in analogy with electric charge. Each particle has a charge number, and its charge equals that number times the electron charge e. Conservation of electric charge was first established experimentally and only later, with the discovery of Maxwell's equations, found to be required by basic principles. Lepton and baryon numbers also describe chargelike quantities; they are examples of *quantum numbers*. We don't yet know if their conservation is required by fundamental principles.

Next let us examine the three types of natural radioactive decay to see how these conservation laws apply.

■ **Figure 36.11**
Cloud chamber photograph of particles produced in the α decay of ^{214}Po to ^{210}Pb. Note the almost constant range of the particles. Each α has an energy of 7.69 MeV. The one long track is due to a rare decay from an excited state of ^{214}Po.

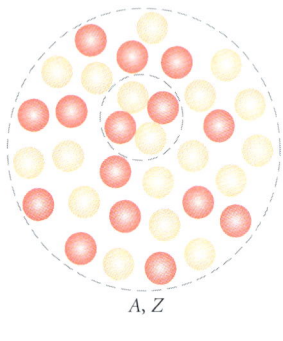

■ **Figure 36.12**
An α decay. Four nucleons bind together and escape as an α-particle. Baryons are conserved, so the total mass number A of the two daughter nuclei equals that of the parent. Electric charge is also conserved, so the sum of the atomic numbers of the daughters equals that of the parent. Kinetic energy of the daughters arises from a reduction in the total mass.

36.2.2 α Decay

In α decay, two neutrons and two protons within a parent nucleus bind together to form a helium nucleus, which then escapes (■ Figures 36.11, 36.12, and 36.13). The potential energy released in forming the α-particle is converted to kinetic energy, which is shared by the daughter particles. The daughter particles have a smaller total mass than the parent nucleus; the loss of rest energy (traditionally given the symbol Q) equals the gain of kinetic energy:

$$K = Q = (\text{mass of parent nucleus} - \text{total mass of daughter particles})c^2.$$

The Q values in α decay are always much less than the rest energy of a single nucleon. This observation is a consequence of baryon conservation. Nucleons cannot disappear in the reaction; they can only be bound more tightly. Thus the total atomic mass number of the daughter particles is the same as that of the parent nucleus. Electrons play no role in α decay.

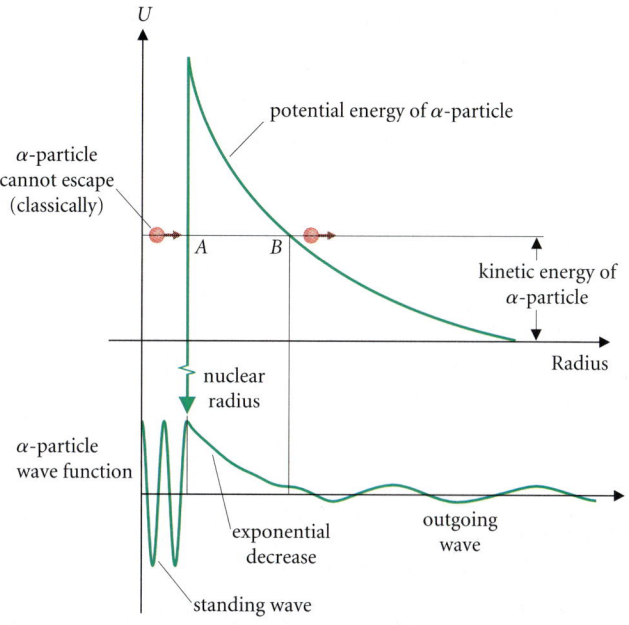

■ FIGURE 36.13

α-decay. Four nucleons associate into an α-particle and escape the nucleus together. Such combinations of nucleons continually form and redissolve within a nucleus. Because α-particles are particularly tightly bound (Figure 36.9), they form with sizable kinetic energy and have a particularly large probability of escape compared with other small substructures that might form. The potential energy function of an α-particle is large and negative from the center of the nucleus to a radius of a few fermis due to the attractive strong forces exerted by the rest of the nucleons. Farther out, its potential energy is positive because of repulsive Coulomb forces. A classical particle could not exist in the region between points A and B on the graph, where it would have to have negative kinetic energy. However, the wave function (cf. §35.3 and §35.4) for an α-particle moving outward is partially transmitted through this region. The outgoing wave that results corresponds to a small probability that the α-particle will escape. (The particle is said to *tunnel* through the classical barrier.) Calculations based on this tunneling model are in excellent agreement with measured α-decay rates.

Conservation of electric charge requires that the number of protons be the same before and after the reaction. That is, the total atomic number of the daughters is the same as that of the parent nucleus.

A general equation for α decay that expresses these conservation laws (■ Figure 36.14) is:

$$^A Z \rightarrow {}^{(A-4)}(Z-2) + {}^4\text{He} + Q. \qquad (36.4)$$

For example:

$$^{144}\text{Nd} \rightarrow {}^{140}\text{Ce} + {}^4\text{He} + Q.$$

(Neodymium decays to cerium plus helium, releasing energy Q.)

The Q value of a reaction can be calculated from tabulated masses. Decay is possible when Q is positive; otherwise, the parent nucleus is stable.

EXAMPLE 36.3 ◆◆ Find the energy released in the α decay of ^{144}Nd.

MODEL The Q value is determined by the decrease of total *nuclear* mass, while tabulated data give atomic masses, which include electrons. However, as many electrons are included in the atomic mass of the parent as in the total atomic mass of the daughters. So, the electron masses cancel in the calculation of Q. The total binding energy of the electrons differs somewhat between parent and daughter atoms, resulting in errors of at most a few keV.

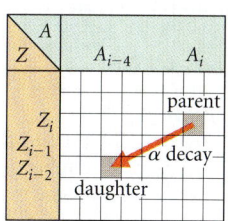

■ FIGURE 36.14

An α decay displayed on a table of atomic number Z versus mass number A. The daughter nucleus is displaced by 4 in mass number A and 2 in atomic number Z.

SETUP AND SOLVE We obtain the masses from Table 36.1. Thus:

$$Q = c^2(M_{Nd} - M_{Ce} - M_{He})$$
$$= c^2(143.9101 - 139.9055 - 4.0026) \text{ u}$$
$$= c^2(0.0020 \text{ u})(931.49 \text{ MeV}/c^2 \cdot \text{u})$$
$$= 1.9 \text{ MeV}.$$

ANALYZE Notice how many significant figures are lost in the subtraction!

36.2.3 β Decay

In β decay, the parent nucleus emits an energetic electron. To conserve charge, the daughter nucleus must have one more proton than the parent and, since the total number of baryons cannot change, one fewer neutron. In effect, a single neutron within the nucleus decays to produce a proton and an electron (■ Figure 36.15). Free neutrons undergo just such a reaction.

A tentative formula for β decay (■ Figure 36.16) is:

$$^AZ \rightarrow {}^A(Z+1) + e^- + Q + X.$$

The reaction product X must be included; otherwise, nearly all the conservation laws are violated. For example, the measured kinetic energies of emitted electrons spread over a continuous range from 0 to Q (■ Figure 36.17). Without X, this could happen only if individual decays violate energy conservation. In cloud chamber photographs (■ Figure 36.18), the electron and daughter nucleus leave tracks that are not in opposite directions. Without X, this would indicate a violation of linear momentum conservation. Nor can the reaction conserve angular momentum. One neutron with spin $\frac{1}{2}\hbar$ is replaced with a proton and electron, each with spin $\frac{1}{2}\hbar$. According to quantum mechanics, these spins either cancel to zero or add to a spin of $1\hbar$. Orbital angular momentum comes in whole multiples of $\hbar$ and so cannot account for the difference of $\frac{1}{2}\hbar$ between final and initial spin angular momentum. Finally, total lepton number would not be the same if X is not present. The parent has zero lepton number (no electrons) while the electron daughter has lepton number 1. Thus X must have lepton number -1. To understand β decay, we must either believe in a new kind of particle or abandon conservation laws altogether.

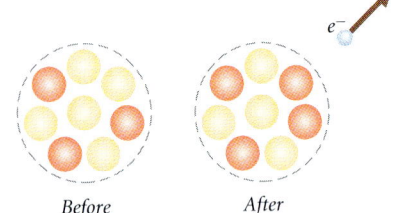

■ **FIGURE 36.15**
A β decay. The daughter nucleus has the same atomic mass number A. It has one fewer neutron but one more proton, so its charge is increased by 1. An energetic electron is ejected from the nucleus.

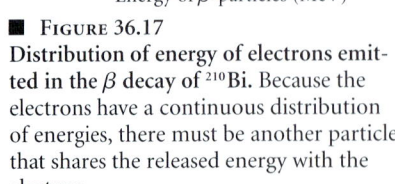

■ **FIGURE 36.16**
A β decay displayed on a table of atomic number Z versus mass number A. The daughter nucleus has the same mass number A but is displaced by 1 in atomic number Z.

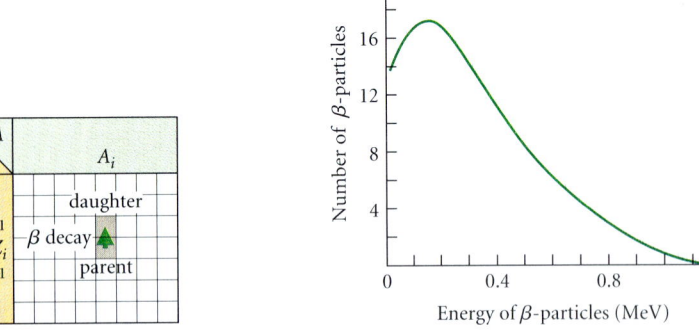

■ **FIGURE 36.17**
Distribution of energy of electrons emitted in the β decay of ^{210}Bi. Because the electrons have a continuous distribution of energies, there must be another particle that shares the released energy with the electron.

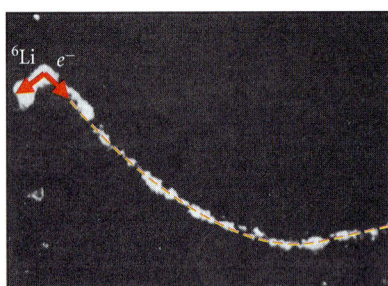

■ **FIGURE 36.18**
Cloud chamber photograph of β decay of ^{6}He. The short, thick track is the ^{6}Li daughter nucleus; the thinner, curved track is the electron. The two particles do not move opposite each other but instead have a net downward momentum in this picture, indicating the presence of a third, upward-moving, particle. Since it leaves no track, the third particle is electrically neutral. The photograph was taken in 1957 by S. Szalay and J. Csikay.

In summary, the new particle has spin $\tfrac{1}{2}\hbar$, no electric charge, and lepton number -1. It is a *neutrino*—a name given by Enrico Fermi meaning "little neutral one" in Italian. It is believed to have zero—or at least a very small—mass. This specific type is called an *electron antineutrino*, symbolized $\bar{\nu}_e$. Neutrinos have a very small probability of interacting with other particles and remained undetected by experiment for 20 years after Fermi proposed their existence. They were first observed in 1953 by Cowan and Reines, who detected reactions of chlorine atoms with the intense neutrino flux from a nuclear reactor.

The general equation for β decay, including the antineutrino, is

$$^AZ \rightarrow\ ^A(Z+1) + e^- + \bar{\nu}_e + Q. \tag{36.5}$$

The Q value for a β decay may also be computed using tabulated atomic masses. The atomic mass of the daughter element *includes* the mass of the emitted electron.

EXERCISE 36.2 ♦♦ What kinetic energy is shared by the electron and neutrino in the β decay of ^{12}B?

36.2.4 Antiparticles and Positron (β^+) Decay

For each kind of particle, a corresponding *antiparticle* exists that has equal mass but opposite values of electric charge and baryon or lepton number. In 1927, Paul Dirac predicted that such antiparticles must exist if both quantum mechanics and special relativity are correct. Antielectrons (*positrons*) were first observed by Carl Anderson in 1932 in a cloud chamber photograph of cosmic rays (■ Figure 36.19). Antiprotons and antineutrons were observed in accelerator experiments during the 1950s (■ Figure 36.20). Since then, Dirac's prediction has been completely verified experimentally.

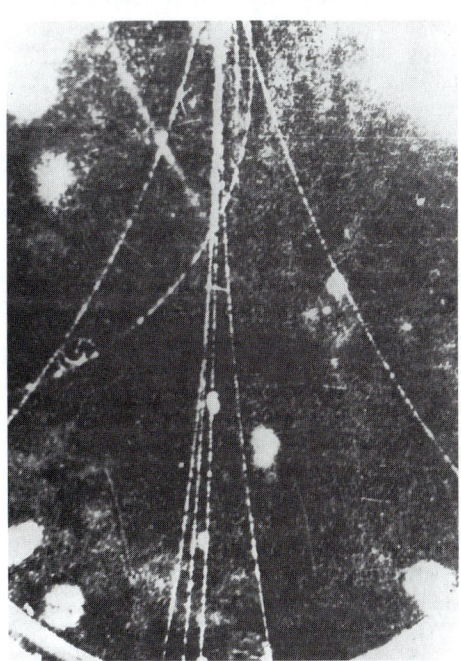

■ **Figure 36.19**
One of the first pictures of a cosmic ray electron shower passing through a cloud chamber. A magnetic field causes the charged particles to follow curved paths (§29.1). Tracks curving to both right and left indicate the presence of both electrons and positrons.

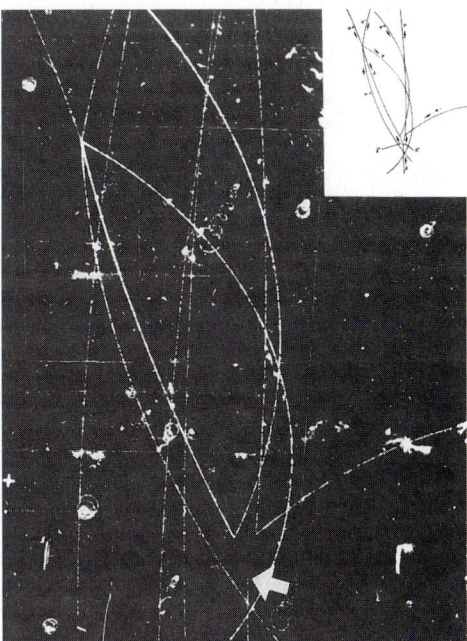

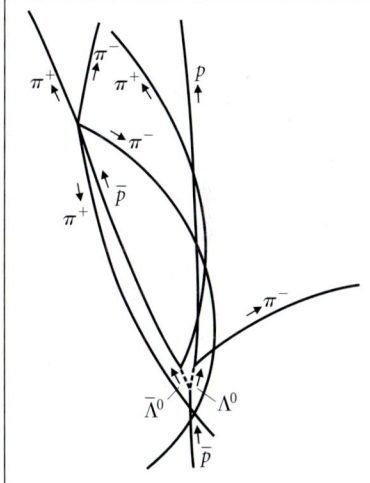

■ **Figure 36.20**
Track of an antiproton in a bubble chamber. Upon colliding with a proton (at arrow), the pair annihilates, producing two neutral Λ-particles that leave no tracks. The Λ^0, on the right, decays into a proton and a π^{-1}. The antiparticle $\bar{\Lambda}^0$ decays into the exact antiparticles, an antiproton and a π^+. The Λ- and π-particles are discussed in Chapter 37.

Digging Deeper

Neutrinos

Neutrinos interact only very weakly with ordinary matter. While most kinds of particle are stopped by a few centimeters thickness of solid material, a typical neutrino could pass through several light-years of solid iron!

Cosmic Gall*

Neutrinos, they are very small
   They have no charge and have no mass
And do not interact at all.
   The earth is just a silly ball
To them, through which they simply pass,
   Like dustmaids down a drafty hall
Or photons through a sheet of glass.
   They snub the most exquisite gas,
Ignore the most substantial wall,
   Cold shoulder steel and sounding brass,
Insult the stallion in his stall,
   And, scorning barriers of class,
Infiltrate you and me! Like tall
   And painless guillotines, they fall
Down through our heads into the grass.
   At night they enter at Nepal
And pierce the lover and his lass
   From underneath the bed—you call
It wonderful; I call it crass.

John Updike

The Sun seems to be producing only a third as many neutrinos as expected from the nuclear reactions that fuel its energy output. For more than 20 years, Raymond Davis has been observing neutrino reactions with the chlorine atoms in a 400 000-L tank of perchlorethylene cleaning fluid (■ Fig-

---

*From *Telephone Poles and Other Poems* by John Updike. Copyright © 1963 by John Updike. Reprinted by permission of Alfred A. Knopf Inc.

■ Figure 36.21
Tank of perchlorethylene used in a neutrino detection experiment. Located deep underground at the Homestake gold mine near Lead, South Dakota, the tank is covered with water that provides even more shielding against stray particles that might mimic neutrino reactions. A small fraction of the neutrinos entering the tank undergo reactions that convert chlorine nuclei to argon. Periodically, the argon is flushed from the system and its amount measured.

ure 36.21). The tank is kept in a deep gold mine to shield it from particles that cannot pass through a few thousand feet of earth. Neutrino reactions produce 30 argon atoms per month, which are flushed from the tank every 3 months and counted! The argon production rate is too small, and we don't understand why. Neutrino detectors using water (■ Figure 36.22) and gallium also fail to detect enough neutrinos.

Neutrinos may be able to change form during their 8-min flight from the Sun; models of the nuclear reactions inside the Sun may be wrong in some unknown way. Some astronomers

---

The particle and antiparticle have exactly opposite values of electric charge, lepton number, and baryon number, so the total value of each is zero. A system consisting of a particle and its antiparticle possesses only mass/energy and linear and angular momentum, precisely the conserved quantities possessed by a photon. Thus, the system can *annihilate* to form two photons. For example:

$$e^+ + e^- \rightarrow 2\gamma.$$

Two photons are required in order to conserve the system's momentum.

Figure 36.22
The Kamioka detector. The detector is at a depth of 2700 m in a lead-zinc mine, about 300 km west of Tokyo. The detector contains a total of 1000 photomultipliers (cf. §35.1) that detect photons produced by interactions of particles with the water in the detector.

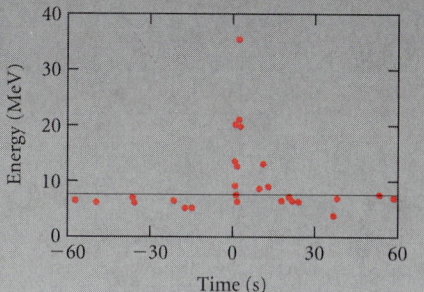

Figure 36.23
Neutrino events observed in the Kamiokande detector on February 23, 1987. The time of the detection coincides with the arrival of light from supernova 1987a in the Large Magellanic Cloud. (This detection was an important confirmation of supernova theory.)

Figure 36.24
The region of the Large Magellanic Cloud surrounding supernova 1987a before (top) and after the supernova outburst. The arrow indicates the star prior to the explosion.

have proposed that the Sun may turn off its energy production periodically. Because the photons produced inside the Sun take about 10 million y to work their way out, we would have to wait that long to detect a reduction in heat and light.

Neutrinos from outside the solar system were detected in early 1987 (■ Figure 36.23), when two laboratories detected a total of 16 neutrinos at the same time as astronomers first observed light from the explosion of supernova SN1987a in a nearby galaxy (■ Figure 36.24). The explosion compressed the star's core, forcing protons and electrons together in reverse β decay, converting the entire core to neutrons. The number of neutrinos observed is consistent with the theoretical expectation.

The inverse reaction is called *pair production*. An energetic photon disappears and produces a particle-antiparticle pair: $\gamma \rightarrow e^- + e^+$. The energy of the photon must exceed the rest energy of the two electrons:

$$E_\gamma > 2m_e c^2 = 1.022 \text{ MeV}.$$

Pair production by a single photon alone does not conserve momentum, but the process can occur near a charged particle or in an intense electromagnetic field, where momentum can be

See Problem 76.

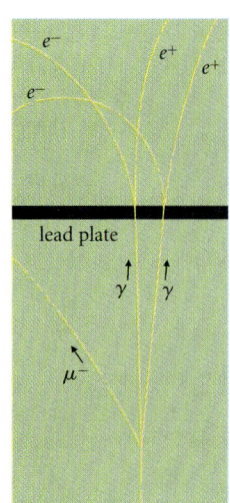
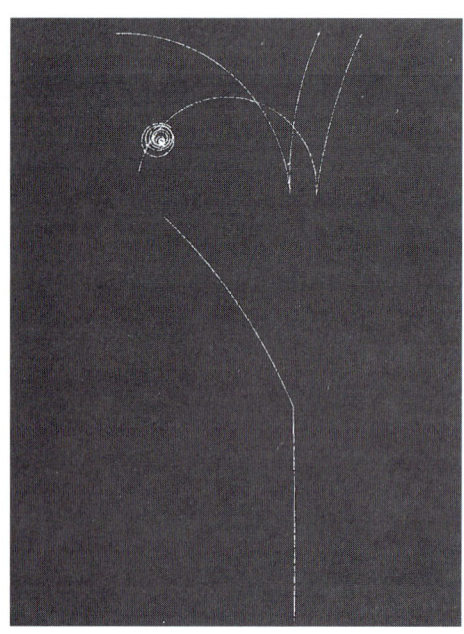

■ FIGURE 36.25
Pair production. The K meson entering the bubble chamber at the bottom decays into an antineutrino, a μ meson, and two γ rays. The γ rays pass through a lead plate placed across the chamber, where they produce electron-positron pairs. (Notice that the tracks curve in opposite directions, indicating the opposite signs of charge.) The antineutrino does not interact at all and its track remains invisible.

transferred to the particle or the field. Pair production is the principal energy loss mechanism for high-energy γ rays passing through matter (■ Figure 36.25).

Some nuclei are unstable against positron emission. The daughter nucleus has one more neutron and one fewer proton than its parent:

THIS PROCESS IS CALLED β^+ DECAY.

$$^A Z \to {}^A(Z-1) + e^+ + \nu_e + Q. \tag{36.6}$$

Here ν_e represents an electron neutrino, which is the antiparticle of the electron antineutrino produced in β decay.

EXERCISE 36.3 ❖ What are the baryon number and lepton number of each particle involved in a positron emission?

36.2.5 γ Decay

In γ decay, a nucleus emits a photon, which carries neither baryon number, lepton number, nor electric charge. Neither the atomic number nor the atomic mass number of the parent nucleus changes in the decay. The parent and daughter in this case are just different energy states of the same nuclear species (■ Figure 36.26). The process is very similar to the emission of photons by transitions of electrons between atomic levels, except that γ emission is only crudely modeled by transitions involving a single nucleon. The general formula for γ emission is:

$$^A Z^* \to {}^A Z + \gamma, \tag{36.7}$$

where * denotes a nucleus excited from its ground state. For example:

$$^{57}\text{Fe}^* \to {}^{57}\text{Fe} + \gamma.$$

To conserve momentum, the daughter nucleus recoils and carries away a small kinetic energy that is usually, but not always, unimportant.

SEE THE ESSAY ON A GEOMETRIC THEORY OF GRAVITY.

If a photon of just the right energy strikes an iron nucleus, γ emission can go in reverse. The γ emission by $^{57}\text{Fe}^*$ is the transition used by Pound and Rebka in their search for the gravitational redshift. With even a small nuclear recoil, an emitted photon would not have enough energy to cause the inverse reaction, and the Pound–Rebka experiment would be impossible. In iron, however, individual atoms are bound in a crystal that absorbs the recoil

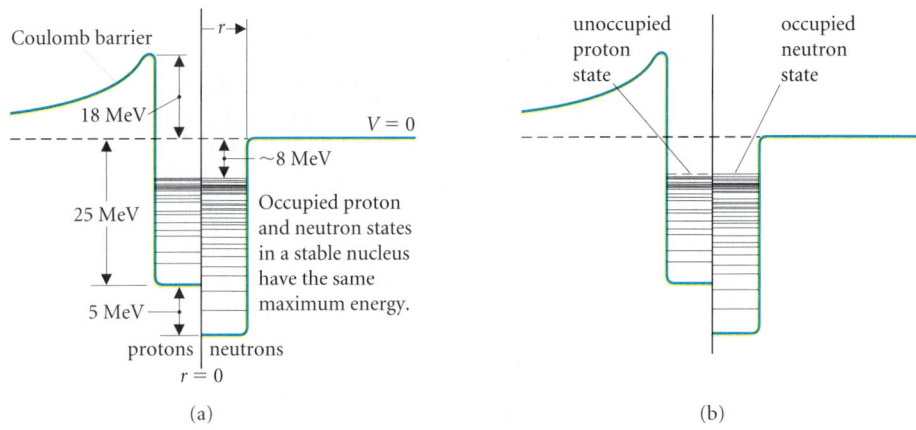

■ **FIGURE 36.26**
The shell model. Individual nucleons move rapidly throughout the volume of the nucleus influenced by an average potential energy function due to all the other nucleons. Neutrons experience only the strong force, whose potential is represented in the right half of the figure. Protons have, in addition, positive electric potential energy. The energy values shown here are estimates for a heavy nucleus like uranium. The energy states for a particle in this collective potential are represented by the horizontal lines in the figure. Individual nucleons fill these states in a manner similar to that of the electron states in a neutral atom. The Pauli exclusion principle allows only two nucleons in each level so neutrons and protons must occupy ever higher quantum number states in larger nuclei.

Gamma ray emission by nuclei occurs for the same reason that electrons in atoms emit photons. A nucleon, excited to a state above the lowest unoccupied state emits a γ ray during transition to a lower energy state. Absorption of a γ ray results in transition of a nucleon to a higher state. (a) In stable nuclei, neutrons and protons occupy states at nearly the same energy. Neutrons cannot decay because the resulting proton would have to enter a state with higher quantum numbers and would require energy exceeding the mass difference between the neutron and proton. (b) A nucleus with an unoccupied proton state close in energy to an occupied neutron state is unstable against β emission.

momentum with negligible change in energy. The resulting photons are precisely tuned to be reabsorbed and so provide a tool for measuring small frequency shifts due to relative motion or different height in a gravitational field.

> **EXERCISE 36.4** ♦♦ What is the kinetic energy of a free ^{57}Fe nucleus after the emission of the 14.4-keV x ray? If the half-life of the γ decay is 10^{-7} s, compare the energy of the nucleus with the uncertainty in the x-ray energy obtained from Heisenberg's principle. How large a crystal must the iron nucleus couple with so that the recoil energy loss is less than the uncertainty in the energy?

36.2.6 The Law of Decay

All three kinds of radiation (α, β, and γ) follow a simple statistical law.

> The number of nuclei per second that decay—that is, emit a particle—is proportional to the number N of nuclei present in the sample:
>
> $$R = \lambda N.$$

RADIOACTIVE DECAY IS GOVERNED BY QUANTUM MECHANICS, WHICH ONLY ALLOWS US TO KNOW THE PROBABILITY THAT ANY FUTURE STATE WILL OCCUR (CF. CHAPTER 35). THUS THE LAW OF DECAY IS A DIRECT CONSEQUENCE OF QUANTUM MECHANICS.

The decay rate R equals the rate of decrease in the number of parent nuclei:

$$\frac{dN}{dt} = -R = -\lambda N. \qquad (36.8)$$

WE'VE SEEN THIS EQUATION BEFORE. SEE, FOR EXAMPLE, §§13.3.4, 14.4, 31.1.

In a given time interval, every parent nucleus in a sample has an equal probability of decaying. In a large sample, the number of nuclei that actually decay equals the number present multiplied by the chance each has of decaying. The constant of proportionality λ, the probability of decay per unit time, depends on the particular parent nucleus and on the particular type of decay. Values of λ are usually measured experimentally and range over more than 40 orders of magnitude. The solution of the differential equation (36.8) is:

$$N(t) = N_o e^{-\lambda t}, \tag{36.9}$$

where N_o is the number of parent nuclei present in the sample at $t = 0$. The rate of particle emission also decreases exponentially:

$$R(t) \equiv -\frac{dN}{dt} = \lambda N_o e^{-\lambda t}. \tag{36.10}$$

Equation (36.10) expresses the characteristic experimental feature of radioactive decay: exponential decrease of the decay rate.

Nuclear decay may also be described by the half-life of a parent nucleus.

> The *half-life* $T_{1/2}$ of a nuclear species is the time required for half of a large sample to decay.

According to eqn. (36.9), the number of parent nuclei remaining after one half-life is:

$$N(T_{1/2}) \equiv N_o/2 = N_o \exp(-\lambda T_{1/2}).$$

Canceling the common factor of N_o and taking natural logarithms, we find:

$$-\ln 2 = -\lambda T_{1/2},$$

or
$$T_{1/2} = (\ln 2)/\lambda. \tag{36.11}$$

EXERCISE 36.5 ♦♦ Show that the law of decay may be expressed in the form:

$$N(t) = N_o 2^{-t/T_{1/2}}.$$

EXAMPLE 36.4 ♦♦ A sample of ^{214}Pb nuclei (half-life = 3.05 min) initially emits 352 β-particles per second. When will the emission rate have decreased to 10 /s?

MODEL The decay rate is proportional to the number of parent nuclei.

SETUP We use the result of Exercise 36.5. At $t = 0$, $R(0) = R_o = \lambda N_o$. Then:

$$\frac{R}{R_o} = \frac{N}{N_o} = 2^{-t/T_{1/2}}.$$

SOLVE Taking logarithms:
$$\ln\left(\frac{R}{R_o}\right) = -\frac{t}{T_{1/2}} \ln 2.$$

So,
$$t = T_{1/2} \frac{-\ln(R/R_o)}{\ln 2}$$

$$= (3.05 \text{ min}) \frac{\ln(35.2)}{0.69315} = 15.7 \text{ min}.$$

ANALYZE The emission rate decreases by a factor of 2 per half-life. In five half-lives, the rate decreases by $2^5 = 32$.

The decay rate R of a radioactive sample is called its *activity*. The SI unit for activity is named after Henri Becquerel.

Digging Deeper

γ-Ray Imaging

Doctors can examine a patient's internal organs by injecting a radioactive substance that is absorbed by the organ and then detecting γ rays from the absorbed substance. The most popular type of detector is the Anger camera (■ Figure 36.27), which measures the location of light pulses (scintillations) produced when γ rays are absorbed in a sodium iodide crystal. The resulting image of the organ is displayed on a television screen (■ Figure 36.28).

The radioisotope most commonly used is ^{99m}Tc, an excited state of ^{99}Tc with a half-life of 6 h. The 140-keV γ ray emitted by ^{99m}Tc is strongly absorbed by NaI, allowing use of a thin detector crystal and a correspondingly light and versatile camera. Furthermore, technetium enters a variety of nontoxic chemical compounds designed for absorption by different organs.

The medical goal of imaging is to make an accurate diagnosis while minimizing the radiation dose to the patient. Trade-offs are inevitable. For example, the radiation dose would be minimized by choosing a radioisotope with a half-life of about 70% of the time for the target organ to absorb the isotope. In thyroid examinations, 2 mCi of sodium pertechnetate ($Na^{99m}TcO_4$) are injected 2 h before the image is taken. The 6-h half-life is not the optimum value, but this disadvantage is much less important than the advantages of efficient camera design and effective medical chemistry.

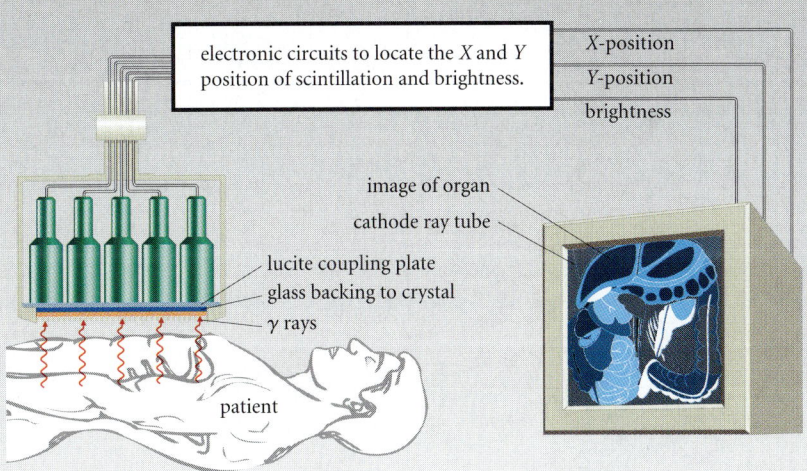

■ **Figure 36.27**
Schematic of an Anger-type camera. The γ rays from nuclear decay interact with the sodium iodide crystal, producing visible light that is detected by the array of photomultiplier tubes (cf. §35.1) and converted to an image on the cathode ray tube.

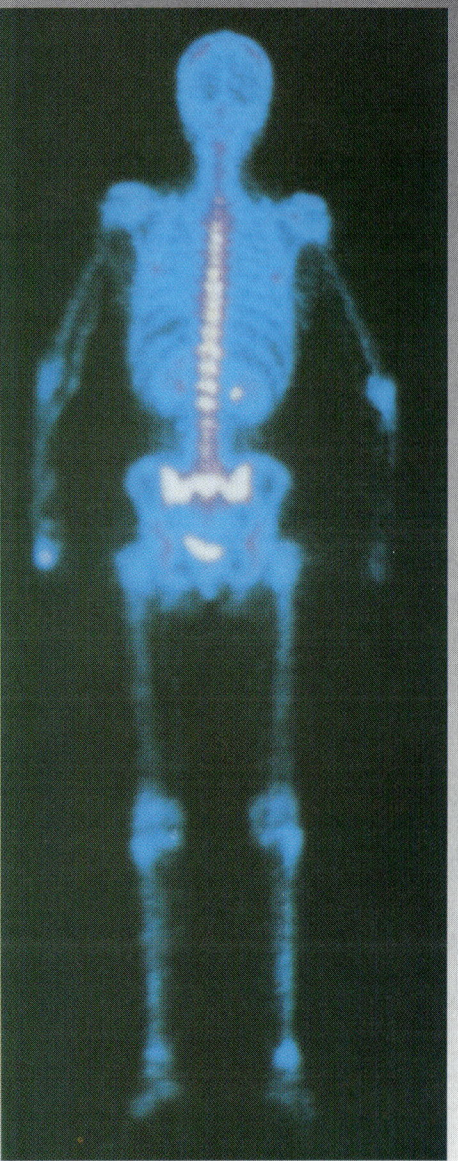

■ **Figure 36.28**
Whole skeleton image produced by an Anger camera using ^{99}Tc.

One *becquerel* (1 Bq) is an activity of one disintegration per second.

$$1 \text{ Bq} \equiv 1 \text{ s}^{-1}.$$

A unit still in common use is named after Marie Curie. The curie (Ci), originally the activity of 1 gram of pure radium metal, is now defined in terms of the becquerel:

$$1 \text{ Ci} \equiv 3.7 \times 10^{10} \text{ Bq}.$$

■ **FIGURE 36.29**
Radioactive series. (a) The $4n$ series begins with ^{232}Th, which has a half-life of 14 billion years. It passes through elements with mass number $A = 4n$, $n = 58-52$, ending with ^{208}Pb, which is stable. The series is a sequence of α decays (orange lines) and β decays (green lines). In some instances, two possible paths are available. For example, ^{212}Bi α-decays to ^{208}Tl 34% of the time and β-decays to ^{212}Po 66% of the time. The half-life of each element in the series is indicated on the chart.

(b) The $4n + 1$ series begins with ^{237}Np. Its half-life of 2 million years is so short that there is essentially none left in nature. The series ends at the stable isotope ^{209}Bi.

(c) The $4n + 2$ series begins with ^{238}U and ends with ^{206}Pb; ^{234}Th β-decays to an excited state of ^{234}Pa, which can itself β-decay or γ-decay to the ground state, which subsequently β-decays.

(d) The $4n + 3$ series begins with ^{235}U and ends with ^{207}Pb.

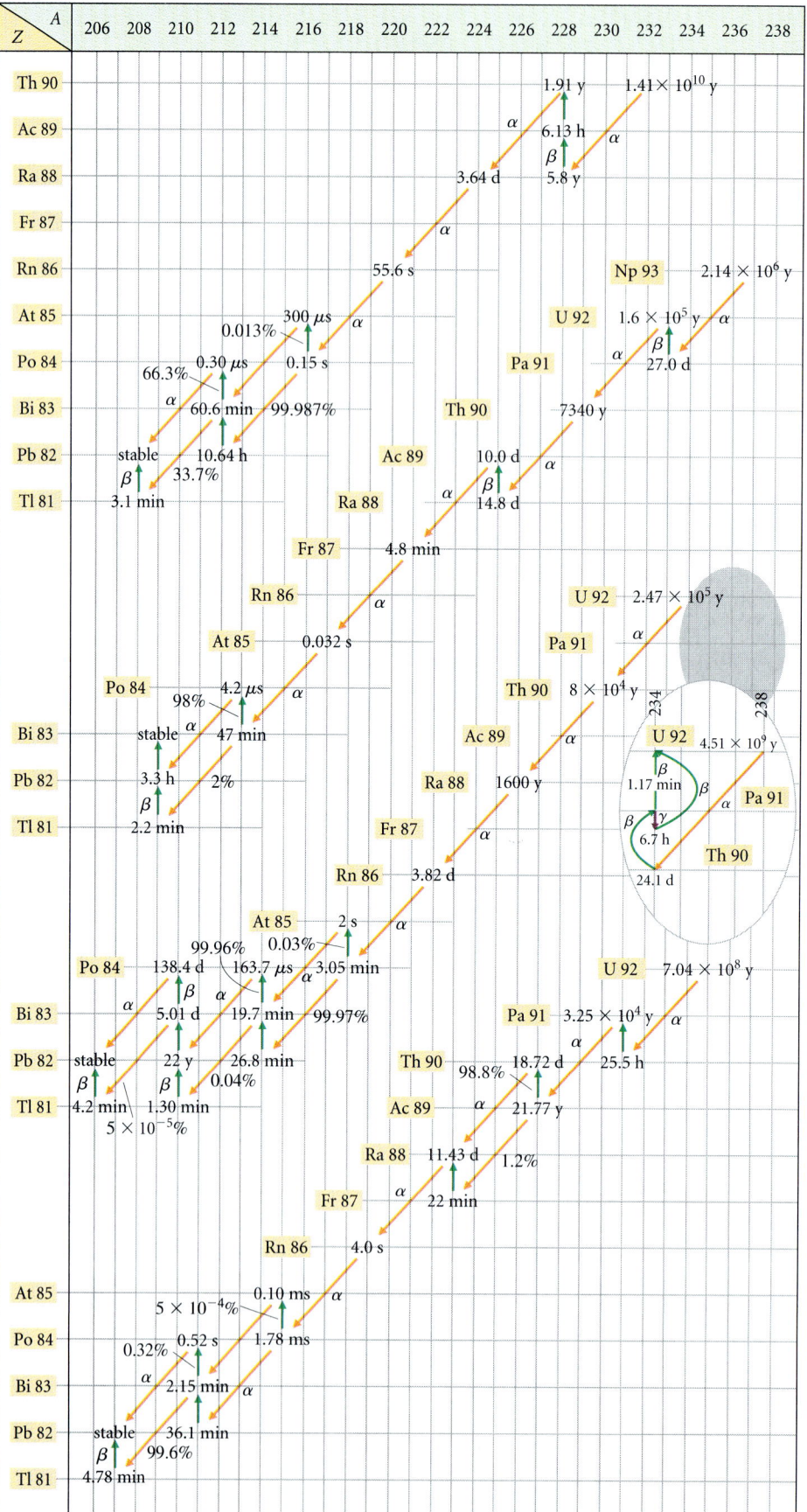

36.2.7 Radioactive Series

Among the heavier nuclei, radioactive decay frequently results in an unstable daughter nucleus, which then itself decays to form granddaughter nuclei, and so forth, until at last a stable form is produced. In such a series, β decays do not alter the atomic mass number and α decays each reduce A by 4. So all the nuclear species in a series have mass numbers that differ from each other by whole multiples of 4, and four distinct series exist (■ Figure 36.29). Three of the series are found in nature. In each case, the heaviest isotope has a half-life of a billion or so years. These heavy elements are thought to have been produced by exploding stars shortly before the solar system formed, some 4.5 billion years ago. Elements heavier than the series leader have long since decayed. The fourth possible series begins with the isotope ^{237}Np, whose half-life is too short for a noticeable quantity to remain in nature.

The elements lighter than the series leader have much shorter half-lives. They are continually formed from earlier elements in the series and decay into later members. In a particular sample of material, the series reaches an equilibrium in which each species is formed at the same rate it decays. The series is like a sequence of lakes connected by a mountain stream. Water melts from the parent snowpack and flows through the lakes. The level of each lake adjusts until water flowing in is balanced by water flowing out (■ Figure 36.30).

Radioactive series provide a useful tool for dating geological samples. As ^{238}U in the rock decays, ^{206}Pb is produced at the end of the series. Similarly, ^{235}U decays to ^{207}Pb. Assuming that no lead was present originally, the ratios of ^{238}U/^{206}Pb or ^{235}U/^{207}Pb determine the age of the sample via the law of decay. Of the naturally occurring lead isotopes, only ^{204}Pb is not a product of radioactivity. This isotope may be used to check the assumption of zero initial lead.

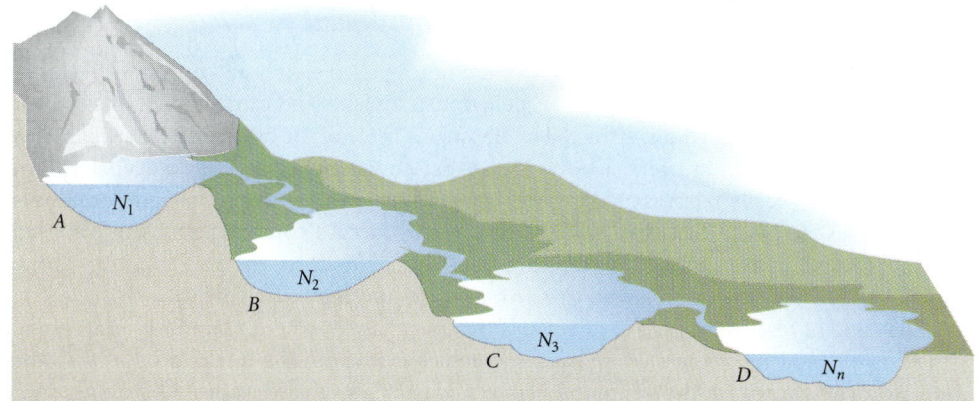

■ **Figure 36.30**
Water analogy for radioactive series. The decay of the parent element is like water pouring from a large reservoir, through a small outlet into a series of lakes and streams. The level of each lake adjusts until water flowing out of it just balances the water flowing in. The water ultimately enters a reservoir with no outlet, which corresponds to the stable end product of the series.

EXAMPLE 36.5 ♦♦ A 1.0-g sample of granite contains 0.16 μg of ^{238}U and 0.13 μg of ^{206}Pb. No ^{204}Pb was found. Estimate the age of the rock.

MODEL The ^{206}Pb is produced by the decay of ^{238}U. Because there is no ^{204}Pb, we assume there was no lead in the rock when it was formed.

SETUP If the sample initially contained N_o atoms of uranium, after a time t, $N(t) = N_o e^{-\lambda t}$ would remain. We assume all the decayed atoms have reached the end of the series. Thus:

$$\frac{N_{Pb}}{N_U} = \frac{N_o(1 - e^{-\lambda t})}{N_o e^{-\lambda t}} = e^{\lambda t} - 1.$$

The ratio of masses is:
$$\frac{M_{Pb}}{M_U} = \frac{A_{Pb} N_{Pb}}{A_U N_U} = \frac{206}{238}(e^{\lambda t} - 1).$$

SOLVE
$$e^{\lambda t} = 1 + \frac{238 M_{Pb}}{206 M_U} = 1 + \frac{238(0.13 \ \mu g)}{206(0.16 \ \mu g)} = 1.94.$$

$$\lambda t = \frac{\ln(2)}{T_{1/2}} t = \ln(1.94).$$

$$t = T_{1/2} \frac{\ln(1.94)}{\ln(2)} = (4.51 \times 10^9 \text{ y})(0.96) = 4.3 \times 10^9 \text{ y}.$$

ANALYZE The result is independent of N_o, the original amount of ^{238}U present. A check on the answer could be obtained from the ratio ^{235}U/^{207}Pb in the same sample. ∎

Uranium is present in small quantities in the ground in most parts of the country. One of the products of uranium decay is radon gas (^{222}Rn). With a half-life of about 4 d, radon can diffuse out of the ground before it decays. The gas mixes into the atmosphere and disperses. However, if the gas diffuses through the foundations into a modern home that has been tightly sealed for fuel efficiency, radon can build up within the house, forming an air pollutant that can be a health hazard. Not all radon is of natural origin. Uranium was used in a bright orange paint during the 1930s and 1940s. Some dishes made with this paint have become collector's items that can contribute to the radon level within a home. The hazard can be readily avoided by ensuring that your home is adequately ventilated.

36.3 NUCLEAR REACTIONS

Interactions between nuclei depend strongly on the energy available. At energies of ≲ 1 MeV, the nuclei collide elastically, while at energies of several GeV per nucleon, they disintegrate into a spray of small fragments. The subject of reactions in general is much too vast to be dealt with here, so we shall discuss only a few specific reactions of particular interest. The same conservation laws we used in the study of radioactive decay apply to nuclear reactions, although the reactions often involve more complex initial and final states than occur in simple decay.

36.3.1 Transmutation by Neutron Bombardment

Because neutrons have no electric charge and are not repelled electrically by a nucleus, even neutrons moving with thermal speeds can penetrate a nucleus and cause a reaction. Once it gets close enough, the neutron is attracted by the strong force of the other nucleons and forms a *compound nucleus* with the same Z but larger A than the original target nucleus. Absorption of the neutron typically releases about 8 MeV of binding energy, leaving the compound nucleus in an excited state that then undergoes γ decay. Decays continue until the resulting product is stable. For example, the resulting nucleus may β-decay to form an atom of one higher Z. The original element is said to be *transmuted* into the new element.

This kind of reaction is used in neutron activation analysis, where a sample is bombarded with neutrons and its original composition is determined from the nuclei produced in the bombardment. One use of the technique is in testing whether valuable paintings are authentic. The induced radiation is short lived, and the painting need not be damaged to remove a sample for chemical analysis. Many of the chemical elements in the solar system were formed by neutron-capture reactions in the interior of earlier stars.

EXAMPLE 36.6 ♦♦♦ Neglecting any kinetic energy of the incident neutron, what are the energies of the γ ray and the electron + neutrino produced in the transmutation of ^{13}C into ^{14}N?

MODEL When the neutron is absorbed by the carbon nucleus, its mass number increases by 1 to form ^{14}C, which then β-decays to ^{14}N, releasing energy Q. The formation of the compound nucleus and its γ decay follow eqn. (36.7). Equation (36.5) describes the subsequent β decay. We neglect the initial kinetic energy of the neutron and any recoil kinetic energy of the ^{14}C nucleus.

SETUP Applying eqn. (36.7):

$$n + {}^{13}C \to {}^{14}C^* \to {}^{14}C + \gamma.$$

and eqn. (36.5): $\quad {}^{14}C \to {}^{14}N + e^- + \bar{\nu}_e + Q.$

Then, the energy of the photon is:

$$E_\gamma = c^2(\text{neutron mass} + \text{mass of } {}^{13}C - \text{mass of } {}^{14}C).$$

SOLVE We take the mass values from Tables 36.1 and 36.2.

$$E_\gamma = c^2(1.00866 + 13.00336 - 14.00324)(u)(931.5 \text{ MeV}/c^2 \cdot u)$$
$$= 8.2 \text{ MeV}.$$

SETUP Neglecting recoil of the daughter nucleus, the energy Q released in the β decay is transformed into kinetic energy of the electron and the neutrino.

SOLVE
$$Q = c^2(\text{mass of } {}^{14}\text{C} - \text{mass of } {}^{14}\text{N})$$
$$= c^2(14.00324 - 14.00307)(u)(931.5 \text{ MeV}/c^2 \cdot u)$$
$$= 0.16 \text{ MeV}.$$

ANALYZE Once again, subtraction of numbers that are almost equal greatly reduces the number of significant figures in the results. Note that ^{14}N is ordinary nitrogen, the major constituent of air.

REMEMBER: THE TABULATED MASSES ARE ATOMIC MASSES, SO THE EXTRA ELECTRON IS INCLUDED IN THE MASS OF ^{14}N.

36.3.2 Energy Generation in Stars

A star has to have a central temperature of millions of kelvin so that gas pressure can support its material against its own gravitational attraction. At the same time, the star cools by radiation so that an intense energy source is needed to maintain its temperature. Nuclear reactions provide that energy source. Stars form out of material that is primarily hydrogen, and the most important reaction is the fusion of hydrogen to form helium. In a star like the Sun, the reaction begins with the fusion of two protons to form deuterium:

ARTHUR EDDINGTON WAS THE FIRST TO SUGGEST THAT NUCLEAR REACTIONS COULD POWER THE SUN. CRITICS ASSERTED THAT THE SUN WAS NOT HOT ENOUGH, TO WHICH HE REPLIED "THEN FIND ME A HOTTER PLACE!"

The Proton–Proton Cycle

Reaction	Energy Release (MeV)
$p + p \rightarrow {}^2\text{H} + e^+ + \nu_e$	1.442
${}^2\text{H} + p \rightarrow {}^3\text{He} + \gamma$	5.493
${}^3\text{He} + {}^3\text{He} \rightarrow {}^4\text{He} + 2p$	12.859

Stars more massive than the Sun have higher central temperatures at which a sequence of reactions catalyzed by carbon, nitrogen, and oxygen synthesize helium more rapidly than in the proton–proton cycle.

EVEN IN THE SUN, SOME He IS SYNTHESIZED VIA THE CNO CYCLE.

The CNO Cycle

Reaction	Energy Release (MeV)
$p + {}^{12}\text{C} \rightarrow {}^{13}\text{N} + \gamma$	1.944
${}^{13}\text{N} \rightarrow {}^{13}\text{C} + e^+ + \nu_e$	2.221
$p + {}^{13}\text{C} \rightarrow {}^{14}\text{N} + \gamma$	7.550
$p + {}^{14}\text{N} \rightarrow {}^{15}\text{O} + \gamma$	7.293
${}^{15}\text{O} \rightarrow {}^{15}\text{N} + e^+ + \nu_e$	2.761
$p + {}^{15}\text{N} \rightarrow {}^{12}\text{C} + {}^4\text{He}$	4.965

The carbon, nitrogen, and oxygen nuclei are not consumed by the reaction cycle so that a relatively small number of these nuclei can process a large number of the star's protons into helium.

Because a proton needs a large amount of kinetic energy to overcome the Coulomb repulsion between nuclei, both of these reaction cycles are extremely temperature-sensitive. Under conditions in stellar interiors, the rate of the proton–proton cycle is proportional to T^{12} and the rate of the CNO cycle to T^{20}. It is much easier for two protons to come close enough to react than for a proton to approach a more heavily charged CNO nucleus. However, a compound nucleus formed from two protons is much more likely to separate again rather than become ^{2}H via β decay. The compound nuclei in the CNO cycle tend to β-decay rather than dissociate. Consequently, the CNO is the more efficient cycle at temperatures where it occurs at all.

Digging Deeper

WE ARE CHILDREN OF THE STARS

During their lives, stars serve as factories for fusing hydrogen into heavier elements (■ Figure 36.31a). Helium formation releases the greatest binding energy and lasts for the greatest part of the star's lifetime (about 10^{10} y for the Sun but as short as 10^6 y for the biggest and brightest stars). To maintain energy production as its core is converted to helium, the star contracts, causing its temperature to increase and the zone of hydrogen fusion to move outward in a shell surrounding the helium core. Gravitational contraction raises the central temperature until helium can fuse into carbon:

Triple α Process

$$^4\text{He} + {}^4\text{He} \rightarrow {}^8\text{Be} + \gamma.$$
$$^4\text{He} + {}^8\text{Be} \rightarrow {}^{12}\text{C} + \gamma.$$

This reaction stabilizes the star for a few thousand years until a carbon core builds up, requiring further contraction.

In small stars like the Sun, the core stops contracting because its electrons cannot move closer together without sharing quantum numbers and thus violating the Pauli exclusion principle. The core no longer has to remain hot to support itself, and the star ends its life as a slowly cooling object about the size of the Earth. In larger stars, the core becomes hot enough to start a carbon-burning reaction. Eventually, such stars build an onionlike structure, ■ Figure 36.31b, adding layers ever faster until an iron core forms at the center.

At this point, the star runs out of nuclear resources. Very high temperatures are required to overcome the Coulomb repulsion between large nuclei, and a large flux of γ rays is associated with matter at high temperatures (see §35.1.3). These γ rays drive inverse nuclear reactions that dissociate the large nuclei. The balance of fusion and fission tends to produce a majority of ^{56}Fe nuclei, which have the maximum binding energy per nucleon. The continuing loss of energy to neutrinos, which escape the star, causes its ultimate demise. Contraction proceeds rapidly, and catastrophe occurs in a few hours as the temperature of the core continues to increase and dissociation of iron nuclei by γ rays dominates fusion.

The star is forced to return the nuclear energy it has used for millions of years. The core collapses in less than a millisecond, squeezing itself into a ball of neutrons and emitting a flash of some 10^{57} neutrinos (Figure 36.23). Unsupported by the collapsed core, the outer layers fall inward and heat up. Nuclear reactions in the still fresh outer material, together with the neutrino flash, produce a vast explosion that reverses the infall and hurls the outer layers back into space. This material, much of it now processed into heavy elements, mixes with the interstellar gas. Millions of years later, these elements join in the formation of new stars and provide the stuff to make planets and people. We are the product of several generations of stars that preceded the Sun.

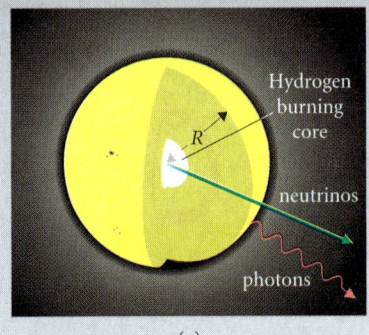

(a)

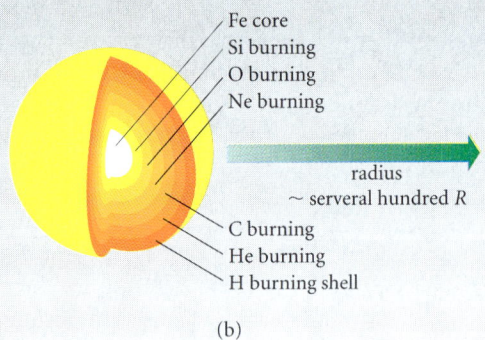

(b)

■ **Figure 36.31**
Structure of a star at various stages of its evolution. (a) During most of its lifetime, a star burns H to He in its inner core. (b) Late red giant phase. Shells have developed around the core. In each shell, a different nuclear reaction occurs, the heavier elements being produced closer to the center. The process terminates when an iron core is formed because elements heavier than iron are less tightly bound (Figure 36.9). If the star is sufficiently massive, it collapses violently, followed by a supernova explosion (Figure 36.24). Heavy elements can be formed by neutron bombardment during the explosion.

The difficulty in forming deuterium (^{2}H) from protons is equally significant for fusion power research, where the chief problem is to confine a dense, hot plasma long enough for the plasma nuclei to form helium. Current designs use deuterium and tritium (^{3}H) as fuel and avoid the troublesome proton-fusion step.

36.3.3 Fission

Nuclear *fission* is the breakup of a nucleus into smaller fragments.

Fission reactions are possible because the binding energy per nucleon for elements heavier than iron decreases with increasing mass number. Massive nuclei can thus release potential energy by reorganizing into smaller nuclei (■ Figure 36.32). Spontaneous fission, a common mode of decay for artificially produced nuclei more massive than uranium, also occurs in uranium, though at a much lower rate than α decay. In nuclear power plants, induced fission results when a free neutron is absorbed by a uranium (235) nucleus. The 8 MeV released when the neutron is absorbed leaves an excited compound nucleus that has a high probability of fission.

The fission reaction is useful because it is a messy reaction that releases neutrons as well as daughter nuclei. If these neutrons strike other uranium nuclei and result in further fissions, it is possible for a *chain reaction* to release energy from every atom in a macroscopic quantity of material. The uranium sample need only be large enough to give a daughter neutron a high probability of causing fission before escaping the sample or being absorbed.

The probability of a neutron causing fission is described by the cross section, or effective target area, σ, of a uranium nucleus. We showed in §22.7 that a molecule traveling through a gas with n molecules per cubic meter, each with cross-sectional area σ, travels an average distance $\ell = 1/n\sigma$ before striking another molecule. A similar formula describes the average distance a neutron travels through a sample of uranium before causing a fission reaction. Thus ℓ gives the size of a uranium sample that can chain-react. The criterion for a chain reaction is usually described by a critical mass. Roughly speaking, this is the mass in a sphere of radius ℓ. The critical mass depends on the concentration of uranium in the sample, but it is typically of the order of a kilogram.

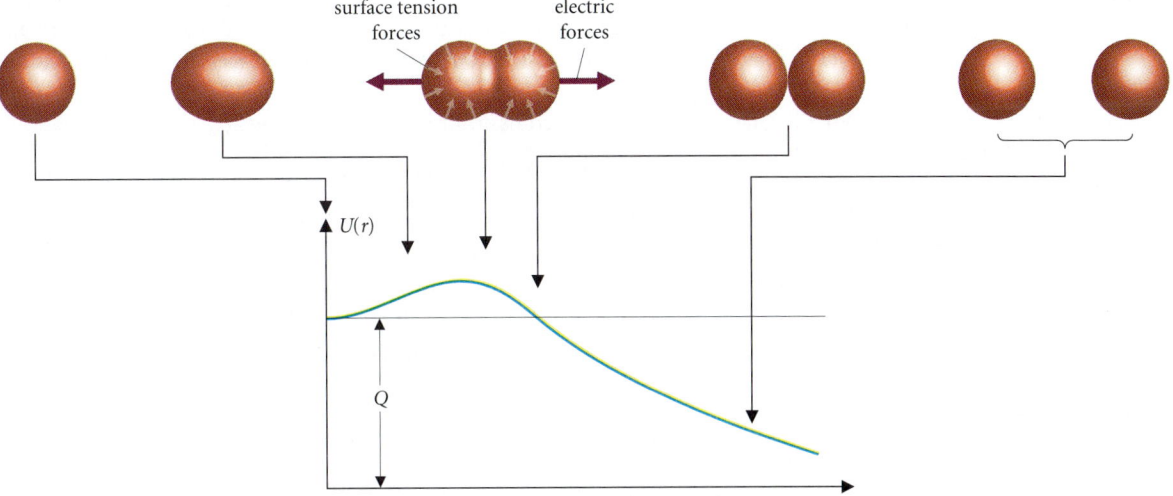

■ **Figure 36.32**
The liquid drop model. Nuclear fission may be understood with a model that, in contrast to the shell model, ignores the behavior of individual nucleons and treats the nucleus as a drop of liquid. The nucleons play the role of molecules of the liquid, and the nuclear forces are treated like the surface tension in a liquid. The fission instability arises when the energy released by absorption of a neutron leaves the nucleus rotating and thus with a larger diameter perpendicular to the rotation axis. Electrical repulsion tends to stretch the nucleus. Nuclear forces are less effective in the distorted, rotating nucleus. As the two parts of the nucleus separate, the surface tension tends to pull them into separate spheres rather than opposing the electric force and holding them together.

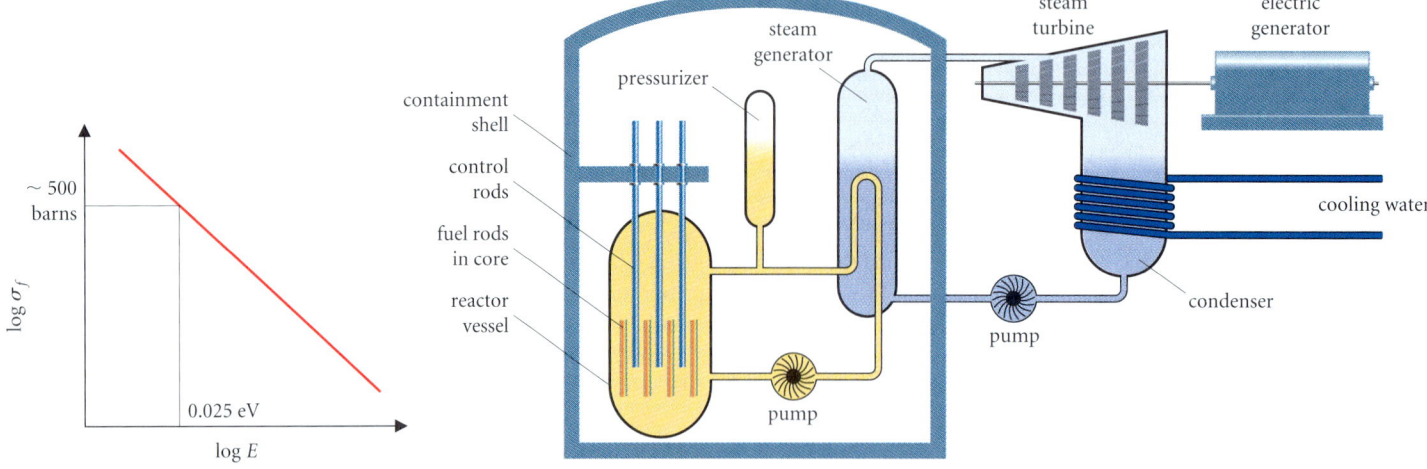

■ **FIGURE 36.33**
Fission cross section of ^{235}U as a function of neutron energy. The unit of nuclear cross section, the *barn*, is equal to 10^{-28} m².

■ **FIGURE 36.34**
Schematic of a pressurized water reactor. Pellets of uranium fuel are placed in fuel rods and immersed in water, which serves both as coolant and moderator of neutrons. Control rods absorb neutrons and stop the reaction when inserted into the reactor core.

PLUTONIUM AND THORIUM ARE ALSO USED. THORIUM IS A NATURALLY OCCURRING ELEMENT THAT PLAYS A MINOR ROLE IN TECHNOLOGY. PLUTONIUM IS PRODUCED IN URANIUM REACTORS AND THEN ITSELF USED AS FUEL.

The fission cross section of the fissionable isotope of uranium, ^{235}U, decreases rapidly at high neutron energy (■ Figure 36.33). For the neutrons released in fission of uranium, the cross section is so low that an impractical amount of material would be needed for a reactor, unless the neutrons can be slowed down, or *moderated*. ■ Figure 36.34 shows the schematic of a typical reactor operated in the United States. The uranium is concentrated in fuel pellets that are much too small to sustain a chain reaction. The water surrounding the pellets serves both to transport the heat released by the reactor and to moderate the neutrons. Collisions with the nearly equal-mass hydrogen nuclei in the water efficiently transfer the neutrons' kinetic energy to the water. The control rods are usually made of cadmium, which absorbs neutrons efficiently. The reactor is designed so that, with the control rods completely inserted, any neutron is almost certainly absorbed before causing a fission. In operation, the rods are adjusted so that the neutrons from each fission, on average, produce precisely one new fission.

EXAMPLE 36.7 ♦♦ An induced fission reaction is described by:

$$n + {}^{235}U \rightarrow {}^{236}U^* \rightarrow {}^{141}Ba + {}^{92}Kr + 3n + Q.$$

Estimate the energy of each neutron produced in this reaction.

MODEL Since the neutrons are much less massive than the two nuclei that are produced, nearly all the energy released is shared by the three neutrons. For this estimate, we may assume that, on average, they share the energy equally.

SETUP The energy released is due to the mass difference between the reacting particles and the fission products.

$$Q = c^2[\text{mass of } {}^{235}U - (\text{mass of } {}^{141}Ba + {}^{92}Kr + 2n)]$$
$$= c^2[235.0439 - 140.9141 - 91.9262 - 2(1.0087)](u)(931.49 \text{ MeV}/c^2 \cdot u)$$
$$= (0.1862)(931.49) \text{ MeV}$$
$$= 173.4 \text{ MeV}.$$

SOLVE The energy per neutron is $Q/3 = 57.8$ MeV.

ANALYZE This energy is off scale in Figure 36.33. The cross section for one of these unmoderated neutrons to produce another fission is tiny.

Chapter Summary

Where Are We Now?

In this chapter we have delved into the atomic nucleus to find that it is composed of more fundamental particles—nucleons—and that these particles can be split apart and rearranged in nuclear reactions, thus changing one chemical species into another. In our efforts to understand the reactions between the nucleons, we have identified the strong and weak nuclear forces and the quantum mechanical quantities of lepton number and baryon number. In the next chapter we shall pursue our quest to even smaller sizes and observe that the nucleons themselves are composite structures.

What Did We Do?

Atomic nuclei are made up of nucleons: neutrons and protons. A nuclear species is described by its atomic number Z and atomic mass number A. A nucleus contains Z protons and $N \equiv A - Z$ neutrons. The two kinds of nucleon have similar masses. Protons each have electric charge $+e$ and neutrons are electrically neutral.

Besides the electrical force we have encountered previously, two kinds of force act between nucleons: the strong force and the weak force. The strong force has short range and cannot bind nuclei larger than a few fermis in radius. Stable nuclei generally contain more neutrons than protons since neutrons enhance the stabilizing strong force but do not share the electrical repulsion of the protons. The weak force is responsible for reactions that change neutrons to protons (β decay), and vice versa. Nuclei with too large an excess of neutrons are unstable against β decay.

There are three principal categories of natural radioactivity. In α decay, a parent nucleus decays to a daughter with the emission of an α-particle (helium nucleus). Both the charge and mass of the nucleus are changed.

$$\alpha \text{ decay: } {}^A Z \rightarrow {}^{(A-4)}(Z-2) + {}^4\text{He} + Q.$$

In β decay, the weak force causes the decay of a neutron within the nucleus, changing the charge of the nucleus but not its mass number A.

$$\beta \text{ decay: } {}^A Z \rightarrow {}^A(Z+1) + e^- + \bar{\nu}_e + Q.$$

In γ decay, a nucleus in an excited state decays to a lower energy state with the emission of a photon (γ ray). All such reactions may be analyzed using conservation laws for energy, charge, lepton and baryon number, and linear and angular momentum. To conserve energy, nuclei always decay to a combination of particles whose total mass is less than or equal to that of the original nucleus. Applying the conservation laws to β decay, we find that a particle called the neutrino must exist. The neutrino is a lepton with zero charge and zero (or very small) mass.

The rate of particle emission by a sample of radioactive material decreases exponentially according to the *law of decay*:

$$\frac{dN}{dt} = -\lambda N.$$

The half-life of a nuclear species is the time required for half of a large sample to decay.

Two important classes of nuclear reactions are fusion, in which elements lighter than iron combine to form more massive nuclei with the release of energy, and fission, in which nuclei heavier than iron split into smaller nuclei, again releasing energy. Fusion of hydrogen to helium provides the energy source in the Sun and other stars, while fission of uranium and related elements is used in nuclear power plants.

Practical Applications

Nuclear power plants have become a major source of energy worldwide and are also used to power some ships and submarines. Nuclear decay has applications in the diagnosis and treatment of various ailments, including thyroid disease and cancer. Radioactive decay provides a reliable method for finding the ages of archaeological and geological samples. Nuclear fusion of deuterium and tritium into helium is being developed as a possible power source for the next millennium.

Solutions to Exercises

36.1 From Figure 36.9, the binding energy per nucleon for $A = 24$ (^{24}Mg) is $B/A \approx 8.3$ MeV, while $B/A \approx 7.1$ MeV at $A = 4$. The energy released by fusing six helium nuclei into one ^{24}Mg nucleus is:
$$\Delta E = 24(8.3 \text{ MeV}) - 24(7.1 \text{ MeV}) = 24(1.2 \text{ MeV}) = 29 \text{ MeV}.$$
Equivalently, 1.2 MeV/nucleon equals:
$$\frac{(1.2 \text{ MeV/nucleon})(1.6 \times 10^{-13} \text{ J/MeV})}{1.7 \times 10^{-24} \text{ g/nucleon}} = 1.1 \times 10^{11} \text{ J/g}.$$

36.2 The equation for β decay of ^{12}B is:
$$^{12}\text{B} \rightarrow {}^{12}\text{C} + e^- + \bar{\nu}_e + Q.$$
The tabulated mass for ^{12}C includes a carbon nucleus and six electrons, while the tabulated value for ^{12}B includes a nucleus and only five electrons. The neutrino is assumed massless. Therefore, the energy released in the reaction is:
$$Q = [M(^{12}\text{B}) - M(^{12}\text{C})]c^2$$
$$= [(12.0144 - 12.0000) \text{ u}](931.49 \text{ MeV/u}) = 13.4 \text{ MeV}.$$

36.3 In the general case of positron emission, the two nuclei ^AZ and $^A(Z-1)$ have zero lepton number and baryon number A. The positron and the electron neutrino have zero baryon number. The positron, being the antiparticle of the electron, has the opposite lepton number of the electron—that is, -1. The total lepton number equals that of the parent nucleus, zero; so the neutrino has lepton number $+1$.

36.4 The photon has momentum $p_\gamma = E/c$. The momentum of the nucleus is equal and opposite to that of the γ, and its kinetic energy is given by the nonrelativistic expression:

$$E_{\text{Fe}} = \frac{p^2}{2M}$$
$$= \frac{[(14.4 \text{ keV})(1.60 \times 10^{-16} \text{ J/keV})/(3.00 \times 10^8 \text{ m/s})]^2}{2(57)(1.66 \times 10^{-27} \text{ kg})}$$
$$= 3.12 \times 10^{-22} \text{ J}$$
$$= 1.95 \times 10^{-3} \text{ eV}.$$

From Heisenberg's principle, the uncertainty in the photon's energy is:
$$\Delta E = \frac{\hbar}{\Delta t} = \frac{1.05 \times 10^{-34} \text{ J}\cdot\text{s}}{(10^{-7} \text{ s})(1.6 \times 10^{-19} \text{ J/eV})} \approx 7 \times 10^{-9} \text{ eV}.$$

For the photon to be reabsorbed by another iron nucleus, its energy (in the rest frame of the absorbing nucleus) must not be shifted by more than the Heisenberg uncertainty. The recoil energy of the nucleus is large enough to prevent reabsorption. To suppress the recoil energy to an amount less than ΔE requires that M be increased by a factor of 3×10^5. This can be achieved if the nucleus couples to a crystal containing 3×10^5 atoms.

36.5 Using eqn. (36.11) in eqn. (36.9), we have:
$$N(t) = N_o \exp[-(\ln 2)t/T_{1/2}].$$
Since $e^{ab} = (e^a)^b$, we may write the law of decay as:
$$N(t) = N_o(e^{\ln 2})^{-t/T_{1/2}}$$
$$= N_o 2^{-t/T_{1/2}}.$$

Basic Skills

Review Questions

§36.1 BASIC NUCLEAR STRUCTURE

- What is the *atomic number Z* of an element?
- What is the *atomic mass number A* of an element? What rule gives the relation of A to Z?
- Which nuclear property determines the chemical behavior of an atom? How do *isotopes* of an element differ?
- How is the *atomic mass unit* defined?
- How is the radius of a nucleus related to its mass number A?
- Approximately how big is a carbon nucleus?
- What are the three basic constituents of atoms?
- What are *nucleons*?

- What are the two kinds of nuclear force? Which is responsible for holding nuclei together?
- What is β decay? Which nuclear force is responsible for this process?
- What is the *binding energy* of a nucleus?

§36.2 NATURAL RADIOACTIVITY

- List three types of natural radioactivity.
- Which quantities are conserved in radioactive decay?
- Describe α decay. How are the mass and charge of the daughter nucleus related to those of the parent?
- What particles emerge from a nucleus that undergoes β decay?
- Which nuclear quantities change as a result of γ decay?
- What is a *positron*? Which of its parameters differ from those of an electron?
- What happens when a particle and its antiparticle collide?
- What is the *law of decay*?
- What is the *half-life* $T_{1/2}$ of an atomic species? How is it related to the decay probability λ?
- What is the *activity* of a radioactive sample? In which units is it measured?
- How many radioactive series are there? How many are found in nature?

§36.3 NUCLEAR REACTIONS

- What happens when neutrons bombard a relatively small nucleus like carbon? What happens if the target nucleus is large, like uranium?
- What is the proton–proton cycle? Why is it important?
- Why does a nuclear reactor need water to moderate the neutrons?

Basic Skill Drill

§36.1 BASIC NUCLEAR STRUCTURE

1. Write the standard symbols for nuclei of yttrium containing 90 nucleons and for aluminum nuclei with 27 nucleons.
2. Natural lithium has a chemical atomic mass of 6.941 u. It consists of isotopes with atomic masses of 6.015123 u and 7.016005 u. Find the percentages of the two isotopes in the natural mixture.
3. Calculate the binding energy of a deuterium nucleus (^{2}H).
4. Some nuclear theorists speculate that a nucleus with atomic number $Z = 110$ and atomic mass number $A = 294$ might be stable. What would be the radius of such a nucleus?

§36.2 NATURAL RADIOACTIVITY

5. Write a formula for the α decay of ^{183}Au.
6. Write a formula for the β decay of ^{8}Li.
7. What is the energy of γ rays produced in the annihilation of a proton/antiproton pair?
8. By how much does the mass of a ^{60}Co nucleus change when it emits a 58.6-keV x ray?
9. ^{23}Ne has a half-life of 37.6 s against β decay. What is the activity of 15 mg of ^{23}Ne?
10. A natural sample of ^{235}U ore contains 0.056 mg of ^{227}Th. How much ^{235}U is in the sample?

§36.3 NUCLEAR REACTIONS

11. What nucleus is produced when a ^{59}Co nucleus captures a neutron? What is the Q value for this reaction?
12. How much energy is liberated in the triple α process in which three α-particles combine to produce a ^{12}C nucleus?
13. When ^{235}U absorbs a neutron, one possible reaction produces ^{143}La and ^{87}Br. How many neutrons are produced? How much energy is liberated?

Questions and Problems

§36.1 BASIC NUCLEAR STRUCTURE

14. ❖ A stable isotope is known to have a mass number $A = 45$. Within roughly what range of atomic number Z would you expect the isotope to be?
15. ❖ Discuss the factors that place lower and upper limits on the size of nuclei with a given atomic number. Which nucleus would you expect to be more stable, ^{7}Li or ^{8}Li?
16. ♦ Natural boron is a mixture of ^{10}B and ^{11}B and has a chemical atomic mass of 10.82 u. What is the isotope ratio of natural boron?
17. ♦ Natural lead contains the isotopes ^{204}Pb (1.4%), ^{206}Pb (24.1%), ^{207}Pb (22.1%), and ^{208}Pb (52.4%). Calculate the chemical atomic mass, and compare with the value in the periodic table.
18. ♦ Calculate the binding energy of a ^{12}C nucleus.
19. ♦ Use eqn. (36.1) to estimate the size of a californium nucleus ($Z = 98$) with $A = 245$. (This artificially produced nucleus has a lifetime of 8500 y.)

20. ♦♦ A deuteron (^{2}H) consists of a neutron and proton confined within a distance of approximately 1.5 fm. Use Heisenberg's uncertainty principle to estimate the kinetic energy of the proton and neutron. Compare your result with the binding energy of the deuteron.
21. ♦♦ Calculate the binding energy per nucleon for tritium (^{3}H), and compare with the result for ^{3}He.

§36.2 NATURAL RADIOACTIVITY

22. ❖ Explain what is wrong with each of the following reaction equations. There may be more than one error in each.

$$^{227}\text{Th} \rightarrow {}^{223}\text{Ca} + \alpha.$$
$$^{15}\text{C} \rightarrow {}^{15}\text{B} + e^+ + \bar{\nu}_e.$$
$$^{3}\text{H} \rightarrow {}^{3}\text{He} + e^+ + \nu_e.$$

23. ❖ It is never possible for the activity of a radioactive substance to increase with time. Is this statement true or false? Explain your reasoning.

24. ❖ A great, great ... granddaughter species in a radioactive series can never have an activity greater than that of the parent species. Is this statement true or false? Explain your reasoning.

25. ❖ Under what circumstances can a great, great ... granddaughter species in a radioactive series have an activity less than that of the parent species?

26. ❖ The oxygen isotopes ^{14}O and ^{19}O both decay by β emission. Which do you expect to emit an electron, and which a positron? Why?

27. ❖ State the rule for computing Q, the energy released by a positron decay, in terms of tabulated atomic masses. Show explicitly how the rule accounts for the extra electrons included in the atomic masses.

28. ❖ Recent observations of the central regions of our galaxy show strong γ-ray emission at an energy of 0.511 MeV. How do you suppose these photons may be produced?

29. ❖ Suppose the reaction n → γ were possible, in violation of baryon conservation. Would it be possible to have nuclear species heavier than hydrogen? Why or why not?

30. ❖ Since antielectrons, antiprotons, and antineutrons all exist, why do we not encounter antinuclei, or even antipeople? (*Hint:* What would be the major experimental problem with maintaining a macroscopic quantity of antimatter?)

31. ❖ Referring to Figure 36.26, explain why the shell model requires large, stable nuclei to contain more neutrons than protons.

32. ♦ What energy photons would result from the annihilation of a neutron and an antineutron?

33. ♦ Write a formula for the α decay of each of the following nuclear species: ^{224}Th, ^{211}Bi, ^{243}Cm, ^{221}Fr, ^{215}At.

34. ♦ Write a formula for the β decay of each of the following nuclear species. Determine from Table 36.1 and/or Figure 36.3 which are electron emitters and which are positron emitters: ^{12}B, ^{34}Cl, ^{140}La, ^{209}Pb, ^{138}Pr.

35. ♦♦ Compute the activity of a 1.0-g sample of ^{226}Ra, whose half-life is 1622 y.

36. ♦♦ Use tabulated data to show that ^{56}Fe is stable against α decay.

37. ♦♦ A ^{60}Co source has a half-life of 5.2 y and an initial activity of 1100 Ci. Express its activity in becquerels. What is its activity after 40 y? How many cobalt atoms does it contain initially?

38. ♦♦ A sample of ^{131}I has a half-life of 8.06 d and an initial activity of 2.5×10^8 Bq. What will be the total number of disintegrations from the sample?

39. ♦♦ The mass of ^{64}Cu is 63.929759 u. It decays by positron emission to ^{64}Ni, which has mass 63.92796 u. What is the maximum energy of positrons emitted by a ^{64}Cu sample?

40. ♦♦ Find the number of γ rays emitted inside a patient given a 2-mCi standard dose of ^{99m}Tc ($T_{1/2} = 6$ h) for a thyroid examination.

41. ♦♦ What is the energy of α-particles emitted by ^{152}Sm?

42. ♦♦ The half-life of ^{239}Pu is 2.439×10^4 y. By how much is the activity of a sample decreased after 500 y?

43. ♦♦ Find the energy released in the α decay of ^{212}Rn.

44. ♦♦ Find the energy released in the β decay of ^{16}N.

45. ♦♦ A ^{226}U nucleus undergoes five successive α decays before reaching a stable form. What is the final nucleus?

46. ♦♦ The ^{235}U in a sample of ore has an activity of 3.4×10^5 Bq. How much ^{227}Th is in the sample?

47. ♦♦ A sample of thorium ore emits 1.0×10^7 Bq of β radiation. How much thorium is contained in the sample?

48. ♦♦ A sample of ^{65}Ni decays by β emission with a half-life of 25.20 h. At $t = 0$, a sample of ^{78}As ($T_{1/2} = 1.515$ h) has 4.00 times the β activity of the nickel sample. At what time will the two samples have equal activities?

49. ♦♦♦ A technician wishes to assay a prepared sample of ^{112}Ag but knows it is contaminated with ^{111}Pd. Both nuclides undergo β decay, the silver with a half-life of 3.13 h and the palladium with a half-life of 22 min. The initial activity of the sample is 6.35×10^7 Bq. After 1.00 h, the activity has decreased to 4.50×10^7 Bq. What are the initial activities of the silver and the palladium? How long must the technician wait for the palladium contaminant to contribute less than 0.1% of the activity?

50. ♦♦♦ A subdivision is to be built on land where the waste rock from a uranium mine was discarded 10 y ago. Assume that all of the original 30 g·m^{-3} of uranium was removed and all of the gaseous daughter elements escaped from the waste. Estimate the remaining activity per cubic meter of the waste rock.

51. ♦♦♦ A geochemist wants to know the age of a granite rock sample. Analysis shows the sample to contain 600 g of ^{206}Pb for every kilogram of ^{238}U. We may assume the sample has been undisturbed since its formation, at which time it contained no ^{206}Pb. How old is the sample?

52. ♦♦♦ A sample initially contains a mixture of two isotopes with decay constants λ_1 and λ_2 and initially equal activities. Find an expression for the ratio of the activities of the two isotopes at later times. After what time will the shortest-lived isotope contribute less than 1% of the total activity?

53. ♦♦♦ Show that the minimum number of parent nuclei N_o required to obtain a given decay rate R after a time t_a occurs when the half-life of each nucleus is $T_{1/2} = t_a \ln 2$. Find the minimum number N_o. (This result has application in medicine, where a sufficient decay rate R for a procedure should be obtained with a minimum total radioactive dose to the patient.)

§36.3 NUCLEAR REACTIONS

54. ❖ Give reasons why the following reactions cannot occur:

$$^{23}\text{Na} + \text{n} \rightarrow {}^{19}\text{F} + \alpha.$$

$$^{3}\text{H} + {}^{2}\text{H} \rightarrow {}^{4}\text{He} + \text{p}.$$

55. ❖ Suppose you wish to investigate the amounts of copper and zinc in the paint of a supposed masterpiece using neutron activation analysis. What radioactive elements would you expect to produce? (Use Figure 36.3.)

56. ❖ Using Table 36.3, determine which elements are likely to be produced when ^{238}U is exposed to an intense neutron flux in a reactor. Do you see why plutonium is important in nuclear technology?

57. ❖ Consider some of the fission products that are likely to be produced in the fission reactions ^{235}U → X + Y + 3n. Are the products stable? Do you see why a fission reactor continues to require a large flow of cooling water even after the fission reaction is shut down?

58. ♦ Calculate the energy released in forming a ^{16}O nucleus from a ^{12}C nucleus plus an α-particle.

59. ♦ A neutron is absorbed by an unknown target atom. After a γ decay and a β^+ decay, the product is observed to be ^{65}Cu. What was the target species?

60. ♦ What is the unknown product X in each of the following reactions?

$$^{14}\text{N} + \alpha \rightarrow X + \text{n}.$$

$$^{12}\text{C} + \gamma \rightarrow X + \alpha.$$

$$^{7}\text{Li} + \text{p} \rightarrow X + \text{n}.$$

$$^{31}\text{P} + {}^{2}\text{H} \rightarrow X + \text{p}.$$

61. ♦ Find the energy Q released by the fusion reaction:

$$^{2}\text{H} + {}^{2}\text{H} \rightarrow {}^{3}\text{He} + \text{n} + Q.$$

TABLE 36.3 Half-lives of Heavy Elements Near ^{238}U

Z \ A	237	238	239	240	241	242
91 Pa	8.7 min	2.3 min				
92 U	6.75 d	4.5×10^9 y	23.5 min	14.1 h		
93 Np	2×10^6 y	2.12 d	2.35 d	7.5 min	16.0 min	
94 Pu	1.1 μs	87.8 y	8 μs	6540 y	23 μs	4×10^5 y
95 Am	75 min	0.06 s	11.9 h	0.9 ms	1.5 μs	14 ms

62. ◆◆ Each of the following reactions leads to the same compound nucleus X. What is that nucleus?

$$^{17}O + {}^3He \rightarrow X.$$
$$^{14}N + {}^6Li \rightarrow X.$$
$$^{10}B + {}^{10}B \rightarrow X.$$

What is the remaining reaction product when nucleus X emits each of the following: α, ^{2}H, ^{3}H, ^{9}Be?

63. ◆◆ What compound nucleus results when an α-particle is incident on a ^{19}F nucleus? What is the remaining reaction product when the compound nucleus emits each of the following: α, ^{2}H, ^{3}H, ^{9}Be?

64. ◆◆ Assume that a star consuming helium by the triple α process produces ten times the power of the Sun and can consume 0.3 solar masses of helium. How long will the star's helium-burning phase last? You will need the result of Problem 12.

65. ◆◆ What is the minimum antineutrino energy necessary to cause the reaction:

$$\bar{\nu}_e + p \rightarrow n + e^+?$$

66. ◆◆ What is the minimum-energy photon necessary to remove a neutron from magnesium in the reaction $^{24}Mg + \gamma \rightarrow {}^{23}Mg + n$? Neglect any kinetic energy of the reaction products.

67. ◆◆ Find the value of Q for the reaction $^8Be \rightarrow {}^4He + {}^4He$. Compare your result with the electrostatic energy of two α-particles, modeled as uniformly charged spheres barely in contact.

68. ◆◆ The wear of automobile parts is frequently tested by neutron activation. For example, a 30-g steel piston ring is exposed to neutrons to form ^{59}Fe with a half-life of 45 d and an activity of 15 μCi. The ring is then installed in a car, and the oil is removed after 30 d of testing. If the oil then shows an activity of 6.3 Bq, how much steel has worn from the ring?

69. ◆◆ The remains of a hominid woman, named Lucy by her discoverer, were found in Africa in 1974, forcing anthropologists to re-evaluate their theories of human origins. The bones were embedded in rock containing potassium. A fraction $f = 1.18 \times 10^{-5}$ of the potassium was ^{40}K, which decays to ^{40}Ar with a half-life of 1.28×10^9 y. For each 0.1 g of potassium, the rock contained 2.39×10^{-9} g of ^{40}Ar. Contamination by the atmosphere accounts for 7.25×10^{-10} g of the argon. What is the age of the fossils?

70. ◆◆ How much energy is released when an α-particle combines with ^{52}Cr to form ^{56}Fe? What is the Q value for the reaction $^{54}Cr + {}^{55}Mn \rightarrow {}^{109}In$ (mass 108.9071 u)?

71. ◆◆ Find the energy released in the reaction $^{12}C + {}^{12}C \rightarrow {}^{24}Mg$.

72. ◆◆ Calculate the energy per gram released in the fusion reaction sequence:

$$^2D + {}^2D \rightarrow {}^3T + p,$$
$$^3T + {}^2D \rightarrow {}^4He + n.$$

Compare your answer with the energy per gram released in a fission reaction such as that in Example 36.7.

73. ◆◆◆ At what temperature is the mean kinetic energy of two protons equal to their electrical potential energy at a separation of 2.0 fm? This greatly overestimates the temperature required for fusion, because the protons can tunnel through the Coulomb barrier in a reverse version of α decay. (See Figure 36.13.)

74. ◆◆◆ The neutron's mass cannot be measured in a mass spectrometer, and so it is obtained indirectly. From the following data find the neutron mass:

$$^1H + {}^1H \rightarrow {}^2H + e^+ + \nu_e + 1.444 \text{ MeV}$$
$$^2H + {}^2H \rightarrow {}^3H + {}^1H + 4.031 \text{ MeV}$$
$$^2H + {}^2H \rightarrow {}^3He + n + 3.269 \text{ MeV}$$
$$^3H \rightarrow {}^3He + e^- + \bar{\nu}_e + 0.0186 \text{ MeV}$$
$$^1H + n \rightarrow {}^2H + \gamma \text{ with } E_\gamma = 2.225 \text{ MeV}$$
$$M({}^1H)c^2 = 938.78 \text{ MeV}.$$

Additional Problems

75. ❖ Explain why a free neutron can undergo β decay while a free proton cannot emit a positron. Why can a proton emit a positron when it is inside a nucleus? (Refer to Figure 36.26.)

76. ❖ By comparing initial and final states of the pair production and annihilation reactions in the CM reference frame of an $e^+ e^-$ pair, show that a single photon cannot by itself decay into a pair, nor can a pair annihilate to form a single photon.

77. ◆ In neutron stars, which contain approximately one solar mass (2×10^{30} kg) within a radius of 10 km, gravitation is intense enough to prevent β decay and allow the existence of a single nucleus containing some 10^{57} neutrons! Compute the density of a typical nucleus and compare your result with the density of a neutron star.

78. ◆◆ Compare the de Broglie wavelength of an α-particle emitted by ^{217}Fr with the diameter of the nucleus.

79. ◆◆ Natural uranium is 0.72% ^{235}U and 99.28% ^{238}U. If the two isotopes were equally abundant when the Earth was formed, how old is the Earth?

80. ◆◆ In the decay of ^{144}Nd (Example 36.3), what fraction of the total energy release is in the form of kinetic energy of the cerium daughter nucleus?

81. ◆◆◆ A 5.000-MeV γ ray dissociates a stationary deuteron (binding energy 2.226 MeV). Find the kinetic energies of the resulting neutron and proton if the neutron emerges at an angle of 45° to the direction of the γ ray.

82. ♦♦♦ Compare the fission cross section of ^{235}U (Figure 36.33) with (a) the geometrical cross section of the uranium nucleus, and (b) a circle with radius equal to the de Broglie wavelength of the neutron. In this case, compare both the magnitude and dependence on neutron energy.

83. ♦♦♦ The unstable nucleus ^{243}Am emits an α-particle with kinetic energy 5.3 MeV. Assuming the α has the same kinetic energy inside the nucleus as outside, compute the rate at which the α undergoes collisions with the inner surface of the nucleus. If the half-life for the decay is 8000 y, what is the probability of escape at each collision?

84. ♦♦♦ A patient's body excretes a particular chemical element with a biological half-life T_b of 2 d and is given a dose of a radioisotope of that element with a decay half-life $T_{1/2} = 8$ h. Derive a formula for the fraction of the original dose remaining in the patient's body as a function of time and compute the fraction remaining after 1.5 d.

85. ♦♦♦ A natural sample of ^{238}U ore emits 1.26×10^4 α-particles per second from decays to ^{234}Th. How many ^{238}U nuclei are contained in the sample? How many decays per second of ^{226}Ra are expected? What mass of radium is present? How many β decays of ^{210}Tl are expected? (Refer to Figure 36.29c.)

86. ♦♦♦ Assuming a neutron loses one-half its kinetic energy in each collision with a thermal proton, what number N of such collisions is necessary to reduce a 2-MeV neutron to thermal energies of 0.04 eV? Assuming the cross section for such collisions equals the geometrical cross section of a proton (radius 1.2 fm), what is the average distance ℓ the neutron travels between collisions in water? Neglect any collisions with oxygen nuclei. Estimate the root mean square distance the neutron travels during this process, taking $d_{rms} \approx \ell\sqrt{N}$.

87. ♦♦♦ Use the results of Problem 90 to show: If $T_2 \ll t \ll T_1$, then N_2 attains a value proportional to N_1 such that both decay with the same activity. Show that, in the case $T_1 \ll t \ll T_2$, species 1 has decayed away, that nearly all the atoms are daughters, and that almost none are granddaughters. Discuss how these results relate to the hydrodynamic analogy for radioactive series (Figure 36.30).

Computer Problems

88. The activity of a sample of ^{55}Cr after successive intervals of 5 min is found to be 9.60, 3.57, 1.33, 0.495, and 0.185 mCi. Plot the logarithm of the activity versus time. What is the half-life of ^{55}Cr?

89. One of the fission products of uranium is ^{149}Nd, which decays with a 1.8-h half-life to ^{149}Pm. It subsequently decays with a 54-h half-life to ^{149}Sm, which is stable. Assuming there is 1 g of Nd at $t = 0$, use a spreadsheet program to calculate the activity of the Nd, the Pm, and the whole sample as a function of time. Plot your results.

Challenge Problems

90. Suppose that initially we have a pure sample containing N_0 atoms of a radioactive species with half-life T_1. The daughter species decays with half-life T_2 to a stable granddaughter species. Show that the numbers of atoms of the three species are given by:

$$N_1(t) = N_0 2^{-t/T_1},$$

$$N_2(t) = N_0 \left(\frac{T_2}{T_1 - T_2}\right)(2^{-t/T_1} - 2^{-t/T_2}),$$

and
$$N_3(t) = N_0 \left(\frac{T_2 2^{-t/T_2} - T_1 2^{-t/T_1}}{T_1 - T_2} + 1\right).$$

91. Use the results of Problem 90 to find the time when the maximum number of atoms are in the daughter form and the value of that maximum number. A medical radiology department receives a "source" containing 100 mCi of ^{99}Mo Monday morning at 10:00 A.M. The molybdenum decays to ^{99m}Tc with a half-life of 66.02 h. The technetium decays with a half-life of 6.02 h. When should technicians milk the source to obtain the maximum quantity of ^{99m}Tc? All the technetium is used for scheduled tests, but a new patient urgently requires a test 3 h after the previous milking. How many curies of technetium are available for the test?

92. Use Heisenberg's uncertainty principle and the observed Q value of the reaction $n \rightarrow p + e^+ + \nu_e$ to argue that we *cannot* think of a neutron as a bound state of proton, electron, and neutrino that somehow splits up. How does the reaction $p + e^- \rightarrow n + \nu_e$ support this conclusion?

93. After a long time has passed and essentially all the parent nuclei in a large sample have decayed, show that the average time a parent nucleus survived before decaying is λ^{-1}. [*Hint:* Calculate the average as $\int tN(t)\,dt/N_0$, where $N(t)$ is the number of nuclei that reach age t and then decay before $t + dt$, and N_0 is the initial number of nuclei.] If a given atom has two distinct modes of decay with decay constants λ_1 and λ_2, find an expression for the number of atoms in a sample remaining after time t, and show that the mean lifetime of an atom is $t_{mean} = (\lambda_1 + \lambda_2)^{-1}$.

94. In Chadwick's original experiment, 5.30-MeV α-particles from a polonium source strike ^{9}Be nuclei. Supposing that the resulting reaction is: ^{9}Be $+ \alpha \rightarrow {}^{13}$C $+ \gamma$, what is the energy of the emerging γ rays? Assuming that these hypothetical γ rays Compton-scatter from the protons in Chadwick's target, what is the minimum-energy γ necessary to produce the observed 5.7-MeV recoil protons? Supposing instead that the reaction is ^{9}Be $+ \alpha \rightarrow {}^{12}$C $+ n$, what is the energy of the resulting neutron? What energy would it transfer to a stationary proton in a head-on elastic collision?

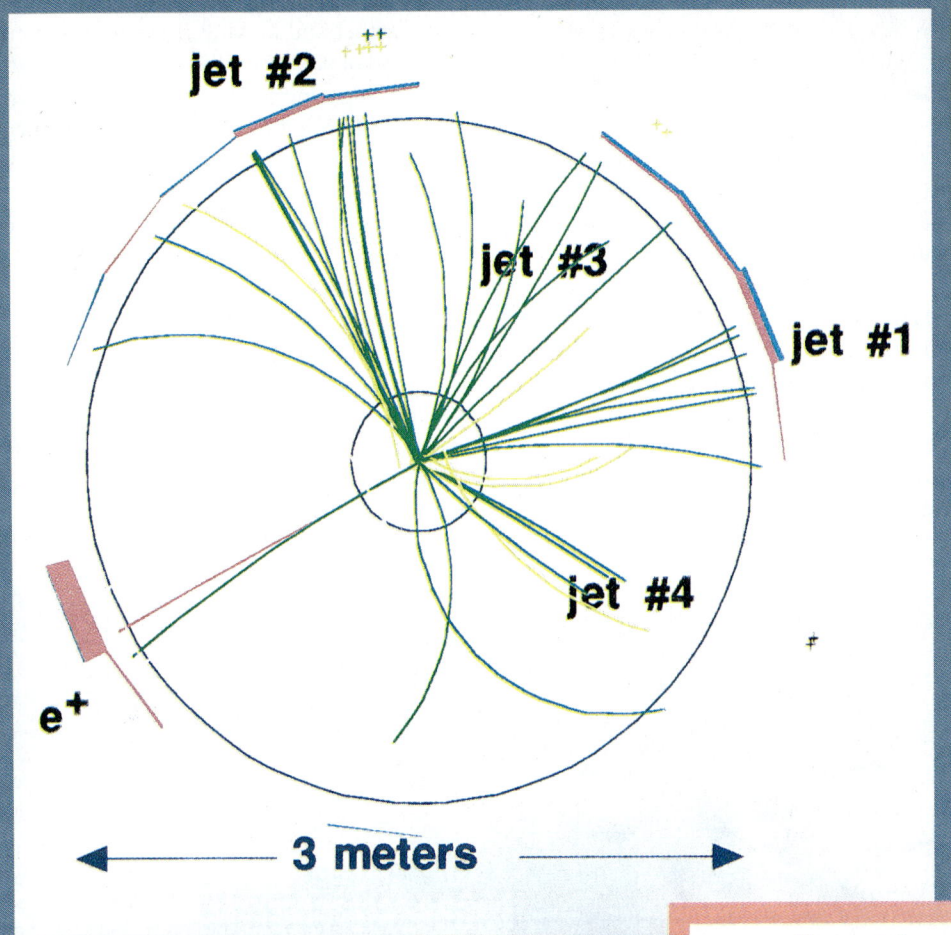

Result of a proton/antiproton collision in the Tevatron at Fermi National Accelerator Laboratory. A top quark and a top antiquark were created in the collision. Each quark decayed into a W boson and a bottom quark. One W boson decayed into a positron and a neutrino, while the other decayed into a light quark and a light antiquark. The particle tracks form four jets, showing the tracks of particles formed in the decay of the four quarks (two bottom quarks, and the light quark/antiquark pair). The positron caused the red track at the left. The neutrino left no track. Its presence is inferred from momentum conservation. The bottom panel shows the energy deposited in detectors (calorimeters) around the experiment chamber, as a function of angle. The position and energy of the neutrino are deduced, not directly observed.

... three quarks for Muster Mark ...

JAMES JOYCE

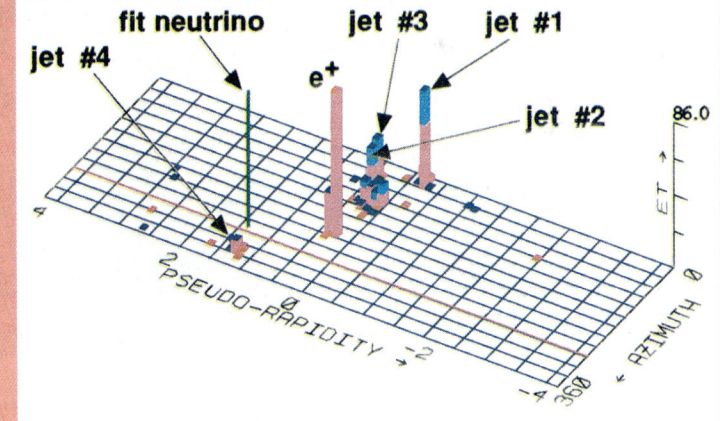

CHAPTER 37
Particle Physics

CONCEPTS
Virtual particles
Quarks
Color charge
Unification of forces

GOALS
Be able to:
Describe qualitatively the electroweak and quantum chromodynamic models of fundamental particles.

The European Laboratory for Particle Physics, recently renamed, is still known by the initials of its parent organization, Conseil Européen de Recherche Nucléaire (European Council for Nuclear Research).

Remember: The symbol T stands for tera = 10^{12} (Table 1.1).

Richard Feynman once remarked that doing particle physics is like studying Swiss watches by smashing them together and looking at the broken pieces that fly out. The structure of nuclei and of particles once called "elementary" is determined by accelerating them to high energy, smashing them together, and looking at the resulting debris. Understanding the smallest constituents of matter requires enormous energies and correspondingly gigantic machines. Since the 1950s, the size and cost of particle accelerators have grown rapidly, and they are now major national facilities such as the Fermi National Accelerator Laboratory (■ Figure 37.1) or international collaborations such as the CERN laboratory in Geneva, Switzerland.

As the energy available for experimental reactions has increased from a few GeV to several TeV per particle, a wealth of subnuclear particles has been revealed. Between 1940 and 1970 physicists gradually developed a new concept of force and discovered the principles that govern the subnuclear world. The resulting *standard model* explains subnuclear behavior with a small number of *fundamental* particle types and three rules of behavior. In this chapter we describe the standard model, which takes us to the frontiers of research. It currently provides our best answers to the questions the ancient Greeks asked at the birth of physics: What are the constituents of the universe and what are the rules they follow?

■ Figure 37.1
Aerial photograph of Fermilab. The circular structure in the photograph follows the path of the underground tunnel where particles are accelerated. Counterrotating beams of particles meet in experiment chambers where they collide within elaborate detectors. Computer analysis of the detector readings (see frontispiece) determines which particles are produced in the collisions.

37.1 Particle Creation and Fundamental Forces

37.1.1 Creation and Destruction

We are used to a stable world. Everyday objects neither appear nor disappear spontaneously. Even in nuclear reactions the particles convert only a small fraction of their mass to energy. The subnuclear world, however, could not be more different. An energetic photon disappears and its energy reappears as a positron and electron, or a neutron is destroyed and a proton, an electron, and an antineutrino are created in its place. Such creation and destruction of subnuclear particles occurs incessantly and is limited only by the need to satisfy conservation laws. This chaos is not apparent in our macroscopic world because the electron and proton appear to be stable. If an electron were to decay, the resulting particles would have to have a total mass less than m_e, total charge $-e$, and lepton number $+1$. We know of no such set of particles. Proton decay, if it occurs at all, has a half-life greater than 10^{31} y.

Despite intensive searches, proton decay has never been observed.

37.1.2 Virtual Particles and Fundamental Forces

Particle creations may appear to violate energy conservation, but only for limited periods of time! For example, a photon with energy E can come into existence for a time Δt limited by the Heisenberg uncertainty principle (■ Figure 37.2):

$$\Delta t \approx \hbar/2E.$$

Such a particle, existing for a limited time on borrowed energy, is called a *virtual particle*.

Virtual particles are the mechanism by which forces are transmitted (■ Figure 37.3). For example, when a virtual photon emitted by one electron is absorbed by another, the photon transfers momentum between the electrons. The continuous exchange of such photons produces a continuous momentum transfer—the repulsive Coulomb force between the electrons. We introduced the idea of fields, such as the electric field (Chapter 23), to explain the interaction of separated particles. On the subatomic level, we recognize the electromagnetic field as a swarm of virtual photons.

During its brief life, a virtual photon can travel a maximum distance:

$$c\,\Delta t \sim \frac{\hbar c}{2E} = \frac{\hbar c}{2hf} = \frac{c}{4\pi f} = \frac{\lambda}{4\pi}.$$

Because photons have no rest mass, they can have arbitrarily small energies, and the distance $c\,\Delta t$ can be arbitrarily large. For this reason, the Coulomb force acts over distances much larger than atomic dimensions.

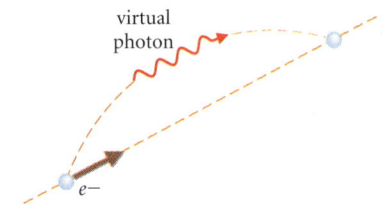

■ **FIGURE 37.2**
An electron emits and reabsorbs a virtual photon. According to Heisenberg's uncertainty principle, a photon of energy E can exist for a time $\Delta t \approx \hbar/(2E)$ before being reabsorbed.

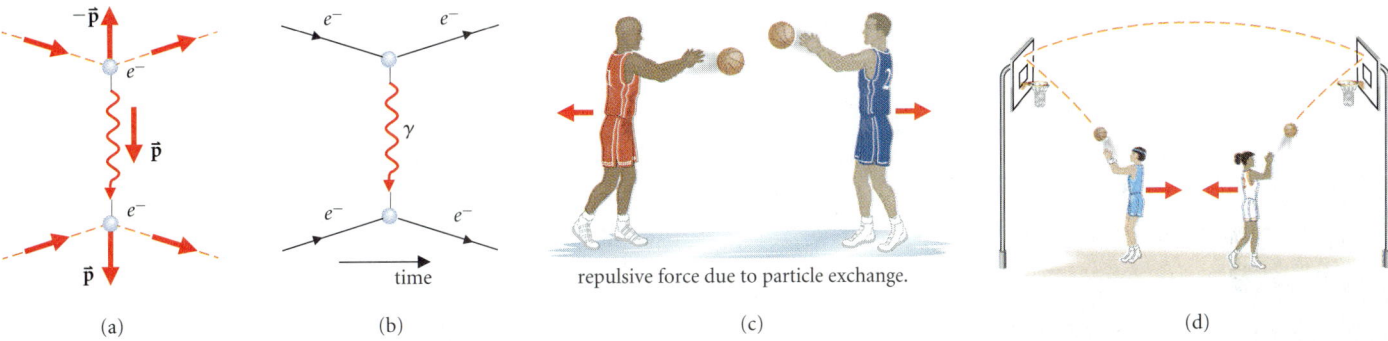

■ **FIGURE 37.3**
(a) A virtual photon transmits electromagnetic force between two electrons. An electron emits a photon with momentum $\vec{p}$, changing its own momentum by $-\vec{p}$. The photon is reabsorbed by a second electron, which gains momentum $\vec{p}$. A continual exchange of photons results in the Coulomb repulsion of the electrons. (b) A diagram showing these processes is called a Feynman diagram. The photon is labeled γ. (c) In a classical analog of particle exchange, two students standing on ice throw basketballs back and forth. As the student on the left throws his ball, he gives it momentum $\vec{p}$ to the right and he recoils with momentum $-\vec{p}$ to the left. Upon catching the ball thrown by his partner, he gains additional leftward momentum. Similarly, the student on the right gains rightward momentum. (d) A similar mechanism produces an attractive Coulomb force between oppositely charged particles, but there is no obvious classical analog. If the two students stand back to back, they can bounce their balls off walls before catching them, and hence be pushed together. But there are no walls in the particle world.

37.1.3 Feynman Diagrams

Richard Feynman (1918–1988) invented a diagram that illustrates the role of virtual particles. A *Feynman diagram* (Figure 37.3b) is a schematic of the paths of the real and virtual particles involved in an interaction. Particles whose paths cross the boundaries of the diagram are real. A point where the paths join is called a vertex. At each vertex all the conservation laws we have discussed so far must be obeyed: the sum of energy, momentum, charge, lepton number, and baryon number of the particles entering the vertex must equal the sum for the particles leaving. In this example time increases from left to right. It is conventional to draw antiparticles moving backward in time.

THESE DIAGRAMS ARE USEFUL GUIDES FOR CALCULATING THE OUTCOME OF A PARTICLE INTERACTION.

SEE §36.2.1 FOR DEFINITIONS OF THESE QUANTUM NUMBERS.

The theory of quantum electrodynamics (QED), based on virtual photons, allows us to calculate interactions between charged particles with great accuracy. Experimental tests currently match theory to better than 1 part in 10^7. Because of this success, physicists now believe that all fundamental forces act by the exchange of virtual particles. Thus to understand subatomic physics we need a catalog of the particles that participate in the interactions that govern the behavior of matter.

37.1.4 The π Meson

In 1935 Hideki Yukawa (■ Figure 37.4) proposed that a virtual particle carries the strong force (§36.1.4). This Yukawa particle is now known as the π meson or *pion*. The name *meson*, or middle particle, reflects the fact that the pion mass is between those of electrons and of nucleons. Yukawa predicted its mass from the short range (10^{-15} m) of the strong force. The energy of a virtual Yukawa particle is at least as great as its rest energy mc^2. This minimum energy corresponds to the maximum time the particle may exist and so to the maximum distance it can travel:

$$c \, \Delta t_{max} = \hbar c/(2mc^2) \approx \text{range of strong force} \approx 10^{-15} \text{ m}.$$

NOTE THAT THE DISTANCE $c \, \Delta t \sim \hbar/mc$ IS APPROXIMATELY THE PION COMPTON WAVELENGTH.

So:

$$mc^2 \approx \frac{\hbar c}{2 \times 10^{-15} \text{ m}} = \frac{(1.0 \times 10^{-34} \text{ J} \cdot \text{s})(3 \times 10^8 \text{ m/s})}{(2 \times 10^{-15} \text{ m})(1.6 \times 10^{-19} \text{ J/eV})}$$

$$= 100 \text{ MeV}.$$

MESON NOW MEANS ANY STRONGLY INTERACTING PARTICLE WITH SPIN ANGULAR MOMENTUM EQUAL TO AN INTEGER TIMES $\hbar$.

This estimate is very close to the measured value of 135 MeV for the neutral pion. Pions were among the first of the new kinds of particle detected in accelerator experiments in the 1950s (■ Figure 37.5).

■ FIGURE 37.4
Hideki Yukawa won the Nobel Prize for physics in 1949 for predicting the existence of mesons.

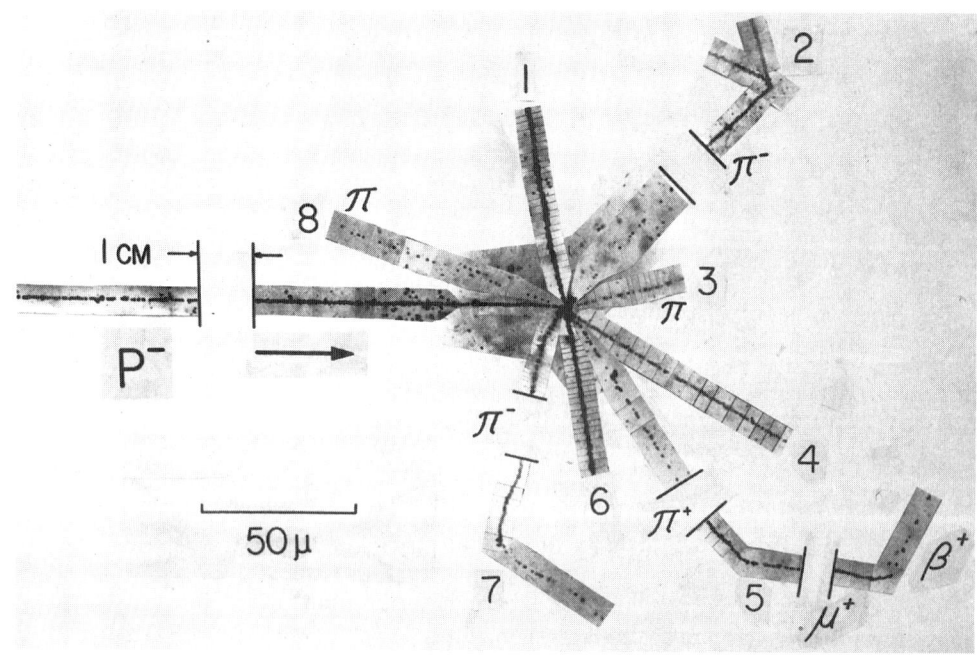

■ FIGURE 37.5
Proton/antiproton collision observed at the Bevatron, University of California, Berkeley, in 1956. The two particles annihilate, forming a spray of pions. One pion at the lower right later decays into a positively charged muon and a neutral particle that leaves no track. Several other decays are also visible.

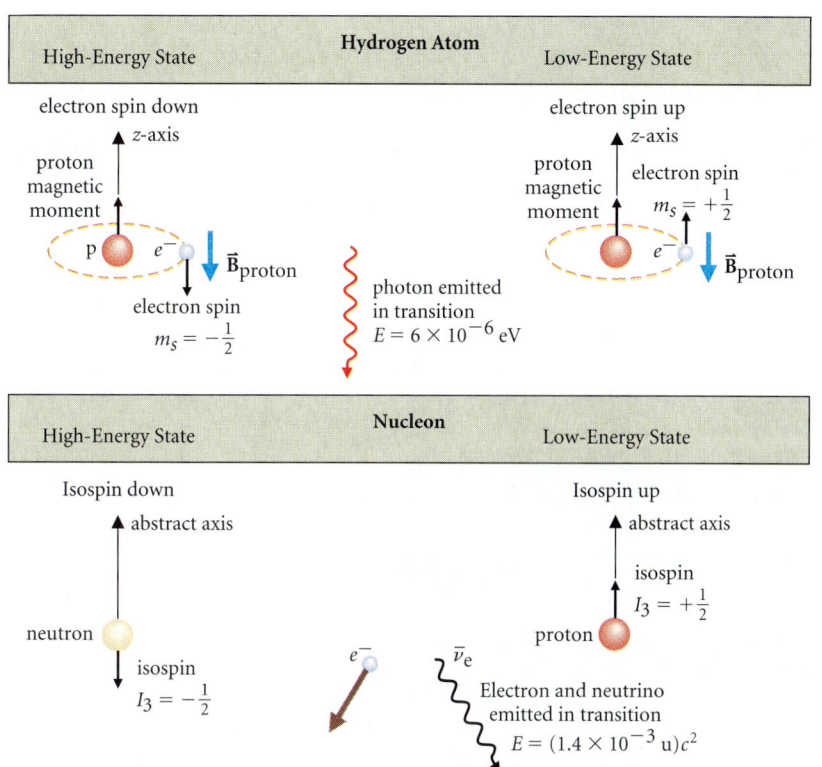

■ FIGURE 37.6
Analogy between electron spin and the isotopic spin of a nucleus. An atom emits a photon as the electron makes a transition from the high-energy, "spin-down" state to the low-energy, "spin-up" state. The neutron ("isospin-down" state) is the high-energy state of the nucleon. In the transition to the low-energy, "isospin-up" state (the proton), an electron (e^-) and an antineutrino ($\bar{\nu}_e$) are emitted.

37.1.5 Isospin

Discovery of the pion confirmed the idea that forces occur because of particle exchange, but also introduced a new puzzle. The π meson occurs in three forms (π^0, π^+, and π^-) with different electric charges (0, $+e$, and $-e$) but very nearly the same mass and identical strong interactions. Neutrons and protons behave similarly: they have nearly equal masses and experience the strong force in exactly the same way. In an imaginary world where only strong nuclear forces act, neutrons and protons would be identical. We could think of them as being two different quantum states of one particle, the nucleon. The observed differences in mass and electric charge arise because the weak and electric forces act differently on the two quantum states.

The two possible spin states of the electron in a hydrogen atom (§35.3) provide an analogy for the two nucleon states (■ Figure 37.6). If there were no magnetic fields, both electron spin states would have the same energy and be indistinguishable. In the actual world, magnetic fields cause the two spin states to have different energies. This analogy between spin and nucleon states is not just superficial: as we shall see, it proves very important in understanding the strong force (§37.2). Because of this analogy and because *iso*topes differ in neutron number, the property that distinguishes protons from neutrons was named *isospin*.

The neutron and proton have different names because they have different charges and respond differently to electromagnetic fields. But electric charge is irrelevant to the strong force. Had we understood atoms from the inside out instead of from the outside in, we would have recognized a single particle, the nucleon, with two isospin states.

Isospin is described by a quantum number I. States with different isospin have different energies (i.e., different masses) and different electric charges. For nucleons $I = \frac{1}{2}$, and there are two charge states. The pions have $I = 1$, and three charge states. In both cases, the number of charge states equals $2I + 1$.

THIS NUMERICAL RELATION HOLDS FOR PARTICLES CONTAINING FIRST-GENERATION QUARKS. SEE §37.2 AND §37.3.

37.2 SUBNUCLEAR PARTICLES AND THE QUARK MODEL

37.2.1 The Population Explosion

Machines capable of accelerating proton beams to energies of several GeV per particle were first built in the 1950s. Experiments with these machines soon discovered evidence for numerous kinds of strongly interacting particle whose existence was unexpected from experience with nuclear reactions (■ Figure 37.7). These particles exist for very short times before decaying into less massive products (■ Figure 37.8). If their lifetime is too short, the particles do not travel a detectable distance from their origin. The existence of such particles is inferred from a phenomenon called *resonance:* The rate at which particles interact in a collision experiment suddenly increases when their total energy in the CM frame is equal to the new particle's rest

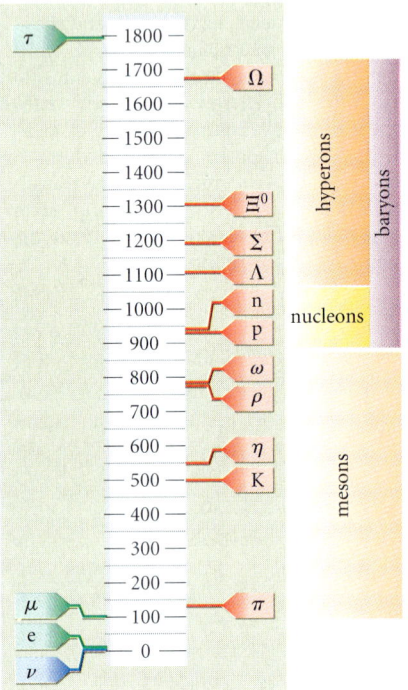

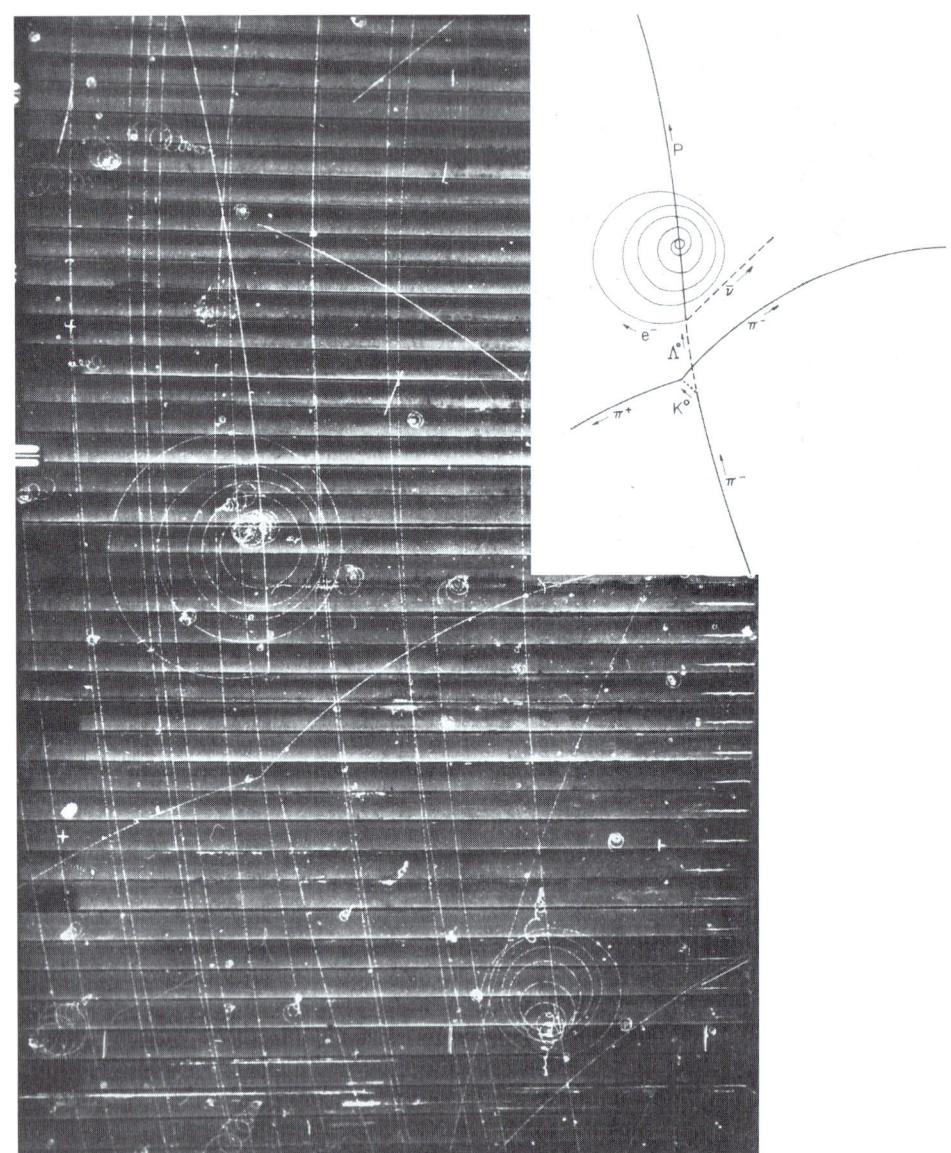

■ **FIGURE 37.7**
Some of the particles that have been discovered at high-energy particle accelerators. The initial classification scheme was based on particle mass: *leptons,* or light particles; *mesons,* intermediate mass particles; and *baryons,* or heavy particles. As more particles were discovered, the mass scheme was found to be inappropriate—notice the mass of the τ—and was replaced by a classification in terms of particle spin and response to forces. Leptons do not respond to the strong force, and they have spin $\frac{1}{2}$. Hadrons—mesons and baryons—do respond to the strong force; mesons have integer spin, hyperons and nucleons have half-integer spin. The forces felt by each particle are indicated by the color code. Red: strong, electromagnetic, and weak. Green: electromagnetic and weak. Blue: weak only.

■ **FIGURE 37.8**
Several particle decays are evident in this bubble chamber photograph. A π^- collides with a proton in the chamber, producing a Λ^0 and a K^0. The K^0 subsequently decays into a π^- and a π^+. The Λ^0 decays into a proton, an electron, and an antineutrino, which leaves no track. The distance traveled by each particle between production and decay is indicative of its lifetime. The magnetic field in the chamber causes charged particle tracks to curve: note the spiral followed by the electron.

energy (■ Figure 37.9). The colliding particles form the new particle, which then decays into particles of lesser mass. The population explosion of such particles demolished the tidy explanation of matter as electrons, protons, and neutrons exerting forces via virtual photons and pions. The new kinds of particle seemed to have no role in nuclear phenomena, yet their existence demanded explanation.

37.2.2 Strangeness and Quarks

Our understanding of subnuclear particles has come with the recognition of several new properties that are important in subnuclear reactions but not apparent in nuclear or atomic structure. The first of these properties, recognized in 1953, is called *strangeness* because particles possessing it seemed at the time to behave strangely—they decay too slowly. Nuclei and other strongly interacting systems have radii of the order of 10^{-15} m. Strong interactions require about 10^{-22} s to occur, roughly the time for light to travel this typical distance of 10^{-15} m. The strange particles ultimately decay to stable particles but require the much longer time, about 10^{-10} s, typical of weak interactions.

Apparently the strange particles are unable to decay via strong interactions, but can decay via weak interactions. This behavior can be understood if strong interactions obey a conservation law that weak interactions may violate. Strangeness, the new partially conserved quantity, occurs only in multiples of a basic unit, just as electric charge occurs in multiples of the electron charge. Thus particles are assigned an integer S, which gives the number of strangeness units they possess. The Λ^0 in Figure 37.8, for example, has a strangeness of -1. Strangeness may change by one unit during weak interactions.

The recognition of strangeness led to a powerful classification scheme for strongly interacting particles. ■ Figure 37.10 shows how particles of roughly similar mass fall into groups that show great regularity in their properties. When this classification was proposed, the

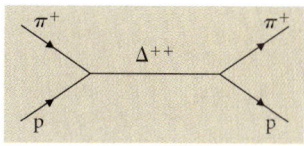

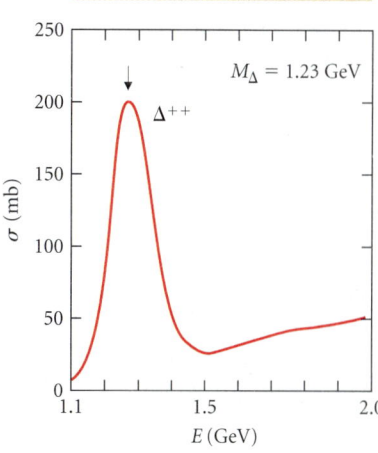

■ **FIGURE 37.9**
A Δ^{++} particle with mass 1.23 GeV/c^2 is formed in collisions between protons and pions. The particle itself decays rapidly and no track is seen. Its existence is inferred from its effect on the scattering of the π^+ and p. The rate of interactions (measured by the cross section) shows a marked peak as the total beam energy approaches 1.23 GeV. (1 mb = 10^{-31} m², cf. Figure 36.33.)

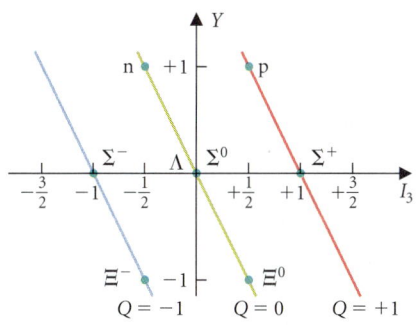

(a) Baryon octet.

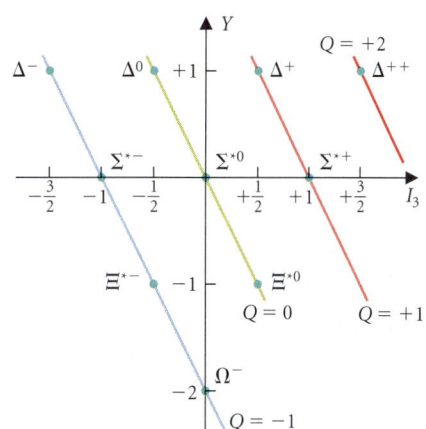

(b) Baryon decuplet.

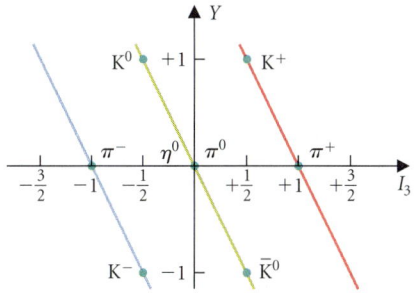

(c) Meson octet.

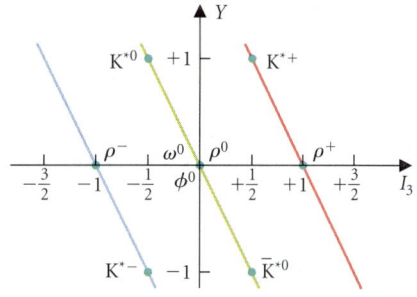

(d) Meson nonet.

■ **FIGURE 37.10**
Isospin/hypercharge diagrams. Hypercharge, $Y = B + S$, is the sum of baryon number plus strangeness. When we plot particle properties in a diagram of Y versus I_3, the z-component of isospin, an amazing regularity emerges. Particle families of roughly similar mass form regular geometrical figures. Lines of constant electric charge are diagonal lines in these plots. The notation K* denotes an excited state of the K particle. These patterns correspond to symmetries in the underlying mathematical description of the particles (§37.4).

■ **FIGURE 37.11**
Discovery of the Ω^-. The K^- incident at the bottom of the photograph collides with a proton in the chamber, producing an Ω^-, a K^0, and a K^+. The Ω^- subsequently decays into a Ξ^0 and a π^-.

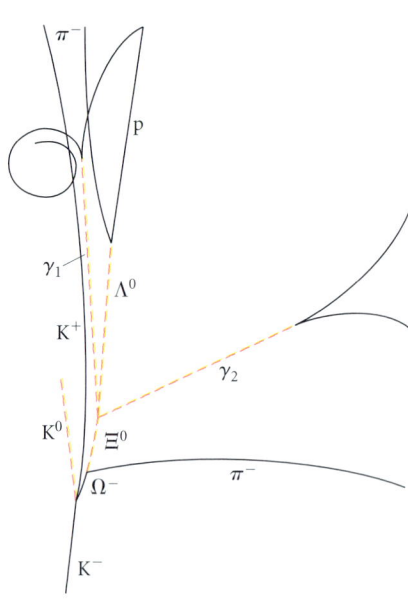

Ω^- particle in Figure 37.10b had not been observed. Its subsequent discovery (■ Figure 37.11) is convincing evidence for the reality of strangeness.

Just as the periodic table of chemical elements reflects the behavior of electrons in atoms, this classification of subnuclear particles hints at an underlying structure. In 1964, Murray Gell-Mann and Stephan Zweig pointed out that the classification is explained if the strongly interacting particles are composites of three kinds of particles they dubbed *quarks*. Each quark has spin $\frac{1}{2}$, baryon number $\frac{1}{3}$, and charge that is a multiple of $e/3$. The different kinds of quark are referred to as *flavors*. The first two flavors, *up* and *down,* refer to the isotopic spin: up

THE NAME IS GERMAN FOR MILK CURDS AND IS TAKEN FROM A PASSAGE IN *FINNIGAN'S WAKE* BY JAMES JOYCE. (SEE CHAPTER QUOTE.)

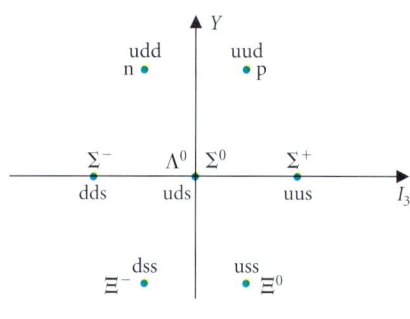

(a) Baryon octet.

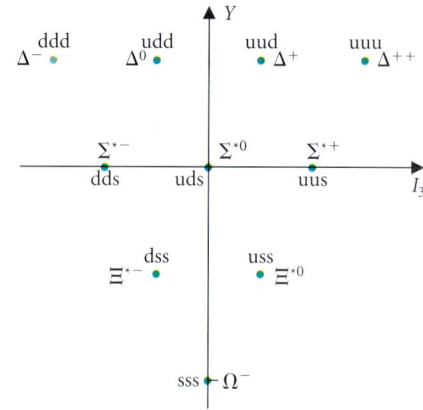

(b) Baryon decuplet.

■ **FIGURE 37.12**
Quark theory explains the hypercharge/isospin diagram. Each particle is described by the *quarks* that compose it. The flavorless particles at the centers of diagrams (c) and (d) are superpositons of combinations of quark/antiquark pairs. As in Figure 37.10, the * denotes an excited state.

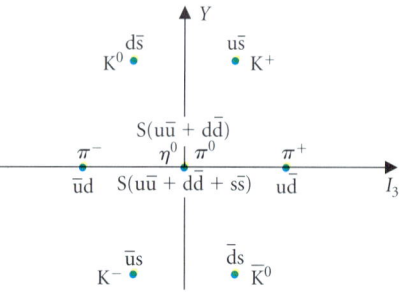

(c) Meson octet.

(d) Meson nonet.

quarks have isospin up ($I_3 = +\frac{1}{2}$), while the down quark has isospin down ($I_3 = -\frac{1}{2}$). The strange quark is assigned $S = -1$ and $I = 0$. ■ Figure 37.12 illustrates how the model explains particle properties. As more massive particles were discovered, additional quarks with additional quantum numbers were needed to explain them. The other three flavors are distinguished by quantum numbers called C (charm), T (topness), and B (bottomness) (● Table 37.1). In ● Table 37.2 we list properties of some mesons and baryons, including their masses and quark

REMEMBER: YOU MIGHT FIND IT USEFUL TO THINK OF THESE NEW QUANTUM NUMBERS IN ANALOGY WITH ELECTRIC CHARGE (CF. §36.2.1). TOPNESS IS SOMETIMES CALLED TRUTH AND BOTTOMNESS, BEAUTY.

TABLE 37.1 Properties of Quarks

Flavor	Symbol	Electric Charge	Isospin	I_3	S	C	B	T	Mass[a] (MeV)
Up	u	$2e/3$	$\frac{1}{2}$	$+\frac{1}{2}$	0	0	0	0	330
Down	d	$-e/3$	$\frac{1}{2}$	$-\frac{1}{2}$	0	0	0	0	333
Strange	s	$-e/3$	0	0	-1	0	0	0	486
Charm	c	$2e/3$	0	0	0	$+1$	0	0	1650
Bottom	b	$-e/3$	0	0	0	0	-1	0	4500
Top	t	$2e/3$	0	0	0	0	0	$+1$	180 000

[a]Masses are not directly observable. Values are derived from an analysis of hadron masses.

TABLE 37.2 A Selection of Mesons and Baryons

Name	Valence Quarks	Mass (MeV)	Isospin	I_3
π^+	$u\bar{d}$	139.57	1	$+1$
π^-	$\bar{u}d$	139.57	1	-1
π^0	$S(\bar{u}u,\bar{d}d)$[a]	134.97	1	0
K^+	$u\bar{s}$	493.65	$\frac{1}{2}$	$\frac{1}{2}$
K^-	$\bar{u}s$	493.65	$\frac{1}{2}$	$-\frac{1}{2}$
K^0	$d\bar{s}$	497.7	$\frac{1}{2}$	$-\frac{1}{2}$
η	$S(\bar{u}u,\bar{d}d,\bar{s}s)$[a]	549	0	0
p	uud	938.272	$\frac{1}{2}$	$+\frac{1}{2}$
n	udd	939.566	$\frac{1}{2}$	$-\frac{1}{2}$
Λ	uds	1115.6	0	0
Σ^+	uus	1189.4	1	$+1$
Σ^0	uds	1192.6	1	0
Σ^-	dds	1197.4	1	-1
Δ^{++}	uuu	1230	$\frac{3}{2}$	$+\frac{3}{2}$
Δ^+	uud	1230	$\frac{3}{2}$	$+\frac{1}{2}$
Δ^0	udd	1230	$\frac{3}{2}$	$-\frac{1}{2}$
Δ^-	ddd	1230	$\frac{3}{2}$	$-\frac{3}{2}$
Ξ^0	uss	1315	$\frac{1}{2}$	$+\frac{1}{2}$
Ξ^-	dss	1321	$\frac{1}{2}$	$-\frac{1}{2}$
Ω^-	sss	1672	0	0
Λ_c^+	udc	2285	0	0
B^+	$u\bar{b}$	5271	$\frac{1}{2}$	$\frac{1}{2}$
B^0	$d\bar{b}$	5275	$\frac{1}{2}$	$-\frac{1}{2}$

[a]The notation $S(x, y)$ means that the particle is a superposition of states x and y.

TABLE 37.3 Leptons

Name		Mass (MeV)	Charge (e)
Electron	e	0.511	−1
Muon	μ	105.658	−1
Tau	τ	1784 ± 3	−1
Electron neutrino	ν_e	$< 1.8 \times 10^{-5}$	0
Muon neutrino	ν_μ	< 0.25	0
Tau neutrino	ν_τ	< 35	0

content. All the mesons are quark/antiquark pairs. Each baryon contains three quarks. In • Table 37.3 we list the leptons. Each particle has an antiparticle with opposite quantum numbers.

37.2.3 Proton Structure and the Reality of Quarks

No experiment has found evidence for an isolated quark, and experiments fail to detect objects with fractional charge. A quark, it seems, cannot be separated from another quark by more than the 10^{-15}-m scale typical of strong interactions. Since they cannot be isolated, we must look elsewhere for proof of their existence. Rutherford detected atomic nuclei by showing that charged particles are scattered by something much smaller than a single atom. If protons really consist of three quarks, we should be able to observe scattering by individual quarks. Doing such an experiment is like using a microscope; one must use a wavelength small enough to resolve what one wants to see. By 1970, the Stanford Linear Accelerator Center had produced electron beams with de Broglie wavelengths of the order of 10^{-16} m, and had obtained definite evidence of pointlike particles within protons. The continued experimental and theoretical success of the quark model has overcome initial skepticism among physicists, who now generally accept it as the working model of reality.

IN 1977, WILLIAM FAIRBANK AND CO-WORKERS AT STANFORD UNIVERSITY REPORTED OBSERVING FRACTIONALLY CHARGED NIOBIUM SPHERES. THE RESULT REMAINS UNCONFIRMED AND IN CONTRADICTION WITH SEVERAL OTHER EXPERIMENTS.

THESE SUBPARTICLES WERE ORIGINALLY NAMED PARTONS TO AVOID PREMATURE IDENTIFICATION WITH THE QUARKS.

EXERCISE 37.1 ♦♦ What is the energy of an electron that has a de Broglie wavelength of 10^{-16} m?

EXAMPLE 37.1 ♦♦ One possible decay mode of the Λ^0 is: $\Lambda^0 \rightarrow p + \pi^-$ (Figure 36.24). Verify that the masses of the particles permit this decay, and determine which quantum numbers are conserved.

MODEL This is a conservation law problem for which we may use the standard plan.

SETUP The particle properties are given in Table 37.2 and Figure 37.12, and the quantum numbers for the quarks are in Table 37.1. The charge of the Λ^0 is the sum of the charges of the quarks that compose it: $q = 2e/3 - e/3 - e/3 = 0$. The baryon number is 1 for each baryon and zero for each meson or lepton.

	BEFORE	AFTER
	Λ^0	$p + \pi^-$
Mass (MeV)	1115.6	938.3 + 139.6 = 1077.9
Charge	0	$+e + -e =$ 0
Baryon number	+1	+1 + 0 = +1
Strangeness	−1	0 + 0 = 0
Isospin I_3	0	$\frac{1}{2} - 1 = -\frac{1}{2}$

SOLVE The mass of the two decay products is less than that of the original particle, as required. The Q value for the reaction is 1115.6 MeV − 1077.9 MeV = 37.7 MeV. Charge and baryon number are conserved; strangeness and isospin are not.

ANALYZE The Λ^0 is composed of the quarks u, d, and s. The proton is uud and the pion is $d\bar{u}$. Thus in this interaction the strange quark s converted to a down quark d with the production of a $u\bar{u}$ pair. Since the strangeness changes by one unit, this decay is governed by the weak interaction. We may conclude that isospin, like strangeness, is not conserved under the weak interaction.

37.3 THE STANDARD MODEL

The quark model of the 1960s offered no explanation of the forces that hold quarks together or of why the composites they form interact with each other. Since then, the idea behind quantum electrodynamics—that forces are transmitted by virtual particles—has been applied successfully to the strong force and to a unified version of the weak and electromagnetic forces known as the electroweak interaction. Together, these ideas form the standard model, which we describe in this section.

37.3.1 Electroweak Unification

In a weak interaction, the charge of a particle may change (as in neutron decay) or it may remain unchanged (■ Figure 37.13). Thus three different species of virtual particle with different electric charges are necessary to transmit weak forces. These particles are given the names $W^\pm$ and Z^0, with the superscripts indicating their electric charges. Furthermore, under weak interactions, the electron and the neutrino act as different states of the same kind of particle—just as the neutron and proton are the same when only strong forces are involved. To accommodate these features, the electroweak theory requires *four* force-transmitting particles, called *gauge bosons*, instead of three (● Table 37.4). The fourth has neither mass nor electric charge, and transmits forces between electric charges—it is the photon. Weak and electromagnetic

WE INTRODUCED THE NEUTRINO IN THE LAST CHAPTER; SEE §36.2.3. THINK OF THIS IN ANALOGY WITH THE ENERGY STATES OF AN ATOM. THE STATE OF THE ATOM CHANGES WHEN A PHOTON IS EMITTED OR ABSORBED. SIMILARLY, THE STATE OF A PARTICLE (AND ITS NAME!) CHANGE WHEN ONE OF THE GAUGE BOSONS IS EMITTED OR ABSORBED.

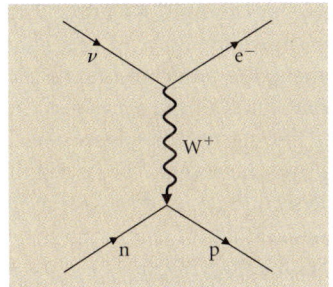
(a) $n + \nu_e \rightarrow p + e^-$.

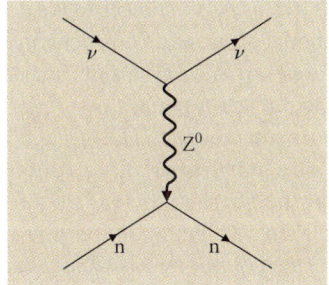
(b) Elastic collision between a ν and n.

■ FIGURE 37.13
Examples of weak interactions. (a) A neutrino emits a W^+ and becomes an electron. A neutron absorbs the W^+ and becomes a proton. (b) Elastic collision between a neutrino and a neutron. The Z^0 exchanges energy and momentum between the particles but does not alter the character of the neutron or the neutrino.

TABLE 37.4 Gauge Bosons

Name	Force	Mass (MeV)	Charge
Photon γ	electromagnetic	0	0
W^+	weak	80	$+e$
W^-	weak	80	$-e$
Z	weak	91	0

GAUGE REFERS TO A MATHEMATICAL FEATURE OF THE THEORY. BOSONS, AFTER SATYENDA NATH BOSE (1894–1974), ARE PARTICLES WITH INTEGER SPIN. THE QUARKS AND LEPTONS ARE FERMIONS, WITH HALF-INTEGER SPIN.

■ FIGURE 37.14
The UA1 detector at CERN, Switzerland. This detector was used in the detection of the Z^0 (Figure 10.16).

ELECTRICITY AND MAGNETISM OFFER A SIMILAR EXAMPLE OF UNIFICATION. IN SPECIAL RELATIVITY, THE DISTINCTION BETWEEN ELECTRIC AND MAGNETIC FIELDS BREAKS DOWN UNDER A CHANGE OF REFERENCE FRAME. THE FIELDS ARE TWO ASPECTS OF ONE EFFECT.

forces turn out to be the same interaction acting under different conditions. The great apparent difference between the two arises because massless virtual photons are easily produced and can act over macroscopic distances, while the W and Z bosons, with masses of about 80 GeV/c^2, are produced rarely and can act only over very short distances. In an (imaginary!) experiment in which 80 GeV were a negligible particle energy, the W, Z, and γ particles would produce forces of equal strength. This electroweak theory, developed by Sheldon Glashow, Steven Weinberg, and Abdus Salaam in the late 1960s, was confirmed in 1983 by the production of free W and Z bosons at the CERN accelerator (■ Figure 37.14).

The quarks are also subject to weak interactions. For example, ■ Figure 37.15 illustrates the quark model of the neutron–neutrino interaction shown in Figure 37.13a. The interaction changes a down quark into an up quark and also changes a neutrino into an electron. The weak interaction sees up and down quarks as different states of the same particle, as it does the electron/neutrino pair. The strange quark alone doesn't fit this scheme, since it doesn't have a pair, so a fourth quark flavor, dubbed *charm*, proves necessary. The last significant skepticism about the quark model faded quickly after the simultaneous discovery at the Stanford and Brookhaven accelerators of both ground and excited states of charmonium—a charmed and anticharmed quark in orbit about each other! (See ■ Figure 37.16.)

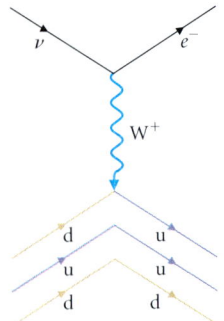

■ FIGURE 37.15
Quark model of the reaction $n + \nu_e \rightarrow p + e^-$. The W^+ boson emitted by the neutrino is absorbed by one of the d quarks in the neutron, converting it into a u quark. The resulting combination of one d and two u quarks forms a proton.

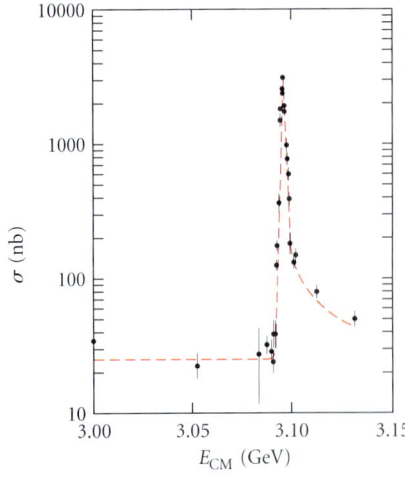

■ FIGURE 37.16
Discovery of the charmed quark. The graph shows the probability for an electron/positron pair to annihilate and produce strongly interacting particles. The peak is a resonance due to the formation of a charm/anticharm pair (J/ψ particle), which then decays to other particles. This peak occurs when the total energy of the interacting particles (in the CM frame) equals 3.1 GeV, corresponding to the 3.1 GeV/c^2 mass of the ψ. (One nanobarn is an effective cross-sectional area of 10^{-37} m^2. The probability of a reaction, and hence the number of observed events, is proportional to this effective area.)

TABLE 37.5	The World According to the Weak Interaction		
Family	Leptons		Quarks
1	electron	e^-, ν_e	u, d up, down
2	muon	μ^-, ν_μ	s, c strange, charm
3	tau	τ^-, ν_τ	t, b top, bottom

The weak interaction organizes the fundamental particles into the families, or *generations,* shown in ● Table 37.5. Particles from family 1 appear to be stable for indefinite periods under present conditions and form the macroscopic world we observe about us. Family 2 is involved in the particle reactions we have discussed so far. Recent discovery of the τ lepton and the top and bottom species of quark suggests a complete third family, though the τ neutrino has not been observed as of 1996. We have no theory that explains why there should be more than one generation nor how many generations there should be altogether.

The W boson can change a quark within a generation or between generations. Such interactions, involving a charged W boson, are called charged-current interactions. The Z boson participates in neutral current interactions and does not change quark flavor.

EXAMPLE 37.2 ❖ Draw a diagram like Figure 37.15 showing the decay $\Lambda^0 \rightarrow p + \pi^-$ discussed in Example 37.1.

MODEL From Table 37.2, the quark compositions are:

$$\Lambda^0 = uds, \quad p = uud, \quad \text{and} \quad \pi^- = \bar{u}d.$$

Thus this is a weak decay in which a strange quark is converted to a down plus a $u\bar{u}$ pair. The Z boson cannot change quark flavors, so the interaction involves a W.

SETUP The s (charge $-e/3$, Table 37.1) converts to a u (charge $2e/3$) by emitting a W^- with charge $-e$: $-e/3 - (-e) = 2e/3$. The W^- then decays to a $\bar{u}$ and a d.

SOLVE ■ Figure 37.17 is the diagram for this interaction.

ANALYZE We shall call these diagrams *quark line diagrams* because they trace the paths of the quarks as they rearrange into new particles. It is customary to draw the arrows on antiparticle lines showing them moving backward in time, as we have done in Figure 37.17 for the $\bar{u}$.

PERHAPS THE BEST STATEMENT OF OUR PUZZLEMENT OVER THE PARTICLE FAMILIES WAS I. I. RABI'S COMMENT ON THE DISCOVERY OF THE MUON: "WHO ORDERED THAT?" COSMOLOGY PROVIDES A LIMIT ON THE MENU'S SIZE. THE AMOUNT OF ^{4}He PRODUCED IN THE EARLY UNIVERSE DEPENDS STRONGLY ON THE NUMBER OF NEUTRINO SPECIES. THE EXISTENCE OF MORE THAN FOUR SPECIES IS INCONSISTENT WITH THE OBSERVED AMOUNT OF ^{4}He.

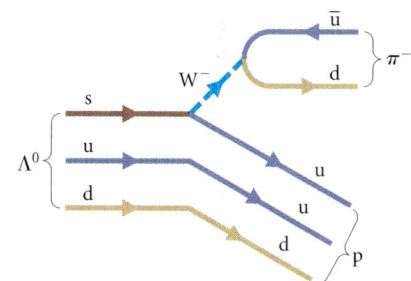

■ **FIGURE 37.17**
Feynman diagram for the weak hadronic decay of a Λ^0 to a proton plus π^-. The up and the down quark remain unchanged. The strange quark converts to a u by emitting a W^-, which then decays to the $\bar{u}$ and d comprising the π^-. Check the conservation of charge at each vertex of the diagram.

37.3.2 Quantum Chromodynamics

Quarks are also bound together by force-transmitting particles, named *gluons* in the whimsical tradition of high-energy physics names. The same scattering experiments that detect quarks inside protons also show that they do not carry all of a proton's momentum. Something else—the gluons—must exist inside a proton to carry the remaining momentum. A proton, then, is a little ball containing three quarks together with a cloud of virtual quark/antiquark pairs and virtual gluons which continually transmit the forces that hold the ball together.

In quantum electrodynamics, the electromagnetic field is represented by a cloud of virtual photons. We also know that the electromagnetic field is produced by, and acts on, the electric charge carried by particles. We say that the photons *couple with* electric charge. By analogy, we would expect quarks to possess a property, a strong force charge, by which they couple with the gluons. None of the observed particles, which are combinations of quarks, show any net amount of this strong force charge, just as neutral atoms contain positive and negative electric charge with a total of zero.

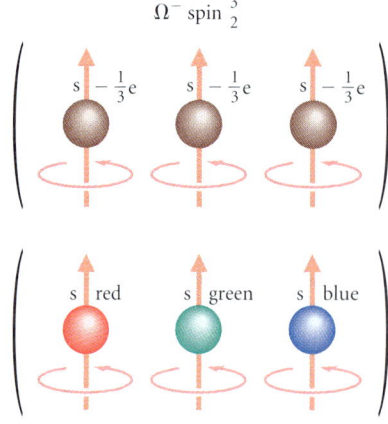
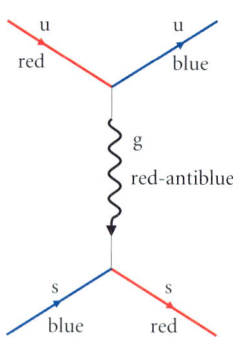

FIGURE 37.18
The Ω^- consists of three apparently identical quarks, which seems to contradict the Pauli exclusion principle. The introduction of color allows us to distinguish the quarks and preserve the principle.

FIGURE 37.19
The chromodynamic force is transmitted by gluons. Each gluon carries a combination of color charges. In this example a red, up quark emits a red-antiblue gluon. The up quark changes from red to blue. The blue, strange quark absorbs the gluon and becomes red.

A GOOD REASON FOR USING THESE NAMES IS THE ANALOGY WITH PRIMARY COLORS OF LIGHT. IF RED, GREEN, AND BLUE LIGHT BEAMS WITH EQUAL INTENSITY SHINE ON A REFLECTING SURFACE, THE RESULT IS SEEN AS WHITE. THIS MIMICS THE FACT THAT EQUAL AMOUNTS OF THE THREE COLOR CHARGES COMBINE TO BE NEUTRAL.

A TOTAL OF EIGHT GLUONS ARE NEEDED IN THE THEORY. IN ADDITION TO THE SIX COMBINATIONS LIKE BLUE-ANTIRED, THERE ARE TWO THAT CARRY MIXTURES SUCH AS BLUE-ANTIBLUE WITH RED-ANTIRED.

We can pull atoms apart and observe electric charge directly, but since we cannot isolate individual quarks, evidence for the existence of a strong force charge is indirect. The Ω^- particle gives an example of such evidence and shows that strong force charge has to have three different forms. The quark model of an Ω^- (■ Figure 37.18) consists of three s quarks with parallel spins to account for the particle's total spin of $3\hbar/2$. According to the Pauli exclusion principle (§35.3), no two quarks can be in identical states. Each of the three quarks in the Ω^- must have a different value of some quantum number, which we recognize as the value of their strong force charge. The accepted name for strong force charge is *color* and the three forms are *red, green,* and *blue.* Each of the observed particles has zero *net* color. Any names could have been chosen, but these have the virtue of being easy to remember. Of course, color charge has no relation to light, which is an electromagnetic wave produced by electric charge.

A theory modeled on quantum electrodynamics but involving *colors* of strong charge is called *quantum chromodynamics*. ■ Figure 37.19 illustrates a typical chromodynamic interaction and shows the crucial new feature that the gluons themselves carry color charge.

■ Figure 37.20 shows why isolated quarks cannot exist. When quarks are separated by distances of the order of 10^{-15} m, one can think of the chromodynamic force between them, like the electric force, in terms of field lines. The field lines represent gluon fields that attract each other and so concentrate in a tube between the quarks, rather than spreading out as electric field lines do. As a result, the chromodynamic force between quarks does not decrease at large separations. Separating two quarks would require an arbitrarily large amount of work, which would be stored as energy in the gluon field. Once this stored energy exceeds the rest energy of a quark/antiquark pair, such a pair materializes, producing two mesons rather than two separated quarks.

37.3.3 Conservation Laws for the Strong and the Weak Forces

• Table 37.6 shows the quantum numbers we have discussed, and their conservation properties. The traditional conserved quantities—energy, momentum, angular momentum, and charge—are also conserved in particle interactions. Quark number and baryon number are also absolutely conserved in the standard model. Certain theories that have been developed in an attempt to unify the strong and the electroweak forces predict nonconservation of quark and baryon number. At this time (1996), there is no experimental support for this idea. Quark flavor is conserved by the strong and electromagnetic forces but not by the weak force.

TABLE 37.6	The Conservation Laws of Particle Physics		
Quantity	Strong	Weak	Electromagnetic
Energy	✓	✓	✓
Momentum	✓	✓	✓
Angular momentum	✓	✓	✓
Electric charge	✓	✓	✓
Quark number	✓	✓	✓
Baryon number	✓	✓	✓
Lepton number	✓	✓	✓
Quark flavor	✓	×	✓
Isotopic spin	✓	×	×
Quark color	×	✓	✓
Lepton generation	✓	✓	✓

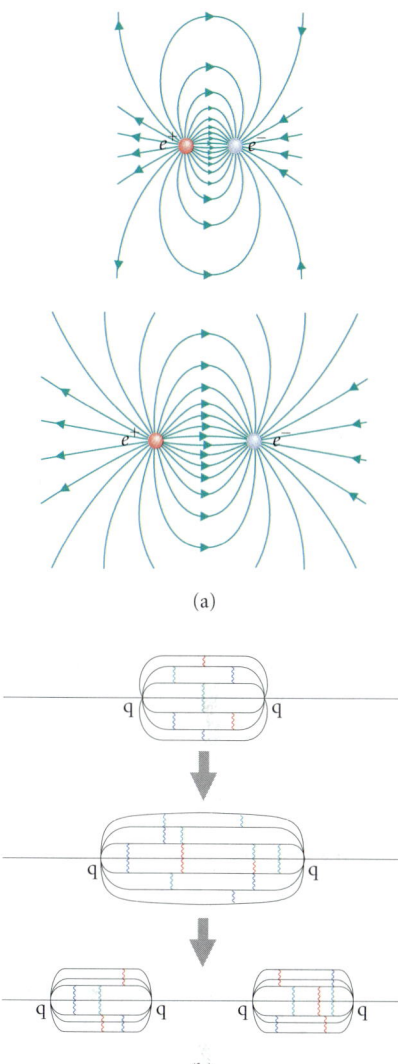

EXAMPLE 37.3 ♦♦ Compare the reactions $\pi^- + p \rightarrow K^0 + \Lambda^0$ (Figure 37.8) and $\Lambda^0 \rightarrow p + e^- + \bar{\nu}_e$. Use the conservation laws to determine whether each reaction involves the strong or the weak interaction.

MODEL The conserved quantities that distinguish strong hadronic interactions from weak ones are quark flavor and isotopic spin (I_3). (*Remember:* The hadrons are the baryons plus the mesons. See Figure 37.7.)

SETUP We investigate the values of these two quantities before and after the interaction. We find the quark composition of each particle from Table 37.2.

$$\pi^-: \bar{u}d, \quad \pi^+: u\bar{d}, \quad p: uud, \quad K^0: d\bar{s}, \quad \Lambda^0: uds.$$

	BEFORE $\pi^- + p$	→	AFTER $K^0 + \Lambda^0$	BEFORE Λ^0	→	AFTER $p + e^- + \bar{\nu}_e$
Quark number	$0 + 3 = 3$		$0 + 3 = 3$	3		3
Quark flavor	$\bar{u}d + uud = udd$		$d\bar{s} + uds = udd$	uds		uud
I_3	$-1 + \frac{1}{2} = -\frac{1}{2}$		$-\frac{1}{2} + 0 = -\frac{1}{2}$	0		$\frac{1}{2}$
Lepton number				0		$0 + 1 - 1 = 0$

SOLVE Neither quark flavor nor isospin change in the $p + \pi^-$ interaction, but both change in the decay of the Λ^0. Thus the first interaction is a strong interaction and the second is a weak interaction.

ANALYZE Both interactions conserve quark number, lepton number, and charge. Production of leptons in the Λ decay is a telltale sign of a weak process. ∎

EXERCISE 37.2 ♦♦ Is the decay $\rho^0 \rightarrow \pi^+ + \pi^-$ governed by the strong or weak interaction? What about $K^0 \rightarrow \pi^+ + \pi^-$? ∎

■ **FIGURE 37.20**
Isolated quarks do not exist because the color force does not decrease with distance. (a) Electric field lines between oppositely charged particles spread out as the separation of the charges increases, illustrating the decrease of the Coulomb force with distance (cf. §23.1 and §23.2). (b) In contrast, color force lines form a tubelike structure and do not separate as the quarks separate. Eventually, when the applied force has done enough work, a single tube splits into two, forming a quark/antiquark pair. The result is two mesons rather than isolated quarks.

37.3.4 Limitations of the Standard Model

The standard model has been completely successful in explaining qualitative features of subnuclear particles. Quantitative calculations with the model are often extremely intricate, but those that have been done are in good agreement with experiment. Nevertheless, no one accepts the model as a final description of nature.

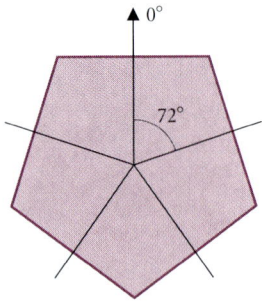

Figure 37.21
A pentagon illustrates the classical idea of symmetry. Rotation of the figure through any multiple of 72° leaves its appearance unchanged.

- Its own great success argues against it.

  The model describes electroweak and chromodynamic interactions by very similar theories. If such theories show weak and electromagnetic forces to be unified, should we not look for a more comprehensive model that includes the strong force and perhaps even gravity in the unified theory?

- Important phenomena lie beyond the scope of the theory.

  Why, for example, are there three apparently different types of force rather than some other number? What determines the number of generations of particles? Why do the numerical parameters in the theory have the values they do?

In the last section of this chapter, we describe some of the ideas that led to discovery of the standard model and that are guiding attempts to extend it.

37.4 Characteristics of Modern Particle Theories

37.4.1 Symmetries and Groups

The pentagon in ■ Figure 37.21 illustrates the classical, geometrical idea of symmetry—a regularity of shape that is pleasing to the eye. A precise statement of this regularity is that a rotation of the figure through 72° results in an identical shape. This statement describes the pentagon's symmetry in a way that generalizes to much more abstract *symmetries* of physical systems: a set of operations (rotation by 72°, 144°, etc.), and a property (shape) that remains unchanged when any of the operations is performed. A closely related example, important in physics, is the fact that all the laws of physics remain unchanged if the coordinate system is rotated in any way. Equivalently, the behavior of a physical system does not depend on spatial orientation.

FOR EXAMPLE, A PENDULUM SWINGING IN EARTH'S GRAVITY IS PART OF A SYSTEM THAT INCLUDES THE EARTH. THE PENDULUM'S BEHAVIOR DOESN'T VARY DURING THE COURSE OF A DAY, AS THE EARTH ROTATES.

Symmetries are closely related to the mathematical idea of groups and to the physical idea of conservation laws. Any set of symmetry operations corresponds to the elements of a group, and every symmetry of physical laws corresponds to a conserved physical property. For example, the symmetry of physical law under rotations corresponds to conservation of angular momentum.

THIS RESULT, WHILE TRUE, IS CERTAINLY NOT OBVIOUS AND WE DO NOT CLAIM TO HAVE PROVED IT.

The group properties can be satisfied only by special sets of elements, which in turn means that only special kinds of symmetry are possible for physical laws. As a result, the mathematics of groups has proved a powerful guide in the discovery of the unified electroweak and quantum chromodynamic theories. One approach to further unification is the attempt to determine which group describes the symmetry of the unified physical laws.

The power of group theory is well illustrated by isotopic spin. We could imagine changing each proton in the universe into a neutron, and vice versa. This operation is a symmetry of the strong force, since it leaves the strong forces acting on any object unchanged. The group of such particle-exchange symmetries of the strong force is the same as the group of spatial rotations. Thus the corresponding quantity, conserved by strong interactions, should follow the same rules as angular momentum. This quantity is the isotopic spin.

WE SAY THAT THE ROTATIONS AND THE STRONG FORCE SYMMETRIES ARE DIFFERENT REPRESENTATIONS OF THE SAME GROUP.

Math Topic

GROUP PROPERTIES ILLUSTRATED BY AN EXAMPLE

PROPERTIES OF A GROUP	ROTATIONS OF A PENTAGON
A group is collection of elements with an operation $*$.	Rotations through 0°, 72°, 144°, 216°, 288° (360° ≡ 0°) follow one rotation by another.
If a and b are in the group, then so is $a*b$.	$72°*144°$ = rotation through $144° + 72° = 216°$.
There is an identity element e.	$e = 0°$—no rotation.
$e*a = a$ for every element a in the group.	$0°*144°$ = rotation through $144° + 0° = 144°$.
For any element a in the group, there is an inverse element a^{-1} such that $a^{-1}*a = e$.	$(144°)^{-1} = 360° - 144° = 216°$ $216°*144°$ = rotation through $144° + 216° = 360° \equiv 0°$.

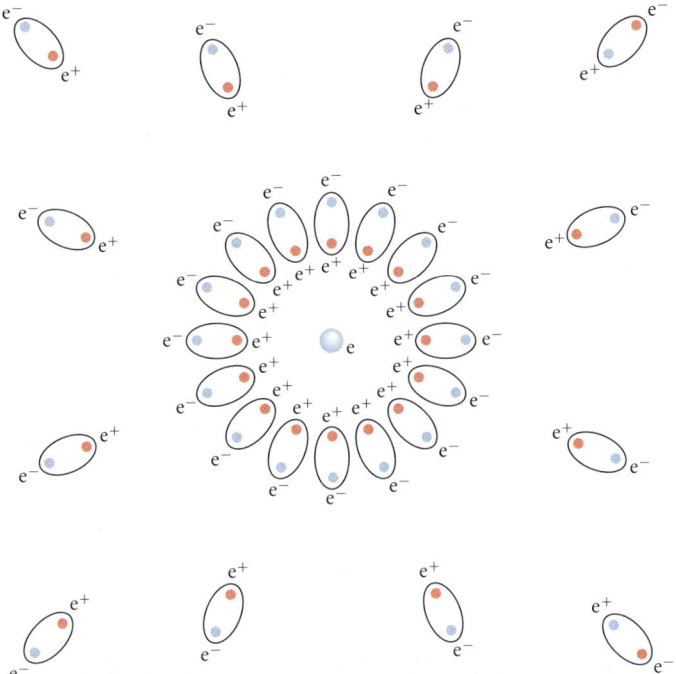

■ **FIGURE 37.22**
An electron is surrounded by a sea of virtual electron/positron pairs. Like the molecules in a dielectric, the pairs are polarized by the central charge and shield the *bare* charge from outside observers. The charge we observe and call *e* is the shielded charge. Both the bare charge and the shielding are infinite. Fortunately, the shielded charge can be calculated. Predictions made using this theory are extremely well supported by experiment.

37.4.2 Renormalization

In Chapter 25 we found that the total electrostatic energy of a point charge is infinite. Though startling, this infinite energy does not cause trouble, since we only observe energy changes. In the classical theory of the electron, we neglect the self-energy, which does not change during classical electrical processes. Such neglect of constant, infinite quantities is an example of *renormalization*.

Infinities appear in the quantum theory via the existence of virtual particles. An electron, ■ Figure 37.22, produces a cloud of virtual photons. According to the Heisenberg principle, the most energetic particles have the shortest time of existence and so remain close to the electron. These virtual photons can themselves produce virtual electron/positron pairs. It seems we can't have a single isolated particle; particles are necessarily surrounded by a virtual particle/photon cloud. The cloud possesses energy and, because the central electron attracts virtual positrons and repels virtual electrons, contributes to the total observed charge. The effects of the cloud are infinite. The observed mass and charge of the electron are the renormalized differences between infinite *bare* mass and charge and the infinite effects of the virtual cloud. Renormalization must result in finite quantities if a quantum theory is to make sensible statements about observed reality. This is a trivial requirement for classical electromagnetism but is a strong restriction on quantum theories. Indeed, one of the strongest reasons for believing the electroweak and quantum chromodynamic theories is that they are renormalizable.

37.4.3 Spontaneous Symmetry Breaking

The symmetry of a physical theory is often disguised by the phenomena we observe. One example is the magnetism of an iron sample. The magnetization of the sample points in some specific direction, yet Maxwell's equations show no preference for any particular direction over another. The theory is symmetric, but the iron sample is not. If the sample is heated above its Curie temperature T_C, the iron atoms no longer align, the magnetization disappears, and the sample regains the symmetry of electromagnetic theory. This behavior of the iron sample is called *spontaneous symmetry breaking*. The rotational symmetry of electromagnetism is *broken* by the iron sample once the typical energy of an atom becomes less than approximately kT_C. The breaking is spontaneous in that random fluctuations in the sample will cause it to magnetize even in the absence of external fields. Furthermore, the loss of symmetry is not complete: the direction in which an isolated sample will magnetize is completely random.

The differences between electromagnetism and the weak force are another example of symmetry breaking. The energy at which the breaking occurs is set by the masses of the W and Z bosons that transmit the weak force. The quest for further unification of physics presumes that electroweak and chromodynamic forces differ only because some, as yet undemonstrated, symmetry remains broken at energies currently achievable by accelerators.

37.5 Conclusion

Quantum chromodynamics and the unified electroweak theory are as far as we can go in describing fundamental particle phenomena, based on correspondence between theory and experimental data. Together with general relativity, they form a working set of answers to the ancient Greek questions about the constituents of the universe and the rules it follows. Three rules (for the three fundamental forces) and 24 basic constituents suffice. Furthermore, each of the three theories shows a high degree of mathematical elegance and economy of physical assumptions. We have every reason to be pleased with the achievements of physics in explaining the world, but no excuse for being smug about them. It remains for the physicists of the future to show how quantum principles apply to gravity and to continue the quest for unity in physical law.

SIX QUARKS, SIX LEPTONS, AND THEIR ANTIPARTICLES. WE HAVE NOT INCLUDED THE FORCE-CARRYING GAUGE BOSONS IN THIS COUNT.

Chapter Summary

Where Are We Now?

We have completed our survey of physical theory, including results discovered centuries ago and those discovered only shortly before the publication of this book. Theories of atomic and subatomic physics involve complex and abstract mathematics, which we have described more qualitatively than our treatment of classical ideas. The quest for knowledge goes on, and the physical models we have described will be modified and refined as research progresses.

What Did We Do?

In the standard model of particle physics, matter is composed of 24 fundamental quantities: 6 quarks, 6 leptons, and their antiparticles. Quarks have a property, akin to electric charge, called color charge, which comes in three kinds: red, green, and blue. All forces result from the exchange of particles: exchange of photons causes the electromagnetic force between electrically charged particles. Similarly, the weak force is due to the exchange of W and Z bosons, and the strong force results from the exchange of gluons between particles possessing color charge. All known particles are accounted for as either individual leptons or as combinations of 2 or 3 of the 12 quark types, bound by a cloud of gluons.

Quarks have fractional electromagnetic charge, in multiples of $\frac{1}{3}e$. Familiar nuclear particles such as the proton are combinations of quarks with an integer total electric charge and no net color.

Conserved quantities in physics are related to underlying symmetries in the physical laws. For example, space-time is invariant under spatial rotations; all forces respect this symmetry, and the corresponding conserved quantity is angular momentum. Symmetry is described by the mathematical idea of a group. The groups describing the symmetries of electromagnetic and weak interactions are combined into one larger group to describe the unified electroweak force. Physicists hope to find a theory that unifies the strong force with the electroweak force.

Practical Applications

The theory presented in this chapter is so new that practical applications, if any, remain for the future.

Solutions to Exercises

37.1 A particle's de Broglie wavelength is given by its momentum: $\lambda = h/p$. For so small a wavelength as we desire, we expect to require highly relativistic electrons, each with an energy much greater than its rest mass. So:
$$p^2c^2 = E^2 - m_0^2 c^4 \approx E^2.$$
Then,
$$E \approx pc = \frac{hc}{\lambda}$$
$$= \frac{(6.64 \times 10^{-34} \text{ J·s})(3 \times 10^8 \text{ m/s})}{10^{-16} \text{ m}}$$
$$= 2 \times 10^{-9} \text{ J}$$
$$= 10 \text{ GeV}.$$

37.2 The ρ^0, at the center of the meson nonet (Figure 37.12d), is composed of equal numbers of quarks and their corresponding antiquarks, and has no net quark flavor. The π^+ and π^-, being the antiparticle of each other, also have no net quark flavor. $I_3 = 0$ for the ρ^0, -1 for the π^-, and $+1$ for the π^+. Quark flavor and the total value of I_3 are both zero before and after the decay, which is therefore governed by the strong interaction.

The K^0 has quark flavor $d\bar{s}$, so quark flavor is not conserved in its decay to two pions. Additionally, I_3 changes from $-\frac{1}{2}$ to 0. This is a weak decay.

Basic Skills

Review Questions

§37.1 PARTICLE CREATION AND FUNDAMENTAL FORCES

- Why does the macroscopic world appear stable?
- What is a *virtual particle*? How long can it exist?
- Which particle did Yukawa first propose? Why?
- To what does *isospin* refer? Explain why the name is appropriate.

§37.2 SUBNUCLEAR PARTICLES AND THE QUARK MODEL

- What were the three original classes of particles? What distinguished the classes?
- Why are *strange* particles strange?
- What is a *quark*? What is the fundamental unit of charge for quarks?
- How many original quark flavors were there?

§37.3 THE STANDARD MODEL

- Which particles transmit the weak force? Are any of them charged? If so, which?
- Which bosons can change quark flavor?
- How many generations of particles are there?
- What name is given to the strong force charge? How many kinds are there and what are their names?
- How does the chromodynamic force vary with quark separation?
- State two limitations of the standard model.

§37.4 CHARACTERISTICS OF MODERN PARTICLE THEORIES

- Which mathematical structure is important for understanding particle theory?
- State a symmetry of the strong force and the corresponding conserved quantity.
- What is *renormalization*?
- What is *spontaneous symmetry breaking*?

Basic Skill Drill

§37.1 PARTICLE CREATION AND FUNDAMENTAL FORCES

1. For how long can a virtual electron/positron pair exist?
2. What is the energy of each photon produced in the decay of a π^0: $\pi^0 \rightarrow \gamma + \gamma$? (The two photons have equal energy in the CE frame. Do you recall why?)

§37.2 SUBNUCLEAR PARTICLES AND THE QUARK MODEL

3. Find the Q value for the reaction:
$$p + \bar{p} \rightarrow \pi^+ + \pi^+ + \pi^- + \pi^-.$$
4. Draw a quark line diagram for the reaction:
$$\pi^- + p \rightarrow \Lambda^0 + K^0.$$
What is the Q value for this reaction?

§37.3 THE STANDARD MODEL

5. Can the reaction
$$K^0 + p \rightarrow K^+ + n$$
proceed via the strong interaction? Why or why not?
6. Draw a Feynman diagram for the interaction:
$$K^- \rightarrow \mu^- + \bar{\nu}_\mu.$$
Which boson is involved?

§37.4 CHARACTERISTICS OF MODERN PARTICLE THEORIES

7. The Δ particle has isospin $I = -\frac{3}{2}$. How many different Δs are there? What are the corresponding values of I_3?

Questions and Problems

§37.1 PARTICLE CREATION AND FUNDAMENTAL FORCES

8. ❖ Discuss the reasons why one must increase the particle energy available from accelerators in order to study phenomena on smaller spatial scales or on shorter time scales.

9. ◆ Calculate the minimum energy of a photon that can cause the reaction:
$$\gamma + p \to \pi^0 + p.$$

10. ◆ Calculate the minimum energy in the CE frame for each incident proton in the reaction:
$$p + p \to p + p + \Lambda + \bar{\Lambda}.$$

11. ◆◆ For an electric motor to function according to Maxwell's equations, virtual photons must be able to travel at least as far as the dimensions of the motor. Derive a value for the maximum possible rest mass of a photon from the dimensions of a large industrial motor, approximately 1 m. What limit on the photon rest mass follows from the fact that Maxwell's equations describe the Earth's magnetic field over distances of the order of 10^7 m?

§37.2 SUBNUCLEAR PARTICLES AND THE QUARK MODEL

12. ❖ Strangeness is conserved in some processes and not in others. In what sense is this similar to your experience with *liquid* water. In what ways is it different?

13. ❖ Mesons have integer spin and zero baryon number. Explain why only combinations of a quark and an antiquark can form a meson. Similarly, explain why only combinations of three quarks may form a baryon.

14. ❖ The baryon octet in Figures 37.10 and 37.12 consists of spin $\frac{1}{2}$ particles, while the baryon decuplet contains spin $\frac{3}{2}$ particles. How are the quarks arranged differently in an octet particle and in a decuplet particle with the same isospin and hypercharge?

15. ◆◆ Determine which quantum numbers are conserved in the decay:
$$\overline{\Sigma^+} \to \bar{p} + \pi^0. \quad (\overline{\Sigma^+} \text{ is the antiparticle of the } \Sigma^+.)$$

§37.3 THE STANDARD MODEL

16. ❖ Draw diagrams showing the exchange of bosons in each of the following reactions:
(a) $p + \bar{\nu}_e \to n + e^+$.
(b) $e^- + \nu_\mu \to \nu_e + \mu^-$.
(c) $d + e^- \to u + \nu_e$.
(d) $\nu_e + \nu_\mu \to \nu_e + \nu_\mu$.
(e) $\bar{\nu}_e + p \to e^+ + \Lambda$.

17. ❖ Determine which of these reactions cannot occur at all and which cannot occur by the strong interaction. In each case, indicate the reasons for your decision.
(a) $K^- + p \to \overline{K^0} + n$.
(b) $\pi^- + p \to K^- + \Sigma^+$.
(c) $\pi^- + p \to K^+ + \Sigma^0 + \pi^-$.
(d) $K^- + p \to \Sigma^+ + n + \pi^-$.
(e) $K^- + p \to K^0 + \pi^+ + e^-$.
(f) $\pi^- + p \to \pi^0 + \Sigma^+$.

Draw quark line diagrams for those reactions that can occur by the strong interaction.

18. ◆◆ Draw a quark line diagram for the reaction:
$$p + \pi^- \to n + \pi^0.$$
What is the Q value for this reaction?

19. ◆◆ Draw a quark line diagram for the reaction:
$$\Xi^- + p \to \Lambda + \Lambda.$$
Find the Q value.

§37.4 CHARACTERISTICS OF MODERN PARTICLE THEORIES

20. ✽ ❖ Do the positive integers (0, 1, 2, 3, . . .) form a group? If so, what are the operation and the identity element of the group? Answer the same question for the following sets:

- All integers (0, ±1, ±2, ±3, . . .).
- All rational fractions (a/b, where a and b are positive integers).
- Translations of the origin of coordinates along a single axis.

21. ❖ At what temperature would you expect the weak interaction to attain comparable strength to electromagnetism? Explain your reasoning.

Additional Problems

22. ◆◆ Given that the Λ_c^+ charmed baryon can decay weakly into $\Delta^{++} + K^-$ or $p + \overline{K^0} + \pi^+ + \pi^-$, determine a limit to the mass of Λ_c^+.

23. ◆◆ Calculate the Q value for the weak decay:
$$\Omega^- \to \Lambda + K^-.$$
Verify that quark number is conserved.

24. ◆◆ The Δ^0 and the Λ both decay to $p + \pi^-$, but their half-lives are very different. Explain why. Which half-life is the shorter? (The values are approximately 10^{-23} s and 3×10^{-10} s, respectively.)

25. ◆◆ Both the ρ^0 and the K^0 decay to $\pi^+ + \pi^-$, but the ρ^0 has a half-life of about 10^{-23} s, while the half-life of the K^0 is about 10^{-10} s. Explain the difference. What can you conclude about the quark flavor of the ρ^0? Determine the Q value for each decay [$M(\rho^0) = 770$ MeV].

26. ◆◆ What is the minimum energy of each proton in the CE frame for the reaction
$$p + p \to \Lambda^0 + K^0 + p + \pi^+$$
to occur? Determine whether this is a strong or weak interaction and draw a quark line diagram or a Feynman diagram for it.

27. ◆◆ Draw a Feynman diagram at the quark level for the following reactions:
(a) $\Omega^- \to \Xi^0 + \pi^-$.
(b) $\Lambda_c^+ \to p + \overline{K^0}$.

Find the Q value for each reaction.

Part VIII Problems

1. ❖ Tachyons are imaginary particles that move faster than light. Just as ordinary particles cannot move faster than light, tachyons cannot go slower. On a space-time diagram, draw the world line of a tachyon emitted from the origin in the positive x-direction at speed $2.2c$, toward an observer moving in the x-direction at speed $0.8c$. Upon receiving the signal, the observer transmits a tachyon signal at $2.2c$ (in his own frame) in the negative x'-direction. Show that this signal arrives at the unprime time-axis *before* the original signal was transmitted. (See Greg Benford's novel *Timescape* for some consequences of tachyon communication!)

2. ❖ Explain, using the shell model (Figure 36.26), why the spins of individual nucleons do *not* add to produce nuclei with large total spin.

3. ❖❖ Show that when a particle's energy is much greater than its rest energy, its de Broglie wavelength is approximately the same as the wavelength of a photon with the same energy.

4. ❖❖ In an accelerator experiment, a K^0 particle decays into two π^0 mesons. In the lab frame, the π^0 mesons' velocities make equal angles $\theta = 27°$ with the direction of motion of the K^0. Find the speed of the K^0. Mass of $K^0 = 497.71$ MeV; mass of $\pi^0 = 134.97$ MeV. (*Hint:* It is easiest to work in the rest frame of the K^0 and then transform to the lab frame.)

5. ❖❖❖ A quasar 3×10^9 light-years away emits a blob of luminous gas moving with $\gamma = 10$ at an angle of 0.01 rad from the line of sight to the Earth. Show that the blob will *appear* to move away from the quasar at greater than the speed of light. Such *superluminal motion* is observed in many quasars. (*Hint:* Assume the blob emits light pulses at regular intervals, and find the separation of the pulses in space and time as seen from Earth. Only distance perpendicular to the line of sight can be measured!)

6. ❖❖❖ In 1959, Pound and Rebka measured the effects of gravitational time dilation on γ rays propagating from the top to the bottom of a 22.5-m-high tower. In the experiment, nuclei in a sample of ^{57}Co emit γ rays with precisely determined frequency as they decay to ^{57}Fe. The γ rays are absorbed in a sample of ^{57}Fe at the bottom of the tower. The detector absorbs the γ rays only when the Doppler effect due to its motion cancels the frequency shift due to time dilation. What fractional change in frequency of the γ rays results from gravitational time dilation? What velocity of the detector results in maximum absorption?

7. ❖❖❖ In an inertial reference frame at rest with respect to the center of a rotating disk, events A and B occur simultaneously at the opposite ends of a diameter. At the same time observers O and O' on the disk arrive at opposite ends of the perpendicular diameter (■ Figure VIII.1). Observers O and O' do not agree that A and B are simultaneous in their respective instantaneous rest frames. They do not even agree as to which event occurs first. By how much time do they disagree? This illustrates that observers in a noninertial reference frame, such as the surface of the Earth, cannot consistently synchronize their clocks. How big is the effect on Earth? Atomic clocks are stable to 1 part in 10^{13}. How much time must pass before the uncertainty in the clock reading is larger than the synchronization error?

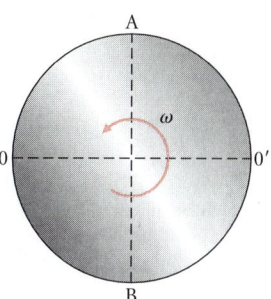

■ Figure VIII.1

8. ❖❖❖ An astronaut jokes that NASA should pay overtime to space shuttle crews because of relativistic effects on the experience of time. Which is the more important effect in a space shuttle orbit, special relativistic time dilation due to the shuttle's speed or general relativistic effects due to the different gravitational potential in orbit as compared with the Earth's surface? Should the astronauts be paid overtime or have their pay docked due to the dominant effect?

9. ❖❖❖ Use Heisenberg's uncertainty principle to estimate the minimum kinetic energy of an electron confined within a nucleus with radius 4.2×10^{-15} m. If the nucleus is a uniformly charged sphere with atomic number $Z = 14$, *estimate* the electric potential energy of the electron. From your results, is it plausible that electric forces can bind the electron within the nucleus?

10. ❖❖❖ Consider a 10-MeV proton incident on a ^{12}C nucleus. Compare the speed of the proton with the speed of an electron in the $n = 1$ Bohr orbit. Also compute the maximum energy the proton can transfer to the electron in a head-on collision. Comment on why the chemical state of an atom is unimportant in understanding its nuclear reactions.

11. ❖❖❖ (a) Apply Bohr's atomic theory to the orbital motion of an electron and a positron around each other. Show that the energy of the nth orbit is given by: $E_n = -mc^2\alpha^2/(4n^2)$, where $\alpha = ke^2/(\hbar c) \approx \frac{1}{137}$ is the fine-structure constant, which measures the strength of the electromagnetic force. (b) An early model for charmonium assumed that the force between a c and a $\bar{c}$ quark follows an inverse square law, but with strength α_s instead of α. Assume that the J/Ψ particle (mass = 3.096 GeV) and the Ψ' particle (mass = 3.687 GeV) are the $n = 1$ and $n = 2$ Bohr states of charmonium. Assume also that the mass of a charmed quark is $m_c \approx 1.5$ GeV. Find the ratio α_s/α.

Challenge Problems

12. A photon with initial frequency f Compton-scatters from an electron moving relativistically in the same direction as the photon. Find the frequency of the scattered photon as a function of scattering

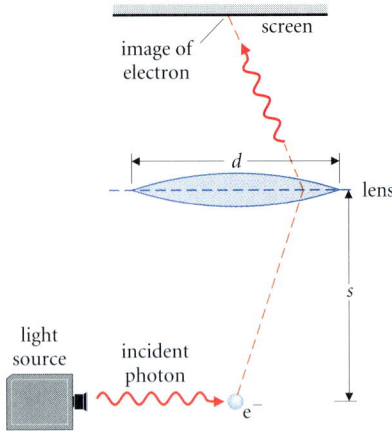

Figure VIII.2

angle and the initial momentum of the electron. Repeat the calculation for an electron moving in the opposite direction from the photon. Comment on the reasons why your results differ from the Compton formula for a stationary electron. A beam of 400-keV x rays is scattered by a gas of singly ionized helium atoms. Estimate the error made in assuming the electrons are stationary rather than in orbit about the helium nuclei.

13. Suppose we attempt to measure the position of an electron with the apparatus shown in ■ Figure VIII.2. A photon from the light source scatters from the electron and is focused by the lens on a photographic plate. The accuracy of the measured position is limited by the resolution of the lens. Show that the uncertainty in the measured position is $\delta x \sim s\lambda/d$. In principle, the accuracy of the measurement can be improved indefinitely by decreasing the wavelength of the photons used. However, the photon transfers momentum to the electron as it scatters. Show that the resulting uncertainty in the momentum of the electron after the measurement is $\delta p_x = (h/\lambda)(d/s)$ so Heisenberg's uncertainty principle is satisfied.

*This is not the end.
It is not even the beginning of the end.
It is perhaps, the end of the beginning.*

Winston Churchill

EPILOGUE

If history is a guide, today's fundamental understanding will ultimately be recognized as an approximation. The standard model succeeds in explaining physics with only three interactions and a small number of particle types. Yet, as we noted in Chapter 37, it contains many arbitrary features. Following the spirit of Einstein, physicists now have appetites whetted for a theory with a single interaction in which all details arise as necessary consequences. With little experimental evidence, the search for such a theory is highly speculative, and we shall take only a quick glance at current ideas.

The logic of the standard model suggests that the electroweak and strong interactions appear different only because we cannot yet observe phenomena in which their unity is apparent. Pursuing this logic leads to a class of Grand Unified Theories, or GUTs, in which the photons, W and Z bosons, and gluons of the standard model are seen as special cases of particles that carry the grand unified interaction. The quest for unity is also a quest for higher experimental energy. Energies of 80 GeV are required to produce the W and Z bosons that carry the weak interaction; much higher energies are needed to test GUTs. The Superconducting Supercollider would have produced particle energies of approximately 40 TeV, at a cost of billions. That massive an enterprise is needed to search for phenomena predicted by GUTs.

The universe provides the ultimate high-energy physics laboratory. It is expanding away from an immensely hot, dense origin some 10–20 billion years ago. Near this beginning, temperatures were high enough to produce any kind of elementary particle. Cosmologists have shown that the current state of the universe may be the result of particle reactions in the first fraction of the first second. An exciting consequence of some GUT models is that universes arise spontaneously from a quantum vacuum. Fundamental particle dynamics may explain the existence of our universe!

Beyond GUTs comes the unification of gravity with the other interactions. Gravity is intimately related to the geometry of space-time, and abstract geometrical notions play a large role in such theories. For example, *superstring* theories involve a space-time with ten dimensions. Four are the usual space and time dimensions. Space in the remaining six dimensions is curved so strongly that the entire six-dimensional space we usually think of as a single space-time event is approximately 10^{-35} m in extent. In these theories, the fundamental type of object is not a point particle, but a one-dimensional *string* that wraps around the *compactified* six-space. The possible vibrations of such strings give rise to the different kinds of particle observed at the level of the standard model.

We do not know now which, if any, of these speculations will prove correct. The only certain thing is that physics will remain exciting and continue to stretch our imagination for some time to come.

The earliest and simplest GUT models predicted decay of protons with a lifetime of $10^{32\pm2}$ y. Several experiments have ruled out a lifetime this short.

The Superconducting Supercollider, already under construction in Texas, was canceled by Congress in 1994.

APPENDIXES

Appendix I

MATHEMATICS
A. Scientific Notation
B. Arithmetic, Algebra, Geometry and Trigonometry
C. Vectors
D. Coordinate Systems
E. Derivatives
F. Integrals
G. Rules of Probability

Appendix II

SYMBOLS
A. The Greek Alphabet
B. Symbols Used in the Text
C. Mathematical Symbols
D. Symbols for Units
E. Roman Numerals

Appendix III

ANSWERS TO SELECTED PROBLEMS

APPENDIX I

Mathematics

A: SCIENTIFIC NOTATION

In problems of interest to physicists, quantities may vary over a very large range of magnitudes. For example, consider the concept of length; sizes much smaller than single atoms and as large as the visible universe may be important in the same discussion. Scientific discussions thus require a convenient and efficient way to describe these magnitudes and to calculate with the very large or very small numerical ratios that arise from comparing them.

Common decimal notation certainly allows the expression of such numbers. A ratio of one-millionth can be written 0.000001, but already the number of zeros makes the expression difficult to read quickly. Most people would be unable to read a much smaller number directly and would have to resort to counting zeros. Scientific notation is a way of using exponents to express numbers by counting zeros and reporting the count rather than writing the zeros. Consider the following examples.

$$0.001 = 1/1000 = 1/(10 \times 10 \times 10) = (1/10)^3 = 10^{-3}$$

Or, $\quad 1000. = 10 \times 10 \times 10 = 10^3.$

If the leading digit is to the right of the decimal, the exponent is negative; if the leading digit is to the left, the exponent is positive.

$$33\,760\,000 = 3.376 \times 10\,000\,000 = 3.376 \times 10^7.$$
$$0.0000462 = 4.62 \times 0.00001 = 4.62 \times 10^{-5}.$$

The number that multiplies the power of 10 is chosen by convention. One convention, which we generally use in this text, is to express numbers greater than 1000 or less than 0.001 as a value between 0.1 and 10 multiplied by the appropriate power of 10. Another convention, popular among engineers, is to make the exponent of 10 be a multiple of 3. Thus we might write the numbers above as:

$$3.376 \times 10^7 = 0.3376 \times 10^8$$

and $\quad 4.62 \times 10^{-5} = 0.462 \times 10^{-4},$

while the engineering convention would be

$$33.76 \times 10^6 \quad \text{and} \quad 46.2 \times 10^{-6}.$$

Multiplication and division of numbers in scientific notation rely on the distributive law and the rules for multiplying or dividing exponents. For example:

$$\begin{aligned}
(3.376 \times 10^7) \times (4.62 \times 10^{-5}) &= (3.376 \times 4.62) \\
&\quad \times (10^7 \times 10^{-5}) \\
&= 15.6 \times 10^{7-5} \\
&= 15.6 \times 10^2 \\
&= 1.56 \times 10^3.
\end{aligned}$$

$$\frac{3.376 \times 10^7}{4.62 \times 10^{-5}} = \left(\frac{3.376}{4.62}\right)\left(\frac{10^7}{10^{-5}}\right) = 0.731 \times 10^{12}.$$

To add and subtract numbers in scientific notation, first express the numbers as multiples of the *same* power of 10. Then the numbers can be added or subtracted in the normal manner:

$$\begin{aligned}
3.376 \times 10^7 + 6.417 \times 10^5 &= 3.376 \times 10^7 + 0.06417 \times 10^7 \\
&= 3.440 \times 10^7.
\end{aligned}$$

$$\begin{aligned}
4.62 \times 10^{-5} - 3.71 \times 10^{-4} &= 0.462 \times 10^{-4} - 3.71 \times 10^{-4} \\
&= -3.25 \times 10^{-4}.
\end{aligned}$$

B: Arithmetic, Algebra, Geometry, and Trigonometry

Arithmetic and Algebra

Associative law of multiplication:	$a(b + c) = ab + ac$
Distributive law of addition:	$(a + b) + c = a + (b + c)$
Distributive law of multiplication:	$(ab)c = a(bc)$
Logarithm of a product:	$\log(ab) = \log(a) + \log(b)$
Change of base of a logarithm:	$\log_a x = (\log_a b)(\log_b x)$

A-1

Conversion of the natural logarithm to base 10: $\log_e 10 \approx 2.303$

Conversion of base 10 to the natural logarithm: $\log_{10} e \approx 0.434$

Multiplication of exponents: $(a^x)(a^y) = a^{x+y}$

Powers of exponents: $(a^x)^y = a^{xy}$

Solution of a quadratic, $ax^2 + bx + c = 0$: $x = \dfrac{-b \pm \sqrt{b^2 - 4ac}}{2a}$

Factors: $a^2 + 2ab + b^2 = (a + b)^2$
$a^2 - b^2 = (a - b)(a + b)$

Factorial: $n! = n(n - 1)(n - 2) \cdots (2)1$

Taylor series:
$$f(x_o + x) = f(x_o) + f'(x_o)x + f''(x_o)x^2/2! + \cdots$$

Binomial expansion:
$$(1 + x)^n = 1 + nx + n(n - 1)x^2/2! + \cdots$$

Series expansions of common functions:
$$e^x = 1 + x + x^2/2! + x^3/3! + \cdots$$
$$\ln(1 + x) = x - x^2/2 + x^3/3 + \cdots$$
$$\sin(x) = x - x^3/3! + x^5/5! + \cdots$$
$$\cos(x) = 1 - x^2/2! + x^4/4! + \cdots$$
$$\tan(x) = x + x^3/3 + 2x^5/15 + \cdots$$

Geometry

Circumference of circle of radius r: $2\pi r$

Area of circle of radius r: πr^2

Surface area of a sphere of radius r: $4\pi r^2$

Volume of a sphere of radius r: $4\pi r^3/3$

Volume of a circular cylinder of radius r and height h: $\pi r^2 h$

Area of a triangle: $bh/2$

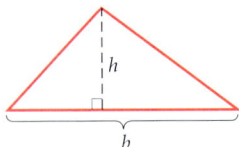

Pythagorean theorem: $a^2 + b^2 = c^2$

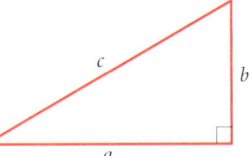

Straight line: $y = mx + c$
$m = \tan\theta = \text{rise/run}$

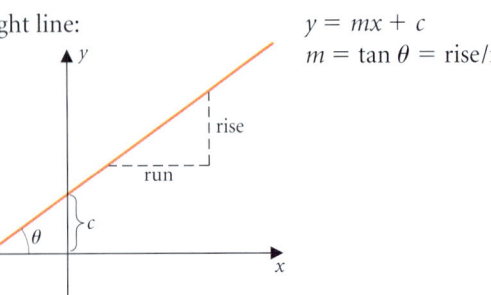

Circle: $(x - x_o)^2 + (y - y_o)^2 = r^2$

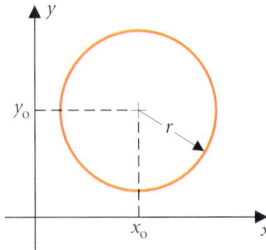

Ellipse: $\dfrac{(x - x_o)^2}{a^2} + \dfrac{(y - y_o)^2}{b^2} = 1$

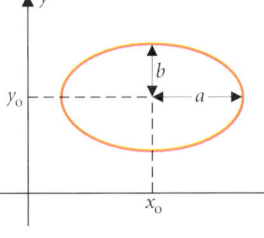

Hyperbola: $\dfrac{(x - x_o)^2}{a^2} - \dfrac{(y - y_o)^2}{b^2} = 1$

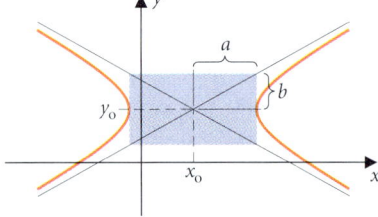

Parabola: $(y - y_o)^2 = 4a(x - x_o)$

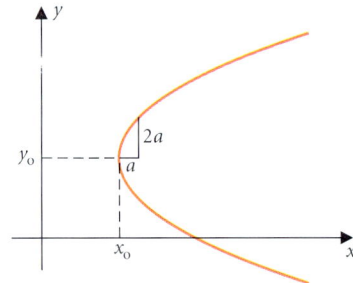

Trigonometry

$\sin\theta = \dfrac{b}{c}$ $\csc\theta = \text{cosec}\,\theta = \dfrac{1}{\sin\theta} = \dfrac{c}{b}$

$\cos\theta = \dfrac{a}{c}$ $\sec\theta = \dfrac{1}{\cos\theta} = \dfrac{c}{a}$

$\tan\theta = \dfrac{b}{a}$ $\cot\theta = \text{cotan}\,\theta = \dfrac{1}{\tan\theta} = \dfrac{a}{b}$

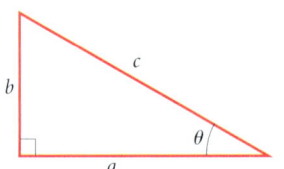

$$\sin^2\theta + \cos^2\theta = 1 \qquad 1 + \tan^2\theta = \sec^2\theta$$

$$\sin(2\theta) = 2\sin\theta\cos\theta \qquad \sin(\theta/2) = \sqrt{(1-\cos\theta)/2}$$

$$\cos(2\theta) = \cos^2\theta - \sin^2\theta \qquad \cos(\theta/2) = \sqrt{(1+\cos\theta)/2}$$

$$= 2\cos^2\theta - 1$$

$$= 1 - 2\sin^2\theta$$

$$\tan(2\theta) = \frac{2\tan\theta}{1-\tan^2\theta} \qquad \tan(\theta/2) = \frac{\sin\theta}{1+\cos\theta} = \frac{1-\cos\theta}{\sin\theta}$$

$$\sin A + \sin B = 2\sin\left(\frac{A+B}{2}\right)\cos\left(\frac{A-B}{2}\right)$$

$$\cos A + \cos B = 2\cos\left(\frac{A+B}{2}\right)\cos\left(\frac{A-B}{2}\right)$$

$$\sin(A+B) = \sin A\cos B + \cos A\sin B$$

$$\cos(A+B) = \cos A\cos B - \sin A\sin B$$

$$\tan(A+B) = \frac{\tan A + \tan B}{1 - \tan A\tan B}$$

Triangle Rules

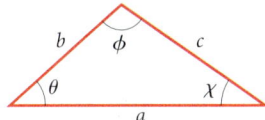

Sine rule: $\quad \dfrac{\sin\theta}{c} = \dfrac{\sin\phi}{a} = \dfrac{\sin\chi}{b}$

Cosine rule: $\quad c^2 = a^2 + b^2 - 2ab\cos\theta$

C: Vectors

Name of vector: $\vec{v}$

Magnitude of vector: $|\vec{v}|$ or v

$|\vec{v}| = \sqrt{v_x^2 + v_y^2 + v_z^2}$

Components of vectors: $\quad \vec{v} = (v_x, v_y, v_z)$

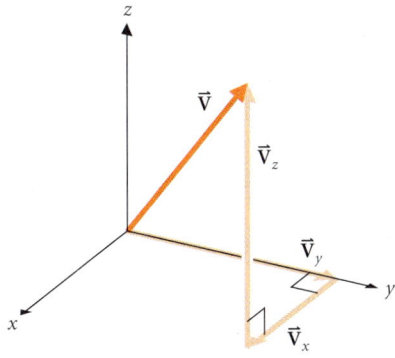

Unit vectors: $\quad \hat{v} \equiv \dfrac{\vec{v}}{|\vec{v}|} \quad$ and $\quad \vec{v} = \vec{v}_x + \vec{v}_y + \vec{v}_z = v_x\hat{i} + v_y\hat{j} + v_z\hat{k}$

Addition of vectors: $\quad S_x = a_x + b_x;\; S_y = a_y + b_y;\; S_z = a_z + b_z$

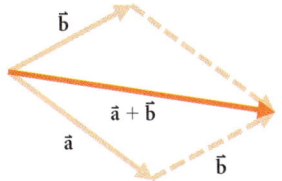

Subtraction of vectors: $\quad D_x = a_x - b_x;\; D_y = a_y - b_y;\; D_z = a_z - b_z$

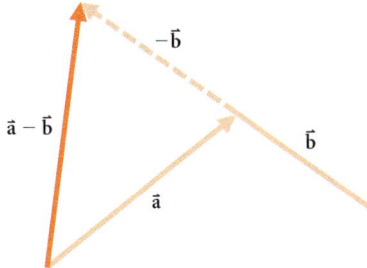

Dot product of two vectors: $\quad \vec{a}\cdot\vec{b} = ab\cos\theta = \vec{b}\cdot\vec{a}$

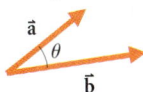

Distributive rule for dot products:
$$\vec{a}\cdot(\vec{b}+\vec{c}) = \vec{a}\cdot\vec{b} + \vec{a}\cdot\vec{c}$$

Dot product in terms of components:
$$\vec{a}\cdot\vec{b} = a_xb_x + a_yb_y + a_zb_z$$

Cross product of two vectors:
$$|\vec{a}\times\vec{b}| = ab\sin\theta,\; \text{direction given by right-hand rule}$$
$$\vec{a}\times\vec{b} = -\vec{b}\times\vec{a}$$

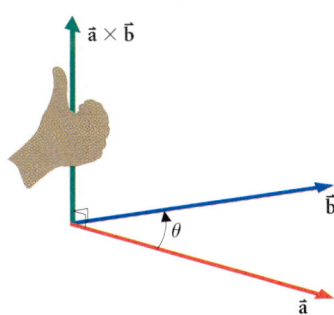

Distributive rule for cross products:
$$\vec{a}\times(\vec{b}+\vec{c}) = \vec{a}\times\vec{b} + \vec{a}\times\vec{c}$$

Cross product in terms of components:
$$\vec{a}\times\vec{b} = \hat{i}(a_yb_z - a_zb_y) + \hat{j}(a_zb_x - a_xb_z) + \hat{k}(a_xb_y - a_yb_x)$$

Identities: $\quad \vec{a}\cdot(\vec{b}\times\vec{c}) = \vec{b}\cdot(\vec{c}\times\vec{a}) = \vec{c}\cdot(\vec{a}\times\vec{b})$
$$\vec{a}\times(\vec{b}\times\vec{c}) = (\vec{a}\cdot\vec{c})\vec{b} - (\vec{a}\cdot\vec{b})\vec{c}$$
$$(\vec{a}\times\vec{b})\cdot(\vec{c}\times\vec{d}) = (\vec{a}\cdot\vec{c})(\vec{b}\cdot\vec{d}) - (\vec{a}\cdot\vec{d})(\vec{b}\cdot\vec{c})$$
$$\vec{a} = (\vec{a}\cdot\hat{b})\hat{b} + \hat{b}\times(\vec{a}\times\hat{b})$$
$$|\vec{a}\times\vec{b}|^2 = a^2b^2 - (\vec{a}\cdot\vec{b})^2$$

D: Coordinate Systems

Rectangular: (x,y,z)
Unit vectors: $\hat{i} = \hat{x}; \hat{j} = \hat{y}; \hat{k} = \hat{z}$
For a right-handed system: $\hat{i} \times \hat{j} = \hat{k}$

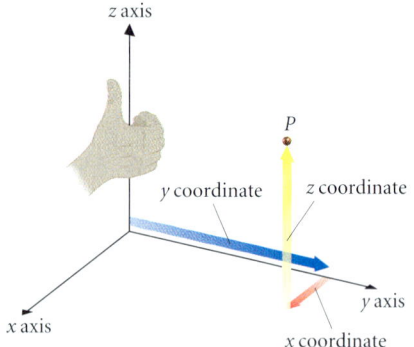

Plane polar: (r, θ)
Unit vectors: $\hat{r}; \hat{\theta}$

$x = r \cos \theta \qquad y = r \sin \theta$
$r = \sqrt{x^2 + y^2} \qquad \theta = \tan^{-1}(y/x)$

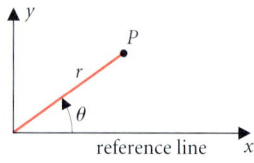

Cylindrical polar: (r, ϕ, z)
Unit vectors: $\hat{r}; \hat{\phi}; \hat{z}$
$\hat{r} \times \hat{\phi} = \hat{z}$

$x = r \cos \phi; \qquad y = r \sin \phi; \qquad z = z$
$r = \sqrt{x^2 + y^2} \qquad \phi = \tan^{-1}(y/x)$

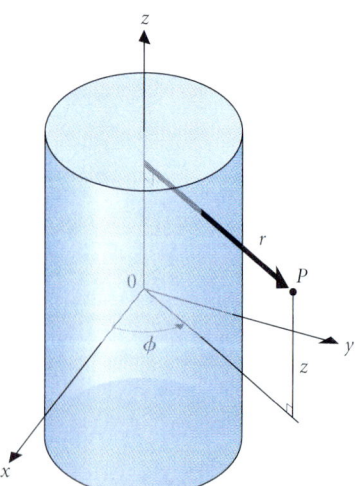

Spherical polar: (r, θ, ϕ)

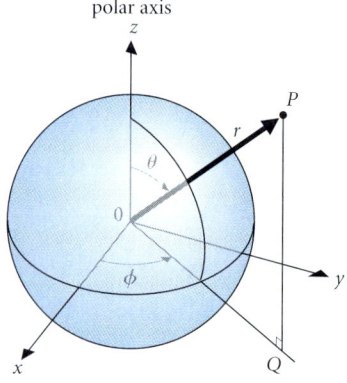

Note: Mathematics texts often define θ and ϕ to have reverse meanings. We have given the definitions that are standard in physics.

Unit vectors: $\hat{r}; \hat{\theta}; \hat{\phi}$
$\hat{r} \times \hat{\theta} = \hat{\phi}$

$x = r \sin \theta \cos \phi; \qquad y = r \sin \theta \sin \phi; \qquad z = r \cos \theta$
$r = \sqrt{x^2 + y^2 + z^2};$
$\theta = \cos^{-1}[z/\sqrt{x^2 + y^2 + z^2}];$
$\phi = \tan^{-1}(y/x)$

E: Derivatives

Let f be a function of variable x.

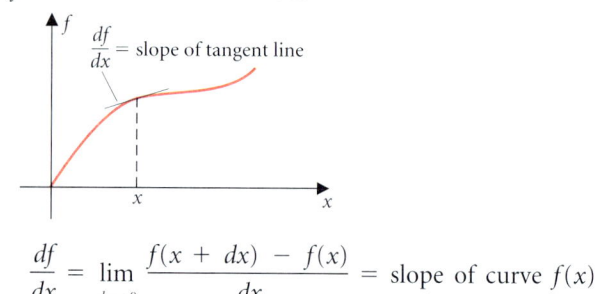

$\dfrac{df}{dx} = \lim_{dx \to 0} \dfrac{f(x + dx) - f(x)}{dx} = $ slope of curve $f(x)$

f	df/dx
x^n	nx^{n-1}
e^{ax}	ae^{ax}
$\sin(kx)$	$k\cos(kx)$
$\cos(kx)$	$-k\sin(kx)$
$\tan(kx)$	$k\sec^2(kx)$
$\sec(kx)$	$k\sec(kx)\tan(kx)$
$\ln(kx)$	$\dfrac{1}{x}$

Chain rule: Let g be a function of $f(x)$, then:

$\dfrac{dg}{dx} = \left(\dfrac{dg}{df}\right)\left(\dfrac{df}{dx}\right)$

Example: $g = \ln(\sin x)$

$\dfrac{dg}{dx} = \dfrac{1}{\sin x} \cos x = \cot x$

Partial derivatives: Let $f = f(x, t)$, then:

$$\frac{\partial f}{\partial x} = \frac{df(x, t)}{dx}, \quad \text{holding } t \text{ constant}$$

Example: $f = \sin(kx - \omega t)$ $\partial f/\partial x = k\cos(kx - \omega t)$

Product rule: Let $h(x) = f(x)g(x)$, then:

$$\frac{dh}{dx} = f(x)\frac{dg}{dx} + \frac{df}{dx}g(x)$$

Let $h(x) = \frac{f(x)}{g(x)}$, then:

$$\frac{dh}{dx} = -\left(\frac{f(x)}{g(x)^2}\right)\left(\frac{dg}{dx}\right) + \left(\frac{1}{g(x)}\right)\left(\frac{df}{dx}\right)$$

$$= \frac{g(x)\, df/dx - f(x)\, dg/dx}{g(x)^2}$$

Inverse functions: $\dfrac{dg}{dx} = \dfrac{1}{dx/dg}$

Example: $\dfrac{d(\cos^{-1} x)}{dx} = \dfrac{d\theta}{d(\cos\theta)}$ (letting $x = \cos\theta$)

$$= \frac{1}{d(\cos\theta)/d\theta} = \frac{1}{-\sin\theta} = \frac{-1}{\sqrt{1 - x^2}}$$

F: INTEGRALS

$$\int_a^b f(x)\, dx = \text{the area under the curve of } f(x) \text{ between } a \text{ and } b$$

$$= \lim_{\Delta x_i \to 0} \sum_a^b f(x_i)\, \Delta x_i$$

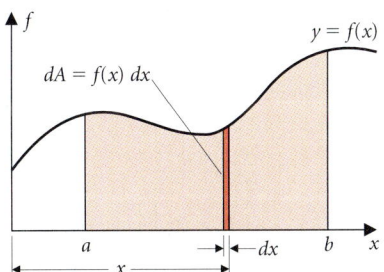

Integration is the inverse of differentiation:

$$\int_a^b \frac{df}{dx}\, dx = f(b) - f(a) \quad \text{and} \quad \frac{d}{dx}\int_a^x f(x')\, dx' = f(x)$$

Substitution

Consider $f(x)$ where $x = g(u)$. Then $dx = (dg/du)\, du$.

$$\int_a^b f(x)\, dx = \int_{g^{-1}(a)}^{g^{-1}(b)} f(g(u))\, \frac{dg}{du}\, du$$

Example: Evaluate

$$\int_0^a \frac{dx}{\sqrt{a^2 - x^2}}.$$

Letting $u = \theta$, $g(u) = a\sin\theta$, $dx = a\cos\theta\, d\theta$, the limits then become:

$$g^{-1}(0) = \sin^{-1}(0) = 0, \; g^{-1}(a) = \sin^{-1}(1) = \pi/2.$$

So, performing the indicated substitutions, we get:

$$\int_0^{\pi/2} \frac{a\cos\theta}{\sqrt{a^2 - a^2\sin^2\theta}}\, d\theta = \int_0^{\pi/2} d\theta = \frac{\pi}{2}.$$

Integration by Parts

Let $h(x) = f(x)\dfrac{dg}{dx}$

$$\int_a^b h(x)\, dx = f(x)g(x)\bigg|_a^b - \int_a^b \frac{df}{dx}g(x)\, dx$$

Some Indefinite Integrals

f	$\int f(x)\, dx$
$x^p \; (p \neq -1)$	$\dfrac{x^{p+1}}{p+1}$
$\sin(kx)$	$\dfrac{-\cos(kx)}{k}$
$\dfrac{1}{x + a}$	$\ln(x + a)$
$\dfrac{1}{1 + x^2}$	$\tan^{-1}(x)$
$\sin^2(x)$	$\dfrac{x}{2} - \dfrac{\sin(2x)}{4}$
$\tan^2(x)$	$\tan(x) - x$
$\dfrac{1}{\sqrt{1 - x^2}}$	$\sin^{-1}(x)$
e^{ax}	$\dfrac{e^{ax}}{a}$
$\cos(kx)$	$\dfrac{\sin(kx)}{k}$
$\tan(kx)$	$\dfrac{-\ln[|\cos(kx)|]}{k}$
$\sec(x)$	$\ln[|\sec(x) + \tan(x)|]$
$\cos^2(x)$	$\dfrac{x}{2} + \dfrac{\sin(2x)}{4}$
$\sec^2(x)$	$\tan(x)$
$\sqrt{1 - x^2}$	$\tfrac{1}{2}\left[x\sqrt{1 - x^2} + \sin^{-1}(x)\right]$

Some Definite Integrals

$$\int_0^{2\pi} \sin^2(x)\, dx = \int_0^{2\pi} \cos^2(x)\, dx = \pi$$

$$\int_0^{2\pi} \sin(x)\, dx = \int_0^{2\pi} \cos(x)\, dx = 0$$

$$\int_0^{\infty} e^{-a^2 x^2}\, dx = \frac{\sqrt{\pi}}{2a}$$

$$\int_0^{\infty} x^{2n+1} e^{-a^2 x^2}\, dx = \frac{n!}{2a^{2(n+1)}}$$

$$\int_0^{\infty} e^{-ax}\, dx = \frac{1}{a} \quad (a > 0)$$

$$\int_0^{\infty} x^2 e^{-a^2 x^2}\, dx = \frac{\sqrt{\pi}}{4a^3}$$

$$\int_0^{\infty} x^4 e^{-a^2 x^2}\, dx = \frac{3\sqrt{\pi}}{8a^5}$$

G: Rules of Probability

If events A and B are independent, then:

$$p(A \text{ and } B) = p(A)p(B)$$
$$\begin{aligned}p(A \text{ or } B) &= 1 - p(\text{not } A \text{ and not } B) \\ &= 1 - [1 - p(A)][1 - p(B)] \\ &= p(A) + p(B) - p(A)p(B)\end{aligned}$$

Appendix II

Symbols

A: The Greek Alphabet

A	α	alpha
B	β	beta
Γ	γ	gamma
Δ	δ	delta
E	ϵ	epsilon
Z	ζ	zeta
H	η	eta
Θ	θ	theta
I	ι	iota
K	κ	kappa
Λ	λ	lambda
M	μ	mu
N	ν	nu
Ξ	ξ	xi
O	o	omicron
Π	π	pi
P	ρ	rho
Σ	σ	sigma
T	τ	tau
Υ	υ	upsilon
Φ	ϕ	phi
X	χ	chi
Ψ	ψ	psi
Ω	ω	omega

B: Symbols Used in the Text

The entry under "Chapter" is the chapter in which the symbol is first used.

Chapter	Symbol	Definition
0	a	semimajor axis of an ellipse
2	$\vec{a}$	acceleration
2	$\vec{a}_o$	specific constant value of acceleration
17	a	slit width
20	a	van der Waals constant
35	a_o	Bohr radius
1	A	area
14	A	amplitude
1	AB	straight line distance between A and B
1	$\overrightarrow{AB}$	vector between points A and B
2	$\overset{\frown}{AB}$	path length along a circular arc from A to B
28	$\vec{B}$	magnetic field (magnetic induction)
36	B	binding energy
37	B	baryon number
37	B	bottomness or beauty
14	b	damping constant
20	b	van der Waals constant
37	b	bottom quark
1	c	speed of light in vacuum
12	c	compression of a spring
37	c	charmed quark
27	C	capacitance
37	C	charm
0	C	a constant
19	c_p	specific heat at constant pressure
19	c'_p	molar specific heat at constant pressure
19	c_v	specific heat at constant volume
19	c'_v	molar specific heat at constant volume
28	$\mathscr{C}$	circulation
0	d, D	distance
3	d, D	diameter
17	d	slit separation
37	d	down quark
2	$\vec{D}$	displacement = change in position
27	$\vec{D}$	electric displacement
17	D	dispersion
36	D	deuterium
2	dx	differential amount of x
0	e	eccentricity of an ellipse

A-7

Chapter	Symbol	Definition
13	e	2.7182818
7	e	efficiency
23	e	electron charge = 1.6×10^{-19} C
24	e, e^-	electron
36	e^+	positron
1	E	east
8	E	energy
23	$\vec{E}$	electric field
26	$\mathcal{E}$	electromotive force
3	f	frequency (Hz)
4	$\vec{f}$	friction force
4	$\vec{F}$	force
0	F_1, F_2	foci of an ellipse
29	g	electron g factor = 2.002 ...
2	$\vec{g}$	acceleration due to gravity
E2	$\vec{g}$	gravitational field strength
5	G	gravitational constant
0	h, H	height
21	H	rate of heat flow
29	$\vec{H}$	magnetic field vector
35	h	Planck's constant
35	$\hbar = h/(2\pi)$	Planck's constant
6	$\vec{I}$	impulse
12	I	rotational inertia
16	I	intensity
26	I	electric current
30	I_d	displacement current
37	I	isospin
37	I_3	z-component of isospin
1	$\hat{i}$	unit vector in the x-direction
26	$\vec{j}$	current density
1	$\hat{j}$	unit vector in the y-direction
1	$\hat{k}$	unit vector in the z-direction
4	k	spring constant
15	k	wave number = $2\pi/\lambda$
33	$\vec{k}$	wave vector
19	k	Boltzmann's constant
21	k	thermal conductivity
23	k	electric force constant
28	k_m	$= \mu_o/4\pi$, magnetic force constant
7	K	kinetic energy
22	K	coefficient of performance
2	ℓ	distance
3	ℓ	length
35	ℓ	angular momentum quantum number
1	L	length
1	L	dimension of length
31	L	(self) inductance
9	$\vec{L}$	angular momentum
15	m	an integer
35	m	quantum number for z-component of angular momentum
19	m	mass of a molecule
28	$\vec{m}$	magnetic moment
1	M, m	mass
1	M	dimension of mass
31	M	mutual inductance
29	$\vec{M}$	magnetization
4	$\vec{n}$	normal force
23	$\hat{n}$	unit normal vector
13	n	number density (particles per unit volume)
16	n	refractive index
15	n	any integer
36	n	neutron
28	n	number of turns per unit length in a coil
1	N	north
19	N	number of molecules of substance
28	N	integer
19	N_A	Avogadro's number
19	$\mathcal{N}$	number of moles of substance
1	O	origin of coordinates
36	p	proton
6	$\vec{p}$	momentum
24	$\vec{p}$	electric dipole moment
1	P	a point
7	P	power
13	P	pressure
33	P	radiation pressure
16	P_*	pressure amplitude in a sound wave
6	$\vec{P}$	momentum of a system
27	$\vec{P}$	electric polarization
19	Q	heat transfer
23	Q, q	electric charge
32	Q	quality factor for an LRC circuit
36	Q	energy produced in a nuclear or particle reaction
3	r, R	radius
1	$\vec{r}$	position vector
3	$\hat{r}$	unit vector in the radial direction
3	R	range of a projectile
17	R	resolving power
19	R	ideal gas constant
21	R	thermal resistance
26	R	electric resistance
21	R_f	R factor for thermal resistance = RA
35	R	Rydberg constant
4	$\vec{s}$	displacement
1	s	arc length
4	s	stretch or compression of a spring
35	s	spin
37	s	strange quark
1	S	south
2	S	speed

A-8 APPENDIX II • SYMBOLS

Chapter	Symbol	Definition
22	S	entropy
37	S	strangeness
33	$\vec{S}$	Poynting vector
2	t	time variable
37	t	top quark
1	T	dimension of time
0	T	period
4	T	tension
19	T	temperature
36	T	tritium
27	u	energy density
37	u	up quark
10	$\vec{u}$	velocity
8	U	potential energy
8	U	internal energy
2	$\vec{v}$	velocity
26	$\vec{v}_d$	drift velocity
2	$\vec{v}_i$	initial velocity
2	$\vec{v}_f$	final velocity
2	$\vec{v}_o$	specific constant value of velocity
6	V	volume
25	V	electric potential
25	V_{AB}	potential difference between A and B
1	$\vec{V}$	arbitrary vector
1	V_x, V_y, V_z	components of $\vec{V}$
4	$\vec{W}$	weight
7	W	work
1	W	west
32	X	reactance
1	x	cartesian coordinate
33	$\hat{x}$	unit vector in the x-direction
1	y	cartesian coordinate
28	$\hat{y}$	unit vector in the y-direction
37	Y	hypercharge
1	z	cartesian coordinate
28	$\hat{z}$	unit vector in the z-direction
32	Z	impedance
10	α	alpha-particle or helium nucleus
15	α	angle
20	α	coefficient of linear thermal expansion
35	α	fine structure constant
3	$\vec{\alpha}$	angular acceleration
16	β	v/c
21	β	coefficient of volume thermal expansion
36	β	beta-particle, or electron
19	γ	ratio of specific heats
29	γ	$1/\sqrt{1-\beta^2}$, relativistic speed factor
36	γ	photon
2	δx	small amount of x, or uncertainty in x
1	Δx	change in x = final value − initial value
37	Δ	delta-particle
10	ϵ	coefficient of restitution
21	ϵ	emissivity
23	ϵ_0	$=1/(4\pi k)$, permittivity of the vacuum
1	Θ, θ	angle
28	$\hat{\theta}$	unit vector in θ-direction
27	κ	dielectric constant
3	λ	latitude
5	λ	mass per unit length
15	λ	wavelength
22	λ	mean free path
24	λ	charge per unit length
35	λ_C	Compton wavelength
36	Λ	lambda particle
15	μ	mass per unit length
34	μ, μ^-	muon
4	μ_s	coefficient of static friction
4	μ_k	coefficient of kinetic friction
28	μ_o	permeability of the vacuum
29	μ	magnetic permeability
36	ν	neutrino
36	$\bar{\nu}$	antineutrino
1	π	3.14159
36	π	π meson, pion
6	ρ	mass density
24	ρ	charge density
26	ρ	resistivity
37	ρ	rho particle
11	σ	surface mass density
21	σ	Stephan–Boltzmann constant
22	σ	collision cross section
24	σ	surface charge density
26	σ	conductivity
37	Σ	sigma particla
14	τ	time interval, characteristic time
37	τ	tau-particle
9	$\vec{\tau}$	torque
1	ϕ	angle
14	ϕ	phase
E2	ϕ	gravitational potential
23	Φ_E	electric flux
28	Φ_B	magnetic flux
27	χ	electric susceptibility
29	χ_m	magnetic susceptibility
4	ψ, Ψ	angle
35	ψ	Schrödinger wave function
3	ω, Ω	angular frequency (rad/sec)
12	$\vec{\omega}$	angular velocity
12	Ω	precession frequency
22	Ω	number of microstates
37	Ω	omega-particle

C: Mathematical Symbols

$=$	equals; set equal to	$\gtrsim$	greater than or about equal to	$\angle$	angle
$\approx$	is approximately equal to	$\lesssim$	less than or about equal to	$\dfrac{d}{dx}$	derivative with respect to x
$\neq$	is not equal to	$\ll$	much less than		
$\sim$	of the order of	$\gg$	much greater than	$\dfrac{\partial}{\partial x}$	partial derivative with respect to x
$\equiv$	is defined as; is equivalent to	$\pm$	plus or minus		
$\Rightarrow$	implies	$\lvert x \rvert$	absolute value of x	$\int$	integral
$\propto$	proportional to	$\lvert \vec{x} \rvert$	magnitude of the vector $\vec{x}$	$\oint$	integral over a closed path or surface
$\rightarrow$	tends to, becomes	$[x]$	physical dimensions of x	lim	limit
$<$	less than	$<x>$	average value of x		
$>$	greater than	$\parallel$	parallel	$\sum\limits_{i=n}^{m} x_i$	sum over x_i, $i = n$ through m, that is, $x_n + x_{n+1} + x_{n+2} + \cdots + x_m$
$\geq$	greater than or equal to	$\perp$	perpendicular		
$\leq$	less than or equal to	$n!$	n factorial		

D: Symbols for Units

kg	kilogram	} Fundamental SI units	ft	foot	N	newton
m	meter		g	gram	nm	nautical mile = 1.15 statute mile
s	second		G	gauss	Oe	oersted
A	ampere		gal	gallon	oz	ounce
Å	angstrom		H	henry	Pa	pascal
atm	atmosphere		h	hour	qt	quart
AU	astronomical unit		hp	horse power	sr	steradian
Bq	bequerel		Hz	hertz	rad	radian
Btu	British thermal unit		in.	inch	rev	revolution
C	coulomb		J	joule	rpm	revolutions per minute
°C	degree celsius		K	kelvin	T	tesla
cal	calorie		kt	knot = nautical mile per hour	ton	ton
Cal	Calorie = kilocalorie		kWh	kilowatt-hour	u	atomic mass unit
cm Hg	centimeters of mercury		lb	pound	V	volt
Ci	curie		ly	light-year	W	watt
d	day		L	liter	Wb	Weber
dyn	dyne		mi	mile	y	year
erg	erg		min	minute	Ω	ohm
eV	electron volt		mol	mole	°	degree
°F	degree Fahrenheit		mph	miles per hour	′	minute of arc
F	farad		$M_\odot$	solar mass	″	second of arc

E: Roman Numerals

i	I	1	xi	XI	11
ii	II	2	xii	XII	12
iii	III	3	xiii	XIII	13
iv	IV	4	xiv	XIV	14
v	V	5	xv	XV	15
vi	VI	6	xvi	XVI	16
vii	VII	7	xvii	XVII	17
viii	VIII	8	xviii	XVIII	18
ix	IX	9	xix	XIX	19
x	X	10	xx	XX	20

APPENDIX III

Answers to Selected Problems

Chapter 0

0.1 Brand X **0.3** 7.2 cm; 31.2 cm
0.5 The comet's speed is greatest at point A and smallest at point C. **0.7** straight down **0.9** all the way to the bottom
0.11 GH **0.15** No; The difference is only 0.2 mm

Chapter 1

1.1 ± 0.5 mm
1.3(a) 5×10^{-8} s; 4 significant figures
(b) 50 kg; 1 significant figure
(c) 5×10^{-6}; 3 significant figures
(d) 50 μm; 3 significant figures
1.5 $10^{18} \mu$s
1.7(a) $\dfrac{M}{L^3}$ **(b)** dimensionless
(c) L^2 **(d)** L^3 **1.9** The vector would have a magnitude of 15 m and point north.
1.11

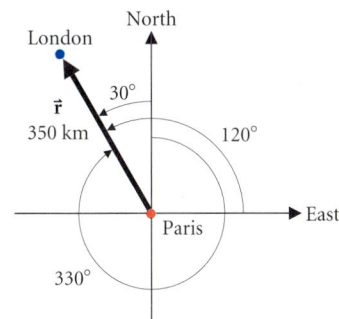

1.13 $a_x = -1, a_y = 2$
$b_x = 3, b_y = 2$
$c_x = 1, c_y = -3$
$d_x = 2, d_y = 2$
$e_x = -2, e_y = -1$
1.15 (2, 3) **1.19(a)** ± 0.5 mm; four decimal places **(b)** the uncertainty is higher; ± 0.5 mm **(c)** ± 0.5 sec; 4 digits

(d) same as part (c); precision of the measurement is totally unsuited for its purpose **1.23** 4×10^4 km
1.27(a) ± 0.8 m **(b)** unnecessary
(c) $0.458°$ **(d)** Ridiculous **(e)** no
1.31 67 in. = 1.7 m
1.35 1.8×10^{12} furlong/fortnight
1.39 $\dfrac{1}{(\text{length})^2}$; 1.5×10^7 m^{-2}; no
1.43(a) $3x^2(\Delta x)$ **(b)** $|\cos(x)|(\Delta x)$
(c) $\left|-\dfrac{22x}{(2x^2-1)^2}\right|(\Delta x)$;
$6.6 \times 10^{-2}; 1.0 \times 10^{-2}; 0.32$ **1.47** (38 m, 23° from the ground) **1.51** (e)
1.55(a) $|\vec{s}| = 2|\vec{b}| \cos\left[\dfrac{1}{2}(\theta_b - \theta_a)\right]$
(b) $|\vec{d}| = 2|\vec{a}| \sin\left(\dfrac{\theta_b - \theta_a}{2}\right)$;
its direction from north is $\dfrac{\theta_a + \theta_b}{2} - \dfrac{\pi}{2}$
1.59 (53 Nmi, 5° N of E) **1.63** $\dfrac{1}{2}\vec{A} + \vec{B}$:
(1 m, 35 m); (35 m, 1.6° from y-axis);
$\vec{B} - \vec{A}$: $(-68$ m, -4 m); (68 m, 3° from x-axis)
1.67 $\sqrt{14}$ m; $\sqrt{12}$ m; $\sqrt{2}$ m
1.71 6.62 cm$^3 \pm 0.17$ cm^3; 6.62 ± 0.05 cm^3; Implied uncertainty is $\approx \frac{1}{3}$ true uncertainty.
1.75 7600 s; 4×10^{-8} sec/day; 5×10^{-13}
1.79 9×10^7 y

Chapter 2

2.1 85 km/h **2.3** -2.6 cm/s; $+1.5$ cm/s
2.5 A: (1.1 m/s, parallel to the x-axis)
B: (1.8 m/s, parallel to the y-axis)
C: (0.77 m/s, toward 30° clockwise from the y-axis) **2.7** b **2.9** only when $v_i = 0$

2.11

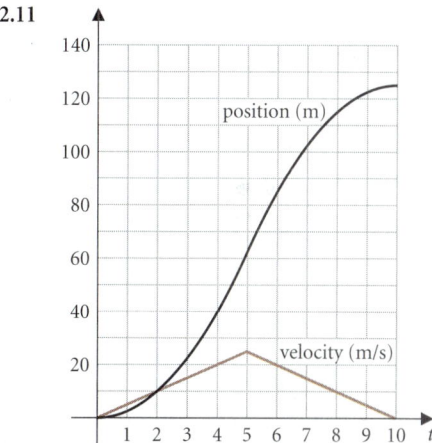

2.13 $(9.3 \times 10^{-2}$ m/s^2, in the direction from D to A) **2.17** (10 m/s, 30° north of west) **2.21** (8.8 m/s, 59° north of east)
2.25 5.0 m/s^2 **2.29** a
2.33 Slope $= \dfrac{\Delta \text{ position}}{\Delta \text{ time}} =$ velocity; no
2.37 (290 km/h, parallel to the track)
2.41 4.00 m/s^2 **2.45** 0.39 s; 3.1 m/s
2.49 24 s **2.53** 78 m; 16 m
2.57(a) 1.2 m; **(b)** (0.29 m/s, downward)
(c) $\alpha = 0.90$ m/s^4 **2.61** 18 m/s
2.65 Modeling the spacecraft as a particle is an excellent approximation
2.69 $g = 980.6$ cm/s^2; $\alpha = 3.5 \times 10^2$ cm/s^4
2.73 $\theta = 57°$

Chapter 3

3.1 d **3.3(a)** 6.28 m/s; 0.100 Hz; 0.628 rad/s; **(b)** 3.95 m/s^2
3.5 $\vec{a}(C) = 0$ m/s^2; $\vec{a}(A) = 20$ m/s$^2\,\hat{\imath}$
$\vec{a}(B) = -20$ m/s$^2\,\hat{\jmath}$
$\vec{a}(D) = 20$ m/s$^2\,\hat{\jmath}$
$\vec{a}(E) = -20$ m/s$^2\,\hat{\imath}$
3.7(a) $(4.4$ mi$)\hat{\jmath} + (1.4$ mi$)\hat{\imath}$
(b) $(3.7$ mi$)\hat{\jmath} - (0.71$ mi$)\hat{\imath}$

A-11

(c) $|\vec{D}_{rel}| = (0.71 \text{ mi})(-3\hat{i} - \hat{j})$
3.9 23°   **3.13** 15°, or 75°
3.17 4.5 s; 45 m away; yes
3.21 19.8 m/s
3.25(a) $y = -\frac{1}{2}gt^2$;
  $x = v_o t$
(b) $+\frac{gD^2}{2v_o^2}; -g\frac{D}{v_o}$
(c) $2D - v_o\sqrt{\frac{2h}{g}}; D\sqrt{\frac{2g}{h}}$ **3.29** 52°
3.33 $\left(\frac{v_c}{v_i}\right)^2 \left[\frac{2\sqrt{3}}{3} \pm \sqrt{\frac{4}{3} - \frac{1}{3}\left(\frac{v_i}{v_c}\right)^2 \left(1 + 3\left(\frac{v_i}{v_c}\right)^2\right)}\right]$;
yes   **3.37** $\Delta\vec{v} = \vec{v}_f - \vec{v}_i = 0$; therefore $\vec{a}_{avg}$ must be 0   **3.41** b   **3.45** 7.3 km
3.49 286 m/s; close to, but not quite equal to, the speed of sound   **3.53** 2.00
3.57 Its speed *relative to the air* is greater.
3.61 1.2 km; 0.67 km; $\frac{1}{3}$ h   **3.65** 3.14 h
3.69 She's cutting it close; her round trip will take 1.9 hours, just 6 minutes short of her limit.   **3.73** speed = 4.0 m/s
3.77 $(7.1 \text{ m/s})\sqrt{1 - \sin[(10 \text{ rad/s})t]}$
3.81 13.9 m/s²; 17.1 m/s²; 17.3 m/s²
3.85 $2h \sin 2\phi [\cos 2\phi + \sqrt{\cos^2 2\phi + 1}]$
3.89 Camelot   **3.93** Air resistance reduces both maximum height and range. Air resistance does not noticeably affect the path unless $\alpha > 0.01$ **3.97(a)** 35°
(b) 4.8 m   **(c)** 5.2 m
(d) $\vec{a}_{AV,land} = (53 \text{ m/s}^2, 28°$ right of upward); $\vec{a}_{AV,horse} = (33 \text{ m/s}^2, 5°$ right of upward);
(e) Leap to horse: 1.5 rad/s; Horse to landing: 1.6 rad/s

Chapter 4

4.1 rock climber: normal forces, frictional forces, gravitational force; buoy: tension force, drag force, gravitational force, buoyant force
4.3 (5.6 N, 60° below the *x*-axis)
4.5 force; 1×10^{28} N   **4.7** 15 boxes
4.9 If I weigh 840 N, I am able to exert half of this on the lever: 420 N. Then $s = 15$ cm;
$\vec{F}_{on\ spring} = +(420 \text{ N})\hat{i}; = +k\vec{s}$
$\vec{F}_{spring\ on\ you} = -(420 \text{ N})\hat{i}; = -k\vec{s}$
$\vec{F}_{on\ mounting} = +(420 \text{ N})\hat{i}$
4.11 $f_s = F$ **4.13** $F_{bump} = Ma$
4.15 $n = mg = 1.5 \times 10^4$ N; friction = $(2.1 \times 10^3$ N, inward)   **4.19** $2M$; $2Mg \sin \theta$ **4.23** 1.8×10^3 N
4.27 6.00 m/s²   **4.31** k = (1 lb-force)/[(1 lb-mass)(32 ft/s²)]; 1 slug = 32 lb-mass = 14.5 kg   **4.35** $n = Mg$; the normal force increases by an amount mg.

4.39 The springs all have the same spring constant.   **4.43** $\frac{2k}{3}$ **4.47** By pulling harder on the string, the person reduces the normal force. If the person releases the tension in the string, n increases to mg.
4.51(a) 640 N   **(b)** 770 N
(c) 510 N   **(d)** 640 N   **(e)** zero
4.55 19 N   **4.59** 1900 N
4.63 $F_D = 120$ N; 0.34 m/s²
4.67 Frictional force; $(1.22 \times 10^4$ N, opposite the car's velocity)   **4.71** $\frac{mg}{\mu_s d}$;
$\frac{2mg}{\mu_s d}$; mg **4.75** $\frac{1}{2}\ell\left(1 + \frac{Mg}{k\ell}\right)$ below the ceiling   **4.79** the plane's path will curve upward   **4.83** 1.0×10^5 N
4.87(a) 29 m/s   **(b)** 7.0×10^1 m/s; 63 m/s   **(c)** If $v < v_1$, friction acts up the slope; there is a minimum speed if $\tan \theta > \mu$.   **4.93** If the rope doesn't stretch the climber suffers a major injury.
4.97(a) 830 N   **(b)** 2800 N
4.101 $F = \frac{\mu mg}{\cos \theta + \mu \sin \theta}$; 22°; 0.40 kN; the slope doesn't change the value of θ at the limit; 850 N   **4.105** 57°

Chapter 5

5.1 1700 N; 3400 N   **5.3** When the cable does not hang in a curve there is no force to balance the cable's weight.
5.5 Assume you and your friend have a mass of 65 kg and 75 kg respectively, then; 4×10^{-8} N; $2 \times 10^{10} = \frac{W}{Fg}$
5.7 2.3×10^{13} kg; 1.5×10^{-3} m/s²; 400 m from the asteroid's surface
5.9 Mg; $2Mg$; $3Mg$ **5.13(a)** Mg
(b) $\frac{Mg}{k} + \ell$ **5.17** $s = \frac{M_2 g}{k}$;
$g(M_1 + M_2) = T$ **5.21** 3.13 m/s
5.25 the blocks do slide with respect to each other; 2.9 m/s²; 5.3 m/s²
5.29 $a = g \sin \theta - \frac{g}{2}\cos\theta(\mu_1 + \mu_2)$;
$s = \frac{Mg \cos\theta}{2k}(\mu_1 + \mu_2)$ **5.33** Since the measured tension differs from the ideal value by much more than the accuracy of the data, we conclude that the pulley system is not ideal.   **5.37** $\frac{2Mg}{5}$ **5.41** $\frac{1}{4}Mg$
5.45 $\theta = \phi$; $\sin\theta = \frac{1}{1+\alpha}$;
$M = \frac{2k\ell}{g}\alpha\sqrt{1 - \left(\frac{1}{1+\alpha}\right)^2}$; $\theta = 49° = \phi$;

$M = 7.7 \times 10^3$ kg   **5.49** The jar would remain right by her hand when released.
5.53 2.0×10^{30} kg   **5.57** 1.4 h
5.61 The length of the Lunar beanstalk is about $\frac{1}{4}$ of the distance from the Earth to the Moon; the lunar beanstalk will not function properly.   **5.65(a)** The monkey accelerates at 1.96 m/s² downward while the bananas accelerate at 1.96 m/s² upward.
(b) The bananas accelerate at 4.9 m/s²
(c) The bananas accelerate at 9.8 m/s²
(d) The monkey will minimize the time it takes for him to reach the bananas by accelerating up the rope; a, 2.26s; b, 2.02s; c, 1.75s   **(e)** no   **5.69** $\frac{1}{3}Mg \sin\theta$
5.73 3.70 m/s²; Mars
5.77(a) $g(r) = \frac{g_0 r}{R_E}$ **5.81(a)** 1.0×10^5 N
(b) $T = 9.3 \times 10^4$ N   **(c)** 9.0×10^4 N
5.85 $\frac{dy}{dx} = \tan\theta = \frac{g\lambda x}{C}$; $y(x) = \frac{g\lambda x^2}{2C}$;
sag $\propto D^2$

Chapter 6

6.1 $\frac{p_1}{p_2} = \frac{3}{8}$ **6.3** c
6.5 2×10^2 N·s   **6.7** 12×10^3 N·s $\hat{i}$
6.9 12 m/s   **6.11** 0.15 m/s
6.13 $\frac{p_{ms}}{p_{vw}} = 0.20$ **6.17** (50 N·s, "down-alley")   **6.21** (12.5 N·s, opposite the ball's initial velocity)   **6.25** 30 m/s
6.29 $(3 \text{ m/s})\hat{i} + (8 \text{ m/s})\hat{j}$
6.33 3×10^4 N   **6.37(a)** (6 N·s, upward)   **(b)** 3.0×10^3 N   **6.41** zero
6.45 less than $m_{exhaust} v_{ex}$ **6.49** It slides off the ice in a direction opposite the dumb dog.   **6.53** astronaut: 0.72 m/s; toolkit: 4.3 m/s   **6.57** 8.3×10^{-2} m/s
6.61 $(0.14 \text{ N·s})\hat{i}$; 2.9 m/s   **6.65** 58 N
6.69 $-0.69 v_{ex}$ **6.73** The speed will be increased by a factor of 1.6
6.77

6.81 $\frac{5h}{4}$ **6.85(a)** $v_1 = 0.27$ m/s, $v_2 = 0$
(b) $v_1 = 0.27$ m/s, $v_2 = 0.26$ m/s
(c) The skater's gliding speeds are not changed. But the velocity components perpendicular to the gliding direction are as given in (a) and (b) above.
6.89 $\frac{v_0^2}{2\mu_r g}$; Back of car #2 is $\frac{v_0^2}{2\mu_r g} - \frac{7L}{4}$ from point A; $2.3\sqrt{\mu_r gL}$ **6.93** $\pi F_0 t_0 \hat{i}$

6.97

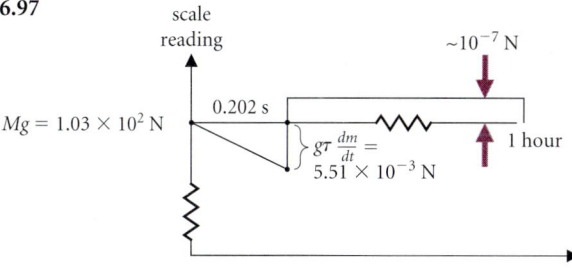

CHAPTER 7

7.1 $K_{car 1} = 7.2 \times 10^4$ J; $K_{truck} = 7.2 \times 10^5$ J; $\frac{K_{car 1}}{K_{truck}} = 10^{-1}$; $\frac{K_{car 1}}{K_{car 2}} = 0.12$

7.3 $W_{haystack\ on\ car} = -4.7 \times 10^5$ J; the car does $+4.7 \times 10^5$ J of work on the hay.
7.5 $\vec{a} \cdot \vec{b} = -3.1$ 7.7 zero
7.9 2.8 kN   7.13 100 J   7.17 d
7.21 98 J   7.25 9.0×10^1 J
7.29 0   7.33 26   7.37 70°
7.41(a) $W_{pirate} = 3 \times 10^{11}$ J; $W_{carrier} = -2 \times 10^{11}$ J; $v_f = 450$ m/s
(b) Yes; -2×10^{10} J
7.43(a) 1.1×10^6 m/s
(b) $\frac{\Delta v}{v} = 4 \times 10^{-3}$ 7.49 2×10^9 J; the same   7.53(a) $\hat{i}_2 = \cos\theta\hat{i}_1 + \sin\theta\hat{j}_1$; $\hat{j}_2 = -\sin\theta\hat{i}_1 + \cos\theta\hat{j}_1$
7.55(a) $\mu Mg2\pi R$ 7.59 By turning East, the rocket is thrusting in the same direction that it's already moving in—increasing its K as efficiently as possible.

7.63 10; 50 N   7.67 $5; \frac{5}{6}$
7.71 9.2×10^5 W   7.75 11
7.79 0.857; 50 J   7.83 True
7.87(a) 16 m/s   (b) 78 km/h
(c) 32 m/s²   (d) fine for backs of envelopes but not OK for a detailed accident report.   (e) 2.0×10^5 J   7.91 $2\sin\theta$
7.95(a) $s_1 = \frac{v_0^2}{2\mu g}\left(1 - \left(\frac{\alpha}{1+\alpha}\right)^2\right)$; $s_2 = \frac{v_0^2}{2\mu g(1+\alpha)^2}$; $W = -\frac{1}{2}\frac{mv_0^2}{(1+\alpha)}$
(b) $\sqrt{2\mu gL(1+\alpha)} = v_{max}$;
$v_m = v_0\frac{\alpha}{1+\alpha} + \frac{\sqrt{v_0^2 - 2\mu gL(1+\alpha)}}{1+\alpha}$
$v_M = \frac{\alpha}{1+\alpha}v_0 - \frac{\alpha}{1+\alpha}\sqrt{v_0^2 - 2\mu gL(1+\alpha)}$;
$W = -\mu_k mgL$

CHAPTER 8

8.1 $\frac{U_1}{U_2} = \frac{1}{4}$ 8.3 3 cm   8.5 Reference level; floor; $U_i = 1400$ J; $U_f = 0$; Reference level: initial position; $U_i = 0$ J; $U_f = -1400$ J; Reference level: highest shelf; $U_i = -4200$ J; $U_f = -5600$ J; $\Delta U = U_f - U_i = -1400$ J; $v = 7.9$ m/s
8.7 choice d) is correct   8.9 1×10^7 J
8.11 700 W   8.13 The spring energy decreases by a factor of 2.
8.17 $\frac{mg}{\mu(\ell - L)}$; $\frac{mg(\ell - L)}{2\mu}$
8.21 Same; The smaller mass will have most of the energy.   8.25 Our exact formula applies to all planets. The practical formula applies to all planets as long as you use the correct value of the acceleration due to gravity.
8.29(a) 4×10^4 W   (b) The KE given to the people requires an additional power of 7 W   8.33 10^{14} kg/s
8.37 $\frac{3\ell}{2} + \frac{2Mg}{k}$ 8.41 4.1 kJ
8.45 42 m/s   8.49 $x = \frac{\ell[k_1 + 3k_2]}{2(k_1 + k_2)}$
8.53 The decrease of TME is due to friction forces within the car's brakes dissipating energy as heat.   8.57 normal force; none; forces that occur in the muscles in your legs.
8.61 Normal and weight forces don't do any work on the truck. Friction and tension do work. The work Isaac does ends up as thermal energy   8.65 2.4 J; normal and weight force, tension; none do work   8.69 5 m/s; probably a large overestimate   8.73 No
8.77 $\Delta E = -40$ J   8.81 2.0×10^5 J; 240 N   8.85 $U = -GMm\left[\frac{1}{R+h} + \frac{1}{4R+h}\right]$; Yes; $g = \frac{17GM}{16R^2}$ for $h \ll R$
8.89 $s_0 = \sqrt{\frac{5mRg}{k}}$
8.93 $k = 1140$ N/m, (stretch ≤ 0.9 m). For larger stretches, the two estimates differ increasingly with $k_1 > k_2$.
8.97 $v = \sqrt{\frac{2g(d - \ell)\sin\theta\,[1 + (1+\alpha)^2\tan^2\theta]}{(1+\alpha)\,[1 + (1+\alpha)\tan^2\theta]}}$

CHAPTER 9

9.1(a) 1.5×10^7 J·s   (b) $\frac{L_{O,car}}{L_{O,truck}} = 0.20$;
9.3 $-(22\ m^2)\hat{j}$ 9.5 $-(200\ N\cdot m)\hat{k}$
9.7 The remaining person's weight force will exert a torque on the bench that will cause the bench to rotate.   9.9 from above
9.11 $\vec{r}_{cm} = (30\ cm)\hat{i}$ 9.13 $(-1.0\ J\cdot s)\hat{k}$
9.15 $\frac{L_{O,1}}{L_{O,2}} = 0.73$ 9.19(a) for any point O on a line through the particle's position parallel to its velocity.   (b) points O in the x-y plane above the $\vec{L} = 0$ line   (c) O must be in the x-y plane below the $\vec{L} = 0$ line   (d) O on the line parallel to the z-axis through the particle's position   (e) a straight path at constant velocity
9.25 $[(2.9)(-\hat{k}) + (1.6)\hat{i} + (6.1)\hat{j}]$ J·s; $\vec{C}$ describes an angular momentum vector.
9.29 $-3\hat{k}$ 9.33 3.14×10^{43} J·s; Jupiter contributes the most
9.37 31 m²; 126°   9.41 $m(\vec{r}_o \times \vec{v}_o)$; $|\vec{L}_O| = mr_o v_o$; $t_\perp = -\frac{\vec{r}_o \cdot \vec{v}_o}{v_o}$;
$|\vec{L}_O| = mv_o\,r(t_\perp)$ 9.45 $(5 \times 10^3\ J\cdot s)\hat{k}$
9.49 20 N·m   9.53 $F = \frac{rMg}{R}$;
$P = rMg\omega$ 9.57(a) $\vec{L}_i = 2(\ell^2 m\omega_0\hat{k})$;
$\vec{\tau} = -\ell\mu mg\hat{k}$; $t = \frac{\ell\omega_0}{\mu_k g}$;   (b) $m\omega_0^2\ell^2$;
same time   9.61 The observed motion could result from the star along with a massive companion traveling through space and orbiting about the system CM as they go.
9.65 You should support the stick at its center.
9.69 $\vec{r}_{CM} = \left(\frac{7}{4}\ m\right)\hat{i} + \left(\frac{7}{4}\ m\right)\hat{j}$;
$\vec{v}_{CM} = \left(\frac{1}{2}\ m/s\right)\hat{i} + \left(\frac{1}{4}\ m/s\right)\hat{j}$;
$K_{int} = 22.9$ J; $\vec{L}_{int} = (8.8\ J\cdot s)\hat{k}$
9.73 $\frac{1}{18}L$ below the square's geometric center.
9.77 -40 m/s $\hat{i} + 60$ m/s $\hat{j} + 0.4$ m/s $\hat{k}$
9.81 The satellite begins to rotate. The astronaut rotates in the opposite direction.
9.85 $(4.9\ m/s)\hat{i}$; $-(3.1\ m/s)\hat{i}$
9.89 The other half ends up with twice its initial velocity and escapes from the Earth.
9.93(a) 0.14 rad/s; 2.5×10^{10} J·s
(b) The velocity of the CM is $-\frac{\omega\ell}{6}\hat{j}$;
The rotational speed about the CM is ω
9.101 The period is 34.3 in scaled units.
9.103 $\frac{\sqrt{2}\,\alpha\,m^2}{36g_o}$

Chapter 10

10.3 The initial velocity of the target ball is 0.1 m/s in the direction of the incident ball.
10.5 $\vec{v}_{CM} = (1 \text{ m/s})\hat{i}$; $\vec{v}_{astr} = 5 \text{ m/s}\hat{i}$; $\vec{v}_{sat} = -1 \text{ m/s }\hat{i}$; Both zero   **10.7** inelastic; $\frac{Mv_0^2}{4}$ **10.9** No   **10.11** only if the particles have equal mass   **10.15** 0.28 m/s; 6.2 m/s   **10.19** 330 kg   **10.23** Yes; $\frac{1}{9}h$
10.27 $\vec{v}_f = -\frac{v_0}{2}[2.5\hat{i} + \hat{j}]$; 8.8 mv_0^2
10.31 $\frac{1}{2}f$ **10.35** 450 m/s
10.39 43 kg·m/s; 79 J
10.43 $v_f = 0.36v_i$
$\vec{u} = (0.49v_i$ at 66° to original axis of molecule)
$\omega = 0.89\frac{v_i}{\ell}$
where ℓ = length of molecule
$\vec{\omega}$ is clockwise in Fig. 10.31
10.47(a) $\vec{v}_{CM} = \frac{\Delta \vec{p}}{2M}$; $\omega = \frac{\Delta p}{M\ell}$
(b) The dumbbell travels along a straight path without rotating. Its CM velocity is the same as in (a).   **(c)** $\frac{(\Delta p)^2}{2M}$; $\frac{(\Delta p)^2}{4M}$
10.51 570 m/s; 2400 J   **10.55** the outgoing velocities are perpendicular;
$\vec{v}_1 = \left(\frac{\sqrt{3}}{2}v_0, \text{ at } 30°\right)$
$\vec{v}_2 = \left(\frac{v_0}{2}, \text{ at } 60°\right)$
10.59 $\frac{1}{3}\vec{v}_0$; $\vec{\omega} = \left(\frac{v_0}{4\ell}, \text{ clockwise}\right)$; $\frac{5}{16}mv_0^2$
10.63(b) $\sqrt{v_n^2 - \mu_k g d}$ **(e)** u_{n+1} increases if $v_1 > 3.0$ m/s and u_{n+1} decreases if $v_1 < 3.0$ m/s; 6   **10.67** No; Yes

Chapter 11

11.1 b, c, and d   **11.3** 0.21 m
11.5 The CM is higher   **11.7** No; No   **11.9** No   **11.11** $(0, \frac{3}{2}\ell)$
11.13 590 N   **11.15** $\frac{1}{2\sqrt{3}}L$ from the bottom of the crate.   **11.19** 4.0×10^2 N;
$\ell_{Mike} = \frac{M_{Jen}}{M_{Mike}}\ell_{Jen}$; Mike must sit at a maximum distance 0.94 m from the center.
11.23 1.2×10^5 N; 1.2×10^4 N·m; (1.6×10^5 N, rightward); (0.4×10^5 N, leftward)   **11.27** 9   **11.33** Vertical force = 95 N. Horizontal force = 110 N. Put bottom screws in first.   **11.37(a)** at her CM   **(b)** None   **(c)** 15°   **(d)** 6°   **(e)** 0.25   **11.41** We want a larger force to be exerted by the shear's blades than the tailor's blades.   **11.45** 20 N
11.49 0.80 g   **11.53** Equilibrium is not possible unless θ_1 and $\theta_2 \leq 45°$ and $M_1 \sin\theta_1 = M_2 \sin\theta_2$.   **11.57** 4500 N;
Force on each front tire = 4.0×10^3 N
Force on each back tire = 8.0×10^3 N
11.61 Top barrel: $\sqrt{2}Mg$ on middle barrel, Mg on wall; Middle barrel: $2Mg\sqrt{2}$ on lowest barrel, $3Mg$ on wall; Bottom barrel: $3Mg$ on floor, $2Mg$ on wall; the value of μ_s is irrelevant
11.65 $m_{max}(\theta) = \frac{Mr_1}{2r_2} \times \frac{[\cos^2\theta(1+\mu^2) + 4\mu^2\sin^2\theta + 4\mu\sin\theta\cos\theta]^{1/2}}{\cos\theta + 2\mu\sin\theta}$;
$m_{min}(\theta)$ can be found by replacing μ by $-\mu$ in the above formula.   **11.69** 56 kg
11.73 at its geometric center   **11.77** $\frac{4R}{3\pi}$ from straight edge   **11.81** 8.5×10^4 N
11.87 2.5×10^3 N; 1.2×10^4 N
11.91 (670 N @ 66° from horizontal)
11.95 11.3 cm   **11.99(b)** 0.53 m
11.103 26.1%

Chapter 12

12.1 centered on module so that the whole station rotates on the $\hat{k}$ axis
12.3 8.0×10^5 **12.5** 60 J·s
12.7 7.6 N   **12.9** 8.96 rad/s; 94.1 J
12.11 0.31 rad/s   **12.19** 0.8 rad/s; 0.3 m/s   **12.23** The square has the greatest rotational inertia, the rod has the smallest
12.27 490 W   **12.31** 5.4 N; 4.9 W
12.35 $\sqrt{\frac{3L}{g}}$; $\sqrt{\frac{4gL}{3}}$ downward; $\frac{1}{R}\sqrt{\frac{4gL}{3}}$;
$\frac{2}{3}MgL$; $\frac{1}{3}MgL$; MgL **12.39** 57.7 N;
57.0 N   **12.43** at the bottom of the barrel; At angle $\sin^{-1}\left[+\frac{2\alpha_{barrel,z}R}{5g}\right]$ from the bottom; $\frac{5\mu g}{2R\sqrt{1+\mu^2}}$ **12.49** 2.7 m/s
12.53 $2.7R - 1.7r$; $2.5R - 1.5r$
12.57 $\frac{2\pi}{1+\frac{I_c}{I_y}}$; Only if $\frac{I_c}{I_y}$ is irrational
12.61 No; When the object rotates about the z-axis $\vec{L}$ and $\vec{\omega}$ are $\parallel$; Neither
12.65 $\alpha = \frac{\alpha}{4} + \frac{1}{8}$ **12.69** $\frac{3MR^2}{2}$
12.73 $\frac{1}{2}M\ell^2$; $\frac{1}{4}M\ell^2$ **12.77** 5.5×10^{22} N·m; 1.4×10^{22} kg $\approx 2 \times 10^{-3} M_E$
12.79 vertically
12.83(a) $\vec{\omega} = (5.2 \text{ rad/s})\hat{k}$
(b) $-(0.12 \text{ rad/s}^2)\hat{k}$
(c) 420 J; (160 J·s)$\hat{k}$
(d) $-(3.8 \text{ N·m})\hat{k}$; 18; -420 N·m
(e) -0.87 rad/s$\hat{k}$; 18 J   **12.87** $\frac{2g}{5}$;
$\frac{2g}{5R}$ **12.91** $3m\ell^2\omega$; $\frac{3}{2}m\ell^2\omega^2$
12.95 8.3 cm from the junction, on the steel side; $I_{CM} = 0.91$ kg·m²; I(steel end) = 3.5 kg·m²; I(Al end) = 6.0 kg·m²
12.99 0.31 m   **12.103(a)** No
(b) $N_i = (23.87 \text{ s})\omega(t_i)$; $b = 3.79 \times 10^{-5}$ /s
12.107(a)
$\vec{v}(t) = [\omega \sin\omega t(v_0 t - R_0) - v_0\cos\omega t]\hat{i}$
$+ [\omega \cos\omega t(R_0 - v_0 t) - v_0\sin\omega t]\hat{j}$;
$\vec{a}(t) = [\omega^2 \cos\omega t(v_0 t - R_0)$
$+ 2\omega v_0 \sin\omega t]\hat{i}$
$- [\omega^2 \sin\omega t(R_0 - v_0 t)$
$+ 2\omega v_0 \cos\omega t]\hat{j}$
(b) $\frac{1}{2}m\omega^2[R_0^2 - r^2]$ **(d)** $\sqrt{7}\omega R_0/2$

Chapter 13

13.1 The beam is primarily supported by shear.   **13.3** 37 atm; 6.3×10^8 N
13.5 the larger piston; 110 N
13.7 3×10^5 Pa   **13.9** 750 m³
13.11 5.85 m/s   **13.13** 14 m
13.17 No; No   **13.21** .2 kg
13.25 3 atm   **13.29** .0207 m²
13.33 1×10^{-3} m²   **13.37** No changes need to be made.   **13.41** 13 m
13.45 1400 m   **13.49** 0.5 N
13.53 neither   **13.59** 41.5%
13.63 448 bags   **13.67** 1.8×10^3 kg/m³
13.73 pushed   **13.77** greater
13.81 60 m   **13.85** 9.7 atm; 0.98 MW
13.89 2.1 cm   **13.91** $\frac{3}{2}\rho v_0^2 LW$
13.95 greater; the same
13.97 1.02×10^5 Pa   **13.99** angular momentum   **13.103** 9.3 km; 140 m/s
13.105 5.05×10^7 N; 1.68×10^8 N·m; 6.67 m below the surface of the water
13.109 6.26 rad/s
13.113(a) 4×10^5 Pa   **(b)** a parabola

Chapter 14

14.3 $x(t) = -(7.5 \text{ cm})\cos[(2.6 \text{ rad/s})t]$;
$v(t) = (19.5 \text{ m/s})\sin[(2.6 \text{ rad/s})t]$
14.5 The period of pendulum A is $\sqrt{2}$ times the period of pendulum B.
14.7 $\theta(t) = 3.0° \cos[(3.6 \text{ rad/s})t]$
14.9 8.7°   **14.11** 0.3 m   **14.13** 0.7 s
14.17 $x(t) = (+1.1 \text{ m})\cos\left([2.6 \text{ /s}]t + \frac{3\pi}{2}\right)$
14.21 4.5 s   **14.25** 0.55 m
14.29 towards the support

14.33 2.005 s
14.37 $\theta(t) = (0.081 \text{ rad}) \sin[(1.98 \text{ rad/s})t]$
14.41 0.64 Hz   **14.45** 1.9 s
14.49 $\frac{E_{2°}}{E_{1°}} = 4$ **14.53** 8.0 rad/s; 0.056 m/s
14.57 3.3×10^3 N/m
14.61 $Z_{res} = -(0.08 \text{ m}) \cos[(6 \text{ rad/s})t]$
14.65 5.0 s; 0.11 rad/s
14.69 $2\pi \sqrt{\frac{m}{k_1 + k_2}}$
14.73 .023 m/s; $+\frac{1}{2} \pi r^2 \rho_w g z_0^2$
14.77 2.10 s, 5% larger than for the simple pendulum.   **14.81** $7.8 \sqrt{\frac{r}{g}}$

CHAPTER 15

15.1 $\sqrt{3}$ **15.3** 440 Hz
15.5 0.48 s   **15.7** $\sqrt{\frac{7}{5} \frac{P}{\rho}} = 330$ m/s at 0° C   **15.9** 2:1   **15.11** 43 W
15.13
$y(x, t) = (5.0 \text{ cm}) \cos \begin{bmatrix} (3.0 \text{ rad/m})x \\ -(5.0 \text{ rad/s})t + \pi \end{bmatrix}$
For a free boundary, the phase difference would be zero rather than π
15.15 9.9 g/m   **15.19** No.
15.23 440 Hz   **15.27** $v = 5.00$ m/s in $-x$ direction; $\lambda = 0.100$ m; $T = 0.0200$ s; $f = 50.0$ Hz; A = 1.00 mm   **15.31** $y(x, t) = (0.750 \text{ mm}) \times \cos[(31.4 \text{ rad/m})x - (314 \text{ rad/s})t + \pi]$ **15.35** 0.6 μm
15.39 No; Yes   **15.43** 2.5 mm
15.47 The wave displacements will not always be equal and opposite at *any* point.
15.51(a) $\frac{1}{2}$ wavelength   **(b)** $\frac{1}{2}$ wavelength
(c) $\frac{1}{4}$ wavelength   **15.55** 13 cm
15.59 increase by 2.3%   **15.63** $13 \frac{g}{m}$; 0.060 mm   **15.67** The path of an element is elliptical.   **15.69** $\frac{m}{2\ell} \sqrt{\frac{T}{\mu}}$, m = integer
15.73 One pulse will appear to stand still and the other will race around at $2v$.
15.77 The superposition is a pulse that travels at 330 m/s, the speed of each individual wave.
15.83 $P = \frac{A^2 T k \omega}{4} \sin(2\pi x) \cos(400\pi t)$

CHAPTER 16

16.1 0.75 m; 0.19 m   **16.3** 6 displacement nodes; 7 pressure nodes; 1.0 kHz
16.5 3.0×10^8 m/s   **16.7(a)** 33 ns
(b) 0.13 s   **(c)** 4 y
16.9 $3 \times 10^{-9} \frac{W}{m^2}$ **16.11** 440 Hz
16.13 7×10^{-5} **16.15** 7.0 m
16.17 24.4°   **16.19** 131 Hz
16.23(a) 5.2 m   **(b)** 2.6 m
16.27 49 Hz; 97 Hz   **16.31** 8.3 min
16.35 No   **16.39** 3.2
16.43 33 W/m²; 140 dB   **16.49** 3×10^8 m/s   **16.53** -9.17×10^{-4}; blueshifted   **16.57** $f = 9.0 \times 10^2$ Hz
16.65 You should point the laser at the apparent position of the object.   **16.69** at 0° to the normal of side BC   **16.73** θ_i;
$\frac{2t}{\sqrt{\left(\frac{n_{glass}}{n_{air}}\right)^2 \csc^2 \theta_i - 1}}$ **16.79(a)** 33μs
(b) 1.3 ms; No; estimates v as $\frac{\Delta d}{\Delta t}$
16.83(a) 3   **(b)** 2.5; a node at each wall
16.87 7.4×10^3 m/s; Yes
16.91 $\theta_1 = \sin^{-1} \frac{d}{r}, \theta_2 = \sin^{-1}\left(\frac{d}{nr}\right)$;
maximum deflection $\phi \approx 138°$ occurs for $\frac{d}{r} \sim 0.9°$ **16.95(a)** 1.3×10^{-5}
(b) Yes; No

CHAPTER 17

17.1 1.20 mm; 2.40 mm; 3.60 mm; 1.20 mm; The separation is not constant except near the center of the screen.
17.3 0.10 μm   **17.5** green; 17.7 mm
17.7(a) $\theta = 6.89°, 13.9°, 21.1°, 28.7°, 36.9° \ldots 73.7°$ **(b)** $\theta = 36.9°$
17.9 $y = (3.9 \text{ mm}) \times \cos\left[\left(3 \frac{rad}{m}\right) x - \left(10^3 \frac{rad}{s}\right) t + 0.14 \text{ rad}\right]$
17.11 $\frac{2\pi}{7}$ **17.13** $\frac{0.215 \text{ nm}}{m}$,
$m = 1, 2, 3 \ldots$ **17.15** The film deforms into a *wedge* shape that is thicker at the bottom.   **17.19** orange   **17.23** 19; 6.1×10^{-5} m from apex   **17.27** $y = 0$, ±1.3 mm, ±2.5 mm, ±3.8 mm etc; 1.52
17.31 67 mm; 44 mm   **17.35** The 600 kHz signals, with their longer wavelengths are bent further into the tunnel.
17.41 $y = 0.74$ m   **17.45** 0.02"
17.49 $d = 2a = 4.0$ μm
17.53 Maximum intensity is 9 times that of the fork. Minimum intensity equals that of the fork.   **17.57** $\theta \approx 3.2°, 6.4°, 9.6°, 12.8°, 16.1°, 22.9°, 26.4°, 30°$ etc.; $\theta = 0, 19.5°, 41.8°, 90°; 4.5°, 8°, 11°, 14.5°$ etc.
17.61(b) $1.2 \times 10^5; 1.7 \times 10^{-6}$ m; $1.2 \times 10^6 \text{ m}^{-1}$ **17.65** 4 Hz; Maximum intensity is ≈ 7.5 times that of the fork. Minimum intensity is ≈ 0.54 that of the fork.
17.69 2.6 nm; 5.1 nm; 7.2 nm
17.73 $\lambda = \frac{2t \sin \theta}{m}; \theta + 2 \tan^{-1}\left(\frac{t}{5d}\right)$

17.79(a) 530 nm green, 410 nm violet
(b) 580 nm yellow, 490 nm blue, 425 nm violet   **(c)** 456 nm blue, 638 nm red
17.83 $\langle u \rangle_{short} = \frac{2I_0}{v_s}[1 + \cos(\Delta kx - \Delta \omega t)]$;
$\frac{2\pi}{\Delta k}; \frac{1}{f_b}; \langle u \rangle_{long} = \frac{2I_0}{v_s} = 2\langle u \rangle_1$
17.91 $d = \frac{\lambda L}{\ell}; \frac{N_A}{N_B} = 3$

CHAPTER 18

18.1

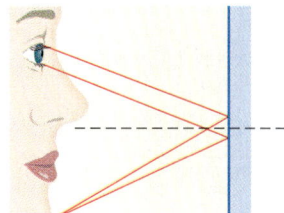

18.3 3 m   **18.5** $x_i = +6.7$ cm; real image; $x_i = -12$ cm; virtual image
18.7(a) Virtual image 2.7 m behind the glass.
(b) Virtual image 0.79 m behind glass.
18.9 2.0   **18.11** A virtual image is formed 17 cm from the lens, on the same side as the object.   **18.13** 10.1 cm
18.15 $x_i = +40$ cm; real; $m = -4.0$; inverted   **18.17** 52   **18.19** 6.137 cm @ 480.0 nm; 6.246 cm @ 656.3 nm
18.21 5   **18.25** twice the separation of the 2 mirrors   **18.29** 0.686 cm
18.31 $\theta = 2\phi$ **18.35** No
18.39 real, inverted image, 0.33 m from the mirror   **18.43** 5.2 cm behind the surface of the bottle.   **18.47** 5.13 cm behind the lens.   **18.53** 0.70 m from the outside surface of the window.   **18.55** The image is virtual and is located inside the bubble, 0.92 cm from the center, on the same side as the crystal. The bubble behaves like a diverging lens.   **18.59** $f = \frac{Rn}{2(n-1)}$
18.63 $x_o > 0$ **18.67** -11 cm; $x_i = -6.4$ cm; $m = -0.40$; virtual
18.71 $m = 0.63$; on axis magnification would be 0.60   **18.73** same; converging
18.77 image moves 20 cm away from object
18.81 +1.0   **18.85(a)** $x_i = \frac{5f}{4}$
(b) $x_i = f$ **(c)** $x_i = +\frac{f}{2}$
(d) Image formed at ∞.   **(e)** $x_i = -f$
(f) $x_i = -\frac{5}{6}f$ **(g)** $x_i = \frac{13}{8}f$
(h) $x_i = \frac{5}{3}f$ **(i)** $x_i = 1.7f$
18.89 0.6 cm   **18.93** The travel time of light along any path through the lens

from O to I is the same. **18.97** The mirror should be at 45°; parallel light **18.99** 33 cm **18.103** The object must be virtual and closer to the surface than $\dfrac{Rn_1}{(n_2-n_1)}$. **18.111** 10 km

Chapter 19

19.1 Yes; No **19.3** 351.7 K; 173° F
19.5(a) 6.2×10^{-3} mol **(b)** 4
(c) helium **19.7** no
19.9(a)

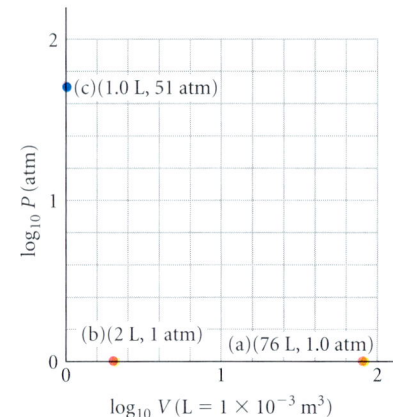

19.11 1030 J/kg·K **19.13** 1.1 atm
19.15 $\dfrac{7}{5}$ **19.17** 740 J/kg·K
19.19(a) Yes. Time scale is months to years.
(b) Yes. Time scale is on the order of hours.
(c) No. **(d)** Yes. Minutes time scale.
19.23 $-37.97°$ F; 234.28 K **19.27** each molecule has the same kinetic energy; less massive molecules have larger rms speed
19.31 The 2 L container
19.35(a) 1.6×10^6 m/s
(b) 6.7×10^7 m/s **19.39** 740 J
19.41(a) 24 J **(b)** $+63$ J **(c)** The block rises 5.3 m. **19.45(a)** 1.8 mol
(b) 0.063 atm **(c)** No
19.49 0.080 mol **19.53** 1010 J/kg·K; 1680 J/kg·K; $\gamma = 1.66$; no change
19.57 330 m/s; $v_{rms} = 490$ m/s
19.61 $W = 1300$ J **19.65(a)** 0.20 mol
(b) 0.028 mol **(c)** nV_p
(d) $n = 22.5$; 0.64 mol **(e)** 475 K
(f) 3.2×10^5 Pa **19.69** 1850 J/K·kg
19.73 8×10^{-23} J; No
19.77 $\dfrac{v_{rms}\text{ electron}}{v_{rms}\text{ ion}} = 205$;
The contributions to the pressure are equal.
19.81 1.47 cm **19.83** 245 K
19.87 No; Yes; work is done on the system
19.91 6.2×10^{-5} m

Chapter 20

20.1 7.5×10^5 Pa **20.3** 2.89×10^6 J
20.5 1.9×10^{-5} K^{-1}

20.7 Heat capacity of water is much greater
20.9 48° C **20.13** 1.06×10^6 Pa
20.17 0.59 mol **20.21** 2.89×10^5 kJ
20.25 No **20.29** 0.2%
20.33 3.8×10^{-4} K^{-1}
20.37 $T = \dfrac{C_A T_A + C_B T_B}{C_A + C_B}$
20.41 246 $\dfrac{J}{kg\cdot K}$ **20.43** 13.5° C
20.47 5.012 J/K; 1.22×10^5 Pa; 232 K; The thermometer *lowered* the system temperature by $\sim$1.5 K **20.51** 0.59 kg
20.55 7.6 kg **20.59** 5.2° C
20.63 9×10^{-3} R *above* the equator.

Chapter 21

21.1 3100 W **21.3** concrete: 7.3 W/m²·K; oak: 4.9 W/m²·K; oak
21.5 b; c **21.7** 8.85 μm **21.9** 8 min
21.13 skewer conducts heat to interior of potato **21.17** $R_{plywood} = 3.4 \times 10^{-3}$ K/W; $R_{window} = 0.058$ K/W; $R_{wall} = 3.2 \times 10^{-3}$ K/W
21.21 360° C; 410 W
21.25 $H = \dfrac{k\pi rR}{h(T_2 - T_1)}$; The heat flow is unchanged. **21.29** Buoyant force accelerates the hot gases upward; no.
21.33 It is very small.
21.37 3×10^{-5} m²; No **21.39** 10 min
21.43 The stove **21.47(a)** 47 W
(b) 7 W; $H_{conv} < H_{rad}$ **21.51** $\alpha \approx 0.01$

Chapter 22

22.1 melting of ice cubes; free expansion of gas; an inelastic collision
22.3(a)

Process:	Q(J)	W(J)	ΔU(J)
AB	3030	0	3030
BC	10100	4040	6060
CD	-6060	0	-6060
DA	-5050	-2020	-3030
Cycle	2020	2020	0

(b) 0.154 **22.5** 1.4×10^3 J; 1.6×10^3 J **22.7** No
22.9 $e < 50\%$ **22.11** 2.4×10^{-5}
22.13 1×10^{14} m
22.21(a) $T_A = \dfrac{P_A V_A}{\mathcal{N}R}$; $T_B = \dfrac{1}{8}\dfrac{P_A V_A}{\mathcal{N}R}$
(b) $W_{AB} = -\dfrac{7}{8}P_A V_A$; $Q_{AB} = -\dfrac{35}{16}P_A V_A$;
(c) $T_C = 4\dfrac{P_A V_A}{\mathcal{N}R}$; $Q_{BC} = \dfrac{93}{16}P_A V_A$
(d) $W_{CA} = \dfrac{9}{2}P_A V_A$ **(e)** 0.6

22.25 The pressure and volume throughout the cycle increase; decrease **22.29** by a factor of 2.25 or by 95 W
22.37 $\dfrac{dS}{dt} = -6.2 \times 10^{22}$ W/K
22.39 850 J/K **22.43** 4.0×10^2 K
22.47 decreases if $\dfrac{v}{v_{mp}} < 1.02$;
increases for $\dfrac{v}{v_{mp}} > 1.02$
22.51 $\lambda_2 = \dfrac{1}{2}\lambda_1$; $\dfrac{\tau_1}{\tau_2} = \sqrt{2}$
22.55 40 nm **22.59** The temperature dependence of the entropy is given by: $\dfrac{3}{2}Nk \ln T$ or $3Nk \ln v_{rms}$.
22.63 $10^{-3 \times 10^8}$; $N(10^{-3 \times 10^8})$; $10^{3 \times 10^8}$ y
22.67 0° C; $\Delta S_{ice} = 12.26$ J/K; $\Delta S_{water} = -10.76$ J/K; $\Delta S_{net} = +1.5$ J/K
22.71 $e_B = 1 - \dfrac{T_A}{T_B}$; $e_C = 1 - \dfrac{T_A}{T_C}$; $e_C > e_B$
22.75(a) 0.42 **(b)** 6×10^{-6}
22.79(a) $\dfrac{V_G}{V_F} = \left[\dfrac{\Delta P}{P_C} + 1\right]^{\frac{1}{\gamma - 1}}$
22.83(c) $\dfrac{P_A + P_B}{2}(V_B - V_B) + \dfrac{P_B V_B + P_A V_A}{\gamma - 1}$
(d) $\dfrac{P_A + P_D}{2}(V_D - V_A) + \dfrac{P_D V_D - P_A V_A}{\gamma - 1}$
(e) 0.366 **(f)** $1.691 V_A$; $(1.201)\dfrac{P_A V_A}{\mathcal{N}R}$
(g) 0.5997

Chapter 23

23.1 zero **23.3** positive charge on the sweater and negative charge on the cotton T-shirt **23.5** $9.0 \times 10^9 \dfrac{N}{C}$ towards Q
23.7 a
23.9

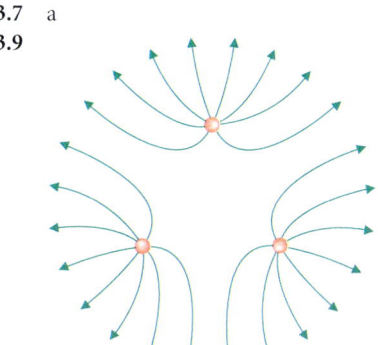

23.11 $3.00 \times 10^6 \dfrac{N\cdot m^2}{C}$; $1.00 \times 10^6 \dfrac{N\cdot m^2}{C}$; $-2.00 \times 10^6 \dfrac{N\cdot m^2}{C}$
23.15 2×10^{-8} N **23.19** $\left(\dfrac{kQ^2}{k_s}\right)^{1/3}$
23.23 0.18 nC

23.27 $\vec{E} = (5.58 \times 10^{-11} \frac{N}{C}$, downward)
23.31 At R, $\vec{E}$ is upward. At P, $\vec{E}$ is to the right.
23.35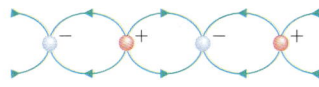

23.39 zero; directly away from the empty corner 23.43(a) 3.4 mN
(b) 4.8 mN 23.47(a) zero
(b) $\left(\frac{kQ}{2a^2}(1 + 2\sqrt{2})\right.$, outward along the diagonal$\left.\right)$ (c) $\left(\frac{16}{25}\sqrt{5}\frac{kQ}{a^2}, \perp$ to the side$\right)$
23.51 $\frac{2\sqrt{2}kq}{a^2}\hat{i}$ 23.55 $\frac{Q}{6\epsilon_0}$; No
23.59 47 N·m²/C 23.63 Unstable; same; no 23.67 $y = 3.3D$
23.71 $y = \pm 1.3 L$
23.75 $\frac{dy}{dx} = \frac{y}{x + a\left(\frac{r_-^3 - r_+^3}{r_-^3 + r_+^3}\right)}$ where $r_- = \sqrt{(x - a)^2 + y^2}$ is the distance from the charge at $+a$, and r_+ is defined similarly.

Chapter 24

24.1 $(3.3 \times 10^2 \text{ N/C})\hat{r}$
24.3 $(6.3 \times 10^6 \text{ N/C}$, radially outward from axis along which filament lies)
24.5 $(2.6 \times 10^8 \text{ N/C}$, normal to the plate)
24.7 9.47×10^{14} m/s²; 1.45×10^{-8} s
24.9 $0.17 \mu C$ 24.13 $-.5$ MC; $-.9$ nC/m² 24.17 zero
24.21 4.7×10^4 N/C; 2.7×10^4 N/C
24.25 $\frac{k\lambda\ell}{x_0^2 - \frac{1}{4}\ell^2}\hat{i}$ 24.29 $-\frac{\pi k\lambda_0}{2R}\hat{j}$
24.33 b 24.37 240 N/C
24.41 $\vec{E}_{net}$ will always point 45° from each plane at every point in space. $|\vec{E}_{net}| = \frac{\sigma\sqrt{2}}{2\epsilon_0}$
24.45 $\frac{2\pi k\sigma_0 z\hat{k}}{a^2}\left[\frac{2z^2 + a^2}{\sqrt{z^2 + a^2}} - 2z\right]$
24.49 $\frac{\rho}{3\epsilon_0}$ 24.53 e
24.57 5.2×10^{-2} m/s²
24.61 -56 nC 24.65 0.42 m/s
24.71 at points A and C the field points in the $-y$-direction; The field is in the y-direction at both points B and D 24.73 1.0 N/C
24.77 zero; $\frac{kQ}{9D^2}$ 24.81 $\frac{L^2}{mke^2}$
24.85 the rod hangs at 50° with the vertical. 11 N 24.89(a) $F_x = -9.3 \mu N$; $F_y = 38 \mu N$;
(b) $F_x = -59 \mu N$; $F_y = -28 \mu N$;
(c) $F_x = 0$; $F_y = -86 \mu N$
24.93 $\frac{\pi k\sigma_0 q}{2g}$; Yes

Chapter 25

25.1 4.6×10^{-20} J 25.3 600 J
25.5 (a)
25.7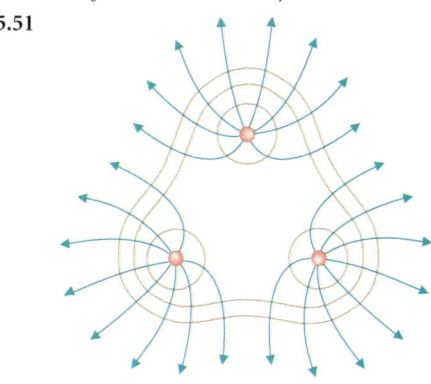

25.9 $\frac{kQ}{a}$ 25.11 (c) 25.13 (b)
25.15 zero 25.19 2.0×10^{-11} m
25.23 same 25.25 zero
25.29 -54 V 25.33 -4.91×10^4 V
25.37 10 25.41 $-\frac{kQ^2}{a}$
25.45 $2\frac{kQq}{\ell}$ 25.47 $k\frac{p\cos\theta}{r^2}$
25.51

25.55(a) $y = \frac{2}{3}$ cm, -2.0 cm
(b) $y = -4.8$ cm; -1.2×10^5 V
25.59 $0.3 \mu C$
25.63(a) $V(y) = 2k\lambda \ln\left(\frac{y_0}{y}\right)$ where y_0 is a suitably chosen reference point
(b) $V(y) = \frac{k\lambda L}{y}$
(c) $V(y) = k\lambda \ln\left(\frac{\sqrt{(L^2/4) + y^2} + L/2}{\sqrt{(L^2/4) + y^2} - L/2}\right)$
25.67 $V = \pi k\rho(b^2 - a^2)$
25.71(a) a (b) d
25.75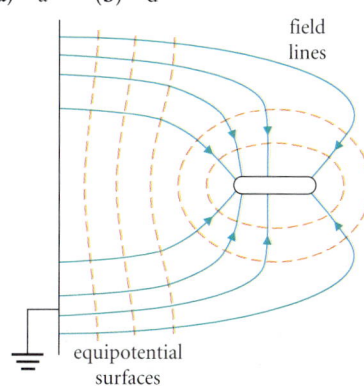

25.79 88 kV/m; The field strength becomes larger 25.81 44 V
25.83 163×10^6 C; 900 25.87 190 V
25.91 39 s; 51 W
25.95 $V(P_1) = 1.45 \times 10^5$ V
  $V(P_2) = -2.0 \times 10^5$ V
25.99(a) Point outside: $\frac{kQ}{d}$ where $d = $ distance from center of sphere;
  Point inside: $V(P) = \frac{kQ}{2}\frac{(3R^2 - d^2)}{R^3}$
(b) $\frac{3}{5}\frac{kQ^2}{R}$ 25.103 $\frac{\pi}{4q}\sqrt{\frac{md^3}{k}}$

Chapter 26

26.1 1200 C 26.3 3 A; 1.7 Ω
26.5 $\frac{1}{3}$ 26.7 4.4 kΩ 26.9 11.2 V
26.11 2.26 Ω 26.13 $\lambda\omega r$
26.17 1.44 V 26.21 1.9 V
26.25 ≥ 3.3 mm 26.29 1.5 mm; 1.7×10^{-4} V/m 26.33 closed
26.37 21.7 kΩ 26.41 0.4 Ω; 49.0 V
26.45 16 Ω; 0.50 A; 0.33 A top arm; 0.17 A through 8 Ω resistor; 0.057 A through each 12 Ω resistor; $V_A = 6.7$ V; $V_B = 6.6$ V; $V_C = 6.0$ V 26.49 both decrease
26.53 14 A through 3 Ω resistor; 4 A through 6 Ω resistor; 18 A through 24 V battery
26.59 top arm: $I = \mathcal{E}/R$ to the left,
  2nd arm: $I = 0$,
  3rd arm: $I = 0$,
  Bottom arm: $I = \mathcal{E}/R$ to the right
26.63 2.5 A; $\frac{2}{3}$ A; 1.8 A; 1.3 V
26.67 2.49 V; 2.50 V 26.71 0.500 mA; less sensitive; $R_2 = 0.20$ Ω; $R_3 = 0.020$ Ω
26.75 0.03 Ω 26.79 $2R$
26.83 0.13 A
26.87 $i_1 = \frac{(R_3 + R_2)\mathcal{E}_1 - R_3\mathcal{E}_2}{R_1R_2 + R_1R_3 + R_2R_3}$;
  $i_2 = \frac{[\mathcal{E}_2(R_1 + R_3) - R_3\mathcal{E}_1]}{(R_1R_2 + R_1R_3 + R_2R_3)}$
26.91 $I_N = 25.3$ A; $R_N = 0.26$ Ω
26.95 5.00 V; 9.99 Ω; 5.00 V; 0.477 A; Case A gives the closer answer for the resistance. Case B gives the closer answer for the emf.

Chapter 27

27.1 120 pC 27.3 $d = 3.1$ mm
27.5 7.0×10^{-8} J 27.7 6.5 pF
27.9 33 pF 27.11 2.1×10^{-6} J·m^{-3}
27.15 (e) 27.19 0.90 kV/m
27.23 $V_1/V_2 = \frac{1}{3}$; $U_1/U_2 = \frac{1}{3}$
27.27 3 μF 27.29 in series; $\frac{8}{7}$ μF

27.33 $Q_i = C\mathcal{E}$; $U = \frac{3}{2}C\mathcal{E}^2$; $3\mathcal{E} = \Delta V_{tot}$; $C\mathcal{E} = Q_f$; $U_f = \frac{3}{2}C\mathcal{E}^2$

27.37 $|\sigma_{bound}| < |\sigma_{free}|$; (b)

27.41 20 pF **27.45** 2.2×10^{-11} F

27.47 $C_1 = 22$ pF, $C_2 = 4.7$ pF, $Q = 320$ pC, $U = 2.0$ nJ, $C_1' = 7.1$ pF, $C_2' = 8.6$ pF, $Q_1 = 0.14$ nC, $Q_2 = 0.18$ nC, $U' = 3.3$ nJ, Work ($W = 1.3$ nJ) is done *on* the system.

27.51 $U_E = 100$ J/m³ **27.55** $r = 2R$

27.57(a) $\frac{1}{2\pi}kQ^2\frac{(\ell^2 + x^2)^2}{(\ell^2 - x^2)^4}$

(b) $\frac{2}{\pi}kQ^2\frac{(x\ell)^2}{(x^2 - \ell^2)^4}$; $\frac{8}{81\pi}\frac{kQ^2}{\ell^4}$

27.61 The Forsterite-filled capacitor can hold 2.7 times as much charge as the paraffin.

27.65 $V_{min} = 10$ m³

27.69 $|x| > 8.25$ cm

27.73 $R = \frac{3}{5}\frac{ke^2}{mc^2} = 1.7 \times 10^{-15}$ m

Chapter 28

28.1 C: zero; D: into the page

28.3 along the y-axis or the z-axis; 1.2×10^{-4} N **28.5** 3.4×10^{-9} T

28.7 4.1×10^{-3} A·m² **28.9** $0; \mu_0 I$; $2\mu_0 I$ **28.11** 1.5×10^{-7} T; zero

28.13 negative **28.17** 6×10^{-7} N

28.21 $(2.2 \times 10^{-16}$ N, out of the page)

28.25(a) a circle **(b)** a single turn

28.29 7.9×10^{-5} T **28.33** 0; 1.1×10^{-4} T **28.37** 3.02×10^{-6} T; using magnetic moment $B_z = 3.41 \times 10^{-6}$ T which gives a 13% error

28.41 7% **28.45** zero

28.49 $B_o = \frac{1}{2}\mu_0 jr$; Within the hole, $\vec{B}$ is perpendicular to the line from the center of the cylinder to the center of the hole and has magnitude $\frac{1}{2}\mu_0 jc$. **28.53(b),(c)** $B = \frac{\mu_0 I}{4\pi r}$

(d) $0; \frac{\mu_0 I}{2\pi r}$ **28.57** yes **28.61** zero

28.67(a) above both sheets $\vec{B} = (\mu_0 K$, perpendicular to $\vec{K}$); between the sheets $\vec{B} = 0$; below the sheets $\vec{B} = (\mu_0 K$, opposite $\vec{B}_{above}$)

(b) Above and below the sheets $\vec{B} = 0$; Between the sheets $\vec{B} = \mu_0 K$, perpendicular to $\vec{K}$) **(c)** Above both sheets $\vec{B} = \frac{\mu_0}{2}K(\hat{i} - \hat{j})$ where $\hat{i}$ and $\hat{j}$ are the directions of the two currents; Between the sheets $\vec{B} = \frac{-\mu_0}{2}K(\hat{i} + \hat{j})$; Below the sheets $\vec{B} = \frac{\mu_0}{2}K(-\hat{i} + \hat{j})$

28.69 7.9×10^{22} A·m²; 2×10^9 A

28.73

x(cm)	B(x) (T)
0.0	5.49×10^{-5}
2.0	5.73×10^{-5}
4.0	5.80×10^{-5}
6.0	5.81×10^{-5}

Chapter 29

29.1 9×10^3 m; 53 rad/s

29.3 1.1 T **29.5** $(0.090$ N$)(\hat{k} - \hat{j})$

29.7 ($1.5\ \mu$N/m, to the right)

29.9 0.30 nV **29.11** 0.028 A/m; 1.25699×10^{-4} T **29.13** clockwise circle

29.17 At the poles, an incoming charged particle will be undeflected. **29.21** Proton 3.9 km, 12 Hz; Electron 2.2 m, 22 kHz

29.25 23 kV **29.29** 1.0×10^7 m/s

29.33 There is no y-component of force regardless of orientation. **29.37** e

29.41 1.2×10^{-2} N·m

29.45 $\vec{F}_1 = \frac{1}{2}ILB\hat{k}$ ($\hat{k}$ is ⊥ to plane of loop)

$\vec{F}_2 = \frac{1}{2}ILB\hat{k}$

$\vec{F}_3 = -ILB\hat{k}$

$\vec{F}_{net} = 0$;

$\vec{\tau} = \left(\frac{\sqrt{3}\ L^2\ IB}{4}\right)$, in plane of triangle and ⊥ to $\vec{B}$) **29.49** ΔV_H is proportional to V, w, and B; ΔV_H is inversely proportional to ρ and ℓ; ΔV_H has no dependence on d

29.53 corner to corner, 21 pV.

29.55 1.4×10^{-14} W **29.59** $\frac{B\sin\theta}{(L/m)}$

29.63 No; Forces attempt to expand the solenoid. **29.67** 3×10^9 A; 10^{19} W

29.71 $\tan\theta = \frac{\mu_0 NI}{2a\ B_E}$

29.75 $\frac{-e(\vec{v}_d \times \vec{B})}{\text{(mass per atom)}}$ (electrons per atom)

For copper: The correction is 2 parts in 10^5

Chapter 30

30.1 0.80 kV

30.3 $(4$ mV$)\{(100)\sin[(400$ rad/s$)t] - (10^{-4}/\text{s}^3)t^3\}$ **30.5** Smaller loop: 0.5 V, 0.025 W. Larger loop: 1 V, 0.1 W; The larger loop. **30.7** 9.4×10^{-3} N·m

30.9 0.24 mV/m **30.11** 60 N

30.13 0.22 nT **30.17** c

30.21 0.599 mV **30.25** 1.0×10^{-9} A in the sense opposite the current in the large loop.

30.27 $(0.13$ A$)\{[t/(1.0$ s$)] - 2[t/(1.0$ s$)]^3\}$; $2.5\ \mu$W; 0.71 s **30.31** 344 turns

30.35 $\frac{\omega a^2}{2}B$ $0 < t < \frac{\pi}{\omega}$; $\frac{-\omega a^2}{2}B$ $\frac{\pi}{\omega} < t < \frac{2\pi}{\omega}$

30.39 zero; 0.1 mV; 0.1 mV; zero

30.41 (b) **30.45** 140 V; satellite end

30.49 $(385$ V/m$)\hat{i} + (165$ V/m$)(\hat{j} - \hat{k})$

30.53 2.0 mV/m; 1.9 mV/m

30.57 $y > 0$ $\vec{F} = 0$; $y < 0$ $\vec{F} = -qvB\hat{k}$

30.61(b) 400 N·m **(c)** 1 ms

30.65 $0.17\ \mu$A; $2.7\ \mu$A

30.69 $\frac{\mu_0 I}{24\pi R}\left(1 - \frac{1}{\sqrt{2}}\right)\hat{\theta}$;

$\frac{\mu_0 I}{24\pi R}\left(1 + \frac{1}{\sqrt{2}}\right)\hat{\theta}$; $0.32\ \mu_0 I\ \hat{\theta}$

30.73 $Wv\left[B(0) + 2t\frac{dB}{dt}\right]$; 75.8 mV (clockwise); 75.8 mV (counterclockwise)

30.77(a) 7.3 V/m **(b)** 7.3×10^{-4} V

(c) 25 V/m **(d)** 18 V/m

(e) 4.4 pC, Yes **30.81** $\frac{mgR\sin\theta}{B^2\ell^2\cos^2\theta}$

30.85 -7.80 Wb

30.89 $\vec{F} = -\hat{k}\frac{\mu_0^2\omega}{16R}\frac{m_0^2 a^3 \sin 2\omega t \sin 2\theta}{(z_0^2 + a^2)^3}$

where $\hat{k}$ is parallel to $\vec{m}$; zero

Chapter 31

31.1 $V = 0.10$ kV; $I = 15$ mA

31.3 For large t $Q = 0.56$ mC, $I = 0$; 1.7 s; $I_{initial} = 1.0$ mA **31.5** $2.0\ \mu$J/m³

31.7 1.0 A **31.9** 390 Hz; 0.43 μJ

31.13 0.69 RC

31.17 $Q(t) = C\mathcal{E}(1 + e^{-t/(RC)})$;

$I(t) = -\frac{\mathcal{E}}{R}e^{-t/RC}$

31.21 capacitors in parallel, resistors in parallel **31.25** solenoid with length 2ℓ will have half the magnetic energy of one with length ℓ **31.29** 0.6 mJ/m³; 7×10^{-5}

31.33 0.57 H **31.35(a)** same

(b) $\frac{N^2\pi\mu_0}{\ell}(a_2^2 + 3a_1^2)$

31.41 Doubling the resistance halves the time constant; ω_1 decreases

31.43 doubles; ω_1 will increase if $L < \frac{3}{8}R^2C$ otherwise, ω_1 decreases

31.47 1.0×10^{-2} A clockwise

31.49 0; 1.00 mA; 0.090 ms

31.53 all in series or all in parallel

31.57 $\tau = \frac{\mu_0\pi rd}{8\rho}$ **31.61** 0.32 mC, 0; 0.51 A, 1.0 ms; 0.57 mC, 2.7 ms

31.65 steady current $\mathcal{E}/R$ through the resistor; capacitor is uncharged

31.69 $Q_1 = \mathcal{E} C_1$; $Q_2 = \mathcal{E} C_2$;
$\dfrac{\mathcal{E}}{R_1} e^{-t/(R_1 C_1)} + \dfrac{\mathcal{E}}{R_2} e^{-t/(R_2 C_2)}$; Nothing

31.73(a) $\tau = \dfrac{L_1 L_2 (R + r)}{(L_1 + L_2) R r}$;

$I_1 = \dfrac{\mathcal{E}}{R} \dfrac{L_2}{(L_1 + L_2)} (1 - e^{-t/\tau})$;

$I_2 = \dfrac{\mathcal{E}}{R} \left(\dfrac{L_1}{L_1 + L_2} \right) (1 - e^{-t/\tau})$;

$I_3 = \dfrac{\mathcal{E}}{R + r} e^{-t/\tau}$; $I_4 = \dfrac{\mathcal{E}}{R} \left[1 - \dfrac{r}{R + r} e^{-t/\tau} \right]$

(b) $\tau = L \left(\dfrac{1}{r_1} + \dfrac{1}{r_2} + \dfrac{1}{R} \right)$;

$I_1 = \dfrac{\mathcal{E} r_2}{(r_2 R + r_1 R + r_1 r_2)} e^{-t/\tau}$;

$I_2 = \dfrac{\mathcal{E} r_1}{r_2 R + r_1 R + r_1 r_2} e^{-t/\tau}$;

$I_3 = \dfrac{\mathcal{E}}{R} (1 - e^{-t/\tau})$;

$I_4 = \dfrac{\mathcal{E}}{R} \left[1 - \dfrac{r_1 r_2}{R(r_1 + r_2) + r_1 r_2} e^{-t/\tau} \right]$

31.77 $L_1 + L_2 \pm 2M$; Yes; $\dfrac{L_1 L_2 - M^2}{L_1 + L_2 \mp 2M}$

31.81 $\dfrac{\mathcal{E}}{R} (1 + e^{-t/RC} - e^{-tR/L})$

31.85

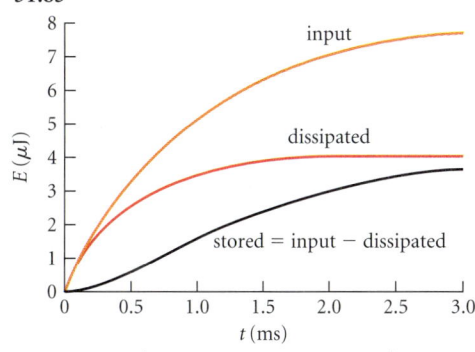

Stored: $\mathcal{E}^2 C \left(\dfrac{1}{2} + \dfrac{1}{2} e^{-2t/(RC)} - e^{-t/(RC)} \right)$;

dissipated: $\dfrac{1}{2} \mathcal{E}^2 C (1 - e^{-2t/(RC)})$;

drawn from the battery: $\mathcal{E}^2 C (1 - e^{-t/(RC)})$

31.89 $\dfrac{\mathcal{E}}{R} \left[1 - \dfrac{1}{2} e^{-2Rt/3L} - \dfrac{1}{2} e^{-2Rt/L} \right]$,

$\dfrac{\mathcal{E}}{2R} [e^{-2Rt/L} - e^{-2Rt/3L}]$; 3.6 mA, -1.3 mA

Chapter 32

32.1 470 Ω **32.3** 240 Ω
32.5 0.54 A; 1.4 W
32.7 $(30 \text{ A}) \cos(\omega t - 0.147 \text{ rad})$. The amplitude has 2 significant figures.

32.9 5.4 W

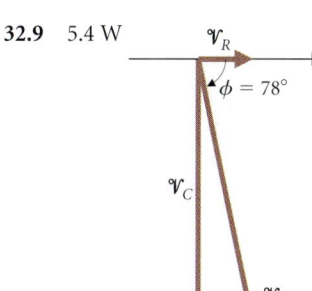

32.11 0.078 Ω; 1.2 μH; 8.7 mA
32.13 The inductance $2L$ has the larger reactance by a factor of 2. **32.17** United States, 60 Hz; Denver, 10 kHz; supercomputer, 150 MHz; chip, 30 GHz

32.21 12 Ω   **32.25** $V_{\text{rms}} = \dfrac{V_0}{\sqrt{3}}; \dfrac{V_0}{\sqrt{3}}$

32.29 capacitor   **32.33** 0.88 μF
32.37 $f = 0.50$ kHz; The woofer should be connected to the capacitor and the tweeter should be connected to the inductor
32.41 $I_o = 1.1$ mA

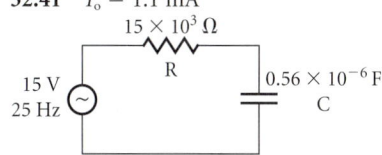

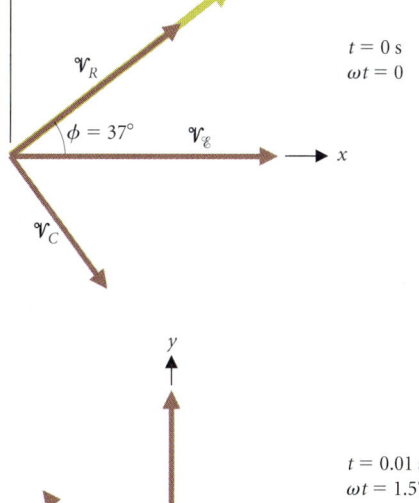

32.45 420 W; 40% TV set; 60% vacuum cleaner

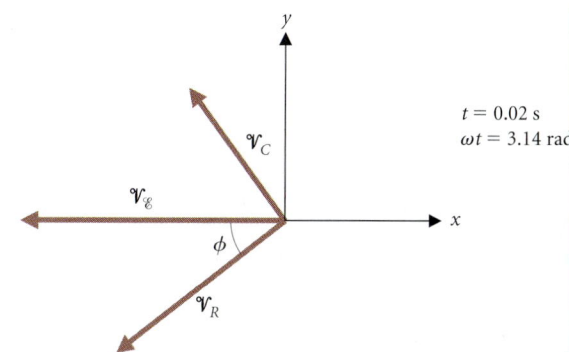

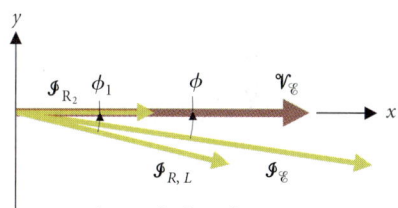

$\phi_1 = 13°$, $\phi = 8°$
32.49 ω_o and Q are each halved, τ is unchanged
32.53(a) 0.78 MHz, 3900;   **(b)** 6600, 0.45 MHz   **32.55** $R \leq 0.18$ Ω
32.59 1.04×10^{-10} F; 1100; 0.55%

32.63 $W_{\text{generator}} = \dfrac{V_0^2}{\omega Z} \left[\dfrac{\pi}{4} \cos \phi + \dfrac{1}{2} \sin \phi \right]$; The first term is the work dissipated by the resistor. The second term is $U_{L,\max} - U_{C,\max}$

32.67 In the steady state no potential difference occurs across the capacitor.
32.71 20 Ω   **32.75(a)** The capacitance is $C \equiv \dfrac{1}{\omega_0^2 L} = \dfrac{\epsilon_o w}{d} [\ell + (\kappa - 1)x]$
(b) $Q_{\min} = 7.1$; $R_{\max} = 24$ Ω
32.79 $V_C / V_{\text{ant}} = 0.17$

32.83 $I_r = \dfrac{X_M I_o}{\sqrt{R_r^2 + (\omega L_r)^2}} \cos \left[\omega t + \dfrac{\pi}{2} - \tan^{-1} \left(\dfrac{\omega L_r}{R_r} \right) \right]$; $\langle F_{\text{up}} \rangle \propto \dfrac{R_r}{R_r^2 + \omega^2 L_r^2}$

Chapter 33

33.1 4.68 V/m

33.3 $\vec{B} = (2.8 \times 10^{-8} \text{ T}) \dfrac{(-\hat{x} + \hat{y})}{\sqrt{2}}$

33.5 $E_0 = 4 \times 10^{-3}$ V/m
33.7 6.0×10^{-11} N   **33.9** 6.84 W/m²

33.11 67.5° **33.13** 4.21 × 10⁹ Hz
33.15 the −z-direction
33.19 $\vec{k} = (0.346 \text{ rad/m})\hat{x}$
$B_0 = 2.64 \times 10^{-9}$ T
33.23 $\vec{E}_0 = (1.92 \times 10^{-2} \text{ V/m})(-2\hat{x} + \hat{y} + 5\hat{z}); f = 107$ MHz **33.27** 7 × 10⁷ m²; 4 × 10⁴ W/m²; 5.5 × 10³ V/m
33.31 9 × 10¹¹ W
33.35 $R_c = 0.19\ \mu$m **33.39** Yes; set the filter's transmission axis horizontal
33.43 1.6 **33.47(a)** 56°
(b) 100% **(c)** 10.5% **33.51** 16°
33.55

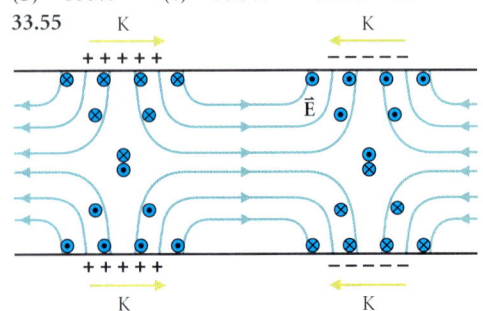

33.59 1 GHz
33.63(a)
$$B_x(y) = -\mu_0\epsilon_0 E_0 \frac{\omega a}{\pi} \cos(\omega t) \sin\left(\frac{\pi y}{a}\right)$$
(b) $\vec{B}$ is zero within the conductor; The surface current per unit length is $2\epsilon_0 E_0 \frac{\omega a}{\pi} \cos \omega t$ in the −z-direction. The displacement current is equal in magnitude to, but opposite, the surface current.
33.67 $\frac{1}{2}$ W; 2 × 10⁻⁹ N **33.73** 6 cm; 10 cm; polarized parallel to the 3-cm sides
33.75 $\alpha = \frac{\mu_0 c}{\rho\sqrt{2}}\left[1 + \sqrt{1 + (\epsilon_0\omega\rho)^{-2}}\right]^{-1/2}$
$k = \frac{\omega}{c\sqrt{2}}\left[1 + \sqrt{1 + (\epsilon_0\omega\rho)^{-2}}\right]^{1/2}$
$\phi_0 = \tan^{-1}\left[\frac{1}{\epsilon_0\omega\rho + \sqrt{(\epsilon_0\omega\rho)^2 + 1}}\right]$
$\frac{B_0}{E_0} = \frac{1}{c}[1 + (\epsilon_0\omega\rho)^{-2}]^{1/4}$

Chapter 34

34.1 71 m **34.3** 2.7 × 10⁷ m/s
34.5 −18 m²; spacelike **34.7** 0.32 c
34.9 1.48; 1020 MeV/c; (938 MeV)²
34.13 The astronaut at the front must leap first, $\frac{\gamma\beta x}{c}$ seconds earlier. **34.17** The minimum number of observers is three. To test the symmetry of time-dilation four observers are needed. **34.21** 0.26 m
34.25 1.4 × 10⁸ m/s, 470 m
34.29 $\Delta\ell/\ell_0 = -6 \times 10^{-10}$
34.33 No **34.37** $\Delta s = 36.4$ m; No. In no frame do both events occur at the same time.

34.43 The fractional error is 2.8 × 10⁻¹⁰%
34.47 $x'_A = 0; x'_B = -5\sqrt{3}$ m;
$x'_C = 4\sqrt{3}$ m; $x'_D = \frac{23\sqrt{3}}{3}$ m.
$ct'_A = \sqrt{3}$ m; $ct'_B = 7\sqrt{3}$ m;
$ct'_C = -4\sqrt{3}$ m; $ct'_D = \frac{-10\sqrt{3}}{3}$ m.
$\Delta s_{AB} = \sqrt{33}$ m; $\Delta s_{CD} = \sqrt{-39}$ m.
34.49 $x_{1A} = x_{1C} = -\frac{5}{3}$ km, $x_{1B} = x_{1D} = \frac{5}{3}$ km; $t_{1A} = t_{1C} = 4.4 \times 10^{-6}$ s, $t_{1B} = t_{1D} = -4.4 \times 10^{-6}$ s; $y_{2A} = y_{2B} = \frac{5}{3}$ km; $y_{2C} = y_{2D} = -\frac{5}{3}$ km; $y_{2E} = 0$; $t_{2A} = t_{2B} = -4.44 \times 10^{-6}$ s; $t_{2C} = t_{2D} = +4.44 \times 10^{-6}$ s
34.53 1.5 cm; To represent a given space time interval requires a geometrical length on the prime axes that is longer by a factor $\sqrt{(1 + \beta^2)/(1 - \beta^2)}$.
34.59 $a'_\parallel = \frac{a_\parallel}{\gamma^3(1 - \beta u_\parallel/c)^3}$
$a'_\perp = \frac{[(1 - \beta u_\parallel/c)a_\perp + (\beta u_\perp/c)a_\parallel]}{\gamma^2(1 - \beta u_\parallel/c)^3}$
34.63 All of the energy in two colliding particles is available for producing massive reaction products. **34.67** The fractional change = 1.44 × 10⁻⁸ **34.69** 6.72 × 10⁻²⁷ J², the same, 4.68 × 10⁻²⁴ kg·m/s, 3.7 × 10⁻¹⁸ kg·m/s **34.73** 1.0 × 10⁻⁸ m/s **34.79** Yes. The same argument applies as in the non-relativistic case.
34.83 −2.6 × 10⁻¹⁵ K⁻¹
34.87 Light requires different time intervals to arrive from different events. 7.5 × 10⁻⁷ s, 5.6 × 10⁻⁴ s, the aircraft's motion
34.91 $u_0 = \gamma c; \vec{u} = \gamma\vec{v}$
34.95

	2.000	10.000
p/p_N	0.866	0.995
K/K_N	0.786	0.985

Chapter 35

35.1 1.2 nm; 2.4 × 10¹⁷ Hz
35.3 10.4 eV **35.5** 0.434 μm
35.7 870 eV **35.9** 9; 18
35.11 0.3 nm **35.13** ≥ 5 × 10⁻²¹ eV
35.15 170 nm **35.19** 0.24 eV
35.23 6.08 × 10⁻⁴ **35.27** 0.8 photons/s
35.31 5.9 × 10⁻³ nm **35.35** 1.8 V
35.39 0.98 V; +0.2 V **35.43** lithium
35.47 $n^1 = 5 \to n = 3$ **35.51** Yes; $n \geq 3$
35.55 8 × 10⁶ **35.59** 6480 nm;
$d = 3$ nm; $d_i = 10^8$ nm; $r_{350} \gg d$, $r_{350} \ll d_i$ **35.65** $z_{\text{eff}} = 25$
35.67 7 nm; 4 × 10⁻¹² m; 2 × 10⁻³⁶ m
35.71 5.9 × 10⁻¹⁵ m, about $\frac{1}{3}$ the nuclear radius **35.75** 0.723 MeV **35.79** 7 fm

35.83 7%; 200% **35.89** 1 eV/molecule; 1.2 μm **35.93** 0.106 nm; 1.24 × 10¹⁵ Hz
35.97 $\frac{Q}{2M}$; 4.63 × 10⁻²⁴ A·m²;
9.26 × 10⁻²⁴ A·m²; 1.8 × 10⁻⁴ eV;
$E_{\text{hyperfine}} \approx 5 \times 10^{-8}$ eV

Chapter 36

36.1 ⁹⁰Y; ²⁷Al **36.3** 1.8 MeV
36.5 ¹⁸³Au → ¹⁷⁹Ir + ⁴He + Q
36.7 γ ray has energy 938 MeV
36.9 7.2 × 10¹⁸ Bq **36.11** ⁶⁰Co;
7.42 MeV **36.13** 6; 153.5 MeV
36.17 207.241 u **36.21** 2.82 MeV;
2.57 MeV **36.25** If sample has been decaying undisturbed for less than $t \sim 1/(\lambda d)$
36.29 No **36.33** ²²⁴Th → ²²⁰Ra + ⁴He + Q; ²¹¹Bi → ²⁰⁷Tl + ⁴He + Q; ²⁴³Cm → ²³⁹Pu + ⁴He + Q; ²²¹Fr → ²¹⁷At + ⁴He + Q; ²¹⁵At → ²¹¹Bi + ⁴He + Q
36.37 4.1 × 10¹³ Bq; 2 × 10¹¹ Bq; 9.6 × 10²¹
36.41 0.186 MeV **36.45** ²⁰⁶Pb
36.49 silver 5.45 × 10⁷ Bq; palladium 0.90 × 10⁷ Bq; 3.1 h **36.53** $\frac{eRT_{1/2}}{\ln 2}$
36.59 ⁶⁴Zn **36.61** 3.27 MeV
36.65 1.805 MeV **36.69** 2.6 × 10⁶ y
36.73 2.8 × 10⁹ K **36.77** $\rho_* \approx 5 \times 10^{17}$ kg/m³; $\rho_{\text{nuc}} = 2.3 \times 10^{17}$ kg/m³
36.81 neutron 1.479 MeV; proton 1.295 MeV
36.85 2.59 × 10²¹; 1.26 × 10⁴ decays/sec; 3.4 × 10⁻¹⁸ kg; 10¹⁸
36.89

36.91 $\left(\frac{T_1 T_2}{T_1 - T_2}\right) \ln\left(\frac{T_1}{T_2}\right) \frac{1}{\ln 2}$;
$N_0 \frac{T_2}{T_1 - T_2}\{2^{-[T_2/(T_1 - T_2)]\ln(T_1/T_2)/\ln 2} - 2^{-[T_1/(T_1 - T_2)]\ln(T_1/T_2)/\ln 2}\}$; Tues at 8ʰ53ᵐ am; 22.6 mCi

Chapter 37

37.1 3 × 10⁻²² s **37.3** 1.3183 GeV
37.5 Yes; quark flavor and isospin are unchanged **37.7** 4; $-\frac{3}{2}, -\frac{1}{2}, +\frac{1}{2}, +\frac{3}{2}$
37.11 10⁻⁷ eV; 10⁻¹⁴ eV **37.15** charge, baryon number, energy, linear momentum, angular momentum **37.19** 27 MeV
37.23 62 MeV **37.27(a)** 217 MeV
(b) 849 MeV

INDEX

Major discussions and definitions are in boldface.

3C catalogue, 546, 1039

A

Aberration of starlight, 115, 535, 1072
Aberration
   optical (chromatic and spherical), 624
   spherical, 606, 637–8
Absolute temperature. See Kelvin, 648
Absolute zero, 648, 734
AC circuit, 1017
   analysis using phasors, 1027
AC power, rating of, 1019
Acceleration, 58–63, 1072
   angular, 392
   average, 59
   due to gravity, 64–67, 196, 290–291
     as a unit, 73
   in circular motion, 97–98, 175
   in components, 62
   in simple harmonic motion, 472
   in special relativity, 1087–8, 1090
   instantaneous, 60
   parallel/perpendicular to velocity, 61
   of a reference frame, 149, 1103–4
   of a rocket, 212
   uniform, 69–73
   versus time, graph, 69
Accelerator, 926, 928, 970–2
Accommodation, 618
Accuracy
   of a measurement, 21, 1132
   of a number, 26
Action and reaction. See force pairs, 130–1
Activity, 1162
Adatom, 1143
Adiabat, 661
Adiabatic process, **661–3**, 721, 723, 728
Advanced Light Source, 900, 1070
Advantage, mechanical, 237, 438
Aether. See ether, 574
Age
   of the Earth, 1175
   of the Sun, 1092
   of the universe, 1109, 1199
Air
   dielectric strength, 885
   as an ideal gas, 649
   incompressible flow of, 454–6
   properties of, 433, 434

   sound waves in, 527
   as a thermal insulator, 702
Air resistance, 276, 286, 430
Aircraft
   compass, 755
   dynamics, xxviii, 147–148, 225, 239, 241,
     390–1, 412–3
   generation of lift, 454–5
   heading indicator, 414
   rotational inertia, 412–3
   stability, 388
Alcohol, in a thermometer, 646
Algebra, of vectors, 41
Alnico, 941
Alpha-decay, 1154–6
Alpha particle, **1150**, 1153
   in nuclear decay, 345, 814, 1154–6
   as probe of atomic structure, 1119, 1149
Alternating current, 1017
Altimeter, 445
Aluminum
   atomic structure, 1131
   and Compton scattering, 1118
   diffraction pattern, 1127
   photoelectric properties, 1114–5
AM radio, 533, 1030
Amber, 758
American concert pitch, 525
Ammeter, 862–863, 908
Ammonia, molecular structure, 665
Ampère, Andre Marie, 954
Ampère's law, 909, 975–7
Ampere (unit) definition, 762, **933**
Ampère-Maxwell law, 976
Amperian curve, 909
Amplification, 1030
Amplitude, **474**, 485, 502
   current, 1017
   electric field, 1041, 1043–4
   and energy, 480, 1044–5
   in interference, 563
   magnetic field, 1041, 1043–4
   of reflected and transmitted waves, 514–5
   splitting, 573
Anderson, Carl, 1157
Angle
   bank, 147
   Brewster's, 1053–4
   critical, 550

   of attack, 455
   of incidence, 548
     and radiation pressure, 1049
   of reflection, 548
   of refraction, 549
   solid, 782
   SI unit, 24
Angstrom unit, 45
Angular acceleration, 94, 98, 392
Angular frequency, 473. See also angular
   speed.
Angular momentum, 294–301, 397–400
   of a system, 313
   internal, 312, 315
   and magnetic moment, 941
   in atoms, 1122, 1128
Angular momentum quantum number, 1128
Angular size, 618
Angular speed, 94
Angular velocity, 392
Annihilation, 1158
Anode, 1116
Antenna circuit, 1030–1, 1037
Antenna, helical, 1056
Antineutrino, 1156–7
Antinode, 511
Antiparticle, 1157, 1179
Antiproton, 347, 1093–4, 1157, 1177, 1180
Aperture, of a lens, 621
Apollo 15, 3
Application, point of, 363
Approximating, 29–31, 121–2
   capacitance, 878
   in cooling calculations, 711
   in dipole problems, 804
   in electric field calculations, 793
   forces on wires, 934
   in magnetic field calculations, 905
   in pendulum analysis, 478
   plane wave approximation, 1041
   small angle approximation, 96–7
   See also *back-of-the-envelope*.
Archimedes, 432, 445, 447
   and pi, 256
Archimedes' principle, **445–9**, 476
Area
   and integrals, 202–4, 249, 251, 656
   and Kepler's laws, 8, 295–6
   as a vector cross product, 299, 906

I-1

Argon
  on the moon, 737
  and search for solar neutrinos, 1158
  specific heat of, 665
Aristotle
  and elements, 641
  and the nature of motion, 5, 259
Arm, of a circuit, 858
Arrow, 10, 86–87
ASA number, 621
Asteroid, 198
Atmosphere, 31, 435
  composition of, 737
  stability of, 708
  structure of, 443–5, 675
  and transmission of EM waves, 533, 622
  U.S. Standard, 444
  and winds, 663
Atmosphere (unit) definition, 436
Atom, 1110, 1119–32, 1181
  Bohr's model, 1122–5
  and definition of units, 23
  and electrical structure of matter, 759
  and friction, 140
  highly excited, 1126
  hydrogenic, 1125
  and magnetic properties of matter, 938–40
  multielectron, 1131
  Schrödinger's model, 1128
  spectra of, 1119
  standing waves in, 511, 1126
  structure of, 1150–1
  and thermal energy, 652
  world as, xxx–xxxi
Atomic clock, 23, 1074
Atomic mass, chemical, 1147
Atomic mass number, 1147
Atomic mass unit, 435, 650, **1147**
Atomic number, 1147
Atomic weight. See atomic mass, 1147
Attitude control system, of a spacecraft, 936
Atwood's machine, 403–4, 407
Aurora, 347, 757, 923, 929
Automobile
  acceleration of, 72–3
  accidents, 73, 339–40
  engine, 278, 721–2
  starter circuit, 858–860
Avogadro's number, **651,** 667
Axes, coordinate, 20, 35, 1081
Axis
  optical, 603
  rotation, 392
  transmission, 1051
    parallel axis theorem, 410
    perpendicular axis theorem, 422

B
Back-of-the-envelope, 29
  Caesar's last breath, 31

Graphical analysis in kinematics, 68
A car accident, 73, 76
Mars or Bust!, 122–3
Spring constant of a scale, 138, 153
Saving the Earth, 202
Timescale for mass-on-spring, 263
Average air drag, 276
A rotating barbell, 397
Biplane versus monoplane, 412–3
Modeling air, 435
Horse force, 437
A pendulum clock, 488
Wave speed, 498, 500
Exposing the film, 622
Thunderstorm, 683
A hot water pipe, 704
Radiation from you, 710
Collisions in air, 740
Charge in your body, 765
Approximating electric field, 796
Needle in an electric field, 830
Estimating capacitance, 878, 879
Charge on the ends of a car axle, 965
Time constant for an LR circuit, 999
Making the spark, 1000
The lifetime of the Sun, 1092
Observing a star, 1115
The plum-pudding model, 1120
Ballet, 310
Balmer, Johann, 1121
Balmer lines, 1121
  in stars, 1106, 1130
Bank angle, 147
Bank of capacitors, 875–6, 883
Banked turn, 147
Barbell, 397–8, 410–1
Barn, 1170
Barometer
  aneroid, 445
  Torricelli's, 432, 443
Barrier, potential, 1142, 1155
Baryon, 1154, **1182,** 1184–5, 1191
Base of support, 365–6
Baseball, 119–121, 199, 201, 271–272
Basilar membrane, 540
Bathysphere, 442, 704–5
Battery, 845
  charging, 859, 862
  circuit symbol, 845
  dead, 847
Beam, electron, 899, 914–916
Beamsplitter, 573
Bean, Alan, 133
Beanstalk, 164–5, 183–185
Beats, 564
Beauty, 1185
Becquerel, Henri, 345, 533, 1153, 1162
Becquerel, unit, 1163
Bell, Alexander Graham, 539
Belt and pulley system, 427

Bending
  of a beam, 372–3, 378
  of light, 1106
Benzene, xxx
Bernoulli's law, 451–3
Beryl, 588
Beryllium
  atomic structure, 1131
  in Chadwick's experiment, 1151
Beta (relativistic factor) defined, 1073
Beta ray, 1153
Beta-decay, 1152, 1156–7
Betatron, 970–2
Bevatron, 1093, 1180
Bias, 855
Bicycle, **426–30,** 673
Billiards, 202, 343
Bimetallic switch, 686–7
Bin, 736
Binnig, Gerd, 1142, 1144
Binoculars, 621
Biot, Jean Baptiste, 901
Biot-Savart law, 901–903
Biplane, 390–1, 413
BIPM, **22,** 688
Birefringence, 1056
Blackbody, 709–10, 1109, 1118–9
Blackett, 1150
Black hole, 1108–9
Blazing, 586
Blimp, 462
Bocci, 311, 335
Body coordinates, 391
Body, rigid, 361–3
  limitations of, 372, 406
Bohr, Niels, 1111, 1121, 1124
Bohr radius, 1124
Boiling, 680–1
Boltzmann factor, **735,** 1119
Boltzmann, Ludwig, 735
Boltzmann relation, **735,** 1119
Boltzmann's constant, 650
Bose, Satyenda Nath, 1187
Boson, 1187
Bottle, magnetic, 899
Bottomness, 1185
Boundary conditons, 509, 907
Boyle, Robert and Boyle's law, 649
Brackett series, 1121
Bradley, James, 115, 534
Bragg, Sir William Henry, 588
Bragg diffraction (reflection), 588
Brahe, Tycho, 8
Brain, MRI scan of, 943
Brayton cycle, 746
Brewster's angle, 1054
Bridge circuits, 864–865
Bridges
  foot, 172
  simple, 376

suspension, 172
thermal expansion in, 684
truss, 363, 376–8
British Thermal Unit, 688
British unit system, 156
Brix scale, 462
Brookhaven, 1188
Brown, Robert, 667
Brownian motion, 667
Brushes, 963
Bubble chamber, 945, 1157, 1182
Building materials, 705
Bungee jumping, 482–4
Buoyant force, 128, **445**
Burbidge, Margaret, 623

C

Cable, coaxial, 996
Calcite, 1056, 1065, 1117
Calculus
  in circular motion, 98
  in kinematics, 58
  differentiation, 56, 60, 62
  See also derivative.
  See also integration, 250
Calibration, 22, 690
Caloric theory, 641–2
Calorie, 688
Calorimetry, 689–690, 1177
Cambridge University, 975
Camera, 621–2
  Anger, 1163
Capacitance
  calculating, 877–9
  definition, 876
  estimating, 878
  per unit length, 879
Capacitor
  in AC circuits, 1018
  charging, 880, 975–7, **992–3**
  circuit symbol, 880
  commercial, 884
  cylindrical, 879
  discharging, 988–90
  parallel plate, 876–7
  polar, 884
  spherical, 878
  variable, 897
Capacitor combinations, 880–884
Capacity, heat, 689
Carbon
  atomic structure, 1131
  in nuclear reactions, 1167–8
  graphite—in resistors, 699, 852
Carbon dating, 928
Carnot cycle, **723–724**, 727, 732–4
Carnot, Sadi, 641, **723**
Carroll, Lewis, 613
Cartesian coordinates, 19–20
Cat, Schrödinger's, 1134

Catastrophe, ultraviolet, 1119
Cathode, 1112, 1116
Cathode ray tube, 930
Causality—principle of, 20
Cause, 4–5
  and effect and the Lorentz transformation, 1085
Cavendish, Henry, 181, 766
Cavitation, 466
Cavity oscillator, 1057–8
Cavity radiation. See Blackbody radiation, 710
Cell, blood, 1070
Celsius, Anders, 646
Center of energy reference frame, 1093
Center of gravity, 306–7
Center of mass, 307–8, 363–6, 369, 391
  motion of, 309–10
  of extended bodies, 374–5
Center of mass reference frame, 310–2, 317
  angular momentum in, 312, 315–7
  collisions in, 337–8, 1093
  energy in, 312
  momentum in, 311
Center of momentum, 311
Ceramic, dielectric strength of, 885
Cerenkov radiation, 556
CERN, 347, 1188
Cesium, 23, 471
Cesium antimonide, 1114
cgs units, 1113
Chadwick, James, 1150–1
Chain reaction, 1169
Characteristic curve, 854–855
Charge
  bound, 884–5, 887
  electric, 757–8, 765–6, 1072
  electron, 759, 765, 772
  free, 888
  of nucleus, 1146
  point, 765, 766
  test, 769
  variable, 988, 990
Charge carriers, 937–8
Charge-to-mass-ratio, 800, 925–6, 930
Charles, Jacques, 649
Charm, 1185, 1188–9
Charmonium, 1188
Chemical bonding and friction, 140
Chemistry
  and atomic structure, 760, 1130–2, 1150
  influence on Newton, 163
  and magnetic moment, 941
  use of moles in, 651
  and MRI, 943
Chernobyl, 1146
Chiang, Shirley, 1142
Chlorine, 1148
  in neutrino detectors, 1158
Chord, of a wing, 147, 455
Chromodynamics, quantum, 1189

Circle, great, 1108
Circuit
  AC, 1017
  bridge, 699, 864–5
  DC, 964, 988–993, 999–1007, 1049–50
  energy transport in, 1048–50
  filter, 1023
  household, 1036
  LC, 1001–2, 1058
  LR, 999–1000, 1022–3
  LRC, 1003–5, 1029–32
  multiloop, 858–62, 1006–7
  radio tuning, 1004–5, 1015, 1030–1, 1037
  RC, 988, 991, 1021–2
  RCL, 1003–5, 1029–32
Circular motion, 93–101, 370–1, 413
  analogy with SHM, 472–4
  dynamics of, 145–8
  in a magnetic field, 924–7
  vectors in, 396
Circulation
  of $\vec{B}$, 907–909, 975–7
  of $\vec{E}$, 959, 969
Classical electron radius, 891
Clausius cycle, 747
Clausius, Rudolf, 720
Clocks
  atomic, 23
  and coordinates of events, 21, 1078
  and oscillations, 469–71, 478–9, 488
  and time in relativity, 1073–4, 1078
Cloud
  electron, 759, 1068, 1128
  Large Magellanic, 1159
  virtual particle, 1193
Cloud chamber, **1150**, 1154, 1157
Clouds and convection, 707
Cluster
  of galaxies, 1070, 1125
  globular, 215
CNO cycle, 1167
Coaxial cable, 879, 911
Cobalt
  and magnetism, 760, 938
  and nuclear medicine, 1146
  nuclei and parity, 613
COBE, 709
Cochlea, 540
Coefficient
  of friction. See friction, 139
  of lift, 455, 465
  of performance, 725
  of restitution, 289, 341
  temperature, of resistivity, 851–2
  of thermal expansion, 683, 684
Coherence, 570
Coil
  simple, 905
  spark, 1000, 1012, 1013
  toroidal, 912, 917, 1011

Collision
  in CM reference frame, 337–338
  Compton, 1117
  definition, 333
  elastic, 334, 341–343
  gravitational, 349–350
  and heat transfer, 653
  inelastic, 338–341, 343–345
  in one dimension, 334–335
  in two dimensions, 336–337, 343
  involving angular momentum, 347–349
  of molecules in a gas, 347–349, 439, 666, 739–41
  subatomic particle, 345–347, 1177–8
  of vehicles, 207–208, 339–340
Collision time, 201, 740, 849–850
Color, and light, 536, 572
Color (color charge), 1190
Coma, 624
Comet, Vomit, 196, 1104
Comets, 198, 202–203, 360
Commutative rule for vector addition, 41
Compass, magnetic, 3, **755**
Components, of a vector, 37–38
Compound circuit, 860–862
Compressibility, **433**, 528
Compression, 167–168, 363, 377, 433
  of a beam, 371
  of a gas, 656, 663, 721, 723
Compton, Arthur Holley, 1116
Compton effect, 1116–8
Compton wavelength, **1118**, 1180
Computer programs
  Kepler, 183, 327
  Trace, 635
  Magmot, 951
Concepts, identification, 117
Conduction, thermal, 701–5
Conductivity
  electrical, 850
  thermal, 701–2
Conductor, electrical, behavior of, 768–9, 829–833
  and electric field, 829–830, 832–33
  and electric potential, 831
  and surface charge density, 830, 832
Cone (in the eye), 536, 579
Conservation law
  for angular momentum, 197, 313–317, 1192
  for charge, **765,** 976
  for energy, 197, 269–273, 275, 452, 959
  for momentum, 197, 205–215
    in relativity, 1092
  in nuclear physics, 1153–4
  in particle physics, 1191
  problem solving plan, **207,** 448
Conservation laws
  and rigid bodies, 406–9
  and symmetry, 1192
Conservation of mass, 450

Conservation of string, 168
Conservative force, 269–271, 966
Consistency check, 118
Constant
  Boltzmann's, 650
  Coulomb, 766
  fine structure, 1124
  phase, 502
  solar, 557, 1045
  Stefan-Boltzmann, 710
  time, 989
Constraints, 167, 401
Construction of equations, 117
Contact electricity. See triboelectricity, 768
Contact, thermal, 645
Context, 117
Continuity, equation of, 450
Continuous systems, 361
Contour map, 292
Control rods, 1170
Convection, 707–8
Conveyor belt, 210
Cooper pairs. See electron pairs, 854
Coordinate axes, 20
Coordinate system, 20
  choice of, 19, 40, 63–64, 250, 403
  origin, 19–20
  right handed, 20
Coordinates
  body, 391
  Cartesian, 20
  spherical, 25
  cylindrical, 25
  in special relativity, 1078
  object and image, 601
Copernicus, Nicholas, 6, 196
Copper
  conduction electrons in, 850
  size of nucleus, 1149
  surface image, 1110
Cornea, 536, 607, 618
Corner reflector, 558
Correspondence principle, 1126
Corrosive gas, 1131
Cosmic microwave background radiation, 709
Cosmic ray, 1074, 1088, 1091, 1157
Cosmology, and particle physics, 1189, 1199
Coulomb constant, 766
Coulomb's law, 766, 770, 901
Coulomb (unit), definition, 762
Couple, 367–8
Cowan, 1157
Crest of a wave, 567
Critical damping, 487
Critical mass, 1169
Critical point, 679
Crommelin, Andrew, 1069–70
Cross product
  and area, 299, 906
  definition, 297

  Math Toolbox, 299–300
  in terms of components, 299–300
Cross section, 740, 1169–70
Crossover network, 1035
Cryogenics, 698
Crystal, 1161
  birefringent, 1056, 1065
  and magnetic domains, 941
  and nuclear recoil, 1160–1
  piezoelectric, 1142
  salt, 587
  specific heat, 667
  and x-ray diffraction, 587–8
Curie, Marie, 345, 1153, 1163
Curie, Pierre, 940, 1153
Curie, unit, 1163
Curling, 336–337
Current
  asymmetry, in diodes, 855
  conduction, 975
  continuity of, 976
  displacement, 975–6
  eddy, 973–4
  electrical, 761, 844, 931
    and energy transport, 1048–50
  magnetization, 939
  variable, 990
Current balance, 934
Current density, 850–851, 910
Current sheet, 912
Curvature
  of a mathematical curve, 68
  of an optical surface, 634
  of space-time, 1107–8, 1199
  of a wave front, 606
Curve, Ampèrian, 909
Cutoff wavelength, 1059
Cycle, 473
Cycle, thermodynamic, 663–5, 720–4
Cyclotron, 926, 928
Cylindrical polar coordinates, 25

D
Damping, 486, 1003
  critical, 487, 1005
Damping time, 487
Dating, radioactive, 928
Daughter (nucleus), 1153
Davis, Raymond, 1158
Day, 23
  sidereal, 418
De Broglie, Louis, 469, 1126
Debye, Pieter, 689
Decay
  alpha, 1154
  beta, 1156
  gamma, 1160
  proton, 556, 1178, 1199
Decay, law of, 1161–2
Decay rate, 1161

Deceleration, 60
Decibel, 539
Decuplet, baryon, 1184
Dees, 926
Defect, mass, 1152
Degree
    of angular measure, 24
    of temperature, 647
    why a?, 647
Degree of freedom, 665
Density, 30, **434–5**
    charge, linear, 793
        surface, 797
        volume, 799
    current, 850
    electron, 1128
    energy, 889, 996, 1044
    in sound waves, 527–8
    surface, 374
Depth of field, 622
Depth, variation of pressure with, 442
Derivative
    of cross product, 302, 320
    of dot product, 230
    of a vector, 62
    partial, **505,** 659, 1042–3
Descartes, Rene, 20, 466
Detector, particle, 1188
Determinant, and cross products, 300
Deuterium, 346, 1048, 1152, 1167
Diagram
    energy-level, 1123, 1129
    Feynman, 1179
    field line, 773–4
    force-torque, 306
    free-body, 140–8, 166–75
    phase, 680
    P-V, 656
    quark-line, 1189
Diamagnetism, 939–40
Diamond, 885
    specific heat, 688
    thermal conductivity, 702
Dielectric, 884–9
    constant, 885
    strength, 885
Diesel engine, 746
Differential element
    in analyzing continuous systems, 361, 375, 409, 440–1
    and the Biot-Savart law, 901
    in electric field calculations, 794–5
    in integration, 249–50
    in kinematics, 58
    of string, 171, 506–7
    and work, 232
Diffraction, 576–78
    circular aperture, 578
    combined with interference, 579
    electron, 1127, 1132, 1150

Fresnel vs Fraunhofer, 576
    intensity in, 586–7
    rectangular aperture, 577–8
    x ray, 588, 1127
Diffusion, 741
Dimension, physical, 29
Dimensional analysis, 29, 498
Diode, 855
Diopter, 611
Dipole
    electric, 677, 760, **802–5,** 885, 887
    magnetic, 756, 903, 905, 937
    moment, 803
Dirac, Paul, 1157
Direction, of a vector, 32
Discrimination, 1030–2
Disk around black hole, 1109
Disorder, 719–20, 726–7, 741–2
Dispersion, 552–3
    of a grating, 585
Displaced fluid, 445–6
Displacement
    definition, 32
    and change in position, 36
    and distance, 34
    electrical, 888–9
    of fluid, 446
    relative, 104
    in sound waves, 527–8, 539, 540
Displacement current, 975
Dissipation, 238, 990
Distance, 34
    proper, 1080
Distribution
    Maxwell-Boltzmann, 736–9
    what is?, 736
Distributive law
    for cross product, 299
    for dot product, 229
    for scalar multiplication of vectors, 41
Domain, magnetic, 940
Doping, of semiconductors, 855
Doppler effect, 542–7
    relativistic, 546, 1099
    transverse, 1099
Doppler, Johann Christian, 542
Dot product
    definition, 228
    Math Toolbox, 229–230
    in terms of components, 230
Down (quark), 1185
Drag, 128, 276. See also air resistance
Drift, $\vec{E} \times \vec{B}$, 930–1
Drift speed in electric current, 849–50
Driver, mass, 973
Driving, 52–7, 72–3, 146–7, 207–8, 278
Dulong-Petit law, 667, 689
Dust, interstellar, 1055
Dynamics, 126
    of circular motion, 145–8

of fluids, 449–54
of particles, 140–8
of rigid bodies, 400–6
of systems of particles, 166–9
relativistic, 1088–90
Dynode, 1112

E

Ear, 524, **540**
Earth, xxix, 181
    atmosphere, 443–5, 737
    electric field of, 808, 931
    gravitational field of, 290–2
    magnetic field of, **756–7,** 936
    mass of, 181
    precession of, 414
    as a reference frame, 35
    rotation of, 39–40, 414
Earth-moon system, 187, 267, 293
Earthquake, 181
Eccentricity, 9, 331
Eclipse, 23, 1069–70, 1107
Eddington, Arthur, 1069–70, 1167
Eddy currents, 973–4
Efficiency
    limit on, 732–3
    of a Carnot cycle, 724
    of an engine, 720–1
    of a machine, 238
    of the Otto cycle, 722
    quantum, 1114
Einstein, Albert, xxvi, 3, **954,** 1067
    and Brownian motion, 667
    and electromagnetism, 754, 762–3, 891, 966, 1070
    and gravitational field, 66, 293, 1104
    and light, 469, 532, 576, 1043, 1113
    and the mass-energy relation, 891
    and quantum physics, 1111, 1113, 1125
    and special relativity, 1068, 1072
Einstein cross, the, 1107
Einstein Observatory, 588
Elastic energy. See energy, 259
Electric charge. See charge, 757
Electric field due to
    a charged sheet, 798
    charged sphere, 806
    dipole, 804, 812
    infinite line, 794
    spherical shell, 792
Electric field strength, 769, 774
Electrical safety, 848
Electrode, 844
Electron, xxx, 4, 1151, 1186, 1193
    and atomic structure, 1126–8
    charge of, 759, **765**
    density, 1128
    and electric current, 844, 849
    and electrical properties of matter, 759, 765
    and ionization, 347, 1123

Electron, (*continued*)
  model of, 891
  oscillating, and EM waves, 1040
  pairs, 854
  wave behavior of, 1110–1, 1127, 1134
Electron-positron pairs, 945, 1158–9
Electron volt definition, 819
Elements, 1150
  creation of, 1168
  in Aristotle's physics, 5, 641
  See also Periodic Table.
Ellipses (*Digging Deeper*), 9
Ellipsoid, 9, 634
Elliptical orbits, 163, 202, 329–31, 349
Elsewhere, 1079
Emf, 845, 964–6, 1017
  circuit symbol
    AC, 1017
    DC, 845
  induced, 956–959, 966
  motional, 961
  and potential difference, 964–5
  and reference frame, 966
  and work, 966
Emissivity, 710
Energy, 197
  binding, 1125, 1152–3
  chemical, 275
  conservation of. See conservation law.
  elastic, 253, 258
  electric potential, 814–6, 821–5, 890–1
  equipartition of, 666
  and equilibrium, 366–7, 481
  forms of, 225
  gravitational, 263–9
  in oscillatory motion, 480–2, 665, 684
  interaction, 891
  internal, 274–8, 312, 334, 338, 651–3, 659
  kinetic, 226, 311, 1072, 1090
  magnetic, 995–6, 1001
  in orbits, 329–30
  potential, 259–69
  rest, 1090
  rotational, 396–7, 666
  self, 890
  thermal, 274, 345, 652–3
  total mechanical, 271
  unit, 226
  and work, 226, 275, 654
Energy density
  electric, 889–92, 1044–5
  magnetic, 996, 1044–5
Energy level diagram, 1123, 1129
Energy stored
  in a capacitor, 880–1, 990, 1001
  in an inductor, 995, 1001
  in a spring, 253
Energy transmission
  by machines, 238–239
  by waves, 507–8, 566
  by EM waves, 1044–6, 1048–50

Energy-momentum invariant, 1092
Engine
  airplane, 675
  automobile, 721–2
  Diesel, 746
  Hero's, 641
  steam, 641
Entropy, 726–32, 741–2
  and the arrow of time, 751–2
  and equilibrium, 742–3
  and reversibility, 729
Envelope, back-of-the, 29. See also Back of the envelope.
Environment, 644
Eötvös, Roland von, 1106
Equation
  differential, 1005
  exponential, 444, 486, 989, 999
  simple harmonic, 472
  of state, 678
  wave, 505–6
Equations, construction, 117
Equilibrium, 166
  3 forces in, 368–9
  dynamic, 369–71
  and entropy, 742
  establishment of, 740
  hydrostatic, 440–5
  neutral, 366
  of nuclear series, 1165
  and potential energy, 366
  of rigid bodies, 363
  of systems of objects, 371–2, 471, 496
  stability of, 366–7, 448
  static, 166
  thermal, 645
Equipartition of energy, 665–6
Equipotential surface, 291–3, 825–8
  and conductors, 831
  and field lines, 825–6
  and Poynting flux, 1049
Equivalence principle, 1104, 1106
Equivalent resistance defined, 853
Escape speed, 184–5, 268, 737
Escapement, 488
Estimation, 29, 56, 73
Ether, 574–5, 1072
Event, 20, 88, 1075, 1077
Everest, Mount, 17, 20, 1105
Everyday scale, xxviii
Evolution, 751–2
Excited state, 1123
Exclusion principle, 1130
Exit pupil, 620–1
Expansion
  free, 730
  thermal, 683–7
  of the universe, 1109
Experiment
  and theory, 14
  Cavendish, 181

Eötvös, 1106
  inclined plane, 11, 197, 257–8
  Joule's, 654
  Michelson-Morley, **574–5,** 1072
  Millikan oil-drop, 772
  nature of, 12
  Pound-Rebka, 1107, 1133, 1160, 1197
  thought, 11, 132, 445
  Young's, 570
Exponential behavior, 444, 487, 712, 989, 999, 1142, 1162
Exposure, of film, 621–2
Extreme Ultraviolet Explorer, 936
Eye, 533, **536,** 579, 599, 607–8, **618,** 633
Eyeglasses, 611, 632
Eyepiece, 619, 620
Eye relief, 620

F
f-stop or f-ratio, 621–2
Fahrenheit, Gabriel, 646
Fairbank, William, 1186
Family, of particles, 1183, **1189**
Farad (unit), definition, 877
Faraday, Michael, 248, 762, 773, 877, **956,** 985
Faraday rotation, 1064
Faraday (unit), 877
Faraday's law, 956–958, 967, 975
  plan for using, 968
Far point, 632
Fast (of a camera), 622, 623
Fechner, 846
Fermat, Pierre, 559
Fermi, Enrico, 1130, 1149
  and estimation, 29
  and neutrinos, 1157
Fermi (unit), 1149
Fermi National Accelerator Laboratory, 929, 1177, 1178
Fermi's problem, 29, 50
Fermion, 1130, 1187
Ferrites, 941
Ferromagnetism, 940–1
Feynman, Richard P., 1110, 1178, 1179
Fiber-optic, 551
Field
  Coulomb, 770, 964
  electric, 760, **769–77,** 1072
    and reference frame, 763
    induced, 959, 967–70
  gravitational, 290–3, 760
  magnetic, 755–7, 899–900, 1130
Field line, 291–3, 756, 760, 773, 1040
  relation to charge, 777–9
Field line diagram, 760, 773–7
  rules for constructing, 774
Film, thin and interference, 571–3
Filter, polarizing, 1051
Fine structure, 1129
Fine structure constant, 1124

Fish, 602, 631, 686
Fish-eye lens, 558, 621, 629
Fission, nuclear, 1152, 1169–70
Fitzgerald contraction. See Lorentz contraction, 1075
Fizeau, Hippolyte Louis, 557
Flasher, highway emergency, 987–8
Flashgun, 988
Flashlight, 859
Flavor (of quarks), 1191
Flow, fluid, 449
Fluid, definition, 433
Fluid dynamics, 449–54
Fluorescence, 1139
Fluorine, 1131
Flux
   of current density, 910
   of electric field, 779
   of fluid, 450, 779
   of heat, 701, 703–5
   of magnetic field, 907, 957, 960
Flying, 88, 102–4, 147–8, 160, 454–5
FM radio, 533, 1031–2
Focal coordinate, 604, 611
Focal length, 605, 611
Focal point
   first and second, 608
   of a lens, 611
   of a mirror, 604–5, 623
   of a refracting surface, 608
Focus
   of an ellipse, 9
   in optics, 604, 636
   prime, 623
Force
   what is it?, 126–7
   buoyant, 445
   centrifugal, 145
   centripetal, 145
   chromodynamic (color), 1190–1
   conservative, 269–71
   contact (See normal force), 128
   Coulomb, 766, 1179
   driving, 486
   due to eddy currents, 974
   elastic, 127–8, 136–8
   electric, 766, 770
   electroweak, 1187–8
   external, 215, 309, 484
   fictitious (see inertial force), 1103
   fluid, 128–9, 432
   frictional, 128, 1144
   fundamental, xxxi, 197
   grand unified, xxxi, 1199
   gravitational, 127, 175–8, 329, 1104
   impulsive, 200
   inertial, 1103
   input, 237
   internal, 215, 406–7, 433, 449, 452
   line of action of, 364
   Lorentz, 965
   magnetic, on particles, 761, 899–900, 924
   on wires, 761, 931–4
   and emf, 961, 966
   molecular, 433, 677
   net, 127
   nonconservative, 274
   normal, 128, 435
   nuclear, xxxi, 1149, 1151–2
   output, 238
   pairs, 130–1, 149, 204–5, 215, 313–5
   and particle exchange, 1179
   per unit length, 933
   periodic, 484
   restoring, 471, 485, 497, 528
   shear, 433
   in special relativity, 1090
   spring, 127, **136–138**
   strong, xxxi, 1151–2, 1183, 1189–90
   tension, 128, 170, 363
   transmission of, 130
   van der Waals, 677, 802, 1144
   vector nature of, 129–30
   weak, xxxi, 613, 1152, 1183, 1187–8, 1190
Force unit, 134
Force at a distance, 179
Force-balance, 127. See also "Equilibrium".
Forced oscillations, 484–6
Forces that occur in classical mechanics, 128
Fourier, Joseph, 502
Fourier's theorem, 502, 548, 1038, 1044
Fovea, 579
Fowler, William, xxv
Fractal, 667
Franklin, Benjamin, 641, **758**, 832
Fraunhofer, Joseph, 576
Free body diagram, 141
   method of, 141, 167
Free Fall, 177–8
   Digging Deeper, 90
Frequency
   angular, 473, 503
   beat, 565
   in circular motion, 95
   crossover, 1036
   cyclotron, 925
   fundamental, 510, 511
   Larmor, 943
   natural, 484, 1029
   in oscillations, 473
   precession, 413–4
   resonant, 1030
   in wave motion, 499, 542–6
Fresnel, Augustine, 576
Friction
   as a cause, 5, 394
   coefficient of, 139
   definition, 128
   kinetic, 139, 274, 404–6
   as a nonconservative force, 274
   properties of, 139
   rough surface model, 139–40, 274
   static, 139, 146, 278, 370, 372, 402, 403
Friedmann, Aleksandr, 1068
Fringes, interference, 569
Full width half maximum (fwhm), 1037
Function
   harmonic, 474
   phase, 502
   wave, 501–2, 1133–5
Fusion, latent heat of, 682
Fusion, nuclear, 346, 816–7, 1152
   experiments, 876, 898–9, 929, 1046, 1048
   in the Sun, 1091, 1167–9
Future, 20, 751–2, 1079, 1109

G

Galaxy, 1109
   cluster, 1107, 1125
   Large Magellanic Cloud, 1159
   M87, 1109
   magnetic field of, 1055
Galilean relativity, 104, 148, 1070
Galilei, Galileo, 1, 3, **10–12**, 18
   influence on Newton, 162
   and oscillations, 471
   and projectile motion, 10–11
   and relative motion, 104, 1070
   and the speed of light, 534
Galileo spacecraft, 332–3, 349–50
Galileo's inclined plane experiments, 11, 197
Galileo's law of fall, 12, 64–66, 181–183
Galvanometer, 862–5, 934
Gamma (ratio of specific heats), 529, **660–1**
Gamma (relativistic factor), 928, **1073**
Gamma decay, 1160–1
Gamma ray, 533, **1153**
   and Compton scattering, 1118
   imaging, 1163
Garriott, Owen, 136
Gas, 433, 438
   corrosive, 1131
   ideal, 438–9, 649–53
   inert, 1131
   laws, 649
Gauge pressure, 459
Gauss, Carl Friedrich, 777, **781**
Gauss' law for electric field, 778, **781–3**, 976
   use in calculating $\vec{E}$, 791
   for magnetic field, 907
Gauss (unit), 900
Gaussian surface, **792**, 909
Gay-Lussac, Joseph Louis, 649
Gear ratio, 426, 428
Geiger counter, 533, **793**
Geiger, Hans, 1119
Gell-Mann, Murray, 1184
Generation, of quarks, 1189
Generator, 955, **962–3**, 1017
   homopolar, 985
Geomagnetism, 756–7

Geometry
   distortion of space-time, 1087
   and gravity, 1108
Geosynchronous orbit, 17, 184, 557
Gerber, Christoph, 1144
Germanium thermometer, 699
Giant, red, 1168
Gilbert, William, 756, 757
Giotto spacecraft, 198
Glashow, Sheldon, 1188
Glass
   dispersion of light by, 552
   speed of light in, 537
   in thermometers, 691
   total internal reflection in, 551
Gliding, 196
Globular cluster, 215
Gluon, 1189
Goddard, Robert H., 222
Goeppert-Meyer, Maria, xxv
Gold, 768, 1143
   and Archimedes' problem, 447
   photoelectric properties, 1114
   and Rutherford's experiments, 1119, 1121, 1149
Gossamer Albatross, 427
Gradient
   Math Topic, 826
   temperature, 701
Gram, 26, 434
Grand Tour, 350
Granules, solar, 707
Graphical methods, 68–69
Graphite
   in resistors, 699, 852
   x-ray scattering by, 1117
Grating, diffraction, 585–6
Gravimeter, 65, 67
Gravitational energy. See energy, 263
Gravitational field, 290–3
Gravitational force law (Newtonian), 126, 175–183, 329
   due to a uniform sphere, 178
Gravity, 3, 1104, 1199
   center of, 306
   Digging Deeper, 66
   specific, 434
Great circle route, 1108
Greek physics, 4, 641, 755, 758, 1150
Grey body, 710
Ground
   electrical, 818
   circuit symbol, 845
Ground state, 1123
Grounding a circuit, 879
Group speed, 566
Groups, and particle physics, 1192
GUTs, 1199
Gyro-radius, 925
Gyroscope, 413–4

H

Hadron, 1182, 1191
Half-life, 1162
Hall effect, 937–8
Hall, E.H., 937
Hall field, 966
Halley's comet, 10, 16, 198
Hard, magnetically, 941
Harmonic function, 474
Harmonic wave, 502
Harmonics, 511–2, 532
Hat Creek Radio Observatory, 698
Hawking, Stephen, 752
Heading indicator for aircraft, 414
Heat, 274, 653
   latent, 681
   specific, 658
Heat capacity, 689
Heat death, 752
Heat engine, 720–2
Heat reservoir, 723
Heat transfer, 653, 657–8, 700–12
   and entropy, 727–8
Height, scale, 444
Heisenberg, Werner, 1132
Helium
   atomic structure, 1125, 1131
   binding energy, 1153
   constraints on neutrino number, 1189
   as an ideal gas, 649
   at low temperatures, 698–9
   in nuclear reactions, 345, 346, 1091, 1167–8
   specific heat, 659
   spectrum, 1119
   See also *alpha particle*.
Helmholtz coils, 906, 920, 922
Helmholtz, Hermann von, 906
Henry, Joseph, 995
Henry (unit), 995
Hero of Alexandria, 641
Herschel, Sir William, 533
Hertz, Heinrich, 95
   and the discovery of radio waves, 533, 762, 970, 1044
   and the photoelectric effect, 1112
Hertz (unit), definition, 95
Hipparchus, 22
Hole
   black, 1108–9
   modeling holes, 374–5
   in semiconductors, 855
Hooke, Robert, 137, 534
   and Newton, 163
Hooke's law, 137, 169
Horsepower (HP), 235
Hubble, Edwin, 1068
Hubble telescope, 624, 637–8
Hulse, Richard, 1109
Human body, 710–1, 765, 848, 879, 943, 1163

Huygens, Christiaan, 548
   and circular motion, 162–3
   and the speed of light, 534
Huygens Principle, 548, 549, 576
Hydraulic machinery. See machines, 438
Hydroelectric power, 956
Hydrogen
   atomic structure, 1128, 1130
   Bohr model, 1122
   energy level diagram, 1123, 1129
   in interstellar space, 1124
   in nuclear reactions, 1167–8
   nuclear structure, 1147
   phase diagram, 680
   specific heat, 665
   spectrum, 1121
Hydrogenic atom, 1125
Hydrometer, 462
Hyperbola, 656
Hyperboloid, 634
Hypercharge, 1183
Hyperfine structure, 1130
Hyperon, 1182
Hysteresis, 942

I

IBM, 1110
Ice
   as frictionless surface, 138, 140
   how fish live beneath, 686
   phase transitions, 681
Iceberg, 446
Ideal gas
   constant, 651
   law, 650, 651
   model, 438–9, 649
   monatomic, 649–653, 658–61
   polyatomic, 666–7
   specific heats of, 658–61, 666–7
Images
   angular size, 618
   definitions, 600–1
   distortion, 617
   erect, 612
   formed by a curved mirror, 603–6, 613–5
   formed by a lens, 610–11, 616–7
   inverted, 612
   in a plane mirror, 599–600, 612
   real, 600
   in a refracting surface, 602, 607–8, 615
   virtual, 600
   visibility of, 617
Impact parameter, 343
Impedance, 1022–3, 1029
Impulse, 200–204, 248–251, 333
   angular, 304
Impulse-momentum theorem, 202
Incidence
   angle of, 548
   plane of, 548

Inclined plane experiments, 11, 197
Incompressibility, 433, 440, 528, 679
Index of refraction, 536
Induced electric field, 959, 967
Induced emf, 956–60
Inductance, 993–5
    mutual, 997–8
    self, 993–5
    of a solenoid, 994
Induction, magnetic, 900
Inductor
    circuit symbol, 995
    in AC circuits, 1018
    energy stored in, 995
    variable, 1015
Inert gas, 1130
Inertia, 528
    moment of, 398
    principle of, 11
    rotational, 396–400
Infeld, Leopold, 955
Infrared, 533
Infrasonic, 525
Insulator
    electrical, 768–9, 884
    thermal, 702, 705–6
Integral laws in E&M, 914
Integrals as areas, 204, 251
Integration, 66–67, 248–253, 409–10, 794–5
Integration method, 250
Intensity
    definition, 539
    in diffraction patterns, 586–7
    in interference patterns, 580–81
    in light waves, 1045
    in sound waves, 539
    magnetic field, 941–2
Interaction
    charged-current, 1189
    neutral current, 1189
Interference
    closely spaced sources, 581
    constructive, 564
    destructive, 564
    electron, 1134
    and image formation, 606
    multiple sources, 583–4
    spatially separated sources, 567–8, 1134
    thin film, 571–3
Interferometer, 573
    LIGO, 1103
    Michelson, 573
    wave-front splitting, 571
International Practical Temperature Scale, 691
International System of Units, 21–31
Interstellar space, 1126, 1130
Interval, space-time, 1079
Invariance, 41
Invariant, 41, **1071**, 1092
    interval, 1080

    electromagnetic, 1095
    energy-momentum, 1092
Inverse square law, 541
Ionization, 347, 841, 925, 1123
Ionization potential, 1123
Ions, in batteries, 845
Iron
    atoms on copper surface, 1110
    binding energy, 1153
    and Compton scattering, 1118
    in intergalactic space, 1125
    magnetic domains in, 940, 942
    and magnetism, 760, 938, 941
    in nuclear reactions, 1160
    and the Pound-Rebka experiment, 1133, 1160–1
    in stars, 1168
Irreversible process, 719–20, 729–31, 751
Irrotational flow, 449
Isentropic, 727
Isobar, 680
Isobaric (constant pressure) process, 656
Isochoric (constant volume) process, 656
Isolated system, 205, 215, 314, 644
Isospin, 1181, 1192
Isotherm, 656
    of the van der Waals equation, 679
Isothermal (constant temperature) process, 655, 661, 728
Isotope, 760, **1147**
Isotropy, 436, 439

J
Jerk, 62
Jet, of particles, 1177
Joliot-Curie, Irene, 1153
Joule (unit) definition, 226
Joule heating, 847, 974
Joule, James Prescott, 226, 654
Junction
    in a circuit, 858
    in a mechanical system, 377
    p-n, 855
Jupiter
    comet collision with, 199
    and Galileo spacecraft, 332–3, 349–50
    intensity of sunlight at, 541
    and Newton's discovery of gravity, 163
    and Roemer/the speed of light, 535

K
K-meson, 751, 1182, 1184
Kamioka detector, 1159
Keck telescope, 624
Kelvin (unit), definition, 648
Kelvin, Lord, **648,** 720
Kepler, Johannes, 1, 6–9, 122
Kepler's laws, 8, 30, 122–3, 183
    and angular momentum, 295–6
    and the law of gravity, 162–3, 175

Kerst, D. W., 971
Kilogram, definition, 26
Kilometer. See meter, 23
Kinematics, 51–106
    of rotation, 393–6
Kinetic energy. See energy, 226
Kinetic theory of gases. See statistical mechanics.
Kirchhoff, Gustav, 858
Kirchhoff's rules, 859, 910
    in AC circuits, 1021, 1022
    in time-dependent circuits, 988, 991, 1001, 1003, 1006
    and phasors, 1025, 1027
Klystron tube, 1058

L
Laboratory reference frame, 310
Ladders, 372–3
Lagrange points, 180
Lambda particle, 1157, 1182, 1185, 1186, 1189
Lampton, Michael, 634
Lande g factor, 941
Laplace, Pierre Simon de, 1108
Larmor frequency, 943
Larmor radius, 925
Laser, NOVA, 1048
Laser interferometer, 1103
Laser ranger for mapping, 24
Lasers, 1046, 1047–8, 1051, 1065
    coherence length for, 571
Latent heat, 681–3
Lavoisier, Antoine, and mass, 150, 641
Lawrence Berkeley Laboratory, 875–6, 881, 883, 1070, 1180
Lawrence, Ernest O., 926
Lawrence Livermore National Laboratory, 929, 1048
Lead, 1160, 1162
    lead-acid battery, 845
    and nuclear series, 1164–5
Lee, T.D., 613
Leidenfrost effect, 702
Length contraction, 1074–5
Length
    focal, 605
    rest, 1080
    unit, 23
Lens, 608–9
    achromatic, 632
    converging or diverging, 611
    fish-eye, 558, 621, 629
    gravitational, 1107
    meniscus, 611
    plano-convex or concave, 611
    spectacle, 611
    thick, 636
    thin, 610–11
    ray-tracing rules, 615
    telephoto, 631

Lens-maker's equation, 610
Lenz, Heinrich, 958
Lenz's law, 958–9, 973
Lepton, 1154, 1182, 1186, 1191
Levitation, magnetic, 973–4
Lick Observatory, 623, 632
Lifetime
   of a state, 1133
   of the Sun, 1092
Lift, 128, 147, 225, 388, **454–5**
Light
   bending of, 549–50, 1107
   dispersion of, 552
   as an EM wave, 1040
   monochromatic, 533, 1114
   as particles, 532, 1112–8
   pipe. See fiber optic, 551
   polarization of, 1050–7
   speed of, 23, 28, 534–5, 915, 1043, 1070, 1072, 1092
   unpolarized, 1050
   visible, 533, 536
Light bulb, 715, 854
Light cone, 1079
Light source, and coherence, 570
Light year, 557
Lightning, 754, 757, 814, 868
Lightning rod, 758
LIGO, 1103
Limit, in kinematics, 55–57
Limits of visibility, 617
Line
   nodal, 568
   spectral, 1119–30
   straight, in curved space, 1108
   world, 1077
Line of action (of a force), 364
Linear induction motor, 973
Liquid, 433, 440
   sound speed in, 528
Liquid drop model of nuclear structure, 1169
Lithium, 1153
   atomic structure, 1131
   and β-decay, 1156
   photoemission from, 1113
   spectrum, 1119
Load, 376
Lodestone, 755
Long straight wire, magnetic field, 903
Loop, in a circuit, 858
Loop currents, 873
LORAN, 592
Lorentz contraction, see length contraction, 1075
Lorentz force law, 927, 967
Lorentz, Hendrik A., 927, 1072
Lorentz transformation, 1081–2
Loudness, 539
Loudspeaker, 592, 900, 932, 1023
Lubrication, 140
Lucid, Shannon, 597, 615
Luminosity, 1045
   Eddington, 1066
Lyman series, 1121

M

M87, 1109
Machines, 237–9
   hydraulic, 438
Macrostate, 656, 734, 741
MAGLEV, 973–4
Magnet
   bar, 755–6, 899
   permanent, 938, 940–2
Magnetic field
   in a coaxial cable, 910–11
   due to a current loop, 905
   due to a current sheet, 913
   of the Earth, 756–7
   of the galaxy, 1055
   due to a long wire, 903
   in a solenoid, 911–2
   of the Sun, 757
   units for, 900
Magnetic moment, **905–6**, 935–7, 938–41
   and angular momentum, 941
   of nuclei, 1151
Magnetic pole, 755
   of the Earth, 756–7
Magnetic resonance imaging, 943
Magnetism, 755
Magnetite, 755, 760
Magnetization, 938
Magnification
   angular, 619, 621
   linear, **612**, 614, 616–7
   longitudinal, 630
   of a microscope, 619
Magnifying glass (magnifier), 616, 618–9
Magnitude, of a vector, 32
Major axis, 9
Malaria, 1070
Malus, Etienne, 1051
Malus, law of, 1051
Manometer, 461
Mars, 6–8, 10, 13, 196, 199, 541
   trip to, 122–3
Marsden, Ernest, 1119
Maser, 1054–5
Mass
   atomic, 1147
   critical, 1169
   defect, 1152
   driver, 213, 269, 973
   electron, 1151
   flow, 209–12
   gravitational, 182–3
   inertial, 182–3
   "negative," 374
   in Newtonian physics, **133**, 150, 182–3, 434, 1071
   nuclear, 1148–9
   proton, 1151
   reduced, 665, 1140
   in special relativity, 1088, 1090–91
   and energy, 891, 1090–1
   unit, 26, 1147
   what is it?, 150
Mass-on-spring system, 261–3, 341–3
   and SHM, 472–6
Mass spectrometer, 925–6
Mästlin, Michael, 6
Material, magnetic, 938–40
Math Toolboxes
   Properties of the scalar product, 229–30
   Properties of the cross product, 299–300
   Harmonic functions, 474
   Sign conventions, 990
   How to solve a linear differential equation, 1005
Math Topic Boxes
   Formal evaluation of the limit, 97
   Use of calculus in circular motion, 98
   Integrals as areas, 251
   Work done around a closed path, 271
   General proof of the parallel axis theorem, 411
   Demonstration that $f(x \pm vt)$ solves the wave equation, 506
   Electric field as the gradient of potential, 826
   Group properties illustrated by an example, 1192
Matrix method for solving linear equations, 873
Maximum, principal, 584–5
Maxwell-Boltzmann distribution, 738
Maxwell, James Clerk, 754, 762–3, **954**, 956 975, 1040, 1067
Maxwell's equations, 954, **977**, 1070, 1072, 1193
McKee, Suzanne, 536
Mean free path, 740, 850
Measurement, 1134
Mechanical advantage, 237
Mechanical energy, 270–1
Mechanical equivalent of heat, 654
Mechanics
   classical (see Newtonian), 1, 150, 1067
   quantum, 1132–6
   statistical, 665–7, 734–743
   structure of Newtonian, 150
Mercury (element), 443
   inch or cm of, 443
   spectrum, 1119
   superconducting transition, 854
   in thermometers, 646, 685, 691
Mercury (planet), 1108
Meson, 1180, 1182, 1185
Metals
   electric current in, 849–50, 937
   heat conduction in, 703
   photoelectric properties, 1114–5
   properties, 769
Meter, definition, 23

Meters, electrical, 862–4
Metric system, 22. See also Units; SI.
Metrology, 21
Michelson, Albert, 573, 1072
Michelson-Morley experiment, 574–5, 1072
Micrometer, 22
Microphone, 765, 798
Microscopes, 619–20
   atomic force, 1144
   electron, 1127
   scanning tunneling, 1110–1, **1142–4**
   x-ray, 1070
Microstate, 656, 735, 741–2
Microwaves, 533, 1051, 1057–9
   cosmic, 709
Milky Way, xxix, 1055, 1124
Millikan, Robert
   and the electron charge, 772
   and the photoelectric effect, 1111, 1113
Minkowski, Hermann, 1069, 1077
Minor axis, 9
Mirror
   magnetic, 929
   parabolic, 606
   plane, 599–600
   primary, 623
   spherical
     concave, 603–4
     convex, 604–5
Mirror heating facility, 9
Modeling
   in kinematics, 58, 469
   and the role of mathematics, 15
   in problem solving, 117
Model
   ideal gas, 438–9
   liquid drop, 1169
   plum-pudding, 1119
   shell, 1161
   of space and time, 19–21
   standard, 1187–92
Moderation, in nuclear reactions, 1170
Modes
   of energy storage, 665–6
   in a wave guide, 1058–9
   wave, on a string, 512
Molecules
   angular momentum, 312
   in a breath, 31
   collisions of, 347–9, 739–41
   diatomic and polyatomic, 665–6
   in dielectrics, 887
   in a fluid, 433, 440
   in an ideal gas, 438–9, 651
   water, as dipoles, 804
Moles, 651
Moment of inertia. See rotational inertia, 397
Moment, magnetic, 905
Momentum
   angular, 294–301, 312
     in atoms, 1122, 1128

     and orbits, 329–31
     of a rigid body, 400
   and impulse, 200
   linear, 162, 199–215
   relativistic, 928, 972, **1088–90**
   of light, 1046–7, 1117
   of a system, 205–6, 311, 1093
   and wavelength, 1127
Momentum transfer, 200–4
Monochromatic light, 533, 1114
Monoplane, 413
Monopole. See magnetic pole, 755
Moon, xxix, 178, 199
   atmosphere, 737
   Galileo and the, 13
   gravitational field, 293
   gravity, 65, 135, 182
   phases of, 3
   physics on the, 3
   A trip around the, 179–80
Morley, Edward, 573
Motion
   Brownian, 667
   of charged particles, 800–2, 924–31
   circular, dynamics of, 145–8
   linear, dynamics of, 141–4
   linear, graphical analysis, 64–65, 68–69
   general, modeled with linear and circular, 101
   projectile, 84–93
   relative, 102–4
   of rigid bodies, 363, 391–6
   simple harmonic, 472
   simultaneous, 89
   thermal, 651, 734–41, 1088
   uniform circular, 93–100
   uniformly accelerated linear, 69
Motional emf, 961
Motions, coupled, 496
Motorcycle, 362–3, 371
Motor, electric, 761, **963–4**, 982, 986
MRI, 943
Muller, Richard, 928
Multiloop circuits, 859–62, 1006
Muon, 1074, 1141, 1160, 1182, 1186, 1189
Music, 495–6, 510, 512, 524, 525
Mutual inductance, 997–8

N

n-type semiconductor, 855
National Institute of Standards and Technology, 22, 65, 691
National standard kilogram, 26
Navigation, 102–4, 118–9
Near point, 618
Nebula
   horsehead, 1055
   keyhole, 1125
Neon
   atomic structure, 1131
   in stars, 1168
Neptune, 350

Network, of capacitors, 882–3
Neutrino, 1156–9, 1186
Neutron, xxxi, **759, 1150–1**
   discovery of, 1151
   and fission of uranium, 1169–70
   in nuclear reactions, 346
   quark content, 1185, 1188
   star, 1109
   and the strong force, 1181
Neutron activation analysis, 1166
Neutron bombardment, 1166
Newcomen, 641
Newton (unit), definition, 134
Newton, Isaac, xxix, 1, 9, 19
   and the apple, 134
   discoveries, 162–3
   and force, 131
   and the gravitational force, 178, 290
   and the laws of motion, 66, 126, 132, 199
   and light, 532–3, 534
   and sound waves, 528
   and telescopes, 623
Newtonian mechanics, structure, 150
Newton's first law, 126, 148–50
Newton's law of cooling, 711–2
Newton's laws, 126, 148
Newton's rings, 571
Newton's second law, **132–4,** 148–50, 248, 1071
   applied to a system, 209, 214–5, 309
   in special relativity, 1090
   in terms of momentum, 199
   in terms of torque, 303, 400, 413
Newton's theorem, 178, 195
Newton's third law, 130–1, 149–50, 454
   and conservation of angular momentum, 313–5
   and conservation of momentum, 204–5, 215
   and the electric force, 771
   and magnetic forces, 936
   strong form, 314–5
Nickel
   and magnetism, 941
   size of nucleus, 1150
Nitrogen
   molecules and specific heat, 665
   in stars (nuclear reactions), 1167–8
Nobel prize, 772, 938, 954, 1109, 1142, 1153, 1180
Noble gas, 1130
Nodal line, 563, 568
Node, **511,** 530–1, 1128
Noise, 541
Nonconservative force, 271, 275–8
Nonet, meson, 1183, 1184
Nonohmic devices, 854–5
Normal, to a surface, 548
Normal force. See force, 128
Northern lights. See Aurora, 757
Norton's theorem, 874
Nuclear decay, 345, 1154–63
Nuclear magnetic resonance. See MRI, 943

Nucleon, 1150, 1181
Nucleus
   atomic, xxxi, 943, 1121, 1146–53
      compound, 1166
      size, 1121, 1150
   comet, 198
   galaxy, 1039, 1109
Number
   atomic, 1147
   Avogadro's, 651
   quantum, 1128, 1154
   wave, 502

O

O'Neill, Gerard, 973
Objective, 619, 620
Objects, in optics, 600–1
Ocean, 442
Octet, baryon, 1183, 1184
Octet, meson, 1183, 1184
Oersted, Christian, 760, 901
Oersted's rule, 901, 903, 908
Ohm, Georg Simon, 846
Ohm (unit), 846
Ohm's law, 852
Ohmic losses, 847
Ohmmeter, 864
Omega particle, 1183–4, 1190
Onnes, H. Kamerlingh, 854
Optical axis, 603
Optical path, 559
Orbits, 8–10, 202, 329–31, 349, 1108
   atomic, 1122
   geosynchronous, 17, 184, 557
   of the space shuttle, 330–1
Order of a diffraction peak, 585
Order of magnitude, 30
Origin
   and angular momentum, 298, 300
   and torque balance, 364, 369–70
   of a reference frame, 35–36
Oscillation, 471–84, 1001–3
   coupled, 496
   damped, 487, 1003–5
   of an electron, 1040
   electromagnetic, 1057
   forced, 484–6
Oscilloscope, 801
Otto cycle, 721–2
Overdamped, 487, 1005
Overtone, 532
Oxides, metal, as superconductors, 854
Oxygen
   isotherms of, 679–70
   molecules, 665
   in stars (nuclear reactions), 1167–8

P

p-type semiconductor, 855
P-T diagram, 672, 680
P-V diagram, 656

Pair creation, 945, 1158–60
Palomar, Mount, 623
Parabola, 91–92, 482, 820
Paraboloid, 634
Paradox
   hydrostatic, 443
   Olbers', 561
   twin, 1087–8
Paraffin
   and Chadwick's experiment, 1151
   and Compton scattering, 1118
Parallel Axis Theorem, 410–11
Parallel connection in circuits, 856
Parallelogram rule, 35
Parallel plate capacitor, 876–7
Paramagnetism, 938–9
Parent (in nuclear decay), 1153
Parity, 613
Partial derivative, 505
Particle, 52, 150, 165, 199, 312, 334
   charged, 765
   in continuous systems, 361, 432
   "elementary," 892
   fundamental, 1178
   validity of model, 215, 309
   virtual, 1179
Parton, 1186
Pascal, Blaise, 432
Pascal (unit), definition, 436
Pascal's Principle, 438, 442
Paschen series, 1121
Past, 751, 1079
Path difference, 567
Path length, 32
   and interference, 571
   optical, 559
Pauli, Wolfgang, 1130
Pendulum
   physical, 478–9
   simple, 302, 469, **477–8**
      large amplitude, 494
Performance, coefficient of, 725
Period
   of circular motion, 94
   of an oscillation, 473
   of a planet's orbit, 9
   of a wave, 499
Periodic table, 1130–2, back cover
   analog with particle physics, 1184
Permeability, magnetic, 901, 942
Permittivity of free space ϵ_0, 766
Perpendicular axis theorem, 422
Perrin, Jean Baptiste, 667
Perspective, 13
Phase, 474
   change, 509, 571, 680–1, 1059
   constant, 474, 502
   diagram, 680
   difference, 563
   function, 502
   shift, 474, 571

   in AC circuits, 1020, 1022–3, 1026, 1027, 1029
   speed, 566
   transition, 680–1
   of a wave, 502–4
Phasor, **581**–3
   in AC circuits, 1025–9
Photocathode, 1112
Photoelectric effect, 1112–6
Photoelectron, 1112
Photography, 621–2, 1053, 1054, 1140
Photomultiplier tube, **1112**, 1159, 1163
Photon, 1111, 1113
   virtual, 1179
Physical integral, 249
Physics
   why do?, xxv
   what is?, xxv
Piano, 511, 565
Pion, 1157, **1180**
   in collisions, 1182
Pioneer spacecraft, 350
Pipe, organ, 530–2
Piston, 655
Pitch, American concert, 525
Pith, 759
Pitot tube, 456
Plan, 117
Planck, Max, 1067
   and blackbody radiation, 1118
   and quantization, 1111
   and the 2nd law of thermodynamics, 720
Planck radiation law, 709, 1118–9
Planck's constant, 1113, 1116
Plane of incidence, 548
Planets, 5, 6, 350
Plasma, 1118
   propagation of EM waves through, 1057
Plastic, dielectric strength, 885
Platinum, in thermometers, 699
Plutonium, 1146, 1170, 1175
Point
   critical, 679
   triple, 690
Point charge, 766
   electric field, 770
   potential, 818, 825
Point of application (of a force), 363
Polar coordinates, 25
Polarization of EM waves
   linear, 1050–3
   circular, 1056–7
   in nature, 1054–5
   by reflection, 1053–6
   by scattering, 1054
   of starlight, 1054–5
Polarization vector, 887–8
Polaroid, 1055. See also filter, polarizing.
Pole, magnetic, 755–6, 899
Polestar, 414
Position vector, 35–6

Position versus time graph, 68–9
Positron, 347, 1157–60
Postulate, quantum, 1111
Postulates of special relativity, 1072
Potential difference, 818, 876, 964–6
Potential, electric, 817–820
    due to a continuous charge distribution, 828–9
    in circuits, 860
    gravitational, 290, 1104
    Hall, 937
    stopping, 1116
Potential energy. See also energy, 259
    of α-particle in nucleus, 1155
    and equilibrium, 366
    and oscillations, 480–1
    and phase transitions, 680–1
    of proton and neutron, 1161
    and thermal expansion, 684
Potential, ionization, 1123
Potentiometer, 864
Pound, R.V., 1087, 1107
Power, 234–9, 1090
    in AC circuits, 1019, 1024, 1026
    in DC circuits, 847–8, 966
    in EM waves, 1045–6
    of an optical element, 606, 611
    and phasors, 1026
    in reflected and transmitted waves, 514
    in relativity, 1090
    in a rotating system, 304–5, 400
    in sound waves, 538
    in waves on strings, 507–8
Power curve, 235–7, 428–30
Power factor, 1024
Power plant, nuclear, 1145–6, 1170
Power transmission, 900, 956, 998, 1016–7
Poynting, John Henry, 1045
Poynting vector, 1045
Precession, 413–4
Precision
    and measurement, 22
    and significant figures, 26
Prefixes, SI, 28
Pressure, 435–40, 453
    atmospheric, 436–7
    gauge, 459
    in an ideal gas, 439, 649
    radiation, 1046–9
    ram, 455–6
    in sound waves, 527, 529, 530, 564
    in a van der Waals gas, 678
    variation in a fluid at rest, 440–2
Primary coil, of a transformer, 998
Primary mirror, 623
Principal maximum, 584–5
Principal quantum number, 1128
Principia, 19
Principle
    Archimedes', **445–9**, 476
    correspondence, 1126
    of equipartition, 666
    equivalence, 1104
    exclusion (Pauli), 1130
    Fermat's, 559
    Huygens', 548
    Pascal's, 438, 442
    uncertainty (Heisenberg), 1132–3, 1179
Prism, 533, 552
    erecting, 621
Probability
    and quantum theory, 1128, 1133–4, 1142, 1161
    and statistical mechanics, 735–6, 738
Problem solving steps, 117
Problems, ratings, xxvii
Process, thermodynamic, 656–8
Projectile, Galileo's work on, 10
Projectile motion, 84–93
    in Aristotle's physics, 5
    and orbits, 92
    range, 87
    shape of trajectory, 91
Projectiles and orbits (*Digging Deeper*), 92
Prominence, solar, 757
Proper distance, 1080
Proper time, 1080
Properties of potential energy, 261
Proportion, 9–10, 12
Proton, xxxi, 346, 759, 1091, 1181, 1185–6
    collisions of, 1093, 1146, 1177, 1180, 1182
    decay of, 556, 1178
Proton–proton cycle, 1092–3, **1167**
Ptolemy, 6
Pulleys, 173–5, 238, 371, 403–4
Pulse, 497
Pump, suction, 432
Pupil
    exit, 620–1
    of the eye, 536, 621
Purcell, Edward M., xxv
Pyramids, 369, 414
Pythagoras, and stringed instruments, 4
Pythagoras' theorem, 39, 49, 1080
Pythagorean solids, 7

Q

QED, 1180, 1189
Quality factor, 1030
Quantity of motion, 12, **199**. See also energy, momentum.
Quantization, 666, **1111**, 1122
Quantum mechanics, 1132
    origins of, 1111
Quantum number, 1123, 1128, 1154, 1185
    principal, 1128
Quark, xxxi, 1177, 1184–5
Quark-search experiment, 772
Quarter-wave plate, 1056
Quasar, 1039–40, 1107, 1197
    and redshift, 546
Quate, Calvin, 1144

R

R-factor, 705
Rabi, I. I., 1145, **1189**
Radar, 547, 1058
Radian
    and circular motion, 94, 393
    definition, 24
Radiation
    cosmic, 709
    thermal (blackbody or cavity), 709–11, 1118
Radiation pressure, 1046
Radio
    AM, 533
    FM, 533
Radio source, 546, 1055
Radioactive dating, 1165
Radioactivity, 1153
Radium, 345–6, 1163
Radius
    Bohr, 1124
    classical electron, 891
    of curvature, 101, 146
    Larmor, 925
    of nucleus, 1121, **1150**
    Schwarzschild, 1109
Radon, 345–6, 814
    and air pollution, 1166
Rainbow, 523, 553, 560
Ramp, as a machine, 237–9
Random motion, 439, 741
Random walk, 741
Range
    of penetrating radiation, 1153
    of a projectile, 87
    of the strong force, 1180
Rapidity, 1100
Rate, decay, 1161
Ray, 538, 547, 598, 617
    α, β, γ, 1153
    construction, 603, 613–7, 619, 620
    paraxial, **603**, 613, 624
Ray tracing
    rules for lenses, 615
    rules for mirrors, 614
    with a computer, 615, 624, **634–8**
Rayleigh, Lord John William Strutt, 578
Rayleigh's criterion, 579
Rayleigh-Jeans law, 1119, 1126
Reactance, 882, **1020**, 1024
    in parallel, 1027–8
    in series, 1024, 1026
Reactions, nuclear, 1091
    chain, 1169
Reactor, nuclear, 929, 1146, 1152
    pressurized, 1169–70
Rebka, G. A., 1087, 1107, 1197
Recoil, nuclear, 1160
Redshift, 546
    gravitational, 1106
Reference frame, **19**, 35
    accelerated, 149, 1103–4

Reference frame, (*continued*)
  body, 391
  center of mass, 310
  and emf, 966
  inertial, 148–50, 1072, 1104
  noninertial, 149, 1103–4
  relatively moving, **104,** 118–9, 310, 763, 1070–71
  rest frame, 1094
  which is easiest, 105
Reference level, 264, 266
Reference point for torques, choice of, 364
Reference state, 261, 267, 816
Reflection
  law of, 548
  of microwaves in a guide, 1059
  and polarization, 1053–6
  of sound and light, 547–9
  total internal, 550–2
  of waves at a boundary, 508–10, 513–4, 548
Refraction of sound and light, 549–50
Refraction, index of, 536–7, 550, 571
Refractive index. See index of refraction, 536
Refrigerator, 724–6
Reines, 1157
Relativity
  Galilean, 20, 148, 1070–72
  General, 1069, 1103–9
  Special, 545, 1072–9
    postulates, 1072
  special principle of, 1071
Renormalization, 892, 1193
Resistance
  electrical, 846
    in AC circuits, 1017
    internal, 847
    thermal, 705–6
Resistivity, 851–2
Resistors
  circuit symbol, 845
  commercial, 852
  in parallel, 856–8
  in series, 853–5
Resolution, 578–9
Resolving power, 585, 594
Resonance
  in oscillations, 484–5, 1029
  particle, 1182–3
Retina, 536, 599, 618
Retrograde motion, 5
Reversibility, 72, 719, 726, 729, 751
Right hand rule
  and angular momentum, 296
  and angular velocity, 392
  and cross products, 297, 901
  and magnetic fields, 901, 905
  and torque, 301
Rigid body
  definition, 361, 363

  dynamics, 400–406
  limitations, 372–3, 406
  rotation, 99–100, 391–5
Rigidity
  of a body, 363
  of a particle, 1101
Ring, Rowland, 912
Rings
  Newton's, 571
  Saturn's, 329
Ripples, 547, 1110
Ritter, 533
Rms, 651, 1019
Road race, 52–7, 61–2, 101
Rock climbing, 142–3, 165, 168–9, 384
Rock, speed of sound in, 560
Rockets, 210–2
Rod
  in the eye, 536
  control, 1170
  fuel, 1170
Roemer, Olaus, 534
Rohrer, Heinrich, 1142
Roller coaster, 257–8, 265–6
Rolling without slipping, 106, 394–5, 402–3
Röntgen, 533
Root mean square. See rms, 651, 1019
Rope. See string, 170
Rotational inertia, 396–400
  calculation of, 409–13
  definition, 397
  table, 398–9
Rowland ring, 912, 917, 1011
Rumford, Count, 642, 653
Rutherford, Ernest, 1068, 1119, 1149, 1186
Rydberg, 1124

S
Safety considerations in electricity, 848
Sailing, 34, 39, 40, 431–2, 464
Salaam, Abdus, 1188
Salt, 587
San Francisco State University, 772
Santa Ana winds, 643–644, 663
Satellites, 92
  de-spinning of, 407–8
  geosynchronous, 184
  pointing of, 936
  tethered, 983
Saturation, of magnetic field, 942
Saturn, 163, 329, 350
Saturn V rocket, 28
Savart, Felix, 901
Scalar, 32, 41
Scalar product. See dot product, 228
Scalar triple product, 327
Scale height of the atmosphere, 444
Scales, calibration of, 136
Scanning tunneling microscope, 1110–1, **1142–4**

Scattering
  Compton, 1117
  Rutherford, 1121
  of sunlight, 1054
Schrödinger, Erwin, 1128, **1134**
Schwarzschild, Karl, 708, 1109
Scott, David, and hammer experiment, 3
Seat (source) of emf. See emf, 845
Second, definition, 23
Secondary coil, of a transformer, 998
Self-energy, 890–2
Self-inductance. See inductance, 993
Semiconductors, 769, **854–5,** 1143
  devices in electrical meters, 862–4
Semimajor axis, 9, 330
Sequence, triboelectric, 769
Series connection in circuits, 853
Series, radioactive, 1164–6
Setup of solution, 117
Shane telescope, 632
Shell, closed, 1131
Sherwin, Chalmers, 1087
Shock absorber, 486–7
SI, 21–31
Sidereal period, 10
Sigma particle, 1080, 1185
Sign conventions
  in Ampere's law, 909
  in circuits, 859, 990
  in Faraday's law, 960
  in optics, 601
  in thermodynamics, 654
Significant figures, 26–27
Silicon
  crystallization experiment, 9
  in semiconductors, 855
  in stars, 1168
  STM image, 1142–3
Silver, size of nucleus, 1149
Simple harmonic motion, 472–7, 503
Simultaneity, 1075–7
Siphon, 453–4
Size, angular, 618
Skating, 206, 294–5, 315–7
Skiing, 84, 91–3, 143–5
Skylab, 133
Slope of a graph, 68
Slug, 156
Small angle approximations, **97,** 478, 490, 506
Snell, Willebrord, 550
Snell's law, 550, 607
Snider, 1107
Soap, 571–2, 588
  solution, flow of, 449
Sodium chloride. See salt, 587
Sodium D lines, 585–6, **1128**
Soft, magnetically, 941
Solar constant, 1045
Solar system, xxix, 5, 6, 350
Solar wind, 347, 757, 923–4

Solenoid
  inductance of, 994
  magnetic field inside, 911–2
Solid
  versus liquid, 433
  specific heat, 667, 688–9
  sound in, 524, 529
Solution, trial, 117, 120
Sound, 524
  intensity level, 539
  pressure level, 560
  quality, 512
  speed, 528–9
Sound waves
  basic relations, 525–7
  standing, 530–2
Space, definition, 19
Space colonies, 180, 183–5, 973
Space shuttle, 212, 443
  acceleration of, 59
  experiments on, xxix, 9, 136
  orbit, 330–1
  speed of, 176
  tiles, 702
Spacelab I, 9, 136
Spacelike, 1079
Space-time, 1077–80
  diagram, 1077–8
  interval, 1079–80
Span, 376
Spark coil, 1000
Spark, electric, 765
Spark plug, 1000
Species, nuclear, 1147
Specific gravity, 434–5
Specific heat
  definition, 658, 687
  at constant pressure, 659–60
  at constant volume, 658–9
  and internal energy, 659
  molar, 661, 665, 667, 669, **688**
  ratio, 660
  of a polyatomic gas, 666
  table of values, 688
Spectrometer, 588
Spectrum
  acoustic, 525
  atomic, 1119
  electromagnetic, 532–3, 585
  of hydrogen, 1121
  produced by a prism, 552
Speed
  angular, 94, 392
  average, definition, 53
  in circular motion, 95
  drift, 849
  escape, 184–5, 268, 1108
  group, 566
  instantaneous, 56
  molecular, 651–2

phase, 537, 566
  of light, 534–7, **1043**
  of rocket, 211–2
  of sound, 528–9
  wave, 499
Spherical polar coordinates, 25
Spin, 941, 1129, 1181
  isotopic (see also isospin), 1192
  nuclear, 943
Spring constant, 137
Spring force. See Hooke's Law, 137
SQUID, 699
Stability. See also equilibrium, 366–7, 448, 471
  of the atmosphere, 683, 708
  nuclear, 1147, 1152
Standard, and units, 21
Standard Model, 1187–92
Standards, temperature, 646–7, 690–1
Standing waves
  and atoms, 1126, 1128
  electromagnetic, 1057
  sound, 530–2
  on a string, 510–3
Stanford Linear Accelerator Center, 1186
Stapp, John P., 59, 132–3
Star
  neutron, 468, 900, 1108
  red giant, 1168
  white dwarf, 1106, 1130
Star-delta transformation, 874
Stark, Johannes, 1111
Stars
  energy generation in, 1167–8
  radiation pressure in, 1046
  universe of, xxix, 1067
State
  excited, 1123, 1183
  ground, 1123
  nuclear, 1161
  thermodynamic, 656
State variable, 656
  entropy as, 728–9
Statics
  of particles, 166
  of rigid bodies, 363–9
Statistical mechanics, 665–7, 734–743
Steady flow
  of fluids, 449
  of heat, 703–5
Steady state response of AC circuit, 1021–4
Stefan-Boltzmann law, 710
Step-up and step-down transformers, 998, 1016
Steradian, 782
Stokes, Sir George, 975
Stonehenge, 3
Strangeness, 1183
Streamline, 449–50, 454
Stream tube, 450

String
  conservation of, 168
  ideal, 170–1
  massive, 172–3
  in particle physics, 1199
  waves on, 497, 505–7, 510–515
String tension, 128, 170–5
Stromer, 647
Study problems,
  An Accident in the Lock, 448–9
  The Betatron, 970–2
  Bungee Jumping, 482–4
  The Collapsing Square, 824–5
  A Compound Circuit, 860–2
  A Cycle in the P-V diagram, 663–5
  Delivering the Mail, 212–4
  The Egg factory, 276–9
  An Electron Beam, 914–6
  F-Stops, 621–2
  Hotdog Skiing on Spring Snow, 143–5
  Lunch at Noon?, 118–9
  Mars or Bust!, 122–3
  Rockfall!, 404–6
  Slugger Jose's Pop Fly, 119–22
  The student's revenge, 1083–5
  A Thermal Switch, 686–7
  A Trip around the Moon, 179–80
  The Two Skaters, 315–7
  Two Unequal Charges, 776–7
Subatomic world, xxxi, 1110 et. seq.
Subroutine
  in solution plans, 116
Sun, xxix
  convection in, 707
  density, 434
  eclipse of, 1069–70
  effect on Earth, 347, 707
  energy production in, 1091, 1158–9, 1167–8
  intensity, 541, 710, 1045
  magnetic field of, 757
  and planetary orbits, 163, 298, 349
  radius, 710
  Schwarzschild radius, 1109
  spectrum, 533, 709
  temperature, 709
Sunspot, 900
Superconductivity, 698, **854**
Superfluid, 698
Superluminal motion, 1197
Supernova, 1159, 1168
Superposition
  of particle states, 1184, 1185
  principle of
    for waves, 508, 1132–3
    for electric field, 771
    for magnetic field, 902
    for potential, 821
Superstring, 892, 1199
Surface
  closed, 780

Surface, (*continued*)
 Gaussian, 792
 optical, 600
 refracting, 607–8
Susceptibility
 dielectric, 887–8
 magnetic, 941–2
Switch
 circuit symbol, 845
 thermal, 686–7
Symmetry
 and angular momentum of bodies, 400
 and Gauss' law, 791–2
 in physical theories, 613, 1192
 time-reversal, 72, 751
Symmetry breaking, 613, **1193**
Synchronization of clocks, 1075–6, 1078
Synchrotron, 347, 928, 1070
Synodic period, 10
System
 of particles,
  angular momentum of, 313–5
  center of mass, 307–8, 374
  dynamics of, 165–6
  electric potential energy of, 821–5
  kinetic energy of, 233–4
  momentum of, 205–15
  potential energy of, 273–4
 of rigid bodies, 371–2, 406
 thermodynamic, 644

T

Tachyon, 1092, 1197
Tail, exponential, 989
Tangent line, and velocity, 68
Tangent substitution, 795
Tau, 1182, 1186
Taylor, Joseph, 1109
TE mode, 1059
Technetium, 1163
Telephoto lens, 631
Telescope
 Cassegrain, 623
 Galilean, 631
 Hubble, 624, 637–8
 Keck, 624
 Keplerian, 620
 Newtonian, 623
 radio, 598
 reflecting, 622–4
 Shane, 632
Television
 brightness control, 820
 electron beam in, 914–5, 1091, 1127
 electric fields in, 790–1, 801
 frequency of transmission, 533
 magnetic fields in, 899, 900
TEM mode, 1058
Temperature
 as scalar, 32
 brightness, 710

color, 709
critical, 679
Curie, 940, 1193
Debye, 667, **688–9**
definition, 646
during a phase transition, 680–1
in an ideal gas, 650
low, 698–9
measurement of. See thermometry, 690
and nuclear reactions, 1167
Temperature scale, 646–8
 Celsius (centigrade), 646–8
 Fahrenheit, 646–8
 International practical, 691
 Kelvin, 648
 Rankine, 648
 Thermodynamic, 733–4
Tension
 in beams, 363, 377–8
 in strings, 128, **170–1**, 497, 506–7
Terminal voltage, 846–847
Tesla, Nikola, 900, 1017
Tesla (unit), 900
Test charge, 769
Tevatron, 1177
Text, suggestions for using, xxvi
Thales, and Ionian philosophy, 4, 758
Theodolite, 24
Theory
 and experiment, 14
 physical, 150
Thermocouple, 692, 696
Thermodynamics
 first law, 654
 second law, 719–20, 726
 third law, 734
 zeroth law, 645
Thermometer, 646
 constant-volume gas, 691–2, 699
 germanium semiconductor, 699
 mercury-in-glass, 646, 685, 691, 696
Thermometry, 690–2, 698–9
Thevenin's theorem, 873–4
Thompson, Benjamin. See Rumford, 642
Thompson, J.J., 930
Thomson, William. See Kelvin, 648
Threshold of hearing, 539
Thrust, 128
Thunderstorm, 676–7, 683, 754
Tides, 23, 195, 360
Time
 arrow of, 20, 743, 751–2
 Newton's definition, 20
 proper, 1080
 unit, 22
Time average, 508, 1019
Time constant, 989, 999
Time dilation
 gravitational, 1104–5
 in special relativity, 1073–4
Timelike, 1079

TM mode, 1059
Tokamak, 898, 900, 929
Top, motion of, 413
Top-down reasoning, 116
Topness, 1185
Torque, 301–4
 and angular acceleration, 400, 477
 and angular momentum, 303, 315
 as a cross product, 302
 on a current loop, 934–7, 963–4
 on a dipole, 803
 on a wire segment, 931
Torque balance, 306–7, 363, 368, 445
Torque ratio, 428
Torricelli, Evangelista, 432
Torricelli's law, 454
Total internal reflection, 550–2
Tracks, particle, 347, 1146, 1150, 1154, 1156, 1157, 1177, 1182, 1184
Trains, 105, 166–8, 209, 252–3, 718–9, 973
Trajectory
 parabolic, 91
Transformation law
 for electric field, 966, 1095
 Lorentz, 1081–2
 for momentum and energy, 1095
 for velocity, Galilean, 1071
 for velocity, relativistic, 1085–6
Transformer, 974, 998, 1016
Transient response, 1025
Transition, in an atom, 1123
Transmission axis of polarizer, 1051
Transmission of waves across a boundary, 513–4
Transmutation, 1166
Trapezoidal rule, 223
Trial solutions, 117
Triboelectricity, 768
Triple alpha process, 1168
Triple point, 690–1
Tritium, 346, 1048, 1152, 1169
Truth
 and physical theory, 14
 quantum number, 1185
Tug-of-war, 129
Tunneling, quantum mechanical, 1142, 1155
Turbulence in fluid flow, 449
Turn, standard rate, 147
Tweeter, 1023

U

UHF, 533
Ultrasound, 524
Ultraviolet, 533, 1119
Uncertainty in a measurement or number, 26–7, 57, 1132
Uncertainty principle, 1132–3, 1179
Underdamped, 487
Unification, xxxi, 754, 763, 1187, 1188, 1199
Uniform circular motion, 93–100
Uniformly accelerated linear motion, 69–73, 119–21

Unit vector, **39–40,** 62, 248, 302
Units
   cgs, 1113
   conversion, 28
   derived, 134
   International System, 21
Universal law, 12
Universe
   Aristotle's model, 5–6
   expansion of, 546, 752, 1068, 1109, 1199
   overview, xxvii–xxxi
   Ptolemy's model, 6
   as a reference frame, 149
Unpolarized light, 1050
Up (quark), 1185
Uranium, 741, 1153, 1166, 1169–70, 1175
Uranus, 350

V

Van Allen belts, 757, 925, 944
Van de Graaff, 841, 922
Van der Waals, Johannes, 677
Van der Waals model for gases, 677–80, 802, 1144
Vaporization, latent heat of, 682
Variable mass systems, 208–212
Vector
   addition, 33
   algebra, 41
   components of, 37
   cross product, 297–300, 899
   differentiation, 62–63, 98
   dot product, 228–30
   four-dimensional, 1092, 1102
   integration, 66–7, 233
   magnitude, 32
   multiplication by a scalar, 34
   notation, 32, 38, 40
   position, 35
   preliminary definition, 31
   subtraction, 34
   unit, 39
   zero vector, 33
Velocity
   angular, 392
   average, definition, 54
   drift, 849, 901
   instantaneous, 55–58
   instantaneous, in circular motion, 96
   and power, 237
   relative, 104
   in sound waves, 527
Velocity selector, 927–8
Velocity transformation
   Galilean, 104
   special relativistic, 1086
Velocity versus time graph, 69
Venus
   and the Galileo spacecraft, 333, 350
   and tests of GR, 1108

Verne, Jules, 179
Vertex, 1179
Very Large Array (VLA), 595
Vibration, 540, 665, 1199. See also oscillation.
Violin, 512
Visibility, limits of, 617
Vision, 536
Visualization, 117
VLA, 595
VLF, 533
Volt (unit), definition, 818
Volta, Alessandro, 818
Voltage
   high, 848
   household, 1019
Voltage divider, 863–864
Voltmeter, 863–864
Volume, 434, 446, 448, 649
   as a vector product, 327
   displaced, 445
Vomit comet, 196, 1104
Von Drais, Karl, 426
Von Guericke, Otto, 437
Von Klitzing, Klaus, 938
VOR, 595
Vortex, 465
Voyager spacecraft, 350

W

W particle, 1187–9
Walk, drunkard's or random, 741
Water
   analogy for electric circuits, 844–5
   analogy for radioactive series, 1165
   as a conserved substance, 197
   as a particle detector, 1158–9
   and definition of gram, 26
   dielectric properties, 885
   in Greek physics, 4
   molecules, 804
   in nuclear reactors, 1170
   ocean, 442
   phase transitions in, 681
   and specific gravity, 434
   specific heat, 688
   supercooled, 742–3
   and temperature scales, 647–8
   thermal expansion of, 686
   triple point, 691
   waves on, 496, 500, 521
Watt, James, 234–235, 641
Watt (unit), definition, 234
Waves
   De Broglie, 1126–7
   earthquake, 181
   electromagnetic, 762–3, 1039–60
   electron, 511, 1126, 1134
   gravitational, 1109
   harmonic, 502–4
   heat, 698
   longitudinal, 500

   mathematical description of, 501–2
   matter, 1128
   on springs, 500
   on strings, 497–500
   on water, 496, 500
   periodic, 498
   plane, 547, 1041
   sound, 525–7, 574
   standing, 510–2, 530–2, 1057, 1126
   transverse, 500
Wave equation, 505–6
   for EM waves, 1041–3
   for sound waves, 531
   for waves on a string, 506–7
Waveform, 498
Wave front, 537–8, 542, 544, 547, 606
Wave function, 501–2, 1133–5
   for α particle, 1155
   atomic, 1128
Waveguide, 1058–60
Wavelength, 499, 542–7, 550, 598, 1059
   Compton, 1118
   cutoff, 1059–60
   de Broglie, 1126–7, 1186
Wave number, 502
Wave-particle duality, 1118, 1132
Wave pulse, 497, 1133
Wave speed, 497
Wave vector, 1040
Weber, Wilhelm, 957
Weber (unit), 957
Weight
   atomic, see atomic mass, 1147
   definition, 127
   distinction from mass, 135
   perception of, 147–8
   practical formula for, 135
Weightlessness, 178
Weinberg, Stephen, 1188
Wheatstone Bridge, 864–5
Wheatstone, Charles, 846, **864**
White dwarf star, 1106, 1130
Whitehead, Alfred, 3
Wien displacement law, 709
Wind, 545
Wind, solar, 347, 757
Window, 610, 705, 1053
Wollaston, 533
Woofer, 1023
Work, 226–33
   and emf, 966
   formal definition, 233
   and potential energy, 261
   unit, 226
Work done
   by the Coulomb force, 814–5
   by a fluid, 451
   by the gravitational force, 263, 266–7
   by an ideal gas, 655
   by nonconservative forces, 275
   by a spring, 252–3, 260–1

by a variable force, 232, 252–3
in an adiabatic process, 662
in an isothermal process, 655
in a rotating system, 304–5
in thermodynamics, 654–8
Work-energy theorem, 226, 233
   applied to fluids, 451
   in special relativity, 1090
   in thermodynamics, 653
Work function, 1113, 1115–6
World line, 1077
Wu, C. S., 613

X

x rays, 533, 1070
   from a black hole, 1109
   and the Compton effect, 1116–8
   diffraction, 587–8
   source, 1109
   speed in glass, 566

Y

Yang, C. N., 613
Young, Thomas, 570
Yukawa, Hideki, 1180

Z

Z particle, 347, 1187–9
Zeeman effect, 1130
Zeeman, Pieter, 1130
Zero
   absolute, 648, 691, 734
   of electric potential, 816, 818
   of potential energy, 266
   vector, 33
Zook, Alma, 512
Zweig, Stephan, 1184

PHOTO CREDITS

PROLOGUE xxv Brookhaven National Laboratory; **xxviii (top left)** © Brian Yarvin 1994/Photo Researchers, Inc.; **(top right)** © Renee Lynn/Photo Researchers, Inc.; **(bottom)** © David Nunuk/Science Photo Library/Photo Researchers, Inc.; **xxix (top left & top middle)** NASA; **(bottom left)** NOAO; **(bottom middle)** Painting by Alfred Kamajian, from "How the Milky Way Formed" by Sidney Van den Bergh and James E. Hesser. © 1993 by Scientific American, Inc. All Rights Reserved; **(bottom right)** Alan Dressler (Carnegie Institution) and NASA; **xxx (top)** M. J. Geller, J. P. Huchva and V. deLapparent (graphics by M. J. Geller and E. E. Falco) © Smithsonian Astrophysical Observatory; **(bottom left)** © A. B. Dowsett/Science Photo Library/Photo Researchers, Inc.; **(bottom right)** IBM Corp., Research Division, Almaden Research Center.

PART I 1 Rockwell International/NASA; **(inset)** NASA.

CHAPTER 0 2 NASA; 3 **(top)** Agence Photographique de la Réunion des Musées Nationaux; **(bottom)** © English Heritage/Photographic Library; 5 Erich Lessing/Art Resource, New York; 6 © John W. Allen; 7 **(top)** The Bettmann Archive, Inc.; **(left)** Science Photo Library/Photo Researchers, Inc.; **(right)** © Owen Gingerich. By permission of the Houghton Library, Harvard University; 8 © Owen Gingerich. By permission of the Houghton Library, Harvard University; 9 Deutsche Forschungsanstalt für Luft-und Raumfahrt e.V.; 10 Science Photo Library/Photo Researchers, Inc.; 13 **(top left)** © Owen Gingerich, 1981, by permission of the Houghton Library, Harvard University; 13 **(top right)** Authors; **(bottom left & right)** Scala/Art Resource, NY.

CHAPTER 1 18 By Chesley Bonestell/Space Art International; 20 The Granger Collection, New York; 22 **(top)** © Charles M. Falco/Photo Researchers, Inc.; **(bottom)** © Courtesy of the National Institute of Standards and Technology; 23 **(both)** © Courtesy of the National Institute of Standards and Technology; 24 Photo courtesy of the design consulting firm Greenhorne & O'Mara, Inc.; 26 National Institute of Standards; 29 **(top)** NASA; **(bottom)** AIP Emilio Segrè Visual Archives; 47 Appeared in "Why Round the Answer?" by D. L. Mathieson, *The Physics Teacher*, Oct. 1990, p. 471; 48 © Dorval/Explorer/Photo Researchers, Inc.

CHAPTER 2 51 © Jo Lillini, Agence Vandystadt/Photo Researchers, Inc.; 59 **(left)** Smithsonian; **(right)** © Deborah Davis/PhotoEdit; 65 James E. Faller/NIST; 67 © 1994 Ken M. Johns/Photo Researchers, Inc.; 71 © The Harold E. Edgerton 1992 Trust, courtesy of Palm Press, Inc.

CHAPTER 3 83 © Jean Marc Barey/Agence Vandystadt/Photo Researchers, Inc.; 84 **(top)** © Richard Megna/Fundamental Photographs; **(bottom)** David Young Wolff/PhotoEdit; 93 **(top)** © Mike Fizer. All rights reserved; **(bottom)** Tom Nebbia/Palm Press, Inc.; 99 Authors/Scottish Heavyweight Champion Francis Brebner throwing the hammer; 102 © 1990, Richard Megna/Fundamental Photographs.

CHAPTER 4 125 © Classic Bike/EMAP Archives; 126 Courtesy of the Trustees of the Portsmouth Estates; 129 © Stephen Dalton/NHPA; 132 AP/Wide World Photos; 133 NASA, Johnson Space Center; 135 NIST; 136 **(both)** NASA; 139 F. P. Bowden, D. Tabor, *Friction and Lubrication of Solids*, Oxford University Press © 1950. By permission of Oxford University Press; 140 Reuters/Bettmann.

ESSAY 1 162 Courtesy of the Royal Ontario Museum.

CHAPTER 5 172 **(left)** Jonathan Blair, © National Geographic Society; **(right)** © Mirror Syndication International; 173 West Publishing; 179 Bettmann Archive; 191 © Earl Zwicker, from *The Physics Teacher*, January 1987, p. 59; 196 Michael Lampton.

PART II 197 © The Stock Market/Globus Brothers, 1989.

CHAPTER 6 198 Roger Ressmeyer, © 1990 Corbis; **(inset)** © 1986 Max-Planck-Institut für Aeronomie, Lindau/Harz, Germany, by courtesy of Dr. H. U. Keller; 200 © The Harold E. Edgerton 1992 Trust, courtesy of Palm Press, Inc.; 212 NASA; 215 NOAO; 220 © Dennis MacDonald/PhotoEdit.

CHAPTER 7 224 Boeing Aircraft Corp.; 228 © Tony Freeman/PhotoEdit; 235 © The Granger Collection, New York; 237 © Tom Pantages; 242 NASA.

CHAPTER 8 257 © Busch Entertainment Corp.; 274 © 1992 Richard Megna, Fundamental Photographs.

ESSAY 2 290 © Giraudon/Art Resource, NY; 292 USGS.

CHAPTER 9 294 © David Madison 1994; 295 © David Madison 1994; 302 John Burke; 310 © 1993 Barry Rosenthal/FPG; 324 © Vandystadt/Photo Researchers, Inc.

ESSAY 3 329 NASA; 331 NASA.

CHAPTER 10 335 © 1995 Comstock; 336 © Bernard Asset, Agence Vandystadt/Photo Researchers, Inc.; 345 Photograph of model of a solid, assembled for "inelastic" collisions. Figure 2 in "Elastic and Inelastic Collisions" by Uri Ganiel, *The Physics Teacher*, January 1992, p. 19; 347 **(top)** Photo courtesy of CERN, Geneva; **(bottom)** Pekka Parviainen/Science Photo Library/Photo Researchers, Inc.; 359 English Heritage; 360 Space Telescope Science Institute/NASA.

PART III   **361** © Tad Stamm. All Rights Reserved. Picturesque.

CHAPTER 11   **362** West Publishing; **(inset)** © Tom Pantages; **363** Susan Lea; **382** © Reuters/Bettmann.

CHAPTER 12   **390** © 1990 Comstock; **394** © FPG International 1936; **407** Permission granted by the Courier-Journal, Louisville, KY; **413** © Tom Pantages; **414 (left)** © Jean Kugler/1986 FPG International; **(right)** Reprinted by permission from page 24 of *The Dynamic Universe: An Introduction to Astronomy*, 4/e, by Theodore P. Snow, Copyright © 1991 by West Publishing Company, All rights reserved; **424** © David Young-Wolff/PhotoEdit.

ESSAY 4   **426** © Scott Markewitz 1992/FPG International; **427 (top)** © FPG International; **(middle)** © C. Michael Lewis; **(bottom)** Peter Marlow/SYGMA; **428** Scott Markewitz 1991/FPG International.

CHAPTER 13   **431** © Merpool/Photo Researchers, Inc.; **432** © The Granger Collection, New York; **437** Photo Deutsches Museum Munchen; **446** © Hubertus Kanus/Photo Researchers, Inc.; **449 (top)** © Maarten Rutgers and Xiao-lun Wu **(bottom)** ONERA; **454** U.S. Navy; **455 (left)** P. Bradshaw, Stanford University; **(right)** Courtesy Thomas J. Mueller/University of Notre Dame; **456** ONERA; **462 (top)** © Tom Pantages; **(bottom)** © Robert Brenner/PhotoEdit; **465 (both)** ONERA.

PART IV   **469** © Farley Lewis/Photo Researchers, Inc.

CHAPTER 14   **470** © Tom Pantages; **482** © Reuters/Bettmann; **486** UPI/Bettmann; **488** © Tom Pantages; **493** © Deborah Davis, PhotoEdit.

CHAPTER 15   **495** © Tony Freeman/Photo Edit; **496** Alan Oddie/PhotoEdit; **498** From *Inquiry Into Physics*, by Ostdiek and Bord, © 1995 West Publishing Company, All rights reserved; **500** © Tom Pantages; **511** © Tom Pantages; **512 (both)** © Tom Pantages.

CHAPTER 16   **523** Jane Preston; **524** Shuleen Martin; **532** From "The Physics of Organ Pipes" by N. H. Fletcher and S. Thwaites. *Scientific American* January 1983, p. 97; **533** The Granger Collection, New York; **536** The Smith-Kettlewell Eye Research Institute pamphlet; **537** Alan Oddie/PhotoEdit; **538** © Jack Stein Grove/PhotoEdit; **540** A. J. Hudspeth and R. A. Jacobs; **546** Alan Bridle/NRAO; **547** © Erich Schrempp/Photo Researchers, Inc.; **548** © Tom Pantages; **549** © Tom Pantages; **551 (top)** © 1990 Richard Megna/Fundamental Photographs; **(bottom)** Amnon Yariv; **552** © David Parker/Science Photo Library/Photo Researchers, Inc.; **558** © 1994 Comstock, Inc.

CHAPTER 17   **562** © Raymond F. Newell; **563** © 1986 Richard Megna/Fundamental Photographs; **569** © Tom Pantages; **571** © 1987 Ken Kay/Fundamental Photographs; **573** Picture from article by G. Ramme: "Colors on Soap Films—An Interference Phenomenon" in *The Physics Teacher*, October 1990, p. 480; **574** Special Collections Division, Nimitz Library, United States Naval Academy, Courtesy Hale Observatories; **576 (all)** © Tom Pantages; **578** © Tom Pantages; **579 (all)** © Tom Pantages; **583 (both)** © Tom Pantages; **585** © Tom Pantages; **587** PSSC PHYSICS, 2/e, 1965; D. C. Heath & Company with Educational Development Center, Inc., Newton, MA. By permission of D. C. Heath & Company; **588 (left)** Kodak RL/Visuals Unlimited; **(right)** © C. Canizares, MIT; **592** © Dr. Brian J. Thompson, Provost Emeritus; Distinguished University Professor; **594** The Observatories of the Carnegie Institute of Washington; **595 (left)** R. J. Nelmes and M. I. McMahon, "High-Pressure Powder Diffraction on Synchrotron Sources", *Journal of Synchrotron Radiation*, Vol. I, Part 1, October 1994, p. 71, Figure 2; **(right)** NRAO.

CHAPTER 18   **597** NASA; **598 (both left)** © Tom Pantages; **(right)** NRAO; **599** © Jeff Isaac Greenberg/Photo Researchers, Inc.; **600** © Tom Pantages; **603** © Tom Pantages; **604** © Tom Pantages; **605 (left)** © Tom Pantages; **(right)** © 1993 Richard Megna/Fundamental Photographs; **613** © 1992 Martin Hough/Fundamental Photographs; **620** © Tom Pantages; **623** From *Science Year*, The 1969 World Book Science Annual. © Field Enterprises Educational Corporation. By permission of World Book, Inc.; **624** Roger Ressmeyer, © 1992 Corbis; **629** © Robert Greenler/Robert Greenler Sky Photos.

ESSAY 5   **634** NASA.

PART V   **640** © Bonnie Sue Rauch/Photo Researchers, Inc.; **641 (top)** Courtesy of Central Scientific Company; **(bottom)** The Bettmann Archive; **642** © The Granger Collection, New York.

CHAPTER 19   **643** © Will & Demi McIntyre. All Rights Reserved. Picturesque; **644** © Tom Pantages; **648** Science Photo Library/Photo Researchers, Inc.; **654** The Science Museum/Science & Society Picture Library; **667** From "Brownian Motion" by Bernard H. Lavenda, *Scientific American*, February 1985, p. 71.

CHAPTER 20   **676** © Uniphoto Inc.; **684** © Tom Pantages; **686** © Bill Terr. All Rights Reserved. Picturesque Stock Photos; **691** Courtesy of the National Institute of Standards and Technology; **692** Authors.

ESSAY 6   **698 (top)** Jim Lockhart; **(bottom)** R. Plambeck, Radio Astronomy Lab, U.C. Berkeley; **699 (right)** after data courtesy of Lakeshore Cryotronics.

CHAPTER 21   **700** © Tom Pantages; **702 (left)** © 1993 James L. Amos, All Rights Reserved; **(right)** © Tom Pantages; **707 (top)** NASA; **(bottom left)** © Hans Namuth/Science Photo Source/Photo Researchers; **(bottom middle)** Courtesy of E. L. Kohlschmieder; **(bottom right)** NASA; **716** © Tom Pantages.

CHAPTER 22   **718** Tadashi Hachimandai; **723** © Photo Deutsches Museum Munchen.

ESSAY 7   **751 (both)** © Tom Pantages.

PART VI   **754** © Murray & Associates, Inc. 1987, Picturesque Stock Photo; **756** © Tom Pantages; **757** NASA; **758 (left)** © The Granger Collection, New York; **(right)**; Tom Pantages; **759 (all)** Authors; **760** © Tom Pantages; **761** Authors; **762** Reprinted from *The Physics Teacher*, Dec. 1978, p. 631. Illustration from Faraday's "Diary" Vol. I. London, 1932.

CHAPTER 23   **764** © 1994 Comstock; **768** Courtesy of IBM Corp., Research Division, Almaden Research Center; **772** Dr. Roger Bland, San Francisco State University; **781** © The Bettmann Archive.

CHAPTER 24   **790** AP/Wide World Photos; **(inset)** © Tom Pantages; **793** © Tom Pantages; **797** © Tom Pantages; **801 (left)** Jim Lockhart, San Francisco State University; **(right)** © 1980 Kip Peticolas/Fundamental Photographs.

CHAPTER 25   813 © 1993 Peter Menzel Photography/Boston Museum of Science; 826 Christopher Johnson, The University of Utah.

CHAPTER 26   843 © Tom Pantages; 844 **(left)** © Tom Pantages; **(right)** Authors; 846 The Bettmann Archive; 848 **(left)** © Tom Pantages; **(right)** © Kip Peticolas/Fundamental Photographs; 850 © Tom Pantages; 852 © Tom Pantages; 862 © Tom Pantages.

CHAPTER 27   875 Lawrence Berkeley Laboratory; 883 Lawrence Berkeley Laboratory; 884 © Tom Pantages.

CHAPTER 28   898 Courtesy of Princeton University; 899 Authors; 900 © The Bettmann Archive; 908 Chip Kozy; 911 © Tom Pantages.

CHAPTER 29   923 Mark Marten/Science Source/Photo Researchers, Inc.; 925 Courtesy of Finnigan MAT, San Jose, CA; 926 Lawrence Berkeley Laboratory; 928 Courtesy of Mallinckrodt Institute of Radiology at Washington University; 930 The Science Museum/Science & Society Picture Library; 934 © Tom Pantages; 940 K. Koike and K. Hayakawa from *Applied Physics Letters* vol. 45, No. 5, Sept. 1984; 943 Paul Shambroom/Photo Researchers, Inc.; 945 Lawrence Berkeley Laboratory.

PART VII   953 Courtesy of GE Research and Development Center; 954 **(left)** The Bettmann Archive; **(right)** Courtesy of the Archives, California Institute of Technology.

CHAPTER 30   955 Bureau of Reclamation; 956 AIP Emilio Segrè Visual Archives; 957 © Tom Pantages; 962 © 1996 by Joseph Somsel. Photograph courtesy of Pacific Gas and Electric Company; 971 University of Illinois/Courtesy AIP Emilio Segrè Visual Archives; 973 **(right)** Courtesy of the Japanese Railway Technical Research Institute; **(both on bottom)** Authors.

CHAPTER 31   987 © Tom Pantages.

CHAPTER 32   1016 © 1996 by Joseph Somsel. Photograph courtesy of Pacific Gas and Electric Company.

CHAPTER 33   1039 Maarten Schmidt/California Institute of Technology; 1048 Courtesy of Lawrence Livermore National Laboratory; 1053 **(both)** © Tom Pantages; 1054 © Tom Pantages; 1055 **(top)** Royal Observatory, Edinburgh/Science Photo Library/Photo Researchers, Inc.; **(bottom)** Dr. W. J. Welch, Radio Astronomy Lab, U.C. Berkeley; 1056 Russ Underwood, Lockheed Martin Missiles & Space; 1058 © Tom Pantages.

PART VIII   1067 photo courtesy of Lockheed Corporation.

CHAPTER 34   1069 © courtesy of Roland and Marjorie Christen, taken with Astro-Physics 130mm StarFire EDT refractor; 1070 Werner Meyer-Ilse, Center for X-ray Optics, Lawrence Berkeley National Laboratory; 1091 © 1977 by Sidney Harris—"What's So Funny About Science?", William Kaufmann, Inc; 1093 Lawrence Berkeley Laboratory.

ESSAY 8   1103 The LIGO Project, California Institute of Technology; 1106 From data courtesy of Iain F. Reid, California Institute of Technology, taken at the Keck Telescope; 1107 **(left)** © Royal Astronomical Society Library; **(right)** NASA/STSI; 1109 **(top)** Dana Berry/STScI (Hubble Space Telescope Science Institute); **(bottom)** NASA/STSI.

CHAPTER 35   1110 IBM Corporation, Research Division, Almaden Research Center; 1112 Hamamatsu Corporation; 1113 Permission to copy granted from *Physical Review* Vol. VII, No. 3, 2nd series, March 1916; 1118 Permission to copy granted from *Physical Review* Vol. VII, 2nd series, March 1916; 1119 © 1994 Wabash Instrument Corp./Fundamental Photographs; 1121 David Gauntt and A. O. Schawlow/Stanford University; 1124 Bettmann Archive; 1125 NOAO; 1127 PSSC Physics 2/e, 1965 D. C. Heath and Company Educational and Development Center; 1128 © 1994 Wabash Instrument Corp./Fundamental Photographs; 1130 From data courtesy of Gary D. Schmidt, Steward observatory, University of Arizona; 1134 **(left)** Akira Tonomura/Hitachi, Ltd.; **(right)** Bettmann Archive.

ESSAY 9   1142 Shirley Chiang; 1143 **(both)** courtesy of IBM Corp., Reasearch Division, Almaden Research Center.

CHAPTER 36   1145 © 1996 by Joseph Somsel. Photograph courtesy of Pacific Gas and Electric Company; 1146 **(left)** © Tim Bedow/Science Photo Library/Photo Researchers, Inc.; **(right)** Lawrence Berkeley National Laboratory; 1150 Blackett and Lees, Plate 7 from the Proceeding of the Royal Society, A136, pp. 325–338, 1932; 1153 © The Bettmann Archive; 1154 K. Phillipp, Naturwissenschaft, Vol. 14, p. 1204 (1926). © Springer-Verlag; 1156 copyright Società Italiana di Fisica; 1157 **(left)** © California Institute of Technology; **(right)** Lawrence Berkeley Laboratory; 1158 photo courtesy of Brookhaven National Laboratory; 1159 **(both left)** Kamiokande; **(right)** © Anglo-Australian Observatory/Royal Observatory Edinburgh. Photographs made from UK Schmidt plates by David Malin; 1160 Lawrence Berkeley Laboratory; 1163 © 1994 Peter Berndt, M.D., P.A. All Rights Reserved/Custom Medical Stock Photo.

CHAPTER 37   1177 CDF Collaboration; 1178 Fermi National Accelator Laboratory/Science Photo Library/Photo Researchers, Inc.; 1180 **(left)** AP/Wide World Photos; **(right)** Lawrence Berkeley Laboratory; 1182 Lawrence Berkeley Laboratory; 1184 Courtesy Brookhaven National Laboratory; 1188 Photo courtesy of CERN Geneva.

QC 21.2 .L43 1997

Lea, Susan.

Physics

Solution Plans

Subject		Page
General problem solving	Interlude 1	117
Integration in physics	Interlude 2	250
Projectile motion	§3.1.2	88
Free-body diagrams	§4.6.1	141
Dynamic systems	Figure 5.4	167
Conservation laws	Figure 6.10	207
Rigid-body dynamics	Figure 12.12	401
Steady state fluid flow	Figure 13.32	453
Thin-film interference	Figure 17.12	572
Using PV diagram	Figure 19.18	664
Using Gauss' law	Figure 24.1	791
Kirchhoff's rules for circuits	Figure 26.27	859
Ampère's law	Figure 28.19	910
Faraday's law	Figure 30.19	968
Time-dependent circuits	Figure 31.4	991
How to solve a linear differential eqn.	§31.3.3	1005
Circuit analysis using phasors	Figure 32.13	1027

List of Tabulated Data

Quantity	Table #	Page
Atomic constituents, properties	36.2	115
Conductivity, thermal	21.1	70
Curie temperatures	29.1	94
Densities	13.1	43
Dielectric constant	27.1	88
Dynamic relations/rotational motion	12.4	40
Electromagnetic spectrum	16.2	53
Electron distribution	35.2	113
Forces that occur in classical mechanics	4.1	12
Gauge bosons	37.4	118
International Practical Temperature Scale	20.5	69
Invariant and relative quantities in Galilean relativity	34.1	107
Invariant and relative quantities in special relativity	34.2	109
Kinematic quantities in rotation/fixed axis	12.1	39
Kinematic quantities/uniform acceleration	12.2	39
Latent heats of fusion and vaporization	20.2	68
Leptons	37.3	118
Magnetic fields	28.1	90
Magnetic susceptibility	29.2	94
Mesons and Baryons	37.2	118
Nuclei, selected masses	36.1	114
Objects and Images, Types of	18.1	60
Particle physics, conservation laws of	37.6	119
Periodic table	inside back cover	
Phase relations in AC circuits	32.1	102
Photoelectric properties of selected metals	35.1	111
Quarks, properties of	37.1	118
Refractive index	16.3	53
Resistance, thermal, building materials	21.2	70
Resistivity, electrical	26.1	85
Rotational inertia, selected objects	12.3	39
SI prefixes	1.1	2
Sound, speed	16.1	52
Sound, speed in rock	16.4	56
Specific heats	20.4	68
Specific heats, molar, for gases	19.2	66
Temperatures	19.1	64
Triboelectric sequence	23.1	76
Thermal expansion, coefficient of	20.3	68
Uniformly accelerated linear motion, equations	2.4	7
van der Waals constants	20.1	67
Wavelengths of hydrogen lines	35.18	112
Weak interaction, world according to	37.5	118